Fundamental Constants

Avogadro's number (N_A)	6.0221418×10^{23}
Electron charge (e)	1.6022×10^{-19} C
Electron mass	9.109387×10^{-28} g
Faraday constant (F)	96,485.3 C/mol e^-
Gas constant (R)	0.08206 L $\cdot$ atm/K $\cdot$ mol
	8.314 J/K $\cdot$ mol
	62.36 L $\cdot$ torr/K $\cdot$ mol
	1.987 cal/K $\cdot$ mol
Planck's constant (h)	6.6256×10^{-34} J $\cdot$ s
Proton mass	1.672623×10^{-24} g
Neutron mass	1.674928×10^{-24} g
Speed of light in a vacuum	2.99792458×10^{8} m/s

Some Prefixes Used with SI Units

tera (T)	10^{12}	centi (c)	10^{-2}
giga (G)	10^{9}	milli (m)	10^{-3}
mega (M)	10^{6}	micro (μ)	10^{-6}
kilo (k)	10^{3}	nano (n)	10^{-9}
deci (d)	10^{-1}	pico (p)	10^{-12}

Useful Conversion Factors and Relationships

$$1 \text{ lb} = 453.6 \text{ g}$$

$$1 \text{ in} = 2.54 \text{ cm (exactly)}$$

$$1 \text{ mi} = 1.609 \text{ km}$$

$$1 \text{ km} = 0.6215 \text{ mi}$$

$$1 \text{ pm} = 1 \times 10^{-12} \text{ m} = 1 \times 10^{-10} \text{ cm}$$

$$1 \text{ atm} = 760 \text{ mmHg} = 760 \text{ torr} = 101,325 \text{ N/m}^2 = 101,325 \text{ Pa}$$

$$1 \text{ cal} = 4.184 \text{ J (exactly)}$$

$$1 \text{ L} \cdot \text{atm} = 101.325 \text{ J}$$

$$1 \text{ J} = 1 \text{ C} \times 1 \text{ V}$$

$$?^\circ\text{C} = (^\circ\text{F} - 32^\circ\text{F}) \times \frac{5^\circ\text{C}}{9^\circ\text{F}}$$

$$?^\circ\text{F} = \frac{9^\circ\text{F}}{5^\circ\text{C}} \times (^\circ\text{C}) + 32^\circ\text{F}$$

$$?\text{K} = (^\circ\text{C} + 273.15^\circ\text{C}) \left(\frac{1\text{K}}{1^\circ\text{C}} \right)$$

Periodic Table of the Elements

Key

Atomic number	6
Symbol	C
Name	Carbon
Average atomic mass	12.01

An element

Main group

Period number	Group number	1A 1	2A 2		3B 3	4B 4	5B 5	6B 6	7B 7	8B 8	8B 9	8B 10	1B 11	2B 12	3A 13	4A 14	5A 15	6A 16	7A 17	8A 18
1		1 H Hydrogen 1.008																		2 He Helium 4.003
2		3 Li Lithium 6.941	4 Be Beryllium 9.012												5 B Boron 10.81	6 C Carbon 12.01	7 N Nitrogen 14.01	8 O Oxygen 16.00	9 F Fluorine 19.00	10 Ne Neon 20.18
3		11 Na Sodium 22.99	12 Mg Magnesium 24.31												13 Al Aluminum 26.98	14 Si Silicon 28.09	15 P Phosphorus 30.97	16 S Sulfur 32.07	17 Cl Chlorine 35.45	18 Ar Argon 39.95
4		19 K Potassium 39.10	20 Ca Calcium 40.08		21 Sc Scandium 44.96	22 Ti Titanium 47.87	23 V Vanadium 50.94	24 Cr Chromium 52.00	25 Mn Manganese 54.94	26 Fe Iron 55.85	27 Co Cobalt 58.93	28 Ni Nickel 58.69	29 Cu Copper 63.55	30 Zn Zinc 65.41	31 Ga Gallium 69.72	32 Ge Germanium 72.64	33 As Arsenic 74.92	34 Se Selenium 78.96	35 Br Bromine 79.90	36 Kr Krypton 83.80
5		37 Rb Rubidium 85.47	38 Sr Strontium 87.62		39 Y Yttrium 88.91	40 Zr Zirconium 91.22	41 Nb Niobium 92.91	42 Mo Molybdenum 95.94	43 Tc Technetium (98)	44 Ru Ruthenium 101.1	45 Rh Rhodium 102.9	46 Pd Palladium 106.4	47 Ag Silver 107.9	48 Cd Cadmium 112.4	49 In Indium 114.8	50 Sn Tin 118.7	51 Sb Antimony 121.8	52 Te Tellurium 127.6	53 I Iodine 126.9	54 Xe Xenon 131.3
6		55 Cs Cesium 132.9	56 Ba Barium 137.3		57 La Lanthanum 138.9	72 Hf Hafnium 178.5	73 Ta Tantalum 180.9	74 W Tungsten 183.8	75 Re Rhenium 186.2	76 Os Osmium 190.2	77 Ir Iridium 192.2	78 Pt Platinum 195.1	79 Au Gold 197.0	80 Hg Mercury 200.6	81 Tl Thallium 204.4	82 Pb Lead 207.2	83 Bi Bismuth 209.0	84 Po Polonium (209)	85 At Astatine (210)	86 Rn Radon (222)
7		87 Fr Francium (223)	88 Ra Radium (226)		89 Ac Actinium (227)	104 Rf Rutherfordium (267)	105 Db Dubnium (268)	106 Sg Seaborgium (271)	107 Bh Bohrium (272)	108 Hs Hassium (270)	109 Mt Meitnerium (276)	110 Ds Darmstadtium (281)	111 Rg Roentgenium (280)	112 Cn Copernicium (285)	113 Nh Nihonium (286)	114 Fl Flerovium (289)	115 Mc Moscovium (289)	116 Lv Livermorium (293)	117 Ts Tennessine (293)	118 Og Oganesson (294)

Transition metals

Lanthanides 6

58 Ce Cerium 140.1	59 Pr Praseodymium 140.9	60 Nd Neodymium 144.2	61 Pm Promethium (145)	62 Sm Samarium 150.4	63 Eu Europium 152.0	64 Gd Gadolinium 157.3	65 Tb Terbium 158.9	66 Dy Dysprosium 162.5	67 Ho Holmium 164.9	68 Er Erbium 167.3	69 Tm Thulium 168.9	70 Yb Ytterbium 173.0	71 Lu Lutetium 175.0

Actinides 7

90 Th Thorium 232.0	91 Pa Protactinium 231.0	92 U Uranium 238.0	93 Np Neptunium (237)	94 Pu Plutonium (244)	95 Am Americium (243)	96 Cm Curium (247)	97 Bk Berkelium (247)	98 Cf Californium (251)	99 Es Einsteinium (252)	100 Fm Fermium (257)	101 Md Mendelevium (258)	102 No Nobelium (259)	103 Lr Lawrencium (262)

Metals

Nonmetals

Metalloids

List of the Elements with Their Symbols and Atomic Masses*

Element	Symbol	Atomic Number	Atomic Mass†	Element	Symbol	Atomic Number	Atomic Mass†
Actinium	Ac	89	(227)	Mendelevium	Md	101	(258)
Aluminum	Al	13	26.9815386	Mercury	Hg	80	200.59
Americium	Am	95	(243)	Molybdenum	Mo	42	95.94
Antimony	Sb	51	121.760	Moscovium	Mc	115	(289)
Argon	Ar	18	39.948	Neodymium	Nd	60	144.242
Arsenic	As	33	74.92160	Neon	Ne	10	20.1797
Astatine	At	85	(210)	Neptunium	Np	93	(237)
Barium	Ba	56	137.327	Nickel	Ni	28	58.6934
Berkelium	Bk	97	(247)	Nihonium	Nh	113	(286)
Beryllium	Be	4	9.012182	Niobium	Nb	41	92.90638
Bismuth	Bi	83	208.98040	Nitrogen	N	7	14.0067
Bohrium	Bh	107	(272)	Nobelium	No	102	(259)
Boron	B	5	10.811	Oganesson	Og	118	(294)
Bromine	Br	35	79.904	Osmium	Os	76	190.23
Cadmium	Cd	48	112.411	Oxygen	O	8	15.9994
Calcium	Ca	20	40.078	Palladium	Pd	46	106.42
Californium	Cf	98	(251)	Phosphorus	P	15	30.973762
Carbon	C	6	12.0107	Platinum	Pt	78	195.084
Cerium	Ce	58	140.116	Plutonium	Pu	94	(244)
Cesium	Cs	55	132.9054519	Polonium	Po	84	(209)
Chlorine	Cl	17	35.453	Potassium	K	19	39.0983
Chromium	Cr	24	51.9961	Praseodymium	Pr	59	140.90765
Cobalt	Co	27	58.933195	Promethium	Pm	61	(145)
Copernicium	Cn	112	(285)	Protactinium	Pa	91	231.03588
Copper	Cu	29	63.546	Radium	Ra	88	(226)
Curium	Cm	96	(247)	Radon	Rn	86	(222)
Darmstadtium	Ds	110	(281)	Rhenium	Re	75	186.207
Dubnium	Db	105	(268)	Rhodium	Rh	45	102.90550
Dysprosium	Dy	66	162.500	Roentgenium	Rg	111	(280)
Einsteinium	Es	99	(252)	Rubidium	Rb	37	85.4678
Erbium	Er	68	167.259	Ruthenium	Ru	44	101.07
Europium	Eu	63	151.964	Rutherfordium	Rf	104	(267)
Fermium	Fm	100	(257)	Samarium	Sm	62	150.36
Flerovium	Fl	114	(289)	Scandium	Sc	21	44.955912
Fluorine	F	9	18.9984032	Seaborgium	Sg	106	(271)
Francium	Fr	87	(223)	Selenium	Se	34	78.96
Gadolinium	Gd	64	157.25	Silicon	Si	14	28.0855
Gallium	Ga	31	69.723	Silver	Ag	47	107.8682
Germanium	Ge	32	72.64	Sodium	Na	11	22.98976928
Gold	Au	79	196.966569	Strontium	Sr	38	87.62
Hafnium	Hf	72	178.49	Sulfur	S	16	32.065
Hassium	Hs	108	(270)	Tantalum	Ta	73	180.94788
Helium	He	2	4.002602	Technetium	Tc	43	(98)
Holmium	Ho	67	164.93032	Tellurium	Te	52	127.60
Hydrogen	H	1	1.00794	Tennessine	Ts	117	(293)
Indium	In	49	114.818	Terbium	Tb	65	158.92535
Iodine	I	53	126.90447	Thallium	Tl	81	204.3833
Iridium	Ir	77	192.217	Thorium	Th	90	232.03806
Iron	Fe	26	55.845	Thulium	Tm	69	168.93421
Krypton	Kr	36	83.798	Tin	Sn	50	118.710
Lanthanum	La	57	138.90547	Titanium	Ti	22	47.867
Lawrencium	Lr	103	(262)	Tungsten	W	74	183.84
Lead	Pb	82	207.2	Uranium	U	92	238.02891
Lithium	Li	3	6.941	Vanadium	V	23	50.9415
Livermorium	Lv	116	(293)	Xenon	Xe	54	131.293
Lutetium	Lu	71	174.967	Ytterbium	Yb	70	173.04
Magnesium	Mg	12	24.3050	Yttrium	Y	39	88.90585
Manganese	Mn	25	54.938045	Zinc	Zn	30	65.409
Meitnerium	Mt	109	(276)	Zirconium	Zr	40	91.224

*These atomic masses show as many significant figures as are known for each element. The atomic masses in the periodic table are shown to four significant figures, which is sufficient for solving the problems in this book.

†Approximate values of atomic masses for radioactive elements are given in parentheses.

Chemistry

Julia Burdge

COLLEGE OF WESTERN IDAHO

CHEMISTRY, SIXTH EDITION

Published by McGraw Hill LLC, 1325 Avenue of the Americas, New York, NY 10019. Copyright ©2023 by McGraw Hill LLC. All rights reserved. Printed in the United States of America. Previous editions ©2020, 2017, and 2014. No part of this publication may be reproduced or distributed in any form or by any means, or stored in a database or retrieval system, without the prior written consent of McGraw Hill LLC, including, but not limited to, in any network or other electronic storage or transmission, or broadcast for distance learning.

Some ancillaries, including electronic and print components, may not be available to customers outside the United States.

This book is printed on acid-free paper.

1 2 3 4 5 6 7 8 9 LWI 27 26 25 24 23 22

ISBN 978-1-264-08577-4 (bound edition)
MHID 1-264-08577-X (bound edition)
ISBN 978-1-264-50191-5 (loose-leaf edition)
MHID 1-264-50191-9 (loose-leaf edition)

Executive Portfolio Manager: *Michelle Hentz*
Senior Product Developer: *Mary Hurley*
Senior Marketing Manager: *Cassie Cloutier*
Content Project Managers: *Jane Mohr and Rachael Hillebrand*
Buyer: *Sandy Ludovissy*
Designer: *David Hash*
Content Licensing Specialist: *Melissa Homer*
Cover Image: *Stockbyte/Alamy Stock Photo*
Compositor: *Aptara®, Inc.*

All credits appearing on page or at the end of the book are considered to be an extension of the copyright page.

Library of Congress Cataloging-in-Publication Data

Names: Burdge, Julia, author.
Title: Chemistry / Julia Burdge.
Description: Sixth edition. | New York : McGraw Hill Education, [2023] |
   Includes index.
Identifiers: LCCN 2021031239 (print) | LCCN 2021031240 (ebook) | ISBN
   9781264085774 (hardcover) | ISBN 9781264501915 (spiral bound) | ISBN
   9781264507795 (pdf) | ISBN 9781264494576 (ebook other)
Subjects: LCSH: Chemistry—Textbooks.
Classification: LCC QD33.2 .B865 2023 (print) | LCC QD33.2 (ebook) | DDC
   540—dc23
LC record available at https://lccn.loc.gov/2021031239
LC ebook record available at https://lccn.loc.gov/2021031240

The Internet addresses listed in the text were accurate at the time of publication. The inclusion of a website does not indicate an endorsement by the authors or McGraw Hill LLC, and McGraw Hill LLC does not guarantee the accuracy of the information presented at these sites.

Dedication

To the people who will always matter the most: Katie, Beau, and Sam.

About the Author

Courtesy of Julia Burdge

Julia Burdge received her Ph.D. (1994) from the University of Idaho in Moscow, Idaho. Her research and dissertation focused on instrument development for analysis of trace sulfur compounds in air and the statistical evaluation of data near the detection limit.

In 1994, she accepted a position at The University of Akron in Akron, Ohio, as an assistant professor and director of the Introductory Chemistry program. In the year 2000, she was tenured and promoted to associate professor at The University of Akron on the merits of her teaching, service, and research in chemistry education. In addition to directing the general chemistry program and supervising the teaching activities of graduate students, she helped establish a future-faculty development program and served as a mentor for graduate students and post-doctoral associates. In 2008, Julia relocated back to the northwest to be near family. She lives in Boise, Idaho, and holds an adjunct faculty position at the College of Western Idaho in Nampa.

In her free time, Julia enjoys the company of her children and Erik Nelson, her husband and best friend.

Brief Contents

Contents

Bernhard Staehli/Shutterstock

Zoonar/O Popova/age fotostock

Zigy Kaluzny/The Image Bank/Getty Images

4 REACTIONS IN AQUEOUS SOLUTIONS 140

Sara Stathas/Alamy Stock Photo

Pixtal/age fotostock

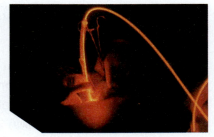
Source: John Crawford/National Cancer Institute

7 ELECTRON CONFIGURATION AND THE PERIODIC TABLE 312

Yoshikazu Tsuno/AFP/Getty Images

8   CHEMICAL BONDING I: BASIC CONCEPTS   358

Dinodia Photos/Alamy Stock Photo

10 GASES 464

Corbis/VCG/Image 100/Getty Images

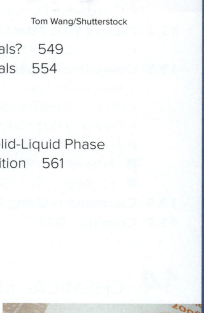

Tom Wang/Shutterstock

Jonas Ekstromer/AFP/Getty Images

Purestock/SuperStock

Chalabala/iStock/Getty Images

15 CHEMICAL EQUILIBRIUM   726

Buena Vista Images/DigitalVision/Getty Images

16 ACIDS AND BASES 786

Environmental Images/Universal Images Group/Shutterstock

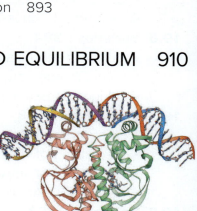

Margouillat photo/Shutterstock

LAGUNA DESIGN/Science Photo Library/ Science Source

19 ELECTROCHEMISTRY 958

Science Photo Library/Alamy Stock Photo

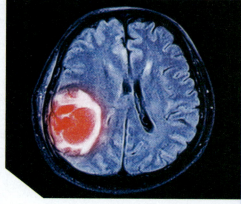

Puwadol Jaturawutthichai/Alamy Stock Photo

Digital Vision/Getty Images

David Kay/Shutterstock

Courtesy of Julia Burdge

24 METALLURGY AND THE CHEMISTRY OF METALS   (ONLINE ONLY)

David A. Tietz/Editorial Image, LLC

25 NONMETALLIC ELEMENTS AND THEIR COMPOUNDS (ONLINE ONLY)

M. Brodie/Alamy Stock Photo

Preface

Welcome to the exciting and dynamic world of Chemistry! My desire to create a general chemistry textbook grew out of my concern for the interests of students and faculty alike. Having taught general chemistry for many years, and having helped new teachers and future faculty develop the skills necessary to teach general chemistry, I believe I have developed a distinct perspective on the common problems and misunderstandings that students encounter while learning the fundamental concepts of chemistry—and that professors encounter while teaching them. I believe that it is possible for a textbook to address many of these issues while conveying the wonder and possibilities that chemistry offers. With this in mind, I have tried to write a text that balances the necessary fundamental concepts with engaging real-life examples and applications, while utilizing a consistent, step-by-step problem-solving approach and an innovative art and media program.

Key Features

Problem-Solving Methodology

Sample Problems are worked examples that guide the student step-by-step through the process of solving problems. Each Sample Problem follows the same four-step method: Strategy, Setup, Solution, and Think About It (check).

SAMPLE PROBLEM 4.8

For an aqueous solution of glucose ($C_6H_{12}O_6$), determine (a) the molarity of 2.00 L of a solution that contains 50.0 g of glucose, (b) the volume of this solution that would contain 0.250 mol of glucose, and (c) the number of moles of glucose in 0.500 L of this solution.

Strategy Convert the mass of glucose given to moles, and use the equations for interconversions of M, liters, and moles to calculate the answers.

Setup The molar mass of glucose is 180.2 g.

$$\text{moles of glucose} = \frac{50.0 \text{ g}}{180.2 \text{ g/mol}} = 0.277 \text{ mol}$$

Solution (a) molarity $= \dfrac{0.227 \text{ mol } C_6H_{12}O_6}{2.00 \text{ L solution}} = 0.139 \text{ } M$

A common way to state the concentration of this solution is to say, "This solution is 0.139 M in glucose."

(b) volume $= \dfrac{0.250 \text{ mol } C_6H_{12}O_6}{0.139 \text{ } M} = 1.80 \text{ } L$

(c) moles of $C_6H_{12}O_6$ in 0.500 L = 0.500 L $\times$ 0.139 M = 0.0695 mol

THINK ABOUT IT

Check to see that the magnitudes of your answers are logical. For example, the mass given in the problem corresponds to 0.277 mol of solute. If you are asked, as in part (b), for the volume that contains a number of moles smaller than 0.277, make sure your answer is smaller than the original volume.

Practice Problem ATTEMPT For an aqueous solution of sucrose ($C_{12}H_{22}O_{11}$), determine (a) the molarity of 5.00 L of a solution that contains 235 g of sucrose, (b) the volume of this solution that would contain 1.26 mol of sucrose, and (c) the number of moles of sucrose in 1.89 L of this solution.

Practice Problem BUILD For an aqueous solution of sodium chloride (NaCl), determine (a) the molarity of 3.75 L of a solution that contains 155 g of sodium chloride, (b) the volume of this solution that would contain 4.58 mol of sodium chloride, and (c) the number of moles of sodium chloride in 22.75 L of this solution.

Practice Problem CONCEPTUALIZE The diagrams represent solutions of two different concentrations. What volume of solution 2 contains the same amount of solute as 5.00 mL of solution 1? What volume of solution 1 contains the same amount of solute as 30.0 mL of solution 2?

solution 1   solution 2

Strategy: plan is laid out for solving the problem.

Setup: necessary information is gathered and organized.

Solution: problem is worked out.

Think About It:
– Assess the result.
– Provides information that shows the relevance of the result or the technique.
– Sometimes shows an alternate route to the same answer.

Each Sample Problem is followed by my ABC approach of three Practice Problems: **A**ttempt, **B**uild, and **C**onceptualize.

Practice Problem **A** (or "**A**ttempt") asks the student to apply the same Strategy to solve a problem very similar to the Sample Problem. In general, the same Setup and series of steps in the Solution can be used to solve Practice Problem A.

Practice Problem **B** (or "**B**uild") assesses mastery of the same skills as those required for the Sample Problem and Practice Problem A, but everywhere possible; Practice Problem B cannot be solved using the same Strategy used for the Sample Problem and for Practice Problem A. This provides the student an opportunity to develop a strategy independently, and combats the tendency that some students have to want to apply a "template" approach to solving chemistry problems. Practice Problems "Attempt" and "Build" have been incorporated into the problems available in Connect (R) and can be used in online homework and/or quizzing.

Practice Problem **C** (or "**C**onceptualize") provides an exercise that probes the student's conceptual understanding of the material. Practice Problems C often include concept and molecular art.

Applying What You've Learned
Sports drinks typically contain sucrose ($C_{12}H_{22}O_{11}$), fructose ($C_6H_{12}O_6$), sodium citrate ($Na_3C_6H_5O_7$), potassium citrate ($K_3C_6H_5O_7$), and ascorbic acid ($H_2C_6H_6O_6$), among other ingredients. (a) Classify each of these ingredients as a nonelectrolyte, a weak electrolyte, or a strong electrolyte [◄◄ Sample Problem 4.1]. (b) If a sports drink is 0.0015 *M* in both potassium citrate and potassium phosphate, what is the overall concentration of potassium in the drink [◄◄ Sample Problem 4.11]? (c) The aqueous iodine used to determine vitamin C content in sports drinks can be prepared by combining aqueous solutions of iodic acid (HIO_3) and hydroiodic acid (HI). (The products are aqueous iodine and liquid water.) Write a balanced equation for this reaction [◄◄ Sample Problem 3.3], (d) Write the net ionic equation for the reaction [◄◄ Sample Problem 4.3], (e) Determine the oxidation number for each element in the net ionic equation [◄◄ Sample Problem 4.5].

Each chapter's end-of-chapter questions and problems begin with an **Integrative Problem,** titled *Applying What You've Learned*. These integrative problems incorporate multiple concepts from the chapter, with each step of the problem providing a specific reference to the appropriate Sample Problem in case the student needs direction.

Key Skills

Newly located immediately before the end-of-chapter problems, Key Skills pages are modules that provide a review of specific problem-solving techniques from that particular chapter. These are techniques the author knows are vital to success in later chapters. The Key Skills pages are designed to be easy-to-find touchstones to hone specific skills from earlier chapters—in the context of later chapters. The answers to the Key Skills Problems can be found in the Answer Appendix in the back of the book.

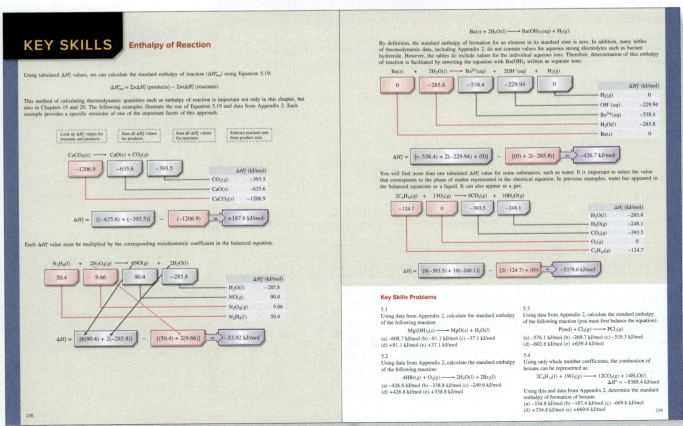

New to the Sixth Edition

- Updated periodic-table numbering scheme.
- **New chapter openers,** with emphasis on the chemistry associated with global climate change.
- **New End-of-Chapter Problems** have been added in response to user comments. These include additional conceptual problems, and updates of information in topical questions.
- **Specific references to Key Skills pages** in the "Before You Begin, Review These Skills" sections.
- **New figures** to help students develop conceptual understanding.
- **Continued development of truly comprehensive and consistent problem-solving.** Hundreds of worked examples (Sample Problems) help students get started learning how to approach and solve problems.

New and updated chapter content includes:
Incorporation of essential information from student notes into the main flow of text in each chapter. The remaining student notes are designed to help students over a variety of stumbling blocks. They include timely warnings about common errors, reminders of important information from previous chapters, and general information that helps place the material in an easily understood context.

Chapter 1—New and updated end-of-chapter problems and a new figure illustrating intensive and extensive properties

Chapter 2—Updated end-of-chapter problems

Chapter 4—New and updated conceptual end-of-chapter problems

Chapter 5—New and updated conceptual end-of-chapter problems

Chapter 7—New conceptual checkpoint questions

Chapter 9—New chapter opener and Applying-What-You've-Learned problems

Chapter 10—Updated end-of-chapter problems

Chapter 11—New Sample and Practice Problems

Chapter 13—New chapter opener and new end-of-chapter problems

Chapter 14—New and updated conceptual end-of-chapter problems

Chapter 17—New conceptual end-of-chapter problems

Chapter 19—New conceptual end-of-chapter problems

Instructor and Student Resources

Instructor Resources

ALEKS (Assessment and LEarning in Knowledge Spaces) is a web-based system for individualized assessment and learning available 24/7 over the Internet. ALEKS uses artificial intelligence to accurately determine a student's knowledge and then guides her to the material that she is most ready to learn. ALEKS offers immediate feedback and access to ALEKSPedia—an interactive text that contains concise entries on chemistry topics. ALEKS is also a full-featured course management system with rich reporting features that allow instructors to monitor individual and class perfor-mance, set student goals, assign/grade online quizzes, and more. ALEKS allows instructors to spend more time on concepts while ALEKS teaches students practical problem-solving skills. And with ALEKS 360, your student also has access to this text's eBook. Learn more at **www.aleks.com/highered/science.**

Instructors have access to the following instructor resources:

- **Instructor's Manual** This supplement contains Learning Objectives; Applications, Demonstrations, Tips and References; a list of End-of-Chapter Problems sorted by dif-ficulty; and a list of End-of-Chapter Problems sorted by type for each chapter of the text.

- **Art** Full-color digital files of all illustrations, photos, and tables in the book can be readily incorporated into lecture presentations, exams, or custom-made classroom materials. In addition, all files have been inserted into PowerPoint slides for ease of lecture preparation.
- **Animations** Numerous full-color animations illustrating important processes are also provided. Harness the visual impact of concepts in motion by importing these files into classroom presentations or online course materials.
- **PowerPoint Lecture Outlines** Ready-made presentations that combine art and lecture notes are provided for each chapter of the text.
- **Computerized Test Bank** Test questions that accompany *Chemistry* are available for creating exams or quizzes.
- **Instructor's Solutions Manual** This supplement contains complete, worked-out solutions for *all* the end-of-chapter problems in the text.

McGraw Hill Virtual Labs is a must-see, outcomes-based lab simulation. It assesses a student's knowledge and adaptively corrects deficiencies, allowing the student to learn faster and retain more knowledge with greater success. First, a student's knowledge is adaptively leveled on core learning outcomes: Questioning reveals knowledge deficiencies that are corrected by the delivery of content that is conditional on a student's response. Then, a simulated lab experience requires the student to think and act like a scientist: recording, interpreting, and analyzing data using simulated equipment found in labs and clinics. The student is allowed to make mistakes—a powerful part of the learning experience! A virtual coach provides subtle hints when needed, asks questions about the student's choices, and allows the student to reflect on and correct those mistakes. Whether your need is to overcome the logistical challenges of a traditional lab, provide better lab prep, improve student performance, or make your online experience one that rivals the real world, McGraw Hill Virtual Labs accomplishes it all.

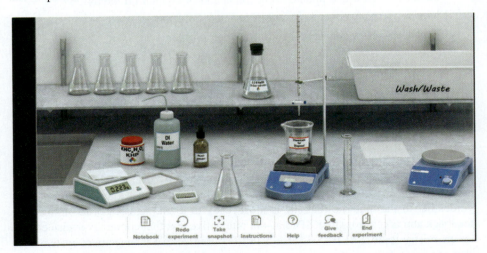

McGraw Hill Create™

With **McGraw Hill Create,** you can easily rearrange chapters, combine material from other content sources, and quickly upload content you have written, like your course syllabus or teaching notes. Find the content you need in Create by searching through thousands of leading McGraw Hill textbooks. Arrange your book to fit your teaching style. Create even allows you to personalize your book's appearance by selecting the cover and adding your name, school, and course information. Order a Create book and you'll receive a complimentary print review copy in three to five business days or a complimentary electronic review copy (eComp) via email in minutes. Go to www.mcgrawhillcreate.com today and register to experience how McGraw Hill Create empowers you to teach *your* students *your* way. **www.mcgrawhillcreate.com**

Additional Student Resources

All students will have access to **chemistry animations** for the animated Visualizing Chemistry figures as well as other chemistry animations. Within the text, the animations are mapped to the appropriate content.

Additionally, students can purchase a Student Solution Manual that contains detailed solutions and explanations for the odd-numbered problems in the main text.

For me, this text will always remain a work in progress. I encourage you to contact me with any comments or questions.

Julia Burdge
juliaburdge@cwidaho.cc

Acknowledgments

I wish to thank the many people who have contributed to the continued development of this text. Raymond Chang's lifetime commitment and Jason Overby's tireless work on the development and demonstration of the book's digital content continue to ensure and augment the quality of this endeavor.

My family, as always, continues to be there for me—no matter what.

Finally, I wish to thank my McGraw Hill family, for their continued confidence and support. This family consists of Vice President, Science, Engineering, and Math Portfolio Kathleen McMahon, Executive Portfolio Manager Michelle Hentz, Senior Marketing Manager Cassie Cloutier, Senior Product Developer Mary Hurley, Senior Content Project Manager Jane Mohr, Product Development Manager Robin Reed, and Lead Designer David Hash.

Create More Lightbulb Moments.

Every student has different needs and enters your course with varied levels of preparation. ALEKS® pinpoints what students already know, what they don't and, most importantly, what they're ready to learn next. Optimize your class engagement by aligning your course objectives to ALEKS® topics and layer on our textbook as an additional resource for students.

ALEKS® Creates a Personalized and Dynamic Learning Path

ALEKS® creates an optimized path with an ongoing cycle of learning and assessment, celebrating students' small wins along the way with positive real-time feedback. Rooted in research and analytics, ALEKS® improves student outcomes by fostering better preparation, increased motivation and knowledge retention.

*visit **bit.ly/whatmakesALEKSunique** to learn more about the science behind the most powerful adaptive learning tool in education!

Preparation & Retention

The more prepared your students are, the more effective your instruction is. Because ALEKS® understands the prerequisite skills necessary for mastery, students are better prepared when a topic is presented to them. ALEKS® provides personalized practice and guides students to what they need to learn next to achieve mastery. ALEKS® improves knowledge and student retention through periodic knowledge checks and personalized learning paths. This cycle of learning and assessment ensures that students remember topics they have learned, are better prepared for exams, and are ready to learn new content as they continue into their next course.

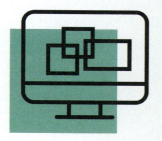

Flexible Implementation:
Your Class Your Way!

ALEKS® enables you to structure your course regardless of your instruction style and format. From a traditional classroom, to various co-requisite models, to an online prep course before the start of the term, ALEKS® can supplement your instruction or play a lead role in delivering the content.

*visit **bit.ly/ALEKScasestudies** to see how your peers are delivering better outcomes across various course models!

Outcomes & Efficacy

Our commitment to improve student outcomes services a wide variety of implementation models and best practices, from lecture-based to labs and co-reqs to summer prep courses. Our case studies illustrate our commitment to help you reach your course goals and our research demonstrates our drive to support all students, regardless of their math background and preparation level.

*visit **bit.ly/outcomesandefficacy** to review empirical data from ALEKS® users around the country

Turn Data Into Actionable Insights

ALEKS® Reports are designed to inform your instruction and create more meaningful interactions with your students when they need it the most. ALEKS® Insights alert you when students might be at risk of falling behind so that you can take immediate action. Insights summarize students exhibiting at least one of four negative behaviors that may require intervention including Failed Topics, Decreased Learning, Unusual Learning and Procrastination & Cramming.

Winner of 2019 Digital Edge 50 Award for Data Analytics!

bit.ly/ALEKS_MHE

Chemistry

CHAPTER 1

Chemistry: The Central Science

Calving glacier.

Bernhard Staehli/Shutterstock

In This Chapter, You Will Learn

Some of what chemistry is and how it is studied using the scientific method. You will learn about the system of units used by scientists and about expressing and dealing with the numbers that result from scientific measurements.

Before You Begin, Review These Skills

- Basic algebra
- Scientific notation [⏭ Appendix 1]

Global Climate Change and the Scientific Method

To advance understanding of science, researchers use a set of guidelines known as the *scientific method.* The guidelines involve careful observations, educated reasoning, and the development and experimental testing of hypotheses and theories. One field of study in which the scientific method has informed our understanding of the world is that of *global climate change*.

Late in the nineteenth century, Swedish chemist Svante Arrhenius used the principles of chemistry to describe the "greenhouse effect," the process by which certain components of the atmosphere absorb some of the energy radiating from Earth's surface and prevent it from escaping into space—thereby warming the planet. The greenhouse effect is a natural phenomenon, responsible in part for Earth's average global temperature being hospitable to humans and other forms of life. But Arrhenius also predicted what he perceived to be an inevitable, eventual consequence of the burning of coal and other fossil fuels, which increased significantly during the industrial revolution. He believed that, unchecked, the dramatic increase in atmospheric CO_2 caused by human activities would cause a potentially dangerous increase in global temperature via the "enhanced greenhouse effect."

Several groups of climate scientists, including those at the National Aeronautics and Space Administration's Goddard Institute for Space Studies (NASA/GISS) at Columbia University, study global temperature trends by analyzing observations from many thousands of data sets gathered using a variety of different measurement techniques over the course of more than a century. Their findings have consistently validated Arrhenius's prediction. There is no doubt that the temperature of our planet is increasing. Moreover, the connection between global temperature change and human activities—most importantly the burning of fossil fuels—is undeniable.

The issue of global climate change is one that appears frequently in the popular press. Unfortunately, it has become something of a political issue, with some people dismissing its importance or denying its existence outright. As a student of science, you will want to develop an informed perspective. To do this, you must understand how observations, hypotheses, theories, and experimentation contribute to a self-correcting scientific narrative; and how they have given rise to the current scientific consensus regarding climate change and humankind's role in it.

At the end of this chapter, you will be able to answer several questions related to the study of global climate change [⏭ Applying What You've Learned, page 34].

1.1   The Study of Chemistry

Chemistry often is called the *central science* because knowledge of the principles of chemistry can facilitate understanding of other sciences, including physics, biology, geology, astronomy, oceanography, engineering, and medicine. ***Chemistry*** is the study of *matter* and the *changes* that matter undergoes. Matter is what makes up our bodies, our belongings, our physical environment, and in fact our universe. ***Matter*** is anything that has mass and occupies space.

Although it can take many different forms, all matter consists of various combinations of atoms of only a relatively small number of simple substances called *elements*. The properties of matter depend on which of these elements it contains and on how the atoms of those elements are arranged.

Chemistry You May Already Know

You may already be familiar with some of the terms used in chemistry. Even if this is your first chemistry course, you may have heard of *molecules* and know them to be tiny pieces of a substance—much too tiny to see. Further, you may know that molecules are made up of *atoms,* even smaller pieces of matter. And even if you don't know what a chemical formula is, you probably know that H_2O is water and CO_2 is carbon dioxide. You may have used, or at least heard, the term *chemical reaction;* and you are undoubtedly familiar with a variety of chemical reactions, such as those shown in Figure 1.1.

Familiar chemical reactions, such as those shown in Figure 1.1, are all things that you can observe at the *macroscopic level.* In other words, these processes and their results are visible to the human eye. In studying chemistry, you will learn to understand and visualize many of these processes at the *molecular level.*

Because atoms and molecules are far too small to observe directly, we need a way to visualize them. One way is through the use of molecular models. Throughout

(a)

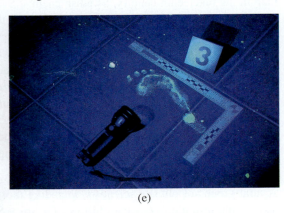
(c)

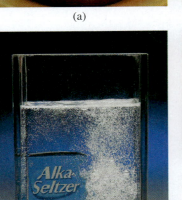
(b)

(d)

(e)

Figure 1.1 Many familiar processes are chemical reactions: (a) The flame of a creme brulee torch is the combustion of butane. (b) The bubbles produced when Alka-Seltzer dissolves in water are carbon dioxide, produced by a chemical reaction between two ingredients in the tablets. (c) The formation of rust is a chemical reaction that occurs when iron, water, and oxygen are all present. (d) Many baked goods "rise" as the result of a chemical reaction that produces carbon dioxide. (e) The glow produced when luminol is used to detect traces of blood in crime-scene investigations is the result of a chemical reaction.

a: Mike Liu/Shutterstock; b: Charles D. Winters/McGraw Hill; c: Danie van Niekerk/Shutterstock; d: Marie C Fields/Shutterstock; e: Couperfield/Shutterstock

How Can I Enhance My Chances of Success in Chemistry Class?

Success in a chemistry class depends largely on problem-solving ability. The Sample Problems throughout this text are designed to help you develop problem-solving skills. Each is divided into four steps: Strategy, Setup, Solution, and Think About It.

Strategy: Read the problem carefully and determine what is being asked and what information is provided. The Strategy step is where you should think about what skills are required and lay out a plan for solving the problem. Give some thought to what you expect the result to be. If you are asked to determine the number of atoms in a sample of matter, for example, you should expect the answer to be a whole number. Determine what, if any, units should be associated with the result. When possible, make a ballpark estimate of the magnitude of the correct result, and make a note of your estimate.

Setup: Next, gather the information necessary to solve the problem. Some of the information will have been given in the problem itself. Other information, such as equations, constants, and tabulated data (including atomic masses), should also be brought together in this step. Write down and label clearly all of the information you will use to solve the problem. Be sure to write appropriate units with each piece of information.

Solution: Using the necessary equations, constants, and other information, calculate the answer to the problem. Pay particular attention to the units associated with each number, tracking and canceling units throughout the calculation. In the event that multiple calculations are required, carefully label any intermediate results.

Think About It: Consider your calculated result and ask yourself whether or not it makes sense. Compare the units and the magnitude of your result with your ballpark estimate from the Strategy step. If your result does not have the appropriate units, or if its magnitude or sign is not reasonable, check your solution for possible errors. A very important part of problem solving is being able to judge whether the answer is reasonable. It is relatively easy to spot a wrong sign or incorrect units, but you should also develop a sense of magnitude and be able to tell when an answer is either way too big or way too small. For example, if a problem asks how many molecules are in a sample and you calculate a number that is less than 1, you should know that it cannot be correct.

For additional practice, each Sample Problem is followed by three Practice Problems: A, B, and C. Practice Problem A, "Attempt," typically is very similar to the Sample Problem and can be solved using the same strategy. Practice Problem B, "Build," generally tests the same skills as Practice Problem A, but usually requires a slightly different approach. Practice Problem B lets you practice devising your own problem-solving strategy—an indispensable skill in any science curriculum. Practice Problem C, "Conceptualize," specifically probes your understanding of the underlying chemical concepts associated with the Sample Problem.

Regular use of the Sample Problems and Practice Problems A, B, and C in this text can help you develop an effective set of problem-solving skills. They can also help you assess whether you are ready to move on to the next new concepts. If you struggle with the Practice Problems, then you probably need to review the corresponding Sample Problem and the concepts that led up to it.

this book, we will represent matter at the molecular level using *molecular art*, the two-dimensional equivalent of molecular models. In these pictures, atoms are represented as spheres, and atoms of particular elements are represented using specific colors. Table 1.1 lists some of the elements that you will encounter most often and the colors used to represent them in this book.

Molecular art can be of *ball-and-stick* models, in which the bonds connecting atoms appear as sticks [Figure 1.2(b)], or of *space-filling* models, in which the atoms appear to overlap one another [Figure 1.2(c)]. Ball-and-stick and space-filling models illustrate the specific, three-dimensional arrangement of the atoms. The ball-and-stick model does a good job of illustrating the arrangement of atoms, but exaggerates the distances between atoms, relative to their sizes. The space-filling model gives a more accurate picture of these *interatomic* distances but can obscure the details of the three-dimensional arrangement.

The Scientific Method

Experiments are the key to advancing our understanding of chemistry—or any science. Although not all scientists will necessarily take the same approach to experimentation, they all follow a set of guidelines known as the ***scientific method*** to add their results

H_2O

(a)

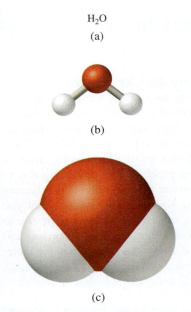

(b)

(c)

Figure 1.2 Water represented with a (a) molecular formula, (b) ball-and-stick model, and (c) space-filling model.

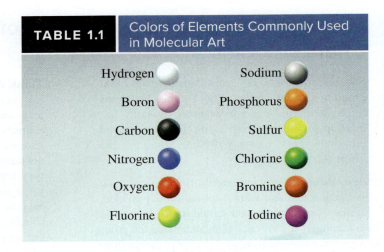

TABLE 1.1	Colors of Elements Commonly Used in Molecular Art	
Hydrogen		Sodium
Boron		Phosphorus
Carbon		Sulfur
Nitrogen		Chlorine
Oxygen		Bromine
Fluorine		Iodine

to the larger body of knowledge within a given field. The flowchart in Figure 1.3 illustrates this basic process. The method begins with the gathering of data via observations and experiments. Scientists study these data and try to identify *patterns* or *trends*. When they find a pattern or trend, they may summarize their findings with a *law,* a concise verbal or mathematical statement of a reliable relationship between phenomena. Scientists may then formulate a *hypothesis,* a tentative explanation for their observations. Further experiments are designed to test the hypothesis. If experiments indicate that the hypothesis is incorrect, the scientists go back to the drawing board, try to come up with a different interpretation of their data, and formulate a new hypothesis. The new hypothesis will then be tested by experiment. When a hypothesis stands the test of extensive experimentation, it may evolve into a theory. A *theory* is a unifying principle that explains a body of experimental observations and the laws that are based on them. Theories can also be used to predict related phenomena, so theories are constantly being tested. If a theory is disproved by experiment, then it must be discarded or modified so that it becomes consistent with experimental observations.

A fascinating example of the use of the scientific method is the story of how smallpox was eradicated. Late in the eighteenth century, an English doctor named Edward Jenner observed that even during outbreaks of smallpox in Europe, milkmaids seldom contracted the disease. He reasoned that when people who had frequent contact with cows contracted *cowpox,* a similar but far less harmful disease, they developed a natural immunity to smallpox. He predicted that intentional exposure to the cowpox virus would produce the same immunity. In 1796, Jenner exposed an 8-year-old boy to the cowpox virus using pus from the cowpox lesions of an infected milkmaid. Six weeks later, he exposed the boy to the *smallpox* virus and, as Jenner had predicted, the boy did *not* contract the disease. Subsequent experiments using the same technique (later dubbed *vaccination* from the Latin *vacca* meaning *cow*) confirmed that immunity to smallpox could be induced.

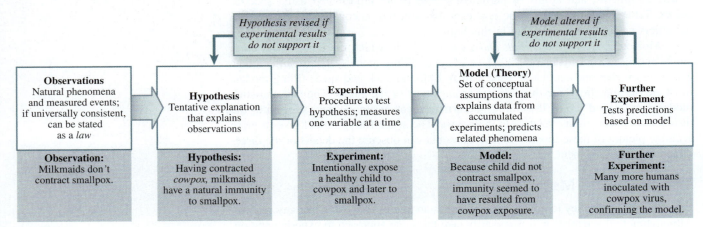

Figure 1.3 Flowchart of the scientific method.

A superbly coordinated international effort on the part of healthcare workers was successful in eliminating smallpox worldwide. In 1980, the World Health Organization declared smallpox officially eradicated in nature. This historic triumph over a dreadful disease, one of the greatest medical advances of the twentieth century, began with Jenner's astute observations, inductive reasoning, and careful experimentation—the essential elements of the *scientific method.*

1.2   Classification of Matter

Chemists classify matter as either a *substance* or a *mixture* of substances. A **substance** is a form of matter that has a specific composition and distinct properties. Examples are salt (sodium chloride), iron, water, mercury, carbon dioxide, and oxygen. Substances can be further classified as either *elements* (such as iron, mercury, and oxygen) or *compounds* (such as salt, water, and carbon dioxide). Different substances differ from one another in composition and properties, and each can be identified by its appearance, taste, smell, or other properties.

Student Note: Some books refer to substances as *pure substances.* These two terms generally mean the same thing although the adjective *pure* is unnecessary in this context because a substance is, by definition, pure.

States of Matter

Every substance can, in principle, exist as a solid, a liquid, and a gas, the three physical states depicted in Figure 1.4. Solids and liquids sometimes are referred to collectively as the *condensed phases.* Liquids and gases sometimes are referred to collectively as *fluids.* In a solid, particles are held close together in an orderly fashion with little freedom of motion. As a result, a solid does not conform to the shape of its container. Particles in a liquid are close together but are not held rigidly in position; they are free to move past one another. Thus, a liquid conforms to the shape of the part of the container it fills. In a gas, the particles are separated by distances that are very large compared to the size of the particles. A sample of gas assumes both the shape and the volume of its container.

The three states of matter can be interconverted without changing the chemical composition of the substance. Upon heating, a solid (e.g., ice) will melt to form a liquid (water). Further heating will vaporize the liquid, converting it to a gas (water vapor). Conversely, cooling a gas will cause it to condense into a liquid. When the liquid is cooled further, it will freeze into the solid form. Figure 1.5 shows the three physical states of water.

Animation
Matter—three states of matter.

Elements

An *element* is a substance that cannot be separated into simpler substances by chemical means. Iron, mercury, oxygen, and hydrogen are just 4 of the 118 elements that have

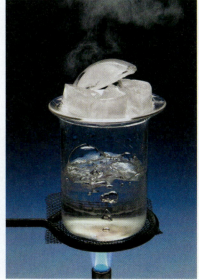

Figure 1.5 Water as a solid (ice), liquid, and gas. (We can't actually see water vapor, any more than we can see the nitrogen and oxygen that make up most of the air we breathe. When we see steam or clouds, what we are actually seeing is water vapor that has condensed upon encountering cold air.)
Charles D. Winters/Timeframe Photography/ McGraw Hill

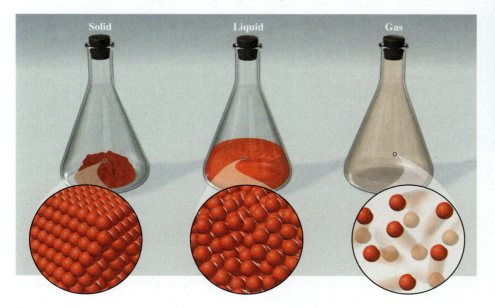

Figure 1.4 Molecular-level illustrations of a solid, liquid, and gas.

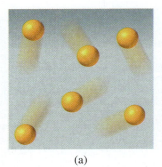

(a)

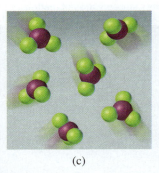

(b)

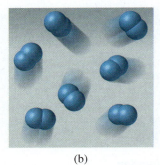

(c)

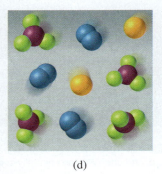

(d)

Figure 1.6 (a) Isolated atoms of an element. (b) Molecules of an element. (c) Molecules of a compound, consisting of more than one element. (d) A mixture of atoms of an element and molecules of an element and a compound.

been identified. Most of the known elements occur naturally on Earth. The others have been produced by scientists via nuclear processes, which are discussed in Chapter 20. As shown in Figure 1.6(a) and (b), an element may consist of atoms or molecules.

For convenience, chemists use symbols of one or two letters to represent the elements. Only the first letter of an element's chemical symbol is capitalized. A list of the elements and their symbols appears at the beginning of this book. The symbols of some elements are derived from their Latin names—for example, Ag from *argentum* (silver), Pb from *plumbum* (lead), and Na from *natrium* (sodium)—while most of them come from their English names—for example, H for hydrogen, Co for cobalt, and Br for bromine.

Compounds

Most elements can combine with other elements to form compounds. Hydrogen gas, for example, burns in the presence of oxygen gas to form water, which has properties that are distinctly different from those of either hydrogen or oxygen. Thus, water is a *compound,* a substance composed of atoms of two or more elements chemically united in fixed proportions [Figure 1.6(c)]. The elements that make up a compound are called the compound's *constituent elements.* For example, the constituent elements of water are hydrogen and oxygen; and water always contains twice as many hydrogen atoms as oxygen atoms (fixed proportions).

A compound cannot be separated into simpler substances by any physical process. (A physical process [▶▶ Section 1.4] is one that does not change the identity of the matter. Examples of physical processes include boiling, freezing, and filtering.) Instead, the separation of a compound into its constituent elements requires a *chemical reaction.*

Mixtures

A *mixture* is a combination of two or more substances [Figure 1.6(d)] in which the substances retain their distinct identities. Like pure substances, mixtures can be solids, liquids, or gases. Some familiar examples are mixed nuts, 14-carat gold, apple juice, salt water, and air. Unlike compounds, mixtures do not have a universal constant composition. Therefore, samples of air collected in different locations will differ in composition because of differences in altitude, pollution, and other factors. The ratio of salt to water in different samples of salt water will vary depending on how they were prepared.

Mixtures are either *homogeneous,* having uniform composition throughout; or *heterogeneous,* having variable composition. When we dissolve a teaspoon of sugar in a glass of water, we get a *homogeneous mixture.* However, if we mix sand with iron filings, we get a *heterogeneous mixture* in which the two substances remain distinct and discernible from each other (Figure 1.7).

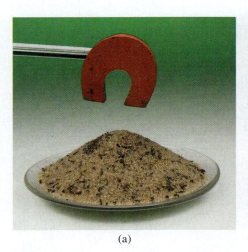

(a)

(b)

Figure 1.7 (a) A heterogeneous mixture contains iron filings and sand. (b) A magnet is used to separate the iron filings from the mixture.

a: Charles D. Winters/McGraw Hill; b: Charles D. Winters/Timeframe Photography/McGraw Hill

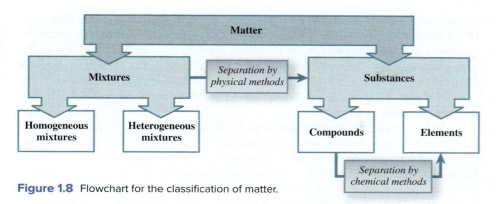

Figure 1.8 Flowchart for the classification of matter.

Mixtures, whether homogeneous or heterogeneous, can be separated into pure components by physical means—without changing the identities of the components. Thus, sugar can be recovered from a water solution by evaporating the solution to dryness. Condensing the vapor will give us back the water component. To separate the sand–iron mixture, we can use a magnet to remove the iron filings from the sand, because sand is not attracted to the magnet [see Figure 1.7(b)]. After separation, the components of the mixture will have the same composition and properties as they did prior to being combined. The relationships among substances, elements, compounds, and mixtures are summarized in Figure 1.8.

1.3 Scientific Measurement

Scientists use a variety of devices to measure the properties of matter. A meterstick is used to measure length; a burette, pipette, graduated cylinder, and volumetric flask are used to measure volume (Figure 1.9); a balance is used to measure mass; and a

Figure 1.9 (a) A volumetric flask is used to prepare a precise volume of a solution for use in the laboratory. (b) A graduated cylinder is used to measure a volume of liquid. It is less precise than the volumetric flask. (c) A volumetric pipette is used to deliver a precise amount of liquid. (d) A burette is used to measure the volume of a liquid that has been added to a container. A reading is taken before and after the liquid is delivered, and the volume delivered is determined by subtracting the first reading from the second.

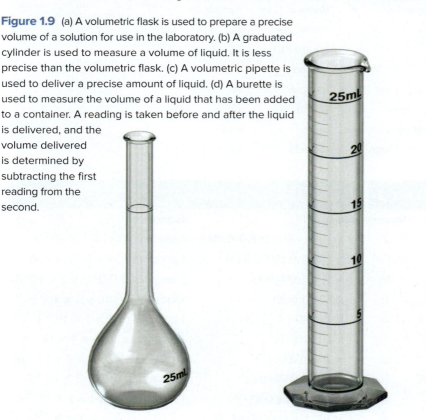

Volumetric flask
(a)

Graduated cylinder
(b)

Pipette
(c)

Burette
(d)

thermometer is used to measure temperature. Properties that can be measured are called *quantitative* properties because they are expressed using numbers. When we express a measured quantity with a number, though, we must always include the appropriate unit; otherwise, the measurement is meaningless. For example, to say that the depth of a swimming pool is 3 is insufficient to distinguish between one that is 3 *feet* (0.9 meter) and one that is 3 *meters* (9.8 feet) deep. Units are essential to reporting measurements correctly.

Student Note: According to the U.S. Metric Association (USMA), the United States is "the only significant holdout" with regard to adoption of the metric system. The other countries that continue to use traditional units are Myanmar (formerly Burma) and Liberia.

The two systems of units with which you are probably most familiar are the *English system* (foot, gallon, pound, etc.) and the *metric system* (meter, liter, kilogram, etc.). Although there has been an increase in the use of metric units in the United States in recent years, English units still are used commonly. For many years, scientists recorded measurements in metric units, but in 1960, the General Conference on Weights and Measures, the international authority on units, proposed a revised metric system for universal use by scientists. We use both metric and revised metric (SI) units in this book.

SI Base Units

The revised metric system is called the ***International System of Units*** (abbreviated SI, from the French *Système Internationale d'Unités*). Table 1.2 lists the seven SI base units. All other units of measurement can be derived from these base units. The ***SI unit*** for *volume,* for instance, is derived by cubing the SI base unit for *length*. The prefixes listed in Table 1.3 are used to denote decimal fractions and multiples of SI units. This enables scientists to tailor the magnitude of a unit to a particular application. For example, the meter (m) is appropriate for describing the dimensions of a classroom, but the kilometer (km), 1000 m, is more appropriate for describing the

TABLE 1.2	Base SI Units	
Base Quantity	**Name of Unit**	**Symbol**
Length	meter	m
Mass	kilogram	kg
Time	second	s
Electric current	ampere	A
Temperature	kelvin	K
Amount of substance	mole	mol
Luminous intensity	candela	cd

Student Note: Only one of the seven SI base units, the kilogram, itself contains a prefix.

TABLE 1.3	Prefixes Used with SI Units		
Prefix	**Symbol**	**Meaning**	**Example**
Tera-	T	1×10^{12} (1,000,000,000,000)	1 teragram (Tg) = 1×10^{12} g
Giga-	G	1×10^{9} (1,000,000,000)	1 gigawatt (GW) = 1×10^{9} W
Mega-	M	1×10^{6} (1,000,000)	1 megahertz (MHz) = 1×10^{6} Hz
Kilo-	k	1×10^{3} (1,000)	1 kilometer (km) = 1×10^{3} m
Deci-	d	1×10^{-1} (0.1)	1 deciliter (dL) = 1×10^{-1} L
Centi-	c	1×10^{-2} (0.01)	1 centimeter (cm) = 1×10^{-2} m
Milli-	m	1×10^{-3} (0.001)	1 millimeter (mm) = 1×10^{-3} m
Micro-	μ	1×10^{-6} (0.000001)	1 microliter (μL) = 1×10^{-6} L
Nano-	n	1×10^{-9} (0.000000001)	1 nanosecond (ns) = 1×10^{-9} s
Pico-	p	1×10^{-12} (0.000000000001)	1 picogram (pg) = 1×10^{-12} g

distance between two cities. Units that you will encounter frequently in the study of chemistry include those for mass, temperature, volume, and density.

Mass

Although the terms *mass* and *weight* often are used interchangeably, they do not mean the same thing. Strictly speaking, weight is the force exerted by an object or sample due to gravity. **Mass** is a measure of the amount of matter in an object or sample. Because gravity varies from location to location (gravity on the moon is only about one-sixth that on Earth), the weight of an object varies depending on where it is measured. The mass of an object remains the same regardless of where it is measured. The SI base unit of mass is the kilogram (kg), but in chemistry the smaller gram (g) often is more convenient and is more commonly used:

$$1 \text{ kg} = 1000 \text{ g} = 1 \times 10^3 \text{ g}$$

Temperature

There are two temperature scales used in chemistry. Their units are degrees Celsius (°C) and kelvin (K). The Celsius scale was originally defined using the freezing point (0°C) and the boiling point (100°C) of pure water at sea level. As Table 1.2 shows, the SI base unit of temperature is the **kelvin.** Kelvin is known as the *absolute* temperature scale, meaning that the lowest temperature possible is 0 K, a temperature referred to as "absolute zero." No *degree* sign (°) is used to represent a temperature on the Kelvin scale. The theoretical basis of the Kelvin scale has to do with the behavior of gases and is discussed in Chapter 10.

Units of the Celsius and Kelvin scales are equal in magnitude, so *a degree Celsius is equivalent to a kelvin.* Thus, if the temperature of an object increases by 5°C, it also increases by 5 K. Absolute zero on the Kelvin scale is equivalent to −273.15°C on the Celsius scale. We use the following equation to convert a temperature from units of degrees Celsius to kelvin:

$$K = {}^\circ C + 273.15 \qquad\qquad \textbf{Equation 1.1}$$

Depending on the precision required, the conversion from degrees Celsius to kelvin often is done simply by adding 273, rather than 273.15.

Sample Problem 1.1 illustrates conversions between these two temperature scales.

SAMPLE PROBLEM 1.1

Normal human body temperature can range over the course of the day from about 36°C in the early morning to about 37°C in the afternoon. Express these two temperatures and the range that they span using the Kelvin scale.

Strategy Use Equation 1.1 to convert temperatures from the Celsius scale to the Kelvin scale. Then convert the range of temperatures from degrees Celsius to kelvin, keeping in mind that 1°C is equivalent to 1 K.

Setup Equation 1.1 is already set up to convert the two temperatures from degrees Celsius to kelvin. No further manipulation of the equation is needed. The range in kelvin will be the same as the range in degrees Celsius.

Solution 36°C + 273 = 309 K, 37°C + 273 = 310 K, and the range of 1°C is equal to a range of 1 K.

THINK ABOUT IT

Check your math and remember that converting a temperature from degrees Celsius to kelvin is different from converting a *difference* in temperature from degrees Celsius to kelvin.

(Continued on next page)

Practice Problem ●**A**TTEMPT Express the freezing point of water (0°C), the boiling point of water (100°C), and the range spanned by the two temperatures using the Kelvin scale.

Practice Problem ●**B**UILD According to the website of the National Aeronautics and Space Administration (NASA), the average temperature of the universe is 2.7 K. Convert this temperature to degrees Celsius.

Practice Problem ●**C**ONCEPTUALIZE If a single degree on the Celsius scale is represented by the rectangle on the left, which of the rectangles on the right best represents a single kelvin?

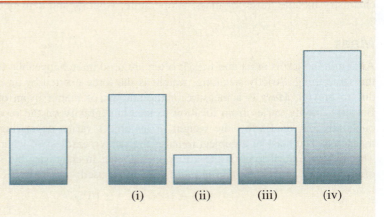

(i)   (ii)   (iii)   (iv)

Bringing Chemistry to Life

Fahrenheit Temperature Scale

Outside of scientific circles, the Fahrenheit temperature scale is the one most used in the United States. Before the work of Daniel Gabriel Fahrenheit (German physicist, 1686–1736), there were numerous different, arbitrarily defined temperature scales, none of which gave consistent measurements. Accounts of exactly how Fahrenheit devised his temperature scale vary from source to source. In one account, in 1724, Fahrenheit labeled as 0° the lowest artificially attainable temperature at the time (the temperature of a mixture of ice, water, and ammonium chloride). Using a traditional scale consisting of 12 degrees, he labeled the temperature of a healthy human body as the twelfth degree. On this scale, the freezing point of water occurred at the fourth degree. For better resolution, each degree was further divided into eight smaller degrees. This convention makes the freezing point of water 32° and normal body temperature 96°. Today we consider normal body temperature to be somewhat higher than 96°F.

The boiling point of water on the Fahrenheit scale is 212°, meaning that there are 180 degrees (212° − 32°) between the freezing and boiling points. This separation is considerably more than the 100 degrees between the freezing point and boiling point of water on the Celsius scale [named after Swedish physicist Ander Celsius (1701–1744)]. Thus, the size of a degree on the Fahrenheit scale is only 100/180 or five-ninths of a degree on the Celsius scale. Equations 1.2 and 1.3 give the relationship between Fahrenheit and Celsius temperatures.

Equation 1.2

$$\text{temp in °C} = (\text{temp in °F} - 32\text{°F}) \times \frac{5\text{°C}}{9\text{°F}}$$

and

Equation 1.3

$$\text{temp in °F} = \frac{9\text{°F}}{5\text{°C}} \times (\text{temp in °C}) + 32\text{°F}$$

Sample Problem 1.2 illustrates the conversion between Celsius and Fahrenheit scales.

SAMPLE PROBLEM (1.2)

A body temperature below 35.0°C constitutes hypothermia, whereas one above 39.0°C constitutes a high fever. Convert each of these temperatures to the Fahrenheit scale.

Strategy We are given temperatures in Celsius and are asked to convert them to Fahrenheit.

Setup We use Equation 1.3:

$$\text{temp in }°F = \frac{9°F}{5°C} \times (\text{temp in }°C) + 32°F \qquad \textbf{Equation 1.3}$$

Solution

$$\text{temp in }°F = \frac{9°F}{5°C} \times 35.0°C + 32°F = 95.0°F$$

$$\text{temp in }°F = \frac{9°F}{5°C} \times 39.0°C + 32°F = 102.2°F$$

THINK ABOUT IT

"Normal" body temperature on the Fahrenheit scale is generally considered to be 98.6°F. The temperatures of hypothermia and high fever should be *below* and *above* that number, respectively. Therefore, 95.0°F and 102.2°F seem like reasonable results.

Practice Problem ATTEMPT Convert the temperatures 45.0°C and 90.0°C, and the difference between them, to degrees Fahrenheit.

Practice Problem BUILD In Ray Bradbury's 1953 novel *Fahrenheit 451*, 451°F is said to be the temperature at which books, which have been banned in the story, ignite. Convert 451°F to the Celsius scale.

Practice Problem CONCEPTUALIZE If a single degree on the Fahrenheit scale is represented by the rectangle on the left, which of the rectangles on the right best represents a single degree on the Celsius scale? Which best represents a single kelvin?

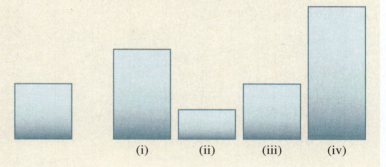

(i)   (ii)   (iii)   (iv)

Derived Units: Volume and Density

There are many quantities, such as volume and density, that require units not included in the base SI units. In these cases, we must combine base units to *derive* appropriate units for the quantity.

The derived SI unit for volume, the meter cubed (m^3), is a larger volume than is practical in most laboratory settings. The more commonly used metric unit, the *liter* (L), is derived by cubing the *decimeter* (one-tenth of a meter) and is therefore also referred to as the cubic decimeter (dm^3). Another commonly used metric unit of volume is the *milliliter* (mL), which is derived by cubing the centimeter (1/100 of a meter). The milliliter is also referred to as the cubic centimeter (cm^3). Figure 1.10 illustrates the relationship between the liter (or dm^3) and the milliliter (or cm^3).

Density is the ratio of mass to volume. Oil floats on water, for example, because, in addition to not mixing with water, oil has a lower density than water. That is, given *equal volumes* of the two liquids, the oil will have a *smaller mass* than the water. Density is calculated using the following equation:

$$d = \frac{m}{V} \qquad \textbf{Equation 1.4}$$

Oil floating on water is a familiar demonstration of density differences.

David A. Tietz/Editorial Image, LLC

where *d, m,* and *V* denote density, mass, and volume, respectively. The SI-derived unit for density is the kilogram per cubic meter (kg/m^3). This unit is too large for most common uses, however, so grams per cubic centimeter (g/cm^3) and its equivalent,

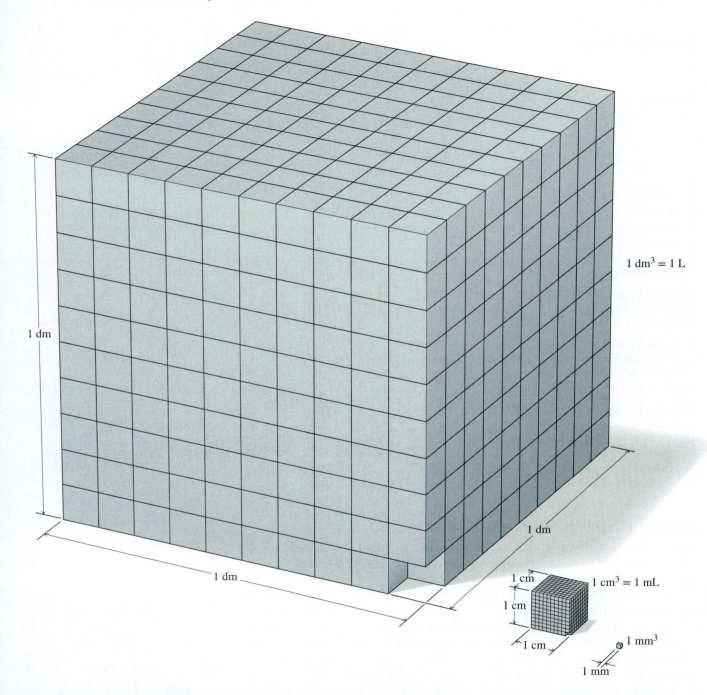

1 dm

1 dm

1 dm

$1 \text{ dm}^3 = 1 \text{ L}$

1 cm

1 cm

1 cm

$1 \text{ cm}^3 = 1 \text{ mL}$

1 mm^3

1 mm

Figure 1.10 The larger cube has 1-dm (10-cm) sides and a volume of 1 L. The next smaller cube has 1-cm (10-mm) sides and a volume of 1 cm^3 or 1 mL. The smallest cube has 1-mm sides and a volume of 1 mm^3. Note that although there are 10 cm in a decimeter, there are 1000 cm^3 in a cubic decimeter. This figure is drawn to scale to give you a sense of the actual dimensions of liters and cubic centimeters.

grams per milliliter (g/mL), are used to express the densities of most solids and liquids. Water, for example, has a density of 1.00 g/cm^3 at 4°C. Because gas densities generally are very low, we typically express them in units of grams per liter (g/L):

$$1 \text{ g/cm}^3 = 1 \text{ g/mL} = 1000 \text{ kg/m}^3$$

$$1 \text{ g/L} = 0.001 \text{ g/mL}$$

Sample Problem 1.3 illustrates density calculations.

SAMPLE PROBLEM 1.3

Ice cubes float in a glass of water because solid water is less dense than liquid water. (a) Calculate the density of ice given that, at 0°C, a cube that is 2.0 cm on each side has a mass of 7.36 g, and (b) determine the volume occupied by 23 g of ice at 0°C.

Strategy (a) Determine density by dividing mass by volume (Equation 1.4), and (b) use the calculated density to determine the volume occupied by the given mass.

Setup (a) We are given the mass of the ice cube, but we must calculate its volume from the dimensions given. The volume of the ice cube is $(2.0 \text{ cm})^3$, or 8.0 cm^3. (b) Rearranging Equation 1.4 to solve for volume gives $V = m/d$.

Solution (a) $d = \dfrac{7.36 \text{ g}}{8.0 \text{ cm}^3} = 0.92 \text{ g/cm}^3$ or 0.92 g/mL (b) $V = \dfrac{23 \text{ g}}{0.92 \text{ g/cm}^3} = 25 \text{ cm}^3$ or 25 mL

THINK ABOUT IT

For a sample with a density *less* than 1 g/cm^3, the number of cubic centimeters should be *greater* than the number of grams. In this case, 25 (cm^3) > 23 (g).

Practice Problem ATTEMPT Given that 25.0 mL of mercury has a mass of 340.0 g, calculate (a) the density of mercury and (b) the volume of 155 g of mercury.

Practice Problem BUILD Calculate (a) the density of a solid substance if a cube measuring 2.33 cm on one side has a mass of 117 g and (b) the mass of a cube of the same substance measuring 7.41 cm on one side.

Practice Problem CONCEPTUALIZE Using the picture of the graduated cylinder and its contents, arrange the following in order of increasing density: blue liquid, pink liquid, yellow liquid, grey solid, blue solid, green solid.

The following box illustrates the importance of using units carefully in scientific work.

Why Are Units So Important?

On December 11, 1998, NASA launched the 125-million-dollar Mars Climate Orbiter, which was intended to be the Red Planet's first weather satellite. After a 416-million-mile (mi) journey, the spacecraft was supposed to go into Mars's orbit on September 23, 1999. Instead, it entered Mars's atmosphere about 100 km (62 mi) lower than planned and was destroyed by heat. Mission controllers later determined that the spacecraft was lost because English measurement units were not converted to metric units in the navigation software.

Engineers at Lockheed Martin Corporation, who built the spacecraft, specified its thrust in pounds, which is an English unit of force. Scientists at NASA's Jet Propulsion Laboratory, on the other hand, who were responsible for deployment, had assumed that the thrust data they were given were expressed in *newtons,* a metric unit. To carry out the conversion between pound and newton, we would start with 1 lb = 0.4536 kg and, from Newton's second law of motion,

force = (mass)(acceleration) = $(0.4536 \text{ kg})(9.81 \text{ m/s}^2)$
= $4.45 \text{ kg} \cdot \text{m/s}^2 = 4.45 \text{ N}$

because 1 newton (N) = 1 kg · m/s^2. Therefore, instead of converting 1 lb of *force* to 4.45 N, the scientists treated it as a force of 1 N. The considerably smaller engine thrust employed because of the engineers' failure to convert from English to metric units resulted in a lower orbit and the ultimate destruction of the spacecraft.

Commenting on the failure of the Mars mission, one scientist said, "This is going to be the cautionary tale that will be embedded into introduction to the metric system in elementary school, high school, and college science courses until the end of time."

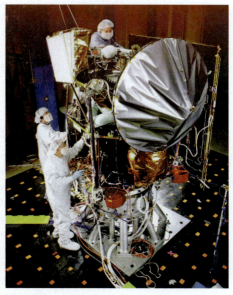

Mars Climate Orbiter during preflight tests.
Source: *NASA Image Collection/Alamy Stock Photo*

CHECKPOINT – SECTION 1.3   Scientific Measurement

1.3.1 The coldest temperature ever recorded on Earth was −128.6°F (recorded at Vostok Station, Antarctica, on July 21, 1983). Express this temperature in degrees Celsius and in kelvins.

   a) −89.2°C, −89.2 K   d) −173.9°C, 99.3 K

   b) −289.1°C, −15.9 K   e) −7.0°C, 266.2 K

   c) −89.2°C, 183.9 K

1.3.2 What is the density of an object that has a volume of 34.2 cm^3 and a mass of 19.6 g?

   a) 0.573 g/cm^3 d) 53.8 g/cm^3

   b) 1.74 g/cm^3 e) 14.6 g/cm^3

   c) 670 g/cm^3

1.3.3 A sample of water is heated from room temperature to just below the boiling point. The overall change in temperature is 72°C. Express this temperature change in kelvins.

   a) 345 K   d) 201 K

   b) 72 K   e) 273 K

   c) 0 K

1.3.4 Given that the density of gold is 19.3 g/cm^3, calculate the volume (in cm^3) of a gold nugget with a mass of 5.98 g.

   a) 3.23 cm^3 d) 0.310 cm^3

   b) 5.98 cm^3 e) 13.3 cm^3

   c) 115 cm^3

1.4   The Properties of Matter

Substances are identified by their properties as well as by their composition. Properties of a substance may be *quantitative* (measured and expressed with a number) or *qualitative* (not requiring explicit measurement).

Physical Properties

Color, melting point, boiling point, and physical state are all physical properties. A *physical property* is one that can be observed and measured without changing the *identity* of a substance. For example, we can determine the melting point of ice by heating a block of ice and measuring the temperature at which the ice is converted to water. Liquid water differs from ice in appearance but not in composition; both liquid water and ice are H_2O. Melting is a *physical change:* one in which the state of matter changes, but the identity of the matter does not change. We can recover the original ice by cooling the water until it freezes. Therefore, the melting point of a substance is a *physical* property. Similarly, when we say that nitrogen dioxide gas is brown, we are referring to the physical property of color.

Chemical Properties

The statement "Hydrogen gas burns in oxygen gas to form water" describes a *chemical property* of hydrogen, because to observe this property we must carry out a *chemical change*—burning in oxygen (combustion), in this case. After a chemical change, the original substance (hydrogen gas in this case) will no longer exist. What remains is a different substance (water, in this case). We *cannot* recover the hydrogen gas from the water by means of a physical process, such as boiling or freezing.

     Every time we bake cookies, we bring about a chemical change. When heated, the sodium bicarbonate (baking soda) in cookie dough undergoes a chemical change that produces carbon dioxide gas. The gas forms numerous little bubbles in the dough during the baking process, causing the cookies to "rise." Once the cookies are baked, we cannot recover the sodium bicarbonate by cooling the cookies, or by *any* physical

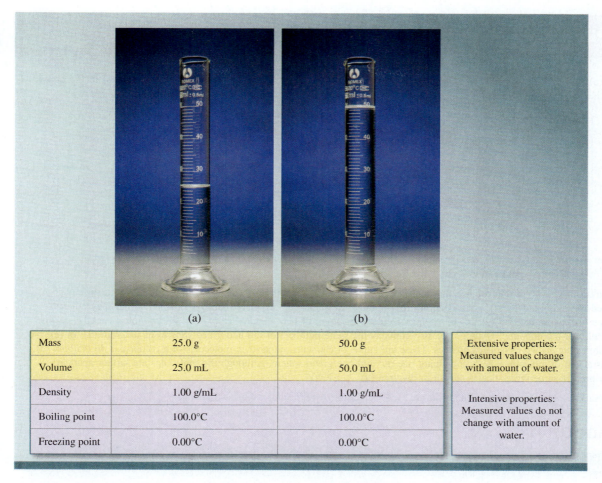

	(a)	(b)	
Mass	25.0 g	50.0 g	Extensive properties: Measured values change with amount of water.
Volume	25.0 mL	50.0 mL	
Density	1.00 g/mL	1.00 g/mL	Intensive properties: Measured values do not change with amount of water.
Boiling point	100.0°C	100.0°C	
Freezing point	0.00°C	0.00°C	

Figure 1.11 Some extensive properties (mass and volume) and intensive properties (density, boiling point, and freezing point) of water. The measured values of the extensive properties depend on the amount of water. The measured values of the intensive properties are independent of the amount of water.

(Photos): *H.S. Photos/Alamy Stock Photo*

process. When we eat the cookies, we cause further chemical changes that occur during digestion and metabolism.

Extensive and Intensive Properties

All properties of matter are either *extensive* or *intensive*. The measured value of an **extensive property** depends on the amount of matter. *Mass* is an extensive property. More matter means more mass. Values of the same extensive property can be added together. For example, two gold nuggets will have a combined mass that is the sum of the masses of each nugget, and the length of two city buses is the sum of their individual lengths. The value of an extensive property depends on the amount of matter.

The value of an **intensive property** does *not* depend on the amount of matter. *Density* and *temperature* are intensive properties. Suppose that we have two beakers of water at the same temperature and we combine them to make a single quantity of water in a larger beaker. The density and the temperature of the water in the larger combined quantity will be the same as they were in the two separate beakers. Unlike mass and length, which are additive, temperature, density, and other intensive properties are not additive. Figure 1.11 illustrates some of the extensive and intensive properties of water.

Sample Problem 1.4 shows you how to differentiate chemical and physical processes.

SAMPLE PROBLEM   1.4

The diagram in (a) shows a compound made up of atoms of two elements (represented by the green and red spheres) in the liquid state. Which of the diagrams in (b) to (d) represent a physical change, and which diagrams represent a chemical change?

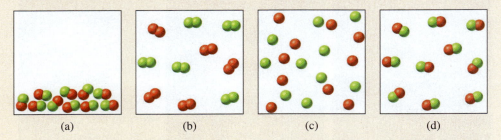

(a)   (b)   (c)   (d)

Strategy We review the discussion of physical and chemical changes. A physical change does not change the *identity* of a substance, whereas a chemical change *does* change the identity of a substance.

Setup The diagram in (a) shows a substance that consists of molecules of a compound, each of which contains two different atoms, represented by green and red spheres. Diagram (b) contains the same number of red and green spheres, but they are not arranged the same way as in diagram (a). In (b), each molecule is made up of two identical atoms. These are molecules of *elements*, rather than molecules of a compound. Diagram (c) also contains the same numbers of red and green spheres as diagram (a). In (c), however, all the atoms are shown as isolated spheres. These are atoms of elements, rather than molecules of a compound. In diagram (d), the spheres are arranged in molecules, each containing one red and one green sphere. Although the molecules are farther apart in diagram (d), they are the same molecules as shown in diagram (a).

Solution Diagrams (b) and (c) represent chemical changes. Diagram (d) represents a physical change.

THINK ABOUT IT

A chemical change changes the *identity* of matter. A physical change does not.

Practice Problem **A**TTEMPT Which of the following processes is a physical change? (a) evaporation of water; (b) combination of hydrogen and oxygen gas to produce water; (c) dissolution of sugar in water; (d) separation of sodium chloride (table salt) into its constituent elements, sodium and chlorine; (e) combustion of sugar to produce carbon dioxide and water.

Practice Problem **B**UILD The diagram on the left shows a system prior to a process taking place. Which of the other diagrams [(i) to (iv)] could represent the system after a *physical* process; which could represent the system after a *chemical* process; and which could not represent either?

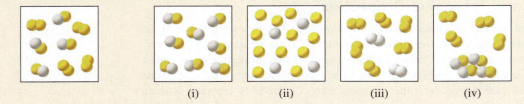

(i)   (ii)   (iii)   (iv)

Practice Problem **C**ONCEPTUALIZE The diagram on the right represents the result of a process. Which of the diagrams [(i) to (iii)] could represent the starting material if the process were physical, and which could represent the starting material if the change were chemical?

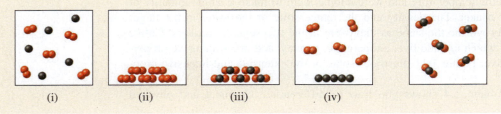

(i)   (ii)   (iii)   (iv)

CHECKPOINT – SECTION 1.4   The Properties of Matter

1.4.1   Which of the following [(a)–(f)] represents a physical change? (Select all that apply.)

1.4.2   Which of the following [(a)–(f)] represents a chemical change? (Select all that apply.)

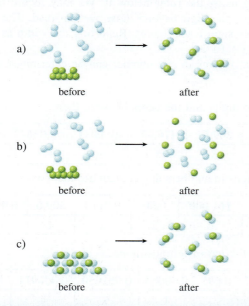

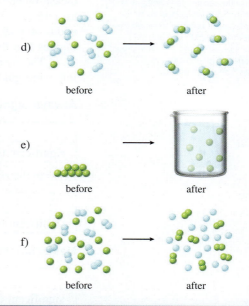

1.5   Uncertainty in Measurement

Chemistry makes use of two types of numbers: exact and inexact. *Exact* numbers include numbers with defined values, such as 2.54 in the definition 1 inch (in) = 2.54 cm, 1000 in the definition 1 kg = 1000 g, and 12 in the definition 1 dozen = 12 objects. (The number 1 in each of these definitions is also an exact number.) Exact numbers also include those that are obtained by counting. Numbers measured by any method other than counting are *inexact*.

Measured numbers are inexact because of the measuring devices that are used, the individuals who use them, or both. For example, a ruler that is poorly calibrated will result in measurements that are in error—no matter how carefully it is used. Another ruler may be calibrated properly but have insufficient resolution for the necessary measurement. Finally, whether or not an instrument is properly calibrated or has sufficient resolution, there are unavoidable differences in how different people see and interpret measurements.

Significant Figures

An inexact number must be reported in such a way as to indicate the uncertainty in its value. This is done using significant figures. *Significant figures* are the *meaningful digits* in a reported number. Consider the measurement of the USB plug in Figure 1.12 using the ruler above it. The plug's width is slightly greater than 1 cm. We may record the width as 1.2 cm, but because there are no gradations between 1 and 2 cm on this ruler, we are *estimating* the second digit. Although we are certain about the 1 in 1.2, we are *not* certain about the 2. The last digit in a measured number is referred to as the *uncertain digit;* and the uncertainty associated with a measured number is generally considered to be ±1 in the place of the last digit. Thus, when we report the width of the USB plug to be 1.2 cm, we are implying that its width is 1.2 ± 0.1 cm, meaning

Figure 1.12 The width we report for the USB plug depends on which ruler we use to measure it.

Mega Pixel/Shutterstock

that its actual width may be as low as 1.1 cm or as high as 1.3 cm. Each of the digits in a measured number, including the uncertain digit, is a significant figure. The reported width of the USB plug, 1.2 cm, contains *two* significant figures.

A ruler with millimeter gradations would enable us to be certain about the second digit in this measurement and to estimate a third digit. Now consider the measurement of the USB plug using the ruler below it. We may record the width as 1.15 cm. Again, we estimate one digit beyond those we can read. The reported width of 1.15 cm contains *three* significant figures. Reporting the width as 1.15 cm implies that the width is 1.15 ± 0.01 cm.

The number of significant figures in any number can be determined using the following guidelines:

Always Significant   Nonzero digits and the zeros between them:

	137.1	209.51	410.05	10.0011	0.036	0.00501
significant figures	4	5	5	6	2	3

Zeros to the *right* of nonzero digits in numbers that contain decimal points:

	8.300	161.000	0.50	0.0113	309.0	0.0052500
significant figures	4	6	2	3	4	5

Never Significant   Zeros to the *left* of leftmost nonzero digit:

	0.00137	0.695	0.00008	0.051050	0.006011	0.00090
significant figures	3	3	1	5	4	2

Sometimes Significant   Zeros to the *right* of the rightmost nonzero digit in a number that does *not* contain a decimal point may or may not be considered significant, depending on circumstance. For example, the number 1000 may have anywhere from one to four significant figures. Without additional information, it is not possible to know. To avoid ambiguity in such cases, it is best to express such numbers using scientific notation [▶ Appendix 1].

	1×10^3	1.0×10^3	1.00×10^3	1.000×10^3
significant figures	1	2	3	4

Sample Problem 1.5 lets you practice determining the number of significant figures in a number.

SAMPLE PROBLEM (1.5)

Determine the number of significant figures in the following measurements: (a) 443 cm, (b) 15.03 g, (c) 0.0356 kg, (d) 3.000×10^{-7} L, (e) 50 mL, (f) 0.9550 m.

Strategy   All nonzero digits are significant, so the goal will be to determine which of the zeros is significant.

Setup   Zeros are significant if they appear between nonzero digits or if they appear after a nonzero digit in a number that contains a decimal point. Zeros may or may not be significant if they appear to the right of the last nonzero digit in a number that does not contain a decimal point.

Solution   (a) 3; (b) 4; (c) 3; (d) 4; (e) 1 or 2, an ambiguous case; (f) 4.

THINK ABOUT IT

Be sure that you have identified zeros correctly as either significant or not significant. They are significant in (b), (d), and (f); they are not significant in (c); and it is not possible to tell in (e).

Practice Problem Ⓐ**TTEMPT** Determine the number of significant figures in the following measurements: (a) 1129 m, (b) 0.0003 kg, (c) 1.094 cm, (d) 3.5×10^{12} atoms, (e) 150 mL, (f) 9.550 km.

Practice Problem Ⓑ**UILD** For each of the following numbers, determine the number of significant figures it contains, rewrite it without using scientific notation, and determine the number of significant figures in the result. (a) 3.050×10^{-4}, (b) 4.3200×10^{2}, (c) 8.001×10^{-7}, (d) 2.006080×10^{5}, (e) 1.503×10^{-5}, (f) 6.07510×10^{4}.

 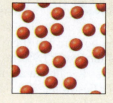

Practice Problem Ⓒ**ONCEPTUALIZE** Report the number of colored objects contained within each square and, in each case, indicate the number of significant figures in the number you report.

Calculations with Measured Numbers

Because we often use one or more measured numbers to calculate a desired result, a second set of guidelines specifies how to handle significant figures in calculations.

1. In addition and subtraction, the answer cannot have more digits to the right of the decimal point than the original number with the smallest number of digits to the right of the decimal point. For example:

$$
\begin{array}{ll}
102.50 & \leftarrow \text{two digits after the decimal point} \\
+\,0.231 & \leftarrow \text{three digits after the decimal point} \\
\hline
102.731 & \leftarrow \text{round to } 102.73
\end{array}
$$

$$
\begin{array}{ll}
143.29 & \leftarrow \text{two digits after the decimal point} \\
-\,20.1 & \leftarrow \text{one digit after the decimal point} \\
\hline
123.19 & \leftarrow \text{round to } 123.2
\end{array}
$$

The rounding procedure works as follows. Suppose we want to round 102.13 and 54.86 each to one digit to the right of the decimal point. To begin, we look at the digit(s) that will be dropped. If the leftmost digit to be dropped is less than 5, as in 102.13, we *round down* (to 102.1), meaning that we simply drop the digit(s). If the leftmost digit to be dropped is equal to or greater than 5, as in 54.86, we *round up* (to 54.9), meaning that we add 1 to the preceding digit.

2. In multiplication and division, the number of significant figures in the final product or quotient is determined by the original number that has the smallest number of significant figures. The following examples illustrate this rule:

$1.4 \times 8.011 = 11.2154$ $\leftarrow$ round to 11 (limited by 1.4 to *two* significant figures)

$\dfrac{11.57}{305.88} = 0.037825290964$ $\leftarrow$ round to 0.03783 (limited by 11.57 to *four* significant figures)

3. *Exact numbers* can be considered to have an infinite number of significant figures and do not limit the number of significant figures in a calculated result. For example, a penny minted after 1982 has a mass of 2.5 g. If we have three such pennies, the total mass is

$$3 \times 2.5\ g = 7.5\ g$$

The answer should *not* be rounded to one significant figure because 3 is an *exact* number.

Student Note: Note that it is the number of pennies (3), not the mass, that is an exact number.

What's Significant About Significant Figures?

The rules regarding significant figures in calculations may seem arbitrary, but their purpose is to ensure that scientific information is not reported with more certainty than is appropriate.

Consider the following measurements of the dimensions of a box. The length and width of the box are 10.10 cm and 10.15 cm, respectively. The depth of the box is 1.95 cm.

Recall that the implied uncertainty in a measured number lies in the *last* significant figure, and that it is generally considered to be ± 1 (in the last digit). Therefore, the uncertainty implied by each of these measured numbers is ± 0.01 cm. We can express this uncertainty as a percentage of each number as follows:

$$\frac{0.01 \text{ cm}}{10.10 \text{ cm}} \times 100\% \approx 0.099\%$$

$$\frac{0.01 \text{ cm}}{10.15 \text{ cm}} \times 100\% \approx 0.099\%$$

$$\frac{0.01 \text{ cm}}{1.95 \text{ cm}} \times 100\% \approx 0.51\%$$

Multiplying these dimensions to calculate the volume of the box gives:

$$10.10 \text{ cm} \times 10.15 \text{ cm} \times 1.95 \text{ cm} = 199.90425 \text{ cm}^3$$

The implied uncertainty in the resulting number, expressed as a percentage, is

$$\frac{0.00001 \text{ cm}^3}{199.90425 \text{ cm}^3} \times 100\% \approx 5.0 \times 10^{-6}\%$$

This implied uncertainty is far too small. Our result cannot be *more* certain than the numbers that were used to calculate it. Rounding to just *three* significant figures, as per the rule for multiplication, gives a volume of $2.00 \times 10^2 \text{ cm}^3$, with an implied uncertainty of

$$\frac{1 \text{ cm}^3}{2.00 \times 10^2 \text{ cm}^3} \times 100\% \approx 0.50\%$$

This uncertainty is comparable to that of the box depth—the *least* certain of the measured dimensions.

---

4. In calculations with multiple steps, rounding the result of each step can result in "rounding error." Consider the following two-step calculation:

First step: $A \times B = C$
Second step: $C \times D = E$

Suppose that $A = 3.66$, $B = 8.45$, and $D = 2.11$. The value of E depends on whether we round the value of C prior to using it in the second step of the calculation (Method 1) or not (Method 2).

Method 1	Method 2
$C = 3.66 \times 8.45 = 30.9$	$C = 3.66 \times 8.45 = 30.93$
$E = 30.9 \times 2.11 = 65.2$	$E = 30.93 \times 2.11 = 65.3$

In general, it is best to retain at least one extra digit until the end of a multistep calculation, as shown by Method 2, to minimize rounding error.

Sample Problems 1.6 and 1.7 show how significant figures are handled in arithmetic operations.

SAMPLE PROBLEM   1.6

Perform the following arithmetic operations and report the result to the proper number of significant figures: (a) 317.5 mL + 0.675 mL, (b) 47.80 L − 2.075 L, (c) 13.5 g ÷ 45.18 L, (d) 6.25 cm × 1.175 cm, (e) 5.46×10^2 g + 4.991×10^3 g.

Strategy Apply the rules for significant figures in calculations, and round each answer to the appropriate number of digits.

Setup (a) The answer will contain one digit to the right of the decimal point to match 317.5, which has the fewest digits to the right of the decimal point. (b) The answer will contain two digits to the right of the decimal point to match 47.80. (c) The answer will contain three significant figures to match 13.5, which has the fewest number of significant figures in the calculation. (d) The answer will contain three significant figures to match 6.25. (e) To add numbers expressed in scientific notation, first write both numbers to the same power of 10. That is, $4.991 \times 10^3 = 49.91 \times 10^2$, so the answer will contain two digits to the right of the decimal point (when multiplied by 10^2) to match both 5.46 and 49.91.

Solution

(a) 317.5 mL
    + 0.675 mL
    ―――――――
     318.175 mL   ← round to 318.2 mL

(b) 47.80 mL
    −2.075 mL
    ―――――――
     45.725 L   ← round to 45.73 L

(c) $\dfrac{13.5\ g}{45.18\ L} = 0.298804781$ g/L   ← round to 0.299 g/L

(d) 6.25 cm × 1.175 cm = 7.34375 cm² ← round to 7.34 cm²

(e) 5.46×10^2 g
     +49.91×10^2 g
     ――――――――――――
      55.37×10^2 g = 5.537×10^3 g

THINK ABOUT IT

It may look as though the rule of addition has been violated in part (e) because the final answer (5.537×10^3 g) has three places past the decimal point, not two. However, the rule was applied to get the answer 55.37×10^2 g, which has *four* significant figures. Changing the answer to correct scientific notation doesn't change the number of significant figures, but in this case it changes the number of places past the decimal point.

Practice Problem **TTEMPT** Perform the following arithmetic operations, and report the result to the proper number of significant figures: (a) 105.5 L + 10.65 L, (b) 81.058 m − 0.35 m, (c) 3.801×10^{21} atoms + 1.228×10^{19} atoms, (d) 1.255 dm × 25 dm, (e) 139 g ÷ 275.55 mL.

Practice Problem **B**UILD Perform the following arithmetic operations, and report the result to the proper number of significant figures: (a) 1.0267 cm × 2.508 cm × 12.599 cm, (b) 15.0 kg ÷ 0.036 m³, (c) 1.113×10^{10} kg − 1.050×10^9 kg, (d) 25.75 mL + 15.00 mL, (e) 46 cm³ + 180.5 cm³.

Practice Problem **C**ONCEPTUALIZE A citrus dealer in Florida sells boxes of 100 oranges at a roadside stand. The boxes routinely are packed with one to three extra oranges to help ensure that customers are happy with their purchases. The average weight of an orange is 7.2 ounces, and the average weight of the boxes in which the oranges are packed is 3.2 pounds. Determine the total weight of five of these 100-orange boxes.

SAMPLE PROBLEM   1.7

An empty container with a volume of 9.850×10^2 cm³ is weighed and found to have a mass of 124.6 g. The container is filled with a gas and reweighed. The mass of the container and the gas is 126.5 g. Determine the density of the gas to the appropriate number of significant figures.

Strategy This problem requires two steps: subtraction to determine the mass of the gas, and division to determine its density. Apply the corresponding rule regarding significant figures to each step.

Setup In the subtraction of the container mass from the combined mass of the container and the gas, the result can have only one place past the decimal point: 126.5 g − 124.6 g = 1.9 g. Thus, in the division of the mass of the gas by the volume of the container, the result can have only two significant figures.

Solution

$$
\text{mass of gas} = \begin{array}{r} 126.5\ g \\ -124.6\ g \\ \hline 1.9\ g \end{array} \quad \leftarrow \text{one place past the decimal point (two significant figures)}
$$

$$
\text{density} = \frac{1.9\ g}{9.850 \times 10^2\ cm^3} = 0.00193\ g/cm^3 \leftarrow \text{round to } 0.0019\ g/cm^3
$$

The density of the gas is 1.9×10^{-3} g/cm³.

THINK ABOUT IT

In this case, although each of the three numbers we started with has *four* significant figures, the solution has only *two* significant figures.

(Continued on next page)

Practice Problem **A**TTEMPT An empty container with a volume of 150.0 cm³ is weighed and found to have a mass of 72.5 g. The container is filled with a liquid and reweighed. The mass of the container and the liquid is 194.3 g. Determine the density of the liquid to the appropriate number of significant figures.

Practice Problem **B**UILD Another empty container with an unknown volume is weighed and found to have a mass of 81.2 g. The container is then filled with a liquid with a density of 1.015 g/cm³ and reweighed. The mass of the container and the liquid is 177.9 g. Determine the volume of the container to the appropriate number of significant figures.

Practice Problem **C**ONCEPTUALIZE A solid block of a heat-resistant plastic with a total mass of 8.172 g is dropped into a graduated cylinder of water to determine its volume. The graduated cylinder is shown before and after the plastic has been added. Use the information shown here to determine the density of the plastic. Be sure to report your answer to the appropriate number of significant figures.

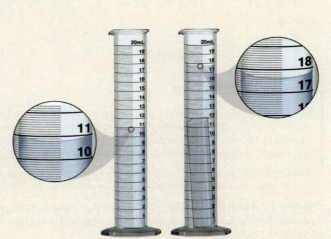

Accuracy and Precision

Accuracy and precision are two ways to gauge the quality of a set of measured numbers. Although the difference between the two terms may be subtle, it is important. *Accuracy* tells us how close a measurement is to the *true* value. *Precision* tells us how close multiple measurements of the same thing are to one another (Figure 1.13).

Suppose that three students are asked to determine the mass of an aspirin tablet. Each student weighs the aspirin tablet three times. The results (in grams) are

	Student A	Student B	Student C
	0.335	0.357	0.369
	0.331	0.375	0.373
	0.333	0.338	0.371
Average value	0.333	0.357	0.371

The true mass of the tablet is 0.370 g. Student A's results are more precise than those of student B, but neither set of results is very accurate. Student C's results are both precise (very small deviation of individual masses from the average mass) and accurate (average value very close to the true value). Figure 1.14 shows all three students' results in relation to the true mass of the tablet. Highly accurate measurements

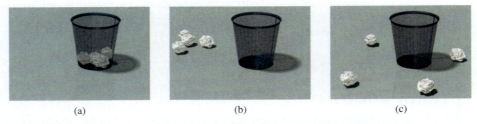

(a)   (b)   (c)

Figure 1.13 The distribution of papers shows the difference between accuracy and precision. (a) Good accuracy and good precision. (b) Poor accuracy but good precision. (c) Poor accuracy and poor precision.

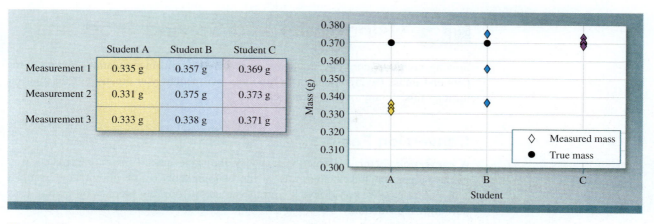

	Student A	Student B	Student C
Measurement 1	0.335 g	0.357 g	0.369 g
Measurement 2	0.331 g	0.375 g	0.373 g
Measurement 3	0.333 g	0.338 g	0.371 g

Figure 1.14 Graphing the students' data illustrates the difference between precision and accuracy. Student A's results are precise (values are close to one another) but not accurate because the average value is far from the true value. Student B's results are neither precise nor accurate. Student C's results are both precise and accurate.

are usually precise, as well, although highly precise measurements do not necessarily guarantee accurate results. For example, an improperly calibrated meterstick or a faulty balance may give precise readings that are significantly different from the correct value.

CHECKPOINT – SECTION 1.5   Uncertainty in Measurement

1.5.1 What volume of water does the graduated cylinder contain (to the proper number of significant figures)?

a) 32.2 mL

b) 30.25 mL

c) 32.5 mL

d) 32.50 mL

e) 32.500 mL

1.5.2 Which of the following is the sum of the following numbers to the correct number of significant figures?

$$3.115 + 0.2281 + 712.5 + 45 =$$

a) 760.8431

b) 760.843

c) 760.84

d) 760.8

e) 761

1.5.3 The true dependence of y on x is represented by the line. Three students measured y as a function of x and plotted their data on the graph. Which set of data has the best accuracy and which has the best precision, respectively?

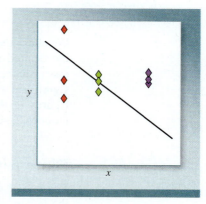

a) red, green

b) green, green

c) green, purple

d) purple, purple

e) purple, green

1.5.4 What is the result of the following calculation to the correct number of significant figures?

$$(6.266 - 6.261) \div 522.0 =$$

a) 9.5785×10^{-6}

b) 9.579×10^{-6}

c) 9.58×10^{-6}

d) 9.6×10^{-6}

e) 1×10^{-5}

1.6   Using Units and Solving Problems

Solving problems correctly in chemistry requires careful manipulation of both numbers and units. Paying attention to the units will benefit you greatly as you proceed through this, or any other, science course.

Conversion Factors

A *conversion factor* is a fraction in which the same quantity is expressed one way in the numerator and another way in the denominator. By definition, for example, 1 in = 2.54 cm. We can derive a conversion factor from this equality by writing it as the following fraction:

$$\frac{1 \text{ in}}{2.54 \text{ cm}}$$

Because the numerator and denominator express the same length, this fraction is equal to 1; as a result, we can equally well write the conversion factor as

$$\frac{2.54 \text{ cm}}{1 \text{ in}}$$

Because both forms of this conversion factor are equal to 1, we can multiply a quantity by either form without changing the value of that quantity. This is useful for changing the units in which a given quantity is expressed—something you will do often throughout this text. For instance, if we need to convert a length from inches to centimeters, we multiply the length in inches by the appropriate conversion factor.

$$12.00 \text{ in} \times \frac{2.54 \text{ cm}}{1 \text{ in}} = 30.48 \text{ cm}$$

We chose the form of the conversion factor that cancels the unit inches and produces the desired unit, centimeters. The result contains four significant figures because exact numbers, such as those obtained from definitions, do not limit the number of significant figures in the result of a calculation. Thus, the number of significant figures in the answer to this calculation is based on the number 12.00, not the number 2.54.

Dimensional Analysis—Tracking Units

The use of conversion factors in problem solving is called *dimensional analysis* or the *factor-label method*. Many problems require the use of more than one conversion factor. The conversion of 12.00 inches into meters, for example, takes two steps: one to convert inches to centimeters, which we have already demonstrated, and one to convert centimeters to meters. The additional conversion factor required is derived from the equality

$$1 \text{ m} = 100 \text{ cm}$$

and is expressed as either

$$\frac{100 \text{ cm}}{1 \text{ m}} \quad \text{or} \quad \frac{1 \text{ m}}{100 \text{ cm}}$$

We must choose the conversion factor that will introduce the unit meter and cancel the unit centimeter (i.e., the one on the right). We can set up a problem of this type as the following series of unit conversions so that it is unnecessary to calculate an intermediate answer at each step:

$$12.00 \text{ in} \times \frac{2.54 \text{ cm}}{1 \text{ in}} \times \frac{1 \text{ m}}{100 \text{ cm}} = 0.3048 \text{ m}$$

Careful tracking of units and their cancellation can be a valuable tool in checking your work. If we had accidentally used the *reciprocal* of one of the conversion factors, the resulting units would have been something other than meters. Unexpected or nonsensical units can reveal an error in your problem-solving strategy.

Sample Problem 1.8 shows how to derive conversion factors and use them to do unit conversions.

Student Note: If we had accidentally used the reciprocal of the conversion from centimeters to meters, the result would have been 3048 cm²/m, which would make no sense—both because the units are nonsensical and because the numerical result is not reasonable. You know that 12 inches is a foot and that a foot is not equal to *thousands* of meters!

SAMPLE PROBLEM   1.8

The Food and Drug Administration (FDA) recommends that dietary sodium intake be no more than 2400 mg per day. What is this mass in pounds (lb)? (1 lb = 453.6 g)

Strategy   This problem requires a two-step dimensional analysis, because we must convert milligrams to grams and then grams to pounds. Assume the number 2400 has four significant figures.

Setup   The necessary conversion factors are derived from the equalities 1 g = 1000 mg and 1 lb = 453.6 g.

$$\frac{1 \text{ g}}{1000 \text{ mg}} \quad \text{or} \quad \frac{1000 \text{ mg}}{1 \text{ g}} \quad \text{and} \quad \frac{1 \text{ lb}}{453.6 \text{ g}} \quad \text{or} \quad \frac{453.6 \text{ g}}{1 \text{ lb}}$$

From each pair of conversion factors, we select the one that will result in the proper unit cancellation.

Solution

$$2400 \text{ mg} \times \frac{1 \text{ g}}{1000 \text{ mg}} \times \frac{1 \text{ lb}}{453.6 \text{ g}} = 0.005291 \text{ lb}$$

THINK ABOUT IT

Make sure that the magnitude of the result is reasonable and that the units have canceled properly. Because pounds are much larger than milligrams, a given mass will be a much smaller number of pounds than of milligrams. If we had mistakenly multiplied by 1000 and 453.6 instead of dividing by them, the result (2400 mg × 1000 mg/g × 453.6 g/lb = 1.089×10^9 mg²/lb) would be unreasonably large—and the units would not have canceled properly.

Practice Problem A TTEMPT   The American Heart Association recommends that healthy adults limit dietary cholesterol to no more than 300 mg per day. Convert this mass of cholesterol to ounces (1 oz = 28.3459 g). Assume 300 mg has just one significant figure.

Practice Problem B UILD   An object has a mass of 24.98 oz. What is its mass in grams?

Practice Problem C ONCEPTUALIZE   The diagram contains several objects that are constructed using colored blocks and grey connectors. Note that each of the objects is essentially identical, consisting of the same number and arrangement of blocks and connectors. Give the appropriate conversion factor for each of the specified operations.

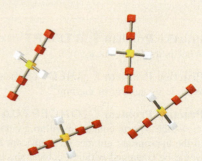

(a) We know the number of objects and wish to determine the number of red blocks.

(b) We know the number of yellow blocks and wish to determine the number of objects.

(c) We know the number of yellow blocks and wish to determine the number of white blocks.

(d) We know the number of grey connectors and wish to determine the number of yellow blocks.

Many familiar quantities require units raised to specific powers. For example, an area may be expressed in units of *length* squared (e.g., square meters, m^2, or square inches, in^2). Volumes sometimes are expressed in units of length *cubed* (e.g., cubic feet, ft^3, or cubic centimeters, cm^3). More often, though, volumes are expressed in liters (L) or milliliters (mL). It's important to remember that these are the common names given to specific units of length cubed. The liter is defined as a decimeter (dm) cubed: $1 \text{ L} = 1 \text{ dm}^3$; and the milliliter is defined as the centimeter cubed: $1 \text{ mL} = 1 \text{ cm}^3$. (See Figure 1.10.) When units are squared or cubed, special care must be taken when using them in dimensional analysis. For example, converting from cubic meters to cubic centimeters requires the following operation:

$$1 \text{ m}^3 \times \frac{100 \text{ cm}}{1 \text{ m}} \times \frac{100 \text{ cm}}{1 \text{ m}} \times \frac{100 \text{ cm}}{1 \text{ m}} = 1.00 \times 10^6 \text{ cm}^3$$

or

$$1 \text{ m}^3 \times \left(\frac{100 \text{ cm}}{1 \text{ m}}\right)^3 = 1.00 \times 10^6 \text{ cm}^3$$

Failing to raise the conversion factor to the same power as the unit itself is a common error—and one that can happen easily when the units L or mL appear—because they do not explicitly show the power 3.

Sample Problem 1.9 shows how to handle problems in which conversion factors are squared or cubed in dimensional analysis.

SAMPLE PROBLEM 1.9

An average adult has 5.2 L of blood. What is the volume of blood in cubic meters?

Strategy There are several ways to solve a problem such as this. One way is to convert liters to cubic centimeters and then cubic centimeters to cubic meters.

Setup $1 \text{ L} = 1000 \text{ cm}^3$ and $1 \text{ cm} = 1 \times 10^{-2}$ m. When a unit is raised to a power, the corresponding conversion factor must also be raised to that power in order for the units to cancel appropriately.

Solution

$$5.2 \text{ L} \times \frac{1000 \text{ cm}^3}{1 \text{ L}} \times \left(\frac{1 \times 10^{-2} \text{ m}}{1 \text{ cm}}\right)^3 = 5.2 \times 10^{-3} \text{ m}^3$$

THINK ABOUT IT

Based on the preceding conversion factors, $1 \text{ L} = 1 \times 10^{-3} \text{ m}^3$. Therefore, 5 L of blood would be equal to $5 \times 10^{-3} \text{ m}^3$, which is very close to the calculated answer.

Practice Problem A TTEMPT The density of silver is 10.5 g/cm^3. What is its density in kg/m^3?

Practice Problem B UILD The density of mercury is 13.6 g/cm^3. What is its density in mg/mm^3?

Practice Problem C ONCEPTUALIZE Each diagram [(i) or (ii)] shows the objects contained within a cubical space. In each case, determine to the appropriate number of significant figures the number of objects that would be contained within a cubical space in which the length of the cube's edge is exactly five times that of the cube shown in the diagram.

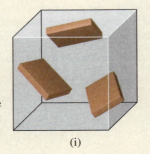
(i)

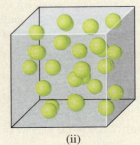
(ii)

CHECKPOINT – SECTION 1.6   Using Units and Solving Problems

1.6.1 The density of lithium metal is 535 kg/m³. What is this density in g/cm³?

a) 0.000535 g/cm³

b) 0.535 g/cm³

c) 0.0535 g/cm³

d) 0.54 g/cm³

e) 53.5 g/cm³

1.6.2 Convert 43.1 cm³ to liters.

a) 43.1 L

b) 43,100 L

c) 0.0431 L

d) 4310 L

e) 0.043 L

1.6.3 What is the volume of a 5.75-g object that has a density of 3.97 g/cm³?

a) 1.45 cm³

b) 0.690 cm³

c) 22.8 cm³

d) 0.0438 cm³

e) 5.75 cm³

1.6.4 How many cubic centimeters are there in a cubic meter?

a) 10

b) 100

c) 1000

d) 1×10^4

e) 1×10^6

Chapter Summary

Section 1.1

- *Chemistry* is the study of *matter* and the changes matter undergoes.

- Chemists go about research using a set of guidelines and practices known as the *scientific method,* in which observations give rise to *laws,* data give rise to *hypotheses,* hypotheses are tested with experiments, and successful hypotheses give rise to *theories,* which are further tested by experiment.

Section 1.2

- All matter exists either as a *substance* or as a mixture of substances. Substances may be *elements* (containing only one kind of atom) or *compounds* (containing two or more kinds of atoms). A *mixture* may be *homogeneous* (a solution) or *heterogeneous.* Mixtures may be separated using physical processes. Compounds can be separated into their constituent elements using chemical processes. Elements cannot be separated into simpler substances.

Section 1.3

- Scientists use a system of units referred to as the *International System of Units* or *SI units.*

- There are seven *base* SI units including the kilogram (for *mass*) and the *kelvin* (for temperature). SI units for such quantities as volume and *density* are derived from the base units.

Section 1.4

- Substances are identified by their *quantitative* (involving numbers) and *qualitative* (not involving numbers) properties.

- *Physical properties* are those that can be determined without the matter in question undergoing a chemical change. A *physical change* is one in which the identity of the matter involved does not change.

- *Chemical properties* are determined only as the result of a *chemical change,* in which the original substance is converted to a different substance. Physical and chemical properties may be *extensive* (dependent on the amount of matter) or *intensive* (independent of the amount of matter).

Section 1.5

- Measured numbers are *inexact.* Numbers obtained by counting or that are part of a definition are *exact* numbers.

- *Significant figures* are used to specify the uncertainty in a measured number or in a number calculated using measured numbers. Significant figures must be carried through calculations such that the implied uncertainty in the final answer is reasonable.

- *Accuracy* refers to how close measured numbers are to a *true* value. *Precision* refers to how close measured numbers are to *one another*.

Section 1.6

- A *conversion factor* is a fraction in which the numerator and denominator are the same quantity expressed in different units. Multiplying by a conversion factor is *unit conversion.*

- *Dimensional analysis* is a series of unit conversions used in the solution of a multistep problem.

Key Words

Key Equations

1.1 $K = °C + 273.15$	Temperature in kelvins is determined by adding 273.15 to the temperature in Celsius. Often we simply add 273, depending on the precision with which the Celsius temperature is known.
1.2 temp in $°C = (\text{temp } °F - 32°F) \times \dfrac{5°C}{9°F}$	Temperature in Fahrenheit is used to determine temperature in Celsius.
1.3 temp in $°F = \dfrac{9°F}{5°C} \times (\text{temp in } °C) + 32°F$	Temperature in Celsius is used to determine temperature in Fahrenheit.
1.4 $d = \dfrac{m}{V}$	Density is the ratio of mass to volume. For liquids and solids, densities are typically expressed in g/cm^3.

KEY SKILLS

Dimensional Analysis

Solving problems in chemistry often involves mathematical combinations of measured values and constants. A conversion factor is a fraction (equal to one) derived from an equality. For example, 1 inch is, by definition, equal to 2.54 centimeters:

$$\boxed{1 \text{ in}} = \boxed{2.54 \text{ cm}}$$

We can derive two different conversion factors from this equality:

$$\boxed{\dfrac{1 \text{ in}}{2.54 \text{ cm}}} \quad \text{or} \quad \boxed{\dfrac{2.54 \text{ cm}}{1 \text{ in}}}$$

Which fraction we use depends on what units we start with, and what units we expect our result to have. If we are converting a distance given in centimeters to inches, we multiply by the first fraction.

$$\boxed{37.6 \text{ cm}} \times \boxed{\dfrac{1 \text{ in}}{2.54 \text{ cm}}} = \boxed{14.8 \text{ in}}$$

If we are converting a distance given in inches to centimeters, we multiply by the second fraction.

$$\boxed{5.23 \text{ in}} \times \boxed{\dfrac{2.54 \text{ cm}}{1 \text{ in}}} = \boxed{13.3 \text{ cm}}$$

In each case, the units cancel to give the desired units in the result.

When a unit is raised to a power to express, for example, an area (cm^2) or a volume (cm^3), the conversion factor must be raised to the same power. For example, converting an area expressed in square centimeters to square inches requires that we square the conversion factor; converting a volume expressed in cubic centimeters to cubic meters requires that we cube the conversion factor. The following individual flowcharts converting an area in cm^2 to m^2 show why this is so:

$$\boxed{48.5 \text{ cm}^2} = \boxed{48.5 \text{ cm}} \times \boxed{\text{cm}} \times \boxed{\left(\dfrac{1 \text{ in}}{2.54 \text{ cm}}\right)^2} = \boxed{\dfrac{1 \text{ in}}{2.54 \text{ cm}}} \times \boxed{\dfrac{1 \text{ in}}{2.54 \text{ cm}}}$$

$$\boxed{48.5 \text{ cm}} \times \boxed{\text{cm}} \times \boxed{\dfrac{1 \text{ in}}{2.54 \text{ cm}}} \times \boxed{\dfrac{1 \text{ in}}{2.54 \text{ cm}}} = \boxed{7.52 \text{ in}^2}$$

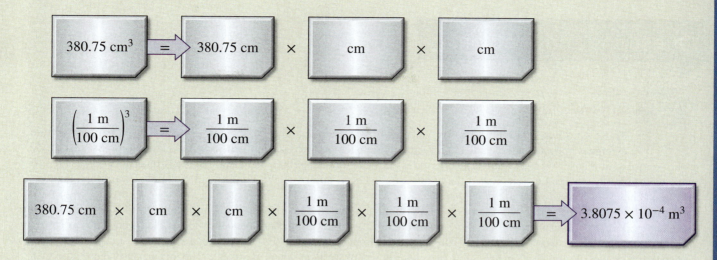

Failure to raise the conversion factor to the appropriate power would result in units not canceling properly.

Often the solution to a problem requires several different conversions, which can be combined on a single line. For example: If we know that a 157-lb athlete running at 7.09 miles per hour consumes 55.8 cm³ of oxygen per kilogram of body weight for every minute spent running, we can calculate how many liters of oxygen this athlete consumes by running 10.5 miles (1 kg = 2.2046 lb, 1 L = 1 dm³).

$$157 \text{ lb} \times \frac{1 \text{ kg}}{2.2046 \text{ lb}} \times \frac{1 \text{ h}}{7.09 \text{ mi}} \times \frac{55.8 \text{ cm}^3}{\text{kg} \cdot \text{min}} \times \frac{60 \text{ min}}{1 \text{ h}} \times \left(\frac{1 \text{ dm}}{10 \text{ cm}}\right)^3 \times 10.5 \text{ mi} = 353 \text{ dm}^3$$

$$353 \text{ dm}^3 = 353 \text{ L}$$

Key Skills Problems

1.1
Given that the density of gold is 19.3 g/cm³, calculate the volume (in cm³) of a gold nugget with a mass of 5.98 g.

(a) 3.23 cm³ (b) 5.98 cm³ (c) 115 cm³ (d) 0.310 cm³ (e) 13.3 cm³

1.2
The SI unit for energy is the joule (J), which is equal to the kinetic energy possessed by a 2.00-kg mass moving at 1.00 m/s. Convert this velocity to mph (1 mi = 1.609 km).

(a) 4.47×10^{-7} mph (b) 5.79×10^6 mph (c) 5.79 mph (d) 0.0373 mph (e) 2.24 mph

1.3
Determine the density of the following object in g/cm³. A cube with edge length = 0.750 m and mass = 14.56 kg.

(a) 0.0345 g/cm³ (b) 1.74 g/cm³ (c) 670 g/cm³ (d) 53.8 g/cm³ (e) 14.6 g/cm³

1.4
A 28-kg child can consume a maximum of 23 children's acetaminophen tablets in an 8-h period without exceeding the safety-limit maximum allowable dose. Given that each children's tablet contains 80 mg of acetaminophen, determine the maximum allowable dose in mg per pound of body weight for one day.

(a) 80 mg/lb (b) 90 mg/lb (c) 430 mg/lb (d) 720 mg/lb (e) 3.7 mg/lb

Questions and Problems

Applying What You've Learned

Those who describe themselves as "skeptical" about climate change sometimes posit that global temperature change is normal, and that any observed increase in temperature is simply the result of natural processes—outside the control of humans. However, there is an enormous body of climate research that clearly demonstrates otherwise. One line of inquiry that has helped to established the connection between human activity and so-called "global warming" involves what is known as *vertical structure of temperature*.

Earth's atmosphere is divided into a series of altitudinal layers: the troposphere (ground-level to 8–14.5 km), the stratosphere (top of the troposphere–50 km), the mesosphere (50–80 km), the thermosphere (80–700 km), and the exosphere (700–10,000 km). The two layers closest to the planet, the troposphere and the stratosphere, are those most important to our discussion of global climate change. The troposphere is where we live, where weather events occur, and where nearly all human activity takes place. When we burn fossil fuels, we increase the amount of CO_2 in the *troposphere*.

In 1988, atmospheric scientist V. Ramanathan, now of the Scripps Institution of Oceanography at the University of California, San Diego, proposed that global temperature change caused by the anthropogenic increase in atmospheric CO_2 could be readily distinguished from that caused by *natural* events, such as increased solar activity. Global temperature increase caused by the sun, he reasoned, would occur in both the troposphere *and* the stratosphere. Conversely, changes caused by the enhanced greenhouse effect (the result of increased atmospheric CO_2 concentration) would cause warming of the troposphere; but *cooling* of the stratosphere—because more of the heat radiating from Earth's surface would be trapped by greenhouse gases in the troposphere, thus never reaching the stratosphere. Indeed, temperature monitoring over several decades has demonstrated an *increase* in tropospheric temperature, and a *decrease* in stratospheric temperature. This is one of the observations that climate scientists refer to as a *human fingerprint* on global climate change.

The National Weather Service recently reported that the average temperature at the tropopause, the boundary between the troposphere and the stratosphere, is 51°C. (a) Convert this temperature to the Kelvin scale [◄◄ Sample Problem 1.1]. (b) Temperature data compiled since the 1950s indicate that the troposphere has warmed by 0.14°C per decade, and that the lower stratosphere has cooled by as much as 0.20°C per decade. Express these temperature changes in degrees Fahrenheit [◄◄ Sample Problem 1.2]. (c) A 1.250-L sample of air collected at sea level has a mass of 1.53 g. Determine the density of this air sample in g/L, in g/cm³, and in kg/m³ [◄◄ Sample Problem 1.7]. (Be sure to report your answer to the proper number of significant figures.) (d) In the lowest part of the troposphere, near sea level, there are roughly 4.4×10^{20} molecules per cubic inch of air. Convert this figure to molecules per cubic centimeter. [◄◄ Sample Problem 1.9].

SECTION 1.1: THE STUDY OF CHEMISTRY

Review Questions

1.1   Define the terms *chemistry* and *matter*.
1.2   Explain what is meant by the scientific method.
1.3   What is the difference between a hypothesis and a theory?

Conceptual Problems

1.4   Classify each of the following statements as a hypothesis, law, or theory. (a) All matter is composed of very small particles called atoms. (b) English Romantic poet John Keats would have started writing novels had he not died at age 25. (c) The planets in our solar system move about the sun in elliptical orbits.

1.5   Classify each of the following statements as a hypothesis, law, or theory. (a) Chimpanzees can be taught to communicate using human languages. (b) The force acting on an object is equal to its mass times its acceleration. (c) An individual with a trait that gives it a reproductive advantage will pass that trait on to its offspring, increasing the frequency of the trait in subsequent generations.

1.6   Identify the elements present in the following molecules (see Table 1.1).

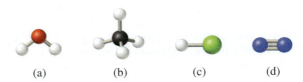

(a)   (b)   (c)   (d)

1.7   Identify the elements present in the following molecules (see Table 1.1).

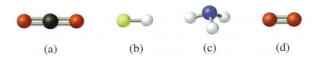

(a)   (b)   (c)   (d)

SECTION 1.2: CLASSIFICATION OF MATTER

Review Questions

1.8   Give an example for each of the following terms: (a) matter, (b) substance, (c) mixture.
1.9   Give an example of a homogeneous mixture and an example of a heterogeneous mixture.

1.10 Give an example of an element and a compound. How do elements and compounds differ?

1.11 What is the number of known elements?

Computational Problems

1.12 Give the names of the elements represented by the chemical symbols Li, F, P, Cu, As, Zn, Cl, Pt, Mg, U, Al, Si, Ne (see the table at the beginning of the book).

1.13 Give the chemical symbols for the following elements: (a) potassium, (b) tin, (c) chromium, (d) boron, (e) barium, (f) plutonium, (g) sulfur, (h) argon, (i) mercury.

1.14 Classify each of the following substances as an element or a compound: (a) hydrogen, (b) water, (c) gold, (d) sugar.

1.15 Classify each of the following as an element, a compound, a homogeneous mixture, or a heterogeneous mixture: (a) seawater, (b) helium gas, (c) sodium chloride (salt), (d) a bottle of soft drink, (e) a milkshake, (f) air in a bottle, (g) concrete.

Conceptual Problems

1.16 Identify each of the diagrams shown here as a solid, liquid, gas, or mixture of two substances.

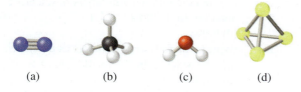

(a)   (b)   (c)   (d)

1.17 Identify each of the diagrams shown here as an element or a compound.

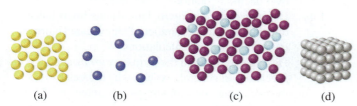

(a)   (b)   (c)   (d)

SECTION 1.3: SCIENTIFIC MEASUREMENT

Review Questions

1.18 Name the SI base units that are important in chemistry, and give the SI units for expressing the following: (a) length, (b) volume, (c) mass, (d) time, (e) temperature.

1.19 Write the numbers represented by the following prefixes: (a) mega-, (b) kilo-, (c) deci-, (d) centi-, (e) milli-, (f) micro-, (g) nano-, (h) pico-.

1.20 What units do chemists normally use for the density of liquids and solids? For the density of gas? Explain the differences.

1.21 What is the difference between mass and weight? If a person weighs 168 lb on Earth, about how much would the person weigh on the moon?

1.22 Describe the three temperature scales used in the laboratory and in everyday life: the Fahrenheit, Celsius, and Kelvin scales.

Computational Problems

1.23 Mercury is the only metal that is a liquid at room temperature. Calculate its density (in g/mL) if a 349.2-g sample of mercury occupies 25.75 mL.

1.24 The density of a particular brand of olive oil is 0.925 g/mL. Calculate the mass of 515 mL of the liquid.

1.25 Convert the following temperatures to degrees Celsius or Fahrenheit: (a) 95°F, the temperature on a hot summer day; (b) 12°F, the temperature on a cold winter day; (c) a 102°F fever; (d) a furnace operating at 1852°F; (e) −273.15°C (theoretically the lowest attainable temperature).

1.26 (a) Normally the human body can endure a temperature of 105°F for only short periods of time without permanent damage to the brain and other vital organs. What is this temperature in degrees Celsius? (b) Ethylene glycol is a liquid organic compound that is used as an antifreeze in car radiators. It freezes at −11.5°C. Calculate its freezing temperature in degrees Fahrenheit. (c) The temperature on the surface of the sun is about 6300°C. What is this temperature in degrees Fahrenheit?

1.27 The density of water at 84°C is 0.969 g/mL. What is the volume of 67.0 g of water at this temperature?

1.28 The density of copper (Cu) is 8.96 g/cm^3 at 25°C. What is the mass of a piece of copper that occupies 35.3 cm^3 at this temperature?

1.29 Convert the following temperatures to kelvin: (a) 115.21°C, the melting point of sulfur; (b) 37°C, the normal body temperature; (c) 357°C, the boiling point of mercury.

1.30 Convert the following temperatures to degrees Celsius: (a) 77 K, the boiling point of liquid nitrogen, (b) 4.22 K, the boiling point of liquid helium, (c) 600.61 K, the melting point of lead.

Conceptual Problems

1.31 Which of the following illustrations best represents the measurement of the temperature of boiling water using Celsius (blue) and Kelvin (red) scales? Explain.

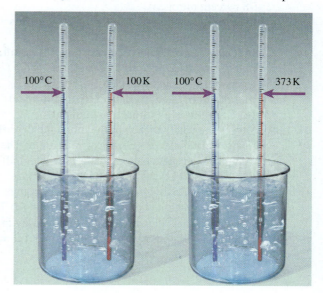

1.32   The diagram shows balls of aluminum foil dropped into water. In one case, the ball floats on the water; in the other case, it has been squeezed into a smaller ball and sinks. What does this indicate about the relative densities of the two aluminum-foil balls? Does this mean that the density of aluminum is not constant? Explain.

SECTION 1.4: THE PROPERTIES OF MATTER

Review Questions

1.33   What is the difference between qualitative data and quantitative data?

1.34   Using examples, explain the difference between a physical property and a chemical property.

1.35   How does an intensive property differ from an extensive property?

1.36   Determine which of the following properties are intensive and which are extensive: (a) length, (b) volume, (c) temperature, (d) mass.

Conceptual Problems

1.37   Classify the following as qualitative or quantitative statements, giving your reasons. (a) Food tastes better with salt and pepper. (b) The world's deepest lake, Lake Baikal, has a maximum depth of 1642 m. (c) The most abundant element in the universe is hydrogen, which accounts for about 75 percent of all matter. (d) Water fluoridation helps prevent tooth decay. (e) Earth's atmosphere is approximately 78 percent nitrogen.

1.38   Determine whether the following statements describe chemical or physical properties: (a) Oxygen gas supports combustion. (b) Fertilizers help to increase agricultural production. (c) Water boils below 100°C on top of a mountain. (d) Lead is denser than aluminum. (e) Uranium is a radioactive element.

1.39   Determine whether each of the following describes a physical change or a chemical change: (a) The helium gas inside a balloon tends to leak out after a few hours. (b) A flashlight beam slowly gets dimmer and finally goes out. (c) Frozen orange juice is reconstituted by adding water to it. (d) The growth of plants depends on the sun's energy in a process called photosynthesis. (e) A spoonful of salt dissolves in a bowl of soup.

1.40   A student pours 44.3 g of water at 10°C into a beaker containing 115.2 g of water at 10°C. What are the final mass, temperature, and density of the combined water? The density of water at 10°C is 1.00 g/mL.

1.41   A 37.2-g sample of lead (Pb) pellets at 20°C is mixed with a 62.7-g sample of lead pellets at the same temperature. What are the final mass, temperature, and density of the combined sample? The density of Pb at 20°C is 11.35 g/cm³.

SECTION 1.5: UNCERTAINTY IN MEASUREMENT

Review Questions

1.42   Comment on whether each of the following statements represents an exact number: (a) 50,247 tickets were sold at a sporting event, (b) 509.2 mL of water was used to make a birthday cake, (c) 3 dozen eggs were used to make a breakfast, (d) 0.41 g of oxygen was inhaled in each breath, (e) One inch is equal to 2.54 cm.

1.43   What is the advantage of using scientific notation over decimal notation?

1.44   Define *significant figure*. Discuss the importance of using the proper number of significant figures in measurements and calculations.

1.45   Distinguish between the terms *accuracy* and *precision*. In general, explain why a precise measurement does not always guarantee an accurate result.

Computational Problems

1.46   Express the following numbers in scientific notation: (a) 0.000000027, (b) 356, (c) 47,764, (d) 0.096.

1.47   Express the following numbers as decimals: (a) 1.52×10^{-2}, (b) 7.78×10^{-8}, (c) 1×10^{-6}, (d) 1.6001×10^{-3}.

1.48   Express the answers to the following calculations in scientific notation:
(a) $145.75 + (2.3 \times 10^{-1})$
(b) $79,500 \div (2.5 \times 10^{2})$
(c) $(7.0 \times 10^{-3}) - (8.0 \times 10^{-4})$
(d) $(1.0 \times 10^{4}) - (9.9 \times 10^{6})$

1.49   Express the answers to the following calculations in scientific notation:
(a) $0.0095 + (8.5 \times 10^{-3})$
(b) $653 \div (5.75 \times 10^{-8})$
(c) $850,000 - (9.0 \times 10^{5})$
(d) $(3.6 \times 10^{-4}) \times (3.6 \times 10^{6})$

1.50   Determine the number of significant figures in each of the following measurements: (a) 4867 mi, (b) 56 mL, (c) 60,104 tons, (d) 2900 g, (e) 40.2 g/cm³, (f) 0.0000003 cm, (g) 0.7 min, (h) 4.6×10^{19} atoms.

1.51   Determine the number of significant figures in each of the following measurements: (a) 0.006 L,

(b) 0.0605 dm, (c) 60.5 mg, (d) 605.5 cm^2, (e) 9.60×10^3 g, (f) 6 kg, (g) 60 m.

1.52   Carry out the following operations as if they were calculations of experimental results, and express each answer in the correct units with the correct number of significant figures:
(a) 5.6792 m + 0.6 m + 4.33 m
(b) 3.70 g − 2.9133 g
(c) 4.51 cm × 3.6666 cm

1.53   Carry out the following operations as if they were calculations of experimental results, and express each answer in the correct units with the correct number of significant figures:
(a) 7.310 km ÷ 5.70 km
(b) $(3.26 \times 10^{-3}$ mg$) - (7.88 \times 10^{-5}$ mg$)$
(c) $(4.02 \times 10^6$ dm$) + (7.74 \times 10^7$ dm$)$

1.54   Three students (A, B, and C) are asked to determine the volume of a sample of ethanol. Each student measures the volume three times with a graduated cylinder. The results in milliliters are: A (87.1, 88.2, 87.6); B (86.9, 87.1, 87.2); C (87.6, 87.8, 87.9). The true volume is 87.0 mL. Comment on the precision and the accuracy of each student's results.

1.55   Three apprentice tailors (X, Y, and Z) are assigned the task of measuring the seam of a pair of trousers. Each one makes three measurements. The results in inches are X (31.5, 31.6, 31.4); Y (32.8, 32.3, 32.7); Z (31.9, 32.2, 32.1). The true length is 32.0 in. Comment on the precision and the accuracy of each tailor's measurements.

SECTION 1.6: USING UNITS AND SOLVING PROBLEMS

Computational Problems

1.56   Carry out the following conversions: (a) 22.6 m to decimeters, (b) 25.4 mg to kilograms, (c) 556 mL to liters, (d) 10.6 kg/m^3 to g/cm^3.

1.57   Carry out the following conversions: (a) 242 lb to milligrams, (b) 68.3 cm^3 to cubic meters, (c) 7.2 m^3 to liters, (d) 28.3 μg to pounds.

1.58   The average speed of helium at 25°C is 1255 m/s. Convert this speed to miles per hour (mph).

1.59   How many minutes are there in a solar year (365.24 days)?

1.60   How many minutes does it take light from the sun to reach Earth? (The distance from the sun to Earth is 93 million mi; the speed of light is 2.99792458×10^8 m/s.)

1.61   A slow jogger runs a mile in 13 min. Calculate the speed in (a) in/s, (b) m/min, (c) km/h (1 mi = 1609 m; 1 in = 2.54 cm).

1.62   A 6.0-ft person weighs 168 lb. Express this person's height in meters and weight in kilograms (1 lb = 453.6 g; 1 m = 3.28 ft).

1.63   In 2015, a Japanese high-speed train set a record—traveling at 374 mph during a test run near Mt. Fuji.

What is the speed in kilometers per hour (1 mi = 1609 m)? Report your answer as a whole number.

1.64   For a fighter jet to take off from the deck of an aircraft carrier, it must reach a speed of 62 m/s. Calculate the speed in miles per hour.

1.65   The "normal" lead content in human blood is about 0.40 part per million (i.e., 0.40 g of lead per million grams of blood). A value of 0.80 part per million (ppm) is considered to be dangerous. How many grams of lead are contained in 6.0×10^3 g of blood (the amount in an average adult) if the lead content is 0.62 ppm?

1.66   Carry out the following conversions: (a) 32.4 yd to centimeters, (b) 3.0×10^{10} cm/s to ft/s, (c) 1.42 light-years to miles (a light-year is an astronomical measure of distance—the distance traveled by light in a year, or 365 days; the speed of light is 3.00×10^8 m/s).

1.67   Carry out the following conversions: (a) 185 nm to meters, (b) 4.5 billion years (roughly the age of Earth) to seconds (assume 365 days in a year), (c) 71.2 cm^3 to cubic meters, (d) 88.6 m^3 to liters.

1.68   The term *heavy metal* generally refers to a metal with a density greater than 5 g/cm^3. What is this density in kg/m^3?

1.69   The density of ammonia gas under certain conditions is 0.625 g/L. Calculate its density in g/cm^3.

1.70   (a) Carbon monoxide (CO) is a poisonous gas because it binds very strongly to the oxygen carrier hemoglobin in blood. A concentration of 8.00×10^2 ppm by volume of carbon monoxide is considered lethal to humans. Calculate the volume in liters occupied by carbon monoxide in a room that measures 17.6 m long, 8.80 m wide, and 2.64 m high at this concentration. (b) Prolonged exposure to mercury (Hg) vapor can cause neurological disorders and respiratory problems. For safe air quality control, the concentration of mercury vapor must be under 0.050 mg/m^3. Convert this number to g/L. (c) The general test for type II diabetes is that the blood sugar (glucose) level should be below 120 mg per deciliter (mg/dL). Convert this number to micrograms per milliliter (μg/mL).

1.71   The average time it takes for a molecule to diffuse a distance of x cm is given by

$$t = \frac{x^2}{2D}$$

where t is the time in seconds and D is the diffusion coefficient. Given that the diffusion coefficient of glucose is 5.7×10^{-7} cm^2/s, calculate the time it would take for a glucose molecule to diffuse 10 μm, which is roughly the size of a cell.

1.72   A human brain weighs about 1 kg and contains about 10^{11} cells. Assuming that each cell is completely filled with water (density = 1 g/mL), calculate the length of one side of such a cell if it were a cube. If the cells are spread out into a thin layer that is a single cell thick, what is the surface area in square meters?

ADDITIONAL PROBLEMS

1.73 Using the appropriate number of significant figures, report the length of the blue rectangle (a) using the ruler shown above the rectangle and (b) using the ruler shown below the rectangle.

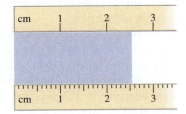

1.74 A piece of metal with a mass of 13.2 g was dropped into a graduated cylinder containing 17.00 mL of water. The graduated cylinder after the addition of the metal is shown. Determine the density of the metal to the appropriate number of significant figures.

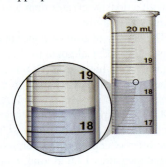

1.75 Which of the following statements describe physical properties and which describe chemical properties? (a) Iron has a tendency to rust. (b) Rainwater in industrialized regions tends to be acidic. (c) Hemoglobin molecules have a red color. (d) When a glass of water is left out in the sun, the water gradually disappears. (e) Carbon dioxide in air is converted to more complex molecules by plants during photosynthesis.

1.76 In determining the density of a rectangular metal bar, a student made the following measurements: length, 8.53 cm; width, 2.4 cm; height, 1.0 cm; mass, 52.7064 g. Calculate the density of the metal to the correct number of significant figures.

1.77 Calculate the mass of each of the following: (a) a sphere of gold with a radius of 10.0 cm (volume of a sphere with a radius r is $V = \frac{4}{3}\pi r^3$; density of gold = 19.3 g/cm^3), (b) a cube of platinum of edge length 0.040 mm (density = 21.4 g/cm^3), (c) 50.0 mL of ethanol (density = 0.798 g/mL).

1.78 A cylindrical glass tube 12.7 cm in length is filled with mercury (density = 13.6 g/mL). The mass of mercury needed to fill the tube is 105.5 g. Calculate

the inner diameter of the tube (volume of a cylinder of radius r and length h is $V = \pi r^2 h$).

1.79 The following procedure was used to determine the volume of a flask. The flask was weighed dry and then filled with water. If the masses of the empty flask and filled flask were 37.25 g and 91.03 g, respectively, and the density of water is 0.9976 g/cm^3, calculate the volume of the flask in cubic centimeters.

1.80 The speed of sound in air at room temperature is about 343 m/s. Calculate this speed in miles per hour (1 mi = 1609 m).

1.81 A piece of silver (Ag) metal weighing 194.3 g is placed in a graduated cylinder containing 242.0 mL of water. The volume of water now reads 260.5 mL. From these data calculate the density of silver.

1.82 The experiment described in Problem 1.81 is a crude but convenient way to determine the density of some solids. Describe a similar experiment that would enable you to measure the density of ice. Specifically, what would be the requirements for the liquid used in your experiment?

1.83 A lead sphere has a mass of 1.20×10^4 g, and its volume is 1.05×10^3 cm^3. Calculate the density of lead.

1.84 Lithium is the least dense metal known (density = 0.53 g/cm^3). What is the volume occupied by 3.15×10^3 g of lithium?

1.85 At what temperature does the numerical reading on a Celsius thermometer equal that on a Fahrenheit thermometer?

1.86 Suppose that a new temperature scale has been devised on which the melting point of ethanol (−117.3°C) and the boiling point of ethanol (78.3°C) are taken as 0°S and 100°S, respectively, where S is the symbol for the new temperature scale. Derive an equation relating a reading on this scale to a reading on the Celsius scale. What would this thermometer read at 25°C?

1.87 The total volume of seawater is 1.5×10^{21} L. Assume that seawater contains 3.1 percent sodium chloride by mass and that its density is 1.03 g/mL. Calculate the total mass of sodium chloride in kilograms and in tons (1 ton = 2000 lb; 1 lb = 453.6 g).

1.88 A sheet of aluminum (Al) foil has a total area of 1.000 ft^2 and a mass of 3.636 g. What is the thickness of the foil in millimeters (density of Al = 2.699 g/cm^3)?

1.89 A student is given a crucible and asked to prove whether it is made of pure platinum. She first weighs the crucible in air and then weighs it suspended in water (density = 0.9986 g/mL). The readings are 860.2 g and 820.2 g, respectively. Based on these measurements and given that the density of platinum is 21.45 g/cm^3, what should her conclusion be? (*Hint:* An object suspended in a fluid is buoyed up by the mass of the fluid displaced by the object. Neglect the buoyancy of air.)

1.90   The surface area and average depth of the Pacific Ocean are 1.8×10^8 km^2 and 3.9×10^3 m, respectively. Calculate the volume of water in the ocean in liters.

1.91   The unit "troy ounce" is often used for precious metals such as gold (Au) and platinum (Pt) (1 troy ounce = 31.103 g). (a) A gold coin weighs 2.41 troy ounces. Calculate its mass in grams. (b) Is a troy ounce heavier or lighter than an ounce (1 lb = 16 oz; 1 lb = 453.6 g)?

1.92   Osmium (Os) is the densest element known (density = 22.57 g/cm^3). Calculate the mass in pounds and in kilograms of an Os sphere 15 cm in diameter (about the size of a grapefruit) (volume of a sphere of radius r is $\frac{4}{3}\pi r^3$).

1.93   Calculate the percent error for the following measurements: (a) The density of alcohol (ethanol) is found to be 0.802 g/mL (true value = 0.798 g/mL). (b) The mass of gold in an earring is analyzed to be 0.837 g (true value = 0.864 g).

1.94   In water conservation, chemists spread a thin film of a certain inert material over the surface of water to cut down on the rate of evaporation of water in reservoirs. This technique was pioneered by Benjamin Franklin three centuries ago. Franklin found that 0.10 mL of oil could spread over the surface of water about 40 m^2 in area. Assuming that the oil forms a *monolayer,* that is, a layer that is only one molecule thick, estimate the length of each oil molecule in nanometers (1 nm = 1×10^{-9} m).

1.95   You are given a liquid. Briefly describe the steps you would take to show whether it is a pure substance or a homogeneous mixture.

1.96   A gas company in Massachusetts charges $1.30 for 15.0 ft^3 of natural gas. (a) Convert this rate to dollars per liter of gas. (b) If it takes 0.304 ft^3 of gas to boil a liter of water, starting at room temperature (25°C), how much would it cost to boil a 2.1-L kettle of water?

1.97   A 250-mL glass bottle was filled with 242 mL of water at 20°C and tightly capped. It was then left outdoors overnight, where the average temperature was −5°C. Predict what would happen. The density of water at 20°C is 0.998 g/cm^3 and that of ice at −5°C is 0.916 g/cm^3.

1.98   A bank teller is asked to assemble $1 sets of coins for his clients. Each set is made up of three quarters, one nickel, and two dimes. The masses of the coins are quarter, 5.645 g; nickel, 4.967 g; and dime, 2.316 g. What is the maximum number of sets that can be assembled from 33.871 kg of quarters, 10.432 kg of nickels, and 7.990 kg of dimes? What is the total mass (in grams) of the assembled sets of coins?

1.99   The men's world record for running a mile outdoors (set in 1999) is 3 min 43.13 s. At this rate, how long would it take to run a 2-km race (1 mi = 1609 m)?

1.100   Venus, the second closest planet to the sun, has a surface temperature of 7.3×10^2 K. Convert this temperature to degrees Celsius and degrees Fahrenheit.

1.101   Comment on whether each of the following is a homogeneous mixture or a heterogeneous mixture: (a) air in a closed bottle, (b) air over New York City.

1.102   It has been estimated that 8.0×10^4 tons of gold (Au) have been mined. Assume gold costs $1350 per troy ounce. What is the total worth of this quantity of gold? (1 troy ounce = 31.103 g)

1.103   A 1.0-mL volume of seawater contains about 4.0×10^{-12} g of gold. The total volume of ocean water is 1.5×10^{21} L. Calculate the total amount of gold (in grams) that is present in seawater and the worth of the gold in dollars (see Problem 1.102). With so much gold out there, why hasn't someone become rich by mining gold from the ocean?

1.104   Measurements show that 1.0 g of iron (Fe) contains 1.1×10^{22} Fe atoms. How many Fe atoms are in 4.9 g of Fe, which is the total amount of iron in the body of an average adult?

1.105   The thin outer layer of Earth, called the crust, contains only 0.50 percent of Earth's total mass and yet is the source of almost all the elements (the atmosphere provides elements such as oxygen, nitrogen, and a few other gases). Silicon (Si) is the second most abundant element in Earth's crust (27.2 percent by mass). Calculate the mass of silicon in kilograms in Earth's crust (mass of Earth = 5.9×10^{21} tons; 1 ton = 2000 lb; 1 lb = 453.6 g).

1.106   The radius of a copper (Cu) atom is roughly 1.3×10^{-10} m. How many times can you divide evenly a 10-cm-long piece of copper wire until it is reduced to two separate copper atoms? (Assume there are appropriate tools for this procedure and that copper atoms are lined up in a straight line, in contact with each other. Round off your answer to an integer.)

1.107   A graduated cylinder is filled to the 40.00-mL mark with a mineral oil. The masses of the cylinder before and after the addition of the mineral oil are 124.966 g and 159.446 g, respectively. In a separate experiment, a metal ball bearing of mass 18.713 g is placed in the cylinder and the cylinder is again filled to the 40.00-mL mark with the mineral oil. The combined mass of the ball bearing and mineral oil is 50.952 g. Calculate the density and radius of the ball bearing (volume of a sphere of radius r is $\frac{4}{3}\pi r^3$).

1.108   A chemist mixes two liquids A and B to form a homogeneous mixture. The densities of the liquids are 2.0514 g/mL for A and 2.6678 g/mL for B. When she drops a small object into the mixture, she finds that the object becomes suspended in the liquid; that is, it neither sinks nor floats. If the mixture is made of 41.37 percent A and 58.63 percent B by volume, what is the density of the object? Can this procedure be used in general to determine the densities of solids? What assumptions must be made in applying this method?

1.109 A chemist in the nineteenth century prepared an unknown substance. In general, do you think it would be more difficult to prove that it is an element or a compound? Explain.

Industrial Problems

1.110 Chlorine is used to disinfect swimming pools. The accepted concentration for this purpose is 1 ppm chlorine, or 1 g of chlorine per million grams of water. Calculate the volume of a chlorine solution (in milliliters) a homeowner should add to her swimming pool if the solution contains 6.0 percent chlorine by mass and there are 2.0×10^4 gallons (gal) of water in the pool (1 gal = 3.79 L; density of liquids = 1.0 g/mL).

1.111 The world's total petroleum reserve is estimated at 2.0×10^{22} joules [a joule (J) is the unit of energy, where $1 \text{ J} = 1 \text{ kg} \cdot \text{m}^2/\text{s}^2$]. At the present rate of consumption, 1.8×10^{20} joules per year (J/yr), how long would it take to exhaust the supply?

1.112 Bronze is an alloy made of copper (Cu) and tin (Sn). Calculate the mass of a bronze cylinder of radius 6.44 cm and length 44.37 cm. The composition of the bronze is 79.42 percent Cu and 20.58 percent Sn and the densities of Cu and Sn are 8.94 g/cm^3 and 7.31 g/cm^3, respectively. What assumption should you make in this calculation?

1.113 Chalcopyrite, the principal ore of copper (Cu), contains 34.63 percent Cu by mass. How many grams of Cu can be obtained from 7.35×10^3 kg of the ore?

Engineering Problems

1.114 Vanillin (used to flavor vanilla ice cream and other foods) is the substance whose aroma the human nose detects in the smallest amount. The threshold limit is 2.0×10^{-11} g per liter of air. If the current price of 50 g of vanillin is $112, determine the cost to supply enough vanillin so that the aroma could be detected in a large aircraft hangar with a volume of 5.0×10^7 ft^3.

1.115 One gallon of gasoline in an automobile's engine produces on the average 9.5 kg of carbon dioxide, which is a greenhouse gas; that is, it promotes the warming of Earth's atmosphere. Calculate the annual production of carbon dioxide in kilograms if there are 40 million cars in the United States and each car covers a distance of 5000 mi at a consumption rate of 20 miles per gallon.

1.116 Magnesium (Mg) is a valuable metal used in alloys, in batteries, and in the manufacture of chemicals. It is obtained mostly from seawater, which contains about 1.3 g of Mg for every kilogram of seawater. Referring to Problem 1.87, calculate the volume of

seawater (in liters) needed to extract 8.0×10^4 tons of Mg, which is roughly the annual production in the United States.

1.117 Fluoridation is the process of adding fluorine compounds to drinking water to help fight tooth decay. A concentration of 1 ppm of fluorine is sufficient for the purpose (1 ppm means one part per million, or 1 g of fluorine per 1 million g of water). The compound normally chosen for fluoridation is sodium fluoride, which is also added to some toothpastes. Calculate the quantity of sodium fluoride in kilograms needed per year for a city of 50,000 people if the daily consumption of water per person is 150 gal. What percent of the sodium fluoride is "wasted" if each person uses only 6.0 L of water a day for drinking and cooking (sodium fluoride is 45.0 percent fluorine by mass; 1 gal = 3.79 L; 1 year = 365 days; 1 ton = 2000 lb; 1 lb = 453.6 g; density of water = 1.0 g/mL)?

Biological Problems

1.118 The natural abundances of elements in the human body, expressed as percent by mass, are oxygen (O), 65 percent; carbon (C), 18 percent; hydrogen (H), 10 percent; nitrogen (N), 3 percent; calcium (Ca), 1.6 percent; phosphorus (P), 1.2 percent; all other elements, 1.2 percent. Calculate the mass in grams of each element in the body of a 62-kg person.

1.119 A resting adult requires about 240 mL of pure oxygen per minute and breathes about 12 times every minute. If inhaled air contains 20 percent oxygen by volume and exhaled air 16 percent, what is the volume of air per breath? (Assume that the volume of inhaled air is equal to that of exhaled air.)

1.120 (a) Referring to Problem 1.119, calculate the total volume (in liters) of air an adult breathes in a day. (b) In a city with heavy traffic, the air contains 2.1×10^{-6} L of carbon monoxide (a poisonous gas) per liter. Calculate the average daily intake of carbon monoxide in liters by a person.

1.121 The medicinal thermometer commonly used in homes can be read to $\pm 0.1°F$, whereas those in the doctor's office may be accurate to $\pm 0.1°C$. Percent error is often expressed as the absolute value of the difference between the true value and the experimental value, divided by the true value:

$$\text{percent error} = \frac{|\text{true value} - \text{experimental value}|}{\text{true value}} \times 100\%$$

The vertical lines indicate absolute value. In degrees Celsius, express the percent error expected from each of these thermometers in measuring a person's body temperature of 38.9°C.

1.122 TUMS is a popular remedy for acid indigestion. A typical TUMS tablet contains calcium carbonate plus some inert substances. When ingested, it reacts with the gastric juice (hydrochloric acid) in the stomach to give off carbon dioxide gas. When a 1.328-g tablet reacted with 40.00 mL of hydrochloric acid (density = 1.140 g/mL), carbon dioxide gas was given off and the resulting solution weighed 46.699 g. Calculate the number of liters of carbon dioxide gas released if its density is 1.81 g/L.

1.123 Pheromones are compounds secreted by females of many insect species to attract mates. Typically, 1.0×10^{-8} g of a pheromone is sufficient to reach all targeted males within a radius of 0.50 mi. Calculate the density of the pheromone (in grams per liter) in a cylindrical air space having a radius of 0.50 mi and a height of 40 ft (volume of a cylinder of radius r and height h is $\pi r^2 h$).

Standardized-Exam Practice Problems

Verbal Reasoning

English writer and essayist Lady Mary Wortley Montagu (1689–1762) traveled extensively and was fascinated by the customs in other countries. While in Turkey, she observed the practice of "engrafting" wherein people were inoculated against smallpox by intentional exposure to a mild form of the disease. She was so convinced of the efficacy and the safety of engrafting, that she had both of her children inoculated. She herself had survived smallpox as a child. Lady Montagu campaigned for the practice when she returned to England, and despite opposition from doctors and religious leaders, inoculation came into common use. It remained the primary defense against the scourge of smallpox for decades—until Jenner developed the practice of vaccination.

1. The main point of the passage is that

   a) Lady Montagu survived smallpox as a child.
   b) Lady Montagu brought the practice of engrafting from Turkey to England.
   c) doctors in eighteenth-century England were opposed to the practice of engrafting.
   d) Jenner developed the practice of vaccination.

2. Based on the passage, Lady Montagu was most likely

   a) a doctor.
   b) Turkish.
   c) severely scarred by smallpox.
   d) a member of a prominent British family.

3. The author refers to Lady Montagu having survived smallpox to

   a) explain why Lady Montagu was fascinated by the practice of engrafting.
   b) compare Lady Montagu to the doctors and religious leaders in England.
   c) explain why Lady Montagu herself did not undergo the engrafting procedure.
   d) emphasize Lady Montagu's fascination with other cultures.

4. Based on the passage, the author most likely thinks that Lady Montagu was

   a) educated and influential.
   b) inconsequential in the prevention of smallpox in England.
   c) trained in science and medicine.
   d) married to the British ambassador to Turkey.

Answers to In-Chapter Materials

Practice Problems

1.1A 273 K and 373 K, range = 100 K. **1.1B** −270.5°C. **1.2A** 113°F, 194°F; difference = 81°F. **1.2B** 233°C. **1.3A** (a) 13.6 g/mL, (b) 11.4 mL. **1.3B** (a) 9.25 g/cm^3, (b) 3.76 × 10^3 g. **1.4A** (a) and (c). **1.4B** Physical: iv, chemical: ii and iii, neither: i. **1.5A** (a) 4, (b) 1, (c) 4, (d) 2, (e) 2 or 3, (f) 4. **1.5B** (a) 4, 0.0003050, 4; (b) 5, 432.00, 5; (c) 4, 0.0000008001, 4; (d) 7, 200608.0, 7; (e) 4, 0.00001503, 4; (f) 6, 60751.0, 6. **1.6A** (a) 116.2 L, (b) 80.71 m, (c) 3.813 × 10^{21} atoms, (d) 31 dm^2, (e) 0.504 g/mL. **1.6B** (a) 32.44 cm^3, (b) 4.2 × 10^2 kg/m^3, (c) 1.008 × 10^{10} kg, (d) 40.75 mL, (e) 227 cm^3. **1.7A** 0.8120 g/cm^3. **1.7B** 95.3 cm^3. **1.8A** 0.01 oz. **1.8B** 708.1 g. **1.9A** 1.05 × 10^4 kg/m^3. **1.9B** 13.6 mg/mm^3.

Checkpoints

1.3.1 c. **1.3.2** a. **1.3.3** b. **1.3.4** d. **1.4.1** b, c, e. **1.4.2** a, d, f. **1.5.1** c. **1.5.2** e. **1.5.3** c. **1.5.4** e. **1.6.1** b. **1.6.2** c. **1.6.3** a. **1.6.4** e.

CHAPTER 2

Atoms, Molecules, and Ions

Legumes, including peas and beans, are a good source of dietary iron. Other iron-rich foods include meat, eggs, some vegetables, and fortified cereals. When diet alone does not provide an adequate supply, iron supplements can be taken.

Zoonar/O Popova/age fotostock

In This Chapter, You Will Learn

What atoms are made of and how they are arranged in molecules and ions that make up the substances that we encounter every day. You will also learn how to associate the name of a substance with its chemical formula.

Before You Begin, Review These Skills

- Significant figures [◄◄ Section 1.5]
- Dimensional analysis [◄◄ Section 1.6]
- Chapter 1 Key Skills [◄◄ pages 32–33]

How Certain Atoms, Molecules, and Ions Can Affect Human Health

Atoms, molecules, and ions make up the substances we encounter every day. Some of these substances are important components of a balanced diet. An estimated 25 percent of the world's population suffers from iron deficiency, the most common nutritional deficiency in the world. Iron is necessary for the production of hemoglobin, the component in red blood cells responsible for the transport of oxygen. An inadequate supply of iron and the resulting shortage of hemoglobin can cause iron deficiency anemia (IDA). Some of the symptoms of IDA are fatigue, weakness, pale color, poor appetite, headache, and light-headedness.

Although IDA can be caused by loss of blood or by poor absorption of iron, the most common cause is insufficient iron in the diet. Dietary iron comes from such sources as meat, eggs, leafy green vegetables, dried beans, and dried fruits. Some breakfast cereals, such as Cream of Wheat, are fortified with iron in the form of iron metal, also known as *elemental* or *reduced* iron. The absorption of dietary iron can be enhanced by the intake of vitamin C (ascorbic acid). When the diet fails to provide enough iron, a nutritional supplement may be necessary to prevent a deficiency. Many supplements provide iron in the form of a compound called *ferrous sulfate.*

Elemental iron, ascorbic acid, and the iron in ferrous sulfate are examples of some of the *atoms, molecules,* and *ions* that are essential for human health.

Student Note: Iron absorption can be diminished by certain disorders, such as Crohn's disease, and by some medications.

David A. Tietz/Editorial Image, LLC

At the end of this chapter, you will be able to solve a series of problems involving iron, iron sulfate, and ascorbic acid [►► Applying What You've Learned, page 82].

Figure 2.1 John Dalton.

Georgios Kollidas/GeorgiosArt/iStock/ Getty Images

2.1 The Atomic Theory

In the fifth century B.C., the Greek philosopher Democritus proposed that all matter consists of very small, indivisible particles, which he named *atomos* (meaning uncuttable or indivisible). Although Democritus's idea was not accepted by many of his contemporaries (notably Plato and Aristotle), somehow it endured. Experimental evidence from early scientific investigations provided support for the notion of "atomism" and gradually gave rise to the modern definitions of elements and compounds. In 1808, an English scientist and schoolteacher, John Dalton[1] (Figure 2.1), formulated a precise definition of the indivisible building blocks of matter that we call atoms.

Dalton's work marked the beginning of the modern era of chemistry. The hypotheses about the nature of matter on which Dalton's atomic theory is based can be summarized as follows:

1. Elements are composed of extremely small particles called *atoms*. All atoms of a given element are identical, having the same size, mass, and chemical properties. The atoms of one element are different from the atoms of all other elements.
2. *Compounds* are composed of atoms of more than one element. In any given compound, the same types of atoms are always present in the same relative numbers.
3. A chemical reaction *rearranges* atoms; it does not create or destroy them.

Figure 2.2 is a schematic representation of these hypotheses.

Dalton's concept of an atom was far more detailed and specific than that of Democritus. The first hypothesis states that atoms of one element are different from atoms of all other elements. Dalton made no attempt to describe the structure or composition of atoms—he had no idea what an atom was really like. He did realize, though, that the different properties shown by elements such as hydrogen and oxygen could be explained by assuming that hydrogen atoms were not the same as oxygen atoms.

The second hypothesis suggests that, to form a certain compound, we not only need atoms of the right *kinds* of elements, but specific *numbers* of these atoms as well. This idea is an extension of a law published in 1799 by Joseph Proust, a French chemist. According to Proust's **law of definite proportions,** different samples of a given compound always contain the same elements in the same mass *ratio*. Thus, if we were to analyze samples of carbon dioxide gas obtained from different sources, such as the exhaust from a car in Mexico City or the air above a pine forest in northern

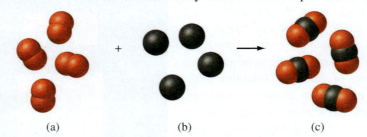

(a)   (b)   (c)

Figure 2.2 This represents a chemical reaction between the elements oxygen and carbon. (a) Oxygen does not exist as isolated atoms under ordinary conditions, but rather as *molecules,* each of which consists of *two* oxygen atoms. Note that the oxygen atoms (red spheres) appear all to be identical to one another (Dalton's 1st hypothesis). (b) Likewise, the carbon atoms (black spheres) all appear to be identical to one another. Carbon also exists in the form of molecules that are more varied and complex than those of oxygen, but the carbon has been represented here as isolated atoms to simplify the figure. (c) The compound CO_2 forms when each carbon atom combines with two oxygen atoms (Dalton's 2nd hypothesis). Finally, the reaction results in the rearrangement of the atoms, but all the atoms present before the reaction (left of the arrow) are also present after the reaction (right of the arrow) (Dalton's 3rd hypothesis).

---

1. John Dalton (1766–1844). English chemist, mathematician, and philosopher. In addition to the atomic theory, Dalton formulated several gas laws and gave the first detailed description of the type of color blindness, now called "Daltonism," from which he suffered. He also assigned relative weights to the elements, many of which differ considerably from those that we use today. He was described by his friends as awkward and without social grace.

Maine, each sample would contain the same ratio by mass of oxygen to carbon. Consider the following results of the analysis of three samples of carbon dioxide, each from a different source:

Sample	Mass of O (g)	Mass of C (g)	Ratio (g O : g C)
123 g carbon dioxide	89.4	33.6	2.66:1
50.5 g carbon dioxide	36.7	13.8	2.66:1
88.6 g carbon dioxide	64.4	24.2	2.66:1

In any sample of pure carbon dioxide, there are 2.66 g of oxygen for every gram of carbon present. This constant mass ratio can be explained by assuming that the elements exist in tiny particles of fixed mass (atoms), and that compounds are formed by the combination of fixed numbers of each type of particle.

Dalton's second hypothesis also supports the *law of multiple proportions*. According to this law, if two elements can combine to form more than one compound with each other, the masses of one element that combine with a fixed mass of the other element are in ratios of small whole numbers. That is, different compounds made up of the same elements differ in the number of atoms of each kind that combine. For example, carbon combines with oxygen to form carbon dioxide and carbon monoxide. In any sample of pure carbon monoxide, there are 1.33 g of oxygen for every gram of carbon.

Sample	Mass of O (g)	Mass of C (g)	Ratio (g O : g C)
16.3 g carbon monoxide	9.31	6.99	1.33:1
25.9 g carbon monoxide	14.8	11.1	1.33:1
88.4 g carbon monoxide	50.5	37.9	1.33:1

Thus, the ratio of oxygen to carbon in carbon *di*oxide is 2.66; and the ratio of oxygen to carbon in carbon *mon*oxide is 1.33. According to the law of multiple proportions, the ratio of two such ratios can be expressed as small whole numbers.

$$\frac{\text{ratio of O to C in carbon dioxide}}{\text{ratio of O to C in carbon monoxide}} = \frac{2.66}{1.33} = 2:1$$

For samples containing equal masses of carbon, the ratio of oxygen in carbon dioxide to oxygen in carbon monoxide is 2:1. Modern measurement techniques indicate that one atom of carbon combines with two atoms of oxygen in carbon dioxide and with one atom of oxygen in carbon monoxide. This result is consistent with the law of multiple proportions (Figure 2.3).

Dalton's third hypothesis is another way of stating the *law of conservation of mass,*[2] which is that matter can be neither created nor destroyed. Because matter is made up of atoms that are unchanged in a chemical reaction, it follows that mass must be conserved as well. Dalton's brilliant insight into the nature of matter was the main stimulus for the rapid progress of chemistry during the nineteenth century.

Animation
Law of conservation of mass.

Figure 2.3 An illustration of the law of multiple proportions.

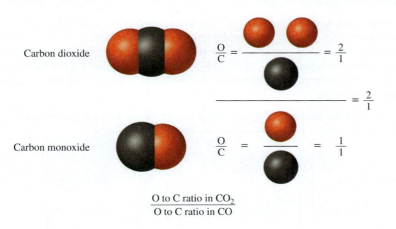

Carbon dioxide

$$\frac{\text{O}}{\text{C}} = \frac{}{} = \frac{2}{1}$$

$$\frac{}{} = \frac{2}{1}$$

Carbon monoxide

$$\frac{\text{O}}{\text{C}} = \frac{}{} = \frac{1}{1}$$

$$\frac{\text{O to C ratio in CO}_2}{\text{O to C ratio in CO}}$$

---

2. According to Albert Einstein, mass and energy are alternate aspects of a single entity called mass-energy. Chemical reactions usually involve a gain or loss of heat and other forms of energy. Thus, when energy is lost in a reaction, for example, mass is also lost. Except for nuclear reactions (see Chapter 20), however, changes of mass in chemical reactions are far too small to detect. Therefore, for all practical purposes mass is conserved.

Sample Problem 2.1 shows how some common compounds obey the law of multiple proportions.

SAMPLE PROBLEM 2.1

(a) Both water (H_2O) and hydrogen peroxide (H_2O_2) are composed of hydrogen and oxygen. When water is decomposed into its constituent elements, it produces 0.125 g hydrogen for every gram of oxygen. When hydrogen peroxide is decomposed, it produces 0.063 g hydrogen for every gram of oxygen. Determine the whole number ratio of g H : 1.00 g O in water to g H : 1.00 g O in hydrogen peroxide to show how these data illustrate the law of multiple proportions. (b) Sulfur and oxygen can combine to form several compounds including sulfur dioxide (SO_2) and sulfur trioxide (SO_3). Sulfur dioxide contains 0.9978 g oxygen for every gram of sulfur. Sulfur trioxide contains 1.497 g oxygen for every gram of sulfur. Determine the whole number ratio of g O : 1.00 g S in sulfur dioxide to g O : 1.00 g S in sulfur trioxide.

Strategy For two compounds, each consisting of just two different elements, we are given the mass ratio of one element to the other. In each case, to show how the information given illustrates the law of multiple proportions, we divide the larger ratio by the smaller ratio.

Setup (a) The mass ratio of hydrogen to oxygen is higher in water than it is in hydrogen peroxide. Therefore, we divide the number of grams of hydrogen per gram of oxygen given for *water* by that given for *hydrogen peroxide*. (b) The mass ratio of oxygen to sulfur is higher in sulfur trioxide than it is in sulfur dioxide. Therefore, we divide the number of grams of oxygen per gram of sulfur given for *sulfur trioxide* by that given for *sulfur dioxide*.

Solution

(a) $$\frac{\text{g H : 1.00 g O in water}}{\text{g H : 1.00 g O in hydrogen peroxide}} = \frac{0.125}{0.063} = 1.98{:}1 \approx 2{:}1$$

(b) $$\frac{\text{g O : 1.00 g S in sulfur trioxide}}{\text{g O : 1.00 g S in sulfur dioxide}} = \frac{1.497}{0.9978} = 1.50{:}1. \text{ Multiplying through by 2 gives 3:2.}$$

THINK ABOUT IT

When the result of such calculations is not a whole number, we must decide whether the result is close enough to *round* to a whole number as in part (a), or whether to multiply through to get whole numbers as in part (b). Only numbers that are *very* close to whole can be rounded. For example, numbers ending in approximately .25, .33, or .5 should be multiplied by 4, 3, or 2, respectively, to give a whole number.

Practice Problem A TTEMPT In each case, calculate the appropriate ratio to show that the information given is consistent with the law of multiple proportions. (a) Both ammonia (NH_3) and hydrazine (N_2H_4) are composed of nitrogen and hydrogen. Ammonia contains 0.2158 g hydrogen for every gram of nitrogen. Hydrazine contains 0.1439 g hydrogen for every gram of nitrogen. (b) Two of the compounds that consist of nitrogen and oxygen are nitric oxide, also known as nitrogen monoxide (NO) and nitrous oxide (N_2O), which is also known as dinitrogen monoxide. Nitric oxide contains 1.142 g oxygen for every gram of nitrogen. Nitrous oxide contains 0.571 g oxygen for every gram of nitrogen.

Practice Problem B UILD (a) Two of the simplest compounds containing just carbon and hydrogen are methane and ethane. Given that methane contains 0.3357 g hydrogen for every 1.00 g carbon and that the ratio of g hydrogen/1.00 g carbon in methane to g hydrogen/1.00 g carbon in ethane is 4:3, determine the number of grams of hydrogen per gram of carbon in ethane. (b) Xenon (Xe) and fluorine (F) can combine to form several different compounds, including XeF_2, which contains 0.2894 g fluorine for every gram of xenon. Use the law of multiple proportions to determine n, which represents the number of F atoms in another compound, XeF_n, given that it contains 0.8682 g of F for every gram of Xe.

Practice Problem C ONCEPTUALIZE Which of the following diagrams illustrates the law of multiple proportions?

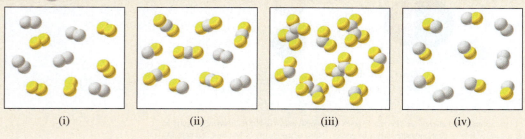

(i)   (ii)   (iii)   (iv)

CHECKPOINT – SECTION 2.1   The Atomic Theory

2.1.1

For the two compounds pictured, evaluate the following ratio:

$$\frac{\text{g yellow : 1.00 g blue (right)}}{\text{g yellow : 1.00 g blue (left)}}$$

a) 2:1

b) 3:2

c) 3:4

d) 5:2

e) 5:4

2.1.2 For which of the following pairs of compounds is the ratio

$$\frac{\text{g yellow : 1.00 g blue (right)}}{\text{g yellow : 1.00 g blue (left)}}$$

equal to 4:1?

a)

b)

c)

d)

e)

2.2   The Structure of the Atom

On the basis of Dalton's atomic theory, we can define an **atom** as the basic unit of an element that can enter into a chemical combination. Dalton imagined an atom that was both extremely small and indivisible. However, a series of investigations that began in the 1850s and extended into the twentieth century clearly demonstrated that atoms actually possess internal structure; that is, they are made up of even smaller particles, which are called *subatomic particles.* This research led to the discovery of electrons, protons, and neutrons.

Discovery of the Electron

Many scientists in the 1890s studied ***radiation,*** the emission and transmission of energy through space in the form of waves. Information gained from this research contributed greatly to our understanding of atomic structure. One device used to investigate this phenomenon was a cathode ray tube, the forerunner of the tubes used in older televisions and computer monitors (Figure 2.4).

A cathode ray tube consists of two metal plates sealed inside a glass tube from which most of the air has been evacuated. When the metal plates are connected to a high-voltage source, the negatively charged plate, called the *cathode,* emits an invisible ray. The cathode ray is drawn to the positively charged plate, called the *anode,* where it passes through a hole and continues traveling to the other end of the tube. When the ray strikes the specially phosphor-coated surface, it produces a bright light.

Because consistent results are observed regardless of the composition of the cathode, cathode rays were presumed to be a component of all matter. Furthermore, because the path of the cathode rays could be deflected by magnetic and electric fields, as shown in Figure 2.4, they must be streams of charged *particles.* According to electromagnetic theory, a moving charged body behaves like a magnet and can interact with electric and magnetic fields through which it passes. Because the cathode ray is attracted by the plate bearing positive charges and repelled by the plate bearing

Animation
Cathode ray tube experiment.

Figure 2.4 A cathode ray tube with an electric field perpendicular to the direction of the cathode rays and an external magnetic field. The symbols N and S denote the north and south poles of the magnet. The cathode rays will strike the end of the tube at point A in the presence of a magnetic field and at point B in the presence of an electric field. (In the absence of any external field—or when the effects of the electric field and magnetic field cancel each other—the cathode rays will not be deflected but will travel in a straight line and strike the middle of the circular screen.)

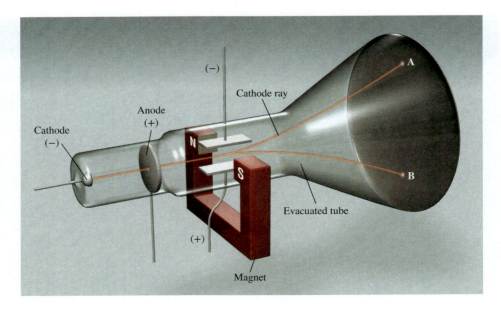

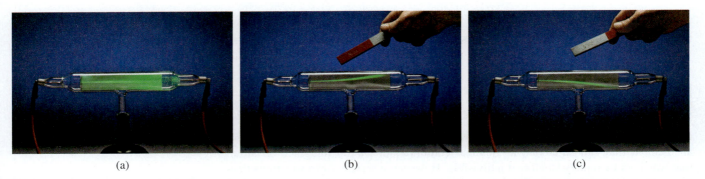

(a)   (b)   (c)

Figure 2.5 (a) A cathode ray produced in a discharge tube. The ray itself is invisible, but the fluorescence of a zinc sulfide coating on the glass causes it to appear green. (b) The cathode ray bends toward one pole of a magnet and (c) bends away from the opposite pole.
(a), (c): Charles D. Winters/Timeframe Photography/McGraw Hill; (b): Charles D. Winters/McGraw Hill

negative charges, it must consist of negatively charged particles. We now know these negatively charged particles as *electrons.* Figure 2.5 shows the deflection of a stream of electrons away from the south pole of a bar magnet. This effect on the path of electrons is what causes a television picture to become distorted temporarily when a magnet is brought close to the screen.

Late in the nineteenth century, English physicist J. J. Thomson used a cathode ray tube and his knowledge of electromagnetic theory to determine the ratio of electric charge to the mass of an individual electron. The number he calculated was 1.76×10^8 C/g, where C stands for *coulomb,* which is the derived SI unit of electric charge. (In SI base units, 1 C = 1 A · s. Recall that A and s are the SI base units *ampere* and *second,* respectively [◄◄ Section 1.3, Table 1.2].)

Early in the twentieth century, American physicist R. A. Millikan conducted an ingenious experiment that enabled him to calculate precisely the charge of the electron. In his experiment, Millikan examined the motion of tiny droplets of oil as they fell through an electric field in the space between two electrically charged plates. An X-ray source was used to knock electrons off of molecules in the air, and the electrons became attached to the oil droplets, giving each droplet a negative charge. He then varied the strength of the electric field and was able to slow, stop, or even *reverse* the downward motion of the droplets. By precisely varying the electric field, and knowing the mass of the droplets, he was able to calculate the

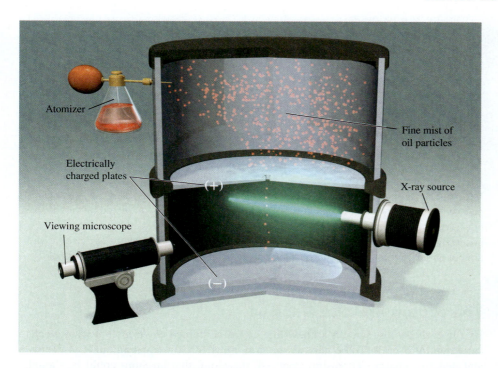

Figure 2.6 Schematic diagram of Millikan's oil-drop experiment.

Atomizer

Fine mist of oil particles

Electrically charged plates

(+)

X-ray source

Viewing microscope

(−)

Animation
Millikan oil-drop experiment.

charge on each droplet (Figure 2.6). His work proved that the charge on each droplet was an integral multiple of -1.6022×10^{-19} C, which he deduced was the charge on a single electron. Using Thomson's charge-to-mass ratio and the charge from his own experiments, Millikan calculated the mass of an electron as follows:

$$\text{mass of an electron} = \frac{\text{charge}}{\text{charge/mass}} = \frac{-1.6022 \times 10^{-19} \text{ C}}{-1.76 \times 10^8 \text{ C/g}} = 9.10 \times 10^{-28} \text{ g}$$

This is an *extremely* small mass.

Student Note: Millikan actually determined the charge on droplets with multiple electrons attached. The charges he determined were always *multiples* of -1.6022×10^{-19} C.

Radioactivity

In 1895, the German physicist Wilhelm Röntgen noticed that cathode rays caused glass and metals to emit yet another type of ray. This highly energetic radiation penetrated matter, darkened covered photographic plates, and caused a variety of substances to *fluoresce* (give off light). These rays were *not* deflected by a magnet, however, so unlike cathode rays, they could not contain charged particles. Röntgen called them X rays because of their mysterious nature.

Not long after Röntgen's discovery, Antoine Becquerel, a professor of physics in Paris, began to study the fluorescent properties of substances. Purely by accident, he found that exposing thickly wrapped photographic plates (so no light could get in) to a uranium compound caused them to darken, even without the stimulation of cathode rays. Like X rays, the rays from the uranium compound were highly energetic and could not be deflected by a magnet, but they differed from X rays because they arose spontaneously. One of Becquerel's students, Marie Curie, suggested the name ***radioactivity*** to describe this spontaneous emission of particles and/or radiation. Today, we use the term *radioactive* to describe any element that spontaneously emits radiation.

Three types of rays are produced by the breakdown, or *decay,* of radioactive substances such as uranium. Two of the three are deflected by oppositely charged metal plates (Figure 2.7). ***Alpha (α) rays*** consist of positively charged particles, called ***α particles,*** that are deflected *away* from the positively charged plate. ***Beta (β) rays,*** or ***β particles,*** are *electrons,* so they are deflected away from the *negatively* charged plate. The third type of radioactive radiation consists of high-energy ***gamma (γ) rays.*** Like X rays, γ rays have no charge and are unaffected by external electric or magnetic fields.

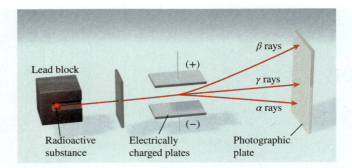

Figure 2.7 Three types of rays emitted by radioactive elements. So-called β rays actually consist of negatively charged particles (electrons) and are therefore attracted by the positively charged plate. The opposite holds true for so-called α rays—they are actually positively charged particles and are drawn to the negatively charged plate. Because γ rays are not particles and have no charge, their path is unaffected by an external electric field.

The Proton and the Nucleus

By the early 1900s, scientists knew that atoms contained electrons but were electrically neutral overall. To be neutral, an atom must contain equal amounts of positive and negative charge. Thomson proposed, therefore, that an atom could be thought of as a sphere of positively charged matter in which negatively charged electrons were embedded uniformly, like the chocolate chips in a scoop of mint chocolate chip ice cream. This so-called plum-pudding model was the accepted theory for a number of years.

In 1910, New Zealand physicist Ernest Rutherford used α particles to probe the structure of the atom. He carried out a series of experiments using very thin foils of gold and other metals as targets for α particles from a radioactive source. He observed that the majority of the α particles penetrated the foil either completely undeflected or with only a small angle of deflection. Every now and then, however, an α particle was scattered (or deflected) at a large angle. In some instances, the α particle actually bounced back in the direction of the radioactive source. This was an extraordinarily surprising result. In Thomson's model of the atom, the positive charge was so diffuse that the α particles, with their relatively high mass, should all have passed through the foil with little or no deflection. To quote Rutherford's initial reaction to this discovery: "It was as incredible as if you had fired a 15-inch shell at a piece of tissue paper and it came back and hit you." The result of Rutherford's gold-foil experiment necessitated the development of a new model of the atom. Figure 2.8 illustrates the results of Rutherford's α-scattering experiment.

Nuclear Model of the Atom

Rutherford later explained the results of the α-scattering experiment by proposing a new model for the atom. According to Rutherford, most of the atom must be empty space. This would explain why the majority of α particles passed through the gold foil with little or no deflection. The atom's positive charges, Rutherford proposed, were all concentrated in the ***nucleus,*** which is an extremely dense central *core* within the atom. Whenever an α particle came close to a nucleus in the scattering experiment, it experienced a large repulsive force and therefore a large deflection. Moreover, an α particle traveling directly toward a nucleus would be completely repelled and its direction would be reversed.

The positively charged particles in the nucleus are called ***protons.*** In separate experiments, it was found that each proton carried the same *quantity* of charge as an electron (just opposite in sign) but had a mass of 1.67262×10^{-24} g. Although this is an extremely small value, it is nearly 2000 times the mass of an electron.

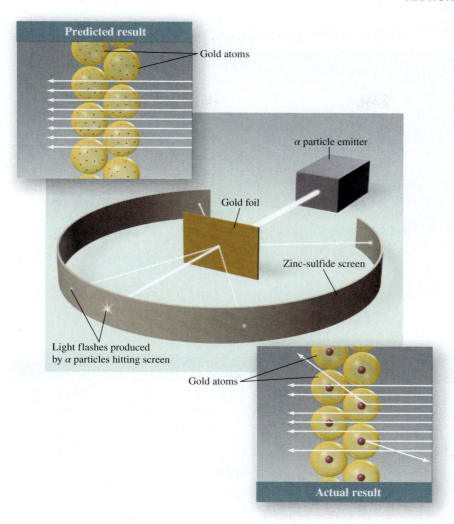

Figure 2.8 Rutherford's experimental design for measuring the scattering of α particles by a piece of gold foil. The plum-pudding model predicted that the α particles would all pass through the gold foil undeflected. The actual result: Most of the α particles do pass through the gold foil with little or no deflection, but a few are deflected at large angles. Occasionally an α particle bounces off the foil back toward the source. The nuclear model explains the results of Rutherford's experiments.

Based on these data, the atom was believed to consist of a nucleus that accounted for most of the mass of the atom, but which occupied only a tiny fraction of its volume. We express atomic (and molecular) dimensions using the SI unit *picometer* (*pm*), where

$$1 \text{ pm} = 1 \times 10^{-12} \text{ m}$$

A typical atomic radius is about 100 pm, whereas the radius of an atomic nucleus is only about 5×10^{-3} pm. You can appreciate the relative sizes of an atom and its nucleus by imagining that if an atom were the size of the New Orleans Superdome, the volume of its nucleus would be comparable to that of a marble. While the protons are confined to the nucleus of the atom, the electrons are distributed *around* the nucleus at relatively large distances from it.

> **Student Note:** Atomic radii are sometimes given in *angstroms,* where 1 angstrom (Å) = 1×10^{-10} m. Using angstroms, a typical atomic radius is about 1 Å and a typical nuclear radius is about 5×10^{-5} Å.

The concept of atomic radius is useful experimentally, but you should not get the impression that atoms have well-defined boundaries or surfaces. We explain in Chapter 6 that the outer regions of atoms are actually relatively "fuzzy," rather than sharply defined.

The Neutron

Rutherford's model of atomic structure left one major problem unsolved. It was known that hydrogen, the simplest atom, contained only one proton and that the helium atom contained two protons. Therefore, the ratio of the mass of a helium atom to that of a

Figure 2.9 The protons and neutrons in an atom are contained in the tiny volume of the nucleus. Electrons are distributed within the sphere surrounding the nucleus.

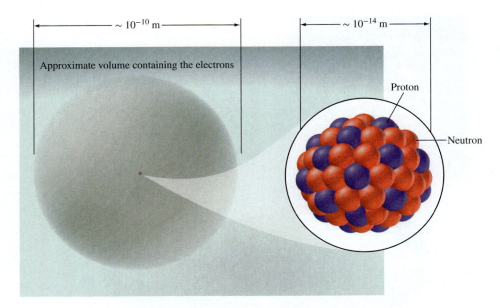

TABLE 2.1	Masses and Charges of Subatomic Particles		
Particle	**Mass (g)**	**Charge (C)**	**Charge Unit**
Electron*	9.10938×10^{-28}	-1.6022×10^{-19}	-1
Proton	1.67262×10^{-24}	$+1.6022 \times 10^{-19}$	$+1$
Neutron	1.67493×10^{-24}	0	0

*More refined measurements have resulted in a small change to Millikan's original value.

hydrogen atom should be 2:1. (Because electrons are much lighter than protons, their contribution to atomic mass is insignificant.) In reality, however, the mass ratio of helium to hydrogen is 4:1. Rutherford and others postulated that there must be another type of subatomic particle in the atomic nucleus, the proof of which was provided in 1932 by James Chadwick, an English physicist. Chadwick's experiments established the existence of a third subatomic particle, the ***neutron.*** Neutrons are so-named because they are electrically neutral, and have a mass slightly greater than that of a proton. The mystery of the mass ratio could now be explained. A typical helium nucleus consists of two protons and two neutrons, whereas a typical hydrogen nucleus contains only a proton; the mass ratio, therefore, is 4:1.

Figure 2.9 shows the location of the elementary particles (protons, neutrons, and electrons) in an atom. There are other subatomic particles, but the electron, the proton, and the neutron are the three fundamental components of the atom that are important in chemistry. Table 2.1 lists the masses and charges of these three elementary particles.

2.3   Atomic Number, Mass Number, and Isotopes

All atoms can be identified by the number of protons and neutrons they contain. The ***atomic number (Z)*** is the number of protons in the nucleus of each atom of an element. It also indicates the number of *electrons* in the atom—because atoms are *neutral* and contain the same number of protons and electrons. The chemical identity of an atom can be determined solely from its atomic number. For example, the atomic number of nitrogen is 7. Thus, each nitrogen atom has seven protons and seven electrons. Or, viewed another way, every atom in the universe that contains seven protons is a nitrogen atom.

The *mass number (A)* is the total number of neutrons *and* protons present in the nucleus of an atom of an element. Except for the most common form of hydrogen, which has one proton and no neutrons, all atomic nuclei contain both protons and neutrons. Collectively, protons and neutrons are called *nucleons.* A nucleon is a particle within the nucleus. In general, the mass number is given by

mass number (A) = number of protons (Z) + number of neutrons

The number of neutrons in an atom equals the difference between the mass number and the atomic number, or $(A - Z)$. For example, the mass number of fluorine is 19 and the atomic number is 9 (indicating 9 protons in the nucleus). Thus, the number of neutrons in an atom of fluorine is $19 - 9 = 10$. The atomic number, number of neutrons, and mass number all must be positive integers (whole numbers).

The accepted way to denote the atomic number and mass number of an atom of an element (X) is as follows:

Mass number
(number of protons + neutrons)

$$_{Z}^{A}\text{X} \longleftarrow \text{Element symbol}$$

Atomic number
(number of protons)

Contrary to the first hypothesis of Dalton's atomic theory, atoms of a given element do not all have the same mass. Most elements have two or more *isotopes,* atoms that have the same atomic number (Z) but different mass numbers (A). For example, there are three isotopes of hydrogen, called *hydrogen* (or *protium*), *deuterium,* and *tritium.* Hydrogen has one proton and no neutrons in its nucleus, deuterium has one proton and one neutron, and tritium has one proton and two neutrons. Thus, to represent the isotopes of hydrogen, we write

$$_{1}^{1}\text{H} \qquad _{1}^{2}\text{H} \qquad _{1}^{3}\text{H}$$
$$\text{hydrogen} \qquad \text{deuterium} \qquad \text{tritium}$$

Similarly, the two common isotopes of uranium $(Z = 92)$, which have mass numbers of 235 and 238, respectively, can be represented as follows:

$$_{92}^{235}\text{U} \qquad _{92}^{238}\text{U}$$

The first isotope, with $235 - 92 = 143$ neutrons in its nucleus, is used in nuclear reactors and atomic bombs, whereas the second isotope, with 146 neutrons, lacks the properties necessary for these applications. With the exception of hydrogen, which has different names for each of its isotopes, the isotopes of other elements are identified by their mass numbers. The two isotopes of uranium are called uranium-235 (pronounced "uranium two thirty-five") and uranium-238 (pronounced "uranium two thirty-eight"). Because the subscripted atomic number can be determined from the elemental symbol, it may be omitted from these representations without the loss of any information. The symbols ^{3}H and ^{235}U are sufficient to specify the isotopes tritium and uranium-235, respectively.

The chemical properties of an element are determined primarily by the protons and electrons in its atoms; neutrons do not take part in chemical changes under normal conditions. Therefore, isotopes of the same element exhibit similar chemical properties, forming the same types of compounds and displaying similar reactivities.

Sample Problem 2.2 shows how to calculate the number of protons, neutrons, and electrons using atomic numbers and mass numbers.

SAMPLE PROBLEM   2.2

Determine the numbers of protons, neutrons, and electrons in each of the following species: (a) $^{35}_{17}Cl$, (b) $^{37}_{17}Cl$, (c) ^{41}K, and (d) carbon-14.

Strategy Recall that the superscript denotes the mass number (A), and the subscript denotes the atomic number (Z). In the case where no subscript is shown, as in parts (c) and (d), the atomic number can be deduced from the elemental symbol or name. For the purpose of determining the number of electrons, remember that atoms are neutral, so the number of electrons equals the number of protons.

Setup Number of protons = Z, number of neutrons = $A - Z$, and number of electrons = number of protons. Recall that the 14 in carbon-14 is the mass number.

Solution

(a) The atomic number is 17, so there are 17 protons. The mass number is 35, so the number of neutrons is $35 - 17 = 18$. The number of electrons equals the number of protons, so there are 17 electrons.

(b) Again, the atomic number is 17, so there are 17 protons. The mass number is 37, so the number of neutrons is $37 - 17 = 20$. The number of electrons equals the number of protons, so there are 17 electrons, too.

(c) The atomic number of K (potassium) is 19, so there are 19 protons. The mass number is 41, so there are $41 - 19 = 22$ neutrons. There are 19 electrons.

(d) Carbon-14 can also be represented as ^{14}C. The atomic number of carbon is 6, so there are 6 protons and 6 electrons. There are $14 - 6 = 8$ neutrons.

THINK ABOUT IT

Verify that the number of protons and the number of neutrons for each example sum to the mass number that is given. In part (a), for example, there are 17 protons and 18 neutrons, which sum to give a mass number of 35, the value given in the problem. In part (b), 17 protons + 20 neutrons = 37. In part (c), 19 protons + 22 neutrons = 41. In part (d), 6 protons + 8 neutrons = 14.

Practice Problem **A**TTEMPT How many protons, neutrons, and electrons are there in an atom of (a) $^{10}_{5}B$, (b) ^{36}Ar, (c) $^{85}_{38}Sr$, and (d) carbon-11?

Practice Problem **B**UILD Give the correct symbols to identify an atom that contains (a) 4 protons, 4 electrons, and 5 neutrons; (b) 23 protons, 23 electrons, and 28 neutrons; (c) 54 protons, 54 electrons, and 70 neutrons; and (d) 31 protons, 31 electrons, and 38 neutrons.

Practice Problem **C**ONCEPTUALIZE Based on the numbers of nucleons, write the nuclear symbol for each of the following diagrams:

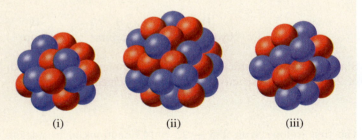

(i)   (ii)   (iii)

CHECKPOINT – SECTION 2.3   Atomic Number, Mass Number, and Isotopes

2.3.1 How many neutrons are there in an atom of ^{60}Ni?

    a) 60

    b) 30

    c) 28

    d) 32

    e) 29

2.3.2 What is the *mass number* of an oxygen atom with nine neutrons in its nucleus?

    a) 8

    b) 9

    c) 17

    d) 16

    e) 18

2.4 The Periodic Table

More than half of the elements known today were discovered between 1800 and 1900. During this period, chemists noted that the physical and chemical properties of certain groups of elements were similar to one another. These similarities, together with the need to organize the large volume of available information about the structure and properties of elemental substances, led to the development of the ***periodic table,*** a chart in which elements having similar chemical and physical properties are grouped together. Figure 2.10 shows the modern periodic table in which the elements are arranged by atomic number (shown above the element symbol) in horizontal rows called ***periods*** and in vertical columns called ***groups*** or ***families.*** Elements in the same *group* tend to have similar physical and chemical properties.

The elements can be categorized as metals, nonmetals, or metalloids. A ***metal*** is a good conductor of heat and electricity, whereas a ***nonmetal*** is usually a poor conductor of heat and electricity. A ***metalloid*** has properties that are intermediate between those of metals and nonmetals. Figure 2.10 shows that the majority of known elements are metals; only 17 elements are nonmetals, and fewer than 10 elements are metalloids. Although most sources, including this text, designate the elements B, Si, Ge, As, Sb, and Te as metalloids, sources vary for the elements Po and At. In this text, we classify both Po and

1 1A	2 2A	3 3B	4 4B	5 5B	6 6B	7 7B	8 8B	9 8B	10 8B	11 1B	12 2B	13 3A	14 4A	15 5A	16 6A	17 7A	18 8A
1 H Hydrogen																	2 He Helium
3 Li Lithium	4 Be Beryllium											5 B Boron	6 C Carbon	7 N Nitrogen	8 O Oxygen	9 F Fluorine	10 Ne Neon
11 Na Sodium	12 Mg Magnesium											13 Al Aluminum	14 Si Silicon	15 P Phosphorus	16 S Sulfur	17 Cl Chlorine	18 Ar Argon
19 K Potassium	20 Ca Calcium	21 Sc Scandium	22 Ti Titanium	23 V Vanadium	24 Cr Chromium	25 Mn Manganese	26 Fe Iron	27 Co Cobalt	28 Ni Nickel	29 Cu Copper	30 Zn Zinc	31 Ga Gallium	32 Ge Germanium	33 As Arsenic	34 Se Selenium	35 Br Bromine	36 Kr Krypton
37 Rb Rubidium	38 Sr Strontium	39 Y Yttrium	40 Zr Zirconium	41 Nb Niobium	42 Mo Molybdenum	43 Tc Technetium	44 Ru Ruthenium	45 Rh Rhodium	46 Pd Palladium	47 Ag Silver	48 Cd Cadmium	49 In Indium	50 Sn Tin	51 Sb Antimony	52 Te Tellurium	53 I Iodine	54 Xe Xenon
55 Cs Cesium	56 Ba Barium	57 La Lanthanum	72 Hf Hafnium	73 Ta Tantalum	74 W Tungsten	75 Re Rhenium	76 Os Osmium	77 Ir Iridium	78 Pt Platinum	79 Au Gold	80 Hg Mercury	81 Tl Thallium	82 Pb Lead	83 Bi Bismuth	84 Po Polonium	85 At Astatine	86 Rn Radon
87 Fr Francium	88 Ra Radium	89 Ac Actinium	104 Rf Rutherfordium	105 Db Dubnium	106 Sg Seaborgium	107 Bh Bohrium	108 Hs Hassium	109 Mt Meitnerium	110 Ds Darmstadtium	111 Rg Roentgenium	112 Cn Copernicium	113 Nh Nihonium	114 Fl Flerovium	115 Mc Moscovium	116 Lv Livermorium	117 Ts Tennessine	118 Og Oganesson

58 Ce Cerium	59 Pr Praseodymium	60 Nd Neodymium	61 Pm Promethium	62 Sm Samarium	63 Eu Europium	64 Gd Gadolinium	65 Tb Terbium	66 Dy Dysprosium	67 Ho Holmium	68 Er Erbium	69 Tm Thulium	70 Yb Ytterbium	71 Lu Lutetium
90 Th Thorium	91 Pa Protactinium	92 U Uranium	93 Np Neptunium	94 Pu Plutonium	95 Am Americium	96 Cm Curium	97 Bk Berkelium	98 Cf Californium	99 Es Einsteinium	100 Fm Fermium	101 Md Mendelevium	102 No Nobelium	103 Lr Lawrencium

Figure 2.10 The modern periodic table. The elements are arranged according to atomic number (see Section 2.3), which is shown above each element's symbol. With the exception of hydrogen (H), nonmetals appear at the far right of the table. The two rows of metals beneath the main body of the table are set apart to keep the table from being too wide. Actually, cerium (58) should follow lanthanum (57), and thorium (90) should follow actinium (89). In this book, we use the 1–18 group designation, which has been recommended by the International Union of Pure and Applied Chemistry (IUPAC). A and B group designations were standard U.S. notation until recently and still appear on some periodic tables.

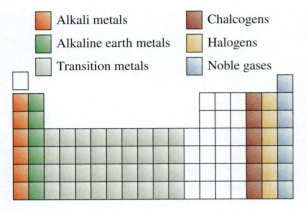

- ▨ Alkali metals
- ▨ Alkaline earth metals
- ▨ Transition metals
- ▨ Chalcogens
- ▨ Halogens
- ▨ Noble gases

At as metalloids. From left to right across any period, the physical and chemical properties of the elements change gradually from metallic to nonmetallic.

Elements are often referred to collectively by their periodic table group number (Group 1, Group 2, and so on). For convenience, however, some element groups have been given special names. The Group 1 elements, with the exception of H (i.e., Li, Na, K, Rb, Cs, and Fr), are called *alkali metals,* and the Group 2 elements (Be, Mg, Ca, Sr, Ba, and Ra) are called *alkaline earth metals.* Elements in Group 16 (O, S, Se, Te, and Po) are sometimes referred to as the *chalcogens.* Elements in Group 17 (F, Cl, Br, I, and At) are known as *halogens,* and elements in Group 18 (He, Ne, Ar, Kr, Xe, and Rn) are called *noble gases,* or rare gases. The elements in Groups 3–11 collectively are called the *transition elements* or *transition metals.*

The periodic table is a handy tool that correlates the properties of the elements in a systematic way and helps us to predict chemical behavior. At the turn of the twentieth century, the periodic table was deemed "the most predictive tool in all of science." We take a more detailed look at this keystone of chemistry in Chapter 7.

The following Bringing Chemistry to Life box describes the distribution of the elements in Earth's crust.

Bringing Chemistry to Life

Distribution of Elements on Earth

Earth's crust extends from the surface to a depth of about 40 km (about 25 mi). Because of technical difficulties, scientists have been unable to study the inner portions of Earth as easily and as thoroughly as the crust. Nevertheless, it is believed that there is a solid core consisting mostly of iron at the center of Earth. Surrounding the core is a layer called the *mantle,* which consists of hot fluid containing iron, carbon, silicon, and sulfur.

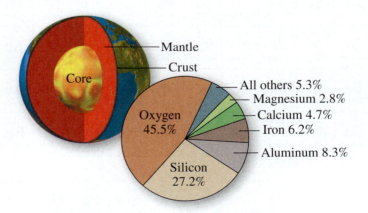

Of the 83 elements that are found in nature, 12 make up 99.7 percent of Earth's crust by mass. They are, in decreasing order of natural abundance, oxygen (O), silicon (Si), aluminum (Al), iron (Fe), calcium (Ca), magnesium (Mg), sodium (Na), potassium (K), titanium (Ti), hydrogen (H), phosphorus (P), and manganese (Mn). When discussing the natural abundance of the elements, keep in mind that the elements are unevenly distributed throughout Earth's crust, and most elements are combined chemically with other elements. Some of the elements that can be found in nature in pure form include gold, copper, and sulfur.

CHECKPOINT – SECTION 2.4   **The Periodic Table**

2.4.1 Which of the following series of elemental symbols lists a nonmetal, a metal, and a metalloid?

a) Ca, Cu, Si

b) K, Mg, B

c) Br, Ba, Ge

d) O, Na, S

e) Ag, Cr, As

2.4.2 Which of the following elements would you expect to have properties most similar to those of chlorine (Cl)?

a) Cu

b) F

c) Na

d) Cr

e) S

2.5   The Atomic Mass Scale and Average Atomic Mass

In some experimental work, it is important to know the masses of individual atoms, which depend on the number of protons, neutrons, and electrons they contain. However, even the smallest speck of dust that our unaided eyes can perceive consists of as many as 1×10^{16} atoms! Although we cannot weigh a single atom, it is possible to determine experimentally the mass of one atom *relative* to another. The first step is to assign a value to the mass of one atom of a given element so that it can be used as a standard.

According to international agreement, one **atomic mass unit (amu)** is defined as a mass exactly equal to one-twelfth the mass of one carbon-12 atom. (Carbon-12 is the carbon isotope that has six protons and six neutrons.) Setting the atomic mass of carbon-12 at 12 amu provides the standard for measuring the **atomic mass** of the other elements. For example, experiments have shown that a hydrogen atom (^1H) is only 8.3985 percent as massive as the carbon-12 atom. Thus, if the mass of one carbon-12 atom is exactly 12 amu, the atomic mass of hydrogen must be 0.083985×12 amu, or 1.0078 amu. (Because the mass of the carbon atom is an exact number [◀◀ Section 1.5], it does not limit the number of significant figures in the calculated result.) Similar calculations show that the atomic mass of fluorine-19 is 18.9984 amu and that of oxygen-16 is 15.9949 amu. Thus, although we cannot measure the mass of a single oxygen-16 atom, we know that it is approximately 16 times as massive as a hydrogen-1 atom.

When you look up the atomic mass of carbon in a table such as the one at the beginning of this book, you will find that its value is 12.01 amu, not 12.00 amu. The difference arises because most naturally occurring elements (including carbon) have more than one isotope. This means that when we measure the atomic mass of an element, we must generally settle for the average mass of the naturally occurring mixture of isotopes. For example, the natural abundances of carbon-12 and carbon-13 are 98.93 percent and 1.07 percent, respectively. The atomic mass of carbon-13 has been determined to be 13.003355 amu. Thus, the average atomic mass of natural carbon can be calculated as follows:

$$\left(\frac{98.93}{100}\right)(12.00000 \text{ amu}) + \left(\frac{1.07}{100}\right)(13.003355 \text{ amu}) = 12.01 \text{ amu}$$

Note that this is a *weighted* average. Because there are many more carbon-12 atoms than carbon-13 atoms in naturally occurring carbon, the average atomic mass is much closer to the mass of carbon-12 than to that of carbon-13.

When we say that the atomic mass of carbon is 12.01 amu, we are referring to the *average* value. If we could examine an individual atom of naturally occurring carbon, we would never find one of atomic mass 12.01 amu. We would find either an atom of atomic mass of exactly 12 amu or one of atomic mass 13.003355 amu—although those of atomic mass 12 amu would be far more common. The atomic

Figure 2.11 Schematic diagram of one type of mass spectrometer.

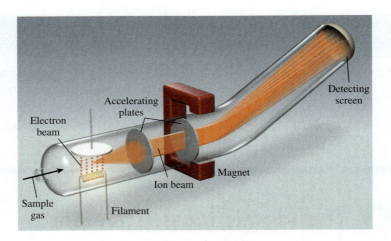

masses in the periodic table are *average atomic masses.* The term ***atomic weight*** is sometimes used to mean average atomic mass.

The atomic masses of many isotopes have been accurately determined to five or six significant figures. For most purposes, though, we will use average atomic masses, which are generally given to four significant figures (see the table at the beginning of this book). For simplicity, we will omit the word *average* when we discuss the atomic masses of the elements.

The most direct and most accurate method for determining atomic and molecular masses is mass spectrometry. In a *mass spectrometer,* such as that depicted in Figure 2.11, a gaseous sample is bombarded by a stream of high-energy electrons. Collisions between the electrons and the gaseous atoms (or molecules) produce positively charged species, called *ions,* by dislodging an electron from the atoms or molecules. These positive ions (of mass m and charge e) are accelerated as they pass through two oppositely charged plates. The emerging ions are deflected into a circular path by a magnet. The radius of the path depends on the charge-to-mass ratio (i.e., e/m). Ions with a small e/m ratio trace a wider arc than those having a larger e/m ratio, so ions with equal charges but different masses are separated from one another. The mass of each ion (and hence its parent atom or molecule) is determined from the magnitude of its deflection. Eventually the ions arrive at the detector, which registers a current for each type of ion. The amount of current generated is directly proportional to the number of ions, so it enables us to determine the relative abundance of isotopes.

The first mass spectrometer, developed in the 1920s by the English physicist F. W. Aston, was crude by today's standards. Nevertheless, it provided indisputable evidence of the existence of isotopes, such as neon-20 (natural abundance 90.48 percent) and neon-22 (natural abundance 9.25 percent). When more sophisticated and sensitive mass spectrometers became available, scientists identified a *third* isotope (neon-21) with natural abundance 0.27 percent (Figure 2.12). This example illustrates

Figure 2.12 Mass spectrum of neon.

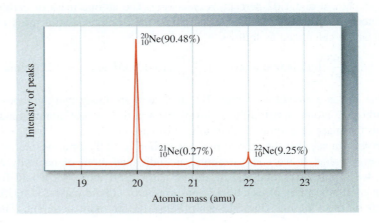

how very important experimental accuracy is to a quantitative science like chemistry. Early experiments failed to detect neon-21 because its natural abundance was so small. Only 27 in 10,000 Ne atoms are neon-21.

Sample Problem 2.3 shows how to calculate the average atomic mass of oxygen.

SAMPLE PROBLEM 2.3

Oxygen is the most abundant element in both Earth's crust and the human body. The atomic masses of its three stable isotopes, $^{16}_{8}O$ (99.757 percent), $^{17}_{8}O$ (0.038 percent), and $^{18}_{8}O$ (0.205 percent), are 15.9949, 16.9991, and 17.9992 amu, respectively. Calculate the average atomic mass of oxygen using the relative abundances given in parentheses. Report the result to four significant figures.

Strategy Each isotope contributes to the average atomic mass based on its relative abundance. Multiplying the mass of each isotope by its fractional abundance (percent value divided by 100) will give its contribution to the average atomic mass.

Setup Each percent abundance can be converted to a fractional abundance: 99.757 percent to 99.757/100 or 0.99757, 0.038 percent to 0.038/100 or 0.00038, and 0.205 percent to 0.205/100 or 0.00205. Once we find the contribution to the average atomic mass for each isotope, we can then add the contributions together to get the average atomic mass.

Solution

$$(0.99757)(15.9949 \text{ amu}) + (0.00038)(16.9991 \text{ amu}) + (0.00205)(17.9992) = 15.999 \text{ amu} \approx 16.00 \text{ amu}$$

THINK ABOUT IT

The average atomic mass should be closest to the atomic mass of the most abundant isotope (oxygen-16, in this case) and should, to the appropriate number of significant figures, be the same number that appears in the periodic table (16.00 amu, in this case).

Practice Problem ATTEMPT The atomic masses of the two stable isotopes of copper, $^{63}_{29}Cu$ (69.17 percent) and $^{65}_{29}Cu$ (30.83 percent), are 62.929599 and 64.927793 amu, respectively. Calculate the average atomic mass of copper.

Practice Problem BUILD The average atomic mass of nitrogen is 14.0067. The atomic masses of the two stable isotopes of nitrogen, ^{14}N and ^{15}N, are 14.003074002 and 15.00010897 amu, respectively. Use this information to determine the percent abundance of each nitrogen isotope.

Practice Problem CONCEPTUALIZE The following diagrams show collections of metal spheres. Each collection consists of two or more different types of metal—represented here by different colors. The masses of metal spheres are as follows: white = 2.3575 g, black = 3.4778 g, and blue = 5.1112 g. For each diagram, determine the average mass of a single metal sphere.

 (i)    (ii)   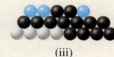 (iii)

CHECKPOINT – SECTION 2.5   The Atomic Mass Scale and Average Atomic Mass

2.5.1 Boron has two naturally occurring isotopes, ^{10}B and ^{11}B, which have masses 10.0129 and 11.0093 amu, respectively. Given the average atomic mass of boron (10.81 amu), determine the percent abundance of each isotope.

a) 50% ^{10}B, 50% ^{11}B

b) 20% ^{10}B, 80% ^{11}B

c) 98% ^{10}B, 2% ^{11}B

d) 93% ^{10}B, 7% ^{11}B

e) 22% ^{10}B, 78% ^{11}B

2.5.2 The two naturally occurring isotopes of antimony, ^{121}Sb (57.21 percent) and ^{123}Sb (42.79 percent), have masses of 120.904 and 122.904 amu, respectively. What is the average atomic mass of Sb?

a) 121.90 amu

b) 122.05 amu

c) 121.76 amu

d) 121.34 amu

e) 122.18 amu

2.6 Ions and Ionic Compounds

Out of all the elements, only the six noble gases in Group 18 of the periodic table (He, Ne, Ar, Kr, Xe, and Rn) exist in the form of isolated atoms under ordinary conditions. For this reason, they are called *monatomic* (meaning a single atom) gases. Most matter is composed of *ions,* which we discuss in this section, or *molecules,* which we discuss in Section 2.7.

The number of positively charged protons in the nucleus of an atom remains the same during ordinary chemical reactions, but negatively charged electrons may be lost or gained—resulting in the formation of *ions.* An **ion** is an atom (or a *group* of atoms) that has a net positive or negative charge. The ions that make up an ionic compound are held together by strong electrostatic forces known as *ionic bonds* [▸▸ Section 8.2].

Atomic Ions

An **atomic ion** or **monatomic ion** is one that consists of just *one* atom with a positive or negative charge. The *loss* of one or more electrons from an atom yields a **cation,** an ion with a net *positive* charge. For example, a sodium atom (Na) can readily lose an electron to become a sodium cation, which is represented by Na^+:

Na Atom	Na$^+$ Ion
11 protons	11 protons
11 electrons	10 electrons

An **anion** is an ion whose net charge is *negative* due to an *increase* in the number of electrons. A chlorine atom (Cl), for instance, can gain an electron to become a chloride ion (Cl^-):

Cl Atom	Cl$^-$ Ion
17 protons	17 protons
17 electrons	18 electrons

Sodium chloride (NaCl), ordinary table salt, is called an **ionic compound** because it consists of *cations* (Na^+) and *anions* (Cl^-).

An atom can lose or gain more than one electron. Examples include Mg^{2+}, Fe^{3+}, S^{2-}, and N^{3-}. Figure 2.13 shows the charges of many more monatomic ions from across the periodic table. With very few exceptions, metals tend to form cations and nonmetals form anions. The charges on monatomic ions of elements in Groups 1 and 2 are predictable. Elements in these groups form cations with charges equal to their

Student Note: Note that a multiple charge is denoted with the number followed by the sign; thus, 2+ not +2.

Figure 2.13 Common monatomic ions arranged by their positions in the periodic table. Note that mercury(I), Hg_2^{2+}, is actually a *poly*atomic ion.

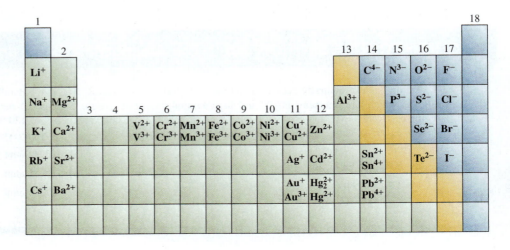

group numbers. Charges on monatomic ions of elements in Group 13 are also predictable, forming cations with charges equal to the second digit of their group number. Aluminum, for example, in Group 13, forms a cation with a charge of +3. Most of the anions that form from elements of Groups 14 through 17 have charges equal to the corresponding group number minus 18. For example, the monatomic anion formed by oxygen (Group 16) has a charge of $16 - 18 = -2$. You should be able to determine the charges on ions of elements in Groups 1 and 2, and elements in Groups 13 through 17 (for any element that forms only one common ion) using only a periodic table.

A monatomic cation is named simply by adding the word *ion* to the name of the element. Thus, the ion of potassium (K^+) is known as potassium ion. Similarly, the cations formed by the elements magnesium and aluminum (Mg^{2+} and Al^{3+}) are called magnesium ion and aluminum ion, respectively. It is not necessary for the name to specify the charge on these ions because their charges are equal to their group numbers.

Certain metals, especially the *transition metals,* can form cations of more than one possible charge. Iron, for example, can form Fe^{2+} and Fe^{3+}. An older nomenclature system that is still in limited use assigns the ending *–ous* to the cation with the *smaller* positive charge and the ending *–ic* to the cation with the *greater* positive charge:

$$Fe^{2+}: \text{ferrous ion}$$
$$Fe^{3+}: \text{ferric ion}$$

This method of naming ions has some distinct limitations. First, the *–ous* and *–ic* suffixes indicate the *relative* charges of the two cations involved, not the *actual* charges. Thus, Fe^{3+} is the ferric ion, but Cu^{2+} is the cupric ion. In addition, the *–ous* and *–ic* endings make it possible to name only two cations with different charges. Some metals, such as manganese (Mn), can form cations with three or more different charges.

Therefore, it has become increasingly common to designate different cations with Roman numerals, using the Stock[3] system. In this system, the Roman numeral I indicates a positive charge of one, II means a positive charge of two, and so on, as shown for manganese:

$$Mn^{2+}: \text{manganese(II) ion}$$
$$Mn^{3+}: \text{manganese(III) ion}$$
$$Mn^{4+}: \text{manganese(IV) ion}$$

These names are pronounced "manganese-two ion," "manganese-three ion," and "manganese-four ion," respectively. Using the Stock system, the ferrous and ferric ions are iron(II) and iron(III), respectively. To avoid confusion, and in keeping with modern practice, we use the Stock system to name compounds in this textbook.

A monatomic anion is named by changing the ending of the element's name to *–ide,* and adding the word *ion.* Thus, the anion of chlorine (Cl^-), is called *chloride ion.* The anions of carbon, nitrogen, and oxygen (C^{4-}, N^{3-}, and O^{2-}) are called *carbide, nitride,* and *oxide,* respectively. Because there is only one possible charge for an ion formed from a nonmetal, it is unnecessary for the ion's name to specify its charge. Table 2.2 lists alphabetically a number of common monatomic ions.

Polyatomic Ions

Ions that consist of a combination of two or more atoms are called **polyatomic ions.** The atoms that make up a polyatomic ion are held together by covalent chemical bonds [▶ Section 8.3]. Because these ions are encountered frequently throughout general chemistry, it is important that you learn and commit to memory the names, formulas, and charges of the polyatomic ions listed in Table 2.3. Although most of the common polyatomic ions are anions, a few are cations.

TABLE 2.2	Names and Formulas of Some Common Monatomic Ions

Name	Formula
Cations	
aluminum	Al^{3+}
barium	Ba^{2+}
cadmium	Cd^{2+}
calcium	Ca^{2+}
cesium	Cs^+
chromium(III)	Cr^{3+}
cobalt(II)	Co^{2+}
copper(I)	Cu^+
copper(II)	Cu^{2+}
hydrogen	H^+
iron(II)	Fe^{2+}
iron(III)	Fe^{3+}
lead(II)	Pb^{2+}
lithium	Li^+
magnesium	Mg^{2+}
manganese(II)	Mn^{2+}
mercury(II)	Hg^{2+}
potassium	K^+
silver	Ag^+
sodium	Na^+
strontium	Sr^{2+}
tin(II)	Sn^{2+}
zinc	Zn^{2+}
Anions	
bromide	Br^-
chloride	Cl^-
fluoride	F^-
hydride	H^-
iodide	I^-
nitride	N^{3-}
oxide	O^{2-}
sulfide	S^{2-}

---

3. Alfred E. Stock (1876–1946). German chemist. Stock did most of his research in the synthesis and characterization of boron, beryllium, and silicon compounds. He was a pioneer in the study of mercury poisoning.

TABLE 2.3	Common Polyatomic Ions
Name	**Formula/Charge**
Cations	
ammonium	NH_4^+
hydronium	H_3O^+
mercury(I)	Hg_2^{2+}
Anions	
acetate	$C_2H_3O_2^-$
azide	N_3^-
carbonate	CO_3^{2-}
chlorate	ClO_3^-
chlorite	ClO_2^-
chromate	CrO_4^{2-}
cyanide	CN^-
dichromate	$Cr_2O_7^{2-}$
dihydrogen phosphate	$H_2PO_4^-$
hydrogen carbonate or bicarbonate	HCO_3^-
hydrogen phosphate	HPO_4^{2-}
hydrogen sulfate or bisulfate	HSO_4^-
hydroxide	OH^-
hypochlorite	ClO^-
nitrate	NO_3^-
nitrite	NO_2^-
oxalate	$C_2O_4^{2-}$
perchlorate	ClO_4^-
permanganate	MnO_4^-
peroxide	O_2^{2-}
phosphate	PO_4^{3-}
phosphite	PO_3^{3-}
sulfate	SO_4^{2-}
sulfite	SO_3^{2-}
thiocyanate	SCN^-

Student Note: Some oxoanions occur in series of ions that contain the same central atom and have the same charge, but contain different numbers of oxygen atoms.

perchlorate	ClO_4^-
chlorate	ClO_3^-
chlorite	ClO_2^-
hypochlorite	ClO^-
nitrate	NO_3^-
nitrite	NO_2^-
phosphate	PO_4^{3-}
phosphite	PO_3^{3-}
sulfate	SO_4^{2-}
sulfite	SO_3^{2-}

Formulas of Ionic Compounds

The formulas of ionic compounds indicate the smallest whole number ratio in which the ions combine to form an electrically neutral substance. For example, in sodium chloride, sodium ions and chloride ions combine in a 1:1 ratio, making the formula NaCl. In magnesium chloride, magnesium ions and chloride ions are combined in a 1:2 ratio, making the formula $MgCl_2$. Formulas such as these, that indicate the ratio of combination, are called *empirical formulas.* The word *empirical* means "from experience" or, in the context of chemical formulas, "from experiment."

An ionic compound does not exist as a collection of discrete units—there is no such thing as an NaCl *particle.* Rather, it consists of a vast, highly ordered array of interspersed cations and anions called a *lattice.* For example, solid sodium chloride (NaCl) consists of equal numbers of Na^+ and Cl^- ions arranged in a three-dimensional network of alternating cations and anions (Figure 2.14). As you can see in Figure 2.14, no Na^+ ion in NaCl is associated with any particular Cl^- ion. In fact, each Na^+ ion

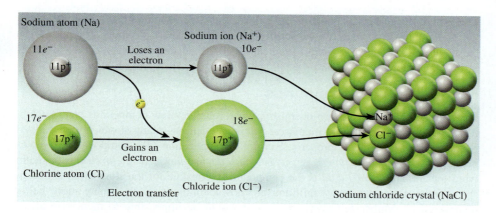

Figure 2.14 An electron is transferred from the sodium atom to the chlorine atom, giving a sodium ion and a chloride ion. The oppositely charged ions are attracted to each other electrostatically and form a solid lattice.

is surrounded by six Cl^- ions and vice versa. In other ionic compounds, the actual structure may be different, but the arrangement of cations and anions is such that the compounds are all electrically neutral. The charges on the cation and anion are not shown in the formula for an ionic compound.

For ionic compounds to be electrically neutral, the sum of the charges on the cations and anions in each formula unit must be zero. When the charges on the cations and anions are numerically *equal,* the ions will combine in a 1:1 ratio. If the charges on the cations and anions are numerically *different,* you can apply the following guideline to make the formula electrically neutral (and thus obtain the empirical formula): Write a subscript for the cation that is numerically equal to the charge on the anion and a subscript for the anion that is numerically equal to the charge on the cation. As we have seen in the cases of NaCl and $MgCl_2$, when the appropriate subscript would be 1, we do not include it in the formula.

Let's consider some examples.

Potassium Bromide The potassium ion (K^+) and the bromide ion (Br^-) combine to form the ionic compound *potassium bromide*. The sum of the charges is $1 + (-1) = 0$, so no subscripts are necessary. The formula is KBr.

Zinc Iodide The zinc ion (Zn^{2+}) and the iodide ion (I^-) combine to form *zinc iodide*. The sum of the charges of one Zn^{2+} ion and one I^- ion is $+2 + (-1) = +1$. To make the charges add up to zero, we multiply the -1 charge of the anion by 2 and add the subscript "2" to the symbol for iodine. Thus, the formula for zinc iodide is ZnI_2.

Ammonium Chloride The cation is NH_4^+ and the anion is Cl^-. The sum of the charges is $1 + (-1) = 0$, so the ions combine in a 1:1 ratio and the resulting formula is NH_4Cl.

Aluminum Oxide The cation is Al^{3+} and the anion is O^{2-}. The following diagram can be used to determine the subscripts for this compound:

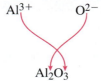

The sum of the charges for aluminum oxide is $2(+3) + 3(-2) = 0$. Thus, the formula is Al_2O_3.

Calcium Phosphate The cation is Ca^{2+} and the anion is PO_4^{3-}. The following diagram can be used to determine the subscripts:

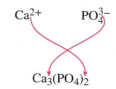

The sum of the charges is $3(+2) + 2(-3) = 0$. Thus, the formula for calcium phosphate is $Ca_3(PO_4)_2$. When we add a subscript to a polyatomic ion, we must first put parentheses around the ion's formula to indicate that the subscript applies to *all* the atoms in the polyatomic ion.

Naming Ionic Compounds

Early in the history of modern chemistry, when relatively few compounds were known, it was possible for chemists to memorize compound names—many of which were derived from physical appearance, properties, origin, or application of a compound. Examples include *milk of magnesia, laughing gas, baking soda,* and *formic acid* (*formica* is the Latin word for ant, and *formic acid* is the compound responsible for the sting of an ant bite).

Today there are many millions of known compounds, and many more being discovered or synthesized every year, so it would be impossible to memorize all of their names. Fortunately, it is unnecessary, because over the years chemists have devised a convenient system for naming chemical substances. The rules are the same worldwide, facilitating communication among scientists and providing a useful way of labeling an overwhelming variety of substances. Mastering these rules now will benefit you tremendously as you progress through your chemistry course. You must be able to name a compound, given its chemical formula, and you must be able to write the chemical formula of a compound, given its name.

An ionic compound is named simply by using the name of the cation followed by the name of the anion, eliminating the word *ion* from each. Several examples were given earlier in the Formulas of Ionic Compounds section. Other examples are sodium cyanide (NaCN), potassium permanganate ($KMnO_4$), and ammonium sulfate [$(NH_4)_2SO_4$]. Note that the names of these ionic compounds do not indicate the ratios of combination explicitly—although the chemical formulas do. There is only one neutral combination possible for each of these examples. Lithium ion has a charge of +1, always. Cyanide ion has a charge of −1, always. Their ratio of combination can only be 1:1. The same is true of potassium permanganate, in which the charges are also +1 and −1. Ammonium ion has a charge of +1 and sulfate ion has a charge of −2, always—making the only possible neutral combination of ammonium and sulfate ions 2:1. With ionic compounds, it is unnecessary to include the ratio of combination in the *name* because the charges on the ions are known.

In cases where a metal cation can have more than one possible charge, recall that the charge is indicated in the name of the ion with a Roman numeral in parentheses. Thus, the compounds $FeCl_2$ and $FeCl_3$ are named *iron(II) chloride* and *iron(III) chloride,* respectively. (These are pronounced "iron-two chloride" and "iron-three chloride.")

Sample Problems 2.4 and 2.5 illustrate how to name ionic compounds and write formulas for ionic compounds based on the information given in Tables 2.2 and 2.3.

SAMPLE PROBLEM 2.4

Name the following ionic compounds: (a) MgO, (b) Al(OH)$_3$, and (c) Fe$_2$(SO$_4$)$_3$.

Strategy Begin by identifying the cation and the anion in each compound, and then combine the names for each, eliminating the word *ion.*

Setup MgO contains Mg^{2+} and O^{2-}, the magnesium ion and the oxide ion; Al(OH)$_3$ contains Al^{3+} and OH^-, the aluminum ion and the hydroxide ion; and Fe$_2$(SO$_4$)$_3$ contains Fe^{3+} and SO_4^{2-}, the iron(III) ion and the sulfate ion. We know that the iron in Fe$_2$(SO$_4$)$_3$ is iron(III), Fe^{3+}, because it is combined with the sulfate ion in a 2:3 ratio.

Solution (a) Combining the cation and anion names, and eliminating the word *ion* from each of the individual ions' names, we get *magnesium oxide* as the name of MgO; (b) Al(OH)$_3$ is *aluminum hydroxide;* and (c) Fe$_2$(SO$_4$)$_3$ is *iron(III) sulfate.*

THINK ABOUT IT

Be careful not to confuse the subscript in a formula with the charge on the metal ion. In part (c), for example, the subscript on Fe is 2, but this is an iron(III) compound.

Practice Problem **A**TTEMPT Name the following ionic compounds: (a) Na_2SO_4, (b) $Cu(NO_3)_2$, (c) $Fe_2(CO_3)_3$.

Practice Problem **B**UILD Name the following ionic compounds: (a) $K_2Cr_2O_7$, (b) $Li_2C_2O_4$, (c) $CuNO_3$.

Practice Problem **C**ONCEPTUALIZE The diagram represents a small sample of an ionic compound where red spheres represent nitrate ions and grey spheres represent iron ions. Deduce the correct formula and name of the compound.

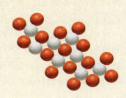

SAMPLE PROBLEM 2.5

Deduce the formulas of the following ionic compounds: (a) mercury(I) chloride, (b) lead(II) chromate, and (c) potassium hydrogen phosphate.

Strategy Identify the ions in each compound, and determine their ratios of combination using the charges on the cation and anion in each.

Setup (a) Mercury(I) chloride is a combination of Hg_2^{2+} and Cl^-. [Mercury(I) is one of the few cations listed in Table 2.3.] In order to produce a neutral compound, these two ions must combine in a 1:2 ratio. (b) Lead(II) chromate is a combination of Pb^{2+} and CrO_4^{2-}. These ions combine in a 1:1 ratio. (c) Potassium hydrogen phosphate is a combination of K^+ and HPO_4^{2-}. These ions combine in a 2:1 ratio.

Solution The formulas are (a) Hg_2Cl_2, (b) $PbCrO_4$, and (c) K_2HPO_4.

> ### THINK ABOUT IT
>
> Make sure that the charges sum to zero in each compound formula. In part (a), for example, $Hg_2^{2+} + Cl^- = (2+) + 2(-1) = 0$; in part (b), $(+2) + (-2) = 0$; and in part (c), $2(+1) + (-2) = 0$.

Practice Problem **A**TTEMPT Deduce the formulas of the following ionic compounds: (a) lead(II) chloride, (b) magnesium carbonate, and (c) ammonium phosphate.

Practice Problem **B**UILD Deduce the formulas of the following ionic compounds: (a) iron(III) sulfide, (b) mercury(II) nitrate, and (c) potassium sulfite.

Practice Problem **C**ONCEPTUALIZE The diagram represents a small sample of an ionic compound where yellow spheres represent sulfite ions and blue spheres represent copper ions. Deduce the correct formula and name of the compound.

Oxoanions

Oxoanions are polyatomic anions that contain one or more oxygen atoms and one atom (the "central atom") of another element. Examples include the chlorate (ClO_3^-), nitrate (NO_3^-), and sulfate (SO_4^{2-}) ions. Often, two or more oxoanions have the same central atom but different numbers of O atoms (e.g., NO_3^- and NO_2^-). Starting with the oxoanions whose names end in *–ate,* we can name these ions as follows:

1. The ion with one *more* O atom than the *–ate* ion is called the *per . . . ate* ion. Thus, ClO_3^- is the chlorate ion, so ClO_4^- is the *perchlorate ion.*
2. The ion with one *less* O atom than the *–ate* anion is called the *–ite* ion. Thus, ClO_2^- is the *chlorite ion.*
3. The ion with *two* fewer O atoms than the *–ate* ion is called the *hypo . . . ite* ion. Thus, ClO^- is the *hypochlorite ion.*

At a minimum, you must commit to memory the formulas and charges of the oxoanions whose names end in *–ate* so that you can apply these guidelines when necessary.

Sample Problem 2.6 tests your ability to name and identify oxoanions.

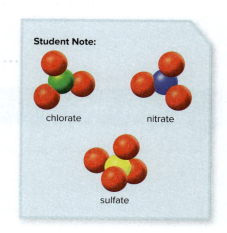

Student Note:

chlorate   nitrate

sulfate

SAMPLE PROBLEM 2.6

Name the following species: (a) BrO_4^-, (b) HCO_3^-, and (c) SO_3^{2-}.

Strategy Each species is an oxoanion. Identify the "reference oxoanion" (the one with the *–ate* ending) for each, and apply the rules to determine appropriate names.

Setup (a) Chlorine, bromine, and iodine (members of Group 17) all form analogous series of oxoanions with one to four oxygen atoms. Thus, the reference oxoanion is bromate (BrO_3^-), which is analogous to chlorate (ClO_3^-). In part (b), HCO_3^- has one more hydrogen than the carbonate ion (CO_3^{2-}). In part (c), the reference ion is sulfate (SO_4^{2-}).

Solution (a) BrO_4^- has one more O atom than the bromate ion (BrO_3^-), so BrO_4^- is the *perbromate* ion. (b) CO_3^{2-} is the carbonate ion. Because HCO_3^- has one ionizable hydrogen atom, it is called the *hydrogen carbonate ion*. (c) SO_3^{2-} has one less oxygen atom than the reference ion, so it is the *sulfite* ion.

THINK ABOUT IT

Remembering all these names and formulas is greatly facilitated by memorizing the common ions that end in *–ate*.

chlorate	ClO_3^-	nitrate	NO_3^-	bromate	BrO_3^-	oxalate	$C_2O_4^{2-}$	phosphate	PO_4^{3-}
iodate	IO_3^-	carbonate	CO_3^{2-}	sulfate	SO_4^{2-}	chromate	CrO_4^{2-}	permanganate	MnO_4^-

Practice Problem (A)TTEMPT Name the following species: (a) BrO^-, (b) HSO_4^-, and (c) $C_2O_4^{2-}$.

Practice Problem (B)UILD Name the following species: (a) IO_2^-, (b) $HCrO_4^-$, and (c) $HC_2O_4^-$.

Practice Problem (C)ONCEPTUALIZE The diagrams show models of a series of oxoanions. Which of the models represent an anion whose name ends in *–ate*?

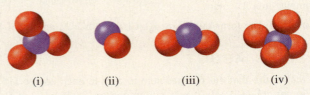

(i)   (ii)   (iii)   (iv)

Hydrates

Figure 2.15 $CuSO_4$ is white. The pentahydrate, $CuSO_4 \cdot 5H_2O$, is blue.

Charles D. Winters/Timeframe Photography/ McGraw Hill

Hydrates are *ionic* compounds that also contain water *molecules* within their solid structure. In its normal state, for example, each unit of copper(II) sulfate has five water molecules associated with it. The systematic name for this compound is copper(II) sulfate pentahydrate, and its formula is written as $CuSO_4 \cdot 5H_2O$. The water molecules can be driven off by heating. When this occurs, the resulting compound is $CuSO_4$, which is sometimes called anhydrous copper(II) sulfate; *anhydrous* means that the compound no longer has water molecules associated with it. Hydrates and the corresponding anhydrous compounds often have distinctly different physical and chemical properties (Figure 2.15).

Some other hydrates are

$BaCl_2 \cdot 2H_2O$	barium chloride dihydrate
$LiCl \cdot H_2O$	lithium chloride monohydrate
$MgSO_4 \cdot 7H_2O$	magnesium sulfate heptahydrate
$Sr(NO_3)_2 \cdot 4H_2O$	strontium nitrate tetrahydrate

CHECKPOINT – SECTION 2.6   Ions and Ionic Compounds

2.6.1 What is the correct name of the compound $PbSO_4$?

a) Lead sulfate

b) Lead(I) sulfate

c) Lead(II) sulfate

d) Monolead sulfate

e) Lead monosulfate

2.6.2 What is the correct formula for the compound iron(III) carbonate?

a) $FeCO_3$

b) Fe_3CO_3

c) Fe_2CO_3

d) $Fe_2(CO_3)_3$

e) $Fe_3(CO_3)_2$

2.6.3 Which of the following is the correct formula for nitrous acid?

a) HNO

b) HN_2O

c) N_2O

d) HNO_2

e) HNO_3

2.6.4 What is the formula of nickel(II) nitrate hexahydrate?

a) $NiNO_3 \cdot 6H_2O$

b) $Ni_2NO_3 \cdot 6H_2O$

c) $Ni(NO_3)_2 \cdot 6H_2O$

d) $NiNO_3 \cdot 12H_2O$

e) $Ni(NO_3)_2 \cdot 12H_2O$

2.6.5 What is the correct formula for sodium nitride?

a) NaN

b) NaN_3

c) Na_3N

d) $NaNO_3$

e) $NaNO_2$

2.6.6 What is the correct name of the compound Hg_2CrO_4?

a) Mercury(I) chromate

b) Mercury(II) chromate

c) Mercury dichromate

d) Dimercury chromate

e) Monomercury chromate

2.7 Molecules and Molecular Compounds

A *molecule* is an electrically neutral combination of at least two atoms in a specific arrangement held together by electrostatic forces known as *covalent chemical bonds* [▶▶ Section 8.3]. A molecule may contain atoms of the same element, or it may contain atoms of two or more elements joined in a fixed ratio, in accordance with the law of definite proportions [◀◀ Section 2.1]. Thus, a molecule can be an *element* or it can be a *compound,* which, by definition, is made up of two or more elements [◀◀ Section 1.2]. Hydrogen gas, for example, is an element, but it consists of molecules, each of which is made up of two H atoms. Water, on the other hand, is a compound that consists of molecules, each of which contains two H atoms and one O atom.

The hydrogen molecule, symbolized as H_2, is called a *diatomic molecule* because it contains *two* atoms. Other elements that normally exist as diatomic molecules are nitrogen (N_2), oxygen (O_2), and the Group 17 elements—fluorine (F_2), chlorine (Cl_2), bromine (Br_2), and iodine (I_2). These are known as *homonuclear* diatomic molecules because both atoms in each molecule are of the same element. A diatomic molecule can also contain atoms of different elements. Examples of these *heteronuclear* diatomic molecules include hydrogen chloride (HCl) and carbon monoxide (CO).

Most molecules contain more than two atoms. They can all be atoms of the same element, as in ozone (O_3) and white phosphorus (P_4), or they can be combinations of two or more different elements, as in water (H_2O) and methane (CH_4). Molecules containing more than two atoms are called *polyatomic molecules.*

Homonuclear diatomic

Heteronuclear diatomic

Polyatomic

Molecular Formulas

A *chemical formula* denotes the composition of the substance. A *molecular formula* shows the exact number of atoms of each element in a molecule. In our discussion of molecules, each example was given with its molecular formula in parentheses. Thus, H_2 is the molecular formula for hydrogen, O_2 is that for oxygen, O_3 is that for ozone, and H_2O is that of water. The subscript numeral indicates the number of atoms of an element present in the molecule. There is no subscript for O in H_2O because there is only one atom of oxygen in a molecule of water. Just as with the formulas of ionic compounds, we do not write 1 as a subscript. Oxygen (O_2) and ozone (O_3) are allotropes of oxygen. An *allotrope* is one of two or more distinct forms of an

H₂O

H:O:H

H—O—H

Figure 2.16 Several ways to represent the water molecule.

element. Two of the allotropic forms of the element carbon—diamond and graphite—have dramatically different properties (and *prices*).

We can also represent molecules with *structural formulas*. The **structural formula** shows not only the elemental composition, but also the general arrangement of atoms within the molecule. In the case of water, each of the hydrogen atoms is connected to the oxygen atom. Figure 2.16 shows the molecular formula, structural formula, and molecular models (both ball-and-stick and space-filling variety) for water. Note that the chemical bond between two atoms can be represented with either a pair of dots or a line.

In Chapters 8 and 9, we explain how to use the molecular formula to deduce the structural formula and the three-dimensional arrangement of a molecule. We use all of these methods for representing molecules throughout the book, so you should be familiar with each method and with the information it provides.

Sample Problem 2.7 shows how to write a molecular formula from the corresponding molecular model.

SAMPLE PROBLEM   2.7

Write the molecular formula of ethanol based on its ball-and-stick model, shown here.

Strategy See Table 1.1.

Setup There are *two* carbon atoms, *six* hydrogen atoms, and *one* oxygen atom, so the subscript on C will be 2 and the subscript on H will be 6, and there will be no subscript on O.

Solution C_2H_6O

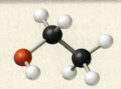

Ethanol

THINK ABOUT IT

Often the molecular formula for an organic compound such as ethanol is written so that the formula more closely resembles the actual arrangement of atoms in the molecule. Thus, the molecular formula for ethanol is commonly written as C_2H_5OH.

Practice Problem **A** **TTEMPT** Chloroform was used as an anesthetic for childbirth and surgery during the nineteenth century. Write the molecular formula for chloroform based on the molecular model shown here.

Chloroform

Practice Problem **B** **UILD** Write the molecular formula for acetone based on the molecular model shown below.

Student Note: Note that acetone contains a *double* bond. Multiple bonds between atoms are discussed in Chapters 8 and 9.

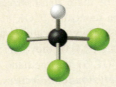

Acetone

Practice Problem **C** **ONCEPTUALIZE** How many ball-and-stick models of ethanol molecules can be constructed using the collection of balls shown here? How many of each color ball will be left over?

TABLE 2.4	Greek Prefixes		
Prefix	**Meaning**	**Prefix**	**Meaning**
Mono–	1	Hexa–	6
Di–	2	Hepta–	7
Tri–	3	Octa–	8
Tetra–	4	Nona–	9
Penta–	5	Deca–	10

TABLE 2.5	Some Compounds Named Using Greek Prefixes		
Compound	**Name**	**Compound**	**Name**
CO	Carbon monoxide	SO_3	Sulfur trioxide
CO_2	Carbon dioxide	NO_2	Nitrogen dioxide
SO_2	Sulfur dioxide	N_2O_5	Dinitrogen pentoxide

Naming Molecular Compounds

Most molecular substances are **binary compounds,** meaning that they are composed of atoms of two different elements. Binary molecular compounds contain two different *nonmetals* (see Figure 2.10). (The term binary can also refer to ionic compounds composed of two different elements—a metal and a nonmetal.) To name a binary molecular compound, we first name the element that appears first in the formula. For HCl that would be hydrogen. We then name the second element, changing the ending of its name to *–ide.* For HCl, the second element is chlorine, so we would change chlorine to chloride. Thus, the systematic name of HCl is *hydrogen chloride.* Similarly, HI is hydrogen iodide (iod*ine* → iod*ide*) and SiC is silicon carbide (carb*on* → carb*ide*).

Unlike the case with binary ionic compounds, which can only combine in one specific ratio, it is quite common for one pair of nonmetallic elements to form several *different* binary molecular compounds. In these cases, confusion in naming the compounds is avoided by the use of Greek prefixes to denote the number of atoms of each element present. Some of the Greek prefixes are listed in Table 2.4, and several compounds named using prefixes are listed in Table 2.5.

The prefix *mono–* is generally omitted for the first element. SO_2, for example, is named *sulfur dioxide,* not *monosulfur dioxide.* Thus, the absence of a prefix for the first element usually means there is only one atom of that element present in the molecule. In addition, for ease of pronunciation, we usually eliminate the last letter of a prefix that ends in "o" or "a" when naming an oxide. Thus, N_2O_5 is *dinitrogen pentoxide,* rather than *dinitrogen pentaoxide.*

Sample Problem 2.8 gives you some practice naming binary molecular compounds from their formulas.

SAMPLE PROBLEM (2.8)

Name the following binary molecular compounds: (a) NF_3 and (b) N_2O_4.

Strategy Each compound will be named using the systematic nomenclature including, where necessary, appropriate Greek prefixes.

Setup With binary compounds, we start with the name of the element that appears *first* in the formula, and we change the ending of the *second* element's name to *–ide.* We use prefixes, where appropriate, to indicate the number of atoms of each element. In part (a) the molecule contains one nitrogen atom and three fluorine atoms. We will omit the prefix *mono–* for nitrogen because it is the first element listed in the formula, and we will use the prefix *tri–* to denote the number of fluorine atoms. In part (b) the molecule contains two nitrogen atoms and four oxygen atoms, so we will use the prefixes *di–* and *tetra–* in naming the compound. Recall that in naming an oxide, the last letter of a prefix that ends in "a" or "o" is omitted.

Solution (a) nitrogen trifluoride and (b) dinitrogen tetroxide

THINK ABOUT IT

Make sure that the prefixes match the subscripts in the molecular formulas and that the word *oxide* is not preceded immediately by an "a" or an "o."

(Continued on next page)

Practice Problem (A)**TTEMPT** Name the following binary molecular compounds: (a) Cl_2O and (b) $SiCl_4$.

Practice Problem (B)**UILD** Name the following binary molecular compounds: (a) ClO_2 and (b) CBr_4.

Practice Problem (C)**ONCEPTUALIZE** Name the binary molecular compound shown.

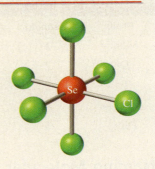

Writing the formula for a molecular compound, given its systematic name, is usually straightforward. For instance, the name *boron trichloride* indicates the presence of one boron atom (no prefix) and three chlorine atoms (*tri*–), so the corresponding molecular formula is BCl_3. Note once again that the order of the elements is the same in both the name and the formula.

Sample Problem 2.9 gives you some practice determining the formulas of binary molecular compounds from their names.

SAMPLE PROBLEM 2.9

Write the chemical formulas for the following binary molecular compounds: (a) sulfur tetrafluoride and (b) tetraphosphorus decasulfide.

Strategy The formula for each compound will be deduced using the systematic nomenclature guidelines.

Setup In part (a) there is no prefix for sulfur, so there is only one sulfur atom in a molecule of the compound. Therefore, we will use no prefix for the S in the formula. The prefix *tetra*– means that there are four fluorine atoms. In part (b) the prefixes *tetra*– and *deca*– denote four and ten, respectively.

Solution (a) SF_4 and (b) P_4S_{10}

THINK ABOUT IT

Double-check that the subscripts in the formulas match the prefixes in the compound names: (a) 4 = *tetra* and (b) 4 = *tetra* and 10 = *deca*.

Practice Problem (A)**TTEMPT** Give the molecular formula for each of the following compounds: (a) carbon disulfide and (b) dinitrogen trioxide.

Practice Problem (B)**UILD** Give the molecular formula for each of the following compounds: (a) sulfur hexafluoride and (b) disulfur decafluoride.

Practice Problem (C)**ONCEPTUALIZE** Draw a molecular model of sulfur trioxide.

Student Note: Binary compounds containing carbon and hydrogen are *organic* compounds and do not follow the same naming conventions as other molecular compounds. Organic compounds and their nomenclature are discussed in detail in Chapter 25.

The names of molecular compounds containing hydrogen do not usually conform to the systematic nomenclature guidelines. Traditionally, many of these compounds are called either by their common, nonsystematic names or by names that do not indicate explicitly the number of H atoms present:

B_2H_6	Diborane	PH_3	Phosphine
SiH_4	Silane	H_2O	Water
NH_3	Ammonia	H_2S	Hydrogen sulfide

Even the order in which the elements are written in these hydrogen-containing compounds is irregular. In water and hydrogen sulfide, H is written first, whereas it is written last in the other compounds.

TABLE 2.6	Some Simple Acids	
Formula	**Binary Compound Name**	**Acid Name**
HF	Hydrogen fluoride	Hydrofluoric acid
HCl	Hydrogen chloride	Hydrochloric acid
HBr	Hydrogen bromide	Hydrobromic acid
HI	Hydrogen iodide	Hydroiodic acid
HCN*	Hydrogen cyanide	Hydrocyanic acid

*Although HCN is not a *binary* compound, it is included in this table because it is similar chemically to HF, HCl, HBr, and HI.

Simple Acids

Acids make up another important class of molecular compounds. One definition of an ***acid*** is a substance that produces hydrogen ions (H^+) when dissolved in water. Several binary molecular compounds produce hydrogen ions when dissolved in water and are, therefore, acids. In these cases, two different names can be assigned to the same chemical formula. For example, HCl, *hydrogen chloride,* is a gaseous compound. When it is dissolved in water, however, we call it *hydrochloric acid.* The rules for naming simple acids of this type are as follows: remove the *–gen* ending from hydrogen (leaving *hydro–*), change the *–ide* ending on the second element to *–ic,* combine the two words, and add the word *acid.*

$$\text{hydro}gen \text{ chlor}ide + -ic \text{ acid} \longrightarrow \text{hydrochloric acid}$$

Likewise, hydrogen fluoride (HF) becomes *hydrofluoric acid.* Table 2.6 lists these and other examples.

For a compound to produce hydrogen ions upon dissolving, it must contain at least one ***ionizable hydrogen atom.*** An ionizable hydrogen atom is one that separates from the molecule upon dissolving and becomes a hydrogen ion (H^+).

Oxoacids

In addition to the simple acids, there is another important class of acids known as ***oxoacids,*** which ionize to produce hydrogen ions and the corresponding oxoanions. The formula of an oxoacid can be determined by adding enough H^+ ions to the corresponding oxoanion to yield a formula with no net charge. For example, the formulas of oxoacids based on the nitrate (NO_3^-) and sulfate (SO_4^{2-}) ions are HNO_3 and H_2SO_4, respectively. The names of oxoacids are derived from the names of the corresponding oxoanions using the following guidelines:

1. An acid based on an *–ate* ion is called . . . *ic* acid. Thus, $HClO_3$ is called *chloric acid.*
2. An acid based on an *–ite* ion is called . . . *ous* acid. Thus, $HClO_2$ is called *chlorous acid.*
3. Prefixes in oxoanion names are retained in the names of the corresponding oxoacids. Thus, $HClO_4$ and $HClO$ are called *perchloric acid* and *hypochlorous acid,* respectively.

Many oxoacids, such as H_2SO_4 and H_3PO_4, are ***polyprotic acid***—meaning that they have more than one ionizable hydrogen atom. In these cases, the names of anions in which one or more (but not all) of the hydrogen ions have been removed must indicate the number of H ions that remain, as shown for the anions derived from phosphoric acid:

H_3PO_4	phosphoric acid	HPO_4^{2-}	hydrogen phosphate ion
$H_2PO_4^-$	dihydrogen phosphate ion	PO_4^{3-}	phosphate ion

Student Note: In Chapter 16, we explore acids and bases in greater detail; and we explain that there are other ways to define the terms *acid* and *base.*

Student Note: An ion is a *charged* species.

Sample Problem 2.10 lets you practice identifying and naming oxoacids.

SAMPLE PROBLEM 2.10

Determine the formula of sulfurous acid.

Strategy The *–ous* ending in the name of an acid indicates that the acid is derived from an oxoanion ending in *–ite.* Determine the formula and charge of the oxoanion, and add enough hydrogens to make a neutral formula.

Setup The sulfite ion is SO_3^{2-}.

Solution The formula of sulfurous acid is H_2SO_3.

THINK ABOUT IT

Note that none of the oxoacids' names begins with the prefix *hydro–*. The prefix *hydro–* in an acid's name indicates that the acid is *binary.*

Practice Problem A TTEMPT Determine the formula of perbromic acid. (Refer to the information in Sample Problem 2.6.)

Practice Problem B UILD Determine the formula of chromic acid.

Practice Problem C ONCEPTUALIZE Referring to the diagrams in Practice Problem 2.6C, which of the ions shown would be part of an acid whose name begins with a prefix?

So far, our discussion of nomenclature has focused on **inorganic compounds,** which generally are defined as compounds that do not contain carbon—although some carbon-containing species such as CN^- and CO_3^{2-} are considered inorganic. Another important class of molecular substances is *organic* compounds, which have their own system of nomenclature. **Organic compounds** contain carbon and hydrogen, sometimes in combination with other elements such as oxygen, nitrogen, sulfur, and the halogens. The simplest organic compounds are those that contain only carbon and hydrogen and are known as **hydrocarbons.** Among hydrocarbons, the simplest examples are compounds known as **alkanes.** The name of an alkane depends on the number of carbon atoms in the molecule. Table 2.7 gives the molecular formulas, systematic names, and ball-and-stick models of some of the simplest alkanes.

Many organic compounds are derivatives of alkanes in which one of the H atoms has been replaced by a group of atoms known as a **functional group.** The functional group determines many of the chemical properties of a compound because it typically is where a chemical reaction *occurs.* Table 2.8 lists the names and provides ball-and-stick models of several important functional groups.

Ethanol, for example, the alcohol in alcoholic beverages, is ethane (C_2H_6) with one of the hydrogen atoms replaced by an alcohol (—OH) group. Its name is derived from that of *ethane,* indicating that it contains *two* carbon atoms.

The molecular formula of ethanol can also be written C_2H_6O, but C_2H_5OH conveys more information about the structure of the molecule. Organic compounds and several functional groups are discussed in greater detail in Chapter 25.

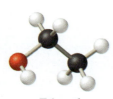

Ethanol

Empirical Formulas of Molecular Substances

In addition to the ways we have learned so far, like ionic substances, molecular substances can be represented using *empirical formulas* [◀◀ Section 2.6]. The empirical formula tells what elements are present in a molecule and in what whole number ratio they are combined. For example, the molecular formula of hydrogen peroxide is H_2O_2, but its empirical formula is simply HO. Hydrazine, which has

TABLE 2.7	Formulas, Names, and Models of Some Simple Alkanes	
Formula	Name	Model
CH_4	Methane	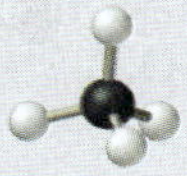
C_2H_6	Ethane	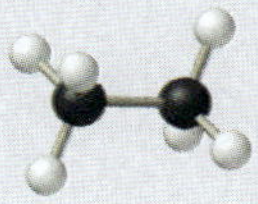
C_3H_8	Propane	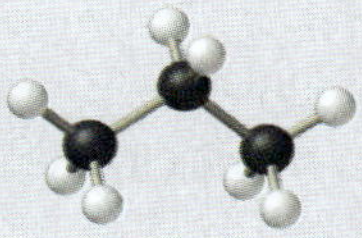
C_4H_{10}	Butane	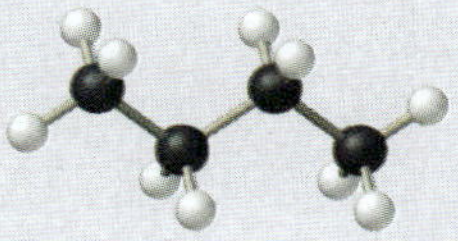
C_5H_{12}	Pentane	
C_6H_{14}	Hexane	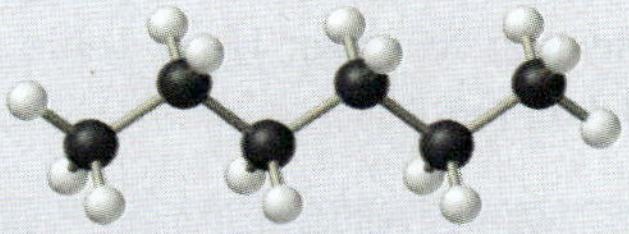
C_7H_{16}	Heptane	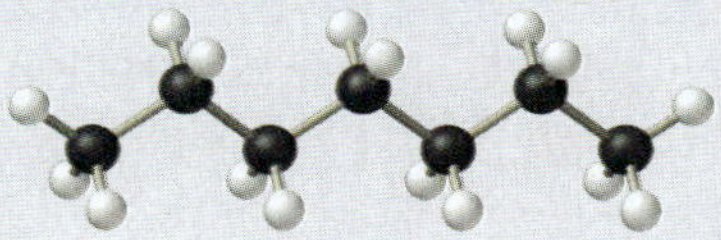
C_8H_{18}	Octane	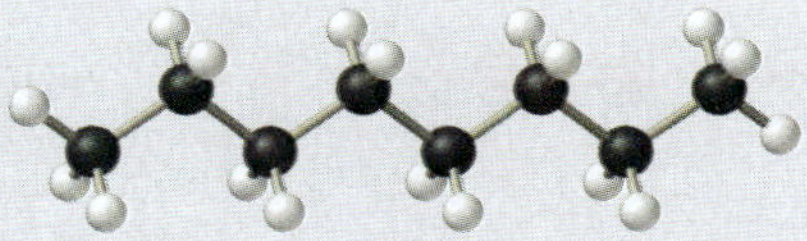
C_9H_{20}	Nonane	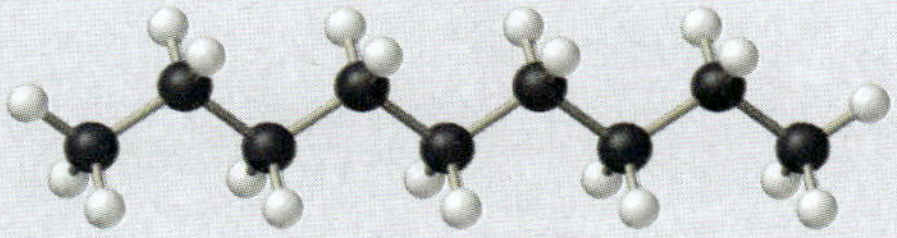
$C_{10}H_{22}$	Decane	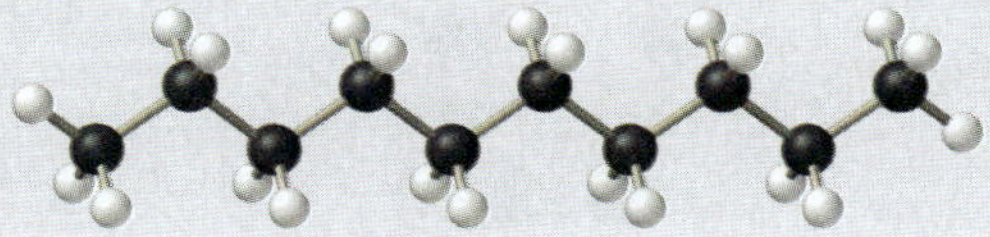

been used as a rocket fuel, has the molecular formula N_2H_4, so its empirical formula is NH_2. Although the ratio of nitrogen to hydrogen is 1:2 in both the molecular formula (N_2H_4) and the empirical formula (NH_2), only the molecular formula tells us the actual number of N atoms (two) and H atoms (four) present in a hydrazine molecule.

TABLE 2.8	Organic Functional Groups	
Name	**Functional Group**	**Model**
Alcohol	—OH	
Aldehyde	—CHO	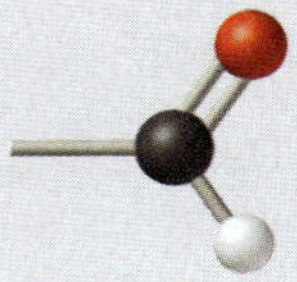
Carboxylic acid	—COOH	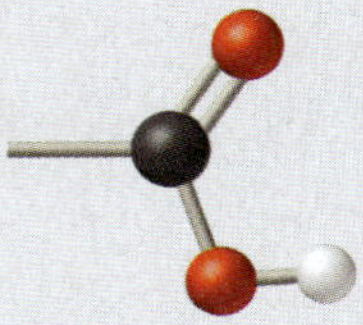
Amine	—NH$_2$	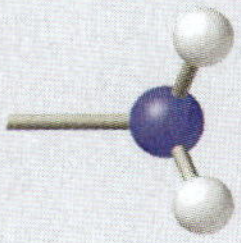

In many cases, the empirical and molecular formulas are identical. In the case of water, for example, there is no combination of smaller whole numbers that can convey the ratio of two H atoms for every one O atom, so the empirical formula is the same as the molecular formula: H$_2$O. Table 2.9 lists the molecular and empirical formulas for several compounds.

TABLE 2.9	Molecular and Empirical Formulas			
Compound	**Molecular Formula**	**Model**	**Empirical Formula**	**Model**
Water	H$_2$O		H$_2$O	
Hydrogen peroxide	H$_2$O$_2$	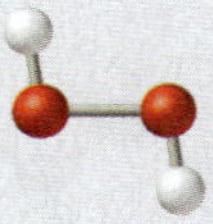	HO	
Ethane	C$_2$H$_6$	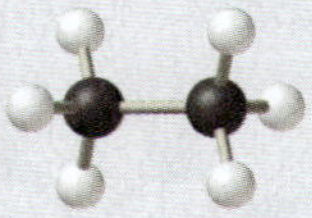	CH$_3$	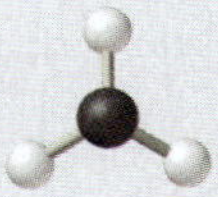
Propane	C$_3$H$_8$		C$_3$H$_8$	
Acetylene	C$_2$H$_2$		CH	
Benzene	C$_6$H$_6$		CH	

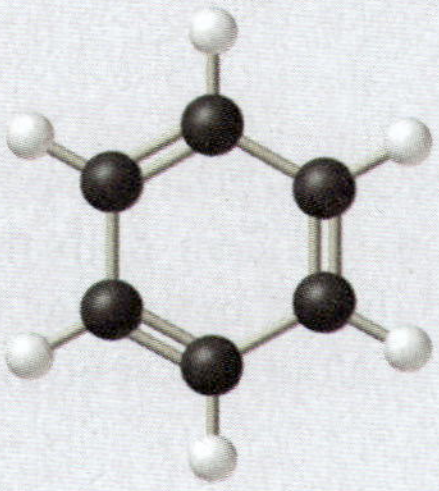

Empirical formulas are the *simplest* chemical formulas; they are written by reducing the subscripts in molecular formulas to the smallest possible whole numbers (without altering the relative numbers of atoms). Molecular formulas are the *true* formulas of molecules. As we discuss in Chapter 3, when chemists analyze an unknown compound, the first step is usually the determination of the compound's empirical formula.

Sample Problem 2.11 lets you practice determining empirical formulas from molecular formulas.

SAMPLE PROBLEM 2.11

Write the empirical formulas for the following molecules: (a) glucose ($C_6H_{12}O_6$), a substance known as blood sugar; (b) adenine ($C_5H_5N_5$), also known as vitamin B$_4$; and (c) nitrous oxide (N_2O), a gas that is used as an anesthetic ("laughing gas") and as an aerosol propellant for whipped cream.

Strategy To write the empirical formula, the subscripts in the molecular formula must be reduced to the smallest possible whole numbers (without altering the relative numbers of atoms).

Setup The molecular formulas in parts (a) and (b) each contain subscripts that are divisible by common numbers. Therefore, we will be able to express the formulas with smaller whole numbers than those in the molecular formulas. In part (c), the molecule has only one O atom, so it is impossible to simplify this formula further.

Solution (a) Dividing each of the subscripts in the molecular formula for glucose by 6, we obtain the empirical formula, CH_2O. If we had divided the subscripts by 2 or 3, we would have obtained the formulas $C_3H_6O_3$ and $C_2H_4O_2$, respectively. Although the ratio of carbon to hydrogen to oxygen atoms in each of these formulas is correct (1:2:1), neither is the simplest formula because the subscripts are not in the smallest possible whole number ratio. (b) Dividing each subscript in the molecular formula of adenine by 5, we get the empirical formula, CHN. (c) Because the subscripts in the formula for nitrous oxide are already the smallest possible whole numbers, its empirical formula is the same as its molecular formula, N_2O.

THINK ABOUT IT

Make sure that the *ratio* in each empirical formula is the same as that in the corresponding molecular formula and that the subscripts are the smallest possible whole numbers. In part (a), for example, the ratio of C:H:O in the molecular formula is 6:12:6, which is equal to 1:2:1, the ratio expressed in the empirical formula.

Practice Problem ATTEMPT Write empirical formulas for the following molecules: (a) caffeine ($C_8H_{10}N_4O_2$), a stimulant found in tea and coffee, (b) butane (C_4H_{10}), which is used in cigarette lighters, and (c) glycine ($C_2H_5NO_2$), an amino acid.

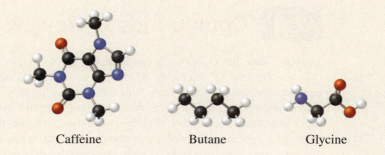

Caffeine   Butane   Glycine

Practice Problem BUILD For which of the following molecular formulas is the formula shown in parentheses the correct empirical formula? (a) $C_{12}H_{22}O_{11}$ ($C_{12}H_{22}O_{11}$), (b) $C_8H_{12}O_4$ ($C_4H_6O_2$), (c) H_2O_2 (H_2O)?

(Continued on next page)

Practice Problem **C**ONCEPTUALIZE Which of the following molecules has/have the same empirical formula as acetic acid ($HC_2O_2H_3$)?

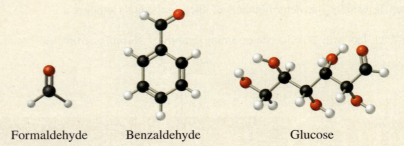

Formaldehyde   Benzaldehyde   Glucose

CHECKPOINT – SECTION 2.7   Molecules and Molecular Compounds

2.7.1 What is the correct systematic
name of PCl_5?

a) Phosphorus chloride

b) Phosphorus pentachloride

c) Monophosphorus chloride

d) Pentachlorophosphorus

e) Pentaphosphorus chloride

2.7.2 What is the name of the compound
shown?

a) Methane

b) Carbon tetrahydrogen monoxide

c) Methanol

d) Methane monoxide

e) Tetrahydrogen carbon monoxide

2.7.3 What is the correct formula for the
compound carbon tetrachloride?

a) C_4Cl_4

b) C_4Cl

c) CCl_4

d) CCl_2

e) CCl

2.7.4 What is the empirical formula of the compound
shown?

a) C_6H_6

b) C_6H_4

c) C_2H_2

d) CH_2

e) CH

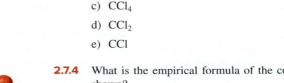

2.8   Compounds in Review

Student Note: In general, we
determine whether a compound is
molecular or ionic by asking these
questions: Does the compound
consist of only nonmetals? Then it
is probably molecular. Does the
compound contain either a metal
cation or the NH_4^+ ion? Then it is
probably ionic. (Note that it is
possible for an ionic compound to
contain only nonmetals if it contains
the NH_4^+ ion.)

We have seen that compounds can be made up of ions, or molecules. It is important
that you be able to determine whether a compound is ionic or molecular from its
chemical formula. In many cases, this will be possible only if you can easily recognize
the common polyatomic ions. Further, we have learned a systematic approach to naming compounds. Remember that by convention, we use Greek prefixes to indicate
ratios of combination in *molecular* compounds, but *not* in ionic compounds. Prefixes
are unnecessary in the names of ionic compounds because the charges on the constituent ions determine the ratio of combination. Figure 2.17 summarizes the steps for
naming molecular and ionic compounds.

Some compounds are better known by their common names than by their systematic chemical names. Familiar examples are listed in Table 2.10.

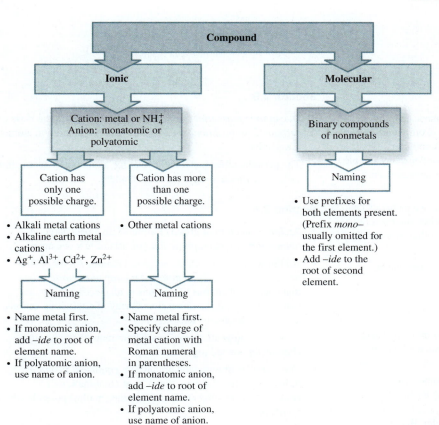

Figure 2.17 Steps for naming molecular and ionic compounds.

TABLE 2.10	Common and Systematic Names of Some Familiar Inorganic Compounds	
Formula	**Common Name**	**Systematic Name**
$CaCO_3$	Marble, chalk, limestone	Calcium carbonate
$NaHCO_3$	Baking soda	Sodium hydrogen carbonate
$Mg(OH)_2$	Milk of magnesia	Magnesium hydroxide
H_2O	Water	Dihydrogen monoxide
NH_3	Ammonia	Trihydrogen nitride
CO_2	Dry ice	Solid carbon dioxide
$NaCl$	Salt	Sodium chloride
N_2O	Nitrous oxide, laughing gas	Dinitrogen monoxide
$MgSO_4 \cdot 7H_2O$	Epsom salt	Magnesium sulfate heptahydrate

Chapter Summary

Section 2.1

- Dalton's atomic theory states that all matter is made up of tiny indivisible, immutable particles called *atoms.* Compounds form, moreover, when atoms of different elements combine in fixed ratios. According to the *law of definite proportions,* any sample of a given compound will always contain the same elements in the same mass ratio.

- The *law of multiple proportions* states that if two elements can form more than one compound with one another, the mass ratio of one will be related to the mass ratio of the other by a small whole number.

- The *law of conservation of mass* states that matter can be neither created nor destroyed.

Section 2.2

- On the basis of Dalton's atomic theory, the *atom* is the basic unit of an element. Studies with *radiation* indicated that atoms contained subatomic particles, one of which was the *electron.*

- Experiments with *radioactivity* have shown that some atoms give off different types of radiation, called *alpha (α) rays, beta (β) rays,* and *gamma (γ) rays.* Alpha rays are composed of *α particles,* which are actually helium nuclei. Beta rays are composed of *β particles,* which are actually electrons. Gamma rays are high-energy radiation.

- Most of the mass of an atom resides in a tiny, dense region known as the *nucleus.* The nucleus contains positively charged particles called *protons* and electrically neutral particles called *neutrons.* The charge on a proton is equal in magnitude but opposite in sign to the charge on an electron. The electrons occupy the relatively large volume around the nucleus. A neutron has a slightly greater mass than a proton, but each is almost 2000 times as massive as an electron.

Section 2.3

- The *atomic number (Z)* is the number of protons in the nucleus of an atom. The atomic number determines the identity of the atom. The *mass number (A)* is the sum of the protons and neutrons in the nucleus.

- Protons and neutrons are referred to collectively as *nucleons.*

- Atoms with the same atomic number but different mass numbers are called *isotopes.*

Section 2.4

- The *periodic table* arranges the elements in rows (*periods*) and columns (*groups* or *families*). Elements in the same group exhibit similar properties.

- All elements fall into one of three categories: *metal, nonmetal,* or *metalloid.*

- Some of the groups have special names including *alkali metals* (Group 1, except hydrogen), *alkaline earth metals* (Group 2), *chalcogens* (Group 16), *halogens* (Group 17), *noble gases* (Group 18), and *transition elements* or *transition metals* (Groups 3–11).

Section 2.5

- One *atomic mass unit (amu)* is exactly one-twelfth the mass of a carbon-12 atom. *Atomic mass* is the mass of an atom in atomic mass units.

- The periodic table contains the *average* atomic mass (sometimes called the *atomic weight*) of each element.

Section 2.6

- An *ion* is an atom or group of atoms with a net charge. An *atomic ion* or a *monatomic ion* consists of just one atom.

- An ion with a net positive charge is a *cation.* An ion with a net negative charge is an *anion.* An *ionic compound* is one that consists of cations and anions in an electrically neutral combination. A three-dimensional array of alternating cations and anions is called a *lattice.*

- *Polyatomic ions* are those that contain more than one atom chemically bonded together.

- The formulas of ionic compounds are *empirical formulas,* indicating with subscripts the ratio of combination of a compound's constituent elements in the smallest possible whole numbers.

- Ionic compounds' names are written as the name of the cation followed by the name of the anion—each with the word "ion" removed. The *name* of an ionic compound need not indicate the ratio of combination of elements because the charges on ions are known.

- *Oxoanions* are polyatomic ions that contain one or more oxygen atoms.

- *Hydrates* are compounds whose formulas include a specific number of water molecules.

Section 2.7

- A *molecule* is an electrically neutral group of two or more atoms. Molecules consisting of just two atoms are called *diatomic.* Diatomic molecules may be *homonuclear* (just one kind of atom) or *heteronuclear* (two kinds of atoms). In general, molecules containing more than two atoms are called *polyatomic.*

- A *chemical formula* denotes the composition of a substance. A *molecular formula* specifies the exact numbers of atoms in a molecule of a compound. A *structural formula* shows the arrangement of atoms in a substance.

- An *allotrope* is one of two or more different forms of an element.

- Molecular compounds are named according to a set of rules, including the use of Greek prefixes to specify the number of each kind of atom in the molecule.

- *Binary* compounds are those that consist of *two* elements. An *acid* is a substance that generates hydrogen ions when it dissolves in water. An *ionizable hydrogen atom* is one that can be removed in water to become a hydrogen ion (H^+).

- *Oxoacids* are acids based on oxoanions. Acids with more than one ionizable hydrogen atom are called *polyprotic.*

- *Inorganic compounds* are generally those that do not contain carbon. *Organic compounds* contain carbon and hydrogen, sometimes in combination with other elements. *Hydrocarbons* contain only carbon and hydrogen. The simplest hydrocarbons are the *alkanes*. A *functional group* is a group of atoms that determines the chemical properties of an organic compound.

- The formulas of molecular compounds may be expressed as molecular formulas, or as empirical formulas. The empirical and molecular formulas of a given compound may or may not be identical.

Key Words

Acid, 71

Alkali metal, 56

Alkaline earth metal, 56

Alkane, 72

Allotrope, 67

α particle, 49

Alpha (α) ray, 49

Anion, 60

Atom, 47

Atomic ion, 60

Atomic mass, 57

Atomic mass unit (amu), 57

Atomic number (Z), 52

Atomic weight, 58

β particle, 49

Beta (β) ray, 49

Binary compound, 69

Cation, 60

Chalcogens, 56

Chemical formula, 67

Diatomic molecule, 67

Electron, 48

Empirical formula, 62

Family, 55

Functional group, 72

Gamma (γ) rays, 49

Group, 55

Halogens, 56

Heteronuclear, 67

Homonuclear, 67

Hydrate, 66

Hydrocarbon, 72

Inorganic compounds, 72

Ion, 60

Ionic compound, 60

Ionizable hydrogen atom, 71

Isotope, 53

Lattice, 62

Law of conservation of mass, 45

Law of definite proportions, 44

Law of multiple proportions, 45

Mass number (A), 53

Metal, 55

Metalloid, 55

Molecular formula, 67

Molecule, 67

Monatomic ion, 60

Neutron, 52

Noble gases, 56

Nonmetal, 55

Nucleons, 53

Nucleus, 50

Organic compounds, 72

Oxoacid, 71

Oxoanion, 65

Period, 55

Periodic table, 55

Polyatomic ion, 61

Polyatomic molecule, 67

Polyprotic acid, 71

Proton, 50

Radiation, 47

Radioactivity, 49

Structural formula, 68

Transition elements, 56

Transition metals, 56

The process of naming binary molecular compounds follows the procedure outlined in Section 2.7. The element that appears first in the formula is named first, followed by the name of the second element—with its ending changed to –*ide*. Greek prefixes are used to indicate numbers of atoms, but the prefix *mono*– is *not* used when there is only one atom of the *first* element in the formula.

Examples:

N_2O	NO_2	Cl_2O_7	P_4O_6
dinitrogen monoxide	nitrogen dioxide	dichlorine heptoxide	tetraphosphorus hexoxide

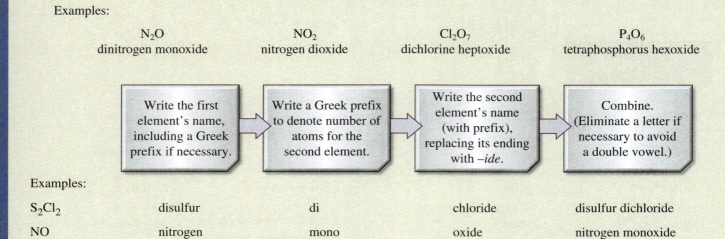

Examples:

S_2Cl_2	disulfur	di	chloride	disulfur dichloride
NO	nitrogen	mono	oxide	nitrogen monoxide

The process of naming binary ionic compounds follows the simple procedure outlined in Section 2.6. Naming compounds that contain polyatomic ions follows essentially the same procedure; but it does require you to recognize the common polyatomic ions [◄◄ Table 2.3]. Because many ionic compounds contain polyatomic ions, it is important that you know their names, formulas, and charges—well enough that you can identify them readily.

In ionic compounds with ratios of combination other than 1:1, subscript numbers are used to denote the number of each ion in the formula.

Examples: $CaBr_2$, Na_2S, $AlCl_3$, Al_2O_3, FeO, Fe_2O_3

Recall that because the common ions of main group elements have predictable charges, it is unnecessary to use prefixes to denote their numbers when naming compounds that contain them. Thus, the names of the first four examples above are calcium bromide, sodium sulfide, aluminum chloride, and aluminum oxide. The last two contain transition metal ions, many of which have more than one possible charge. In these cases, in order to avoid ambiguity, the charge on the metal ion is designated with a Roman numeral in parentheses. The names of these two compounds are iron(II) oxide and iron(III) oxide, respectively.

When a subscript number is required for a polyatomic ion, the ion's formula must first be enclosed in parentheses.

Examples: $Ca(NO_3)_2$, $(NH_4)_2S$, $Ba(C_2H_3O_2)_2$, $(NH_4)_2SO_4$, $Fe_3(PO_4)_2$, $Co_2(CO_3)_3$

Names: calcium nitrate, ammonium sulfide, barium acetate, ammonium sulfate, iron(II) phosphate, cobalt(III) carbonate

The process of naming ionic compounds given their formulas can be summarized with the following flowchart:

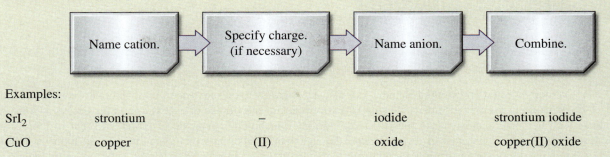

Examples:

SrI_2	strontium	–	iodide	strontium iodide
CuO	copper	(II)	oxide	copper(II) oxide

It is equally important that you be able to write the formula of an ionic compound given its name. Again, knowledge of the common polyatomic ions is critical. The process of writing an ionic compound's formula given its name is summarized as follows:

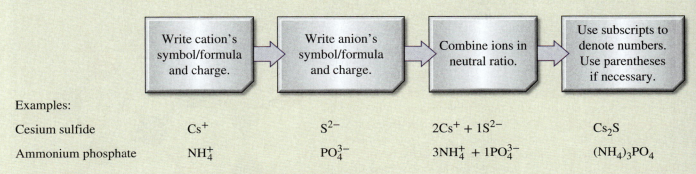

Examples:

Cesium sulfide	Cs^+	S^{2-}	$2Cs^+ + 1S^{2-}$	Cs_2S
Ammonium phosphate	NH_4^+	PO_4^{3-}	$3NH_4^+ + 1PO_4^{3-}$	$(NH_4)_3PO_4$

The process of writing a molecular compound's formula given its name is summarized as follows:

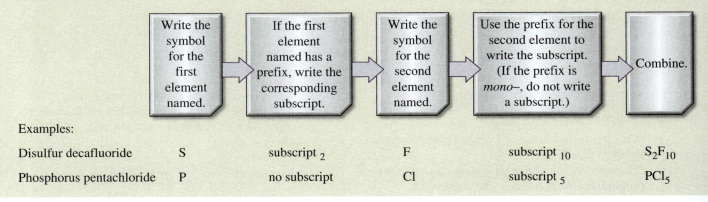

Examples:

Disulfur decafluoride	S	subscript $_2$	F	subscript $_{10}$	S_2F_{10}
Phosphorus pentachloride	P	no subscript	Cl	subscript $_5$	PCl_5

Key Skills Problems

2.1
What is the correct name for $CaSO_4$?

(a) calcium sulfoxide (b) calcium sulfite (c) calcium sulfur oxide (d) calcium sulfate (e) calcium sulfide tetroxide

2.2
What is the correct formula for nickel(II) perchlorate?

(a) $NiClO_4$ (b) Ni_2ClO_4 (c) $Ni(ClO_4)_2$ (d) $NiClO_3$ (e) $Ni(ClO_3)_2$

2.3
What is the correct name for NCl_3?

(a) trinitrogen chloride (b) mononitrogen chloride (c) nitrogen trichloride (d) nitride trichloride (e) mononitride chloride

2.4
What is the correct formula for phosphorus pentachloride?

(a) PCl_5 (b) P_5Cl (c) $P(ClO)_5$ (d) PO_4Cl (e) $PClO$

Questions and Problems

Applying What You've Learned

Although iron is an essential element, it is also a potentially toxic substance. Hemochromatosis is one of the most common hereditary disorders, causing "iron overload" or the storage of excess iron in the tissues and organs. Individuals with hemochromatosis often must undergo periodic phlebotomy (removal of blood) in order to remove excess stored iron, which would otherwise cause irreversible damage to internal organs, including the liver and kidneys. Those who have a tendency to store too much iron are advised to avoid combining iron-rich foods with substances that enhance iron absorption, such as ascorbic acid (vitamin C).

Because of iron's toxicity, iron supplements are potentially dangerous, especially to children. In fact, iron poisoning is the most common toxicological emergency in young children—due in part to the resemblance many iron supplements bear to candy. Most vitamins that contain iron are sold with childproof caps to help prevent accidental overdose. The Food and Drug Administration (FDA) recommends supplements containing more than 30 mg of iron per dose to be sold in single-dose blister packs to make it more difficult for a child to consume a dangerous amount. (a) Iron has four naturally occurring isotopes: ^{54}Fe (53.9396 amu), ^{56}Fe (55.9349 amu), ^{57}Fe (56.9354 amu), and ^{58}Fe (57.9333 amu). For each isotope, determine the number of neutrons in the nucleus [◄◄ Sample Problem 2.2]. (b) Calculate the average atomic mass of iron given that the natural abundances of the four isotopes are 5.845, 91.754, 2.119, and 0.282 percent, respectively [◄◄ Sample Problem 2.3]. (c) Write the formula for ferrous sulfate [iron(II) sulfate] [◄◄ Sample Problem 2.5]. (d) Write the molecular formula for ascorbic acid (see the ball-and-stick model at right) [◄◄ Sample Problem 2.7]. (e) Determine the empirical formula of ascorbic acid [◄◄ Sample Problem 2.11].

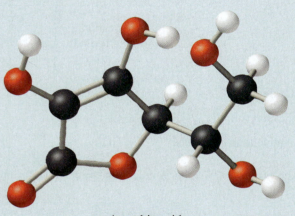

Ascorbic acid

SECTION 2.1: THE ATOMIC THEORY

Review Questions

2.1   What are the hypotheses on which Dalton's atomic theory is based?

2.2   State the laws of definite proportions and multiple proportions. Illustrate each with an example.

Computational Problems

2.3   The elements sulfur and oxygen can form a variety of different compounds, the most common of which are SO_2 and SO_3. Samples of these two compounds were decomposed into their constituent elements. One produced 1.002 g S for every gram of O; the other produced 0.668 g S for every gram of O. Show that these results are consistent with the law of multiple proportions.

2.4   Two different compounds, each containing only phosphorus and chlorine, were decomposed into their constituent elements. One produced 0.2912 g P for every gram of Cl; the other produced 0.1747 g P for every gram of Cl. Show that these results are consistent with the law of multiple proportions.

2.5   Sulfur reacts with fluorine to produce three different compounds. The mass ratio of fluorine to sulfur for each compound is given in the following table:

Compound	mass F : mass S
S_2F_{10}	2.962
SF_4	2.370
SF_6	3.555

Show that these data are consistent with the law of multiple proportions.

2.6   Both FeO and Fe_2O_3 contain only iron and oxygen. The mass ratio of oxygen to iron for each compound is given in the following table:

Compound	mass O : mass Fe
FeO	0.2865
Fe_2O_3	0.4297

Show that these data are consistent with the law of multiple proportions.

Conceptual Problems

2.7 For the two compounds pictured, evaluate the following ratio:

$$\frac{\text{g blue} : 1.00 \text{ g red(right)}}{\text{g blue} : 1.00 \text{ g red(left)}}$$

2.8 For the two compounds pictured, evaluate the following ratio:

$$\frac{\text{g green} : 1.00 \text{ g yellow(right)}}{\text{g green} : 1.00 \text{ g yellow(left)}}$$

SECTION 2.2: THE STRUCTURE OF THE ATOM

Review Questions

2.9 Define the following terms: (a) α particle, (b) β particle, (c) γ ray, (d) X ray.

2.10 Name the types of radiation known to be emitted by radioactive elements.

2.11 Compare the properties of the following: α particles, cathode rays, protons, neutrons, and electrons.

2.12 Describe the contributions of the following scientists to our knowledge of atomic structure: J. J. Thomson, R. A. Millikan, Ernest Rutherford, and James Chadwick.

2.13 Describe the experimental basis for believing that the nucleus occupies a very small fraction of the volume of the atom.

Problems

2.14 The diameter of a neutral helium atom is about 1×10^2 pm. Suppose that we could line up helium atoms side by side in contact with one another. Approximately how many atoms would it take to make the distance 1 in from end to end?

2.15 Roughly speaking, the radius of an atom is about 10,000 times greater than that of its nucleus. If an atom were magnified so that the radius of its nucleus became 2.0 cm, about the size of a marble, what would be the radius of the atom in miles (1 mi = 1609 m)?

SECTION 2.3: ATOMIC NUMBER, MASS NUMBER, AND ISOTOPES

Review Questions

2.16 Use the helium-4 isotope to define atomic number and mass number. Why does knowledge of the atomic number enable us to deduce the number of electrons present in an atom?

2.17 Why do all atoms of an element have the same atomic number, although they may have different mass numbers?

2.18 What do we call atoms of the same elements with different mass numbers?

2.19 Explain the meaning of each term in the symbol $_Z^A X$.

Computational Problems

2.20 What is the mass number of an iron atom that has 28 neutrons?

2.21 Calculate the number of neutrons of ^{239}Pu.

2.22 For each of the following species, determine the number of protons and the number of neutrons in the nucleus: $_2^3$He, $_2^4$He, $_{12}^{24}$Mg, $_{12}^{25}$Mg, $_{22}^{48}$Ti, $_{35}^{79}$Br, $_{78}^{195}$Pt.

2.23 Indicate the number of protons, neutrons, and electrons in each of the following species: $_7^{15}$N, $_{16}^{33}$S, $_{29}^{63}$Cu, $_{38}^{84}$Sr, $_{56}^{130}$Ba, $_{74}^{186}$W, $_{80}^{202}$Hg.

2.24 Write the appropriate symbol for each of the following isotopes: (a) $Z = 11$, $A = 23$; (b) $Z = 28$, $A = 64$; (c) $Z = 50$, $A = 115$; (d) $Z = 20$, $A = 42$.

2.25 Write the appropriate symbol for each of the following isotopes: (a) $Z = 74$, $A = 186$; (b) $Z = 80$, $A = 201$; (c) $Z = 34$, $A = 76$; (d) $Z = 94$, $A = 239$.

2.26 Determine the mass number of (a) a boron atom with 6 neutrons, (b) a magnesium atom with 13 neutrons, (c) a bromine atom with 44 neutrons, and (d) a mercury atom with 119 neutrons.

2.27 Determine the mass number of (a) a fluorine atom with 11 neutrons, (b) a sulfur atom with 16 neutrons, (c) an arsenic atom with 45 neutrons, and (d) a platinum atom with 120 neutrons.

2.28 The following radioactive isotopes are used in medicine for imaging organs, studying blood circulation, treating cancer, and so on. Give the number of neutrons present in each isotope: ^{198}Au, ^{47}Ca, ^{60}Co, ^{18}F, ^{125}I, ^{131}I, ^{42}K, ^{43}K, ^{24}Na, ^{32}P, ^{85}Sr, ^{99}Tc.

SECTION 2.4: THE PERIODIC TABLE

Review Questions

2.29 What is the periodic table, and what is its significance in the study of chemistry?

2.30 State two differences between a metal and a nonmetal.

2.31 Write the names and symbols for four elements in each of the following categories: (a) nonmetal, (b) metal, (c) metalloid.

2.32 Give two examples of each of the following: (a) alkali metals, (b) alkaline earth metals, (c) halogens, (d) noble gases, (e) chalcogens, (f) transition metals.

2.33 The explosion of an atomic bomb in the atmosphere releases many radioactive isotopes into the environment. One of the isotopes is ^{90}Sr. Via a relatively short food chain, it can enter the human body. Considering the position of strontium in the periodic table, explain why it is particularly harmful to humans.

Computational Problems

2.34 Elements whose names end with –*ium* are usually metals; sodium is one example. Identify a nonmetal whose name also ends with –*ium*.

2.35 Describe the changes in properties (from metals to nonmetals or from nonmetals to metals) as we move (a) down a periodic group and (b) across the periodic table from left to right.

2.36 Consult webelements.com to find (a) two metals less dense than water, (b) two metals more dense than mercury, (c) the densest known solid metallic element, and (d) the densest known solid nonmetallic element.

2.37 Group the following elements in pairs that you would expect to show similar chemical properties: K, F, P, Na, Cl, and N.

2.38 Group the following elements in pairs that you would expect to show similar chemical properties: I, Ba, O, Br, S, and Ca.

2.39 Write the symbol for each of the following biologically important elements in the given periodic table: iron (present in hemoglobin for transporting oxygen), iodine (present in the thyroid gland), sodium (present in intracellular and extracellular fluids), phosphorus (present in bones and teeth), sulfur (present in proteins), and magnesium (present in chlorophyll molecules).

SECTION 2.5: THE ATOMIC MASS SCALE AND AVERAGE ATOMIC MASS

Review Questions

2.40 What is an atomic mass unit? Why is it necessary to introduce such a unit?

2.41 What is the mass (in amu) of a carbon-12 atom? Why is the atomic mass of carbon listed as 12.01 amu in the table at the beginning of this book?

2.42 Explain clearly what is meant by the statement "The atomic mass of gold is 197.0 amu."

2.43 What information would you need to calculate the average atomic mass of an element?

Problems

2.44 The atomic masses of $^{79}_{35}Br$ (50.69 percent) and $^{81}_{35}Br$ (49.31 percent) are 78.9183361 and 80.916289 amu, respectively. Calculate the average atomic mass of bromine. The percentages in parentheses denote the relative abundances.

2.45 The atomic masses of ^{204}Pb (1.4 percent), ^{206}Pb (24.1 percent), ^{207}Pb (22.1 percent), and ^{208}Pb (52.4 percent) are 203.973020, 205.974440, 206.975872, and 207.976627 amu, respectively. Calculate the average atomic mass of lead. The percentages in parentheses denote the relative abundances.

2.46 The atomic masses of ^{24}Mg (78.99 percent), ^{25}Mg (10.00 percent), and ^{26}Mg (11.01 percent) are 23.9850423, 24.9858374, and 25.9825937 amu, respectively. Calculate the average atomic mass of magnesium. The percentages in parentheses denote the relative abundances.

2.47 The atomic masses of ^{191}Ir (37.3 percent) and ^{193}Ir (62.7 percent) are 190.960584 and 192.962917 amu, respectively. Calculate the average atomic mass of iridium. The percentages in parentheses denote the relative abundances.

2.48 The atomic masses of ^{203}Tl and are ^{205}Tl are 202.972320 and 204.974401, respectively. Calculate the natural abundances of these two isotopes. The average atomic mass of thallium is 204.4 amu.

2.49 The atomic masses of ^{6}Li and ^{7}Li are 6.0151 amu and 7.0160 amu, respectively. Calculate the natural abundances of these two isotopes. The average atomic mass of lithium is 6.941 amu.

SECTION 2.6: IONS AND IONIC COMPOUNDS

Review Questions

2.50 Give an example of each of the following: (a) a monatomic cation, (b) a monatomic anion, (c) a polyatomic cation, (d) a polyatomic anion.

2.51 What is an ionic compound? How is electrical neutrality maintained in an ionic compound?

2.52 Explain why the chemical formulas of ionic compounds are usually the same as their empirical formulas.

2.53 What is the Stock system? What are its advantages over the older system of naming cations?

2.54 Give an example of an ionic compound that does not contain a metal.

Conceptual Problems

2.55 Give the number of protons and electrons in each of the following common ions: Na^+, Ca^{2+}, Al^{3+}, Fe^{2+}, I^-, F^-, S^{2-}, O^{2-}, N^{3-}.

2.56 Give the number of protons and electrons in each of the following common ions: K^+, Mg^{2+}, Fe^{3+}, Br^-, Mn^{2+}, C^{4-}, Cu^{2+}.

2.57 Write the formulas for the following ionic compounds: (a) sodium oxide, (b) iron sulfide (containing the Fe^{2+} ion), (c) cobalt sulfate (containing the Co^{3+} and SO_4^{2-} ions), (d) barium fluoride.

2.58 Write the formulas for the following ionic compounds: (a) copper bromide (containing the Cu^+ ion), (b) manganese oxide (containing the

Mn³⁺ ion), (c) mercury iodide (containing the Hg₂²⁺ ion), (d) magnesium phosphate (containing the PO₄³⁻ ion).

2.59 Which of the following compounds are likely to be ionic? Which are likely to be molecular? $SiCl_4$, LiF, $BaCl_2$, B_2H_6, KCl, C_2H_4.

2.60 Which of the following compounds are likely to be ionic? Which are likely to be molecular? CH_4, NaBr, BaF_2, CCl_4, ICl, CsCl, NF_3.

2.61 Name the following compounds: (a) KH_2PO_4, (b) K_2HPO_4, (c) HBr (gas), (d) HBr (in water), (e) Li_2CO_3, (f) $K_2Cr_2O_7$, (g) NH_4NO_2, (h) HIO_3, (i) PF_5, (j) P_4O_6, (k) CdI_2, (l) $SrSO_4$, (m) $Al(OH)_3$.

2.62 Name the following compounds: (a) KClO, (b) Ag_2CO_3, (c) HNO_2, (d) $KMnO_4$, (e) $CsClO_3$, (f) KNH_4SO_4, (g) FeO, (h) Fe_2O_3, (i) $TiCl_4$, (j) NaH, (k) Li_3N, (l) Na_2O, (m) Na_2O_2.

2.63 Write the formulas for the following compounds: (a) rubidium nitrite, (b) potassium sulfide, (c) sodium hydrogen sulfide, (d) magnesium phosphate, (e) calcium hydrogen phosphate, (f) lead(II) carbonate, (g) tin(II) fluoride, (h) ammonium sulfate, (i) silver perchlorate, (j) boron trichloride.

2.64 Write the formulas for the following compounds: (a) copper(I) cyanide, (b) strontium chlorite, (c) perbromic acid, (d) hydroiodic acid, (e) disodium ammonium phosphate, (f) potassium dihydrogen phosphate, (g) iodine heptafluoride, (h) tetraphosphorus decasulfide, (i) mercury(II) oxide, (j) mercury(I) iodide, (k) selenium hexafluoride.

2.65 In the diagrams shown here, match each of the drawings with the following ionic compounds: Al_2O_3, LiH, Na_2S, $Mg(NO_3)_2$. (Green spheres represent cations and red spheres represent anions.)

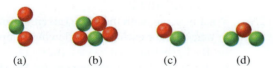

    (a)   (b)   (c)   (d)

2.66 Given the formulas for the ionic compounds, draw the correct ratio of cations to anions as shown in Problem 2.65: (a) $BaSO_4$, (b) CaF_2, (c) Mg_3N_2, (d) K_2O.

SECTION 2.7: MOLECULES AND MOLECULAR COMPOUNDS

Review Questions

2.67 What is the difference between an atom and a molecule?

2.68 What are allotropes? Give an example. How are allotropes different from isotopes?

2.69 What does a chemical formula represent? Determine the ratio of the atoms in the following molecular formulas: (a) NO, (b) NCl_3, (c) N_2O_4, (d) P_4O_6.

2.70 Define molecular formula and empirical formula. What are the similarities and differences between the empirical formula and molecular formula of a compound?

2.71 Give an example of a case in which two molecules have different molecular formulas but the same empirical formula.

2.72 What is the difference between inorganic compounds and organic compounds?

2.73 Give one example each for a binary compound and a ternary compound. (A ternary compound is one that contains three different elements.)

2.74 Explain why the formula HCl can represent two different chemical systems.

Conceptual Problems

2.75 For each of the following diagrams, determine whether it represents diatomic molecules, polyatomic molecules, molecules that are not compounds, molecules that are compounds, or an elemental form of the substance.

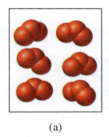  

    (a)   (b)   (c)

2.76 For each of the following diagrams, determine whether it represents diatomic molecules, polyatomic molecules, molecules that are not compounds, molecules that are compounds, or an elemental form of the substance.

 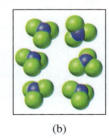 

    (a)   (b)   (c)

2.77 Identify the following as elements or compounds: NH_3, N_2, S_8, NO, CO, CO_2, H_2, SO_2.

2.78 Give two examples of each of the following: (a) a diatomic molecule containing atoms of the same element, (b) a diatomic molecule containing atoms of different elements, (c) a polyatomic molecule containing atoms of the same element, (d) a polyatomic molecule containing atoms of different elements.

2.79 Write the empirical formulas of the following compounds: (a) C_2N_2, (b) C_6H_6, (c) C_9H_{20}, (d) P_4O_{10}, (e) B_2H_6.

2.80 Write the empirical formulas of the following compounds: (a) Al_2Br_6, (b) $Na_2S_2O_4$, (c) N_2O_5, (d) $K_2Cr_2O_7$, (e) $H_2C_2O_4$.

2.81 Write the molecular formula of alanine, an amino acid used in protein synthesis. The color codes are black (carbon), blue (nitrogen), red (oxygen), and white (hydrogen).

2.82 Write the molecular formula of ethanol. The color codes are black (carbon), red (oxygen), and white (hydrogen).

2.83 Name the following binary molecular compounds: (a) NCl_3, (b) IF_7, (c) P_4O_6, (d) S_2Cl_2.

2.84 Write chemical formulas for the following molecular compounds: (a) phosphorus tribromide, (b) dinitrogen tetrafluoride, (c) xenon tetroxide, (d) selenium trioxide.

2.85 Write the molecular formulas and names of the following compounds.

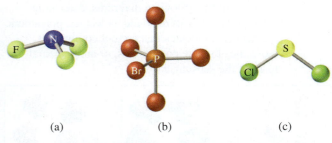

(a)   (b)   (c)

2.86 Write the molecular formulas and names of the following compounds.

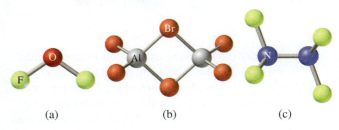

(a)   (b)   (c)

ADDITIONAL PROBLEMS

2.87 Define the following terms: acids, bases, oxoacids, oxoanions, and hydrates.

2.88 A sample of a uranium compound is found to be losing mass gradually. Explain what is happening to the sample.

2.89 In which one of the following pairs do the two species resemble each other most closely in chemical properties: (a) $_1^1H$ and $_1^1H^+$, (b) $_7^{14}N$ and $_7^{14}N^{3-}$, (c) $_6^{12}C$ and $_6^{13}C$? Explain.

2.90 One isotope of a metallic element has mass number 137 and 82 neutrons in the nucleus. The cation derived from the isotope has 54 electrons. Write the chemical symbol for this cation.

2.91 One isotope of a nonmetallic element has mass number 31 and 16 neutrons in the nucleus. The anion derived from the isotope has 18 electrons. Write the chemical symbol for this anion.

2.92 The following table gives numbers of electrons, protons, and neutrons in atoms or ions of a number of elements. Answer the following: (a) Which of the species are neutral? (b) Which are negatively charged? (c) Which are positively charged? (d) What are the conventional symbols for all the species?

	Atom or Ion of Element						
	A	B	C	D	E	F	G
Number of electrons	5	10	18	28	36	5	9
Number of protons	5	7	19	30	35	5	9
Number of neutrons	5	7	20	36	46	6	10

2.93 What is wrong with or ambiguous about the phrase "four molecules of NaCl"?

2.94 The following phosphorus sulfides are known: P_4S_3, P_4S_7, and P_4S_{10}. Do these compounds obey the law of multiple proportions?

2.95 Which of the following are elements, which are molecules but not compounds, which are compounds but not molecules, and which are both compounds and molecules? (a) SO_2, (b) S_8, (c) Cs, (d) N_2O_5, (e) O, (f) O_2, (g) O_3, (h) CH_4, (i) KBr, (j) S, (k) P_4, (l) LiF.

2.96 What is wrong with the name (given in parentheses or brackets) for each of the following compounds: (a) $BaCl_2$ (barium dichloride), (b) Fe_2O_3 [iron(II) oxide], (c) $CsNO_2$ (cesium nitrate), (d) $Mg(HCO_3)_2$ [magnesium(II) bicarbonate]?

2.97 Discuss the significance of assigning an atomic mass of exactly 12 amu to the carbon-12 isotope.

2.98 Determine what is wrong with the chemical formula and write the correct chemical formula for each of the following compounds: (a) $(NH_3)_2CO_3$ (ammonium carbonate), (b) $CaOH_2$ (calcium hydroxide), (c) $CdSO_3$ (cadmium sulfide), (d) $ZnCrO_4$ (zinc dichromate).

2.99 Fill in the blanks in the table.

Symbol		$_{26}^{54}Fe^{2+}$			
Protons	5			79	86
Neutrons	6		16	117	136
Electrons	5		18	79	
Net charge			−3		0

2.100   (a) Which elements are most likely to form ionic compounds? (b) Which metallic elements are most likely to form cations with different charges?

2.101   Write the formula of the common ion derived from each of the following: (a) Li, (b) S, (c) I, (d) N, (e) Al, (f) Cs, (g) Mg.

2.102   Which of the following symbols provides more information about the atom: ^{23}Na or $_{11}$Na? Explain.

2.103   Write the chemical formulas and names of the binary acids and oxoacids that contain Group 17 elements. Do the same for elements in Groups 13, 14, 15, and 16.

2.104   Determine the molecular and empirical formulas of the compounds shown here. (Black spheres are carbon, and white spheres are hydrogen.)

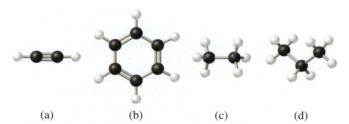

(a)   (b)   (c)   (d)

2.105   For the noble gases (the Group 18 elements) $_{2}^{4}$He, $_{10}^{20}$Ne, $_{18}^{40}$Ar, $_{36}^{84}$Kr, and $_{54}^{132}$Xe, (a) determine the number of protons and neutrons in the nucleus of each atom, and (b) determine the ratio of neutrons to protons in the nucleus of each atom. Describe any general trend you discover in the way this ratio changes with increasing atomic number.

2.106   A monatomic ion has a charge of +2. The nucleus of the parent atom has a mass number of 55. If the number of neutrons in the nucleus is 1.2 times that of the number of protons, what is the name and symbol of the element?

2.107   The Group 11 metals, Cu, Ag, and Au, are called coinage metals. What chemical properties make them especially suitable for making coins and jewelry?

2.108   The elements in Group 18 of the periodic table are called noble gases. Can you suggest what "noble" means in this context?

2.109   The formula for calcium oxide is CaO. What are the formulas for magnesium oxide and strontium oxide?

2.110   A common mineral of barium is barytes, or barium sulfate (BaSO$_4$). Because elements in the same periodic group have similar chemical properties, we might expect to find some radium sulfate (RaSO$_4$) mixed with barytes because radium is the last member of Group 2. However, the only source of radium compounds in nature is in uranium minerals. Explain why this is so.

2.111   Two elements form a compound that can be represented as ⚫🔴. The same two elements also combine to form several other compounds, which can be represented with the diagrams shown here.

For each of these, determine the ratio of g red : 1.00 g blue in this compound to g red : 1.00 g blue in ⚫🔴.

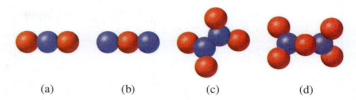

(a)   (b)   (c)   (d)

2.112   Which of the diagrams can be used to illustrate the law of constant composition? Which can be used to illustrate the law of multiple proportions? In each case, for a diagram that cannot be used to illustrate the law, explain why.

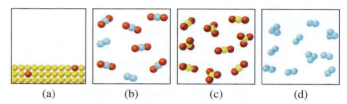

(a)   (b)   (c)   (d)

2.113   Fluorine reacts with hydrogen (H) and deuterium (D) to form hydrogen fluoride (HF) and deuterium fluoride (DF), where deuterium ($_{1}^{2}$H) is an isotope of hydrogen. Would a given amount of fluorine react with different masses of the two hydrogen isotopes? Does this violate the law of definite proportions? Explain.

2.114   Predict the formula and name of a binary compound formed from the following elements: (a) Na and H, (b) B and O, (c) Na and S, (d) Al and F, (e) F and O, (f) Sr and Cl.

2.115   Identify each of the following elements: (a) a halogen whose anion contains 36 electrons, (b) a radioactive noble gas with 86 protons, (c) a Group 16 element whose anion contains 36 electrons, (d) an alkali metal cation that contains 36 electrons, (e) a Group 14 cation that contains 80 electrons.

2.116   Show the locations of (a) alkali metals, (b) alkaline earth metals, (c) the halogens, and (d) the noble gases in the given outline of a periodic table. Also draw dividing lines between metals and metalloids and between metalloids and nonmetals.

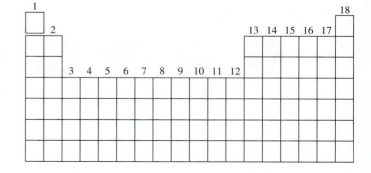

2.117   Fill in the blanks in the table.

Cation	Anion	Formula	Name
			Magnesium bicarbonate
		$SrCl_2$	
Fe^{3+}	NO_2^-		
			Manganese(II) chlorate
		$SnBr_4$	
Co^{2+}	PO_4^{3-}		
Hg_2^{2+}	I^-		
		Cu_2CO_3	
			Lithium nitride
Al^{3+}	S^{2-}		

2.118   Some compounds are better known by their common names than by their systematic chemical names. Give the chemical formulas of the following substances: (a) dry ice, (b) salt, (c) laughing gas, (d) marble (chalk, limestone), (e) baking soda, (f) ammonia, (g) water, (h) milk of magnesia, (i) epsom salt.

2.119   The relationship between mass and energy is expressed by Einstein's equation, $E = mc^2$, where E is energy, m is mass, and c is the speed of light. In a combustion experiment, it was found that 12.096 g of hydrogen molecules combined with 96.000 g of oxygen molecules to form water and released 1.715×10^3 kJ of heat. Use Einstein's equation to calculate the corresponding mass change in this process, and comment on whether or not the law of conservation of mass holds for ordinary chemical processes.

2.120   (a) Describe Rutherford's experiment and how the results revealed the nuclear structure of the atom. (b) Consider the ^{23}Na atom. Given that the radius and mass of the nucleus are 3.04×10^{-15} m and 3.82×10^{-23} g, respectively, calculate the density of the nucleus in g/cm^3. The radius of a ^{23}Na atom is 186 pm. Calculate the density of the space occupied by the electrons outside the nucleus in the sodium atom. Do your results support Rutherford's model of an atom? [The volume of a sphere of radius r is $\frac{4}{3}\pi r^3$.]

2.121   Name the given acids.

2.122   Draw two different structural formulas based on the molecular formula C_2H_6O. Is the fact that you can have more than one compound with the same molecular formula consistent with Dalton's atomic theory?

2.123   Ethane and acetylene are two gaseous hydrocarbons. Chemical analyses show that in one sample of ethane, 2.65 g of carbon are combined with 0.665 g of hydrogen, and in one sample of acetylene, 4.56 g of carbon are combined with 0.383 g of hydrogen. (a) Are these results consistent with the law of multiple proportions? (b) Write reasonable molecular formulas for these compounds.

Multiconcept Problems

2.124   A cube made of platinum (Pt) has an edge length of 1.0 cm. (a) Calculate the number of Pt atoms in the cube. (b) Atoms are spherical in shape. Therefore, the Pt atoms in the cube cannot fill all the available space. If only 74 percent of the space inside the cube is taken up by Pt atoms, calculate the radius in picometers of a Pt atom. The density Pt is 21.45 g/cm^3, and the mass of a single Pt atom is 3.240×10^{-22} g. [The volume of a sphere of radius r is $\frac{4}{3}\pi r^3$.]

2.125   (a) Assuming an atomic nucleus is spherical in shape, show that its radius r is proportional to the cube root of the mass number (A). (b) In general, the radius of a nucleus is given by $r = r_0 A^{1/3}$, where r_0 is a proportionality constant given by 1.2×10^{-15} m. Calculate the volume of the Li nucleus. (c) Given that the radius of $^{7}_{3}Li$ atom is 152 pm, calculate what fraction of the atom's volume is occupied by its nucleus. Does your result support Rutherford's model of the atom?

2.126   A distributor of ball bearings sells packages of "remnant" ball bearings that contain a mixture of copper bearings and titanium bearings—along with bearings made of a proprietary alloy of several different metals. Although the bearings all have the same diameter, 4.175 mm, the metals from which they are made all have different densities. The density of the proprietary alloy is 5.659 g/cm^3. Packages of the remnant bearings are, on average, 30.4 percent copper bearings and 51.2 percent titanium bearings, the remainder being proprietary alloy. Using the formula for the volume of a sphere, $V = \frac{4}{3}\pi r^3$, and the densities of the metals from webelements.com, determine the average mass of a ball bearing in one of the packages.

Standardized-Exam Practice Problems

Physical and Biological Sciences

Carbon-14, a radioactive isotope of carbon, is used to determine the ages of fossils in a technique called *carbon dating*. Carbon-14 is produced in the upper atmosphere when nitrogen-14 atoms are bombarded by neutrons from cosmic rays. ^{14}C undergoes a process called β emission in which a neutron in the nucleus decays to form a proton and an electron. The electron, or β particle, is ejected from the nucleus. Because the production and decay of ^{14}C occur simultaneously, the total amount of ^{14}C in the atmosphere is constant. Plants absorb ^{14}C in the form of CO_2 and animals consume plants and other animals. Thus, all living things contain a constant ratio of ^{12}C to ^{14}C. When a living thing dies, the ^{14}C it contains continues to decay, but because replenishment ceases, the ratio of ^{12}C to ^{14}C changes over time. Scientists use the ^{12}C to ^{14}C ratio to determine the age of material that was once living.

1. If atmospheric conditions were to change such that ^{14}C were produced at twice the current rate,

   a) the world's supply of ^{14}N would be consumed completely.
   b) the ^{12}C to ^{14}C ratio in living things would increase.
   c) the ^{12}C to ^{14}C ratio in living things would decrease.
   d) the ^{12}C to ^{14}C ratio in living things would not change.

2. When a ^{14}N nucleus is bombarded by a neutron to produce a ^{14}C nucleus, what else is produced?

   a) Nothing
   b) Another neutron
   c) An electron
   d) A proton

3. Based on the description of β emission in the passage, what nucleus results from the decay of a ^{14}C nucleus by β emission?

   a) ^{14}N
   b) ^{13}N
   c) ^{12}C
   d) ^{13}C

4. The accuracy of carbon dating depends on the assumption that

   a) ^{14}C is the only radioactive species in the material being tested.
   b) the rate of decay of ^{14}C is constant.
   c) ^{12}C and ^{14}C undergo radioactive decay at the same rate.
   d) each ^{14}C nucleus decays to give a ^{12}C nucleus.

Answers to In-Chapter Materials

Practice Problems

2.1A (a) 3:2, (b) 2:1. **2.1B** (a) 0.2518 g, (b) $n = 6(XeF_6)$. **2.2A** (a) p = 5, n = 5, e = 5; (b) p = 18, n = 18, e = 18; (c) p = 38, n = 47, e = 38; (d) p = 6, n = 5, e = 6. **2.2B** (a) $^{9}_{4}Be$, (b) $^{51}_{23}V$, (c) $^{124}_{54}Xe$, (d) $^{69}_{31}Ga$. **2.3A** 63.55 amu. **2.3B** 99.64% ^{14}N, 0.36% ^{15}N. **2.4A** (a) sodium sulfate, (b) copper(II) nitrate, (c) iron(III) carbonate. **2.4B** (a) potassium dichromate, (b) lithium oxalate, (c) copper(I) nitrate. **2.5A** (a) $PbCl_2$, (b) $MgCO_3$, (c) $(NH_4)_3PO_4$. **2.5B** (a) Fe_2S_3, (b) $Hg(NO_3)_2$, (c) K_2SO_3. **2.6A** (a) hypobromite ion, (b) hydrogen sulfate ion, (c) oxalate ion. **2.6B** (a) iodite ion, (b) hydrogen chromate ion, (c) hydrogen oxalate ion. **2.7A** $CHCl_3$. **2.7B** C_3H_6O. **2.8A** (a) dichlorine monoxide, (b) silicon tetrachloride. **2.8B** (a) chlorine dioxide, (b) carbon tetrabromide. **2.9A** (a) CS_2, (b) N_2O_3. **2.9B** (a) SF_6, (b) S_2F_{10}. **2.10A** $HBrO_4$. **2.10B** (a) H_2CrO_4. **2.11A** (a) $C_4H_5N_2O$, (b) C_2H_5, (c) $C_2H_5NO_2$. **2.11B** (a).

Checkpoints

2.1.1 e. **2.1.2** b. **2.3.1** d. **2.3.2** c. **2.4.1** c. **2.4.2** b. **2.5.1** b. **2.5.2** c. **2.6.1** c. **2.6.2** d. **2.6.3** d. **2.6.4** c. **2.6.5** c. **2.6.6** a. **2.7.1** b. **2.7.2** c. **2.7.3** c. **2.7.4** e.

Stoichiometry: Ratios of Combination

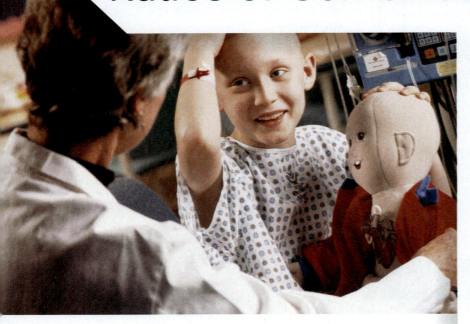

Happy outcomes for young cancer patients and their families are far more common than they used to be. St. Jude Children's Research Hospital has reported dramatic increases in the survival rates of the ten most common pediatric cancers over the last 50 years. Increased survival rates are attributed both to earlier detection and to improvements in treatment options. The drug cisplatin is an important component of many such treatments.

Zigy Kaluzny/The Image Bank/Getty Images

Stoichiometry's Importance in the Manufacture of Drugs

Stoichiometry refers to the quantitative relationships between the substances that are consumed and produced by chemical reactions. These quantitative relationships are important in the development of large-scale production of such things as chemotherapeutic drugs for the treatment of cancer.

One of cancer chemotherapy's greatest success stories began with an accidental discovery. In 1964, Barnett Rosenberg and his research group at Michigan State University were studying the effect of an electric field on the growth of bacteria. Using platinum electrodes, they passed an electric current through a bacterial culture. To their surprise, the cells in the culture stopped dividing. The researchers determined that cisplatin, $Pt(NH_3)_2Cl_2$, a compound containing platinum from the electrodes, was responsible. Furthermore, they reasoned that because cancer is the result of the uncontrolled division of abnormal cells, the compound might be useful as an anticancer drug.

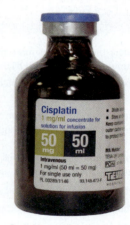

Dr. P. Marazzi/Science Source

Platinol, the name under which cisplatin is marketed, was approved by the FDA in 1978 for the treatment of metastatic testicular and ovarian cancers. Today it is one of the most commonly prescribed cancer drugs—often being used as part of a chemotherapeutic combination for the treatment of a wide variety of cancers. Because it plays an important role in the treatment of so many different cancers, it is sometimes called "the penicillin of chemotherapy." Cisplatin works by attaching itself to the DNA of cancer cells and preventing their replication. The damaged cells are then destroyed by the body's immune system. Unfortunately, cisplatin can cause serious side effects, including severe kidney damage. Ongoing research efforts are directed toward finding related compounds that are both effective and less toxic.

In 1964, cisplatin was produced accidentally when platinum electrodes reacted with ammonia molecules and chloride ions that were present in a bacterial culture. Today, manufacturers use the principles of stoichiometry to produce cisplatin in the most efficient, economical way possible.

At the end of this chapter, you will be able to solve several problems related to the drug cisplatin [▶▶ Applying What You've Learned, page 128].

3.1 Molecular and Formula Masses

Using atomic masses from the periodic table and a molecular formula, we can determine the ***molecular mass,*** which is the mass in atomic mass units (amu) of an individual molecule. The molecular mass is simply the sum of the atomic masses of the atoms that make up the molecule. We multiply the atomic mass of each element by the number of atoms of that element in the molecule and then sum the masses for each element present. For example,

$$\text{molecular mass of } H_2O = 2(\text{atomic mass of H}) + \text{atomic mass of O}$$
$$= 2(1.008 \text{ amu}) + 16.00 \text{ amu} = 18.02 \text{ amu}$$

Because the atomic masses on the periodic table are average atomic masses, the result of such a determination is an average molecular mass, sometimes referred to as the ***molecular weight.*** As with the term *atomic* mass, we will use the term *molecular* mass in this text.

Although an ionic compound does not have a molecular mass, we can use its empirical formula to determine its ***formula mass*** (the mass of a "formula unit"), sometimes called the ***formula weight.*** Sample Problem 3.1 illustrates how to determine molecular mass and formula mass.

SAMPLE PROBLEM   3.1

Calculate the molecular mass or the formula mass, as appropriate, for each of the following compounds: (a) propane, (C_3H_8), (b) lithium hydroxide, (LiOH), and (c) barium acetate, $[Ba(C_2H_3O_2)_2]$.

Strategy Determine the molecular mass (for each molecular compound) or formula mass (for each ionic compound) by summing all the atomic masses.

Setup Using the formula for each compound, determine the number of atoms of each element present. A molecule of propane contains three C atoms and eight H atoms. The compounds in parts (b) and (c) are ionic and will therefore have formula masses rather than molecular masses. A formula unit of lithium hydroxide contains one Li atom, one O atom, and one H atom. A formula unit of barium acetate contains one Ba atom, four C atoms, six H atoms, and four O atoms. (Remember that the subscript after the parentheses means that there are two acetate ions, each of which contains two C atoms, three H atoms, and two O atoms.)

Solution For each compound, multiply the number of atoms by the atomic mass of each element and then sum the calculated values.

(a) The molecular mass of propane is 3(12.01 amu) + 8(1.008 amu) = 44.09 amu.

(b) The formula mass of lithium hydroxide is 6.941 amu + 16.00 amu + 1.008 amu = 23.95 amu.

(c) The formula mass of barium acetate is 137.3 amu + 4(12.01 amu) + 6(1.008 amu) + 4(16.00 amu) = 255.4 amu.

THINK ABOUT IT

Double-check that you have counted the number of atoms correctly for each compound and that you have used the proper atomic masses from the periodic table.

Practice Problem **A**TTEMPT Calculate the molecular or formula mass of each of the following compounds: (a) magnesium chloride ($MgCl_2$), (b) sulfuric acid (H_2SO_4), and (c) oxalic acid ($H_2C_2O_4$).

Practice Problem **B**UILD Calculate the molecular or formula mass of each of the following compounds: (a) calcium carbonate ($CaCO_3$), (b) nitrous acid (HNO_2), and (c) ammonium sulfide $[(NH_4)_2S]$.

Practice Problem **C**ONCEPTUALIZE Some over-the-counter medicines for migraines contain both ibuprofen and caffeine, molecular models of which are shown here. Write a molecular formula for each of these compounds.

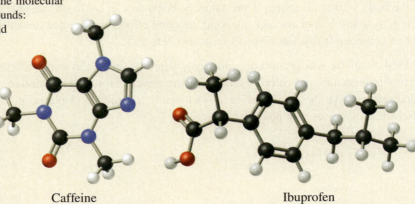

Caffeine   Ibuprofen

CHECKPOINT – SECTION 3.1   Molecular and Formula Masses

3.1.1 What is the molecular mass of citric acid ($H_3C_6H_5O_7$)?

a) 192.12 amu

b) 189.10 amu

c) 132.07 amu

d) 29.02 amu

e) 89.07 amu

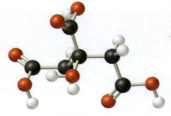

Citric acid

3.1.2 What is the formula mass of calcium citrate $[Ca_3(C_6H_5O_7)_2]$?

a) 309.34 amu

b) 69.10 amu

c) 229.18 amu

d) 498.44 amu

e) 418.28 amu

3.2 Percent Composition of Compounds

The formula of a compound indicates the number of atoms of each element in a unit of the compound. From a molecular or empirical formula, we can calculate what percent of the total mass is contributed by each element in a compound. A list of the percent by mass of each element in a compound is known as the compound's *percent composition by mass*. One way that the purity of a compound can be verified is by comparing its percent composition by mass, determined experimentally, with its calculated percent composition. Percent composition is calculated by dividing the mass of each element in a unit of the compound by the molecular or formula mass of the compound and then multiplying by 100 percent. Mathematically, the percent by mass of an element in a compound is expressed as

$$\text{percent by mass of an element} = \frac{n \times \text{atomic mass of element}}{\text{molecular or formula mass of compound}} \times 100\% \qquad \textbf{Equation 3.1}$$

where n is the number of atoms of the element in a molecule or formula unit of the compound. For example, in a molecule of hydrogen peroxide (H_2O_2), there are two H atoms and two O atoms. The atomic masses of H and O are 1.008 and 16.00 amu, respectively, so the molecular mass of H_2O_2 is 34.02 amu. Therefore, the percent composition of H_2O_2 is calculated as follows:

$$\%H = \frac{2 \times 1.008 \text{ amu H}}{34.02 \text{ amu } H_2O_2} \times 100\% = 5.926\%$$

$$\%O = \frac{2 \times 16.00 \text{ amu O}}{34.02 \text{ amu } H_2O_2} \times 100\% = 94.06\%$$

The sum of percentages is 5.296% + 94.06% = 99.99%. The small discrepancy from 100 percent is due to rounding of the atomic masses of the elements. We could equally well have used the empirical formula of hydrogen peroxide (HO) for the calculation. In this case, we would have used the *empirical formula mass,* 17.01 amu, in place of the molecular mass.

$$\%H = \frac{1.008 \text{ amu H}}{17.01 \text{ amu}} \times 100\% = 5.926\%$$

$$\%O = \frac{16.00 \text{ amu O}}{17.01 \text{ amu}} \times 100\% = 94.06\%$$

Because both the molecular formula and the empirical formula tell us the composition of the compound, they both give the same percent composition by mass. Sample Problem 3.2 shows how to calculate percent composition by mass.

SAMPLE PROBLEM   3.2

Lithium carbonate (Li_2CO_3) was the first "mood-stabilizing" drug approved by the FDA for the treatment of mania and manic-depressive illness, also known as bipolar disorder. Calculate the percent composition by mass of lithium carbonate.

Strategy Use Equation 3.1 to determine the percent by mass contributed by each element in the compound.

Setup Lithium carbonate is an ionic compound that contains Li, C, and O. In a formula unit, there are two Li atoms, one C atom, and three O atoms with atomic masses 6.941, 12.01, and 16.00 amu, respectively. The formula mass of (Li_2CO_3) is 2(6.941 amu) + 12.01 amu + 3(16.00 amu) = 73.89 amu.

Solution For each element, multiply the number of atoms by the atomic mass, divide by the formula mass, and multiply by 100 percent.

$$\%Li = \frac{2 \times 6.941 \text{ amu Li}}{73.89 \text{ amu Li}_2\text{CO}_3} \times 100\% = 18.79\%$$

$$\%C = \frac{12.01 \text{ amu C}}{73.89 \text{ amu Li}_2\text{CO}_3} \times 100\% = 16.25\%$$

$$\%O = \frac{3 \times 16.00 \text{ amu O}}{73.89 \text{ amu Li}_2\text{CO}_3} \times 100\% = 64.96\%$$

THINK ABOUT IT

Make sure that the percent composition results for a compound sum to approximately 100. (In this case, the results sum to exactly 100 percent—18.79% + 16.25% + 64.96% = 100.00%—but remember that because of rounding, the percentages may sum to very slightly more or very slightly less.)

Practice Problem **A**TTEMPT Determine the percent composition by mass of the artificial sweetener aspartame ($C_{14}H_{18}N_2O_5$).

Practice Problem **B**UILD Determine the simplest molecular formula for a compound that is 62.04 percent carbon, 10.41 percent hydrogen, and 27.55 percent oxygen by mass.

Practice Problem **C**ONCEPTUALIZE Determine the percent composition by mass of acetaminophen, the active ingredient in over-the-counter pain relievers such as Tylenol.

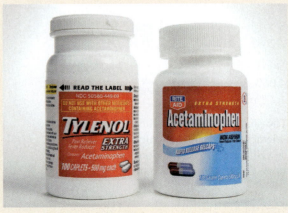

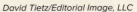
David Tietz/Editorial Image, LLC

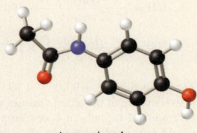
Acetaminophen

CHECKPOINT – SECTION 3.2   **Percent Composition of Compounds**

3.2.1 What is the percent composition by mass of aspirin ($C_9H_8O_4$)?

a) 44.26% C, 3.28% H, 52.46% O

b) 60.00% C, 4.48% H, 35.53% O

c) 41.39% C, 3.47% H, 55.14% O

d) 42.86% C, 6.35% H, 50.79% O

e) 42.86% C, 38.09% H, 19.05% O

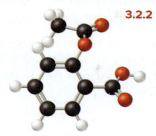

Aspirin

3.2.2 What is the percent composition by mass of sodium bicarbonate ($NaHCO_3$)?

a) 20.89% Na, 2.75% H, 32.74% C, 43.62% O

b) 44.20% Na, 1.94% H, 23.09% C, 30.76% O

c) 21.28% Na, 0.93% H, 33.35% C, 44.43% O

d) 44.20% Na, 1.94% H, 23.09% C, 30.76% O

e) 27.37% Na, 1.20% H, 14.30% C, 57.14% O

3.3   Chemical Equations

A *chemical reaction,* as described in the third hypothesis of Dalton's atomic theory [◄◄ Section 2.1], is the rearrangement of atoms in a sample of matter. Examples include the rusting of iron and the explosive combination of hydrogen and oxygen gases to produce water. A **chemical equation** uses chemical symbols to denote what occurs in a chemical reaction. We have seen how chemists represent elements and compounds using chemical symbols. We now look at how chemists represent chemical reactions using chemical equations.

Interpreting and Writing Chemical Equations

A chemical equation represents a *chemical statement.* When you encounter a chemical equation, you may find it useful to read it as though it were a sentence.

Read

$$NH_3 + HCl \longrightarrow NH_4Cl$$

as "Ammonia and hydrogen chloride react to produce ammonium chloride."

Read

$$CaCO_3 \longrightarrow CaO + CO_2$$

as "Calcium carbonate reacts to produce calcium oxide and carbon dioxide." Thus, the plus signs can be interpreted simply as the word *and,* and the arrows can be interpreted as the phrase "react(s) to produce."

In addition to interpreting chemical equations, you must be able to write chemical equations to represent reactions. For example, the equation for the process by which sulfur and oxygen react to produce sulfur dioxide is written as

$$S + O_2 \longrightarrow SO_2$$

Likewise, we write the equation for the reaction of sulfur trioxide and water to produce sulfuric acid as

$$SO_3 + H_2O \longrightarrow H_2SO_4$$

Each chemical species that appears to the left of the arrow is called a **reactant.** Reactants are those substances that are *consumed* in the course of a chemical reaction. Each species that appears to the right of the arrow is called a **product.** Products are the substances that *form* during the course of a chemical reaction.

Chemists usually indicate the physical states of reactants and products with italicized letters in parentheses following each species in the equation. Gases, liquids, and solids are labeled with (*g*), (*l*), and (*s*), respectively. Chemical species that are dissolved

Student Note: Students sometimes fear that they will be asked to determine the products of an unfamiliar chemical reaction—and will be unable to do so. In this chapter and in Chapter 4, we explain how to deduce the products of several different types of reactions.

in water are said to be *aqueous* and are labeled (*aq*). The equation examples given previously can be written as follows:

$$NH_3(g) + HCl(g) \longrightarrow NH_4Cl(s) \qquad S(s) + O_2(g) \longrightarrow SO_2(g)$$

$$CaCO_3(s) \longrightarrow CaO(s) + CO_2(g) \qquad SO_3(g) + H_2O(l) \longrightarrow H_2SO_4(aq)$$

In some cases, the physical state must be expressed more specifically. Carbon, for example, exists in two different solid forms: diamond and graphite. Rather than simply write C(*s*), we must specify the form of solid carbon by writing C(diamond) or C(graphite).

$$C(graphite) + O_2(g) \longrightarrow CO_2(g)$$

$$C(diamond) + O_2(g) \longrightarrow CO_2(g)$$

Some problems in later chapters can be solved only if the states of reactants and products are specified. It is a good idea to get in the habit now of including the physical states of reactants and products in the chemical equations that you write.

Chemical equations are also used to represent *physical* processes. Sucrose ($C_{12}H_{22}O_{11}$) dissolving in water, for example, is a physical process [◄◄ Section 1.2] that can be represented with the following chemical equation:

$$C_{12}H_{22}O_{11}(s) \xrightarrow{H_2O} C_{12}H_{22}O_{11}(aq)$$

The H_2O over the arrow in the equation denotes the process of dissolving a substance in water. Although formulas or symbols often are omitted for simplicity, they can be written over an arrow in a chemical equation to indicate the conditions under which the reaction takes place. For example, in the chemical equation

$$2KClO_3(s) \xrightarrow{\Delta} 2KCl(s) + 3O_2(g)$$

the symbol Δ indicates that the addition of heat is necessary to make $KClO_3$ react to form KCl and O_2.

Balancing Chemical Equations

Based on what you have learned so far, the chemical equation for the explosive reaction of hydrogen gas with oxygen gas to form liquid water would be

$$H_2(g) \quad + \quad O_2(g) \quad \longrightarrow \quad H_2O(l)$$

This equation as written violates the law of conservation of mass, however, because four atoms (two H and two O) react to produce only three atoms (two H and one O).

The equation must be *balanced* so that the same number of each kind of atom appears on both sides of the reaction arrow. Balancing is achieved by writing appropriate *stoichiometric coefficients* (often referred to simply as *coefficients*) to the left of the chemical formulas. In this case, we write a coefficient of 2 to the left of both the $H_2(g)$ and the $H_2O(l)$:

$$2H_2(g) \quad + \quad O_2(g) \quad \longrightarrow \quad 2H_2O(l)$$

There are now four H atoms and two O atoms on each side of the arrow. When balancing a chemical equation, we can change only the coefficients that precede the chemical formulas, *not* the subscripts within the chemical formulas. Changing the subscripts would change the formulas for the species involved in the reaction. For example, changing the product from H_2O to H_2O_2 would result in equal numbers of each kind of atom on both sides of the equation, but the equation we set out to balance represented the combination of hydrogen gas and oxygen gas to form water, not hydrogen peroxide. Additionally, we cannot add reactants or products to the chemical equation for the purpose of balancing it. To do so would result in an equation that represents the wrong reaction. The chemical equation must be made quantitatively correct without changing its qualitative chemical statement.

Balancing a chemical equation requires something of a trial-and-error approach. You may find that you change the coefficient for a particular reactant or product, only to have to change it again later in the process. In general, it will facilitate the balancing process if you do the following:

1. Change the coefficients of compounds (e.g., CO_2) before changing the coefficients of elements (e.g., O_2).
2. Treat polyatomic ions that appear on both sides of the equation (e.g., CO_3^{2-}) as units, rather than counting their constituent atoms individually.
3. Count atoms and/or polyatomic ions carefully, and track their numbers each time you change a coefficient.

Combustion refers to burning in the presence of oxygen. Combustion of a hydrocarbon such as butane produces carbon dioxide and water. To balance the chemical equation for the combustion of butane, we first take an inventory of the numbers of each type of atom on each side of the arrow.

$$C_4H_{10}(g) + O_2(g) \longrightarrow CO_2(g) + H_2O(l)$$

$$4-C-1$$
$$10-H-2$$
$$2-O-3$$

Initially, there are four C atoms on the left and one on the right; ten H atoms on the left and two on the right; and two O atoms on the left with three on the right. As a first step, we place a coefficient of 4 in front of $CO_2(g)$ on the product side.

$$C_4H_{10}(g) + O_2(g) \longrightarrow \mathbf{4}CO_2(g) + H_2O(l)$$

$$4-C-4$$
$$10-H-2$$
$$2-O-9$$

This changes the tally of atoms as shown. Thus, the equation is balanced for carbon, but not for hydrogen or oxygen. Next, we place a coefficient of 5 in front of $H_2O(l)$ on the product side and tally the atoms on both sides again.

$$C_4H_{10}(g) + O_2(g) \longrightarrow \mathbf{4}CO_2(g) + \mathbf{5}H_2O(l)$$

$$4-C-4$$
$$10-H-10$$
$$2-O-13$$

Now the equation is balanced for carbon and hydrogen. Only oxygen remains to be balanced. There are 13 O atoms on the product side of the equation (eight in CO_2 molecules and another five in H_2O molecules), so we need 13 O atoms on the reactant side. Because each oxygen molecule contains two O atoms, we have to place a coefficient of $\frac{13}{2}$ in front of $O_2(g)$:

$$C_4H_{10}(g) + \tfrac{13}{2}O_2(g) \longrightarrow \mathbf{4}CO_2(g) + \mathbf{5}H_2O(l)$$

$$4-C-4$$
$$10-H-10$$
$$13-O-13$$

Student Note: When we study electrochemistry in detail, we present a method for balancing certain equations that *does* allow the addition of H_2O, H^+, and OH^- [▶ Section 19.1].

Student Note: The water produced by combustion can also be represented by $H_2O(g)$.

Student Note: A balanced equation is, in a sense, a mathematical equality. We can multiply or divide through by any number, and the equality will still be valid.

With equal numbers of each kind of atom on both sides of the equation, this equation is now balanced. For now, however, you should practice balancing equations with the smallest possible *whole* number coefficients. Multiplying each coefficient by 2 gives all whole numbers and a final balanced equation:

$$\mathbf{2}C_4H_{10}(g) + \mathbf{13}O_2(g) \longrightarrow \mathbf{8}CO_2(g) + \mathbf{10}H_2O(l)$$

8—C—8

20—H—20

26—O—26

Sample Problem 3.3 lets you practice writing and balancing a chemical equation.

SAMPLE PROBLEM 3.3

Write and balance the chemical equation for the aqueous reaction of barium hydroxide and perchloric acid to produce aqueous barium perchlorate and water.

Strategy Determine the formulas and physical states of all reactants and products, and use them to write a chemical equation that makes the correct chemical statement. Finally, adjust coefficients in the resulting chemical equation to ensure that there are identical numbers of each type of atom on both sides of the reaction arrow.

Setup The reactants are $Ba(OH)_2$ and $HClO_4$, and the products are $Ba(ClO_4)_2$ and H_2O [⏮ Sections 2.6 and 2.7]. Because the reaction is aqueous, all species except H_2O will be labeled (*aq*) in the equation. Being a liquid, H_2O will be labeled (*l*).

Solution The chemical statement "barium hydroxide and perchloric acid react to produce barium perchlorate and water" can be represented with the following unbalanced equation:

$$Ba(OH)_2(aq) + HClO_4(aq) \longrightarrow Ba(ClO_4)_2(aq) + H_2O(l)$$

Perchlorate ions (ClO_4^-) appear on both sides of the equation, so count them as units, rather than count the individual atoms they contain. Thus, the tally of atoms and polyatomic ions is

1—Ba—1

2—O—1 (not including O atoms in ClO_4^- ions)

3—H—2

1—ClO_4^-—2

The barium atoms are already balanced, and placing a coefficient of 2 in front of $HClO_4(aq)$ balances the number of perchlorate ions.

$$Ba(OH)_2(aq) + 2HClO_4(aq) \longrightarrow Ba(ClO_4)_2(aq) + H_2O(l)$$

1—Ba—1

2—O—1 (not including O atoms in ClO_4^- ions)

4—H—2

2—ClO_4^-—2

Placing a coefficient of 2 in front of $H_2O(l)$ balances both the O and H atoms, giving us the final balanced equation:

$$Ba(OH)_2(aq) + 2HClO_4(aq) \longrightarrow Ba(ClO_4)_2(aq) + 2H_2O(l)$$

1—Ba—1

2—O—2 (not including O atoms in ClO_4^- ions)

4—H—4

2—ClO_4^-—2

THINK ABOUT IT

Check to be sure the equation is balanced by counting all the atoms individually.

1—Ba—1

10—O—10

4—H—4

2—Cl—2

Practice Problem ATTEMPT Write and balance the chemical equation that represents the combustion of propane (i.e., the reaction of propane gas with oxygen gas to produce carbon dioxide gas and liquid water).

Practice Problem BUILD Write and balance the chemical equation that represents the reaction of sulfuric acid with sodium hydroxide to form water and sodium sulfate.

Practice Problem CONCEPTUALIZE Write a balanced equation for the reaction shown here.

Bringing Chemistry to Life

The Stoichiometry of Metabolism

The carbohydrates and fats we eat are broken down into small molecules in the digestive system. Carbohydrates are broken down into simple sugars such as glucose ($C_6H_{12}O_6$), and fats are broken down into fatty acids (carboxylic acids that contain hydrocarbon chains) and glycerol ($C_3H_8O_3$). The small molecules produced in the digestion process are subsequently consumed by a series of complex biochemical reactions. Although the metabolism of simple sugars and fatty acids involves relatively complex processes, the results are essentially the same as that of combustion—that is, simple sugars and fatty acids react with oxygen to produce carbon dioxide, water, and energy. The balanced chemical equation for the metabolism of glucose is

Glycerol

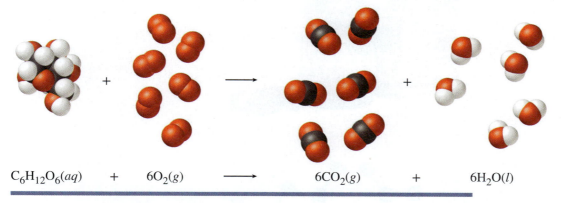

$$C_6H_{12}O_6(aq) \quad + \quad 6O_2(g) \quad \longrightarrow \quad 6CO_2(g) \quad + \quad 6H_2O(l)$$

Sample Problem 3.4 shows how to balance and use the equation for metabolism.

SAMPLE PROBLEM 3.4

Butyric acid (also known as butanoic acid, $C_4H_8O_2$) is one of many compounds found in milk fat. First isolated from rancid butter in 1869, butyric acid has received a great deal of attention in recent years as a potential anticancer agent. Assuming that the only products are CO_2 and H_2O, write and balance the equation for the metabolism of butyric acid.

Strategy Write an unbalanced equation to represent the combination of reactants and formation of products as stated in the problem, and then balance the equation.

Setup Metabolism in this context refers to the combination of $C_4H_8O_2$ with O_2 to produce CO_2 and H_2O.

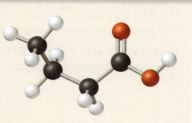

Butyric acid

(Continued on next page)

Solution

$$C_4H_8O_2(aq) + O_2(g) \longrightarrow CO_2(g) + H_2O(l)$$

Balance the number of C atoms by changing the coefficient for CO_2 from 1 to 4.

$$C_4H_8O_2(aq) + O_2(g) \longrightarrow \mathbf{4}CO_2(g) + H_2O(l)$$

Balance the number of H atoms by changing the coefficient for H_2O from 1 to 4.

$$C_4H_8O_2(aq) + O_2(g) \longrightarrow \mathbf{4}CO_2(g) + \mathbf{4}H_2O(l)$$

Finally, balance the number of O atoms by changing the coefficient for O_2 from 1 to 5.

$$C_4H_8O_2(aq) + \mathbf{5}O_2(g) \longrightarrow \mathbf{4}CO_2(g) + \mathbf{4}H_2O(l)$$

THINK ABOUT IT

Count the number of each type of atom on each side of the reaction arrow to verify that the equation is properly balanced. There are 4 C, 8 H, and 12 O in the reactants and in the products, so the equation is balanced.

Practice Problem **A**TTEMPT Another compound found in milk fat that appears to have anticancer and antiobesity properties is conjugated linoleic acid (CLA; $C_{18}H_{32}O_2$). Assuming again that the only products are CO_2 and H_2O, write and balance the equation for the metabolism of CLA.

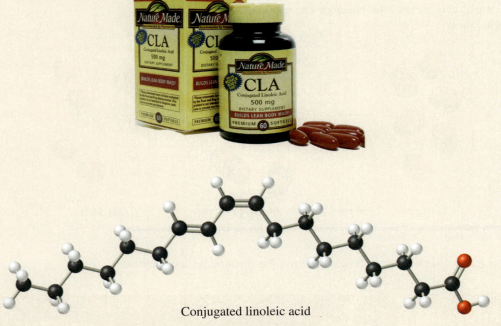

Conjugated linoleic acid

John Flournoy/McGraw Hill

Practice Problem **B**UILD Write and balance the equation for the combination of ammonia gas with solid copper(II) oxide to produce copper metal, nitrogen gas, and liquid water.

Practice Problem **C**ONCEPTUALIZE The compound shown on the left reacts with nitrogen dioxide to form the compound shown on the right and iodine. Write a balanced equation for the reaction.

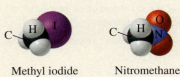

Methyl iodide   Nitromethane

CHECKPOINT – SECTION 3.3   **Chemical Equations**

3.3.1 What are the stoichiometric coefficients in the following equation when it is balanced?

$$CH_4(g) + H_2O(g) \longrightarrow H_2(g) + CO_2(g)$$

a) 1, 2, 2, 2

b) 2, 1, 1, 2

c) 1, 2, 2, 1

d) 2, 2, 2, 1

e) 1, 2, 4, 1

3.3.2 Which chemical equation represents the reaction shown? (Blue spheres are nitrogen, and white spheres are hydrogen.)

a) $6N(g) + 18H(g) \longrightarrow 6NH_3(g)$

b) $3N_2(g) + 3H_2(g) \longrightarrow 2NH_3(g)$

c) $2N_2(g) + 3H_2(g) \longrightarrow 2NH_3(g)$

d) $N_2(g) + 3H_2(g) \longrightarrow 2NH_3(g)$

e) $3N_2(g) + 6H_2(g) \longrightarrow 6NH_3(g)$

3.3.3 Which is the correctly balanced form of the given equation?

$$S(s) + O_3(g) \longrightarrow SO_2(g)$$

a) $S(s) + O_3(g) \longrightarrow SO_3(g)$

b) $3S(s) + 6O_3(g) \longrightarrow 3SO_2(g)$

c) $3S(s) + O_3(g) \longrightarrow 3SO_2(g)$

d) $3S(s) + 2O_3(g) \longrightarrow SO_2(g)$

e) $3S(s) + 2O_3(g) \longrightarrow 3SO_2(g)$

3.3.4 Carbon monoxide reacts with oxygen to produce carbon dioxide according to the following balanced equation:

$$2CO(g) + O_2(g) \longrightarrow 2CO_2(g)$$

A reaction vessel containing the reactants is pictured.

Which of the following represents the contents of the reaction vessel when the reaction is complete?

a)

b)

c)

d)

3.4   The Mole and Molar Masses

A balanced chemical equation tells us not only what species are consumed and produced in a chemical reaction, but also in what relative amounts. In the combination of hydrogen and oxygen to produce water, the balanced chemical equation tells us that two H_2 molecules react with one O_2 molecule to produce two H_2O molecules. When carrying out a reaction such as this, however, chemists do not work with individual molecules. Rather, they work with macroscopic quantities that contain enormous numbers of molecules. Regardless of the number of molecules involved, though, the molecules combine in the ratio specified in the balanced equation. Twenty molecules of hydrogen would combine with 10 molecules of oxygen, 10 million molecules of hydrogen would combine with 5 million molecules of oxygen, and so on. Clearly, it would not be convenient to express the quantities used in the laboratory in terms of the numbers of molecules involved! Instead, chemists use a unit of measurement called the *mole*.

Figure 3.1 Bulk nails are sold by the pound. How many nails there are in a pound depends on the size and type of nail.

Ryan McVay/Getty Images

Type of nail	Size (in)	Nails/lb
3d box	1.5	635
6d box	2	236
10d box	3	94
4d casing	1.5	473
8d casing	2.5	145
2d common	1	876
4d common	1.5	316
6d common	2	181
8d common	2.5	106

The Mole

If you're getting doughnuts for yourself, you probably buy them individually, but if you're getting them for your entire chemistry class, you had better buy them by the dozen. One dozen doughnuts contains exactly 12 doughnuts. In fact, a dozen of anything contains exactly 12 of that thing. Pencils typically come 12 to a box, and 12 such boxes may be shipped to the campus bookstore in a bigger box. The bigger box contains one gross of pencils (144). Whether a dozen or a gross, each is a convenient quantity that contains a reasonable, specific, exact number of items.

Chemists, too, have adopted such a number to make it easy to express the number of molecules (or atoms or ions) in a typical macroscopic sample of matter. Atoms and molecules are so much smaller than doughnuts or pencils, though, that the number used by chemists is significantly bigger than a dozen or a gross. The quantity used by chemists is the ***mole (mol),*** which is defined as the amount of a substance that contains as many elementary entities (atoms, molecules, formula units, etc.) as there are atoms in exactly 0.012 kg (12 g) of carbon-12. The number of atoms in exactly 12 g of carbon-12, which is determined experimentally, is known as ***Avogadro's number (N_A),*** in honor of the Italian scientist Amedeo Avogadro.[1] The currently accepted value of Avogadro's number is 6.0221418×10^{23}, although we usually round it to 6.022×10^{23}. Thus, a dozen doughnuts contains 12 doughnuts; a gross of pencils contains 144 pencils; and a mole of O_2 gas, an amount that at room temperature and ordinary pressure would fill slightly more than half of a 10-gallon fish tank, contains 6.022×10^{23} O_2 molecules. Unlike the dozen and the gross, which are arrived at by counting objects, the mole is not a quantity that can be determined by counting. The number of atoms or molecules in a macroscopic sample of matter is simply too big to be counted. Instead, the number of atoms or molecules in a quantity of substance is determined by weighing—like nails in a hardware store (Figure 3.1). The number of nails needed to build a fence or a house is fairly large. Therefore, when nails are purchased for such a project, they are not *counted;* but rather, they are *weighed* to determine their number. How many nails there are to a pound depends on the size and type of nail.

For example, if we want to buy 1000 1.5-in 4d common nails, we would not count out a thousand nails. Instead, we would weigh out an amount just over three pounds. Using data from the table in Figure 3.1:

$$1000 \ \text{4d common nails} \times \frac{1 \ \text{lb}}{316 \ \text{4d common nails}} = 3.16 \ \text{lb}$$

Student Note: Avogadro's number is almost unimaginably big. If 6.022×10^{23} pennies were distributed equally among every inhabitant of the United States, each man, woman, and child would have over 20 trillion dollars! Furthermore, each person's share of pennies, neatly stacked, would occupy approximately the same volume as 760 Empire State Buildings.

---

1. Lorenzo Romano Amedeo Carlo Avogadro di Quaregua e di Cerret (1776–1856). Italian mathematical physicist. He practiced law for many years before he became interested in science. His most famous work, now known as Avogadro's law (see Chapter 10), was largely ignored during his lifetime, although in the late nineteenth century it became the basis for determining atomic masses.

If we were to weigh out something *other* than the calculated amount, we could determine the number of nails using the same conversion factor:

$$5.00 \text{ lb} \times \frac{316 \text{ 4d common nails}}{1 \text{ lb}} = 1580 \text{ 4d common nails}$$

Note that the same mass of a different type of nail would contain a different number of nails. For example, we might weigh out 5.00 lb of 3-in 10d box nails:

$$5.00 \text{ lb} \times \frac{94 \text{ 10d box nails}}{1 \text{ lb}} = 470 \text{ 10d box nails}$$

The number of nails in each case is determined by weighing a sample of nails, and the number of nails in a given mass depends on the type of nail. Numbers of elementary entities, too, are determined by *weighing* a sample; and the number of elementary entities per unit mass also depends on the *type* of elementary entities in the sample. Figure 3.2 shows samples containing one mole each of several common substances.

Consider again the formation of water from hydrogen and oxygen gases. No matter how large the number of molecules involved, the ratio is always *two* H_2 combining with *one* O_2 to form *two* H_2O. Therefore, just as two molecules of hydrogen combine with one molecule of oxygen to form two molecules of water, two moles of hydrogen molecules ($2 \times N_A$ or 12.044×10^{23} H_2 molecules) combine with one mole of oxygen molecules (6.022×10^{23} O_2 molecules) to form two moles of water molecules (12.044×10^{23} H_2O molecules). Thus, we can now interpret the chemical equation for this reaction in terms of moles. Figure 3.3 depicts this reaction on both the molecular and macroscopic scales.

Note also that the subscripts in a chemical formula denote the ratio of combination of atoms in a substance. Just as a water molecule contains two H atoms and one O atom, a mole of water molecules contains two moles of H atoms and one mole of O atoms. The ratio of combination stays the same, regardless of the size of the sample.

Figure 3.2 One mole each of several familiar substances. Left to right: copper, water, salt, sugar, and aluminum.

Charles D. Winters/McGraw Hill

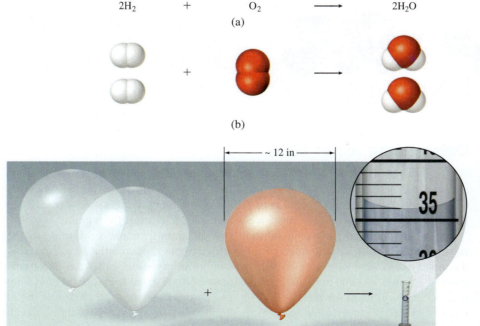

$$2H_2 \quad + \quad O_2 \quad \longrightarrow \quad 2H_2O$$

(a)

(b)

(c)

Figure 3.3 The reaction of H_2 and O_2 to form H_2O shown as (a) a balanced chemical equation, (b) molecular models, and (c) an experiment (at 0°C and atmospheric pressure) in which 44.8 L of H_2 are combined with 22.4 L of O_2 to produce 36.0 mL of liquid H_2O.

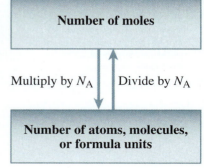

Number of moles

Multiply by N_A | Divide by N_A

Number of atoms, molecules, or formula units

N_A = Avogadro's number

Sample Problem 3.5 lets you practice converting between moles and atoms.

SAMPLE PROBLEM 3.5

Calcium is the most abundant metal in the human body. A typical human body contains roughly 30 moles of calcium. Determine (a) the number of Ca atoms in 30.00 moles of calcium and (b) the number of moles of calcium in a sample containing 1.00×10^{20} atoms.

Strategy Use Avogadro's number to convert from moles to atoms and from atoms to moles.

Setup When the number of moles is known, we multiply by Avogadro's number to convert to atoms. When the number of atoms is known, we divide by Avogadro's number to convert to moles.

Solution

(a) $30.00 \text{ mol Ca} \times \dfrac{6.022 \times 10^{23} \text{ Ca atoms}}{1 \text{ mol Ca}} = 1.807 \times 10^{25} \text{ Ca atoms}$

(b) $1.00 \times 10^{20} \text{ Ca atoms} \times \dfrac{1 \text{ mol Ca}}{6.022 \times 10^{23} \text{ Ca atoms}} = 1.66 \times 10^{-4} \text{ mol Ca}$

THINK ABOUT IT

Make sure that units cancel properly in each solution and that the result makes sense. In part (a), for example, the number of moles (30) is greater than one, so the number of atoms is greater than Avogadro's number. In part (b), the number of atoms (1×10^{20}) is less than Avogadro's number, so there is less than a mole of substance.

Practice Problem **A**TTEMPT Potassium is the second most abundant metal in the human body. Calculate (a) the number of atoms in 7.31 moles of potassium and (b) the number of moles of potassium that contains 8.91×10^{25} atoms.

Practice Problem **B**UILD Calculate (a) the number of atoms in 1.05×10^{-6} mole of helium and (b) the number of moles of helium that contains 2.33×10^{21} atoms.

Practice Problem **C**ONCEPTUALIZE These diagrams show collections of objects. For each diagram, express the number of objects using units of *dozen* and using units of *gross*. (Report each answer to four significant figures but explain why the answers to this problem actually have more than four significant figures.)

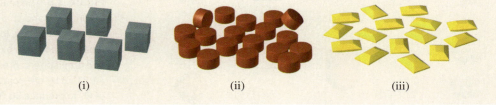

(i)   (ii)   (iii)

Student Note: Because the number of moles specifies the number of particles (atoms, molecules, or ions), using molar mass as a conversion factor, in effect, allows us to count the particles in a sample of matter by weighing the sample.

Determining Molar Mass

Although chemists often wish to combine substances in specific mole ratios, there is no direct way to measure the number of moles in a sample of matter. Instead, chemists determine how many moles there are of a substance by measuring its mass (usually in grams). The molar mass of the substance is then used to convert from grams to moles.

The ***molar mass*** ($\mathscr{M}$) of a substance is the mass in grams of 1 mole of the substance. By definition, the mass of a mole of carbon-12 is exactly 12 g. Note that the molar mass of carbon is numerically equal to its atomic mass. Likewise, the atomic mass of calcium is 40.08 amu and its molar mass is 40.08 g, the atomic mass of sodium is 22.99 amu and its molar mass is 22.99 g, and so on. In general, an element's molar mass in grams is numerically equal to its atomic mass in atomic mass units. Recall from Section 2.5 that

$$1 \text{ amu} = 1.661 \times 10^{-24} \text{ g}$$

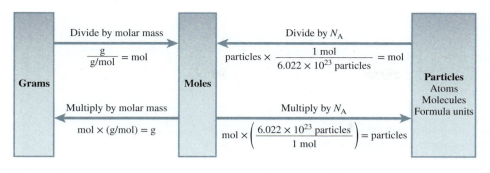

Figure 3.4 Flowchart for conversions among mass, moles, and number of particles.

This is the reciprocal of Avogadro's number. Expressed another way:

$$1 \text{ g} = 6.022 \times 10^{23} \text{ amu}$$

In effect, there is 1 mole of atomic mass units in a gram. The molar mass (in grams) of any compound is numerically equal to its molecular or formula mass (in amu). The molar mass of water, for example, is 18.02 g, and the molar mass of sodium chloride (NaCl) is 58.44 g.

When it comes to expressing the molar mass of elements such as oxygen and hydrogen, we have to be careful to specify what form of the element we mean. For instance, the element oxygen exists predominantly as diatomic molecules (O_2). Thus, if we say 1 mole of oxygen and by *oxygen* we mean O_2, the molecular mass is 32.00 amu and the molar mass is 32.00 g. If on the other hand we mean a mole of atomic oxygen (O), then the molar mass is only 16.00 g, which is numerically equal to the atomic mass of O (16.00 amu). You should be able to tell from the context which form of an element is intended, as the following examples illustrate:

Context	*Oxygen* Means	Molar Mass
How many moles of oxygen react with 2 moles of hydrogen to produce water?	O_2	32.00 g
How many moles of oxygen are there in 1 mole of water?	O	16.00 g
Air is approximately 21% oxygen.	O_2	32.00 g
Many organic compounds contain oxygen.	O	16.00 g

Although the term molar mass specifies the mass of 1 mole of a substance, making the appropriate units simply grams (g), we usually express molar masses in units of grams per mole (g/mol) to facilitate calculations involving moles.

Interconverting Mass, Moles, and Numbers of Particles

Molar mass is the conversion factor that we use to convert from mass (m) to moles (n), and vice versa. We use Avogadro's number to convert from number of moles to number of particles (N), and vice versa. *Particles* in this context may refer to atoms, molecules, ions, or formula units. Figure 3.4 summarizes the operations involved in these conversions.

Sample Problems 3.6 and 3.7 illustrate how the conversions are done.

SAMPLE PROBLEM **3.6**

Determine (a) the number of moles of C in 10.00 g of naturally occurring carbon and (b) the mass of 0.905 mole of sodium chloride.

Strategy Use molar mass to convert from mass to moles and to convert from moles to mass.

Setup The molar mass of carbon is 12.01 g/mol. The molar mass of a compound is numerically equal to its formula mass. The molar mass of sodium chloride (NaCl) is 58.44 g/mol.

(Continued on next page)

Solution

(a) $10.00 \text{ g C} \times \dfrac{1 \text{ mol C}}{12.01 \text{ g C}} = 0.8326 \text{ mol C}$ 　　　　(b) $0.905 \text{ mol NaCl} \times \dfrac{58.44 \text{ g NaCl}}{1 \text{ mol NaCl}} = 52.9 \text{ g NaCl}$

THINK ABOUT IT

Always double-check unit cancellations in problems such as these—errors are common when molar mass is used as a conversion factor. Also make sure that the results make sense. In both cases, a mass smaller than the molar mass corresponds to less than a mole of substance.

Practice Problem A**TTEMPT**　(a) Determine the mass in grams of 2.75 moles of glucose ($C_6H_{12}O_6$). (b) Determine the number of moles in 59.8 g of sodium nitrate ($NaNO_3$).

Practice Problem B**UILD**　(a) Determine the mass of sodium metal that contains the same number of moles as 87.2 g of copper metal. (b) Determine the number of moles of helium that has the same mass as 4.505 moles of neon.

Practice Problem C**ONCEPTUALIZE**　Plain doughnuts from a particular bakery have an average mass of 32.6 g, whereas jam-filled doughnuts from the same bakery have an average mass of 40.0 g. (a) Determine the mass of a dozen plain doughnuts and the mass of a dozen jam-filled doughnuts. (b) Determine the number of doughnuts in a kilogram of plain and the number in a kilogram of jam-filled. (c) Determine the mass of plain doughnuts that contains the same number of doughnuts as a kilogram of jam-filled. (d) Determine the total mass of a dozen doughnuts consisting of three times as many plain as jam-filled.

SAMPLE PROBLEM (3.7)

(a) Determine the number of water molecules and the numbers of H and O atoms in 3.26 g of water. (b) Determine the mass of 7.92×10^{19} carbon dioxide molecules.

Strategy　Use molar mass and Avogadro's number to convert from mass to molecules, and vice versa. Use the molecular formula of water to determine the numbers of H and O atoms.

Setup　(a) Starting with mass (3.26 g of water), we use molar mass (18.02 g/mol) to convert to moles of water. From moles, we use Avogadro's number to convert to number of water molecules. In part (b), we reverse the process in part (a) to go from number of molecules to mass of carbon dioxide.

Solution

(a) $3.26 \text{ g H}_2\text{O} \times \dfrac{1 \text{ mol H}_2\text{O}}{18.02 \text{ g H}_2\text{O}} \times \dfrac{6.022 \times 10^{23} \text{ H}_2\text{O molecules}}{1 \text{ mol H}_2\text{O}} = 1.09 \times 10^{23} \text{ H}_2\text{O molecules}$

Using the molecular formula, we can determine the number of H and O atoms in 3.26 g of H_2O as follows:

$$1.09 \times 10^{23} \text{ H}_2\text{O molecules} \times \dfrac{2 \text{ H atoms}}{1 \text{ H}_2\text{O molecule}} = 2.18 \times 10^{23} \text{ H atoms}$$

$$1.09 \times 10^{23} \text{ H}_2\text{O molecules} \times \dfrac{1 \text{ O atom}}{1 \text{ H}_2\text{O molecule}} = 1.09 \times 10^{23} \text{ O atoms}$$

(b) $7.92 \times 10^{19} \text{ CO}_2 \text{ molecules} \times \dfrac{1 \text{ mol CO}_2}{6.022 \times 10^{23} \text{ CO}_2 \text{ molecules}} \times \dfrac{44.01 \text{ g CO}_2}{1 \text{ mol CO}_2} = 5.79 \times 10^{-3} \text{ g CO}_2$

THINK ABOUT IT

Again, check the cancellation of units carefully and make sure that the magnitudes of your results are reasonable.

Practice Problem (A)**TTEMPT**   (a) Calculate the number of oxygen molecules and the number of oxygen atoms in 35.5 g of O_2. (b) Calculate the mass of 9.95×10^{14} SO_3 molecules.

Practice Problem (B)**UILD**   (a) Determine the number of oxygen atoms and hydrogen atoms in 1.00 kg of water. (b) Calculate the mass of calcium carbonate that contains 1.00 mol of oxygen atoms.

Practice Problem (C)**ONCEPTUALIZE**   A particular commemorative set of coins contains two 1.00-oz silver coins and three 0.500-oz gold coins. (a) How many gold coins are there in 49.0 lb of coin sets? (b) How many silver coins are there in a collection of sets that has a total mass of 63.0 lb? (c) What is the total mass (in lb) of a collection of sets that contains 93 silver coins? (d) What is the mass of silver coins (in lb) in a collection of coin sets that contains 9.00 lb of gold coins?

Empirical Formula from Percent Composition

In Section 3.2, we learned how to use the chemical formula (either molecular or empirical) to determine the percent composition by mass. With the concepts of the mole and molar mass, we can now use the experimentally determined percent composition to determine the empirical formula of a compound. Sample Problem 3.8 shows how to do this.

SAMPLE PROBLEM (3.8)

Determine the empirical formula of a compound that is 30.45 percent nitrogen and 69.55 percent oxygen by mass.

Strategy   Assume a 100-g sample so that the mass percentages of nitrogen and oxygen given in the problem statement correspond to the masses of N and O in the compound. Then, using the appropriate molar masses, convert the grams of each element to moles. Use the resulting numbers as subscripts in the empirical formula, reducing them to the lowest possible whole numbers for the final answer.

Setup   The empirical formula of a compound consisting of N and O is N_xO_y. The molar masses of N and O are 14.01 and 16.00 g/mol, respectively. One hundred grams of a compound that is 30.45 percent nitrogen and 69.55 percent oxygen by mass contains 30.45 g N and 69.55 g O.

Solution

$$30.45 \ \cancel{g \ N} \times \frac{1 \ mol \ N}{14.01 \ \cancel{g \ N}} = 2.173 \ mol \ N$$

$$69.55 \ \cancel{g \ O} \times \frac{1 \ mol \ O}{16.00 \ \cancel{g \ O}} = 4.347 \ mol \ O$$

This gives a formula of $N_{2.173}O_{4.347}$. Dividing both subscripts by the smaller of the two to get the smallest possible whole numbers (2.173/2.173 = 1, 4.347/2.173 ≈ 2) gives an empirical formula of NO_2. This may or may not be the molecular formula of the compound because both NO_2 and N_2O_4 have this empirical formula. Without knowing the molar mass, we cannot be sure which one it is.

THINK ABOUT IT

Use the method described in Sample Problem 3.2 to calculate the percent composition of the empirical formula NO_2 and verify that it is the same as that given in this problem.

Practice Problem (A)**TTEMPT**   Determine the empirical formula of a compound that is 52.15 percent C, 13.13 percent H, and 34.73 percent O by mass.

Practice Problem (B)**UILD**   Determine the empirical formula of a compound that is 85.63 percent C and 14.37 percent H by mass.

Practice Problem (C)**ONCEPTUALIZE**   What is the smallest collection of coins from Practice Problem 3.7C that would be 60 percent gold and 40 percent silver by mass? (Assume that the gold coins are pure gold and the silver coins are pure silver.)

CHECKPOINT – SECTION 3.4   The Mole and Molar Masses

3.4.1 How many molecules are in 30.1 g of sulfur dioxide (SO_2)?

a) 1.81×10^{25}

b) 2.83×10^{23}

c) 6.02×10^{23}

d) 1.02×10^{24}

e) 5.00×10^{-23}

3.4.2 How many moles of hydrogen are there in 6.50 g of ammonia (NH_3)?

a) 0.382 mol

b) 1.39 mol

c) 0.215 mol

d) 1.14 mol

e) 2.66 mol

3.4.3 Determine the empirical formula of a compound that has the following composition: 92.3 percent C and 7.7 percent H.

a) CH

b) C_2H_3

c) C_4H_6

d) C_6H_7

e) C_4H_3

3.4.4 Determine the empirical formula of a compound that has the following composition: 48.6 percent C, 8.2 percent H, and 43.2 percent O.

a) C_3H_8O

b) C_3H_6O

c) $C_2H_5O_2$

d) C_2H_6O

e) $C_3H_6O_2$

3.5 Combustion Analysis

As we saw in [◄◄ Section 3.4], knowing the mass of each element contained in a sample of a substance enables us to determine the empirical formula of the substance. One common, practical use of this ability is the experimental determination of empirical formula by **combustion analysis.**

Combustion analysis of organic compounds (containing carbon, hydrogen, and sometimes oxygen) is carried out using an apparatus like the one shown in Figure 3.5. A sample of known mass is placed in the furnace and heated in the presence of oxygen. The carbon dioxide and water produced from carbon and hydrogen, respectively, in the combustion reaction are collected in "traps," which are weighed before and after the combustion. The difference in mass of each trap before and after the reaction is the mass of the collected product. Knowing the mass of each product, we can determine the percent composition of the compound. And, from percent composition, we can determine the empirical formula.

Determination of Empirical Formula

When a compound such as glucose is burned in a combustion analysis apparatus, carbon dioxide (CO_2) and water (H_2O) are produced. Because only oxygen gas is added to the reaction, the carbon and hydrogen present in the products must have

Figure 3.5 Schematic of a combustion analysis apparatus. CO_2 and H_2O produced in combustion are trapped and weighed. The amounts of these products are used to determine how much carbon and hydrogen the combusted sample contained. (CuO is used to ensure complete combustion of all carbon to CO_2.)

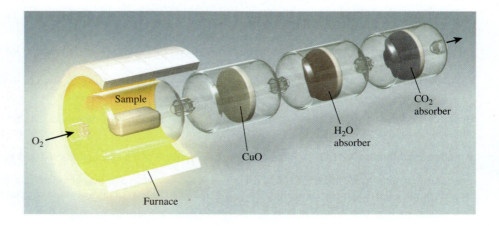

come from the glucose. The oxygen in the products may have come from the glucose, but it may also have come from the added oxygen. Suppose that in one such experiment the combustion of 18.8 g of glucose produced 27.6 g of CO_2 and 11.3 g of H_2O. We can calculate the mass of carbon and hydrogen in the original 18.8-g sample of glucose as follows:

$$\text{mass of C} = 27.6 \text{ g } CO_2 \times \frac{1 \text{ mol } CO_2}{44.01 \text{ g } CO_2} \times \frac{1 \text{ mol C}}{1 \text{ mol } CO_2} \times \frac{12.01 \text{ g C}}{1 \text{ mol C}} = 7.53 \text{ g C}$$

$$\text{mass of H} = 11.3 \text{ g } H_2O \times \frac{1 \text{ mol } H_2O}{18.02 \text{ g } H_2O} \times \frac{2 \text{ mol H}}{1 \text{ mol } H_2O} \times \frac{1.008 \text{ g H}}{1 \text{ mol H}} = 1.26 \text{ g H}$$

Thus, 18.8 g of glucose contains 7.53 g of carbon and 1.26 g of hydrogen. The remaining mass [18.8 g − (7.53 g + 1.26 g) = 10.0 g] is oxygen.

The number of moles of each element present in 18.8 g of glucose is

$$\text{moles of C} = 7.53 \text{ g C} \times \frac{1 \text{ mol C}}{12.01 \text{ g C}} = 0.627 \text{ mol C}$$

$$\text{moles of H} = 1.26 \text{ g H} \times \frac{1 \text{ mol H}}{1.008 \text{ g H}} = 1.25 \text{ mol H}$$

$$\text{moles of O} = 10.0 \text{ g O} \times \frac{1 \text{ mol O}}{16.00 \text{ g O}} = 0.626 \text{ mol O}$$

The empirical formula of glucose can therefore be written $C_{0.627}H_{1.25}O_{0.626}$. Because the numbers in an empirical formula must be integers, we divide each of the subscripts by the smallest subscript, 0.626(0.627/0.626 ≈ 1, 1.25/0.626 ≈ 2, and 0.626/0.626 = 1), and obtain CH_2O for the empirical formula.

Student Hot Spot

Student data indicate you may struggle with combustion analysis. Access the eBook to view additional Learning Resources on this topic.

Student Note: Determination of an empirical formula from combustion data can be especially sensitive to rounding error. When solving problems such as these, don't round until the very end.

Determination of Molecular Formula

The empirical formula gives only the ratio of combination of the atoms in a molecule, so there may be numerous compounds with the same empirical formula. If we know the approximate molar mass of the compound, though, we can determine the molecular formula from the empirical formula. For instance, the molar mass of glucose is about 180 g. The *empirical-formula mass* of CH_2O is about 30 g [12.01 g + 2(1.008 g) + 16.00 g]. To determine the molecular formula, we first divide the molar mass by the empirical-formula mass: 180 g/30 g = 6. This tells us that there are six empirical-formula units per molecule in glucose. Multiplying each subscript by 6 (recall that when none is shown, the subscript is understood to be a 1) gives the molecular formula, $C_6H_{12}O_6$.

Sample Problem 3.9 shows how to determine the molecular formula of a compound from its combustion data and molar mass.

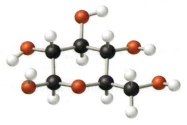

Glucose

SAMPLE PROBLEM 3.9

Combustion of a 5.50-g sample of benzene produces 18.59 g CO_2 and 3.81 g H_2O. Determine the empirical formula and the molecular formula of benzene, given that its molar mass is approximately 78 g/mol.

Strategy From the product masses, determine the mass of C and the mass of H in the 5.50-g sample of benzene. Sum the masses of C and H; the difference between this sum and the original sample mass is the mass of O in the sample (if O is in fact present in benzene). Convert the mass of each element to moles, and use the results as subscripts in a chemical formula. Convert the subscripts to whole numbers by dividing each by the smallest subscript. This gives the empirical formula. To calculate the molecular formula, first divide the molar mass given in the problem statement by the empirical-formula mass. Then, multiply the subscripts in the empirical formula by the resulting number to obtain the subscripts in the molecular formula.

Benzene

(Continued on next page)

These conversion factors enable us to determine how many moles of CO_2 will be produced upon reaction of a given amount of CO, or how much CO is necessary to produce a specific amount of CO_2. Consider the complete reaction of 3.82 moles of CO to form CO_2. To calculate the number of moles of CO_2 produced, we use the conversion factor with moles of CO_2 in the numerator and moles of CO in the denominator.

$$\text{moles } CO_2 \text{ produced} = 3.82 \text{ mol CO} \times \frac{1 \text{ mol } CO_2}{1 \text{ mol CO}} = 3.82 \text{ mol } CO_2$$

Similarly, we can use other ratios represented in the balanced equation as conversion factors. For example, we have 1 mol $O_2 \simeq 2$ mol CO_2 and 2 mol CO $\simeq 1$ mol O_2. The corresponding conversion factors allow us to calculate the amount of CO_2 produced upon reaction of a given amount of O_2, and the amount of one reactant necessary to react completely with a given amount of the other. Using the preceding example, we can determine the *stoichiometric amount* of O_2 (how many moles of O_2 are needed to react with 3.82 moles of CO).

$$\text{moles } O_2 \text{ needed} = 3.82 \text{ mol CO} \times \frac{1 \text{ mol } O_2}{2 \text{ mol CO}} = 1.91 \text{ mol } O_2$$

Sample Problem 3.10 illustrates how to determine reactant and product amounts using a balanced chemical equation.

SAMPLE PROBLEM 3.10

Urea [$(NH_2)_2CO$] is a by-product of protein metabolism. This waste product is formed in the liver and then filtered from the blood and excreted in the urine by the kidneys. Urea can be synthesized in the laboratory by the combination of ammonia and carbon dioxide according to the equation

$$2NH_3(g) + CO_2(g) \longrightarrow (NH_2)_2CO(aq) + H_2O(l)$$

(a) Calculate the amount of urea that will be produced by the complete reaction of 5.25 moles of ammonia. (b) Determine the stoichiometric amount of carbon dioxide required to react with 5.25 moles of ammonia.

Strategy Use the balanced chemical equation to determine the correct stoichiometric conversion factors, and then multiply by the number of moles of ammonia given.

Setup According to the balanced chemical equation, the conversion factor for ammonia and urea is either

$$\frac{2 \text{ mol } NH_3}{1 \text{ mol } (NH_2)_2CO} \quad \text{or} \quad \frac{1 \text{ mol } (NH_2)_2CO}{2 \text{ mol } NH_3}$$

To multiply by moles of NH_3 and have the units cancel properly, we use the conversion factor with moles of NH_3 in the denominator.
   Similarly, the conversion factor for ammonia and carbon dioxide can be written as

$$\frac{2 \text{ mol } NH_3}{1 \text{ mol } CO_2} \quad \text{or} \quad \frac{1 \text{ mol } CO_2}{2 \text{ mol } NH_3}$$

Again, we select the conversion factor with ammonia in the denominator so that moles of NH_3 will cancel in the calculation.

Solution

(a) moles $(NH_2)_2CO$ produced = 5.25 mol NH$_3$ × $\dfrac{1 \text{ mol } (NH_2)_2CO}{2 \text{ mol } NH_3}$ = 2.63 mol $(NH_2)_2CO$

(b) moles CO_2 required = 5.25 mol NH$_3$ × $\dfrac{1 \text{ mol } CO_2}{2 \text{ mol } NH_3}$ = 2.63 mol CO_2

> ## THINK ABOUT IT
>
> As always, check to be sure that units cancel properly in the calculation. Also, the balanced equation indicates that there will be *fewer* moles of urea produced than ammonia consumed. Therefore, your calculated number of moles of urea (2.63) should be *smaller* than the number of moles given in the problem (5.25). Similarly, the stoichiometric coefficients in the balanced equation are the same for carbon dioxide and urea, so your answers to this problem should also be the same for both species.

Practice Problem **A**TTEMPT Nitrogen and hydrogen react to form ammonia according to the following balanced equation: $N_2(g) + 3H_2(g) \longrightarrow 2NH_3(g)$. Calculate the number of moles of hydrogen required to react with 0.0880 mole of nitrogen, and the number of moles of ammonia that will form.

Practice Problem **B**UILD Tetraphosphorus decoxide (P_4O_{10}) reacts with water to produce phosphoric acid. Write and balance the equation for this reaction, and determine the number of moles of each reactant required to produce 5.80 moles of phosphoric acid.

Practice Problem **C**ONCEPTUALIZE The models represent the reaction of nitric acid with tin metal to form metastannic acid (H_2SnO_3), water, and nitrogen dioxide. Determine how many moles of nitric acid must react to produce 8.75 mol H_2SnO_3.

Mass of Reactants and Products

Balanced chemical equations give us the relative amounts of reactants and products in terms of moles. However, because we measure reactants and products in the laboratory by weighing them, most often such calculations start with mass rather than the number of moles. Sample Problem 3.11 illustrates how to determine amounts of reactants and products in terms of grams.

SAMPLE PROBLEM 3.11

Nitrous oxide (N_2O), also known as "laughing gas," is commonly used as an anesthetic in dentistry. It is manufactured by heating ammonium nitrate. The balanced equation is

$$NH_4NO_3(s) \xrightarrow{\Delta} N_2O(g) + 2H_2O(g)$$

(a) Calculate the mass of ammonium nitrate that must be heated in order to produce 10.0 g of nitrous oxide. (b) Determine the corresponding mass of water produced in the reaction.

Strategy For part (a), use the molar mass of nitrous oxide to convert the given mass of nitrous oxide to moles, use the appropriate stoichiometric conversion factor to convert to moles of ammonium nitrate, and then use the molar mass of ammonium nitrate to convert to grams of ammonium nitrate. For part (b), use the molar mass of nitrous oxide to convert the given mass of nitrous oxide to moles, use the stoichiometric conversion factor to convert from moles of nitrous oxide to moles of water, and then use the molar mass of water to convert to grams of water.

(Continued on next page)

Setup The molar masses are as follows: 80.05 g/mol for NH_4NO_3, 44.02 g/mol for N_2O, and 18.02 g/mol for H_2O. The conversion factors from nitrous oxide to ammonium nitrate and from nitrous oxide to water are, respectively:

$$\frac{1 \text{ mol } NH_4NO_3}{1 \text{ mol } N_2O} \quad \text{and} \quad \frac{2 \text{ mol } H_2O}{1 \text{ mol } N_2O}$$

Solution

(a) $\qquad 10.0 \text{ g } N_2O \times \dfrac{1 \text{ mol } N_2O}{44.02 \text{ g } N_2O} = 0.227 \text{ mol } N_2O$

$\qquad 0.227 \text{ mol } N_2O \times \dfrac{1 \text{ mol } NH_4NO_3}{1 \text{ mol } N_2O} = 0.227 \text{ mol } NH_4NO_3$

$\qquad 0.227 \text{ mol } NH_4NO_3 \times \dfrac{80.05 \text{ g } NH_4NO_3}{1 \text{ mol } NH_4NO_3} = 18.2 \text{ g } NH_4NO_3$

Thus, 18.2 g of ammonium nitrate must be heated in order to produce 10.0 g of nitrous oxide.

(b) Starting with the number of moles of nitrous oxide determined in the first step of part (a),

$$0.227 \text{ mol } N_2O \times \frac{2 \text{ mol } H_2O}{1 \text{ mol } N_2O} = 0.454 \text{ mol } H_2O$$

$$0.454 \text{ mol } H_2O \times \frac{18.02 \text{ g } H_2O}{1 \text{ mol } H_2O} = 8.18 \text{ g } H_2O$$

Therefore, 8.18 g of water will also be produced in the reaction.

THINK ABOUT IT

Use the law of conservation of mass to check your answers. Make sure that the combined mass of both products is equal to the mass of reactant you determined in part (a). In this case (rounded to the appropriate number of significant figures), 10.0 g + 8.18 g = 18.2 g. Remember that small differences may arise as the result of rounding.

Practice Problem (A)TTEMPT Calculate the mass of water produced by the metabolism of 56.8 g of glucose. (See the Bringing Chemistry to Life box in Section 3.3 for the necessary equation.)

Practice Problem (B)UILD What mass of glucose must be metabolized in order to produce 175 g of water?

Practice Problem (C)ONCEPTUALIZE The models here represent the reaction of nitrogen dioxide with water to form nitrogen monoxide and nitric acid. What mass of nitrogen dioxide must react for 100.0 g HNO_3 to be produced? (Don't forget to balance the equation.)

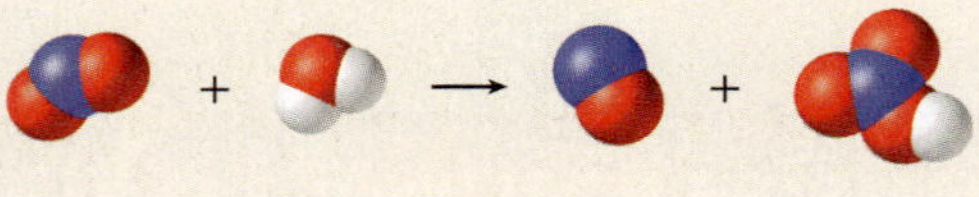

CHECKPOINT – SECTION 3.6   Calculations with Balanced Chemical Equations

3.6.1 How many moles of LiOH will be produced if 0.550 mol Li reacts according to the following equation?

$$2Li(s) + 2H_2O(l) \longrightarrow 2LiOH(aq) + H_2(g)$$

a) 0.550 mol

b) 1.10 mol

c) 0.275 mol

d) 2.20 mol

e) 2.00 mol

3.6.2 Determine the stoichiometric amount (in grams) of O_2 necessary to react with 5.71 g Al according to the following equation:

$$4Al(s) + 3O_2(g) \longrightarrow 2Al_2O_3(s)$$

a) 5.08 g

b) 9.03 g

c) 2.54 g

d) 4.28 g

e) 7.61 g

3.7   Limiting Reactants

When a chemist carries out a reaction, the reactants usually are not present in stoichiometric amounts. Because the goal of a reaction is usually to produce the maximum quantity of a useful compound from the starting materials, an excess of one reactant is commonly supplied to ensure that the more expensive or more important reactant is converted completely to the desired product. Consequently, some of the reactant supplied in excess will be left over at the end of the reaction. The reactant used up first in a reaction is called the ***limiting reactant,*** because the amount of this reactant *limits* the amount of product that can form. When all the limiting reactant has been consumed, no more product can be formed. ***Excess reactants*** are those present in quantities *greater* than necessary to react with the quantity of the limiting reactant.

The concept of a limiting reactant applies to everyday tasks, too, such as making ham sandwiches. Suppose you want to make the maximum number of ham sandwiches possible, each of which will consist of two slices of bread and one slice of ham. If you have eight slices of bread and six slices of ham, how many sandwiches can you make? The answer is four, because after making four sandwiches you will be out of bread. You will have two slices of ham left over, but without additional bread you will be unable to make any more sandwiches. In this case, bread is the limiting reactant and ham is the excess reactant.

Student Note: Limiting reactants and excess reactants are also referred to as *limiting reagents* and *excess reagents*.

Animation
Limiting reagent in reaction of NO and O_2.

Determining the Limiting Reactant

In problems involving limiting reactants, the first step is to determine which is the limiting reactant. After the limiting reactant has been identified, the rest of the problem can be solved using the approach outlined in Section 3.6. Consider the formation of methanol (CH_3OH) from carbon monoxide and hydrogen:

$$CO(g) + 2H_2(g) \longrightarrow CH_3OH(l)$$

Suppose that initially we have 5 moles of CO and 8 moles of H_2, the ratio shown in Figure 3.6(a).

We can use the stoichiometric conversion factors to determine how many moles of H_2 are necessary for all the CO to react. From the balanced equation, we have 1 mol CO $\simeq$ 2 mol H_2. Therefore, the amount of H_2 necessary to react with 5 mol CO is

$$\text{moles of } H_2 = 5 \text{ mol CO} \times \frac{2 \text{ mol } H_2}{1 \text{ mol CO}} = 10 \text{ mol } H_2$$

Because there are only 8 moles of H_2 available, there is insufficient H_2 to react with all the CO. Therefore, H_2 is the limiting reactant and CO is the excess reactant. H_2 will be used up first, and when it is gone, the formation of methanol will cease and there will be some CO left over, as shown in Figure 3.6(b). To determine how much CO will be left over when the reaction is complete, we must first calculate the amount of CO that will react with all 8 moles of H_2:

$$\text{moles of CO} = 8 \text{ mol } H_2 \times \frac{1 \text{ mol CO}}{2 \text{ mol } H_2} = 4 \text{ mol CO}$$

Thus, there will be 4 moles of CO consumed and 1 mole (5 mol − 4 mol) left over. Sample Problem 3.12 illustrates how to combine the concept of a limiting reactant with the conversion between mass and moles. Figure 3.7 illustrates the steps for this type of calculation.

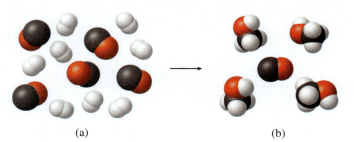

(a)   (b)

Figure 3.6 The reaction of (a) H_2 and CO to form (b) CH_3OH. Each molecule represents 1 mol of substance. In this case, H_2 is the limiting reactant and there is 1 mol of CO remaining when the reaction is complete.

Figure 3.7

Limiting Reactant Problems

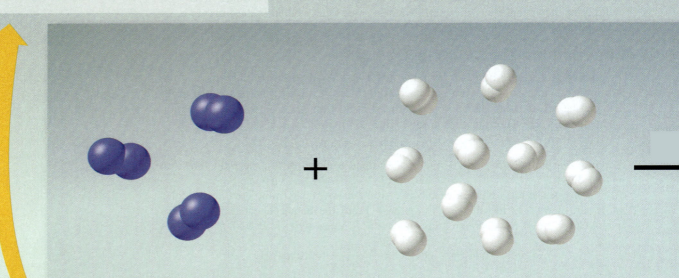

START Determine what mass of NH_3 forms when 84.06 g N_2 and 22.18 g H_2 react according to the equation:

$$N_2 + 3H_2 \longrightarrow 2NH_3$$

Convert to moles.

$$\frac{84.06 \text{ g } N_2}{28.02 \text{ g/mol}} = 3.000 \text{ mol } N_2$$

$$\frac{22.18 \text{ g } H_2}{2.016 \text{ g/mol}} = 11.00 \text{ mol } H_2$$

Determine moles NH_3.

Total mass before reaction:

84.06 g N_2 + 22.18 g H_2 = 106.24 g

+

Compare the total mass *after* the reaction with the total mass *before* the reaction. The small difference between the masses before and after is due to rounding.

Σ after reaction = 102.2 g NH_3 + 4.03 g H_2 = 106.2 g

Add the mass of the product and the mass of leftover excess reactant to get the total mass after reaction.

Use coefficients as conversion factors.

Method 1

$$3.000 \text{ mol N}_2 \times \frac{2 \text{ mol NH}_3}{1 \text{ mol N}_2} = 6.000 \text{ mol NH}_3$$

$$11.00 \text{ mol H}_2 \times \frac{2 \text{ mol NH}_3}{3 \text{ mol H}_2} = 7.333 \text{ mol NH}_3$$

or

Method 2

Rewrite the balanced equation using actual amounts. According to the balanced equation, 3.667 mol N_2 are required to react with 11.00 mol H_2.

$$3.000 \text{ N}_2 + 9.000 \text{ H}_2 \longrightarrow 6.000 \text{ NH}_3$$

$$3.667 \text{ N}_2 + 11.00 \text{ H}_2 \longrightarrow 7.333 \text{ NH}_3$$

Either way, the *smaller* amount of product is correct.

6.000 mol NH_3

Convert to grams.

$$6.000 \text{ mol NH}_3 \times \frac{17.03 \text{ g NH}_3}{1 \text{ mol NH}_3} = 102.2 \text{ g NH}_3$$

CHECK

N_2 was the limiting reactant. Calculate how much H_2 is left over.

$$2.00 \text{ mol H}_2 \times \frac{2.016 \text{ g H}_2}{1 \text{ mol H}_2} = 4.03 \text{ g H}_2$$

Convert to grams.

11.00 mol initially
− 9.00 mol consumed
2.00 mol H_2 remaining

What's the point?

There is more than one correct method for solving many types of problems. This limiting reactant problem shows two different routes to the correct answer, and shows how the result can be compared to the information given in the problem to determine whether or not it is reasonable and correct.

(See Visualizing Chemistry questions VC 3.1–VC 3.4 on page 133.)

SAMPLE PROBLEM 3.12

Alka-Seltzer tablets contain aspirin, sodium bicarbonate, and citric acid. When they come into contact with water, the sodium bicarbonate ($NaHCO_3$) and citric acid ($H_3C_6H_5O_7$) react to form carbon dioxide gas, among other products.

$$3NaHCO_3(aq) + H_3C_6H_5O_7(aq) \longrightarrow 3CO_2(g) + 3H_2O(l) + Na_3C_6H_5O_7(aq)$$

The formation of CO_2 causes the trademark fizzing when the tablets are dropped into a glass of water. An Alka-Seltzer tablet contains 1.700 g of sodium bicarbonate and 1.000 g of citric acid. Determine, for a single tablet dissolved in water, (a) which ingredient is the limiting reactant, (b) what mass of the excess reactant is left over when the reaction is complete, and (c) what mass of CO_2 forms.

Strategy Convert each of the reactant masses to moles. Use the balanced equation to write the necessary stoichiometric conversion factor and determine which reactant is limiting. Again, using the balanced equation, write the stoichiometric conversion factors to determine the number of moles of excess reactant remaining and the number of moles of CO_2 produced. Finally, use the appropriate molar masses to convert moles of excess reactant and moles of CO_2 to grams.

The reaction of sodium bicarbonate and citric acid produces Alka-Seltzer's effervescence.

Charles D. Winters/Timeframe Photography/McGraw Hill

Setup The required molar masses are 84.01 g/mol for $NaHCO_3$, 192.12 g/mol for $H_3C_6H_5O_7$, and 44.01 g/mol for CO_2. From the balanced equation we have 3 mol $NaHCO_3 \simeq$ 1 mol $H_3C_6H_5O_7$, 3 mol $NaHCO_3 \simeq$ 3 mol CO_2, and 1 mol $H_3C_6H_5O_7 \simeq$ 3 mol CO_2. The necessary stoichiometric conversion factors are therefore:

$$\frac{3 \text{ mol } NaHCO_3}{1 \text{ mol } H_3C_6H_5O_7} \qquad \frac{1 \text{ mol } H_3C_6H_5O_7}{3 \text{ mol } NaHCO_3} \qquad \frac{3 \text{ mol } CO_2}{3 \text{ mol } NaHCO_3} \qquad \frac{3 \text{ mol } CO_2}{1 \text{ mol } H_3C_6H_5O_7}$$

Solution

$$1.700 \text{ g } NaHCO_3 \times \frac{1 \text{ mol } NaHCO_3}{84.01 \text{ g } NaHCO_3} = 0.02024 \text{ mol } NaHCO_3$$

$$1.000 \text{ g } H_3C_6H_5O_7 \times \frac{1 \text{ mol } H_3C_6H_5O_7}{192.12 \text{ g } H_3C_6H_5O_7} = 0.005205 \text{ mol } H_3C_6H_5O_7$$

(a) To determine which reactant is limiting, calculate the amount of citric acid necessary to react completely with 0.02024 mol sodium bicarbonate.

$$0.02024 \text{ mol } NaHCO_3 \times \frac{1 \text{ mol } H_3C_6H_5O_7}{3 \text{ mol } NaHCO_3} = 0.006745 \text{ mol } H_3C_6H_5O_7$$

The amount of $H_3C_6H_5O_7$ required to react with 0.02024 mol of $NaHCO_3$ is more than a tablet contains. Therefore, citric acid is the limiting reactant and sodium bicarbonate is the excess reactant.

(b) To determine the mass of excess reactant ($NaHCO_3$) left over, first calculate the amount of $NaHCO_3$ that will react:

$$0.005205 \text{ mol } H_3C_6H_5O_7 \times \frac{3 \text{ mol } NaHCO_3}{1 \text{ mol } H_3C_6H_5O_7} = 0.01562 \text{ mol } NaHCO_3$$

Thus, 0.01562 mol of $NaHCO_3$ will be consumed, leaving 0.00462 mol unreacted. Convert the unreacted amount to grams as follows:

$$0.00462 \text{ mol } NaHCO_3 \times \frac{84.01 \text{ g } NaHCO_3}{1 \text{ mol } NaHCO_3} = 0.388 \text{ g } NaHCO_3$$

(c) To determine the mass of CO_2 produced, first calculate the number of moles of CO_2 produced from the number of moles of limiting reactant ($H_3C_6H_5O_7$) consumed:

$$0.005205 \text{ mol } H_3C_6H_5O_7 \times \frac{3 \text{ mol } CO_2}{1 \text{ mol } H_3C_6H_5O_7} = 0.01562 \text{ mol } CO_2$$

Convert this amount to grams as follows:

$$0.01562 \text{ mol } CO_2 \times \frac{44.01 \text{ g } CO_2}{1 \text{ mol } CO_2} = 0.6874 \text{ g } CO_2$$

To summarize the results: (a) citric acid is the limiting reactant, (b) 0.388 g sodium bicarbonate remains unreacted, and (c) 0.6874 g carbon dioxide is produced.

THINK ABOUT IT

In a problem such as this, it is a good idea to check your work by calculating the amounts of the other products in the reaction. According to the law of conservation of mass, the combined starting mass of the two reactants (1.700 g + 1.000 g = 2.700 g) should equal the sum of the masses of products and leftover excess reactant. In this case, the masses of H_2O and $Na_3C_6H_5O_7$ produced are 0.2815 g and 1.343 g, respectively. The mass of CO_2 produced is 0.6874 g [from part (c)] and the amount of excess $NaHCO_3$ is 0.388 g [from part (b)]. The total, 0.2815 g + 1.343 g + 0.6874 g + 0.388 g, is 2.700 g, identical to the total mass of the reactants.

Practice Problem A TTEMPT Ammonia is produced by the reaction of nitrogen and hydrogen according to the equation, $N_2(g) + 3H_2(g) \longrightarrow 2NH_3(g)$. Calculate the mass of ammonia produced when 35.0 g of nitrogen react with 12.5 g of hydrogen. Which is the excess reactant and how much of it will be left over when the reaction is complete?

Practice Problem B UILD Potassium hydroxide and phosphoric acid react to form potassium phosphate and water according to the equation: $3KOH(aq) + H_3PO_4(aq) \longrightarrow K_3PO_4(aq) + 3H_2O(l)$. Determine the starting mass of each reactant if 55.7 g K_3PO_4 is produced and 89.8 g H_3PO_4 remains unreacted.

Practice Problem C ONCEPTUALIZE The diagrams show a reaction mixture before and after a chemical reaction. Write the balanced equation for the reaction, using the smallest possible whole numbers, and identify the limiting reactant.

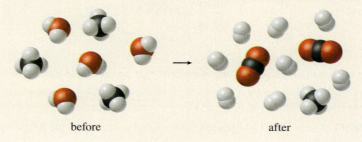

before   after

Reaction Yield

When you use stoichiometry to calculate the amount of product formed in a reaction, you are calculating the *theoretical yield* of the reaction. The theoretical yield is the amount of product that forms when *all* the limiting reactant reacts to form the desired product. It is the *maximum* obtainable yield, predicted by the balanced equation. In practice, the *actual yield*—the amount of product actually obtained from a reaction— is almost always less than the theoretical yield. There are many reasons for the difference between the actual and theoretical yields. For instance, some of the reactants may not react to form the desired product. They may react to form different products, in something known as *side reactions,* or they may simply remain unreacted. In addition, it may be difficult to isolate and recover all the product at the end of the reaction. Chemists often determine the efficiency of a chemical reaction by calculating its *percent yield,* which tells *what percentage the actual yield is of the theoretical yield.* It is calculated as follows:

$$\% \text{ yield} = \frac{\text{actual yield}}{\text{theoretical yield}} \times 100\% \qquad \textbf{Equation 3.2}$$

Percent yields may range from a tiny fraction to 100 percent. (They cannot exceed 100 percent.) Chemists try to maximize percent yield in a variety of ways. Factors that can affect percent yield, including temperature and pressure, are discussed in Chapter 15. Sample Problem 3.13 shows how to calculate the percent yield of a pharmaceutical manufacturing process.

SAMPLE PROBLEM (3.13)

Emergency oxygen supplies aboard aircraft often are generated by the decomposition of sodium chlorate ($NaClO_3$).

$$2NaClO_3(s) \longrightarrow 2NaCl(s) + 3O_2(g)$$

Determine the percent yield if the decomposition of 75.0 g sodium chlorate generates 29.45 g oxygen.

Strategy Convert grams sodium chlorate to moles and use the balanced equation to determine how many moles of oxygen can be produced. Convert moles oxygen to grams. This is the theoretical yield of oxygen. Compare the actual yield (given in the problem) to the theoretical to determine percent yield.

Setup The necessary molar masses are 106.44 g/mol for $NaClO_3$ and 32.00 g/mol for O_2.

Solution

$$75.0 \text{ g } NaClO_3 \times \frac{1 \text{ mol } NaClO_3}{106.44 \text{ g } NaClO_3} = 0.7046 \text{ mol } NaClO_3$$

$$0.7046 \text{ mol } NaClO_3 \times \frac{3 \text{ mol } O_2}{2 \text{ mol } NaClO_3} = 1.057 \text{ mol } O_2$$

$$1.057 \text{ mol } O_2 \times \frac{32.00 \text{ g } O_2}{1 \text{ mol } O_2} = 33.82 \text{ g } O_2$$

$$\% \text{ yield } = \frac{29.45 \text{ g}}{33.82 \text{ g}} \times 100\% = 87.1\% \text{ yield}$$

THINK ABOUT IT

Make sure you have used the correct molar masses and remember that percent yield can never exceed 100 percent.

Practice Problem (A)TTEMPT   Diethyl ether is produced from ethanol according to the following equation:

$$2CH_3CH_2OH(l) \longrightarrow CH_3CH_2OCH_2CH_3(l) + H_2O(l)$$

Calculate the percent yield if 68.6 g of ethanol reacts to produce 16.1 g of ether.

Practice Problem (B)UILD   What mass of ether will be produced if 221 g of ethanol reacts with a 68.9 percent yield?

Practice Problem (C)ONCEPTUALIZE   Consider the reaction pictured, where each red sphere represents an oxygen atom and each yellow sphere represents a sulfur atom. Write the balanced equation and identify the limiting reactant.

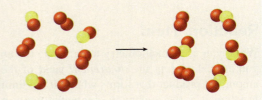

SAMPLE PROBLEM (3.14)

Aspirin, acetylsalicylic acid ($C_9H_8O_4$), is the most commonly used pain reliever in the world. It is produced by the reaction of salicylic acid ($C_7H_6O_3$) and acetic anhydride ($C_4H_6O_3$) according to the following equation:

$C_7H_6O_3$	+	$C_4H_6O_3$	$\longrightarrow$	$C_9H_8O_4$	+	$HC_2H_3O_2$
salicylic acid		acetic anhydride		acetylsalicylic acid		acetic acid

In a certain aspirin synthesis, 104.8 g of salicylic acid and 110.9 g of acetic anhydride are combined. Calculate the percent yield of the reaction if 105.6 g of aspirin are produced.

Strategy Convert reactant grams to moles, and determine which is the limiting reactant. Use the balanced equation to determine the number of moles of aspirin that can be produced, and convert this number of moles to grams for the theoretical yield. Use the actual yield (given in the problem) and the calculated theoretical yield to calculate the percent yield.

Setup The necessary molar masses are 138.12 g/mol for salicylic acid, 102.09 g/mol for acetic anhydride, and 180.15 g/mol for aspirin.

Solution

$$104.8 \text{ g } C_7H_6O_3 \times \frac{1 \text{ mol } C_7H_6O_3}{138.12 \text{ g } C_7H_6O_3} = 0.7588 \text{ mol } C_7H_6O_3$$

$$110.9 \text{ g } C_4H_6O_3 \times \frac{1 \text{ mol } C_4H_6O_3}{102.09 \text{ g } C_4H_6O_3} = 1.086 \text{ mol } C_4H_6O_3$$

Because the two reactants combine in a 1:1 mole ratio, the reactant present in the smallest number of moles (in this case, salicylic acid) is the limiting reactant. According to the balanced equation, one mole of aspirin is produced for every mole of salicylic acid consumed.

$$1 \text{ mol salicylic acid } (C_7H_6O_3) \simeq 1 \text{ mol aspirin } (C_9H_8O_4)$$

Therefore, the theoretical yield of aspirin is 0.7588 mol. We convert this to grams using the molar mass of aspirin:

$$0.7588 \text{ mol } C_9H_8O_4 \times \frac{180.15 \text{ g } C_9H_8O_4}{1 \text{ mol } C_9H_8O_4} = 136.7 \text{ g } C_9H_8O_4$$

Thus, the theoretical yield is 136.7 g. If the actual yield is 105.6 g, the percent yield is

$$\% \text{ yield } = \frac{105.6 \text{ g}}{136.7 \text{ g}} \times 100\% = 77.25\% \text{ yield}$$

THINK ABOUT IT

When there are two or more reactants, start by identifying the *limiting* reactant.

Practice Problem Ⓐ**TTEMPT** Determine the percent yield if 175 g salicylic acid and 125 g acetic anhydride react to produce 157.6 g aspirin.

Practice Problem Ⓑ**UILD** What mass of aspirin will be produced if 75.5 g salicylic acid and 58.0 g acetic anhydride react with a 59.4 percent yield?

Practice Problem Ⓒ**ONCEPTUALIZE** The diagrams show a mixture of reactants and the mixture of recovered products for an experiment using the chemical reaction introduced in Practice Problem 3.12C. Identify the limiting reactant and determine the percent yield of carbon dioxide.

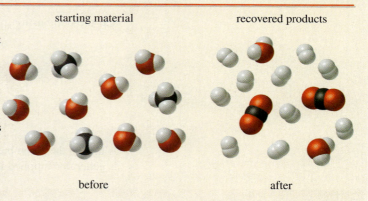

starting material   recovered products

before   after

Types of Chemical Reactions

As you continue to study chemistry, you will encounter a wide variety of chemical reactions. The sheer number of different reactions can seem daunting at times, but most of them fall into a relatively small number of categories. Becoming familiar with several reaction types and learning to recognize patterns of reactivity will help you make sense out of the reactions in this book. Three of the most commonly encountered reaction types are *combination, decomposition,* and *combustion.*

> *Combination.* A reaction in which two or more reactants combine to form a single product is known as a ***combination reaction.*** Examples include the reaction of ammonia and hydrogen chloride to form ammonium chloride,

$$NH_3(g) + HCl(g) \longrightarrow NH_4Cl(s)$$

and the reaction of nitrogen and hydrogen gases to form ammonia,

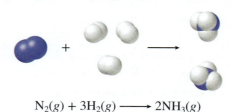

$$N_2(g) + 3H_2(g) \longrightarrow 2NH_3(g)$$

Decomposition. A reaction in which two or more products form from a single reactant is known as a ***decomposition reaction.*** A decomposition reaction is essentially the opposite of a combination reaction. Examples of this type of reaction include the decomposition of calcium carbonate to produce calcium oxide and carbon dioxide gas,

$$CaCO_3(s) \xrightarrow{\Delta} CaO(s) + CO_2(g)$$

and the decomposition of hydrogen peroxide to produce water and oxygen gas,

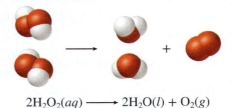

$$2H_2O_2(aq) \longrightarrow 2H_2O(l) + O_2(g)$$

Combustion. As you learned in Section 3.3, a *combustion reaction* is one in which a substance burns in the presence of oxygen. Combustion of a compound that contains C and H (or C, H, and O) produces carbon dioxide gas and water. By convention, we consider the water produced in a combustion reaction to be *liquid* water. Examples of this type of combustion are the combustion of formaldehyde,

$$CH_2O(l) + O_2(g) \longrightarrow CO_2(g) + H_2O(l)$$

and the combustion of methane,

$$CH_4(g) + 2O_2(g) \longrightarrow CO_2(g) + 2H_2O(l)$$

Although these combustion reactions are shown here as balanced equations, oxygen is generally supplied in excess in such processes to ensure complete combustion.

SAMPLE PROBLEM 3.15

Determine whether each of the following equations represents a combination reaction, a decomposition reaction, or a combustion reaction:
(a) $H_2(g) + Br_2(g) \longrightarrow 2HBr(g)$, (b) $2HCO_2H(l) + O_2(g) \longrightarrow 2CO_2(g) + 2H_2O(l)$, (c) $2KClO_3(s) \longrightarrow 2KCl(s) + 3O_2(g)$.

Strategy Look at the reactants and products in each balanced equation to see if two or more reactants combine into one product (a combination reaction), if one reactant splits into two or more products (a decomposition reaction), or if the main products formed are carbon dioxide gas and water (a combustion reaction).

Setup The equation in part (a) depicts *two reactants* and *one product*. The equation in part (b) represents a combination with O_2 of a compound containing C, H, and O to produce CO_2 and H_2O. The equation in part (c) represents *two products* being formed from a *single reactant*.

Solution These equations represent (a) a combination reaction, (b) a combustion reaction, and (c) a decomposition reaction.

THINK ABOUT IT

Make sure that a reaction identified as a combination has only one product [as in part (a)], a reaction identified as a decomposition has only one reactant [as in part (b)], and a reaction identified as a combustion produces only CO_2 and H_2O [as in part (c)].

Practice Problem **A**TTEMPT Identify each of the following as a combination, decomposition, or combustion reaction:
(a) $C_2H_4O_2(l) + 2O_2(g) \longrightarrow 2CO_2(g) + 2H_2O(l)$, (b) $2Na(s) + Cl_2(g) \longrightarrow 2NaCl(s)$, (c) $2NaH(s) \longrightarrow 2Na(s) + H_2(g)$.

Practice Problem **B**UILD Using the chemical species A_2, B, and AB, write a balanced equation for a combination reaction.

Practice Problem 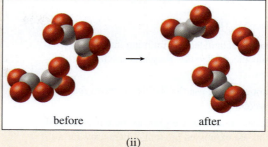 **CONCEPTUALIZE** Each of the diagrams represents a reaction mixture before and after a chemical reaction. Identify each of the reactions shown as *combination, decomposition,* or *combustion.*

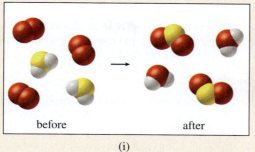

before	after

(i)

before	after

(ii)

CHECKPOINT – SECTION 3.7   Limiting Reactants

3.7.1 What mass of $CaSO_4$ is produced according to the given equation when 5.00 g of each reactant are combined?

$$CaF_2(s) + H_2SO_4(aq) \longrightarrow CaSO_4(s) + 2HF(g)$$

a) 10.0 g

b) 11.6 g

c) 6.94 g

d) 8.72 g

e) 5.02 g

3.7.2 What is the percent yield for a process in which 10.4 g CH_3OH reacts and 10.1 g CO_2 forms according to the following equation?

$$2CH_3OH(l) + 3O_2(g) \longrightarrow 2CO_2(g) + 4H_2O(l)$$

a) 97.1%

b) 70.7%

c) 52.1%

d) 103%

e) 37.9%

3.7.3 How many moles of NH_3 can be produced by the combination of 3.0 mol N_2 and 1.5 mol H_2?

a) 2.0 mol

b) 1.5 mol

c) 0.50 mol

d) 6.0 mol

e) 1.0 mol

3.7.4 What mass of water is produced by the reaction of 50.0 g CH_3OH with an excess of O_2 when the yield is 53.2 percent?

$$2CH_3OH(g) + 3O_2(g) \longrightarrow 2CO_2(g) + 4H_2O(l)$$

a) 28.1 g

b) 56.2 g

c) 29.9 g

d) 15.0 g

e) 26.6 g

3.7.5 Reactants A (red) and B (blue) combine to form a single product C (purple) according to the equation

$2A + B \longrightarrow$ C. What is the limiting reactant in the reaction vessel shown?

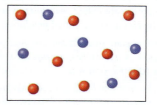

a) A

b) B

c) C

d) None. Reactants are present in stoichiometric amounts.

3.7.6 Which of the following represents the contents of the reaction vessel in Checkpoint 3.7.5 after the reaction is complete?

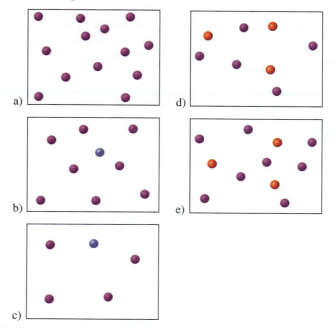

Chapter Summary

Section 3.1

- *Molecular mass* is calculated by summing the masses of all atoms in a molecule. *Molecular weight* is another term for molecular mass.

- For ionic compounds, we use the analogous terms *formula mass* and *formula weight.*

- Molecular masses, molecular weights, formula masses, and formula weights are expressed in atomic mass units (amu).

Section 3.2

- Molecular or formula mass can be used to determine *percent composition by mass* of a compound.

Section 3.3

- A *chemical equation* is a written representation of a chemical reaction or a physical process. Chemical species on the left side of the equation are called *reactants,* whereas those on the right side of the equation are called *products.*

- The physical state of each reactant and product is specified in parentheses as (*s*), (*l*), (*g*), or (*aq*) for *solid, liquid, gas,* and *aqueous* (dissolved in water), respectively.

- Chemical equations are balanced only by changing the *stoichiometric coefficients* of the reactants and/or products, and never by changing the formulas of the reactants and/or products (i.e., by changing their subscripted numbers).

- *Combustion* refers to chemical combination with oxygen. Combustion of hydrocarbons produces carbon dioxide and water.

Section 3.4

- A *mole* is the amount of a substance that contains 6.022×10^{23} [*Avogadro's number* (N_A)] of elementary particles (atoms, molecules, ions, formula units, etc.).

- *Molar mass* ($\mathcal{M}$) is the mass of one mole of a substance, usually expressed in grams. The molar mass of a substance in grams is numerically equal to the *atomic, molecular,* or *formula* mass of the substance in amu.

- Molar mass and Avogadro's number can be used to interconvert among *mass, moles,* and *number of particles* (atoms, molecules, ions, formula units, etc.).

Section 3.5

- *Combustion analysis* is used to determine the empirical formula of a compound. The *empirical formula* can be used to calculate percent composition.

- The empirical formula and molar mass can be used to determine the molecular formula.

Section 3.6

- A balanced chemical equation can be used to determine how much product will form from given amounts of reactants, how much of one reactant is necessary to react with a given amount of another, or how much reactant is required to produce a specified amount of product. Reactants that are combined in exactly the ratio specified by the balanced equation are said to be "combined in *stoichiometric amounts.*"

Section 3.7

- The *limiting reactant* is the reactant that is consumed completely in a chemical reaction. An *excess reactant* is the reactant that is not consumed completely. The maximum amount of product that can form depends on the amount of limiting reactant.

- The *theoretical yield* of a reaction is the amount of product that will form if all the limiting reactant is consumed by the desired reaction.

- The *actual yield* is the amount of product actually recovered.

- *Percent yield* [(actual/theoretical) × 100%] is a measure of the efficiency of a chemical reaction.

- *Combustion* (in which a substance burns in the presence of oxygen), *combination* (in which two or more reactants combine to form a single product), and *decomposition* (in which a reactant splits apart to form two or more products) are three types of *chemical reactions* that are commonly encountered.

Key Words

Actual yield, 119

Aqueous, 96

Avogadro's number (N_A), 102

Chemical equation, 95

Combination reaction, 121

Combustion, 97

Combustion analysis, 108

Decomposition reaction, 122

Excess reactant, 115

Formula mass, 92

Formula weight, 92

Limiting reactant, 115

Molar mass ($\mathcal{M}$), 104

Mole, 102

Molecular mass, 92

Molecular weight, 92

Percent composition by mass, 93

Percent yield, 119

Product, 95

Reactant, 95

Stoichiometric amount, 112

Stoichiometric coefficients, 96

Theoretical yield, 119

Key Equations

3.1 percent by mass of an element =

$$\frac{n \times \text{atomic mass of element}}{\text{molecular or formula mass of compound}} \times 100\%$$

Using a compound's formula (molecular or empirical), we can calculate its percent composition by mass.

3.2 % yield = $\dfrac{\text{actual yield}}{\text{theoretical yield}} \times 100\%$

The amount of product actually produced in a reaction will nearly always be less than that predicted by the balanced equation. We use the actual (measured) amount of product and the calculated amount of product to determine the percent yield of a reaction.

Key Constant

Section 3.4 Avogadro's number 6.022×10^{23}

Avogadro's number is the number of elementary entities (atoms, molecules, formula units) in a mole. It is generally considered to have units of mol^{-1}, and is used in conversions between moles and atoms, moles and molecules, or moles and formula units.

The amount of product that can be produced in a chemical reaction typically is limited by the amount of *one* of the reactants—known as the *limiting* reactant. The practice of identifying the limiting reactant, calculating the maximum possible amount of product, and determining the percent yield and remaining amount of an excess reactant requires several skills:

- Balancing chemical equations [◂◂ Section 3.3]
- Determining molar mass [◂◂ Section 3.4]
- Converting between mass and moles [◂◂ Section 3.4]
- Using stoichiometric conversion factors [◂◂ Section 3.6]

Consider the following example. Hydrazine (N_2H_4) reacts with dinitrogen tetroxide (N_2O_4) to form nitrogen monoxide (NO) and water. Determine the mass of NO that can be produced when 10.45 g of N_2H_4 and 53.68 g of N_2O_4 are combined. The unbalanced equation is

$$N_2H_4 + N_2O_4 \longrightarrow NO + H_2O$$

We first balance the equation.

$$N_2H_4 + 2N_2O_4 \longrightarrow 6NO + 2H_2O$$

Next, we determine the necessary molar masses.

$$N_2H_4: 2(14.01) + 4(1.008) = \boxed{\frac{32.05\ g}{mol}} \qquad N_2O_4: 2(14.01) + 4(16.00) = \boxed{\frac{92.02\ g}{mol}} \qquad NO: 14.01 + 16.00 = \boxed{\frac{30.01\ g}{mol}}$$

We convert the reactant masses given in the problem to moles. Then we determine the mole amount of NO that could be produced from the mole amount of each reactant by multiplying each of the *reactant* mole amounts by the appropriate stoichiometric conversion factor, which we derive from the balanced equation. According to the balanced equation:

$$1\ mol\ N_2H_4 \simeq 6\ mol\ NO \qquad and \qquad 2\ mol\ N_2O_4 \simeq 6\ mol\ NO$$

$$\frac{10.45\ g\ N_2H_4}{32.05\ g/mol} = 0.32605\ mol\ N_2H_4 \qquad 0.32605\ mol\ N_2H_4 \times \frac{6\ mol\ NO}{1\ mol\ N_2H_4} = 1.9563\ mol\ NO$$

$$\frac{53.68\ g\ N_2O_4}{92.02\ g/mol} = 0.58335\ mol\ N_2O_4 \qquad 0.58335\ mol\ N_2O_4 \times \frac{6\ mol\ NO}{2\ mol\ N_2O_4} = 1.7501\ mol\ NO$$

The reactant that produces the smaller amount of product is the limiting reactant; in this case, N_2O_4.

We continue the problem using the mole amount of NO produced by reaction of the given amount of N_2O_4. To convert from moles to mass (grams), we multiply the number of moles NO by the molar mass of NO:

$$1.7501\ mol\ NO \times \frac{30.01\ g}{mol} = 52.52\ g\ NO$$

Thus, 52.52 g NO can be produced by the reaction. Note that we retained an extra significant figure until the end of the calculation.

To determine the mass of remaining excess reactant, we must first determine what amount was consumed in the reaction. To do this, we multiply the mole amount of limiting reactant (N_2O_4) by the appropriate stoichiometric conversion factor. According to the balanced equation:

$$1 \text{ mol } N_2H_4 \rightleftharpoons 2 \text{ mol } N_2O_4$$

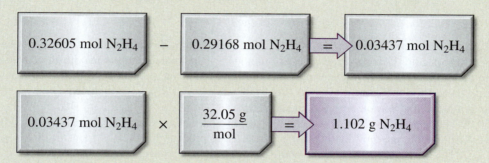

$$\boxed{0.58335 \text{ mol } N_2O_4} \times \boxed{\dfrac{1 \text{ mol } N_2H_4}{2 \text{ mol } N_2O_4}} = \boxed{0.29168 \text{ mol } N_2H_4}$$

This is the amount of N_2H_4 consumed. The amount remaining is the difference between this and the original amount. We convert the remaining mole amount to grams using the molar mass of N_2H_4.

$$\boxed{0.32605 \text{ mol } N_2H_4} - \boxed{0.29168 \text{ mol } N_2H_4} = \boxed{0.03437 \text{ mol } N_2H_4}$$

$$\boxed{0.03437 \text{ mol } N_2H_4} \times \boxed{\dfrac{32.05 \text{ g}}{\text{mol}}} = \boxed{1.102 \text{ g } N_2H_4}$$

Thus, 1.102 g N_2H_4 remain when the reaction is complete.

We can check our work in a problem such as this by also calculating the mass of the other product: in this case, water. The mass of all products plus the mass of any remaining reactant must equal the sum of starting reactant masses.

Key Skills Problems

3.1
Calculate the mass of water produced in the example.

(a) 21.02 g (b) 10.51 g (c) 11.61 g (d) 11.75 g (e) 5.400 g

Use the following information to answer questions 3.2, 3.3, and 3.4.

Calcium phosphide (Ca_3P_2) and water react to form calcium hydroxide and phosphine (PH_3). In a particular experiment, 225.0 g Ca_3P_2 and 125.0 g water are combined.

$$Ca_3P_2(s) + H_2O(l) \longrightarrow Ca(OH)_2(aq) + PH_3(g)$$

(Don't forget to balance the equation.)

3.2
How much PH_3 can be produced?

(a) 350.0 g (b) 235.0 g (c) 78.59 g (d) 83.96 g (e) 41.98 g

3.3
How much $Ca(OH)_2$ can be produced?

(a) 91.51 g (b) 274.5 g (c) 513.8 g (d) 85.63 g (e) 257.0 g

3.4
How much of the excess reactant remains when the reaction is complete?

(a) 14.37 g (b) 235.0 g (c) 78.56 g (d) 83.96 g (e) 41.98 g

Questions and Problems

Applying What You've Learned

Cisplatin [Pt(NH$_3$)$_2$Cl$_2$] is sometimes called the *penicillin* of cancer drugs because it is effective in the treatment of so many different cancers. It is prepared by the reaction of ammonium tetrachloroplatinate(II) with ammonia (note that the equation is not balanced):

$$(NH_4)_2PtCl_4(aq) \quad + \quad NH_3(aq) \quad \longrightarrow \quad Pt(NH_3)_2Cl_2(s) \quad + \quad NH_4Cl(aq)$$

This is a very expensive process because (NH$_4$)$_2$PtCl$_4$ contains platinum, a precious metal that historically has sold for roughly twice the price of gold. Using a large excess of ammonia helps manufacturers maximize conversion of the high-priced reactant to the desired product. (a) Determine the molecular mass and percent composition by mass for ammonium tetrachloroplatinate(II) and for cisplatin [◄◄ Sample Problems 3.1 and 3.2]. (b) Balance the equation for the production of cisplatin from ammonium tetrachloroplatinate(II) and ammonia [◄◄ Sample Problem 3.3]. (c) Determine the number of each type of atom in 50.00 g of cisplatin [◄◄ Sample Problem 3.7]. (d) In a particular process, 172.5 g (NH$_4$)$_2$PtCl$_4$ is combined with an excess of NH$_3$. Assuming all the limiting reactant is converted to product, how many grams of Pt(NH$_3$)$_2$Cl$_2$ will be produced [◄◄ Sample Problem 3.11]? (e) If the actual amount of Pt(NH$_3$)$_2$Cl$_2$ produced in part (d) is 129.6 g, what is the percent yield [◄◄ Sample Problem 3.13].?

SECTION 3.1: MOLECULAR AND FORMULA MASSES

Review Questions

3.1 What is meant by the term *molecular mass,* and why is the molecular mass that we calculate generally an average molecular mass?

3.2 Explain the difference between the terms *molecular mass* and *formula mass.* To what type of compound does each term refer?

Computational Problems

3.3 Calculate the molecular mass (in amu) of each of the following substances: (a) CH$_3$Cl, (b) N$_2$O$_4$, (c) SO$_2$, (d) C$_6$H$_{12}$, (e) H$_2$O$_2$, (f) C$_{12}$H$_{22}$O$_{11}$, (g) NH$_3$.

3.4 Calculate the molecular mass (in amu) of each of the following substances: (a) C$_6$H$_6$O, (b) H$_2$SO$_4$, (c) C$_6$H$_6$, (d) C$_6$H$_{12}$O$_6$, (e) BCl$_3$, (f) N$_2$O$_5$, (g) H$_3$PO$_4$.

3.5 Calculate the molecular mass or formula mass (in amu) of each of the following substances: (a) CH$_4$, (b) NO$_2$, (c) SO$_3$, (d) C$_6$H$_6$, (e) NaI, (f) K$_2$SO$_4$, (g) Ca$_3$(PO$_4$)$_2$.

3.6 Calculate the molecular mass or formula mass (in amu) of each of the following substances: (a) Li$_2$CO$_3$, (b) C$_2$H$_6$, (c) NF$_2$, (d) Al$_2$O$_3$, (e) Fe(NO$_3$)$_3$, (f) PCl$_5$, (g) Mg$_3$N$_2$.

SECTION 3.2: PERCENT COMPOSITION OF COMPOUNDS

Review Questions

3.7 Use ammonia (NH$_3$) to explain what is meant by the percent composition by mass of a compound.

3.8 Describe how the knowledge of the percent composition by mass of an unknown compound can help us identify the compound.

Computational Problems

3.9 Tin (Sn) exists in Earth's crust as SnO$_2$. Calculate the percent composition by mass of Sn and O in SnO$_2$.

3.10 For many years, chloroform (CHCl$_3$) was used as an inhalation anesthetic in spite of the fact that it is also a toxic substance that may cause severe liver, kidney, and heart damage. Calculate the percent composition by mass of this compound.

3.11 All the substances listed here are fertilizers that contribute nitrogen to the soil. Which of these is the richest source of nitrogen on a mass percentage basis?
(a) Urea [(NH$_2$)$_2$CO]
(b) Ammonium nitrate (NH$_4$NO$_3$)
(c) Guanidine [HNC(NH$_2$)$_2$]
(d) Ammonia (NH$_3$)

3.12 Limonene, shown here, is a by-product of the commercial processing of citrus. It has a pleasant orange scent and is a critical ingredient in some popular "Earth-friendly" cleaning products. Calculate the percent composition of limonene.

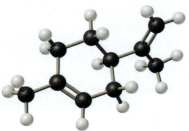

3.13 Tooth enamel is $Ca_5(PO_4)_3(OH)$. Calculate the percent composition of the elements present.

3.14 A four-pack of Red Bull Energy Drink consists of four cans of Red Bull and one cardboard holder. How many cans are there in 112 four-packs? How many four-packs would contain 68 cans?

3.15 A "variety pack" of ramen noodles consists of a dozen individual packs of noodles: six packs of chicken flavor, three packs of beef flavor, and three packs of vegetable flavor. (a) How many vegetable noodle packs are in the following numbers of variety packs: 20, 4.667, 0.25? (b) How many variety packs are necessary to provide the following numbers of beef noodle packs: 72, 3, 10? (c) How many vegetable noodle packs are there in the number of variety packs that contain each of the following numbers of the other flavors: 30 chicken flavor, 2 chicken flavor, 25 beef flavor? (For any inexact numbers, report your answers to four significant figures.)

SECTION 3.3: CHEMICAL EQUATIONS

Review Questions

3.16 Use the formation of water from hydrogen and oxygen to explain the following terms: *chemical reaction, reactant,* and *product.*

3.17 What is the difference between a chemical reaction and a chemical equation?

3.18 Why must a chemical equation be balanced? What law is obeyed by a balanced chemical equation?

3.19 Write the symbols used to represent gas, liquid, solid, and the aqueous phase in chemical equations.

Conceptual Problems

3.20 Write an unbalanced equation to represent each of the following reactions: (a) nitrogen and oxygen react to form nitrogen dioxide, (b) dinitrogen pentoxide reacts to form dinitrogen tetroxide and oxygen, (c) ozone reacts to form oxygen, (d) chlorine and sodium iodide react to form iodine and sodium chloride, and (e) magnesium and oxygen react to form magnesium oxide. (f) Balance equations (a)–(e).

3.21 Write an unbalanced equation to represent each of the following reactions: (a) potassium hydroxide and phosphoric acid react to form potassium phosphate and water; (b) zinc and silver chloride react to form zinc chloride and silver; (c) sodium hydrogen carbonate reacts to form sodium carbonate, water, and carbon dioxide; (d) ammonium nitrite reacts to form nitrogen and water; and (e) carbon dioxide and potassium hydroxide react to form potassium carbonate and water. (f) Balance equations (a)–(e).

3.22 For each of the following unbalanced chemical equations, write the corresponding chemical statement.
(a) $S_8 + O_2 \longrightarrow SO_2$
(b) $CH_4 + O_2 \longrightarrow CO_2 + H_2O$
(c) $N_2 + H_2 \longrightarrow NH_3$
(d) $P_4O_{10} + H_2O \longrightarrow H_3PO_4$
(e) $S + HNO_3 \longrightarrow H_2SO_4 + NO_2 + H_2O$

3.23 For each of the following unbalanced chemical equations, write the corresponding chemical statement.
(a) $K + H_2O \longrightarrow KOH + H_2$
(b) $Ba(OH)_2 + HCl \longrightarrow BaCl_2 + H_2O$
(c) $Cu + HNO_3 \longrightarrow Cu(NO_3)_2 + NO + H_2O$
(d) $Al + H_2SO_4 \longrightarrow Al_2(SO_4)_3 + H_2$
(e) $HI \longrightarrow H_2 + I_2$

3.24 Balance the following equations using the method outlined in Section 3.3.
(a) $C + O_2 \longrightarrow CO$
(b) $CO + O_2 \longrightarrow CO_2$
(c) $H_2 + Br_2 \longrightarrow HBr$
(d) $K + H_2O \longrightarrow KOH + H_2$
(e) $Mg + O_2 \longrightarrow MgO$
(f) $O_3 \longrightarrow O_2$
(g) $H_2O_2 \longrightarrow H_2O + O_2$
(h) $N_2 + H_2 \longrightarrow NH_3$
(i) $Zn + AgCl \longrightarrow ZnCl_2 + Ag$
(j) $S_8 + O_2 \longrightarrow SO_2$
(k) $NaOH + H_2SO_4 \longrightarrow Na_2SO_4 + H_2O$
(l) $Cl_2 + NaI \longrightarrow NaCl + I_2$
(m) $KOH + H_3PO_4 \longrightarrow K_3PO_4 + H_2O$
(n) $CH_4 + Br_2 \longrightarrow CBr_4 + HBr$

3.25 Balance the following equations using the method outlined in Section 3.3.
(a) $N_2O_5 \longrightarrow N_2O_4 + O_2$
(b) $KNO_3 \longrightarrow KNO_2 + O_2$
(c) $NH_4NO_3 \longrightarrow N_2O + H_2O$
(d) $NH_4NO_2 \longrightarrow N_2 + H_2O$
(e) $NaHCO_3 \longrightarrow Na_2CO_3 + H_2O + CO_2$
(f) $P_4O_{10} + H_2O \longrightarrow H_3PO_4$
(g) $HCl + CaCO_3 \longrightarrow CaCl_2 + H_2O + CO_2$
(h) $Al + H_2SO_4 \longrightarrow Al_2(SO_4)_3 + H_2$
(i) $CO_2 + KOH \longrightarrow K_2CO_3 + H_2O$
(j) $CH_4 + O_2 \longrightarrow CO_2 + H_2O$
(k) $Be_2C + H_2O \longrightarrow Be(OH)_2 + CH_4$
(l) $Cu + HNO_3 \longrightarrow Cu(NO_3)_2 + NO + H_2O$

(m) $S + HNO_3 \longrightarrow H_2SO_4 + NO_2 + H_2O$

(n) $NH_3 + CuO \longrightarrow Cu + N_2 + H_2O$

3.26   Which of the following equations best represents the reaction shown in the diagram?

(a) $8A + 4B \longrightarrow C + D$

(b) $4A + 8B \longrightarrow 4C + 4D$

(c) $2A + B \longrightarrow C + D$

(d) $4A + 2B \longrightarrow 4C + 4D$

(e) $2A + 4B \longrightarrow C + D$

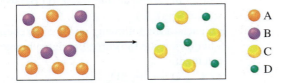

3.27   Which of the following equations best represents the reaction shown in the diagram?

(a) $A + B \longrightarrow C + D$

(b) $6A + 4B \longrightarrow C + D$

(c) $A + 2B \longrightarrow 2C + D$

(d) $3A + 2B \longrightarrow 2C + D$

(e) $3A + 2B \longrightarrow 4C + 2D$

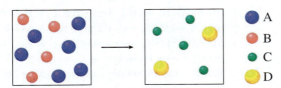

SECTION 3.4: THE MOLE AND MOLAR MASSES

Review Questions

3.28   Define the term *mole*. What is the unit for mole in calculations? What does the mole have in common with the pair, the dozen, and the gross? What does Avogadro's number represent?

3.29   What is the molar mass of an atom? What are the commonly used units for molar mass?

3.30   What does the word *empirical* in empirical formula mean?

3.31   If we know the empirical formula of a compound, what additional information do we need to determine its molecular formula?

Computational Problems

3.32   Earth's population is about 7.0 billion. Suppose that every person on Earth participates in a process of counting identical particles at the rate of two particles per second. How many years would it take to count 6.0×10^{23} particles? Assume that there are 365 days in a year.

3.33   The thickness of a piece of paper is 0.0036 in. Suppose a certain book has an Avogadro's number of pages; calculate the thickness of the book in light-years. (*Hint:* See Problem 1.66 for the definition of light-year.)

3.34   How many atoms are there in 5.10 moles of sulfur (S)?

3.35   How many moles of cobalt (Co) atoms are there in 6.00×10^9 (6 billion) Co atoms?

3.36   How many moles of calcium (Ca) atoms are in 77.4 g of Ca?

3.37   How many grams of gold (Au) are there in 15.3 moles of Au?

3.38   What is the mass in grams of a single atom of each of the following elements: (a) Ag, (b) K?

3.39   What is the mass in grams of a single atom of each of the following elements: (a) Si, (b) Fe?

3.40   What is the mass in grams of 1.00×10^{12} lead (Pb) atoms?

3.41   How many atoms are present in 25.85 g of copper (Cu)?

3.42   Which of the following has more atoms: 0.302 g of hydrogen atoms or 14.7 g of chromium atoms?

3.43   Which of the following has a greater mass: two atoms of lead or 5.1×10^{-23} mole of helium?

3.44   Calculate the molar mass of the following substances: (a) Li_2CO_3, (b) CS_2, (c) $CHCl_3$ (chloroform), (d) $C_6H_8O_6$ (ascorbic acid, or vitamin C), (e) KNO_3, (f) Mg_3N_2.

3.45   Calculate the molar mass of a compound if 0.372 mol of it has a mass of 152 g.

3.46   How many molecules of ethane (C_2H_6) are present in 0.334 g of C_2H_6?

3.47   Calculate the number of C, H, and O atoms in 1.50 g of glucose ($C_6H_{12}O_6$), a sugar.

3.48   The density of water is 1.00 g/mL at 4°C. How many water molecules are present in 15.78 mL of water at this temperature?

3.49   How many grams of sulfur (S) are needed to react completely with 246 g of mercury (Hg) to form HgS?

3.50   Calculate the mass in grams of iodine (I_2) that will react completely with 20.4 g of aluminum (Al) to form aluminum iodide (AlI_3).

3.51   Tin(II) fluoride (SnF_2) is often added to toothpaste as an ingredient to prevent tooth decay. What is the mass of F in grams in 24.6 g of the compound?

3.52   Determine the empirical formulas of the compounds with the following compositions: (a) 2.1 percent H, 65.3 percent O, 32.6 percent S; (b) 20.2 percent Al, 79.8 percent Cl.

3.53   Determine the empirical formulas of the compounds with the following compositions: (a) 40.1 percent C, 6.6 percent H, 53.3 percent O; (b) 18.4 percent C, 21.5 percent N, 60.1 percent K.

3.54   The empirical formula of a compound is CH. If the molar mass of this compound is about 78 g, what is its molecular formula?

3.55   The molar mass of caffeine is 194.19 g. Is the molecular formula of caffeine $C_4H_5N_2O$ or $C_8H_{10}N_4O_2$?

3.56   Monosodium glutamate (MSG), a food-flavor enhancer, has been blamed for "Chinese restaurant syndrome," the symptoms of which are headaches and chest pains. MSG has the following composition by mass: 35.51 percent C, 4.77 percent H, 37.85 percent O, 8.29 percent N, and 13.60 percent Na. What is its molecular formula if its molar mass is about 169 g?

3.57 Toxicologists use the term LD_{50} to describe the number of grams of a substance per kilogram of body weight that is a lethal dose for 50 percent of test animals. Calculate the number of arsenic(VI) oxide molecules corresponding to an LD_{50} value of 0.015 for a 184-lb man, assuming that the test animals and humans have the same LD_{50}.

3.58 Chemical analysis shows that the oxygen-carrying protein hemoglobin is 0.34 percent Fe by mass. What is the minimum possible molar mass of hemoglobin? The actual molar mass of hemoglobin is about 65,000 g. How would you account for the discrepancy between your minimum value and the experimental value?

Conceptual Problems

3.59 In response to invasion by a microorganism, the cells of some plants release azelaic acid, shown here, as a molecular "distress flare" to help other cells prepare for and build immunity against the invasion.

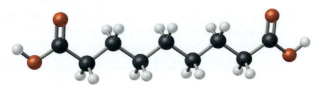

Write the molecular and empirical formulas for azelaic acid and calculate its percent composition by mass.

3.60 Researchers recently reported that the compound in stale beer that attracts cockroaches is DDMP, shown here.

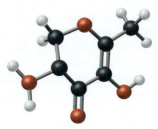

Write the molecular and empirical formulas for DDMP and calculate its percent composition by mass.

SECTION 3.5: COMBUSTION ANALYSIS

Review Questions

3.61 In combustion analysis, is the combined mass of the products (CO_2 and H_2O) less than, equal to, or greater than the combined mass of the compound that is combusted and the O_2 that reacts with it? Explain.

3.62 Explain why, in combustion analysis, we cannot determine the amount of oxygen in the sample directly from the amount of oxygen in the products H_2O and CO_2.

Computational Problems

3.63 Menthol is a flavoring agent extracted from peppermint oil. It contains C, H, and O. In one combustion analysis, 10.00 mg of the substance yields 11.53 mg H_2O and 28.16 mg CO_2. What is the empirical formula of menthol?

3.64 Ascorbic acid (vitamin C) contains C, H, and O. In one combustion analysis, 5.24 g of ascorbic acid yields 7.86 g CO_2 and 2.14 g H_2O. Calculate the empirical formula and molecular formula of ascorbic acid given that its molar mass is about 176 g.

3.65 The amino acid cysteine plays an important role in the three-dimensional structure of proteins by forming "disulfide bridges." The percent composition of cysteine is 29.74 percent C, 5.82 percent H, 26.41 percent O, 11.56 percent N, and 26.47 percent S. What is the molecular formula if its molar mass is approximately 121 g?

3.66 The diagram shows the products of a combustion analysis. Determine the empirical formula of the compound being analyzed if (a) it is a hydrocarbon and (b) it is a compound containing C, H, and O, with a formula weight of approximately 92.

3.67 Which of the following diagrams could represent the products of combustion of a sample of (a) acetylene (C_2H_2) and (b) ethylene (C_2H_4)?

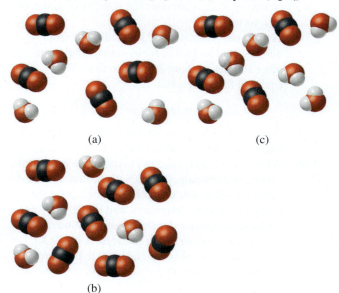

(a)   (c)

(b)

SECTION 3.6: CALCULATIONS WITH BALANCED CHEMICAL EQUATIONS

Review Questions

3.68   On what law is stoichiometry based? Why is it essential to use balanced equations in solving stoichiometric problems?

3.69   Describe the steps involved in balancing a chemical equation.

Computational Problems

3.70   Consider the combustion of carbon monoxide (CO) in oxygen gas:

$$2CO(g) + O_2(g) \longrightarrow 2CO_2(g)$$

Starting with 3.60 moles of CO, calculate the number of moles of CO_2 produced if there is enough oxygen gas to react with all the CO.

3.71   Silicon tetrachloride ($SiCl_4$) can be prepared by heating Si in chlorine gas:

$$Si(s) + 2Cl_2(g) \longrightarrow SiCl_4(l)$$

In one reaction, 0.507 mol of $SiCl_4$ is produced. How many moles of molecular chlorine were used in the reaction?

3.72   Ammonia is a principal nitrogen fertilizer. It is prepared by the reaction between hydrogen and nitrogen:

$$3H_2(g) + N_2(g) \longrightarrow 2NH_3(g)$$

In a particular reaction, 6.0 mol of NH_3 were produced. How many moles of H_2 and how many moles of N_2 were consumed to produce this amount of NH_3?

3.73   Consider the combustion of butane (C_4H_{10}):

$$2C_4H_{10}(g) + 13O_2(g) \longrightarrow 8CO_2(g) + 10H_2O(l)$$

In a particular reaction, 5.0 mol of C_4H_{10} react with an excess of O_2. Calculate the number of moles of CO_2 formed.

3.74   The annual production of sulfur dioxide from burning coal and fossil fuels, auto exhaust, and other sources is about 26 million tons. The equation for the reaction is

$$S(s) + O_2(g) \longrightarrow SO_2(g)$$

How much sulfur (in tons), present in the original materials, would result in that quantity of SO_2?

3.75   When baking soda (sodium bicarbonate or sodium hydrogen carbonate, $NaHCO_3$) is heated, it releases carbon dioxide gas, which is responsible for the rising of cookies, doughnuts, and bread. (a) Write a balanced equation for the decomposition of the compound (one of the products is Na_2CO_3). (b) Calculate the mass of $NaHCO_3$ required to produce 20.5 g of CO_2.

3.76   When potassium cyanide (KCN) reacts with acids, a deadly poisonous gas, hydrogen cyanide (HCN), is given off. Here is the equation:

$$KCN(aq) + HCl(aq) \longrightarrow KCl(aq) + HCN(g)$$

If a sample of 0.140 g of KCN is treated with an excess of HCl, calculate the amount of HCN formed, in grams.

3.77   Fermentation is a complex chemical process of winemaking in which glucose is converted into ethanol and carbon dioxide:

$$\underset{\text{glucose}}{C_6H_{12}O_6} \longrightarrow \underset{\text{ethanol}}{2C_2H_5OH} + 2CO_2$$

Starting with 500.4 g of glucose, what is the maximum amount of ethanol in grams and in liters that can be obtained by this process (density of ethanol = 0.789 g/mL)?

3.78   Each copper(II) sulfate unit is associated with five water molecules in crystalline copper(II) sulfate pentahydrate ($CuSO_4 \cdot 5H_2O$). When this compound is heated in air above 100°C, it loses the water molecules and also its blue color:

$$CuSO_4 \cdot 5H_2O \longrightarrow CuSO_4 + 5H_2O$$

If 9.60 g of $CuSO_4$ is left after heating 15.01 g of the blue compound, calculate the number of moles of H_2O originally present in the compound.

3.79   For many years, the extraction of gold—that is, the separation of gold from other materials—involved the use of potassium cyanide:

$$4Au + 8KCN + O_2 + 2H_2O \longrightarrow 4KAu(CN)_2 + 4KOH$$

What is the minimum amount of KCN in moles needed to extract 29.0 g (about an ounce) of gold?

3.80   Limestone ($CaCO_3$) is decomposed by heating to quicklime (CaO) and carbon dioxide. Calculate how many grams of quicklime can be produced from 1.0 kg of limestone.

3.81   Nitrous oxide (N_2O) is also called "laughing gas." It can be prepared by the thermal decomposition of ammonium nitrate (NH_4NO_3). The other product is H_2O. (a) Write a balanced equation for this reaction. (b) How many grams of N_2O are formed if 0.46 mol of NH_4NO_3 is used in the reaction?

3.82   The fertilizer ammonium sulfate [$(NH_4)_2SO_4$] is prepared by the reaction between ammonia (NH_3) and sulfuric acid:

$$2NH_3(g) + H_2SO_4(aq) \longrightarrow (NH_4)_2SO_4(aq)$$

How many kilograms of NH_3 are needed to produce 1.00×10^5 kg of $(NH_4)_2 SO_4$?

3.83   A common laboratory preparation of oxygen gas is the thermal decomposition of potassium chlorate ($KClO_3$). Assuming complete decomposition, calculate the number of grams of O_2 gas that can be obtained from 182.5 g of $KClO_3$. (The products are KCl and O_2.)

SECTION 3.7: LIMITING REACTANTS

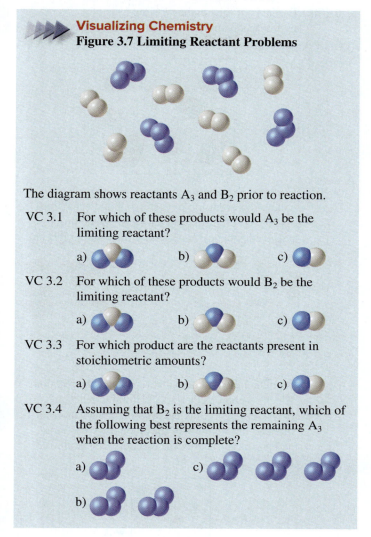

▶▶▶ **Visualizing Chemistry**
Figure 3.7 Limiting Reactant Problems

The diagram shows reactants A_3 and B_2 prior to reaction.

VC 3.1   For which of these products would A_3 be the limiting reactant?

   a)   b)   c)

VC 3.2   For which of these products would B_2 be the limiting reactant?

   a)   b)   c)

VC 3.3   For which product are the reactants present in stoichiometric amounts?

   a)   b)   c)

VC 3.4   Assuming that B_2 is the limiting reactant, which of the following best represents the remaining A_3 when the reaction is complete?

   a)   c)

   b)

Review Questions

3.84   Define *limiting reactant* and *excess reactant*. What is the significance of the limiting reactant in predicting the amount of the product obtained in a reaction? Can there be a limiting reactant if only one reactant is present?

3.85   Give an everyday example that illustrates the limiting reactant concept.

3.86   Why is the theoretical yield of a reaction determined only by the amount of the limiting reactant?

3.87   Why is the actual yield of a reaction almost always smaller than the theoretical yield?

Computational Problems

3.88   Nitric oxide (NO) reacts with oxygen gas to form nitrogen dioxide (NO_2), a dark-brown gas:

$$2NO(g) + O_2(g) \longrightarrow 2NO_2(g)$$

In one experiment, 0.886 mol of NO is mixed with 0.503 mol of O_2. Determine which of the two

reactants is the limiting reactant. Calculate also the number of moles of NO_2 produced.

3.89   Consider the reaction

$$MnO_2 + 4HCl \longrightarrow MnCl_2 + Cl_2 + 2H_2O$$

If 1.15 mol of MnO_2 and 175.4 g of HCl react, which reactant will be used up first? How many grams of Cl_2 will be produced?

3.90   Nickel carbonyl can be prepared by the direct combination of nickel metal with carbon monoxide gas according to the following chemical equation:

$$Ni(s) + 4CO(g) \longrightarrow Ni(CO)_4(s)$$

Determine the mass of nickel carbonyl that can be produced by the combination of 50.03 g Ni(s) with 78.25 g CO(g). Which reactant is consumed completely? How much of the other reactant remains when the reaction is complete?

3.91   Phosgene and ammonia gases can react to produce urea and ammonium chloride solids according to the following chemical equation:

$$COCl_2(g) + 4NH_3(g) \longrightarrow CO(NH_2)_2(s) + 2NH_4Cl(s)$$

Determine the mass of each product formed when 52.68 g $COCl_2(g)$ and 35.50 g $NH_3(g)$ are combined. Which reactant is consumed completely? How much of the other reactant remains when the reaction is complete?

3.92   Zinc metal reacts with aqueous silver nitrate to produce silver metal and aqueous zinc nitrate according to the following equation (unbalanced):

$$Zn(s) + AgNO_3(aq) \longrightarrow Ag(s) + Zn(NO_3)_2(aq)$$

What mass of silver metal is produced when 25.00 g Zn is added to a beaker containing 105.5 g $AgNO_3$ dissolved in 250 mL of water. Determine the mass amounts of each substance present in the beaker when the reaction is complete.

3.93   When combined, aqueous solutions of sulfuric acid and potassium hydroxide react to form water and aqueous potassium sulfate according to the following equation (unbalanced):

$$H_2SO_4(aq) + KOH(aq) \longrightarrow H_2O(l) + K_2SO_4(aq)$$

Determine what mass of water is produced when a beaker containing 100.0 g H_2SO_4 dissolved in 250 mL water is added to a larger beaker containing 100.0 g KOH dissolved in 225 mL water. Determine the mass amounts of each substance (other than water) present in the large beaker when the reaction is complete.

3.94   Hydrogen fluoride is used in the manufacture of Freons (which destroy ozone in the stratosphere) and in the production of aluminum metal. It is prepared by the reaction

$$CaF_2 + H_2SO_4 \longrightarrow CaSO_4 + 2HF$$

In one process, 6.00 kg of CaF_2 is treated with an excess of H_2SO_4 and yields 2.86 kg of HF. Calculate the percent yield of HF.

3.95   Nitroglycerin ($C_3H_5N_3O_9$) is a powerful explosive. Its decomposition may be represented by

$$4C_3H_5N_3O_9 \longrightarrow 6N_2 + 12CO_2 + 10H_2O + O_2$$

This reaction generates a large amount of heat and gaseous products. It is the sudden formation of these gases, together with their rapid expansion, that produces the explosion. (a) What is the maximum amount of O_2 in grams that can be obtained from 2.00×10^2 g of nitroglycerin? (b) Calculate the percent yield in this reaction if the amount of O_2 generated is found to be 6.55 g.

3.96   Titanium(IV) oxide (TiO_2) is a white substance produced by the action of sulfuric acid on the mineral ilmenite ($FeTiO_3$):

$$FeTiO_3 + H_2SO_4 \longrightarrow TiO_2 + FeSO_4 + H_2O$$

Its opaque and nontoxic properties make it suitable as a pigment in plastics and paints. In one process, 8.00×10^3 kg of $FeTiO_3$ yielded 3.67×10^3 kg of TiO_2. What is the percent yield of the reaction?

3.97   Ethylene (C_2H_4), an important industrial organic chemical, can be prepared by heating hexane (C_6H_{14}) at 800°C:

$$C_6H_{14} \xrightarrow{\Delta} C_2H_4 + \text{other products}$$

If the yield of ethylene production is 31.7 percent, what mass of hexane must be used to produce 397 g of ethylene?

3.98   When heated, lithium reacts with nitrogen to form lithium nitride:

$$6Li(s) + N_2(g) \xrightarrow{\Delta} 2Li_3N(s)$$

What is the theoretical yield of Li_3N in grams when 12.3 g of Li is heated with 33.6 g of N_2? If the actual yield of Li_3N is 5.89 g, what is the percent yield of the reaction?

3.99   Disulfide dichloride (S_2Cl_2) is used in the vulcanization of rubber, a process that prevents the slippage of rubber molecules past one another when stretched. It is prepared by heating sulfur in an atmosphere of chlorine:

$$S_8(l) + 4Cl_2(g) \xrightarrow{\Delta} 4S_2Cl_2(l)$$

What is the theoretical yield of S_2Cl_2 in grams when 4.06 g of S_8 is heated with 6.24 g of Cl_2? If the actual yield of S_2Cl_2 is 6.55 g, what is the percent yield?

Conceptual Problems

3.100   Products of the combustion analysis of a hydrocarbon are represented as shown. Determine the empirical formula of the hydrocarbon.

3.101   Consider the reaction pictured, where each red sphere represents an oxygen atom and each blue sphere represents a nitrogen atom. Write the balanced equation and identify the limiting reactant.

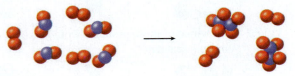

3.102   Consider the reaction

$$2A + B \longrightarrow C$$

(a) In the diagram here that represents the reaction, which reactant, A or B, is the limiting reactant? (b) Assuming a complete reaction, draw a molecular-model representation of the amounts of reactants and products left after the reaction. The atomic arrangement in C is ABA.

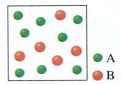

● A
● B

3.103   Consider the reaction

$$N_2 + 3H_2 \longrightarrow 2NH_3$$

Assuming each model represents one mole of the substance, show the number of moles of the product and the excess reactant left after the complete reaction.

○ H_2
● N_2
◐ NH_3

3.104   Determine whether each of the following equations represents a combination reaction, a decomposition reaction, or a combustion reaction: (a) $2NaHCO_3 \longrightarrow Na_2CO_3 + CO_2 + H_2O$, (b) $NH_3 + HCl \longrightarrow NH_4Cl$, (c) $2CH_3OH + 3O_2 \longrightarrow 2CO_2 + 4H_2O$.

3.105 Determine whether each of the following equations represents a combination reaction, a decomposition reaction, or a combustion reaction: (a) $C_3H_8 + 5O_2 \longrightarrow 3CO_2 + 4H_2O$, (b) $2NF_2 \longrightarrow N_2F_4$, (c) $CuSO_4 \cdot 5H_2O \longrightarrow CuSO_4 + 5H_2O$.

ADDITIONAL PROBLEMS

3.106 The diagram represents the products (CO_2 and H_2O) formed after the combustion of a hydrocarbon (a compound containing only C and H atoms). Write an equation for the reaction. (*Hint:* The molar mass of the hydrocarbon is about 30 g.)

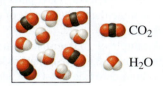

3.107 Consider the reaction of hydrogen gas with oxygen gas:

$$2H_2(g) + O_2(g) \longrightarrow 2H_2O(g)$$

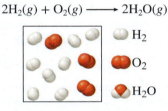

Assuming a complete reaction, which of the diagrams (a–d) shown here represents the amounts of reactants and products left after the reaction?

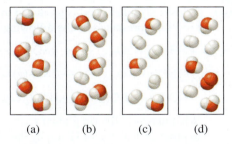

(a)   (b)   (c)   (d)

3.108 The atomic mass of element X is 33.42 amu. A 27.22-g sample of X combines with 84.10 g of another element Y to form a compound XY. Calculate the atomic mass of Y.

3.109 How many moles of O are needed to combine with 0.212 mol of C to form (a) CO and (b) CO_2?

3.110 The aluminum sulfate hydrate [$Al_2(SO_4)_3 \cdot xH_2O$] contains 8.10 percent Al by mass. Calculate x, that is, the number of water molecules associated with each $Al_2(SO_4)_3$ unit.

3.111 A sample of a compound of Cl and O reacts with an excess of H_2 to give 0.233 g of HCl and 0.403 g of H_2O. Determine the empirical formula of the compound.

3.112 The carat is the unit of mass used by jewelers. One carat is exactly 200 mg. How many carbon atoms are present in a 2-carat diamond?

3.113 An iron bar weighed 664 g. After the bar had been standing in moist air for a month, exactly one-eighth of the iron turned to rust (Fe_2O_3.) Calculate the final mass of the iron bar and rust.

3.114 A certain metal oxide has the formula MO where M denotes the metal. A 39.46-g sample of the compound is strongly heated in an atmosphere of hydrogen to remove oxygen as water molecules. At the end, 31.70 g of the metal is left over. If O has an atomic mass of 16.00 amu, calculate the atomic mass of M and identify the element.

3.115 Suppose you are given a cube made of magnesium (Mg) metal of edge length 1.0 cm. (a) Calculate the number of Mg atoms in the cube. (b) Atoms are spherical in shape. Therefore, the Mg atoms in the cube cannot fill all the available space. If only 74 percent of the space inside the cube is taken up by Mg atoms, calculate the radius in picometers of an Mg atom. (The density of Mg is 1.74 g/cm^3, and the volume of a sphere of radius r is $\frac{4}{3}\pi r^3$.)

3.116 Carbohydrates are compounds containing carbon, hydrogen, and oxygen in which the hydrogen to oxygen ratio is 2:1. A certain carbohydrate contains 40.0 percent carbon by mass. Calculate the empirical and molecular formulas of the compound if the approximate molar mass is 178 g.

3.117 Which of the following has the greater mass: 0.72 g of O_2 or 0.0011 mol of chlorophyll ($C_{55}H_{72}MgN_4O_5$)?

3.118 Analysis of a metal chloride XCl_3 shows that it contains 67.2 percent Cl by mass. Calculate the molar mass of X, and identify the element.

3.119 Calculate the number of cations and anions in each of the following compounds: (a) 8.38 g of KBr, (b) 5.40 g of Na_2SO_4, (c) 7.45 g of $Ca_3(PO_4)_2$.

3.120 A mixture of NaBr and Na_2SO_4 contains 29.96 percent Na by mass. Calculate the percent by mass of each compound in the mixture.

3.121 Avogadro's number has sometimes been described as a conversion factor between amu and grams. Use the fluorine atom (19.00 amu) as an example to show the relationship between the atomic mass unit and the gram.

3.122 The natural abundances of the two stable isotopes of hydrogen (hydrogen and deuterium) are 99.99 percent ^1_1H and 0.01 percent ^2_1H. Assume that water exists as either H_2O or D_2O. Calculate the number of D_2O molecules in exactly 400 mL of water (density 1.00 g/mL).

3.123 In the formation of carbon monoxide, CO, it is found that 2.445 g of carbon combine with 3.257 g of oxygen. What is the atomic mass of oxygen if the atomic mass of carbon is 12.01 amu?

3.124 What mole ratio of molecular chlorine (Cl_2) to molecular oxygen (O_2) would result from the breakup of the compound Cl_2O_7 into its constituent elements?

3.125 Which of the following substances contains the greatest mass of chlorine: (a) 5.0 g Cl_2, (b) 60.0 g $NaClO_3$, (c) 0.10 mol KCl, (d) 30.0 g $MgCl_2$, (e) 0.50 mol Cl_2?

3.126 A compound made up of C, H, and Cl contains 55.0 percent Cl by mass. If 9.00 g of the compound contain 4.19×10^{23} H atoms, what is the empirical formula of the compound?

3.127 Platinum forms two different compounds with chlorine. One contains 26.7 percent Cl by mass, and the other contains 42.1 percent Cl by mass. Determine the empirical formulas of the two compounds.

3.128 Heating 2.40 g of the oxide of metal X (molar mass of X = 55.9 g/mol) in carbon monoxide (CO) yields the pure metal and carbon dioxide. The mass of the metal product is 1.68 g. From the data given, show that the simplest formula of the oxide is X_2O_3 and write a balanced equation for the reaction.

3.129 A compound X contains 63.3 percent manganese (Mn) and 36.7 percent O by mass. When X is heated, oxygen gas is evolved and a new compound Y containing 72.0 percent Mn and 28.0 percent O is formed. (a) Determine the empirical formulas of X and Y. (b) Write a balanced equation for the conversion of X to Y.

3.130 A mixture of $CuSO_4 \cdot 5H_2O$ and $MgSO_4 \cdot 7H_2O$ is heated until all the water is lost. If 5.020 g of the mixture gives 2.988 g of the anhydrous salts, what is the percent by mass of $CuSO_4 \cdot 5H_2O$ in the mixture?

3.131 When 0.273 g of Mg is heated strongly in a nitrogen (N_2) atmosphere, a chemical reaction occurs. The product of the reaction weighs 0.378 g. Calculate the empirical formula of the compound containing Mg and N. Name the compound.

3.132 A mixture of methane (CH_4) and ethane (C_2H_6) of mass 13.43 g is completely burned in oxygen. If the total mass of CO_2 and H_2O produced is 64.84 g, calculate the fraction of CH_4 in the mixture.

3.133 Air is a mixture of many gases. However, in calculating its molar mass we need consider only the three major components: nitrogen, oxygen, and argon. Given that one mole of air at sea level is made up of 78.08 percent nitrogen, 20.95 percent oxygen, and 0.97 percent argon, what is the molar mass of air?

3.134 A die has an edge length of 1.5 cm. (a) What is the volume of one mole of such dice? (b) Assuming that the mole of dice could be packed in such a way that they were in contact with one another, forming stacking layers covering the entire surface of Earth, calculate the height in meters the layers would extend outward. [The radius (r) of Earth is 6371 km, and the area of a sphere is $4\pi r^2$.]

3.135 A certain metal M forms a bromide containing 53.79 percent Br by mass. What is the chemical formula of the compound?

3.136 A sample of iron weighing 15.0 g was heated with potassium chlorate ($KClO_3$) in an evacuated container. The oxygen generated from the decomposition of $KClO_3$ converted some of the Fe to Fe_2O_3. If the combined mass of Fe and Fe_2O_3 was 17.9 g, calculate the mass of Fe_2O_3 formed and the mass of $KClO_3$ decomposed.

3.137 A sample containing NaCl, Na_2SO_4, and $NaNO_3$ gives the following elemental analysis: 32.08 percent Na, 36.01 percent O, 19.51 percent Cl. Calculate the mass percent of each compound in the sample.

3.138 A sample of 10.0 g of sodium reacts with oxygen to form 13.83 g of sodium oxide (Na_2O) and sodium peroxide (Na_2O_2). Calculate the percent composition of the product mixture.

3.139 Propane (C_3H_8) is a minor component of natural gas and is used in domestic cooking and heating. (a) Balance the following equation representing the combustion of propane in air:

$$C_3H_8 + O_2 \longrightarrow CO_2 + H_2O$$

(b) How many grams of carbon dioxide can be produced by burning 3.65 mol of propane? Assume that oxygen is the excess reactant in this reaction.

Industrial Problems

3.140 Industrially, nitric acid is produced by the Ostwald process, represented by the following equations:

$$4NH_3(g) + 5O_2(g) \longrightarrow 4NO(g) + 6H_2O(l)$$
$$2NO(g) + O_2(g) \longrightarrow 2NO_2(g)$$
$$2NO_2(g) + H_2O(l) \longrightarrow HNO_3(aq) + HNO_2(aq)$$

What mass of NH_3 (in grams) must be used to produce 1.00 ton of HNO_3 by the Ostwald process, assuming an 80 percent yield in each step (1 ton = 2000 lb; 1 lb = 453.6 g)?

3.141 An impure sample of zinc (Zn) is treated with an excess of sulfuric acid (H_2SO_4) to form zinc sulfate ($ZnSO_4$) and molecular hydrogen (H_2). (a) Write a balanced equation for the reaction. (b) If 0.0764 g of H_2 is obtained from 3.86 g of the sample, calculate the percent purity of the sample. (c) What assumptions must you make in part (b)?

3.142 One of the reactions that occurs in a blast furnace, where iron ore is converted to cast iron, is

$$Fe_2O_3 + 3CO \longrightarrow 2Fe + 3CO_2$$

Suppose that 1.64×10^3 kg of Fe is obtained from a 2.62×10^3-kg sample of Fe_2O_3. Assuming that the reaction goes to completion, what is the percent purity of Fe_2O_3 in the original sample?

3.143 Industrially, hydrogen gas can be prepared by combining propane gas (C_3H_8) with steam at about 400°C. The products are carbon monoxide (CO) and hydrogen gas (H_2). (a) Write a balanced equation for the reaction. (b) How many kilograms of H_2 can be obtained from 2.84×10^3 kg of propane?

Engineering Problems

3.144 A reaction having a 90 percent yield may be considered a successful experiment. However, in the synthesis of complex molecules such as chlorophyll and many anticancer drugs, a chemist often has to carry out multiple-step syntheses. What is the overall percent yield for such a synthesis, assuming it is a 30-step reaction with a 90 percent yield at each step?

3.145 A certain sample of coal contains 1.6 percent sulfur by mass. When the coal is burned, the sulfur is converted to sulfur dioxide. To prevent air pollution, this sulfur dioxide is treated with calcium oxide (CaO) to form calcium sulfite ($CaSO_3$). Calculate the daily mass (in kilograms) of CaO needed by a power plant that uses 6.60×10^6 kg of coal per day.

Biological Problems

3.146 Aspirin or acetylsalicylic acid is synthesized by combining salicylic acid with acetic anhydride:

$$C_7H_6O_3 + C_4H_6O_3 \longrightarrow C_9H_8O_4 + HC_2H_3O_2$$
salicylic acid   acetic anhydride   aspirin   acetic acid

(a) How much salicylic acid is required to produce 0.400 g of aspirin (about the content in a tablet), assuming acetic anhydride is present in excess? (b) Calculate the amount of salicylic acid needed if only 74.9 percent of salicylic is converted to aspirin. (c) In one experiment, 9.26 g of salicylic acid reacts with 8.54 g of acetic anhydride. Calculate the theoretical yield of aspirin and the percent yield if only 10.9 g of aspirin is produced.

3.147 Lactic acid, which consists of C, H, and O, has long been thought to be responsible for muscle soreness following strenuous exercise. Determine the empirical formula of lactic acid given that combustion of a 10.0-g sample produces 14.7 g CO_2 and 6.00 g H_2O.

3.148 Calculate the percent composition by mass of all the elements in calcium phosphate [$Ca_3(PO_4)_2$], a major component of bone.

3.149 Lysine, an essential amino acid in the human body, contains C, H, O, and N. In one experiment, the complete combustion of 2.175 g of lysine gave 3.94 g CO_2 and 1.89 g H_2O. In a separate experiment, 1.873 g of lysine gave 0.436 g NH_3. (a) Calculate the empirical formula of lysine. (b) The approximate molar mass of lysine is 150 g. What is the molecular formula of the compound?

3.150 The compound 2,3-dimercaptopropanol ($HSCH_2CHSHCH_2OH$), commonly known as British Anti-Lewisite (BAL), was developed during World War I as an antidote to arsenic-containing poison gas. (a) If each BAL molecule binds one arsenic (As) atom, how many As atoms can be removed by 1.0 g of BAL? (b) BAL can also be used to remove poisonous heavy metals like mercury (Hg) and lead (Pb). If each BAL binds one Hg atom, calculate the mass percent of Hg in a BAL-Hg complex. (A H atom is removed when a BAL molecule binds a Hg atom.)

3.151 Mustard gas ($C_4H_8Cl_2S$) is a poisonous gas that was used in World War I and banned afterward. It causes general destruction of body tissues, resulting in the formation of large water blisters. There is no effective antidote. Calculate the percent composition by mass of the elements in mustard gas.

3.152 Myoglobin stores oxygen for metabolic processes in muscle. Chemical analysis shows that it contains 0.34 percent Fe by mass. What is the molar mass of myoglobin? (There is one Fe atom per molecule.)

3.153 Hemoglobin ($C_{2952}H_{4664}N_{812}O_{832}S_8Fe_4$) is the oxygen carrier in blood. (a) Calculate its molar mass. (b) An average adult has about 5.0 L of blood. Every milliliter of blood has approximately 5.0×10^9 erythrocytes, or red blood cells, and every red blood cell has about 2.8×10^8 hemoglobin (HG) molecules. Calculate the mass of hemoglobin molecules in grams in an average adult.

3.154 Cysteine, shown here, is one of the 20 amino acids found in proteins in humans. Write the molecular formula of cysteine, and calculate its molar mass.

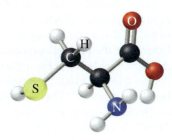

3.155 Isoflurane, shown here, is a common inhalation anesthetic. Write its molecular formula, and calculate its molar mass.

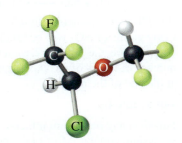

Environmental Problems

3.156 Carbon dioxide (CO_2) is the gas that is mainly responsible for global warming (the greenhouse effect). The burning of fossil fuels is a major cause of the increased concentration of CO_2 in the atmosphere. Carbon dioxide is also the end product of metabolism (see Sample Problem 3.4). Using glucose as an example of food, calculate the annual human production of CO_2 in grams, assuming that each person consumes 5.0×10^2 g of glucose per day, that the world's population is 6.5 billion, and that there are 365 days in a year.

3.157 It is estimated that the day Mt. St. Helens erupted (May 18, 1980), about 4.0×10^5 tons of SO_2 were released into the atmosphere. If all the SO_2 were eventually converted to sulfuric acid, how many tons of H_2SO_4 were produced?

3.158 Leaded gasoline contains an additive to prevent engine "knocking." On analysis, the additive compound is found to contain carbon, hydrogen, and lead (Pb) (hence, "leaded gasoline"). When 51.36 g of this compound is burned in an apparatus such as that shown in Figure 3.5, 55.90 g of CO_2 and 28.61 g of H_2O are produced. Determine the empirical formula of the gasoline additive. Because of its detrimental effect on the environment, the original lead additive has been replaced in recent years by methyl *tert*-butyl ether (a compound of C, H, and O) to enhance the performance of gasoline. (As of 1999, this compound is also being phased out because of its contamination of drinking water.) When 12.1 g of the compound is burned in an apparatus like the one shown in Figure 3.5, 30.2 g of CO_2 and 14.8 g of H_2O are formed. What is the empirical formula of this compound?

3.159 Peroxyacylnitrate (PAN) is one of the components of smog. It is a compound of C, H, N, and O. Determine the percent composition of oxygen and the empirical formula from the following percent composition by mass: 19.8 percent C, 2.50 percent H, 11.6 percent N. What is its molecular formula given that its molar mass is about 120 g?

3.160 The depletion of ozone (O_3) in the stratosphere has been a matter of great concern among scientists in recent years. It is believed that ozone can react with nitric oxide (NO) that is discharged from high-altitude jet planes. The reaction is

$$O_3 + NO \longrightarrow O_2 + NO_2$$

If 0.740 g of O_3 reacts with 0.670 g of NO, how many grams of NO_2 will be produced? Which compound is the limiting reactant? Calculate the number of moles of the excess reactant remaining at the end of the reaction.

Multiconcept Problems

3.161 Potash is any potassium mineral that is used for its potassium content. Most of the potash produced in the United States goes into fertilizer. The major sources of potash are potassium chloride (KCl) and potassium sulfate (K_2SO_4). Potash production is often reported as the potassium oxide (K_2O) equivalent or the amount of K_2O that could be made from a given mineral. (a) If KCl costs $0.55 per kg, for what price (dollar per kg) must K_2SO_4 be sold to supply the same amount of potassium on a per dollar basis? (b) What mass (in kg) of K_2O contains the same number of moles of K atoms as 1.00 kg of KCl?

3.162 Octane (C_8H_{18}) is a component of gasoline. Complete combustion of octane yields H_2O and CO_2. Incomplete combustion produces H_2O and CO, which not only reduces the efficiency of the engine using the fuel but is also toxic. In a certain test run, 1.000 gallon (gal) of octane is burned in an engine. The total mass of CO, CO_2, and H_2O produced is 11.53 kg. Calculate the efficiency of the process; that is, calculate the fraction of octane converted to CO_2. The density of octane is 2.650 kg/gal.

3.163 The following is a crude but effective method for estimating the *order of magnitude* of Avogadro's number using stearic acid ($C_{18}H_{36}O_2$). When stearic acid is added to water, its molecules collect at the surface and form a monolayer; that is, the layer is only one molecule thick. The cross-sectional area of each stearic acid molecule has been measured to be 0.21 nm². In one experiment, it is found that 1.4×10^{-4} g of stearic acid is needed to form a monolayer over water in a dish of diameter 20 cm. Based on these measurements, what is Avogadro's number? (The area of a circle of radius r is πr^2.)

3.164 The photograph at the beginning of Chapter 2 shows a bottle of iron supplements in the form of ferrous fumarate ($FeC_4H_2O_4$). (a) Determine the molar mass of ferrous fumarate and calculate its percent composition. (b) Write the empirical formula of ferrous fumarate. (c) What mass of each CO_2 and H_2O would be produced by the combustion of 1.000 g $FeC_4H_2O_4$? (d) The FDA recommends that iron supplements containing more than 30 mg iron be sold in unit-dose blister packaging, of the type shown here, to help prevent accidental overdose—especially in children. Explain why it is not necessary for the tablets in the photograph (according to the label, 65 mg) to be in unit-dose blister packaging, in accordance with the FDA recommendation.

David A. Tietz/Editorial Image, LLC

Standardized-Exam Practice Problems

Physical and Biological Sciences

The first step in producing phosphorus fertilizer is the treatment of fluorapatite, a phosphate rock, with sulfuric acid to yield calcium dihydrogen phosphate, calcium sulfate, and hydrogen fluoride gas. In one experiment, a chemist combines 1.00 kg of each reactant.

1. Select the correct balanced equation to represent the reaction.

    a) $2CaPO_4F(s) + 2H_2SO_4(aq)$
    $$\longrightarrow CaH_2PO_4(aq) + CaSO_4(aq) + HF(g)$$
    b) $2CaPO_4F(s) + H_2SO_4(aq)$
    $$\longrightarrow CaH_2PO_4(aq) + CaSO_4(aq) + HF(g)$$
    c) $Ca_5(PO_4)_3F(s) + 3H_2SO_4(aq)$
    $$\longrightarrow 3Ca(H_2PO_4)_2(aq) + 2CaSO_4(aq) + HF(g)$$
    d) $2Ca_5(PO_4)_3F(s) + 7H_2SO_4(aq)$
    $$\longrightarrow 3Ca(H_2PO_4)_2(aq) + 7CaSO_4(aq) + 2HF(g)$$

2. Assuming that all the limiting reactant is converted to products, what mass of calcium dihydrogen phosphate is produced?

    a) 464 g   b) 696 g   c) 92.8 g   d) 199 g

3. What mass of the excess reactant remains unreacted?

    a) 319 g   b) 681 g   c) 406 g   d) 490 g

4. What is the percent yield of the reaction if 197 g of calcium dihydrogen phosphate is produced?

    a) 99.0%   b) 42.5%   c) 28.3%   d) 212%

Answers to In-Chapter Materials

Answers to Practice Problems

3.1A (a) 95.21 amu, (b) 98.09 amu, (c) 90.04 amu.
3.1B (a) 100.09 amu, (b) 47.02 amu, (c) 68.15 amu. **3.2A** 57.13% C, 6.165% H, 9.521% N, 27.18% O. **3.2B** C_3H_6O. **3.3A** $C_3H_8(g) +$ $5O_2(g) \longrightarrow 3CO_2(g) + 4H_2O(l)$. **3.3B** $H_2SO_4(aq) + 2NaOH(aq)$ $\longrightarrow Na_2SO_4(aq) + 2H_2O(l)$. **3.4A** $C_{18}H_{32}O_2(aq) + 25O_2(g)$ $\longrightarrow 18CO_2(g) + 16H_2O(l)$. **3.4B** $2NH_3(g) + 3CuO(s) \longrightarrow$ $3Cu(s) + N_2(g) + 3H_2O(l)$. **3.5A** (a) 4.40×10^{24} atoms K, (b) 1.48×10^2 mol K. **3.5B** (a) 6.32×10^{17} atoms He, (b) 3.87×10^{-3} mol He. **3.6A** (a) 495 g, (b) 0.704 mol. **3.6B** (a) 31.5 g, (b) 22.71 mol. **3.7A** (a) 6.68×10^{23} O_2 molecules, 1.34×10^{24} O atoms, (b) 1.32×10^{-7} g. **3.7B** (a) 3.34×10^{25} O atoms, $6.68 \times$ 10^{25} H atoms, (b) 33.4 g. **3.8A** C_2H_6O. **3.8B** CH_2. **3.9A** $C_3H_4O_3$ and $C_6H_8O_6$. **3.9B** 1.44 g CO_2, 1.18 g H_2O. **3.10A** 0.264 mol H_2 and 0.176 mol NH_3. **3.10B** $P_4O_{10} + 6H_2O \longrightarrow 4H_3PO_4$; 1.45 mol P_4O_{10}, 8.70 mol H_2O. **3.11A** 34.1 g. **3.11B** 292 g. **3.12A** 42.6 g ammonia; nitrogen is limiting reactant and 4.95 g is hydrogen left over. **3.12B** 44.2 g KOH, 115.5 g H_3PO_4. **3.13A** 29.2%. **3.13B** 122 g. **3.14A** 71.5% **3.14B** 58.5 g. **3.15A** (a) combustion, (b) combination, (c) decomposition. **3.15B** $A_2 + 2B \longrightarrow 2AB$.

Answers to Checkpoints

3.1.1 a. **3.1.2** d. **3.2.1** b. **3.2.2** e. **3.3.1** e. **3.3.2** d. **3.3.3** e. **3.3.4** d. **3.4.1** b. **3.4.2** d. **3.4.3** a. **3.4.4** e. **3.5.1** e. **3.5.2** d. **3.5.3** a. **3.5.4** d. **3.6.1** a. **3.6.2** a. **3.7.1** c. **3.7.2** b. **3.7.3** e. **3.7.4** c. **3.7.5** a. **3.7.6** c.

CHAPTER 4

Reactions in Aqueous Solutions

Gatorade, the original sports beverage, was developed to remedy depleted blood sugar, blood volume, and electrolyte balance in college athletes.

Sara Stathas/Alamy Stock Photo

In This Chapter, You Will Learn

About some of the properties of aqueous solutions and about several different types of reactions that can occur between dissolved substances. You will also learn how to express the concentration of a solution and how concentration can be useful in solving quantitative problems.

Before You Begin, Review These Skills

- Identifying compounds as either *ionic* or *molecular* [◄◄ Sections 2.6 and 2.7]
- Names, formulas, and charges of the common polyatomic ions [◄◄ Table 2.3]
- Chapter 3 Key Skills [◄◄ pages 126–127]

How Aqueous Solutions Impact Athletic Stamina and Performance

In 1965, University of Florida (UF) assistant coach Dwayne Douglas was concerned about the health of Gators football players. He noted that during practices and games in hot weather the players (1) lost a great deal of weight, (2) seldom needed to urinate, and (3) had limited stamina, especially during the second half of a practice or game. He consulted Dr. Robert Cade, researcher and kidney-disease specialist at UF's medical college, who embarked on a project to identify the cause of the athletes' lack of endurance. It was found that after a period of intense activity accompanied by profuse sweating, the players had low blood sugar, low blood volume, and an imbalance of electrolytes—all of which contributed to heat exhaustion. Cade and his research fellows theorized that the depletion of sugar, water, and electrolytes might be remedied by having the athletes drink a solution containing just the right amounts of each. Using this theory, they developed a beverage containing water, sugar, and sodium and potassium salts similar to those present in sweat. By all accounts, the beverage tasted so bad that no one would drink it. Mary Cade, Robert Cade's wife, suggested adding lemon juice to make the concoction more palatable—and the drink that would become Gatorade was born. In their 1966 season, the Gators earned a reputation as the "second-half" team, often coming from behind in the third or fourth quarter. Gators coach Ray Graves attributed his team's newfound late-in-the-game strength to the newly developed sideline beverage that replenished blood sugar, blood volume, and electrolyte balance. Sports drinks are now a multibillion-dollar industry, and there are several popular brands, although Gatorade still maintains a large share of the market.

Development of beverages that help replenish blood sugar and restore electrolyte balance requires comprehension of the properties of *aqueous solutions*.

David A. Tietz/Editorial Image, LLC

At the end of this chapter, you will be able to solve several problems related to typical sports drinks and an aqueous reaction used to analyze their contents [►► Applying What You've Learned, page 191].

Animation
Solutions—strong, weak, and nonelectrolytes.

4.1 General Properties of Aqueous Solutions

A **solution** is a homogeneous mixture of two or more substances [◀◀ Section 1.2]. Solutions may be gaseous (such as air), solid (such as brass), or liquid (such as salt-water). Usually, the substance present in the largest amount is referred to as the **solvent** and any substance present in a smaller amount is called the **solute.** For example, if we dissolve a teaspoon of sugar in a glass of water, water is the solvent and sugar is the solute. In this chapter, we will focus on the properties of aqueous solutions—those in which water is the solvent. Throughout the remainder of this chapter, unless otherwise noted, *solution* will refer specifically to an aqueous solution.

Electrolytes and Nonelectrolytes

You have probably heard of electrolytes in the context of sports drinks such as Gatorade. Electrolytes in body fluids are necessary for the transmission of electrical impulses, which are critical to physiological processes such as nerve impulses and muscle contractions. In general, an **electrolyte** is a substance that dissolves in water to yield a solution that conducts electricity. By contrast, a **nonelectrolyte** is a substance that dissolves in water to yield a solution that does *not* conduct electricity. Every water-soluble substance fits into one of these two *categories*.

> **Student Note:** A substance that *dissolves* in a particular solvent is said to be "soluble" in that solvent. In this chapter, we use the word *soluble* to mean "water-soluble."

The difference between an aqueous solution that conducts electricity and one that does not is the presence or absence of *ions*. As an illustration, consider solutions of sugar and salt. The physical processes of sugar (sucrose, $C_{12}H_{22}O_{11}$) dissolving in water and salt (sodium chloride, NaCl) dissolving in water can be represented with the following chemical equations:

$$C_{12}H_{22}O_{11}(s) \xrightarrow{H_2O} C_{12}H_{22}O_{11}(aq) \quad \text{and} \quad NaCl(s) \xrightarrow{H_2O} Na^+(aq) + Cl^-(aq)$$

Note that while the sucrose molecules remain intact upon dissolving, becoming aqueous sucrose molecules, the sodium chloride dissociates, producing aqueous sodium ions and aqueous chloride ions. **Dissociation** is the process by which an ionic compound, upon dissolution, breaks apart into its constituent ions. It is the presence of ions that allows the solution of sodium chloride to conduct electricity. Thus, sodium chloride is an *electrolyte* and sucrose is a *nonelectrolyte*.

Like sucrose, which is a molecular compound [◀◀ Section 2.7], many water-soluble molecular compounds are nonelectrolytes. Some molecular compounds are electrolytes, however, because they ionize on dissolution. **Ionization** is the process by which a molecular compound forms ions when it dissolves. Recall from Chapter 2 that acids are compounds that dissolve in water to produce hydrogen ions (H^+) [◀◀ Section 2.7]. HCl, for example, ionizes to produce H^+ ions and Cl^- ions.

$$HCl(g) \xrightarrow{H_2O} H^+(aq) + Cl^-(aq)$$

Acids constitute one of two important classes of molecular compounds that are electrolytes. Molecular *bases* constitute the other one. A **base** is a compound that dissolves in water to produce hydroxide ions (OH^-). Ammonia (NH^3), for example, ionizes in water to produce ammonium (NH_4^+) and hydroxide (OH^-) ions.

> **Student Note:** Bases may be *molecular*, like ammonia (NH_3), or *ionic*, like sodium hydroxide (NaOH).

$$NH_3(g) + H_2O(l) \rightleftharpoons NH_4^+(aq) + OH^-(aq)$$

Strong Electrolytes and Weak Electrolytes

In a solution of sodium chloride, *all* the dissolved compound exists in the form of ions. Thus, NaCl, which is an ionic compound [◀◀ Section 2.6], is said to have *dissociated completely*. An electrolyte that dissociates completely is known as a **strong electrolyte.** All water-soluble ionic compounds dissociate completely upon dissolving, so all water-soluble ionic compounds are strong electrolytes.

The list of molecular compounds that are strong electrolytes is fairly short. It comprises the seven strong acids, which are listed in Table 4.1. A strong acid ionizes

TABLE 4.1	The Strong Acids
Acid	**Ionization Equation**
Hydrochloric acid	$HCl(aq) \longrightarrow H^+(aq) + Cl^-(aq)$
Hydrobromic acid	$HBr(aq) \longrightarrow H^+(aq) + Br^-(aq)$
Hydroiodic acid	$HI(aq) \longrightarrow H^+(aq) + I^-(aq)$
Nitric acid	$HNO_3(aq) \longrightarrow H^+(aq) + NO_3^-(aq)$
Chloric acid	$HClO_3(aq) \longrightarrow H^+(aq) + ClO_3^-(aq)$
Perchloric acid	$HClO_4(aq) \longrightarrow H^+(aq) + ClO_4^-(aq)$
Sulfuric acid*	$H_2SO_4(aq) \longrightarrow H^+(aq) + HSO_4^-(aq)$
	$HSO_4^-(aq) \rightleftharpoons H^+(aq) + SO_4^{2-}(aq)$

*Note that although each sulfuric acid molecule has two ionizable hydrogen atoms, it only undergoes the first ionization completely, effectively producing one H^+ ion and one HSO_4^- ion per H_2SO_4 molecule. The second ionization happens only to a very small extent.

completely, resulting in a solution that contains hydrogen ions and the corresponding anions but essentially no acid molecules.

Most of the molecular compounds that are electrolytes are weak electrolytes. A **weak electrolyte** is a compound that produces ions upon dissolving but exists in solution *predominantly* as molecules that are *not* ionized. Most acids (except those listed in Table 4.1) are weak electrolytes. Acetic acid ($HC_2H_3O_2$) is not one of the strong acids listed in Table 4.1, so it is a weak acid. Its ionization in water is represented by the following chemical equation:

$$HC_2H_3O_2(l) \rightleftharpoons H^+(aq) + C_2H_3O_2^-(aq)$$

Note the use of the double arrow, $\rightleftharpoons$, in this equation and in two earlier equations, including one in Table 4.1. This denotes a reaction that occurs in both directions and does not result in *all* the reactant(s) (e.g., acetic acid) being converted permanently to product(s) (e.g., hydrogen ions and acetate ions). Instead, forward and reverse reactions both occur, and a state of dynamic chemical equilibrium is established.

Although acetic acid molecules ionize, the resulting ions have a strong tendency to recombine to form acetic acid molecules again. Eventually, the ions produced by the ionization will be recombining at the same rate at which they are produced, and there will be no further change in the numbers of acetic acid molecules, hydrogen ions, or acetate ions. Because there is a stronger tendency for the ions to recombine than for the molecules to *ionize,* at any given point in time, most of the dissolved acetic acid exists as molecules that are not ionized (reactant). Only a very small percentage exists in the form of hydrogen ions and acetate ions (products).

The ionization of a weak base, while similar in many ways to the ionization of a weak acid, requires some additional explanation. Ammonia (NH_3) is a common weak base. The ionization of ammonia in water is represented by the equation

$$NH_3(g) + H_2O(l) \rightleftharpoons NH_4^+(aq) + OH^-(aq)$$

Note that the ammonia molecule does not ionize by breaking apart into ions. Rather, it does so by ionizing a *water* molecule. The H^+ ion from a water molecule attaches to an ammonia molecule, producing an ammonium ion (NH_4^+) and leaving what remains of the water molecule, the OH^- ion, in solution.

Student Note: In a state of dynamic *chemical equilibrium,* or simply *equilibrium,* both forward and reverse reactions continue to occur. However, because they are occurring at the same *rate,* no net change is observed over time in the amounts of reactants or products. Chemical equilibrium is the subject of Chapters 15, 16, and 17.

$$NH_3(g) \quad + \quad H_2O(l) \quad \rightleftharpoons \quad NH_4^+(aq) \quad + \quad OH^-(aq)$$

As with the ionization of a weak acid, the reverse process predominates, and at any given point in time there will be far more NH_3 molecules present than there will be NH_4^+ and OH^- ions.

We can distinguish between electrolytes and nonelectrolytes experimentally using an apparatus like the one pictured in Figure 4.1. A lightbulb is connected to a battery using a circuit that includes the contents of the beaker. For the bulb to light, electric current must flow from one electrode to the other. Pure water is a very poor conductor of electricity because H_2O ionizes to only an extremely small extent. There are virtually no ions in pure water to conduct the current, so H_2O is considered a nonelectrolyte. If we add a small amount of salt (sodium chloride), however, the lightbulb will begin to glow as soon as the salt dissolves in the water. Sodium chloride dissociates completely in water to give Na^+ and Cl^- ions. Because the NaCl solution conducts electricity, we say that NaCl is an electrolyte.

If the solution contains a nonelectrolyte, as it does in Figure 4.1(a), the bulb will not light. If the solution contains an electrolyte, as it does in Figure 4.1(b) and (c), the bulb will light. The cations in solution are attracted to the negative electrode, and the anions are attracted to the positive electrode. This movement sets up an electric current that is equivalent to the flow of electrons along a metal wire. How brightly the bulb burns depends upon the number of ions in solution. In Figure 4.1(b), the solution contains a *weak* electrolyte and therefore a relatively small number of ions, so the bulb lights only weakly. The solution in Figure 4.1(c) contains a *strong* electrolyte, which produces a relatively large number of ions, so the bulb lights brightly.

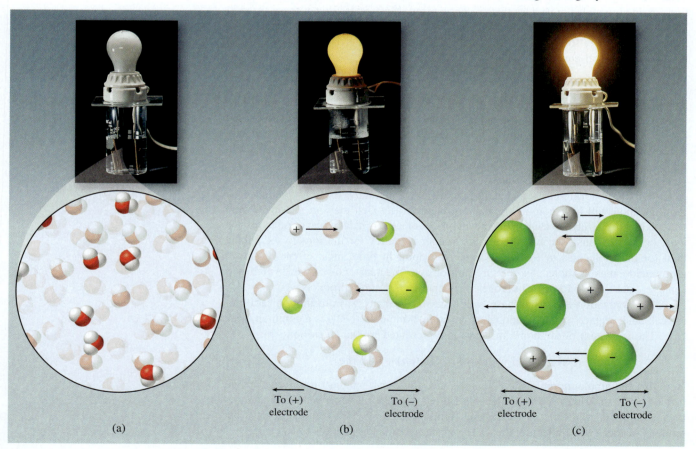

(a)   (b)   (c)

Figure 4.1 An apparatus for distinguishing between electrolytes and nonelectrolytes and between weak electrolytes and strong electrolytes. A solution's ability to conduct electricity depends on the number of ions it contains. (a) Pure water contains almost no ions and does not conduct electricity; therefore, the lightbulb is not lit. (b) A weak electrolyte solution such as HF(aq) contains a small number of ions, and the lightbulb is dimly lit. (c) A strong electrolyte solution such as NaCl(aq) contains a large number of ions, and the lightbulb is brightly lit. The molar amounts of dissolved substances in the beakers in (b) and (c) are equal.

(a), (b), (c): Stephen Frisch/McGraw Hill

Identifying Electrolytes

While the experimental method described in Figure 4.1 can be useful, often you will have to characterize a compound as a non-electrolyte, a weak electrolyte, or a strong electrolyte just by looking at its formula. A good first step is to determine whether the compound is *ionic* or *molecular*.

An ionic compound contains a *cation* (which is either a metal ion or the ammonium ion) and an *anion* (which may be atomic or polyatomic). A binary compound that contains a metal and a nonmetal is almost always ionic. This is a good time to review the polyatomic anions in Table 2.3 [◄◄ Section 2.6]. You will need to be able to recognize them in the formulas of compounds. Any ionic compound that dissolves in water is a strong electrolyte.

If a compound does not contain a metal cation or the ammonium cation, it is molecular. In this case, you will need to determine whether or not the compound is an acid. Acids generally can be recognized by the way their formulas are written, with the ionizable hydrogens written first. $HC_2H_3O_2$, H_2CO_3, and H_3PO_4 are acetic acid, carbonic acid, and phosphoric acid, respectively. Formulas of carboxylic acids, such as acetic acid, often are written with their ionizable hydrogen atoms *last* to keep the functional group together in the formula. Thus, either $HC_2H_3O_2$ or CH_3COOH is correct for acetic acid. To make it easier to identify compounds as acids, in this chapter we write all acid formulas with the ionizable H atom(s) first. If a compound is an acid, it is an electrolyte. If it is one of the acids listed in Table 4.1, it is a strong acid and therefore a strong electrolyte. Any acid not listed in Table 4.1 is a weak acid and therefore a weak electrolyte.

If a molecular compound is not an acid, you must then consider whether or not it is a weak base. Many weak bases are related to ammonia in that they consist of a nitrogen atom bonded to hydrogen and/or carbon atoms. Examples include methylamine (CH_3NH_2), pyridine (C_5H_5N), and hydroxylamine (NH_2OH). Weak bases are weak electrolytes.

If a molecular compound is neither an acid nor a weak base, it is a nonelectrolyte. The flowchart in Figure 4.2 can be useful for classification of water-soluble compounds.

Sample Problem 4.1 lets you practice using chemical formulas to classify compounds as electrolytes and nonelectrolytes.

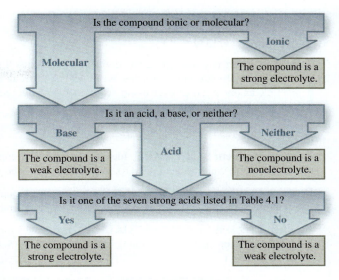

Figure 4.2 Flowchart for determining if a compound is a strong electrolyte, a weak electrolyte, or a nonelectrolyte.

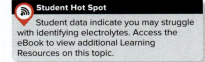

Student Hot Spot

Student data indicate you may struggle with identifying electrolytes. Access the eBook to view additional Learning Resources on this topic.

SAMPLE PROBLEM 4.1

Classify each of the following compounds as a nonelectrolyte, a weak electrolyte, or a strong electrolyte: (a) methanol (CH_3OH), (b) sodium hydroxide (NaOH), (c) ethylamine ($C_2H_5NH_2$), and (d) hydrofluoric acid (HF).

Strategy Classify each compound as ionic or molecular. Soluble ionic compounds are strong electrolytes. Classify each molecular compound as an acid, base, or neither. Molecular compounds that are neither acids nor bases are nonelectrolytes. Molecular compounds that are bases are weak electrolytes. Finally, classify acids as either strong or weak. Strong acids are strong electrolytes, and weak acids are weak electrolytes.

Setup (a) Methanol contains neither a metal cation nor the ammonium ion. It is therefore molecular. Its formula does not begin with H, so it is probably not an acid, and it does not contain a nitrogen atom, so it is not a weak base. Molecular compounds that are neither acids nor bases are *nonelectrolytes*. (b) Sodium hydroxide contains a metal cation (Na^+) and is therefore ionic. It is also one of the strong bases. (c) Ethylamine contains no cations and is therefore molecular. It is also a nitrogen-containing base, similar to ammonia. (d) Hydrofluoric acid is, as its name suggests, an acid. However, it is not on the list of strong acids in Table 4.1 and is, therefore, a weak acid.

Solution (a) Nonelectrolyte   (b) Strong electrolyte   (c) Weak electrolyte   (d) Weak electrolyte

(Continued on next page)

THINK ABOUT IT

Make sure that you have correctly identified compounds that are ionic and compounds that are molecular. Remember that strong acids are strong electrolytes, weak acids and weak bases are weak electrolytes, and strong bases are strong electrolytes (by virtue of their being soluble ionic compounds). Molecular compounds, with the exceptions of acids and weak bases, are nonelectrolytes.

Practice Problem (A)TTEMPT Identify the following compounds as nonelectrolytes, weak electrolytes, or strong electrolytes: ethanol (C_2H_5OH), nitrous acid (HNO_2), and sodium hydrogen carbonate ($NaHCO_3$, also known as *bicarbonate*).

Practice Problem (B)UILD Identify the following compounds as nonelectrolytes, weak electrolytes, or strong electrolytes: phosphorous acid (H_3PO_3), hydrogen peroxide (H_2O_2), and ammonium sulfate [$(NH_4)_2SO_4$].

Practice Problem (C)ONCEPTUALIZE Determine which diagram, if any, could represent an aqueous solution of each of the following compounds: LiCl, $CuSO_4$, K_2SO_4, H_2CO_3, $Al_2(SO_4)_3$, $AlCl_3$, Na_3PO_4. (Red and blue spheres represent different chemical species.)

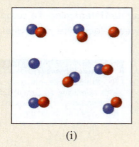
(i)

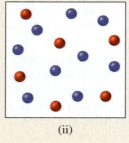
(ii)

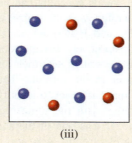
(iii)

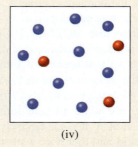
(iv)

CHECKPOINT – SECTION 4.1   **General Properties of Aqueous Solutions**

4.1.1   Soluble ionic compounds are _____.

   a) always nonelectrolytes

   b) always weak electrolytes

   c) always strong electrolytes

   d) never strong electrolytes

   e) sometimes nonelectrolytes

4.1.2   Soluble molecular compounds are _____.

   a) always nonelectrolytes

   b) always weak electrolytes

   c) always strong electrolytes

   d) never strong electrolytes

   e) sometimes strong electrolytes

4.1.3   Which of the following compounds is a weak electrolyte?

   a) LiCl   d) NaI

   b) $(C_2H_5)_2NH$ e) HNO_3

   c) KNO_3

4.1.4   Which of the following compounds is a strong electrolyte?

   a) HF   d) NH_3

   b) H_2CO_3 e) H_2O

   c) NaF

4.2   Precipitation Reactions

When an aqueous solution of lead(II) nitrate [$Pb(NO_3)_2$] is added to an aqueous solution of sodium iodide (NaI), a yellow insoluble solid—lead(II) iodide (PbI_2)—forms. Sodium nitrate ($NaNO_3$), the other reaction product, remains in solution. Figure 4.3 shows this reaction in progress. An insoluble solid product that separates from a solution is called a *precipitate,* and a chemical reaction in which a precipitate forms is called a *precipitation reaction.*

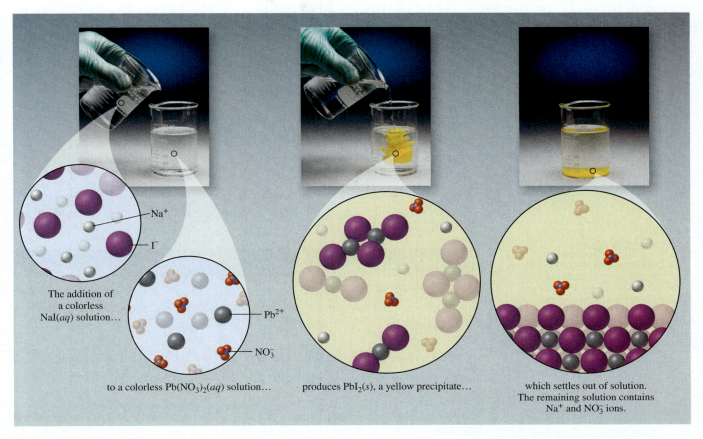

Figure 4.3 A colorless aqueous solution of NaI is added to a colorless aqueous solution of $Pb(NO_3)_2$. A yellow precipitate, PbI_2, forms. Na^+ and NO_3^- ions remain in solution.

Photos: *(left and middle): Charles D. Winters/Timeframe Photography/McGraw Hill; (right): Charles D. Winters/McGraw Hill*

Precipitation reactions usually involve ionic compounds, but a precipitate does not form every time two solutions of electrolytes are combined. Instead, whether or not a precipitate forms when two solutions are mixed depends on the solubility of the products.

Solubility Guidelines for Ionic Compounds in Water

When an ionic substance such as sodium chloride dissolves in water, the water molecules remove individual ions from the three-dimensional solid structure and surround them. This process, called *hydration,* is shown in Figure 4.4. Water is an excellent

Animation
Precipitation of $BaSO_4$.

Figure 4.4 Hydration of anions and cations of a soluble ionic compound. Water molecules surround each anion with their partial positive charges (H atoms) oriented toward the negatively charged anion; and they surround each cation with their partial negative charges (O atoms) oriented toward the positively charged cation.

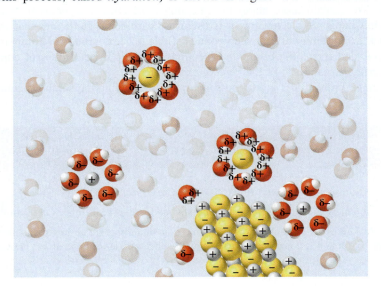

TABLE 4.2	Solubility Guidelines: Soluble Compounds
Water-Soluble Compounds	**Insoluble Exceptions**
Compounds containing an alkali metal cation (Li^+, Na^+, K^+, Rb^+, Cs^+) or the ammonium ion (NH_4^+)	
Compounds containing the nitrate ion (NO_3^-), acetate ion ($C_2H_3O_2^-$), or chlorate ion (ClO_3^-)	
Compounds containing the chloride ion (Cl^-), bromide ion (Br^-), or iodide ion (I^-)	Compounds containing Ag^+, Hg_2^{2+}, or Pb^{2+}
Compounds containing the sulfate ion (SO_4^{2-})	Compounds containing Ag^+, Hg_2^{2+}, Pb^{2+}, Ca^{2+}, Sr^{2+}, or Ba^{2+}

Student Note: Ionic compounds often are classified according to the anions they contain. Compounds that contain the chloride ion are called *chlorides*, compounds containing the nitrate ion are called *nitrates*, and so on.

TABLE 4.3	Solubility Guidelines: Insoluble Compounds
Water-Insoluble Compounds	**Soluble Exceptions**
Compounds containing the carbonate ion (CO_3^{2-}), phosphate ion (PO_4^{3-}), chromate ion (CrO_4^{2-}), or sulfide ion (S^{2-})	Compounds containing Li^+, Na^+, K^+, Rb^+, Cs^+, or NH_4^+
Compounds containing the hydroxide ion (OH^-)	Compounds containing Li^+, Na^+, K^+, Rb^+, Cs^+, or Ba^{2+}

Student Note: Note the recurrence of the same three groups of ions in the exceptions columns in Tables 4.2 and 4.3: Group 1 or the ammonium cation; Ag^+, Hg_2^{2+}, or Pb^{2+}; and the heavier Group 2 cations.

Student Note: The partial charges on the oxygen atom and the hydrogen atoms sum to zero. Water molecules, although polar, have no *net* charge. Chapters 8 and 9 cover partial charges and molecular polarity in more detail.

solvent for ionic compounds because H_2O is a *polar* molecule; that is, its electrons are distributed such that there is a partial negative charge on the oxygen atom, denoted by the δ− symbol, and partial positive charges, denoted by the δ+ symbol, on each of the *hydrogen atoms*. The oxygen atoms in the surrounding water molecules are attracted to the cations, while the hydrogen atoms are attracted to the anions. These attractions explain the orientation of water molecules around each of the ions in solution. The surrounding water molecules prevent the cations and anions from recombining.

Solubility is defined as the maximum amount of solute that will dissolve in a given quantity of solvent at a specific temperature. Not all ionic compounds dissolve in water. Whether or not an ionic compound is water-soluble depends on the relative magnitudes of the water molecules' attraction to the ions, and the ions' attraction to one another. If the water molecules' attraction for the ions exceeds the ions' attraction to one another, then the ionic compound will dissolve. If the ions' attraction to each other exceeds the water molecules' attraction to the ions, then the compound won't dissolve. We explain more about the magnitudes of attractive forces in ionic compounds in Chapter 8, but for now it is useful to learn some guidelines that enable us to predict the solubility of ionic compounds. Table 4.2 lists groups of compounds that are *soluble* and shows the *insoluble* exceptions. Table 4.3 lists groups of compounds that are *insoluble* and shows the *soluble* exceptions.

Sample Problem 4.2 gives you some practice applying the solubility guidelines.

SAMPLE PROBLEM 4.2

Classify each of the following compounds as soluble or insoluble in water: (a) $AgNO_3$, (b) $CaSO_4$, (c) K_2CO_3.

Strategy Use the guidelines in Tables 4.2 and 4.3 to determine whether or not each compound is expected to be water-soluble.

Setup (a) $AgNO_3$ contains the nitrate ion (NO_3^-). According to Table 4.2, *all* compounds containing the nitrate ion are soluble. (b) $CaSO_4$ contains the sulfate ion (SO_4^{2-}). According to Table 4.2, compounds containing the sulfate ion are soluble unless the cation is Ag^+, Hg_2^{2+}, Pb^{2+}, Ca^{2+}, Sr^{2+}, or Ba^{2+}. Thus, the Ca^{2+} ion is one of the insoluble exceptions. (c) K_2CO_3 contains an alkali metal cation (K^+) for which, according to Table 4.2, there are no insoluble exceptions. Alternatively, Table 4.3 shows that most compounds containing the carbonate ion (CO_3^{2-}) are insoluble—but compounds containing a Group 1 cation such as K^+ are soluble exceptions.

Solution (a) soluble, (b) insoluble, (c) soluble

> ## THINK ABOUT IT
>
> Check the ions in each compound against the information in Tables 4.2 and 4.3 to confirm that you have drawn the right conclusions.

Practice Problem ATTEMPT Classify each of the following compounds as soluble or insoluble in water: (a) $PbCl_2$, (b) $(NH_4)_3PO_4$, (c) $Fe(OH)_3$.

Practice Problem BUILD Classify each of the following compounds as soluble or insoluble in water: (a) $MgBr_2$, (b) $Ca_3(PO_4)_2$, (c) $KClO_3$.

Practice Problem CONCEPTUALIZE Using Tables 4.2 and 4.3, identify a compound that will cause precipitation of two different insoluble ionic compounds when an aqueous solution of it is added to an aqueous solution of iron(III) sulfate.

Molecular Equations

The reaction shown in Figure 4.3 can be represented with the chemical equation

$$Pb(NO_3)_2(aq) + 2NaI(aq) \longrightarrow 2NaNO_3(aq) + PbI_2(s)$$

Based on this chemical equation, the metal cations seem to exchange anions. That is, the Pb^{2+} ion, originally paired with NO_3^- ions, ends up paired with I^- ions; similarly, each Na^+ ion, originally paired with an I^- ion, ends up paired with an NO_3^- *ion*. This equation, as written, is called a ***molecular equation,*** which is a chemical equation written with all compounds represented by their chemical formulas, making it look as though they exist in solution as molecules or formula units.

> **Student Note:** Reactions in which compounds exchange ions are sometimes called *metathesis* or *double replacement* reactions.

You now know enough chemistry to predict the products of this type of chemical reaction! Simply write the formulas for the reactants, and then write formulas for the compounds that would form if the cations in the reactants were to trade anions. For example, if you want to write the equation for the reaction that occurs when solutions of sodium sulfate and barium hydroxide are combined, you would first write the formulas of the reactants [◄◄ Section 2.6]:

$$Na_2SO_4(aq) + Ba(OH)_2(aq) \longrightarrow$$

Then you would write the formula for one product by combining the cation from the first reactant (Na^+) with the anion from the second (OH^-); and write the formula for the other product by combining the cation from the second reactant (Ba^{2+}) with the anion from the first (SO_4^{2-}). Thus, the equation is

$$Na_2SO_4(aq) + Ba(OH)_2(aq) \longrightarrow 2NaOH + BaSO_4$$

Although we have balanced the equation [◄◄ Section 3.3], we have not yet put phases in parentheses for the products.

The final step in predicting the outcome of such a reaction is to determine which of the products, if any, will precipitate from solution. We do this using the solubility guidelines for ionic compounds (Tables 4.2 and 4.3). The first product (NaOH) contains a Group 1 cation (Na^+) and will therefore be soluble. We indicate its phase as (*aq*). The second product ($BaSO_4$) contains the sulfate ion (SO_4^{2-}). Sulfate compounds are soluble unless the cation is Ag^+, Hg_2^{2+}, Pb^{2+}, Ca^{2+}, Sr^{2+}, or Ba^{2+}. $BaSO_4$ is therefore insoluble and will precipitate. We indicate its phase as (*s*):

$$Na_2SO_4(aq) + Ba(OH)_2(aq) \longrightarrow 2NaOH(aq) + BaSO_4(s)$$

Ionic Equations

Although molecular equations are useful, especially from the standpoint of knowing which solutions to combine in the laboratory, they are in a sense unrealistic. Soluble ionic compounds are *strong electrolytes* [◀◀ Section 4.1]. As such, they exist in solution as hydrated *ions,* rather than as formula units. Thus, it would be more realistic to represent the aqueous species in the reaction of $Na_2SO_4(aq)$ with $Ba(OH)_2(aq)$ as follows:

$$Na_2SO_4(aq) \longrightarrow 2Na^+(aq) + SO_4^{2-}(aq)$$

$$Ba(OH)_2(aq) \longrightarrow Ba^{2+}(aq) + 2OH^-(aq)$$

$$NaOH(aq) \longrightarrow Na^+(aq) + OH^-(aq)$$

If we were to rewrite the equation, representing the dissolved compounds as hydrated ions, it would be

$$2Na^+(aq) + SO_4^{2-}(aq) + Ba^{2+}(aq) + 2OH^-(aq) \longrightarrow$$
$$2Na^+(aq) + 2OH^-(aq) + BaSO_4(s)$$

This version of the equation is called an **ionic equation,** a chemical equation in which any compound that exists completely or predominantly as ions in solution is represented as those ions. Species that are insoluble or that exist in solution completely or predominantly as molecules are represented with their chemical formulas, as they were in the molecular equation.

Net Ionic Equations

$Na^+(aq)$ and $OH^-(aq)$ both appear as reactants and products in the ionic equation for the reaction of $Na_2SO_4(aq)$ with $Ba(OH)_2(aq)$. Ions that appear on both sides of the equation arrow are called **spectator ions** because they do not participate in the reaction. Spectator ions cancel one another, just as identical terms on both sides of an algebraic equation cancel one another, so we need not show spectator ions in chemical equations.

$$2\cancel{Na^+(aq)} + SO_4^{2-}(aq) + Ba^{2+}(aq) + 2\cancel{OH^-(aq)} \longrightarrow$$
$$2\cancel{Na^+(aq)} + 2\cancel{OH^-(aq)} + BaSO_4(s)$$

Eliminating the spectator ions yields the following equation:

$$Ba^{2+}(aq) + SO_4^{2-}(aq) \longrightarrow BaSO_4(s)$$

This version of the equation is called a **net ionic equation,** which is a chemical equation that includes only the species that are actually involved in the reaction. The net ionic equation, in effect, tells us what actually happens when we combine solutions of sodium sulfate and barium hydroxide.

The steps necessary to determine the molecular, ionic, and net ionic equations for a precipitation reaction are as follows:

1. Write and balance the molecular equation, predicting the products by assuming that the cations trade anions.
2. Write the ionic equation by separating strong electrolytes into their constituent ions.
3. Write the net ionic equation by identifying and canceling spectator ions on both sides of the equation.

If both products of a reaction are strong electrolytes, all the ions in solution are spectator ions. In this case, there is no net ionic equation and no reaction takes place.

Sample Problem 4.3 illustrates the stepwise determination of molecular, ionic, and net ionic equations.

Student Note: Although the reactants may be written in either order in the net ionic equation, it is common for the cation to be shown first and the anion second.

SAMPLE PROBLEM 4.3

Write the molecular, ionic, and net ionic equations for the reaction that occurs when aqueous solutions of lead acetate [$Pb(C_2H_3O_2)_2$] and calcium chloride ($CaCl_2$) are combined.

Strategy Predict the products by exchanging ions and balance the equation. Determine which product will precipitate based on the solubility guidelines in Tables 4.2 and 4.3. Rewrite the equation showing strong electrolytes as ions. Identify and cancel spectator ions.

Setup The products of the reaction are $PbCl_2$ and $Ca(C_2H_3O_2)_2$. $PbCl_2$ is insoluble, because Pb^{2+} is one of the insoluble exceptions for chlorides, which are generally soluble. $Ca(C_2H_3O_2)_2$ is soluble because all acetates are soluble.

Solution Molecular equation:

$$Pb(C_2H_3O_2)_2(aq) + CaCl_2(aq) \longrightarrow PbCl_2(s) + Ca(C_2H_3O_2)_2(aq)$$

Ionic equation:

$$Pb^{2+}(aq) + 2C_2H_3O_2^-(aq) + Ca^{2+}(aq) + 2Cl^-(aq) \longrightarrow PbCl_2(s) + Ca^{2+}(aq) + 2C_2H_3O_2^-(aq)$$

Net ionic equation:

$$Pb^{2+}(aq) + 2Cl^-(aq) \longrightarrow PbCl_2(s)$$

THINK ABOUT IT

Remember that the charges on ions in a compound must sum to zero. Make sure that you have written correct formulas for the products and that each of the equations you have written is balanced. If you find that you are having trouble balancing an equation, check to make sure you have correct formulas for the products.

Practice Problem **A**TTEMPT Write the molecular, ionic, and net ionic equations for the combination of $Sr(NO_3)_2(aq)$ and $Li_2SO_4(aq)$.

Practice Problem **B**UILD Write the molecular, ionic, and net ionic equations for the combination of $KNO_3(aq)$ and $BaCl_2(aq)$.

Practice Problem **C**ONCEPTUALIZE Which diagram best represents the result when equal volumes of equal-concentration aqueous solutions of barium nitrate and potassium phosphate are combined?

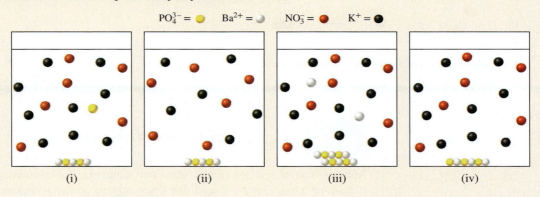

PO_4^{3-} = ⬤ Ba^{2+} = ⬤ NO_3^- = ⬤ K^+ = ⬤

(i)   (ii)   (iii)   (iv)

CHECKPOINT – SECTION 4.2   Precipitation Reactions

4.2.1 Which of the following are water-soluble? (Choose all that apply.)

a) Na_2S

b) $Ba(C_2H_3O_2)_2$

c) $CaCO_3$

d) $CuBr_2$

e) Hg_2Cl_2

4.2.2 Which of the following are water-insoluble? (Choose all that apply.)

a) Ag_2CrO_4

b) Li_2CO_3

c) $Ca_3(PO_4)_2$

d) $BaSO_4$

e) $ZnCl_2$

4.2.3 What are the spectator ions in the ionic equation for the combination of $Li_2CO_3(aq)$ and $Ba(OH)_2(aq)$?

a) CO_3^{2-} and OH^-

b) Li^+ and OH^-

c) Li^+ and Ba^{2+}

d) Ba^{2+} and OH^-

e) Ba^{2+} and CO_3^{2-}

4.2.4 Select the correct net ionic equation for the combination of $Fe(NO_3)_2(aq)$ and $Na_2CO_3(aq)$.

a) $Na^+(aq) + CO_3^{2-}(aq) \longrightarrow NaCO_3(s)$

b) $Fe^{2+}(aq) + CO_3^{2-}(aq) \longrightarrow FeCO_3(s)$

c) $2Na^+(aq) + CO_3^{2-}(aq) \longrightarrow Na_2CO_3(s)$

d) $Fe^{2+}(aq) + 2NO_3^-(aq) \longrightarrow Fe(NO_3)_2(s)$

e) $Na^+(aq) + NO_3^-(aq) \longrightarrow NaNO_3(s)$

4.2.5 Which reaction is represented by the net ionic equation for the combination of aqueous solutions of LiOH and $Cu(NO_3)_2$?

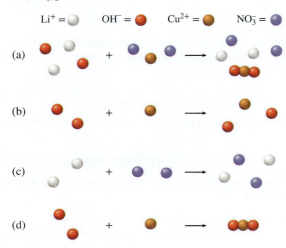

(e)   There is no net ionic equation. No reaction occurs.

4.2.6 Which reaction is represented by the net ionic equation for the combination of aqueous solutions of $LiNO_3$ and $NaC_2H_3O_2$?

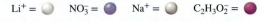

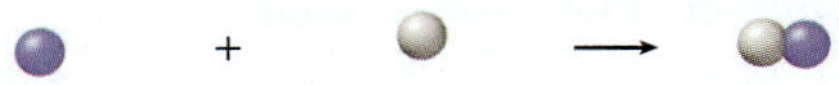

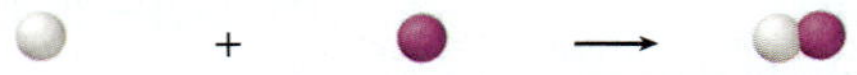

(e)   There is no net ionic equation. No reaction occurs.

4.3   # Acid-Base Reactions

Another type of reaction occurs when two solutions, one containing an acid and one containing a base, are combined. We frequently encounter acids and bases in everyday life (Figure 4.5). Ascorbic acid, for instance, is also known as vitamin C; acetic acid is the component responsible for the sour taste and characteristic smell of vinegar; and hydrochloric acid is the acid in muriatic acid and is also the principal ingredient in gastric juice (stomach acid). Ammonia, found in many cleaning products, and sodium hydroxide, found in drain cleaner, are common bases. Acid-base chemistry is extremely important to biological processes. Let's look again at the properties of acids and bases, and then look at acid-base reactions.

Strong Acids and Bases

As we saw in Section 4.1, the seven strong acids—those that ionize completely in solution—are listed in Table 4.1. All other acids are weak acids. The strong bases are the hydroxides of Group 1 and heavy Group 2 metals. These are soluble ionic compounds,

Figure 4.5 Some common acids and bases. From left to right: Sodium hydroxide (NaOH), ascorbic acid ($C_6H_8O_6$ or, with its ionizable hydrogens written first, $H_2C_6H_6O_6$), hydrochloric acid (HCl), acetic acid ($HC_2H_3O_2$), and ammonia (NH_3). HCl and NaOH are both strong electrolytes and exist in solution entirely as ions. Water molecules are not shown.

David A. Tietz/Editorial Image, LLC

TABLE 4.4	Strong Acids and Strong Bases	
Strong Acids		**Strong Bases**
HCl		LiOH
HBr		NaOH
HI		KOH
HNO_3		RbOH
$HClO_3$		CsOH
$HClO_4$		$Ca(OH)_2$
H_2SO_4		$Sr(OH)_2$
		$Ba(OH)_2$

which dissociate completely and exist entirely as ions in solution. Thus, both strong acids and strong bases are strong electrolytes. Table 4.4 lists the strong acids and strong bases. It is important that you know these compounds.

Student Note: Although three of the Group 2 hydroxides [$Ca(OH)_2$, $Sr(OH)_2$, and $Ba(OH)_2$] are typically classified as strong bases, only $Ba(OH)_2$ is sufficiently soluble to be used commonly in the laboratory. For any ionic compound, what does dissolve—even if it is only a tiny amount—dissociates completely.

Brønsted Acids and Bases

In Section 2.7, we defined an acid as a substance that ionizes in water to produce H^+ ions, and a base as a substance that ionizes (or dissociates, in the case of an ionic base)

Student Note: Acid-base neutralization reactions, like precipitation reactions [◄◄ Section 4.2], are *metathesis* reactions, where two species exchange ions.

The following are also examples of acid-base neutralization reactions, represented by molecular equations:

$$HNO_3(aq) + KOH(aq) \longrightarrow H_2O(l) + KNO_3(aq)$$

$$H_2SO_4(aq) + 2NaOH(aq) \longrightarrow 2H_2O(l) + Na_2SO_4(aq)$$

$$2HC_2H_3O_2(aq) + Ba(OH)_2(aq) \longrightarrow 2H_2O(l) + Ba(C_2H_3O_2)_2(aq)$$

$$HCl(aq) + NH_3(aq) \longrightarrow NH_4Cl(aq)$$

The last equation looks different because it does not show water as a product. Recall, however, that $NH_3(aq)$ ionizes to give $NH_4^+(aq)$ and $OH^-(aq)$. If we include these two species as reactants in place of $NH_3(aq)$, the equation becomes

$$HCl(aq) + NH_4^+(aq) + OH^-(aq) \longrightarrow H_2O(l) + NH_4Cl(aq)$$

Sample Problem 4.4 involves an acid-base neutralization reaction.

SAMPLE PROBLEM 4.4

Milk of magnesia, an over-the-counter laxative, is a mixture of magnesium hydroxide [$Mg(OH)_2$] and water. Because $Mg(OH)_2$ is insoluble in water (see Table 4.3), milk of magnesia is a *suspension* rather than a solution. The undissolved solid is responsible for the milky appearance of the product. When acid such as HCl is added to milk of magnesia, the suspended $Mg(OH)_2$ dissolves, and the result is a clear, colorless solution. Write and balance the molecular equation, and then give the ionic and net ionic equations for this reaction.

Student Note: Most suspended solids will settle to the bottom of the bottle, making it necessary to "shake well before using." Shaking redistributes the solid throughout the liquid.

  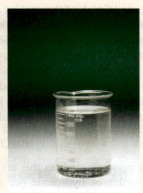

(a) Milk of magnesia   (b) Addition of HCl   (c) Resulting clear solution

(a), (b), (c): Charles D. Winters/McGraw Hill

Strategy Determine the products of the reaction; then write and balance the equation. Remember that one of the reactants, $Mg(OH)_2$, is a solid. Identify any strong electrolytes and rewrite the equation showing strong electrolytes as ions. Identify and cancel the spectator ions.

Setup Because this is an acid-base neutralization reaction, one of the products is water. The other product is a salt comprising the cation from the base, Mg^{2+}, and the anion from the acid, Cl^-. For the formula to be neutral, these ions combine in a 1:2 ratio, giving $MgCl_2$ as the formula of the salt.

Solution

$$Mg(OH)_2(s) + 2HCl(aq) \longrightarrow 2H_2O(l) + MgCl_2(aq)$$

Of the species in the molecular equation, only HCl and $MgCl_2$ are strong electrolytes. Therefore, the ionic equation is

$$Mg(OH)_2(s) + 2H^+(aq) + 2Cl^-(aq) \longrightarrow 2H_2O(l) + Mg^{2+}(aq) + 2Cl^-(aq)$$

Cl^- is the only spectator ion. The net ionic equation is

$$Mg(OH)_2(s) + 2H^+(aq) \longrightarrow 2H_2O(l) + Mg^{2+}(aq)$$

THINK ABOUT IT

Make sure your equation is balanced and that you only show strong electrolytes as ions. $Mg(OH)_2$ is *not* shown as aqueous ions because it is insoluble.

Practice Problem **A**TTEMPT Write and balance the molecular equation and then give the ionic and net ionic equations for the neutralization reaction between $Ba(OH)_2(aq)$ and $HF(aq)$.

Practice Problem **B**UILD Write and balance the molecular equation and then give the ionic and net ionic equations for the neutralization reaction between $NH_3(aq)$ and $H_2SO_4(aq)$.

Practice Problem **C**ONCEPTUALIZE Which diagram best represents the ions remaining in solution after stoichiometric amounts of aqueous barium hydroxide and hydrobromic acid are combined?

$OH^- = $ ⬤ $Ba^{2+} = $ ⬤ $H^+ = $ ⚪ $Br^- = $ ⬤

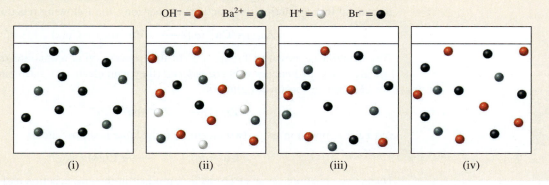

(i)　　　　(ii)　　　　(iii)　　　　(iv)

CHECKPOINT – SECTION 4.3　　Acid-Base Reactions

4.3.1 Identify the Brønsted acid in the following equation:

$$H_2SO_4(aq) + 2NH_3(aq) \longrightarrow (NH_4)_2SO_4(aq)$$

a) $H_2SO_4(aq)$

b) $NH_3(aq)$

c) $H_2O(l)$

d) $(NH_4)_2SO_4$

e) This equation does not contain a Brønsted acid.

4.3.2 Identify the Brønsted base in the following equation:

$$HCl(aq) + NO_2^-(aq) \longrightarrow HNO_2(aq) + Cl^-(aq)$$

a) $HCl(aq)$

b) $NO_2^-(aq)$

c) $HNO_2(aq)$

d) $Cl^-(aq)$

e) $H_2O(l)$

4.3.3 Which of the following is the correct net ionic equation for the reaction of H_2SO_4 and KOH?

a) $H_2(aq) + 2OH^-(aq) \longrightarrow 2H_2O(l)$

b) $2H^+(aq) + 2OH^-(aq) \longrightarrow 2H_2O(l)$

c) $2H^+(aq) + OH^-(aq) \longrightarrow H_2O(l)$

d) $H_2SO_4(aq) + 2OH^-(aq) \longrightarrow 2H_2O(l) + SO_4^{2-}(aq)$

e) $H^+(aq) + HSO_4^-(aq) + 2OH^-(aq) \longrightarrow$ $2H_2O(l) + SO_4^{2-}(aq)$

4.3.4 Which of the following is the correct net ionic equation for the reaction of HF and LiOH?

a) $HF(aq) + LiOH(aq) \longrightarrow H_2O(aq) + LiF(aq)$

b) $H^+(aq) + F^-(aq) + Li^+(aq) + OH^-(aq) \longrightarrow$ $H_2O(aq) + Li^+(aq) + F^-(aq)$

c) $H^+(aq) + OH^-(aq) \longrightarrow H_2O(l)$

d) $HF(aq) + OH^-(aq) \longrightarrow H_2O(l) + F^-(aq)$

e) $H^+(aq) + Li^+(aq) + OH^-(aq) \longrightarrow$ $H_2O(l) + Li^+(aq)$

Use the following diagrams to answer Checkpoint questions 4.3.5 and 4.3.6.

$OH^- = $ ⬤ $Ba^{2+} = $ ⬤ $H^+ = $ ⚪ $Cl^- = $ 🟢

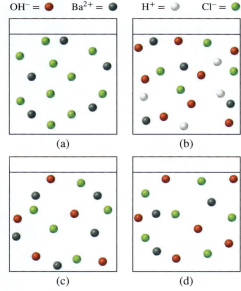

(a)　　　　(b)

(c)　　　　(d)

4.3.5 Which diagram best represents the ions remaining in solution when equimolar amounts of aqueous barium hydroxide and hydrochloric acid are combined?

4.3.6 Which diagram best represents the ions remaining in solution when stoichiometric amounts of aqueous barium hydroxide and hydrochloric acid are combined?

4.4   Oxidation-Reduction Reactions

In Sections 4.2 and 4.3, we encountered two types of chemical reactions that can occur when two electrolyte solutions are combined: *precipitation,* in which ionic compounds exchange ions, and *acid-base neutralization,* in which a proton is transferred from an acid to a base. In this section, we introduce **oxidation-reduction reactions,** commonly called *redox* reactions. A **redox reaction** is a chemical reaction in which *electrons* are transferred from one reactant to another. For example, if we place a piece of zinc metal into a solution that contains copper ions, the following reaction will occur:

$$Zn(s) + Cu^{2+}(aq) \longrightarrow Zn^{2+}(aq) + Cu(s)$$

This reaction is shown in Figure 4.6. In this process, zinc atoms are *oxidized* (they lose electrons) and copper ions are *reduced* (they gain electrons). Each zinc atom loses two electrons to become a zinc ion,

$$Zn(s) \longrightarrow Zn^{2+}(aq) + 2e^-$$

and each copper ion gains two electrons to become a copper atom.

$$Cu^{2+}(aq) + 2e^- \longrightarrow Cu(s)$$

These two equations show electrons as a product in the zinc reaction and as a reactant in the copper reaction. Each of these two equations represents a **half-reaction,** the oxidation or the reduction reaction in a redox reaction. The sum of the two half-reaction equations is the overall equation for the redox reaction:

$$Zn(s) \longrightarrow Zn^{2+}(aq) + 2e^-$$
$$+ \; Cu^{2+}(aq) + 2e^- \longrightarrow Cu(s)$$
$$\overline{Zn(s) + Cu^{2+}(aq) + \cancel{2e^-} \longrightarrow Zn^{2+}(aq) + Cu(s) + \cancel{2e^-}}$$

Figure 4.6 Oxidation of zinc in a solution of copper(II) sulfate.

(a), (b): Charles D. Winters/McGraw Hill

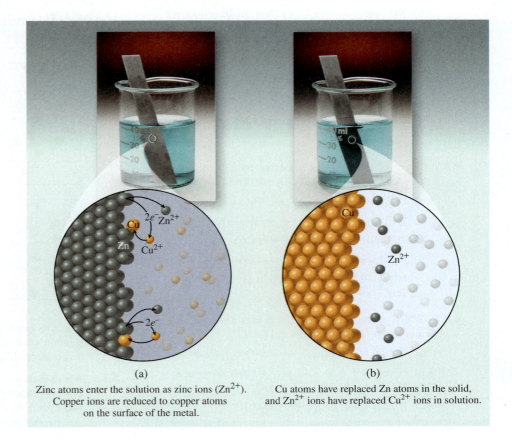

(a)

Zinc atoms enter the solution as zinc ions (Zn^{2+}). Copper ions are reduced to copper atoms on the surface of the metal.

(b)

Cu atoms have replaced Zn atoms in the solid, and Zn^{2+} ions have replaced Cu^{2+} ions in solution.

Although these two processes can be represented by separate equations, they cannot occur separately. For one species to gain electrons, another must lose them, and vice versa.

   Oxidation is the *loss* of electrons. The opposite process, the *gain* of electrons, is called **reduction.** In the reaction of Zn with Cu^{2+}, Zn is called the **reducing agent** because it donates electrons, causing Cu^{2+} to be reduced. Cu^{2+} is called the **oxidizing agent,** on the other hand, because it accepts electrons, causing Zn to be oxidized.

   Another example of a redox reaction is the formation of calcium oxide (CaO) from its constituent elements.

$$2Ca(s) + O_2(g) \longrightarrow 2CaO(s)$$

In this reaction, each calcium atom loses two electrons (is oxidized) and each oxygen atom gains two electrons (is reduced). The corresponding half-reactions are

$$2Ca \longrightarrow 2Ca^{2+} + 4e^-$$
$$O_2 + 4e^- \longrightarrow 2O^{2-}$$

The resulting Ca^{2+} and O^{2-} ions combine to form CaO.

   Redox reactions take place because atoms of different elements have different tendencies to gain electrons. Oxygen, for instance, has a much greater tendency to gain electrons than does calcium. Calcium, being a metal, has a significant tendency to lose electrons. Compounds that form between elements with significantly different tendencies to gain electrons generally are ionic. By knowing the charges on the monatomic ions in such compounds, we can keep track of the electrons that have been lost and gained.

Oxidation Numbers

When elements of similar abilities to gain electrons combine, they tend to form molecular compounds, as in the formation of HF and NH_3 from their respective elements:

$$H_2(g) + F_2(g) \longrightarrow 2HF(g)$$
$$N_2(g) + 3H_2(g) \longrightarrow 2NH_3(g)$$

In the formation of hydrogen fluoride (HF), therefore, fluorine does not gain an electron per se—and hydrogen does not lose one. Experimental evidence shows, however, that there is a partial transfer of electrons from H to F. Oxidation numbers provide us with a way to "balance the books" with regard to electrons in a chemical equation. The **oxidation number,** also called the **oxidation state,** is the charge an atom would have *if* electrons were transferred completely. For example, we can rewrite the preceding equations for the formation of HF and NH_3 as follows:

$H_2(g)$	+	$F_2(g)$	$\longrightarrow$	$2HF(g)$
0		0		+1 −1

$N_2(g)$	+	$3H_2(g)$	$\longrightarrow$	$2NH_3(g)$
0		0		−3 +1

The numbers below each element are the oxidation numbers. In both of the reactions shown, the reactants are all homonuclear diatomic molecules. Thus, we would expect *no* transfer of electrons from one atom to the other and the oxidation number of each is zero. For the product molecules, however, for the sake of determining oxidation numbers, we assume that *complete* electron transfer has taken place and that each atom has either gained or lost one or more electrons. The oxidation numbers reflect the number of electrons assumed to have been transferred.

   Oxidation numbers enable us to identify elements that are oxidized and reduced at a glance. The elements that show an *increase* in oxidation number—hydrogen in the preceding examples—are oxidized, whereas the elements that show a *decrease* in oxidation number—fluorine and nitrogen—are reduced.

Student Note: The term *oxidation* originally was used by chemists to mean "reaction with oxygen." Like definitions of acids and bases, though, it has been redefined to include any reaction in which electrons are lost.

Animation
Chemical reactions—formation of Ag_2S by oxidation reduction.

In a sense, each atom's oxidation number makes a contribution to the overall charge on the species. Note that the oxidation numbers in both HF [(+1) + (−1) = 0] and NH_3 [(−3) + 3(+1) = 0] sum to zero. Because compounds are electrically neutral, the oxidation numbers in any compound will sum to zero. For a polyatomic ion, oxidation numbers must sum to the charge on the ion. (The oxidation number of a monatomic ion is equal to its charge.)

The following guidelines will help you assign oxidation numbers. There are essentially two rules:

1. The oxidation number of any element, in its elemental form, is zero.
2. The oxidation numbers in any chemical species must sum to the overall charge on the species. That is, oxidation numbers must sum to zero for any molecule and must sum to the charge on any polyatomic ion. The oxidation number of a monatomic ion is equal to the charge on the ion.

In addition to these two rules, it is necessary to know the elements that always, or nearly always, have the same oxidation number. Table 4.5 lists elements whose oxidation numbers are "reliable," in order of decreasing reliability.

To determine oxidation numbers in a compound or a polyatomic ion, you must use a stepwise, systematic approach. Draw a circle under each element's symbol in the chemical formula. Then draw a square under each circle. In the circle, write the oxidation number of the element; in the square, write the total contribution to charge by that element. Start with the oxidation numbers you know, and use them to figure out the ones you don't know. Here is an example:

$$KMnO_4$$

Oxidation number
Total contribution to charge

Fill in the oxidation number first for the element that appears highest on the list in Table 4.5. Potassium (K) is a Group 1 metal. In its compounds, it always has the oxidation number +1. We write +1 in the circle beneath the K. Because there is only one K atom in this formula, the total contribution to charge is also +1, so we also write +1 in the square beneath the K.

$$KMnO_4$$

Oxidation number
Total contribution to charge

TABLE 4.5	Elements with Reliable Oxidation Numbers in Compounds or Polyatomic Ions	

Element	Oxidation Number	Exceptions
Fluorine	−1	
Group 1 or 2 metal	+1 or +2, respectively	
Hydrogen	+1	Any combination with a Group 1 or 2 metal to form a metal hydride. Examples: LiH and CaH_2—the oxidation number of H is −1 in both examples.
Oxygen	−2	Any combination with something higher on the list that necessitates its having a different oxidation number (see rule 2 for assigning oxidation numbers). Examples: H_2O_2 and KO_2—the oxidation number of O for H_2O_2 is −1 and for KO_2 is $-\frac{1}{2}$.
Group 17 (other than fluorine)	−1	Any combination with something higher on the list that necessitates its having a different oxidation number (see rule 2 for assigning oxidation numbers). Examples: ClF, BrO_4^-, and IO_3^-—the oxidation numbers of Cl, Br, and I are +1, +7, and +5, respectively. Remember that these exceptions do not apply to fluorine, which *always* has an oxidation state of −1 when it is part of a compound.

Next on the list is oxygen (O). In compounds, O usually has the oxidation number −2, so we assign it −2. Because there are four O atoms in the formula, the total contribution to charge by O atoms is 4(−2) = −8.

	$KMnO_4$		
Oxidation number	+1		−2
Total contribution to charge	+1		−8

The numbers in the squares, all the contributions to overall charge, must sum to zero. This requires putting +7 in the box beneath the Mn atom. Because there is just one Mn atom in this formula, the contribution to charge is the same as the oxidation number. Thus, (+1) + (+7) + (−8) = 0.

	$KMnO_4$		
Oxidation number	+1	+7	−2
Total contribution to charge	+1	+7	−8

Student Hot Spot

Student data indicate you may struggle with oxidation numbers. Access the eBook to view additional Learning Resources on this topic.

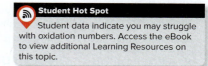

Sample Problem 4.5 lets you determine oxidation numbers in three more compounds and a polyatomic ion.

SAMPLE PROBLEM 4.5

Determine the oxidation number of each atom in the following compounds and ion: (a) SO_2, (b) NaH, (c) CO_3^{2-}, (d) N_2O_5.

Strategy For each compound, assign an oxidation number first to the element that appears higher in Table 4.5. Then use rule 2 to determine the oxidation number of the other element.

Setup (a) O appears in Table 4.5 but S does not, so we assign oxidation number −2 to O. Because there are two O atoms in the molecule, the total contribution to charge by O is 2(−2) = −4. The lone S atom must therefore contribute +4 to the overall charge. (b) Both Na and H appear in Table 4.5, but Na appears higher in the table, so we assign the oxidation number +1 to Na. This means that H must contribute −1 to the overall charge. (H^- is the hydride ion.) (c) We assign the oxidation number −2 to O. Because there are three O atoms in the carbonate ion, the total contribution to charge by O is −6. To have the contributions to charge sum to the charge on the ion (−2), the C atom must contribute +4. (d) We assign the oxidation number −2 to O. Because there are five O atoms in the N_2O_5 molecule, the total contribution to charge by O is −10. To have the contributions to charge sum to zero, the contribution by N must be +10, and because there are two N atoms, each one must contribute +5. Therefore, the oxidation number of N is +5.

Solution

(a) In SO_2, the oxidation numbers of S and O are +4 and −2, respectively.

SO_2	
+4	−2
+4	−4

(b) In NaH, the oxidation numbers of Na and H are +1 and −1, respectively.

NaH	
+1	−1
+1	−1

(c) In CO_3^{2-}, the oxidation numbers of C and O are +4 and −2, respectively.

CO_3^{2-}	
+4	−2
+4	−6

(Continued on next page)

(d) In N_2O_5, the oxidation numbers of N and O are +5 and −2, respectively.

$$N_2O_5$$

$$\boxed{\overset{(+5)}{+10}} \boxed{\overset{(-2)}{-10}}$$

THINK ABOUT IT

Use the circle and square system to verify that the oxidation numbers you have assigned do indeed sum to the overall charge on each species.

Practice Problem ATTEMPT Assign oxidation numbers to each atom in the following compounds: H_2O_2, MnO_2, H_2SO_4.

Practice Problem BUILD Assign oxidation numbers to each atom in the following polyatomic ions: O_2^{2-}, ClO^-, ClO_3^-.

Practice Problem CONCEPTUALIZE Write the balanced equation for the reaction represented by the models and determine oxidation states for each element before and after the reaction.

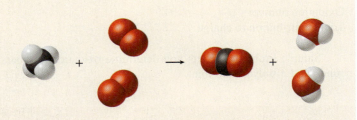

Oxidation of Metals in Aqueous Solutions

Recall from the beginning of this section that zinc metal reacts with aqueous copper ions to form aqueous zinc ions and copper metal. One way that this reaction might be carried out is for zinc metal to be immersed in a solution of copper(II) sulfate ($CuSO_4$), as depicted in Figure 4.6. The molecular equation for this reaction is

$$Zn(s) \quad + \quad CuCl_2(aq) \quad \longrightarrow \quad ZnCl_2(aq) \quad + \quad Cu(s)$$

$$\boxed{\overset{(0)}{0}} \qquad \boxed{\overset{(+2)}{+2}}\boxed{\overset{(-1)}{-2}} \qquad \boxed{\overset{(+2)}{+2}}\boxed{\overset{(-1)}{-2}} \qquad \boxed{\overset{(0)}{0}}$$

This is an example of a **displacement reaction.** Zinc *displaces,* or *replaces,* copper in the dissolved salt by being oxidized from Zn to Zn^{2+}. Copper is displaced from the salt (and removed from solution) by being reduced from Cu^{2+} to Cu. Chloride (Cl^-), which is neither oxidized nor reduced, is a spectator ion in this reaction.

What would happen, then, if we placed copper metal into a solution containing zinc chloride ($ZnCl_2$)? Would $Cu(s)$ be oxidized to $Cu^{2+}(aq)$ by $Zn^{2+}(aq)$ the way $Zn(s)$ is oxidized to $Zn^{2+}(aq)$ by $Cu^{2+}(aq)$? The answer is no. In fact, no reaction would occur if we were to immerse copper metal into an aqueous solution of $ZnCl_2$.

$$Cu(s) + ZnCl_2(aq) \longrightarrow \text{no reaction}$$

No reaction occurs between $Cu(s)$ and $Zn^{2+}(aq)$, whereas a reaction does occur between $Zn(s)$ and $Cu^{2+}(aq)$ because zinc is more easily oxidized than copper.

The **activity series** (Table 4.6) is a list of metals (and hydrogen) arranged from top to bottom in order of decreasing ease of oxidation. The second column shows the oxidation half-reaction corresponding to each element in the first column. Note the positions of zinc and copper in the table. Zinc appears higher in the table and is therefore oxidized more easily. In fact, an element in the series will be oxidized by the ions of any element that appears below it. According to Table 4.6, therefore, zinc metal will be oxidized by a solution containing any of the following ions: Cr^{3+}, Fe^{2+}, Cd^{2+}, Co^{2+}, Ni^{2+}, Sn^{2+}, H^+, Cu^{2+}, Ag^+, Hg^{2+}, Pt^{2+}, or Au^{3+}. On the other hand, zinc will not be *oxidized* by a solution containing Mn^{2+}, Al^{3+}, Mg^{2+}, Na^+, Ca^{2+}, Ba^{2+}, K^+, or Li^+ ions.

Some metals such as copper are so unreactive that they are found in nature in the uncombined state.

Charles D. Winters/McGraw Hill

Student Note: When a metal is *oxidized* by an aqueous solution, it becomes an aqueous ion.

TABLE 4.6	Activity Series	
Element	**Oxidation Half-Reaction**	
Lithium	$Li \longrightarrow Li^+ + e^-$	
Potassium	$K \longrightarrow K^+ + e^-$	
Barium	$Ba \longrightarrow Ba^{2+} + 2e^-$	
Calcium	$Ca \longrightarrow Ca^{2+} + 2e^-$	
Sodium	$Na \longrightarrow Na^+ + e^-$	
Magnesium	$Mg \longrightarrow Mg^{2+} + 2e^-$	
Aluminum	$Al \longrightarrow Al^{3+} + 3e^-$	
Manganese	$Mn \longrightarrow Mn^{2+} + 2e^-$	
Zinc	$Zn \longrightarrow Zn^{2+} + 2e^-$	
Chromium	$Cr \longrightarrow Cr^{3+} + 3e^-$	
Iron	$Fe \longrightarrow Fe^{2+} + 2e^-$	
Cadmium	$Cd \longrightarrow Cd^{2+} + 2e^-$	
Cobalt	$Co \longrightarrow Co^{2+} + 2e^-$	
Nickel	$Ni \longrightarrow Ni^{2+} + 2e^-$	
Tin	$Sn \longrightarrow Sn^{2+} + 2e^-$	
Lead	$Pb \longrightarrow Pb^{2+} + 2e^-$	
Hydrogen	$H_2 \longrightarrow 2H^+ + 2e^-$	
Copper	$Cu \longrightarrow Cu^{2+} + 2e^-$	
Silver	$Ag \longrightarrow Ag^+ + e^-$	
Mercury	$Hg \longrightarrow Hg^{2+} + 2e^-$	
Platinum	$Pt \longrightarrow Pt^{2+} + 2e^-$	
Gold	$Au \longrightarrow Au^{3+} + 3e^-$	

(Left margin, vertical arrow pointing up: *Increasing ease of oxidation*)

   Metals listed at the top of the activity series are called the *active metals.* These include the alkali and alkaline earth metals. These metals are so reactive that they are not found in nature in their elemental forms. Metals at the bottom of the series, such as copper, silver, platinum, and gold, are called the *noble metals* because they have very little tendency to react. These are the metals most often used for jewelry and coins. Reactions in which hydrogen ion is reduced to hydrogen gas are known as *hydrogen displacement* reactions.

Balancing Simple Redox Equations

To learn how to balance redox equations, let's revisit the practice of balancing equations. In Chapter 3, you learned to balance equations by counting the number of each kind of atom on each side of the equation arrow. For the purpose of balancing redox equations, it is also necessary to count electrons. For example, consider the net ionic equation for the reaction of chromium metal with nickel ion:

$$Cr(s) + Ni^{2+}(aq) \longrightarrow Cr^{3+}(aq) + Ni(s)$$

Although this equation has equal numbers of each type of atom on both sides, it is not balanced because there is a charge of $+2$ on the reactant side and a charge of $+3$ on the product side. To balance it, we can separate it into its half-reactions.

$$Cr(s) \longrightarrow Cr^{3+}(aq) + 3e^-$$
$$Ni^{2+}(aq) + 2e^- \longrightarrow Ni(s)$$

When we add half-reactions to get the overall reaction, the electrons must cancel. Because any electrons lost by one species must be gained by the other, electrons may *not* appear in an overall chemical equation. Therefore, prior to adding these two half-reactions, we must multiply the chromium half-reaction by 2

$$2[Cr(s) \longrightarrow Cr^{3+}(aq) + 3e^-]$$

and the nickel half-reaction by 3.

$$3[Ni^{2+}(aq) + 2e^- \longrightarrow Ni(s)]$$

Then when we add the half-reactions, the electrons cancel and we get the balanced overall equation.

$$2Cr(s) \longrightarrow 2Cr^{3+}(aq) + 6e^-$$
$$+\ 3Ni^{2+}(aq) + 6e^- \longrightarrow 3Ni(s)$$
$$\overline{2Cr(s) + 3Ni^{2+}(aq) \longrightarrow 2Cr^{3+}(aq) + 3Ni(s)}$$

This is known as the **half-reaction method** of balancing redox equations. We use this method extensively when we examine more complex redox reactions in Chapter 19.

The activity series enables us to predict whether or not a metal will be oxidized by a solution containing a particular salt or by an acid. Sample Problems 4.6 and 4.7 give you more practice making such predictions and balancing redox equations.

SAMPLE PROBLEM   4.6

Predict which of the following reactions will occur, and for those that will occur, write the net ionic equation and indicate which element is oxidized and which is reduced: (a) $Fe(s) + PtCl_2(aq) \longrightarrow$? (b) $Cr(s) + AuCl_3(aq) \longrightarrow$? (c) $Pb(s) + Zn(NO_3)_2(aq) \longrightarrow$?

Strategy The salt in each equation (the compound on the reactant side) is a strong electrolyte. What is important is the identity of the metal cation *in* the salt. For each equation, compare the positions in Table 4.6 of the solid metal and the metal cation from the salt to determine whether or not the solid metal will be oxidized. If the cation appears lower in the table, the solid metal will be oxidized (i.e., the reaction will occur). If the cation appears higher in the table, the solid metal will not be oxidized (i.e., no reaction will occur).

Setup (a) The cation in $PtCl_2$ is Pt^{2+}. Platinum appears lower in Table 4.6 than iron, so $Pt^{2+}(aq)$ will oxidize $Fe(s)$.

(b) The cation in $AuCl_3$ is Au^{3+}. Gold appears lower in Table 4.6 than chromium, so $Au^{3+}(aq)$ will oxidize $Cr(s)$.

(c) The cation in $Zn(NO_3)_2$ is Zn^{2+}. Zinc appears higher in Table 4.6 than lead, so $Zn^{2+}(aq)$ will not oxidize $Pb(s)$.

Solution (a) $Fe(s) + Pt^{2+}(aq) \longrightarrow Fe^{2+}(aq) + Pt(s)$; iron is oxidized (0 to +2) and platinum is reduced (+2 to 0).

(b) $Cr(s) + Au^{3+}(aq) \longrightarrow Cr^{3+}(aq) + Au(s)$; chromium is oxidized (0 to +3) and gold is reduced (+3 to 0).

(c) No reaction.

THINK ABOUT IT

Check your conclusions by working each problem backward. For part (b), for example, write the net ionic equation in reverse, using the products as the reactants: $Au(s) + Cr^{3+}(aq) \longrightarrow$? Now compare the positions of gold and chromium in Table 4.6 again. Chromium is higher, so chromium(III) ions cannot oxidize gold. This confirms your conclusion that the forward reaction (the oxidation of chromium by gold ions) will occur.

Practice Problem ATTEMPT Predict which of the following reactions will occur, and for those that will occur, write the net ionic equation and indicate which element is oxidized and which is reduced: (a) $Co(s) + BaI_2(aq) \longrightarrow$? (b) $Sn(s) + CuBr_2(aq) \longrightarrow$? (c) $Ag(s) + NaCl(aq) \longrightarrow$?

Practice Problem BUILD Predict which of the following reactions will occur, and for those that will occur, write the net ionic equation and indicate which element is oxidized and which is reduced: (a) $Ni(s) + Cu(NO_3)_2(aq) \longrightarrow$? (b) $Ag(s) + KCl(aq) \longrightarrow$? (c) $Al(s) + AuCl_3(aq) \longrightarrow$?

Practice Problem **C**ONCEPTUALIZE Given the following data, construct an activity series similar to Table 4.6 for five metals: A, B, C, D, and E. The data indicate the results of specific combinations of metals and metal ions.

Experiment 1: $A(s) + D^+(aq) \longrightarrow A^+(aq) + D(s)$

Experiment 2: $C(s) + B^+(aq) \longrightarrow C^+(aq) + B(s)$

Experiment 3: $D(s) + B^+(aq) \longrightarrow$ no reaction

Experiment 4: $C(s) + A^+(aq) \longrightarrow$ no reaction

Experiment 5: $B(s) + E^+(aq) \longrightarrow B^+(aq) + E(s)$

Experiment 6: $D(s) + E^+(aq) \longrightarrow$ no reaction

SAMPLE PROBLEM 4.7

Predict which of the following reactions will occur, and for those that will occur, balance the equation and indicate which element is oxidized and which is reduced: (a) $Al(s) + CaCl_2(aq) \longrightarrow$? (b) $Cr(s) + Pb(C_2H_3O_2)_2(aq) \longrightarrow$? (c) $Sn(s) + HI(aq) \longrightarrow$?

Strategy As in Sample Problem 4.6, identify the cation in the aqueous species and for each equation, compare the positions in Table 4.6 of the solid metal and the cation to determine whether or not the solid metal will be oxidized. If the cation appears lower in the table, the reaction will occur.

Setup (a) The cation in $CaCl_2$ is Ca^{2+}. Calcium appears higher in Table 4.6 than aluminum, so $Ca^{2+}(aq)$ will not oxidize $Al(s)$. (b) The cation in $Pb(C_2H_3O_2)_2$ is Pb^{2+}. Lead appears lower in Table 4.6 than chromium, so $Pb^{2+}(aq)$ will oxidize $Cr(s)$. (c) The cation in HI is H^+. Hydrogen appears lower in Table 4.6 than tin, so $H^+(aq)$ will oxidize $Sn(s)$.

Solution (a) No reaction.

(b) The two half-reactions are represented by the following:

Oxidation: $Cr(s) \longrightarrow Cr^{3+}(aq) + 3e^-$

Reduction: $Pb^{2+}(aq) + 2e^- \longrightarrow Pb(s)$

To balance the charges, we must multiply the oxidation half-reaction by 2 and the reduction half-reaction by 3:

$$2 \times [Cr(s) \longrightarrow Cr^{3+}(aq) + 3e^-] = 2Cr(s) \longrightarrow 2Cr^{3+}(aq) + 6e^-$$

$$3 \times [Pb^{2+}(aq) + 2e^- \longrightarrow Pb(s)] = 3Pb^{2+}(aq) + 6e^- \longrightarrow 3Pb(s)$$

We can then add the two half-reactions, canceling the electrons on both sides to get

$$2Cr(s) + 3Pb^{2+}(aq) \longrightarrow 2Cr^{3+}(aq) + 3Pb(s)$$

The overall, balanced molecular equation is

$$2Cr(s) + 3Pb(C_2H_3O_2)_2(aq) \longrightarrow 2Cr(C_2H_3O_2)_3(aq) + 3Pb(s)$$

Chromium is oxidized (0 to +3) and lead is reduced (+2 to 0).

(c) The two half-reactions are as follows:

Oxidation: $Sn(s) \longrightarrow Sn^{2+}(aq) + 2e^-$

Reduction: $2H^+(aq) + 2e^- \longrightarrow H_2(g)$

Adding the two half-reactions and canceling the electrons on both sides yields

$$Sn(s) + 2H^+(aq) \longrightarrow Sn^{2+}(aq) + H_2(g)$$

The overall, balanced molecular equation is

$$Sn(s) + 2HI(aq) \longrightarrow SnI_2(aq) + H_2(g)$$

Tin is oxidized (0 to +2) and hydrogen is reduced (+1 to 0).

THINK ABOUT IT

Check your conclusions by working each problem backward. Write each equation in reverse and compare the positions of the elements in the activity series.

(Continued on next page)

Practice Problem Ⓐ**TTEMPT** Predict which of the following reactions will occur, and for those that will occur, give the overall, balanced molecular equation and indicate which element is oxidized and which is reduced.
(a) $Mg(s) + Cr(C_2H_3O_2)_3(aq) \longrightarrow$? (b) $Cu(s) + HBr(aq) \longrightarrow$?
(c) $Cd(s) + AgNO_3(aq) \longrightarrow$?

Practice Problem Ⓑ**UILD** Predict which of the following reactions will occur, and for those that will occur, indicate which element is oxidized and which is reduced. (a) $Pt(s) + Cu(NO_3)_2(aq) \longrightarrow$?
(b) $Ag(s) + AuCl_3(aq) \longrightarrow$? (c) $Sn(s) + HNO_3(aq) \longrightarrow$?

Practice Problem Ⓒ**ONCEPTUALIZE** Metals M and N are represented by yellow and white spheres, respectively. Based on the diagrams before and after the reaction, write the corresponding balanced equation and assign oxidation numbers to the metals and their ions.

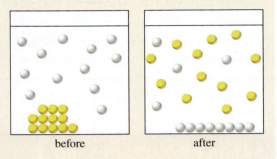

before   after

Other Types of Redox Reactions

Several of the reaction types that you have already encountered are also redox reactions.

Combination Reactions

Combination reactions such as the formation of ammonia from its constituent elements can involve oxidation and reduction.

$$N_2(g) \quad + \quad 3H_2(g) \quad \longrightarrow \quad 2NH_3(g)$$

In this reaction, nitrogen is reduced from 0 to −3, while hydrogen is oxidized from 0 to +1. Other examples of combination reactions include those shown in Figure 4.7.

Decomposition

Decomposition can also be a redox reaction, as illustrated by the following examples:

$$2NaH(s) \quad \longrightarrow \quad 2Na(s) \quad + \quad H_2(g)$$

$$2KClO_3(s) \quad \longrightarrow \quad 2KCl(s) \quad + \quad 3O_2(g)$$

$$2H_2O_2(aq) \quad \longrightarrow \quad 2H_2O(l) \quad + \quad O_2(g)$$

$$2Na \quad + \quad Cl_2 \quad \longrightarrow \quad 2NaCl$$
(a)

$$2H_2 \quad + \quad O_2 \quad \longrightarrow \quad 2H_2O$$
(b)

Figure 4.7 (a) Reaction between sodium and chlorine to form sodium chloride, and (b) reaction between hydrogen and oxygen to form water. For each element, the oxidation number appears in the circle and the total contribution to charge appears in the square below it.

The decomposition of hydrogen peroxide, shown in the preceding equation, is an example of a ***disproportionation reaction,*** in which one element undergoes both oxidation and reduction. In the case of H_2O_2, the oxidation number of O is initially -1. In the products of the decomposition, O has an oxidation number of -2 in H_2O and of 0 in O_2.

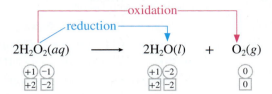

$$2H_2O_2(aq) \longrightarrow 2H_2O(l) + O_2(g)$$

Finally, ***combustion*** [◄◄ Section 3.3] is a redox process.

$$CH_4(g) + 2O_2(g) \longrightarrow CO_2(g) + 2H_2O(l)$$

Figure 4.8 shows the known oxidation numbers of elements in compounds—arranged according to their positions in the periodic table.

1	2	3	4	5	6	7	8	9	10	11	12	13	14	15	16	17	18
1 **H** +1, −1																	2 **He**
3 **Li** +1	4 **Be** +2											5 **B** +3	6 **C** +4, +2, −4	7 **N** +5, +4, +3, +2, +1, −3	8 **O** +2, 2½, 21, −2	9 **F** −1	10 **Ne**
11 **Na** +1	12 **Mg** +2											13 **Al** +3	14 **Si** +4, −4	15 **P** +5, +3, −3	16 **S** +6, +4, +2, −2	17 **Cl** +7, +6, +5, +4, +3, +1, −1	18 **Ar**
19 **K** +1	20 **Ca** +2	21 **Sc** +3	22 **Ti** +4, +3, +2	23 **V** +5, +4, +3, +2	24 **Cr** +6, +5, +4, +3, +2	25 **Mn** +7, +6, +4, +3, +2	26 **Fe** +3, +2	27 **Co** +3, +2	28 **Ni** +2	29 **Cu** +2, +1	30 **Zn** +2	31 **Ga** +3	32 **Ge** +4, −4	33 **As** +5, +3, −3	34 **Se** +6, +4, −2	35 **Br** +5, +3, +1, −1	36 **Kr** +4, +2
37 **Rb** +1	38 **Sr** +2	39 **Y** +	40 **Zr** +4	41 **Nb** +5, +4	42 **Mo** +6, +4, +3	43 **Tc** +7, +6, +4	44 **Ru** +8, +6, +4, +3	45 **Rh** +4, +3, +2	46 **Pd** +4, +2	47 **Ag** +1	48 **Cd** +2	49 **In** +3	50 **Sn** +4, +2	51 **Sb** +5, +3, −3	52 **Te** +6, +4, −2	53 **I** +7, +5, +1, −1	54 **Xe** +6, +4, +2
55 **Cs** +1	56 **Ba** +2	71 **Lu** +3	72 **Hf** +4	73 **Ta** +5	74 **W** +6, +4	75 **Re** +7, +6, +4	76 **Os** +8, +4	77 **Ir** +4, +3	78 **Pt** +4, +2	79 **Au** +3, +1	80 **Hg** +2, +1	81 **Tl** +3, +1	82 **Pb** +4, +2	83 **Bi** +5, +3	84 **Po** +2	85 **At** −1	86 **Rn**

Figure 4.8 Periodic table showing oxidation numbers for each element. The most common oxidation numbers are shown in red.

CHECKPOINT – SECTION 4.4   Oxidation-Reduction Reactions

4.4.1 Determine the oxidation number of sulfur in each of the following species: H_2S, HSO_3^-, SCl_2, and S_8.

a) $+2$, $+6$, -2, $+\frac{1}{4}$

b) -2, $+3$, $+2$, 0

c) -2, $+5$, $+2$, $-\frac{1}{4}$

d) -1, $+4$, $+2$, 0

e) -2, $+4$, $+2$, 0

4.4.2 What species is the reducing agent in the following equation?

$$Mg(s) + 2HCl(aq) \longrightarrow MgCl_2(aq) + H_2(g)$$

a) $Mg(s)$ d) $Mg^{2+}(aq)$

b) $H^+(aq)$ e) $H_2(g)$

c) $Cl^-(aq)$

4.4.3 Which of the following equations represents a redox reaction? (Choose all that apply.)

a) $2Mg(s) + O_2(g) \longrightarrow 2MgO(s)$

b) $Cu(s) + PtCl_2(aq) \longrightarrow CuCl_2(aq) + Pt(s)$

c) $NH_4Cl(aq) + AgNO_3(aq) \longrightarrow NH_4NO_3(aq) + AgCl(s)$

d) $2NaN_3(s) \longrightarrow 2Na(s) + 3N_2(g)$

e) $CaCO_3(s) \longrightarrow CaO(s) + CO_2(g)$

4.4.4 According to the activity series, which of the following redox reactions will occur? (Choose all that apply.)

a) $Fe(s) + NiBr_2(aq) \longrightarrow FeBr_2(aq) + Ni(s)$

b) $Sn(s) + Pb(NO_3)_2(aq) \longrightarrow Sn(NO_3)_2(aq) + Pb(s)$

c) $Mg(s) + BaI_2(aq) \longrightarrow MgI_2(aq) + Ba(s)$

d) $Pb(s) + PtCl_2(aq) \longrightarrow PbCl_2(aq) + Pt(s)$

e) $Zn(s) + CaBr_2(aq) \longrightarrow ZnBr_2(aq) + Ca(s)$

4.5   Concentration of Solutions

One of the factors that can influence reactions in aqueous solution is concentration. The **concentration** of a solution is the amount of solute dissolved in a given quantity of solvent or solution. Consider the two solutions of iodine pictured in Figure 4.9. The solution on the left is more concentrated than the one on the right—that is, it contains a higher ratio of solute to solvent. By contrast, the solution on the right is more dilute. (The qualitative terms *concentrated* and *dilute* are relative terms, like *expensive* and *cheap*.) The color is more intense in the more concentrated solution. Often the concentrations of reactants determine how fast a chemical reaction occurs. For example, the reaction of magnesium metal and acid [◄◄ Section 4.4] happens faster if the concentration of acid is greater. As we explain in Chapter 13, there are several different ways to express the concentration of a solution. In this chapter, we introduce only molarity, which is one of the most commonly used units of concentration.

Figure 4.9 Two solutions of iodine in benzene. The solution on the left is more concentrated. The solution on the right is more dilute.

Photos: *Charles D. Winters/McGraw Hill*

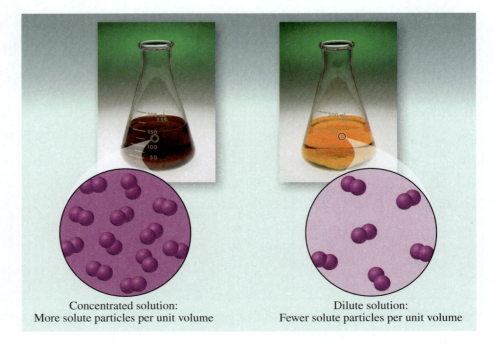

Concentrated solution: More solute particles per unit volume

Dilute solution: Fewer solute particles per unit volume

Molarity

Molarity, or *molar concentration,* symbolized *M,* is defined as the number of moles of solute per liter of solution. Thus, 1 L of a 1.5 molar solution of glucose ($C_6H_{12}O_6$), written as 1.5 M $C_6H_{12}O_6$, contains 1.5 mol of dissolved glucose. Half a liter of the same solution would contain 0.75 mol of dissolved glucose, a milliliter of the solution would contain 1.5×10^{-3} mol of dissolved glucose, and so on.

Student Note: Molarity can equally well be defined as millimoles per milliliter (mmol/mL), which can simplify some calculations.

$$\text{molarity} = \frac{\text{moles solute}}{\text{liters solution}} \qquad \textbf{Equation 4.1}$$

To calculate the molarity of a solution, we divide the number of moles of solute by the volume of the solution in liters.

Equation 4.1 can be rearranged in three ways to solve for any of the three variables: molarity (*M*), moles of solute (mol), or volume of solution in liters (L).

Student Note: Students sometimes have difficulty seeing how units cancel in these equations. It may help to write *M* as mol/L until you become completely comfortable with these equations.

$$(1) \; M = \frac{\text{mol}}{L} \qquad\qquad (2) \; L = \frac{\text{mol}}{M} \qquad\qquad (3) \; mol = M \times L$$

Sample Problem 4.8 illustrates how to use these equations to solve for molarity, volume of solution, and moles of solute.

SAMPLE PROBLEM 4.8

For an aqueous solution of glucose ($C_6H_{12}O_6$), determine (a) the molarity of 2.00 L of a solution that contains 50.0 g of glucose, (b) the volume of this solution that would contain 0.250 mol of glucose, and (c) the number of moles of glucose in 0.500 L of this solution.

Strategy Convert the mass of glucose given to moles, and use the equations for interconversions of *M*, liters, and moles to calculate the answers.

Setup The molar mass of glucose is 180.2 g.

$$\text{moles of glucose} = \frac{50.0 \text{ g}}{180.2 \text{ g/mol}} = 0.277 \text{ mol}$$

Solution (a) molarity $= \dfrac{0.227 \text{ mol } C_6H_{12}O_6}{2.00 \text{ L solution}} = 0.139 \; M$

A common way to state the concentration of this solution is to say, "This solution is 0.139 *M* in glucose."

(b) volume $= \dfrac{0.250 \text{ mol } C_6H_{12}O_6}{0.139 \; M} = 1.80 \; L$

(c) moles of $C_6H_{12}O_6$ in 0.500 L $= 0.500$ L $\times 0.139 \; M = 0.0695$ mol

THINK ABOUT IT

Check to see that the magnitudes of your answers are logical. For example, the mass given in the problem corresponds to 0.277 mol of solute. If you are asked, as in part (b), for the volume that contains a number of moles smaller than 0.277, make sure your answer is smaller than the original volume.

Practice Problem **A**TTEMPT For an aqueous solution of sucrose ($C_{12}H_{22}O_{11}$), determine (a) the molarity of 5.00 L of a solution that contains 235 g of sucrose, (b) the volume of this solution that would contain 1.26 mol of sucrose, and (c) the number of moles of sucrose in 1.89 L of this solution.

Practice Problem **B**UILD For an aqueous solution of sodium chloride (NaCl), determine (a) the molarity of 3.75 L of a solution that contains 155 g of sodium chloride, (b) the volume of this solution that would contain 4.58 mol of sodium chloride, and (c) the number of moles of sodium chloride in 22.75 L of this solution.

Practice Problem **C**ONCEPTUALIZE The diagrams represent solutions of two different concentrations. What volume of solution 2 contains the same amount of solute as 5.00 mL of solution 1? What volume of solution 1 contains the same amount of solute as 30.0 mL of solution 2?

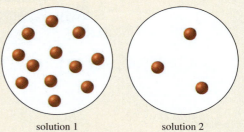

solution 1   solution 2

Figure 4.10

Preparing a Solution from a Solid

The mass likely will not be exactly the calculated number.

3.896 g

Transfer the weighed $KMnO_4$ to the volumetric flask.

Weigh out the solid $KMnO_4$. (The tare function on a digital balance automatically subtracts the mass of the weighing paper.)

2.003 g

Calculate the mass of $KMnO_4$ necessary for the target concentration of 0.1 M.

$$\frac{0.1 \text{ mol}}{L} \times 0.2500 \text{ L} = 0.02500 \text{ mol}$$

$$0.02500 \text{ mol} \times \frac{158.04 \text{ g}}{\text{mol}} = 3.951 \text{ g } KMnO_4$$

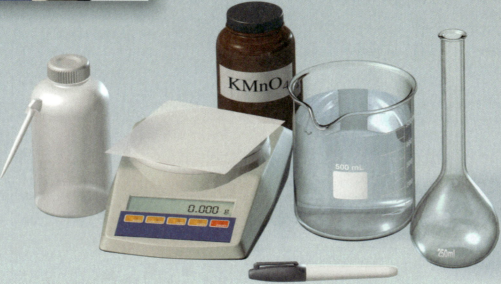

Add water sufficient to dissolve the KMnO₄.

Swirl the flask to dissolve the solid.

Add more water.

Fill exactly to the calibration mark using a wash bottle or eyedropper.

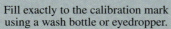

0.09861 N

(See Visualizing Chemistry questions VC 4.1–VC 4.4 on pages 193–194.)

After capping and inverting the flask to ensure complete mixing, we calculate the actual concentration of the prepared solution.

$$3.896 \text{ g KMnO}_4 \times \frac{1 \text{ mol}}{158.04 \text{ g}} = 0.024652 \text{ mol}$$

$$\frac{0.024652 \text{ mol}}{0.2500 \text{ L}} = 0.09861 \ M$$

What's the point?

The goal is to prepare a solution of precisely known concentration, with that concentration being very close to the target concentration of 0.1 *M*. Note that because 0.1 is a *specified* number, it does not limit the number of significant figures in our calculations.

Animation
Figure 4.10, Preparing a Solution
from a Solid pp. 170–171.

Student Note: It is important to remember that molarity is defined in terms of the volume of *solution,* not the volume of solvent. In many cases, these two are not the same.

Animation
Dilution.

The procedure for preparing a solution of known molarity is shown in Figure 4.10. First, the solute is weighed accurately and transferred, often with a funnel, to a volumetric flask of the desired volume. Next, water is added to the flask, which is then swirled to dissolve the solid. After all the solid has dissolved, more water is added slowly to bring the level of solution exactly to the volume mark. Finally, the flask is capped and inverted to ensure thorough mixing. Knowing the volume of the solution in the flask and the quantity of compound dissolved, we can determine the *molarity of the solution* using Equation 4.1. Note that this procedure does not require that we know the exact amount of water added. Because of the way molarity is defined, it is important only that we know the final volume of the *solution.*

Dilution

Concentrated "stock" solutions of commonly used substances typically are kept in the laboratory stockroom. Often we need to dilute these stock solutions before using them. *Dilution* is the process of preparing a less concentrated solution from a more concentrated one. Suppose that we want to prepare 1.00 L of a 0.400 M $KMnO_4$ solution from a solution of 1.00 M $KMnO_4$. For this purpose, we need 0.400 mol of $KMnO_4$. Because there is 1.00 mol of $KMnO_4$ in 1.00 L of a 1.00 M $KMnO_4$ solution, there is 0.400 mol of $KMnO_4$ in 0.400 L of the same solution:

$$\frac{1.00 \text{ mol } KMnO_4}{1.00 \text{ L of solution}} = \frac{0.400 \text{ mol } KMnO_4}{0.400 \text{ L of solution}}$$

Therefore, we must withdraw precisely 400 mL (0.400 L) from the 1.00 M $KMnO_4$ solution and dilute it to 1.00 L by adding water (in a 1.00-L volumetric flask). This method gives us 1.00 L of the desired 0.400 M $KMnO_4$.

In carrying out a dilution process, it is useful to remember that adding more solvent to a given amount of the stock solution changes (decreases) the concentration of the solution without changing the number of moles of solute present in the solution (Figure 4.11).

Equation 4.2   moles of solute before dilution = moles of solute after dilution

Using arrangement (3) of Equation 4.1, we can calculate the number of moles of solute:

$$\text{moles of solute} = \frac{\text{moles of solute}}{\text{liters of solution}} \times \text{liters of solution}$$

Figure 4.11 Dilution changes the concentration of a solution; it does *not* change the number of moles of solute in the solution.

Photos: *David A. Tietz/Editorial Image, LLC*

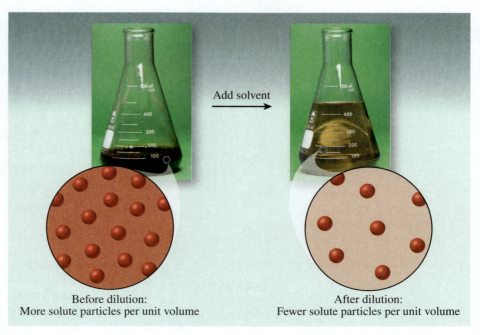

Add solvent

Before dilution:	After dilution:
More solute particles per unit volume	Fewer solute particles per unit volume

Because the number of moles of solute before the dilution is the same as that after dilution, we can write

$$M_c \times L_c = M_d \times L_d \qquad \textbf{Equation 4.3}$$

where the subscripts c and d stand for *concentrated* and *dilute*, respectively. Thus, by knowing the molarity of the concentrated stock solution (M_c) and the desired final molarity (M_d) and volume (L_d) of the dilute solution, we can calculate the volume of stock solution required for the dilution (L_c).

Because most volumes measured in the laboratory are in milliliters rather than liters, it is worth pointing out that Equation 4.3 can also be written with volumes of the concentrated and dilute solutions in milliliters.

Student Note: Students sometimes resist using the unit *millimole*. However, using the $M_c \times mL_c = M_d \times mL_d$ form in Equation 4.4 often reduces the number of steps in a problem, thereby reducing the number of opportunities to make calculation errors.

$$M_c \times mL_c = M_d \times mL_d \qquad \textbf{Equation 4.4}$$

In this form of the equation, the product of each side is in *millimoles* (mmol) rather than moles. We apply Equation 4.4 in Sample Problem 4.9.

SAMPLE PROBLEM 4.9

What volume of 12.0 *M* HCl, a common laboratory stock solution, must be used to prepare 250.0 mL of 0.125 *M* HCl?

Strategy Use Equation 4.4 to determine the volume of 12.0 *M* HCl required for the dilution.

Setup $M_c = 12.0$ *M*, $M_d = 0.125$ *M*, $mL_d = 250.0$ mL.

Solution

Student Note: It is very important to note that, for safety, when diluting a concentrated acid, the acid must be added to the water, and *not* the other way around.

$$12.0 \ M \times mL_c = 0.125 \ M \times 250.0 \ \text{mL}$$

$$mL_c = \frac{0.125 \ M \times 250.0 \ \text{mL}}{12.0 \ M} = 2.60 \ \text{mL}$$

THINK ABOUT IT

Plug the answer into Equation 4.4, and make sure that the product of concentration and volume is the same on both sides of the equation.

Practice Problem ATTEMPT What volume of 6.0 *M* H$_2$SO$_4$ is needed to prepare 500.0 mL of a solution that is 0.25 *M* in H$_2$SO$_4$?

Practice Problem BUILD What volume of 0.20 *M* H$_2$SO$_4$ can be prepared by diluting 125 mL of 6.0 *M* H$_2$SO$_4$?

Practice Problem CONCEPTUALIZE The diagrams represent a concentrated stock solution (left) and a dilute solution (right) that can be prepared by dilution of the stock solution. How many milliliters of the concentrated stock solution are needed to prepare solutions of the same concentration as the dilute solution of each of the following final volumes? (a) 50.0 mL, (b) 100.0 mL, (c) 250.0 mL

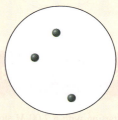

Serial Dilution

A series of dilutions may be used in the laboratory to prepare a number of increasingly dilute solutions from a stock solution. The method involves preparing a solution as described previously and diluting a portion of the prepared solution to make a *more* dilute solution. For example, we could use our 0.400 *M* KMnO$_4$ solution, to prepare a series of five increasingly dilute solutions—with the concentration decreasing by a factor of 10

Solution The percent Cl in the unknown compound is the mass of Cl in the precipitate divided by the mass of the original sample:

$$\frac{0.3862 \text{ g}}{0.8633 \text{ g}} \times 100\% = 44.73\% \text{ Cl}$$

THINK ABOUT IT

Pay close attention to which numbers correspond to which quantities. It is easy in this type of problem to lose track of which mass is the precipitate and which is the original sample. Dividing by the wrong mass at the end will result in an incorrect answer.

Practice Problem Ⓐ**TTEMPT** A 0.5620-g sample of an ionic compound containing the bromide ion (Br⁻) is dissolved in water and treated with an excess of AgNO₃. If the mass of the AgBr precipitate that forms is 0.8868 g, what is the percent by mass of Br in the original compound?

Practice Problem Ⓑ**UILD** A sample that is 63.9 percent chloride by mass is dissolved in water and treated with an excess of AgNO₃. If the mass of the AgCl precipitate that forms is 1.085 g, what was the mass of the original sample?

Practice Problem Ⓒ**ONCEPTUALIZE** Which diagram best represents the solution (originally containing sodium chloride) from which the chloride has been removed by the addition of excess silver nitrate?

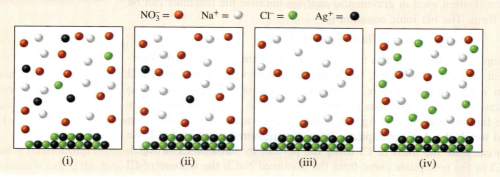

$$NO_3^- = \bullet \qquad Na^+ = \bullet \qquad Cl^- = \bullet \qquad Ag^+ = \bullet$$

(i)	(ii)	(iii)	(iv)

Gravimetric analysis is a quantitative method, not a qualitative one, so it does not establish the identity of the unknown substance. Thus, the results in Sample Problem 4.12 do *not* identify the cation. However, knowing the percent by mass of Cl greatly helps us narrow the possibilities. Because no two compounds containing the same anion (or cation) have the same percent composition by mass, comparison of the percent by mass obtained from gravimetric analysis with that calculated from a series of known compounds could reveal the identity of the unknown compounds.

Acid-Base Titrations

Quantitative studies of acid-base neutralization reactions are most conveniently carried out using a technique known as titration. In **titration,** a solution of accurately known concentration, called a **standard solution,** is added gradually to another solution of unknown concentration, until the chemical reaction between the two solutions is complete, as shown in Figure 4.15. If we know the volumes of the standard and unknown solutions used in the titration, along with the concentration of the standard solution, we can calculate the concentration of the unknown solution.

A solution of the strong base sodium hydroxide can be used as the standard solution in a titration, but it must first be *standardized,* because sodium hydroxide in solution reacts with carbon dioxide in the air, making its concentration unstable over time. We can *standardize* the sodium hydroxide solution by titrating it against an acid solution of accurately known concentration. The acid often chosen for this task is a monoprotic acid called potassium hydrogen phthalate (KHP), for which the molecular formula is

Student Note: *Standardization* in this context is the meticulous determination of concentration.

Figure 4.15 Apparatus for titration.

David A. Tietz/Editorial Image, LLC

KHC$_8$H$_4$O$_4$. KHP is a white, soluble solid that is commercially available in highly pure form. The reaction between KHP and sodium hydroxide is

$$\text{KHC}_8\text{H}_4\text{O}_4(aq) + \text{NaOH}(aq) \longrightarrow \text{KNaC}_8\text{H}_4\text{O}_4(aq) + \text{H}_2\text{O}(l)$$

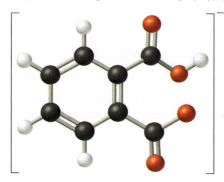

HC$_8$H$_4$O$_4^-$

and the net ionic equation is

$$\text{HC}_8\text{H}_4\text{O}_4^-(aq) + \text{OH}^-(aq) \longrightarrow \text{C}_8\text{H}_4\text{O}_4^{2-}(aq) + \text{H}_2\text{O}(l)$$

Note that KHP is a *monoprotic* acid, so it reacts in a 1:1 ratio with hydroxide ion.

To standardize a solution of NaOH with KHP, a known amount of KHP is transferred to an Erlenmeyer flask and some distilled water is added to make up a solution. Next, NaOH solution is carefully added to the KHP solution from a burette until all the acid has reacted with the base. This point in the titration, where the acid has been completely neutralized, is called the ***equivalence point.*** It is usually signaled by the ***endpoint,*** where an indicator causes a sharp change in the color of the solution. In acid-base titrations, ***indicators*** are substances that have distinctly different colors in acidic and basic media. One commonly used indicator is phenolphthalein, which is colorless in acidic and neutral solutions but reddish pink in basic solutions. At the equivalence point, all the KHP present has been neutralized by the added NaOH and the solution is still colorless. However, if we add just one more drop of NaOH solution from the burette, the solution will be basic and will immediately turn pink. Sample Problem 4.13 illustrates just such a titration.

Student Hot Spot

Student data indicate you may struggle with titrations. Access the eBook to view additional Learning Resources on this topic.

Student Note: The endpoint in a titration is used to approximate the equivalence point. A careful choice of indicators, which we discuss in Chapter 16, helps make this approximation reasonable. Phenolphthalein, although very common, is not appropriate for every acid-base titration.

(i)	(ii)	(iii)	(iv)

Sample Problem 4.15 shows how titration with a standard base can be used to determine the molar mass of an unknown acid.

SAMPLE PROBLEM 4.15

A 0.1216-g sample of a monoprotic acid is dissolved in 25 mL water, and the resulting solution is titrated with 0.1104 *M* NaOH solution. A 12.5-mL volume of the base is required to neutralize the acid. Calculate the molar mass of the acid.

Strategy Using the concentration and volume of the base, we can determine the number of moles of base required to neutralize the acid. We then determine the number of moles of acid and divide the mass of the acid by the number of moles to get the molar mass.

Setup Because the acid is monoprotic, it will react in a 1:1 ratio with the base; therefore, the number of moles of acid will be equal to the number of moles of base. The volume of base in liters is 0.0125 L.

Solution

$$\text{moles of base} = 0.0125 \text{ L} \times 0.1104 \text{ mol/L} = 0.00138$$

Because moles of base = moles of acid, the moles of acid = 0.00138 mol. Therefore,

$$\text{molar mass of the acid} = \frac{0.1216 \text{ g}}{0.00138 \text{ mol}} = 88.1 \text{ g/mol}$$

THINK ABOUT IT

For this technique to work, we must know whether the acid is monoprotic, diprotic, or polyprotic. A diprotic acid, for example, would combine in a 1:2 ratio with the base, and the result would have been a molar mass twice as large.

Practice Problem **A** **TTEMPT** What is the molar mass of a monoprotic acid if 28.1 mL of 0.0788 *M* NaOH is required to neutralize a 0.205-g sample?

Practice Problem **B** **UILD** What is the molar mass of a diprotic acid if 30.5 mL of 0.1112 *M* NaOH is required to neutralize a 0.1365-g sample?

Practice Problem **C** **ONCEPTUALIZE** Consider aqueous solutions of two different acids. Each contains the same mass of acid, and each requires the same volume of 0.10 *M* NaOH for complete neutralization—and yet the two acids do not have the same molar mass. Explain how this is possible.

Figure 4.16 A redox titration of oxalate using KMnO₄(*aq*) as the oxidizing agent and the indicator. Prior to the equivalence point, the solution in the flask is nearly colorless. At the endpoint, all of the reducing agent has been consumed and the excess permanganate ion causes the solution to turn purple.

Stephen Frisch/McGraw Hill

KMnO₄(*aq*)

Na₂C₂O₄(*aq*)

Redox Titration

Another quantitative-analysis method is ***redox titration.*** Redox titration involves the use of an oxidation-reduction reaction, with one reactant being delivered via a burette. In one common type of redox titration, the titrant is a solution of potassium permanganate, which serves as both the oxidizing agent and the indicator. In the analysis of oxalate ion, for example, permanganate ion reacts with oxalate ion according to the equation

$$2MnO_4^-(aq) + 5C_2O_4^{2-}(aq) + 16H^+(aq) \longrightarrow 2Mn^{2+}(aq) + 10CO_2(aq) + 8H_2O(aq)$$

Prior to the equivalence point, the solution is nearly colorless. When all of the oxalate ion has been consumed, one additional drop of potassium permanganate titrant will impart a purple color to the solution—indicating the endpoint. (See Figure 4.16.)

In some redox titrations, a separate indicator is used. For example, several common redox-titration methods use an iodine (I₂) solution as the oxidizing agent, and starch as the indicator. When all of the reducing agent has been consumed, any excess iodine combines with the starch indicator to produce an intensely colored blue species. When the iodine solution is used as the titrant, the appearance of blue indicates that the reaction is complete.

Sample Problem 4.16 illustrates the use of redox titration.

SAMPLE PROBLEM 4.16

The vitamin C (ascorbic acid, $C_6H_8O_6$) content of Gatorade and other sports beverages can be measured by titration with iodine solution. The reaction can be represented with the equation

$$I_2(aq) + C_6H_8O_6(aq) \longrightarrow 2I^-(aq) + C_6H_6O_6(aq) + 2H^+(aq)$$

Determine the mass of vitamin C (in mg) contained in a 350-mL bottle of Gatorade if a 25.0-mL sample requires 29.25 mL of 0.00125 M I_2 solution to reach the endpoint.

Strategy Use the volume and concentration of the iodine solution to determine the number of moles of iodine reacted; then use the balanced equation to determine the number of moles of vitamin C reacted. (In this case, the ratio of combination is 1:1.) Use this number of moles and the molar mass of vitamin C to determine the mass of vitamin C in the 25.0-mL sample; and then determine the mass of vitamin C in the total volume (350 mL).

Setup The molar mass of vitamin C is 176.1 g/mol. The volume of I_2 solution in liters is 0.02925 L.

Solution

$$0.02925 \text{ L} \times 0.00125 \ M = 3.656 \times 10^{-5} \text{ mol } I_2$$
$$= 3.656 \times 10^{-5} \text{ mol vitamin C}$$

$$3.656 \times 10^{-5} \text{ mol} \times \frac{176.1 \text{ g vitamin C}}{\text{mol}} = 6.44 \times 10^{-3} \text{ g}$$

$$6.44 \times 10^{-3} \text{ g} \times \frac{1000 \text{ mg}}{1 \text{ g}} = 6.440 \text{ mg vitamin C in 25.0 mL}$$

$$\frac{6.440 \text{ mg}}{25.0 \text{ mL}} \times 350 \text{ mL} = 90 \text{ mg}$$

THINK ABOUT IT

This problem could also be solved using fewer steps by using *millimoles* instead of *moles*.

Practice Problem Ⓐ**TTEMPT** Iodine is also used to analyze the sulfur dioxide content in wine. The species that reacts with iodine is actually sulfurous acid (H_2SO_3), and the reaction is represented by the equation

$$I_2(aq) + H_2SO_3(aq) \longrightarrow 2I^-(aq) + HSO_3^-(aq) + 3H^+(aq)$$

Determine the amount of sulfurous acid (in mg) in a 750-mL bottle of wine if a 50.0-mL sample requires 14.75 mL of 0.00115 M aqueous iodine to reach the endpoint.

Practice Problem Ⓑ**UILD** The iron content of drinking water can be measured by titration with potassium permanganate. The reaction is represented by the equation

$$5Fe^{2+}(aq) + KMnO_4^-(aq) + 8H^+(aq) \longrightarrow 5Fe^{3+}(aq) + Mn^{2+}(aq) + 4H_2O(l)$$

Determine the concentration of iron in ppm (mg/L) of a sample of water if 25.0 mL of the water requires 21.30 mL of 2.175×10^{-5} M $KMnO_4$ to reach the endpoint in a titration.

Practice Problem Ⓒ**ONCEPTUALIZE** Because iodine itself is not very soluble in water, "iodine" solutions used in redox titrations generally contain the triiodide ion (I_3^-). Thus, the equation for the redox titration of vitamin C with iodine can be written as

$$C_6H_8O_6(aq) + I_3^-(aq) \longrightarrow C_6H_6O_6(aq) + 3I^-(aq) + 2H^+(aq)$$

Which diagram best represents the ions remaining in solution at the equivalence point in a titration of vitamin C with triiodide? (Spectator ions are not shown.)

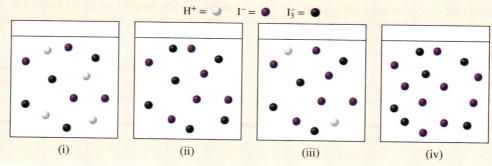

H$^+$ = ⚪   I$^-$ = 🟣   I$_3^-$ = ⚫

(i)   (ii)   (iii)   (iv)

5.1 Energy and Energy Changes

Matter can undergo physical changes and chemical changes [◂◂ Section 1.4]. The melting of ice, for example, is a physical change that can be represented by the following equation:

$$H_2O(s) \longrightarrow H_2O(l)$$

The formation of water from its constituent elements, represented by the following equation, is an example of a chemical change:

$$2H_2(g) + O_2(g) \longrightarrow 2H_2O(l)$$

In each case, there is energy involved in the change. Energy (in the form of heat) must be *supplied* to melt ice, whereas energy (in the form of heat and light) is *produced* by the explosive combination of hydrogen and oxygen gases. In fact, every change that matter undergoes is accompanied by either the absorption or the release of energy. In this chapter, we focus on the energy changes associated with physical and chemical processes.

Forms of Energy

Energy is usually defined as the capacity to do work or transfer heat. All forms of energy are either kinetic or potential. *Kinetic energy* is the energy that results from *motion*. It is calculated with the equation

Equation 5.1
$$E_k = \tfrac{1}{2}mu^2$$

Student Note: Some textbooks use the letter *v* to denote velocity.

where m is the mass of the object and u is its velocity. One form of kinetic energy of particular interest to chemists is *thermal energy,* which is the energy associated with the random motion of atoms and molecules. We can monitor changes in thermal energy by measuring temperature changes.

Potential energy is the energy possessed by an object by virtue of its position. The two forms of potential energy of greatest interest to chemists are chemical energy and electrostatic energy. *Chemical energy* is energy stored within the structural units (molecules or polyatomic ions) of chemical substances. The amount of chemical energy in a sample of matter depends on the types and arrangements of atoms in the structural units that make up the sample.

Electrostatic energy is potential energy that results from the interaction of charged particles. Oppositely charged particles *attract* each other, whereas particles of like charges *repel* each other. The magnitude of the resulting electrostatic potential energy is proportional to the product of the two charges (Q_1 and Q_2) divided by the distance (d) between them.

Equation 5.2
$$E_{el} \propto \frac{Q_1 Q_2}{d}$$

If the charges Q_1 and Q_2 are opposite (i.e., one positive and one negative), the result is a negative value for E_{el}, which indicates *attraction*. Like charges (i.e., both positive or both negative) result in a positive value for E_{el}, indicating *repulsion*.

Kinetic and potential energy are interconvertible—that is, one form can be converted to the other. For example, dropping an object and allowing it to fall converts potential energy to kinetic energy. Likewise, a chemical reaction that gives off heat converts chemical energy (potential) to thermal energy (kinetic). Although energy can assume many different forms that are interconvertible, the total amount of energy in the universe is constant. When energy of one form disappears, the same amount of energy must appear in another form or forms. This principle is known as the *law of conservation of energy.*

Energy Changes in Chemical Reactions

To analyze energy changes associated with chemical reactions, we must first define the *system,* the specific part of the universe that is of interest to us. For chemists, systems usually include the substances involved in chemical and physical changes. In an acid-base neutralization experiment, for example, the system may be the contents of a beaker in which HCl and NaOH react with each other. The rest of the universe outside the system, including the beaker itself, constitutes the *surroundings.*

Many chemical reactions are carried out for the purpose of exploiting the associated energy change, rather than for the purpose of obtaining the products of the reactions. For example, combustion reactions involving fossil fuels are carried out for the thermal energy they produce, not for their products, which are carbon dioxide and water.

It is important to distinguish between thermal energy and heat. *Heat* is the transfer of thermal energy between two bodies that are at different temperatures. Although the term *heat* by itself implies the transfer of energy, we customarily talk of "heat flow," meaning "heat absorbed" or "heat released," when describing the energy changes that occur during a process. *Thermochemistry* is the study of heat (the transfer of thermal energy) in chemical reactions.

The combustion of hydrogen gas in oxygen is one of many chemical reactions that release considerable quantities of energy (Figure 5.1):

$$2H_2(g) + O_2(g) \longrightarrow 2H_2O(l) + \text{energy}$$

In this case, we label the mixture of reactants and product (hydrogen, oxygen, and water molecules) the *system.* Because energy cannot be created or destroyed, any energy released by the system must be gained by the surroundings. Thus, the heat generated by the combustion process is transferred from the system to its surroundings. This reaction is an example of an *exothermic process,* which is any process that gives off heat—that is, transfers thermal energy *from* the system *to* the surroundings. Figure 5.2(a) shows the energy change for the combustion of hydrogen gas.

Next, consider the decomposition of mercury(II) oxide (HgO) at high temperatures:

$$\text{energy} + 2HgO(s) \longrightarrow 2Hg(l) + O_2(g)$$

This reaction is an *endothermic process* because heat has to be supplied to the system (i.e., to HgO) by the surroundings [Figure 5.2(b)] for the reaction to occur. Thus, thermal energy is transferred *from* the surroundings *to* the system in an endothermic process.

According to Figure 5.2, the energy of the products of an exothermic reaction is lower than the energy of the reactants. The difference in energy between the reactants H_2 and O_2 and the product H_2O is the heat released by the system to the

Figure 5.1 The Hindenburg, a German airship filled with hydrogen gas, was destroyed in a horrific fire as it landed at Lakehurst, New Jersey, in 1937.
Everett Historical/Shutterstock

Animation
Thermochemistry—heat flow in endothermic and exothermic reactions.

Figure 5.2 (a) An exothermic reaction. Heat is given off by the system. (b) An endothermic reaction. Heat is absorbed by the system.

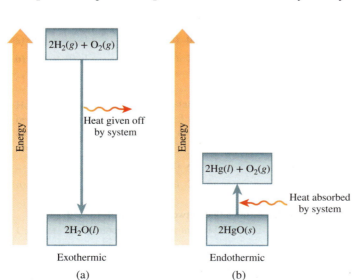

a) i < ii < iii < iv

d) ii < i = iii < iv

b) iv < iii < ii < i

e) iv < i = ii < iii

c) i = iii < ii < iv

5.1.6   Which of the following changes would double the magnitude of the electrostatic attraction between two oppositely charged particles? (Select all that apply.)

a) Doubling both charges

b) Doubling one of the charges

c) Doubling the distance between the charges

d) Reducing the distance between the charges by half

e) Doubling both charges and doubling the distance between the charges

---

5.2   Introduction to Thermodynamics

Thermochemistry is part of a broader subject called *thermodynamics,* which is the scientific study of the interconversion of heat and other kinds of energy. The laws of thermodynamics provide useful guidelines for understanding the energetics and directions of processes. In this section, we introduce the first law of thermodynamics, which is particularly relevant to the study of thermochemistry. We continue our discussion of thermodynamics in Chapter 18.

We have defined a system as the part of the universe we are studying. There are three types of systems. An *open system* can exchange mass and energy with its surroundings. For example, an open system may consist of a quantity of water in an open container, as shown in Figure 5.3(a). If we close the flask, as in Figure 5.3(b), so that no water vapor can escape from or condense into the container, we create a *closed system,* which allows the transfer of *energy* but not *mass.* By placing the water in an insulated container, as shown in Figure 5.3(c), we can construct an *isolated system,* which does not exchange either mass or energy with its surroundings.

> **Student Note:** The energy exchanged between open systems or closed systems and their surroundings is usually in the form of heat.

States and State Functions

In thermodynamics, we study changes in the *state of a system,* which is defined by the values of all relevant macroscopic properties, such as composition, energy, temperature, pressure, and volume. Energy, pressure, volume, and temperature are said to be *state functions*—properties that are determined by the state of the system, regardless of how that condition was achieved. In other words, when the state of a system changes, the magnitude of change in any state function depends only on the initial and final states of the system and not on how the change is accomplished.

Consider, for example, your position in a six-story building. Your elevation depends upon which floor you are on. If you change your elevation by taking the stairs from the ground floor up to the fourth floor, the change in your elevation depends only upon your initial state (the ground floor—the floor you started on) and your final state (the fourth floor—the floor you went to). It does not depend on whether you went directly to the fourth floor or up to the sixth and then down to the fourth floor. Your

Figure 5.3 (a) An open system allows exchange of both energy and matter with the surroundings. (b) A closed system allows exchange of energy but not matter. (c) An isolated system does not allow exchange of energy or matter. (This flask is enclosed by an insulating vacuum jacket.)

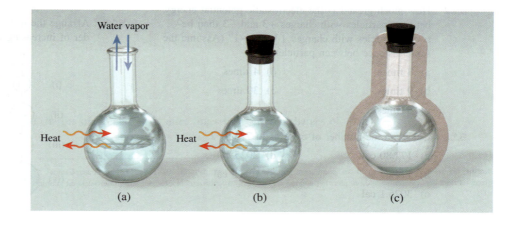

Water vapor

Heat   Heat

(a)   (b)   (c)

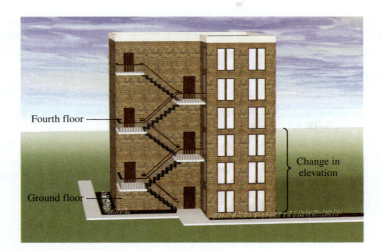

Figure 5.4 The change in elevation that occurs when a person goes from the ground floor to the fourth floor in a building does not depend on the path taken.

overall change in elevation is the same either way because it depends only on your initial and final elevations. Thus, elevation is a state function.

The amount of effort it takes to get from the ground floor to the fourth floor, on the other hand, depends on how you get there. More effort has to be exerted to go from the ground floor to the sixth floor and back down to the fourth floor than to go from the ground floor to the fourth floor directly. The effort required for this change in elevation is *not* a state function. Furthermore, if you subsequently return to the ground floor, your overall change in elevation will be zero, because your initial and final states are the same, but the amount of effort you exerted going from the ground floor to the fourth floor and back to the ground floor is *not* zero. Even though your initial and final states are the same, you do not get back the effort that went into climbing up and down the stairs.

Energy is a state function, too. Using potential energy as an example, your net increase in gravitational potential energy is always the same, regardless of how you get from the ground floor to the fourth floor of a building (Figure 5.4).

The First Law of Thermodynamics

The *first law of thermodynamics,* which is based on the law of conservation of energy, states that energy can be converted from one form to another but cannot be created or destroyed. It would be impossible to demonstrate this by measuring the total amount of energy in the universe; in fact, just determining the total energy content of a small sample of matter would be extremely difficult. Fortunately, because energy is a state function, we can demonstrate the first law by measuring the change in the energy of a system between its initial state and its final state in a process. The change in internal energy, ΔU, is given by

$$\Delta U = U_f - U_i$$

where U_i and U_f are the internal energies of the system in the initial and final states, respectively; and the symbol Δ means *final* minus *initial*.

The internal energy of a system has two components: kinetic energy and potential energy. The kinetic energy component consists of various types of molecular motion and the movement of electrons within molecules. Potential energy is determined by the attractive interactions between electrons and nuclei and by repulsive interactions between electrons and between nuclei in individual molecules, as well as by interactions between molecules. It is impossible to measure all these contributions accurately, so we cannot calculate the total energy of a system with any certainty. Changes in energy, on the other hand, can be determined experimentally.

Consider the reaction between 1 mole of sulfur and 1 mole of oxygen gas to produce 1 mole of sulfur dioxide:

$$S(s) + O_2(g) \longrightarrow SO_2(g)$$

Student Hot Spot

Student data indicate you may struggle with state functions. Access the eBook to view additional Learning Resources on this topic.

Student Note: Elemental sulfur exists as S_8 molecules, but we typically represent it as S to simplify chemical equations.

In this case, our system is composed of the reactant molecules and the product molecules. We do not know the internal energy content of either the reactants or the product, but we can accurately measure the *change* in energy content, ΔU, given by

$$\Delta U = U(\text{products}) - U(\text{reactants})$$
$$= \text{energy content of 1 mol } SO_2(g) - \text{energy content of 1 mol } S(s) \text{ and 1 mol } O_2(g)$$

This reaction gives off heat. Therefore, the energy of the product is less than that of the reactants, and ΔU is negative.

The release of heat that accompanies this reaction indicates that some of the chemical energy contained in the system has been converted to thermal energy. Furthermore, the thermal energy released by the system is absorbed by the surroundings. The transfer of energy from the system to the surroundings does not change the total energy of the universe. That is, the sum of the energy changes is zero:

$$\Delta U_{sys} + \Delta U_{surr} = 0$$

where the subscripts "sys" and "surr" denote system and surroundings, respectively. Thus, if a system undergoes an energy change ΔU_{sys}, the rest of the universe, or the surroundings, must undergo a change in energy that is equal in magnitude but opposite in sign:

$$\Delta U_{sys} = -\Delta U_{surr}$$

Energy released in one place must be gained somewhere else. Furthermore, because energy can be changed from one form to another, the energy lost by one system can be gained by another system in a different form. For example, the energy released by burning coal in a power plant may ultimately turn up in our homes as electric energy, heat, light, and so on.

Work and Heat

Student Note: The units for heat and work are the same as those for energy: joules, kilojoules, or calories.

Recall from Section 5.1 that energy is defined as the capacity to do work or transfer heat. When a system releases or absorbs heat, its internal energy changes. Likewise, when a system does work on its surroundings, or when the surroundings do work on the system, the system's internal energy also changes. The overall change in the system's internal energy is given by

Equation 5.3	$\Delta U = q + w$

where q is heat (released or absorbed by the system) and w is work (done *on* the system or done *by* the system). Note that it is possible for the heat and work components to cancel each other out and for there to be no change in the system's internal energy. Interestingly, although neither q nor w is a state function (each depends on the path between the initial and final states of the system), their sum, ΔU, does *not* depend on the path between initial and final states because U *is* a state function.

In chemistry, we are normally interested in the energy changes associated with the system rather than the surroundings. Therefore, unless otherwise indicated, ΔU will refer specifically to ΔU_{sys}. The sign conventions for q and w are as follows: q is positive for an endothermic process and negative for an exothermic process, and w is positive for work done on the system by the surroundings and negative for work done by the system on the surroundings. Table 5.1 summarizes the sign conventions for q and w.

TABLE 5.1	Sign Conventions for Heat (q) and Work (w)	
Process		**Sign**
Heat absorbed by the system (endothermic process)		q is positive
Heat released by the system (exothermic process)		q is negative
Work done on the system by the surroundings (e.g., a volume decrease)		w is positive
Work done by the system on the surroundings (e.g., a volume increase)		w is negative

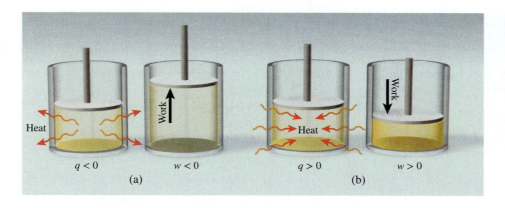

Figure 5.5 (a) When heat is released by the system (to the surroundings), q is negative. When work is done by the system (on the surroundings), w is negative. (b) When heat is absorbed by the system (from the surroundings), q is positive. When work is done on the system (by the surroundings), w is positive.

The drawings in Figure 5.5 illustrate the logic behind the sign conventions for q and w. If a system releases heat to the surroundings or does work on the surroundings [Figure 5.5(a)], we would expect its internal energy to decrease because they are energy-depleting processes. For this reason, both q and w are negative. Conversely, if heat is added to the system or if work is done on the system [Figure 5.5(b)], then the internal energy of the system increases. In this case, both q and w are positive.

Sample Problem 5.2 shows how to determine the overall change in the internal energy of a system.

SAMPLE PROBLEM 5.2

Calculate the overall change in internal energy, ΔU, (in joules) for a system that absorbs 188 J of heat and does 141 J of work on its surroundings.

Strategy Combine the two contributions to internal energy using Equation 5.3 and the sign conventions for q and w.

Setup The system absorbs heat, so q is *positive*. The system does work on the surroundings, so w is negative.

Solution

$$\Delta U = q + w = 188 \text{ J} + (-141 \text{ J}) = 47 \text{ J}$$

> ### THINK ABOUT IT
> Consult Table 5.1 to make sure that you have used the proper sign conventions for q and w.

Practice Problem ATTEMPT Calculate the change in total internal energy for a system that releases 1.34×10^4 kJ of heat and does 2.98×10^4 kJ of work on the surroundings.

Practice Problem BUILD Calculate the magnitude of q for a system that does 7.05×10^5 kJ of work on its surroundings and for which the change in total internal energy is -9.55×10^3 kJ. Indicate whether heat is absorbed or released by the system.

Practice Problem CONCEPTUALIZE
The diagram on the far left in the row shows a system before a process. Which of the diagrams on its right could represent the system after it undergoes a process in which the system absorbs heat and ΔU is positive?

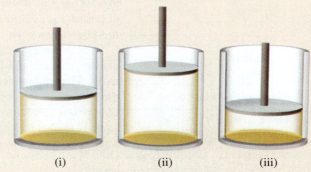

(i)   (ii)   (iii)

CHECKPOINT – SECTION 5.2   Introduction to Thermodynamics

5.2.1 Calculate the overall change in internal energy for a system that releases 43 J in heat in a process in which no work is done.

a) 43 J

b) -2.3×10^{-2} J

c) 0 J

d) 2.3×10^{-2} J

e) -43 J

5.2.2 Calculate w, and determine whether work is done by the system or on the system when 928 kJ of heat is released and $\Delta U = -1.47 \times 10^3$ kJ.

a) $w = -1.36 \times 10^6$ kJ, done by the system

b) $w = 1.36 \times 10^6$ kJ, done on the system

c) $w = -5.4 \times 10^2$ kJ, done by the system

d) $w = 2.4 \times 10^3$ kJ, done on the system

e) $w = -2.4 \times 10^3$ kJ, done by the system

5.2.3 Which of the following is a closed system? (Select all that apply.)

a) Pot of boiling water on a stove

b) Pot of cold water on a countertop

c) Helium-filled Mylar balloon

d) Styrofoam cup full of hot coffee (without lid)

e) Glass of iced tea

5.2.4 Which of the following could be described as a state function? (Select all that apply.)

a) Dieter's weight

b) Calories burned by a person walking to the health club

c) Distance traveled by a person walking to the health club

d) Distance between two points on a map

e) Energy required to heat a given amount of water from 25°C to the boiling point

5.3   Enthalpy

To calculate ΔU, we must know the values and signs of both q and w. As we will see in Section 5.4, we determine q by measuring temperature changes. To determine w, we need to know whether the reaction occurs under constant-volume or constant-pressure conditions.

Reactions Carried Out at Constant Volume or at Constant Pressure

Student Note: The explosive decomposition of NaN_3 is the reaction that inflates air bags in cars.

Imagine carrying out the decomposition of sodium azide (NaN_3) in two different experiments. In the first experiment, the reactant is placed in a metal cylinder with a fixed volume. When detonated, the NaN_3 reacts, generating a large quantity of N_2 gas inside the closed, fixed-volume container.

$$2NaN_3(s) \longrightarrow 2Na(s) + 3N_2(g)$$

The effect of this reaction will be an increase in the pressure inside the container, similar to what happens if you shake a bottle of soda vigorously prior to opening it. (The concept of pressure is examined in detail in Chapter 10. However, if you have ever put air in the tire of an automobile or a bicycle, you are familiar with the concept.)

Now imagine carrying out the same reaction in a metal cylinder with a movable piston. As this explosive decomposition proceeds, the piston in the metal cylinder will move. The gas produced in the reaction pushes the cylinder upward, thereby increasing the volume of the container and preventing any increase in pressure. This is a simple example of mechanical work done by a chemical reaction. Specifically, this type of work is known as *pressure-volume*, or *PV*, work. The amount of work done by such a process is given by

Student Note: Pressure-volume work and electrical work are two important types of work done by chemical reactions. Electrical work is discussed in detail in Chapter 19.

Equation 5.4	$w = -P\Delta V$

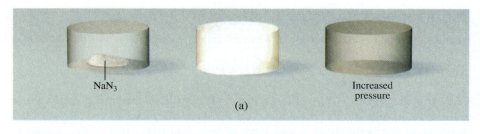

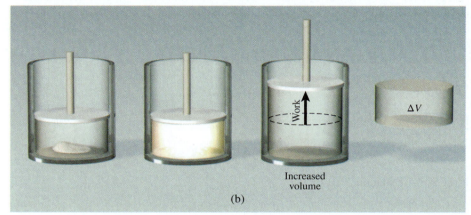

where P is the external, opposing pressure and ΔV is the change in the volume of the container as the result of the piston being pushed upward. In keeping with the sign conventions in Table 5.1, an increase in volume results in a negative value for w, whereas a decrease in volume results in a positive value for w. Figure 5.6 illustrates this reaction (a) being carried out at a constant volume, and (b) at a constant pressure.

When a chemical reaction is carried out at constant volume, then no PV work can be done because $\Delta V = 0$ in Equation 5.4. From Equation 5.3 it follows that

$$\Delta U = q - P\Delta V \qquad \textbf{Equation 5.5}$$

and, because $P\Delta V = 0$ at constant volume,

$$q_V = \Delta U \qquad \textbf{Equation 5.6}$$

We add the subscript "V" to indicate that this is a constant-volume process. This equality may seem strange at first. We said earlier that q is *not* a state function. However, for a process carried out under constant-volume conditions, q can have only one specific value, which is equal to ΔU. In other words, while q is *not* a state function, q_V *is* one.

Constant-volume conditions are often inconvenient and sometimes impossible to achieve. Most reactions occur in open containers, under conditions of constant pressure (usually at whatever the atmospheric pressure happens to be where the experiments are conducted). In general, for a constant-pressure process, we write

$$\Delta U = q + w$$
$$= q_P - P\Delta V$$

or

$$q_P = \Delta U + P\Delta V \qquad \textbf{Equation 5.7}$$

where the subscript "P" denotes constant pressure.

Enthalpy and Enthalpy Changes

There is a thermodynamic function of a system called ***enthalpy* (*H*)**, which is defined as

Equation 5.8 $$H = U + PV$$

Student Note: The SI unit of pressure is the *pascal* (Pa), which in SI base units is 1 kg/(m · s^2). Volume in SI base units is cubic meters (m^3). Therefore, multiplying units of *pressure* by units of *volume* gives (1 kg/m · s^2) × m^3 = 1 kg · m^2/s^2, which is the definition of the *joule* (J). Thus, $P\Delta V$ has units of *energy*.

where U is the internal energy of the system and P and V are the pressure and volume of the system, respectively. Because U and PV have energy units, enthalpy also has energy units. Furthermore, U, P, and V are all state functions—that is, the changes in $(U + PV)$ depend only on the initial and final states. It follows, therefore, that the change in H, or ΔH, also depends only on the initial and final states. Thus, H is a state function.

For any process, the *change* in enthalpy is given by

Equation 5.9 $$\Delta H = \Delta U + \Delta(PV)$$

If the pressure is held constant, then

Equation 5.10 $$\Delta H = \Delta U + P\Delta V$$

If we solve Equation 5.10 for ΔU,

$$\Delta U = \Delta H - P\Delta V$$

Then, substituting the result for ΔU into Equation 5.7, we obtain

$$q_P = (\Delta H - P\Delta V) + P\Delta V$$

The $P\Delta V$ terms cancel, and for a constant-pressure process, the heat exchanged between the system and the surroundings is equal to the enthalpy change:

Equation 5.11 $$q_P = \Delta H$$

Again, q is *not* a state function, but q_P *is;* that is, the heat change at constant pressure can have only one specific value and is equal to ΔH.

We now have two quantities—ΔU and ΔH—that can be associated with a reaction. If the reaction occurs under constant-volume conditions, then the heat change, q_V, is equal to ΔU. If the reaction is carried out at constant pressure, on the other hand, the heat change, q_P, is equal to ΔH.

Student Note: The enthalpy of reaction is often symbolized by ΔH_{rxn}. The subscript can be changed to denote a specific type of reaction or physical process: ΔH_{vap} can be used for the enthalpy of *vaporization*, for example.

Because most laboratory reactions are constant-pressure processes, the heat exchanged between the system and surroundings is equal to the change in enthalpy for the process. For any reaction, we define the change in enthalpy, called the ***enthalpy of reaction* (ΔH_{rxn})**, as the difference between the enthalpies of the products and the enthalpies of the reactants:

Equation 5.12 $$\Delta H = H(\text{products}) - H(\text{reactants})$$

The enthalpy of reaction can be positive or negative, depending on the process. For an endothermic process (where heat is absorbed by the system from the surroundings), ΔH is positive (i.e., $\Delta H > 0$). For an exothermic process (where heat is released by the system to the surroundings), ΔH is negative (i.e., $\Delta H < 0$).

We now apply the idea of enthalpy changes to two common processes, the first involving a physical change and the second involving a chemical change.

Thermochemical Equations

Under ordinary atmospheric conditions at sea level, ice melts to form liquid water when exposed to temperatures above 0°C. Measurements show that for every mole of ice converted to liquid water under these conditions, 6.01 kJ of heat energy is absorbed by the

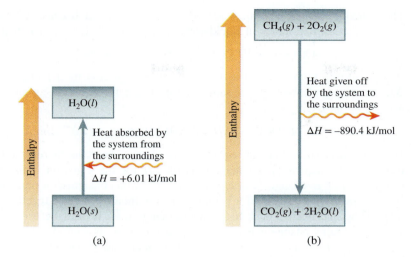

Figure 5.7 (a) Melting 1 mole of ice at 0°C, an endothermic process, results in an enthalpy increase of 6.01 kJ (ΔH = +6.01 kJ/mol). (b) The burning of 1 mole of methane in oxygen gas, an exothermic process, results in an enthalpy decrease in the system of 890.4 kJ (ΔH = −890.4 kJ/mol). The enthalpy diagrams of these two processes are not shown to the same scale.

system (the ice). Because the pressure is constant, the heat change is equal to the enthalpy change, ΔH. This is an *endothermic* process ($\Delta H > 0$), because heat is absorbed by the ice from its surroundings [Figure 5.7(a)]. The equation for this physical change is

$$H_2O(s) \longrightarrow H_2O(l) \qquad \Delta H = +6.01 \text{ kJ/mol}$$

The "per mole" in the unit for ΔH means that this is the enthalpy change *per mole of the reaction (or process) as it is written*—that is, when 1 mole of ice is converted to 1 mole of liquid water.

> **Student Note:** Although, strictly speaking, it is unnecessary to include the sign of a positive number, we will include the sign of all positive ΔH values to emphasize the thermochemical sign convention.

Now consider the combustion of methane (CH_4), the principal component of natural gas:

$$CH_4(g) + 2O_2(g) \longrightarrow CO_2(g) + 2H_2O(l) \qquad \Delta H = -890.4 \text{ kJ/mol}$$

From experience we know that burning natural gas releases heat to the surroundings, so it is an exothermic process. Under constant-pressure conditions, this heat change is equal to the enthalpy change and ΔH must have a negative sign [Figure 5.7(b)]. Again, the "per mole" in the units for ΔH means that when 1 mole of CH_4 reacts with 2 moles of O_2 to yield 1 mole of CO_2 and 2 moles of liquid H_2O, 890.4 kJ of heat is released to the surroundings. Note that when you specify that a particular amount of heat is released, it is not necessary to include a negative sign.

It is important to keep in mind that the ΔH value in kJ/mol does not mean *per mole* of a particular reactant or product. It refers to all the species in a reaction in the molar amounts specified by the coefficients in the balanced equation. Thus, for the combustion of methane, the ΔH value of −890.4 kJ/mol can be expressed in any of the following ways:

$$\frac{-890.4 \text{ kJ}}{1 \text{ mol } CH_4} \qquad \frac{-890.4 \text{ kJ}}{2 \text{ mol } O_2} \qquad \frac{-890.4 \text{ kJ}}{1 \text{ mol } CO_2} \qquad \frac{-890.4 \text{ kJ}}{2 \text{ mol } H_2O}$$

Although the importance of expressing ΔH in units of kJ/mol (rather than just kilojoules) will become apparent when we study thermodynamics in greater detail [▶▶ Chapter 18], you should learn and become comfortable with this convention now.

The equations for the melting of ice and the combustion of methane are examples of ***thermochemical equations,*** which are chemical equations that show the enthalpy changes as well as the mass relationships. It is essential to specify a balanced chemical equation when quoting the enthalpy change of a reaction. The following guidelines are helpful in interpreting, writing, and manipulating thermochemical equations:

1. When writing thermochemical equations, we must always specify the physical states of all reactants and products, because they help determine the actual enthalpy changes. In the equation for the combustion of methane, for example, changing the liquid water product to water vapor changes the value of ΔH:

$$CH_4(g) + 2O_2(g) \longrightarrow CO_2(g) + 2H_2O(g) \qquad \Delta H = -802.4 \text{ kJ/mol}$$

Student Note: Remember that the "per mole" in this context refers to per mole of reaction—not, in this case, per mole of water.

The enthalpy change is −802.4 kJ rather than −890.4 kJ because 88.0 kJ is needed to convert 2 moles of liquid water to 2 moles of water vapor; that is,

$$2H_2O(l) \longrightarrow 2H_2O(g) \qquad \Delta H = +88.0 \text{ kJ/mol}$$

2. If we multiply both sides of a thermochemical equation by a factor n, then ΔH must also change by the same factor. Thus, for the melting of ice, if $n = 2$, we have

$$2H_2O(s) \longrightarrow 2H_2O(l) \qquad \Delta H = 2(6.01 \text{ kJ/mol}) = +12.02 \text{ kJ/mol}$$

3. When we reverse a chemical equation, we change the roles of reactants and products. Consequently, the magnitude of ΔH for the equation remains the same, but its sign changes. For example, if a reaction consumes thermal energy from its surroundings (i.e., if it is endothermic), then the reverse reaction must release thermal energy back to its surroundings (i.e., it must be exothermic) and the enthalpy change expression must also change its sign. Thus, reversing the melting of ice and the combustion of methane, the thermochemical equations become

$$H_2O(l) \longrightarrow H_2O(s) \qquad\qquad \Delta H = −6.01 \text{ kJ/mol}$$
$$CO_2(g) + 2H_2O(l) \longrightarrow CH_4(g) + 2O_2(g) \qquad \Delta H = +890.4 \text{ kJ/mol}$$

What was an endothermic process becomes an exothermic process when reversed, and vice versa.

Sample Problem 5.3 illustrates the use of a thermochemical equation to relate the mass of a product to the energy consumed in the reaction.

SAMPLE PROBLEM 5.3

Given the thermochemical equation for photosynthesis,

$$6H_2O(l) + 6CO_2(g) \longrightarrow C_6H_{12}O_6(s) + 6O_2(g) \qquad \Delta H = +2803 \text{ kJ/mol}$$

calculate the solar energy required to produce 75.0 g of $C_6H_{12}O_6$.

Strategy The thermochemical equation shows that for every mole of $C_6H_{12}O_6$ produced, 2803 kJ is absorbed. We need to find out how much energy is absorbed for the production of 75.0 g of $C_6H_{12}O_6$. We must first find out how many moles there are in 75.0 g of $C_6H_{12}O_6$.

Setup The molar mass of $C_6H_{12}O_6$ is 180.2 g/mol, so 75.0 g of $C_6H_{12}O_6$ is

$$75.0 \text{ g} \times \frac{1 \text{ mol } C_6H_{12}O_6}{180.2 \text{ g}} = 0.416 \text{ mol}$$

We will multiply the thermochemical equation, including the enthalpy change, by 0.416, in order to write the equation in terms of the appropriate amount of $C_6H_{12}O_6$.

Solution

$$(0.416 \text{ mol})[6H_2O(l) + 6CO_2(g) \longrightarrow C_6H_{12}O_6(s) + 6O_2(g)]$$

and $(0.416 \text{ mol})(\Delta H) = (0.416 \text{ mol})(2803 \text{ kJ/mol})$ gives

$$2.50 \text{ } H_2O(l) + 2.50 \text{ } CO_2(g) \longrightarrow 0.416 \text{ } C_6H_{12}O_6(s) + 2.50 \text{ } O_2(g) \qquad \Delta H = +1.17 \times 10^3 \text{ kJ}$$

Therefore, 1.17×10^3 kJ of energy in the form of sunlight is consumed in the production of 75.0 g of $C_6H_{12}O_6$. Note that the "per mole" units in ΔH are canceled when we multiply the thermochemical equation by the number of moles of $C_6H_{12}O_6$.

THINK ABOUT IT

The specified amount of $C_6H_{12}O_6$ is less than half a mole. Therefore, we should expect the associated enthalpy change to be less than half that specified in the thermochemical equation for the production of 1 mole of $C_6H_{12}O_6$.

Practice Problem ⒶTTEMPT Calculate the solar energy required to produce 5255 g of $C_6H_{12}O_6$.

Practice Problem ⒷUILD Calculate the mass (in grams) of O_2 that is produced by photosynthesis when 2.49×10^4 kJ of solar energy is consumed.

Practice Problem ⒸONCEPTUALIZE The diagrams represent systems before and after reaction for two related chemical processes. ΔH for the first reaction is 1755.0 kJ/mol. Determine the value of ΔH for the second reaction.

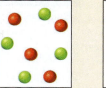

before   after

$\Delta H = 1755.0$ kJ/mol

before   after

$\Delta H = ?$

CHECKPOINT – SECTION 5.3   Enthalpy

5.3.1 Given the thermochemical equation: $H_2(g) + Br_2(l) \longrightarrow 2HBr(g)$, $\Delta H = -72.4$ kJ/mol, calculate the amount of heat released when a kilogram of $Br_2(l)$ is consumed in this reaction.

a) 7.24×10^4 kJ

b) 453 kJ

c) 906 kJ

d) 227 kJ

e) 724 kJ

5.3.2 Given the thermochemical equation: $2Cu_2O(s) \longrightarrow 4Cu(s) + O_2(g)$, $\Delta H = +333.8$ kJ/mol, calculate the mass of copper produced when 1.47×10^4 kJ is consumed in this reaction.

a) 11.2 kg

b) 176 kg

c) 44.0 kg

d) 334 kg

e) 782 kg

5.4 | Calorimetry

In the study of thermochemistry, heat changes that accompany physical and chemical processes are measured with a *calorimeter,* a closed container designed specifically for this purpose. We begin our discussion of *calorimetry,* the measurement of heat changes, by defining two important terms: *specific heat* and *heat capacity.*

Specific Heat and Heat Capacity

The *specific heat* (s) of a substance is the amount of heat required to raise the temperature of 1 g of the substance by 1°C. The *heat capacity* (C) is the amount of heat required to raise the temperature of an *object* by 1°C. We can use the specific heat of a substance to determine the heat capacity of a specified quantity of that substance. For example, we can use the specific heat of water, 4.184 J/(g · °C), to determine the heat capacity of a kilogram of water:

> **Student Note:** Although heat capacity is typically given for an object rather than for a substance— the "object" may be a given quantity of a particular substance.

$$\text{heat capacity of 1 kg of water} = \times 1000 \text{ g} = 4184 \frac{4.184 \text{ J}}{1 \text{ g} \cdot °\text{C}} \quad \text{or} \quad 4.184 \times 10^3 \text{ J/°C}$$

Note that specific heat has the units J/(g·°C) and heat capacity has the units J/°C. Table 5.2 shows the specific heat values of some common substances. If we know the specific heat and the amount of a substance, then the change in the sample's temperature (ΔT) will tell us the amount of heat (q) that has been absorbed or released

> 📶 **Student Hot Spot**
>
> Student data indicate you may struggle with specific heat and heat capacity. Access the eBook to view additional Learning Resources on this topic.

Animation

Figure 5.9, Determination of ΔH°_{rxn} by Constant-Pressure Calorimetry.

TABLE 5.2	Specific Heat Values of Some Common Substances		
Substance	**Specific Heat (J/g · °C)**	**Substance**	**Specific Heat (J/g · °C)**
Al(*s*)	0.900	Fe(*s*)	0.444
Au(*s*)	0.129	Hg(*l*)	0.139
C(graphite)	0.720	$H_2O(l)$	4.184
C(diamond)	0.502	Ni(*s*)	0.440
Cu(*s*)	0.385	$C_2H_5OH(l)$ (ethanol)	2.46

in a particular process. One equation for calculating the heat associated with a temperature change is given by

Equation 5.13	$q = sm\Delta T$

where s is the specific heat, m is the mass of the substance undergoing the temperature change, and ΔT is the temperature change: $\Delta T = T_{final} - T_{initial}$. Another equation for calculating the heat associated with a temperature change is given by

Equation 5.14	$q = C\Delta T$

Student Note: Note that $C = sm$. Although specific heat is a property of substances, and heat capacity is a property of objects, we can define a specified quantity of a substance as an "object" and determine its heat capacity using its mass and its specific heat.

where C is the heat capacity and ΔT is the temperature change. The sign convention for q is the same as that for an enthalpy change: q is positive for endothermic processes and negative for exothermic processes.

Sample Problem 5.4 shows how to use the specific heat of a substance to calculate the amount of heat needed to raise the temperature of the substance by a particular amount.

SAMPLE PROBLEM (5.4)

Calculate the amount of heat (in kJ) required to heat 255 g of water from 25.2°C to 90.5°C.

Strategy Use Equation 5.13 ($q = sm\Delta T$) to calculate q.

Setup $m = 255$ g, $s = 4.184$ J/g · °C, and $\Delta T = 90.5°C - 25.2°C = 65.3°C$

Solution

$$q = \frac{4.184 \text{ J}}{\text{g} \cdot °C} \times 255 \text{ g} \times 65.3°C = 6.97 \times 10^4 \text{ J} \quad \text{or} \quad 69.7 \text{ kJ}$$

THINK ABOUT IT

Look carefully at the cancellation of units and make sure that the number of kilojoules is smaller than the number of joules. It is a common error to multiply by 1000 instead of dividing in conversions of this kind.

Practice Problem ATTEMPT Calculate the amount of heat (in kJ) required to heat 1.01 kg of water from 0.05°C to 35.81°C.

Practice Problem BUILD What will be the final temperature of a 514-g sample of water, initially at 10.0°C, after 90.8 kJ have been added to it?

Practice Problem CONCEPTUALIZE Shown here are two samples of the same substance. When equal amounts of heat are added to both samples, the temperature of the sample on the left increases by 15.3°C. Determine the increase in temperature of the sample on the right.

Constant-Pressure Calorimetry

A crude constant-pressure calorimeter can be constructed from two Styrofoam coffee cups, as shown in Figure 5.8. This device, called a coffee-cup calorimeter, can be used to measure the heat exchanged between the system and surroundings for a variety of reactions, such as acid-base neutralization, heat of solution, and heat of dilution. Because the pressure is constant, the heat change for the process (q) is equal to the enthalpy change (ΔH). In such experiments, we consider the reactants and products to be the system, and the water in the calorimeter to be the surroundings. We neglect the small heat capacity of the Styrofoam cups in our calculations. In the case of an exothermic reaction, the heat released by the system is absorbed by the water (surroundings), thereby increasing its temperature. Knowing the mass of the water in the calorimeter, the specific heat of water, and the change in temperature, we can calculate q_P of the system using the equation

$$q_{sys} = -sm\Delta T \qquad \textbf{Equation 5.15}$$

Note that the minus sign makes q_{sys} a negative number if ΔT is a positive number (i.e., if the temperature goes up). This is in keeping with the sign conventions listed in Table 5.1. A negative ΔH or a negative q indicates an exothermic process, whereas a positive ΔH or a positive q indicates an endothermic process. Table 5.3 lists some of the reactions that can be studied with a constant-pressure calorimeter. Figure 5.9 shows how constant-pressure calorimetry can be used to determine ΔH for a reaction.

Constant-pressure calorimetry can also be used to determine the heat capacity of an object or the specific heat of a substance. Suppose, for example, that we have a lead pellet with a mass of 26.47 g originally at 89.98°C. We drop the pellet into a constant-pressure calorimeter containing 100.0 g of water at 22.50°C. The temperature of the water increases to 23.17°C. In this case, we consider the pellet to be the system and the water to be the surroundings. Because it is the temperature of the surroundings that we measure and because $q_{sys} = -q_{surr}$, we can rewrite Equation 5.15 as

$$q_{surr} = sm\Delta T$$

Thus, q_{water} of the water is

$$q_{water} = \frac{4.184 \text{ J}}{\text{g} \cdot °\text{C}} \times 100.0 \text{ g} \times (23.17°\text{C} - 22.50°\text{C}) = 280 \text{ J}$$

and q_{pellet} is −280 J. The negative sign indicates that heat is *released* by the pellet. Dividing q_{pellet} by the temperature change (ΔT) gives us the heat capacity of the pellet (C_{pellet}).

$$C_{pellet} = \frac{-280 \text{ J}}{23.17°\text{C} - 89.95°\text{C}} = 4.19 \text{ J/°C}$$

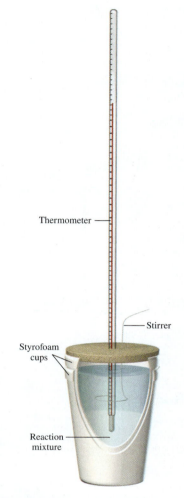

Figure 5.8 A constant-pressure coffee-cup calorimeter made of two Styrofoam cups. The nested cups help to insulate the reaction mixture from the surroundings. Two solutions of known volume containing the reactants at the same temperature are carefully mixed in the calorimeter. The heat produced or absorbed by the reaction can be determined by measuring the temperature change.

Student Note: The temperature of the water stops changing when it and the temperature of the lead pellet are equal. Therefore, the final temperature of the pellet is also 23.17°C.

TABLE 5.3	Heats of Some Typical Reactions and Physical Processes Measured at Constant Pressure		
Type of Reaction	**Example**		**ΔH (kJ/mol)**
Heat of neutralization	$HCl(aq) + NaOH(aq) \longrightarrow H_2O(l) + NaCl(aq)$		−56.2
Heat of ionization	$H_2O(l) \longrightarrow H^+(aq) + OH^-(aq)$		+56.2
Heat of fusion	$H_2O(s) \longrightarrow H_2O(l)$		+6.01
Heat of vaporization	$H_2O(l) \longrightarrow H_2O(g)$		+40.0*

*Measured at 25°C. At 100°C, the value is +40.79 kJ.

Figure 5.9

Determination of ΔH°_{rxn} by Constant-Pressure Calorimetry

One at a time, we pour the solutions into the calorimeter.

We start with 50.0 mL each of 1.00 M HCl and 1.00 M NaOH. Both solutions are at room temperature, which in this example is 25.0°C. The net ionic equation that represents the reaction is

$$H^{+}(aq) + OH^{-}(aq) \longrightarrow H_2O(l)$$

We have 0.0500 L × 1.00 M = 0.0500 mol of each reactant.

When both solutions have been added, we cap the calorimeter to prevent loss of energy to the environment and use the stirrer to ensure that the solutions are mixed thoroughly.

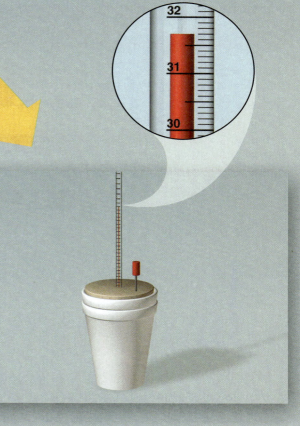

As the reaction proceeds, the temperature of the water increases as it absorbs the energy given off by the reaction. We record the maximum water temperature as 31.7°C.

Assuming the density and specific heat of the solution to be the same as those of water (1 g/mL, 4.184 J/g · °C), we calculate q_{soln} as follows:

q_{soln} = specific heat of water × mass of water × temperature change

$$= \frac{4.184 \text{ J}}{1 \text{ g} \cdot °C} \times 100.0 \text{ g} \times (31.7 - 25.0)°C = 2803 \text{ J}$$

Assuming that the heat capacity of the calorimeter is negligible, we know that $q_{soln} = -q_{rxn}$, and we can write

$q_{rxn} = -2803 \text{ J}$

This is the heat of reaction when 0.0500 mol H^+ reacts with 0.0500 mol OH^-. To determine ΔH_{rxn}, we divide q_{rxn} by the number of moles. (These reactants are present in stoichiometric amounts. If there were a *limiting* reactant, we would divide by the number of moles of limiting reactant.)

$$\Delta H_{rxn} = \frac{-2803 \text{ J}}{0.0500 \text{ mol}} = -5.61 \times 10^4 \text{ J/mol or } -56.1 \text{ kJ/mol}$$

This result is very close to the number we get using Equation 5.12 and the data in Appendix 2.

(See Visualizing Chemistry questions VC 5.1–VC 5.4 on page 242.)

What's the point?

Constant-pressure calorimetry can be used to determine ΔH_{rxn}—the heat of reaction for the reactant quantities specified by the balanced equation. However, when we carry out calorimetry experiments in the laboratory, we typically use much smaller quantities of reactants than those represented in a chemical equation. By measuring the temperature change of the surroundings (a known quantity of water in which the reactants are dissolved), we can determine q_{rxn} for the reactant quantities in the experiment. We can then divide q_{rxn} by the number of moles of reactant to determine ΔH_{rxn}.

Furthermore, because we know the *mass* of the pellet, we can determine the specific heat of lead (s_{lead}):

$$S_{Pb} = \frac{C_{pellet}}{m_{pellet}} = \frac{4.19 \text{ J/°C}}{26.47 \text{ g}} = 0.158 \text{ J/g} \cdot °C \quad \text{or} \quad 0.16 \text{ J/g} \cdot °C$$

Animation

Figure 5.10, Determination of Specific Heat by Constant-Pressure Calorimetry.

Sample Problem 5.5 shows how to use constant-pressure calorimetry to calculate the heat capacity (C) of an object. Figure 5.10 illustrates the process of determining the specific heat of a metal using constant-pressure calorimetry.

SAMPLE PROBLEM   5.5

A metal pellet with a mass of 100.0 g, originally at 88.4°C, is dropped into 125 g of water originally at 25.1°C. The final temperature of both the pellet and the water is 31.3°C. Calculate the heat capacity C (in J/°C) of the pellet.

Strategy Use Equation 5.13 ($q = sm\Delta T$) to determine the heat absorbed by the water; then use Equation 5.14 ($q = C\Delta T$) to determine the heat capacity of the metal pellet.

Setup $m_{water} = 125$ g, $s_{water} = 4.184$ J/g · °C, and $\Delta T_{water} = 31.3°C - 25.1°C = 6.2°C$. The heat absorbed by the water must be released by the pellet: $q_{water} = -q_{pellet}$, $m_{pellet} = 100.0$ g, and $\Delta T_{pellet} = 31.3°C - 88.4°C = -57.1°C$.

Solution From Equation 5.13, we have

$$q_{water} = \frac{4.184 \text{ J}}{\text{g} \cdot °C} \times 125 \text{ g} \times 6.2°C = 3242.6 \text{ J}$$

Thus,

$$q_{pellet} = -3242.6 \text{ J}$$

From Equation 5.14, we have

$$-3242.6 \text{ J} = C_{pellet} \times (-57.1°C)$$

Thus,

$$C_{pellet} = 57 \text{ J/°C}$$

THINK ABOUT IT

The units cancel properly to give appropriate units for heat capacity. Moreover, ΔT_{pellet} is a negative number because the temperature of the pellet decreases.

Practice Problem ATTEMPT What would the final temperature be if the pellet from Sample Problem 5.5, initially at 95°C, were dropped into a 218-g sample of water, initially at 23.8°C?

Practice Problem BUILD What mass of water could be warmed from 23.8°C to 46.3°C by the pellet in Sample Problem 5.5 initially at 116°C?

Practice Problem CONCEPTUALIZE Two samples of the same substance are shown. The temperatures of the two samples are indicated on the thermometer. Which of the final temperatures shown (i, ii, or iii) best represents the final temperature when the two samples are combined?

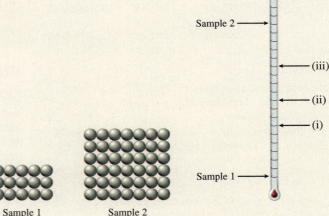

Sample 1   Sample 2

Bringing Chemistry to Life

Heat Capacity and Hypothermia

Like a warm metal pellet, the human body loses heat when it is immersed in cold water. Because we are warm-blooded animals, our body temperature is maintained at around 37°C. The human body is about 70 percent water by mass and water has a very high specific heat, so fluctuations in body temperature normally are very small. An air temperature of 25°C (often described as "room temperature") feels warm to us because air has a small specific heat (about 1 J/g · °C) and a low density. Consequently, very little heat is lost from the body to the surrounding air. The situation is drastically different if the body is immersed in water at 25°C. The heat lost by the human body when immersed in water can be thousands of times greater than that lost to air of the same temperature.

Hypothermia occurs when the body's mechanisms for producing and conserving heat are exceeded by loss of heat to the surroundings. Although hypothermia is dangerous and potentially deadly, there are certain circumstances under which it may actually be beneficial. A colder body temperature slows down all the normal biochemical processes, reducing the brain's need for oxygen, and prolonging the time period during which resuscitation efforts can be effective. Occasionally we hear about a seemingly miraculous recovery of a near-drowning victim who was submerged for a long period of time. These victims are usually small children who were submerged in icy water. The small size and, therefore, small heat capacity of a child allows for rapid cooling and may afford some protection from hypoxia—the lack of oxygen that causes death in drowning victims.

The lowest body temperature ever recorded for a human (who survived) was 13.7°C. Anna Bagenholm, 29, spent over an hour submerged after falling headfirst through the ice on a frozen river. Although she was clinically dead when she was pulled from the river, she has made a full recovery. There's a saying among doctors who treat hypothermia patients: "No one is dead until they're *warm* and dead."

> **Student Note:** Hypothermia routinely is induced in patients undergoing open heart surgery, drastically reducing the body's need for oxygen. Under these conditions, the heart can be stopped for the duration of the surgery.

Constant-Volume Calorimetry

The heat of combustion is usually measured using constant-volume calorimetry. Typically, a known mass of the compound to be analyzed is placed in a steel container called a *constant-volume bomb*, or simply a *bomb*, which is pressurized with oxygen. The closed bomb is then immersed in a known amount of water in an insulated

Figure 5.10

Determination of Specific Heat by Constant-Pressure Calorimetry

We add metal shot (125.0 g at 100.0°C) to the water, and we cap the calorimeter to prevent loss of energy to the environment.

We place 100.0 mL (100.0 g) of water in the calorimeter. The temperature of this water is 25.0°C.

We place 125.0 g of metal shot in a test tube and immerse it in boiling water long enough to heat all of the metal to the boiling point of water (100.0°C).

As energy is transferred from the metal shot to the water, the temperature of the water increases and the temperature of the metal shot decreases. We use the stirrer to ensure thorough mixing. The thermometer measures the temperature of the water.

When the temperature of the metal shot and the water are equal, the temperature of the water has reached a maximum value. We record this temperature as 34.1°C.

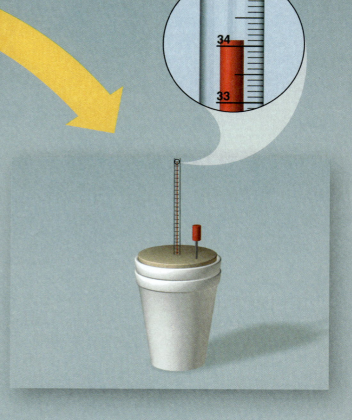

We know that $q_{water} = -q_{metal}$.
Substituting in the information given, we write:

q_{water} = specific heat of water × mass of water × temperature change

$$= \frac{4.184 \text{ J}}{1 \text{ g} \cdot {}^\circ\text{C}} \times 100.0 \text{ g} \times (34.1 - 25.0){}^\circ\text{C} = 3807 \text{ J}$$

q_{metal} = specific heat of metal × mass of metal × temperature change

$$= x \times 125.0 \text{ g} \times (34.1 - 100.0){}^\circ\text{C} = -8238 \, x \text{ g} \cdot {}^\circ\text{C}$$

and

$$3807 \text{ J} = -(-8238 \, x) \text{ g} \cdot {}^\circ\text{C}$$

$$x = \frac{3807 \text{ J}}{8238 \text{ g} \cdot {}^\circ\text{C}} = 0.46 \text{ J/g} \cdot {}^\circ\text{C}$$

The specific heat of the metal is therefore 0.46 J/g · °C.

(See Visualizing Chemistry questions VC 5.5–VC 5.8 on page 242.)

What's the point?

We can determine the specific heat of a metal by combining a known mass of the metal at a known temperature with a known mass of water at a known temperature. Assuming the calorimeter has a negligible heat capacity, the amount of energy lost by the hotter metal is equal to the amount of energy gained by the cooler water.

Figure 5.11 A constant-volume bomb calorimeter. The calorimeter is filled with oxygen gas at high pressure before it is placed in the bucket. The sample is ignited electrically, and the heat produced by the reaction is determined by measuring the temperature increase in the known amount of water surrounding the bomb.

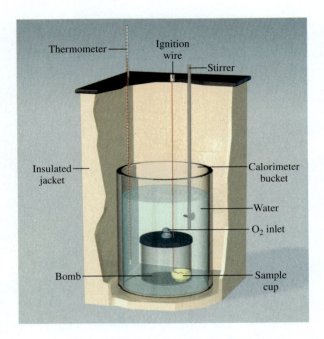

container, as shown in Figure 5.11. (Together, the steel bomb and the water in which it is submerged constitute the *calorimeter*.) The sample is ignited electrically, and the heat released by the combustion of the sample is absorbed by the bomb and the water and can be determined by measuring the increase in temperature of the water. The special design of this type of calorimeter allows us to assume that no heat (or mass) is lost to the surroundings during the time it takes to carry out the reaction and measure the temperature change. Therefore, we can call the bomb and the water in which it is submerged an *isolated* system. Because no heat enters or leaves the system during the process, the heat change of the system overall (q_{system}) is zero and we can write

$$q_{cal} = -q_{rxn}$$

where q_{cal} and q_{rxn} are the heat changes for the calorimeter and the reaction, respectively. Thus,

$$q_{rxn} = -q_{cal}$$

To calculate q_{cal}, we need to know the heat capacity of the calorimeter (C_{cal}) and the change in temperature, that is,

Equation 5.16 $$q_{cal} = C_{cal}\Delta T$$

And, because $q_{rxn} = -q_{cal}$,

Equation 5.17 $$q_{rxn} = -C_{cal}\Delta T$$

The heat capacity of the calorimeter (C_{cal}) is determined by burning a substance with an accurately known heat of combustion. For example, it is known that the combustion of a 1.000-g sample of benzoic acid (C_6H_5COOH) releases 26.38 kJ of heat. If the measured temperature increase is 4.673°C, then the heat capacity of the calorimeter is given by

$$C_{cal} = \frac{q_{cal}}{\Delta T} = \frac{26.38 \text{ kJ}}{4.673°C} = 5.645 \text{ kJ/°C}$$

Once C_{cal} has been determined, the calorimeter can be used to measure the heat of combustion of other substances. Because a reaction in a bomb calorimeter occurs under constant-volume rather than constant-pressure conditions, the measured heat

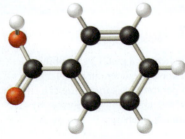

Benzoic acid

change corresponds to the *internal energy* change (ΔU) rather than to the *enthalpy* change (ΔH) (see Equations 5.6 and 5.11). It is possible to correct the measured heat changes so that they correspond to ΔH values, but the corrections usually are quite small, so we will not concern ourselves with the details here.

Sample Problem 5.6 shows how to use constant-volume calorimetry to determine the energy content per gram of a substance.

SAMPLE PROBLEM 5.6

A Famous Amos bite-sized chocolate chip cookie weighing 7.25 g is burned in a bomb calorimeter to determine its energy content. The heat capacity of the calorimeter is 39.97 kJ/°C. During the combustion, the temperature of the water in the calorimeter increases by 3.90°C. Calculate the energy content (in kJ/g) of the cookie.

Strategy Use Equation 5.17 ($q_{rxn} = -C_{cal}\Delta T$) to calculate the heat released by the combustion of the cookie. Divide the heat released by the mass of the cookie to determine its energy content per gram.

Setup $C_{cal} = 39.97$ kJ/°C and $\Delta T = 3.90$°C

Solution From Equation 5.17 we have

$$q_{rxn} = -C_{cal}\Delta T = -(39.97 \text{ kJ/°C})(3.90\text{°C}) = -1.559 \times 10^2 \text{ kJ}$$

The negative sign in the result indicates that heat is released by the combustion. Because energy content is a positive quantity, we write

$$\text{energy content per gram} = \frac{1.559 \times 10^2 \text{ kJ}}{7.25 \text{ g}} = 21.5 \text{ kJ/g}$$

Nutrition facts label on a box of Famous Amos chocolate chip cookies.

Brian Moeskau

THINK ABOUT IT

According to the label on the cookie package, a serving size is four cookies, or 29 g, and each serving contains 150 Cal. Convert the energy per gram to Calories per serving to verify the result.

$$\frac{21.5 \text{ kJ}}{g} \times \frac{1 \text{ Cal}}{4.184 \text{ kJ}} \times \frac{29 \text{ g}}{\text{serving}} = 1.5 \times 10^2 \text{ Cal/serving}$$

Practice Problem A TTEMPT A serving of Grape-Nuts cereal (5.80 g) is burned in a bomb calorimeter with a heat capacity of 43.7 kJ/°C. During the combustion, the temperature of the water in the calorimeter increased by 1.92°C. Calculate the energy content (in kJ/g) of Grape-Nuts.

Practice Problem B UILD The energy content of raisin bread is 13.1 kJ/g. Calculate the temperature increase when a slice of raisin bread (32.0 g) is burned in the calorimeter in Sample Problem 5.6.

Practice Problem C ONCEPTUALIZE Suppose an experiment to determine the energy content of food used a calorimeter that contained less water than it did when it was calibrated. Explain how this would affect the result of the experiment.

CHECKPOINT – SECTION 5.4   Calorimetry

5.4.1 A 1.000-g sample of benzoic acid is burned in a calorimeter to determine its heat capacity, C_{cal}. The reaction gives off 26.42 kJ of heat and the temperature of the water in the calorimeter increases from 23.40°C to 27.20°C. What is the heat capacity of the calorimeter?

a) 3.80 kJ/°C

b) 6.95 kJ/°C

c) 100 kJ/°C

d) 0.144 kJ/°C

e) 7.81 kJ/°C

5.4.2 One-gram samples of Al, Fe, and Au are each heated from 40°C to 75°C. Arrange the metals in order of increasing amount of heat absorbed in the process.

a) Al < Fe < Au

b) Al < Au < Fe

c) Fe < Au < Al

d) Au < Fe < Al

e) All of the metals absorbed the same amount of heat.

5.4.3 A reaction, carried out in a bomb calorimeter with $C_{cal} = 5.01$ kJ/°C, gives off 318 kJ of heat. The initial temperature of the water is 24.8°C. What is the final temperature of the water in the calorimeter?

a) 88.3°C   d) 162°C

b) 63.5°C   e) 76.7°C

c) 29.8°C

5.4.4 Quantities of 50.0 mL of 1.00 M HCl and 50.0 mL of 1.00 M NaOH are combined in a constant-pressure calorimeter. Both solutions are initially at 24.4°C. Calculate the final temperature of the combined solutions. (Use the data from Table 5.3. Assume that the mass of the combined solutions is 100.0 g and that the solution's specific heat is the same as that for water, 4.184 J/g · °C.) The heat capacity of the calorimeter is negligibly small.

a) 31.1°C   d) 91.8°C

b) 29.0°C   e) 35.7°C

c) 44.2°C

What if the Heat Capacity of the Calorimeter Isn't Negligible?

Although we usually assume that no heat is absorbed by the Styrofoam cups we use in the laboratory for constant-pressure calorimetry, in reality, the calorimeter generally does absorb a small portion of the heat produced by a chemical reaction. We can determine the heat capacity of a coffee-cup calorimeter by combining reactant solutions with precisely known concentrations and masses. Once we have determined the heat capacity, we can correct for the heat absorbed by the calorimeter when we carry out other experiments.

Consider an experiment in which we combine 50.0 mL of 0.250 M HCl(aq) with 50.0 mL of 0.250 M NaOH. The calorimeter and both solutions are initially at 23.50°C, and the density and specific heat of the combined solution are the same as that of water (1.000 g/mL and 4.184 J/g · °C, respectively). We can determine the amount of heat the reaction will generate using moles of reactants and the heat of neutralization value from Table 5.3. We have 0.0500 L × 0.250 M = 0.0125 mol of each reactant. According to Table 5.3, the heat of neutralization is −56.2 kJ/mol. Thus, we expect the enthalpy change of the system to be 0.0125 mol × (−56.2 kJ/mol) = −0.703 kJ. Converting this to joules and using Equation 5.15, we calculate the temperature change we expect from the combination of these reactants.

$$q_{sys} = -703 \text{ J} = -4.184 \text{ J/g} \cdot {}^\circ\text{C} \times 100.0 \text{ g} - \Delta T$$

$$\Delta T = \frac{-703 \text{ J}}{(-100.0 \text{ g})(4.184 \text{ J/g} \cdot {}^\circ\text{C})} = 1.68 {}^\circ\text{C}$$

Thus, we expect the temperature to increase by 1.68°C. However, the measured final temperature is 25.09°C, an increase of only 1.59°C. The water temperature increased by less than expected because the calorimeter absorbed part of the heat produced by the neutralization reaction. We can determine how much heat the calorimeter absorbed by using Equation 5.15 again. This time, we use the measured temperature change to calculate the amount of heat absorbed by the water.

$$q_{surr} = 4.184 \text{ J/g} \cdot {}^\circ\text{C} \times 100.0 \text{ g} \times 1.59 {}^\circ\text{C} = 665 \text{ J}$$

We know that the reaction produced 703 J but the water absorbed only 665 J. The remaining energy, 703 − 665 = 38 J, is $q_{calorimeter}$, the heat absorbed by the calorimeter itself. We use Equation 5.14 to calculate the heat capacity of the calorimeter.

$$38 \text{ J} = C_{calorimeter} \times 1.59 {}^\circ\text{C}$$

$$C_{calorimeter} = 23.9 \text{ J/}{}^\circ\text{C}$$

Remember that the calorimeter had the same initial temperature and the same final temperature as the solutions.

5.5   Hess's Law

Because enthalpy is a state function, the change in enthalpy that occurs when reactants are converted to products in a reaction is the same whether the reaction takes place in one step or in a series of steps. This observation is called **Hess's law.**[2] An analogy for Hess's law can be made to the floors in a building. Suppose, for example, that you take the elevator from the first floor to the sixth floor of the building. The net

---

2. Germain Henri Hess (1802–1850). Swiss chemist. Hess was born in Switzerland but spent most of his life in Russia. For formulating Hess's law, he is called the father of thermochemistry.

gain in your gravitational potential energy (which is analogous to the enthalpy change for the overall process) is the same whether you go directly there or stop at each floor on your way up (breaking the trip into a series of steps).

Recall from Section 5.3 that the enthalpy change for the combustion of a mole of methane depends on whether the product water is liquid or gas. More heat is given off by the reaction that produces liquid water. We can use this example to illustrate Hess's law by envisioning the first of these reactions happening in two steps. In step 1, methane and oxygen are converted to carbon dioxide and liquid water, releasing heat.

$$CH_4(g) + 2O_2(g) \longrightarrow CO_2(g) + 2H_2O(l) \qquad \Delta H = -890.4 \text{ kJ/mol}$$

In step 2, the liquid water is vaporized, which requires an input of heat.

$$2H_2O(l) \longrightarrow 2H_2O(g) \qquad \Delta H = +88.0 \text{ kJ/mol}$$

We can add balanced chemical equations just as we can add algebraic equalities, canceling identical items on opposite sides of the equation arrow:

$$
\begin{array}{ll}
CH_4(g) + 2O_2(g) \longrightarrow CO_2(g) + 2H_2O(l) & \Delta H = -890.4 \text{ kJ/mol} \\
+ \qquad 2H_2O(l) \longrightarrow 2H_2O(g) & \Delta H = +88.0 \text{ kJ/mol} \\
\hline
CH_4(g) + 2O_2(g) \longrightarrow CO_2(g) + 2H_2O(g) & \Delta H = -802.4 \text{ kJ/mol}
\end{array}
$$

When we add thermochemical equations, we add the ΔH values as well. This gives us the overall enthalpy change for the net reaction. Using this method, we can deduce the enthalpy changes for many reactions, some of which may not be possible to carry out directly. In general, we apply Hess's law by arranging a series of chemical equations (corresponding to a series of steps) in such a way that they sum to the desired overall equation. Often, in applying Hess's law, we must manipulate the equations involved, multiplying by appropriate coefficients, reversing equations, or both. It is important to follow the guidelines [◄◄ Section 5.3] for the manipulation of thermochemical equations and to make the corresponding change to the enthalpy change of each step.

Sample Problem 5.7 illustrates the use of this method for determining ΔH.

Student Hot Spot

Student data indicate you may struggle with Hess's law. Access the eBook to view additional Learning Resources on this topic.

SAMPLE PROBLEM 5.7

Given the following thermochemical equations,

$$NO(g) + O_3(g) \longrightarrow NO_2(g) + O_2(g)$$
$$\Delta H = -198.9 \text{ kJ/mol}$$

$$O_3(g) \longrightarrow O_2(g)\tfrac{3}{2}$$
$$\Delta H = -142.3 \text{ kJ/mol}$$

$$O_2(g) \longrightarrow 2O(g)$$
$$\Delta H = +495 \text{ kJ/mol}$$

determine the enthalpy change for the reaction

$$NO(g) + O(g) \longrightarrow NO_2(g)$$

Strategy Arrange the given thermochemical equations so that they sum to the desired equation. Make the corresponding changes to the enthalpy changes, and add them to get the desired enthalpy change.

Setup The first equation has NO as a reactant with the correct coefficient, so we use it as is.

$$NO(g) + O_3(g) \longrightarrow NO_2(g) + O_2(g) \qquad \Delta H = -198.9 \text{ kJ/mol}$$

The second equation must be reversed so that the O_3 introduced by the first equation will cancel (O_3 is not part of the overall chemical equation). We also must change the sign on the corresponding ΔH value.

$$\tfrac{3}{2}O_2(g) \longrightarrow O_3(g) \qquad \Delta H = +142.3 \text{ kJ/mol}$$

These two steps sum to give the following:

$$
\begin{array}{ll}
NO(g) + O_3(g) \longrightarrow NO_2(g) + O_2(g) & \Delta H = -198.9 \text{ kJ/mol} \\
+ \tfrac{1}{2}O_2(g) \tfrac{3}{2}O_2(g) \longrightarrow O_3(g) & \Delta H = +142.3 \text{ kJ/mol} \\
\hline
NO(g) + \tfrac{1}{2}O_2(g) \longrightarrow NO_2(g) & \Delta H = -56.6 \text{ kJ/mol}
\end{array}
$$

(Continued on next page)

We then replace the $\frac{1}{2}O_2$ on the left with O by incorporating the last equation. To do so, we divide the third equation by 2 and reverse its direction. As a result, we must also divide its ΔH value by 2 and change its sign.

$$O(g) \longrightarrow \tfrac{1}{2}O_2(g) \qquad \Delta H = -247.5 \text{ kJ/mol}$$

Finally, we sum all the steps and add their enthalpy changes.

Solution

$$
\begin{array}{ll}
NO(g) + \cancel{O_3(g)} \longrightarrow NO_2(g) + \cancel{O_2(g)} & \Delta H = -198.9 \text{ kJ/mol} \\
\cancel{\tfrac{3}{2}O_2(g)} \longrightarrow \cancel{O_3(g)} & \Delta H = +142.3 \text{ kJ/mol} \\
+ O(g) \longrightarrow \cancel{\tfrac{1}{2}O_2(g)} & \Delta H = -274.5 \text{ kJ/mol} \\
\hline
NO(g) + O(g) \longrightarrow NO_2(g) & \Delta H = -304 \text{ kJ/mol}
\end{array}
$$

THINK ABOUT IT

Recall that when you first learned to balance chemical equations, you were advised to use the smallest possible *whole* numbers as coefficients [◄◄ Section 3.3]. Using Hess's law often requires you to use *fractions* as coefficients. Remember that a balanced chemical equation can be thought of as an algebraic equality— and that you can multiply through the entire equation, including the enthalpy of reaction, by any number you choose.

before   after
$\Delta H = 25 \text{ kJ/mol}$

before   after
$\Delta H = -60 \text{ kJ/mol}$

before   after
$\Delta H = 100 \text{ kJ/mol}$

before   after
$\Delta H = ?$

Practice Problem ATTEMPT Use the thermochemical equations provided in Sample Problem 5.7 to determine the enthalpy change for the reaction $2NO(g) + 4O(g) \longrightarrow 2NO_2(g) + O_2(g)$.

Practice Problem BUILD Use the thermochemical equations provided in Sample Problem 5.7 to determine the enthalpy change for the reaction $2NO_2(g) \longrightarrow 2NO(g) + O_2(g)$.

Practice Problem CONCEPTUALIZE The diagrams shown are representations of four systems before and after reactions involving five different chemical species—each represented by a different color sphere. The ΔH values are given for the first three. Determine ΔH for the last reaction.

CHECKPOINT – SECTION 5.5   Hess's Law

5.5.1 Given the following information:

$$2H_2(g) + O_2(g) \longrightarrow 2H_2O(g) \qquad \Delta H = -483.6 \text{ kJ/mol}$$
$$3O_2(g) \longrightarrow 2O_3(g) \qquad \Delta H = +284.6 \text{ kJ/mol}$$

what is ΔH for $3H_2(g) + O_3(g) \longrightarrow 3H_2O(g)$?

a) −199 kJ/mol

b) −1010 kJ/mol

c) −867.7 kJ/mol

d) +768.2 kJ/mol

e) −440.8 kJ/mol

5.5.2 Given the following information:

$$P_4(s) + 3O_2(g) \longrightarrow P_4O_6(s) \qquad \Delta H = -1640.1 \text{ kJ/mol}$$
$$P_4(s) + 5O_2(g) \longrightarrow P_4O_{10}(s) \qquad \Delta H = -2940.1 \text{ kJ/mol}$$

what is the value of ΔH_{rxn} for $P_4O_6(s) + 2O_2(g) \longrightarrow P_4O_{10}(s)$?

a) −1300.0 kJ/mol

b) +4580.2 kJ/mol

c) −4580.2 kJ/mol

d) +982.6 kJ/mol

e) −982.6 kJ/mol

5.5.3 Each diagram shows a system before and after a chemical reaction along with the corresponding enthalpy of reaction. Determine the enthalpy of reaction for the last reaction represented.

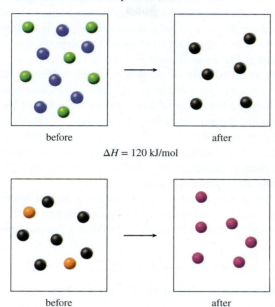

before   after

$\Delta H = 120$ kJ/mol

before   after

$\Delta H = 45$ kJ/mol

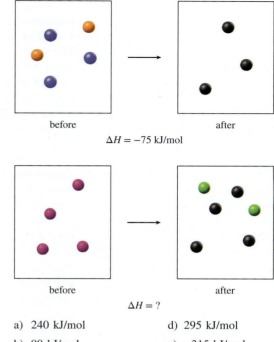

before   after

$\Delta H = -75$ kJ/mol

before   after

$\Delta H = ?$

a) 240 kJ/mol   d) 295 kJ/mol

b) 90 kJ/mol   e) −315 kJ/mol

c) −120 kJ/mol

5.5.4 Given the thermochemical data,

$$A + B \longrightarrow C \qquad \Delta H_1 = -100 \text{ kJ/mol}$$
$$A + \tfrac{3}{2}B \longrightarrow D \qquad \Delta H_2 = -150 \text{ kJ/mol}$$

determine the enthalpy of reaction for

$$2C + B \longrightarrow 2D \qquad \Delta H_3 = ?$$

a) 500 kJ/mol   d) −100 kJ/mol

b) −500 kJ/mol   e) 200 kJ/mol

c) −50 kJ/mol

5.6 Standard Enthalpies of Formation

So far we have learned that we can determine the enthalpy change that accompanies a reaction by measuring the heat absorbed or released (at constant pressure). According to Equation 5.12, ΔH can also be calculated if we know the enthalpies of all reactants and products. However, there is no way to measure the *absolute* value of the enthalpy of a substance. Only values *relative* to an arbitrary reference can be determined. This problem is similar to the one geographers face in expressing the elevations of specific mountains or valleys. Rather than trying to devise some type of "absolute" elevation scale (perhaps based on the distance from the center of Earth), by common agreement all geographical heights and depths are expressed relative to sea level, an arbitrary reference with a defined elevation of "zero" meters or feet. Similarly, chemists have agreed on an arbitrary reference point for enthalpy.

The "sea level" reference point for all enthalpy expressions is called the ***standard enthalpy of formation*** (ΔH_f°), which is defined as the heat change that results when 1 mole of a compound is formed from its constituent elements in their standard states. The superscripted degree sign denotes standard-state conditions, and the subscripted f stands for *formation*. The phrase "in their standard states" refers to the most stable form of an element under standard conditions, meaning at ordinary atmospheric pressure. The element oxygen, for example, can exist as atomic oxygen (O), diatomic oxygen (O_2), or ozone (O_3). By far the most stable form at ordinary atmospheric pressure, though, is diatomic oxygen. Thus, the standard state of oxygen is $O_2(g)$. Although the standard state does not specify a temperature, we will always use ΔH_f° values measured at 25°C.

Appendix 2 lists the standard enthalpies of formation for a number of elements and compounds. By convention, the standard enthalpy of formation of any element in its most stable form is zero. Again, using the element oxygen as an example, we can write $\Delta H_f^\circ(O_2) = 0$, but $\Delta H_f^\circ(O_3) \neq 0$ and $\Delta H_f^\circ(O) \neq 0$. Similarly, graphite is a more stable allotropic form of carbon than diamond under standard conditions and 25°C, so we have ΔH_f° (graphite) = 0 and ΔH_f° (diamond) $\neq 0$.

The importance of the standard enthalpies of formation is that once we know their values, we can readily calculate the ***standard enthalpy of reaction*** (ΔH_{rxn}°), defined as the enthalpy of a reaction carried out under standard conditions. For example, consider the hypothetical reaction

$$a\text{A} + b\text{B} \longrightarrow c\text{C} + d\text{D}$$

where *a*, *b*, *c*, and *d* are stoichiometric coefficients. For this reaction, ΔH_{rxn}° is given by

Equation 5.18 $\Delta H_{rxn}^\circ = [c\Delta H_f^\circ(\text{C}) + d\Delta H_f^\circ(\text{D})] - [a\Delta H_f^\circ(\text{A}) + b\Delta H_f^\circ(\text{B})]$

We can generalize Equation 5.18 as

Equation 5.19 $\Delta H_{rxn}^\circ = \Sigma n\Delta H_f^\circ(\text{products}) - \Sigma m\Delta H_f^\circ(\text{reactants})$

where *m* and *n* are the stoichiometric coefficients for the reactants and products, respectively, and Σ (sigma) means "the sum of." In these calculations, the stoichiometric coefficients are treated as numbers without units. Thus, the result has units of kJ/mol, where again, "per mole" means per mole of reaction as written. To use Equation 5.19 to calculate ΔH_{rxn}°, we must know the ΔH_f° values of the compounds that take part in the reaction. These values, tabulated in Appendix 2, are determined by either the direct method or the indirect method.

The *direct* method of measuring ΔH_f° works for compounds that can be synthesized from their elements easily and safely. Suppose we want to know the enthalpy of formation of carbon dioxide. We must measure the enthalpy of the reaction when carbon (graphite) and molecular oxygen in their standard states are converted to carbon dioxide in its standard state:

$$\text{C(graphite)} + O_2(g) \longrightarrow CO_2(g) \qquad \Delta H_{rxn}^\circ = -393.5 \text{ kJ/mol}$$

We know from experience that this combustion goes to completion. Thus, from Equation 5.19 we can write

$$\Delta H_{rxn}^\circ = \Delta H_f^\circ(CO_2) - [\Delta H_f^\circ(\text{graphite}) + \Delta H_f^\circ(O_2)] = -393.5 \text{ kJ/mol}$$

Because graphite and O_2 are the most stable allotropic forms of their respective elements, ΔH_f°(graphite) and $\Delta H_f^\circ(O_2)$ are both zero. Therefore,

$$\Delta H_{rxn}^\circ = \Delta H_f^\circ(CO_2) - 393.5 \text{ kJ/mol}$$

or

$$\Delta H_f^\circ(CO_2) - 393.5 \text{ kJ/mol}$$

Arbitrarily assigning a value of zero to ΔH_f° for each element in its standard state does *not* affect the outcome of these calculations. Remember, in thermochemistry we are interested only in enthalpy changes because they can be determined experimentally, whereas the absolute enthalpy values cannot. The choice of a zero "reference level" for enthalpy is intended to simplify the calculations. Referring again to the terrestrial altitude analogy, we find that Mt. Everest (the highest peak in the world) is 8708 ft higher than Mt. Denali (the highest peak in North America). This difference in altitude would be the same whether we had chosen sea level or the center of Earth as our reference elevation.

Other compounds that can be studied by the direct method are SF_6, P_4O_{10}, and CS_2. The equations representing their syntheses are

$$S(\text{rhombic}) + 3F_2(g) \longrightarrow SF_6(g)$$
$$P_4(\text{white}) + 5O_2(g) \longrightarrow P_4O_{10}(s)$$
$$C(\text{graphite}) + 2S(\text{rhombic}) \longrightarrow CS_2(l)$$

S(rhombic) and P(white) are the most stable allotropes of sulfur and phosphorus, respectively, at 1 atm and 25°C, so their ΔH_f° values are zero.

Sample Problem 5.8 shows how ΔH_f° values can be used to determine ΔH_{rxn}°.

SAMPLE PROBLEM 5.8

Using data from Appendix 2, calculate ΔH_{rxn}° for $Ag^+(aq) + Cl^-(aq) \longrightarrow AgCl(s)$.

Strategy Use Equation 5.19 [$\Delta H_{rxn}^\circ = \Sigma n \Delta H_f^\circ(\text{products}) - \Sigma m \Delta H_f^\circ(\text{reactants})$] and ΔH_f° values from Appendix 2 to calculate ΔH_{rxn}°.

Setup The ΔH_f° values for $Ag^+(aq)$, $Cl^-(aq)$, and $AgCl(s)$ are +105.9, −167.2, and −127.0 kJ/mol, respectively.

Solution Using Equation 5.19,

$$\Delta H_{rxn}^\circ = \Delta H_f^\circ(AgCl) - [\Delta H_f^\circ(Ag^+) + \Delta H_f^\circ(Cl^-)]$$
$$= -127.0 \text{ kJ/mol} - [(+105.9 \text{ kJ/mol}) + (-167.2 \text{ kJ/mol})]$$
$$= -127.0 \text{ kJ/mol} - (-61.3 \text{ kJ/mol}) = -65.7 \text{ kJ/mol}$$

THINK ABOUT IT
Watch out for misplaced or missing minus signs. This is an easy place to lose track of them.

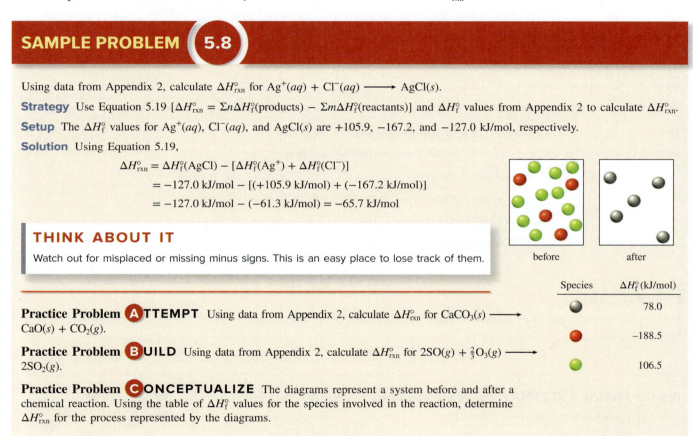

before   after

Species	ΔH_f° (kJ/mol)
(gray)	78.0
(red)	−188.5
(green)	106.5

Practice Problem Ⓐ**TTEMPT** Using data from Appendix 2, calculate ΔH_{rxn}° for $CaCO_3(s) \longrightarrow CaO(s) + CO_2(g)$.

Practice Problem Ⓑ**UILD** Using data from Appendix 2, calculate ΔH_{rxn}° for $2SO(g) + \frac{2}{3}O_3(g) \longrightarrow 2SO_2(g)$.

Practice Problem Ⓒ**ONCEPTUALIZE** The diagrams represent a system before and after a chemical reaction. Using the table of ΔH_f° values for the species involved in the reaction, determine ΔH_{rxn}° for the process represented by the diagrams.

Many compounds cannot be synthesized from their elements directly. In some cases, the reaction proceeds too slowly, or side reactions produce substances other than the desired compound. In these cases, ΔH_f° can be determined by an *indirect* approach, using Hess's law. If we know a series of reactions for which ΔH_{rxn}° can be measured, and we can arrange them in such a way as to have them sum to the equation corresponding to the formation of the compound of interest, we can calculate ΔH_f° for the compound.

Sample Problem 5.9 shows how to use Hess's law to calculate the ΔH_f° value by the indirect method for a compound that cannot be produced easily from its constituent elements.

SAMPLE PROBLEM (5.9)

Given the following information, calculate the standard enthalpy of formation of acetylene (C_2H_2) from its constituent elements:

$$C(graphite) + O_2(g) \longrightarrow CO_2(g) \qquad \Delta H^\circ_{rxn} = -393.5 \text{ kJ/mol} \qquad (1)$$

$$H_2(g) + \tfrac{1}{2}O_2(g) \longrightarrow H_2O(l) \qquad \Delta H^\circ_{rxn} = -285.8 \text{ kJ/mol} \qquad (2)$$

$$2C_2H_2(g) + 5O_2(g) \longrightarrow 4CO_2(g) + 2H_2O(l) \qquad \Delta H^\circ_{rxn} = -2598.8 \text{ kJ/mol} \qquad (3)$$

Strategy Arrange the equations that are provided so that they will sum to the desired equation. This may require reversing or multiplying one or more of the equations. For any such change, the corresponding change must also be made to the ΔH°_{rxn} value.

Setup The equation corresponding to the standard enthalpy of formation of acetylene is

$$2C(graphite) + H_2(g) \longrightarrow C_2H_2(g)$$

We multiply Equation (1) and its ΔH°_{rxn} value by 2:

$$2C(graphite) + 2O_2(g) \longrightarrow 2CO_2(g) \qquad \Delta H^\circ_{rxn} = 2(-393.5 \text{ kJ/mol}) = -787.0 \text{ kJ/mol}$$

We include Equation (2) and its ΔH°_{rxn} value as is:

$$H_2(g) + \tfrac{1}{2}O_2(g) \longrightarrow H_2O(l) \qquad \Delta H^\circ_{rxn} = -285.8 \text{ kJ/mol}$$

We reverse Equation (3) and divide it by 2 (i.e., multiply through by $\tfrac{1}{2}$):

$$2CO_2(g) + H_2O(l) \longrightarrow C_2H_2(g) + \tfrac{5}{2}O_2(g) \qquad \Delta H^\circ_{rxn} = -(-2598.8 \text{ kJ/mol})/2 = +1299.4 \text{ kJ/mol}$$

The original ΔH°_{rxn} value of Equation (3) has its sign reversed and it is divided by 2.

Solution Summing the resulting equations and the corresponding ΔH°_{rxn} values:

$$2C(graphite) + 2\cancel{O_2(g)} \longrightarrow 2\cancel{CO_2(g)} \qquad \Delta H^\circ_{rxn} = -787.0 \text{ kJ/mol}$$
$$H_2(g) + \tfrac{1}{2}\cancel{O_2(g)} \longrightarrow \cancel{H_2O(l)} \qquad \Delta H^\circ_{rxn} = -285.8 \text{ kJ/mol}$$
$$+\; 2\cancel{CO_2(g)} + \cancel{H_2O(l)} \longrightarrow C_2H_2(g) + \tfrac{5}{2}\cancel{O_2(g)} \qquad \Delta H^\circ_{rxn} = +1299.4 \text{ kJ/mol}$$
$$\overline{2C(graphite) + H_2(g) \longrightarrow C_2H_2(g) \qquad\qquad \Delta H^\circ_f = +226.6 \text{ kJ/mol}}$$

THINK ABOUT IT

Watch out for misplaced or missing minus signs. This is an easy place to lose track of them.

Practice Problem **A**TTEMPT Use the following data to calculate ΔH°_f for $CS_2(l)$:

$$C(graphite) + O_2(g) \longrightarrow CO_2(g) \qquad \Delta H^\circ_{rxn} = -393.5 \text{ kJ/mol}$$
$$S(rhombic) + O_2(g) \longrightarrow SO_2(g) \qquad \Delta H^\circ_{rxn} = -296.4 \text{ kJ/mol}$$
$$CS_2(l) + 3O_2(g) \longrightarrow CO_2(g) + 2SO_2(g) \qquad \Delta H^\circ_{rxn} = -1073.6 \text{ kJ/mol}$$

Practice Problem **B**UILD ΔH°_f of hydrogen chloride [HCl(g)] is −92.3 kJ/mol. Given the following data, determine the identity of the two missing products and calculate ΔH°_{rxn} for Equation 3. [*Hint:* Start by writing the chemical equation that corresponds to ΔH°_f for HCl(g).]

$$N_2(g) + 4H_2(g) + Cl_2(g) \longrightarrow 2NH_4Cl(s) \qquad \Delta H^\circ_{rxn} = -630.78 \text{ kJ/mol} \qquad (1)$$
$$N_2(g) + 3H_2(g) \longrightarrow 2NH_3(g) \qquad \Delta H^\circ_{rxn} = -92.6 \text{ kJ/mol} \qquad (2)$$
$$NH_4Cl(s) \longrightarrow \qquad\qquad\qquad\qquad\qquad (3)$$

Practice Problem **C**ONCEPTUALIZE The diagrams represent a system before and after a chemical reaction for which ΔH°_{rxn} is −2624.9 kJ/mol. Use this information to complete the table of ΔH°_f values for the species involved in the reaction.

	before	after

Species	ΔH°_f (kJ/mol)
(blue)	−148.7
(red)	?
(black)	255.1

CHECKPOINT – SECTION 5.6   Standard Enthalpies of Formation

5.6.1 Using data from Appendix 2, calculate ΔH°_{rxn} for
$H_2(g) + F_2(g) \longrightarrow 2HF(g)$.

a) -271.6 kJ/mol

b) -543.2 kJ/mol

c) $+271.6$ kJ/mol

d) 0 kJ/mol

e) -135.8 kJ/mol

5.6.2 Using data from Appendix 2, calculate ΔH°_{rxn} for
$2NO_2(g) \longrightarrow N_2O_4(g)$.

a) -24.19 kJ/mol

b) -33.85 kJ/mol

c) $+9.66$ kJ/mol

d) $+67.7$ kJ/mol

e) -58.04 kJ/mol

5.6.3 Which of the following ΔH°_{rxn} values is a ΔH°_f value?
(Select all that apply.)

a) $H_2(g) + Br_2(l) \longrightarrow 2HBr(g)$
$\Delta H^\circ_{rxn} = -72.4$ kJ/mol

b) $4Al(s) + 3O_2(g) \longrightarrow 2Al_2O_3(s)$
$\Delta H^\circ_{rxn} = -3339.6$ kJ/mol

c) $Ag(s) + \frac{1}{2}Cl_2(g) \longrightarrow AgCl(s)$
$\Delta H^\circ_{rxn} = -127.0$ kJ/mol

d) $Cu^{2+}(aq) + SO_4^{2-}(aq) \longrightarrow CuSO_4(s)$
$\Delta H^\circ_{rxn} = +73.25$ kJ/mol

e) $\frac{1}{2}H_2(g) + \frac{1}{2}N_2(g) + \frac{3}{2}O_2(g) \longrightarrow HNO_3(l)$
$\Delta H^\circ_{rxn} = -173.2$ kJ/mol

5.6.4 Using the following data, calculate ΔH°_f for $CO(g)$:

$$C(graphite) + O_2(g) \longrightarrow CO_2(g)$$
$$\Delta H^\circ_{rxn} = -393.5 \text{ kJ/mol}$$

$$CO(g) + \frac{1}{2}O_2(g) \longrightarrow CO_2(g)$$
$$\Delta H^\circ_{rxn} = -283.0 \text{ kJ/mol}$$

a) -393.5 kJ/mol

b) -676.5 kJ/mol

c) $+676.5$ kJ/mol

d) $+110.5$ kJ/mol

e) -110.5 kJ/mol

Chapter Summary

Section 5.1

- *Energy* is the capacity to do work or transfer heat. Energy may be *kinetic energy* (the energy associated with *motion*) or *potential energy* (energy possessed by virtue of *position*). *Thermal energy* is a form of kinetic energy. *Chemical energy* and *electrostatic energy* are forms of potential energy.

- The *law of conservation of energy* states that energy can neither be created nor destroyed. The SI unit of energy is the *joule* (J).

- The *system* is the particular part of the universe that we are interested in studying—such as the reactants and products in a chemical reaction. The term *surroundings* refers to the rest of the universe. System + surroundings = universe.

- *Heat* refers to the flow of thermal energy between two bodies at different temperatures. *Thermochemistry* is the study of the heat associated with chemical reactions and physical processes.

- In an *exothermic process,* heat is released to the surroundings, so the energy of the system decreases. In an *endothermic process,* heat is absorbed from the surroundings, so the energy of the system increases.

Section 5.2

- *Thermodynamics* is the study of the conversions among different types of energy. Thermochemistry is a branch of thermodynamics.

- An *open system* is one that can exchange both matter and energy with its surroundings. A *closed system* is one that can exchange energy but not matter with its surroundings. An *isolated system* is one that cannot exchange either energy or matter with its surroundings.

- The *state of a system* is defined by the values of all relevant macroscopic properties, such as temperature, volume, and pressure. A *state function* is one whose value depends only on the state of the system and not on how that state was achieved. State functions include energy, pressure, volume, and temperature.

- The *first law of thermodynamics* states that energy cannot be created or destroyed, but it can be changed from one form to another. The first law of thermodynamics is based on the law of conservation of energy.

Section 5.3

- *Enthalpy (H)* is the heat exchanged between the system and surroundings at constant pressure. It is a state function. *Enthalpy of reaction (ΔH_{rxn})* is the heat exchanged at constant pressure for a specific reaction.

- A *thermochemical equation* is a balanced chemical equation for which the enthalpy change (ΔH_{rxn}) is given.

Section 5.4

- *Calorimetry* is the science of measuring temperature changes to determine heats associated with chemical reactions. Calorimetry may be carried out at constant pressure (in a coffee-cup calorimeter) or at constant volume (in a bomb calorimeter).

- The *specific heat (s)* of a substance is the amount of heat required to increase the temperature of 1 g of the substance by 1°C. The *heat capacity (C)* of an object is the amount of heat required to increase the temperature of the object by 1°C.

Section 5.5

- *Hess's law* states that the enthalpy change for a reaction that occurs in a series of steps is equal to the sum of the enthalpy changes of the individual steps. Hess's law is valid because enthalpy is a state function.

Section 5.6

- The *standard enthalpy of formation (ΔH_f°)* is the enthalpy change associated with the formation of 1 mole of a substance from its constituent elements, each in their standard states. The *standard enthalpy of reaction (ΔH_{rxn}°)* can be calculated for any reaction using tabulated standard enthalpies of formation (ΔH_f°) of the products and reactants.

Key Words

Key Equations

5.1 $E_k = \frac{1}{2}mu^2$	The kinetic energy of a moving object is calculated using the mass (m) and velocity (u) of the object.
5.2 $E_{el} \propto \dfrac{Q_1 Q_2}{d}$	The electrostatic potential energy (E_{el}) between two charged objects is calculated using the magnitudes of charge (Q_1 and Q_2) and the distance (d) between the charges.
5.3 $\Delta U = q + w$	The change in internal energy of a system (ΔU) is the sum of heat (q) and the work (w) associated with a process. Proper sign conventions must be used for heat and work (Table 5.1).
5.4 $w = -P\Delta V$	Pressure-volume work done by (or on) a system is calculated using the external pressure (P) and the change in volume (ΔV).
5.5 $\Delta U = q - P\Delta V$	The change in internal energy of a system (ΔU) is equal to heat (q) minus pressure-volume work ($P\Delta V$).
5.6 $q_V = \Delta U$	Heat given off (or absorbed) by a system at constant volume (q_V) is equal to the change in internal energy (ΔU).
5.7 $q_P = \Delta U + P\Delta V$	Heat given off (or absorbed) by a system at constant pressure (q_P) is equal to the sum of change in internal energy (ΔU) and pressure-volume work ($P\Delta V$).
5.8 $H = U + PV$	Enthalpy (H) is equal to the sum of internal energy (U) and pressure-volume work ($P\Delta V$).
5.9 $\Delta H = \Delta U + \Delta(PV)$	The change in enthalpy (ΔH) is equal to the sum of change in internal energy (ΔU) and change in the product of pressure and volume [$\Delta(PV)$].
5.10 $\Delta H = \Delta U + P\Delta V$	The change in enthalpy (ΔH) is equal to the sum of change in internal energy (ΔU) and the product of external pressure (P) and change in volume (ΔV).
5.11 $q_P = \Delta H$	Heat given off (or absorbed) by a process at constant pressure (q_P) is equal to the change in enthalpy (ΔH).
5.12 $\Delta H = H(\text{products}) - H(\text{reactants})$	The enthalpy change for a reaction (ΔH) is the difference between the enthalpy of products [$H(\text{products})$] and enthalpy of reactants [$H(\text{reactants})$], although this is not the equation generally used to calculate enthalpy changes because the absolute values of enthalpy are not known.
5.13 $q = sm\Delta T$	Heat given off (or absorbed) by a substance (q) is equal to the product of specific heat of the substance (s), mass of the substance (m), and the change in temperature (ΔT).
5.14 $q = C\Delta T$	Heat given off (or absorbed) by an object (q) is equal to the product of specific heat of the object (C) and the change in temperature (ΔT).
5.15 $q_{sys} = -sm\Delta T$	Heat given off (or absorbed) by a system (q_{sys}) is equal in magnitude and opposite in sign to the heat given off or absorbed by the surroundings.
5.16 $q_{cal} = C_{cal}\Delta T$	Heat given off (or absorbed) by a calorimeter (q_{cal}) is equal to the product of heat capacity of the calorimeter (C_{cal}) and the change in temperature (ΔT).
5.17 $q_{rxn} = -C_{cal}\Delta T$	Heat of reaction (q_{rxn}) is equal in magnitude and opposite in sign to heat of the calorimeter (q_{cal}).
5.18 $\Delta H^{\circ}_{rxn} = [c\Delta H^{\circ}_f(C) + d\Delta H^{\circ}_f(D)]$ $- [a\Delta H^{\circ}_f(A) + b\Delta H^{\circ}_f(B)]$	Standard enthalpy change for a reaction (ΔH°_{rxn}) can be calculated by multiplying the coefficient of each species in the reaction by the corresponding standard enthalpy of formation (ΔH°_f).
5.19 $\Delta H^{\circ}_{rxn} = \Sigma n\Delta H^{\circ}_f\,(\text{products})$ $- \Sigma m\Delta H^{\circ}_f\,(\text{reactants})$	Standard enthalpy change for a reaction (ΔH°_{rxn}) is the difference between the sum of standard enthalpies of formation of products $\Sigma\Delta H^{\circ}_f\,(\text{products})$ and the sum of standard enthalpies of formation of reactants $\Sigma\Delta H^{\circ}_f\,(\text{reactants})$.

Enthalpy of Reaction

Using tabulated ΔH_f° values, we can calculate the standard enthalpy of reaction (ΔH_{rxn}°) using Equation 5.19:

$$\Delta H_{rxn}^\circ = \Sigma n \Delta H_f^\circ \text{ (products)} - \Sigma m \Delta H_f^\circ \text{ (reactants)}$$

This method of calculating thermodynamic quantities such as enthalpy of reaction is important not only in this chapter, but also in Chapters 19 and 20. The following examples illustrate the use of Equation 5.19 and data from Appendix 2. Each example provides a specific reminder of one of the important facets of this approach.

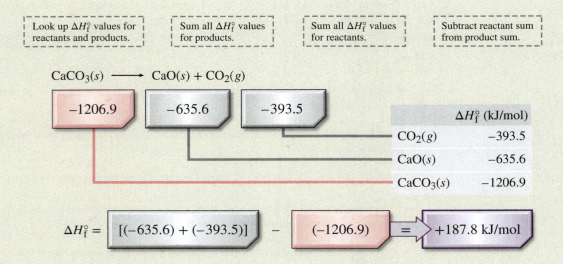

Each ΔH_f° value must be multiplied by the corresponding stoichiometric coefficient in the balanced equation.

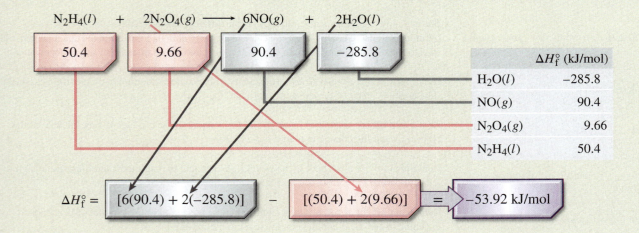

$$Ba(s) + 2H_2O(l) \longrightarrow Ba(OH)_2(aq) + H_2(g)$$

By definition, the standard enthalpy of formation for an element in its standard state is zero. In addition, many tables of thermodynamic data, including Appendix 2, do not contain values for aqueous strong electrolytes such as barium hydroxide. However, the tables do include values for the individual aqueous ions. Therefore, determination of this enthalpy of reaction is facilitated by rewriting the equation with $Ba(OH)_2$ written as separate ions:

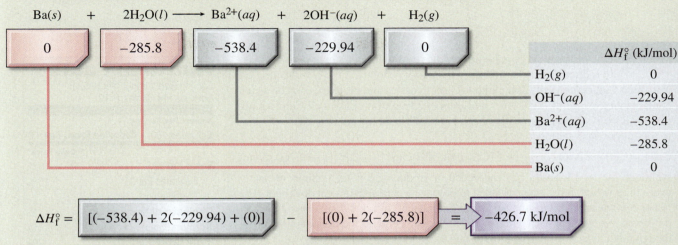

You will find more than one tabulated ΔH_f° value for some substances, such as water. It is important to select the value that corresponds to the phase of matter represented in the chemical equation. In previous examples, water has appeared in the balanced equations as a liquid. It can also appear as a gas.

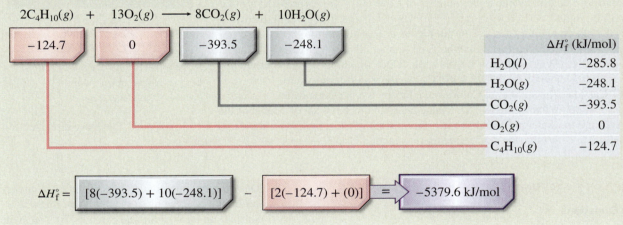

Key Skills Problems

5.1

Using data from Appendix 2, calculate the standard enthalpy of the following reaction:

$$Mg(OH)_2(s) \longrightarrow MgO(s) + H_2O(l)$$

(a) −608.7 kJ/mol (b) −81.1 kJ/mol (c) −37.1 kJ/mol
(d) +81.1 kJ/mol (e) +37.1 kJ/mol

5.2

Using data from Appendix 2, calculate the standard enthalpy of the following reaction:

$$4HBr(g) + O_2(g) \longrightarrow 2H_2O(l) + 2Br_2(l)$$

(a) −426.8 kJ/mol (b) −338.8 kJ/mol (c) −249.6 kJ/mol
(d) +426.8 kJ/mol (e) +338.8 kJ/mol

5.3

Using data from Appendix 2, calculate the standard enthalpy of the following reaction (you must first balance the equation):

$$P(red) + Cl_2(g) \longrightarrow PCl_3(g)$$

(a) −576.1 kJ/mol (b) −269.7 kJ/mol (c) −539.3 kJ/mol
(d) −602.6 kJ/mol (e) +639.4 kJ/mol

5.4

Using only whole number coefficients, the combustion of hexane can be represented as:

$$2C_6H_{14}(l) + 19O_2(g) \longrightarrow 12CO_2(g) + 14H_2O(l)$$
$$\Delta H^\circ = -8388.4 \text{ kJ/mol}$$

Using this and data from Appendix 2, determine the standard enthalpy of formation of hexane.
(a) −334.8 kJ/mol (b) −167.4 kJ/mol (c) −669.6 kJ/mol
(d) +334.8 kJ/mol (e) +669.6 kJ/mol

239

Questions and Problems

Applying What You've Learned

One of the most popular approaches to dieting in recent years has been to reduce dietary fat. One reason many people want to avoid eating fat is its high Calorie content. Compared to carbohydrates and proteins, each of which contains an average of 4 Calories per gram (17 kJ/g), fat contains 9 Calories per gram (38 kJ/g). Tristearin, a typical fat, is metabolized (or combusted) according to the following equation:

$$C_{57}H_{110}O_6(s) + 81.5O_2(g) \longrightarrow 57CO_2(g) + 55H_2O(l) \qquad \Delta H° = -37,760 \text{ kJ/mol}$$

Although the food industry has succeeded in producing low-fat versions of nearly everything we eat, it has thus far failed to produce a palatable low-fat doughnut. The flavor, texture, and what the industry calls "mouth feel" of a doughnut depends largely on the process of deep-fat frying. Fortunately for people in the doughnut business, though, high fat content has not diminished the popularity of doughnuts.

According to information obtained from www.krispykreme.com, a Krispy Kreme original glazed doughnut weighs 52 g and contains 200 Cal and 12 g of fat.

(a) Assuming that the fat in the doughnut is metabolized according to the given equation for tristearin, calculate the number of Calories in the reported 12 g of fat in each doughnut [◀◀ Sample Problem 5.3]. (b) If all the energy contained in a Krispy Kreme doughnut (not just in the fat) were transferred to 6.00 kg of water originally at 25.5°C, what would be the final temperature of the water [◀◀ Sample Problem 5.4]? (c) When a Krispy Kreme apple fritter weighing 101 g is burned in a bomb calorimeter with C_{cal} = 95.3 kJ/°C, the measured temperature increase is 16.7°C. Calculate the number of Calories in a Krispy Kreme apple fritter [◀◀ Sample Problem 5.6]. (d) What would the $\Delta H°$ value be for the metabolism of 1 mole of the fat tristearin if the water produced by the reaction were gaseous instead of liquid [◀◀ Sample Problem 5.7]? [*Hint:* Use data from Appendix 2 to determine the $\Delta H°$ value for the reaction $H_2O(l) \longrightarrow H_2O(g)$ [◀◀ Sample Problem 5.8]].

Nutrition Facts

Serving Size 1 donut (about 52g)
Servings Per Container 12

Amount Per Serving

Calories 200   **Calories From Fat** 100

	%Daily Value*
Total Fat 12g	18%
Saturated Fat 3g	15%
Trans Fat 4g	
Cholesterol 5mg	1%
Sodium 95mg	4%
Total Carbohydrate 22g	7%
Dietary Fiber <1g	1%
Sugars 10g	
Protein 2g	

Vitamin A	0%	•	Vitamin C	2%
Calcium	6%	•	Iron	4%

*Percent of Daily Values (DV) are based on a 2,000 calorie diet.

Nutrition facts label for Krispy Kreme original glazed doughnuts.

David A. Tietz/Editorial Image, LLC

SECTION 5.1: ENERGY AND ENERGY CHANGES

Review Questions

5.1   Define these terms: *system, surroundings, thermal energy, chemical energy, potential energy, kinetic energy, law of conservation of energy.*

5.2   What is heat? How does heat differ from thermal energy? Under what condition is heat transferred from one system to another?

5.3   What are the units for energy commonly employed in chemistry?

5.4   A truck initially traveling at 60 km/h is brought to a complete stop at a traffic light. Does this change violate the law of conservation of energy? Explain.

5.5   These are various forms of energy: chemical, heat, light, mechanical, and electrical. Suggest several ways of converting one form of energy to another.

5.6   Define these terms: *thermochemistry, exothermic process, endothermic process.*

Conceptual Problems

5.7   Stoichiometry is based on the law of conservation of mass. On what law is thermochemistry based?

5.8   Describe the interconversions of forms of energy occurring in these processes: (a) You throw a softball up into the air and catch it. (b) You switch on a flashlight. (c) You ride the ski lift to the top of the hill and then ski down. (d) You strike a match and let it burn completely.

5.9   Decomposition reactions are usually endothermic, whereas combination reactions are usually exothermic. Give a qualitative explanation for these trends.

5.10   For charges of +1 and −1, separated by a distance of d, the electrostatic potential energy is E. In terms of E, determine the electrostatic potential energy between each of the pairs of charges shown.

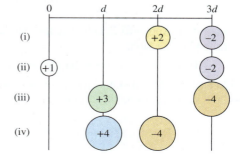

SECTION 5.2: INTRODUCTION TO THERMODYNAMICS

Review Questions

5.11 On what law is the first law of thermodynamics based? Explain the sign conventions in the equation

$$\Delta U = q + w$$

5.12 Explain what is meant by a state function. Give two examples of quantities that are state functions and two that are not state functions.

Computational Problems

5.13 The work done to compress a gas is 47 J. As a result, 93 J of heat is given off to the surroundings. Calculate the change in internal energy of the gas.

5.14 In a gas expansion, 87 J of heat is released to the surroundings and the energy of the system decreases by 128 J. Calculate the work done.

5.15 Calculate w, and determine whether work is done *by* the system or *on* the system when 415 J of heat is released and $\Delta U = 510$ J.

5.16 Calculate q, and determine whether heat is absorbed or released when a system does work on the surroundings equal to 64 J and $\Delta U = 213$ J.

Conceptual Problems

Use the following diagrams for Problems 5.17 and 5.18.

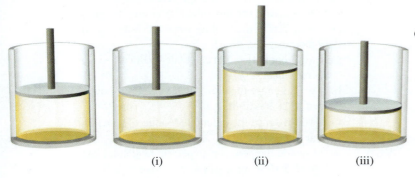

(i)   (ii)   (iii)

5.17 The diagram on the far left shows a system before a process. Determine which of the diagrams on the right could represent the system after it undergoes a process in which (a) the system absorbs heat and ΔU is negative; (b) the system absorbs heat and does work on the surroundings; (c) the system releases heat and does work on the surroundings.

5.18 The diagram on the far left shows a system before a process. Determine which of the diagrams on the right could represent the system after it undergoes a process in which (a) work is done on the system and ΔU is negative; (b) the system releases heat and ΔU is positive; (c) the system absorbs heat and ΔU is positive.

SECTION 5.3: ENTHALPY

Review Questions

5.19 Consider these changes.
(a) $Hg(l) \longrightarrow Hg(g)$
(b) $3O_2(g) \longrightarrow 2O_3(g)$
(c) $CuSO_4 \cdot 5H_2O(s) \longrightarrow CuSO_4(s) + 5H_2O(g)$
(d) $H_2(g) + F_2(g) \longrightarrow 2HF(g)$
At constant pressure, in which of the reactions is work done by the system on the surroundings? By the surroundings on the system? In which of them is no work done?

5.20 Define these terms: *enthalpy* and *enthalpy of reaction*. Under what condition is the heat of a reaction equal to the enthalpy change of the same reaction?

5.21 In writing thermochemical equations, why is it important to indicate the physical state (i.e., gaseous, liquid, solid, or aqueous) of each substance?

5.22 Explain the meaning of this thermochemical equation:

$$4NH_3(g) + 5O_2(g) \longrightarrow 4NO(g) + 6H_2O(g)$$
$$\Delta H = -904 \text{ kJ/mol}$$

5.23 Consider this reaction:

$$2CH_3OH(l) + 3O_2(g) \longrightarrow 4H_2O(l) + 2CO_2(g)$$
$$\Delta H = -1452.8 \text{ kJ/mol}$$

What is the value of ΔH if (a) the equation is multiplied throughout by 2; (b) the direction of the reaction is reversed so that the products become the reactants, and vice versa; (c) water vapor instead of liquid water is formed as the product?

Computational Problems

5.24 A sample of nitrogen gas expands in volume from 1.6 to 5.4 L at constant temperature. Calculate the work done in joules if the gas expands (a) against a vacuum, (b) against a constant pressure of 0.80 atm, and (c) against a constant pressure of 3.7 atm. See Equation 5.4. (1 L·atm = 101.3 J)

5.25 A gas expands in volume from 26.7 to 89.3 mL at constant temperature. Calculate the work done (in joules) if the gas expands (a) against a vacuum, (b) against a constant pressure of 1.5 atm, and (c) against a constant pressure of 2.8 atm. (1 L·atm = 101.3 J)

5.26 A gas expands and does *PV* work on the surroundings equal to 325 J. At the same time, it absorbs 127 J of heat from the surroundings. Calculate the change in energy of the gas.

5.27 The first step in the industrial recovery of zinc from the zinc sulfide ore is roasting; that is, the conversion of ZnS to ZnO by heating:

$$2ZnS(s) + 3O_2(g) \longrightarrow 2ZnO(s) + 2SO_2(g)$$
$$\Delta H = -879 \text{ kJ/mol}$$

Calculate the heat evolved (in kJ) per gram of ZnS roasted.

5.28   Determine the amount of heat (in kJ) given off when 1.26×10^4 g of NO_2 are produced according to the equation

$$2NO(g) + O_2(g) \longrightarrow 2NO_2(g) \quad \Delta H = -114.6 \text{ kJ/mol}$$

5.29   Consider the reaction

$$2H_2O(g) \longrightarrow 2H_2(g) + O_2(g) \quad \Delta H = +483.6 \text{ kJ/mol}$$

at a certain temperature. If the increase in volume is 32.7 L against an external pressure of 1.00 atm, calculate ΔU for this reaction. (1 L·atm = 101.3 J)

5.30   Consider the reaction

$$H_2(g) + Cl_2(g) \longrightarrow 2HCl(g) \quad \Delta H = -184.6 \text{ kJ/mol}$$

If 3 moles of H_2 react with 3 moles of Cl_2 to form HCl, calculate the work done (in joules) against a pressure of 1.0 atm. What is ΔU for this reaction? Assume the reaction goes to completion and that $\Delta V = 0$. (1 L·atm = 101.3 J)

Conceptual Problems

5.31   The following diagrams represent systems before and after reaction for two related chemical processes. ΔH for the first reaction is −595.8 kJ/mol. Determine the value of ΔH for the second reaction.

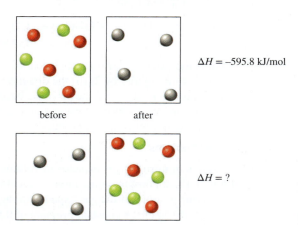

$\Delta H = -595.8$ kJ/mol

before   after

$\Delta H = ?$

before   after

5.32   For most biological processes, the changes in internal energy are approximately equal to the changes in enthalpy. Explain.

SECTION 5.4: CALORIMETRY

▶▶▶ **Visualizing Chemistry**
Figure 5.9 and Figure 5.10

VC 5.1   Referring to Figure 5.9, which of the following would result in the calculated value of ΔH_{rxn} being too high?
a) Spilling some of one of the reactant solutions before adding it to the calorimeter.
b) Reading the final temperature before it reached its maximum value.
c) Misreading the thermometer at the beginning of the experiment and recording too low an initial temperature.

VC 5.2   How would the ΔH_{rxn} calculated in Figure 5.9 be affected if the concentration of one of the reactant solutions were twice as high as it was supposed to be?
a) The calculated ΔH_{rxn} would not be affected.
b) The calculated ΔH_{rxn} would be too low.
c) The calculated ΔH_{rxn} would be too high.

VC 5.3   For an exothermic reaction like the one depicted in Figure 5.9, if the heat capacity of the calorimeter is not negligibly small, the heat absorbed by the water will be _____ the heat given off by the reaction.
a) greater than
b) less than
c) equal to

VC 5.4   Referring to Figure 5.9, how would the results of the experiment have been different if the reaction had been endothermic?
a) The results would have been the same.
b) There would have been a smaller temperature increase.
c) There would have been a temperature decrease.

VC 5.5   What would happen to the specific heat calculated in Figure 5.10 if some of the warm metal shot were lost during the transfer to the calorimeter?
a) It would not affect the calculated value of specific heat.
b) It would cause the calculated value of specific heat to be too high.
c) It would cause the calculated value of specific heat to be too low.

VC 5.6   What would happen to the specific heat calculated in Figure 5.10 if the test tube containing the metal shot were left in the boiling water for longer than the recommended time?
a) It would not affect the calculated value of specific heat.
b) It would cause the calculated value of specific heat to be too high.
c) It would cause the calculated value of specific heat to be too low.

VC 5.7   What would happen to the specific heat calculated in Figure 5.10 if some of the water were spilled prior to being added to the calorimeter?
a) It would not affect the calculated value of specific heat.
b) It would cause the calculated value of specific heat to be too high.
c) It would cause the calculated value of specific heat to be too low.

VC 5.8   Referring to the process depicted in Figure 5.10, which of the following must be known precisely for the calculated specific heat to be accurate?
a) The mass of the boiling water.
b) The temperature of the metal shot before it is immersed in the boiling water.
c) The mass of the water that is added to the calorimeter.

Review Questions

5.33 What is the difference between specific heat and heat capacity? What are the units for these two quantities? Which is the intensive property and which is the extensive property?

5.34 Define *calorimetry* and describe two commonly used calorimeters. In a calorimetric measurement, why is it important that we know the heat capacity of the calorimeter? How is this value determined?

Computational Problems

5.35 A 6.22-kg piece of copper metal is heated from 20.5°C to 324.3°C. Calculate the heat absorbed (in kJ) by the metal.

5.36 Calculate the amount of heat liberated (in kJ) from 366 g of mercury when it cools from 77.0°C to 12.0°C.

5.37 A sheet of gold weighing 10.0 g and at a temperature of 18.0°C is placed flat on a sheet of iron weighing 20.0 g and at a temperature of 55.6°C. What is the final temperature of the combined metals? Assume that no heat is lost to the surroundings. (*Hint:* The heat gained by the gold must be equal to the heat lost by the iron. The specific heats of the metals are given in Table 5.2.)

5.38 A 0.1375-g sample of solid magnesium is burned in a constant-volume bomb calorimeter that has a heat capacity of 3024 J/°C. The temperature increases by 1.126°C. Calculate the heat given off by the burning Mg, in kJ/g and in kJ/mol.

5.39 A quantity of 2.00×10^2 mL of 0.862 M HCl is mixed with 2.00×10^2 mL of 0.431 M $Ba(OH)_2$ in a constant-pressure calorimeter of negligible heat capacity. The initial temperature of the HCl and $Ba(OH)_2$ solutions is the same at 20.48°C. For the process

$$H^+(aq) + OH^-(aq) \longrightarrow H_2O(l)$$

the heat of neutralization is −56.2 kJ/mol. What is the final temperature of the mixed solution? Assume the specific heat of the solution is the same as that for pure water.

5.40 A 50.75-g sample of water at 75.6°C is added to a sample of water at 24.1°C in a constant-pressure calorimeter. If the final temperature of the combined water is 39.4°C and the heat capacity of the calorimeter is 26.3 J/°C, calculate the mass of the water originally in the calorimeter.

5.41 A 25.95-g sample of methanol at 35.6°C is added to a 38.65-g sample of ethanol at 24.7°C in a constant-pressure calorimeter. If the final temperature of the combined liquids is 28.5°C and the heat capacity of the calorimeter is 19.3 J/°C, determine the specific heat of methanol.

5.42 75.0 mL of 1.75 M HCl and 125.0 mL of 1.50 M NaOH are combined in a constant-pressure calorimeter. Both solutions are initially at 23.9°C.

Calculate the final temperature of the combined solutions. (Use the data from Table 5.3. Assume that the mass of the combined solutions is 200.0 g and that the solution's specific heat is the same as that for water, 4.184 J/g·°C.) The heat capacity of the calorimeter is negligibly small.

Conceptual Problems

5.43 Consider two metals, A and B, each having a mass of 100 g and an initial temperature of 20°C. The specific heat of A is larger than that of B. Under the same heating conditions, which metal would take longer to reach a temperature of 21°C?

5.44 Consider the following data:

Metal	Al	Cu
Mass (g)	10	30
Specific heat (J/g · °C)	0.900	0.385
Temperature (°C)	40	60

When these two metals are placed in contact, which of the following will take place?
(a) Heat will flow from Al to Cu because Al has a larger specific heat.
(b) Heat will flow from Cu to Al because Cu has a larger mass.
(c) Heat will flow from Cu to Al because Cu has a larger heat capacity.
(d) Heat will flow from Cu to Al because Cu is at a higher temperature.
(e) No heat will flow in either direction.

SECTION 5.5: HESS'S LAW

Review Questions

5.45 State Hess's law. Explain, with one example, the usefulness of Hess's law in thermochemistry.

5.46 Describe how chemists use Hess's law to determine the ΔH_f° of a compound by measuring its heat (enthalpy) of combustion.

Computational Problems

5.47 Given the thermochemical data,

$$A + 6B \longrightarrow 4C \qquad \Delta H_1 = -1200 \text{ kJ/mol}$$
$$C + B \longrightarrow D \qquad \Delta H_1 = -150 \text{ kJ/mol}$$

Determine the enthalpy change for each of the following:
a) $D \longrightarrow C + B$ d) $2D \longrightarrow 2C + 2B$
b) $2C \longrightarrow \frac{1}{2}A + 3B$ e) $6D + A \longrightarrow 10C$
c) $3D + \frac{1}{2}A \longrightarrow 5C$

5.48 Given the thermochemical data,

$$A + B \longrightarrow 2C \qquad \Delta H_1 = 600 \text{ kJ/mol}$$
$$2C + D \longrightarrow 2E \qquad \Delta H_1 = 210 \text{ kJ/mol}$$

Determine the enthalpy change for each of the following:
a) $4E \longrightarrow 4C + 2D$ d) $2C + 2E \longrightarrow 2A + 2B + D$
b) $A + B + D \longrightarrow 2E$ e) $E \longrightarrow \frac{1}{2}A + \frac{1}{2}B + \frac{1}{2}D$
c) $C \longrightarrow \frac{1}{2}A + \frac{1}{2}B$

5.49 From these data,

$$S(\text{rhombic}) + O_2(g) \longrightarrow SO_2(g)$$
$$\Delta H_{rxn}^{\circ} = -296.4 \text{ kJ/mol}$$

$$S(\text{monoclinic}) + O_2(g) \longrightarrow SO_2(g)$$
$$\Delta H_{rxn}^{\circ} = -296.7 \text{ kJ/mol}$$

calculate the enthalpy change for the transformation

$$S(\text{rhombic}) \longrightarrow S(\text{monoclinic})$$

(Monoclinic and rhombic are different allotropic forms of elemental sulfur.)

5.50 From the following data,

$$C(\text{graphite}) + O_2(g) \longrightarrow CO_2(g)$$
$$\Delta H_{rxn}^{\circ} = -393.5 \text{ kJ/mol}$$

$$H_2(g) + \tfrac{1}{2}O_2(g) \longrightarrow H_2O(l)$$
$$\Delta H_{rxn}^{\circ} = -285.8 \text{ kJ/mol}$$

$$2C_2H_6(g) + 7O_2(g) \longrightarrow 4CO_2(g) + 6H_2O(l)$$
$$\Delta H_{rxn}^{\circ} = -3119.6 \text{ kJ/mol}$$

calculate the enthalpy change for the reaction

$$2C(\text{graphite}) + 3H_2(g) \longrightarrow C_2H_6(g)$$

5.51 From the following heats of combustion,

$$CH_3OH(l) + \tfrac{3}{2}O_2(g) \longrightarrow CO_2(g) + 2H_2O(l)$$
$$\Delta H_{rxn}^{\circ} = -726.4 \text{ kJ/mol}$$

$$C(\text{graphite}) + O_2(g) \longrightarrow CO_2(g)$$
$$\Delta H_{rxn}^{\circ} = -393.5 \text{ kJ/mol}$$

$$H_2(g) + \tfrac{1}{2}O_2(g) \longrightarrow H_2O(l)$$
$$\Delta H_{rxn}^{\circ} = -285.8 \text{ kJ/mol}$$

calculate the enthalpy of formation of methanol (CH_3OH) from its elements:

$$C(\text{graphite}) + 2H_2(g) + \tfrac{1}{2}O_2(g) \longrightarrow CH_3OH(l)$$

5.52 Calculate the standard enthalpy change for the reaction

$$2Al(s) + Fe_2O_3(s) \longrightarrow 2Fe(s) + Al_2O_3(s)$$

given that

$$2Al(s) + \tfrac{3}{2}O_2(g) \longrightarrow Al_2O_3(s)$$
$$\Delta H_{rxn}^{\circ} = -1669.8 \text{ kJ/mol}$$

$$2Fe(s) + \tfrac{3}{2}O_2(g) \longrightarrow Fe_2O_3(s)$$
$$\Delta H_{rxn}^{\circ} = -822.2 \text{ kJ/mol}$$

5.53 Determine the enthalpy change for the gaseous reaction of sulfur dioxide with ozone to form sulfur trioxide given the following thermochemical data:

$$2SO(g) + O_2(g) \longrightarrow 2SO_2(g)$$
$$\Delta H^{\circ} = -602.8 \text{ kJ/mol}$$

$$3SO(g) + 2O_3(g) \longrightarrow 3SO_3(g)$$
$$\Delta H_{rxn}^{\circ} = -1485.03 \text{ kJ/mol}$$

$$\tfrac{3}{2}O_2(g) \longrightarrow O_3(g)$$
$$\Delta H_{rxn}^{\circ} = 142.2 \text{ kJ/mol}$$

5.54 Determine the enthalpy change for the reaction of carbon disulfide and chlorine gases to form carbon

tetrachloride and disulfur dichloride liquids given the following thermochemical data:

$$2S(\text{rhombic}) + C(\text{graphite}) \longrightarrow CS_2(g)$$
$$\Delta H^{\circ} = 115.3 \text{ kJ/mol}$$

$$C(\text{graphite}) + 2Cl_2(g) \longrightarrow CCl_4(l)$$
$$\Delta H_{rxn}^{\circ} = -128.2 \text{ kJ/mol}$$

$$2S(\text{rhombic}) + O_2(g) \longrightarrow SO_2(g)$$
$$\Delta H_{rxn}^{\circ} = -296.4 \text{ kJ/mol}$$

$$2S(\text{rhombic}) + Cl_2(g) \longrightarrow S_2Cl_2(l)$$
$$\Delta H_{rxn}^{\circ} = -58.2 \text{ kJ/mol}$$

$$3H_2(g) + S_2Cl_2(l) \longrightarrow 2H_2S(g) + 2HCl(g)$$
$$\Delta H_{rxn}^{\circ} = -166.7 \text{ kJ/mol}$$

Conceptual Problems

5.55 Each diagram shows a system before and after a chemical reaction along with the corresponding enthalpy of reaction. Determine the enthalpy of reaction for the last reaction represented.

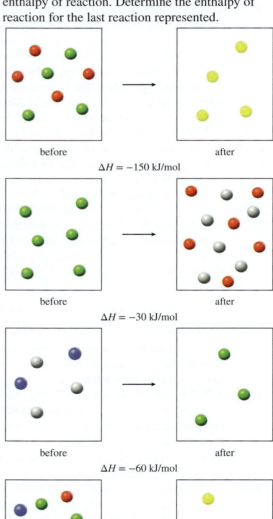

before   after

$$\Delta H = -150 \text{ kJ/mol}$$

before   after

$$\Delta H = -30 \text{ kJ/mol}$$

before   after

$$\Delta H = -60 \text{ kJ/mol}$$

before   after

$$\Delta H = ?$$

The following diagrams depict four chemical reactions involving five different chemical species—each represented by a different color sphere. Use this information to solve Problems 5.56–5.59.

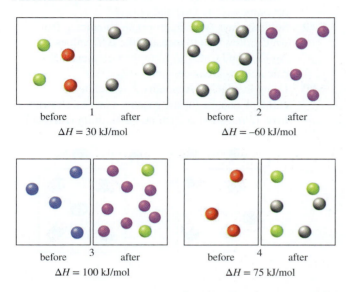

before 1 after
$\Delta H = 30$ kJ/mol

before 2 after
$\Delta H = -60$ kJ/mol

before 3 after
$\Delta H = 100$ kJ/mol

before 4 after
$\Delta H = 75$ kJ/mol

5.56 Determine the value of ΔH for the following reaction:

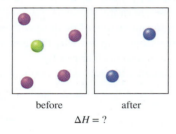

before   after
$\Delta H = ?$

5.57 Determine the value of ΔH for the following reaction:

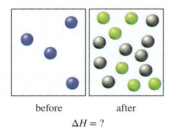

before   after
$\Delta H = ?$

5.58 Determine the value of ΔH for the following reaction:

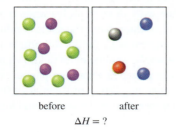

before   after
$\Delta H = ?$

5.59 Determine the value of ΔH for the following reaction:

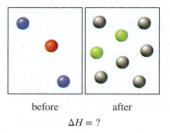

before   after
$\Delta H = ?$

SECTION 5.6: STANDARD ENTHALPIES OF FORMATION

Review Questions

5.60 What is meant by the standard-state condition?

5.61 How are the standard enthalpies of an element and of a compound determined?

5.62 What is meant by the standard enthalpy of reaction? Write the equation for calculating the standard enthalpy of reaction. Define all the terms.

Computational Problems

5.63 Which of the following standard enthalpy of formation values is not zero at 25°C: Na(monoclinic), Ne(g), CH$_4$(g), S$_8$(monoclinic), Hg(l), H(g)?

5.64 The ΔH_f° values of the two allotropes of oxygen, O$_2$ and O$_3$, are 0 and 142.2 kJ/mol, respectively, at 25°C. Which is the more stable form at this temperature?

5.65 Which is the more negative quantity at 25°C: ΔH_f° for H$_2$O(l) or ΔH_f° for H$_2$O(g)?

5.66 The standard enthalpies of formation of ions in aqueous solutions are obtained by arbitrarily assigning a value of zero to H$^+$ ions; that is, $\Delta H_f^\circ [H^+(aq)] = 0$. (a) For the following reaction,

$$HCl(g) \xrightarrow{\text{H}_2\text{O}} H^+(aq) + Cl^-(aq) \qquad \Delta H^\circ = -74.9 \text{ kJ/mol}$$

calculate ΔH_f° for the Cl$^-$ ions. (b) Given that ΔH_f° for OH$^-$ ions is -229.6 kJ/mol, calculate the enthalpy of neutralization when 1 mole of a strong monoprotic acid (such as HCl) is titrated by 1 mole of a strong base (such as KOH) at 25°C.

5.67 Calculate the heats of reaction for the following reactions from the standard enthalpies of formation listed in Appendix 2:
(a) CO(g) + H$_2$O(g) $\longrightarrow$ CO$_2$(g) + H$_2$(g)
(b) 3SO(g) + O$_3$(g) $\longrightarrow$ 3SO$_2$(g)

5.68 Calculate the heats of combustion for the following reactions from the standard enthalpies of formation listed in Appendix 2:
(a) C$_2$H$_4$(g) + 3O$_2$(g) $\longrightarrow$ 2CO$_2$(g) + 2H$_2$O(l)
(b) 2H$_2$S(g) + 3O$_2$(g) $\longrightarrow$ 2H$_2$O(l) + 2SO$_2$(g)

5.69 Methanol, ethanol, and *n*-propanol are three common alcohols. When 1.00 g of each of these alcohols is burned in air, heat is liberated as follows: (a) methanol (CH_3OH), −22.6 kJ; (b) ethanol (C_2H_5OH), −29.7 kJ; (c) *n*-propanol (C_3H_7OH), −33.4 kJ. Calculate the heats of combustion of these alcohols in kJ/mol.

5.70 The standard enthalpy change for the following reaction is 436.4 kJ/mol:

$$H_2(g) \longrightarrow H(g) + H(g)$$

Calculate the standard enthalpy of formation of atomic hydrogen (H).

5.71 From the standard enthalpies of formation, calculate ΔH°_{rxn} for the reaction

$$C_6H_{12}(l) + 9O_2(g) \longrightarrow 6CO_2(g) + 6H_2O(l)$$

For $C_6H_{12}(l)$, $\Delta H^\circ_f = -151.9$ kJ/mol.

5.72 Calculate the heat of decomposition for this process at constant pressure and 25°C:

$$CaCO_3(s) \longrightarrow CaO(s) + CO_2(g)$$

(Look up the standard enthalpy of formation of the reactant and products in Appendix 2.)

5.73 Consider the reaction

$$N_2(g) + 3H_2(g) \longrightarrow 2NH_3(g) \qquad \Delta H = -92.6 \text{ kJ/mol}$$

When 2 mol of N_2 react with 6 mol of H_2 to form 4 mol of NH_3 at 1 atm and a certain temperature, there is a decrease in volume equal to 98 L. Calculate ΔU for this reaction. (The conversion factor is $1 L \cdot atm = 101.3$ J.)

5.74 Calculate the heat released when 2.00 L of $Cl_2(g)$ with a density of 1.88 g/L reacts with an excess of sodium metal at 25°C and 1 atm to form sodium chloride.

5.75 Pentaborane-9 (B_5H_9) is a colorless, highly reactive liquid that will burst into flames when exposed to oxygen. The reaction is

$$2B_5H_9(l) + 12O_2(g) \longrightarrow 5B_2O_3(s) + 9H_2O(l)$$

Calculate the kilojoules of heat released per gram of the compound reacted with oxygen. The standard enthalpy of formation of B_5H_9 is 73.2 kJ/mol.

5.76 Determine the amount of heat (in kJ) given off when 1.26×10^4 g of ammonia is produced according to the equation

$$N_2(g) + 3H_2(g) \longrightarrow 2NH_3(g) \quad \Delta H^\circ_{rxn} = -92.6 \text{ kJ/mol}$$

Assume that the reaction takes place under standard-state conditions at 25°C.

Conceptual Problems

5.77 Predict the value of ΔH°_f (greater than, less than, or equal to zero) for these elements at 25°C (a) $Br_2(g)$, $Br_2(l)$; (b) $I_2(g)$, $I_2(s)$.

5.78 In general, compounds with negative ΔH°_f values are more stable than those with positive ΔH°_f values. $H_2O_2(l)$ has a negative ΔH°_f (see Appendix 2). Why, then, does $H_2O_2(l)$ have a tendency to decompose to $H_2O(l)$ and $O_2(g)$?

5.79 Suggest ways (with appropriate equations) that would allow you to measure the ΔH°_f values of $Ag_2O(s)$ and $CaCl_2(s)$ from their elements. No calculations are necessary.

5.80 Using the data in Appendix 2, calculate the enthalpy change for the gaseous reaction shown here. (*Hint:* First determine the limiting reactant.)

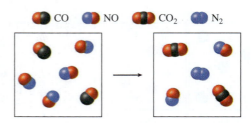

ADDITIONAL PROBLEMS

5.81 The convention of arbitrarily assigning a zero enthalpy value for the most stable form of each element in the standard state at 25°C is a convenient way of dealing with enthalpies of reactions. Explain why this convention cannot be applied to nuclear reactions.

5.82 Consider the following two reactions:

$$A \longrightarrow 2B \qquad \Delta H^\circ_{rxn} = H_1$$
$$A \longrightarrow C \qquad \Delta H^\circ_{rxn} = H_2$$

Determine the enthalpy change for the process

$$2B \longrightarrow C$$

5.83 The standard enthalpy change ΔH° for the thermal decomposition of silver nitrate according to the following equation is +78.67 kJ:

$$AgNO_3(s) \longrightarrow AgNO_2(s) + \tfrac{1}{2}O_2(g)$$

The standard enthalpy of formation of $AgNO_3(s)$ is −123.02 kJ/mol. Calculate the standard enthalpy of formation of $AgNO_2(s)$.

5.84 Consider the reaction:

$$2Na(s) + 2H_2O(l) \longrightarrow 2NaOH(aq) + H_2(g)$$

When 2 moles of Na react with water at 25°C and 1 atm, the volume of H_2 formed is 24.5 L. Calculate the work done in joules when 0.34 g of Na reacts with water under the same conditions. (The conversion factor is $1 L \cdot atm = 101.3$ J.)

5.85 A 44.0-g sample of an unknown metal at 99.0°C was placed in a constant-pressure calorimeter containing 80.0 g of water at 24.0°C. The final temperature of the system was found to be 28.4°C.

Calculate the specific heat of the metal. (The heat capacity of the calorimeter is 12.4 J/°C.)

5.86 A student mixes 88.6 g of water at 74.3°C with 57.9 g of water at 24.8°C in an insulated flask. What is the final temperature of the combined water?

5.87 You are given the following data:

$$H_2(g) \longrightarrow 2H(g) \qquad \Delta H° = 436.4 \text{ kJ/mol}$$
$$Br_2(g) \longrightarrow 2Br(g) \qquad \Delta H° = 192.5 \text{ kJ/mol}$$
$$H_2(g) + Br_2(g) \longrightarrow 2HBr(g) \qquad \Delta H° = -72.4 \text{ kJ/mol}$$

Calculate $\Delta H°$ for the reaction

$$H(g) + Br(g) \longrightarrow HBr(g)$$

5.88 Compare the heat produced by the complete combustion of 1 mole of methane (CH_4) with a mole of water gas (0.50 mol H_2 and 0.50 mol CO) under the same conditions. On the basis of your answer, would you prefer methane over water gas as a fuel? Can you suggest two other reasons why methane is preferable to water gas as a fuel?

5.89 Ethanol (C_2H_5OH) and gasoline (assumed to be all octane, C_8H_{18}) are both used as automobile fuel. If gasoline is selling for $2.75/gal, what would the price of ethanol have to be in order to provide the same amount of heat per dollar? The density and $\Delta H_f°$ of octane are 0.7025 g/mL and -249.9 kJ/mol, respectively, and of ethanol are 0.7894 g/mL and -277.0 kJ/mol, respectively (1 gal = 3.785 L).

5.90 The combustion of how many moles of ethane (C_2H_6) would be required to heat 371 g of water from 55.0°C to 98.0°C?

5.91 The heat of vaporization of a liquid (ΔH_{vap}) is the energy required to vaporize 1.00 g of the liquid at its boiling point. In one experiment, 60.0 g of liquid nitrogen (boiling point = -196°C) is poured into a Styrofoam cup containing 2.00×10^2 g of water at 55.3°C. Calculate the molar heat of vaporization of liquid nitrogen if the final temperature of the water is 41.0°C.

5.92 Explain the cooling effect experienced when ethanol is rubbed on your skin, given that

$$C_2H_5OH(l) \longrightarrow C_2H_5OH(g) \qquad \Delta H° = 42.2 \text{ kJ/mol}$$

5.93 For which of the following reactions does $\Delta H_{rxn}° = \Delta H_f°$?
(a) $H_2(g) + S(\text{rhombic}) \longrightarrow H_2S(g)$
(b) $C(\text{diamond}) + O_2(g) \longrightarrow CO_2(g)$
(c) $H_2(g) + CuO(s) \longrightarrow H_2O(l) + Cu(s)$
(d) $O(g) + O_2(g) \longrightarrow O_3(g)$

5.94 Calculate the work done (in joules) when 1.0 mole of water is frozen at 0°C and 1.0 atm. The volumes of 1 mole of water and ice at 0°C are 0.0180 and 0.0196 L, respectively. (The conversion factor is 1 L · atm = 101.3 J.)

5.95 A certain gas initially at 0.050 L undergoes expansion until its volume is 0.50 L. Calculate the work done (in joules) by the gas if it expands (a) against a vacuum and (b) against a constant pressure of 0.20 atm. (The conversion factor is 1 L · atm = 101.3 J.)

5.96 Calculate the standard enthalpy of formation for diamond, given that

$$C(\text{graphite}) + O_2(g) \longrightarrow CO_2(g)$$
$$\Delta H° = -393.5 \text{ kJ/mol}$$
$$C(\text{diamond}) + O_2(g) \longrightarrow CO_2(g)$$
$$\Delta H° = -395.4 \text{ kJ/mol}$$

5.97 The enthalpy of combustion of benzoic acid (C_6H_5COOH) is commonly used as the standard for calibrating constant-volume bomb calorimeters; its value has been accurately determined to be -3226.7 kJ/mol. When 1.9862 g of benzoic acid are burned in a calorimeter, the temperature rises from 21.84°C to 25.67°C. What is the heat capacity of the bomb? (Assume that the quantity of water surrounding the bomb is exactly 2000 g.)

5.98 At 25°C, the standard enthalpy of formation of HF(aq) is -320.1 kJ/mol; of OH$^-$(aq), it is -229.6 kJ/mol; of F$^-$(aq), it is -329.1 kJ/mol; and of $H_2O(l)$, it is -285.8 kJ/mol.
(a) Calculate the standard enthalpy of neutralization of HF(aq):

$$HF(aq) + OH^-(aq) \longrightarrow F^-(aq) + H_2O(l)$$

(b) Using the value of -56.2 kJ as the standard enthalpy change for the reaction

$$H^+(aq) + OH^-(aq) \longrightarrow H_2O(l)$$

calculate the standard enthalpy change for the reaction

$$HF(aq) \longrightarrow H^+(aq) + F^-(aq)$$

5.99 From the enthalpy of formation for CO_2 and the following information, calculate the standard enthalpy of formation for carbon monoxide (CO).

$$CO(g) + \tfrac{1}{2}O_2(g) \longrightarrow CO_2(g) \qquad \Delta H° = -283.0 \text{ kJ/mol}$$

Why can't we obtain the standard enthalpy of formation directly by measuring the enthalpy of the following reaction?

$$C(\text{graphite}) + \tfrac{1}{2}O_2(g) \longrightarrow CO(g)$$

5.100 In the nineteenth century, two scientists named Dulong and Petit noticed that for a solid element, the product of its molar mass and its specific heat is approximately 25 J/°C. This observation, now called Dulong and Petit's law, was used to estimate the specific heat of metals. Verify the law for the metals listed in Table 5.2. The law does not apply to one of the metals. Which one is it? Why?

5.101 Determine the standard enthalpy of formation of ethanol (C_2H_5OH) from its standard enthalpy of combustion (-1367.4 kJ/mol).

5.102 Acetylene (C_2H_2) and benzene (C_6H_6) have the same empirical formula. In fact, benzene can be made from acetylene as follows:

$$3C_2H_2(g) \longrightarrow C_6H_6(l)$$

The enthalpies of combustion for C_2H_2 and C_6H_6 are -1299.4 and -3267.4 kJ/mol, respectively. Calculate the standard enthalpies of formation of C_2H_2 and C_6H_6 and hence the enthalpy change for the formation of C_6H_6 from C_2H_2.

5.103 Ice at $0°C$ is placed in a Styrofoam cup containing 361 g of a soft drink at $23°C$. The specific heat of the drink is about the same as that of water. Some ice remains after the ice and soft drink reach an equilibrium temperature of $0°C$. Determine the mass of ice that has melted. Ignore the heat capacity of the cup. (*Hint:* It takes 334 J to melt 1 g of ice at $0°C$.)

5.104 A quantity of 85.0 mL of 0.600 M HCl is mixed with 85.0 mL of 0.600 M KOH in a constant-pressure calorimeter. The initial temperature of both solutions is the same at $17.35°C$, and the final temperature of the mixed solution is $19.02°C$. What is the heat capacity of the calorimeter? Assume that the specific heat of the solutions is the same as that of water and the molar heat of neutralization is -56.2 kJ/mol.

5.105 When 1.034 g of naphthalene ($C_{10}H_8$) is burned in a constant-volume bomb calorimeter at 298 K, 41.56 kJ of heat is evolved. Calculate ΔU and w for the reaction on a molar basis.

5.106 From a thermochemical point of view, explain why a carbon dioxide fire extinguisher or water should not be used on a magnesium fire.

5.107 A 4.117-g impure sample of glucose ($C_6H_{12}O_6$) was burned in a constant-volume calorimeter having a heat capacity of 19.65 kJ/$°C$. If the rise in temperature is $3.134°C$, calculate the percent by mass of the glucose in the sample. Assume that the impurities are unaffected by the combustion process and that $\Delta U = \Delta H$. See Appendix 2 for thermodynamic data.

5.108 The combustion of 0.4196 g of a hydrocarbon releases 17.55 kJ of heat. The masses of the products are $CO_2 = 1.419$ g and $H_2O = 0.290$ g. (a) What is the empirical formula of the compound? (b) If the approximate molar mass of the compound is 76 g/mol, calculate its standard enthalpy of formation.

5.109 In a constant-pressure calorimetry experiment, a reaction gives off 21.8 kJ of heat. The calorimeter contains 150 g of water, initially at $23.4°C$. What is the final temperature of the water? The heat capacity of the calorimeter is negligibly small.

5.110 At $850°C$, $CaCO_3$ undergoes substantial decomposition to yield CaO and CO_2. Assuming that the ΔH_f° values of the reactant and products are the same at $850°C$ as they are at $25°C$, calculate the enthalpy change (in kJ) if 66.8 g of CO_2 is produced in one reaction.

5.111 Give an example for each of the following situations: (a) adding heat to a system raises its temperature, (b) adding heat to a system does not change its temperature, and (c) a system's temperature changes despite no heat being added to it or removed from it.

5.112 Which of the constant-pressure processes given here has the smallest difference between ΔH and ΔU: (a) water $\longrightarrow$ water vapor, (b) water $\longrightarrow$ ice, (c) ice $\longrightarrow$ water vapor? Explain.

5.113 Construct a table with the headings q, w, ΔU, and ΔH. For each of the following processes, deduce whether each of the quantities listed is positive (+), negative (−), or zero (0): (a) freezing of benzene, (b) reaction of sodium with water, (c) boiling of liquid ammonia, (d) melting of ice, (e) expansion of a gas at constant temperature.

5.114 A 3.52-g sample of ammonium nitrate (NH_4NO_3) was added to 80.0 mL of water in a constant-pressure calorimeter of negligible heat capacity. As a result, the temperature of the solution decreased from $21.6°C$ to $18.1°C$. Calculate the heat of solution (ΔH_{soln}) in kJ/mol:

$$NH_4NO_3(s) \longrightarrow NH_4^+(aq) + NO_3^-(aq)$$

Assume the specific heat of the solution is the same as that of water.

5.115 A quantity of 50.0 mL of 0.200 M Ba(OH)$_2$ is mixed with 50.0 mL of 0.400 M HNO$_3$ in a constant-pressure calorimeter having a heat capacity of 496 J/$°C$. The initial temperature of both solutions is the same at $22.4°C$. What is the final temperature of the mixed solution? Assume that the specific heat of the solutions is the same as that of water and the molar heat of neutralization is -56.2 kJ/mol.

Industrial Problems

5.116 Methanol (CH_3OH) is an organic solvent and is also used as a fuel in some automobile engines. From the following data, calculate the standard enthalpy of formation of methanol:

$$2CH_3OH(l) + 3O_2(g) \longrightarrow 2CO_2(g) + 4H_2O(l)$$
$$\Delta H_{rxn}^\circ = -1452.8 \text{ kJ/mol}$$

5.117 Producer gas (carbon monoxide) is prepared by passing air over red-hot coke:

$$C(s) + \tfrac{1}{2}O_2(g) \longrightarrow CO(g)$$

Water gas (a mixture of carbon monoxide and hydrogen) is prepared by passing steam over red-hot coke:

$$C(s) + H_2O(g) \longrightarrow CO(g) + H_2(g)$$

For many years, both producer gas and water gas were used as fuels in industry and for domestic cooking. The large-scale preparation of these gases was carried out alternately; that is, first producer gas, then water gas, and so on. Using thermochemical reasoning, explain why this procedure was chosen.

Engineering Problems

5.118 Glauber's salt, sodium sulfate decahydrate ($Na_2SO_4 \cdot 10H_2O$), undergoes a phase transition (i.e., melting or freezing) at a convenient temperature of about 32°C:

$$Na_2SO_4 \cdot 10H_2O(s) \longrightarrow Na_2SO_4 \cdot 10H_2O(l)$$
$$\Delta H° = 74.4 \text{ kJ/mol}$$

As a result, this compound is used to regulate the temperature in homes. It is placed in plastic bags in the ceiling of a room. During the day, the endothermic melting process absorbs heat from the surroundings, cooling the room. At night, it gives off heat as it freezes. Calculate the mass of Glauber's salt in kilograms needed to lower the temperature of air in a room by 8.2°C. The mass of air in the room is 605.4 kg; the specific heat of air is 1.2 J/g · °C.

5.119 An excess of zinc metal is added to 50.0 mL of a 0.100 M $AgNO_3$ solution in a constant-pressure calorimeter like the one pictured in Figure 5.8. As a result of the reaction

$$Zn(s) + 2Ag^+(aq) \longrightarrow Zn^{2+}(aq) + 2Ag(s)$$

the temperature rises from 19.25°C to 22.17°C. If the heat capacity of the calorimeter is 98.6 J/°C, calculate the enthalpy change for the given reaction on a molar basis. Assume that the density and specific heat of the solution are the same as those for water, and ignore the specific heats of the metals.

5.120 A driver's manual states that the stopping distance quadruples as the speed doubles; that is, if it takes 30 ft to stop a car moving at 25 mph, then it would take 120 ft to stop a car moving at 50 mph. Justify this statement by using mechanics and the first law of thermodynamics. [Assume that when a car is stopped, its kinetic energy ($\frac{1}{2}mu^2$) is totally converted to heat.]

5.121 A gas company in Massachusetts charges 27 cents for a mole of natural gas (CH_4). Calculate the cost of heating 200 mL of water (enough to make a cup of coffee or tea) from 20°C to 100°C. Assume that only 50 percent of the heat generated by the combustion is used to heat the water; the rest of the heat is lost to the surroundings.

5.122 Portable hot packs are available for skiers and people engaged in other outdoor activities in a cold climate. The air-permeable paper packet contains a mixture of powdered iron, sodium chloride, and other components, all moistened by a little water. The exothermic reaction that produces the heat is a very common one—the rusting of iron:

$$4Fe(s) + 3O_2(g) \longrightarrow 2Fe_2O_3(s)$$

When the outside plastic envelope is removed, O_2 molecules penetrate the paper, causing the reaction to begin. A typical packet contains 250 g of iron to warm your hands or feet for up to 4 hours. How much heat (in kJ) is produced by this reaction? (*Hint:* See Appendix 2 for $\Delta H_f°$ values.)

5.123 For reactions in condensed phases (liquids and solids), the difference between ΔH and ΔU is usually quite small. This statement holds for reactions carried out under atmospheric conditions. For certain geochemical processes, however, the external pressure may be so great that ΔH and ΔU can differ by a significant amount. A well-known example is the slow conversion of graphite to diamond under Earth's surface. Calculate $\Delta H - \Delta U$ for the conversion of 1 mole of graphite to 1 mole of diamond at a pressure of 50,000 atm. The densities of graphite and diamond are 2.25 g/cm^3 and 3.52 g/cm^3, respectively.

5.124 Consider the reaction

$$2H_2(g) + O_2(g) \longrightarrow 2H_2O(l)$$

Under atmospheric conditions (1.00 atm) it was found that the formation of water resulted in a decrease in volume equal to 73.4 L. Calculate ΔU for the process. $\Delta H = -571.6$ kJ/mol. (The conversion factor is 1 L · atm = 101.3 J.)

5.125 The total volume of the Pacific Ocean is estimated to be 7.2 × 10^8 km^3. A medium-sized atomic bomb produces 1.0 × 10^{15} J of energy upon explosion. Calculate the number of atomic bombs needed to release enough energy to raise the temperature of the water in the Pacific Ocean by 1°C.

5.126 The so-called hydrogen economy is based on hydrogen produced from water using solar energy. The gas is then burned as a fuel:

$$2H_2(g) + O_2(g) \longrightarrow 2H_2O(l)$$

A primary advantage of hydrogen as a fuel is that it is nonpolluting. A major disadvantage is that it is a gas and therefore is harder to store than liquids or solids. Calculate the number of moles of H_2 required to produce an amount of energy equivalent to that produced by the combustion of a gallon of octane (C_8H_{18}). The density of octane is 2.66 kg/gal, and its standard enthalpy of formation is −249.9 kJ/mol.

Biological Problems

5.127 Photosynthesis produces glucose ($C_6H_{12}O_6$) and oxygen from carbon dioxide and water:

$$6CO_2(g) + 6H_2O(l) \longrightarrow C_6H_{12}O_6(s) + 6O_2(g)$$

(a) How would you determine experimentally the ΔH_{rxn}° value for this reaction? (b) Solar radiation produces about 7.0×10^{14} kg of glucose a year on Earth. What is the corresponding ΔH° change?

5.128 Calculate the standard enthalpy change for the fermentation process, in which glucose ($C_6H_{12}O_6$) is converted to ethanol (C_2H_5OH) and carbon dioxide.

5.129 A 46-kg person drinks 500 g of milk, which has a "caloric" value of approximately 3.0 kJ/g. If only 17 percent of the energy in milk is converted to mechanical work, how high (in meters) can the person climb based on this energy intake? [*Hint:* The work done in ascending is given by *mgh*, where *m* is the mass (in kg), *g* is the gravitational acceleration (9.8 m/s^2), and *h* is the height (in meters).]

5.130 A man ate 0.50 pound of cheese (an energy intake of 4×10^3 kJ). Suppose that none of the energy was stored in his body. What mass (in grams) of water would he need to perspire in order to maintain his original temperature? (It takes 44.0 kJ to vaporize 1 mole of water.)

5.131 Why are cold, damp air and hot, humid air more uncomfortable than dry air at the same temperatures? [The specific heats of water vapor and air are approximately 1.9 J/(g·°C) and 1.0 J/(g·°C), respectively.]

5.132 A woman expends 95 kJ of energy walking a kilometer. The energy is supplied by the metabolic breakdown of food, which has an efficiency of 35 percent. How much energy does she save by walking the kilometer instead of driving a car that gets 8.2 km per liter of gasoline (approximately 20 mi/gal)? The density of gasoline is 0.71 g/mL, and its enthalpy of combustion is −49 kJ/g.

5.133 The carbon dioxide exhaled by sailors in a submarine is often removed by reaction with an aqueous lithium hydroxide solution. (a) Write a balanced equation for this process. (*Hint:* The products are water and a soluble salt.) (b) If every sailor consumes 1.2×10^4 kJ of energy every day and assuming that this energy is totally supplied by the metabolism of glucose ($C_6H_{12}O_6$), calculate the amounts of CO_2 produced and LiOH required to purify the air.

5.134 How much metabolic energy must a 5.2-g hummingbird expend to fly to a height of 21 m? (See the hint in Problem 5.129.)

5.135 Acetylene (C_2H_2) can be made by combining calcium carbide (CaC_2) with water. (a) Write an equation for the reaction. (b) What is the maximum amount of heat (in joules) that can be obtained from the combustion of acetylene, starting with 74.6 g of CaC_2?

5.136 (a) A person drinks four glasses of cold water (3.0°C) every day. The volume of each glass is 2.5×10^2 mL. How much heat (in kJ) does the body have to supply to raise the temperature of the water to 37°C, the body temperature? (b) How much heat would your body lose if you were to ingest 8.0×10^2 g of snow at 0°C to quench your thirst? (The amount of heat necessary to melt snow is 6.01 kJ/mol.)

5.137 Both glucose and fructose are simple sugars with the same molecular formula of $C_6H_{12}O_6$. Sucrose ($C_{12}H_{22}O_{11}$), or table sugar, consists of a glucose molecule bonded to a fructose molecule (a water molecule is eliminated in the formation of sucrose). (a) Calculate the energy released when a 2.0-g glucose tablet is burned in air. (b) To what height can a 65-kg person climb after ingesting such a tablet, assuming only 30 percent of the energy released is available for work? (See the hint for Problem 5.129.) Repeat the calculations for a 2.0-g sucrose tablet.

5.138 Metabolic activity in the human body releases approximately 1.0×10^4 kJ of heat per day. Assume that a 55-kg body has the same specific heat as water; how much would the body temperature rise if it were an isolated system? How much water must the body eliminate as perspiration to maintain the normal body temperature (98.6°F)? Comment on your results. (The heat of vaporization of water is 2.41 kJ/g.)

Environmental Problems

5.139 Calcium oxide (CaO) is used to remove sulfur dioxide generated by coal-burning power stations:

$$2CaO(s) + 2SO_2(g) + O_2(g) \longrightarrow 2CaSO_4(s)$$

Calculate the enthalpy change if 6.6×10^5 g of SO_2 is removed by this process.

5.140 About 6.0×10^{13} kg of CO_2 is fixed (converted to more complex organic molecules) by photosynthesis every year. (a) Assuming all the CO_2 ends up as glucose ($C_6H_{12}O_6$), calculate the energy (in kJ) stored by photosynthesis per year. (b) A typical coal-burning electric power station generates about 2.0×10^6 W per year. How many such stations are needed to generate the same amount of energy as that captured by photosynthesis (1 W = 1 J/s)?

5.141 The average temperature in deserts is high during the day but quite cool at night, whereas that in regions along the coastline is more moderate. Explain.

Multiconcept Problems

5.142   *Lime* is a term that includes calcium oxide (CaO, also called quicklime) and calcium hydroxide [$Ca(OH)_2$, also called slaked lime]. It is used in the steel industry to remove acidic impurities, in air-pollution control to remove acidic oxides such as SO_2, and in water treatment. Quicklime is made industrially by heating limestone ($CaCO_3$) above 2000°C:

$$CaCO_3(s) \longrightarrow CaO(s) + CO_2(g)$$
$$\Delta H° = 177.8 \text{ kJ/mol}$$

Slaked lime is produced by treating quicklime with water:

$$CaO(s) + H_2O(l) \longrightarrow Ca(OH)_2(s)$$
$$\Delta H° = -65.2 \text{ kJ/mol}$$

The exothermic reaction of quicklime with water and the rather small specific heats of both quicklime [0.946 J/(g·°C)] and slaked lime [1.20 J/(g·°C)] make it hazardous to store and transport lime in vessels made of wood. Wooden sailing ships carrying lime would occasionally catch fire when water leaked into the hold. (a) If a 500.0-g sample of water reacts with an equimolar amount of CaO (both at an initial temperature of 25°C), what is the final temperature of the product, $Ca(OH)_2$? Assume that the product absorbs all the heat released in the reaction. (b) Given that the standard enthalpies of formation of CaO and H_2O are -635.6 and -285.8 kJ/mol, respectively, calculate the standard enthalpy of formation of $Ca(OH)_2$.

5.143   Hydrazine (N_2H_4) decomposes to form ammonia and nitrogen gases. (a) Write a balanced chemical equation for this process. (b) Given that the standard enthalpy of formation of hydrazine is 50.42 kJ/mol, calculate $\Delta H°_{rxn}$ for its decomposition. (c) Both hydrazine and ammonia will burn in oxygen to produce $H_2O(l)$ and $N_2(g)$. Write balanced equations for these processes and determine $\Delta H°_{rxn}$ for each process. (d) If equal masses of hydrazine and ammonia were burned in separate bomb calorimeter experiments, which would cause the greater increase in temperature?

5.144   Flameless ration heaters are used to warm military MREs (meals ready-to-eat). A heater consists of a pouch containing 7.5 g of a metal alloy powder that is 95% Mg. According to the instructions, 30 mL of water should be added to the pouch. The resulting reaction produces aqueous magnesium hydroxide and hydrogen gas. Over the course of 12 minutes, the temperature of an 8.0-oz. meal in contact with the heater increases by 100°F. (a) Write a balanced chemical equation for the reaction that occurs when water is added to the pouch. (b) Determine $\Delta H°$ for the reaction. (c) Calculate q of the reaction. (d) Calculate the heat capacity of the food. (Assume that the density of water is 0.998 g/cm³; and that the reaction is the system, and the water and food together constitute the surroundings.)

5.145   A piece of metal with a mass of 5.05 g originally at 25.5°C is dropped into 25 g of water originally at 82.7°C. The final temperature of the metal and the water is 81.5°C. Determine the specific heat of the metal and consult Table 5.2 to determine its possible identity.

Standardized-Exam Practice Problems

Physical and Biological Sciences

A bomb calorimeter was calibrated by burning 1.013 g of benzoic acid ($C_7H_6O_2$) ($\Delta U_{comb} = 3.221 \times 10^3$ kJ/mol). The temperature change in the calorimeter during the calibration combustion was 5.19°C. A nutritional chemist then used the calibrated calorimeter to determine the energy content of food. The chemist carefully dried a sample of food and placed 0.8996 g of the sample in the calorimeter with sufficient oxygen for the combustion to go to completion. Combustion of the food sample caused the temperature of the calorimeter to increase by 4.42°C.

1. Approximately how many moles of O_2 gas were consumed in the calibration combustion?

   a) 0.008
   b) 0.1
   c) 0.2
   d) 0.06

2. What is the heat capacity (C_V) of the calorimeter?

   a) 5.15 kJ/°C
   b) 5.08 kJ/°C
   c) 5.12 kJ/°C
   d) 4.97 kJ/°C

3. What is the energy content of the food?

   a) 22.8 kJ/g
   b) 4.97 kJ/g
   c) 25.3 kJ/g
   d) 0.201 kJ/g

4. What would be the effect on the result if the food sample were not completely dried prior to being placed in the calorimeter?

   a) The combustion of the sample would be incomplete.
   b) The calculated energy content per gram would be too low.
   c) The calculated energy content per gram would be too high.
   d) There would be no effect on the result.

Answers to In-Chapter Materials

Answers to Practice Problems

5.1A (a) 1.13×10^3 J, (b) 4. **5.1B** (a) 372 m/s, (b) neither. **5.2A** -4.32×10^4 kJ. **5.2B** 6.95×10^5 kJ, heat is absorbed. **5.3A** 8.174×10^4 kJ. **5.3B** 1.71×10^3 g. **5.4A** 151 kJ. **5.4B** 52.2°C. **5.5A** 28°C. **5.5B** 42 g. **5.6A** 14.5 kJ/g. **5.6B** 10.5°C rise. **5.7A** −1103 kJ/mol. **5.7B** 113.2 kJ/mol. **5.8A** 177.8 kJ/mol. **5.8B** −697.6 kJ/mol. **5.9A** 87.3 kJ/mol. **5.9B** $NH_3(g) + HCl(g)$, 176.8 kJ/mol.

Answers to Checkpoints

5.1.1 a. **5.1.2** b. **5.1.3** a. **5.1.4** e. **5.1.5** c. **5.1.6** b, d, e. **5.2.1** e. **5.2.2** c. **5.2.3** c. **5.2.4** a, d, e. **5.3.1** b. **5.3.2** a. **5.4.1** b. **5.4.2** d. **5.4.3** a. **5.4.4** a. **5.5.1** c. **5.5.2** a. **5.5.3** c. **5.5.4** d. **5.6.1** b. **5.6.2** e. **5.6.3** c, e. **5.6.4** e.

Design Icon Credits: Animation icon: ©McGraw Hill Education; Student Hot Spot icon: ©LovArt/Shutterstock.com; Satellite view of planet Earth: Stockbyte/Alamy

CHAPTER 6

Quantum Theory and the Electronic Structure of Atoms

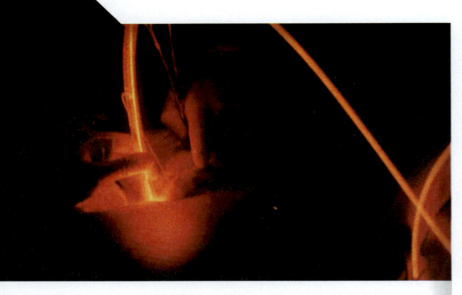

Light is used in a variety of medical procedures, including some cancer treatments. Photodynamic therapy, shown here, is done using visible light.

Source: *John Crawford/National Cancer Institute*

In This Chapter, You Will Learn

About some of the properties of electromagnetic radiation or *light* and how these properties have been used to study and elucidate the electronic structure of atoms. You will also learn how to determine the arrangement of electrons in a particular atom.

Before You Begin, Review These Skills

- Tracking units [◄◄ Section 1.6]
- The nuclear model of the atom [◄◄ Section 2.2]

How Our Understanding of Electronic Structure Has Advanced Medicine

For more than a century, radiation has been used in the treatment of cancer. Traditional therapies involve the use of powerful *ionizing radiation* [►► Chapter 20], such as X rays or gamma rays. Ionizing radiation damages the DNA of cancer cells, preventing their continued replication—although it does the same to healthy, noncancerous cells. Over time, technicians and physicians have developed methods that minimize the damage done to healthy tissue, but there remains an inherent risk in the use of ionizing radiation, and stringent safety protocols are required for treatment.

In recent decades, a new type of radiation therapy has emerged and is becoming increasingly popular: *photodynamic therapy* (PDT). PDT works by irradiation with *visible* light, which is far less energetic than X-ray or gamma radiation, and thus safer to use. Treatment begins with the administration of a *photosensitizer* (a substance that undergoes reaction when irradiated) that is selectively taken up by cancerous tissue. Subsequent irradiation with a specific wavelength of visible light causes the photosensitizer to produce an especially reactive form of oxygen known as *singlet oxygen,* which damages the cancerous cells and prevents further replication.

The visible light used in photodynamic therapy results from electron transitions within atoms. Although the nuclear model that Rutherford proposed based on his gold-foil experiment specified the location of the protons and the neutrons in an atom, it failed to describe the location or behavior of the *electrons.* Early in the twentieth century, the application of a radical new theory in physics called *quantum theory,* and the ingenious interpretation of experimental evidence by Max Planck, Albert Einstein, and others, led to our current understanding of the *electronic structure of atoms.* This understanding of electronic structure is what makes such things as photodynamic therapy possible.

At the end of this chapter, you will be able to answer a variety of questions about the nature of light and the electronic structure of the atom [►► Applying What You've Learned, page 302].

6.1   The Nature of Light

Student Note: It's useful to have a sense of the relative positions of the types of radiation in the electromagnetic spectrum. A mnemonic device for remembering the regions in order of *increasing wavelength* is: "Great Xylophones Use Very Important Musical Records." (Gamma, X ray, Ultraviolet, Visible, Infrared, Microwave, and Radio waves.) As we explain in Section 6.2, this also lists the types of radiation in order of *decreasing energy*.

When we say "light," we generally mean *visible* light, which is the light we can detect with our eyes. Visible light, however, is only a small part of the continuum of radiation that makes up the **electromagnetic spectrum.** In addition to visible light, the electromagnetic spectrum includes radio waves, microwave radiation, infrared and ultraviolet radiation, X rays, and gamma rays, as shown in Figure 6.1. Some of these terms may be familiar to you. For instance, the danger of exposure to ultraviolet radiation is why you need to use sunscreen. You may have used microwave radiation from a microwave oven to reheat food or to pop popcorn; you may have had X rays during a routine dental checkup or after breaking a bone; and you may recall from Chapter 2 that gamma rays are emitted from some radioactive materials. Although these phenomena may seem very different from each other and from visible light, they all are the transmission of energy in the form of *waves*.

Properties of Waves

Student Note: The speed of light is *defined* to be 2.99792458 × 10⁸ m/s. It is, therefore, an *exact* number and usually does not limit the number of significant figures in a calculated result. In most calculations, however, the speed of light is rounded to three significant figures: $c = 3.00 \times 10^8$ m/s.

The fundamental properties of waves are illustrated in Figure 6.2. Waves are characterized by their wavelength, frequency, and amplitude. **Wavelength** λ (lambda) is the distance between identical points on successive waves (e.g., successive peaks or successive troughs). The **frequency** ν (nu) is the *number* of waves that pass through a particular point in 1 second. **Amplitude** is the vertical distance from the midline of a wave to the top of the peak or the bottom of the trough.

The speed of a wave depends on the type of wave and the nature of the medium through which the wave is traveling (e.g., air, water, or a vacuum). The speed of light through a vacuum, c, is 2.99792458×10^8 m/s. The speed, wavelength, and frequency of a wave are related by the equation

Equation 6.1 $c = \lambda\nu$

Student Note: Frequency is expressed as cycles per second, or simply *reciprocal seconds* (s⁻¹), which is also known as *hertz* (Hz).

where λ and ν are expressed in meters (m) and reciprocal seconds (s⁻¹), respectively. Although wavelength in meters is convenient for this equation, the units customarily used

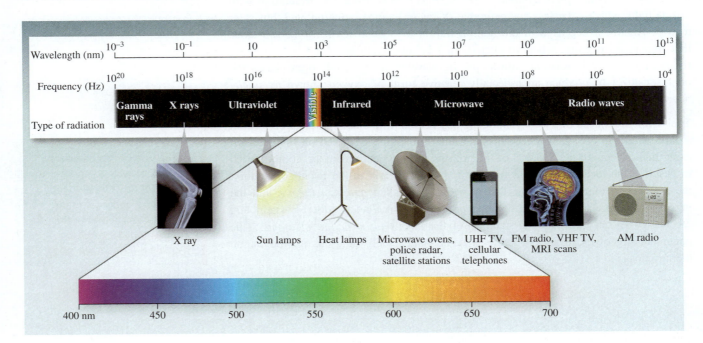

Figure 6.1 Electromagnetic spectrum. Each type of radiation is spread over a specific range of wavelengths (and frequencies). Visible light ranges from 400 nm (violet) to 700 nm (red).

(knee joint X ray): *kaling2100/Shutterstock;* (MRI): *Don Farrall/Photodisc/Getty Images*

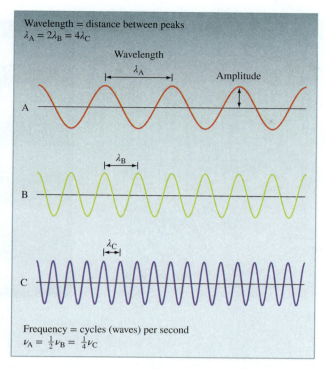

Wavelength = distance between peaks
$\lambda_A = 2\lambda_B = 4\lambda_C$

Wavelength
λ_A

Amplitude

A

λ_B

B

λ_C

C

Frequency = cycles (waves) per second
$\nu_A = \frac{1}{2}\nu_B = \frac{1}{4}\nu_C$

Figure 6.2 Characteristics of waves: wavelength, amplitude, and frequency.

to express the wavelength of electromagnetic radiation depend on the type of radiation and the magnitude of the corresponding wavelength. The wavelength of visible light, for instance, is on the order of nanometers (nm, or 10^{-9} m), and that of microwave radiation is on the order of centimeters (cm, or 10^{-2} m).

The Electromagnetic Spectrum

In 1873, James Clerk Maxwell proposed that visible light consisted of electromagnetic waves. According to Maxwell's theory, an ***electromagnetic wave*** has an electric field component and a magnetic field component. These two components have the same wavelength and frequency, and hence the same speed, but they travel in mutually perpendicular planes (Figure 6.3). The significance of Maxwell's theory is that it provides a mathematical description of the general behavior of light. In particular, his model accurately describes how energy in the form of radiation can be propagated through space as oscillating electric and magnetic fields.

The Double-Slit Experiment

A simple yet convincing demonstration of the wave nature of light is the phenomenon of *interference.* When a light source passes through a narrow opening, called a *slit,* a bright line is generated in the path of the light through the slit. When the same light source passes through two closely spaced slits, however, as shown in Figure 6.4, the result is not two bright lines, one in the path of each slit, but rather a series of light and dark lines known as an *interference pattern.* When the light sources recombine after passing through the slits, they do so *constructively* where the two waves are *in phase* (giving rise to the light lines) and *destructively* where the waves are *out of phase* (giving rise to the dark lines). Constructive interference and destructive interference are properties of waves.

The various types of electromagnetic radiation in Figure 6.1 differ from one another in wavelength and frequency. Radio waves, which have long wavelengths and low frequencies, are emitted by large antennas, such as those used by broadcasting stations. The shorter, visible light waves are produced by the motions of electrons within atoms and molecules. The shortest waves, which also have the highest frequency, are γ (gamma) rays, which result from nuclear processes [◄◄ Section 2.2]. The higher the frequency, the more energetic the radiation. Thus, ultraviolet radiation, X rays, and γ rays are high-energy radiation, whereas infrared radiation, microwave radiation, and radio waves are low-energy radiation.

Sample Problem 6.1 illustrates the conversion between wavelength and frequency.

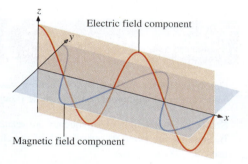

Electric field component

Magnetic field component

Figure 6.3 Electric field and magnetic field components of an electromagnetic wave. These two components have the same wavelength, frequency, and amplitude, but they vibrate in two mutually perpendicular planes.

Figure 6.4 Double-slit experiment. (a) Red lines correspond to the maximum intensity resulting from constructive interference. Dashed blue lines correspond to the minimum intensity resulting from destructive interference. (b) Interference pattern with alternating bright and dark lines.

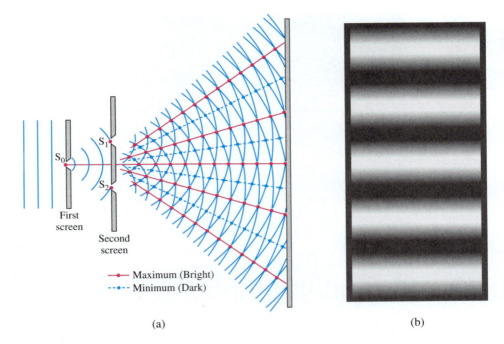

First screen

Second screen

•—•— Maximum (Bright)
--•-- Minimum (Dark)

(a)   (b)

SAMPLE PROBLEM   6.1

One type of laser used in the treatment of vascular skin lesions is a neodymium-doped yttrium aluminum garnet or Nd:YAG laser. The wavelength commonly used in these treatments is 532 nm. What is the frequency of this radiation?

Strategy Wavelength and frequency are related by Equation 6.1 ($c = \lambda\nu$), so we must rearrange Equation 6.1 to solve for frequency. Because we are given the wavelength of the electromagnetic radiation in nanometers, we must convert this wavelength to meters and use $c = 3.00 \times 10^8$ m/s.

Setup Solving for frequency gives $\nu = c/\lambda$. Next we convert the wavelength to meters:

$$\lambda(\text{in meters}) = 532 \text{ nm} \times \frac{1 \times 10^{-9} \text{ m}}{1 \text{ nm}} = 5.32 \times 10^{-7} \text{ m}$$

Solution

$$\nu = \frac{3.00 \times 10^8 \text{ m/s}}{5.32 \times 10^{-7} \text{ m}} = 5.64 \times 10^{14} \text{ s}^{-1}$$

THINK ABOUT IT

Make sure your units cancel properly. A common error in this type of problem is neglecting to convert wavelength to meters.

Practice Problem ATTEMPT What is the frequency (in reciprocal seconds) of electromagnetic radiation with a wavelength of 1.03 cm?

Practice Problem BUILD What is the wavelength (in meters) of an electromagnetic wave whose frequency is 1.61×10^{12} s^{-1}?

Practice Problem CONCEPTUALIZE Which of the following sets of waves best represents the relative wavelengths/frequencies of visible light of the colors shown?

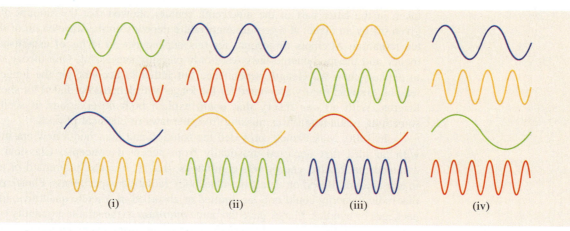

(i)   (ii)   (iii)   (iv)

CHECKPOINT – SECTION 6.1   The Nature of Light

6.1.1   Calculate the wavelength of light with frequency 3.45×10^{14} s^{-1}.

  a) 1.15×10^{-6} nm

  b) 1.04×10^{23} nm

  c) 8.70×10^{2} nm

  d) 115 nm

  e) 9.66×10^{-24} nm

6.1.2   Calculate the frequency of light with wavelength 126 nm.

  a) 2.38×10^{15} s^{-1}

  b) 4.20×10^{-16} s^{-1}

  c) 37.8 s^{-1}

  d) 2.65×10^{-2} s^{-1}

  e) 3.51×10^{19} s^{-1}

6.1.3   When traveling through a translucent medium, such as glass, light travels more slowly than it travels through a vacuum. Red light with a wavelength of 684 nm travels through Pyrex glass with a frequency of 2.92×10^{14} s^{-1}. Calculate the speed of this light.

  a) 3.00×10^{8} m/s

  b) 2.00×10^{8} m/s

  c) 2.92×10^{6} m/s

  d) 4.23×10^{7} m/s

  e) 2.23×10^{8} m/s

6.1.4   Of the waves pictured, _____ has the greatest frequency, _____ has the greatest wavelength, and _____ has the greatest amplitude.

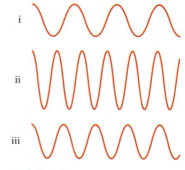

  a) i, ii, iii

  b) i, iii, ii

  c) ii, i, ii

  d) ii, i, iii

  e) ii, iii, ii

6.2   Quantum Theory

Early attempts by nineteenth-century physicists to figure out the structure of the atom met with only limited success. This was largely because they were attempting to understand the behavior of subatomic particles using the laws of classical physics that govern the behavior of macroscopic objects. It took a long time to realize—and an even longer time to accept—that the properties of atoms are *not* governed by the same physical laws as larger objects.

Quantization of Energy

When a solid is heated above 800 K, it emits electromagnetic radiation, including visible wavelengths, known as **blackbody radiation.** As the temperature of the solid increases, the light emitted goes from red, to orange, to yellow, to white. Measurements taken in the latter part of the nineteenth century showed that the amount of energy given off by an object at a certain temperature depends on the wavelength of the emitted radiation. Attempts to account for this dependence in terms of established wave theory and thermodynamic laws were only partially successful. One theory was able to explain short-wavelength dependence but failed to account for the longer wavelengths. Another theory accounted for the longer wavelengths but failed for short wavelengths. With no *one* theory that could explain both observations, it seemed that something fundamental was missing from the laws of classical physics.

In 1900, Max Planck[1] provided the solution and launched a new era in physics with an idea that departed drastically from accepted concepts. Classical physics assumed that radiant energy was continuous; that is, it could be emitted or absorbed in any amount. Based on data from blackbody radiation experiments, Planck proposed that radiant energy could only be emitted or absorbed in discrete quantities, like small packages or bundles. Planck gave the name **quantum** to the smallest quantity of energy that can be emitted (or absorbed) in the form of electromagnetic radiation. The energy E of a single quantum of energy is given by

$$E = h\nu$$

Equation 6.2

where h is called *Planck's constant* and ν is the *frequency* of the radiation. The value of Planck's constant is 6.63×10^{-34} J $\cdot$ s.

Student Note: The National Institute of Standards and Technology (NIST) gives a value of 6.626070040 × 10^{-34} J · s for Planck's constant. Typically, three significant figures are sufficient for solving problems.

According to quantum theory, energy is always emitted in whole-number multiples of $h\nu$. At the time Planck presented his theory, he could not explain why energies should be fixed or quantized in this manner. Starting with this hypothesis, however, he had no difficulty correlating the experimental data for the emission by solids over the entire range of wavelengths; the experimental data supported his new *quantum theory*.

The idea that energy is *quantized* rather than *continuous* may seem strange, but the concept of quantization has many everyday analogies. For example, vending machines dispense cans or bottles of soft drinks only in whole numbers (you can't buy part of a can or bottle from a machine). Each can or bottle is a quantum of its soft drink. Even processes in living systems involve quantized phenomena. The eggs laid by hens are quanta (hens lay only whole eggs). Similarly, when a dog or cat gives birth to a litter, the number of offspring is always an integer. Each puppy or kitten is a quantum of that animal. Planck's quantum theory revolutionized physics. Indeed, the flurry of research that ensued altered our concept of nature forever.

Sample Problem 6.2 shows how to compare energy per photon for different wavelengths.

Bringing Chemistry to Life

Laser Pointers

The laser pointers that have become so common typically emit radiation in the red region of the visible spectrum with output wavelengths ranging from 630 to 680 nm. Low prices and availability have made the devices popular not only with instructors and other speakers, but also with teenagers and even children—raising some significant safety concerns. Although the human blink reflex generally is sufficient to protect against serious or permanent injury to the eye by one of these devices, intentional prolonged exposure of the eye to the beam from a laser pointer can be dangerous. Of particular concern are the new green laser pointers that emit a wavelength of 532 nm. The lasers in these devices also produce radiation in the infrared region of the electromagnetic spectrum (1064 nm), but they are equipped

---

1. Max Karl Ernst Ludwig Planck (1858–1947). German physicist. Planck received the Nobel Prize in Physics in 1918 for his quantum theory. He also made significant contributions in thermodynamics and other areas of physics.

with filters to prevent the emission of infrared radiation. However, some of the inexpensive imported lasers do not bear adequate safety labeling and the filters are easily removed—potentially resulting in the emission of dangerous radiation. Although the 1064-nm laser beam is less energetic than the 532-nm beam, it poses a greater danger to the eye because it is not visible and does not evoke the blink response that visible wavelengths do. Despite not being visible, a 1064-nm beam can pass through the anterior structures of the eye and injure the retina, although the damage may not be immediately apparent.

SAMPLE PROBLEM   6.2

How much more energy per photon is there in green light of wavelength 532 nm than in red light of wavelength 635 nm?

Strategy Convert each wavelength to frequency using Equation 6.1, and then calculate the energy per photon using Equation 6.2.

Setup Convert the wavelengths to meters:

$$532 \text{ nm} \times \frac{1 \times 10^{-9} \text{ m}}{1 \text{ nm}} = 5.32 \times 10^{-7} \text{ m}$$

$$635 \text{ nm} \times \frac{1 \times 10^{-9} \text{ m}}{1 \text{ nm}} = 6.35 \times 10^{-7} \text{ m}$$

Planck's constant, h, is 6.63×10^{-34} J · s.

Solution For 532 nm,

$$\nu = \frac{c}{\lambda} = \frac{3.00 \times 10^8 \text{ m/s}}{5.32 \times 10^{-7} \text{ m}} = 5.64 \times 10^{14} \text{ s}^{-1}$$

and

$$E = h\nu = (6.63 \times 10^{-34} \text{ J} \cdot \text{s})(5.64 \times 10^{14} \text{ s}^{-1}) = 3.74 \times 10^{-19} \text{ J}$$

The energy of a single photon with wavelength 532 nm is 3.74×10^{-19} J.

Following the same procedure, the energy of a photon of wavelength 635 nm is 3.13×10^{-19} J. The difference between them is $(3.74 \times 10^{-19}$ J$) - (3.13 \times 10^{-19}$ J$) = 6.1 \times 10^{-20}$ J. Therefore, a photon of green light ($\lambda = 532$ nm) has 6.1×10^{-20} J *more* energy than a photon of red light ($\lambda = 635$ nm).

Brian Moeskau

THINK ABOUT IT

As wavelength decreases, frequency increases. Energy, being directly proportional to frequency, also increases.

Practice Problem ATTEMPT Calculate the difference in energy (in joules) between a photon with $\lambda = 680$ nm (red) and a photon with $\lambda = 442$ nm (blue).

Practice Problem BUILD In what region of the electromagnetic spectrum is a photon found that possesses twice as much energy as one in the blue region ($\lambda = 442$ nm) of the visible spectrum?

Practice Problem CONCEPTUALIZE Shown here are waves of electromagnetic radiation of two different frequencies and two different amplitudes. Assume that intensity of radiation (photons/s) is directly proportional to amplitude. Which of the waves is made up of photons of greater energy? Which wave delivers more photons during a given period of time? Which wave delivers more total energy during a given time period?

(i)

(ii)

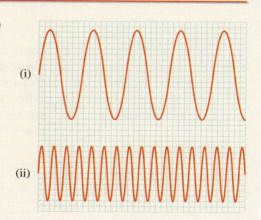

Figure 6.5 The photoelectric effect. (a) Light of frequency lower than the threshold frequency does not cause electrons to be ejected. (b) Light with the threshold frequency causes electrons to be ejected. (c) Light with greater than the threshold frequency causes electrons to be ejected with additional kinetic energy (moving faster).

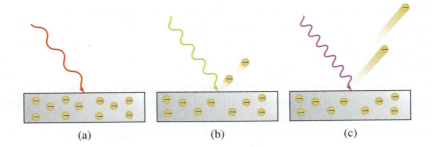

(a)   (b)   (c)

Photons and the Photoelectric Effect

In 1905, only 5 years after Planck presented his quantum theory, Albert Einstein[2] used the theory to explain another mysterious physical phenomenon, the ***photoelectric effect,*** in which electrons are ejected from the surface of a metal exposed to light of at least a certain minimum frequency, called the *threshold frequency* (Figure 6.5). The number of electrons ejected was proportional to the intensity (or brightness) of the light, but the energies of the ejected electrons were not. Below the threshold frequency no electrons were ejected no matter how intense the light.

The photoelectric effect could not be explained by the wave theory of light, which associated the energy of light with its intensity. Einstein, however, made an extraordinary assumption. He suggested that a beam of light is really a stream of particles. These *particles of light* are now called ***photons.*** Using Planck's quantum theory of radiation as a starting point, Einstein deduced that each *photon* must possess energy E given by the equation

$$E_{photon} = h\nu$$

where h is Planck's constant and ν is the frequency of the light. Electrons are held in a metal by attractive forces, and so removing them from the metal requires light of a sufficiently high frequency (which corresponds to a sufficiently high energy) to break them free. Shining a beam of light onto a metal surface can be thought of as shooting a beam of particles—*photons*—at the metal atoms. If the frequency of the photons is such that $h\nu$ exactly equals the energy that binds the electrons in the metal, then the light will have just enough energy to knock the electrons loose. If we use light of a higher frequency, then not only will the electrons be knocked loose, but they will also acquire some kinetic energy. This situation is summarized by the equation

Equation 6.3 $h\nu = KE + W$

where KE is the kinetic energy of the ejected electron and W is the binding energy of the electron in the metal. [Binding energies are typically given in units of electron volts (eV), where 1 eV = 1.602×10^{-19} J.] Rewriting Equation 6.3 as

$$KE = h\nu - W$$

shows that the more energetic the photon (i.e., the higher its frequency), the greater the kinetic energy of the ejected electron. If the frequency of light is below the threshold frequency, the photon will simply bounce off the surface and no electrons will be ejected. If the frequency is equal to the threshold frequency, it will dislodge the most loosely held electron. Above the threshold frequency, it will not only dislodge the electron, but also impart some kinetic energy to the ejected electron.

---

2. Albert Einstein (1879–1955). German-born American physicist. Regarded by many as one of the two greatest physicists the world has known (the other is Isaac Newton). The three papers (on special relativity, Brownian motion, and the photoelectric effect) that he published in 1905 while employed as a technical assistant in the Swiss patent office in Berne have profoundly influenced the development of physics. He received the Nobel Prize in Physics in 1921 for his explanation of the photoelectric effect.

Where Have I Encountered the Photoelectric Effect?

Chances are good that you encounter the photoelectric effect regularly. Some everyday applications include the type of device that prevents a garage door from closing when something is in the door's path and motion-detection systems used in museums and other high-security environments. These sorts of devices work simply by responding to an interruption in a beam of light. In each case, a beam of light normally shines on a *photocathode,* a surface that emits photoelectrons, and the photoelectrons, accelerated toward an anode by high voltage, constitute a current. When the beam of light is interrupted, the flow of photoelectrons stops and the current is cut off. In the case of the garage-door safety device, when the current stops, the movement of the door stops. In the case of the motion-detection systems, when the current stops, an alarm may sound or a light may turn on.

One of the more exotic uses of the photoelectric effect is in night-vision goggles. Although you may never have looked through night-vision goggles, you have probably seen night-vision images on the news or in a suspense-filled movie such as *Silence of the Lambs.* Typically, the night-vision images we see are from what are known as third-generation night-vision devices. These devices use a photocathode material (gallium arsenide) that emits photoelectrons when struck by photons in the *infrared* region of the electromagnetic spectrum. (First- and second-generation devices used different photocathode materials and relied more on the amplification of low-level visible light.) The photoelectrons emitted by the photocathode enter a microchannel plate (MCP), an array of tiny parallel tubes, where each strikes the internal surface of a tube causing many more electrons to be ejected—a process called *secondary emission.* This effectively amplifies the current generated by each photoelectron. The amplified current is then accelerated by high voltage toward a phosphorus screen, where incident electrons cause the emission of visible light, generating the familiar green glow of night vision.

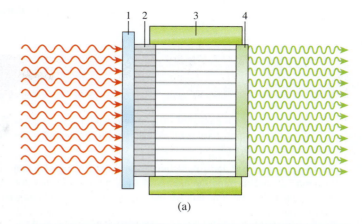

(a)

(b)

(a) Night-vision goggles schematic. 1. Photocathode (gallium arsenide). 2. Microchannel plate (MCP). 3. High-voltage source. 4. Phosphorus screen. (b) The green glow of night vision.

Official Marine Corps photo by Sgt. Brian A. Tuthill

Now consider two beams of light having the same frequency (greater than the threshold frequency) but different intensities. The more intense beam of light consists of a larger number of photons, so it ejects more electrons from the metal's surface than the weaker beam of light. Thus, the more intense the light, the greater the number of electrons emitted by the target metal; the higher the frequency of the light, the greater the kinetic energy of the ejected electrons.

Sample Problem 6.3 shows how to determine the energy of a single photon of light of a given wavelength and how to determine the maximum kinetic energy of an electron ejected via the photoelectric effect.

Einstein's theory of light posed a dilemma for scientists. On the one hand, it explains the photoelectric effect. On the other hand, the particle theory of light is inconsistent with the known wavelike properties of light. The only way to resolve the dilemma is to accept the idea that light possesses properties characteristic of both particles *and* waves. Depending on the experiment, light behaves either as a wave or as a stream of particles. This concept was totally alien to the way physicists had thought about matter and radiation, and it took a long time for them to accept it. We will see in Section 6.4 that possessing properties of both particles and waves is not unique to light but ultimately is characteristic of all matter, including electrons.

SAMPLE PROBLEM 6.3

Calculate the energy (in joules) of (a) a photon with a wavelength of 5.00×10^4 nm (infrared region) and (b) a photon with a wavelength of 52 nm (ultraviolet region). (c) Calculate the kinetic energy of an electron ejected by the photon in part (b) from a metal with a binding energy of 3.7 eV.

Strategy In parts (a) and (b), we are given the wavelength of light. Use Equation 6.1 to convert wavelength to frequency; then use Equation 6.2 to determine the energy of the photon for each wavelength. In part (c), we are asked to determine the kinetic energy of an ejected electron. For this we use Equation 6.3. The binding energy, given in electron volts, must be converted to joules for units to cancel.

Setup The wavelengths must be converted from nanometers to meters:

(a) $5.00 \times 10^4 \; \cancel{nm} \times \dfrac{1 \times 10^{-9} \; m}{1 \; \cancel{nm}} = 5.00 \times 10^{-5} \; m$
   (b) $52 \; \cancel{nm} \times \dfrac{1 \times 10^{-9} \; m}{1 \; \cancel{nm}} = 5.2 \times 10^{-8} \; m$

Planck's constant, h, is 6.63×10^{-34} J · s.

(c) $W = 3.7 \; \cancel{eV} \times \dfrac{1.602 \times 10^{-19} \; J}{1 \; \cancel{eV}} = 5.9 \times 10^{-19} \; J$

Solution (a) $\nu = \dfrac{c}{\lambda} = \dfrac{3.00 \times 10^8 \; m/s}{5.00 \times 10^{-5} \; m} = 6.00 \times 10^{12} \; s^{-1}$

and

$E = h\nu = (6.63 \times 10^{-34} \; J \cdot s)(6.00 \times 10^{12} \; s^{-1}) = 3.98 \times 10^{-21} \; J$

This is the energy of a single photon with wavelength 5.00×10^4 nm.

(b) Following the same procedure as in part (a), the energy of a photon of wavelength 52 nm is 3.8×10^{-18} J.

(c) $KE = h\nu - W = 3.8 \times 10^{-18} \; J - 5.9 \times 10^{-19} \; J = 3.2 \times 10^{-18} \; J$

THINK ABOUT IT

Remember that frequency and wavelength are inversely proportional (Equation 6.1). Thus, as wavelength *decreases*, frequency and energy *increase*. Note that the binding energy becomes less significant as the energy of the incident photon increases.

Practice Problem **A**TTEMPT Calculate the energy (in joules) of (a) a photon with wavelength 2.11×10^2 nm, and (b) a photon with frequency 1.78×10^8 s^{-1}. (c) Calculate the kinetic energy of an electron ejected by the photon in part (a) from a metal with a binding energy of 4.66 eV.

Practice Problem **B**UILD (a) Calculate the wavelength (in nm) of light with energy 1.89×10^{-20} J per photon. (b) For light of wavelength 410 nm, calculate the number of photons per joule. (c) Determine the binding energy (in eV) of a metal if the kinetic energy possessed by an ejected electron [using one of the photons in part (b)] is 2.93×10^{-19} J.

Practice Problem **C**ONCEPTUALIZE A blue billiard ball with a mass of 165 g rests in a shallow well on an otherwise flat surface. When a red billiard ball with the same mass moving at any velocity less than 1.20 m/s strikes the blue ball, the blue ball does not move (i). When the red ball strikes the blue ball moving at exactly 1.20 m/s, the blue ball is just barely dislodged from the well (ii). What will be the velocity (iii) of the blue ball when it is struck by the red ball moving at 1.75 m/s?

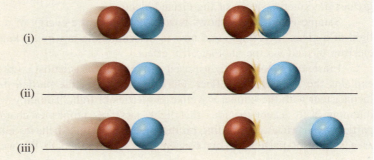

(i)

(ii)

(iii)

CHECKPOINT – SECTION 6.2   Quantum Theory

6.2.1 Calculate the energy per photon of light with wavelength 650 nm.

   a) 1.29×10^{-31} J   d) 1.44×10^{-48} J

   b) 4.31×10^{-40} J   e) 3.06×10^{-19} J

   c) 1.02×10^{-27} J

6.2.2 Calculate the wavelength of light that has energy 1.32×10^{-23} J/photon.

   a) 5.02×10^{-9} cm   d) 1.51 cm

   b) 6.64×10^{3} cm   e) 66.4 cm

   c) 2.92×10^{-63} cm

6.2.3 Which of the following is characteristic of high-energy radiation?

   a) High wavelength   d) High amplitude

   b) High velocity   e) All of the above

   c) High frequency

6.2.4 A clean metal surface is irradiated with light of three different wavelengths: λ_1, λ_2, and λ_3. The kinetic energies of the ejected electrons are as follows: λ_1: 2.9×10^{-20} J; λ_2: approximately zero; λ_3: 4.2×10^{-19} J. Arrange the light in order of increasing wavelength.

   a) $\lambda_1 < \lambda_2 < \lambda_3$ d) $\lambda_3 < \lambda_1 < \lambda_2$

   b) $\lambda_2 < \lambda_1 < \lambda_3$ e) $\lambda_2 < \lambda_3 < \lambda_1$

   c) $\lambda_3 < \lambda_2 < \lambda_1$

6.3   Bohr's Theory of the Hydrogen Atom

In addition to explaining the photoelectric effect, Planck's quantum theory and Einstein's ideas made it possible for scientists to unravel another nineteenth-century mystery in physics: atomic line spectra.

In the seventeenth century, Newton had shown that sunlight is composed of various color components that can be recombined to produce *white light*. Since that time, chemists and physicists have studied the characteristics of such *emission spectra.* The emission spectrum of a substance can be seen by energizing a sample of material with either thermal energy or some other form of energy (such as a high-voltage electrical discharge if the substance is a gas). A "red-hot" or "white-hot" iron bar freshly removed from a fire produces a characteristic glow. The glow is the visible portion of its emission spectrum. The heat given off by the same iron bar is another portion of its emission spectrum—the infrared region. A feature common to the emission spectrum of the sun and that of a heated solid is that both are continuous; that is, all wavelengths of visible light are present in each spectrum (Figure 6.6).

> **Student Note:** If you have ever seen a rainbow, you are familiar with this phenomenon. The rainbow is the visible portion of the sun's emission spectrum.

(a)

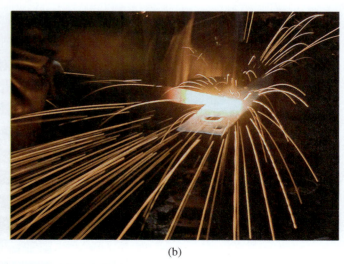

(b)

Figure 6.6 The visible white light emitted by (a) the sun and (b) a white-hot iron bar. In each case, the white light is the combination of all visible wavelengths (see Figure 6.1).

(a): Doug Menuez/Photodisc/Getty Images; (b): Charles D. Winters/McGraw Hill

Atomic Line Spectra

Unlike those of the sun or a white-hot iron bar, the emission spectra of atoms in the gas phase do not show a continuous spread of wavelengths from red to violet; rather, the atoms produce bright lines in distinct parts of the visible spectrum. These *line spectra* are the emission of light only at *specific wavelengths*. Figure 6.7 is a schematic diagram of a discharge tube that is used to study emission spectra.

Every element has a unique emission spectrum, so the characteristic lines in atomic spectra can be used in chemical analysis to identify elements, much as finger-prints are used to identify people. When the lines of the emission spectrum of a known element exactly match the lines of the emission spectrum of an unknown sample, the identity of the element in the sample is established. Although the procedure of iden-tifying elements by their line spectra had been used for many years in chemical analy-sis, the origin of the spectral lines was not understood until early in the twentieth century. Figure 6.8 shows the emission spectra of several elements.

Figure 6.7 (a) Experimental arrangement for studying the emission spectra of atoms and molecules. The gas being studied is in a discharge tube containing two electrodes. As electrons flow from the negative electrode to the positive electrode, they collide with the gas particles. The collisions lead to the emission of light by the atoms (or molecules). The emitted light is separated into its components by a prism. (b) Line emission spectrum of hydrogen.

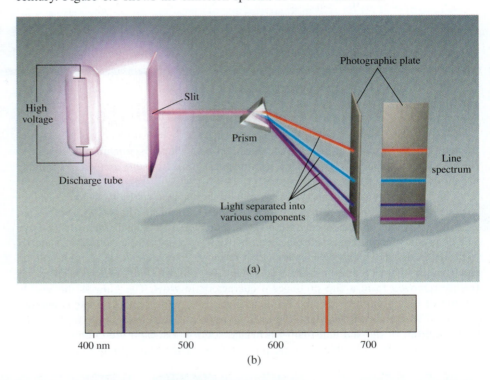

Figure 6.8 Emission spectra of several elements.

McGraw Hill

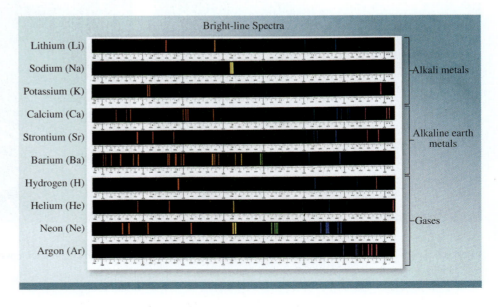

In 1885, Johann Balmer[3] developed a remarkably simple equation that could be used to calculate the wavelengths of the four visible lines in the emission spectrum of hydrogen. Johannes Rydberg[4] developed Balmer's equation further, yielding an equation that could calculate not only the *visible* wavelengths, but those of *all* hydrogen's spectral lines:

Equation 6.4

$$\frac{1}{\lambda} = R_\infty \left(\frac{1}{n_1^2} - \frac{1}{n_2^2} \right)$$

Student Note: The Rydberg equation is a mathematical relationship that was derived from experimental data. Although it predates quantum theory by decades, it agrees remarkably well with it for one-electron systems such as the hydrogen atom.

In Equation 6.4, now known as the *Rydberg equation,* λ is the wavelength of a line in the spectrum; R_∞ is the Rydberg constant (1.09737316×10^7 m^{-1}); and n_1 and n_2 are positive integers, where $n_2 > n_1$.

The Line Spectrum of Hydrogen

In 1913, not long after Planck's and Einstein's discoveries, a theoretical explanation of the emission spectrum of the hydrogen atom was presented by the Danish physicist Niels Bohr.[5] Bohr's treatment is very complex and is no longer considered to be correct in all its details. We concentrate only on his important assumptions and final results, which account for the observed spectral lines and provide an important step toward the understanding of quantum theory.

When Bohr first approached this problem, physicists already knew that the atom contains electrons and protons. They thought of an atom as an entity in which electrons whirled around the nucleus in circular orbits at high velocities. This was an appealing description because it resembled the familiar model of planetary motion around the sun. However, according to the laws of classical physics, an electron moving in an orbit of a hydrogen atom would experience an acceleration toward the nucleus by radiating away energy in the form of electromagnetic waves. Thus, such an electron would quickly spiral into the nucleus and annihilate itself with the proton. To explain why this does not happen, Bohr postulated that the electron is allowed to occupy only certain orbits of specific energies. In other words, the energies of the electron are quantized. An electron in any of the allowed orbits will not radiate energy and therefore will not spiral into the nucleus.

Bohr attributed the emission of radiation by an energized hydrogen atom to the electron dropping from a higher-energy orbit to a lower one and giving up a quantum of energy (a photon) in the form of light (Figure 6.10). Using arguments based on electrostatic interaction and Newton's laws of motion, Bohr showed that the energies that the electron in the hydrogen atom can possess are given by

Animation
Figure 6.10, Emission Spectrum of Hydrogen.

Equation 6.5

$$E_n = -2.18 \times 10^{-18} \text{ J} \left(\frac{1}{n^2} \right)$$

where n is an integer with values $n = 1, 2, 3$, and so on. The negative sign in Equation 6.5 is an arbitrary convention, signifying that the energy of the electron in the atom is *lower* than the energy of a *free electron,* which is an electron that is infinitely far from the nucleus. The energy of a free electron is arbitrarily assigned a value of zero. Mathematically, this corresponds to setting n equal to infinity in Equation 6.5:

$$E_\infty = -2.18 \times 10^{-18} \text{ J} \left(\frac{1}{\infty^2} \right) = 0$$

3. Johann Jakob Balmer (1825–1898). Swiss mathematician. From 1859 until his death in 1898 Balmer taught math at a secondary school for girls in Basel, Switzerland. Although physicists did not understand why his equation worked until long after his death, the visible series of lines in the spectrum of hydrogen is named for him.

4. Johannes Robert Rydberg (1854–1919). Swedish mathematician and physicist. Rydberg analyzed many atomic spectra in an effort to understand the periodic properties of elements. Although he was nominated twice for the Nobel Prize in Physics, he never received it.

5. Niels Henrik David Bohr (1885–1962). Danish physicist. One of the founders of modern physics, he received the Nobel Prize in Physics in 1922 for his theory explaining the line spectrum of hydrogen.

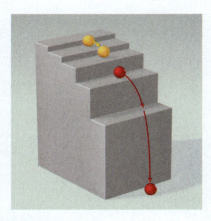

Figure 6.9 Mechanical analogy for the emission processes. The ball can rest on any step but not between steps.

As the electron gets closer to the nucleus (as n decreases), E_n becomes larger in absolute value, but also more negative. The most negative value, then, is reached when $n = 1$, which corresponds to the most stable energy state. We call this the **ground state,** the *lowest* energy state of an atom. The stability of the electron diminishes as n increases. Each energy state in which $n > 1$ is called an **excited state.** Each excited state is higher in energy than the ground state. In the hydrogen atom, an electron for which n is greater than 1 is said to be in an excited state.

The radius of each circular orbit in Bohr's model depends on n^2. Thus, as n increases from 1 to 2 to 3, the orbit radius increases very rapidly. The higher the excited state, the farther away the electron is from the nucleus (and the less tightly held it is by the nucleus).

Bohr's theory enables us to explain the line spectrum of the hydrogen atom. Radiant energy absorbed by the atom causes the electron to move from the ground state ($n = 1$) to an excited state ($n > 1$). Conversely, radiant energy (in the form of a photon) is *emitted* when the electron moves from a higher-energy excited state to a lower-energy excited state or the ground state.

The quantized movement of the electron from one energy state to another is analogous to the movement of a tennis ball either up or down a set of stairs (Figure 6.9). The ball can be on any of several steps but never between steps. The journey from a lower step to a higher one is an energy-requiring process, whereas movement from a higher step to a lower step is an energy-releasing process. The quantity of energy involved in either type of change is determined by the distance between the beginning and ending steps. Similarly, the amount of energy needed to move an electron in the Bohr atom depends on the difference in energy levels between the initial and final states.

To apply Equation 6.5 to the emission process in a hydrogen atom, let us suppose that the electron is initially in an excited state characterized by n_i. During emission, the electron drops to a lower-energy state characterized by n_f (the subscripts i and f denote the *initial* and *final* states, respectively). This lower-energy state may be the ground state, but it can be any state lower than the initial excited state. The difference between the energies of the initial and final states is

$$\Delta E = E_f - E_i$$

From Equation 6.5,

$$E_f = -2.18 \times 10^{-18}\,\text{J}\left(\frac{1}{n_f^2}\right)$$

and

$$E_i = -2.18 \times 10^{-18}\,\text{J}\left(\frac{1}{n_i^2}\right)$$

Therefore,

$$\Delta E = \left(\frac{-2.18 \times 10^{-18}\,\text{J}}{n_f^2}\right) - \left(\frac{-2.18 \times 10^{-18}\,\text{J}}{n_i^2}\right)$$

$$= -2.18 \times 10^{-18}\,\text{J}\left(\frac{1}{n_f^2} - \frac{1}{n_i^2}\right)$$

Student Note: When $n_i > n_f$, ΔE is *negative*, indicating energy is *emitted*. When $n_f > n_i$, ΔE is *positive*, indicating energy is *absorbed*.

Because this transition results in the emission of a photon of frequency ν and energy $h\nu$, we can write

$$\Delta E = h\nu = -2.18 \times 10^{-18}\,\text{J}\left(\frac{1}{n_f^2} - \frac{1}{n_i^2}\right) \qquad \textbf{Equation 6.6}$$

A photon is emitted when $n_i > n_f$. Consequently, the term in parentheses is *positive*, making ΔE *negative* (energy is lost to the surroundings). A photon is absorbed when $n_f > n_i$, making the term in parentheses *negative,* so ΔE is *positive*. Each spectral line in the emission spectrum of hydrogen corresponds to a particular transition in a hydrogen atom. When we study a large number of hydrogen atoms, we observe all possible transitions and hence the corresponding spectral lines. The brightness of a spectral line depends on how many photons of the same wavelength are emitted.

To calculate the wavelength of an emission line, we substitute c/λ for ν and then divide both sides of Equation 6.6 by hc. In addition, because wavelength (and frequency) can only have positive values, we take the absolute value of the right side of the *equation*. (In this case, we do so simply by eliminating the negative sign.)

Student Note: Because 2.18×10^{-18} J/$hc = 1.096 \times 10^7$ m^{-1}, which to three significant figures is equal to R_∞, this equation is essentially the same as the Rydberg equation (Equation 6.4).

$$\frac{1}{\lambda} = \frac{2.18 \times 10^{-18}\,\text{J}}{hc}\left(\frac{1}{n_f^2} - \frac{1}{n_i^2}\right)$$ **Equation 6.7**

The emission spectrum of hydrogen includes a wide range of wavelengths from the infrared to the ultraviolet. Table 6.1 lists the series of transitions in the hydrogen spectrum, each with a different value of n_f. The series are named after their discoverers (Lyman, Balmer, Paschen, and Brackett). The Balmer series was the first to be studied because some of its lines occur in the visible region.

Figure 6.11 shows transitions associated with spectral lines in each of the emission series. Each horizontal line represents one of the allowed energy levels for the electron in a hydrogen atom. The energy levels are labeled with their n values.

Sample Problem 6.4 illustrates the use of Equation 6.7.

TABLE 6.1	**Emission Series in the Hydrogen Spectrum**		
Series	**n_f**	**n_i**	**Spectrum Region**
Lyman	1	2, 3, 4, . . .	Ultraviolet
Balmer	2	3, 4, 5, . . .	Visible and ultraviolet
Paschen	3	4, 5, 6, . . .	Infrared
Brackett	4	5, 6, 7, . . .	Infrared

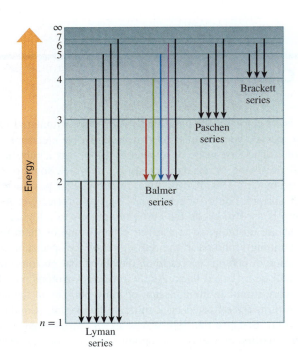

Figure 6.11 Energy levels in the hydrogen atom and the various emission series. Each series terminates at a different value of n.

SAMPLE PROBLEM 6.4

Calculate the wavelength (in nm) of the photon emitted when an electron transitions from the $n = 4$ state to the $n = 2$ state in a hydrogen atom.

Strategy Use Equation 6.7 to calculate λ.

Setup According to the problem, the transition is from $n = 4$ to $n = 2$, so $n_i = 4$ and $n_f = 2$. The required constants are $h = 6.63 \times 10^{-34}$ J · s and $c = 3.00 \times 10^8$ m/s.

Solution

$$\frac{1}{\lambda} = \frac{2.18 \times 10^{-18}\,\mathrm{J}}{hc}\left(\frac{1}{n_f^2} - \frac{1}{n_i^2}\right)$$

$$= \frac{2.18 \times 10^{-18}\,\mathrm{J}}{(6.63 \times 10^{-34}\,\mathrm{J \cdot s})(3.00 \times 10^8\,\mathrm{m/s})}\left(\frac{1}{2^2} - \frac{1}{4^2}\right)$$

$$= 2.055 \times 10^6\,\mathrm{m^{-1}}$$

Remember to keep at least one extra digit in intermediate answers to avoid rounding error in the final result [◄◄ Section 1.5]. Therefore,

$$\lambda = 4.87 \times 10^{-7}\,\mathrm{m} = 487\,\mathrm{nm}$$

THINK ABOUT IT

Look again at the line spectrum of hydrogen in Figure 6.7 and make sure that your result matches one of them. Note that for an emission, n_i is always greater than n_f, and Equation 6.7 gives a positive result.

Practice Problem Ⓐ**TTEMPT** What is the wavelength (in nm) of a photon emitted during a transition from the $n = 3$ state to the $n = 1$ state in the H atom?

Practice Problem Ⓑ**UILD** What is the value of n_i for an electron that emits a photon of wavelength 93.14 nm when it returns to the ground state in the H atom?

Practice Problem Ⓒ**ONCEPTUALIZE** For each pair of transitions, determine which one results in emission of the larger amount of energy.

(a) $n = 6$ to $n = 3$ $n = 3$ to $n = 2$ (b) $n = 3$ to $n = 1$ $n = 10$ to $n = 2$ (c) $n = 2$ to $n = 1$ $n = 99$ to $n = 2$

Bringing Chemistry to Life

Lasers

The word *laser* is an acronym for *l*ight *a*mplification by *s*timulated *e*mission of *r*adiation. It is a special type of emission that may involve electronic transitions in *atoms* or *molecules*. The first laser, developed in 1960, was a ruby laser. Ruby is a deep-red mineral consisting of corundum (Al_2O_3), in which some of the Al^{3+} ions have been replaced by Cr^{3+} ions. A cylindrical ruby crystal is positioned between two perfectly parallel mirrors, one of which is only partially reflective. A light source called a flashlamp is used to excite the chromium atoms to a higher energy level. The excited atoms are unstable, so at a given instant some of them return to the ground state by emitting a photon in the red region of the spectrum ($\lambda = 694.3$ nm). Spontaneous emission of photons occurs in all directions, but photons emitted directly at either mirror will be reflected back. As the reflected photons pass back through the ruby crystal, they stimulate the emission of more photons in the same direction. These photons, in turn, are reflected back through the crystal, stimulating still more emissions, and so on. Because the light waves are *in phase*—that is, their maxima coincide and their minima coincide—the photons enhance one another, increasing their power with each passage between the mirrors. When the light reaches a certain intensity, it emerges from the partially reflective mirror as a laser beam. Laser light is characterized by three properties: It is *intense,* it has a precisely known wavelength and therefore *energy,* and it is *coherent.*

Student Note: *Coherent* means that the light waves are all in phase.

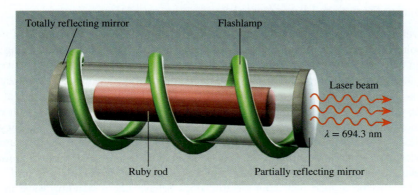

The uses of lasers are quite numerous, including many medical applications. Their high intensity and ease of focus make them suitable for drilling holes in metals, welding, and carrying out nuclear fusion. The fact that they are highly directional and have precisely known wavelengths makes them very useful for telecommunications. Lasers are also used in isotope separation, in chemical analysis, in holography (three-dimensional photography), in compact disc players, and in supermarket scanners.

CHECKPOINT – SECTION 6.3   Bohr's Theory of the Hydrogen Atom

6.3.1 Calculate the energy of an electron in the $n = 3$ state in a hydrogen atom.

a) 2.42×10^{-19} J

b) -2.42×10^{-19} J

c) 7.27×10^{-19} J

d) -7.27×10^{-19} J

e) -6.54×10^{-18} J

6.3.2 Calculate ΔE of an electron that goes from $n = 1$ to $n = 5$.

a) 8.72×10^{-20} J

b) -8.72×10^{-20} J

c) 5.45×10^{-17} J

d) 2.09×10^{-18} J

e) -2.09×10^{-18} J

6.3.3 What is the wavelength of light emitted when an electron in a hydrogen atom goes from $n = 5$ to $n = 3$?

a) 4.87×10^{-7} m

b) 6.84×10^{-7} m

c) 1.28×10^{-6} m

d) 3.65×10^{-7} m

e) 1.02×10^{-7} m

6.3.4 Which wavelength corresponds to the transition of an electron in a hydrogen atom from $n = 2$ to $n = 1$?

a) 182 nm

b) 91.2 nm

c) 724 nm

d) 812 nm

e) 122 nm

6.4   Wave Properties of Matter

Bohr's theory was both fascinating and puzzling. It fit the experimental data for hydrogen, but physicists did not understand the underlying principle. Why, for example, was an electron restricted to orbiting the nucleus at certain fixed distances? For a decade no one, not even Bohr himself, could offer a logical explanation. In 1924, Louis de Broglie[6] provided a solution to this puzzle. De Broglie reasoned that if

6. Louis Victor Pierre Raymond Duc de Broglie (1892–1987). French physicist. A member of an old and noble family in France, he held the title of a prince. In his doctoral dissertation, he proposed that matter and radiation have the properties of both wave and particle. For this work, de Broglie was awarded the Nobel Prize in Physics in 1929.

energy (light) can, under certain circumstances, behave like a stream of particles (photons), then perhaps particles such as electrons can, under certain circumstances, exhibit wavelike properties.

The de Broglie Hypothesis

In developing his revolutionary theory, de Broglie incorporated his observations of macroscopic phenomena that exhibited quantized behavior. For example, a guitar string has certain discrete frequencies of vibration, like those shown in Figure 6.12(a). The waves generated by plucking a guitar string are *standing* or *stationary waves* because they do not travel along the string. Some points on the string, called **nodes,** do not move at all; that is, the amplitude of the wave at these points is *zero*. There is a node at each end, and there may be one or more nodes between the ends. The greater the frequency of vibration, the shorter the wavelength of the standing wave and the greater the number of nodes. According to de Broglie, an electron in an atom behaves like a *standing wave*. However, as Figure 6.12 shows, only certain wavelengths are possible or *allowed*.

De Broglie argued that if an electron does behave like a standing wave in the hydrogen atom, the wavelength must fit the circumference of the orbit exactly; that is, the circumference of the orbit must be an integral multiple of the wavelength, as shown in Figure 6.12(b). Otherwise, the wave would partially cancel itself by destructive interference on each successive orbit, quickly reducing its amplitude to zero.

The relationship between the circumference of an allowed orbit ($2\pi r$) and the wavelength (λ) of the electron is given by

Equation 6.8	$2\pi r = n\lambda$

where r is the radius of the orbit, λ is the wavelength of the electron wave, and n is a positive integer (1, 2, 3, …). Because n is an integer, r can have only certain values (integral multiples of λ) as n increases from 1 to 2 to 3 and so on. And, because the

Figure 6.12 (a) Standing waves of a vibrating guitar string. The length of the string must be equal to a whole number times one-half the wavelength ($\lambda/2$). (b) In a circular orbit, only whole-number multiples of wavelengths are allowed. Any fractional number of wavelengths would result in cancellation of the wave due to destructive interference.

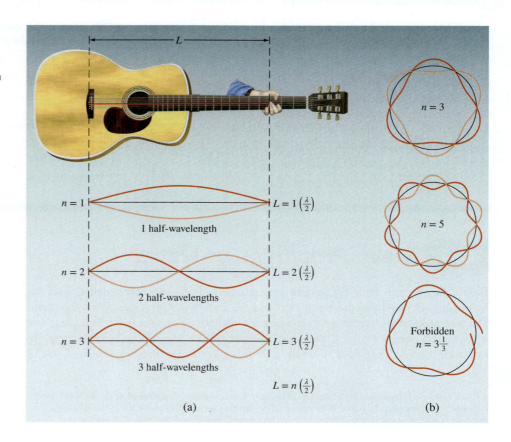

energy of the electron depends on the size of the orbit (or the value of r), the energy can have only certain values, too. Thus, the energy of the electron in a hydrogen atom, if it behaves like a standing wave, must be quantized.

De Broglie's reasoning led to the conclusion that waves can behave like particles and particles can exhibit wavelike properties. De Broglie deduced that the particle and wave properties are related by the following expression:

$$\lambda = \frac{h}{mu}$$

Equation 6.9

where λ, m, and u are the wavelength associated with a moving particle, its *mass*, and its velocity, respectively. Equation 6.9 implies that a particle in motion can be treated as a wave, and a wave can exhibit the properties of a particle. To help you remember this important point, notice that the left side of Equation 6.9 involves the wavelike property of wavelength, whereas the right side involves mass, a property of particles. A wavelength calculated using Equation 6.9 is usually referred to specifically as a *de Broglie wavelength*. Likewise, we refer to a *mass* calculated using Equation 6.9 as a *de Broglie mass*.

Sample Problem 6.5 illustrates how de Broglie's theory and Equation 6.9 can be applied.

Student Note: Mass (m) must be expressed in kilograms for units to cancel properly in Equation 6.9.

SAMPLE PROBLEM 6.5

Calculate the de Broglie wavelength of the "particle" in the following two cases: (a) a 25-g bullet traveling at 612 m/s, and (b) an electron ($m = 9.109 \times 10^{-31}$ kg) moving at 63.0 m/s.

Strategy Use Equation 6.9 to calculate the de Broglie wavelengths. Remember that the mass in Equation 6.9 must be expressed in kilograms for the units to cancel properly.

Setup Planck's constant, h, is 6.63×10^{-34} J · s or, for the purpose of making the unit cancellation obvious, 6.63×10^{-34} kg · m^2/s. Remember that 1 J = 1 kg · m^2/s^2.

Solution

(a) $25 \text{ g} \times \dfrac{1 \text{ kg}}{1000 \text{ g}} = 0.025 \text{ kg}$

$$\lambda = \frac{h}{mu} = \frac{6.63 \times 10^{-34} \text{ kg} \cdot \text{m}^2/\text{s}}{(0.025 \text{ kg})(612 \text{ m/s})} = 4.3 \times 10^{-35} \text{ m}$$

(b) $\lambda = \dfrac{h}{mu} = \dfrac{6.63 \times 10^{-34} \text{ kg} \cdot \text{m}^2/\text{s}}{(9.109 \times 10^{-31} \text{ kg})(63.0 \text{ m/s})} = 1.16 \times 10^{-5} \text{ m}$

THINK ABOUT IT

While you are new at solving these problems, always write out the units of Planck's constant (J · s) as kg · m^2/s. This will enable you to check your unit cancellations and detect common errors such as expressing mass in grams rather than kilograms. Note that the calculated wavelength of a macroscopic object, even one as small as a bullet, is extremely small. An object must be at least as small as a subatomic particle for its wavelength to be large enough for us to observe.

Practice Problem ATTEMPT Calculate the de Broglie wavelength (in nm) of a hydrogen atom ($m = 1.674 \times 10^{-27}$ kg) moving at 1500 cm/s.

Practice Problem BUILD Use Equation 6.9 to calculate the *momentum*, p (defined as mass times velocity, $m \times u$) associated with a photon of radiation of wavelength 810 nm. The velocity of a photon is the speed of light, c.

Student Note: Momentum has units of kg · m/s or N · s, where N is the *newton*, the SI unit of *force*. The newton is a derived SI unit: 1 N = 1 kg · m/s^2.

(Continued on next page)

Practice Problem **C**ONCEPTUALIZE Consider the impact of early electron diffraction experiments on scientists' understanding of the behavior of matter. Which of the following imaginary macroscopic experiments most closely corresponds to the remarkable outcome of electron diffraction?

(a) Combining one marble with another by one method yields two marbles; but combining the two marbles by another method yields four marbles.

(b) Combining one marble with another by one method yields two marbles; but combining the two marbles by another method yields zero marbles.

(c) Combining one marble with another by any method yields two marbles.

Diffraction of Electrons

Shortly after de Broglie introduced his equation and predicted that electrons should exhibit wave properties, successful electron diffraction experiments were carried out by Clinton Davisson[7] and Lester Germer[8] in the United States and G. P. Thomson[9] in England. These experiments demonstrated that electrons do indeed possess wavelike properties. By directing a beam of electrons (which are most definitely particles) through a thin piece of gold foil, Thomson obtained a set of concentric rings on a screen, similar to the diffraction pattern observed when X rays (which are most definitely waves) were used. Figure 6.13 shows X-ray and electron diffraction patterns for aluminum.

Figure 6.13 (a) X-ray diffraction pattern and (b) electron diffraction of titanium compounds. The similarity of these two patterns shows that electrons can behave like X rays and display wave properties.

(a): Omikron/Science Source; (b): Dr David Wexler, colored by Dr Jeremy Burgess/ Science Source

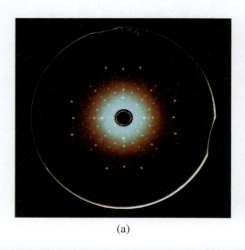
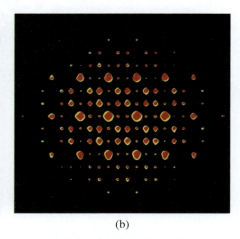

(a)   (b)

CHECKPOINT – SECTION 6.4   Wave Properties of Matter

6.4.1 Calculate the de Broglie wavelength associated with a helium-4 atom (4.00 amu) moving at 3.0×10^6 m/s.

a) 2.0×10^{-20} m

b) 3.3×10^{-11} m

c) 3.3×10^{-14} m

d) 1.8×10^{-19} m

e) 6.6×10^{-27} m

6.4.2 What is the momentum of a photon of light with $\lambda = 122$ nm? (The velocity of a photon is the speed of light, c.)

a) 5.43×10^{-27} kg · m/s

b) 6.63×10^{-34} kg · m/s

c) 6.00×10^{-17} kg · m/s

d) 8.64×10^{-11} kg · m/s

e) 2.71×10^{-27} kg · m/s

7. Clinton Joseph Davisson (1881–1958). American physicist. He and G. P. Thomson shared the Nobel Prize in Physics in 1937 for demonstrating the wave properties of electrons.

8. Lester Halbert Germer (1896–1972). American physicist. Discoverer (with Davisson) of the wave properties of electrons.

9. George Paget Thomson (1892–1975). English physicist. Son of J. J. Thomson, he received the Nobel Prize in Physics in 1937, along with Clinton Davisson, for demonstrating the wave properties of electrons.

6.5 Quantum Mechanics

The discovery that waves could have matterlike properties and that matter could have wavelike properties was revolutionary. Although scientists had long believed that energy and matter were distinct entities, the distinction between them, at least at the atomic level, was no longer clear. Bohr's theory was tremendously successful in explaining the line spectrum of hydrogen, but it failed to explain the spectra of atoms with more than one electron. The electron appeared to behave as a particle in some circumstances and as a wave in others. Neither description could completely explain the behavior of electrons in atoms. This left scientists frustrated in their quest to understand exactly where the electrons in an atom are.

The Uncertainty Principle

To describe the problem of trying to locate a subatomic particle that behaves like a wave, Werner Heisenberg[10] formulated what is now known as the **Heisenberg uncertainty principle:** It is impossible to know simultaneously both the *momentum p* and the *position x* of a particle with certainty. Stated mathematically,

$$\Delta x \cdot \Delta p \geq \frac{h}{4\pi}$$ **Equation 6.10**

For a particle of mass *m*,

$$\Delta x \cdot m\Delta u \geq \frac{h}{4\pi}$$ **Equation 6.11**

> **Student Note:** Like the de Broglie wavelength equation, Equation 6.11 requires that mass be expressed in kilograms. Unit cancellation will be more obvious if you express Planck's constant in $kg \cdot m^2/s$ rather than $J \cdot s$.

where Δx and Δu are the uncertainties in measuring the position and velocity of the particle, respectively. The $\geq$ signs have the following meaning. If the measured uncertainties of position and velocity are large (say, in a crude experiment), their product can be substantially greater than $h/4\pi$ (hence the > sign). The significance of Equation 6.11 is that even in the most favorable conditions for measuring position and velocity, the product of the uncertainties can never be less than $h/4\pi$ (hence the = sign). Thus, making measurement of the velocity of a particle *more* precise (i.e., making Δu a *small* quantity) means that the position must become correspondingly *less* precise (i.e., Δx will become *larger*). Similarly, if the position of the particle is known more precisely, its velocity measurement must become less precise.

If the Heisenberg uncertainty principle is applied to the hydrogen atom, we find that the electron cannot orbit the nucleus in a well-defined path, as Bohr thought. If it did, we could determine precisely both the position of the electron (from the radius of the orbit) and its speed (from its kinetic energy) at the same time. This would violate the uncertainty principle.

Sample Problem 6.6 shows how the Heisenberg uncertainty principle can be applied.

SAMPLE PROBLEM 6.6

An electron in a hydrogen atom is known to have a velocity of 5×10^6 m/s ± 1 percent. Using the uncertainty principle, calculate the minimum uncertainty in the position of the electron and, given that the diameter of the hydrogen atom is less than 1 angstrom (Å), comment on the magnitude of this uncertainty compared to the size of the atom.

(Continued on next page)

---

10. Werner Karl Heisenberg (1901–1976). German physicist. One of the founders of modern quantum theory. Heisenberg received the Nobel Prize in Physics in 1932.

Strategy The uncertainty in the velocity, 1 percent of 5×10^6 m/s, is Δu. Using Equation 6.11, calculate Δx and compare it with the diameter of the hydrogen atom. Recall that 1 Å is equal to 1×10^{-10} m [◀◀ Section 2.2].

Setup The mass of an electron (from Table 2.1, rounded to three significant figures and converted to kilograms) is 9.11×10^{-31} kg. Planck's constant, h, is 6.63×10^{-34} kg · m²/s.

Solution

$$\Delta u = 0.01 \times 5 \times 10^6 \text{ m/s} = 5 \times 10^4 \text{ m/s}$$

$$\Delta x = \frac{h}{4\pi \cdot m\Delta u}$$

Therefore,

$$\Delta x = \frac{6.63 \times 10^{-34} \text{ kg} \cdot \text{m}^2/\text{s}}{4\pi(9.11 \times 10^{-31} \text{ kg})(5 \times 10^4 \text{ m/s})} \geq 1 \times 10^{-9} \text{ m}$$

The *minimum* uncertainty in the position x is 1×10^{-9} m $= 10$ Å. The uncertainty in the electron's position is 10 times larger than the atom!

THINK ABOUT IT

A common error is expressing the mass of the particle in grams instead of kilograms, but you should discover this inconsistency if you check your unit cancellation carefully. Remember that if one uncertainty is small, the other must be large. The uncertainty principle applies in a practical way only to submicroscopic particles. In the case of a macroscopic object, where the mass is much larger than that of an electron, small uncertainties, relative to the size of the object, are possible for both position and velocity.

Practice Problem Ⓐ**TTEMPT** Calculate the minimum uncertainty in the position of the 25-g bullet from Sample Problem 6.5 if the uncertainty in its velocity is (a) ±1 percent and (b) ±0.01 percent.

Practice Problem Ⓑ**UILD** (a) Calculate the minimum uncertainty in the momentum of an object for which the uncertainty in position is 3 Å. To what minimum uncertainty in velocity does this correspond if the particle is (b) a neutron (mass = 1.0087 amu) and (c) an electron (mass = 5.486×10^{-4} amu)?

Practice Problem Ⓒ**ONCEPTUALIZE** Using Equation 6.11, we can calculate the minimum uncertainty in the position or the velocity of any moving particle, including a macroscopic object such as a marble. Calculate the uncertainty in position of a 10-g marble moving at 2.5 m/s (±5 percent) and comment on the significance of your result.

The Schrödinger Equation

Bohr made a significant contribution to our understanding of atoms, and his suggestion that the energy of an electron in an atom is quantized remains unchallenged, but his theory did not provide a complete description of the behavior of electrons in atoms. In 1926, the Austrian physicist Erwin Schrödinger,[11] using a complex mathematical technique, formulated an equation that describes the behavior and energies of submicroscopic particles in general, an equation analogous to Newton's laws of motion for macroscopic objects. The *Schrödinger equation* requires advanced calculus to solve, and we do not discuss it here. The equation, however, incorporates both particle behavior, in terms of mass m, and wave behavior, in terms of a *wave function* ψ (psi), which depends on the location in space of the system (such as an electron in an atom).

The wave function itself has no direct physical meaning. However, the probability of finding the electron in a certain region in space is proportional to the square of the wave function, ψ^2. The idea of relating ψ^2 to probability stemmed from a wave theory analogy. According to wave theory, the intensity of light is proportional to the

---

11. Erwin Schrödinger (1887–1961). Austrian physicist. Schrödinger formulated wave mechanics, which laid the foundation for modern quantum theory. He received the Nobel Prize in Physics in 1933.

square of the amplitude of the wave, or ψ^2. The most likely place to find a photon is where the intensity is greatest—that is, where the value of ψ^2 is greatest. A similar argument associates ψ^2 with the likelihood of finding an electron in regions surrounding the nucleus.

Schrödinger's equation launched an entirely new field, called *quantum mechanics* (or *wave mechanics*), and began a new era in physics and chemistry. We now refer to the developments in quantum theory from 1913—when Bohr presented his model of the hydrogen atom—to 1926 as "old quantum theory."

The Quantum Mechanical Description of the Hydrogen Atom

The Schrödinger equation specifies the possible energy states the electron can occupy in a hydrogen atom and identifies the corresponding wave functions (ψ). These energy states and wave functions are characterized by a set of *quantum numbers* (to be discussed shortly), with which we can construct a comprehensive model of the hydrogen atom.

Although quantum mechanics does not allow us to specify the exact location of an electron in an atom, it does define the region where the electron is most likely to be at a given time. The concept of **electron density** gives the probability that an electron will be found in a particular region of an atom. The square of the wave function, ψ^2, defines the distribution of electron density in three-dimensional space around the nucleus. Regions of high electron density represent a high probability of locating the electron (Figure 6.14).

To distinguish the quantum mechanical description of an atom from Bohr's model, we speak of an atomic *orbital,* rather than an orbit. An **atomic orbital** can be thought of as the wave function of an electron in an atom. When we say that an electron is in a certain orbital, we mean that the distribution of the electron density or the probability of locating the electron in space is described by the square of the wave function associated with that orbital. An atomic orbital, therefore, has a characteristic energy, as well as a characteristic distribution of electron density.

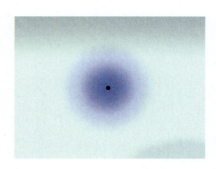

Figure 6.14 Representation of the electron density distribution surrounding the nucleus in the hydrogen atom. It shows a higher probability of finding the electron closer to the nucleus.

CHECKPOINT – SECTION 6.5   Quantum Mechanics

6.5.1 What is the minimum uncertainty in the position of an electron moving at a speed of 4×10^6 m/s ± 1 percent? (The mass of an electron is 9.11×10^{-31} kg.)

a) 2×10^{-8} m

b) 1×10^{-9} m

c) 6×10^{-9} m

d) 7×10^{-8} m

e) 1×10^{-12} m

6.5.2 What is the minimum uncertainty in the position of a proton moving at a speed of 4×10^6 m/s ± 1 percent? (The mass of a proton is 1.67×10^{-27} kg.)

a) 1×10^{-13} m

b) 8×10^{-10} m

c) 4×10^{-11} m

d) 3×10^{-12} m

e) 8×10^{-13} m

6.6   Quantum Numbers

In Bohr's model of the hydrogen atom, only one number, n, was necessary to describe the location of the electron. In quantum mechanics, three **quantum numbers** are required to describe the *distribution of electron density* in an atom. These numbers are derived from the mathematical solution of Schrödinger's equation for the hydrogen atom. They are called the *principal* quantum number, the *angular momentum* quantum number, and the *magnetic* quantum number. Each atomic orbital in an atom is characterized by a unique set of these three quantum numbers.

Student Note: The three quantum numbers n, ℓ, and m_ℓ specify the *size*, *shape*, and *orientation* of an orbital, respectively.

Principal Quantum Number (n)

The *principal quantum number (n)* designates the *size* of the orbital. The larger n is, the greater the average distance of an electron in the orbital from the nucleus and therefore the larger the orbital. The principal quantum number can have integral values of 1, 2, 3, and so forth, and it corresponds to the quantum number in Bohr's model of the hydrogen atom. Recall from Equation 6.5 that in a hydrogen atom, the value of n alone determines the energy of an orbital. (As we explain shortly, this is *not* the case for an atom that contains more than one electron.)

Angular Momentum Quantum Number (ℓ)

The *angular momentum quantum number (ℓ)* describes the *shape* of the atomic orbital (see Section 6.7). The values of ℓ are integers that depend on the value of the principal quantum number, n. For a given value of n, the possible values of ℓ range from 0 to $n - 1$. If $n = 1$, there is only one possible value of ℓ; that is, 0 ($n - 1$ where $n = 1$). If $n = 2$, there are two values of ℓ: 0 and 1. If $n = 3$, there are three values of ℓ: 0, 1, and 2. The value of ℓ is designated by the letters s, p, d, and f as follows:[12]

ℓ	0	1	2	3
Orbital designation	s	p	d	f

Thus, if $\ell = 0$, we have an s orbital; if $\ell = 1$, we have a p orbital; and so on.

A collection of orbitals with the same value of n is frequently called a *shell*. One or more orbitals with the same n and ℓ values are referred to as a *subshell*. For example, the shell designated by $n = 2$ is composed of two subshells: $\ell = 0$ and $\ell = 1$ (the allowed values of ℓ for $n = 2$). These subshells are called the 2s and 2p subshells where 2 denotes the value of n, and s and p denote the values of ℓ.

Magnetic Quantum Number (m_ℓ)

The *magnetic quantum number (m_ℓ)* describes the *orientation* of the orbital in space (see Section 6.7). Within a subshell, the value of m_ℓ depends on the value of ℓ. For a certain value of ℓ, there are ($2\ell + 1$) integral values of m_ℓ as follows:

$$-\ell, \ldots 0, \ldots, +\ell$$

If $\ell = 0$, there is only one possible value of m_ℓ: 0. If $\ell = 1$, then there are *three* values of m_ℓ: -1, 0, and $+1$. If $\ell = 2$, there are *five* values of m_ℓ, namely, -2, -1, 0, $+1$, and $+2$, and so on. The number of m_ℓ values indicates the number of *orbitals* in a subshell with a particular ℓ value; that is, each m_ℓ value refers to a different orbital.

Table 6.2 summarizes the allowed values of the three quantum numbers, n, ℓ, and m_ℓ, and Figure 6.15 illustrates schematically how the allowed values of quantum numbers give rise to the number of subshells and orbitals in each shell of an atom.

Sample Problem 6.7 gives you some practice with the allowed values of quantum numbers.

Student Note: The number of subshells in a shell is equal to n. The number of orbitals in a shell is equal to n^2. The number of orbitals in a subshell is equal to $2\ell + 1$.

---

12. The unusual sequence of letters (s, p, d, and f) has an historical origin. Physicists who studied atomic emission spectra tried to correlate their observations of spectral lines with the energy states involved in the transitions. They described the emission lines as *s*harp, *p*rincipal, *d*iffuse, and *f*undamental.

TABLE 6.2	Allowed Values of the Quantum Numbers n, ℓ, and m_ℓ		
When n is	ℓ can be	When ℓ is	m_ℓ can be
1	only 0	0	only 0
2	0 or 1	0	only 0
		1	-1, 0, or $+1$
3	0, 1, or 2	0	only 0
		1	-1, 0, or $+1$
		2	-2, -1, 0, $+1$, or $+2$
4	0, 1, 2, or 3	0	only 0
		1	-1, 0, or $+1$
		2	-2, -1, 0, $+1$, or $+2$
		3	-3, -2, -1, 0, $+1$, $+2$, or $+3$
.	.	.	.
.	.	.	.
.	.	.	.

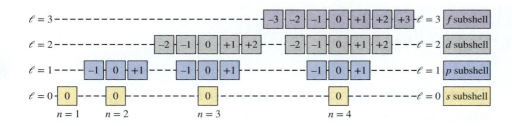

Figure 6.15 Illustration of how quantum numbers designate shells, subshells, and orbitals.

SAMPLE PROBLEM 6.7

What are the possible values for the magnetic quantum number (m_ℓ) when the principal quantum number (n) is 3 and the angular momentum quantum number (ℓ) is 1?

Strategy Use the rules governing the allowed values of m_ℓ. Recall that the possible values of m_ℓ depend on the value of ℓ, not on the value of n.

Setup The possible values of m_ℓ are $-\ell, \ldots, 0, \ldots, +\ell$.

Solution The possible values of m_ℓ are -1, 0, and $+1$.

THINK ABOUT IT

Consult Table 6.2 to make sure your answer is correct. Table 6.2 confirms that it is the value of ℓ, not the value of n, that determines the possible values of m_ℓ.

Practice Problem **A**TTEMPT (a) What are the possible values for m_ℓ when the principal quantum number (n) is 2 and the angular momentum quantum number (ℓ) is 0? (b) What are the possible values for m_ℓ when the principal quantum number (n) is 3 and the angular momentum quantum number (ℓ) is 2?

(Continued on next page)

Practice Problem **B**UILD (a) What is the lowest possible value of the principal quantum number (n) when the angular momentum quantum number (ℓ) is 1? (b) What are the possible values of the angular momentum quantum number (ℓ) when the principal quantum number (n) is 4 and the magnetic quantum number (m_ℓ) is 0?

Practice Problem **C**ONCEPTUALIZE Imagine a cobbler's business (where shoes are repaired) with trendy, V-shaped cabinets, four of which are shown here. When a pair of shoes is brought in for repair, it is kept in a shoebox in one of these cabinets. The location of each pair of shoes is recorded using a set of numbers that designate the cabinet (C), the shelf (S), and the specific box (B).

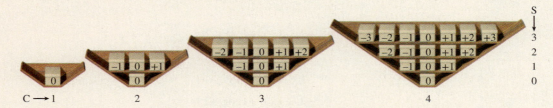

Each cabinet has a number, corresponding to the number of shelves it has. Thus, for the cabinets shown here, the value of C can be 1, 2, 3, or 4. Shelves within each cabinet are numbered sequentially from the bottom, starting with zero. For the smallest cabinet, with just one shelf, 0 is the only shelf designation. For the other cabinets, shelf designations can have integer values of 0 through C − 1. In addition, each individual box has a number on it. Boxes in the bottom row (row 0) all have the number 0 on them. Any box that resides directly above a box labeled 0, is also labeled 0. Boxes to the right or the left of the zero box on each shelf are numbered sequentially, starting with +1 (for boxes on the right), and starting with −1 (for boxes on the left). Using this numbering system, the cobbler can specify the location of a pair of shoes by designating three numbers: C, S, and B. For each of the following sets of numbers (C, S, B) determine whether or not they designate a box in one of the cabinets. For a set of numbers that does *not* designate a box in one of the cabinets, explain why.

(a) (1, 0, 0); (b) (0, 0, 0); (c) (3, 2, −2); (d) (2, 0, 0); (e) (4, 3, +1); (f) (2, 2, +2).

Electron Spin Quantum Number (m_s)

Whereas three quantum numbers are sufficient to describe an atomic orbital, an additional quantum number becomes necessary to describe an electron that occupies the orbital.

Experiments on the emission spectra of hydrogen and sodium atoms indicated that each line in the emission spectra could be split into two lines by the application of an external magnetic field. The only way physicists could explain these results was to assume that electrons act like tiny magnets. If electrons are thought of as spinning on their own axes, as Earth does, their magnetic properties can be accounted for. According to electromagnetic theory, a spinning charge generates a magnetic field, and it is this motion that causes an electron to behave like a magnet. Figure 6.16 shows the two possible spinning motions of an electron, one clockwise and the other counterclockwise. To specify the electron's spin, we use the ***electron spin quantum number (m_s)***. Because there are two possible directions of spin, opposite each other, m_s has *two* possible values: $+\frac{1}{2}$ and $-\frac{1}{2}$. Two electrons in the same orbital with opposite spins are referred to as "*paired*."

Conclusive proof of electron spin was established by Otto Stern[13] and Walther Gerlach[14] in 1924. Figure 6.17 shows the basic experimental arrangement. A beam of gaseous atoms generated in a hot furnace passes through a nonuniform magnetic field. The interaction between an electron and the magnetic field causes the atom to be deflected from its straight-line path. Because the direction of spin is random, the electrons in *half* of the atoms will be spinning in one direction. Those atoms will be deflected in one way. The electrons in the other half of the atoms will be spinning in

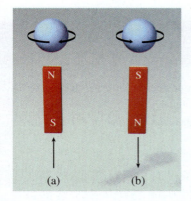

Figure 6.16 (a) Clockwise and (b) counterclockwise spins of an electron. The magnetic fields generated by these two spinning motions are analogous to those from the two magnets. The upward and downward arrows are used to denote the direction of spin.

13. Otto Stern (1888–1969). German physicist. He made important contributions to the study of the magnetic properties of atoms and the kinetic theory of gases. Stern was awarded the Nobel Prize in Physics in 1943.

14. Walther Gerlach (1889–1979). German physicist. Gerlach's main area of research was in quantum theory.

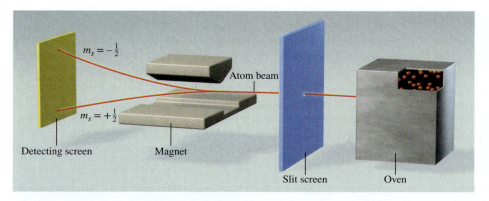

Figure 6.17 Experimental arrangement for demonstrating electron spin. A beam of atoms is directed through a magnetic field. When a hydrogen atom, with a single electron, passes through the field, it is deflected in one direction or the other, depending on the direction of the electron's spin. In a stream consisting of many atoms, there will be equal distributions of the two kinds of spins, so two spots of equal intensity are detected on the screen.

the *opposite* direction. Those atoms will be deflected in the other direction. Thus, two spots of equal intensity are observed on the detecting screen.

To summarize, we can designate an orbital in an atom with a set of *three* quantum numbers. These three quantum numbers indicate the size (n), shape (ℓ), and orientation (m_ℓ) of the orbital. A fourth quantum number (m_s) is necessary to designate the spin of an electron in the orbital.

CHECKPOINT – SECTION 6.6   Quantum Numbers

6.6.1 Which of the following is a legitimate set of three quantum numbers: n, ℓ, and m_ℓ? (Select all that apply.)

a) 1, 0, 0

b) 2, 0, 0

c) 1, 0, +1

d) 2, 1, +1

e) 2, 2, −1

6.6.2 How many orbitals are there in a subshell designated by the quantum numbers $n = 3$, $\ell = 2$?

a) 2

b) 3

c) 5

d) 7

e) 10

6.6.3 How many subshells are there in the shell designated by $n = 3$?

a) 1

b) 2

c) 3

d) 6

e) 9

6.6.4 What is the total number of orbitals in the shell designated by $n = 3$?

a) 1

b) 2

c) 3

d) 6

e) 9

6.7   Atomic Orbitals

Strictly speaking, an atomic orbital does not have a well-defined shape because the wave function characterizing the orbital extends from the nucleus to infinity. In that sense, it is difficult to say what an orbital looks like. On the other hand, it is certainly useful to think of orbitals as having specific shapes. Being able to visualize atomic orbitals is essential to understanding the formation of chemical bonds and molecular geometry, which are discussed in Chapters 8 and 9. In this section, we look at each type of orbital separately.

s Orbitals

For any value of the principal quantum number (n), the value 0 is possible for the angular momentum quantum number (ℓ), corresponding to an s subshell. Furthermore, when $\ell = 0$, the magnetic quantum number (m_ℓ) has only one possible value, 0, corresponding to an s orbital. Therefore, there is an s subshell in every shell, and each s subshell contains just one orbital, an *s orbital.*

Figure 6.18 illustrates three ways to represent the distribution of electrons: the probability density, the spherical distribution of electron density, and the *radial probability distribution* (the probability of finding the electron as a function

Student Note: The radial probability distribution can be thought of as a map of "where an electron spends most of its time."

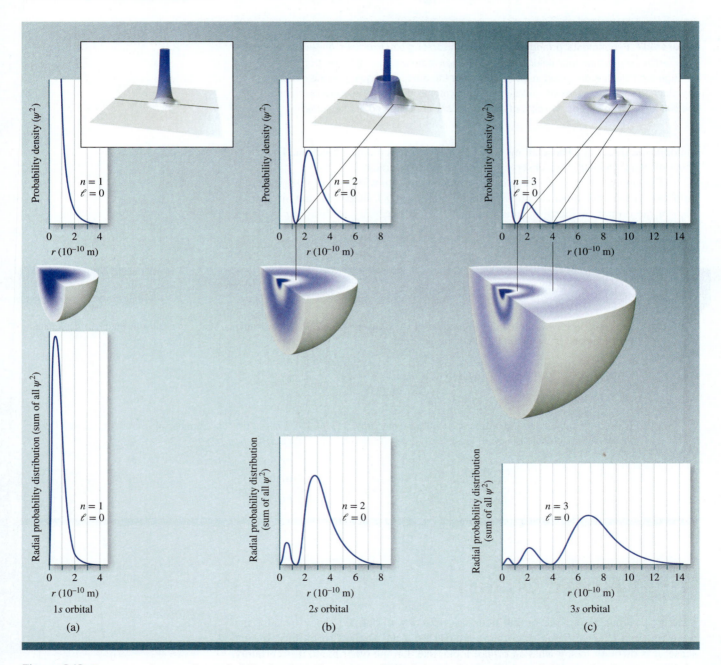

Figure 6.18 From top to bottom, the *probability density* and corresponding relief map, the distribution of electron density represented spherically with shading corresponding to the relief map, and the *radial probability distribution* for (a) the 1s, (b) the 2s, and (c) the 3s orbitals of hydrogen.

of distance from the nucleus) for the 1s, 2s, and 3s orbitals of hydrogen. The boundary surface (the outermost surface of the spherical representation) is a common way to represent atomic orbitals, incorporating the volume in which there is about a 90 percent probability of finding the electron at any given time.

All s orbitals are spherical in shape but differ in size, which increases as the principal quantum number increases. The radial probability distribution for the 1s orbital exhibits a maximum at 52.9 pm (0.529 Å) from the nucleus. The radial probability distribution plots for the 2s and 3s orbitals exhibit two and three maxima, respectively, with the greatest probability occurring at a greater distance from the nucleus as n increases. Between the two maxima for the 2s orbital there is a point on the plot where the probability drops to zero. This corresponds to a *node* in the electron density, where the standing wave has zero amplitude. There are two such nodes in the radial probability distribution plot of the 3s orbital.

Although the boundary surface diagram of an s orbital does not show the number of nodes, the most important features of atomic orbitals, for our purposes, are their overall shapes and *relative* sizes. These features are adequately represented by boundary surface diagrams.

> **Student Note:** Interestingly, this distance is equal to the radius of the $n = 1$ orbit in the Bohr model of the hydrogen atom and is defined as the *Bohr radius.*

p Orbitals

When the principal quantum number (n) is 2 or greater, the value 1 is possible for the angular momentum quantum number (ℓ), corresponding to a p subshell. And, when $\ell = 1$, the magnetic quantum number (m_ℓ) has three possible values: -1, 0, and $+1$, each corresponding to a different **p orbital.** Therefore, there is a p subshell in every shell for which $n \geq 2$, and each p subshell contains three p orbitals. These three p orbitals are labeled p_x, p_y, and p_z (Figure 6.19), with the subscripted letters indicating the axis along which each orbital is oriented. These three p orbitals are identical in size, shape, and energy; they differ from one another only in orientation. Note, however, that there is no simple relation between the values of m_ℓ and the x, y, and z directions. For our purpose, you need only remember that because there are three possible values of m_ℓ, there are three p orbitals with different orientations.

The boundary surface diagrams of p orbitals in Figure 6.19 show that each p orbital can be thought of as two lobes on opposite sides of the nucleus. Like s orbitals, p orbitals increase in size from 2p to 3p to 4p orbital and so on.

d Orbitals and Other Higher-Energy Orbitals

When the principal quantum number (n) is 3 or greater, the value 2 is possible for the angular momentum quantum number (ℓ), corresponding to a d subshell. When $\ell = 2$, the magnetic quantum number (m_ℓ) has *five* possible values, -2, -1, 0, $+1$, and $+2$, each corresponding to a different **d orbital.** Again there is no direct correspondence between a given orientation and a particular m_ℓ value. All the 3d orbitals in an atom are identical in energy and are labeled with subscripts denoting their orientation with respect to the x, y, and z axes and to the planes defined by them. The

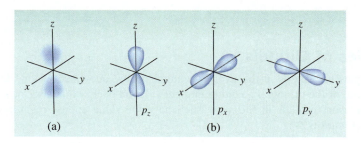

(a)   (b)

Figure 6.19 (a) Electron distribution in a p orbital. (b) Boundary surfaces for the p_x, p_y, and p_z orbitals.

Figure 6.20 Boundary surfaces for the *d* orbitals.

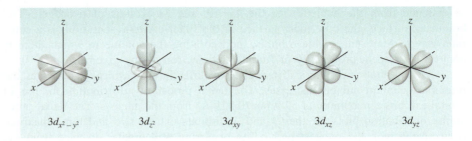

d orbitals that have higher principal quantum numbers (4*d*, 5*d*, etc.) have shapes similar to those shown for the 3*d* orbitals in Figure 6.20.

The *f orbitals* are important when accounting for the behavior of elements with atomic numbers greater than 57, but their shapes are difficult to represent. In this text, we do not concern ourselves with the shapes of orbitals having ℓ values greater than 2. Sample Problem 6.8 shows how to label orbitals with quantum numbers.

SAMPLE PROBLEM   6.8

List the values of *n*, ℓ, and m_ℓ for each of the orbitals in a 4*d* subshell.

Strategy Consider the significance of the number and the letter in the 4*d* designation and determine the values of *n* and ℓ. There are multiple possible values for m_ℓ which will have to be deduced from the value of ℓ.

Setup The integer at the beginning of an orbital designation is the principal quantum number (*n*). The letter in an orbital designation gives the value of the angular momentum quantum number (ℓ). The magnetic quantum number (m_ℓ) can have integral values of $-\ell, \ldots, 0, \ldots, +\ell$.

Solution The values of *n* and ℓ are 4 and 2, respectively, so the possible values of m_ℓ are $-2, -1, 0, +1,$ and $+2$.

THINK ABOUT IT

Consult Figure 6.15 to verify your answers.

Practice Problem (A)**TTEMPT** Give the values of *n*, ℓ, and m_ℓ for the orbitals in a 3*d* subshell.

Practice Problem (B)**UILD** Using quantum numbers, explain why there is no 2*d* subshell.

Practice Problem (C)**ONCEPTUALIZE** Recall the cabinets, shelves, and shoeboxes in Practice Problem 6.7C, which are reproduced here. Write the set of three numbers that specifies each of the highlighted boxes.

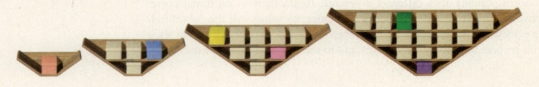

Energies of Orbitals

The energies of orbitals in the hydrogen atom, or any one-electron ion, depend only on the value of the principal quantum number (*n*), and energy increases as *n* increases. For this reason, orbitals in the same shell have the same energy regardless of their subshell (Figure 6.21).

$$1s < 2s = 2p < 3s = 3p = 3d < 4s = 4p = 4d = 4f$$

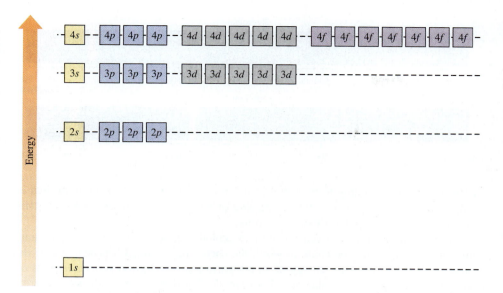

Figure 6.21 Orbital energy levels in the hydrogen atom. Each box represents one orbital. Orbitals with the same principal quantum number (*n*) all have the same energy.

Thus, all four orbitals (one 2*s* and three 2*p*) in the second shell have the *same energy;* all nine orbitals (one 3*s*, three 3*p*, and five 3*d*) in the third shell have the same energy; and all sixteen orbitals (one 4*s*, three 4*p*, five 4*d*, and seven 4*f*) in the fourth shell have the same energy. The energy picture is more complex for many-electron atoms than it is for hydrogen, as is discussed in Section 6.8.

Student Note: Recall that Bohr, whose model only *had* one quantum number, was able to calculate the energies of electrons in the hydrogen atom accurately.

CHECKPOINT – SECTION 6.7   Atomic Orbitals

6.7.1 How many orbitals are there in the 5*f* subshell?

a) 5   b) 7   c) 14   d) 16   e) 28

6.7.2 The energy of an orbital in the hydrogen atom depends on _____.

a) *n*, *ℓ*, and *m*$_ℓ$ c) *n* only   e) *m*$_ℓ$ only

b) *n* and *ℓ*   d) *ℓ* only

6.7.3 In a hydrogen atom, which orbitals are higher in energy than a 3*s* orbital? (Select all that apply.)

a) 3*p*   b) 4*s*   c) 2*p*   d) 3*d*   e) 4*p*

6.7.4 Which of the following sets of quantum numbers, *n*, *ℓ*, and *m*$_ℓ$, corresponds to a 3*p* orbital?

a) 3, 0, 0   c) 3, 2, −1   e) 1, 3, 1

b) 3, 1, 0   d) 1, 1, −2

6.8 Electron Configuration

The hydrogen atom is a particularly simple system because it contains only one electron. The electron may reside in the 1*s* orbital (the *ground state*), or it may be found in some higher-energy orbital (an *excited state*). With many-electron systems, we need to know the ground-state **electron configuration**—that is, how the electrons are distributed in the various atomic orbitals. To do this, we need to know the relative energies of atomic orbitals in a many-electron system, which differ from those in a one-electron system such as hydrogen.

Energies of Atomic Orbitals in Many-Electron Systems

Consider the two emission spectra shown in Figure 6.22. The spectrum of helium contains more lines than that of hydrogen. This indicates that there are more possible

Figure 6.22 Comparison of the emission spectra of H and He.

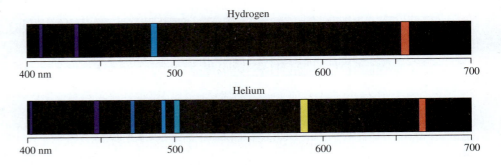

transitions, corresponding to emission in the visible range, in a helium atom than in a hydrogen atom. This is due to the *splitting* of energy levels caused by electrostatic interactions between helium's two electrons.

Figure 6.23 shows the general order of orbital energies in a many-electron atom. In contrast to the hydrogen atom, in which the energy of an orbital depends only on the value of n (Figure 6.21), the energy of an orbital in a many-electron system depends on both the value of n and the value of ℓ. For example, $3p$ orbitals all have the same energy, but they are higher in energy than the $3s$ orbital and lower in energy than the $3d$ orbitals. In a many-electron atom, for a given value of n, the energy of an orbital increases with increasing value of ℓ. One important consequence of the splitting of energy levels is the relative energies of d orbitals in one shell and the s orbital in the next higher shell. As Figure 6.23 shows, the $4s$ orbital is lower in energy than the $3d$ orbitals. Likewise, the $5s$ orbital is lower in energy than the $4d$ orbital, and so on. This fact becomes important when we determine how the electrons in an atom populate the atomic orbitals.

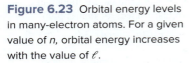

Student Note: "Splitting" of energy levels refers to the splitting of a shell into subshells of different energies, as shown in Figure 6.23.

The Pauli Exclusion Principle

According to the ***Pauli***[15] ***exclusion principle,*** no two electrons in the same atom can have the same four quantum numbers. If two electrons in an atom have the

Figure 6.23 Orbital energy levels in many-electron atoms. For a given value of n, orbital energy increases with the value of ℓ.

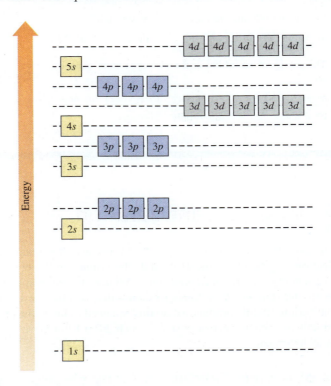

15. Wolfgang Pauli (1900–1958). Austrian physicist. One of the founders of quantum mechanics, Pauli was awarded the Nobel Prize in Physics in 1945.

same n, ℓ, and m_ℓ values (meaning that they occupy the same *orbital*), then they must have different values of m_s; that is, one must have $m_s = +\frac{1}{2}$ and the other must have $m_s = -\frac{1}{2}$. Because there are only two possible values for m_s, and no two electrons in the same orbital may have the same value for m_s, a maximum of *two* electrons may occupy an atomic orbital, and these two electrons must have opposite spins. Two electrons in the same orbital with opposite spins are said to have *paired spins*.

We can indicate the arrangement of electrons in atomic orbitals with labels that identify each orbital (or subshell) and the number of electrons in it. Thus, we could describe a hydrogen atom in the ground state using $1s^1$.

Student Note: $1s^1$ is read as "one s one."

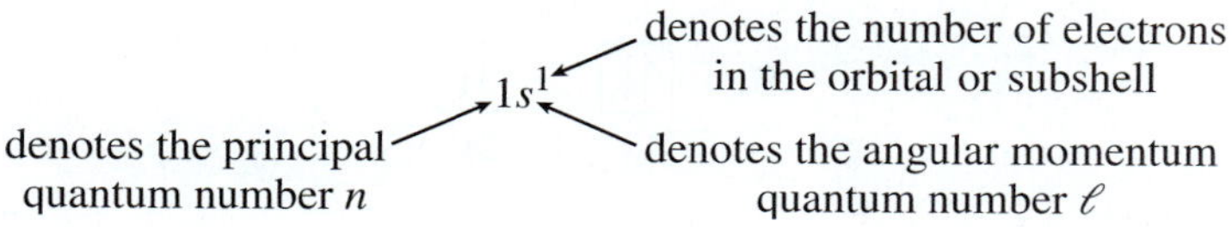

denotes the principal quantum number n ⟶ $1s^1$ ⟵ denotes the number of electrons in the orbital or subshell
⟵ denotes the angular momentum quantum number ℓ

We can also represent the arrangement of electrons in an atom using *orbital diagrams,* in which each orbital is represented by a labeled box. The orbital diagram for a hydrogen atom in the ground state is

H [↑]
$1s^1$

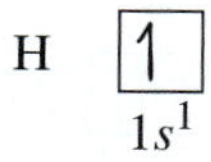

The upward arrow denotes one of the two possible spins (one of the two possible m_s values) of the electron in the hydrogen atom (the other possible spin is indicated with a downward arrow). Under certain circumstances, it is useful to indicate the explicit locations of electrons.

The orbital diagram for a helium atom in the *ground state* is

Student Note: The *ground state* for a many-electron atom is the one in which all the electrons occupy orbitals of the lowest possible energy.

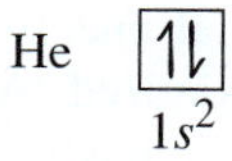

He [↑↓]
$1s^2$

The label $1s^2$ indicates there are *two* electrons in the $1s$ orbital. Note also that the arrows in the box point in opposite directions, representing opposite electron spins. Generally when an orbital diagram includes an orbital with a single electron, we represent it with an upward arrow—although we could represent it equally well with a downward arrow. The choice is arbitrary and has no effect on the energy of the electron.

Student Note: $1s^2$ is read as "one s two," *not* as "one s squared."

The Aufbau Principle

We can continue the process of writing electron configurations for elements based on the order of orbital energies and the Pauli exclusion principle. This process is based on the **Aufbau principle,** which makes it possible to "build" the periodic table of the elements and determine their electron configurations by steps. Each step involves adding one proton to the nucleus and one electron to the appropriate atomic orbital. Through this process we gain a detailed knowledge of the electron configurations of the elements. As is discussed in later chapters, knowledge of electron configurations helps us understand and predict the properties of the elements. It also explains why the elements fit into the periodic table the way they do.

After helium, the next element in the periodic table is lithium, which has three electrons. Because of the restrictions imposed by the Pauli exclusion principle, an orbital can accommodate no more than two electrons. Thus, the third electron cannot reside in the $1s$ orbital. Instead, it must reside in the next available orbital with the lowest possible energy. According to Figure 6.23, this is the $2s$ orbital. Therefore, the electron configuration of lithium is $1s^2 2s^1$, and the orbital diagram is

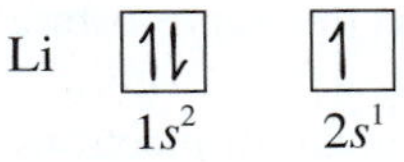

Li   [↑↓]   [↑]
      $1s^2$ $2s^1$

Similarly, we can write the electron configuration of beryllium as $1s^2 2s^2$ and represent it with the orbital diagram

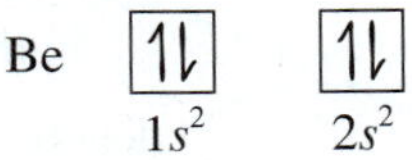

Be   [↑↓]   [↑↓]
      $1s^2$ $2s^2$

With both the $1s$ and the $2s$ orbitals filled to capacity, the next electron, which is needed for the electron configuration of boron, must reside in the $2p$ subshell. Because all three $2p$ orbitals are of equal energy, or **degenerate**, the electron can occupy any one of them. By convention, we usually show the first electron to occupy the p subshell in the first empty box in the orbital diagram.

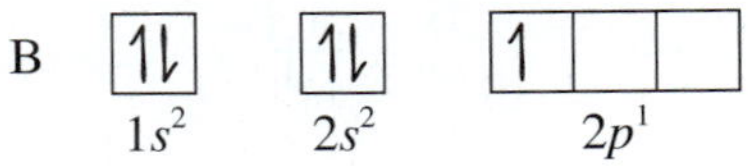

B   [↑↓]   [↑↓]   [↑][ ][ ]
     $1s^2$ $2s^2$ $2p^1$

Hund's Rule

Student Note: Electrons with the same spin are said to have *parallel spins*.

Will the sixth electron, which is needed to represent the electron configuration of carbon, reside in the $2p$ orbital that is already half occupied, or will it reside in one of the other, empty $2p$ orbitals? According to **Hund's**[16] **rule,** the most stable arrangement of electrons in orbitals of equal energy is the one in which the number of electrons with the *same spin* is maximized. As we have seen, no two electrons in any orbital may have the same spin, so maximizing the number of electrons with the same spin requires putting the electrons in separate orbitals. Accordingly, in any subshell, an electron will occupy an empty orbital rather than one that already contains an electron. Because electrons are negatively charged, they repel one another. Maximizing parallel spins in separate orbitals minimizes the electron-electron repulsions in a subshell.

The electron configuration of carbon is, therefore, $1s^2 2s^2 2p^2$, and its orbital diagram is

C   [↑↓]   [↑↓]   [↑][↑][ ]
     $1s^2$ $2s^2$ $2p^2$

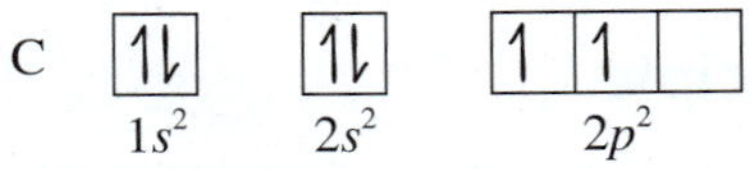

Similarly, the electron configuration of nitrogen is $1s^2 2s^2 2p^3$, and its orbital diagram is

N   [↑↓]   [↑↓]   [↑][↑][↑]
     $1s^2$ $2s^2$ $2p^3$

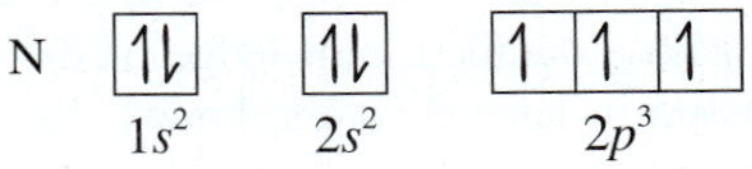

16. Frederick Hund (1896–1997). German physicist. Hund's work was mainly in quantum mechanics. He also helped to develop the molecular orbital theory of chemical bonding.

Once all the $2p$ orbitals are singly occupied, additional electrons will have to pair with those already in the orbitals. Thus, the electron configurations and orbital diagrams for O, F, and Ne are as follows:

O $1s^2 2s^2 2p^4$ $\boxed{\uparrow\downarrow}$ $\boxed{\uparrow\downarrow}$ $\boxed{\uparrow\downarrow\,|\,\uparrow\,|\,\uparrow}$
                          $1s^2$ $2s^2$ $2p^4$

F $1s^2 2s^2 2p^5$ $\boxed{\uparrow\downarrow}$ $\boxed{\uparrow\downarrow}$ $\boxed{\uparrow\downarrow\,|\,\uparrow\downarrow\,|\,\uparrow}$
                          $1s^2$ $2s^2$ $2p^5$

Ne $1s^2 2s^2 2p^6$ $\boxed{\uparrow\downarrow}$ $\boxed{\uparrow\downarrow}$ $\boxed{\uparrow\downarrow\,|\,\uparrow\downarrow\,|\,\uparrow\downarrow}$
                          $1s^2$ $2s^2$ $2p^6$

General Rules for Writing Electron Configurations

Based on the preceding examples we can formulate the following general rules for determining the electron configuration of an element in the ground state:

1. Electrons will reside in the available orbitals of the lowest possible energy.
2. Each orbital can accommodate a maximum of two electrons.
3. Electrons will not pair in *degenerate* orbitals if an empty orbital is available.
4. Orbitals will fill in the order indicated in Figure 6.23. Figure 6.24 provides a simple way for you to remember the proper order.

Sample Problem 6.9 illustrates the procedure for determining the ground-state electron configuration of an atom.

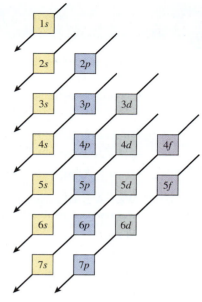

Figure 6.24 A simple way to remember the order in which orbitals fill with electrons.

Student Note: Remember that in this context, *degenerate* means "of equal energy." Orbitals in the same subshell are degenerate.

SAMPLE PROBLEM (6.9)

Write the electron configuration and give the orbital diagram of a calcium (Ca) atom ($Z = 20$).

Strategy Use the general rules given and the Aufbau principle to "build" the electron configuration of a calcium atom and represent it with an orbital diagram.

Setup Because $Z = 20$, we know that a Ca atom has 20 electrons. They will fill orbitals in the order designated in Figure 6.23, obeying the Pauli exclusion principle and Hund's rule. Orbitals will fill in the following order: $1s$, $2s$, $2p$, $3s$, $3p$, $4s$. Each s subshell can contain a maximum of two electrons, whereas each p subshell can contain a maximum of six electrons.

Solution

Ca $1s^2 2s^2 2p^6 3s^2 3p^6 4s^2$ $\boxed{\uparrow\downarrow}$ $\boxed{\uparrow\downarrow}$ $\boxed{\uparrow\downarrow\,|\,\uparrow\downarrow\,|\,\uparrow\downarrow}$ $\boxed{\uparrow\downarrow}$ $\boxed{\uparrow\downarrow\,|\,\uparrow\downarrow\,|\,\uparrow\downarrow}$ $\boxed{\uparrow\downarrow}$
                                          $1s^2$ $2s^2$ $2p^6$ $3s^2$ $3p^6$ $4s^2$

Animation
Atomic structure—electron configurations.

THINK ABOUT IT

Look at Figure 6.23 again to make sure you have filled the orbitals in the right order and that the sum of electrons is 20. Remember that the $4s$ orbital fills before the $3d$ orbitals.

Practice Problem **A**TTEMPT Write the electron configuration and give the orbital diagram of a rubidium (Rb) atom ($Z = 37$).

Practice Problem **B**UILD Write the electron configuration and give the orbital diagram of a bromine (Br) atom ($Z = 35$).

Practice Problem **C**ONCEPTUALIZE Imagine an alternate universe in which the allowed values of the magnetic quantum number, m_ℓ, can have values of $-(\ell + 1) \ldots 0 \ldots +(\ell + 1)$. In this alternate universe, what would be the maximum number of electrons that could have the principal quantum number 3 in a given atom?

CHECKPOINT – SECTION 6.8   Electron Configuration

6.8.1 Which of the following electron configurations correctly represents the Ti atom?

a) $1s^2 2s^2 2p^6 3s^2 3p^6 3d^4$

b) $1s^2 2s^2 2p^6 3s^2 3p^6 4s^2 3d^2$

c) $1s^2 2s^2 2p^6 3s^2 3p^6 4s^2 3d^{10}$

d) $1s^2 2s^2 2p^6 3s^2 3p^6 3d^{10}$

e) $1s^2 2s^2 2p^6 3s^2 3p^6 4s^4$

6.8.2 What element is represented by the following electron configuration? $1s^2 2s^2 2p^6 3s^2 3p^6 4s^2 3d^{10} 4p^4$

a) Br

b) As

c) S

d) Se

e) Te

6.8.3 Which orbital diagram is correct for the ground-state S atom?

a) ⇅ $1s^2$ | ⇅ $2s^2$ | ⇅ ⇅ ⇅ $2p^6$ | ⇅ $3s^2$ | ⇅ ⇅ ☐ $3p^4$

b) ⇅ $1s^2$ | ⇅ $2s^2$ | ⇅ ⇅ ⇅ $2p^6$ | ⇅ $3s^2$ | ⇅ ↑ ↑ $3p^4$

c) ⇅ $1s^2$ | ⇅ $2s^2$ | ⇅ ⇅ ☐ $2p^4$

d) ⇅ $1s^2$ | ⇅ $2s^2$ | ⇅ ↑ ↑ $2p^4$

e) ⇅ $1s^2$ | ⇅ $2s^2$ | ⇅ ⇅ ⇅ $2p^6$

6.9 Electron Configurations and the Periodic Table

The electron configurations of all elements except hydrogen and helium can be represented using a ***noble gas core,*** which shows in brackets the electron configuration of the noble gas element that most recently precedes the element in question, followed by the electron configuration in the outermost occupied subshells. Figure 6.25 gives the outermost ground-state electron configurations of elements from H ($Z = 1$) through Rg ($Z = 111$). Notice the similar pattern of electron configurations in the elements lithium ($Z = 3$) through neon ($Z = 10$) and those of sodium ($Z = 11$) through argon ($Z = 18$). Both Li and Na, for example, have the configuration ns^1 in their outermost occupied subshells. For Li, $n = 2$; for Na, $n = 3$. Both F and Cl have electron configuration $ns^2 np^5$, where $n = 2$ for F and $n = 3$ for Cl, and so on.

As mentioned in Section 6.8, the $4s$ subshell is filled before the $3d$ subshell in a many-electron atom (see Figure 6.23). Thus, the electron configuration of potassium ($Z = 19$) is $1s^2 2s^2 2p^6 3s^2 3p^6 4s^1$. Because $1s^2 2s^2 2p^6 3s^2 3p^6$ is the electron configuration of argon, we can simplify the electron configuration of potassium by writing $[\text{Ar}]4s^1$, where [Ar] denotes the "argon core."

$$\text{K} \qquad \underbrace{1s^2 2s^2 2p^6 3s^2 3p^6}_{[\text{Ar}]}4s^1 \qquad \longrightarrow \qquad [\text{Ar}]\,4s^1$$

The placement of the outermost electron in the $4s$ orbital (rather than in the $3d$ orbital) of potassium is strongly supported by experimental evidence. The physical and chemical properties of potassium are very similar to those of lithium and sodium, the first two alkali metals. In both lithium and sodium, the outermost electron is in an s orbital (there is no doubt that their outermost electrons occupy s orbitals because there is no $1d$ or $2d$ subshell). Based on its similarities to the other alkali metals, we expect potassium to have an analogous electron configuration; that is, we expect the last electron in potassium to occupy the $4s$ rather than the $3d$ orbital.

The elements from Group 3 through Group 11 are *transition metals* [◀◀ Section 2.4]. Transition metals either have incompletely filled d subshells or readily give rise to

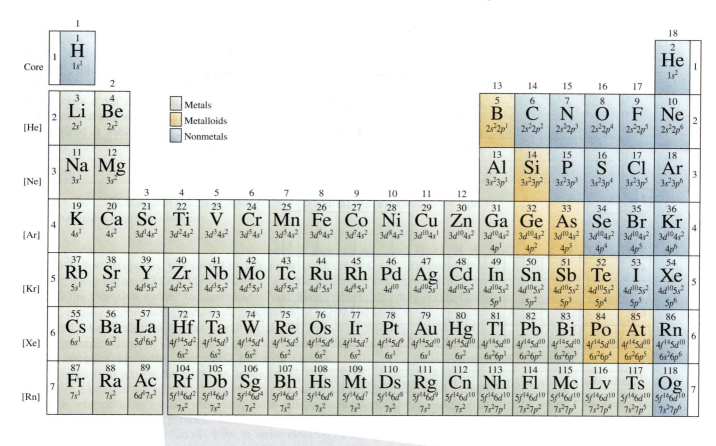

Figure 6.25 Outermost ground-state electron configurations for the known elements.

cations that have incompletely filled d subshells. In the first transition metal series, from scandium ($Z = 21$) through *copper* ($Z = 29$), additional electrons are placed in the $3d$ orbitals according to Hund's rule. However, there are two anomalies. The electron configuration of chromium ($Z = 24$) is $[Ar]4s^1 3d^5$ and not $[Ar]4s^2 3d^4$, as we might expect. A similar break in the pattern is observed for copper, whose electron configuration is $[Ar]4s^1 3d^{10}$ rather than $[Ar]4s^2 3d^9$. The reason for these anomalies is that a slightly greater stability is associated with the half-filled ($3d^5$) and completely filled ($3d^{10}$) subshells.

Cr [Ar] | 1 | | 1 | 1 | 1 | 1 | 1 |
         | 4s | | 3d | 3d | 3d | 3d | 3d |

Cu [Ar] | 1 | | 1↓ | 1↓ | 1↓ | 1↓ | 1↓ |
         | 4s | | 3d | 3d | 3d | 3d | 3d |

For elements Zn ($Z = 30$) through Kr ($Z = 36$), the $3d$, $4s$, and $4p$ subshells fill in a straightforward manner. With rubidium ($Z = 37$), electrons begin to enter the $n = 5$ energy level.

Some of the electron configurations in the second transition metal series [yttrium ($Z = 39$) through silver ($Z = 47$)] are also irregular, but the details of many of these

Student Note: Although zinc and the other elements in Group 12 sometimes are included under the heading "transition metals," they neither have nor readily acquire partially filled d subshells. Strictly speaking, they are *not* transition metals.

Student Note: Electron configurations such as these may also be written with the d subshell first. For example, $[Ar]4s^1 3d^{10}$ can also be written as $[Ar]3d^{10}4s^1$. Either way is acceptable.

CHECKPOINT – SECTION 6.9   Electron Configurations and the Periodic Table

6.9.1 Which of the following electron configurations correctly represents the Ag atom?

a) $[Kr]5s^2 4d^9$

b) $[Kr]5s^2 4d^{10}$

c) $[Kr]5s^1 4d^{10}$

d) $[Xe]5s^2 4d^9$

e) $[Xe]5s^1 4d^{10}$

6.9.2 What element is represented by the following electron configuration: $[Kr]5s^2 4d^{10} 5p^5$?

a) Tc

b) Br

c) I

d) Xe

e) Te

6.9.3 Which of the following is a *d*-block element? (Select all that apply.)

a) Sb

b) Au

c) Ca

d) Zn

e) U

6.9.4 Which of the following is a *p*-block element? (Select all that apply.)

a) Pb

b) C

c) Sr

d) Xe

e) Na

Chapter Summary

Section 6.1

- What we commonly refer to as "light" is actually the visible portion of the *electromagnetic spectrum*. All light has certain common characteristics including wavelength, frequency, and amplitude.

- *Wavelength* (λ) is the distance between two crests or two troughs of a wave. *Frequency* (ν) is the number of waves that pass a point per unit time. *Amplitude* is the distance between the midpoint and crest or trough of a wave.

- *Electromagnetic waves* have both electric and magnetic components that are both mutually perpendicular and in phase.

Section 6.2

- *Blackbody radiation* is the electromagnetic radiation given off by a solid when it is heated.

- Max Planck proposed that energy, like matter, was composed of tiny, indivisible "packages" called *quanta*. *Quanta* is the plural of *quantum*.

- Albert Einstein used Planck's revolutionary quantum theory to explain the *photoelectric effect,* in which electrons are emitted when light of a certain minimum frequency shines on a metal surface.

- A *quantum* of light is referred to as a *photon.*

Section 6.3

- An *emission spectrum* is the light given off by an object when it is excited thermally. Emission spectra may be *continuous,* including all the wavelengths within a particular range, or they may be *line spectra,* consisting only of certain discrete wavelengths.

- The *ground state* is the lowest possible energy state for an atom. An *excited state* is any energy level higher than the ground state.

Section 6.4

- A *node* is a point at which a standing wave has zero amplitude.

- Having observed that light could exhibit particle-like behavior, de Broglie proposed that matter might also exhibit wavelike behavior. The *de Broglie wavelength* is the wavelength associated with a particle of very small mass. Soon after de Broglie's proposal, experiments showed that electrons could exhibit diffraction—a property of waves.

Section 6.5

- According to the *Heisenberg uncertainty principle,* the product of the uncertainty of the *location* and the uncertainty of the *momentum* of a very small particle must have a certain minimum value. It is thus impossible to know simultaneously both the location and momentum of an electron.

- The *electron density* gives the probability of finding an electron in a particular region in an atom. An *atomic orbital* is the region of three-dimensional space, defined by ψ^2 (the square of the wave function, ψ), where the probability of finding an electron is high. An atomic orbital can accommodate a maximum of *two* electrons.

Section 6.6

- An atomic orbital is defined by three *quantum numbers:* the *principal quantum number (n),* the *angular momentum quantum number (ℓ),* and the *magnetic quantum number (m_ℓ).*

- The principal quantum number (n) indicates distance from the nucleus. Possible values of n are (1, 2, 3, . . .). The angular momentum quantum number (ℓ) indicates the shape of the orbital. Possible values of ℓ are (0, 1, . . . , $n - 1$). The magnetic quantum number (m_ℓ) indicates the orbital's orientation in space. Possible values of m_ℓ are ($-\ell$, . . . , 0, . . . , $+\ell$).

- Two electrons that occupy the same atomic orbital in the ground state must have different *electron spin quantum numbers (m_s),* either $+\frac{1}{2}$ or $-\frac{1}{2}$.

Section 6.7

- The value of the angular momentum quantum number (ℓ) determines the type of the atomic orbital: $\ell = 0$ corresponds to an *s orbital,* $\ell = 1$ corresponds to a *p orbital,* $\ell = 2$ corresponds to a *d orbital,* and $\ell = 3$ corresponds to an *f orbital.*

Section 6.8

- The *electron configuration* specifies the arrangement of electrons in the atomic orbitals of an atom.

- According to the *Pauli exclusion principle,* no two electrons in an atom in the ground state can have the same four quantum numbers, n, ℓ, m_ℓ, and m_s.

- The *Aufbau principle* describes the theoretical, sequential building up of the elements in the periodic table by the stepwise addition of protons and electrons.

- Atomic orbitals that have the same energy are called *degenerate*. According to *Hund's rule,* degenerate orbitals must all contain one electron before any can contain two electrons.

Section 6.9

- The *noble gas core* makes it possible to abbreviate the writing of electron configurations.

- The *lanthanide (rare earth) series* and *actinide series* appear at the bottom of the periodic table. They represent the filling of *f* orbitals.

Key Words

Actinide series, 294

Amplitude, 256

Angular momentum quantum number (ℓ), 280

Atomic orbital, 279

Aufbau principle, 289

Blackbody radiation, 259

d orbital, 285

de Broglie wavelength, 275

Degenerate, 290

Electromagnetic spectrum, 256

Electromagnetic wave, 257

Electron configuration, 287

Electron density, 279

Electron spin quantum number (m_s), 282

Emission spectra, 265

Excited state, 270

f orbital, 286

Frequency (ν), 256

Ground state, 270

Heisenberg uncertainty principle, 277

Hund's rule, 290

Lanthanide (rare earth) series, 294

Line spectra, 266

Magnetic quantum number (m_ℓ), 280

Noble gas core, 292

Node, 274

p orbital, 285

Pauli exclusion principle, 288

Photoelectric effect, 262

Photons, 262

Principal quantum number (n), 280

Quantum, 260

Quantum numbers, 279

s orbital, 284

Wavelength (λ), 256

Key Equations

6.1 $c = \lambda\nu$	The wavelength (λ) and frequency (ν) of electromagnetic radiation are related to one another through the speed of light (c). If wavelength is known, frequency can be determined, and vice versa.
6.2 $E = h\nu$	The energy of a photon (E) is equal to the product of Planck's constant (h) and frequency (ν) of the photon.
6.3 $h\nu = \mathrm{KE} + W$	The energy ($h\nu$) of a photon used to eject electrons from a metal surface via the photoelectric effect is equal to the sum of kinetic energy of the ejected electron (KE) and the binding energy (W).
6.4 $\dfrac{1}{\lambda} = R_\infty\left(\dfrac{1}{n_1^2} - \dfrac{1}{n_2^2}\right)$	When an electron transitions from one quantum state to another (n_i to n_f), the difference in energy between the two states is emitted (or absorbed) in the form of light. The wavelength of the emitted/absorbed light can be calculated using Equation 6.4.
6.5 $E_n = -2.18 \times 10^{-18}\,\mathrm{J}\left(\dfrac{1}{n^2}\right)$	The energy of an electron for a given value of n (E_n) is inversely proportional to the square of n—and is by convention a negative number.
6.6 $\Delta E = h\nu = -2.18 \times 10^{-18}\,\mathrm{J}\left(\dfrac{1}{n_f^2} - \dfrac{1}{n_i^2}\right)$	The difference in energy between two quantum states (n_i to n_f) is calculated using Equation 6.6.
6.7 $\dfrac{1}{\lambda} = \dfrac{2.18 \times 10^{-18}\,\mathrm{J}}{hc}\left(\dfrac{1}{n_f^2} - \dfrac{1}{n_i^2}\right)$	Similar to Equation 6.4, Equation 6.7 allows calculation of the wavelength of emitted/absorbed light when n_i and n_f are known.
6.8 $2\pi r = n\lambda$	This is the relationship between the allowed orbit ($2\pi r$) and wavelength (λ) of an electron behaving as a standing wave.
6.9 $\lambda = \dfrac{h}{mu}$	The de Broglie wavelength (λ) of a particle can be calculated using Planck's constant (h), the mass of the particle in kilograms (m), and velocity (u) of the particle.

6.10 $\Delta x \cdot \Delta p \geq \dfrac{h}{4\pi}$

The Heisenberg uncertainty principle states that the product of uncertainties in position (Δx) and momentum (Δp) of a particle cannot be less than Planck's constant (h) over 4π. Knowing the uncertainty in one (position or momentum) allows us to calculate the minimum uncertainty in the other.

6.11 $\Delta x \cdot m\Delta u \geq \dfrac{h}{4\pi}$

Similar to Equation 6.10, when mass of the particle is known, knowing the uncertainty in position (Δx) allows us to calculate the minimum uncertainty in its velocity (Δu).

Key Constants

Section 6.1 speed of light, c
2.99792458×10^8 m/s

Used to convert between frequency and wavelength.

Section 6.2 Planck's constant, h
$6.626070040 \times 10^{-34}$ J · s

Used in a wide variety of calculations, including to determine energy of a photon, de Broglie wavelength, and uncertainty in position or momentum of particle.

Section 6.3 Rydberg constant, R_∞
1.09737316×10^7 m^{-1}

Used to calculate wavelength of light emitted during an electron transition.

Determining Ground-State Valence Electron Configurations Using the Periodic Table

An easy way to determine the electron configuration of an element is by using the periodic table. Although the table is arranged by atomic number, it is also divided into blocks that indicate the type of orbital occupied by an element's outermost electrons. Outermost valence electrons of elements in the s-block (shown in yellow) reside in s orbitals; those of elements in the p-block (blue) reside in p orbitals; and so on.

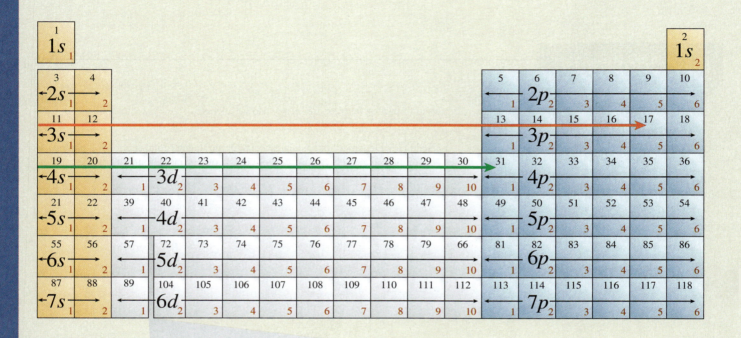

To determine the ground-state electron configuration of any element, we start with the most recently completed noble gas core, and count across the following period to determine the valence electron configuration. Consider the example of Cl, which has atomic number 17. The noble gas that precedes Cl is Ne, with atomic number 10. Therefore, we begin by writing [Ne]. The noble gas symbol in square brackets represents the core electrons—with a completed p subshell. To complete the electron configuration, we count from the left of period 3 as shown by the red arrow, adding the last (rightmost) configuration label from each block the arrow touches:

There are seven electrons in addition to the noble gas core. Two of them reside in an s subshell, and five of them reside in a p subshell. By simply counting across the third period, we can determine the specific subshells that contain the valence electrons, and arrive at the correct ground-state electron configuration: $[Ne]3s^23p^5$.

For Ga, with atomic number 31, the preceding noble gas is Ar, with atomic number 18. Counting across the fourth period (green arrow) gives the ground-state electron configuration: $[Ar]4s^23d^{10}4p^1$.

There are a few elements for which this method will not give the correct configuration. For example, there is no element with a ground-state valence electron configuration ending in $3d^4$ or $3d^9$. Instead, Cr and Cu are $[Ar]4s^13d^5$ and $[Ar]4s^13d^{10}$, respectively. Remember that this is the result of the unusual stability of either a *half-filled* or a *filled d* subshell [◂◂ Section 6.9].

We can also use the periodic table to determine the identity of an element, given its ground-state electron configuration. For example, given the configuration $[Ne]3s^23p^4$, we focus on the last entry in the configuration: $3p^4$. This tells us that the element is in the *third* period (3), in the *p*-block (p), and that it has *four* electrons in its p subshell (superscript 4). This corresponds to atomic number 16, which is the element sulfur (S).

Key Skills Problems

6.1

What is the noble gas core for Mo?

(a) Ar (b) Kr (c) Xe (d) Ne (e) Rn

6.2

Which of the following electron configurations correctly represents the V atom?

(a) $[Ar]3d^5$ (b) $[Ar]4s^23d^2$ (c) $[Ar]4s^23d^4$ (d) $[Ar]4s^23d^3$
(e) $[Kr]4s^23d^3$

6.3

What element is represented by the electron configuration $[Kr]5s^24d^{10}5p^1$?

(a) Sn (b) Ga (c) In (d) Tl (e) Zr

6.4

What is the electron configuration of the Lu atom?

(a) $[Xe]6s^24f^{14}$ (b) $[Xe]6s^25d^1$ (c) $[Xe]6s^24f^{13}$
(d) $[Xe]6s^24f^{14}5d^1$ (e) $[Xe]4f^{14}$

Questions and Problems

Applying What You've Learned

The two most commonly used lasers in photodynamic therapy are the Ar ion laser, which emits beams at 488 nm (blue) and 514 nm (green), and a 635-nm (red) diode laser. These wavelengths do not penetrate biological tissue very deeply, and currently are useful only for tumors that are on or just beneath the surface of the skin—or of the lining of an organ that can be exposed to light, such as the bladder or esophagus. (a) Calculate the frequency and the energy per photon of the light emitted by a 635-nm diode laser [◄◄ Sample Problems 6.1 and 6.2]. (b) Calculate the difference in energy per photon in the two wavelengths emitted by the Ar laser. Which of the wavelengths is more energetic [◄◄ Sample Problem 6.3]? (c) Calculate the de Broglie wavelength of an argon atom moving at 1000 m/s [◄◄ Sample Problem 6.5]. (d) Give the values of n, ℓ, and m_ℓ for the $3p$ orbitals in an Ar atom [◄◄ Sample Problem 6.8]. (e) Write the electron configuration of Ar [◄◄ Sample Problems 6.9 and 6.10].

SECTION 6.1: THE NATURE OF LIGHT

Review Questions

6.1 What is a wave? Using a diagram, define the following terms associated with waves: wavelength, frequency, amplitude.

6.2 What are the units for wavelength and frequency of electromagnetic waves? What is the speed of light in meters per second and miles per hour?

6.3 List the types of electromagnetic radiation, starting with the radiation having the longest wavelength and ending with the radiation having the shortest wavelength.

6.4 Give the high and low wavelength values that define the visible region of the electromagnetic spectrum.

Computational Problems

6.5 (a) What is the wavelength (in nm) of light having a frequency of 8.6×10^{13} Hz? (b) What is the frequency (in Hz) of light having a wavelength of 566 nm?

6.6 (a) What is the frequency of light having a wavelength of 456 nm? (b) What is the wavelength (in nm) of radiation having a frequency of 2.45×10^9 Hz? (This is the type of radiation used in microwave ovens.)

6.7 The SI unit of time is the second, which is defined as 9,192,631,770 cycles of radiation associated with a certain emission process in the cesium atom. Calculate the wavelength of this radiation (to three significant figures). In which region of the electromagnetic spectrum is this wavelength found?

Conceptual Problems

6.8 How many minutes would it take a radio wave to travel from the planet Venus to Earth? (The average distance from Venus to Earth = 28 million miles.) How long would it take an infrared wave to travel the same distance?

6.9 The average distance between Mars and Earth is about 1.3×10^8 miles. How long would it take video

images transmitted from the Mars Spirit rover on Mars' surface to reach Earth (1 mile = 1.61 km)?

6.10 Four waves represent light in four different regions of the electromagnetic spectrum: visible, microwave, infrared, and ultraviolet. Determine the best match of regions to the waves shown here. Explain your choices.

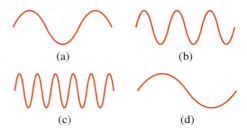

(a)   (b)

(c)   (d)

SECTION 6.2: QUANTUM THEORY

Review Questions

6.11 Briefly explain Planck's quantum theory and explain what a quantum is. What are the units for Planck's constant?

6.12 Give two everyday examples that illustrate the concept of quantization.

6.13 Explain what is meant by the photoelectric effect.

6.14 What are photons? What role did Einstein's explanation of the photoelectric effect play in the development of the particle-wave interpretation of the nature of electromagnetic radiation?

Computational Problems

6.15 A photon has a wavelength of 705 nm. Calculate the energy of the photon in joules.

6.16 The blue color of the sky results from the scattering of sunlight by molecules in the air. The blue light has a frequency of about 7.5×10^{14} Hz. (a) Calculate the wavelength (in nm) associated with this radiation, and (b) calculate the energy (in joules) of a single photon associated with this frequency.

6.17 A photon has a frequency of 6.5×10^9 Hz.
(a) Convert this frequency into wavelength (nm). Does this frequency fall in the visible region?
(b) Calculate the energy (in joules) of this photon.
(c) Calculate the energy (in joules) of 1 mole of photons all with this frequency.

6.18 What is the wavelength (in nm) of radiation that has an energy content of 2.13×10^3 kJ/mol? In which region of the electromagnetic spectrum is this radiation found?

6.19 When copper is bombarded with high-energy electrons, X rays are emitted. Calculate the energy (in joules) associated with the photons if the wavelength of the X rays is 0.154 nm.

6.20 A particular form of electromagnetic radiation has a frequency of 9.87×10^{15} Hz. (a) What is its wavelength in nanometers? In meters? (b) To what region of the electromagnetic spectrum would you assign it? (c) What is the energy (in joules) of one quantum of this radiation?

6.21 The retina of a human eye can detect light when radiant energy incident on it is at least 4.0×10^{-17} J. For light of 585-nm wavelength, how many photons does this energy correspond to?

6.22 The radioactive ^{60}Co isotope is used in nuclear medicine to treat certain types of cancer. Calculate the wavelength and frequency of an emitted gamma particle having the energy of 1.29×10^{11} J/mol.

Conceptual Problems

6.23 Photosynthesis makes use of visible light to bring about chemical changes. Explain why heat energy in the form of infrared radiation is ineffective for photosynthesis.

6.24 A red light was shined onto a metal sample and the result shown in (i) was observed. When the light source was changed to a blue light, the result shown in (ii) was observed. Explain how these results can be interpreted with respect to the photoelectric effect.

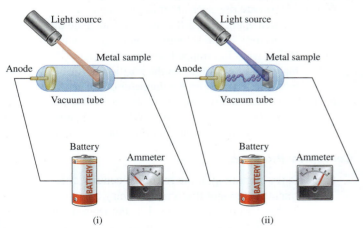

(i)   (ii)

Describe the result you would expect for each of the following: (a) The intensity of red light is increased. (b) The intensity of blue light is increased. (c) Violet light is used.

6.25 A photoelectric experiment was performed by separately shining a laser at 450 nm (blue light) and a laser at 560 nm (yellow light) on a clean metal surface and measuring the number and kinetic energy of the ejected electrons. Which light would generate more electrons? Which light would eject electrons with greater kinetic energy? Assume that the same amount of energy is delivered to the metal surface by each laser and that the frequencies of the laser lights exceed the threshold frequency.

SECTION 6.3: BOHR'S THEORY OF THE HYDROGEN ATOM

Visualizing Chemistry
Figure 6.10

VC 6.1 Which of the following best explains why we see only four lines in the emission spectrum of hydrogen?
a) Hydrogen has only four different electronic transitions.
b) Only four of hydrogen's electronic transitions correspond to visible wavelengths.
c) The other lines in hydrogen's emission spectrum can't be seen easily against the black background.

VC 6.2 One way to see the emission spectrum of hydrogen is to view the hydrogen in an electric discharge tube through a *spectroscope,* a device that separates the wavelengths. Why can we not view the emission spectrum simply by pointing the spectroscope at a sample of hydrogen confined in a glass tube or flask?
a) Without the electrons being in excited states, there would be no emission of light.
b) The glass would make it impossible to see the emission spectrum.
c) Hydrogen alone does not exhibit an emission spectrum—it must be combined with oxygen.

VC 6.3 How many lines would we see in the emission spectrum of hydrogen if the downward transitions from excited states all ended at $n = 1$ and no transitions ended at $n = 2$?
a) We would still see four lines.
b) We would see five lines.
c) We would not see any lines.

VC 6.4 For a hydrogen atom in which the electron has been excited to $n = 4$, how many different transitions can occur as the electron eventually returns to the ground state?
a) 1
b) 3
c) 6

Review Questions

6.26 What are emission spectra? How do line spectra differ from continuous spectra?

6.27 What is an energy level? Explain the difference between ground state and excited state.

6.28   Briefly describe Bohr's theory of the hydrogen atom and how it explains the appearance of an emission spectrum. How does Bohr's theory differ from concepts of classical physics?

Computational Problems

6.29   The first line of the Balmer series occurs at a wavelength of 656.3 nm. What is the energy difference between the two energy levels involved in the emission that results in this spectral line?

6.30   Calculate the wavelength (in nm) of a photon emitted by a hydrogen atom when its electron drops from the $n = 7$ state to the $n = 2$ state.

6.31   Calculate the frequency (Hz) and wavelength (nm) of the emitted photon when an electron drops from the $n = 4$ to the $n = 3$ level in a hydrogen atom.

6.32   Careful spectral analysis shows that the familiar yellow light of sodium lamps (such as street lamps) is made up of photons of two wavelengths, 589.0 nm and 589.6 nm. What is the difference in energy (in joules) between photons with these wavelengths?

6.33   An electron in the hydrogen atom makes a transition from an energy state of principal quantum number n_i to the $n = 1$ state. If the photon emitted has a wavelength of 94.9 nm, what is the value of n_i?

6.34   Consider the following energy levels of a hypothetical atom:

E_4: -1.0×10^{-19} J E_2: -10×10^{-19} J
E_3: -5.0×10^{-19} J E_1: -15×10^{-19} J

(a) What is the wavelength of the photon needed to excite an electron from E_1 to E_4? (b) What is the energy (in joules) a photon must have to excite an electron from E_2 to E_3? (c) When an electron drops from the E_3 level to the E_1 level, the atom is said to undergo emission. Calculate the wavelength of the photon emitted in this process.

Conceptual Problems

6.35   Some copper compounds emit green light when they are heated in a flame. How would you determine whether the light is of one wavelength or a mixture of two or more wavelengths?

6.36   Is it possible for a fluorescent material to emit radiation in the ultraviolet region after absorbing visible light? Explain your answer.

6.37   Explain how astronomers are able to tell which elements are present in distant stars by analyzing the electromagnetic radiation emitted by the stars.

SECTION 6.4: WAVE PROPERTIES OF MATTER

Review Questions

6.38   How does de Broglie's hypothesis account for the fact that the energies of the electron in a hydrogen atom are quantized?

6.39   Why is Equation 6.9 meaningful only for submicroscopic particles, such as electrons and atoms, and not for macroscopic objects?

6.40   Does a baseball in flight possess wave properties? If so, why can we not determine its wave properties?

Computational Problems

6.41   Thermal neutrons are neutrons that move at speeds comparable to those of air molecules at room temperature. These neutrons are most effective in initiating a nuclear chain reaction among ^{235}U isotopes. Calculate the wavelength (in nm) associated with a beam of neutrons moving at 7.00×10^2 m/s (mass of a neutron = 1.675×10^{-27} kg).

6.42   Protons can be accelerated to speeds near that of light in particle accelerators. Estimate the wavelength (in nm) of such a proton moving at 2.90×10^8 m/s (mass of a proton = 1.673×10^{-27} kg).

6.43   What is the de Broglie wavelength (in cm) of a 12.4-g hummingbird flying at 1.20×10^2 mph (1 mile = 1.61 km)?

6.44   What is the de Broglie wavelength (in nm) associated with a 2.50-g Ping-Pong ball traveling at 15.0 mph?

SECTION 6.5: QUANTUM MECHANICS

Review Questions

6.45   What are the inadequacies of Bohr's theory?

6.46   What is the Heisenberg uncertainty principle? What is the Schrödinger equation?

6.47   What is the physical significance of the wave function?

6.48   How is the concept of electron density used to describe the position of an electron in the quantum mechanical treatment of an atom?

6.49   What is an atomic orbital? How does an atomic orbital differ from an orbit?

Computational Problems

6.50   Alveoli are tiny sacs of air in the lungs. Their average diameter is 5.0×10^{-5} m. Calculate the uncertainty in the velocity of an oxygen molecule (5.3×10^{-26} kg) trapped within a sac. (*Hint:* The maximum uncertainty in the position of the molecule is given by the diameter of the sac.)

6.51   The speed of a thermal neutron (see Problem 6.41) is known to within 2.0 km/s. What is the minimum uncertainty in the position of the thermal neutron?

Conceptual Problems

6.52   In the beginning of the twentieth century, some scientists thought that a nucleus may contain both electrons and protons. Use the Heisenberg uncertainty principle to show that an electron cannot be confined within a nucleus. Repeat the calculation for a proton. Comment on your results. Assume the radius of a nucleus to be 1.0×10^{-15} m.

The masses of an electron and a proton are 9.109×10^{-31} kg and 1.673×10^{-27} kg, respectively. (*Hint:* Treat the radius of the nucleus as the uncertainty in position.)

6.53 Suppose that photons of blue light (430 nm) are used to locate the position of a 2.80-g Ping-Pong ball in flight and that the uncertainty in the position is equal to one wavelength. What is the minimum uncertainty in the speed of the Ping-Pong ball? Comment on the magnitude of your result.

SECTION 6.6: QUANTUM NUMBERS

Review Questions

6.54 Describe the four quantum numbers used to characterize an electron in an atom.

6.55 Which quantum number defines a shell? Which quantum numbers define a subshell?

6.56 Which of the four quantum numbers (n, ℓ, m_ℓ, m_s) determine (a) the energy of an electron in a hydrogen atom and in a many-electron atom, (b) the size of an orbital, (c) the shape of an orbital, (d) the orientation of an orbital in space?

Conceptual Problems

6.57 An electron in a certain atom is in the $n = 2$ quantum level. List the possible values of ℓ and m_ℓ that it can have.

6.58 An electron in an atom is in the $n = 3$ quantum level. List the possible values of ℓ and m_ℓ that it can have.

6.59 List all the possible subshells and orbitals associated with the principal quantum number n, if $n = 4$.

6.60 List all the possible subshells and orbitals associated with the principal quantum number n, if $n = 5$.

SECTION 6.7: ATOMIC ORBITALS

Review Questions

6.61 Describe the shapes of s, p, and d orbitals. How are these orbitals related to the quantum numbers n, ℓ, and m_ℓ?

6.62 List the hydrogen orbitals in increasing order of energy.

6.63 Describe the characteristics of an s orbital, p orbital, and d orbital. Which of the following orbitals do not exist: $1p$, $2s$, $2d$, $3p$, $3d$, $3f$, $4s$, $4f$?

6.64 Why is a boundary surface diagram useful in representing an atomic orbital?

Conceptual Problems

6.65 Give the values of the quantum numbers associated with the following orbitals: (a) $2p$, (b) $3s$, (c) $5d$.

6.66 Give the values of the four quantum numbers of an electron in the following orbitals: (a) $3s$, (b) $4p$, (c) $3d$.

6.67 Discuss the similarities and differences between a $1s$ and a $2s$ orbital.

6.68 What is the difference between a $2p_x$ and a $2p_y$ orbital?

6.69 Why do the $3s$, $3p$, and $3d$ orbitals have the same energy in a hydrogen atom but different energies in a many-electron atom?

6.70 Make a chart of all allowable orbitals in the first four principal energy levels of the hydrogen atom. Designate each by type (e.g., s, p), and indicate how many orbitals of each type there are.

6.71 For each of the following pairs of hydrogen orbitals, indicate which is higher in energy: (a) $1s$, $2s$; (b) $2p$, $3p$; (c) $3d_{xy}$, $3d_{yz}$; (d) $3s$, $3d$; (e) $4f$, $5s$.

6.72 Which orbital in each of the following pairs is lower in energy in a many-electron atom: (a) $2s$, $2p$; (b) $3p$, $3d$; (c) $3s$, $4s$; (d) $4d$, $5f$?

6.73 A $3s$ orbital is illustrated here. Using this as a reference to show the relative size of the other four orbitals, answer the following questions.

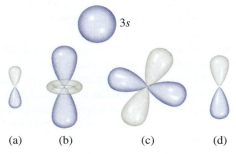

(a)   (b)   (c)   (d)

(a) Which orbital has the greatest value of n?
(b) How many orbitals have a value of $\ell = 1$?
(c) How many other orbitals with the same value of n would have the same general shape as orbital (b)?

SECTION 6.8: ELECTRON CONFIGURATION

Review Questions

6.74 What is electron configuration? Describe the roles that the Pauli exclusion principle and Hund's rule play in writing the electron configuration of elements.

6.75 Explain the meaning of the symbol $4d^6$.

6.76 State the Aufbau principle, and explain the role it plays in classifying the elements in the periodic table.

Computational Problems

6.77 Calculate the total number of electrons that can occupy (a) one s orbital, (b) three p orbitals, (c) five d orbitals, (d) seven f orbitals.

6.78 What is the total number of electrons that can be held in all orbitals having the same principal quantum number n?

6.79 Determine the maximum number of electrons that can be found in each of the following subshells: $3s$, $3d$, $4p$, $4f$, $5f$.

6.80   Indicate the total number of (a) p electrons in N ($Z = 7$), (b) s electrons in Si ($Z = 14$), and (c) $3d$ electrons in S ($Z = 16$).

Conceptual Problems

6.81   Indicate which of the following sets of quantum numbers in an atom are unacceptable and explain why: (a) $(1, 1, +\frac{1}{2}, -\frac{1}{2})$, (b) $(3, 0, -1, +\frac{1}{2})$, (c) $(2, 0, +1, +\frac{1}{2})$, (d) $(4, 3, -2, +\frac{1}{2})$, (e) $(3, 2, +1, 1)$.

6.82   The ground-state electron configurations listed here are incorrect. Explain what mistakes have been made in each and write the correct electron configurations.
Al: $1s^2 2s^2 2p^4 3s^2 3p^3$
B: $1s^2 2s^2 2p^5$
F: $1s^2 2s^2 2p^6$

6.83   Indicate the number of unpaired electrons present in each of the following atoms: B, Ne, P, Sc, Mn, Se, Kr, Fe, Cd, I, Pb.

6.84   The electron configuration of a neutral atom is $1s^2 2s^2 2p^6 3s^2$. Write a complete set of quantum numbers for each of the electrons. Name the element.

6.85   Which of the following species has the greatest number of unpaired electrons: S^+, S, or S^-?

6.86   Portions of orbital diagrams representing the ground-state electron configurations of certain elements are shown here. Which of them violate the Pauli exclusion principle? Which violate Hund's rule?

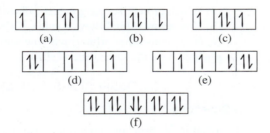

SECTION 6.9: ELECTRON CONFIGURATIONS AND THE PERIODIC TABLE

Review Questions

6.87   Describe the characteristics of transition metals.
6.88   What is the noble gas core? How does it simplify the writing of electron configurations?
6.89   What are the group and period of the element osmium?
6.90   Define the following terms and give an example of each: lanthanides, actinides.
6.91   Explain why the ground-state electron configurations of Cr and Cu are different from what we might expect.
6.92   Write the electron configuration of a xenon core.
6.93   Comment on the correctness of the following statement: The probability of finding two electrons with the same four quantum numbers in an atom is zero.

Conceptual Problems

6.94   Use the Aufbau principle to obtain the ground-state electron configuration of selenium.
6.95   Use the Aufbau principle to obtain the ground-state electron configuration of technetium.
6.96   Write the ground-state electron configurations for the following elements: B, V, C, As, I, Au.
6.97   Write the ground-state electron configurations for the following elements: Ge, Fe, Zn, Ni, W, Tl.

ADDITIONAL PROBLEMS

6.98   Spectral lines of the Lyman and Balmer series do not overlap. Verify this statement by calculating the longest wavelength associated with the Lyman series and the shortest wavelength associated with the Balmer series (in nm).

6.99   Discuss the current view of the correctness of the following statements. (a) The electron in the hydrogen atom is in an orbit that never brings it closer than 100 pm to the nucleus. (b) Atomic absorption spectra result from transitions of electrons from lower to higher energy levels. (c) A many-electron atom behaves somewhat like a solar system that has a number of planets.

6.100   Distinguish carefully between the following terms: (a) wavelength and frequency, (b) wave properties and particle properties, (c) quantization of energy and continuous variation in energy.

6.101   What is the maximum number of electrons in an atom that can have the following quantum numbers? Specify the orbitals in which the electrons would be found.
(a) $n = 2$, $m_s = +\frac{1}{2}$; (b) $n = 4$, $m_\ell = +1$; (c) $n = 3$, $\ell = 2$; (d) $n = 2$, $\ell = 0$, $m_s = -\frac{1}{2}$; (e) $n = 4$, $\ell = 3$, $m_\ell = -2$.

6.102   Identify the following individuals and their contributions to the development of quantum theory: Bohr, de Broglie, Einstein, Planck, Heisenberg, Schrödinger.

6.103   Consider the graph here.

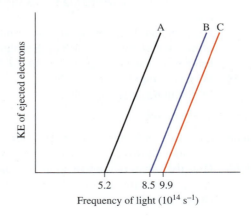

(a) Calculate the binding energy (W) of each metal. Which metal has the highest binding energy? (b) A photon with a wavelength of 333 nm is fired at the three metals. Which, if any, of the metals will eject an electron?

6.104 A baseball pitcher's fastball has been clocked at about 100 mph. (a) Calculate the wavelength of a 0.141-kg baseball (in nm) at this speed. (b) What is the wavelength of a hydrogen atom at the same speed (1 mile = 1609 m)?

6.105 The He$^+$ ion contains only one electron and is therefore a hydrogen-like ion. Calculate the wavelengths, in increasing order, of the first four transitions in the Balmer series of the He$^+$ ion. Compare these wavelengths with the same transitions in an H atom. Comment on the differences. (The Rydberg constant for He is 4.39×10^7 m^{-1}.)

6.106 Draw the shapes (boundary surfaces) of the following orbitals: (a) $2p_y$, (b) $3d_{z^2}$, (c) $3d_{x^2-y^2}$. (Show coordinate axes in your sketches.)

6.107 Draw orbital diagrams for atoms with the following electron configurations:
(a) $1s^2 2s^2 2p^5$
(b) $1s^2 2s^2 2p^6 3s^2 3p^3$
(c) $1s^2 2s^2 2p^6 3s^2 3p^6 4s^2 3d^7$

6.108 Ionization energy is the minimum energy required to remove an electron from an atom. It is usually expressed in units of kJ/mol, that is, the energy in kilojoules required to remove one mole of electrons from one mole of atoms. (a) Calculate the ionization energy for the hydrogen atom. (b) Repeat the calculation, assuming in this second case that the electrons are removed from the $n = 2$ state, instead of from the ground state.

6.109 An electron in a hydrogen atom is excited from the ground state to the $n = 4$ state. Comment on the correctness of the following statements (true or false).
(a) $n = 4$ is the first excited state.
(b) It takes more energy to ionize (remove) the electron from $n = 4$ than from the ground state.
(c) The electron is farther from the nucleus (on average) in $n = 4$ than in the ground state.
(d) The wavelength of light emitted when the electron drops from $n = 4$ to $n = 1$ is longer than that from $n = 4$ to $n = 2$.
(e) The wavelength the atom absorbs in going from $n = 1$ to $n = 4$ is the same as that emitted as it goes from $n = 4$ to $n = 1$.

6.110 The ionization energy of a certain element is 412 kJ/mol (see Problem 6.108). However, when the atoms of this element are in the first excited state, the ionization energy is only 126 kJ/mol. Based on this information,

calculate the wavelength of light emitted in a transition from the first excited state to the ground state.

6.111 The electron configurations described in this chapter all refer to gaseous atoms in their ground states. An atom may absorb a quantum of energy and promote one of its electrons to a higher-energy orbital. When this happens, we say that the atom is in an excited state. The electron configurations of some excited atoms are given. Identify these atoms and write their ground-state configurations:
(a) $1s^1 2s^1$
(b) $1s^2 2s^2 2p^2 3d^1$
(c) $1s^2 2s^2 2p^6 4s^1$
(d) $[Ar]4s^1 3d^{10} 4p^4$
(e) $[Ne]3s^2 3p^4 3d^1$

6.112 All molecules undergo vibrational motions. Quantum mechanical treatment shows that the vibrational energy E_{vib} of a diatomic molecule such as HCl is given by

$$E_{vib} = \left(n + \frac{1}{2} \right) n\nu$$

where n is a quantum number ($n = 0, 1, 2, 3, \ldots$) and ν is the fundamental frequency of vibration. (a) Sketch the first three vibrational energy levels for HCl. (b) Calculate the energy required to excite an HCl molecule from the ground level to the first excited level. The fundamental frequency of vibration for HCl is 8.66×10^{13} s^{-1}. (c) The fact that the lowest vibrational energy in the ground level is not zero but equal to $\frac{1}{2}h\nu$ means that molecules will vibrate at all temperatures, including absolute zero. Use the Heisenberg uncertainty principle to justify this prediction. (*Hint:* Consider a molecule that is not vibrating and start by predicting the uncertainty in its momentum.)

6.113 When an electron makes a transition between energy levels of a hydrogen atom, there are no restrictions on the initial and final values of the principal quantum number n. However, there is a quantum mechanical rule that restricts the initial and final values of the orbital angular momentum ℓ. This is the *selection rule,* which states that $\Delta \ell = \pm 1$; that is, in a transition, the value of ℓ can only increase or decrease by 1. According to this rule, which of the following transitions are allowed: (a) $1s \longrightarrow 2s$, (b) $2p \longrightarrow 1s$, (c) $1s \longrightarrow 3d$, (d) $3d \longrightarrow 4f$, (e) $4d \longrightarrow 3s$? In view of this selection rule, explain why it is possible to observe the various emission series shown in Figure 6.11.

6.114 In 1996, physicists created an anti-atom of hydrogen. In such an atom, which is the antimatter equivalent of an ordinary atom, the electric charges of all the component particles are reversed. Thus the nucleus of an anti-atom is made of an antiproton, which has the same mass as a proton but bears a negative charge, while the electron is replaced by an anti-electron (also called a positron) with the same mass as an electron, but bearing a positive charge. Would you expect the energy levels, emission spectra, and atomic orbitals of an antihydrogen atom to be different from those of a hydrogen atom? What would happen if an anti-atom of hydrogen collided with a hydrogen atom?

6.115 An electron in an excited state in a hydrogen atom can return to the ground state in two different ways: (a) via a direct transition in which a photon of wavelength λ_1 is emitted and (b) via an intermediate excited state reached by the emission of a photon of wavelength λ_2. This intermediate excited state then decays to the ground state by emitting another photon of wavelength λ_3. Derive an equation that relates λ_1 to λ_2 and λ_3.

6.116 (a) An electron in the ground state of the hydrogen atom moves at an average speed of 5×10^6 m/s. If the speed is known to an uncertainty of 20 percent, what is the minimum uncertainty in its position? Given that the radius of the hydrogen atom in the ground state is 5.29×10^{-11} m, comment on your result. The mass of an electron is 9.1094×10^{-31} kg. (b) A 0.15-kg baseball thrown at 100 mph has a momentum of 6.7 kg · m/s. If the uncertainty in measuring the momentum is 1.0×10^{-7} of the momentum, calculate the uncertainty in the baseball's position.

6.117 The wave function for the 2s orbital in the hydrogen atom is

$$\psi_{2s} = \frac{1}{\sqrt{2a_0^3}}\left(1 - \frac{\rho}{2}\right) e^{-\rho/2}$$

where a_0 is the value of the radius of the first Bohr orbit, equal to 0.529 nm; ρ is $Z(r/a_0)$; and r is the distance from the nucleus in meters. Calculate the distance from the nucleus (in nm) of the node of the 2s wave function.

6.118 Calculate the energies needed to remove an electron from the $n = 1$ state and the $n = 5$ state in the Li^{2+} ion. What is the wavelength (in nm) of the emitted photon in a transition from $n = 5$ to $n = 1$? Solving Equation 6.4 for energy gives $\Delta E = R_\infty hc \left(\dfrac{1}{n_1^2} - \dfrac{1}{n_2^2}\right)$, where the Rydberg constant R_∞ for hydrogen-like atoms is 1.097×10^7 m^{-1} · Z^2, and Z is the atomic number.

6.119 According to Einstein's special theory of relativity, the mass of a moving particle, m_{moving}, is related to its mass at m_{rest}, by the following equation

$$m_{\text{moving}} = \frac{m_{\text{rest}}}{\sqrt{1 - (u/c)^2}}$$

where u and c are the speeds of the particle and light, respectively. (a) In particle accelerators, protons, electrons, and other charged particles are often accelerated to speeds close to the speed of light. Calculate the wavelength (in nm) of a proton moving at 50.0 percent the speed of light. The mass of a proton is 1.67×10^{-27} kg. (b) Calculate the mass of a 6.0×10^{-2} kg tennis ball moving at 63 m/s. Comment on your results.

6.120 The mathematical equation for studying the photoelectric effect is

$$h\nu = W + \tfrac{1}{2}m_e u^2$$

where ν is the frequency of light shining on the metal; W is the energy needed to remove an electron from the metal; and m_e and u are the mass and speed of the ejected electron, respectively. In an experiment, a student found that a maximum wavelength of 351 nm is needed to just dislodge electrons from a zinc metal surface. Calculate the velocity (in m/s) of an ejected electron when the student employed light with a wavelength of 313 nm.

6.121 Calculate the wavelength and frequency of an emitted gamma particle having the energy of 3.14×10^{11} J/mol.

6.122 The figure illustrates a series of transitions that occur in a hydrogen atom.

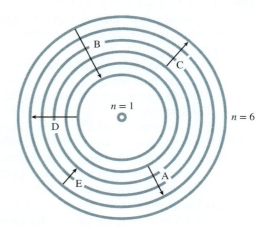

(a) Which transitions are absorptions and which are emissions?
(b) Rank the emissions in order of increasing energy.
(c) Rank the emissions in order of increasing wavelength of light emitted.

Engineering Problems

6.123 In a photoelectric experiment a student uses a light source whose frequency is greater than that needed to eject electrons from a certain metal. However, after continuously shining the light on the same area of the metal for a long period of time the student notices that the maximum kinetic energy of ejected electrons begins to decrease, even though the frequency of the light is held constant. How would you account for this behavior?

6.124 When a compound containing cesium ion is heated in a Bunsen burner flame, photons with an energy of 4.30×10^{-19} J are emitted. What color is the cesium flame?

6.125 In an electron microscope, electrons are accelerated by passing them through a voltage difference. The kinetic energy thus acquired by the electrons is equal to the voltage times the charge on the electron. Thus a voltage difference of 1 volt imparts a kinetic energy of 1.602×10^{-19} volt-coulomb or 1.602×10^{-19} J. Calculate the wavelength associated with electrons accelerated by 5.00×10^3 volts.

6.126 The sun is surrounded by a white circle of gaseous material called the corona, which becomes visible during a total eclipse of the sun. The temperature of the corona is in the millions of degrees Celsius, which is high enough to break up molecules and remove some or all of the electrons from atoms. One way astronomers have been able to estimate the temperature of the corona is by studying the emission lines of ions of certain elements. For example, the emission spectrum of Fe^{14+} ions has been recorded and analyzed. Knowing that it takes 3.5×10^4 kJ/mol to convert Fe^{13+} to Fe^{14+} estimate the temperature of the sun's corona. (*Hint:* The average kinetic energy of one mole of a gas is $\frac{3}{2}RT$.)

6.127 Scientists have found interstellar hydrogen atoms with quantum number n in the hundreds. Calculate the wavelength of light emitted when a hydrogen atom undergoes a transition from $n = 236$ to $n = 235$. In what region of the electromagnetic spectrum does this wavelength fall?

6.128 Only a fraction of the electric energy supplied to a tungsten lightbulb is converted to visible light. The rest of the energy shows up as infrared radiation (i.e., heat). A 75-W lightbulb converts 15.0 percent of the energy supplied to it into visible light (assume the wavelength to be 550 nm). How many photons are emitted by the lightbulb per second (1 W = 1 J/s)?

6.129 Certain sunglasses have small crystals of silver chloride (AgCl) incorporated in the lenses. When the lenses are exposed to light of the appropriate wavelength, the following reaction occurs:

$$AgCl \longrightarrow Ag + Cl$$

The Ag atoms formed produce a uniform grey color that reduces the glare. If ΔH for the preceding reaction is 248 kJ/mol, calculate the maximum wavelength of light that can induce this process.

6.130 A ruby laser produces radiation of wavelength 633 nm in pulses whose duration is 1.00×10^{-9} s. (a) If the laser produces 0.376 J of energy per pulse, how many photons are produced in each pulse? (b) Calculate the power (in watts) delivered by the laser per pulse (1 W = 1 J/s).

Biological Problems

6.131 The UV light that is responsible for tanning the skin falls in the 320- to 400-nm region. Calculate the total energy (in joules) absorbed by a person exposed to this radiation for 2.5 h, given that there are 2.0×10^{16} photons hitting Earth's surface per square centimeter per second over a 80-nm (320 to 400 nm) range and that the exposed body area is 0.45 m^2. Assume that only half of the radiation is absorbed and the other half is reflected by the body. (*Hint:* Use an average wavelength of 360 nm in calculating the energy of a photon.)

6.132 The retina of a human eye can detect light when radiant energy incident on it is at least 4.0×10^{-17} J. For light of 575-nm wavelength, how many photons does this correspond to?

6.133 Blackbody radiation is the term used to describe the dependence of the radiation energy emitted by an object on wavelength at a certain temperature. Planck proposed the quantum theory to account for the dependence. Shown in the figure is a plot of the radiation energy emitted by our sun versus wavelength. This curve is characteristic of objects at about 6000 K, which is the temperature at the surface of the sun. At a higher temperature, the curve has a similar shape but the maximum will shift to a shorter wavelength. (a) What does this curve reveal about two consequences of great biological significance on Earth? (b) How are astronomers able to determine the temperature at the surface of stars in general?

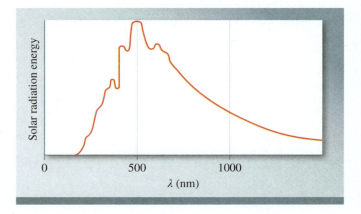

Multiconcept Problems

6.134   Photodissociation of water

$$H_2O(l) + h\nu \longrightarrow H_2(g) + \tfrac{1}{2}O_2(g)$$

has been suggested as a source of hydrogen. The ΔH_{rxn}° for the reaction, calculated from thermochemical data, is 285.8 kJ per mole of water decomposed. Calculate the maximum wavelength (in nm) that would provide the necessary energy. In principle, is it feasible to use sunlight as a source of energy for this process?

6.135   A microwave oven operating at 1.22×10^8 nm is used to heat 150 mL of water (roughly the volume of a tea cup) from 20°C to 100°C. Calculate the number of photons needed if 92.0 percent of microwave energy is converted to the thermal energy of water.

6.136   How many photons at 586 nm must be absorbed to melt 5.0×10^2 g of ice? On average, how many H_2O molecules does one photon convert from ice to water? (*Hint:* It takes 334 J to melt 1 g of ice at 0°C.)

6.137   A 368-g sample of water absorbs infrared radiation at 1.06×10^4 nm from a carbon dioxide laser. Suppose all the absorbed radiation is converted to heat. Calculate the number of photons at this wavelength required to raise the temperature of the water by 5.00°C.

Standardized-Exam Practice Problems

Physical and Biological Sciences

According to Wien's law, the wavelength of maximum intensity in blackbody radiation, λ_{max}, is inversely proportional to the temperature of the radiating body. Mathematically,

$$\lambda_{max} = \frac{b}{T}$$

where b is Wien's displacement constant (2.898×10^6 mm · K) and T is the temperature of the radiating body in kelvins. The sun, composed primarily of hydrogen, emits a continuous spectrum from the region known as the photosphere, with the most intense emission occurring at approximately 500 nm.

1. What is the approximate surface temperature of the sun?

    a) 60,000 K   b) 6000 K   c) 1500 K   d) 500 K

2. What is the frequency of the sun's λ_{max}?

    a) 6.0×10^{14} s^{-1} c) 2.5×10^2 s^{-1}
    b) 5.0×10^{14} s^{-1} d) 5.9×10^{-14} s^{-1}

3. What is the energy of a photon with wavelength λ_{max} for a blackbody at 5200 K?

    a) 5.4×10^{-14} J   c) 3.6×10^{-19} J
    b) 5.4×10^{14} J   d) 5.6×10^{-7} J

4. The visible region of the electromagnetic spectrum includes wavelengths from about 400 nm to about 700 nm. What is the minimum approximate temperature that would be required for the maximum intensity wavelength emitted by a blackbody to occur in the ultraviolet region of the spectrum?

    a) 71,000 K   b) 9100 K   c) 7300 K   d) 64,000 K

Answers to In-Chapter Materials

Answers to Practice Problems

6.1A 2.91×10^{10} s^{-1}. **6.1B** 1.86×10^{-4} m. **6.2A** 1.58×10^{-19} J. **6.2B** UV. **6.3A** (a) 9.43×10^{-19} J, (b) 1.18×10^{-25} J, (c) 1.96×10^{-19} J. **6.3B** (a) 1.05×10^{4} nm, (b) 2.1×10^{18}, (c) 1.2 eV. **6.4A** 103 nm. **6.4B** 7. **6.5A** 26 nm. **6.5B** 8.2×10^{-28} kg · m/s. **6.6A** (a) 3×10^{-34} m, (b) 3×10^{-32} m. **6.6B** (a) $\pm 2 \times 10^{-25}$ kg · m/s, (b) $\pm 1 \times 10^{2}$ m/s, (c) $\pm 2 \times 10^{5}$ m/s. **6.7A** Only 0, (b) $-2, -1, 0, +1, +2$. **6.7B** (a) 2, (b) 0, 1, 2, 3. **6.8A** $n = 3$, $\ell = 2$, $m_\ell = -2, -1, 0, +1, +2$. **6.8B** For a d orbital, $\ell = 2$, but when $n = 2$, ℓ cannot be 2. **6.9A** $1s^2 2s^2 2p^6 3s^2 3p^6 4s^2 3d^{10} 4p^6 5s^1$. **6.9B** $1s^2 2s^2 2p^6 3s^2 3p^6 4s^2 3d^{10} 4p^5$. **6.10A** [Rn]$7s^2$. **6.10B** At.

Answers to Checkpoints

6.1.1 c. **6.1.2** a. **6.1.3** b. **6.1.4** c. **6.2.1** e. **6.2.2** d. **6.2.3** c. **6.2.4** d. **6.3.1** b. **6.3.2** d. **6.3.3** c. **6.3.4** e. **6.4.1** c. **6.4.2** a. **6.5.1** b. **6.5.2** e. **6.6.1** a, b, d. **6.6.2** c. **6.6.3** c. **6.6.4** e. **6.7.1** b. **6.7.2** c. **6.7.3** b, e. **6.7.4** b. **6.8.1** b. **6.8.2** d. **6.8.3** b. **6.9.1** c. **6.9.2** c. **6.9.3** b, d. **6.9.4** a, b, d.

CHAPTER 7

Electron Configuration and the Periodic Table

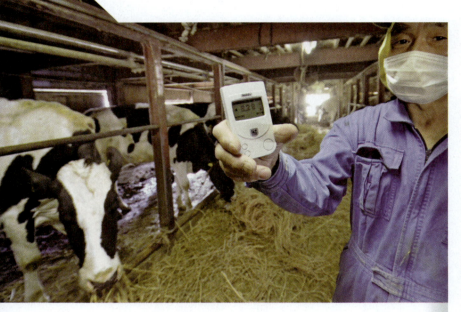

One of many cows *abandoned in the evacuation region surrounding the Fukushima Daiichi Nuclear Power Plant in Japan. Measurement of the levels of a Group 2 element in the teeth of these cattle has contributed to our understanding of the environmental impact of the Fukushima disaster in 2011.*

Yoshikazu Tsuno/AFP/Getty Images

In This Chapter, You Will Learn

Some of the chemical and physical properties of the elements and how these properties are related to each element's position in the periodic table.

Before You Begin, Review These Skills

- Atomic ions [⏮ Section 2.6]
- Electron configurations and the periodic table [⏮ Section 6.9]
- Chapter 6 Key Skills [⏮ pages 300–301]

What an Element's Position in the Periodic Table Can Tell Us

Elements in the same group of the periodic table tend to exhibit similar physical and chemical properties. The Group 2 metals, for example, magnesium, calcium, and strontium, share many similar properties. Strontium, for example, is sufficiently similar to calcium for it to be incorporated into the developing teeth of animals—*in place* of calcium. In fact, measurement of the levels of ^{90}Sr in human teeth has been used for decades to assess the range and extent of radioactive pollution in areas of nuclear testing or nuclear accident.

In 2011, a tsunami resulting from an earthquake caused catastrophic damage to the Fukushima Daiichi Nuclear Power Plant (FNPP) in Japan. This was the first significant incident at a nuclear power facility since the Chernobyl disaster in 1986. The damage to the facility allowed the release of significant quantities of radioactive material, including ^{90}Sr, into the surrounding area. Humans were evacuated from a large area surrounding FNPP, but thousands of cattle were abandoned there. Recent studies of the teeth of these cattle have enabled researchers to study the environmental and ecological impact of this event.

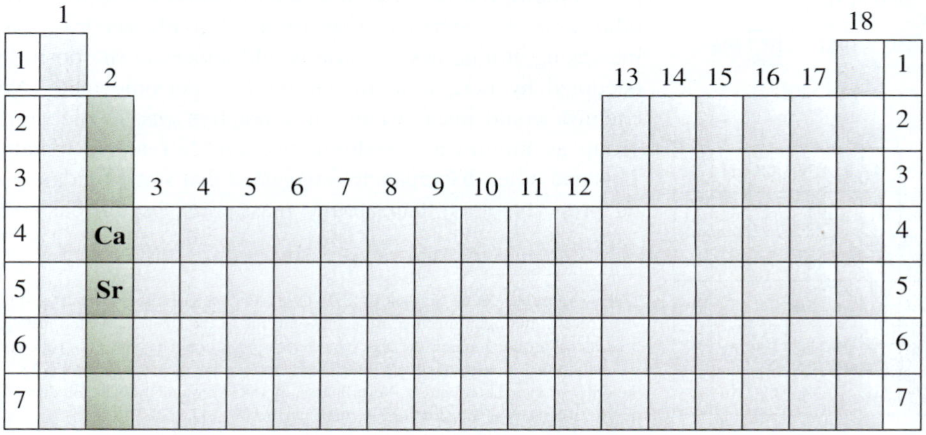

The physical and chemical properties of the Group 2 metals are quite similar. This is why they were placed in the same group in the periodic table in the first place. It turns out they belong in the same group because they have the same valence electron configuration.

At the end of this chapter, you will be able to answer a series of questions about the Group 2 metals and the ions that they form [⏭ Applying What You've Learned, page 350].

7.1 Development of the Periodic Table

In the nineteenth century, chemists had only a vague idea of atoms and molecules and did not yet know about electrons and protons. Nevertheless, they devised the periodic table using their knowledge of atomic masses. Accurate measurements of the atomic masses of many elements had already been made. Arranging elements according to their atomic masses in a periodic table seemed logical to those chemists, who believed that chemical behavior should somehow be related to atomic mass.

In 1864, the English chemist John Newlands[1] noticed that when the elements were arranged in order of atomic mass, every eighth element had similar properties. Newlands referred to this peculiar relationship as the *law of octaves*. However, this "law" turned out to be inadequate for elements beyond calcium, and Newlands's work was not accepted by the scientific community.

In 1869, the Russian chemist Dmitri Mendeleev[2] and the German chemist Lothar Meyer[3] independently proposed a much more extensive tabulation of the elements based on the regular, periodic recurrence of properties—a phenomenon known as *periodicity*.

Mendeleev's classification system was a great improvement over Newlands's for two reasons. First, it grouped the elements together more accurately, according to their properties. Second, and equally important, it made it possible to predict the properties of several elements that had not yet been discovered. For example, Mendeleev proposed the existence of an unknown element that he called eka-aluminum and predicted a number of its properties. (*Eka* is a Sanskrit word meaning "first"; thus, eka-aluminum would be the first element under aluminum in the same group.) When gallium was discovered 4 years later, its properties matched the predicted properties of eka-aluminum remarkably well:

	Eka-Aluminum (Ea)	Gallium (Ga)
Atomic mass	68 amu	69.9 amu
Melting point	Low	30.15°C
Density	5.9 g/cm^3	5.94 g/cm^3
Formula of oxide	Ea_2O_3	Ga_2O_3

Mendeleev's periodic table included 66 known elements. By 1900, some 30 more had been added to the list, filling in some of the empty spaces. Figure 7.1 gives the time period during which each element was discovered.

Although this periodic table was remarkably successful, the early versions had some inconsistencies that were impossible to overlook. For example, the atomic mass of argon (39.95 amu) is greater than that of potassium (39.10 amu), but argon comes before potassium in the periodic table. If elements were arranged solely according to increasing atomic mass, argon would appear in the position occupied by potassium in our modern periodic table. No chemist would place argon, an unreactive *gas,* in the same group as lithium and sodium, two *highly* reactive metals. This and other discrepancies suggested that some fundamental property other than atomic mass must be the basis of

Figure 7.1 Periodic table of elements classified by dates of discovery.

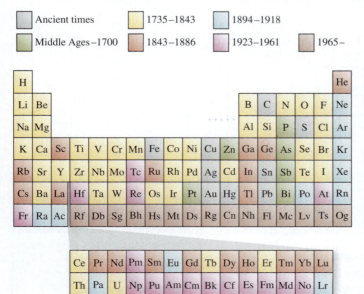

1. John Alexander Reina Newlands (1838–1898). English chemist. Newlands's work was a step in the right direction in the classification of the elements. Unfortunately, because of its shortcomings, he was subjected to much criticism, and even ridicule. At one meeting he was asked if he had ever examined the elements according to the order of their initial letters! Nevertheless, in 1887 Newlands was honored by the Royal Society of London for his contribution.

2. Dmitri Ivanovich Mendeleev (1836–1907). Russian chemist. His work on the periodic classification of elements is regarded by many as the most significant achievement in chemistry in the nineteenth century.

3. Julius Lothar Meyer (1830–1895). German chemist. In addition to his contribution to the periodic table, Meyer discovered the chemical affinity of hemoglobin for oxygen.

periodicity. The fundamental property turned out to be the number of protons in an atom's nucleus, something that could not have been known by Mendeleev and his contemporaries.

In 1913, a young English physicist, Henry Moseley,[4] discovered a correlation between what he called *atomic number* and the frequency of X rays generated by bombarding an element with high-energy electrons. Moseley noticed that, in general, the frequencies of X rays emitted from the elements increased with increasing atomic mass. Among the few exceptions he found were argon and potassium. Although argon has a greater atomic mass than potassium, the X-ray emission from potassium indicated that it has the greater atomic number. Ordering the periodic table using atomic number enabled scientists to make sense out of the discrepancies that had puzzled them earlier. Moseley concluded that the atomic number was equal to the number of protons in the nucleus and to the number of electrons in an atom.

Entries in modern periodic tables usually include an element's atomic number along with its symbol. Electron configurations of elements help to explain the periodic recurrence of physical and chemical properties. The importance and usefulness of the periodic table lie in the fact that we can use our understanding of the general properties and trends within a group or a period to predict with considerable accuracy the properties of any element, even though that element may be unfamiliar to us.

Sample Problem 7.1 shows how the periodic table can be used to predict similarities and differences in the properties of elements.

SAMPLE PROBLEM 7.1

What elements would you expect to exhibit properties most similar to those of chlorine?

Strategy Because elements in the same group tend to have similar properties, you should identify elements in the same group as chlorine.

Setup Chlorine is a member of Group 17.

Solution Fluorine, bromine, and iodine, the other nonmetals in Group 17, should have properties most similar to those of chlorine.

THINK ABOUT IT

Astatine (At) is also in Group 17. Astatine, though, is classified as a metalloid, and we have to be careful comparing nonmetals to *metalloids* (or to *metals*). As a metalloid, the properties of astatine should be less similar to those of chlorine than the other members of Group 17. (Actually, astatine is radioactive and very little is known about its properties.)

Practice Problem Ⓐ**TTEMPT** What element(s) would you expect to exhibit properties most similar to those of silicon (Si)?

Practice Problem Ⓑ**UILD** Arrange the following Group 15 elements in order of increasing similarity of properties to N: As, Bi, and P.

Practice Problem Ⓒ**ONCEPTUALIZE** Three different groups are highlighted in the periodic table shown here. Which of the three highlighted groups contains the largest number of elements with similar properties? Explain.

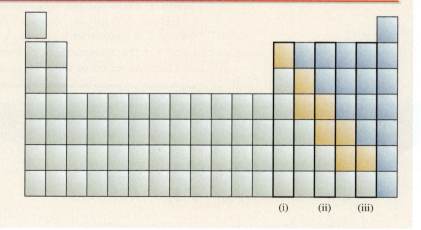

(i)   (ii)   (iii)

---

4. Henry Gwyn-Jeffreys Moseley (1887–1915). English physicist. Moseley discovered the relationship between X-ray spectra and atomic number. A lieutenant in the Royal Engineers, he was killed in action at the age of 28 during the British campaign in Gallipoli, Turkey.

Bringing Chemistry to Life

The Chemical Elements of Life

Of the 117 elements currently known, relatively few are essential to living systems. Actually, only six nonmetallic elements form the building blocks of cells: C, H, O, N, P, and S. They are the constituent elements of proteins, nucleic acids, and carbohydrates. Although the natural abundance of carbon is quite low (about 0.1 percent by mass of Earth's crust), it is present in nearly all biological molecules. Carbon is perhaps the most versatile element because it has the ability to form various types of chemical bonds. Carbon atoms can form bonds to each other, linking up to form an enormous variety of chains and ring structures.

The metals play several different roles in living systems. As cations (Na^+, K^+, Ca^{2+}, and Mg^{2+}), they serve to maintain the balance between intracellular and extracellular fluids, nerve transmissions, and other activities. They are also needed for protein functions. For example, the Fe^{2+} ion binds oxygen in hemoglobin molecules, and Cu^{2+}, Zn^{2+}, and Mg^{2+} ions are essential for enzyme activity. In addition, calcium in the form of $Ca_5(PO_4)_3(OH)$ and $Ca_3(PO_4)_2$ is an essential component of teeth and bones.

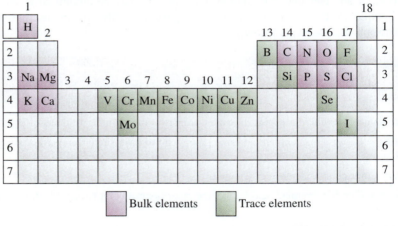

Bulk elements   Trace elements

The periodic table shown here highlights the essential elements in living systems. Of special interest are the *trace elements,* such as iron (Fe), copper (Cu), zinc (Zn), iodine (I), cobalt (Co), selenium (Se), and fluorine (F), which together make up about 0.1 percent of the human body's mass. Although the trace elements are present in very small amounts, they are crucial for our health. In many cases, however, their exact biological role is still not fully understood.

These elements are necessary for biological functions such as growth, the transport of oxygen for metabolism, and defense against disease. There is a delicate balance in the amounts of these elements in our bodies. Too much or too little over an extended period of time can lead to serious illness or even death.

CHECKPOINT – SECTION 7.1   Development of the Periodic Table

7.1.1   Which of the following elements would you expect to have chemical properties most similar to those of S?

a) P   b) Cl   c) Se   d) Na   e) Sr

7.1.2   Which of the following elements would you expect to have properties similar to those of Ba?

a) Sr   b) Rb   c) Na   d) K   e) B

7.1.3   The first synthesis of element 117 was reported in 2010. Based on its position in the periodic table, which of the following properties would you expect it to exhibit?

(i) gaseous at room temperature
(ii) unreactive
(iii) properties similar to At
(iv) properties similar to Ra

a) i and iv   c) iii only   e) i and iii

b) i, ii, and iv   d) iv only

7.2 The Modern Periodic Table

Figure 7.2 shows the modern periodic table together with the outermost ground-state electron configurations of the elements. (The electron configurations of the elements are also given in Figure 6.25.) Starting with hydrogen, the electronic subshells are filled in the order shown in Figure 6.23 [⏮ Section 6.8].

Classification of Elements

Based on the type of subshell containing the outermost electrons, the elements can be divided into categories—the main group elements, the noble gases, the transition elements (or transition metals), the lanthanides, and the actinides. The ***main group elements*** (also called the *representative elements*) are the elements in Groups 1, 2, and 13 through 17. With the exception of helium, each of the *noble gases* (the Group 18 elements) has a completely filled *p* subshell. The outermost electron configurations are $1s^2$ for helium and ns^2np^6 for the other noble gases, where *n* is the principal quantum number for the outermost shell.

Student Note: In this context, *outermost* electrons refers to those that are placed in orbitals *last* using the Aufbau principle [⏮ Section 6.8].

The transition metals are the elements in Groups 3 through 11. Transition metals either have incompletely filled *d* subshells or readily produce cations with incompletely filled *d* subshells. According to this definition, the elements of Group 12 are *not* transition metals. They typically form +2 ions, although they can also form +1 ions. In either case, the electron configuration includes a completed *d* subshell [⏭ Section 7.6].

	1	2	3	4	5	6	7	8	9	10	11	12	13	14	15	16	17	18	
1	1 **H** $1s^1$																	2 **He** $1s^2$	1
2	3 **Li** $2s^1$	4 **Be** $2s^2$											5 **B** $2s^22p^1$	6 **C** $2s^22p^2$	7 **N** $2s^22p^3$	8 **O** $2s^22p^4$	9 **F** $2s^22p^5$	10 **Ne** $2s^22p^6$	2
3	11 **Na** $3s^1$	12 **Mg** $3s^2$											13 **Al** $3s^23p^1$	14 **Si** $3s^23p^2$	15 **P** $3s^23p^3$	16 **S** $3s^23p^4$	17 **Cl** $3s^23p^5$	18 **Ar** $3s^23p^6$	3
4	19 **K** $4s^1$	20 **Ca** $4s^2$	21 **Sc** $4s^23d^1$	22 **Ti** $4s^23d^2$	23 **V** $4s^23d^3$	24 **Cr** $4s^13d^5$	25 **Mn** $4s^23d^5$	26 **Fe** $4s^23d^6$	27 **Co** $4s^23d^7$	28 **Ni** $4s^23d^8$	29 **Cu** $4s^13d^{10}$	30 **Zn** $4s^23d^{10}$	31 **Ga** $4s^24p^1$	32 **Ge** $4s^24p^2$	33 **As** $4s^24p^3$	34 **Se** $4s^24p^4$	35 **Br** $4s^24p^5$	36 **Kr** $4s^24p^6$	4
5	37 **Rb** $5s^1$	38 **Sr** $5s^2$	39 **Y** $5s^24d^1$	40 **Zr** $5s^24d^2$	41 **Nb** $5s^14d^4$	42 **Mo** $5s^14d^5$	43 **Tc** $5s^24d^5$	44 **Ru** $5s^14d^7$	45 **Rh** $5s^14d^8$	46 **Pd** $4d^{10}$	47 **Ag** $5s^14d^{10}$	48 **Cd** $5s^24d^{10}$	49 **In** $5s^25p^1$	50 **Sn** $5s^25p^2$	51 **Sb** $5s^25p^3$	52 **Te** $5s^25p^4$	53 **I** $5s^25p^5$	54 **Xe** $5s^25p^6$	5
6	55 **Cs** $6s^1$	56 **Ba** $6s^2$	57 **La** $6s^25d^1$	72 **Hf** $6s^25d^2$	73 **Ta** $6s^25d^3$	74 **W** $6s^25d^4$	75 **Re** $6s^25d^5$	76 **Os** $6s^25d^6$	77 **Ir** $6s^25d^7$	78 **Pt** $6s^15d^9$	79 **Au** $6s^15d^{10}$	80 **Hg** $6s^25d^{10}$	81 **Tl** $6s^26p^1$	82 **Pb** $6s^26p^2$	83 **Bi** $6s^26p^3$	84 **Po** $6s^26p^4$	85 **At** $6s^26p^5$	86 **Rn** $6s^26p^6$	6
7	87 **Fr** $7s^1$	88 **Ra** $7s^2$	89 **Ac** $7s^26d^1$	104 **Rf** $7s^26d^2$	105 **Db** $7s^26d^3$	106 **Sg** $7s^26d^4$	107 **Bh** $7s^26d^5$	108 **Hs** $7s^26d^6$	109 **Mt** $7s^26d^7$	110 **Ds** $7s^26d^8$	111 **Rg** $7s^26d^9$	112 **Cn** $7s^26d^{10}$	113 **Nh** $7s^27p^1$	114 **Fl** $7s^27p^2$	115 **Mc** $7s^27p^3$	116 **Lv** $7s^27p^4$	117 **Ts** $7s^27p^5$	118 **Og** $7s^27p^6$	7

58 **Ce** $6s^24f^15d^1$	59 **Pr** $6s^24f^3$	60 **Nd** $6s^24f^4$	61 **Pm** $6s^24f^5$	62 **Sm** $6s^24f^6$	63 **Eu** $6s^24f^7$	64 **Gd** $6s^24f^75d^1$	65 **Tb** $6s^24f^9$	66 **Dy** $6s^24f^{10}$	67 **Ho** $6s^24f^{11}$	68 **Er** $6s^24f^{12}$	69 **Tm** $6s^24f^{13}$	70 **Yb** $6s^24f^{14}$	71 **Lu** $6s^24f^{14}5d^1$
90 **Th** $7s^26d^2$	91 **Pa** $7s^25f^26d^1$	92 **U** $7s^25f^36d^1$	93 **Np** $7s^25f^46d^1$	94 **Pu** $7s^25f^6$	95 **Am** $7s^25f^7$	96 **Cm** $7s^25f^76d^1$	97 **Bk** $7s^25f^9$	98 **Cf** $7s^25f^{10}$	99 **Es** $7s^25f^{11}$	100 **Fm** $7s^25f^{12}$	101 **Md** $7s^25f^{13}$	102 **No** $7s^25f^{14}$	103 **Lr** $7s^25f^{14}6d^1$

Figure 7.2 Valence electron configurations of the elements. For simplicity, the filled *f* subshells are not shown in elements 72 through 86 and 104 through 118.

Figure 7.3 Periodic table with color-coding of main group elements, noble gases, transition metals, Group 12 metals, lanthanides, and actinides.

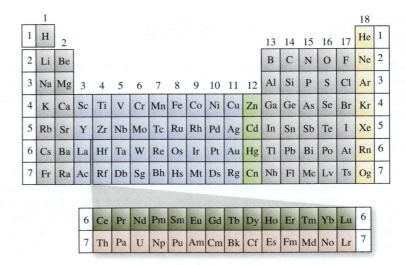

Zinc, cadmium, and mercury are *d*-block elements, though, so they generally are included in the discussion of transition metals.

The lanthanides and actinides are sometimes called *f*-block transition elements because they have incompletely filled *f* subshells. Figure 7.3 distinguishes the groups of elements discussed here.

There is a distinct pattern to the electron configurations of the elements in a particular group. See, for example, the electron configurations of Groups 1 and 2 in Table 7.1. Each member of Group 1 has a noble gas core plus one additional electron, giving each alkali metal the general electron configuration of [noble gas]ns^1. Similarly, the Group 2 alkaline earth metals have a noble gas core and an outer electron configuration of ns^2.

The outermost electrons of an atom are called **valence electrons,** which are the ones involved in the formation of chemical bonds between atoms. The similarity of the valence electron configurations (i.e., they have the same number and type of valence electrons) is what makes the elements in the same group resemble one another chemically. This observation holds true for the other main group elements as well. For instance, the halogens (Group 17) all have outer electron configurations of ns^2np^5, and they have similar properties.

In predicting properties for Groups 13 through 17, we must take into account that each of these groups contains elements on both sides of the line that divides metals and nonmetals. For example, the elements in Group 14 all have the same outer electron configuration, ns^2np^2, but there is considerable variation in chemical properties among these elements because carbon is a nonmetal, silicon and germanium are metalloids, and tin and lead are metals.

As a group, the noble gases behave very similarly. The noble gases are generally unreactive because they all have completely filled outer *ns* and *np* subshells, a condition that imparts unusual stability.

Student Note: Although hydrogen's electron configuration is 1s^1 [◄◄ Section 6.9, Figure 6.25], it is a nonmetal and is not really a member of Group 1.

TABLE 7.1	Electron Configurations of Group 1 and Group 2 Elements		
Group 1		**Group 2**	
Li	$[He]2s^1$	Be	$[He]2s^2$
Na	$[Ne]3s^1$	Mg	$[Ne]3s^2$
K	$[Ar]4s^1$	Ca	$[Ar]4s^2$
Rb	$[Kr]5s^1$	Sr	$[Kr]5s^2$
Cs	$[Xe]6s^1$	Ba	$[Xe]6s^2$
Fr	$[Rn]7s^1$	Ra	$[Rn]7s^2$

Although the outer electron configuration of the transition metals is not always the same within a group and there is often no regular pattern in the way the electron configuration changes from one metal to the next in the same period, all transition metals share many characteristics (multiple oxidation states, richly colored compounds, magnetic properties, and so on) that set them apart from other elements. These properties are similar because all these metals have incompletely filled d subshells. Likewise, the lanthanide and actinide elements resemble one another because they have incompletely filled f subshells.

Sample Problem 7.2 shows how to determine the electron configuration from the number of electrons in an atom.

SAMPLE PROBLEM 7.2

Without using a periodic table, give the ground-state electron configuration and block designation (s-, p-, d-, or f-block) of an atom with (a) 17 electrons, (b) 37 electrons, and (c) 22 electrons. Classify each atom as a main group element or transition metal.

Strategy Use the Aufbau principle discussed in Section 6.8. Start writing each electron configuration with principal quantum number $n = 1$, and then continue to assign electrons to orbitals in the order presented in Figure 6.23 until all the electrons have been accounted for.

Setup According to Figure 6.23, orbitals fill in the following order: $1s$, $2s$, $2p$, $3s$, $3p$, $4s$, $3d$, $4p$, $5s$, $4d$, $5p$, $6s$, and so on. Recall that an s subshell contains one orbital, a p subshell contains three orbitals, and a d subshell contains five orbitals. Remember, too, that each orbital can accommodate a maximum of two electrons. The block designation of an element corresponds to the type of subshell occupied by the last electrons added to the configuration according to the Aufbau principle.

Solution

(a) $1s^2 2s^2 2p^6 3s^2 3p^5$, p-block, main group

(b) $1s^2 2s^2 2p^6 3s^2 3p^6 4s^2 3d^{10} 4p^6 5s^1$, s-block, main group

(c) $1s^2 2s^2 2p^6 3s^2 3p^6 4s^2 3d^2$, d-block, transition metal

THINK ABOUT IT

Consult Figure 6.25 to confirm your answers.

Practice Problem **A**TTEMPT Without using a periodic table, give the ground-state electron configuration and block designation (s-, p-, d-, or f-block) of an atom with (a) 15 electrons, (b) 20 electrons, and (c) 35 electrons.

Practice Problem **B**UILD Identify the elements represented by (a) $1s^2 2s^2 2p^6 3s^2 3p^1$, (b) $1s^2 2s^2 2p^6 3s^2 3p^6 4s^2 3d^{10}$, and (c) $1s^2 2s^2 2p^6 3s^2 3p^6 4s^2 3d^{10} 4p^6 5s^2$.

Practice Problem **C**ONCEPTUALIZE Determine the *total* number of electrons and the number of *valence* electrons for each of the indicated elements.

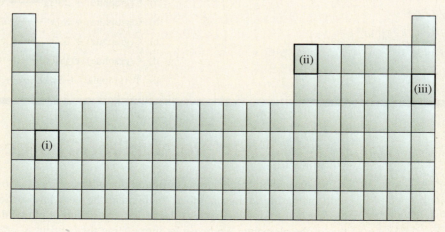

Representing Free Elements in Chemical Equations

Having classified the elements according to their ground-state electron configurations, we can now learn how chemists represent elements in chemical equations.

Metals   Because metals typically do not exist in discrete molecular units but rather in complex, three-dimensional networks of atoms, we always use their empirical formulas in chemical equations. The empirical formulas are the same as the symbols that represent the elements. For example, the empirical formula for iron is Fe, the same as the symbol for the element.

Nonmetals   There is no single rule regarding the representation of nonmetals in chemical equations. Carbon, for example, exists in several allotropic forms. Regardless of the allotrope, we use its empirical formula C to represent elemental carbon in chemical equations. Often the symbol C will be followed by the specific allotrope in parentheses as in the equation representing the conversion of graphite to diamond, two of carbon's allotropic forms:

$$C(\text{graphite}) \longrightarrow C(\text{diamond})$$

> **Student Note:** Recall that allotropes are different forms of the same element [◄◄ Section 2.6].

For nonmetals that exist as polyatomic molecules, we generally use the molecular formula in equations: H_2, N_2, O_2, F_2, Cl_2, Br_2, I_2, and P_4, for example. In the case of sulfur, however, we usually use the empirical formula S rather than the molecular formula S_8. Thus, instead of writing the equation for the combustion of sulfur as

$$S_8(s) + 8O_2(g) \longrightarrow 8SO_2(g)$$

we usually write

$$S(s) + O_2(g) \longrightarrow SO_2(g)$$

although, technically, both ways are correct.

Noble Gases   All the noble gases exist as isolated atoms, so we use their symbols: He, Ne, Ar, Kr, Xe, and Rn.

Metalloids   The metalloids, like the metals, all have complex three-dimensional networks, so we also represent them with their empirical formulas—that is, their symbols: B, Si, Ge, and so on.

CHECKPOINT – SECTION 7.2   The Modern Periodic Table

7.2.1   Which electron configuration is correct for a germanium (Ge) atom in the ground state?

a) $1s^2 2s^2 2p^6 3s^2 3p^6 4s^2 4p^2$

b) $1s^2 2s^2 2p^6 3s^2 3p^6 4s^2 3d^{10} 4p^2$

c) $1s^2 2s^2 2p^6 3s^2 3p^2$

d) $1s^2 2s^2 2p^6 3s^2 3p^6 4s^2 4p^2 4d^{10}$

e) $1s^2 2s^2 2p^6 3s^2 3p^2 3d^{10}$

7.2.2   Which of the following equations correctly represent the chemical reaction in which graphite combines with sulfur to form carbon disulfide gas [$CS_2(g)$]? (Select all that apply.)

a) $C(\text{graphite}) + 2S(s) \longrightarrow CS_2(g)$

b) $C(\text{graphite}) + S_2(s) \longrightarrow CS_2(g)$

c) $C(\text{graphite}) + S_8(s) \longrightarrow CS_8(g)$

d) $C(\text{graphite}) + \frac{1}{4}S_8(s) \longrightarrow CS_2(g)$

e) $4C(\text{graphite}) + S_8(s) \longrightarrow 4CS_2(g)$

7.3   Effective Nuclear Charge

As we have seen, the electron configurations of the elements show a periodic variation with increasing atomic number. In this and the next few sections, we examine how electron configuration explains the periodic variation of physical and chemical properties of the elements. We begin by introducing the concept of *effective nuclear charge*.

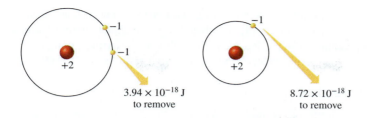

Figure 7.4 Removal of the first electron in He requires less energy than removal of the second electron because of shielding.

Nuclear charge (Z) is simply the number of protons in the nucleus of an atom. *Effective nuclear charge* (Z_{eff}) is the actual magnitude of positive charge that is "experienced" by an electron in the atom. The only atom in which the nuclear charge and effective nuclear charge are the same is hydrogen, which has only one electron. In all other atoms, the electrons are simultaneously attracted to the nucleus and repelled by one another. This results in a phenomenon known as *shielding*. An electron in a many-electron atom is partially shielded from the positive charge of the nucleus by the other electrons in the atom.

Student Note: *Shielding* is also known as *screening*.

One way to illustrate how electrons in an atom shield one another is to consider the amounts of energy required to remove the two electrons from a helium atom, shown in Figure 7.4. Experiments show that it takes 3.94×10^{-18} J to remove the first electron but 8.72×10^{-18} J to remove the second one. There is no shielding once the first electron is removed, so the second electron feels the full effect of the +2 nuclear charge and is more difficult to remove.

Although all the electrons in an atom shield one another to some extent, those that are most effective at shielding are the *core* electrons. As a result, the value of Z_{eff} increases steadily from left to right across a period of the periodic table because the number of core electrons remains the same (only the number of protons, Z, and the number of *valence* electrons increases).

Student Note: Core electrons are those in the completed inner shells.

As we move to the right across period 2, the nuclear charge increases by 1 with each new element, but the *effective* nuclear charge increases only by an average of 0.64. (If the valence electrons did *not* shield one another, the effective nuclear charge would also increase by 1 each time a proton was added to the nucleus.)

	Li	Be	B	C	N	O	F
Z	3	4	5	6	7	8	9
Z_{eff} (felt by valence electrons)	1.28	1.91	2.42	3.14	3.83	4.45	5.10

In general, the effective nuclear charge is given by

$$Z_{eff} = Z - \sigma \qquad \textbf{Equation 7.1}$$

Student Hot Spot

Student data indicate you may struggle with effective nuclear charge. Access the eBook to view additional Learning Resources on this topic.

where σ is the shielding constant. The shielding constant is greater than zero but smaller than Z.

The change in Z_{eff} as we move from the top of a group to the bottom is generally less significant than the change as we move across a period. Although each step down a group represents a large increase in the nuclear charge, there is also an additional shell of core electrons to shield the valence electrons from the nucleus. Consequently, the *effective* nuclear charge changes less than the nuclear charge as we move down a column of the periodic table.

7.4 Periodic Trends in Properties of Elements

Several physical and chemical properties of the elements depend on effective nuclear charge. To understand the trends in these properties, it is helpful to visualize the electrons of an atom in *shells*. Recall that the value of the principal quantum number (n) increases as the distance from the nucleus increases [◄◄ Section 6.7]. If we take this statement

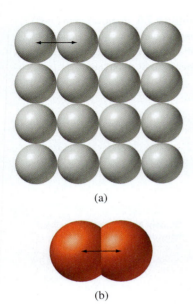

Figure 7.5 (a) Atomic radius in metals is defined as half the distance between adjacent metal atoms. (b) Atomic radius in nonmetals is defined as half the distance between bonded identical atoms in a molecule.

literally, and picture all the electrons in a shell at the same distance from the nucleus, the result is a sphere of uniformly distributed negative charge, with its distance from the nucleus depending on the value of n. With this as a starting point, we examine the periodic trends in atomic radius, ionization energy, and electron affinity.

Atomic Radius

Intuitively, we think of the *atomic radius* as the distance between the nucleus of an atom and its valence shell (i.e., the outermost shell that is occupied by one or more electrons), because we usually envision atoms as spheres with discrete boundaries. According to the quantum mechanical model of the atom, though, there is no specific distance from the nucleus beyond which an electron may not be found [◄◄ Section 6.7]. Therefore, the atomic radius requires a specific definition.

There are two ways in which the atomic radius is commonly defined. One is the *metallic radius,* which is half the distance between the nuclei of two adjacent, identical metal atoms [Figure 7.5(a)]. The other is the *covalent radius,* which is half the distance between adjacent, identical nuclei in a molecule [Figure 7.5(b)].

Figure 7.6 shows the atomic radii of the main group elements according to their positions in the periodic table. There are two distinct trends. The atomic radius *decreases* as we move from left to right across a period and *increases* from top to bottom as we move down within a group. The increase down a group is fairly easily explained. As we step down a column, the outermost occupied shell has an ever-increasing value of n, so it lies farther from the nucleus, making the radius bigger.

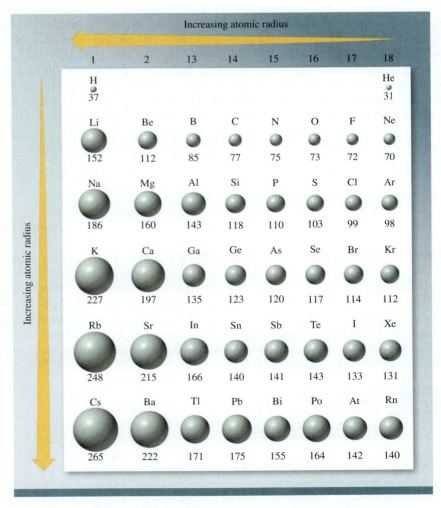

Figure 7.6 Atomic radii of main group elements.

Animation
Periodic table—atomic radius.

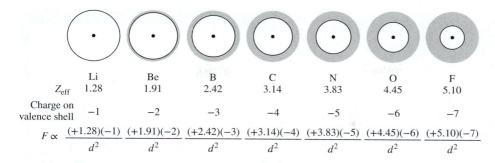

	Li	Be	B	C	N	O	F
Z_{eff}	1.28	1.91	2.42	3.14	3.83	4.45	5.10
Charge on valence shell	−1	−2	−3	−4	−5	−6	−7

$$F \propto \quad \frac{(+1.28)(-1)}{d^2} \quad \frac{(+1.91)(-2)}{d^2} \quad \frac{(+2.42)(-3)}{d^2} \quad \frac{(+3.14)(-4)}{d^2} \quad \frac{(+3.83)(-5)}{d^2} \quad \frac{(+4.45)(-6)}{d^2} \quad \frac{(+5.10)(-7)}{d^2}$$

Figure 7.7 Atomic radius decreases from left to right across a period because of the increased electrostatic attraction between the effective nuclear charge and the charge on the valence shell. The white circle shows the atomic size in each case. The comparison of attractive forces between the nuclei and valence shells is done using Coulomb's law [▶▶ Explaining Periodic Trends].

Now let's try to understand the decrease in radius from left to right across a period. Although this trend may at first seem counterintuitive, given that the number of valence electrons is increasing with each new element, consider the shell model in which all the electrons in a shell form a uniform sphere of negative charge around the nucleus at a distance specified by the value of n. As we move from left to right across a period, the effective nuclear charge increases and each step to the right adds another electron to the valence shell. Coulomb's law dictates that there will be a more powerful attraction between the nucleus and the valence shell when the magnitudes of both charges increase. The result is that as we step across a period the valence shell is drawn closer to the nucleus, making the atomic radius smaller. Figure 7.7 shows how the effective nuclear charge, charge on the valence shell, and atomic radius vary across period 2. We can picture the valence shells in all the atoms as being initially at the same distance (determined by n) from the nuclei, but being pulled closer by a larger attractive force resulting from increases in both Z_{eff} and the number of valence electrons.

Sample Problem 7.3 shows how to use these trends to compare the atomic radii of different elements.

Student Note: Although the overall trend in atomic size for transition elements is also to decrease from left to right and increase from top to bottom, the observed radii do not vary in as regular a way as do the main group elements.

SAMPLE PROBLEM 7.3

Referring only to a periodic table, arrange the elements P, S, and O in order of increasing atomic radius.

Strategy Use the left-to-right (decreasing) and top-to-bottom (increasing) trends to compare the atomic radii of two of the three elements at a time.

Setup Sulfur is to the right of phosphorus in the third row, so sulfur should be smaller than phosphorus. Oxygen is above sulfur in Group 16, so oxygen should be smaller than sulfur.

Solution O < S < P

THINK ABOUT IT

Consult Figure 7.6 to confirm the order. Note that there are circumstances under which the trends alone will be insufficient to compare the radii of two elements. Using only a periodic table, for example, it would not be possible to determine that chlorine ($r = 99$ pm) has a larger radius than oxygen ($r = 73$ pm).

Practice Problem ATTEMPT Referring only to a periodic table, arrange the elements Ge, Se, and F in order of increasing atomic radius.

Practice Problem BUILD For which of the following pairs of elements can the atomic radii *not* be compared using the periodic table alone: P and Se, Se and Cl, or P and O?

Practice Problem CONCEPTUALIZE Based on size and using only a periodic table, identify the colored spheres as Al, B, Mg, and Sr.

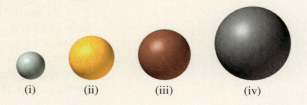

(i)   (ii)   (iii)   (iv)

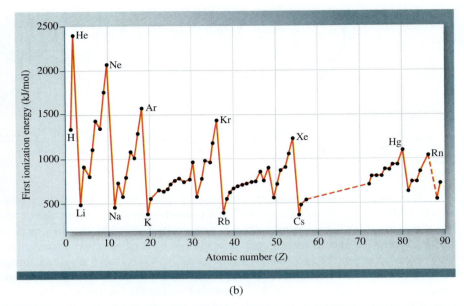

Figure 7.8 (a) First ionization energies (in kJ/mol) of the main group elements. (b) First ionization energy as a function of atomic number.

Ionization Energy

Ionization energy (IE) is the minimum energy required to remove an electron from an atom in the gas phase. Typically, we express ionization energy in kJ/mol, the number of kilojoules required to remove a mole of electrons from a mole of gaseous atoms. Sodium, for example, has an ionization energy of 495.8 kJ/mol, meaning that the energy input required to drive the process

$$Na(g) \longrightarrow Na^+(g) + e^-$$

is 495.8 kJ/mol. Specifically, this is the *first* ionization energy of sodium, $IE_1(Na)$, which corresponds to the removal of the most loosely held electron. Figure 7.8(a) shows the first ionization energies of the main group elements according to their positions in the periodic table. Figure 7.8(b) shows a graph of IE_1 as a function of atomic number.

In general, as effective nuclear charge increases, ionization energy also increases. Thus, IE_1 increases from left to right across a period. Despite this trend, the graph in Figure 7.8(b) shows that IE_1 for a Group 13 element is smaller than that for the corresponding Group 2 element. Likewise, IE_1 for a Group 16 element is smaller than that for the corresponding Group 15 element. Both of these *interruptions* of the upward trend in IE_1 can be explained by using electron configuration.

Recall that the energy of an electron in a many-electron system depends not only on the principal quantum number (n), but also on the angular momentum quantum number (ℓ) [◄◄ Section 6.8, Figure 6.23]. Within a given shell, electrons with the higher value of ℓ have a higher energy (are less tightly held by the nucleus) and are therefore *easier* to remove. Figure 7.9(a) shows the relative energies of an *s* subshell ($\ell = 0$) and a *p* subshell ($\ell = 0$). Ionization of an element in Group 2 requires the removal of an electron from an *s* orbital, whereas ionization of an element in Group 13 requires the removal of an electron from a *p* orbital; therefore, the element in Group 13 has a lower ionization energy than the element in Group 2.

As for the decrease in ionization energy in elements of Group 16 compared to those in Group 15, both ionizations involve the removal of a *p* electron, but the ionization of an atom in Group 16 involves the removal of a *paired* electron. The repulsive force between two electrons in the same orbital makes it easier to remove one of them, making the ionization energy for the Group 16 element actually lower than that for the Group 15 element. [See Figure 7.9(b).]

Student Note: As with atomic radius, ionization energy changes in a similar but somewhat less regular way among the transition elements.

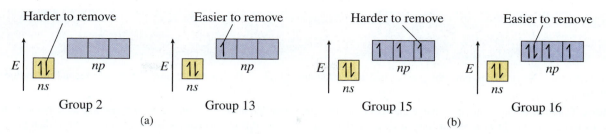

Figure 7.9 (a) It is harder to remove an electron from an *s* orbital than it is to remove an electron from a *p* orbital with the same principal quantum number. (b) Within a *p* subshell, it is easier to remove an electron from a doubly occupied orbital than from a singly occupied orbital.

The first ionization energy IE_1 decreases as we move from top to bottom within a group due to the increasing atomic radius. Although the effective nuclear charge does not change significantly as we step down a group, the atomic radius increases because the value of *n* for the valence shell increases. According to Coulomb's law, the attractive force between a valence electron and the effective nuclear charge gets *weaker* as the distance between them increases. This makes it easier to remove an electron, and so IE_1 decreases.

It is possible to remove additional electrons in subsequent ionizations, giving IE_2, IE_3, and so on. The second and third ionizations of sodium, for example, can be represented, respectively, as

$$Na^+(g) \longrightarrow Na^{2+}(g) + e^- \qquad \text{and} \qquad Na^{2+}(g) \longrightarrow Na^{3+}(g) + e^-$$

However, the removal of successive electrons requires ever-increasing amounts of energy because it is harder to remove an electron from a cation than from an atom (and it gets even harder as the charge on the cation increases). Table 7.2 lists the ionization energies of the elements in period 2 and of sodium. These data show that it takes much more energy to remove core electrons than to remove valence electrons. There are two reasons for this. First, core electrons are closer to the nucleus, and second, core electrons experience a greater effective nuclear charge because there are fewer filled shells shielding them from the nucleus. Both of these factors contribute to a greater attractive force between the electrons and the nucleus, which must be overcome to remove the electrons.

Sample Problem 7.4 shows how to use these trends to compare first ionization energies, and subsequent ionization energies, of specific atoms.

TABLE 7.2	Ionization Energies (in kJ/mol) for Elements 3 Through 11*									
	Z	IE_1	IE_2	IE_3	IE_4	IE_5	IE_6	IE_7	IE_8	IE_9	IE_{10}
Li	3	520	7,298	11,815							
Be	4	899	1,757	14,848	21,007						
B	5	800	2,427	3,660	25,026	32,827					
C	6	1,086	2,353	4,621	6,223	37,831	47,277				
N	7	1,402	2,856	4,578	7,475	9,445	53,267	64,360			
O	8	1,314	3,388	5,301	7,469	10,990	13,327	71,330	84,078		
F	9	1,681	3,374	6,050	8,408	11,023	15,164	17,868	92,038	106,434	
Ne	10	2,080	3,952	6,122	9,371	12,177	15,238	19,999	23,069	115,380	131,432
Na	11	496	4,562	6,910	9,543	13,354	16,613	20,117	25,496	28,932	141,362

*Cells shaded with blue represent the removal of core electrons.

SAMPLE PROBLEM   7.4

Would you expect Na or Mg to have the greater first ionization energy (IE_1)? Which should have the greater second ionization energy (IE_2)?

Strategy Consider effective nuclear charge and electron configuration to compare the ionization energies. Effective nuclear charge increases from left to right in a period (thus increasing IE), and it is more difficult to remove a paired core electron than an unpaired valence electron.

Setup Na is in Group 1, and Mg is beside it in Group 2. Na has one valence electron, and Mg has two valence electrons.

Solution $IE_1(Mg) > IE_1(Na)$ because Mg is to the right of Na in the periodic table (i.e., Mg has the greater effective nuclear charge, so it is more difficult to remove its electron). $IE_2(Na) > IE_2(Mg)$ because the second ionization of Mg removes a valence electron, whereas the second ionization of Na removes a core electron.

THINK ABOUT IT

The first ionization energies of Na and Mg are 496 and 738 kJ/mol, respectively. The second ionization energies of Na and Mg are 4562 and 1451 kJ/mol, respectively.

Practice Problem **A**TTEMPT   Which element, Mg or Al, will have the higher first ionization energy and which will have the higher third ionization energy?

Practice Problem **B**UILD   Explain why Rb has a lower IE_1 than Sr, but Sr has a lower IE_2 than Rb.

Practice Problem **C**ONCEPTUALIZE   Imagine an arrangement of atomic orbitals in an alternate universe, in which the s subshell contains *two* orbitals instead of one, and the p subshell contains *four* orbitals rather than three. Under these circumstances, in which groups would you expect the anomalously low first ionization energies to occur?

Electron Affinity

Student Note: Some books define electron affinity as ΔH, rather than the negative of ΔH for the process of adding an electron. This simply changes the sign of *EA* relative to what we show here.

Electron affinity (EA) is the energy released (the negative of the enthalpy change ΔH) when an atom in the gas phase accepts an electron. Consider the process in which a gaseous chlorine atom accepts an electron:

$$Cl(g) + e^- \longrightarrow Cl^-(g) \qquad \Delta H = -349.0 \text{ kJ/mol}$$

A negative value of ΔH indicates an exothermic process [◄◄ Section 5.3], so 349.0 kJ/mol of energy is released (the definition of electron affinity) when a mole of gaseous chlorine atoms accepts a mole of electrons. A positive electron affinity indicates a process that is energetically favorable. In general, the larger and more positive the *EA* value, the more favorable the process and the more apt it is to occur. Figure 7.10 shows electron affinities for the main group elements.

Student Note: Although *IE* and *EA* both increase from left to right across a period, an increase in *IE* means that it is *less* likely that an electron will be *removed* from an atom. An increase in *EA*, on the other hand, means that it is *more* likely that an electron will be *accepted* by an atom.

Like ionization energy, electron affinity increases from left to right across a period. This trend in *EA* is due to the increase in effective nuclear charge from left to right (i.e., it becomes progressively easier to add a negatively charged electron as the positive charge of the element's nucleus increases). There are also periodic interruptions of the upward trend of *EA* from left to right, similar to those observed for IE_1, although they do *not* occur for the same elements. For example, the *EA* of a Group 2 element is lower than that for the corresponding Group 1 element, and the *EA* of a Group 15 element is lower than that for the corresponding Group 14 element. These exceptions to the trend are due to the electron configurations of the elements involved.

It is harder to add an electron to a Group 2 element (ns^2) than to the Group 1 element (ns^1) in the same period because the electron added to the Group 2 element is placed in an orbital of higher energy (a p orbital versus an s orbital). Likewise, it is harder to add an electron to a Group 15 element (ns^2np^3) than to the corresponding Group 14 element (ns^2np^2) because the electron added to the Group 15 element must

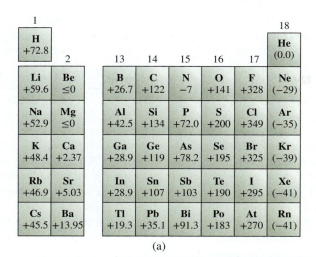

1							18
H +72.8							**He** (0.0)
	2	13	14	15	16	17	
Li +59.6	**Be** ≤0	**B** +26.7	**C** +122	**N** −7	**O** +141	**F** +328	**Ne** (−29)
Na +52.9	**Mg** ≤0	**Al** +42.5	**Si** +134	**P** +72.0	**S** +200	**Cl** +349	**Ar** (−35)
K +48.4	**Ca** +2.37	**Ga** +28.9	**Ge** +119	**As** +78.2	**Se** +195	**Br** +325	**Kr** (−39)
Rb +46.9	**Sr** +5.03	**In** +28.9	**Sn** +107	**Sb** +103	**Te** +190	**I** +295	**Xe** (−41)
Cs +45.5	**Ba** +13.95	**Tl** +19.3	**Pb** +35.1	**Bi** +91.3	**Po** +183	**At** +270	**Rn** (−41)

(a)

Figure 7.10 (a) Electron affinities (kJ/mol) of the main group elements. (b) Electron affinity as a function of atomic number.

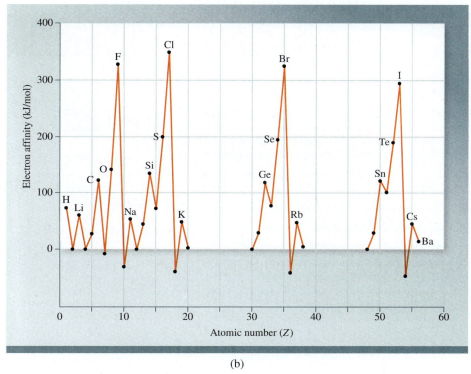

(b)

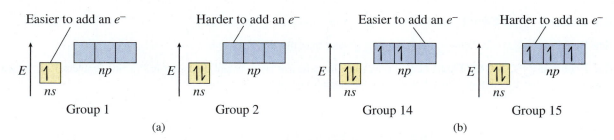

Easier to add an e^- Group 1

Harder to add an e^- Group 2

Easier to add an e^- Group 14

Harder to add an e^- Group 15

(a)   (b)

Figure 7.11 (a) It is easier to add an electron to an s orbital than to add one to a p orbital with the same principal quantum number. (b) Within a p subshell, it is easier to add an electron to an empty orbital than to add one to an orbital that already contains an electron.

be placed in an orbital that already contains an electron. Figure 7.11 illustrates these points. Note that there is a much less significant and less regular variation in electron affinities from top to bottom within a group [Figure 7.10(a)].

Just as more than one electron can be removed from an atom, more than one electron can also be added to an atom. While many first electron affinities are positive, subsequent electron affinities are always negative. Considerable energy is required to overcome the repulsive forces between the electron and the negatively charged ion. The addition of two electrons to a gaseous oxygen atom can be represented as:

Process	ΔH (kJ/mol)	Electron Affinity
$O(g) + e^- \longrightarrow O^-(g)$	-141	$EA_1 = 141$ kJ/mol
$O^-(g) + e^- \longrightarrow O^{2-}(g)$	744	$EA_2 = -744$ kJ/mol

The term *second electron affinity* may seem like something of a misnomer, because an anion in the gas phase has no real "affinity" for an electron. As is discussed in Chapter 8, a significantly *endothermic* process such as the addition of an electron to a gaseous O^- ion happens only in concert with one or more *exothermic* processes that more than compensate for the required energy input.

Sample Problem 7.5 lets you practice using the periodic table to compare the electron affinities of elements.

SAMPLE PROBLEM 7.5

For each pair of elements, indicate which one you would expect to have the greater first electron affinity, EA_1: (a) Al or Si, (b) Si or P.

Strategy Consider the effective nuclear charge and electron configuration to compare the electron affinities. The effective nuclear charge increases from left to right in a period (thus generally increasing EA), and it is more difficult to add an electron to a partially occupied orbital than to an empty one. Writing out orbital diagrams for the valence electrons is helpful for this type of problem.

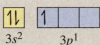
$3s^2$ $3p^1$
Valence orbital diagram for Al

Setup (a) Al is in Group 13 and Si is beside it in Group 14. Al has three valence electrons ($[Ne]3s^23p^1$), and Si has four valence electrons ($[Ne]3s^23p^2$).

(b) P is in Group 15 (to the right of Si), so it has five valence electrons ($[Ne]3s^23p^3$).

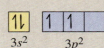
$3s^2$ $3p^2$
Valence orbital diagram for Si

Solution (a) $EA_1(Si) > EA_1(Al)$ because Si is to the right of Al and therefore has a greater effective nuclear charge.

(b) $EA_1(Si) > EA_1(P)$ because although P is to the right of Si in the third period of the periodic table (giving P the larger Z_{eff}), adding an electron to a P atom requires placing it in a $3p$ orbital that is partially occupied. The energy cost of *pairing* electrons outweighs the energy advantage of adding an electron to an atom with a larger effective nuclear charge.

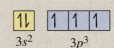
$3s^2$ $3p^3$
Valence orbital diagram for P

THINK ABOUT IT
The first electron affinities of Al, Si, and P are 42.5, 134, and 72.0 kJ/mol, respectively.

Practice Problem ATTEMPT Would you expect Mg or Al to have the higher EA_1?

Practice Problem BUILD Why is the EA_1 for Ge greater than the EA_1 for As?

Practice Problem CONCEPTUALIZE In the same hypothetical arrangement described in Practice Problem 7.4C, in which groups would you expect the anomalously low electron affinities to occur?

Metallic Character

Metals tend to

- Be shiny, lustrous, and malleable
- Be good conductors of both heat and electricity
- Have low ionization energies (so they commonly form *cations*)
- Form ionic compounds with chlorine (metal chlorides)
- Form basic, ionic compounds with oxygen (metal oxides)

Nonmetals, on the other hand, tend to

- Vary in color and lack the shiny appearance associated with metals
- Be brittle, rather than malleable
- Be poor conductors of both heat and electricity
- Form acidic, molecular compounds with oxygen
- Have high electron affinities (so they commonly form *anions*)

Metallic character increases from top to bottom in a group and decreases from left to right within a period.

Metalloids are elements with properties intermediate between those of metals and nonmetals. Because the definition of metallic character depends on a combination of properties, there may be some variation in the elements identified as metalloids in different sources. Astatine (At), for example, is listed as a metalloid in some sources and a nonmetal in others.

Explaining Periodic Trends

Many of the periodic trends in properties of the elements can be explained using *Coulomb's law,* which states that the force (F) between two charged objects (Q_1 and Q_2) is directly proportional to the product of the two charges and *inversely* proportional to the distance (d) between the objects squared. Recall that the *energy* between two oppositely charged particles (E_{el}) is inversely proportional to d [◂ Section 5.1, Equation 5.2]. The SI unit of force is the newton (1 N = 1 kg · m/s^2), and the SI unit of energy is the joule (1 J = 1 kg · m^2/s^2).

$$F \propto \frac{Q_1 \times Q_2}{d^2}$$

When the charges have opposite signs, F is negative—indicating an *attractive* force between the objects. When the charges have the same sign, F is positive—indicating a *repulsive* force. Table 7.3 shows how the magnitude of the attractive force between two oppositely charged objects at a fixed distance from each other varies with changes in the magnitudes of the charges.

Sample Problem 7.6 illustrates how Coulomb's law can be used to compare the magnitudes of attractive forces between charged objects.

TABLE 7.3	Attractive Force Between Oppositely Charged Objects at a Fixed Distance ($d = 1$) from Each Other		
Q_1	Q_2	Attractive force is proportional to	
+1	−1	1	
+2	−2	4	
+3	−3	9	

SAMPLE PROBLEM 7.6

For carbon and nitrogen, use the effective nuclear charges given in Figure 7.7 and the atomic radii given in Figure 7.6 to compare the attractive force between the nucleus in each atom and the valence electron that would be removed by the first ionization.

Strategy Use Coulomb's law to calculate a number to which the attractive force will be proportional in each case.

Setup From Figure 7.7, the effective nuclear charges of C and N are 3.14 and 3.83, respectively, and the radii of C and N are 77 pm and 75 pm, respectively. The first ionization energies are 1086 kJ/mol (C) and 1402 kJ/mol (N). The charge on the valence electron in each case is −1.

(Continued on next page)

Solution For C: $F \propto \dfrac{3.14 \times (-1)}{(77 \text{ pm})^2} = -5.3 \times 10^{-4}$

For N: $F \propto \dfrac{3.83 \times (-1)}{(75 \text{ pm})^2} = -6.8 \times 10^{-4}$

Note that in this type of comparison, it doesn't matter what units we use for the distance between the charges. We are not trying to calculate a particular attractive force, only to compare the magnitudes of these two attractive forces.

THINK ABOUT IT

Remember that the negative sign simply indicates that the force is attractive rather than repulsive. The calculated number for nitrogen is about 28 percent larger than that for carbon.

Practice Problem **A**TTEMPT Between which two charges is the attractive force larger: +3.26 and −1.15 separated by a distance of 1.5 pm, or +2.84 and −3.63 separated by a distance of 2.5 pm?

Practice Problem **B**UILD What must the distance be between charges of +2.25 and −1.86 for the attractive force between them to be the same as that between charges of +4.06 and −2.11 separated by a distance of 2.16 pm?

Practice Problem **C**ONCEPTUALIZE Rank these pairs of charged objects in order of increasing magnitude of the attractive force between them.

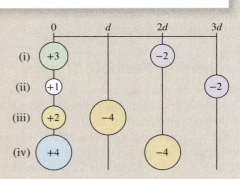

CHECKPOINT – SECTION 7.4   Periodic Trends in Properties of Elements

7.4.1 Arrange the elements Ca, Sr, and Ba in order of increasing IE_1.

a) Ca < Sr < Ba

b) Ba < Sr < Ca

c) Ba < Ca < Sr

d) Sr < Ba < Ca

e) Sr < Ca < Ba

7.4.2 Arrange the elements Li, Be, and B in order of increasing IE_2.

a) Li < Be < B

b) Li < B < Be

c) Be < B < Li

d) Be < Li < B

e) B < Be < Li

7.4.3 For each of the following pairs of elements, indicate which will have the greater EA_1: Rb or Sr, C or N, O or F.

a) Rb, C, O

b) Sr, N, F

c) Sr, C, F

d) Sr, N, O

e) Rb, C, F

7.4.4 Which element, K or Ca, will have the greater IE_1, which will have the greater IE_2, and which will have the greater EA_1?

a) Ca, K, K

b) K, K, Ca

c) K, Ca, K

d) Ca, Ca, K

e) Ca, Ca, Ca

7.4.5 Which best represents the third ionization of aluminum?

a) $2+$ + e^- ⟶ $3+$

b) ◯ ⟶ $3+$ + e^- + e^- + e^-

c) ◯ + e^- + e^- + e^- ⟶ $3+$

d) $2+$ ⟶ $3+$ + e^-

e) $3+$ + e^- + e^- + e^- ⟶ ◯

7.4.6 Which best represents the second electron affinity of nitrogen?

a) $-$ + e^- ⟶ $2-$

b) ◯ + e^- + e^- ⟶ $2-$

c) $2-$ ⟶ ◯ + e^- + e^-

d) $-$ ⟶ ◯ + e^-

e) ◯ ⟶ $2-$ + e^- + e^-

7.5 Electron Configuration of Ions

Because many ionic compounds are made up of monatomic anions and cations, it is helpful to know how to write the electron configurations of these ionic species. Just as for atoms, we use the Pauli exclusion principle and Hund's rule to write the ground-state electron configurations of cations and anions.

Recall from Chapter 2 that we can use the periodic table to predict the charges on many of the ions formed by main group elements. Elements in Groups 1 and 2, for example, form ions with charges of +1 and +2, respectively. Elements in Groups 16 and 17 form ions with charges of −2 and −1, respectively. Knowing something about electron configurations enables us to explain these charges. (See Figure 2.13.)

Ions of Main Group Elements

In Section 7.4, we learned about the tendencies of atoms to lose or gain electrons. In every period of the periodic table, the element with the highest IE_1 is the Group 18 element, the noble gas. [See Figure 7.8(b).] Also, Group 18 is the only group in which none of the members has *any* tendency to accept an electron; that is, they all have negative EA values. [See Figure 7.10(b).] High ionization energies and low electron affinities make the noble gases almost completely unreactive. Ultimately, the cause of this lack of reactivity is electron configuration. The $1s^2$ configuration of He and the $ns^2np^6 (n \geq 2)$ valence electron configurations of the other noble gases are extraordinarily stable. Other main group elements tend to either lose or gain the number of electrons needed to achieve the same number of electrons as the nearest noble gas. Species with identical electron configurations are called *isoelectronic*.

> **Student Note:** It is a common error to mistake species with the same valence electron configuration for isoelectronic species. For example, F^- and Ne are isoelectronic. F^- and Cl^- are not.

To write the electron configuration of an ion formed by a main group element, we first write the configuration for the atom and either add or remove the appropriate number of electrons. Electron configurations for the sodium and chloride ions are

Na: $1s^2 2s^2 2p^6 3s^1 \longrightarrow$ Na$^+$: $1s^2 2s^2 2p^6$ (10 electrons total, isoelectronic with Ne)

Cl: $1s^2 2s^2 2p^6 3s^2 3p^5 \longrightarrow$ Cl$^-$: $1s^2 2s^2 2p^6 3s^2 3p^6$ (18 electrons total, isoelectronic with Ar)

We can also write electron configurations for ions using the noble gas core.

Na: [Ne]$3s^1 \longrightarrow$ Na$^+$: [Ne]

Cl: [Ne]$3s^2 3p^5 \longrightarrow$ Cl$^-$: [Ne]$3s^2 3p^6$ or [Ar]

Sample Problem 7.7 gives you some practice writing electron configurations for the ions of main group elements.

SAMPLE PROBLEM 7.7

Write electron configurations for the following ions of main group elements: (a) N^{3-}, (b) Ba^{2+}, and (c) Be^{2+}.

Strategy First write electron configurations for the atoms. Then add electrons (for anions) or remove electrons (for cations) to account for the charge.

Setup (a) N^{3-} forms when N ($1s^2 2s^2 2p^3$ or [He]$2s^2 2p^3$), a main group nonmetal, gains three electrons.

(b) Ba^{2+} forms when Ba ($1s^2 2s^2 2p^6 3s^2 3p^6 4s^2 3d^{10} 4p^6 5s^2 4d^{10} 5p^6 6s^2$ or [Xe]$6s^2$) loses two electrons.

(c) Be^{2+} forms when Be ($1s^2 2s^2$ or [He]$2s^2$) loses two electrons.

Solution (a) [He]$2s^2 2p^6$ or [Ne]

(b) [Kr]$5s^2 4d^{10} 5p^6$ or [Xe]

(c) $1s^2$ or [He]

(Continued on next page)

THINK ABOUT IT

Be sure to add electrons to form an anion, and remove electrons to form a cation.

Practice Problem **A**TTEMPT Write electron configurations for (a) O^{2-}, (b) Ca^{2+}, and (c) Se^{2-}.

Practice Problem **B**UILD List all the species (atoms and/or ions) that are likely to have the following electron configuration: $1s^2 2s^2 2p^6$.

Practice Problem **C**ONCEPTUALIZE Select the correct valence orbital diagram for the Mg^{2+} ion and for the S^{2-} ion.

Mg^{2+}

(i) [Ne] $3s$ ⇅ $3p$ ↑ ↑

(ii) [Ne] $3s$ □ $3p$ □ □

(iii) [Ne] $3s$ ⇅ $3p$ ⇅ □

(iv) [Ne] $3s$ □ $3p$ ↑ ↑

S^{2-}

(i) [Ne] $3s$ ⇅ $3p$ ↑ ↑

(ii) [Ne] $3s$ □ $3p$ ⇅ ↑ ↑

(iii) [Ne] $3s$ ⇅ $3p$ ⇅ ⇅ ⇅

(iv) [Ne] $3s$ ↑ $3p$ ↑ ↑ ↑

Ions of *d*-Block Elements

Recall from Section 6.8 that the $4s$ orbital fills before the $3d$ orbitals for the elements in the first row of the *d*-block (Sc to Zn) [◄ Section 6.8]. Following the pattern for writing electron configurations for main group ions, then, we might expect the two electrons lost in the formation of the Fe^{2+} ion to come from the $3d$ subshell. It turns out, though, that an atom always loses electrons first from the shell with the *highest* value of *n*. In the case of Fe, that would be the $4s$ subshell.

$$Fe: [Ar]4s^2 3d^6 \longrightarrow Fe^{2+}: [Ar]3d^6$$

Iron can also form the Fe^{3+} ion, in which case the third electron is removed from the $3d$ subshell.

$$Fe: [Ar]4s^2 3d^6 \longrightarrow Fe^{3+}: [Ar]3d^5$$

In general, when a *d*-block element becomes an ion, it loses electrons first from the ns subshell and then from the $(n-1)d$ subshell. This explains, in part, why many of the transition metals can form ions with a +2 charge.

Sample Problem 7.8 gives you some practice writing electron configurations for the ions of *d*-block elements.

SAMPLE PROBLEM   7.8

Write electron configurations for the following ions of *d*-block elements: (a) Zn^{2+}, (b) Mn^{2+}, and (c) Cr^{3+}.

Strategy First write electron configurations for the atoms. Then add electrons (for anions) or remove electrons (for cations) to account for the charge. The electrons removed from a *d*-block element must come first from the outermost *s* subshell, not the partially filled *d* subshell.

Setup (a) Zn^{2+} forms when Zn ($1s^2 2s^2 2p^6 3s^2 3p^6 4s^2 3d^{10}$ or $[Ar]4s^2 3d^{10}$) loses two electrons.

(b) Mn^{2+} forms when Mn ($1s^2 2s^2 2p^6 3s^2 3p^6 4s^2 3d^5$ or $[Ar]4s^2 3d^5$) loses two electrons.

(c) Cr^{3+} forms when Cr ($1s^2 2s^2 2p^6 3s^2 3p^6 4s^1 3d^5$ or $[Ar]4s^1 3d^5$) loses three electrons—one from the $4s$ subshell and two from the $3d$ subshell. Remember that the electron configuration of Cr is anomalous in that it has only one $4s$ electron, making its *d* subshell half filled [◄ Section 6.9].

Solution (a) $[Ar]3d^{10}$ (b) $[Ar]3d^5$ (c) $[Ar]3d^3$

THINK ABOUT IT

Be sure to add electrons to form an anion and remove electrons to form a cation. Also, double-check to make sure that electrons removed from a *d*-block element come first from the *ns* subshell and then, if necessary, from the $(n - 1)d$ subshell.

Practice Problem **A**TTEMPT Write electron configurations for (a) Co^{3+}, (b) Cu^{2+}, and (c) Ag^+.

Practice Problem **B**UILD What common *d*-block ion (see Figure 2.13) is isoelectronic with Zn^{2+}?

Practice Problem **C**ONCEPTUALIZE Select the correct valence orbital diagram for the Fe^{2+} ion and for the Fe^{3+} ion.

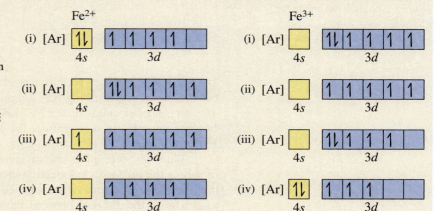

CHECKPOINT – SECTION 7.5   Electron Configuration of Ions

7.5.1 Which of the following ions are isoelectronic with a noble gas? (Select all that apply.)

a) Mn^{2+}

b) Ca^{2+}

c) Br^-

d) O^{2+}

e) F^-

7.5.2 Which of the following pairs are isoelectronic with each other? (Select all that apply.)

a) Ca^{2+} and Sr^{2+}

b) O^{2-} and Mg^{2+}

c) I^- and Kr

d) S^{2-} and Cl^-

e) He and H^+

7.5.3 Select the correct ground-state electron configuration for Ti^{2+}.

a) $[Ar]4s^23d^2$

b) $[Ar]4s^23d^4$

c) $[Ar]4s^2$

d) $[Ar]3d^2$

e) $[Ar]4s^13d^1$

7.5.4 Select the correct ground-state electron configuration for S^{2-}.

a) $[Ne]3p^4$

b) $[Ne]3s^23p^6$

c) $[Ne]3s^23p^2$

d) $[Ne]3p^6$

e) $[Ne]$

7.6 Ionic Radius

When an atom gains or loses one or more electrons to become an ion, its radius changes. The ***ionic radius,*** the radius of a cation or an anion, affects the physical and chemical properties of an ionic compound. The three-dimensional structure of an ionic compound, for example, depends on the relative sizes of its cations and anions.

Comparing Ionic Radius with Atomic Radius

When an atom loses an electron and becomes a cation, its radius decreases due in part to a reduction in electron-electron repulsions (and consequently a reduction in shielding) in the valence shell. A significant decrease in radius occurs when *all* of an atom's valence electrons are removed. This is the case with ions of most main group elements, which are isoelectronic with the noble gases preceding them. Consider Na, which loses its $3s$ electron to become Na^+:

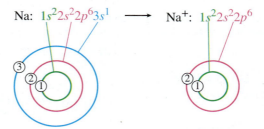

$$Na:\ 1s^2 2s^2 2p^6 3s^1 \longrightarrow Na^+:\ 1s^2 2s^2 2p^6$$

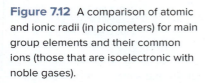

Valence orbitals in Na and in Na^+

The valence electron of Na has a principal quantum number of $n = 3$. When it has been removed, the resulting Na^+ ion no longer has any electrons in the $n = 3$ shell. The outermost electrons of the Na^+ ion have a principal quantum number of $n = 2$. Because the value of n determines the distance from the nucleus, this corresponds to a smaller radius.

When an atom gains one or more electrons and becomes an anion, its radius increases due to increased electron-electron repulsions. Adding an electron causes the rest of the electrons in the valence shell to spread out and take up more space to maximize the distance between them.

Figure 7.12 shows the ionic radii for those ions of main group elements that are isoelectronic with noble gases and compares them to the radii of the parent atoms. Note that the ionic radius, like the atomic radius, increases from top to bottom in a group.

Isoelectronic Series

An *isoelectronic series* is a series of two or more species that have identical electron configurations, but different nuclear charges. For example, O^{2-}, F^-, and Ne constitute

Figure 7.12 A comparison of atomic and ionic radii (in picometers) for main group elements and their common ions (those that are isoelectronic with noble gases).

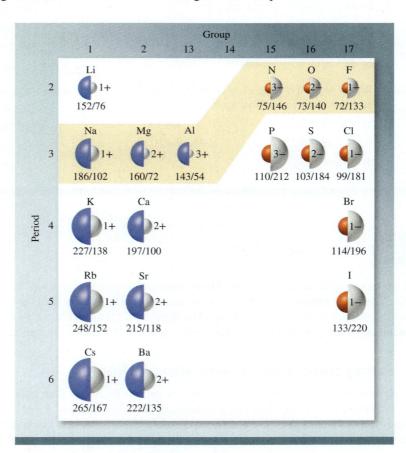

an isoelectronic series. Although these three species have identical electron configurations, they have different radii. In an isoelectronic series, the species with the smallest nuclear charge (i.e., the smallest atomic number Z) will have the largest radius. The species with the largest nuclear charge (i.e., the largest Z) will have the smallest radius.

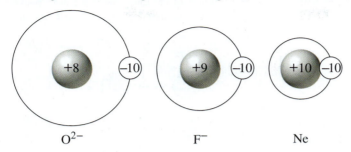

O^{2-} F^- Ne

Student Hot Spot

Student data indicate you may struggle with ionic radii. Access the eBook to view additional Learning Resources on this topic.

Sample Problem 7.9 shows how to identify members of an isoelectronic series and how to arrange them according to radius.

SAMPLE PROBLEM   7.9

Identify the isoelectronic series in the following group of species, and arrange the ions in order of increasing radius: K^+, Ne, Ar, Kr, P^{3-}, S^{2-}, and Cl^-.

Strategy Isoelectronic series are species with identical electron configurations but different nuclear charges. Determine the number of electrons in each species. The radii of isoelectronic ions within a series decrease with increasing nuclear charge.

Setup The number of electrons in each species is as follows: 18 (K^+), 10 (Ne), 18 (Ar), 36 (Kr), 18 (P^{3-}), 18 (S^{2-}), and 18 (Cl^-). The species with 18 electrons constitute the isoelectronic series. The nuclear charges of the ions with 18 electrons are +19 (K^+), +15 (P^{3-}), +16 (S^{2-}), and +17 (Cl^-).

Solution The isoelectronic series includes K^+, Ar, P^{3-}, S^{2-}, and Cl^-. The ions, in order of increasing radius, are: $K^+ < Cl^- < S^{2-} < P^{3-}$.

THINK ABOUT IT

Consult Figure 7.12 to check your result. With identical electron configurations, the attractive force between the valence electrons and the nucleus will be strongest for the largest nuclear charge. Thus, the larger the nuclear charge, the closer in the valence electrons will be pulled and the smaller the radius of an ion will be.

Practice Problem **A**TTEMPT Arrange the following isoelectronic series in order of increasing radius: Se^{2-}, Br^-, and Rb^+.

Practice Problem **B**UILD List all the common ions that are isoelectronic with Ne.

Practice Problem **C**ONCEPTUALIZE Which periodic table's highlighted portion includes elements that can form an isoelectronic series? (Select all that apply.)

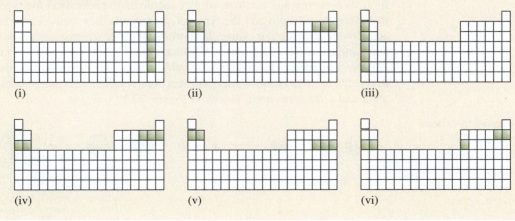

CHECKPOINT – SECTION 7.6   Ionic Radius

7.6.1   Which of the following species are isoelectronic with Kr? (Select all that apply.)

   a) He

   b) Ne

   c) Ar

   d) Br^-

   e) Rb^+

7.6.2   Which of the following are arranged correctly in order of increasing radius? (Select all that apply.)

   a) $F^- < Cl^- < Br^- < I^-$

   b) $O^{2-} < F^- < Na^+ < Mg^{2+}$

   c) $Ca^{2+} < K^+ < S^{2-} < Cl^-$

   d) $Rb^+ < K^+ < Na^+ < Li^+$

   e) $Sr^{2+} < Ca^{2+} < Mg^{2+} < Be^{2+}$

7.6.3   Which of the following is the most realistic representation of an atom from Group 16 becoming an ion?

7.6.4   Which of the following is the most realistic representation of an atom from Group 1 becoming an ion?

7.7   Periodic Trends in Chemical Properties of the Main Group Elements

Ionization energy and electron affinity enable us to understand the types of reactions that elements undergo and the types of compounds they form. These two parameters actually measure similar things. Ionization energy is a measure of how powerfully an atom attracts its own electrons, while electron affinity is a measure of how powerfully an atom can attract electrons from another source. As a very simple example of how this helps us understand a chemical reaction, consider the combination of a sodium atom and a chlorine atom, shown in Figure 7.13.

Figure 7.13 Formation of NaCl from its constituent elements. Note that although the charges are not all shown, the solid consists of a three-dimensional array of alternating oppositely charged ions.

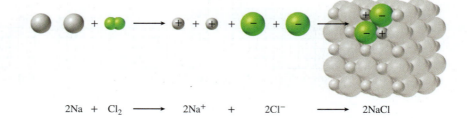

$$2Na + Cl_2 \longrightarrow 2Na^+ + 2Cl^- \longrightarrow 2NaCl$$

Sodium, with its low ionization energy, has a relatively weak attraction for its one valence electron. Chlorine, with its energetically favorable electron affinity, has the ability to attract electrons from another source. In this case, the electron that is loosely held by the Na atom, and powerfully attracted by the Cl atom, is transferred from Na to Cl, thus producing a sodium ion (Na^+) and a chloride ion (Cl^-). According to Coulomb's law, oppositely charged objects attract each other. The positively charged sodium ion and the negatively charged chloride ion are drawn together by electrostatic attraction, and the result is the formation of the solid ionic compound sodium chloride (NaCl).

Animation
Periodic table—properties of alkali and alkaline earth metals.

General Trends in Chemical Properties

Before we examine the elements in individual groups, let's identify some overall trends. We have said that elements in the same group resemble one another in chemical behavior because they have similar valence electron configurations. This statement, although correct in the general sense, must be applied with caution. Chemists have long known that the properties of the first member of each group (Li, Be, B, C, N, O, and F) are different from those of the rest of the members of the same group. Lithium, for example, exhibits many, but not all, of the properties characteristic of the Group 1 (alkali) metals. For example, unlike the other Group 1 elements, Li reacts with the O_2 and N_2 in air to form a simple oxide (Li_2O) and nitride (Li_3N), respectively. Similarly, beryllium is a somewhat atypical member of Group 2 (alkaline earth metals) in that it forms covalent compounds, and so on. The differences can be attributed to the unusually small size of the first element in each group (see Figure 7.6).

Another trend in the chemical behavior of main group elements is the diagonal relationship. ***Diagonal relationships*** refer to similarities between pairs of elements in different groups and periods of the periodic table. Specifically, the first three members of the second period (Li, Be, and B) exhibit many similarities to the elements located diagonally below them in the periodic table (Mg, Al, and Si). The reason for this phenomenon is the similarity of charge densities of their cations. (Charge density is the charge on an ion divided by its volume.) Cations with comparable charge densities react similarly with anions and therefore form the same types of compounds. Thus, the chemistry of lithium resembles that of magnesium in some ways; the same holds for beryllium and aluminum and for boron and silicon. Each of these pairs is said to exhibit a diagonal relationship.

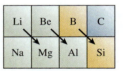

Diagonal relationships

A comparison of the properties of elements in the same group is most valid if the elements in question have a similar metallic (or nonmetallic) character. The elements in Groups 1 and 2, for example, are all metals, whereas those in Groups 17 and 18 are all nonmetals. We have to be more careful when comparing the elements of Groups 13 through 16, though, because a single group may contain metals, metalloids, and nonmetals. In these groups, we should expect a greater variation in chemical properties even though all group members have similar valence electron configurations.

Hydrogen ($1s^1$)

There is no completely suitable position for hydrogen in the periodic table (it really belongs in a group by itself). Traditionally hydrogen is shown at the top of Group 1, because, like the alkali metals, it has a single s valence electron and forms a cation with a charge of $+1$ (H^+), which is hydrated in solution. On the other hand, hydrogen also forms the *hydride* ion (H^-) in ionic compounds such as NaH and CaH_2. In this respect, hydrogen resembles the members of Group 17 (halogens), all of which form -1 anions (F^-, Cl^-, Br^-, and I^-) in ionic compounds. Ionic hydrides react with water to produce hydrogen gas and the corresponding metal hydroxides:

$$2NaH(s) + 2H_2O(l) \longrightarrow 2NaOH(aq) + 2H_2(g)$$

$$CaH_2(s) + 2H_2O(l) \longrightarrow Ca(OH)_2(aq) + 2H_2(g)$$

The most important compound of hydrogen is water, which forms when hydrogen burns in air:

$$2H_2(g) + O_2(g) \longrightarrow 2H_2O(l)$$

Figure 7.14 Group 1 elements.

(lithium, sodium, potassium):
Charles D. Winters/McGraw Hill;
(rubidium, cesium): *Stephen Frisch/McGraw Hill*

| Lithium | Sodium | Potassium |

| Rubidium | Cesium |

Properties of the Active Metals

Group 1 Elements (ns^1, $n \geq 2$)

Figure 7.14 shows samples of the Group 1 elements. These elements all have low ionization energies, making it easy for them to become M^+ ions. In fact, these metals are so reactive that they are never found in nature in the pure elemental state. They react with water to produce hydrogen gas and the corresponding metal hydroxide:

$$2M(s) + 2H_2O(l) \longrightarrow 2MOH(aq) + H_2(g)$$

where M denotes an alkali metal. When exposed to air, they gradually lose their shiny appearance as they react with oxygen to form metal oxides. Lithium forms lithium oxide (containing the oxide ion, O^{2-}):

$$4Li(s) + O_2(g) \longrightarrow 2Li_2O(s)$$

The other alkali metals all form oxides or *peroxides* (containing the peroxide ion, O_2^{2-}):

$$2Na(s) + O_2(g) \longrightarrow Na_2O_2(s)$$

Potassium, rubidium, and cesium also form *superoxides* (containing the superoxide ion, O_2^-):

$$K(s) + O_2(g) \longrightarrow KO_2(s)$$

The type of oxide that forms when an alkali metal reacts with oxygen has to do with the stability of the various oxides. Because these oxides are all ionic compounds, their stability depends on how strongly the cations and anions attract one another. Lithium tends to form predominantly the oxide because lithium oxide is more stable than lithium peroxide.

Group 2 Elements (ns^2, $n \geq 2$)

Student Note: Because they have less metallic character than the other Group 2 elements, beryllium and magnesium form some molecular compounds such as BeH_2 and MgH_2.

Figure 7.15 shows samples of the Group 2 elements. As a group, the alkaline earth metals are somewhat less reactive than the alkali metals. Both the first and the second ionization energies decrease (and metallic character increases) from beryllium to barium. Group 2 elements tend to form M^{2+} ions, where M denotes an alkaline earth metal atom.

Beryllium

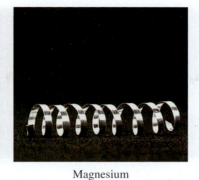

Magnesium

Figure 7.15 Group 2 elements.

(beryllium, magnesium, barium):
Stephen Frisch/McGraw Hill;
(calcium, strontium): *Charles D. Winters/
McGraw Hill*

Calcium

Strontium   Barium

The reactions of alkaline earth metals with water vary considerably. Beryllium does not react with water; magnesium reacts slowly with steam; and calcium, strontium, and barium react vigorously with cold water.

$$Ca(s) + 2H_2O(l) \longrightarrow Ca(OH)_2(s) + H_2(g)$$

$$Sr(s) + 2H_2O(l) \longrightarrow Sr(OH)_2(s) + H_2(g)$$

$$Ba(s) + 2H_2O(l) \longrightarrow Ba(OH)_2(aq) + H_2(g)$$

The reactivity of the alkaline earth metals toward oxygen also increases from Be to Ba. Beryllium and magnesium form oxides (BeO and MgO) only at elevated temperatures, whereas CaO, SrO, and BaO form at room temperature.

Magnesium reacts with aqueous acid to produce hydrogen gas:

$$Mg(s) + 2H^+(aq) \longrightarrow Mg^{2+}(aq) + H_2(g)$$

Calcium, strontium, and barium also react with aqueous acid solutions to produce hydrogen gas. However, because the metals also react with water, the two different reactions (with H^+ and with H_2O) occur simultaneously.

Properties of Other Main Group Elements

Group 13 Elements (ns^2np^1, $n \geq 2$)

Figure 7.16 shows samples of the first four Group 13 elements. Boron, the first member of the group, is a metalloid; the others (Al, Ga, In and Tl) are metals. Boron does not form binary ionic compounds and is unreactive toward both oxygen and water. Aluminum, the next element in the group, readily forms aluminum oxide when exposed to air:

$$4Al(s) + 3O_2(g) \longrightarrow 2Al_2O_3(s)$$

The aluminum oxide forms a protective coating, preventing the underlying metal from reacting further. This fact makes it possible to use aluminum for structural materials, such as aluminum siding and the shells of airplanes. Without the protective coating, layer after layer of Al atoms would become oxidized, and the structure would eventually crumble.

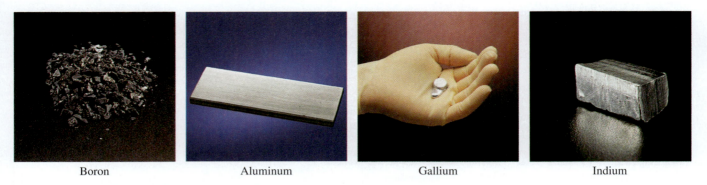

Boron   Aluminum   Gallium   Indium

Figure 7.16 Group 13 elements.

(boron, gallium, indium): *Stephen Frisch/McGraw Hill;* (aluminum): *Charles D. Winters/McGraw Hill*

Aluminum forms the Al^{3+} ion. It reacts with hydrochloric acid according to the equation:

$$2Al(s) + 6H^+(aq) \longrightarrow 2Al^{3+}(aq) + 3H_2(g)$$

The other Group 13 metals (Ga, In, and Tl) can form both M^+ and M^{3+} ions. As we move down the group, the M^+ ion becomes the more stable of the two.

The metallic elements in Group 13 also form many molecular compounds. For example, aluminum reacts with hydrogen to form AlH_3, which has properties similar to those of BeH_2. The progression of properties across the second row of the periodic table illustrates the gradual shift from metallic to nonmetallic character in the main group elements.

Group 14 Elements (ns^2np^2, $n \geq 2$)

Figure 7.17 shows samples of the Group 14 elements. Carbon, the first member of the group, is a nonmetal, whereas silicon and germanium, the next two members, are metalloids. Tin and lead, the last two members of the group, are metals. They do not react with water, but they do react with aqueous acid to produce hydrogen gas:

$$Sn(s) + 2H^+(aq) \longrightarrow Sn^{2+}(aq) + H_2(g)$$
$$Pb(s) + 2H^+(aq) \longrightarrow Pb^{2+}(aq) + H_2(g)$$

The Group 14 elements form compounds in both the +2 and +4 oxidation states. For carbon and silicon, the +4 oxidation state is the more stable one. For example, CO_2

Figure 7.17 Group 14 elements.

(graphite, silicon, germanium, lead): *Charles D. Winters/McGraw Hill;* (tin): *Charles D. Winters/ Timeframe Photography/McGraw Hill;* (diamond): *Steve Hamblin/Alamy Stock Photo*

Carbon (graphite)   Carbon (diamond)   Silicon

Germanium   Tin   Lead

Nitrogen   Phosphorus (white)   Phosphorus (red)

Arsenic   Antimony   Bismuth

Figure 7.18 Group 15 elements.
(nitrogen, antimony): *Charles D. Winters/ Timeframe Photography/McGraw Hill;* (white phosphorus, red phosphorus): *Charles D. Winters/McGraw Hill;* (arsenic, bismuth): *Stephen Frisch/McGraw Hill*

is more stable than CO, and SiO_2 is a stable compound, but SiO does not exist under ordinary conditions. As we move down the group, however, the relative stability of the two oxidation states is reversed. In tin compounds the +4 oxidation state is only slightly more stable than the +2 oxidation state. In lead compounds the +2 oxidation state is the more stable one. The outer electron configuration of lead is $6s^2 6p^2$, and lead tends to lose only the $6p$ electrons to form Pb^{2+} rather than both the $6p$ and $6s$ electrons to form Pb^{4+}.

Student Note: Carbon, being a nonmetal, achieves its +4 oxidation state without actually losing four electrons, which would be very "expensive" in terms of the energy required. For the same reason, compounds containing metals in very high oxidation states also tend to be molecular rather than ionic.

Group 15 Elements ($ns^2 np^3$, $n \geq 2$)

Figure 7.18 shows samples of the Group 15 elements. Nitrogen and phosphorus are nonmetals, arsenic and antimony are metalloids, and bismuth is a metal. Because Group 15 contains elements in all three categories, we expect greater variation in their chemical properties.

Elemental nitrogen is a diatomic gas (N_2). It forms a variety of oxides (NO, N_2O, NO_2, N_2O_4, and N_2O_5), all of which are gases except for N_2O_5, which is a solid at room temperature. Nitrogen has a tendency to accept three electrons to form the nitride ion (N^{3-}). Most metal nitrides, such as Li_3N and Mg_3N_2, are ionic compounds. Phosphorus exists as individual P_4 molecules (white phosphorus) or chains of P_4 molecules (red phosphorus). It forms two solid oxides with the formulas P_4O_6 and P_4O_{10}. The industrially important oxoacids nitric acid and phosphoric acid form when N_2O_5 and P_4O_{10}, respectively, react with water:

$$N_2O_5(s) + H_2O(l) \longrightarrow 2HNO_3(aq)$$

$$P_4O_{10}(s) + 6H_2O(l) \longrightarrow 4H_3PO_4(aq)$$

Arsenic, antimony, and bismuth have extensive three-dimensional structures. Bismuth is far less reactive than metals in the preceding groups.

Student Note: Elemental bismuth appears colorful because of a thin surface layer of bismuth oxide (Bi_2O_3).

Credit: *Charles D. Winters/McGraw Hill*

Group 16 Elements ($ns^2 np^4$, $n \geq 2$)

Figure 7.19 shows samples of the first three Group 16 elements. The first three members of the group (oxygen, sulfur, and selenium) are nonmetals, whereas the last two (tellurium and polonium) are metalloids. Oxygen is a colorless, odorless, diatomic gas; elemental sulfur and selenium exist as the molecules S_8 and Se_8, respectively;

Figure 7.19 Group 16 elements.

(oxygen): *David A. Tietz/Editorial Image, LLC;* (sulfur): *Charles D. Winters/McGraw Hill;* (selenium): *Charles D. Winters/Timeframe Photography/McGraw Hill;* (tellurium): *Stephen Frisch/McGraw Hill*

and tellurium and polonium have more extensive three-dimensional structures. (Polonium is a radioactive element that is difficult to study in the laboratory.) Oxygen has a tendency to accept two electrons to form the oxide ion (O^{2-}) in many compounds. Sulfur, selenium, and tellurium also form ions by accepting two electrons: S^{2-}, Se^{2-}, and Te^{2-}. The elements in Group 16 (especially oxygen) form a large number of molecular compounds with nonmetals. Some of the important compounds of sulfur are SO_2, SO_3, and H_2S. Sulfuric acid, an oxoacid, forms when sulfur trioxide reacts with water:

$$SO_3(g) + H_2O(l) \longrightarrow H_2SO_4(aq)$$

Group 17 Elements (ns^2np^5, $n \geq 2$)

Figure 7.20 shows samples of the first four Group 17 elements. All the halogens are nonmetals with the general formula X_2, where X denotes a halogen element. Like the Group 1 metals, the Group 17 nonmetals are too reactive to be found in nature in the elemental form. (Astatine, the last member of Group 17, is radioactive. Very little is known about its properties.)

The halogens have high ionization energies and large, energetically favorable electron affinities. Anions derived from the halogens (F^-, Cl^-, Br^-, and I^-) are called *halides*. The vast majority of alkali metal halides are ionic compounds. The halogens also form many molecular compounds among themselves, such as ICl and BrF_3, and with nonmetals in other groups, such as NF_3, PCl_5, and SF_6. The halogens react with hydrogen to form hydrogen halides:

$$H_2(g) + X_2(g) \longrightarrow 2HX(g)$$

This reaction is explosive when it involves fluorine, but it becomes less and less violent as we substitute chlorine, bromine, and iodine. The hydrogen halides dissolve in water to form hydrohalic acids. Hydrofluoric acid (HF) is a weak acid (meaning it is a weak electrolyte), but the other hydrohalic acids (HCl, HBr, and HI) are all strong acids (strong electrolytes).

Fluorine

Chlorine

Bromine

Iodine

Figure 7.20 Group 17 elements.

(fluorine, bromine): *Stephen Frisch/McGraw Hill;* (chlorine): *Rvkamalov gmail.com/Shutterstock;* (iodine): *Charles D. Winters/Timeframe Photography/McGraw Hill*

Group 18 Elements (ns^2np^6, $n \geq 2$)

Figure 7.21 shows samples of the Group 18 elements. All the noble gases exist as monatomic species. With the exception of helium, which has the electron configuration $1s^2$, their atoms have completely filled outer ns and np subshells. Their electron configurations give the noble gases their great stability. The Group 18 ionization energies are among the highest of all the elements (see Figure 7.8). Their electron affinities are all less than zero (Figure 7.10), so they have no tendency to accept extra electrons.

For years the noble gases were called *inert gases* because they were not known to react with anything. Beginning in 1963, however, compounds were prepared from the heavier members of the group by exposing them to very strong oxidizing agents such as fluorine and oxygen. Some of the compounds that have been prepared are XeF_4, XeO_3, $XeOF_4$, KrF_2, and most recently, HArF. Although the chemistry of the noble gases is interesting, their compounds are not involved in any natural biological processes and they currently have no major commercial applications.

Student Note: Note that the common ions formed by the other main group elements are those that make them isoelectronic with a noble gas. In Group 1 elements, for example, each atom *loses* one electron to become isoelectronic with the noble gas that immediately precedes it; in Group 17 elements, each atom *gains* one electron to become isoelectronic with the noble gas that immediately follows it; and so on.

Comparison of Group 1 and Group 11 Elements

Although the outer electron configurations of Groups 1 and 11 are similar (members of both groups have a single valence electron in an *s* orbital), their chemical properties are very different.

Helium

Neon

Figure 7.21 Discharge tubes containing Group 18 elements.

(helium, neon, argon, krypton, xenon): *Stephen Frisch/McGraw Hill*

Argon

Krypton

Xenon

The first ionization energies of Cu, Ag, and Au are 745, 731, and 890 kJ/mol, respectively. Because these values are considerably larger than those of the alkali metals, the Group 11 elements are much less reactive. The higher ionization energies of the Group 11 elements result from incomplete shielding of the nucleus by the inner *d* electrons (compared with the more effective shielding by the completely filled noble gas cores). Consequently, the outer *s* electrons of the Group 11 elements are more strongly attracted by the nucleus. In fact, copper, silver, and gold are so unreactive that they are usually found in the uncombined state in nature. The inertness, rarity, and attractive appearance of these metals make them valuable in the manufacture of coins and jewelry. For this reason, these metals are also known as "coinage metals." The differences in the chemistry of the elements in Group 2 from that of the elements in Group 12 can be explained in a similar way.

David A. Tietz/Editorial Image, LLC

David A. Tietz/Editorial Image, LLC

Bringing Chemistry to Life

Salt Substitutes

Sodium chloride (NaCl), one of the world's most common seasonings, contains the Group 1 metal cation, Na^+. Because of their similarities to sodium, the other Group 1 metals also form chlorides. Lithium chloride (LiCl) and potassium chloride (KCl) are particularly similar to sodium chloride in appearance and taste, and both have been used as salt substitutes for people on low-sodium diets. Lithium chloride, used briefly as a salt substitute in the 1940s, turned out to be dangerous and its use resulted in cases of severe lithium toxicity—some of which were fatal. Today, salt substitutes such as Morton Salt Substitute and Nu-Salt contain *potassium* chloride.

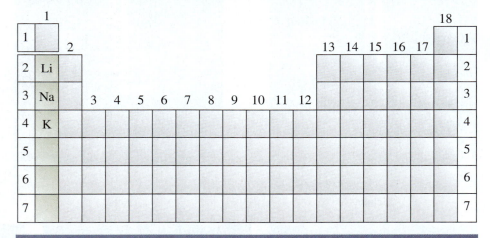

Variation in Properties of Oxides Within a Period

One way to compare the properties of the main group elements across a period is to examine the properties of a series of similar compounds. Because oxygen combines with almost all elements, we will compare the properties of oxides of the third-period elements to see how metals differ from metalloids and nonmetals. Some elements in the third period (P, S, and Cl) form several types of oxides, but for simplicity we will consider only those oxides in which the elements have the highest oxidation number. Table 7.4 lists a few general characteristics of these oxides and some specific physical properties of the oxides of third-period elements.

The tendency of oxygen to form the oxide ion is greatly favored when oxygen combines with metals that have low ionization energies, such as those in Groups 1 and 2 and aluminum. Thus, Na_2O, MgO, and Al_2O_3 are ionic compounds, as evidenced by their high melting points and boiling points. They have extensive three-dimensional structures in which each cation is surrounded by a specific number of anions, and vice versa. As the ionization energies of the elements increase from left to right, so does the molecular nature

TABLE 7.4	Some Properties of Oxides of the Third-Period Elements						
	Na_2O	MgO	Al_2O_3	SiO_2	P_4O_{10}	SO_3	Cl_2O_7
Type of compound	← Ionic →			← Molecular →			
Structure	← Extensive three-dimensional →			← Discrete molecular units →			
Melting point (°C)	1275	2800	2045	1610	580	16.8	−91.5
Boiling point (°C)	?	3600	2980	2230	?	44.8	82
Acid-base nature	Basic	Basic	Amphoteric	← Acidic →			

of the oxides that form. Silicon is a metalloid; its oxide (SiO_2) also has a huge three-dimensional network, although it is not an ionic compound. The oxides of phosphorus, sulfur, and chlorine are molecular compounds composed of small discrete units. The weak attractions among these molecules result in relatively low melting points and boiling points.

Most oxides can be classified as acidic or basic depending on whether they produce acidic or basic solutions when dissolved in water (or whether they *react* as acids or bases). Some oxides are **amphoteric,** which means that they display both acidic and basic properties. The first two oxides of the third period, Na_2O and MgO, are basic oxides. For example, Na_2O reacts with water to form the base sodium hydroxide:

$$Na_2O(s) + H_2O(l) \longrightarrow 2NaOH(aq)$$

Magnesium oxide is quite insoluble; it does not react with water to any appreciable extent. However, it does react with acids in a manner that resembles an acid-base reaction:

$$MgO(s) + 2HCl(aq) \longrightarrow MgCl_2(aq) + H_2O(l)$$

The products of this reaction are a salt ($MgCl_2$) and water, the same kind of products that are obtained in an acid-base neutralization.

Aluminum oxide is even less soluble than magnesium oxide. It, too, does not react with water, but it exhibits the properties of a base by reacting with acids:

$$Al_2O_3(s) + 6HCl(aq) \longrightarrow 2AlCl_3(aq) + 3H_2O(l)$$

It also exhibits acidic properties by reacting with bases:

$$Al_2O_3(s) + 2NaOH(aq) + 3H_2O(l) \longrightarrow 2NaAl(OH)_4(aq)$$

Thus, Al_2O_3 is classified as an amphoteric oxide because it has properties of both acids and bases. Other amphoteric oxides are ZnO, BeO, and Bi_2O_3.

Silicon dioxide is insoluble and does not react with water. It has acidic properties, however, because it reacts with a very concentrated aqueous base:

$$SiO_2(s) + 2OH^-(aq) \longrightarrow SiO_3^{2-}(aq) + H_2O(l)$$

For this reason, concentrated aqueous, strong bases such as sodium hydroxide ($NaOH$) should *not* be stored in Pyrex glassware, which is made of SiO_2.

The remaining third-period oxides (P_4O_{10}, SO_3, and Cl_2O_7) are acidic. They react with water to form phosphoric acid, sulfuric acid, and perchloric acid, respectively:

$$P_4O_{10}(s) + 6H_2O(l) \longrightarrow 4H_3PO_4(aq)$$

$$SO_3(g) + H_2O(l) \longrightarrow 2H_2SO_4(aq)$$

$$Cl_2O_7(l) + H_2O(l) \longrightarrow 2HClO_4(aq)$$

Student Note: Certain oxides such as CO and NO are neutral; that is, they do not react with water to produce acidic or basic solutions. In general, oxides of nonmetals are either acidic or neutral.

This brief examination of oxides of the third-period elements shows that as the metallic character of the elements decreases from left to right across the period, their oxides change from basic to amphoteric to acidic. Metal oxides are usually basic, and most oxides of nonmetals are acidic. The intermediate properties of the oxides (as demonstrated by the amphoteric oxides) are exhibited by elements whose positions are intermediate within the period. Because the metallic character of the elements increases from top to bottom within a group of main group elements, the oxides of elements with higher atomic numbers are more basic than the lighter elements.

Chapter Summary

Section 7.1

- The modern periodic table was devised independently by Dmitri Mendeleev and Lothar Meyer in the nineteenth century. The elements that were known at the time were grouped based on their physical and chemical properties. Using his arrangements of the elements, Mendeleev successfully predicted the existence of elements that had not yet been discovered.

- Early in the twentieth century, Henry Moseley refined the periodic table with the concept of the *atomic number,* thus resolving a few inconsistencies in the tables proposed by Mendeleev and Meyer.

- Elements in the same group of the periodic table tend to have similar physical and chemical properties.

Section 7.2

- The periodic table can be divided into the *main group elements* (also known as the *representative elements*) and the *transition metals.* It is further divided into smaller groups or *columns* of elements that all have the same configuration of *valence electrons*.

- The 18 columns of the periodic table are labeled 1 through 18. Groups 1 and 2 contain the *s*-block elements; Groups 3 through 12 contain the *d*-block elements; and Groups 13 through 18 contain the *p*-block elements.

Section 7.3

- *Effective nuclear charge* (Z_{eff}) is the nuclear charge that is "felt" by the valence electrons. It is usually lower than the nuclear charge due to *shielding* by the core electrons.

Section 7.4

- *Atomic radius* is the distance between an atom's nucleus and its valence shell. The atomic radius of a metal atom is defined as the *metallic radius,* which is one-half the distance between adjacent, identical nuclei in a metal solid. The atomic radius of a nonmetal is defined as the *covalent radius,* which is one-half the distance between adjacent, identical nuclei in a molecule. In general, atomic radii *decrease* from left to right across a *period* of the periodic table and *increase* from top to bottom down a *group.*

- *Ionization energy (IE)* is the energy required to remove an electron from a gaseous atom. The first ionization energy (IE_1) is smaller than subsequent ionization energies [e.g., second (IE_2), third (IE_3), and so on]. The first ionization of any atom removes a valence electron. Ionization energies increase dramatically when core electrons are being removed.

- First ionization energies (IE_1 values) tend to increase across a period and decrease down a group. Exceptions to this trend can be explained based upon the electron configuration of the element.

- *Electron affinity (EA)* is the energy released when an atom in the gas phase accepts an electron. *EA* is equal to $-\Delta H$ for the process $A(g) + e^- \longrightarrow A^-(g)$

- Electron affinities tend to increase across a period. As with first ionization energies, exceptions to the trend can be explained based on the electron configuration of the element.

- Metals tend to be shiny, lustrous, malleable, ductile, and conducting (for both heat and electricity). Metals typically lose electrons to form cations, and they tend to form ionic compounds (including basic oxides).

- Nonmetals tend to be brittle and not good conductors (for either heat or electricity). They can gain electrons to form anions but they commonly form molecular compounds (including acidic oxides).

- In general, metallic character decreases across a period and increases down a group of the periodic table. *Metalloids* are elements with properties intermediate between metals and nonmetals.

- According to *Coulomb's law,* the attractive force (F) between two oppositely charged particles (Q_1 and Q_2) is directly proportional to the product of the charges and inversely proportional to the distance (d) between the objects squared: ($F \propto Q_1 \cdot Q_2/d^2$).

Section 7.5

- The common ions of main group elements are *isoelectronic* with noble gases. When a *d*-block element loses one or more electrons, it loses them first from the shell with the highest principal quantum number (e.g., electrons in the 4*s* subshell are lost before electrons in the 3*d* subshell).

Section 7.6

- *Ionic radius* is the distance between the nucleus and valence shell of a cation or an anion. A cation is smaller than its parent atom. An anion is larger than its parent atom.

- An *isoelectronic series* consists of one or more ions and sometimes an atom, all of which have identical electron configurations. Within an isoelectronic series of ions, the greater the nuclear charge, the smaller the radius.

Section 7.7

- A *diagonal relationship* describes similarities in the chemical properties of elements that are in different groups, but that are positioned diagonally from each other in the periodic table.

- Although members of a group in the periodic table exhibit similar chemical and physical properties, the first member of each group tends to be significantly different from the other members. Hydrogen is essentially a group unto itself.

- The alkali metals (Group 1) tend to be highly reactive toward oxygen, water, and acid. Group 2 metals are less reactive than Group 1 metals, but the heavier members all react with water to produce metal hydroxides and hydrogen gas. Groups that contain both metals and nonmetals (e.g., Groups 14, 15, and 16) tend to show greater variability in their physical and chemical properties.

- *Amphoteric* oxides, such as Al_2O_3, are those that exhibit both acidic and basic behavior.

Key Words

Key Equation

7.1 $Z_{eff} = Z - \sigma$

Effective nuclear charge (Z_{eff}) is equal to *nuclear* charge (*Z*) minus the shielding constant (σ).

The properties of an element are determined in large part by the *size* (radius) and the valence-shell *electron configuration* of its atoms and ions. Together, principal quantum number (n), effective nuclear charge (Z_{eff}), and charge on the valence shell (number of valence electrons) determine atomic radius.

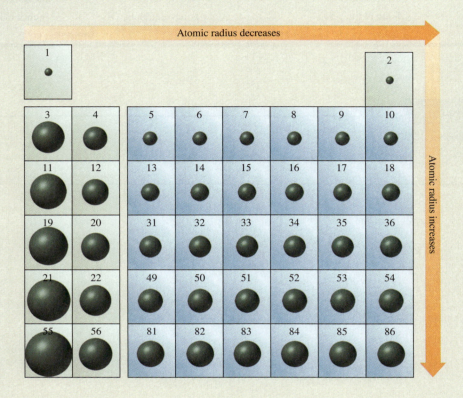

From left to right across a period, Z_{eff} and charge on the valence shell both *increase*. Each step to the right involves the addition of a proton, which increases Z, and the addition of an electron. Each additional electron resides in the same shell. Remember that electrons in the same shell do not shield one another well. The result is that Z_{eff} increases by nearly as much as Z. As the magnitude of opposite charges increases, coulombic attraction between them increases, and they are drawn closer together, thus reducing the atomic radius.

From top to bottom within a group, valence electrons reside in shells with increasingly larger values of n, putting them farther away from the nucleus—thereby increasing the atomic radius. Valence-shell electron configuration and, to a large extent, Z_{eff} remain the same from top to bottom in a group. Variations in atomic radii of the *d*-block elements vary less regularly than the main group elements.

From left to right across a period, Z_{eff} *increases* and radius *decreases*. Both factors contribute to stronger coulombic attraction between the positively charged nucleus and the valence electrons—making it harder to remove an electron; therefore, ionization energy (IE_1) *increases*. From top to bottom within a group, although Z_{eff} remains fairly constant, radius increases. The increased distance between the nucleus and the valence electrons corresponds to weaker coulombic attraction, making it easier to remove an electron; therefore, IE_1 *decreases*.

Because Z_{eff} *increases* and radius *decreases* from left to right across a period, the coulombic attraction between the positively charged nucleus and an added electron gets stronger, making it easier to add an electron; therefore, electron affinity (EA) *increases*. And because Z_{eff} remains fairly constant and radius *increases* from top to bottom within a group, there is a smaller coulombic attraction between the nucleus and an added electron—making it harder to add an electron; therefore, EA decreases.

Specific anomalies in the general trends of IE_1 and EA are determined by valence-electron configuration.

Ionization energy (IE_1) increases →

1 1312							2 2372
3 520	4 899	5 800	6 1086	7 1402	8 1314	9 1681	10 2080
11 496	12 738	13 577	14 786	15 1012	16 999	17 1256	18 1520
19 419	20 590	31 579	32 761	33 947	34 941	35 1143	36 1351
21 403	22 549	49 558	50 708	51 834	52 869	53 1009	54 1170
55 376	56 503	81 589	82 715	83 703	84 813	85 (926)	86 1037

(Ionization energy (IE_1) decreases ↓)

Electron affinity (EA) increases →

1 +72.8							2 (0.0)
3 +59.6	4 ≤0	5 +26.7	6 +122	7 −7	8 +141	9 +328	10 (−29)
11 +52.9	12 ≤0	13 +42.5	14 +134	15 +72.0	16 +200	17 +349	18 (−35)
19 +48.4	20 +2.37	31 +28.9	32 +119	33 +78.2	34 +195	35 +325	36 (−39)
21 +46.9	22 +5.03	49 +28.9	50 +107	51 +103	52 +190	53 +295	54 (−41)
55 +45.5	56 +13.95	81 +19.3	82 +35.1	83 +91.3	84 +183	85 +270	86 (−41)

(Electron affinity (EA) decreases ↓)

For example, although the trend is an increase in IE_1 from left to right, IE_1 for Group 13 is lower than IE_1 for Group 2—within a period. This is because the electron removed by ionization comes from the *p* subshell, which is higher in energy than the *s* subshell [◄◄ Figure 7.9(a)]. Likewise, IE_1 for Group 16 is lower than IE_1 for Group 15. In this case, the electron removed by ionization is one of a pair of electrons in a *p* orbital. Because paired electrons in a single orbital repel one another, removing one of them is relatively easy [◄◄ Figure 7.9(b)].

Similarly, despite the general trend, EA for Group 2 is lower than EA for Group 1. In this case, the electron added by electron affinity goes into the *p* subshell, which is higher in energy than the *s* subshell [◄◄ Figure 7.11(a)]. And EA for Group 15 is lower than that for Group 14 because the added electron must go into an already occupied *p* orbital [◄◄ Figure 7.11(b)].

Each of these periodic properties can be explained and understood using Coulomb's law: $F \propto \dfrac{Q_1 \times Q_2}{d^2}$.

Key Skills Problems

7.1

Often we can compare properties of two elements based solely on periodic trends. For which pair of elements is the periodic trend not sufficient to determine which has the higher first ionization energy?

(a) C and Si   (b) Al and Ga   (c) Ga and Si   (d) Tl and Sn   (e) B and Si

7.2

The colored spheres represent the ions Ca^{2+}, Cl^-, K^+, P^{3-}, and S^{2-}. Based on size and using only a periodic table, determine which of the following has the ions in the same order as the diagram.

(a) Ca^{2+}, Cl^-, K^+, P^{3-}, S^{2-}
(b) Ca^{2+}, K^+, P^{3-}, S^{2-}, Cl^-
(c) P^{3-}, S^{2-}, Cl^-, K^+, Ca^{2+}
(d) Ca^{2+}, K^+, Cl^-, S^{2-}, P^{3-}
(e) P^{3-}, S^{2-}, Cl^-, Ca^{2+}, K^+

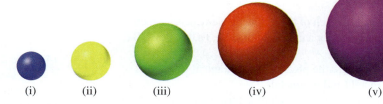

(i)   (ii)   (iii)   (iv)   (v)

7.3

Group 18 exhibits the highest first ionization energy. Which group do you expect to exhibit the highest *second* ionization energy (EA_2)?

(a) 17   (b) 16   (c) 15   (d) 2   (e) 1

7.4

Which of the following best describes why Z_{eff} does not change significantly from top to bottom within a group?

(a) There are fewer completed electron shells.
(b) There are fewer valence electrons.
(c) There are more valence electrons.
(d) There are more completed electron shells.
(e) The number of protons in the nucleus does not change significantly.

Questions and Problems

Applying What You've Learned

As we have seen, one of the consequences of two chemical species having similar properties is that human physiology sometimes mistakes one species for the other. Healthy bones require constant replenishment of calcium. Our dietary calcium comes primarily from dairy products but can also be found in some vegetables—especially dark, leafy greens such as spinach and kale. Strontium-90 is one of the radioactive isotopes released into the atmosphere as the result of nuclear testing or an accident like the one at Fukushima. The strontium will eventually settle on land and water, and it can enter our bodies through ingestion—especially of vegetation—and through the inhalation of airborne particles. The body mistakes Sr^{2+} ions for Ca^{2+} ions and incorporates them into otherwise healthy bone. Constant exposure to the radiation emitted by strontium-90 affects not only the bone and surrounding soft tissue, but also the bone marrow, damaging and destroying stem cells vital to the immune system. Such long-term exposure leads to an increased risk of leukemia and other cancers. (a) Without referring to a periodic table, write the electron configuration of strontium ($Z = 38$) [◄◄ Sample Problem 7.2]. (b) Referring only to a periodic table, arrange Sr and the other Group 2 metals (not including Ra) in order of increasing atomic radius [◄◄ Sample Problem 7.3]. (c) Again referring only to the periodic table, arrange the members of Group 2 (not including Ra) in order of increasing ionization energy (IE_1) [◄◄ Sample Problem 7.4]. (d) Write the electron configuration for each of the Group 2 metal cations [◄◄ Sample Problem 7.7]. (e) For each metal cation in part (d), identify an isoelectronic series consisting of a noble gas and one or more common ions (see Figure 2.13) [◄◄ Sample Problem 7.8]. List the ions of each isoelectronic series in order of increasing radius [◄◄ Sample Problem 7.9].

SECTION 7.1: DEVELOPMENT OF THE PERIODIC TABLE

Review Questions

7.1   Briefly describe the significance of Mendeleev's periodic table.

7.2   What is Moseley's contribution to the modern periodic table?

7.3   Describe the general layout of a modern periodic table.

7.4   What is the most important relationship among elements in the same group in the periodic table?

SECTION 7.2: THE MODERN PERIODIC TABLE

Review Questions

7.5   Classify each of the following elements as a metal, a nonmetal, or a metalloid: As, Xe, Fe, Li, B, Cl, Ba, P, I, Si.

7.6   Compare the physical and chemical properties of metals and nonmetals.

7.7   Draw a rough sketch of a periodic table (no details are required). Indicate regions where metals, nonmetals, and metalloids are located.

7.8   What is a main group element? Give names and symbols of four main group elements.

7.9   Without referring to a periodic table, write the name and give the symbol for one element in each of the following groups: 1, 2, 13, 14, 15, 16, 17, 18, transition metals.

7.10 Indicate whether the following elements exist as atomic species, molecular species, or extensive three-dimensional structures in their most stable states at room temperature, and write the molecular or empirical formula for each one: phosphorus, iodine, magnesium, neon, carbon, sulfur, cesium, and oxygen.

7.11 You are given a sample of a dark, shiny solid and asked to determine whether it is the nonmetal *iodine* or a *metallic* element. What test could you do that would enable you to answer the question without destroying the sample?

7.12 What are valence electrons? For main group elements, the number of valence electrons of an element is equal to its group number. Show that this is true for the following elements: Al, Sr, K, Br, P, S, C.

7.13 Write the outer electron configurations for the (a) alkali metals, (b) alkaline earth metals, (c) halogens, (d) noble gases.

7.14 Use the first-row transition metals (Sc to Cu) as an example to illustrate the characteristics of the electron configurations of transition metals.

7.15 Arsenic is not an essential element for the human body. Based on its position in the periodic table, suggest a reason for its toxicity.

Conceptual Problems

7.16 In the periodic table, the element hydrogen is sometimes grouped with the alkali metals and sometimes with the halogens. Explain why hydrogen can resemble the Group 1 and the Group 17 elements.

7.17 A neutral atom of a certain element has 34 electrons. Consulting only the periodic table, identify the element and write its ground-state electron configuration.

7.18 Group the following electron configurations in pairs that would represent elements with similar chemical properties:
(a) $1s^2 2s^2 2p^6 3s^2$
(b) $1s^2 2s^2 2p^3$
(c) $1s^2 2s^2 2p^6 3s^2 3p^6 4s^2 3d^{10} 4p^6$
(d) $1s^2 2s^2$
(e) $1s^2 2s^2 2p^6$
(f) $1s^2 2s^2 2p^6 3s^2 3p^3$

7.19 Group the following electron configurations in pairs that would represent elements with similar chemical properties:
(a) $1s^2 2s^2 2p^5$
(b) $1s^2 2s^1$
(c) $1s^2 2s^2 2p^6$
(d) $1s^2 2s^2 2p^6 3s^2 3p^5$
(e) $1s^2 2s^2 2p^6 3s^2 3p^6 4s^1$
(f) $1s^2 2s^2 2p^6 3s^2 3p^6 4s^2 3d^{10} 4p^6$

7.20 Without referring to a periodic table, write the electron configuration of elements with the following atomic numbers: (a) 9, (b) 20, (c) 26, (d) 33.

7.21 Specify the group of the periodic table in which each of the following elements is found:
(a) $[Ne]3s^1$, (b) $[Ne]3s^2 3p^3$, (c) $[Ne]3s^2 3p^6$,
(d) $[Ar]4s^2 3d^8$.

SECTION 7.3: EFFECTIVE NUCLEAR CHARGE

Review Questions

7.22 Explain the term *effective nuclear charge.*

7.23 Explain why the atomic radius of Be is smaller than that of Li.

Computational Problems

7.24 The electron configuration of B is $1s^2 2s^2 2p^1$. (a) If each core electron (i.e., the $1s$ electrons) were totally effective in shielding the valence electrons (i.e., the $2s$ and $2p$ electrons) from the nucleus and the valence electrons did not shield one another, what would be the shielding constant (σ) and the effective nuclear charge (Z_{eff}) for the $2s$ and $2p$ electrons? (b) In reality, the shielding constants for the $2s$ and $2p$ electrons in B are slightly different. They are 2.42 and 2.58, respectively. Calculate Z_{eff} for these electrons, and explain the differences from the values you determined in part (a).

7.25 The electron configuration of C is $1s^2 2s^2 2p^2$. (a) If each core electron (i.e., the $1s$ electrons) were totally effective in shielding the valence electrons (i.e., the $2s$ and $2p$ electrons) from the nucleus and the valence electrons did not shield one another, what would be the shielding constant (σ) and the effective nuclear charge (Z_{eff}) for the $2s$ and $2p$ electrons? (b) In reality, the shielding constants for the $2s$ and $2p$ electrons in C are slightly different. They are 2.78 and 2.86, respectively. Calculate Z_{eff} for these electrons, and explain the differences from the values you determined in part (a).

SECTION 7.4: PERIODIC TRENDS IN PROPERTIES OF ELEMENTS

Review Questions

7.26 Define *atomic radius.* Does the size of an atom have a precise meaning? Explain.

7.27 How does atomic radius change (a) from left to right across a period and (b) from top to bottom in a group?

7.28 Define *ionization energy.* Explain why ionization energy measurements are usually made when atoms are in the gaseous state. Why is the second ionization energy always greater than the first ionization energy for any element?

7.29 Sketch the outline of the periodic table, and show group and period trends in the first ionization energy of the elements. What types of elements have the highest ionization energies and what types have the lowest ionization energies?

7.30 (a) Define *electron affinity.* (b) Explain why electron affinity measurements are made with gaseous atoms. (c) Ionization energy is always a positive quantity, whereas electron affinity may be either positive or negative. Explain.

7.31 Explain the trends in electron affinity from aluminum to chlorine (see Figure 7.10).

Computational Problems

7.32 A hydrogen-like ion is an ion containing only one electron. The energies of the electron in a hydrogen-like ion are given by

$$E_n = -(2.18 \times 10^{-18} \text{ J})Z^2 \left(\frac{1}{n^2}\right)$$

where n is the principal quantum number and Z is the atomic number of the element. Calculate the ionization energy (in kJ/mol) of the He^+ ion.

7.33 Plasma is a state of matter consisting of positive gaseous ions and electrons. In the plasma state, a mercury atom could be stripped of its 80 electrons and therefore would exist as Hg^{80+}. Use the equation in Problem 7.32 to calculate the energy required for the last ionization step, that is,

$$Hg^{79+}(g) \longrightarrow Hg^{80+}(g) + e^-$$

Conceptual Problems

7.34 On the basis of their positions in the periodic table, select the atom with the larger atomic radius in each of the following pairs: (a) Na, Si; (b) Ba, Be; (c) N, F; (d) Br, Cl; (e) Ne, Kr.

7.35 Arrange the following atoms in order of increasing atomic radius: Na, Al, P, Cl, Mg.

7.36 Which is the largest atom in the third period of the periodic table?

7.37 Which is the smallest atom in Group 17?

7.38 Based on size, identify the spheres shown as Na, Mg, O, and S.

7.39 Based on size, identify the spheres shown as K, Ca, S, and Se.

7.40 Why is the radius of the lithium atom considerably larger than the radius of the hydrogen atom?

7.41 Use the second period of the periodic table as an example to show that the size of atoms decreases as we move from left to right. Explain the trend.

7.42 Arrange the following in order of increasing first ionization energy: Na, Cl, Al, S, and Cs.

7.43 Arrange the following in order of increasing first ionization energy: F, K, P, Ca, and Ne.

7.44 Use the third period of the periodic table as an example to illustrate the change in first ionization energies of the elements as we move from left to right. Explain the trend.

7.45 In general, the first ionization energy increases from left to right across a given period. Aluminum, however, has a lower first ionization energy than magnesium. Explain.

7.46 The first and second ionization energies of K are 419 and 3052 kJ/mol, and those of Ca are 590 and 1145 kJ/mol, respectively. Compare their values and comment on the differences.

7.47 Two atoms have the electron configurations $1s^2 2s^2 2p^6$ and $1s^2 2s^2 2p^6 3s^1$. The first ionization energy of one is 2080 kJ/mol, and that of the other is 496 kJ/mol. Match each ionization energy with one of the given electron configurations. Justify your choice.

7.48 Arrange the elements in each of the following groups in order of increasing electron affinity: (a) Li, Na, K; (b) F, Cl, Br, I.

7.49 Specify which of the following elements you would expect to have the greatest electron affinity: He, K, Co, S, Cl.

7.50 Considering their electron affinities, do you think it is possible for the alkali metals to form an anion like M^-, where M represents an alkali metal?

7.51 Explain why alkali metals have a greater affinity for electrons than alkaline earth metals.

SECTION 7.5: ELECTRON CONFIGURATION OF IONS

Review Questions

7.52 How does the electron configuration of ions derived from main group elements give them stability?

7.53 What do we mean when we say that two ions or an atom and an ion are *isoelectronic*?

7.54 Is it possible for the atoms of one element to be isoelectronic with the atoms of another element? Explain.

7.55 Give three examples of first-row transition metal (Se to Cu) ions that are isoelectronic with argon.

Conceptual Problems

7.56 A M^{2+} ion derived from a metal in the first transition metal series has four electrons in the $3d$ subshell. What element might M be?

7.57 A metal ion with a net +3 charge has five electrons in the $3d$ subshell. Identify the metal.

7.58 Write the ground-state electron configurations of the following ions: (a) Li^+, (b) H^-, (c) N^{3-}, (d) F^-, (e) S^{2-}, (f) Al^{3+}, (g) Se^{2-}, (h) Br^-, (i) Rb^+, (j) Sr^{2+}, (k) Sn^{2+}, (l) Te^{2-}, (m) Ba^{2+}, (n) Pb^{2+}, (o) In^{3+}, (p) Tl^+, (q) Tl^{3+}.

7.59 Group the species that are isoelectronic: Be^{2+}, F^-, Fe^{2+}, N^{3-}, He, S^{2-}, Co^{3+}, Ar.

7.60 Write the ground-state electron configurations of the following transition metal ions: (a) Sc^{3+}, (b) Ti^{4+}, (c) V^{5+}, (d) Cr^{3+}, (e) Mn^{2+}, (f) Fe^{2+}, (g) Fe^{3+}, (h) Co^{2+}, (i) Ni^{2+}, (j) Cu^+, (k) Cu^{2+}, (l) Ag^+, (m) Au^+, (n) Au^{3+}, (o) Pt^{2+}.

7.61 Name the ions with three charges that have the following electron configurations: (a) $[Ar]3d^3$, (b) $[Ar]$, (c) $[Kr]4d^6$, (d) $[Xe]4f^{14}5d^6$.

7.62 Which of the following species are isoelectronic with each other: C, Cl^-, Mn^{2+}, B^-, Ar, Zn, Fe^{3+}, Ge^{2+}?

SECTION 7.6: IONIC RADIUS

Review Questions

7.63 Define *ionic radius*. How does the size of an atom change when it is converted to (a) an anion and (b) a cation?

7.64 Explain why, for isoelectronic ions, the anions are larger than the cations.

Conceptual Problems

7.65 Indicate which one of the two species in each of the following pairs is smaller: (a) Cl or Cl^-, (b) Na or Na^+, (c) O^{2-} or S^{2-}, (d) Mg^{2+} or Al^{3+}, (e) Au^+ or Au^{3+}.

7.66 List the following ions in order of increasing ionic radius: N^{3-}, Na^+, F^-, Mg^{2+}, O^{2-}.

7.67 Explain which of the following cations is larger, and why: Cu^+ or Cu^{2+}.

7.68 Explain which of the following anions is larger, and why: Se^{2-} or Te^{2-}.

SECTION 7.7: PERIODIC TRENDS IN CHEMICAL PROPERTIES OF THE MAIN GROUP ELEMENTS

Review Questions

7.69 Why do members of a group exhibit similar chemical properties?

7.70 Explain why hydrogen belongs in its own group.

7.71 Which elements are more likely to form acidic oxides? Basic oxides? Amphoteric oxides?

Conceptual Problems

7.72 Give the physical states (gas, liquid, or solid) of the main group elements in the fourth period (K, Ca, Ga, Ge, As, Se, Br) at room temperature.

7.73 The boiling points of neon and krypton are −246.1°C and −153.2°C, respectively. Using these data, estimate the boiling point of argon.

7.74 Use the alkali metals and alkaline earth metals as examples to show how we can predict the chemical properties of elements simply from their electron configurations.

7.75 Based on your knowledge of the chemistry of the alkali metals, predict some of the chemical properties of francium, the last member of the group.

7.76 As a group, the noble gases are very stable chemically. Why?

7.77 Why are Group 11 elements more stable than Group 1 elements even though they seem to have the same outer electron configuration, ns^1, where n is the principal quantum number of the outermost shell?

7.78 How do the chemical properties of oxides change from left to right across a period? How do they change from top to bottom within a particular group?

7.79 Write balanced equations for the reactions between each of the following oxides and water: (a) Li_2O, (b) CaO, (c) SO_3.

7.80 Write formulas for and name the binary hydrogen compounds of the second-period elements (Li to F). Describe how the physical and chemical properties of these compounds change from left to right across the period.

7.81 Which oxide is more basic, MgO or BaO? Why?

ADDITIONAL PROBLEMS

7.82 State whether each of the following properties of the main group elements generally increases or decreases (a) from left to right across a period and (b) from top to bottom within a group: metallic character, atomic size, ionization energy, acidity of oxides.

7.83 Referring to the periodic table, name (a) the halogen in the fourth period, (b) an element similar to phosphorus in chemical properties, (c) the most reactive metal in the fifth period, (d) an element that has an atomic number smaller than 20 and is similar to strontium.

7.84 Write equations representing the following processes:
(a) The electron affinity of S^-
(b) The third ionization energy of titanium
(c) The electron affinity of Mg^{2+}
(d) The ionization energy of O^{2-}

7.85 Arrange the following isoelectronic species in order of increasing ionization energy: O^{2-}, F^-, Na^+, Mg^{2+}.

7.86 Write the empirical (or molecular) formulas of compounds that the elements in the third period (sodium to chlorine) should form with (a) molecular oxygen and (b) molecular chlorine. In each case indicate whether you would expect the compound to be ionic or molecular in character.

7.87 Arrange the following species in isoelectronic pairs: O^+, Ar, S^{2-}, Ne, Zn, Cs^+, N^{3-}, As^{3+}, N, Xe.

7.88 In which of the following are the species written in decreasing order by size of radius: (a) Be, Mg, Ba, (b) N^{3-}, O^{2-}, F^-, (c) Tl^{3+}, Tl^{2+}, Tl^+?

7.89 Which of the following is the most realistic representation of an atom from Group 2 becoming an ion?

7.90 Which of the following is the most realistic representation of an atom from Group 17 becoming an ion?

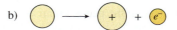

7.91 You are given four substances: a fuming red liquid, a dark metallic-looking solid, a pale-yellow gas, and a yellow-green gas that attacks glass. You are told that these substances are the first four members of Group 17, the halogens. Name each one.

7.92 For each pair of elements listed, give three properties that show their chemical similarity: (a) sodium and potassium and (b) chlorine and bromine.

7.93   What is the most reactive element on the periodic table?

7.94   Explain why the first electron affinity of sulfur is 200 kJ/mol but the second electron affinity is −649 kJ/mol.

7.95   The H^- ion and the He atom have two $1s$ electrons each. Which of the two species is larger? Explain.

7.96   Predict the products of the following oxides with water: Na_2O, BaO, CO_2, N_2O_5, P_4O_{10}, SO_3. Write an equation for each of the reactions. Specify whether the oxides are acidic, basic, or amphoteric.

7.97   Write the formulas and names of the oxides of the second-period elements (Li to N). Identify the oxides as acidic, basic, or amphoteric. Use the highest oxidation state of each element.

7.98   State whether each of the following elements is a gas, liquid, or solid under atmospheric conditions. Also state whether it exists in the elemental form as atoms, molecules, or a three-dimensional network: Mg, Cl, Si, Kr, O, I, Hg, Br.

7.99   The formula for calculating the energies of an electron in a hydrogen-like ion is given in Problem 7.32. This equation can be applied only to one-electron atoms or ions. One way to modify it for more complex species is to replace Z with $Z - \sigma$ or Z_{eff}. Calculate the value of σ if the first ionization energy of helium is 3.94×10^{-18} J per atom. (Disregard the minus sign in the given equation in your calculation.)

7.100 Why do noble gases have negative electron affinity values?

7.101 The atomic radius of K is 227 pm and that of K^+ is 138 pm. Calculate the percent decrease in volume that occurs when $K(g)$ is converted to $K^+(g)$. (The volume of a sphere is $\frac{4}{3}\pi r^3$, where r is the radius of the sphere.)

7.102 The atomic radius of F is 72 pm and that of F^- is 133 pm. Calculate the percent increase in volume that occurs when $F(g)$ is converted to $F^-(g)$. (See Problem 7.101 for the volume of a sphere.)

7.103 Match each of the elements on the right with its description on the left:

(a) A dark-red liquid   Calcium (Ca)
(b) A colorless gas that burns in oxygen gas   Gold (Au)
(c) A metal that reacts violently with water   Hydrogen (H_2)
(d) A shiny metal that is used in jewelry   Argon (Ar)
(e) An inert gas   Bromine (Br_2)

7.104 The energy needed for the following process is 1.96×10^4 kJ/mol:

$$Li(g) \longrightarrow Li^{3+}(g) + 3e^-$$

If the first ionization energy of lithium is 520 kJ/mol, calculate the second ionization energy of lithium, that is, the energy required for the process

$$Li^+(g) \longrightarrow Li^{2+}(g) + e^-$$

(*Hint:* You need the equation in Problem 7.32.)

7.105 A student is given samples of three elements, X, Y, and Z which could be an alkali metal, a member of Group 14, or a member of Group 15. She makes the following observations: Element X has a metallic luster and conducts electricity. It reacts slowly with hydrochloric acid to produce hydrogen gas. Element Y is a light yellow solid that does not conduct electricity. Element Z has a metallic luster and conducts electricity. When exposed to air, it slowly forms a white powder. A solution of the white powder in water is basic. What can you conclude about the elements from these observations?

7.106 What is the electron affinity (in kJ/mol) of the Na^+ ion?

7.107 The ionization energies of sodium (in kJ/mol), starting with the first and ending with the eleventh, are 496, 4562, 6910, 9543, 13,354, 16,613, 20,117, 25,496, 28,932, 141,362, 159,075. Plot the log of ionization energy (y axis) versus the number of ionization (x axis); for example, log 496 is plotted versus 1 (labeled IE_1, the first ionization energy), log 4562 is plotted versus 2 (labeled IE_2, the second ionization energy), and so on. (a) Label IE_1 through IE_{11} with the electrons in orbitals such as $1s$, $2s$, $2p$, and $3s$. (b) What can you deduce about electron shells from the breaks in the curve?

7.108 Explain, in terms of their electron configurations, why Fe^{2+} is more easily oxidized to Fe^{3+} than Mn^{2+} is to Mn^{3+}.

7.109 Write the formulas and names of the hydrides of the following second-period elements: Li, C, N, O, F. Predict their reactions with water.

7.110 In halogen displacement reactions a halogen element can be generated by oxidizing its anions with a halogen element that lies above it in the periodic table. This means that there is no way to prepare elemental fluorine, because it is the first member of Group 17. Indeed, for years the only way to prepare elemental fluorine was to oxidize F^- ions by electrolytic means. Then, in 1986, a chemist reported that by combining potassium hexafluoromanganate(IV) (K_2MnF_6) with antimony pentafluoride (SbF_5) at 150°C, he had generated elemental fluorine. Balance the following equation representing the reaction:

$$K_2MnF_6 + SbF_5 \longrightarrow KSbF_6 + MnF_3 + F_2$$

7.111 Write a balanced equation for the preparation of (a) molecular oxygen, (b) ammonia, (c) carbon dioxide, (d) molecular hydrogen, (e) calcium oxide. Indicate the physical state of the reactants and products in each equation.

7.112 Write chemical formulas for oxides of nitrogen with the following oxidation numbers: +1, +2, +3, +4, +5. (*Hint:* There are two oxides of nitrogen with a +4 oxidation number.)

7.113 Most transition metal ions are colored. For example, a solution of $CuSO_4$ is blue. How would you show that the blue color is due to the hydrated Cu^{2+} ions and not the SO_4^{2-} ions?

7.114 In general, atomic radius and ionization energy have opposite periodic trends. Why?

7.115 Explain why the electron affinity of nitrogen is approximately zero, while the elements on either side, carbon and oxygen, have substantial positive electron affinities.

7.116 Consider the halogens chlorine, bromine, and iodine. The melting point and boiling point of chlorine are $-101.5°C$ and $-34.0°C$ and those of iodine are $113.7°C$ and $184.3°C$, respectively. Thus chlorine is a gas and iodine is a solid under room conditions. Estimate the melting point and boiling point of bromine. Compare your values with those from the webelements.com website.

7.117 Although it is possible to determine the second, third, and higher ionization energies of an element, the same cannot usually be done with the electron affinities of an element. Explain.

7.118 Why do elements that have high ionization energies also have more positive electron affinities? Which group of elements would be an exception to this generalization?

7.119 The first four ionization energies of an element are approximately 738, 1450, 7.7×10^3, and 1.1×10^4 kJ/mol. To which periodic group does this element belong? Explain your answer.

7.120 Predict the atomic number and ground-state electron configuration of the next member of the alkali metals after francium.

7.121 (a) The formula of the simplest hydrocarbon is CH_4 (methane). Predict the formulas of the simplest compounds formed between hydrogen and the following elements: silicon, germanium, tin, and lead. (b) Sodium hydride (NaH) is an ionic compound. Would you expect rubidium hydride (RbH) to be more or less ionic than NaH? (c) Predict the reaction between radium (Ra) and water. (d) When exposed to air, aluminum forms a tenacious oxide (Al_2O_3) coating that protects the metal from corrosion. Which metal in Group 2 would you expect to exhibit similar properties? (See Section 7.7.)

7.122 Match each of the elements on the right with its description on the left:

(a) A pale yellow gas that reacts with water   Nitrogen (N_2)

Boron (B)

(b) A soft metal that reacts with water to produce hydrogen   Fluorine (F_2)

Aluminum (Al)

(c) A metalloid that is hard and has a high melting point   Sodium (Na)

(d) A colorless, odorless gas

(e) A metal that is more reactive than iron, but does not corrode in air

7.123 One way to estimate the effective charge (Z_{eff}) of a many-electron atom is to use the equation $IE_1 = (1312 \text{ kJ/mol}) (Z_{eff}^2/n^2)$, where IE_1 is the first ionization energy and n is the principal quantum number of the shell in which the electron resides. Use this equation to calculate the effective nuclear charges of Li, Na, and K. Also calculate Z_{eff}/n for each metal. Comment on your results.

7.124 Use your knowledge of thermochemistry to calculate the ΔH for the following processes: (a) $Cl^-(g) \longrightarrow Cl^+(g) + 2e^-$ and (b) $K^+(g) + 2e^- \longrightarrow K^-(g)$.

7.125 To prevent the formation of oxides, peroxides, and superoxides, alkali metals are sometimes stored in an inert atmosphere. Which of the following gases should not be used for lithium: Ne, Ar, N_2, Kr? Explain. (*Hint:* As mentioned in the chapter, Li and Mg exhibit a diagonal relationship. Compare the common compounds of these two elements.)

7.126 On one graph, plot the effective nuclear charge (shown in parentheses) and atomic radius (see Figure 7.6) versus atomic number for the second-period elements: Li(1.28), Be(1.91), B(2.42), C(3.14), N(3.83), O(4.45), F(5.10), Ne(5.76). Comment on the trends.

7.127 One allotropic form of an element X is a colorless crystalline solid. The reaction of X with an excess amount of oxygen produces a colorless gas. This gas dissolves in water to yield an acidic solution. Choose one of the following elements that matches X: (a) sulfur, (b) phosphorus, (c) carbon, (d) boron, (e) silicon.

Engineering Problems

7.128 Calculate the maximum wavelength of light (in nm) required to ionize a single sodium atom.

7.129 A technique called photoelectron spectroscopy is used to measure the ionization energy of atoms. A gaseous sample is irradiated with UV light, and electrons are ejected from the valence shell. The kinetic energies of the ejected electrons are measured. Because the energy of the UV photon and the kinetic energy of the ejected electron are known, we can write

$$h\nu = IE + \tfrac{1}{2}mu^2$$

where ν is the frequency of the UV light, and m and u are the mass and velocity of the electron, respectively. In one experiment the kinetic energy of the ejected electron from potassium is found to be 5.34×10^{-19} J using a UV source of wavelength 162 nm. Calculate the ionization energy of potassium. How can you be sure that this ionization energy corresponds to the electron in the valence shell (i.e., the most loosely held electron)?

7.130 Element M is a shiny and highly reactive metal (melting point 63°C), and element X is a highly reactive nonmetal (melting point −7.2°C). They react to form a compound with the empirical formula MX, a colorless, brittle white solid that melts at −734°C. When dissolved in water or when in the molten state, the substance conducts electricity. When chlorine gas is bubbled through an aqueous solution containing MX, a reddish-brown liquid appears and Cl^- ions are formed. From these observations, identify M and X. (You may need to consult a handbook of chemistry for the melting-point values.)

Biological Problems

7.131 Write the ground-state electron configurations of the following ions, which play important roles in biochemical processes in our bodies: (a) Na^+, (b) Mg^{2+}, (c) Cl^-, (d) K^+, (e) Ca^{2+}, (f) Fe^{2+}, (g) Cu^{2+}, (h) Zn^{2+}.

7.132 Thallium (Tl) is a neurotoxin and exists mostly in the Tl(I) oxidation state in its compounds. Aluminum (Al), which causes anemia and dementia, is only stable in the Al(III) form. The first, second, and third ionization energies of Tl are 589, 1971, and 2878 kJ/mol, respectively. The first, second, and third ionization energies of Al are 577.5, 1817, and 2745 kJ/mol, respectively. Plot the ionization energies of Al and Tl versus the number of electrons removed and explain the trends.

7.133 Both Mg^{2+} and Ca^{2+} are important biological ions. One of their functions is to bind to the phosphate group of ATP molecules or amino acids of proteins. For Group 2 metals in general, the tendency for binding to the anions increases in the order $Ba^{2+} < Sr^{2+} < Ca^{2+} < Mg^{2+}$. Explain this trend.

7.134 The air in a spacecraft or submarine needs to be purified of exhaled carbon dioxide. Write equations for the reactions between carbon dioxide and (a) lithium oxide (Li_2O), (b) sodium peroxide (Na_2O_2), and (c) potassium superoxide (NO_2).

Multiconcept Problems

7.135 As discussed in the chapter, the atomic mass of argon is greater than that of potassium. This observation created a problem in the early development of the periodic table because it meant that argon should be placed after potassium. (a) How was this difficulty resolved? (b) From the following data, calculate the average atomic masses of argon and potassium: Ar-36 (35.9675 amu, 0.337 percent), Ar-38 (37.9627 amu, 0.063 percent), Ar-40 (39.9624 amu, 99.60 percent), K-39 (38.9637 amu, 93.258 percent), K-40 (39.9640 amu, 0.0117 percent), K-41 (40.9618 amu, 6.730 percent).

7.136 Little is known of the chemistry of astatine, the last member of Group 17. Describe the physical characteristics that you would expect this halogen to have. Predict the products of the reaction between sodium astatide (NaAt) and sulfuric acid. (*Hint:* Sulfuric acid is an oxidizing agent.)

7.137 Based on knowledge of the electronic configuration of titanium, state which of the following compounds of titanium is unlikely to exist: K_3TiF_6, $K_2Ti_2O_5$, $TiCl_3$, K_2TiO_4, K_2TiF_6.

7.138 The ionization energy of a certain element is 412 kJ/mol. When the atoms of this element are in the first excited state, however, the ionization energy is only 126 kJ/mol. Based on this information, calculate the wavelength of light emitted in a transition from the first excited state to the ground state.

7.139 Experimentally, the electron affinity of an element can be determined by using a laser light to ionize the anion of the element in the gas phase:

$$X^-(g) + h\nu \longrightarrow X(g) + e^-$$

Referring to Figure 7.10, calculate the photon wavelength (in nm) corresponding to the electron affinity for chlorine. In what region of the electromagnetic spectrum does this wavelength fall?

Standardized-Exam Practice Problems

Physical and Biological Sciences

These questions are not based on a descriptive passage.

1. A halogen has valence electrons in which orbitals?

   a) *s*   b) *s* and *p*   c) *p*   d) *s, p,* and *d*

2. How many subshells does a shell with principal quantum number *n* contain?

   a) *n*   b) n^2 c) $n - 1$ d) $2n - 1$

3. In a shell that contains an *f* subshell, what is the ratio of *f* orbitals to *s* orbitals?

   a) 14:1   b) 7:1   c) 7:3   d) 7:5

4. What is the maximum number of electrons that can be in the $n = 3$ shell?

   a) 2   b) 6   c) 8   d) 18

Answers to In-Chapter Materials

Answers to Practice Problems

7.1A Ge. **7.1B** Bi < As < P. **7.2A** (a) $1s^22s^22p^63s^23p^3$, p-block,
(b) $1s^22s^22p^63s^23p^64s^2$, s-block, (c) $1s^22s^22p^63s^23p^64s^23d^{10}4p^5$,
p-block. **7.2B** (a) Al, (b) Zn, (c) Sr. **7.3A** F < Se < Ge.
7.3B P and Se. **7.4A** Mg, Mg. **7.4B** Rb has a smaller Z_{eff}; IE_2
for Rb corresponds to the removal of a core electron. **7.5A** Al.
7.5B Adding an electron to As involves pairing. **7.6A** The attractive
force is slightly larger between +3.26 and −1.15 separated by
1.5 pm. **7.6B** 1.51 pm. **7.7A** (a) [Ne], (b) [Ar], (c) [Kr].

7.7B N^{3-}, O^{2-}, F^-, Ne, Na^+, Mg^{2+}, Al^{3+}. **7.8A** (a) $[Ar]3d^6$,
(b) $[Ar]3d^9$, (c) $[Kr]4d^{10}$. **7.8B** Cu^+. **7.9A** Rb^+ < Br^- < Se^{2-}.
7.9B F^-, O^{2-}, N^{3-}, Na^+, Mg^{2+}, Al^{3+}.

Answers to Checkpoints

7.1.1 c. **7.1.2** a. **7.1.3** c. **7.2.1** b. **7.2.2** a, d, e. **7.4.1** b. **7.4.2** c.
7.4.3 e. **7.4.4** a. **7.4.5** d. **7.4.6** a. **7.5.1** b, c, e. **7.5.2** b, d. **7.5.3** d.
7.5.4 b. **7.6.1** d, e. **7.6.2** a. **7.6.3** c. **7.6.4** d.

CHAPTER 8

Chemical Bonding I: Basic Concepts

Dynamite, a stabilized form of the explosive nitroglycerin, is used to blast through solid rock.

Dinodia Photos/Alamy Stock Photo

In This Chapter, You Will Learn

About two different types of chemical bonds, called *ionic* and *covalent*. You will also learn how to represent atoms and atomic ions using Lewis dot symbols and how to represent molecules and polyatomic ions using Lewis *structures*.

Before You Begin, Review These Skills

- Atomic radius [◄◄Section 7.4]
- Ions of main group elements [◄◄ Section 7.5]

What Chemical Bonds Have to Do with Explosives

When the Swedish chemist Alfred Nobel died in 1896, his will specified that the bulk of his considerable fortune was to be used to establish the prizes that bear his name. The prizes, given annually in five categories (Chemistry, Physics, Physiology or Medicine, Literature, and Peace), are intended to recognize significant contributions to the betterment of humankind. In life, Nobel had been a prolific scientist and entrepreneur. He held more than 300 patents, including the one for dynamite, a stabilized form of the explosive nitroglycerin. His extensive work on the development and manufacture of explosives earned him the title "merchant of death" and was the cause of personal tragedy when his younger brother was killed in an explosion at one of the family's factories.

Ironically, toward the end of his life, Nobel developed heart disease–related chest pain (angina pectoris) and was directed by his physician to take nitroglycerin orally, which Nobel refused to do. Glyceryl trinitrate, the name used by the medical community—perhaps to avoid the impression that doctors are prescribing explosives to patients—still is widely used to treat the symptoms of heart disease and other medical conditions. One of the more recently developed and intriguing uses of nitroglycerin is its placement in the tip of a condom to stimulate an erection.

Both the explosive nature of nitroglycerin and its effectiveness in treating such conditions as heart disease and erectile dysfunction can be illuminated by an understanding of the basic concepts of *chemical bonding*.

Student Note: A sixth prize, the *Nobel Memorial Prize in Economics,* is awarded along with the others but was not specified in Nobel's will, and it is not really a Nobel Prize.

At the end of this chapter, you will be able to answer a series of questions about nitric oxide and nitroglycerin [▶▶ Applying What You've Learned, page 399].

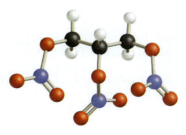

Nitroglycerin

8.1   Lewis Dot Symbols

The development of the periodic table and the concept of electron configuration gave chemists a way to explain the formation of compounds. The explanation, formulated by Gilbert Lewis,[1] is that atoms combine to achieve a more stable electron configuration. Maximum stability results when an atom is isoelectronic with a noble gas.

When atoms interact to form compounds, it is their *valence electrons* that actually interact. Therefore, it is helpful to have a method for depicting the valence electrons of the atoms involved. This can be done using Lewis dot symbols. A **Lewis dot symbol** consists of the element's symbol surrounded by dots, where each dot represents a valence electron. For the main group elements, the number of dots in the Lewis dot symbol is the same as the group number for Groups 1 and 2, and the same as the *second digit* in the group number for Groups 13 through 18, as shown in Figure 8.1. (Because they have incompletely filled inner shells, transition metals typically are not represented with Lewis dot symbols.)

Figure 8.1 shows that dots are placed above and below as well as to the left and right of the symbol. The exact order in which the dots are placed around the element symbol is not important, but the number of dots is. Thus, any of the following would be correct for the Lewis dot symbol for boron:

When writing Lewis dot symbols, though, we do not "pair" dots until absolutely necessary. Thus, we would *not* represent boron with a pair of dots on one side and a single dot on another.

For main group metals such as Na or Mg, the number of dots in the Lewis dot symbol is the number of electrons that are lost when the atom forms a cation that is isoelectronic with the preceding noble gas. For nonmetals of the second period (B through F), the number of unpaired dots is the number of bonds the atom can form. (As we will see shortly [►► Section 8.8], a larger nonmetal atom [one in the third period or beyond] can actually form as many bonds as the total number of dots in its Lewis dot symbol.)

In addition to atoms, we can also represent atomic *ions* with Lewis dot symbols. To do so, we simply add (for anions) or subtract (for cations) the appropriate number of dots from the Lewis dot symbol of the atom and include the ion's charge.

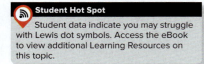

Figure 8.1 Lewis dot symbols of the main group elements.

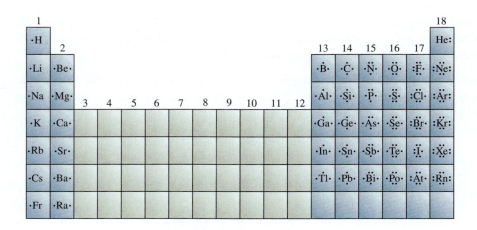

---

1. Gilbert Newton Lewis (1875–1946). American chemist. Lewis made many important contributions in the areas of chemical bonding, thermodynamics, acids and bases, and spectroscopy. Despite the significance of Lewis's work, he was never awarded a Nobel Prize.

Sample Problem 8.1 shows how to use Lewis dot symbols to represent atomic ions.

SAMPLE PROBLEM 8.1

Write Lewis dot symbols for (a) fluoride ion (F^-), (b) potassium ion (K^+), and (c) sulfide ion (S^{2-}).

Strategy Starting with the Lewis dot symbol for each element, add dots (for anions) or remove dots (for cations) as needed to achieve the correct charge on each ion. Don't forget to include the appropriate charge on the Lewis dot symbol.

Setup The Lewis dot symbols for F, K, and S are $:\ddot{F}\cdot$, $K\cdot$, and $\cdot\ddot{S}\cdot$, respectively.

Solution (a) $\left[:\ddot{F}:\right]^-$

(b) K^+

(c) $\left[:\ddot{S}:\right]^{2-}$

THINK ABOUT IT

For ions that are isoelectronic with noble gases, cations should have no dots remaining around the element symbol, whereas anions should have eight dots around the element symbol. Note, too, that we put square brackets around the Lewis dot symbol for an anion and place the negative charge outside the brackets.

Practice Problem **A**TTEMPT Write Lewis dot symbols for (a) Ca^{2+}, (b) N^{3-}, and (c) I^-.

Practice Problem **B**UILD Indicate the charge on each of the ions represented by the following Lewis dot symbols: (a) $\left[:\ddot{O}:\right]^?$, (b) $H^?$, and (c) $\left[:\ddot{P}:\right]^?$.

Practice Problem **C**ONCEPTUALIZE For each of the highlighted positions on the periodic table, write a Lewis structure for an atom of the element and for the common ion that it forms. Use the generic symbol X for each element. For example, rather than writing ·Na and [Na]$^+$ for sodium and sodium ion, respectively, you should write ·X and [X]$^+$.

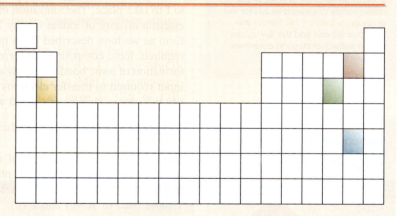

CHECKPOINT – SECTION 8.1   Lewis Dot Symbols

8.1.1 Using only a periodic table, determine the correct Lewis dot symbol for a silicon (Si) atom.

   a) $:Si:$ d) $\cdot Si\cdot$

   b) $\cdot\dot{S}i\cdot$ e) $:Si$

   c) $:\dot{S}i\cdot$

8.1.2 Using only a periodic table, determine the correct Lewis dot symbol for the bromide ion (Br^-).

   a) $\left[:\ddot{B}r\cdot\right]^-$ d) $\left[Br\right]^-$

   b) $\left[:\ddot{B}r\right]^-$ e) $\left[:\ddot{B}r:\right]^-$

   c) $\left[\cdot\ddot{B}r\cdot\right]^-$

8.1.3 To which group does element X belong if its Lewis symbol is $\cdot\ddot{X}\cdot$?

   a) 1   d) 14   g) 17

   b) 2   e) 15   h) It is not possible to tell.

   c) 13   f) 16

8.1.4 To which group does element Y belong if the Lewis symbol for its anion is $\left[:\ddot{Y}:\right]^{m-}$, where m represents the charge?

   a) 14   c) 16   e) It is not possible to tell.

   b) 15   d) 17

Figure 8.6

Born-Haber Cycle

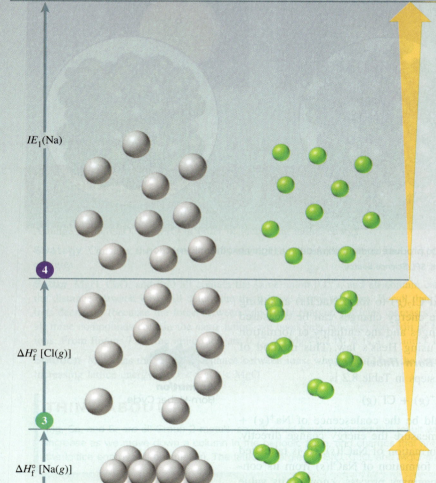

IE_1(Na)

ΔH_f° [Cl(g)]

ΔH_f° [Na(g)]

ΔH_f° [NaCl(s)]

4

3

2

1

Step 4

The first ionization energy of sodium, IE_1 (Na), gives the amount of energy required to convert 1 mole of Na(g) to 1 mole of Na$^+$(g):

$$Na(g) \longrightarrow Na^+(g) + e^-$$

Step 3

The tabulated value of ΔH_f° for Cl(g) gives the amount of energy needed to convert $\frac{1}{2}$ mole of Cl$_2$(g) to 1 mole of Cl(g):

$$\tfrac{1}{2}Cl_2(g) \longrightarrow Cl(g)$$

Step 2

The tabulated value of ΔH_f° for Na(g) gives the amount of energy needed to convert 1 mole of Na(s) to 1 mole of Na(g):

$$Na(s) \longrightarrow Na(g)$$

Step 1

The tabulated value of ΔH_f° for NaCl(s) gives us the energy produced when 1 mole of Na and $\frac{1}{2}$ mole of Cl$_2$ combine to form 1 mole of NaCl:

$$Na(s) + \tfrac{1}{2}Cl_2(g) \longrightarrow NaCl(s)$$

Step 1 in the Born-Haber cycle involves converting 1 mole of NaCl into 1 mole of Na and $\frac{1}{2}$ mole of Cl$_2$ (the reverse of the ΔH_f° reaction):

$$NaCl(s) \longrightarrow Na(s) + \tfrac{1}{2}Cl_2(g)$$

Therefore, ΔH for step 1 is $-\Delta H_f^\circ$ [NaCl(s)].

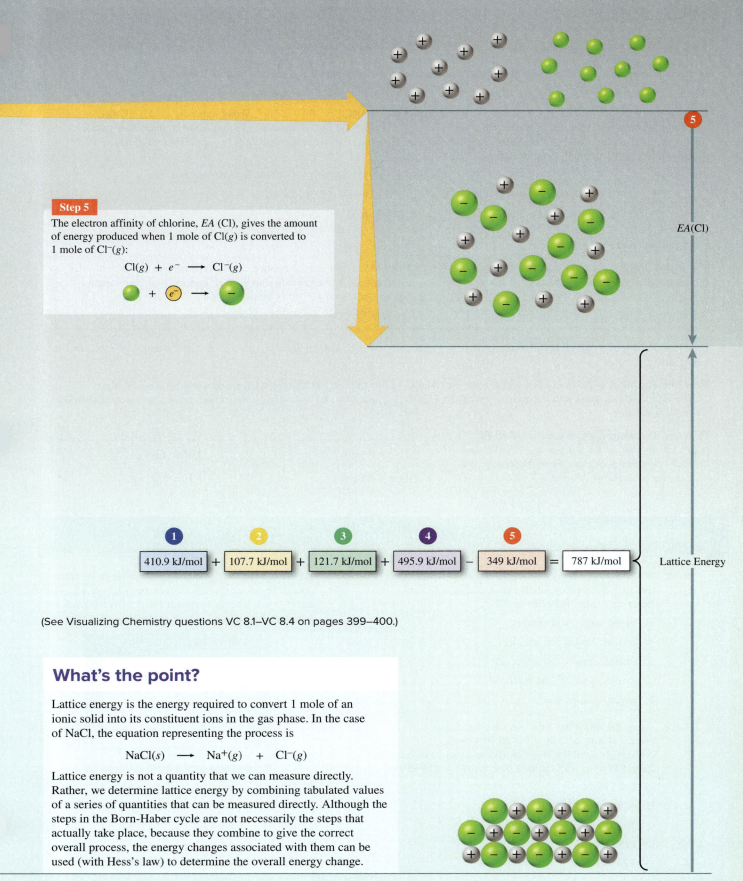

(See Visualizing Chemistry questions VC 8.1–VC 8.4 on pages 399–400.)

What's the point?

Lattice energy is the energy required to convert 1 mole of an ionic solid into its constituent ions in the gas phase. In the case of NaCl, the equation representing the process is

$$NaCl(s) \longrightarrow Na^+(g) + Cl^-(g)$$

Lattice energy is not a quantity that we can measure directly. Rather, we determine lattice energy by combining tabulated values of a series of quantities that can be measured directly. Although the steps in the Born-Haber cycle are not necessarily the steps that actually take place, because they combine to give the correct overall process, the energy changes associated with them can be used (with Hess's law) to determine the overall energy change.

Step 5

The electron affinity of chlorine, *EA* (Cl), gives the amount of energy produced when 1 mole of Cl(*g*) is converted to 1 mole of Cl⁻(*g*):

$$Cl(g) + e^- \longrightarrow Cl^-(g)$$

EA(Cl)

410.9 kJ/mol + 107.7 kJ/mol + 121.7 kJ/mol + 495.9 kJ/mol − 349 kJ/mol = 787 kJ/mol

Lattice Energy

SAMPLE PROBLEM 8.3

Using data from Figures 7.8 and 7.10 and Appendix 2, calculate the lattice energy of cesium chloride (CsCl).

Strategy Using Figure 8.6 as a guide, combine the pertinent thermodynamic data and use Hess's law to calculate the lattice energy.

Setup From Figure 7.8, $IE_1(\text{Cs}) = 376$ kJ/mol. From Figure 7.10, $EA_1(\text{Cl}) = 349.0$ kJ/mol. From Appendix 2, $\Delta H_f^\circ[\text{Cs}(g)] = 76.50$ kJ/mol, $\Delta H_f^\circ[\text{Cl}(g)] = 121.7$ kJ/mol, and $\Delta H_f^\circ[\text{CsCl}(s)] = -442.8$ kJ/mol. Because we are interested in magnitudes only, we can use the absolute values of the thermodynamic data. And, because only the standard heat of formation of CsCl(s) is a negative number, it is the only one for which the sign changes.

Solution
$$\{\Delta H_f^\circ[\text{Cs}(g)] + \Delta H_f^\circ[\text{Cl}(g)] + IE_1(\text{Cs}) + |\Delta H_f^\circ[\text{CsCl}(s)]|\} - EA_1(\text{Cl}) = \text{lattice energy}$$

$$= (76.50 \text{ kJ/mol} + 121.7 \text{ kJ/mol} + 376 \text{ kJ/mol} + 442.8 \text{ kJ/mol}) - 349.0 \text{ kJ/mol}$$

$$= 668 \text{ kJ/mol}$$

THINK ABOUT IT

Compare this value to that for NaCl in Figure 8.6 (787 kJ/mol). Both compounds contain the same anion (Cl⁻) and both have cations with the same charge (+1), so the relative sizes of the cations will determine the relative strengths of their lattice energies. Because Cs⁺ is larger than Na⁺, the lattice energy of CsCl is smaller than that of NaCl.

Practice Problem A TTEMPT Using data from Figures 7.8 and 7.10 and Appendix 2, calculate the lattice energy of rubidium iodide (RbI).

Practice Problem B UILD The lattice energy of MgO is 3890 kJ/mol, and the second ionization energy (IE_2) of Mg is 1450.6 kJ/mol. Using these data, as well as data from Figures 7.8 and 7.10 and Appendix 2, determine the second electron affinity for oxygen, $EA_2(\text{O})$.

Practice Problem C ONCEPTUALIZE Five points (A through E) lie along a line. The known distances between points are given. Determine the distance between points A and C.

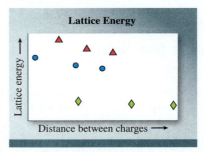

AB = 5.05 cm   DE = 4.65 cm

BD = 7.65 cm   CE = 6.27 cm

CHECKPOINT – SECTION 8.2   Ionic Bonding

8.2.1 Will the lattice energy of KF be larger or smaller than that of LiF, larger or smaller than that of KCl, and larger or smaller than that of KI?

   a) larger, larger, and smaller

   b) smaller, larger, and smaller

   c) smaller, larger, and larger

   d) smaller, smaller, and smaller

   e) larger, smaller, and larger

8.2.2 Using the following data, calculate the lattice energy of KF: $\Delta H_f^\circ[\text{K}(g)] = 89.99$ kJ/mol, $\Delta H_f^\circ[\text{F}(g)] = 80.0$ kJ/mol, $IE_1(\text{K}) = 418.8$ kJ/mol, $\Delta H_f^\circ[\text{KF}(s)] = -547$ kJ/mol, and $EA_1(\text{F}) = 328$ kJ/mol.

   a) 808 kJ/mol

   b) −286 kJ/mol

   c) 261 kJ/mol

   d) 1355 kJ/mol

   e) −261 kJ/mol

8.2.3 Lattice energies are graphed for three series of compounds in which the ion charges are +2, −2; +2, −1; and +1, −1. The ions in each series of compounds are separated by different distances. Identify the series.

   a) red +2, −2; blue +1, −1; green +2, −1

   b) red +2, −2; blue +2, −1; green +1, −1

   c) red +2, −1; blue +2, −2; green +1, −1

   d) red +1, −1; blue +2, −1; green +2, −2

   e) red +2, −1; blue +1, −1; green +2, −2

8.3 Covalent Bonding

We learned in Section 8.2 that ionic compounds tend to form between metals and nonmetals when electrons are transferred from an element with a low ionization energy (the metal) to one with a high electron affinity (the nonmetal). When compounds form between elements with more similar properties, electrons are not transferred from one element to another but instead are shared to give each atom a noble gas electron configuration. It was Gilbert Lewis who first suggested that a chemical bond involves atoms sharing electrons, and this approach is known as the ***Lewis theory of bonding.***

Lewis theory depicts the formation of the bond in H_2 as

$$H\cdot \; + \; \cdot H \; \longrightarrow \; H\!:\!H$$

In essence, two H atoms move close enough to each other to *share* the electron pair. Although there are still two atoms and just two electrons, this arrangement allows each H atom to "count" both electrons as its own and to "feel" as though it has the noble gas electron configuration of helium. This type of arrangement, where two atoms share a pair of electrons, is known as ***covalent bonding,*** and the shared pair of electrons constitutes the ***covalent bond.*** (For the sake of simplicity, the shared pair of electrons can be represented by a dash, rather than by two dots: H—H.) In a covalent bond, each electron in a shared pair is attracted to the nuclei of both atoms. It is this attraction that holds the two atoms together.

Lewis summarized much of his theory of chemical bonding with the octet rule. According to the ***octet rule,*** atoms will lose, gain, or share electrons to achieve a *noble gas electron configuration*. This rule enables us to predict many of the formulas for compounds consisting of specific elements. The octet rule holds for nearly all the compounds made up of second-period elements and is therefore especially important in the study of organic compounds, which contain mostly C, N, and O atoms.

> **Student Note:** For nearly all elements, achieving a noble gas electron configuration results in eight electrons around each atom—hence the name *octet* rule. For H, the octet rule dictates that it have two electrons, giving it the electron configuration of the noble gas He.

As with ionic bonding, covalent bonding of many-electron atoms involves only the valence electrons. Consider the fluorine molecule (F_2). The electron configuration of F is $1s^2 2s^2 2p^5$. The $1s$ electrons are low in energy and stay near the nucleus most of the time, so they do not participate in bond formation. Thus, each F atom has seven valence electrons (the two $2s$ and five $2p$ electrons). According to Figure 8.1, there is only one unpaired electron on F, so the formation of the F_2 molecule can be represented as follows:

$$:\!\ddot{F}\!\cdot \; + \; \cdot\ddot{F}\!: \; \longrightarrow \; :\!\ddot{F}\!:\!\ddot{F}\!: \quad \text{or} \quad :\!\ddot{F}\!-\!\ddot{F}\!:$$

Only two valence electrons participate in the bond that forms F_2. The other, nonbonding electrons, are called ***lone pairs***—pairs of valence electrons that are not involved in covalent bond formation. Thus, each F in F_2 has three lone pairs of electrons.

$$\longrightarrow :\!\ddot{F}\!-\!\ddot{F}\!:\longleftarrow \text{lone pair}$$

Lewis Structures

The structures used to represent molecules held together by covalent bonds, such as H_2 and F_2, are called *Lewis structures*. A ***Lewis structure*** is a representation of covalent bonding in which shared electron pairs are shown either as dashes or as pairs of dots between two atoms, and lone pairs are shown as pairs of dots on individual atoms. Only valence electrons are shown in a *Lewis structure*.

> **Student Note:** Lewis structures are also referred to as Lewis dot structures. In this book, we will use the term *Lewis structure* to avoid confusion with the term *Lewis dot symbol.*

To draw the Lewis structure of the water molecule, recall that the Lewis dot symbol for oxygen has two unpaired dots (Figure 8.1), meaning that it has two unpaired electrons and can form two bonds. Because hydrogen has only one electron, it can form only one covalent bond. Thus, the Lewis structure for water is

$$H\!:\!\ddot{O}\!:\!H \quad \text{or} \quad H\!-\!\ddot{O}\!-\!H$$

In this case, the O atom has two lone pairs. The hydrogen atom has no lone pairs because its only valence electron is used to form a covalent bond.

In the F_2 and H_2O molecules, the F, H, and O atoms each achieve a stable noble gas configuration by sharing electrons, thus illustrating the octet rule:

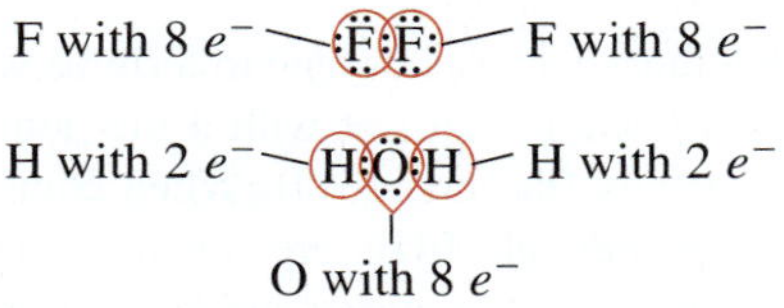

$$\text{F with } 8\,e^- \quad \text{\textcircled{F}\textcircled{F}} \quad \text{F with } 8\,e^-$$

$$\text{H with } 2\,e^- \quad \text{H\textcircled{O}H} \quad \text{H with } 2\,e^-$$

$$\text{O with } 8\,e^-$$

The octet rule works best for elements in the second period of the periodic table. These elements have only $2s$ and $2p$ valence subshells, which can hold a total of eight electrons. When an atom of one of these elements forms a covalent compound, it can attain the noble gas electron configuration [Ne] by sharing electrons with other atoms in the same compound. In Section 8.8, we will discuss some important exceptions to the octet rule.

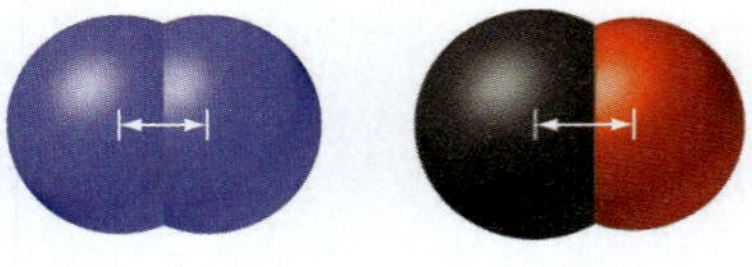

Multiple Bonds

Atoms can form several different types of covalent bonds, such as single bonds and multiple bonds. In a **single bond**, two atoms are held together by one electron pair. **Multiple bonds** form, on the other hand, when two atoms share two or more pairs of electrons. A multiple bond in which the atoms share two pairs of electrons is called a **double bond.** Double bonds are found in molecules such as carbon dioxide (CO_2) and ethylene (C_2H_4):

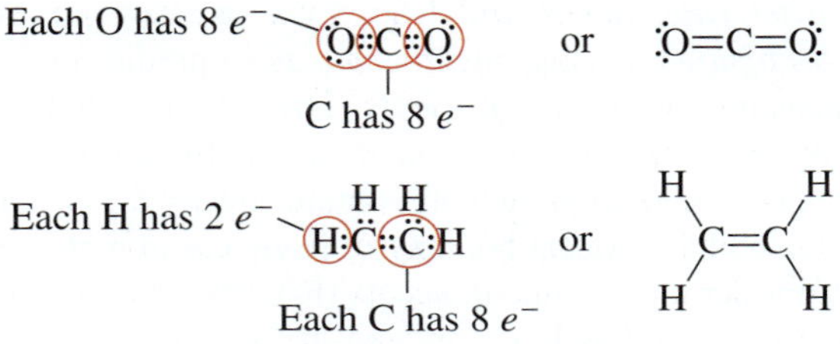

Each O has $8\,e^-$ — O:C:O or $\ddot{O}=C=\ddot{O}$
C has $8\,e^-$

Each H has $2\,e^-$ — H:C::C:H or
Each C has $8\,e^-$

A **triple bond** arises when two atoms share three pairs of electrons, as in molecules such as nitrogen (N_2) and acetylene (C_2H_2):

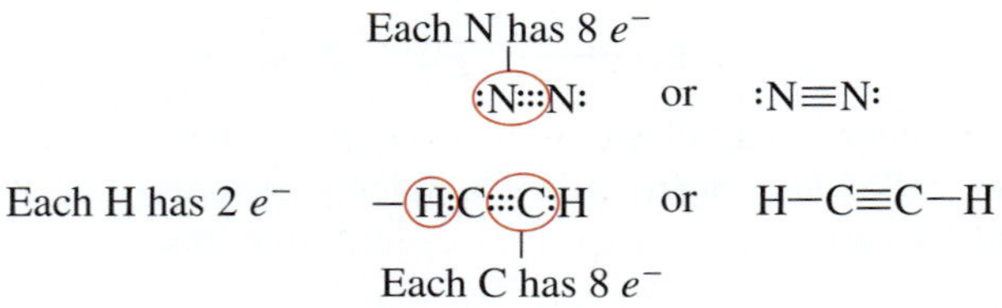

Each N has $8\,e^-$
$:\text{N:::N:}$ or $:N\equiv N:$

Each H has $2\,e^-$ H:C:::C:H or $H{-}C\equiv C{-}H$
Each C has $8\,e^-$

In ethylene and acetylene, all the valence electrons are used in bonding; there are no lone pairs on the carbon atoms. In fact, with the important exception of carbon monoxide (CO), most stable molecules containing carbon do not have lone pairs on the carbon atoms.

Multiple bonds are shorter than single bonds. *Bond length* is defined as the distance between the nuclei of two covalently bonded atoms in a molecule (Figure 8.7). Table 8.3 lists some experimentally determined bond lengths. For a given pair of atoms, such as carbon and nitrogen, triple bonds are shorter than double bonds, and double bonds are shorter than single bonds. The shorter multiple bonds are also stronger than single bonds, as we will see in Section 8.9.

Comparison of Ionic and Covalent Compounds

Ionic and covalent compounds differ markedly in their general physical properties because of differences in the nature of their bonds. There are two types of attractive forces in covalent compounds, the *intramolecular* bonding force that holds the atoms together in

N_2
Bond length 1.10 Å

CO
Bond length 1.13 Å

Figure 8.7 Bond length is the distance between the nuclei of two bonded atoms.

TABLE 8.3	Average Bond Lengths of Some Common Single, Double, and Triple Bonds

Bond Type	Bond Length (pm)
C—H	107
O—H	96
C—O	143
C=O	121
C≡O	113
C—C	154
C=C	133
C≡C	120
C—N	143
C=N	138
C≡N	116
N—N	147
N=N	124
N≡N	110
N—O	136
N=O	122
O—O	148
O=O	121

TABLE 8.4	Comparison of Some Properties of an Ionic Compound (NaCl) and a Covalent Compound (CCl_4)	
Property	**NaCl**	**CCl_4**
Appearance	White solid	Colorless liquid
Melting point (°C)	801	−23
Molar heat of fusion* (kJ/mol)	30.2	2.5
Boiling point (°C)	1413	76.5
Molar heat of vaporization* (kJ/mol)	600	30
Density (g/cm³)	2.17	1.59
Solubility in water	High	Very low
Electrical conductivity		
Solid	Poor	Poor
Liquid	Good	Poor
Aqueous	Good	Poor

*The *molar heat of fusion* and *molar heat of vaporization* are the amounts of heat needed to melt 1 mole of the solid and to vaporize 1 mole of the liquid, respectively.

a molecule, and the *intermolecular* forces between molecules. Bond enthalpy, discussed in Section 8.9, can be used to quantify the intramolecular bonding force. Intermolecular forces are usually quite weak compared to the forces holding atoms together within a molecule, so molecules of a covalent compound are not held together tightly. As a result, covalent compounds are usually gases, liquids, or low-melting solids.

On the other hand, the electrostatic forces holding ions together in an ionic compound are usually very strong, so ionic compounds are solids at room temperature and have high melting points. Many ionic compounds are soluble in water, and the resulting aqueous solutions conduct electricity because the compounds are strong electrolytes. Most covalent compounds are insoluble in water, or if they do dissolve, their aqueous solutions generally do not conduct electricity, because the compounds are nonelectrolytes. Molten ionic compounds conduct electricity because they contain mobile cations and anions; liquid or molten covalent compounds do *not* conduct electricity because no ions are present. Table 8.4 compares some of the properties of a typical ionic compound, sodium chloride (NaCl), with those of a covalent compound, carbon tetrachloride (CCl_4).

8.4 Electronegativity and Polarity

So far, we have described chemical bonds as either *ionic,* when they occur between a metal and a nonmetal, or *covalent,* when they occur between nonmetals. In fact, ionic and covalent bonds are simply the extremes in a spectrum of bonding. Bonds that fall between these two extremes are *polar,* meaning that electrons are shared but are not shared equally. Such bonds are referred to as *polar covalent bonds.* The following shows a comparison of the different types of bonds, where M and X represent two different elements:

M:X	$M^{\delta+}X^{\delta-}$	M^+X^-
Pure covalent bond	**Polar covalent bond**	**Ionic bond**
Neutral atoms held together by *equally* shared electrons	Partially charged atoms held together by *unequally* shared electrons	Oppositely charged ions held together by electrostatic attraction

To illustrate the spectrum of bonding, let's consider three substances: H_2, HF, and NaF. In the H_2 molecule, where the two bonding atoms are identical, the electrons are shared equally. That is, the electrons in the covalent bond spend roughly the same amount of time in the vicinity of each H atom. In the HF molecule, on the other hand, where the two bonding atoms are different, the electrons are *not* shared equally. They spend more time in the vicinity of the F atom than in the vicinity of the H atom. (The δ symbol is used to denote partial charges on the atoms.) In NaF, the electrons are not shared at all but rather are transferred from sodium to fluorine.

One way to visualize the distribution of electrons in species such as H_2, HF, and NaF is to use electrostatic potential models (Figure 8.8). These models show regions where electrons spend a lot of time in red, and regions where electrons spend very little time in blue. (Regions where electrons spend a moderate amount of time appear in green.)

Student Note: In fact, the transfer of electrons in NaF is *nearly* complete. Even in an ionic bond, the electrons in question spend a small amount of time near the cation.

Figure 8.8 Electron density maps show the distribution of charge in a covalent species (H_2), a polar covalent species (HF), and an ionic species (NaF). The most electron-rich regions are red; the most electron-poor regions are blue.

H₂   HF   NaF

Animation
Periodic table—electronegativity.

Electronegativity

Electronegativity is the ability of an atom in a compound to draw electrons to itself. It determines where electrons in a compound spend most of their time. Elements with high electronegativity have a greater tendency to attract electrons than do elements with low electronegativity. Electronegativity is related to electron affinity and ionization energy. An atom such as fluorine, which has a high electron affinity (tends to accept electrons) and a high ionization energy (does not lose electrons easily), has a high electronegativity. Sodium, on the other hand, has a low electron affinity, a low ionization energy, and therefore a low electronegativity.

Electronegativity is a relative concept, meaning that an element's electronegativity can be measured only in relation to the electronegativity of other elements. Linus Pauling[2] devised a method for calculating the relative electronegativities of most elements. These values are shown in Figure 8.9. In general, electronegativity increases

Figure 8.9 Electronegativities of common elements.

Increasing electronegativity →

Increasing electronegativity ↑

1	2	3	4	5	6	7	8	9	10	11	12	13	14	15	16	17	18
H 2.1																	
Li 1.0	**Be** 1.5											**B** 2.0	**C** 2.5	**N** 3.0	**O** 3.5	**F** 4.0	
Na 0.9	**Mg** 1.2											**Al** 1.5	**Si** 1.8	**P** 2.1	**S** 2.5	**Cl** 3.0	
K 0.8	**Ca** 1.0	**Sc** 1.3	**Ti** 1.5	**V** 1.6	**Cr** 1.6	**Mn** 1.5	**Fe** 1.8	**Co** 1.9	**Ni** 1.9	**Cu** 1.9	**Zn** 1.6	**Ga** 1.6	**Ge** 1.8	**As** 2.0	**Se** 2.4	**Br** 2.8	**Kr** 3.0
Rb 0.8	**Sr** 1.0	**Y** 1.2	**Zr** 1.4	**Nb** 1.6	**Mo** 1.8	**Tc** 1.9	**Ru** 2.2	**Rh** 2.2	**Pd** 2.2	**Ag** 1.9	**Cd** 1.7	**In** 1.7	**Sn** 1.8	**Sb** 1.9	**Te** 2.1	**I** 2.5	**Xe** 2.6
Cs 0.7	**Ba** 0.9	**La** 1.1	**Hf** 1.3	**Ta** 1.5	**W** 1.7	**Re** 1.9	**Os** 2.2	**Ir** 2.2	**Pt** 2.2	**Au** 2.4	**Hg** 1.9	**Tl** 1.8	**Pb** 1.9	**Bi** 1.9	**Po** 2.0	**At** 2.2	
Fr 0.7	**Ra** 0.9																

2. Linus Carl Pauling (1901–1994). American chemist. Regarded by many as the most influential chemist of the twentieth century. Pauling received the Nobel Prize in Chemistry in 1954 for his work on protein structure, and the Nobel Peace Prize in 1962 for his tireless campaign against the testing and proliferation of nuclear arms. He is the only person ever to have received two unshared Nobel Prizes.

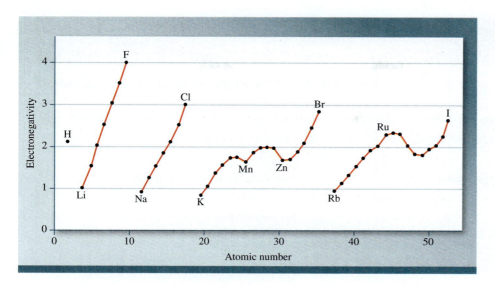

Figure 8.10 Variation of electronegativity with atomic number.

from left to right across a period in the periodic table, as the metallic character of the elements decreases. Within each group, electronegativity decreases with increasing atomic number and increasing metallic character. The transition metals do not follow these trends. The most electronegative elements (the halogens, oxygen, nitrogen, and sulfur) are found in the upper right-hand corner of the periodic table, and the least electronegative elements (the alkali and alkaline earth metals) are clustered near the lower left-hand corner. These trends are readily apparent in the graph in Figure 8.10.

Electronegativity and electron affinity are related but distinct concepts. Both indicate the tendency of an atom to attract electrons. Electron affinity, however, refers to an isolated atom's ability to attract an additional electron in the gas phase, whereas electronegativity refers to the ability of an atom in a chemical bond (with another atom) to attract the shared electrons. Electron affinity, moreover, is an experimentally measurable quantity, whereas electronegativity is an estimated number that cannot be measured directly.

Atoms of elements with widely different electronegativities tend to form ionic compounds with each other because the atom of the less electronegative element gives up its electrons to the atom of the more electronegative element. Atoms of elements with comparable electronegativities tend to form polar covalent bonds, or simply polar bonds, with each other because the shift in electron density from one atom to the other is usually small. Only atoms of the same element, which have the same electronegativity, can be joined by a pure covalent bond.

There is no sharp distinction between *nonpolar covalent* and *polar covalent* or between *polar covalent* and *ionic,* but the following guidelines can help distinguish among them:

- A bond between atoms whose electronegativities differ by less than 0.5 is generally considered purely covalent or **nonpolar.**
- A bond between atoms whose electronegativities differ by the range of 0.5 to 2.0 is generally considered *polar covalent.*
- A bond between atoms whose electronegativities differ by 2.0 or more is generally considered *ionic.*

Sometimes chemists describe bonds using the term *percent ionic character* [▶▶ Equation 8.2]. A purely ionic bond would have 100 percent ionic character (although no such bond is known). A purely covalent, nonpolar bond has 0 percent ionic character.

Sample Problem 8.4 shows how to use electronegativities to identify a chemical bond as nonpolar, polar, or ionic.

SAMPLE PROBLEM 8.4

Classify the following bonds as nonpolar, polar, or ionic: (a) the bond in ClF, (b) the bond in CsBr, and (c) the carbon-carbon double bond in C_2H_4.

Strategy Using the information in Figure 8.9, determine which bonds have identical, similar, and widely different electronegativities.

Setup Electronegativity values from Figure 8.9 are: Cl (3.0), F (4.0), Cs (0.7), Br (2.8), C (2.5).

Solution

(a) The difference between the electronegativities of F and Cl is $4.0 - 3.0 = 1.0$, making the bond in ClF polar.

(b) In CsBr, the difference is $2.8 - 0.7 = 2.1$, making the bond ionic.

(c) In C_2H_4, the two atoms are identical. (Not only are they the same element, but each C atom is bonded to two H atoms.) The carbon-carbon double bond in C_2H_4 is nonpolar.

THINK ABOUT IT

By convention, the difference in electronegativity is always calculated by subtracting the smaller number from the larger one, so the result is always positive.

Practice Problem ATTEMPT Classify the following bonds as nonpolar, polar, or ionic: (a) the bonds in H_2S, (b) the H—O bonds in H_2O_2, and (c) the O—O bond in H_2O_2.

Practice Problem BUILD In order of increasing polarity, list the bonds between carbon and each of the Group 16 elements.

Practice Problem CONCEPTUALIZE Electrostatic potential maps are shown for HCl and LiH. Determine which diagram is which. (The H atom is shown on the left in both.)

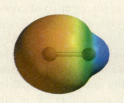 

Dipole Moment, Partial Charges, and Percent Ionic Character

The shift of electron density in a polar bond is symbolized by placing a crossed arrow (a dipole arrow) above the Lewis structure to indicate the direction of the shift. For example,

$$\overset{\longmapsto}{\text{H}}-\overset{\cdot\cdot}{\underset{\cdot\cdot}{\text{F}}}:$$

The consequent charge separation can be represented as

$$\overset{\delta+}{\text{H}}-\overset{\cdot\cdot}{\underset{\cdot\cdot}{\text{F}}}\overset{\delta-}{:}$$

A quantitative measure of the polarity of a bond is its ***dipole moment*** (μ), which is calculated as the product of the charge (Q) and the distance (r) between the charges:

Equation 8.1 $\mu = Q \times r$

The distance, r, between partial charges in a polar diatomic molecule is the bond length expressed in meters. Bond lengths are usually given in angstroms (Å) or picometers (pm), so it is generally necessary to convert to meters. For a diatomic molecule containing a polar bond to be electrically neutral, the partial positive and partial negative charges must have the same magnitude. Therefore, the Q term in Equation 8.1 refers to the magnitude of the partial charges and the calculated value of μ is always positive.

TABLE 8.5	Bond Lengths and Dipole Moments of the Hydrogen Halides		
Molecule	**Bond Length (Å)**	**Dipole Moment (D)**	
HF	0.92	1.82	
HCl	1.27	1.08	
HBr	1.41	0.82	
HI	1.61	0.44	

Dipole moments are usually expressed in debye units (D), named for Peter Debye.[3] In terms of more familiar SI units,

$$1 \text{ D} = 3.336 \times 10^{-30} \text{ C} \cdot \text{m}$$

where *C is coulombs* and m is meters. Table 8.5 lists several polar diatomic molecules, their bond lengths, and their experimentally measured dipole moments.

Sample Problem 8.5 shows how to use bond lengths and dipole moments to determine the magnitude of the partial charges in a polar diatomic molecule.

Student Note: We usually express the charge on an electron as −1. This refers to *units of electronic charge*. However, remember that the charge on an electron can also be expressed in *coulombs* [◄◄ Section 2.2]. The conversion factor between the two is necessary to calculate dipole moments: $1e^- = 1.6022 \times 10^{-19}$ C.

SAMPLE PROBLEM 8.5

Hydrofluoric acid [HF(*aq*)] has several important industrial applications, including the etching of glass and the manufacture of electronic components. Burns caused by hydrofluoric acid are unlike any other acid burns and present unique medical complications. HF solutions typically penetrate the skin and damage internal tissues, including bone, often with minimal surface damage. Less concentrated solutions actually can cause greater injury than more concentrated ones by penetrating more deeply before causing injury, thus delaying the onset of symptoms and preventing timely treatment. Determine the magnitude of the partial positive and partial negative charges in the HF molecule.

Strategy Rearrange Equation 8.1 to solve for *Q*. Convert the resulting charge in coulombs to charge in units of electronic charge.

Setup According to Table 8.5, $\mu = 1.82$ D and $r = 0.92$ Å for HF. The dipole moment must be converted from debye to C · m and the distance between the ions must be converted to meters.

$$\mu = 1.82 \text{ D} \times \frac{3.336 \times 10^{-30} \text{ C} \cdot \text{m}}{1 \text{ D}} = 6.07 \times 10^{-30} \text{ C} \cdot \text{m}$$

$$r = 0.92 \text{ Å} \times \frac{1 \times 10^{-10} \text{ m}}{1 \text{ Å}} = 9.2 \times 10^{-11} \text{ m}$$

Solution In coulombs:

$$Q = \frac{\mu}{r} = \frac{6.07 \times 10^{-30} \text{ C} \cdot \text{m}}{9.2 \times 10^{-11} \text{ m}} = 6.598 \times 10^{-20} \text{ C}$$

In units of electronic charge:

$$6.598 \times 10^{-20} \text{ C} \times \frac{1e^-}{1.6022 \times 10^{-19} \text{ C}} = 0.41e^-$$

Therefore, the partial charges in HF are +0.41 and −0.41 on H and F, respectively.

$$^{+0.41}\text{H} - \overset{..}{\underset{..}{\text{F}}}:^{-0.41}$$

(Continued on next page)

---

3. Peter Joseph William Debye (1884–1966). American chemist and physicist of Dutch origin. Debye made many significant contributions to the study of molecular structure, polymer chemistry, X-ray analysis, and electrolyte solutions. He was awarded the Nobel Prize in Chemistry in 1936.

THINK ABOUT IT

Calculated partial charges should always be less than 1. If a "partial" charge were 1 or greater, it would indicate that at least one electron had been transferred from one atom to the other. Remember that polar bonds involve unequal sharing of electrons, not a complete transfer of electrons.

Practice Problem A TTEMPT Using data from Table 8.5, determine the magnitude of the partial charges in HBr.

Practice Problem B UILD Given that the partial charges on C and O in carbon monoxide are +0.020 and −0.020, respectively, calculate the dipole moment of CO. (The distance between the partial charges, r, is 113 pm.)

Practice Problem C ONCEPTUALIZE Based on the answers to Sample Problem 8.5 and Practice Problem 8.5A, estimate the magnitude of the partial charges in HCl.

Although the designations "covalent," "polar covalent," and "ionic" can be useful, sometimes chemists wish to describe and compare chemical bonds with more precision. For this purpose, we can use Equation 8.1 to calculate the dipole moment we would expect if the charges on the atoms were discrete instead of partial; that is, if an electron had actually been transferred from one atom to the other. Comparing this calculated dipole moment with the measured value gives us a quantitative way to describe the nature of a bond using the term ***percent ionic character,*** which is defined as the ratio of observed μ to calculated μ, multiplied by 100.

Equation 8.2 $$\text{percent ionic character} = \frac{\mu(\text{observed})}{\mu(\text{calculated assuming discrete charges})} \times 100\%$$

Figure 8.11 illustrates the relationship between percent ionic character and the electronegativity difference in a heteronuclear diatomic molecule.

Sample Problem 8.6 shows how to calculate percent ionic character using Equation 8.2.

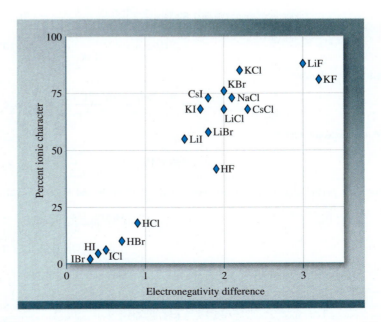

Figure 8.11 Relationship between percent ionic character and electronegativity difference.

SAMPLE PROBLEM 8.6

Using data from Table 8.5, calculate the percent ionic character of the bond in HI.

Strategy Use Equation 8.1 to calculate the dipole moment in HI assuming that the charges on H and I are +1 and −1, respectively; and use Equation 8.2 to calculate percent ionic character. The magnitude of the charges must be expressed as coulombs ($1\ e^- = 1.6022 \times 10^{-19}$ C); the bond length (r) must be expressed as meters ($1\ \text{Å} = 1 \times 10^{-10}$ m); and the calculated dipole moment should be expressed as debyes ($1\ \text{D} = 3.336 \times 10^{-30}$ C · m).

Setup From Table 8.5, the bond length in HI is 1.61 Å (1.61×10^{-10} m) and the measured dipole moment of HI is 0.44 D.

Solution The dipole moment we would expect if the magnitude of charges were 1.6022×10^{-19} C is

$$\mu = Q \times r = (1.6022 \times 10^{-19}\ \text{C}) \times (1.61 \times 10^{-10}\ \text{m}) = 2.58 \times 10^{-29}\ \text{C} \cdot \text{m}$$

Converting to debyes gives

$$2.58 \times 10^{-29}\ \cancel{\text{C} \cdot \text{m}} \times \frac{1\ \text{D}}{3.336 \times 10^{-30}\ \cancel{\text{C} \cdot \text{m}}} = 7.73\ \text{D}$$

The percent ionic character of the H—I bond is

$$\frac{0.44\ \text{D}}{7.73\ \text{D}} \times 100\% = 5.7\%$$

THINK ABOUT IT

A purely covalent bond (in a homonuclear diatomic molecule such as H_2) would have 0 percent ionic character. In theory, a purely ionic bond would be expected to have 100 percent ionic character, although no such bond is known.

Practice Problem ATTEMPT Using data from Table 8.5, calculate the percent ionic character of the bond in HF.

Practice Problem BUILD Using information from Figure 7.12, and given that the NaI bond has 59.7 percent ionic character, determine the measured dipole moment of NaI.

Practice Problem CONCEPTUALIZE One metal and three nonmetals are highlighted on the periodic table shown here. List the nonmetal elements in order of increasing ionic character of the bond each might form with the highlighted metal.

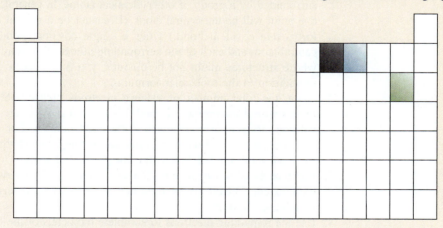

CHECKPOINT – SECTION 8.4   Electronegativity and Polarity

8.4.1 In which of the following molecules are the bonds *most* polar?

    a) H_2Se c) CO_2 e) PCl_5

    b) H_2O d) BCl_3

8.4.2 Using data from Table 8.5, calculate the magnitude of the partial charges in HI.

    a) 0.39   c) 0.057   e) 0.15

    b) 1.8   d) 0.60

8.4.3 Arrange molecules A through E in order of increasing percent ionic character.

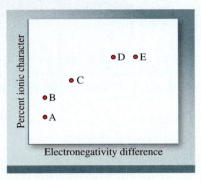

a) A < B < C < D < E

b) A = B < C < D < E

c) A = B < C < D = E

d) A < B < C < D = E

e) A < B = C = D < E

8.4.4 Using data from Table 8.5, calculate the percent ionic character of HCl.

a) 85.0 percent

b) 6.10 percent

c) 17.7 percent

d) 2.03 percent

e) 20.3 percent

8.5   Drawing Lewis Structures

Although the octet rule and Lewis structures alone do not present a complete picture of covalent bonding, they do help us account for some of the properties of molecules. In addition, Lewis structures provide a starting point for the bonding theories that we examine in Chapter 9. It is crucial, therefore, that you learn a system for drawing correct Lewis structures for molecules and polyatomic ions. The basic steps, illustrated in Figure 8.12, are as follows:

1. From the molecular formula, draw the skeletal structure of the compound, using chemical symbols and placing bonded atoms next to one another. For simple compounds, this step is fairly easy. Often there will be a unique central atom surrounded by a group of other identical atoms. In general, the *least* electronegative atom will be the central atom. (H cannot be a central atom because it only forms *one* covalent bond.) Draw a single covalent bond (dash) between the central atom and each of the surrounding atoms. (For more complex compounds whose structures might not be obvious, you may need to be given information in addition to the molecular formula.)

2. Count the total number of valence electrons present. Remember that an element's *group number* (1, 2, and 13–18) is used to determine the number of valence electrons it contributes to the total number. For polyatomic ions, add electrons to the total number to account for negative charges; subtract electrons from the total number to account for positive charges.

3. For each bond (dash) in the skeletal structure, subtract two electrons from the total valence electrons (determined in step 2) to determine the number of remaining electrons.

4. Use the remaining electrons to complete the octets of the terminal atoms (those bonded to the central atom) by placing pairs of electrons on each atom. (Remember that an H atom only requires two electrons to complete its valence shell.) If there is more than one type of terminal atom, complete the octets of the most electronegative atoms first.

5. If any electrons remain after step 4, place them in pairs on the central atom.

6. If the central atom has fewer than eight electrons after completing steps 1 to 5, move one or more pairs from the terminal atoms to form multiple bonds between the central atom and the terminal atoms. (Unless the central atom is a Group 13 element.) Like Lewis dot symbols for atomic anions, Lewis structures for polyatomic anions are enclosed by square brackets.

Steps for Drawing Lewis Structures

Step	CH$_4$	CCl$_4$	H$_2$O	O$_2$	CN$^-$
1	H–C–H (with H above and H below)	Cl–C–Cl (with Cl above and Cl below)	H–O–H	O–O	C–N
2	8	32	8	12	10
3	$8 - 8 = 0$	$32 - 8 = 24$	$8 - 4 = 4$	$12 - 2 = 10$	$10 - 2 = 8$
4	H–C–H (with H above and H below)	:Cl–C–Cl: (with Cl above and Cl below)	H–O–H	:Ö–Ö:	:C–N̈:
5	—	—	H–Ö–H	—	—
6	—	—	—	:O=O:	[:C≡N:]$^-$

Student Note: To summarize the process of drawing Lewis structures:

1. Draw the skeletal structure.
2. Count the valence electrons.
3. Subtract two electrons for each bond.
4. Distribute the remaining electrons.
5. Complete the octets of all atoms.
6. Use multiple bonds if necessary.

Figure 8.12 Drawing Lewis structures.

Sample Problem 8.7 shows how to draw a Lewis structure.

SAMPLE PROBLEM 8.7

Draw the Lewis structure for carbon disulfide (CS$_2$).

Strategy Use the procedure described in steps 1 through 6 in Figure 8.12 for drawing Lewis structures.

Setup

Step 1:   C and S have identical electronegativities. We will draw the skeletal structure with the unique atom, C, at the center.

$$S–C–S$$

Step 2:   The total number of valence electrons is 16, six from each S atom and four from the C atom [2(6) + 4 = 16].

Step 3:   Subtract four electrons to account for the bonds in the skeletal structure, leaving us 12 electrons to distribute.

Step 4:   Distribute the 12 remaining electrons as three lone pairs on each S atom.

$$:S̈–C–S̈:$$

Step 5:   There are no electrons remaining after step 4, so step 5 does not apply.

Step 6:   To complete carbon's octet, use one lone pair from each S atom to make a double bond to the C atom.

Solution

$$:S̈=C=S̈:$$

THINK ABOUT IT

Counting the total number of valence electrons should be relatively simple to do, but it is often done hastily and is therefore a potential source of error in this type of problem. Remember that the number of valence electrons for each element is equal to the group number for elements in Groups 1 and 2, and to the second *digit* of the group number for elements in Groups 13 through 18.

(Continued on next page)

Practice Problem Ⓐ**TTEMPT** Draw the Lewis structure for NF_3.

Practice Problem Ⓑ**UILD** Draw the Lewis structure for ClO_3^-.

Practice Problem Ⓒ**ONCEPTUALIZE** Of the three Lewis structures shown here, identify any that are not correct and specify what is wrong.

$$[:C≡N:]^-   :O=\ddot{C}=O:$$
(i)   (ii)

$$:\ddot{O}-\ddot{S}-\ddot{O}:$$
(iii)

CHECKPOINT – SECTION 8.5   Drawing Lewis Structures

8.5.1 Identify the correct Lewis structure for formic acid (HCOOH).

a) $H-\ddot{C}-\ddot{O}-\ddot{O}-H$

b) $H-\ddot{O}=C=\ddot{O}-H$

c) $H-\ddot{O}-\ddot{C}-\ddot{O}-H$

d) $H-\overset{\displaystyle :O:}{\overset{\|}{C}}-\ddot{O}-H$

e) $:\ddot{O}-\overset{\displaystyle :O:}{\underset{\displaystyle H}{C}}-H$

8.5.2 Identify the correct Lewis structure for hydrogen peroxide (H_2O_2).

a) $H-\ddot{O}=\ddot{O}-H$

b) $H-O≡O-H$

c) $H-\ddot{O}-\ddot{O}-H$

d) $H=O=O=H$

e) $H-H-\ddot{O}-\ddot{O}:$

8.6 Lewis Structures and Formal Charge

So far you have learned two different methods of electron "bookkeeping." In Chapter 4, you learned about oxidation numbers [◄◄ Section 4.4], and in Section 8.4, you learned how to calculate partial charges. There is one additional commonly used method of electron bookkeeping—namely, *formal charge,* which can be used to determine the most plausible Lewis structures when more than one possibility exists for a compound. Formal charge is determined by comparing the number of electrons associated with an atom in a Lewis structure with the number of electrons that would be associated with the isolated atom. In an isolated atom, the number of electrons associated with the atom is simply the number of valence electrons. (As usual, we need not be concerned with the core electrons.)

To determine the number of electrons associated with an atom in a Lewis structure, keep in mind the following:

- All the atom's nonbonding electrons are associated with the atom.
- Half of the atom's bonding electrons are associated with the atom.

Animation

Chemical Bonding—formal charge calculations.

| **Equation 8.3** | formal charge = valence electrons − associated electrons |

An atom's formal charge is calculated as follows: We can illustrate the concept of formal charge using the ozone molecule (O_3). Use the step-by-step method for drawing Lewis structures to draw the Lewis structure for ozone, and then determine the formal charge on each O atom by subtracting the number of associated electrons from the number of valence electrons.

$$2 \text{ unshared} + \frac{6 \text{ shared}}{2} = 5 \, e^-$$

$$4 \text{ unshared} + \frac{4 \text{ shared}}{2} = 6 \, e^- \qquad\qquad 6 \text{ unshared} + \frac{2 \text{ shared}}{2} = 7 \, e^-$$

Valence e^-	6	6	6
e^- associated with atom	6	5	7
Difference (formal charge)	0	+1	−1

$$:\!\ddot{O}\!=\!\ddot{O}\!-\!\ddot{O}\!:$$

....

As with oxidation numbers, the sum of the formal charges must equal the overall charge on the species [◀ Section 4.4]. Because O_3 is a molecule, its formal charges must sum to zero. For ions, the formal charges must sum to the overall charge on the ion.

Formal charges do *not* represent actual charges on atoms in a molecule. In the O_3 molecule, for example, there is no evidence that the central atom bears a net +1 charge or that one of the terminal atoms bears a −1 charge. Assigning formal charges to the atoms in the Lewis structure merely helps us keep track of the electrons involved in bonding in the molecule.

Sample Problem 8.8 lets you practice determining formal charges.

Student Note: While you are new at determining formal charges, it may be helpful to draw Lewis structures with all dots, rather than dashes. This can make it easier to see how many electrons are associated with each atom.

Remember that for the purpose of counting associated electrons, those shared by two atoms are evenly split between them.

SAMPLE PROBLEM 8.8

The widespread use of fertilizers has resulted in the contamination of some groundwater with nitrates, which are potentially harmful. Nitrate toxicity is due primarily to its conversion in the body to nitrite (NO_2^-), which interferes with the ability of hemoglobin to transport oxygen. Determine the formal charges on each atom in the nitrate ion (NO_3^-).

Strategy Use steps 1 through 6 in Figure 8.12 for drawing Lewis stuctures to draw the Lewis structure of NO_3^-. For each atom, subtract the associated electrons from the valence electrons.

Setup

$$\left[\begin{array}{c} :\!\ddot{O}\!: \\ | \\ :\!\ddot{O}\!-\!N\!=\!\ddot{O}\!: \end{array}\right]^-$$

The N atom has five valence electrons and four associated electrons (one from each single bond and two from the double bond). Each singly bonded O atom has six valence electrons and seven associated electrons (six in three lone pairs and one from the single bond). The doubly bonded O atom has six valence electrons and six associated electrons (four in two lone pairs and two from the double bond).

Solution The formal charges are as follows: +1 (N atom), −1 (singly bonded O atoms), and 0 (doubly bonded O atom).

THINK ABOUT IT

The sum of formal charges (+1) + (−1) + (−1) + (0) = −1 is equal to the overall charge on the nitrate ion.

(Continued on next page)

Practice Problem Ⓐ**TTEMPT** Determine the formal charges on each atom in the carbonate ion (CO_3^{2-}).

Practice Problem Ⓑ**UILD** Determine the formal charges and use them to determine the overall charge, if any, on the species represented by the following Lewis structure:

$$\begin{bmatrix} & \ddot{\ddot{O}}: & \\ :\ddot{O}- & \!\!S\!\!-\ddot{O}: \end{bmatrix}^{?}$$

Practice Problem Ⓒ**ONCEPTUALIZE** The hypothetical element A is shown here in three different partial Lewis structures. For each structure, determine what the formal charge is on A (a) if it is a member of Group 17, (b) if it is a member of Group 15, and (c) if it is a member of Group 13.

$$-\ddot{A}: \qquad -\ddot{A}- \qquad -\overset{|}{\underset{|}{\ddot{A}}}-$$
$$\text{(i)} \qquad\quad \text{(ii)} \qquad\quad \text{(iii)}$$

Sometimes, there is more than one possible skeletal arrangement of atoms for the Lewis structure for a given species. In such cases, we often can select the best skeletal arrangement by using formal charges and the following guidelines:

- For molecules, a Lewis structure in which all formal charges are zero is preferred to one in which there are nonzero formal charges.
- Lewis structures with small formal charges (0 and ±1) are preferred to those with large formal charges (±2, ±3, and so on).
- The best skeletal arrangement of atoms will give rise to Lewis structures in which the formal charges are consistent with electronegativities. For example, the more electronegative atoms should have the more negative formal charges.

Sample Problem 8.9 shows how formal charge can be used to determine the best skeletal arrangement of atoms for the Lewis structure of a molecule or polyatomic ion.

SAMPLE PROBLEM   8.9

Formaldehyde (CH_2O), which can be used to preserve biological specimens, is commonly sold as a 37% aqueous solution. Use formal charges to determine which skeletal arrangement of atoms shown here is the best choice for the Lewis structure of CH_2O.

$$\text{H}-\text{C}-\text{O}-\text{H} \qquad \text{H}-\overset{\overset{\displaystyle O}{|}}{\text{C}}-\text{H}$$

Strategy Complete the Lewis structures for each of the CH_2O skeletons shown and determine the formal charges on the atoms in each one.

Setup The completed Lewis structures for the skeletons shown are:

$$\text{H}-\ddot{\text{C}}=\ddot{\text{O}}-\text{H} \qquad \text{H}-\overset{\overset{\displaystyle \cdot\ddot{O}\cdot}{\|}}{\text{C}}-\text{H}$$

In the structure on the left, the formal charges are as follows:

Both H atoms: 1 valence e^- – 1 associated e^- (from single bond) = 0

C atom: 4 valence e^- – 5 associated e^- (two in the lone pair, one from the single bond, and two from the double bond) = –1

O atom: 6 valence e^- – 5 associated e^- (two from the lone pair, one from the single bond, and two from the double bond) = +1

$$\text{H}-\ddot{\text{C}}=\ddot{\text{O}}-\text{H}$$
$$\text{Formal charges} \quad 0 \quad -1 \quad +1 \quad 0$$

In the structure on the right, the formal charges are as follows:

Both H atoms: 1 valence e^- − 1 associated e^- (from single bond) = 0

C atom: 4 valence e^- − 4 associated e^- (one from each single bond, and two from the double bond) = 0

O atom: 6 valence e^- − 6 associated e^- (four from the two lone pairs and two from the double bond) = 0

$$\ddot{O}$$
$$\|$$
$$H-C-H$$

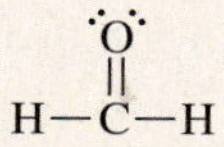

Formal charges all zero

Solution Of the two possible arrangements, the structure on the left has an O atom with a positive formal charge, which is inconsistent with oxygen's high electronegativity. Therefore, the structure on the right, in which both H atoms are attached directly to the C atom and all atoms have a formal charge of zero, is the better choice for the Lewis structure of CH_2O.

THINK ABOUT IT

For a molecule, formal charges of zero are preferred. When there are nonzero formal charges, they should be consistent with the electronegativities of the atoms in the molecule. A positive formal charge on oxygen, for example, is inconsistent with oxygen's high electronegativity.

Practice Problem Ⓐ**TTEMPT** Two possible arrangements are shown for the Lewis structure of a carboxyl group, −COOH. Use formal charges to determine which of the two arrangements is better.

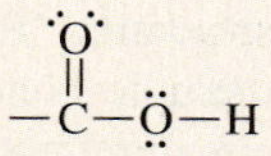

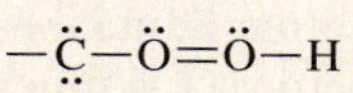

$$\ddot{O}$$
$$\|$$
$$-C-\ddot{O}-H \qquad -\ddot{C}-\ddot{O}=\ddot{O}-H$$

Practice Problem Ⓑ**UILD** Use Lewis structures and formal charges to determine the best skeletal arrangement of atoms in NCl_2^-.

Practice Problem Ⓒ**ONCEPTUALIZE** For each partial Lewis structure shown here, determine what group element A must belong to in order for its formal charge to be zero.

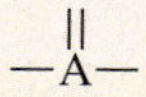

$$\begin{array}{cccc} \| & & & \\ -A- & -\ddot{A}- & -\ddot{A}- & =A= \\ & & | & \\ (i) & (ii) & (iii) & (iv) \end{array}$$

CHECKPOINT − SECTION 8.6   Lewis Structures and Formal Charge

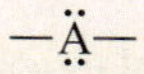

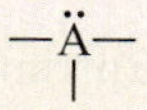

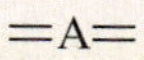

8.6.1 Determine the formal charges on H, C, and N, respectively, in HCN.

a) 0, +1, and −1

b) −1, +1, and 0

c) 0, −1, and +1

d) 0, +1, and +1

e) 0, 0, and 0

8.6.2 Which of the Lewis structures shown is most likely preferred for NCO^-?

a) 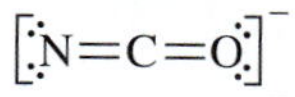 $\left[:N=C=\ddot{O}: \right]^-$

b) $\left[:\ddot{N}-C\equiv O: \right]^-$

c) $\left[:N\equiv C-\ddot{O}: \right]^-$

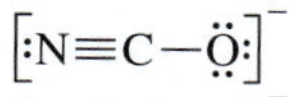

d) 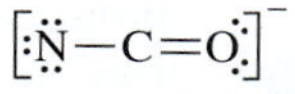 $\left[:\ddot{N}-C=\ddot{O}: \right]^-$

e) 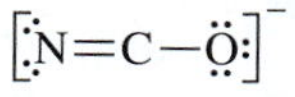 $\left[:N=C-\ddot{O}: \right]^-$

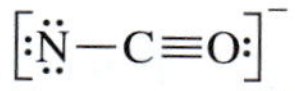

8.7   Resonance

Our drawing of the Lewis structure for ozone (O_3) satisfied the octet rule for the central O atom because we placed a double bond between it and one of the two terminal O atoms. In fact, we can put the double bond at either end of the molecule, as shown by the following two equivalent Lewis structures:

$$\ddot{O}=\ddot{O}-\ddot{O}: \longleftrightarrow :\ddot{O}-\ddot{O}=\ddot{O}$$

A single bond between O atoms should be longer than a double bond between O atoms, but experimental evidence indicates that both of the bonds in O_3 are equal in length (128 pm). Because neither one of these two Lewis structures accounts for the known bond lengths in O_3, we use both Lewis structures to represent the ozone molecule.

Each of the Lewis structures is called a resonance structure. A **resonance structure** is one of two or more Lewis structures for a single molecule that cannot be represented accurately by only one Lewis structure. The double-headed arrow indicates that the structures shown are resonance structures. Like the medieval European traveler to Africa who described a rhinoceros as a cross between a griffin and a unicorn (two familiar but imaginary animals), we describe ozone, a real molecule, in terms of two familiar but nonexistent structures.

A common misconception about resonance is that a molecule such as ozone somehow shifts quickly back and forth from one resonance structure to the other. Neither resonance structure, though, adequately represents the actual molecule, which has its own unique, stable structure. "Resonance" is a human invention, designed to address the limitations of a simple bonding model. To extend the animal analogy, a rhinoceros is a distinct, real creature, not some oscillation between the mythical griffin and unicorn!

The carbonate ion provides another example of resonance:

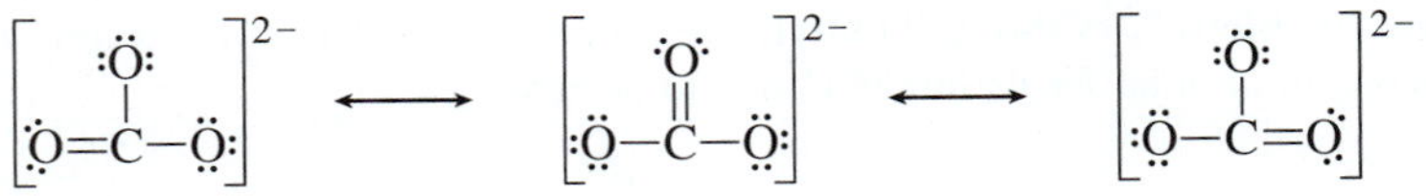

$$\left[\begin{array}{c} :\ddot{O}: \\ | \\ \ddot{O}=C-\ddot{O}: \end{array}\right]^{2-} \longleftrightarrow \left[\begin{array}{c} \dot{O} \\ || \\ :\ddot{O}-C-\ddot{O}: \end{array}\right]^{2-} \longleftrightarrow \left[\begin{array}{c} :\ddot{O}: \\ | \\ :\ddot{O}-C=\ddot{O} \end{array}\right]^{2-}$$

According to experimental evidence, all three carbon-oxygen bonds in CO_3^{2-} are equivalent. Therefore, the properties of the carbonate ion are best explained by considering its resonance structures together.

The concept of resonance applies equally well to organic systems. A good example is the benzene molecule (C_6H_6):

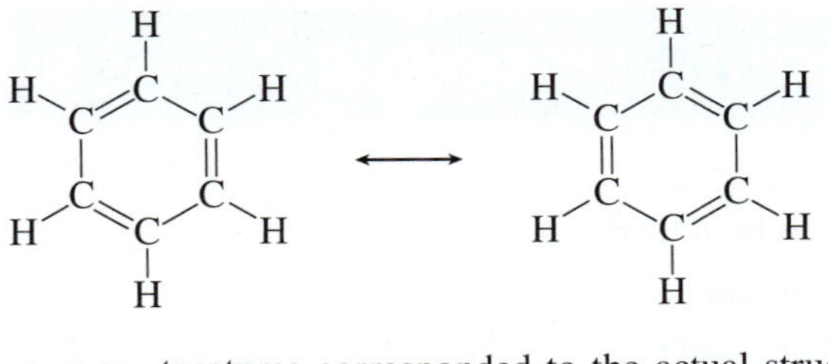

If one of these resonance structures corresponded to the actual structure of benzene, there would be two different bond lengths between adjacent C atoms, one with the properties of a single bond and the other with the properties of a double bond. In fact, the distance between all adjacent C atoms in benzene is 140 pm, which is shorter than a C—C bond (154 pm) and longer than a C=C bond (133 pm).

Resonance structures differ only in the positions of their *electrons*—not in the positions of their atoms. Thus, $:\ddot{N}=N=\ddot{O}$ and $:N\equiv N-\ddot{O}:$ are resonance structures of each other, whereas $:\ddot{N}=N=\ddot{O}$ and $:\ddot{N}=O=\ddot{N}:$ are not.

Sample Problem 8.10 shows how to draw resonance structures.

SAMPLE PROBLEM 8.10

High oil and gasoline prices have renewed interest in alternative methods of producing energy, including the "clean" burning of coal. Part of what makes "dirty" coal *dirty* is its high sulfur content. Burning dirty coal produces sulfur dioxide (SO_2), among other pollutants. Sulfur dioxide is oxidized in the atmosphere to form sulfur trioxide (SO_3), which subsequently combines with water to produce sulfuric acid—a major component of acid rain. Draw all possible resonance structures of sulfur trioxide.

Strategy Draw two or more Lewis structures for SO_3 in which the atoms are arranged the same way but the electrons are arranged differently.

Setup Following the steps for drawing Lewis structures, we determine that a correct Lewis structure for SO_3 contains two sulfur-oxygen single bonds and one sulfur-oxygen double bond.

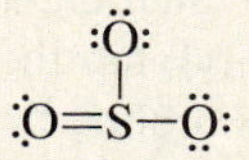

$$:\ddot{O}: \\ | \\ :\ddot{O}=S-\ddot{O}:$$

But the double bond can be put in any one of three positions in the molecule.

Solution

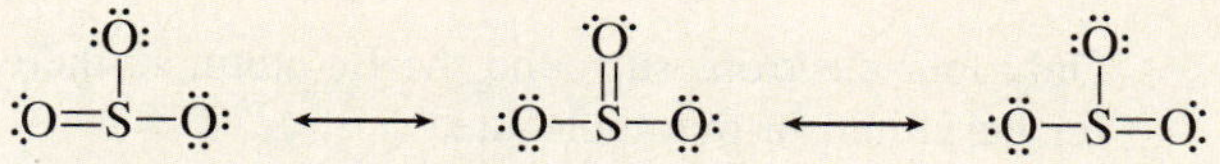

$$:\ddot{O}: \qquad\qquad :\ddot{O}: \qquad\qquad :\ddot{O}:$$
$$:\ddot{O}=S-\ddot{O}: \quad\longleftrightarrow\quad :\ddot{O}-S-\ddot{O}: \quad\longleftrightarrow\quad :\ddot{O}-S=\ddot{O}:$$

THINK ABOUT IT

Always make sure that resonance structures differ only in the positions of the electrons, *not* in the positions of the atoms.

Practice Problem **A**TTEMPT Draw all possible resonance structures for the nitrate ion (NO_3^-).

Practice Problem **B**UILD Draw three resonance structures for the thiocyanate ion (NCS^-), and determine the formal charges in each resonance structure. Based on formal charges, list the resonance structures in order of increasing relative importance.

Practice Problem **C**ONCEPTUALIZE The Lewis structure of a molecule consisting of the hypothetical elements A, B, and C is shown here. Of the four other structures, identify any that is *not* a resonance structure of the original and explain why it is not a resonance structure.

$$:\ddot{B}: \\ | \\ :\ddot{C}-A=\ddot{C}:$$

$:\ddot{A}:$	$\dot{B}$	$:\ddot{B}:$	$:\ddot{B}:$
$:\ddot{C}=B-\ddot{C}:$	$:\ddot{C}-A-\ddot{C}:$	$:\ddot{C}-A-\ddot{C}:$	$:\ddot{C}-A-\ddot{C}:$
(i)	(ii)	(iii)	(iv)

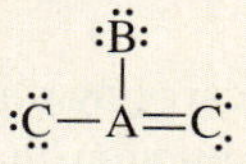

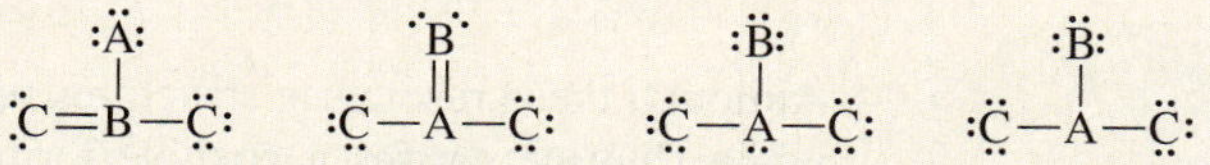

CHECKPOINT – SECTION 8.7   Resonance

8.7.1 Indicate which of the following are resonance structures of $:\ddot{C}l-Be-\ddot{C}l:$ (select all that apply).

a) $:\ddot{C}l=Be=\ddot{C}l:$

b) $:\ddot{C}l\equiv Be-\ddot{C}l:$

c) $:Be\equiv Cl-\ddot{C}l:$

d) $:\ddot{C}l-Be\equiv Cl:$

e) $:Be=Cl=\ddot{C}l:$

8.7.2 How many resonance structures can be drawn for the nitrite ion (NO_2^-)? (N and O must obey the octet rule.)

a) 1

b) 2

c) 3

d) 4

e) 5

8.8   Exceptions to the Octet Rule

The octet rule almost always holds for second-period elements. Exceptions to the octet rule fall into three categories:

1. The central atom has fewer than eight electrons due to a shortage of electrons.
2. The central atom has fewer than eight electrons due to an odd number of electrons.
3. The central atom has more than eight electrons.

Incomplete Octets

In some compounds, the number of electrons surrounding the central atom in a stable molecule is fewer than eight. Beryllium, for example, which is the Group 2 element in the second period, has the electron configuration $[He]2s^2$. Thus, it has two valence electrons in the $2s$ orbital. In the gas phase, beryllium hydride (BeH_2) exists as discrete molecules. The Lewis structure of BeH_2 is

$$H-Be-H$$

Only four electrons surround the Be atom, so there is no way to satisfy the octet rule for beryllium in this molecule.

Elements in Group 13 also tend to form compounds in which they are surrounded by fewer than eight electrons. Boron, for example, has the electron configuration $[He]2s^22p^1$, so it has only three valence electrons. Boron reacts with the halogens to form a class of compounds having the general formula BX_3, where X is a halogen atom. Thus, there are only six electrons around the boron atom in boron trifluoride:

$$:\ddot{F}-\overset{\displaystyle :\ddot{F}:}{\underset{}{B}}-\ddot{F}:$$

We actually *can* satisfy the octet rule for boron in BF_3 by using a lone pair on one of the F atoms to form a double bond between the F atom and boron. This gives rise to three additional resonance structures:

$$\ddot{F}=\overset{:\ddot{F}:}{B}-\ddot{F}: \quad\longleftrightarrow\quad :\ddot{F}-\overset{\cdot\dot{F}}{B}-\ddot{F}: \quad\longleftrightarrow\quad :\ddot{F}-\overset{:\ddot{F}:}{B}=\ddot{F}:$$

Although these resonance structures result in boron carrying a negative formal charge while fluorine carries a positive formal charge, a situation that is inconsistent with the electronegativities of the atoms involved, the experimentally determined bond length in BF_3 (130.9 pm) is shorter than a single bond (137.3 pm). The shorter bond length would appear to support the idea behind the three resonance structures.

On the other hand, boron trifluoride combines with ammonia in a reaction that is better represented using the Lewis structure in which boron has only six valence electrons around it:

$$\overset{:\ddot{F}:}{\underset{:\ddot{F}:}{:\ddot{F}-B}} \;+\; \overset{H}{\underset{H}{:N-H}} \;\longrightarrow\; \overset{:\ddot{F}:\;\;H}{\underset{:\ddot{F}:\;\;H}{:\ddot{F}-B-N-H}}$$

It seems, then, that the properties of BF_3 are best explained by all four resonance structures.

The B—N bond in F_3B-NH_3 is different from the covalent bonds discussed so far in the sense that both electrons are contributed by the N atom. This type of bond is called a ***coordinate covalent bond*** (also referred to as a ***dative bond***), which is defined as a covalent bond in which one of the atoms donates both electrons. Although the properties of a coordinate covalent bond do not differ from those of a normal covalent bond (i.e., the electrons are shared in both cases), the distinction is useful for keeping track of valence electrons and assigning formal charges.

Odd Numbers of Electrons

Some molecules, such as nitrogen dioxide (NO_2), contain an odd number of electrons.

$$\ddot{O}=\dot{N}-\ddot{O}:$$

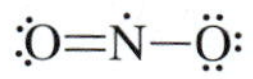

Because we need an even number of electrons for every atom in a molecule to have a complete octet, the octet rule cannot be obeyed for all the atoms in these molecules. Molecules with an odd number of electrons are sometimes referred to as *free radicals* (or just *radicals*). Many radicals are highly reactive, because there is a tendency for the unpaired electron to form a covalent bond with an unpaired electron on another molecule. When two nitrogen dioxide molecules collide, for example, they form dinitrogen tetroxide, a molecule in which the octet rule *is* satisfied for both the N and O atoms.

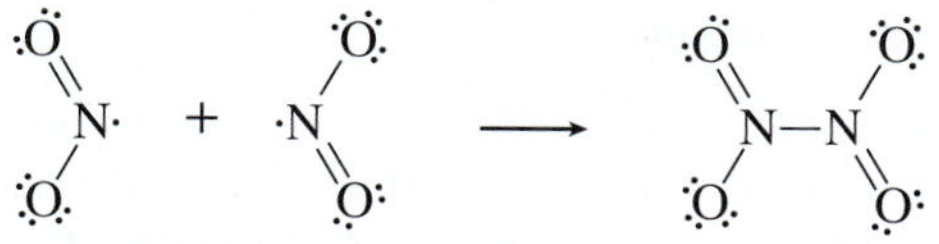

Bringing Chemistry to Life

The Power of Radicals

Beginning about a week after the September 11, 2001, attacks, letters containing anthrax bacteria were mailed to several news media offices and to two U.S. senators. Of the 22 people who subsequently contracted anthrax, five died. Anthrax, a spore-forming bacterium (*Bacillus anthracis*), is classified by the CDC as a *Category A bioterrorism agent*. Spore-forming bacteria are notoriously difficult to kill, making the cleanup of the buildings contaminated by anthrax costly and time-consuming. The American Media Inc. (AMI) building in Boca Raton, Florida, was not deemed safe to enter until July 2004, after it had been treated with chlorine dioxide (ClO_2), the only structural fumigant approved by the Environmental Protection Agency (EPA) for anthrax decontamination. The effectiveness of ClO_2 in killing anthrax and other hardy biological agents stems in part from its being a *radical*, meaning that it contains an odd number of electrons.

The American Media Inc. building in Boca Raton, Florida.

Greg Mathieson/Shutterstock

Sample Problem 8.11 lets you practice drawing Lewis structures for species with odd numbers of electrons.

SAMPLE PROBLEM (8.11)

Draw the Lewis structure of chlorine dioxide (ClO_2).

Strategy The skeletal structure is

$$O-Cl-O$$

This puts the unique atom, Cl, in the center and puts the more electronegative O atoms in terminal positions.

(Continued on next page)

Setup There are a total of 19 valence electrons (seven from the Cl and six from each of the two O atoms). We subtract four electrons to account for the two bonds in the skeleton, leaving us with 15 electrons to distribute as follows: three lone pairs on each O atom, one lone pair on the Cl atom, and the last remaining electron also on the Cl atom.

Solution

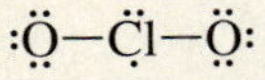

$$:\ddot{O}—\ddot{C}l—\ddot{O}:$$

THINK ABOUT IT

CIO_2 is used primarily to bleach wood pulp in the manufacture of paper, but it is also used to bleach flour, disinfect drinking water, and deodorize certain industrial facilities. In recent years, it has become popular for use in the disinfection of produce—particularly sprouts, which have been implicated in numerous bacterial-infection outbreaks.

Practice Problem Ⓐ**TTEMPT** Draw the Lewis structure for the OH species. [The OH species is a *radical*, not to be confused with the hydroxide *ion* (OH⁻).]

Practice Problem Ⓑ**UILD** Draw the Lewis structure for the NS_2 molecule.

Practice Problem Ⓒ**ONCEPTUALIZE** Hypothetical elements A and B combine to form a number of molecules and polyatomic ions. (Element A is a member of Group 15; element B is a member of Group 16.) Using the chemical formulas, determine which of the following species must be represented by a Lewis structure with an unpaired electron.

AB	AB_2	A_2B	AB_3	AB_3^-	AB_2^-

Expanded Octets

Atoms of the second-period elements cannot have more than eight valence electrons around them, but atoms of elements in and beyond the third period of the periodic table can. In addition to the $3s$ and $3p$ orbitals, elements in the third period also have $3d$ orbitals that can be used in bonding. These orbitals enable an atom to form an *expanded octet*. One compound in which there is an expanded octet is sulfur hexafluoride, a very stable compound. The electron configuration of sulfur is $[Ne]3s^23p^4$. In SF_6, each of sulfur's 6 valence electrons forms a covalent bond with a fluorine atom, so there are 12 electrons around the central sulfur atom:

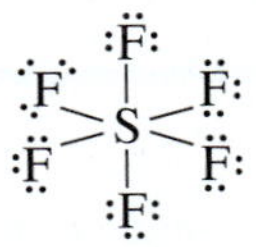

$$
\begin{array}{c}
:\ddot{F}: \\
\ddot{F}\diagdown \mid \diagup \ddot{F}: \\
\quad S \\
:\ddot{F}\diagup \mid \diagdown \ddot{F}: \\
:\ddot{F}:
\end{array}
$$

In Chapter 9, we will see that these 12 electrons, or six bonding pairs, are accommodated in six orbitals that originate from the one $3s$, the three $3p$, and two of the five $3d$ orbitals. Sulfur also forms many compounds in which it does obey the octet rule. In sulfur dichloride, for instance, S is surrounded by only eight electrons:

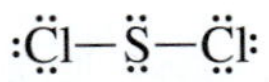

$$:\ddot{C}l—\ddot{S}—\ddot{C}l:$$

When drawing Lewis structures of compounds containing a central atom from the third period and beyond, the octet rule may be satisfied for all the atoms before all the valence electrons have been used up. When this happens, the extra electrons should be placed as lone pairs on the central atom.

Sample Problem 8.12 involves compounds that do not obey the octet rule.

Which Is More Important: Formal Charge or the Octet Rule?

We have learned that although central atoms from the third period (and beyond) often obey the octet rule, it is not necessary for them to do so. Rather, they can have what we call "expanded octets," meaning they are surrounded by more than eight electrons.

The sulfate ion (SO_4^{2-}), for example, can be represented by a resonance structure that obeys the octet rule (structure I), or by a structure that does not (structure II):

$$
\begin{bmatrix} \text{:Ö:}^{-1} \\ \text{}^{-1}\text{:Ö} \!-\! \overset{+2}{\text{S}} \!-\! \text{Ö:}^{-1} \\ \text{:Ö:}_{-1} \end{bmatrix}^{2-} \qquad \begin{bmatrix} \text{:Ö:}^{-1} \\ \text{:O} \!=\! \text{S} \!=\! \text{Ö:} \\ \text{:Ö:}_{-1} \end{bmatrix}^{2-}
$$

Structure I   Structure II

In structure I, although the octet rule is obeyed for the central atom, there are nonzero formal charges (shown in blue) on all the atoms. In structure II, by relocating a lone pair from each of two oxygen atoms and creating two double bonds, we change three of the formal charges to zero.

In some species, including the sulfate ion, it is possible to incorporate too many double bonds. Structures with *three* and *four* double bonds to sulfur would give formal charges on S and O that are inconsistent with the electronegativities of these elements. In general, if you are trying to minimize formal charges by expanding the central atom's octet, only add enough double bonds to make the formal charge on the central atom zero.

Whether structure I or structure II is better (or "more important") has been the subject of some debate among educators over the past two decades. Although some chemists may have a strong preference for one or the other, it is important for you to understand that *both* are valid Lewis structures and you should be able to draw both types of structures.

(See end-of-chapter problems 8.90, 8.102, and 8.103.)

SAMPLE PROBLEM 8.12

Draw the Lewis structures of (a) boron triiodide (BI_3), (b) arsenic pentafluoride (AsF_5), and (c) xenon tetrafluoride (XeF_4).

Strategy Follow the step-by-step procedure for drawing Lewis structures. The skeletal structures are

$$
\text{(a)} \ \overset{\overset{\textstyle I}{|}}{I \!-\! B \!-\! I} \qquad \text{(b)} \ \overset{\overset{\textstyle F}{|}}{\underset{F}{\overset{F\diagdown}{\diagup}} As \!-\! F} \qquad \text{(c)} \ \overset{\overset{\textstyle F}{|}}{\underset{F}{F \!-\! Xe \!-\! F}}
$$

Note that the skeletal structure already has more than an octet around the As atom.

Setup (a) There are a total of 24 valence electrons in BI_3 (three from the B and seven from each of the three I atoms). We subtract six electrons to account for the three bonds in the skeleton, leaving 18 electrons to distribute as three lone pairs on each I atom.
(b) There are 40 total valence electrons [five from As (Group 15) and seven from each of the five F atoms (Group 17)]. We subtract 10 electrons to account for the five bonds in the skeleton, leaving 30 to be distributed. Next, place three lone pairs on each F atom, thereby completing all their octets and using up all the electrons. (c) There are 36 total valence electrons (eight from Xe and seven from each of the four F atoms). We subtract eight electrons to account for the bonds in the skeleton, leaving 28 to distribute. We first complete the octets of all four F atoms. When this is done, four electrons remain, so we place two lone pairs on the Xe atom.

Solution

$$
\text{(a)} \ \overset{\overset{\textstyle \text{:Ï:}}{|}}{\text{:Ï} \!-\! B \!-\! \text{Ï:}} \qquad \text{(b)} \ \overset{\overset{\textstyle \text{:F̈:}}{|}}{\underset{\text{:F̈:}}{\overset{\text{:F̈}\diagdown}{\diagup}} As \!-\! \text{F̈:}} \qquad \text{(c)} \ \overset{\overset{\textstyle \text{:F̈:}}{|}}{\underset{\text{:F̈:}}{\text{:F̈} \!-\! \text{Xe} \!-\! \text{F̈:}}}
$$

THINK ABOUT IT

Boron is one of the elements that does not always follow the octet rule. Like BF_3, however, BI_3 can be drawn with a double bond to satisfy the octet of boron. This gives rise to a total of four resonance structures:

$$
\overset{\overset{\textstyle \text{:Ï:}}{|}}{\text{:Ï} \!-\! B \!-\! \text{Ï:}} \longleftrightarrow \overset{\overset{\textstyle \text{:Ï:}}{|}}{\text{:Ï} \!=\! B \!-\! \text{Ï:}} \longleftrightarrow \overset{\overset{\textstyle \text{·Ï·}}{\|}}{\text{:Ï} \!-\! B \!-\! \text{Ï:}} \longleftrightarrow \overset{\overset{\textstyle \text{:Ï:}}{|}}{\text{:Ï} \!-\! B \!=\! \text{Ï·}}
$$

Atoms beyond the second period can accommodate more than an octet of electrons, whether those electrons are used in bonds or reside on the central atom as lone pairs.

(Continued on next page)

Practice Problem ATTEMPT Draw the Lewis structures of (a) beryllium fluoride (BeF_2), (b) phosphorus pentachloride (PCl_5), and (c) the iodine tetrachloride ion (ICl_4^-).

Practice Problem BUILD Draw the Lewis structures of (a) boron trichloride (BCl_3), (b) antimony pentafluoride (SbF_5), and (c) krypton difluoride (KrF_2).

Practice Problem CONCEPTUALIZE Elements in the same group exhibit similar chemistry and sometimes form analogous species. For example, nitrogen and phosphorus (both members of Group 15) can combine with chlorine in a 1:3 ratio to form NCl_3 and PCl_3, respectively. Phosphorus can also combine with chlorine in a 1:5 ratio. Explain why nitrogen cannot.

CHECKPOINT – SECTION 8.8   Exceptions to the Octet Rule

8.8.1 In which of the following species does the central atom *not* obey the octet rule?

a) ClO_2^-

b) CO_2

c) BrO_3^-

d) HCN

e) ICl_4^-

8.8.2 Which elements cannot have more than an octet of electrons? (Select all that apply.)

a) N

b) C

c) S

d) Br

e) O

8.8.3 In which species does the central atom *obey* the octet rule? (Select all that apply.)

a) I_3^-

b) BH_3

c) AsF_6^-

d) NO_2

e) ClO_2^-

8.8.4 How many lone pairs are there on the central atom in the Lewis structure of ICl_2^-?

a) 0

b) 1

c) 2

d) 3

e) 4

8.9   Bond Enthalpy

One measure of the stability of a molecule is its **bond enthalpy,** which is the enthalpy change associated with breaking a particular bond in 1 mole of gaseous molecules. (Bond enthalpies in solids and liquids are affected by neighboring molecules.) The experimentally determined bond enthalpy of the diatomic hydrogen molecule, for example, is

$$H_2(g) \longrightarrow H(g) + H(g) \qquad \Delta H° = 436.4 \text{ kJ/mol}$$

According to this equation, breaking the covalent bonds in 1 mole of gaseous H_2 molecules requires 436.4 kJ of energy. For the less stable chlorine molecule,

$$Cl_2(g) \longrightarrow Cl(g) + Cl(g) \qquad \Delta H° = 242.7 \text{ kJ/mol}$$

Bond enthalpies can also be directly measured for heteronuclear diatomic molecules, such as HCl,

$$HCl(g) \longrightarrow H(g) + Cl(g) \qquad \Delta H° = 431.9 \text{ kJ/mol}$$

as well as for molecules containing multiple bonds:

$$O_2(g) \longrightarrow O(g) + O(g) \qquad \Delta H° = 498.7 \text{ kJ/mol}$$

$$N_2(g) \longrightarrow N(g) + N(g) \qquad \Delta H° = 941.4 \text{ kJ/mol}$$

TABLE 8.6	Bond Enthalpies				
Bond	**Bond Enthalpy (kJ/mol)**	**Bond**	**Bond Enthalpy (kJ/mol)**	**Bond**	**Bond Enthalpy (kJ/mol)**
H−H*	436.4	C≡O	1070	O−O	142
H−N	393	C−P	263	O=O	498.7
H−O	460	C−S	255	O−P	502
H−S	368	C=S	477	O=S	469
H−P	326	C−F	453	O−F	190
H−F	568.2	C−Cl	339	O−Cl	203
H−Cl	431.9	C−Br	276	O−Br	234
H−Br	366.1	C−I	216	O−I	234
H−I	298.3	N−N	193	P−P	197
C−H	414	N=N	418	P=P	489
C−C	347	N≡N	941.4	S−S	268
C=C	620	N−O	176	S=S	352
C≡C	812	N=O	607	F−F	156.9
C−N	276	N−F	272	Cl−Cl	242.7
C=N	615	N−Cl	200	Cl−F	193
C≡N	891	N−Br	243	Br−Br	192.5
C−O	351	N−I	159	I−I	151.0
C=O†	745				

*Bond enthalpies shown in red are for *diatomic molecules*.

†The C=O bond enthalpy in CO_2 is 799 kJ/mol.

Student Note: Bond enthalpies for diatomic molecules have more significant figures than those for polyatomic molecules. Those for polyatomic molecules are average values based on the bonds in more than one compound.

Measuring the strength of covalent bonds in polyatomic molecules is more complicated. For example, measurements show that the energy needed to break the first O−H bond in H_2O is different from that needed to break the second O−H bond:

$$H_2O(g) \longrightarrow H(g) + OH(g) \qquad \Delta H° = 502 \text{ kJ/mol}$$

$$OH(g) \longrightarrow H(g) + O(g) \qquad \Delta H° = 427 \text{ kJ/mol}$$

In each case, an O−H bond is broken, but the first step requires the input of more energy than the second. The difference between the two $\Delta H°$ values suggests that the second O−H bond itself undergoes change, because of the changes in its chemical environment.

We can now understand why the bond enthalpy of the same O−H bond in two different molecules, such as methanol (CH_3OH) and water (H_2O), will not be the same: their environments are different. For polyatomic molecules, therefore, we speak of the *average* bond enthalpy of a particular bond. For example, we can measure the enthalpy of the O−H bond in 10 different polyatomic molecules and obtain the average O−H bond enthalpy by dividing the sum of the bond enthalpies by 10. Table 8.6 lists the average bond enthalpies of a number of diatomic and polyatomic molecules. As we noted earlier, triple bonds are stronger than double bonds, and double bonds are stronger than single bonds.

A comparison of the thermochemical changes that take place during a number of reactions reveals a strikingly wide variation in the enthalpies of different reactions. For example, the combustion of hydrogen gas in oxygen gas is fairly *exothermic:*

$$H_2(g) + \tfrac{1}{2}O_2(g) \longrightarrow H_2O(l) \qquad \Delta H° = -285.8 \text{ kJ/mol}$$

The formation of glucose from carbon dioxide and water, on the other hand, best achieved by photosynthesis, is highly *endothermic:*

$$6CO_2(g) + 6H_2O(l) \longrightarrow C_6H_{12}O_6(s) + 6O_2(g) \qquad \Delta H° = 2801 \text{ kJ/mol}$$

We can account for such variations by looking at the stability of individual reactant and product molecules. After all, most chemical reactions involve the making and breaking of bonds. Therefore, knowing the bond enthalpies and hence the stability of molecules reveals something about the thermochemical nature of the reactions that molecules undergo.

In many cases, it is possible to predict the approximate enthalpy of a reaction by using the average bond enthalpies. Because energy is always required to break chemical bonds and chemical bond formation is always accompanied by a release of energy, we can estimate the enthalpy of a reaction by counting the total number of bonds broken and formed in the reaction and recording all the corresponding enthalpy changes. The enthalpy of reaction in the gas phase is given by

Equation 8.4 $\Delta H^\circ = \Sigma BE(\text{reactants}) - \Sigma BE(\text{products})$

= total energy *input* (to *break* bonds) − total energy *released* (by bond *formation*)

where BE stands for average bond enthalpy and Σ is the summation sign. As written, Equation 8.4 takes care of the sign convention for ΔH°. Thus, if the total energy input needed to break bonds in the reactants is less than the total energy released when bonds are formed in the products, then ΔH° is negative and the reaction is exothermic [Figure 8.13(a)]. On the other hand, if less energy is released (bond making) than absorbed (bond breaking), ΔH° is positive and the reaction is endothermic [Figure 8.13(b)].

If all the reactants and products are diatomic molecules, then the equation for the enthalpy of reaction will yield accurate results because the bond enthalpies of diatomic molecules are accurately known. If some or all of the reactants and products are polyatomic molecules, the equation will yield only approximate results because the bond enthalpies used will be averages.

Sample Problem 8.13 shows how to estimate enthalpies of reaction using bond enthalpies.

Figure 8.13 Enthalpy changes in (a) an exothermic reaction and (b) an endothermic reaction. The ΔH° values are calculated using Equation 5.19 and tabulated ΔH_f° values from Appendix 2.

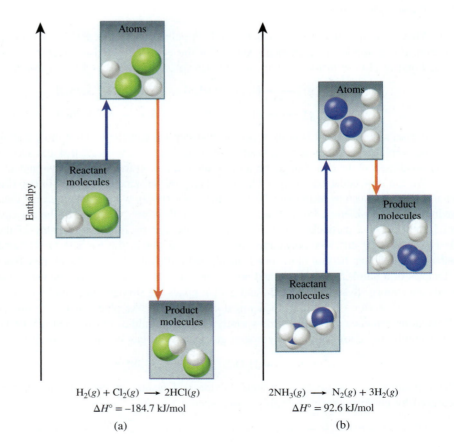

$H_2(g) + Cl_2(g) \longrightarrow 2HCl(g)$
$\Delta H^\circ = -184.7$ kJ/mol
(a)

$2NH_3(g) \longrightarrow N_2(g) + 3H_2(g)$
$\Delta H^\circ = 92.6$ kJ/mol
(b)

SAMPLE PROBLEM 8.13

Use bond enthalpies from Table 8.6 to estimate the enthalpy of reaction for the combustion of methane:

$$CH_4(g) + 2O_2(g) \longrightarrow CO_2(g) + 2H_2O(l)$$

Strategy Draw Lewis structures to determine what bonds are to be broken and what bonds are to be formed. (Don't skip the step of drawing Lewis structures. This is the only way to know for certain what types and numbers of bonds must be broken and formed.)

Setup

$$H-\overset{\overset{\displaystyle H}{|}}{\underset{\underset{\displaystyle H}{|}}{C}}-H \quad + \quad \begin{matrix} \ddot{O}=\ddot{O} \\ \\ \ddot{O}=\ddot{O} \end{matrix} \quad \longrightarrow \quad \ddot{O}=C=\ddot{O} \quad + \quad \begin{matrix} H-\ddot{O}-H \\ \\ H-\ddot{O}-H \end{matrix}$$

Bonds to break: 4 C—H and 2 O=O.

Bonds to form: 2 C=O and 4 H—O.

Bond enthalpies from Table 8.6: 414 kJ/mol (C—H), 498.7 kJ/mol (O=O), 799 kJ/mol (C=O in CO_2), and 460 kJ/mol (H—O).

Solution

$$[4(414 \text{ kJ/mol}) + 2(498.7 \text{ kJ/mol})] - [2(799 \text{ kJ/mol}) + 4(460 \text{ kJ/mol})] = -785 \text{ kJ/mol}$$

Remember that heats of reaction are expressed in kJ/mol, where the "per mole" refers to *per mole of reaction as written* [◄◄ Section 5.3].

THINK ABOUT IT

Use Equation 5.19 [◄◄ Section 5.6] and data from Appendix 2 to calculate this enthalpy of reaction again; then compare your results using the two approaches. The difference in this case is due to two things: Most tabulated bond enthalpies are averages and, by convention, we show the product of combustion as liquid water—but average bond enthalpies apply to species in the gas phase, where there is little or no influence exerted by neighboring molecules.

Practice Problem A TTEMPT Use bond enthalpies from Table 8.6 to estimate the enthalpy of reaction for the combination of carbon monoxide and oxygen to produce carbon dioxide:

$$2CO(g) + O_2(g) \longrightarrow 2CO_2(g)$$

Practice Problem B UILD Using the following chemical equation, data from Table 8.6, and data from Appendix 2, determine the P—Cl bond enthalpy:

$$PH_3(g) + 3HCl(g) \longrightarrow PCl_3(g) + 3H_2(g)$$

Practice Problem C ONCEPTUALIZE Four different chemical reactions are represented here. For each reaction, indicate whether it is endothermic or exothermic—or if there is not enough information to determine.

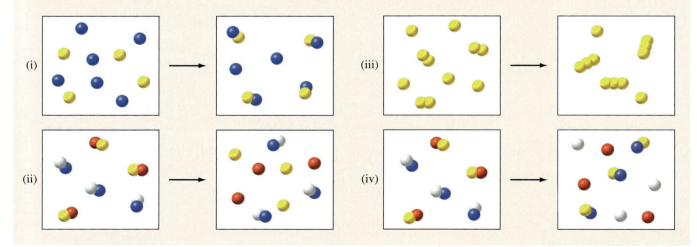

CHECKPOINT – SECTION 8.9   Bond Enthalpy

8.9.1 Use data from Table 8.6 to estimate ΔH_{rxn} for the decomposition of hydrogen peroxide to form water and oxygen.

$$H_2O_2(l) \longrightarrow H_2O(l) + \tfrac{1}{2}O_2(g)$$

a) −71 kJ/mol

b) +357 kJ/mol

c) −357 kJ/mol

d) −107 kJ/mol

e) +71 kJ/mol

8.9.2 Use data from Table 8.6 to estimate ΔH_{rxn} for the reaction of fluorine and chlorine to produce ClF.

$$F_2(g) + Cl_2(g) \longrightarrow 2ClF(g)$$

a) −77.5 kJ/mol

b) −206.6 kJ/mol

c) 206.6 kJ/mol

d) −13.6 kJ/mol

e) 13.6 kJ/mol

8.9.3 Use bond enthalpies to determine ΔH_{rxn} for the reaction shown.

a) −1139 kJ/mol

b) +1139 kJ/mol

c) −346 kJ/mol

d) +114 kJ/mol

e) −114 kJ/mol

8.9.4 Use bond enthalpies to determine ΔH_{rxn} for the reaction shown.

a) −1028 kJ/mol

b) −200 kJ/mol

c) −392 kJ/mol

d) +392 kJ/mol

e) +200 kJ/mol

Chapter Summary

Section 8.1

- A *Lewis dot symbol* depicts an atom or an atomic ion of a main group element with dots (representing the valence electrons) arranged around the element's symbol. Main group atoms lose or gain one or more electrons to become isoelectronic with noble gases.

Section 8.2

- The electrostatic attraction that holds ions together in an ionic compound is referred to as *ionic bonding*.

- *Lattice energy* is the amount of energy required to convert a mole of ionic solid to its constituent ions in the gas phase. Lattice energy cannot be measured directly, but is determined using the *Born-Haber cycle* and thermodynamic quantities that can be measured directly.

Section 8.3

- According to the *Lewis theory of bonding, covalent bonding* results when atoms *share* valence electrons. The atoms in molecules and those in polyatomic ions are held together by *covalent bonds*.

- According to the *octet rule,* atoms will lose, gain, or share electrons to achieve a noble gas configuration. Pairs of valence electrons that are *not* involved in the covalent bonding in a molecule or polyatomic ion (i.e., valence electrons that are not shared) are called *lone pairs*.

- *Lewis structures* are drawn to represent molecules and polyatomic ions, showing the arrangement of atoms and the positions of all valence electrons. Lewis structures represent the shared pairs of valence electrons either as two dots, ··, or as a single dash, —. Any *unshared* electrons are represented as dots.

- One shared pair of electrons between atoms constitutes a *single bond. Multiple bonds* form between atoms that share more than one pair of electrons. Two shared pairs constitute a *double bond,* and three shared pairs constitute a *triple bond*.

Section 8.4

- Bonds in which electrons are not shared equally are *polar* and are referred to as *polar covalent bonds*.

- *Electronegativity* is an atom's ability to draw shared electrons toward itself. Bonds between elements of widely different electronegativities ($\Delta \geq 2.0$) are *ionic*. Covalent bonds between atoms with significantly different electronegativities ($0.5 \leq \Delta < 2.0$) are *polar*. Bonds between atoms with very similar electronegativities ($\Delta < 0.5$) are *nonpolar*.

- *Percent ionic character* quantifies the polarity of a bond, and is determined by comparing the measured dipole moment to the one predicted by assuming that the bonded atoms have discrete charges.

- The *dipole moment (μ)* is a quantitative measure of the polarity of a bond.

Section 8.5

- Lewis structures of molecules or polyatomic ions can be drawn using the following step-by-step procedure:
  1. Use the molecular formula to draw the skeletal structure.
  2. Count the total number of valence electrons, adding electrons to account for a negative charge and subtracting electrons to account for a positive charge.
  3. Subtract two electrons for each bond in the skeletal structure.
  4. Distribute the remaining valence electrons to complete octets, completing the octets of the more electronegative atoms first.
  5. Place any remaining electrons on the central atom.
  6. Include double or triple bonds, if necessary, to complete the octets of all atoms.

Section 8.6

- *Formal charge* is a way of keeping track of the valence electrons in a species. Formal charges should be consistent with electronegativities and can be used to determine the best arrangement of atoms and electrons for a Lewis structure.

Section 8.7

- *Resonance structures* are two or more equally correct Lewis structures that differ in the positions of the electrons but *not* in the positions of the atoms. Different resonance structures of a compound can be separated by a resonance arrow, $\longleftrightarrow$.

Section 8.8

- In an ordinary covalent bond, each atom contributes one electron to the shared pair of electrons. In cases where just one of the atoms contributes *both* of the electrons, the bond is called a *coordinate covalent bond* or a *dative bond*.

- A species that contains an odd number of electrons is called a *free radical*.

Section 8.9

- *Bond enthalpy* is the energy required to break 1 mole of a particular type of bond. Bond enthalpies are a measure of the stability of covalent bonds and can be used to estimate the enthalpy change for a reaction.

The first step in solving many problems is drawing a correct Lewis structure. The process of drawing a Lewis structure was first described in Figure 8.12. The steps, which are summarized in the flowchart below, are:

1. Count the total valence electrons. Recall that each atom contributes a number equal to its group number; and remember to add or subtract valence electrons to account for charge on a polyatomic ion.

2. Use the chemical formula to draw a skeletal structure. Usually, the central atom is less electronegative than the terminal atoms; although hydrogen cannot be a central atom because it can form only one bond.

3. For each bond in the skeletal formula, subtract two from the total number of valence electrons.

4. Distribute the remaining electrons, satisfying first the octets of the more electronegative (usually terminal) atoms.

5. If all terminal atoms have complete octets, and there are valence electrons still to be distributed, place them on the central atom as lone pairs.

6. If the valence electrons run out before all octets are satisfied, use multiple bonds to complete the octets of all atoms.

7. For a charged species, enclose the Lewis structure in square brackets and add a superscript charge.

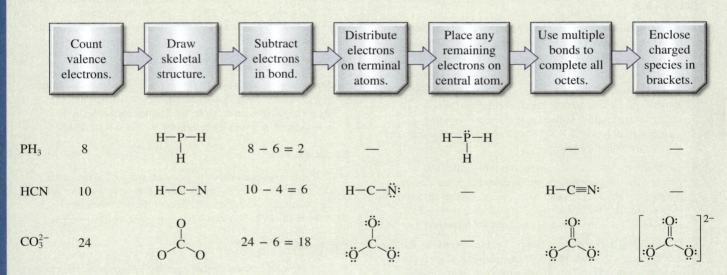

There are exceptions to the octet rule.

- Be and B, small atoms with low electronegativity, need not obey the octet rule.
- Species with an odd number of valence electrons cannot obey the octet rule.
- Elements in the third period and beyond need not obey the octet rule.
- A larger central atom (from the third period or beyond) can accommodate more than eight electrons and can have an "expanded" octet.

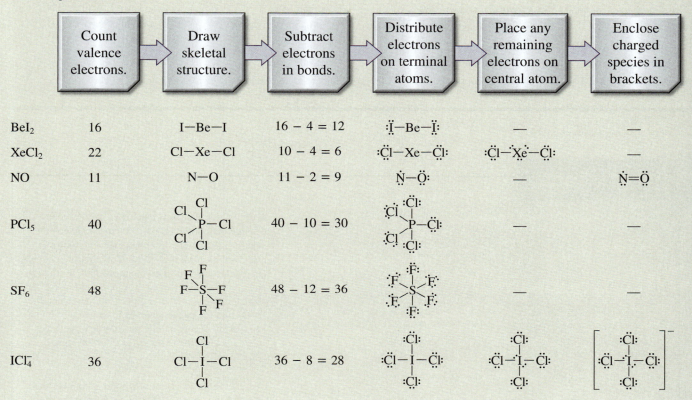

Key Skills Problems

8.1
Which of the following atoms must always obey the octet rule? (Select all that apply.)

(a) C   (b) N   (c) S   (d) Br   (e) Xe

8.2
Which of the following species has an odd number of electrons? (Select all that apply.)

(a) N_2O (b) NO_2 (c) NO_2^- (d) NO_3^- (e) NS

8.3
How many lone pairs are on the central atom in $XeOF_2$?

(a) 0   (b) 1   (c) 2   (d) 3   (e) 4

8.4
How many lone pairs are on the central atom in the perchlorate ion?

(a) 0   (b) 1   (c) 2   (d) 3   (e) 4

Key Words

Bond enthalpy, 390
Born-Haber cycle, 365
Coordinate covalent bond, 386
Covalent bond, 369
Covalent bonding, 369
Dative bond, 386
Dipole moment (μ), 374
Double bond, 370
Electronegativity, 372
Formal charge, 380
Free radical, 387
Ionic bonding, 362
Lattice energy, 362
Lewis dot symbol, 360
Lewis structure, 369
Lewis theory of bonding, 369
Lone pair, 369
Multiple bond, 370
Nonpolar, 373
Octet rule, 369
Percent ionic character, 376
Polar, 371
Polar covalent bond, 371
Resonance structure, 384
Single bond, 370
Triple bond, 370

Key Equations

8.1 $\mu = Q \times r$	Dipole moment (μ) is calculated as the product of charge magnitude (Q) and distance between the charges (bond length, r) in a diatomic molecule. Because molecules are neutral, the partial charges in a heteronuclear diatomic molecule are equal in magnitude and opposite in sign. Equation 8.1 can be used to calculate the dipole moment when the magnitude of partial charges is known—or it can be used to determine the magnitude of partial charges when the experimentally determined dipole moment is known.
8.2 $\text{percent ionic character} = \dfrac{\mu(\text{observed})}{\mu(\text{calculated assuming discrete charges})} = 100\%$	Percent ionic character of a bond is equal to the ratio of the observed dipole moment to the dipole moment calculated, assuming discrete charges on the atoms.
8.3 formal charge = valence electrons − associated electrons	Formal charge on an atom in a Lewis structure is equal to the number of valence electrons (group number) minus half of the electrons it shares with other atoms in the structure.
8.4 $\Delta H^\circ = \Sigma\text{BE(reactants)} - \Sigma\text{BE(products)}$	The enthalpy change of a reaction can be estimated by subtracting the sum of bond enthalpies in products from the sum of bond enthalpies in reactants.

Questions and Problems

Applying What You've Learned

Researchers in the early 1990s made the sensational announcement that nitric oxide (NO), which had long been thought of only as a component of air pollution, turns out to play an important role in human physiology. They found that NO serves as a *signal* molecule, being produced *in vivo* and regulating a wide variety of cell functions in the body including in the cardiovascular, nervous, and immune systems.

The discovery of the biological role of nitric oxide has shed light on how nitroglycerin works as a drug. For many years, nitroglycerin tablets have been prescribed for cardiac patients to relieve the pain caused by brief interruptions in the flow of blood to the heart, although how it worked was not understood. We now know that nitroglycerin produces nitric oxide in the body, which causes muscles to relax and allows the arteries to dilate.

Research continues to uncover the role nitric oxide plays in biological processes, and medicine continues to find new uses for this molecule.

Problems:

(a) Without consulting Figure 8.1, give the Lewis dot symbols for N and O [◄◄ Sample Problem 8.1]. (b) Classify the bond in NO as nonpolar, polar, or ionic [◄◄ Sample Problem 8.4]. (c) Given the experimentally determined dipole moment (0.16 D) and the bond length (1.15 Å), determine the magnitude of the partial charges in the NO molecule [◄◄ Sample Problem 8.5]. (d) Draw the Lewis structures for NO and for nitroglycerin ($C_3H_5N_3O_9$) [◄◄ Sample Problem 8.7]. (e) Determine the formal charges on each atom in NO and in nitroglycerin [◄◄ Sample Problem 8.8]. (f) Nitroglycerin decomposes explosively to give carbon dioxide, water, nitrogen, and oxygen. Given the balanced equation for this reaction,

$$C_3H_5(NO_3)_3 \longrightarrow 3CO_2 + 2.5H_2O + 1.5N_2 + 0.25O_2$$

use bond enthalpies to estimate $\Delta H°$ for the reaction [◄◄ Sample Problem 8.13].

SECTION 8.1: LEWIS DOT SYMBOLS

Review Questions

8.1   What is a Lewis dot symbol? What elements do we generally represent with Lewis symbols?

8.2   Use the second member of each group from Group 1 to Group 17 to show that the number of valence electrons on an atom of the element is the same as its group number.

Conceptual Problems

8.3   Without referring to Figure 8.1, write Lewis dot symbols for atoms of the following elements: (a) Be, (b) K, (c) Ca, (d) Ga, (e) O, (f) Br, (g) N, (h) I, (i) As, (j) F.

8.4   Write Lewis dot symbols for the following ions: (a) Li^+, (b) Cl^-, (c) S^{2-}, (d) Sr^{2+}, (e) N^{3-}.

8.5   Write Lewis dot symbols for the following atoms and ions: (a) I, (b) I^-, (c) S, (d) S^{2-}, (e) P, (f) P^{3-}, (g) Na, (h) Na^+, (i) Mg, (j) Mg^{2+}, (k) Al, (l) As^{3+}, (m) Pb, (n) Pb^{2+}.

SECTION 8.2: IONIC BONDING

Visualizing Chemistry
Figure 8.6

VC 8.1   What additional information would you need to calculate the lattice energy for a compound if the charges on the cation and anion were +2 and −1, respectively, rather than +1 and −1?
a) No additional information is needed.
b) IE_2 of the cation.
c) IE_2 of the cation and EA_2 of the anion.

VC 8.2   What additional information would you need to calculate the lattice energy for a compound if the charges on the cation and anion were +2 and −2, respectively, rather than +1 and −1?
a) No additional information is needed.
b) IE_2 of the cation.
c) IE_2 of the cation and EA_2 of the anion.

VC 8.3   How would the magnitude of the lattice energy calculated using the Born-Haber cycle change if the charges on the cation and anion were +2 and −2, respectively, rather than +1 and −1?
a) Lattice energy would increase.
b) Lattice energy would decrease.
c) Whether lattice energy would increase or decrease depends on the relative magnitudes of IE_2 of the cation and EA_2 of the anion.

VC 8.4   What law enables us to use the Born-Haber cycle to calculate lattice energy?
a) Coulomb's law
b) Hess's law
c) Law of multiple proportions

Review Questions

8.6   Explain what *ionic bonding* is.

8.7   Explain how ionization energy and electron affinity determine whether atoms of elements will combine to form ionic compounds.

8.8   Name five metals and five nonmetals that are very likely to form ionic compounds. Write formulas for compounds that might result from the combination of these metals and nonmetals. Name these compounds.

8.9   Name one ionic compound that contains only nonmetallic elements.

8.10 Name one ionic compound that contains a polyatomic cation and a polyatomic anion (see Table 2.3).

8.11 Explain why ions with charges greater than ±3 are seldom found in ionic compounds.

8.12 The term *molar mass* was introduced in Chapter 3. Molar mass is numerically equivalent to molecular mass, although the units are different, for a covalent compound. What is the advantage of using the term molar mass when we discuss ionic compounds?

8.13 In which of the following states would NaCl be electrically conducting: (a) solid, (b) molten (i.e., melted), (c) dissolved in water? Explain.

8.14 Beryllium forms a compound with chlorine that has the empirical formula $BeCl_2$. How would you determine whether it is an ionic compound? (The compound is not soluble in water.)

8.15 What is *lattice energy,* and what does it indicate about the stability of an ionic compound?

8.16 Explain how the lattice energy of an ionic compound such as KCl can be determined using the Born-Haber cycle. On what law is this procedure based?

8.17 Specify which compound in each of the following pairs of ionic compounds should have the higher lattice energy: (a) KCl or MgO, (b) LiF or LiBr, (c) Mg_3N_2 or NaCl. Explain your choice.

8.18 Specify which compound in each of the following pairs of ionic compounds should have the higher lattice energy: (a) AlN or CaO, (b) NaF or CsF, (c) $MgCl_2$ or MgF_2. Explain your choice.

Computational Problems

8.19 Use the Born-Haber cycle outlined in Section 8.2 for NaCl to calculate the lattice energy of LiCl. Use data from Figures 7.8 and 7.10 and Appendix 2.

8.20 Calculate the lattice energy of $CaCl_2$. Use data from Figures 7.8 and 7.10 and Appendix 2. (The second ionization energy of Ca, IE_2, is 1145 kJ/mol.)

Conceptual Problems

8.21 An ionic bond is formed between a cation A^+ and an anion B^-. Based on Coulomb's law

$$E \propto \frac{Q_1 \times Q_2}{d}$$

how would the energy of the ionic bond be affected by the following changes: (a) doubling the radius of A^+, (b) tripling the charge on A^+, (c) doubling the charges on A^+ and B^-, (d) decreasing the radii of A^+ and B^- to half their original values?

8.22 Give the empirical formulas and names of the compounds formed from the following pairs of ions: (a) Rb^+ and I^-, (b) Cs^+ and SO_4^{2-}, (c) Sr^{2+} and N^{3-}, (d) Al^{3+} and S^{2-}.

8.23 Use Lewis dot symbols to show the transfer of electrons between the following atoms to form cations and anions: (a) Na and F, (b) K and S, (c) Ba and O, (d) Al and N.

8.24 Write the Lewis dot symbols of the reactants and products in the following reactions. (First balance the equations.)
(a) $Sr + Se \longrightarrow SrSe$ (c) $Li + N_2 \longrightarrow Li_3N$
(b) $Ca + H_2 \longrightarrow CaH_2$ (d) $Al + S \longrightarrow Al_2S_3$

SECTION 8.3: COVALENT BONDING

Review Questions

8.25 Describe Lewis's contribution to our understanding of the covalent bond.

8.26 Use an example to illustrate each of the following terms: *lone pair, Lewis structure, octet rule, bond length.*

8.27 What is the difference between a Lewis *symbol* and a Lewis *structure*?

8.28 How many lone pairs are on the underlined atoms in these compounds: H<u>Br</u>, H<u>S</u>, C<u>H</u>₄?

8.29 Compare single, double, and triple bonds in a molecule, and give an example of each. For the same bonding atoms, how does the bond length change from single bond to triple bond?

8.30 Compare the properties of ionic compounds and covalent compounds.

8.31 Summarize the essential features of the Lewis octet rule. The octet rule applies mainly to the second-period elements. Explain.

Conceptual Problems

8.32   For each of the following pairs of elements, state whether the binary compound they form is likely to be ionic or covalent. Write the empirical formula and name of the compound: (a) I and Cl, (b) Mg and F.

8.33   For each of the following pairs of elements, state whether the binary compound they form is likely to be ionic or covalent. Write the empirical formula and name of the compound: (a) B and F, (b) K and Br.

SECTION 8.4: ELECTRONEGATIVITY AND POLARITY

Review Questions

8.34   Define *electronegativity,* and explain the difference between electronegativity and electron affinity. Describe in general how the electronegativities of the elements change according to their position in the periodic table.

8.35   What is a *polar covalent bond*? Name two compounds that contain one or more polar covalent bonds.

Computational Problems

8.36   Four atoms are arbitrarily labeled D, E, F, and G. Their electronegativities are as follows: D = 3.8, E = 3.3, F = 2.8, and G = 1.3. If the atoms of these elements form the molecules DE, DG, EG, and DF, how would you arrange these molecules in order of increasing covalent bond character?

8.37   Using information in Table 8.5, calculate the magnitude of the partial charges on the atoms in HI.

8.38   Using information in Table 8.5, calculate the percent ionic character of the bond in HBr.

Conceptual Problems

8.39   List the following bonds in order of increasing ionic character: cesium to fluorine, chlorine to chlorine, bromine to chlorine, silicon to carbon.

8.40   Classify the following bonds as covalent, polar covalent, or ionic, and explain: (a) the CC bond in H_3CCH_3, (b) the KI bond in KI, (c) the NB bond in H_3NBCl_3, (d) the CF bond in CF_4.

8.41   Classify the following bonds as covalent, polar covalent, or ionic, and explain: (a) the SiSi bond in $Cl_3SiSiCl_3$, (b) the SiCl bond in $Cl_3SiSiCl_3$, (c) the CaF bond in CaF_2, (d) the NH bond in NH_3.

8.42   List the following bonds in order of increasing ionic character: the lithium-to-fluorine bond in LiF, the potassium-to-oxygen bond in K_2O, the nitrogen-to-nitrogen bond in N_2, the sulfur-to-oxygen bond in SO_2, the chlorine-to-fluorine bond in ClF_3.

8.43   Arrange the following bonds in order of increasing ionic character: carbon to hydrogen, fluorine to hydrogen, bromine to hydrogen, sodium to chlorine, potassium to fluorine, lithium to chlorine.

8.44   Two pairs of elements are highlighted in the periodic table shown here. Consider two binary compounds, one consisting of the two elements highlighted in yellow and one consisting of the two elements highlighted in blue. For which pair of elements will the partial charges be largest? Explain.

8.45   Considering all four highlighted elements in the periodic table shown here, which pair of elements would produce a binary compound with bonds of the greatest percent ionic character? (The binary compound may consist of two yellow elements, two blue elements, or one of each.) Explain.

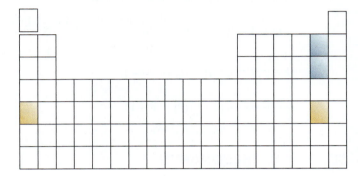

SECTION 8.5: DRAWING LEWIS STRUCTURES

Conceptual Problems

8.46   Draw Lewis structures for the following molecules and ions: (a) NCl_3, (b) OCS, (c) H_2O_2, (d) CH_3COO^-, (e) CN^-, (f) $CH_3CH_2NH_3^+$.

8.47   Draw Lewis structures for the following molecules and ions: (a) OF_2, (b) N_2F_2, (c) Si_2H_6, (d) OH^-, (e) CH_2ClCOO^-, (f) $CH_3NH_3^+$.

8.48   Draw Lewis structures for the following molecules: (a) ICl, (b) PH_3, (c) P_4 (each P is bonded to three other P atoms), (d) H_2S, (e) N_2H_4, (f) $HClO_3$.

8.49   Draw Lewis structures for the following molecules: (a) $COBr_2$ (C is bonded to O and Br atoms), (b) H_2Se, (c) NH_2OH, (d) CH_3NH_2, (e) CH_3CH_2Br, (f) NCl_3.

SECTION 8.6: LEWIS STRUCTURES AND FORMAL CHARGE

Review Question

8.50   Explain the concept of *formal charge*. Do formal charges represent an actual separation of charges?

Conceptual Problems

8.51   Draw Lewis structures for the following ions: (a) NO_2^+, (b) SCN^-, (c) S_2^{2-}, (d) ClF_2^+. Show formal charges.

8.52   Draw Lewis structures for the following ions: (a) O_2^{2-}, (b) C_2^{2-}, (c) NO^+, (d) NH_4^+. Show formal charges.

8.53   The skeletal structure of acetic acid shown here
is correct, but some of the bonds are wrong.
(a) Identify the incorrect bonds and explain
what is wrong with them. (b) Write the correct
Lewis structure for acetic acid.

$$\begin{array}{ccc} H & :O: \\ | & | \\ H=C-C-O-H \\ | \\ H \end{array}$$

8.54   The following Lewis structures are incorrect.
Explain what is wrong with each one, and give a
correct Lewis structure for the molecule. (Relative
positions of atoms are shown correctly.)

(a) $H-\ddot{C}=\ddot{N}$ (d) $\begin{array}{c} :\ddot{F}\ddot{F}: \\ \diagdown B \diagup \\ :\ddot{F}: \end{array}$ (g) $\begin{array}{c} :\ddot{F}\ddot{F}: \\ \diagdown N \diagup \\ :\ddot{F}: \end{array}$

(b) $H=C=C=H$ (e) $H-\ddot{O}=\ddot{F}:$

(c) $\ddot{O}-Sn-\ddot{O}$ (f) $\begin{array}{c} H \\ | \\ C-\ddot{F}: \\ | \\ :\ddot{O}: \end{array}$

SECTION 8.7: RESONANCE

Review Questions

8.55   What is a *resonance structure*? Is it possible to
isolate one resonance structure of a compound for
analysis? Explain.
8.56   What are the rules for writing resonance structures?

Conceptual Problems

8.57   Draw Lewis structures for the following species,
including all resonance forms, and show formal
charges: (a) HCO_2^-, (b) $CH_2NO_2^-$, The relative
positions of the atoms are as follows:

$$\begin{array}{cccccc} & O & & & H & O \\ H & C & & & C & N \\ & O & & & H & O \end{array}$$

8.58   Draw three resonance structures for the chlorate
ion (ClO_3^-). Show formal charges.
8.59   Draw three resonance structures for hydrazoic acid
(HN_3). The atomic arrangement is HNNN. Show
formal charges.
8.60   Draw two resonance structures for diazomethane
(CH_2N_2). Show formal charges. The skeletal
structure of the molecule is

$$\begin{array}{c} H \\ C \ N \ N \\ H \end{array}$$

8.61   Draw three reasonable resonance structures for
the OCN^- ion. Show formal charges and rank the
importance of the structures.
8.62   Draw three resonance structures for the molecule
N_2O in which the atoms are arranged in the order

NNO. Indicate formal charges and arrange the
resonance structures in order of increasing relative
importance.
8.63   Draw a resonance structure of the adenine
molecule shown here, which is part of the DNA
structure. Show all the lone pairs and label the
formal charges.

SECTION 8.8: EXCEPTIONS TO THE OCTET RULE

Review Questions

8.64   Why does the octet rule not hold for many
compounds containing elements in the third period
of the periodic table and beyond?
8.65   Give three examples of compounds that do not
satisfy the octet rule. Write a Lewis structure for
each.
8.66   Because fluorine has seven valence electrons
($2s^2 2p^5$), seven covalent bonds in principle could
form around the atom. Such a compound might be
FH_7 or FCl_7. These compounds have never been
prepared. Why?
8.67   What is a *coordinate covalent bond*? Is it different
from an ordinary covalent bond?
8.68   Because the central atom in each case is from
the same group of the periodic table, the Lewis
structures we draw for SO_2 and O_3 are essentially
the same. Explain why we can draw a resonance
structure for SO_2 in which the formal charge on the
central atom is zero, but we cannot do this for O_3.
8.69   What is the advantage of drawing the Lewis
structures of oxoanions and oxoacids using
expanded octets?

Conceptual Problems

8.70   The AlI_3 molecule has an incomplete octet around
Al. Draw three resonance structures of the
molecule in which the octet rule is satisfied for
both the Al and the I atoms. Show formal charges.
8.71   In the vapor phase, beryllium chloride consists of
discrete $BeCl_2$ molecules. Is the octet rule satisfied
for Be in this compound? If not, can you form an
octet around Be by drawing another resonance
structure? How plausible is this structure?
8.72   Of the noble gases, only Kr, Xe, and Rn are known
to form a few compounds with O and/or F. Write
Lewis structures for the following molecules:
(a) XeF_2, (b) XeF_4, (c) XeF_6, (d) $XeOF_4$,
(e) XeO_2F_2. In each case Xe is the central atom.

8.73 Write a Lewis structure for $SbCl_5$. Does this molecule obey the octet rule?

8.74 Write Lewis structures for SeF_4 and SeF_6. Is the octet rule satisfied for Se?

8.75 Write Lewis structures for the reaction

$$AlCl_3 + Cl^- \longrightarrow AlCl_4^-$$

What kind of bond joins Al and Cl in the product?

8.76 Draw two resonance structures for the bromate ion (BrO_3^-), one that obeys the octet rule and one in which the formal charge on the central atom is zero.

8.77 Draw two resonance structures for the sulfite ion (SO_3^{2-}), one that obeys the octet rule and one in which the formal charge on the central atom is zero.

SECTION 8.9: BOND ENTHALPY

Review Questions

8.78 What is *bond enthalpy*? Bond enthalpies of polyatomic molecules are average values, whereas those of diatomic molecules can be accurately determined. Why?

8.79 Explain why the bond enthalpy of a molecule is usually defined in terms of a gas-phase reaction. Why are bond-breaking processes always endothermic and bond-forming processes always exothermic?

Computational Problems

8.80 From the following data, calculate the average bond enthalpy for the NH bond:

$$NH_3(g) \longrightarrow NH_2(g) + H(g) \quad \Delta H° = 435 \text{ kJ/mol}$$
$$NH_2(g) \longrightarrow NH(g) + H(g) \quad \Delta H° = 381 \text{ kJ/mol}$$
$$NH(g) \longrightarrow N(g) + H(g) \quad \Delta H° = 360 \text{ kJ/mol}$$

8.81 For the reaction

$$O(g) + O_2(g) \longrightarrow O_3(g) \quad \Delta H° = -107.2 \text{ kJ/mol}$$

Calculate the average bond enthalpy in O_3.

8.82 The bond enthalpy of $F_2(g)$ is 156.9 kJ/mol. Calculate $\Delta H_f°$ for $F(g)$.

8.83 For the reaction

$$2C_2H_6(g) + 7O_2(g) \longrightarrow 4CO_2(g) + 6H_2O(g)$$

(a) Predict the enthalpy of reaction from the average bond enthalpies in Table 8.6. (b) Calculate the enthalpy of reaction from the standard enthalpies of formation (see Appendix 2) of the reactant and product molecules, and compare the result with your answer for part (a).

Conceptual Problems

8.84 Use average bond enthalpies from Table 8.6 to estimate ΔH_{rxn} for the following reaction.

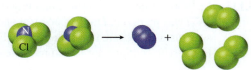

8.85 Use average bond enthalpies from Table 8.6 to estimate ΔH_{rxn} for the following reaction.

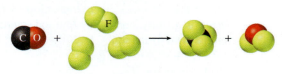

8.86 For each reaction represented, indicate whether it is endothermic, exothermic, or if there is not enough information to determine.

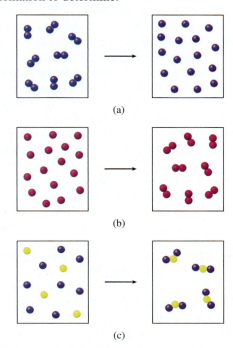

(a)

(b)

(c)

8.87 For each reaction represented, indicate whether it is endothermic, exothermic, or if there is not enough information to deterimine.

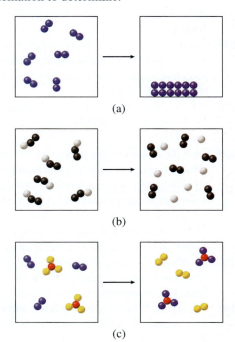

(a)

(b)

(c)

ADDITIONAL PROBLEMS

8.88 Match each of the following energy changes with one of the processes given: ionization energy, electron affinity, bond enthalpy, and standard enthalpy of formation.
(a) $F(g) + e^- \longrightarrow F^-(g)$
(b) $F_2(g) \longrightarrow 2F(g)$
(c) $Na(g) \longrightarrow Na^+(g) + e^-$
(d) $Na(s) + \frac{1}{2}F_2(g) \longrightarrow NaF(s)$

8.89 The formulas for the fluorides of the third-period elements are NaF, MgF_2, AlF_3, SiF_4, PF_5, SF_6, and ClF_3. Classify these compounds as covalent or ionic.

8.90 Use ionization energy (see Figure 7.8) and electron affinity (see Figure 7.10) values to calculate the energy change (in kJ/mol) for the following reactions:
(a) $Li(g) + I(g) \longrightarrow Li^+(g) + I^-(g)$
(b) $Na(g) + F(g) \longrightarrow Na^+(g) + F^-(g)$
(c) $K(g) + Cl(g) \longrightarrow K^+(g) + Cl^-(g)$

8.91 Describe some characteristics of an ionic compound such as KF that would distinguish it from a covalent compound such as benzene (C_6H_6).

8.92 Write Lewis structures for BrF_3, ClF_5, and IF_7. Identify those in which the octet rule is not obeyed.

8.93 Write three reasonable resonance structures for the azide ion N_3^- in which the atoms are arranged as NNN. Show formal charges.

8.94 Draw two resonance structures for sulfurous acid (H_2SO_3): one that obeys the octet rule for the central atom, and one that minimizes the formal charges. Determine the formal charge on each atom in both structures.

8.95 Give an example of an ion or molecule containing Al that (a) obeys the octet rule, (b) has an expanded octet, and (c) has an incomplete octet.

8.96 Draw four reasonable resonance structures for the PO_3F^{2-} ion. The central P atom is bonded to the three O atoms and to the F atom. Show formal charges.

8.97 Attempts to prepare the compounds CF_2, LiO_2, $CsCl_2$, and PI_5 as stable species under atmospheric conditions have failed. Suggest possible reasons for the failure.

8.98 Draw reasonable resonance structures for the following ions: (a) HSO_4^-, (b) PO_4^{3-}, (c) HPO_3^{2-}, (d) IO_3^-.

8.99 Are the following statements true or false? (a) Formal charges represent an actual separation of charges. (b) ΔH_{rxn}° can be estimated from the bond enthalpies of reactants and products. (c) All second-period elements obey the octet rule in their compounds. (d) The resonance structures of a molecule can be separated from one another in the laboratory.

8.100 A rule for drawing plausible Lewis structures is that the central atom is generally less electronegative than the surrounding atoms. Explain why this is so.

8.101 Using the following information and the fact that the average C—H bond enthalpy is 414 kJ/mol, estimate the standard enthalpy of formation of methane (CH_4).

$$C(s) \longrightarrow C(g) \qquad \Delta H_{rxn}^\circ = 716 \text{ kJ/mol}$$
$$2H_2(g) \longrightarrow 4H(g) \qquad \Delta H_{rxn}^\circ = 872.8 \text{ kJ/mol}$$

8.102 Based on changes in enthalpy, which of the following reactions will occur more readily?
(a) $Cl(g) + CH_4(g) \longrightarrow CH_3Cl(g) + H(g)$
(b) $Cl(g) + CH_4(g) \longrightarrow CH_3(g) + HCl(g)$

8.103 Which of the following molecules has the shortest nitrogen-to-nitrogen bond: N_2H_4, N_2O, N_2, N_2O_4? Explain.

8.104 Most organic acids can be represented as RCOOH, where COOH is the carboxyl group and R is the rest of the molecule. [For example, R is CH_3 in acetic acid (CH_3COOH).] (a) Draw a Lewis structure for the carboxyl group. (b) Upon ionization, the carboxyl group is converted to the carboxylate group (COO^-). Draw resonance structures for the carboxylate group.

8.105 Which of the following species are isoelectronic: NH_4^+, C_6H_6, CO, CH_4, N_2, $B_3N_3H_6$?

8.106 Draw three resonance structures for the hydrogen sulfite ion (HSO_3^-)—one that obeys the octet rule for the central atom, and two that expand the octet of the central atom. Calculate the formal charges on all atoms in each structure and determine which, if any, of the resonance structures has formal charges that are inconsistent with the elements' electronegativities.

8.107 Draw two resonance structures for each species— one that obeys the octet rule, and one in which the formal charge on the central atom is zero: PO_4^{3-}, $HClO_3$, SO_3, SO_2.

8.108 The following species have been detected in interstellar space: (a) CH, (b) OH, (c) C_2, (d) HNC, (e) HCO. Draw Lewis structures for these species.

8.109 The amide ion (NH_2^-) is a Brønsted base. Use Lewis structures to represent the reaction between the amide ion and water.

8.110 Draw Lewis structures for the following organic molecules: (a) tetrafluoroethylene (C_2F_4), (b) propane (C_3H_8), (c) butadiene ($CH_2CHCHCH_2$), (d) propyne (CH_3CCH), (e) benzoic acid (C_6H_5COOH). (To draw C_6H_5COOH, replace an H atom in benzene with a COOH group.)

8.111 The triiodide ion (I_3^-) in which the I atoms are arranged in a straight line is stable, but the corresponding F_3^- ion does not exist. Explain.

8.112 Compare the bond enthalpy of F_2 with the overall energy change for the following process:

$$F_2(g) \longrightarrow F^+(g) + F^-(g)$$

Which is the preferred dissociation for F_2, energetically speaking?

8.113 In 1999, an unusual cation containing only nitrogen (N_5^+) was prepared. Draw three resonance structures of the ion, showing formal charges. (*Hint:* The N atoms are joined in a linear fashion.)

8.114   Write the formulas of the binary hydrides for the second-period elements from Li to F. Identify the bonding in each as covalent, polar covalent, or ionic.

8.115   Several resonance structures for the molecule CO_2 are shown here. Explain why some of them are likely to be of little importance in describing the bonding in this molecule.
(a) $:\ddot{O}=C=\ddot{O}:$ (c) $:O\equiv\ddot{C}-\ddot{O}:$
(b) $:O\equiv C-\ddot{O}:$ (d) $:\ddot{O}-C-\ddot{O}:$

8.116   For each of the following organic molecules, draw a Lewis structure in which the carbon atoms are bonded to each other by single bonds: (a) C_2H_6, (b) C_4H_{10}, (c) C_5H_{12}. For parts (b) and (c), show only structures in which each C atom is bonded to no more than two other C atoms.

8.117   In the gas phase, aluminum chloride exists as a dimer (a unit of two) with the formula Al_2Cl_6. Its skeletal structure is given by

$$\begin{array}{ccccc} Cl & & Cl & & Cl \\ & \diagdown & & \diagup & \\ & Al & & Al & \\ & \diagup & & \diagdown & \\ Cl & & Cl & & Cl \end{array}$$

Complete the Lewis structure and indicate the coordinate covalent bonds in the molecule. Does this dimer possess a dipole moment? Explain.

8.118   Draw Lewis structures for the following organic molecules: C_2H_3F, C_3H_6, C_4H_8. In each there is one C=C bond, and the rest of the carbon atoms are joined by C—C bonds.

8.119   Calculate $\Delta H°$ for the reaction

$$H_2(g) + I_2(g) \longrightarrow 2HI(g)$$

using (a) Equation 8.4 and (b) Equation 5.19, given that $\Delta H_f°$ for $I_2(g)$ is 61.0 kJ/mol.

8.120   Draw Lewis structures for the following organic molecules: (a) methanol (CH_3OH); (b) ethanol (CH_3CH_2OH); (c) tetraethyl lead [$Pb(CH_2CH_3)_4$], which is used in "leaded gasoline"; (d) methylamine (CH_3NH_2), which is used in tanning; (e) mustard gas ($ClCH_2CH_2SCH_2CH_2Cl$), a poisonous gas used in World War 1; (f) urea [$(NH_2)_2CO$], a fertilizer; and (g) glycine (NH_2CH_2COOH), an amino acid.

8.121   Write Lewis structures for the following four isoelectronic species: (a) CO, (b) NO^+, (c) CN^-, (d) N_2. Show formal charges.

8.122   Oxygen forms three types of ionic compounds in which the anions are oxide (O^{2-}), peroxide (O_2^{2-}), and superoxide (O_2^-). Draw Lewis structures of these ions.

8.123   Comment on the correctness of the statement, "All compounds containing a noble gas atom violate the octet rule."

8.124   Write three resonance structures for (a) the cyanate ion (NCO^-) and (b) the isocyanate ion (CNO^-). In each case, rank the resonance structures in order of increasing importance.

8.125   (a) From the following data, calculate the bond enthalpy of the F_2^- ion.

$F_2(g) \longrightarrow 2F(g)$	$\Delta H_{rxn}° = 156.9$ kJ/mol	
$F^-(g) \longrightarrow F(g) + e^-$	$\Delta H_{rxn}° = 333$ kJ/mol	
$F_2^-(g) \longrightarrow F_2(g) + e^-$	$\Delta H_{rxn}° = 290$ kJ/mol	

(b) Explain the difference between the bond enthalpies of F_2 and F_2^-.

8.126   The resonance concept is sometimes described by analogy to a mule, which is a cross between a horse and a donkey. Compare this analogy with the one used in this chapter, that is, the description of a rhinoceros as a cross between a griffin and a unicorn. Which description is more appropriate? Why?

8.127   The N—O bond distance in nitric oxide is 115 pm, which is intermediate between a triple bond (106 pm) and a double bond (120 pm). (a) Draw two resonance structures for NO, and comment on their relative importance. (b) Is it possible to draw a resonance structure having a triple bond between the atoms?

8.128   Vinyl chloride (C_2H_3Cl) differs from ethylene (C_2H_4) in that one of the H atoms is replaced with a Cl atom. Vinyl chloride is used to prepare poly(vinyl chloride), which is an important polymer used in pipes. (a) Draw the Lewis structure of vinyl chloride. (b) The repeating unit in poly(vinyl chloride) is $-CH_2-CHCl-$. Draw a portion of the molecule showing three such repeating units. (c) Calculate the enthalpy change when 1.0×10^3 kg of vinyl chloride forms poly(vinyl chloride).

8.129   Experiments show that it takes 1656 kJ/mol to break all the bonds in methane (CH_4) and 4006 kJ/mol to break all the bonds in propane (C_3H_8). Based on these data, calculate the average bond enthalpy of the C—C bond.

8.130   Draw a Lewis structure for nitrogen pentoxide (N_2O_5) in which each N is bonded to three O atoms.

8.131   The American chemist Robert S. Mulliken suggested a different definition for the electronegativity (EN) of an element, given by

$$EN = \frac{IE_1 + EA}{2}$$

where IE_1 is the first ionization energy and EA is the electron affinity of the element. Calculate the electronegativities of O, F, and Cl using the preceding equation. Compare the electronegativities of these elements on the Mulliken and Pauling scales. (To convert to the Pauling scale, divide each EN value by 230 kJ/mol.)

8.132   Among the common inhaled anesthetics are:

Halothane ($CF_3CHClBr$)
Isoflurane ($CF_3CHClOCHF_2$)
Enflurane ($CHFClCF_2OCHF_2$)
Methoxyflurane ($CHCl_2CF_2OCH_3$)

Draw Lewis structures of these molecules.

8.133 Using Table 8.6, compare the following bond enthalpies: $C-C$ in C_2H_6, $N-N$ in N_2H_4, and $O-O$ in H_2O_2. What effect do lone pairs on adjacent atoms appear to have on bond enthalpy?

8.134 From the lattice energy of KCl in Table 8.6, and the ionization energy of K and electron affinity on Cl in Figures 7.8 and 7.10, respectively, calculate the $\Delta H°$ for the reaction

$$K(g) + Cl(g) \longrightarrow KCl(s)$$

Engineering Problems

8.135 In 1998, scientists using a special type of electron microscope were able to measure the force needed to break a *single* chemical bond. If 2.0×10^{-9} N was needed to break a $C-Si$ bond, estimate the bond enthalpy in kJ/mol. Assume that the bond has to be stretched by a distance of 2 Å (2×10^{-10} m) before it is broken.

8.136 $\Delta H_f°$ [$MgF_2(g)$] $= -1123$ kJ/mol. Using this and data from Appendix 2, Figures 7.8 and 7.10, and Practice Problem 8.3B, calculate the lattice energy of MgF_2.

Biological Problems

8.137 A student in your class claims that magnesium oxide actually consists of Mg^+ and O^- ions, not Mg^{2+} and O^{2-} ions. Suggest some experiments one could do to show that your classmate is wrong.

8.138 The following is a simplified (skeletal) structure of the amino acid histidine. Draw a complete Lewis structure of the molecule.

8.139 The following is a simplified (skeletal) structure of the amino acid tryptophan. Draw a complete Lewis structure of the molecule.

8.140 Do a Web search of the following ionic compounds and give brief descriptions of their medical uses: $AgNO_3$, $BaSO_4$, $CaSO_4$, KI, Li_2CO_3, $Mg(OH)_2$, $NaHCO_3$, NaF, TiO_2, ZnO.

8.141 Methyl isocyanate (CH_3NCO) is used to make certain pesticides. In December 1984, water leaked into a tank containing this substance at a chemical plant, producing a toxic cloud that killed thousands of people in Bhopal, India. Draw Lewis structures for CH_3NCO, showing formal charges.

8.142 The amide group plays an important role in determining the structure of proteins:

Draw another resonance structure for this group. Show formal charges.

Environmental Problems

8.143 Draw Lewis structures for the following chlorofluorocarbons (CFCs), which are partly responsible for the depletion of ozone in the stratosphere: (a) $CFCl_3$, (b) CF_2Cl_2, (c) CHF_2Cl, (d) CF_3CHF_2.

8.144 Although nitrogen dioxide (NO_2) is a stable compound, there is a tendency for two such molecules to combine to form dinitrogen tetroxide (N_2O_4). Why? Draw four resonance structures of N_2O_4, showing formal charges.

8.145 The chlorine nitrate ($ClONO_2$) molecule is believed to be involved in the destruction of ozone in the Antarctic stratosphere. Draw a plausible Lewis structure for this molecule.

Multiconcept Problems

8.146 The hydroxyl radical (OH) plays an important role in atmospheric chemistry. It is highly reactive and has a tendency to combine with an H atom from other compounds, causing them to break up. Thus OH is sometimes called a "detergent" radical because it helps to clean up the atmosphere. (a) Draw the Lewis structure for the radical. (b) Refer to Table 8.6 and explain why the radical has a high affinity for H atoms. (c) Estimate the enthalpy change for the following reaction:

$$OH(g) + CH_4(g) \longrightarrow CH_3(g) + H_2O(g)$$

(d) The radical is generated when sunlight hits water vapor. Calculate the maximum wavelength (in nm) required to break an $O-H$ bond in H_2O.

8.147 The species H_3^+ is the simplest polyatomic ion. The geometry of the ion is that of an equilateral triangle. (a) Draw three resonance structures to represent the ion. (b) Given the following information

$$2H + H^+ \longrightarrow H_3^+ \qquad \Delta H° = -849 \text{ kJ/mol}$$
$$H_2 \longrightarrow 2H \qquad \Delta H° = 436.4 \text{ kJ/mol}$$

calculate $\Delta H°$ for the reaction

$$H^+ + H_2 \longrightarrow H_3^+$$

8.148 The bond enthalpy of the $C-N$ bond in the amide group of proteins (see Problem 8.142) can be treated as an average of $C-N$ and $C=N$ bonds. Calculate the maximum wavelength of light needed to break the bond.

Standardized-Exam Practice Problems

Physical and Biological Sciences

Nitrous oxide (N_2O) is an anesthetic commonly used for dental procedures. Because of the euphoria caused by inhaling it, N_2O is commonly known as "laughing gas." It is licensed for use as a food additive and as an aerosol propellant. It is used to displace air from potato chip bags to extend shelf life and as the propellant in whipped cream canisters. In recent years N_2O has become popular as a recreational drug, due in part to its ready availability to consumers. Although N_2O is legal, it is regulated by the FDA; its sale and distribution for the purpose of human consumption are not permitted.

Bond	Bond Enthalpy (kJ/mol)
N$-$O	176
N$=$O	607
N$-$N	193
N$=$N	418
N$\equiv$N	941.4
O$=$O	498.7

1. Which of the following Lewis structures are possible for N_2O?

$:N\equiv N-\ddot{O}:$ $\dot{\ddot{N}}=N=\ddot{O}$ $\dot{\ddot{N}}=O=\dot{\ddot{N}}$ $:\ddot{N}-O-\ddot{N}:$

    I   II   III   IV

a) I only   b) I and II   c) I, II, and III   d) I, II, III, and IV

2. Use formal charges to choose the best of the resonance structures shown.

a) I   b) II   c) III   d) IV

3. Using the best resonance structure and the average bond enthalpies given, calculate ΔH_f° for N_2O.

a) 73 kJ/mol   b) -73 kJ/mol   c) 166 kJ/mol   d) -166 kJ/mol

4. Why does the calculated ΔH_f° value differ from the tabulated value of 81.56 kJ/mol?

a) The tabulated value is wrong.
b) None of the resonance structures depicts the bonds realistically.
c) To a reasonable number of significant figures, the calculated and tabulated values are the same.
d) Using bond enthalpies gives only an estimate of ΔH_{rxn}°.

Answers to In-Chapter Materials

Practice Problems

8.1A (a) Ca^{2+}, (b) $[:\ddot{N}:]^{3-}$, (c) $[:\ddot{I}:]^-$. **8.1B** (a) 2$-$, (b) $+$, (c) 3$-$.

8.2A $MgCl_2$. **8.2B** NaF < MgO < AlN. **8.3A** 629 kJ/mol.

8.3B -841 kJ/mol. **8.4A** (a) nonpolar, (b) polar, (c) nonpolar.

8.4B C$-$S < C$-$Se < C$-$Te < C$-$Po < C$-$O. **8.5A** 0.12.

8.5B 0.11 D. **8.6A** 41 percent. **8.6B** 9.23 D.

8.7A $:\ddot{F}-\overset{\displaystyle :\ddot{F}:}{\underset{}{N}}-\ddot{F}:$ **8.7B** $\left[:\ddot{O}-\overset{\displaystyle :\ddot{O}:}{\underset{}{Cl}}-\ddot{O}:\right]^-$ **8.8A** C atom = 0, double-bonded O atom = 0, single-bonded O atoms = -1.

8.8B S atom = $+1$, O atoms = -1, overall charge = -2.

8.9A $-\overset{\displaystyle \ddot{O}}{\underset{}{\overset{\|}{C}}}-\ddot{O}-H$ **8.9B** Cl$-$N$-$Cl.

8.10A $\left[:\ddot{O}=N-\ddot{O}:\right]^- \longleftrightarrow \left[:\ddot{O}-\overset{\displaystyle \ddot{O}}{\underset{}{\overset{\|}{N}}}-\ddot{O}:\right]^- \longleftrightarrow \left[:\ddot{O}-N=\ddot{O}:\right]^-$

8.10B $\left[:\overset{-2}{\ddot{N}}-\overset{0}{C}\equiv\overset{+1}{S}:\right]^- \longleftrightarrow \left[:\overset{-1}{\ddot{N}}=\overset{0}{C}=\overset{0}{\ddot{S}}:\right]^- \longleftrightarrow \left[:\overset{0}{N}\equiv\overset{0}{C}-\overset{-1}{\ddot{S}}:\right]^-$

8.11A $\cdot\ddot{O}-H$ **8.11B** $:S=\dot{N}-\ddot{S}:$

8.12A (a) $:\ddot{F}-Be-\ddot{F}:$ (b) $\overset{\displaystyle :\ddot{Cl}:}{\underset{\displaystyle :\ddot{Cl}:}{\ddot{Cl}-P-\ddot{Cl}:}}$ (c) $\left[:\ddot{Cl}-\overset{\displaystyle :\ddot{Cl}:}{\underset{\displaystyle :\ddot{Cl}:}{\overset{|}{\underset{|}{I}}}-\ddot{Cl}:}\right]^-$

8.12B (a) $:\ddot{Cl}-\overset{\displaystyle :\ddot{Cl}:}{\underset{}{B}}-\ddot{Cl}:$ (b) $\overset{\displaystyle :\ddot{F}:}{\underset{\displaystyle :\ddot{F}:}{\ddot{F}-Sb-\ddot{F}:}}$ (c) $:\ddot{F}-\ddot{Kr}-\ddot{F}:$

8.13A -557 kJ/mol. **8.13B** 328 kJ/mol.

Answers to Checkpoints

8.1.1 b. **8.1.2** e. **8.1.3** e. **8.1.4** e. **8.2.1** c. **8.2.2** a. **8.2.3** b.
8.4.1 b. **8.4.2** c. **8.4.3** d. **8.4.4** c. **8.5.1** d. **8.5.2** c. **8.6.1** e.
8.6.2 c. **8.7.1** a, b, d. **8.7.2** b. **8.8.1** e. **8.8.2** a, b, e. **8.8.3** e.
8.8.4 d. **8.9.1** d. **8.9.2** e. **8.9.3** e. **8.9.4** b.

CHAPTER 9

Chemical Bonding II: Molecular Geometry and Bonding Theories

Folic acid is a ubiquitous ingredient in "fortified" processed foods, such as flour, white rice, and cereal. The naturally occurring form of this vitamin, *folate,* is found in a variety of foods, including many fruits and vegetables.

Editorial Image, LLC

In This Chapter, You Will Learn

How to determine the three-dimensional shape of a molecule and how the interactions of atomic orbitals give rise to chemical bonds.

Before You Begin, Review These Skills

- Shapes of atomic orbitals [⏮ Section 6.7]
- Electron configurations of atoms [⏮ Section 6.8]
- Drawing Lewis structures [⏮ Section 8.5]
- Chapter 8 Key Skills [⏮ pages 397–398]

How Molecular Shape Affects the Function of Vitamins and Drugs

The properties and reactivity of a substance depend significantly on the shape, or *geometry,* of its molecules. Nowhere is this more evident than in the function of enzymes.

Student Note: Enzymes are biological catalysts [⏭ Chapter 14] that facilitate specific biochemical reactions in living systems.

Folic acid, also known as vitamin B-9, is vitally important for the production of new cells. For decades, folic acid supplements have been recommended for women of childbearing age for prevention of neural tube birth defects such as spina bifida and anencephaly. Although folic acid itself is not biochemically active, it is converted to the active form, tetrahydrofolate, by the enzyme *dihydrofolate reductase* (DHFR). The specificity of this enzyme's function is due, in large part, to the complementary shapes of the folic acid molecule and the active site where it binds to the enzyme.

During the first half of the twentieth century, oncologists observed that administration of folic acid appeared to worsen leukemia, and that restriction of folic acid intake could cause improvement in leukemia patients. Although the role of folic acid in leukemia was not understood at the time, cancer is, broadly speaking, the uncontrolled growth and division of cells. We know now that because folic acid enhances cell production, it can accelerate the progression of some cancers.

The observations of folic acid's effect on leukemia led to the development of the drug methotrexate. Methotrexate reduces folic acid absorption by competing for the DHFR enzyme, a phenomenon known as *competitive enzyme inhibition.* This drug is able to compete for this very specific enzyme because its molecular shape is so similar to that of folic acid. In fact, design and manipulation of molecular shape is a critical part of the development of new pharmaceuticals.

At the end of this chapter, you will be able to answer several questions about the molecular geometry of folic acid and the drug, methotrexate, that blocks its absorption **[⏭ Applying What You've Learned, page 456]**.

Many biochemical processes are specific in that they depend on the shapes of the molecules involved. *Chemical bonding theories* help us to predict and/or explain these shapes.

Editorial Image, LLC

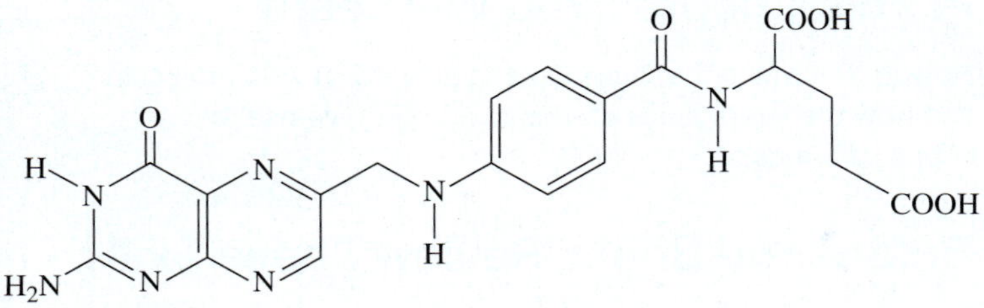

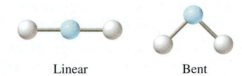

Editorial Image, LLC

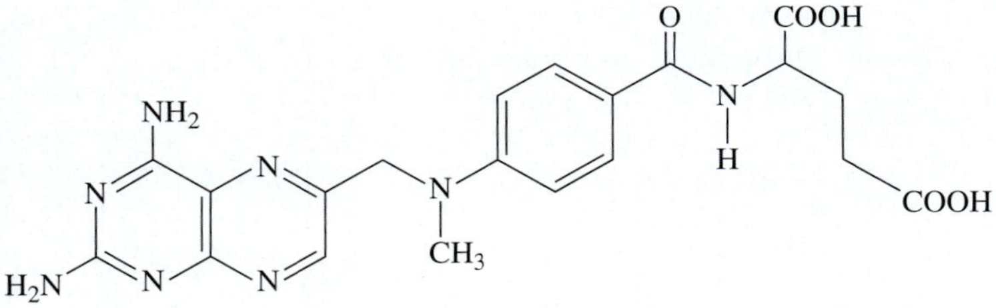

9.1 Molecular Geometry

Many familiar chemical and biochemical processes depend heavily on the three-dimensional shapes of the molecules and/or ions involved. Our sense of smell is one example; the effectiveness of a particular drug is another. Although the shape of a molecule or poly-atomic ion must be determined experimentally, we can predict their shapes reasonably well using Lewis structures [◄◄ Section 8.5] and the ***valence-shell electron-pair repulsion (VSEPR)*** model. In this section, we focus primarily on determining the shapes of mole-cules of the general type AB_x, where A is a central atom surrounded by x B atoms and x can have integer values of 2 to 6. (Any atom that is bonded to two or more other atoms can be considered a "central" atom.) For example, NH_3 is an AB_3 molecule in which A is nitrogen, B is hydrogen, and $x = 3$. Its Lewis structure is shown here.

Table 9.1 lists examples of each type of AB_x molecule and polyatomic ion that we will consider. Throughout this chapter, we discuss concepts that apply both to molecules and to polyatomic ions, but we usually refer to them collectively as "molecules."

Having the molecular formula alone is insufficient to predict the shape of a molecule. For instance, AB_2 molecules may be linear or bent:

$$H—\overset{..}{N}—H$$
$$|$$
$$H$$

Lewis structure of NH_3

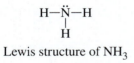

Animation
Chemical bonding—valence-shell electron-pair repulsion theory.

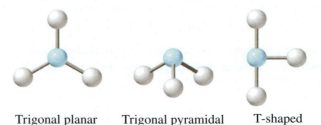

Linear   Bent

Moreover, AB_3 molecules may be planar, pyramidal, or T-shaped:

Trigonal planar   Trigonal pyramidal   T-shaped

To determine shape, we must start with a correct Lewis structure and apply the VSEPR model.

TABLE 9.1	Examples of AB_x Molecules and Polyatomic Ions
AB_2	$BeCl_2$, SO_2, H_2O, NO_2^-
AB_3	BF_3, NH_3, ClF_3, SO_3^{2-}
AB_4	CCl_4, NH_4^+, SF_4, XeF_4, ClO_4^-
AB_5	PCl_5, IF_5, SbF_5, BrF_5
AB_6	SF_6, UF_6, $TiCl_6^{3-}$

The VSEPR Model

Recall that the electrons in the valence shell are the ones involved in chemical bonding [◄◄ Section 8.1]. The basis of the VSEPR model is that electron pairs in the valence shell of an atom *repel* one another. As we learned in Chapter 8, there are two types of electron pairs: bonding pairs and nonbonding pairs (also known as lone pairs). Furthermore, bonding pairs may be found in single bonds or in multiple bonds. For clarity, we refer to electron *domains* instead of electron pairs when we use the VSEPR model. An ***electron domain*** in this context is a lone pair or a bond, regardless of whether the bond is single, double, or triple. Consider the following examples:

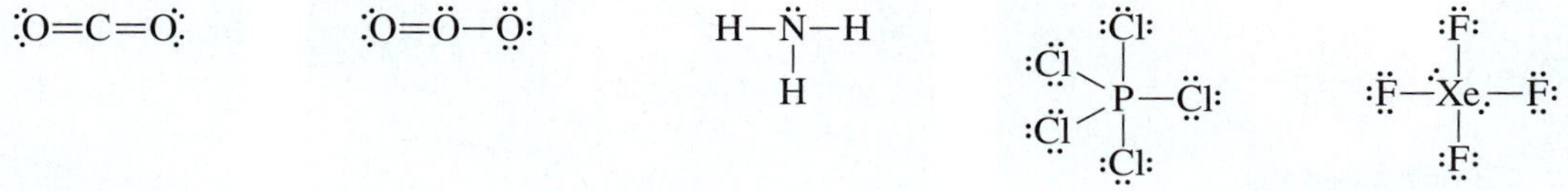

	CO₂	O₃	NH₃	PCl₅	XeF₄
	$\ddot{O}=C=\ddot{O}$	$\ddot{O}=\ddot{O}-\ddot{O}$	H—N̈—H, H	(Lewis structure)	(Lewis structure)
	2 double bonds	1 single bond 1 double bond + 1 lone pair	3 single bonds + 1 lone pair	5 single bonds	4 single bonds + 2 lone pairs
Total number of electron domains on central atom	2 electron domains	3 electron domains	4 electron domains	5 electron domains	6 electron domains

Note the number of electron domains on the central atom in each molecule. The VSEPR model predicts that because these electron domains repel one another, they will arrange themselves to be as far apart as possible, thus minimizing the repulsive interactions between them. It is important to understand that you cannot tell the shape of a molecule or ion simply from its formula—you must apply VSEPR theory.

   We can visualize the arrangement of electron domains using balloons, as shown in Figure 9.1. Like the B atoms in our AB_x molecules, the balloons are all connected to a central, fixed point, which represents the central atom (A). When they are as far apart as possible, they adopt the five geometries shown in the figure. When there are only two balloons, they orient themselves to point in opposite directions [Figure 9.1(a)]. With three balloons the arrangement is a trigonal plane [Figure 9.1(b)]. With four balloons the arrangement adopted is a tetrahedron [Figure 9.1(c)]. With five balloons, three of them adopt positions in a trigonal plane whereas the other two point opposite to each other, forming an axis that is perpendicular to the trigonal plane [Figure 9.1(d)]. This geometry is called a trigonal bipyramid. Finally, with six balloons the arrangement is an octahedron, which is essentially a square bipyramid [Figure 9.1(e)]. Each of the AB_x molecules we consider will have one of these five electron-domain geometries: linear, trigonal planar, tetrahedral, trigonal bipyramidal, or octahedral.

Figure 9.1 The arrangements adopted by (a) two, (b) three, (c) four, (d) five, and (e) six balloons.

(all) *Stephen Frisch/McGraw Hill*

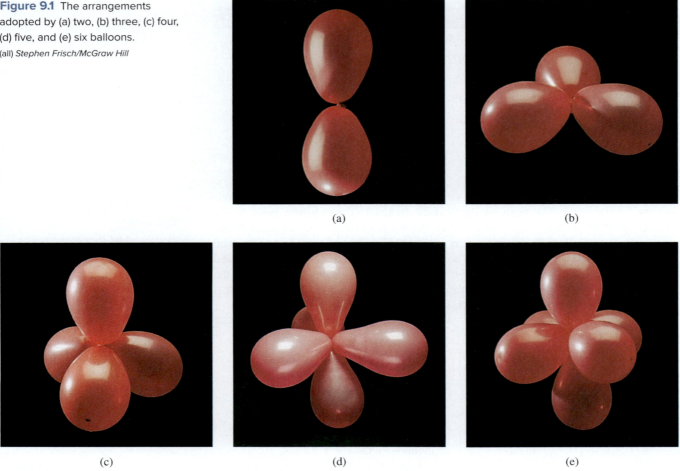

(a)

(b)

(c)

(d)

(e)

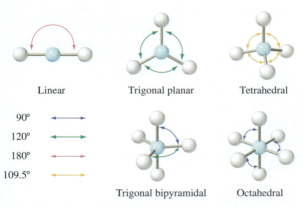

Linear

Trigonal planar

Tetrahedral

90°

120°

180°

109.5°

Trigonal bipyramidal

Octahedral

Figure 9.2 Five geometries of AB_x molecules in which all the electron domains are bonds.

Electron-Domain Geometry and Molecular Geometry

It is important to distinguish between the *electron-domain geometry,* which is the arrangement of electron domains (bonds and lone pairs) around the central atom, and the *molecular geometry,* which is the arrangement of bonded *atoms*. Figure 9.2 illustrates the molecular geometries of AB_x molecules in which all the electron domains are bonds—that is, there are no lone pairs on any of the central atoms. In these cases, the molecular geometry is the same as the electron-domain geometry.

In an AB_x molecule, a *bond angle* is the angle between two adjacent A—B bonds. In an AB_2 molecule there are only two bonds and therefore only one bond angle, and, provided that there are no lone pairs on the central atom, the bond angle is 180°. AB_3 and AB_4 molecules have three and four bonds, respectively. However, in each case there is only one bond angle possible between any two A—B bonds. In an AB_3 molecule, the bond angle is 120°, and in an AB_4 molecule, the bond angle is 109.5°—again, provided that there are no lone pairs on the central atoms. Similarly, in an AB_6 molecule the bond angles between adjacent bonds are all 90°. (The angle between any two A—B bonds that point in opposite directions is 180°.)

AB_5 molecules contain two different bond angles between adjacent bonds. The reason for this is that, unlike those in the other

AB_x molecules, the positions occupied by bonds in a trigonal bipyramid are not all equivalent. The three bonds that are arranged in a trigonal plane are referred to as **equatorial.** The bond angle between any two of the three equatorial bonds is 120°. The two bonds that form an axis perpendicular to the trigonal plane are referred to as **axial.** The bond angle between either of the axial bonds and any one of the equatorial bonds is 90°. (As in the case of the AB_6 molecule, the angle between any two A—B bonds that point in opposite directions is 180°.) Figure 9.2 illustrates all these bond angles. The angles shown in the figure are the bond angles that are observed when all the electron domains on the central atom are identical. As we explain later in this section, the bond angles in many molecules differ slightly from these *ideal* values.

When the central atom in an AB_x molecule bears one or more lone pairs, the electron-domain geometry and the molecular geometry are no longer the same. However, we still use the electron-domain geometry as a first step in determining the molecular geometry. The first step in determining the molecular geometry of O_3 (or any species), for example, is to draw its Lewis structure. Two different resonance structures can be drawn for O_3:

$$\ddot{\text{O}}{=}\ddot{\text{O}}{-}\ddot{\ddot{\text{O}}}: \quad\longleftrightarrow\quad :\ddot{\ddot{\text{O}}}{-}\ddot{\text{O}}{=}\ddot{\text{O}}:$$

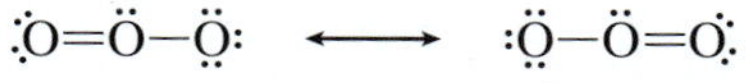

Either one can be used to determine its geometry.

The next step is to count the electron domains on the central atom. In this case, there are three: one single bond, one double bond, and one lone pair. Using the VSEPR model, we first determine the electron-domain geometry. According to the information in Figure 9.2, three electron domains on the central atom will be arranged in a trigonal plane. Molecular geometry, however, is dictated by the arrangement of *atoms.* If we consider only the positions of the three atoms in this molecule, the molecular geometry (the molecule's shape) is *bent.*

Electron-domain geometry: Molecular geometry:
trigonal planar   bent

In addition to the five basic geometries depicted in Figure 9.2, you must be familiar with how molecular geometry can differ from electron-domain geometry. Table 9.2 shows the common molecular geometries where there are one or more lone pairs on the central atom. Note the positions occupied by the lone pairs in the trigonal bipyramidal electron-domain geometry. When there are lone pairs on the central atom in a trigonal bipyramid, the lone pairs preferentially occupy equatorial positions because repulsion is greater when the angle between electron domains is 90° or less. Placing a lone pair in an *axial* position would put it at 90° to three other electron domains. Placing it in an *equatorial* position puts it at 90° to only two other domains, thus minimizing the number of strong repulsive interactions.

All positions are equivalent in the octahedral geometry, so one lone pair on the central atom can occupy any of the positions. If there is a second lone pair in this geometry, though, it must occupy the position opposite the first. This arrangement minimizes the repulsive forces between the two lone pairs (they are 180° apart instead of 90° apart).

In summary, the steps to determine the electron-domain and molecular geometries are as follows:

1. Draw the Lewis structure of the molecule or polyatomic ion.
2. Count the number of electron domains on the central atom.
3. Determine the electron-domain geometry by applying the VSEPR model.
4. Determine the molecular geometry by considering the positions of the atoms only.

TABLE 9.2	Electron-Domain and Molecular Geometries of Molecules with Lone Pairs on the Central Atom

Total Number of Electron Domains	Type of Molecule	Electron-Domain Geometry	Number of Lone Pairs	Placement of Lone Pairs	Molecular Geometry	Example
3	AB_2	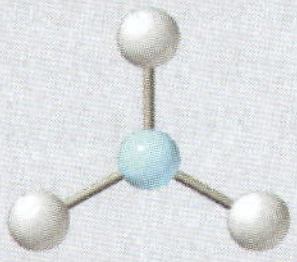 Trigonal planar	1		 Bent	SO_2
4	AB_3	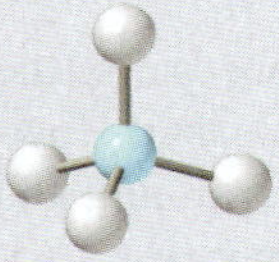 Tetrahedral	1	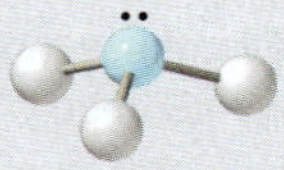	 Trigonal pyramidal	NH_3
4	AB_2	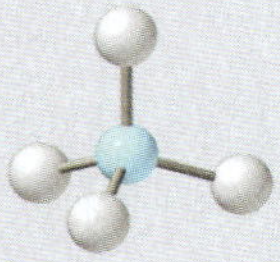 Tetrahedral	2		 Bent	H_2O
5	AB_4	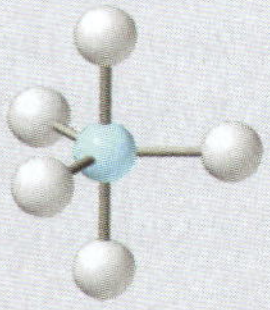 Trigonal bipyramidal	1	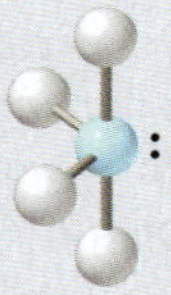	 Seesaw-shaped	SF_4
5	AB_3	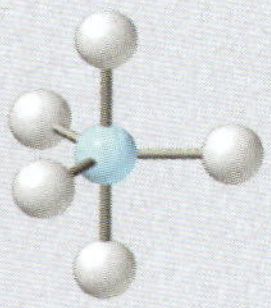 Trigonal bipyramidal	2	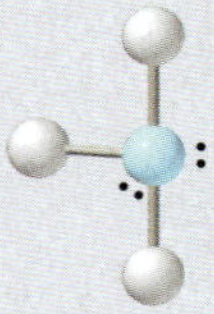	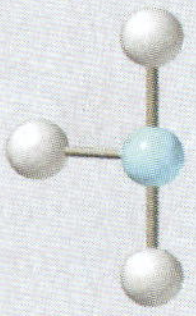 T-shaped	ClF_3
5	AB_2	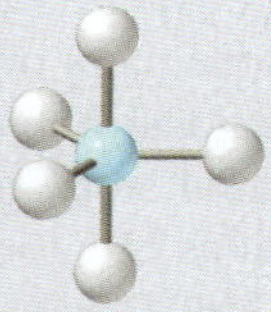 Trigonal bipyramidal	3		 Linear	IF_2^-
6	AB_5	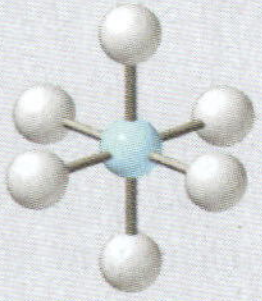 Octahedral	1	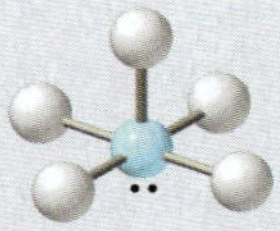	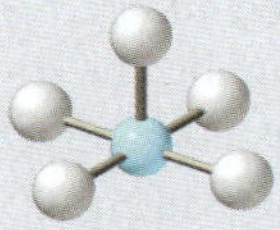 Square pyramidal	BrF_5
6	AB_4	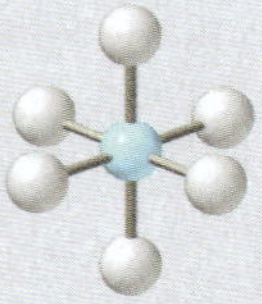 Octahedral	2	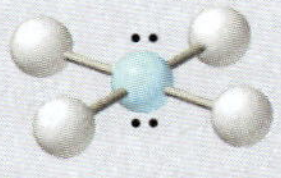	 Square planar	XeF_4

Sample Problem 9.1 shows how to determine the shape of a molecule or poly-atomic ion.

SAMPLE PROBLEM 9.1

Determine the shapes of (a) SO_3 and (b) ICl_4^-.

Strategy Use Lewis structures and the VSEPR model to determine first the electron-domain geometry and then the molecular geometry (shape).

Setup (a) The Lewis structure of SO_3 is:

$$\ddot{O}=\overset{\displaystyle :\ddot{O}:}{\underset{\displaystyle |}{S}}-\ddot{O}:$$

There are three electron domains on the central atom: one double bond and two single bonds.

(b) The Lewis structure of ICl_4^- is:

$$\left[\begin{array}{c} :\ddot{Cl}: \\ | \\ :\ddot{Cl}-\ddot{I}.-\ddot{Cl}: \\ | \\ :\ddot{Cl}: \end{array} \right]^-$$

There are six electron domains on the central atom in ICl_4^-: four single bonds and two lone pairs.

Solution

(a) According to the VSEPR model, three electron domains will be arranged in a trigonal plane. Since there are no lone pairs on the central atom in SO_3, the molecular geometry is the same as the electron-domain geometry. Therefore, the shape of SO_3 is trigonal planar.

Electron-domain geometry: trigonal planar ⟶ Molecular geometry: trigonal planar

(b) Six electron domains will be arranged in an octahedron. Two lone pairs on an octahedron will be located on opposite sides of the central atom, making the shape of ICl_4^- square planar.

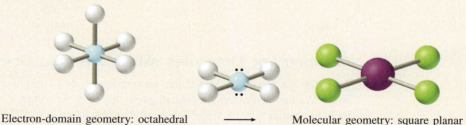

Electron-domain geometry: octahedral ⟶ Molecular geometry: square planar

THINK ABOUT IT

Compare these results with the information in Figure 9.2 and Table 9.2. Make sure that you can draw Lewis structures correctly. Without a correct Lewis structure, you will be unable to determine the shape of a molecule.

Practice Problem ATTEMPT Determine the shapes of (a) CO_2 and (b) SCl_2.

Practice Problem BUILD (a) From what group must the terminal atoms come in an AB_x molecule where the central atom is from Group 16, for the electron-domain geometry and the molecular geometry both to be trigonal planar? (b) From what group must the terminal atoms come in an AB_x molecule where the central atom is from Group 17, for the electron-domain geometry to be octahedral and the molecular geometry to be square pyramidal?

(Continued on next page)

Practice Problem **C**ONCEPTUALIZE
These four models may represent molecules or polyatomic ions. Lone pairs on the central atom, if any, are not shown. Which of these could represent a species in which there are lone pairs on the central atom? Which could represent a species in which there are no lone pairs on the central atom?

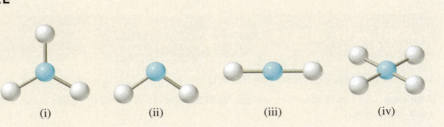

(i)   (ii)   (iii)   (iv)

Deviation from Ideal Bond Angles

Some electron domains are better than others at repelling neighboring domains. As a result, the bond angles may be slightly different from those shown in Figure 9.2. For example, the electron-domain geometry of ammonia (NH_3) is tetrahedral, so we might predict the H—N—H bond angles to be 109.5°. In fact, the bond angles are about 107°, slightly smaller than predicted. The lone pair on the nitrogen atom repels the N—H bonds more strongly than the bonds repel one another. It therefore "squeezes" them closer together than the ideal tetrahedral angle of 109.5°.

In effect, a lone pair takes up more *space* than the bonding pairs. This can be understood by considering the attractive forces involved in determining the location of the electron pairs. A lone pair on a central atom is attracted only to the nucleus of that atom. A bonding pair of electrons, on the other hand, is simultaneously attracted by the nuclei of both of the bonding atoms. As a result, the lone pair has more freedom to spread out and greater capacity to repel other electron domains. Also, because they contain more electron density, multiple bonds repel more strongly than single bonds. Consider the bond angles in each of the following examples:

Student Hot Spot

Student data indicate you may struggle with determining molecular shape and bond angles using VSEPR. Access the eBook to view additional Learning Resources on this topic.

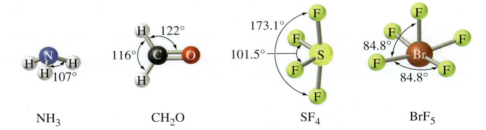

NH_3 CH_2O SF_4 BrF_5

Geometry of Molecules with More than One Central Atom

Thus far, we have considered the geometries of molecules having only one central atom. We can determine the overall geometry of more complex molecules by treating them as though they have multiple central atoms. Methanol (CH_3OH), for example, has a central C atom and a central O atom, as shown in the following Lewis structure:

$$\begin{array}{c} H \\ | \\ H-C-\ddot{O}-H \\ | \\ H \end{array}$$

Student Note: When we specify the geometry of a particular portion of a molecule, we refer to it as the geometry "about" a particular atom. In methanol, for example, we say that the geometry is *tetrahedral about the C atom* and *bent about the O atom*.

Both the C and the O atoms are surrounded by four electron domains. In the case of C, they are three C—H bonds and one C—O bond. In the case of O, they are one O—C bond, one O—H bond, and two lone pairs. In each case, the electron-domain geometry is tetrahedral. However, the molecular geometry of the C part of the molecule is *tetrahedral,* whereas the molecular geometry of the O part of the molecule is *bent.* Note that although the Lewis structure makes it appear as though there is a 180° angle between the O—C and O—H bonds, the angle is actually approximately 109.5°, the angle in a tetrahedral arrangement of electron domains.

Sample Problem 9.2 shows how to determine when bond angles differ from ideal values.

SAMPLE PROBLEM 9.2

Acetic acid, the substance that gives vinegar its characteristic smell and sour taste, is sometimes used in combination with corticosteroids to treat certain types of ear infections. Its Lewis structure is:

$$\begin{array}{ccccc} & H & \ddot{O} & & \\ & | & \| & & \\ H\!-\!C\!-\!C\!-\!\ddot{O}\!-\!H & & & \\ & | & & & \\ & H & & & \end{array}$$

Determine the molecular geometry about each of the central atoms, and determine the approximate value of each of the bond angles in the molecule. Which if any of the bond angles would you expect to be smaller than the ideal values?

Strategy Identify the central atoms and count the number of electron domains around each of them. Use the VSEPR model to determine each electron-domain geometry, and the information in Table 9.2 to determine the molecular geometry about each central atom.

Setup The leftmost C atom is surrounded by four electron domains: one C—C bond and three C—H bonds. The middle C atom is surrounded by three electron domains: one C—C bond, one C—O bond, and one C=O (double) bond. The O atom is surrounded by four electron domains: one O—C bond, one O—H bond, and two lone pairs.

Solution The electron-domain geometry of the leftmost C is tetrahedral. Because all four electron domains are bonds, the molecular geometry of this part of the molecule is also tetrahedral. The electron-domain geometry of the middle C is trigonal planar. Again, because all the domains are bonds, the molecular geometry is also trigonal planar. The electron-domain geometry of the O atom is tetrahedral. Because two of the domains are lone pairs, the molecular geometry about the O atom is bent.

Bond angles are determined using electron-domain geometry. Therefore, the approximate bond angles about the leftmost C are 109.5°, those about the middle C are 120°, and those about the O are 109.5°. The angle between the two single bonds on the middle carbon will be *less* than 120° because the double bond repels the single bonds more strongly than they repel each other. Likewise, the bond angle between the two bonds on the O will be less than 109.5° because the lone pairs on O repel the single bonds more strongly than they repel each other and push the two bonding pairs closer together. The angles are labeled as follows:

$$\begin{array}{ccccc} \sim 109.5° & H & \ddot{O} & > 120° & \\ H\!-\!C\!-\!C\!-\!\ddot{O}\!-\!H & & & \\ & H & & & \\ & <120° & <109.5° & & \end{array}$$

THINK ABOUT IT

Compare these answers with the information in Figure 9.2 and Table 9.2.

Practice Problem **A**TTEMPT Ethanolamine ($HOCH_2CH_2NH_2$) has a smell similar to ammonia and is commonly found in biological tissues. Its Lewis structure is:

$$\begin{array}{ccccc} & H & H & & \\ & | & | & & \\ H\!-\!\ddot{O}\!-\!C\!-\!C\!-\!\ddot{N}\!-\!H & & & \\ & | & | & | & \\ & H & H & H & \end{array}$$

Determine the molecular geometry about each central atom and label all the bond angles. Cite any expected deviations from ideal bond angles.

Practice Problem **B**UILD The bond angle in NH_3 is significantly smaller than the ideal bond angle of 109.5° because of the lone pair on the central atom. Explain why the bond angle in SO_2 is very close to 120° despite there being a lone pair on the central atom.

Practice Problem **C**ONCEPTUALIZE Which of these models represents a species in which there is deviation from ideal bond angles?

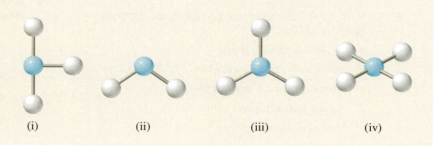

(i)   (ii)   (iii)   (iv)

How Are Larger, More Complex Molecules Represented?

Organic molecules, which can be quite large and complex, commonly are represented using *bond-line* structures. As the name suggests, each line in a bond-line structure represents a bond. Because organic compounds all contain carbon, most of the lines represent bonds between carbon atoms. In fact, with only a few exceptions, the carbon atoms in an organic molecule are not shown explicitly in bond-line structures—rather, they are understood to be at the end of each line, unless an atom *other* than carbon is shown. Also not shown in bond-line structures are the hydrogen atoms that are bonded to carbon. They are understood to be attached as appropriate to complete the octet of each carbon atom. For example, the bond-line structure of propane (C_3H_8) is:

Each line represents a carbon-carbon bond; there are carbon atoms at the ends of each line; and each carbon atom is bonded to the number of hydrogen atoms necessary to satisfy its octet. Thus, the bond-line structure shown here corresponds to the molecular model

Atoms other than carbon and hydrogen are shown explicitly in bond-line structures. For example, the bond-line structure of ethanol (C_2H_5OH) is

Note how similar this structure appears to the structure of propane. In this case, though, one of the lines represents a carbon-carbon bond and the other represents a carbon-oxygen bond. This structure corresponds to the molecular model

Bond-line structures make it possible to represent large, complex molecules, such as those found in biological systems. You may have noticed that the benzaldehyde molecule shown at the beginning of this chapter is represented with a bond-line structure. Some other examples are shown here along with their molecular formulas.

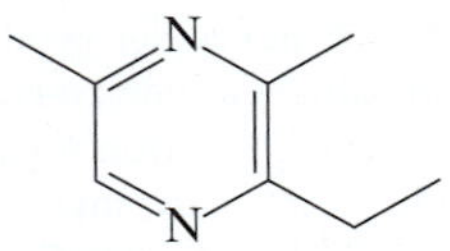

2-ethyl-3,5-dimethypyrazine,   3-methylbutanal,
$C_8H_{12}N_2$ $C_5H_{10}O$

(Two of the many volatile organic compounds responsible for the aroma of coffee.)

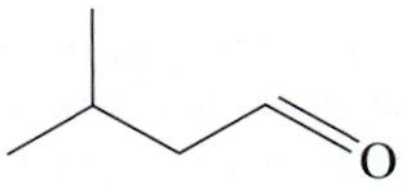

CHECKPOINT – SECTION 9.1   Molecular Geometry

9.1.1 What are the electron-domain geometry and molecular geometry of CO_3^{2-}?

a) tetrahedral, trigonal planar

b) tetrahedral, trigonal pyramidal

c) trigonal pyramidal, trigonal pyramidal

d) trigonal planar, trigonal planar

e) tetrahedral, tetrahedral

9.1.2 What are the electron-domain geometry and molecular geometry of ClO_3^-?

a) tetrahedral, trigonal planar

b) tetrahedral, trigonal pyramidal

c) trigonal pyramidal, trigonal pyramidal

d) trigonal planar, trigonal planar

e) tetrahedral, tetrahedral

9.1.3 What is the approximate value of the bond angle indicated?

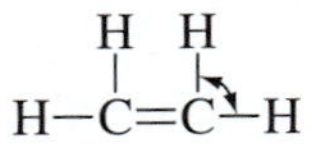

$$\begin{array}{c} \text{H}\quad\text{H} \\ |\quad\ \ \nwarrow \\ \text{H}-\text{C}=\text{C}-\text{H} \end{array}$$

a) < 90°   c) > 109.5°   e) < 120°

b) < 109.5°   d) > 120°

9.1.4 What is the approximate value of the bond angle indicated?

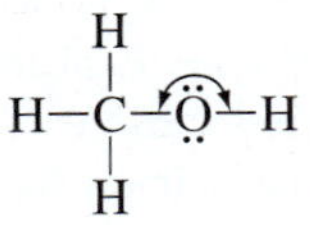

$$\begin{array}{c} \text{H} \\ | \\ \text{H}-\text{C}-\overset{\frown}{\text{O}}-\text{H} \\ | \\ \text{H} \end{array}$$

a) < 180°   c) < 109.5°   e) < 90°

b) > 180°   d) > 109.5°

9.1.5 Which of the following shows a deviation from ideal bond angles that is not possible for an AB_x molecule?

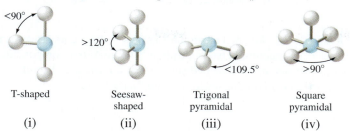

T-shaped

(i)

Seesaw-shaped

(ii)

Trigonal pyramidal

(iii)

Square pyramidal

(iv)

a) ii only

b) i and iii

c) i, ii, and iv

d) ii and iv

e) All these deviations from ideal bond angles are possible.

9.2 Molecular Geometry and Polarity

Molecular geometry is tremendously important in understanding the physical and chemical behavior of a substance. *Molecular polarity,* for example, is one of the most important consequences of molecular geometry, because molecular polarity influences physical, chemical, and biological properties. Recall from Section 8.4 that a bond between two atoms of different electronegativities is polar and that a diatomic molecule containing a polar bond is a *polar molecule.* Whether a molecule made up of three or more atoms is polar depends not only on the polarity of the individual bonds, but also on its molecular geometry.

Each of the CO_2 and H_2O molecules contains two identical atoms bonded to a central atom and two polar bonds. However, only one of these molecules is polar. To understand why, think of each individual bond dipole as a vector. The overall dipole moment of the molecule is determined by vector addition of the individual bond dipoles.

In the case of CO_2, we have two identical vectors pointing in opposite directions. When the vectors are placed on a Cartesian coordinate system, they have no y component and their x components are equal in magnitude but opposite in sign. The sum of these two vectors is zero in both the x and y directions. Thus, although the *bonds* in CO_2 are polar, the *molecule* is nonpolar.

Animation
Chemical bonding—molecular geometry and polarity.

Student Note: Recall that we can represent an individual bond dipole using a crossed arrow that points toward the more electronegative atom [◄◄ Section 8.4].

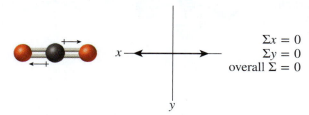

$$\Sigma x = 0$$
$$\Sigma y = 0$$
overall $\Sigma = 0$

The vectors representing the bond dipoles in water, although equal in magnitude and opposite in the x direction, are not opposite in the y direction. Therefore, although their x components sum to zero, their y components do not. This means that there is a net resultant dipole and H_2O is *polar.*

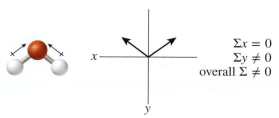

$$\Sigma x = 0$$
$$\Sigma y \neq 0$$
overall $\Sigma \neq 0$

Can More Complex Molecules Contain Polar Bonds and Still Be Nonpolar?

In AB_x molecules where $x \geq 3$, it may be less obvious whether the individual bond dipoles cancel one another. Consider the molecule BF_3, for example, which has a trigonal planar geometry:

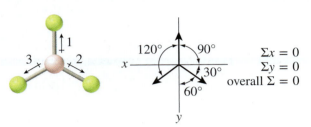

We will simplify the math in this analysis by assigning the vectors representing the three identical B—F bonds an arbitrary magnitude of 1.00. The x, y coordinates for the end of arrow 1 are (0, 1.00). Determining the coordinates for the ends of arrows 2 and 3 requires the use of trigonometric functions. You may have learned the mnemonic SOH CAH TOA, where the letters stand for

$$\text{Sin} = \text{Opposite over Hypotenuse}$$

$$\text{Cos} = \text{Adjacent over Hypotenuse}$$

$$\text{Tan} = \text{Opposite over Adjacent}$$

The x coordinate for the end of arrow 2 corresponds to the length of the line *opposite* the 60° angle. The hypotenuse of the triangle has a length of 1.00 (the arbitrarily assigned value). Therefore, using SOH,

$$\sin 60° = 0.866 = \frac{\text{opposite}}{\text{hypotenuse}} = \frac{\text{opposite}}{1}$$

so the x coordinate for the end of arrow 2 is 0.866.

The magnitude of the y coordinate corresponds to the length of the line *adjacent* to the 60° angle. Using TOA,

$$\tan 60° = 1.73 = \frac{\text{opposite}}{\text{adjacent}} = \frac{0.866}{\text{adjacent}}$$

$$\text{adjacent} = \frac{0.866}{1.73} = 0.500$$

so the y coordinate for the end of arrow 2 is −0.500. (The trigonometric formula gives us the length of the side. We know from the diagram that the sign of this y component is negative.)

Arrow 3 is similar to arrow 2. Its x component is equal in magnitude but opposite in sign, and its y component is the same magnitude and sign as that for arrow 2. Therefore, the x and y coordinates for all three vectors are

	x	y
Arrow 1	0	1
Arrow 2	0.866	−0.500
Arrow 3	−0.866	−0.500
Sum =	0	0

Because the individual bond dipoles (represented here as the vectors) sum to zero, the molecule is nonpolar overall.

Although it is somewhat more complicated, a similar analysis can be done to show that all x, y, and z coordinates sum to zero when there are four identical polar bonds arranged in a tetrahedron about a central atom. In fact, any time there are identical bonds symmetrically distributed around a central atom, with no lone pairs on the central atom, the molecule will be nonpolar overall, even if the bonds themselves are polar.

In cases where the bonds are distributed symmetrically around the central atom, the atoms surrounding the central atom determine whether the molecule is polar overall. For example, CCl_4 and $CHCl_3$ have the same molecular geometry (tetrahedral), but CCl_4 is nonpolar because all four bonds are identical and their dipoles cancel one another. In $CHCl_3$, however, the bonds are not all identical, one of the C—Cl bonds has been replaced by a C—H bond, and therefore the bond dipoles do *not* sum to zero. The $CHCl_3$ molecule is *polar*.

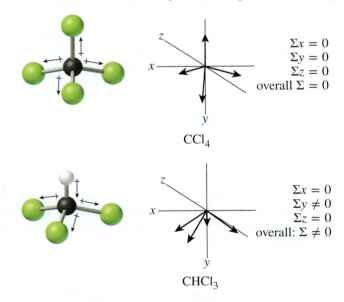

(See end-of-chapter problems 9.79, 9.85, and 9.87.)

Dipole moments can be used to distinguish between molecules that have the same chemical formula but different arrangements of atoms. Such compounds are called **structural isomers.** For example, there are two structural isomers of dichloroethylene ($C_2H_2Cl_2$). Because the individual bond dipoles sum to zero in *trans*-dichloroethylene, the *trans* isomer is nonpolar:

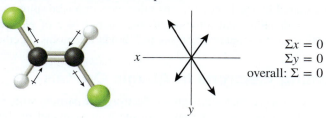

$$\Sigma x = 0$$
$$\Sigma y = 0$$
$$\text{overall: } \Sigma = 0$$

The bond dipoles in the *cis* isomer do not cancel one another, so *cis*-dichloroethylene is polar:

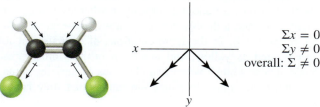

$$\Sigma x = 0$$
$$\Sigma y \neq 0$$
$$\text{overall: } \Sigma \neq 0$$

Because of the difference in polarity, these two isomers can be distinguished experimentally by measuring the dipole moment.

CHECKPOINT – SECTION 9.2   Molecular Geometry and Polarity

9.2.1 Identify the polar molecules in the following group: HBr, CH_4, CS_2.

a) HBr only

b) HBr and CS_2

c) HBr, CH_4, and CS_2

d) CH_4 and CS_2

e) CH_4 only

9.2.2 Identify the nonpolar molecules in the following group: SO_2, NH_3, XeF_2.

a) SO_2, NH_3, and XeF_2

b) SO_2 only

c) XeF_2 only

d) SO_2 and XeF_2

e) SO_2 and NH_3

9.3 Valence Bond Theory

The Lewis theory of chemical bonding provides a relatively simple way for us to visualize the arrangement of electrons in molecules. It is insufficient, however, to explain the differences between the covalent bonds in compounds such as H_2, F_2, and HF. Although Lewis theory describes the bonds in these three molecules in exactly the same way, they really are quite different from one another, as evidenced by their bond lengths and bond enthalpies listed in Table 9.3. Understanding these differences and why covalent bonds form in the first place requires a bonding model that combines

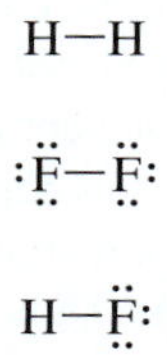

H—H

:$\overset{..}{\underset{..}{F}}$—$\overset{..}{\underset{..}{F}}$:

H—$\overset{..}{\underset{..}{F}}$:

Lewis dot structures of H_2, F_2, and HF

TABLE 9.3	Bond Lengths and Bond Enthalpies of H_2, F_2, and HF	
	Bond Length (Å)	**Bond Enthalpy (kJ/mol)**
H_2	0.74	436.4
F_2	1.42	150.6
HF	0.92	568.2

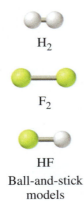

H₂

F₂

HF

Ball-and-stick
models

Lewis's notion of atoms sharing electron pairs and the quantum mechanical descriptions of atomic orbitals.

According to *valence bond theory*, atoms share electrons when an atomic orbital on one atom overlaps with an atomic orbital on the other. Each of the overlapping atomic orbitals must contain a single, unpaired electron. Furthermore, the two electrons shared by the bonded atoms must have opposite spins [◄◄ Section 6.6]. The nuclei of both atoms are attracted to the shared pair of electrons. It is this mutual attraction for the shared electrons that holds the atoms together.

Representing Electrons in Atomic Orbitals

In Chapter 6, we learned that although an electron is a particle with a known mass, it exhibits wavelike properties. The quantum mechanical model of the atom, which gives rise to the familiar shapes of *s* and *p* atomic orbitals, treats electrons in atoms as waves, rather than particles. Therefore, rather than use arrows to denote the locations and spins of electrons, we adopt a convention whereby a singly occupied orbital appears as a light color and a doubly occupied orbital appears as a darker version of the same color (Figure 9.3). In the representations of orbitals that follow, atomic *s* orbitals are represented as yellow, and atomic *p* orbitals are represented as blue. (Empty *p* orbitals appear white.) When two electrons occupy the same atomic orbital in the ground state, their spins are paired—meaning that they have opposite spins [◄◄ Section 6.6].

The H—H bond in H_2 forms when the singly occupied 1*s* orbitals of the two H atoms overlap:

Student Note: Keep in mind that although there are still just two electrons, each atom "thinks" it owns them both, so when the singly occupied orbitals overlap, both orbitals end up doubly occupied.

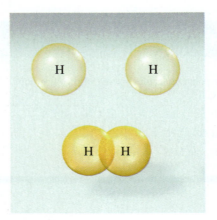

Figure 9.3 Representations of atomic orbitals. Singly occupied *s* orbitals appear light yellow. Doubly occupied *s* orbitals appear darker yellow. Singly and doubly occupied *p* orbitals appear lighter blue and darker blue, respectively.

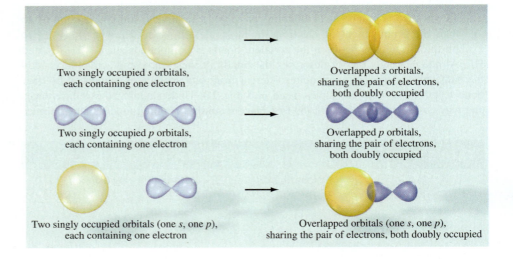

Two singly occupied *s* orbitals, each containing one electron

Overlapped *s* orbitals, sharing the pair of electrons, both doubly occupied

Two singly occupied *p* orbitals, each containing one electron

Overlapped *p* orbitals, sharing the pair of electrons, both doubly occupied

Two singly occupied orbitals (one *s*, one *p*), each containing one electron

Overlapped orbitals (one *s*, one *p*), sharing the pair of electrons, both doubly occupied

Similarly, the F—F bond in F_2 forms when the singly occupied $2p$ orbitals of the two F atoms overlap:

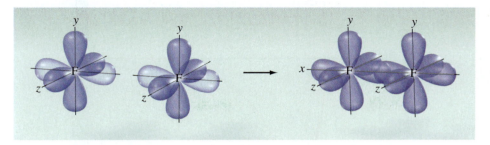

Recall that the ground-state electron configuration of the F atom is $[He]2s^22p^5$ [◄◄ Section 6.8]. (The ground-state orbital diagram of F is shown here.)

We can also depict the formation of an H—F bond using the valence bond model. In this case, the singly occupied $1s$ orbital of the H atom overlaps with the singly occupied $2p$ orbital of the F atom:

$2s^2$ $2p^5$

Orbital diagram for F

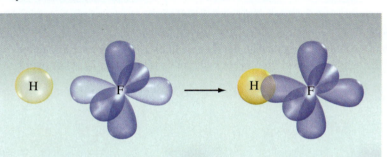

According to the quantum mechanical model, the sizes, shapes, and energies of the $1s$ orbital of H and the $2p$ orbital of F are different. Therefore, it is not surprising that the bonds in H_2, F_2, and HF vary in strength and length.

Energetics and Directionality of Bonding

Why do covalent bonds form? According to valence bond theory, a covalent bond will form between two atoms if the potential energy of the resulting molecule is lower than that of the isolated atoms. Simply put, this means that the formation of covalent bonds is exothermic. While this fact may not seem intuitively obvious, you know that energy must be supplied to a molecule to *break* covalent bonds [◄◄ Section 8.9]. Because the formation of a bond is the reverse process, we should expect energy to be given off. (Recall that the enthalpy change for a forward process and that for the reverse process differ only in sign: $\Delta H_{\text{forward}} = -\Delta H_{\text{reverse}}$ [◄◄ Section 5.3].) Figure 9.4 illustrates how the potential energy of two hydrogen atoms varies with distance between the nuclei.

Valence bond theory also introduces the concept of *directionality* to chemical bonds. For example, we expect the bond formed by the overlap of a p orbital to coincide with the axis along which the p orbital lies. Consider the molecule H_2S. Unlike the other molecules that we have encountered, H_2S does not have the bond angle that Lewis theory and the VSEPR model would lead us to predict. (With four electron domains on the central atom, we would expect the bond angle to be on the order of 109.5°.) In fact, the H—S—H bond angle is 92°.

Looking at this in terms of valence bond theory, the central atom (S) has two unpaired electrons, each of which resides in a $3p$ orbital. The orbital diagram for the ground-state electron configuration of the S atom is

92°

H_2S

Student Note: For you to understand the material in this section and Section 9.4, you must be able to draw orbital diagrams for ground-state electron configurations [◄◄ Section 6.8].

S   [Ne]

$3s^2$ $3p^4$

Figure 9.4 The change in potential energy of two hydrogen atoms as a function of internuclear distance. The minimum potential energy (−436 kJ/mol) occurs when the distance between the nuclei is 74 pm. The yellow spheres represent the 1s orbitals of hydrogen.

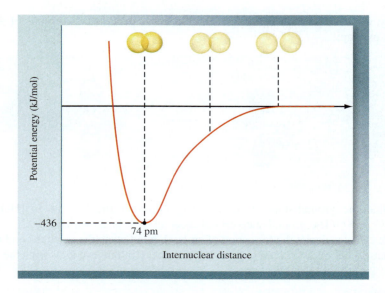

Remember that p orbitals are mutually perpendicular, lying along the x, y, and z axes [◄◄ Section 6.7]. We can rationalize the observed bond angle by envisioning the overlap of each of the singly occupied $3p$ orbitals with the $1s$ orbital of a hydrogen atom:

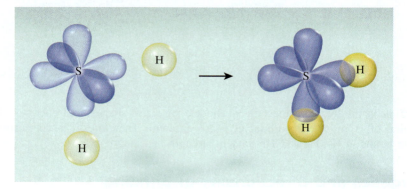

In summary, the important features of valence bond theory are as follows:

- A bond forms when singly occupied atomic orbitals on two atoms overlap.
- The two electrons shared in the region of orbital overlap must be of opposite spin.
- Formation of a bond results in a lower potential energy for the system.

Sample Problem 9.3 shows how to use valence bond theory to explain the bonding in a molecule.

SAMPLE PROBLEM (9.3)

Hydrogen selenide (H_2Se) is a foul-smelling gas that can cause eye and respiratory tract inflammation. The H−Se−H bond angle in H_2Se is approximately 92°. Use valence bond theory to describe the bonding in this molecule.

Strategy Consider the central atom's ground-state electron configuration, and determine what orbitals are available for bond formation.

Setup The ground-state electron configuration of Se is $[Ar]4s^23d^{10}4p^4$. Its orbital diagram (showing only the $4p$ orbitals) is

$$\boxed{\uparrow\downarrow} \boxed{\uparrow} \boxed{\uparrow}$$

$$4p^4$$

Solution Two of the 4*p* orbitals are singly occupied and therefore available for bonding. The bonds in H_2Se form as the result of the overlap of a hydrogen 1*s* orbital with each of these orbitals on the Se atom.

THINK ABOUT IT

Because the 4*p* orbitals on the Se atom are all mutually perpendicular, we should expect the angles between bonds formed by their overlap to be approximately 90°.

Practice Problem **A**TTEMPT Use valence bond theory to describe the bonding in phosphine (PH_3), which has H—P—H bond angles of approximately 94°.

Practice Problem **B**UILD For which molecule(s) can we *not* use valence bond theory to explain the bonding: SO_2 (O—S—O bond angle ~120°), CH_4 (H—C—H bond angles = 109.5°), AsH_3 (H—As—H bond angles = 92°)? Explain.

Practice Problem **C**ONCEPTUALIZE Which of these models could represent a species for which valence bond theory is sufficient to explain the observed bond angle? Explain.

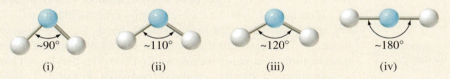

(i) ~90°   (ii) ~110°   (iii) ~120°   (iv) ~180°

CHECKPOINT – SECTION 9.3 Valence Bond Theory

9.3.1 Which of the following atoms, in its ground state, does not have unpaired electrons? (Select all that apply.)

a) O

b) Be

c) B

d) F

e) Ne

9.3.2 According to valence bond theory, how many bonds would you expect a nitrogen atom (in its ground state) to form?

a) 2

b) 3

c) 4

d) 5

e) 6

9.4 Hybridization of Atomic Orbitals

Although valence bond theory is useful and can explain more of our experimental observations than Lewis bond theory, it fails to explain the bonding in many of the molecules that we encounter. According to valence bond theory, for example, an atom must have a singly occupied atomic orbital to form a bond with another atom. How then do we explain the bonding in $BeCl_2$? The central atom, Be, has a ground-state electron configuration of [He]2*s*^{2}, so it has no unpaired electrons. With no singly occupied atomic orbitals in its ground state, how does Be form two bonds?

Furthermore, in cases where the ground-state electron configuration of the central atom *does* have the required number of unpaired electrons, how do we explain the observed bond angles? Carbon, like sulfur, has two unpaired electrons in its ground state. Using valence bond theory as our guide, we might envision the formation of two covalent bonds with oxygen, as in CO_2. If the two unpaired electrons on C (each residing in a 2*p* orbital) were to form bonds, however, the O—C—O bond angle

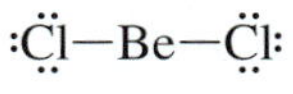

:C̈l—Be—C̈l:

$BeCl_2$

Student Note: The ground-state orbital diagram for C is:

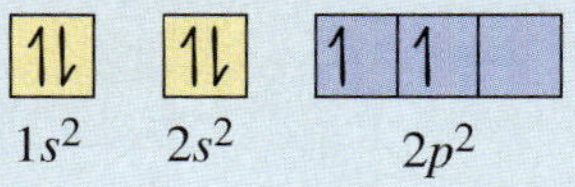

$1s^2$ $2s^2$ $2p^2$

should be on the order of 90°, like the bond angle in H_2S. In fact, the bond angle in CO_2 is 180°:

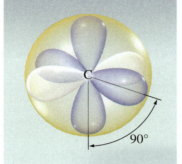

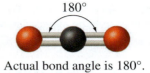

Actual bond angle is 180°.

Bond angle should be 90°.

Animation

Hybrid orbitals—orbital hybridization and valence bond theory.

Student Note: Remember that Be is one of the atoms that does not obey the octet rule [◄◄ Section 8.1]. The Lewis structure of $BeCl_2$ is

:C̈l—Be—C̈l:

To explain these and other observations, we need to extend our discussion of orbital overlap to include the concept of **hybridization** or *mixing* of atomic orbitals.

The idea of hybridization of atomic orbitals begins with the molecular geometry and works backward to explain the bonds and the observed bond angles in a molecule. To extend our discussion of orbital overlap and introduce the concept of hybridization of atomic orbitals, we first consider beryllium chloride ($BeCl_2$), which has two electron domains on the central atom. Using its Lewis structure (shown in the margin) and the VSEPR model, we predict that $BeCl_2$ will have a Cl—Be—Cl bond angle of 180°. If this is true, though, how does Be form two bonds with no unpaired electrons, and why is the angle between the two bonds 180°?

To answer the first part of the question, we envision the *promotion* of one of the electrons in the 2s orbital to an empty 2p orbital. Recall that electrons can be promoted from a lower atomic orbital to a higher one [◄◄ Section 6.3]. The ground-state electron configuration is the one in which all the electrons occupy orbitals of the lowest possible energy. A configuration in which one or more electrons occupy a *higher* energy orbital is called an *excited* state. An excited state generally is denoted with an asterisk (e.g., Be* for an excited-state Be atom). Showing only the valence orbitals, we can represent the promotion of one of the valence electrons of beryllium as

$3s^2$ $3p^5$

Orbital diagram for Cl

With one of its valence electrons promoted to the 2p subshell, the Be atom now has two unpaired electrons and therefore can form two bonds. However, the orbitals in which the two unpaired electrons reside are different from each other, so we would expect bonds formed as a result of the overlap of these two orbitals (each with a 3p orbital on a Cl atom) to be different:

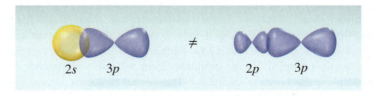

Experimentally, though, the bonds in $BeCl_2$ are identical in length and strength.

Hybridization of *s* and *p* Orbitals

To explain how beryllium forms two identical bonds, we must mix the orbitals in which the unpaired electrons reside, thus yielding two equivalent orbitals. The mixing of beryllium's 2s orbital with one of its 2p orbitals, a process known as *hybridization,* yields two *hybrid orbitals* that are neither *s* nor *p,* but have some character of each. The **hybrid orbitals** are designated 2*sp* or simply *sp.*

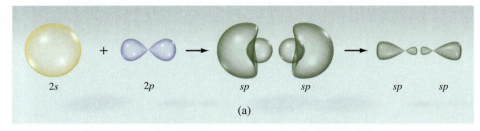

Mixing of one *s* orbital and one *p* orbital to yield two *sp* orbitals

The mathematical combination of the quantum mechanical wave functions for an *s* orbital and a *p* orbital gives rise to two new, equivalent wave functions. As shown in Figure 9.5(a), each *sp* hybrid orbital has one small lobe and one large lobe and, like any two electron domains on an atom, they are oriented in opposite directions with a 180° angle between them. The figure shows the atomic and hybrid orbitals separately for clarity. Note that the hybrid orbitals are shown in two ways: the first is a more realistic shape, whereas the second is a simplified shape that we use to keep the figures clear and make the visualization of orbitals easier. Note also that the representations of hybrid orbitals are green. Figure 9.5(b) shows the locations of the atomic orbitals and the hybrid orbitals, respectively, relative to the beryllium nucleus.

With two *sp* hybrid orbitals, each containing a single unpaired electron, we can see how the Be atom is able to form two identical bonds with two Cl atoms [Figure 9.5(c)]. Each of the singly occupied *sp* hybrid orbitals on the Be atom overlaps with the singly occupied 3*p* atomic orbital on a Cl atom. The energy required to promote an electron in an atom is more than compensated for by the energy given off when a bond forms.

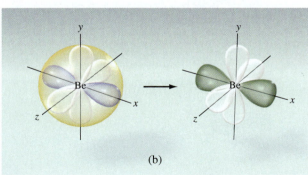

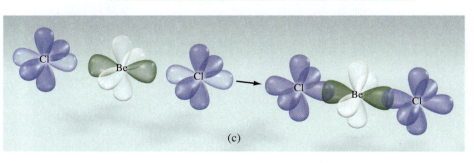

Figure 9.5 (a) An atomic *s* orbital (yellow) and one atomic *p* orbital (blue) combine to form two *sp* hybrid orbitals (green). The realistic hybrid orbital shapes are shown first. The thinner representations are used to keep diagrams clear. (b) The 2*s* orbital and one of the 2*p* orbitals on Be combine to form two *sp* hybrid orbitals. Unoccupied orbitals are shown in white. Like any two electron domains, the hybrid orbitals on Be are 180° apart. (c) The hybrid orbitals on Be each overlap with a singly occupied 3*p* orbital on a Cl atom.

BF$_3$

Student Note: A hybrid orbital analysis starts with a known molecular geometry and known bond angles. It is not used to predict geometries.

We can do a similar analysis of the bonds and the trigonal-planar geometry of boron trifluoride (BF$_3$). The ground-state electron configuration of the B atom is [He]$2s^2 2p^1$, containing just one unpaired electron. Promotion of one of the $2s$ electrons to an empty $2p$ orbital gives the three unpaired electrons needed to explain the formation of *three* bonds. The ground-state and excited-state electron configurations can be represented by

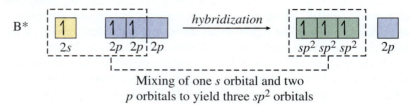

Because the three bonds in BF$_3$ are identical, we must hybridize the three singly occupied atomic orbitals (the one s and two p orbitals) to give three singly occupied hybrid orbitals:

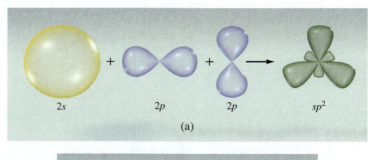

Mixing of one s orbital and two
p orbitals to yield three sp^2 orbitals

Figure 9.6 illustrates the hybridization and bond formation in BF$_3$.

Figure 9.6 (a) An s atomic orbital and two p atomic orbitals combine to form three sp^2 hybrid orbitals. (b) The three sp^2 hybrid orbitals on B are arranged in a trigonal plane. (Empty atomic orbitals are shown in white.) (c) Hybrid orbitals on B overlap with $2p$ orbitals on F.

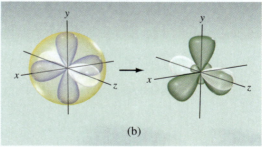

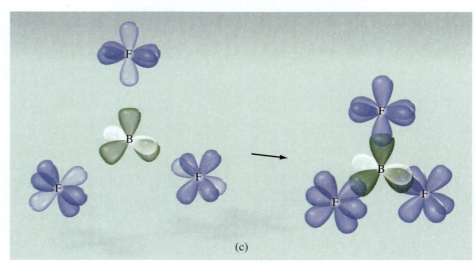

In both cases (i.e., for $BeCl_2$ and BF_3), some but not *all* of the p orbitals are hybridized. When the remaining unhybridized atomic p orbitals do not contain electrons, as in the case of BF_3, they will not be part of the discussion of bonding in this chapter. As we will see in Section 9.5, though, unhybridized atomic orbitals that *do* contain electrons are important in our description of the bonding in a molecule.

We can now apply the same kind of analysis to the methane molecule (CH_4). The Lewis structure of CH_4 has four electron domains around the central carbon atom. This means that we need four hybrid orbitals, which in turn means that four atomic orbitals must be hybridized. The ground-state electron configuration of the C atom contains two unpaired electrons. Promotion of one electron from the $2s$ orbital to the empty $2p$ orbital yields the four unpaired electrons needed for the formation of four bonds:

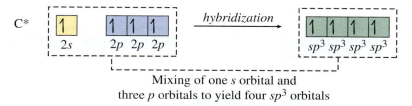

Hybridization of the s orbital and the three p orbitals yields four hybrid orbitals designated sp^3. We can then place the electrons that were originally in the s and p atomic orbitals into the sp^3 hybrid orbitals:

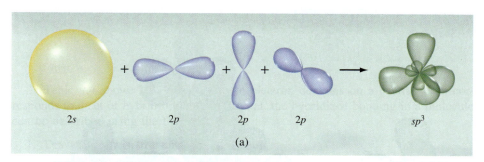

Mixing of one s orbital and three p orbitals to yield four sp^3 orbitals

The set of four sp^3 hybrid orbitals on carbon, like any four electron domains on a central atom, assumes a tetrahedral arrangement. Figure 9.7 illustrates how the hybridization of the C atom results in the formation of the four bonds and the 109.5° bond angles observed in CH_4.

H—C—H (with H above and H below C)

CH_4

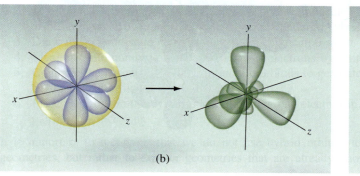

(b)

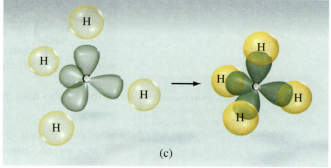

(c)

Figure 9.7 (a) An *s* atomic orbital and three *p* atomic orbitals combine to form four sp^3 hybrid orbitals. (b) The four sp^3 hybrid orbitals on C are arranged in a tetrahedron. (c) Hybrid orbitals on C overlap with 1s orbitals on H. For clarity, the small lobes of the hybrid orbitals are not shown.

THINK ABOUT IT

The Lewis structure given for thalidomide is one of two possible resonance structures. Draw the other resonance structure, and count sigma and pi bonds again. Make sure you get the same answer.

Practice Problem **A**TTEMPT The active ingredient in Tylenol and a host of other over-the-counter pain relievers is acetaminophen ($C_8H_9NO_2$). Determine the total number of sigma and pi bonds in the acetaminophen molecule.

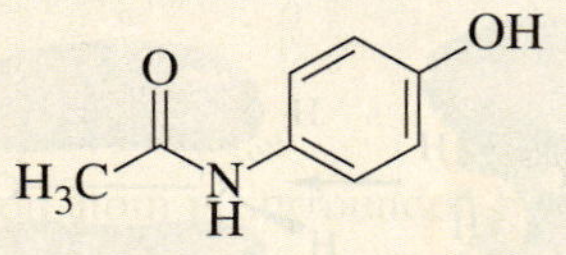

Acetaminophen

Practice Problem **B**UILD Determine the total number of sigma and pi bonds in a molecule of aspirin ($C_9H_8O_4$).

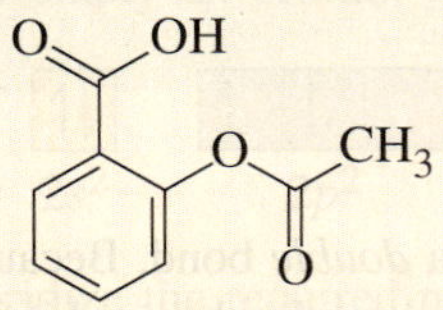

Aspirin

Practice Problem **C**ONCEPTUALIZE In terms of valence bond theory and hybrid orbitals, explain why C_2H_2 and C_2H_4 contain pi bonds, whereas C_2H_6 does not.

Student Note: Remember that *isomers* are molecules with the same chemical formula but different structural arrangements of atoms [◄ Section 9.2].

Because the *p* orbitals that form pi bonds must be parallel to each other, pi bonds restrict the rotation of a molecule in a way that sigma bonds do not. For example, the molecule 1,2-dichloroethane exists as a single isomer. Although we can draw the molecule in several different ways, including the two shown in Figure 9.9(a), all of them are equivalent because the molecule can rotate freely about the sigma bond between the two carbon atoms.

On the other hand, 1,2-dichloroethylene exists as two distinct isomers—*cis* and *trans*—as shown in Figure 9.9(b). The double bond between the carbon atoms consists of one sigma bond and one pi bond. The pi bond restricts rotation about the sigma bond, making the molecules rigid, planar, and not interchangeable. To change one isomer into the other, the pi bond would have to be broken and rotation would have to occur about the sigma bond and the pi bond. This process would require a significant input of energy.

The acetylene molecule (C_2H_2) is linear. Because each carbon atom has two electron domains around it in the Lewis structure, the carbon atoms are *sp*-hybridized. As before, promotion of an electron first maximizes the number of unpaired electrons:

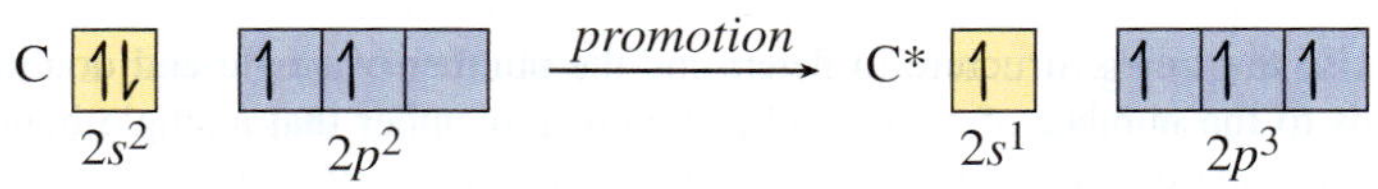

$$H-C\equiv C-H$$
$$C_2H_2$$

The 2s orbital and one of the 2p orbitals then mix to form two *sp* hybrid orbitals:

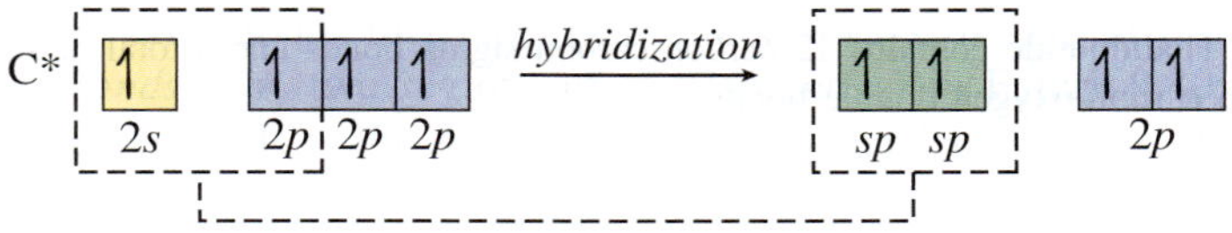

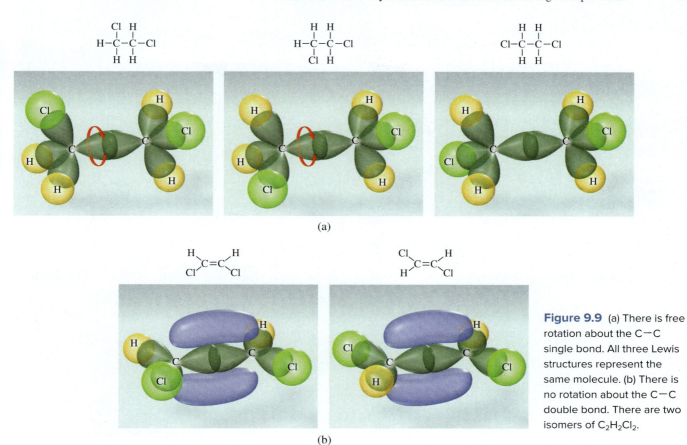

(a)

(b)

Figure 9.9 (a) There is free rotation about the C—C single bond. All three Lewis structures represent the same molecule. (b) There is no rotation about the C—C double bond. There are two isomers of $C_2H_2Cl_2$.

This leaves two unhybridized p orbitals (each containing an electron) on each C atom. Figure 9.10 shows the sigma and pi bonds in the acetylene molecule (also known as *ethyne*). Just as one sigma bond and one pi bond make up a *double* bond, one sigma bond and two pi bonds make up a *triple* bond. Figure 9.11 summarizes the formation of bonds in ethane, ethylene, and acetylene.

Animation
Figure 9.11, Formation of Pi Bonds in Ethylene.

Figure 9.10 (a) Formation of the sigma bond in acetylene. (b) Formation of the pi bonds in acetylene.

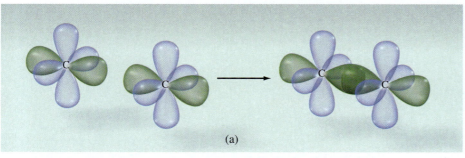

(a)

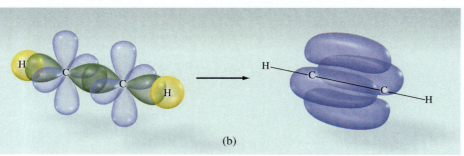

(b)

Figure 9.11

Formation of Pi Bonds in Ethylene and Acetylene

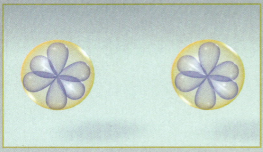

$2s$ and $2p$ atomic orbitals on two C atoms.

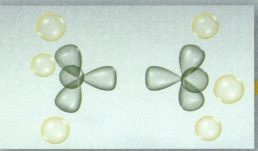

In ethane (C_2H_6) the C atoms are sp^3-hybridized.

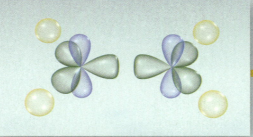

In ethylene (C_2H_4) the C atoms are sp^2-hybridized.

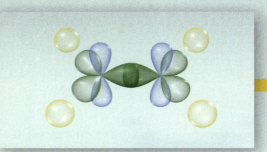

Overlap of sp^2 hybrid orbitals forms a sigma bond between the two C atoms.

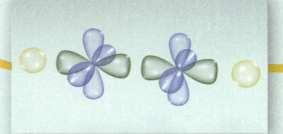

In acetylene (C_2H_2) the C atoms are sp-hybridized.

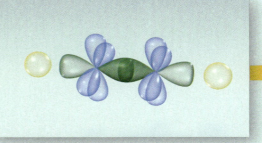

Overlap of sp hybrid orbitals forms a sigma bond between the two C atoms.

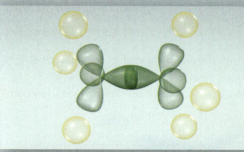

Overlap of sp^3 hybrid orbitals forms a sigma bond between the two C atoms.

Each C atom forms three sigma bonds with H atoms.

Each C atom forms two sigma bonds with H atoms. Each C atom has one leftover unhybridized p orbital.

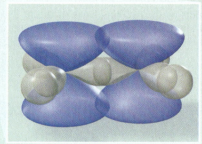

Using the more realistic shape shows how p orbitals overlap.

The parallel, unhybridized p orbitals overlap to form a pi bond with two lobes.

Each C atom forms one sigma bond with an H atom. Each C atom has two leftover unhybridized p orbitals.

Each pair of parallel, unhybridized p orbitals overlaps to form a pi bond with two lobes.

(See Visualizing Chemistry Questions VC 9.1–VC 9.4 on pages 457–458.)

What's the point?

When carbon is sp^2- or sp-hybridized, parallel, unhybridized p orbitals interact to form pi bonds. Each pi bond consists of two lobes as a result of the overlap.

Sample Problem 9.6 shows how hybrid orbitals and pi bonds can be used to explain the bonding in formaldehyde, a molecule with a carbon-oxygen double bond.

Animation
Figure 9.11, Formation of Pi Bonds
in Acetylene.

SAMPLE PROBLEM   9.6

In addition to its use in aqueous solution as a preservative for laboratory specimens, formaldehyde gas is used as an antibacterial fumigant. Use hybridization to explain the bonding in formaldehyde (CH_2O).

Strategy Draw the Lewis structure of formaldehyde, determine the hybridization of the C and O atoms, and describe the formation of the sigma and pi bonds in the molecule.

Setup The Lewis structure of formaldehyde is

$$\ddot{O} \atop \| \atop H{-}\overset{}{C}{-}H$$

The C and O atoms each have three electron domains around them. [Carbon has two single bonds (C—H) and a double bond (C=O); oxygen has a double bond (O=C) and two lone pairs.]

Solution Three electron domains correspond to sp^2 hybridization. For carbon, promotion of an electron from the $2s$ orbital to the empty $2p$ orbital is necessary to maximize the number of unpaired electrons. For oxygen, no promotion is necessary. Each undergoes hybridization to produce sp^2 hybrid orbitals; and each is left with a singly occupied, unhybridized p orbital:

A sigma bond is formed between the C and O atoms by the overlap of one of the sp^2 hybrid orbitals from each of them. Two more sigma bonds form between the C atom and the H atoms by the overlap of carbon's remaining sp^2 hybrid orbitals with the $1s$ orbital on each H atom. Finally, the remaining p orbitals on C and O overlap to form a pi bond:

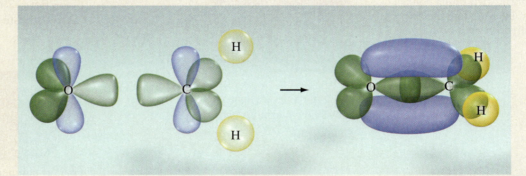

THINK ABOUT IT

Our analysis describes the formation of both a sigma bond and a pi bond between the C and O atoms. This corresponds correctly to the double bond predicted by the Lewis structure. (The two lone pairs on the O atom are the electrons in the doubly occupied sp^2 hybrid orbitals.)

Practice Problem Ⓐ**TTEMPT** Use valence bond theory and hybrid orbitals to explain the bonding in hydrogen cyanide (HCN).

Practice Problem Ⓑ**UILD** Use valence bond theory and hybrid orbitals to explain the bonding in diatomic nitrogen N_2.

Practice Problem Ⓒ**ONCEPTUALIZE** Explain why hybrid orbitals are necessary to explain the bonding in N_2 and O_2, but not to explain the bonding in H_2 or Br_2.

CHECKPOINT – SECTION 9.5   Hybridization in Molecules Containing Multiple Bonds

9.5.1 Which of the following molecules contain one or more pi bonds? (Select all that apply.)

a) N_2

b) Cl_2

c) CO_2

d) CH_3OH

e) CCl_4

9.5.2 From left to right, give the hybridization of each carbon atom in the allene molecule ($H_2C{=}C{=}CH_2$).

a) sp^2, sp^2, sp^2

b) sp^3, sp^2, sp^3

c) sp^2, sp, sp^2

d) sp^3, sp, sp^3

e) sp^3, sp^3, sp^3

9.5.3 Which of the following pairs of atomic orbitals on adjacent nuclei can overlap to form a sigma bond? Consider the x axis to be the internuclear axis.

a) $1s$ and $2s$

b) $1s$ and $2p_x$

c) $2p_y$ and $2p_y$

d) $3p_y$ and $3p_z$

e) $2p_x$ and $3p_x$

9.5.4 Which of the following pairs of atomic orbitals on adjacent nuclei can overlap to form a pi bond? Consider the x axis to be the internuclear axis.

a) $1s$ and $2s$

b) $2s$ and $2s$

c) $2p_y$ and $2p_y$

d) $3p_y$ and $3p_z$

e) $2p_x$ and $3p_x$

9.6   Molecular Orbital Theory

Although the bonding theories we have seen thus far provide simple and effective ways for us to visualize molecules and to predict their shapes and bond angles, Lewis structures and valence bond theory do not enable us to describe or predict some important properties of molecules. Diatomic oxygen, for example, exhibits a property called *paramagnetism*. **Paramagnetic** species are attracted by magnetic fields, whereas **diamagnetic** species are weakly repelled by them. Such magnetic properties are the result of a molecule's electron configuration. Species in which all the electrons are *paired* are diamagnetic, whereas species that contain one or more *unpaired* electrons are paramagnetic. Because O_2 exhibits paramagnetism, it must contain unpaired electrons. According to the Lewis structure of O_2 (shown in the margin) and the valence bond theory description of O_2, however, all the electrons in O_2 are paired. Another bonding theory, called molecular orbital theory, is needed to describe the paramagnetism of O_2 and other important molecular properties.

According to **molecular orbital theory,** the atomic orbitals involved in bonding actually combine to form new orbitals that are the "property" of the entire molecule, rather than of the atoms forming the bonds. These new orbitals are called **molecular orbitals.** In molecular orbital theory, electrons shared by atoms in a molecule reside in the molecular orbitals.

Molecular orbitals are like atomic orbitals in several ways: they have specific shapes and specific energies, and they can each accommodate a maximum of two electrons. As was the case with atomic orbitals, two electrons residing in the same molecular orbital must have opposite spins, as required by the Pauli exclusion principle. And, like hybrid orbitals, the number of molecular orbitals we get is equal to the number of atomic orbitals we combine.

Our treatment of molecular orbital theory in this book is limited to descriptions of bonding in diatomic molecules consisting of elements from the first two periods of the periodic table (H through Ne).

Animation

Chemical bonding—paramagnetic liquid oxygen.

O_2

Bonding and Antibonding Molecular Orbitals

To begin our discussion, we consider H_2, the simplest homonuclear diatomic molecule. According to *valence* bond theory, an H_2 molecule forms when two H atoms are

Figure 9.12 (a) Two *s* atomic orbitals combine to give two sigma molecular orbitals. (b) One of the molecular orbitals is lower in energy than the original atomic orbitals (darker), and (c) one is higher in energy (lighter). The two light yellow lobes make up one molecular orbital. (d) Atomic and molecular orbitals shown relative to the H nuclei.

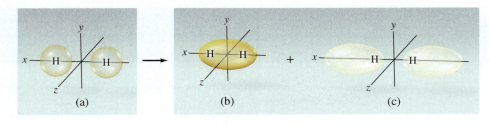

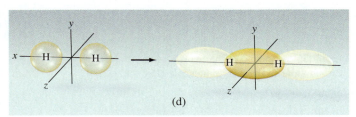

Student Note: Remember that the quantum mechanical approach treats atomic orbitals as *wave functions* [◄ Section 6.5], and that one of the properties of waves is their capacity for both constructive combination and destructive combination [◄ Section 6.1].

close enough for their 1*s* atomic orbitals to overlap. According to *molecular* orbital theory, two H atoms come together to form H_2 when their 1*s* atomic orbitals combine to give molecular orbitals. Figure 9.12 shows the 1*s* atomic orbitals of the isolated H atoms and the molecular orbitals that result from their constructive and destructive combinations. The *constructive* combination of the two 1*s* orbitals gives rise to a molecular orbital [Figure 9.12(b)] that lies along the internuclear axis, directly between the two H nuclei. Just as electron density shared between two nuclei in overlapping atomic orbitals drew the nuclei together, electron density in a molecular orbital that lies between two nuclei will draw them together, too. Thus, this molecular orbital is referred to as a ***bonding molecular orbital.***

The *destructive* combination of the 1*s* atomic orbitals also gives rise to a molecular orbital that lies along the internuclear axis, but, as Figure 9.12(c) shows, this molecular orbital, which consists of two lobes, does not lie in between the two nuclei. Electron density in this molecular orbital would actually pull the two nuclei in opposite directions, rather than toward each other. This is referred to as an ***antibonding molecular orbital.***

σ Molecular Orbitals

Molecular orbitals that lie along the internuclear axis (such as the bonding and antibonding molecular orbitals in H_2) are referred to as σ molecular orbitals. Specifically, the *bonding* molecular orbital formed by the combination of two 1*s* atomic orbitals is designated σ_{1s} and the *antibonding* orbital is designated σ_{1s}^*, where the asterisk distinguishes an antibonding orbital from a bonding orbital. Figure 9.12(d) summarizes the combination of two 1*s* atomic orbitals to yield two molecular orbitals: one bonding and one antibonding.

Animation
Chemical bonding—formation of molecular orbitals.

Like atomic orbitals, molecular orbitals have specific energies. The combination of two atomic orbitals of equal energy, such as two 1*s* orbitals on two H atoms, yields one molecular orbital that is lower in energy (bonding) and one molecular orbital that is higher in energy (antibonding) than the original atomic orbitals. The bonding molecular orbital in H_2 is concentrated between the nuclei, along the internuclear axis. Electron density in this molecular orbital both attracts the nuclei and shields them from each other, stabilizing the molecule. Thus, the bonding molecular orbital is lower in energy than the isolated atomic orbitals. In contrast, the antibonding molecular orbital has most of its electron density outside the internuclear region. Electron density in this orbital does not shield one nucleus from the other, which increases the nuclear repulsions and makes the antibonding molecular orbital higher in energy than the isolated atomic orbitals.

Student Note: The designations σ and π are used in molecular orbital theory just as they are in valence bond theory: σ refers to electron density along the internuclear axis, and π refers to electron density that influences both nuclei but that does not lie directly along the internuclear axis.

Showing all the molecular orbitals in a molecule can make for a very complicated picture. Rather than represent molecules with pictures of their molecular orbitals, we generally use diagrams in which molecular orbitals are represented with boxes placed at the appropriate relative energy levels. Figure 9.13 shows the energies of the

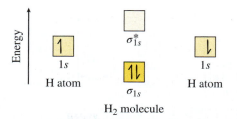

Figure 9.13   Relative energies of atomic orbitals in H and molecular orbitals in H_2.

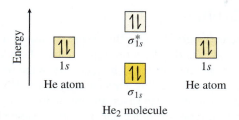

Figure 9.14   Relative energies of atomic orbitals in He and molecular orbitals in He_2.

σ_{1s} and σ_{1s}^*, molecular orbitals in H_2, relative to the energy of the original $1s$ orbitals on two isolated H atoms. Like atomic orbitals, molecular orbitals fill in order of increasing energy. Note that the electrons that originally resided in the atomic orbitals both occupy the lowest-energy molecular orbital, σ_{1s}, with opposite spins.

We can construct a similar molecular orbital diagram for the hypothetical molecule He_2. Like the H atom, the He atom has a $1s$ orbital (unlike H, though, the $1s$ orbital on He has two electrons, not one). The combination of $1s$ orbitals to form molecular orbitals in He_2 is essentially the same as what we have described for H_2. The placement of electrons is shown in Figure 9.14.

Bond Order

With molecular orbital diagrams such as those for H_2 and He_2, we can begin to see the power of molecular orbital theory. For a diatomic molecule described using molecular orbital theory, we can calculate the ***bond order***. The value of the bond order indicates, qualitatively, how *stable* a molecule is. The higher the bond order, the more stable the molecule. Bond order is calculated in the following way:

$$\text{bond order} = \frac{\begin{array}{c}\text{number of electrons} \\ \text{in bonding} \\ \text{molecular orbitals}\end{array} - \begin{array}{c}\text{number of electrons} \\ \text{in antibonding} \\ \text{molecular orbitals}\end{array}}{2} \qquad \textbf{Equation 9.1}$$

In the case of H_2, where both electrons reside in the σ_{1s} orbital, the bond order is $[(2 - 0)/2] = 1$. In the case of He_2, where the two additional electrons reside in the σ_{1s}^* orbital, the bond order is $[(2 - 2)/2] = 0$. Molecular orbital theory predicts that a molecule with a bond order of zero will not exist and He_2 in fact, does *not* exist under ordinary conditions.

We can do similar analyses of the molecules Li_2 and Be_2. (The Li and Be atoms have ground-state electron configurations of $[He]2s^1$ and $[He]2s^2$, respectively.) The $2s$ atomic orbitals also combine to form the corresponding σ and σ^* molecular orbitals. Figure 9.15 shows the molecular orbital diagrams and bond orders for Li_2 and Be_2.

As predicted by molecular orbital theory, Li_2, with a bond order of 1, is a stable molecule, whereas Be_2, with a bond order of 0, does not exist.

Figure 9.15 Bond order determination for Li_2 and Be_2.

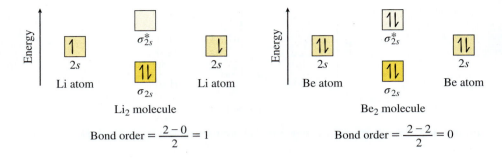

$$\text{Bond order} = \frac{2-0}{2} = 1 \qquad\qquad \text{Bond order} = \frac{2-2}{2} = 0$$

π Molecular Orbitals

To consider diatomic molecules beyond Be_2, we must also consider the combination of p atomic orbitals. Like s orbitals, p orbitals combine both constructively, to give bonding molecular orbitals that are lower in energy than the original atomic orbitals, and destructively, to give antibonding molecular orbitals that are higher in energy than the original atomic orbitals. However, the orientations of p_x, p_y, and p_z orbitals give rise to two different types of molecular orbitals: σ molecular orbitals, in which the regions of electron density in the bonding and antibonding molecular orbitals lie along the internuclear axis, and π molecular orbitals, in which the regions of electron density affect both nuclei but do *not* lie along the internuclear axis.

Orbitals that lie along the internuclear axis, as the $2p_x$ orbitals do in Figure 9.16(a), point directly toward each other and combine to form σ molecular orbitals. Figure 9.16(b) shows the combination of two $2p_x$ atomic orbitals to give two molecular orbitals designated σ_{2p_x} and $\sigma^*_{2p_x}$. Figure 9.16(c) shows the relative energies of these molecular orbitals.

Orbitals that are aligned parallel to each other, like the $2p_y$ and $2p_z$ orbitals shown in Figure 9.16(a), combine to form π molecular orbitals. These bonding molecular orbitals are designated π_{2p_y} and π_{2p_z}; the corresponding antibonding molecular orbitals are designated $\pi^*_{2p_y}$ and $\pi^*_{2p_z}$. Often we refer to the molecular orbitals collectively using the designations $\pi_{2p_{y,z}}$ and $\pi^*_{2p_{y,z}}$. Figure 9.17(a) shows the constructive and destructive combination of parallel p orbitals. Figure 9.17(b) shows the locations of the molecular orbitals resulting from the combination of p_y, and p_z orbitals relative to the two atomic nuclei. Again, electron density in the resulting *bonding* molecular

Figure 9.16 (a) Two sets of $2p$ orbitals. (b) The p atomic orbitals that point toward each other (p_x) combine to give bonding and antibonding σ molecular orbitals. (c) The antibonding σ molecular orbital is higher in energy than the corresponding bonding σ molecular orbital.

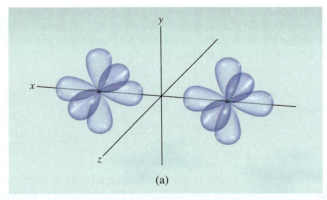

(a)

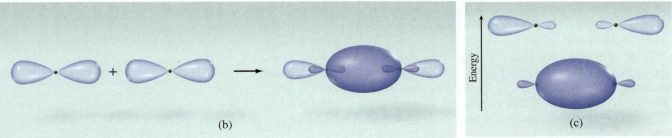

(b)

(c)

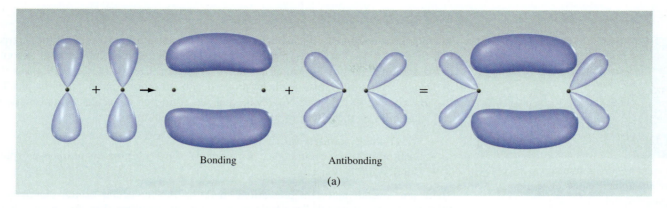

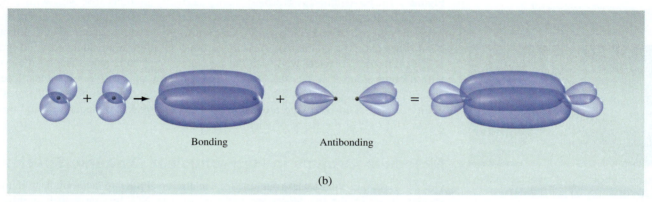

Figure 9.17   Parallel *p* atomic orbitals combine to give *π* molecular orbitals. (a) Bonding and antibonding molecular orbitals shown separately. (b) Bonding and antibonding molecular orbitals shown together relative to the two nuclei.

orbitals serves to hold the nuclei together, whereas electron density in the *antibonding* molecular orbitals does not.

Just as the *p* atomic orbitals within a particular shell are higher in energy than the *s* orbital in the same shell, all the molecular orbitals resulting from the combination of *p* atomic orbitals are higher in energy than the molecular orbitals resulting from the combination of *s* atomic orbitals. To understand better the relative energy levels of the molecular orbitals resulting from *p*-orbital combinations, consider the fluorine molecule (F_2).

In general, molecular orbital theory predicts that the more effective the interaction or overlap of the atomic orbitals, the lower in energy will be the resulting bonding molecular orbital and the higher in energy will be the resulting antibonding molecular orbital. Thus, the relative energy levels of molecular orbitals in F_2 can be represented by the diagram in Figure 9.18(a). The p_x orbitals, which lie along the internuclear axis, overlap most effectively, giving the lowest-energy bonding molecular orbital and the highest-energy antibonding molecular orbital.

The order of orbital energies shown in Figure 9.18(a) assumes that *p* orbitals interact only with other *p* orbitals and *s* orbitals interact only with other *s* orbitals—that there is no significant interaction between *s* and *p* orbitals. In fact, the relatively smaller nuclear charges of boron, carbon, and nitrogen atoms cause their atomic orbitals to be held less tightly than those of atoms with larger nuclear charges, and some *s-p* interaction does take place. This results in a change in the relative energies of the σ_{2p_x} and $\pi_{2p_{y,z}}$ molecular orbitals. Although energies of several of the resulting molecular orbitals change, the most important of these changes is the energy of the σ_{2p} orbital, making it higher than the $\pi_{2p_{y,z}}$ orbitals. The relative energy levels of molecular orbitals in the B_2, C_2, and N_2 molecules can be represented by the diagram in Figure 9.18(b).

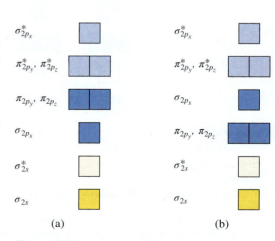

Figure 9.18   (a) Ordering of molecular orbital energies for O_2 and F_2. (b) Ordering of molecular orbital energies for Li_2, B_2, C_2, and N_2. Bonding orbitals are darker; antibonding orbitals are lighter.

Molecular Orbital Diagrams

Liquid oxygen is attracted to the poles of a magnet because O_2 is paramagnetic.

Charles D. Winters/McGraw Hill

Beginning with oxygen, the nuclear charge is sufficiently large to prevent the interaction of s and p orbitals. Thus, for O_2 and Ne_2, the order of molecular orbital energies is the same as that for F_2, which is shown in Figure 9.18(a). Figure 9.19 gives the molecular orbital diagrams, magnetic properties, bond orders, and bond enthalpies for Li_2, B_2, C_2, N_2, O_2, F_2, and Ne_2. Note that the filling of molecular orbitals follows the same rules as the filling of atomic orbitals [◀◀ Section 6.8]:

- Lower-energy orbitals fill first.
- Each orbital can accommodate a maximum of two electrons with opposite spins.
- Hund's rule is obeyed.

There are several important predictions made by the molecular orbital diagrams in Figure 9.19. First, molecular orbital theory correctly predicts that Ne_2, with a bond order of 0, does not exist. Second, it correctly predicts the magnetic properties of the molecules that do exist. Both B_2 and O_2 are known to be paramagnetic. Third, although bond order is only a qualitative measure of bond strength, the calculated bond orders of the molecules correlate well with the measured bond enthalpies. The N_2 molecule, with a bond order of 3, has the largest bond enthalpy of the five molecules. The B_2 and F_2 molecules, each with a bond order of 1, have the smallest bond enthalpies. Its ability to predict correctly the properties of molecules makes molecular orbital theory a powerful tool in the study of chemical bonding.

Liquid nitrogen is not attracted to the poles of a magnet because N_2 is diamagnetic.

Charles D. Winters/McGraw Hill

Molecular Orbitals in Heteronuclear Diatomic Species

Our description of molecular orbital theory can also be applied to heteronuclear diatomic species, in which the two atoms are different, such as NO. In a case such as this, our description of the molecular orbitals involved in bonding must be modified slightly.

The atomic orbitals of a more electronegative atom are lower in energy than the corresponding atomic orbitals of a less electronegative atom. The $2s$ and $2p$ atomic

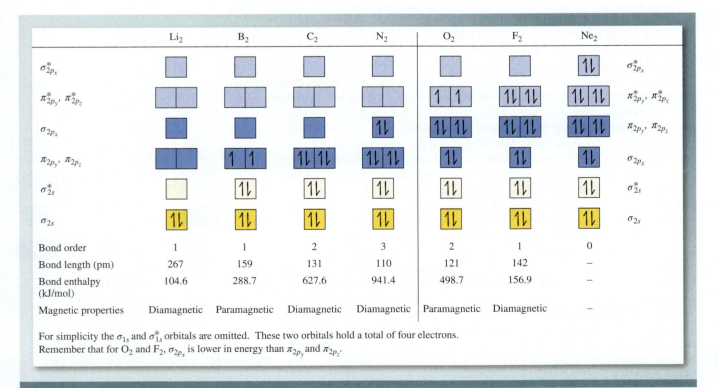

Figure 9.19 Molecular orbital diagrams for second-period homonuclear diatomic molecules.

orbitals are lower in energy for oxygen, which is more electronegative than nitrogen [◄◄ Figure 8.9]. When atomic orbitals of different energies interact to form molecular orbitals, the lower-energy atomic orbital contributes more to the *bonding* molecular orbital; and the higher-energy atomic orbital contributes more to the *antibonding* molecular orbital. The result is that the bonding molecular orbital more closely resembles the atomic orbital of the *more* electronegative atom, and the antibonding molecular orbital more closely resembles the atomic orbital of the *less* electronegative atom. The result of this is that electron density in the resulting bonding molecular orbital is *greater* in the vicinity of the *more* electronegative atom.

Recall that the order of energies of molecular orbitals in second-period homonuclear diatomic molecules is different for B_2, C_2, and N_2, than it is for O_2, F_2, and Ne_2 (Figure 9.19). For second-period heteronuclear diatomic species containing one atom from each group, such as NO and CO, there is no simple rule that can tell us which order the molecular orbitals follow. Note that the molecular orbitals in NO (Figure 9.20) follow the same order as those in O_2.

Often the bond order determined from a molecular orbital diagram corresponds to the number of bonds in the Lewis structure of the molecule. In the case of NO, though, the Lewis structure contains a double bond whereas the molecular orbital approach gives a bond order of 2.5. In fact, the molecular orbital approach gives a bond order that is more consistent with experimental data. The experimentally determined strength of the bond in NO (631 kJ/mol) is greater than that of the average nitrogen-oxygen double bond (607 kJ/mol) [◄◄ Table 8.6].

Sample Problem 9.7 shows how to use molecular orbital diagrams to determine the magnetic properties and bond order of the superoxide ion.

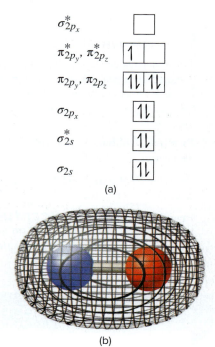

Figure 9.20 (a) Molecular orbital diagram for NO. (b) Molecular orbital representation of NO molecule.

SAMPLE PROBLEM 9.7

The superoxide ion (O_2^-) has been implicated in a number of degenerative conditions, including aging and Alzheimer's disease. Using molecular orbital theory, determine whether O_2^- is paramagnetic or diamagnetic, and then calculate its bond order.

Strategy Start with the molecular orbital diagram for O_2, add an electron, and then use the resulting diagram to determine the magnetic properties and bond order.

Setup The molecular orbital diagram for O_2 is shown in Figure 9.19. The additional electron must be added to the lowest-energy molecular orbital available.

$$\sigma_{2p_x}^* \quad \square$$
$$\pi_{2p_y}^*, \pi_{2p_z}^* \quad \boxed{\uparrow\downarrow}\,\boxed{\uparrow}$$
$$\pi_{2p_y}, \pi_{2p_z} \quad \boxed{\uparrow\downarrow}\,\boxed{\uparrow\downarrow}$$
$$\sigma_{2p_x} \quad \boxed{\uparrow\downarrow}$$
$$\sigma_{2s}^* \quad \boxed{\uparrow\downarrow}$$
$$\sigma_{2s} \quad \boxed{\uparrow\downarrow}$$

Molecular orbital diagram for O_2^-

Solution In this case, either of the two singly occupied π_{2p}^* orbitals can accommodate an additional electron. This gives a molecular orbital diagram in which there is one unpaired electron, making O_2^- paramagnetic. The new diagram has six electrons in bonding molecular orbitals and three in antibonding molecular orbitals. We can ignore the electrons in the σ_{2s} and σ_{2s}^* orbitals because their contributions to the bond order cancel each other. The bond order is $(6 - 3)/2 = 1.5$.

(Continued on next page)

THINK ABOUT IT

Experiments confirm that the superoxide ion is paramagnetic. Also, any time we add one or more electrons to an antibonding molecular orbital, as we did in this problem, we should expect the bond order to decrease. Electrons in antibonding orbitals cause a bond to be less stable.

Practice Problem (A)**TTEMPT** Use molecular orbital theory to determine whether N_2^{2-} is paramagnetic or diamagnetic, and then calculate its bond order.

Practice Problem (B)**UILD** Use molecular orbital theory to determine whether F_2^{2+} is paramagnetic or diamagnetic, and then calculate its bond order.

Practice Problem (C)**ONCEPTUALIZE** For most of the homonuclear diatomic species shown in Figure 9.19, *addition* and *removal* of one or more electrons (to form polyatomic ions) have opposite effects on the bond order. For some species, addition and removal of electrons have the *same* effect on bond order. Identify the species for which this is true and explain how it can be so.

CHECKPOINT – SECTION 9.6   Molecular Orbital Theory

9.6.1 Calculate the bond order of N_2^{2+}, and determine whether it is paramagnetic or diamagnetic.

  a) 2, paramagnetic   d) 3, paramagnetic

  b) 2, diamagnetic   e) 1, paramagnetic

  c) 3, paramagnetic

9.6.2 Which of the following species is paramagnetic? (Select all that apply.)

  a) C_2^{2-} b) O_2^{2+} c) F_2^{2+} d) F_2^{2-} e) C_2^{2+}

9.6.3 Calculate the bond order of He_2^+.

  a) 0   b) 0.5   c) 1.0   d) 1.5   e) 2

9.6.4 Which if any of the following species has a bond order of 0? (Select all that apply.)

  a) B_2^{2+} b) Ne_2^{2+} c) F_2^{2-} d) He_2^{2+} e) H_2^{2-}

9.7 Bonding Theories and Descriptions of Molecules with Delocalized Bonding

The progression of bonding theories in this chapter illustrates the importance of model development. Scientists use models to understand experimental results and to predict future observations. A model is useful as long as it agrees with observation. When it fails to do so, it must be replaced with a new model. What follows is a synopsis of the strengths and weaknesses of the bonding theories presented in Chapters 8 and 9:

Lewis Theory

Strength: The Lewis theory of bonding enables us to make qualitative predictions about bond strengths and bond lengths. Lewis structures are easy to draw and are widely used by chemists [◄◄ Section 8.9].

Weakness: Lewis structures are two dimensional, whereas molecules are three dimensional. In addition, Lewis theory fails to account for the differences in bonds in compounds such as H_2, F_2, and HF. It also fails to explain *why* bonds form.

The Valence-Shell Electron-Pair Repulsion Model

Strength: The VSEPR model enables us to predict the shapes of many molecules and polyatomic ions.

Weakness: Because the VSEPR model is based on the Lewis theory of bonding, it also fails to explain why bonds form.

Valence Bond Theory

Strength: Valence bond theory describes the formation of covalent bonds as the overlap of atomic orbitals. Bonds form because the resulting molecule has a lower potential energy than the original, isolated atoms.

Weakness: Valence bond theory alone fails to explain the bonding in many molecules such as $BeCl_2$, BF_3, and CH_4, in which the central atom in its ground state does not have enough unpaired electrons to form the observed number of bonds.

Hybridization of Atomic Orbitals

Strength: The hybridization of atomic orbitals is not a separate bonding theory; rather, it is an *extension* of valence bond theory. Using hybrid orbitals, we can understand the bonding and geometry of more molecules, including $BeCl_2$, BF_3, and CH_4.

Weakness: Valence bond theory and hybrid orbitals fail to predict some of the important properties of molecules, such as the paramagnetism of O_2.

Molecular Orbital Theory

Strength: Molecular orbital theory enables us to predict accurately the magnetic and other properties of molecules and ions.

Weakness: Pictures of molecular orbitals can be very complex.

Although molecular orbital theory is in many ways the most powerful of the bonding models, it is also the most complex, so we continue to use the other models when they do an adequate job of explaining or predicting the properties of a molecule. For example, if you need to predict the three-dimensional shape of an AB_x molecule on an exam, you should draw its Lewis structure and apply the VSEPR model. Don't try to draw its molecular orbital diagram. On the other hand, if you need to determine the bond order of a diatomic molecule or ion, you should draw a molecular orbital diagram. In general chemistry, it is best to use the *simplest* theory that can answer a particular question.

Because they remain useful, we don't discard the old models when we develop new ones. In fact, the bonding in some molecules, such as benzene (C_6H_6), is best described using a combination of models. Benzene can be represented with two resonance structures [◄◄ Section 8.7].

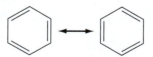

According to its Lewis structure and valence bond theory, the benzene molecule contains twelve σ bonds (six carbon-carbon, and six carbon-hydrogen) and three π bonds. From experimental evidence, however, we know that benzene does not have three single bonds and three double bonds between carbon atoms. Rather, there are six equivalent carbon-carbon bonds. This is precisely the reason that two different Lewis structures are necessary to represent the molecule. Neither one alone accurately depicts the nature of the carbon-carbon bonds. In fact, the π bonds in benzene are ***delocalized,*** meaning that they are spread out over the entire molecule, rather than confined between two specific atoms. (Bonds that are confined between two specific atoms are called ***localized*** bonds.) Valence bond theory does a good job of describing the localized σ bonds in benzene, but molecular orbital theory does a better job of using delocalized π bonds to describe the bonding scheme in benzene.

To describe the σ bonds in benzene, begin with a Lewis structure and count the electron domains on the carbon atoms. (Either resonance structure will give the same result.) Each C atom has three electron domains around it (two single bonds and one double bond). Recall from Table 9.4 that an atom that has three electron domains is sp^2-hybridized. To obtain the three unpaired electrons necessary on each C atom, one electron from each C atom must be promoted from the doubly occupied $2s$ orbital to an empty $2p$ orbital:

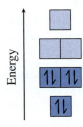

This actually creates four unpaired electrons. Next, the orbitals are sp^2-hybridized, leaving one singly occupied, unhybridized $2p$ orbital on each C atom:

The sp^2 hybrid orbitals adopt a trigonal planar arrangement and overlap with one another (and with $1s$ orbitals on H atoms) to form the σ bonds in the molecule.

The remaining unhybridized $2p$ orbitals (one on each C atom) combine to form molecular orbitals. Because the p orbitals are all *parallel* to one another, only π_{2p} and π_{2p}^* molecular orbitals form. The combination of these six $2p$ atomic orbitals forms six molecular orbitals: three bonding and three antibonding. These molecular orbitals are delocalized over the entire benzene molecule:

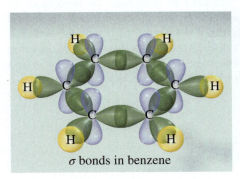

σ bonds in benzene

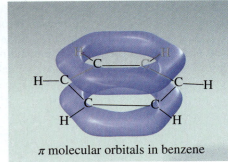

π molecular orbitals in benzene

In the ground state, the lower-energy *bonding* molecular orbitals contain all six electrons. The electron density in the delocalized π molecular orbitals lies above and below the plane that contains all the atoms and the σ bonds in the molecule.

Sample Problem 9.8 shows how to combine valence bond theory and molecular orbital theory to explain the bonding in the carbonate ion.

Animation
Chemical bonding—sigma and pi bonding in benzene.

SAMPLE PROBLEM   9.8

It takes three resonance structures to represent the carbonate ion (CO_3^{2-}):

None of the three, though, is a completely accurate depiction. As with benzene, the bonds that are shown in the Lewis structure as one double and two single are actually three equivalent bonds. Use a combination of valence bond theory and molecular orbital theory to explain the bonding in CO_3^{2-}.

Strategy Starting with the Lewis structure, use valence bond theory and hybrid orbitals to describe the σ bonds. Then use molecular orbital theory to describe the delocalized π bonding.

Setup The Lewis structure of the carbonate ion shows three electron domains around the central C atom, so the carbon must be sp^2-hybridized.

Solution Each of the sp^2 hybrid orbitals on the C atom overlaps with a singly occupied p orbital on an O atom, forming the three σ bonds. Each O atom has an additional, singly occupied p orbital, perpendicular to the one involved in σ bonding. The unhybridized p orbital on C overlaps with the p orbitals on O to form π bonds, which have electron densities above and below the plane of the molecule. Because the species can be represented with resonance structures, we know that the π bonds are delocalized.

THINK ABOUT IT

Double bonds that appear in different places in different resonance structures represent delocalized π bonds.

Practice Problem (A)TTEMPT Use a combination of valence bond theory and molecular orbital theory to describe the bonding in ozone (O_3).

Practice Problem (B)UILD Use a combination of valence bond theory and molecular orbital theory to describe the bonding in the nitrite ion (NO_2^-).

Practice Problem (C)ONCEPTUALIZE For which of the following species is the bonding best described as delocalized?

$$SO_3 \qquad SO_3^{2-} \qquad S_8 \qquad O_2$$

CHECKPOINT – SECTION 9.7   Bonding Theories and Descriptions of Molecules with Delocalized Bonding

9.7.1 Which of the following contain one or more delocalized π bonds? (Select all that apply.)

a) O_2

b) CO_2

c) NO_2^-

d) CH_4

e) CH_2Cl_2

9.7.2 Which of the atoms in BCl_3 need hybrid orbitals to describe the bonding in the molecule?

a) all four atoms

b) only the B atom

c) only the three Cl atoms

d) only the B atom and one Cl atom

e) only the B atom and two Cl atoms

9.7.3 Which of the following can hybrid orbitals be used for? (Select all that apply.)

a) to explain the geometry of a molecule

b) to explain how a central atom can form more bonds than the number of unpaired electrons in its ground-state configuration

c) to predict the geometry of a molecule

d) to explain the magnetic properties of a molecule

e) to predict the magnetic properties of a molecule

9.7.4 Which of the following enables us to explain the paramagnetism of O_2?

a) Lewis theory

b) valence bond theory

c) valence-shell electron-pair repulsion

d) hybridization of atomic orbitals

e) molecular orbital theory

Chapter Summary

Section 9.1

- According to the *valence-shell electron-pair repulsion (VSEPR)* model, electron pairs in the valence shell of an atom repel one another. An *electron domain* is a lone pair or a bond. Any bond (single, double, or triple) constitutes one electron domain.

- The arrangement of electron domains about a central atom, determined using the VSEPR model, is called the *electron-domain geometry*. The arrangement of *atoms* in a molecule is called the *molecular geometry*. The basic molecular geometries are linear, bent, trigonal planar, tetrahedral, trigonal pyramidal, trigonal bipyramidal, seesaw-shaped, T-shaped, octahedral, square pyramidal, and square planar.

- The *bond angle* is the angle between two adjacent bonds in a molecule or polyatomic ion. A trigonal bipyramid contains two types of bonds: *axial* and *equatorial.*

Section 9.2

- The polarity of a molecule depends on the polarity of its individual bonds and on its molecular geometry. Even a molecule containing polar bonds may be nonpolar overall if the bonds are distributed symmetrically.

- *Structural isomers* are molecules with the same chemical formula, but different structural arrangements.

Section 9.3

- According to *valence bond theory,* bonds form between atoms when atomic orbitals overlap, thus allowing the atoms to share valence electrons. A bond forms, furthermore, when the resulting molecule is lower in energy than the original, isolated atoms.

Section 9.4

- To explain the bonding in some molecules, we need to employ the concept of *hybridization,* in which atomic orbitals mix to form hybrid orbitals.

- To use hybrid-orbital analysis, we must already know the molecular geometry and bond angles in a molecule. Hybrid orbitals are not used to predict molecular geometries.

Section 9.5

- *Sigma (σ) bonds* form when the region of orbital overlap lies directly between the two atoms. *Pi (π) bonds* form when parallel, unhybridized *p* orbitals interact. A *double* bond consists of one sigma bond and one pi bond. A *triple* bond consists of one sigma bond and two pi bonds.

Section 9.6

- A *paramagnetic* species is one that contains unpaired electrons. A *diamagnetic* species is one in which there are no unpaired electrons. Paramagnetic species are weakly attracted by a magnetic field, whereas diamagnetic species are weakly repelled by a magnetic field.

- According to *molecular orbital theory,* atomic orbitals combine to form new *molecular orbitals* that are associated with the molecule, rather than with individual atoms. Molecular orbitals may be *sigma,* if the orbital lies directly along the internuclear axis, or *pi,* if the orbital does not lie directly along the internuclear axis.

- Molecular orbitals may be *bonding* or *antibonding*. A *bonding molecular orbital* is lower in energy than the isolated atomic orbitals that combined to form it. The corresponding *antibonding molecular orbital* is higher in energy than the isolated atomic orbitals. *Bond order* is a measure of the *strength* of a bond and can be determined using a molecular orbital diagram.

Section 9.7

- It is generally best to use the bonding theory that most easily describes the bonding in a particular molecule or polyatomic ion. In species that can be represented by two or more resonance structures, the pi bonds are *delocalized,* meaning that they are spread out over the molecule and not constrained to just two atoms. *Localized* bonds are those constrained to two atoms. Many species are best described using a combination of valence bond theory and molecular orbital theory.

Key Words

Antibonding molecular orbital, 442

Axial, 413

Bond angle, 412

Bond order, 443

Bonding molecular orbital, 442

Delocalized, 449

Diamagnetic, 441

Electron domain, 411

Electron-domain geometry, 412

Equatorial, 413

Hybridization, 426

Localized, 449

Molecular geometry, 412

Molecular orbital, 441

Molecular orbital theory, 441

Paramagnetic, 441

Pi (π) bond, 434

Sigma (σ) bond, 434

Structural isomer, 421

Valence bond theory, 422

Valence-shell electron-pair repulsion (VSEPR), 410

Key Equation

9.1 $\text{bond order} = \dfrac{\begin{array}{c}\text{number of electrons}\\\text{in bonding}\\\text{molecular orbitals}\end{array} - \begin{array}{c}\text{number of electrons}\\\text{in antibonding}\\\text{molecular orbitals}\end{array}}{2}$

Bond order is calculated by subtracting the number of electrons in antibonding molecular orbitals from the number in bonding molecular orbitals, and dividing the result by 2.

Questions and Problems

Applying What You've Learned

The structures of folic acid, also known as vitamin B-9, and methotrexate, the drug developed to block folic acid absorption, are shown here.

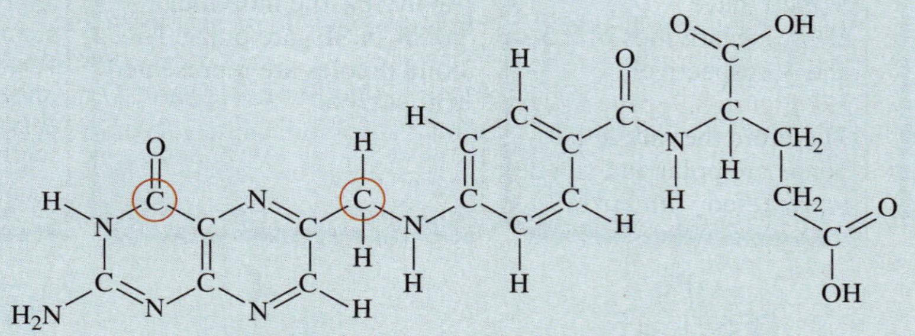

folic acid

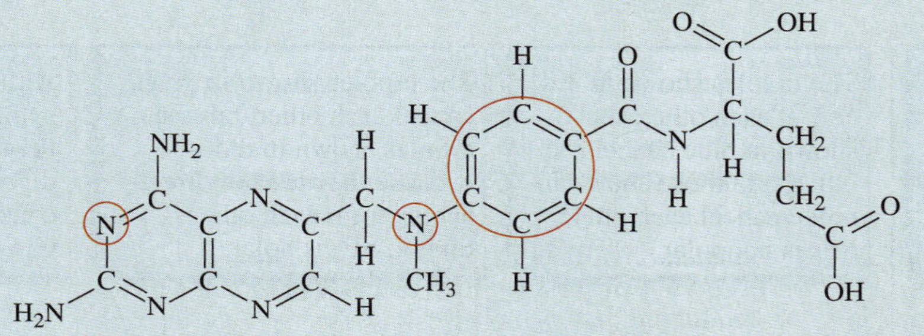

methotrexate

Problems:

(a) How many carbon-carbon sigma bonds are there in folic acid and how many carbon-carbon, carbon-nitrogen, or carbon-oxygen pi bonds are there [◀◀ Sample Problem 9.5]? (b) Determine the hybridization of the carbon atoms circled in red in the folic acid molecule and of the nitrogen atoms circled in red in the methotrexate molecule, and describe the geometry about each circled atom [◀◀ Sample Problem 9.4]. (c) Which of the pi bonds in folic acid are delocalized? (d) Describe the bonding in the circled six-membered ring in methotrexate using a combination of valence bond theory and molecular orbital theory [◀◀ Sample Problem 9.8].

SECTION 9.1: MOLECULAR GEOMETRY

Review Questions

9.1   How is the geometry of a molecule defined, and why is the study of molecular geometry important?

9.2   Sketch the shape of a linear triatomic molecule, a trigonal planar molecule containing four atoms, a tetrahedral molecule, a trigonal bipyramidal molecule, and an octahedral molecule. Give the bond angles in each case.

9.3   How many atoms are directly bonded to the central atom in a tetrahedral molecule, a trigonal bipyramidal molecule, and an octahedral molecule?

9.4   Discuss the basic features of the VSEPR model. Explain why the magnitude of repulsion decreases in the following order: lone pair–lone pair > lone pair–bonding pair > bonding pair–bonding pair.

9.5   In the trigonal bipyramidal arrangement, why does a lone pair occupy an equatorial position rather than an axial position?

9.6   Explain why the CH_4 molecule is not square planar, although its Lewis structure makes it look as though it could be.

Conceptual Problems

9.7   Predict the geometries of the following species using the VSEPR method: (a) PCl_3, (b) $CHCl_3$, (c) SiH_4, (d) $TeCl_4$.

9.8   Predict the geometries of the following species:
(a) $AlCl_3$, (b) $ZnCl_2$, (c) $HgBr_2$, (d) N_2O
(arrangement of atoms is NNO).

9.9   Predict the geometry of the following molecules
and ion using the VSEPR model: (a) CBr_4,
(b) BCl_3, (c) NF_3, (d) H_2Se, (e) NO_2^-.

9.10   Predict the geometry of the following molecules
and ion using the VSEPR model: (a) CH_3I,
(b) ClF_3, (c) H_2S, (d) SO_3, (e) SO_4^{2-}.

9.11   Predict the geometry of the following ions using
the VSEPR method: (a) SCN^- (arrangement of
atoms is SCN), (b) AlH_4^-, (c) $SnCl_5^-$, (d) H_3O^+,
(e) BeF_4^{2-}.

9.12   Predict the geometries of the following ions:
(a) NH_4^+, (b) NH_2^-, (c) CO_3^{2-}, (d) ICl_2^-, (e) ICl_4^-.

9.13   Describe the geometry around each of the three
central atoms in the CH_3COOH molecule.

9.14   Which of the following species are tetrahedral:
$SiCl_4$, SeF_4, XeF_4, CI_4, $CdCl_4^{2-}$?

SECTION 9.2: MOLECULAR GEOMETRY AND POLARITY

Review Questions

9.15   Explain why an atom cannot have a permanent
dipole moment.

9.16   The bonds in beryllium hydride (BeH_2) molecules
are polar, and yet the dipole moment of the
molecule is zero. Explain.

Conceptual Problems

9.17   Determine whether (a) BrF_5 and (b) BCl_3 are polar.

9.18   Determine whether (a) OCS and (b) XeF_4 are polar.

9.19   Which of the molecules shown is polar?

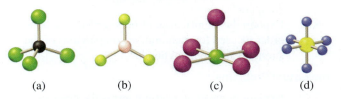

(a)   (b)   (c)   (d)

9.20   Which of the molecules shown is polar?

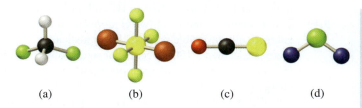

(a)   (b)   (c)   (d)

SECTION 9.3: VALENCE BOND THEORY

Review Questions

9.21   What is valence bond theory? How does it differ
from the Lewis concept of chemical bonding?

9.22   Use valence bond theory to explain the bonding in
Cl_2 and HCl. Show how the atomic orbitals overlap
when a bond is formed.

9.23   According to valence bond theory, how many
bonds would you expect each of the following
atoms (in the ground state) to form: Be, C?

9.24   According to valence bond theory, how many
bonds would you expect each of the following
atoms (in the ground state) to form: P, S?

SECTION 9.4: HYBRIDIZATION OF ATOMIC ORBITALS

Review Questions

9.25   What is the hybridization of atomic orbitals? Why
is it impossible for an isolated atom to exist in the
hybridized state?

9.26   How does a hybrid orbital differ from a pure
atomic orbital? Can two $2p$ orbitals of an atom
hybridize to give two hybridized orbitals?

9.27   What is the angle between the following two
hybrid orbitals on the same atom: (a) sp and sp
hybrid orbitals, (b) sp^2 and sp^2 hybrid orbitals,
(c) sp^3 and sp^3 hybrid orbitals?

Conceptual Problems

9.28   Describe the bonding scheme of the AsH_3
molecule in terms of hybridization.

9.29   What is the hybridization state of Si (a) in SiH_4
and (b) in $H_3Si{-}SiH_3$?

9.30   Describe the change in hybridization (if any) of
the Al atom in the following reaction:

$$AlCl_3 + Cl^- \longrightarrow AlCl_4^-$$

9.31   Consider the reaction

$$BF_3 + NH_3 \longrightarrow F_3B{-}NH_3$$

Describe the changes in hybridization (if any) of
the B and N atoms as a result of this reaction.

9.32   What hybrid orbitals are used by nitrogen atoms in the
following species: (a) NH_3, (b) $H_2N{-}NH_2$, (c) NO_3^-?

9.33   Describe the hybridization of phosphorus in PF_5.

SECTION 9.5: HYBRIDIZATION IN MOLECULES CONTAINING MULTIPLE BONDS

 **Visualizing Chemistry**
Figure 9.11

VC 9.1   How is a sigma bond different from a pi bond?
a) A sigma bond is a bonding molecular orbital;
a pi bond is an antibonding molecular orbital.
b) A sigma bond is a single bond, whereas a pi
bond is a double bond.
c) The electron density in a sigma bond lies along
the internuclear axis; that of a pi bond does not.

VC 9.2   Pi bonds form when _____ atomic orbitals
on _____ atom(s) overlap.
a) perpendicular, adjacent
b) parallel, adjacent
c) parallel, the same

VC 9.3 Formation of two pi bonds requires the combination of _____ atomic orbitals.
a) two   b) four   c) six

VC 9.4 Why are there no pi bonds in ethane (C_2H_6)?
a) The remaining unhybridized p orbitals do not contain any electrons.
b) There are no unhybridized p orbitals remaining on either C atom.
c) The remaining unhybridized p orbitals are not parallel to each other.

Review Questions

9.34 How would you distinguish between a sigma bond and a pi bond?

9.35 Which of the following pairs of atomic orbitals of adjacent nuclei can overlap to form a sigma bond? Which overlap to form a pi bond? Which cannot overlap (no bond)? Consider the x axis to be the internuclear axis, that is, the line joining the nuclei of the two atoms. (a) $1s$ and $1s$, (b) $1s$ and $2p_x$, (c) $2p_x$ and $2p_y$, (d) $3p_y$ and $3p_y$, (e) $2p_x$ and $2p_x$, (f) $1s$ and $2s$.

Conceptual Problems

9.36 What are the hybrid orbitals of the carbon atoms in the following molecules?
(a) H_3C-CH_3
(b) $H_3C-CH=CH_2$
(c) $CH_3-C\equiv C-CH_2OH$
(d) $CH_3CH=O$
(e) CH_3COOH

9.37 Specify which hybrid orbitals are used by carbon atoms in the following species: (a) CO, (b) CO_2, (c) CN^-.

9.38 The allene molecule ($H_2C=C=CH_2$) is linear (the three C atoms lie on a straight line). What are the hybridization states of the carbon atoms? Draw diagrams to show the formation of sigma bonds and pi bonds in allene.

9.39 What is the hybridization of the central N atom in the azide ion (N_3^-)? (The arrangement of atoms is NNN.)

9.40 How many sigma bonds and pi bonds are there in each of the following molecules?

(a)   (b)   (c)

9.41 How many pi bonds and sigma bonds are there in the tetracyanoethylene molecule?

9.42 Tryptophan is one of the 20 amino acids in the human body. Describe the hybridization state of the C and N atoms, and determine the number of sigma and pi bonds in the molecule.

9.43 Benzo(*a*)pyrene is a potent carcinogen found in coal and cigarette smoke. Determine the number of sigma and pi bonds in the molecule.

SECTION 9.6: MOLECULAR ORBITAL THEORY

Review Questions

9.44 What is molecular orbital theory? How does it differ from valence bond theory?

9.45 Define the following terms: bonding molecular orbital, antibonding molecular orbital, pi molecular orbital, sigma molecular orbital.

9.46 Sketch the shapes of the following molecular orbitals: σ_{1s}, σ_{1s}^*, π_{2p}, π_{2p}^*. How do their energies compare?

9.47 Explain the significance of bond order. Can bond order be used for quantitative comparisons of the strengths of chemical bonds?

Conceptual Problems

9.48 Explain in molecular orbital terms the changes in H—H internuclear distance that occur as the molecular H_2 is ionized first to H_2^+ and then to H_2^{2+}.

9.49 The formation of H_2 from two H atoms is an energetically favorable process. Yet statistically there is less than a 100 percent chance that any two H atoms will undergo the reaction. Apart from energy considerations, how would you account for this observation based on the electron spins in the two H atoms?

9.50 Draw a molecular orbital energy level diagram for each of the following species: He_2, HHe, He_2^+. Compare their relative stabilities in terms of bond orders. (Treat HHe as a diatomic molecule with three electrons.)

9.51 Arrange the following species in order of increasing stability: Li_2, Li_2^+, Li_2^-. Justify your choice with a molecular orbital energy level diagram.

9.52 Use molecular orbital theory to explain why the Be_2 molecule does not exist.

9.53 Which of these species has a longer bond, B_2 or B_2^+? Explain in terms of molecular orbital theory.

9.54 Acetylene (C_2H_2) has a tendency to lose two protons (H^+) and form the carbide ion (C_2^{2-}), which is present in a number of ionic compounds, such as CaC_2 and MgC_2. Describe the bonding scheme in the C_2^{2-} ion in terms of molecular orbital theory. Compare the bond order in C_2^{2-} with that in C_2.

9.55 Compare the Lewis and molecular orbital treatments of the oxygen molecule.

9.56 Explain why the bond order of N_2 is greater than that of N_2^+, but the bond order of O_2 is less than that of O_2^+.

9.57 Compare the relative bond orders of the following species and indicate their magnetic properties (i.e., diamagnetic or paramagnetic): O_2, O_2^+, O_2^- (superoxide ion), O_2^{2-} (peroxide ion).

9.58 Use molecular orbital theory to compare the relative stabilities of F_2 and F_2^+.

9.59 A single bond is almost always a sigma bond, and a double bond is almost always made up of a sigma bond and a pi bond. There are very few exceptions to this rule. Show that the B_2 and C_2 molecules are examples of the exceptions.

SECTION 9.7: BONDING THEORIES AND DESCRIPTIONS OF MOLECULES WITH DELOCALIZED BONDING

Review Questions

9.60 How does a delocalized molecular orbital differ from a molecular orbital such as that found in H_2 or C_2H_4? What do you think are the minimum conditions (e.g., number of atoms and types of orbitals) for forming a delocalized molecular orbital?

9.61 In Chapter 8, we saw that the resonance concept is useful for dealing with species such as the benzene molecule and the carbonate ion. How does molecular orbital theory deal with these species?

Conceptual Problems

9.62 Both ethylene (C_2H_4) and benzene (C_6H_6) contain the $C=C$ bond. The reactivity of ethylene is greater than that of benzene. For example, ethylene readily reacts with molecular bromine, whereas benzene is normally quite inert toward molecular bromine and many other compounds. Explain this difference in reactivity.

9.63 Explain why the symbol on the left is a better representation of benzene molecules than that on the right.

9.64 Determine which of these molecules has a more delocalized orbital, and justify your choice. (*Hint:* Both molecules contain two benzene rings. In naphthalene, the two rings are fused together. In biphenyl, the two rings are joined by a single bond, about which the two rings can rotate.)

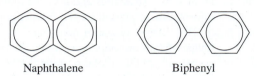

      Naphthalene   Biphenyl

9.65 Nitryl fluoride (FNO_2) is very reactive chemically. The fluorine and oxygen atoms are bonded to the nitrogen atom. (a) Write a Lewis structure for FNO_2. (b) Indicate the hybridization of the nitrogen atom. (c) Describe the bonding in terms of molecular orbital theory. Where would you expect delocalized molecular orbitals to form?

9.66 Describe the bonding in the nitrate ion NO_3^- in terms of delocalized molecular orbitals.

9.67 What is the state of hybridization of the central O atom in O_3? Describe the bonding in O_3 in terms of delocalized molecular orbitals.

ADDITIONAL PROBLEMS

9.68 Which of the following species is not likely to have a tetrahedral shape: (a) $SiBr_4$, (b) NF_4^+, (c) SF_4, (d) $BeCl_4^{2-}$, (e) BF_4^-, (f) $AlCl_4^-$?

9.69 Draw the Lewis structure of mercury(II) bromide. Is this molecule linear or bent? How would you establish its geometry?

9.70 Although both carbon and silicon are in Group 14, very few $Si=Si$ bonds are known. Account for the instability of silicon-to-silicon double bonds in general. (*Hint:* Compare the atomic radii of C and Si in Figure 7.6, page 322. What effect would the larger size have on pi bond formation?)

9.71 Predict the geometry of sulfur dichloride (SCl_2) and the hybridization of the sulfur atom.

9.72 Antimony pentafluoride (SbF_5) reacts with XeF_4 and XeF_6 to form ionic compounds, $XeF_3^+ SbF_6^-$ and $XeF_5^+ SbF_6^-$. Describe the geometries of the cations and anions in these two compounds.

9.73 Assume that the third-period element phosphorus forms a diatomic molecule, P_2, in an analogous way as nitrogen does to form N_2. (a) Write the electronic configuration for P_2. Use [Ne_2] to represent the electron configuration for the first two periods. (b) Calculate its bond order. (c) What are its magnetic properties (diamagnetic or paramagnetic)?

9.74 The molecule benzyne (C_6H_4) is a very reactive species. It resembles benzene in that it has a six-membered ring of carbon atoms. Draw a Lewis structure of the molecule and account for the molecule's high reactivity.

9.75   Predict the bond angles for the following molecules:
(a) $BeCl_2$, (b) BCl_3, (c) CCl_4, (d) CH_3Cl, (e) Hg_2Cl_2
(arrangement of atoms: ClHgHgCl), (f) $SnCl_2$,
(g) H_2O_2, (h) SnH_4.

9.76   Briefly compare the VSEPR and hybridization
approaches to the study of molecular geometry.

9.77   Draw Lewis structures and give the other
information requested for the following molecules:
(a) BF_3. Shape: planar or nonplanar? (b) ClO_3^-.
Shape: planar or nonplanar? (c) HCN. Polar or
nonpolar? (d) OF_2. Polar or nonpolar? (e) NO_2.
Estimate the ONO bond angle.

9.78   Describe the hybridization state of arsenic in
arsenic pentafluoride (AsF_5).

9.79   Determine whether (a) PCl_5 and (b) H_2CO
(C double bonded to O) are polar.

9.80   Draw Lewis structures and give the other
information requested for the following: (a) SO_3.
Polar or nonpolar molecule? (b) PF_3. Polar or
nonpolar? (c) F_3SiH. Polar or nonpolar? (d) SiH_3^-.
Shape: planar or pyramidal? (e) Br_2CH_2. Polar or
nonpolar molecule?

9.81   Which of the following molecules are linear:
ICl_2^-, IF_2^+, OF_2, SnI_2, $CdBr_2$?

9.82   Draw the Lewis structure for the $BeCl_4^{2-}$ ion.
Predict its geometry, and describe the hybridization
state of the Be atom.

9.83   The N_2F_2 molecule can exist in either of the
following two forms:

(a) What is the hybridization of N in the molecule?
(b) Which structure is polar?

9.84   Cyclopropane (C_3H_6) has the shape of a triangle in
which a C atom is bonded to two H atoms and two
other C atoms at each corner. Cubane (C_8H_8) has the
shape of a cube in which a C atom is bonded to one H
atom and three other C atoms at each corner. (a) Draw
Lewis structures of these molecules. (b) Compare the
CCC angles in these molecules with those predicted
for an sp^3-hybridized C atom. (c) Would you expect
these molecules to be easy to make?

9.85   Determine whether (a) CH_2Cl_2 and (b) XeF_4 are
polar.

9.86   Does the following molecule have a dipole
moment? Explain.

9.87   For which molecular geometries (linear, bent,
trigonal planar, trigonal pyramidal, tetrahedral,
square planar, T-shaped, seesaw-shaped, trigonal

bipyramidal, square pyramidal, octahedral) can an
AB_x molecule be nonpolar if there are (a) two
different types of terminal atoms and (b) three
different types of terminal atoms?

9.88   (a) From what group must the terminal atoms
come in an AB_x molecule where the central atom
is from Group 15, for both the electron-domain
geometry and the molecular geometry to be
trigonal bipyramidal? (b) From what group must
the terminal atoms come in an AB_x molecule
where the central atom is from Group 16, for the
electron-domain geometry to be tetrahedral and
the molecular geometry to be bent?

9.89   Carbon suboxide (C_3O_2) is a colorless pungent-
smelling gas. Does this molecule possess a dipole
moment? Explain.

9.90   The following molecules (AX_4Y_2) all have an
octahedral geometry. Group the molecules that are
equivalent to each other.

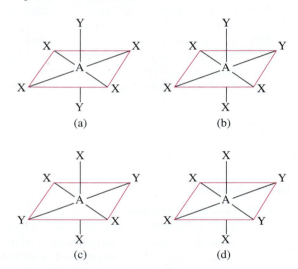

9.91   The compounds carbon tetrachloride (CCl_4)
and silicon tetrachloride ($SiCl_4$) are similar in
geometry and hybridization. However, CCl_4
does not react with water but $SiCl_4$ does. Explain
the difference in their chemical reactivities.
(*Hint:* The first step of the reaction is believed
to be the addition of a water molecule to the
Si atom in $SiCl_4$.)

9.92   Write the ground-state electron configuration for
B_2. Is the molecule diamagnetic or paramagnetic?

9.93   What is the hybridization of C and of N in this
molecule?

9.94 The stable allotropic form of phosphorus is P_4, in which each P atom is bonded to three other P atoms. Draw a Lewis structure of this molecule and describe its geometry. At high temperatures, P_4 dissociates to form P_2 molecules containing a P=P bond. Explain why P_4 is more stable than P_2.

9.95 Use molecular orbital theory to explain the difference between the bond enthalpies of F_2 and F_2^-. (See Problem 8.125.)

9.96 Use molecular orbital theory to explain the bonding in the azide ion (N_3^-). (The arrangement of atoms is NNN.)

9.97 Carbon dioxide has a linear geometry and is nonpolar. Yet we know that the molecule exhibits bending and stretching motions that create a dipole moment. How would you reconcile these seemingly conflicting descriptions of CO_2?

9.98 Draw three Lewis structures for compounds with the formula $C_2H_2F_2$. Indicate which of the compounds are polar.

9.99 Write the electron configuration of the cyanide ion (CN^-). Name a stable molecule that is isoelectronic with the ion.

9.100 Aluminum trichloride ($AlCl_3$) is an electron-deficient molecule. It has a tendency to form a dimer (a molecule made up of two $AlCl_3$ units):

$$AlCl_3 + AlCl_3 \longrightarrow Al_2Cl_6.$$

(a) Draw a Lewis structure for the dimer. (b) Describe the hybridization state of Al in $AlCl_3$ and Al_2Cl_6. (c) Sketch the geometry of the dimer. (d) Do these molecules possess a dipole moment?

9.101 The Lewis structure for O_2 is

$$\ddot{\text{O}}=\ddot{\text{O}}$$

Use molecular orbital theory to show that the structure actually corresponds to an excited state of the oxygen molecule.

9.102 Draw the Lewis structure of ketene (C_2H_2O) and describe the hybridization states of the C atoms. The molecule does not contain O—H bonds. On separate diagrams, sketch the formation of the sigma and pi bonds.

9.103 Which of the following geometries has a greater stability for tin(IV) hydride (SnH_4)?

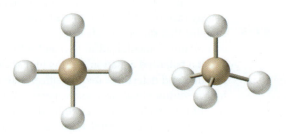

9.104 Which of the following ions possess a dipole moment: (a) ClF_2^+, (b) ClF_2^-, (c) IF_4^+, (d) IF_4^-?

Biological Problems

9.105 The molecular model of vitamin C is shown here. (a) Write the molecular formula of the compound. (b) What is the hybridization of each C and O atom? (c) Describe the geometry about each C and O atom.

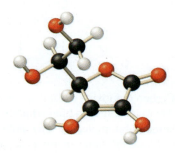

9.106 The molecular model of nicotine (a stimulant) is shown here. (a) Write the molecular formula of the compound. (b) What is the hybridization of each C and N atom? (c) Describe the geometry about each C and N atom.

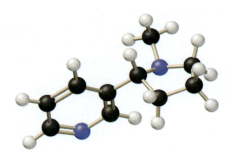

9.107 The compound TCDD, or 2,3,7,8-tetrachlorodibenzo-*p*-dioxin, is highly toxic:

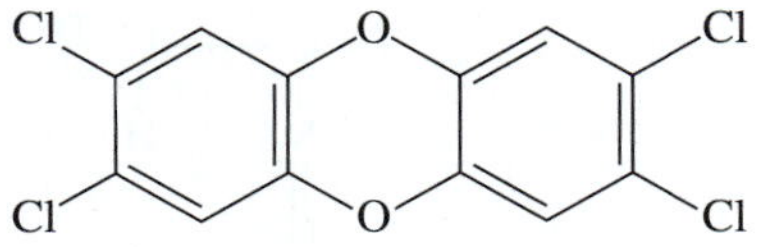

Cl, O, Cl
Cl, O, Cl

It gained considerable notoriety in 2004 when it was implicated in the attempted murder of a Ukrainian politician. (a) Describe its geometry, and state whether the molecule has a dipole moment. (b) How many pi bonds and sigma bonds are there in the molecule?

9.108 Progesterone is a hormone responsible for female sex characteristics. In the usual shorthand structure, each point where lines meet represents a C atom, and most H atoms are not shown. Draw the complete structure of the molecule, showing all C and H atoms. Indicate which C atoms are sp^2- and sp^3-hybridized.

9.109 Carbon monoxide (CO) is a poisonous compound due to its ability to bind strongly to Fe^{2+} in the hemoglobin molecule. The molecular orbitals of CO have the same energy order as those of the N_2 molecule. (a) Draw a Lewis structure of CO and assign formal charges. Explain why CO has a rather small dipole moment of 0.12 D. (b) Compare the bond order of CO with that from molecular orbital theory. (c) Which of the atoms (C or O) is more likely to form bonds with the Fe^{2+} ion in hemoglobin?

9.110 The compound 3′-azido-3′-deoxythymidine, commonly known as AZT, is one of the drugs used to treat AIDS. What are the hybridization states of the C and N atoms in this molecule?

9.111 The disulfide bond, $-S-S-$, plays an important role in determining the three-dimensional structure of proteins. Describe the nature of the bond and the hybridization state of the S atoms.

Environmental Problems

9.112 Greenhouse gases absorb (and trap) outgoing infrared radiation (heat) from Earth and contribute to global warming. A molecule of a greenhouse gas

either possesses a permanent dipole moment or has a changing dipole moment during its vibrational motions. Consider three of the vibrational modes of carbon dioxide

where the arrows indicate the movement of the atoms. (During a complete cycle of vibration, the atoms move toward one extreme position and then reverse their direction to the other extreme position.) Which of the preceding vibrations are responsible for CO_2 behaving as a greenhouse gas? Which of the following gases are greenhouse gases: N_2, O_2, O_3, CO, NO_2, N_2O, CH_4, $CFCl_3$?

9.113 The compound 1,2-dichloroethane ($C_2H_4Cl_2$) is nonpolar, while *cis*-dichloroethylene ($C_2H_2Cl_2$) has a dipole moment: The reason for the difference is that groups connected by a single bond can rotate with respect to each other, but no rotation occurs when a double bond connects the groups. On the basis of bonding considerations, explain why rotation occurs in 1,2-dichloroethane but not in *cis*-dichloroethylene.

1,2-dichloroethane   *cis*-dichloroethylene

Multiconcept Problems

9.114 Consider an N_2 molecule in its first excited electronic state, that is, when an electron in the highest occupied molecular orbital is promoted to the lowest empty molecular orbital. (a) Identify the molecular orbitals involved, and sketch a diagram to show the transition. (b) Compare the bond order and bond length of N_2^* with N_2, where the asterisk denotes the excited molecule. (c) Is N_2^* diamagnetic or paramagnetic? (d) When N_2^* loses its excess energy and converts to the ground state N_2, it emits a photon of wavelength 470 nm, which makes up part of the auroras' lights. Calculate the energy difference between these levels.

9.115 Imagine that sulfur dioxide can be prepared by combining elemental sulfur (S_8) and sulfur trioxide. Write a balanced equation for this hypothetical reaction and determine the oxidation state and the hybridization of sulfur in each species. What mass of sulfur trioxide would be required to combine with 1.00 kg of elemental sulfur; and what total mass of sulfur dioxide would result—assuming the reaction goes to completion?

Standardized-Exam Practice Problems

Physical Sciences

These questions are not based on a descriptive passage.

1. What is the shape of the ICl_3 molecule?

   a) Trigonal planar   c) T-shaped
   b) Trigonal pyramidal   d) Tetrahedral

2. Which of the following molecules is nonpolar?

   a) NCl_3 c) PCl_3
   b) BCl_3 d) $BrCl_3$

3. Which of the following has a bond order of 2?

   $$N_2^{2-} \qquad N_2 \qquad N_2^{2+}$$
   $$\text{I} \qquad\quad \text{II} \qquad \text{III}$$

   a) I only   c) III only
   b) II only   d) I and III

4. Which of the following are paramagnetic?

   $$O_2^{2-} \qquad O_2^{-} \qquad O_2^{+} \qquad O_2^{2+}$$
   $$\text{I} \qquad\quad \text{II} \qquad\quad \text{III} \qquad\quad \text{IV}$$

   a) I and II   c) III and IV
   b) II and III   d) I and III

Answers to In-Chapter Materials

Answers to Practice Problems

9.1A (a) linear, (b) bent. **9.1B** (a) Group 16, (b) Group 17.
9.2A Bent about O, tetrahedral about each C, trigonal pyramidal about N. All bond angles are ~109.5°. Angles labeled in blue are <109.5°.

$$\underset{\text{H}\ \ \text{H}\ \ \text{H}}{\overset{\text{H}\ \ \text{H}}{\text{H}-\ddot{\text{O}}-\text{C}-\text{C}-\ddot{\text{N}}-\text{H}}}$$

9.2B SO_2 contains double bonds, which are not pushed together by the central atom's lone pair as easily as the single bonds in NH_3. **9.3A** Singly occupied $3p$ orbitals from the P atom overlap with s orbitals from H atoms. **9.3B** We cannot use valence bond theory to explain the bonding in SO_2 or CH_4. In the case of SO_2, although the central atom has two unpaired electrons and can form two bonds, the unpaired electrons on S are in $3p$ orbitals. Formation of two bonds by the overlap of two $3p$ orbitals on S would be expected to result in a bond angle of approximately 90°. In the case of CH_4, the central atom does not have enough unpaired electrons to form four bonds. **9.4A** Two of the $4p$ electrons in Br are promoted to empty d orbitals. The s orbital, all three p orbitals, and two of the d orbitals hybridize to form six sp^3d^2 hybrid orbitals. One of the hybrid orbitals contains the lone pair. Each of the remaining hybrid orbitals contains one electron and overlaps with a singly occupied $2p$ orbital on an F atom. The arrangement of hybrid orbitals is octahedral and the bond angles are ~90°. **9.4B** One of the $2s$ electrons in Be is promoted to an empty p orbital. The s orbital and one p orbital hybridize to form two sp hybrid orbitals. Each hybrid orbital contains one electron and overlaps with a singly occupied $2p$ orbital on an F atom. The arrangement is linear with a bond angle of ~180°. **9.5A** 16 σ bonds and 4 π bonds. **9.5B** 17 σ bonds and 5 π bonds. **9.6A** C and N atoms are sp-hybridized. The triple bond between C and N is composed of one sigma bond (from overlap of hybrid orbitals) and two pi bonds (from interaction of remaining p orbitals). The single bond between H and C is the result of an sp orbital from C overlapping with an s orbital from H. **9.6B** Each N atom is sp-hybridized. One sp orbital on each is singly occupied and one contains a lone pair. The singly occupied sp orbitals overlap to form a sigma bond between the N atoms. The remaining unhybridized p orbitals interact to form two pi bonds. **9.7A** paramagnetic; bond order = 2. **9.7B** paramagnetic; bond order = 2. **9.8A** Two different resonance structures are possible; therefore we consider all three atoms to be sp^2-hybridized. One of the hybrid orbitals on the central O atom contains the lone pair, the other two form sigma bonds to the terminal O atoms. Each atom has one remaining unhybridized p orbital. The p orbitals combine to form π molecular orbitals. **9.8B** Two different resonance structures are possible; therefore we consider all three atoms to be sp^2-hybridized. One of the hybrid orbitals on the central N atom contains the lone pair, the other two form sigma bonds to the terminal O atoms. Each atom has one remaining unhybridized p orbital. The p orbitals combine to form π molecular orbitals.

Answers to Checkpoints

9.1.1 d. **9.1.2** b. **9.1.3** e. **9.1.4** c. **9.1.5** d. **9.2.1** a. **9.2.2** c.
9.3.1 b, e. **9.3.2** b. **9.4.1** b. **9.4.2** d. **9.5.1** a, c. **9.5.2** c. **9.5.3** a, b, e.
9.5.4 c. **9.6.1** b. **9.6.2** c, e. **9.6.3** b. **9.6.4** a, c, e. **9.7.1** c. **9.7.2** b.
9.7.3 a, b. **9.7.4** e.

CHAPTER 10

Gases

Scuba divers breathe a compressed mixture of gases. For shallow recreational diving, compressed air is generally used. For dives to greater depths, various mixtures of helium, nitrogen, and oxygen are used.

Corbis/VCG/Image 100/Getty Images

How the Properties of Gases Contribute to the Hazards of Scuba Diving

One of the first lessons taught in scuba certification is that divers must never hold their breath during ascent to the surface. Failure to heed this warning can result in serious injury or death. During underwater ascent, the air in a diver's lungs expands. If the air is not expelled, it causes overexpansion and rupture of alveoli—the tiny sacks that normally fill with air on inhalation. This condition, known as "burst lung," can cause air to escape into the chest cavity, where further expansion can collapse the ruptured lung. The potential for these catastrophic injuries is not limited to deep-sea divers, though, as burst lung can occur during a rapid ascent of as little as 3 meters, a depth common in public swimming pools. Furthermore, *spontaneous pneumothorax,* the medical term for this injury, can be caused even without the failure to exhale if there is an air-filled cyst in the diver's lung, or if there is a region of lung tissue blocked by phlegm due to a respiratory tract infection. In addition to the risk of spontaneous pneumothorax, burst lung can result in gas bubbles entering the bloodstream. Expansion of these bubbles during rapid underwater ascent can block circulation, a condition known as *gas embolism,* which can lead to heart attack or stroke.

Because the cause of burst lung in divers is rapid *de*compression, treatment for the resulting conditions, especially gas embolism, usually includes *re*compression in a hyperbaric chamber. Hyperbaric chambers are cylindrical enclosures built to withstand pressures significantly above atmospheric pressure. With the victim of a rapid decompression and, in some cases, medical personnel inside the enclosure, the door is sealed and high-pressure *air* is pumped in until the interior pressure is high enough to compress the gas bubbles in the victim's system. The victim usually breathes pure oxygen through a mask to help purge the undesirable gases during treatment. The pressure in a typical medical hyperbaric chamber can be increased to as many as six times atmospheric pressure.

Understanding the risks associated with scuba diving and the efforts to prevent and treat related injuries requires knowledge of the behavior of *gases.*

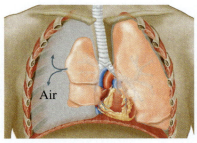

Pneumothorax

Monoplace hyperbaric chamber
ERproductions Ltd/Blend Images LLC

Student Note: "Monoplace" hyperbaric chambers, which are large enough to accommodate only one person, typically are pressurized with pure oxygen, although monoplace chambers are not used to treat decompression injuries.

At the end of this chapter, you will be able to solve a series of problems involving the properties of gases as they relate to the safety of athletes [►► Applying What You've Learned, page 514].

Figure 10.1 Solid, liquid, and gaseous states of a substance.

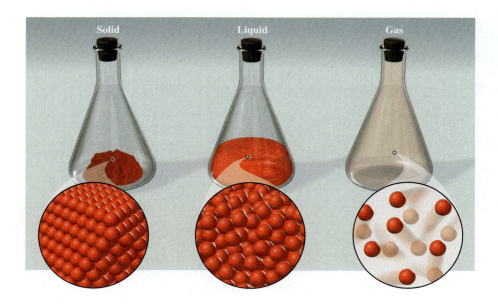

10.1   Properties of Gases

Recall from Chapter 1 that matter exists in one of three states: solid, liquid, or gas. In fact, most substances that are solid or liquid at room temperature (25°C) *can* exist as gases under appropriate conditions. Water, for instance, *evaporates* under the right conditions. Water vapor is a gas. (In general, the term *vapor* is used to refer to the gaseous state of a substance that is a liquid or solid at room temperature.) In this chapter we explore the nature of gases and how their properties at the molecular level give rise to the macroscopic properties that we observe. Figure 10.1 illustrates the three states of matter at the macroscopic level and at the molecular level.

Relatively few elements exist as gases at room temperature. Those that do are hydrogen, nitrogen, oxygen, fluorine, chlorine, and the noble gases. Of these, the noble gases exist as isolated atoms, whereas the others exist as diatomic molecules [◄◄ Section 2.7]. Figure 10.2 shows where the gaseous elements appear in the periodic table.

Many molecular compounds, most often those with low molar masses, exist as gases at room temperature. Table 10.1 lists some gaseous compounds that may be familiar to you.

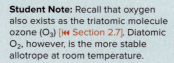

Student Note: Recall that oxygen also exists as the triatomic molecule ozone (O_3) [◄◄ Section 2.7]. Diatomic O_2, however, is the more stable allotrope at room temperature.

Characteristics of Gases

Gases differ from the condensed phases (solids and liquids) in the following important ways:

1. *A sample of gas assumes both the shape and volume of its container.* Like a liquid, a gas consists of particles (molecules or atoms) that do not have fixed positions in the sample [◄◄ Section 1.2]. As a result, both liquids and gases are

Figure 10.2 Elements that exist as gases at room temperature.

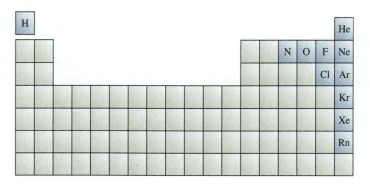

TABLE 10.1	Molecular Compounds That Are Gases at Room Temperature
Molecular Formula	**Compound Name**
HCl	Hydrogen chloride
NH_3	Ammonia
CO_2	Carbon dioxide
N_2O	Dinitrogen monoxide or nitrous oxide
CH_4	Methane
HCN	Hydrogen cyanide

able to *flow.* (Recall from Chapter 1 that we refer to liquids and gases collectively as *fluids.*) While a sample of liquid will assume the shape of the part of its container that it occupies, a sample of *gas* will expand to fill the entire *volume* of its container.

2. *Gases are compressible.* Unlike a solid or a liquid, a gas consists of particles with relatively large distances between them; that is, the distance between any two particles in a gas is much larger than the size of a molecule or atom. Because gas particles are far apart, it is possible to move them closer together by confining them to a smaller volume.

3. *The densities of gases are much smaller than those of liquids and solids and are highly variable depending on temperature and pressure.* The densities of gases are typically expressed in g/L, whereas those of liquids and solids are typically expressed in g/mL or g/cm^3.

    When we compress a sample of gas, we decrease its volume. Because its mass remains the same, the ratio of mass to volume (density) increases. Conversely, if we increase the volume to which a sample of gas is confined, we decrease its density.

    If you have ever seen a hot-air balloon aloft, you have seen a demonstration of how the density of a gas varies with temperature. Hot air is less dense than cold air, so hot air "floats" on cold air, much like oil floats on water.

4. *Gases form homogeneous mixtures (solutions) with one another in any proportion.* Some liquids (e.g., oil and water) do not mix with one another. Gases, on the other hand, because their particles are so far apart, do not interact with one another to any significant degree unless a chemical reaction takes place between them. This allows molecules of different gases to mix uniformly. That is, gases that don't react with each other are mutually *miscible.*

Each of these four characteristics is the result of the properties of gases at the molecular level.

Student Note: 1 mL = 1 cm^3 [◄◄ Section 1.3].

Gas Pressure: Definition and Units

A sample of gas confined to a container exerts a pressure on the walls of its container. For example, the air in the tires of your car exerts pressure on the inside walls of the tires. In fact, gases exert pressure on everything they touch. Thus, while you may add enough air to increase the pressure inside your tire to 32 pounds per square inch (psi), there is also a pressure of approximately 14.7 psi, called *atmospheric pressure,* acting on the outside of the tire—and on everything else, including your body. The reason you don't feel the pressure of the atmosphere pushing on the outside of your body is that an equal pressure exists inside your body so that there is no net pressure on you.

Atmospheric pressure, the pressure exerted by Earth's atmosphere, can be demonstrated using the empty metal container shown in Figure 10.3(a). Because the container is open to the atmosphere, atmospheric pressure acts on both the internal and external walls of the container. When we attach a vacuum pump to the opening of

Student Note: Common pencil-type pressure gauges actually measure the difference between internal and external pressure. Thus, if the tire is completely flat, the reading of 0 psi means that the pressure *inside* the tire is the same as that *outside* the tire.

Pencil-type tire gauge

David Tietz/Editorial Image, LLC

Figure 10.3 (a) An empty metal can.
(b) When the air is removed by a
vacuum pump, atmospheric pressure
crushes the can.

(both): *Charles D. Winters/Timeframe
Photography/McGraw Hill*

Student Note: Force is
mass × acceleration.

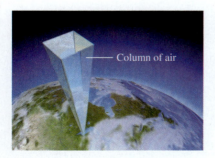

Figure 10.4 A column of air
1 cm × 1 cm from Earth's surface to
the top of the atmosphere weighs
approximately 1 kg.

Student Note: Remember that when
a unit is raised to a power, any
conversion factor you use must also be
raised to that power [◄◄ Section 1.6].

TABLE 10.2	Units of Pressure Commonly Used in Chemistry	
Unit	**Origin**	**Definition**
standard atmosphere (atm)	Pressure at sea level	$1\text{ atm} = 101{,}325\text{ Pa}$
mmHg	Barometer measurement	$1\text{ mmHg} = 133.222\text{ Pa}$
torr	Name given to mmHg in honor of Torricelli, the inventor of the barometer	$1\text{ torr} = 133.322\text{ Pa}$
bar	Same order of magnitude as atm, but a decimal multiple of Pa	$1\text{ bar} = 1 \times 10^5\text{ Pa}$

the container and draw air out of it, however, we reduce the pressure inside the container. When the pressure against the interior walls is reduced, atmospheric pressure crushes the container [Figure 10.3(b)].

   Pressure is defined as the force applied per unit area:

$$\text{pressure} = \frac{\text{force}}{\text{area}}$$

The SI unit of force is the ***newton (N),*** where

$$1\text{ N} = 1\text{ kg} \cdot \text{m/s}^2$$

The SI unit of pressure is the ***pascal (Pa),*** defined as 1 newton per square meter.

$$1\text{ Pa} = 1\text{ N/m}^2$$

In SI base units, $1\text{ Pa} = 1\text{ kg/m} \cdot \text{s}^2$. Although the pascal is the SI unit of pressure, there are other units of pressure that are more commonly used. Table 10.2 lists the units in which pressure is most commonly expressed in chemistry and their definitions in terms of pascals. Which of these units you encounter most often will depend on your specific field of study. Use of *atmospheres* (atm) is common in chemistry, although use of the *bar* is becoming increasingly common. Use of *millimeters mercury* (mmHg) is common in medicine and meteorology. We use all these units in this text.

Calculation of Pressure

The force experienced by an area exposed to Earth's atmosphere is equal to the weight of the column of air above it. For example, the mass of air above a spot on the ground near sea level, with an area of 1 cm², is approximately 1 kg (Figure 10.4).

   The weight of an object that is subject to Earth's gravitational pull is equal to its mass times the gravitational constant, 9.80665 m/s². Thus, the force exerted by this column of air is

$$1\text{ kg} \times \frac{9.80665\text{ m}}{\text{s}^2} \approx 10\text{ kg} \cdot \text{m/s}^2 = 10\text{ N}$$

Pressure, though, is force per unit area. Specifically, pressure in *pascals* is equal to force in *newtons* per *square meter*. We must first convert area from cm² to m²,

$$1\text{ cm}^2 \times \left(\frac{1\text{ m}}{100\text{ m}}\right)^2 = 0.0001\text{ m}^2$$

and then divide force by area,

$$\frac{10\text{ N}}{0.0001\text{ m}^2} = 1 \times 10^5\text{ Pa}$$

This pressure is roughly equal to 1 atm (~1×10^5 Pa), which we would expect at sea level.

We can calculate the pressure exerted by a column of any fluid (gas or liquid) in the same way. In fact, this is how atmospheric pressure is commonly measured—by determining the height of a column of mercury it can support.

Measurement of Pressure

A simple *barometer,* an instrument used to measure atmospheric pressure, consists of a long glass tube, closed at one end and filled with mercury. The tube is carefully inverted in a container of mercury so that no air enters the tube. When the tube is inverted, and the open end is submerged in the mercury in the container, some of the mercury in the tube will flow out into the container, creating an empty space at the top (closed end) of the tube (Figure 10.5). The weight of the mercury remaining in the tube is supported by atmospheric pressure pushing down on the surface of the mercury in the container. In other words, the pressure exerted by the column of mercury is *equal* to the pressure exerted by the atmosphere. *Standard atmospheric pressure* (1 atm) was originally defined as the pressure that would support a column of mercury exactly 760 mm high at 0°C at sea level. The mmHg unit is also called the torr, after the Italian scientist Evangelista Torricelli,[1] who invented the barometer. Standard atmospheric pressure is given in all the common units of pressure in the Student Note.

A *manometer* is a device used to measure pressures other than atmospheric pressure. The principle of operation of a manometer is similar to that of a barometer. There are two types of manometers, both of which are shown in Figure 10.6. The closed-tube manometer [Figure 10.6(a)] is normally used to measure pressures below atmospheric pressure, whereas the open-tube manometer [Figure 10.6(b)] is generally used to measure pressures equal to or greater than atmospheric pressure.

The pressure exerted by a column of fluid, such as that in a barometer (Figure 10.5), is given by Equation 10.1.

$$P = hdg \qquad \text{**Equation 10.1**}$$

where h is the height of the column in meters, d is the density of the fluid in kg/m^3, and g is the gravitational constant equal to 9.80665 m/s^2. This equation explains why barometers historically have been constructed using mercury. The height of a column

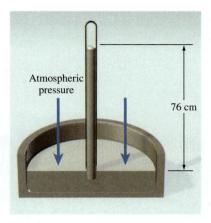

Figure 10.5 Barometer.

Student Note:
1 atm*
101,325 Pa
760 mmHg*
760 torr*
1.01325 bar
14.7 psi
*These are exact numbers.

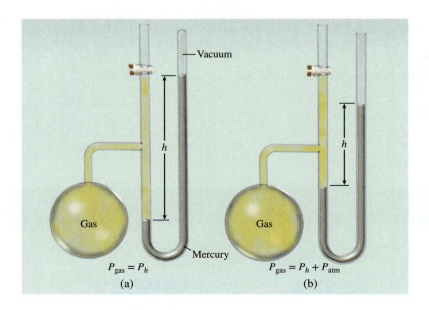

Vacuum

Gas

h

$P_{gas} = P_h$

Mercury

(a)

Gas

h

$P_{gas} = P_h + P_{atm}$

(b)

Figure 10.6 (a) Closed-tube manometer. The space labeled "vacuum" actually contains a small amount of mercury vapor. (b) Open-tube manometer.

1. Evangelista Torricelli (1608–1647). Italian mathematician. Torricelli was supposedly the first person to recognize the existence of atmospheric pressure.

of fluid supported by a given pressure is inversely proportional to the density of the fluid. (At a given *P*, as *d* goes up, *h* must go down—and vice versa.) Mercury's high density made it possible to construct barometers and manometers of manageable size. For example, a barometer filled with mercury that stands 1 m tall would have to be over 13 m tall if it were filled with water.

Sample Problem 10.1 shows how to calculate the pressure exerted by a column of fluid.

SAMPLE PROBLEM 10.1

Calculate the pressure exerted by a column of mercury 70.0 cm high. Express the pressure in pascals and in atmospheres. The density of mercury is 13.5951 g/cm^3.

Strategy Use Equation 10.1 to calculate pressure. Remember that height must be expressed in meters and density must be expressed in kg/m^3.

Setup

$$h = 70.0 \text{ cm} \times \frac{1 \text{ m}}{100 \text{ cm}} = 0.700 \text{ m}$$

$$d = \frac{13.5951 \text{ g}}{\text{cm}^3} \times \frac{1 \text{ kg}}{1000 \text{ g}} \times \left(\frac{100 \text{ cm}}{1 \text{ m}}\right)^3 = 1.35951 \times 10^4 \text{ kg/m}^3$$

$$g = 9.80665 \text{ m/s}^2$$

Solution

$$\text{pressure} = 0.700 \text{ m} \times \frac{1.35951 \times 10^4 \text{ kg}}{\text{m}^3} \times \frac{9.80665 \text{ m}}{\text{s}^2} = 9.33 \times 10^4 \text{ kg/m} \cdot \text{s}^2 = 9.33 \times 10^4 \text{ Pa}$$

$$9.33 \times 10^4 \text{ Pa} \times \frac{1 \text{ atm}}{101,325 \text{ Pa}} = 0.921 \text{ atm}$$

THINK ABOUT IT

Make sure your units cancel properly in this type of problem. Common errors include forgetting to express height in meters and density in kg/m^3. You can avoid these errors by becoming familiar with the value of atmospheric pressure in the various units. A column of mercury slightly less than 760 mm is equivalent to slightly less than 101,325 Pa and slightly less than 1 atm.

Practice Problem **A**TTEMPT What pressure (in atm) is exerted by a column of mercury exactly 1 m high?

Practice Problem **B**UILD What would be the height of a column of water supported by the pressure you calculated in Sample Problem 10.1? Assume that the density of the water is 1.00 g/cm^3.

Practice Problem **C**ONCEPTUALIZE Arrange the four columns of liquid in order of increasing pressure they exert.

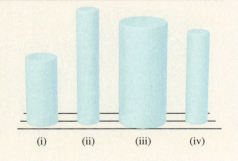

(i)   (ii)   (iii)   (iv)

CHECKPOINT – SECTION 10.1   Properties of Gases

10.1.1 Express a pressure of 1.15 atm in units of bar.

a) 1.17 bar

b) 1.13 bar

c) 0.881 bar

d) 874 bar

e) 1.51 × 10^{-3} bar

10.1.2 Which of the following is true? (Select all that apply.)

a) 0.80 atm = 0.80 torr

b) 4180 mmHg = 5.573 × 10^5 Pa

c) 433 torr = 433 mmHg

d) 2.300 atm = 1748 torr

e) 5.5 atm = 1.0 × 10^5 Pa

10.1.3 Calculate the height of a column of ethanol that would be supported by atmospheric pressure (1 atm). The density of ethanol is 0.789 g/cm³.

a) 1.31×10^4 m

d) 13.1 m

b) 0.789 m

e) 780 mm

c) 600 mm

10.1.4 What pressure is exerted by a column of water 50.0 m high? Assume the density of the water is 1.00 g/cm³.

a) 490 atm

b) 4.84 atm

c) 0.087 atm

d) 50 atm

e) 1.62 atm

10.1.5 How do the pressures exerted by the following columns of water compare?

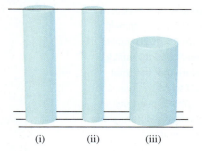

(i)   (ii)   (iii)

a) i = ii < iii

d) ii < i = iii

b) i < iii < ii

e) iii < i = ii

c) iii < i < ii

10.2   The Gas Laws

Animation
Gas laws.

In contrast to the condensed phases, all gases, even those with vastly different chemical compositions, exhibit remarkably similar physical behavior. Numerous experiments carried out in the seventeenth and eighteenth centuries showed that the physical state of a sample of gas can be described completely with just four parameters: temperature (T), pressure (P), volume (V), and number of moles (n). Knowing any three of these parameters enables us to calculate the fourth. The relationships between these parameters are known as the **gas laws**.

Student Note: Because they arise from experiment, these laws are referred to as the *empirical* gas laws.

Boyle's Law: The Pressure-Volume Relationship

Imagine that you have a plastic syringe filled with air. If you hold your finger tightly against the tip of the syringe and push the plunger with your other hand, decreasing the volume of the air, you will increase the pressure in the syringe. During the seventeenth century Robert Boyle[2] conducted systematic studies of the relationship between gas volume and pressure using a simple apparatus like the one shown in Figure 10.7. The J-shaped tube contains a sample of gas confined by a column of mercury. The apparatus functions as an open-end manometer. When the mercury levels on both sides are equal [Figure 10.7(a)], the pressure of the confined gas is equal to atmospheric pressure. When more mercury is added through the open end, the pressure of the confined gas is increased by an amount proportional to the height of the added mercury—and the volume of the gas decreases. If, for example, as shown in Figure 10.7(b), we *double* the pressure on the confined gas by adding enough mercury to make the difference in mercury levels on the left and right 760 mm (the height of a mercury column that exerts a pressure equal to 1 atm), the volume of the gas is reduced by *half*. If we triple the original pressure on the confined gas by adding more mercury, the volume of the gas is reduced to one-third of its original volume [Figure 10.7(c)].

Table 10.3 gives a set of data typical of Boyle's experiments. Figure 10.8 shows some of the volume data plotted (a) as a function of pressure and (b) as a function of the inverse of pressure, respectively. These data illustrate **Boyle's law,** which states that the pressure of a fixed amount of gas at a constant temperature is inversely proportional to the volume of the gas. This inverse relationship between pressure and volume can be expressed mathematically as follows:

$$V \propto \frac{1}{P}$$

Student Note: Remember that the symbol $\propto$ means "is proportional to."

---

2. Robert Boyle (1627–1691). British chemist and natural philosopher. Although Boyle is commonly associated with the gas law that bears his name, he made many other significant contributions to the fields of chemistry and physics.

Figure 10.7 Demonstration of Boyle's law. The volume of a sample of gas is inversely proportional to its pressure. (a) $P = 760$ mmHg, $V = 100$ mL. (b) $P = 1520$ mmHg, $V = 50$ mL. (c) $P = 2280$ mmHg, $V = 33$ mL. Note that the total pressure exerted on the gas is the sum of atmospheric pressure (760 mmHg) and the difference in height of the mercury.

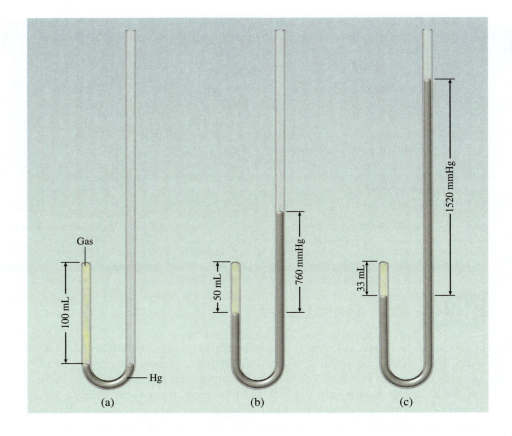

TABLE 10.3	Typical Data from Experiments with the Apparatus of Figure 10.7									
P(mmHg)	760	855	950	1045	1140	1235	1330	1425	1520	2280
V(mL)	100	89	78	72	66	59	55	54	50	33
	Shown in Figure 10.7(a)								Shown in Figure 10.7(b)	Shown in Figure 10.7(c)

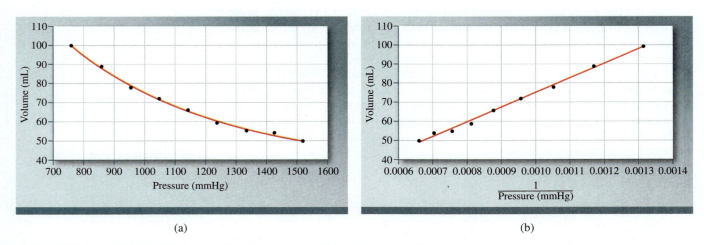

(a)   (b)

Figure 10.8 Plots of volume (a) as a function of pressure and (b) as a function of 1/pressure.

or

$$V = k_1 \frac{1}{P} \quad \text{(at constant temperature)} \qquad \textbf{Equation 10.2(a)}$$

where k_1 is a *proportionality constant*. We can rearrange Equation 10.2(a) to get

$$PV = k_1 \quad \text{(at constant temperature)} \qquad \textbf{Equation 10.2(b)}$$

According to this form of Boyle's law, the *product* of the pressure and the volume of a given sample of gas (at constant temperature) is a *constant*.

Although the individual values of pressure and volume can vary greatly for a given sample of gas, the product of P and V is always equal to the same constant as long as the temperature is held constant and the amount of gas does not change. Therefore, for a given sample of gas under two different sets of conditions at constant temperature, we can write

$$P_1 V_1 = k_1 = P_2 V_2$$

or

$$P_1 V_1 = P_2 V_2 \quad \text{(at constant temperature)} \qquad \textbf{Equation 10.3}$$

where V_1 is the volume at pressure P_1 and V_2 is the volume at pressure P_2.

Sample Problem 10.2 illustrates the use of Boyle's law.

SAMPLE PROBLEM 10.2

If a skin diver takes a breath at the surface, filling his lungs with 5.82 L of air, what volume will the air in his lungs occupy when he dives to a depth where the pressure is 1.92 atm? (Assume constant temperature and that the pressure at the surface is exactly 1 atm.)

Strategy Use Equation 10.3 to solve for V_2.

Setup $P_1 = 1.00$ atm, $V_1 = 5.82$ L, and $P_2 = 1.92$ atm.

Solution

$$V_2 = \frac{P_1 \times V_1}{P_2} = \frac{1.00 \text{ atm} \times 5.82 \text{ L}}{1.92 \text{ atm}} = 3.03 \text{ L}$$

THINK ABOUT IT

At higher pressure, the volume should be smaller. Therefore, the answer makes sense.

Practice Problem A TTEMPT Calculate the volume of a sample of gas at 5.75 atm if it occupies 5.14 L at 2.49 atm. (Assume constant temperature.)

Practice Problem B UILD At what pressure would a sample of gas occupy 7.86 L if it occupies 3.44 L at 4.11 atm? (Assume constant temperature.)

Practice Problem C ONCEPTUALIZE Which of the following diagrams could represent a gas sample in a balloon at constant temperature before and after an increase in external pressure?

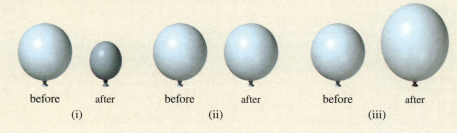

before   after   before   after   before   after

(i)   (ii)   (iii)

Animation
Gas laws—Charles's law.

(a)

(b)

Figure 10.9 (a) Air-filled balloon. (b) Lowering the temperature with liquid nitrogen causes a volume decrease. The pressure inside the balloon, which is roughly equal to the external pressure, remains constant in this process.

(both): *Charles D. Winters/Timeframe Photography/McGraw Hill*

Student Note: Remember that a kelvin and a degree Celsius have the same magnitude. Thus, while we add 273.15 to the temperature in °C to get the temperature in K, a *change* in temperature in Celsius is *equal* to the change in temperature in K. A temperature of 20°C is the same as 293.15 K. A *change* in temperature of 20°C, however, is the same as a *change* in temperature of 20 K.

Charles's and Gay-Lussac's Law: The Temperature-Volume Relationship

If you took a helium-filled Mylar balloon outdoors on a cold day, the balloon would shrink somewhat when it came into contact with the cold air. This would occur because the volume of a sample of gas depends on the temperature. A more dramatic illustration is shown in Figure 10.9, where liquid nitrogen is being poured over an air-filled balloon. The large drop in temperature of the air in the balloon (the boiling liquid nitrogen has a temperature −196°C) results in a significant decrease in its volume, causing the balloon to shrink. Note that the pressure inside the balloon is roughly equal to the external pressure.

The first to study the relationship between gas volume and temperature were French scientists Jacques Charles[3] and Joseph Gay-Lussac.[4] Their studies showed that, at constant pressure, the volume of a gas sample increases when heated and decreases when cooled. Figure 10.10(a) shows a plot of data typical of Charles's and Gay-Lussac's experiments. Note that with pressure held constant, the volume of a sample of gas plotted as a function of temperature yields a straight line. These experiments were carried out at several different pressures [Figure 10.10(b)], each yielding a different straight line. Interestingly, if the lines are extrapolated to zero volume, they all meet at the *x* axis at the temperature −273.15°C. The implication is that a gas sample occupies zero volume at −273.15°C. This is not observed in practice, however, because all gases condense to form liquids or solids before −273.15°C is reached.

In 1848, Lord Kelvin[5] realized the significance of the extrapolated lines all meeting at −273.15°C. He identified −273.15°C as **absolute zero,** theoretically the lowest attainable temperature. Then he set up an **absolute temperature scale,** now called the **Kelvin temperature scale,** with absolute zero as the lowest point [◄◄ Section 1.3]. On the Kelvin scale, 1 kelvin (K) is equal in magnitude to 1 degree Celsius. The difference is simply an offset of 273.15. We obtain the absolute temperature by adding 273.15 to the temperature expressed in Celsius, although we often use simply 273 instead of *273.15*. Several important points on the two scales match up as follows:

	Kelvin Scale (K)	**Celsius Scale (°C)**
Absolute zero	0 K	−273.15°C
Freezing point of water	273.15 K	0°C
Boiling point of water	373.15 K	100°C

The dependence of the volume of a sample of gas on temperature is given by

$$V \propto T$$

or

Equation 10.4(a) $V = k_2 T$ (at constant pressure)

where k_2 is the proportionality constant. We can rearrange Equation 10.4(a) to get

Equation 10.4(b) $\dfrac{V}{T} = k_2$ (at constant pressure)

---

3. Jacques Alexandre Cesar Charles (1746–1823). French physicist. Charles was a gifted lecturer, an inventor of scientific apparatus, and the first person to use hydrogen to inflate balloons.

4. Joseph Louis Gay-Lussac (1778–1850). French chemist and physicist. Like Charles, Gay-Lussac was a balloon enthusiast. He once ascended to an altitude of 20,000 ft to collect air samples for analysis.

5. William Thomson, Lord Kelvin (1824–1907). Scottish mathematician and physicist. Kelvin did important work in many branches of physics.

Student Note: Don't forget that volume is proportional to *absolute* temperature. The volume of a sample of gas at constant pressure doubles if the temperature increases from 100 K to 200 K—but *not* if the temperature increases from 100°C to 200°C!

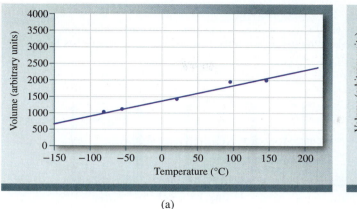
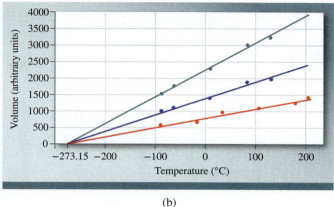

(a)   (b)

Figure 10.10 (a) Plot of the volume of a sample of gas as a function of temperature. (b) Plot of the volume of a sample of gas as a function of temperature at three different pressures.

Equations 10.4(a) and (b) are expressions of *Charles's and Gay-Lussac's law,* often referred to simply as **Charles's law,** which states that the volume of a fixed amount of gas maintained at constant pressure is directly proportional to the absolute temperature of the gas.

Just as we did with the pressure-volume relationship at constant temperature, we can compare two sets of volume-temperature conditions for a given sample of gas at constant pressure. From Equation 10.4 we can write

$$\frac{V_1}{T_1} = k_2 = \frac{V_2}{T_2}$$

or

$$\frac{V_1}{T_1} = \frac{V_2}{T_2} \quad \text{(at constant pressure)} \qquad \textbf{Equation 10.5}$$

where V_1 is the volume of the gas at T_1 and V_2 is the volume of the gas at T_2.

Sample Problem 10.3 shows how to use Charles's law.

SAMPLE PROBLEM 10.3

A sample of argon gas that originally occupied 14.6 L at 25.0°C was heated to 50.0°C at constant pressure. What is its new volume?

Strategy Use Equation 10.5 to solve for V_2. Remember that temperatures must be expressed in kelvin.

Setup $T_1 = 298.15$ K, $V_1 = 14.6$ L, and $T_2 = 323.15$ K.

Solution

$$V_2 = \frac{V_1 \times T_2}{T_1} = \frac{14.6 \text{ L} \times 323.15 \text{ K}}{298.15 \text{ K}} = 15.8 \text{ L}$$

THINK ABOUT IT

When temperature increases at constant pressure, the volume of a gas sample increases.

Practice Problem A TTEMPT A sample of gas originally occupies 29.1 L at 0.0°C. What is its new volume when it is heated to 15.0°C? (Assume constant pressure.)

Practice Problem B UILD At what temperature (in °C) will a sample of gas occupy 82.3 L if it occupies 50.0 L at 75.0°C? (Assume constant pressure.)

(Continued on next page)

Practice Problem C ONCEPTUALIZE A sample of gas at 50°C is contained in a cylinder with a movable piston shown below (far left). Which of the other diagrams [(i)–(iv)] best represents the system when the temperature of the sample has been increased to 100°C?

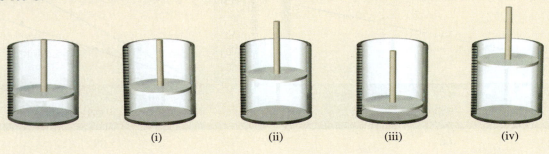

(i)   (ii)   (iii)   (iv)

Avogadro's Law: The Amount-Volume Relationship

In 1811, the Italian scientist Amedeo Avogadro proposed that equal volumes of different gases contain the same number of particles (molecules or atoms) at the same temperature and pressure. This hypothesis gave rise to *Avogadro's law,* which states that the volume of a sample of gas is directly proportional to the number of moles in the sample at constant temperature and pressure:

$$V \propto n$$

or

Equation 10.6(a) $V = k_3 n$ (at constant temperature and pressure)

Rearranging Equation 10.6(a) gives

Equation 10.6(b) $\dfrac{V}{n} = k_3$

Equations 10.6(a) and (b) are expressions of Avogadro's law.

As with the other gas laws, we can compare two sets of conditions using Avogadro's law where n and V both change at constant pressure and temperature and write

$$\frac{V_1}{n_1} = k_3 = \frac{V_2}{n_2}$$

or

Equation 10.7 $\dfrac{V_1}{n_1} = \dfrac{V_2}{n_2}$

where V_1 is the volume of a sample of gas consisting of n_1 moles and V_2 is the volume of a sample consisting of n_2 moles—under conditions of constant temperature and pressure. Coupled with a balanced chemical equation, Avogadro's law enables us to predict the volumes of gaseous reactants and products. Consider the reaction of H_2 and N_2 to form NH_3:

$$3H_2(g) + N_2(g) \longrightarrow 2NH_3(g)$$

The balanced equation reveals the ratio of combination of reactants in terms of *moles* [◄◄ Section 3.4]. However, because the volume of a gas (at a given temperature and pressure) is directly *proportional* to the number of moles, the balanced equation also reveals the ratio of combination in terms of *volume*. Thus, if we were to combine three volumes (liters, milliliters, etc.) of hydrogen gas with one volume of nitrogen gas, assuming they react completely according to the balanced equation, we would expect two volumes of ammonia gas to be produced (Figure 10.11). The ratio of combination of H_2 and N_2 (and production of NH_3), whether expressed in moles or units of volume, is 3:1:2.

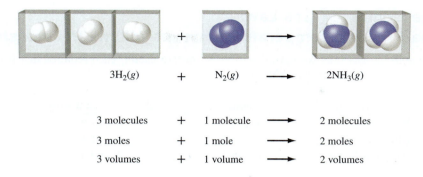

Figure 10.11 Illustration of Avogadro's law. The volume of a sample of gas is directly proportional to the number of moles.

$$3H_2(g) + N_2(g) \longrightarrow 2NH_3(g)$$

3 molecules	+	1 molecule	$\longrightarrow$	2 molecules
3 moles	+	1 mole	$\longrightarrow$	2 moles
3 volumes	+	1 volume	$\longrightarrow$	2 volumes

Sample Problem 10.4 shows how to apply Avogadro's law.

SAMPLE PROBLEM 10.4

If we combine 3.0 L of NO and 1.5 L of O_2, and they react according to the balanced equation $2NO(g) + O_2(g) \longrightarrow 2NO_2(g)$, what volume of NO_2 will be produced? (Assume that the reactants and product are all at the same temperature and pressure.)

Strategy Apply Avogadro's law to determine the volume of a gaseous product.

Setup Because volume is proportional to the number of moles, the balanced equation determines in what volume ratio the reactants combine and the ratio of product volume to reactant volume. The amounts of reactants given are stoichiometric amounts [◄◄ Section 3.6].

Solution According to the balanced equation, the volume of NO_2 formed will be equal to the volume of NO that reacts. Therefore, 3.0 L of NO_2 will form.

THINK ABOUT IT

Remember that the coefficients in balanced chemical equations indicate ratios in molecules or moles. Under conditions of constant temperature and pressure, the volume of a gas is proportional to the number of moles. Therefore, the coefficients in balanced equations containing only gases also indicate ratios in *liters,* provided the reactions occur at constant temperature and pressure. It is important to recognize that coefficients indicate ratios in liters only in balanced equations in which all the reactants and products are gases. We cannot apply the same approach to reactions in which there are solid, liquid, or aqueous species.

Practice Problem **A**TTEMPT What volume (in liters) of water vapor will be produced when 34 L of H_2 and 17 L of O_2 react according to the equation $2H_2(g) + O_2(g) \longrightarrow 2H_2O(g)$?

Practice Problem **B**UILD What volumes (in liters) of carbon monoxide and oxygen gas must react according to the equation $2CO(g) + O_2(g) \longrightarrow 2CO_2(g)$ to form 3.16 L of carbon dioxide?

Practice Problem **C**ONCEPTUALIZE A hypothetical gaseous reaction is depicted here with molecular models. Imagine that this reaction takes place in a cylinder with a movable piston at constant temperature and pressure, and that the diagram below (on the far left) represents the reaction mixture before the reaction. Which of the diagrams [(i)–(iv)] best represents the system when the reaction is complete? (Assume that reactants are combined in stoichiometric amounts. Note that the reaction depicted is not balanced.)

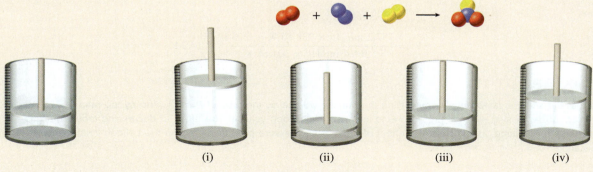

The Combined Gas Law:
The Pressure-Temperature-Amount-Volume Relationship

Although the gas laws we have discussed so far are useful, each requires that two of the system's parameters be held constant.

Problem Type	Relates	Requires Constant
Boyle's law	P and V	n and T
Charles's law	T and V	n and P
Avogadro's law	n and V	P and T

Many of the problems we will encounter involve changes to P, T, and V, and, in some cases, also to n. To solve such problems, we need a gas law that relates all the variables. By combining Equations 10.3, 10.5, and 10.7, we obtain the *combined gas law:*

Equation 10.8(a)
$$\frac{P_1 V_1}{n_1 T_1} = \frac{P_2 V_2}{n_2 T_2}$$

The combined gas law can be used to solve problems where any or all of the variables change. Note that when a problem involves a fixed quantity of gas, Equation 10.8(a) reduces to the more common form of the combined gas law:

Equation 10.8(b)
$$\frac{P_1 V_1}{T_1} = \frac{P_2 V_2}{T_2}$$

Sample Problem 10.5 illustrates the use of the combined gas law.

SAMPLE PROBLEM 10.5

If a child releases a 6.25-L helium balloon in the parking lot of an amusement park where the temperature is 28.50°C and the air pressure is 757.2 mmHg, what will the volume of the balloon be when it has risen to an altitude where the temperature is −34.35°C and the air pressure is 366.4 mmHg?

Strategy In this case, because there is a fixed amount of gas, we use Equation 10.8(b). The only value we don't know is V_2. Temperatures must be expressed in kelvins. We can use any units of pressure, as long as we are consistent.

Setup $T_1 = 301.65$ K, $T_2 = 238.80$ K. Solving Equation 10.8(b) V_2 gives

$$V_2 = \frac{P_1 T_2 V_1}{P_2 T_1}$$

Solution

$$V_2 = \frac{757.2 \text{ mmHg} \times 238.80 \text{ K} \times 6.25 \text{ L}}{366.4 \text{ mmHg} \times 301.65 \text{ K}} = 10.2 \text{ L}$$

THINK ABOUT IT

Note that the solution is essentially multiplying the original volume by the ratio of P_1 to P_2, and by the ratio of T_2 to T_1. The effect of decreasing external pressure is to increase the balloon volume. The effect of decreasing *temperature* is to *decrease* the volume. In this case, the effect of decreasing pressure predominates and the balloon volume increases significantly.

Practice Problem **A**TTEMPT   What would be the volume of the balloon in Sample Problem 10.5 if, instead of being released to rise in the atmosphere, it were submerged in a swimming pool to a depth where the pressure is 922.3 mmHg and the temperature is 26.35°C?

Practice Problem **B**UILD   The volume of a bubble that starts at the bottom of a lake at 4.55°C increases by a factor of 10 as it rises to the surface where the temperature is 18.45°C and the air pressure is 0.965 atm. Assuming that the density of the lake water is 1.00 g/cm³, determine the depth of the lake. (*Hint:* You will need to use Equation 10.1.)

Practice Problem **C**ONCEPTUALIZE   Which of the following diagrams could represent a gas sample in a balloon before and after an increase in temperature and an increase in external pressure?

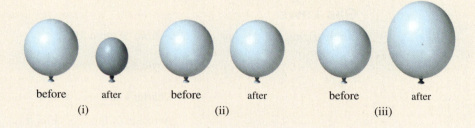

before	after	before	after	before	after
(i)		(ii)		(iii)	

CHECKPOINT – SECTION 10.2   The Gas Laws

10.2.1   Given $P_1 = 1.50$ atm, $V_1 = 37.3$ mL, and $P_2 = 1.18$ atm, calculate V_2. Assume that n and T are constant.

a) 0.0211 mL

b) 0.0341 mL

c) 29.3 mL

d) 12.7 mL

e) 47.4 mL

10.2.2   Given $T_1 = 21.5°C$, $V_1 = 50.0$ mL, and $T_2 = 316°C$, calculate V_2. Assume that n and P are constant.

a) 100 mL

b) 73.5 mL

c) 25.0 mL

d) 3.40 mL

e) 26.5 mL

10.2.3   At what temperature will a gas sample occupy 100.0 L if it originally occupies 76.1 L at 89.5°C? Assume constant P.

a) 276°C

b) 118°C

c) 203°C

d) 68.1°C

e) 99.6°C

10.2.4   What volume of NH_3 will be produced when 180 mL of H_2 reacts with 60.0 mL of N_2 according to the following equation:

$$3H_2(g) + N_2(g) \longrightarrow 2NH_3(g)$$

Assume constant T and P for reactants and products.

a) 120 mL

b) 60 mL

c) 180 mL

d) 240 mL

e) 220 mL

10.2.5   Which diagram could represent the result of increasing the temperature and decreasing the external pressure on a fixed amount of gas in a balloon?

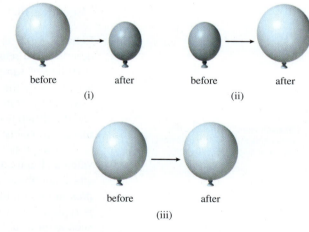

before	after	before	after
(i)		(ii)	

before	after
(iii)	

a) i only

b) ii only

c) i and ii

d) i, ii, and iii

e) iii only

10.2.6   Which diagram in question 10.2.5 could represent the result of decreasing both the temperature and the pressure?

a) i only

b) ii only

c) i and ii

d) i, ii, and iii

e) iii only

10.3   The Ideal Gas Equation

Recall that the state of a sample of gas is described completely using the four variables *T*, *P*, *V*, and *n*. Each of the gas laws introduced in Section 10.2 relates one variable of a sample of gas to another while the *other* two variables are held constant. In experiments with gases, however, there are usually changes in more than just two of the variables. Therefore, it is useful for us to combine the equations representing the gas laws into a single equation that will enable us to account for changes in any or all of the four variables.

Deriving the Ideal Gas Equation from the Empirical Gas Laws

Summarizing the gas law equations from Section 10.2:

$$\text{Boyle's law:} \quad V \propto \frac{1}{P}$$

$$\text{Charles's law:} \quad V \propto T$$

$$\text{Avogadro's law:} \quad V \propto n$$

We can combine these equations into the following general equation that describes the physical behavior of all gases:

$$V \propto \frac{nT}{P}$$

or

$$V = R\frac{nT}{P}$$

where *R* is the proportionality constant. This equation can be rearranged to give:

Equation 10.9	$PV = nRT$

Equation 10.9 is the most commonly used form of the **ideal gas equation,** which describes the relationship among the four variables *P*, *V*, *n*, and *T*. An **ideal gas** is a hypothetical sample of gas whose pressure-volume-temperature behavior is predicted accurately by the ideal gas equation. Although the behavior of *real* gases generally differs slightly from that predicted by Equation 10.9, in most of the cases we will encounter, the differences are usually small enough for us to use the ideal gas equation to make reasonably good predictions about the behavior of *gases*.

Student Note: We discuss the conditions that result in deviation from ideal behavior in Section 10.7.

The proportionality constant, *R*, in Equation 10.9 is called the **gas constant.** Its value and units depend on the units in which *P* and *V* are expressed. (The variables *n* and *T* are always expressed in mol and K, respectively.) Recall from Section 10.1 that pressure is commonly expressed in atmospheres, mmHg (torr), pascals, or bar. Volume is typically expressed in liters or milliliters, but can also be expressed in other units, such as m^3. Table 10.4 lists several different expressions of the gas constant, *R*.

TABLE 10.4	Various Equivalent Expressions of the Gas Constant, *R*
Numerical Value	**Unit**
0.08206	$L \cdot atm/K \cdot mol$
62.36	$L \cdot torr/K \cdot mol$
0.08314	$L \cdot bar/K \cdot mol$
8.314	$m^3 \cdot Pa/K \cdot mol$
8.314	$J/K \cdot mol$
1.987	$cal/K \cdot mol$

Note that the product of volume and pressure gives units of *energy* (i.e., joules and calories).

Keep in mind that all these expressions of R are equal to one another, just as 1 yard is equal to 3 ft. They are simply expressed in different units.

One of the simplest uses of the ideal gas equation is the calculation of one of the variables when the other three are already known. For example, we can calculate the volume of 1 mole of an ideal gas at 0°C and 1 atm, conditions known as ***standard temperature and pressure (STP).*** In this case, n, T, and P are given. R is a constant, leaving V as the only unknown. We can rearrange Equation 10.9 to solve for V,

$$V = \frac{nRT}{P}$$

enter the information that is given, and calculate V. Remember that in calculations using the ideal gas equation, temperature must *always* be expressed in kelvins.

$$V = \frac{(1 \text{ mol})(0.08206 \text{ L} \cdot \text{atm/K} \cdot \text{mol})(273.15 \text{ K})}{1 \text{ atm}} = 22.41 \text{ L}$$

Thus, the volume occupied by 1 mole of an ideal gas at STP is 22.41 L, a volume slightly less than 6 gal.

Sample Problem 10.6 shows how to calculate the molar volume of a gas at a temperature other than 0°C.

> **Student Note:** In thermochemistry we often used 25°C as the "standard" temperature—although temperature is *not* actually part of the definition of the standard state [◄◄ Section 5.6]. The standard temperature for gases is defined specifically as 0°C.

> **Student Note:** In this problem, because they are specified rather than measured, 0°C, 1 mole, and 1 atm are *exact* numbers and do not affect the number of significant figures in the result [◄◄ Section 1.5].

SAMPLE PROBLEM (10.6)

Calculate the volume of a mole of ideal gas at room temperature (25°C) and 1 atm.

Strategy Convert the temperature in °C to temperature in kelvins, and use the ideal gas equation to solve for the unknown volume.

Setup The data given are $n = 1$ mol, $T = 298.15$ K, and $P = 1$ atm. Because the pressure is expressed in atmospheres, we use $R = 0.08206$ L · atm/K · mol to solve for volume in liters.

Solution

$$V = \frac{(1 \text{ mol})(0.08206 \text{ L} \cdot \text{atm/K} \cdot \text{mol})(298.15 \text{ K})}{1 \text{ atm}} = 24.5 \text{ L}$$

THINK ABOUT IT

With the pressure held constant, we should expect the volume to increase with increased temperature. Room temperature is higher than the standard temperature for gases (0°C), so the molar volume at room temperature (25°C) *should* be higher than the molar volume at 0°C—and it is.

Practice Problem ATTEMPT What is the volume of 5.12 moles of an ideal gas at 32°C and 1.00 atm?

Practice Problem BUILD At what temperature (in °C) would 1 mole of ideal gas occupy 50.0 L ($P = 1.00$ atm)?

Practice Problem CONCEPTUALIZE The diagram shown below represents a sample of an ideal gas at STP in a container whose volume is not fixed. Which of the diagrams [(i)–(iv)] best represents the sample after the absolute temperature has been doubled and the external pressure has been increased by a factor of 3?

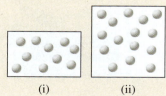

         (i)   (ii)   (iii)   (iv)

Applications of the Ideal Gas Equation

Using some simple algebraic manipulation, we can solve for variables other than those that appear explicitly in the ideal gas equation. For example, if we know the molar mass of a gas (g/mol), we can determine its density at a given temperature and pressure. Recall from Section 10.1 that the density of a gas is generally expressed in units of g/L. We can rearrange the ideal gas equation to solve for mol/L:

$$\frac{n}{V} = \frac{P}{RT}$$

If we then multiply both sides by the molar mass, $\mathcal{M}$, we get

$$\mathcal{M} \times \frac{n}{V} = \frac{P}{RT} \times \mathcal{M}$$

where $\mathcal{M} \times n/V$ gives g/L *or* density, *d*. *Therefore,*

> **Student Note:** $n \times \mathcal{M} = m$, where m is mass in grams.

Equation 10.10	$d = \dfrac{P\mathcal{M}}{RT}$

Conversely, if we know the density of a gas, we can determine its molar mass:

> **Student Note:** Another way to arrive at Equation 10.10 is to substitute $m/\mathcal{M}$ for n in the ideal gas equation and rearrange to solve for m/V (density):
>
> $PV = \dfrac{m}{\mathcal{M}}RT$ and $\dfrac{m}{V} = d = \dfrac{P\mathcal{M}}{RT}$

Equation 10.11	$\mathcal{M} = \dfrac{dRT}{P}$

In a typical experiment, in which the molar mass of a gas is determined, a flask of known volume is evacuated and weighed [Figure 10.12(a)]. It is then filled (to a known pressure) with the gas of unknown molar mass and reweighed [Figure 10.12(b)]. The difference in mass is the mass of the gas sample. Dividing by the known volume of the flask gives the density of the gas, and the molar mass can then be determined using Equation 10.11.

Similarly, the molar mass of a volatile liquid can be determined by placing a small volume of it in the bottom of a flask, the mass and volume of which are known. The flask is then immersed in a hot-water bath, causing the volatile liquid to completely evaporate and its vapor to fill the flask. Because the flask is open, some of the excess vapor escapes. When no more vapor escapes, the flask is capped and *removed from the water bath*. The flask is then weighed to determine the mass of the vapor. (At this point, some or all of the vapor has condensed but the mass remains the same.) The density of the vapor is determined by dividing the mass of the vapor by the volume of the flask. Equation 10.11 is then used to calculate the molar mass of the volatile liquid.

Sample Problems 10.7 and 10.8 illustrate the use of Equations 10.10 and 10.11.

> 🛜 **Student Hot Spot**
>
> Student data indicate you may struggle with the ideal gas equation. Access the eBook to view additional Learning Resources on this topic.

> **Student Note:** Because the flask is open to the atmosphere while the volatile liquid vaporizes, we can use atmospheric pressure as P. Also, because the flask is capped at the water bath temperature, we can use the water bath temperature as T.

Figure 10.12 (a) Evacuated flask. (b) Flask filled with gas. The mass of the gas is the difference between the two masses. The density of the gas is determined by dividing mass by volume.

(a)   (b)

SAMPLE PROBLEM 10.7

Carbon dioxide is effective in fire extinguishers partly because its density is greater than that of air, so CO_2 can smother the flames by depriving them of oxygen. (Air has a density of approximately 1.2 g/L at room temperature and 1 atm.) Calculate the density of CO_2 at room temperature (25°C) and 1.0 atm.

Strategy   Use Equation 10.10 to solve for density. Because the pressure is expressed in atm, we should use $R = 0.08206$ L · atm/K · mol. Remember to express temperature in kelvins.

Setup   The molar mass of CO_2 is 44.01 g/mol.

Solution

$$d = \frac{P\mathcal{M}}{RT} = \frac{(1 \text{ atm})\left(\dfrac{44.01 \text{ g}}{\text{mol}}\right)}{\left(\dfrac{0.08206 \text{ L} \cdot \text{atm}}{\text{K} \cdot \text{mol}}\right)(298.15 \text{ K})} = 1.8 \text{ g/L}$$

THINK ABOUT IT

The calculated density of CO_2 is greater than that of air under the same conditions (as expected). Although it may seem tedious, it is a good idea to write units for each and every entry in a problem such as this. Unit cancellation is very useful for detecting errors in your reasoning or your solution setup.

Practice Problem ATTEMPT   Calculate the density of air at 0°C and 1 atm. (Assume that air is 80 percent N_2 and 20 percent O_2.)

Practice Problem BUILD   What pressure would be required for helium at 25°C to have the same density as carbon dioxide at 25°C and 1 atm?

Practice Problem CONCEPTUALIZE   Two samples of gas are shown at the same temperature and pressure. Which sample has the greater density? Which exerts the greater pressure?

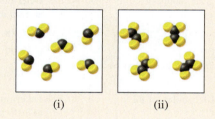

(i)   (ii)

SAMPLE PROBLEM 10.8

A company has just patented a new synthetic alcohol for alcoholic beverages. The new product is said to have all the pleasant properties associated with ethanol but none of the undesirable effects such as hangover, impairment of motor skills, and risk of addiction. The chemical formula is proprietary. You analyze a sample of the new product by placing a small volume of it in a round-bottomed flask with a volume of 511.0 mL and an evacuated mass of 131.918 g. You submerge the flask in a water bath at 100.0°C and allow the volatile liquid to vaporize. You then cap the flask and remove it from the water bath. You weigh it and determine the mass of the vapor in the flask to be 0.768 g. What is the molar mass of the volatile liquid, and what does it mean with regard to the new product? (Assume the pressure in the laboratory is 1 atm.)

Strategy   Use the measured mass of the vapor and the given volume of the flask to determine the density of the vapor at 1 atm and 100.0°C, and then use Equation 10.8 to determine molar mass.

Setup $P = 1$ atm, $V = 0.5110$ L, $R = 0.8206$ L · atm/K · mol, and $T = 373.15$ K.

(Continued on next page)

Solution

$$d = \frac{0.768 \text{ g}}{0.5110 \text{ L}} = 1.5029 \text{ g/L}$$

$$\mathcal{M} = \frac{\left(\dfrac{1.5029 \text{ g}}{\text{L}}\right)\left(\dfrac{0.08206 \text{ L} \cdot \text{atm}}{\text{K} \cdot \text{mol}}\right)(373.15 \text{ K})}{1 \text{ atm}} = 46.02 \text{ g/mol}$$

The result is a molar mass suspiciously close to that of ethanol!

> **THINK ABOUT IT**
>
> Because more than one compound can have a particular molar mass, this method is not definitive for identification. However, in this circumstance, further testing of the proprietary formula certainly would be warranted.

Practice Problem Ⓐ**TTEMPT** Determine the molar mass of a gas with a density of 1.905 g/L at 80.0°C and 1.00 atm.

Practice Problem Ⓑ**UILD** A sample of the volatile liquid propyl acetate ($C_5H_{10}O_2$) is analyzed using the procedure and equipment described in Sample Problem 10.8. What will the mass of the 511.0-mL flask be after evaporation of the propyl acetate?

Practice Problem Ⓒ**ONCEPTUALIZE** These models represent two compounds that contain different amounts of the same two elements. Both compounds are liquids at room temperature. If the compound represented on the left is analyzed by the method described in Sample Problem 10.8 and adds 0.412 g to the mass of the evacuated flask, what mass will be added to the flask when the compound represented on the right is analyzed under the same experimental conditions?

CHECKPOINT – SECTION 10.3   The Ideal Gas Equation

10.3.1 Calculate the volume occupied by 8.75 moles of an ideal gas at STP.

  a) 196 L   d) 18.0 L

  b) 268 L   e) 2.56 L

  c) 0.718 L

10.3.2 Calculate the pressure exerted by 10.2 moles of an ideal gas in a 7.5-L vessel at 150°C.

  a) 17 atm   d) 1.3 atm

  b) 31 atm   e) 47 atm

  c) 0.72 atm

10.3.3 Determine the density of a gas with $\mathcal{M} = 146.07$ g/mol at 1.00 atm and 100.0°C.

  a) 6.85×10^{-3} g/L   d) 30.6 g/L

  b) 4.77 g/L   e) 17.8 g/L

  c) 146 g/L

10.3.4 Determine the molar mass of a gas with $d = 1.963$ g/L at 1.00 atm and 100.0°C.

  a) 0.0166 g/mol   d) 6.09×10^3 g/mol

  b) 60.1 g/mol   e) 1.63×10^3 g/mol

  c) 16.1 g/mol

10.4   Reactions with Gaseous Reactants and Products

In Chapter 3, we used balanced chemical equations to calculate amounts of reactants and/or products in chemical reactions—expressing those amounts in mass (usually grams). However, in the case of reactants and products that are gases, it is more practical to measure and express amounts in volume (liters or milliliters). This makes the ideal gas equation useful in the stoichiometric analysis of chemical reactions that involve gases.

Calculating the Required Volume of a Gaseous Reactant

According to Avogadro's law, the volume of a gas at a given temperature and pressure is proportional to the number of moles. Moreover, balanced chemical equations give the ratio of combination of gaseous reactants in both moles and volume (see Figure 10.11). Therefore, if we know the volume of one reactant in a gaseous reaction, we can determine the required amount of another reactant (at the same temperature and pressure). For example, consider the reaction of carbon monoxide and oxygen to yield carbon dioxide:

$$2CO(g) + O_2(g) \longrightarrow 2CO_2(g)$$

The ratio of combination of CO and O_2 is 2:1, whether we are talking about moles or units of volume. Thus, if we want to determine the stoichiometric amount [◄◄ Section 3.6] of O_2 required to combine with a particular volume of CO, we simply use the conversion factor provided by the balanced equation, which can be expressed as any of the following:

$$\frac{1 \text{ mol } O_2}{2 \text{ mol CO}} \quad \text{or} \quad \frac{1 \text{ L } O_2}{2 \text{ L CO}} \quad \text{or} \quad \frac{1 \text{ mL } O_2}{2 \text{ mL CO}}$$

Let's say we want to determine what volume of O_2 is required to react completely with 65.8 mL of CO at STP. We could use the ideal gas equation to convert the volume of CO to moles, use the stoichiometric conversion factor to convert to moles O_2, and then use the ideal gas equation again to convert moles O_2 to volume. But this method involves several unnecessary steps. We get the same result simply by using the conversion factor expressed in milliliters:

$$65.8 \text{ mL CO} \times \frac{1 \text{ mL } O_2}{2 \text{ mL CO}} = 32.9 \text{ mL } O_2$$

In cases where only one of the reactants is a gas, we *do* need to use the ideal gas equation in our analysis. Recall, for example, the reaction of sodium metal and chlorine gas used to illustrate the Born-Haber cycle [◄◄ Section 8.2]:

$$2Na(s) + Cl_2(g) \longrightarrow 2NaCl(s)$$

Given moles (or more commonly the *mass*) of Na, and information regarding temperature and pressure, we can determine the volume of Cl_2 required to react completely:

$$\times \frac{1 \text{ mol Na}}{22.99 \text{ g Na}} = \qquad \times \frac{1 \text{ mol } Cl_2}{2 \text{ mol Na}} = \qquad \times \frac{RT}{P} =$$

| grams Na | → | moles Na | → | moles Cl_2 | → | liters Cl_2 |

Sample Problem 10.9 shows how to use the ideal gas equation in a stoichiometric analysis.

SAMPLE PROBLEM (10.9)

Sodium peroxide (Na_2O_2) is used to remove carbon dioxide from (and add oxygen to) the air supply in spacecrafts. It works by reacting with CO_2 in the air to produce sodium carbonate (Na_2CO_3) and O_2.

$$2Na_2O_2(s) + 2CO_2(g) \longrightarrow 2Na_2CO_3(s) + O_2(g)$$

What volume (in liters) of CO_2 (at STP) will react with a kilogram of Na_2O_2?

(Continued on next page)

Strategy Convert the given mass of Na_2O_2 to moles, use the balanced equation to determine the stoichiometric amount of CO_2, and then use the ideal gas equation to convert moles of CO_2 to liters.

Setup The molar mass of Na_2O_2 is 77.98 g/mol (1 kg = 1000 g). (Treat the specified mass of Na_2O_2 as an exact number.)

Solution

$$1000 \text{ g Na}_2\text{O}_2 \times \frac{1 \text{ mol Na}_2\text{O}_2}{77.98 \text{ g Na}_2\text{O}_2} = 12.82 \text{ mol Na}_2\text{O}_2$$

$$12.82 \text{ mol Na}_2\text{O}_2 \times \frac{2 \text{ mol CO}_2}{2 \text{ mol Na}_2\text{O}_2} = 12.82 \text{ mol CO}_2$$

$$V_{CO_2} = \frac{(12.82 \text{ mol CO}_2)(0.08206 \text{ L} \cdot \text{atm/K} \cdot \text{mol})(273.15 \text{ K})}{1 \text{ atm}} = 287.4 \text{ L CO}_2$$

THINK ABOUT IT

The answer may seem like an enormous volume of CO_2. If you check the cancellation of units carefully in ideal gas equation problems, however, with practice you will develop a sense of whether such a calculated volume is reasonable.

Practice Problem **A**TTEMPT   What volume (in liters) of CO_2 can be consumed at STP by 525 g Na_2O_2?

Practice Problem **B**UILD   What mass (in grams) of Na_2O_2 is necessary to consume 1.00 L CO_2 at STP?

Practice Problem **C**ONCEPTUALIZE   The decomposition reactions of two solid compounds are represented here.

If an equal number of moles of each solid reactant were to decompose, how would the volume of products of the second decomposition compare to the volume of products of the first decomposition?

Determining the Amount of Reactant Consumed Using Change in Pressure

Although none of the empirical gas laws focuses on the relationship between n and P explicitly, we can rearrange the ideal gas equation to show that n is directly proportional to P at constant V and T:

Equation 10.12(a) $n = P \times \left(\dfrac{V}{RT}\right)$ (at constant V and T)

Therefore, we can use the change in pressure in a reaction vessel to determine how many moles of a gaseous reactant are consumed in a chemical reaction:

Equation 10.12(b) $\Delta n = \Delta P \times \left(\dfrac{V}{RT}\right)$ (at constant V and T)

Student Note: This refers to a reaction in which there is only *one* gaseous reactant and in which none of the products is a gas, such as the reaction described in Sample Problem 10.10. In reactions involving multiple gaseous species, Δn refers to the *net* change in number of moles of gas—and the analysis gets somewhat more complicated.

where Δn is the number of moles of gas consumed and ΔP is the change in pressure in the reaction vessel.

Sample Problem 10.10 shows how to use Equation 10.12(b).

SAMPLE PROBLEM (10.10)

Another air-purification method for enclosed spaces involves the use of "scrubbers" containing aqueous lithium hydroxide, which reacts with carbon dioxide to produce lithium carbonate and water:

$$2LiOH(aq) + CO_2(g) \longrightarrow Li_2CO_3(s) + H_2O(l)$$

Consider the air supply in a submarine with a total volume of 2.5×10^5 L. The pressure is 0.9970 atm, and the temperature is 25°C. If the pressure in the submarine drops to 0.9891 atm as the result of carbon dioxide being consumed by an aqueous lithium hydroxide scrubber, how many moles of CO_2 are consumed?

Strategy Use Equation 10.12(b) to determine Δn, the number of moles of CO_2 consumed.

Setup $\Delta P = 0.9970$ atm $- 0.9891$ atm $= 7.9 \times 10^{-3}$ atm. According to the problem statement, $V = 2.5 \times 10^5$ L and $T = 298.15$ K. For problems in which P is expressed in atmospheres and V in liters, use $R = 0.08206$ L · atm/K · mol.

Solution

$$\Delta n_{CO_2} = 7.9 \times 10^{-3} \text{ atm} \times \frac{2.5 \times 10^5 \text{ L}}{(0.08206 \text{ L} \cdot \text{atm/K} \cdot \text{mol}) \times (298.15 \text{ K})} = 81 \text{ moles } CO_2 \text{ consumed}$$

THINK ABOUT IT

Careful cancellation of units is *essential*. Note that this amount of CO_2 corresponds to 162 moles or 3.9 kg of LiOH. (It's a good idea to verify this yourself.)

Practice Problem **ATTEMPT** Using all the same conditions as those described in Sample Problem 10.10, calculate the number of moles of CO_2 consumed if the pressure drops by 0.010 atm.

Practice Problem **BUILD** By how much would the pressure in the submarine drop if 2.55 kg of LiOH were completely consumed by reaction with CO_2? (Assume the same starting P, V, and T as in Sample Problem 10.10.)

Practice Problem **CONCEPTUALIZE** The diagrams represent a reaction in which all of the species (reactants and products) are gases. If stoichiometric amounts of reactants are combined in a reaction vessel of fixed volume, how will the pressure after the reaction compare to the pressure before the reaction? (Assume that temperature is constant.)

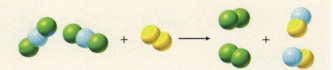

Predicting the Volume of a Gaseous Product

Using a combination of stoichiometry and the ideal gas equation, we can calculate the volume of gas that we expect to be produced in a chemical reaction. We first use stoichiometry to determine the number of moles produced, and then apply the ideal gas equation to determine what volume will be occupied by that number of moles under the specified conditions.

Sample Problem 10.11 shows how to predict the volume of a gaseous product.

SAMPLE PROBLEM (10.11)

The air bags in cars are inflated when a collision triggers the explosive, highly exothermic decomposition of sodium azide (NaN_3):

$$2NaN_3(s) \longrightarrow 2Na(s) + 3N_2(g)$$

A typical driver-side air bag contains about 50 g of NaN_3. Determine the volume of N_2 gas that would be generated by the decomposition of 50.0 g of sodium azide at 85.0°C and 1.00 atm.

Strategy Convert the given mass of NaN_3 to moles, use the ratio of the coefficients from the balanced chemical equation to determine the corresponding number of moles of N_2 produced, and then use the ideal gas equation to determine the volume of that number of moles at the specified temperature and pressure.

(Continued on next page)

Setup The molar mass of NaN_3 is 65.02 g/mol.

Solution

$$\text{mol NaN}_3 = \frac{50.0 \text{ g NaN}_3}{65.02 \text{ g/mol}} = 0.769 \text{ mol NaN}_3$$

$$0.769 \text{ mol NaN}_3 \times \left(\frac{3 \text{ mol N}_2}{2 \text{ mol NaN}_3}\right) = 1.15 \text{ mol N}_2$$

$$V_{N_2} = \frac{(1.15 \text{ mol N}_2)(0.08206 \text{ L} \cdot \text{atm/K} \cdot \text{mol})(358.15 \text{ K})}{1 \text{ atm}} = 33.9 \text{ L N}_2$$

THINK ABOUT IT

The calculated volume represents the space between the driver and the steering wheel and dashboard that must be filled by the air bag to prevent injury. Air bags also contain an oxidant that consumes the sodium metal produced in the reaction.

Practice Problem A **TTEMPT** The chemical equation for the metabolic breakdown of glucose ($C_6H_{12}O_6$) is the same as that for the combustion of glucose [◄◄ Section 3.3—Bringing Chemistry to Life box]:

$$C_6H_{12}O_6(aq) + 6O_2(g) \longrightarrow 6CO_2(g) + 6H_2O(l)$$

Calculate the volume of CO_2 produced at normal human body temperature (37°C) and 1.00 atm when 10.0 g of glucose is consumed in the reaction.

Practice Problem B **UILD** The passenger-side air bag in a typical car must fill a space approximately four times as large as the driver-side air bag to be effective. Calculate the mass of sodium azide required to fill a 125-L air bag at 85.0°C and 1.00 atm.

Practice Problem C **ONCEPTUALIZE** The unbalanced decomposition reactions of two solid compounds are represented here. Both decompose to form the same two gaseous products, in different amounts. Which compound will produce the greater volume of products when equal numbers of moles decompose? Which compound will produce the greater volume of products when equal numbers of grams decompose?

(i)   (ii)

CHECKPOINT – SECTION 10.4   Reactions with Gaseous Reactants and Products

10.4.1 Determine the volume of Cl_2 gas at STP that will react with 1.00 mole of Na solid to produce NaCl according to the equation,

$$2Na(s) + Cl_2(g) \longrightarrow 2NaCl(s)$$

a) 22.4 L

b) 44.8 L

c) 15.3 L

d) 11.2 L

e) 30.6 L

10.4.2 Determine the mass of NaN_3 required for an air bag to produce 100.0 L of N_2 gas at 85.0°C and 1.00 atm according to the equation,

$$2NaN_3(s) \longrightarrow 2Na(s) + 3N_2(g)$$

a) 332 g

b) 148 g

c) 221 g

d) 664 g

e) 442 g

10.5   Gas Mixtures

So far our discussion of the physical properties of gases has focused on the behavior of *pure* gaseous substances, even though the gas laws were all developed based on observations of samples of air, which is a *mixture* of gases. In this section, we consider gas mixtures and their physical behavior. We restrict our discussion in this section to gases that behave ideally and that do not react with one another.

Dalton's Law of Partial Pressures

When two or more gaseous substances are placed in a container, each gas behaves as though it occupies the container alone. For example, if we place 1.00 mole of N_2 gas in a 5.00-L container at 0°C, it exerts a pressure of

$$P = \frac{(1.00 \text{ mol}(0.08206 \text{ L} \cdot \text{atm/K} \cdot \text{mol}(273.15 \text{ K})}{5.00 \text{ L}} = 4.48 \text{ atm}$$

If we then add a mole of another gas, such as O_2, the pressure exerted by N_2 does not change. It remains at 4.48 atm. The O_2 gas exerts its own pressure, also 4.48 atm. Neither gas is affected by the presence of the other. In a mixture of gases, the pressure exerted by each gas is known as the ***partial pressure*** (P_i) of the gas. We use subscripts to denote partial pressures:

$$P_{N_2} = \frac{(1.00 \text{ mol}(0.08206 \text{ L} \cdot \text{atm/K} \cdot \text{mol}(273.15 \text{ K})}{5.00 \text{ L}} = 4.48 \text{ atm}$$

$$P_{O_2} = \frac{(1.00 \text{ mol}(0.08206 \text{ L} \cdot \text{atm/K} \cdot \text{mol}(273.15 \text{ K})}{5.00 \text{ L}} = 4.48 \text{ atm}$$

and we can solve the ideal gas equation for each component of any gas mixture:

$$P_i = \frac{n_i RT}{V}$$

Dalton's law of partial pressures states that the total pressure exerted by a gas mixture is the sum of the partial pressures exerted by each component of the mixture:

$$P_{\text{total}} = \Sigma P_i$$

Thus, the total pressure exerted by a mixture of 1.00 mol N_2 and 1.00 mol O_2 in a 5.00-L vessel at 0°C is

$$P_{\text{total}} = P_{N_2} + P_{O_2} = 4.48 \text{ atm} + 4.48 \text{ atm} = 8.96 \text{ atm}$$

Figure 10.13 illustrates Dalton's law of partial pressures.

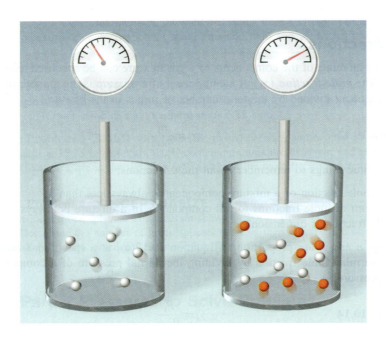

Figure 10.13 Schematic illustration of Dalton's law of partial pressures. Total pressure is equal to the sum of partial pressures.

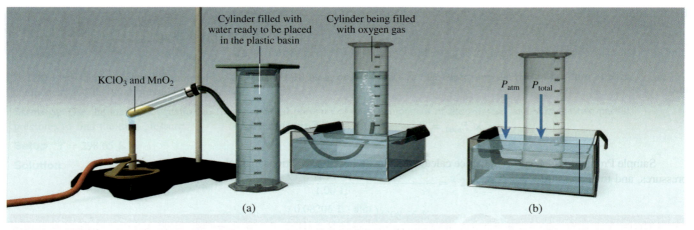

Figure 10.14 (a) Apparatus for measuring the amount of gas produced in a chemical reaction. (b) When the water levels inside and outside the collection vessel are the same, the pressure inside the vessel is equal to atmospheric pressure.

TABLE 10.5	Vapor Pressure of Water (P_{H_2O}) as a Function of Temperature				
T (°C)	**P (torr)**	**T (°C)**	**P (torr)**	**T (°C)**	**P (torr)**
0	4.6	35	42.2	70	233.7
5	6.5	40	55.3	75	289.1
10	9.2	45	71.9	80	355.1
15	12.8	50	92.5	85	433.6
20	17.5	55	118.0	90	525.8
25	23.8	60	149.4	95	633.9
30	31.8	65	187.5	100	760.0

Student Hot Spot

Student data indicate you may struggle with mole fractions. Access the eBook to view additional Learning Resources on this topic.

of potassium chlorate ($KClO_3$), the reaction used to generate emergency oxygen supplies on airplanes, produces potassium chloride and oxygen:

$$2KClO_3(s) \longrightarrow 2KCl(s) + 3O_2(g)$$

The oxygen gas is collected over water, as shown in Figure 10.14(a). The volume of water displaced by the gas is equal to the volume of gas produced. (Prior to reading the volume of the gas, the level of the graduated cylinder must be adjusted such that the water levels inside and outside the cylinder are the *same*. This ensures that the *pressure* inside the graduated cylinder is the same as atmospheric pressure [Figure 10.14(b)].) However, because the measured volume contains both the oxygen produced by the reaction *and* water vapor, the pressure exerted inside the graduated cylinder is the sum of the two partial pressures:

$$P_{total} = P_{O_2} + P_{H_2O}$$

By subtracting the partial pressure of water from the total pressure, which is equal to atmospheric pressure, we can determine the partial pressure of oxygen—and thereby determine how many moles are produced by the reaction. We get the partial pressure of water, which depends on temperature, from a table of values. Table 10.5 lists the partial pressure (also known as the *vapor pressure*) of water at various temperatures.

Sample Problem 10.14 shows how to use Dalton's law of partial pressures to determine the amount of gas produced in a chemical reaction and collected over water.

SAMPLE PROBLEM 10.14

Calcium metal reacts with water to produce hydrogen gas [◀◀ Section 7.7]:

$$Ca(s) + 2H_2O(l) \longrightarrow Ca(OH)_2(aq) + H_2(g)$$

Determine the mass of H_2 produced at 25°C and 0.967 atm when 525 mL of the gas is collected over water as shown in Figure 10.14.

Strategy Use Dalton's law of partial pressures to determine the partial pressure of H_2, use the ideal gas equation to determine moles of H_2, and then use the molar mass of H_2 to convert to mass. (Pay careful attention to units. Atmospheric pressure is given in atmospheres, whereas the vapor pressure of water is tabulated in torr.)

Setup $V = 0.525$ L and $T = 298.15$ K. The partial pressure of water at 25°C is 23.8 torr (Table 10.5) or 23.8 torr (1 atm/760 torr) = 0.0313 atm. The molar mass of H_2 is 2.016 g/mol.

Solution

$$P_{H_2} = P_{total} - P_{H_2O} = 0.967 \text{ atm} - 0.0313 \text{ atm} = 0.936 \text{ atm}$$

$$\text{moles of } H_2 = \frac{(0.9357 \text{ atm})(0.525 \text{ L})}{\left(\dfrac{0.08206 \text{ L} \cdot \text{atm}}{\text{K} \cdot \text{mol}}\right)(298.15 \text{ K})} = 2.01 \times 10^{-2} \text{ mol}$$

$$\text{mass of } H_2 = (2.008 \times 10^{-2} \text{ mol})(2.016 \text{ g/mol}) = 0.0405 \text{ g } H_2$$

THINK ABOUT IT

Check unit cancellation carefully, and remember that the densities of gases are relatively low. The mass of approximately half a liter of hydrogen at or near room temperature and 1 atm should be a very small number.

Practice Problem **A**TTEMPT Calculate the mass of O_2 produced by the decomposition of $KClO_3$ when 821 mL of O_2 is collected over water at 30.0°C and 1.015 atm.

Practice Problem **B**UILD Determine the volume of gas collected over water when 0.501 g O_2 is produced by the decomposition of $KClO_3$ at 35.0°C and 1.08 atm.

Practice Problem **C**ONCEPTUALIZE The first diagram represents the result of an experiment in which the oxygen gas produced by a chemical reaction is collected over water at typical room temperature. Which of the diagrams [(i)–(iv)] best represents the result of the same experiment on a day when the temperature in the laboratory is significantly warmer?

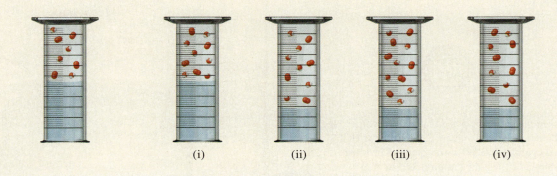

(i)   (ii)   (iii)   (iv)

Collection over water can also be used to determine the molar volume of a gas generated in a chemical reaction as shown in Figure 10.15.

Animation
Molar Volume of a Gas

Figure 10.15

Molar Volume of a Gas

The reaction between zinc and hydrochloric acid produces hydrogen gas. The net ionic equation for the reaction is

$$Zn(s) + 2H^+(aq) \longrightarrow Zn^{2+}(aq) + H_2(g)$$

The H_2 gas evolved in the reaction is collected in the inverted graduated cylinder. Using the balanced equation, we determine how much H_2 will be produced.

$$0.072 \text{ g Zn} \times \frac{1 \text{ mol Zn}}{65.41 \text{ g Zn}} = 0.0011 \text{ mol Zn}$$

$$0.0011 \text{ mol Zn} \times \frac{1 \text{ mol } H_2}{1 \text{ mol Zn}} = 0.0011 \text{ mol } H_2$$

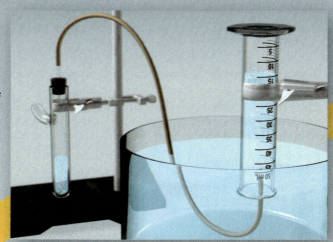

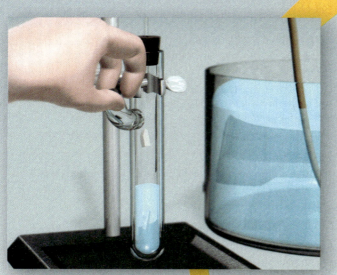

The reservoir is inverted and the zinc drops into the 1.0 M HCl.

START

A 0.072-g sample of zinc is placed in the reservoir. The vessel contains approximately 5 mL of 1.0 M HCl. A graduated cylinder is filled with water and inverted over the tubing to collect the gas produced by the reaction of zinc and acid. The temperature of the water is 25.0°C and the pressure in the room is 748.0 torr.

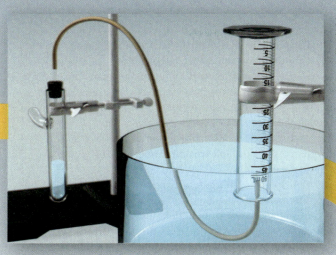

Zn is the limiting reactant. When all of the zinc has been consumed, the reaction is complete and no more gas is evolved.

We adjust the level of the graduated cylinder to make the water level the same inside and outside of the cylinder. This lets us know that the pressure inside the cylinder is the same as the pressure in the room. When the water levels are the same, we can read the volume of gas collected, which is 26.5 mL.

The pressure inside the cylinder is the sum of two partial pressures: that of the collected H_2 and that of water vapor. To determine the partial pressure of H_2, which is what we want, we must subtract the vapor pressure of water from the total pressure. Table 10.5 gives the vapor pressure of water at 25.0°C as 23.8 torr. Therefore, the pressure of H_2 is 748.0 − 23.8 = 724.2 torr. Using Equation 10.8(b),

$$\frac{P_1 V_1}{T_1} = \frac{P_2 V_2}{T_2}$$

we calculate the volume the H_2 gas would occupy at STP.

$$\frac{(724.2 \text{ torr})(26.5 \text{ mL})}{298.15 \text{ K}} = \frac{(760 \text{ torr})V_2}{273.15 \text{ K}}$$

$$V_2 = \frac{(724.2 \text{ torr})(26.5 \text{ mL})(273.15 \text{ K})}{(760 \text{ torr})(298.15 \text{ K})} = 23.13 \text{ mL}$$

This is the volume of 0.0011 mol H_2. The molar volume is

$$\frac{23.13 \text{ mL}}{0.0011 \text{ mol}} = 2.1 \times 10^4 \text{ mL/mol or 21 L/mol}$$

(See Visualizing Chemistry questions VC 10.1–VC 10.4 on page 517.)

What's the point?

The gas collected in the graduated cylinder is a mixture of the gas produced by the reaction and water vapor. We determine the pressure of the gas produced by the reaction by subtracting the tabulated partial pressure of water from the total pressure. Knowing the volume, pressure, and temperature of a sample of gas, in this case H_2, we can determine what volume the same sample of gas would occupy at STP. This enables us to determine experimentally the molar volume of H_2 at STP, which turns out to be fairly close to the accepted value of 22.4 L. Several sources of error, including uncertainty in the volume of the collected gas, contribute to the result not being exactly 22.4 L.

Bringing Chemistry to Life

Hyperbaric Oxygen Therapy

In 1918, during the Spanish flu epidemic that claimed tens of millions of lives worldwide, physician Orville Cunningham noted that people living at lower elevations appeared to have a greater chance of surviving the flu than those living at higher elevations. Believing this to be the result of increased air pressure, he developed a hyperbaric chamber to treat flu victims. One of Cunningham's earliest and most notable successes was the recovery of a flu-stricken colleague who had been near death. Cunningham subsequently built a hyperbaric chamber large enough to accommodate dozens of patients and treated numerous flu victims, most with success.

In the decades following the Spanish flu epidemic, hyperbaric therapy fell out of favor with the medical community and was largely discontinued. Interest in it was revived when the U.S. military ramped up its underwater activities in the 1940s and hyperbaric chambers were constructed to treat military divers suffering from decompression sickness (DCS), also known as "the bends." Significant advancement in hyperbaric methods began in the 1970s when the Undersea Medical Society (renamed the Undersea and Hyperbaric Medical Society in 1976) became involved in the clinical use of hyperbaric chambers. Today, hyperbaric oxygen therapy (HBOT) is used to treat a wide variety of conditions, including carbon monoxide poisoning, anemia caused by critical blood loss, severe burns, and life-threatening bacterial infections. Once considered an "alternative" therapy and viewed with skepticism, HBOT is now covered by most insurance plans.

Student Note: One unfortunate group of patients was undergoing treatment when the power to the chamber was shut off accidentally. All the patients died. At the time, their deaths were attributed to influenza, but they almost certainly died as the result of the unintended rapid decompression.

Sample Problem 10.15 illustrates the importance of mole fractions and partial pressures in hyperbaric oxygen therapy.

SAMPLE PROBLEM (10.15)

Large hyperbaric chambers are used mainly to treat victims of diving accidents. In one treatment protocol at a typical facility, the chamber is pressurized to 6.0 atm with compressed air and the patient breathes a mixture of gases that is 47 percent O_2 by volume. In another protocol, the chamber is pressurized with compressed air to 2.8 atm and the patient breathes pure O_2. Determine the partial pressure of O_2 in each treatment protocol and compare the results.

Strategy Because the gas the patient breathes is inside the hyperbaric chamber, its total pressure is the same as the chamber pressure. To obtain the partial pressure of O_2, multiply the mole fraction of O_2 in the breathing gas in each protocol by the total pressure (Equation 10.16).

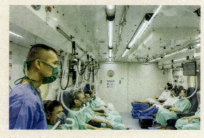

Hyperbaric chamber
narin phapnam/Shutterstock

Setup In the first protocol, the breathing gas is 47 percent O_2, so the mole fraction of O_2 is 0.47. In the second protocol, where pure O_2 is used, the mole fraction of O_2 is 1. Use Equation 10.16 to calculate the partial pressure of O_2 in each protocol.

$$\chi_i \times P_{total} = P_i$$

Solution In the first protocol, the pressure of O_2 is

$$0.47 \times 6.0 \text{ atm} = 2.8 \text{ atm}$$

In the second protocol, the pressure of O_2 is

$$1 \times 2.8 \text{ atm} = 2.8 \text{ atm}$$

Both protocols produce the same pressure of O_2.

THINK ABOUT IT

Monoplace hyperbaric chambers, which are large enough to accommodate only one person, commonly are pressurized to 2.8 atm with pure O_2.

Practice Problem **A**TTEMPT   What mole fraction of O_2 is necessary for the partial pressure of O_2 to be 2.8 atm when the total pressure is 4.6 atm?

Practice Problem **B**UILD   What chamber pressure would be required for a patient to receive the therapeutic partial pressure of O_2 (2.8 atm) without breathing a special mixture of gases through a mask? Assume that the air used to pressurize the chamber is 21 percent O_2 by volume.

Practice Problem **C**ONCEPTUALIZE   The diagram on the top shows a gas mixture at a particular temperature. Which of the diagrams [(i)–(iv)] represents a mixture with the same mole fraction of red? Which represents a mixture with the same partial pressure of red? Which represents a mixture with the same total pressure?

(i)   (ii)   (iii)   (iv)

CHECKPOINT – SECTION 10.5   Gas Mixtures

10.5.1 What is the partial pressure of He in a 5.00-L vessel at 25°C that contains 0.0410 mole of He, 0.121 mole of Ne, and 0.0922 mole of Ar?

a) 1.24 atm   d) 2.87 atm

b) 0.248 atm   e) 0.201 atm

c) 0.117 atm

10.5.2 What is the mole fraction of CO_2 in a mixture of 0.756 mole of N_2, 0.189 mole of O_2, and 0.0132 mole of CO_2?

a) 0.789   d) 1.003

b) 0.0138   e) 0.798

c) 0.0140

10.5.3 What is the partial pressure of oxygen in a gas mixture that contains 4.10 moles of oxygen, 2.38 moles of nitrogen, and 0.917 mole of carbon dioxide and that has a total pressure of 2.89 atm?

a) 1.60 atm   d) 0.705 atm

b) 3.59 atm   e) 0.624 atm

c) 0.391 atm

10.5.4 What mass of acetylene (C_2H_2) is produced by the reaction of calcium carbide (CaC_2) and water,

$$CaC_2(s) + 2H_2O(l) \longrightarrow C_2H_2(g) + Ca(OH)_2(aq)$$

if 425 mL of the gas is collected over water at 30°C and a pressure of 0.996 atm?

a) 0.016 g   d) 0.424 g

b) 16.3 g   e) 0.443 g

c) 0.019 g

10.5.5 In the following diagram, each color represents a different gas molecule. Calculate the mole fraction of each gas.

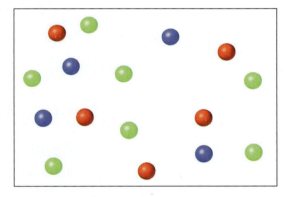

a) $\chi_{red} = 0.5$, $\chi_{blue} = 0.4$, $\chi_{green} = 0.7$

b) $\chi_{red} = 0.05$, $\chi_{blue} = 0.07$, $\chi_{green} = 0.07$

c) $\chi_{red} = 0.3125$, $\chi_{blue} = 0.25$, $\chi_{green} = 0.4375$

d) $\chi_{red} = 0.4167$, $\chi_{blue} = 0.333$, $\chi_{green} = 0.5833$

e) $\chi_{red} = 0.333$, $\chi_{blue} = 0.333$, $\chi_{green} = 0.333$

10.5.6 Calculate the partial pressure of each gas in the diagram in question 10.5.5 if the total pressure is 8.21 atm.

a) $P_{red} = 0.5$ atm, $P_{blue} = 0.25$ atm, $P_{green} = 4.5$ atm

b) $P_{red} = 2.57$ atm, $P_{blue} = 2.05$ atm, $P_{green} = 3.59$ atm

c) $P_{red} = 3.13$ atm, $P_{blue} = 2.50$ atm, $P_{green} = 2.58$ atm

d) $P_{red} = 2.74$ atm, $P_{blue} = 2.74$ atm, $P_{green} = 2.74$ atm

e) $P_{red} = 3.125$ atm, $P_{blue} = 2.500$ atm, $P_{green} = 4.375$ atm

10.6 The Kinetic Molecular Theory of Gases

The gas laws were derived empirically, and they enable us to predict the macroscopic behavior of gases. They do not explain, however, *why* gases behave as they do. The **kinetic molecular theory,** which was put forth in the nineteenth century by a number of physicists, notably Ludwig Boltzmann[6] and James Maxwell,[7] explains how the molecular nature of gases gives rise to their macroscopic properties. The basic assumptions of the kinetic molecular theory are as follows:

1. A gas is composed of particles that are separated by relatively large distances. The volume occupied by individual molecules is negligible.
2. Gas molecules are constantly in random motion, moving in straight paths, colliding with the walls of their container and with one another in perfectly elastic collisions. (Energy is *transferred* but not *lost* in the collisions.)
3. Gas molecules do not exert attractive or repulsive forces on one another.
4. The average kinetic energy, $\overline{E_k}$, of gas molecules in a sample is proportional to the absolute temperature:

$$\overline{E_k} \propto T$$

Recall that kinetic energy is the energy associated with motion [◄◄ Section 5.1]:

$$E_k = \frac{1}{2}mu^2$$

Thus, the kinetic energy of an individual gas molecule is proportional to its mass and its velocity squared. When we talk about a group of gas molecules, we determine the average kinetic energy using the *mean square speed, $\overline{u^2}$*, which is the average of the speed squared for all the molecules in the sample:

$$\overline{u^2} = \frac{u_1^2 + u_2^2 + u_3^2 + \cdots + u_N^2}{N}$$

where N is the number of molecules in the sample.

Application to the Gas Laws

Kinetic molecular theory enables us to understand some of the properties and behavior of gases in the following ways.

Compressibility

Gases are compressible because molecules in the gas phase are separated by large distances (assumption 1) and can be moved closer together by decreasing the volume occupied by a sample of gas (Figure 10.16).

Boyle's Law ($V \propto 1/P$)

The pressure exerted by a gas is the result of the collisions of gas molecules with the walls of their container (assumption 2). The magnitude of the pressure depends on

---

6. Ludwig Eduard Boltzmann (1844–1906). Austrian physicist. Although Boltzmann was one of the greatest theoretical physicists of all time, his work was not recognized by other scientists in his own lifetime. He suffered from poor health and severe depression and committed suicide in 1906.

7. James Clerk Maxwell (1831–1879). Scottish physicist. Maxwell was one of the great theoretical physicists of the nineteenth century. His work covered many areas in physics, including the kinetic theory of gases, thermodynamics, and electricity and magnetism.

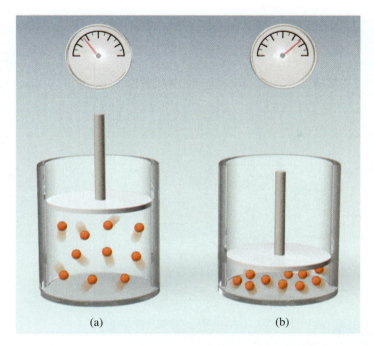

(a)   (b)

Figure 10.16 Gases can be compressed by decreasing their volume. (a) Before volume decrease. (b) After volume decrease, the increased frequency of collisions between molecules and the walls of their container constitutes higher pressure.

both the frequency of collision and the speed of molecules when they collide with the walls. Decreasing the volume occupied by a sample of gas increases the frequency of these collisions, thus increasing the pressure (Figure 10.16).

Charles's Law ($V \propto T$)

Heating a sample of gas increases its average kinetic energy (assumption 4). Because the masses of the molecules do not change, an increase in average kinetic energy must be accompanied by an increase in the mean square speed of the molecules. In other words, heating a sample of gas makes the gas molecules move faster. Faster-moving molecules collide more frequently and with greater speed at impact, thus increasing the pressure. If the container can expand (as is the case with a balloon or a cylinder with a movable piston), the volume of the gas sample will increase, thereby decreasing the frequency of collisions until the pressure inside the container and the pressure outside are again equal (Figure 10.17).

Avogadro's Law ($V \propto n$)

Because the magnitude of the pressure exerted by a sample of gas depends on the frequency of the collisions with the container wall, the presence of more molecules would cause an increase in pressure. Again, the container will expand if it can. Expansion of the container will decrease the frequency of collisions until the pressures inside and outside the container are once again equal (Figure 10.18).

Dalton's Law of Partial Pressures ($P_{total} = \Sigma P_i$)

Gas molecules do not attract or repel one another (assumption 3), so the pressure exerted by one gas is unaffected by the presence of another gas. Consequently, the

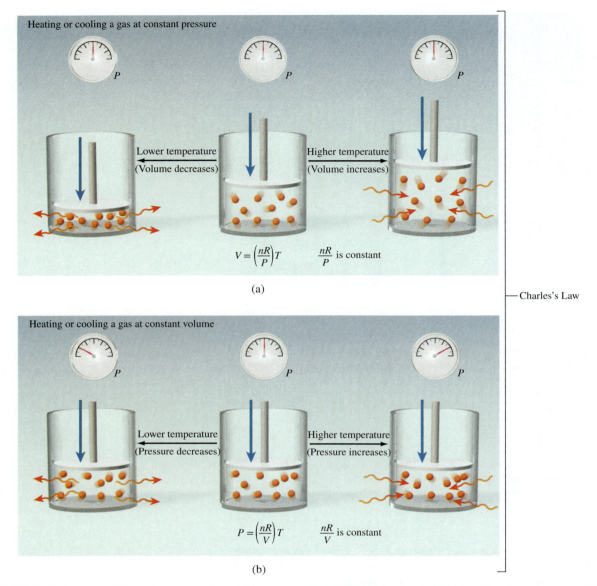

Heating or cooling a gas at constant pressure

Lower temperature
(Volume decreases)

Higher temperature
(Volume increases)

$$V = \left(\frac{nR}{P}\right)T \qquad \frac{nR}{P} \text{ is constant}$$

(a)

Charles's Law

Heating or cooling a gas at constant volume

Lower temperature
(Pressure decreases)

Higher temperature
(Pressure increases)

$$P = \left(\frac{nR}{V}\right)T \qquad \frac{nR}{V} \text{ is constant}$$

(b)

Figure 10.17 Charles's law. (a) The volume of a sample of gas at constant pressure is proportional to its absolute temperature. (b) The pressure of a sample of gas at constant volume is proportional to its absolute temperature.

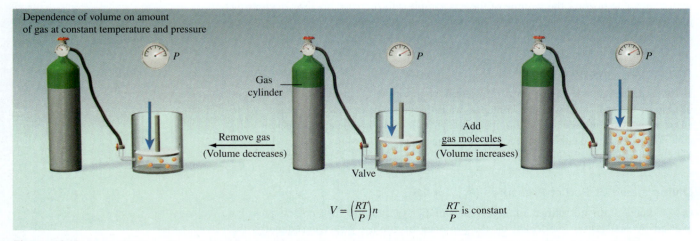

Dependence of volume on amount
of gas at constant temperature and pressure

Gas
cylinder

Remove gas
(Volume decreases)

Valve

Add
gas molecules
(Volume increases)

$$V = \left(\frac{RT}{P}\right)n \qquad \frac{RT}{P} \text{ is constant}$$

Figure 10.18 Avogadro's law. The volume of a gas at constant temperature and pressure is proportional to the number of moles.

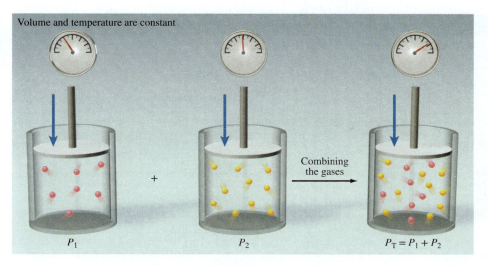

Figure 10.19 Each component of a gas mixture exerts a pressure independent of the other components. The total pressure is the sum of the individual components' partial pressures.

total pressure exerted by a mixture of gases is simply the sum of the partial pressures of the individual components in the mixture (Figure 10.19).

Molecular Speed

One of the important outcomes of the kinetic molecular theory is that the total kinetic energy of a mole of gas (any gas) is equal to $\frac{3}{2}RT$. With assumption 4 we saw that the average kinetic energy of one molecule is $\frac{1}{2}m\overline{u^2}$. For 1 mole of the gas we write

$$N_A\left(\frac{1}{2}m\overline{u^2}\right) = \frac{3}{2}RT$$

where N_A is Avogadro's number and R is the gas constant expressed as 8.314 J/K · mol. Because $m \times N_A = \mathcal{M}$, we can rearrange the preceding equation as follows:

$$\overline{u^2} = \frac{3RT}{\mathcal{M}}$$

Taking the square root of both sides gives

$$\sqrt{\overline{u^2}} = \sqrt{\frac{3RT}{\mathcal{M}}}$$

where $\sqrt{\overline{u^2}}$ is the ***root-mean-square (rms) speed (u_{rms})***. The result,

$$u_{\text{rms}} = \sqrt{\frac{3RT}{\mathcal{M}}} \qquad\qquad \textbf{Equation 10.17}$$

gives us the root-mean-square speed, which is the speed of a molecule with the average kinetic energy in a gas sample. Equation 10.17 indicates two important things: (1) The root-mean-square speed is directly proportional to the square root of the absolute temperature, and (2) the root-mean-square speed is inversely proportional to the square root of $\mathcal{M}$. Thus, for any two samples of gas at the same temperature, the gas with the larger molar mass will have the lower root-mean-square speed, u_{rms}. Remember that the average kinetic energy of a gas depends on its absolute temperature. Therefore, any two gas samples at the same temperature have the same average kinetic energy.

Keep in mind that most molecules will have speeds either higher or lower than u_{rms}—and that u_{rms} is temperature dependent. James Maxwell studied extensively the behavior of gas molecules at various temperatures. Figure 10.20(a) shows typical Maxwell speed distribution curves for nitrogen gas at three different temperatures. At a given temperature, the distribution curve tells us the number of molecules moving at a certain speed. The maximum of each curve represents the most probable speed—

Student Note: In Equation 10.17, R must be expressed as 8.314 J/K · mol and $\mathcal{M}$ must be expressed in kg/mol.

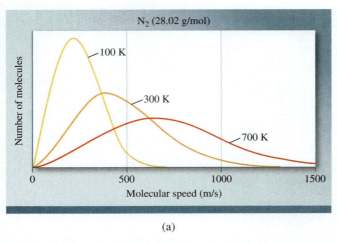

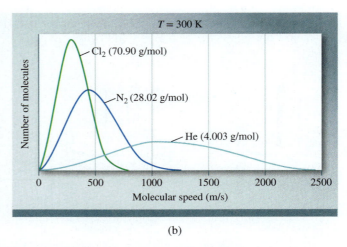

(a)   (b)

Figure 10.20 (a) The distribution of speeds for nitrogen gas at three different temperatures. At higher temperatures more molecules are moving faster. (b) The distribution of speeds for three different gases at the same temperature. On average, lighter molecules move faster than heavier molecules.

that is, the speed of the largest number of molecules. Note that the most probable speed increases as temperature increases [the maximum shifts toward the right in Figure 10.20(a)]. Furthermore, the curve also begins to flatten out with increasing temperature, indicating that larger numbers of molecules are moving faster.

Figure 10.20(b) shows the speed distributions of three different gases (Cl_2, N_2, and He) at the same temperature (300 K). The difference in these curves can be explained by noting that lighter molecules, on average, move faster than heavier ones.

Although we can use Equation 10.18 to calculate u_{rms} of a molecule in a particular sample, we will generally find it more useful to compare the u_{rms} values of molecules in different gas samples. For example, we can write Equation 10.18 for two different gases:

$$u_{rms}(1) = \sqrt{\frac{3RT}{\mathcal{M}_1}} \quad \text{and} \quad u_{rms}(2) = \sqrt{\frac{3RT}{\mathcal{M}_2}}$$

We can then determine the u_{rms} of a molecule in one gas relative to that in the other gas:

$$\frac{u_{rms}(1)}{u_{rms}(2)} = \frac{\sqrt{\dfrac{3RT}{\mathcal{M}_1}}}{\sqrt{\dfrac{3RT}{\mathcal{M}_2}}}$$

Canceling identical terms, when both gases are at the same temperature, we can write

Student Note: Because Equation 10.18 contains the ratio of two molar masses, we can express the molar masses as g/mol or kg/mol. (In Equation 10.17, we had to express molar mass as kg/mol for the units to cancel properly.)

Equation 10.18

$$\frac{u_{rms}(1)}{u_{rms}(2)} = \sqrt{\frac{\mathcal{M}_2}{\mathcal{M}_1}}$$

Using Equation 10.18 we can compare u_{rms} values of molecules with different molar masses (at a given temperature).

Animation
Diffusion of gases.

Diffusion and Effusion

The random motion of gas molecules gives rise to two readily observable phenomena: diffusion and effusion. ***Diffusion*** is the mixing of gases as the result of random motion and frequent collisions (Figure 10.21), while ***effusion*** is the escape of gas molecules from a container to a region of vacuum (Figure 10.22). One of the earliest successes of the kinetic molecular theory was its ability to explain diffusion and effusion.

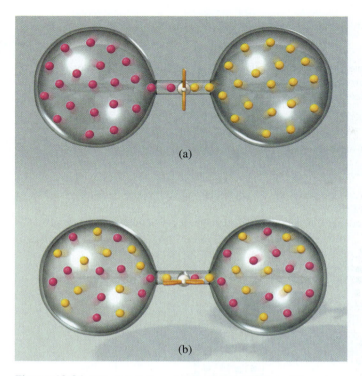

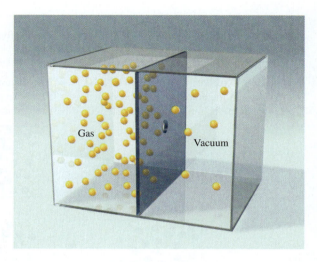

Figure 10.21 Diffusion is the mixing of gases. (a) Two different gases in separate containers. (b) When the stopcock is opened, the gases mix by diffusion.

Figure 10.22 Effusion is the escape of a gas into a vacuum.

Graham's law states that the rate of diffusion or effusion of a gas is inversely proportional to the square root of its molar mass:

$$\text{Rate} \propto \frac{1}{\sqrt{M}}$$

This is essentially a restatement of Equation 10.18. Thus, lighter gases diffuse and effuse more rapidly than heavier gases.

Sample Problem 10.16 shows how to compare the speeds of molecules, knowing their masses.

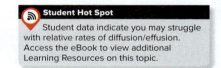

Student Hot Spot

Student data indicate you may struggle with relative rates of diffusion/effusion. Access the eBook to view additional Learning Resources on this topic.

SAMPLE PROBLEM 10.16

Determine how much faster a helium atom moves, on average, than a carbon dioxide molecule at the same temperature.

Strategy Use Equation 10.18 and the molar masses of He and CO_2 to determine the ratio of their root-mean-square speeds. When solving a problem such as this, it is generally best to label the lighter of the two molecules as molecule 1 and the heavier molecule as molecule 2. This ensures that the result will be greater than 1, which is relatively easy to interpret.

Setup The molar masses of He and CO_2 are 4.003 and 44.01 g/mol, respectively.

Solution

$$\frac{u_{rms}(\text{He})}{u_{rms}(CO_2)} = \frac{\sqrt{\dfrac{44.01 \text{ g}}{\text{mol}}}}{\sqrt{\dfrac{4.003 \text{ g}}{\text{mol}}}} = 3.316$$

On average, He atoms move 3.316 times faster than CO_2 molecules at the same temperature.

(Continued on next page)

THINK ABOUT IT

Remember that the relationship between molar mass and molecular speed (Equation 10.18) is reciprocal. A CO_2 molecule has approximately 10 times the mass of an He atom. Therefore, we should expect an He atom, on average, to be moving approximately $\sqrt{10}$ times (~3.2 times) as fast as a CO_2 molecule.

Practice Problem ATTEMPT Determine the relative root-mean-square speeds of O_2 and SF_6 at a given temperature.

Practice Problem BUILD Determine the molar mass and identity of a gas that moves 4.67 times as fast as CO_2.

Practice Problem CONCEPTUALIZE The diagram on the top represents an equimolar mixture of two gases prior to effusion into the adjoining evacuated chamber. The molar mass of the brown gas is significantly larger than the molar mass of the yellow gas. Which of the diagrams [(i)–(iii)] best represents the contents of the two chambers after a period of time has passed?

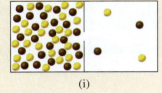
(i)

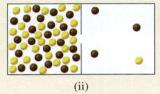
(ii)

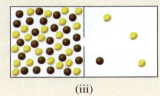
(iii)

CHECKPOINT – SECTION 10.6   The Kinetic Molecular Theory of Gases

10.6.1 Methane (CH_4) diffuses approximately 2.4 times as fast as a certain unknown gas. Which of the following could be the unknown gas?

a) O_2

b) C_2H_6

c) CH_3Br

d) CH_3I

e) F_2

10.6.2 Which gas effuses faster, He or Ar, and how much faster does it effuse?

a) He effuses 9.98 times as fast as Ar.

b) He effuses 3.16 times as fast as Ar.

c) Ar effuses 9.98 times as fast as He.

d) Ar effuses 3.16 times as fast as He.

e) He and Ar effuse at the same rate.

10.7   Deviation from Ideal Behavior

The gas laws and the kinetic molecular theory assume that molecules in the gas phase occupy negligible volume (assumption 1) and that they do not exert any force on one another, either attractive or repulsive (assumption 3). Gases that behave as though these assumptions were strictly true are said to exhibit *ideal behavior.* Many gases do exhibit ideal or nearly ideal behavior under ordinary conditions. Figure 10.23 shows the molar volumes of some common gases at STP. All are remarkably close to the ideal value of 22.41 L. Although we generally assume that real gases behave ideally, there are conditions, namely, high pressure and low temperature, under which the behavior of a real gas deviates from ideal.

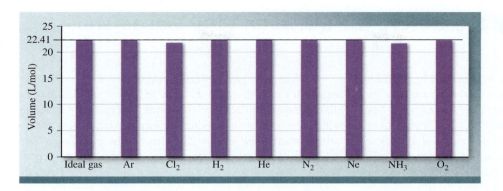

Figure 10.23 Molar volumes of some common gases at STP.

Factors That Cause Deviation from Ideal Behavior

At high pressures, gas molecules are relatively close together. We can assume that gas molecules occupy no volume only when the distances between molecules are large. When the distances between molecules are reduced, the volume occupied by each individual molecule becomes more significant.

At low temperatures, gas molecules are moving more slowly. We can assume that there are no intermolecular forces between gas molecules, either attractive or repulsive, when the gas molecules are moving very fast and the magnitude of their kinetic energies is much larger than the magnitude of any intermolecular forces. When molecules move more slowly, they have lower kinetic energies and the magnitude of the forces between them becomes more significant.

The van der Waals Equation

Because there are conditions under which use of the ideal gas equation would result in large errors (i.e., high pressure and/or low temperature), we must use a slightly different approach when gases do not behave ideally. Analyses of real gases that took into account nonzero molecular volumes and intermolecular forces were first carried out by J. D. van der Waals[8] in 1873. Van der Waals's treatment provides us with an interpretation of the behavior of real gases at the molecular level.

Consider the approach of a particular molecule toward the wall of its container (Figure 10.24). The intermolecular attractions exerted by neighboring molecules prevent the molecule from hitting the wall as hard as it otherwise would. This results in the pressure exerted by a real gas being lower than that predicted by the ideal gas equation. Van der Waals suggested that the pressure exerted by an ideal gas, P_{ideal}, is related to the experimentally measured pressure, P_{real}, by the equation

$$P_{ideal} = P_{real} + \frac{an^2}{V^2}$$

where a is a constant and n and V are the number of moles and volume of the gas, respectively. The correction term for pressure (an^2/V^2) can be understood as follows. The intermolecular interaction that gives rise to nonideal behavior depends on how frequently any two molecules encounter each other. The number of such encounters increases with the square of the number of molecules per unit volume (n/V), and a is a proportionality constant. The quantity P_{ideal} is the pressure we would measure if there were no intermolecular attractions.

The other correction concerns the volume occupied by the gas molecules. In the ideal gas equation, V represents the volume of the container. However, each molecule actually occupies a very small but nonzero volume. We can correct for the volume occupied by the gas molecules by subtracting a term, nb, from the volume of the container:

$$V_{real} = V_{ideal} - nb$$

where n and b are the number of moles and the proportionality constant, respectively.

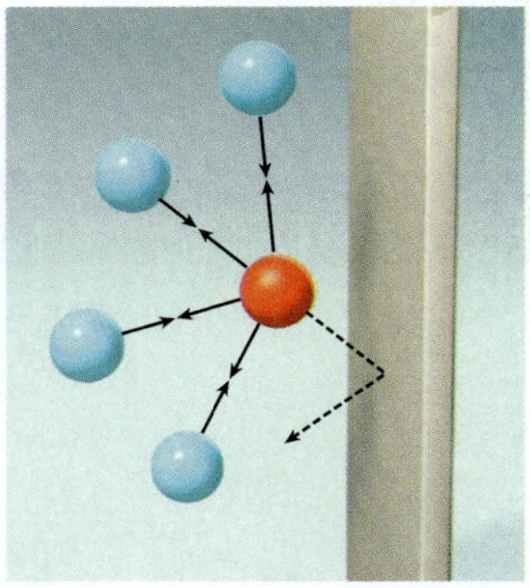

Figure 10.24 The effect of intermolecular attractions on the pressure exerted by a gas.

8. Johannes Diderik van der Waals (1837–1923). Dutch physicist. Van der Waals received the Nobel Prize in Physics in 1910 for his work on the properties of gases and liquids.

What's *Really* the Difference Between Real Gases and Ideal Gases?

As illustrated by the molar volumes in Figure 10.23, the ideal gas equation is remarkably accurate for common gases at or near STP. However, at low temperatures and high pressures, the assumptions that enable us to treat gases as ideal are no longer valid. In these cases, we must consider the effects of attractive forces between the molecules and the nonzero volume that the molecules occupy.

When a gas is cooled and/or compressed, it condenses to a liquid—indicating that there are attractive forces between the molecules [▶ Chapter 11]. Even in the gas phase, the attractive forces between molecules can impact the observed behavior of a substance. As illustrated in Figure 10.24, a molecule that is attracted to other molecules in a sample of gas will not strike the wall of the container with as high a velocity as it would if there were no such intermolecular attractions. The pressure term in the van der Waals equation, $P + a(n/V)^2$, is the experimentally determined pressure, P, plus a correction for the pressure that we do *not* observe because of attractive forces between the gas molecules. Note that the correction factor depends on the moles-per-unit-volume (n/V) squared. The value of the constant a is specific to a particular gas.

When two gas molecules (assumed to be spherical) approach each other, the distance of closest approach is the sum of their radii ($2r$). The volume around each molecule into which the center of another molecule cannot penetrate is called the *excluded volume*. The effect of the excluded volume is to limit the fraction of the container volume actually available for molecules to move about in a gas sample. Thus, the volume term in the van der Waals equation, $V - nb$, is the container volume V minus the correction for the excluded volume, nb, where n is the number of moles of the gas and b is the excluded volume per mole of the gas.

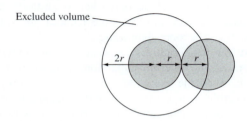
Excluded volume

(See end-of-chapter problems 10.117 and 10.118.)

Incorporating both corrections into the ideal gas equation gives us the **van der Waals equation,** with which we can analyze gases under conditions where ideal behavior is not expected.

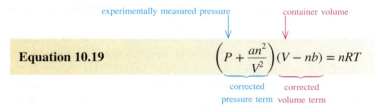

Equation 10.19

experimentally measured pressure   container volume

$$\left(P + \frac{an^2}{V^2}\right)(V - nb) = nRT$$

corrected   corrected
pressure term   volume term

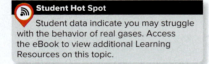
The van der Waals constants a and b for a number of gases are listed in Table 10.6. The magnitude of a indicates how strongly molecules of a particular type of gas attract one another. The magnitude of b is related to molecular (or atomic) size, although the relationship is not a simple one.

TABLE 10.6   Van der Waals Constants of Some Common Gases

Gas	$a\left(\dfrac{atm \cdot L^2}{mol^2}\right)$	$b\left(\dfrac{L}{mol}\right)$	Gas	$a\left(\dfrac{atm \cdot L^2}{mol^2}\right)$	$b\left(\dfrac{L}{mol}\right)$
He	0.034	0.0237	O_2	1.36	0.0318
Ne	0.211	0.0171	Cl_2	6.49	0.0562
Ar	1.34	0.0322	CO_2	3.59	0.0427
Kr	2.32	0.0398	CH_4	2.25	0.0428
Xe	4.19	0.0510	CCl_4	20.4	0.138
H_2	0.244	0.0266	NH_3	4.17	0.0371
N_2	1.39	0.0391	H_2O	5.46	0.0305

Sample Problem 10.17 shows how to use the van der Waals equation.

SAMPLE PROBLEM 10.17

A sample of 3.50 moles of NH_3 gas occupies 5.20 L at 47°C. Calculate the pressure of the gas (in atm) using (a) the ideal gas equation and (b) the van der Waals equation.

Strategy (a) Use the ideal gas equation, $PV = nRT$.

(b) Use Equation 10.19 and a and b values for NH_3 from Table 10.6.

Setup $T = 320.15$ K, $a = 4.17$ atm · L/mol², and $b = 0.0371$ L/mol.

Solution (a) $P = \dfrac{nRT}{V} = \dfrac{(3.50 \text{ mol})\left(\dfrac{0.08206 \text{ L} \cdot \text{atm}}{\text{K} \cdot \text{mol}}\right)(320.15 \text{ K})}{5.20 \text{ L}} = 17.7$ atm

(b) Evaluating the correction terms in the van der Waals equation, we get

$$\frac{an^2}{V^2} = \frac{\left(\dfrac{4.17 \text{ atm} \cdot \text{L}^2}{\text{mol}^2}\right)(3.50 \text{ mol})^2}{(5.20 \text{ L})^2} = 1.89 \text{ atm}$$

$$nb = (3.50 \text{ mol})\left(\frac{0.0371 \text{ L}}{\text{mol}}\right) = 0.130 \text{ L}$$

Finally, substituting these results into Equation 10.19, we have

$$(P + 1.89 \text{ atm})(5.20 \text{ L} - 0.130 \text{ L}) = (3.50 \text{ mol})\left(\frac{0.08206 \text{ L} \cdot \text{atm}}{\text{K} \cdot \text{mol}}\right)(320.15 \text{ K})$$

$$P = 16.2 \text{ atm}$$

THINK ABOUT IT

As is often the case, the pressure exerted by the real gas sample is lower than predicted by the ideal gas equation.

Practice Problem **A**TTEMPT Using data from Table 10.6, calculate the pressure exerted by 11.9 mol of neon gas in a volume of 5.75 L at 25°C using (a) the ideal gas equation and (b) the van der Waals equation (Equation 10.19). Compare your results.

Practice Problem **B**UILD Calculate the pressure exerted by 0.3500 mol of oxygen gas in a volume of 6.500 L at 32.0°C using (a) the ideal gas equation and (b) the van der Waals equation.

Practice Problem **C**ONCEPTUALIZE What properties of real gases prevent them from exhibiting ideal behavior? Explain why gases exhibit a greater degree of ideal behavior at very high temperatures and/or at very low pressures.

One way to measure a gas's deviation from ideal behavior is to determine its compressibility factor, Z, where $Z = PV/RT$. For one mole of an ideal gas, Z is equal to 1 at all pressures and temperatures. For real gases, the factors that contribute to nonideal behavior cause the value of Z to deviate from 1. Intermolecular forces help to account for the plots in Figure 10.25(a). Molecules exert both attraction and repulsion on one another. At large separations (low pressures), attraction predominates. In this region, the gas is more compressible than an ideal gas and the curve dips below the horizontal line ($Z < 1$). As molecules are brought closer together under pressure, repulsion begins to play an important role. If the pressure continues to increase, a point is reached when the gas becomes less compressible than an ideal gas because the molecules repel one another and the curve rises above the horizontal line ($Z > 1$). Figure 10.25(b) shows the plots of Z versus pressure at different temperatures. We see that as the temperature increases, the gas behaves more like an ideal gas (the curves become closer to the horizontal line). The increase in the molecules' kinetic energy makes molecular attraction less important.

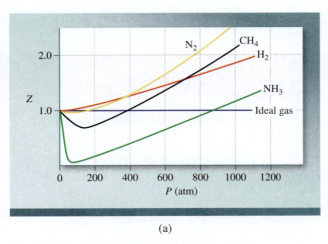

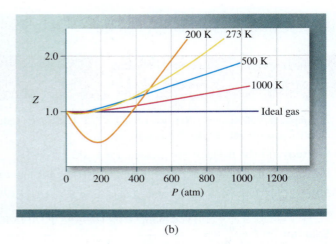

(a)   (b)

Figure 10.25 (a) Compressibility factor (Z) as a function of pressure for several gases at 0°C. (b) Compressibility factor as a function of pressure for N_2 gas at several different temperatures.

CHECKPOINT – SECTION 10.7   Deviation from Ideal Behavior

10.7.1 Which of the following conditions cause deviation from ideal behavior? (Select all that apply.)

a) High pressure

b) Low pressure

c) High temperature

d) Low temperature

e) High volume

10.7.2 Using the van der Waals equation, calculate the pressure exerted by 1.5 moles of carbon dioxide in a 3.75-L vessel at 10°C.

a) 9.3 atm

b) 8.9 atm

c) 10 atm

d) 8.6 atm

e) –2.4 atm

Chapter Summary

Section 10.1

- A gas assumes the volume and shape of its container and is compressible. Gases generally have low densities (expressed in g/L) and will mix in any proportions to give homogeneous solutions.
- Gases exert *pressure,* which is the force per unit area. The SI units of force and pressure are the *newton (N)* and the *pascal (Pa),* respectively. Other commonly used units of pressure are atmosphere (atm), mmHg, torr, and bar.
- Pressure can be measured using a *barometer* or a *manometer.*
- *Standard atmospheric pressure* (1 atm) is the pressure exerted by the atmosphere at sea level.

Section 10.2

- The physical state of a sample of gas can be described using four parameters: temperature (T), pressure (P), volume (V), and number of moles (n). Equations relating these parameters are called the *gas laws.*
- *Boyle's law* states that the volume of a sample of gas at constant temperature is inversely proportional to pressure.
- Experiments done by Charles and Gay-Lussac showed that the volume of a gas at constant pressure is directly proportional to temperature. Lord Kelvin used Charles's and Gay-Lussac's data to propose that *absolute zero* is the lowest theoretically attainable temperature. The *absolute temperature scale,* also known as the *Kelvin temperature scale,* is used for all calculations involving gases.
- *Charles's and Gay-Lussac's law,* commonly known as *Charles's law,* states that the volume of a sample of gas at constant pressure is directly proportional to its absolute temperature.
- *Avogadro's law* states that the volume of a sample of gas at constant temperature and pressure is directly proportional to the number of moles.
- The *combined gas law* combines the laws of Boyle, Charles, and Avogadro and relates pressure, volume, temperature, and number of moles without assuming that any of the parameters is constant.

Section 10.3

- The *ideal gas equation,* $PV = nRT$, makes it possible to predict the behavior of gases. An *ideal gas* is one that behaves in a way predicted by the ideal gas equation. R is the *gas constant,* which may be expressed in a variety of units. The units used to express R depend on the units used to express P and V.
- *Standard temperature and pressure* (STP) is defined as 0°C and 1 atm.
- The ideal gas equation can be used to calculate the density of a gas and to interconvert between density and molar mass.

Section 10.4

- For a reaction occurring at constant temperature and pressure, and involving only gases, the coefficients in the balanced chemical equation apply to units of volume, as well as to numbers of molecules or moles.
- A balanced chemical equation and the ideal gas equation can be used to determine volumes of gaseous reactants and/or products in a reaction.

Section 10.5

- Each component in a mixture of gases exerts a *partial pressure* (P_i) independent of the other mixture components. *Dalton's law of partial pressures* states that the total pressure exerted by a gas *mixture* is the sum of the partial pressures of the components.
- *Mole fraction* (χ_i) is the unitless quotient of the number of moles of a mixture component and the total number of moles in the mixture, n_i/n_{total}.

Section 10.6

- According to the *kinetic molecular theory,* gases are composed of particles with negligible volume that are separated by large distances; the particles are in constant, random motion, and collisions between the particles and between the particles and their container walls are perfectly elastic; there are no attractive or repulsive forces between the particles; and the average kinetic energy of particles in a sample is proportional to the absolute temperature of the sample.
- Kinetic molecular theory can be used to explain the compressibility of gases and the empirical gas laws.
- The *root-mean-square (rms) speed* (u_{rms}) of gas molecules in a sample at a given temperature is inversely proportional to the molecular mass.
- According to *Graham's law,* the rates of *diffusion* (mixing of gases) and *effusion* (escape of a gas from a container into a vacuum) are inversely proportional to the square root of the molar mass of the gas.

Section 10.7

- Deviation from ideal behavior is observed at high pressure and/or low temperature. The *van der Waals equation* makes corrections for the nonzero volume of gas molecules and the attractive forces between molecules.

Key Words

Key Equations

10.1 $P = hdg$	The pressure exerted by a column of fluid is calculated as the product of the column height (in m), the density of the fluid (in kg/m^3), and the gravitational constant (9.80665 m/s^2).
10.2(a) $V = k_1 \dfrac{1}{P}$ (at constant temperature) **10.2(b)** $PV = k_1$ (at constant temperature)	Boyle's law: At constant temperature, (a) the volume of a gas is inversely proportional to pressure; and (b) the product of volume and pressure for a sample of gas is constant.
10.3 $P_1V_1 = P_2V_2$ (at constant temperature)	For a sample of gas at constant temperature, because the product of pressure and volume is constant, we can calculate the change in volume for a given change in pressure—or vice versa.
10.4(a) $V = k_2T$ (at constant pressure) **10.4(b)** $\dfrac{V}{T} = k_2$ (at constant pressure)	Charles's law: At constant pressure, (a) the volume of a sample of gas is directly proportional to absolute temperature; and (b) the ratio of volume to absolute temperature is constant.
10.5 $\dfrac{V_1}{T_1} = \dfrac{V_2}{T_2}$ (at constant pressure)	For a sample of gas at constant pressure, because the ratio of volume to absolution temperature is constant, we can calculate the change in volume for a given change in temperature.
10.6(a) $V = k_3n$ (at constant temperature and pressure) **10.6(b)** $\dfrac{V}{n} = k_3$	Avogadro's law: At constant temperature and pressure, (a) the volume of a sample of gas is directly proportional to the number of moles; and (b) the ratio of volume to number of moles is constant.
10.7 $\dfrac{V_1}{n_1} = \dfrac{V_2}{n_2}$	For a sample of gas at constant temperature and pressure, because the ratio of volume to number of moles is constant, we can calculate the change in volume for a given change in number of moles.
10.8(a) $\dfrac{P_1V_1}{n_1T_1} = \dfrac{P_2V_2}{n_2T_2}$ **10.8(b)** $\dfrac{P_1V_1}{T_1} = \dfrac{P_2V_2}{T_2}$	Combined gas law: (a) The quantity $\dfrac{PV}{nT}$ is constant; and (b) because the quantity $\dfrac{PV}{T}$ is constant for a given amount of gas, we can calculate the changes in pressure, volume, and/or temperature as the other parameters change.
10.9 $PV = nRT$	Ideal gas equation: The product of pressure and volume is directly proportional to the product of number of moles and absolute temperature. R is the gas constant. Units of R depend on the units used for the other parameters.

10.10 $d = \dfrac{P\mathcal{M}}{RT}$	By rearranging the ideal gas equation, we can calculate the density of a gas using its molar mass.
10.11 $\mathcal{M} = \dfrac{dRT}{P}$	By rearranging the ideal gas equation, we can calculate the molar mass of a gas using its density.
10.12(a) $n = P \times \left(\dfrac{V}{RT}\right)$ (at constant V and T) **10.12(b)** $\Delta n = \Delta P \times \left(\dfrac{V}{RT}\right)$ (at constant V and T)	The net number of gaseous moles consumed or produced in a reaction can be calculated using the measured change in pressure.
10.13 $\chi_i = \dfrac{n_i}{n_{\text{total}}}$	The mole fraction of a component in a mixture is the ratio of number of moles of the component to the total number of moles.
10.14 $\chi_i = \dfrac{P_i}{P_{\text{total}}}$	Mole fractions in a mixture of gases can also be calculated as the ratio of partial pressure of a component to total pressure.
10.15 $\chi_i \times n_{\text{total}} = n_i$	The number of moles of a component in a gaseous mixture is the product of the component's mole fraction and the total number of moles.
10.16 $\chi_i \times P_{\text{total}} = P_i$	The partial pressure of a component of a gaseous mixture is the product of the component's mole fraction and the total pressure.
10.17 $u_{\text{rms}} = \sqrt{\dfrac{3RT}{\mathcal{M}}}$	The root-mean-square speed of gas molecules in a sample is inversely proportional to the square root of the molar mass of the gas.
10.18 $\dfrac{u_{\text{rms}}(1)}{u_{\text{rms}}(2)} = \sqrt{\dfrac{\mathcal{M}_2}{\mathcal{M}_1}}$	Rates of effusion/diffusion of gases of different molar masses can be compared using the square root of the ratio of their molar masses, with the rate of each gas being inversely proportional to the square root of its molar mass.
10.19 $\left(P + \dfrac{an^2}{V^2}\right)(V - nb) = nRT$	The ideal gas equation is modified for real gases by applying a correction to both the pressure term and the volume term. The constants a and b depend on the identity of the gas.

Key Constants

Section 10.1	Gravitational constant $g = 9.80665$ m/s^2	Used to calculate the pressure exerted by of a column of fluid of known density.
Section 10.3	Gas constant $R = 0.08206$ L $\cdot$ atm/K $\cdot$ mol	Used with the ideal gas equation.

Most of the gases that we encounter are mixtures of two or more different gases. The concentrations of gases in a mixture are typically expressed using mole fractions, which are calculated using Equation 10.13:

$$\chi_i = \frac{n_i}{n_{total}}$$

Depending on the information given in a problem, calculating mole fractions may require you to determine molar masses and carry out mass-to-mole conversions [◀◀ Section 3.4].

For example, consider a mixture that consists of known masses of three different gases: 5.50 g He, 7.75 g N_2O, and 10.00 g SF_6. Molar masses of the components are

He: 4.003 = $\dfrac{4.003\ g}{mol}$ N_2O: 2(14.01) + (16.00) = $\dfrac{44.02\ g}{mol}$ SF_6: 32.07 + 6(19.00) = $\dfrac{146.1\ g}{mol}$

We convert each of the masses given in the problem to moles by dividing each by the corresponding molar mass:

$\dfrac{5.50\ g\ He}{4.003\ g/mol}$ ⟹ 1.374 mol He $\dfrac{7.75\ g\ N_2O}{44.02\ g/mol}$ ⟹ 0.1761 mol N_2O $\dfrac{10.00\ g\ SF_6}{146.1\ g/mol}$ ⟹ 0.06846 mol SF_6

We then determine the total number of moles in the mixture:

1.374 mol He + 0.1761 mol N_2O + 0.06846 mol SF_6 = 1.619 mol

We divide the number of moles of each component by the total number of moles to get each component's mole fraction.

$\chi_{He} = \dfrac{1.374\ mol\ He}{1.619\ mol}$ ⟹ 0.849 $\chi_{N_2O} = \dfrac{0.1761\ mol\ N_2O}{1.619\ mol}$ ⟹ 0.109 $\chi_{SF_6} = \dfrac{0.06846\ mol\ SF_6}{1.619\ mol}$ = 0.0423

The resulting mole fractions have no units; and for any mixture, the sum of mole fractions of all components is 1. Rounding error may result in the overall sum of mole fractions not being exactly 1. In this case, to the appropriate number of significant figures [◀◀ Section 1.5], the sum is 1.00. (Note that we kept an extra digit throughout the calculations.)

Because at a given temperature, pressure is proportional to the number of moles, mole fractions can also be calculated using the partial pressures of the gaseous components using Equation 10.14:

$$\chi_i = \frac{P_i}{P_{total}}$$

Because gases that do not react with one another are all mutually miscible [◀◀ Section 10.1], gas mixtures are homogeneous—and can also be referred to as *solutions*. And although we first encounter mole fractions in the context of gases, they are also used extensively in the context of other solutions—including *aqueous* solutions [▶▶ Section 13.5]. Determination of mole fraction is done the same way, regardless of the nature of the solution. When liquids are involved, it is sometimes necessary to convert from volume to mass using the liquid's *density* [◀◀ Section 1.3].

volume of liquid (mL) × density of liquid (g/mL) ⟹ mass of liquid (g)

Consider the following example: 5.75 g of sugar (sucrose, $C_{12}H_{22}O_{11}$) is dissolved in 100.0 mL of water at 25°C. We first determine the molar masses of sucrose and water.

$$H_2O: 2(1.008) + 16.00 = \boxed{\frac{18.02 \text{ g}}{\text{mol}}} \qquad C_{12}H_{22}O_{11}: 12(12.01) + 22(1.008) + 11(16.00) = \boxed{\frac{342.3 \text{ g}}{\text{mol}}}$$

Then we use the density of water to convert the volume given to a mass. The density of water at 25°C is 0.9970 g/mL.

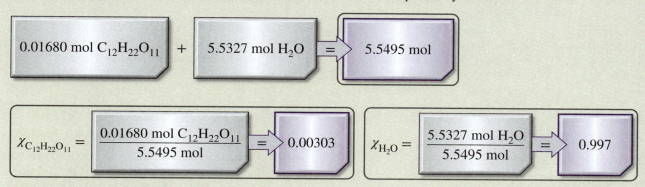

$$100.0 \text{ mL H}_2O \times \frac{0.9970 \text{ g}}{\text{mL}} = 99.70 \text{ g H}_2O$$

We convert the masses of both solution components to moles:

$$\frac{5.75 \text{ g C}_{12}H_{22}O_{11}}{342.3 \text{ g/mol}} = 0.01680 \text{ mol C}_{12}H_{22}O_{11} \qquad \frac{99.70 \text{ g H}_2O}{18.02 \text{ g/mol}} = 5.5327 \text{ mol H}_2O$$

We then sum the number of moles and divide moles of each individual component by the total.

$$0.01680 \text{ mol C}_{12}H_{22}O_{11} + 5.5327 \text{ mol H}_2O = 5.5495 \text{ mol}$$

$$\chi_{C_{12}H_{22}O_{11}} = \frac{0.01680 \text{ mol C}_{12}H_{22}O_{11}}{5.5495 \text{ mol}} = 0.00303 \qquad \chi_{H_2O} = \frac{5.5327 \text{ mol H}_2O}{5.5495 \text{ mol}} = 0.997$$

To the appropriate number of significant figures, the mole fractions sum to 1.

Key Skills Problems

10.1
Determine the mole fraction of helium in a gaseous mixture consisting of 0.524 g He, 0.275 g Ar, and 2.05 g CH₄.

(a) 0.0069 (b) 0.0259 (c) 0.481 (d) 0.493 (e) 0.131

10.2
Determine the mole fraction of argon in a gaseous mixture in which the partial pressures of H₂, N₂, and Ar are 0.01887 atm, 0.3105 atm, and 1.027 atm, respectively.

(a) 0.01391 (b) 0.2289 (c) 0.7572 (d) 0.01887 (e) 1.027

10.3
Determine the mole fraction of *water* in a solution consisting of 5.00 g glucose ($C_6H_{12}O_6$) and 250.0 g water.

(a) 0.00200 (b) 0.998 (c) 0.0278 (d) 1.00 (e) 0.907

10.4
Determine the mole fraction of ethanol in a solution containing 15.50 mL ethanol (C_2H_5OH) and 110.0 mL water. (The density of ethanol is 0.789 g/mL; the density of water is 0.997 g/mL.)

(a) 0.0436 (b) 6.08 (c) 0.265 (d) 0.958 (e) 0.0418

Questions and Problems

Applying What You've Learned

Scuba divers are not the only athletes who can suffer the detrimental effects of sudden changes in pressure. Mountain climbers, too, are susceptible to the dangers of rapid ascent. At high elevation, air pressure is significantly lower than at sea level. A lower total pressure means a lower partial pressure of oxygen, and insufficient oxygen or *hypoxia* can cause altitude sickness. Early symptoms of altitude sickness include headache, dizziness, and nausea. In severe cases, climbers may suffer hallucinations, seizure, coma, and even death.

In 1990, Igor Gamow, a professor of microbiology at the University of Colorado, patented a portable device for high-altitude treatment of altitude sickness. The Gamow Bag is an inflatable cylinder large enough to accommodate an adult mountain climber. The bag is inflated and pressurized with a foot pump, and the afflicted climber remains sealed inside the pressurized bag until symptoms subside enough to begin descent. Although the bag is pressurized to only about 0.14 atm above atmospheric pressure, at very high altitudes this corresponds to a simulated descent on the order of 10,000 ft!

Gamow Bag
Pascal Boegli/Alamy Stock Photo

Problems:
(a) If a Gamow Bag is pressurized to 0.14 atm above atmospheric pressure at an altitude of 25,000 ft, where atmospheric pressure is 0.37 atm, what height column of mercury would be supported by the pressure inside the pressurized bag [◄◄ Sample Problem 10.1]? (b) The Gamow Bag inflates to a volume of 4.80×10^2 L. What volume would be occupied by the air in the pressurized bag in part (a) at standard atmospheric pressure? (Assume no change in temperature.) [◄◄ Sample Problem 10.2] (c) Calculate the density of the air in the pressurized bag in part (a) at 0°C. (Assume that air is 80 percent N_2 and 20 percent O_2 by volume.) [◄◄ Sample Problem 10.7] (d) LiOH scrubbers are sometimes used to prevent the buildup of CO_2 during use of the bag. What volume of CO_2 (at 0°C) can be removed from the pressurized bag in part (a) by 0.50 kg LiOH [◄◄ Sample Problem 10.9]? (e) Calculate the number of moles of each gas in the pressurized bag in part (a) at 0°C [◄◄ Sample Problem 10.13].

SECTION 10.1: PROPERTIES OF GASES

Review Questions

10.1   Name five elements and five compounds that exist as gases at room temperature.

10.2   List the physical characteristics of gases.

10.3   Define *pressure* and give the common units for pressure.

10.4   Describe how a barometer and a manometer are used to measure gas pressure.

10.5   Why is mercury a more suitable substance to use in a barometer than water?

10.6   Explain why the height of mercury in a barometer is independent of the cross-sectional area of the tube.

10.7   Would it be easier to drink water with a straw on top of Mt. Everest or at the foot? Explain.

10.8   Is the atmospheric pressure in a mine that is 500 m below sea level greater or less than 1 atm?

10.9   What is the difference between the terms *gas* and *vapor*? At 25°C, which of the following substances in the gas phase should be properly called a gas and which should be called a vapor: molecular chlorine (Cl_2), molecular iodine (I_2)?

10.10   If the maximum distance that water may be brought up a well by a suction pump is 34 ft (10.3 m), how is it possible to obtain water and oil from hundreds of feet below the surface of Earth?

10.11   Why is it that if the barometer reading falls in one part of the world, it must rise somewhere else?

10.12   Why do astronauts have to wear protective suits when they are on the surface of the moon?

Computational Problems

10.13   Convert 375 mmHg to atmospheres, bar, torr, and pascals.

10.14   The atmospheric pressure at the summit of Mt. McKinley is 581 mmHg on a certain day. What is the pressure in atmospheres and in kilopascals?

10.15   Calculate the height of a column of methanol (CH_3OH) that would be supported by atmospheric pressure. The density of methanol is 0.787 g/cm³.

10.16   Calculate the height of a column of ethylene glycol [$CH_2(OH)CH_2(OH)$] that would be supported by atmospheric pressure (1 atm). The density of ethylene glycol is 1.12 g/cm³.

10.17   What pressure (in atm) is exerted by a column of toluene (C_7H_8) 87 m high? The density of toluene is 0.867 g/cm³.

10.18 What pressure (in atm) is exerted by a column of isopropanol (C_3H_7OH) 264 m high? The density of isopropanol is 0.785 g/cm³.

SECTION 10.2: THE GAS LAWS

Review Questions

10.19 State the following gas laws in words and also in the form of an equation: Boyle's law, Charles's law, Avogadro's law. In each case, indicate the conditions under which the law is applicable, and give the units for each quantity in the equation.

10.20 Explain why a helium weather balloon expands as it rises in the air. Assume that the temperature remains constant.

Computational Problems

10.21 A gas sample occupying a volume of 25.6 mL at a pressure of 0.970 atm is allowed to expand at constant temperature until its pressure reaches 0.541 atm. What is its final volume?

10.22 At 46°C a sample of ammonia gas exerts a pressure of 5.3 atm. What is the pressure when the volume of the gas is reduced to one-fourth of the original value at the same temperature?

10.23 The volume of a gas is 7.15 L, measured at 1.00 atm. What is the pressure of the gas in mmHg if the volume is changed to 9.25 L? (The temperature remains constant.)

10.24 A sample of air occupies 3.8 L when the pressure is 1.2 atm. (a) What volume does it occupy at 6.6 atm? (b) What pressure is required to compress it to 0.075 L? (The temperature is kept constant.)

10.25 A 28.4-L volume of methane gas is heated from 35°C to 72°C at constant pressure. What is the final volume of the gas?

10.26 Under constant-pressure conditions a sample of hydrogen gas initially at 88°C and 9.6 L is cooled until its final volume is 3.4 L. What is its final temperature?

Conceptual Problems

10.27 Which diagram(s) could represent the result of increasing temperature at constant pressure?

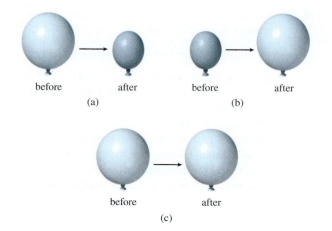

(a) (b)

(c)

10.28 Which of the diagrams in Problem 10.27 could represent the result of decreasing temperature and increasing external pressure?

10.29 A gaseous sample of a substance is cooled at constant pressure. Which of the following diagrams best represents the situation if the final temperature is (a) above the boiling point of the substance and (b) below the boiling point but above the freezing point of the substance?

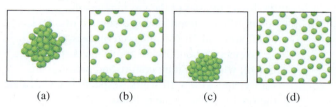

(a) (b) (c) (d)

10.30 Consider the following gaseous sample in a cylinder fitted with a movable piston. Initially there are n moles of the gas at temperature T, pressure P, and volume V.

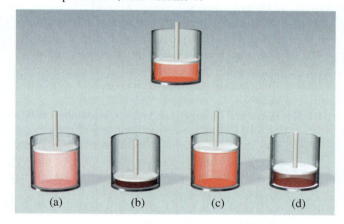

(a) (b) (c) (d)

Choose the cylinder that correctly represents the gas after each of the following changes. (1) The pressure on the piston is tripled at constant n and T. (2) The absolute temperature is doubled at constant n and P. (3) n more moles of the gas are added at constant T and P. (4) Absolute temperature is halved at constant P.

SECTION 10.3: THE IDEAL GAS EQUATION

Review Questions

10.31 List the characteristics of an ideal gas.

10.32 What are standard temperature and pressure (STP)? What is the significance of STP in relation to the volume of 1 mole of an ideal gas?

10.33 Why is the density of a gas much lower than that of a liquid or solid under atmospheric conditions? What units are normally used to express the density of gases?

Computational Problems

10.34 A sample of nitrogen gas in a 4.5-L container at a temperature of 27°C exerts a pressure of 4.1 atm. Calculate the number of moles of gas in the sample.

10.35 Given that 6.9 moles of carbon monoxide gas are present in a container of volume 30.4 L, what is the pressure of the gas (in atm) if the temperature is 82°C?

10.36 What volume will 9.8 moles of sulfur hexafluoride (SF_6) gas occupy if the temperature and pressure of the gas are 105°C and 9.4 atm, respectively?

10.37 The temperature of 1.75 L of a gas initially at STP is raised to 313°C at constant volume. Calculate the final pressure of the gas in atmospheres.

10.38 A gas-filled balloon having a volume of 2.50 L at 1.2 atm and 20°C is allowed to rise to the stratosphere (about 30 km above the surface of Earth), where the temperature and pressure are −23°C and 3.00×10^{-3} atm, respectively. Calculate the final volume of the balloon.

10.39 A gas evolved during the fermentation of glucose (wine making) has a volume of 0.67 L at 22.5°C and 1.00 atm. What was the volume of this gas at the fermentation temperature of 36.5°C and 1.00 atm pressure?

10.40 An ideal gas originally at 0.85 atm and 66°C was allowed to expand until its final volume, pressure, and temperature were 94 mL, 0.60 atm, and 45°C, respectively. What was its initial volume?

10.41 Calculate the volume (in liters) of 75.5 g of CO_2 at STP.

10.42 A gas at 572 mmHg and 35.0°C occupies a volume of 6.15 L. Calculate its volume at STP.

10.43 Dry ice is solid carbon dioxide. A 0.071-g sample of dry ice is placed in an evacuated 1.8-L vessel at 35°C. Calculate the pressure inside the vessel after all the dry ice has been converted to CO_2 gas.

10.44 At STP, 0.7678 L of a gas weighs 5.005 g. Calculate the molar mass of the gas.

10.45 At 914.3 torr and 72.5°C, 14.85 g of a gas occupies a volume of 3.787 L. What is the molar mass of the gas?

10.46 Ozone molecules in the stratosphere absorb much of the harmful radiation from the sun. Typically, the temperature and pressure of ozone in the stratosphere are 250 K and 1.0×10^{-3} atm, respectively. How many ozone molecules are present in 1.0 L of air under these conditions?

10.47 Assuming that air contains 78 percent N_2, 21 percent O_2, and 1.0 percent Ar, all by volume, how many molecules of each type of gas are present in 1.0 L of air at STP?

10.48 A 2.10-L vessel contains 4.65 g of a gas at 1.00 atm and 27.0°C. (a) Calculate the density of the gas in g/L. (b) What is the molar mass of the gas?

10.49 Calculate the density of hydrogen bromide (HBr) gas in g/L at 733 mmHg and 46°C.

10.50 A certain anesthetic contains 64.9 percent C, 13.5 percent H, and 21.6 percent O by mass. At 120°C and 750 mmHg, 1.00 L of the gaseous

compound weighs 2.30 g. What is the molecular formula of the compound?

10.51 A compound has the empirical formula SF_4. At 20°C, 0.100 g of the gaseous compound occupies a volume of 22.1 mL and exerts a pressure of 1.02 atm. What is the molecular formula of the gas?

Conceptual Problems

10.52 The pressure of 6.0 L of an ideal gas in a flexible container is decreased to one-third of its original pressure, and its absolute temperature is decreased by one-half. What is the final volume of the gas?

10.53 A certain amount of gas at 25°C and at a pressure of 0.800 atm is contained in a vessel. Suppose that the vessel can withstand a pressure no higher than 5.00 atm. How high can you raise the temperature of the gas without bursting the vessel?

SECTION 10.4: REACTIONS WITH GASEOUS REACTANTS AND PRODUCTS

Problems

10.54 Consider the formation of nitrogen dioxide from nitric oxide and oxygen:

$$2NO(g) + O_2(g) \longrightarrow 2NO_2(g)$$

If 9.0 L of NO is combined with excess O_2 at STP, what is the volume in liters of the NO_2 produced?

10.55 Methane, the principal component of natural gas, is used for heating and cooking. The combustion process is:

$$CH_4(g) + 2O_2(g) \longrightarrow CO_2(g) + 2H_2O(l)$$

If 15.0 moles of CH_4 react with oxygen, what is the volume of CO_2 (in liters) produced at 23.0°C and 0.985 atm?

10.56 When coal is burned, the sulfur present in coal is converted to sulfur dioxide (SO_2), which is responsible for the acid rain phenomenon:

$$S(s) + O_2(g) \longrightarrow SO_2(g)$$

If 3.15 kg of S reacts with oxygen, calculate the volume of SO_2 gas (in mL) formed at 30.5°C and 1.04 atm.

10.57 In alcohol fermentation, yeast converts glucose to ethanol and carbon dioxide:

$$C_6H_{12}O_6(s) \longrightarrow 2C_2H_5OH(l) + 2CO_2(g)$$

If 5.97 g of glucose reacts and 1.44 L of CO_2 gas is collected at 293 K and 0.984 atm, what is the percent yield of the reaction?

10.58 The diagram shows the contents of a vessel before and after the reaction of two gaseous species. Write a balanced equation to represent the reaction. If the pressure before the reaction is 1.20 atm, what will be the pressure when the reaction is complete? (Assume the temperature and volume of the vessel are constant.) A = White, B = Blue, C = Striped.

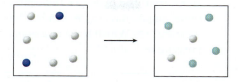

10.59 The diagram shows the contents of a flexible vessel before and after the reaction of the gaseous species XY and Y_3. Write a balanced equation to represent the reaction and determine the final volume when 50.0 mL of XY and 20.0 mL of Y_3 are combined at constant temperature and pressure.

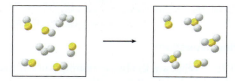

10.60 What is the mass of the solid NH_4Cl formed when 73.0 g of NH_3 is mixed with an equal mass of HCl? What is the volume of the gas remaining, measured at 14.0°C and 752 mmHg? What gas is it?

10.61 Dissolving 3.00 g of an impure sample of calcium carbonate in hydrochloric acid produced 0.656 L of carbon dioxide (measured at 20.0°C and 792 mmHg). Calculate the percent by mass of calcium carbonate in the sample. State any assumptions.

10.62 Calculate the mass in grams of hydrogen chloride produced when 5.6 L of molecular hydrogen measured at STP react with an excess of molecular chlorine gas.

10.63 Ethanol (C_2H_5OH) burns in air:

$$C_2H_5OH(l) + O_2(g) \longrightarrow CO_2(g) + H_2O(l)$$

Balance the equation and determine the volume of air in liters at 52.0°C and 825 mmHg required to burn 235 g of ethanol. Assume that air is 21.0 percent O_2 by volume.

SECTION 10.5: GAS MIXTURES

 Visualizing Chemistry
Figure 10.15

VC 10.1 The molar volume of hydrogen can be determined using the reaction of zinc metal and acid,

$$Zn(s) + 2H^+(aq) \longrightarrow Zn^{2+}(aq) + H_2(g)$$

as shown in Figure 10.15. When the reaction is complete, what does the space above the water in the graduated cylinder contain?
a) $H_2(g)$, $Zn^{2+}(aq)$, and $H_2O(g)$
b) $H_2(g)$ and $H_2O(g)$
c) $H_2(g)$, $H_2O(g)$, and air

VC 10.2 How would the calculated molar volume be affected if we neglected to subtract the partial pressure of water vapor from the total pressure?
a) It would be greater.
b) It would be smaller.
c) It would not change.

VC 10.3 How would the calculated molar volume be affected if we neglected to adjust the level of the graduated cylinder prior to reading the volume of gas collected? Assume that the level of water inside the graduated cylinder is higher than the level outside.
a) It would be greater.
b) It would be smaller.
c) It would not change.

VC 10.4 How would the calculated molar volume be affected if some of the zinc metal failed to drop into the aqueous acid?
a) It would be greater.
b) It would be smaller.
c) It would not change.

Review Questions

10.64 State Dalton's law of partial pressures and explain what *mole fraction* is. Does mole fraction have units?

10.65 What are the approximate partial pressures of N_2 and O_2 in air at the top of a mountain where atmospheric pressure is 0.8 atm? (See Problem 10.47.)

Computational Problems

10.66 The diagram represents a mixture of three different gases. The partial pressure of the gas represented by yellow spheres is 0.385 atm. Determine the mole fractions of each gas, the total pressure, and the partial pressures of the gases represented by blue and green spheres.

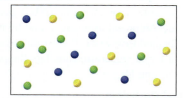

10.67 The diagram represents a mixture of three different gases. The partial pressure of the gas represented by purple spheres is 0.220 atm. Determine the mole fractions of each gas, the total pressure, and the partial pressures of the gases represented by red and yellow spheres.

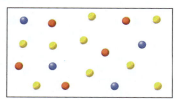

10.68   Dry air near sea level has the following composition by volume: N_2, 78.08 percent; O_2, 20.94 percent; Ar, 0.93 percent; CO_2, 0.05 percent. The atmospheric pressure is 1.00 atm. Calculate (a) the partial pressure of each gas in atmospheres and (b) the concentration of each gas in mol/L at 0°C. (*Hint:* Because volume is proportional to the number of moles present, mole fractions of gases can be expressed as ratios of volumes at the same temperature and pressure.)

10.69   A mixture of helium and neon gases is collected over water at 28.0°C and 745 mmHg. If the partial pressure of helium is 368 mmHg, what is the partial pressure of neon? (Vapor pressure of water at 28°C = 28.3 mmHg.)

10.70   A piece of sodium metal reacts completely with water as follows:

$$2Na(s) + 2H_2O(l) \longrightarrow 2NaOH(aq) + H_2(g)$$

The hydrogen gas generated is collected over water at 25.0°C. The volume of the gas is 246 mL measured at 1.00 atm. Calculate the number of grams of sodium used in the reaction. (Vapor pressure of water at 25°C = 0.0313 atm.)

10.71   A sample of zinc metal reacts completely with an excess of hydrochloric acid:

$$Zn(s) + 2HCl(aq) \longrightarrow ZnCl_2(aq) + H_2(g)$$

The hydrogen gas produced is collected over water at 25.0°C using an arrangement similar to that shown in Figure 10.14(a). The volume of the gas is 7.80 L, and the pressure is 0.980 atm. Calculate the amount of zinc metal in grams consumed in the reaction. (Vapor pressure of water at 25°C = 28.3 mmHg.)

10.72   Helium is mixed with oxygen gas for deep-sea divers. Calculate the percent by volume of oxygen gas in the mixture if the diver has to submerge to a depth where the total pressure is 5.2 atm. The partial pressure of oxygen is maintained at 0.20 atm at this depth.

10.73   A sample of ammonia (NH_3) gas is completely decomposed to nitrogen and hydrogen gases over heated iron wool. If the total pressure is 866 mmHg after the reaction, calculate the partial pressures of N_2 and H_2.

Conceptual Problems

10.74   Consider the three containers shown, all of which have the same volume and are at the same temperature. (a) Which container has the smallest mole fraction of gas A (red)? (b) Which container has the highest partial pressure of gas B (green)? (c) Which container has the highest total pressure?

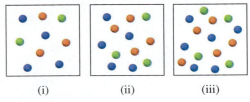

(i)   (ii)   (iii)

10.75   The volume of the box on the right is twice that of the box on the left. The boxes contain helium atoms (red) and hydrogen molecules (green) at the same temperature. (a) Which box has a higher total pressure? (b) Which box has a higher partial pressure of helium?

SECTION 10.6: THE KINETIC MOLECULAR THEORY OF GASES

Review Questions

10.76   What are the basic assumptions of the kinetic molecular theory of gases?

10.77   How does the kinetic molecular theory explain Boyle's law, Charles's law, Avogadro's law, and Dalton's law of partial pressures?

10.78   What does the Maxwell speed distribution curve tell us? Does Maxwell's theory work for a sample of 200 molecules? Explain.

10.79   Which of the following statements is correct? (a) Heat is produced by the collision of gas molecules against one another. (b) When a gas is heated at constant volume, the molecules collide with one another more often.

10.80   Uranium hexafluoride (UF_6) is a much heavier gas than helium, yet at a given temperature, the average kinetic energies of the samples of the two gases are the same. Explain.

10.81   What is the difference between gas diffusion and effusion?

Computational Problems

10.82   Compare the root-mean-square speeds of O_2 and UF_6 at 65°C.

10.83   The temperature in the stratosphere is −23°C. Calculate the root-mean-square speeds of N_2, O_2, and O_3 molecules in this region.

10.84   Nickel forms a gaseous compound of the formula $Ni(CO)_x$. What is the value of x given the fact that under the same conditions of temperature and pressure, methane (CH_4) effuses 3.3 times faster than the compound?

10.85   At a certain temperature the speeds of six gaseous molecules in a container are 2.0, 2.2, 2.6, 2.7, 3.3, and 3.5 m/s. Calculate the root-mean-square speed and the average speed of the molecules. These two average values are close to each other, but the root-mean-square value is always the larger of the two. Why?

10.86   The ^{235}U isotope undergoes fission when bombarded with neutrons. However, its natural abundance is only 0.72 percent. To separate it from the more abundant ^{238}U isotope, uranium is first converted to UF_6, which is easily vaporized above room temperature. The mixture of the $^{235}UF_6$ and $^{238}UF_6$ gases is then subjected to many stages of effusion. Calculate how much faster $^{235}UF_6$ effuses than $^{238}UF_6$.

10.87   An unknown gas evolved from the fermentation of glucose is found to effuse through a porous barrier in 15.0 min. Under the same conditions of temperature and pressure, it takes an equal volume of N_2 12.0 min to effuse through the same barrier. Calculate the molar mass of the unknown gas, and suggest what the gas might be.

Conceptual Problems

10.88   The average distance traveled by a molecule between successive collisions is called *mean free path*. For a given amount of a gas, how does the mean free path of a gas depend on (a) density, (b) temperature at constant volume, (c) pressure at constant temperature, (d) volume at constant temperature, and (e) size of the atoms?

10.89   Each pair of diagrams represents a mixture of gases before and after effusion. In each case, determine how the molar masses of the two gases compare.

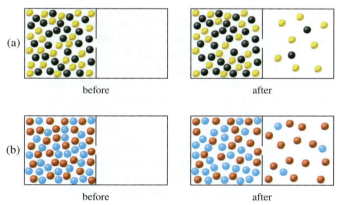

SECTION 10.7: DEVIATION FROM IDEAL BEHAVIOR

Review Questions

10.90   Cite two pieces of evidence to show that gases do not behave ideally under all conditions. Under what set of conditions would a gas be expected to behave most ideally: (a) high temperature and low pressure, (b) high temperature and high pressure, (c) low temperature and high pressure, or (d) low temperature and low pressure?

10.91   Figure 10.25(a) shows that at 0°C, with the exception of H_2, each of the gases has a pressure at which its compressibility factor is equal to 1—the point at which the curve crosses the ideal gas line. What is the significance of this point? Does each of these gases have a pressure at which the assumptions of ideal behavior (negligible molecular volume and no intermolecular attractions) are valid? Explain.

10.92   Write the van der Waals equation for a real gas. Explain the corrective terms for pressure and volume.

10.93   (a) A real gas is introduced into a flask of volume V. Is the corrected volume of the gas greater or less than V? (b) Ammonia has a larger a value than neon does (see Table 10.6). What can you conclude about the relative strength of the attractive forces between molecules of ammonia and between atoms of neon?

Computational Problems

10.94   Using the data shown in Table 10.6, calculate the pressure exerted by 2.50 moles of CO_2 confined in a volume of 5.00 L at 450 K. Compare the pressure with that predicted by the ideal gas equation.

10.95   At 27°C, 10.0 moles of a gas in a 1.50-L container exert a pressure of 130 atm. Is this an ideal gas?

ADDITIONAL PROBLEMS

10.96   Discuss the following phenomena in terms of the gas laws: (a) the pressure increase in an automobile tire on a hot day, (b) the "popping" of a paper bag, (c) the expansion of a weather balloon as it rises in the air, (d) the loud noise heard when a lightbulb shatters.

10.97   Under the same conditions of temperature and pressure, which of the following gases would behave most ideally: Ne, N_2, or CH_4? Explain.

10.98   Nitroglycerin, an explosive compound, decomposes according to the equation

$$4C_3H_5(NO_3)_3(s) \longrightarrow 12CO_2(g) + 10H_2O(g) + 6N_2(g) + O_2(g)$$

Calculate the total volume of gases when collected at 1.2 atm and 25°C from 2.6×10^2 g of nitroglycerin. What are the partial pressures of the gases under these conditions?

10.99   The empirical formula of a compound is CH. At 200°C, 0.145 g of this compound occupies 97.2 mL at a pressure of 0.74 atm. What is the molecular formula of the compound?

10.100   When ammonium nitrite (NH_4NO_2) is heated, it decomposes to give nitrogen gas. This property is used to inflate some tennis balls. (a) Write a balanced equation for the reaction. (b) Calculate the quantity (in grams) of NH_4NO_2 needed to inflate a tennis ball to a volume of 86.2 mL at 1.20 atm and 22°C.

10.101   Three flasks containing gases A (red) and B (blue) are shown here. (a) If the total pressure in (i) is 2.0 atm, what are the pressures in (ii) and (iii)? (b) Calculate the total pressure and the partial pressure of each gas after the valves are opened. The volumes of (i) and (iii) are 2.0 L each, and the volume of (ii) is 1.0 L. The temperature is the same throughout.

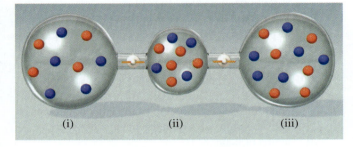

(i)   (ii)   (iii)

10.102   The boiling point of liquid nitrogen is –196°C. On the basis of this information alone, do you think nitrogen is an ideal gas at STP?

10.103   On heating, potassium chlorate ($KClO_3$) decomposes to yield potassium chloride and oxygen gas. In one experiment, a student heated 20.4 g of $KClO_3$ until the decomposition was complete. (a) Write a balanced equation for the reaction. (b) Calculate the volume of oxygen (in liters) if it was collected at 0.962 atm and 18.3°C.

10.104   The volume of a sample of pure HCl gas was 189 mL at 25°C and 108 mmHg. It was completely dissolved in about 60 mL of water and titrated with an NaOH solution; 15.7 mL of the NaOH solution was required to neutralize the HCl. Calculate the molarity of the NaOH solution.

10.105   Propane (C_3H_8) burns in oxygen to produce carbon dioxide gas and water vapor. (a) Write a balanced equation for this reaction. (b) Calculate the number of liters of carbon dioxide measured at STP that could be produced from 7.45 g of propane.

10.106   Consider the following apparatus. Calculate the partial pressures of helium and neon after the stopcock is opened. The temperature remains constant at 16°C.

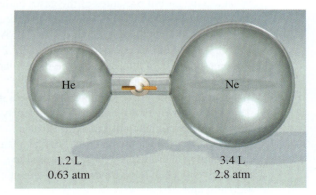

1.2 L   3.4 L
0.63 atm   2.8 atm

10.107   Nitric oxide (NO) reacts with molecular oxygen as follows:

$$2NO(g) + O_2(g) \longrightarrow 2NO_2(g)$$

Initially NO and O_2 are separated as shown here. When the valve is opened, the reaction quickly goes to completion. Determine what gases remain at the end and calculate their partial pressures. Assume that the temperature remains constant at 25°C.

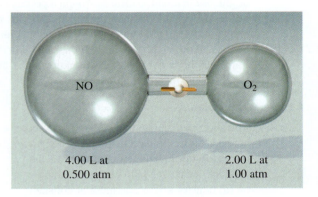

4.00 L at   2.00 L at
0.500 atm   1.00 atm

10.108   Nitrous oxide (N_2O) can be obtained by the thermal decomposition of ammonium nitrate (NH_4NO_3). (a) Write a balanced equation for the reaction. (b) In a certain experiment, a student obtains 0.340 L of the gas at 718 mmHg and 24°C. If the gas weighs 0.580 g, calculate the value of the gas constant.

10.109   Describe how you would measure, by either chemical or physical means, the partial pressures of a mixture of gases of the following composition: (a) CO_2 and H_2, (b) He and N_2.

10.110   A certain hydrate has the formula $MgSO_4 \cdot xH_2O$. A quantity of 54.2 g of the compound is heated in an oven to drive off the water. If the steam generated exerts a pressure of 24.8 atm in a 2.00-L container at 120°C, calculate x.

10.111   A mixture of Na_2CO_3 and $MgCO_3$ of mass 7.63 g is combined with an excess of hydrochloric acid. The CO_2 gas generated occupies a volume of 1.67 L at 1.24 atm and 26°C. From these data, calculate the percent composition by mass of Na_2CO_3 in the mixture.

10.112   Interstellar space contains mostly hydrogen atoms at a concentration of about 1 atom/cm^3. (a) Calculate the pressure of the H atoms. (b) Calculate the volume (in liters) that contains 1.0 g of H atoms. The temperature is 3 K.

10.113   If 10.00 g of water is introduced into an evacuated flask of volume 2.500 L at 65°C, calculate the mass of water vaporized. (*Hint:* Assume that the volume of the remaining liquid water is negligible; the vapor pressure of water at 65°C is 187.5 mmHg.)

10.114 Two vessels are labeled A and B. Vessel A contains NH_3 gas at 70°C, and vessel B contains Ne gas at the same temperature. If the average kinetic energy of NH_3 is 7.1×10^{-21} J/molecule, calculate the root-mean-square speed of Ne atoms in m^2/s^2.

10.115 Which of the following molecules has the largest a value: CH_4, F_2, C_6H_6, or Ne?

10.116 The following procedure is a simple though somewhat crude way to measure the molar mass of a gas. A liquid of mass 0.0184 g is introduced into a syringe like the one shown here by injection through the rubber tip using a hypodermic needle. The syringe is then transferred to a temperature bath heated to 45°C, and the liquid vaporizes. The final volume of the vapor (measured by the outward movement of the plunger) is 5.58 mL, and the atmospheric pressure is 760 mmHg. Given that the compound's empirical formula is CH_2, determine the molar mass of the compound.

Rubber tip

10.117 Consider a gas sample consisting of molecules with radius r. (a) Determine the excluded volume defined by two molecules and (b) calculate the excluded volume per mole (b) for the gas. Compare the excluded volume per mole with the volume actually occupied by a mole of the molecules.

10.118 Determine the excluded volume per mole and the total volume of the molecules in a mole for a gas consisting of molecules with radius 165 picometers (pm). [Note: To obtain the volume in liters, we must express the radius in decimeters (dm).]

10.119 Because the van der Waals constant b is the excluded volume per mole of a gas, we can use the value of b to estimate the radius of a molecule or atom. Consider a gas that consists of molecules, for which the van der Waals constant b is 0.0315 L/mol. Estimate the molecular radius in pm. Assume that the molecules are spherical.

10.120 A gaseous reaction takes place at constant volume and constant pressure in a cylinder as shown here. Which of the following equations best describes the reaction? The initial temperature (T_1) is twice that of the final temperature (T_2).
a) $A + B \longrightarrow C$
b) $AB \longrightarrow C + D$
c) $A + B \longrightarrow C + D$
d) $A + B \longrightarrow 2C + D$

10.121 The partial pressure of carbon dioxide varies with seasons. Would you expect the partial pressure in the Northern Hemisphere to be higher in the summer or winter? Explain.

10.122 (a) What volume of air at 1.0 atm and 22°C is needed to fill a 0.98-L bicycle tire to a pressure of 5.0 atm at the same temperature? (Note that the 5.0 atm is the gauge pressure, which is the difference between the pressure in the tire and atmospheric pressure. Before filling, the pressure in the tire was 1.0 atm.) (b) What is the total pressure in the tire when the gauge pressure reads 5.0 atm? (c) The tire is pumped by filling the cylinder of a hand pump with air at 1.0 atm and then, by compressing the gas in the cylinder, adding all the air in the pump to the air in the tire. If the volume of the pump is 33 percent of the tire's volume, what is the gauge pressure in the tire after three full strokes of the pump? Assume constant temperature.

10.123 At what temperature will He atoms have the same u_{rms} value as N_2 molecules at 25°C?

10.124 Estimate the distance (in nm) between molecules of water vapor at 100°C and 1.0 atm. Assume ideal behavior. Repeat the calculation for liquid water at 100°C, given that the density of water is 0.96 g/cm^3 at that temperature. Comment on your results. (Assume each water molecule to be a sphere with a diameter of 0.3 nm.) (*Hint:* First calculate the number density of water molecules. Next, convert the number density to linear density, that is, the number of molecules in one direction.)

10.125 Which of the noble gases would not behave ideally under any circumstance? Why?

10.126 A 5.72-g sample of graphite was heated with 68.4 g of O_2 in a 8.00-L flask. The reaction that took place was

$$C(graphite) + O_2(g) \longrightarrow CO_2(g)$$

After the reaction was complete, the temperature in the flask was 182°C. What was the total pressure inside the flask?

10.127 A 6.11-g sample of a Cu-Zn alloy reacts with HCl acid to produce hydrogen gas. If the hydrogen gas has a volume of 1.26 L at 22°C and 728 mmHg, what is the percent of Zn in the alloy? (*Hint:* Cu does not react with HCl.)

10.128   Nitrogen forms several gaseous oxides. One of them has a density of 1.33 g/L measured at 764 mmHg and 150°C. Write the formula of the compound.

10.129   Nitrogen dioxide (NO_2) cannot be obtained in a pure form in the gas phase because it exists as a mixture of NO_2 and N_2O_4. At 25°C and 0.98 atm, the density of this gas mixture is 2.7 g/L. What is the partial pressure of each gas?

10.130   Lithium hydride reacts with water as follows:

$$LiH(s) + H_2O(l) \longrightarrow LiOH(aq) + H_2(g)$$

During World War II, U.S. pilots carried LiH tablets. In the event of a crash landing at sea, the LiH would react with the seawater and fill their life jackets and lifeboats with hydrogen gas. How many grams of LiH are needed to fill a 4.1-L life jacket at 0.97 atm and 12°C?

10.131   Assuming ideal behavior, which of the following gases will have the greatest volume at STP? (a) 0.82 mole of He, (b) 24 g of N_2, or (c) 5.0×10^{23} molecules of Cl_2.

10.132   Calculate the density of helium in a helium balloon at 25.0°C. (Assume that the pressure inside the balloon is 1.10 atm.)

10.133   Helium atoms in a closed container at room temperature are constantly colliding with one another and with the walls of their container. Does this "perpetual motion" violate the law of conservation of energy? Explain.

10.134   Uranium hexafluoride (UF_6) is a much heavier gas than hydrogen, yet at a given temperature, the average kinetic energies of these two gases are the same. Explain.

10.135   Consider the molar volumes shown in Figure 10.23. (a) Explain why Cl_2 and NH_3 have molar volumes significantly smaller from that of an ideal gas. (b) Explain why H_2, He, and Ne have molar volumes greater than that of an ideal gas. (*Hint:* Look up the boiling points of the gases shown in the figure.)

10.136   The plot of Z versus P for a gas at 0°C is shown. Explain the causes of the negative deviation from ideal behavior at lower pressures and the positive deviation from ideal behavior at higher pressures.

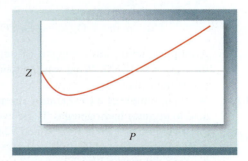

10.137   In 2.00 min, 29.7 mL of He effuses through a small hole. Under the same conditions of pressure and temperature, 10.0 mL of a mixture of CO and CO_2 effuses through the hole in the same amount of time. Calculate the percent composition by volume of the mixture.

10.138   A mixture of methane (CH_4) and ethane (C_2H_6) is stored in a container at 294 mmHg. The gases are burned in air to form CO_2 and H_2O. If the pressure of CO_2 is 356 mmHg, measured at the same temperature and volume as the original mixture, calculate the mole fractions of the gases.

10.139   Use the kinetic theory of gases to explain why hot air rises.

10.140   Given that the van der Waals constant b is the excluded volume and that the excluded volume is four times the volume actually occupied by the gas molecules in a sample, determine what percentage of the container volume is actually occupied by $CCl_4(g)$ molecules at (a) STP, (b) 10.0 atm and 273 K, and (c) 50.0 atm and 273 K.

Engineering Problems

10.141   The running engine of an automobile produces carbon monoxide (CO), a toxic gas, at the rate of about 188 g CO per hour. A car is left idling in a poorly ventilated garage that is 6.0 m long, 4.0 m wide, and 2.2 m high at 20°C. (a) Calculate the rate of CO production in mol/min. (b) How long would it take to build up a lethal concentration of CO of 1000 ppmv (parts per million by volume)?

10.142   Consider the apparatus shown here. When a small amount of water is introduced into the flask by squeezing the bulb of the medicine dropper, water is squirted upward out of the long glass tubing. Explain this observation. (*Hint:* Hydrogen chloride gas is soluble in water.)

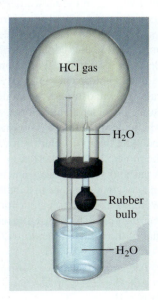

10.143 About 8.0×10^6 tons of urea [$(NH_2)_2CO$] is used annually as a fertilizer. The urea is prepared at 200°C and under high-pressure conditions from carbon dioxide and ammonia (the products are urea and steam). Calculate the volume of ammonia (in liters) measured at 150 atm needed to prepare 1.0 ton of urea.

10.144 Some ballpoint pens have a small hole in the main body of the pen. What is the purpose of this hole?

10.145 A student breaks a thermometer and spills most of the mercury (Hg) onto the floor of a laboratory that measures 15.2 m long, 6.6 m wide, and 2.4 m high. (a) Calculate the mass of mercury vapor (in grams) in the room at 20°C. The vapor pressure of mercury at 20°C is 1.7×10^{-6} atm. (b) Does the concentration of mercury vapor exceed the air quality regulation of 0.050 mg Hg/m³ of air? (c) One way to deal with small quantities of spilled mercury is to spray sulfur powder over the metal. Suggest a physical and a chemical reason for this action.

10.146 The apparatus shown here can be used to measure atomic and molecular speeds. Suppose that a beam of metal atoms is directed at a rotating cylinder in a vacuum. A small opening in the cylinder allows the atoms to strike a target area. Because the cylinder is rotating, atoms traveling at different speeds will strike the target at different positions. In time, a layer of the metal will deposit on the target area, and the variation in its thickness is found to correspond to Maxwell's speed distribution. In one experiment it is found that at 850°C some bismuth (Bi) atoms struck the target at a point 2.80 cm from the spot directly opposite the slit. The diameter of the cylinder is 15.0 cm, and it is rotating at 130 revolutions per second. (a) Calculate the speed (in m/s) at which the target is moving. (*Hint:* The circumference of a circle is given by $2\pi r$, where r is the radius.) (b) Calculate the time (in seconds) it takes for the target to travel 2.80 cm. (c) Determine the speed of the Bi atoms. Compare your result in part (c) with the u_{rms} of Bi at 850°C. Comment on the difference.

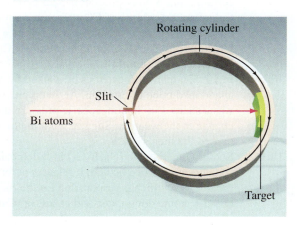

Rotating cylinder

Slit

Bi atoms

Target

10.147 A student tries to determine the volume of a flask like the one shown in Figure 10.12. These are her results: mass of the bulb filled with dry air at 23°C and 744 mmHg = 91.6843 g; mass of evacuated bulb = 91.4715 g. Assume the composition of air is 78 percent N_2, 21 percent O_2, and 1 percent argon by volume. What is the volume (in mL) of the bulb? (*Hint:* First calculate the average molar mass of air, as shown in Problem 3.133.)

10.148 Apply your knowledge of the kinetic theory of gases to the following situations. (a) Two flasks of volumes V_1 and V_2 ($V_2 > V_1$) contain the same number of helium atoms at the same temperature. (i) Compare the root-mean-square (rms) speeds and average kinetic energies of the helium (He) atoms in the flasks. (ii) Compare the frequency and the force with which the He atoms collide with the walls of their containers. (b) Equal numbers of He atoms are placed in two flasks of the same volume at temperatures T_1 and T_2 ($T_2 > T_1$). (i) Compare the rms speeds of the atoms in the two flasks. (ii) Compare the frequency and the force with which the He atoms collide with the walls of their containers. (c) Equal numbers of He and neon (Ne) atoms are placed in two flasks of the same volume, and the temperature of both gases is 74°C. Comment on the validity of the following statements: (i) The rms speed of He is equal to that of Ne. (ii) The average kinetic energies of the two gases are equal. (iii) The rms speed of each He atom is 1.47×10^3 m/s.

10.149 A 5.00-mol sample of NH_3 gas is kept in a 1.92-L container at 300 K. If the van der Waals equation is assumed to give the correct answer for the pressure of the gas, calculate the percent error made in using the ideal gas equation to calculate the pressure.

10.150 In the metallurgical process of refining nickel, the metal is first combined with carbon monoxide to form tetracarbonylnickel, which is a gas at 43°C:

$$Ni(s) + 4CO(g) \longrightarrow Ni(CO)_4(g)$$

This reaction separates nickel from other solid impurities. (a) Starting with 86.4 g of Ni, calculate the pressure of $Ni(CO)_4$ in a container of volume 4.00 L. (Assume the preceding reaction goes to completion.) (b) At temperatures above 43°C, the pressure of the gas is observed to increase much more rapidly than predicted by the ideal gas equation. Explain.

10.151 Some commercial drain cleaners contain a mixture of sodium hydroxide and aluminum powder. When the mixture is poured down a clogged drain, the following reaction occurs:

$$2NaOH(aq) + 2Al(s) + 6H_2O(l) \longrightarrow$$
$$2NaAl(OH)_4(aq) + 3H_2(g)$$

The heat generated in this reaction helps melt away obstructions such as grease, and the hydrogen gas released stirs up the solids clogging the drain. Calculate the volume of H_2 formed at 23°C and 1.00 atm if 3.12 g of Al are treated with an excess of NaOH.

10.152   A stockroom supervisor measured the contents of a 25.0-gal drum partially filled with acetone on a day when the temperature was 18.0°C and atmospheric pressure was 750 mmHg, and found that 15.4 gal of the solvent remained. After tightly sealing the drum, an assistant dropped the drum while carrying it upstairs to the organic laboratory. The drum was dented, and its internal volume was decreased to 20.4 gal. What is the total pressure inside the drum after the accident? The vapor pressure of acetone at 18.0°C is 400 mmHg. (*Hint:* At the time the drum was sealed, the pressure inside the drum, which is equal to the sum of the pressures of air and acetone, was equal to the atmospheric pressure.)

10.153   Commercially, compressed oxygen is sold in metal cylinders. If a 120-L cylinder is filled with oxygen to a pressure of 132 atm at 22°C, what is the mass of O_2 present? How many liters of O_2 gas at 1.00 atm and 22°C could the cylinder produce? (Assume ideal behavior.)

10.154   The root-mean-square speed of a certain gaseous oxide is 493 m/s at 20°C. What is the molecular formula of the compound?

10.155   Referring to Figure 10.20, we see that the maximum of each speed distribution plot is called the most probable speed (u_{mp}) because it is the speed possessed by the largest number of molecules. It is given by $u_{mp} = \sqrt{2RT/\mathcal{M}}$.
(a) Compare u_{mp} with u_{rms} for nitrogen at 25°C.
(b) The following diagram shows the Maxwell speed distribution curves for an ideal gas at two different temperatures T_1 and T_2. Calculate the value of T_2.

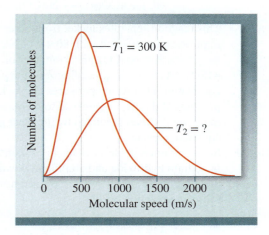

Biological Problems

10.156   Air entering the lungs ends up in tiny sacs called alveoli. It is from the alveoli that oxygen diffuses into the blood. The average radius of the alveoli is 0.0050 cm, and the air inside contains 14 percent oxygen. Assuming that the pressure in the alveoli is 1.0 atm and the temperature is 37°C, calculate the number of oxygen molecules in one of the alveoli. (*Hint:* The volume of a sphere of radius r is $\frac{4}{3}\pi r^3$.)

10.157   The shells of hard-boiled eggs sometimes crack due to the rapid thermal expansion of the shells at high temperatures. Suggest another reason why the shells may crack.

10.158   Ethylene gas (C_2H_4) is emitted by fruits and is known to be responsible for their ripening. Based on this information, explain why a bunch of bananas ripens faster in a closed paper bag than in an open bowl.

10.159   The gas laws are vitally important to scuba divers. The pressure exerted by 33 ft of seawater is equivalent to 1 atm pressure. (a) A diver ascends quickly to the surface of the water from a depth of 36 ft without exhaling gas from his lungs. By what factor will the volume of his lungs increase by the time he reaches the surface? Assume that the temperature is constant. (b) The partial pressure of oxygen in air is about 0.20 atm. (Air is 20 percent oxygen by volume.) In deep-sea diving, the composition of air the diver breathes must be changed to maintain this partial pressure. What must the oxygen content (in percent by volume) be when the total pressure exerted on the diver is 4.0 atm? (At constant temperature and pressure, the volume of a gas is directly proportional to the number of moles of gases.)

10.160   A healthy adult exhales about 5.0×10^2 mL of a gaseous mixture with each breath. Calculate the number of molecules present in this volume at 37°C and 1.1 atm. List the major components of this gaseous mixture.

10.161   The percent by mass of bicarbonate (HCO_3^-) in a certain Alka-Seltzer product is 32.5 percent. Calculate the volume of CO_2 generated (in mL) at 37°C and 1.00 atm when a person ingests a 3.29-g tablet. (*Hint:* The reaction is between HCO_3^- and HCl acid in the stomach.)

10.162   In 1995, a man suffocated as he walked by an abandoned mine in England. At that moment there was a sharp drop in atmospheric pressure due to a change in the weather. Suggest what might have caused the man's death.

10.163   Sodium bicarbonate ($NaHCO_3$) is called baking soda because, when heated, it releases carbon dioxide gas, which is responsible for the rising of cookies, some doughnuts, and cakes. (a) Calculate the volume (in liters) of CO_2 produced by heating 5.0 g of $NaHCO_3$ at 180°C and 1.3 atm. (b) Ammonium bicarbonate (NH_4HCO_3) has also been used for the same purpose. Suggest one advantage and one disadvantage of using NH_4HCO_3 instead of $NaHCO_3$ for baking.

Environmental Problems

10.164   Under the same conditions of temperature and pressure, why does 1 L of moist air weigh less than 1 L of dry air?

10.165   Atop Mt. Everest, the atmospheric pressure is 210 mmHg and the air density is 0.426 kg/m³. (a) Calculate the air temperature, given that the molar mass of air is 29.0 g/mol. (b) Assuming no change in air composition, calculate the percent decrease in oxygen gas from sea level to the top of Mt. Everest.

10.166   Relative humidity is defined as the ratio (expressed as a percentage) of the partial pressure of water vapor in the air to the equilibrium vapor pressure (see Table 10.5) at a given temperature. On a certain summer day in North Carolina the partial pressure of water vapor in the air is 3.9×10^3 Pa at 30°C. Calculate the relative humidity.

10.167   The atmosphere on Mars is composed mainly of carbon dioxide. The surface temperature is 220 K, and the atmospheric pressure is about 6.0 mmHg. Taking these values as Martian "STP," calculate the molar volume in liters of an ideal gas on Mars.

10.168   Venus's atmosphere is composed of 96.5 percent CO_2, 3.5 percent N_2, and 0.015 percent SO_2 by volume. Its standard atmospheric pressure is 9.0×10^6 Pa. Calculate the partial pressures of the gases in pascals.

Multiconcept Problems

10.169   Acidic oxides such as carbon dioxide react with basic oxides like calcium oxide (CaO) and barium oxide (BaO) to form salts (metal carbonates). (a) Write equations representing these two reactions. (b) A student placed a mixture of CaO and BaO of combined mass 4.88 g in a 1.46-L flask containing carbon dioxide gas at 35°C and 746 mmHg. After the reactions were complete, she found that the CO_2 pressure had dropped to 252 mmHg. Calculate the percent composition by mass of the mixture. Assume that the volumes of the solids are negligible.

10.170   Sulfur dioxide reacts with oxygen to form sulfur trioxide. (a) Write the balanced equation and use data from Appendix 2 to calculate $\Delta H°$ for this reaction. (b) At a given temperature and pressure, what volume of oxygen is required to react with 1 L of sulfur dioxide? What volume of sulfur trioxide will be produced? (c) The diagram at right represents the combination of equal volumes of the two reactants. Which of the following diagrams [(i)–(iv)] best represents the result?

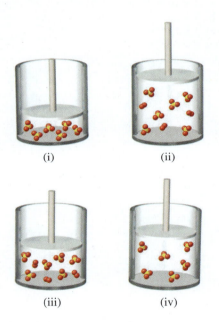

(i)   (ii)

(iii)   (iv)

10.171   In a constant-pressure calorimetry experiment, a 2.675-g piece of zinc metal is dropped into 100.0 mL of 1.75 M hydrochloric acid in a closed vessel with a movable piston. The pressure and temperature in the laboratory are 769 torr and 23.8°C, respectively. Calculate the work done by the system.

Standardized-Exam Practice Problems

Physical and Biological Sciences

Every breath we take, on average, contains molecules that were once exhaled by Wolfgang Amadeus Mozart (1756–1791).

1. Calculate the total number of molecules in the atmosphere. (Assume that the total mass of the atmosphere is 6×10^{18} kg and the average molar mass of air is 29.0 g/mol.)

   a) 1×10^{23}
   b) 1×10^{26}
   c) 1×10^{29}
   d) 1×10^{18}

2. Assuming the volume of every breath (inhale or exhale) is 0.5 L, calculate the number of molecules exhaled in each breath at body temperature (37°C) and 1 atm.

   a) 1×10^{22}
   b) 1×10^{21}
   c) 1×10^{23}
   d) 6×10^{23}

3. Calculate the mass of air exhaled with each breath.

   a) 0.02 g
   b) 0.6 g
   c) 0.2 g
   d) 6 g

4. If Mozart's life span was exactly 35 years, what is the number of molecules he exhaled in that period (given that an average person breathes 12 times per minute)?

   a) 2×10^{8}
   b) 2×10^{29}
   c) 1×10^{29}
   d) 3×10^{30}

Answers to In-Chapter Materials

Answers to Practice Problems

10.1A 1.32 atm. **10.1B** 9.52 m. **10.2A** 2.23 L. **10.2B** 1.80 atm. **10.3A** 30.7 L. **10.3B** 300°C. **10.4A** 34 L. **10.4B** 3.16 L CO, 1.58 L O_2. **10.5A** 5.09 L. **10.5B** 85.0 m. A common mistake in this problem is failure to subtract the atmospheric pressure (0.965 atm) from the total pressure at the bottom of the lake (9.19 atm). The pressure due to the *water* is only 8.23 atm. **10.6A** 128 L. **10.6B** 336°C. **10.7A** 1.29 g/L. **10.7B** 11 atm. **10.8A** 55.2 g/mol. **10.8B** 133.622 g. **10.9A** 151 L. **10.9B** 3.48 g. **10.10A** 1.0×10^2 mol. **10.10B** 0.0052 atm. **10.11A** 8.48 L. **10.11B** 184 g. **10.12A** $P_{He} = 0.184$ atm, $P_{H_2} = 0.390$ atm, $P_{Ne} = 1.60$ atm, $P_{total} = 2.18$ atm. **10.12B** 0.032 mol CH_4, 0.091 mol C_2H_6, 0.123 mol total. **10.13A** $\chi_{CO_2} = 0.0495$, $\chi_{CH_4} = 0.255$, $\chi_{He} = 0.695$, $P_{CO_2} = 0.286$ atm, $P_{CH_4} = 1.47$ atm, $P_{He} = 4.02$ atm. **10.13B** $P_{Xe} = 4.95$ atm, $P_{Ne} = 1.55$ atm, $n_{Xe} = 3.13$, $n_{Ne} = 0.984$. **10.14A** 1.03 g. **10.14B** 0.386 L. **10.15A** 0.61. **10.15B** 13 atm. **10.16A** 2.137. **10.16B** 2.02 g/mol, H_2. **10.17A** 50.6 atm, 51.6 atm. **10.17B** 1.348 atm, 1.347 atm.

Answers to Checkpoints

10.1.1 a. **10.1.2** b, c, d. **10.1.3** d. **10.1.4** b. **10.1.5** e. **10.2.1** e. **10.2.2** a. **10.2.3** c. **10.2.4** a. **10.2.5** b. **10.2.6** d. **10.3.1** a. **10.3.2** e. **10.3.3** b. **10.3.4** b. **10.4.1** d. **10.4.2** b. **10.5.1** e. **10.5.2** b. **10.5.3** a. **10.5.4** d. **10.5.5** c. **10.5.6** b. **10.6.1** c. **10.6.2** b. **10.7.1** a, d. **10.7.2** b.

CHAPTER 11

Intermolecular Forces and the Physical Properties of Liquids and Solids

Late in 2007, reports began to surface that infants in China were falling ill after being fed formula contaminated with melamine. The contamination was the result of criminal activity.

Tom Wang/Shutterstock

In This Chapter, You Will Learn

About the types of forces that hold the molecules (or atoms or ions) in substances together and about the properties that arise because of these forces.

Before You Begin, Review These Skills

- Molecular geometry and polarity [◀◀ Section 9.2]
- Basic trigonometry [▶▶ Appendix 1]
- Chapter 9 Key Skills [◀◀ pages 454–455]

How Intermolecular Forces Contributed to the Contaminated Infant Formula Tragedy in China

Life as we know it depends heavily on intermolecular forces—the attractions that keep molecules together. Without them, water—perhaps the most essential ingredient in life—would not be a liquid. On this occasion, though, intermolecular forces caused two contaminants in infant formula, each relatively innocuous by itself, to form a highly toxic complex that caused many thousands of babies in China to develop acute kidney failure.

The contamination turned out to have been the result of foul play. Unscrupulous vendors of dairy products sought to increase their profits by diluting the milk they sold. However, because its protein content would have appeared too low, diluted milk would not have passed the industry quality control tests. To boost the apparent protein content, the vendors added melamine ($C_3H_6N_6$) to the diluted milk. Although melamine is not a protein, its high nitrogen content (66.6 percent by mass) made the milk appear to contain adequate protein. The industry test for protein does not measure protein specifically but rather nitrogen content, which in unadulterated milk would give a measure of the protein content. In addition to being added to the milk illegally, the melamine was itself contaminated with another compound, cyanuric acid ($C_3H_3N_3O_3$). At sufficiently high concentrations, these two contaminants form an insoluble complex that precipitates as an unusual type of stone, causing severe kidney damage. The molecules in the insoluble complex are held together by powerful attractive forces known as *hydrogen bonds* (shown as dashed red lines).

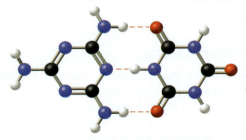

Secrecy surrounding the scandal has made it impossible to know for certain how many babies were poisoned—and how many died. The formation of the toxic melamine-cyanuric acid complex is the result of particularly strong attractive forces between the melamine and cyanuric acid molecules. Attractive forces between molecules (intermolecular forces) and within molecules (intramolecular forces) determine many of the physical properties of liquids and solids.

At the end of this chapter, you will be able to answer questions about the physical properties of substances used in the treatment and prevention of kidney stones [▶▶ Applying What You've Learned, page 572].

11.1   Intermolecular Forces

In Chapter 10, we learned that gases consist of rapidly moving particles (molecules or atoms), separated by relatively large distances. Liquids and solids, the *condensed phases* [◄◄ Section 1.2], consist of particles (molecules, atoms, or ions) that are touching one another. The attractive forces that hold particles together in the condensed phases are called **intermolecular forces.** The *magnitude* of intermolecular forces is what determines whether the particles that make up a substance are a gas, liquid, or solid.

Gas

Particles in a gas are separated by large distances and free to move entirely independently of one another.

Liquid

Particles in a liquid are touching one another but free to move about.

Solid

Particles in a solid are essentially locked in place with respect to one another.

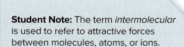

Student Note: The term *intermolecular* is used to refer to attractive forces between molecules, atoms, or ions.

We have already encountered an example of "intermolecular" forces in the form of ionic bonding [◄◄ Section 8.2], where the magnitude of attraction between oppositely charged particles is governed by Coulomb's law [◄◄ Section 7.3]. Because the particles that make up an ionic compound have discrete (*full*) charges, the attractive forces that hold them together are especially powerful. In fact, that's the reason that they're solids at room temperature. The intermolecular forces that we discuss in this chapter are also the result of Coulombic attractions, but the attractions involve only partial charges [◄◄ Section 8.4] rather than discrete charges and are therefore weaker than the forces involved in ionic bonding.

We begin our discussion of intermolecular forces with the attractive forces that act between atoms or molecules in a pure substance. These forces are known collectively as **van der Waals forces,** and they include *dipole-dipole interactions,* including *hydrogen bonding* and *dispersion forces.*

Dipole-Dipole Interactions

Dipole-dipole interactions are attractive forces that act between *polar molecules.* Recall that a diatomic molecule containing elements of significantly different electronegativities, such as HCl, has an unequal distribution of electron density and therefore has partial charges, positive ($\delta+$) at one end and negative ($\delta-$) at the other.

The partial positive charge on one molecule is attracted to the partial negative charge on a neighboring molecule. Figure 11.1 shows the orientation of polar molecules in a solid and in a liquid. The arrangement is somewhat less orderly in the liquid than it is in the solid.

Because this attractive force between polar molecules is Coulombic, the magnitude of the attractive forces depends on the magnitude of the dipole. In general, the larger the dipole, the larger the attractive force. Certain physical properties such as

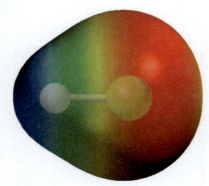

HCl

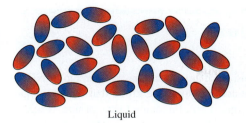

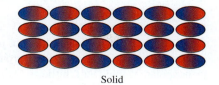

Liquid   Solid

Figure 11.1 Arrangement of polar molecules in a liquid and in a solid.

TABLE 11.1	Dipole Moments and Boiling Points of Compounds with Similar Molecular Masses		
Compound	**Structural Formula**	**Dipole Moment (D)**	**Boiling Point (°C)**
Propane	$CH_3CH_2CH_3$	0.1	−42
Dimethyl ether	CH_3OCH_3	1.3	−25
Methyl chloride	CH_3Cl	1.9	−24
Acetaldehyde	CH_3CHO	2.7	21
Acetonitrile	CH_3CN	2.9	82

boiling point *reflect* the magnitude of intermolecular forces. A substance in which the particles are held together by larger intermolecular attractions will require more energy to *separate* the particles and will therefore boil at a higher temperature. Table 11.1 lists several compounds with similar molar masses along with their dipole moments and boiling points.

Hydrogen Bonding

Hydrogen bonding is a special type of dipole-dipole interaction. But, whereas dipole-dipole interactions act between any polar molecules, hydrogen bonding occurs only in molecules that contain H bonded to a small, highly electronegative atom, such as *N, O, or F*. In a molecule such as HF, which is shown in Figure 11.2, the F atom to which H is bonded draws electron density toward itself. Being small and very electronegative, the F atom draws electron density away from H very effectively. Because H has only one electron, this leaves the hydrogen nucleus practically unshielded— giving H a very large partial positive charge. This large partial positive charge is powerfully attracted to the large partial negative charge (lone pairs) on the small, highly electronegative F atom of a neighboring HF molecule. The result is an especially strong dipole-dipole attraction.

In Sample Problem 8.5, we calculated the partial charges on H and F as +0.41 and −0.41, respectively. The partial charges on H and Cl, by contrast, are only +0.18 and −0.18, respectively. They are smaller in HCl because Cl is larger and less electronegative than F. Because of the very high electronegativity of F, and the high partial positive charges that result, HF exhibits significant hydrogen bonding, whereas HCl does not. (HCl does exhibit hydrogen bonding to a very small degree. But only molecules containing an N—H, O—H, or F—H bond exhibit significant hydrogen bonding.)

Student Note: It is a common error to assume that hydrogen bonding occurs in any molecule that contains hydrogen. It *only* occurs to a significant degree in molecules with N—H, O—H, or F—H bonds.

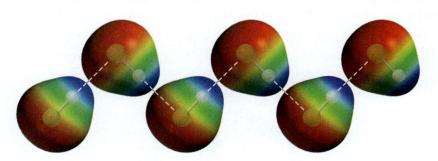

Figure 11.2 Hydrogen bonds between HF molecules.

Bringing Chemistry to Life

Sickle Cell Disease

Amino acids are the building blocks of proteins. Each amino acid has both amino ($-NH_2$) and carboxy ($-COOH$) functional groups, in addition to a side group that may be polar or nonpolar. Valine (Val), histidine (His), leucine (Leu), threonine (Thr), proline (Pro), and glutamic acid (Glu) are 6 of the 20 amino acids [▶▶| Section 25.6] that make up human proteins. The characteristic side group of each amino acid is shaded to identify it:

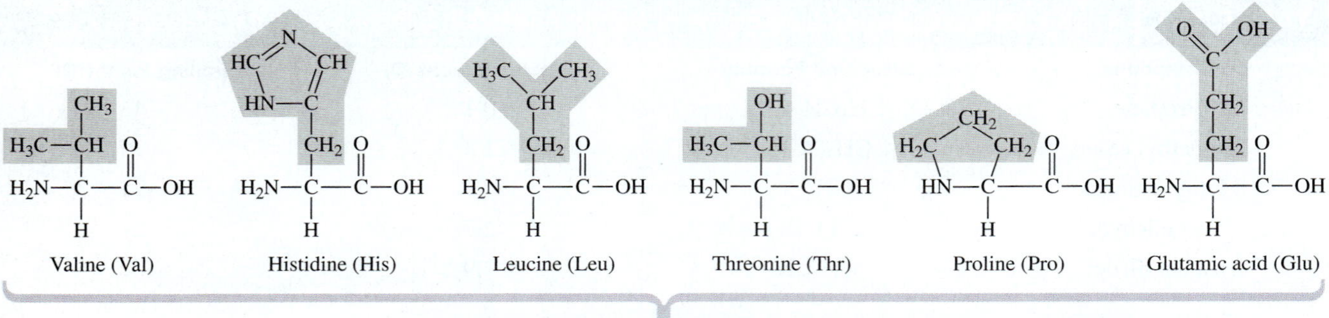

Valine (Val)   Histidine (His)   Leucine (Leu)   Threonine (Thr)   Proline (Pro)   Glutamic acid (Glu)

Proteins form when amino acids are joined together with peptide bonds, also known as *amide linkages,* in which the carboxy group on one amino acid connects to the amino group of the next. The carboxy group of the second amino acid connects to the amino group of another, and so on. A water molecule is eliminated with the formation of each peptide bond. What remains of each amino acid after the formation of peptide bonds (and the corresponding loss of water molecules) is referred to as an *amino acid residue,* or simply a *residue.* (Peptide bonds are shown in blue.)

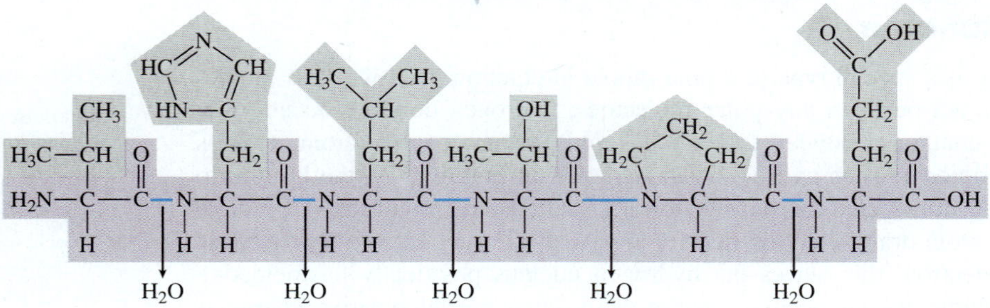

H_2O H_2O H_2O H_2O H_2O

The primary structure of a protein refers to the sequence of amino acids that make up the protein chain.

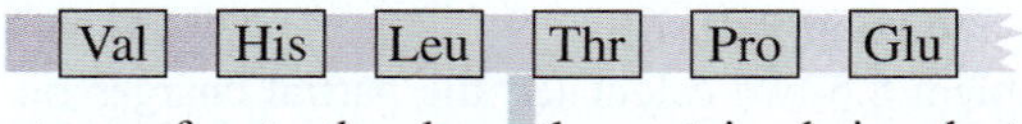

Val   His   Leu   Thr   Pro   Glu

Secondary structure refers to the shape the protein chain adopts as the result of hydrogen bonding between nearby residues. One of the possible secondary structures is a helix.

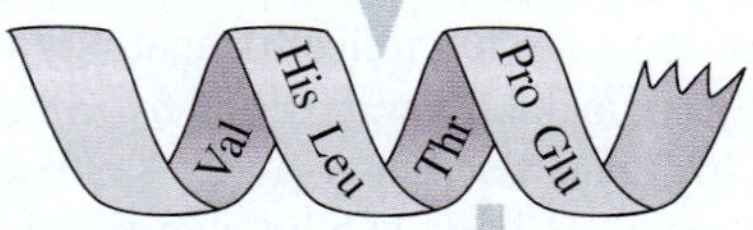

Val His Leu Thr Pro Glu

Tertiary structure refers to the folding of the protein into a characteristic shape, which is stabilized by attractive forces between more distant residues.

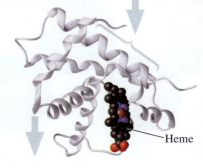

Heme

Quaternary structure refers to the shape resulting from the assembly of two or more proteins, called subunits, to form a larger protein complex. In the case of hemoglobin, four subunits [two α subunits (purple), each consisting of 141 amino acids; and two β subunits (pink), each consisting of 146 amino acids] make up the larger complex.

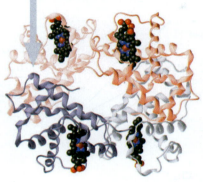

Hemoglobin is a *globular* protein, meaning that it assembles into a roughly spherical shape with polar residues covering most of the sphere's surface and nonpolar residues largely confined to the interior of the complex. Distribution of polar residues on the outside of the sphere enhances a globular protein's aqueous solubility.

In normal hemoglobin, the sixth amino acid residue (of 146) in each β subunit is a glutamic acid residue, which is polar. In hemoglobin S, the type associated with sickle cell disease, the glutamic acid residue in the sixth position of each β subunit is replaced by a valine residue, which is nonpolar. This results in a nonpolar group on the surface of the protein complex, where there should be a polar group.

Although most of the surface of the spherical hemoglobin complex is polar, when it is not oxygenated, one of hemoglobin's β subunits has a small nonpolar region on the exterior of the complex. Ordinarily, this is of no consequence.

Nonpolar region

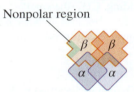

In hemoglobin S, however, the nonpolar valine residue on the surface of one complex is attracted to the nonpolar region on a nearby complex.

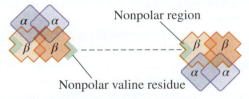

This results in hemoglobin forming long chains that precipitate from solution, causing the deformation of red blood cells that is characteristic of sickle cell disease.

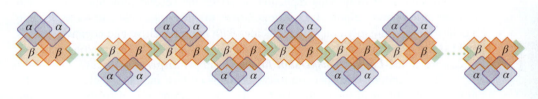

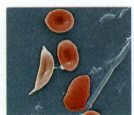

Source: *Janice Haney Carr/CDC*

These deformed blood cells clog narrow capillaries, restricting blood flow to the body's organs. The reduced blood flow gives rise to anemia, episodes of debilitating pain, susceptibility to infections, stroke, and eventually premature death.

Figure 11.3 Boiling points of the hydrogen compounds of elements from Groups 14 through 17. Although normally the boiling point increases with increasing mass within a group, the *lightest* compound has the *highest* boiling point in Groups 15 through 17. This departure from the observed trend is due to hydrogen bonding.

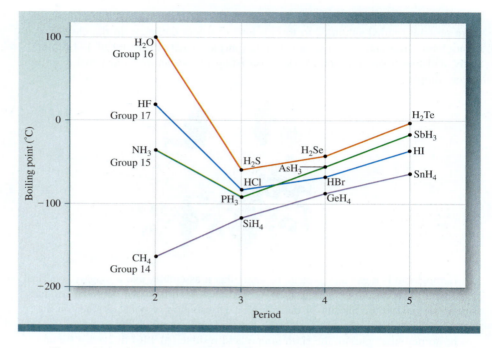

Figure 11.3 shows the boiling points of the binary hydrogen compounds of Groups 14 through 17. Within the series of hydrogen compounds of Group 14, the boiling point increases with increasing molar mass. For Groups 15 through 17, the same trend is observed for all but the smallest member of each series, which has what appears to be an unexpectedly high boiling point. This departure from the trend in boiling points illustrates how powerful hydrogen bonding can be. NH_3 (Group 15), HF (Group 17), and H_2O (Group 16) have anomalously high boiling points because they exhibit strong hydrogen bonding. CH_4 (the smallest member of the Group 14 series) does not exhibit hydrogen bonding because it does not contain an N—H, O—H, or F—H bond; hence, it has the lowest boiling point of that group.

Student Note: Hydrogen bonding also occurs in mixtures, between solute and solvent molecules that contain N—H, F—H, or O—H bonds.

Dispersion Forces

Nonpolar gases, such as N_2 and O_2, can be liquefied under the right conditions of pressure and temperature, so nonpolar molecules must also exhibit attractive intermolecular forces. These intermolecular forces are Coulombic in nature (as are all intermolecular forces), but they differ from other intermolecular forces because they arise from the movement of electrons in nonpolar molecules.

On average, the distribution of electron density in a nonpolar molecule is uniform and symmetrical (which is why a nonpolar molecule has no dipole moment). However, because electrons in a molecule have some freedom to move about, at any given point in time the molecule may have a nonuniform distribution of electron density, giving it a fleeting, temporary dipole—called an ***instantaneous dipole.*** An instantaneous dipole in one molecule can induce dipoles in neighboring molecules. For example, the temporary partial negative charge on a molecule repels the electrons in a molecule next to it. This repulsion polarizes the second molecule, which then acquires a temporary dipole. It in turn polarizes the next molecule and so on, leaving a collection of ordinarily nonpolar molecules with partial positive and negative charges and Coulombic attractions between them as shown in Figure 11.4. The resulting attractive forces are called ***London[1] dispersion forces*** or simply ***dispersion forces.***

The magnitude of dispersion forces depends on how mobile the electrons in the molecule are. In small molecules, such as F_2, the electrons are relatively close to the nuclei and cannot move about very freely; thus, the electron distribution in F_2 is not

---

1. Fritz London (1900–1954). German physicist. London was a theoretical physicist whose major work was on superconductivity in liquid helium.

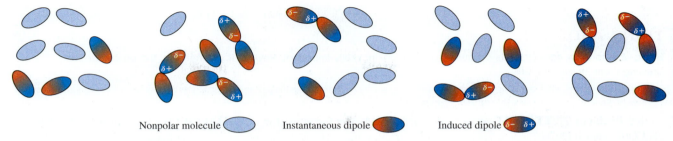

Nonpolar molecule   Instantaneous dipole   Induced dipole

Figure 11.4 Instantaneous dipoles in ordinarily nonpolar molecules can induce temporary dipoles in neighboring molecules, causing the molecules to be attracted to one another. This type of interaction is responsible for the condensation of nonpolar gases.

TABLE 11.2	Molar Masses, Boiling Points, and States of the Halogens at Room Temperature		
Molecule	**Molar Mass (g/mol)**	**Boiling Point (°C)**	**State (Room Temp.)**
F_2	38.0	−188	Gas
Cl_2	70.9	−34	Gas
Br_2	159.8	59	Liquid
I_2	253.8	184	Solid

easily *polarized*. In larger molecules, such as Cl_2, the electrons are somewhat farther away from the nuclei and therefore move about more freely. The electron density in Cl_2 is more easily polarized than that in F_2, resulting in larger instantaneous dipoles, larger induced dipoles, and larger intermolecular attractions overall. The trend in magnitude of dispersion forces with increasing molecular size is illustrated by the data in Table 11.2. (It is also illustrated by the correlation between molar mass and boiling point in Figure 11.3.) Dispersion forces act not only between nonpolar molecules, but between *all* molecules.

Sample Problem 11.1 lets you practice determining what kinds of forces exist between particles in liquids and solids.

Student Note: Molecules that more readily acquire an instantaneous dipole are said to be *polarizable*. In general, larger molecules have greater *polarizability* than small ones.

SAMPLE PROBLEM 11.1

What kind(s) of intermolecular forces exist in (a) $CCl_4(l)$, (b) $CH_3COOH(l)$, (c) $CH_3COCH_3(l)$, and (d) $H_2S(l)$?

Strategy Draw Lewis dot structures and apply VSEPR theory [◀◀ Section 9.2] to determine whether each molecule is polar or nonpolar. Nonpolar molecules exhibit dispersion forces only. Polar molecules exhibit both dipole-dipole interactions and dispersion forces. Polar molecules with N—H, F—H, or O—H bonds exhibit dipole-dipole interactions (including hydrogen bonding) and dispersion forces.

Setup The Lewis dot structures for molecules (a) to (d) are:

(a)   (b)   (c)   (d)

Solution (a) CCl_4 is nonpolar, so the only intermolecular forces are dispersion forces.

(b) CH_3COOH is polar and contains an O—H bond, so it exhibits dipole-dipole interactions (including hydrogen bonding) and dispersion forces.

(c) CH_3COCH_3 is polar but does *not* contain N—H, O—H, or F—H bonds, so it exhibits dipole-dipole interactions and dispersion forces.

(d) H_2S is polar but does not contain N—H, O—H, or F—H bonds, so it exhibits dipole-dipole interactions and dispersion forces.

(Continued on next page)

THINK ABOUT IT

Being able to draw correct Lewis structures is, once again, vitally important. Review, if you need to, the procedure for drawing them [◄◄ Section 8.5].

Practice Problem **A**TTEMPT   What kind(s) of intermolecular forces exist in (a) $CH_3CH_2CH_2CH_2CH_3(l)$, (b) $CH_3CH_2OH(l)$, (c) $H_2CO(l)$, and (d) $O_2(l)$?

Practice Problem **B**UILD   What kind(s) of intermolecular forces exist in (a) $CH_2Cl_2(l)$, (b) $CH_3CH_2CH_2OH(l)$, (c) $H_2O_2(l)$, and (d) $N_2(l)$?

Practice Problem **C**ONCEPTUALIZE   Using the molecular formula C_2H_6O, draw two Lewis structures that exhibit different intermolecular forces. List the types of intermolecular forces exhibited by each structure.

Ion-Dipole Interactions

Student Note: Hydration [◄◄ Section 4.2] is one example of an ion-dipole interaction.

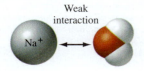

Weak interaction

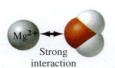

Strong interaction

Figure 11.5 Ion-dipole interactions between ions and water molecules. In each case the positive charge on the ion is attracted to the partial negative charge on the oxygen atom in the water molecule. The attraction is stronger when the distance between the two species is smaller.

Ion-dipole interactions are Coulombic attractions between ions (either positive or negative) and polar molecules. These interactions occur in mixtures of ionic and polar species such as an aqueous solution of sodium chloride. The magnitude of ion-dipole interactions depends on the charge and the size of the ion, and on the dipole moment and size of the polar molecule. Cations generally interact more strongly with dipoles than anions (of the same magnitude charge) because they tend to be smaller.

Figure 11.5 shows the ion-dipole interaction between the Na^+ and Mg^{2+} ions with a water molecule, which has a large dipole moment (1.87 D). Because the Mg^{2+} ion has a *higher charge* and a *smaller size* [◄◄ Section 7.6] than the Na^+ ion (the ionic radii of Mg^{2+} and Na^+ are 78 and 98 pm, respectively), Mg^{2+} interacts more strongly with water molecules. The properties of solutions are discussed in Chapter 13.

Although the intermolecular forces discussed so far are all *attractive* forces, molecules also exert repulsive forces on one another (when *like* charges approach one another). When two molecules approach each other closely, the repulsions between electrons and between nuclei in the molecules become significant. The magnitude of these repulsive forces rises very steeply as the distance separating the molecules in a condensed phase decreases. This is the reason that liquids and solids are so hard to compress. In these phases, the molecules are already in close contact with one another, and so they greatly resist being compressed further.

CHECKPOINT – SECTION 11.1   Intermolecular Forces

11.1.1 What kind(s) of intermolecular forces exist between benzene molecules (C_6H_6)? (Select all that apply.)

a) Dispersion forces

b) Dipole-dipole interactions

c) Hydrogen bonding

d) Ion-dipole interactions

e) Ionic bonding

11.1.2 Which of the following exhibits significant hydrogen bonding? (Select all that apply.)

a) HBr

b) H_2CF_2

c) H_2

d) H_2O_2

e) CH_3CN

11.2 | Properties of Liquids

Several of the physical properties of a liquid depend on the magnitude of its intermolecular forces. In this section we consider three such properties: surface tension, viscosity, and vapor pressure.

Surface Tension

A molecule within a liquid is pulled in all directions by the intermolecular forces between it and the other molecules that surround it. There is no *net* pull in any one direction. A molecule at the surface of the liquid is similarly pulled down and to the sides by neighboring molecules, but Figure 11.6 shows there is no upward pull to balance the downward or inward pull (into the bulk of the liquid). This results in a net pull inward on molecules at the surface, causing the surface of a liquid to tighten like an elastic film, thus minimizing its surface area. The "beading" of water on the leaves of plants, shown in Figure 11.7, is an illustration of this phenomenon.

A quantitative measure of the elastic force in the surface of a liquid is the ***surface tension,*** the amount of energy required to stretch or increase the surface of a liquid by a unit area (e.g., by 1 cm^2). A liquid with strong intermolecular forces has a high surface tension. Water, for instance, with its strong hydrogen bonds, has a very high surface tension.

Another illustration of surface tension is the *meniscus,* the curved surface of a liquid contained in a narrow tube. Figure 11.8(a) shows the concave surface of water in a graduated cylinder. (You probably know from your laboratory class that you are

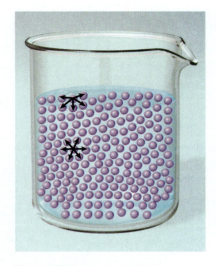

Figure 11.6 Intermolecular forces acting on a molecule in the surface layer of a liquid and in the interior region of the liquid.

Figure 11.7 Surface tension causes rainwater to bead on leaves.
David Planchet/McGraw Hill

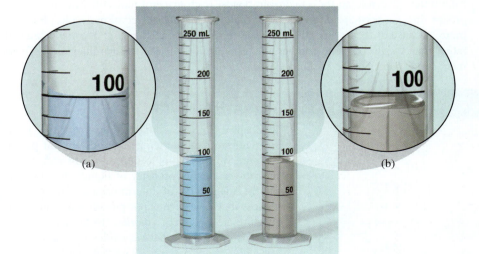

(a)   (b)

Figure 11.8 (a) When water is measured in a glass graduated cylinder, the adhesive forces are greater than the cohesive forces and the meniscus is concave. (b) When mercury is measured in a glass graduated cylinder, cohesive forces are greater than adhesive forces and the meniscus is convex.

Blood being drawn from a finger into a capillary tube.

pittawut/Shutterstock

Student Note: *Cohesion* and *adhesion* are also called *cohesive forces* and *adhesive forces.*

to read the volume level with the bottom of the meniscus.) This is caused by a thin film of water adhering to the wall of the glass cylinder. The surface tension of water causes this film to contract, and as it does, it pulls the water up the cylinder. This effect, known as **capillary action,** is more pronounced in a cylinder with a very small diameter, such as a capillary tube used to draw a small amount of blood. Two types of forces bring about capillary action. One is **cohesion,** the attractions between *like* molecules (in this case, between water molecules). The other is **adhesion,** the attractions between *unlike* molecules (in this case, between water molecules and the molecules that make up the interior surface of the graduated cylinder). If adhesion is stronger than cohesion, as it is in Figure 11.8(a), the contents of the tube will be pulled upward. The upward movement is limited by the weight of the liquid in the tube. In the case of mercury, shown in Figure 11.8(b), the cohesive forces are stronger than the adhesive forces, resulting in a convex meniscus in which the liquid level at the glass wall is lower than that in the middle.

Viscosity

Another property determined by the magnitude of intermolecular forces in a liquid is viscosity. **Viscosity,** with units of $N \cdot s/m^2$, is a measure of a fluid's resistance to flow. The higher the viscosity, the more slowly a liquid flows. The viscosity of a liquid typically decreases with increasing temperature. The phrase "slow as molasses in winter" refers to the fact that molasses pours more slowly (has a higher viscosity) in cold weather. You may also have noticed that honey and maple syrup seem thinner when they are heated.

Liquids that have strong intermolecular forces have higher viscosities than those that have weaker intermolecular forces. Table 11.3 lists the viscosities of some liquids that may be familiar to you. Water's high viscosity, like its high surface tension, is the result of hydrogen bonding. Note how large the viscosity of glycerol is compared to the other liquids listed in Table 11.3. The structure of glycerol, which is a sweet tasting, syrupy liquid used for a wide variety of things, including the manufacture of candy and antibiotics, is

$$
\begin{array}{c}
\text{H} \\
| \\
\text{H}-\text{C}-\text{O}-\text{H} \\
| \\
\text{H}-\text{C}-\text{O}-\text{H} \\
| \\
\text{H}-\text{C}-\text{O}-\text{H} \\
| \\
\text{H}
\end{array}
$$

Glycerol

Like water, glycerol can form hydrogen bonds. Each glycerol molecule has three —OH groups that can participate in hydrogen bonding with other glycerol molecules. Furthermore, because of their shape, the molecules have a great tendency to become entangled rather than to slip past one another as the molecules of less viscous liquids do. These interactions contribute to its high viscosity.

TABLE 11.3	Viscosities of Some Familiar Liquids at 20°C
Liquid	**Viscosity ($N \cdot s/m^2$)**
Acetone (C_3H_6O)	3.16×10^{-4}
Water (H_2O)	1.01×10^{-3}
Ethanol (C_2H_5OH)	1.20×10^{-3}
Mercury (Hg)	1.55×10^{-3}
Blood	4×10^{-3}
Glycerol ($C_3H_8O_3$)	1.49

Vapor Pressure

In Chapter 10, we encountered the term *vapor pressure,* referring to the temperature-dependent partial pressure of water [◄◄ Section 10.5]. In fact, vapor pressure is another property of liquids that depends on the magnitude of intermolecular forces. Substances that have high vapor pressures at room temperature are said to be **volatile.** (Note that this does not mean that a substance is explosive—only that it has a high vapor pressure.) The molecules in a liquid

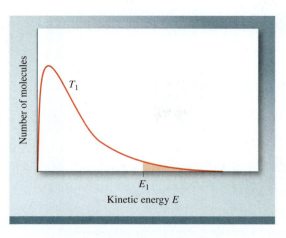
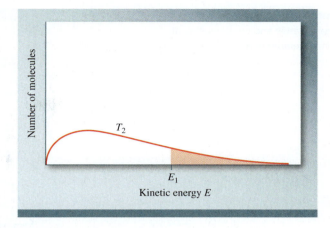

Figure 11.9 Kinetic energy distribution curves for molecules in a liquid (a) at temperature T_1 and (b) at a higher temperature T_2. Note that at the higher temperature, the curve flattens out. The shaded areas represent the number of molecules possessing kinetic energy equal to or greater than a certain kinetic energy E_1. The higher the temperature, the greater the number of molecules with high kinetic energy.

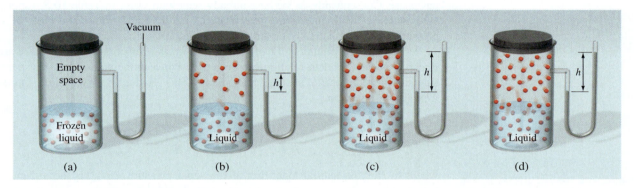

Figure 11.10 Establishment of equilibrium vapor pressure. (a) Initially, there are no molecules in the gas phase. (b) Molecules enter the gas phase, increasing the total pressure above the liquid. (c) The partial pressure of the liquid continues to increase until the rates of vaporization and condensation are equal. (d) Vaporization and condensation continue to occur at the same rate, and there is no further net change in pressure.

are in constant motion, and, like the molecules in a gas, they have a distribution of kinetic energies. The most probable kinetic energy for molecules in a sample of liquid increases with increasing temperature, as shown in Figure 11.9. If a molecule at the surface of a liquid has sufficient kinetic energy, it can escape from the liquid phase into the gas phase. This phenomenon is known as ***evaporation*** or ***vaporization.*** Consider the apparatus shown in Figure 11.10. As a liquid begins to evaporate, molecules leave the liquid phase and become part of the gas phase in the space above the liquid. Molecules in the gas phase can return to the liquid phase if they strike the liquid surface and again become trapped by intermolecular forces, a process known as ***condensation.*** Initially, evaporation occurs more rapidly than condensation. As the number of molecules in the gas phase increases, however, so does the rate of condensation. The vapor pressure over the liquid increases until the rate of condensation is equal to the rate of evaporation, which is constant at any given temperature (Figure 11.11). This (or any other) situation, wherein a forward process and reverse process are occurring at the same rate, is called a ***dynamic equilibrium.*** Although both processes are ongoing (*dynamic*), the number of molecules in the gas phase at any given point in time does not change (*equilibrium*). The pressure exerted by the molecules that have escaped to the gas phase, once the pressure has stopped increasing, is the ***equilibrium vapor pressure,*** or simply the *vapor pressure.*

   The average kinetic energy of molecules in a liquid increases with increasing temperature (see Figure 11.9). At a higher temperature, therefore, a greater percentage of molecules at the liquid surface will possess sufficient kinetic energy to escape into the

Animation
Equilibrium vapor pressure.

Student Hot Spot

   Student data indicate you may struggle with the concept of vapor pressure. Access the eBook to view additional Learning Resources on this topic.

Student Note: The processes of vaporization and condensation are examples of *phase changes*. These and other phase changes are discussed in detail in Section 11.6.

Figure 11.11 Comparison of the rates of vaporization and condensation at constant temperature.

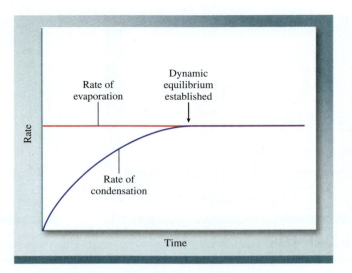

gas phase. Consequently, as we have already seen with water in Table 10.5, vapor pressure increases with increasing temperature. Figure 11.12 shows the plots of vapor pressure versus temperature for three different liquids.

The plots in Figure 11.12 of vapor pressure as a function of temperature are not linear. However, a linear relationship does exist between the natural log of vapor pressure and the reciprocal of absolute temperature. This relationship is called the ***Clausius*[2]-*Clapeyron*[3] *equation:***

Equation 11.1

$$\ln P = -\frac{\Delta H_{vap}}{RT} + C$$

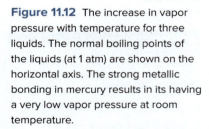

Student Note: The units of R are those that enable us to cancel the units of J/mol (or kJ/mol) associated with ΔH_{vap} [◄ Section 10.3, Table 10.4].

where $\ln P$ is the natural logarithm of the vapor pressure, ΔH_{vap} is the molar heat of vaporization (kJ/mol), R is the gas constant (8.314 J/K · mol), and C is a constant that must be determined experimentally for each different compound. The Clausius-Clapeyron equation has the form of the general linear equation $y = mx + b$:

$$\ln P = \left(-\frac{\Delta H_{vap}}{R}\right)\left(\frac{1}{T}\right) + C$$
$$y = \qquad mx \qquad + b$$

Figure 11.12 The increase in vapor pressure with temperature for three liquids. The normal boiling points of the liquids (at 1 atm) are shown on the horizontal axis. The strong metallic bonding in mercury results in its having a very low vapor pressure at room temperature.

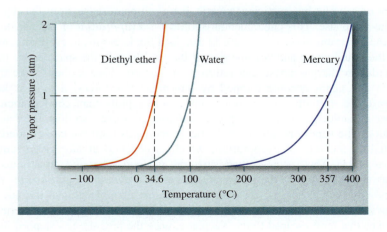

2. Rudolf Julius Emanuel Clausius (1822–1888). German physicist. Clausius's work was mainly in electricity, kinetic theory of gases, and thermodynamics.

3. Benoit Paul Emile Clapeyron (1799–1864). French engineer. Clapeyron made contributions to the thermodynamics of steam engines.

By measuring the vapor pressure of a liquid at several different temperatures and plotting $\ln P$ versus $1/T$, we can determine the slope of the line, which is equal to $-\Delta H_{vap}/R$. (ΔH_{vap} is assumed to be independent of temperature.)

If we know the value of ΔH_{vap} and the vapor pressure of a liquid at one temperature, we can use the Clausius-Clapeyron equation to calculate the vapor pressure of the liquid at a different temperature. At temperatures T_1 and T_2, the vapor pressures are P_1 and P_2. From Equation 11.1 we can write

Student Hot Spot

Student data indicate you may struggle with Clausius-Clapeyron equation. Access the eBook to view additional Learning Resources on this topic.

$$\ln P_1 = -\frac{\Delta H_{vap}}{RT_1} + C \qquad \textbf{Equation 11.2}$$

$$\ln P_2 = -\frac{\Delta H_{vap}}{RT_2} + C \qquad \textbf{Equation 11.3}$$

Subtracting Equation 11.3 from Equation 11.2, we get

$$\ln P_1 - \ln P_2 = -\frac{\Delta H_{vap}}{RT_1} - \left(-\frac{\Delta H_{vap}}{RT_2}\right)$$

$$= \frac{\Delta H_{vap}}{R}\left(\frac{1}{T_2} - \frac{1}{T_1}\right)$$

and finally

$$\ln \frac{P_1}{P_2} = \frac{\Delta H_{vap}}{R}\left(\frac{1}{T_2} - \frac{1}{T_1}\right) \qquad \textbf{Equation 11.4}$$

Sample Problem 11.2 shows how to use Equation 11.4.

SAMPLE PROBLEM 11.2

Diethyl ether is a volatile, highly flammable organic liquid that today is used mainly as a solvent. (It was used as an anesthetic during the nineteenth century and as a recreational intoxicant early in the twentieth century during prohibition, when ethanol was difficult to obtain.) The vapor pressure of diethyl ether is 401 mmHg at 18°C, and its molar heat of vaporization is 26 kJ/mol. Calculate its vapor pressure at 32°C.

Strategy Given the vapor pressure at one temperature, P_1, use Equation 11.4 to calculate the vapor pressure at a second temperature, P_2.

$$\ln \frac{P_1}{P_2} = \frac{\Delta H_{vap}}{R}\left(\frac{1}{T_2} - \frac{1}{T_1}\right)$$

Setup Temperature must be expressed in kelvins, so $T_1 = 291.15$ K and $T_2 = 305.15$ K. Because the molar heat of vaporization is given in kJ/mol, we will have to convert it to J/mol for the units of R to cancel properly: $\Delta H_{vap} = 2.6 \times 10^4$ J/mol. The inverse function of $\ln x$ is e^x.

Solution

$$\ln \frac{P_1}{P_2} = \frac{2.6 \times 10^4 \text{ J/mol}}{8.314 \text{ J/K} \cdot \text{mol}}\left(\frac{1}{305.15 \text{ K}} - \frac{1}{291.15 \text{ K}}\right)$$

$$= -0.4928$$

$$\frac{P_1}{P_2} = e^{-0.4928} = 0.6109$$

$$\frac{P_1}{0.6109} = P_2$$

$$P_2 = \frac{401 \text{ mmHg}}{0.6109} = 6.6 \times 10^2 \text{ mmHg}$$

(Continued on next page)

THINK ABOUT IT

It is easy to switch P_1 and P_2 or T_1 and T_2 accidentally and get the wrong answer to a problem such as this. One way to help safeguard against this common error is to verify that the vapor pressure is *higher* at the higher temperature.

Practice Problem **A**TTEMPT   The vapor pressure of ethanol is 1.00×10^2 mmHg at 34.9°C. What is its vapor pressure at 55.8°C? (ΔH_{vap} for ethanol is 39.3 kJ/mol.)

Practice Problem **B**UILD   Estimate the molar heat of vaporization of a liquid whose vapor pressure doubles when the temperature is raised from 85°C to 95°C. At what temperature will the vapor pressure be five times the value at 85°C?

Practice Problem **C**ONCEPTUALIZE   The diagram on the left depicts a system at room temperature. (The space above the liquid contains air and the vapor of the liquid in the container, and is sealed with a movable piston, making the pressure inside the vessel equal to atmospheric pressure.) Which of the diagrams [(i)–(v)] could represent the same system at a higher temperature? Assume that atmospheric pressure is constant.

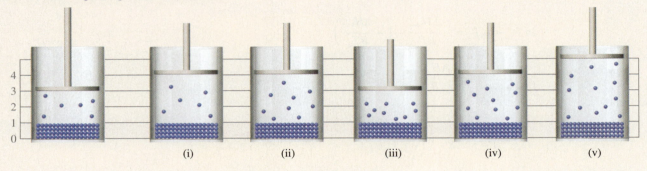

(i)	(ii)	(iii)	(iv)	(v)

CHECKPOINT – SECTION 11.2   Properties of Liquids

11.2.1 At what temperature would diethyl ether have a vapor pressure of 250 mmHg? Use the vapor pressure at 18°C and ΔH_{vap} given in Sample Problem 11.2.

a) 5.6°C

b) 280°C

c) 17°C

d) 6.5°C

e) −270°C

11.2.2 Given the following information for C_6F_6, calculate its ΔH_{vap}. At 300 K, the vapor pressure is 92.47 mmHg, and at 320 K, the vapor pressure is 225.1 mmHg.

a) 208 kJ/mol

b) 411 kJ/mol

c) 16.4 kJ/mol

d) 10.3 kJ/mol

e) 35.5 kJ/mol

11.2.3 Using the graph, estimate the vapor pressure of the liquid at 100°C.

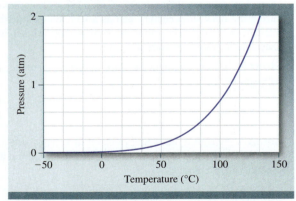

a) 100 atm

d) 110 atm

b) 0.75 atm

e) 1.50 atm

c) 1.00 atm

11.2.4 Using the result from question 11.2.3 and another point from the graph, estimate ΔH_{vap} for the liquid.

a) 15 kJ/mol

d) 75 kJ/mol

b) 35 kJ/mol

e) 95 kJ/mol

c) 55 kJ/mol

11.3 Crystal Structure

Solids can be categorized as either crystalline or amorphous. A ***crystalline solid*** possesses rigid and long-range order; its atoms, molecules, or ions occupy specific positions. The arrangement of the particles in a crystalline solid, which we call the ***lattice structure,*** depends on the nature and the size of the particles involved. The forces responsible for the stability of a crystal can be ionic forces, covalent bonds, van der Waals forces, hydrogen bonds, or a combination of some of these forces. Amorphous solids lack a well-defined arrangement and long-range molecular order. In this section, we concentrate on the nature of crystalline solids. (Amorphous solids are discussed in Section 11.5.)

Student Note: Ice is a crystalline solid. Glass is an amorphous solid.

Unit Cells

A ***unit cell*** is the basic repeating structural unit of a crystalline solid. Figure 11.13 shows a unit cell and its extension in three dimensions. Each sphere represents an atom, ion, or molecule and is called a ***lattice point.*** For the purpose of clarity, we limit our discussion in this section to metal crystals in which each lattice point is occupied by an atom.

Every crystalline solid can be described in terms of one of the seven types of unit cells shown in Figure 11.14. The geometry of the cubic unit cell is particularly simple because all sides and all angles are equal. Any of the unit cells, when repeated in space in all three dimensions, forms the lattice structure characteristic of a crystalline solid.

Animation
Cubic unit cells and their origins.

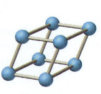

Figure 11.13 (a) A single unit cell and (b) a lattice (three-dimensional array) made up of many unit cells. Each sphere represents a lattice point, which may be an atom, a molecule, or an ion.

(a)   (b)

Figure 11.14 The seven types of unit cells. Angle α is defined by edges b and c, angle β by edges a and c, and angle γ by edges a and b.

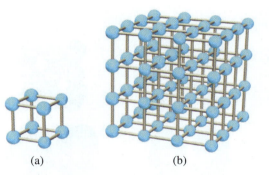

Simple cubic
$a = b = c$
$\alpha = \beta = \gamma = 90°$

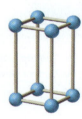

Tetragonal
$a = b \neq c$
$\alpha = \beta = \gamma = 90°$

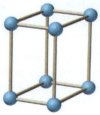

Orthorhombic
$a \neq b \neq c$
$\alpha = \beta = \gamma = 90°$

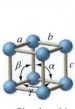

Rhombohedral
$a = b = c$
$\alpha = \beta = \gamma \neq 90°$

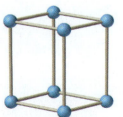

Monoclinic
$a \neq b \neq c$
$\gamma \neq \alpha = \beta = 90°$

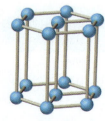

Triclinic
$a \neq b \neq c$
$\alpha \neq \beta \neq \gamma \neq 90°$

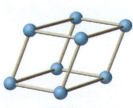

Hexagonal
$a = b \neq c$
$\alpha = \beta = 90°; \gamma = 120°$

Packing Spheres

We can understand the geometric requirements for crystal formation by considering the different ways of packing a number of identical atoms to form an ordered three-dimensional structure. The way the atoms are arranged in layers determines the type of unit cell.

In the simplest case, a layer of atoms can be arranged as shown in Figure 11.15(a). The three-dimensional structure can be generated by placing a layer above and below this layer in such a way that atoms in one layer are directly over the atoms in the layer below. This procedure can be extended to generate many, many layers, as in the case of a crystal. Focusing on the atom labeled with an *"x,"* we see that it is in contact with four atoms in its own layer, one atom in the layer above it, and one atom in the layer below it. Each atom in this arrangement is said to have a coordination number of 6 because it has six immediate neighbors. The ***coordination number*** is defined as the number of atoms surrounding an atom in a crystal lattice. The value of the coordination number indicates how tightly the atoms are packed together—the larger the coordination number, the closer the atoms are to one another. The basic repeating unit in the array of atoms is called a ***simple cubic cell*** (scc) [Figure 11.15(b)]. (The simple cubic cell is also called the *primitive* cubic cell.)

The other types of cubic cells, shown in Figure 11.16, are the ***body-centered cubic cell*** (bcc) and the ***face-centered cubic cell*** (fcc). Unlike the simple cube, the second layer of atoms in the body-centered cubic arrangement fits into the depressions of the first layer and the third layer fits into the depressions of the second layer (Figure 11.17).

Student Note: Depending on the type of particles that make up a solid, the coordination number can also refer to the number of surrounding molecules or ions.

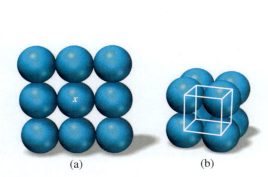

(a)   (b)

Figure 11.15 Arrangement of identical spheres in a simple cubic cell. (a) Top view of one layer of spheres. (b) Definition of a simple cubic cell.

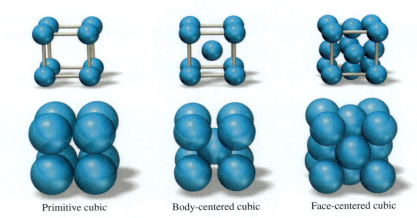

Primitive cubic   Body-centered cubic   Face-centered cubic

Figure 11.16 Three types of cubic cells. The top view makes it easier to see the locations of the lattice points, but the bottom view is more realistic, with the spheres touching one another.

Figure 11.17 In the body-centered cubic arrangement, the spheres in each layer rest in the depressions between spheres in the previous layer.

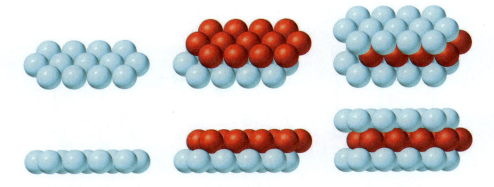

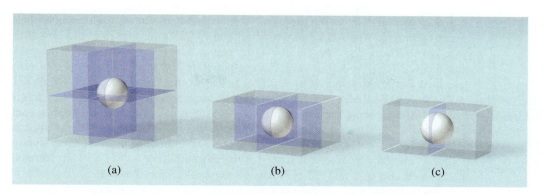

(a)   (b)   (c)

Figure 11.18 (a) A corner
atom in any cell is shared by
eight unit cells. (b) An edge
atom is shared by four unit
cells. (c) A face-centered
atom in a cubic cell is shared
by two unit cells.

The coordination number of each atom in the bcc structure is 8 (each sphere is in contact with four others in the layer above and four others in the layer below). In the face-centered cubic cell, there are atoms at the center of each of the six faces of the cube, in addition to the eight corner atoms. The coordination number in the face-centered cubic cell is 12 (each sphere is in contact with four others in its own layer, four others in the layer above, and four others in the layer below).

Because every unit cell in a crystalline solid is adjacent to other unit cells, most of a cell's atoms are shared by neighboring cells. (The atom at the center of the body-centered cubic cell is an exception.) In all types of cubic cells, for example, each corner atom belongs to eight unit cells whose corners all touch [Figure 11.18(a)]. An atom that lies on an edge, on the other hand, is shared by four unit cells [Figure 11.18(b)], and a face-centered atom is shared by two unit cells [Figure 11.18(c)]. Because a simple cubic cell has lattice points only at each of the eight corners, and because each corner atom is shared by eight unit cells, there will be the equivalent of only *one* complete atom contained within a simple cubic unit cell (Figure 11.19). A body-centered cubic cell contains the equivalent of two complete atoms, one in the center and eight shared corner atoms. A face-centered cubic cell contains the equivalent of four complete atoms—three from the six face-centered atoms and one from the eight shared corner atoms.

Figure 11.19 Because each sphere
is shared by eight unit cells and there
are eight corners in a cube, there is
the equivalent of one complete sphere
inside a simple cubic unit cell.

Closest Packing

There is more empty space in the simple cubic and body-centered cubic cells than in the face-centered cubic cell. Closest packing, the most efficient arrangement of atoms, starts with the structure shown in Figure 11.20(a), which we call layer A. Focusing on the only atom that is surrounded completely by other atoms, we see that it has six immediate neighbors in its own layer. In the second layer, which we call layer B, atoms are packed into the depressions between the atoms in the first layer so that all the atoms are as close together as possible [Figure 11.20(b)].

There are two ways that a third layer of atoms can be arranged. They may sit in the depressions between second-layer atoms such that the third-layer atoms lie directly over atoms in the first layer [Figure 11.20(c)]. In this case, the third layer is also labeled A. Alternatively, atoms in the third layer may sit in a *different* set of depressions such that they do not lie directly over atoms in the first layer [Figure 11.20(d)]. In this case, we label the third layer C.

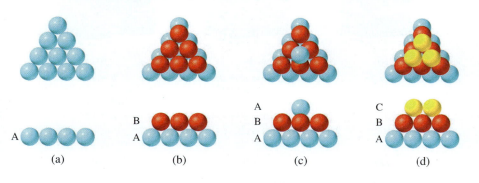

(a)   (b)   (c)   (d)

Figure 11.20 (a) In a close-packed
layer, each sphere is in contact with six
others. (b) Spheres in the second layer
fit into the depressions between the
first-layer spheres. (c) In the hexagonal
close-packed structure, each third-layer
sphere is directly over a first-layer
sphere. (d) In the cubic close-packed
structure, each third-layer sphere fits
into a depression that is directly over a
depression in the first layer.

Figure 11.21 shows the exploded views and the structures resulting from these two arrangements. The ABA arrangement [Figure 11.21(a)] is known as the *hexagonal close-packed (hcp) structure,* and the ABC arrangement [Figure 11.21(b)] is the *cubic close-packed (ccp) structure,* which corresponds to the face-centered cube already described. In the hcp structure, the spheres in every other layer occupy the same vertical position (ABABAB . . .), while in the ccp structure, the spheres in every fourth layer occupy the same vertical position (ABCABCA . . .). In both structures, each sphere has a coordination number of 12 (each sphere is in contact with six spheres in its own layer, three spheres in the layer above, and three spheres in the layer below). Both the hcp and ccp structures represent the most efficient way of packing identical spheres in a unit cell, and the coordination number cannot exceed 12.

Many metals form crystals with hcp or ccp structures. For example, magnesium, titanium, and zinc crystallize with their atoms in an hcp array, while aluminum, nickel, and silver crystallize in the ccp arrangement. A substance will crystallize with the arrangement that maximizes the stability of the solid.

Figure 11.22 summarizes the relationship between the atomic radius *r* and the edge length *a* of a simple cubic cell, a body-centered cubic cell, and a face-centered cubic cell. This relationship can be used to determine the atomic radius of a sphere in which the density of the crystal is known.

Student Note: The noble gases, which are monatomic, crystallize in the ccp structure, with the exception of helium, which crystallizes in the hcp structure.

Figure 11.21 Exploded views of (a) a hexagonal close-packed structure and (b) a cubic close-packed structure. This view is tilted to show the face-centered cubic unit cell more clearly. Note that the cubic close-packed arrangement is the same as the face-centered unit cell.

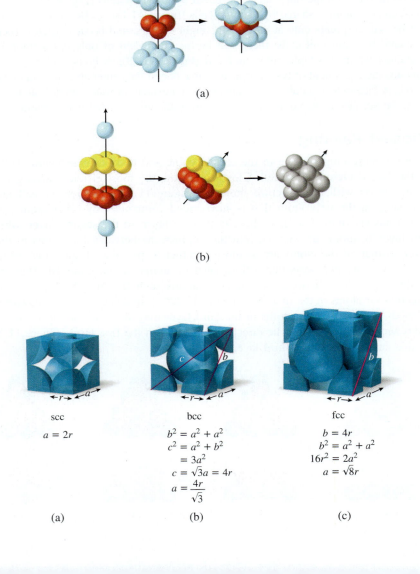

(a)

(b)

Figure 11.22 The relationship between the edge length (*a*) and radius (*r*) of atoms in the (a) simple cubic cell, (b) body-centered cubic cell, and (c) face-centered cubic cell.

scc

$a = 2r$

(a)

bcc

$$b^2 = a^2 + a^2$$
$$c^2 = a^2 + b^2$$
$$= 3a^2$$
$$c = \sqrt{3}a = 4r$$
$$a = \frac{4r}{\sqrt{3}}$$

(b)

fcc

$$b = 4r$$
$$b^2 = a^2 + a^2$$
$$16r^2 = 2a^2$$
$$a = \sqrt{8}r$$

(c)

Sample Problem 11.3 illustrates the relationships between the unit cell type, cell dimensions, and density of a metal.

SAMPLE PROBLEM 11.3

Gold crystallizes in a cubic close-packed structure (face-centered cubic unit cell) and has a density of 19.3 g/cm^3. Calculate the atomic radius of an Au atom in angstroms (Å).

Strategy Using the given density and the mass of gold contained within a face-centered cubic unit cell, determine the volume of the unit cell. Then, use the volume to determine the value of a, and use the equation supplied in Figure 11.22(c) to find r. Be sure to use consistent units for mass, length, and volume.

Setup The face-centered cubic unit cell contains a total of four atoms of gold [six faces, each shared by two unit cells, and eight corners, each shared by eight unit cells—Figure 11.22(c)]. $d = m/V$ and $V = a^3$.

Solution First, we determine the mass of gold (in grams) contained within a unit cell:

$$m = \frac{4 \text{ atoms}}{\text{unit cell}} \times \frac{1 \text{ mol}}{6.022 \times 10^{23} \text{ atoms}} \times \frac{197.0 \text{ g Au}}{1 \text{ mol Au}} = 1.31 \times 10^{-21} \text{ g/unit cell}$$

Then we calculate the volume of the unit cell in cm^3:

$$V = \frac{m}{d} = \frac{1.31 \times 10^{-21} \text{ g}}{19.3 \text{ g/cm}^3} = 6.78 \times 10^{-23} \text{ cm}^3$$

Using the calculated volume and the relationship $V = a^3$ (rearranged to solve for a), we determine the length of a side of a unit cell:

$$a = \sqrt[3]{V} = \sqrt[3]{6.78 \times 10^{-23} \text{ cm}^3} = 4.08 \times 10^{-8} \text{ cm}$$

Using the relationship provided in Figure 11.22(c) (rearranged to solve for r), we determine the radius of a gold atom in centimeters:

$$r = \frac{a}{\sqrt{8}} = \frac{4.08 \times 10^{-8} \text{ cm}}{\sqrt{8}} = 1.44 \times 10^{-8} \text{ cm}$$

Finally, we convert centimeters to angstroms:

$$1.44 \times 10^{-8} \text{ cm} \times \frac{1 \times 10^{-2} \text{ m}}{1 \text{ cm}} \times \frac{1 \text{ Å}}{1 \times 10^{-10} \text{ m}} = 1.44 \text{ Å}$$

THINK ABOUT IT

Atomic radii tend to be on the order of 1 Å, so this answer is reasonable.

Practice Problem A TTEMPT When silver crystallizes, it forms face-centered cubic cells. The unit cell edge length is 4.087 Å. Calculate the density of silver.

Practice Problem B UILD The density of sodium metal is 0.971 g/cm^3, and the unit cell edge length is 4.285 Å. Determine the unit cell (simple, body-centered, or face-centered cubic) of sodium metal.

Practice Problem C ONCEPTUALIZE The diagram shows two different arrangements of circles. Using the areas defined by the red rectangles, determine the two-dimensional "density" (ratio of area occupied by circles to total area) for each arrangement; and determine the ratio of densities for the two arrangements. (Report the ratio of densities to two significant figures.)

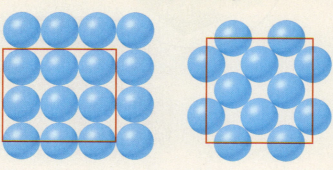

CHECKPOINT – SECTION 11.3   Crystal Structure

11.3.1 Nickel has a face-centered cubic unit cell with an edge length of 352.4 pm. Calculate the density of nickel.

a) 2.227 g/cm^3

b) 4.455 g/cm^3

c) 38.99 g/cm^3

d) 8.908 g/cm^3

e) 11.14 g/cm^3

11.3.2 A metal crystallizes in a body-centered cubic unit cell with an edge length of 5.065×10^{-8} cm. Given that the approximate density of the metal is 3.51 g/cm^3, determine its molar mass and its probable identity.

a) 68.5 g/mol, Ga

b) 137 g/mol, Ba

c) 34.2 g/mol, K

d) 205 g/mol, Tl

e) 274 g/mol, Mt

11.4   Types of Crystals

The structures and properties of crystalline solids, such as melting point, density, and hardness, are determined by the kinds of forces that hold the particles together. We can classify any crystal as one of four types: ionic, covalent, molecular, or metallic.

Ionic Crystals

Ionic crystals are composed of charged spheres (cations and anions) that are held together by Coulombic attraction. Anions typically are considerably bigger than cations [◄◄ Section 7.6], and the relative sizes and relative numbers of the ions in a compound determine how the ions are arranged in the solid lattice. NaCl adopts a face-centered cubic arrangement as shown in Figure 11.23. Note the positions of ions within the unit cell, and within the lattice overall. Both the Na$^+$ ions and the Cl$^-$ ions adopt face-centered cubic arrangements, and the unit cell defined by the arrangement of cations overlaps with the unit cell defined by the arrangement of anions. Look closely at the unit cell shown in Figure 11.23(a). It is defined as fcc by the positions of the Cl$^-$ ions. Recall that there is the equivalent of four spheres contained in the fcc unit cell (half a sphere at each of six faces and one-eighth of a sphere at each of eight corners). In this case the spheres are Cl$^-$ ions, so the unit cell of NaCl contains four Cl$^-$ ions. Now look at the positions of the Na$^+$ ions. There are Na$^+$ ions centered

Student Note: The lattice points used to define a unit cell must all be identical. We can define the unit cell of NaCl based on the positions of the Na$^+$ ions or the positions of the Cl$^-$ ions.

Figure 11.23 The unit cell of an ionic compound can be defined by either (a) the positions of anions or (b) the positions of cations.

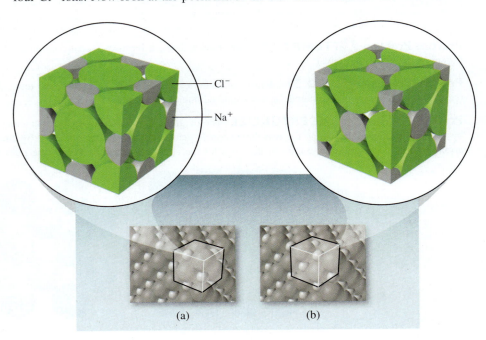

Cl$^-$

Na$^+$

(a)   (b)

How Do We Know the Structures of Crystals?

Virtually all we know about crystal structure has been learned from X-ray diffraction studies. **X-ray diffraction** is the scattering of X rays by the units of a crystalline solid. The scattering, or *diffraction patterns,* produced are used to deduce the arrangement of particles in the solid lattice.

In Section 6.1, we discussed the interference phenomenon associated with waves (see Figure 6.4). Because X rays are a form of electromagnetic radiation (i.e., they are *waves*), they exhibit interference phenomena under suitable conditions. In 1912, Max von Laue[4] correctly suggested that, because the wavelength of X rays is comparable in magnitude to the distances between lattice points in a crystal, the lattice should be able to diffract X rays. An X-ray diffraction pattern is the result of interference in the waves associated with X rays.

Figure 11.24 shows a typical X-ray diffraction setup. A beam of X rays is directed at a mounted crystal. When X-ray

photons encounter the electrons in the atoms of a crystalline solid, some of the incoming radiation is reflected, much as visible light is reflected by a mirror; the process is called the scattering of X rays.

To understand how a diffraction pattern arises, consider the scattering of X rays by the atoms in two parallel planes (Figure 11.25). Initially, the two incident rays are in phase with each other (their maxima and minima occur at the same positions). The upper wave is scattered, or reflected, by an atom in the first layer, while the lower wave is scattered by an atom in the second layer. For these two scattered waves to be in phase again, the extra distance traveled by the lower wave (the sum of the distance between points B and C and the distance between points C and D) must be an integral multiple of the wavelength (λ) of the X ray; that is,

Equation 11.5 $BC + CD = 2d \sin \theta = n\lambda$
$$n = 1, 2, 3, \ldots$$

4. Max Theodor Felix von Laue (1879–1960). German physicist. Von Laue received the Nobel Prize in Physics in 1914 for his discovery of X-ray diffraction.

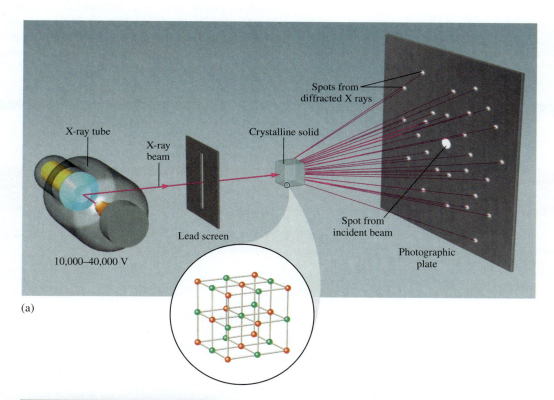

(a)

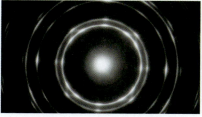

(b)

Figure 11.24 (a) Arrangement for obtaining the X-ray diffraction pattern of a crystal. The shield prevents the intense beam of undiffracted X rays from damaging the photographic plate. (b) X-ray diffraction pattern.
(b) Science History Images/Alamy Stock Photo

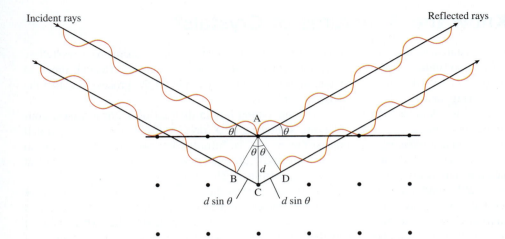

Incident rays

Reflected rays

Figure 11.25 Reflection of X rays from two layers of atoms. The lower wave travels a distance $2d \sin \theta$ longer than the upper wave does. For the two waves to be in phase again after reflection, it must be true that $2d \sin \theta = n\lambda$, where λ is the wavelength of the X ray and $n = 1, 2, 3, \ldots$ The sharply defined spots in Figure 11.24 are observed only if the crystal is large enough to consist of hundreds of parallel layers.

where θ is the angle between the X rays and the plane of the crystal and d is the distance between adjacent planes. Equation 11.5 is known as the ***Bragg equation,*** after William H. Bragg and Sir William L. Bragg.[5] The reinforced waves produce a dark spot on a photographic film for each value of θ that satisfies the Bragg equation.

The X-ray diffraction technique offers the most accurate method for determining bond lengths and bond angles in molecules in solids. Because X rays are scattered by electrons, chemists can construct an electron-density contour map from the diffraction patterns by using a complex mathematical procedure. Basically, an electron-density contour map tells us the relative electron densities at various locations in a molecule. The densities reach a maximum near the center of each atom. In this manner, we can determine the positions of the nuclei and hence the geometric parameters of the molecule.

(See end-of-chapter problems 11.56, 11.57, 11.108, and 11.109.)

---

5. William Henry Bragg (1862–1942) and Sir William Lawrence Bragg (1890–1972). English physicists, father and son. Both worked on X-ray crystallography. The younger Bragg formulated the fundamental equation for X-ray diffraction. The two shared the Nobel Prize in Physics in 1915.

SAMPLE PROBLEM 11.4

When X rays of wavelength 0.122 nm are diffracted by a metallic crystal, the angle of first-order diffraction ($n = 1$) is measured to be 16.8°. What is the distance (in pm) between the layers of atoms responsible for the diffraction?

Strategy We rearrange Equation 11.5 to solve for d.

$$d = \frac{n\lambda}{2 \sin \theta}$$

Setup The values given in the problem are $n = 1$, $\lambda = 0.122$ nm, and $\theta = 16.8°$. To get d in pm, we must express λ in pm.

$$0.122 \text{ nm} \times \frac{1000 \text{ pm}}{1 \text{ nm}} = 122 \text{ pm}$$

Solution

$$d = \frac{122 \text{ pm}}{2 \sin (16.8°)} = 211 \text{ pm}$$

THINK ABOUT IT

A common error in calculations such as this is to neglect to convert the units of λ to match the desired units of d. Also, if you find that your calculation yields a result different from this one, check to make sure that your calculator's mode is set to *degrees,* not radians.

Practice Problem **A**TTEMPT When X rays of wavelength 0.0831 nm are diffracted by a metallic crystal, the angle of first-order diffraction ($n = 1$) is measured to be 13.4°. What is the distance (in pm) between the layers of atoms responsible for the diffraction?

Practice Problem **B**UILD What will be the measured angle of first-order diffraction (θ) when a crystalline substance with a distance between layers of 273 pm is analyzed by X-ray diffraction using X rays of wavelength 0.115 nm?

Practice Problem **C**ONCEPTUALIZE Would you expect the angle of diffraction (θ) to increase or decrease as (a) distance between layers in a crystal (d) increases with X-ray wavelength held constant; (b) as X-ray wavelength increases with distance between layers in a crystal (d) held constant?

on each edge of the cube, in addition to one Na^+ at the center. Each sphere on the cube's edge is shared by four unit cells, and there are 12 such edges. Thus, the unit cell in Figure 11.23(a) also contains four Na^+ ions (one-quarter sphere at each of 12 edges, giving three spheres, and one sphere at the center). The unit cell of an ionic compound always contains the same ratio of cations to anions as the empirical formula of the compound.

Figure 11.26 shows the crystal structures of three ionic compounds: CsCl, ZnS, and CaF_2. Cesium chloride [Figure 11.26(a)] has the simple cubic lattice. Despite the apparent similarity of the formulas of CsCl and NaCl, CsCl adopts a different arrangement because the Cs^+ ion is much larger than the Na^+ ion. Zinc sulfide [Figure 11.26(b)] has the *zincblende* structure, which is based on the face-centered cubic lattice. If the S^{2-} ions occupy the lattice points, the smaller Zn^{2+} ions are arranged tetrahedrally about each S^{2-} ion. Other ionic compounds that have the zincblende structure include CuCl, BeS, CdS, and HgS. Calcium fluoride [Figure 11.26(c)] has the *fluorite* structure. The unit cell in Figure 11.26(c) is defined based on the positions of the cations, rather than the positions of the anions. The Ca^{2+} ions occupy the lattice points, and each F^- ion is surrounded tetrahedrally by four Ca^{2+} ions. The compounds SrF_2, BaF_2, $BaCl_2$, and PbF_2 also have the fluorite structure.

> **Student Note:** It is a common mistake to identify the CsCl structure as body-centered cubic. Remember that the lattice points used to define a unit cell must all be identical. In this case, they are all Cl^- ions. CsCl has a simple cubic unit cell.

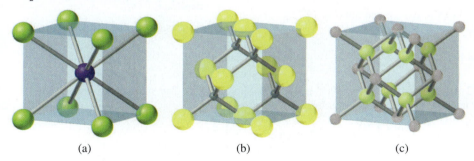

(a) (b) (c)

Figure 11.26 Crystal structures of (a) CsCl, (b) ZnS, and (c) CaF_2. In each case, the smaller sphere represents the cation.

Sample Problems 11.5 and 11.6 show how to determine the number of ions in a unit cell and the density of an ionic crystal, respectively.

SAMPLE PROBLEM 11.5

How many of each ion are contained within a unit cell of ZnS?

Strategy Determine the contribution of each ion in the unit cell based on its position.

Setup Referring to Figure 11.26, the unit cell has four Zn^{2+} ions completely contained within the unit cell, and S^{2-} ions at each of the eight corners and at each of the six faces. Interior ions (those completely contained within the unit cell) contribute one, those at the corners each contribute one-eighth, and those on the faces each contribute one-half.

Solution The ZnS unit cell contains four Zn^{2+} ions (interior) and four S^{2-} ions [$8 \times \frac{1}{8}$ (corners) and $6 \times \frac{1}{2}$ (faces)].

(Continued on next page)

THINK ABOUT IT

Make sure that the ratio of cations to anions that you determine for a unit cell matches the ratio expressed in the compound's empirical formula.

Practice Problem **A**TTEMPT   Referring to Figure 11.26, determine how many of each ion are contained within a unit cell of CaF_2.

Practice Problem **B**UILD   Referring to Figure 11.26, determine how many of each ion are contained within a unit cell of CsCl.

Practice Problem **C**ONCEPTUALIZE   Explain why the CsCl structure [Figure 11.26(a)] is not considered body-centered cubic.

SAMPLE PROBLEM   11.6

The edge length of the NaCl unit cell is 564 pm. Determine the density of NaCl in g/cm^3.

Strategy Use the number of Na^+ and Cl^- ions in a unit cell (four of each) to determine the mass of a unit cell. Calculate volume using the edge length given in the problem statement. Density is mass divided by volume ($d = m/V$). Be careful to use units consistently.

Setup The masses of Na^+ and Cl^- ions are 22.99 amu and 35.45 amu, respectively. The conversion factor from amu to grams is

> **Student Note:** The mass of an atomic ion is treated the same as the mass of the parent atom. In these cases, the mass of an electron is not significant [◄◄ Section 2.2, Table 2.1].

$$\frac{1 \text{ g}}{6.022 \times 10^{23} \text{ amu}}$$

so the masses of the Na^+ and Cl^- ions are 3.818×10^{-23} g and 5.887×10^{-23} g, respectively. The unit cell length is

$$564 \text{ pm} \times \frac{1 \times 10^{-12} \text{ m}}{1 \text{ pm}} \times \frac{1 \text{ cm}}{1 \times 10^{-2} \text{ m}} = 5.64 \times 10^{-8} \text{ cm}$$

Solution The mass of a unit cell is 3.882×10^{-22} g ($4 \times 3.818 \times 10^{-23}$ g $+ 4 \times 5.887 \times 10^{-23}$ g). The volume of a unit cell is 1.794×10^{-22} cm^3 [$(5.64 \times 10^{-8}$ cm$)^3$]. Therefore, the density is given by

$$d = \frac{3.882 \times 10^{-22} \text{ g}}{1.794 \times 10^{-22} \text{ cm}^3} = 2.16 \text{ g/cm}^3$$

THINK ABOUT IT

If you were to hold a cubic centimeter (1 cm^3) of salt in your hand, how heavy would you expect it to be? Common errors in this type of problem include errors of unit conversion—especially with regard to length and volume. Such errors can lead to results that are off by many orders of magnitude. Often you can use common sense to gauge whether or not a calculated answer is reasonable. For instance, simply getting the centimeter-meter conversion upside down would result in a calculated density of 2.16×10^{12} g/cm^3! You *know* that a cubic centimeter of salt doesn't have a mass that large. (That's billions of kilograms!) If the magnitude of a result is not reasonable, go back and check your work.

Practice Problem **A**TTEMPT   LiF has the same unit cell as NaCl (fcc). The edge length of the LiF unit cell is 402 pm. Determine the density of LiF in g/cm^3.

Practice Problem **B**UILD   NiO also adopts the face-centered cubic arrangement. Given that the density of NiO is 6.67 g/cm^3, calculate the length of the edge of its unit cell (in pm).

Practice Problem **C**ONCEPTUALIZE   Referring to the cesium chloride structure in Figure 11.26(a), determine the density of a hypothetical ionic compound with the same unit cell, given that the radii of the anions and cations are 150 pm and 92 pm, respectively; and that their average masses are 98 amu and 192 amu, respectively.

Most ionic crystals have high melting points, which is an indication of the strong cohesive forces holding the ions together. A measure of the stability of ionic crystals is the lattice energy [◄◄ Section 8.2]; the higher the lattice energy, the more stable the compound. Ionic solids do not conduct electricity because the ions are fixed in position. In the molten (melted) state or when dissolved in water, however, the compound's ions are free to move and the resulting liquid conducts electricity.

Covalent Crystals

In covalent crystals, atoms are held together in an extensive three-dimensional network entirely by covalent bonds. Well-known examples are two of carbon's allotropes: diamond and graphite. In diamond, each carbon atom is sp^3-hybridized and bonded to four other carbon atoms [Figure 11.27(a)]. The strong covalent bonds in three dimensions contribute to diamond's unusual hardness (it is the hardest material known) and very high melting point (3550°C). In graphite, carbon atoms are arranged in six-membered rings [Figure 11.27(b)]. The atoms are all sp^2-hybridized, and each atom is bonded to three other atoms. The remaining unhybridized $2p$ orbital on each carbon atom is used in pi bonding. In fact, each layer of graphite has the kind of delocalized molecular orbital that is present in benzene [◄◄ Section 9.7]. Because electrons are free to move around in this extensively delocalized molecular orbital, graphite is a good conductor of electricity in directions along the planes of the carbon atoms. The layers are held together by weak van der Waals forces. The covalent bonds in graphite account for its hardness; however, because the layers can slide past one another, graphite is slippery to the touch and is effective as a lubricant. It is also used as the "lead" in pencils.

Another covalent crystal is quartz (SiO_2). The arrangement of silicon atoms in quartz is similar to that of carbon in diamond, but in quartz there is an oxygen atom between each pair of Si atoms. Because Si and O have different electronegativities, the Si—O bond is polar. Nevertheless, SiO_2 is similar to diamond in many respects, such as being very hard and having a high melting point (1610°C).

Figure 11.27 Structures of (a) diamond and (b) graphite. In diamond, each carbon atom is bonded in a tetrahedral arrangement to four other carbon atoms. In graphite, each carbon atom is bonded in a trigonal planar arrangement to three other carbon atoms. The distance between layers in graphite is 335 pm.

(both) *Charles D. Winters/Timeframe Photography/McGraw Hill*

335 pm

(a)   (b)

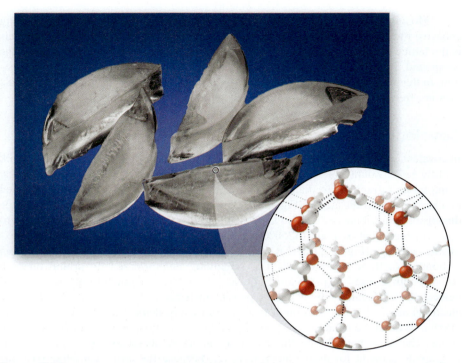

Figure 11.28 The three-dimensional structure of ice. The covalent bonds are shown by short solid lines and the weaker hydrogen bonds by long dotted lines between O and H. The empty space in the structure accounts for the low density of ice, relative to liquid water.
Charles D. Winters/Timeframe Photography/McGraw Hill

Molecular Crystals

In a molecular crystal, the lattice points are occupied by molecules, so the attractive forces between them are van der Waals forces and/or hydrogen bonding. An example of a molecular crystal is solid sulfur dioxide (SO_2), in which the predominant attractive force is a dipole-dipole interaction. Intermolecular hydrogen bonding is mainly responsible for maintaining the three-dimensional lattice of ice (Figure 11.28). Other examples of molecular crystals are I_2, P_4, and S_8.

Except in ice, molecules in molecular crystals are generally packed together as closely as their size and shape allow. Because van der Waals forces and hydrogen bonding are usually quite weak compared with covalent and ionic bonds, molecular crystals are more easily broken apart than ionic and covalent crystals. Indeed, most molecular crystals melt at temperatures below 100°C.

Metallic Crystals

Every lattice point in a metallic crystal is occupied by an atom of the same metal. Metallic crystals are generally body-centered cubic, face-centered cubic, or hexagonal close-packed. Consequently, metallic elements are usually very dense.

The bonding in metals is quite different from that in other types of crystals. In a metal, the bonding electrons are delocalized over the entire crystal. In fact, metal atoms in a crystal can be imagined as an array of positive ions immersed in a sea of delocalized valence electrons (Figure 11.29). The great cohesive force resulting from delocalization is responsible for a metal's strength, whereas the mobility of the delocalized electrons makes metals good conductors of heat and electricity. Table 11.4 summarizes the properties of the four different types of crystals discussed. Note that the data in Table 11.4 refer to the solid phase of each substance listed.

Sample Problem 11.7 lets you practice relating the cell dimensions in a solid to the density of the solid.

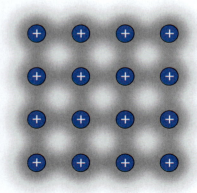

Figure 11.29 A cross section of a metallic crystal. Each circled positive charge represents the nucleus and inner electrons of a metal atom. The grey area surrounding the positive metal ions indicates the mobile "sea" of electrons.

TABLE 11.4	Types of Crystals and Their General Properties		
Type of Crystal	**Cohesive Forces**	**General Properties**	**Examples**
Ionic	Coulombic attraction and dispersion forces	Hard, brittle, high melting point, poor conductor of heat and electricity	NaCl, LiF, MgO, $CaCO_3$
Covalent	Covalent bonds	Hard, brittle, high melting point, poor conductor of heat and electricity	C (diamond),* SiO_2 (quartz)
Molecular[†]	Dispersion and dipole-dipole forces, hydrogen bonds	Soft, low melting point, poor conductor of heat and electricity	Ar, CO_2, I_2, H_2O, $C_{12}H_{22}O_{11}$
Metallic	Metallic bonds	Variable hardness and melting point, good conductor of heat and electricity	All metallic elements, such as Na, Mg, Fe, Cu

*Diamond is a good conductor of heat.

[†]Included in this category are crystals made up of individual atoms.

SAMPLE PROBLEM 11.7

The metal iridium (Ir) crystallizes with a face-centered cubic unit cell. Given that the length of the edge of a unit cell is 383 pm, determine the density of iridium in g/cm³.

Strategy A face-centered metallic crystal contains four atoms per unit cell [$8 \times \frac{1}{8}$ (corners) and $6 \times \frac{1}{2}$ (faces)]. Use the number of atoms per cell and the atomic mass to determine the mass of a unit cell. Calculate volume using the edge length given in the problem statement. Density is then mass divided by volume ($d = m/V$). Be sure to make all necessary unit conversions.

Setup The mass of an Ir atom is 192.2 amu. The conversion factor from amu to grams is

$$\frac{1 \text{ g}}{6.022 \times 10^{23} \text{ amu}}$$

so the mass of an Ir atom is 3.192×10^{-22} g. The unit cell length is

$$383 \text{ pm} \times \frac{1 \times 10^{-12} \text{ m}}{1 \text{ pm}} \times \frac{1 \text{ cm}}{1 \times 10^{-2} \text{ m}} = 3.83 \times 10^{-8} \text{ cm}$$

Solution The mass of a unit cell is 1.277×10^{-21} g ($4 \times 3.192 \times 10^{-22}$ g). The volume of a unit cell is 5.618×10^{-23} cm³ [$(3.83 \times 10^{-8}$ cm$)^3$]. Therefore, the density is given by

$$d = \frac{1.277 \times 10^{-21} \text{ g}}{5.62 \times 10^{-23} \text{ cm}^3} = 22.7 \text{ g/cm}^3$$

THINK ABOUT IT
Metals typically have high densities, so common sense can help you decide whether or not your calculated answer is reasonable.

Practice Problem A TTEMPT Aluminum metal crystallizes in a face-centered cubic unit cell. If the length of the cell edge is 404 pm, what is the density of aluminum in g/cm³?

Practice Problem B UILD Copper crystallizes in a face-centered cubic lattice. If the density of the metal is 8.96 g/cm³, what is the length of the unit cell edge in picometers?

Practice Problem C ONCEPTUALIZE Given that the diameter and average mass of a billiard ball are 5.72 cm and 165 g, respectively, determine the density of a billiard ball. Assuming that they can be packed like atoms in a metal, determine the density of a collection of billiard balls packed with a simple cubic unit cell, and those packed with a face-centered unit cell. Explain why the three densities are different despite all referring to the same objects.

CHECKPOINT – SECTION 11.4   Types of Crystals

11.4.1 The diagram here shows the anions in the hypothetical edge-centered unit cell of an ionic compound in which the ions combine in a 1:1 ratio. Use the diagram to determine the total number of ions contained within a unit cell of the compound.

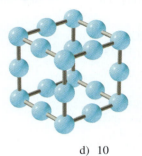

a) 2   d) 10

b) 4   e) 20

c) 8

11.4.2 At what angle would you expect X rays of wavelength 0.154 nm to be reflected from a crystal in which the distance between layers is 312 pm? (Assume $n = 1$.)

a) 1.6°

b) 29.6°

c) 0.25°

d) 6.8°

e) 14.3°

11.5 Amorphous Solids

Solids are most stable in crystalline form. However, if a solid is formed rapidly (e.g., when a liquid is cooled quickly), its atoms or molecules do not have time to align themselves and may become locked in positions other than those of a regular crystal. The resulting solid is said to be amorphous. *Amorphous solids,* such as glass, lack a regular three-dimensional arrangement of atoms. In this section, we will briefly discuss the properties of glass.

Glass is one of civilization's most valuable and versatile materials. It is also one of the oldest—glass articles date back as far as 1000 B.C. *Glass* commonly refers to an optically transparent fusion product of inorganic materials that has cooled to a rigid state without crystallizing. By fusion product we mean that the glass is formed by mixing molten silicon dioxide (SiO_2), its chief component, with compounds such as sodium oxide (Na_2O), boron oxide (B_2O_3), and certain transition metal oxides for color and other properties. In some respects, glass behaves more like a liquid than a solid.

There are about 800 different types of glass in common use today. Figure 11.30 shows two-dimensional schematic representations of crystalline quartz and amorphous quartz glass. Table 11.5 lists the composition and properties of quartz, Pyrex, and

Figure 11.30 Two-dimensional representation of (a) crystalline quartz and (b) noncrystalline (amorphous) quartz glass. The small spheres represent silicon. In reality, the structure of quartz is three-dimensional. Each Si atom is bonded in a tetrahedral arrangement to four O atoms.

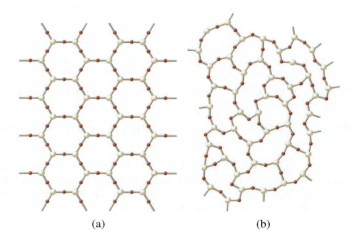

(a)   (b)

TABLE 11.5	Composition and Properties of Three Types of Glass	
Pure quartz glass	100% SiO_2	Low thermal expansion, transparent to a wide range of wavelengths. Used in optical research.
Pyrex glass	60%–80% SiO_2, 10%–25% B_2O_3, some Al_2O_3	Low thermal expansion; transparent to visible and infrared, but not to ultraviolet light. Used in cookware and laboratory glassware.
Soda-lime glass	75% SiO_2, 15% Na_2O, 10% CaO	Easily attacked by chemicals and sensitive to thermal shocks. Transmits visible light but absorbs ultraviolet light. Used in windows and bottles.

soda-lime glass. The color of glass is due largely to the presence of oxides of metal ions—mostly transition metal ions. For example, green glass contains iron(III) oxide (Fe_2O_3) or copper(II) oxide (CuO), yellow glass contains uranium(IV) oxide (UO_2), blue glass contains cobalt(II) and copper(II) oxides (CoO) and (CuO), and red glass contains small particles of gold and copper.

11.6 Phase Changes

A *phase* is a homogeneous part of a system that is separated from the rest of the system by a well-defined boundary. When an ice cube floats in a glass of water, for example, the liquid water is one phase and the solid water (the ice cube) is another. Although the chemical properties of water are the same in both phases, the physical properties of a solid are different from those of a liquid.

When a substance goes from one phase to another phase, we say that it has undergone a ***phase change.*** Phase changes in a system are generally caused by the addition or removal of energy, usually in the form of heat. Familiar examples of phase changes include the following:

Student Note: Phase changes are *physical* changes [◄◄ Section 1.4].

Example	Phase Change
Freezing of water	$H_2O(l) \longrightarrow H_2O(s)$
Evaporation (or vaporization) of water	$H_2O(l) \longrightarrow H_2O(g)$
Melting (fusion) of ice	$H_2O(s) \longrightarrow H_2O(l)$
Condensation of water vapor	$H_2O(g) \longrightarrow H_2O(l)$
Sublimation of dry ice	$CO_2(s) \longrightarrow CO_2(g)$

The establishment of an equilibrium vapor pressure, as described in Section 11.2, involved two of these phase changes: vaporization and condensation. Figure 11.31 summarizes the various types of phase changes.

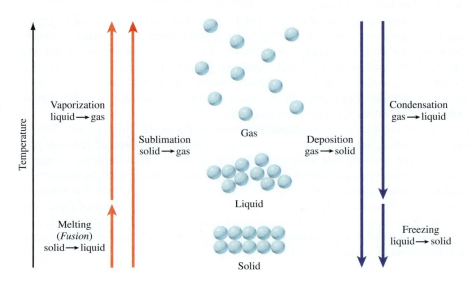

Figure 11.31 The six possible phase changes: melting (fusion), vaporization, sublimation, deposition, condensation, and freezing.

TABLE 11.6	Molar Heats of Vaporization for Selected Liquids	
Substance	Boiling Point (°C)	ΔH_{vap} (kJ/mol)
Argon (Ar)	−186	6.3
Benzene (C_6H_6)	80.1	31.0
Ethanol (C_2H_5OH)	78.3	39.3
Diethyl ether ($C_2H_5OC_2H_5$)	34.6	26.0
Mercury (Hg)	357	59.0
Methane (CH_4)	−164	9.2
Water (H_2O)	100	40.79

Liquid-Vapor Phase Transition

In Section 11.2, we learned that the vapor pressure of a liquid increases with increasing temperature. When the vapor pressure reaches the external pressure, the liquid boils. In fact, the **boiling point** of a substance is defined as the temperature at which its vapor pressure equals the external, atmospheric pressure. As a result, the boiling point of a substance varies with the external pressure. At the top of a mountain, for example, where the atmospheric pressure is lower than that at sea level, the vapor pressure of water (or any liquid) reaches the external pressure at a lower temperature. Thus, the boiling point is lower than it would be at sea level. The temperature at which the vapor pressure of a liquid is equal to 1 atm is called the **normal boiling point**.

Because the boiling point is defined in terms of the vapor pressure of the liquid, the boiling point is related to the **molar heat of vaporization (ΔH_{vap})**, the amount of heat required to vaporize a mole of substance at its boiling point. Indeed, the data in Table 11.6 show that the boiling point generally increases as ΔH_{vap} increases. Ultimately, both the boiling point and ΔH_{vap} are determined by the strength of intermolecular forces. For example, argon (Ar) and methane (CH_4), which have only relatively weak dispersion forces, have low boiling points and small molar heats of vaporization. Diethyl ether ($C_2H_5OC_2H_5$) has a dipole moment, and the dipole-dipole forces account for its moderately high boiling point and ΔH_{vap}. Both ethanol (C_2H_5OH) and water have strong hydrogen bonding, which accounts for their high boiling points and large ΔH_{vap} values. Strong metallic bonding causes mercury to have the highest boiling point and ΔH_{vap} of the liquids in Table 11.6. Interestingly, benzene (C_6H_6), although nonpolar, has a high *polarizability* due to the distribution of its electrons in delocalized π molecular orbitals. The dispersion forces that result can be as strong as (or even stronger than) dipole-dipole forces and/or hydrogen bonds.

The opposite of vaporization is condensation. In principle, a gas can be liquefied (made to condense) either by cooling or by applying pressure. Cooling a sample of gas decreases the kinetic energy of its molecules, so eventually the molecules aggregate to form small drops of liquid. Applying pressure to the gas (compression), on the other hand, reduces the distance between molecules, so they can be pulled together by intermolecular attractions. Many liquefaction processes use a combination of reduced temperature and increased pressure.

Student Note: T_c is the *highest* temperature at which a substance can exist as a liquid.

Every substance has a **critical temperature (T_c)** above which its gas phase cannot be liquefied, no matter how great the applied pressure. **Critical pressure (P_c)** is the minimum pressure that must be applied to liquefy a substance *at* its critical temperature. At temperatures above the critical temperature, there is no fundamental distinction between a liquid and a gas—we simply have a fluid. A fluid at a temperature and pressure that exceed T_c and P_c, respectively, is called a **supercritical fluid.** Supercritical fluids have some remarkable properties and are used as solvents in a wide variety of industrial applications. The first such large-scale industrial use was the decaffeination of coffee with supercritical CO_2.

TABLE 11.7	Critical Temperatures and Critical Pressures of Selected Substances	
Substance	T_c (°C)	P_c (atm)
Ammonia (NH₃)	132.4	111.5
Argon (Ar)	−122.2	6.3
Benzene (C_6H_6)	288.9	47.9
Carbon dioxide (CO_2)	31.0	73.0
Ethanol (C_2H_5OH)	243	63.0
Diethyl ether ($C_2H_5OC_2H_5$)	192.6	35.6
Mercury (Hg)	1462	1036
Methane (CH_4)	−83.0	45.6
Molecular hydrogen (H_2)	−239.9	12.8
Molecular nitrogen (N_2)	−147.1	33.5
Molecular oxygen (O_2)	−118.8	49.7
Sulfur hexafluoride (SF_6)	45.5	37.6
Water (H_2O)	374.4	219.5

Table 11.7 lists the critical temperatures and critical pressures of a number of common substances. The critical temperature of a substance reflects the strength of its intermolecular forces. Benzene, ethanol, mercury, and water, which have strong intermolecular forces, also have high critical temperatures compared with the other substances listed in the table.

Solid-Liquid Phase Transition

The transformation of liquid to solid is called *freezing,* and the reverse process is called *melting,* or **fusion.** The **melting point** of a solid or the **freezing point** of a liquid is the temperature at which solid and liquid phases coexist in equilibrium. The normal melting (or freezing) point of a substance is the temperature at which it melts (or freezes) at 1 atm.

The most familiar liquid-solid equilibrium is probably that of water and ice. At 0°C and 1 atm, the dynamic equilibrium is represented by

$$\text{ice} \rightleftharpoons \text{water}$$

or

$$H_2O(s) \rightleftharpoons H_2O(l)$$

A glass of ice water at 0°C provides a practical illustration of this dynamic equilibrium. As the ice cubes melt to form water, some of the water between ice cubes may freeze, thus joining the cubes together. Remember that in a dynamic equilibrium, forward and reverse processes are occurring at the same rate [◄◄ Section 4.1].

Because molecules are more strongly held in the solid phase than in the liquid phase, heat is required to melt a solid into a liquid. The heating curve in Figure 11.32 shows that when a solid is heated, its temperature increases gradually until point A is reached. At this point, the solid begins to melt. During the melting period (A ⟶ B), the first flat portion of the curve in Figure 11.32, heat is being absorbed by the system, yet its temperature remains constant. The heat helps the molecules overcome the attractive forces in the solid. Once the sample has melted completely (point B), the heat absorbed increases the average kinetic energy of the liquid molecules and the liquid temperature rises (B ⟶ C). The vaporization process (C ⟶ D) can be explained similarly. The temperature remains constant during the period when the increased kinetic energy is used to overcome the cohesive forces in the liquid. When all molecules are in the gas phase, the temperature rises again.

Student Note: In most cases a glass of ice water would not be a *true* example of a dynamic equilibrium because it would not be kept at 0°C. At room temperature, all the ice eventually melts.

Figure 11.32 A typical heating curve, from the solid phase through the liquid phase to the gas phase of a substance. Because ΔH_{fus} is smaller than ΔH_{vap}, a substance melts in less time than it takes to boil. This explains why AB is shorter than CD. The steepness of the solid, liquid, and vapor heating lines is determined by the specific heat of the substance in each state.

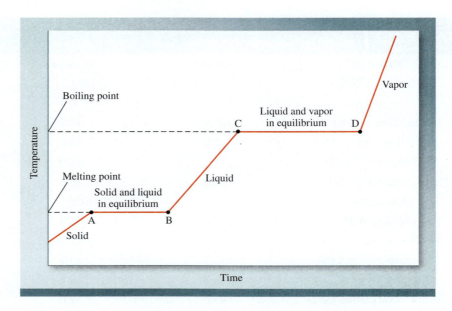

The **_molar heat of fusion_ (ΔH_{fus})** is the energy, usually expressed in kJ/mol, required to melt 1 mole of a solid. Table 11.8 lists the molar heats of fusion for the substances in Table 11.6. A comparison of the data in the two tables shows that ΔH_{fus} is smaller than ΔH_{vap} for each substance. This is consistent with the fact that molecules in a liquid are still fairly closely packed together, so some energy (but not a lot of energy, relatively speaking) is needed to bring about the rearrangement from solid to liquid. When a liquid is vaporized, on the other hand, its molecules become completely separated from one another, so considerably more energy is required to overcome the intermolecular attractive forces.

Cooling a substance has the opposite effect of heating it. If we remove heat from a gas sample at a steady rate, its temperature decreases. As the liquid is being formed, heat is given off by the system, because its potential energy is decreasing. For this reason, the temperature of the system remains constant over the condensation period (D $\longrightarrow$ C). After all the vapor has condensed, the temperature of the liquid begins to drop again. Continued cooling of the liquid finally leads to freezing (B $\longrightarrow$ A).

Supercooling is a phenomenon in which a liquid can be temporarily cooled to below its freezing point. Supercooling occurs when heat is removed from a liquid so rapidly that the molecules literally have no time to assume the ordered structure of a solid. A supercooled liquid is unstable. Gentle stirring or the addition to it of a small "seed" crystal of the same substance will cause it to solidify quickly.

TABLE 11.8	Molar Heats of Fusion for Selected Substances	
Substance	**Melting Point (°C)**	**ΔH_{fus} (kJ/mol)**
Argon (Ar)	−190	1.3
Benzene (C_6H_6)	5.5	10.9
Ethanol (C_2H_5OH)	−117.3	7.61
Diethyl ether ($C_2H_5OC_2H_5$)	−116.2	6.90
Mercury (Hg)	−39	23.4
Methane (CH_4)	−183	0.84
Water (H_2O)	0	6.01

Solid-Vapor Phase Transition

Solids can be vaporized, so solids, too, have a vapor pressure. **Sublimation** is the process by which molecules go directly from the solid phase to the vapor phase. The reverse process, in which molecules go directly from the vapor phase to the solid phase, is called **deposition.** Naphthalene, which is the substance used to make mothballs, has a fairly high vapor pressure for a solid (1 mmHg at 53°C); thus, its pungent vapor quickly permeates an enclosed space. Iodine also sublimes. At room temperature, the violet color of iodine vapor is easily visible in a closed container.

Because molecules are more tightly held in a solid, the vapor pressure of a solid is generally much less than that of the corresponding liquid. The **molar enthalpy of sublimation (ΔH_{sub})** of a substance is the energy, usually expressed in kilojoules, required to sublime 1 mole of a solid. It is equal to the sum of the molar enthalpies of fusion and vaporization:

$$\Delta H_{sub} = \Delta H_{fus} + \Delta H_{vap} \qquad \textbf{Equation 11.6}$$

Equation 11.6 is an illustration of Hess's law [◄◄ Section 5.5]. The enthalpy, or heat change, for the overall process is the same whether the substance changes directly from the solid to the vapor phase or if it changes from the solid to the liquid and then to the vapor phase.

Solid iodine in equilibrium with its vapor.

Turtle Rock Scientific/Science Source

Bringing Chemistry to Life

The Dangers of Phase Changes

If you have ever suffered a steam burn, you know that it can be far more serious than a burn caused simply by boiling water—even though steam and boiling water are both at the same temperature. A heating curve helps explain why this is so (see Figure 11.33). When boiling water touches your skin, it is cooled to body temperature because it deposits the heat it contains on your skin. The heat deposited on your skin by a sample of boiling water at 100°C can be represented by the orange line under the curve. When an equivalent mass of steam contacts your skin, it first deposits heat as it condenses and *then* cools to body temperature. The heat deposited on your skin by a sample of steam is represented by the red line under the curve. Notice how much more heat is deposited by steam than by liquid water at the same temperature. The steam contains more heat because it has been heated *and* vaporized. The additional heat that was absorbed by the water to vaporize it is what makes a steam burn worse than a burn from boiling water.

A heating curve can also be used to explain why hikers stranded by blizzards are warned not to consume snow in an effort to stay hydrated. When you drink cold

Student Note: Equation 11.6 is generally used to approximate ΔH_{sub}. It only holds strictly when all the phase changes occur at the same temperature.

Figure 11.33 Heating curve of water.

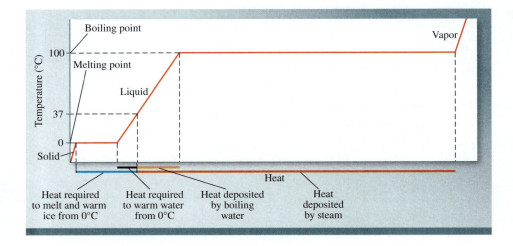

water, your body expends energy to warm the water you consume to body temperature. If you consume snow, your body must first expend the energy necessary to melt the snow, and then to warm it. Because a phase change is involved, the amount of energy required to assimilate snow is much greater than the amount necessary to assimilate an equal mass of water—even if the water is ice-cold. This can contribute to *hypothermia,* a potentially dangerous drop in body temperature.

Sample Problem 11.8 shows several applications of calculations of the energy transferred between system and surroundings.

SAMPLE PROBLEM 11.8

(a) Calculate the amount of heat deposited on the skin of a person burned by 1.00 g of liquid water at 100.0°C and (b) the amount of heat deposited by 1.00 g of steam at 100.0°C. (c) Calculate the amount of energy necessary to warm 100.0 g of water from 0.0°C to body temperature and (d) the amount of heat required to melt 100.0 g of ice at 0.0°C and then warm it to body temperature. (Assume that body temperature is 37.0°C.) You may want to review the calculation of the heat exchanged between the system and surroundings for temperature changes and phase changes [◄◄ Sections 5.3 and 5.4].

Strategy For the purpose of following the sign conventions, we can designate the water as the *system* and the body as the *surroundings.* (a) Heat is transferred from hot water to the skin in a single step: a temperature change. (b) The transfer of heat from steam to the skin takes place in two steps: a phase change and a temperature change. (c) Cold water is warmed to body temperature in a single step: a temperature change. (d) The melting of ice and the subsequent warming of the resulting liquid water takes place in two steps: a phase change and a temperature change. In each case, the heat transferred during a temperature change depends on the mass of the water, the specific heat of water, and the change in temperature. For the phase changes, the heat transferred depends on the amount of water (in moles) and the molar heat of vaporization (ΔH_{vap}) or molar heat of fusion (ΔH_{fus}). In each case, the total energy transferred or required is the sum of the energy changes for the individual steps.

Setup The required specific heats (*s*) are 4.184 J/g · °C for water and 1.99 J/g · °C for steam. (Assume that the specific heat values do not change over the range of temperatures in the problem.) From Table 11.6, the molar heat of vaporization (ΔH_{vap}) of water is 40.79 kJ/mol, and from Table 11.8, the molar heat of fusion (ΔH_{fus}) of water is 6.01 kJ/mol. The molar mass of water is 18.02 g/mol. *Note:* The ΔH_{vap} of water is the amount of heat required to vaporize a mole of water. In this problem, however, we want to know how much heat is deposited when water vapor *condenses,* so we must use the negative, −40.79 kJ/mol.

Solution (a) $\Delta T = 37.0°C - 100.0°C = -63.0°C$

From Equation 5.13, we write

$$q = ms\Delta T = 1.00 \text{ g} \times \frac{4.184 \text{ J}}{\text{g} \cdot °\text{C}} \times -63.0°\text{C} = -2.64 \times 10^2 \text{ J} = -0.264 \text{ kJ}$$

Thus, 1.00 g of water at 100.0°C deposits 0.264 kJ of heat on the skin. (The negative sign indicates that heat is given off by the system and absorbed by the surroundings.)

(b) $\dfrac{1.00 \text{ g}}{18.02 \text{ g/mol}} = 0.0555 \text{ mol water}$

$$q_1 = n\Delta H_{vap} = 0.0555 \text{ mol} \times \frac{-40.79 \text{ kJ}}{\text{mol}} = -2.26 \text{ kJ}$$

$$q_2 = ms\Delta T = 1.00 \text{ g} \times \frac{4.184 \text{ J}}{\text{g} \cdot °\text{C}} \times -63.0°\text{C} = -2.64 \times 10^2 \text{ J} = -0.264 \text{ kJ}$$

The overall energy deposited on the skin by 1.00 g of steam is the sum of q_1 and q_2:

$$-2.26 \text{ kJ} + (-0.264 \text{ kJ}) = -2.53 \text{ kJ}$$

The negative sign indicates that the system (steam) gives off the energy.

(c) $\Delta T = 37.0°C - 0.0°C = 37.0°C$

$$q = ms\Delta T = 100.0 \text{ g} \times \frac{4.184 \text{ J}}{\text{g} \cdot °\text{C}} \times 37.0°\text{C} = 1.55 \times 10^4 \text{ J} = 15.5 \text{ kJ}$$

The energy required to warm 100.0 g of water from 0.0°C to 37.0°C is 15.5 kJ.

(d) $\dfrac{100.0 \text{ g}}{18.02 \text{ g/mol}} = 5.55 \text{ mol}$

$$q_1 = n\Delta H_{\text{fus}} = 5.55 \text{ mol} \times \dfrac{6.01 \text{ kJ}}{\text{mol}} = 33.4 \text{ kJ}$$

$$q_2 = ms\Delta T = 100.0 \text{ g} \times \dfrac{4.184 \text{ J}}{\text{g} \cdot {}^\circ\text{C}} \times 37.0{}^\circ\text{C} = 1.55 \times 10^4 \text{ J} = 15.5 \text{ kJ}$$

The energy required to melt 100.0 g of ice at 0.0°C and warm it to 37.0°C is the sum of q_1 and q_2:

$$33.4 \text{ kJ} + 15.5 \text{ kJ} = 48.9 \text{ kJ}$$

THINK ABOUT IT

In problems that include phase changes, the q values corresponding to the phase-change steps will be the largest contributions to the total. If you find that this is not the case in your solution, check to see if you have made the common error of neglecting to convert the q values corresponding to temperature changes from J to kJ.

Practice Problem Ⓐ**TTEMPT** Calculate the amount of energy (in kJ) necessary to convert 346 g of liquid water from 0°C to water vapor at 182°C.

Practice Problem Ⓑ**UILD** Determine the final state and temperature of 100 g of water originally at 25.0°C after 50.0 kJ of heat have been added to it.

Practice Problem Ⓒ**ONCEPTUALIZE** Two samples of the same pure liquid are represented here. If adding 753 J to the sample on the left causes its temperature to increase by 41.2°C, by how much will the temperature of the sample on the right increase when an equal amount of energy is added to it?

CHECKPOINT – SECTION 11.6   Phase Changes

11.6.1 How much energy (in kJ) is required to convert 25.0 g of liquid water at room temperature (25°C) to steam at 110°C?

a) 64.9 kJ

b) 562 kJ

c) 1339 kJ

d) 1.34 kJ

e) 26.9 kJ

11.6.2 How much energy (in kJ) is given off when 1.0 g of steam at 100°C cools to room temperature (25°C)?

a) 0.326 kJ

b) 316 kJ

c) 2.58 kJ

d) 48.9 kJ

e) 22.1 kJ

11.7 Phase Diagrams

The relationships between the phases of a substance can be represented in a single graph known as a phase diagram. A ***phase diagram*** summarizes the conditions (temperature and pressure) at which a substance exists as a solid, liquid, or gas. Figure 11.34(a) shows the phase diagram of CO_2, which is typical of many substances. The graph is divided into three regions, each of which represents a pure phase. The line separating any two regions, called a *phase boundary line,* indicates conditions under which these two phases can exist in equilibrium. The point at which

Figure 11.34 (a) The phase diagram of carbon dioxide. Note that the solid-liquid boundary line has a positive slope. There is no liquid phase below 5.2 atm, so only the solid and vapor phases can exist under ordinary atmospheric conditions. (b) Heating solid CO_2 initially at −100°C and 1 atm (point 1) causes it to sublime when it reaches −78°C (point 2). At 25°C, increasing the pressure from 1 atm (point 3) to about 70 (point 4) will cause CO_2 to condense to a liquid.

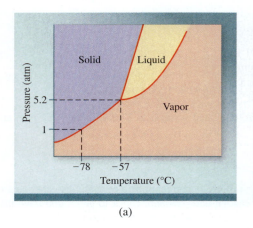

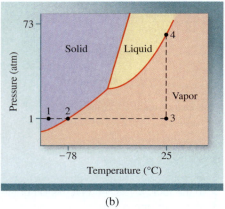

(a)   (b)

all three phase boundary lines meet is called the **triple point.** The triple point is the only combination of temperature and pressure at which all three phases of a substance can be in equilibrium with one another. The point at which the liquid-vapor phase boundary line abruptly ends is the critical point, corresponding to the critical temperature (T_c) and the critical pressure (P_c).

In order to understand the information in a phase diagram, consider the dashed lines between numbered points in Figure 11.34(b). If we start with a sample of CO_2 at 1 atm and −100°C, the sample is initially a solid (point 1). If we then add heat to the sample at 1 atm, its temperature increases until it reaches −78°C, the sublimation point of CO_2 at 1 atm (point 2). When the entire sample has sublimed at −78°C, the temperature of the resulting vapor will begin to increase. We continue adding heat until the temperature of the vapor is −25°C (point 3). At this point, we maintain the temperature at −25°C and begin increasing the pressure until the vapor condenses, which would occur at a pressure of about 70 atm (point 4). With a phase diagram, we can tell in what phase a substance will exist at any given temperature and pressure. Furthermore, we can tell what phase changes will occur as the result of increases or decreases in temperature, pressure, or both.

One of the interesting things about the phase diagram of CO_2 is that its triple point occurs above atmospheric pressure −57°C, 5.2 atm. This means that there is no liquid phase at atmospheric pressure, the condition that makes dry ice "dry." At elevated pressure, solid CO_2 does melt. In fact, liquid CO_2 is used as the solvent in many dry-cleaning operations.

The phase diagram of water (Figure 11.35) is unusual because the solid-liquid phase-boundary line has a negative slope. (Compare this with the solid-liquid phase-boundary line of CO_2 in Figure 11.34.) As a result, ice can be liquefied within a narrow temperature range by applying pressure.

🔊 Student Hot Spot

Student data indicate you may struggle with interpreting phase diagrams. Access the eBook to view additional Learning Resources on this topic.

Student Note: Bismuth is another example of a substance whose solid-liquid phase boundary (under ordinary atmospheric conditions) has a negative slope. Like water, bismuth has a liquid density (1.005 g/cm³ at room temperature) that is higher than its solid density (0.9780 g/cm³ at its melting point of 271°C).

Figure 11.35 The phase diagram of water. Each solid line between two phases specifies the conditions of pressure and temperature under which the two phases can exist in equilibrium. The point at which all three phases can exist in equilibrium (0.006 atm and 0.01°C) is called the triple point.

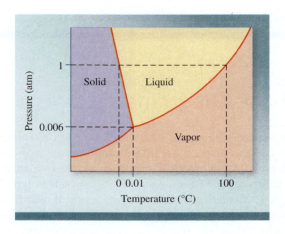

Sample Problem 11.9 lets you practice interpreting the information in a phase diagram.

SAMPLE PROBLEM   11.9

Using the following phase diagram, (a) determine the normal boiling point and the normal melting point of the substance, (b) determine the physical state of the substance at 2 atm and 110°C, and (c) determine the pressure and temperature that correspond to the triple point of the substance.

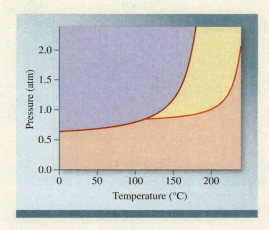

Strategy Each point on the phase diagram corresponds to a pressure-temperature combination. The normal boiling and melting points are the temperatures at which the substance undergoes phase changes. These points fall on the phase boundary lines. The triple point is where the three phase boundaries meet.

Setup By drawing lines corresponding to a given pressure and/or temperature, we can determine the temperature at which a phase change occurs, or the physical state of the substance under specified conditions.

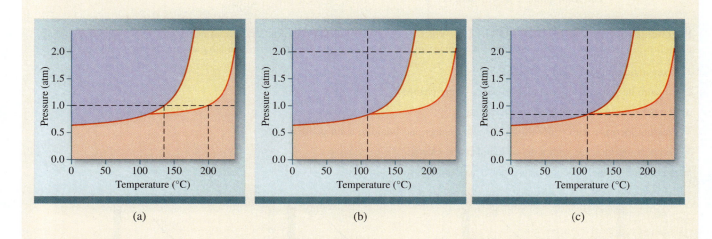

(a)   (b)   (c)

Solution (a) The normal boiling and melting points are ~200°C and ~135°C, respectively.

(b) At 2 atm and 110°C the substance is a solid.

(c) The triple point occurs at ~0.8 atm and ~115°C.

(Continued on next page)

THINK ABOUT IT

The triple point of this substance occurs at a pressure below atmospheric pressure. Therefore, it will melt rather than sublime when it is heated under ordinary conditions.

Practice Problem **A**TTEMPT Use the following phase diagram to determine (a) the normal boiling point and melting point of the substance, and (b) the physical state of the substance at 1.2 atm and 100°C.

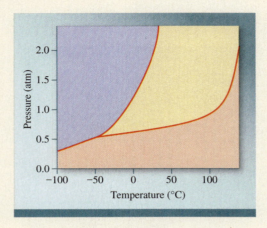

Practice Problem **B**UILD Sketch the phase diagram of a substance using the following data:

Pressure (atm)	Melting Point (°C)	Boiling Point (°C)	Sublimation Point (°C)
0.5	—	—	0
1.0	60	110	—
1.5	75	200	—
2.0	105	250	—
2.5	125	275	—

The triple point is at 0.75 atm and 45°C.

Practice Problem **C**ONCEPTUALIZE Which of the phase diagrams shown here has an arrow that traces a path including the following changes? For the other two phase diagrams, write a numbered list of changes in temperature, pressure, and phase.

1. Temperature increase with no phase change

2. Pressure decrease causing a solid-to-vapor phase change

3. Temperature increase with no phase change

4. Pressure increase with vapor-to-liquid phase change and liquid-to-solid phase change

5. Temperature increase with solid-to-liquid phase change

6. Pressure decrease with liquid-to-vapor phase change

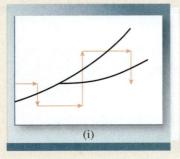

(i)

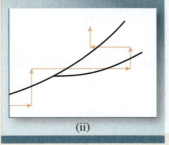

(ii)

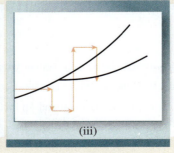

(iii)

CHECKPOINT – SECTION 11.7   Phase Diagrams

Refer to the following phase diagrams to answer questions 11.7.1 and 11.7.2.

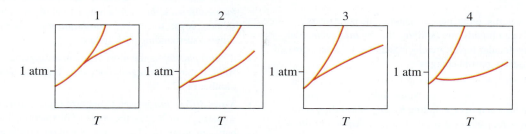

11.7.1 Which phase diagram corresponds to a substance that will sublime rather than melt as it is heated at 1 atm?

a) 1

b) 1 and 2

c) 1, 2, and 4

d) 3

e) none

11.7.2 Which phase diagram corresponds to a substance that will liquefy when pressure is increased at a temperature below its freezing point?

a) 1

b) 1 and 2

c) 1, 2, and 4

d) 3

e) none

Chapter Summary

Section 11.1

- The particles (atoms, molecules, or ions) in the condensed phases (solids and liquids) are held together by ***intermolecular forces.*** Intermolecular forces are electrostatic attractions between opposite charges or partial charges.

- Intermolecular forces acting between atoms or molecules in a pure substance are called ***van der Waals forces*** and include ***dipole-dipole interactions*** (including ***hydrogen bonding***) and ***London dispersion forces*** (also called simply ***dispersion forces***).

- Dipole-dipole interactions exist between *polar* molecules, whereas *nonpolar* molecules are held together by *dispersion* forces alone. Dispersion forces are those between ***instantaneous dipoles*** and *induced* dipoles. When a nonpolar molecule acquires an instantaneous dipole, it is said to be ***polarized.*** Dispersion forces act between all molecules, nonpolar and polar.

- Hydrogen bonding is an especially strong type of dipole-dipole interaction that occurs in molecules that contain $H-N$, $H-O$, or $H-F$ bonds.

- ***Ion-dipole interactions*** are those that occur (in solutions) between ions and polar molecules.

Section 11.2

- Physical properties of liquids that result from intermolecular forces include surface tension, viscosity, and vapor pressure.

- ***Surface tension*** is the net pull inward on molecules at the surface of a liquid. Surface tension is related to ***cohesion,*** the attractive forces between molecules within a substance, and ***adhesion,*** the attractive forces between molecules in a substance and their container. The balance between cohesion and adhesion determines whether a liquid meniscus is concave or convex. It also gives rise to ***capillary action,*** in which liquid is drawn upward into a narrow tube against gravity.

- ***Viscosity*** is resistance to flow, reflecting how easily molecules move past one another.

- ***Evaporation*** (also known as ***vaporization***) is the phase change from liquid to vapor. ***Condensation*** is the phase change from vapor to liquid. In a closed system, when vaporization and condensation are occurring at the same rate, a state of ***dynamic equilibrium*** exists and the *vapor pressure* is equal to the ***equilibrium vapor pressure.*** Vapor pressure measures how easily molecules escape to the vapor phase. A ***volatile*** substance has a high vapor pressure.

- Surface tension, viscosity, and vapor pressure are all temperature dependent. The ***Clausius-Clapeyron equation*** relates the vapor pressure of a substance to its absolute temperature.

Section 11.3

- The spheres (molecules, atoms, or ions) in a ***crystalline solid*** are arranged in a three-dimensional ***lattice structure*** consisting of a repeating pattern of ***unit cells.*** The type of unit cell is determined by the positions of the ***lattice points.***

- Cubic unit cells may be ***simple cubic, body-centered cubic,*** or ***face-centered cubic,*** containing a total of one, two, or four spheres, respectively. The ***coordination number*** is the number of spheres surrounding each sphere.

- *Closest packing* is the most efficient arrangement of spheres in a solid. It may be *hexagonal* or *cubic*. Each has a coordination number of 12.

- ***X-ray diffraction*** is used to analyze crystal structure. The wavelength of X rays, the angle of diffraction, and the spacing between atoms in a lattice are related by the ***Bragg equation.***

Section 11.4

- Crystals may be ionic, covalent, molecular, or metallic. Each type of crystalline solid has characteristics determined in part by the types of interactions holding it together.

Section 11.5

- ***Amorphous solids*** such as ***glass*** lack regular three-dimensional structure.

Section 11.6

- The possible ***phase changes*** are melting or ***fusion*** ($s \longrightarrow l$), freezing ($l \longrightarrow s$), vaporization ($l \longrightarrow g$), condensation ($g \longrightarrow l$), ***sublimation*** ($s \longrightarrow g$), and ***deposition*** ($g \longrightarrow s$).

- The ***boiling point*** of a substance is the temperature at which its vapor pressure equals the external pressure. The ***normal boiling point*** is the temperature at which its vapor pressure equals 1 atm. Boiling point is pressure dependent. The ***molar heat of vaporization*** (ΔH_{vap}) is the amount of heat required to vaporize 1 mole of a substance at its boiling point.

- The ***critical temperature*** (T_c) is the temperature above which a gas cannot be liquefied by applying pressure. The ***critical pressure*** (P_c) is the pressure necessary to liquefy a gas at its critical temperature. A substance above its critical temperature and pressure is a ***supercritical fluid.***

- The ***melting point*** or ***freezing point*** is the temperature at which the solid and liquid phases are in equilibrium. The ***molar heat of fusion*** (ΔH_{fus}) is the amount of heat required to melt 1 mole of a substance at its melting point. ***Supercooling*** is the process of rapidly lowering a liquid's temperature below its freezing point.

- The ***molar heat of sublimation*** (ΔH_{sub}) is equal to the sum of the molar heats of fusion and vaporization: $\Delta H_{sub} = \Delta H_{fus} + \Delta H_{vap}$

Section 11.7

- A ***phase diagram*** indicates the phase of a substance under any combination of temperature and pressure. Lines between phases are called *phase boundaries.*

- The ***triple point*** is where all three phase boundaries meet. This is the temperature and pressure combination at which all three phases are in equilibrium.

Key Words

Adhesion, 538

Amorphous solid, 556

Body-centered cubic cell, 544

Boiling point, 558

Bragg equation, 550

Capillary action, 538

Clausius-Clapeyron equation, 540

Cohesion, 538

Condensation, 539

Coordination number, 544

Critical pressure (P_c), 558

Critical temperature (T_c), 558

Crystalline solid, 543

Deposition, 561

Dipole-dipole interactions, 530

Dispersion forces, 534

Dynamic equilibrium, 539

Equilibrium vapor pressure, 539

Evaporation, 539

Face-centered cubic cell, 544

Freezing point, 559

Fusion, 559

Glass, 556

Hydrogen bonding, 531

Instantaneous dipole, 534

Intermolecular forces, 530

Ion-dipole interactions, 536

Lattice point, 543

Lattice structure, 543

London dispersion forces, 534

Melting point, 559

Molar enthalpy of sublimation (ΔH_{sub}), 561

Molar heat of fusion (ΔH_{fus}), 560

Molar heat of vaporization (ΔH_{vap}), 558

Normal boiling point, 558

Phase change, 557

Phase diagram, 563

Polarized, 535

Simple cubic cell, 544

Sublimation, 561

Supercooling, 560

Supercritical fluid, 558

Surface tension, 537

Triple point, 564

Unit cell, 543

van der Waals forces, 530

Vaporization, 539

Viscosity, 538

Volatile, 538

X-ray diffraction, 549

Key Equations

11.1 $\ln P = -\dfrac{\Delta H_{vap}}{RT} + C$	The Clausius-Clapeyron equation relates the natural logarithm of the vapor pressure of a substance to its heat of vaporization (ΔH_{vap}) and the absolute temperature.
11.2 $\ln P_1 = -\dfrac{\Delta H_{vap}}{RT_1} + C$	Equations 11.2 and 11.3 are the Clausius-Clapeyron equation written for one substance at two different temperatures, T_1 and T_2.
11.3 $\ln P_2 = -\dfrac{\Delta H_{vap}}{RT_2} + C$	
11.4 $\ln \dfrac{P_1}{P_2} = \dfrac{\Delta H_{vap}}{R}\left(\dfrac{1}{T_2} - \dfrac{1}{T_1}\right)$	Subtracting Equation 11.3 from Equation 11.2 gives an equation that can be used to determine vapor pressure at a new temperature—provided that vapor pressure at one temperature and ΔH_{vap} for the substance are known. This equation can also be rearranged to solve for ΔH_{vap} if the vapor pressure is known at two different temperatures.
11.5 $BC + CD = 2d \sin \theta = n\lambda \qquad n = 1, 2, 3, \ldots$	The Bragg equation is used with X-ray diffraction data to determine the structure of a crystalline solid.
11.6 $\Delta H_{sub} = \Delta H_{fus} + \Delta H_{vap}$	The heat of sublimation (ΔH_{sub}) is the sum of heat of fusion (ΔH_{fus}) and heat of vaporization (ΔH_{vap}). This is a consequence of Hess's law. Sublimation (the phase change from solid to vapor) can be thought of as a two-step process in which a solid becomes a liquid, and the liquid becomes a vapor. Because the two steps sum to the overall process, their ΔH values sum to the overall ΔH.

Questions and Problems

Applying What You've Learned

The melamine tragedy in China caused an epidemic of unusual kidney stones in infants, a condition that is ordinarily very rare. In the general population of adults, though, kidney stones are relatively common with roughly 10 percent of adults experiencing symptoms during their lifetimes. The most common type of kidney stone is composed of calcium oxalate (CaC_2O_4), and several different approaches can be used for treatment and/or prevention. One method of treatment involves the use of diuretics—drugs that cause elimination of water from the body in the form of urine. Increasing the volume of fluid that passes through the kidneys can help to reduce the amount of calcium available for stone formation. It can also facilitate the passing of a stone that has already formed. Two of the drugs that are prescribed for the treatment and prevention of kidney stones are hydrochlorothiazide and chlorthalidone.

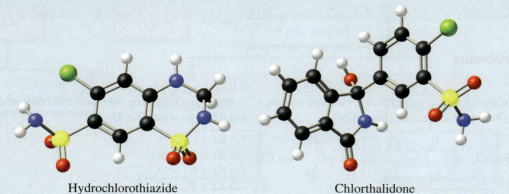

Hydrochlorothiazide   Chlorthalidone

Problems:

(a) What types of intermolecular forces exist between chlorthalidone molecules and between hydrochlorothiazide molecules [◄◄ Sample Problem 11.1]? (b) Calcium is one of the components of the most common type of kidney stone. Calcium metal crystallizes in a face-centered cubic unit cell with a cell edge length of 558.84 pm. Calculate the radius of a calcium atom in angstroms (Å) [◄◄ Sample Problem 11.3]. (c) How many Ca atoms are contained within a unit cell [◄◄ Sample Problem 11.5]? (d) Calculate the density of calcium metal in g/cm^3 [◄◄ Sample Problem 11.6]. (e) Patients who have suffered from kidney stones often are advised to drink extra water to help prevent the formation of additional stones. An article on WebMD.com recommends drinking at least three quarts (2.84 L) of water every day—nearly 50 percent more than the amount recommended for healthy adults. How much energy must the body expend to warm this amount of water consumed at 10°C to body temperature (37°C)? How much *more* energy would have to be expended if the same quantity of water were consumed as ice at 0°C? ΔH_{fus} for water is 6.01 kJ/mol. Assume the density and specific heat of water are 1.00 g/cm^3 and 4.184 J/g · °C, respectively, and that both quantities are independent of temperature [◄◄ Sample Problem 11.7].

SECTION 11.1: INTERMOLECULAR FORCES

Review Questions

11.1 Give an example for each type of intermolecular force:
(a) dipole-dipole interaction, (b) ion-dipole interaction,
(c) dispersion forces, (d) van der Waals forces.

11.2 Explain the term *polarizability*. What kind of molecules tend to have high polarizabilities? What is the relationship between polarizability and intermolecular forces?

11.3 Explain the difference between a temporary dipole moment and a permanent dipole moment.

11.4 What evidence can you cite that all molecules exert attractive forces on one another?

11.5 What physical properties are determined by the strength of intermolecular forces in solids and in liquids?

11.6 Explain why hydrogen bonding is exhibited by some hydrogen-containing compounds and not by others.

11.7 Describe the types of intermolecular forces that govern the folding of a protein molecule into its physiologically functioning three-dimensional state.

Conceptual Problems

11.8 The compounds Br_2 and ICl are isoelectronic (have the same number of electrons) and have similar molar masses, yet Br_2 melts at −7.2°C and ICl melts at 27.2°C. Explain.

11.9 If you lived in Alaska, which of the following natural gases could you keep in an outdoor storage tank in winter: methane (CH_4), propane (C_3H_8), or butane (C_4H_{10})? Explain why.

11.10 The binary hydrogen compounds of the Group 14 elements and their boiling points are: CH_4, −162°C; SiH_4, −112°C; GeH_4, −88°C; and SnH_4, −52°C. Explain the increase in boiling points from CH_4 to SnH_4.

11.11 List the types of intermolecular forces that exist between molecules (or atoms or ions) in each of the following species: (a) benzene (C_6H_6), (b) CH_3Cl, (c) PF_3, (d) NaCl, (e) CS_2.

11.12 Ammonia is both a donor and an acceptor of hydrogen in hydrogen-bond formation. Draw a diagram showing the hydrogen bonding of an ammonia molecule with two other ammonia molecules.

11.13 Which of the following species are capable of hydrogen-bonding among themselves: (a) C_2H_6, (b) HI, (c) KF, (d) BeH_2, (e) CH_3COOH?

11.14 Arrange the following in order of increasing boiling point: RbF, CO_2, CH_3OH, CH_3Br. Explain your reasoning.

11.15 Diethyl ether has a boiling point of 34.5°C, and 1-butanol has a boiling point of 117°C:

Diethyl ether   1-Butanol

Both of these compounds have the same numbers and types of atoms. Explain the difference in their boiling points.

11.16 Which member of each of the following pairs of substances would you expect to have a higher boiling point: (a) O_2 and Cl_2, (b) SO_2 and CO_2, (c) HF and HI?

11.17 Which substance in each of the following pairs would you expect to have the higher boiling point: (a) Ne or Xe, (b) CO_2 or CS_2, (c) CH_4 or Cl_2, (d) F_2 or LiF, (e) NH_3 or PH_3? Explain why.

11.18 Explain in terms of intermolecular forces why (a) NH_3 has a higher boiling point than CH_4 and (b) KCl has a higher melting point than I_2.

11.19 What kind of attractive forces must be overcome to (a) melt ice, (b) boil molecular bromine, (c) melt solid iodine, and (d) dissociate F_2 into F atoms?

11.20 The following compounds have the same molecular formula (C_4H_{10}). Which one would

you expect to have a higher boiling point, (i) or (ii)?

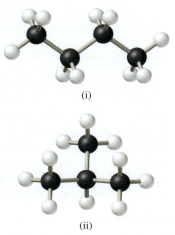

(i)

(ii)

11.21 Explain the difference in the melting points of the following compounds:

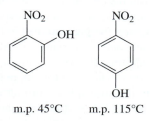

m.p. 45°C   m.p. 115°C

(*Hint:* One of the two can form *intra*molecular hydrogen bonds.)

SECTION 11.2: PROPERTIES OF LIQUIDS

Review Questions

11.22 Explain why liquids, unlike gases, are virtually incompressible.

11.23 What is surface tension? What is the relationship between intermolecular forces and surface tension? How does surface tension change with temperature?

11.24 Despite the fact that stainless steel is much denser than water, a stainless-steel razor blade can be made to float on water. Why?

11.25 Use water and mercury as examples to explain adhesion and cohesion.

11.26 A glass can be filled slightly above the rim with water. Explain why the water does not overflow.

11.27 Draw diagrams showing the capillary action of (a) water and (b) mercury in three tubes of different radii.

11.28 What is viscosity? What is the relationship between intermolecular forces and viscosity?

11.29 Why does the viscosity of a liquid decrease with increasing temperature?

11.30 Why is ice less dense than water?

11.31 Outdoor water pipes have to be drained or insulated in winter in a cold climate. Why?

Computational Problems

11.32 The vapor pressure of benzene (C_6H_6) is 40.1 mmHg at 7.6°C. What is its vapor pressure at 60.6°C? The molar heat of vaporization of benzene is 31.0 kJ/mol.

11.33 The vapor pressure of phosphorus trichloride (PCl_3) is 119.7 mmHg at 25.0°C. What will the vapor pressure be at 50.0°C? The molar heat of vaporization of PCl_3 is 31.63 kJ/mol.

11.34 Estimate the molar heat of vaporization of a liquid whose vapor pressure doubles when the temperature is raised from 75°C to 100°C.

11.35 Determine the normal boiling point of a liquid with vapor pressure 63.8 mmHg at 32.5°C given that its molar heat of vaporization is 43.5 kJ/mol.

Conceptual Problems

11.36 The diagram on the left depicts a system at room temperature. (The space above the liquid contains air and the vapor of the liquid in the container, and is sealed with a movable piston, making the pressure inside the vessel equal to atmospheric pressure.) Which of the other diagrams, [(i)–(v)], could represent the same system where (a) the temperature is the same, but the atmospheric pressure is lower; and (b) both the temperature and the atmospheric pressure are lower?

11.40 Vapor pressure measurements at several different temperatures are shown for mercury. Determine graphically the molar heat of vaporization for mercury.

T(°C)	200	250	300	320	340
P(mmHg)	17.3	74.4	246.8	376.3	557.9

11.41 The vapor pressure of liquid X is lower than that of liquid Y at 20°C, but higher at 60°C. What can you deduce about the relative magnitude of the molar heats of vaporization of X and Y?

SECTION 11.3: CRYSTAL STRUCTURE

Review Questions

11.42 Define the following terms: *crystalline solid, lattice point, unit cell, coordination number, closest packing*.

11.43 Describe the geometries of the following cubic cells: simple cubic, body-centered cubic, face-centered cubic. Which of these structures would give the highest density for the same type of atoms? Which the lowest?

11.44 Classify the solid states in terms of crystal types of the elements in the third period of the periodic table. Predict the trends in their melting points and boiling points.

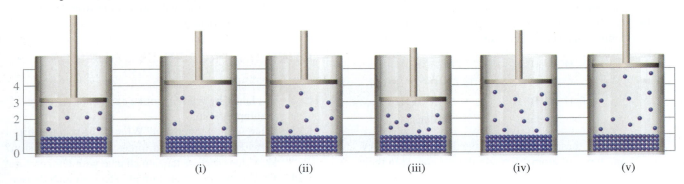

(i)   (ii)   (iii)   (iv)   (v)

11.37 Which of the diagrams in Problem 11.36 could represent the system where (a) the temperature is the same, but the atmospheric pressure is higher; and (b) both the temperature and the atmospheric pressure are higher?

11.38 Predict which of the following liquids has greater surface tension: ethanol (C_2H_5OH) or dimethyl ether (CH_3OCH_3).

11.39 Predict the viscosity of ethylene glycol relative to that of ethanol and glycerol (see Table 11.3).

$$CH_2-OH$$
$$|$$
$$CH_2-OH$$

Ethylene glycol

11.45 The melting points of the oxides of the third-period elements given in parentheses: Na_2O (1275°C), MgO (2800°C), Al_2O_3 (2045°C), SiO_2 (1610°C), P_4O_{10} (580°C), SO_3 (16.8°C), Cl_2O_7 (−91.5°C). Classify these solids in terms of crystal types.

11.46 Define X-ray diffraction. What are the typical wavelengths (in nm) of X rays? (See Figure 6.1.)

11.47 Write the Bragg equation. Define every term and describe how this equation can be used to measure interatomic distances.

Computational Problems

11.48 What is the coordination number of each sphere in (a) a simple cubic cell, (b) a body-centered cubic cell, and (c) a face-centered cubic cell? Assume the spheres are all the same.

11.49 Calculate the number of spheres that would be found within a simple cubic cell, body-centered cubic cell, and face-centered cubic cell. Assume that the spheres are the same.

11.50 Metallic iron crystallizes in a cubic lattice. The unit cell edge length is 287 pm. The density of iron is 7.87 g/cm^3. How many iron atoms are within a unit cell?

11.51 Barium metal crystallizes in a body-centered cubic lattice (the Ba atoms are at the lattice points only). The unit cell edge length is 502 pm, and the density of the metal is 3.50 g/cm^3. Using this information, calculate Avogadro's number. [*Hint:* First calculate the volume (in cm^3) occupied by 1 mole of Ba atoms in the unit cells. Next calculate the volume (in cm^3) occupied by one Ba atom in the unit cell. Assume that 68 percent of the unit cell is occupied by Ba atoms.]

11.52 Vanadium crystallizes in a body-centered cubic lattice (the V atoms occupy only the lattice points). How many V atoms are present in a unit cell?

11.53 Europium crystallizes in a body-centered cubic lattice (the Eu atoms occupy only the lattice points). The density of Eu is 5.26 g/cm^3. Calculate the unit cell edge length in picometers.

11.54 Crystalline silicon has a cubic structure. The unit cell edge length is 543 pm. The density of the solid is 2.33 g/cm^3. Calculate the number of Si atoms in one unit cell.

11.55 A face-centered cubic cell contains 8 X atoms at the corners of the cell and 6 Y atoms at the faces. What is the empirical formula of the solid?

11.56 When X rays of wavelength 0.0900 nm are diffracted by a metallic crystal, the angle of first-order diffraction (n = 1) is measured to be 15.2°. What is the distance (in pm) between the layers of atoms responsible for the diffraction?

11.57 The distance between layers in an NaCl crystal is 282 pm. X rays are diffracted from these layers at an angle of 23.0°. Assuming that n = 1, calculate the wavelength of the X rays in nanometers.

Conceptual Problems

11.58 Identify the unit cell of molecular iodine (I$_2$) shown here. (*Hint:* Consider the position of iodine molecules, not individual iodine atoms.)

11.59 Shown here is a zinc oxide unit cell. What is the formula of zinc oxide?

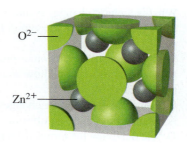

SECTION 11.4: TYPES OF CRYSTALS

Review Questions

11.60 Describe and give examples of the following types of crystals: (a) ionic crystals, (b) covalent crystals, (c) molecular crystals, (d) metallic crystals.

11.61 Why are metals good conductors of heat and electricity? Why does the ability of a metal to conduct electricity decrease with increasing temperature?

Conceptual Problems

11.62 A solid is hard, brittle, and electrically nonconducting. Its melt (the liquid form of the substance) and an aqueous solution containing the substance conduct electricity. Classify the solid.

11.63 A solid is soft and has a low melting point (below 100°C). The solid, its melt, and an aqueous solution containing the substance are all nonconductors of electricity. Classify the solid.

11.64 A solid is very hard and has a high melting point. Neither the solid nor its melt conducts electricity. Classify the solid.

11.65 Which of the following are molecular solids and which are covalent solids: Se$_8$, HBr, Si, CO$_2$, C, P$_4$O$_6$, SiH$_4$?

11.66 Classify the solid state of the following substances as ionic crystals, covalent crystals, molecular crystals, or metallic crystals: (a) CO$_2$, (b) B$_{12}$, (c) S$_8$, (d) KBr, (e) Mg, (f) SiO$_2$, (g) LiCl, (h) Cr.

11.67 Explain why diamond is harder than graphite. Why is graphite an electrical conductor but diamond is not?

SECTION 11.5: AMORPHOUS SOLIDS

Review Questions

11.68 What is an amorphous solid? How does it differ from a crystalline solid?

11.69 Define glass. What is the chief component of glass? Name three types of glass.

SECTION 11.6: PHASE CHANGES

Review Questions

11.70   What is a phase change? Name all possible changes that can occur among the vapor, liquid, and solid phases of a substance.

11.71   What is the equilibrium vapor pressure of a liquid? How is it measured, and how does it change with temperature?

11.72   Use any one of the phase changes to explain what is meant by dynamic equilibrium.

11.73   Define the following terms: (a) molar heat of vaporization, (b) molar heat of fusion, (c) molar heat of sublimation. What are their typical units?

11.74   How is the molar heat of sublimation related to the molar heats of vaporization and fusion? On what law are these relationships based?

11.75   What can we learn about the intermolecular forces in a liquid from the molar heat of vaporization?

11.76   The greater the molar heat of vaporization of a liquid, the greater its vapor pressure. True or false?

11.77   Define boiling point. How does the boiling point of a liquid depend on external pressure? Referring to Table 10.5, what is the boiling point of water when the external pressure is 187.5 mmHg?

11.78   As a liquid is heated at constant pressure, its temperature rises. This trend continues until the boiling point of the liquid is reached. No further rise in temperature of the liquid can be induced by heating. Explain.

11.79   What is critical temperature? What is the significance of critical temperature in condensation of gases?

11.80   What is the relationship between intermolecular forces in a liquid and the liquid's boiling point and critical temperature? Why is the critical temperature of water greater than that of most other substances?

11.81   How do the boiling points and melting points of water and carbon tetrachloride vary with pressure? Explain any difference in behavior of these two substances.

11.82   Why is solid carbon dioxide called dry ice?

11.83   The vapor pressure of a liquid in a closed container depends on which of the following: (a) the volume above the liquid, (b) the amount of liquid present, (c) temperature, (d) intermolecular forces between the molecules in the liquid?

11.84   Wet clothes dry more quickly on a hot, dry day than on a hot, humid day. Explain.

11.85   Which of the following phase transitions gives off more heat: (a) 1 mole of steam to 1 mole of water at 100°C, or (b) 1 mole of water to 1 mole of ice at 0°C?

11.86   A beaker of water is heated to boiling by a Bunsen burner. Would adding another burner raise the temperature of the boiling water? Explain.

11.87   Explain why splashing a small amount of liquid nitrogen (b.p. 77 K) is not as harmful as splashing boiling water on your skin.

Computational Problems

11.88   Calculate the amount of heat (in kJ) required to convert 150.2 g of water to steam at 100°C.

11.89   How much heat (in kJ) is needed to convert 866 g of ice at −15°C to steam at 146°C? (The specific heats of ice and steam are 2.03 and 1.99 J/g · °C, respectively.)

11.90   The molar heats of fusion and sublimation of molecular iodine are 15.27 and 62.30 kJ/mol, respectively. Estimate the molar heat of vaporization of liquid iodine.

Conceptual Problems

11.91   How is the rate of evaporation of a liquid affected by (a) temperature, (b) the surface area of a liquid exposed to air, (c) intermolecular forces?

11.92   The following compounds, listed with their boiling points, are liquid at −10°C: butane, −0.5°C; ethanol, 78.3°C; toluene, 110.6°C. At −10°C, which of these liquids would you expect to have the highest vapor pressure? Which the lowest? Explain.

11.93   Freeze-dried coffee is prepared by freezing brewed coffee and then removing the ice component with a vacuum pump. Describe the phase changes taking place during these processes.

11.94   A student hangs wet clothes outdoors on a winter day when the temperature is −15°C. After a few hours, the clothes are found to be fairly dry. Describe the phase changes in this drying process.

11.95   Explain why steam at 100°C causes more serious burns than water at 100°C.

SECTION 11.7: PHASE DIAGRAMS

Review Questions

11.96   What is a phase diagram? What useful information can be obtained from studying a phase diagram?

11.97   Explain how water's phase diagram differs from those of most substances. What property of water causes the difference?

Conceptual Problems

11.98   The blades of ice skates are quite thin, so the pressure exerted on ice by a skater can be substantial. Explain how this facilitates skating on ice.

11.99   A length of wire is placed on top of a block of ice. The ends of the wire extend over the edges of the ice, and a heavy weight is attached to each end. It is found that the ice under the wire gradually melts, so the wire slowly moves through the ice block. At the same time, the water above the wire refreezes. Explain the phase changes that accompany this phenomenon.

11.100 The boiling point and freezing point of sulfur dioxide are $-10°C$ and $-72.7°C$ (at 1 atm), respectively. The triple point is $-75.5°C$ and 1.65×10^{-3} atm, and its critical point is at $157°C$ and 78 atm. On the basis of this information, draw a rough sketch of the phase diagram of SO_2.

11.101 A phase diagram of water is shown. Label the regions. Predict what would happen as a result of the following changes: (a) Starting at A, we raise the temperature at constant pressure. (b) Starting at B, we lower the pressure at constant temperature. (c) Starting at C, we lower the temperature at constant pressure.

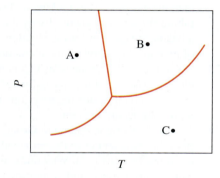

ADDITIONAL PROBLEMS

11.102 Name the kinds of attractive forces that must be overcome to (a) boil liquid ammonia, (b) melt solid phosphorus (P_4), (c) dissolve CsI in liquid HF, (d) melt potassium metal.

11.103 Which of the following properties indicates very strong intermolecular forces in a liquid: (a) very low surface tension, (b) very low critical temperature, (c) very low boiling point, (d) very low vapor pressure?

11.104 At $-35°C$, liquid HI has a higher vapor pressure than liquid HF. Explain.

11.105 Based on the following properties of elemental boron, classify it as one of the crystalline solids discussed in Section 11.4: high melting point ($2300°C$), poor conductor of heat and electricity, insoluble in water, very hard substance.

11.106 Referring to Figure 11.34, determine the stable phase of CO_2 at (a) 4 atm and $-60°C$ and (b) 0.5 atm and $-20°C$.

11.107 Which of the following substances has the highest polarizability: CH_4, H_2, CCl_4, SF_6, H_2S?

11.108 X rays of wavelength 0.154 nm are diffracted from a crystal at an angle of $14.17°$. Assuming that $n = 1$, calculate the distance (in pm) between layers in the crystal.

11.109 At what angle will X rays of wavelength 0.154 nm be diffracted from a crystal if the distance (in pm) between layers in the crystal is 188 pm? (Assume $n = 1$.)

11.110 A CO_2 fire extinguisher is located on the outside of a building in Massachusetts. During the winter months, one can hear a sloshing sound when the extinguisher is gently shaken. In the summertime there is often no sound when it is shaken. Explain. Assume that the extinguisher has no leaks and that it has not been used.

11.111 What is the vapor pressure of mercury at its normal boiling point ($357°C$)?

11.112 A flask of water is connected to a powerful vacuum pump. When the pump is turned on, the water begins to boil. After a few minutes, the same water begins to freeze. Eventually, the ice disappears. Explain what happens at each step.

11.113 The liquid-vapor boundary line in the phase diagram of any substance always stops abruptly at a certain point. Why?

11.114 The interionic distances of several alkali halide crystals are as follows:

Crystal	NaCl	NaBr	NaI	KCl	KBr	KI
Interionic distance (pm)	282	299	324	315	330	353

Plot lattice energy versus the reciprocal interionic distance. How would you explain the plot in terms of the dependence of lattice energy on the distance of separation between ions? What law governs this interaction? (For lattice energies, see Table 8.1.)

11.115 Which has a greater density, crystalline SiO_2 or amorphous SiO_2? Why?

11.116 A student is given four solid samples labeled W, X, Y, and Z. All have a metallic luster. She is told that the solids could be gold, lead sulfide, mica (which is quartz, or SiO_2), and iodine. The results of her investigations are: (a) W is a good electrical conductor; X, Y, and Z are poor electrical conductors. (b) When the solids are hit with a hammer, W flattens out, X shatters into many pieces, Y is smashed into a powder, and Z is not affected. (c) When the solids are heated with a Bunsen burner, Y melts with some sublimation, but X, W, and Z do not melt. (d) In treatment with 6 M HNO_3, X dissolves; there is no effect on W, Y, or Z. On the basis of these test results, identify the solids.

11.117 Which of the following statements are false? (a) Dipole-dipole interactions between molecules are greatest if the molecules possess only temporary dipole moments. (b) All compounds containing hydrogen atoms can participate in hydrogen-bond formation. (c) Dispersion forces exist between all atoms, molecules, and ions.

11.118   The diagram shows a kettle of boiling water. Identify the phases in regions A and B.

11.119   X rays of wavelength 0.154 nm strike an aluminum crystal; the rays are reflected at an angle of 19.3°. Assuming that $n = 1$, calculate the spacing between the planes of aluminum atoms (in pm) that is responsible for this angle of reflection.

11.120   The properties of gases, liquids, and solids differ in a number of respects. How would you use the kinetic molecular theory (see Section 10.6) to explain the following observations? (a) Ease of compressibility decreases from gas to liquid to solid. (b) Solids retain a definite shape, but gases and liquids do not. (c) For most substances, the volume of a given amount of material increases as it changes from solid to liquid to gas.

11.121   Select the substance in each pair that should have the higher boiling point. In each case identify the principal intermolecular forces involved and account briefly for your choice: (a) K_2S or $(CH_3)_3N$, (b) Br_2 or $CH_3CH_2CH_2CH_3$.

11.122   A small drop of oil in water assumes a spherical shape. Explain. (*Hint:* Oil is made up of nonpolar molecules, which tend to avoid contact with water.)

11.123   Under the same conditions of temperature and density, which of the following gases would you expect to behave less ideally: CH_4 or SO_2? Explain.

11.124   The fluorides of the second-period elements and their melting points are: LiF, 845°C; BeF_2, 800°C; BF_3, −126.7°C; CF_4, −184°C; NF_3, −206.6°C; OF_2, −223.8°C; F_2, −219.6°C. Classify the type(s) of intermolecular forces present in each compound.

11.125   The standard enthalpy of formation of gaseous molecular iodine is 62.4 kJ/mol. Use this information to calculate the molar heat of sublimation of molecular iodine at 25°C.

11.126   The distance between Li^+ and Cl^- is 257 pm in solid $LiCl$ and 203 pm in an $LiCl$ unit in the gas phase. Explain the difference in the bond lengths.

11.127   Heat of hydration, that is, the heat change that occurs when ions become hydrated in solution, is largely due to ion-dipole interactions. The heats of hydration for the alkali metal ions are Li^+, −520 kJ/mol; Na^+, −405 kJ/mol; K^+, −321 kJ/mol. Account for the trend in these values.

11.128   If water were a linear molecule, (a) would it still be polar, and (b) would the water molecules still be able to form hydrogen bonds with one another?

11.129   Calculate the $\Delta H°$ for the following processes at 25°C: (a) $Br_2(l) \longrightarrow Br_2(g)$ and (b) $Br_2(g) \longrightarrow 2Br(g)$. Comment on the relative magnitudes of these $\Delta H°$ values in terms of the forces involved in each case. (*Hint:* See Table 8.6.)

11.130   Which liquid would you expect to have a greater viscosity, water or diethyl ether? The structural formula of diethyl ether is shown in Problem 11.15.

11.131   A beaker of water is placed in a closed container. Predict the effect on the vapor pressure of the water when (a) its temperature is lowered, (b) the volume of the container is doubled, (c) more water is added to the beaker.

11.132   Ozone (O_3) is a strong oxidizing agent that can oxidize all the common metals except gold and platinum. A convenient test for ozone is based on its action on mercury. When exposed to ozone, mercury becomes dull looking and sticks to glass tubing (instead of flowing freely through it). Write a balanced equation for the reaction. What property of mercury is altered by its interaction with ozone?

11.133   A sample of limestone ($CaCO_3$) is heated in a closed vessel until it is partially decomposed. Write an equation for the reaction, and state how many phases are present.

11.134   Carbon and silicon belong to Group 14 of the periodic table and have the same valence electron configuration (ns^2np^2). Why does silicon dioxide (SiO_2) have a much higher melting point than carbon dioxide (CO_2)?

11.135   Provide an explanation for each of the following phenomena: (a) Solid argon (m.p. −189.2°C; b.p. −185.7°C) can be prepared by immersing a flask containing argon gas in liquid nitrogen (b.p. −195.8°C) until it liquefies and then connecting the flask to a vacuum pump. (b) The melting point of cyclohexane (C_6H_{12}) increases with increasing pressure exerted on the solid cyclohexane. (c) Certain high-altitude clouds contain water droplets at −10°C. (d) When a piece of dry ice is added to a beaker of water, fog forms above the water.

11.136   Argon crystallizes in the face-centered cubic arrangement at 40 K. Given that the atomic radius of argon is 191 pm, calculate the density of solid argon.

11.137   A pressure cooker is a sealed container that allows steam to escape when it exceeds a predetermined pressure. How does this device reduce the time needed for cooking?

11.138   A 1.20-g sample of water is injected into an evacuated 5.00-L flask at 65°C. What percentage of the water will be vapor when the system reaches equilibrium? Assume ideal behavior of water vapor and that the volume of liquid water is negligible. The vapor pressure of water at 65°C is 187.5 mmHg.

11.139   What are the advantages of cooking the vegetable broccoli with steam instead of boiling it in water?

11.140   A quantitative measure of how efficiently spheres pack into unit cells is called *packing efficiency,* which is the percentage of the cell space occupied by the spheres. Calculate the packing efficiencies of a simple cubic cell, a body-centered cubic cell, and a face-centered cubic cell. (*Hint:* Refer to Figure 11.22 and use the relationship that the volume of a sphere is $\frac{4}{3}\pi r^3$, where r is the radius of the sphere.)

11.141 A student heated a beaker of cold water (on a tripod) with a Bunsen burner. When the gas was ignited, she noticed that there was water condensed on the outside of the beaker. Explain what happened.

11.142 The compound dichlorodifluoromethane (CCl_2F_2) has a normal boiling point of $-30°C$, a critical temperature of $112°C$, and a corresponding critical pressure of 40 atm. If the gas is compressed to 18 atm at $20°C$, will the gas condense? Your answer should be based on a graphical interpretation.

11.143 Iron crystallizes in a body-centered cubic lattice. The cell length as determined by X-ray diffraction is 286.7 pm. Given that the density of iron is 7.874 g/cm^3, calculate Avogadro's number.

11.144 Sketch the cooling curves of water from about $110°C$ to about $-10°C$. How would you also show the formation of supercooled liquid below $0°C$ that then freezes to ice? The pressure is at 1 atm throughout the process. The curves need not be drawn quantitatively.

11.145 The boiling point of methanol is $65.0°C$, and the standard enthalpy of formation of methanol vapor is -201.2 kJ/mol. Calculate the vapor pressure of methanol (in mmHg) at $25°C$. (*Hint:* See Appendix 2 for other thermodynamic data of methanol.)

11.146 A sample of water shows the following behavior as it is heated at a constant rate:

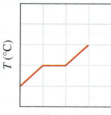

If twice the mass of water has the same amount of heat transferred to it, which of the following graphs best describes the temperature variation? Note that the scales for all the graphs are the same.

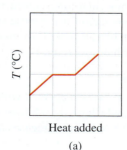

(a)

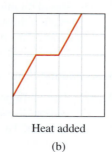

(b)

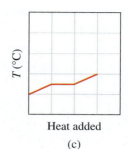

(c)

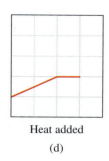

(d)

11.147 A closed vessel of volume 9.6 L contains 2.0 g of water. Calculate the temperature (in °C) at which only half of the water remains in the liquid phase. (See Table 10.5 for vapor pressures of water at different temperatures.)

11.148 The electrical conductance of copper metal decreases with increasing temperature, but that of a $CuSO_4$ solution increases with increasing temperature. Explain.

11.149 Assuming ideal behavior, calculate the density of gaseous HF at its normal boiling point ($19.5°C$). The experimentally measured density under the same conditions is 3.10 g/L. Account for the difference between your calculated result and the experimental value.

11.150 Referring to the cesium chloride structure in Figure 11.26(a), determine the density of a hypothetical ionic compound with the same unit cell, given that the radii of the anions and cations are 150 pm and 92 pm, respectively, and that their average masses are 98 amu and 192 amu, respectively.

11.151 Which of the following compounds is most likely to exist as a liquid at room temperature: ethane (C_2H_6), hydrazine (N_2H_4), fluoromethane (CH_3F)? Explain.

11.152 A chemistry instructor performed the following mystery demonstration. Just before the students arrived in class, she heated some water to boiling in an Erlenmeyer flask. She then removed the flask from the flame and closed the flask with a rubber stopper. After the class commenced, she held the flask in front of the students and announced that she could make the water boil simply by rubbing an ice cube on the outside walls of the flask. To the amazement of everyone, it worked. Give an explanation for this phenomenon.

Industrial Problems

11.153 Given the phase diagram of carbon, answer the following questions: (a) How many triple points are there and what are the phases that can coexist at each triple point? (b) Which has a higher density, graphite or diamond? (c) Synthetic diamond can be made from graphite. Using the phase diagram, how would you go about making diamond?

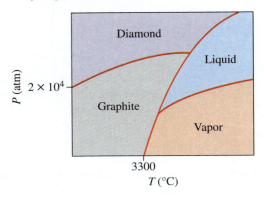

11.154   What is the origin of dark spots on the inner glass walls of an old tungsten lightbulb? What is the purpose of filling these lightbulbs with argon gas?

Engineering Problems

11.155   The phase diagram of helium is shown. Helium is the only known substance that has two different liquid phases: helium-I and helium-II. (a) What is the maximum temperature at which helium-II can exist? (b) What is the minimum pressure at which solid helium can exist? (c) What is the normal boiling point of helium-I? (d) Can solid helium sublime?

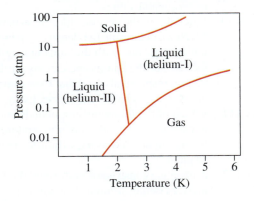

11.156   The phase diagram of sulfur is shown. (a) How many triple points are there? (b) Which is the more stable allotrope under ordinary atmospheric conditions? (c) Describe what happens when sulfur at 1 atm is heated from 80°C to 200°C.

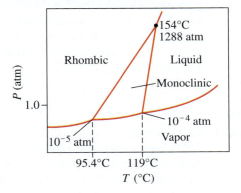

Biological Problems

11.157   Swimming coaches sometimes suggest that a drop of alcohol (ethanol) placed in an ear plugged with water "draws out the water." Explain this action from a molecular point of view.

11.158   Given the general properties of water and ammonia, comment on the problems that a biological system (as we know it) would have developing in an ammonia medium.

	H_2O	NH_3
Boiling point	373.15 K	239.65 K
Melting point	273.15 K	195.3 K
Molar heat capacity	75.3 J/K · mol	8.53 J/K · mol
Molar heat of vaporization	40.79 kJ/mol	23.3 kJ/mol
Molar heat of fusion	6.0 kJ/mol	5.9 kJ/mol
Viscosity	0.001 N · s/m^2	0.0254 N · s/m^2 (at 240 K)
Dipole moment	1.82 D	1.46 D
Phase at 300 K	Liquid	Gas

11.159   Given two complementary strands of DNA containing 100 base pairs each, calculate the ratio of two separate strands to hydrogen-bonded double helix in solution at 300 K. (*Hint:* The formula for calculating this ratio is $e^{-\Delta E/RT}$, where ΔE is the energy difference between hydrogen-bonded double-strand DNAs and single-strand DNAs and R is the gas constant.) Assume the energy of hydrogen bonds per base pair to be 10 kJ/mol.

11.160   The average distance between base pairs measured parallel to the axis of a DNA molecule is 3.4 Å. The average molar mass of a pair of nucleotides is 650 g/mol. Estimate the length in centimeters of a DNA molecule of molar mass 5.0×10^9 g/mol. Roughly how many base pairs are contained in this molecule?

11.161   Gaseous or highly volatile liquid anesthetics are often preferred in surgical procedures because once inhaled, these vapors can quickly enter the bloodstream through the alveoli and then enter the brain. Several common gaseous anesthetics are shown with their boiling points:

Halothane
50°C

Isoflurane
48.5°C

Enflurane
56.5°C

Based on intermolecular force considerations, explain the advantages of using these anesthetics. (*Hint:* The brain barrier is made of membranes that have a nonpolar interior region.)

Environmental Problems

11.162   The south pole of Mars is covered with solid carbon dioxide, which partly sublimes during the summer. The CO_2 vapor recondenses in the winter when the temperature drops to 150 K. Given that the heat of sublimation of CO_2 is 25.9 kJ/mol, calculate the atmospheric pressure on the surface of Mars. [*Hint:* Use Figure 11.34 to determine the

normal sublimation temperature of dry ice and Equation 11.4, which also applies to sublimations.]

11.163 Why do citrus growers spray their trees with water to protect them from freezing in very cold weather?

Standardized-Exam Practice Problems

Physical and Biological Sciences

Silicon used in computer chips must have an impurity level below 10^{-9} (i.e., fewer than one impurity atom for every 10^9 Si atoms). Silicon is prepared by the reduction of quartz (SiO_2) with coke (a form of carbon made by the destructive distillation of coal) at about 2000°C:

$$SiO_2(s) + 2C(s) \xrightarrow{\Delta} Si(l) + 2CO(g)$$

Next, solid silicon is separated from other solid impurities by treatment with hydrogen chloride at 350°C to form gaseous trichlorosilane ($SiCl_3H$):

$$Si(s) + 3HCl(g) \xrightarrow{\Delta} SiCl_3H(g) + H_2(g)$$

Finally, ultrapure Si can be obtained by reversing the preceding reaction at 1000°C:

$$SiCl_3H(g) + H_2(g) \xrightarrow{\Delta} Si(s) + 3HCl(g)$$

1. The molar heat of vaporization of trichlorosilane is 28.8 kJ/mol. Using this information and the equation

$$\ln \frac{P_1}{P_2} = \frac{\Delta H_{vap}}{R} \left(\frac{1}{T_2} - \frac{1}{T_1} \right)$$

determine the normal boiling point of trichlorosilane.

    a) −28.0°C   b) −276°C   c) 30.1°C   d) 275°C

2. What kind(s) of intermolecular forces exist between trichlorosilane molecules?

    a) Dispersion, dipole-dipole, and hydrogen bonding
    b) Dispersion and dipole-dipole
    c) Ion-ion
    d) Dipole-dipole

3. Each cubic unit cell (edge length $a = 543$ pm) contains eight Si atoms. If there are 1.0×10^{13} boron atoms per cubic centimeter in a sample of pure silicon, how many Si atoms are there for every B atom in the sample?

    a) 5×10^{22}
    b) 2×10^{-10}
    c) 2×10^{-22}
    d) 5×10^9

4. Calculate the density of pure silicon.

    a) 2.33 g/cm³
    b) 0.292 g/cm³
    c) 4.67 g/cm³
    d) 3.72 × 10⁻²² g/cm³

Answers to In-Chapter Materials

Answers to Practice Problems

11.1A (a) Dispersion forces, (b) dipole-dipole forces, H bonding and dispersion forces, (c) dipole-dipole and dispersion forces, (d) dispersion forces. **11.1B** (a) Dipole-dipole and dispersion forces; (b) dipole-dipole forces, H bonding, and dispersion forces; (c) dipole-dipole forces, H bonding, and dispersion forces; (d) dispersion forces. **11.2A** 265 mmHg. **11.2B** 76.0 kJ/mol, 109°C. **11.3A** 10.5 g/cm³. **11.3B** Body-centered cubic. **11.4A** 179 pm. **11.4B** 12.2° **11.5A** 4 Ca, 8 F. **11.5B** 1 Cs, 1 Cl. **11.6A** 2.65 g/cm³. **11.6B** 421 pm. **11.7A** 2.72 g/cm³. **11.7B** 361 pm. **11.8A** 984 kJ. **11.8B** 100°C, liquid and vapor in equilibrium. **11.9A** (a) ~110°C, ~−10°C, (b) liquid.

11.9B

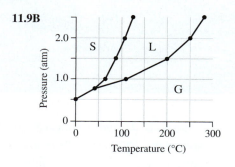

Answers to Checkpoints

11.1.1 a. **11.1.2** d. **11.2.1** a. **11.2.2** e. **11.2.3** b. **11.2.4** b. **11.3.1** d. **11.3.2** b. **11.4.1** c. **11.4.2** e. **11.6.1** a. **11.6.2** c. **11.7.1** a. **11.7.2** e.

Modern Materials

Left to right: Richard R. Schrock (USA), Yves Chauvin (France), and Robert H. Grubbs (USA) shared the 2005 Nobel Prize in Chemistry for their work on the metathesis method of organic synthesis.

Jonas Ekstromer/AFP/Getty Images

In This Chapter, You Will Learn

About some of the chemistry involved in the development of modern materials.

Before You Begin, Review These Skills

- Functional groups [◀◀ Section 2.7]
- Molecular polarity [◀◀ Section 9.2]

Modern Materials Chemistry and the 2005 Nobel Prize

The 2005 Nobel Prize in Chemistry was awarded to Yves Chauvin, Robert H. Grubbs, and Richard R. Schrock for the development of the metathesis method in organic synthesis. In the metathesis method, double bonds in organic molecules are broken, and then new double bonds are formed. In olefin metathesis, a carbon-carbon double bond is broken in each of two olefins. The subsequent formation of two new carbon-carbon double bonds results in products in which the compounds have exchanged substituent groups with each other:

Student Note: *Metathesis* is a word meaning "an exchange of position," and *olefin* is an old-fashioned word for an alkene, a molecule with a carbon-carbon double bond.

Note that each reactant molecule has one R group and one R′ group, whereas each product molecule has two *identical* groups (i.e., two R groups or two R′ groups). This process requires a metal catalyst and is believed to occur by the formation of a four-membered ring:

The metathesis method can be applied to many industrial chemical processes, including those that potentially can produce pharmaceuticals for the treatment of such conditions as bacterial and viral infections, cancer, and Alzheimer's disease.

At the end of this chapter, you will be able to answer questions about modern polymeric materials [▶▶ Applying What You've Learned, page 611].

Bottles that can be recycled are made of thermoplastic polymers.

Thinkstock/SuperStock

12.1 | Polymers

Many biological molecules, such as DNA, starch, and proteins, have very large molecular masses. These molecules are *polymers* because they are made up of many smaller parts linked together. The prefix *poly* comes from a Greek root meaning "many," and the small molecules that make up the individual building blocks of polymers are called *monomers*. Many natural polymers, such as silk and rubber, now have synthetic analogues that are manufactured by the chemical industry.

Polymers are used in many medical applications, such as flexible wound dressings, surgical implants, and prosthetics. Some of these polymers are *thermoplastic,* which means that they can be melted and reshaped, or heated and bent. Others are *thermosetting,* which means that their shape is determined as part of the chemical process that formed the polymer. Thermosetting polymers cannot be reshaped easily and are not easily recycled, whereas thermoplastic polymers can be melted down and cast into new shapes in different products.

Addition Polymers

Student Note: The systematic name of C_2H_4 is *ethene*, but the common name *ethylene* is still widely used.

Student Note: Recall that species with an unpaired electron are called *free radicals* [◄◄ Section 8.8].

The simplest type of polymerization reaction involves the bonding of monomer molecules by movement of electrons from a multiple bond into new single bonds between molecules. This type of polymerization is called *addition polymerization.* The synthesis of polyethylene from ethylene (C_2H_4) illustrates an addition polymerization. The process of "splitting open" the double bond is initiated by a molecule or an atom with an unpaired electron—known as a *free radical*. The free radical initiator attaches to one of the ethylene carbon atoms and breaks the pi bond between the two carbon atoms in ethylene. Figure 12.1 shows the scheme of addition polymerization with respect to the double bonds opening up to form single bonds between monomer units to form polyethylene from ethylene. (The arrows with half-pointed heads show movement of a single electron.)

The structure of polyethylene can be represented as $\text{-}(CH_2\text{-}CH_2)_n$, where n is the number of ethylene monomer molecules that reacted (which could be hundreds, thousands, or more). The end of the chain contains a carbon atom with an unpaired electron (i.e., $-CH_2CH_2\cdot$), though, so it must be closed by bonding with another atom. This might be a hydrogen atom (e.g., $-CH_2CH_2-H$), but it might also be part of the initiator molecule (e.g., $-CH_2CH_2-OR$). Although the residue of the initiator molecule has very different chemical properties than the main polymer chain, it will not significantly affect the polymer's properties because the polyethylene chain is so long.

Animation
Organic and biochemistry—natural and synthetic polymers.

Although the structure of a polyethylene chain is drawn in Figure 12.1 as a straight line, the bond angles around a central carbon atom with four bonds are approximately 109.5° each [◄◄ Section 9.1]. Thus, the "straight" chain really zigzags, instead. The polymer chain bends, too, because there is free rotation about carbon-carbon single bonds. These molecular-level features manifest themselves in such macroscopic properties as flexibility.

It is possible, depending on the reaction conditions, for branches to form off the main polyethylene chain during polymerization. Polymer chains with many branches cannot pack together as efficiently as "straight" or unbranched chains, so branched-chain polyethylene is more flexible and lower in density than straight-chain polyethylene. The difference between branched and unbranched polyethylene is shown schematically in Figure 12.2.

Polyethylene consisting primarily of unbranched chains is known as high-density polyethylene (HDPE), whereas polyethylene that consists primarily of branched

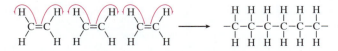

Figure 12.1 Addition polymerization to form polyethylene from ethylene. Each curved, single-headed arrow represents the movement of one electron.

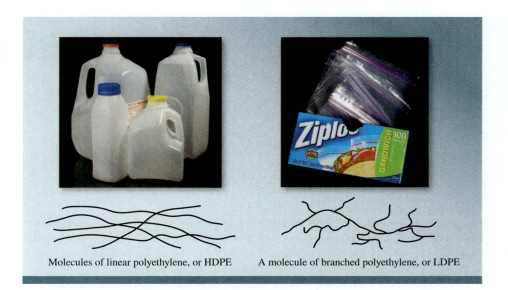

Figure 12.2 Linear polyethylene is used to make plastic milk bottles. Plastic bags and wrap like Glad Cling Wrap are made from branched polyethylene.

(both): *David A. Tietz/Editorial Image, LLC*

chains is known as low-density polyethylene (LDPE). HDPE is used in plastic bottles and containers where it is important to maintain shape, whereas LDPE is used in plastic food bags and wraps.

Natural rubber formed from latex is an addition polymer. The monomer unit is isoprene, a molecule with two C=C double bonds. In the reaction that forms *polyisoprene,* a pair of electrons moves from one double bond to form a new double bond between the middle two carbon atoms, and the other double bond opens up to form a carbon-carbon single bond to the next monomer unit.

Isoprene

cis-Polyisoprene

Natural rubber is not very durable—it only retains its useful properties in a relatively narrow temperature range. To address this problem, Charles Goodyear[1] developed industrial *vulcanization* in 1839, a process in which sulfur is used to create linkages between individual polymer chains in the rubber. By heating the rubber with sulfur (S_8) under carefully controlled conditions, some of the carbon-hydrogen bonds are broken and replaced with carbon-sulfur bonds. The sulfur atoms are bonded to one or more additional sulfur atoms and then to another polymer chain through another sulfur-carbon bond (Figure 12.3). These linkages, called **cross-links,** make the rubber

1. Charles Goodyear (1800–1860). American chemist. Goodyear was the first person to realize the potential of natural rubber. His vulcanization process made the use of rubber practical, contributing greatly to the development of the automobile industry.

Figure 12.3 Vulcanization.

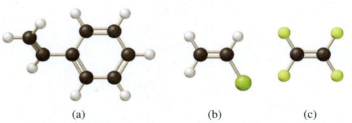

much stronger and more rigid than natural polyisoprene. Rubber is an ***elastomer,*** a material that can stretch or bend and then return to its original shape as long as the limits of its elasticity are not exceeded. The cross-linking of the polyisoprene chains plays an important role in maintaining rubber's elastomer properties.

Goodyear, the current supplier of tires to NASCAR teams, varies the composition of the rubber in its tires for each race depending on the expected tire wear and banking of the track. The left- and right-side tires can even be of different compositions depending on the track layout and conditions. Softer tire compounds offer more grip but wear out faster, whereas harder compounds last longer but grip less.

Many other molecules containing carbon-carbon double bonds can form addition polymers. Styrene [Figure 12.4(a)] is an analogue of ethylene in which a hydrogen atom has been replaced with a C_6H_5 (a *phenyl*) group. The polymerization of styrene forms *polystyrene* (Figure 12.5). As a foam (with air or some other gas blown into the solid), polystyrene is used in disposable cups and plates as well as Styrofoam insulation and craft materials. Vinyl chloride [Figure 12.4(b)], an analogue of

Student Note: Each tire used in the NASCAR race series costs approximately $400. A single race car can go through as many as 32 tires in one 500-mile race, depending on track conditions. There are dozens of races throughout the year, and even more tires are used up during practices and prerace qualifying. It takes a lot of tires to compete in NASCAR!

Robert E. Klein/AP Images

Student Note: This figure shows the structure of a general "vinyl" compound. Styrene and vinyl chloride are vinyl compounds in which R is a phenyl group and a chlorine atom, respectively.

(a)   (b)   (c)

Figure 12.4 (a) Styrene, (b) vinyl chloride, and (c) tetrafluoroethylene (the monomer of Teflon).

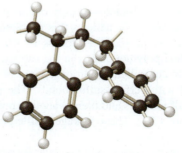

Figure 12.5 Structure of polystyrene.

ethylene in which one of the hydrogen atoms has been replaced by a chlorine atom, is another monomer that can be used to form a polymer with many common uses. Poly(vinyl chloride) (PVC) is used in pipes and fittings, some plastic wraps, and the classic "vinyl" records that were the predecessors to musical recordings on cassettes and CDs.

Polytetrafluoroethylene, more commonly known as the trademarked brand name Teflon, is formed from the addition polymerization of tetrafluoroethylene. Tetrafluoroethylene, shown in Figure 12.4(c), is an analogue of ethylene in which fluorine atoms have replaced all four of the hydrogen atoms. Teflon is a very good electrical insulator, so it is commonly used to coat wires. It is probably best known, though, as a nonstick substance used to coat bakeware, frying pans, and pots. It is also used in films that can be inserted into threaded joints between metal pipes to make it easier to unscrew the connection when necessary. Because Teflon is chemically inert, it is not possible to cross-link the chains like an elastomer. The structures of some addition polymers, including the structures of their respective monomers and what they are typically used for, are summarized in Table 12.1.

The diversity of polymers increases greatly when different monomers are used in the same polymer chain. *Copolymers* are polymers made from two or more different monomers. Different combinations of the two or more monomers give rise to different types of copolymers. These different copolymers are summarized in Table 12.2.

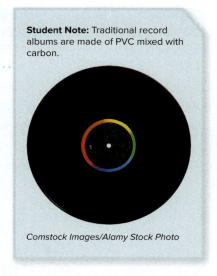

Student Note: Traditional record albums are made of PVC mixed with carbon.

Comstock Images/Alamy Stock Photo

TABLE 12.1	Some Common Addition Polymers		
Polymer	**Structure**	**Monomer**	**Uses**
Polyethylene	$-(CH_2-CH_2)_n-$	$H_2C=CH_2$	Plastic bags and wraps, bottles, toys, Tyvek wrap
Polystyrene	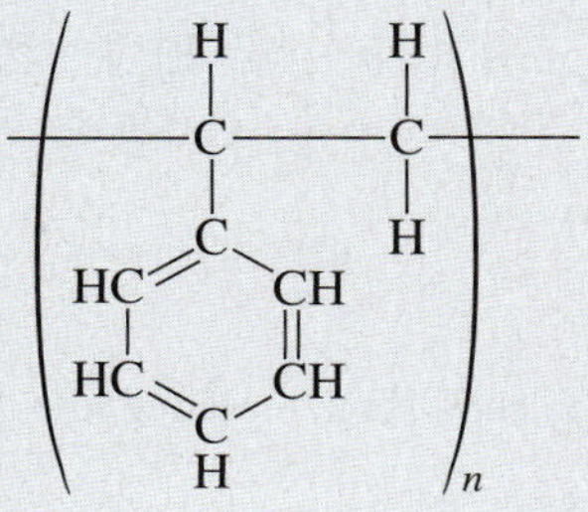 *(structure shown)*	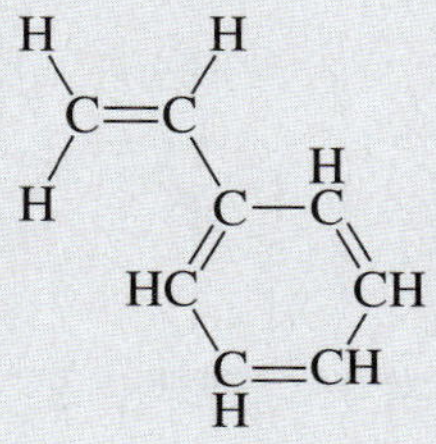 *(monomer shown)*	Disposable cups and plates, insulation, packing materials
Polyvinylchloride	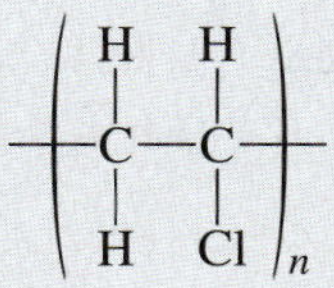 *(structure shown)*	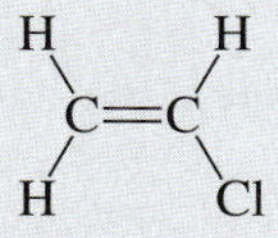 *(monomer shown)*	Pipes and fittings, plastic wraps, records
Polyacrylonitrile	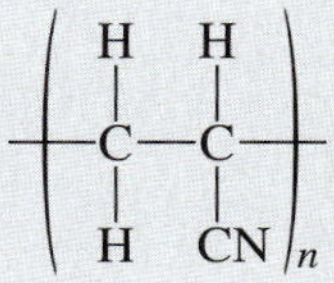 *(structure shown)*	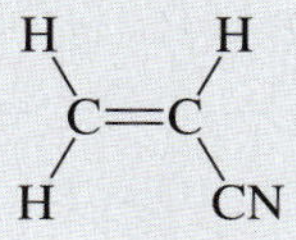 *(monomer shown)*	Fibers for carpeting and clothing (Orlon)
Polybutadiene	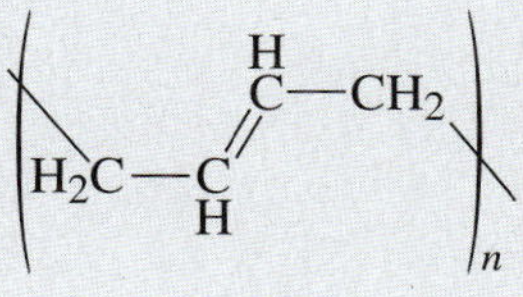 *(structure shown)*	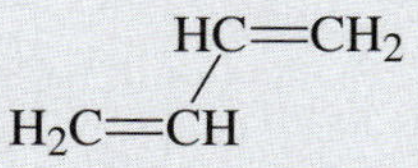 $HC=CH_2$ / $H_2C=CH$	Synthetic rubber

TABLE 12.2	Types of Copolymers
Copolymer Type	**Representation***
Random	... −A−A−B−A−B−B−B−A−A−B−A−B−A−B−A−B−B−A− ...
Alternating	... −A−B−A−B−A−B−A−B−A−B−A−B−A−B−A−B− ...
Block	... −A−A−A−A−A−A−B−B−B−B−B−B−A−A−A−A−A−A− ...
Graft	A−A−A−A−A−A−A−A−A−A−A−A−A−A−A−A−A−A ... | B−B−B−B−B−B−B−B−B ...

*A and B represent different monomers (e.g., ethylene and styrene).

Sample Problem 12.1 lets you practice deducing the structure of an addition polymer based on the structure of its monomer.

SAMPLE PROBLEM 12.1

Polypropylene is used in applications that range from indoor-outdoor carpeting to soda bottles to bank notes. It is produced by the addition polymerization of propylene, $H_2C=CH-CH_3$. Draw the structure of polypropylene, showing at least two repeating monomer units, and write its general formula.

Student Note: An Australian banknote issued in 1988 is printed on polypropylene for increased durability and security.

David A. Tietz/Editorial Image, LLC

Strategy Addition polymers form via a free-radical reaction in which the pi electrons in the carbon-carbon double bond of a monomer molecule are used to form carbon-carbon single bonds to other monomer molecules. Draw the structural formula of propylene such that the double bond can be "opened up" to form single bonds between consecutive monomer units.

Setup The structural formula of propylene is

By drawing three adjacent propylene molecules, we can rearrange the bonds to show three repeating units.

Solution The structure of polypropylene showing three monomer units is

Thus, the general formula of polypropylene is

THINK ABOUT IT

The general formula looks like the structure of the monomer, except that the carbon-carbon double bond has been opened up to form bonds to monomer units on each side. This means the answer is reasonable. In fact, the opposite process (i.e., restoring the carbon-carbon double bond) can be used to determine the structure of the monomer used to create an addition polymer.

Practice Problem Ⓐ**TTEMPT** Saran Wrap, the original plastic film, is polyvinylidene chloride, which is formed by the addition polymerization of vinylidene chloride, $H_2C{=}CCl_2$. Draw the structure of polyvinylidene chloride, showing at least two repeating units, and write its general structure.

Practice Problem Ⓑ**UILD** The general structure of polymethyl methacrylate, an addition polymer used in Plexiglas and Lucite, is as follows:

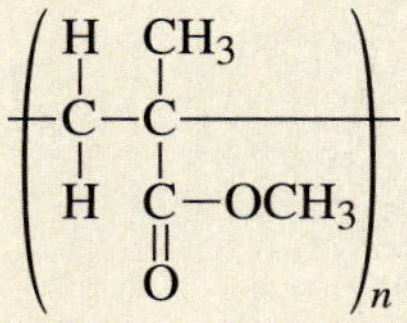

$$\left(\begin{array}{c} \text{H} \quad \text{CH}_3 \\ | \quad \ | \\ -\text{C}-\text{C}- \\ | \quad \ | \\ \text{H} \quad \text{C}-\text{OCH}_3 \\ \quad\ \ \| \\ \quad\ \ \text{O} \end{array}\right)_n$$

Draw the structure of the monomer used to make this polymer.

Practice Problem Ⓒ**ONCEPTUALIZE** Which diagram best represents the movement of electrons in the formation of an addition polymer? (Electrons in carbon-carbon double bonds are color coded for identification.)

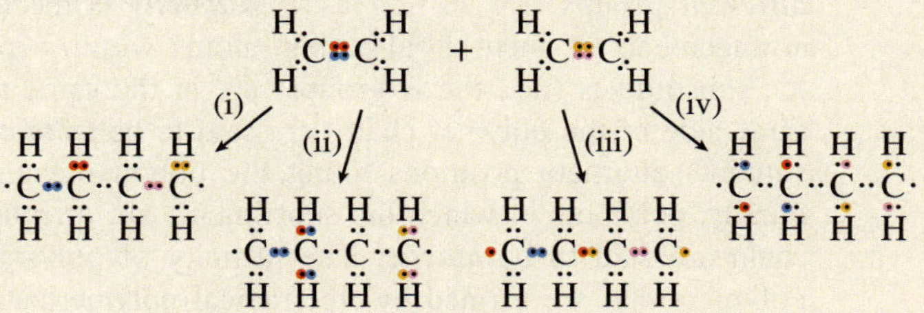

Butadiene ($CH_2{=}CH{-}CH{=}CH_2$) and styrene, two monomers listed in Table 12.1, can form a block copolymer called poly(styrene-butadiene-styrene) or SBS rubber [Figure 12.6(a)], in which there are alternating blocks of polybutadiene and polystyrene. They also form a graft polymer known as high-impact polystyrene (HIPS) [Figure 12.6(b)], in which chains of polybutadiene branch off of the main polystyrene chain.

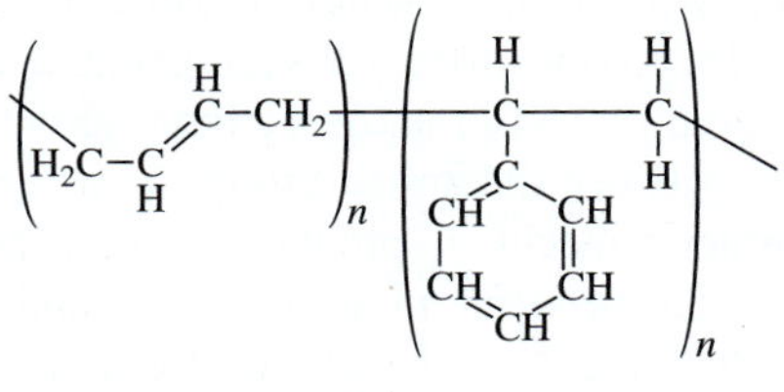

(a)

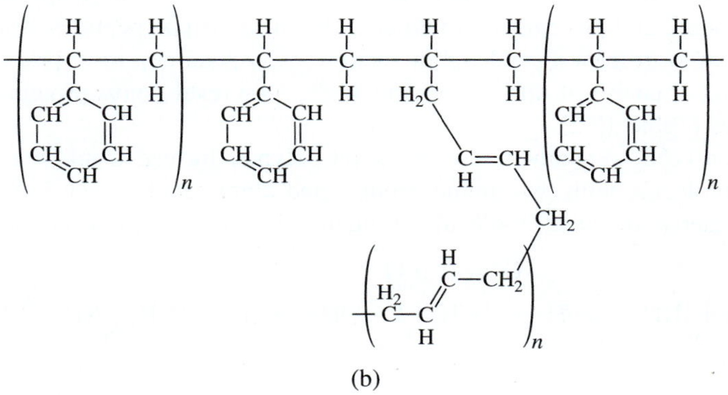

(b)

Figure 12.6 Structures of (a) SBS rubber and (b) HIPS. SBS consists of alternating blocks of polybutadiene and polystyrene, whereas HIPS consists of a polystyrene backbone with polybutadiene branches.

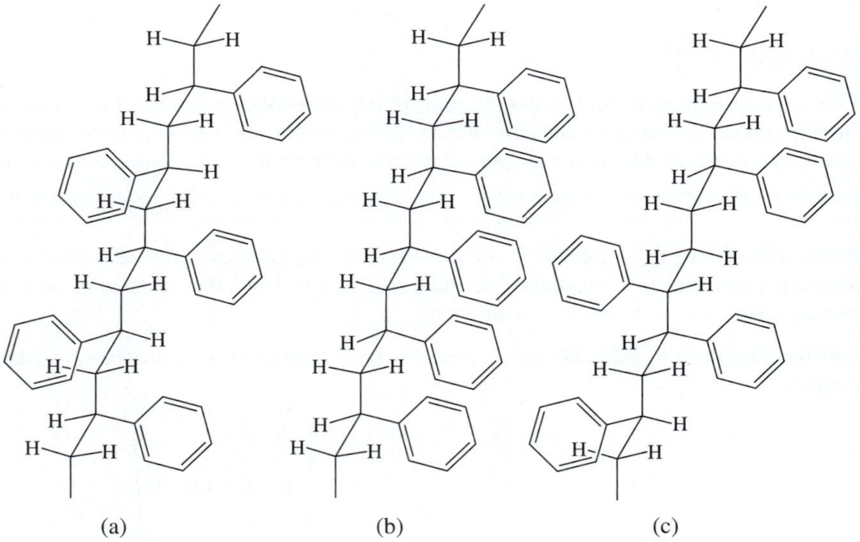

Figure 12.7 Structures of (a) isotactic, (b) syndiotactic, and (c) atactic polymers.

(a)   (b)   (c)

Polymers containing a vinyl group ($-CH_2-CHR-$) can exist as optical isomers because the carbon atom bonded to the R group is tetrahedral and bonded to four *different* groups [▶▶ Section 25.4]. **Tacticity** is the term that describes the relative arrangements of such chiral carbon atoms within a polymer. Polymers in which all the substituents (i.e., the R groups) are in the same relative orientation (i.e., on the same side of the polymer chain) are said to be **isotactic**. Polymers in which the substituents alternate positions along the polymer chain are said to be **syndiotactic**. Finally, polymers in which the substituents are oriented randomly along the polymer chain are said to be **atactic**. The tacticity of polystyrene is shown in Figure 12.7. Polymers that are formed by free-radical polymerization processes, such as polystyrene and PVC, are usually atactic. Special catalysts have been developed, though, that can be used to synthesize isotactic and syndiotactic polymers.

Condensation Polymers

Condensation polymers form when monomers with two different functional groups combine, resulting in the elimination of a small molecule (usually water). Proteins are biological condensation polymers that can have molecular weights in the thousands or millions of grams per mole. Proteins form between amino acid monomers. Each amino acid contains an amino group ($-NH_2$) and a carboxy group ($-COOH$), and the condensation reaction occurs between the amino group of one amino acid and the carboxy group of another. One water molecule is produced as each new C$-$N bond is formed.

The C$-$N bond formed between an amino group and a carboxy group is called an amide bond because RCONR′R″ is the general form of an amide [▶▶ Section 25.2, Table 25.2]. Polymers in which the monomers are connected by amide linkages are called **polyamides**. The amide bond between amino acids is more specifically called a peptide bond, and proteins are routinely referred to as polypeptides. Although polypeptides can form from a single amino acid (e.g., polyvaline), most proteins in organisms consist of chains of different amino acids. As a result, most proteins are random copolymers (Table 12.2).

Nylon 6,6 is a synthetic condensation polymer formed from hexamethylenediamine, a molecule with two amino groups, and adipic acid, which has two carboxy groups. Water is the small molecule eliminated in this condensation reaction.

$$n H_2N\!\!-\!\!\left(CH_2\right)_{\!6}\!\!-\!\!NH_2 + n OH\!-\!\overset{\overset{O}{\|}}{C}\!\!-\!\!\left(CH_2\right)_{\!4}\!\!-\!\!\overset{\overset{O}{\|}}{C}\!\!-\!\!OH \longrightarrow \left[NH\!\!-\!\!\left(CH_2\right)_{\!6}\!\!-\!\!NH\!-\!\overset{\overset{O}{\|}}{C}\!\!-\!\!\left(CH_2\right)_{\!4}\!\!-\!\!\overset{\overset{O}{\|}}{C}\right]_{\!n}$$

$$+\, 2n H_2O$$

Both hexamethylenediamine and adipic acid contain six carbon atoms, which is why the resulting polymer is called nylon 6,6. Like proteins, nylon 6,6 contains amide linkages between monomers, so nylon 6,6 is a polyamide, too. Also like proteins, nylon 6,6 is a copolymer, although nylon 6,6 is an *alternating* copolymer (Table 12.2), not a *random* copolymer.

Dacron is the trade name for the condensation copolymer formed from ethylene glycol (a diol—a molecule with two alcohol groups) and *para*-terephthalic acid (a diacid—a molecule with two carboxy groups). The condensation reaction removes the hydrogen atom from the alcohol on one monomer and the OH group from the carboxy group on a second monomer, producing water.

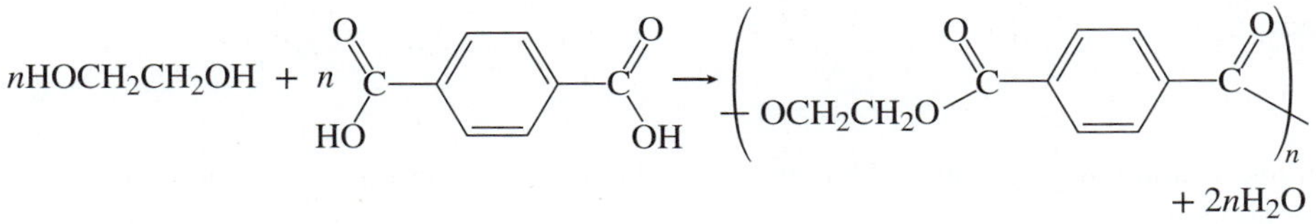

$$n\text{HOCH}_2\text{CH}_2\text{OH} + n \text{ (diacid)} \longrightarrow (\text{polyester})_n + 2n\text{H}_2\text{O}$$

The bond that forms between each ethylene glycol monomer and each *para*-terephthalic acid monomer is called an ester linkage because RCOOR′ is the general form of an ester [▶▶ Section 25.2, Table 25.2]. Polymers in which the monomers are connected by ester linkages are called **polyesters**.

Deoxyribonucleic acid (DNA; see Figure 12.8), which stores the genetic information for every known organism and has a molecular weight as large as billions of grams per mole, is a biological condensation copolymer. If your body contained anything but a tiny fraction of a mole of DNA, you would be extremely heavy! DNA is a copolymer of a five-carbon sugar called deoxyribose, four different molecular bases (adenine, thymine, guanine, and cytosine), and phosphoric acid (H_3PO_4). Ribonucleic acid (RNA) is a biological condensation copolymer analogous to DNA, except that RNA is made from a five-carbon sugar called ribose and the four different molecular bases are adenine, uracil, guanine, and cytosine. RNA molecules vary greatly in size. They are large molecules, but still much smaller than DNA molecules.

Sample Problem 12.2 shows how to determine the monomers in a copolymer.

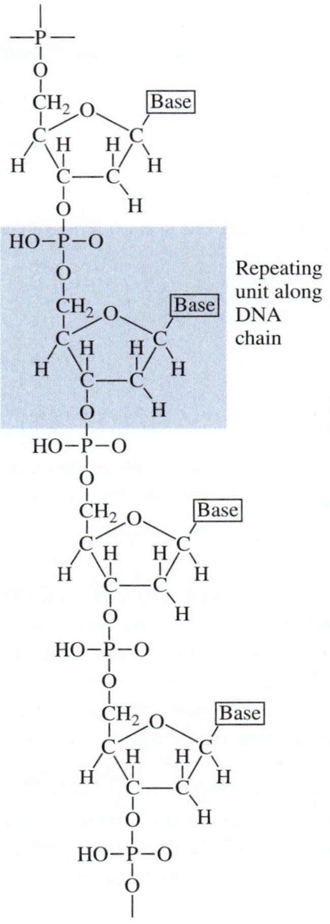

Figure 12.8 Structure of DNA.

SAMPLE PROBLEM 12.2

Kevlar, a condensation polymer used in bulletproof vests, has the following general structure:

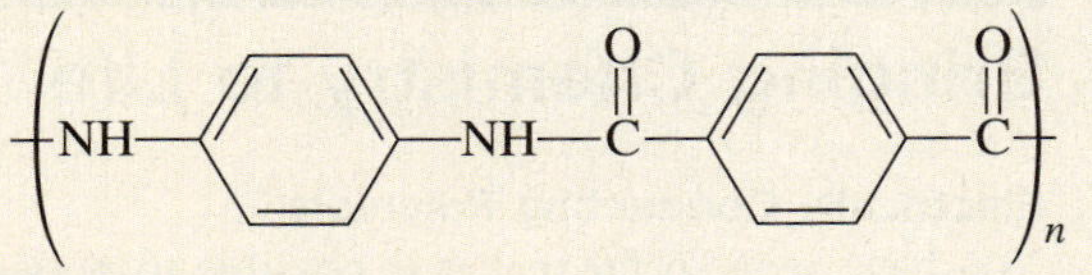

Write the structure of the two monomers that form Kevlar.

Strategy Identify the condensation linkages, split them apart, and then add H to one of the resulting exposed bonds and OH to the other.

Setup Kevlar contains an *amide* (linkage between C and N). We split the C—N bond, adding an H to the exposed N and an OH to the exposed C to produce an amine and a carboxylic acid, respectively.

Solution To determine what the monomers are, first remove the C—N bond:

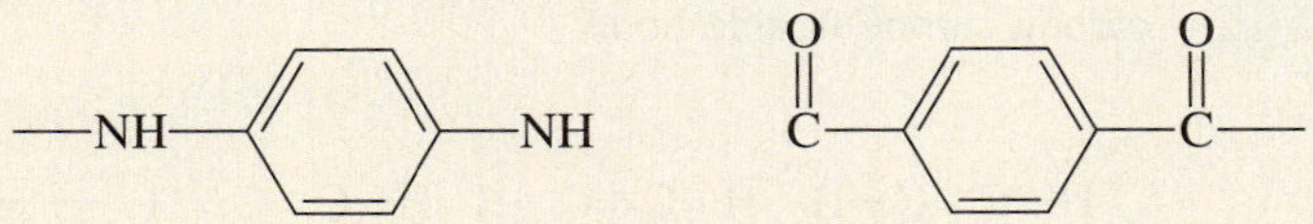

Next, add H_2O as H to the N and OH to the C:

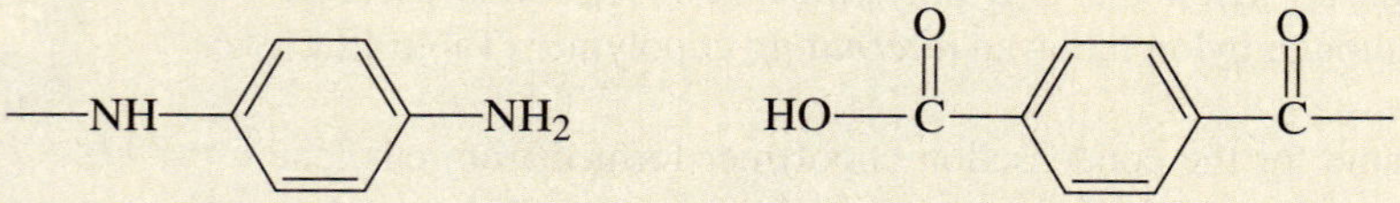

Finally, add the components of another water molecule to the open ends of the molecules—H to N and OH to C, as before:

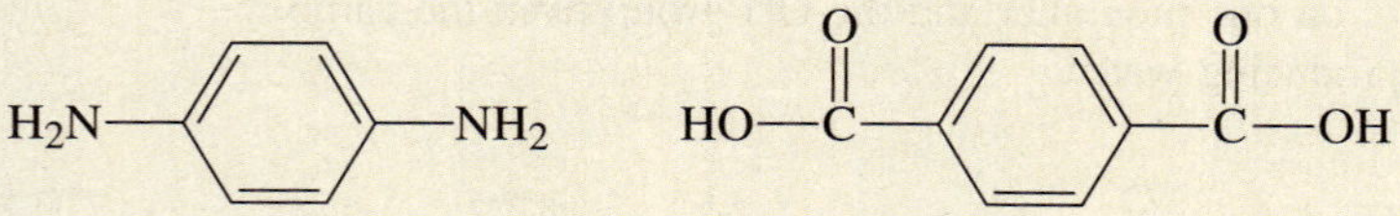

THINK ABOUT IT

To determine whether the monomers in the final answer are reasonable, recombine them in a condensation reaction to see if they re-form Kevlar, the polymer in the problem.

Practice Problem (A)**TTEMPT** Nylon 6, like nylon 6,6, is a condensation polymer (a polyamide). Unlike nylon 6,6, however, nylon 6 is *not* a copolymer because it is synthesized from a single monomer called 6-aminohexanoic acid. Draw the structure of at least three repeating units of nylon 6.

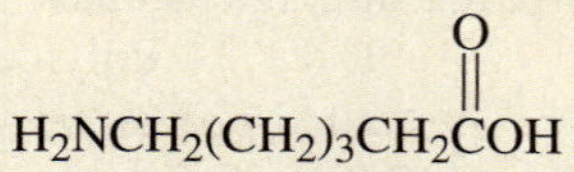

$$H_2NCH_2(CH_2)_3CH_2\overset{\displaystyle O}{\overset{\|}{C}}OH$$

Practice Problem (B)**UILD** Kodel is a polymer used to make stain-resistant carpets. Draw the structure of the monomer. Is this an addition polymer or a condensation polymer?

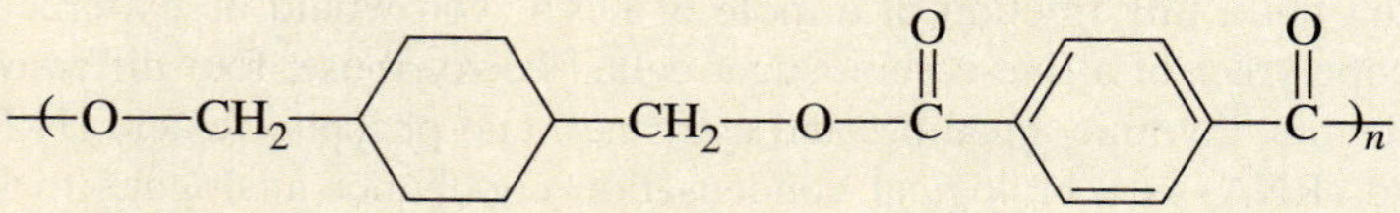

Practice Problem (C)**ONCEPTUALIZE** Draw the structure of the substance that would be eliminated if the molecule shown here were to form a condensation polymer.

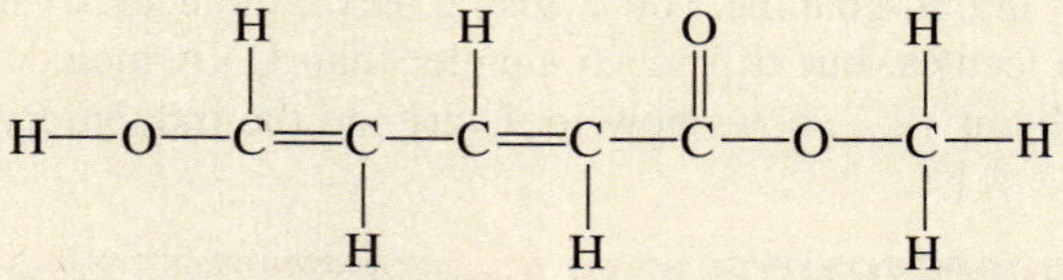

Bringing Chemistry to Life

Electrically Conducting Polymers

You have seen so far that it is possible to polymerize organic compounds with carbon-carbon double bonds (alkenes). The resulting addition polymers (e.g., polyethylene) have carbon-carbon single bonds. It is also possible, though, to polymerize organic compounds with carbon-carbon triple bonds. The simplest of these compounds (known collectively as *alkynes*) is acetylene (C_2H_2), which is commonly used in welding torches. The polymerization of acetylene is very similar to the addition polymerization of ethylene. That is, a free-radical initiator molecule attaches to one of the carbon atoms and breaks one of the pi bonds between the two carbon atoms in acetylene. The resulting polymer, polyacetylene, thus retains a carbon-carbon double bond:

$$H-C\equiv C-H \quad H-C\equiv C-H \quad H-C\equiv C-H \longrightarrow$$

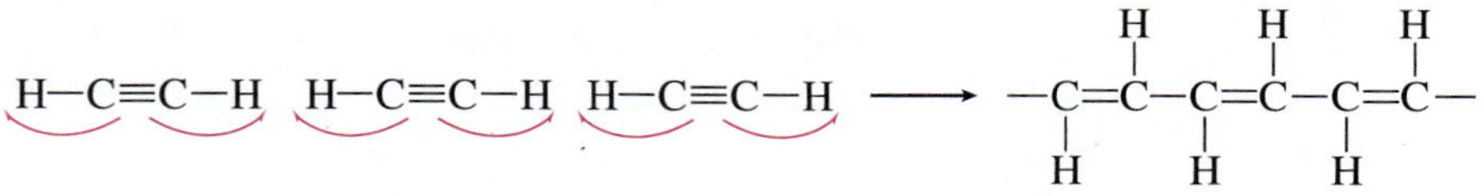

The generalized form of the reaction can be represented as follows:

$$n\ H{-}C{\equiv}C{-}H \longrightarrow \left(\begin{array}{c} H \\ | \\ C{=}C \\ | \\ H \end{array}\right)_n$$

The structure of polyacetylene contains alternating carbon-carbon single and carbon-carbon double bonds. Recall from Chapter 9 that the carbon atoms in a $C{=}C$ double bond are sp^2-hybridized and that it is the leftover p orbital on each carbon atom that overlaps to form the pi bond. Each carbon atom in polyacetylene is identical (structurally and electronically), so there is extended overlap of the p orbitals throughout the polymer chain. The p electrons are *delocalized*—that is, they can move throughout the network of overlapping orbitals that extend the length of the polymer chain. By being able to move electrons from one end of the polymer chain to the other, polyacetylene can conduct electricity much like a wire. That is, polyacetylene is a plastic that conducts electricity!

For comparison, consider polyethylene and Teflon. Both are polyalkenes in which the polymer chain contains only carbon-carbon single bonds. Because all the electrons are localized in sigma bonds between the carbon atoms, there are no delocalized electrons available to carry charge throughout the polymer chain. The presence of electrons in *delocalized* orbitals, such as those in polyacetylene, is the key to electrical conductivity.

Sample Problem 12.3 shows how to determine the structure of an electrically conducting polymer.

SAMPLE PROBLEM 12.3

Propyne ($HC{\equiv}CCH_3$) can be used to form an electrically conducting polymer. Draw the structure of polypropyne, showing at least three repeating units, and write the general formula for the polymer.

Strategy Addition polymers, whether they are synthesized from alkenes or alkynes, form via a free-radical reaction in which one pair of pi electrons in the carbon-carbon multiple bond of a monomer molecule is used to form carbon-carbon single bonds to other monomer molecules. Draw the structural formula of propyne such that the triple bond can be "opened up" to form single bonds between consecutive monomer units.

Setup The structural formula of propyne can be determined from the formula given and by analogy to acetylene. The only difference between propyne and acetylene is the CH_3 group in place of one of the hydrogen atoms:

$$H{-}C{\equiv}C{-}CH_3$$

By drawing three adjacent propyne molecules, we can rearrange the bonds to show three repeating units:

$$\text{---}H{-}C{\equiv}C{-}CH_3\text{---}H{-}C{\equiv}C{-}CH_3\text{---}H{-}C{\equiv}C{-}CH_3\text{---}$$

Solution The structure of polypropylene showing three monomer units is

$$-\underset{\underset{H}{|}}{C}{=}\underset{\underset{H}{|}}{\overset{\overset{CH_3}{|}}{C}}-\underset{\underset{H}{|}}{C}{=}\underset{\underset{H}{|}}{\overset{\overset{CH_3}{|}}{C}}-C{=}\overset{\overset{CH_3}{|}}{C}-$$

The repeating unit is $-CH{=}C(CH_3)-$, so the general structure is

$$\left(\begin{array}{c} CH_3 \\ | \\ C{=}C \\ | \\ H \end{array}\right)_n$$

THINK ABOUT IT

The repeating polymer unit in the general structure resembles the original monomer except that there is a carbon-carbon double bond in place of a carbon-carbon triple bond. If each of the carbon-carbon single bonds at the ends of the repeating unit in parentheses were split open, then one electron from each bond could be combined to re-form the second pi bond in the carbon-carbon triple bond of the monomer.

Practice Problem **A**TTEMPT Draw the structures of poly(1-butyne) and poly(1-butene), showing three or more repeating units. How are they different? How are they similar? Which of them, if any, do you think will conduct electricity?

$$H—C≡C—CH_2CH_3 \qquad H_2C=CH—CH_2CH_3$$
$$\text{1-butyne} \qquad\qquad \text{1-butene}$$

Practice Problem **B**UILD Explain why many conducting polymers form from monomers that contain triple bonds.

Practice Problem **C**ONCEPTUALIZE Which best represents the movement of electrons in the formation of an electrically conducting polymer from an alkyne monomer? (Electrons in carbon-carbon triple bonds are color coded for identification.)

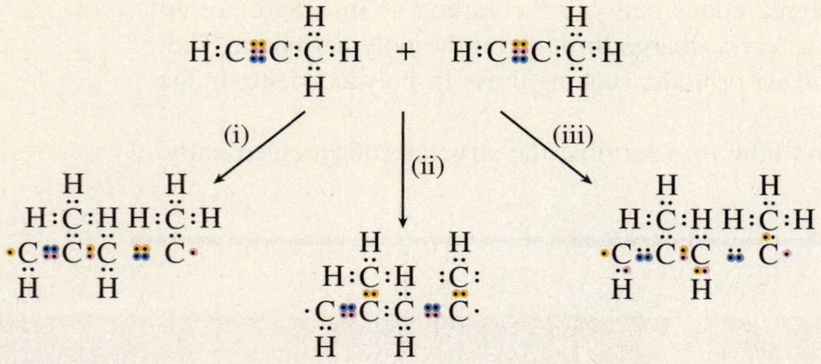

CHECKPOINT – SECTION 12.1   Polymers

12.1.1 Classify the following copolymer:

$$-B-B-B-B-B-A-A-A-A-A-B-B-B-B-B-A-A-A-A-A-$$

(Select all that apply.)

a) Block   b) Random   c) Graft   d) Alternating   e) All of the above

12.1.2 What feature is common to molecules that can undergo polymerization?

a) Fluorine   b) Hydrogen bonds   c) Sulfur   d) Multiple bonds   e) Lone pairs

12.2   Ceramics and Composite Materials

Ceramics

Student Note: The larger the lattice energy of an ionic compound, the higher the melting point [◄◄ Section 8.2, Table 8.1]. Ceramics all have large lattice energies.

The use of ceramics in the form of pottery dates back to antiquity. Along with common ceramic substances like brick and cement, modern advanced ceramics are found in electronic devices and on the exterior of spaceships. All of these *ceramics* are polymeric inorganic compounds that share the properties of hardness, strength, and high melting points. Ceramics are usually formed by melting and then solidifying inorganic substances (including clays).

Figure 12.9 Captain Wendy Lawrence inspects the ceramic tiles on the exterior of the space shuttle.
Source: *NASA Kennedy Space Center (NASA-KSC)*

Most ceramic materials are electrical insulators, but some conduct electricity very well at very cold temperatures. Ceramics are also good heat insulators, which is why the outer layer of the space shuttle (shown in Figure 12.9) was made of ceramic tiles. These ceramic tiles could withstand the very hot temperatures of reentry into the atmosphere (where temperatures can reach 1650°C), while keeping the underlying metal structural body of the orbiter (and the astronauts within) relatively cool.

Ceramics can be prepared by heating a slurry of a powder of the inorganic substance in water to a very high temperature under high pressure. This process, called *sintering*, bonds the particles to each other, thus producing the finished ceramic. Sintering is a relatively easy process to perform, but the resulting particle size is irregular, so the solid may contain cracks, spaces, and imperfections. Imperfections in the interior of a solid ceramic can be impossible to detect, but they make weak spots in the substance that can cause the ceramic piece to fail. To avoid these problems, the *sol-gel process* is frequently employed for modern ceramics that are used in structural applications. The sol-gel process produces particles of nearly uniform size that are much more likely to produce a solid ceramic without gaps or cracks.

The first step in the sol-gel process is the preparation of an alkoxide of the metal or metalloid that is going to be made into the ceramic. This can be illustrated with the yttrium(III) ion, which is used to prepare a yttrium-oxygen ceramic:

$$Y + 3CH_3CH_2OH \longrightarrow Y(OCH_3CH_2)_3 + 3H^+$$
yttrium alkoxide

The resulting metal alkoxide is dissolved in alcohol. Water is then added to generate the metal hydroxide and regenerate the alcohol:

$$Y(OCH_3CH_2)_3 + 3H_2O \longrightarrow Y(OH)_3 + 3CH_3CH_2OH$$
yttrium alkoxide   yttrium hydroxide   ethanol

The metal hydroxide, once formed, undergoes condensation polymerization to form a chain with bridging oxygen atoms between the metal atoms:

$$\underset{\displaystyle HO-Y-OH}{\overset{\displaystyle OH}{|}} \quad \underset{\displaystyle HO-Y-OH}{\overset{\displaystyle OH}{|}} \longrightarrow \underset{\displaystyle HO-Y-O-Y-OH}{\overset{\displaystyle OH \quad OH}{|\qquad|}} + H_2O \text{ (water)}$$

Just as in condensation polymerization between organic molecules, a small molecule (water, in this case) is produced by the combination of atoms from the two monomers. Repeated condensation produces an insoluble chain of metal atoms connected by oxygen

Student Note: The sintering of bronze in the manufacture of bearings takes advantage of the inherent imperfections produced by the process. Porous bearings are sometimes preferred because the porosity allows a lubricant to flow throughout the material.

Student Note: The "sol" in sol-gel refers to a colloidal suspension of individual particles. The "gel" refers to the suspension of the resulting polymer.

atoms. The polymer chain may contain branches (i.e., it may be three dimensional) because all three of the hydroxide groups attached to the central metal ion are identical (and are thus equally likely to react). The suspension of the metal oxide-hydroxide polymer is called a *gel*. The gel is carefully heated to remove the liquid, and what remains is a collection of tiny, remarkably uniformly sized particles. Sintering of material produced by the sol-gel process produces a ceramic with relatively few imperfections.

Composite Materials

A *composite material* is made from two or more substances with different properties that remain separate in the bulk material. Each contributes properties to the overall material, though, such that the composite exhibits the best properties of each of its components. One of the oldest examples of a human-made composite material is the formation of bricks from mud and straw, a process that dates back to biblical times.

Modern composite materials commonly include reinforcing fibers, analogous to the straw in mud bricks, in a polymer matrix called a resin. Fiberglass, Kevlar, or carbon fibers can be used to give strength to the composite, whereas polyester, polyamide, or epoxy forms the matrix that holds the fibers together. These types of composites are called *polymer matrix composites*. Composites made from a metal and a ceramic, organic polymer, or another metal are called *metal matrix composites*. Carbide drill tips are made with a combination of softer cobalt and tougher tungsten carbide. Toyota has used a metal matrix composite in the engine block of some of its cars, and some bicycles are made with aluminum metal matrix composites. Composites made of a ceramic as the primary reinforcing material accompanied by organic polymers are called *ceramic matrix composites*. A common biological ceramic matrix composite is bone. Bones consist of reinforcing fibers made of collagen (a protein) surrounded by a matrix of hydroxyapatite $[Ca_5(PO_4)_3OH]$.

Another type of composite material that does not fit neatly into one of these categories is a reinforced carbon-carbon composite (RCC), which consists of carbon fibers in a graphite matrix. To prevent oxidation, a silicon carbide coating is applied. RCC is used on the nose cones of missiles, but it is most commonly known as the coating on the nose of the space shuttle. It is believed that the Space Shuttle *Columbia* disaster in 2003 occurred because the ceramic tiles and RCC panel material were both damaged on liftoff by the impact of falling foam insulation. The extreme heat of reentry into the atmosphere damaged the orbiter's structure through the breaches in the ceramic tiles and the RCC panel, and the orbiter and all seven astronauts on board were lost.

Carbon fibers can be woven into fabrics and threads to be used as the structural components of vehicles and for sporting equipment such as bicycles, tennis rackets, and skateboards.

12.3   Liquid Crystals

Liquid crystals are substances that exhibit properties of both liquids, such as the ability to flow and to take on the shape of a container, and those of crystals, such as a regular arrangement of particles in a lattice. Some substances exhibit liquid crystal behavior when they are melted from a solid. As the temperature increases, the solid liquefies but retains some order in one or two dimensions. At a higher temperature, the ordered liquid becomes a more conventional liquid in which there is no consistent orientation of the molecules.

Liquids are *isotropic,* because their properties are independent of the direction of testing. Because particles in a liquid are free to rotate, there are molecules present in every possible orientation, so the bulk sample gives the same measurements regardless of the direction in which the measurements are taken or observations are made. Liquid crystals are *anisotropic* because the properties they display depend on the direction (orientation) of the measurement. A box full of pencils, all neatly aligned, is an analogy for anisotropy, because the pencils are long and narrow. Looking at the pencils end on is different than looking at them from the side. What you see depends on the direction from which you view the pencils. Similarly, what you see when you look at anisotropic molecules depends on the direction from which you view them.

Frederick Reinitzer discovered the first compound to exhibit liquid crystal behavior in 1888. Cholesteryl benzoate was observed to form an ordered liquid crystalline phase when melted, and this liquid crystalline phase became an ordinary liquid at higher temperatures:

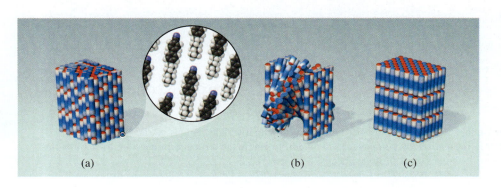

Cholesteryl benzoate

The structure of cholesteryl benzoate is fairly rigid due to the presence of the fused rings and sp^2-hybridized carbon atoms. The molecule is relatively long compared to its width, too. Rigidity and this particular shape makes it possible for the cholesteryl benzoate molecules to arrange themselves in an orderly manner, in much the way that pencils, chopsticks, or tongue depressors can be arranged.

There are three different types of ordering (called nematic, smectic, and cholesteric) that a liquid crystal can adopt (see Figure 12.10). A ***nematic*** liquid crystal contains molecules ordered in one dimension only. Nematic molecules are all aligned parallel to one another, but there is no organization into layers or rows. A ***smectic*** liquid crystal is ordered in two dimensions. Smectic molecules are aligned parallel to one other and are further arranged in layers that are parallel to one another. ***Cholesteric*** liquid crystal molecules are parallel to each other within each layer, but each layer is rotated with respect to the layers above and below it. The layers are rotated because there are repulsive intermolecular forces between layers.

Liquid crystalline substances were curiosities for many years but have found applications in many fields. Liquid crystal displays (LCDs) in calculators, watches, and laptop computers are possible, for example, because polarized light can be transmitted through liquid crystals in one phase but not transmitted through another (Figure 12.11). Polarized light is produced by passing ordinary light (which oscillates in all directions perpendicular to the beam) through a filter that allows only the light waves oscillating in one particular plane to pass through. This plane-polarized light can be rotated by a twisted liquid crystal such that it is allowed to pass through another filter when a voltage is applied to the liquid. If the light passes through the liquid when the voltage is off, though, the light will *not* pass through. This principle can be used in calculator and watch displays because each digit is made up of no more than the seven segments shown in Figure 12.11, and the specific digit depends on which segments are bright and which are dark. LCDs in laptop computers and televisions are more sophisticated, but still operate on the same basic principle.

Student Note: Recall from Chapter 9 that there is free rotation about carbon-carbon single bonds (i.e., bonds between sp^3-hybridized carbon atoms), but not about carbon-carbon double bonds (i.e., bonds between sp^2-hybridized carbon atoms).

Figure 12.10 (a) Nematic, (b) cholesteric, and (c) smectic liquid crystals.

Figure 12.11 Liquid crystal display. When the current is on, the liquid crystal molecules are aligned and polarized light cannot pass through. When the current is off, the molecules are not aligned and light can pass through.

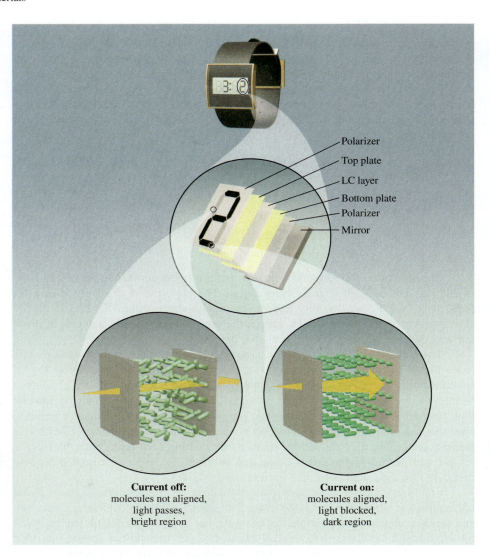

Polarizer
Top plate
LC layer
Bottom plate
Polarizer
Mirror

Current off:
molecules not aligned,
light passes,
bright region

Current on:
molecules aligned,
light blocked,
dark region

Figure 12.12 LCD thermometer.

Stockbyte/Getty Images

Student Note: A material that changes color as the temperature changes is called *thermochromic*.

Liquid crystals can also be used in thermometers (Figure 12.12). The spacing between crystal layers depends on temperature, and the wavelength of light reflected by the crystal depends on this spacing. Thus, the color reflected by the liquid crystal will change with temperature, and this color can be used to indicate the temperature to which the liquid crystal is exposed.

Sample Problem 12.4 lets you practice determining whether or not a molecule might exhibit liquid crystal properties.

SAMPLE PROBLEM 12.4

Would the following molecule make a good liquid crystal? Why or why not?

$$CH_3-\overset{\overset{O}{\|}}{C}-\underset{}{\bigcirc}-\overset{\overset{O}{\|}}{C}-\underset{}{\bigcirc}-CH_2-CH_2-CH_2-CH_2-CH_2-CH_3$$

Strategy Cholesteryl benzoate exhibits liquid crystal properties because it has structurally rigid regions (fused rings and sp^2-hybridized carbon atoms) and because it is relatively long compared to its width. Examine the structure in question to see if it has rigid regions and to see if it is longer in one dimension than in another.

Setup Carbon atoms that are sp^2-hybridized contribute to the rigidity of a molecule's shape.

Solution The left-hand portion of the molecule contains an sp^2-hybridized carbon atom, a benzene ring, another sp^2-hybridized carbon atom, and another benzene ring. These features are relatively rigid and should allow the molecule to maintain its shape when heated. The chain of CH_2 groups ending in a CH_3 group is not rigid due to free rotation about the carbon-carbon single bonds (i.e., sp^3-hybridized carbon atoms). The overall shape of the molecule is longer than it is wide, though, which combined with a large portion that is rigid should make the molecule a good candidate for liquid crystal behavior.

THINK ABOUT IT

Double bonds in a molecular structure indicate the presence of pi bonds. Recall that it is the pi bonds that restrict rotation about bonds in a molecule [◄◄ Section 9.5] and lend rigidity to the structure.

Practice Problem **A**TTEMPT Based on its structure, explain why you would expect this molecule to exhibit liquid crystal properties.

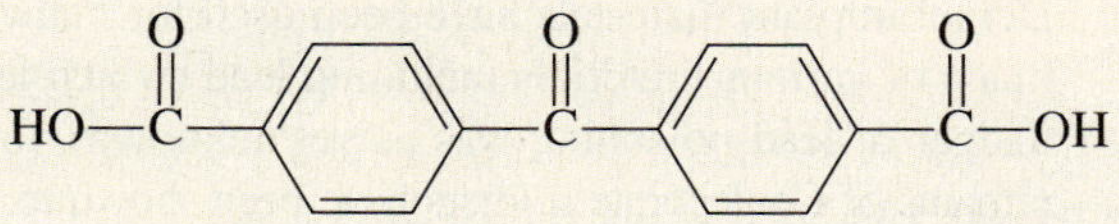

Practice Problem **B**UILD Would you expect this molecule to exhibit liquid crystal properties? Explain your answer.

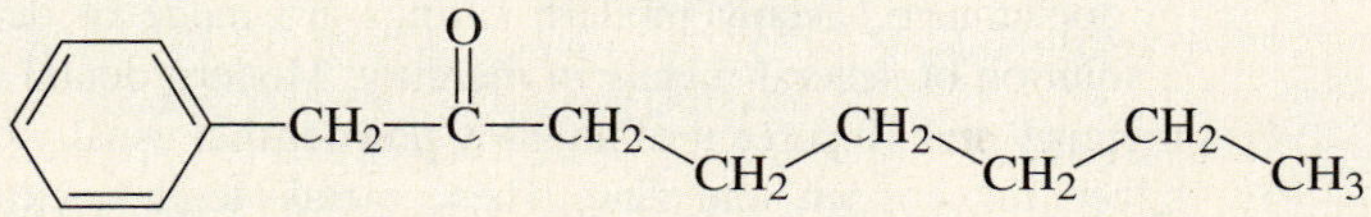

Practice Problem **C**ONCEPTUALIZE This compound does not possess liquid crystal properties. What changes are needed (e.g., addition of a group) to convert it to a liquid crystal?

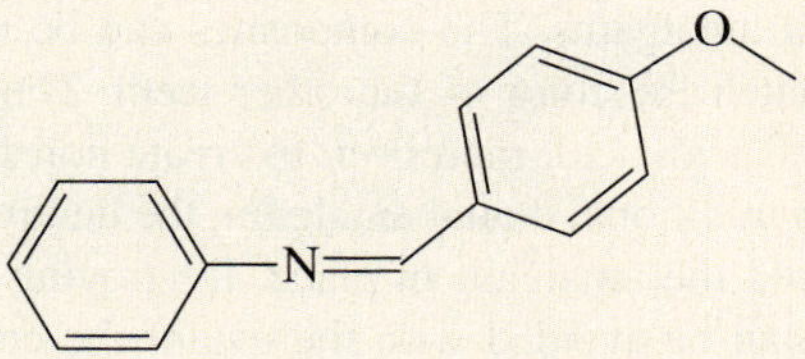

CHECKPOINT – SECTION 12.3   Liquid Crystals

12.3.1 Which of the following is a good analogy for anisotropy? (Select all that apply.)

a) Glass of water   d) Box of rice

b) Bucket of ice cubes   e) Box of macaroni

c) Box of spaghetti

12.3.2 What characteristics make a molecule likely to exhibit liquid crystal properties? (Select all that apply.)

a) Long narrow shape   d) Low molar mass

b) Flexibility   e) High molar mass

c) Rigidity

12.4 Biomedical Materials

Many modern materials are finding uses in medical applications. Replacement joints, dental implants, and artificial organs all contain modern polymers, composites, and ceramics. To function successfully in a biomedical application, though, a material must first be compatible with the living system. The human body very easily recognizes foreign substances and attacks them in an effort to rid them from the body. Thus, a biomaterial must be designed with enough similarity to the body's own systems that the body will accept the material as its own. Additionally, if the polymer, ceramic, or composite contains leftover chemicals from its manufacture, these contaminants may lead to adverse reactions with the body over the lifetime of the medical implant. It is important, therefore, that the substance be as pure as possible.

The physical properties of a biomedical material are important, too, because the longer the material lasts, the less often the medical device has to be replaced during the lifetime of the patient. Most biomedical materials must possess great strength and flexibility to perform in the body. For example, the materials in artificial joints and heart valves must be able to flex many times without breaking. Materials used in dental implants must show great hardness, moreover, so as not to crack during biting and chewing. It takes years of research to develop successful biomedical materials that meet these needs.

Dental Implants

Dental implant materials have been used for many years. The oldest dental fillings were made of various materials including lead (which fell out of favor due to its softness—the danger of lead poisoning was as yet unknown), tin, platinum, gold, and aluminum. The remains of Confederate soldiers have been shown to include fillings made of a lead-tungsten mixture, probably from shotgun pellets; tin-iron; a mercury amalgam; gold; and even radioactive thorium. (The dentist probably thought he was using tin, which is similar in appearance.) Many modern fillings are made of dental *amalgam* [▶▶ Chapter 19], a solution of several metals in mercury. Modern dental amalgam consists of 50 percent mercury and 50 percent of an alloy powder that usually contains (in order of abundance) silver, tin, copper, and zinc. These metals tend to expand slightly with age, causing fissures and cracks in the tooth that may require further intervention (e.g., crowns, root canals, or tooth replacement). Some people consider amalgam fillings to be unsightly, too.

Dental composite materials are now used that have several advantages over traditional amalgams. The composites can be made in a wide range of colors, for example, to match the color of the other teeth. The existing healthy portion of the tooth can be etched with acid, moreover, to create pores into which the composite material can bond. With traditional dental amalgam, the dentist must create indentations in the healthy tooth to hold the amalgam in place. Destroying a portion of the healthy tooth is undesirable and can be avoided with the use of the composite.

The dental composite consists of a matrix (made from a methacrylate resin) and a silica filler. The composite material is applied to the cavity in a putty-like consistency and then is dried and polymerized by a photochemical reaction initiated by light of a particular wavelength. Because the light does not penetrate very far into the composite, the thickness of the applied composite is critical. If the layer of composite is too thick, some of the composite will remain soft. A properly constructed dental composite filling will last more than 10 years. Over time, though, composite fillings tend to shrink, leaving breaches that can cause leakage, a situation that must be addressed to prevent further tooth decay.

Porcelain ceramic fillings and crowns are very common, but the ceramic has two disadvantages: it is very hard and can wear on the opposing teeth, and it is brittle and may crack if subjected to great force. Most dentists, therefore, do not recommend porcelain ceramic crowns and fillings for the molar teeth, which do the bulk of the grinding work during chewing.

A material used in dentistry must have properties that maximize both patient comfort and the lifetime of the implant. Dental fillings and tooth replacements must be resistant to acids, for example, because many foods (such as citrus fruits and soft drinks) contain acids directly. Any food containing carbohydrates can produce acids, though, if traces are left in the mouth because the bacteria that reside there consume carbohydrates, producing acids in the process. These acids eat away at the natural hydroxyapatite in teeth, creating caries (cavities). Materials like dental amalgam, gold, and dental composite resist attack by acids in the mouth.

A dental material should also have low thermal conductivity; that is, it should conduct heat poorly. This is especially important in applications where the implant material is in contact with the nerve inside the tooth. If the implant material conducts heat well, then whenever hot or cold foods come in contact with the implant, the hot or cold will be transmitted through the material to the nerve, causing pain. Metals are good thermal conductors, so dental amalgams must not touch nerves.

Finally, dental materials should resist wear (so they last a long time) and they should resist expansion and contraction as the temperature fluctuates. The close spacing and precision fits involved in dental fillings and implants would result in discomfort and possibly failure of the implant if the material expanded or contracted appreciably with changes in temperature.

Soft Tissue Materials

Burn victims who have lost large amounts of skin do not have enough cells to grow new skin, so a synthetic substitute must be used. The most promising artificial skin material is based on a polymer of lactic acid and glycolic acid. Both of these compounds contain an alcohol group ($-OH$) and a carboxy group ($-COOH$), so they can form a polyester copolymer in a condensation reaction that mirrors the formation of Dacron polyester (Section 12.1):

glycolic acid   lactic acid

This copolymer forms the structural mesh that supports the growth of skin tissue cells from a source other than the patient. Once the skin cells grow on the structural mesh, the artificial skin is applied to the patient, and the copolymer mesh eventually disappears as the ester linkages are hydrolyzed. During this treatment, the patients must take drugs that will suppress their immune systems, which ordinarily would attack these substances as foreign to prevent their bodies from rejecting the new skin.

Sutures, commonly known as stitches, are now made of the same lactic acid–glycolic acid copolymer as the structural mesh of artificial skin. Before, sutures were made of materials that did not degrade so they eventually had to be removed. This type of suture is still used in cases where the degrading suture is deemed a risk or there might be a need to reopen the incision at a later time. Biodegradable sutures are routinely used during operations on animals, too, which decreases the number of trips to the veterinarian, thereby reducing the stress on the pet and its owner.

When portions of blood vessels need to be replaced or circumvented in the body, it may or may not be possible to obtain a suitable vascular graft from another part of the patient's body. Dacron polyester (Section 12.1) can be woven into a tubular shape for this purpose, and pores can be worked into the material so that the graft can integrate itself into the tissue around it, as well as to allow capillaries to connect. Unfortunately, the body's immune system recognizes this as a foreign substance and blood platelets stick to the walls of the Dacron graft, causing clotting.

Artificial hearts and heart valves are potential solutions to the shortage of donor organs. Artificial hearts have not yet been perfected to the point that they can work on a permanent basis, but valve replacement is a common procedure that can successfully prolong a patient's life. The most common replacement heart valve is fixed in position by surrounding the valve with lactic acid–glycolic acid copolymer, the same polyester that is used in skin grafts and biodegradable sutures. The polymer is woven into a mesh ring that allows the body tissue to grow into the ring and hold it in place.

Researchers today are using *electrospinning* to place nanofibers directly onto wounds. (Nanofibers are products of nanotechnology; see Section 12.5.) Electrospinning uses a high-voltage electric field to create a jet of polymer solution that

Student Note: Hydrolysis is essentially the opposite of a condensation reaction.

transforms into a spiraling dry fiber only nanometers in diameter. The wound dressings applied by direct electrospinning have high rates of moisture transmission and bacterial resistance. Enzymes can be added to the solution containing the nanofiber polymer, and these enzymes can be released into the skin after the dressing is applied. Proteins can also be cross-linked onto the nanofiber, making possible many applications of the nanofiber wound dressing.

Artificial Joints

Joint replacements, such as artificial knees, elbows, and hips, require major surgery but have become relatively commonplace because suitable materials have become available for not only the bone replacement but also the contact surfaces. In knee replacement, for example, the goal is usually to replace worn-out cartilage surfaces (which may have resulted from arthritis or injury). It is possible to replace shattered bones that cannot heal back to their original state as well.

The materials originally used in joint replacements were ivory, metals, and glass, but they are fast being replaced by polymeric plastics and ceramics. Polymethyl methacrylate (PMMA), which was the subject of Practice Problem 12.1B in Section 12.1, and polyethylene have been used in many total joint replacements. However, these joints tend to fail over time when they are implanted in younger, more active patients, so efforts are under way to improve the wear quality of the PMMA and polyethylene parts.

Although metallurgical research provided new substances that were much stronger and less likely to break, including titanium-based alloys, total joint replacements were predicted to last no more than 20 years in a patient before needing to be replaced again. Ultrahigh-molecular-weight polyethylene, treated with radiation to initiate cross-linking of the polymer chains and stored under inert nitrogen gas to stop free-radical damage to the chain from the radiation, is now commonly used in many replacement joints, with the expectation of improved wear. It remains to be seen whether ultrahigh-molecular-weight polyethylene will prolong the life of these devices, since many of the joints have been in service in patients for less than a decade.

Other surfaces have also been researched to reduce wear. Metal-on-metal bearing surfaces suffered early setbacks due to the manufacturing tolerances of the cobalt-based alloy used. As the bearing surfaces wore, they produced metal sludge, and high friction between components caused loosening of the joint. Modern metal-on-metal bearing surfaces have improved as the manufacturing has become more consistent and produced components lower in friction.

Ceramic-on-ceramic bearing surfaces have also been developed. Some ceramics have been prone to cracking, but many have shown great promise for long-lasting use. When a ceramic component in a replacement joint does crack, however, it can be disastrous because the entire joint may need to be replaced and further healthy bone may need to be sacrificed to install the new replacement joint.

Despite the progress made so far on developing safe, durable, and useful artificial joints, much work remains to be done. One long-term issue that has yet to be researched extensively, for example, is the effect of the release of metal cations from the joint into the body over the lifetime of the joint replacement. This may become a more urgent health issue as the lifetimes of the artificial joints increase and the ages at which patients get the replacements decrease.

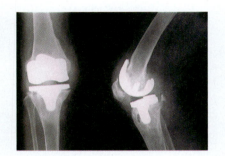

Jarva Jar/Shutterstock

12.5 Nanotechnology

Atoms and molecules have sizes on the order of a tenth of a nanometer (where 1 nm = 10^{-9} m) [◄◄ Section 2.2]. *Nanotechnology* is the development and study of such small-scale materials and objects. An atomic force microscope [Figure 12.13(a)] is one way to image the surface of a material at the atomic or molecular level.

A sharp-pointed stylus is moved over the surface of the substance under study, and the intermolecular forces between the material and the stylus force the stylus up

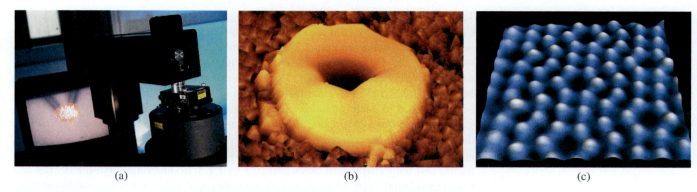

(a)   (b)   (c)

Figure 12.13 (a) Atomic force microscope (AFM). (b) AFM image of a yeast cell. (c) STM image of iron atoms arranged to display the Chinese characters for *atom* on a copper surface.

a: *Tek Image/Science Source;* b: *Dr. Andre Kempe/Oxford Scientific/Getty Images;* c: *Andrew Dunn/Alamy Stock Photo*

and down. The stylus is mounted on a probe that reflects a laser beam into a detector. From the differences in deflection of the laser beam, the electronics attached to the probe can create an image of the surface of the object [Figure 12.13(b)].

The scanning tunneling microscope (STM) works on a similar principle but only for samples that conduct electricity. The STM measures the peak and valley heights of the sample from the flow of electric current. The tip of the STM must move up and down to keep the current constant, and the up-and-down movements are translated into an image of the atomic and molecular structure of the substance [Figure 12.13(c)].

Graphite, Buckyballs, and Nanotubes

Carbon exists as several allotropes, one of which is graphite. Graphite consists of sheets of carbon atoms that are all sp^2-hybridized [◄◄ Section 9.4]:

> **Student Note:** Recall that *allotropes* are different forms of the same element [◄◄ Section 2.6].

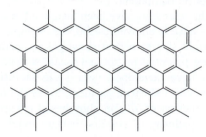

Intermolecular forces [◄◄ Chapter 11] hold the sheets together. Because of the delocalization of the electrons in the extended network of *p* orbitals perpendicular to the graphite sheet, graphite is an electrical conductor within its sheets, but not between them. Electrical conductivity is much more common for metals than it is for nonmetals like graphite.

Graphite and diamond were long thought to be the only two allotropes of carbon. In 1985, though, a compound consisting of 60 carbon atoms was isolated from graphite rods that had been vaporized with an intense pulse of laser light. The structure of the compound (C_{60}) was identical to that of a soccer ball, with 12 pentagonal and 20 hexagonal faces and carbon atoms at the corners of each face:

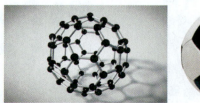 

(left): Angel Soler Gollonet/Shutterstock; (right): Jules Frazier/Photodisc/Getty Images

This pattern also resembles an architectural structure called a geodesic dome, so the C_{60} molecule was named *buckminsterfullerene* in honor of R. Buckminster Fuller, the designer of the geodesic dome. The nickname for C_{60} is the *buckyball.*

Since the discovery of C_{60}, elongated and elliptical cages of 70 and 80 carbon atoms have also been discovered. These molecules, called **fullerenes** as a group, have been proposed to exist in such exotic places as stars and interstellar media and have been observed in such mundane places as deposits of chimney soot.

A single sheet of graphite is known as **graphene.** A graphene is essentially a two-dimensional carbon crystal consisting of an array of six-membered rings. Graphenes are extraordinarily strong (hundreds of times stronger than steel), they conduct electricity, and their shapes can be customized by the introduction of five-membered and seven-membered rings at specific locations. Because of their unique combination of properties, graphenes currently are the subject of a great deal of research. Proposed uses of graphenes include replacement of copper interconnects in integrated circuits, production of transparent conduction electrodes—useful for touchscreens and liquid crystal displays—and ultrahigh-sensitivity detection devices.

Unless otherwise constrained, graphene sheets will wrap around on themselves forming cylinders with diameters on the scale of nanometers. Such cylinders are called **carbon nanotubes.** Two types of nanotube structures are known, as shown in Figure 12.14. The differences in arrangement of hexagons give the two kinds of nanotubes different electrical conductivities.

Carbon nanotubes of different diameters can nest inside each other, producing multiple-walled nanotubes. Most nanotubes are single-walled, though, and they can be as narrow as 1 nm in diameter. These nanotubes are remarkably strong—stronger, in fact, than a comparably sized piece of *steel*! Because nanotubes consist only of carbon atoms, they can come into close contact, thereby allowing intermolecular forces to hold large numbers of tubes together. This, in turn, makes it possible to align groups of tubes, potentially forming long strands and fibers. Lightweight fibers and strands with tremendous strength have many applications in such things as auto parts, medical implants, and sports equipment.

Besides carbon, other elements have been fashioned into nanotubes, too. Molybdenum sulfide nanotubes have been synthesized, for example, with openings between atoms that are just the right size to permit some gas molecules to pass through, but not others. This may have potential for the nanoscale storage of gases for fuel cells.

Many universities have developed large-scale facilities for nanotechnology research in collaboration with industrial partners. Since the lower limits of size have been reached in the production of silicon chip–based electronic circuits, nanotechnology may provide the pathway to circuits that are much smaller than have ever been imagined.

Student Note: An analogy is to take a piece of paper and roll it so that you have a tube.

Figure 12.14 Carbon nanotubes.

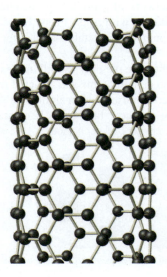

12.6   Semiconductors

Recall from Section 9.6 that bonds between atoms can be described as resulting from the combination of *atomic* orbitals to form *molecular* orbitals. In a bulk sample of metal containing many, many atoms, there are many, many molecular orbitals. These molecular orbitals are so close in energy that instead of forming individual bonding and antibonding orbitals for each pair of atomic orbitals that combine, they form a band of bonding levels and a band of antibonding levels. The energies of the bonding and antibonding bands depend on the energies of the atomic orbitals that combined to form them in the first place. As a result, the band structure of a bulk sample depends on the original atoms' energy levels. The band energies and the gaps (or lack of gaps) between the bonding band (called the **valence band**) and the antibonding band (called the **conduction band**) make it possible to classify a substance as an electrical conductor, a semiconductor, or an electrical insulator based on the behavior of the electrons in the bands (Figure 12.15).

The valence band is filled with the valence (bonding) electrons, whereas the conduction band is empty or only partially filled with electrons. It is the conduction band that allows electrons to move between atoms. The gap between the valence band and the conduction band can vary from nothing (for an electrical conductor) to a small amount (for a semiconductor) to a very large amount (for an electrical insulator). Thus, the size of the band gap determines the conduction behavior of the material.

Metals have no band gap, so they are good conductors of electricity. The valence band and conduction band actually overlap or are one and the same in metallic conductors, so there is no energy barrier to the movement of electrons from one atom to another in a metal. **Semiconductors** have a band gap, but it is relatively small, so there is limited movement of electrons from the valence band to the conduction band. **Nonconductors** (electrical insulators) have large band gaps, so it is nearly impossible to promote electrons from the valence band to the conduction band.

Silicon, germanium, and carbon in the form of graphite are the only elemental semiconductors at room temperature. All three of these elements are in Group 14 and have four valence electrons. (Tin and lead are also in Group 14, but they are metals, not semiconductors.) Other semiconductors consist of combinations of elements whose valence electron count totals 8. For example, gallium (Group 13) and phosphorus (Group 15) form a semiconductor because gallium contributes three valence electrons and phosphorus contributes five, giving a total of eight valence electrons. Semiconductors have also been formed between Group 12 elements (particularly Zn and Cd) and Group 16 elements.

The conductivity of a semiconductor can be enhanced greatly by **doping**, the addition of very small quantities of an element with one more or one fewer valence electron than the natural semiconductor. For the purpose of simplicity, we will consider a pure silicon semiconductor. Silicon is a Group 14 element, so it has four valence electrons per atom. A small (parts per million) amount of phosphorus (Group 15, five valence electrons) can be added, thus doping the silicon with phosphorus. Since each

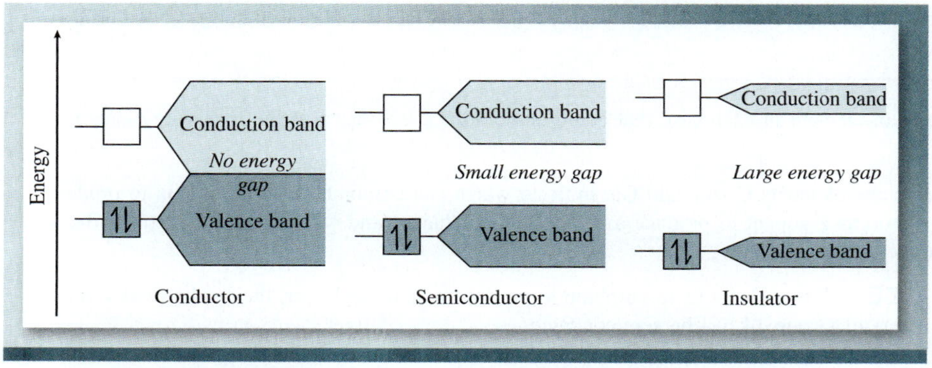

Student Note: The group number for a main group element is used to determine the number of valence electrons that the atom has [◄◄ Section 8.5].

Figure 12.15 Energy bands in metals (conductors), semiconductors, and insulators.

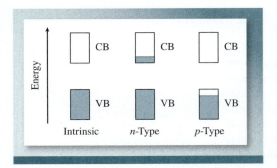

Figure 12.16 The effects of doping on the conduction of semiconductors.

phosphorus atom has an "extra" electron relative to the pure semiconductor, these extra electrons must reside in the conduction band, where they increase its conductivity. This type of doped semiconductor is called an ***n-type semiconductor*** because the semiconductivity has been enhanced by the addition of negative particles, the extra electrons.

It is also possible to dope a semiconductor with an element that has fewer valence electrons than the semiconductor itself. For example, a silicon semiconductor could be doped with small amounts (again, parts per million) of gallium (Group 13, three valence electrons). Now, instead of an "extra" electron, there is a "hole" (one less electron) in every place that a gallium atom has replaced a silicon atom. These holes are effectively positive charges (because each is the absence of an electron), so this type of material is called a ***p-type semiconductor.*** A *p*-type semiconductor has increased conductivity because the holes (which are in the valence band) move through the silicon rather than electrons. The energy required to move an electron from the valence band into a hole (also in the valence band) is considerably less than the energy needed to promote an electron from the valence band to the conduction band of a semiconductor (Figure 12.16). Thus, the movement of the holes results from the movement of charge, and conductivity is the movement of charge.

A combination of *p*-type and *n*-type semiconductors can be used to create a solar cell in which the holes and extra electrons in the semiconductors are at equilibrium in the dark. At equilibrium, the positive holes in the *n*-type layer have offset the movement of electrons into the *p*-type layer. Light falling on the solar cell causes the electrons that have entered the *p*-type layer to go back to the *n*-type layer, and other electrons to pass through a wire from the *n*-type layer to the *p*-type layer. This movement of electrons constitutes an electric current.

Sample Problem 12.5 lets you practice identifying combinations of elements that can exhibit semiconductor properties.

SAMPLE PROBLEM 12.5

State whether each of the following combinations of elements could form a semiconductor: (a) Ga-Se, (b) In-P, (c) Cd-Te.

Strategy Count the valence electrons in each type of atom. If they total eight for the two elements, then the combination will probably form a semiconductor.

Setup Locate each element in the periodic table. All of them are main group elements, so we can use their group numbers to determine the number of valence electrons each element contributes: (a) Ga is in Group 13, and Se is in Group 16. (b) In is in Group 13, and P is in Group 15. (c) Cd is in Group 12, and Te is in Group 16.

Solution (a) Ga has three valence electrons, and Se has six. This gives a total of nine, which is too many to form a semiconductor.

(b) In (three valence electrons) and P (five valence electrons) combine for a total of eight, so In-P should be a semiconductor.

(c) Cd (two valence electrons) and Te (six valence electrons) also combine for a total of eight, so Cd-Te should be a semiconductor, too.

THINK ABOUT IT

Semiconductors consisting of Group 13 and Group 15 elements are used in the contact layers of light-emitting diodes (LEDs).

Practice Problem Ⓐ**TTEMPT** Name an element in each case that could be combined with each of the following elements to form a semiconductor: (a) O, (b) Sb, (c) Zn.

Practice Problem Ⓑ**UILD** For each of the elements C, As, and Ga, indicate which can be combined with silicon to produce a semiconductor, which can be added to silicon as a dopant to produce a *p*-type semiconductor, and which can be added to silicon as a dopant to produce an *n*-type semiconductor.

Practice Problem Ⓒ**ONCEPTUALIZE** For two elements to combine to produce a semiconductor, their valence electrons must sum to eight. Explain how pure members of Group 14 can be semiconductors.

Diodes, which are electronic devices that restrict the flow of electrons in a circuit to one direction, work in essentially the opposite direction from solar cells. A light-emitting diode (LED) consists of *n*-type and *p*-type semiconductor layers placed in contact. When a small voltage is applied, the "extra" electrons from the *n*-type side combine with the holes in the *p*-type side, thus emitting energy (light) whose wavelength depends on the band gap. The band gap is different for different semiconductors (e.g., Ga-As versus Ga-P), so the specific combination of semiconductor materials can be chosen to make the band gap correspond to a desired color of light. LEDs are finding increasing use in applications previously dominated by traditional incandescent lightbulbs such as emergency exit signs, car tail lights, and traffic signals (Figure 12.17).

Figure 12.17 LED arrays in traffic signals. Many communities have replaced the red and green signals with LED arrays that are much more energy efficient than the incandescent bulbs used previously. The yellow lights are being replaced more slowly, however, because LED arrays are expensive and the short duration of yellow lights would not necessarily allow an overall cost saving when the energy efficiency is factored in.

David A. Tietz/Editorial Image, LLC

CHECKPOINT – SECTION 12.6   Semiconductors

12.6.1 Semiconductors are substances that

a) have a large band gap.

b) have a small band gap.

c) have no band gap.

d) do not conduct electricity.

e) conduct electricity only when combined with other elements.

12.6.2 Elements that are semiconductors at room temperature are

a) members of Group 2.

b) members of Group 13.

c) members of Group 14.

d) combinations of elements from Group 13 and Group 15.

e) combinations of elements from Group 2 and Group 16.

12.7   Superconductors

No ordinary electrical conductor is perfect. Even metals have some resistance to the flow of electrons, which wastes energy in the form of heat. *Superconductors* have no resistance to the flow of electrons and thus could be very useful for the transmission of electricity over the long distances between power plants and cities and towns. The first superconductor was discovered in 1911 by the Dutch physicist H. Kamerlingh-Onnes. He received the 1913 Nobel Prize in Physics for showing that mercury was a superconductor at 4 K, the temperature of liquid helium. Liquid helium is very expensive, however, so there were few feasible applications of a mercury superconductor.

Since that time, more superconductors have been discovered including a lanthanum-, barium-, copper-, and oxygen-containing *ceramic* compound, and a series of copper oxides ("cuprates"), all of which become superconducting below 77 K, the temperature of liquid nitrogen (a much more common and less expensive refrigerant than liquid helium). The temperature below which an element, compound, or material becomes superconducting is called the *superconducting transition temperature, T_c.* The higher the T_c, the more useful the superconductor.

By 1987, superconductivity was observed in another ceramic, yttrium barium copper oxide (YBCO) ($YBa_2Cu_3O_7$). Commonly called a "1-2-3 compound" because of the 1:2:3 ratio of yttrium to barium to copper, YBCO has a T_c of 93 K (above the temperature of liquid nitrogen). Although 93 K is still cold, it qualifies as quite warm in the world of superconductors, so YBCO is considered to be the first *high-temperature superconductor.* YBCO was first prepared by combining the three metal carbonates at high temperature (1000 to 1300 K). The copper ions are present in a mixture of +2 and +3 oxidation states:

$$4BaCO_3 + Y_2(CO_3)_3 + 4CuCO_3 + Cu_2(CO_3)_3 \longrightarrow 2YBa_2Cu_3O_7 + 14CO_2$$

YBCO must be sintered carefully to obtain the optimal superconductivity.

Student Note: The decomposition of a metal carbonate to the metal oxide and carbon dioxide gas is a very common reaction.

Figure 12.18 The Meissner effect. A magnet levitates above a YBCO pellet at the temperature of liquid nitrogen (−196°C).

ktsdesign/Shutterstock

Superconducting wires were first made from $Bi_2Sr_2CaCu_2O_8$, known as BSCCO-2212, a high-temperature superconductor with $T_c = 95$ K. It can be prepared in wire forms because it has layers of bismuth and oxygen atoms that YBCO does not have. BSCCO-2212 is so named because its formula consists of two bismuth, two strontium, one calcium, and two copper atoms. Other BSCCO compounds exist, too, and they all have a general formula of $Bi_2Sr_2Ca_nCu_{n+1}O_{2n+6}$. BSCCO-2223, for example, which has the formula $Bi_2Sr_2Ca_2Cu_3O_{10}$ (where $n = 2$), is a superconductor with $T_c = 107$ K.

Superconductors at or below T_c exhibit the ***Meissner effect***, the exclusion of magnetic fields. In Figure 12.18, for example, a magnet is shown levitating above a YBCO pellet, a superconductor at or below −180°C, its T_c.

Magnetic levitation is being researched for use in trains, because the amount of friction between the train and the tracks would be drastically reduced in a magnetically levitated ("maglev") train versus one running on ordinary tracks and wheels. The engineering challenge to a maglev train is that the superconductor must be kept at or below its T_c for the magnetic levitation to occur, and the highest temperatures obtained for superconducting substances are approximately 140 K—warm on the superconductivity scale but still very cold by ordinary standards.

Superconductivity in metals can be explained satisfactorily by BCS theory, first proposed by John Bardeen, Leon Neil Cooper, and John Robert Schrieffer in 1957. They received the Nobel Prize in Physics in 1972 for their work. BCS theory treats superconductivity using quantum mechanical effects, proposing that electrons with opposite spin can pair due to fundamental attractive forces between the electrons. At temperatures below T_c, the paired electrons resist energetic interference from other atoms and experience no resistance to flow. Superconductivity in ceramics has yet to be satisfactorily explained.

Student Note: You might be skeptical that two negatively charged electrons would attract each other, but there is evidence that all subatomic particles do exert some attractive forces on each other.

Chapter Summary

Section 12.1

- *Polymers* are made up of many repeating units, called *monomers,* that are linked together by covalent bonds.

- *Thermoplastic* polymers can be melted and reshaped, whereas *thermosetting* polymers assume their final shape as part of the chemical reaction that forms them in the first place.

- *Addition polymers* are formed via a radical reaction in which a pi bond in an alkene or alkyne is opened up to form new bonds to adjacent monomers.

- Polymer chains can be *cross-linked* to provide added durability and flexibility.

- *Elastomers,* such as natural rubber, can stretch and bend and then return to their original shape.

- *Copolymers* contain more than one type of monomer. Copolymers can be random, alternating, block, or graft, depending on the sequence of the different monomers and the way that the monomers are attached in the polymer.

- *Tacticity* describes the relative geometric arrangements of the groups attached to the chiral carbon atoms in vinyl polymers. In *isotactic* polymers, all the groups attached to the chiral carbon atoms have identical arrangements. In *syndiotactic* polymers, they alternate positions along the polymer chain. In *atactic* polymers, they have random arrangements.

- *Condensation polymers* form when two different groups on monomers react, thus producing a covalent bond between monomers and a small molecule (usually water) that is expelled. The two reacting groups may be on the same monomer (as in the amino acids that form polypeptides and proteins), or they may be on two different monomers (as in hexamethylenediamine and adipic acid, which form nylon 6,6). Most condensation polymers are really copolymers.

- Amines and organic acids form condensation polymers known as *polyamides*. Alcohols and organic acids form condensation polymers known as *polyesters*. Amino acids form polyamide polymers known as polypeptides. Long, straight-chain polypeptides are known as proteins.

- DNA and RNA are condensation polymers consisting of sugars, organic bases, and phosphate ions.

Section 12.2

- *Ceramics* are polymeric inorganic compounds that are hard and strong and have high melting points.

- Ceramics are *sintered* to bond the particles to each other and to close any gaps that may exist in the material.

- The sol-gel process is used to prepare ceramics for structural applications because it produces particles of nearly uniform size that are much more likely to produce a solid ceramic without gaps or cracks.

- *Composite materials* contain several components each of which contributes properties to the overall material. Modern composites include reinforcing fibers and a surrounding matrix. The reinforcing fibers may be made of polymers, metals, or ceramics.

Section 12.3

- *Liquid crystals* are substances that maintain an ordered arrangement in the liquid phase. Molecules containing structurally rigid regions (e.g., rings and multiple bonds) and molecules that are longer in one dimension than in another seem to form the best liquid crystals. Although ordinary liquids are *isotropic,* liquid crystals are *anisotropic,* meaning that their apparent properties depend on the direction from which we view them.

- Liquid crystals may be *nematic, smectic,* or *cholesteric,* depending on the type of ordering the molecules can adopt. Nematic liquid crystals contain order in only one dimension (parallel molecules), smectic liquid crystals contain order in two dimensions (parallel molecules in parallel layers), and cholesteric liquid crystals contain layers that are rotated with respect to one another. Within each cholesteric layer, the molecules are parallel.

Section 12.4

- Dental amalgam consists of several metals dissolved in an alloy with mercury. Dental composite materials consist of a methacrylate polymer in a silica matrix. The composite can be varied in color to match existing teeth, making the repair more aesthetically pleasing than metallic amalgam.

- Biodegradable *sutures* and artificial skin are made of synthetic polyesters that support the growth of tissue around the polymer and then hydrolyze into the body.

- Artificial joint replacements can be made of metal, ceramics, and organic polymers.

Section 12.5

- *Nanotechnology* is the study and development of objects with sizes on the order of a nanometer (one billionth of a meter).

- Pure carbon exists as diamond, graphite, *fullerenes,* and *carbon nanotubes.*

- Two-dimensional sheets of sp^2-hybridized carbon atoms are called *graphenes.*

Section 12.6

- Electrical conductors (e.g., metals) have no gap between their *valence band* and *conduction band,* so electrons are easily promoted from the valence to the conduction band. Semiconductors, which conduct electricity more than nonconductors (insulators) but less than metallic conductors, have a small gap. The gap in **nonconductors** is too large for electrons to be promoted from the valence to the conduction band.

- **Semiconductors** generally have a total of eight valence electrons per formula unit. Thus, they typically form between two Group 14 elements, one Group 13 element and one Group 15 element, or one Group 12 element and one Group 16 element.
- Semiconductors can be **doped** with an element with fewer valence electrons to produce a **p-type semiconductor,** or with an element with more valence electrons to produce an **n-type semiconductor.**
- p-Type and n-type semiconductors can be used together in solar cells and **diodes.**

Section 12.7

- **Superconductors** offer no resistance to the flow of electrons. The earliest known superconductors only exhibited the behavior at liquid helium temperature (4 K), but modern **high-temperature superconductors** conduct at temperatures as high as 140 K.
- The highest temperature at which a superconductor exhibits superconductivity is called the **superconducting transition temperature (T_c).**
- Superconducting substances exhibit the **Meissner effect,** the exclusion of magnetic fields in the substance's volume. This can be demonstrated by levitating a magnet above a superconductor.

Key Words

Addition polymerization, 584

Anisotropic, 596

Atactic, 590

Carbon nanotube, 604

Ceramic, 594

Cholesteric, 597

Composite material, 596

Condensation polymer, 590

Conduction band, 605

Copolymer, 587

Cross-link, 585

Diode, 607

Doping, 605

Elastomer, 586

Fullerene, 604

Graphene, 604

High-temperature superconductor, 607

Isotactic, 590

Isotropic, 596

Liquid crystal, 596

Meissner effect, 608

Monomer, 584

Nanotechnology, 602

Nematic, 597

Nonconductor, 605

n-Type semiconductor, 606

Polyamide, 590

Polyester, 591

Polymer, 584

p-Type semiconductor, 606

Semiconductor, 605

Sintering, 595

Smectic, 597

Superconducting transition temperature (T_c), 607

Superconductor, 607

Suture, 601

Syndiotactic, 590

Tacticity, 590

Thermoplastic, 584

Thermosetting, 584

Valence band, 605

Questions and Problems

Applying What You've Learned

Now that you know more about polymers, we can apply some of the concepts you've learned to ring-opening metathesis polymerization (ROMP), a polymerization reaction that utilizes the metathesis reaction introduced at the beginning of the chapter. The general mechanism for ROMP is as follows:

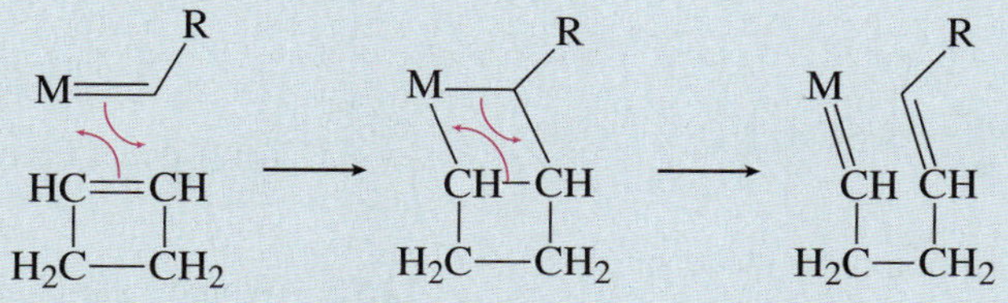

The pi bond from the carbon-carbon double bond in the ring of the monomer reacts with the metal-carbon double bond in the transition metal catalyst (M=CHR) to first form a metallacyclobutane complex. The metallacyclobutane complex then breaks up to form a new carbon-carbon double bond and a new metal-carbon double bond. The new carbon-carbon double bond is the first in the polymer chain, and the new metal-carbon double bond is available to react with additional monomers and thus grow the chain. As long as the metal from the catalyst is attached to the polymer via this metal-carbon double bond, the polymer is said to be "living" because it will continue to grow as more monomer is added. The process we consider is the polymerization of two cyclic alkenes (cyclobutene and 3-chlorocyclobutene) using ROMP with an appropriate transition metal catalyst (abbreviated as M=CHR):

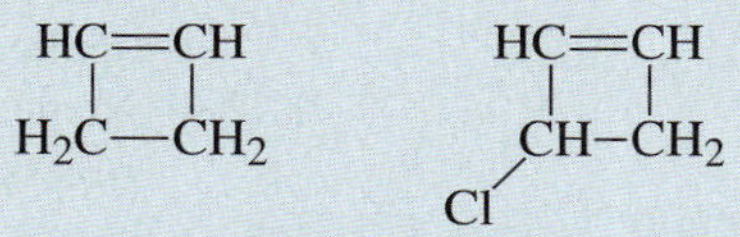

Cyclobutene   3-Chlorocyclobutene

Problems:

(a) What is the formula of the polymer formed by polymerizing first fifty cyclobutene molecules and then seventy 3-chloro-cyclobutene molecules [◀◀ Sample Problem 12.1]? (b) What type of polymer (see Table 12.2) is this [◀◀ Sample Problem 12.2]? (c) Would this polymer conduct electricity [◀◀ Sample Problem 12.3]?

SECTION 12.1: POLYMERS

Review Questions

12.1 Bakelite, the first commercially produced polymer, contains monomer units of phenol and formaldehyde. If an item made of Bakelite were broken, it could not be melted down and re-formed. Is Bakelite a thermoplastic or thermosetting polymer? Can Bakelite be recycled?

12.2 How are SBS rubber [Figure 12.6(a)] and nylon 6,6 similar? How are they different?

Conceptual Problems

12.3 What type of monomer unit is needed to make an electrically conducting organic polymer?

12.4 Draw the structure of the addition polymer formed by 1-butene:

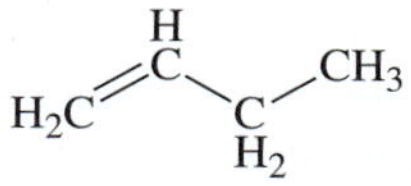

Show at least two repeating units.

12.5   Draw the structure of the alternating copolymer of these two compounds, showing at least two repeating units.

$$H_3C \quad CH_2 \qquad H-C\equiv C-H$$

(structures shown: a branched alkene and acetylene)

12.6   Draw the structures of the two monomer units in this polymer:

$$\left[-NH-(CH_2)_8-NH-\overset{O}{\overset{\|}{C}}-(CH_2)_4-\overset{O}{\overset{\|}{C}}- \right]_n$$

Is this polymer a condensation polymer or an addition polymer? Is it a polyester, a polyamide, or something else?

SECTION 12.2: CERAMICS AND COMPOSITE MATERIALS

Review Questions

12.7   What types of atoms and bonds are present in the polymer "backbone" in a ceramic containing a transition metal?

12.8   Describe two natural types of composite materials and the substances that they contain.

Conceptual Problems

12.9   What steps would be used to form a ceramic material from scandium and ethanol (CH_3CH_2OH) using the sol-gel process?

12.10   Amorphous silica (SiO_2) can be formed in uniform spheres from a sol-gel process using methanol (CH_3OH) solvent. List the chemical reactions involved in this process.

12.11   Bakelite, described in Review Question 12.1, is manufactured using materials like cellulose paper or glass fibers that are compressed and heated among the monomer units. Is Bakelite a simple polymer? If not, what classification best fits Bakelite?

SECTION 12.3: LIQUID CRYSTALS

Review Questions

12.12   Is a normal liquid isotropic or anisotropic? How is an anisotropic material different from an isotropic material?

12.13   Describe how nematic, smectic, and cholesteric liquid crystals are similar and different.

Conceptual Problems

12.14   Would each of these molecules be likely to form a liquid crystal? If not, why not?

(a) (structure shown)

(b) (structure shown)

(c) $H_2C=C=C$ (structure shown)

12.15   Would long-chain hydrocarbons or addition polymers of alkene molecules form liquid crystals? Why or why not?

12.16   Would an ionic compound form a liquid crystal? Why or why not?

SECTION 12.4: BIOMEDICAL MATERIALS

Review Questions

12.17   Name three major concerns that must be addressed when proposing any material to be used in medical applications inside the body.

12.18   Name a necessary chemical property and a necessary physical property of dental materials.

Conceptual Problems

12.19   Is the polymer commonly used in artificial skin material (shown here) an addition polymer or a condensation polymer? Is it a block, alternating, graft, or random copolymer?

$$\left(-OCH_2\overset{O}{\overset{\|}{C}}-OCH\overset{O}{\overset{\|}{C}}- \right)_n \quad \overset{|}{CH_3}$$

12.20   What are some advantages and disadvantages of the use of ceramics in replacement joint bearing surfaces relative to metals?

12.21   How do amalgam and composite tooth fillings differ in the ways they fail over time?

SECTION 12.5: NANOTECHNOLOGY

Review Questions

12.22   How does an STM measure the peak and valley heights along the surface of a sample?

12.23   Name four allotropic forms of carbon.

Conceptual Problems

12.24 How does an AFM differ from an STM? Why can graphite possibly be investigated with either instrument whereas diamond can only be investigated with an AFM?

12.25 What is the hybridization of the carbon atoms in a single-walled nanotube?

12.26 What structural or bonding aspect of graphite and carbon nanotubes allows their use as semiconductors, whereas diamond is an electrical insulator?

12.27 What type of intermolecular forces holds the sheets of carbon atoms together in graphite? What type of intermolecular forces hold carbon nanotubes together in strands and fibers?

SECTION 12.6: SEMICONDUCTORS

Review Questions

12.28 How do the band gaps differ among conductors, semiconductors, and insulators?

12.29 The only elemental semiconductors (Si, Ge, and graphite) are members of Group 14. Why are Sn and Pb, which are also in Group 14, not classified as semiconductors?

Problems

12.30 State whether each combination of elements would be expected to produce a semiconductor: (a) Ga and Sb, (b) As and N, (c) B and P, (d) Zn and Sb, (e) Cd and S.

12.31 What type of semiconductor would be formed (a) if Si were doped with Sb and (b) if Ge were doped with B?

12.32 Describe the movement of electrons in a solar battery that uses *p*-type and *n*-type semiconductors when light shines on the solar cell.

SECTION 12.7: SUPERCONDUCTORS

Review Questions

12.33 Why is YBCO called a "high-temperature" semiconductor when it exhibits superconductivity only below −180°C, which is very cold?

12.34 How might the Meissner effect be used in a real-world application? What are the engineering challenges to this application with the current superconductors?

Conceptual Problems

12.35 What is the molecular formula of BSCCO-2201?

12.36 Superconductivity at room temperature is sometimes a subject of science fiction or fantasy. For example, in the movie *Terminator 2: Judgment Day,* a cyborg has a CPU that contains a room-temperature superconductor. How might a computer with such a CPU differ in design from current computers?

12.37 What types of bonding (covalent, ionic, network, metallic) are present in a plastic polymer? In a ceramic material?

12.38 Draw representations of isotactic, syndiotactic, and atactic polyacrylonitrile (for the monomer unit, see Table 12.1).

12.39 What are the electron configurations of the copper ions in YBCO? Are both ions expected oxidation states for copper?

ADDITIONAL PROBLEMS

12.40 Draw representations of block copolymers and graft copolymers using the symbols A and B for two types of monomer units. Are the categories exclusive? Using a third monomer type, see if you can draw a representation of a polymer that is both a block and a graft copolymer.

12.41 Would the metal alkoxide in the sol-gel process be acidic or basic?

Engineering Problems

12.42 Would the compound shown form a liquid crystal? Why or why not?

12.43 The semiconductor band gap energy determines the color of an LED. Knowing what you learned in Chapter 6 about the energies of photons of different colors of light, which semiconductor would have the larger band gap: the one in a green exit sign's LEDs, or the one in a red traffic signal?

Biological Problems

12.44 Draw the structure of an alternating copolymer of these two amino acids, showing at least two repeating units:

12.45 Fluoride ion is commonly used in drinking water supplies and in toothpaste (in the form of sodium fluoride) to prevent tooth decay. The fluoride ions replace hydroxide ions in the enamel mineral hydroxyapatite [$Ca_5(PO_4)_3OH$, the same mineral as in bone], forming fluoroapatite [$Ca_5(PO_4)_3F$]. Why does replacing the hydroxide ion with fluoride ion prevent tooth decay? How does this relate to the properties of materials used in artificial fillings?

Standardized-Exam Practice Problems

Physical and Biological Sciences
These questions are not based on a passage.

1. Which of the following would form an *n*-type semiconductor: (i) doping Ge with As, (ii) doping Si with B, (iii) doping graphite with P?

   a) i, ii, and iii
   b) i and iii
   c) i and ii
   d) iii only

2. What is unusual about the copper in YBCO?

   a) It exists as a mixture of +1 and +2 oxidation states.
   b) It exists as a mixture of +1 and 0 oxidation states.
   c) It exists as a mixture of +1 and +3 oxidation states.
   d) It exists as a mixture of +2 and +3 oxidation states.

3. A semiconductor can be produced by the combination of two elements with a total of how many valence electrons?

   a) 6
   b) 8
   c) 12
   d) 16

4. Metals are good conductors of electricity because they

   a) have no band gap.
   b) have a large band gap.
   c) have a small band gap.
   d) combine with nonmetals to form semiconductors.

Answers to In-Chapter Materials

Answers to Practice Problems

12.1A   12.1B   12.2A

–(CH$_2$CCl$_2$)–

12.2B Condensation.

12.3A —CH=C(CH₂CH₃)—CH=C(CH₂CH₃)—CH=C(CH₂CH₃)—

—CH₂—CH(CH₂CH₃)—CH₂—CH(CH₂CH₃)—CH₂—CH(CH₂CH₃)—

The two structures are similar in that they both have ethyl groups on alternating carbons. They are different in that poly(1-butyne) is a conjugated system and poly(1-butene) is not. The first structure is more likely to be an electrically conducting polymer. **12.3B** Addition polymerization of alkynes results in conjugated double bonds—giving the polymer the rigidity of structure necessary for it to be electrically conducting. **12.4A** This molecule would likely exhibit liquid crystal behavior because it is long, thin, and has conjugated rings that limit free rotation. **12.4B** This molecule lacks the rigidity needed for liquid crystal behavior. **12.5A** (a) Elements with two valence electrons (e.g., Be, Sr, Cd), (b) elements with three valence electrons (e.g., B, Ga, In), (c) elements with six valence electrons (e.g., O, Se, Te). **12.5B** Carbon can be combined with silicon to produce a semiconductor. Arsenic can be used to dope silicon to produce an *n*-type semiconductor, and gallium can be used to dope silicon to produce a *p*-type semiconductor.

Answers to Checkpoints

12.1.1 a. **12.1.2** d. **12.3.1** c. **12.3.2** a, c. **12.6.1** b. **12.6.2** c.

CHAPTER 13

Physical Properties of Solutions

Seawater covers most of Earth's surface, but the salt in seawater makes it unsuitable for drinking.

Purestock/SuperStock

Desalination of Seawater

Roughly 71 percent of Earth's surface is covered by water, but more than 99 percent of that is seawater and is not suitable for most uses by humans. In recent decades, with global population growth and increasing freshwater scarcity, desalination of seawater has become an important part of maintaining an adequate freshwater supply in densely populated coastal regions. Historically, desalination has been done by thermal distillation, a practice that requires a great deal of energy. In recent decades, though, reverse osmosis has become an economically competitive alternative. Reverse osmosis uses high pressure to push seawater through a semipermeable membrane, through which water molecules pass, but salts and other dissolved substances do not.

Raquel Lonas/Moment/Getty Images

A relatively new technology, *forward* osmosis, uses the osmotic pressure of a very concentrated solution to draw water across a membrane, with no need for high-pressure pumping. The concentrated solution is known as the draw solution or simply the *draw*. For purposes of desalination, the purified water must then be separated from the draw—and the draw recovered for subsequent reuse. Although forward osmosis was initially thought to be a method of desalination that required far less energy than reverse osmosis, it turns out that a great deal of energy is required to separate the water from the draw—something typically done by reverse osmosis.

The latest news of forward osmosis is the development of an "osmotic sponge." Instead of using a concentrated aqueous solution as the draw, the sponges use *hydrogels*—polymers that absorb volumes of water many times their own. One of the most promising features of the osmotic sponges is that certain hydrogels can be made to release the water they have absorbed by being warmed to just above room temperature.

The development of technologies such as reverse osmosis and forward osmosis for desalination of water requires an understanding of the *properties of solutions*.

At the end of this chapter, you will be able to solve a series of problems related to concentrations of aqueous solutions [►► Applying What You've Learned, page 654].

13.1   Types of Solutions

As we noted in Section 1.2, a solution is a homogeneous mixture of two or more substances. Recall that a solution consists of a *solvent* and one or more *solutes* [◄◄ Section 4.1]. Although many of the most familiar solutions are those in which a solid is dissolved in a liquid (e.g., saltwater or sugar water), the components of a solution may be solid, liquid, or gas. The possible combinations give rise to seven distinct types of solutions, which we classify by the original states of the solution components. Table 13.1 gives an example of each type.

In this chapter, we focus on solutions in which the solvent is a liquid, and the liquid solvent we encounter most often is water. Recall that solutions in which water is the solvent are called *aqueous* solutions [◄◄ Section 4.1].

Solutions can also be classified by the amount of solute dissolved relative to the maximum amount that can be dissolved. A ***saturated solution*** is one that contains the maximum amount of a solute that will dissolve in a solvent at a specific temperature. The amount of solute dissolved in a given volume of a saturated solution is called the **solubility**. It is important to realize that solubility refers to a specific solute, a specific solvent, and a specific *temperature*. For example, the solubility of NaCl in water at 20°C is 36 g per 100 mL. The solubility of NaCl at another temperature, or in another solvent, would be different. An ***unsaturated solution*** is one that contains less solute than it has the capacity to dissolve. A ***supersaturated solution,*** on the other hand, contains *more* dissolved solute than is present in a saturated solution (Figure 13.1). It is generally not stable, and eventually the dissolved solute will come out of solution. An example of this phenomenon is shown in Figure 13.2.

> **Student Note:** The term *solubility* was also defined in Section 4.2.

TABLE 13.1	Types of Solutions		
Solute	**Solvent**	**State of Resulting Solution**	**Example**
Gas	Gas	Gas*	Air
Gas	Liquid	Liquid	Carbonated water
Gas	Solid	Solid	H_2 gas in palladium
Liquid	Liquid	Liquid	Ethanol in water
Liquid	Solid	Solid	Mercury in silver
Solid	Liquid	Liquid	Saltwater
Solid	Solid	Solid	Brass (Cu/Zn)

*Gaseous solutions can only contain gaseous solutes

(a)   (b)   (c)   (d)   (e)

Figure 13.1 (a) Many solutions consist of a solid dissolved in water. (b) When all the solid dissolves, the solution is unsaturated. (c) If more solid is added than will dissolve, the solution is saturated. (d) A saturated solution is, by definition, in contact with undissolved solid. (e) Some saturated solutions can be made into supersaturated solutions by heating to dissolve more solid, and cooling carefully to prevent crystallization.
(a-e) Charles D. Winters/McGraw Hill

(a)   (b)   (c)   (d)   (e)

Figure 13.2 In a supersaturated solution, (a) addition of a tiny seed crystal initiates crystallization of excess solute. (b)–(e) Crystallization proceeds rapidly to give a *saturated* solution and the crystallized solid.

(a-e) Charles D. Winters/McGraw Hill

13.2   The Solution Process

In Chapter 4, we learned guidelines that helped us predict whether or not an ionic solid is soluble in water. We now take a more general look at the factors that determine solubility. This discussion enables us to understand why so many ionic substances are soluble in water, which is a polar solvent, and it helps us to predict the solubility of ionic and molecular compounds in both polar and nonpolar solvents.

Intermolecular Forces and Solubility

The intermolecular forces that hold molecules together in liquids and solids play a central role in the solution process. When the solute dissolves in the solvent, molecules of the solute disperse throughout the solvent. They are, in effect, *separated* from one another and each solute molecule is *surrounded* by solvent molecules—a process known as ***solvation.*** The ease with which solute molecules are separated from one another and surrounded by solvent molecules depends on the relative strengths of three types of interactions:

1. Solute-solute interactions
2. Solvent-solvent interactions
3. Solute-solvent interactions

Unlike the intermolecular forces covered in Chapter 11, most of which were between the molecules, atoms, or ions of a pure substance, solute-solvent interactions are those that exist in a mixture of different substances. Because the components of a mixture can have different properties, there is a greater variety of intermolecular forces to consider. In addition to dispersion forces, which exist between all substances, dipole-dipole forces, which exist between polar molecules, and ion-ion forces, which exist between ions, solutions can exhibit the following intermolecular forces:

Intermolecular Forces	Example	Model
Ion-dipole. The charge of an ion is attracted to the partial charge on a polar molecule.	NaCl or KI in H_2O	
Dipole-induced dipole. The partial charge on a polar molecule induces a temporary partial charge on a neighboring nonpolar molecule or atom.	He or CO_2 in H_2O	
Ion-induced dipole. The charge of an ion induces a temporary partial charge on a neighboring nonpolar molecule or atom.	Fe^{2+} and O_2	

Student Note: A hemoglobin molecule contains four Fe^{2+} ions. In the early stages of O_2 binding, oxygen molecules are attracted to the Fe^{2+} ions by an ion-induced dipole interaction.

Figure 13.3 A molecular view of the solution process portrayed as taking place in three steps: First the solute and solvent molecules are separated (steps 1 and 2, respectively—both endothermic). Then the solvent and solute molecules mix (step 3—exothermic).

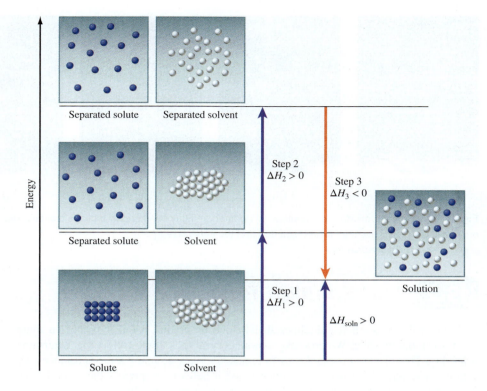

Student Note: Like the formation of chemical bonds [◀◀ Section 8.9], the formation of intermolecular attractions is exothermic. If that isn't intuitively obvious, think of it this way: It would require energy to *separate* molecules that are attracted to each other. The reverse process, the *combination* of molecules that attract each other, would give off an equal amount of energy [◀◀ Section 5.3].

For simplicity, we can imagine the solution process taking place in the three distinct steps shown in Figure 13.3. Step 1 is the separation of solute molecules from one another, and step 2 is the separation of solvent molecules from one another. Both of these steps require an input of energy to overcome intermolecular attractions, so they are *endothermic*. In step 3 the solvent and solute molecules mix. This process is usually *exothermic*. The enthalpy change for the overall process, ΔH_{soln}, is given by

$$\Delta H_{soln} = \Delta H_1 + \Delta H_2 + \Delta H_3$$

The overall solution-formation process is exothermic ($\Delta H_{soln} < 0$) when the energy given off in step 3 is greater than the sum of energy required for steps 1 and 2. The overall process is endothermic ($\Delta H_{soln} > 0$) when the energy given off in step 3 is less than the total required for steps 1 and 2. (Figure 13.3 depicts a solution formation that is endothermic overall.)

The saying "like dissolves like" is useful in predicting the solubility of a substance in a given solvent. What this expression means is that two substances with intermolecular forces of similar type and magnitude are likely to be soluble in each other. For example, both carbon tetrachloride (CCl_4) and benzene (C_6H_6) are nonpolar liquids. The only intermolecular forces present in these substances are dispersion forces [◀◀ Section 11.1]. When these two liquids are mixed, they readily dissolve in each other, because the attraction between CCl_4 and C_6H_6 molecules is comparable in magnitude to the forces between molecules in pure CCl_4 and those between molecules in pure C_6H_6. Two liquids are said to be ***miscible*** if they are completely soluble in each other in all proportions. Alcohols such as methanol, ethanol, and 1,2-ethylene glycol are miscible with water because they can form hydrogen bonds with water molecules:

| Methanol | Ethanol | 1,2-Ethylene glycol |

The guidelines listed in Tables 4.2 and 4.3 enable us to predict the solubility of a particular ionic compound in water. When sodium chloride dissolves in water, the ions are stabilized in solution by *hydration,* which involves ion-dipole interactions. In general, ionic compounds are much more soluble in polar solvents, such as water, liquid ammonia, and liquid hydrogen fluoride, than in nonpolar solvents. Because the molecules of nonpolar solvents, such as benzene and carbon tetrachloride, do not have a dipole moment, they cannot effectively solvate the Na^+ and Cl^- ions. The predominant intermolecular interaction between ions and nonpolar compounds is an ion-induced dipole interaction, which typically is much weaker than ion-dipole interactions. Consequently, ionic compounds usually have extremely low solubility in nonpolar solvents.

Sample Problem 13.1 lets you practice predicting relative solubilities using the principle of "like dissolves like."

> **Student Note:** *Solvation* refers in a general way to solute particles being surrounded by solvent molecules. When the solvent is water, we use the more specific term *hydration* [◄◄ Section 4.1].

SAMPLE PROBLEM 13.1

Determine for each solute whether the solubility will be greater in water, which is polar, or in benzene (C_6H_6), which is nonpolar: (a) bromine (Br_2), (b) sodium iodide (NaI), (c) carbon tetrachloride (CCl_4), and (d) formaldehyde (CH_2O).

Strategy Consider the structure of each solute to determine whether or not it is polar. For molecular solutes, start with a Lewis structure and apply the VSEPR theory [◄◄ Section 9.1]. We expect polar solutes, including ionic compounds, to be more soluble in water. Nonpolar solutes will be more soluble in benzene.

Setup

(a) Bromine is a homonuclear diatomic molecule and is nonpolar.

(b) Sodium iodide is ionic.

(c) Carbon tetrachloride has the following Lewis structure:

$$:\ddot{C}l:$$
$$:\ddot{C}l-C-\ddot{C}l:$$
$$:\ddot{C}l:$$

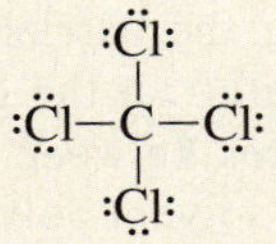

With four electron domains around the central atom, we expect a tetrahedral arrangement. A symmetrical arrangement of identical bonds results in a nonpolar molecule.

(d) Formaldehyde has the following Lewis structure:

$$\ddot{O}$$
$$H-C-H$$

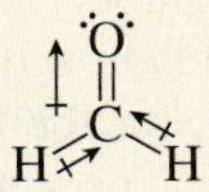

Crossed arrows can be used to represent the individual bond dipoles [◄◄ Section 9.2]. This molecule is polar and can form hydrogen bonds with water.

Solution

(a) Bromine is more soluble in benzene.

(b) Sodium iodide is more soluble in water.

(c) Carbon tetrachloride is more soluble in benzene.

(d) Formaldehyde is more soluble in water.

THINK ABOUT IT

Remember that molecular formula alone is not sufficient to determine the shape or the polarity of a polyatomic molecule. It must be determined by starting with a correct Lewis structure and applying the VSEPR theory [◄◄ Section 9.2].

(Continued on next page)

Practice Problem **A**TTEMPT Predict whether iodine (I_2) is more soluble in liquid ammonia (NH_3) or in carbon disulfide (CS_2).

Practice Problem **B**UILD Which of the following should you expect to be more soluble in benzene than in water: C_3H_8, HCl, I_2, CS_2?

Practice Problem **C**ONCEPTUALIZE The first diagram represents a closed system consisting of water and a water-soluble gas in a container fitted with a movable piston. Which of the other diagrams best represents the system when the piston is moved downward, decreasing the volume of the gas over the water? (Each diagram includes a thermometer indicating the temperature of the system.)

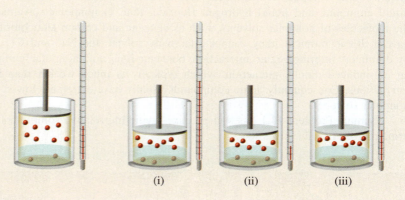

(i)   (ii)   (iii)

The Driving Force for Dissolution

Previously, we learned that the driving force behind some processes is the lowering of the system's potential energy. Recall the minimization of potential energy when two hydrogen atoms are 74 pm apart [◂◂ Section 9.3]. However, because there are substances that dissolve endothermically, meaning that the process increases the system's potential energy, something else must be involved in determining whether or not a substance will dissolve. That something else is *entropy*.

The ***entropy*** of a system is a measure of how *dispersed* or *spread out* its energy is. Consider two samples of different gases separated by a physical barrier. When we remove the barrier, the gases mix, forming a solution. Under ordinary conditions, we can treat the gases as ideal, meaning we can assume that there are no attractive forces between the molecules in either sample before they mix (no solute-solute or solvent-solvent attractions to break)—and no attractive forces between the molecules in the mixture (no solute-solvent attractions form). The energy of the system does not change—and yet the gases mix spontaneously. The reason such a solution forms is that, although there is no change in the energy of either of the original samples of gas, the energy possessed by each sample of gas spreads out into a larger volume. This increased dispersal of the system's energy is an increase in the *entropy* of the system. There is a natural tendency for entropy to increase; that is, for the energy of a system to become more dispersed—unless there is something preventing that dispersal. Initially, the physical barrier between the two gases prevented their energy from spreading out into the larger volume. It is the increase in entropy that drives the formation of this solution—and many others.

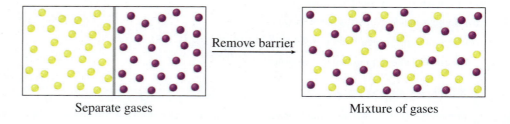

Separate gases   Remove barrier   Mixture of gases

Now consider the case of solid ammonium nitrate (NH_4NO_3), which dissolves in water in an *endothermic* process. In this case, the dissolution *increases* the potential energy of the system. However, the energy possessed by the ammonium nitrate solid

Why Are Vitamins Referred to as Water Soluble and Fat Soluble?

Vitamins can be categorized as either water soluble or fat soluble. An excess of a fat-soluble vitamin can be stored in the body, whereas excesses of most water-soluble vitamins are eliminated in the urine.

The difference between water-soluble and fat-soluble vitamins is their molecular structures. Water-soluble vitamins, such as vitamin C ($C_6H_8O_6$), have multiple polar groups that can interact with water to form hydrogen bonds:

Vitamin C

In general, fats are nonpolar molecules containing hydrocarbon chains. Fat-soluble vitamins, such as vitamin E, tend to be more hydrocarbon-like. They may have a polar group or two, but the molecule is predominantly nonpolar:

Vitamin E

Elimination of vitamin C causes the urine to be more acidic, which can inhibit the growth of some bacteria responsible for urinary tract infections. Taking larger doses than the official recommended daily amount of 60 mg is recommended by many physicians to lower the incidence of such infections. Fat-soluble vitamins cannot be eliminated in this way, and taking too much can lead to vitamin toxicity.

(See end-of-chapter problems 13.95, 13.103, and 13.105.)

---

spreads out to occupy the volume of the resulting solution causing the entropy of the system to increase. Ammonium nitrate dissolves in water because the favorable increase in the system's entropy outweighs the unfavorable increase in its potential energy. Although the process being endothermic is a barrier to solution formation, it is not enough of a barrier to prevent it.

In some cases, a process is so endothermic that even an increase in entropy is not enough to allow it to happen spontaneously. Sodium chloride (NaCl) is not soluble in a nonpolar solvent such as benzene, for example, because the solvent-solute interactions that would result are too weak to compensate for the energy required to separate the network of positive and negative ions in sodium chloride. The magnitude of the exothermic step in solution formation (ΔH_3) is so small compared to the combined magnitude of the endothermic steps (ΔH_1 and ΔH_2) that the overall process is *very* endothermic and does not happen to any significant degree despite the increase in entropy that would result.

CHECKPOINT – SECTION 13.2   The Solution Process

13.2.1 Which compounds do you expect to be more soluble in benzene than in water? (Select all that apply.)

a) SO_2

b) CO_2

c) Na_2SO_4

d) C_2H_6

e) Br_2

13.2.2 Which of the following solutions exhibits hydrogen bonding between the solute and solvent? (Select all that apply.)

a) $H_2(g)$ in $H_2O(l)$

b) $CH_3OH(l)$ in $H_2O(l)$

c) $CO_2(g)$ in $H_2O(l)$

d) $NH_3(g)$ in $H_2O(l)$

e) $NaCl(s)$ in $H_2O(l)$

13.3   Concentration Units

We learned in Chapter 4 that chemists often express concentration of solutions in units of *molarity*. Recall that molarity, *M*, is defined as the number of moles of solute divided by the number of liters of solution [◄◄ Section 4.5]:

$$\text{molarity} = M = \frac{\text{moles of solute}}{\text{liters of solution}}$$

Mole fraction, χ, which is defined as the number of moles of solute divided by the total number of moles, is also an expression of concentration [◄◄ Section 10.5]:

$$\text{mole fraction of component A} = \chi_A = \frac{\text{moles of A}}{\text{sum of moles of all components}}$$

Student Note: We have already used percent by mass to describe the composition of a pure substance [◄◄ Section 3.2]. In this chapter, we use percent by mass to describe *solutions*.

In this section, we learn about *molality* and *percent by mass,* two additional ways to express the concentration of a mixture component. How a chemist expresses concentration depends on the type of problem being solved.

Molality

Molality (m) is the number of moles of solute dissolved in 1 kg (1000 g) of solvent:

Equation 13.1	$\text{molality} = m = \dfrac{\text{moles of solute}}{\text{mass of solvent (in kg)}}$

Student Note: It is a common mistake to confuse molarity and molality. Molarity depends on the *volume* of the *solution*. Molality depends on the *mass* of the *solvent*.

For example, to prepare a 1-molal (1-*m*) aqueous sodium sulfate solution, we must dissolve 1 mol (142.0 g) of Na_2SO_4 in 1 kg of water.

Percent by Mass

The *percent by mass* (also called *percent by weight*) is the ratio of the mass of a solute to the mass of the solution, multiplied by 100 percent. Because the units of mass cancel on the top and bottom of the fraction, any units of mass can be used—provided they are used consistently:

Equation 13.2	$\text{percent by mass} = \dfrac{\text{mass of solute}}{\text{mass of solute + mass of solvent}} \times 100\%$

For example, we can express the concentration of the 1-*m* aqueous sodium sulfate solution as:

$$\text{percent by mass } Na_2SO_4 = \frac{\text{mass of } Na_2SO_4}{\text{mass of } Na_2SO_4 + \text{mass of water}} \times 100\%$$

$$= \frac{142.0 \text{ g}}{1142.0 \text{ g}} \times 100\% = 12.4\%$$

The term *percent* literally means "parts per hundred." If we were to use Equation 13.2 but multiply by 1000 instead of 100, we would get "parts per thousand"; multiplying by 1,000,000 would give "parts per million" or *ppm;* and so on. Parts per million, parts per billion, parts per trillion, and so forth, usually are used to express very low concentrations, such as those of some pollutants in the atmosphere or a body of water. For example, if a 1-kg sample of water is found to contain 3 μg (3×10^{-6} g) of arsenic, its concentration can be expressed in parts per billion (ppb) as follows:

$$\frac{3 \times 10^{-6} \text{g}}{1000 \text{ g}} \times 10^9 = 3 \text{ ppb}$$

Student Hot Spot

Student data indicate you may struggle with percent by mass. Access the eBook to view additional Learning Resources on this topic.

Sample Problem 13.2 shows how to calculate the concentration in molality and in percent by mass.

SAMPLE PROBLEM 13.2

A solution is made by dissolving 170.1 g of glucose ($C_6H_{12}O_6$) in enough water to make a liter of solution. The density of the solution is 1.062 g/mL. Express the concentration of the solution in (a) molality, (b) percent by mass, and (c) parts per million.

Strategy Use the molar mass of glucose to determine the number of moles of glucose in a liter of solution. Use the density (in g/L) to calculate the mass of a liter of solution. Subtract the mass of glucose from the mass of solution to determine the mass of water. Use Equation 13.1 to determine the molality. Knowing the mass of glucose and the total mass of solution in a liter, use Equation 13.2 to calculate the percent concentration by mass.

Setup The molar mass of glucose is 180.2 g/mol; the density of the solution is 1.062 g/mL.

Solution

(a)
$$\frac{170.1 \text{ g}}{180.2 \text{ g/mol}} = 0.9440 \text{ mol glucose per L of solution}$$

$$1 \text{ L of solution} \times \frac{1062 \text{ g}}{L} = 1062 \text{ g}$$

$$1062 \text{ g solution} - 170.1 \text{ g glucose} = 892 \text{ g water} = 0.892 \text{ kg water}$$

$$\frac{0.9440 \text{ mol glucose}}{0.892 \text{ kg water}} = 1.06 \ m$$

(b) $\dfrac{170.1 \text{ g glucose}}{1062 \text{ g solution}} \times 100\% = 16.02\%$ glucose by mass

(c) $\dfrac{170.1 \text{ g glucose}}{1062 \text{ g solution}} \times 1,000,000 = 1.602 \times 10^5$ ppm glucose

THINK ABOUT IT

Pay careful attention to units in problems such as this. Most require conversions between grams and kilograms and/or liters and milliliters.

Practice Problem **A**TTEMPT   Determine (a) the molality and (b) the percent by mass of urea for a solution prepared by dissolving 5.46 g urea [$(NH_2)_2CO$] in 215 g of water.

Practice Problem **B**UILD   Determine the molality of an aqueous solution that is 4.5 percent urea by mass.

Practice Problem **C**ONCEPTUALIZE   For a given solute/solvent pair at a given temperature, which graph best depicts the relationship between percent composition by mass and molality of the solute?

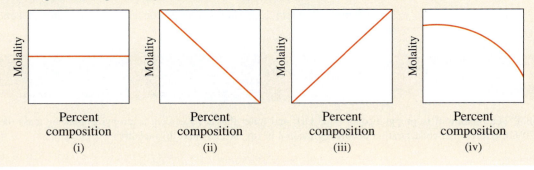

Comparison of Concentration Units

The choice of a concentration unit is based on the purpose of the experiment. For instance, we typically use molarity to express the concentrations of solutions for titrations and gravimetric analyses. *Mole fractions* are used to express the concentrations of gases—and of solutions when we are working with vapor pressures, which we will discuss in Section 13.5.

The advantage of molarity is that it is generally easier to measure the volume of a solution, using precisely calibrated volumetric flasks, than to weigh the solvent. *Molality*, on the other hand, has the advantage of being temperature independent. The volume of a solution typically increases slightly with increasing temperature, which would change the molarity. The mass of solvent in a solution, however, does *not* change with temperature.

Student Note: For very dilute aqueous solutions, molarity and molality have the same value. The mass of a liter of water is 1 kg, and in a very dilute solution, the mass of solute is negligible compared to that of the solvent.

Percent by mass is similar to molality in that it is independent of temperature. Furthermore, because it is defined in terms of the ratio of the mass of solute to the mass of solution, we do not need to know the molar mass of the solute to calculate the percent by mass.

Often it is necessary to convert the concentration of a solution from one unit to another. For example, the same solution may be used for different experiments that require different concentration units for calculations. Suppose we want to express the concentration of a 0.396-m aqueous glucose ($C_6H_{12}O_6$) solution (at 25°C) in molarity. We know there is 0.396 mole of glucose in 1000 g of the solvent. We need to determine the *volume* of this solution to calculate molarity. To determine volume, we must first calculate its mass:

$$0.396 \text{ mol } C_6H_{12}O_6 \times \frac{180.2 \text{ g}}{1 \text{ mol } C_6H_{12}O_6} = 71.4 \text{ g } C_6H_{12}O_6$$

$$71.4 \text{ g } C_6H_{12}O_6 + 1000 \text{ g } H_2O = 1071 \text{ g solution}$$

Student Note: Problems that require conversions between molarity and molality must provide density or sufficient information to determine density.

Once we have determined the mass of the solution, we use the density *of the solution,* which is typically determined experimentally, to determine its volume. The density of a 0.396 m glucose solution is 1.16 g/mL at 25°C. Therefore, its volume is:

$$\text{volume} = \frac{\text{mass}}{\text{density}}$$

$$= \frac{1071 \text{ g}}{1.16 \text{ g/mL}} \times \frac{1 \text{ L}}{1000 \text{ mL}}$$

$$= 0.923 \text{ L}$$

Having determined the volume of the solution, the molarity is given by:

$$\text{molarity} = \frac{\text{moles of solute}}{\text{liters of solution}}$$

$$= \frac{0.396 \text{ mol}}{0.923 \text{ L}}$$

$$= 0.429 \text{ mol/L} = 0.429 \text{ M}$$

Sample Problem 13.3 shows how to convert from one unit of concentration to another.

SAMPLE PROBLEM   13.3

"Rubbing alcohol" is a mixture of isopropyl alcohol (C_3H_7OH) and water that is 70 percent isopropyl alcohol by mass (density = 0.79 g/mL at 20°C). Express the concentration of rubbing alcohol in (a) molarity and (b) molality.

Strategy
(a) Use density to determine the total mass of a liter of solution, and use percent by mass to determine the mass of isopropyl alcohol in a liter of solution. Convert the mass of isopropyl alcohol to moles, and divide moles by liters of solution to get molarity. We can choose to start with any volume in a problem like this. Choosing 1 L simplifies the math.

(b) Subtract the mass of C_3H_7OH from the mass of solution to get the mass of water. Divide moles of C_3H_7OH by the mass of water (in kg) to get molality.

Setup The mass of a liter of rubbing alcohol is 790 g, and the molar mass of isopropyl alcohol is 60.09 g/mol.

Solution

(a)
$$\frac{790 \text{ g solution}}{\text{L solution}} \times \frac{70 \text{ g C}_3\text{H}_7\text{OH}}{100 \text{ g solution}} = \frac{553 \text{ g C}_3\text{H}_7\text{OH}}{\text{L solution}}$$

$$\frac{553 \text{ g C}_3\text{H}_7\text{OH}}{\text{L solution}} \times \frac{1 \text{ mol}}{60.09 \text{ g C}_3\text{H}_7\text{OH}} = \frac{9.20 \text{ mol C}_3\text{H}_7\text{OH}}{\text{L solution}} = 9.2 \text{ } M$$

(b) 790 g solution − 553 g C$_3$H$_7$OH = 237 g water = 0.237 kg water

$$\frac{9.20 \text{ mol C}_3\text{H}_7\text{OH}}{0.237 \text{ kg water}} = 39 \text{ } m$$

Rubbing alcohol is 9.2 M and 39 m in isopropyl alcohol.

THINK ABOUT IT

Note the large difference between molarity and molality in this case. Molarity and molality are the same (or similar) only for very dilute aqueous solutions.

Practice Problem A TTEMPT An aqueous solution that is 16 percent sulfuric acid (H$_2$SO$_4$) by mass has a density of 1.109 g/mL at 25°C. Determine (a) the molarity and (b) the molality of the solution at 25°C.

Practice Problem B UILD Determine the percent sulfuric acid by mass of a 1.49-m aqueous solution of H$_2$SO$_4$.

Practice Problem C ONCEPTUALIZE The diagrams represent solutions of a solid substance that is soluble in both water (density 1 g/cm^3) and chloroform (density 1.5 g/cm^3). For which of these solutions will the numerical value of molarity be closest to that of the molality? For which will the values of molarity and molality be most different?

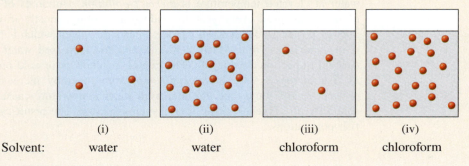

	(i)	(ii)	(iii)	(iv)
Solvent:	water	water	chloroform	chloroform

CHECKPOINT – SECTION 13.3   Concentration Units

13.3.1 A solution of a particular concentration is prepared at 20°C. Which of the following does not change when the solution is heated to 75°C? (Select all that apply.)

a) Molarity

b) Molality

c) Percent by mass

d) Mole fraction

e) All of these change.

13.3.2 What is the molality of a solution prepared by dissolving 6.44 g of naphthalene (C$_{10}$H$_8$) in 80.1 g benzene?

a) 1.13 m

b) 80.4 m

c) 0.804 m

d) 0.627 m

e) 11.7 m

13.3.3 At 20.0°C, a 0.258-m aqueous solution of glucose (C$_6$H$_{12}$O$_6$) has a density of 1.0173 g/mL. Calculate the molarity of this solution.

a) 0.258 M

b) 0.300 M

c) 0.456 M

d) 0.251 M

e) 0.448 M

13.3.4 At 25.0°C, an aqueous solution that is 25.0 percent H$_2$SO$_4$ by mass has a density of 1.178 g/mL. Calculate the molarity and the molality of this solution.

a) 3.00 M and 3.40 m

b) 3.40 M and 3.40 m

c) 3.00 M and 3.00 m

d) 3.00 M and 2.98 m

e) 3.44 M and 3.14 m

13.4 Factors That Affect Solubility

Recall that solubility is defined as the maximum amount of solute that will dissolve in a given quantity of solvent at a specific *temperature*. Temperature affects the solubility of most substances. In this section we consider the effects of temperature on the aqueous solubility of solids and gases, and the effect of *pressure* on the aqueous solubility of gases.

Temperature

More sugar dissolves in hot tea than in iced tea because the aqueous solubility of sugar, like that of most solid substances, increases as the temperature increases. Figure 13.4 shows the solubility of some common solids in water as a function of temperature. Note how the solubility of a solid and the *change* in solubility over a particular temperature range vary considerably. The relationship between temperature and solubility is complex and often nonlinear.

The relationship between temperature and the aqueous solubility of gases is somewhat simpler than that of solids. Most gaseous solutes become less soluble in water as temperature increases. If you get a glass of water from your faucet and leave it on the kitchen counter for a while, you will see bubbles forming in the water as it warms to room temperature. As the temperature of the water increases, dissolved gases become less soluble and come out of solution, resulting in the formation of bubbles.

One of the more important consequences of the reduced solubility of gases in water at elevated temperature is *thermal pollution.* Hundreds of billions of gallons of water are used every year for industrial cooling, mostly in electric power and nuclear power production. This process heats the water, which is then returned to the rivers and lakes from which it was taken. The increased water temperature has a twofold impact on aquatic life. The rate of metabolism of cold-blooded species such as fish increases with increasing temperature, thereby increasing their need for oxygen. At the same time, the increased water temperature causes a decrease in the solubility of oxygen, making less oxygen available. The result can be disastrous for fish populations.

Figure 13.4 Temperature dependence of the solubility of glucose and several ionic compounds in water.

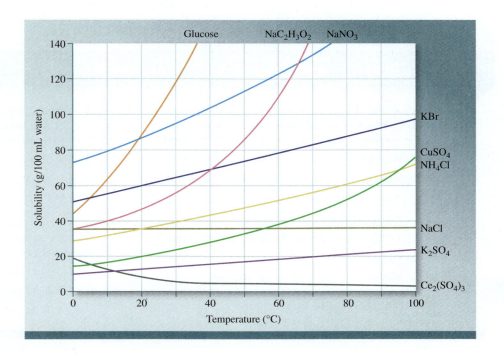

Pressure

Although pressure does not influence the solubility of a liquid or a solid significantly, it does greatly affect the solubility of a gas. The quantitative relationship between gas solubility and pressure is given by **Henry's**[1] **law,** which states that the solubility of a gas in a liquid is *proportional* to the pressure of the gas over the solution:

$$c \propto P$$

and is expressed as:

$$c = kP \qquad \qquad \textbf{Equation 13.3}$$

where c is the molar concentration (mol/L) of the dissolved gas, P is the pressure (in atm) of the gas over the solution, and k is a proportionality constant called the **Henry's law constant.** Henry's law constants are specific to the gas-solvent combination and vary with temperature. The units of k are mol/L · atm. If there is a mixture of gases over the solution, then P in Equation 13.3 is the partial pressure of the gas in question.

Henry's law can be understood qualitatively in terms of the kinetic molecular theory [◀◀ Section 10.6]. The amount of gas that will dissolve in a solvent depends on how frequently the gas molecules collide with the liquid surface and become trapped by the condensed phase. Suppose we have a gas in dynamic equilibrium [◀◀ Section 11.2] with a solution, as shown in Figure 13.5(a). At any point in time, the number of gas molecules entering the solution is equal to the number of dissolved gas molecules leaving the solution and entering the vapor phase. If the partial pressure of the gas is increased [Figure 13.5(b)], more molecules strike the liquid surface, causing more of them to dissolve. As the concentration of dissolved gas increases, the number of gas molecules leaving the solution also increases. These processes continue until the concentration of dissolved gas in the solution reaches the point again where the number of molecules leaving the solution per second equals the number entering the solution per second.

One interesting application of Henry's law is the production of carbonated beverages. Manufacturers put the "fizz" in soft drinks using pressurized carbon dioxide. The pressure of CO_2 applied (typically on the order of 5 atm) is many thousands of times greater than the partial pressure of CO_2 in the air. Thus, when a can or bottle of soda is opened, the CO_2 dissolved under high-pressure conditions comes out of solution—resulting in the bubbles that make carbonated drinks appealing.

Sample Problem 13.4 illustrates the use of Henry's law.

Student Note: Henry's law means that if we double the pressure of a gas over a solution (at constant temperature), we double the concentration of gas dissolved in the solution; triple the pressure, triple the concentration; and so on.

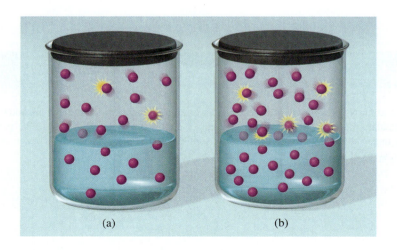

(a) (b)

Figure 13.5 A molecular view of Henry's law. When the partial pressure of the gas over the solution increases from (a) to (b), the concentration of the dissolved gas also increases according to Equation 13.3.

---

1. William Henry (1775–1836). English chemist. Henry's major contribution to science was his discovery of the law describing the solubility of gases, which now bears his name.

SAMPLE PROBLEM 13.4

Calculate the concentration of carbon dioxide in a soft drink that was bottled under a partial pressure of 5.0 atm CO_2 at 25°C (a) before the bottle is opened and (b) after the soda has gone "flat" at 25°C. The Henry's law constant for CO_2 in water at this temperature is 3.1×10^{-2} mol/L · atm. Assume that the partial pressure of CO_2 in air is 0.0004 atm and that the Henry's law constant for the soft drink is the same as that for water.

Strategy Use Equation 13.3 and the given Henry's law constant to solve for the molar concentration (mol/L) of CO_2 at 25°C and the two CO_2 pressures given.

Setup At 25°C, the Henry's law constant for CO_2 in water is 3.1×10^{-2} mol/L · atm.

Solution (a) $c = (3.1 \times 10^{-2}$ mol/L · atm$)(5.0$ atm$) = 1.6 \times 10^{-1}$ mol/L

(b) $c = (3.1 \times 10^{-2}$ mol/L · atm$)(0.0004$ atm$) = 1 \times 10^{-5}$ mol/L

THINK ABOUT IT

With a pressure on the order of 13,000 times smaller in part (b) than in part (a), we expect the concentration of CO_2 to be approximately 13,000 times smaller—and it is.

Practice Problem (A)TTEMPT Calculate the concentration of CO_2 in water at 25°C when the pressure of CO_2 over the solution is 4.0 atm.

Practice Problem (B)UILD Calculate the pressure of O_2 necessary to generate an aqueous solution that is 3.4×10^{-2} M in O_2 at 25°C. The Henry's law constant for O_2 in water at 25°C is 1.3×10^{-3} mol/L · atm.

Practice Problem (C)ONCEPTUALIZE The first diagram represents a closed system with two different gases dissolved in water. Which of the other diagrams could represent a closed system consisting of the same two gases at the same temperature?

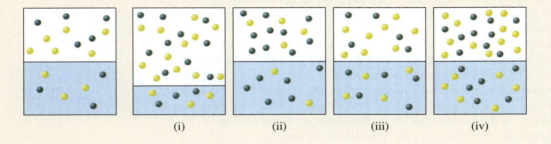

(i)   (ii)   (iii)   (iv)

CHECKPOINT – SECTION 13.4   Factors That Affect Solubility

13.4.1 The solubility of N_2 in water at 25°C and an N_2 pressure of 1 atm is 6.8×10^{-4} mol/L. Calculate the concentration of dissolved N_2 in water under atmospheric conditions where the partial pressure of N_2 is 0.78 atm.

a) 6.8×10^{-4} M

b) 8.7×10^{-4} M

c) 5.3×10^{-4} M

d) 1.5×10^{-4} M

e) 3.1×10^{-3} M

13.4.2 Calculate the molar concentration of O_2 in water at 25°C under atmospheric conditions where the partial pressure of O_2 is 0.22 atm. The Henry's law constant for O_2 is 1.3×10^{-3} mol/L · atm.

a) 2.9×10^{-4} M

b) 5.9×10^{-3} M

c) 1.7×10^{-3} M

d) 1.0×10^{-3} M

e) 1.3×10^{-3} M

13.5 Colligative Properties

Colligative properties are properties that depend on the number of solute particles in solution but do not depend on the nature of the solute particles. That is, colligative properties depend on the concentration of solute particles regardless of whether those particles are atoms, molecules, or ions. The colligative properties are vapor-pressure lowering, boiling-point elevation, freezing-point depression, and osmotic pressure. We begin by considering the colligative properties of relatively dilute solutions ($\leq 0.2\ M$) of nonelectrolytes.

Vapor-Pressure Lowering

We have seen that a liquid exerts a characteristic vapor pressure [◄◄ Section 11.2]. When a **nonvolatile** solute (one that does *not* exert a vapor pressure) is dissolved in a liquid, the vapor pressure exerted by the liquid decreases. The difference between the vapor pressure of a pure solvent and that of the corresponding solution depends on the concentration of the solute in the solution. This relationship is expressed by **Raoult's**[2] **law,** which states that the partial pressure of a solvent over a solution, P_1, is given by the *vapor pressure of the pure solvent,* P_1°, times the mole fraction of the solvent in the solution, χ_1:

$$P_1 = \chi_1 P_1^\circ \qquad \textbf{Equation 13.4}$$

Student Note: Table 10.5 gives the vapor pressure of water at various temperatures.

In a solution containing only one solute, $\chi_1 = 1 - \chi_2$, where χ_2 is the mole fraction of the solute. Equation 13.4 can therefore be rewritten as:

$$P_1 = (1 - \chi_2)P_1^\circ$$

or:

$$P_1 = P_1^\circ - \chi_2 P_1^\circ$$

so that:

$$P_1^\circ - P_1 = \Delta P = \chi_2 P_1^\circ \qquad \textbf{Equation 13.5}$$

Thus, the decrease in vapor pressure, ΔP, is directly proportional to the solute concentration expressed as a *mole fraction.*

To understand the phenomenon of vapor-pressure lowering, we must understand the degree of *order* associated with the states of matter involved. As we saw in Section 13.2, molecules in the liquid state are fairly highly ordered; that is, they have low entropy. Molecules in the gas phase have significantly less order—they have high entropy. Because there is a natural tendency toward increased entropy, molecules have a certain tendency to leave the region of lower entropy and enter the region of higher entropy. This corresponds to molecules leaving the liquid and entering the gas phase. As we have seen, when a solute is added to a liquid, the liquid's order is disrupted. Thus, the solution has greater entropy than the pure liquid. Because there is a smaller difference in entropy between the solution and the gas phase than there was between the pure liquid and the gas phase, there is a decreased tendency for molecules to leave the solution and enter the gas phase—resulting in a lower vapor pressure exerted by the solvent. This qualitative explanation of vapor-pressure lowering is illustrated in Figure 13.6. The smaller difference in entropy between the solution and gas phases, relative to that between the pure liquid and gas phases, results in a decreased tendency for solvent molecules to enter the gas phase. This results in a lowering of vapor pressure. The solvent in a solution will always exert a lower vapor pressure than the pure solvent.

---

2. François Marie Raoult (1839–1901). French chemist. Raoult's work was mainly in solution properties and electrochemistry.

Figure 13.6 The smaller difference in entropy between the solution and gas phases, relative to that between the pure liquid and gas phases, results in a decreased tendency for solvent molecules to enter the gas phase. This results in a lowering of vapor pressure. The solvent in a solution always exerts a lower vapor pressure than the pure solvent.

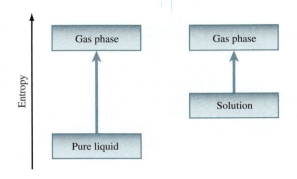

Sample Problem 13.5 shows how to use Raoult's law.

SAMPLE PROBLEM 13.5

Calculate the vapor pressure of water over a solution made by dissolving 225 g of glucose in 575 g of water at 35°C. (At 35°C, $P°_{H_2O}$ = 42.2 mmHg.)

Strategy Convert the masses of glucose and water to moles, determine the mole fraction of water, and use Equation 13.4 to find the vapor pressure over the solution.

Setup The molar masses of glucose and water are 180.2 and 18.02 g/mol, respectively.

Solution

$$\frac{225 \text{ g glucose}}{180.2 \text{ g/mol}} = 1.25 \text{ mol glucose} \qquad \frac{575 \text{ g water}}{18.02 \text{ g/mol}} = 31.9 \text{ mol water}$$

$$\chi_{water} = \frac{31.9 \text{ mol water}}{1.25 \text{ mol glucose} + 31.9 \text{ mol water}} = 0.962$$

$$P_{H_2O} = \chi_{water} \times P°_{H_2O} = 0.962 \times 42.2 \text{ mmHg} = 40.6 \text{ mmHg}$$

The vapor pressure of water over the solution is 40.6 mmHg.

THINK ABOUT IT

This problem can also be solved using Equation 13.5 to calculate the vapor-pressure lowering, ΔP.

Practice Problem ATTEMPT Calculate the vapor pressure of a solution made by dissolving 115 g of urea [$(NH_2)_2CO$; molar mass = 60.06 g/mol] in 485 g of water at 25°C. (At 25°C, $P°_{H_2O}$ = 23.8 mmHg.)

Practice Problem BUILD Calculate the mass of urea that should be dissolved in 225 g of water at 35°C to produce a solution with a vapor pressure of 37.1 mmHg. (At 35°C, $P°_{H_2O}$ = 42.2 mmHg.)

Practice Problem CONCEPTUALIZE The diagrams represent four closed systems containing aqueous solutions of the same nonvolatile solute at the same temperature. Over which solution is the vapor pressure of water the highest? Over which solution is it the lowest? Over which two solutions is the vapor pressure the same?

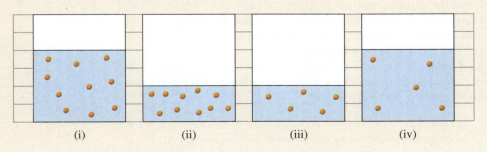

(i)   (ii)   (iii)   (iv)

If both components of a solution are **volatile** (i.e., have measurable vapor pressure), the vapor pressure of the solution is the sum of the individual partial pressures exerted by the solution components. Raoult's law holds equally well in this case:

$$P_A = \chi_A P_A^\circ$$
$$P_B = \chi_B P_B^\circ$$

where P_A and P_B are the partial pressures over the solution for components A and B, P_A° and P_B° are the vapor pressures of the pure substances A and B, and χ_A and χ_B are their mole fractions. The total pressure is given by Dalton's law of partial pressures [◀◀ Section 10.5]:

$$P_T = P_A + P_B$$

or:

$$P_T = \chi_A P_A^\circ + \chi_B P_B^\circ$$

For example, benzene and toluene are volatile components that have similar structures and therefore similar intermolecular forces:

$$\text{CH}_3$$

Benzene   Toluene

In a solution of benzene and toluene, the vapor pressure of each component obeys Raoult's law. Figure 13.7 shows the dependence of the total vapor pressure (P_T) in a benzene-toluene solution on the composition of the solution. Because there are only two components in the solution, we need only express the composition of the solution in terms of the mole fraction of *one* component. For any value of $\chi_{benzene}$, the mole fraction of toluene, $\chi_{toluene}$, is given by the equation $(1 - \chi_{benzene})$. The benzene-toluene solution is an example of an **ideal solution,** which is simply a solution that *obeys* Raoult's law.

Note that for a mixture in which the mole fractions of benzene and toluene are both 0.5, although the liquid mixture is equimolar, the vapor above the solution is not. Because pure benzene has a higher vapor pressure (75 mmHg at 20°C) than pure toluene (22 mmHg at 20°C), the vapor phase over the mixture will contain a higher concentration of the more volatile benzene molecules than it will the less volatile toluene molecules.

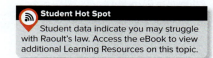

Student Hot Spot

Student data indicate you may struggle with Raoult's law. Access the eBook to view additional Learning Resources on this topic.

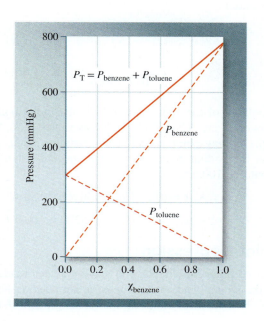

Figure 13.7 The dependence of partial pressures of benzene and toluene on their mole fractions in a benzene-toluene solution ($\chi_{toluene} = 1 - \chi_{benzene}$) at 80°C. This solution is said to be ideal because the vapor pressures obey Raoult's law.

Figure 13.8 Phase diagram illustrating the boiling-point elevation and freezing-point depression of aqueous solutions. The dashed curves pertain to the solution, and the solid curves to the pure solvent. As this diagram shows, the boiling point of the solution is higher than that of water, and the freezing point of the solution is lower than that of water.

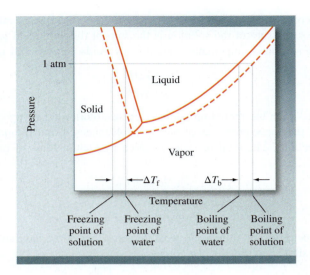

Boiling-Point Elevation

Recall that the *boiling point* of a substance is the temperature at which its vapor pressure equals the external atmospheric pressure [◄◄ Section 11.6]. Because the presence of a nonvolatile solute lowers the vapor pressure of a solution, it also affects the boiling point of the solution—relative to that of the pure liquid. Figure 13.8 shows the phase diagram of water and the changes that occur when a nonvolatile solute is added to it. At any temperature the vapor pressure over a solution is lower than that over the pure liquid, so the liquid-vapor curve for the solution lies below that for the pure solvent. Consequently, the dashed solution curve intersects the horizontal line that marks $P = 1$ atm at a higher temperature than the normal boiling point of the pure solvent. That is, a higher temperature is needed to make the solvent's vapor pressure equal to atmospheric pressure.

The boiling-point elevation (ΔT_b) is defined as the difference between the boiling point of the solution (T_b) and the boiling point of the pure solvent (T_b°):

Student Note: Rearranging this equation, we get the boiling point of the solution by adding ΔT_b to the boiling point of the pure solvent: $T_b = T_b^\circ + \Delta T_b$.

$$\Delta T_b = T_b - T_b^\circ$$

Because $T_b > T_b^\circ$, ΔT_b is a positive quantity.

The value of ΔT_b is proportional to the concentration, expressed in molality, of the solute in the solution:

$$\Delta T_b \propto m$$

or:

Equation 13.6 $$\Delta T_b = K_b m$$

where m is the molality of the solution and K_b is the molal boiling-point elevation constant. The units of K_b are °C/m. Table 13.2 lists values of K_b for several common solvents. Using the boiling-point elevation constant for water and Equation 13.6, you can show that the boiling point of a 1.00-m aqueous solution of a nonvolatile, nonelectrolyte would be 100.5°C:

$$\Delta T_b = K_b m = (0.52°C/m)(1.00\ m) = 0.52°C$$

$$T_b = T_b^\circ + \Delta T_b = 100.0°C + 0.52°C = 100.5°C$$

Freezing-Point Depression

If you have ever lived in a cold climate, you may have seen roads and sidewalks that were "salted" in the winter. The application of a salt such as NaCl or $CaCl_2$ thaws ice (or prevents its formation) by lowering the freezing point of water.

TABLE 13.2	Molal Boiling-Point Elevation and Freezing-Point Depression Constants of Several Common Solvents			
Solvent	Normal Boiling Point (°C)	K_b (°C/m)	Normal Freezing Point (°C)	K_f (°C/m)
Water	100.0	0.52	0.0	1.86
Benzene	80.1	2.53	5.5	5.12
Ethanol	78.4	1.22	−117.3	1.99
Acetic acid	117.9	2.93	16.6	3.90
Cyclohexane	80.7	2.79	6.6	20.0

The phase diagram in Figure 13.8 shows that in addition to shifting the liquid-vapor phase boundary down, the addition of a nonvolatile solute also shifts the solid-liquid phase boundary to the left. Consequently, this dashed line intersects the solid horizontal line at 1 atm at a temperature lower than the freezing point of pure water. The freezing-point depression (ΔT_f) is defined as the difference between the freezing point of the pure solvent and the freezing point of the solution:

$$\Delta T_f = T_f^\circ - T_f$$

Because $T_f^\circ > T_f$, ΔT_f is a positive quantity. Again, the change in temperature is proportional to the molal concentration of the solution:

$$\Delta T_f \propto m$$

or:

$$\Delta T_f = K_f m \qquad \textbf{Equation 13.7}$$

where m is the concentration of solute expressed in molality and K_f is the molal freezing-point depression constant (see Table 13.2). Like K_b, K_f has units of °C/m.

Like boiling-point elevation, freezing-point depression can be explained in terms of differences in entropy. Freezing involves a transition from the more disordered liquid state to the more ordered solid state. For this to happen, energy must be removed from the system. Because a solution has greater disorder than the solvent, there is a bigger difference in entropy between the solution and the solid than there is between the pure solvent and the solid (Figure 13.9). The larger difference in entropy means that more energy must be removed for the liquid-solid transition to happen. Thus, *the solution freezes* at a lower temperature than does the pure solvent. Boiling-point elevation occurs only when the solute is nonvolatile. Freezing-point depression occurs regardless of the solute's volatility.

Sample Problem 13.6 demonstrates a practical application of freezing-point depression and boiling-point elevation.

Student Note: We get the freezing point of the solution by *subtracting* ΔT_f from the freezing point of the pure solvent: $\Delta T_f = T_f^\circ - T_f$.

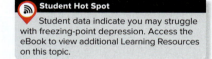
Student Hot Spot
Student data indicate you may struggle with freezing-point depression. Access the eBook to view additional Learning Resources on this topic.

Student Note: When a solution freezes, the solid that separates out is actually pure solvent. The solute remains in the liquid solution.

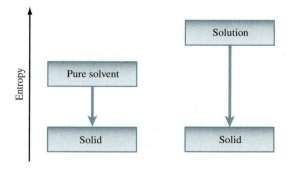

Figure 13.9 The solution has greater entropy than the pure solvent. The bigger difference in entropy between the solution and the solid means that more energy must be removed from the solution for it to freeze. Thus, the solution freezes at a lower temperature than the pure solvent.

SAMPLE PROBLEM 13.6

Ethylene glycol [$CH_2(OH)CH_2(OH)$] is a common automobile antifreeze. It is water soluble and fairly nonvolatile (b.p. 197°C). Calculate (a) the freezing point and (b) the boiling point of a solution containing 685 g of ethylene glycol in 2075 g of water.

Strategy Convert grams of ethylene glycol to moles, and divide by the mass of water in kilograms to get molal concentration. Use molal concentration in Equations 13.7 and 13.6 to determine ΔT_f and ΔT_b, respectively.

Setup The molar mass of ethylene glycol ($C_2H_6O_2$) is 62.07 g/mol. K_f and K_b for water are 1.86°C/m and 0.52°C/m, respectively.

Solution

$$\frac{685 \text{ g } C_2H_6O_2}{62.07 \text{ g/mol}} = 11.04 \text{ mol } C_2H_6O_2 \quad \text{and} \quad \frac{11.04 \text{ mol } C_2H_6O_2}{2.075 \text{ kg } H_2O} = 5.32 \text{ m } C_2H_6O_2$$

(a) $\Delta T_f = K_f m = (1.86°C/m)(5.32 \ m) = 9.89°C$

The freezing point of the solution is $(0 - 9.89)°C = -9.89°C$.

(b) $\Delta T_b = K_b m = (0.52°C/m)(5.32 \ m) = 2.8°C$

The boiling point of the solution is $(100.0 + 2.8)°C = 102.8°C$.

THINK ABOUT IT

Because it both lowers the freezing point and raises the boiling point, antifreeze is useful at both temperature extremes.

Practice Problem **A**TTEMPT Calculate the freezing point and boiling point of a solution containing 268 g of ethylene glycol and 1015 g of water.

Practice Problem **B**UILD What mass of ethylene glycol must be added to 1525 g of water to raise the boiling point to 103.9°C?

Practice Problem **C**ONCEPTUALIZE The diagrams represent four different aqueous solutions of the same solute. Which of the solutions has the lowest freezing point? Which has the highest boiling point?

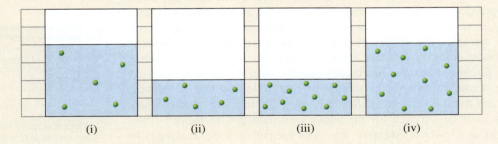

(i)   (ii)   (iii)   (iv)

Osmotic Pressure

Many chemical and biological processes depend on *osmosis*, the selective passage of solvent molecules through a porous membrane from a more dilute solution to a more concentrated one. Figure 13.10 illustrates osmosis. The left compartment of the apparatus contains pure solvent; the right compartment contains a solution made with the same solvent. The two compartments are separated by a *semipermeable membrane*, which allows the passage of solvent molecules but blocks the passage of solute molecules. At the beginning, the liquid levels in the two tubes are equal [Figure 13.10(a)]. As time passes, the level in the right tube rises. It continues to rise until equilibrium is reached, after which no further net change in levels is observed. The *osmotic pressure* (π) of a solution is the pressure required to stop osmosis. As shown in Figure 13.10(b), this pressure can be measured directly from the difference in the final liquid levels.

Animation
Osmosis.

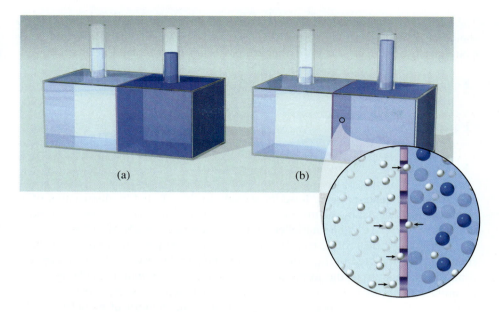

The osmotic pressure of a solution is directly proportional to the concentration, expressed in *molarity,* of the solute in solution:

$$\pi \propto M$$

and is given by:

$$\pi = MRT \qquad \textbf{Equation 13.8}$$

where M is the molarity of the solution, R is the gas constant (0.08206 L · atm/K · mol), and T is the absolute temperature. The osmotic pressure (π) is typically expressed in atmospheres.

Like boiling-point elevation and freezing-point depression, osmotic pressure is directly proportional to the concentration of the solution. This is what we would expect, though, because all colligative properties depend only on the number of solute particles in solution, not on the identity of the solute particles. Two solutions of equal concentration have the same osmotic pressure and are said to be *isotonic* to each other.

Electrolyte Solutions

So far we have discussed the colligative properties of nonelectrolyte solutions. Because electrolytes undergo *dissociation* when dissolved in water [◄◄ Section 4.1], we must consider them separately. Recall, for example, that when NaCl dissolves in water, it dissociates into $Na^+(aq)$ and $Cl^-(aq)$. For every mole of NaCl dissolved, we get two moles of ions in solution. Similarly, when a formula unit of $CaCl_2$ dissolves, we get three ions: one Ca^{2+} ion and two Cl^- ions. Thus, for every mole of $CaCl_2$ dissolved, we get three moles of ions in solution. Colligative properties depend only on the number of dissolved particles—not on the type of particles. This means that a 0.1-*m* solution of NaCl will exhibit a freezing-point depression twice that of a 0.1-*m* solution of a nonelectrolyte, such as sucrose. Similarly, we expect a 0.1-*m* solution of $CaCl_2$ to depress the freezing point of water three times as much as a 0.1 *m* sucrose solution. To account for this effect, we introduce and define a quantity called the ***van't Hoff*[3] *factor (i),*** which is given by:

$$i = \frac{\text{actual number of particles in solution after dissociation}}{\text{number of formula units initially dissolved in solution}}$$

3. Jacobus Henricus van't Hoff (1852–1911). Dutch chemist. One of the most prominent chemists of his time, van't Hoff did significant work in thermodynamics, molecular structure and optical activity, and solution chemistry. In 1901 he received the first Nobel Prize in Chemistry.

Thus, i is 1 for all nonelectrolytes. For strong electrolytes such as NaCl and KNO$_3$, i should be 2, and for strong electrolytes such as Na$_2$SO$_4$ and CaCl$_2$, i should be 3. Consequently, the equations for colligative properties must be modified as follows:

Equation 13.9	$\Delta T_b = iK_b m$

Equation 13.10	$\Delta T_f = iK_f m$

Equation 13.11	$\pi = iMRT$

The amount of vapor-pressure lowering would also be affected by dissociation of an electrolyte. In calculating the mole fraction of solute and solvent, the number of moles of solute would have to be multiplied by the appropriate van't Hoff factor.

In reality, the colligative properties of electrolyte solutions are usually smaller than predicted by Equations 13.9 through 13.11, especially at higher concentrations, because of the formation of *ion pairs*. An **ion pair** is made up of one or more cations and one or more anions held together by electrostatic forces (Figure 13.11). The presence of an ion pair reduces the number of particles in solution, thus reducing the observed colligative properties. Tables 13.3 and 13.4 list experimentally measured values of i and those calculated using the *van't Hoff factor*.

Student Note: Calculated van't Hoff factors are exact numbers [◄◄ Section 1.5].

Figure 13.11 (a) Free ions and (b) ion pairs in solution. Ion pairing reduces the number of dissolved particles in a solution, causing a decrease in the observed colligative properties. Furthermore, an ion pair bears no net charge and therefore cannot conduct electricity in solution.

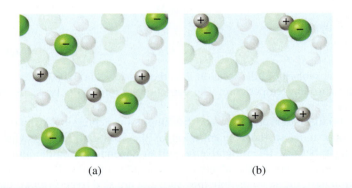

(a)   (b)

TABLE 13.3	Calculated and Measured van't Hoff Factors of 0.0500-*M* Electrolyte Solutions at 25°C		
Electrolyte	***i* (Calculated)**		***i* (Measured)**
Sucrose*	1		1.0
HCl	2		1.9
NaCl	2		1.9
MgSO$_4$	2		1.3
MgCl$_2$	3		2.7
FeCl$_3$	4		3.4

*Sucrose is a nonelectrolyte. It is listed here for comparison only.

TABLE 13.4	Experimentally Measured van't Hoff Factors of Sucrose and NaCl Solutions at 25°C		
	Concentration		
Compound	**0.100 *m***	**0.00100 *m***	**0.000100 *m***
Sucrose	1.00	1.00	1.00
NaCl	1.87	1.94	1.97

Sample Problem 13.7 demonstrates the experimental determination of a van't Hoff factor.

SAMPLE PROBLEM (13.7)

The osmotic pressure of a 0.0100 M potassium iodide (KI) solution at 25°C is 0.465 atm. Determine the experimental van't Hoff factor for KI at this concentration.

Strategy Use osmotic pressure to calculate the molar concentration of KI, and divide by the *nominal* concentration of 0.0100 M.

Setup $R = 0.08206$ L · atm/K · mol, and $T = 298$ K.

Solution Solving Equation 13.8 for M,

$$M = \frac{\pi}{RT} = \frac{0.465 \text{ atm}}{(0.08206 \text{ L} \cdot \text{atm/K} \cdot \text{mol})(298 \text{ K})} = 0.0190 \ M$$

$$i = \frac{0.0190 \ M}{0.0100 \ M} = 1.90$$

The experimental van't Hoff factor for KI at this concentration is 1.90.

THINK ABOUT IT

The calculated van't Hoff factor for KI is 2. The experimentally determined van't Hoff factor must be less than or equal to the calculated value.

Practice Problem A TTEMPT The freezing-point depression of a 0.100 m MgSO$_4$ solution is 0.225°C. Determine the experimental van't Hoff factor of MgSO$_4$ at this concentration.

Practice Problem B UILD Using the experimental van't Hoff factor from Table 13.4, determine the freezing point of a 0.100-m aqueous solution of NaCl. (Assume that the van't Hoff factors do not change with temperature.)

Practice Problem C ONCEPTUALIZE The diagram represents an aqueous solution of an electrolyte. Determine the experimental van't Hoff factor for the solute.

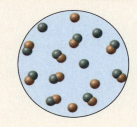

Bringing Chemistry to Life

Intravenous Fluids

Human blood consists of red blood cells (*erythrocytes*), white blood cells (*leukocytes*), and platelets (*thrombocytes*) suspended in plasma, an aqueous solution containing a variety of solutes including salts and proteins. Each red blood cell is surrounded by a protective semipermeable membrane. Inside this membrane, the concentration of dissolved substances is about 0.3 M. Likewise, the concentration of dissolved substances in plasma is also about 0.3 M. Having the same concentration (and therefore the same osmotic pressure of ~7.6 atm at 37°C) inside and outside the red blood cell prevents a net movement of water into or out of the cell through the protective semipermeable membrane [Figure 13.12(a)]. To maintain this balance of osmotic pressure, fluids that are given intravenously must be *isotonic* to plasma. Five percent dextrose (sugar) and normal saline, which is 0.9% sodium chloride, are two of the most commonly used isotonic intravenous fluids.

A solution that has a lower concentration of dissolved substances than plasma is said to be **hypotonic** to plasma. If a significant volume of pure water were administered intravenously, it would dilute the plasma, lowering its concentration and making it hypotonic to the solution inside the red blood cells. If this were to happen, water would enter the red blood cells via osmosis [Figure 13.12(b)]. The

Figure 13.12 A cell in (a) a hypotonic solution, (b) an isotonic solution, and (c) a hypertonic solution. The cell swells and may eventually burst in (a); it shrinks in (c).

(photos): *David M. Phillips/Science Source*

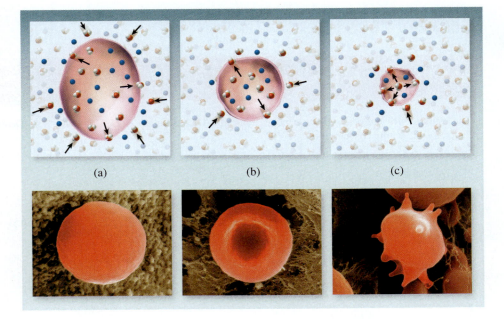

(a)   (b)   (c)

cells would swell and could potentially burst, a process called *hemolysis*. On the other hand, if red blood cells were placed in a solution with a higher concentration of dissolved substances than plasma, a solution said to be **hypertonic** to plasma, water would leave the cells via osmosis [Figure 13.12(c)]. The cells would shrink, a process called *crenation,* which is also potentially dangerous. The osmotic pressure of human plasma must be maintained within a very narrow range to prevent damage to red blood cells. Hypotonic and hypertonic solutions can be administered intravenously to treat specific medical conditions, but the patient must be carefully monitored throughout the treatment.

Interestingly, humans have for centuries exploited the sensitivity of cells to osmotic pressure. The process of "curing" meat with salt or with sugar causes crenation of the bacteria cells that would otherwise cause spoilage.

Sample Problem 13.8 explores the relationship between concentration and osmotic pressure.

SAMPLE PROBLEM 13.8

Ringer's lactate, a solution containing several different salts, is often administered intravenously for the initial treatment of trauma patients. One liter of Ringer's lactate contains 0.102 mol of sodium chloride, 4×10^{-3} mol of potassium chloride, 1.5×10^{-3} mol of calcium chloride, and 2.8×10^{-2} mol of sodium lactate. Determine the osmotic pressure of this solution at normal body temperature (37°C). Assume no ion pairing. (The formula of the lactate ion is $CH_3CH_2COO^-$.)

Strategy Because the solutes are all ionic, each solute concentration will have to be multiplied by the appropriate van't Hoff factor. (Because we are assuming no ion pairing, we can use calculated van't Hoff factors.) Sum the concentrations for all solutes to determine the total concentration of dissolved particles, and use Equation 13.8 to calculate the osmotic pressure.

Setup Because the volume of the solution described is 1 L, the number of moles is also the molarity for each solute. $R = 0.08206 \; L \cdot atm/K \cdot mol$, $T = 310$ K, and the van't Hoff factors for the solutes in Ringer's lactate are as follows:

$$NaCl(s) \longrightarrow Na^+(aq) + Cl^-(aq) \qquad\qquad i = 2$$

$$KCl(s) \longrightarrow K^+(aq) + Cl^-(aq) \qquad\qquad i = 2$$

$$CaCl_2(s) \longrightarrow Ca^{2+}(aq) + 2Cl^-(aq) \qquad\qquad i = 3$$

$$NaCH_3CH_2COO(s) \longrightarrow Na^+(aq) + CH_3CH_2COO^-(aq) \qquad\qquad i = 2$$

Solution The total concentration of ions in solution is the sum of the individual concentrations.

$$\text{total concentration} = 2[\text{NaCl}] + 2[\text{KCl}] + 3[\text{CaCl}_2] + 2[\text{NaCH}_3\text{CH}_2\text{COO}]$$

$$= 2(0.102\ M) + 2(4 \times 10^{-3}\ M) + 3(1.5 \times 10^{-3}\ M) + 2(2.8 \times 10^{-2}\ M)$$

$$= 2.73 \times 10^{-1}\ M$$

$$\pi = MRT = (0.273\ M)(0.08206\ \text{L} \cdot \text{atm/K} \cdot \text{mol})(310\ \text{K}) = 6.93\ \text{atm}$$

THINK ABOUT IT

Ringer's lactate is isotonic with human plasma, which is often the case with fluids administered intravenously.

Practice Problem Ⓐ**TTEMPT** Determine the osmotic pressure of a solution that is 0.200 *M* in glucose and 0.100 *M* in sodium chloride at 37.0°C. Assume no ion pairing.

Practice Problem Ⓑ**UILD** Determine the concentration of an aqueous solution that has an osmotic pressure of 5.6 atm at 37°C if (a) the solute is glucose, and (b) the solute is sodium chloride. Assume no ion pairing.

Practice Problem Ⓒ**ONCEPTUALIZE** The first diagram represents an aqueous solution. Which of the other pictures represents a solution that is isotonic with the first?

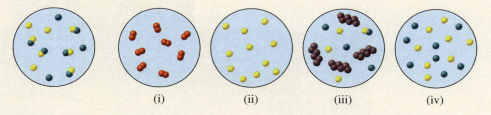

(i)	(ii)	(iii)	(iv)

To review:

- Vapor-pressure lowering depends on concentration expressed as *mole fraction, χ*.
- Boiling-point elevation depends on concentration expressed as *molality, m*.
- Freezing-point depression depends on concentration expressed as *molality, m*.
- Osmotic pressure depends on concentration expressed as *molarity, M*.

CHECKPOINT – SECTION 13.5   Colligative Properties

13.5.1 A solution contains 75.0 g of glucose (molar mass 180.2 g/mol) in 425 g of water. Determine the vapor pressure of water over the solution at 35°C. ($P^\circ_{\text{H}_2\text{O}}$ = 42.2 mmHg at 35°C.)

  a) 0.732 mmHg   d) 41.5 mmHg

  b) 42.9 mmHg   e) 42.2 mmHg

  c) 243 mmHg

13.5.2 Determine the boiling point and the freezing point of a solution prepared by dissolving 678 g of glucose in 2.0 kg of water. For water, K_b = 0.52°C/*m* and K_f = 1.86°C/*m*.

  a) 101°C and 3.5°C   d) 112°C and 6.2°C

  b) 99°C and −3.5°C   e) 88°C and −6.2°C

  c) 101°C and −3.5°C

13.5.3 Calculate the osmotic pressure of a solution prepared by dissolving 65.0 g of Na_2SO_4 in enough water to make 500 mL of solution at 20°C. (Assume no ion pairing.)

  a) 0.75 atm   d) 1×10^{-2} atm

  b) 66 atm   e) 22 atm

  c) 44 atm

13.5.4 A 1.00-*m* solution of HCl has a freezing point of −3.30°C. Determine the experimental van't Hoff factor for HCl at this concentration.

  a) 1.77   d) 2

  b) 2.01   e) 1

  c) 1.90

Bringing Chemistry to Life

Hemodialysis

Osmosis refers to the movement of solvent through a membrane from the side where the solute concentration is lower to the side where the solute concentration is higher. Hemodialysis involves a more porous membrane, through which both solvent (water) and small solute particles can pass. The size of the membrane pores is such that only small waste products such as excess potassium ion, creatinine, urea, and extra fluid can pass through. Larger components in blood, such as blood cells and proteins, are too large to pass through the membrane. A solute will pass through the membrane from the side where its concentration is higher to the side where its concentration is lower. The composition of the dialysate ensures that the necessary solutes in the blood (e.g., sodium and calcium ions) are not removed.

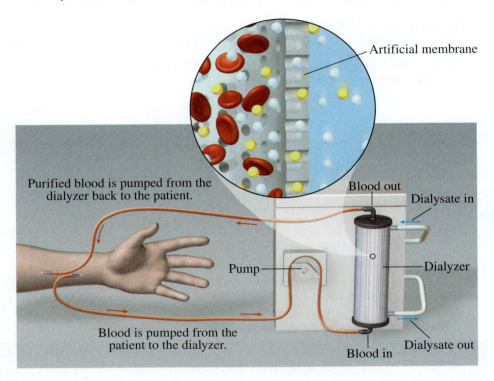

Because it is not normally found in blood, fluoride ion, if present in the dialysate, will flow across the membrane into the blood. In fact, this is true of any sufficiently small solute that is not normally found in blood—necessitating requirements for the purity of water used to prepare dialysate solutions that far exceed those for drinking water.

13.6 Calculations Using Colligative Properties

The colligative properties of nonelectrolyte solutions provide a means of determining the molar mass of a solute. Although any of the four colligative properties can be used in theory for this purpose, only freezing-point depression and osmotic pressure are used in practice because they show the most pronounced, and therefore the most easily measured, changes. From the experimentally determined freezing-point depression or osmotic pressure, we can calculate the solution's *molality or molarity,* respectively. Knowing the mass of dissolved solute, we can readily determine its molar mass.

Sample Problems 13.9 and 13.10 illustrate this technique.

Student Note: These calculations require Equations 13.7 and 13.8, respectively.

SAMPLE PROBLEM 13.9

Quinine was the first drug widely used to treat malaria, and it remains the treatment of choice for severe cases. A solution prepared by dissolving 10.0 g of quinine in 50.0 mL of ethanol has a freezing point 1.55°C below that of pure ethanol. Determine the molar mass of quinine. (The density of ethanol is 0.789 g/mL.) Assume that quinine is a nonelectrolyte.

Strategy Use Equation 13.7 to determine the molal concentration of the solution. Use the density of ethanol to determine the mass of solvent. The molal concentration of quinine multiplied by the mass of ethanol (in kg) gives moles of quinine. The mass of quinine (in grams) divided by moles of quinine gives the molar mass.

Setup

$$\text{mass of ethanol} = 50.0 \text{ mL} \times 0.789 \text{ g/mL} = 39.5 \text{ g} \quad \text{or} \quad 3.95 \times 10^{-2} \text{ kg}$$

K_f for ethanol (from Table 13.2) is 1.99°C/*m*.

Solution Solving Equation 13.7 for molal concentration,

$$m = \frac{\Delta T_f}{K_f} = \frac{1.55°C}{1.99°C/m} = 0.779 \ m$$

The solution is 0.779 *m* in quinine (i.e., 0.779 mol quinine/kg ethanol solvent).

$$\left(\frac{0.779 \text{ mol quinine}}{\text{kg ethanol}}\right)(3.95 \times 10^{-2} \text{ kg ethanol}) = 0.0308 \text{ mol quinine}$$

$$\text{molar mass of quinine} = \frac{10.0 \text{ g quinine}}{0.0308 \text{ mol quinine}} = 325 \text{ g/mol}$$

THINK ABOUT IT

Check the result using the molecular formula of quinine: $C_{20}H_{24}N_2O_2$ (324.4 g/mol). Multistep problems such as this one require careful tracking of units at each step.

Practice Problem ATTEMPT Calculate the molar mass of naphthalene, the organic compound in "mothballs," if a solution prepared by dissolving 5.00 g of naphthalene in exactly 100 g of benzene has a freezing point 2.00°C below that of pure benzene.

Practice Problem BUILD What mass of naphthalene must be dissolved in 2.00×10^2 g of benzene to give a solution with a freezing point 2.50°C below that of pure benzene?

Practice Problem CONCEPTUALIZE The first diagram represents an aqueous solution that has a freezing point of −3.5°C. Which of the other diagrams could represent the system after the temperature has been lowered to −4.0°C? Liquid water is shown as blue; ice is shown as white. (To keep the diagrams uncluttered, individual water molecules are not shown.)

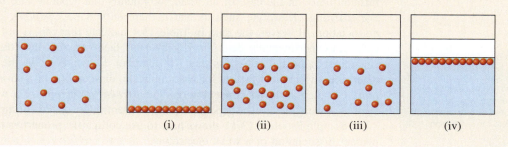

(i)   (ii)   (iii)   (iv)

SAMPLE PROBLEM 13.10

A solution is prepared by dissolving 50.0 g of hemoglobin (Hb) in enough water to make 1.00 L of solution. The osmotic pressure of the solution is measured and found to be 14.3 mmHg at 25°C. Calculate the molar mass of hemoglobin. (Assume that there is no change in volume when the hemoglobin is added to the water.)

Strategy Use Equation 13.8 to calculate the molarity of the solution. Because the solution volume is 1 L, the molarity is equal to the number of moles of hemoglobin. Dividing the mass of hemoglobin, which is given in the problem statement, by the number of moles gives the molar mass.

Setup $R = 0.08206 \text{ L} \cdot \text{atm/K} \cdot \text{mol}$, $T = 298 \text{ K}$, and $\pi = 14.3 \text{ mmHg}/(760 \text{ mmHg/atm}) = 1.88 \times 10^{-2} \text{ atm}$.

Solution Rearranging Equation 13.8 to solve for molarity, we get:

$$M = \frac{\pi}{RT} = \frac{1.88 \times 10^{-2} \text{ atm}}{(0.08206 \text{ L} \cdot \text{atm/K} \cdot \text{mol})(298 \text{ K})} = 7.69 \times 10^{-4} \, M$$

Thus, the solution contains 7.69×10^{-4} mole of hemoglobin:

$$\text{molar mass of hemoglobin} = \frac{50.0 \text{ g}}{7.69 \times 10^{-4} \text{ mol}} = 6.50 \times 10^{4} \text{ g/mol}$$

THINK ABOUT IT

Biological molecules can have *very* high molar masses.

Practice Problem 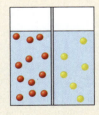TTEMPT A solution made by dissolving 25 mg of insulin in 5.0 mL of water has an osmotic pressure of 15.5 mmHg at 25°C. Calculate the molar mass of insulin. (Assume that there is no change in volume when the insulin is added to the water.)

Practice Problem BUILD What mass of insulin must be dissolved in 50.0 mL of water to produce a solution with an osmotic pressure of 16.8 mmHg at 25°C?

Practice Problem CONCEPTUALIZE The first diagram represents one aqueous solution separated from another by a semipermeable membrane. Which of the other diagrams could represent the same system after the passage of some time?

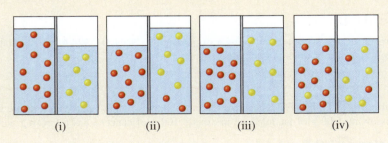

(i)   (ii)   (iii)   (iv)

The colligative properties of an electrolyte solution can be used to determine percent dissociation. **Percent dissociation** is the percentage of dissolved molecules (or formula units, in the case of an ionic compound) that separate into ions in solution. For a strong electrolyte such as NaCl, there should be *complete,* or 100 percent, dissociation. However, the data in Table 13.4 indicate that this is not necessarily the case. An experimentally determined van't Hoff factor smaller than the corresponding calculated value indicates less than 100 percent dissociation. As the experimentally determined van't Hoff factors for NaCl indicate, dissociation of a strong electrolyte is more complete at lower concentration. The **percent ionization** of a weak electrolyte, such as a weak acid, also depends on the concentration of the solution.

Sample Problem 13.11 shows how to use colligative properties to determine the percent dissociation of a weak electrolyte.

> **Student Note:** Recall that the term *dissociation* is used for ionic electrolytes and the term *ionization* is used for molecular electrolytes. In this context, they mean essentially the same thing.

SAMPLE PROBLEM (**13.11**)

A solution that is 0.100 *M* in hydrofluoric acid (HF) has an osmotic pressure of 2.64 atm at 25°C. Calculate the percent ionization of HF at this concentration.

Strategy Use the osmotic pressure and Equation 13.8 to determine the molar concentration of the particles in solution. Compare the concentration of particles to the *nominal* concentration (0.100 *M*) to determine what percentage of the original HF molecules are ionized.

Setup $R = 0.08206$ L · atm/K · mol, and $T = 298$ K.

Solution Rearranging Equation 13.8 to solve for molarity,

$$M = \frac{\pi}{RT} = \frac{2.64 \text{ atm}}{(0.08206 \text{ L} \cdot \text{atm/K} \cdot \text{mol})(298 \text{ K})} = 0.108 \text{ } M$$

The concentration of dissolved particles is 0.108 *M*. Consider the ionization of HF [◄◄ Section 4.3]:

$$\text{HF}(aq) \rightleftharpoons \text{H}^+(aq) + \text{F}^-(aq)$$

According to this equation, if *x* HF molecules ionize, we get *x* H⁺ ions and *x* F⁻ ions. Thus, the total concentration of particles in solution will be the original concentration of HF minus *x*, which gives the concentration of intact HF molecules, plus 2*x*, which is the concentration of ions (H⁺ and F⁻):

$$(0.100 - x) + 2x = 0.100 + x$$

Therefore, $0.108 = 0.100 + x$ and $x = 0.008$. Because we earlier defined *x* as the amount of HF ionized, the percent ionization is given by:

$$\text{percent ionization} = \frac{0.008 \text{ } M}{0.100 \text{ } M} \times 100\% = 8\%$$

At this concentration, HF is 8 percent ionized.

THINK ABOUT IT

For weak acids, the lower the concentration, the greater the percent ionization. A 0.010-*M* solution of HF has an osmotic pressure of 0.30 atm, corresponding to 23 percent ionization. A 0.0010-*M* solution of HF has an osmotic pressure of 3.8×10^{-2} atm, corresponding to 56 percent ionization.

Practice Problem (A)TTEMPT An aqueous solution that is 0.0100 *M* in acetic acid ($\text{HC}_2\text{H}_3\text{O}_2$) has an osmotic pressure of 0.255 atm at 25°C. Calculate the percent ionization of acetic acid at this concentration.

Practice Problem (B)UILD An aqueous solution that is 0.015 *M* in acetic acid ($\text{HC}_2\text{H}_3\text{O}_2$) is 3.5 percent ionized at 25°C. Calculate the osmotic pressure of this solution.

Practice Problem (C)ONCEPTUALIZE The diagrams represent aqueous solutions of weak electrolytes. List the solutions in order of increasing percent ionization.

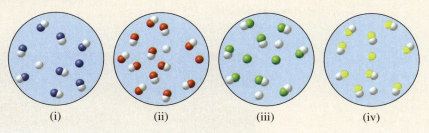

(i)   (ii)   (iii)   (iv)

CHECKPOINT – SECTION 13.6   Calculations Using Colligative Properties

13.6.1 A solution made by dissolving 14.2 g of sucrose in 100 g of water exhibits a freezing-point depression of 0.77°C. Calculate the molar mass of sucrose.

a) 34 g/mol   c) 2.4 g/mol   e) 68 g/mol

b) 3.4×10^2 g/mol   d) 1.8×10^2 g/mol

13.6.2 A 0.010-*M* solution of the weak electrolyte HA has an osmotic pressure of 0.27 atm at 25°C. What is the percent ionization of the electrolyte at this concentration?

a) 27%   c) 15%   e) 90%

b) 10%   d) 81%

| 13.7 | **Colloids** |

The solutions discussed so far in this chapter are true homogeneous mixtures. Now consider what happens if we add fine sand to a beaker of water and stir. The sand particles are suspended at first but gradually settle to the bottom of the beaker. This is an example of a heterogeneous mixture. Between the two extremes of homogeneous and heterogeneous mixtures is an intermediate state called a colloidal suspension, or simply, a colloid. A **colloid** is a dispersion of particles of one substance throughout another *substance*. Colloidal particles are much larger than the normal solute molecules; they range from 1×10^3 pm to 1×10^6 pm. Also, a colloidal suspension lacks the homogeneity of a true solution.

Student Note: The substance dispersed is called the *dispersed phase*; the substance in which it is dispersed is called the *dispersing medium*.

Colloids can be further categorized as aerosols (liquid or solid dispersed in gas), foams (gas dispersed in liquid or solid), emulsions (liquid dispersed in another liquid), sols (solid dispersed in liquid or in another solid), and gels (liquid dispersed in a solid). Table 13.5 lists the different types of colloids and gives one or more examples of each.

One way to distinguish a solution from a colloid is by the **Tyndall[4] effect.** When a beam of light passes through a colloid, it is scattered by the dispersed phase (Figure 13.13).

TABLE 13.5	Types of Colloids		
Dispersing Medium	**Dispersed Phase**	**Name**	**Example**
Gas	Liquid	Aerosol	Fog, mist
Gas	Solid	Aerosol	Smoke
Liquid	Gas	Foam	Whipped cream, meringue
Liquid	Liquid	Emulsion	Mayonnaise
Liquid	Solid	Sol	Milk of magnesia
Solid	Gas	Foam	Styrofoam*
Solid	Liquid	Gel	Jelly, butter
Solid	Solid	Solid sol	Alloys such as steel, gemstones (glass with dispersed metal)

*Styrofoam is a registered trademark of the Dow Chemical Company. It refers specifically to extruded polystyrene used for insulation in home construction. "Styrofoam" cups, coolers, and packing peanuts are not really made of Styrofoam. They are made of expanded polystyrene or EPS foam.

Figure 13.13 The Tyndall effect. Light is scattered by colloidal particles (left and right) but not by dissolved particles (center).

Charles D. Winters/McGraw Hill

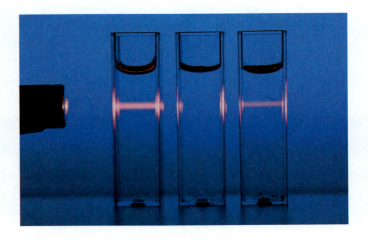

4. John Tyndall (1820–1893). Irish physicist. Tyndall did important work in magnetism and also explained glacier motion.

Figure 13.14 A familiar example of the Tyndall effect: headlights illuminating fog.

logoboom/Shutterstock

No such scattering is observed with true solutions because the solute molecules are too small to interact with visible light. Another demonstration of the Tyndall effect is the scattering of light from automobile headlights in fog (Figure 13.14).

Among the most important colloids are those in which the dispersing medium is water. Such colloids can be categorized as **hydrophilic** (*water loving*) or **hydrophobic** (*water fearing*). Hydrophilic colloids contain extremely large molecules such as proteins. In the aqueous phase, a protein like hemoglobin folds in such a way that the hydrophilic parts of the molecule, the parts that can interact favorably with water molecules by ion-dipole forces or hydrogen-bond formation, are on the outside surface (Figure 13.15).

A hydrophobic colloid normally would not be stable in water, and the particles would clump together, like droplets of oil in water merging to form a film at the water's surface. They can be *stabilized*, however, by the adsorption of ions on their surface (Figure 13.16). Material that collects on the surface is *adsorbed*, whereas material that passes to the interior is *absorbed*. The adsorbed ions are hydrophilic and can interact with water to stabilize the colloid. In addition, because adsorption of ions leaves the colloid particles *charged*, electrostatic repulsion prevents them from clumping together. Soil particles in rivers and streams are hydrophobic particles that are stabilized in this way. When river water enters the sea, the charges on the dispersed particles are neutralized by the high-salt medium. With the charges on their surfaces neutralized, the particles no longer repel one another and they clump together to form the silt that is seen at the mouth of the river.

Another way hydrophobic colloids can be stabilized is by the presence of other hydrophilic groups on their surfaces. Consider sodium stearate, a soap molecule that has a polar group at one end, often called the "head," and a long hydrocarbon "tail"

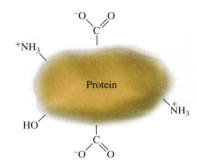

Figure 13.15 Hydrophilic groups on the surface of a large molecule such as a protein stabilize the molecule in water. Note that all the hydrophilic groups can form hydrogen bonds with water.

Student Note: A hydrophobic colloid must be *stabilized* to remain suspended in water.

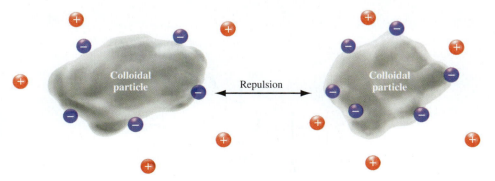

Figure 13.16 Diagram showing the stabilization of hydrophobic colloids. Negative ions are adsorbed onto the surface, and the repulsion between like charges prevents aggregation of the particles.

that is nonpolar (Figure 13.17). The cleansing action of soap is due to the dual nature of the hydrophobic tail and the hydrophilic head. The hydrocarbon tail is readily soluble in oily substances, which are also nonpolar, while the ionic $-COO^-$ group remains outside the oily surface. When enough soap molecules have surrounded an oil droplet, as shown in Figure 13.18, the entire system becomes stabilized in water because the exterior portion is now largely hydrophilic. This is how greasy substances are removed by the action of soap. In general, the process of stabilizing a colloid that would otherwise not stay dispersed is called *emulsification,* and a substance used for such stabilization is called an *emulsifier* or *emulsifying agent.*

A mechanism similar to that involving sodium stearate makes it possible for us to digest dietary fat. When we ingest fat, the gallbladder excretes a substance known as bile. Bile contains a variety of substances including bile salts. A *bile salt* is a derivative of cholesterol with an attached amino acid. Like sodium stearate, a bile salt has both a hydrophobic end and a hydrophilic end. (Figure 13.19 shows the bile salt sodium glycocholate.) The bile salts surround fat particles with their hydrophobic ends oriented toward the fat and their hydrophilic ends facing the water, emulsifying the fat in the aqueous medium of the digestive system. This process allows fats to be digested and other nonpolar substances such as *fat-soluble vitamins* to be absorbed through the wall of the small intestine.

Student Note: It is being nonpolar that makes some vitamins soluble in fat. Remember the axiom "like dissolves like."

Figure 13.17 (a) A sodium stearate molecule. (b) The simplified representation of the molecule that shows a hydrophilic head and a hydrophobic tail.

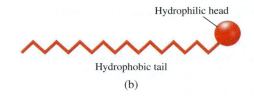

Sodium stearate ($C_{17}H_{35}COO^-Na^+$)

(a)

Hydrophilic head

Hydrophobic tail

(b)

Figure 13.18 The mechanism by which soap removes grease. (a) Grease (oily substance) is not soluble in water. (b) When soap is added to water, the nonpolar tails of soap molecules dissolve in grease. (c) The grease can be washed away when the polar heads of the soap molecules stabilize it in water.

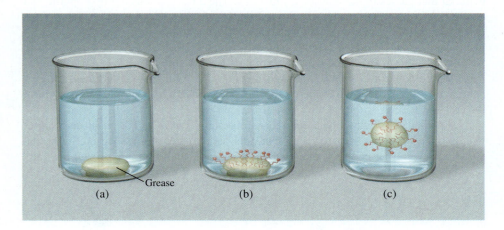

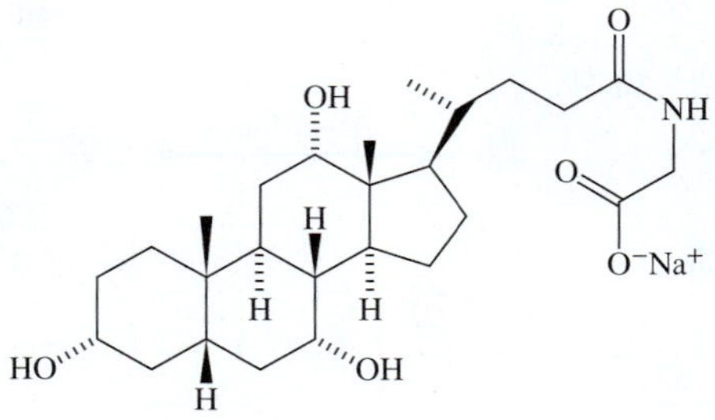

Grease

(a)   (b)   (c)

Figure 13.19 Structure of sodium glycocholate. The hydrophobic tail of sodium glycocholate dissolves in ingested fats, stabilizing them on the aqueous medium of the digestive system.

Chapter Summary

Section 13.1

- Solutions are homogeneous mixtures of two or more substances, which may be solids, liquids, or gases.

- *Saturated solutions* contain the maximum possible amount of dissolved solute.

- The amount of solute dissolved in a saturated solution is the *solubility* of the solute in the specified solvent at the specified temperature.

- *Unsaturated solutions* contain less than the maximum possible amount of solute.

- *Supersaturated solutions* contain more solute than specified by the solubility.

Section 13.2

- Substances with similar intermolecular forces tend to be soluble in one another. "Like dissolves like." Two liquids that are soluble in each other are called *miscible.*

- Solution formation may be endothermic or exothermic overall. An increase in *entropy* is the driving force for solution formation. Solute particles are surrounded by solvent molecules in a process called *solvation.*

Section 13.3

- In addition to molarity (M) and mole fraction (χ), *molality (m)* and *percent by mass* are used to express the concentrations of solutions.

- *Molality* is defined as the number of moles of solute per kilogram of solvent. *Percent by mass* is defined as the mass of solute divided by the total mass of the solution, all multiplied by 100 percent.

- Molality and percent by mass have the advantage of being temperature independent. Conversion among molarity, molality, and percent by mass requires solution *density.*

- The units of concentration used depend on the type of problem to be solved.

Section 13.4

- Increasing the temperature *increases* the solubility of most solids in water and *decreases* the solubility of most gases in water.

- Increasing the pressure increases the solubility of gases in water but does not affect the solubility of solids.

- According to *Henry's law,* the solubility of a gas in a liquid is directly proportional to the partial pressure of the gas over the solution: $c = kP$.

- The proportionality constant k is the *Henry's law constant.* Henry's law constants are specific to the gas and solvent, and they are temperature dependent.

Section 13.5

- *Colligative properties* depend on the *number* (but not on the *type*) of dissolved particles. The colligative properties are *vapor-pressure lowering, boiling-point elevation, freezing-point depression,* and *osmotic pressure.*

- A *volatile* substance is one that has a measurable vapor pressure. A *nonvolatile* substance is one that does not have a measurable vapor pressure.

- According to *Raoult's law,* the partial pressure of a substance over a solution is equal to the *mole fraction (χ)* of the substance times its *pure vapor pressure ($P°$)*. An *ideal solution* is one that obeys Raoult's law.

- *Osmosis* is the flow of solvent through a *semipermeable membrane,* one that allows solvent molecules but not solute particles to pass, from a more dilute solution to a more concentrated one.

- *Osmotic pressure (π)* is the pressure required to prevent osmosis from occurring.

- Two solutions with the same osmotic pressure are called *isotonic. Hypotonic* refers to a solution with a *lower* osmotic pressure. *Hypertonic* refers to a solution with a *higher* osmotic pressure. These terms are often used in reference to human plasma, which has an osmotic pressure of 7.6 atm.

- In electrolyte solutions, the number of dissolved particles is increased by dissociation or ionization. The magnitudes of colligative properties are increased by the *van't Hoff factor (i),* which indicates the degree of dissociation or ionization.

- The experimentally determined van't Hoff factor is generally smaller than the calculated value due to the formation of *ion pairs*—especially at high concentrations. Ion pairs are oppositely charged ions that are attracted to each other and effectively become a single "particle" in solution.

Section 13.6

- Experimentally determined colligative properties can be used to calculate the molar mass of a nonelectrolyte or the *percent dissociation* (or *percent ionization*) of a weak electrolyte.

Section 13.7

- A *colloid* is a dispersion of particles (about 1×10^3 pm to 1×10^6 pm) of one substance in another substance.

- Colloids can be distinguished from true solutions by the *Tyndall effect,* which is the scattering of visible light by colloidal particles.

- Colloids are classified as either *hydrophilic* (water loving) or *hydrophobic* (water fearing).

- Hydrophobic colloids can be stabilized in water by surface interactions with ions or polar molecules.

Key Words

Colligative properties, 631	Hypotonic, 639	Osmotic pressure (π), 636	Solvation, 619
Colloid, 646	Ideal solution, 633	Percent by mass, 624	Supersaturated solution, 618
Entropy, 622	Ion pair, 638	Percent dissociation, 644	Tyndall effect, 646
Henry's law, 629	Isotonic, 637	Percent ionization, 644	Unsaturated solution, 618
Henry's law constant (k), 629	Miscible, 620	Raoult's law, 631	van't Hoff factor (i), 637
Hydrophilic, 647	Molality (m), 624	Saturated solution, 618	Volatile, 633
Hydrophobic, 647	Nonvolatile, 631	Semipermeable membrane, 636	
Hypertonic, 640	Osmosis, 636	Solubility, 618	

Key Equations

13.1 $\text{molality} = m = \dfrac{\text{moles of solute}}{\text{mass of solvent (in kg)}}$

Molality (m) of a particular solute in a solution is calculated by dividing the number of moles of that solute by the number of kilograms of solvent.

13.2 percent by mass
$= \dfrac{\text{mass of solute}}{\text{mass of solute + mass of solvent}} \times 100\%$

Percent mass of a particular solute in a solution is calculated by dividing the mass of that solute by the total mass of solvent and solute.

13.3 $c = kP$

The concentration of gas dissolved in a liquid at a particular temperature is equal to the product of that gas's partial pressure over the solution (P) and the Henry's law constant (k). The value of the Henry's law constant is specific to the gas-solvent-temperature combination.

13.4 $P_1 = \chi_1 P_1^\circ$

The pressure exerted by a solvent over a solution (P_1) is equal to the product of that solvent's mole fraction (χ_1) and the pressure exerted by the pure solvent (P_1°).

13.5 $P_1^\circ - P_1 = \Delta P = \chi_2 P_1^\circ$

The difference between the pressure exerted by a solvent over a solution (P_1) and the pressure exerted by the pure solvent (P_1°) is the vapor-pressure lowering (ΔP) caused by the presence of solute. In the case where only one solute is present, the mole fraction of the solute can be calculated from the value of ΔP. ΔP is the product of the solute's mole fraction (χ_2) and the pressure exerted by the pure solvent (P_1°).

13.6 $\Delta T_b = K_b m$

The amount by which a solvent's boiling point is increased by the presence of a solute (ΔT_b) is calculated as the product of the boiling-point elevation constant (K_b) and the molality of the solute (m). The value of the boiling-point elevation constant depends on the identity of the solvent.

13.7 $\Delta T_f = K_f m$

The amount by which a solvent's boiling point is decreased by the presence of a solute (ΔT_f) is calculated as the product of the freezing-point depression constant (K_f) and the molality of the solute (m). The value of the freezing-point depression constant depends on the identity of the solvent.

13.8 $\pi = MRT$

Osmotic pressure of a solution (π) is the product of solute molarity (M), the gas constant (R) expressed in units of pressure—typically 0.0823 L · atm/K · mol, and absolute temperature (T).

13.9 $\Delta T_b = iK_b m$

For the purpose of calculating the boiling-point elevation of a solution in which the solute is an electrolyte, the solute's molality (m) is multiplied by its van't Hoff factor (i). The value of i depends on the percent dissociation (or ionization) of the solute.

13.10 $\Delta T_f = iK_f m$

For the purpose of calculating the freezing-point depression of a solution in which the solute is an electrolyte, the solute's molality (m) is multiplied by its van't Hoff factor (i). The value of i depends on the percent dissociation (or ionization) of the solute.

13.11 $\pi = iMRT$

For the purpose of calculating the osmotic pressure of a solution in which the solute is an electrolyte, the solute's molarity (M) is multiplied by its van't Hoff factor (i). The value of i depends on the percent dissociation (or ionization) of the solute.

KEY SKILLS

Entropy as a Driving Force

We have seen that although a decrease in system energy can be the driving force for a process [◄◄ Section 9.3], entropy also plays a role in determining whether or not a process will occur. Recall that entropy is a measure of how spread out a system's energy is. The simplest way to interpret this is to consider how spread out a system's energy is in *space*. Consider the example of a compressed gas in one side of a divided container. If the barrier between the two compartments is removed, the compressed gas will expand to fill the new, larger volume:

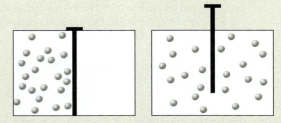

The energy possessed by the gas molecules was originally contained within a smaller volume. After the expansion, the energy possessed by the molecules occupies a larger volume, meaning that the energy is more spread out in space. This spreading out in space of the system's energy is an increase in entropy.

In addition to applying this interpretation of entropy to solution formation and gas expansion [◄◄ Section 13.2], we can apply it to phase changes, such as the sublimation of dry ice [$CO_2(s)$]:

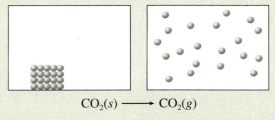

$$CO_2(s) \longrightarrow CO_2(g)$$

The CO_2 molecules in a sample of dry ice possess energy that is confined to the volume of the solid. Sublimation of the sample results in the molecules (and the energy they possess) occupying a much larger volume—corresponding to an increase in entropy. Because of this increase in entropy, although it is endothermic, the sublimation of dry ice does happen spontaneously.

For any process to happen, it must be exothermic, or be accompanied by an entropy increase, or both. An endothermic process (one in which system energy increases) may occur if there is a sufficient increase in the system's entropy. For example, although the dissolution of sucrose ($C_{12}H_{22}O_{11}$) is endothermic, sucrose dissolves in water because of the resulting increase in entropy:

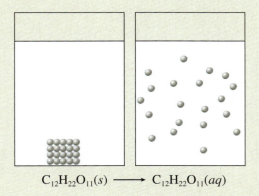

$$C_{12}H_{22}O_{11}(s) \longrightarrow C_{12}H_{22}O_{11}(aq)$$

Likewise, a process that results in an entropy *decrease* may occur if it is sufficiently exothermic. An example of this is the condensation of water vapor on a cool surface. Although the energy of the water molecules is less spread out when it condenses, the condensation process is *exothermic* enough to compensate for the entropy decrease—and it does happen:

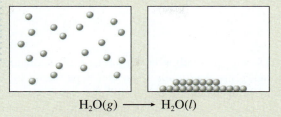

Student Note: Condensation is the reverse of vaporization. The molar enthalpy of vaporization of water (ΔH_{vap}) is 40.79 kJ/mol [◀ Section 11.6]. Therefore, the molar enthalpy of condensation is −40.79 kJ/mol.

$$H_2O(g) \longrightarrow H_2O(l)$$

Processes that are neither exothermic nor accompanied by an entropy increase do not occur.

Being able to make a qualitative assessment of the entropy change associated with a process, and being able to determine whether a process is exothermic or endothermic, can help us predict which processes are likely to happen, and which are not.

Key Skills Problems

13.1

Which of the following processes is accompanied by an increase in entropy? (Select all that apply.)

(a) $Br_2(l) \longrightarrow Br_2(g)$
(b) $NH_3(g) + HCl(g) \longrightarrow NH_4Cl(s)$
(c) $NaCl(s) \longrightarrow Na^+(g) + Cl^-(g)$
(d) $H_2(g) + O_2(g) \longrightarrow 2H_2O(l)$

13.2

For each of the processes depicted here, determine if it is endothermic or exothermic, or if there is not enough information to determine.

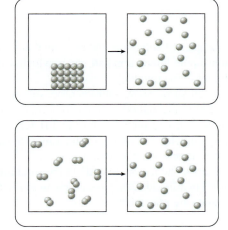

(a) Endothermic, exothermic
(b) Exothermic, endothermic
(c) Endothermic, not enough information to determine
(d) Endothermic, endothermic
(e) Exothermic, not enough information to determine

13.3

For each of the processes depicted here, determine if it is endothermic or exothermic, or if there is not enough information to determine.

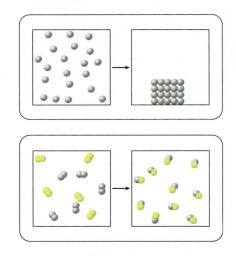

(a) Endothermic, exothermic
(b) Exothermic, endothermic
(c) Endothermic, not enough information to determine
(d) Endothermic, endothermic
(e) Exothermic, not enough information to determine

13.4

For some ionic solutes, dissolution actually causes an overall decrease in entropy. In order for such a dissolution to occur spontaneously, it must be _____.

(a) Exothermic
(b) Endothermic
(c) There is not enough information to determine this.

Questions and Problems

Applying What You've Learned

Although forward osmosis is not yet widely used for large-scale desalination, its use is quite common in *emergency hydration bags*. One type, made by Fluid Technology Solutions (FTSH$_2$O) encloses an aqueous draw "syrup" of electrolytes and other nutrients inside a semipermeable-membrane bag. The bag is designed to be immersed in seawater or other non-potable water. The high concentration of solutes inside the bag draws in water, via forward osmosis, and the semipermeable membrane excludes salt and/or pathogens. The passage of water into the bag dilutes the syrup. When the bag reaches capacity, a process that can take several hours, the diluted syrup is safe to consume and contains lifesaving purified water, calories, and electrolytes.

(a) Emergency hydration bag.

(b) Emergency hydration bags before and after forward osmosis.

(a-b) Fluid Technology Solutions

In one type of emergency hydration bag, the draw syrup consists of 28.0 g dextrose (C$_6$H$_{12}$O$_6$), 22.0 g fructose (C$_6$H$_{12}$O$_6$), 0.59 g KH$_2$PO$_4$, 0.25 g NaCl, and a trace amount of malic acid (C$_4$H$_6$O$_5$) dissolved in enough water to make 100 mL of solution. The density of the solution is 1.308 g/mL. (a) Express the concentration of each of the four major solutes in molality and percent by mass [◄◄ Sample Problem 13.2]. (b) Sodium chloride is an important electrolyte that can be replenished by drinking the contents of the diluted syrup from an emergency hydration bag. Determine the boiling point and freezing point of a solution prepared by dissolving 25 g NaCl in 200.0 mL water [◄◄ Sample Problem 13.6]. (For water, $d = 1$ g/mL.) Explain why the experimentally determined boiling point and freezing point will be significantly different from the calculated values. (c) The draw syrup contains a small amount of malic acid for flavor. Determine the van't Hoff factor of malic acid in a 0.0100-M solution if the osmotic pressure of the solution is 296 mmHg at 80°C [◄◄ Sample Problem 13.7]. (d) Another flavorant used in emergency hydration bags is tartaric acid. A 1.30-M solution of tartaric acid has an osmotic pressure of 32.8 atm at 25°C. Determine the percent ionization of the acid at this concentration [◄◄ Sample Problem 13.11].

SECTION 13.1: TYPES OF SOLUTIONS

Review Questions

13.1 Describe and give examples of an unsaturated solution, a saturated solution, and a supersaturated solution.

13.2 Describe the different types of solutions that can be formed by the combination of solids, liquids, and gases. Give examples of each type of solution.

SECTION 13.2: THE SOLUTION PROCESS

Review Questions

13.3 Briefly describe the solution process at the molecular level. Use the dissolution of a solid in a liquid as an example.

13.4 Basing your answer on intermolecular force considerations, explain what "like dissolves like" means.

13.5 What is solvation? What factors influence the extent to which solvation occurs? Give two examples of solvation; include one that involves ion-dipole interaction and one in which dispersion forces come into play.

13.6 As you know, some solution processes are endothermic and others are exothermic. Provide a molecular interpretation for the difference.

13.7 Explain why dissolving a solid almost always leads to an increase in disorder.

13.8 Describe the factors that affect the solubility of a solid in a liquid. What does it mean to say that two liquids are miscible?

Conceptual Problems

13.9 Why is naphthalene (C$_{10}$H$_8$) more soluble than CsF in benzene?

13.10 Explain why ethanol (C$_2$H$_5$OH) is not soluble in cyclohexane (C$_6$H$_{12}$).

13.11 Arrange the following compounds in order of increasing solubility in water: O$_2$, LiCl, Br$_2$, methanol (CH$_3$OH).

13.12   Explain the variations in solubility in water of the listed alcohols:

Compound	Solubility in Water (g/100 g) at 20°C
CH_3OH	∞
CH_3CH_2OH	∞
$CH_3CH_2CH_2OH$	∞
$CH_3CH_2CH_2CH_2OH$	9
$CH_3CH_2CH_2CH_2CH_2OH$	2.7

Note: ∞ means that the alcohol and water are completely miscible in all proportions.

SECTION 13.3: CONCENTRATION UNITS

Review Questions

13.13   Define the following concentration terms and give their units: percent by mass, mole fraction, molarity, molality. Compare their advantages and disadvantages.

13.14   Outline the steps required for conversion between molarity, molality, and percent by mass.

Computational Problems

13.15   Calculate the percent by mass of the solute in each of the following aqueous solutions: (a) 5.75 g of NaBr in 67.9 g of solution, (b) 24.6 g of KCl in 114 g of water, (c) 4.8 g of toluene in 39 g of benzene.

13.16   Calculate the amount of water (in grams) that must be added to (a) 5.00 g of urea ($(NH_2)_2CO$ in the preparation of a 16.2 percent by mass solution, and (b) 26.2 g of $MgCl_2$ in the preparation of a 1.5 percent by mass solution.

13.17   Calculate the molality of each of the following solutions: (a) 14.3 g of sucrose ($C_{12}H_{22}O_{11}$) in 685 g of water, (b) 7.15 moles of ethylene glycol ($C_2H_6O_2$) in 3505 g of water.

13.18   Calculate the molality of each of the following aqueous solutions: (a) 2.55 M NaCl solution (density of solution = 1.08 g/mL), (b) 45.2 percent by mass KBr solution.

13.19   Calculate the molalities of the following aqueous solutions: (a) 1.22 M sugar ($C_{12}H_{22}O_{11}$) solution (density of solution = 1.12 g/mL), (b) 0.87 M NaOH solution (density of solution = 1.04 g/mL), (c) 5.24 M $NaHCO_3$ solution (density of solution = 1.19 g/mL).

13.20   Calculate the molality of a 0.015-M aqueous solution of NaCl, and the molality of a 0.015-M solution of CS_2 in n-hexane (d = 0.6606 g/mL). Use the comparison of molarity and molality of each solution to explain why the two are equal for dilute aqueous solutions but that, in general, molarity and molality are different numbers.

13.21   The alcohol content of hard liquor is normally given in terms of the "proof," which is defined as twice the percentage by volume of ethanol (C_2H_5OH) present.

Calculate the number of grams of alcohol present in 1.00 L of 80 proof vodka. The density of ethanol is 0.798 g/mL.

13.22   The concentrated sulfuric acid we use in the laboratory is 98.0 percent H_2SO_4 by mass. Calculate the molality and molarity of the acid solution. The density of the solution is 1.83 g/mL.

13.23   Calculate the molarity and molality of an NH_3 solution made up of 35.0 g of NH_3 in 75.0 g of water. The density of the solution is 0.982 g/mL.

13.24   The density of an aqueous solution containing 15.0 percent of ethanol (C_2H_5OH) by mass is 0.984 g/mL. (a) Calculate the molality of this solution. (b) Calculate its molarity. (c) What volume of the solution would contain 0.250 mole of ethanol?

13.25   Fish breathe the dissolved air in water through their gills. Assuming the partial pressures of oxygen and nitrogen in air to be 0.20 and 0.80 atm, respectively, calculate the mole fractions of oxygen and nitrogen in the air dissolved in water at 298 K. The solubilities of O_2 and N_2 in water at 298 K are 1.3×10^{-3} mol/L · atm and 6.8×10^{-4} mol/L · atm, respectively. Comment on your results.

SECTION 13.4: FACTORS THAT AFFECT SOLUBILITY

Review Questions

13.26   How do the solubilities of most ionic compounds in water change with temperature? With pressure?

13.27   Discuss the factors that influence the solubility of a gas in a liquid.

13.28   What is thermal pollution? Why is it harmful to aquatic life?

13.29   What is Henry's law? Define each term in the equation, and give its units. How would you account for the law in terms of the kinetic molecular theory of gases? Give two exceptions to Henry's law.

13.30   A student is observing two beakers of water. One beaker is heated to 30°C, and the other is heated to 100°C. In each case, bubbles form in the water. Are these bubbles of the same origin? Explain.

13.31   A man bought a goldfish in a pet shop. Upon returning home, he put the goldfish in a bowl of recently boiled water that had been cooled quickly. A few minutes later the fish was found dead. Explain what happened to the fish.

Computational Problems

13.32   A 3.20-g sample of a salt dissolves in 9.10 g of water to give a saturated solution at 25°C. What is the solubility (in g salt/100 g of H_2O) of the salt?

13.33   The solubility of KNO_3 is 155 g per 100 g of water at 75°C and 38.0 g at 25°C. What mass (in grams) of KNO_3 will crystallize out of solution if exactly 100 g of its saturated solution at 75°C is cooled to 25°C?

13.34   A 50-g sample of impure $KClO_3$ (solubility = 7.1 g per 100 g H_2O at 20°C) is contaminated with 10 percent of KCl (solubility = 25.5 g per 100 g of H_2O at 20°C). Calculate the minimum quantity of 20°C water needed to dissolve all the KCl from the sample. How much $KClO_3$ will be left after this treatment? (Assume that the solubilities are unaffected by the presence of the other compound.)

13.35   The solubility of CO_2 in water at 25°C and 1 atm is 0.034 mol/L. What is its solubility under atmospheric conditions? (The partial pressure of CO_2 in air is 0.0003 atm.) Assume that CO_2 obeys Henry's law.

13.36   The solubility of N_2 in blood at 37°C and at a partial pressure of 0.80 atm is 5.6×10^{-4} mol/L. A deep-sea diver breathes compressed air with the partial pressure of N_2 equal to 4.0 atm. Assume that the total volume of blood in the body is 5.0 L. Calculate the amount of N_2 gas released (in liters at 37°C and 1 atm) when the diver returns to the surface of the water, where the partial pressure of N_2 is 0.80 atm.

Conceptual Problems

13.37   A miner working 260 m below sea level opened a carbonated soft drink during a lunch break. To his surprise, the soft drink tasted rather "flat." Shortly afterward, the miner took an elevator to the surface. During the trip up, he could not stop belching. Why?

13.38   A beaker of water is initially saturated with dissolved air. Explain what happens when He gas at 1 atm is bubbled through the solution for a long time.

13.39   A student carried out the following experiment to measure the pressure of carbon dioxide in the space above the carbonated soft drink in a bottle. First, she weighed the bottle (853.5 g). Next, she carefully removed the cap to let the CO_2 gas escape. She then reweighed the bottle with the cap (851.3 g). Finally, she measured the volume of the soft drink (452.4 mL). Given that the Henry's law constant for CO_2 in water at 25°C is 3.4×10^{-2} mol/L · atm, calculate the pressure of CO_2 over the soft drink in the bottle before it was opened. Explain why this pressure is only an estimate of the true value.

13.40   The first diagram represents an open system with two different gases dissolved in water. Which of the diagrams (a)–(c) could represent the same system at a higher temperature?

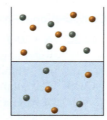

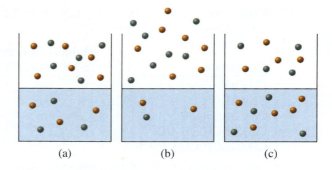

(a)   (b)   (c)

13.41   The following diagrams represent an aqueous solution at two different temperatures. In both diagrams, the solution is saturated in two different solutes, each of which is represented by a different color. Using the diagrams, determine for each solute whether the dissolution process is endothermic or exothermic. For which of the solutes do you think the numerical value of ΔH_{soln} is greater? Explain.

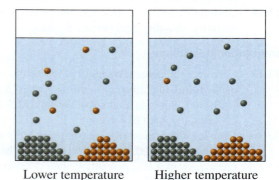

Lower temperature   Higher temperature

SECTION 13.5: COLLIGATIVE PROPERTIES

Review Questions

13.42   What are colligative properties? What is the meaning of the word *colligative* in this context?

13.43   Give two examples of (a) a volatile liquid and (b) a nonvolatile liquid.

13.44   Write the equation representing Raoult's law, and express it in words.

13.45   Use a solution of benzene in toluene to explain what is meant by an ideal solution.

13.46   Write the equations relating boiling-point elevation and freezing-point depression to the concentration of the solution. Define all the terms, and give their units.

13.47   How is vapor-pressure lowering related to a rise in the boiling point of a solution?

13.48   Use a phase diagram to show the difference in freezing points and boiling points between an aqueous urea solution and pure water.

13.49   What is osmosis? What is a semipermeable membrane?

13.50 Write the equation relating osmotic pressure to the concentration of a solution. Define all the terms, and specify their units.

13.51 What does it mean when we say that the osmotic pressure of a sample of seawater is 25 atm at a certain temperature?

13.52 Explain why molality is used for boiling-point elevation and freezing-point depression calculations and molarity is used in osmotic pressure calculations.

13.53 Why is the discussion of the colligative properties of electrolyte solutions more involved than that of nonelectrolyte solutions?

13.54 What are ion pairs? What effect does ion-pair formation have on the colligative properties of a solution? How does the ease of ion-pair formation depend on (a) charges on the ions, (b) size of the ions, (c) nature of the solvent (polar versus nonpolar), (d) concentration?

13.55 What is the van't Hoff factor? What information does it provide?

13.56 For most intravenous injections, great care is taken to ensure that the concentration of solutions to be injected is comparable to that of blood plasma. Explain.

Computational Problems

13.57 A solution is prepared by dissolving 396 g of sucrose ($C_{12}H_{22}O_{11}$) in 624 g of water. What is the vapor pressure of this solution at 30°C? (The vapor pressure of water is 31.8 mmHg at 30°C.)

13.58 How many grams of sucrose ($C_{12}H_{22}O_{11}$) must be added to 552 g of water to give a solution with a vapor pressure 2.0 mmHg less than that of pure water at 20°C? (The vapor pressure of water at 20°C is 17.5 mmHg.)

13.59 The vapor pressure of benzene is 100.0 mmHg at 26.1°C. Calculate the vapor pressure of a solution containing 24.6 g of camphor ($C_{10}H_{16}O$) dissolved in 98.5 g of benzene. (Camphor is a low-volatility solid.)

13.60 The vapor pressures of ethanol (C_2H_5OH) and 1-propanol (C_3H_7OH) at 35°C are 100 and 37.6 mmHg, respectively. Assume ideal behavior and calculate the partial pressures of ethanol and 1-propanol at 35°C over a solution of ethanol in 1-propanol, in which the mole fraction of ethanol is 0.455.

13.61 How many grams of urea [($NH_2)_2CO$] must be added to 658 g of water to give a solution with a vapor pressure 2.50 mmHg lower than that of pure water at 30°C? (The vapor pressure of water at 30°C is 31.8 mmHg.)

13.62 What are the boiling point and freezing point of a 3.12-m solution of naphthalene in benzene? (The

boiling point and freezing point of benzene are 80.1°C and 5.5°C, respectively.)

13.63 An aqueous solution contains the amino acid glycine (NH_2CH_2COOH). Assuming that the acid does not ionize in water, calculate the molality of the solution if it freezes at −1.1°C.

13.64 How many liters of the antifreeze ethylene glycol [$CH_2(OH)CH_2(OH)$] would you add to a car radiator containing 6.50 L of water if the coldest winter temperature in your area is −20°C? Calculate the boiling point of this water-ethylene glycol mixture. (The density of ethylene glycol is 1.11 g/mL.)

13.65 A solution is prepared by condensing 4.00 L of a gas, measured at 27°C and 748 mmHg pressure, into 75.0 g of benzene. Calculate the freezing point of this solution.

13.66 What is the osmotic pressure (in atm) of a 1.57-M aqueous solution of urea [($NH_2)_2CO$] at 27.0°C?

13.67 What are the normal freezing points and boiling points of the following solutions: (a) 21.2 g NaCl in 135 mL of water and (b) 15.4 g of urea in 66.7 mL of water?

13.68 At 25°C, the vapor pressure of pure water is 23.76 mmHg and that of seawater is 22.98 mmHg. Assuming that seawater contains only NaCl, estimate its molal concentration.

13.69 Both NaCl and $CaCl_2$ are used to melt ice on roads and sidewalks in winter. What advantages do these substances have over sucrose or urea in lowering the freezing point of water?

13.70 A 0.86 percent by mass solution of NaCl is called "physiological saline" because its osmotic pressure is equal to that of the solution in blood cells. Calculate the osmotic pressure of this solution at normal body temperature (37°C). Note that the density of the saline solution is 1.005 g/mL.

13.71 The osmotic pressure of 0.010-M solutions of $CaCl_2$ and urea at 25°C are 0.605 and 0.245 atm, respectively. Calculate the van't Hoff factor for the $CaCl_2$ solution.

13.72 Calculate the osmotic pressure of a 0.0500 M $MgSO_4$ solution at 25°C. (*Hint:* See Table 13.3.)

13.73 The tallest trees known are the redwoods in California. Assuming the height of a redwood to be 105 m (about 350 ft), estimate the osmotic pressure required to push water up to the treetop.

13.74 Calculate the difference in osmotic pressure (in atm) at the normal body temperature between the blood plasma of a diabetic patient and that of a healthy adult. Assume that the sole difference between the two people is due to the higher glucose level in the diabetic patient. The glucose levels are 1.75 and 0.84 g/L, respectively. Based on your result, explain why such a patient frequently feels thirsty.

Conceptual Problems

13.75 Which of the following aqueous solutions has (a) the higher boiling point, (b) the higher freezing point, and (c) the lower vapor pressure: 0.35 *m* CaCl$_2$ or 0.90 *m* urea? Explain. Assume complete dissociation.

13.76 Consider two aqueous solutions, one of sucrose (C$_{12}$H$_{22}$O$_{11}$) and the other of nitric acid (HNO$_3$). Both solutions freeze at −1.5°C. What other properties do these solutions have in common?

13.77 Arrange the following solutions in order of decreasing freezing point: 0.10 *m* Na$_3$PO$_4$, 0.35 *m* NaCl, 0.20 *m* MgCl$_2$, 0.15 *m* C$_6$H$_{12}$O$_6$, 0.15 *m* CH$_3$COOH.

13.78 Arrange the following aqueous solutions in order of decreasing freezing point, and explain your reasoning: 0.50 *m* HCl, 0.50 *m* glucose, 0.50 *m* acetic acid.

13.79 Indicate which compound in each of the following pairs is more likely to form ion pairs in water: (a) NaCl or Na$_2$SO$_4$, (b) MgCl$_2$ or MgSO$_4$, (c) LiBr or KBr.

13.80 The first diagram represents an aqueous solution of a weak electrolyte. Indicate which of the other diagrams, (a) through (c), could represent an aqueous solution in which the solute exhibits the same van't Hoff factor as the original.

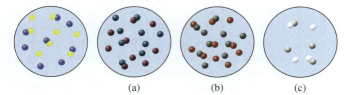

(a)   (b)   (c)

13.81 The diagram at right represents a system consisting of an aqueous solution separated from pure water by a semipermeable membrane. Indicate which of the other diagrams, (a) through (d), could represent the system after the passage of time.

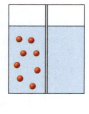

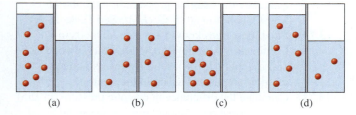

(a)   (b)   (c)   (d)

SECTION 13.6: CALCULATIONS USING COLLIGATIVE PROPERTIES

Review Questions

13.82 Describe how you would use freezing-point depression and osmotic pressure measurements to determine the molar mass of a compound. Why are boiling-point elevation and vapor-pressure lowering normally not used for this purpose?

13.83 Describe how you would use the osmotic pressure to determine the percent ionization of a weak, monoprotic acid.

Computational Problems

13.84 The elemental analysis of an organic solid extracted from gum arabic (a gummy substance used in adhesives, inks, and pharmaceuticals) showed that it contained 40.0 percent C, 6.7 percent H, and 53.3 percent O. A solution of 0.650 g of the solid in 27.8 g of the solvent diphenyl gave a freezing-point depression of 1.56°C. Calculate the molar mass and molecular formula of the solid. (*K*$_f$ for diphenyl is 8.00°C/*m*.)

13.85 A solution of 2.50 g of a compound having the empirical formula C$_6$H$_5$P in 25.0 g of benzene is observed to freeze at 4.3°C. Calculate the molar mass of the solute and its molecular formula.

13.86 The molar mass of benzoic acid (C$_6$H$_5$COOH) determined by measuring the freezing-point depression in benzene is twice what we would expect for the molecular formula, C$_7$H$_6$O$_2$. Explain this apparent anomaly.

13.87 A solution containing 0.8330 g of a polymer of unknown structure in 170.0 mL of an organic solvent was found to have an osmotic pressure of 5.20 mmHg at 25°C. Determine the molar mass of the polymer.

13.88 A quantity of 7.480 g of an organic compound is dissolved in water to make 300.0 mL of solution. The solution has an osmotic pressure of 1.43 atm at 27°C. The analysis of this compound shows that it contains 41.8 percent C, 4.7 percent H, 37.3 percent O, and 16.3 percent N. Calculate the molecular formula of the compound.

13.89 A solution of 6.85 g of a carbohydrate in 100.0 g of water has a density of 1.024 g/mL and an osmotic pressure of 4.61 atm at 20.0°C. Calculate the molar mass of the carbohydrate.

13.90 A 0.036-*M* aqueous nitrous acid (HNO$_2$) solution has an osmotic pressure of 0.93 atm at 25°C. Calculate the percent ionization of the acid.

13.91 A 0.100-*M* aqueous solution of the base HB has an osmotic pressure of 2.83 atm at 25°C. Calculate the percent ionization of the base.

SECTION 13.7: COLLOIDS

Review Questions

13.92 What are colloids? Referring to Table 13.5, why is there no colloid in which both the dispersed phase and the dispersing medium are gases?

13.93 Describe how hydrophilic and hydrophobic colloids are stabilized in water.

13.94 Describe and give an everyday example of the Tyndall effect.

ADDITIONAL PROBLEMS

13.95 Predict whether each vitamin will be water soluble or fat soluble.

Vitamin D

Vitamin B$_2$ (riboflavin)

13.96 Solutions A and B have osmotic pressures of 2.4 and 4.6 atm, respectively, at a certain temperature. What is the osmotic pressure of a solution prepared by mixing equal volumes of A and B at the same temperature?

13.97 Acetic acid is a polar molecule and can form hydrogen bonds with water molecules. Therefore, it has a high solubility in water. Yet acetic acid is also soluble in benzene (C_6H_6), a nonpolar solvent that lacks the ability to form hydrogen bonds. A solution of 3.8 g of CH_3COOH in 80 g C_6H_6 has a freezing point of 3.5°C. Calculate the molar mass of the solute, and suggest what its structure might be. (*Hint:* Acetic acid molecules can form hydrogen bonds between themselves.)

13.98 A 2.6-L sample of water contains 192 μg of lead. Does this concentration of lead exceed the safety limit of 0.050 ppm of lead per liter of drinking water?

13.99 Expanded polystyrene foams (EPS) are commonly used to make disposable cups, food containers, and packing peanuts. When 1.00 g of a particular EPS is dissolved in enough benzene to make 500.0 mL of solution, the experimentally measured osmotic pressure of the solution at 25°C is 4.76 mmHg. Determine the molar mass of this EPS.

13.100 Two liquids A and B have vapor pressures of 76 and 132 mmHg, respectively, at 25°C. What is the total vapor pressure of the ideal solution made up of (a) 1.00 mol of A and 1.00 mol of B and (b) 2.00 mol of A and 5.00 mol of B?

13.101 Determine the van't Hoff factor of Na_3PO_4 in a 0.40-m solution whose freezing point is −2.6°C.

13.102 A 262-mL sample of a sugar solution containing 1.22 g of the sugar has an osmotic pressure of 30.3 mmHg at 35°C. What is the molar mass of the sugar?

13.103 Predict whether vitamin B$_6$, also known as pyridoxine, is water soluble or fat soluble.

Vitamin B$_6$

13.104 A forensic chemist is given a white powder for analysis. She dissolves 0.50 g of the substance in 8.0 g of benzene. The solution freezes at 3.9°C. Can the chemist conclude that the compound is cocaine ($C_{17}H_{21}NO_4$)? What assumptions are made in the analysis?

13.105 Predict whether vitamin A is water soluble or fat soluble.

Vitamin A

13.106 A solution of 1.00 g of anhydrous aluminum chloride ($AlCl_3$) in 50.0 g of water freezes at −1.11°C. Does the molar mass determined from this freezing point agree with that calculated from the formula? Why?

13.107 Explain why reverse osmosis is (theoretically) more desirable as a desalination method than distillation or freezing. What minimum pressure must be applied to seawater at 25°C for reverse osmosis to occur? (Treat seawater as a 0.70 M NaCl solution.)

13.108 A 1.32-g sample of a mixture of cyclohexane (C_6H_{12}) and naphthalene ($C_{10}H_8$) is dissolved in 18.9 g of benzene (C_6H_6). The freezing point of the solution is 2.2°C. Calculate the mass percent of the mixture. (See Table 13.2 for constants.)

13.109 How does each of the following affect the solubility of an ionic compound: (a) lattice energy, (b) solvent (polar versus nonpolar), (c) enthalpies of hydration of cation and anion?

13.110 A solution contains two volatile liquids A and B. Complete the following table, in which the symbol $\longleftrightarrow$ indicates attractive intermolecular forces.

Attractive Forces	Deviation from Raoult's Law	ΔH_{soln}
A $\longleftrightarrow$ A, B $\longleftrightarrow$ B > A $\longleftrightarrow$ B		
	Negative	
		Zero

13.111 The concentration of commercially available concentrated nitric acid is 70.0 percent by mass, or 15.9 M. Calculate the density and the molality of the solution.

13.112 A mixture of ethanol and 1-propanol behaves ideally at 36°C and is in equilibrium with its vapor. If the mole fraction of ethanol in the solution is 0.62, calculate its mole fraction in the vapor phase at this temperature. (The vapor pressures of pure ethanol and 1-propanol at 36°C are 108 and 40.0 mmHg, respectively.)

13.113 Ammonia (NH_3) is very soluble in water, but nitrogen trichloride (NCl_3) is not. Explain.

13.114 For ideal solutions, the volumes are additive. This means that if 5 mL of A and 5 mL of B form an ideal solution, the volume of the solution is 10 mL. Provide a molecular interpretation for this observation. When 500 mL of ethanol (C_2H_5OH) is mixed with 500 mL of water, the final volume is less than 1000 mL. Why?

13.115 Acetic acid is a weak acid that ionizes in solution as follows:

$$CH_3COOH(aq) \rightleftharpoons CH_3COO^-(aq) + H^+(aq)$$

If the freezing point of a 0.106 m CH_3COOH solution is −0.203°C, calculate the percent of the acid that has undergone ionization.

13.116 Iodine (I_2) is only sparingly soluble in water (left photo). Yet upon the addition of iodide ions (e.g., from KI), iodine is converted to the triiodide ion, which readily dissolves (right photo):

$$I_2(s) + I^-(aq) \rightleftharpoons I_3^-(aq)$$

Describe the change in solubility of I_2 in terms of the change in intermolecular forces.

13.117 Concentrated hydrochloric acid is usually available at a concentration of 37.7 percent by mass. What is its molar concentration? (The density of the solution is 1.19 g/mL.)

13.118 Explain each of the following statements: (a) The boiling point of seawater is higher than that of pure water. (b) Carbon dioxide escapes from the solution when the cap is removed from a carbonated soft drink bottle. (c) Molal and molar concentrations of dilute aqueous solutions are approximately equal. (d) In discussing the colligative properties of a solution (other than osmotic pressure), it is preferable to express the concentration in units of molality rather than in molarity. (e) Methanol (b.p. 65°C) is useful as an antifreeze, but it should be removed from the car radiator during the summer season.

13.119 A mixture of NaCl and sucrose ($C_{12}H_{22}O_{12}$) of combined mass 10.2 g is dissolved in enough water to make up a 250-mL solution. The osmotic pressure of the solution is 7.32 atm at 23°C. Calculate the mass percent of NaCl in the mixture.

13.120 (a) Derive the equation relating the molality (m) of a solution to its molarity (M)

$$m = \frac{M}{d - \dfrac{M\mathcal{M}}{1000}}$$

where d is the density of the solution (g/mL) and $\mathcal{M}$ is the molar mass of the solute (g/mol). (*Hint:* Start by expressing the solvent in kilograms in terms of the difference between the mass of the solution and the mass of the solute.) (b) Show that, for dilute aqueous solutions, m is approximately equal to M.

13.121 At 27°C, the vapor pressure of pure water is 23.76 mmHg and that of an aqueous solution of urea is 22.98 mmHg. Calculate the molality of urea in the solution.

13.122 A nonvolatile organic compound Z was used to make up two solutions. Solution A contains 5.00 g of Z dissolved in 100 g of water, and solution B contains 2.31 g of Z dissolved in 100 g of benzene. Solution A has a vapor pressure of 754.5 mmHg at the normal boiling point of water, and solution B has the same vapor pressure at the normal boiling point of benzene. Calculate the molar mass of Z in solutions A and B, and account for the difference.

13.123 The diagrams below represent aqueous solutions of various solutes. Arrange them in order of increasing boiling point; increasing freezing point; and increasing van't Hoff factor value.

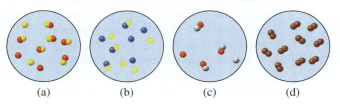

(a)   (b)   (c)   (d)

13.124   The diagram below on the left represents an aqueous solution with a boiling point of 106.5°C. Determine the boiling point of the aqueous solution represented by the diagram on the right.

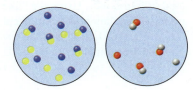

13.125   Hydrogen peroxide with a concentration of 3.0% (3.0 g of H_2O_2 in 100 mL of solution) is sold in drugstores for use as an antiseptic. For a 10.0-mL 3.0% H_2O_2 solution, calculate (a) the oxygen gas produced (in liters) at STP when the compound undergoes complete decomposition and (b) the ratio of the volume of O_2 collected to the initial volume of the H_2O_2 solution.

13.126   State which of the alcohols listed in Problem 13.12 you would expect to be the best solvent for each of the following substances, and explain why: (a) I_2, (b) KBr, (c) $CH_3CH_2CH_2CH_2CH_3$.

13.127   The diagram shows vapor pressure curves for pure benzene and a solution of a nonvolatile solute in benzene. Estimate the molality of the solution.

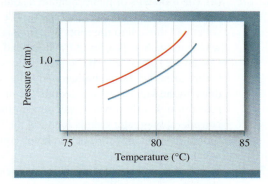

13.128   A mixture of liquids A and B exhibits ideal behavior. At 84°C, the total vapor pressure of a solution containing 1.2 moles of A and 2.3 moles of B is 331 mmHg. Upon the addition of another mole of B to the solution, the vapor pressure increases to 347 mmHg. Calculate the vapor pressure of pure A and B at 84°C.

13.129   Use Henry's law and the ideal gas equation to prove the statement that the volume of a gas that dissolves in a given amount of solvent is *independent* of the pressure of the gas. (*Hint:* Henry's law can be modified as $n = kP$, where n is the number of moles of the gas dissolved in the solvent.)

13.130   At 298 K, the osmotic pressure of a glucose solution is 10.50 atm. Calculate the freezing point of the solution. The density of the solution is 1.16 g/mL.

13.131   Two beakers are placed in a closed container. Beaker A initially contains 0.15 mole of naphthalene ($C_{10}H_8$) in 100 g of benzene (C_6H_6), and beaker B

initially contains 31 g of an unknown compound dissolved in 100 g of benzene. At equilibrium, beaker A is found to have lost 7.0 g of benzene. Assuming ideal behavior, calculate the molar mass of the unknown compound. State any assumptions made.

Engineering Problems

13.132   Consider the three mercury manometers shown in the diagram at right. One of them has 1 mL of water on top of the mercury, another has 1 mL of a 1 *m* urea solution on top of the mercury, and the third one has 1 mL of a 1 *m* NaCl solution placed on top of the mercury. Which of these solutions is in the tube labeled X, which is in Y, and which is in Z?

13.133   Before a carbonated beverage bottle is sealed, it is pressurized with a mixture of air and carbon dioxide. (a) Explain the effervescence that occurs when the cap of the bottle is removed. (b) What causes the fog to form near the mouth of the bottle right after the cap is removed?

13.134   Aluminum sulfate [$Al_2(SO_4)_3$] is sometimes used in municipal water treatment plants to remove undesirable particles. Explain how this process works.

13.135   Two beakers, one containing a 50-mL aqueous 1.0 *M* glucose solution and the other a 50-mL aqueous 2.0 *M* glucose solution, are placed under a tightly sealed bell jar at room temperature. What are the volumes in these two beakers at equilibrium?

13.136   In the apparatus shown, what will happen if the membrane is (a) permeable to both water and the Na^+ and Cl^- ions, (b) permeable to water and the Na^+ ions but not to the Cl^- ions, (c) permeable to water but not to the Na^+ and Cl^- ions?

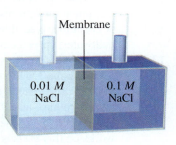

Biological Problems

13.137   Making mayonnaise involves beating oil into small droplets in water, in the presence of egg yolk. What is the purpose of the egg yolk? (*Hint:* Egg yolk contains lecithins, which are molecules with a polar head and a long nonpolar hydrocarbon tail.)

13.138   Fish in the Antarctic Ocean swim in water at about −2°C. (a) To prevent their blood from freezing, what must be the concentration (in molality) of the blood? Is this a reasonable physiological concentration? (b) In recent years, scientists have discovered a special type of

Dmytro Pylypenko/Shutterstock

protein in the blood of these fish that, although present in quite low concentrations ($\leq 0.001\ m$), has the ability to prevent the blood from freezing. Suggest a mechanism for its action.

13.139 Lysozyme is an enzyme that cleaves bacterial cell walls. A sample of lysozyme extracted from egg white has a molar mass of 13,930 g. A quantity of 0.100 g of this enzyme is dissolved in 150 g of water at 25°C. Calculate the vapor-pressure lowering, the depression in freezing point, the elevation in boiling point, and the osmotic pressure of this solution. (The vapor pressure of water at 25°C is 23.76 mmHg.)

13.140 The blood sugar (glucose) level of a diabetic patient is approximately 0.140 g of glucose/100 mL of blood. Every time the patient ingests 40 g of glucose, her blood glucose level rises to approximately 0.240 g/100 mL of blood. Calculate the number of moles of glucose per milliliter of blood and the total number of moles and grams of glucose in the blood before and after consumption of glucose. (Assume that the total volume of blood in her body is 5.0 L.)

13.141 Trees in cold climates may be subjected to temperatures as low as −60°C. Estimate the concentration of an aqueous solution in the body of the tree that would remain unfrozen at this temperature. Is this a reasonable concentration? Comment on your result.

13.142 A cucumber placed in concentrated brine (saltwater) shrivels into a pickle. Explain.

13.143 "Time-release" drugs have the advantage of releasing the drug to the body at a constant rate so that the drug concentration at any time is not too high as to have harmful side effects or too low as to be ineffective. A schematic diagram of a pill that works on this basis is shown. Explain how it works.

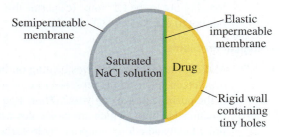

13.144 Valinomycin is an antibiotic. It functions by binding K^+ ions and transporting them across the membrane into cells to offset the ionic balance. The molecule is represented here by its skeletal structure in which

the end of each straight line corresponds to a carbon atom (unless a different atom is shown at the end of the line). There are as many H atoms attached to each C atom as necessary to give each C atom a total of four bonds. Using the "like dissolves like" principle, explain how the molecule functions. (*Hint:* The $-CH_3$ groups at the two ends of each Y shape are nonpolar.)

13.145 (a) The root cells of plants contain a solution that is hypertonic in relation to water in the soil. Thus, water can move into the roots by osmosis. Explain why salts such as NaCl and $CaCl_2$ spread on roads to melt ice can be harmful to nearby trees. (b) Just before urine leaves the human body, the collecting ducts in the kidney (which contain the urine) pass through a fluid whose salt concentration is considerably greater than is found in the blood and tissues. Explain how this action helps conserve water in the body.

13.146 What masses of sodium chloride, magnesium chloride, sodium sulfate, calcium chloride, potassium chloride, and sodium bicarbonate are needed to produce 1 L of artificial seawater for an aquarium? The required ionic concentrations are $[Na^+] = 2.56\ M$, $[K^+] = 0.0090\ M$, $[Mg^{2+}] = 0.054\ M$, $[Ca^{2+}] = 0.010\ M$, $[HCO_3^-] = 0.0020\ M$, $[Cl^-] = 2.60\ M$, $[SO_4^{2-}] = 0.051\ M$.

13.147 The osmotic pressure of blood plasma is approximately 7.5 atm at 37°C. Estimate the total concentration of dissolved species and the freezing point of blood.

13.148 The antibiotic gramicidin A can transport Na^+ ions into a certain cell at the rate of 5.0×10^7 Na^+ ions/channel · s. Calculate the time in seconds to transport enough Na^+ ions to increase its concentration by 8.0×10^{-3} M in a cell whose intracellular volume is 2.0×10^{-10} mL.

13.149 A protein has been isolated as a salt with the formula $Na_{20}P$ (this notation means that there are 20 Na^+ ions associated with a negatively charged protein P^{20-}). The osmotic pressure of a 10.0-mL solution containing 0.225 g of the protein is 0.257 atm at 25.0°C. (a) Calculate the molar mass of the protein from these data. (b) Calculate the actual molar mass of the protein.

Multiconcept Problems

13.150   Hemoglobin, the oxygen-transport protein, binds about 1.35 mL of oxygen per gram of the protein. The concentration of hemoglobin in normal blood is 150 g/L blood. Hemoglobin is about 95 percent saturated with O_2 in the lungs and only 74 percent saturated with O_2 in the capillaries. Calculate the volume of O_2 released by hemoglobin when 100 mL of blood flows from the lungs to the capillaries.

13.151   Pheromones are compounds secreted by the females of many insect species to attract males. One of these compounds contains 80.78 percent C, 13.56 percent H, and 5.66 percent O. A solution of 1.00 g of this pheromone in 8.50 g of benzene freezes at 3.37°C. What are the molecular formula

and molar mass of the compound? (The normal freezing point of pure benzene is 5.50°C.)

13.152   The vapor pressure of ethanol (C_2H_5OH) at 20°C is 44 mmHg, and the vapor pressure of methanol (CH_3OH) at the same temperature is 94 mmHg. A mixture of 30.0 g of methanol and 45.0 g of ethanol is prepared (and can be assumed to behave as an ideal solution). (a) Calculate the vapor pressure of methanol and ethanol above this solution at 20°C. (b) Calculate the mole fraction of methanol and ethanol in the vapor above this solution at 20°C. (c) Suggest a method for separating the two components of the solution.

13.153   A very long pipe is capped at one end with a semipermeable membrane. How deep (in meters) must the pipe be immersed into the sea for fresh water to begin to pass through the membrane? Assume the water to be at 20°C, and treat it as a 0.70 M NaCl solution. The density of seawater is 1.03 g/cm^3, and the acceleration due to gravity is 9.81 m/s^2.

Standardized-Exam Practice Problems

Physical and Biological Sciences

A mixture of two volatile liquids is said to be ideal if each component obeys Raoult's law:

$$P_i = \chi_i P_i^\circ$$

Two volatile liquids A (molar mass 100 g/mol) and B (molar mass 110 g/mol) form an ideal solution. At 55°C, A has a vapor pressure of 98 mmHg and B has a vapor pressure of 42 mmHg. A solution is prepared by mixing equal masses of A and B.

1. Calculate the mole fraction of each component in the solution.

    a) $\chi_A = 0.50$, $\chi_B = 0.50$ c) $\chi_A = 1.0$, $\chi_B = 0.91$
    b) $\chi_A = 0.52$, $\chi_B = 0.48$ d) $\chi_A = 0.09$, $\chi_B = 0.91$

2. Calculate the partial pressures of A and B over the solution at 55°C.

    a) $P_A = 98$ mmHg, $P_B = 42$ mmHg
    b) $P_A = 49$ mmHg, $P_B = 21$ mmHg
    c) $P_A = 70$ mmHg, $P_B = 70$ mmHg
    d) $P_A = 51$ mmHg, $P_B = 20$ mmHg

3. Suppose that some of the vapor over the solution at 55°C is condensed to a liquid. Calculate the mole fraction of each component in the condensed liquid.

    a) $\chi_A = 0.50$, $\chi_B = 0.50$
    b) $\chi_A = 0.52$, $\chi_B = 0.48$
    c) $\chi_A = 0.72$, $\chi_B = 0.28$
    d) $\chi_A = 1.0$, $\chi_B = 0.86$

4. Calculate the partial pressures of the components above the condensed liquid at 55°C.

    a) $P_A = 98$ mmHg, $P_B = 42$ mmHg
    b) $P_A = 30$ mmHg, $P_B = 27$ mmHg
    c) $P_A = 71$ mmHg, $P_B = 12$ mmHg
    d) $P_A = 72$ mmHg, $P_B = 28$ mmHg

Answers to In-Chapter Materials

Answers to Practice Problems

13.1A CS_2. **13.1B** C_3H_8, I_2, CS_2. **13.2A** (a) 0.423 m, (b) 2.48%. **13.2B** 0.78 m. **13.3A** (a) 1.8 M, (b) 1.9 m. **13.3B** 12.8%. **13.4A** 0.12 M. **13.4B** 26 atm. **13.5A** 22.2 mmHg. **13.5B** 103 g. **13.6A** f.p. = −7.91°C, b.p. = 102.2°C. **13.6B** 710 g. **13.7A** $i = 1.21$. **13.7B** f.p. = −0.35°C. **13.8A** 10.2 atm. **13.8B** (a) 0.22 M, (b) 0.11 M. **13.9A** 128 g/mol. **13.9B** 12.5 g. **13.10A** 6.0 × 10^3 g/mol. **13.10B** 0.27 g. **13.11A** 4%. **13.11B** 0.38 atm.

Answers to Checkpoints

13.2.1 b, d, e. **13.2.2** b, d. **13.3.1** b, c, d. **13.3.2** d. **13.3.3** d. **13.3.4** a. **13.4.1** c. **13.4.2** a. **13.5.1** d. **13.5.2** c. **13.5.3** b. **13.5.4** a. **13.6.1** b. **13.6.2** b.

CHAPTER 14

Chemical Kinetics

Methanol is a component of some de-icing fluids.

Chalabala/iStock/Getty Images

In This Chapter, You Will Learn

How rates of chemical reaction are determined and expressed and about the factors that influence the rates of reactions.

Before You Begin, Review These Skills

- The value of R expressed in J/K · mol [◀◀ Section 10.3, Table 10.4]
- Use of logarithms [▶▶ Appendix 1]

How Chemical Kinetics Can Be Useful in Medicine

The different rates at which specific biochemical reactions occur can be exploited to save the life of a patient who has ingested a poison, such as methanol. Accidental or intentional ingestion of methanol can cause headache, nausea, blindness, seizures, and even death. In the liver, methanol is metabolized by the enzyme alcohol dehydrogenase (ADH) to yield formaldehyde:

$$CH_3OH + NAD^+ \longrightarrow HCHO + NADH + H^+$$
methanol   formaldehyde

The formaldehyde is subsequently converted to formic acid by another enzyme, aldehyde dehydrogenase (ALDH):

$$HCHO + NADP^+ + H_2O \longrightarrow HCOOH + NADPH + H^+$$
formaldehyde   formic acid

Formaldehyde and formic acid are the species responsible for the toxic effects of methanol poisoning—including metabolic acidosis, in which the blood becomes dangerously acidic.

To treat this kind of poisoning, ADH must be prevented from metabolizing any more methanol. In years past, this was done by administering large quantities of ethanol, which has an affinity for ADH roughly 100 times that of methanol, making the reaction of ADH with ethanol significantly faster than the reaction with methanol. In this treatment, while the body's supply of ADH was busy converting ethanol to acetaldehyde (a reaction analogous to the conversion of methanol to formaldehyde), the methanol was removed from the body by dialysis. This treatment was not without complications, however, because the quantity of ethanol required typically caused significant inebriation and central nervous system (CNS) depression. The metabolic products of ethanol (acetaldehyde and acetic acid) are also toxic—although generally to a lesser degree than those of methanol. In fact, acetaldehyde is the species responsible for the constellation of symptoms known collectively as a hangover.

In 2000, the FDA approved the drug fomepizole, marketed under the name Antizol, for the treatment of methanol poisoning. Fomepizole ($C_4H_6N_2$) has an affinity for ADH approximately 8000 times that of methanol and is used to treat methanol toxicity without the intoxication and CNS depression caused by ethanol. As before, dialysis is used to remove methanol from the blood.

Understanding chemical kinetics can make it possible to minimize the damage done by undesirable reactions—and to increase the speed of desirable reactions.

At the end of this chapter, you will be able to solve a series of problems related to the production of methanol [▶▶ Applying What You've Learned, page 714].

14.1   Reaction Rates

Chemical kinetics is the study of how fast reactions take place. Many familiar reactions, such as the initial steps in vision and photosynthesis, happen almost instantaneously, whereas others, such as the rusting of iron or the conversion of diamond to graphite, take place on a time scale of days or even millions of years.

Knowledge of kinetics is important to many scientific endeavors, including drug design, pollution control, and food processing. The job of an industrial chemist often is to work on increasing the *rate* of a reaction rather than maximizing its yield or developing a new process.

A chemical reaction can be represented by the general equation:

$$\text{reactants} \longrightarrow \text{products}$$

This equation tells us that during the course of a reaction, reactants are consumed while products are formed. We can follow the progress of a reaction by monitoring either the decrease in concentration of the reactants or the increase in the concentrations of the products. The method used to monitor changes in reactant or product concentrations depends on the specific reaction. In a reaction that either consumes or produces a colored species, we can measure the intensity of the color over time with a spectrometer. In a reaction that either consumes or produces a gas, we can measure the change in pressure over time with a manometer. Electrical conductance measurement can be used to monitor the progress if ionic species are consumed or produced.

Average Reaction Rate

Consider the hypothetical reaction represented by:

$$A \longrightarrow B$$

in which A molecules are converted to B molecules. Figure 14.1 shows the progress of this reaction as a function of time.

The decrease in the number of A molecules and the increase in the number of B molecules with time are shown graphically in Figure 14.2. It is generally convenient to express the rate in terms of the change in concentration with time. Thus, for the reaction A ⟶ B, we can express the rate as:

$$\text{rate} = -\frac{\Delta[A]}{\Delta t} \quad \text{or} \quad \text{rate} = \frac{\Delta[B]}{\Delta t}$$

where $\Delta[A]$ and $\Delta[B]$ are the changes in concentration (molarity) over a time period Δt. The rate expression containing $\Delta[A]$ has a minus sign because the concentration of A decreases during the time interval—that is, $\Delta[A]$ is a negative quantity. The rate expression containing $\Delta[B]$ does not have a minus sign because the concentration of B increases during the time interval. Rate is always a positive quantity, so when it is

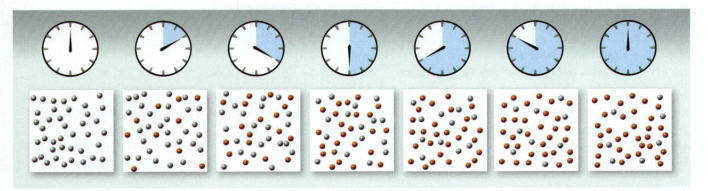

Figure 14.1 The progress of the reaction A ⟶ B. Initially, only A molecules (grey spheres) are present. As time progresses, there are more and more B molecules (red spheres).

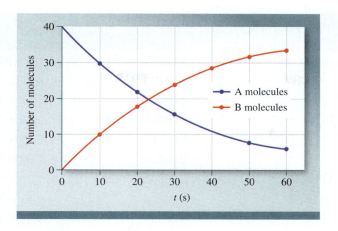

Figure 14.2 The rate of reaction A ⟶ B represented as the decrease of A molecules with time and as the increase of B molecules with time.

expressed in terms of the change in a reactant concentration, a minus sign is needed in the rate expression to make the rate positive. When the rate is expressed in terms of the change in a product concentration, no negative sign is needed to make the rate positive because the product concentration increases with time. Rates calculated in this way are average rates over the time period Δt.

To understand rates of chemical reactions and how they are determined, it is useful to consider some specific reactions. First, we consider the aqueous reaction of molecular bromine (Br_2) with formic acid (HCOOH):

$$Br_2(aq) + HCOOH(aq) \longrightarrow 2Br^-(aq) + 2H^+(aq) + CO_2(g)$$

Molecular bromine is reddish-brown, whereas all the other species in the reaction are colorless. As the reaction proceeds, the concentration of bromine decreases and its color fades (Figure 14.3). The decrease in intensity of the color (and therefore in the concentration of bromine) can be monitored with a spectrometer, which registers the amount of visible light absorbed by bromine (Figure 14.4).

Figure 14.3 From left to right: The decrease in bromine concentration as time elapses is indicated by the loss of color.

Ken Karp/McGraw Hill

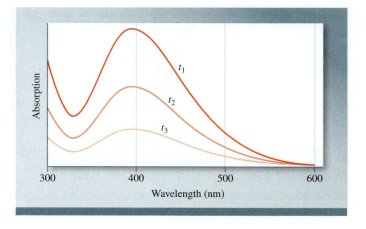

Figure 14.4 Plot of the absorption of bromine versus wavelength. The maximum absorption of visible light by bromine occurs at 393 nm. As the reaction progresses (t_1 to t_3), the absorption, which is proportional to [Br_2], decreases.

TABLE 14.1	Rates of the Reaction of Molecular Bromine and Formic Acid at 25°C	
Time (s)	$[Br_2]$ (M)	Rate (M/s)
0.0	0.0120	4.20×10^{-5}
50.0	0.0101	3.52×10^{-5}
100.0	0.00846	2.96×10^{-5}
150.0	0.00710	2.49×10^{-5}
200.0	0.00596	2.09×10^{-5}
250.0	0.00500	1.75×10^{-5}
300.0	0.00420	1.48×10^{-5}
350.0	0.00353	1.23×10^{-5}
400.0	0.00296	1.04×10^{-5}

Measuring the bromine concentration at some initial time and then at some final time enables us to determine the average rate of the reaction during that time interval:

$$\text{average rate} = -\frac{\Delta[Br_2]}{\Delta t}$$

$$= -\frac{[Br_2]_{\text{final}} - [Br_2]_{\text{initial}}}{t_{\text{final}} - t_{\text{initial}}}$$

Using data from Table 14.1, we can calculate the average rate over the first 50-s time interval as follows:

$$\text{average rate} = -\frac{(0.0101 - 0.0120)\,M}{50.0} = 3.80 \times 10^{-5}\ M/s$$

If we had chosen the first 100 s as our time interval, the average rate would then be given by:

$$\text{average rate} = -\frac{(0.00846 - 0.0120)\,M}{100.0\ s} = 3.54 \times 10^{-5}\ M/s$$

These calculations demonstrate that the average rate of this reaction depends on the time interval we choose. In other words, the rate changes over time. This is why a plot of the concentration of a reactant or product as a function of time is a curve rather than a straight line [Figure 14.5(a)].

Instantaneous Rate

If we were to calculate the average rate over shorter and shorter time intervals, we could obtain the *instantaneous rate,* which is the rate for a specific instant in time.

Figure 14.5 (a) The plot of $[Br_2]$ against time is a curve because the reaction rate changes as $[Br_2]$ changes with time. (b) The plot of the reaction rate against $[Br_2]$ is a straight line because the rate is proportional to $[Br_2]$.

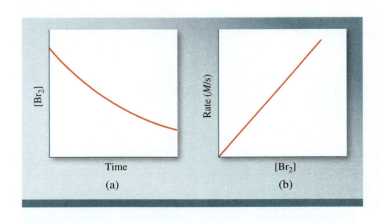

(a)   (b)

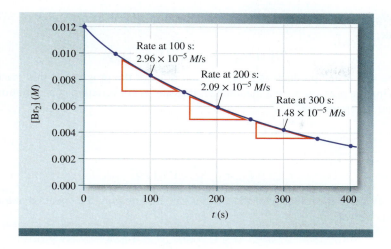

Figure 14.6 The instantaneous rates
of the reaction between molecular
bromine and formic acid at $t = 100$,
200, and 300 s are given by the
slopes of the tangents at these times.

Figure 14.6 shows the plot of $[Br_2]$ versus time based on the data from Table 14.1.
The instantaneous rate is equal to the slope of a tangent to the curve at any particular
time. Note that we can pick *any* two points along a tangent to calculate its slope. For
a chemist, the instantaneous rate is generally a more useful quantity than the average
rate. For the remainder of this chapter, therefore, the term *rate* is used to mean "instantaneous rate" (unless otherwise stated).

The slope of the tangent, and therefore the reaction *rate,* diminishes with time
because the concentration of bromine decreases with time [Figure 14.5(a)]. The data
in Table 14.1 show how the rate of this reaction depends on the concentration of
bromine. At 50.0 s, for example, when the concentration of bromine is 0.0101 *M,* the
rate is 3.52×10^{-5} *M*/s. When the concentration of bromine has been reduced by half
(i.e., reduced to 0.00500 *M*), at 250 s, the rate is also reduced by half (i.e., reduced
to 1.75×10^{-5} *M*/s):

$$\frac{[Br_2]_{50\,s}}{[Br_2]_{250\,s}} \approx 2 \quad \text{and} \quad \frac{\text{rate at 50.0 s}}{\text{rate at 250.0 s}} = \frac{3.52 \times 10^{-5}\ M/s}{1.75 \times 10^{-5}\ M/s} \approx 2$$

Thus, the rate is directly proportional to the concentration of bromine:

$$\text{rate} \propto [Br_2]$$

$$\text{rate} = k[Br_2]$$

where **k,** the proportionality constant, is called the **rate constant.**
Rearranging the preceding equation gives:

$$k = \frac{\text{rate}}{[Br_2]}$$

We can use the concentration and rate data from Table 14.1 for any value of t to calculate the value of k for this reaction. For example, using the data for $t = 50.0$ s gives:

$$k = \frac{3.52 \times 10^{-5}\ M/s}{0.0101\ M} = 3.49 \times 10^{-3}\ s^{-1} \qquad (\text{at } t = 50.0 \text{ s})$$

Likewise, using the data for $t = 300.0$ s gives:

$$k = \frac{1.48 \times 10^{-5}\ M/s}{0.00420\ M} = 3.52 \times 10^{-3}\ s^{-1} \qquad (\text{at } t = 300 \text{ s})$$

Slight variations in the calculated values of k are due to experimental deviations in rate
measurements. To two significant figures, we get $k = 3.5 \times 10^{-3}\ s^{-1}$ for this reaction,
regardless of which line of data we chose from Table 14.1. It is important to note that
the value of k does not depend on the concentration of bromine. The rate constant is
constant at constant temperature. (In Section 14.4 we discuss how k depends on
temperature.)

We now consider another specific reaction, the decomposition of hydrogen peroxide:

$$2H_2O_2(aq) \longrightarrow 2H_2O(l) + O_2(g)$$

Because one of the products is a gas, we can monitor the progress of this reaction by measuring the pressure with a manometer. Pressure is converted to concentration using the ideal gas equation:

$$PV = nRT$$

or

$$P_{O_2} = \frac{n}{V} RT = [O_2]\, RT$$

where n/V gives the molarity of the oxygen gas. Rearranging the equation and solving for $[O_2]$, we get:

$$[O_2] = \frac{1}{RT} P_{O_2}$$

The reaction rate, which is expressed as the rate of oxygen production, can now be written as:

Student Note: Many students prefer simply to convert pressures to concentrations using the ideal gas equation [◄◄ Section 10.3] and then determine the rate using the familiar form of the equation: rate = Δ[O₂]/Δt.

$$\text{rate} = \frac{\Delta[O_2]}{\Delta t} = \frac{1}{RT}\frac{\Delta P_{O_2}}{\Delta t}$$

Determining the rate from pressure data is possible only if the temperature (in kelvins) at which the reaction is carried out is known. Figure 14.7(a) shows the apparatus used to monitor the pressure change in the decomposition of hydrogen peroxide. Figure 14.7(b) shows the plot of oxygen pressure, P_{O_2}, versus time. As we did in Figure 14.6, we can draw a tangent to the curve in Figure 14.7(b) to determine the instantaneous rate at any point.

Stoichiometry and Reaction Rate

For stoichiometrically simple reactions of the type A ⟶ B, the rate can be expressed either in terms of the decrease in reactant concentration with time, $-\Delta[A]/\Delta t$, or in terms of the increase in product concentration with time, $\Delta[B]/\Delta t$—both expressions give the same result. For reactions in which one or more of the stoichiometric coefficients is something other than 1, we must take extra care in expressing the rate. For example, consider again the reaction between molecular bromine and formic acid:

$$Br_2(aq) + HCOOH(aq) \longrightarrow 2Br^-(aq) + 2H^+(aq) + CO_2(g)$$

Figure 14.7 (a) The rate of hydrogen peroxide decomposition can be measured with a manometer, which (b) shows the increase in the oxygen gas pressure with time.

a: *Ken Karp/McGraw Hill*

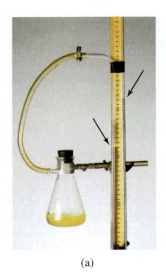

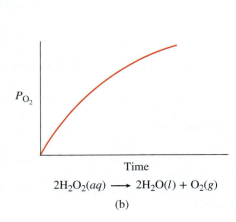

P_{O_2}

Time

$$2H_2O_2(aq) \longrightarrow 2H_2O(l) + O_2(g)$$

(a)   (b)

We have expressed the rate of this reaction in terms of the disappearance of bromine. But what if we chose instead to express the rate in terms of the appearance of bromide ion? According to the balanced equation, 2 moles of Br^- are generated for each mole of Br_2 consumed. Thus, Br^- appears at twice the rate that Br_2 disappears. To avoid the potential ambiguity of reporting the rate of disappearance or appearance of a specific chemical *species,* we report the **rate of reaction.** We determine the rate of reaction such that the result is the same regardless of which species we monitor. For the hypothetical reaction:

$$A \longrightarrow 2B$$

the rate of reaction can be written as either

$$rate = -\frac{\Delta[A]}{\Delta t} \quad \text{or} \quad rate = \frac{1}{2}\frac{\Delta[B]}{\Delta t}$$

both of which give the same result. For the bromine and formic acid reaction, we can write the rate of reaction as either:

$$rate = -\frac{\Delta[Br_2]}{\Delta t}$$

as we did earlier, or:

$$rate = \frac{1}{2}\frac{\Delta[Br^-]}{\Delta t}$$

In general, for the reaction:

$$aA + bB \longrightarrow cC + dD$$

the rate is given by:

$$rate = -\frac{1}{a}\frac{\Delta[A]}{\Delta t} = -\frac{1}{b}\frac{\Delta[B]}{\Delta t} = \frac{1}{c}\frac{\Delta[C]}{\Delta t} = \frac{1}{d}\frac{\Delta[D]}{\Delta t} \qquad \text{**Equation 14.1**}$$

Expressing the rate in this fashion ensures that the rate of reaction is the same regardless of which species we measure to monitor the reaction's progress.

Sample Problems 14.1 and 14.2 show how to write expressions for reaction rates and how to take stoichiometry into account in rate expressions.

Student Note: The rate of change in concentration of each species is divided by the coefficient of that species in the balanced equation.

SAMPLE PROBLEM 14.1

Write the rate expressions for each of the following reactions:

(a) $I^-(aq) + OCl^-(aq) \longrightarrow Cl^-(aq) + OI^-(aq)$

(b) $2O_3(g) \longrightarrow 3O_2(g)$

(c) $4NH_3(g) + 5O_2(g) \longrightarrow 4NO(g) + 6H_2O(g)$

Strategy Use Equation 14.1 to write rate expressions for each of the reactions.

Setup For reactions containing gaseous species, progress is generally monitored by measuring pressure. Pressures are converted to molar concentrations using the ideal gas equation, and rate expressions are written in terms of molar concentrations.

Solution (a) All the coefficients in this equation are 1. Therefore:

$$rate = -\frac{\Delta[I^-]}{\Delta t} = -\frac{\Delta[OCl^-]}{\Delta t} = \frac{\Delta[Cl^-]}{\Delta t} = \frac{\Delta[OI^-]}{\Delta t}$$

(b) $rate = -\frac{1}{2}\frac{\Delta[O_3]}{\Delta t} = \frac{1}{3}\frac{\Delta[O_2]}{\Delta t}$

(c) $rate = -\frac{1}{4}\frac{\Delta[NH_3]}{\Delta t} = -\frac{1}{5}\frac{\Delta[O_2]}{\Delta t} = \frac{1}{4}\frac{\Delta[NO]}{\Delta t} = \frac{1}{6}\frac{\Delta[H_2O]}{\Delta t}$

(Continued on next page)

THINK ABOUT IT

Make sure that the change in concentration of each species is divided by the corresponding coefficient in the balanced equation. Also make sure that the rate expressions written in terms of reactant concentrations have a negative sign to make the resulting rate positive.

Practice Problem ATTEMPT   Write the rate expressions for each of the following reactions:

(a) $CO_2(g) + 2H_2O(g) \longrightarrow CH_4(g) + 2O_2(g)$

(b) $3O_2(g) \longrightarrow 2O_3(g)$

(c) $2NO(g) + O_2(g) \longrightarrow 2NO_2(g)$

Practice Problem BUILD   Write the balanced equation corresponding to the following rate expressions:

(a) $\text{rate} = -\dfrac{1}{3}\dfrac{\Delta[CH_4]}{\Delta t} = -\dfrac{1}{2}\dfrac{\Delta[H_2O]}{\Delta t} = -\dfrac{\Delta[CO_2]}{\Delta t} = \dfrac{1}{4}\dfrac{\Delta[CH_3OH]}{\Delta t}$

(b) $\text{rate} = -\dfrac{1}{2}\dfrac{\Delta[N_2O_5]}{\Delta t} = \dfrac{1}{2}\dfrac{\Delta[N_2]}{\Delta t} = \dfrac{1}{5}\dfrac{\Delta[O_2]}{\Delta t}$

(c) $\text{rate} = -\dfrac{\Delta[H_2]}{\Delta t} = -\dfrac{\Delta[CO]}{\Delta t} = -\dfrac{\Delta[O_2]}{\Delta t} = \dfrac{\Delta[H_2CO_3]}{\Delta t}$

Practice Problem CONCEPTUALIZE
The diagrams represent a system that initially consists of reactants A (red) and B (blue), which react to form product C (purple). Write the balanced chemical equation that corresponds to the reaction.

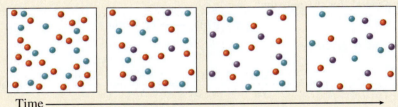

Time ⟶

SAMPLE PROBLEM   14.2

Consider the reaction:

$$4NO_2(g) + O_2(g) \longrightarrow 2N_2O_5(g)$$

At a particular time during the reaction, nitrogen dioxide is being consumed at the rate of 0.00130 M/s. (a) At what rate is molecular oxygen being consumed? (b) At what rate is dinitrogen pentoxide being produced?

Strategy Determine the rate of reaction using Equation 14.1, and, using the stoichiometry of the reaction, convert to rates of change for the specified individual species.

Setup

$$\text{rate} = -\frac{1}{4}\frac{\Delta[NO_2]}{\Delta t} = -\frac{\Delta[O_2]}{\Delta t} = \frac{1}{2}\frac{\Delta[N_2O_5]}{\Delta t}$$

We are given:

$$\frac{\Delta[NO_2]}{\Delta t} = -0.00130 \ M/s$$

where the minus sign indicates that the concentration of NO_2 is decreasing with time. The rate of reaction, therefore, is:

$$\text{rate} = -\frac{1}{4}\frac{\Delta[NO_2]}{\Delta t} = -\frac{1}{4}(-0.00130 \ M/s)$$

$$= 3.25 \times 10^{-4} \ M/s$$

Solution (a) $3.25 \times 10^{-4} \ M/s = -\dfrac{\Delta[O_2]}{\Delta t}$

$$\frac{\Delta[O_2]}{\Delta t} = -3.25 \times 10^{-4} \ M/s$$

Molecular oxygen is being consumed at a rate of $3.25 \times 10^{-4} \ M/s$.

(b) $3.25 \times 10^{-4}\ M/s = \dfrac{1}{2}\dfrac{\Delta[N_2O_5]}{\Delta t}$

$2(3.25 \times 10^{-4}\ M/s) = \dfrac{\Delta[N_2O_5]}{\Delta t}$

$\dfrac{\Delta[N_2O_5]}{\Delta t} = 6.50 \times 10^{-4}\ M/s$

Dinitrogen pentoxide is being produced at a rate of $6.50 \times 10^{-4}\ M/s$.

THINK ABOUT IT

Remember that the negative sign in a rate expression indicates that a species is being consumed rather than produced. Rates are always expressed as positive quantities.

Practice Problem ATTEMPT Consider the reaction:

$$4PH_3(g) \longrightarrow P_4(g) + 6H_2(g)$$

At a particular point during the reaction, molecular hydrogen is being formed at the rate of 0.168 M/s. (a) At what rate is P_4 being produced? (b) At what rate is PH_3 being consumed?

Practice Problem BUILD Consider the following unbalanced equation:

$$A + B \longrightarrow C$$

When C is being formed at the rate of 0.086 M/s, A is being consumed at a rate of 0.172 M/s and B is being consumed at a rate of 0.258 M/s. Balance the equation based on the relative rates of formation and consumption of products and reactants.

Practice Problem CONCEPTUALIZE Consider the reaction $2A + B \longrightarrow 2C$. Which graph could represent the concentrations of A, B, and C as the reaction progresses from $t = 0$ s?

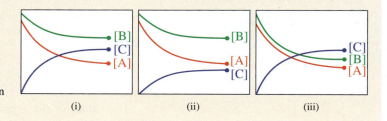

(i) (ii) (iii)

CHECKPOINT – SECTION 14.1   Reaction Rates

14.1.1 Which expressions are correct for the rate of the following reaction?

$$F_2(g) + 2ClO_2(g) \longrightarrow 2FClO_2(g)$$

(Select all that apply.)

a) $\text{rate} = -\dfrac{\Delta[F_2]}{\Delta t}$

b) $\text{rate} = -\dfrac{\Delta[ClO_2]}{\Delta t}$

c) $\text{rate} = -\dfrac{1}{2}\dfrac{\Delta[F_2]}{\Delta t}$

d) $\text{rate} = -\dfrac{1}{2}\dfrac{\Delta[ClO_2]}{\Delta t}$

e) $\text{rate} = -\dfrac{\Delta[FClO_2]}{\Delta t}$

14.1.2 In the same reaction:

$$F_2(g) + 2ClO_2(g) \longrightarrow 2FClO_2(g)$$

if the concentration of F_2 is changing at a rate of 0.026 M/s, what is the rate of change in the concentration of ClO_2?

a) 0.026 M/s

b) 0.052 M/s

c) 0.013 M/s

d) 0.078 M/s

e) 0.004 M/s

14.2 Dependence of Reaction Rate on Reactant Concentration

We saw in Section 14.1 that the rate of reaction between bromine and formic acid is proportional to the concentration of bromine and that the proportionality constant k is the rate constant. We now explore how the rate, the rate constant, and the reactant concentrations are related.

The Rate Law

The equation relating the rate of reaction to the concentration of molecular bromine:

$$\text{rate} = k[Br_2]$$

is an example of a ***rate law.*** The rate law is an equation that relates the rate of reaction to the concentrations of reactants. For the general reaction:

$$aA + bB \longrightarrow cC + dD$$

the rate law is:

Equation 14.2 $$\text{rate} = k[A]^x[B]^y$$

Student Note: It is important to emphasize that the exponents in a rate law must be determined from experimental data and that, in general, they are *not* equal to the coefficients from the chemical equation.

where k is the rate constant and the exponents x and y are numbers that must be determined experimentally. When we know the values of k, x, and y, we can use Equation 14.2 to calculate the rate of the reaction, given the concentrations of A and B.

In the case of the reaction of molecular bromine and formic acid, the rate law is:

$$\text{rate} = k(Br_2)^x(HCOOH)^y$$

where $x = 1$ and $y = 0$.

The values of the exponents in the rate law indicate the *order* of the reaction with respect to each reactant. In the reaction of bromine and formic acid, for example, the exponent for the bromine concentration, $x = 1$, means that the reaction is *first order* with respect to bromine. The exponent of 0 for the formic acid concentration indicates that the reaction is *zeroth order* with respect to formic acid. The sum of x and y is called the overall ***reaction order.*** Thus, the reaction of bromine and formic acid is first order in bromine, zeroth order in formic acid, and first order $(1 + 0 = 1)$ overall.

Student Hot Spot

Student data indicate you may struggle with rate laws. Access the eBook to view additional Learning Resources on this topic.

Experimental Determination of the Rate Law

To see how a rate law is determined from experimental data, consider the following reaction between fluorine and chlorine dioxide:

$$F_2(g) + 2ClO_2(g) \longrightarrow 2FClO_2(g)$$

Rate laws are commonly determined using a table of starting reactant concentrations and initial rates. The ***initial rate*** is the instantaneous rate at the beginning of the reaction. By varying the starting concentrations of reactants and observing the changes that result in the initial rate, we can determine how the rate depends on each reactant concentration.

Table 14.2 shows the initial rate data for the reaction of fluorine and chlorine dioxide being carried out three different times. Each time, the combination of reactant concentrations is different, and each time the initial rate is different.

To determine the values of the exponents x and y in the rate law:

$$\text{rate} = k[F_2]^x[ClO_2]^y$$

we must compare two experiments in which one reactant concentration *changes,* and the other remains constant. For example, we first compare the data from experiments 1 and 3:

Experiment	$[F_2]$ (*M*)	$[ClO_2]$ (*M*)	Initial Rate (*M*/s)
1	0.10	0.010	1.2×10^{-3}
2	[F₂] doubles { 0.10 [ClO₂] unchanged { 0.040	Rate doubles { 4.8 \times 10^{-3}	
3	0.20	0.010	2.4×10^{-3}

When the concentration of fluorine doubles, with the chlorine dioxide concentration held constant, the rate doubles:

$$\frac{[F_2]_3}{[F_2]_1} = \frac{0.20\ M}{0.10\ M} = 2 \qquad \frac{rate_3}{rate_1} = \frac{2.4 \times 10^{-3}\ M/s}{1.2 \times 10^{-3}\ M/s} = 2$$

This indicates that the rate is directly proportional to the concentration of fluorine, and the value of x is 1:

$$rate = k[F_2][ClO_2]^y$$

Student Note: Recall that when an exponent is 1, it need not be shown [◄◄ Section 3.3].

Similarly, we can compare experiments 1 and 2:

Experiment	$[F_2]$ (*M*)	$[ClO_2]$ (*M*)	Initial Rate (*M*/s)
1	[F₂] unchanged { 0.10	[ClO₂] quadruples { 0.010	Rate quadruples { 1.2 × 10⁻³
2	0.10	0.040	4.8×10^{-3}
3	0.20	0.010	2.4×10^{-3}

We find that the rate quadruples when the concentration of chlorine dioxide is quadrupled, but the fluorine concentration is held constant:

$$\frac{[ClO_2]_2}{[ClO_2]_1} = \frac{0.040\ M}{0.010\ M} = 4 \qquad \frac{rate_2}{rate_1} = \frac{4.8 \times 10^{-3}\ M/s}{1.2 \times 10^{-3}\ M/s} = 4$$

This indicates that the rate is also directly proportional to the concentration of chlorine dioxide, so the value of y is also 1. Thus, we can write the rate law as follows:

$$rate = k[F_2][ClO_2]$$

Because the concentrations of F_2 and ClO_2 are each raised to the first power, we say that the reaction is first order in F_2 and first order in ClO_2. The reaction is second order overall.

Knowing the rate law, we can then use the data from any one of the experiments to calculate the rate constant. Using the data for the first experiment in Table 14.2, we can write:

$$k = \frac{rate}{[F_2][ClO_2]} = \frac{1.2 \times 10^{-3}\ M/s}{(0.10\ M)(0.010\ M)} = 1.2\ M^{-1} \cdot s^{-1}$$

Table 14.3 contains initial rate data for the hypothetical reaction:

$$aA + bB \longrightarrow cC + dD$$

which has the general rate law:

$$rate = k[A]^x[B]^y$$

TABLE 14.2	Initial Rate Data for the Reaction Between F_2 and ClO_2		
Experiment	$[F_2]$ (*M*)	$[ClO_2]$ (*M*)	Initial Rate (*M*/s)
1	0.10	0.010	1.2×10^{-3}
2	0.10	0.040	4.8×10^{-3}
3	0.20	0.010	2.4×10^{-3}

TABLE 14.3	Initial Rate Data for the Reaction Between A and B		
Experiment	**[A] (M)**	**[B] (M)**	**Initial Rate (M/s)**
1	0.10	0.015	2.1×10^{-4}
2	0.20	0.015	4.2×10^{-4}
3	0.10	0.030	8.4×10^{-4}

Comparing experiments 1 and 2, we see that when [A] doubles, with [B] unchanged, the rate also doubles.

Experiment	**[A] (M)**		**[B] (M)**		**Initial Rate (M/s)**
1	[A] doubles $\begin{cases} 0.10 \\ 0.20 \end{cases}$		[B] unchanged $\begin{cases} 0.015 \\ 0.015 \end{cases}$		Rate doubles $\begin{cases} 2.1 \times 10^{-4} \\ 4.2 \times 10^{-4} \end{cases}$
2					
3	0.10		0.030		8.4×10^{-4}

Thus, $x = 1$.

Comparing experiments 1 and 3, when [B] doubles with [A] unchanged, we see that the rate quadruples.

Experiment	**[A] (M)**		**[B] (M)**		**Initial Rate (M/s)**
1	$\begin{cases} 0.10 \end{cases}$		$\begin{cases} 0.015 \end{cases}$		$\begin{cases} 2.1 \times 10^{-4} \end{cases}$
2	[A] unchanged $\begin{cases} 0.20 \end{cases}$		[B] doubles $\begin{cases} 0.015 \end{cases}$	Rate quadruples	$\begin{cases} 4.2 \times 10^{-4} \end{cases}$
3	$\begin{cases} 0.10 \end{cases}$		$\begin{cases} 0.030 \end{cases}$		$\begin{cases} 8.4 \times 10^{-4} \end{cases}$

Thus, the rate is *not* directly proportional to [B] to the first power, but rather it is directly proportional to [B] to the second power (i.e., $y = 2$):

$$\text{rate} \propto [B]^2$$

The overall rate law is:

$$\text{rate} = k[A][B]^2$$

This reaction is therefore first order in A, second order in B, and third order overall.

Once again, knowing the rate law, we can use data from any of the experiments in the table to calculate the rate constant. Using the data from experiment 1, we get:

$$k = \frac{\text{rate}}{[A][B]^2} = \frac{2.1 \times 10^{-4} \ M/s}{(0.10 \ M)(0.015 \ M)^2} = 9.3 \ M^{-2} \cdot s^{-1}$$

Note that the units of this rate constant are different from those for the rate constant we calculated for the F_2-ClO_2 reaction and the bromine reaction. In fact, the units of a rate constant depend on the *overall* order of the reaction. Table 14.4 compares the units of the rate constant for reactions that are zeroth, first, second, and third order overall.

TABLE 14.4	Units of the Rate Constant k for Reactions of Various Overall Orders	
Overall Reaction Order	**Sample Rate Law**	**Units of k**
0	rate $= k$	$M \cdot s^{-1}$
1	rate $= k[A]$ or rate $= k[B]$	s^{-1}
2	rate $= k[A]^2$, rate $= k[B]^2$, or rate $= k[A][B]$	$M^{-1} \cdot s^{-1}$
3*	rate $= k[A]^2[B]$ or rate $= k[A][B]^2$	$M^{-2} \cdot s^{-1}$

*Another possibility for a third-order reaction is rate $= k[A][B][C]$, although such reactions are very rare.

The following are three important things to remember about the rate law:

1. The exponents in a rate law must be determined from a table of experimental data—in general, they are not related to the stoichiometric coefficients in the balanced chemical equation.
2. Comparing changes in individual reactant concentrations with changes in rate shows how the rate depends on each reactant concentration.
3. Reaction order is generally defined in terms of reactant concentrations rather than product concentrations.

Sample Problem 14.3 shows how to use initial rate data to determine a rate law.

SAMPLE PROBLEM 14.3

The gas-phase reaction of nitric oxide with hydrogen at 1280°C is:

$$2NO(g) + 2H_2(g) \longrightarrow N_2(g) + 2H_2O(g)$$

From the following data collected at 1280°C, determine (a) the rate law, (b) the rate constant, including units, and (c) the rate of the reaction when $[NO] = 4.8 \times 10^{-3}$ M and $[H_2] = 6.2 \times 10^{-3}$ M.

Experiment	[NO] (M)	[H$_2$] (M)	Initial Rate (M/s)
1	5.0×10^{-3}	2.0×10^{-3}	1.3×10^{-5}
2	1.0×10^{-2}	2.0×10^{-3}	5.0×10^{-5}
3	1.0×10^{-2}	4.0×10^{-3}	1.0×10^{-4}

Strategy Compare two experiments at a time to determine how the rate depends on the concentration of each reactant.

Setup The rate law is rate = $k[NO]^x[H_2]^y$. Comparing experiments 1 and 2, we see that the rate increases by approximately a factor of 4 when [NO] is doubled but [H$_2$] is held constant. Comparing experiments 2 and 3 shows that the rate doubles when [H$_2$] doubles but [NO] is held constant.

Solution (a) Dividing the rate from experiment 2 by the rate from experiment 1, we get:

$$\frac{\text{rate}_2}{\text{rate}_1} = \frac{5.0 \times 10^{-5} \, M \cdot s^{-1}}{1.3 \times 10^{-5} \, M \cdot s^{-1}} \approx 4 = \frac{k(1.0 \times 10^{-2} \, M)^x(2.0 \times 10^{-3} \, M)^y}{k(5.0 \times 10^{-3} \, M)^x(2.0 \times 10^{-3} \, M)^y}$$

A quotient of numbers, each raised to the same power, is equal to the quotient raised to that power: $x^n/y^n = (x/y)^n$. Canceling identical terms in the numerator and denominator gives:

$$\frac{(1.0 \times 10^{-2} \, M)^x}{(5.0 \times 10^{-3} \, M)^x} = 2^x = 4$$

Therefore, $x = 2$. The reaction is second order in NO.

Dividing the rate from experiment 3 by the rate from experiment 2, we get:

$$\frac{\text{rate}_3}{\text{rate}_2} = \frac{1.0 \times 10^{-4} \, M \cdot s^{-1}}{5.0 \times 10^{-5} \, M \cdot s^{-1}} = 2 = \frac{k(1.0 \times 10^{-2} \, M)^x(4.0 \times 10^{-3} \, M)^y}{k(1.0 \times 10^{-2} \, M)^x(2.0 \times 10^{-3} \, M)^y}$$

Canceling identical terms in the numerator and denominator gives:

$$\frac{(4.0 \times 10^{-3} \, M)^y}{(2.0 \times 10^{-3} \, M)^y} = 2^y = 2$$

Therefore, $y = 1$. The reaction is first order in H$_2$. The overall rate law is:

$$\text{rate} = k[NO]^2[H_2]$$

(b) We can use data from any of the experiments to calculate the value and units of k. Using the data from experiment 1 gives:

$$k = \frac{\text{rate}}{[NO]^2[H_2]} = \frac{1.3 \times 10^{-5} \, M/s}{(5.0 \times 10^{-3} \, M)^2(2.0 \times 10^{-3} \, M)} = 2.6 \times 10^2 \, M^{-2} \cdot s^{-1}$$

(c) Using the rate constant determined in part (b) and the concentrations of NO and H$_2$ given in the problem statement, we can determine the reaction rate as follows:

$$\text{rate} = (2.6 \times 10^2 \, M^{-2} \cdot s^{-1})(4.8 \times 10^{-3} \, M)^2(6.2 \times 10^{-3} \, M)$$
$$= 3.7 \times 10^{-5} \, M/s$$

(Continued on next page)

THINK ABOUT IT

The exponent for the concentration of H_2 in the rate law is 1, whereas the coefficient for H_2 in the balanced equation is 2. It is a common error to try to write a rate law using the stoichiometric coefficients as the exponents. Remember that, in general, the exponents in the rate law are *not* related to the coefficients in the balanced equation. Rate laws must be determined by examining a table of experimental data.

Practice Problem ATTEMPT The reaction of peroxydisulfate ion $(S_2O_8^{2-})$ with iodide ion (I^-) is:

$$S_2O_8^{2-}(aq) + 3I^-(aq) \longrightarrow 2SO_4^{2-}(aq) + I_3^-(aq)$$

From the following data collected at a certain temperature, determine the rate law and calculate the rate constant, including its units.

Experiment	$[S_2O_8^{2-}]$ (M)	$[I^-]$ (M)	Initial Rate (M/s)
1	0.080	0.034	2.2×10^{-4}
2	0.080	0.017	1.1×10^{-4}
3	0.16	0.017	2.2×10^{-4}

Practice Problem BUILD For the following general reaction, rate $= k[A]^2$ and $k = 1.3 \times 10^{-2}$ $M^{-1} \cdot s^{-1}$:

$$A + B \longrightarrow 2C$$

Use this information to fill in the missing table entries.

Experiment	$[A]$ (M)	$[B]$ (M)	Initial Rate (M/s)
1	0.013	0.250	2.20×10^{-6}
2	0.026	0.250	
3		0.500	2.20×10^{-6}

Practice Problem CONCEPTUALIZE Three initial-rate experiments are shown here depicting the reaction of X (red) and Y (yellow) to form Z (green). Using the diagrams, determine the rate law for the reaction X + Y $\longrightarrow$ Z.

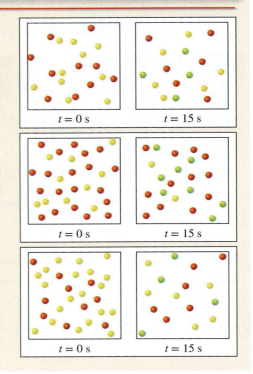

$t = 0$ s $t = 15$ s

$t = 0$ s $t = 15$ s

$t = 0$ s $t = 15$ s

CHECKPOINT – SECTION 14.2   Dependence of Reaction Rate on Reactant Concentration

Answer questions 14.2.1 through 14.2.4 using the table of initial rate data for the reaction:

$$A + 2B \longrightarrow 2C + D$$

Experiment	$[A]$ (M)	Initial $[B]$ (M)	Rate (M/s)
1	0.12	0.010	2.2×10^{-3}
2	0.36	0.010	6.6×10^{-3}
3	0.12	0.020	2.2×10^{-3}

14.2.1 What is the rate law for the reaction?

a) rate $= k[A][B]^2$

b) rate $= k[A]^2[B]$

c) rate $= k[A]^3$

d) rate $= k[A]^2$

e) rate $= k[A]$

14.2.2 Calculate the rate constant.

a) 0.15 $M^{-1} \cdot s^{-1}$ d) 0.018 s^{-1}

b) 0.15 $M \cdot s^{-1}$ e) 0.018 $M^{-1} \cdot s^{-1}$

c) 0.15 s^{-1}

14.2.3 What is the overall order of the reaction?

a) 0   d) 3

b) 1   e) 4

c) 2

14.2.4 Determine the rate when $[A] = 0.50$ M and $[B] = 0.25$ M.

a) 9.2×10^{-3} M/s

b) 2.3×10^{-3} M/s

c) 4.5×10^{-3} M/s

d) 5.0×10^{-3} M/s

e) 1.3×10^{-2} M/s

14.2.5 The diagrams represent three experiments in which the reaction A + B ⟶ C is carried out with varied initial concentrations of A and B. Determine the rate law for the reaction. (A = green, B = yellow, C = red.)

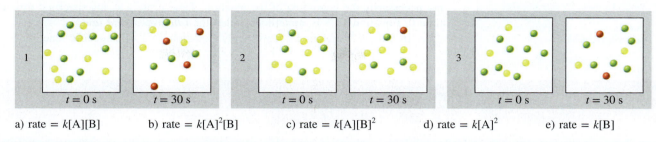

$t = 0$ s $\qquad$ $t = 30$ s	$t = 0$ s $\qquad$ $t = 30$ s	$t = 0$ s $\qquad$ $t = 30$ s

a) rate = $k[A][B]$ b) rate = $k[A]^2[B]$ c) rate = $k[A][B]^2$ d) rate = $k[A]^2$ e) rate = $k[B]$

14.3 Dependence of Reactant Concentration on Time

We can use the rate law to determine the rate of a reaction using the rate constant and the reactant concentrations:

$$\overset{\text{rate law}}{\underset{\underset{\text{rate}\qquad\text{rate constant}}{}}{\text{rate} = k[A]^x[B]^y}}$$

A rate law can also be used to determine the remaining concentration of a reactant at a specific time during a reaction. We can illustrate this use of rate laws using reactions that are first order overall and reactions that are second order overall.

First-Order Reactions

A *first-order reaction* is a reaction whose rate depends on the concentration of one of the reactants raised to the first power. Two examples are the decomposition of ethane (C_2H_6) into highly reactive fragments called methyl radicals ($\cdot CH_3$), and the decomposition of dinitrogen pentoxide (N_2O_5) into nitrogen dioxide (NO_2) and molecular oxygen (O_2):

$$C_2H_6 \longrightarrow 2 \cdot CH_3 \qquad\qquad \text{rate} = k[C_2H_6]$$
$$2N_2O_5(g) \longrightarrow 4NO_2(g) + O_2(g) \qquad \text{rate} = k[N_2O_5]$$

In a first-order reaction of the type:

$$A \longrightarrow \text{product}$$

the rate can be expressed as the rate of change in reactant concentration:

$$\text{rate} = -\frac{\Delta[A]}{\Delta t}$$

as well as in the form of the rate law:

$$\text{rate} = k[A]$$

Setting these two expressions of the rate equal to each other we get:

$$-\frac{\Delta[A]}{\Delta t} = k[A]$$

Applying calculus to the preceding equation, we can show that:

$$\ln \frac{[A]_t}{[A]_0} = -kt \qquad\qquad \textbf{Equation 14.3}$$

where ln is the natural logarithm, and $[A]_0$ and $[A]_t$ are the concentrations of A at times 0 and t, respectively. In general, time 0 refers to any specified time during a reaction—not necessarily the beginning of the reaction. Time t refers to any specified time *after* time 0. Equation 14.3 is sometimes called the ***integrated rate law***.

In Sample Problem 14.4, we apply Equation 14.3 to a specific reaction.

SAMPLE PROBLEM 14.4

The decomposition of hydrogen peroxide is first order in H_2O_2:

$$2H_2O_2(aq) \longrightarrow 2H_2O(l) + O_2(g)$$

The rate constant for this reaction at 20°C is $1.8 \times 10^{-5}\ s^{-1}$. If the starting concentration of H_2O_2 is 0.75 M, determine (a) the concentration of H_2O_2 remaining after 3 h and (b) how long it will take for the H_2O_2 concentration to drop to 0.10 M.

Strategy Use Equation 14.3 to find $[H_2O_2]_t$, where $t = 3$ h, and then solve Equation 14.3 for t to determine how much time must pass for $[H_2O_2]_t$ to equal 0.10 M.

Setup $[H_2O_2]_0 = 0.75\ M$; time t for part (a) is (3 h)(60 min/h)(60 s/min) = 10,800 s.

Solution (a) $\ln \dfrac{[H_2O_2]_t}{[H_2O_2]_0} = -kt$

$$\ln \frac{[H_2O_2]_t}{0.75\ M} = -(1.8 \times 10^{-5}\ s^{-1})(10,800\ s) = -0.1944$$

Take the inverse natural logarithm of both sides of the equation (the inverse of ln x is e^x [▸▸ Appendix 1]) to get

$$\frac{[H_2O_2]_t}{0.75\ M} = e^{-0.1944} = 0.823$$

$$[H_2O_2]_t = (0.823)(0.75\ M) = 0.62\ M$$

The concentration of H_2O_2 after 3 h is 0.62 M.

(b) $\ln\left(\dfrac{0.10\ M}{0.75\ M}\right) = -2.015 = -(1.8 \times 10^{-5}\ s^{-1})\,t$

$$\frac{2.015}{1.8 \times 10^{-5}\ s^{-1}} = t = 1.12 \times 10^5\ s$$

The time required for the peroxide concentration to drop to 0.10 M is 1.1×10^5 s or about 31 h.

THINK ABOUT IT

Don't forget the minus sign in Equation 14.3. If you calculate a reactant concentration at time t that is greater than the concentration at time 0 (or if you get a negative time required for the concentration to drop to a specified level), check your solution for this common error.

Practice Problem ATTEMPT The rate constant for the reaction 2A $\longrightarrow$ B is $7.5 \times 10^{-3}\ s^{-1}$ at 110°C. The reaction is first order in A. How long (in seconds) will it take for [A] to decrease from 1.25 M to 0.71 M?

Practice Problem BUILD Refer again to the reaction 2A $\longrightarrow$ B, for which $k = 7.5 \times 10^{-3}\ s^{-1}$ at 110°C. With a starting concentration of [A] = 2.25 M, what will [A] be after 2.0 min?

Practice Problem CONCEPTUALIZE The diagrams on the right illustrate the first-order reaction of B (blue) to form C (yellow). Use the information given in the first set of diagrams to determine how much time is required for the change depicted in the second set of diagrams to take place.

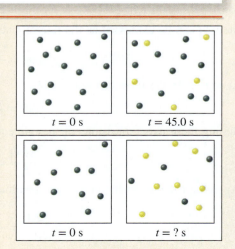

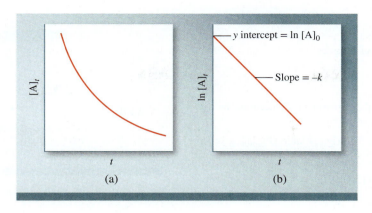

Figure 14.8 First-order reaction characteristics: (a) Decrease of reactant concentration with time. (b) A plot of ln $[A]_t$ versus t. The slope of the line is equal to $-k$ and the y intercept is equal to ln $[A]_0$.

Equation 14.3 can be rearranged as follows:

$$\ln [A]_t = -kt + \ln [A]_0 \qquad \textbf{Equation 14.4}$$

Equation 14.4 has the form of the linear equation $y = mx + b$:

$$\ln [A]_t = (-k)(t) + \ln [A]_0$$
$$y \quad = \quad m \ \ x \ + \quad b$$

Figure 14.8(a) shows the decrease in concentration of reactant A during the course of the reaction. As we saw in Section 14.1, the plot of reactant concentration as a function of time is not a straight line. For a first-order reaction, however, we do get a straight line if we plot the natural log of reactant concentration (ln $[A]_t$) versus time (y versus x). The slope of the line is equal to $-k$ [Figure 14.8(b)], so we can determine the rate constant from the slope of this plot.

Sample Problem 14.5 shows how a rate constant can be determined from experimental data.

> **Student Note:** Because pressure is proportional to concentration, for gaseous reactions [◄◄ Section 10.4] Equations 14.3 and 14.4 can be written as
>
> $$\ln \frac{P_t}{P_0} = -kt$$
>
> and
>
> $$\ln P_t = -kt + \ln P_0$$
>
> respectively, where P_0 and P_t are the pressures of reactant A at times 0 and t, respectively.

> **Student Note:** This graphical determination is an alternative to using the method of initial rates to determine the value of k.

SAMPLE PROBLEM 14.5

The rate of decomposition of azomethane is studied by monitoring the partial pressure of the reactant as a function of time:

$$CH_3-N{=}N-CH_3(g) \longrightarrow N_2(g) + C_2H_6(g)$$

The data obtained at 300°C are listed in the following table:

Time (s)	$P_{\text{azomethane}}$ (mmHg)
0	284
100	220
150	193
200	170
250	150
300	132

Determine the rate constant of the reaction at this temperature.

Strategy We can use Equation 14.3 only for first-order reactions, so we must first determine if the decomposition of azomethane is first order. We do this by plotting ln P against time. If the reaction is first order, we can use Equation 14.3 and the data at any two of the times in the table to determine the rate constant.

Setup The table expressed as ln P is:

Time (s)	ln P
0	5.649
100	5.394
150	5.263
200	5.136
250	5.011
300	4.883

(Continued on next page)

Plotting these data gives a straight line, indicating that the reaction is indeed first order. Thus, we can use Equation 14.3 expressed in terms of pressure:

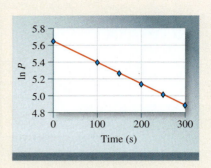

$$\ln \frac{P_t}{P_0} = -kt$$

P_t and P_0 can be pressures at any two times during the experiment. P_0 need not be the pressure at 0 s—it need only be at the earlier of the two times.

Solution Using data from times 100 s and 250 s of the original table ($P_{azomethane}$ versus t), we get:

$$\ln \frac{150 \text{ mmHg}}{220 \text{ mmHg}} = -k(150 \text{ s})$$

$$\ln 0.682 = -k(150 \text{ s})$$

$$k = 2.55 \times 10^{-3} \text{ s}^{-1}$$

THINK ABOUT IT

We could equally well have determined the rate constant by calculating the slope of the plot of ln P versus t. Using the two points labeled on the plot, we get

$$\text{slope} = \frac{5.011 - 5.394}{250 - 100}$$

$$= -2.55 \times 10^{-3} \text{ s}^{-1}$$

Remember that slope = $-k$, so $k = 2.55 \times 10^{-3} \text{ s}^{-1}$.

Practice Problem **A**TTEMPT Ethyl iodide (C_2H_5I) decomposes at a certain temperature in the gas phase as follows:

$$C_2H_5I(g) \longrightarrow C_2H_4(g) + HI(g)$$

From the following data, determine the rate constant of this reaction. Begin by constructing a plot to verify that the reaction is first order:

Time (min)	$[C_2H_5I]$ (M)
0	0.36
15	0.30
30	0.25
48	0.19
75	0.13

Practice Problem **B**UILD Use the calculated k from Practice Problem A to fill in the missing values in the following table:

Time (min)	$[C_2H_5I]$ (M)
0	0.45
10	___
20	___
30	___
40	___

Practice Problem **C**ONCEPTUALIZE Use the graph in the Setup section of Sample Problem 14.5 to estimate the pressure of azomethane (in mmHg) at $t = 50$ s.

We often describe the rate of a reaction using the half-life. The **half-life ($t_{1/2}$)** is the time required for the reactant concentration to drop to *half* its original value. We obtain an expression for $t_{1/2}$ for a first-order reaction as follows:

$$t = \frac{1}{k} \ln \frac{[A]_0}{[A]_t}$$

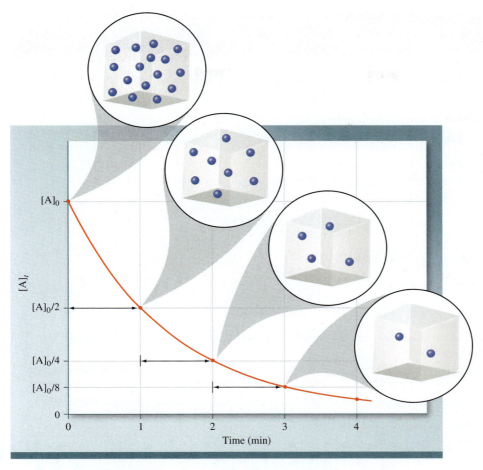

Figure 14.9 A plot of [A] versus time for the first-order reaction A ⟶ products. The half-life of the reaction is 1 min. The concentration of A is halved every half-life.

According to the definition of half-life, $t = t_{1/2}$ when $[A]_t = \frac{1}{2}[A]_0$, so:

$$t_{1/2} = \frac{1}{k} \ln \frac{[A]_0}{\frac{1}{2}[A]_0}$$

Because $\dfrac{[A]_0}{\frac{1}{2}[A]_0} = 2$, and $\ln 2 = 0.693$, the expression for $t_{1/2}$ simplifies to:

$$t_{1/2} = \frac{0.693}{k} \qquad\qquad \textbf{Equation 14.5}$$

According to Equation 14.5, the half-life of a first-order reaction is independent of the initial concentration of the reactant. Thus, it takes the same time for the concentration of the reactant to decrease from 1.0 M to 0.50 M as it does for the concentration to decrease from 0.10 M to 0.050 M (Figure 14.9). Measuring the half-life of a reaction is one way to determine the rate constant of a first-order reaction.

The half-life of a first-order reaction is inversely proportional to its rate constant, so a short half-life corresponds to a large rate constant. Consider, for example, two radioactive isotopes used in nuclear medicine: ^{24}Na ($t_{1/2} = 14.7$ h) and ^{60}Co ($t_{1/2} = 5.3$ yr). Sodium-24, with the shorter half-life, decays faster. If we started with an equal number of moles of each isotope, most of the sodium-24 would be gone in a week whereas most of the cobalt-60 would remain unchanged.

Sample Problem 14.6 shows how to calculate the half-life of a first-order reaction, given the rate constant.

SAMPLE PROBLEM 14.6

The decomposition of ethane (C_2H_6) to methyl radicals (CH_3) is a first-order reaction with a rate constant of $5.36 \times 10^{-4}\,s^{-1}$ at 700°C:

$$C_2H_6 \longrightarrow 2CH_3$$

Calculate the half-life of the reaction in minutes.

Strategy Use Equation 14.5 to calculate $t_{1/2}$ in seconds, and then convert to minutes.

Setup

$$\text{seconds} \times \frac{1\text{ minute}}{60\text{ seconds}} = \text{minutes}$$

Solution

$$t_{1/2} = \frac{0.693}{k} = \frac{0.693}{5.36 \times 10^{-4}\,s^{-1}} = 1293\text{ s}$$

$$1293\text{ s} \times \frac{1\text{ min}}{60\text{ s}} = 21.5\text{ min}$$

The half-life of ethane decomposition at 700°C is 21.6 min.

THINK ABOUT IT

Half-lives and rate constants can be expressed using any units of time and reciprocal time, respectively. Track units carefully when you convert from one unit of time to another.

Practice Problem ATTEMPT   Calculate the half-life of the decomposition of azomethane, discussed in Sample Problem 14.5.

Practice Problem BUILD   Calculate the rate constant for the first-order decay of ^{24}Na ($t_{1/2} = 14.7$ h).

Practice Problem CONCEPTUALIZE The diagrams show a system in which A (red) is reacting to form B (blue) over the course of time. Use the information in the diagrams to determine the half-life of the reaction.

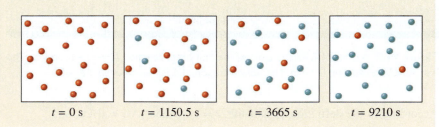

$t = 0$ s $t = 1150.5$ s $t = 3665$ s $t = 9210$ s

Second-Order Reactions

A *second-order reaction* is a reaction whose rate depends on the concentration of one reactant raised to the second power or on the product of the concentrations of two different reactants (each raised to the first power). For simplicity, we consider only the first type of reaction:

$$A \longrightarrow \text{product}$$

where the rate can be expressed as:

$$\text{rate} = -\frac{\Delta[A]}{\Delta t}$$

or as:

$$\text{rate} = k[A]^2$$

As before, we can combine the two expressions of the rate:

$$-\frac{\Delta[A]}{\Delta t} = k[A]^2$$

Again, using calculus, we obtain the following integrated rate law:

Equation 14.6 $$\frac{1}{[A]_t} = kt + \frac{1}{[A]_0}$$

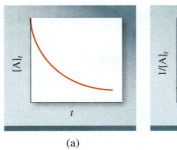

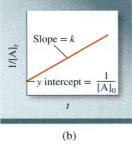

(a)   (b)

Thus, for a second-order reaction, we obtain a straight line when we plot the reciprocal of concentration ($1/[A]_t$) against time (Figure 14.10), and the slope of the line is equal to the rate constant, k. As before, we can obtain the expression for the half-life by setting $[A]_t = \frac{1}{2}[A]_0$ in Equation 14.6.

$$\frac{1}{\frac{1}{2}[A]_0} = k t_{1/2} + \frac{1}{[A]_0}$$

Solving for $t_{1/2}$, we obtain:

$$t_{1/2} = \frac{1}{k[A]_0} \qquad \text{Equation 14.7}$$

Note that unlike the half-life of a first-order reaction, which is independent of the starting concentration, the half-life of a second-order reaction is inversely proportional to the initial reactant concentration. Determining the half-life at several different initial concentrations is one way to distinguish between first-order and second-order reactions.

Sample Problem 14.7 shows how to use Equations 14.6 and 14.7 to calculate reactant concentrations and the half-life of a second-order reaction.

SAMPLE PROBLEM 14.7

Iodine atoms combine to form molecular iodine in the gas phase:

$$I(g) + I(g) \longrightarrow I_2(g)$$

This equation can also be written as $2I(g) \longrightarrow I_2(g)$. The reaction is second order and has a rate constant of $7.0 \times 10^9\ M^{-1} \cdot s^{-1}$ at 23°C. (a) If the initial concentration of I is 0.086 M, calculate the concentration after 2.0 min. (b) Calculate the half-life of the reaction when the initial concentration of I is 0.60 M and when the initial concentration of I is 0.42 M.

Strategy Use Equation 14.6 to determine $[I]_t$ at $t = 2.0$ min; use Equation 14.7 to determine $t_{1/2}$ when $[I]_0 = 0.60\ M$ and when $[I]_0 = 0.42\ M$.

Setup
$$t = (2.0\ \text{min})(60\ \text{s/min}) = 120\ \text{s}$$

Solution (a) $\dfrac{1}{[A]_t} = kt + \dfrac{1}{[A]_0}$

$$= (7.0 \times 10^9\ M^{-1} \cdot s^{-1})(120\ s) + \frac{1}{0.086\ M}$$

$$= 8.4 \times 10^{11}\ M^{-1}$$

$$[A]_t = \frac{1}{8.4 \times 10^{11}\ M^{-1}} = 1.2 \times 10^{-12}\ M$$

The concentration of atomic iodine after 2.0 min is $1.2 \times 10^{-12}\ M$.

(b) When $[I]_0 = 0.60\ M$,

$$t_{1/2} = \frac{1}{k[A]_0} = \frac{1}{(7.0 \times 10^9\ M^{-1} \cdot s^{-1})(0.60\ M)} = 2.4 \times 10^{-10}\ s$$

When $[I]_0 = 0.42\ M$,

$$t_{1/2} = \frac{1}{k[A]_0} = \frac{1}{(7.0 \times 10^9\ M^{-1} \cdot s^{-1})(0.42\ M)} = 3.4 \times 10^{-10}\ s$$

(Continued on next page)

THINK ABOUT IT

(a) Iodine, like the other halogens, exists as diatomic molecules at room temperature. It makes sense, therefore, that atomic iodine would react quickly, and essentially completely, to form I_2 at room temperature. The very low remaining concentration of I after 2.0 min makes sense. (b) As expected, the half-life of this second-order reaction is not constant. (A constant half-life is a characteristic of first-order reactions.)

Practice Problem ATTEMPT   The reaction $2A \longrightarrow B$ is second order in A with a rate constant of $32.0\ M^{-1} \cdot s^{-1}$ at 25.0°C. (a) Starting with $[A]_0 = 0.0075\ M$, how long will it take for the concentration of A to drop to $0.0018\ M$? (b) Calculate the half-life of the reaction for $[A]_0 = 0.0075\ M$ and for $[A]_0 = 0.0025\ M$.

Practice Problem BUILD   Determine the initial concentration, $[A]_0$, for the reaction in Practice Problem A necessary for the half-life to be (a) 1.50 s, (b) 25.0 s, and (c) 175 s.

Practice Problem CONCEPTUALIZE   The diagrams below show three different experiments with the reaction of A (red) and B (blue) to form C (purple). The reaction $A + B \longrightarrow C$ is second order in A. In the last experiment, the red spheres at $t = 0$ s are not shown. Determine how many red spheres must be included for the diagram to be correct.

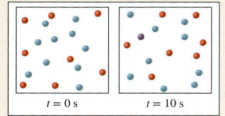

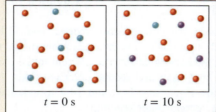

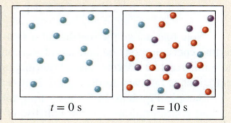

First- and second-order reactions are the most common reaction types. Reactions of overall order zero exist but are relatively rare. For a zeroth-order reaction:

$$A \longrightarrow product$$

the rate law is given by:

$$rate = k[A]^0 = k$$

Thus, the rate of a ***zeroth-order reaction*** is a constant, independent of reactant concentration. Third-order and higher-order reactions are quite rare and too complex to be covered in this book. Table 14.5 summarizes the kinetics for first-order and second-order reactions of the type $A \longrightarrow product$.

TABLE 14.5	Summary of the Kinetics of Zeroth-Order, First-Order, and Second-Order Reactions		
Order	**Rate Law**	**Integrated Rate Law**	**Half-Life**
0	$rate = k$	$[A]_t = -kt + [A]_0$	$\dfrac{[A]_0}{2k}$
1	$rate = k[A]$	$\ln\dfrac{[A]_t}{[A]_0} = -kt$	$\dfrac{0.693}{k}$
2	$rate = k[A]^2$	$\dfrac{1}{[A]_t} = kt + \dfrac{1}{[A]_0}$	$\dfrac{1}{k[A]_0}$

CHECKPOINT – SECTION 14.3   Dependence of Reactant Concentration on Time

The first-order decomposition of dinitrogen pentoxide (N_2O_5) is represented by

$$2N_2O_5(g) \longrightarrow 4NO_2(g) + O_2(g)$$

Use the table of data to answer questions 14.3.1 and 14.3.2:

$t(s)$	$[N_2O_5]$ (M)
0	0.91
300	0.75
600	0.64
1200	0.44
3000	0.16

14.3.1 What is the rate constant for the decomposition of N_2O_5?

a) 9×10^{-4} s^{-1}

b) 4×10^{-4} s^{-1}

c) 6×10^{-4} s^{-1}

d) 5×10^{-3} s^{-1}

e) 1×10^{-3} s^{-1}

14.3.2 Approximately how long will it take for $[N_2O_5]$ to fall from 0.62 M to 0.10 M?

a) 1000 s   d) 4000 s

b) 2000 s   e) 5000 s

c) 3000 s

14.3.3 Consider the first-order reaction A ⟶ B in which A molecules (blue spheres) are converted to B molecules (yellow spheres). What is the half-life for the reaction?

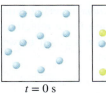

$t = 0$ s $t = 10$ s

a) 5 s   d) 20 s

b) 10 s   e) 30 s

c) 15 s

14.3.4 Which figure below represents the numbers of molecules present after 20 s for the reaction in question 14.3.3?

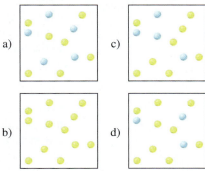

a)   c)

b)   d)

14.3.5 Of the plots shown here, ____ corresponds to a zeroth-order reaction, ____ corresponds to a first-order reaction, and ____ corresponds to a second-order reaction.

a) ii, iii, i

b) i, ii, iii

c) iii, i, ii

d) ii, i, iii

e) i, iii, ii

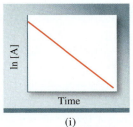

(i)

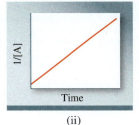

(ii)

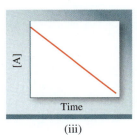

(iii)

14.4   Dependence of Reaction Rate on Temperature

Nearly all reactions happen faster at higher temperatures. For example, the time required to cook food depends largely on the boiling point of water. Cookbooks sometimes give alternate directions for cooking at high altitudes, where the lower atmospheric pressure results in water boiling at a lower temperature [◂◂ Section 11.6]. The reaction involved in hard-boiling an egg happens faster at 100°C (about 10 min) than at 80°C (about 30 min). The dependence of reaction rate on temperature is the reason we keep food in a refrigerator—and why food keeps even longer in a freezer. The lower the temperature, the slower the processes that cause food to spoil.

Collision Theory

Chemical reactions generally occur as a result of collisions between reacting molecules. A greater frequency of collisions usually leads to a higher reaction rate. According to the **collision theory** of chemical kinetics, the reaction rate is directly proportional to the number of molecular collisions per second:

$$\text{rate} \propto \frac{\text{number of collisions}}{\text{s}}$$

Consider the reaction of A molecules with B molecules to form some product. Suppose that each product molecule is formed by the direct combination of an A molecule and a B molecule. If we doubled the concentration of A, then the number of A-B collisions would also double, because there would be twice as many A molecules that could collide with B molecules in any given volume. Consequently, the rate would increase by a factor of 2. Similarly, doubling the concentration of B molecules would increase the rate twofold. Thus, we can express the rate law as

$$\text{rate} = k[\text{A}][\text{B}]$$

The reaction is first order in both A and B and is second order overall.

This view of collision theory is something of a simplification, though, because not every collision between molecules results in a reaction. A collision that *does* result in a reaction is called an **effective collision.** A molecule in motion possesses kinetic energy; the faster it is moving, the greater its kinetic energy. When molecules collide, part of their kinetic energy is converted to vibrational energy. If the initial kinetic energies are large, then the colliding molecules will vibrate so strongly as to break some of the chemical bonds. This bond breaking is the first step toward product formation. If the initial kinetic energies are small, the molecules will merely bounce off of each other intact. There is a minimum amount of energy, the **activation energy (Eₐ),** required to initiate a chemical reaction. Without this minimum amount of energy at impact, a collision will be *ineffective;* that is, it will not result in a reaction.

When molecules react (as opposed to when *atoms* react), having sufficient kinetic energy is not the only requirement for a collision to be effective. Molecules must also be oriented in a way that favors reaction. The reaction between chlorine atoms and nitrosyl chloride (NOCl) illustrates this point:

$$\text{Cl} + \text{NOCl} \longrightarrow \text{Cl}_2 + \text{NO}$$

This reaction is most favorable when a free Cl atom collides directly with the Cl atom in the NOCl molecule [Figure 14.11(a)]. Otherwise, the reactants simply bounce off of each other and no reaction occurs [Figure 14.11(b)].

> **Student Note:** Kinetic energy is the result of motion of the whole molecule, relative to its surroundings. Vibrational energy is the result of motion of the atoms in a molecule, relative to one another.

Figure 14.11 (a) For an effective collision to take place, the free Cl atom must collide directly with the Cl atom in NOCl. (b) Otherwise, the reactants bounce off of one another and the collision is ineffective—no reaction takes place.

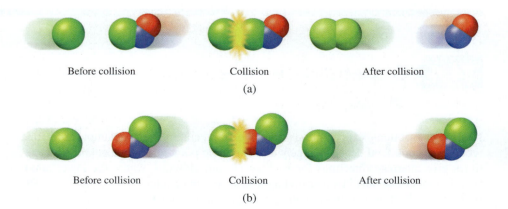

Before collision   Collision   After collision

(a)

Before collision   Collision   After collision

(b)

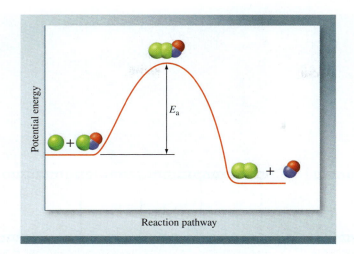

Figure 14.12 Energy profile for the reaction of Cl with NOCl. In addition to being oriented properly, reactant molecules must possess sufficient energy to overcome the activation energy.

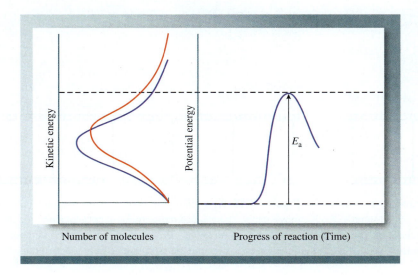

Figure 14.13 Kinetic molecular theory shows that the average speed and therefore average kinetic energy of a collection of molecules increases with increasing temperature. The blue line represents a collection of molecules at lower temperature; the red line represents the collection of molecules at higher temperature. At higher temperature, more molecules have sufficient kinetic energy to exceed the activation energy and undergo effective collision.

When molecules collide (in an effective collision), they form an ***activated complex*** (also called the ***transition state***), a temporary species formed by the reactant molecules as a result of the collision. Figure 14.12 shows a potential-energy profile for the reaction between Cl and NOCl.

We can think of the activation energy as an energy *barrier* that prevents less energetic molecules from reacting. Because the number of reactant molecules in an ordinary reaction is very large, the speeds, and therefore also the kinetic energies of the molecules, vary greatly. Normally, only a small fraction of the colliding molecules—the fastest-moving ones—have sufficient kinetic energy to exceed the activation energy. These molecules can therefore take part in the reaction. The relationship between rate and temperature should now make sense. According to kinetic molecular theory, the average kinetic energy of a sample of molecules increases as the temperature increases [◀◀ Section 10.6, Figure 10.20]. Thus, a higher percentage of the molecules in the sample have sufficient kinetic energy to exceed the activation energy (Figure 14.13), and the reaction rate increases.

Collision theory gives us a way to visualize how reactant concentration and temperature affect reaction rate. Figures 14.14 and 14.15 illustrate these effects at the molecular level.

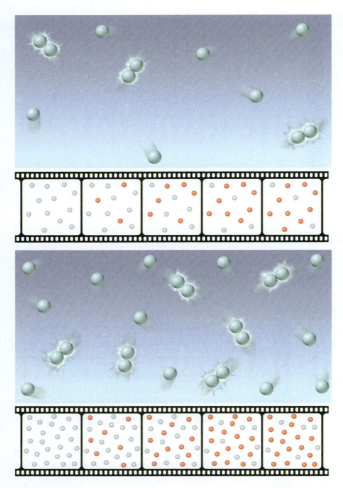

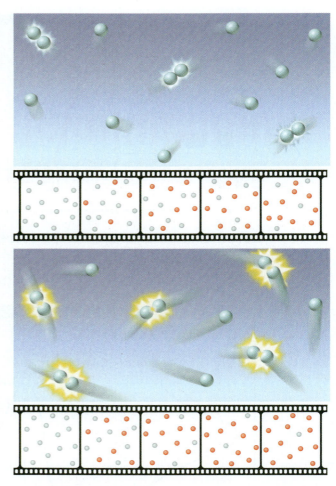

Figure 14.14 At higher concentration, reactant molecules collide more often, giving rise to a greater number of collisions overall, which increases the number of effective collisions. An increase in the number of effective collisions increases the reaction rate.

Figure 14.15 When temperature increases, reactant molecules move faster. This causes more frequent collisions and results in greater energy at impact. Both factors increase the number of effective collisions and cause an increase in reaction rate.

The Arrhenius Equation

The dependence of the rate constant of a reaction on temperature can be expressed by the *Arrhenius equation:*

Equation 14.8
$$k = Ae^{-E_a/RT}$$

Student Note: Absolute temperature is expressed in *kelvins* [◀◀ Section 1.3].

where E_a is the activation energy of the reaction (in kJ/mol), R is the gas constant (8.314 J/K · mol), T is the *absolute temperature,* and e is the base of the natural logarithm [▶▶ Appendix 1]. The quantity A represents the collision frequency and is called the frequency factor. It can be treated as a constant for a given reaction over a reasonably wide temperature range. Equation 14.8 shows that the rate constant decreases with increasing activation energy and increases with increasing temperature. This equation can be expressed in a more useful form by taking the natural logarithm of both sides:

$$\ln k = \ln Ae^{-E_a/RT}$$

or:

Equation 14.9
$$\ln k = \ln A - \frac{E_a}{RT}$$

which can be rearranged to give the following linear equation:

$$\ln k = \left(-\frac{E_a}{R}\right)\left(\frac{1}{T}\right) + \ln A$$

$$y \ = \quad m \quad x \ + \ b$$

Equation 14.10

Thus, a plot of ln k versus $1/T$ gives a straight line whose slope is equal to $-E_a/R$ and whose y intercept (b) is equal to ln A.

Sample Problem 14.8 demonstrates a graphical method for determining the activation energy of a reaction.

SAMPLE PROBLEM (14.8)

Rate constants for the reaction:

$$CO(g) + NO_2(g) \longrightarrow CO_2(g) + NO(g)$$

were measured at four different temperatures. The data are shown in the table. Plot ln k versus $1/T$, and determine the activation energy (in kJ/mol) for the reaction:

k ($M^{-1} \cdot s^{-1}$)	T (K)
0.0521	288
0.101	298
0.184	308
0.332	318

Strategy Plot ln k versus $1/T$, and determine the slope of the resulting line. According to Equation 14.10, slope = $-E_a/R$.

Setup $R = 8.314$ J/K · mol. Taking the natural log of each value of k and the inverse of each value of T gives:

ln k	$1/T$ (K^{-1})
−2.95	3.47×10^{-3}
−2.29	3.36×10^{-3}
−1.69	3.25×10^{-3}
−1.10	3.14×10^{-3}

Solution A plot of these data yields the following graph:

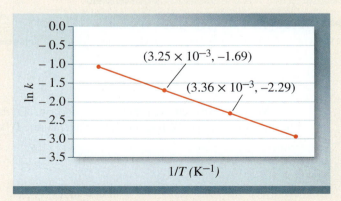

The slope is determined using the x and y coordinates of any two points on the line. Using the points that are labeled on the graph gives:

$$\text{slope} = \frac{-1.69 - (-2.29)}{3.25 \times 10^{-3} \text{ K}^{-1} - 3.36 \times 10^{-3} \text{ K}^{-1}} = -5.5 \times 10^3 \text{ K}$$

(Continued on next page)

The value of the slope is -5.5×10^3 K. Because slope $= -E_a/R$,

$$E_a = -(\text{slope})(R)$$

$$= -(-5.5 \times 10^3 \text{ K})(8.314 \text{ J/K} \cdot \text{mol})$$

$$= 4.6 \times 10^4 \text{ J/mol} \quad \text{or} \quad 46 \text{ kJ/mol}$$

The activation energy is 46 kJ/mol.

THINK ABOUT IT

Note that while k has units of $M^{-1} \cdot s^{-1}$, ln k has no units.

Practice Problem **A**TTEMPT   The second-order rate constant for the decomposition of nitrous oxide to nitrogen molecules and oxygen atoms has been determined at various temperatures:

k $(M^{-1} \cdot s^{-1})$	T (°C)
1.87×10^{-3}	600
0.0113	650
0.0569	700
0.244	750

Determine the activation energy graphically.

Practice Problem **B**UILD   Use the graph to determine the value of k at 475 K.

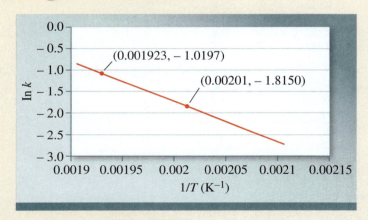

Practice Problem **C**ONCEPTUALIZE   Each line in the graph represents a different reaction. List the reactions in order of increasing activation energy.

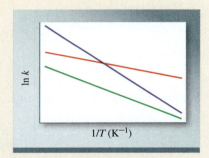

Student Note: Note that for two different temperatures, the only thing in Equation 14.9 that changes is k. The other variables, A and E_a (and R), are constant.

We can derive an even more useful form of the Arrhenius equation starting with Equation 14.9 written for two different temperatures, T_1 and T_2:

$$\ln k_1 = \ln A - \frac{E_a}{RT_1}$$

$$\ln k_2 = \ln A - \frac{E_a}{RT_2}$$

Subtracting ln k_2 from ln k_1 gives:

$$\ln k_1 - \ln k_2 = \left(\ln A - \frac{E_a}{RT_1}\right) - \left(\ln A - \frac{E_a}{RT_2}\right)$$

$$\ln \frac{k_1}{k_2} = \frac{E_a}{R}\left(-\frac{1}{T_1} + \frac{1}{T_2}\right)$$

$$\ln \frac{k_1}{k_2} = \frac{E_a}{R}\left(\frac{1}{T_2} - \frac{1}{T_1}\right) \qquad \textbf{Equation 14.11}$$

Equations 14.8 through 14.11 are all "Arrhenius equations," but Equation 14.11 is the form you will use most often to solve kinetics problems. Equation 14.11 enables us to do two things:

1. If we know the rate constant at two different temperatures, we can calculate the activation energy.
2. If we know the activation energy and the rate constant at one temperature, we can determine the value of the rate constant at any other temperature.

Sample Problems 14.9 and 14.10 show how to use Equation 14.11.

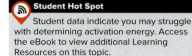

Student Hot Spot

Student data indicate you may struggle with determining activation energy. Access the eBook to view additional Learning Resources on this topic.

SAMPLE PROBLEM 14.9

The rate constant for a particular first-order reaction is given for three different temperatures:

T (K)	k (s^{-1})
400	2.9×10^{-3}
450	6.1×10^{-2}
500	7.0×10^{-1}

Using these data, calculate the activation energy of the reaction.

Strategy Use Equation 14.11 to solve for E_a.

Setup Solving Equation 14.11 for E_a gives:

$$E_a = R\left(-\frac{\ln\dfrac{k_1}{k_2}}{\dfrac{1}{T_2} - \dfrac{1}{T_1}}\right)$$

Solution Using the rate constants for 400 K (T_1) and 450 K (T_2), we get:

$$E_a = \frac{8.314\,\text{J}}{\text{K}\cdot\text{mol}}\left(\frac{\ln\dfrac{2.9 \times 10^{-3}\,\text{s}^{-1}}{6.1 \times 10^{-2}\,\text{s}^{-1}}}{\dfrac{1}{450\,\text{K}} - \dfrac{1}{400\,\text{K}}}\right)$$

$$= 91{,}173\,\text{J/mol} = 91\,\text{kJ/mol}$$

The activation energy of the reaction is 91 kJ/mol.

THINK ABOUT IT

A good way to check your work is to use the value of E_a that you calculated (and Equation 14.11) to determine the rate constant at 500 K. Make sure it agrees with the value in the table.

Practice Problem **ATTEMPT** Use the data in the following table to determine the activation energy of the reaction:

T (K)	k (s^{-1})
625	1.1×10^{-4}
635	1.5×10^{-4}
645	2.0×10^{-4}

(Continued on next page)

Practice Problem **B**UILD   Based on the data shown in Practice Problem A, what will be the value of k at 655 K?

Practice Problem **C**ONCEPTUALIZE   According to the Arrhenius equation, which graph best represents the relationship between temperature and rate constant?

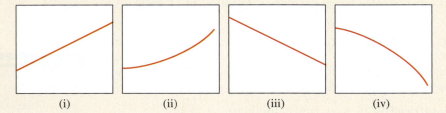

(i)   (ii)   (iii)   (iv)

SAMPLE PROBLEM (14.10)

A certain first-order reaction has an activation energy of 83 kJ/mol. If the rate constant for this reaction is 2.1×10^{-2} s^{-1} at 150°C, what is the rate constant at 300°C?

Strategy   Use Equation 14.11 to solve for k_2. Pay particular attention to units in this type of problem.

Setup   Solving Equation 14.11 for k_2 gives:

$$k_2 = \frac{k_1}{e^{\left(\frac{E_a}{R}\right)\left(\frac{1}{T_2} - \frac{1}{T_1}\right)}}$$

$E_a = 8.3 \times 10^4$ J/mol, $T_1 = 423$ K, $T_2 = 573$ K, $R = 8.314$ J/K · mol, and
$k_1 = 2.1 \times 10^{-2}$ s^{-1}.

> **Student Note:** E_a is converted to joules so that the units will cancel properly. Alternatively, R could be expressed as 0.008314 kJ/K · mol. Also note again that T must be expressed in kelvins.

Solution

$$k_2 = \frac{2.1 \times 10^{-2} \text{ s}^{-1}}{e^{\left(\frac{8.3 \times 10^4 \text{ J/mol}}{8.314 \text{ J/K} \cdot \text{mol}}\right)\left(\frac{1}{573 \text{ K}} - \frac{1}{423 \text{ K}}\right)}}$$

$$= 1.0 \times 10^1 \text{ s}^{-1}$$

The rate constant at 300°C is 10 s^{-1}.

THINK ABOUT IT

Make sure that the rate constant you calculate at a higher temperature is in fact higher than the original rate constant. According to the Arrhenius equation, the rate constant always increases with increasing temperature. If you get a smaller k at a higher temperature, check your solution for mathematical errors.

Practice Problem **A**TTEMPT   Calculate the rate constant at 200°C for a reaction that has a rate constant of 8.1×10^{-4} s^{-1} at 90°C and an activation energy of 99 kJ/mol.

Practice Problem **B**UILD   Calculate the rate constant at 200°C for a reaction that has a rate constant of 8.1×10^{-4} s^{-1} at 90°C and an activation energy of 59 kJ/mol.

Practice Problem **C**ONCEPTUALIZE   According to the Arrhenius equation, which graph best represents the relationship between temperature and activation energy?

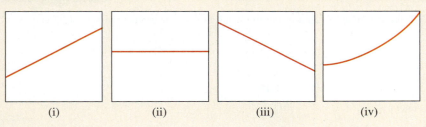

(i)   (ii)   (iii)   (iv)

CHECKPOINT – SECTION 14.4 Dependence of Reaction Rate on Temperature

Use the table of data collected for a first-order reaction to answer questions 14.4.1 and 14.4.2:

T (K)	*k* (s^{-1})
300	3.9×10^{-2}
310	1.1×10^{-1}
320	2.8×10^{-1}

14.4.1 What is the activation energy of the reaction?

a) 88 kJ/mol
b) 8.8×10^4 kJ/mol
c) 8.0×10^4 kJ/mol
d) 8.8×10^{-3} kJ/mol
e) 80 kJ/mol

14.4.2 What is the rate constant at 80°C?

a) $5.8 \times 10^{-40}\ s^{-1}$
b) $4.8\ s^{-1}$
c) $2.8\ s^{-1}$
d) $5.2 \times 10^{-3}\ s^{-1}$
e) $1.7 \times 10^2\ s^{-1}$

14.4.3 Which of the following changes when temperature changes?

a) Rate
b) Rate law
c) Rate constant
d) Activation energy
e) Reaction order

14.5 Reaction Mechanisms

A balanced chemical equation does not tell us much about *how* a reaction actually takes place. In many cases, the balanced equation is simply the sum of a series of steps. Consider the following hypothetical example. In the first step of a reaction, a molecule of reactant A combines with a molecule of reactant B to form a molecule of C:

$$A + B \longrightarrow C$$

In the second step, the molecule of C combines with another molecule of B to produce D:

$$C + B \longrightarrow D$$

The overall balanced equation is the sum of these two equations:

$$\begin{array}{ll} \textit{Step 1:} & A + B \longrightarrow \cancel{C} \\ \textit{Step 2:} & \underline{\cancel{C} + B \longrightarrow D} \\ & A + 2B \longrightarrow D \end{array}$$

The sequence of steps that sum to give the overall reaction is called the ***reaction mechanism.*** A reaction mechanism is comparable to the *route* traveled during a trip, whereas the overall balanced chemical equation specifies only the *origin* and the *destination.*

For a specific example of a reaction mechanism, we consider the reaction between nitric oxide and oxygen:

$$2NO(g) + O_2(g) \longrightarrow 2NO_2(g)$$

We know that NO_2 does not form as the result of a single collision between two NO molecules and one O_2 molecule because N_2O_2 is detected during the course of the reaction. We can envision the reaction taking place via the following two steps:

$$2NO(g) \longrightarrow N_2O_2(g)$$

$$N_2O_2(g) + O_2(g) \longrightarrow 2NO_2(g)$$

In the first step, two NO molecules collide to form an N_2O_2 molecule. This is followed by the step in which N_2O_2 and O_2 combine to give two molecules of NO_2. The net chemical equation, which represents the overall change, is given by the sum of the first and second steps:

$$\begin{array}{lr} \textit{Step 1:} & NO + NO \longrightarrow N_2O_2 \\ \textit{Step 2:} & \underline{N_2O_2 + O_2 \longrightarrow 2NO_2} \\ \textit{Overall reaction:} & 2NO(g) + O_2(g) \longrightarrow 2NO_2(g) \end{array}$$

Species such as N_2O_2 (and C in the hypothetical equation) are called ***intermediates*** because they appear in the mechanism of the reaction but not in the overall balanced equation. An intermediate is produced in an early step in the reaction and consumed in a later step.

Elementary Reactions

Student Note: Termolecular reactions actually are quite rare simply because the simultaneous encounter of three molecules is far less likely than that of two molecules.

Each step in a reaction mechanism represents an ***elementary reaction,*** one that occurs in a single collision of the reactant molecules. The ***molecularity*** of an elementary reaction is essentially the *number* of reactant molecules involved in the collision. Elementary reactions may be ***unimolecular*** (one reactant molecule), ***bimolecular*** (two reactant molecules), or ***termolecular*** (three reactant molecules). These molecules may be of the same or different types. Each of the elementary steps in the formation of NO_2 from NO and O_2 is bimolecular, because there are *two* reactant molecules in each step. Likewise, each of the steps in the hypothetical $A + 2B \longrightarrow D$ reaction is bimolecular. The overall equation has three reactant molecules. Because it is not an elementary reaction, though, we do not classify it as a termolecular reaction. Rather, it is the combination of two bimolecular reactions (steps 1 and 2).

Knowing the steps of a reaction enables us to deduce the rate law. Suppose we have the following elementary reaction:

$$A \longrightarrow \text{products}$$

Because there is only one molecule present, this is a unimolecular reaction. It follows that the larger the number of A molecules present, the faster the rate of product formation. Thus, the rate of a unimolecular reaction is directly proportional to the concentration of A—the reaction is first order in A:

$$\text{rate} = k[A]$$

For a bimolecular elementary reaction involving A and B molecules:

$$A + B \longrightarrow \text{products}$$

the rate of product formation depends on how frequently A and B collide, which in turn depends on the concentrations of A and B. Thus, we can express the rate as:

$$\text{rate} = k[A][B]$$

Therefore, this is a second-order reaction. Similarly, for a bimolecular elementary reaction of the type:

$$A + A \longrightarrow \text{products}$$

or:

$$2A \longrightarrow \text{products}$$

the rate becomes:

$$\text{rate} = k[A]^2$$

Student Note: It is important to remember that this is not the case in general. It only applies to elementary reactions—and whether or not a reaction is elementary must be determined experimentally.

which is also a second-order reaction. The preceding examples show that the reaction order for each reactant in an elementary reaction is equal to its stoichiometric coefficient in the chemical equation for that step. In general, we cannot tell just by looking at the balanced equation whether the reaction occurs as shown or in a series of steps. This determination must be made using data obtained experimentally.

Rate-Determining Step

In a reaction mechanism consisting of more than one elementary step, the rate law for the overall process is given by the ***rate-determining step,*** which is the *slowest* step in the sequence. An analogy for a process in which there is a rate-determining step is the amount of time required to buy stamps at the post office when there is a long line of customers. The process consists of several steps: waiting in line, requesting

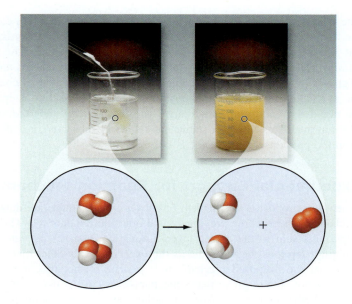

Figure 14.16 The decomposition of hydrogen peroxide is catalyzed by the addition of an iodide salt. Some of the iodide ions are oxidized to molecular iodine, which then reacts with iodide ions to form the brown triiodide ion (I_3^-).
(photos): *Charles D. Winters/Timeframe Photography/McGraw Hill*

the stamps, receiving the stamps, and paying for the stamps. At a time when the line is very long, the amount of time spent waiting in line (step 1) largely determines how much time the overall process takes.

To study reaction mechanisms, we first do a series of experiments to establish initial rates at various reactant concentrations. We then analyze the data to determine the rate constant and overall order of the reaction, and we write the rate law. Finally, we propose a plausible mechanism for the reaction in terms of logical elementary steps. The steps of the proposed mechanism must satisfy two requirements:

1. The sum of the elementary reactions must be the overall balanced equation for the reaction.
2. The rate-determining step must have the same rate law as that determined from the experimental data.

The decomposition of hydrogen peroxide can be facilitated by iodide ions (Figure 14.16). The overall reaction is:

$$2H_2O_2(aq) \longrightarrow 2H_2O(l) + O_2(g)$$

By experiment, we find the rate law to be:

$$\text{rate} = k[H_2O_2][I^-]$$

Thus, the reaction is first order with respect to both H_2O_2 and I^-.

The decomposition of H_2O_2 is not an elementary reaction, because it does not occur in a single step. If it did, the reaction would be second order in H_2O_2 (as a result of the collision of two H_2O_2 molecules). What's more, the I^- ion, which is not even part of the overall equation, would not appear in the rate law expression. How can we reconcile these facts? First, we can account for the observed rate law by assuming that the reaction takes place in two separate elementary steps, each of which is bimolecular:

$$\textit{Step 1:} \quad H_2O_2 + I^- \xrightarrow{k_1} H_2O + IO^-$$

$$\textit{Step 2:} \quad H_2O_2 + IO^- \xrightarrow{k_2} H_2O + O_2 + I^-$$

If we further assume that step 1 is the rate-determining step, then the rate of the reaction can be determined from the first step alone:

$$\text{rate} = k_1[H_2O_2][I^-]$$

where $k_1 = k$. The IO^- ion is an intermediate because it is produced in the first step and consumed in the second step. It does not appear in the overall balanced equation.

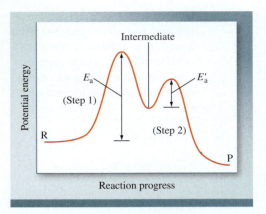

Figure 14.17 Potential-energy profile for a two-step reaction in which the first step is rate determining. R and P represent reactants and products, respectively.

The I^- ion also does not appear in the overall equation, but it is consumed in the first step and then produced in the second step. In other words, it is present at the start of the reaction, and it is present at the end. The I^- ion is a *catalyst,* and its function is to *speed up* the reaction. Catalysts are discussed in greater detail in Section 14.6. Figure 14.17 shows the potential-energy profile for a reaction like the decomposition of H_2O_2. The first step, which is the rate-determining step, has a larger activation energy than the second step. The intermediate, although stable enough to be observed, reacts quickly to form the products. Its existence is only fleeting.

Experimental Support for Reaction Mechanisms

When you propose a mechanism based on experimental rate data, and your mechanism satisfies the previously laid out requirements [(1) The individual steps must sum to the correct overall equation, and (2) the rate law of the rate-determining step must be the same as the experimentally-determined rate law)], you can say that the proposed mechanism is *plausible*—not that it is necessarily *correct*. It is not possible to prove that a mechanism is correct using rate data alone. To determine whether or not a proposed reaction mechanism is actually correct, we must conduct other experiments. In the case of hydrogen peroxide decomposition, we might try to detect the presence of the IO^- ions. If we can detect them, it will support the proposed mechanism. Similarly, for the hydrogen iodide reaction, detection of iodine atoms would lend support to the proposed two-step mechanism. For example, I_2 dissociates into atoms when it is irradiated with visible light. Thus, we might predict that the formation of HI from H_2 and I_2 would speed up as the intensity of light is increased—because that would increase the concentration of I atoms. Indeed, this is just what is observed.

In one case, chemists wanted to know which $C-O$ bond was broken in the reaction between methyl acetate and water to better understand the reaction:

$$CH_3-\overset{\overset{\textstyle O}{\|}}{C}-O-CH_3 \ + \ H_2O \ \longrightarrow \ CH_3-\overset{\overset{\textstyle O}{\|}}{C}-OH \ + \ CH_3OH$$

The two possibilities of bond breaking are:

$$CH_3-\overset{\overset{\textstyle O}{\|}}{C}-\overset{\downarrow}{}-O-CH_3 \quad \text{and} \quad CH_3-\overset{\overset{\textstyle O}{\|}}{C}-O-\overset{\downarrow}{}-CH_3$$
$$\text{(a)} \qquad\qquad\qquad \text{(b)}$$

To distinguish between schemes (a) and (b), chemists used water containing the oxygen-18 isotope instead of ordinary water, which contains the oxygen-16 isotope. When the ^{18}O water was used, only the acetic acid formed contained ^{18}O:

$$CH_3-\overset{\overset{\textstyle O}{\|}}{C}-{^{18}O}-H$$

Thus, the reaction must have occurred via bond-breaking scheme (a), because the product formed via scheme (b) would have retained both of its original oxygen atoms and would contain no ^{18}O.

Another example is photosynthesis, the process by which green plants produce glucose from carbon dioxide and water:

$$6CO_2 + 6H_2O \longrightarrow C_6H_{12}O_6 + 6O_2$$

A question that arose early in the studies of photosynthesis was whether the molecular oxygen produced came from the water, the carbon dioxide, or both. By using water containing only the oxygen-18 isotope, it was concluded that all the oxygen produced by photosynthesis came from the water and none came from the carbon dioxide, because the O_2 produced contained only ^{18}O.

These examples illustrate how creative chemists must be to study reaction mechanisms.

Identifying Plausible Reaction Mechanisms

In assessing the plausibility of a reaction mechanism, we apply the same two criteria that we used to propose a mechanism:

1. The individual steps (elementary processes) must sum to the correct overall reaction.
2. The rate-determining step (the slow step) must give the same rate law as the experimentally determined rate law.

> **Student Note:** Recall that for an elementary process, the exponents in the rate law are simply the coefficients in the balanced equation.

Consider the gas-phase reaction of hydrogen with iodine monochloride to form hydrogen chloride and iodine:

$$H_2 + 2ICl \longrightarrow 2HCl + I_2$$

The experimentally determined rate law for this reaction is rate = $k[H_2][ICl]$. We present four different proposed mechanisms.

Mechanism 1

Step 1: $ICl + ICl \longrightarrow I_2 + Cl_2$ (slow)
Step 2: $Cl_2 + H_2 \longrightarrow 2HCl$

The steps in mechanism 1 sum to the correct overall reaction:

$$ICl + ICl \longrightarrow I_2 + \cancel{Cl_2} \text{ (slow)}$$
$$\cancel{Cl_2} + H_2 \longrightarrow 2HCl$$
$$\overline{\text{Sum: } H_2 + 2ICl \longrightarrow 2HCl + I_2}$$

However, the rate law of the rate-determining step is rate = $k[ICl]^2$. Therefore, mechanism 1 is not plausible. It meets criterion 1, but it does not meet criterion 2.

Mechanism 2

Step 1: $H_2 + ICl \longrightarrow HI + HCl$ (slow)
Step 2: $ICl + HCl \longrightarrow HI + Cl_2$

The rate-determining step in mechanism 2 has the correct rate law:

$$\text{rate} = k[H_2][ICl]$$

However, the steps do not sum to the correct overall reaction:

$$H_2 + ICl \longrightarrow HI + \cancel{HCl} \text{ (slow)}$$
$$ICl + \cancel{HCl} \longrightarrow HI + Cl_2$$
$$\overline{\text{Sum: } H_2 + 2ICl \longrightarrow 2HI + Cl_2}$$

Therefore, mechanism 2 is not plausible. It meets criterion 2, but it does not meet criterion 1.

Mechanism 3

Step 1: $H_2 \longrightarrow 2H$ (slow)
Step 2: $ICl + H \longrightarrow HCl + I$
Step 3: $H + I \longrightarrow HI$

The rate law for the rate-determining step in mechanism 3 (rate = $k[H_2]$) is not correct.

Furthermore, the steps do not sum to the correct overall reaction:

$$H_2 \longrightarrow 2\cancel{H} \text{ (slow)}$$

$$ICl + \cancel{H} \longrightarrow HCl + \cancel{I}$$

$$\cancel{H} + \cancel{I} \longrightarrow HI$$

$$\overline{\text{Sum: } H_2 + ICl \longrightarrow HCl + HI}$$

Therefore, mechanism 3 is not plausible. It meets neither criterion 1 nor criterion 2.

Mechanism 4

Step 1: $H_2 + ICl \longrightarrow HCl + HI$ (slow)
Step 2: $HI + ICl \longrightarrow HCl + I_2$

The rate-determining step in mechanism 4 has the correct rate law:

$$\text{rate} = k[H_2][ICl]$$

Furthermore, the steps in mechanism 4 sum to the correct overall reaction:

$$H_2 + ICl \longrightarrow HCl + \cancel{HI} \text{ (slow)}$$

$$\cancel{HI} + ICl \longrightarrow HCl + I_2$$

$$\overline{\text{Sum: } H_2 + 2ICl \longrightarrow 2HCl + I_2}$$

Therefore, mechanism 4 is plausible. It meets both criteria.

Sample Problem 14.11 lets you practice determining if a proposed reaction mechanism is plausible.

SAMPLE PROBLEM (14.11)

The gas-phase decomposition of nitrous oxide (N_2O) is believed to occur in two steps:

Step 1: $N_2O \xrightarrow{k_1} N_2 + O$

Step 2: $N_2O + O \xrightarrow{k_2} N_2 + O_2$

Experimentally the rate law is found to be rate = $k[N_2O]$. (a) Write the equation for the overall reaction. (b) Identify the intermediate(s). (c) Identify the rate-determining step.

Strategy Add the two equations, canceling identical terms on opposite sides of the arrow, to obtain the overall reaction. The canceled terms will be the intermediates if they were first generated and then consumed. Write rate laws for each elementary step; the one that matches the experimental rate law will be the rate-determining step.

Setup Intermediates are species that are generated in an earlier step and consumed in a later step. We can write rate laws for elementary reactions simply by using the stoichiometric coefficient for each species as its exponent in the rate law.

Step 1: $N_2O \xrightarrow{k_1} N_2 + O$ rate = $k[N_2O]$

Step 2: $N_2O + O \xrightarrow{k_2} N_2 + O_2$ rate = $k[N_2O][O]$

Solution (a) $2N_2O \longrightarrow 2N_2 + O_2$

(b) O (atomic oxygen) is the intermediate.

(c) Step 1 is the rate-determining step because its rate law is the same as the experimental rate law: rate = $k[N_2O]$.

THINK ABOUT IT

A species that gets canceled when steps are added may be an intermediate or a catalyst. In this case, the canceled species is an intermediate because it was first generated and then consumed. A species that is first consumed and then generated, but doesn't appear in the overall equation, is a catalyst.

Practice Problem **A**TTEMPT   The reaction between NO_2 and CO to produce NO and CO_2 is thought to occur in two steps:

$$Step\ 1:\quad NO_2 + NO_2 \xrightarrow{k_1} NO + NO_3$$

$$Step\ 2:\quad NO_3 + CO \xrightarrow{k_2} NO_2 + CO_2$$

The experimental rate law is rate $= k[NO_2]^2$. (a) Write the equation for the overall reaction. (b) Identify the intermediate(s). (c) Identify the rate-determining step.

Practice Problem **B**UILD   Propose a plausible mechanism for the reaction $F_2 + 2ClO_2 \longrightarrow 2FClO_2$, given that the rate law for the reaction is rate $= k[F_2][ClO_2]$.

Practice Problem **C**ONCEPTUALIZE   How many steps are there in the reaction represented by the potential-energy profile? Which step is the rate-determining step? How many intermediates, if any, are there in the reaction mechanism?

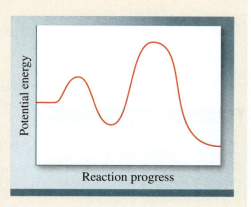

Reaction progress

Mechanisms with a Fast Initial Step

For first- and second-order reactions, it is reasonably straightforward to propose a plausible mechanism. We simply use the experimentally determined rate law to write the rate-determining step as the first step, and then write one or more additional steps such that the appropriate species will cancel to give the correct overall equation. Sometimes, however, we will encounter a reaction with an experimentally determined rate law that suggests an unlikely scenario. For example, consider the gaseous reaction of nitric oxide with chlorine to produce nitrosyl chloride:

$$2NO(g) + Cl_2(g) \longrightarrow 2NOCl(g)$$

for which the experimentally determined rate law is rate $= k[NO]^2[Cl_2]$. Note that the exponents in the rate law are the same as the coefficients in the balanced equation—making this reaction *third order* overall. Although one possibility is that the reaction is an elementary process, this would require the simultaneous collision of three molecules. As was noted earlier, such a termolecular reaction is unlikely. A more reasonable mechanism would be one in which a fast first step is followed by the slower, rate-determining step:

$$Step\ 1:\quad NO(g) + Cl_2(g) \xrightarrow{k_1} NOCl_2(g)\quad (fast)$$

$$Step\ 2:\quad NOCl_2(g) + NO(g) \xrightarrow{k_2} 2NOCl(g)\quad (slow)$$

However, when we write the overall rate law using the equation for the rate-determining step, as we have done previously, the resulting rate law:

$$rate = k_2[NOCl_2][NO]$$

includes the concentration of an *intermediate* ($NOCl_2$). Although this rate law is correct for the proposed mechanism, in this form it does not enable us to determine the plausibility of the mechanism by comparison to the experimentally determined rate law. To do this, we must derive a rate law in which only *reactants* from the overall equation appear.

When the intermediate ($NOCl_2$) is not consumed by step 2 as fast as it is produced by step 1, its concentration builds up, causing step 1 to happen in *reverse;* that is, $NOCl_2$ reacts to produce NO and Cl_2. This results in the establishment of a dynamic equilibrium [◄◄ Section 4.1], where the forward and reverse of step 1 are occurring at the same rate. We denote this by changing the single reaction arrow in step 1 to equilibrium arrows:

$$NO(g) + Cl_2(g) \underset{k_{-1}}{\overset{k_1}{\rightleftarrows}} NOCl_2(g)$$

We can write the rate laws for the forward and reverse of step 1, both elementary processes, as:

$$rate_{forward} = k_1[NO][Cl_2] \quad and \quad rate_{reverse} = k_{-1}[NOCl_2]$$

where k_1 and k_{-1} are the individual rate constants for the forward and reverse processes, respectively. Because the two rates are equal, we can write:

$$k_1[NO][Cl_2] = k_{-1}[NOCl_2]$$

Rearranging to solve for the concentration of $NOCl_2$ gives:

$$\frac{k_1}{k_{-1}}[NO][Cl_2] = [NOCl_2]$$

When we substitute the result into the original rate law for the concentration of $NOCl_2$, we get

$$\text{rate} = k_2[NOCl_2][NO] = k_2\frac{k_1}{k_{-1}}[NO]^2[Cl_2]$$

which agrees with the experimentally determined rate law, rate $= k[NO]^2[Cl_2]$, where k is equal to $\frac{k_2 k_1}{k_{-1}}$.

Sample Problem 14.12 lets you practice relating the experimentally determined rate law to a reaction mechanism in which a fast first step is followed by a slower, rate-determining step.

SAMPLE PROBLEM 14.12

Consider the gas-phase reaction of nitric oxide and oxygen that was described at the beginning of Section 14.5:

$$2NO(g) + O_2(g) \longrightarrow 2NO_2(g)$$

Show that the following mechanism is plausible. The experimentally determined rate law is rate $= k[NO]^2[O_2]$:

Step 1: $NO(g) + NO(g) \underset{k_{-1}}{\overset{k_1}{\rightleftharpoons}} N_2O_2(g)$ (fast)

Step 2: $N_2O_2(g) + O_2(g) \overset{k_2}{\longrightarrow} 2NO_2(g)$ (slow)

Strategy To establish the plausibility of a mechanism, we must compare the rate law of the rate-determining step to the experimentally determined rate law. In this case, the rate-determining step has an intermediate (N_2O_2) as one of its reactants, giving us a rate law of rate $= k_2[N_2O_2][O_2]$. Because we cannot compare this directly to the experimental rate law, we must solve for the intermediate concentration in terms of *reactant* concentrations.

Setup The first step is a rapidly established equilibrium. Both the forward and reverse of step 1 are elementary processes, which enables us to write their rate laws from the balanced equation:

$$\text{rate}_{\text{forward}} = k_1[NO]^2 \quad \text{and} \quad \text{rate}_{\text{reverse}} = k_{-1}[N_2O_2]$$

Solution Because at equilibrium the forward and reverse processes are occurring at the same rate, we can set their rates equal to each other and solve for the intermediate concentration:

$$k_1[NO]^2 = k_{-1}[N_2O_2]$$

$$[N_2O_2] = \frac{k_1[NO]^2}{k_{-1}}$$

Substituting the solution into the original rate law (rate $= k[N_2O_2][O_2]$) gives:

$$\text{rate} = k_2\frac{k_1[NO]^2}{k_{-1}}[O_2] = k[NO]^2[O_2] \quad \text{where } k = \frac{k_2 k_1}{k_{-1}}$$

THINK ABOUT IT

Not all reactions have a single rate-determining step. Analyzing the kinetics of reactions with two or more comparably slow steps is beyond the scope of this book.

Practice Problem Ⓐ**TTEMPT** Show that the following two-step mechanism is consistent with the experimentally determined rate law of rate $= k[NO]^2[Br_2]$ for the reaction of nitric oxide and bromine: $2NO(g) + Br_2(g) \longrightarrow 2NOBr(g)$.

Step 1: $NO(g) + Br_2(g) \underset{k_{-1}}{\overset{k_1}{\rightleftharpoons}} NOBr_2(g)$ (fast)

Step 2: $NOBr_2(g) + NO(g) \overset{k_2}{\longrightarrow} 2NOBr(g)$ (slow)

Practice Problem (B)**UILD** The reaction $H_2(g) + I_2(g) \longrightarrow 2HI(g)$ proceeds via a two-step mechanism in which the rate law for the rate-determining step is rate = $k[H_2][I]^2$. Write the mechanism and rewrite the rate law using only reactant concentrations.

Practice Problem (C)**ONCEPTUALIZE** The reaction of $A + B \longrightarrow C + D$ is believed to proceed via a two-step mechanism in which the second step is rate-limiting and the first step produces an intermediate (I). Which graph could represent the concentrations of the reactants and the intermediate as the reaction progresses?

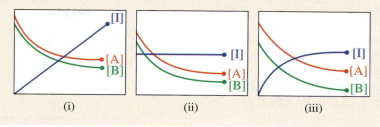

(i)   (ii)   (iii)

CHECKPOINT – SECTION 14.5   Reaction Mechanisms

Use the following information to answer questions 14.5.1 to 14.5.3: For the reaction:

$$A + B \longrightarrow C + D$$

the experimental rate law is rate = $k[B]^2$.

14.5.1 What is the order of the reaction with respect to A and B, respectively?

a) 0 and 1

b) 1 and 0

c) 0 and 2

d) 2 and 1

e) 2 and 0

14.5.2 What is the overall order of the reaction?

a) 0

b) 1

c) 2

d) 3

e) 4

14.5.3 Which of the following is a plausible mechanism for the reaction?

a) Step 1: $A + B \longrightarrow C + E$ (slow)
   Step 2: $E + A \longrightarrow D$ (fast)

b) Step 1: $B + B \longrightarrow C + E$ (slow)
   Step 2: $E + A \longrightarrow D + B$ (fast)

c) Step 1: $A + A \longrightarrow B + D$ (slow)
   Step 2: $B + B \longrightarrow A + C$ (fast)

d) Step 1: $B + B \longrightarrow C + E$ (slow)
   Step 2: $C + A \longrightarrow D + B$ (fast)

e) Step 1: $A + B \longrightarrow D + E$ (slow)
   Step 2: $E + A \longrightarrow C$ (fast)

14.5.4 A plausible mechanism for the reaction:

$$H_2 + 2IBr \longrightarrow I_2 + 2HBr$$

is the following:

Step 1: $H_2 + IBr \xrightarrow{k_1} HI + HBr$ (slow)

Step 2: $HI + IBr \xrightarrow{k_2} I_2 + HBr$ (fast)

Select the rate law that will be determined experimentally.

a) rate = $k[H_2]^2$

b) rate = $k[H_2][IBr]^2$

c) rate = $k[IBr]^2$

d) rate = $k[H_2][IBr]$

e) rate = $k[HI][IBr]$

14.6 Catalysis

Recall from Section 14.5 that the reaction rate for the decomposition of hydrogen peroxide depends on the concentration of iodide ions, even though I^- does not appear in the overall equation. Instead, I^- acts as a catalyst for the reaction. A *catalyst* is a substance that increases the rate of a chemical reaction without itself being consumed. The catalyst may react to form an intermediate, but it is regenerated in a subsequent step of the reaction.

Animation
Catalysis.

Molecular oxygen is prepared in the laboratory by heating potassium chlorate. The reaction is:

$$2KClO_3(s) \longrightarrow 2KCl(s) + 3O_2(g)$$

Figure 14.18 Comparison of the activation energy barriers of (a) an uncatalyzed reaction and (b) the same reaction with a catalyst. A catalyst lowers the energy barrier but does not affect the energies of the reactants or products. Although the reactants and products are the same in both cases, the reaction mechanisms and rate laws are different in (a) and (b).

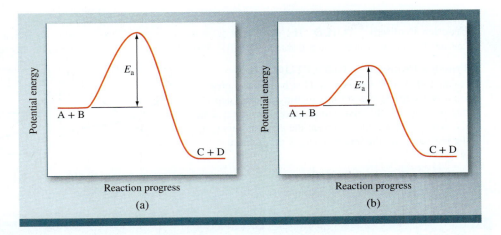

However, this thermal decomposition process is very slow in the absence of a catalyst. The rate of decomposition can be increased dramatically by adding a small amount of manganese(IV) dioxide (MnO_2), a black powdery substance. All the MnO_2 can be recovered at the end of the reaction, just as all the I^- ions remain following the decomposition of H_2O_2.

A catalyst speeds up a reaction by providing a set of elementary steps with more favorable kinetics than those that exist in its absence. From Equation 14.8 we know that the rate constant k (and hence the rate) of a reaction depends on the frequency factor (A) and the activation energy (E_a)—the larger the value of A (or the smaller the value of E_a), the greater the rate. In many cases, a catalyst increases the rate by lowering the activation energy for the reaction.

Let's assume that the following reaction has a certain rate constant k and an activation energy E_a:

$$A + B \xrightarrow{k} C + D$$

In the presence of a catalyst, however, the rate constant is k_c, called the *catalytic rate constant*.

By the definition of a catalyst:

$$\text{rate}_{\text{catalyzed}} > \text{rate}_{\text{uncatalyzed}}$$

Figure 14.18 shows the potential energy profiles for both reactions. The total energies of the reactants (A and B) and those of the products (C and D) are unaffected by the catalyst; the only difference between the two is a lowering of the activation energy from E_a to E_a'. Because the activation energy for the reverse reaction is also lowered, a catalyst enhances the rates of the forward and reverse reactions equally.

There are three general types of catalysis, depending on the nature of the rate-increasing substance: heterogeneous catalysis, homogeneous catalysis, and enzyme catalysis.

Heterogeneous Catalysis

In *heterogeneous catalysis,* the reactants and the catalyst are in different phases. The catalyst is usually a solid, and the reactants are either gases or liquids. Heterogeneous catalysis is by far the most important type of catalysis in industrial chemistry, especially in the synthesis of many important chemicals—including ammonia. Gerhard Ertl's elucidation of the mechanism of heterogeneous catalysis showed that it occurs in four steps: *adsorption,* where reactant molecules bind to the surface of the solid catalyst; *activation,* where reactant molecules become oriented such that reaction is possible; *reaction;* and *desorption,* where product molecules leave the solid surface and the cycle can begin again with new reactants. Figure 14.19 illustrates this process for the Haber-Bosch reaction.

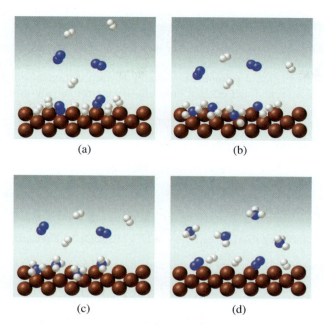

(a)   (b)

(c)   (d)

Figure 14.19 (a) N$_2$ and H$_2$ molecules become attached (adsorption) to the surface of the solid catalyst. (b) The molecules become oriented (activation) so that the distances between atoms facilitate the formation of new bonds. (c) Reaction produces NH$_3$ molecules. (d) NH$_3$ molecules leave the solid surface (desorption), exposing the metal surface for the adsorption of fresh reactant molecules.

Heterogeneous catalysis is also used in the catalytic converters in automobiles. At high temperatures inside a car's engine, nitrogen and oxygen gases react to form nitric oxide:

$$N_2(g) + O_2(g) \longrightarrow 2NO(g)$$

When released into the atmosphere, NO rapidly combines with O$_2$ to form NO$_2$. Nitrogen dioxide and other gases emitted by automobiles, such as carbon monoxide (CO) and various unburned hydrocarbons, make automobile exhaust a major source of air pollution.

New cars are equipped with catalytic converters [Figure 14.20(a)]. An efficient catalytic converter serves two purposes: It oxidizes CO and unburned hydrocarbons to CO$_2$ and H$_2$O, and it converts NO and NO$_2$ to N$_2$ and O$_2$. Hot exhaust gases into which air has been injected are passed through the first chamber of one converter to accelerate the complete burning of hydrocarbons and to decrease CO emissions. Because high temperatures increase NO production, however, a second chamber containing a different catalyst (a transition metal or a transition metal oxide such as CuO or Cr$_2$O$_3$) and operating at a lower temperature is required to dissociate NO into N$_2$ and O$_2$ before the exhaust is discharged through the tailpipe [Figure 14.20(b)].

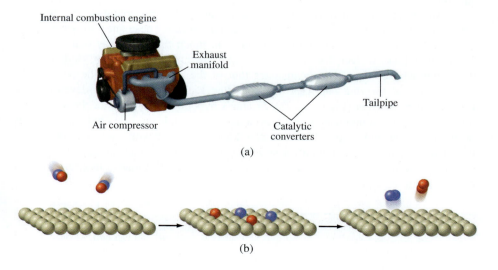

Internal combustion engine

Exhaust manifold

Air compressor

Catalytic converters

Tailpipe

(a)

(b)

Figure 14.20 (a) A two-stage catalytic converter for an automobile. (b) In the second stage, NO molecules bind to the surface of the catalyst. The N atoms bond to each other and the O atoms bond to each other, producing N$_2$ and O$_2$, respectively.

Homogeneous Catalysis

In *homogeneous catalysis,* the reactants and the catalyst are dispersed in a single phase, usually liquid. Acid and base catalyses are the most important types of homogeneous catalysis in liquid solution. For example, the reaction of ethyl acetate with water to form acetic acid and ethanol normally occurs too slowly to be measured:

$$CH_3COOC_2H_5 + H_2O \longrightarrow CH_3COOH + C_2H_5OH$$

In the absence of the catalyst, the rate law is given by:

$$rate = k[CH_3COOC_2H_5]$$

The reaction, however, can be catalyzed by an acid. Often a catalyst is shown above the arrow in a chemical equation:

$$CH_3COOC_2H_5 + H_2O \xrightarrow{\text{H}^+} H^+CH_3COOH + C_2H_5OH$$

In the presence of acid, the rate is faster and the rate law is given by:

$$rate = k_c[CH_3COOC_2H_5][H^+]$$

Because $k_c > k$ in magnitude, the rate is determined solely by the catalyzed portion of the reaction.

Homogeneous catalysis has several advantages over heterogeneous catalysis. For one thing, the reactions can often be carried out under atmospheric conditions, thus reducing production costs and minimizing the decomposition of products at high temperatures. In addition, homogeneous catalysts can be designed to function selectively for particular types of reactions, and homogeneous catalysts cost less than the precious metals (e.g., platinum and gold) used in heterogeneous catalysis.

Enzymes: Biological Catalysts

Of all the intricate processes that have evolved in living systems, none is more striking or more essential than enzyme catalysis. *Enzymes* are *biological* catalysts. The amazing fact about enzymes is that not only can they increase the rate of biochemical reactions by factors ranging from 10^6 to 10^{18}, but they are also highly specific. An enzyme acts only on certain reactant molecules, called *substrates,* while leaving the rest of the system unaffected. It has been estimated that an average living cell may contain some 3000 different enzymes, each of them catalyzing a specific reaction in which a substrate is converted into the appropriate product(s). Enzyme catalysis is usually homogeneous because the substrate and enzyme are both present in an aqueous solution.

An enzyme is typically a large protein molecule that contains one or more active sites where interactions with substrates take place. These sites are structurally compatible with specific substrate molecules, in much the same way that a key fits a particular lock. In fact, the notion of a rigid enzyme structure that binds only to molecules whose shape exactly matches that of the active site was the basis of an early theory of enzyme catalysis, the so-called lock-and-key theory developed by Emil Fischer[1] in 1894 (Figure 14.21). Fischer's hypothesis accounts for the specificity of enzymes, but it contradicts research evidence that a single enzyme binds to substrates of different sizes and shapes. Chemists now know that an enzyme molecule (or at least its active site) has a fair amount of structural flexibility and can modify its shape to accommodate more than one type of substrate. Figure 14.22 shows a molecular model of an enzyme in action.

---

1. Emil Fischer (1852–1919). German chemist. Regarded by many as the greatest organic chemist of the nineteenth century. Fischer made many significant contributions in the synthesis of sugars and other important molecules. He was awarded the Nobel Prize in Chemistry in 1902.

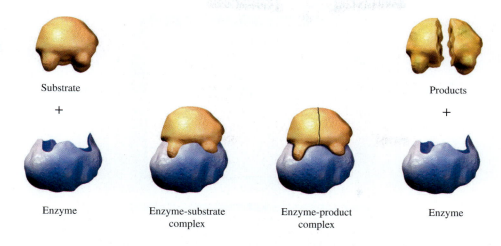

Figure 14.21 The lock-and-key model of an enzyme's specificity for substrate molecules.

Substrate

+

Enzyme

Enzyme-substrate complex

Enzyme-product complex

Products

+

Enzyme

Figure 14.22 Left to right: The binding of glucose molecule (red) to hexokinase (an enzyme in the metabolic pathway). Note how the region at the active site closes around glucose after binding. Often, the geometries of both the substrate and the active site are altered to fit each other.

The mathematical treatment of enzyme kinetics is quite complex, even when we know the basic steps involved in the reaction. A simplified scheme is given by the following elementary steps:

$$E + S \underset{k_{-1}}{\overset{k_1}{\rightleftharpoons}} ES$$

$$ES \overset{k_2}{\longrightarrow} E + P$$

where E, S, and P represent enzyme, substrate, and product, respectively, and ES is the enzyme-substrate intermediate. It is often assumed that the formation of ES and its decomposition back to enzyme and substrate molecules occur rapidly and that the rate-determining step is the formation of product. Figure 14.23 shows the potential-energy profile for the reaction.

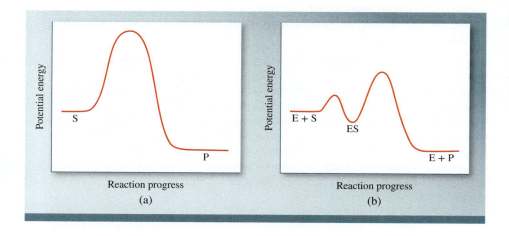

Reaction progress
(a)

Reaction progress
(b)

Figure 14.23 Comparison of (a) an uncatalyzed reaction and (b) the same reaction catalyzed by an enzyme. The plot in (b) assumes that the catalyzed reaction has a two-step mechanism, in which the second step (ES $\longrightarrow$ E + P) is rate determining.

Figure 14.24 Plot of the rate of product formation versus substrate concentration in an enzyme-catalyzed reaction.

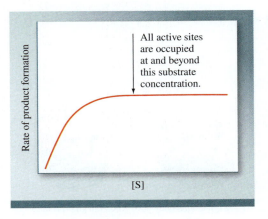

In general, the rate of such a reaction is given by the equation:

$$\text{rate} = \frac{\Delta[P]}{\Delta t}$$

$$= k[ES]$$

The concentration of the ES intermediate is itself proportional to the amount of the substrate present, and a plot of the rate versus the concentration of substrate typically yields a curve like that shown in Figure 14.24. Initially the rate rises rapidly with increasing substrate concentration. Above a certain concentration, however, all the active sites are occupied, and the reaction becomes zeroth order in the substrate. In other words, the rate remains the same even though the substrate concentration increases. At and beyond this point, the rate of formation of product depends only on how fast the ES intermediate breaks down, not on the number of substrate molecules present.

Bringing Chemistry to Life

Catalysis and Hangovers

Alcohol dehydrogenase (ADH) and aldehyde dehydrogenase (ALDH) catalyze the metabolism of methanol. In analogous reactions, ADH and ALDH catalyze the metabolism of *ethanol,* the alcohol in alcoholic beverages.

The first reaction converts ethanol to acetaldehyde, which is far more toxic than ethanol:

$$CH_3CH_2OH + NAD^+ \longrightarrow CH_3CHO + NADH + H^+$$

It is acetaldehyde (CH_3CHO) that causes the misery associated with a hangover—including headache, nausea, and vomiting.

In the second reaction, acetaldehyde is converted to acetic acid, which is harmless:

$$CH_3CHO + NADP^+ + H_2O \longrightarrow CH_3COOH + NADPH + H^+$$

An effective but painful part of the treatment for alcohol abuse is to administer *disulfiram,* marketed under the name Antabuse. Disulfiram blocks the action of ALDH, preventing the conversion of acetaldehyde to acetic acid. The resulting buildup of acetaldehyde causes the patient to feel very sick almost immediately, making the next cocktail *far* less appealing. The action of disulfiram was discovered accidentally when Danish pharmaceutical researchers who were taking the drug as an experimental treatment for parasitic diseases became very ill every time they consumed alcohol.

Q77photo/Shutterstock

Chapter Summary

Section 14.1

- The rate of a chemical reaction is the change in concentration of reactants or products over time. Rates may be expressed as an *average* rate over a given time interval or as an *instantaneous rate.*

- The *rate constant (k)* is a proportionality constant that relates the *rate of reaction* with the concentration(s) of reactant(s). The rate constant k for a given reaction changes only with temperature.

Section 14.2

- The *rate law* is an equation that expresses the relationship between rate and reactant concentration(s). In general, the rate law for the reaction of A and B is rate = $k[A]^x[B]^y$.

- The *reaction order* is the power to which the concentration of a given reactant is raised in the rate law equation. The overall reaction order is the sum of the powers to which reactant concentrations are raised in the rate law.

- The *initial rate* is the instantaneous rate of reaction when the reactant concentrations are *starting* concentrations.

- The rate law and reaction order must be determined by comparing changes in the initial rate with changes in starting reactant concentrations. In general, the rate law cannot be determined solely from the balanced equation.

Section 14.3

- The *integrated rate law* can be used to determine reactant concentrations after a specified period of time. It can also be used to determine how long it will take to reach a specified reactant concentration.

- The rate of a *first-order reaction* is proportional to the concentration of a single reactant. The rate of a *second-order reaction* is proportional to the product of two reactant concentrations ([A][B]), or on the concentration of a single reactant squared ($[A]^2$ or $[B]^2$). The rate of a *zeroth-order reaction* does not depend on reactant concentration.

- The *half-life ($t_{1/2}$)* of a reaction is the time it takes for half of a reactant to be consumed. The half-life is *constant* for first-order reactions, and it can be used to determine the rate constant of the reaction.

Section 14.4

- *Collision theory* explains why the rate constant, and therefore the reaction rate, increases with increasing temperature. The relationship between temperature and the rate constant is expressed by the *Arrhenius equation.*

- Reactions occur when molecules of sufficient energy (and appropriate orientation) collide. *Effective collisions* are those that result in the formation of an *activated complex,* also called a *transition state.* Only *effective* collisions can result in product formation.

- The *activation energy* (E_a) is the minimum energy that colliding molecules must possess in order for the collision to be effective.

Section 14.5

- A *reaction mechanism* may consist of a series of *steps,* called *elementary reactions.* Unlike rate laws in general, the rate law for an elementary reaction can be written from the balanced equation, using the stoichiometric coefficient for each reactant species as its exponent in the rate law.

- A species that is produced in one step of a reaction mechanism and subsequently consumed in another step is called an *intermediate.* A species that is first consumed and later regenerated is called a *catalyst.* Neither intermediates nor catalysts appear in the overall balanced equation.

- The rate law of each step in a reaction mechanism indicates the *molecularity* or overall *order* of the step. A *unimolecular* step is first order, involving just one molecule; a *bimolecular* step is second order, involving the collision of two molecules; and a *termolecular* step is third order, involving the collision of three molecules. Termolecular processes are relatively rare.

- If one step in a reaction is much slower than all the other steps, it is the *rate-determining step.* The rate-determining step has a rate law identical to the experimental rate law.

Section 14.6

- A *catalyst* speeds up a reaction, usually by lowering the value of the activation energy. *Catalysis* refers to the process by which a catalyst increases the reaction rate.

- Catalysis may be *heterogeneous,* in which the catalyst and reactants exist in different phases, or *homogeneous,* in which the catalyst and reactants exist in the same phase.

- *Enzymes* are biological catalysts with high specificity for the reactions that they catalyze.

Key Words

Key Equations

14.1 $\text{rate} = -\dfrac{1}{a}\dfrac{\Delta[A]}{\Delta t} = -\dfrac{1}{b}\dfrac{\Delta[B]}{\Delta t} = \dfrac{1}{c}\dfrac{\Delta[C]}{\Delta t} = \dfrac{1}{d}\dfrac{\Delta[D]}{\Delta t}$	Rate can be expressed in terms of the concentration of any species in a reaction. The minus signs in the equation correct for the fact that although reactant concentrations decrease as a reaction proceeds, rate is always expressed as a positive quantity.
14.2 $\text{rate} = k[A]^x[B]^y$	Rate can also be expressed using the *rate law,* which relates the rate to reactant concentrations and a rate constant, k. Units of the rate constant depend on overall reaction order.
14.3 $\ln\dfrac{[A]_t}{[A]_0} = -kt$	The integrated rate law for a first-order reaction relates natural logarithm (ln) of the ratio of reactant concentration at time zero to reactant concentration at a subsequent time, $\ln\left(\dfrac{[A]_0}{[A]_t}\right)$, to the rate constant, k, and the time elapsed, t.
14.4 $\ln[A]_t = -kt + \ln[A]_0$	The integrated rate law for a first-order reaction can be written in $y = mx + b$ form. This form of the equation indicates that when $\ln[A]_t$, is plotted against time, the slope of the resulting line is the negative of the rate constant, $-k$, and the intercept is $\ln[A]_0$.
14.5 $t_{1/2} = \dfrac{0.693}{k}$	The half-life, $t_{1/2}$, of a first-order reaction is constant and is calculated as the ratio of 0.693 (ln 2) to the rate constant of the reaction.
14.6 $\dfrac{1}{[A]_t} = kt + \dfrac{1}{[A]_0}$	The integrated rate law for a second-order reaction, written in $y = mx + b$ form, indicates that when $\dfrac{1}{[A]_t}$ is plotted against time, the slope of the resulting line is the rate constant and the intercept is $\dfrac{1}{[A]_0}$.
14.7 $t_{1/2} = \dfrac{1}{k[A]_0}$	The half-life of a second-order reaction is not constant, and is calculated as the reciprocal product of rate constant and reactant concentration at time 0.

14.8 $k = Ae^{-E_a/RT}$

The Arrhenius equation relates the rate constant to the frequency factor, A, the activation energy, E_a, and the absolute temperature. For units to cancel properly in this equation, R must be expressed using units of energy (J/K · mol or kJ/K · mol, depending on the units used for E_a).

14.9 $\ln k = \ln A - \dfrac{E_a}{RT}$

Another form of the Arrhenius equation.

14.10 $\ln k = \left(-\dfrac{E_a}{R}\right)\left(\dfrac{1}{T}\right) + \ln A$

The Arrhenius equation written in $y = mx + b$ form indicates that when $\ln k$ is plotted against $1/T$, the slope of the resulting line is $-E_a/R$ and the intercept is $\ln A$.

14.11 $\ln \dfrac{k_1}{k_2} = \dfrac{E_a}{R}\left(\dfrac{1}{T_2} - \dfrac{1}{T_1}\right)$

This is the most useful form of the Arrhenius equation. Subtracting Equation 14.10 at one temperature from Equation 14.10 at another temperature eliminates the frequency factor by cancellation and allows us to calculate the activation energy (E_a) for a reaction if we know the rate constants at two different temperatures. It also allows us to calculate rate constant at any other temperature, provided that we know E_a.

One of the important applications of kinetics is the analysis of radioactive decay, which is first order. Although the equations are the same as those we have already derived for first-order kinetics, the terms and symbols vary slightly from what we have seen to this point. For example, the amount of radioactive material is typically expressed using *activity* (A) in *disintegrations per second* (dps), or in numbers of nuclei (N), rather than in units of concentration. Thus, Equation 14.3 becomes:

$$\ln \frac{A_t}{A_0} = -kt \qquad \text{or} \qquad \ln \frac{N_t}{N_0} = -kt$$

The use of activity is very similar to the use of concentration. As with other kinetics problems, we may be given the initial activity (A_0), the rate constant (k), and the time elapsed (t)—and be asked to determine the new activity (A_t). Or we may be asked to solve for one of the other parameters, which simply requires manipulation of the original equation. For example, we may be given A_0, A_t, and t and be asked to determine the rate constant—or the half-life. (Recall that for any first-order process, the half-life is constant, and is related to the rate constant by Equation 14.5.) In this case, we solve Equation 14.3 for k, or for $t_{1/2}$—or for whichever parameter we need to find.

When the missing parameter is either A_0 or A_t, we must pay special attention to the manipulation of logarithms [▶▶ Appendix 1]. Solving the activity version of Equation 14.3 for A_t, for example, gives:

$$A_t = A_0(e^{-kt})$$

When radioactive decay is quantified using numbers of nuclei (N), some additional thought must go into determining the "known" parameters. Rocks that contain uranium, for example, can be dated by measuring the amounts of uranium-238 and lead-206 they contain. ^{238}U is unstable and undergoes a series of radioactive decay steps [▶▶ Section 20.3], ultimately producing a ^{206}Pb nucleus for every ^{238}U nucleus that decays. Although rocks may contain other isotopes of lead, ^{206}Pb results strictly from the decay of ^{238}U. Therefore, we can assume that every ^{206}Pb nucleus was originally a ^{238}U nucleus. If we know the mass of ^{238}U (N_t) and the mass of ^{206}Pb in the rock, we can determine N_0 as follows:

Student Note: Because mass is proportional to the number of nuclei for a given isotope, we can use masses in the number-of-nuclei version of Equation 14.3—as long as they are mass amounts of the same nucleus.

$$\text{mass of } ^{206}\text{Pb} \quad \times \quad \frac{1}{\text{atomic mass of } ^{206}\text{Pb}} \quad = \quad \text{number of } ^{206}\text{Pb nuclei}$$

Because every ^{206}Pb nucleus was originally a ^{238}U nucleus:

$$\text{number of } ^{206}\text{Pb nuclei} \quad = \quad \text{number of } ^{238}\text{U nuclei}$$

$$\text{number of } ^{238}\text{U nuclei} \quad \times \quad \text{atomic mass of } ^{238}\text{U} \quad = \quad \text{mass of } ^{238}\text{U}$$

These operations condense to give:

$$\boxed{\text{mass of } ^{206}\text{Pb}} \times \boxed{\dfrac{\text{atomic mass of } ^{238}\text{U}}{\text{atomic mass of } ^{206}\text{Pb}}} = \boxed{\text{mass of } ^{238}\text{U}}$$

This gives the mass of ^{238}U *not* accounted for in the mass of ^{238}U found in the rock. Adding this mass to the mass of ^{238}U in the rock gives the mass of ^{238}U *originally* present (N_0). When we have both N_t and N_0, given the rate constant for decay of ^{238}U (1.54×10^{-10} yr^{-1}), we can determine the age of the rock (t).

Consider the following example:

A rock is found to contain 23.17 g ^{238}U and 2.02 g ^{206}Pb. Its age is determined as follows:

$$\boxed{2.02 \text{ g } ^{206}\text{Pb}} \times \boxed{\dfrac{238 \text{ g } ^{238}\text{U}}{206 \text{ g } ^{206}\text{Pb}}} = \boxed{2.334 \text{ g } ^{238}\text{U}}$$

$$\boxed{N_0 = (23.17 + 2.334) = 25.50 \text{ g}} \qquad \boxed{N_t = 23.17 \text{ g}}$$

Solving Equation 14.3 for t gives:

$$t = -\left(\dfrac{\ln \dfrac{N_t}{N_0}}{k} \right)$$

and:

$$t = -\left(\dfrac{\ln \dfrac{23.17 \text{ g}}{25.50 \text{ g}}}{1.54 \times 10^{-10} \text{ yr}^{-1}} \right) = \boxed{6.22 \times 10^8 \text{ yr}}$$

Therefore, the rock is 622 million years old.

Key Skills Problems

14.1
It takes 218 hours for the activity of a certain radioactive isotope to fall to one-tenth of its original value. Calculate the half-life of the isotope.

(a) 3.18×10^{-4} h (b) 21.8 h (c) 0.0152 h (d) 65.6 h (e) 0.0106 h

14.2
^{61}Cu decays with a half-life of 3.35 h. Determine the original mass of a sample of ^{61}Cu if 612.8 mg remains after exactly 24 hours.

(a) 85.5 mg (b) 4.39×10^3 mg (c) 736 mg (d) 6.40×10^3 mg (e) 8.78×10^4 mg

14.3
The rate constant for the radioactive decay of iodine-126 is 0.0533 d^{-1}. How much ^{126}I remains of a 2.55-g sample of ^{126}I after exactly 24 hours?

(a) 0.948 g (b) 2.42 g (c) 0.136 g (d) 0.710 g (e) 0.873 g

14.4
Determine the age of a rock that contains 45.7 mg ^{238}U and 1.02 mg ^{206}Pb.

(a) 165 million years (b) 2.50 billion years (c) 143 million years (d) 6.49 billion years (e) 63.7 million years

Questions and Problems

Applying What You've Learned

It takes as little as 5 mL (1 tsp) of methanol to cause permanent blindness or death; and, unlike ethanol, methanol can be absorbed in toxic amounts not only by ingestion but also by inhalation of vapor or absorption through the skin. Nevertheless, methanol is present in a number of common household products including antifreeze, windshield-washing fluid, and paint remover. Its presence in a large number of such products makes it an important industrial substance. One method used to synthesize methanol is the combination of carbon monoxide and hydrogen gases at 100°C:

$$CO(g) + 2H_2(g) \longrightarrow CH_3OH(g)$$

The reaction is catalyzed by a nickel compound.

Problems:
(a) Write the expression for the rate of this reaction in terms of [CO] [◄◄ Sample Problem 14.1]. (b) Write the expression for the rate of this reaction in terms of [H_2] and in terms of [CH_3OH] [◄◄ Sample Problem 14.2]. (c) Given the following table of experimental data at 100°C, determine the rate law and the rate constant for the reaction. Then, determine the initial rate of the reaction when the starting concentration of CO is 16.5 M [◄◄ Sample Problem 14.3].

Experiment	[CO] (M)	[H_2] (M)	Initial Rate (M/s)
1	5.60	11.2	0.952
2	5.60	22.4	0.952
3	11.2	11.2	1.90

(d) Calculate the time required for the concentration of CO to be reduced from 16.5 M to 1.91 M [◄◄ Sample Problem 14.4].
(e) Calculate $t_{1/2}$ of the reaction [◄◄ Sample Problem 14.6]. (f) Given that k is 3.0 s^{-1} at 200°C, calculate E_a of the reaction [◄◄ Sample Problem 14.9]. (g) Use the calculated value of E_a to determine the value of k at 180°C [◄◄ Sample Problem 14.10].

SECTION 14.1: REACTION RATES

Review Questions

14.1   What is meant by the *rate* of a chemical reaction? What are the units of the rate of a reaction?

14.2   Distinguish between average rate and instantaneous rate. Which of the two rates gives us an unambiguous measurement of reaction rate? Why?

14.3   What are the advantages of measuring the initial rate of a reaction?

14.4   Identify two reactions that are very slow (take days or longer to complete) and two reactions that are very fast (reactions that are over in minutes or seconds).

Problems

14.5   Write the reaction rate expressions for the following reactions in terms of the disappearance of the reactants and the appearance of products:
(a) $H_2(g) + I_2(g) \longrightarrow 2HI(g)$
(b) $5Br^-(aq) + BrO_3^-(aq) + 6H^+(aq) \longrightarrow$ $3Br_2(aq) + 3H_2O(l)$

14.6   Write the reaction rate expressions for the following reactions in terms of the disappearance of the reactants and the appearance of products:
(a) $2H_2(g) + O_2(g) \longrightarrow 2H_2O(g)$
(b) $4NH_3(g) + 5O_2(g) \longrightarrow 4NO(g) + 6H_2O(g)$

14.7   Consider the reaction:

$$2NO(g) + O_2(g) \longrightarrow 2NO_2(g)$$

Suppose that at a particular moment during the reaction nitric oxide (NO) is reacting at the rate of 0.066 M/s. (a) At what rate is NO_2 being formed? (b) At what rate is molecular oxygen reacting?

14.8   Consider the reaction:

$$N_2(g) + 3H_2(g) \longrightarrow 2NH_3(g)$$

Suppose that at a particular moment during the reaction molecular hydrogen is reacting at the rate of 0.082 M/s. (a) At what rate is ammonia being formed? (b) At what rate is molecular nitrogen reacting?

SECTION 14.2: DEPENDENCE OF REACTION RATE ON REACTANT CONCENTRATION

Review Questions

14.9 Explain what is meant by the *rate law* of a reaction.

14.10 Explain what is meant by the *order* of a reaction.

14.11 What are the units for the rate constants of first-order and second-order reactions?

14.12 Consider the zeroth-order reaction: A $\longrightarrow$ product. (a) Write the rate law for the reaction. (b) What are the units for the rate constant? (c) Plot the rate of the reaction versus [A].

14.13 The rate constant of a first-order reaction is 66 s^{-1}. What is the rate constant in units of minutes?

14.14 On which of the following properties does the rate constant of a reaction depend: (a) reactant concentrations, (b) nature of reactants, (c) temperature?

Computational Problems

14.15 The rate law for the reaction:

$$NH_4^+(aq) + NO_2^-(aq) \longrightarrow N_2(g) + 2H_2O(l)$$

is given by rate = $k[NH_4^+][NO_2^-]$. At 25°C, the rate constant is $3.0 \times 10^{-4}/M \cdot s$. Calculate the rate of the reaction at this temperature if $[NH_4^+] = 0.36\ M$ and $[NO_2^-] = 0.075\ M$.

14.16 Use the data in Table 14.2 to calculate the rate of the reaction at the time when $[F_2] = 0.020\ M$ and $[ClO_2] = 0.035\ M$.

14.17 Consider the reaction:

$$A + B \longrightarrow products$$

From the following data obtained at a certain temperature, determine the order of the reaction and calculate the rate constant.

[A] (*M*)	[B] (*M*)	Rate (*M*/s)
1.50	1.50	3.20×10^{-1}
1.50	2.50	3.20×10^{-1}
3.00	1.50	6.40×10^{-1}

14.18 Consider the reaction:

$$X + Y \longrightarrow Z$$

From the following data, obtained at 360 K, (a) determine the order of the reaction, and (b) determine the initial rate of disappearance of X when the concentration of X is 0.30 *M* and that of Y is 0.40 *M*.

Initial Rate of Disappearance of X (*M*/s)	[X] (*M*)	[Y] (*M*)
0.053	0.10	0.50
0.127	0.20	0.30
1.02	0.40	0.60
0.254	0.20	0.60
0.509	0.40	0.30

14.19 Determine the overall orders of the reactions to which the following rate laws apply: (a) rate = $k[NO_2]^2$, (b) rate = k, (c) rate = $k[H_2]^2[Br_2]^{1/2}$, (d) rate = $k[NO]^2[O_2]$.

14.20 Consider the reaction:

$$A \longrightarrow B$$

The rate of the reaction is $1.6 \times 10^{-2}\ M$/s when the concentration of A is 0.15 *M*. Calculate the rate constant if the reaction is (a) first order in A and (b) second order in A.

Conceptual Problems

14.21 Cyclobutane decomposes to ethylene according to the equation:

$$C_4H_8(g) \longrightarrow 2C_2H_4(g)$$

Determine the order of the reaction and the rate constant based on the following pressures, which were recorded when the reaction was carried out at 430°C in a constant-volume vessel.

Time (s)	$P_{C_4H_8}$ (mmHg)
0	400
2,000	316
4,000	248
6,000	196
8,000	155
10,000	122

14.22 The following gas-phase reaction was studied at 290°C by observing the change in pressure as a function of time in a constant-volume vessel:

$$ClCO_2CCl_3(g) \longrightarrow 2COCl_2(g)$$

Determine the order of the reaction and the rate constant based on the following data.

Time (s)	P (mmHg)
0	15.76
181	18.88
513	22.79
1164	27.08

where *P* is the total pressure.

SECTION 14.3: DEPENDENCE OF REACTANT CONCENTRATION ON TIME

Review Questions

14.23 Write an equation relating the concentration of a reactant A at $t = 0$ to that at $t = t$ for a first-order reaction. Define all the terms, and give their units. Do the same for a second-order reaction.

14.24 Define *half-life*. Write the equation relating the half-life of a first-order reaction to the rate constant.

14.25 Write the equation relating the half-life of a second-order reaction to the rate constant. How does it differ from the equation for a first-order reaction?

14.26 For a first-order reaction, how long will it take for the concentration of reactant to fall to one-eighth its original value? Express your answer in terms of the half-life ($t_{1/2}$) and in terms of the rate constant k.

Computational Problems

14.27 What is the half-life of a compound if 75 percent of a given sample of the compound decomposes in 60 min? Assume first-order kinetics.

14.28 The thermal decomposition of phosphine (PH_3) into phosphorus and molecular hydrogen is a first-order reaction:

$$4PH_3(g) \longrightarrow P_4(g) + 6H_2(g)$$

The half-life of the reaction is 35.0 s at 680°C. Calculate (a) the first-order rate constant for the reaction and (b) the time required for 95 percent of the phosphine to decompose.

14.29 The rate constant for the second-order reaction:

$$2NOBr(g) \longrightarrow 2NO(g) + Br_2(g)$$

is 0.80/$M \cdot$ s at 10°C. (a) Starting with a concentration of 0.086 M, calculate the concentration of NOBr after 22 s. (b) Calculate the half-lives when $[NOBr]_0 = 0.072$ M and $[NOBr]_0 = 0.054$ M.

14.30 The rate constant for the second-order reaction:

$$2NO_2(g) \longrightarrow 2NO(g) + O_2(g)$$

is 0.54/$M \cdot$ s at 300°C. How long (in seconds) would it take for the concentration of NO_2 to decrease from 0.65 M to 0.18 M?

14.31 The second-order rate constant for the dimerization of a protein (P)

$$P + P \longrightarrow P_2$$

is $6.2 \times 10^{-3}/M \cdot$ s at 25°C. If the concentration of the protein is 2.7×10^{-4} M, calculate the initial rate (M/s) of formation of P_2. How long (in seconds) will it take to decrease the concentration of P to 2.7×10^{-5} M?

Conceptual Problems

14.32 Consider the first-order reaction X $\longrightarrow$ Y shown here. (a) What is the half-life of the reaction? (b) Draw pictures showing the number of X (red) and Y (blue) molecules at 20 s and at 30 s.

$t = 0$ s $t = 10$ s

14.33 The reaction A $\longrightarrow$ B follows first-order kinetics. Initially different amounts of A molecules, represented here by green spheres, are placed in three containers of equal volume at the same temperature. (a) What are the relative rates of the

reaction in these three containers? (b) How would the relative rates be affected if the volume of each container were doubled? (c) What are the relative half-lives of the reactions in (i) to (iii)?

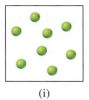 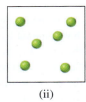 

(i)   (ii)   (iii)

SECTION 14.4: DEPENDENCE OF REACTION RATE ON TEMPERATURE

Review Questions

14.34 Define *activation energy*. What role does activation energy play in chemical kinetics?

14.35 Write the Arrhenius equation, and define all terms.

14.36 Use the Arrhenius equation to show why the rate constant of a reaction (a) decreases with increasing activation energy and (b) increases with increasing temperature.

14.37 The burning of methane in oxygen is a highly exothermic reaction. Yet a mixture of methane and oxygen gas can be kept indefinitely without any apparent change. Explain.

14.38 Sketch a potential-energy versus reaction progress plot for the following reactions:
(a) S(s) + $O_2(g)$ $\longrightarrow$ $SO_2(g)$ $\Delta H° = -296$ kJ/mol
(b) $Cl_2(g)$ $\longrightarrow$ Cl(g) + Cl(g) $\Delta H° = 243$ kJ/mol

14.39 The reaction H + H_2 $\longrightarrow$ H_2 + H has been studied for many years. Sketch a potential-energy versus reaction progress diagram for this reaction.

14.40 Over the range of about ±3°C from normal body temperature, the metabolic rate, M_T, is given by $M_T = M_{37}(1.1)^{\Delta T}$, where M_{37} is the normal rate (at 37°C) and ΔT is the change in T. Discuss this equation in terms of a possible molecular interpretation.

Computational Problems

14.41 For the reaction:

$$NO(g) + O_3(g) \longrightarrow NO_2(g) + O_2(g)$$

the frequency factor A is 8.7×10^{12} s^{-1} and the activation energy is 63 kJ/mol. What is the rate constant for the reaction at 75°C?

14.42 The rate constant of a first-order reaction is 4.60×10^{-4} s^{-1} at 350°C. If the activation energy is 104 kJ/mol, calculate the temperature at which its rate constant is 8.80×10^{-4} s^{-1}.

14.43 The rate constants of some reactions double with every 10° rise in temperature. Assume that a reaction takes place at 295 K and 305 K. What must the activation energy be for the rate constant to double as described?

14.44 The rate at which tree crickets chirp is 2.0×10^2 per minute at 27°C but only 39.6 per minute at 5°C. From these data, calculate the "activation energy" for the chirping process. (*Hint:* The ratio of rates is equal to the ratio of rate constants.)

14.45 The rate of bacterial hydrolysis of fish muscle is twice as great at 2.2°C as at −1.1°C. Estimate an E_a value for this reaction. Is there any relation to the problem of storing fish for food?

14.46 The activation energy for the denaturation of a protein is 396 kJ/mol. At what temperature will the rate of denaturation be 20 percent greater than its rate at 25°C?

14.47 Given the same reactant concentrations, the reaction:

$$CO(g) + Cl_2(g) \longrightarrow COCl_2(g)$$

at 250°C is 1.50×10^3 times as fast as the same reaction at 150°C. Calculate the activation energy for this reaction. Assume that the frequency factor is constant.

Conceptual Problems

14.48 Variation of the rate constant with temperature for the first-order reaction:

$$2N_2O_5(g) \longrightarrow 2N_2O_4(g) + O_2(g)$$

is given in the following table. Determine graphically the activation energy for the reaction.

T (K)	k (s^{-1})
298	1.74×10^{-5}
308	6.61×10^{-5}
318	2.51×10^{-4}
328	7.59×10^{-4}
338	2.40×10^{-3}

14.49 Diagram A describes the initial state of reaction:

$$H_2 + Cl_2 \longrightarrow 2HCl$$

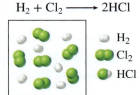

Diagram A

Suppose the reaction is carried out at two different temperatures as shown in diagram B. Which picture represents the result at the higher temperature? (The reaction proceeds for the same amount of time at both *temperatures*.)

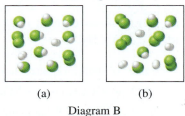

(a)   (b)

Diagram B

SECTION 14.5: REACTION MECHANISMS

Review Questions

14.50 What do we mean by the *mechanism* of a reaction?

14.51 What is an elementary step? What is the molecularity of a reaction?

14.52 Classify the following elementary reactions as unimolecular, bimolecular, or termolecular:
(a) $2NO + Br_2 \longrightarrow 2NOBr$
(b) $CH_3NC \longrightarrow CH_3CN$
(c) $SO + O_2 \longrightarrow SO_2 + O$

14.53 Reactions can be classified as unimolecular, bimolecular, and so on. Why are there no zero-molecular reactions? Explain why termolecular reactions are rare.

14.54 Determine the molecularity, and write the rate law for each of the following elementary steps:
(a) $X \longrightarrow$ products
(b) $X + Y \longrightarrow$ products
(c) $X + Y + Z \longrightarrow$ products
(d) $X + X \longrightarrow$ products
(e) $X + 2Y \longrightarrow$ products

14.55 What is the rate-determining step of a reaction? Give an everyday analogy to illustrate the meaning of *rate determining*.

14.56 The equation for the combustion of ethane (C_2H_6) is:

$$2C_2H_6(g) + 7O_2(g) \longrightarrow 4CO_2(g) + 6H_2O(l)$$

Explain why it is unlikely that this reaction happens in a single step.

14.57 Specify which of the following species cannot be isolated in a reaction: activated complex, product, intermediate.

Conceptual Problems

14.58 Classify each of the following elementary steps as unimolecular, bimolecular, or termolecular.

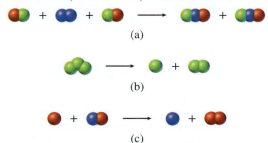

(a)

(b)

(c)

14.59 The rate law for the reaction:

$$2NO(g) + Cl_2(g) \longrightarrow 2NOCl(g)$$

is given by rate = $k[NO][Cl_2]$. (a) What is the order of the reaction? (b) A mechanism involving the following steps has been proposed for the reaction:

$$NO(g) + Cl_2(g) \longrightarrow NOCl_2(g)$$
$$NOCl_2(g) + NO(g) \longrightarrow 2NOCl(g)$$

If this mechanism is correct, what does it imply about the relative rates of these two steps?

14.60   For the reaction $X_2 + Y + Z \longrightarrow XY + XZ$, it is found that doubling the concentration of X_2 doubles the reaction rate, tripling the concentration of Y triples the rate, and doubling the concentration of Z has no effect. (a) What is the rate law for this reaction? (b) Why is it that the change in the concentration of Z has no effect on the rate? (c) Suggest a mechanism for the reaction that is consistent with the rate law.

14.61   The rate law for the reaction:

$$2H_2(g) + 2NO(g) \longrightarrow N_2(g) + 2H_2O(g)$$

is rate = $k[H_2][NO]^2$. Which of the following mechanisms can be ruled out on the basis of the observed rate expression?

Mechanism I

$$H_2 + NO \longrightarrow H_2O + N \quad \text{(slow)}$$
$$N + NO \longrightarrow N_2 + O \quad \text{(fast)}$$
$$O + H_2 \longrightarrow H_2O \quad \text{(fast)}$$

Mechanism II

$$H_2 + 2NO \longrightarrow N_2O + H_2O \quad \text{(slow)}$$
$$N_2O + H_2 \longrightarrow N_2 + H_2O \quad \text{(fast)}$$

Mechanism III

$$2NO \rightleftharpoons N_2O_2 \quad \text{(fast equilibrium)}$$
$$N_2O_2 + H_2 \longrightarrow N_2O + H_2O \quad \text{(slow)}$$
$$N_2O + H_2 \longrightarrow N_2 + H_2O \quad \text{(fast)}$$

14.62   The rate law for the decomposition of ozone to molecular oxygen,

$$2O_3(g) \longrightarrow 3O_2(g)$$

is:

$$\text{rate} = k\frac{[O_3]^2}{[O_2]}$$

The mechanism proposed for this process is:

$$O_3 \underset{k_{-1}}{\overset{k_1}{\rightleftharpoons}} O + O_2$$
$$O + O_3 \overset{k_2}{\longrightarrow} 2O_2$$

Derive the rate law from these elementary steps. Clearly state the assumptions you use in the derivation. Explain why the rate decreases with increasing O_2 concentration.

SECTION 14.6: CATALYSIS

Review Questions

14.63   How does a catalyst increase the rate of a reaction?

14.64   What are the characteristics of a catalyst?

14.65   A certain reaction is known to proceed slowly at room temperature. Is it possible to make the reaction proceed at a faster rate without changing the temperature?

14.66   Most reactions, including enzyme-catalyzed reactions, proceed faster at higher temperatures. However, for a given enzyme, the rate drops off abruptly at a certain temperature. Account for this behavior.

14.67   Are enzyme-catalyzed reactions examples of homogeneous or heterogeneous catalysis? Explain.

14.68   The concentrations of enzymes in cells are usually quite small. What is the biological significance of this fact?

14.69   When fruits such as apples and pears are cut, the exposed areas begin to turn brown. This is the result of an enzyme-catalyzed reaction. Often the browning can be prevented or slowed by adding a few drops of lemon juice. What is the chemical basis of this treatment?

14.70   The first-order rate constant for the dehydration of carbonic acid:

$$H_2CO_3 \longrightarrow CO_2 + H_2O$$

is about $1 \times 10^2 \text{ s}^{-1}$. In view of this rather high rate constant, explain why it is necessary to have the enzyme carbonic anhydrase to enhance the rate of dehydration in the lungs.

Conceptual Problems

14.71   Which two potential-energy profiles represent the same reaction catalyzed and uncatalyzed?

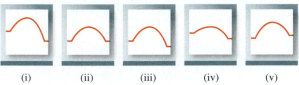

(i)   (ii)   (iii)   (iv)   (v)

14.72   Consider the following mechanism for the enzyme-catalyzed reaction:

$$E + S \underset{k_{-1}}{\overset{k_1}{\rightleftharpoons}} ES \quad \text{(fast equilibrium)}$$
$$ES \overset{k_2}{\longrightarrow} E + P \quad \text{(slow)}$$

Derive an expression for the rate law of the reaction in terms of the concentrations of E and S. (*Hint:* To solve for [ES], make use of the fact that, at equilibrium, the rate of the forward reaction is equal to the rate of the reverse reaction.)

ADDITIONAL PROBLEMS

14.73   List four factors that influence the rate of a reaction.

14.74   Suggest experimental means by which the rates of the following reactions could be followed:
(a) $CaCO_3(s) \longrightarrow CaO(s) + CO_2(g)$
(b) $Cl_2(g) + 2Br^-(aq) \longrightarrow Br_2(aq) + 2Cl^-(aq)$
(c) $C_2H_6(g) \longrightarrow C_2H_4(g) + H_2(g)$
(d) $C_2H_5I(g) + H_2O(l) \longrightarrow$
$$\quad\quad\quad C_2H_5OH(aq) + H^+(aq) + I^-(aq)$$

14.75   "The rate constant for the reaction:

$$NO_2(g) + CO(g) \longrightarrow NO(g) + CO_2(g)$$

is $1.64 \times 10^{-6}/M \cdot s$." What is incomplete about this statement?

14.76 A compound X undergoes two *simultaneous* first-order reactions as follows: X⟶Y with rate constant k_1 and X⟶Z with rate constant k_2. The ratio of k_1/k_2 at 40°C is 8.0. What is the ratio at 300°C? Assume that the frequency factors of the two reactions are the same.

14.77 The following diagrams represent the progress of the reaction A⟶B, where the red spheres represent A molecules and the green spheres represent B molecules. Calculate the rate constant of the reaction.

$t = 0$ s	$t = 20$ s	$t = 40$ s

14.78 The following diagrams show the progress of the reaction 2A ⟶ A₂. Determine whether the reaction is first order or second order, and calculate the rate constant.

$t = 0$ min	$t = 15$ min	$t = 30$ min

14.79 Use the data in Sample Problem 14.5 to determine graphically the half-life of the reaction.

14.80 The following data were collected for the reaction between hydrogen and nitric oxide at 700°C:

$$2H_2(g) + 2NO(g) \longrightarrow 2H_2O(g) + N_2(g)$$

Experiment	[H₂] (M)	[NO] (M)	Initial Rate (M/s)
1	0.010	0.025	2.4×10^{-6}
2	0.0050	0.025	1.2×10^{-6}
3	0.010	0.0125	0.60×10^{-6}

(a) Determine the order of the reaction. (b) Calculate the rate constant. (c) Suggest a plausible mechanism that is consistent with the rate law. (*Hint:* Assume that the oxygen atom is the intermediate.)

14.81 When methyl phosphate is heated in acid solution, it reacts with water:

$$CH_3OPO_3H_2 + H_2O \longrightarrow CH_3OH + H_3PO_4$$

If the reaction is carried out in water enriched with ^{18}O, the oxygen-18 isotope is found in the phosphoric acid product but not in the methanol. What does this tell us about the mechanism of the reaction?

14.82 The rate of the reaction:

$$CH_3COOC_2H_5(aq) + H_2O(l) \longrightarrow$$
$$CH_3COOH(aq) + C_2H_5OH(aq)$$

shows first-order characteristics—that is, rate = $k[CH_3COOC_2H_5]$—even though this is a second-order reaction (first order in $CH_3COOC_2H_5$ and first order in H_2O). Explain.

14.83 Explain why most metals used in catalysis are transition metals.

14.84 The reaction 2A + 3B ⟶ C is first order with respect to A and B. When the initial concentrations are [A] = 1.6×10^{-2} M and [B] = 2.4×10^{-3} M, the rate is 4.1×10^{-4} M/s. Calculate the rate constant of the reaction.

14.85 The bromination of acetone is acid-catalyzed:

$$CH_3COCH_3 + Br_2 \xrightarrow[\text{catalyst}]{H^+} CH_3COCH_2Br + H^+ + Br^-$$

The rate of disappearance of bromine was measured for several different concentrations of acetone, bromine, and H⁺ ions at a certain temperature:

	[CH₃COCH₃] (M)	[Br₂] (M)	[H⁺] (M)	Rate of Disappearance of Br₂ (M/s)
(1)	0.30	0.050	0.050	5.7×10^{-5}
(2)	0.30	0.10	0.050	5.7×10^{-5}
(3)	0.30	0.050	0.10	1.2×10^{-4}
(4)	0.40	0.050	0.20	3.1×10^{-4}
(5)	0.40	0.050	0.050	7.6×10^{-5}

(a) What is the rate law for the reaction?
(b) Determine the rate constant. (c) The following mechanism has been proposed for the reaction:

$$CH_3-\overset{O}{\overset{\|}{C}}-CH_3 + H_3O^+ \rightleftharpoons CH_3-\overset{+OH}{\overset{\|}{C}}-CH_3 + H_2O \quad \text{(fast equilibrium)}$$

$$CH_3-\overset{+OH}{\overset{\|}{C}}-CH_3 + H_2O \longrightarrow CH_3-\overset{OH}{\overset{|}{C}}=CH_2 + H_3O^+ \quad \text{(slow)}$$

$$CH_3-\overset{OH}{\overset{|}{C}}=CH_2 + Br_2 \longrightarrow CH_3-\overset{O}{\overset{\|}{C}}-CH_2Br + HBr \quad \text{(fast)}$$

Show that the rate law deduced from the mechanism is consistent with that shown in part (a).

14.86 The decomposition of N₂O to N₂ and O₂ is a first-order reaction. At 730°C the half-life of the reaction is 3.58×10^3 min. If the initial pressure of N₂O is 2.10 atm at 730°C, calculate the total gas pressure after one half-life. Assume that the volume remains constant.

14.87 The reaction $S_2O_8^{2-} + 2I^- \longrightarrow 2SO_4^{2-} + I_2$ proceeds slowly in aqueous solution, but it can be catalyzed by the Fe³⁺ ion. Given that Fe³⁺ can oxidize I⁻ and Fe²⁺ can reduce $S_2O_8^{2-}$, write a plausible two-step mechanism for this reaction. Explain why the uncatalyzed reaction is slow.

14.88 What are the units of the rate constant for a third-order reaction?

14.89 The integrated rate law for the zeroth-order reaction A ⟶ B is $[A]_t = [A]_0 - kt$. (a) Sketch the following plots: (i) rate versus $[A]_t$ and (ii) $[A]_t$ versus t. (b) Derive an expression for the half-life of the reaction. (c) Calculate the time in half-lives when the integrated rate law is no longer valid, that is, when $[A]_t = 0$.

14.90   A flask contains a mixture of compounds A and B. Both compounds decompose by first-order kinetics. The half-lives are 50.0 min for A and 18.0 min for B. If the concentrations of A and B are equal initially, how long will it take for the concentration of A to be four times that of B?

14.91   Referring to Sample Problem 14.5, explain how you would measure the partial pressure of azomethane experimentally as a function of time.

14.92   The rate law for the reaction $2NO_2(g) \longrightarrow N_2O_4(g)$ is rate = $k[NO_2]^2$. Which of the following changes will change the value of k? (a) The pressure of NO_2 is doubled. (b) The reaction is run in an organic solvent. (c) The volume of the container is doubled. (d) The temperature is decreased. (e) A catalyst is added to the container.

14.93   The reaction of G_2 with E_2 to form 2EG is exothermic, and the reaction of G_2 with X_2 to form 2XG is endothermic. The activation energy of the exothermic reaction is greater than that of the endothermic reaction. Sketch the potential-energy profile diagrams for these two reactions on the same graph.

14.94   The activation energy for the decomposition of hydrogen peroxide:

$$2H_2O_2(aq) \longrightarrow 2H_2O(l) + O_2(g)$$

is 42 kJ/mol, whereas when the reaction is catalyzed by the enzyme catalase, it is 7.0 kJ/mol. Calculate the temperature that would cause the uncatalyzed decomposition to proceed as rapidly as the enzyme-catalyzed decomposition at 20°C. Assume the frequency factor A to be the same in both cases.

14.95   Briefly comment on the effect of a catalyst on each of the following: (a) activation energy, (b) reaction mechanism, (c) enthalpy of reaction, (d) rate of forward reaction, (e) rate of reverse reaction.

14.96   When 6 g of granulated Zn is added to a solution of 2 M HCl in a beaker at room temperature, hydrogen gas is generated. For each of the following changes (at constant volume of the acid), state whether the rate of hydrogen gas evolution will be increased, decreased, or unchanged: (a) 6 g of powdered Zn is used, (b) 4 g of granulated Zn is used, (c) 2 M acetic acid is used instead of 2 M HCl, (d) temperature is raised to 40°C.

14.97   Strictly speaking, the rate law derived for the reaction in Problem 14.80 applies only to certain concentrations of H_2. The general rate law for the reaction takes the form

$$\text{rate} = \frac{k_1[NO]^2[H_2]}{1 + k_2[H_2]}$$

where k_1 and k_2 are constants. Derive rate law expressions under the conditions of very high and very low hydrogen concentrations. Does the result from Problem 14.80 agree with one of the rate expressions here?

14.98   A certain first-order reaction is 35.5 percent complete in 4.90 min at 25°C. What is its rate constant?

14.99   The decomposition of dinitrogen pentoxide has been studied in carbon tetrachloride solvent (CCl_4) at a certain temperature:

$$2N_2O_5 \longrightarrow 4NO_2 + O_2$$

$[N_2O_5]$ (M)	Initial Rate (M/s)
0.92	0.95×10^{-5}
1.23	1.20×10^{-5}
1.79	1.93×10^{-5}
2.00	2.10×10^{-5}
2.21	2.26×10^{-5}

Determine graphically the rate law for the reaction, and calculate the rate constant.

14.100   The thermal decomposition of N_2O_5 obeys first-order kinetics. At 45°C, a plot of ln $[N_2O_5]$ versus t gives a slope of -6.18×10^{-4} min^{-1}. What is the half-life of the reaction?

14.101   When a mixture of methane and bromine is exposed to light, the following reaction occurs slowly:

$$CH_4(g) + Br_2(g) \longrightarrow CH_3Br(g) + HBr(g)$$

Suggest a reasonable mechanism for this reaction. (*Hint:* Bromine vapor is deep red; methane is colorless.)

14.102   The rate of the reaction between H_2 and I_2 to form HI increases with the intensity of visible light. (a) Explain why this fact supports a two-step mechanism. (I_2 vapor is purple.) (b) Explain why the visible light has no effect on the formation of H atoms.

14.103   The rate constant for the gaseous reaction:

$$H_2(g) + I_2(g) \longrightarrow 2HI(g)$$

is $2.42 \times 10^{-2}/M \cdot$ s at 400°C. Initially an equimolar sample of H_2 and I_2 is placed in a vessel at 400°C, and the total pressure is 1658 mmHg. (a) What is the initial rate ($M/$min) of formation of HI? (b) Determine the rate of formation of HI and the concentration of HI (in molarity) after 10.0 min.

14.104   A gas mixture containing CH_3 fragments, C_2H_6 molecules, and an inert gas (He) was prepared at 600 K with a total pressure of 5.42 atm. The elementary reaction

$$CH_3 + C_2H_6 \longrightarrow CH_4 + C_2H_5$$

has a second-order rate constant of $3.0 \times 10^4/M \cdot$ s. Given that the mole fractions of CH_3 and C_2H_6 are 0.00093 and 0.00077, respectively, calculate the initial rate of the reaction at this temperature.

14.105   Consider the following elementary step:

$$X + 2Y \longrightarrow XY_2$$

(a) Write a rate law for this reaction. (b) If the initial rate of formation of XY_2 is 3.8×10^{-3} $M/$s and the initial concentrations of X and Y are 0.26 M and 0.88 M, respectively, what is the rate constant of the reaction?

14.106 The following scheme in which A is converted to B, which is then converted to C, is known as a consecutive reaction:

$$A \longrightarrow B \longrightarrow C$$

Assuming that both steps are first order, sketch on the same graph the variations of [A], [B], and [C] with time.

14.107 (a) Consider two reactions, A and B. If the rate constant for reaction B increases by a larger factor than that of reaction A when the temperature is increased from T_1 to T_2, what can you conclude about the relative values of the activation energies of the two reactions? (b) If a bimolecular reaction occurs every time an A and a B molecule collide, what can you say about the orientation factor and activation energy of the reaction?

14.108 The rate law for the following reaction:

$$CO(g) + NO_2(g) \longrightarrow CO_2(g) + NO(g)$$

is rate $= k[NO_2]^2$. Suggest a plausible mechanism for the reaction, given that the unstable species NO_3 is an intermediate.

14.109 Consider the following elementary steps for a consecutive reaction:

$$A \xrightarrow{k_1} B \xrightarrow{k_2} C$$

(a) Write an expression for the rate of change of B.
(b) Derive an expression for the concentration of B under "steady-state" conditions; that is, when B is decomposing to C at the same rate as it is formed from A.

14.110 Consider the potential-energy profiles for the following three reactions (from left to right). (1) Rank the rates (slowest to fastest) of the reactions. (2) Calculate ΔH for each reaction, and determine which reaction(s) are exothermic and which reaction(s) are endothermic. Assume the reactions have roughly the same frequency factors.

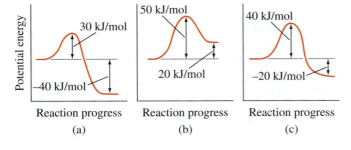

(a)   (b)   (c)

14.111 Consider the following potential-energy profile for the A $\longrightarrow$ D reaction. (a) How many elementary steps are there? (b) How many intermediates are formed? (c) Which step is rate determining? (d) Is the overall reaction exothermic or endothermic?

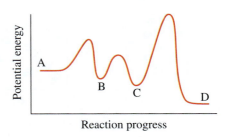

14.112 The rate of a reaction was followed by the absorption of light by the reactants and products as a function of wavelengths (λ_1, λ_2, λ_3) as time progresses. Which of the following mechanisms is consistent with the experimental data?
(a) A $\longrightarrow$ B, A $\longrightarrow$ C
(b) A $\longrightarrow$ B + C
(c) A $\longrightarrow$ B, B $\longrightarrow$ C + D
(d) A $\longrightarrow$ B, B $\longrightarrow$ C

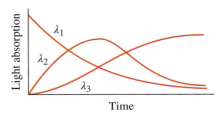

14.113 The first-order rate constant for the decomposition of dimethyl ether:

$$(CH_3)_2O(g) \longrightarrow CH_4(g) + H_2(g) + CO(g)$$

is $3.2 \times 10^{-4} \text{ s}^{-1}$ at 450°C. The reaction is carried out in a constant-volume flask. Initially only dimethyl ether is present and the pressure is 0.350 atm. What is the pressure of the system after 3.5 min? Assume ideal behavior.

14.114 Four experiments with different starting conditions were done using the reaction A $\longrightarrow$ B, where A is represented by blue spheres and B is represented by yellow spheres. Determine the overall order of the reaction.

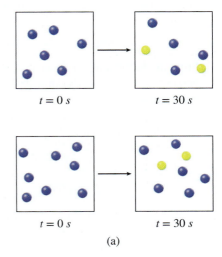

(a)

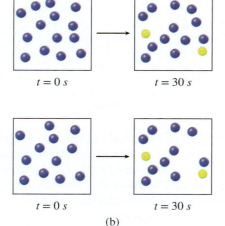

$t = 0\ s$ $t = 30\ s$

$t = 0\ s$ $t = 30\ s$

(b)

14.115   Thallium(I) is oxidized by cerium(IV) as follows:

$$Tl^+ + 2Ce^{4+} \longrightarrow Tl^{3+} + 2Ce^{3+}$$

The elementary steps, in the presence of Mn(II), are as follows:

$$Ce^{4+} + Mn^{2+} \longrightarrow Ce^{3+} + Mn^{3+}$$
$$Ce^{4+} + Mn^{3+} \longrightarrow Ce^{3+} + Mn^{4+}$$
$$Tl^+ + Mn^{4+} \longrightarrow Tl^{3+} + Mn^{2+}$$

(a) Identify the catalyst, intermediates, and the rate-determining step if the rate law is rate = $k[Ce^{4+}][Mn^{2+}]$. (b) Explain why the reaction is slow without the catalyst. (c) Classify the type of catalysis (homogeneous or heterogeneous).

14.116   The activation energy for the reaction:

$$N_2O(g) \longrightarrow N_2(g) + O(g)$$

is 2.4×10^2 kJ/mol at 600 K. Calculate the percentage of the increase in rate from 600 K to 606 K. Comment on your results.

14.117 $\Delta H°$ for the reaction in Problem 14.116 is -164 kJ/mol. What is the activation energy (E_a) for the reverse reaction?

14.118   At a certain elevated temperature, ammonia decomposes on the surface of tungsten metal as follows:

$$2NH_3 \longrightarrow N_2 + 3H_2$$

From the following plot of the rate of the reaction versus the pressure of NH_3, describe the mechanism of the reaction.

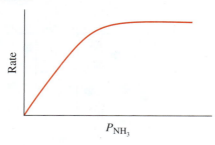

14.119   The following expression shows the dependence of the half-life of a reaction ($t_{1/2}$) on the initial reactant concentration $[A]_0$:

$$t_{1/2} \propto \frac{1}{[A]_0^{n-1}}$$

where n is the order of the reaction. Verify this dependence for zeroth-, first-, and second-order reactions.

Engineering Problems

14.120   In a certain industrial process involving a heterogeneous catalyst, the volume of the catalyst (in the shape of a sphere) is 10.0 cm^3. Calculate the surface area of the catalyst. If the sphere is broken down into eight smaller spheres, each having a volume of 1.25 cm^3, what is the total surface area of the spheres? Which of the two geometric configurations of the catalyst is more effective? (The surface area of a sphere is $4\pi r^2$, where r is the radius of the sphere.) Based on your analysis here, explain why it is sometimes dangerous to work in grain elevators.

14.121   Strontium-90, a radioactive isotope, is a major product of an atomic bomb explosion. It has a half-life of 28.1 yr. (a) Calculate the first-order rate constant for the nuclear decay. (b) Calculate the fraction of ^{90}Sr that remains after 10 half-lives. (c) Calculate the number of years required for 99.0 percent of ^{90}Sr to disappear.

14.122   In the nuclear industry, workers use a rule of thumb that the radioactivity from any sample will be relatively harmless after 10 half-lives. Calculate the fraction of a radioactive sample that remains after this time period. (*Hint:* Radioactive decays obey first-order kinetics.)

14.123   Radioactive plutonium-239 ($t_{1/2} = 2.44 \times 10^5$ yr) is used in nuclear reactors and atomic bombs. If there are 5.0×10^2 g of the isotope in a small atomic bomb, how long will it take for the substance to decay to 1.0×10^2 g, too small an amount for an effective bomb?

14.124   A factory that specializes in the refinement of transition metals such as titanium was on fire. The firefighters were advised not to douse the fire with water. Why?

14.125   When the concentration of A in the reaction A $\longrightarrow$ B was changed from 1.20 *M* to 0.60 *M*, the half-life increased from 2.0 min to 4.0 min at 25°C. Calculate the order of the reaction and the rate constant. (*Hint:* Use the equation in Problem 14.119.)

14.126   The *activity* of a radioactive sample is the number of nuclear disintegrations per second, which is equal to the first-order rate constant times the number of radioactive nuclei present. The fundamental unit of radioactivity is the *curie* (Ci), where 1 Ci corresponds to exactly 3.70×10^{10} disintegrations per second. This decay rate is equivalent to that of 1 g of radium-226. Calculate the rate constant and half-life for the radium decay. Starting with 1.0 g of the radium sample, what is the activity after 500 yr? The molar mass of Ra-226 is 226.03 g/mol.

Biological Problems

14.127 To carry out metabolism, oxygen is taken up by hemoglobin (Hb) to form oxyhemoglobin (HbO_2) according to the simplified equation:

$$Hb(aq) + O_2(aq) \xrightarrow{k} HbO_2(aq)$$

where the second-order rate constant is $2.1 \times 10^6/M \cdot s$ at 37°C. For an average adult, the concentrations of Hb and O_2 in the blood at the lungs are $8.0 \times 10^{-6}\ M$ and $1.5 \times 10^{-6}\ M$, respectively. (a) Calculate the rate of formation of HbO_2. (b) Calculate the rate of consumption of O_2. (c) The rate of formation of HbO_2 increases to $1.4 \times 10^{-4}\ M/s$ during exercise to meet the demand of the increased metabolism rate. Assuming the Hb concentration to remain the same, what must the oxygen concentration be to sustain this rate of HbO_2 formation?

14.128 Sucrose ($C_{12}H_{22}O_{11}$), commonly called table sugar, undergoes hydrolysis (reaction with water) to produce fructose ($C_6H_{12}O_6$) and glucose ($C_6H_{12}O_6$):

$$C_{12}H_{22}O_{11} + H_2O \longrightarrow \underset{\text{fructose}}{C_6H_{12}O_6} + \underset{\text{glucose}}{C_6H_{12}O_6}$$

This reaction is of considerable importance in the candy industry. First, fructose is sweeter than sucrose. Second, a mixture of fructose and glucose, called *invert sugar,* does not crystallize, so the candy containing this sugar would be chewy rather than brittle as candy containing sucrose crystals would be. (a) From the following data determine the order of the reaction. (b) How long does it take to hydrolyze 95 percent of sucrose? (c) Explain why the rate law does not include [H_2O] even though water is a reactant.

Time (min)	[$C_{12}H_{22}O_{11}$] (M)
0	0.500
60.0	0.400
96.4	0.350
157.5	0.280

14.129 To prevent brain damage, a standard procedure is to lower the body temperature of someone who has been resuscitated after suffering cardiac arrest. What is the physiochemical basis for this procedure?

14.130 Ethanol is a toxic substance that, when consumed in excess, can impair respiratory and cardiac functions by interference with the neurotransmitters of the nervous system. In the human body, ethanol

is metabolized by the enzyme alcohol dehydrogenase to acetaldehyde, which causes hangovers. Based on your knowledge of enzyme kinetics, explain why binge drinking (i.e., consuming too much alcohol too fast) can prove fatal.

14.131 A protein molecule P of molar mass $\mathcal{M}$ dimerizes when it is allowed to stand in solution at room temperature. A plausible mechanism is that the protein molecule is first denatured (i.e., loses its activity due to a change in overall structure) before it dimerizes:

$$P \xrightarrow{k} P^* \text{ (denatured)} \quad \text{(slow)}$$
$$2P^* \longrightarrow P_2 \quad \text{(fast)}$$

where the asterisk denotes a denatured protein molecule. Derive an expression for the average molar mass (of P and P_2), $\mathcal{\overline{M}}$, in terms of the initial protein concentration $[P]_0$ and the concentration at time t, $[P]_t$, and $\mathcal{M}$. Describe how you would determine k from molar mass measurements.

Environmental Problems

14.132 At 25°C, the rate constant for the ozone-depleting reaction:

$$O(g) + O_3(g) \longrightarrow 2O_2(g)$$

is 7.9×10^{-15} cm^3/molecule $\cdot$ s. Express the rate constant in units of $1/M \cdot s$.

14.133 Chlorine oxide (ClO), which plays an important role in the depletion of ozone, decays rapidly at room temperature according to the equation:

$$2ClO(g) \longrightarrow Cl_2(g) + O_2(g)$$

From the following data, determine the reaction order and calculate the rate constant of the reaction.

Time (s)	[ClO]
0.12×10^{-3}	8.49×10^{-6}
0.96×10^{-3}	7.10×10^{-6}
2.24×10^{-3}	5.79×10^{-6}
3.20×10^{-3}	5.20×10^{-6}
4.00×10^{-3}	4.77×10^{-6}

14.134 At a certain elevated temperature, ammonia decomposes on the surface of tungsten metal as follows:

$$NH_3 \longrightarrow \tfrac{1}{2}N_2 + \tfrac{3}{2}H_2$$

The kinetic data are expressed as the variation of the half-life with the initial pressure of NH_3:

P (mmHg)	$t_{1/2}$ (s)
264	456
130	228
59	102
16	60

(a) Determine the order of the reaction. (b) How does the order depend on the initial pressure? (c) How does the mechanism of the reaction vary with pressure? (*Hint:* You need to use the equation in Problem 14.119 and plot log $t_{1/2}$ versus log P.)

14.135 Polyethylene is used in many items, including water pipes, bottles, electrical insulation, toys, and mailer envelopes. It is a *polymer,* a molecule with a very high molar mass made by joining many ethylene molecules together. (Ethylene is the basic unit, or monomer, for polyethylene.) The initiation step is:

$$R_2 \xrightarrow{k_1} 2R \cdot \quad \text{(initiation)}$$

The R · species (called a radical) reacts with an ethylene molecule (M) to generate another radical:

$$R \cdot + M \longrightarrow M_1 \cdot$$

The reaction of $M_1 \cdot$ with another monomer leads to the growth or propagation of the polymer chain:

$$M_1 \cdot + M \xrightarrow{k_p} M_2 \cdot \quad \text{(propagation)}$$

This step can be repeated with hundreds of monomer units. The propagation terminates when two radicals combine:

$$M' \cdot + M'' \cdot \xrightarrow{k_t} M' - M'' \quad \text{(termination)}$$

The initiator frequently used in the polymerization of ethylene is benzoyl peroxide [$(C_6H_5COO)_2$]:

$$(C_6H_5COO)_2 \longrightarrow 2C_6H_5COO$$

This is a first-order reaction. The half-life of benzoyl peroxide at 100°C is 19.8 min.

(a) Calculate the rate constant (in min^{-1}) of the reaction. (b) If the half-life of benzoyl peroxide is 7.30 h, or 438 min, at 70°C, what is the activation energy (in kJ/mol) for the decomposition of benzoyl peroxide? (c) Write the rate laws for the elementary steps in the preceding polymerization process, and identify the reactant, product, and intermediates. (d) What condition would favor the growth of long, high-molar-mass polyethylenes?

14.136 In recent years, ozone in the stratosphere has been depleted at an alarmingly fast rate by chlorofluorocarbons (CFCs). A CFC molecule such as $CFCl_3$ is first decomposed by UV radiation:

$$CFCl_3 \longrightarrow CFCl_2 + Cl$$

The chlorine radical then reacts with ozone as follows:

$$Cl + O_3 \longrightarrow ClO + O_2$$
$$ClO + O \longrightarrow Cl + O_2$$

(a) Write the overall reaction for the last two steps. (b) What are the roles of Cl and ClO? (c) Why is the fluorine radical not important in this mechanism? (d) One suggestion to reduce the concentration of chlorine radicals is to add hydrocarbons such as ethane (C_2H_6) to the stratosphere. How will this work? (e) Draw potential-energy versus reaction progress diagrams for the uncatalyzed and catalyzed (by Cl) destruction of ozone: $O_3 + O \longrightarrow 2O_2$. Use the thermodynamic data in Appendix 2 to determine whether the reaction is exothermic or endothermic.

Standardized-Exam Practice Problems

Physical and Biological Sciences

Metastron, an aqueous solution of $^{89}SrCl_2$, is a drug used to alleviate severe bone pain associated with some metastatic cancers. It contains the radioactive isotope ^{89}Sr, which emits β radiation and has a constant half-life of 50.5 days. Because the strontium ion is chemically similar to the calcium ion, ^{89}Sr is incorporated into bone tissue, especially in the bone lesions caused by cancer that has spread. Treatment with Metastron typically involves intravenous injection of 4 mL over the course of about 2 min.

1. Calculate the numerical value of the rate constant (k) for the decay of ^{89}Sr.

    a) 0.0137
    b) 0.0127
    c) 0.693
    d) 0.0198

2. What are the units of the rate constant for the decay of ^{89}Sr?

    a) Days
    b) Reciprocal days
    c) Reciprocal seconds
    d) The rate constant has no units.

3. What percentage of the original activity of ^{89}Sr remains 101 days after an injection of Metastron?

    a) 75 percent   c) 25 percent
    b) 50 percent   d) 12 percent

4. How many days must pass following an injection of Metastron for the β emission to fall to 10 percent of its original value?

    a) 182   c) 505
    b) 168   d) 455

Answers to In-Chapter Materials

Answers to Practice Problems

14.1A (a) Rate $= -\dfrac{\Delta[CO_2]}{\Delta t} = -\dfrac{1}{2}\dfrac{\Delta[H_2O]}{\Delta t} = \dfrac{\Delta[CH_4]}{\Delta t} = \dfrac{1}{2}\dfrac{\Delta[O_2]}{\Delta t}$,

(b) rate $= -\dfrac{1}{3}\dfrac{\Delta[O_2]}{\Delta t} = \dfrac{1}{2}\dfrac{\Delta[O_3]}{\Delta t}$, (c) rate $= -\dfrac{1}{2}\dfrac{\Delta[NO]}{\Delta t} = -\dfrac{\Delta[O_2]}{\Delta t} = $

$\dfrac{1}{2}\dfrac{\Delta[NO_2]}{\Delta t}$. **14.1B** (a) $3CH_4 + 2H_2O + CO_2 \longrightarrow 4CH_3OH$,

(b) $2N_2O_5 \longrightarrow 2N_2 + 5O_2$, (c) $H_2 + CO + O_2 \longrightarrow H_2CO_3$.
14.2A (a) $0.0280\ M/s$. (b) $0.112\ M/s$. **14.2B** $2A + 3B \longrightarrow C$.
14.3A Rate $= k[S_2O_8^{2-}][I^-]$, $k = 8.1 \times 10^{-2}/M \cdot s$. **14.3B** $0.013\ M$,
$8.8 \times 10^{-6}\ M/s$. **14.4A** 75 s. **14.4B** $0.91\ M$.

14.5A

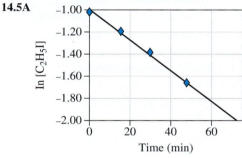

$k = 1.4 \times 10^{-2}/min$. **14.5B** $0.39\ M$, $0.34\ M$, $0.30\ M$, $0.26\ M$.
14.6A $t_{1/2} = 272$ s. **14.6B** $k = 4.71 \times 10^{-2}/h$. **14.7A** (a) 13.2 s,

(b) $t_{1/2} = 4.2$ s, 13 s. **14.7B** (a) $0.0208\ M$, (b) $1.25 \times 10^{-3}\ M$,·
(c) $1.79 \times 10^{-4}\ M$. **14.8A** 241 kJ/mol. **14.8B** $0.0655\ s^{-1}$.
14.9A 1.0×10^2 kJ/mol. **14.9B** $2.7 \times 10^{-4}/s$. **14.10A** $1.7/s$.
14.10B $7.6 \times 10^{-2}/s$. **14.11A** (a) $NO_2 + CO \longrightarrow NO + CO_2$,
(b) NO_3, (c) step 1 is rate determining. **14.11B** $F_2 + ClO_2 \longrightarrow$
$FClO_2 + F$ (slow), $ClO_2 + F \longrightarrow FClO_2$ (fast). This is one
possibility. Any mechanism in which the slow step has the proper
rate law and the steps add to give the correct overall equation
is plausible. **14.12A** The first step is a rapidly established
equilibrium. Setting the rates of forward and reverse reactions
equal to each other gives $k_1[NO][Br_2] = k_{-1}[NOBr_2]$. Solving
for $[NOBr_2]$ gives $k_1[NO][Br_2]/k_{-1}$. Substituting this into the
rate law for the rate-determining step, rate $= k_2[NOBr_2][NO]$,
gives rate $= (k_1k_2/k_{-1})[NO]^2[Br_2]$ or $k[NO]^2[Br_2]$.

14.12B Step 1: $I_2(g) \underset{k_{-1}}{\overset{k_1}{\rightleftharpoons}} 2I(g)$.

Step 2: $H_2(g) + 2I(g) \longrightarrow 2HI(g)$. Rate $= k[H_2][I_2]$.

Answers to Checkpoints

14.1.1 a, d. **14.1.2** b. **14.2.1** e. **14.2.2** d. **14.2.3** b. **14.2.4** a.
14.2.5 b. **14.3.1** c. **14.3.2** c. **14.3.3** b. **14.3.4** d. **14.3.5** c. **14.4.1** e.
14.4.2 b. **14.4.3** a, c. **14.5.1** c. **14.5.2** c. **14.5.3** b. **14.5.4** d.

CHAPTER 15

Chemical Equilibrium

Mountain climbers sometimes fall ill due to the lower oxygen content of air at high altitudes. Long-term exposure to an oxygen-poor environment causes the production of more hemoglobin. The additional red blood cells facilitate the transport of oxygen to the body.

Buena Vista Images/DigitalVision/Getty Images

How the Principles of Equilibrium Favor Athletes Who "Live High and Train Low"

In chemistry, and in biology, when a necessary ingredient for a process is in short supply, nature sometimes responds in a way that compensates for the shortage. For example, in an isolated population of certain types of fish or frogs consisting of a single sex, some of the individuals will *change sexes,* enabling the population to reproduce.

An important example of nature responding to a shortage is the production of extra red blood cells in individuals who reside at high elevation—where the concentration of oxygen is lower than at sea level. Extra red blood cells enable the blood to transport adequate oxygen from the lungs to the rest of the body, despite the low oxygen concentration. Enhanced oxygen-carrying capacity of the blood can give an athlete greater aerobic capacity and stamina (at normal oxygen concentrations). This has given rise to attempts by some athletes to increase their red blood count (RBC) artificially—a practice known as "blood doping."

Historically, blood doping has been done with blood infusions or with the drug *erythropoietin*—both of which have been banned by the World Anti-Doping Agency (WADA). In recent years, however, the use of hypoxic sleeping tents has grown in popularity. These tents are kept filled with a mixture of nitrogen and oxygen in which the oxygen concentration is lower than that in natural air. Just as the bodies of people who reside at high elevations naturally produce additional red blood cells to compensate for the lower partial pressure of oxygen in the air, the body of an athlete who spends his or her sleeping hours in a hypoxic tent adjusts in the same way, increasing the athlete's RBC. Those who "live high and train low" are believed to have an advantage over athletes who reside and train at or near sea level. Although this is a natural phenomenon and does not really constitute blood doping, WADA has issued the opinion that the artificial inducement of this effect is "contrary to the spirit of sport." As of the 2009 edition of the WADA code, the use of hypoxic tents has not been banned, but the issue likely will be revisited in the years to come.

The increase in the RBC in people who live at high altitudes or who sleep in hypoxic tents can be explained using the principles of *chemical equilibrium.*

Student Note: In Michael Crichton's 1990 novel *Jurassic Park*, frog DNA was used to repair ancient dinosaur DNA to facilitate cloning of the extinct animals. In the story, the scientists believed the cloned animals could not reproduce because the population was designed to be entirely female. However, some of the dinosaurs became males—something known to occur in the frogs from which the DNA for repair had been taken—and the population of dinosaurs grew out of control.

At the end of this chapter, you will be able to solve a series of problems related to the detection of hypoxic RBC enhancement [[▶▶ Applying What You've Learned page 772].

Hypoxic sleeping tent
Courtesy of Hypoxico, Inc.

15.1 The Concept of Equilibrium

Until now, for the most part, we have treated chemical reactions as processes that go to completion; that is, we start with only *reactants* and end up with only *products*. In reality, this is not what happens with most chemical reactions. Instead, if we start with only reactants, the typical reaction will proceed, causing reactant concentrations to decrease (as reactants are consumed) and product concentrations to increase (as products are produced). Eventually, though, the concentrations of both reactants and products will stop changing; and we will be left with a *mixture* of reactants and products. A system in which the concentrations of reactants and products remain constant is said to be *at equilibrium*.

We have already encountered several examples of *physical* processes that involve equilibrium—including the establishment of vapor pressure over a liquid in a closed system [◄◄ Section 11.2], and the formation of a saturated solution [◄◄ Section 13.2]. Using the formation of a saturated solution of silver iodide (AgI) as an example (Figure 15.1), let's examine the concept of equilibrium. The dissolution process can be represented with the chemical equation $AgI(s) \rightleftharpoons Ag^+(aq) + I^-(aq)$, where the double arrow [◄◄ Section 4.1] indicates that this is a **reversible process,** meaning that both the forward process and the reverse process can occur. In this case, the *forward* process is the dissolution of AgI and the *reverse* process is the recombination of aqueous Ag^+ and I^- ions to form solid AgI. When we add the solid AgI to the water, initially only the forward process can occur because there are no Ag^+ or I^- ions in solution. When some AgI has dissolved, though, the reverse process can also occur. Initially, the forward process occurs at a higher rate than the reverse process, simply because of the low number of ions in solution. After some time has passed, there are enough ions in solution that the reverse process can occur at the same rate as the forward process—and the concentration of dissolved AgI stops changing. A system in which both forward and reverse processes are occurring at the same rate is said to be *at equilibrium.*

Now let's consider a *chemical* example of equilibrium, the decomposition of dinitrogen tetroxide (N_2O_4) to yield nitrogen dioxide (NO_2). This process, like most chemical reactions, is also reversible:

$$N_2O_4(g) \rightleftharpoons 2NO_2(g)$$

N_2O_4 is a colorless gas, whereas NO_2 is brown. (NO_2 is the cause of the brown appearance of some polluted air.) If we begin by placing a sample of pure N_2O_4 in an evacuated flask, the contents of the flask change from colorless to brown as the decomposition produces NO_2 (Figure 15.2). At first, the brown color intensifies as the concentration of NO_2 increases. Eventually, though, the intensity of the brown color stops increasing, indicating that the concentration of NO_2 has stopped

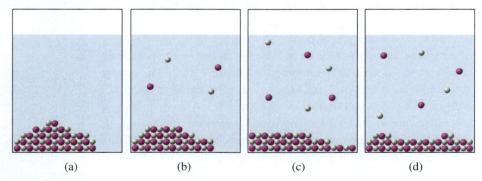

(a)	(b)	(c)	(d)

Figure 15.1 Preparation of a saturated solution of AgI: $AgI(s) \rightleftharpoons Ag^+(aq) + I^-(aq)$. (a) Solid AgI is added to water. (b) Initially, only the forward process (dissolution of solid AgI) can occur, and AgI begins to dissolve. (c) When there are Ag^+ and I^- ions in solution, the reverse process (formation of solid AgI) can also occur. (d) Equilibrium has been achieved when both forward and reverse processes continue to occur at equilibrium; because they are occurring at the same rate, the concentration of dissolved AgI remains constant.

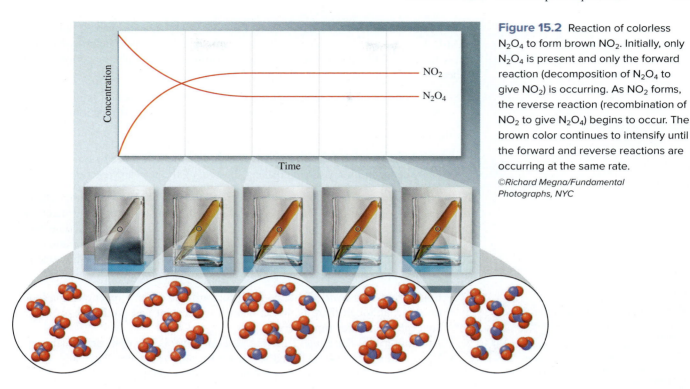

Figure 15.2 Reaction of colorless N_2O_4 to form brown NO_2. Initially, only N_2O_4 is present and only the forward reaction (decomposition of N_2O_4 to give NO_2) is occurring. As NO_2 forms, the reverse reaction (recombination of NO_2 to give N_2O_4) begins to occur. The brown color continues to intensify until the forward and reverse reactions are occurring at the same rate.
©Richard Megna/Fundamental Photographs, NYC

changing. At this point, both the forward and reverse processes are occurring at the same rate, and the system has reached equilibrium.

To understand equilibrium, we must consider the kinetics of the processes involved. In the decomposition of N_2O_4, both the forward and reverse reactions are elementary reactions, so we can write their rate laws from the balanced equation:

$$\text{rate}_{\text{forward}} = k_f[N_2O_4] \qquad \text{and} \qquad \text{rate}_{\text{reverse}} = k_r[NO_2]^2$$

where k_f and k_r are the rate constants for the forward and reverse reactions, respectively [◄◄ Section 14.5]. As always, the square brackets denote *molar concentration*. In the experiment shown in Figure 15.2, initially:

- N_2O_4 concentration is high, and the rate of the forward reaction is high.
- NO_2 concentration is zero, making the rate of the reverse reaction zero.

As the reaction proceeds:

- N_2O_4 concentration falls, decreasing the rate of the forward reaction.
- NO_2 concentration rises, increasing the rate of the reverse reaction.

Figure 15.3(a) shows how the rates of these forward and reverse reactions change over time, and how they eventually become equal as equilibrium is established.

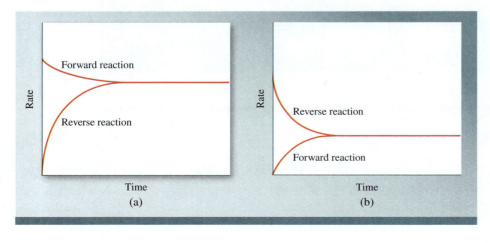

(a)

(b)

Figure 15.3 (a) Starting with just N_2O_4, the rate of the reverse reaction (formation of N_2O_4) is initially zero. The rate of the reverse reaction rises and the rate of the forward reaction falls until both rates are equal. (b) Starting with NO_2, the rate of the forward reaction is initially zero.

We could equally well have started the N_2O_4-NO_2 experiment with pure NO_2 in a flask. Figure 15.4 shows how the brown color, initially intense, fades as NO_2 combines to form N_2O_4. As before, the intensity stops changing after a period of time. When we start the experiment with pure NO_2, initially:

- NO_2 concentration is high, and the rate of the reverse reaction is high.
- N_2O_4 concentration is zero, making the rate of the forward reaction zero.

As the reaction proceeds:

- NO_2 concentration falls, decreasing the rate of the reverse reaction.
- N_2O_4 concentration rises, increasing the rate of the forward reaction.

Figure 15.3(b) shows how the forward and reverse rates change over time when we start with NO_2 instead of N_2O_4. We could also conduct this kind of experiment starting with a mixture of NO_2 and N_2O_4. Again, the forward and reverse reactions would occur, initially at rates determined by the corresponding starting concentrations, and equilibrium would be established when both reactions were occurring at the same rate.

Some important things to remember about equilibrium are:

Student Note: It is a common error to think that equilibrium means equal *concentrations* of reactants and products—it does not. Equilibrium refers to the state in which forward and reverse reactions are occurring at the same rate.

- Equilibrium is a *dynamic* state—both forward and reverse reactions continue to occur, although there is no net change in reactant and product concentrations over time.
- At equilibrium, the rates of the forward and reverse reactions are equal.
- Equilibrium can be established starting with only *reactants,* with only *products,* or with any *mixture* of reactants and products.

For the remainder of this chapter, we limit our discussion to *chemical* equilibria. We explore reversible *physical* processes, such as phase changes, in more detail in Chapter 18.

Figure 15.4 N_2O_4-NO_2 equilibrium starting with NO_2. Initially, only NO_2 is present and only the reverse reaction (recombination of NO_2 to give N_2O_4) is occurring. As N_2O_4 forms, the forward reaction (decomposition of N_2O_4) begins to occur. The brown color continues to fade until the forward and reverse reactions are occurring at the same rate. Note that the experiment in Figure 15.4 is conducted at a lower temperature than the one in Figure 15.2. The decomposition of N_2O_4 is endothermic.

©*Richard Megna/Fundamental Photographs, NYC*

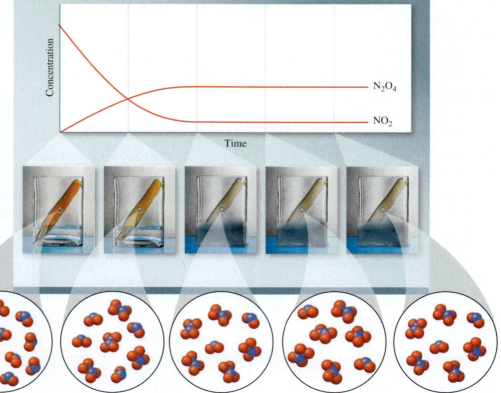

How Do We Know That the Forward and Reverse Processes Are Ongoing in a System at Equilibrium?

Because the concentrations of reactants and products are constant in a system at equilibrium, it may appear that the reaction has simply stopped. In fact, equilibrium is a *dynamic* state in which both forward and reverse processes continue to occur. An experiment that illustrates this involves adding solid silver iodide in which the iodide ion is the radioactive isotope I-131 to a saturated solution of ordinary silver iodide. If the state of equilibrium were the result of the dissolution process having ceased to occur, we would not expect any of the additional solid ($Ag^{131}I$) to dissolve because the solution is already saturated. However, immediately after the addition of solid ($Ag^{131}I$) to a saturated solution of AgI, radioactive iodide ions ($^{131}I^-$) appear in the solution. Moreover, they become distributed throughout the solution and the solid.

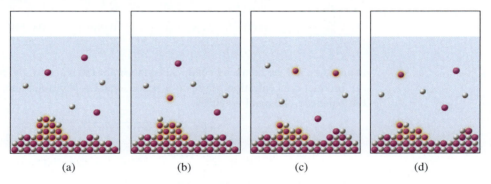

| (a) | (b) | (c) | (d) |

Evidence that equilibrium is dynamic. (a) Solid AgI in which the iodide ion is radioactive isotope I-131 is added to a saturated solution of AgI. (b) $^{131}I^-$ ions immediately appear in the solution. (c) The number of $^{131}I^-$ ions in solution is not constant, although the *total* number of iodide ions in solution *is* constant. (d) $^{131}I^-$ ions become distributed throughout the solution and the solid. If the concentration of dissolved AgI in the original saturated solution were constant because the reaction had stopped occurring, none of the $Ag^{131}I$ would have dissolved.

15.2 The Equilibrium Constant

We have defined equilibrium as the state in which the opposing forward and reverse reactions are occurring at the same rate:

$$rate_{forward} = rate_{reverse}$$

$$k_f[N_2O_4]_{eq} = k_r[NO_2]^2_{eq}$$

where the subscript "eq" denotes a concentration at equilibrium. Rearranging this expression gives

$$\frac{k_f}{k_r} = \frac{[NO_2]^2_{eq}}{[N_2O_4]_{eq}}$$

and because the ratio of two constants (k_f/k_r) is also a constant, we have

equilibrium expression

$$K_c = \frac{[NO_2]^2_{eq}}{[N_2O_4]_{eq}}$$

equilibrium constant

where K_c is the **equilibrium constant** and the equation is known as the **equilibrium expression**. (The subscript "c" stands for *concentration*, referring to the molar concentrations in the expression.) Note the relationship between the equilibrium

Student Note: From this point on, we no longer show the "eq" subscript for concentrations in equilibrium expressions—although you should know that every concentration in an equilibrium expression is the concentration at *equilibrium*.

TABLE 15.1	Initial and Equilibrium Concentrations of N_2O_4 and NO_2 at 25°C				
	Initial Concentrations (M)		**Equilibrium Concentrations (M)**		
Experiment	$[N_2O_4]_i$	$[NO_2]_i$	$[N_2O_4]$	$[NO_2]$	$\dfrac{[NO_2]^2}{[N_2O_4]}$
1	0.670	0.00	0.643	0.0547	4.65×10^{-3}
2	0.446	0.0500	0.448	0.0457	4.66×10^{-3}
3	0.500	0.0300	0.491	0.0475	4.60×10^{-3}
4	0.600	0.0400	0.594	0.0523	4.60×10^{-3}
5	0.000	0.200	0.0898	0.0204	4.63×10^{-3}

expression and the balanced chemical equation for the reaction: the numerator contains the product concentration raised to a power equal to its stoichiometric coefficient in the balanced chemical equation:

$$K_c = \frac{[NO_2]^2_{eq}}{[N_2O_4]_{eq}} \qquad N_2O_4(g) \rightleftharpoons 2NO_2(g)$$

The denominator, moreover, contains the reactant concentration raised to a power equal to its stoichiometric coefficient in the balanced chemical equation. (In this equation the coefficient of N_2O_4 is 1, which generally is not written either as a coefficient or as an exponent.) Table 15.1 lists the starting and equilibrium concentrations of N_2O_4 and NO_2 in a series of experiments carried out at 25°C. Using the equilibrium concentrations from each of the experiments in the table, the value of the expression $[NO_2]^2/[N_2O_4]$ is indeed constant (the average is 4.63×10^{-3}) within the limits of experimental error.

Calculating Equilibrium Constants

In the mid-nineteenth century, Cato Guldberg[1] and Peter Waage[2] studied the equilibrium mixtures of a wide variety of chemical reactions. They observed that at a constant temperature in an equilibrium mixture of reactants and products regardless of the initial concentrations, the *reaction quotient* has a constant value. The **reaction quotient (Q_c)** is a fraction with product concentrations in the numerator and reactant concentrations in the denominator—with each concentration raised to a power equal to the corresponding stoichiometric coefficient in the balanced chemical equation. For the general reaction:

$$aA + bB \rightleftharpoons cC + dD$$

at equilibrium, the reaction quotient Q_c is equal to the equilibrium constant K_c.

Student Note: In our first example, there was just one reactant and one product. When there are multiple species on either side of the equation, the numerator is the *product* of product concentrations and the denominator is the product of *reactant* concentrations—with each concentration raised to the appropriate power, that is, the stoichiometric coefficient.

Equation 15.1	$Q_c = \dfrac{[C]^c[D]^d}{[A]^a[B]^b} = K_c$ (at equilibrium)

This expression is known as the **law of mass action.** Like the equilibrium constant K, Q is subscripted with a "c" to indicate that the quotient is defined in terms of *concentrations*. For the N_2O_4-NO_2 system, this approach gives the same equilibrium

---

1. Cato Maximilian Guldberg (1836–1902). Norwegian chemist and mathematician. Guldberg's research was primarily in the field of thermodynamics.

2. Peter Waage (1833–1900). Norwegian chemist. Like his co-worker, Guldberg, Waage did research in thermodynamics.

expression as we got from our kinetics approach: $[NO_2]_{eq}^2/[N_2O_4]_{eq}$. But the law of mass action was developed *empirically* from countless observations of many different reactions—long before the principles of kinetics were developed. Additionally, it applies not only to elementary reactions, but also to more complex reactions that occur via a series of steps. Furthermore, the law of mass action enables us to write the equilibrium expression for any reaction for which we know the balanced equation. Knowing the equilibrium expression for a reaction, we can use equilibrium concentrations to calculate the value of the equilibrium constant.

Sample Problem 15.1 shows how to use equilibrium concentrations and the law of mass action to calculate the value of an equilibrium constant.

SAMPLE PROBLEM 15.1

Carbonyl chloride ($COCl_2$), also called phosgene, is a highly poisonous gas that was used on the battlefield in World War I. It is produced by the reaction of carbon monoxide with chlorine gas:

$$CO(g) + Cl_2(g) \rightleftharpoons COCl_2(g)$$

In an experiment conducted at 74°C, the equilibrium concentrations of the species involved in the reaction were as follows: $[CO] = 1.2 \times 10^{-2}$ M, $[Cl]_2 = 0.054$ M, and $[COCl_2] = 0.14$ M. (a) Write the equilibrium expression, and (b) determine the value of the equilibrium constant for this reaction at 74°C.

Strategy Use the law of mass action to write the equilibrium expression and plug in the equilibrium concentrations of all three species to evaluate K_c.

Setup The equilibrium expression has the form of concentrations of products over concentrations of reactants, each raised to the appropriate power—in the case of this reaction, all the coefficients are 1, so all the powers will be 1.

Solution

(a) $K_c = \dfrac{[COCl_2]}{[CO][Cl_2]}$ (b) $K_c = \dfrac{(0.14)}{(1.2 \times 10^{-2})(0.054)} = 216$ or 2.2×10^2

K_c for this reaction at 74°C, is 2.2×10^2.

THINK ABOUT IT

When putting the equilibrium concentrations into the equilibrium expression, we leave out the units. It is common practice to express equilibrium constants without units. We examine the reason why in Section 15.5.

Practice Problem 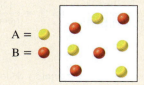**TTEMPT** In an analysis of the following reaction at 100°C,

$$Br_2(g) + Cl_2(g) \rightleftharpoons 2BrCl(g)$$

the equilibrium concentrations were found to be $[Br_2] = 2.3 \times 10^{-3}$ M, $[Cl_2] = 1.2 \times 10^{-2}$ M, and $[BrCl] = 1.4 \times 10^{-2}$ M. Write the equilibrium expression, and calculate the equilibrium constant for this reaction at 100°C.

Practice Problem BUILD In another analysis at 100°C involving the same reaction, the equilibrium concentrations of the reactants were found to be $[Br_2] = 4.1 \times 10^{-3}$ M and $[Cl_2] = 8.3 \times 10^{-3}$ M. Determine the value of $[BrCl]$.

Practice Problem CONCEPTUALIZE Consider the reaction 2A $\rightleftharpoons$ B. The first diagram represents a system at equilibrium where A = ○ and B = ●. Which of the following diagrams [(i)–(iv)] also represents a system at equilibrium? Select all that apply.

A = ○
B = ●

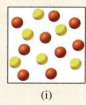
(i)

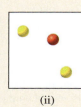
(ii)

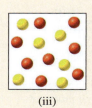
(iii)

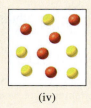
(iv)

Figure 15.5 The value of the reaction quotient, Q, changes as the reaction (experiment 5 from Table 15.1) progresses. When the system reaches equilibrium, Q is equal to the equilibrium constant.

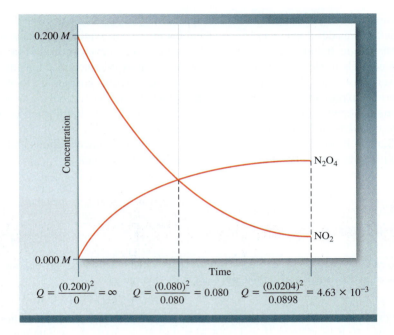

$$Q = \frac{(0.200)^2}{0} = \infty \qquad Q = \frac{(0.080)^2}{0.080} = 0.080 \qquad Q = \frac{(0.0204)^2}{0.0898} = 4.63 \times 10^{-3}$$

Guldberg and Waage's work involved calculating reaction quotients at equilibrium, but the numerical value of the reaction quotient can be determined using starting concentrations or final (equilibrium) concentrations, or concentrations at any time during the reaction (provided that the concentrations are known). The value of Q_c changes as the reaction progresses (Figure 15.5), and Q_c is equal to K_c only when a system is at equilibrium. However, we often find it useful to calculate the value of Q_c for a system that is not at equilibrium. At any point during the progress of a reaction, therefore:

Equation 15.2

$$Q_c = \frac{[C]^c[D]^d}{[A]^a[B]^b}$$

Sample Problem 15.2 lets you practice writing reaction quotients for a variety of balanced equations.

SAMPLE PROBLEM 15.2

Write reaction quotients for the following reactions:

(a) $N_2(g) + 3H_2(g) \rightleftharpoons 2NH_3(g)$

(b) $H_2(g) + I_2(g) \rightleftharpoons 2HI(g)$

(c) $Ag^+(aq) + 2NH_3(aq) \rightleftharpoons Ag(NH_3)_2^+(aq)$

(d) $2O_3(g) \rightleftharpoons 3O_2(g)$

(e) $Cd^{2+}(aq) + 4Br^-(aq) \rightleftharpoons CdBr_4^{2-}(aq)$

(f) $2NO(g) + O_2(g) \rightleftharpoons 2NO_2(g)$

Strategy Use the law of mass action to write reaction quotients.

Setup The reaction quotient for each reaction has the form of the concentrations of *products* over the concentrations of *reactants*, each raised to a power equal to its stoichiometric coefficient in the balanced chemical equation.

Solution

(a) $Q_c = \dfrac{[NH_3]^2}{[N_2][H_2]^3}$ (b) $Q_c = \dfrac{[HI]^2}{[H_2][I_2]}$ (c) $Q_c = \dfrac{[Ag(NH_3)_2^+]}{[Ag^+][NH_3]^2}$ (d) $Q_c = \dfrac{[O_2]^3}{[O_3]^2}$ (e) $Q_c = \dfrac{[CdBr_4^{2-}]}{[Cd^{2+}][Br^-]^4}$ (f) $Q_c = \dfrac{[NO_2]^2}{[NO]^2[O_2]}$

THINK ABOUT IT

With practice, writing reaction quotients becomes second nature. Without sufficient practice, it will seem inordinately difficult. It is important that you become proficient at this. It is very often the first step in solving equilibrium problems.

Practice Problem Ⓐ**TTEMPT** Write the reaction quotient for each of the following reactions:

(a) $2N_2O(g) \rightleftharpoons 2N_2(g) + O_2(g)$

(b) $2NOBr(g) \rightleftharpoons 2NO(g) + Br_2(g)$

(c) $HF(aq) \rightleftharpoons H^+(aq) + F^-(aq)$

(d) $CO(g) + H_2O(g) \rightleftharpoons CO_2(g) + H_2(g)$

(e) $CH_4(g) + 2H_2S(g) \rightleftharpoons CS_2(g) + 4H_2(g)$

(f) $H_2C_2O_4(aq) \rightleftharpoons 2H^+(aq) + C_2O_4^{2-}(aq)$

Practice Problem Ⓑ**UILD** Write the equation for the equilibrium that corresponds to each of the following reaction quotients:

(a) $Q_c = \dfrac{[HCl]^2}{[H_2][Cl_2]}$ (b) $Q_c = \dfrac{[HF]}{[H^+][F^-]}$ (c) $Q_c = \dfrac{[Cr(OH)_4^-]}{[Cr^{3+}][OH^-]^4}$ (d) $Q_c = \dfrac{[H^+][ClO^-]}{[HClO]}$ (e) $Q_c = \dfrac{[H^+][HSO_3^-]}{[H_2SO_3]}$ (f) $Q_c = \dfrac{[NOBr]^2}{[NO]^2[Br_2]}$

Practice Problem Ⓒ**ONCEPTUALIZE**

In principle, in the reaction of A and B to form C, $A(g) + B(g) \rightleftharpoons C(g)$, equilibrium can be established by starting with just a mixture of A and B, starting with just C, or starting with a mixture of A, B, and C. Each of the graphs here depicts the change in the value of the reaction quotient (Q) as equilibrium is established. (The x axis is time.) Identify which graph corresponds to each of the described starting conditions.

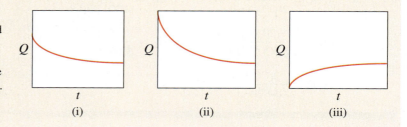

(i)   (ii)   (iii)

Magnitude of the Equilibrium Constant

One of the things the equilibrium constant tells us is the extent to which a reaction proceeds at a particular temperature. If we combine stoichiometric amounts of the reactants in a reaction, three outcomes are possible:

1. The reaction will go essentially to completion, and the equilibrium mixture will consist predominantly of products.
2. The reaction will not occur to any significant degree, and the equilibrium mixture will consist predominantly of reactants.
3. The reaction will proceed to a significant degree but will not go to completion, and the equilibrium mixture will contain comparable amounts of both reactants and products.

When the magnitude of K_c is very large, we expect the first outcome. The formation of the $Ag(NH_3)_2^+$ ion is an example of this possibility:

$$Ag^+(aq) + 2NH_3(aq) \rightleftharpoons Ag(NH_3)_2^+(aq) \qquad K_c = 1.5 \times 10^7 \text{ (at 25°C)}$$

If we were to combine aqueous Ag^+ and aqueous NH_3 in a mole ratio of 1:2, the resulting equilibrium mixture would contain mostly $Ag(NH_3)_2^+$ with only very small amounts of reactants remaining. A reaction with a very large equilibrium constant is sometimes said to "lie to the right" or to "favor products."

When the magnitude of K_c is very small, we expect the second outcome. The chemical combination of nitrogen and oxygen gases to give nitrogen monoxide is an example of this possibility:

$$N_2(g) + O_2(g) \rightleftharpoons 2NO(g) \qquad K_c = 4.3 \times 10^{-25} \text{ (at 25°C)}$$

Nitrogen and oxygen gases do not react to any significant extent at room temperature. If we were to place a mixture of N_2 and O_2 in an evacuated flask and allow the system to reach equilibrium, the resulting mixture would be predominantly N_2 and O_2 with only a *very* small amount of NO. A reaction with a very small equilibrium constant is said to "lie to the left" or to "favor reactants."

The terms *very large* and *very small* are somewhat arbitrary when applied to equilibrium constants. Generally speaking, an equilibrium constant with a magnitude greater than about 1×10^2 can be considered large; one with a magnitude smaller than about 1×10^{-2} can be considered small.

An equilibrium constant that falls between 1×10^2 and 1×10^{-2} indicates that neither products nor reactants are strongly favored. In this case, a system at equilibrium will contain a *mixture* of reactants and products, the exact composition of which will depend on the stoichiometry of the reaction.

CHECKPOINT − SECTION 15.2   The Equilibrium Constant

15.2.1 Select the correct equilibrium expression for the reaction

$$2CO(g) + O_2(g) \rightleftharpoons 2CO_2(g)$$

a) $K_c = \dfrac{2[CO_2]^2}{2[CO]^2[O_2]}$

b) $K_c = \dfrac{(2[CO_2])^2}{(2[CO]^2[O_2])}$

c) $K_c = \dfrac{2[CO_2]}{2[CO][O_2]}$

d) $K_c = \dfrac{2[CO_2]^2}{[2CO]^2[O_2]}$

e) $K_c = \dfrac{[CO_2]^2}{[CO]^2[O_2]}$

15.2.2 Determine the value of the equilibrium constant, K_c, for the reaction

$$A + 2B \rightleftharpoons C + D$$

if the equilibrium concentrations are [A] = 0.0115 M, [B] = 0.0253 M, [C] = 0.109 M, and [D] = 0.0110 M.

a) 163

b) 4.12

c) 2.06

d) 6.14×10^{-3}

e) 0.243

15.3 Equilibrium Expressions

The equilibria we have discussed so far have all been *homogeneous;* that is, the reactants and products have all existed in the same phase—either gaseous or aqueous. In these cases, the equilibrium expression consists of writing the product of the product concentrations at equilibrium over the product of the reactant concentrations at equilibrium, with each concentration raised to a power equal to its stoichiometric coefficient in the balanced chemical equation (Equation 15.2). When the species in a reversible chemical reaction are not all in the same phase, the equilibrium is *heterogeneous.*

Heterogeneous Equilibria

Writing equilibrium expressions for heterogeneous equilibria is also straightforward, but it is slightly different from what we have done so far for homogeneous equilibria. For example, carbon dioxide can combine with elemental carbon to produce carbon monoxide:

$$CO_2(g) + C(s) \rightleftharpoons 2CO(g)$$

The two gases and one solid constitute two separate phases. If we were to write the equilibrium expression for this reaction as we have done previously for homogeneous reactions, including all product concentrations in the numerator and all reactant concentrations in the denominator, we would have:

$$K_c^* = \frac{[CO]^2}{[CO_2][C]}$$

(K_c^* is superscripted with an asterisk to distinguish it from the equilibrium constant that we are about to derive.) The "concentration" of a solid, however, is a constant. If we were to double the number of moles of elemental carbon (C) in the preceding reaction, we would also double its volume. The ratio of moles to volume, which is how we define the concentration, remains the same. Because the concentration of solid carbon is a constant, it is incorporated into the value of the equilibrium constant and does not appear explicitly in the equilibrium expression:

$$K_c^* \times [C] = \frac{[CO]^2}{[CO_2]}$$

The product of K_c^* and $[C]_{eq}$ gives the *real* equilibrium constant for this reaction. The corresponding equilibrium expression is:

$$K_c = \frac{[CO]^2}{[CO_2]}$$

The same argument applies to the concentrations of pure liquids in heterogeneous equilibria. Only gaseous species and aqueous species appear in equilibrium expressions.

Sample Problem 15.3 lets you practice writing equilibrium expressions for heterogeneous equilibria.

SAMPLE PROBLEM   15.3

Write equilibrium expressions for each of the following reactions:

(a) $CaCO_3(s) \rightleftharpoons CaO(s) + CO_2(g)$

(c) $2Fe(s) + 3H_2O(l) \rightleftharpoons Fe_2O_3(s) + 2H_2(g)$

(b) $Hg(l) + Hg^{2+}(aq) \rightleftharpoons Hg_2^{2+}(aq)$

(d) $O_2(g) + 2H_2(g) \rightleftharpoons 2H_2O(l)$

Strategy Use the law of mass action to write the equilibrium expression for each reaction. Only gases and aqueous species appear in the expression.

Setup

(a) Only CO_2 will appear in the expression.

(c) Only H_2 will appear in the expression.

(b) Hg^{2+} and Hg_2^{2+} will appear in the expression.

(d) O_2 and H_2 will appear in the expression.

Solution

(a) $K_c = [CO_2]$

(b) $K_c = \dfrac{[Hg_2^{2+}]}{[Hg^{2+}]}$

(c) $K_c = [H_2]^2$

(d) $K_c = \dfrac{1}{[O_2][H_2]^2}$

THINK ABOUT IT

Like writing equilibrium expressions for homogeneous equilibria, writing equilibrium expressions for heterogeneous equilibria becomes second nature if you practice. The importance of developing this skill now cannot be overstated. Your ability to understand the principles and to solve many of the problems in this chapter, and in Chapters 16 to 19, depends on your ability to write equilibrium expressions *correctly* and *easily*.

Practice Problem **A**TTEMPT Write equilibrium expressions for each of the following reactions:

(a) $SiCl_4(g) + 2H_2(g) \rightleftharpoons Si(s) + 4HCl(g)$

(c) $Ni(s) + 4CO(g) \rightleftharpoons Ni(CO)_4(g)$

(b) $Hg^{2+}(aq) + 2Cl^-(aq) \rightleftharpoons HgCl_2(s)$

(d) $Zn(s) + Fe^{2+}(aq) \rightleftharpoons Zn^{2+}(aq) + Fe(s)$

Practice Problem **B**UILD Which of the following equilibrium expressions corresponds to a heterogeneous equilibrium? How can you tell?

(a) $K_c = [NH_3][HCl]$

(b) $K_c = \dfrac{[H^+][C_2H_3O_2^-]}{[HC_2H_3O_2]}$

(c) $K_c = \dfrac{[Ag(NH_3)_2^+][Cl^-]}{[NH_3]^2}$

(d) $K_c = [Ba^{2+}][F^-]^2$

Practice Problem **C**ONCEPTUALIZE Consider the reaction $A(s) + B(g) \rightleftharpoons C(s)$. Which of the following diagrams could represent a system at equilibrium? Select all that apply. $A = $ ◗, $B = $ ●, and $C = $ ●.

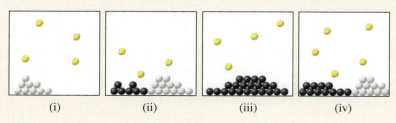

(i)   (ii)   (iii)   (iv)

Manipulating Equilibrium Expressions

In our study of enthalpy, we learned that it is possible to manipulate chemical equations to solve thermochemistry problems [◄◄ Section 5.3]. Recall that when we made a change in a thermochemical equation, we had to make the corresponding change in the ΔH of the reaction. When we reversed a reaction, for example, we had to change the sign of its ΔH. When we change something about how an equilibrium reaction is expressed, we must also make the appropriate changes to the equilibrium expression and the equilibrium constant. Consider the reaction of nitric oxide and oxygen to produce nitrogen dioxide:

$$2NO(g) + O_2(g) \rightleftharpoons 2NO_2(g)$$

The equilibrium expression is:

$$K_c = \frac{[NO_2]^2}{[NO]^2[O_2]}$$

and the equilibrium constant at 500 K is 6.9×10^5. If we were to reverse this equation, writing the equation for the decomposition of NO_2 to produce NO and O_2:

$$2NO_2(g) \rightleftharpoons 2NO(g) + O_2(g)$$

the new equilibrium expression would be the reciprocal of the original equilibrium expression:

$$K_c' = \frac{[NO]^2[O_2]}{[NO_2]^2}$$

Because the equilibrium expression is the reciprocal of the original, the equilibrium constant is also the reciprocal of the original. At 500 K, therefore, the equilibrium constant for the new equation is $1/(6.9 \times 10^5)$ or 1.5×10^{-6}.

Alternatively, if instead of reversing the original equation, we were to multiply it by 2:

$$4NO(g) + 2O_2(g) \rightleftharpoons 4NO_2(g)$$

the new equilibrium expression would be:

$$K_c'' = \frac{[NO_2]^4}{[NO]^4[O_2]^2}$$

which is the original equilibrium expression squared. Again, because the new equilibrium expression is the square of the original, the new equilibrium constant is the square of the original: $K_c = (6.9 \times 10^5)^2$ or 4.8×10^{11}:

Finally, if we were to add the reverse of the original equation:

$$2NO_2(g) \rightleftharpoons 2NO(g) + O_2(g)$$

(for which $K_c = 1.6 \times 10^{-6}$ at 500 K) to the equation:

$$2H_2(g) + O_2(g) \rightleftharpoons 2H_2O(g)$$

(for which $K_c = 2.4 \times 10^{47}$ at 500 K), the sum of the equations and the equilibrium expression (prior to the cancellation of identical terms) would be:

$$2NO_2(g) + 2H_2(g) + O_2(g) \rightleftharpoons 2NO(g) + O_2(g) + 2H_2O(g)$$

and

$$K_c = \frac{[NO]^2[O_2][H_2O]^2}{[NO_2]^2[H_2]^2[O_2]}$$

It may be easier to see before canceling identical terms that for the *sum* of two equations, the equilibrium expression is the *product* of the two corresponding equilibrium expressions:

$$\frac{[NO]^2[O_2][H_2O]^2}{[NO_2]^2[H_2]^2[O_2]} = \frac{[NO]^2[\cancel{O_2}]}{[NO_2]^2} \times \frac{[H_2O]^2}{[H_2]^2[\cancel{O_2}]}$$

TABLE 15.2	Manipulation of Equilibrium Constant Expressions*

$$A(g) + B(g) \rightleftharpoons 2C(g) \quad K_{c_1} = 4.39 \times 10^{-3}$$

$$2C(g) \rightleftharpoons D(g) + E(g) \quad K_{c_2} = 1.15 \times 10^{4}$$

Equation	Equilibrium Expression	Relationship to Original K_c	Equilibrium Constant
$2C(g) \rightleftharpoons A(g) + B(g)$ Original equation is reversed.	$K'_{c_1} = \dfrac{[A][B]}{[C]^2}$	$\dfrac{1}{K_{c_1}}$	2.28×10^2 New constant is the reciprocal of the original.
$2A(g) + 2B(g) \rightleftharpoons 4C(g)$ Original equation is multiplied by a number.	$K''_{c_1} = \dfrac{[C]^4}{[A]^2[B]^2}$	$(K_{c_1})^2$	1.93×10^{-5} New constant is the original raised to the same number.
$\frac{1}{2}A(g) + \frac{1}{2}B(g) \rightleftharpoons C(g)$ Original equation is divided by 2.	$K'''_{c_1} = \dfrac{[C]}{[A]^{1/2}[B]^{1/2}}$	$\sqrt{K_{c_1}}$	6.63×10^{-2} New constant is the square root of the original.
$A(g) + B(g) \rightleftharpoons D(g) + E(g)$ Two equations are added.	$K_{c_3} = \dfrac{[D][E]}{[A][B]}$	$K_{c_1} \times K_{c_2}$	50.5 New constant is the product of the two original constants.

*Temperature is the same for all reactions.

Canceling identical terms on the left and right sides of the equation, and on the top and bottom of the equilibrium expression, gives:

$$2NO_2(g) + 2H_2(g) \rightleftharpoons 2NO(g) + 2H_2O(g) \quad \text{and} \quad K_c = \frac{[NO]^2[H_2O]^2}{[NO_2]^2[H_2]^2}$$

And, because the new equilibrium expression is the product of the individual expressions, the new equilibrium constant is the product of the individual constants. Therefore, for the reaction:

$$2NO_2(g) + 2H_2(g) \rightleftharpoons 2NO(g) + 2H_2O(g)$$

at 500 K, $K_c = (1.6 \times 10^{-6})(2.4 \times 10^{47}) = 3.8 \times 10^{41}$.

Using hypothetical equilibria, Table 15.2 summarizes the various ways that chemical equations can be manipulated and the corresponding changes that must be made to the equilibrium expression and equilibrium constant.

Sample Problem 15.4 shows how to manipulate chemical equations and make the corresponding changes to their equilibrium constants.

SAMPLE PROBLEM 15.4

The following reactions have the indicated equilibrium constants at 100°C:

(1)	$2NOBr(g) \rightleftharpoons 2NO(g) + Br_2(g)$	$K_c = 0.014$
(2)	$Br_2(g) + Cl_2(g) \rightleftharpoons 2BrCl(g)$	$K_c = 7.2$

Determine the value of K_c for the following reactions at 100°C:

(a) $2NO(g) + Br_2(g) \rightleftharpoons 2NOBr(g)$

(b) $4NOBr(g) \rightleftharpoons 4NO(g) + 2Br_2(g)$

(c) $NOBr(g) \rightleftharpoons NO(g) + \frac{1}{2}Br_2(g)$

(d) $2NOBr(g) + Cl_2(g) \rightleftharpoons 2NO(g) + 2BrCl(g)$

(e) $NO(g) + BrCl(g) \rightleftharpoons NOBr(g) + \frac{1}{2}Cl_2(g)$

Strategy Begin by writing the equilibrium expressions for the reactions that are given. Then, determine the relationship of each equation's equilibrium expression to the equilibrium expression of the original equations, and make the corresponding change to the equilibrium constant for each.

(Continued on next page)

Setup The equilibrium expressions for the reactions that are given are:

$$K_c = \frac{[NO]^2[Br_2]}{[NOBr]^2} \quad \text{and} \quad K_c = \frac{[BrCl]^2}{[Br_2][Cl_2]}$$

(a) This equation is the reverse of original equation 1. Its equilibrium expression is the reciprocal of that for the original equation:

$$2NO(g) + Br_2(g) \rightleftharpoons 2NOBr(g) \quad K_c = \frac{[NOBr]^2}{[NO]^2[Br_2]}$$

(b) This is original equation 1 multiplied by a factor of 2. Its equilibrium expression is the original expression squared:

$$4NOBr(g) \rightleftharpoons 4NO(g) + 2Br_2(g) \quad K_c = \left(\frac{[NO]^2[Br_2]}{[NOBr]^2}\right)^2$$

(c) This is original equation 1 multiplied by $\frac{1}{2}$. Its equilibrium expression is the square root of the original:

$$NOBr(g) \rightleftharpoons NO(g) + \tfrac{1}{2}Br_2(g) \quad K_c = \sqrt{\frac{[NO]^2[Br_2]}{[NOBr]^2}} \quad \text{or} \quad K_c = \left(\frac{[NO]^2[Br_2]}{[NOBr]^2}\right)^{1/2}$$

(d) This is the sum of original equations 1 and 2. Its equilibrium expression is the product of the two individual expressions:

$$2NOBr(g) + Cl_2(g) \rightleftharpoons 2NO(g) + 2BrCl(g) \quad K_c = \frac{[NO]^2[BrCl]^2}{[NOBr]^2[Cl_2]}$$

(e) Probably the simplest way to analyze this reaction is to recognize that it is the reverse of the reaction in part (d), multiplied by $\frac{1}{2}$. Its equilibrium expression is the square root of the reciprocal of the expression in part (d):

$$NO(g) + BrCl(g) \rightleftharpoons NOBr(g) + \tfrac{1}{2}Cl_2(g)$$

$$K_c = \sqrt{\frac{[NOBr]^2[Cl_2]}{[NO]^2[BrCl]^2}} \quad \text{or} \quad K_c = \left(\frac{[NOBr]^2[Cl_2]}{[NO]^2[BrCl]^2}\right)^{1/2}$$

Each equilibrium constant will bear the same relationship to the original as the equilibrium expression bears to the original.

Solution

(a) $K_c = 1/0.014 = 71$ (c) $K_c = (0.014)^{1/2} = 0.12$ (e) $K_c = (1/0.10)^{1/2} = 3.2$

(b) $K_c = (0.014)^2 = 2.0 \times 10^{-4}$ (d) $K_c = (0.014)(7.2) = 0.10$

THINK ABOUT IT

The magnitude of an equilibrium constant reveals whether products or reactants are favored, so the reciprocal relationship between K_c values of forward and reverse reactions should make sense. A very large K_c value means that products are favored. In the reaction of hydrogen ion and hydroxide ion to form water, the value of K_c is very large, indicating that the product, water, is favored.

$$H^+(aq) + OH^-(aq) \rightleftharpoons H_2O(l) \qquad K_c = 1.0 \times 10^{14} \text{ (at 25°C)}$$

Simply writing the equation backward doesn't change the fact that water is the predominate species. In the reverse reaction, therefore, the favored species is on the reactant side:

$$H_2O(l) \rightleftharpoons H^+(aq) + OH^-(aq) \qquad K_c = 1.0 \times 10^{-14} \text{ (at 25°C)}$$

As a result, the magnitude of K_c should correspond to reactants being favored; that is, it should be very small.

Practice Problem ⒶTTEMPT The following reactions have the indicated equilibrium constants at a particular temperature:

$$N_2(g) + O_2(g) \rightleftharpoons 2NO(g) \qquad K_c = 4.3 \times 10^{-25}$$
$$2NO(g) + O_2(g) \rightleftharpoons 2NO_2(g) \qquad K_c = 6.4 \times 10^9$$

Determine the values of the equilibrium constants for the following equations at the same temperature: (a) $2NO(g) \rightleftharpoons N_2(g) + O_2(g)$ (b) $\frac{1}{2}N_2(g) + \frac{1}{2}O_2(g) \rightleftharpoons NO(g)$ (c) $N_2(g) + 2O_2(g) \rightleftharpoons 2NO_2(g)$

Practice Problem ⒷUILD The equation $A(g) + B(g) \rightleftharpoons C(g) + D(g)$ represents a reaction for which $K_c = 8$ at 25°C. Write an equation for the related reaction with each of the following K_c values at 25°C: (a) 64, (b) 0.125, (c) 2, (d) 0.3536.

Practice Problem ⒸONCEPTUALIZE Consider a chemical reaction represented by the equation $A(g) + B(g) \rightleftharpoons C(g)$, for which the value of K_c is 100. If we multiply the chemical equation by two, $2A(g) + 2B(g) \rightleftharpoons 2C(g)$, the value of K_c becomes 10,000. Does the larger equilibrium constant indicate that more product [C(g)] will form, given the same starting conditions? Explain.

Equilibrium Expressions Containing Only Gases

When an equilibrium expression contains only gases, we can write an alternate form of the expression in which the concentrations of gases are expressed as partial pressures (atm). Thus, for the equilibrium:

$$N_2O_4(g) \rightleftharpoons 2NO_2(g)$$

we can either write the equilibrium expression as:

$$K_c = \frac{[NO_2]^2}{[N_2O_4]}$$

or as:

$$K_P = \frac{(P_{NO_2})^2}{P_{N_2O_4}}$$

where the subscript "P" in K_P stands for *pressure,* and P_{NO_2} and $P_{N_2O_4}$ are the equilibrium partial pressures of NO_2 and N_2O_4, respectively. In general, K_c is not equal to K_P because the partial pressures of reactants and products expressed in atmospheres are not equal to their concentrations expressed in mol/L. However, a simple relationship between K_P and K_c can be derived using the following equilibrium:

$$aA(g) \rightleftharpoons bB(g)$$

where a and b are the stoichiometric coefficients. The equilibrium constant K_c is given by:

$$K_c = \frac{[B]^b}{[A]^a}$$

and the expression for K_P is:

$$K_P = \frac{(P_B)^b}{(P_A)^a}$$

where P_A and P_B are the partial pressures of A and B. Assuming ideal gas behavior:

$$P_A V = n_A RT$$

$$P_A = \frac{n_A RT}{V} = \left(\frac{n_A}{V}\right)RT$$

and

$$P_A = [A]RT$$

where [A] is the molar concentration of A. Likewise:

$$P_B V = n_B RT$$

$$P_B = \frac{n_B RT}{V} = \left(\frac{n_B}{V}\right)RT$$

and

$$P_B = [B]RT$$

Substituting the expressions for P_A and P_B into the expression for K_P gives:

$$K_P = \frac{(P_B)^b}{(P_A)^a}$$

$$= \frac{[B]^b}{[A]^a}(RT)^{b-a}$$

which simplifies to:

$$K_P = K_c(RT)^{\Delta n}$$

where $\Delta n = b - a$. In general:

Δn = moles of gaseous products − moles of gaseous reactants   **Equation 15.3**

Because pressures are usually expressed in atmospheres, the gas constant R is expressed as 0.08206 L · atm/K · mol, and we can write the relationship between K_P and K_c as:

Equation 15.4 $K_P = K_c[(0.08206 \text{ L} \cdot \text{atm/K} \cdot \text{mol}) \times T]^{\Delta n}$

K_P is equal to K_c only in the special case where $\Delta n = 0$, as in the following equilibrium reaction:

$$H_2(g) + Br_2(g) \rightleftharpoons 2HBr(g)$$

In this case, Equation 15.4 can be written as:

$$K_P = K_c[(0.08206 \text{ L} \cdot \text{atm/K} \cdot \text{mol}) \times T]^0$$
$$= K_c$$

Keep in mind that K_P expressions can only be written for reactions in which every species in the equilibrium expression is a gas. (Remember that solids and pure liquids do not appear in the equilibrium expression.)

Sample Problems 15.5 and 15.6 let you practice writing K_P expressions and illustrate the conversion between K_c and K_P.

SAMPLE PROBLEM 15.5

Write K_P expressions for (a) $PCl_3(g) + Cl_2(g) \rightleftharpoons PCl_5(g)$, (b) $O_2(g) + 2H_2(g) \rightleftharpoons 2H_2O(l)$, and (c) $F_2(g) + H_2(g) \rightleftharpoons 2HF(g)$.

Strategy Write equilibrium expressions for each equation, expressing the concentrations of the gases in partial pressures.

Setup (a) All the species in this equation are gases, so they will all appear in the K_P expression. (b) Only the reactants are gases. (c) All species are gases.

Solution (a) $K_P = \dfrac{(P_{PCl_5})}{(P_{PCl_3})(P_{Cl_2})}$ (b) $K_P = \dfrac{1}{(P_{O_2})(P_{H_2})^2}$ (c) $K_P = \dfrac{(P_{HF})^2}{(P_{F_2})(P_{H_2})}$

THINK ABOUT IT

It isn't necessary for every species in the reaction to be a gas—only those species that appear in the equilibrium expression.

Practice Problem ATTEMPT Write K_P expressions for (a) $2CO(g) + O_2(g) \rightleftharpoons 2CO_2(g)$, (b) $CaCO_3(s) \rightleftharpoons CaO(s) + CO_2(g)$, and (c) $N_2(g) + 3H_2(g) \rightleftharpoons 2NH_3(g)$.

Practice Problem BUILD Write the equation for the gaseous equilibrium corresponding to each of the following K_P expressions:

(a) $K_P = \dfrac{(P_{NO_3})^2}{(P_{NO_2})^2(P_{O_2})}$ (b) $K_P = \dfrac{(P_{CO_2})(P_{H_2})^4}{(P_{CH_4})(P_{H_2O})^2}$ (c) $K_P = \dfrac{(P_{HI})^2}{(P_{I_2})(P_{H_2})}$

Practice Problem CONCEPTUALIZE These diagrams represent closed systems at equilibrium in which red and yellow spheres represent reactants and/or products. For which system(s) can a K_P expression be written? For which can a K_c expression be written? In each case, select all that apply.

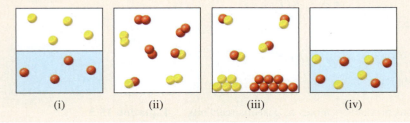

(i)   (ii)   (iii)   (iv)

SAMPLE PROBLEM 15.6

The equilibrium constant, K_c, for the reaction:

$$N_2O_4(g) \rightleftharpoons 2NO_2(g)$$

is 4.63×10^{-3} at 25°C. What is the value of K_P at this temperature?

Strategy Use Equation 15.4 to convert from K_c to K_P. Be sure to convert temperature in degrees Celsius to kelvins.

Setup Using Equation 15.3:

$$\Delta n = 2(NO_2) - 1(N_2O_4) = 1$$

$T = 298$ K.

Solution $K_P = K_c \left(\dfrac{0.08206 \text{ L} \cdot \text{atm}}{\text{K} \cdot \text{mol}} \right) \times T$

$= (4.63 \times 10^{-3})(0.08206 \times 298)$

$= 0.113$

THINK ABOUT IT

Note that we have essentially disregarded the units of R and T so that the resulting equilibrium constant, K_P, is unitless. Equilibrium constants commonly are treated as unitless quantities.

Practice Problem **A**TTEMPT   For the reaction:

$$N_2(g) + 3H_2(g) \rightleftharpoons 2NH_3(g)$$

K_c is 0.31 at 425°C. Calculate K_P for the reaction at this temperature.

Practice Problem **B**UILD $K_P = 2.79 \times 10^{-5}$ for the reaction in Practice Problem A at 472°C. What is K_c for this reaction at 472°C?

Practice Problem **C**ONCEPTUALIZE   Consider the reaction $2A(l) \rightleftharpoons 2B(g)$ at room temperature. Are the values of K_c and K_P numerically equal? Under what conditions might your answer be different?

CHECKPOINT – SECTION 15.3   Equilibrium Expressions

15.3.1 Select the correct equilibrium expression for the reaction:

$$H^+(aq) + OH^-(aq) \rightleftharpoons H_2O(l)$$

a) $K_c = \dfrac{[H_2O]}{[H^+][OH^-]}$

b) $K_c = \dfrac{[H^+][OH^-]}{[H_2O]}$

c) $K_c = [H^+][OH^-]$

d) $K_c = \dfrac{1}{[H^+][OH^-]}$

e) $K_c = [H^+][OH^-][H_2O]$

15.3.2 Select the correct equilibrium expression for the reaction:

$$CaO(s) + CO_2(g) \rightleftharpoons CaCO_3(s)$$

a) $K_c = \dfrac{1}{[CO_2]}$

b) $K_c = \dfrac{[CaCO_3]}{[CaO][CO_2]}$

c) $K_c = [CO_2]$

d) $K_c = \dfrac{[CaO][CO_2]}{[CaCO_3]}$

e) $K_c = \dfrac{[CO_2]}{[CaO][CaCO_3]}$

15.3.3 Given the following information:

$$HF(aq) \rightleftharpoons H^+(aq) + F^-(aq)$$
$$K_c = 6.8 \times 10^{-4} \text{ (at 25°C)}$$
$$H_2C_2O_4(aq) \rightleftharpoons 2H^+(aq) + C_2O_4^{2-}(aq)$$
$$K_c = 3.8 \times 10^{-6} \text{ (at 25°C)}$$

determine the value of K_c for the following reaction at 25°C:

$$C_2O_4^{2-}(aq) + 2HF(aq) \rightleftharpoons 2F^-(aq) + H_2C_2O_4(aq)$$

a) 2.6×10^{-9} c) 1.2×10^{-1} e) 6.8×10^{-4}

b) 1.8×10^{-12} d) 2.6×10^5

15.3.4 K_c for the reaction:

$$Br_2(g) \rightleftharpoons 2Br(g)$$

is 1.1×10^{-3} at 1280°C. Calculate the value of K_P for this reaction at this temperature.

a) 1.1×10^{-3} c) 0.14   e) 8.3×10^{-6}

b) 18   d) 9.1×10^2

15.4 Using Equilibrium Expressions to Solve Problems

We have already used equilibrium expressions to determine the value of an equilibrium constant using equilibrium concentrations. In this section, we explain how to use equilibrium expressions to predict the direction of a reaction and to calculate equilibrium concentrations.

Predicting the Direction of a Reaction

If we start an experiment with only reactants, we know that the reactant concentrations will decrease and the product concentrations will increase; that is, the reaction must proceed in the *forward* direction for equilibrium to be established. Likewise, if we start an experiment with only products, we know that the product concentrations will decrease and the reactant concentrations will increase. In this case, the reaction must proceed in the *reverse* direction to achieve equilibrium. But often we must predict the direction in which a reaction will proceed when we start with a mixture of reactants *and* products. For this situation, we calculate the value of the reaction quotient, Q_c, and compare it to the value of the equilibrium constant, K_c.

The equilibrium constant, K_c, for the gaseous formation of hydrogen iodide from molecular hydrogen and molecular iodine:

$$H_2(g) + I_2(g) \rightleftharpoons 2HI(g)$$

is 54.3 at 430°C. If we were to conduct an experiment starting with a mixture of 0.243 mole of H_2, 0.146 mole of I_2, and 1.98 moles of HI in a 1.00-L container at 430°C, would more HI form or would HI be consumed and more H_2 and I_2 form? Using the starting concentrations, we can calculate the reaction quotient as follows:

$$Q_c = \frac{[HI]_i^2}{[H_2]_i[I_2]_i} = \frac{(1.98)^2}{(0.243)(0.146)} = 111$$

where the subscript "i" indicates *initial* concentration. Because the reaction quotient does not equal K_c ($Q_c = 111$, $K_c = 54.3$), the reaction is not at equilibrium. To establish equilibrium, the reaction will proceed to the left, consuming HI and producing H_2 and I_2, decreasing the value of the numerator and increasing the value in the denominator until the value of the reaction quotient equals that of the equilibrium constant. Thus, the reaction proceeds in the reverse direction (from right to left) to reach equilibrium.

There are three possibilities when we compare Q with K:

$Q < K$ The ratio of initial concentrations of products to reactants is too small. To reach equilibrium, reactants must be converted to products. The system proceeds in the forward direction (from left to right).

$Q = K$ The initial concentrations are equilibrium concentrations. The system is already at equilibrium, and there will be no net reaction in either direction.

$Q > K$ The ratio of initial concentrations of products to reactants is too large. To reach equilibrium, products must be converted to reactants. The system proceeds in the reverse direction (from right to left).

The comparison of Q with K can refer to Q_c and K_c or to Q_P and K_P.

Sample Problem 15.7 shows how the value of Q is used to determine the direction of a reaction that is not at equilibrium.

Student Note: Q_c is calculated using the initial *concentrations of* reactants and products. Similarly, Q_P can be calculated using the initial partial *pressures of* reactants and products.

Student Note: Remember that calculating Q_c is just like calculating K_c: products over reactants, each raised to the appropriate power—except that the concentrations we use to calculate Q_c are the *starting* concentrations. To calculate K_c we must use *equilibrium* concentrations.

SAMPLE PROBLEM   15.7

At 375°C, the equilibrium constant for the reaction

$$N_2(g) + 3H_2(g) \rightleftharpoons 2NH_3(g)$$

is 1.2. At the start of a reaction, the concentrations of N_2, H_2, and NH_3 are 0.071 M, 9.2×10^{-3} M, and 1.83×10^{-4} M, respectively. Determine whether this system is at equilibrium, and if not, determine in which direction it must proceed to establish equilibrium.

Strategy   Use the initial concentrations to calculate Q_c, and then compare Q_c with K_c.

Setup

$$Q_c = \frac{[NH_3]_i^2}{[N_2]_i[H_2]_i^3} = \frac{(1.83 \times 10^{-4})^2}{(0.071)(9.2 \times 10^{-3})^3} = 0.61$$

Solution   The calculated value of Q_c is less than K_c. Therefore, the reaction is not at equilibrium and must proceed to the right to establish equilibrium.

THINK ABOUT IT

In proceeding to the right, a reaction consumes reactants and produces more products. This increases the numerator in the reaction quotient and decreases the denominator. The result is an increase in Q_c until it is equal to K_c, at which point equilibrium will be established.

Practice Problem ATTEMPT   The equilibrium constant, K_c, for the formation of nitrosyl chloride from nitric oxide and chlorine:

$$2NO(g) + Cl_2(g) \rightleftharpoons 2NOCl(g)$$

is 6.5×10^4 at 35°C. In which direction will the reaction proceed to reach equilibrium if the starting concentrations of NO, Cl_2, and NOCl are 1.1×10^{-3} M, 3.5×10^{-4} M, and 1.9 M, respectively?

Practice Problem BUILD   Calculate K_P for the formation of nitrosyl chloride from nitric oxide and chlorine at 35°C, and determine whether the reaction will proceed to the right or the left to achieve equilibrium when the starting pressures are $P_{NO} = 1.01$ atm, $P_{Cl_2} = 0.42$ atm, and $P_{NOCl} = 1.76$ atm.

Practice Problem CONCEPTUALIZE   Consider the reaction 2A $\rightleftharpoons$ B. The first diagram represents a system at equilibrium where A = ● and B = ●. For each of the following diagrams [(i)–(iv)], indicate whether the reaction will proceed to the right, the left, or neither to achieve equilibrium.

A = ●
B = ●

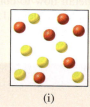
(i)

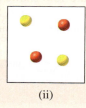
(ii)

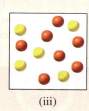
(iii)

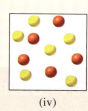
(iv)

Calculating Equilibrium Concentrations

If we know the equilibrium constant for a reaction, we can calculate the concentrations in the equilibrium mixture from the initial reactant concentrations. Consider the following system involving two organic compounds, *cis*- and *trans*-stilbene:

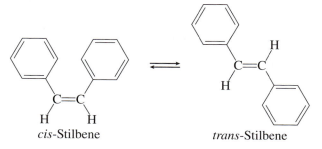

cis-Stilbene   *trans*-Stilbene

give atomic iodine; and the value of x [amount of $I_2(g)$ that reacts] will be very small. In fact, the value of x will be negligible compared to the original concentration of $I_2(g)$. Therefore, $0.00155 - x \approx 0.00155$, and the solution simplifies to:

$$3.39 \times 10^{-12} = \frac{(2x)^2}{0.00155} = \frac{4x^2}{0.00155}$$

$$\frac{3.39 \times 10^{-12}(0.00155)}{4} = 1.31 \times 10^{-15} = x^2$$

$$x = \sqrt{1.31 \times 10^{-15}} = 3.62 \times 10^{-8}\ M$$

According to our ice table, the equilibrium concentration of atomic iodine is $2x$; therefore, $[I(g)] = 2 \times 3.62 \times 10^{-8} = 7.24 \times 10^{-8}\ M$.

THINK ABOUT IT

Having solved for x by this method, we can see that it is indeed insignificant compared to the original concentration of iodine. To an appropriate number of significant figures, $0.00155 - 3.62 \times 10^{-8} = 0.00155$.

Practice Problem Ⓐ**TTEMPT**　Aqueous hydrocyanic acid (HCN) ionizes according to the equation:

$$HCN(aq) \rightleftharpoons H^+(aq) + CN^-(aq)$$

At 25°C, K_c for this reaction is 4.9×10^{-10}. Determine the equilibrium concentrations of all species in a 0.100-M solution of aqueous HCN.

Practice Problem Ⓑ**UILD**　Consider a weak acid, HA, that ionizes according to the equation:

$$HA(aq) \rightleftharpoons H^+(aq) + A^-(aq)$$

At 25°C, a 0.145-M solution of aqueous HA is found to have a hydrogen ion concentration of $2.2 \times 10^{-5}\ M$. Determine the K_c for this reaction at 25°C.

Practice Problem Ⓒ**ONCEPTUALIZE**　Each of the following diagrams shows a system before and after equilibrium is established. Indicate which of the diagrams best represents a system in which you can neglect the x in the solution, as you did in Sample Problem 15.10.

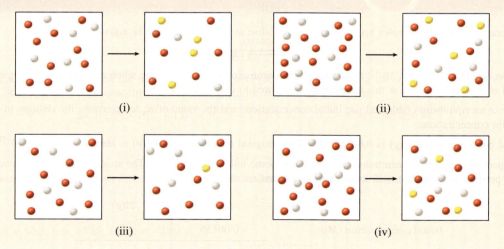

(i)

(ii)

(iii)

(iv)

Here is a summary of the use of initial reactant concentrations to determine equilibrium concentrations:

1. Construct an equilibrium table, and fill in the initial concentrations (including any that are zero).
2. Use initial concentrations to calculate the reaction quotient, Q, and compare Q to K to determine the direction in which the reaction will proceed.
3. Define x as the amount of a particular species consumed, and use the stoichiometry of the reaction to define (in terms of x) the amount of other species consumed or produced.

Student Note: If $Q < K$, the reaction will occur as written. If $Q > K$, the reverse reaction will occur.

4. For each species in the equilibrium, add the change in concentration to the initial concentration to get the equilibrium concentration.
5. Use the equilibrium concentrations and the equilibrium expression to solve for x.
6. Using the calculated value of x, determine the concentrations of all species at equilibrium.
7. Check your work by plugging the calculated equilibrium concentrations into the equilibrium expression. The result should be very close to the K_c stated in the problem.

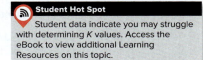

Student Hot Spot

Student data indicate you may struggle with determining K values. Access the eBook to view additional Learning Resources on this topic.

The same procedure applies to K_P.

Sample Problem 15.11 shows how to solve an equilibrium problem using partial pressures.

SAMPLE PROBLEM 15.11

A mixture of 5.75 atm of H_2 and 5.75 atm of I_2 is contained in a 1.0-L vessel at 430°C. The equilibrium constant (K_P) for the reaction:

$$H_2(g) + I_2(g) \rightleftharpoons 2HI(g)$$

at this temperature is 54.3. Determine the equilibrium partial pressures of H_2, I_2, and HI.

Strategy Construct an equilibrium table to determine the equilibrium partial pressures.

Setup The equilibrium table is:

	$H_2(g)$	+	$I_2(g)$	$\rightleftharpoons$	$2HI(g)$
Initial partial pressure (atm):	5.75		5.75		0
Change in partial pressure (atm):	$-x$		$-x$		$+2x$
Equilibrium partial pressure (atm):	$5.75 - x$		$5.75 - x$		$2x$

Solution Setting the equilibrium expression equal to K_P:

$$54.3 = \frac{(2x)^2}{(5.75 - x)^2}$$

Taking the square root of both sides of the equation gives:

$$\sqrt{54.3} = \frac{2x}{5.75 - x}$$

$$7.369 = \frac{2x}{5.75 - x}$$

$$7.369(5.75 - x) = 2x$$

$$42.37 - 7.369x = 2x$$

$$42.37 = 9.369x$$

$$x = 4.52$$

The equilibrium partial pressures are $P_{H_2} = P_{I_2} = 5.75 - 4.52 = 1.23$ atm, and $P_{HI} = 9.04$ atm.

THINK ABOUT IT

Plugging the calculated partial pressures into the equilibrium expression gives:

$$\frac{(P_{HI})^2}{(P_{H_2})(P_{I_2})} = \frac{(9.04)^2}{(1.23)^2} = 54.0$$

The small difference between this result and the equilibrium constant given in the problem statement is due to rounding.

(Continued on next page)

Practice Problem **A**TTEMPT Determine the equilibrium partial pressures of H_2, I_2, and HI if we begin the experiment with 1.75 atm each of H_2 and I_2 at 430°C.

Practice Problem **B**UILD Determine the equilibrium partial pressures of H_2, I_2, and HI (at 430°C) if we begin the experiment with the following conditions: $P_{H_2} = 0.25$ atm, $P_{I_2} = 0.050$ atm, $P_{HI} = 2.5$ atm.

Practice Problem **C**ONCEPTUALIZE Consider the reaction $A(g) + B(g) \rightleftharpoons C(s) + D(s)$. The diagram on the left represents a system at equilibrium. Each of the following diagrams [(i)–(iv)] is missing spheres of a particular color. Indicate how many spheres of the missing color must be included for each diagram to represent a system at equilibrium.

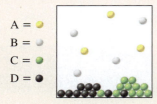

A =
B =
C =
D =

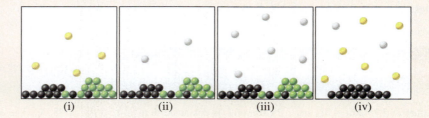

(i)   (ii)   (iii)   (iv)

CHECKPOINT – SECTION 15.4   Using Equilibrium Expressions to Solve Problems

Use the following information to answer questions 15.4.1 and 15.4.2: K_c for the reaction:

$$A + B \rightleftharpoons 2C$$

is 1.7×10^{-2} at 250°C.

15.4.1 What will the equilibrium concentrations of A, B, and C be at this temperature if $[A]_i = [B]_i = 0.750$ M ($[C]_i = 0$)?

a) 6.1×10^{-3} M, 6.1×10^{-3} M, $0.092M$

b) 0.046 M, 0.046 M, 0.092 M

c) 0.70 M, 0.70 M, 0.046 M

d) 0.70 M, 0.70 M, 0.092 M

e) 0.087 M, 0.087 M, 0.66 M

15.4.2 What will the equilibrium concentrations of A, B, and C be at this temperature if $[C]_i = 0.875M$ ($[A]_i = [B]_i = 0$)?

a) 0.41 M, 0.41 M, 0.82 M

b) 0.41 M, 0.41 M, 0.054 M

c) 0.43 M, 0.43 M, 0.0074 M

d) 0.43 M, 0.43 M, 0.44 M

e) 0.43 M, 0.43 M, 0.43 M

15.4.3 If $K_c = 2$ for the reaction $A_2 + B_2 \rightleftharpoons 2AB$ at a certain temperature, which of the following diagrams represents an equilibrium mixture of A_2, B_2, and AB? (Select all that apply.)

● = A   ● = B

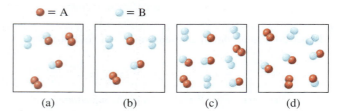

(a)   (b)   (c)   (d)

15.4.4 If $K_c = 3$ for the reaction $X + 2Y \rightleftharpoons Z$ at a certain temperature, then for each of the mixtures of X, Y, and Z shown here, in what direction must the reaction proceed to achieve equilibrium?

X =
Y =
Z =

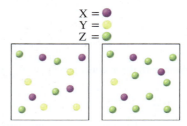

a) Right, right

b) Neither, left

c) Neither, right

d) Left, neither

e) Right, neither

Factors That Affect Chemical Equilibrium

One of the interesting and useful features of chemical equilibria is that they can be manipulated in specific ways to maximize production of a desired substance. Consider, for example, the industrial production of ammonia from its constituent elements by the Haber process:

$$N_2(g) + 3H_2(g) \rightleftharpoons 2NH_3(g)$$

More than 100 million tons of ammonia is produced annually by this reaction, with most of the resulting ammonia being used for fertilizers to enhance crop production. It would be in the best interest of industry to maximize the yield of NH_3. In this section, we explain the various ways in which an equilibrium can be manipulated to accomplish this goal.

Le Châtelier's principle states that when a *stress* is applied to a system at equilibrium, the system will respond by *shifting* in the direction that minimizes the effect of the stress. In this context, "stress" refers to a disturbance of the system at equilibrium by any of the following means:

- The addition of a reactant or product
- The removal of a reactant or product
- A change in volume of the system, resulting in a change in concentration or partial pressure of the reactants and products
- A change in temperature

"Shifting" refers to the occurrence of either the forward or reverse reaction such that the effect of the stress is partially offset as the system reestablishes equilibrium. An equilibrium that shifts to the right is one in which more products are produced by the forward reaction. An equilibrium that shifts to the left is one in which more reactants are produced by the reverse reaction. Using Le Châtelier's principle, we can predict the direction in which an equilibrium will shift, given the specific stress that is applied.

Addition or Removal of a Substance

Again using the Haber process as an example:

$$N_2(g) + 3H_2(g) \rightleftharpoons 2NH_3(g)$$

consider a system at 700 K, in which the equilibrium concentrations are as follows:

$$[N_2] = 2.05\ M \qquad [H_2] = 1.56\ M \qquad [NH_3] = 1.52\ M$$

Using these concentrations in the reaction quotient expression, we can calculate the value of K_c for the reaction at this temperature as follows:

$$Q_c = \frac{[NH_3]^2}{[N_2][H_2]^3} = \frac{(1.52)^2}{(2.05)(1.56)^3} = 0.297 = K_c$$

Student Note: Remember that at equilibrium, the reaction quotient, Q_c, is equal to the equilibrium constant, K_c.

If we were to apply stress to this system by adding more N_2, increasing its concentration from 2.05 M to 3.51 M, the system would no longer be at equilibrium. To see that this is true, use the new concentration of nitrogen in the reaction quotient expression. The new calculated value of Q_c (0.173) is no longer equal to the value of K_c (0.297):

$$Q_c = \frac{[NH_3]^2}{[N_2][H_2]^3} = \frac{(1.52)^2}{(3.51)(1.56)^3} = 0.173 \neq K_c$$

For this system to reestablish equilibrium, the net reaction will have to shift in such a way that Q_c is again equal to K_c, which is constant at a given temperature. Recall from Section 15.4 that when Q is less than K, the reaction proceeds to the right to achieve equilibrium. Likewise, an equilibrium that is stressed in such a way that Q *becomes* less than K will *shift* to the right to reestablish equilibrium. This means that the forward reaction, the consumption of N_2 and H_2 to produce NH_3, will occur. The result will be a net decrease in the concentrations of N_2 and H_2 (thus making the denominator of the reaction quotient smaller), and a net increase in the concentration of NH_3 (thus making the numerator larger). When the concentrations of all species are such that Q_c is again equal to K_c, the system will have established a new *equilibrium position,* meaning that it will have shifted in one direction or the other, resulting in a new equilibrium concentration for each species. Figure 15.7 shows how the concentrations of N_2, H_2, and NH_3 change when N_2 is added to the original equilibrium mixture.

Conversely, if we were to remove N_2 from the original equilibrium mixture, the lower concentration in the denominator of the reaction quotient would result in Q_c being greater than K_c. In this case the reaction will shift to the left. That is, the reverse reaction will take place, thereby increasing the concentrations of N_2 and H_2 and decreasing the concentration of NH_3 until Q_c is once again equal to K_c.

The addition or removal of NH_3 will cause a shift in the equilibrium, too. The addition of NH_3 will cause a shift to the left; the removal of NH_3 will cause a shift to the right. Figure 15.8(a) shows the additions and removals that cause this equilibrium to shift to the right. Figure 15.8(b) shows those that cause it to shift to the left.

In essence, a system at equilibrium will respond to the addition of a species by consuming some of that species, and it will respond to the removal of a species by producing more of that species. It is important to remember that the addition or removal of a species from an equilibrium mixture does not change the value of the equilibrium constant, K. Rather, it changes temporarily the value of the reaction quotient, Q. Furthermore, to cause a shift in the equilibrium, the species added or removed must be one that appears in the reaction quotient expression. In the case of a heterogeneous equilibrium, altering the amount of a solid or liquid species does not change the position of the equilibrium because doing so does not change the value of Q.

Figure 15.7 Adding more of a reactant to a system at equilibrium causes the equilibrium position to shift toward product. The system responds to the addition of N_2 by consuming some of the added N_2 (and some of the other reactant, H_2) to produce more NH_3.

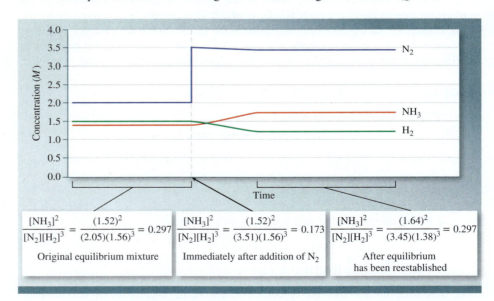

$$\frac{[NH_3]^2}{[N_2][H_2]^3} = \frac{(1.52)^2}{(2.05)(1.56)^3} = 0.297$$

Original equilibrium mixture

$$\frac{[NH_3]^2}{[N_2][H_2]^3} = \frac{(1.52)^2}{(3.51)(1.56)^3} = 0.173$$

Immediately after addition of N_2

$$\frac{[NH_3]^2}{[N_2][H_2]^3} = \frac{(1.64)^2}{(3.45)(1.38)^3} = 0.297$$

After equilibrium has been reestablished

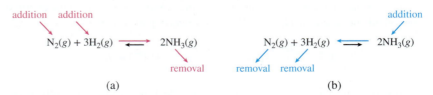

(a)   (b)

Figure 15.8 (a) Addition of a reactant or removal of a product will cause an equilibrium to shift to the right. (b) Addition of a product or removal of a reactant will cause an equilibrium to shift to the left.

Sample Problem 15.12 shows the effects of stress on a system at equilibrium.

SAMPLE PROBLEM (15.12)

Hydrogen sulfide (H_2S) is a contaminant commonly found in natural gas. It is removed by reaction with oxygen to produce elemental sulfur:

$$2H_2S(g) + O_2(g) \rightleftharpoons 2S(s) + 2H_2O(g)$$

For each of the following scenarios, determine whether the equilibrium will shift to the right, shift to the left, or neither: (a) addition of $O_2(g)$, (b) removal of $H_2S(g)$, (c) removal of $H_2O(g)$, and (d) addition of $S(s)$.

Strategy Use Le Châtelier's principle to predict the direction of shift for each case. Remember that the position of the equilibrium is only changed by the addition or removal of a species that appears in the reaction quotient expression.

Setup Begin by writing the reaction quotient expression:

$$Q_c = \frac{[H_2O]^2}{[H_2S]^2[O_2]}$$

Because sulfur is a solid, it does not appear in the expression. Changes in the concentration of any of the other species will cause a change in the equilibrium position. Addition of a reactant or removal of a product that appears in the expression for Q_c will shift the equilibrium to the right:

addition   addition

$$2H_2S(g) \; + \; O_2(g) \; \longrightarrow \; 2S(s) \; + \; 2H_2O(g)$$

removal

Removal of a reactant or addition of a product that appears in the expression for Q_c will shift the equilibrium to the left:

addition

$$N_2(g) \; + \; 3H_2(g) \; \longleftarrow \; 2NH_3(g)$$

removal   removal

Solution

(a) Shift to the right   (c) Shift to the right

(b) Shift to the left   (d) No change

THINK ABOUT IT

In each case, analyze the effect the change will have on the value of Q_c. In part (a), for example, O_2 is added, so its concentration increases. Looking at the reaction quotient expression, we can see that a larger concentration of oxygen corresponds to a larger overall denominator—giving the overall fraction a smaller value. Thus, Q will temporarily be smaller than K and the reaction will have to shift to the right, consuming some of the added O_2 (along with some of the H_2S in the mixture) to reestablish equilibrium.

Practice Problem **A**TTEMPT For each change indicated, determine whether the equilibrium:

$$PCl_3(g) + Cl_2(g) \rightleftharpoons PCl_5(g)$$

will shift to the right, shift to the left, or neither: (a) addition of $PCl_3(g)$, (b) removal of $PCl_3(g)$, (c) removal of $PCl_5(g)$, and (d) removal of $Cl_2(g)$.

Practice Problem **B**UILD What can be added to the equilibrium that will (a) shift it to the left, (b) shift it to the right, (c) not shift it in either direction?

$$AgCl(s) + 2NH_3(aq) \rightleftharpoons Ag(NH_3)_2^+(aq) + Cl^-(aq)$$

(Continued on next page)

Practice Problem 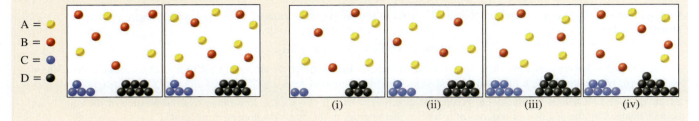 **CONCEPTUALIZE** Consider the reaction A(g) + B(g) $\rightleftharpoons$ C(s) + D(s). The left side of first diagram represents a system at equilibrium where A = 🟡, B = 🔴, C = 🔵, and D = ⚫. The right side of the first diagram represents the system immediately after more A has been added to stress the equilibrium. Which of the following diagrams [(i)–(iv)] could represent the system after equilibrium has been reestablished? Select all that apply.

A = 🟡
B = 🔴
C = 🔵
D = ⚫

(i)   (ii)   (iii)   (iv)

Changes in Volume and Pressure

If we were to start with a gaseous system at equilibrium in a cylinder with a movable piston, we could change the volume of the system, thereby changing the concentrations of the reactants and products.

Consider again the equilibrium between N_2O_4 and NO_2:

$$N_2O_4(g) \rightleftharpoons 2NO_2(g)$$

At 25°C, the equilibrium constant for this reaction is 4.63×10^{-3}. Suppose we have an equilibrium mixture of 0.643 M N_2O_4 and 0.0547 M NO_2 in a cylinder fitted with a movable piston. If we push down on the piston, the equilibrium will be disturbed and will shift in the direction that minimizes the effect of this disturbance. Consider what happens to the concentrations of both species if we decrease the volume of the cylinder by half. Both concentrations are initially *doubled*: [N_2O_4] = 1.286 M and [NO_2] = 0.1094 M. If we plug the new concentrations into the reaction quotient expression, we get:

$$Q_c = \frac{[NO_2]_{eq}^2}{[N_2O_4]_{eq}} = \frac{(0.1094)^2}{1.286} = 9.31 \times 10^{-3}$$

which is not equal to K_c, so the system is no longer at equilibrium. Because Q_c is greater than K_c, the equilibrium will have to shift to the left for equilibrium to be reestablished (Figure 15.9).

In general, a decrease in volume of a reaction vessel will cause a shift in the equilibrium in the direction that minimizes the total number of moles of gas. Conversely, an increase in volume will cause a shift in the direction that maximizes the total number of moles of gas.

Sample Problem 15.13 shows how to predict the equilibrium shift that will be caused by a volume change.

Figure 15.9 The effect of a volume decrease (pressure increase) on the $N_2O_4(g) \rightleftharpoons 2NO_2(g)$ equilibrium. When volume is decreased, the equilibrium is driven toward the side with the smallest number of moles of gas.

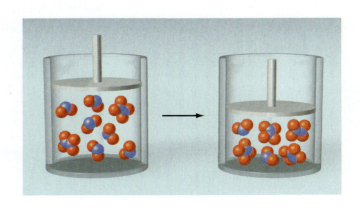

SAMPLE PROBLEM (15.13)

For each reaction, predict in what direction the equilibrium will shift when the volume of the reaction vessel is decreased.

(a) $PCl_5(g) \rightleftharpoons PCl_3(g) + Cl_2(g)$

(b) $2PbS(s) + 3O_2(g) \rightleftharpoons 2PbO(s) + 2SO_2(g)$

(c) $H_2(g) + I_2(g) \rightleftharpoons 2HI(g)$

Strategy Determine which direction minimized the number of moles of gas in the reaction. Count only moles of *gas*.

Setup We have (a) 1 mole of gas on the reactant side and 2 moles of gas on the product side, (b) 3 moles of gas on the reactant side and 2 moles of gas on the product side, and (c) 2 moles of gas on each side.

Solution

(a) Shift to the left   (b) Shift to the right   (c) No shift

---

THINK ABOUT IT

It is a common error to count all the species in a reaction to determine which side has fewer moles. To determine what direction shift a volume change will cause, it is only the number of moles of *gas* that matters. When there is no difference in the number of moles of gas, changing the volume of the reaction vessel will change the concentrations of reactant(s) and product(s)—but the system will remain at equilibrium. (*Q* will remain equal to *K*.)

Practice Problem ATTEMPT For each reaction, predict the direction of shift caused by increasing the volume of the reaction vessel.

(a) $2NOCl(g) \rightleftharpoons 2NO(g) + Cl_2(g)$ (c) $Zn(s) + 2H^+(aq) \rightleftharpoons Zn^{2+}(aq) + H_2(g)$

(b) $CaCO_3(s) \rightleftharpoons CaO(s) + CO_2(g)$

Practice Problem BUILD For the following equilibrium, give an example of a stress that will cause a shift to the right, a stress that will cause a shift to the left, and one that will cause no shift:

$$H_2(g) + F_2(g) \rightleftharpoons 2HF(g)$$

Practice Problem CONCEPTUALIZE Consider the reaction $A(g) + B(g) \rightleftharpoons AB(g)$. The first diagram represents a system at equilibrium. Which of the following diagrams [(i)–(iii)] best represents the system when equilibrium has been reestablished following a volume increase of 50 percent?

A =
B =

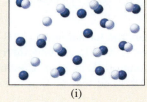
(i)

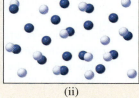
(ii)

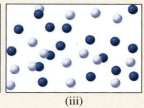
(iii)

It is possible to change the total pressure of a system without changing its volume—by adding an inert gas such as helium to the reaction vessel. Because the total volume remains the same, the concentrations of reactant and product gases do not change. Therefore, the equilibrium is not disturbed and no shift will occur.

Changes in Temperature

A change in concentration or volume may alter the position of an equilibrium (i.e., the relative amounts of reactants and products), but it does not change the value of the equilibrium constant. Only a change in temperature can alter the value of the equilibrium constant. To understand why, consider the following reaction:

$$N_2O_4(g) \rightleftharpoons 2NO_2(g)$$

Figure 15.10 (a) N_2O_4-NO_2 equilibrium. (b) Because the reaction is endothermic, at higher temperature, the $N_2O_4(g) \rightleftharpoons 2NO_2(g)$ equilibrium shifts toward product, making the reaction mixture darker.

Charles D. Winters/Timeframe Photography/ McGraw Hill

(a)   (b)

The forward reaction is endothermic (absorbs heat, $\Delta H > 0$):

$$\text{Heat} + N_2O_4(g) \rightleftharpoons 2NO_2(g) \qquad \Delta H° = 58.0 \text{ kJ/mol}$$

If we treat heat as though it were a reactant, we can use Le Châtelier's principle to predict what will happen if we add or remove heat. Increasing the temperature (adding heat) will shift the reaction in the forward direction because heat appears on the reactant side. Lowering the temperature (removing heat) will shift the reaction in the reverse direction. Consequently, the equilibrium constant, given by:

$$K_c = \frac{[NO_2]^2}{[N_2O_4]}$$

increases when the system is heated and decreases when the system is cooled (Figure 15.10). A similar argument can be made in the case of an exothermic reaction, where heat can be considered to be a product. Increasing the temperature of an exothermic reaction causes the equilibrium constant to decrease, shifting the equilibrium toward reactants.

Another example of this phenomenon is the equilibrium between the following ions:

$$\underset{\text{blue}}{CoCl_4^{2-}} + 6H_2O \rightleftharpoons \underset{\text{pink}}{Co(H_2O)_6^{2+}} + 4Cl^- + \text{heat}$$

The reaction as written, the formation of $Co(H_2O)_6^{2+}$, is exothermic. Thus, the reverse reaction, the formation of $CoCl_4^{2-}$, is endothermic. On heating, the equilibrium shifts to the left and the solution turns blue. Cooling favors the exothermic reaction [the formation of $Co(H_2O)_6^{2+}$] and the solution turns pink (Figure 15.11).

Figure 15.11 (a) An equilibrium mixture of $CoCl_4^{2-}$ ions and $Co(H_2O)_6^{2+}$ ions appears violet. (b) Heating with a Bunsen burner favors the formation of $CoCl_4^{2-}$, making the solution look more blue. (c) Cooling with an ice bath favors the formation of $Co(H_2O)_6^{2+}$, making the solution look more pink.

(all) Charles D. Winters/Timeframe Photography/ McGraw Hill

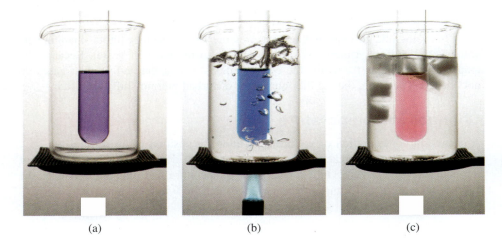

(a)   (b)   (c)

In summary, a temperature increase favors an endothermic reaction, and a temperature decrease favors an exothermic process. Temperature affects the position of an equilibrium by changing the value of the equilibrium constant. Figures 15.12 and 15.13 illustrate the effects of various stresses on systems at equilibrium.

Animation
Le Châtelier's Principle I and II.

SAMPLE PROBLEM 15.14

The Haber process, which is used industrially to generate ammonia—largely for the production of fertilizers—is represented by the equation $N_2(g) + 3H_2(g) \rightleftharpoons 2NH_3(g)$. Using data from Appendix 2, determine the ΔH°_{rxn} for the process and indicate what direction the equilibrium will shift if the temperature is increased. What direction will it shift if the temperature is decreased?

Strategy Determine ΔH°_{rxn} using ΔH°_f values from Appendix 2 and Equation 5.19. Use the sign of ΔH°_{rxn} to determine the direction of shift as temperature changes. For an *endothermic* reaction ($\Delta H^\circ_{rxn} > 0$), equilibrium will shift to the *right* as temperature increases. For an *exothermic* reaction ($\Delta H^\circ_{rxn} < 0$), equilibrium will shift to the *left* as temperature increases.

Setup From Appendix 2, ΔH°_f [$NH_3(g)$] = −46.3 kJ/mol. The reactants are both elements in their standard states. By definition, their ΔH°_f values are zero. Equation 5.19 is:

$$\Delta H^\circ_{rxn} = \Sigma n \Delta H^\circ_f(\text{products}) - \Sigma n \Delta H^\circ_f(\text{reactants})$$

Solution $\Delta H^\circ_{rxn} = 2\Delta H^\circ_f$ [$NH_3(g)$] − {ΔH°_f [$N_2(g)$] + 3ΔH°_f [$H_2(g)$]}

$$= 2(-46.3 \text{ kJ/mol}) - (0 \text{ kJ/mol} + 3 \times 0 \text{ kJ/mol}) = -92.6 \text{ kJ/mol}$$

The reaction is exothermic; therefore, the equilibrium will shift to the left if temperature is increased and to the right if temperature is decreased.

THINK ABOUT IT

Although the Haber process is exothermic, and the equilibrium lies farther to the right at lower temperatures, the industrial process is typically run at about 400°C.

Practice Problem ATTEMPT The reaction of carbon dioxide and calcium hydroxide to produce calcium carbonate and water is represented by the equation $CO_2(g) + Ca(OH)_2(s) \rightleftharpoons CaCO_3(s) + H_2O(l)$. ΔH° for this reaction is −113 kJ/mol. Determine the direction the equilibrium will shift in response to a temperature increase, and a temperature decrease.

Practice Problem BUILD Consider the hypothetical reaction represented by the equation $A(g) + B(g) \rightleftharpoons C(g)$, for which K_c at 100°C = 9.86×10^3 and K_P at 200°C = 1.05×10^3. Determine whether the reaction is endothermic or exothermic.

Practice Problem CONCEPTUALIZE The decomposition of A_2 is represented by the equation $A_2(g) \rightleftharpoons 2A(g)$. Which of the following diagrams best represents the change in equilibrium position as temperature is increased?

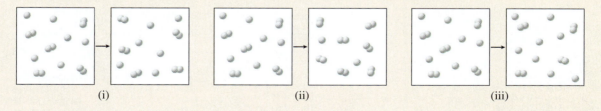

(i)   (ii)   (iii)

Catalysis

A catalyst speeds up a reaction by lowering the reaction's activation energy [◄◄ Section 14.6]. However, a catalyst lowers the activation energy of the forward and reverse reactions to the same extent (see Figure 14.15). The presence of a catalyst, therefore, does not alter the equilibrium constant, nor does it shift the position of an equilibrium system. Adding a catalyst to a reaction mixture that is not at equilibrium will simply cause the mixture to reach equilibrium sooner. The same equilibrium mixture could be obtained without the catalyst, but it might take a much longer time.

Figure 15.12

Effect of Addition to an Equilibrium Mixture

$$H_2(g) + I_2(g) \rightleftharpoons 2HI(g)$$

$[H_2] = 0.112\ M$
$[I_2] = 0.112\ M$
$[HI] = 0.825\ M$

$$Q_c = \frac{[HI]^2}{[H_2][I_2]} = \frac{(0.825)^2}{(0.112)(0.112)} = 54.3$$

$$Q_c = K_c$$

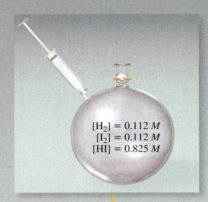

$[H_2] = 0.112\ M$
$[I_2] = 0.112\ M$
$[HI] = 0.825\ M$

$$Q_c = \frac{[HI]^2}{[H_2][I_2]} = \frac{(0.825)^2}{(0.112)(0.112)} = 54.3$$

$$Q_c = K_c$$

$[H_2] = 0.112\ M$
$[I_2] = 0.112\ M$
$[HI] = 0.825\ M$

$$Q_c = \frac{[HI]^2}{[H_2][I_2]} = \frac{(0.825)^2}{(0.112)(0.112)} = 54.3$$

$$Q_c = K_c$$

Add a reactant—$I_2(g)$.

Add a product—$HI(g)$.

Add a species not involved in the equilibrium—$He(g)$.

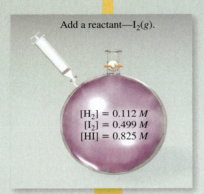

$[H_2] = 0.112\ M$
$[I_2] = 0.499\ M$
$[HI] = 0.825\ M$

$$Q_c = \frac{(0.825)^2}{(0.112)(0.499)} = 12$$

$$Q_c \neq K_c$$

$[H_2] = 0.112\ M$
$[I_2] = 0.112\ M$
$[HI] = 3.17\ M$

$$Q_c = \frac{(3.17)^2}{(0.112)(0.112)} = 801$$

$$Q_c \neq K_c$$

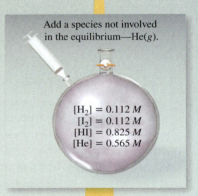

$[H_2] = 0.112\ M$
$[I_2] = 0.112\ M$
$[HI] = 0.825\ M$
$[He] = 0.565\ M$

$$Q_c = \frac{(0.825)^2}{(0.112)(0.112)} = 54.3$$

$$Q_c = K_c$$

Equilibrium shifts toward product.

Equilibrium shifts toward reactant.

Equilibrium does not shift in either direction.

$[H_2] = 0.0404\ M$
$[I_2] = 0.427\ M$
$[HI] = 0.968\ M$

$$Q_c = \frac{(0.968)^2}{(0.0404)(0.427)} = 54.3$$

$$Q_c = K_c$$

$[H_2] = 0.362\ M$
$[I_2] = 0.362\ M$
$[HI] = 2.67\ M$

$$Q_c = \frac{(2.67)^2}{(0.362)(0.362)} = 54.4$$

$$Q_c = K_c$$

$[H_2] = 0.112\ M$
$[I_2] = 0.112\ M$
$[HI] = 0.825\ M$
$[He] = 0.565\ M$

What's the point?

Adding a reactant to an equilibrium mixture shifts the equilibrium toward the product side of the equation. Adding a product shifts the equilibrium toward the reactant side. Adding a species that is neither a reactant nor a product does not cause a shift in the equilibrium.

Effect of Temperature Change

$$N_2O_4(g) \rightleftharpoons 2NO_2(g) \quad \Delta H° = 58.04 \text{ kJ/mol}$$

Temperature decrease drives an endothermic equilibrium toward reactants.

Temperature increase drives an endothermic equilibrium toward products.

$$H_2(g) + I_2(g) \rightleftharpoons 2HI(g) \quad \Delta H° = -9.4 \text{ kJ/mol}$$

Temperature increase drives an exothermic equilibrium toward reactants.

Temperature decrease drives an exothermic equilibrium toward products.

What's the point?

Increasing the temperature of an equilibrium mixture causes a shift toward the product side for an endothermic reaction, and a shift toward the reactant side for an exothermic reaction.

(See Visualizing Chemistry questions VC15.5–VC15.8 on page 776.)

Figure 15.13

Effect of Volume Change

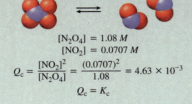

$[N_2O_4] = 1.08\ M$
$[NO_2] = 0.0707\ M$

$$Q_c = \frac{[NO_2]^2}{[N_2O_4]} = \frac{(0.0707)^2}{1.08} = 4.63 \times 10^{-3}$$

$$Q_c = K_c$$

$[N_2O_4] = 0.540\ M$
$[NO_2] = 0.0354\ M$

$$Q_c = \frac{(0.0354)^2}{0.540} = 2.3 \times 10^{-3}$$

$$Q_c \neq K_c$$

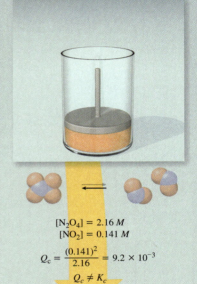

$[N_2O_4] = 2.16\ M$
$[NO_2] = 0.141\ M$

$$Q_c = \frac{(0.141)^2}{2.16} = 9.2 \times 10^{-3}$$

$$Q_c \neq K_c$$

What's the point?

Increasing the volume causes a shift toward the side with the *larger* number of moles of gas. Decreasing the volume of an equilibrium mixture causes a shift toward the side of the equation with the *smaller* number of moles of gas.

$[N_2O_4] = 0.533\ M$
$[NO_2] = 0.0497\ M$

$$Q_c = \frac{(0.0497)^2}{0.533} = 4.6 \times 10^{-3}$$

$$Q_c = K_c$$

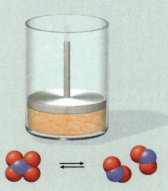

$[N_2O_4] = 2.18\ M$
$[NO_2] = 0.100\ M$

$$Q_c = \frac{(0.100)^2}{2.18} = 4.6 \times 10^{-3}$$

$$Q_c = K_c$$

$[H_2] = 0.112\ M$
$[I_2] = 0.112\ M$
$[HI] = 0.825\ M$

$$Q_c = \frac{[HI]^2}{[H_2][I_2]} = \frac{(0.825)^2}{(0.112)(0.112)} = 54.3$$

$$Q_c = K_c$$

$[H_2] = 0.056\ M$
$[I_2] = 0.056\ M$
$[HI] = 0.413\ M$

$$Q_c = \frac{(0.413)^2}{(0.056)(0.056)} = 54.3$$

$$Q_c = K_c$$

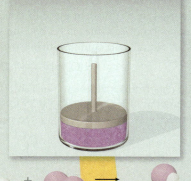

$[H_2] = 0.224\ M$
$[I_2] = 0.224\ M$
$[HI] = 1.65\ M$

$$Q_c = \frac{(1.65)^2}{(0.224)(0.224)} = 54.3$$

$$Q_c = K_c$$

What's the point?

For an equilibrium with equal numbers of gaseous moles on both sides, a change in volume does not cause the equilibrium to shift in either direction.

$[H_2] = 0.056\ M$
$[I_2] = 0.056\ M$
$[HI] = 0.413\ M$

$$Q_c = \frac{(0.413)^2}{(0.056)(0.056)} = 54.3$$

$$Q_c = K_c$$

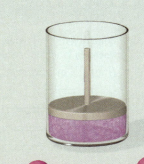

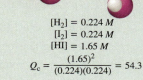

$[H_2] = 0.224\ M$
$[I_2] = 0.224\ M$
$[HI] = 1.65\ M$

$$Q_c = \frac{(1.65)^2}{(0.224)(0.224)} = 54.3$$

$$Q_c = K_c$$

(See Visualizing Chemistry questions VC15.9–VC15.12 on page 776.)

What Happens to the Units in Equilibrium Constants?

We have said that it is common to disregard units and write equilibrium constants as dimensionless quantities. The reason is as follows: Prior to being used in an equilibrium expression, each molar concentration is divided by a reference concentration of 1 M, and each partial pressure is divided by a reference pressure of 1 atm. The reference concentration (1 M) and reference pressure (1 atm) are known as the *standard states* for aqueous and gaseous species, respectively. This cancels the units without changing the value of the concentration or pressure in question. The dimensionless result of dividing a concentration by a reference value is called the *activity*. Activities commonly are used by chemists in place of concentrations, but a detailed discussion of activities is beyond the scope of this book. Using as an example the equilibrium between N_2O_4 and NO_2:

$$N_2O_4(g) \rightleftharpoons 2NO_2(g)$$

and data from experiment 3 in Table 15.1, the equilibrium concentrations of N_2O_4 and NO_2 (at 25°C) are 0.491 M and 0.0475 M, respectively. We can calculate the value of K_c as follows:

$$K_c = \frac{\left(\dfrac{0.0475\ M}{1\ M}\right)^2}{\dfrac{0.491\ M}{1\ M}} = 4.60 \times 10^{-3}$$

We can then convert the molar concentrations to partial pressures using the ideal gas equation:

$$P_{N_2O_4} = \frac{n_{N_2O_4}}{V} RT = [N_2O_4](0.08206)(298) = 12.0\ \text{atm}$$

$$P_{NO_2} = \frac{n_{NO_2}}{V} RT = [NO_2](0.08206)(298) = 1.16\ \text{atm}$$

Finally, we can calculate the value of K_P as:

$$K_P = \frac{\left(\dfrac{1.16\ \text{atm}}{1\ \text{atm}}\right)^2}{\dfrac{12.0\ \text{atm}}{1\ \text{atm}}} = 0.112$$

In each case, dividing by the reference value eliminates the units without changing the numerical value of the result. We do not show the division by reference values explicitly, but you should realize that this is the reason why there are no units associated with equilibrium constants.

Division of each entry in an equilibrium expression by a reference value actually enables us to use molar concentrations and partial pressures in the same equilibrium expression. For example, we can write the equilibrium expression for the reaction:

$$Zn(s) + 2H^+(aq) \rightleftharpoons Zn^{2+}(aq) + H_2(g)$$

as:

$$K = \frac{\left(\dfrac{[Zn^{2+}]}{1\ M}\right)\left(\dfrac{P_{H_2}}{1\ \text{atm}}\right)}{\left(\dfrac{[H^+]}{1\ M}\right)^2}$$

Each of the entries in the expression is divided by the appropriate reference value, making the result dimensionless. Note that this equilibrium constant is neither a K_c nor a K_P. It is referred to simply as K or as K_{eq}, where the subscripted *eq* denotes *equilibrium*. "Mixed" equilibrium constants such as this are used in Chapters 19 and 20.

CHECKPOINT – SECTION 15.5   Factors That Affect Chemical Equilibrium

15.5.1 Which of the following equilibria will shift to the right when H_2 is added? (Select all that apply.)

a) $2H_2 + O_2 \rightleftharpoons 2H_2O$

b) $2HI \rightleftharpoons H_2 + I_2$

c) $H_2 + CO_2 \rightleftharpoons H_2O + CO$

d) $2NaHCO_3 \rightleftharpoons Na_2CO_3 + H_2O + CO_2$

e) $2CO + O_2 \rightleftharpoons 2CO_2$

15.5.2 Which of the following will cause the equilibrium:

$$C(\text{graphite}) + CO_2(g) \rightleftharpoons 2CO(g)$$

to shift to the right? (Select all that apply.)

a) Decreasing the volume

b) Increasing the volume

c) Adding more C(graphite)

d) Adding more $CO_2(g)$

e) Removing CO(g) as it forms

15.5.3 Which of the following equilibria will shift to the left when the temperature is increased? [ΔH (kJ/mol) values are given in parentheses.] (Select all that apply.)

a) $S + H_2 \rightleftharpoons H_2S$ $\Delta H\ (-20)$

b) $C + H_2O \rightleftharpoons CO + H_2$ $\Delta H\ (131)$

c) $H_2 + CO_2 \rightleftharpoons H_2O + CO$ $\Delta H\ (41)$

d) $MgO + CO_2 \rightleftharpoons MgCO_3$ $\Delta H\ (-117)$

e) $2CO + O_2 \rightleftharpoons 2CO_2$ $\Delta H\ (-566)$

15.5.4 For which of the following reactions will a change in volume *not* affect the position of the equilibrium? (Select all that apply.)

a) $MgO(s) + CO_2(g) \rightleftharpoons MgCO_3(s)$

b) $H_2(g) + Cl_2(g) \rightleftharpoons 2HCl(g)$

c) $BaCO_3(s) \rightleftharpoons BaO(s) + CO_2(g)$

d) $Br_2(l) + H_2(g) \rightleftharpoons 2HBr(g)$

e) $C(\text{graphite}) + CO_2(g) \rightleftharpoons 2CO(g)$

15.5.5 The diagram shows the gaseous reaction 2A $\rightleftharpoons$ A$_2$ at equilibrium. How will the numbers of A and A$_2$ change if the volume of the container is increased at constant temperature?

a) A$_2$ will increase and A will decrease.

b) A will increase and A$_2$ will decrease.

c) A$_2$ and A will both decrease.

d) A$_2$ and A will both increase.

e) Neither A$_2$ nor A will change.

15.5.6 The diagrams show equilibrium mixtures of A$_2$, B$_2$, and AB at two different temperatures ($T_2 > T_1$). Is the reaction A$_2$ + B$_2$ $\rightleftharpoons$ 2AB endothermic or exothermic?

a) Endothermic

b) Exothermic

c) Neither

d) There is not enough information to determine.

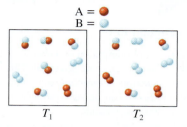

Bringing Chemistry to Life

Hemoglobin Production at High Altitude

As we learned at the end of Chapter 10, rapid ascent to a high altitude can cause altitude sickness. The symptoms of altitude sickness, including dizziness, headache, and nausea, are caused by hypoxia, an insufficient oxygen supply to body tissues. In severe cases, without prompt treatment, a victim may slip into a coma and die. And yet a person staying at a high altitude for weeks or months can recover gradually from altitude sickness, adjust to the low oxygen content in the atmosphere, and live and function normally.

The combination of oxygen with the hemoglobin (Hb) molecule, which carries oxygen through the blood, is a complex reaction, but for our purposes it can be represented by the following simplified equation:*

$$\text{Hb}(aq) + \text{O}_2(aq) \rightleftharpoons \text{HbO}_2(aq)$$

where HbO$_2$ is oxyhemoglobin, the hemoglobin-oxygen complex that actually transports oxygen to tissues. The equilibrium expression for this process is:

$$K_c = \frac{[\text{HbO}_2]}{[\text{Hb}][\text{O}_2]}$$

At an altitude of 3 km, the partial pressure of oxygen is only about 0.14 atm, compared with 0.20 atm at sea level. According to Le Châtelier's principle, a decrease in oxygen concentration will shift the hemoglobin-oxyhemoglobin equilibrium from right to left. This change depletes the supply of oxyhemoglobin, causing hypoxia. Over time, the body copes with this problem by producing more hemoglobin molecules. As the concentration of Hb increases, the equilibrium gradually shifts back toward the right (toward the formation of oxyhemoglobin). It can take several weeks for the increase in hemoglobin production to meet the body's oxygen needs adequately. A return to full capacity may require several years to occur. Studies show that long-time residents of high-altitude areas have high hemoglobin levels in their blood—sometimes as much as 50 percent more than individuals living at sea level. The production of more hemoglobin and the resulting increased capacity of the blood to deliver oxygen to the body have made high-altitude training and hypoxic tents popular among some athletes.

*Biological processes such as this are not true equilibria, but rather they are *steady-state* situations. In a steady state, the constant concentrations of reactants and products are not the result of forward and reverse reactions occurring at the same rate. Instead, reactant concentration is replenished by a previous reaction and product concentration is maintained by a subsequent reaction. Nevertheless, many of the principles of equilibrium, including Le Châtelier's principle, still apply.

Chapter Summary

Section 15.1

- By our definition of a chemical reaction, reactants are consumed and products are produced. This is known as the *forward* reaction. A *reversible process* is one in which the products can also be consumed to produce reactants, a process known as the *reverse* reaction.

- *Equilibrium* is the condition where the forward and reverse reactions are occurring at the same rate and there is no net change in the reactant and product concentrations over time.

- In theory, equilibrium can be established starting with just reactants, with just products, or with a combination of reactants and products.

Section 15.2

- The *reaction quotient* (Q_c) is the product of the *product concentrations* over the product of the *reactant concentrations,* with each concentration raised to a power equal to the corresponding stoichiometric coefficient in the balanced chemical equation. This is known as the *law of mass action.*

- At equilibrium, the reaction quotient is equal to a constant value, the *equilibrium constant* (K_c).

- At equilibrium, concentrations in the reaction quotient are *equilibrium* concentrations and the quotient is called the *equilibrium expression.*

- K_c is constant at constant temperature. The value of K_c can be calculated by plugging equilibrium concentrations into the equilibrium expression.

- A large equilibrium constant ($K_c > 10^3$) indicates that products are favored at equilibrium. A small equilibrium constant ($K_c < 10^{-3}$) indicates that reactants are favored at equilibrium.

Section 15.3

- Solids and liquids do not appear in the equilibrium expression for a heterogeneous reaction.

- When chemical equations that represent equilibria are reversed, multiplied, combined with other equations, or any combination of these processes, the corresponding changes must be made to the equilibrium constants.

- Equilibrium expressions that contain only gases can be written either as K_c expressions or as K_P expressions. K_P expressions have the same form as K_c expressions but contain partial pressures rather than molar concentrations. The reaction quotient, Q, can also be expressed in terms of the pressures of products and reactants. In this case it is labeled Q_P.

- K_c and K_P are not usually equal. The values are the same *only* when the reaction results in no net change in the number of moles of gas.

Section 15.4

- We can predict in which direction a reaction must proceed to achieve equilibrium by comparing the values of Q_c and K_c (or of Q_P and K_P).

- Starting concentrations can be used, along with the equilibrium expression and equilibrium constant, to determine equilibrium concentrations.

Section 15.5

- According to *Le Châtelier's principle,* a system at equilibrium will react to stress by shifting in the direction that will partially offset the effect of the stress.

- The stresses that can be applied to a system at equilibrium include the addition or removal of a substance, changes in the volume of the reaction vessel, and changes in the temperature.

- The addition of a catalyst does not shift the equilibrium in either direction.

Key Words

Key Equations

15.1 $Q_c = \dfrac{[C]^c[D]^d}{[A]^a[B]^b} = K_c$ (at equilibrium)	The reaction quotient, Q, is equal to the product of product concentrations (each raised to the appropriate power) divided by the product of reactant concentrations (each raised to the appropriate power). For each chemical species in the reaction quotient, the "appropriate" power is the coefficient of that species in the balanced chemical equation. When the concentrations used to calculate Q are equilibrium concentrations, $Q = K$. Q and K refer either to Q_c and K_c, or to Q_P and K_P.
15.2 $Q_c = \dfrac{[C]^c[D]^d}{[A]^a[B]^b}$	Reaction quotient, Q, can be calculated using starting concentrations or pressures. In cases where both reactants and products are present initially, we can determine which direction the reaction will proceed to reach equilibrium by comparing the value of Q with the value of K. If Q is less than K, the reaction will proceed to the right. If Q is greater than K, the reaction will proceed to the left. (If Q is equal to K, the system is already at equilibrium.)
15.3 Δn = moles of gaseous products − moles of gaseous reactants	For a reaction in which one or more species is a gas, Δn is the number of gaseous moles on the product side of the equation minus the number of gaseous moles on the reactant side: moles gaseous products − moles gaseous reactants.
15.4 $K_P = K_c[(0.08206\ \text{L} \cdot \text{atm/K} \cdot \text{mol}) \times T]^{\Delta n}$	For a reaction in which all of the species in the equilibrium expression are gases, we can write the expression in terms of pressures (K_P), or we can write the expression in terms of concentrations (K_c). Equation 15.4 is used to convert between the two different equilibrium constants. When there is *no* change in the number of gaseous moles (i.e., when $\Delta n = 0$), $K_P = K_c$.

Questions and Problems

Applying What You've Learned

At present, the World Anti-Doping Agency has no way to detect the use of hypoxic sleeping tents, other than inspection of an athlete's home. Imagine, however, that a biochemical analysis company is developing a way to determine whether an elevated red blood cell count in an athlete's blood is the result of such a practice. A key substance in the detection process is the proprietary compound OD17X, which is produced by the combination of two other proprietary compounds, OD1A and OF2A. The aqueous reaction is represented by the equation:

$$\text{OD1A}(aq) + \text{OF2A}(aq) \rightleftharpoons \text{OD17X}(aq) \qquad \Delta H = 29 \text{ kJ/mol}$$

Problems:

(a) Write the equilibrium expression for the given reaction [◄◄ Sample Problem 15.2]. (b) Given the equilibrium concentrations at room temperature of [OD1A] = 2.12 M, [OF2A] = 1.56 M, and [OD17X] = 1.01 × 10^{-4} M, calculate the value of the equilibrium constant (K_c) at room temperature [◄◄ Sample Problem 15.1]. (c) The OD17X produced is precipitated with another proprietary substance according to the equation:

$$\text{OD17X}(aq) + \text{A771A}(aq) \rightleftharpoons \text{OD17X-A77}(s)$$

K_c for the precipitation equilibrium is 1.00 × 10^6 at room temperature. Write the equilibrium expression for the sum of the two reactions, and determine the value of the overall equilibrium constant [◄◄ Sample Problem 15.4]. (d) Determine whether a mixture with [OD1A] = 3.00 M, [OF2A] = 2.50 M, and [OD17X] = 2.7 × 10^{-4} M is at equilibrium and, if not, in which direction it will have to proceed to achieve equilibrium [◄◄ Sample Problem 15.7]. (e) Because the production of OD17X is endothermic, it can be enhanced by increasing the temperature. At 250°C, the equilibrium constant for the reaction:

$$\text{OD1A}(aq) + \text{OF2A}(aq) \rightleftharpoons \text{OD17X}(aq)$$

is 3.8 × 10^2. If a synthesis at 250°C begins with 1.00 M of each reactant, what will be the equilibrium concentrations of reactants and products [◄◄ Sample Problem 15.8]?

SECTION 15.1: THE CONCEPT OF EQUILIBRIUM

Review Questions

15.1 Define *equilibrium*. Give two examples of a dynamic equilibrium.

15.2 Which of the following statements is correct about a reacting system at equilibrium: (a) the concentrations of reactants are equal to the concentrations of products, (b) the rate of the forward reaction is equal to the rate of the reverse reaction.

15.3 Consider the reversible reaction A $\rightleftharpoons$ B. Explain how equilibrium can be reached by starting with only A, only B, or a mixture of A and B.

SECTION 15.2: THE EQUILIBRIUM CONSTANT

Review Questions

15.4 What is the law of mass action?

15.5 Briefly describe the importance of equilibrium in the study of chemical reactions.

15.6 Define *reaction quotient*. How does it differ from the equilibrium constant?

15.7 Write reaction quotients for the following reactions:
(a) $2\text{NO}(g) + \text{O}_2(g) \rightleftharpoons \text{N}_2\text{O}_4(g)$
(b) $\text{S}(s) + 3\text{F}_2(g) \rightleftharpoons \text{SF}_6(g)$
(c) $\text{Co}^{3+}(aq) + 6\text{NH}_3(aq) \rightleftharpoons \text{Co(NH}_3)_6^{3+}(aq)$
(d) $\text{HCOOH}(aq) \rightleftharpoons \text{HCOO}^-(aq) + \text{H}^+(aq)$

15.8 Write the equation for the reaction that corresponds to each of the following reaction quotients:

(a) $Q_c = \dfrac{[\text{H}_2]^2[\text{S}_2]}{[\text{H}_2\text{S}]^2}$

(c) $Q_c = \dfrac{[\text{HgI}_4^{2-}]}{[\text{Hg}^{2+}][\text{I}^-]^4}$

(b) $Q_c = \dfrac{[\text{NO}_2]^2[\text{Cl}_2]}{[\text{NClO}_2]^2}$

(d) $Q_c = \dfrac{[\text{NO}]^2[\text{Br}_2]}{[\text{NOBr}]^2}$

Computational Problems

15.9 Consider the reaction:

$$2\text{NO}(g) + 2\text{H}_2(g) \rightleftharpoons \text{N}_2(g) + 2\text{H}_2\text{O}(g)$$

At a certain temperature, the equilibrium concentrations are [NO] = 0.71 M, [H$_2$] = 0.10 M, [N$_2$] = 0.15 M, and [H$_2$O] = 2.13 M. (a) Write the equilibrium expression for the reaction. (b) Determine the value of the equilibrium constant.

15.10 The equilibrium constant for the reaction:

$$2SO_2(g) + O_2(g) \rightleftharpoons 2SO_3(g)$$

is 2.8×10^2 at a certain temperature. If $[SO_2] = 0.0124\ M$ and $[O_2] = 0.031\ M$, what is $[SO_3]$?

15.11 Consider the following equilibrium process at 700°C:

$$2H_2(g) + S_2(g) \rightleftharpoons 2H_2S(g)$$

Analysis shows that there are 2.50 moles of H_2, 1.35×10^{-5} mole of S_2, and 8.70 moles of H_2S present in a 12.0-L flask. Calculate the equilibrium constant K_c for the reaction.

15.12 The equilibrium constant for the reaction:

$$2H_2(g) + CO(g) \rightleftharpoons CH_3OH(g)$$

is 1.6×10^{-2} at a certain temperature. If there are 1.17×10^{-2} mole of H_2 and 3.46×10^{-3} mole of CH_3OH at equilibrium in a 5.60-L flask, what is the concentration of CO?

Conceptual Problems

15.13 The first diagram represents a system at equilibrium where A = 🟡 and B = 🔴. How many red spheres must be added to the second diagram for it also to represent a system at equilibrium if the corresponding chemical equation is (a) $2A \rightleftharpoons 2B$ (b) $A \rightleftharpoons B$ (c) $2A \rightleftharpoons B$

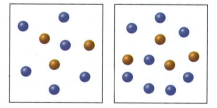

15.14 These two diagrams represent two different instances of a system at equilibrium where X = 🔵 and Z = 🟤. Indicate which chemical equation(s) could represent the corresponding chemical reaction correctly: (a) $X \rightleftharpoons Z$ (b) $2X \rightleftharpoons Z$ (c) $X \rightleftharpoons 2Z$ (d) $2X \rightleftharpoons 2Z$

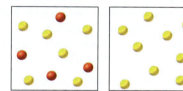

 Visualizing Chemistry
Figure 15.6

VC 15.1 For an equilibrium based on the reaction $A + B \rightleftharpoons C$, where the initial concentrations of A, B, and C are known, what is the correct entry for the highlighted cell in the ice table shown here?

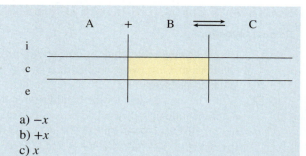

a) $-x$
b) $+x$
c) x

VC 15.2 For an equilibrium based on the reaction $A + B \rightleftharpoons 2C$, where the initial concentrations of A and B are known and the initial concentration of C is zero, what is the correct entry for the highlighted cell in the ice table shown here?

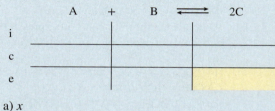

a) x
b) $+2x$
c) $2x$

VC 15.3 For an equilibrium based on the reaction $2A \rightleftharpoons B + C$, where the initial concentration of A is known and the initial concentrations of B and C are zero, what is the correct entry for the highlighted cell in the ice table shown here?

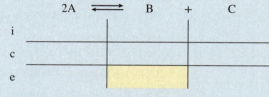

a) $2x$
b) $+x$
c) x

VC 15.4 Of the reactions in questions 15.1–15.3, which has the largest K_c value if this diagram represents the system at equilibrium? A = 🔴, B = 🟡, and C = ⚪.

a) $A + B \rightleftharpoons 2C$
b) $A + B \rightleftharpoons C$
c) $2A \rightleftharpoons B + C$

SECTION 15.3: EQUILIBRIUM EXPRESSIONS

Review Questions

15.15   Define *homogeneous* equilibrium and *heterogeneous* equilibrium. Give two examples of each.

15.16   What do the symbols K_c and K_P represent?

15.17   Write the expressions for the equilibrium constants K_P of the following thermal decomposition reactions:
(a) $2NaHCO_3(s) \rightleftharpoons Na_2CO_3(s) + CO_2(g) + H_2O(g)$
(b) $2CaSO_4(s) \rightleftharpoons 2CaO(s) + 2SO_2(g) + O_2(g)$

15.18   Write equilibrium constant expressions for K_c, and for K_P, if applicable, for the following processes:
(a) $2CO_2(g) \rightleftharpoons 2CO(g) + O_2(g)$
(b) $3O_2(g) \rightleftharpoons 2O_3(g)$
(c) $CO(g) + Cl_2(g) \rightleftharpoons COCl_2(g)$
(d) $H_2O(g) + C(s) \rightleftharpoons CO(g) + H_2(g)$
(e) $HCOOH(aq) \rightleftharpoons H^+(aq) + HCOO^-(aq)$
(f) $2HgO(s) \rightleftharpoons 2Hg(l) + O_2(g)$

15.19   Write the equilibrium constant expressions for K_c and for K_P, if applicable, for the following reactions:
(a) $2NO_2(g) + 7H_2(g) \rightleftharpoons 2NH_3(g) + 4H_2O(l)$
(b) $2ZnS(s) + 3O_2(g) \rightleftharpoons 2ZnO(s) + 2SO_2(g)$
(c) $C(s) + CO_2(g) \rightleftharpoons 2CO(g)$
(d) $C_6H_5COOH(aq) \rightleftharpoons C_6H_5COO^-(aq) + H^+(aq)$

15.20   Write the equation relating K_c to K_P, and define all the terms.

Computational Problems

15.21   The equilibrium constant (K_c) for the reaction:

$$2HCl(g) \rightleftharpoons H_2(g) + Cl_2(g)$$

is 4.17×10^{-34} at 25°C. What is the equilibrium constant for the reaction:

$$H_2(g) + Cl_2(g) \rightleftharpoons 2HCl(g)$$

at the same temperature?

15.22   What is K_P at 1273°C for the reaction

$$2CO(g) + O_2(g) \rightleftharpoons 2CO_2(g)$$

if K_c is 2.24×10^{22} at the same temperature?

15.23   The equilibrium constant K_P for the reaction:

$$2SO_3(g) \rightleftharpoons 2SO_2(g) + O_2(g)$$

is 1.8×10^{-5} at 350°C. What is K_c for this reaction?

15.24   Consider the reaction:

$$N_2(g) + O_2(g) \rightleftharpoons 2NO(g)$$

If the equilibrium partial pressures of N_2, O_2, and NO are 0.15, 0.33, and 0.050 atm, respectively, at 2200°C, what is K_P?

15.25   A reaction vessel contains NH_3, N_2, and H_2 at equilibrium at a certain temperature. The equilibrium concentrations are $[NH_3] = 0.25\ M$, $[N_2] = 0.11\ M$, and $[H_2] = 1.91\ M$. Calculate the equilibrium constant K_c for the synthesis of ammonia if the reaction is represented as:
(a) $N_2(g) + 3H_2(g) \rightleftharpoons 2NH_3(g)$
(b) $\frac{1}{2}N_2(g) + \frac{3}{2}H_2(g) \rightleftharpoons NH_3(g)$

15.26   The equilibrium constant K_c for the reaction:

$$2A(g) \rightleftharpoons A_2(g)$$

is 1.5×10^4 at 155°C. Calculate K_c and K_P for the equilibrium

$$A_2(g) \rightleftharpoons 2A(g)$$

at the same temperature.

15.27   At equilibrium, the pressure of the reacting mixture

$$CaCO_3(s) \rightleftharpoons CaO(s) + CO_2(g)$$

is 9.52 atm at 350°C. Calculate K_P and K_c for this reaction.

15.28   The equilibrium constant K_P for the reaction:

$$PCl_5(g) \rightleftharpoons PCl_3(g) + Cl_2(g)$$

is 1.05 at 250°C. The reaction starts with a mixture of PCl_5, PCl_3, and Cl_2 at pressures of 0.177, 0.223, and 0.111 atm, respectively, at 250°C. When the mixture comes to equilibrium at that temperature, which pressures will have decreased and which will have increased? Explain why.

15.29   Ammonium carbamate ($NH_4CO_2NH_2$) decomposes as follows:

$$NH_4CO_2NH_2(s) \rightleftharpoons 2NH_3(g) + CO_2(g)$$

Starting with only the solid, it is found that when the system reaches equilibrium at 40°C, the total gas pressure (NH_3 and CO_2) is 0.363 atm. Calculate the equilibrium constant K_P.

15.30   The following equilibrium constants were determined at 1123 K:

$$C(s) + CO_2(g) \rightleftharpoons 2CO(g) \qquad K'_P = 1.3 \times 10^{14}$$
$$CO(g) + Cl_2(g) \rightleftharpoons COCl_2(g) \qquad K''_P = 6.0 \times 10^{-3}$$

Write the equilibrium constant expression K_P, and calculate the equilibrium constant at 1123 K for

$$C(s) + CO_2(g) + 2Cl_2(g) \rightleftharpoons 2COCl_2(g)$$

15.31   At a certain temperature, the following reactions have the constants shown:

$$X(s) + O_2(g) \rightleftharpoons XO_2(g) \qquad K'_c = 4.2 \times 10^8$$
$$2X(s) + 3O_2(g) \rightleftharpoons 2XO_3(g) \qquad K''_c = 9.8 \times 10^{23}$$

Calculate the equilibrium constant K_c for the following reaction at that temperature:

$$2XO_2(g) + O_2(g) \rightleftharpoons 2XO_3(g)$$

15.32   Pure phosgene gas ($COCl_2$), 3.00×10^{-2} mol, was placed in a 1.50-L container. It was heated to 800 K, and at equilibrium the pressure of CO was found to be 0.497 atm. Calculate the equilibrium constant K_P for the reaction:

$$CO(g) + Cl_2(g) \rightleftharpoons COCl_2(g)$$

15.33   Consider the equilibrium:

$$2NOBr(g) \rightleftharpoons 2NO(g) + Br_2(g)$$

If nitrosyl bromide (NOBr) is 34 percent dissociated at 25°C and the total pressure is 0.25 atm, calculate K_P and K_c for the dissociation at this temperature.

15.34 The following equilibrium constants have been determined for hydrosulfuric acid at 25°C:

$$H_2S(aq) \rightleftharpoons H^+(aq) + HS^-(aq) \quad K'_c = 9.5 \times 10^{-8}$$

$$HS^-(aq) \rightleftharpoons H^+(aq) + S^{2-}(aq) \quad K''_c = 1.0 \times 10^{-19}$$

Calculate the equilibrium constant for the following reaction at the same temperature:

$$H_2S(aq) \rightleftharpoons 2H^+(aq) + S^{2-}(aq)$$

15.35 The following equilibrium constants have been determined for oxalic acid at 25°C:

$$H_2C_2O_4(aq) \rightleftharpoons H^+(aq) + HC_2O_4^-(aq)$$
$$K'_c = 6.5 \times 10^{-2}$$

$$HC_2O_4^-(aq) \rightleftharpoons H^+(aq) + C_2O_4^{2-}(aq)$$
$$K''_c = 6.1 \times 10^{-5}$$

Calculate the equilibrium constant for the following reaction at the same temperature:

$$H_2C_2O_4(aq) \rightleftharpoons 2H^+(aq) + C_2O_4^{2-}(aq)$$

Conceptual Problems

15.36 The equilibrium constant for the reaction A $\rightleftharpoons$ B is $K_c = 10$ at a certain temperature. (1) Starting with only reactant A, which of the diagrams shown here best represents the system at equilibrium? (2) Which of the diagrams best represents the system at equilibrium if $K_c = 0.10$? Explain why you can calculate K_c in each case without knowing the volume of the container. The grey spheres represent the A molecules, and the green spheres represent the B molecules.

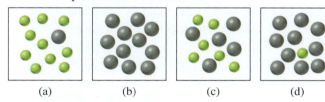

(a)   (b)   (c)   (d)

15.37 The following diagrams represent the equilibrium state for three different reactions of the type:

$$A + X \rightleftharpoons AX(X = B, C, \text{ or } D)$$

A + B ⇌ AB   A + C ⇌ AC   A + D ⇌ AD

(a) Which reaction has the largest equilibrium constant?

(b) Which reaction has the smallest equilibrium constant?

SECTION 15.4: USING EQUILIBRIUM EXPRESSIONS TO SOLVE PROBLEMS

Review Questions

15.38 Outline the steps for calculating the concentrations of species in an equilibrium reaction.

Computational Problems

15.39 The equilibrium constant K_P for the reaction:

$$2SO_2(g) + O_2(g) \rightleftharpoons 2SO_3(g)$$

is 5.60×10^4 at 350°C. The initial pressures of SO_2, O_2, and SO_3 in a mixture are 0.350, 0.762, and 0 atm, respectively, at 350°C. When the mixture reaches equilibrium, is the total pressure less than or greater than the sum of the initial pressures?

15.40 For the synthesis of ammonia:

$$N_2(g) + 3H_2(g) \rightleftharpoons 2NH_3(g)$$

the equilibrium constant K_c at 375°C is 1.2. Starting with $[H_2]_0 = 0.76\ M$, $[N_2]_0 = 0.60\ M$, and $[NH_3]_0 = 0.48\ M$, which gases will have increased in concentration and which will have decreased in concentration when the mixture comes to equilibrium?

15.41 For the reaction:

$$H_2(g) + CO_2(g) \rightleftharpoons H_2O(g) + CO(g)$$

at 700°C, $K_c = 0.534$. Calculate the number of moles of H_2 that are present at equilibrium if a mixture of 0.300 mole of CO and 0.300 mole of H_2O is heated to 700°C in a 10.0-L container.

15.42 At 1000 K, a sample of pure NO_2 gas decomposes:

$$2NO_2(g) \rightleftharpoons 2NO(g) + O_2(g)$$

The equilibrium constant K_P is 158. Analysis shows that the partial pressure of O_2 is 0.25 atm at equilibrium. Calculate the pressure of NO and NO_2 in the mixture.

15.43 The equilibrium constant K_c for the reaction

$$H_2(g) + Br_2(g) \rightleftharpoons 2HBr(g)$$

is 2.18×10^6 at 730°C. Starting with 5.00 mol of HBr in a 22.0-L reaction vessel, calculate the concentrations of H_2, Br_2, and HBr at equilibrium.

15.44 The dissociation of molecular iodine into iodine atoms is represented as:

$$I_2(g) \rightleftharpoons 2I(g)$$

At 1000 K, the equilibrium constant K_c for the reaction is 3.80×10^{-5}. Suppose you start with 0.0456 mole of I_2 in a 2.30-L flask at 1000 K. What are the concentrations of the gases at equilibrium?

15.45 The equilibrium constant K_c for the decomposition of phosgene ($COCl_2$) is 4.63×10^{-3} at 527°C:

$$COCl_2(g) \rightleftharpoons CO(g) + Cl_2(g)$$

Calculate the equilibrium partial pressures of all the components, starting with pure phosgene at 0.760 atm.

15.46   Consider the following equilibrium process at 686°C:

$$CO_2(g) + H_2(g) \rightleftharpoons CO(g) + H_2O(g)$$

The equilibrium concentrations of the reacting species are $[CO] = 0.050\ M$, $[H_2] = 0.045\ M$, $[CO_2] = 0.086\ M$, and $[H_2O] = 0.040\ M$.
(a) Calculate K_c for the reaction at 686°C.
(b) If we add CO_2 to increase its concentration to 0.50 mol/L, what will the concentrations of all the gases be when equilibrium is reestablished?

15.47   Consider the heterogeneous equilibrium process:

$$C(s) + CO_2(g) \rightleftharpoons 2CO(g)$$

At 700°C, the total pressure of the system is found to be 4.50 atm. If the equilibrium constant K_P is 1.52, calculate the equilibrium partial pressures of CO_2 and CO.

15.48   The equilibrium constant K_c for the reaction:

$$H_2(g) + CO_2(g) \rightleftharpoons H_2O(g) + CO(g)$$

is 4.2 at 1650°C. Initially 0.80 mol H_2 and 0.80 mol CO_2 are injected into a 5.0-L flask. Calculate the concentration of each species at equilibrium.

15.49   The aqueous reaction:

L-glutamate + pyruvate $\rightleftharpoons$ α-ketoglutarate + L-alanine

is catalyzed by the enzyme L-glutamate–pyruvate aminotransferase. At 300 K, the equilibrium constant for the reaction is 1.11. Predict whether the forward reaction will occur if the concentrations of the reactants and products are $[\text{L-glutamate}] = 3.0 \times 10^{-5}\ M$, $[\text{pyruvate}] = 3.3 \times 10^{-4}\ M$, $[\alpha\text{-ketoglutarate}] = 1.6 \times 10^{-2}\ M$, and $[\text{L-alanine}] = 6.25 \times 10^{-3}\ M$.

SECTION 15.5: FACTORS THAT AFFECT CHEMICAL EQUILIBRIUM

Visualizing Chemistry
Figure 15.12 and Figure 15.13

Solid calcium carbonate decomposes to form solid calcium oxide and carbon dioxide gas:

$$CaCO_3(s) \rightleftharpoons CaO(s) + CO_2(g) \qquad \Delta H° = 177.8\ \text{kJ/mol}$$

Consider an equilibrium mixture of calcium carbonate and its decomposition products in a container.

VC 15.5   Which of the following changes will result in the formation of more $CaO(s)$?
a) Addition of more $CaCO_3(s)$
b) Addition of more $CO_2(g)$
c) Removal of some of the $CO_2(g)$

VC 15.6   Which of the following changes will result in the formation of more $CaCO_3(s)$?
a) Addition of more $CaO(s)$
b) Addition of more $CO_2(g)$
c) Removal of some of the $CaCO_3(s)$

VC 15.7   Which of the following changes will not cause the equilibrium to shift in either direction? (Select all that apply.)
a) Removal of some of the $CaCO_3(s)$
b) Removal of some of the $CO_2(g)$
c) Addition of more $CaO(s)$

VC 15.8   Which of the following will be affected by a change in the system's temperature? (Select all that apply.)
a) The amount of $CaCO_3(s)$
b) The amount of $CaO(s)$
c) The amount of $CO_2(g)$

A sample of calcium carbonate is in equilibrium with its decomposition products in a closed system:
$$CaCO_3(s) \rightleftharpoons CaO(s) + CO_2(g)$$

VC 15.9   If the volume of the system is increased at constant temperature, relative to the original equilibrium mixture, which of the following will be true when equilibrium is reestablished? (Select all that apply.)
a) There will be a higher concentration of $CO_2(g)$.
b) There will be a lower concentration of $CO_2(g)$.
c) There will be more $CO_2(g)$.

VC 15.10   If the volume of the system is decreased at constant temperature, relative to the original equilibrium mixture, which of the following will be true when equilibrium is reestablished? (Select all that apply.)
a) There will be more $CaCO_3(s)$.
b) There will be more $CaO(s)$.
c) There will be more $CO_2(g)$.

A sample of gaseous hydrogen iodide is in equilibrium with its decomposition products in a closed system:
$$2HI(g) \rightleftharpoons H_2(g) + I_2(g).$$

VC 15.11   If the volume of the system is increased at constant temperature, relative to the original equilibrium mixture, which of the following will be true when equilibrium is reestablished?
a) The concentration of HI will increase and the concentrations of H_2 and I_2 will decrease.
b) The concentration of all three species will be lower.
c) The concentration of all three species will be unchanged.

VC 15.12   If the volume of the system is decreased at constant temperature, relative to the original equilibrium mixture, which of the following will be true when equilibrium is reestablished?
a) There will be more HI.
b) There will be less HI.
c) The amount of HI will be unchanged.

Review Questions

15.50 Explain Le Châtelier's principle. How does this principle enable us to maximize the yields of desirable reactions and minimize the effect of undesirable ones?

15.51 Use Le Châtelier's principle to explain why the equilibrium vapor pressure of a liquid increases with increasing temperature.

15.52 List four factors that can shift the position of an equilibrium. Only one of these factors can alter the value of the equilibrium constant. Which one is it?

15.53 Does the addition of a catalyst have any effects on the position of an equilibrium?

Computational Problems

15.54 Consider the following equilibrium system involving SO_2, Cl_2, and SO_2Cl_2 (sulfuryl dichloride):

$$SO_2(g) + Cl_2(g) \rightleftharpoons SO_2Cl_2(g)$$

Predict how the equilibrium position would change if (a) Cl_2 gas were added to the system, (b) SO_2Cl_2 were removed from the system, (c) SO_2 were removed from the system. The temperature remains constant in each case.

15.55 Heating solid sodium bicarbonate in a closed vessel establishes the following equilibrium:

$$2NaHCO_3(s) \rightleftharpoons Na_2CO_3(s) + H_2O(g) + CO_2(g)$$

What would happen to the equilibrium position if (a) some of the CO_2 were removed from the system, (b) some solid Na_2CO_3 were added to the system, (c) some of the solid $NaHCO_3$ were removed from the system? The temperature remains constant.

15.56 Consider the following equilibrium systems:

(a) $A \rightleftharpoons 2B$ $\Delta H° = 20.0$ kJ/mol
(b) $A + B \rightleftharpoons C$ $\Delta H° = -5.4$ kJ/mol
(c) $A \rightleftharpoons B$ $\Delta H° = 0.0$ kJ/mol

Predict the change in the equilibrium constant K_c that would occur in each case if the temperature of the reacting system were raised.

15.57 What effect does an increase in pressure have on each of the following systems at equilibrium? The temperature is kept constant, and, in each case, the reactants are in a cylinder fitted with a movable piston.

(a) $A(s) \rightleftharpoons 2B(s)$
(b) $2A(l) \rightleftharpoons B(l)$
(c) $A(s) \rightleftharpoons B(g)$
(d) $A(g) \rightleftharpoons B(g)$
(e) $A(g) \rightleftharpoons 2B(g)$

15.58 Consider the equilibrium:

$$2I(g) \rightleftharpoons I_2(g)$$

What would be the effect on the position of equilibrium of (a) increasing the total pressure on the system by decreasing its volume, (b) adding I_2 to the reaction mixture, and (c) decreasing the temperature?

15.59 Consider the following equilibrium process:

$$PCl_5(g) \rightleftharpoons PCl_3(g) + Cl_2(g)$$
$$\Delta H° = 92.5 \text{ kJ/mol}$$

Predict the direction of the shift in equilibrium when (a) the temperature is raised, (b) more chlorine gas is added to the reaction mixture, (c) some PCl_3 is removed from the mixture, (d) the pressure on the gases is increased, (e) a catalyst is added to the reaction mixture.

15.60 Consider the reaction:

$$2SO_2(g) + O_2(g) \rightleftharpoons 2SO_3(g)$$
$$\Delta H° = -198.2 \text{ kJ/mol}$$

Comment on the changes in the concentrations of SO_2, O_2, and SO_3 at equilibrium if we were to (a) increase the temperature, (b) increase the pressure, (c) increase SO_2, (d) add a catalyst, (e) add helium at constant volume.

15.61 In the uncatalyzed reaction:

$$N_2O_4(g) \rightleftharpoons 2NO_2(g)$$

the pressures of the gases at equilibrium are $P_{N_2O_4} = 0.377$ atm and $P_{NO_2} = 1.56$ atm at 100°C. What would happen to these pressures if a catalyst were added to the mixture?

15.62 Consider the gas-phase reaction:

$$2CO(g) + O_2(g) \rightleftharpoons 2CO_2(g)$$

Predict the shift in the equilibrium position when helium gas is added to the equilibrium mixture (a) at constant pressure and (b) at constant volume.

15.63 Consider the following equilibrium reaction in a closed container:

$$CaCO_3(s) \rightleftharpoons CaO(s) + CO_2(g)$$

What will happen if (a) the volume is increased, (b) some CaO is added to the mixture, (c) some $CaCO_3$ is removed, (d) some CO_2 is added to the mixture, (e) a few drops of an NaOH solution are added to the mixture, (f) a few drops of an HCl solution are added to the mixture (ignore the reaction between CO_2 and water), (g) the temperature is increased?

Conceptual Problems

15.64 The following diagrams show the reaction $A + B \rightleftharpoons AB$ at two different temperatures. Is the forward reaction endothermic or exothermic?

      600 K   800 K

15.65   The following diagrams show an equilibrium mixture of O_2 and O_3 at temperatures T_1 and T_2 ($T_2 > T_1$). (a) Write an equilibrium equation showing the forward reaction to be exothermic. (b) Predict how the number of O_2 and O_3 molecules would change if the volume were decreased at constant temperature.

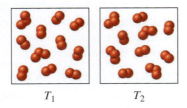

$$T_1 \qquad\qquad T_2$$

ADDITIONAL PROBLEMS

15.66   Consider the reaction $A + B \rightleftharpoons 2C$. The first diagram represents a system at equilibrium where $A = $ ○, $B = $ ●, and $C = $ ●. Which of the following diagrams [(a)–(d)] also represents a system at equilibrium? Select all that apply.

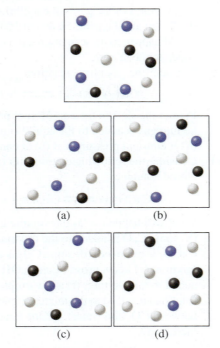

(a)   (b)

(c)   (d)

15.67   Pure nitrosyl chloride (NOCl) gas was heated to 240°C in a 1.00-L container. At equilibrium, the total pressure was 1.00 atm and the NOCl pressure was 0.64 atm:

$$2NOCl(g) \rightleftharpoons 2NO(g) + Cl_2(g)$$

(a) Calculate the partial pressures of NO and Cl_2 in the system. (b) Calculate the equilibrium constant K_P.

15.68   Consider the equilibrium system $3A \rightleftharpoons B$. Sketch the changes in the concentrations of A and B over time for the following situations: (a) initially only A is present, (b) initially only B is present, (c) initially both A and B are present (with A in higher concentration). In each case, assume that the concentration of B is higher than that of A at equilibrium.

15.69   Baking soda (sodium bicarbonate) undergoes thermal decomposition as follows:

$$2NaHCO_3(s) \rightleftharpoons Na_2CO_3(s) + CO_2(g) + H_2O(g)$$

Would we obtain more CO_2 and H_2O by adding extra baking soda to the reaction mixture in (a) a closed vessel or (b) an open vessel?

15.70   Consider the following reaction at equilibrium:

$$A(g) \rightleftharpoons 2B(g)$$

From the data shown here, calculate the equilibrium constant (both K_P and K_c) at each temperature. Is the reaction endothermic or exothermic?

Temperature (°C)	[A] (M)	[B] (M)
200	0.0125	0.843
300	0.171	0.764
400	0.250	0.724

15.71   The equilibrium constant K_P for the reaction:

$$2H_2O(g) \rightleftharpoons 2H_2(g) + O_2(g)$$

is 2×10^{-42} at 25°C. (a) What is K_c for the reaction at the same temperature? (b) The very small value of K_P (and K_c) indicates that the reaction overwhelmingly favors the formation of water molecules. Explain why, despite this fact, a mixture of hydrogen and oxygen gases can be kept at room temperature without any change.

15.72   Consider the following reacting system:

$$2NO(g) + Cl_2(g) \rightleftharpoons 2NOCl(g)$$

What combination of temperature and pressure would maximize the yield of nitrosyl chloride (NOCl)? [*Hint:* ΔH_f° (NOCl) = 51.7 kJ/mol.]

15.73   At a certain temperature and a total pressure of 1.2 atm, the partial pressures of an equilibrium mixture

$$2A(g) \rightleftharpoons B(g)$$

are $P_A = 0.60$ atm and $P_B = 0.60$ atm. (a) Calculate the K_P for the reaction at this temperature. (b) If the total pressure were increased to 1.5 atm, what would be the partial pressures of A and B at equilibrium?

15.74 The decomposition of ammonium hydrogen sulfide:

$$NH_4HS(s) \rightleftharpoons NH_3(g) + H_2S(g)$$

is an endothermic process. A 6.1589-g sample of the solid is placed in an evacuated 4.000-L vessel at exactly 24°C. After equilibrium has been established, the total pressure inside is 0.709 atm. Some solid NH_4HS remains in the vessel. (a) What is the K_P for the reaction? (b) What percentage of the solid has decomposed? (c) If the volume of the vessel were doubled at constant temperature, what would happen to the amount of solid in the vessel?

15.75 Consider the following reaction at a certain temperature:

$$A_2 + B_2 \rightleftharpoons 2AB$$

The mixing of 1 mole of A_2 with 3 moles of B_2 gives rise to x mole of AB at equilibrium. The addition of 2 more moles of A_2 produces another x mole of AB. What is the equilibrium constant for the reaction?

15.76 When heated, ammonium carbamate decomposes as follows:

$$NH_4CO_2NH_2(s) \rightleftharpoons 2NH_3(g) + CO_2(g)$$

At a certain temperature, the equilibrium pressure of the system is 0.318 atm. Calculate K_P for the reaction.

15.77 A mixture of 0.47 mole of H_2 and 3.59 moles of HCl is heated to 2800°C. Calculate the equilibrium partial pressures of H_2, Cl_2, and HCl if the total pressure is 2.00 atm. For the reaction:

$$H_2(g) + Cl_2(g) \rightleftharpoons 2HCl(g)$$

K_P is 193 at 2800°C.

15.78 When heated at high temperatures, iodine vapor dissociates as follows:

$$I_2(g) \rightleftharpoons 2I(g)$$

In one experiment, a chemist finds that when 0.054 mole of I_2 was placed in a flask of volume 0.48 L at 587 K, the degree of dissociation (i.e., the fraction of I_2 dissociated) was 0.0252. Calculate K_c and K_P for the reaction at this temperature.

15.79 One mole of N_2 and three moles of H_2 are placed in a flask at 375°C. Calculate the total pressure of the system at equilibrium if the mole fraction of NH_3 is 0.21. The K_P for the reaction is 4.31×10^{-4}.

15.80 At 1130°C, the equilibrium constant (K_c) for the reaction:

$$2H_2S(g) \rightleftharpoons 2H_2(g) + S_2(g)$$

is 2.25×10^{-4}. If $[H_2S] = 4.84 \times 10^{-3}$ M and $[H_2] = 1.50 \times 10^{-3}$ M, calculate $[S_2]$.

15.81 For the purpose of determining K_P using Equation 15.4, what is Δn for the following equation?

$$C_3H_8(g) + 5O_2(g) \rightleftharpoons 3CO_2(g) + 4H_2O(l)$$

15.82 The following diagram represents a gas-phase equilibrium mixture for the reaction $AB \rightleftharpoons A + B$ at a certain temperature. Describe what would happen to the system after each of the following changes: (a) the temperature is decreased, (b) the volume is increased, (c) He atoms are added to the mixture at constant volume, (d) a catalyst is added to the mixture.

15.83 Consider the following reaction at 1600°C:

$$Br_2(g) \rightleftharpoons 2Br(g)$$

When 1.05 moles of Br_2 are put in a 0.980-L flask, 1.20 percent of the Br_2 undergoes dissociation. Calculate the equilibrium constant K_c for the reaction.

15.84 A quantity of 0.20 mole of carbon dioxide was heated to a certain temperature with an excess of graphite in a closed container until the following equilibrium was reached:

$$C(s) + CO_2(g) \rightleftharpoons 2CO(g)$$

Under these conditions, the average molar mass of the gases was 35 g/mol. (a) Calculate the mole fractions of CO and CO_2. (b) What is K_P if the total pressure is 11 atm? (*Hint:* The average molar mass is the sum of the products of the mole fraction of each gas and its molar mass.)

15.85 When dissolved in water, glucose (corn sugar) and fructose (fruit sugar) exist in equilibrium as follows:

$$fructose \rightleftharpoons glucose$$

A chemist prepared a 0.244 M fructose solution at 25°C. At equilibrium, it was found that its concentration had decreased to 0.113 M. (a) Calculate the equilibrium constant for the reaction. (b) At equilibrium, what percentage of fructose was converted to glucose?

15.86 At room temperature, solid iodine is in equilibrium with its vapor through sublimation and deposition. Describe how you would use radioactive iodine, in either solid or vapor form, to show that there is a dynamic equilibrium between these two phases.

15.87 A student placed a few ice cubes in a drinking glass with water. A few minutes later she noticed that some of the ice cubes were fused together. Explain what happened.

15.88 A mixture containing 3.9 moles of NO and 0.88 mole of CO_2 was allowed to react in a flask at a certain temperature according to the equation:

$$NO(g) + CO_2(g) \rightleftharpoons NO_2(g) + CO(g)$$

At equilibrium, 0.11 mole of CO_2 was present. Calculate the equilibrium constant K_c of this reaction.

15.89 The equilibrium constant K_c for the reaction:

$$H_2(g) + I_2(g) \rightleftharpoons 2HI(g)$$

is 54.3 at 430°C. At the start of the reaction, there are 0.714 mol of H_2, 0.984 mol of I_2, and 0.886 mol of HI in a 2.40-L reaction chamber. Calculate the concentrations of the gases at equilibrium.

15.90 When heated, a gaseous compound A dissociates as follows:

$$A(g) \rightleftharpoons B(g) + C(g)$$

In an experiment, A was heated at a certain temperature until its equilibrium pressure reached 0.14P, where P is the total pressure. Calculate the equilibrium constant K_P of this reaction.

15.91 When a gas was heated under atmospheric conditions, its color deepened. Heating above 150°C caused the color to fade, and at 550°C the color was barely detectable. However, at 550°C, the color was partially restored by increasing the pressure of the system. Which of the following best fits the preceding description: (a) a mixture of hydrogen and bromine, (b) pure bromine, (c) a mixture of nitrogen dioxide and dinitrogen tetroxide. (*Hint:* Bromine has a reddish color, and nitrogen dioxide is a brown gas. The other gases are colorless.) Justify your choice.

15.92 The first diagram represents a system at equilibrium where A = ● and B = ●. The balanced chemical equation for the equilibrium process is A(g) $\rightleftharpoons$ B(g). The equilibrium is stressed by the addition of more reactant. Select a pair of the other diagrams (a–d) that could represent the system (a) immediately following the addition; and (b) after equilibrium has been reestablished.

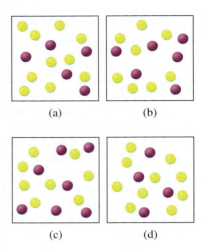

(a)	(b)
(c)	(d)

15.93 A sealed glass bulb contains a mixture of NO_2 and N_2O_4 gases. Describe what happens to the following properties of the gases when the bulb is heated from 20°C to 40°C: (a) color, (b) pressure, (c) average molar mass, (d) degree of dissociation (from N_2O_4 to NO_2), (e) density. Assume that volume remains constant. (*Hint:* NO_2 is a brown gas; N_2O_4 is colorless.)

15.94 At 20°C, the vapor pressure of water is 0.0231 atm. Calculate K_P and K_c for the process:

$$H_2O(l) \rightleftharpoons H_2O(g)$$

15.95 A 2.50-mol sample of NOCl was initially in a 1.50-L reaction chamber at 400°C. After equilibrium was established, it was found that 28.0 percent of the NOCl had dissociated:

$$2NOCl(g) \rightleftharpoons 2NO(g) + Cl_2(g)$$

Calculate the equilibrium constant K_c for the reaction.

15.96 About 75 percent of hydrogen for industrial use is produced by the *steam-reforming* process. This process is carried out in two stages called primary and secondary reforming. In the primary stage, a mixture of steam and methane at about 30 atm is heated over a nickel catalyst at 800°C to give hydrogen and carbon monoxide:

$$CH_4(g) + H_2O(g) \rightleftharpoons CO(g) + 3H_2(g)$$
$$\Delta H° = 206 \text{ kJ/mol}$$

The secondary stage is carried out at about 1000°C, in the presence of air, to convert the remaining methane to hydrogen:

$$CH_4(g) + \tfrac{1}{2}O_2(g) \rightleftharpoons CO(g) + 2H_2(g)$$
$$\Delta H° = 35.7 \text{ kJ/mol}$$

(a) What conditions of temperature and pressure would favor the formation of products in both the primary and secondary stages? (b) The equilibrium constant K_c for the primary stage is 18 at 800°C. (i) Calculate K_P for the reaction. (ii) If the partial pressures of methane and steam were both 15 atm at the start, what are the pressures of all the gases at equilibrium?

15.97 Water is a very weak electrolyte that undergoes the following ionization (called *autoionization*):

$$H_2O(l) \underset{k_{-1}}{\overset{k_1}{\rightleftharpoons}} H^+(aq) + OH^-(aq)$$

(a) If $k_1 = 2.4 \times 10^{-5}$ s^{-1} and $k_{-1} = 1.3 \times 10^{11}/M \cdot$ s, calculate the equilibrium constant K where $K = [H^+]$ $[OH^-]/[H_2O]$. (b) Calculate the product $[H^+][OH^-]$, $[H^+]$, and $[OH^-]$. (*Hint:* Calculate the concentration of liquid water using its density, 1.0 g/mL.)

15.98 Consider the following reaction, which takes place in a single elementary step:

$$2A + B \underset{k_{-1}}{\overset{k_1}{\rightleftharpoons}} A_2B$$

If the equilibrium constant K_c is 12.6 at a certain temperature and if $k_1 = 5.1 \times 10^{-2}$ s^{-1}, calculate the value of k_{-1}.

15.99 The equilibrium constant K_c for the reaction:

$$2NH_3(g) \rightleftharpoons N_2(g) + 3H_2(g)$$

is 0.83 at 375°C. A 14.6-g sample of ammonia is placed in a 4.00-L flask and heated to 375°C. Calculate the concentrations of all the gases when equilibrium is reached.

15.100 At 25°C, a mixture of NO_2 and N_2O_4 gases is in equilibrium in a cylinder fitted with a movable piston. The concentrations are $[NO_2] = 0.0475$ M and $[N_2O_4] = 0.487$ M. The volume of the gas mixture is halved by pushing down on the piston at constant temperature. Calculate the concentrations of the gases when equilibrium is reestablished. Will the color become darker or lighter after the change? [*Hint:* K_c for the dissociation of N_2O_4 is 4.63×10^{-3}. $N_2O_4(g)$ is colorless, and $NO_2(g)$ has a brown color.]

15.101 Consider the reaction between NO_2 and N_2O_4 in a closed container:

$$N_2O_4(g) \rightleftharpoons 2NO_2(g)$$

Initially, 1 mol of N_2O_4 is present. At equilibrium, x mol of N_2O_4 has dissociated to form NO_2. (a) Derive an expression for K_P in terms of x and P, the total pressure. (b) How does the expression in part (a) help you predict the shift in equilibrium due to an increase in P? Does your prediction agree with Le Châtelier's principle?

15.102 In 1899 the German chemist Ludwig Mond developed a process for purifying nickel by converting it to the volatile nickel tetracarbonyl $[Ni(CO)_4]$ (b.p. $= 42.2$°C):

$$Ni(s) + 4CO(g) \rightleftharpoons Ni(CO)_4(g)$$

(a) Describe how you can separate nickel and its solid impurities. (b) How would you recover nickel? $[\Delta H_f^\circ$ for $Ni(CO)_4$ is -602.9 kJ/mol.]

15.103 For which of the following reactions is K_c equal to K_P? For which can we *not* write a K_P expression?
(a) $4NH_3(g) + 5O_2(g) \rightleftharpoons 4NO(g) + 6H_2O(g)$
(b) $CaCO_3(s) \rightleftharpoons CaO(s) + CO_2(g)$
(c) $Zn(s) + 2H^+(aq) \rightleftharpoons Zn^{2+}(aq) + H_2(g)$
(d) $PCl_3(g) + 3NH_3(g) \rightleftharpoons 3HCl(g) +$ $P(NH_2)_3(g)$
(e) $NH_3(g) + HCl(g) \rightleftharpoons NH_4Cl(s)$
(f) $NaHCO_3(s) + H^+(aq) \rightleftharpoons H_2O(l) + CO_2(g) +$ $Na^+(aq)$
(g) $H_2(g) + F_2(g) \rightleftharpoons 2HF(g)$
(h) $C(graphite) + CO_2(g) \rightleftharpoons 2CO(g)$

Industrial Problems

15.104 The equilibrium constant K_P for the following reaction is 4.31×10^{-4} at 375°C:

$$N_2(g) + 3H_2(g) \rightleftharpoons 2NH_3(g)$$

In a certain experiment a student starts with 0.862 atm of N_2 and 0.373 atm of H_2 in a constant-volume vessel at 375°C. Calculate the partial pressures of all species when equilibrium is reached.

15.105 At 1024°C, the pressure of oxygen gas from the decomposition of copper(II) oxide (CuO) is 0.49 atm:

$$4CuO(s) \rightleftharpoons 2Cu_2O(s) + O_2(g)$$

(a) What is K_P for the reaction? (b) Calculate the fraction of CuO that will decompose if 0.16 mol of it is placed in a 2.0-L flask at 1024°C. (c) What would the fraction be if a 1.0-mol sample of CuO were used? (d) What is the smallest amount of CuO (in moles) that would establish the equilibrium?

15.106 The equilibrium constant K_c for the following reaction is 1.2 at 375°C:

$$N_2(g) + 3H_2(g) \rightleftharpoons 2NH_3(g)$$

(a) What is the value of K_P for this reaction?
(b) What is the value of the equilibrium constant K_c for $2NH_3(g) \rightleftharpoons N_2(g) + 3H_2(g)$? (c) What is K_c for $\frac{1}{2}N_2(g) + \frac{3}{2}H_2(g) \rightleftharpoons NH_3(g)$? (d) What are the values of K_P for the reactions described in parts (b) and (c)?

15.107 Industrially, sodium metal is obtained by electrolyzing molten sodium chloride. The reaction at the cathode is $Na^+ + e^- \longrightarrow Na$. We might expect that potassium metal would also be prepared by electrolyzing molten potassium chloride. However, potassium metal is soluble in molten potassium chloride and therefore is hard to recover. Furthermore, potassium vaporizes readily at the operating temperature, creating hazardous conditions. Instead, potassium is prepared by the distillation of molten potassium chloride in the presence of sodium vapor at 892°C:

$$Na(g) + KCl(l) \rightleftharpoons NaCl(l) + K(g)$$

In view of the fact that potassium is a stronger reducing agent than sodium, explain why this approach works. (The boiling points of sodium and potassium are 892°C and 770°C, respectively.)

15.108 Consider the equilibrium reaction described in Problem 15.30. A quantity of 2.50 g of PCl_5 is placed in an evacuated 0.500-L flask and heated to 250°C. (a) Calculate the pressure of PCl_5, assuming it does not dissociate. (b) Calculate the partial pressure of PCl_5 at equilibrium. (c) What is the total pressure at equilibrium? (d) What is the degree of dissociation of PCl_5? (The degree of dissociation is given by the fraction of PCl_5 that has undergone dissociation.)

15.109 The K_P for the reaction:

$$SO_2Cl_2(g) \rightleftharpoons SO_2(g) + Cl_2(g)$$

is 2.05 at 648 K. A sample of SO_2Cl_2 is placed in a container and heated to 648 K, while the total pressure is kept constant at 9.00 atm. Calculate the partial pressures of the gases at equilibrium.

15.110 The "boat" form and the "chair" form of cyclohexane (C_6H_{12}) interconvert as shown here:

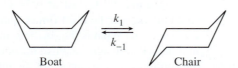

Boat   Chair

In this representation, the H atoms are omitted and a C atom is assumed to be at each intersection of two lines (bonds). The conversion is first order in each direction. The activation energy for the chair $\longrightarrow$ boat conversion is 41 kJ / mol. If the frequency factor is 1.0×10^{12} s^{-1}, what is k_1 at 298 K? The equilibrium constant K_c for the reaction is 9.83×10^3 at 298 K.

15.111 A quantity of 6.75 g of SO_2Cl_2 was placed in a 2.00-L flask. At 648 K, there is 0.0345 mol of SO_2 present. Calculate K_c for the reaction:

$$SO_2Cl_2(g) \rightleftharpoons SO_2(g) + Cl_2(g)$$

15.112 Industrial production of ammonia from hydrogen and nitrogen gases is done using the Haber process.

$$N_2(g) + 3H_2(g) \rightleftharpoons 2NH_3(g) \qquad \Delta H° = -92.6 \text{ kJ/mol}$$

Based on your knowledge of the principles of equilibrium, what would the optimal temperature and pressure conditions be for production of ammonia on a large scale? Are the same conditions also optimal from the standpoint of kinetics? Explain.

Environmental Problems

15.113 The equilibrium constant (K_P) for the formation of the air pollutant nitric oxide (NO) in an automobile engine at 530°C is 2.9×10^{-11}:

$$N_2(g) + O_2(g) \rightleftharpoons 2NO(g)$$

(a) Calculate the partial pressure of NO under these conditions if the partial pressures of nitrogen and oxygen are 3.0 and 0.012 atm, respectively. (b) Repeat the calculation for atmospheric conditions where the partial pressures of nitrogen and oxygen are 0.78 and 0.21 atm and the temperature is 25°C. (The K_P for the reaction is 4.0×10^{-31} at this temperature.) (c) Is the formation of NO endothermic or exothermic? (d) What natural phenomenon promotes the formation of NO? Why?

15.114 Consider the reaction:

$$2NO(g) + O_2(g) \rightleftharpoons 2NO_2(g)$$

At 430°C, an equilibrium mixture consists of 0.020 mol of O_2, 0.040 mol of NO, and 0.96 mol of NO_2. Calculate K_P for the reaction, given that the total pressure is 0.20 atm.

15.115 The formation of SO_3 from SO_2 and O_2 is an intermediate step in the manufacture of sulfuric acid, and it is also responsible for the acid rain phenomenon. The equilibrium constant K_P for the reaction

$$2SO_2(g) + O_2(g) \rightleftharpoons 2SO_3(g)$$

is 0.13 at 830°C. In one experiment, 2.00 mol SO_2 and 2.00 mol O_2 were initially present in a flask. What must the total pressure at equilibrium be to have an 80.0 percent yield of SO_3?

15.116 At 25°C, the equilibrium partial pressures of NO_2 and N_2O_4 are 0.15 atm and 0.20 atm, respectively. If the volume is doubled at constant temperature, calculate the partial pressures of the gases when a new equilibrium is established.

Biological Problems

15.117 The vapor pressure of mercury is 0.0020 mmHg at 26°C. (a) Calculate K_c and K_P for the process $Hg(l) \rightleftharpoons Hg(g)$. (b) A chemist breaks a thermometer and spills mercury onto the floor of a laboratory measuring 6.1 m long, 5.3 m wide, and 3.1 m high. Calculate the mass of mercury (in grams) vaporized at equilibrium and the concentration of mercury vapor (in mg/m^3). Does this concentration exceed the safety limit of 0.05 mg/m^3? (Ignore the volume of furniture and other objects in the laboratory.)

15.118 Both Mg^{2+} and Ca^{2+} are important biological ions. One of their functions is to bind to the phosphate group of ATP molecules or amino acids of proteins. For Group 2A metals in general, the equilibrium constant for binding to the anions increases in the order $Ba^{2+} < Sr^{2+} < Ca^{2+} < Mg^{2+}$. What property of the Group 2 metal cations might account for this trend?

15.119 Photosynthesis can be represented by:

$$6CO_2(g) + 6H_2O(l) \rightleftharpoons C_6H_{12}O_6(s) + 6O_2(g)$$
$$\Delta H° = 2801 \text{ kJ/mol}$$

Explain how the equilibrium would be affected by the following changes: (a) the partial pressure of CO_2 is increased, (b) O_2 is removed from the mixture, (c) $C_6H_{12}O_6$ (glucose) is removed from the mixture, (d) more water is added, (e) a catalyst is added, (f) the temperature is decreased.

15.120 Consider the decomposition of ammonium chloride at a certain temperature:

$$NH_4Cl(s) \rightleftharpoons NH_3(g) + HCl(g)$$

Calculate the equilibrium constant K_P if the total pressure is 2.2 atm at that temperature.

15.121 Eggshells are composed mostly of calcium carbonate ($CaCO_3$) formed by the reaction:

$$Ca^{2+}(aq) + CO_3^{2-}(aq) \rightleftharpoons CaCO_3(s)$$

The carbonate ions are supplied by carbon dioxide produced as a result of metabolism. Explain why eggshells are thinner in the summer when the rate of panting by chickens is greater. Suggest a remedy for this situation.

Multiconcept Problems

15.122 In the gas phase, nitrogen dioxide is actually a mixture of nitrogen dioxide (NO_2) and dinitrogen tetroxide (N_2O_4). If the density of such a mixture is 2.3 g/L at 74°C and 1.3 atm, calculate the partial pressures of the gases and K_P for the dissociation of N_2O_4.

15.123 Consider the potential-energy diagrams for two types of reactions A $\rightleftharpoons$ B. In each case, answer the following questions for the system at equilibrium. (a) How would a catalyst affect the forward and reverse rates of the reaction? (b) How would a catalyst affect the energies of the reactant and product? (c) How would an increase in temperature affect the equilibrium constant?

(d) If the only effect of a catalyst is to lower the activation energies for the forward and reverse reactions, show that the equilibrium constant remains unchanged if a catalyst is added to the reacting mixture.

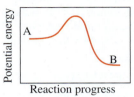
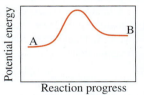

15.124 Iodine is sparingly soluble in water but much more so in carbon tetrachloride (CCl_4). The equilibrium constant, also called the partition coefficient, for the distribution of I_2 between these two phases:

$$I_2(aq) \rightleftharpoons I_2(CCl_4)$$

is 83 at 20°C. (a) A student adds 0.030 L of CCl_4 to 0.200 L of an aqueous solution containing 0.032 g of I_2. The mixture at 20°C is shaken, and the two phases are then allowed to separate. Calculate the fraction of I_2 remaining in the aqueous phase. (b) The student now repeats the extraction of I_2 with another 0.030 L of CCl_4. Calculate the fraction of the I_2 from the original solution that remains in the aqueous phase. (c) Compare the result in part (b) with a single extraction using 0.060 L of CCl_4. Comment on the difference.

15.125 The dependence of the equilibrium constant of a reaction on temperature is given by the van't Hoff equation:

$$\ln K = \frac{-\Delta H°}{RT} + C$$

where C is a constant. The following table gives the equilibrium constant (K_P) for the reaction at various temperatures:

$$2NO(g) + O_2(g) \rightleftharpoons 2NO_2(g)$$

K_P	138	5.12	0.436	0.0626	0.0130
T(K)	600	700	800	900	1000

(a) Determine graphically the $\Delta H°$ for the reaction. (b) Use the van't Hoff equation to derive the following expression, which relates the equilibrium constants at two different temperatures:

$$\ln \frac{K_1}{K_2} = \frac{\Delta H°}{R}\left(\frac{1}{T_2} - \frac{1}{T_1}\right)$$

How does this equation support the prediction based on Le Châtelier's principle about the shift in equilibrium with temperature? (c) The vapor pressures of water are 31.82 mmHg at 30°C and 92.51 mmHg at 50°C. Calculate the molar heat of vaporization of water.

Standardized-Exam Practice Problems

Physical and Biological Sciences

Lime (CaO) is used to prevent SO_2 from escaping from the smokestacks of coal-burning power plants via the formation of solid $CaSO_4 \cdot 2H_2O$ (gypsum). One of the important reactions in the overall process is the decomposition of $CaCO_3$:

$$CaCO_3(s) \rightleftharpoons CaO(s) + CO_2(g)$$

which has an equilibrium constant (K_c) of 3.0×10^{-6} at 725°C.

1. Calculate the value of K_P for the decomposition of $CaCO_3$ at 725°C.

   a) 3.0×10^{-6}
   b) 2.5×10^{-4}
   c) 2.0×10^{-2}
   d) 3.7×10^{-8}

2. If a 12.0-g sample of solid $CaCO_3$ is placed in an evacuated vessel at 725°C, what will the pressure of CO_2 be when the system reaches equilibrium?

   a) 3.0×10^{-6} atm
   b) 3.3×10^{5} atm
   c) 1.7×10^{-3} atm
   d) 1.0 atm

3. Which of the following actions will cause an increase in the pressure of CO_2 in the vessel described in question 2?

   a) Adding He gas
   b) Adding SO_2 gas
   c) Adding more $CaCO_3$ solid
   d) Increasing the volume of the vessel

4. If a 12.0-g sample of solid $CaCO_3$ is placed in a vessel at 725°C in which the pressure of CO_2 is 2.5 atm, what mass of CaO will form?

   a) 6.72 g
   b) 12.0 g
   c) 6.00 g
   d) None

Answers to In-Chapter Materials

Answers to Practice Problems

15.1A $K_c = \dfrac{[BrCl]^2}{[Br_2][Cl_2]}$, $K_c = 7.1$. **15.1B** 0.016 M.

15.2A (a) $Q_c = \dfrac{[N_2]^2[O_2]}{[N_2O]^2}$, (b) $Q_c = \dfrac{[NO]^2[Br_2]}{[NOBr]^2}$, (c) $Q_c = \dfrac{[H^+][F^-]}{[HF]}$,

(d) $Q_c = \dfrac{[CO_2][H_2]}{[CO][H_2O]}$, (e) $Q_c = \dfrac{[CS_2][H_2]^4}{[CH_4][H_2S]^2}$, (f) $Q_c = \dfrac{[H^+]^2[C_2O_4^{2-}]}{[H_2C_2O_4]}$.

15.2B (a) $H_2 + Cl_2 \rightleftharpoons 2HCl$, (b) $H^+ + F^- \rightleftharpoons HF$,
(c) $Cr^{3+} + 4OH^- \rightleftharpoons Cr(OH)_4^-$, (d) $HClO \rightleftharpoons H^+ + ClO^-$,
(e) $H_2SO_3 \rightleftharpoons H^+ + HSO_3^-$, (f) $2NO + Br_2 \rightleftharpoons 2NOBr$.

15.3A (a) $K_c = \dfrac{[HCl]^4}{[SiCl_4][H_2]^2}$, (b) $K_c = \dfrac{1}{[Hg^{2+}][Cl^-]^2}$,

(c) $K_c = \dfrac{[Ni(CO)_4]}{[CO]^4}$, (d) $K_c = \dfrac{[Zn^{2+}]}{[Fe^{2+}]}$. **15.3B** The expressions in
(a), (c), and (d) correspond to heterogeneous equilibria. It would be impossible to write a balanced chemical equation using only the species in each of these expressions, indicating that there are species in each reaction (solids or liquids) that do not appear in the equilibrium expression. **15.4A** (a) 2.3×10^{24}, (b) 6.6×10^{-13}, (c) 2.8×10^{-15}. **15.4B** (a) $2A(g) + 2B(g) \rightleftharpoons 2C(g) + 2D(g)$, (b) $C(g) + D(g) \rightleftharpoons A(g) + B(g)$, (c) $\frac{1}{3}A(g) + \frac{1}{3}B(g) \rightleftharpoons \frac{1}{3}C(g) + \frac{1}{3}D(g)$, (d) $\frac{1}{2}C(g) + \frac{1}{2}D(g) \rightleftharpoons \frac{1}{2}A(g) + \frac{1}{2}B(g)$.

15.5A (a) $K_P = \dfrac{(P_{CO_2})^2}{(P_{CO})^2(P_{O_2})}$, (b) $K_P = P_{CO_2}$, (c) $K_P = \dfrac{(P_{NH_3})^2}{(P_{N_2})(P_{H_2})^3}$.

15.5B (a) $2NO_2 + O_2 \rightleftharpoons 2NO_3$, (b) $CH_4 + 2H_2O \rightleftharpoons CO_2 + 4H_2$, (c) $I_2 + H_2 \rightleftharpoons 2HI$. **15.6A** 9.4×10^{-5}. **15.6B** 0.104.
15.7A Left. **15.7B** Right. **15.8A** $[H_2] = [I_2] = 0.056\ M$; $[HI] = 0.413\ M$. **15.8B** 0.20 M. **15.9A** $[H_2] = [I_2] = 0.081\ M$, $[HI] = 0.594\ M$. **15.9B** $[Br] = 8.4 \times 10^{-3}\ M$, $[Br_2] = 6.5 \times 10^{-2}\ M$.
15.10A $[H^+] = [CN^-] = 7.0 \times 10^{-6}\ M$, $[HCN] \approx 0.100\ M$.
15.10B $K_c = 3.3 \times 10^{-9}$. **15.11A** $[P_{H_2}] = [P_{I_2}] = 0.37$ atm, $P_{HI} = 2.75$ atm. **15.11B** $P_{H_2} = 0.41$ atm, $P_{I_2} = 0.21$ atm, $P_{HI} = 2.2$ atm. **15.12A** (a) Right, (b) left, (c) right, (d) left.
15.12B (a) $Ag(NH_3)_2^+(aq)$ or $Cl^-(aq)$, (b) $NH_3(aq)$, (c) $AgCl(s)$.
15.13A (a) Right, (b) right, (c) right. **15.13B** Right shift: remove HF or add H_2 or add F_2; left shift: remove H_2 or remove F_2 or add HF; no shift: decrease or increase volume of container.
15.14A Temperature increase will cause equilibrium to shift left. Temperature decrease will cause equilibrium to shift right.
15.14B Endothermic.

Answers to Checkpoints

15.2.1 e. **15.2.2** a. **15.3.1** d. **15.3.2** a. **15.3.3** c. **15.3.4** c. **15.4.1** d. **15.4.2** b. **15.4.3** a, b, c. **15.4.4** e. **15.5.1** a, c. **15.5.2** b, d, e. **15.5.3** a, d, e. **15.5.4** b. **15.5.5** b. **15.5.6** b.

CHAPTER 16

Acids and Bases

A helicopter delivers lime to increase the pH of a remote lake acidified by decades of acid rain.

Environmental Images/Universal Images Group/Shutterstock

In This Chapter, You Will Learn

More about the properties of acids and bases and how those properties are related to molecular structure.

Before You Begin, Review These Skills

- The list of strong acids and the list of strong bases [⏮ Section 4.3, Table 4.4]
- How to solve equilibrium problems [⏮ Section 15.4]
- Chapter 15 Key Skills [⏮ pages 770–771]

How Human Activity Has Impacted the Acidity of Rain

Rainwater in unpolluted areas is slightly acidic. The cause is carbon dioxide in the air, which dissolves in raindrops and reacts to form the weak acid, *carbonic* acid (H_2CO_3). In areas where the air is polluted by the burning of fossil fuels, rain can be very acidic. This phenomenon, known as *acid rain,* was first discovered in 1852 by Scottish chemist Robert Angus Smith (1817–1884) in Manchester, England. At that time, the industrial revolution was well under way and the British economy relied heavily on the use of coal to generate steam. The two main atmospheric contributors to the acidity of acid rain are sulfur dioxide (SO_2) and oxides of nitrogen (NO_x), which react to produce, respectively, sulfuric acid (H_2SO_4) and nitric acid (HNO_3)—both *strong* acids. SO_2 and NO_x are both produced by the burning of sulfur-bearing coal.

Student Note: "NO_x" is a collective term used to mean NO and NO_2.

Although acid rain was first discovered in 1852, scientific and societal efforts to understand and remedy its causes did not emerge for another century. U.S. scientists only began to study the causes and effects of acid rain extensively in the late 1960s. Public awareness of the phenomenon was heightened in the 1970s when results of studies at the Hubbard Brook Experimental Forest in New Hampshire were published, detailing the devastating impact of acid rain on the ecology of the region.

Despite the dire circumstance described by atmospheric scientists and ecologists in the last half of the twentieth century, there appears to be reason for optimism with regard to acid rain. The Acid Rain Program, part of the 1990 amendment to the Clean Air Act (originally enacted in 1963), has contributed to reducing the acidity of rainwater in industrial areas of the United States to levels below those seen in the 1960s. Reducing sulfur emissions from coal-burning power plants has involved using coal containing less sulfur, fitting smokestacks with chemical scrubbers to remove SO_2, and developing alternative sources of energy. Significant reductions in NO_x emissions have been achieved through the use of catalytic converters, which have been standard equipment on automobiles in the United States since the mid-1970s.

Understanding the causes of acid rain and working on prevention and remediation requires an understanding of the properties of *acids* and *bases*.

At the end of this chapter, you will be able to solve a series of problems related to the acidity of rainwater [⏭ Applying What You've Learned, page 840].

TABLE 16.1	Conjugate Bases of Some Common Species
Species	**Conjugate Base**
CH_3COOH	CH_3COO^-
H_2O	OH^-
NH_3	NH_2^-
H_2SO_4	HSO_4^-

TABLE 16.2	Conjugate Acids of Some Common Species
Species	**Conjugate Acid**
NH_3	NH_4^+
H_2O	H_3O^+
OH^-	H_2O
H_2NCONH_2 (urea)	$H_2NCONH_3^+$

16.1 Brønsted Acids and Bases

In Chapter 4 we learned that a Brønsted acid is a substance that can donate a proton and a Brønsted base is a substance that can accept a proton [◀◀ Section 4.3]. In this chapter we extend our discussion of Brønsted acid-base theory to include conjugate acids and conjugate bases.

When a Brønsted acid donates a proton, what remains of the acid is known as a *conjugate base.* For example, in the ionization of HCl in water:

$$\underset{\text{acid}}{HCl(aq)} + H_2O(l) \longrightarrow H_3O^+(aq) + \underset{\text{conjugate base}}{Cl^-(aq)}$$

HCl donates a proton to water, producing the hydronium ion (H_3O^+) and the chloride ion (Cl^-), which is the conjugate base of HCl. The two species, HCl and Cl^-, are known as a *conjugate acid-base pair* or simply a *conjugate pair.* Table 16.1 lists the conjugate bases of several familiar species.

Conversely, when a Brønsted base *accepts* a proton, the newly formed *protonated* species is known as a *conjugate acid.* When ammonia (NH_3) ionizes in water:

$$\underset{\text{base}}{NH_3(aq)} + H_2O(l) \rightleftharpoons \underset{\text{conjugate acid}}{NH_4^+(aq)} + OH^-(aq)$$

NH_3 accepts a proton from water to become the ammonium ion (NH_4^+). The ammonium ion is the conjugate acid of ammonia. Table 16.2 lists the conjugate acids of several common species.

Any reaction that we describe using Brønsted acid-base theory involves an acid and a base. The acid donates the proton, and the base accepts it. Furthermore, the products of such a reaction are always a conjugate base and a conjugate acid. It is useful to identify and label each species in a Brønsted acid-base reaction. For the ionization of HCl in water, the species are labeled as follows:

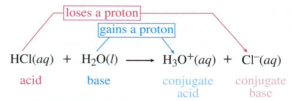

And for the ionization of NH_3 in water:

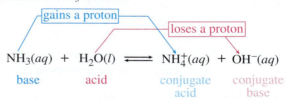

Sample Problems 16.1 and 16.2 let you practice identifying conjugate pairs and the species in a Brønsted acid-base reaction.

Student Hot Spot

Student data indicate you may struggle with conjugate acid-base pairs. Access the eBook to view additional Learning Resources on this topic.

SAMPLE PROBLEM 16.1

What is (a) the conjugate base of HNO_3, (b) the conjugate acid of O^{2-}, (c) the conjugate base of HSO_4^-, and (d) the conjugate acid of HCO_3^-?

Strategy To find the conjugate base of a species, *remove* a proton from the formula. To find the conjugate acid of a species, *add* a proton to the formula.

Setup The word *proton,* in this context, refers to H^+. Thus, the formula and the charge will both be affected by the addition or removal of H^+.

Solution

(a) NO_3^- (b) OH^- (c) SO_4^{2-} (d) H_2CO_3

THINK ABOUT IT

A species does not need to be what we think of as an acid for it to have a conjugate base. For example, we would not refer to the hydroxide ion (OH^-) as an acid—but it does have a conjugate base, the oxide ion (O^{2-}). Furthermore, a species that can either lose or gain a proton, such as HCO_3^-, has both a conjugate base (CO_3^{2-}) and a conjugate acid (H_2CO_3).

Practice Problem **A**TTEMPT What is (a) the conjugate acid of ClO_4^-, (b) the conjugate acid of S^{2-}, (c) the conjugate base of H_2S, and (d) the conjugate base of $H_2C_2O_4$?

Practice Problem **B**UILD HSO_3^- is the conjugate acid of what species? HSO_3^- is the conjugate base of what species?

Practice Problem **C**ONCEPTUALIZE Which of the models represents a species that has a conjugate base? Which represents a species that is the conjugate base of another species?

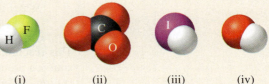

(i)　　(ii)　　(iii)　　(iv)

SAMPLE PROBLEM 16.2

Label each of the species in the following equations as an acid, base, conjugate base, or conjugate acid:

(a) $HF(aq) + NH_3(aq) \rightleftharpoons F^-(aq) + NH_4^+(aq)$　　　　(b) $CH_3COO^-(aq) + H_2O(l) \rightleftharpoons CH_3COOH(aq) + OH^-(aq)$

Strategy In each equation, the reactant that loses a proton is the acid and the reactant that gains a proton is the base. Each product is the conjugate of one of the reactants. Two species that differ only by a proton constitute a conjugate pair.

Setup (a) HF loses a proton and becomes F^-; NH_3 gains a proton and becomes NH_4^+.
(b) CH_3COO^- gains a proton to become CH_3COOH; H_2O loses a proton to become OH^-.

Solution

(a) $HF(aq) + NH_3(aq) \rightleftharpoons F^-(aq) + NH_4^+(aq)$
　　acid　　　　base　　　　conjugate　　conjugate
　　　　　　　　　　　　　　base　　　　acid

(b) $CH_3COO^-(aq) + H_2O(l) \rightleftharpoons CH_3COOH(aq) + OH^-(aq)$
　　　　base　　　　　acid　　　conjugate acid　　conjugate base

THINK ABOUT IT

In a Brønsted acid-base reaction, there is always an acid and a base, and whether a substance behaves as an acid or a base depends on what it is combined with. Water, for example, behaves as a base when combined with HCl but behaves as an acid when combined with NH_3.

Practice Problem **A**TTEMPT Identify and label the species in each reaction:

(a) $NH_4^+(aq) + H_2O(l) \rightleftharpoons NH_3(aq) + H_3O^+(aq)$　　　　(b) $CN^-(aq) + H_2O(l) \rightleftharpoons HCN(aq) + OH^-(aq)$

Practice Problem **B**UILD (a) Write an equation in which HSO_4^- reacts (with water) to form its conjugate base.
(b) Write an equation in which HSO_4^- reacts (with water) to form its conjugate acid.

Practice Problem **C**ONCEPTUALIZE Write the formula and charge for each species in this reaction and identify each as an acid, a base, a conjugate acid, or a conjugate base.

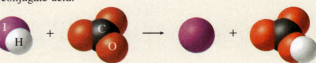

CHECKPOINT – SECTION 16.1　Brønsted Acids and Bases

16.1.1 Which of the following pairs of species are conjugate pairs? (Select all that apply.)

　a) H_2S and S^{2-}　　c) O_2 and H_2O_2　　e) HCl and OH^-

　b) NH_2^- and NH_3　　d) HBr and Br^-

16.1.2 Which of the following species does *not* have a conjugate base? (Select all that apply.)

　a) $HC_2O_4^-$　　c) O^{2-}　　e) HClO

　b) OH^-　　d) CO_3^{2-}

16.2   The Acid-Base Properties of Water

Water is often referred to as the "universal solvent," because it is so common and so important to life on Earth. In addition, most of the acid-base chemistry that we discuss takes place in aqueous solution. In this section, we take a closer look at water's ability to act as either a Brønsted acid (as in the ionization of NH_3) or a Brønsted base (as in the ionization of HCl). A species that can behave as either a Brønsted acid or a Brønsted base is called ***amphoteric.***

Water is a very weak electrolyte, but it does undergo ionization to a small extent:

$$H_2O(l) \rightleftharpoons H^+(aq) + OH^-(aq)$$

This reaction is known as the ***autoionization of water.*** Because we can represent the aqueous proton as either H^+ or H_3O^+ [◄◄ Section 4.3], we can also write the autoionization of water as:

$$2H_2O(l) \rightleftharpoons H_3O^+(aq) + OH^-(aq)$$

| acid | base | conjugate acid | conjugate base |

where one water molecule acts as an acid and the other acts as a base.

As indicated by the double arrow in the equation, the reaction is an equilibrium. The equilibrium expression for the autoionization of water is:

$$K_w = [H_3O^+][OH^-] \quad \text{or} \quad K_w = [H^+][OH^-]$$

Animation
Chemical Equilibrium—equilibrium.

Recall that in a heterogeneous equilibrium such as this, liquids and solids do not appear in the equilibrium expression [◄◄ Section 15.3]. Because the autoionization of water is an important equilibrium that comes up frequently in the study of acids and bases, we use the subscript "w" to indicate that the equilibrium constant is specifically the one for the autoionization of *water*. It is important to realize, though, that K_w, which is sometimes referred to as the ***ion-product constant,*** is simply a K_c for a *specific reaction*. We frequently replace the c in K_c expressions with a letter or a series of letters to indicate the specific type of reaction to which the K_c refers. For example, K_c for the ionization of a weak acid is called K_a, and K_c for the ionization of a weak base is called K_b. In Chapter 17 we introduce K_{sp}, where "sp" stands for "solubility product." Each specially subscripted K is simply a K_c for a specific type of reaction.

In pure water, autoionization is the only source of H_3O^+ and OH^-, and the stoichiometry of the reaction tells us that their concentrations are equal. At 25°C, the concentrations of hydronium and hydroxide ions in pure water are $[H_3O^+] = [OH^-] = 1.0 \times 10^{-7}$ M. Using the equilibrium expression, we can calculate the value of K_w at 25°C as follows:

Student Note: Recall that we disregard the units when we substitute concentrations into an equilibrium expression [◄◄ Section 15.5].

$$K_w = [H_3O^+][OH^-] = (1.0 \times 10^{-7})(1.0 \times 10^{-7}) = 1.0 \times 10^{-14}$$

Furthermore, in any aqueous solution at 25°C, the product of H_3O^+ and OH^- concentrations is equal to 1.0×10^{-14}:

Equation 16.1 $K_w = [H_3O^+][OH^-] = 1.0 \times 10^{-14}$ (at 25°C)

Although their product is a constant, the individual concentrations of hydronium and hydroxide can be influenced by the addition of an acid or a base. However, because the product of H_3O^+ and OH^- concentrations is a constant, we cannot alter the concentrations independently. Any change in one also affects the other. The relative amounts of H_3O^+ and OH^- determine whether a solution is neutral, acidic, or basic:

- When $[H_3O^+] = [OH^-]$, the solution is neutral.
- When $[H_3O^+] > [OH^-]$, the solution is acidic.
- When $[H_3O^+] < [OH^-]$, the solution is basic.

Sample Problem 16.3 shows how to use Equation 16.1.

SAMPLE PROBLEM 16.3

The concentration of hydronium ions in stomach acid is 0.10 M. Calculate the concentration of hydroxide ions in stomach acid at 25°C.

Strategy Use the value of K_w to determine [OH⁻] when [H₃O⁺] = 0.10 M.

Setup $K_w = [H_3O^+][OH^-] = 1.0 \times 10^{-14}$ at 25°C. Rearranging Equation 16.1 to solve for [OH⁻]:

$$[OH^-] = \frac{1.0 \times 10^{-14}}{[H_3O^+]}$$

Solution

$$[OH^-] = \frac{1.0 \times 10^{-14}}{0.10} = 1.0 \times 10^{-13} \, M$$

> **THINK ABOUT IT**
>
> Remember that equilibrium constants are temperature dependent. The value of K_w is 1.0×10^{-14} only at 25°C.

Practice Problem ATTEMPT The concentration of hydroxide ions in the antacid milk of magnesia is $5.0 \times 10^{-4} \, M$. Calculate the concentration of hydronium ions at 25°C.

Practice Problem BUILD The value of K_w at normal body temperature (37°C) is 2.8×10^{-14}. Calculate the concentration of hydroxide ions in stomach acid at body temperature. ([H₃O⁺] = 0.10 M.)

Practice Problem CONCEPTUALIZE The first diagram represents a system consisting of the weak electrolyte AB(l). Like water, liquid AB can autoionize to form A⁺ ions (red) and B⁻ ions (blue). At room temperature, the product of ion concentrations, [A⁺][B⁻], is always equal to 16 for the volume represented here. Also, because every AB molecule that ionizes produces one A⁺ ion and one B⁻ ion, in a pure sample of AB(l), the concentrations [A⁺] and [B⁻] are equal to each other. Which of the other diagrams [(i)–(iii)] best represents the system after enough of the strong electrolyte NaB has been dissolved to increase the number of B⁻ ions to 8? (Although AB is a liquid, the molecules are shown far apart to keep the diagrams from being too crowded. Na⁺ ions also are not shown, in order to keep the diagrams clear.)

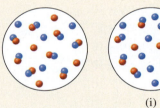
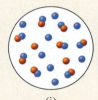
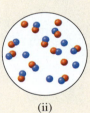
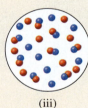

(i)   (ii)   (iii)

CHECKPOINT – SECTION 16.2   The Acid-Base Properties of Water

16.2.1 Calculate [OH⁻] in a solution in which [H₃O⁺] = 0.0012 M at 25°C.

a) $1.2 \times 10^{-3} \, M$

b) $8.3 \times 10^{-17} \, M$

c) $1.0 \times 10^{-14} \, M$

d) $8.3 \times 10^{-12} \, M$

e) $1.2 \times 10^{11} \, M$

16.2.2 Calculate [H₃O⁺] in a solution in which [OH⁻] = 0.25 M at 25°C.

a) $4.0 \times 10^{-14} \, M$

b) $1.0 \times 10^{-14} \, M$

c) $2.5 \times 10^{13} \, M$

d) $1.0 \times 10^{-7} \, M$

e) $4.0 \times 10^{-7} \, M$

16.3 The pH Scale

The acidity of an aqueous solution depends on the concentration of hydronium ions, [H₃O⁺]. This concentration can range over many orders of magnitude, which can make reporting the numbers cumbersome. To describe the acidity of a solution, rather than report the molar concentration of hydronium ions, we typically use the more convenient pH scale. The **pH** of a solution is defined as the negative base-10 logarithm of the hydronium ion concentration (in mol/L):

Equation 16.2	$pH = -\log [H_3O^+]$ or $pH = -\log [H^+]$

Equation 16.2 converts numbers that can span an enormous range ($\sim 10^1$ to 10^{-14}) to numbers generally ranging from ~ 1 to 14. The pH of a solution is a dimensionless quantity, so the units of concentration must be removed from $[H_3O^+]$ before taking the logarithm. Because $[H_3O^+] = [OH^-] = 1.0 \times 10^{-7}$ M in pure water at 25°C, the pH of pure water at 25°C is:

$$-\log (1.0 \times 10^{-7}) = 7.00$$

Student Note: A word about significant figures: When we take the log of a number with two significant figures, we report the result to two places past the decimal point. Thus, pH 7.00 has two significant figures, not three.

Remember, too, that a solution in which $[H_3O^+] = [OH^-]$ is neutral. At 25°C, therefore, a neutral solution has pH 7.00. An acidic solution, one in which $[H_3O^+] > [OH^-]$, has pH < 7.00, whereas a basic solution, in which $[H_3O^+] < [OH^-]$, has pH > 7.00. Table 16.3 shows the calculation of pH for solutions ranging from 0.10 M to 1.0×10^{-14} M.

In the laboratory, pH is measured with a pH meter (Figure 16.1). Table 16.4 lists the pH values of a number of common fluids. Note that the pH of body fluids varies greatly, depending on the location and function of the fluid. The low pH (high acidity) of gastric juices is vital for digestion of food, whereas the higher pH of blood is required to facilitate the transport of oxygen.

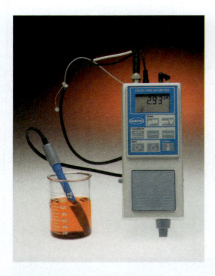

Figure 16.1 A pH meter is commonly used in the laboratory to determine the pH of a solution. Although many pH meters have a range of 1 to 14, pH values can actually be less than 1 and greater than 14.

Charles D. Winters/McGraw Hill

TABLE 16.3	Benchmark pH Values for a Range of Hydronium Ion Concentrations at 25°C		
$[H_3O^+]$ (M)	$-\log [H_3O^+]$	pH	
0.10	$-\log (1.0 \times 10^{-1})$	1.00	
0.010	$-\log (1.0 \times 10^{-2})$	2.00	
1.0×10^{-3}	$-\log (1.0 \times 10^{-3})$	3.00	
1.0×10^{-4}	$-\log (1.0 \times 10^{-4})$	4.00	
1.0×10^{-5}	$-\log (1.0 \times 10^{-5})$	5.00	
1.0×10^{-6}	$-\log (1.0 \times 10^{-6})$	6.00	Acidic
1.0×10^{-7}	$-\log (1.0 \times 10^{-7})$	7.00	Neutral
1.0×10^{-8}	$-\log (1.0 \times 10^{-8})$	8.00	Basic
1.0×10^{-9}	$-\log (1.0 \times 10^{-9})$	9.00	
1.0×10^{-10}	$-\log (1.0 \times 10^{-10})$	10.00	
1.0×10^{-11}	$-\log (1.0 \times 10^{-11})$	11.00	
1.0×10^{-12}	$-\log (1.0 \times 10^{-12})$	12.00	
1.0×10^{-13}	$-\log (1.0 \times 10^{-13})$	13.00	
1.0×10^{-14}	$-\log (1.0 \times 10^{-14})$	14.00	

TABLE 16.4	pH Values of Some Common Fluids		
Fluid	**pH**	**Fluid**	**pH**
Stomach acid	1.5	Saliva	6.4–6.9
Lemon juice	2.0	Milk	6.5
Vinegar	3.0	Pure water	7.0
Grapefruit juice	3.2	Blood	7.35–7.45
Orange juice	3.5	Tears	7.4
Urine	4.8–7.5	Milk of magnesia	10.6
Rainwater (in clean air)	5.5	Household ammonia	11.5

A measured pH can be used to determine experimentally the concentration of hydronium ion in solution. Solving Equation 16.2 for $[H_3O^+]$ gives:

$$[H_3O^+] = 10^{-pH} \qquad \textbf{Equation 16.3}$$

Sample Problems 16.4 and 16.5 illustrate calculations involving pH.

SAMPLE PROBLEM 16.4

Determine the pH of a solution at 25°C in which the hydronium ion concentration is (a) 3.5×10^{-4} M, (b) 1.7×10^{-7} M, and (c) 8.8×10^{-11} M.

Strategy Given $[H_3O^+]$, use Equation 16.2 to solve for pH.

Setup

(a) pH = $-\log (3.5 \times 10^{-4})$ (c) pH = $-\log (8.8 \times 10^{-11})$

(b) pH = $-\log (1.7 \times 10^{-7})$

Solution

(a) pH = 3.46   (c) pH = 10.06

(b) pH = 6.77

THINK ABOUT IT

When a hydronium ion concentration falls between two "benchmark" concentrations in Table 16.3, the pH falls between the two corresponding pH values. In part (c), for example, the hydronium ion concentration (8.8×10^{-11} M) is greater than 1.0×10^{-11} M but less than 1.0×10^{-10} M. Therefore, we expect the pH to be between 11.00 and 10.00.

$[H_3O^+]$ (M)	$-\log [H_3O^+]$	pH
1.0×10^{-10}	$-\log (1.0 \times 10^{-10})$	10.00
$8.8 \times 10^{-11*}$	$-\log (8.8 \times 10^{-11})$	10.06[†]
1.0×10^{-11}	$-\log (1.0 \times 10^{-11})$	11.00

*$[H_3O^+]$ between two benchmark values
[†]pH between two benchmark values

Recognizing the benchmark concentrations and corresponding pH values is a good way to determine whether or not your calculated result is reasonable.

Practice Problem **A**TTEMPT   Determine the pH of a solution at 25°C in which the hydronium ion concentration is (a) 3.2×10^{-9} M, (b) 4.0×10^{-8} M, and (c) 5.6×10^{-2} M.

Practice Problem **B**UILD   Determine the pH of a solution at 25°C in which the hydroxide ion concentration is (a) 8.3×10^{-15} M, (b) 3.3×10^{-4} M, and (c) 1.2×10^{-3} M.

Practice Problem **C**ONCEPTUALIZE   Strong acid is added in 1-mL increments to a liter of water at 25°C. Which of the following graphs best approximates the result of plotting hydronium ion concentration as a function of mL acid added? Which graph best approximates the result of plotting pH as a function of mL acid added?

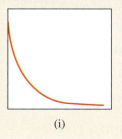
(i)

(ii)

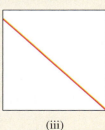
(iii)

(iv)

SAMPLE PROBLEM   16.5

Calculate the hydronium ion concentration in a solution at 25°C in which the pH is (a) 4.76, (b) 11.95, and (c) 8.01.

Strategy Given pH, use Equation 16.3 to calculate $[H_3O^+]$.

Setup

(a) $[H_3O^+] = 10^{-4.76}$ (c) $[H_3O^+] = 10^{-8.01}$

(b) $[H_3O^+] = 10^{-11.95}$

Solution

(a) $[H_3O^+] = 1.7 \times 10^{-5}\ M$ (c) $[H_3O^+] = 9.8 \times 10^{-9}\ M$

(b) $[H_3O^+] = 1.1 \times 10^{-12}\ M$

THINK ABOUT IT

If you use the calculated hydronium ion concentrations to recalculate pH, you will get numbers slightly different from those given in the problem. In part (a), for example, $-\log(1.7 \times 10^{-5}) = 4.77$. The small difference between this and 4.76 (the pH given in the problem) is due to a rounding error. Remember that a concentration derived from a pH with two digits to the right of the decimal point can have only two significant figures. Note also that the benchmarks can be used equally well in this circumstance. A pH between 4 and 5 corresponds to a hydronium ion concentration between $1 \times 10^{-4}\ M$ and $1 \times 10^{-5}\ M$.

Practice Problem ATTEMPT Calculate the hydronium ion concentration in a solution at 25°C in which the pH is (a) 9.90, (b) 1.45, and (c) 7.01.

Practice Problem BUILD Calculate the hydroxide ion concentration in a solution at 25°C in which the pH is (a) 11.89, (b) 2.41, and (c) 7.13.

Practice Problem CONCEPTUALIZE What is the value of the exponent in the hydronium ion concentration for solutions with pH values of 5.90, 10.11, and 1.25?

A *pOH* scale analogous to the pH scale can be defined using the negative base-10 logarithm of the *hydroxide* ion concentration of a solution, $[OH^-]$:

Equation 16.4 $pOH = -\log[OH^-]$

Rearranging Equation 16.4 to solve for hydroxide ion concentration gives:

Equation 16.5 $[OH^-] = 10^{-pOH}$

Now consider again the K_w equilibrium expression for water at 25°C:

$$[H_3O^+][OH^-] = 1.0 \times 10^{-14}$$

Taking the negative logarithm of both sides, we obtain:

$$-\log([H_3O^+][OH^-]) = -\log(1.0 \times 10^{-14})$$

$$-(\log[H_3O^+] + \log[OH^-]) = 14.00$$

$$-\log[H_3O^+] - \log[OH^-] = 14.00$$

$$(-\log[H_3O^+]) + (-\log[OH^-]) = 14.00$$

And from the definitions of pH and pOH, we see that at 25°C:

Equation 16.6 $pH + pOH = 14.00$

TABLE 16.5	Benchmark pOH Values for a Range of Hydroxide Ion Concentrations at 25°C	
[OH⁻] (*M*)	pOH	
0.10	1.00	
1.0×10^{-3}	3.00	
1.0×10^{-5}	5.00	Basic
1.0×10^{-7}	7.00	Neutral
1.0×10^{-9}	9.00	Acidic
1.0×10^{-11}	11.00	
1.0×10^{-13}	13.00	

Equation 16.6 provides another way to express the relationship between the hydronium ion concentration and the hydroxide ion concentration. On the pOH scale, 7.00 is neutral, numbers greater than 7.00 indicate that a solution is acidic, and numbers less than 7.00 indicate that a solution is basic. Table 16.5 lists pOH values for a range of hydroxide ion concentrations at 25°C.

Sample Problems 16.6 and 16.7 illustrate calculations involving pOH.

SAMPLE PROBLEM 16.6

Determine the pOH of a solution at 25°C in which the hydroxide ion concentration is (a) 3.7×10^{-5} *M*, (b) 4.1×10^{-7} *M*, and (c) 8.3×10^{-2} *M*.

Strategy Given [OH⁻], use Equation 16.4 to calculate pOH.

Setup

(a) pOH = $-\log (3.7 \times 10^{-5})$ (b) pOH = $-\log (4.1 \times 10^{-7})$ (c) pOH = $-\log (8.3 \times 10^{-2})$

Solution

(a) pOH = 4.43 (b) pOH = 6.39 (c) pOH = 1.08

THINK ABOUT IT

Remember that the pOH scale is, in essence, the *reverse* of the pH scale. On the pOH scale, numbers below 7 indicate a basic solution, whereas numbers above 7 indicate an acidic solution. The pOH benchmarks (abbreviated in Table 16.5) work the same way the pH benchmarks do. In part (a), for example, a hydroxide ion concentration between 1×10^{-4} *M* and 1×10^{-5} *M* corresponds to a pOH between 4 and 5:

[OH⁻] (*M*)	pOH
1.0×10^{-4}	4.00
3.7×10^{-5}*	4.43†
1.0×10^{-5}	5.00

*[OH⁻] between two benchmark values
†pOH between two benchmark values

Practice Problem (A)TTEMPT Determine the pOH of a solution at 25°C in which the hydroxide ion concentration is (a) 5.7×10^{-12} *M*, (b) 7.3×10^{-3} *M*, and (c) 8.5×10^{-6} *M*.

Practice Problem (B)UILD Determine the pH of a solution at 25°C in which the hydroxide ion concentration is (a) 2.8×10^{-8} *M*, (b) 9.9×10^{-9} *M*, and (c) 1.0×10^{-11} *M*.

Practice Problem (C)ONCEPTUALIZE Without doing any calculations, determine between which two whole numbers the pOH will be for solutions with OH⁻ concentrations of 4.71×10^{-5} *M*, 2.9×10^{-12} *M*, and 7.15×10^{-3} *M*.

SAMPLE PROBLEM 16.7

Calculate the hydroxide ion concentration in a solution at 25°C in which the pOH is (a) 4.91, (b) 9.03, and (c) 10.55.

Strategy Given pOH, use Equation 16.5 to calculate [OH⁻].

Setup

(a) $[OH^-] = 10^{-4.91}$

(b) $[OH^-] = 10^{-9.03}$

(c) $[OH^-] = 10^{-10.55}$

Solution

(a) $[OH^-] = 1.2 \times 10^{-5}\ M$

(b) $[OH^-] = 9.3 \times 10^{-10}\ M$

(c) $[OH^-] = 2.8 \times 10^{-11}\ M$

THINK ABOUT IT

Use the benchmark pOH values to determine whether these solutions are reasonable. In part (a), for example, the pOH between 4 and 5 corresponds to [OH⁻] between $1 \times 10^{-4}\ M$ and $1 \times 10^{-5}\ M$.

Practice Problem **A**TTEMPT Calculate the hydroxide ion concentration in a solution at 25°C in which the pOH is (a) 13.02, (b) 5.14, and (c) 6.98.

Practice Problem **B**UILD Calculate the hydronium ion concentration in a solution at 25°C in which the pOH is (a) 2.74, (b) 10.31, and (c) 12.40.

Practice Problem **C**ONCEPTUALIZE What is the value of the exponent in the hydronium ion concentration for solutions with pOH values of 2.90, 8.75, and 11.86?

Bringing Chemistry to Life

Antacids and the pH Balance in Your Stomach

An average adult produces between 2 and 3 L of gastric juice daily. Gastric juice is an acidic digestive fluid secreted by glands in the mucous membrane that lines the stomach. It contains hydrochloric acid (HCl), among other substances. The pH of gastric juice is about 1.5, which corresponds to a hydrochloric acid concentration of 0.03 M.

The inside lining of the stomach is made up of *parietal* cells, which are fused together to form tight junctions. The interiors of the cells are protected from the surroundings by cell membranes. These membranes allow water and neutral molecules to pass in and out of the stomach, but they usually block the movement of ions such as H_3O^+, Na^+, K^+, and Cl^-. The H_3O^+ ions come from carbonic acid (H_2CO_3) formed as a result of the hydration of CO_2, an end product of metabolism:

$$CO_2(g) + H_2O(l) \rightleftharpoons H_2CO_3(aq)$$

$$H_2CO_3(aq) + H_2O(l) \rightleftharpoons H_3O^+(aq) + HCO_3^-(aq)$$

These reactions take place in the blood plasma bathing the cells in the mucosa. By a process known as *active transport,* H_3O^+ ions move across the membrane into the stomach interior. (Active transport processes are aided by enzymes.) To maintain electrical balance, an equal number of Cl^- ions also move from the blood plasma into the stomach. Once in the stomach, most of these ions are prevented by cell membranes from diffusing back into the blood plasma.

The purpose of the highly acidic medium within the stomach is to digest food and to activate certain digestive enzymes. Eating stimulates H_3O^+ ion secretion. A small fraction of these ions normally are reabsorbed by the mucosa, causing a number of tiny hemorrhages. About half a million cells are shed by the lining every minute, and a healthy stomach is completely relined every few days. However, if the acid content is excessively high, the constant influx of H_3O^+ ions through the membrane back to the blood plasma can cause muscle contraction, pain, swelling, inflammation, and bleeding.

One way to temporarily reduce the H_3O^+ ions concentration in the stomach is to take an antacid. The major function of antacids is to neutralize excess HCl in gastric juice. The following table lists the active ingredients of some popular antacids. The reactions by which these antacids neutralize stomach acid are as follows:

$$NaHCO_3(aq) + HCl(aq) \longrightarrow NaCl(aq) + H_2O(l) + CO_2(g)$$

$$CaCO_3(aq) + 2HCl(aq) \longrightarrow CaCl_2(aq) + H_2O(l) + CO_2(g)$$

$$MgCO_3(aq) + 2HCl(aq) \longrightarrow MgCl_2(aq) + H_2O(l) + CO_2(g)$$

$$Mg(OH)_2(s) + 2HCl(aq) \longrightarrow MgCl_2(aq) + 2H_2O(l)$$

$$Al(OH)_2NaCO_3(s) + 4HCl(aq) \longrightarrow AlCl_3(aq) + NaCl(aq) + 3H_2O(l) + CO_2(g)$$

Active Ingredients in Some Common Antacids

Commercial Name	Active Ingredients
Alka-Seltzer	Aspirin, sodium bicarbonate, citric acid
Milk of magnesia	Magnesium hydroxide
Rolaids	Dihydroxy aluminum sodium carbonate
TUMS	Calcium carbonate
Maalox	Sodium bicarbonate, magnesium carbonate

The CO_2 released by most of these reactions increases gas pressure in the stomach, causing the person to belch. The fizzing that takes place when Alka-Seltzer dissolves in water is caused by carbon dioxide, which is released by the reaction between citric acid and sodium bicarbonate:

$$3NaHCO_3(aq) + H_3C_6H_5O_7(aq) \longrightarrow 3CO_2(g) + 3H_2O(l) + Na_3C_6H_5O_7(aq)$$

This effervescence helps to disperse the ingredients and enhances the palatability of the solution.

CHECKPOINT – SECTION 16.3   The pH Scale

16.3.1 Determine the pH of a solution at 25°C in which $[H_3O^+] = 6.35 \times 10^{-8}\ M$.

a) 7.651

b) 6.803

c) 7.197

d) 6.350

e) 8.000

16.3.2 Determine $[H_3O^+]$ in a solution at 25°C if pH = 5.75.

a) $1.8 \times 10^{-6}\ M$

b) $5.6 \times 10^{-9}\ M$

c) $5.8 \times 10^{-6}\ M$

d) $2.4 \times 10^{-9}\ M$

e) $1.0 \times 10^{-6}\ M$

16.3.3 Determine the pOH of a solution at 25°C in which $[OH^-] = 4.65 \times 10^{-3}\ M$.

a) 11.667

b) 13.687

c) 0.326

d) 4.652

e) 2.333

16.3.4 Determine $[OH^-]$ in a solution at 25°C if pH = 10.50.

a) $3.2 \times 10^{-11}\ M$

b) $3.2 \times 10^{-4}\ M$

c) $1.1 \times 10^{-2}\ M$

d) $7.1 \times 10^{-8}\ M$

e) $8.5 \times 10^{-7}\ M$

16.4 Strong Acids and Bases

Most of this chapter and Chapter 17 deal with equilibrium and the application of the principles of equilibrium to a variety of reaction types. In the context of our discussion of acids and bases, however, it is necessary to review the ionization of strong acids and the dissociation of strong bases. These reactions generally are not treated as *equilibria* but rather as processes that go to completion. This makes the determination of pH for a solution of strong acid or strong base relatively simple.

Student Note: We indicate that ionization of a strong acid is *complete* by using a single arrow ($\longrightarrow$) instead of the double, equilibrium arrow ($\rightleftharpoons$) in the equation.

Strong Acids

There are many different acids, but as we learned in Chapter 4, relatively few qualify as *strong*.

Animation
Acids and Bases—the dissociation of strong and weak acids.

Strong Acid	Ionization Reaction
Hydrochloric acid	$HCl(aq) + H_2O(l) \longrightarrow H_3O^+(aq) + Cl^-(aq)$
Hydrobromic acid	$HBr(aq) + H_2O(l) \longrightarrow H_3O^+(aq) + Br^-(aq)$
Hydroiodic acid	$HI(aq) + H_2O(l) \longrightarrow H_3O^+(aq) + I^-(aq)$
Nitric acid	$HNO_3(aq) + H_2O(l) \longrightarrow H_3O^+(aq) + NO_3^-(aq)$
Chloric acid	$HClO_3(aq) + H_2O(l) \longrightarrow H_3O^+(aq) + ClO_3^-(aq)$
Perchloric acid	$HClO_4(aq) + H_2O(l) \longrightarrow H_3O^+(aq) + ClO_4^-(aq)$
Sulfuric acid	$H_2SO_4(aq) + H_2O(l) \longrightarrow H_3O^+(aq) + HSO_4^-(aq)$

Remember that although sulfuric acid has two ionizable protons, only the first ionization is complete. It is a good idea to commit this short list of strong acids to memory.

Because the ionization of a strong acid is complete, the concentration of hydronium ion at equilibrium is equal to the starting concentration of the strong acid. For instance, if we prepare a 0.10-M solution of HCl, the concentration of hydronium ion in the solution is 0.10 M. All the HCl ionizes, and no HCl molecules remain. Thus, at equilibrium (when the ionization is complete), [HCl] = 0 M and $[H_3O^+]$ = $[Cl^-]$ = 0.10 M. Therefore, the pH of the solution (at 25°C) is:

$$pH = -\log(0.10) = 1.00$$

This is a very low pH, which is consistent with a relatively concentrated solution of a strong acid. As we explain in Section 16.5, a solution of equal concentration but containing a weak acid has a higher pH.

Sample Problems 16.8 and 16.9 let you practice relating the concentration of a strong acid to the pH of an aqueous solution.

SAMPLE PROBLEM   16.8

Calculate the pH of an aqueous solution at 25°C that is (a) 0.035 M in HI, (b) 1.2×10^{-4} M in HNO_3, and (c) 6.7×10^{-5} M in $HClO_4$.

Strategy HI, HNO_3, and $HClO_4$ are all strong acids, so the concentration of hydronium ion in each solution is the same as the stated concentration of the acid. Use Equation 16.2 to calculate pH.

Setup

(a) $[H_3O^+] = 0.035$ M (b) $[H_3O^+] = 1.2 \times 10^{-4}$ M (c) $[H_3O^+] = 6.7 \times 10^{-5}$ M

Solution

(a) pH = $-\log(0.035) = 1.46$ (b) pH = $-\log(1.2 \times 10^{-4}) = 3.92$ (c) pH = $-\log(6.7 \times 10^{-5}) = 4.17$

THINK ABOUT IT

Again, note that when a hydronium ion concentration falls between two of the benchmark concentrations in Table 16.3, the pH falls between the two corresponding pH values. In part (b), for example, the hydronium ion concentration of 1.2×10^{-4} M is greater than 1.0×10^{-4} M and less than 1.0×10^{-3} M. Therefore, we expect the pH to be between 4.00 and 3.00:

$[H_3O^+]$ (M)	$-\log [H_3O^+]$	pH
1.0×10^{-3}	$-\log(1.0 \times 10^{-3})$	3.00
$1.2 \times 10^{-4*}$	$-\log(1.2 \times 10^{-3})$	3.92[†]
1.0×10^{-4}	$-\log(1.2 \times 10^{-4})$	4.00

*$[H_3O^+]$ between two benchmark values
[†]pH between two benchmark values

Being comfortable with the benchmark hydronium ion concentrations and the corresponding pH values will help you avoid some of the common errors in pH calculations.

Practice Problem **A**TTEMPT Calculate the pH of an aqueous solution at 25°C that is (a) 0.081 M in HI, (b) $8.2 \times 10^{-6}\,M$ in HNO_3, and (c) $5.4 \times 10^{-3}\,M$ in $HClO_4$.

Practice Problem **B**UILD Calculate the pOH of an aqueous solution at 25°C that is (a) 0.011 M in HNO_3, (b) $3.5 \times 10^{-3}\,M$ in HBr, and (c) $9.3 \times 10^{-6}\,M$ in HCl.

Practice Problem **C**ONCEPTUALIZE Estimate the pH of a solution prepared by dissolving 1.0×10^{-10} mole of a strong acid in a liter of water at 25°C.

SAMPLE PROBLEM 16.9

Calculate the concentration of HCl in a solution at 25°C that has pH (a) 4.95, (b) 3.45, and (c) 2.78.

Strategy Use Equation 16.3 to convert from pH to the molar concentration of hydronium ion. In a strong acid solution, the molar concentration of hydronium ion is equal to the acid concentration.

Setup

(a) $[HCl] = [H_3O^+] = 10^{-4.95}$ (b) $[HCl] = [H_3O^+] = 10^{-3.45}$ (c) $[HCl] = [H_3O^+] = 10^{-2.78}$

Solution

(a) $1.1 \times 10^{-5}\,M$ (b) $3.5 \times 10^{-4}\,M$ (c) $1.7 \times 10^{-3}\,M$

THINK ABOUT IT

As pH decreases, acid concentration increases. Note that when we take the inverse log of a number with two digits to the right of the decimal point, the result has two significant figures.

Practice Problem **A**TTEMPT Calculate the concentration of HNO_3 in a solution at 25°C that has pH (a) 2.06, (b) 1.77, and (c) 6.01.

Practice Problem **B**UILD Calculate the concentration of HBr in a solution at 25°C that has pOH (a) 9.19, (b) 12.18, and (c) 10.96.

Practice Problem **C**ONCEPTUALIZE Which of the plots best approximates the line that would result if pH were plotted as a function of hydronium ion concentration?

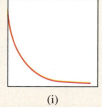 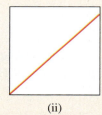  

        (i)   (ii)   (iii)   (iv)

Strong Bases

The list of strong bases is also fairly short. It consists of the hydroxides of alkali metals (Group 1) and the hydroxides of the heaviest alkaline earth metals (Group 2). The dissociation of a strong base is, for practical purposes, complete. Equations representing dissociations of the strong bases are as follows:

Group 1 hydroxides	Group 2 hydroxides
$LiOH(aq) \longrightarrow Li^+(aq) + OH^-(aq)$	$Ca(OH)_2(aq) \longrightarrow Ca^{2+}(aq) + 2OH^-(aq)$
$NaOH(aq) \longrightarrow Na^+(aq) + OH^-(aq)$	$Sr(OH)_2(aq) \longrightarrow Sr^{2+}(aq) + 2OH^-(aq)$
$KOH(aq) \longrightarrow K^+(aq) + OH^-(aq)$	$Ba(OH)_2(aq) \longrightarrow Ba^{2+}(aq) + 2OH^-(aq)$
$RbOH(aq) \longrightarrow Rb^+(aq) + OH^-(aq)$	
$CsOH(aq) \longrightarrow Cs^+(aq) + OH^-(aq)$	

Student Note: Recall that $Ca(OH)_2$ and $Sr(OH)_2$ are not very soluble, but what *does* dissolve dissociates completely [◄◄ Section 4.3, Table 4.4].

Again, because the reaction goes to completion, the pH of such a solution is relatively easy to calculate. In the case of a Group 1 hydroxide, the hydroxide ion concentration is simply the starting concentration of the strong base. In a solution that is 0.018 M

in NaOH, for example, $[OH^-] = 0.018$ M. Its pH can be calculated in two ways. Either we can use Equation 16.1 to determine hydronium ion concentration:

$$[H_3O^+][OH^-] = 1.0 \times 10^{-14}$$

$$[H_3O^+] = \frac{1.0 \times 10^{-14}}{[OH^-]} = \frac{1.0 \times 10^{-14}}{0.018} = 5.56 \times 10^{-13} \, M$$

and then Equation 16.2 to determine pH:

$$pH = -\log(5.56 \times 10^{-13} \, M) = 12.25$$

or we can calculate the pOH with Equation 16.3:

$$pOH = -\log(0.018) = 1.75$$

and use Equation 16.6 to convert to pH:

$$pH + pOH = 14.00$$

$$pH = 14.00 - 1.75 = 12.25$$

Both methods give the same result.

In the case of a Group 2 metal hydroxide, we must be careful to account for the reaction stoichiometry. For instance, if we prepare a solution that is 1.9×10^{-4} M in barium hydroxide, the concentration of hydroxide ion at equilibrium (after complete dissociation) is $2 (1.9 \times 10^{-4} \, M)$ or 3.8×10^{-4} M—twice the original concentration of $Ba(OH)_2$. Once we have determined the hydroxide ion concentration, we can determine pH as before:

$$[H_3O^+] = \frac{1.0 \times 10^{-14}}{[OH^-]} = \frac{1.0 \times 10^{-14}}{3.8 \times 10^{-4}} = 2.63 \times 10^{-11} \, M$$

and

$$pH = -\log(2.63 \times 10^{-11} \, M) = 10.58$$

or

$$pOH = -\log(3.8 \times 10^{-4}) = 3.42$$

$$pH + pOH = 14.00$$

$$pH = 14.00 - 3.42 = 10.58$$

Sample Problems 16.10 and 16.11 illustrate calculations involving hydroxide ion concentration, pOH, and pH.

SAMPLE PROBLEM 16.10

Calculate the pOH of the following aqueous solutions at 25°C: (a) 0.013 M LiOH, (b) 0.013 M $Ba(OH)_2$, (c) 9.2×10^{-5} M KOH.

Strategy LiOH, $Ba(OH)_2$, and KOH are all strong bases. Use reaction stoichiometry to determine hydroxide ion concentration and Equation 16.4 to determine pOH.

Setup (a) The hydroxide ion concentration is simply equal to the concentration of the base. Therefore, $[OH^-] = [LiOH] = 0.013$ M.
(b) The hydroxide ion concentration is twice that of the base:

$$Ba(OH)_2(aq) \longrightarrow Ba^{2+}(aq) + 2OH^-(aq)$$

Therefore, $[OH^-] = 2 \times [Ba(OH)_2] = 2(0.013 \, M) = 0.026$ M.

(c) The hydroxide ion concentration is equal to the concentration of the base. Therefore, $[OH^-] = [KOH] = 9.2 \times 10^{-5}$ M.

Solution (a) $pOH = -\log(0.013) = 1.89$

(b) $pOH = -\log(0.026) = 1.59$

(c) $pOH = -\log(9.2 \times 10^{-5}) = 4.04$

THINK ABOUT IT

These are basic pOH values, which is what we should expect for the solutions described in the problem. Note that while the solutions in parts (a) and (b) have the same base concentration, they do not have the same hydroxide concentration and therefore do not have the same pOH.

Practice Problem **A**TTEMPT   Calculate the pOH of the following aqueous solutions at 25°C: (a) 0.15 M NaOH, (b) 8.4×10^{-3} M RbOH, (c) 1.7×10^{-5} M CsOH.

Practice Problem **B**UILD   Calculate the pH of the following aqueous solutions at 25°C: (a) 9.5×10^{-8} M NaOH, (b) 6.1×10^{-2} M LiOH, (c) 6.1×10^{-2} M Ba(OH)$_2$.

Practice Problem **C**ONCEPTUALIZE   Which of the following plots best represents the relationship between pH and pOH?

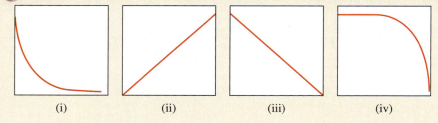

(i)   (ii)   (iii)   (iv)

SAMPLE PROBLEM 16.11

An aqueous solution of a strong base has pH 8.15 at 25°C. Calculate the original concentration of base in the solution (a) if the base is NaOH and (b) if the base is Ba(OH)$_2$.

Strategy Use Equation 16.6 to convert from pH to pOH and Equation 16.5 to determine the hydroxide ion concentration. Consider the stoichiometry of dissociation in each case to determine the concentration of the base itself.

Setup
$$pOH = 14.00 - 8.15 = 5.85$$

(a) The dissociation of 1 mole of NaOH produces 1 mole of OH$^-$. Therefore, the concentration of the base is *equal* to the concentration of hydroxide ion.

(b) The dissociation of 1 mole of Ba(OH)$_2$ produces 2 moles of OH$^-$. Therefore, the concentration of the base is only one-half the concentration of hydroxide ion.

Solution
$$[OH^-] = 10^{-5.85} = 1.41 \times 10^{-6} \, M$$

(a) $[NaOH] = [OH^-] = 1.4 \times 10^{-6} \, M$

(b) $[Ba(OH)_2] = \frac{1}{2}[OH^-] = 7.1 \times 10^{-7} \, M$

THINK ABOUT IT

Alternatively, we could determine the hydroxide ion concentration using Equation 16.3,

$$[H_3O^+] = 10^{-8.15} = 7.1 \times 10^{-9} \, M$$

and Equation 16.1,

$$[OH^-] = \frac{1.0 \times 10^{-14}}{7.1 \times 10^{-9} \, M}$$
$$= 1.4 \times 10^{-6} \, M$$

Once [OH$^-$] is known, the solution is the same as shown previously. Remember to keep an additional significant figure or two until the end of the problem—to avoid *rounding error* [◄◄ Section 1.5].

(Continued on next page)

Practice Problem **A**TTEMPT An aqueous solution of a strong base has pH 8.98 at 25°C. Calculate the concentration of base in the solution (a) if the base is LiOH and (b) if the base is $Ba(OH)_2$.

Practice Problem **B**UILD An aqueous solution of a strong base has pOH 1.76 at 25°C. Calculate the concentration of base in the solution (a) if the base is NaOH and (b) if the base is $Ba(OH)_2$.

Practice Problem **C**ONCEPTUALIZE Which of the following plots best represents pH as a function of concentration for two bases of the same concentration—one monobasic and one dibasic?

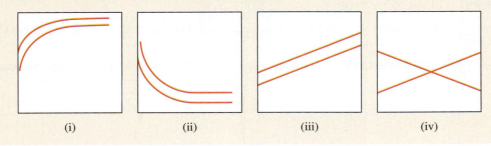

(i)   (ii)   (iii)   (iv)

CHECKPOINT – SECTION 16.4   Strong Acids and Bases

16.4.1 Calculate the pH of a 0.075-M solution of perchloric acid ($HClO_4$) at 25°C.

a) 12.88   d) 1.12

b) 7.75   e) 7.00

c) 6.25

16.4.2 What is the concentration of HBr in a solution with pH 5.89 at 25°C?

a) $7.8 \times 10^{-9}\ M$ d) $8.1 \times 10^{-7}\ M$

b) $1.3 \times 10^{-6}\ M$ e) $1.0 \times 10^{-7}\ M$

c) $5.9 \times 10^{-14}\ M$

16.4.3 What is the pOH of a solution at 25°C that is $1.3 \times 10^{-3}\ M$ in $Ba(OH)_2$?

a) 2.89   d) 11.11

b) 2.59   e) 11.41

c) 3.19

16.4.4 What is the concentration of KOH in a solution at 25°C that has pOH 3.31?

a) $2.0 \times 10^{-11}\ M$ d) $4.5 \times 10^{-4}\ M$

b) $3.3 \times 10^{-1}\ M$ e) $4.9 \times 10^{-4}\ M$

c) $3.3 \times 10^{-7}\ M$

16.4.5 What is the pH of a solution at 25°C that is 0.0095 M in LiOH?

a) 11.68   d) 1.72

b) 2.02   e) 12.28

c) 11.98

16.4.6 What is the concentration of $Ca(OH)_2$ in a solution at 25°C if the pH is 9.01?

a) $1.0 \times 10^{-5}\ M$ d) $9.8 \times 10^{-9}\ M$

b) $5.1 \times 10^{-6}\ M$ e) $4.9 \times 10^{-9}\ M$

c) $2.0 \times 10^{-5}\ M$

16.4.7 Which diagram best represents a solution of sulfuric acid (H_2SO_4)?

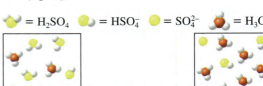

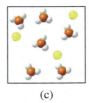

(a)   (b)

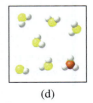

(c)   (d)

16.5 Weak Acids and Acid Ionization Constants

Most acids are **weak acids,** which ionize only to a limited extent in water. At equilibrium, an aqueous solution of a weak acid contains a mixture of aqueous acid molecules, hydronium ions, and the corresponding conjugate base. The degree to which a weak acid ionizes depends on the *concentration* of the acid and the *equilibrium constant* for the ionization.

The Ionization Constant, K_a

Consider a weak monoprotic acid HA. Its ionization in water is represented by:

$$HA(aq) + H_2O(l) \rightleftharpoons H_3O^+(aq) + A^-(aq)$$

or by:

$$HA(aq) \rightleftharpoons H^+(aq) + A^-(aq)$$

The equilibrium expression for this reaction is:

$$K_a = \frac{[H_3O^+][A^-]}{[HA]} \quad \text{or} \quad K_a = \frac{[H^+][A^-]}{[HA]}$$

where K_a is the equilibrium constant for the reaction. More specifically, K_a is called the **acid ionization constant.** Although all weak acids ionize less than 100 percent, they vary in strength. The magnitude of K_a indicates how strong a weak acid is. A large K_a value indicates a stronger acid, whereas a small K_a value indicates a weaker acid. For example, acetic acid (CH_3COOH) and hydrofluoric acid (HF) are both weak acids, but HF is the stronger acid of the two, as evidenced by its larger K_a value. Solutions of equal concentration of the two acids do *not* have the same pH. The pH of the HF solution is lower:

Student Note: Remember that H_3O^+ and H^+ are used interchangeably.

Student Note: Remember that a K_a is a K_c. The subscript "a" simply stands for "acid."

Solution (at 25°C)	K_a	pH
0.10 *M* HF	7.1×10^{-4}	2.09
0.10 *M* CH_3COOH	1.8×10^{-5}	2.87

For comparison, the pH of a 0.1-*M* solution of a strong acid such as HCl or HNO_3 is 1.0. Table 16.6 lists a number of weak acids and their K_a values at 25°C in order of decreasing acid strength.

TABLE 16.6	Ionization Constants of Some Weak Acids at 25°C		
Name of Acid	**Formula**	**Structure**	K_a
Hydrofluoric acid	HF	H—F	7.1×10^{-4}
Nitrous acid	HNO_2	O=N—O—H	4.5×10^{-4}
Formic acid	HCOOH	H—C(=O)—O—H	1.7×10^{-4}
Benzoic acid	C_6H_5COOH	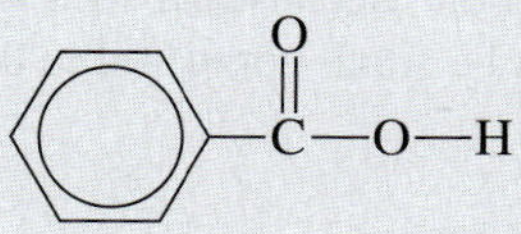 C₆H₅—C(=O)—O—H	6.5×10^{-5}
Acetic acid	CH_3COOH	CH_3—C(=O)—O—H	1.8×10^{-5}
Hydrocyanic acid	HCN	H—C≡N	4.9×10^{-10}
Phenol	C_6H_5OH	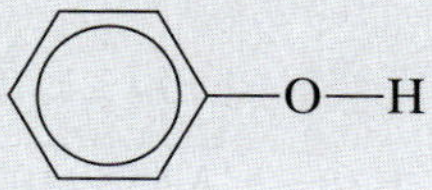 C₆H₅—O—H	1.3×10^{-10}

Animation
Figure 16.2, Using Equilibrium Tables
to Solve Problems.

Calculating pH from K_a

Calculating the pH of a weak acid solution is an equilibrium problem, which we solve using the methods introduced in Chapter 15. Figure 16.2 shows in detail how an equilibrium table is used to determine the pH of a weak acid. Suppose we want to determine the pH of a 0.50 M HF solution at 25°C. The ionization of HF is represented by:

$$HF(aq) + H_2O(l) \rightleftharpoons H_3O^+(aq) + F^-(aq)$$

The equilibrium expression for this reaction is:

$$K_a = \frac{[H_3O^+][F^-]}{[HF]} = 7.1 \times 10^{-4}$$

Student Note: Remember that solids and pure liquids do not appear in the equilibrium expression [◄ Section 15.3].

We construct an equilibrium table and enter the starting concentrations of all species in the equilibrium expression:

$$HF(aq) + H_2O(l) \rightleftharpoons H_3O^+(aq) + F^-(aq)$$

	HF	H_2O	H_3O^+	F^-
Initial concentration (M):	0.50	—		0
Change in concentration (M):		—		
Equilibrium concentration (M):		—		

Using the reaction stoichiometry, we determine the changes in all species:

$$HF(aq) + H_2O(l) \rightleftharpoons H_3O^+(aq) + F^-(aq)$$

	HF	H_2O	H_3O^+	F^-
Initial concentration (M):	0.50	—	0	0
Change in concentration (M):	$-x$	—	$+x$	$+x$
Equilibrium concentration (M):		—		

Finally, we express the equilibrium concentration of each species in terms of x:

$$HF(aq) + H_2O(l) \rightleftharpoons H_3O^+(aq) + F^-(aq)$$

	HF	H_2O	H_3O^+	F^-
Initial concentration (M):	0.50	—	0	0
Change in concentration (M):	$-x$	—	$+x$	$+x$
Equilibrium concentration (M):	$0.50 - x$	—	x	x

These equilibrium concentrations are then entered into the equilibrium expression to give:

$$K_a = \frac{(x)(x)}{0.50 - x} = 7.1 \times 10^{-4}$$

Rearranging this expression, we get:

$$x^2 + 7.1 \times 10^{-4}x - 3.55 \times 10^{-4} = 0$$

This is a quadratic equation, which we can solve using the quadratic formula given in Appendix 1. In the case of a weak acid, however, often we can use a shortcut to simplify the calculation. Because HF is a weak acid, and weak acids ionize only to a slight extent, x must be small compared to 0.50. Therefore, we can make the following approximation:

$$0.50 - x \approx 0.50$$

Now the equilibrium expression becomes:

$$\frac{x^2}{0.50 - x} \approx \frac{x^2}{0.50} = 7.1 \times 10^{-4}$$

Rearranging, we get:

$$x^2 = (0.50)(7.1 \times 10^{-4}) = 3.55 \times 10^{-4}$$

$$x = \sqrt{3.55 \times 10^{-4}} = 1.9 \times 10^{-2} \, M$$

Thus, we have solved for x without having to use the quadratic equation. At equilibrium we have:

$$[HF] = (0.50 - 0.019)\,M = 0.48\,M$$

$$[H_3O^+] = 0.019\,M$$

$$[F^-] = 0.019\,M$$

and the pH of the solution is:

$$pH = -\log(0.019) = 1.72$$

This shortcut gives a good approximation as long as the magnitude of x is significantly smaller than the initial acid concentration. As a rule, it is acceptable to use this short-cut if the calculated value of x is less than 5 percent of the initial acid concentration. In this case, the approximation is acceptable because:

$$\frac{0.019\,M}{0.50\,M} \times 100\% = 3.8\%$$

This is the formula for the *percent ionization* of the acid [◄◄ Section 13.6]. Recall that the percent ionization of a weak electrolyte, such as a weak acid, depends on concentration. Consider a more dilute solution of HF, one that is 0.050 M. Using the preceding procedure to solve for x, we would get $6.0 \times 10^{-3}\,M$. The following test shows, however, that this answer is *not* a valid approximation because it is greater than 5 percent of 0.050 M:

$$\frac{6.0 \times 10^{-3}\,M}{0.050\,M} \times 100\% = 12\%$$

In this case, we must solve for x using the quadratic equation [◄◄ Sample Problem 15.9].

   Sample Problem 16.12 shows how to use K_a to determine the pH of a weak acid solution.

> **Student Hot Spot**
> Student data indicate you may struggle with determining the pH of a weak acid solution. Access the eBook to view additional Learning Resources on this topic.

> **Student Note:** In many cases, use of the quadratic equation can be avoided with a method called *successive approximation*, which is presented in Appendix 1.

SAMPLE PROBLEM   16.12

The K_a of hypochlorous acid (HClO) is 3.5×10^{-8}. Calculate the pH of a solution at 25°C that is 0.0075 M in HClO.

Strategy Construct an equilibrium table, and express the equilibrium concentration of each species in terms of x. Solve for x using the approximation shortcut, and evaluate whether or not the approximation is valid. Use Equation 16.2 to determine pH.

Setup

$$HClO(aq) + H_2O(l) \rightleftharpoons H_3O^+(aq) + ClO^-(aq)$$

	HClO(aq)	H₂O(l)	H₃O⁺(aq)	ClO⁻(aq)
Initial concentration (M):	0.0075	—	0	0
Change in concentration (M):	$-x$	—	$+x$	$+x$
Equilibrium concentration (M):	$0.0075 - x$	—	x	x

Solution These equilibrium concentrations are then substituted into the equilibrium expression to give:

$$K_a = \frac{(x)(x)}{0.0075 - x} = 3.5 \times 10^{-8}$$

Assuming that $0.0075 - x \approx 0.0075$:

$$\frac{x^2}{0.0075} = 3.5 \times 10^{-8}$$

$$x^2 = (3.5 \times 10^{-8})(0.0075)$$

Solving for x, we get:

$$x = \sqrt{2.625 \times 10^{-10}} = 1.62 \times 10^{-5}\,M$$

Applying the 5 percent test indicates that the approximation shortcut is valid in this case: $(1.62 \times 10^{-5}/0.0075) \times 100\% < 5\%$. According to the equilibrium table, $x = [H_3O^+]$. Therefore:

$$pH = -\log(1.62 \times 10^{-5}) = 4.79$$

(Continued on page 808)

Figure 16.2

Using Equilibrium Tables to Solve Problems

Using K_a and Concentration to Determine pH of a Weak Acid

Determine the pH of a weak acid with a concentration of 0.10 M and a K_a of 2.5×10^{-5}. (In this figure, the hydronium ion is represented as H^+.)

The concentrations will change by an unknown amount, x.
- [HA] will decrease by x
- [H^+] and [A^-] will increase by x

We enter the anticipated changes in concentrations in the middle row of the table.

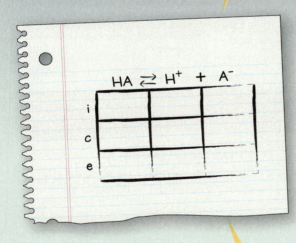

Many pH problems can be solved using an equilibrium table. The table is constructed with a column under each species in the equilibrium equation, and three rows labeled i (initial), c (change), and e (equilibrium). (This is why such a table is sometimes referred to as an "ice" table.)

HA

Using pH and Concentration to Determine K_a of a Weak Acid

Determine the K_a of a weak acid if a 0.12-M solution has a pH of 3.82.

We use pH to determine the equilibrium concentration of H^+:

$$[H^+] = 10^{-pH} = 10^{-3.82} = 1.5 \times 10^{-4} \ M$$

Because the ionization of HA produces equal amounts of H^+ and A^-, the equilibrium concentration of A^- is also $1.5 \times 10^{-4} \ M$. We enter these concentrations in the bottom row of the table.

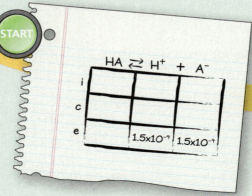

The equilibrium concentrations, which we enter in the last row of the table, are expressed in terms of the unknown x. We write the equilibrium expression for the reaction,

$$K_a = \frac{[H^+][A^-]}{[HA]}$$

and enter the K_a value and the equilibrium concentrations from the last row in the table:

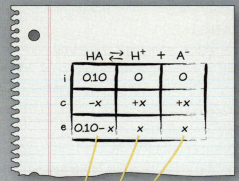

$$2.5 \times 10^{-5} = \frac{(x)(x)}{(0.10 - x)} = \frac{x^2}{(0.10 - x)}$$

Because the magnitude of K_a is small, we expect the amount of HA ionized (x) to be small compared to 0.10. Therefore, we can assume that $0.10 - x \approx 0.10$. This simplifies the solution to

$$2.5 \times 10^{-5} = \frac{x^2}{0.10}$$

Solving for x gives

$$(2.5 \times 10^{-5})(0.10) = x^2$$

$$2.5 \times 10^{-6} = x^2$$

$$\sqrt{2.5 \times 10^{-6}} = x$$

$$x = 1.6 \times 10^{-3}\ M$$

We used x to represent $[H^+]$ at equilibrium. (x is also equal to $[A^-]$ at equilibrium and to the change in $[HA]$.) Therefore, we determine pH by taking the negative log of x.

$$pH = -\log[H^+] = -\log(1.6 \times 10^{-3}) = 2.80$$

 CHECK Verify that $\dfrac{1.6 \times 10^{-3}\ M}{0.10\ M} \times 100\% < 5\%$

and that $\dfrac{(1.6 \times 10^{-3})^2}{(0.10 - 1.6 \times 10^{-3})} \approx K_a$.

$$HA \rightleftharpoons H^+ + A^-$$

To complete the last row of the table, we also need the equilibrium value of $[HA]$. The amount of weak acid that ionized is equal to the amount of H^+ produced by the ionization. Therefore, at equilibrium,

$$[HA] = 0.12\ M - 1.5 \times 10^{-4}\ M \approx 0.12\ M$$

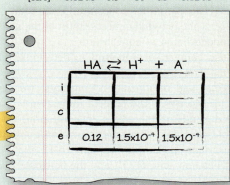

With the bottom row of the table complete, we can use these equilibrium concentrations to calculate the value of K_a.

$$K_a = \frac{[H^+][A^-]}{[HA]} = \frac{(1.5 \times 10^{-4})(1.5 \times 10^{-4})}{(0.12)} = 1.9 \times 10^{-7}$$

 CHECK Verify that when $[HA] = 0.12\ M$ and $K_a = 1.9 \times 10^{-7}$, $[H^+] = \sqrt{(1.9 \times 10^{-7})(0.12)} = 1.5 \times 10^{-4}\ M$ and pH = 3.82.

What's the point?

- Starting with the molar concentration and K_a of a weak monoprotic acid, we can use an equilibrium table to determine pH.
- Starting with pH and molar concentration, we can use an equilibrium table to determine the K_a of a weak monoprotic acid.

(See Visualizing Chemistry questions VC16.1–VC16.4 on page 842.)

THINK ABOUT IT

We learned in Section 16.2 that the concentration of hydronium ion in pure water at 25°C is 1.0×10^{-7} M, yet we use 0 M as the starting concentration to solve for the pH of a solution of weak acid:

$$HA(aq) + H_2O(l) \rightleftharpoons H_3O^+(aq) + A^-(aq)$$

Initial concentration (M):	—	0	0
Change in concentration (M):	—		
Equilibrium concentration (M):	—		

The reason for this is that the *actual* concentration of hydronium ion in pure water is insignificant compared to the amount produced by the ionization of the weak acid. We could use the actual concentration of hydronium as the initial concentration, but doing so would not change the result because $(x + 1.0 \times 10^{-7})$ $M \approx xM$. In solving problems of this type, we neglect the small concentration of H_3O^+ due to the autoionization of water.

Practice Problem Ⓐ **TTEMPT** Calculate the pH at 25°C of a 0.18-M solution of a weak acid that has $K_a = 9.2 \times 10^{-6}$.

Practice Problem Ⓑ **UILD** Calculate the pH at 25°C of a 0.065-M solution of a weak acid that has $K_a = 1.2 \times 10^{-5}$.

Practice Problem Ⓒ **ONCEPTUALIZE** The diagrams show solutions of four different weak acids. In which solution is the concentration of the weak acid highest? Which solution has the highest pH value?

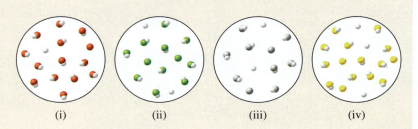

(i)   (ii)   (iii)   (iv)

Percent Ionization

When we first encountered acids [◄◄ Section 4.1], we learned that a strong acid is one that ionizes completely. Weak acids ionize only partially. The degree to which a weak acid ionizes is a measure of its strength, just as the magnitude of its ionization constant (K_a) is a measure of its strength. A quantitative measure of the degree of ionization is *percent ionization,* which, for a weak, monoprotic acid (HA) is calculated as follows:

Equation 16.7
$$\text{percent ionization} = \frac{[H_3O^+]_{eq}}{[HA]_0} \times 100\%$$

Student Note: In fact, a weak acid's original concentration is never *exactly* equal to its equilibrium concentration. However, sometimes the amount that ionizes is so small compared to the original concentration that, to an appropriate number of significant figures, the weak acid concentration does not change and $[HA]_0 \approx [HA]_{eq}$.

in which the hydronium ion concentration, $[H_3O^+]_{eq}$, is the concentration at equilibrium and the weak acid concentration, $[HA]_0$, is the *original* concentration of the acid, which is not necessarily the same as the concentration at equilibrium.

We have learned previously how to use colligative properties to determine the percent ionization of a weak electrolyte, such as a weak acid [◄◄ Section 13.6]. Now that we have an understanding of equilibrium and of ionization constants, we can predict the percent ionization of a weak acid. Moreover, the percent ionization is not constant for a given weak acid, but in fact depends on the acid's concentration.

Consider a 0.10-M solution of benzoic acid, for which the ionization constant (K_a) is 6.5×10^{-5}. Using the procedure described in Sample Problem 16.12, we can determine the equilibrium concentrations of benzoic acid, H_3O^+, and the benzoate ion:

$$C_6H_5COOH(aq) + H_2O(l) \rightleftharpoons H_3O^+(aq) + C_6H_5COO^-(aq)$$

Initial concentration (M):	0.100	—	0	0
Change in concentration (M):	−x	—	+x	+x
Equilibrium concentration (M):	0.100 − x	—	x	x

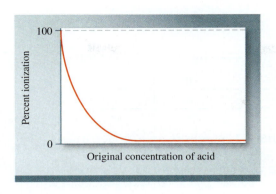

Figure 16.3 Percent ionization of a weak acid depends on the original acid concentration. As concentration approaches zero, ionization approaches 100 percent. The dashed grey line represents 100 percent ionization, which is characteristic of a strong acid.

Solving for x gives 0.0025 M. Therefore, at equilibrium, $[C_6H_5COOH] = 0.097$ M and $[H_3O^+] = [C_6H_5COO^-] = 0.0025$ M. The percent ionization of benzoic acid at this concentration is:

$$\frac{0.0025\ M}{0.100\ M} \times 100\% = 2.5\%$$

Now consider what happens when we dilute this equilibrium mixture by adding enough water to double the volume. The concentrations of all three species are cut in half: $[C_6H_5COOH] = 0.049$ M, $[H_3O^+] = 0.0013$ M, and $[C_6H_5COO^-] = 0.0013$ M. If we plug these new concentrations into the equilibrium expression and calculate the reaction quotient (Q), we get a number different from K_a. In fact, we get $(0.0013^2/0.049) = 3.4 \times 10^{-5}$, which is *smaller* than K_a. When Q is smaller than K, the reaction must proceed to the right to reestablish equilibrium [◀◀ Section 15.4]. Proceeding to the right, in the case of weak acid ionization, corresponds to more of the acid ionizing—meaning that its percent ionization increases.

Student Note: According to Le Châtelier's principle [◀◀ Section 15.5], the reduction in particle concentration caused by dilution stresses the equilibrium. The equilibrium shifts to minimize the effects of the stress by shifting toward the side with more dissolved particles, causing more of the weak acid to ionize.

We can solve again for percent ionization starting with half the original concentration of benzoic acid:

$$C_6H_5COOH(aq) + H_2O(l) \rightleftharpoons H_3O^+(aq) + C_6H_5COO^-(aq)$$

	C_6H_5COOH(aq) +	H_2O(l) ⇌	H_3O^+(aq) +	C_6H_5COO^-(aq)
Initial concentration (M):	0.050	—	0	0
Change in concentration (M):	$-x$	—	$+x$	$+x$
Equilibrium concentration (M):	$0.050 - x$	—	x	x

This time, solving for x gives 0.0018 M. Therefore, at equilibrium, $[C_6H_5COOH] = 0.048$ M, and $[H_3O^+] = [C_6H_5COO^-] = 0.0018$ M. The percent ionization of benzoic acid at this concentration is:

$$\frac{0.0018\ M}{0.050\ M} \times 100\% = 3.6\%$$

Figure 16.3 shows the dependence of percent ionization of a weak acid on concentration. Note that as concentration approaches zero, percent ionization approaches 100.

Sample Problem 16.13 lets you practice calculating percent ionization of weak acid solutions.

SAMPLE PROBLEM 16.13

Determine pH and percent ionization for acetic acid solutions at 25°C with concentrations (a) 0.15 M, (b) 0.015 M, and (c) 0.0015 M.

Strategy Using the procedure described in Sample Problem 16.12, we construct an equilibrium table and for each concentration of acetic acid, we solve for the equilibrium concentration of H_3O^+. We use Equation 16.2 to find pH, and Equation 16.7 to find percent ionization.

Setup From Table 16.6, the ionization constant, K_a, for acetic acid is 1.8×10^{-5}.

(Continued on next page)

Setup

(a)

$$CH_3COOH(aq) + H_2O(l) \rightleftharpoons H_3O^+(aq) + CH_3COO^-(aq)$$

	CH_3COOH	H_2O	H_3O^+	CH_3COO^-
Initial concentration (M):	0.15	—	0	0
Change in concentration (M):	$-x$	—	$+x$	$+x$
Equilibrium concentration (M):	$0.15 - x$	—	x	x

Solving for x gives $[H_3O^+] = 0.0016\ M$ and pH $= -\log(0.0016) = 2.78$.

$$\text{percent ionization} = \frac{0.0016\ M}{0.15\ M} \times 100\% = 1.1\%$$

(b) Solving in the same way as part (a) gives: $[H_3O^+] = 5.2 \times 10^{-4}\ M$ and pH $= 3.28$.

$$\text{percent ionization} = \frac{5.2 \times 10^{-4}\ M}{0.015\ M} \times 100\% = 3.5\%$$

(c) Solving the quadratic equation, or using successive approximation [▶▶ Appendix 1], gives $[H_3O^+] = 1.6 \times 10^{-4}\ M$ and pH $= 3.78$.

$$\text{percent ionization} = \frac{1.6 \times 10^{-4}\ M}{0.0015\ M} \times 100\% = 11\%$$

THINK ABOUT IT

Percent ionization also increases as concentration decreases for weak bases [▶▶ Section 16.6].

Practice Problem **A**TTEMPT   Determine the pH and percent ionization for hydrocyanic acid (HCN) solutions of concentration (a) 0.25 *M*, (b) 0.0075 *M*, and (c) $8.3 \times 10^{-5}\ M$.

Practice Problem **B**UILD   At what concentration does hydrocyanic acid exhibit (a) 0.05 percent ionization, (b) 0.10 percent ionization, and (c) 0.15 percent ionization?

Practice Problem **C**ONCEPTUALIZE   Which of the diagrams shows the weak acid with the highest percent ionization? Is it possible for one of the other weak acids shown here ever to have a higher percent ionization than the one you chose? Explain.

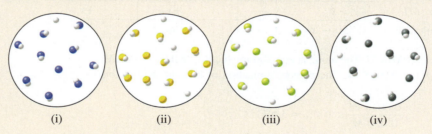

| (i) | (ii) | (iii) | (iv) |

Using pH to Determine K_a

In Chapter 15 we learned that we can determine the value of an equilibrium constant using equilibrium concentrations [◀◀ Section 15.2]. Using a similar approach, we can use the pH of a weak acid solution to determine the value of K_a. Suppose we want to determine the K_a of a weak acid (HA) and we know that a 0.25-*M* solution of the acid has a pH of 3.47 at 25°C. The first step is to use pH to determine the equilibrium hydronium ion concentration. Using Equation 16.3, we get:

$$[H_3O^+] = 10^{-3.47} = 3.39 \times 10^{-4}\ M$$

Student Note: Remember that we keep extra significant figures until the end of a multistep problem to minimize rounding error [◀◀ Section 1.5].

We use the starting concentration of the weak acid and the equilibrium concentration of the hydronium ion to construct an equilibrium table and determine the equilibrium concentrations of all three species:

$$HA(aq) + H_2O(l) \rightleftharpoons H_3O^+(aq) + A^-(aq)$$

Initial concentration (M):	0.25	—	0	0
Change in concentration (M):	-3.39×10^{-4}	—	$+3.39 \times 10^{-4}$	$+3.39 \times 10^{-4}$
Equilibrium concentration (M):	0.2497	—	3.39×10^{-4}	3.39×10^{-4}

These equilibrium concentrations are substituted into the equilibrium expression to give:

$$K_a = \frac{(3.39 \times 10^{-4})^2}{0.2497} = 4.6 \times 10^{-7}$$

Therefore, the K_a of this weak acid is 4.6×10^{-7}:

Sample Problem 16.14 shows how to determine K_a using pH.

SAMPLE PROBLEM 16.14

Aspirin (acetylsalicylic acid, $HC_9H_7O_4$) is a weak acid. It ionizes in water according to the equation:

$$HC_9H_7O_4(aq) + H_2O(l) \rightleftharpoons H_3O^+(aq) + C_9H_7O_4^-(aq)$$

A 0.10-M aqueous solution of aspirin has a pH of 2.27 at 25°C. Determine the K_a of aspirin.

Strategy Determine the hydronium ion concentration from the pH. Use the hydronium ion concentration to determine the equilibrium concentrations of the other species, and plug the equilibrium concentrations into the equilibrium expression to evaluate K_a.

Setup Using Equation 16.3, we have:

$$[H_3O^+] = 10^{-2.27} = 5.37 \times 10^{-3}\ M$$

To calculate K_a, though, we also need the equilibrium concentrations of $C_9H_7O_4^-$ and $HC_9H_7O_4$. The stoichiometry of the reaction tells us that $[C_9H_7O_4^-] = [H_3O^+]$. Furthermore, the amount of aspirin that has *ionized* is equal to the amount of hydronium ion in solution. Therefore, the equilibrium concentration of aspirin is $(0.10 - 5.37 \times 10^{-3})\ M = 0.095\ M$:

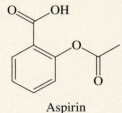

Aspirin

$$HC_9H_7O_4(aq) + H_2O(l) \rightleftharpoons H_3O^+(aq) + C_9H_7O_4^-(aq)$$

Initial concentration (M):	0.10	—	0	0
Change in concentration (M):	−0.005	—	$+5.37 \times 10^{-3}$	$+5.37 \times 10^{-3}$
Equilibrium concentration (M):	0.095	—	5.37×10^{-3}	5.37×10^{-3}

Solution Substitute the equilibrium concentrations into the equilibrium expression as follows:

$$K_a = \frac{[H_3O^+][C_9H_7O_4^-]}{[HC_9H_7O_4]} = \frac{(5.37 \times 10^{-3})^2}{0.095} = 3.0 \times 10^{-4}$$

The K_a of aspirin is 3.0×10^{-4}.

THINK ABOUT IT

Check your work by using the calculated value of K_a to solve for the pH of a 0.10-M solution of aspirin.

Practice Problem **A**TTEMPT Calculate the K_a of a weak acid if a 0.065-M solution of the acid has a pH of 2.96 at 25°C.

Practice Problem **B**UILD Calculate the K_a of a weak acid if a 0.015-M solution of the acid has a pH of 5.03 at 25°C.

Practice Problem **C**ONCEPTUALIZE Calculate K_a values (to two significant figures) for the weak acids represented in the diagrams.

(i)   (ii)   (iii)

CHECKPOINT – SECTION 16.5 Weak Acids and Acid Ionization Constants

16.5.1 The K_a of a weak acid is 5.5×10^{-4}. What is the pH of a 0.63-*M* solution of this acid at 25°C?

a) 3.26
b) 1.83×10^{-2}
c) 1.73
d) 1.63
e) 0.201

16.5.2 A 0.042-*M* solution of a weak acid has pH 4.01 at 25°C. What is the K_a of this acid?

a) 9.5×10^{-9}
b) 9.8×10^{-5}
c) 0.91
d) 4.2×10^{-2}
e) 2.3×10^{-7}

16.5.3 The diagrams show solutions of three different weak acids with the general formula HA. List the acids in order of increasing K_a value.

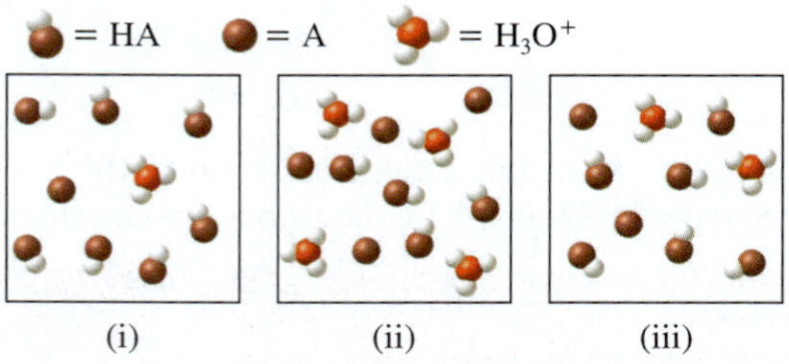

a) i < ii < iii
b) i < iii < ii
c) ii < i < iii
d) ii < iii < i
e) iii < ii < i

16.6 Weak Bases and Base Ionization Constants

Just as most acids are weak, most bases are also weak. The ionization of a ***weak base*** is incomplete and is treated in the same way as the ionization of a weak acid. In this section, we explain how the ionization constant for a weak base, K_b, is related to the pH of an aqueous solution.

The Ionization Constant, K_b

The ionization of a weak base can be represented by the equation:

$$B(aq) + H_2O(l) \rightleftharpoons HB^+(aq) + OH^-(aq)$$

where B is the weak base and HB^+ is its conjugate acid. The equilibrium expression for the ionization is:

$$K_b = \frac{[HB^+][OH^-]}{[B]}$$

where K_b is the equilibrium constant—specifically known as the ***base ionization constant.*** Like weak acids, weak bases vary in strength. Table 16.7 lists a number of common weak bases and their ionization constants. The ability of any one of these substances to act as a base is the result of the lone pair of electrons on the nitrogen atom. The presence of this lone pair is what enables a compound to accept a proton, which is what makes a compound a Brønsted base.

Calculating pH from K_b

Solving problems involving weak bases requires the same approach we used for weak acids. It is important to remember, though, that solving for x in a typical weak base problem gives us the hydroxide ion concentration rather than the hydronium ion concentration.

TABLE 16.7	Ionization Constants of Some Weak Bases at 25°C		
Name of Base	**Formula**	**Structure**	K_b
Ethylamine	$C_2H_5NH_2$	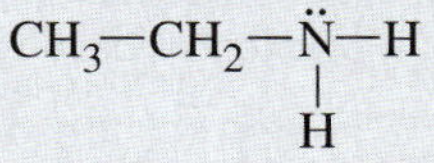 $CH_3-CH_2-\ddot{N}-H$ $\quad\quad\quad\quad\;\; \vert$ $\quad\quad\quad\quad\; H$	5.6×10^{-4}
Methylamine	CH_3NH_2	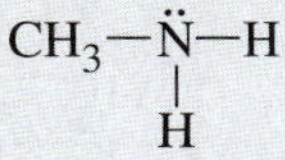 $CH_3-\ddot{N}-H$ $\quad\quad\; \vert$ $\quad\quad H$	4.4×10^{-4}
Ammonia	NH_3	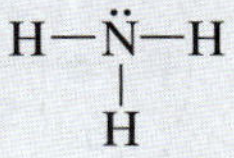 $H-\ddot{N}-H$ $\quad\; \vert$ $\quad H$	1.8×10^{-5}
Pyridine	C_5H_5N	 (ring structure) N:	1.7×10^{-9}
Aniline	$C_6H_5NH_2$	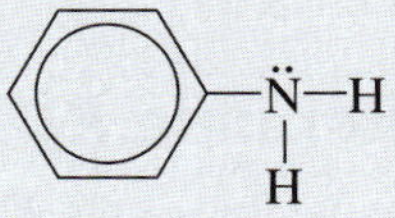 (ring) $-\ddot{N}-H$ $\quad\quad\quad\; \vert$ $\quad\quad\quad H$	3.8×10^{-10}
Urea	H_2NCONH_2	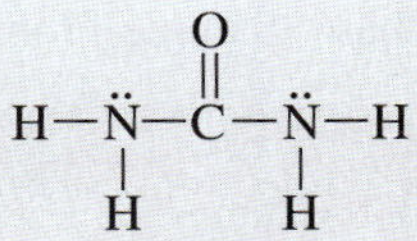 $\quad\quad\quad O$ $\quad\quad\quad \vert\vert$ $H-\ddot{N}-C-\ddot{N}-H$ $\quad\; \vert \quad\quad\; \vert$ $\quad H \quad\quad\; H$	1.5×10^{-14}

Sample Problem 16.15 shows how to use K_b to calculate the pH of a weak base solution.

SAMPLE PROBLEM 16.15

What is the pH of a 0.040 *M* ammonia solution at 25°C?

Strategy Construct an equilibrium table, and express equilibrium concentrations in terms of the unknown *x*. Plug these equilibrium concentrations into the equilibrium expression, and solve for *x*. From the value of *x*, determine the pH.

Setup

$$NH_3(aq) + H_2O(l) \rightleftharpoons NH_4^+(aq) + OH^-(aq)$$

Initial concentration (M):	0.040	—	0	0
Change in concentration (M):	−x	—	+x	+x
Equilibrium concentration (M):	0.040 − x	—	x	x

Solution The equilibrium concentrations are substituted into the equilibrium expression to give:

$$K_b = \frac{[NH_4^+][OH^-]}{[NH_3]} = \frac{(x)(x)}{0.040 - x} = 1.8 \times 10^{-5}$$

Assuming that $0.040 - x \approx 0.040$ and solving for *x* gives:

$$\frac{(x)(x)}{0.040 - x} \approx \frac{(x)(x)}{0.040} = 1.8 \times 10^{-5}$$

$$x^2 = (1.8 \times 10^{-5})(0.040) = 7.2 \times 10^{-7}$$

$$x = \sqrt{7.2 \times 10^{-7}} = 8.5 \times 10^{-4} \; M$$

Applying the 5 percent test indicates that the approximation shortcut is valid in this case: $(8.49 \times 10^{-4}/0.040) \times 100\% = 2\%$. According to the equilibrium table, $x = [OH^-]$. Therefore, $pOH = -\log(x)$:

$$-\log(8.5 \times 10^{-4}) = 3.07$$

and $pH = 14.00 - pOH = 14.00 - 3.07 = 10.93$. The pH of a 0.040-*M* solution of NH_3 at 25°C is 10.93.

(Continued on next page)

THINK ABOUT IT

It is a common error in K_b problems to forget that x is the *hydroxide* ion concentration rather than the *hydronium* ion concentration. Always make sure that the pH you calculate for a solution of base is a basic pH, that is, a pH greater than 7.

Practice Problem **A**TTEMPT Calculate the pH at 25°C of a 0.0028-M solution of a weak base with a K_b of 6.8×10^{-8}.

Practice Problem **B**UILD Calculate the pH at 25°C of a 0.16-M solution of a weak base with a K_b of 2.9×10^{-11}.

Practice Problem **C**ONCEPTUALIZE The diagrams represent solutions of three different weak bases. Arrange the solutions in order of increasing pH. To keep the diagrams from being crowded, hydroxide ions are shown, but water molecules are not.

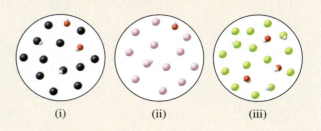

(i)   (ii)   (iii)

Using pH to Determine K_b

Just as we can use pH to determine the K_a of a weak acid, we can also use it to determine the K_b of a weak base. Sample Problem 16.16 demonstrates this procedure.

SAMPLE PROBLEM 16.16

Caffeine, the stimulant in coffee and tea, is a weak base that ionizes in water according to the equation:

$$C_8H_{10}N_4O_2(aq) + H_2O(l) \rightleftharpoons HC_8H_{10}N_4O_2^+(aq) + OH^-(aq)$$

A 0.15-M solution of caffeine at 25°C has a pH of 8.45. Determine the K_b of caffeine.

Strategy Use pH to determine pOH, and pOH to determine the hydroxide ion concentration. From the hydroxide ion concentration, use reaction stoichiometry to determine the other equilibrium concentrations and plug those concentrations into the equilibrium expression to evaluate K_b.

Setup

$$pOH = 14.00 - 8.45 = 5.55$$
$$[OH^-] = 10^{-5.55} = 2.82 \times 10^{-6}\ M$$

Based on the reaction stoichiometry, $[HC_8H_{10}N_4O_2^+] = [OH^-]$, and the amount of hydroxide ion in solution at equilibrium is equal to the amount of caffeine that has ionized. At equilibrium, therefore:

$$[C_8H_{10}N_4O_2] = (0.15 - 2.82 \times 10^{-6})\ M \approx 0.15\ M$$

$$C_8H_{10}N_4O_2(aq) + H_2O(l) \rightleftharpoons HC_8H_{10}N_4O_2^+(aq) + OH^-(aq)$$

Initial concentration (M):	0.15	—	0	0
Change in concentration (M):	-2.82×10^{-6}	—	$+2.82 \times 10^{-6}$	$+2.82 \times 10^{-6}$
Equilibrium concentration (M):	0.15	—	2.82×10^{-6}	2.82×10^{-6}

Caffeine

Solution Plugging the equilibrium concentrations into the equilibrium expression gives:

$$K_b = \frac{[HC_8H_{10}N_4O_2^+][OH^-]}{[C_8H_{10}N_4O_2]} = \frac{(2.82 \times 10^{-6})^2}{0.15} = 5.3 \times 10^{-11}$$

THINK ABOUT IT

Check your answer by using the calculated K_b to determine the pH of a 0.15-M solution.

Practice Problem **A**TTEMPT Determine the K_b of a weak base if a 0.50-M solution of the base has a pH of 9.59 at 25°C.

Practice Problem **B**UILD Determine the K_b of a weak base if a 0.35-M solution of the base has a pH of 11.84 at 25°C.

Practice Problem **C**ONCEPTUALIZE Determine the value of K_b (to two significant figures) for each of the bases represented in Practice Problem 16.15C.

CHECKPOINT – SECTION 16.6   Weak Bases and Base Ionization Constants

16.6.1 What is the pH of a 0.63-M solution of weak base at 25°C if $K_b = 9.5 \times 10^{-7}$?

   a) 3.11

   b) 10.89

   c) 7.00

   d) 1.12

   e) 12.88

16.6.2 A 0.12-M solution of a weak base has a pH of 10.76 at 25°C. Determine K_b.

   a) 2.5×10^{-21}

   b) 8.3×10^{8}

   c) 4.0×10^{-8}

   d) 2.8×10^{-6}

   e) 1.0×10^{-14}

16.6.3 The diagrams show solutions of three different weak bases. List the bases in order of increasing K_b value.

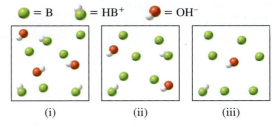

   (i)   (ii)   (iii)

   a) i < ii < iii

   b) i < iii < ii

   c) ii < i < iii

   d) ii < iii < i

   e) iii < ii < i

16.7 Conjugate Acid-Base Pairs

At the beginning of this chapter, we introduced the concept of conjugate acids and conjugate bases. In this section, we examine the properties of conjugate acids and bases, independent of their parent compounds.

The Strength of a Conjugate Acid or Base

When a strong acid such as HCl dissolves in water, it ionizes completely because its conjugate base (Cl^-) has essentially *no* affinity for the H_3O^+ ion in solution:

$$HCl(aq) + H_2O(l) \longrightarrow H_3O^+(aq) + Cl^-(aq)$$

Because the chloride ion has no affinity for the H_3O^+ ion, it does not act as a Brønsted base in water. If we dissolve a chloride salt such as NaCl in water, for example, the Cl^- ions in solution would not accept protons from the water:

$$Cl^-(aq) + H_2O(l) \xrightarrow{\;\;\times\;\;} HCl(aq) + OH^-(aq)$$

The chloride ion, which is the conjugate base of a strong acid, is an example of a **weak conjugate base**.

    Now consider the case of a weak acid. When HF dissolves in water, the ionization happens only to a *limited* degree because the conjugate base, F^-, has a strong affinity for the H_3O^+ ion:

$$HF(aq) + H_2O(l) \rightleftharpoons H_3O^+(aq) + F^-(aq)$$

This equilibrium lies far to the left ($K_a = 7.1 \times 10^{-4}$). Because the fluoride ion has a strong affinity for the H_3O^+ ion, it acts as a Brønsted base in water. If we were to dissolve a fluoride salt, such as NaF, in water, the F^- ions in solution would, to some extent, accept protons from water:

$$F^-(aq) + H_2O(l) \rightleftharpoons HF(aq) + OH^-(aq)$$

The fluoride ion, which is the conjugate base of a weak acid, is an example of a **strong conjugate base**.

    Conversely, a strong base has a **weak conjugate acid** and a weak base has a **strong conjugate acid**. For example, H_2O is the weak conjugate acid of the strong base OH^-, whereas the ammonium ion (NH_4^+) is the strong conjugate acid of the weak base

Student Note: The products HCl and OH$^-$ do *not* actually form when Cl$^-$ and H$_2$O are combined.

Student Note: Note that one of the products of the reaction of a conjugate base with water is always the corresponding weak acid.

Student Note: The Group 1 and heavy Group 2 metal hydroxides are classified as *strong bases*. It is the hydroxide ion itself, however, that accepts a proton and is therefore the Brønsted base. Soluble metal hydroxides are simply sources of the hydroxide ion.

ammonia (NH_3). When an ammonium salt is dissolved in water, the ammonium ions donate protons to the water molecules:

$$NH_4^+(aq) + H_2O(l) \rightleftharpoons NH_3(aq) + H_3O^+(aq)$$

In general, there is a reciprocal relationship between the strength of an acid or base and the strength of its conjugate.

Acid	Example	Conjugate base	Formula
strong	HNO_3	weak conjugate	NO_3^-
weak	HCN	strong conjugate	CN^-

Base	Example	Conjugate acid	Formula
strong	OH^-	weak conjugate	H_2O
weak	NH_3	strong conjugate	NH_4^+

It is important to recognize that the words *strong* and *weak* do not mean the same thing in the context of *conjugate* acids and *conjugate* bases as they do in the context of acids and bases in general. A strong conjugate reacts with water—either accepting a proton from it or donating a proton to it—to a small but measurable extent. A strong conjugate acid acts as a weak Brønsted acid in water; and a strong conjugate base acts as a weak Brønsted base in water. A *weak* conjugate, whether acid or base, does *not* react with water to any measurable extent.

The Relationship Between K_a and K_b of a Conjugate Acid-Base Pair

Because it accepts a proton from water to a small extent, what we refer to as a "strong conjugate base" is actually a weak Brønsted base. Therefore, every strong conjugate base has an ionization constant, K_b. Likewise, every strong conjugate acid, because it acts as a weak Brønsted acid, has an ionization constant, K_a.

A simple relationship between the ionization constant of a weak acid (K_a) and the ionization constant of its conjugate base (K_b) can be derived as follows, using acetic acid as an example:

$$\underset{\text{acid}}{CH_3COOH(aq)} + H_2O(l) \rightleftharpoons H_3O^+(aq) + \underset{\text{conjugate base}}{CH_3COO^-(aq)}$$

$$K_a = \frac{[H_3O^+][CH_3COO^-]}{[CH_3COOH]}$$

Student Note: A conjugate base such as the acetate ion is introduced into a solution by dissolving a soluble salt containing acetate. Sodium acetate, for example, can be used to supply the acetate ion. The sodium ion does not take part in the reaction—it is a spectator ion [◀◀ Section 4.2].

The conjugate base, CH_3COO^-, reacts with water according to the equation:

$$CH_3COO^-(aq) + H_2O(l) \rightleftharpoons CH_3COOH(aq) + OH^-(aq)$$

and the base ionization equilibrium expression is written as:

$$K_b = \frac{[CH_3COOH][OH^-]}{[CH_3COO^-]}$$

As for any chemical equations, we can add these two equilibria and cancel identical terms:

$$\begin{array}{r} \cancel{CH_3COOH(aq)} + H_2O(l) \rightleftharpoons H_3O^+(aq) + \cancel{CH_3COO^-(aq)} \\ + \cancel{CH_3COO^-(aq)} + H_2O(l) \rightleftharpoons \cancel{CH_3COOH(aq)} + OH^-(aq) \\ \hline 2H_2O(l) \rightleftharpoons H_3O^+(aq) + OH^-(aq) \end{array}$$

The sum is the autoionization of water. In fact, this is the case for any weak acid and its conjugate base:

$$\begin{array}{r} \cancel{HA} + H_2O \rightleftharpoons H_3O^+ + \cancel{A^-} \\ + \cancel{A^-} + H_2O \rightleftharpoons \cancel{HA} + OH^- \\ \hline 2H_2O \rightleftharpoons H_3O^+ + OH^- \end{array}$$

or for any weak base and its conjugate acid:

$$B + H_2O \rightleftharpoons HB^+ + OH^-$$
$$+ \underline{HB^+ + H_2O \rightleftharpoons B + H_3O^+}$$
$$2H_2O \rightleftharpoons H_3O^+ + OH^-$$

Recall that when we add two equilibria, the equilibrium constant for the *net* reaction is the product of the equilibrium constants for the individual equations [◄◄ Section 15.3]. Thus, for any conjugate acid-base pair:

$$K_a \times K_b = K_w \qquad \text{Equation 16.8}$$

Equation 16.8 gives the quantitative basis for the reciprocal relationship between the strength of an acid and that of its conjugate base (or between the strength of a base and that of its conjugate acid). Because K_w is a constant, K_b must decrease if K_a increases, and vice versa.

Sample Problem 16.17 shows how to determine ionization constants for conjugates.

SAMPLE PROBLEM 16.17

Determine (a) K_b of the acetate ion (CH_3COO^-), (b) K_a of the methylammonium ion ($CH_3NH_3^+$), (c) K_b of the fluoride ion (F^-), and (d) K_a of the ammonium ion (NH_4^+).

Strategy Each species listed is either a conjugate base or a conjugate acid. Determine the identity of the acid corresponding to each conjugate base and the identity of the base corresponding to each conjugate acid; then, consult Tables 16.6 and 16.7 for their ionization constants. Use the tabulated ionization constants and Equation 16.8 to calculate each indicated K value.

Setup (a) A K_b value is requested, indicating that the acetate ion is a conjugate base. To identify the corresponding Brønsted acid, add a proton to the formula to get CH_3COOH (acetic acid). The K_a of acetic acid (from Table 16.6) is 1.8×10^{-5}.

(b) A K_a value is requested, indicating that the methylammonium ion is a conjugate acid. Determine the identity of the corresponding Brønsted base by removing a proton from the formula to get CH_3NH_2 (methylamine). The K_b of methylamine (from Table 16.7) is 4.4×10^{-4}.

(c) F^- is the conjugate base of HF; $K_a = 7.1 \times 10^{-4}$.

(d) NH_4^+ is the conjugate acid of NH_3; $K_b = 1.8 \times 10^{-5}$. Solving Equation 16.8 separately for K_a and K_b gives, respectively:

$$K_a = \frac{K_w}{K_b} \qquad \text{and} \qquad K_b = \frac{K_w}{K_a}$$

Solution (a) Conjugate base CH_3COO^-: $K_b = \dfrac{1.0 \times 10^{-14}}{1.8 \times 10^{-5}} = 5.6 \times 10^{-10}$

(b) Conjugate acid $CH_3NH_3^+$: $K_a = \dfrac{1.0 \times 10^{-14}}{4.4 \times 10^{-4}} = 2.3 \times 10^{-11}$

(c) Conjugate base F^-: $K_b = \dfrac{1.0 \times 10^{-14}}{7.1 \times 10^{-4}} = 1.4 \times 10^{-11}$

(d) Conjugate acid NH_4^+: $K_a = \dfrac{1.0 \times 10^{-14}}{1.8 \times 10^{-5}} = 5.6 \times 10^{-10}$

THINK ABOUT IT

Because the conjugates of weak acids and bases have ionization constants, salts containing these ions have an effect on the pH of a solution. In Section 16.10 we use the ionization constants of conjugate acids and conjugate bases to calculate pH for solutions containing dissolved salts.

(Continued on next page)

16.9 Molecular Structure and Acid Strength

The strength of an acid is measured by its tendency to ionize:

$$HX \longrightarrow H^+ + X^-$$

Two factors influence the extent to which the acid undergoes ionization. One is the strength of the H—X bond. The stronger the bond, the more difficult it is for the HX molecule to break up and hence the weaker the acid. The other factor is the polarity of the H—X bond. The difference in the electronegativities between H and X results in a polar bond:

$$\overset{\delta+}{H}\text{—}\overset{\delta-}{X}$$

If the bond is highly polarized (i.e., if there is a large accumulation of positive and negative charges on the H and X atoms, respectively), HX will tend to break up into H^+ and X^- ions. A high degree of polarity, therefore, gives rise to a stronger acid. In this section, we consider the roles of bond strength and bond polarity in determining the strength of an acid.

Hydrohalic Acids

Student Note: The polarity of the H—X bond actually decreases from H—F to H—I, largely because F is the most electronegative element. This would suggest that HF would be the strongest of the hydrohalic acids. Based on the data in Table 16.9, however, bond enthalpy is the more important factor in determining the strengths of these acids.

The halogens form a series of binary acids called the hydrohalic acids (HF, HCl, HBr, and HI). Table 16.9 shows that of this series only HF is a weak acid ($K_a = 7.1 \times 10^{-4}$). The data in the table indicate that the predominant factor in determining the strength of the hydrohalic acids is bond strength. HF has the largest bond enthalpy, making its bond the most difficult to break. In this series of binary acids, acid strength increases as bond strength decreases. The strength of the acids increases as follows:

$$HF << HCl < HBr < HI$$

Oxoacids

An oxoacid, as we learned in Chapter 2, contains hydrogen, oxygen, and a central, non-metal atom [◄ Section 2.7]. As the Lewis structures in Figure 16.4 show, oxoacids

TABLE 16.9	Bond Enthalpies for Hydrogen Halides and Acid Strengths for Hydrohalic Acids	
Bond	Bond Enthalpy (kJ/mol)	Acid Strength
H—F	562.8	Weak
H—Cl	431.9	Strong
H—Br	366.1	Strong
H—I	298.3	Strong

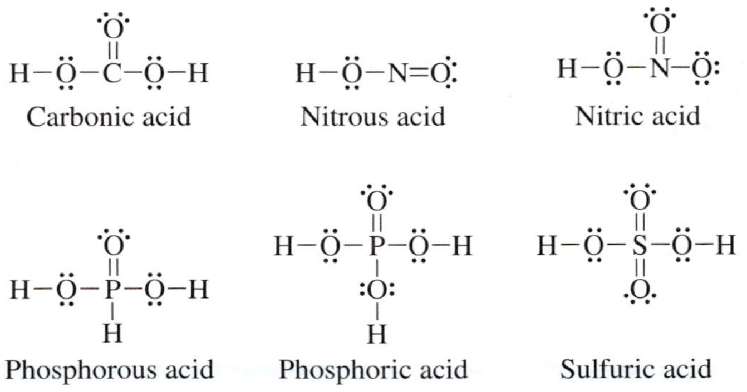

Figure 16.4 Lewis structures of some common oxoacids. Recall that there is more than one possible Lewis structure for an oxoacid in which the central atom is from period 3 or below on the periodic table [◄ Section 8.8].

Carbonic acid   Nitrous acid   Nitric acid

Phosphorous acid   Phosphoric acid   Sulfuric acid

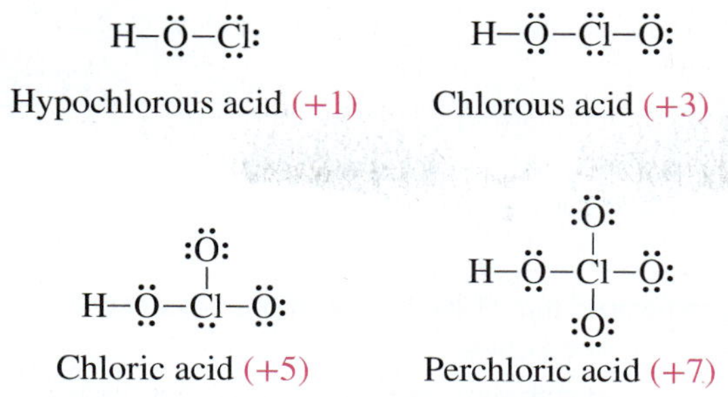

$$H-\overset{..}{\underset{..}{O}}-\overset{..}{\underset{..}{Cl}}:$$
Hypochlorous acid (+1)

$$H-\overset{..}{\underset{..}{O}}-\overset{..}{\underset{..}{Cl}}-\overset{..}{\underset{..}{O}}:$$
Chlorous acid (+3)

$$H-\overset{..}{\underset{..}{O}}-\overset{\overset{\textstyle :\overset{..}{O}:}{|}}{\underset{..}{Cl}}-\overset{..}{\underset{..}{O}}:$$
Chloric acid (+5)

$$H-\overset{..}{\underset{..}{O}}-\overset{\overset{\textstyle :\overset{..}{O}:}{|}}{\underset{\underset{\textstyle :\overset{..}{O}:}{|}}{Cl}}-\overset{..}{\underset{..}{O}}:$$
Perchloric acid (+7)

Figure 16.5 Lewis structures of the oxoacids of chlorine. The oxidation number of the Cl atom is shown in parentheses. Note that although the formulas of these acids are written with the H first, the H atom in each acid is bonded to an O atom, not directly to the Cl atom.

contain one or more O—H bonds. If the central atom is an electronegative element, or is in a high oxidation state, it will attract electrons, causing the O—H bond to be more polar. This makes it easier for the hydrogen to be lost as H^+, making the acid stronger:

To compare their strengths, it is convenient to divide the oxoacids into two groups:

1. *Oxoacids having different central atoms that are from the same group of the periodic table and that have the same oxidation number.* Two examples are:

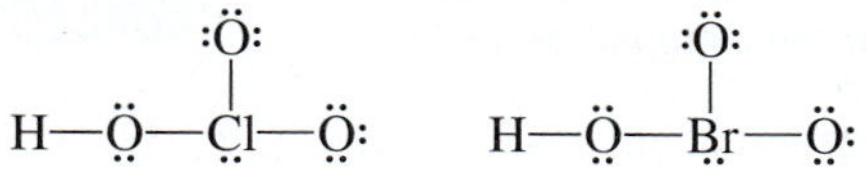

$$H-\overset{..}{\underset{..}{O}}-\overset{\overset{\textstyle :\overset{..}{O}:}{|}}{\underset{..}{Cl}}-\overset{..}{\underset{..}{O}}:$$
HClO$_3$

$$H-\overset{..}{\underset{..}{O}}-\overset{\overset{\textstyle :\overset{..}{O}:}{|}}{Br}-\overset{..}{\underset{..}{O}}:$$
HBrO$_3$

Within this group, acid strength increases with increasing electronegativity of the central atom. Cl and Br have the same oxidation number in these acids, +5. However, because Cl is more electronegative than Br, it attracts the electron pair it shares with oxygen (in the Cl—O—H group) to a greater extent than Br does (in the corresponding Br—O—H group). Consequently, the O—H bond is more polar in chloric acid than in bromic acid and ionizes more readily. The relative acid strengths are:

$$HClO_3 > HBrO_3$$

2. *Oxoacids having the same central atom but different numbers of oxygen atoms.* Within this group, acid strength increases with increasing oxidation number of the central atom. Consider the oxoacids of chlorine shown in Figure 16.5. In this series the ability of chlorine to draw electrons away from the OH group (thus making the O—H bond more polar) increases with the number of electronegative O atoms attached to Cl. Thus, HClO$_4$ is the strongest acid because it has the largest number of oxygen atoms attached to Cl. The acid strength decreases as follows:

$$HClO_4 > HClO_3 > HClO_2 > HClO$$

Sample Problem 16.19 compares acid strengths based on molecular structure:

Student Note: As the number of attached oxygen atoms increases, the oxidation number of the central atom also increases [◄◄ Section 4.4].

SAMPLE PROBLEM 16.19

Predict the relative strengths of the oxoacids in each of the following groups: (a) HClO, HBrO, and HIO; (b) HNO$_3$ and HNO$_2$.

Strategy In each group, compare the electronegativities or oxidation numbers of the central atoms to determine which O—H bonds are the most polar. The more polar the O—H bond, the more readily it is broken and the stronger the acid.

Setup

(a) In a group with different central atoms, we must compare electronegativities. The electronegativities of the central atoms in this group decrease as follows: Cl > Br > I.

(b) These two acids have the same central atom but differ in the number of attached oxygen atoms. In a group such as this, the greater the number of attached oxygen atoms, the higher the oxidation number of the central atom and the stronger the acid.

(Continued on next page)

Solution

(a) Acid strength decreases as follows: HClO > HBrO > HIO.

(b) HNO_3 is a stronger acid than HNO_2.

> **THINK ABOUT IT**
>
> Another way to compare the strengths of these two is to remember that HNO_3 is one of the seven strong acids. HNO_2 is not. Four of the strong acids are oxoacids: HNO_3, $HClO_4$, $HClO_3$, and H_2SO_4.

Practice Problem ATTEMPT Indicate which is the stronger acid: (a) $HBrO_3$ or $HBrO_4$; (b) H_2SeO_4 or H_2SO_4.

Practice Problem BUILD Based on the information in this section, which is the predominant factor in determining the strength of an oxoacid: electronegativity of the central atom or oxidation state of the central atom?

Practice Problem CONCEPTUALIZE The models show formula units of two compounds with the general formula XOH. In the model on the left, X is a metal. In the model on the right, X is a nonmetal. One of these compounds is a weak acid and the other is a strong base. Identify the acid and the base and explain why their properties are so different despite the similarity in their formulas.

Carboxylic Acids

So far our discussion has focused on inorganic acids. A particularly important group of organic acids is the *carboxylic acids,* whose Lewis structures can be represented by:

$$R-\overset{\overset{\textstyle \ddot{O}}{\|}}{C}-\ddot{O}-H$$

where R is part of the acid molecule and the shaded portion represents the carboxyl group, —COOH. You learned in Chapter 4 [◂◂ Section 4.1] that carboxylic acid formulas are often written with the ionizable H atom last, in order to keep the functional group together. You should recognize the formulas for organic acids written either way. For example, acetic acid may be written as $HC_2H_3O_2$ or as CH_3COOH.

The conjugate base of a carboxylic acid, called a carboxylate anion, $RCOO^-$, can be represented by more than one resonance structure:

$$R-\overset{\overset{\textstyle \ddot{O}}{\|}}{C}-\ddot{O}\!:^- \longleftrightarrow R-\overset{\overset{\textstyle :\ddot{O}:^-}{|}}{C}=\ddot{O}:$$

In terms of molecular orbital theory [◂◂ Section 9.6], we attribute the stability of the anion to its ability to spread out or delocalize the electron density over several atoms. The greater the extent of electron delocalization, the more stable the anion and the greater the tendency for the acid to undergo ionization—that is, the stronger the acid.

The strength of carboxylic acids depends on the nature of the R group. Consider, for example, acetic acid and chloroacetic acid:

$$H-\overset{\overset{\textstyle H}{|}}{\underset{\underset{\textstyle H}{|}}{C}}-\overset{\overset{\textstyle \ddot{O}}{\|}}{C}-\ddot{O}-H \qquad\qquad H-\overset{\overset{\textstyle Cl}{|}}{\underset{\underset{\textstyle H}{|}}{C}}-\overset{\overset{\textstyle \ddot{O}}{\|}}{C}-\ddot{O}-H$$

Acetic acid ($K_a = 1.8 \times 10^{-5}$)　　　Chloroacetic acid ($K_a = 1.4 \times 10^{-3}$)

The presence of the electronegative Cl atom in chloroacetic acid shifts the electron density toward the R group, thereby making the O—H bond more polar. Consequently, there is a greater tendency for chloroacetic acid to ionize:

Chloroacetic acid is the stronger of the two acids.

16.10 Acid-Base Properties of Salt Solutions

In Section 16.7, we saw that the conjugate base of a weak acid acts as a weak Brønsted base in water. Consider a solution of the salt sodium fluoride (NaF). (Recall that a *salt* is an ionic compound formed by the reaction between an acid and a base [◄◄ Section 4.3]. Salts are strong electrolytes that dissociate completely into ions.) Because NaF is a strong electrolyte, it dissociates completely in water to give a solution of sodium cations (Na^+) and fluoride anions (F^-). The fluoride ion, which is the conjugate base of hydrofluoric acid, reacts with water to produce hydrofluoric acid and hydroxide ion:

$$F^-(aq) + H_2O(l) \rightleftharpoons HF(aq) + OH^-(aq)$$

This is a specific example of **salt hydrolysis,** in which ions produced by the dissociation of a salt react with water to produce either hydroxide ions or hydronium ions—thus impacting pH. Using our knowledge of how ions from a dissolved salt interact with water, we can determine (based on the identity of the dissolved salt) whether a solution will be neutral, basic, or acidic. Note in the preceding example that sodium ions (Na^+) do not hydrolyze and thus have no impact on the pH of the solution.

Basic Salt Solutions

Sodium fluoride is a salt that dissolves to give a basic solution. In general, an anion that is the conjugate base of a weak acid reacts with water to produce hydroxide ion. Other examples include the acetate ion (CH_3COO^-), the nitrite ion (NO_2^-), the sulfite ion (SO_3^{2-}), and the hydrogen carbonate ion (HCO_3^-). Each of these anions undergoes hydrolysis to produce the corresponding weak acid and hydroxide ion:

$$A^-(aq) + H_2O(l) \rightleftharpoons HA(aq) + OH^-(aq)$$

We can therefore make the qualitative prediction that a solution of a salt in which the anion is the conjugate base of a weak acid will be basic. We calculate the pH of a basic salt solution the same way we calculate the pH of any weak base solution, using the K_b value for the anion. The necessary K_b value is calculated using the tabulated K_a value of the corresponding weak acid (see Table 16.6). Remember that for any conjugate acid-base pair (Equation 16.8):

$$K_a \times K_b = K_w$$

Sample Problem 16.20 shows how to calculate the pH of a basic salt solution.

Student Note: HCO_3^- has an ionizable proton and can also act as a Brønsted acid. However, its tendency to accept a proton is stronger than its tendency to donate a proton:

$$HCO_3^- + H_2O \rightleftharpoons H_2CO_3 + OH^-$$
$$K_b \approx 10^{-8}$$

$$HCO_3^- + H_2O \rightleftharpoons CO_3^{2-} + H_3O^+$$
$$K_a \approx 10^{-11}$$

Student Hot Spot

Student data indicate you may struggle with calculating the pH of a solution containing the anion of a weak acid. Access the eBook to view additional Learning Resources on this topic.

SAMPLE PROBLEM 16.20

Calculate the pH of a 0.10-M solution of sodium fluoride (NaF) at 25°C.

Strategy A solution of NaF contains Na^+ ions and F^- ions. The F^- ion is the conjugate base of the weak acid, HF. Use the K_a value for HF (7.1×10^{-4}, from Table 16.6) and Equation 16.8 to determine K_b for F^-:

$$K_b = \frac{K_w}{K_a} = \frac{1.0 \times 10^{-14}}{7.1 \times 10^{-4}} = 1.4 \times 10^{-11}$$

Then, solve this pH problem like any equilibrium problem, using an equilibrium table.

Setup It's always a good idea to write the equation corresponding to the reaction that takes place along with the equilibrium expression:

$$F^-(aq) + H_2O(l) \rightleftharpoons HF(aq) + OH^-(aq) \quad K_b = \frac{[HF][OH^-]}{[F^-]}$$

Construct an equilibrium table, and determine, in terms of the unknown x, the equilibrium concentrations of the species in the equilibrium expression:

	$F^-(aq)$	$+ H_2O(l)$	$\rightleftharpoons$	$HF(aq)$	$+ OH^-(aq)$
Initial concentration (M):	0.10	—		0	0
Change in concentration (M):	$-x$	—		$+x$	$+x$
Equilibrium concentration (M):	$0.10 - x$	—		x	x

Solution Substituting the equilibrium concentrations into the equilibrium expression and using the shortcut to solve for x, we get:

$$1.4 \times 10^{-11} = \frac{x^2}{0.10 - x} \approx \frac{x^2}{0.10}$$

$$x = \sqrt{(1.4 \times 10^{-11})(0.10)} = 1.2 \times 10^{-6} \, M$$

According to our equilibrium table, $x = [OH^-]$. In this case, the autoionization of water makes a significant contribution to the hydroxide ion concentration so the total concentration will be the sum of $1.2 \times 10^{-6} \, M$ (from the ionization of F^-) and $1.0 \times 10^{-7} \, M$ (from the autoionization of water). Therefore, we calculate the pOH first as:

$$pOH = -\log (1.2 \times 10^{-6} + 1.0 \times 10^{-7}) = 5.95$$

and then the pH:

$$pH = 14.00 - pOH = 14.00 - 5.95 = 8.05$$

The pH of a 0.10-M solution of NaF at 25°C is 8.05.

THINK ABOUT IT

It's easy to mix up pH and pOH in this type of problem. Always make a qualitative prediction regarding the pH of a salt solution first, and then check to make sure that your calculated pH agrees with your prediction. In this case, we would predict a basic pH because the anion in the salt (F^-) is the conjugate base of a weak acid (HF). The calculated pH, 8.05, is indeed basic.

Practice Problem **A**TTEMPT Determine the pH of a 0.15-M solution of sodium acetate (CH_3COONa) at 25°C.

Practice Problem **B**UILD Determine the concentration of a solution of sodium fluoride (NaF) that has pH 8.51 at 25°C.

Practice Problem **C**ONCEPTUALIZE Which of the following graphs best represents the relationship between the pH of a 0.10-M basic salt solution and the K_a of the acid from which the salt is derived?

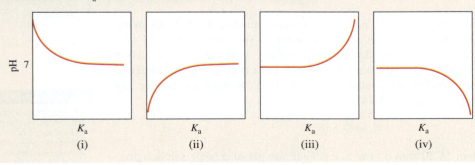

Acidic Salt Solutions

When the cation of a salt is the conjugate acid of a weak base, a solution of the salt will be acidic. For example, when ammonium chloride dissolves in water, it dissociates to give a solution of ammonium ions and chloride ions:

$$NH_4Cl(s) \xrightarrow{H_2O} NH_4^+(aq) + Cl^-(aq)$$

The ammonium ion is the conjugate acid of the weak base ammonia (NH_3). It acts as a weak Brønsted acid, reacting with water to produce hydronium ion:

$$NH_4^+(aq) + H_2O(l) \rightleftharpoons NH_3(aq) + H_3O^+(aq)$$

We would therefore predict that a solution containing the ammonium ion is acidic. To calculate the pH, we must determine the K_a for NH_4^+ using the tabulated K_b value for NH_3 and Equation 16.8. Because Cl^- is the weak conjugate base of the strong acid HCl, Cl^- does not hydrolyze and therefore has no impact on the pH of the solution.

Sample Problem 16.21 shows how to calculate the pH of an acidic salt solution.

SAMPLE PROBLEM 16.21

Calculate the pH of a 0.10-*M* solution of ammonium chloride (NH_4Cl) at 25°C.

Strategy A solution of NH_4Cl contains NH_4^+ cations and Cl^- anions. The NH_4^+ ion is the conjugate acid of the weak base NH_3. Use the K_b value for NH_3 (1.8×10^{-5} from Table 16.7) and Equation 16.8 to determine K_a for NH_4^+.

$$K_a = \frac{K_w}{K_b} = \frac{1.0 \times 10^{-14}}{1.8 \times 10^{-5}} = 5.6 \times 10^{-10}$$

Setup Again, we write the balanced chemical equation and the equilibrium expression:

$$NH_4^+(aq) + H_2O(l) \rightleftharpoons NH_3(aq) + H_3O^+(aq) \qquad K_a = \frac{[NH_3][H_3O^+]}{[NH_4^+]}$$

Next, construct a table to determine the equilibrium concentrations of the species in the equilibrium expression:

$$NH_4^+(aq) + H_2O(l) \rightleftharpoons NH_3(aq) + H_3O^+(aq)$$

	NH_4^+	H_2O	NH_3	H_3O^+
Initial concentration (*M*):	0.10	—	0	0
Change in concentration (*M*):	$-x$	—	$+x$	$+x$
Equilibrium concentration (*M*):	$0.10 - x$	—	x	x

Solution Substituting the equilibrium concentrations into the equilibrium expression and using the shortcut to solve for *x*, we get:

$$5.6 \times 10^{-10} = \frac{x^2}{0.10 - x} \approx \frac{x^2}{0.10}$$

$$x = \sqrt{(5.6 \times 10^{-10})(0.10)} = 7.5 \times 10^{-6}\ M$$

According to the equilibrium table, $x = [H_3O^+]$. The pH can be calculated as follows:

$$pH = -\log(7.5 \times 10^{-6}) = 5.12$$

The pH of a 0.10-*M* solution of ammonium chloride (at 25°C) is 5.12.

THINK ABOUT IT

In this case, we would predict an acidic pH because the cation in the salt (NH_4^+) is the conjugate acid of a weak base (NH_3). The calculated pH is acidic.

(Continued on next page)

Practice Problem (A)**TTEMPT** Determine the pH of a 0.25-M solution of pyridinium nitrate ($C_5H_6NNO_3$) at 25°C. [Pyridinium nitrate dissociates in water to give pyridinium ions ($C_5H_6N^+$), the conjugate acid of *pyridinium* (see Table 16.7), and nitrate ions (NO_3^-).]

Practice Problem (B)**UILD** Determine the concentration of a solution of ammonium chloride (NH_4Cl) that has pH 5.37 at 25°C.

Practice Problem (C)**ONCEPTUALIZE** Which of the following graphs best represents the relationship between the pH of a 0.10-M acidic salt solution and the K_b of the base from which the salt is derived?

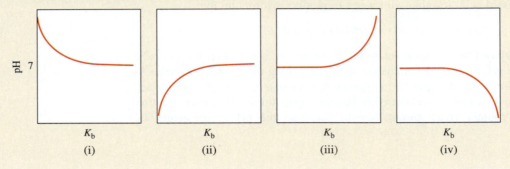

| K_b | K_b | K_b | K_b |
| (i) | (ii) | (iii) | (iv) |

The metal ion in a dissolved salt can also react with water to produce an acidic solution. The extent of hydrolysis is greatest for the small and highly charged metal cations such as Al^{3+}, Cr^{3+}, Fe^{3+}, Bi^{3+}, and Be^{2+}. For example, when aluminum chloride dissolves in water, each Al^{3+} ion becomes associated with six water molecules (Figure 16.6):

Consider one of the bonds that forms between the metal ion and an oxygen atom from one of the six water molecules in $Al(H_2O)_6^{3+}$:

$$Al \rightleftharpoons O \begin{smallmatrix} H \\ \\ H \end{smallmatrix}$$

The positively charged Al^{3+} ion draws electron density toward itself, increasing the polarity of the O—H bonds. Consequently, the H atoms have a greater tendency to ionize than those in water molecules not associated with the Al^{3+} ion. The resulting ionization process can be written as:

$$Al(H_2O)_6^{3+}(aq) + H_2O(l) \rightleftharpoons Al(OH)(H_2O)_5^{2+}(aq) + H_3O^+(aq)$$

The equilibrium constant for the metal cation hydrolysis is given by:

$$K_a = \frac{[Al(OH)(H_2O)_5^{2+}][H_3O^+]}{[Al(H_2O)_6^{3+}]} = 1.3 \times 10^{-5}$$

Figure 16.6 The six H_2O molecules surround the Al^{3+} ion in an octahedral arrangement. The attraction of the small Al^{3+} ion for the lone pairs on the oxygen atoms is so great that the O—H bonds in an H_2O molecule attached to the metal cation are weakened, allowing the loss of a proton (H^+) to an incoming H_2O molecule. This hydrolysis of the metal cation makes the solution acidic.

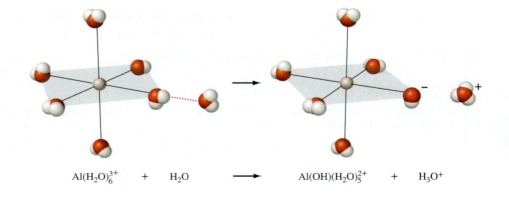

$$Al(H_2O)_6^{3+} \quad + \quad H_2O \quad \longrightarrow \quad Al(OH)(H_2O)_5^{2+} \quad + \quad H_3O^+$$

$Al(OH)(H_2O)_5^{2+}$ can undergo further ionization:

$$Al(OH)(H_2O)_5^{2+}(aq) + H_2O(l) \rightleftharpoons Al(OH)(H_2O)_5^{+}(aq) + H_3O^+(aq)$$

and so on. It is generally sufficient, however, to take into account only the first stage of hydrolysis when determining the pH of a solution that contains metal ions.

Neutral Salt Solutions

The extent of hydrolysis is greatest for the smallest and most highly charged metal ions because a compact, highly charged ion is more effective in polarizing the O—H bond and facilitating ionization. This is why relatively large ions of low charge, including the metal cations of Groups 1 and 2 (the cations of the strong bases), do not undergo significant hydrolysis (Be^{2+} is an exception). Thus, most metal cations of Groups 1 and 2 do not impact the pH of a solution.

Similarly, anions that are conjugate bases of strong acids do not hydrolyze to any significant degree. Consequently, a salt composed of the cation of a strong base and the anion of a strong acid, such as NaCl, produces a neutral solution.

To summarize, the pH of a salt solution can be predicted qualitatively by identifying the ions in solution and determining which of them, if any, undergoes significant hydrolysis.

Student Note: The metal cations of the strong bases are those of the alkali metals (Li^+, Na^+, K^+, Rb^+, and Cs^+) and those of the heavy alkaline earth metals (Sr^{2+} and Ba^{2+}).

	Examples
A cation that will make a solution acidic is	
• the conjugate acid of a weak base.	NH_4^+, $CH_3NH_3^+$, $C_2H_5NH_3^+$
• a small, highly charged metal ion (other than from Group 1 or 2).	Al^{3+}, Cr^{3+}, Fe^{3+}, Bi^{3+}
An anion that will make a solution basic is	
• the conjugate base of a weak acid.	CN^-, NO_2^-, CH_3COO^-
A cation that will not affect the pH of a solution is	
• a Group 1 or heavy Group 2 cation (except Be^{2+}).	Li^+, Na^+, Ba^{2+}
An anion that will not affect the pH of a solution is	
• the conjugate base of a strong acid.	Cl^-, NO_3^-, ClO_4^-

Student Hot Spot

Student data indicate you may struggle with predicting the pH of a salt solution. Access the eBook to view additional Learning Resources on this topic.

Sample Problem 16.22 lets you practice predicting the pH of salt solutions.

SAMPLE PROBLEM 16.22

Predict whether a 0.10-*M* solution of each of the following salts will be basic, acidic, or neutral: (a) LiI, (b) NH_4NO_3, (c) $Sr(NO_3)_2$, (d) KNO_2, (e) NaCN.

Strategy Identify the ions present in each solution, and determine which, if any, will impact the pH of the solution.

Setup (a) Ions in solution: Li^+ and I^-. Li^+ is a Group 1 cation; I^- is the conjugate base of the strong acid HI. Therefore, neither ion hydrolyzes to any significant degree.

(b) Ions in solution: NH_4^+ and NO_3^-. NH_4^+ is the conjugate acid of the weak base NH_3; NO_3^- is the conjugate base of the strong acid HNO_3. In this case, the cation will hydrolyze, making the pH acidic:

$$NH_4^+(aq) + H_2O(aq) \rightleftharpoons NH_3(aq) + H_3O^+(aq)$$

(c) Ions in solution: Sr^{2+} and NO_3^-. Sr^{2+} is a heavy Group 2 cation; NO_3^- is the conjugate base of the strong acid HNO_3. Neither ion hydrolyzes to any significant degree.

(d) Ions in solution: K^+ and NO_2^-. K^+ is a Group 1 cation; NO_2^- is the conjugate base of the weak acid HNO_2. In this case, the anion hydrolyzes, thus making the pH basic:

$$NO_2^-(aq) + H_2O(l) \rightleftharpoons HNO_2(aq) + OH^-(aq)$$

(e) Ions in solution: Na^+ and CN^-. Na^+ is a Group 1 cation; CN^- is the conjugate base of the weak acid HCN. In this case, too, the anion hydrolyzes, thus making the pH basic:

$$CN^-(aq) + H_2O(l) \rightleftharpoons HCN(aq) + OH^-(aq)$$

(Continued on next page)

Solution

(a) Neutral   (d) Basic

(b) Acidic   (e) Basic

(c) Neutral

THINK ABOUT IT

It's very important that you are able to identify the ions in solution correctly. If necessary, review the formulas and charges of the common polyatomic ions [◄◄ Section 2.6, Table 2.3].

Practice Problem ⒶTTEMPT Predict whether a 0.10-M solution of each of the following salts will be basic, acidic, or neutral: (a) CH_3COOLi, (b) C_5H_5NHCl, (c) KF, (d) KNO_3, (e) $KClO_4$.

Practice Problem ⒷUILD In addition to those given in Sample Problem 16.22 and Practice Problem A, identify two salts that will dissolve to give (a) an acidic solution, (b) a basic solution, and (c) a neutral solution.

Practice Problem ⒸONCEPTUALIZE Which of the reactions could correctly illustrate a process by which a salt affects the pH of a solution?

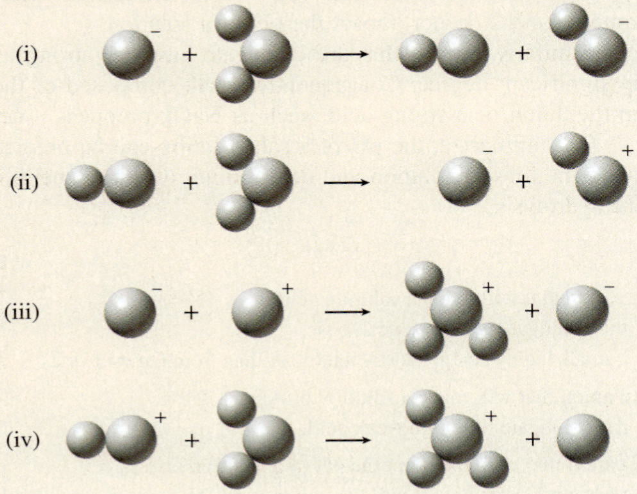

(i)

(ii)

(iii)

(iv)

Salts in Which Both the Cation and the Anion Hydrolyze

So far we have considered salts in which only one ion undergoes hydrolysis. In some salts, both the cation and the anion hydrolyze. Whether a solution of such a salt is basic, acidic, or neutral depends on the relative strengths of the weak acid and the weak base. Although the process of calculating the pH in these cases is more complex than in cases where only one ion hydrolyzes, we can make qualitative predictions regarding pH using the values of K_b (of the salt's anion) and K_a (of the salt's cation):

- When $K_b > K_a$, the solution is basic.
- When $K_b < K_a$, the solution is acidic.
- When $K_b \approx K_a$, the solution is neutral or nearly neutral.

The salt NH_4NO_2, for example, dissociates in solution to give NH_4^+ ($K_a = 5.6 \times 10^{-10}$) and NO_2^- ($K_b = 2.2 \times 10^{-11}$). Because K_a for the ammonium ion is larger than K_b for the nitrite ion, we would expect the pH of an ammonium nitrite solution to be slightly acidic.

CHECKPOINT – SECTION 16.10   Acid-Base Properties of Salt Solutions

16.10.1 Calculate the pH of a 0.075-M solution of NH_4NO_3 at 25°C.

    a) 5.19   d) 2.93

    b) 8.81   e) 11.07

    c) 7.00

16.10.2 Calculate the pH of a 0.082-M solution of NaCN at 25°C.

    a) 5.20   d) 2.89

    b) 8.80   e) 11.11

    c) 7.00

16.10.3 Which of the following salts will produce a basic solution when dissolved in water? (Select all that apply.)

a) Sodium hypochlorite (NaClO)

b) Potassium fluoride (KF)

c) Lithium carbonate (Li_2CO_3)

d) Barium chloride ($BaCl_2$)

e) Ammonium iodide (NH_4I)

16.10.4 Which of the following salts will produce a neutral solution when dissolved in water? (Select all that apply.)

a) Calcium chlorite [$Ca(ClO_2)_2$]

b) Potassium iodide (KI)

c) Lithium nitrate ($LiNO_3$)

d) Barium cyanide [$Ba(CN)_2$]

e) Ammonium iodide (NH_4I)

16.10.5 The diagrams represent solutions of three salts NaX (X = A, B, or C). Arrange the three X^- anions in order of increasing base strength. (The Na^+ ions are not shown.)

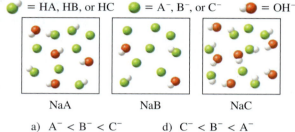

●= HA, HB, or HC   ●= A^-, B^-, or C^- ●= OH^-

NaA   NaB   NaC

a) $A^- < B^- < C^-$ d) $C^- < B^- < A^-$

b) $A^- < C^- < B^-$ e) $C^- < A^- < B^-$

c) $B^- < A^- < C^-$

16.11 Acid-Base Properties of Oxides and Hydroxides

As we saw in Chapter 7, oxides can be classified as acidic, basic, or amphoteric. Thus, our discussion of acid-base reactions would be incomplete if we did not examine the properties of these compounds.

Oxides of Metals and Nonmetals

Figure 16.7 shows the formulas of a number of oxides of the main group elements in their highest oxidation states. All alkali metal oxides and all alkaline earth metal oxides except BeO are basic. Beryllium oxide and several metallic oxides in

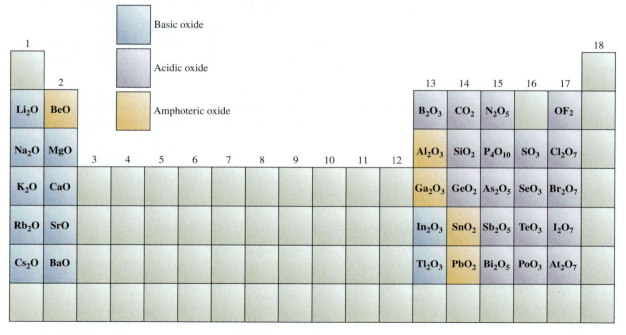

Figure 16.7 Oxides of the main group elements in their highest oxidation states.

Groups 13 and 14 are amphoteric. Nonmetallic oxides in which the oxidation number of the main group element is high are acidic (e.g., N_2O_5, SO_3, and Cl_2O_7), but those in which the oxidation number of the main group element is low (e.g., CO and NO) show no measurable acidic properties. No nonmetallic oxides are known to have basic properties.

The basic metallic oxides react with water to form metal hydroxides:

$$Na_2O(s) + H_2O(l) \longrightarrow 2NaOH(aq)$$

$$BaO(s) + H_2O(l) \longrightarrow Ba(OH)_2(aq)$$

The reactions between acidic oxides and water are as follows:

$$CO_2(g) + H_2O(l) \rightleftharpoons H_2CO_3(aq)$$

$$SO_3(g) + H_2O(l) \rightleftharpoons H_2SO_4(aq)$$

$$N_2O_5(g) + H_2O(l) \rightleftharpoons 2HNO_3(aq)$$

$$P_4O_{10}(g) + 6H_2O(l) \rightleftharpoons 4H_3PO_4(aq)$$

$$Cl_2O_7(g) + H_2O(l) \rightleftharpoons 2HClO_4(aq)$$

The reaction between CO_2 and H_2O explains why pure water gradually becomes acidic when it is exposed to air, which contains CO_2. The pH of rainwater exposed only to unpolluted air is slightly acidic. The reaction between SO_3 and H_2O is largely responsible for acid rain.

Reactions between acidic oxides and bases and those between basic oxides and acids resemble normal acid-base reactions in that the products are a salt and water:

$$CO_2(g) + 2NaOH(aq) \longrightarrow Na_2CO_3(aq) + H_2O(l)$$

$$BaO(s) + 2HNO_3(aq) \longrightarrow Ba(NO_3)_2(aq) + H_2O(l)$$

Aluminum oxide (Al_2O_3) is amphoteric. Depending on the reaction conditions, it can behave either as an acidic oxide or as a basic oxide. For example, Al_2O_3 acts as a base with hydrochloric acid to produce a salt ($AlCl_3$) and water:

$$Al_2O_3(s) + 6HCl(aq) \longrightarrow 2AlCl_3(aq) + 3H_2O(l)$$

and acts as an acid with sodium hydroxide:

$$Al_2O_3(s) + 2NaOH(aq) + 3H_2O \longrightarrow 2NaAl(OH)_4(aq)$$

Only a salt, sodium aluminum hydroxide [$NaAl(OH)_4$, which contains the Na^+ and $Al(OH)_4^-$ ions] is formed in the reaction with sodium hydroxide—no water is produced. Nevertheless, the reaction is still classified as an acid-base reaction because Al_2O_3 neutralizes NaOH.

Some transition metal oxides in which the metal has a high oxidation number act as acidic oxides. Two examples are manganese(VII) oxide (Mn_2O_7) and chromium(VI) oxide (CrO_3), both of which react with water to produce acids:

$$Mn_2O_7(l) + H_2O(l) \longrightarrow 2HMnO_4(aq)$$
permanganic acid

$$CrO_3(s) + H_2O(l) \longrightarrow H_2CrO_4(aq)$$
chromic acid

Basic and Amphoteric Hydroxides

All the alkali and alkaline earth metal hydroxides, except $Be(OH)_2$, are basic. $Be(OH)_2$, $Al(OH)_3$, $Sn(OH)_2$, $Pb(OH)_2$, $Cr(OH)_3$, $Cu(OH)_2$, $Zn(OH)_2$, and $Cd(OH)_2$ are amphoteric. All amphoteric hydroxides are insoluble, but beryllium hydroxide reacts with both acids and bases as follows:

$$Be(OH)_2(s) + 2H_3O^+(aq) \longrightarrow Be^{2+}(aq) + 4H_2O(l)$$

$$Be(OH)_2(s) + 2OH^-(aq) \longrightarrow Be(OH)_4^{2-}(aq)$$

Aluminum hydroxide reacts with both acids and bases in a similar fashion:

$$Al(OH)_3(s) + 3H_3O^+(aq) \longrightarrow Al^{3+}(aq) + 6H_2O(l)$$

$$Al(OH)_3(s) + OH^-(aq) \longrightarrow Al(OH)_4^-(aq)$$

16.12 Lewis Acids and Bases

So far we have discussed acid-base properties in terms of the Brønsted theory. For example, a Brønsted base is a substance that must be able to accept protons. By this definition, both the hydroxide ion and ammonia are bases:

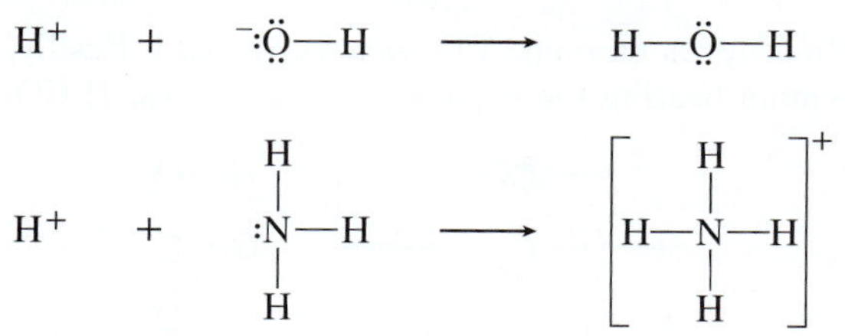

$$H^+ \quad + \quad \,^-\!:\!\ddot{O}\!-\!H \quad \longrightarrow \quad H\!-\!\ddot{O}\!-\!H$$

In each case, the atom to which the proton becomes attached possesses at least one unshared pair of electrons. This characteristic property of OH^-, NH_3, and other Brønsted bases suggests a more general definition of acids and bases.

In 1932 G. N. Lewis defined what we now call a ***Lewis base*** as a substance that can donate a pair of electrons. A ***Lewis acid*** is a substance that can accept a pair of electrons. In the protonation of ammonia, for example, NH_3 acts as a Lewis base because it donates a pair of electrons to the proton H^+, which acts as a Lewis acid by accepting the pair of electrons. A Lewis acid-base reaction, therefore, is one that involves the donation of a pair of electrons from one species to another.

The significance of the Lewis concept is that it is more general than other definitions. Lewis acid-base reactions include many reactions that do not involve Brønsted acids. Consider, for example, the reaction between boron trifluoride (BF_3) and ammonia to form an adduct compound:

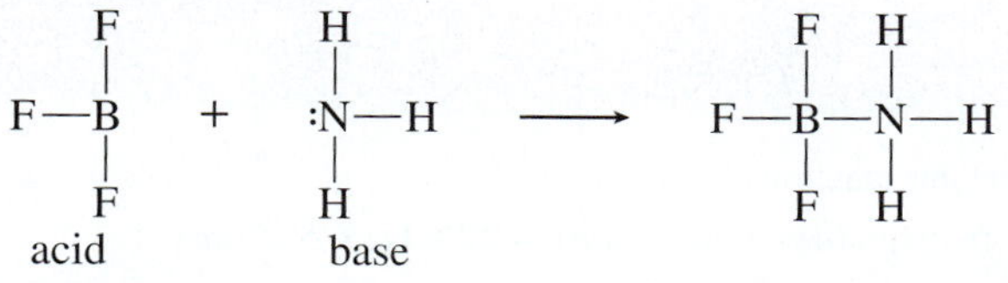

The B atom in BF_3 is sp^2-hybridized [◄◄ Section 9.4]. The vacant, unhybridized $2p_z$ orbital accepts the pair of electrons from NH_3. Thus, BF_3 functions as an acid according to the Lewis definition, even though it does not contain an ionizable proton. A *coordinate covalent* bond [◄◄ Section 8.8] is formed between the B and N atoms. In fact, every Lewis acid-base reaction results in the formation of a coordinate covalent bond.

Boric acid is another Lewis acid containing boron. Boric acid (a weak acid used in eyewash) is an oxoacid with the following structure:

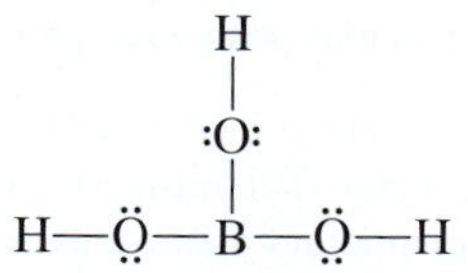

Boric acid itself does not ionize in water to produce H_3O^+. Instead, it produces H_3O^+ in solution by taking a hydroxide ion away from a water molecule:

$$B(OH)_3(aq) + 2H_2O(l) \rightleftharpoons B(OH)_4^-(aq) + H_3O^+(aq)$$

In this Lewis acid-base reaction, boric acid accepts a pair of electrons from the hydroxide ion that is derived from a water molecule, leaving behind a hydronium ion.

The hydration of carbon dioxide to produce carbonic acid:

$$CO_2(g) + H_2O(l) \rightleftharpoons H_2CO_3(aq)$$

can be explained in terms of Lewis acid-base theory as well. The first step involves the donation of a lone pair on the O atom in H_2O to the C atom in CO_2. An orbital is vacated on the C atom to accommodate the lone pair by relocation of the electron pair in one of the C—O pi bonds, changing the hybridization of the oxygen atom from sp^2 to sp^3:

As a result, H_2O is a Lewis base and CO_2 is a Lewis acid. Finally, a proton is transferred onto the O atom bearing the negative charge to form H_2CO_3:

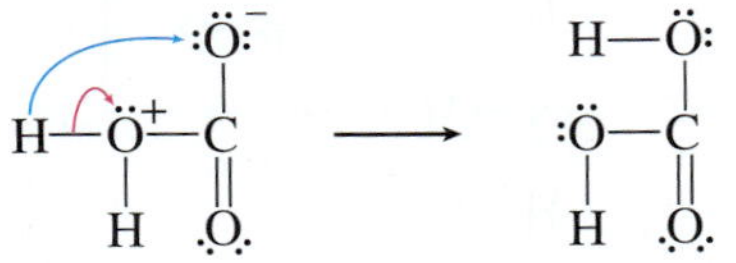

Other examples of Lewis acid-base reactions are:

$$Ag^+(aq) + 2NH_3(aq) \rightleftharpoons Ag(NH_3)_2^+(aq)$$

$$Cd^{2+}(aq) + 4I^-(aq) \rightleftharpoons CdI_4^{2-}(aq)$$

$$Ni(s) + 4CO(g) \rightleftharpoons Ni(CO)_4(g)$$

The hydration of metal ions is in itself a Lewis acid-base reaction. When copper(II) sulfate ($CuSO_4$) dissolves in water, each Cu^{2+} ion becomes associated with six water molecules as $Cu(H_2O)_6^{2+}$. In this case, Cu^{2+} acts as the acid, accepting electrons, whereas H_2O acts as the base, donating electrons.

Sample Problem 16.23 shows how to classify Lewis acids and bases.

SAMPLE PROBLEM 16.23

Identify the Lewis acid and Lewis base in each of the following reactions:

(a) $C_2H_5OC_2H_5 + AlCl_3 \rightleftharpoons (C_2H_5)_2OAlCl_3$

(b) $Hg^{2+}(aq) + 4CN^-(aq) \rightleftharpoons Hg(CN)_4^{2-}(aq)$

Strategy Determine which species in each reaction accepts a pair of electrons (Lewis acid) and which species donates a pair of electrons (Lewis base).

Setup (a) It can be helpful to draw Lewis structures of the species involved:

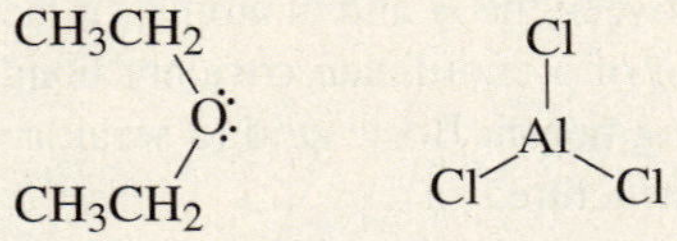

(b) Metal ions act as Lewis acids, accepting electron pairs from anions or molecules with lone pairs.

Solution (a) The Al is sp^2-hybridized in $AlCl_3$ with an empty $2p_z$ orbital. It is electron deficient, sharing only six electrons. Therefore, the Al atom has the capacity to gain two electrons to complete its octet. This property makes $AlCl_3$ a Lewis acid. On the other hand, the lone pairs on the oxygen atom in $C_2H_5OC_2H_5$ make the compound a Lewis base:

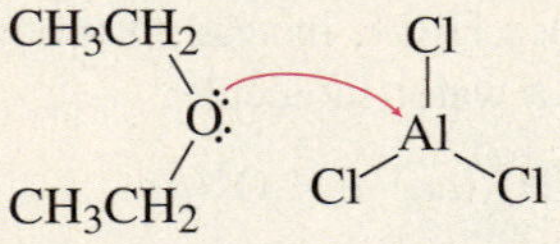

(b) Hg^{2+} accepts four pairs of electrons from the CN^- ions. Therefore, Hg^{2+} is the Lewis acid and CN^- is the Lewis base.

THINK ABOUT IT

In Lewis acid-base reactions, the acid is usually a cation or an electron-deficient molecule, whereas the base is an anion or a molecule containing an atom with lone pairs.

Practice Problem ATTEMPT Identify the Lewis acid and Lewis base in the following reaction:

$$Co^{3+}(aq) + 6NH_3(aq) \rightleftharpoons Co(NH_3)_6^{3+}(aq)$$

Practice Problem BUILD Write formulas for the Lewis acid and Lewis base that react to form H_3NAlBr_3.

Practice Problem CONCEPTUALIZE Which of the diagrams at right best depicts the combination of HCl and water as a Lewis acid-base reaction? Which best depicts the combination as a Brønsted acid-base reaction?

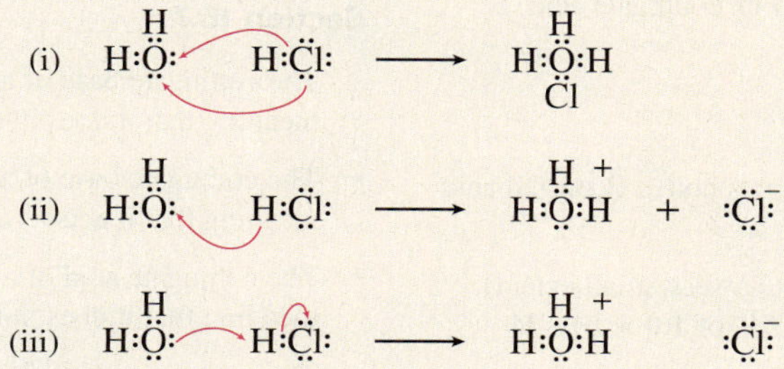

(i)   H:Ö: H:Ċl: ⟶   H:Ö:H
      | |
      H   H
                           Cl

(ii) H:Ö: H:Ċl: ⟶   H:Ö:H + :Ċl:⁻
      | | +
      H   H

(iii) H:Ö: H:Ċl: ⟶   H:Ö:H   :Ċl:⁻
      | | +
      H   H

CHECKPOINT – SECTION 16.12   Lewis Acids and Bases

16.12.1 Which of the following cannot act as a Lewis base? (Select all that apply.)

  a) NH_3

  b) OH^-

  c) CH_4

  d) Fe^{2+}

  e) Al^{3+}

16.12.2 Which of the following is a Lewis acid but not a Brønsted acid? (Select all that apply.)

  a) H_2O

  b) BCl_3

  c) OH^-

  d) Al^{3+}

  e) NH_3

Chapter Summary

Section 16.1

- A Brønsted acid donates a proton; a Brønsted base accepts a proton.

- When a Brønsted acid donates a proton, the anion that remains is a *conjugate base.*

- When a Brønsted base accepts a proton, the resulting cation is a *conjugate acid.*

- The combination of a Brønsted acid and its conjugate base (or the combination of a Brønsted base and its conjugate acid) is called a *conjugate pair.*

Section 16.2

- Water is *amphoteric,* meaning it can act as both a Brønsted acid and a Brønsted base.

- Pure water undergoes *autoionization* (to a very small extent), resulting in concentrations of H^+ and OH^- of 1.0×10^{-7} M at 25°C.

- K_w is the equilibrium constant for the autoionization of water, also called the *ion-product constant.* $K_w = [H^+][OH^-] = 1.0 \times 10^{-14}$ at 25°C.

Section 16.3

- The *pH* scale measures acidity: $pH = -\log [H^+]$.

- $pH = 7.00$ is neutral, $pH < 7.00$ is acidic, and $pH > 7.00$ is basic.

- The *pOH* scale is analogous to the pH scale, but it measures *basicity:* $pOH = -\log [OH^-]$.

- $pOH = 7.00$ is neutral, $pOH < 7.00$ is basic, and $pOH > 7.00$ is acidic.

- $pH + pOH = 14.00$ (at 25°C).

Section 16.4

- There are seven strong acids: HCl, HBr, HI, HNO_3, $HClO_3$, $HClO_4$, and H_2SO_4.

- Strong acids ionize completely in aqueous solution. (Only the first ionization of the diprotic acid H_2SO_4 is complete.)

- The strong bases are the Group 1 and the heaviest Group 2 hydroxides: $LiOH$, $NaOH$, KOH, $RbOH$, $CsOH$, $Ca(OH)_2$, $Sr(OH)_2$, and $Ba(OH)_2$.

Section 16.5

- A *weak acid* ionizes only partially. The *acid ionization constant,* K_a, is the equilibrium constant that indicates to what extent a weak acid ionizes.

- We solve for the pH of a solution of weak acid using the concentration of the acid, the K_a value, and an equilibrium table.

- We can also determine the K_a of a weak acid if we know the initial acid concentration and the pH at equilibrium.

Section 16.6

- A *weak base* ionizes only partially. The *base ionization constant, K_b,* is the equilibrium constant that indicates to what extent a weak base ionizes.

- We solve for the pH of a solution of weak base using the concentration of the base, the K_b value, and an equilibrium table.

- We can also determine the K_b of a weak base if we know the concentration and the pH at equilibrium.

Section 16.7

- The conjugate base of a strong acid is a *weak conjugate base,* meaning that it does not react with water.

- The conjugate base of a weak acid is a *strong conjugate base,* meaning that it acts as a weak Brønsted base in water.

- The conjugate acid of a strong base is a *weak conjugate acid,* meaning that it does not react with water.

- The conjugate acid of a weak base is a *strong conjugate acid,* meaning that it acts as a weak Brønsted acid in water.

- For any conjugate acid-base pair, $K_a \times K_b = K_w$.

Section 16.8

- Diprotic and polyprotic acids have more than one proton to donate. They undergo stepwise ionizations. Each ionization has a K_a value associated with it.

- The K_a values for stepwise ionizations become progressively smaller.

- In most cases, it is only necessary to consider the first ionization of an acid to determine pH. To determine the concentrations of other species at equilibrium, it may be necessary to consider subsequent ionizations.

Section 16.9

- The strength of an acid is affected by molecular structure.

- Polar and weak bonds to the ionizable hydrogen lead to a stronger acid.

- Resonance stabilization of the conjugate base favors the ionization process, resulting in a stronger acid.

Section 16.10

- Salts dissolve in water to give neutral, acidic, or basic solutions depending on their constituent ions. *Salt hydrolysis* is the reaction of an ion with water to produce hydronium or hydroxide ions.

- Cations that are strong conjugate acids such as NH_4^+ make a solution more acidic.

- Anions that are strong conjugate bases such as F^- make a solution more basic. Anions that are conjugate bases of strong acids have no effect on pH.

- Small, highly charged metal ions hydrolyze to give acidic solutions.

Section 16.11

- Oxides of metals generally are basic; oxides of nonmetals generally are acidic.

- Metal hydroxides may be basic or amphoteric.

Section 16.12

- Lewis theory provides more general definitions of acids and bases.

- A *Lewis acid* accepts a pair of electrons; a *Lewis base* donates a pair of electrons.

- A Lewis acid is generally electron-poor and need not have a hydrogen atom.

- A Lewis base is an anion or a molecule with one or more lone pairs of electrons.

Key Words

Acid ionization constant (K_a), 803	Conjugate base, 788	pH, 792	Weak acid, 803
Amphoteric, 790	Conjugate pair, 788	pOH, 794	Weak base, 812
Autoionization of water, 790	Ion-product constant, 790	Salt hydrolysis, 825	Weak conjugate acid, 815
Base ionization constant (K_b), 812	Lewis acid, 833	Strong conjugate acid, 815	Weak conjugate base, 815
Conjugate acid, 788	Lewis base, 833	Strong conjugate base, 815	

Key Equations

16.1 $K_w = [H_3O^+][OH^-] = 1.0 \times 10^{-14}$ (at 25°C)

The equilibrium constant for autoionization of water is K_w. In any aqueous solution, K_w is equal to the product of hydronium ion and hydroxide ion concentrations. At 25°C, the value of K_w is 1.0×10^{-14}.

16.2 $pH = -\log [H_3O^+]$ or $pH = -\log [H^+]$

The pH of an aqueous solution is calculated as minus the base-10 log of hydronium ion concentration.

16.3 $[H_3O^+] = 10^{-pH}$ or $[H^+] = 10^{-pH}$

Hydronium ion concentration can also be calculated from pH.

16.4 $pOH = -\log [OH^-]$

The pOH of an aqueous solution is calculated as minus the base-10 log of hydroxide ion concentration.

16.5 $[OH^-] = 10^{-pOH}$

Hydroxide ion concentration can also be calculated from pOH.

16.6 $pH + pOH = 14.00$ (at 25°C)

The sum of pH and pOH in any aqueous solution at 25°C is 14.00.

16.7 Percent ionization $= \dfrac{[H_3O^+]_{eq}}{[HA]_0} \times 100\%$

Percent ionization of a weak acid is calculated as the ratio of hydronium ion concentration at equilibrium to original weak acid concentration times 100%.

16.8 $K_a \times K_b = K_w$

For any conjugate acid-base pair, the product of K_a for the acid and K_b for the base is K_w.

Salt hydrolysis is critical to the understanding of certain acid-base titrations. When a weak acid is titrated with a strong base, the product of the neutralization is the weak acid's conjugate base:

$$HA(aq) + OH(aq) \longrightarrow A^-(aq) + H_2O(l)$$

The conjugate base of a weak acid behaves as a weak Brønsted base in water:

$$A^-(aq) + H_2O(l) \rightleftharpoons HA(aq) + OH^-(aq)$$

If we know the concentrations of the weak acid and the strong base, we can determine pH at the equivalence point of a weak acid–strong base titration as follows:

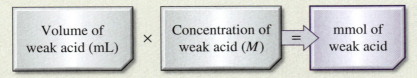

In the case of a monoprotic weak acid such as acetic acid ($HC_2H_2O_2$) and a monobasic strong base such as NaOH, the number of millimoles of base is equal to the number of millimoles of acid at the equivalence point:

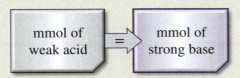

This enables us to determine the volume of strong base necessary to reach the equivalence point:

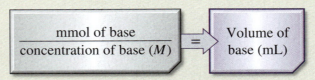

The combination of the original weak acid volume and the volume of base added to reach the equivalence point gives the total volume at the equivalence point. Further, the number of millimoles of conjugate base produced is equal to the number of millimoles of weak acid present at the start of the titration:

Using this information, we find the concentration of conjugate base; and to get K_b, we use the tabulated K_a for the weak acid and Equation 16.8:

$$\frac{\text{mmol of conjugate base}}{\text{Total volume at equivalence point}} = \text{Concentration of conjugate base} \qquad K_b = \frac{K_w}{K_a}$$

Once we have both the concentration and the ionization constant for the conjugate base, we construct an ice table and solve for the equilibrium concentration of hydroxide—and for pH. To illustrate the process, let's determine the pH at the equivalence point of the titration of 30.0 mL 0.15 M acetic acid ($HC_2H_3O_2$, $K_a = 1.8 \times 10^{-5}$) with 0.12 M NaOH at 25°C.

The neutralization net ionic equation is: $HC_2H_3O_2(aq) + OH^-(aq) \longrightarrow C_2H_3O_2^-(aq) + H_2O(l)$

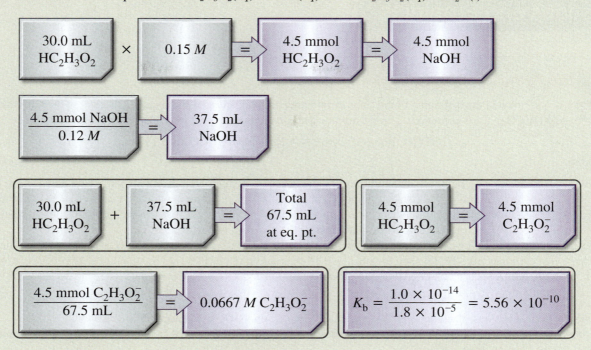

We now construct an equilibrium table and solve for $[OH^-]$. The reaction of acetate ion with water is:

$$C_2H_3O_2^-(aq) + H_2O(l) \rightleftharpoons HC_2H_3O_2(aq) + OH^-(aq)$$

	$C_2H_3O_2^-$		$HC_2H_3O_2$	OH^-
i	0.0667		0	0
c	$-x$		$+x$	$+x$
e	$0.0667 - x$		x	x

$$5.56 \times 10^{-10} = \frac{[HC_2H_3O_2][OH^-]}{[C_2H_3O_2^-]}$$

Solving for x gives 6.09×10^{-6} M. This is the concentration of the OH^- ion:

$$pOH = -\log[OH^-] = -\log(6.09 \times 10^{-6}) = 5.22$$

and $pH = 14.00 - pOH$, which gives:

$$pH = 8.78$$

Key Skills Problems

16.1
Calculate the pH of a solution that is 0.22 M in nitrite ion (NO_2^-) at 25°C. K_a for nitrous acid (HNO_2) is 4.5×10^{-4}.

(a) 11.79   (b) 8.34   (c) 5.65   (d) 7.00   (e) 2.21

16.2
Determine pH at the equivalence point in the titration of 41.0 mL 0.096 M formic acid with 0.108 M NaOH at 25°C.

(a) 12.94   (b) 7.00   (c) 5.76   (d) 8.24   (e) 1.06

16.3
Calculate the pH of a solution that is 0.22 M in pyridinium ion ($C_5H_5NH^+$) at 25°C. K_b for pyridine (C_5H_5N) is 1.7×10^{-9}.

(a) 11.06   (b) 7.00   (c) 7.51   (d) 2.94   (e) 4.19

16.4
Determine pH at the equivalence point in the titration of 26.0 mL 1.12 M pyridine with 0.93 M HCl at 25°C.

(a) 7.00   (b) 2.76   (c) 11.24   (d) 1.73   (e) 12.27

Questions and Problems

Applying What You've Learned

As was noted at the beginning of the chapter, rain in unpolluted areas is slightly acidic because it contains carbonic acid (H_2CO_3). Carbonic acid is a weak, diprotic acid that ionizes to give H_3O^+ and the hydrogen carbonate ion:

$$H_2CO_3(aq) + H_2O(l) \rightleftharpoons H_3O^+(aq) + HCO_3^-(aq)$$

Problems:

(a) Give the formulas for the conjugate acid and the conjugate base of the hydrogen carbonate ion (also known as the bicarbonate ion) [◄◄ Sample Problem 16.1]. (b) Determine $[H_3O^+]$ and pH for a raindrop in which the carbonic acid concentration is 1.8×10^{-5} M. Assume that carbonic acid is the only acid in the raindrop and that the second ionization is negligible [◄◄ Sample Problem 16.12]. (c) Calculate $[OH^-]$ and pOH of the raindrop in part (b) [◄◄ Sample Problems 16.3 and 16.6]. (d) What concentration of HCl would have the same pH as the raindrop in part (b) [◄◄ Sample Problem 16.9]? (e) Determine K_b for the hydrogen carbonate ion (HCO_3^-) and the pH of a 0.10-M solution of $NaHCO_3$ [◄◄ Sample Problems 16.17 and 16.21]. (f) Carbonic acid undergoes a second ionization to produce additional H_3O^+ and the carbonate ion:

$$HCO_3^-(aq) + H_2O(l) \rightleftharpoons H_3O^+(aq) + CO_3^{2-}(aq)$$

Calculate the concentrations of all species (H_2CO_3, H_3O^+, HCO_3^-, and CO_3^{2-}) in the raindrop [◄◄ Sample Problem 16.18]. (g) Sulfuric acid (H_2SO_4) accounts for as much as 80 percent of the acid in acid rain. Its first ionization is complete, producing H_3O^+ and the hydrogen sulfate ion:

$$H_2SO_4(aq) + H_2O(l) \rightleftharpoons H_3O^+(aq) + HSO_4^-(aq)$$

Its second ionization, which produces additional H_3O^+ and the sulfate ion, has an ionization constant (K_{a_2}) of 1.3×10^{-2}:

$$HSO_4^-(aq) + H_2O(l) \rightleftharpoons H_3O^+(aq) + SO_4^{2-}(aq)$$

Calculate the concentrations of all species in a raindrop in which the sulfuric acid concentration is 4.00×10^{-5} M. Assume that sulfuric acid is the only acid present [◄◄ Sample Problem 16.18]. (h) At low concentrations, the percent ionization of a weak acid such as the hydrogen sulfate ion can be quite high. What is the percent ionization of the hydrogen sulfate ion at the concentration calculated in part (g) [◄◄ Sample Problem 16.13]?

SECTION 16.1: BRØNSTED ACIDS AND BASES

Review Questions

16.1   Define Brønsted acids and bases. Give an example of a conjugate pair in an acid-base reaction.

16.2   For a species to act as a Brønsted base, an atom in the species must possess a lone pair of electrons. Explain why this is so.

Conceptual Problems

16.3   Classify each of the following species as a Brønsted acid or base, or both: (a) H_2O, (b) OH^-, (c) H_3O^+, (d) NH_3, (e) NH_4^+, (f) NH_2^-, (g) NO_3^-, (h) CO_3^{2-}, (i) HBr, (j) HCN.

16.4   Identify the acid-base conjugate pairs in each of the following reactions:
(a) $CH_3COO^- + HCN \rightleftharpoons CH_3COOH + CN^-$
(b) $HCO_3^- + HCO_3^- \rightleftharpoons H_2CO_3 + CO_3^{2-}$
(c) $H_2PO_4^- + NH_3 \rightleftharpoons HPO_4^{2-} + NH_4^+$
(d) $HClO + CH_3NH_2 \rightleftharpoons CH_3NH_3^+ + ClO^-$
(e) $CO_3^{2-} + H_2O \rightleftharpoons HCO_3^- + OH^-$

16.5   Write the formulas of the conjugate bases of the following acids: (a) HNO_2, (b) H_2SO_4, (c) H_2S, (d) HCN, (e) HCOOH (formic acid).

16.6   Write the formula for the conjugate acid of each of the following bases: (a) HS^-, (b) HCO_3^-, (c) CO_3^{2-}, (d) $H_2PO_4^-$, (e) HPO_4^{2-}, (f) PO_4^{3-}, (g) HSO_4^-, (h) SO_4^{2-}, (i) SO_3^{2-}.

16.7   Which of the following could represent a Brønsted acid-base reaction?

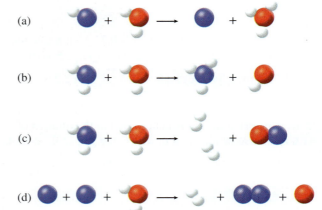

16.8   Oxalic acid ($H_2C_2O_4$) has the following structure:

$$\begin{array}{c} O\!=\!C\!-\!OH \\ | \\ O\!=\!C\!-\!OH \end{array}$$

An oxalic acid solution contains the following species in varying concentrations: $H_2C_2O_4$, $HC_2O_4^-$, $C_2O_4^{2-}$, and H_3O^+. (a) Draw Lewis structures of $HC_2O_4^-$ and $C_2O_4^{2-}$. (b) Which of the four species can act only as acids, which can act only as bases, and which can act as both acids and bases?

SECTION 16.2: THE ACID-BASE PROPERTIES OF WATER

Review Questions

16.9   Write the equilibrium expression for the autoionization of water and an equation relating $[H_3O^+]$ and $[OH^-]$ in solution at 25°C.

16.10   In Section 15.3 we learned that when we multiply a chemical equation by 2, we must square its equilibrium constant. Explain why K_w is the same (1.0×10^{-14} at 25°C) whether we start with one water molecule or two.

16.11   The equilibrium constant for the autoionization of water:

$$2H_2O(l) \rightleftharpoons H_3O^+(aq) + OH^-(aq)$$

is 1.0×10^{-14} at 25°C and 3.8×10^{-14} at 40°C. Is the forward process endothermic or exothermic?

16.12   Define the term *amphoteric*.

16.13   Compare the magnitudes of $[H_3O^+]$ and $[OH^-]$ in aqueous solutions that are acidic, basic, and neutral.

Computational Problems

16.14   Calculate the OH^- concentration in an aqueous solution at 25°C with each of the following H_3O^+ concentrations: (a) 1.13×10^{-4} M, (b) 4.55×10^{-8} M, (c) 7.05×10^{-11} M, (d) 3.13×10^{-2} M.

16.15   Calculate the H_3O^+ concentration in an aqueous solution at 25°C with each of the following OH^- concentrations: (a) 2.50×10^{-2} M, (b) 1.67×10^{-5} M, (c) 8.62×10^{-3} M, (d) 1.75×10^{-12} M.

16.16   The value of K_w at 50°C is 5.48×10^{-14}. Calculate the OH^- concentration in each of the aqueous solutions from Problem 16.14 at 50°C.

16.17   The value of K_w at 100°C is 5.13×10^{-13}. Calculate the H_3O^+ concentration in each of the aqueous solutions from Problem 16.15 at 100°C.

16.18   Which of the following could represent the autoionization of water?

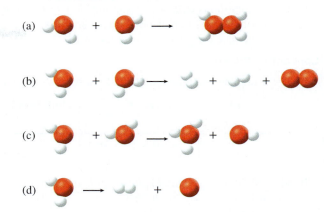

16.19   Indicate which of the following species could, in theory, undergo autoionization: (a) NH_3; (b) NH_4^+; (c) OH^-; (d) O^{2-}; (e) HF; (f) F^-

SECTION 16.3: THE pH SCALE

Review Questions

16.20   Define pH. Why do chemists normally choose to discuss the acidity of a solution in terms of pH rather than hydronium ion concentration $[H_3O^+]$?

16.21   The pH of a solution is 6.7. From this statement alone, can you conclude that the solution is acidic? If not, what additional information would you need? Can the pH of a solution be zero or negative? If so, give examples to illustrate these values.

16.22   Define pOH. Write the equation relating pH and pOH.

Problems

16.23   Calculate the concentration of OH^- ions in a 1.4×10^{-3} M HCl solution.

16.24   Calculate the concentration of H_3O^+ ions in a 0.62 M NaOH solution.

16.25   Calculate the pH of each of the following solutions: (a) 0.0010 M HCl, (b) 0.76 M KOH.

16.26   Calculate the pH of each of the following solutions: (a) 2.8×10^{-4} M $Ba(OH)_2$, (b) 5.2×10^{-4} M HNO_3.

16.27   Calculate the hydronium ion concentration in mol/L for solutions with the following pH values: (a) 2.42, (b) 11.21, (c) 6.96, (d) 15.00.

16.28   Calculate the hydronium ion concentration in mol/L for each of the following solutions: (a) a solution whose pH is 3.11, (b) a solution whose pH is 15.15, (c) a solution whose hydroxide concentration is 2.4×10^{-3} M.

16.29   The pOH of a solution is 9.40 at 25°C. Calculate the hydronium ion concentration of the solution.

16.30   Calculate the number of moles of KOH in 5.50 mL of a 0.360 M KOH solution. What is the pOH of the solution at 25°C?

16.31 How much NaOH (in grams) is needed to prepare 500.0 mL of solution with a pH of 12.00 at 25°C?

16.32 A solution is made by dissolving 18.4 g of HCl in enough water to make 662 mL of solution. Calculate the pH of the solution at 25°C.

Conceptual Problems

16.33 Complete the following table for a solution at 25°C:

pH	$[H_3O^+]$	Solution is
<7		
	$<1.0 \times 10^{-7}\ M$	
		Neutral

16.34 Fill in the word *acidic, basic,* or *neutral* for the following solutions at 25°C:
(a) pOH > 7; solution is _____.
(b) pOH = 7; solution is _____.
(c) pOH < 7; solution is _____.

SECTION 16.4: STRONG ACIDS AND BASES

Review Questions

16.35 Without referring to the text, write the formulas of four strong acids and four strong bases.

16.36 Which of the following statements are true regarding a 1.0-M solution of a strong acid HA at 25°C? (Choose all that apply.)
(a) $[A^-] > [H_3O^+]$
(b) The pH is 0.00.
(c) $[H_3O^+] = 1.0\ M$
(d) $[HA] = 1.0\ M$

16.37 Why are ionizations of strong acids and strong bases generally not treated as equilibria?

Computational Problems

16.38 Calculate the pH of an aqueous solution at 25°C that is (a) 0.12 M in HCl, (b) 2.4 M in HNO$_3$, and (c) 3.2 × 10^{-4} M in HClO$_4$.

16.39 Calculate the pH of an aqueous solution at 25°C that is (a) 1.02 M in HI, (b) 0.035 M in HClO$_4$, and (c) 1.5 × 10^{-6} M in HCl.

16.40 Calculate the concentration of HBr in a solution at 25°C that has a pH of (a) 0.12, (b) 2.46, and (c) 6.27.

16.41 Calculate the concentration of HNO$_3$ in a solution at 25°C that has a pH of (a) 6.13, (b) 4.75, and (c) 1.25.

16.42 Calculate the pOH and pH of the following aqueous solutions at 25°C: (a) 0.066 M KOH, (b) 5.43 M NaOH, (c) 0.74 M Ba(OH)$_2$.

16.43 Calculate the pOH and pH of the following aqueous solutions at 25°C: (a) 1.24 M LiOH, (b) 0.22 M Ba(OH)$_2$, (c) 0.085 M NaOH.

16.44 An aqueous solution of a strong base has a pH of 9.78 at 25°C. Calculate the concentration of the base if the base is (a) LiOH and (b) Ba(OH)$_2$.

16.45 An aqueous solution of a strong base has a pH of 11.04 at 25°C. Calculate the concentration of the base if the base is (a) KOH and (b) Ba(OH)$_2$.

SECTION 16.5: WEAK ACIDS AND ACID IONIZATION CONSTANTS

▶▶▶ **Visualizing Chemistry**
Figure 16.2

Three weak acid solutions are shown here.

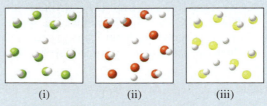

(i)   (ii)   (iii)

VC 16.1 Which weak acid has the largest K_a value?
a) i   b) ii   c) iii

VC 16.2 Which weak acid has the highest pH?
a) i   b) ii   c) iii

VC 16.3 For which weak acid solution can we most likely neglect x in the denominator of the equilibrium expression in the determination of pH?
a) i   b) ii   c) iii

VC 16.4 In the event that we cannot neglect x in the denominator of the equilibrium expression, _____ to solve for pH.
a) we cannot use an equilibrium table
b) we must use the quadratic equation
c) it is unnecessary

Review Questions

16.46 Explain what is meant by the strength of an acid.

16.47 What does the ionization constant tell us about the strength of an acid?

16.48 List the factors on which the K_a of a weak acid depends.

16.49 Why do we normally not quote K_a values for strong acids such as HCl and HNO$_3$? Why is it necessary to specify temperature when giving K_a values?

16.50 Which of the following solutions has the highest pH: (a) 0.40 M HCOOH, (b) 0.40 M HClO$_4$, (c) 0.40 M CH$_3$COOH?

16.51 Without referring to the text, write the formulas of four weak acids.

Computational Problems

16.52 In biological and medical applications, it is often necessary to study the autoionization of water at 37°C instead of 25°C. Given that K_w for water is 2.5 × 10^{-14} at 37°C, calculate the pH of pure water at this temperature.

16.53 The K_a for benzoic acid is 6.5 × 10^{-5}. Calculate the pH of a 0.10-M aqueous solution of benzoic acid at 25°C.

16.54 The K_a for hydrofluoric acid is 7.1 × 10^{-4}. Calculate the pH of a 0.15-M aqueous solution of hydrofluoric acid at 25°C.

16.55 Calculate the pH of an aqueous solution at 25°C that is 0.050 M in hydrocyanic acid (HCN). (K_a for hydrocyanic acid = 4.9×10^{-10}).

16.56 Calculate the pH of an aqueous solution at 25°C that is 0.34 M in phenol (C_6H_5OH). (K_a for phenol = 1.3×10^{-10}).

16.57 Determine the percent ionization of the following solutions of formic acid at 25°C: (a) 0.016 M, (b) 5.7×10^{-4} M, (c) 1.75 M.

16.58 Determine the percent ionization of the following solutions of phenol: (a) 0.56 M, (b) 0.25 M, (c) 1.8×10^{-6} M.

16.59 A 0.015-M solution of a monoprotic acid is 0.92 percent ionized. Calculate the ionization constant for the acid.

16.60 Calculate the concentration at which a monoprotic acid with $K_a = 4.5 \times 10^{-5}$ will be 2.5 percent ionized.

16.61 Calculate the K_a of a weak acid if a 0.19-M aqueous solution of the acid has a pH of 4.52 at 25°C.

16.62 The pH of an aqueous acid solution is 6.20 at 25°C. Calculate the K_a for the acid. The initial acid concentration is 0.010 M.

16.63 What is the original molarity of a solution of formic acid (HCOOH) whose pH is 4.21 at 25°C? (K_a for formic acid = 1.7×10^{-4}.)

16.64 What is the original molarity of a solution of a weak acid whose K_a is 3.5×10^{-5} and whose pH is 5.26 at 25°C?

Conceptual Problems

16.65 Which of the following statements are true for a 0.10-M solution of a weak acid HA? (Choose all that apply.)
(a) The pH is 1.00. (c) $[H_3O^+] = [A^-]$.
(b) $[H_3O^+] \gg [A^-]$. (d) The pH is less than 1.

16.66 Classify each of the following species as a weak or strong acid: (a) HNO_3, (b) HF, (c) H_2SO_4, (d) HSO_4^-, (e) H_2CO_3, (f) HCO_3^-, (g) HCl, (h) HCN, (i) HNO_2.

16.67 Classify each of the following species as a weak or strong base: (a) LiOH, (b) CN^-, (c) H_2O, (d) ClO_4^-, (e) NH_2^-.

SECTION 16.6: WEAK BASES AND BASE IONIZATION CONSTANTS

Review Questions

16.68 Compare the pH values for 0.10-M solutions of NaOH and of NH_3 to illustrate the difference between a strong base and a weak base.

16.69 Which of the following has a higher pH: (a) 1.0 M NH_3, (b) 0.20 M NaOH (K_b for $NH_3 = 1.8 \times 10^{-5}$)?

Computational Problems

16.70 Calculate the pH for each of the following solutions at 25°C: (a) 0.10 M NH_3, (b) 0.050 M C_5H_5N (pyridine). (K_b for pyridine = 1.7×10^{-9}).

16.71 The pH of a 0.30-M solution of a weak base is 10.66 at 25°C. What is the K_b of the base?

16.72 What is the original molarity of an aqueous solution of ammonia (NH_3) whose pH is 11.22 at 25°C (K_b for $NH_3 = 1.8 \times 10^{-5}$)?

16.73 Calculate the pH at 25°C of a 0.61-M aqueous solution of a weak base B with a K_b of 1.5×10^{-4}.

Conceptual Problems

16.74 Determine the K_b of a weak base if a 0.19-M aqueous solution of the base at 25°C has a pH of 10.88.

16.75 The diagrams show aqueous solutions of three different weak bases. Indicate which base has the highest K_b value and which has the lowest K_b value. (Water molecules are not shown.)

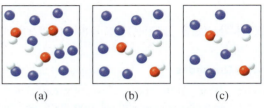

(a)   (b)   (c)

16.76 Rank the solutions in Problem 16.75 in order of decreasing pH.

SECTION 16.7: CONJUGATE ACID-BASE PAIRS

Review Questions

16.77 Write the equation relating K_a for a weak acid and K_b for its conjugate base. Use NH_3 and its conjugate acid NH_4^+ to derive the relationship between K_a and K_b.

16.78 From the relationship $K_aK_b = K_w$, what can you deduce about the relative strengths of a weak acid and its conjugate base?

Computational Problems

16.79 Calculate K_b for each of the following ions: CN^-, F^-, CH_3COO^-, HCO_3^-. (See Table 16.6.)

16.80 Calculate K_a for each of the following ions: NH_4^+, $C_6H_5NH_3^+$, $CH_3NH_3^+$, $C_2H_5NH_3^+$. (See Table 16.7.)

Conceptual Problems

16.81 The following diagrams represent aqueous solutions of three different monoprotic acids: HA, HB, and HC. (a) Which conjugate base (A^-, B^-, or C^-) has the smallest K_b value? (b) Which anion is the strongest base? The water molecules have been omitted for clarity.

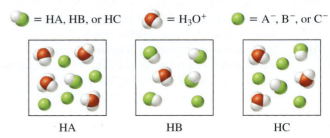

⬤ = HA, HB, or HC   ⬤ = H_3O^+ ⬤ = A^-, B^-, or C^-

HA   HB   HC

16.82 The following diagrams represent solutions of three salts NaX (X = A, B, or C). (a) Which X^- has the weakest conjugate acid? (b) Arrange the three X^- anions in order of decreasing base strength. The Na^+ ion and water molecules have been omitted for clarity.

● = A^-, B^-, or C^- 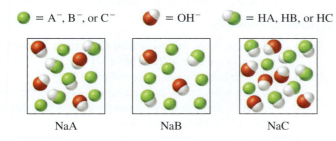 = OH^- ◓ = HA, HB, or HC

NaA	NaB	NaC

SECTION 16.8: DIPROTIC AND POLYPROTIC ACIDS

Review Questions

16.83 Write all the species (except water) that are present in a phosphoric acid solution. Indicate which species can act as a Brønsted acid, which as a Brønsted base, and which as both a Brønsted acid and a Brønsted base.

16.84 Explain why it is generally not necessary to take into account second or third ionization constants when calculating the pH of a polyprotic acid solution.

Computational Problems

16.85 Compare the pH of a 0.040 M HCl solution with that of a 0.040 M H_2SO_4 solution. (*Hint:* H_2SO_4 is a strong acid; K_a for $HSO_4^- = 1.3 \times 10^{-2}$.)

16.86 What are the concentrations of HSO_4^-, SO_4^{2-}, and H_3O^+ in a 0.20 M $KHSO_4$ solution? (*Hint:* H_2SO_4 is a strong acid; K_a for $HSO_4^- = 1.3 \times 10^{-2}$.)

16.87 Calculate the concentrations of H_3O^+, HCO_3^-, and CO_3^{2-} in a 0.050 M H_2CO_3 solution.

16.88 Calculate the pH at 25°C of a 0.25-M aqueous solution of phosphoric acid (H_3PO_4). (K_{a_1}, K_{a_2}, and K_{a_3} for phosphoric acid are 7.5×10^{-3}, 6.25×10^{-8}, and 4.8×10^{-13}, respectively.)

16.89 Calculate the pH at 25°C of a 0.25-M aqueous solution of oxalic acid ($H_2C_2O_4$). (K_{a_1} and K_{a_2} for oxalic acid are 6.5×10^{-2} and 6.1×10^{-5}, respectively.)

Conceptual Problems

16.90 The first and second ionization constants of a diprotic acid H_2A are K_{a_1} and K_{a_2} at a certain temperature. Under what conditions will $[A^{2-}] = K_{a_1}$?

16.91 (a) Which of the following diagrams represents a solution of a weak diprotic acid? (b) Which diagrams represent chemically implausible situations? (The

hydrated proton is shown as a hydronium ion. Water molecules are omitted for clarity.)

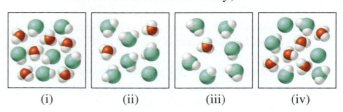

(i)   (ii)   (iii)   (iv)

SECTION 16.9: MOLECULAR STRUCTURE AND ACID STRENGTH

Review Questions

16.92 List four factors that affect the strength of an acid.

16.93 How does the strength of an oxoacid depend on the electronegativity and oxidation number of the central atom?

Conceptual Problems

16.94 Predict the relative acid strengths of the following compounds: H_2O, H_2S, and H_2Se.

16.95 Compare the strengths of the following pairs of acids: (a) H_2SO_4 and H_2SeO_4, (b) H_3PO_4 and H_3AsO_4.

16.96 Which of the following is the stronger acid: $CH_2ClCOOH$ or $CHCl_2COOH$? Explain your choice.

16.97 Consider the following compounds:

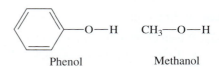

Phenol   Methanol

Experimentally, phenol is found to be a stronger acid than methanol. Explain this difference in terms of the structures of the conjugate bases. (*Hint:* A more stable conjugate base favors ionization. Only one of the conjugate bases can be stabilized by resonance.)

SECTION 16.10: ACID-BASE PROPERTIES OF SALT SOLUTIONS

Review Questions

16.98 Define salt hydrolysis. Categorize salts according to how they affect the pH of a solution.

16.99 Explain why small, highly charged metal ions are able to undergo hydrolysis.

16.100 Al^{3+} is not a Brønsted acid, but $Al(H_2O)_6^{3+}$ is. Explain.

16.101 Specify which of the following salts will undergo hydrolysis: KF, $NaNO_3$, NH_4NO_2, $MgSO_4$, KCN, C_6H_5COONa, RbI, Na_2CO_3, $CaCl_2$, HCOOK.

Computational Problems

16.102   Calculate the pH of a 0.36 M CH_3COONa solution. (K_a for acetic acid = 1.8×10^{-5}).

16.103   Calculate the pH of a 0.42 M NH_4Cl solution. (K_b for ammonia = 1.8×10^{-5}).

16.104   Calculate the pH of a 0.082 M NaF solution. (K_a for HF = 7.1×10^{-4}).

16.105   Calculate the pH of a 0.91 M $C_2H_5NH_3I$ solution. (K_b for $C_2H_5NH_2$ = 5.6×10^{-4}).

Conceptual Problems

16.106   Predict the pH (>7, <7, or ≈7) of aqueous solutions containing the following salts: (a) KBr, (b) $Al(NO_3)_3$, (c) $BaCl_2$, (d) $Bi(NO_3)_3$.

16.107   Predict whether the following solutions are acidic, basic, or nearly neutral: (a) NaBr, (b) K_2SO_3, (c) NH_4NO_2, (d) $Cr(NO_3)_3$.

16.108   A certain salt, MX (containing the M^+ and X^- ions), is dissolved in water, and the pH of the resulting solution is 7.0. What can you say about the strengths of the acid and the base from which the salt is derived?

16.109   In a certain experiment, a student finds that the pHs of 0.10-M solutions of three potassium salts KX, KY, and KZ are 7.0, 9.0, and 11.0, respectively. Arrange the acids HX, HY, and HZ in order of increasing acid strength.

16.110   Predict whether a solution containing the salt K_2HPO_4 will be acidic, neutral, or basic.

16.111   Predict the pH (>7, <7, or ≈7) of a $NaHCO_3$ solution.

SECTION 16.11: ACID-BASE PROPERTIES OF OXIDES AND HYDROXIDES

Review Questions

16.112   Classify the following oxides as acidic, basic, amphoteric, or neutral: (a) CO_2, (b) K_2O, (c) CaO, (d) N_2O_5, (e) CO, (f) NO, (g) SnO_2, (h) SO_3, (i) Al_2O_3, (j) BaO.

16.113   Write equations for the reactions between (a) CO_2 and NaOH(aq), (b) Na_2O and HNO_3(aq).

Conceptual Problems

16.114   Explain why metal oxides tend to be basic if the oxidation number of the metal is low and tend to be acidic if the oxidation number of the metal is high. (*Hint:* Metallic compounds in which the oxidation numbers of the metals are low are more ionic than those in which the oxidation numbers of the metals are high.)

16.115   Arrange the oxides in each of the following groups in order of increasing basicity: (a) K_2O, Al_2O_3, BaO, (b) CrO_3, CrO, Cr_2O_3.

16.116 $Zn(OH)_2$ is an amphoteric hydroxide. Write balanced ionic equations to show its reaction with (a) HCl, (b) NaOH [the product is $Zn(OH)_4^{2-}$].

16.117 $Al(OH)_3$ is insoluble in water. It dissolves in concentrated NaOH solution. Write a balanced ionic equation for this reaction. What type of reaction is this?

SECTION 16.12: LEWIS ACIDS AND BASES

Review Questions

16.118   What are the Lewis definitions of an acid and a base? In what way are they more general than the Brønsted definitions?

16.119   In terms of orbitals and electron arrangements, what must be present for a molecule or an ion to act as a Lewis acid (use H^+ and BF_3 as examples)? What must be present for a molecule or ion to act as a Lewis base (use OH^- and NH_3 as examples)?

Conceptual Problems

16.120   Classify each of the following species as a Lewis acid or a Lewis base: (a) CO_2, (b) H_2O, (c) I^-, (d) SO_2, (e) NH_3, (f) OH^-, (g) H^+, (h) BCl_3.

16.121   Describe the following reaction in terms of the Lewis theory of acids and bases:

$$AlCl_3(s) + Cl^-(aq) \longrightarrow AlCl_4^-(aq)$$

16.122   Which would be considered a stronger Lewis acid: (a) BF_3 or BCl_3, (b) Fe^{2+} or Fe^{3+}? Explain.

16.123   All Brønsted acids are Lewis acids, but the reverse is not true. Give two examples of Lewis acids that are not Brønsted acids.

16.124   Identify the Lewis acid and the Lewis base in the following reactions:
(a) $5CO(g) + Fe(s) \longrightarrow Fe(CO)_5(l)$
(b) $NH_3(g) + BCl_3(g) \longrightarrow Cl_3BNH_3(s)$
(c) $Hg^{2+}(aq) + 4I^-(aq) \longrightarrow HgI_4^{2-}(aq)$

16.125   Identify the Lewis acid and the Lewis base in the following reactions:
(a) $AlBr_3(s) + Br^-(aq) \longrightarrow AlBr_4^-(aq)$
(b) $6CO(g) + Cr(s) \longrightarrow Cr(CO)_6(s)$
(c) $Cu^{2+}(aq) + 4CN^-(aq) \longrightarrow Cu(CN)_4^{2-}(aq)$

ADDITIONAL PROBLEMS

16.126   Predict the direction that predominates in this reaction:

$$F^-(aq) + H_2O(l) \rightleftharpoons HF(aq) + OH^-(aq)$$

16.127   Predict the products and tell whether the following reaction will occur to any measurable extent:

$$CH_3COOH(aq) + Cl^-(aq) \longrightarrow$$

16.128   In a 0.080 M NH_3 solution, what percent of the NH_3 is present as NH_4^+?

16.129 Calculate the pH and percent ionization of a 0.88 M HNO$_2$ solution at 25°C.

16.130 Calculate the pH of a 0.20 M ammonium acetate (CH$_3$COONH$_4$) solution.

16.131 To which of the following would the addition of an equal volume of 0.60 M NaOH lead to a solution having a lower pH: (a) water, (b) 0.30 M HCl, (c) 0.70 M KOH, (d) 0.40 M NaNO$_3$?

16.132 The pH of a 0.0642-M solution of a monoprotic acid is 3.86. Is this a strong acid?

16.133 Like water, liquid ammonia undergoes autoionization:

$$NH_3 + NH_3 \rightleftharpoons NH_4^+ + NH_2^-$$

(a) Identify the Brønsted acids and Brønsted bases in this reaction. (b) What species correspond to H$_3$O$^+$ and OH$^-$, and what is the condition for a neutral solution?

16.134 HA and HB are both weak acids although HB is the stronger of the two. Will it take a larger volume of a 0.10 M NaOH solution to neutralize 50.0 mL of 0.10 M HB than would be needed to neutralize 50.0 mL of 0.10 M HA?

16.135 A solution contains a weak monoprotic acid HA and its sodium salt NaA both at 0.1 M concentration. Show that [OH$^-$] = K_w/K_a.

16.136 The three common chromium oxides are CrO, Cr$_2$O$_3$, and CrO$_3$. If Cr$_2$O$_3$ is amphoteric, what can you say about the acid-base properties of CrO and CrO$_3$?

16.137 Use the data in Table 16.6 to calculate the equilibrium constant for the following reaction:

$$HCOOH(aq) + OH^-(aq) \rightleftharpoons$$
$$HCOO^-(aq) + H_2O(l)$$

16.138 Use the data in Table 16.6 to calculate the equilibrium constant for the following reaction:

$$CH_3COOH(aq) + NO_2^-(aq) \rightleftharpoons$$
$$CH_3COO^-(aq) + HNO_2(aq)$$

16.139 Most of the hydrides of Group 1 and Group 2 metals are ionic (the exceptions are BeH$_2$ and MgH$_2$, which are covalent compounds). (a) Describe the reaction between the hydride ion (H$^-$) and water in terms of a Brønsted acid-base reaction. (b) The same reaction can also be classified as a redox reaction. Identify the oxidizing and reducing agents.

16.140 A 10.0-g sample of white phosphorus was burned in an excess of oxygen. The product was dissolved in enough water to make 500.0 mL of solution. Calculate the pH of the solution at 25°C.

16.141 Use the van't Hoff equation (see Problem 15.125) and the data in Appendix 2 to calculate the pH of water at its normal boiling point.

16.142 Which of the following is the stronger base: NF$_3$ or NH$_3$? (*Hint:* F is more electronegative than H.)

16.143 Which of the following is a stronger base: NH$_3$ or PH$_3$? (*Hint:* The N—H bond is stronger than the P—H bond.)

16.144 The ion product of D$_2$O is 1.35 × 10^{-15} at 25°C. (a) Calculate pD where pD = $-$log [D$_3$O$^+$]. (b) For what values of pD will a solution be acidic in D$_2$O? (c) Derive a relation between pD and pOD.

16.145 Give an example of (a) a weak acid that contains oxygen atoms, (b) a weak acid that does not contain oxygen atoms, (c) a neutral molecule that acts as a Lewis acid, (d) a neutral molecule that acts as a Lewis base, (e) a weak acid that contains two ionizable H atoms, (f) a conjugate acid-base pair, both of which react with HCl to give carbon dioxide gas.

16.146 What is the pH of 250.0 mL of an aqueous solution containing 0.616 g of a strong acid?

16.147 Which of the following diagrams best represents a strong acid, such as HCl, dissolved in water? Which represents a weak acid? Which represents a very weak acid? (The hydrated proton is shown as a hydronium ion. Water molecules are omitted for clarity.)

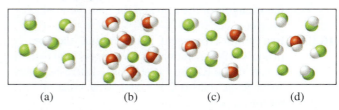

(a)   (b)   (c)   (d)

16.148 HF is a weak acid, but its strength increases with concentration. Explain. (*Hint:* F$^-$ reacts with HF to form HF$_2^-$. The equilibrium constant for this reaction is 5.2 at 25°C.)

16.149 When chlorine reacts with water, the resulting solution is weakly acidic and reacts with AgNO$_3$ to give a white precipitate. Write balanced equations to represent these reactions. Explain why manufacturers of household bleaches add bases such as NaOH to their products to increase their effectiveness.

16.150 When the concentration of a strong acid is not substantially higher than 1.0 × 10^{-7} M, the ionization of water must be taken into account in the calculation of the solution's pH. (a) Derive an expression for the pH of a strong acid solution, including the contribution to [H$_3$O$^+$] from H$_2$O. (b) Calculate the pH of a 1.0 × 10^{-7} M HCl solution.

16.151 Calculate the pH of a 2.00 M NH$_4$CN solution.

16.152 Calculate the concentrations of all species in a 0.100 M H$_3$PO$_4$ solution.

16.153 H$_2$SO$_4$ is a strong acid, but HSO$_4^-$ is a weak acid. Account for the difference in strength of these two related species.

16.154 Calculate the concentrations of all species in a 0.100 M Na$_2$CO$_3$ solution.

16.155 A 20.27-g sample of a metal carbonate (MCO$_3$) is combined with 500 mL of a 1.00 M HCl solution. The excess HCl acid is then neutralized by 32.80 mL of 0.588 M NaOH. Identify M.

16.156 Calculate the pH of a solution that is 1.00 M HCN and 1.00 M HF. Compare the concentration (in molarity) of the CN^- ion in this solution with that in a 1.00 M HCN solution. Comment on the difference.

16.157 How many grams of NaCN would you need to dissolve in enough water to make exactly 250 mL of solution with a pH of 10.00?

16.158 A solution of formic acid (HCOOH) has a pH of 2.53. How many grams of formic acid are there in 100.0 mL of the solution?

16.159 Calculate the pH of a 1-L solution containing 0.150 mole of CH_3COOH and 0.100 mole of HCl.

16.160 A 1.87-g sample of Mg reacts with 80.0 mL of a HCl solution whose pH is −0.544. What is the pH of the solution after all the Mg has reacted? Assume constant volume.

16.161 You are given two beakers, one containing an aqueous solution of strong acid (HA) and the other an aqueous solution of weak acid (HB) of the same concentration. Describe how you would compare the strengths of these two acids by (a) measuring the pH, (b) measuring electrical conductance, and (c) studying the rate of hydrogen gas evolution when these solutions are combined with an active metal such as Mg or Zn.

16.162 Use Le Châtelier's principle to predict the effect of the following changes on the extent of hydrolysis of sodium nitrite ($NaNO_2$) solution: (a) HCl is added, (b) NaOH is added, (c) NaCl is added, (d) the solution is diluted.

16.163 A 0.400 M formic acid (HCOOH) solution freezes at −0.758°C. Calculate the K_a of the acid at that temperature. (*Hint:* Assume that molarity is equal to molality. Carry out your calculations to three significant figures and round off to two for K_a.)

16.164 Which of the following does not represent a Lewis acid-base reaction?
(a) $H_2O + H^+ \longrightarrow H_3O^+$
(b) $NH_3 + BF_3 \longrightarrow H_3NBF_3$
(c) $PF_3 + F_2 \longrightarrow PF_5$
(d) $Al(OH)_3 + OH^- \longrightarrow Al(OH)_4^-$

16.165 A solution of methylamine (CH_3NH_2) as a pH of 10.64. How many grams of methylamine are there in 100.0 mL of the solution?

16.166 Describe the hydration of SO_2 as a Lewis acid-base reaction.

16.167 Both the amide ion (NH_2^-) and the nitride ion (N^{3-}) are stronger bases than the hydroxide ion and hence do not exist in aqueous solutions. (a) Write equations showing the reactions of these ions with water, and identify the Brønsted acid and base in each case. (b) Which of the two is the stronger base?

16.168 Determine whether each of the following statements is true or false. If false, explain why the statement is wrong. (a) All Lewis acids are Brønsted acids. (b) The conjugate base of an acid always carries a negative charge. (c) The percent ionization of a base increases with its concentration in solution. (d) A solution of barium fluoride is acidic.

16.169 How many milliliters of a strong monoprotic acid solution at pH = 4.12 must be added to 528 mL of the same acid solution at pH = 5.76 to change its pH to 5.34? Assume that the volumes are additive.

Biological Problems

16.170 The disagreeable odor of fish is mainly due to organic compounds (RNH_2) containing an amino group, $—NH_2$, where R is the rest of the molecule. Amines are bases just like ammonia. Explain why putting some lemon juice on fish can greatly reduce the odor.

16.171 Explain the action of smelling salt, which is ammonium carbonate [$(NH_4)_2CO_3$]. (*Hint:* The thin film of aqueous solution that lines the nasal passages is slightly basic.)

16.172 A typical reaction between an antacid and the hydrochloric acid in gastric juice is

$$NaHCO_3(s) + HCl(aq) \rightleftharpoons NaCl(aq) + H_2O(l) + CO_2(g)$$

Calculate the volume (in liters) of CO_2 generated from 0.350 g of $NaHCO_3$ and excess gastric juice at 1.00 atm and 37.0°C.

16.173 Novocaine, used as a local anesthetic by dentists, is a weak base ($K_b = 8.91 \times 10^{-6}$). What is the ratio of the concentration of the base to that of its acid in the blood plasma (pH = 7.40) of a patient? (As an approximation, use the K_a values at 25°C.)

16.174 Hemoglobin (Hb) is a blood protein that is responsible for transporting oxygen. It can exist in the protonated form as HbH^+. The binding of oxygen can be represented by the simplified equation:

$$HbH^+ + O_2 \rightleftharpoons HbO_2 + H^+$$

(a) What form of hemoglobin is favored in the lungs where oxygen concentration is highest? (b) In body tissues, where the cells release carbon dioxide produced by metabolism, the blood is more acidic due to the formation of carbonic acid. What form of hemoglobin is favored under this condition? (c) When a person hyperventilates, the concentration of CO_2 in his or her blood decreases. How does this action affect the given equilibrium? Frequently a person who is hyperventilating is advised to breathe into a paper bag. Why does this action help the individual?

16.175 Tooth enamel is largely hydroxyapatite [$Ca_3(PO_4)_3OH$]. When it dissolves in water (a process called *demineralization*), it dissociates as follows:

$$Ca_5(PO_4)_3OH \longrightarrow 5Ca^{2+} + 3PO_4^{3-} + OH^-$$

The reverse process, called *remineralization*, is the body's natural defense against tooth decay. Acids produced from food remove the OH$^-$ ions and thereby weaken the enamel layer. Most toothpastes contain a fluoride compound such as NaF or SnF$_2$. What is the function of these compounds in preventing tooth decay?

Environmental Problems

16.176   Hydrocyanic acid (HCN) is a weak acid and a deadly poisonous compound—in the gaseous form (hydrogen cyanide), it is used in gas chambers. Sodium cyanide is used in the gold-mining industry and has become a significant environmental concern. Why is it dangerous to treat sodium cyanide with acids (such as HCl) without proper ventilation?

16.177   The atmospheric sulfur dioxide (SO$_2$) concentration over a certain region is 0.12 ppm by volume. Calculate the pH of the rainwater due to this pollutant. Assume that the dissolution of SO$_2$ does not affect its pressure. (K_a for H$_2$SO$_3$ = 1.3×10^{-2}.)

Multiconcept Problems

16.178   About half of the hydrochloric acid produced annually in the United States (3.0 billion pounds) is used in metal pickling. This process involves the removal of metal oxide layers from metal surfaces to prepare them for coating. (a) Write the overall and net ionic equations for the reaction between iron(III) oxide, which represents the rust layer over iron, and HCl. Identify the Brønsted acid and base. (b) Hydrochloric acid is also used to remove scale (which is mostly CaCO$_3$) from water pipes. Hydrochloric acid reacts with calcium carbonate in two stages; the first stage forms the bicarbonate ion, which then reacts further to form carbon dioxide. Write equations for these two stages and for the overall reaction. (c) Hydrochloric acid is

used to recover oil from the ground. It dissolves rocks (often CaCO$_3$) so that the oil can flow more easily. In one process, a 15 percent (by mass) HCl solution is injected into an oil well to dissolve the rocks. If the density of the acid solution is 1.073 g/mL, what is the pH of the solution?

16.179   At 28°C and 0.982 atm, gaseous compound HA has a density of 1.16 g/L. A quantity of 2.03 g of this compound is dissolved in water and diluted to exactly 1 L. If the pH of the solution at 25°C is 5.22 (due to ionization of HA), calculate the K_a of the acid.

16.180   In the vapor phase, acetic acid molecules associate to a certain extent to form dimers:

$$2CH_3COOH(g) \rightleftharpoons (CH_3COOH)_2(g)$$

At 51°C, the pressure of a certain acetic acid vapor system is 0.0342 atm in a 360-mL flask. The vapor is condensed and neutralized with 13.8 mL of 0.0568 M NaOH. (a) Calculate the degree of dissociation (α) of the dimer under these conditions:

$$(CH_3COOH)_2 \rightleftharpoons 2CH_3COOH$$

(b) Calculate the equilibrium constant K_P for the reaction in part (a).

16.181   Henry's law constant for CO$_2$ at 38°C is 2.28×10^{-3} mol/L · atm. Calculate the pH of a solution of CO$_2$ at 38°C in equilibrium with the gas at a partial pressure of 3.20 atm.

16.182   (a) Use VSEPR to predict the geometry of the hydronium ion (H$_3$O$^+$). (b) The O atom in H$_2$O has two lone pairs and in principle can accept two H$^+$ ions. Explain why the species H$_4$O^{2+} does not exist. What would be its geometry if it did exist?

Standardized-Exam Practice Problems

Physical and Biological Sciences
The following questions are not based on a descriptive passage.

1. Which of the following best describes the titration of a weak acid with a strong base?

   a) Low starting pH, and pH ≈ 7 at equivalence point
   b) High starting pH, and pH ≈ 7 at equivalence point
   c) High starting pH, and pH < 7 at equivalence point
   d) Low starting pH, and pH > 7 at equivalence point

2. What is the pH of a solution made by combining 150 mL of 0.125 M HCl and 150 mL of 0.120 M Ba(OH)$_2$ at 25°C?

   a) 1.240   c) 2.602
   b) 12.760   d) 11.398

3. What volume of 0.085 M NaOH is necessary to titrate to the equivalence point 25.0 mL of 0.15 M HCN at 25°C?

   a) 44 mL
   b) 88 mL
   c) 22 mL
   d) 25 mL

4. Determine the pH at the equivalence point for the titration in Question 3.

   a) 2.98
   b) 11.02
   c) 8.71
   d) 5.29

Answers to In-Chapter Materials

Answers to Practice Problems

16.1A (a) $HClO_4$, (b) HS^-, (c) HS^-, (d) $HC_2O_4^-$. **16.1B** SO_3^{2-}, H_2SO_3. **16.2A** (a) NH_4^+ acid, H_2O base, NH_3 conjugate base, H_3O^+ conjugate acid; (b) CN^- base, H_2O acid, HCN conjugate acid, OH^- conjugate base. **16.2B** (a) $HSO_4^- + H_2O \longrightarrow H_3O^+ + SO_4^{2-}$, (b) $HSO_4^- + H_2O \longrightarrow H_2SO_4 + OH^-$. **16.3A** 2.0×10^{-11} M. **16.3B** 2.8×10^{-13} M. **16.4A** (a) 8.49, (b) 7.40, (c) 1.25. **16.4B** (a) −0.08, (b) 10.52, (c) 11.08. **16.5A** (a) 1.3×10^{-10} M, (b) 3.5×10^{-2} M, (c) 9.8×10^{-8} M. **16.5B** (a) 7.8×10^{-3} M, (b) 2.6×10^{-12} M, (c) 1.3×10^{-7} M. **16.6A** (a) 11.24, (b) 2.14, (c) 5.07. **16.6B** (a) 6.45, (b) 6.00, (c) 3.00. **16.7A** (a) 9.5×10^{-14} M, (b) 7.2×10^{-6} M, (c) 1.0×10^{-7} M. **16.7B** (a) 5.5×10^{-12} M, (b) 2.0×10^{-4} M, (c) 2.5×10^{-2} M. **16.8A** (a) 1.09, (b) 5.09, (c) 2.27. **16.8B** (a) 12.04, (b) 11.54, (c) 8.97. **16.9A** (a) 8.7×10^{-3} M, (b) 1.7×10^{-2} M, (c) 9.8×10^{-7} M. **16.9B** (a) 1.5×10^{-5} M, (b) 1.5×10^{-2} M, (c) 9.1×10^{-4} M. **16.10A** (a) 0.82, (b) 2.08, (c) 4.77. **16.10B** (a) 6.98, (b) 12.79, (c) 13.09. **16.11A** (a) 9.6×10^{-6} M, (b) 4.8×10^{-6} M. **16.11B** (a) 1.7×10^{-2} M, (b) 8.7×10^{-3} M. **16.12A** 2.89. **16.12B** 3.05. **16.13A** (a) 4.96, 0.0044%, (b) 5.72, 0.026%, (c) 6.70, 0.24%. **16.13B** (a) 2.0×10^{-3} M, (b) 4.9×10^{-4} M, (c) 2.2×10^{-4} M.

16.14A 1.9×10^{-5}. **16.14B** 5.8×10^{-9}. **16.15A** 9.14. **16.15B** 8.33. **16.16A** 3.0×10^{-9}. **16.16B** 1.4×10^{-4}. **16.17A** (a) 1.5×10^{-10}, (b) 1.2×10^{-10}, (c) 1.8×10^{-11}. **16.17B** (a) 1.1×10^{-11}, (b) 4.8×10^{-7}. **16.18A** $[H_2C_2O_4] = 0.11$ M, $[HC_2O_4^-] = 0.086$ M, $[C_2O_4^{2-}] = 6.1 \times 10^{-5}$ M, $[H_3O^+] = 0.086$ M. **16.18B** $[H_2SO_4] = 0$ M, $[HSO_4^-] = 0.13$ M, $[SO_4^{2-}] = 0.011$ M, $[H_3O^+] = 0.15$ M. **16.19A** (a) $HBrO_4$, (b) H_2SO_4. **16.19B** Electronegativity. **16.20A** 8.96. **16.20B** 0.75 M. **16.21A** 2.92. **16.21B** 0.032 M. **16.22A** (a) Basic, (b) acidic, (c) basic, (d) neutral, (e) neutral. **16.22B** (a) NH_4Br and CH_3NH_3I, (b) $NaHPO_4$ and $KOBr$, (c) NaI and KBr. **16.23A** Lewis acid: Co^{3+}, Lewis base: NH_3. **16.23B** NH_3 and $AlBr_3$.

Answers to Checkpoints

16.1.1 b, d. **16.1.2** c, d. **16.2.1** d. **16.2.2** a. **16.3.1** c. **16.3.2** a. **16.3.3** e. **16.3.4** b. **16.4.1** d. **16.4.2** b. **16.4.3** b. **16.4.4** e. **16.4.5** c. **16.4.6** b. **16.4.7** b. **16.5.1** c. **16.5.2** e. **16.5.3** b. **16.6.1** b. **16.6.2** d. **16.6.3** e. **16.7.1** b. **16.7.2** d. **16.7.3** a. **16.8.1** b. **16.8.2** c. **16.8.3** c. **16.8.4** b. **16.10.1** a. **16.10.2** e. **16.10.3** a, b, c. **16.10.4** b, c. **16.10.5** c. **16.12.1** c, d, e. **16.12.2** b, d.

17.1   The Common Ion Effect

Up until now, we have discussed the properties of solutions containing a single solute. In this section, we examine how the properties of a solution change when a second solute is introduced.

Recall that a system at equilibrium will shift in response to being stressed and that stress can be applied in a variety of ways, including the addition of a reactant or a product [◄◄ Section 15.5]. Consider a liter of solution containing 0.10 mole of acetic acid. Using the K_a for acetic acid (1.8×10^{-5}) and an equilibrium table [◄◄ Section 16.5], the pH of this solution at 25°C can be determined:

$$CH_3COOH(aq) + H_2O(l) \rightleftharpoons H_3O^+(aq) + CH_3COO^-(aq)$$

	CH_3COOH	H_2O	H_3O^+	CH_3COO^-
Initial concentration (M):	0.10	—	0	0
Change in concentration (M):	$-x$	—	$+x$	$+x$
Equilibrium concentration (M):	$0.10 - x$	—	x	x

Assuming that $(0.10 - x)\ M \approx 0.10\ M$ and solving for x, we get 1.34×10^{-3}. Therefore, $[CH_3COOH] = 0.09866\ M$, $[H_3O^+] = [CH_3COO^-] = 1.34 \times 10^{-3}\ M$ and pH = 2.87. The percent ionization of acetic acid is:

$$\frac{1.34 \times 10^{-3}\ M}{0.10\ M} \times 100\% = 1.3\%$$

Now consider what happens when we add 0.050 mole of sodium acetate (CH_3COONa) to the solution. Sodium acetate dissociates completely in aqueous solution to give sodium ions and acetate ions:

$$CH_3COONa(aq) \xrightarrow{\ H_2O\ } Na^+(aq) + CH_3COO^-(aq)$$

> **Student Note:** By adding sodium acetate, we also add sodium ions to the solution. However, sodium ions do not interact with water or with any of the other species present [◄◄ Section 16.10].

Thus, by adding sodium acetate, we have increased the concentration of acetate ion. Because acetate ion is a product in the ionization of acetic acid, the addition of acetate ion causes the equilibrium to shift to the left. The net result is a reduction in the percent ionization of acetic acid:

addition

$$CH_3COOH(aq) + H_2O(l) \xleftarrow{\quad} H_3O^+(aq) + CH_3COO^-(aq)$$

Equilibrium is driven toward reactant.

Shifting the equilibrium to the left consumes not only some of the added acetate ion, but also some of the hydronium ion. This causes the pH to change (in this case the pH increases).

Sample Problem 17.1 shows how an equilibrium table can be used to calculate the pH of a solution of acetic acid after the addition of sodium acetate.

SAMPLE PROBLEM (17.1)

Determine the pH at 25°C of a solution prepared by adding 0.050 mole of sodium acetate to 1.0 L of 0.10 *M* acetic acid. (Assume that the addition of sodium acetate does not change the volume of the solution.)

Strategy Construct a new equilibrium table to solve for the hydronium ion concentration. Remember that prior to the ionization of a weak acid, the concentration of hydronium ion in water at 25°C is $1.0 \times 10^{-7}\ M$. However, because this concentration is insignificant compared to the concentration resulting from the ionization, we can neglect it in our equilibrium table.

Setup We use the stated concentration of acetic acid, 0.10 M, and $[H_3O^+] \approx 0\ M$ as the initial concentrations in the table:

$$CH_3COOH(aq) + H_2O(l) \rightleftharpoons H_3O^+(aq) + CH_3COO^-(aq)$$

	CH_3COOH	H_2O	H_3O^+	CH_3COO^-
Initial concentration (M):	0.10	—	0	0.050
Change in concentration (M):	$-x$	—	$+x$	$+x$
Equilibrium concentration (M):	$0.10 - x$	—	x	$0.050 + x$

Solution Substituting the equilibrium concentrations, in terms of the unknown x, into the equilibrium expression gives:

$$1.8 \times 10^{-5} = \frac{(x)(0.050 + x)}{0.10 - x}$$

We expect x to be very small (even smaller than the $1.34 \times 10^{-3}\ M$ concentration of hydronium and acetate ions in 0.10 M acetic acid), because the ionization of CH_3COOH is suppressed by the presence of CH_3COO^-. Therefore, we can assume:

$$(0.10 - x)\ M \approx 0.10\ M \qquad \text{and} \qquad (0.050 + x)\ M \approx 0.050\ M$$

The equilibrium expression simplifies to:

$$1.8 \times 10^{-5} = \frac{(x)(0.050)}{0.10}$$

and $x = 3.6 \times 10^{-5}\ M$. According to the equilibrium table, $[H_3O^+] = x$, so pH $= -\log(3.6 \times 10^{-5}) = 4.44$. In this case, the percent ionization of acetic acid is:

$$\frac{3.6 \times 10^{-5}\ M}{0.10\ M} \times 100\% = 0.036\%$$

This is considerably smaller than the percent ionization prior to the addition of sodium acetate.

THINK ABOUT IT

The equilibrium concentrations of CH_3COOH, CH_3COO^-, and H_3O^+ are the same regardless of whether we add sodium acetate to a solution of acetic acid, add acetic acid to a solution of sodium acetate, or dissolve both species at the same time. We could have constructed an equilibrium table starting with the equilibrium concentrations in the 0.10 M acetic acid solution:

$$CH_3COOH(aq) + H_2O(l) \rightleftharpoons H_3O^+(aq) + CH_3COO^-(aq)$$

	CH_3COOH	H_2O	H_3O^+	CH_3COO^-
Initial concentration (M):	0.09866	—	1.34×10^{-3}	5.134×10^{-2}
Change in concentration (M):	$+y$	—	$-y$	$-y$
Equilibrium concentration (M):	$0.09866 + y$	—	$1.34 \times 10^{-3} - y$	$5.134 \times 10^{-2} - y$

Student Note: This is the sum of the equilibrium concentration of acetate ion in a 0.10-M solution of acetic acid ($1.34 \times 10^{-3}\ M$) and the added acetate ion (0.050 M).

In this case, the reaction proceeds to the left. (The acetic acid concentration increases, and the concentrations of hydronium and acetate ions decrease.) Solving for y gives $1.304 \times 10^{-3}\ M$. $[H_3O^+] = 1.34 \times 10^{-3} - y = 3.6 \times 10^{-5}\ M$ and pH $= 4.44$. We get the same pH either way, although treating the problem as though both CH_3COOH and CH_3COO^- are added at the same time and the reaction proceeds to the right simplifies the solution.

Practice Problem **ⒶTTEMPT** Determine the pH at 25°C of a solution prepared by dissolving 0.075 mole of sodium acetate in 1.0 L of 0.25 M acetic acid. (Assume that the addition of sodium acetate does not change the volume of the solution.)

Practice Problem **ⒷUILD** Determine the pH at 25°C of a solution prepared by dissolving 0.35 mole of ammonium chloride in 1.0 L of 0.25 M aqueous ammonia.

Practice Problem **ⒸONCEPTUALIZE** Which of the following compounds, when added to an aqueous solution of HF, would cause an increase in the pH? Which would cause a decrease in pH? Which would not have an effect on the pH?

NaF	SnF_2	HCl	NaCl	NaOH	H_2O
(i)	(ii)	(iii)	(iv)	(v)	(vi)

An aqueous solution of a weak electrolyte contains both the weak electrolyte and its ionization products, which are ions. If a soluble salt that contains one of those ions is added, the equilibrium shifts to the left, thereby *suppressing* the ionization of the weak electrolyte. In general, when a compound containing an ion in common with a dissolved substance is added to a solution at equilibrium, the equilibrium shifts to the left. This phenomenon is known as the ***common ion effect.*** The *common ion* can also be H_3O^+ or OH^-. For example, addition of a strong acid to a solution of a weak acid suppresses ionization of the weak acid. Similarly, addition of a strong base to a solution of weak base suppresses ionization of the weak base.

CHECKPOINT – SECTION 17.1   The Common Ion Effect

17.1.1 Which of the following would cause a decrease in the percent ionization of nitrous acid (HNO_2) when added to a solution of nitrous acid at equilibrium? (Select all that apply.)

a) $NaNO_2$

b) H_2O

c) $Ca(NO_2)_2$

d) HNO_3

e) $NaNO_3$

17.1.2 What is the pH of a solution prepared by adding 0.05 mole of NaF to 1.0 L of 0.1 M HF at 25°C? (Assume that the addition of NaF does not change the volume of the solution.) (K_a for HF = 7.1×10^{-4}.)

a) 2.1

b) 2.8

c) 1.4

d) 4.6

e) 7.3

17.2   Buffer Solutions

A solution that contains a weak acid and its conjugate base (or a weak base and its conjugate acid) is a *buffer solution* or simply a ***buffer.*** Any solution of a weak acid contains some conjugate base. In a buffer solution, though, the amounts of weak acid and conjugate base must be *comparable,* meaning that the conjugate base must be supplied by a dissolved salt. Buffer solutions, by virtue of their composition, *resist* changes in pH upon addition of small amounts of either an acid or a base. The ability to resist pH change is very important to chemical and biological systems, including the human body. The pH of blood is about 7.4, whereas that of gastric juices is about 1.5. Each of these pH values is crucial for proper enzyme function and the balance of osmotic pressure, and each is maintained within a very narrow pH range by a buffer.

Calculating the pH of a Buffer

Student Note: Remember that sodium acetate is a strong electrolyte [◄ Section 4.1], so it dissociates completely in water to give sodium ions and acetate ions:

$$CH_3COONa \longrightarrow Na^+(aq) + CH_3COO^-(aq)$$

Consider a solution that is 1.0 M in acetic acid and 1.0 M in sodium acetate. If a small amount of acid is added to this solution, it is consumed completely by the acetate ion:

$$H_3O^+(aq) + CH_3COO^-(aq) \longrightarrow CH_3COOH(aq) + H_2O(l)$$

thus converting a strong acid (H_3O^+) to a weak acid (CH_3COOH). Addition of a strong acid lowers the pH of a solution. However, a buffer's ability to convert a strong acid to a weak acid minimizes the effect of the addition on the pH.

Similarly, if a small amount of a base is added, it is consumed completely by the acetic acid:

$$CH_3COOH(aq) + OH^-(aq) \longrightarrow CH_3COO^-(aq) + H_2O(l)$$

thus converting a strong base (OH^-) to a weak base (CH_3COO^-). Addition of a strong base increases the pH of a solution. Again, however, a buffer's ability to convert a strong base to a weak base minimizes the effect of the addition on pH.

To illustrate the function of a buffer, suppose that we have 1 L of the acetic acid–sodium acetate solution described previously. We can calculate the pH of the buffer using the procedure in Section 17.1:

	$CH_3COOH(aq) +$	$H_2O(l) \rightleftharpoons$	$H_3O^+(aq) +$	$CH_3COO^-(aq)$
Initial concentration (M):	1.0	—	0	1.0
Change in concentration (M):	$-x$	—	$+x$	$+x$
Equilibrium concentration (M):	$1.0 - x$	—	x	$1.0 + x$

The equilibrium expression is:

$$K_a = \frac{(x)(1.0 + x)}{1.0 - x}$$

Because the forward reaction is suppressed by the presence of the common ion, CH_3COO^-, and the reverse process is suppressed by the presence of CH_3COOH, it is reasonable to assume that x will be very small:

$$(1.0 - x)\,M \approx 1.0\,M \qquad \text{and} \qquad (1.0 + x)\,M \approx 1.0\,M$$

Thus, the equilibrium expression simplifies to:

$$1.8 \times 10^{-5} = \frac{(x)(1.0)}{1.0} = x$$

At equilibrium, therefore, $[H_3O^+] = 1.8 \times 10^{-5}\,M$ and pH = 4.74.

Now consider what happens when we add 0.10 mole of HCl to the buffer. (We assume that the addition of HCl causes no change in the volume of the solution.) The reaction that takes place when we add a strong acid is the conversion of H_3O^+ to CH_3COOH. The added acid is all consumed, along with an equal amount of acetate ion. We keep track of the amounts of acetic acid and acetate ion when a strong acid (or base) is added by writing the starting amounts above the equation and the final amounts (after the added substance has been consumed) below the equation:

Student Note: As long as the amount of strong acid added to the buffer does not exceed the amount of conjugate base originally present, *all* the added acid will be consumed and converted to weak acid.

Upon addition of H_3O^+:

	1.0 mol	0.1 mol	1.0 mol	
	$CH_3COO^-(aq)$ +	$H_3O^+(aq)$ $\longrightarrow$	$CH_3COOH(aq)$ +	$H_2O(l)$

After H_3O^+ has been consumed: 0.9 mol   0 mol   1.1 mol

We can use the resulting amounts of acetic acid and acetate ion to construct a new equilibrium table:

$$CH_3COOH(aq) + H_2O(l) \rightleftharpoons H_3O^+(aq) + CH_3COO^-(aq)$$

Initial concentration (M):	1.1	—	0	0.9
Change in concentration (M):	$-x$	—	$+x$	$+x$
Equilibrium concentration (M):	$1.1 - x$	—	x	$0.9 + x$

We can solve for pH as we have done before, assuming that x is small enough to be neglected:

$$1.8 \times 10^{-5} = \frac{(x)(0.9 + x)}{1.1 - x} \approx \frac{(x)(0.9)}{1.1}$$

$$x = 2.2 \times 10^{-5}\,M$$

Thus, when equilibrium is reestablished, $[H_3O^+] = 2.2 \times 10^{-5}\,M$ and pH = 4.66—a change of only 0.08 pH unit. Had we added 0.10 mol of HCl to 1 L of pure water, the pH would have gone from 7.00 to 1.00!

In the determination of the pH of a buffer such as the one just described, we always neglect the small amount of weak acid that ionizes (x) because ionization is suppressed by the presence of a common ion. Similarly, we ignore the hydrolysis of the acetate ion because of the presence of acetic acid. This enables us to derive an expression for determining the pH of a buffer. We begin with the equilibrium expression:

$$K_a = \frac{[H_3O^+][A^-]}{[HA]}$$

Rearranging to solve for $[H_3O^+]$ gives:

$$[H_3O^+] = \frac{K_a[HA]}{[A^-]}$$

Taking the negative logarithm of both sides, we obtain:

$$-\log [H_3O^+] = -\log K_a - \log \frac{[HA]}{[A^-]}$$

or:

$$-\log [H_3O^+] = -\log K_a + \log \frac{[A^-]}{[HA]}$$

Thus:

Equation 17.1	$pH = pK_a + \log \dfrac{[A^-]}{[AH]}$

where:

Equation 17.2	$pK_a = -\log K_a$

Equation 17.1 is known as the ***Henderson-Hasselbalch equation.*** Its more general form is:

Equation 17.3	$pH = pK_a + \log \dfrac{[\text{conjugate base}]}{[\text{weak acid}]}$

In the case of our acetic acid (1.0 M) and sodium acetate (1.0 M) buffer, the concentrations of weak acid and conjugate base are *equal*. When this is true, the log term in the Henderson-Hasselbalch equation is zero and the pH is numerically equal to the pK_a. In the case of an acetic acid–acetate ion buffer, $pK_a = -\log 1.8 \times 10^{-5} = 4.74$.

After the addition of 0.10 mole of HCl, we determined that the concentrations of acetic acid and acetate ion were 1.1 M and 0.9 M, respectively. Using these concentrations in the Henderson-Hasselbalch equation gives:

$$pH = 4.74 + \log \frac{[CH_3COO^-]}{[CH_3COOH]}$$

$$= 4.74 + \log \frac{0.9\ M}{1.1\ M}$$

$$= 4.74 + (-0.087) = 4.65$$

The small difference between this pH and the 4.66 calculated using an equilibrium table is due to differences in rounding. Figure 17.1 illustrates how a buffer solution resists drastic changes in pH.

Sample Problem 17.2 shows how the Henderson-Hasselbalch equation is used to determine the pH of a buffer after the addition of a strong base.

Animation
Figure 17.1, Buffer Solutions

SAMPLE PROBLEM 17.2

Starting with 1.00 L of a buffer that is 1.00 M in acetic acid and 1.00 M in sodium acetate, calculate the pH after the addition of 0.100 mole of NaOH. (Assume that the addition does not change the volume of the solution.)

Strategy Added base will react with the acetic acid component of the buffer, converting OH⁻ to CH_3COO^-:

$$CH_3COOH(aq) + OH^-(aq) \longrightarrow H_2O(l) + CH_3COO^-(aq)$$

Write the starting amount of each species above the equation and the final amount of each species below the equation. Use the final amounts as concentrations in Equation 17.1. The volume of the buffer is 1 L in this example, so the number of moles of a substance is equal to the molar concentration. In cases where the buffer volume is something other than 1 L, however, we can still use molar amounts in the Henderson-Hasselbalch equation because the volume would cancel in the top and bottom of the log term.

Setup

Upon addition of OH⁻:

                                1.00 mol   0.10 mol   1.00 mol

$$CH_3COOH(aq) + OH^-(aq) \longrightarrow H_2O(l) + CH_3COO^-(aq)$$

After OH⁻ has been consumed: 0.90 mol   0 mol   1.10 mol

Solution

$$pH = 4.74 + \log \frac{1.10 \ M}{0.90 \ M}$$

$$= 4.74 + \log \frac{1.10 \ M}{0.90 \ M} = 4.83$$

Thus, the pH of the buffer after addition of 0.10 mol of NaOH is 4.83.

THINK ABOUT IT

Always do a "reality check" on a calculated pH. Although a buffer does minimize the effect of added base, the pH does increase. If you find that you've calculated a lower pH after the addition of a base, check for errors like mixing up the weak acid and conjugate base concentrations or losing track of a minus sign.

Practice Problem **A**TTEMPT Calculate the pH of 1 L of a buffer that is 1.0 *M* in acetic acid and 1.0 *M* in sodium acetate after the addition of 0.25 mole of NaOH.

Practice Problem **B**UILD How much HCl must be added to a liter of buffer that is 1.5 *M* in acetic acid and 0.75 *M* in sodium acetate to result in a buffer pH of 4.10?

Practice Problem **C**ONCEPTUALIZE The first diagram represents a buffer solution. Which of the other diagrams [(i)–(iv)] best represents the buffer after strong acid has been added?

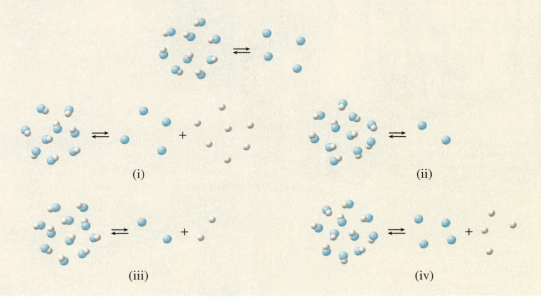

(i)   (ii)

(iii)   (iv)

Preparing a Buffer Solution with a Specific pH

A solution is only a buffer if it has the capacity to resist pH change when either an acid or a base is added. If the concentrations of a weak acid and conjugate base differ by more than a factor of 10, the solution does not have this capacity. Therefore, we consider a solution a buffer, and can use Equation 17.1 to calculate its pH, only if the following condition is met:

$$10 \geq \frac{[\text{conjugate base}]}{[\text{weak acid}]} \geq 0.1$$

Figure 17.1

Buffer Solutions

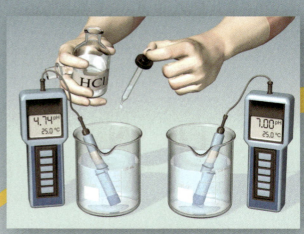

When we add 0.001 mol of strong acid, it is completely consumed by the acetate ion in the buffer.

Before reaction:	0.001 mol	0.010 mol		0.010 mol
	$H_3O^+(aq) + CH_3COO^-(aq)$		$\longrightarrow$	$CH_3COOH(aq) + H_2O(l)$
After reaction:	0 mol	0.009 mol		0.011 mol

0.100 M CH_3COOH
0.100 M CH_3COO^-

START

$pH = 4.74 + \log \dfrac{[CH_3COO^-]}{[CH_3COOH]} = 4.74$

Water
pH = 7.00

The buffer solution is 0.100 M in acetic acid and 0.100 M in sodium acetate. 100 mL of this buffer contains (0.100 mol/L)(0.10 L) = 0.010 mol each acetic acid and acetate ion.

START

When we add 0.001 mol of strong base, it is completely consumed by the acetic acid in the buffer.

Before reaction:	0.001 mol	0.010 mol		0.010 mol
	$OH^-(aq) + CH_3COOH(aq)$		$\longrightarrow$	$H_2O(l) + CH_3COO^-(aq)$
After reaction:	0 mol	0.009 mol		0.011 mol

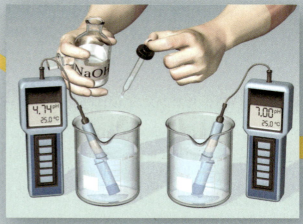

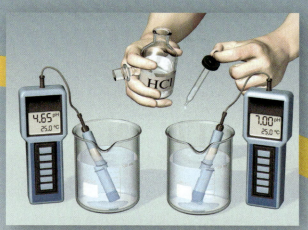

We can calculate the new pH using the Henderson-Hasselbalch equation:

$$pH = 4.74 + \log \frac{0.009}{0.011} = 4.65$$

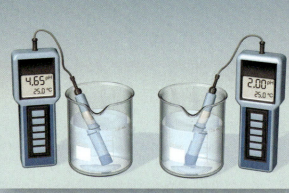

There is nothing in pure water to consume strong acid. Therefore, its pH drops drastically.

$$pH = -\log \frac{0.001 \text{ mol}}{0.10 \text{ L}} = 2.00$$

There is nothing in pure water to consume strong base. Therefore, its pH rises drastically.

$$pOH = -\log \frac{0.001 \text{ mol}}{0.10 \text{ L}} = 2.00, \; pH = 12.00$$

We can calculate the new pH using the Henderson-Hasselbalch equation:

$$pH = 4.74 + \log \frac{0.011}{0.009} = 4.83$$

What's the point?

A buffer contains both a weak acid and its conjugate base.* Small amounts of strong acid or strong base are consumed by the buffer components, thereby preventing drastic pH changes. Pure water does not contain species that can consume acid or base. Even a very small addition of either acid or base causes a large change in pH.

*A buffer could also be prepared using a weak base and its conjugate acid.

(See Visualizing Chemistry questions VC17.1–VC17.4 on page 900.)

Weak Acid	K_a	pK_a
HF	7.1×10^{-4}	3.15
HNO$_2$	4.5×10^{-4}	3.35
HCOOH	1.7×10^{-4}	3.77
C$_6$H$_5$COOH	6.5×10^{-5}	4.19
CH$_3$COOH	1.8×10^{-5}	4.74
HCN	4.9×10^{-10}	9.31
C$_6$H$_5$OH	1.3×10^{-10}	9.89

Consequently, the log term in Equation 17.1 can only have values from -1 to 1, and the pH of a buffer cannot be more than one pH unit different from the pK_a of the weak acid it contains. This is known as the *range* of the buffer, where pH $= pK_a \pm 1$. This enables us to select the appropriate conjugate pair to prepare a buffer with a specific, desired pH.

First, we choose a weak acid whose pK_a is close to the desired pH. Next, we substitute the pH and pK_a values into Equation 17.1 to obtain the necessary ratio of [conjugate base]/[weak acid]. This ratio can then be converted to molar quantities for the preparation of the buffer.

Sample Problem 17.3 demonstrates this procedure.

SAMPLE PROBLEM (17.3)

Select an appropriate weak acid from the table in the margin, and describe how you would prepare a buffer with a pH of 9.50.

Strategy Select an acid with a pK_a within one pH unit of 9.50. Use the pK_a of the acid and Equation 17.1 to calculate the necessary ratio of [conjugate base]/[weak acid]. Select concentrations of the buffer components that yield the calculated ratio.

Setup Two of the acids listed in the table have pK_a values in the desired range: hydrocyanic acid (HCN, $pK_a = 9.31$) and phenol (C$_6$H$_5$OH, $pK_a = 9.89$).

Solution Plugging the values for phenol into Equation 17.1 gives:

$$9.50 = 9.89 + \log \frac{[C_6H_5O^-]}{[C_6H_5OH]}$$

$$9.50 - 9.89 = \log \frac{[C_6H_5O^-]}{[C_6H_5OH]} = -0.39$$

$$\frac{[C_6H_5O^-]}{[C_6H_5OH]} = 10^{-0.39} = 0.41$$

Therefore, the ratio of [C$_6$H$_5$O$^-$] to [C$_6$H$_5$OH] must be 0.41 to 1. One way to achieve this would be to dissolve 0.41 mol of C$_6$H$_5$ONa and 1.00 mol of C$_6$H$_5$OH in 1 L of water.

THINK ABOUT IT

There is an infinite number of combinations of [conjugate base] and [weak acid] that will give the necessary ratio. Note that this pH could also be achieved using HCN and a cyanide salt. For most purposes, it is best to use the least toxic compounds available.

Practice Problem **A**TTEMPT Select an appropriate acid from Table 16.6, and describe how you would prepare a buffer with pH = 4.5.

Practice Problem **B**UILD What range of pH values could be achieved with a buffer consisting of nitrous acid (HNO$_2$) and sodium nitrite (NO$_2^-$)?

Practice Problem **C**ONCEPTUALIZE The diagrams represent three different weak acids, each of which can be combined with its conjugate base to prepare a buffer. Which acid can be used to prepare the buffer with the lowest pH?

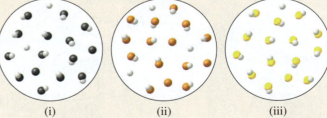

(i)   (ii)   (iii)

Bringing Chemistry to Life

Maintaining the pH of Blood

There are about 5 L of blood in the average adult. Circulating blood keeps cells alive by providing them with oxygen and nutrients and by removing carbon dioxide and other waste materials. The efficiency of this enormously complex system relies on buffers.

The two primary components of blood are *blood plasma* and red blood cells, or *erythrocytes*. Blood plasma contains many compounds, including proteins, metal ions, and inorganic phosphates. The erythrocytes contain hemoglobin molecules, as well as the enzyme carbonic anhydrase, which catalyzes both the formation and the decomposition of carbonic acid (H_2CO_3):

$$CO_2(aq) + H_2O(l) \rightleftharpoons H_2CO_3(aq)$$

The substances inside the erythrocytes are protected from extracellular fluid (blood plasma) by a semipermeable cell membrane that allows only certain molecules to diffuse through it.

The pH of blood plasma is maintained at about 7.40 by several buffer systems, the most important of which is the HCO_3^-/H_2CO_3 system. In the erythrocyte, where the pH is 7.25, the principal buffer systems are HCO_3^-/H_2CO_3 and hemoglobin. The hemoglobin molecule is a complex protein molecule (molar mass of 65,000 g) that contains a number of ionizable protons. As a very rough approximation, we can treat it as a monoprotic acid in the form HHb:

$$HHb(aq) \rightleftharpoons H^+(aq) + Hb^-(aq)$$

where HHb represents the hemoglobin molecule and Hb^- is the conjugate base of HHb. Oxyhemoglobin ($HHbO_2$), formed by the combination of oxygen with hemoglobin, is a stronger acid than HHb:

$$HHbO_2(aq) \rightleftharpoons H^+(aq) + HbO_2^-(aq)$$

Carbon dioxide produced by metabolic processes diffuses into the erythrocyte, where it is rapidly converted to H_2CO_3 by carbonic anhydrase:

$$CO_2(aq) + H_2O(l) \rightleftharpoons H_2CO_3(aq)$$

The ionization of the carbonic acid:

$$H_2CO_3(aq) \rightleftharpoons H^+(aq) + HCO_3^-(aq)$$

has two important consequences. First, the bicarbonate ion diffuses out of the erythrocyte and is carried by the blood plasma to the lungs. This is the major mechanism for removing carbon dioxide. Second, the H^+ ions shift the equilibrium in favor of the un-ionized oxyhemoglobin molecules:

$$H^+(aq) + HbO_2^-(aq) \rightleftharpoons HHbO_2(aq)$$

Because $HHbO_2$ releases oxygen more readily than does its conjugate base (HbO_2^-), the formation of the acid promotes the following reaction from left to right:

$$HHbO_2(aq) \rightleftharpoons HHb(aq) + O_2(aq)$$

The O_2 molecules diffuse out of the erythrocyte and are taken up by other cells to carry out metabolism.

When the venous blood returns to the lungs, the preceding processes are reversed. The bicarbonate ions now diffuse into the erythrocyte, where they react with hemoglobin to form carbonic acid:

$$HHb(aq) + HCO_3^-(aq) \rightleftharpoons Hb^-(aq) + H_2CO_3(aq)$$

Most of the acid is then converted to CO_2 by carbonic anhydrase:

$$H_2CO_3(aq) \rightleftharpoons H_2O(l) + CO_2(aq)$$

The carbon dioxide diffuses to the lungs and is eventually exhaled. The formation of the Hb^- ions (due to the reaction between HHb and HCO_3^-) also favors the uptake of oxygen at the lungs:

$$Hb^-(aq) + O_2(aq) \rightleftharpoons HbO_2^-(aq)$$

because Hb^- has a greater affinity for oxygen than does HHb.

When the arterial blood flows back to the body tissues, the entire cycle is repeated.

CHECKPOINT – SECTION 17.2   Buffer Solutions

17.2.1 Which of the following combinations can be used to prepare a buffer?

a) HCl/NaCl

b) HF/KF

c) NH_3/NH_4Cl

d) HNO_3/HNO_2

e) $NaNO_2/HNO_3$

17.2.2 What is the pH of a buffer that is 0.76 M in HF and 0.98 M in NaF?

a) 3.26

b) 3.04

c) 3.15

d) 10.85

e) 10.74

17.2.3 Consider 1 L of a buffer that is 0.85 M in formic acid (HCOOH) and 1.4 M in sodium formate (HCOONa). Calculate the pH after the addition of 0.15 mol HCl. (Assume the addition causes no volume change.)

a) 4.11

b) 3.99

c) 3.87

d) 10.13

e) 10.01

17.2.4 Consider 1 L of a buffer that is 1.5 M in hydrocyanic acid (HCN) and 1.2 M in sodium cyanide (NaCN). Calculate the pH after the addition of 0.25 mol NaOH. (Assume the addition causes no volume change.)

a) 9.21

b) 9.37

c) 9.04

d) 4.63

e) 4.96

17.2.5 The solutions shown contain one or more of the following compounds: H_2A (a weak diprotic acid), NaHA, and Na_2A (the sodium salts of HA^- and A^{2-}). (For clarity, the sodium ions and water molecules are not shown.) Which solutions are buffers? (Select all that apply.)

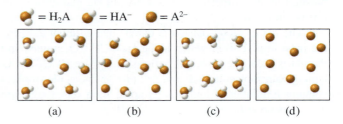

(a)   (b)   (c)   (d)

17.2.6 The solutions shown contain a combination of the weak acid HA and its sodium salt NaA. (For clarity, the sodium ions and water molecules are not shown.) Arrange the solutions in order of increasing pH.

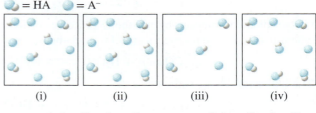

(i)   (ii)   (iii)   (iv)

a) i < iii < iv < ii

b) iv < ii < iii < i

c) i < ii < iv < iii

d) iv < ii < i = iii

e) i = iii < iv < ii

17.3 Acid-Base Titrations

In Section 4.6 we introduced acid-base titrations as a form of chemical analysis. Having discussed buffer solutions, we can now look in more detail at the quantitative aspects of acid-base titrations. We consider three types of reactions: (1) titrations involving a strong acid and a strong base, (2) titrations involving a weak acid and a strong base, and (3) titrations involving a strong acid and a weak base. Titrations involving a weak acid and a weak base are complicated by the hydrolysis of both the cation and the anion of the salt formed. These titrations are not discussed here. Figure 17.2 shows the experimental setup for monitoring the pH over the course of an acid-base titration.

Strong Acid–Strong Base Titrations

The reaction between the strong acid HCl and the strong base NaOH can be represented by:

$$NaOH(aq) + HCl(aq) \longrightarrow NaCl(aq) + H_2O(l)$$

or by the net ionic equation:

$$OH^-(aq) + H_3O^+(aq) \longrightarrow 2H_2O(l)$$

Animation
Acids and Bases—titration of HCl with NaOH.

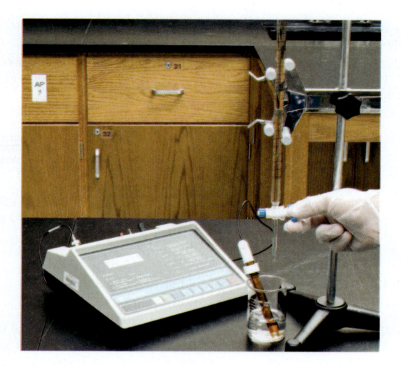

Figure 17.2 A pH meter is used to monitor an acid-base titration.
Matt Meadows/McGraw Hill

Consider the addition of a 0.100 *M* NaOH solution (from a burette) to a vessel containing 25.0 mL of 0.100 *M* HCl. For convenience, we will use only three significant figures for volume and concentration and two significant figures for pH. Figure 17.3 shows the titration curve—the plot of pH as a function of *titrant* volume added.

Before the addition of NaOH begins, the pH of the acid is given by −log(0.100), or 1.00. When NaOH is added, the pH of the solution increases slowly at first. Near the equivalence point, the pH begins to rise steeply, and at the equivalence point, when equimolar amounts of acid and base have reacted, the curve rises almost vertically. In a strong acid–strong base titration, both the hydronium ion and hydroxide ion concentrations are very small at the equivalence point (roughly 1×10^{-7} *M*); consequently, the addition of a single drop of the base causes a large increase in [OH⁻] and a steep rise in the pH of the solution. Beyond the equivalence point, the pH again increases slowly with the continued addition of NaOH.

Student Note: Recall that only the digits to the right of the decimal point are significant in a pH value.

Student Note: The *titrant* is the solution that is added from the burette.

Student Note: Recall that for an acid and base that combine in a 1:1 ratio, the equivalence point is where equal molar amounts of acid and base have been combined [◄◄ Section 4.5].

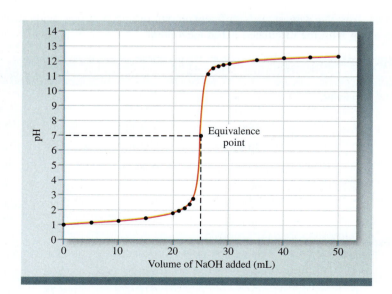

Figure 17.3 Titration curve (pH as a function of volume titrant added) of a strong acid–strong base titration. A 0.100 *M* NaOH solution, the titrant, is added from a burette to 25.0 mL of a 0.100 *M* HCl solution in an Erlenmeyer flask.

Student Note: These calculations could also be done using *moles*, but using millimoles simplifies the calculations. Remember that millimoles = $M \times$ mL [◀◀ Section 4.5].

It is possible to calculate the pH of the solution at every stage of titration. Here are three sample calculations.

1. Consider the addition of 10.0 mL of 0.100 *M* NaOH to 25.0 mL of 0.100 *M* HCl: The total volume of the solution is 35.0 mL. The number of millimoles of NaOH in 10.0 mL is:

$$10.0 \text{ mL} \times \frac{0.100 \text{ mmol NaOH}}{1 \text{ mL}} = 1.00 \text{ mmol}$$

The number of millimoles of HCl originally present in 25.0 mL of solution is:

$$25.0 \text{ mL} \times \frac{0.100 \text{ mmol HCl}}{1 \text{ mL}} = 2.50 \text{ mmol}$$

Thus, the amount of HCl left after partial neutralization is 2.50 − 1.00, or 1.50 mmol. Next, we determine the resulting concentration of H_3O^+. We have 1.50 mmol in 35.0 mL:

$$\frac{1.50 \text{ mmol HCl}}{35.0 \text{ mL}} = 0.0429 \, M$$

Thus $[H_3O^+] = 0.0429 \, M$, and the pH of the solution is:

$$pH = -\log(0.04289) = 1.37$$

2. Consider the addition of 25.0 mL of 0.100 *M* NaOH to 25.0 mL of 0.100 *M* HCl: This is a straightforward calculation, because it involves a complete neutralization reaction and neither ion in the salt (NaCl) undergoes hydrolysis [◀◀ Section 16.10]. At the equivalence point, $[H_3O^+] = [OH^-] = 1.00 \times 10^{-7} \, M$ and the pH of the solution is 7.000.

3. Consider the addition of 35.0 mL of 0.100 *M* NaOH to 25.0 mL of 0.100 *M* HCl: The total volume of the solution is now 60.0 mL. The number of millimoles of NaOH added is:

$$35.0 \text{ mL} \times \frac{0.100 \text{ mmol NaOH}}{1 \text{ mL}} = 3.50 \text{ mmol}$$

There are 2.50 mmol of HCl in 25.0 mL of solution. After complete neutralization of HCl, 2.50 mmol of NaOH have been consumed, and the number of millimoles of NaOH remaining is 3.5 − 2.5 or 1.00 mmol. The concentration of NaOH in 60.0 mL of solution is:

$$\frac{1.00 \text{ mmol NaOH}}{60.0 \text{ mL}} = 0.0167 \, M$$

Thus $[OH^-] = 0.0167 \, M$ and pOH = $-\log(0.0167) = 1.78$. The pH of the solution is 14.00 − 1.78 or 12.22.

Table 17.1 lists the data at nine different points during a strong acid−strong base titration along with the calculated pH at each point.

TABLE 17.1	Determination of pH at Several Different Points in a Strong Acid−Strong Base Titration					
Volume OH⁻ added (mL)	OH⁻ added (mmol)	H_3O^+ remaining (mmol)	Total volume (mL)	$[H_3O^+]$ (mol/L)		pH
0	0	2.5	25.0	0.100		1.000
5.0	0.50	2.0	30.0	0.0667		1.176
10.0	1.0	1.5	35.0	0.0429		1.364
15.0	1.5	1.0	40.0	0.0250		1.602
20.0	2.0	0.5	45.0	0.0111		1.955
25.0	2.5	0	50.0	1.00×10^{-7}		7.000
Volume OH⁻ added (mL)	OH⁻ added (mmol)	Excess OH⁻ (mmol)	Total volume (mL)	$[OH^-]$ (mol/L)	pOH	pH
30.0	3.0	0.5	55.0	0.0091	2.04	11.96
35.0	3.5	1.0	60.0	0.0167	1.78	12.223

Weak Acid–Strong Base Titrations

Consider the neutralization reaction between acetic acid (a weak acid) and sodium hydroxide (a strong base):

$$CH_3COOH(aq) + NaOH(aq) \longrightarrow CH_3COONa(aq) + H_2O(l)$$

This equation can be simplified to:

$$CH_3COOH(aq) + OH^-(aq) \longrightarrow CH_3COO^-(aq) + H_2O(l)$$

The acetate ion that results from this neutralization undergoes hydrolysis [◀ Section 16.10] as follows:

$$CH_3COO^-(aq) + H_2O(l) \rightleftharpoons CH_3COOH(aq) + OH^-(aq)$$

At the equivalence point, therefore, when we only have sodium acetate in solution, the pH will be greater than 7 as a result of the OH^- formed by hydrolysis of the acetate ion.

  The curve for titration of 25.0 mL of 0.1 M acetic acid with 0.10 M sodium hydroxide is shown in Figure 17.4. Note how the shape of the curve differs from the one in Figure 17.3. Compared to the curve for titration of a strong acid with a strong base, the curve for titration of a weak acid with a strong base has a higher initial pH, a more gradual change in pH as base is added, and a shorter vertical region near the equivalence point.

  Again, it is possible to calculate the pH at every stage of the titration. Here are four sample calculations.

1. Prior to the addition of any base, the pH is determined by the ionization of acetic acid. We use its concentration (0.10 M) and its K_a(1.8 × 10^{-5}) to calculate the H_3O^+ concentration using an equilibrium table:

$$CH_3COOH(aq) + H_2O(l) \rightleftharpoons H_3O^+(aq) + CH_3COO^-(aq)$$

	CH₃COOH	H₂O	H₃O⁺	CH₃COO⁻
Initial concentration (M):	0.10	—	0	0
Change in concentration (M):	−x	—	+x	+x
Equilibrium concentration (M):	0.10 − x	—	x	x

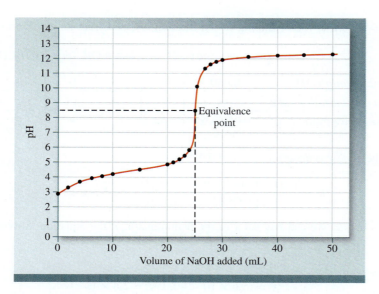

Figure 17.4 Titration curve of a weak acid–strong base titration. A 0.100 M NaOH solution is added from a burette to 25.0 mL of a 0.100 M CH₃COOH solution in an Erlenmeyer flask. Because of the hydrolysis of the salt formed, the pH at the equivalence point is greater than 7.

$$K_a = \frac{[H_3O^+][CH_3COO^-]}{[CH_3COOH]} = \frac{x^2}{0.10 - x} = 1.8 \times 10^{-5}$$

We can neglect x on the bottom of the equation [◄◄ Section 16.5]. Solving for x:

$$\frac{x^2}{0.10} = 1.8 \times 10^{-5}$$

$$x^2 = (1.8 \times 10^{-5})(0.10) = 1.8 \times 10^{-6}$$

$$x = \sqrt{1.8 \times 10^{-6}} = 1.34 \times 10^{-3} \ M$$

gives $[H_3O^+] = 1.34 \times 10^{-3} \ M$ and pH = 2.87.

2. After the first addition of base, some of the acetic acid has been converted to acetate ion via the reaction:

$$CH_3COOH(aq) + OH^-(aq) \longrightarrow CH_3COO^-(aq) + H_2O(l)$$

With significant amounts of both acetic acid and acetate ion in solution, we now treat the solution as a buffer and use the Henderson-Hasselbalch equation to calculate the pH.

After the addition of 10.0 mL of base, the solution contains 1.5 mmol of acetic acid and 1.0 mmol of acetate ion (see Table 17.2):

Student Note: Remember that we can use mol or mmol amounts in place of concentrations in the Henderson-Hasselbalch equation.

$$pH = 4.74 + \log\frac{1.0 \ mmol}{1.5 \ mmol} = 4.56$$

Each of the points between the beginning of the titration and the equivalence point can be calculated in this way.

3. At the equivalence point, all the acetic acid has been neutralized and we are left with acetate ion in solution. (There is also sodium ion, which does not undergo hydrolysis and therefore does not impact the pH of the solution.) At this point, pH is determined by the concentration and the K_b of acetate ion. The equivalence point occurs when 25.0 mL of base has been added, making the total volume 50.0 mL. The 2.5 mmol of acetic acid (see Table 17.2) has all been converted to acetate ion. Therefore, the concentration of acetate ion is:

$$[CH_3COO^-] = \frac{2.5 \ mmol}{50.0 \ mL} = 0.050 \ M$$

As we did at the beginning of the titration, we construct an equilibrium table:

$$CH_3COO^-(aq) + H_2O(l) \rightleftharpoons OH^-(aq) + CH_3COOH(aq)$$

Initial concentration (M):	0.050	—	0	0
Change in concentration (M):	−x	—	+x	+x
Equilibrium concentration (M):	0.050 − x	—	x	x

The K_b for acetate ion is 5.6×10^{-10}:

Student Note: $K_a \times K_b = K_w$ [◄◄ Section 16.7].

$$K_b = \frac{[OH^-][CH_3COOH]}{[CH_3COO^-]} = \frac{x^2}{0.050 - x} = 5.6 \times 10^{-10}$$

As before, we can neglect x in the denominator of the equation. Solving for x:

$$\frac{x^2}{0.050} = 5.6 \times 10^{-10}$$

$$x^2 = (5.6 \times 10^{-10})(0.050) = 2.8 \times 10^{-11}$$

$$x = \sqrt{2.78 \times 10^{-11}} = 5.3 \times 10^{-6} \ M$$

gives $[OH^-] = 5.3 \times 10^{-6} \ M$, pOH = 5.28, and pH = 8.72.

4. After the equivalence point, the curve for titration of a weak acid with a strong base is identical to the curve for titration of a strong acid with a strong base. Because all the acetic acid has been consumed, there is nothing in solution to consume the additional added OH^-, and the pH levels off between 12 and 13.

Table 17.2 lists the data for the titration of 25.0 mL of 0.10 M acetic acid with 0.10 M NaOH.

Sample Problem 17.4 shows how to calculate the pH for the titration of a weak acid with a strong base.

Student Note: Prior to the addition of any base, *and* at the equivalence point, this is an *equilibrium* problem that is solved using a *concentration*, an *ionization constant*, and an *equilibrium table.*

Student Note: *Prior* to the equivalence point, pH is determined using the Henderson-Hasselbalch equation:

$$pH = pK_a + \log \frac{[\text{conjugate base}]}{[\text{weak acid}]}$$

Student Note: After the equivalence point, the titration curve of a weak acid is identical to that of a strong acid.

Student Hot Spot

Student data indicate you may struggle with pH changes during acid-base titration. Access the eBook to view additional Learning Resources on this topic.

TABLE 17.2	Determination of pH at Several Different Points in a CH_3COOH-NaOH Titration			
Volume OH^- added (mL)	OH^- added (mmol)	CH_3COOH (mmol)	CH_3COO^- produced	pH
0	0	2.5	0	2.87*
5.0	0.50	2.0	0.50	4.14
10.0	1.0	1.5	1.0	4.56
15.0	1.5	1.0	1.5	4.92
20.0	2.0	0.5	2.0	5.34
25.0	2.5	0.0	2.5	8.72†

Volume OH^- added (mL)	OH^- added (mmol)	Excess OH^- (mmol)	Total volume (mL)	$[OH^-]$ (mol/L)	pOH	pH
30.0	3.0	0.5	55.0	0.0091	2.04	11.96
35.0	3.5	1.0	60.0	0.0167	1.78	12.22

*$[CH_3COOH] = 0.10\ M$, $K_a = 1.8 \times 10^{-5}$.

†$[CH_3COO^-] = 0.050\ M$, $K_b = 5.6 \times 10^{-10}$.

SAMPLE PROBLEM 17.4

Calculate the pH in the titration of 50.0 mL of 0.120 M acetic acid by 0.240 M sodium hydroxide after the addition of (a) 10.0 mL of base, (b) 25.0 mL of base, and (c) 35.0 mL of base.

Strategy The reaction between acetic acid and sodium hydroxide is:

$$CH_3COOH(aq) + OH^-(aq) \longrightarrow H_2O(l) + CH_3COO^-(aq)$$

Prior to the equivalence point [part (a)], the solution contains both acetic acid and acetate ion, making the solution a buffer. We can solve part (a) using Equation 17.1, the Henderson-Hasselbalch equation. At the equivalence point [part (b)], all the acetic acid has been neutralized and we have only acetate ion in solution. We must determine the concentration of acetate ion and solve part (b) as an equilibrium problem, using the K_b for acetate ion. After the equivalence point [part (c)], all the acetic acid has been neutralized and there is nothing to consume the additional added base. We must determine the concentration of excess hydroxide ion in the solution and solve for pH using Equations 16.4 and 16.6.

Setup Remember that M can be defined as either mol/L or mmol/mL [◄◄ Section 4.5]. For this type of problem, it simplifies the calculations to use millimoles rather than moles. K_a for acetic acid is 1.8×10^{-5}, so $pK_a = 4.74$. K_b for acetate ion is 5.6×10^{-10}.

(a) The solution originally contains (0.120 mmol/mL)(50.0 mL) = 6.00 mmol of acetic acid. A 10.0-mL amount of base contains (0.240 mmol/mL)(10.0 mL) = 2.40 mmol of base. After the addition of 10.0 mL of base, 2.40 mmol of OH^- has neutralized 2.40 mmol of acetic acid, leaving 3.60 mmol of acetic acid and 2.40 mmol acetate ion in solution:

Upon addition of OH^-: 6.00 mmol 2.40 mmol 0 mmol

$$CH_3COOH(aq) + OH^-(aq) \rightleftharpoons H_2O(l) + CH_3COO^-(aq)$$

After OH^- has been consumed: 3.60 mmol 0 mmol 2.40 mmol

(Continued on next page)

(b) After the addition of 25.0 mL of base, the titration is at the equivalence point. We calculate the pH using the concentration and the K_b of acetate ion.

(c) After the addition of 35.0 mL of base, the titration is past the equivalence point and we solve for pH by determining the concentration of excess hydroxide ion.

Solution

(a) $pH = pK_a + \log \dfrac{2.40}{3.60} = 4.74 - 0.18 = 4.56$

(b) At the equivalence point, we have 6.0 mmol of acetate ion in the total volume. We determine the total volume by calculating what volume of 0.24 M base contains 6.0 mmol:

$$(\text{volume})(0.240 \text{ mmol/mL}) = 6.00 \text{ mmol}$$

$$\text{volume} = \frac{6.00 \text{ mmol}}{0.240 \text{ mmol/mL}} = 25.0 \text{ mL}$$

Therefore, the equivalence point occurs when 25.0 mL of base has been added, making the total volume 50.0 mL + 25.0 mL = 75.0 mL. The concentration of acetate ion at the equivalence point is therefore:

$$\frac{6.00 \text{ mmol } CH_3COO^-}{75.0 \text{ mL}} = 0.0800 \ M$$

We can construct an equilibrium table using this concentration and solve for pH using the ionization constant for CH_3COO^- ($K_b = 5.6 \times 10^{-10}$):

$$CH_3COO^-(aq) + H_2O(l) \rightleftharpoons OH^-(aq) + CH_3COOH(aq)$$

Initial concentration (M):	0.0800	—	0	0
Change in concentration (M):	$-x$	—	$+x$	$+x$
Equilibrium concentration (M):	$0.0800 - x$	—	x	x

Using the equilibrium expression and assuming that x is small enough to be neglected:

$$K_b = \frac{[CH_3COOH][OH^-]}{[CH_3COO^-]} = \frac{(x)(x)}{0.0800 - x} \approx \frac{x^2}{0.0800} = 5.6 \times 10^{-10}$$

$$x = \sqrt{4.48 \times 10^{-11}} = 6.7 \times 10^{-6} \ M$$

According to the equilibrium table, $x = [OH^-]$, so $[OH^-] = 6.7 \times 10^{-6} \ M$. At equilibrium, therefore, $pOH = -\log(6.7 \times 10^{-6}) = 5.17$ and $pH = 14.00 \times 5.17 = 8.83$.

(c) After the equivalence point, we must determine the concentration of excess base and calculate pOH and pH using Equations 16.4 and 16.6. A 35.0-mL amount of the base contains $(0.240 \text{ mmol/mL})(35.0 \text{ mL}) = 8.40$ mmol of OH^-. After neutralizing the 6.00 mmol of acetic acid originally present in the solution, this leaves $8.40 - 6.00 = 2.40$ mmol of excess OH^-. The total volume is $50.0 + 35.0 = 85.0$ mL. Therefore, $[OH^-] = 2.40 \text{ mmol/85.0 mL} = 0.0280 \ M$, $pOH = -\log(0.0280) = 1.553$, and $pH = 14.000 - 1.553 = 12.447$.

In summary, (a) $pH = 4.56$, (b) $pH = 8.83$, and (c) $pH = 12.447$.

THINK ABOUT IT

For each point in a titration, decide first what species are in solution and what *type* of problem it is. If the solution contains only a weak acid (or weak base), as is the case before any titrant is added, or if it contains only a conjugate base (or conjugate acid), as is the case at the equivalence point, when pH is determined by salt hydrolysis, it is an *equilibrium* problem that requires a concentration, an ionization constant, and an equilibrium table. If the solution contains comparable concentrations of both members of a conjugate pair, which is the case at points prior to the equivalence point, it is a *buffer* problem and is solved using the Henderson-Hasselbalch equation. If the solution contains excess titrant, either a strong base or strong acid, it is simply a pH problem requiring only a concentration.

Practice Problem Ⓐ**TTEMPT** For the titration of 10.0 mL of 0.15 M acetic acid with 0.10 M sodium hydroxide, determine the pH when (a) 10.0 mL of base has been added, (b) 15.0 mL of base has been added, and (c) 20.0 mL of base has been added.

Practice Problem Ⓑ**UILD** For the titration of 25.0 mL of 0.20 M hydrofluoric acid with 0.20 M sodium hydroxide, determine the volume of base added when pH is (a) 2.85, (b) 3.15, and (c) 11.89. [To solve part (c), you may want to review the approach in Sample Problem 4.9.]

Practice Problem C**ONCEPTUALIZE** Which of the following graphs best represents the plot of pH versus volume of strong base added in the titration of a weak acid?

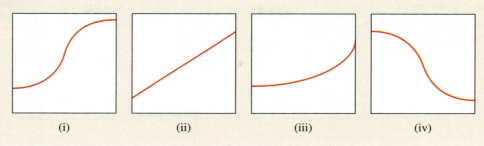

(i)   (ii)   (iii)   (iv)

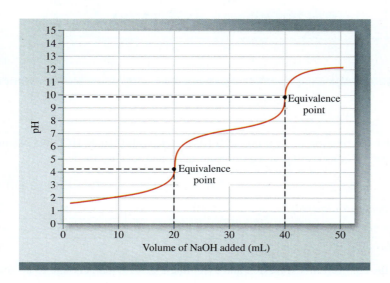

Figure 17.5 Titration curve of a polyprotic acid—strong base titration. A 0.100 *M* NaOH solution is added from a burette to 20.0 mL of a 0.100 *M* H$_2$SO$_3$ solution in an Erlenmeyer flask.

The titration curve for titration of a polyprotic acid (most of which are weak acids) with a strong base will, in theory, exhibit one equivalence point for each proton. Figure 17.5 shows the curve for the titration of 20.0 mL 0.100 *M* H$_2$SO$_3$ with 0.100 *M* NaOH.

Strong Acid–Weak Base Titrations

Consider the titration of HCl, a strong acid, with NH$_3$, a weak base:

$$HCl(aq) + NH_3(aq) \longrightarrow NH_4Cl(aq)$$

or:

$$H_3O^+(aq) + NH_3(aq) \longrightarrow NH_4^+(aq) + H_2O(l)$$

The pH at the equivalence point is less than 7 because the ammonium ion acts as a weak Brønsted acid:

$$NH_4^+(aq) + H_2O(l) \rightleftharpoons NH_3(aq) + H_3O^+(aq)$$

Because of the volatility of an aqueous ammonia solution, it is more convenient to use hydrochloric acid as the titrant (i.e., to add HCl solution from the burette). Figure 17.6 shows the titration curve for this experiment.

Analogous to the titration of a weak acid with a strong base, the initial pH is determined by the concentration and the K_b of ammonia:

$$NH_3(aq) + H_2O(l) \rightleftharpoons NH_4^+(aq) + OH^-(aq)$$

Figure 17.6 Titration curve of a strong acid–weak base titration. A 0.100 M HCl solution is added from a burette to 25.0 mL of a 0.100 M NH$_3$ solution in an Erlenmeyer flask. As a result of salt hydrolysis, the pH at the equivalence point is lower than 7.

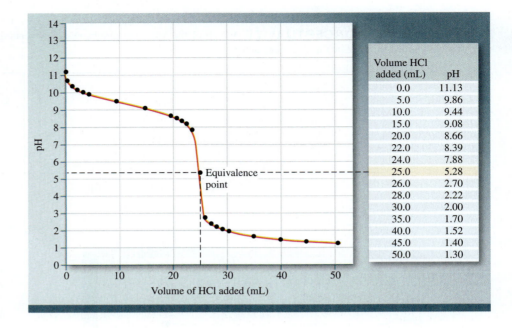

Volume HCl added (mL)	pH
0.0	11.13
5.0	9.86
10.0	9.44
15.0	9.08
20.0	8.66
22.0	8.39
24.0	7.88
25.0	5.28
26.0	2.70
28.0	2.22
30.0	2.00
35.0	1.70
40.0	1.52
45.0	1.40
50.0	1.30

Consider the titration of 25.0 mL of 0.10 M NH$_3$ with 0.10 M HCl. We calculate the initial pH by constructing an equilibrium table and solving for x:

$$NH_3(aq) + H_2O(l) \rightleftharpoons NH_4^+(aq) + OH^-(aq)$$

	NH$_3$(aq) + H$_2$O(l) $\rightleftharpoons$ NH$_4^+$(aq) + OH$^-$(aq)			
Initial concentration (M):	0.10	—	0	0
Change in concentration (M):	$-x$	—	$+x$	$+x$
Equilibrium concentration (M):	$0.10 - x$	—	x	x

Using the equilibrium expression and assuming that x is small enough to be neglected:

$$K_b = \frac{[NH_4^+][OH^-]}{[NH_3]} = \frac{(x)(x)}{0.10 - x} \approx \frac{x^2}{0.10} = 1.8 \times 10^{-5}$$

$$x^2 = 1.8 \times 10^{-6}$$

$$x = \sqrt{1.8 \times 10^{-6}} = 1.3 \times 10^{-3}$$

$$pH = 11.11$$

The pH at the equivalence point is calculated using the concentration and K_a of the conjugate base of NH$_3$, the NH$_4^+$ ion, and an equilibrium table.

Sample Problem 17.5 shows how this is done.

SAMPLE PROBLEM (17.5)

Calculate the pH at the equivalence point when 25.0 mL of 0.100 M NH$_3$ is titrated with 0.100 M HCl.

Strategy The reaction between NH$_3$ and HCl is:

$$NH_3(aq) + H_3O^+(aq) \longrightarrow NH_4^+(aq) + H_2O(l)$$

At the equivalence point, all the NH$_3$ has been converted to NH$_4^+$. Therefore, we must determine the concentration of NH$_4^+$ at the equivalence point and use the K_a for NH$_4^+$ to solve for pH using an equilibrium table.

Setup The solution originally contains $(0.100 \text{ mmol/mL})(25.0 \text{ mL}) = 2.50 \text{ mmol } NH_4^+$. At the equivalence point, 2.50 mmol of HCl has been added. The volume of 0.100 M HCl that contains 2.50 mmol is:

$$(\text{volume})(0.100 \text{ mmol/mL}) = 2.50 \text{ mmol}$$

$$\text{volume} = \frac{2.50 \text{ mmol}}{0.100 \text{ mmol/mL}} = 25.0 \text{ mL}$$

It takes 25.0 mL of titrant to reach the equivalence point, so the total solution volume is $25.0 + 25.0 = 50.0$ mL. At the equivalence point, all the NH_3 originally present has been converted to NH_4^+. The concentration of NH_4^+ is $(2.50 \text{ mmol})/(50.0 \text{ mL}) = 0.0500 \, M$. We must use this concentration as the starting concentration of ammonium ion in our equilibrium table.

Solution

$$NH_4^+(aq) + H_2O(l) \rightleftharpoons NH_3(aq) + H_3O^+(aq)$$

Initial concentration (M):	0.0500	—	0	0
Change in concentration (M):	$-x$	—	$+x$	$+x$
Equilibrium concentration (M):	$0.0500 - x$	—	x	x

The equilibrium expression is:

$$K_a = \frac{[NH_3][H_3O^+]}{[NH_4^+]} = \frac{(x)(x)}{0.0500 - x} \approx \frac{x^2}{0.0500} = 5.6 \times 10^{-10}$$

$$x^2 = 2.8 \times 10^{-11}$$

$$x = \sqrt{2.8 \times 10^{-11}} = 5.3 \times 10^{-6} \, M$$

$[H_3O^+] = x = 5.3 \times 10^6 \, M$. At equilibrium, therefore, pH $= -\log (5.3 \times 10^{-6}) = 5.28$.

THINK ABOUT IT

In the titration of a weak base with a strong acid, the species in solution at the equivalence point is the conjugate acid. Therefore, we should expect an *acidic* pH. Once all the NH_3 has been converted to NH_4^+, there is no longer anything in the solution to consume added acid. Thus, the pH after the equivalence point depends on the number of millimoles of H_3O^+ added and not consumed divided by the new total volume.

Practice Problem **A**TTEMPT Calculate the pH at the equivalence point in the titration of 50.0 mL of 0.10 M methylamine (see Table 16.7) with 0.20 M HCl.

Practice Problem **B**UILD A 50.0-mL quantity of a 0.20-M solution of one of the weak bases in Table 16.7 is titrated with 0.050 M HCl. At the equivalence point, the pH is 2.99. Identify the weak base.

Practice Problem **C**ONCEPTUALIZE Which of the following graphs best represents the plot of pH versus volume of strong acid added in the titration of a weak base?

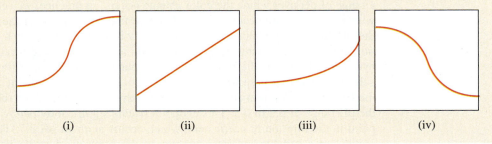

(i)   (ii)   (iii)   (iv)

Acid-Base Indicators

The equivalence point is the point at which the acid has been neutralized completely by the added base. The equivalence point in a titration can be determined by monitoring the pH over the course of the titration, or it can be determined using an *acid-base indicator*. An acid-base indicator is usually a weak organic acid or base for which the ionized and un-ionized forms are different colors.

Figure 17.7 Titration curve of a strong acid with a strong base (blue) and titration curve of a weak acid with a strong base (red). The indicator phenolphthalein can be used to determine the equivalence point of either titration. Methyl red can be used for the strong acid–strong base titration but cannot be used for the weak acid–strong base titration because its color change does not coincide with the steepest part of the curve.

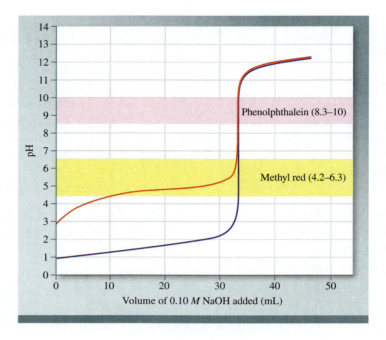

Consider a weak organic acid that we will refer to as HIn. To be an effective acid-base indicator, HIn and its conjugate base, In⁻, must have distinctly different colors. In solution, the acid ionizes to a small extent:

$$HIn(aq) + H_2O(l) \rightleftharpoons H_3O^+(aq) + In^-(aq)$$

In a sufficiently acidic medium, the ionization of HIn is suppressed according to Le Châtelier's principle, and the preceding equilibrium shifts to the left. In this case, the color of the solution will be that of HIn. In a basic medium, on the other hand, the equilibrium shifts to the right and the color of the solution will be that of the conjugate base, In⁻.

The ***endpoint*** of a titration is the point at which the color of the indicator changes. Not all indicators change color at the same pH, however, so the choice of indicator for a particular titration depends on the strength of the acid (and the base) used in the titration. To use the endpoint to determine the equivalence point of a titration, we must select an appropriate indicator.

The endpoint of an indicator does not occur at a specific pH; rather, there is a range of pH over which the color change occurs. In practice, we select an indicator whose color change occurs over a pH range that coincides with the steepest part of the titration curve. Consider the information in Figure 17.7, which shows the titration curves for hydrochloric acid and acetic acid—each being titrated with sodium hydroxide. Either of the indicators shown can be used for the titration of a strong acid with a strong base because both endpoints coincide with the steepest part of the HCl-NaOH titration curve. However, methyl red changes from red to yellow over the pH range of 4.2 to 6.3. This endpoint occurs significantly *before* the equivalence point in the titration of acetic acid, which occurs at about pH 8.7. Therefore, methyl red is *not* a suitable indicator for use in the titration of acetic acid with sodium hydroxide. Phenolphthalein, on the other hand, *is* a suitable indicator for the CH₃COOH-NaOH titration.

Many acid-base indicators are plant pigments. For example, boiling red cabbage in water extracts pigments that exhibit a variety of colors at different pH values (Figure 17.8).

Table 17.3 lists a number of indicators commonly used in acid-base titrations. The choice of indicator for a particular titration depends on the strength of the acid and base to be titrated.

Student Note: The *endpoint* is where the color changes. The *equivalence point* is where neutralization is complete. Experimentally, we use the endpoint to estimate the equivalence point.

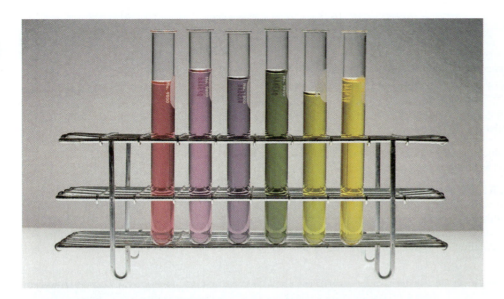

Figure 17.8 Solutions containing extracts of red cabbage (obtained by boiling the cabbage in water) produce different colors when treated with an acid and a base. The pH of the solutions increases from left to right.
Ken Karp/McGraw Hill

TABLE 17.3	Some Common Acid-Base Indicators		

| | Color | | |
Indicator	In Acid	In Base	pH Range
Thymol blue	Red	Yellow	1.2–2.8
Bromophenol blue	Yellow	Bluish purple	3.0–4.6
Methyl orange	Orange	Yellow	3.1–4.4
Methyl red	Red	Yellow	4.2–6.3
Chlorophenol blue	Yellow	Red	4.8–6.4
Bromothymol blue	Yellow	Blue	6.0–7.6
Cresol red	Yellow	Red	7.2–8.8
Phenolphthalein	Colorless	Reddish pink	8.3–10.0

Sample Problem 17.6 illustrates this point.

SAMPLE PROBLEM 17.6

Which indicator (or indicators) listed in Table 17.3 would you use for the acid-base titrations shown in (a) Figure 17.3, (b) Figure 17.4, and (c) Figure 17.6?

Strategy Determine the pH range that corresponds to the steepest part of each titration curve and select an indicator (or indicators) that changes color within that range.

Setup (a) The titration curve in Figure 17.3 is for the titration of a strong acid with a strong base. The steep part of the curve spans a pH range of about 4 to 10.

(b) Figure 17.4 shows the curve for the titration of a weak acid with a strong base. The steep part of the curve spans a pH range of about 7 to 10.

(c) Figure 17.6 shows the titration of a weak base with a strong acid. The steep part of the curve spans a pH range of about 7 to 3.

Solution (a) Most of the indicators listed in Table 17.3, with the exceptions of thymol blue, bromophenol blue, and methyl orange, would work for the titration of a strong acid with a strong base.

(b) Cresol red and phenolphthalein are suitable indicators.

(c) Bromophenol blue, methyl orange, methyl red, and chlorophenol blue are all suitable indicators.

(Continued on next page)

THINK ABOUT IT

If we don't select an appropriate indicator, the endpoint (color change) will not coincide with the equivalence point.

Practice Problem ATTEMPT Referring to Table 17.3, specify at least one indicator that would be suitable for the following titrations: (a) CH_3NH_2 with HBr, (b) HNO_3 with NaOH, (c) HNO_2 with KOH.

Practice Problem BUILD For which of the bases in Table 16.7 could you titrate a 0.1-M solution of base with 0.1 M nitric acid using the indicator thymol blue?

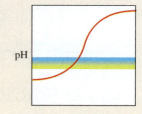

Practice Problem CONCEPTUALIZE The diagram shows the curve for titration of a particular weak acid with a strong base. Also shown is the region of color change for an acid-base indicator that is not a good choice for this titration. Suppose you were to titrate with a known concentration of base to determine the concentration of a sample of this weak acid. How would the acid concentration you determine from the titration be affected by the use of this indicator? Explain.

CHECKPOINT – SECTION 17.3   Acid-Base Titrations

17.3.1 For which of the following titrations will the pH at the equivalence point *not* be neutral? (Select all that apply.)

a) HCN with NaOH

b) HBr with NaOH

c) NH_3 with HNO_3

d) HNO_2 with KOH

e) HF with KOH

17.3.2 Calculate the pH at the equivalence point in the titration of 30 mL of 0.25 M CH_3COOH with 0.25 M KOH.

a) 7.00

b) 5.08

c) 8.92

d) 2.82

e) 11.18

17.3.3 Calculate the pH after the addition of 25 mL of 0.10 M NaOH to 50 mL of 0.10 M HF.

a) 3.15

b) 9.31

c) 12.52

d) 1.48

e) 11.87

17.3.4 Calculate the pH after the addition of 35 mL of 0.10 M NaOH to 30 mL of 0.10 M HCN.

a) 11.89

b) 2.11

c) 12.22

d) 1.78

e) 13.00

For questions 17.3.5 through 17.3.8, refer to the following diagrams. The solutions shown represent various points in the titration of the weak acid HA with NaOH. (For clarity, the sodium ions and water molecules are not shown.)

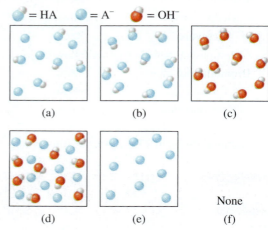

= HA   = A^- = OH^-

(a)   (b)   (c)

(d)   (e)   None (f)

17.3.5 Which diagram corresponds to the beginning of the titration, prior to the addition of any NaOH?

17.3.6 Which diagram corresponds to the equivalence point of the titration? (Ignore salt hydrolysis.)

17.3.7 Which diagram illustrates the solution *after* the equivalence point, when excess NaOH has been added?

17.3.8 Which diagram corresponds to the point at which the pH is equal to the pK_a of the weak acid?

17.3.9 Referring to the titration curve shown in Figure 17.5, what is the predominant species in solution when the pH is 9.86?

a) H_3O^+ c) SO_3^{2-} e) H_2SO_3

b) OH^- d) HSO_3^-

17.4 Solubility Equilibria

The solubility of ionic compounds is important in industry, medicine, and everyday life. For example, barium sulfate ($BaSO_4$), an insoluble compound that is opaque to X rays, is used to diagnose ailments of the digestive tract. Tooth decay begins when tooth enamel, which is mainly made of hydroxyapatite [$Ca_5(PO_4)_3OH$], is made more soluble in saliva by the presence of acid.

The general rules for predicting the solubility of ionic compounds in water were introduced in Section 4.2. The compounds described as "insoluble" in Chapter 4 [◄◄ Section 4.2, Table 4.3] are actually very *slightly* soluble—each to a different degree. While these guidelines are useful, they do not allow us to make quantitative predictions about how much of a given ionic compound will dissolve in water. To develop a quantitative approach, we must start with the principles of chemical equilibrium. Unless otherwise stated, the solvent is water in the following discussion and the temperature is 25°C.

Solubility Product Expression and K_{sp}

Consider a saturated solution of silver chloride that is in contact with undissolved solid silver chloride. The equilibrium can be represented as:

$$AgCl(s) \rightleftharpoons Ag^+(aq) + Cl^-(aq)$$

Although AgCl is not very soluble, *all* the AgCl that does dissolve in water dissociates completely into Ag^+ and Cl^- ions. We can write the equilibrium expression for the dissociation of AgCl as:

$$K_{sp} = [Ag^+][Cl^-]$$

where **K_{sp}** is called the **solubility product constant.** (K_{sp} is just another specially subscripted K_c, where "sp" stands for "solubility product.")

Because each AgCl unit contains only one Ag^+ and one Cl^- ion, its solubility product expression is particularly simple to write. Many ionic compounds dissociate into more than two ions. Table 17.4 lists a number of slightly soluble ionic compounds along with equations representing their dissolution equilibria and their solubility product constants. (Compounds deemed "soluble" by the solubility rules in Chapter 4 are not listed for the same reason we did not list K_a values for the strong acids in Table 16.6.) In general, the magnitude of K_{sp} indicates the solubility of an ionic compound—the smaller the K_{sp} value, the less soluble the compound. To make a direct comparison of K_{sp} values, however, we must compare salts with similar formulas, such as AgCl with ZnS (one cation, one anion) or CaF_2 with $Fe(OH)_2$ (one cation, two anions).

Student Note: A solubility product equilibrium expression is like any other equilibrium expression: K is equal to the concentrations of products over the concentrations of reactants, each raised to its coefficient from the balanced chemical equation.

Thus, for the process

$$MX_n(s) \rightleftharpoons M^{n+}(aq) + nX^-(aq)$$

the K_{sp} expression is

$$K_{sp} = [M^{n+}][X^-]^n$$

MX_n does not appear in the expression because, as for any heterogeneous equilibrium, the equilibrium expression does not include pure liquids or solids [◄◄ Section 15.3].

Calculations Involving K_{sp} and Solubility

There are two ways to express the solubility of a substance: *molar solubility,* which is the number of moles of solute in 1 L of a saturated solution (mol/L), and *solubility,* which is the number of grams of solute in 1 L of a saturated solution (g/L). Both of these expressions refer to concentrations of saturated solutions at a particular temperature (usually 25°C).

Often we know the value of K_{sp} for a compound and are asked to calculate the compound's molar solubility. The procedure for solving such a problem is essentially identical to the procedure for solving weak acid or weak base equilibrium problems:

1. Construct an equilibrium table.
2. Fill in what we know.
3. Figure out what we don't know.

Student Note: Be careful not to confuse the terms *solubility* and K_{sp}. Solubility is the concentration of a saturated solution. K_{sp} is an equilibrium constant.

TABLE 17.4	Solubility Products of Some Slightly Soluble Ionic Compounds at 25°C	
Compound	**Dissolution Equilibrium**	K_{sp}
Aluminum hydroxide	$Al(OH)_3(s) \rightleftharpoons Al^{3+}(aq) + 3OH^-(aq)$	1.8×10^{-33}
Barium carbonate	$BaCO_3(s) \rightleftharpoons Ba^{2+}(aq) + CO_3^{2-}(aq)$	8.1×10^{-9}
Barium fluoride	$BaF_2(s) \rightleftharpoons Ba^{2+}(aq) + 2F^-(aq)$	1.7×10^{-6}
Barium sulfate	$BaSO_4(s) \rightleftharpoons Ba^{2+}(aq) + SO_4^{2-}(aq)$	1.1×10^{-10}
Bismuth sulfide	$Bi_2S_3(s) \rightleftharpoons 2Bi^{3+}(aq) + 3S^{2-}(aq)$	1.6×10^{-72}
Cadmium sulfide	$CdS(s) \rightleftharpoons Cd^{2+}(aq) + S^{2-}(aq)$	8.0×10^{-28}
Calcium carbonate	$CaCO_3(s) \rightleftharpoons Ca^{2+}(aq) + CO_3^{2-}(aq)$	8.7×10^{-9}
Calcium fluoride	$CaF_2(s) \rightleftharpoons Ca^{2+}(aq) + 2F^-(aq)$	4.0×10^{-11}
Calcium hydroxide	$Ca(OH)_2(s) \rightleftharpoons Ca^{2+}(aq) + 2OH^-(aq)$	8.0×10^{-6}
Calcium phosphate	$Ca_3(PO_4)_2(s) \rightleftharpoons 3Ca^{2+}(aq) + 2PO_4^{3-}(aq)$	1.2×10^{-26}
Calcium sulfate	$CaSO_4(s) \rightleftharpoons Ca^{2+}(aq) + SO_4^{2-}(aq)$	2.4×10^{-5}
Chromium(III) hydroxide	$Cr(OH)_3(s) \rightleftharpoons Cr^{3+}(aq) + 3OH^-(aq)$	3.0×10^{-29}
Cobalt(II) sulfide	$CoS(s) \rightleftharpoons Co^{2+}(aq) + S^{2-}(aq)$	4.0×10^{-21}
Copper(I) bromide	$CuBr(s) \rightleftharpoons Cu^+(aq) + Br^-(aq)$	4.2×10^{-8}
Copper(I) iodide	$CuI(s) \rightleftharpoons Cu^+(aq) + I^-(aq)$	5.1×10^{-12}
Copper(II) hydroxide	$Cu(OH)_2(s) \rightleftharpoons Cu^{2+}(aq) + 2OH^-(aq)$	2.2×10^{-20}
Copper(II) sulfide	$CuS(s) \rightleftharpoons Cu^{2+}(aq) + S^{2-}(aq)$	6.0×10^{-37}
Iron(II) hydroxide	$Fe(OH)_2(s) \rightleftharpoons Fe^{2+}(aq) + 2OH^-(aq)$	1.6×10^{-14}
Iron(III) hydroxide	$Fe(OH)_3(s) \rightleftharpoons Fe^{3+}(aq) + 3OH^-(aq)$	1.1×10^{-36}
Iron(III) phosphate	$FePO_4(s) \rightleftharpoons Fe^{3+}(aq) + PO_4^{3-}(aq)$	1.3×10^{-22}
Iron(II) sulfide	$FeS(s) \rightleftharpoons Fe^{2+}(aq) + S^{2-}(aq)$	6.0×10^{-19}
Lead(II) bromide	$PbBr_2(s) \rightleftharpoons Pb^{2+}(aq) + 2Br^-(aq)$	6.6×10^{-6}
Lead(II) carbonate	$PbCO_3(s) \rightleftharpoons Pb^{2+}(aq) + CO_3^{2-}(aq)$	3.3×10^{-14}
Lead(II) chloride	$PbCl_2(s) \rightleftharpoons Pb^{2+}(aq) + 2Cl^-(aq)$	2.4×10^{-4}
Lead(II) chromate	$PbCrO_4(s) \rightleftharpoons Pb^{2+}(aq) + CrO_4^{2-}(aq)$	2.0×10^{-14}
Lead(II) fluoride	$PbF_2(s) \rightleftharpoons Pb^{2+}(aq) + 2F^-(aq)$	4.0×10^{-8}
Lead(II) iodide	$PbI_2(s) \rightleftharpoons Pb^{2+}(aq) + 2I^-(aq)$	1.4×10^{-8}
Lead(II) sulfate	$PbSO_4(s) \rightleftharpoons Pb^{2+}(aq) + SO_4^-(aq)$	1.8×10^{-8}
Lead(II) sulfide	$PbS(s) \rightleftharpoons Pb^{2+}(aq) + S^{2-}(aq)$	3.4×10^{-28}
Magnesium carbonate	$MgCO_3(s) \rightleftharpoons Mg^{2+}(aq) + CO_3^{2-}(aq)$	4.0×10^{-5}
Magnesium hydroxide	$Mg(OH)_2(s) \rightleftharpoons Mg^{2+}(aq) + 2OH^-(aq)$	1.2×10^{-11}
Manganese(II) sulfide	$MnS(s) \rightleftharpoons Mn^{2+}(aq) + S^{2-}(aq)$	3.0×10^{-14}
Mercury(I) bromide	$Hg_2Br_2(s) \rightleftharpoons Hg_2^{2+}(aq) + 2Br^-(aq)$	6.4×10^{-23}
Mercury(I) chloride	$Hg_2Cl_2(s) \rightleftharpoons Hg_2^{2+}(aq) + 2Cl^-(aq)$	3.5×10^{-18}
Mercury(I) sulfate	$Hg_2SO_4(s) \rightleftharpoons Hg_2^{2+}(aq) + SO_4^{2-}(aq)$	6.5×10^{-7}
Mercury(II) sulfide	$HgS(s) \rightleftharpoons Hg^{2+}(aq) + S^{2-}(aq)$	4.0×10^{-54}

TABLE 17.4	(Continued)		
Compound	**Dissolution Equilibrium**		K_{sp}
Nickel(II) sulfide	$NiS(s) \rightleftharpoons Ni^{2+}(aq) + S^{2-}(aq)$		1.4×10^{-24}
Silver bromide	$AgBr(s) \rightleftharpoons Ag^{+}(aq) + Br^{-}(aq)$		7.7×10^{-13}
Silver carbonate	$Ag_2CO_3(s) \rightleftharpoons 2Ag^{+}(aq) + CO_3^{2-}(aq)$		8.1×10^{-12}
Silver chloride	$AgCl(s) \rightleftharpoons Ag^{+}(aq) + Cl^{-}(aq)$		1.6×10^{-10}
Silver chromate	$Ag_2CrO_4(s) \rightleftharpoons 2Ag^{+}(aq) + CrO_4^{2-}(aq)$		1.2×10^{-12}
Silver iodide	$AgI(s) \rightleftharpoons Ag^{+}(aq) + I^{-}(aq)$		8.3×10^{-17}
Silver sulfate	$Ag_2SO_4(s) \rightleftharpoons 2Ag^{+}(aq) + SO_4^{2-}(aq)$		1.5×10^{-5}
Silver sulfide	$Ag_2S(s) \rightleftharpoons 2Ag^{+}(aq) + S^{2-}(aq)$		6.0×10^{-51}
Strontium carbonate	$SrCO_3(s) \rightleftharpoons Sr^{2+}(aq) + CO_3^{2-}(aq)$		1.6×10^{-9}
Strontium hydroxide	$Sr(OH)_2(s) \rightleftharpoons Sr^{2+}(aq) + 2OH^{-}(aq)$		3.2×10^{-4}
Strontium sulfate	$SrSO_4(s) \rightleftharpoons Sr^{2+}(aq) + SO_4^{2-}(aq)$		3.8×10^{-7}
Tin(II) sulfide	$SnS(s) \rightleftharpoons Sn^{2+}(aq) + S^{2-}(aq)$		1.0×10^{-26}
Zinc hydroxide	$Zn(OH)_2(s) \rightleftharpoons Zn^{2+}(aq) + 2OH^{-}(aq)$		1.8×10^{-14}
Zinc sulfide	$ZnS(s) \rightleftharpoons Zn^{2+}(aq) + S^{2-}(aq)$		3.0×10^{-23}

For example, the K_{sp} of silver bromide (AgBr) is 7.7×10^{-13}. We can construct an equilibrium table and fill in the starting concentrations of Ag^{+} and Br^{-} ions:

$$AgBr(s) \rightleftharpoons Ag^{+}(aq) + Br^{-}(aq)$$

Initial concentration (M):	—	0	0
Change in concentration (M):	—		
Equilibrium concentration (M):	—		

Let s be the molar solubility (in mol/L) of AgBr. Because one unit of AgBr yields one Ag^{+} cation and one Br^{-} anion, both $[Ag^{+}]$ and $[Br^{-}]$ are equal to s at equilibrium:

$$AgBr(s) \rightleftharpoons Ag^{+}(aq) + Br^{-}(aq)$$

Initial concentration (M):	—	0	0
Change in concentration (M):	—	$+s$	$+s$
Equilibrium concentration (M):	—	s	s

The equilibrium expression is:

$$K_{sp} = [Ag^{+}][Br^{-}]$$

Therefore:

$$7.7 \times 10^{-13} = (s)(s)$$

and:

$$s = \sqrt{7.7 \times 10^{-13}} = 8.8 \times 10^{-7} \, M$$

Thus, the molar solubility of AgBr is $8.8 \times 10^{-7} \, M$. Furthermore, we can express this solubility in g/L by multiplying the molar solubility by the molar mass of AgBr:

$$\frac{8.8 \times 10^{-7} \text{ mol AgBr}}{1 \text{ L}} \times \frac{187.8 \text{ g}}{1 \text{ mol AgBr}} = 1.7 \times 10^{-4} \text{ g/L}$$

Sample Problem 17.7 demonstrates this approach.

SAMPLE PROBLEM 17.7

Calculate the solubility of copper(II) hydroxide [$Cu(OH)_2$] in g/L.

Strategy Write the dissociation equation for $Cu(OH)_2$, and look up its K_{sp} value in Table 17.4. Solve for molar solubility using the equilibrium expression. Convert molar solubility to solubility in g/L using the molar mass of $Cu(OH)_2$.

Setup The equation for the dissociation of $Cu(OH)_2$ is:

$$Cu(OH)_2(s) \rightleftharpoons Cu^{2+}(aq) + 2OH^-(aq)$$

and the equilibrium expression is $K_{sp} = [Cu^{2+}][OH^-]^2$. According to Table 17.4, K_{sp} for $Cu(OH)_2$ is 2.2×10^{-20}. The molar mass of $Cu(OH)_2$ is 97.57 g/mol.

Solution The stoichiometry of the balanced dissociation equation indicates that the concentration of OH^- increases by twice as much as that of Cu^{2+}:

$$Cu(OH)_2(s) \rightleftharpoons Cu^{2+}(aq) + 2OH^-(aq)$$

		Cu^{2+}	OH^-
Initial concentration (M):	—	0	0
Change in concentration (M):	—	$+s$	$+2s$
Equilibrium concentration (M):	—	s	$2s$

Therefore:

$$2.2 \times 10^{-20} = (s)(2s)^2 = 4s^3$$

$$s = \sqrt[3]{\frac{2.2 \times 10^{-20}}{4}} = 1.8 \times 10^{-7} \, M$$

The molar solubility of $Cu(OH)_2$ is $1.8 \times 10^{-7} \, M$. Multiplying by its molar mass gives:

$$\text{solubility of } Cu(OH)_2 = \frac{1.8 \times 10^{-7} \, \text{mol } Cu(OH)_2}{1 \, L} \times \frac{97.57 \, g \, Cu(OH)_2}{1 \, \text{mol } Cu(OH)_2}$$

$$= 1.7 \times 10^{-5} \, g/L$$

THINK ABOUT IT

Common errors arise in this type of problem when students neglect to raise an entire term to the appropriate power. For example, $(2s)^2$ is equal to $4s^2$ (not $2s^2$).

Practice Problem A TTEMPT Calculate the molar solubility and the solubility in g/L of each salt at 25°C: (a) AgCl, (b) SnS, (c) $SrCO_3$.

Practice Problem B UILD Calculate the molar solubility and the solubility in g/L of each salt at 25°C: (a) PbF_2, (b) Ag_2CO_3, (c) Bi_2S_3.

Practice Problem C ONCEPTUALIZE The following diagrams represent solutions saturated with three different sparingly soluble ionic compounds. Which compound has the greatest molar solubility?

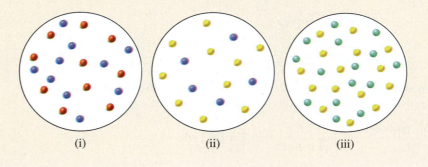

(i)   (ii)   (iii)

We can also use molar solubility to determine the value of K_{sp}. Sample Problem 17.8 illustrates this procedure.

SAMPLE PROBLEM 17.8

The solubility of calcium sulfate ($CaSO_4$) is measured experimentally and found to be 0.67 g/L. Calculate the value of K_{sp} for calcium sulfate.

Strategy Convert solubility to molar solubility using the molar mass of $CaSO_4$, and substitute the molar solubility into the equilibrium expression to determine K_{sp}.

Setup The molar mass of $CaSO_4$ is 136.2 g/mol. The molar solubility of $CaSO_4$ is:

$$\text{molar solubility of } CaSO_4 = \frac{0.67 \text{ g } CaSO_4}{1 \text{ L}} \times \frac{1 \text{ mol } CaSO_4}{136.2 \text{ g } CaSO_4}$$

$$s = 4.9 \times 10^{-3} \text{ mol/L}$$

The equation and the equilibrium expression for the dissociation of $CaSO_4$ are:

$$CaSO_4(s) \rightleftharpoons Ca^{2+}(aq) + SO_4^{2-}(aq) \quad \text{and} \quad K_{sp} = [Ca^{2+}][SO_4^{2-}]$$

Solution Substituting the molar solubility into the equilibrium expression gives:

$$K_{sp} = (s)(s) = (4.9 \times 10^{-3})^2 = 2.4 \times 10^{-5}$$

THINK ABOUT IT

The K_{sp} for $CaSO_4$ is relatively large (compared to many of the K_{sp} values in Table 17.4). In fact, sulfates are listed as soluble compounds in Table 4.2, but calcium sulfate is listed as an insoluble exception. Remember that the term *insoluble* really refers to compounds that are *slightly* soluble, and that different sources may differ with regard to how soluble a compound must be to be considered "soluble."

Practice Problem ATTEMPT Given the solubility, calculate the solubility product constant (K_{sp}) of each salt at 25°C: (a) $PbCrO_4$, $s = 4.0 \times 10^{-5}$ g/L; (b) BaC_2O_4, $s = 0.29$ g/L; (c) $MnCO_3$, $s = 4.2 \times 10^{-6}$ g/L.

Practice Problem BUILD Given the solubility, calculate the solubility product constant (K_{sp}) of each salt at 25°C: (a) Ag_2SO_3, $s = 4.6 \times 10^{-3}$ g/L; (b) Hg_2I_2, $s = 1.5 \times 10^{-7}$ g/L; (c) $Zn_3(PO_4)_2$, $s = 5.9 \times 10^{-5}$ g/L.

Practice Problem CONCEPTUALIZE Which compound in Practice Problem 17.7C has the largest K_{sp}?

Predicting Precipitation Reactions

For the dissociation of an ionic solid in water, any one of the following conditions may exist: (1) the solution is unsaturated, (2) the solution is saturated, or (3) the solution is supersaturated. For concentrations of ions that do not correspond to equilibrium conditions, we use the reaction quotient (Q) [◄◄ Section 15.2] to predict when a precipitate will form. Note that Q has the same form as K_{sp} except that the concentrations of ions are not equilibrium concentrations. For example, if we mix a solution containing Ag^+ ions with one containing Cl^- ions, we write:

$$Q = [Ag^+]_i[Cl^-]_i$$

The subscript "i" denotes that these are *initial* concentrations and do not necessarily correspond to those at equilibrium. If Q is less than or equal to K_{sp}, no precipitate will form. If Q is greater than K_{sp}, AgCl will precipitate out. (Precipitation will continue until the product of the ion concentrations is equal to K_{sp}.)

The ability to predict whether precipitation will occur often has practical value. In industrial and laboratory preparations, we can adjust the concentrations of ions until the ion product exceeds K_{sp} to obtain a desired ionic compound (in the form of a precipitate). The ability to predict precipitation reactions is also useful in medicine. Kidney stones, which can be extremely painful, consist largely of calcium oxalate

Student Hot Spot

Student data indicate you may struggle with determining K_{sp} values. Access the eBook to view additional Learning Resources on this topic.

(CaC_2O_4, $K_{sp} = 2.3 \times 10^{-9}$). The normal physiological concentration of calcium ions in blood plasma is about 5×10^{-3} M. Oxalate ions ($C_2O_4^{2-}$), derived from oxalic acid present in vegetables such as rhubarb and spinach, react with the calcium ions to form insoluble calcium oxalate, which can gradually build up in the kidneys. Proper adjustment of a patient's diet can help to reduce precipitate formation.

Sample Problem 17.9 demonstrates the steps involved in predicting precipitation reactions.

SAMPLE PROBLEM (17.9)

Predict whether a precipitate will form when each of the following is added to 650 mL of 0.0080 M K_2SO_4: (a) 250 mL of 0.0040 M $BaCl_2$; (b) 175 mL of 0.15 M $AgNO_3$; (c) 325 mL of 0.25 M $Sr(NO_3)_2$. (Assume volumes are additive.)

Strategy For each part, identify the compound that might precipitate and look up its K_{sp} value in Table 17.4 or Appendix 3. Determine the concentrations of each compound's constituent ions, and use them to determine the value of the reaction quotient, Q_{sp}; then compare each reaction quotient with the value of the corresponding K_{sp}. If the reaction quotient is greater than K_{sp}, a precipitate will form.

Setup The compounds that might precipitate and their K_{sp} values are (a) $BaSO_4$, $K_{sp} = 1.1 \times 10^{-10}$; (b) Ag_2SO_4, $K_{sp} = 1.5 \times 10^{-5}$; (c) $SrSO_4$, $K_{sp} = 3.8 \times 10^{-7}$.

Solution (a) Concentrations of the constituent ions of $BaSO_4$ are:

$$[Ba^{2+}] = \frac{250 \text{ mL} \times 0.0040 \ M}{650 \text{ mL} + 250 \text{ mL}} = 0.0011 \ M \quad \text{and} \quad [SO_4^{2-}] = \frac{650 \text{ mL} \times 0.0080 \ M}{650 \text{ mL} + 250 \text{ mL}} = 0.0058 \ M$$

Using these concentrations in the equilibrium expression, $[Ba^{2+}][SO_4^{2-}]$, gives a reaction quotient of $(0.0011)(0.0058) = 6.4 \times 10^{-6}$, which is greater than the K_{sp} of $BaSO_4(1.1 \times 10^{-10})$. Therefore, $BaSO_4$ will precipitate.

(b) Concentrations of the constituent ions of Ag_2SO_4 are:

$$[Ag^+] = \frac{175 \text{ mL} \times 0.15 \ M}{650 \text{ mL} + 175 \text{ mL}} = 0.0032 \ M \quad \text{and} \quad [SO_4^{2-}] = \frac{650 \text{ mL} \times 0.0080 \ M}{650 \text{ mL} + 175 \text{ mL}} = 0.0063 \ M$$

Using these concentrations in the equilibrium expression, $[Ag^+]^2[SO_4^{2-}]$, gives a reaction quotient of $(0.032)^2(0.0063) = 6.5 \times 10^{-6}$, which is less than the K_{sp} of $Ag_2SO_4(1.5 \times 10^{-5})$. Therefore, Ag_2SO_4 will not precipitate.

(c) Concentrations of the constituent ions of $SrSO_4$ are:

$$[Sr^{2+}] = \frac{325 \text{ mL} \times 0.25 \ M}{650 \text{ mL} + 325 \text{ mL}} = 0.083 \ M \quad \text{and} \quad [SO_4^{2-}] = \frac{650 \text{ mL} \times 0.0080 \ M}{650 \text{ mL} + 325 \text{ mL}} = 0.0053 \ M$$

Using these concentrations in the equilibrium expression, $[Sr^{2+}][SO_4^{2-}]$, gives a reaction quotient of $(0.083)(0.0053) = 4.4 \times 10^{-4}$, which is greater than the K_{sp} of $SrSO_4(3.8 \times 10^{-7})$. Therefore, $SrSO_4$ will precipitate.

THINK ABOUT IT

Students sometimes have difficulty deciding what compound might precipitate. Begin by writing down the constituent ions in the two solutions before they are combined. Consider the two possible combinations: the cation from the first solution and the anion from the second, or vice versa. You can consult the information in Tables 4.2 and 4.3 to determine whether one of the combinations is insoluble. Also keep in mind that only an *insoluble* salt will have a tabulated K_{sp} value.

Practice Problem ATTEMPT Predict whether a precipitate will form from each of the following combinations: (a) 25 mL of 1×10^{-5} M $Co(NO_3)_2$ and 75 mL of 5×10^{-4} M Na_2S; (b) 500 mL of 7.5×10^{-4} M $AlCl_3$ and 100 mL of 1.7×10^{-5} M $Hg_2(NO_3)_2$; (c) 1.5 L of 0.025 M $BaCl_2$ and 1.25 L of 0.014 M $Pb(NO_3)_2$.

Practice Problem BUILD What is the maximum mass (in grams) of each of the following soluble salts that can be added to 150 mL of 0.050 M $BaCl_2$ without causing a precipitate to form: (a) $(NH_4)_2SO_4$, (b) $Pb(NO_3)_2$, (c) NaF? (Assume that the addition of solid causes no change in volume.)

Practice Problem **C**ONCEPTUALIZE The first two diagrams represent a saturated solution of the slightly soluble salt MA and a solution of the soluble salt NH_4A, respectively. Which of the solutions [(i)–(iv)] of the soluble salt MNO_3 can be added to the solution of NH_4A without causing a precipitate to form? Assume that the volumes of all solutions are equal, and that they are additive when combined. (For clarity, the water molecules, ammonium ions, and nitrate ions are not shown.)

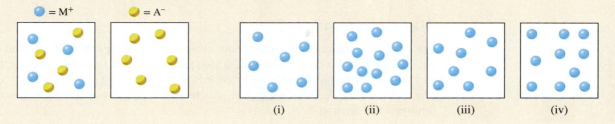

CHECKPOINT – SECTION 17.4   Solubility Equilibria

17.4.1 Using the K_{sp} for aluminum hydroxide $[Al(OH)_3]$, calculate its molar solubility.

a) $2.9 \times 10^{-9}\ M$

b) $7.7 \times 10^{-12}\ M$

c) $4.2 \times 10^{-17}\ M$

d) $8.4 \times 10^{-12}\ M$

e) $3.8 \times 10^{-9}\ M$

17.4.2 What precipitate will form if $0.10\text{-}M$ solutions of $Pb(NO_3)_2$ and NaI are mixed?

a) $Pb(NO_3)_2$

b) NaI

c) PbI_2

d) $NaNO_3$

e) None

17.4.3 The diagrams represent saturated solutions of MX, MY, and MZ, each in equilibrium with its solid. Each solution also contains the soluble nitrate salt of M^{n+}. Arrange the solutions in order of increasing K_{sp}. (For clarity, the water molecules and nitrate ions are not shown.)

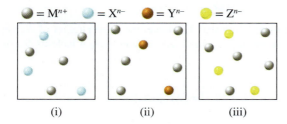

a) i < iii < ii

b) ii < iii < i

c) i = ii < iii

d) iii < ii = i

e) ii < i < iii

17.4.4 The diagram represents a saturated solution of the salt MA_2. Which soluble salt will cause the precipitation of the greatest quantity of MA_2 when 0.1 mol is dissolved in the saturated solution? (For clarity, the water molecules are not shown.)

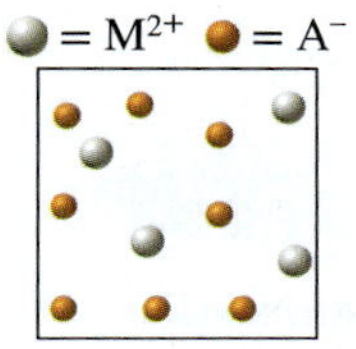

a) NaA   c) AlA_3 e) $M_3(PO_4)_2$

b) BaA_2 d) $M(NO_3)_2$

17.4.5 Diagrams (1) and (2) represent a saturated solution of the slightly soluble salt M_2B and a solution of the soluble salt MNO_3, respectively. Which of the solutions of the soluble salt Na_2B can be added to the solution of MNO_3 without causing a precipitate to form? Select all that apply. Assume that the solution volumes are equal, and that they are additive when combined. (For clarity, the water molecules, nitrate ions, and sodium ions are not shown.)

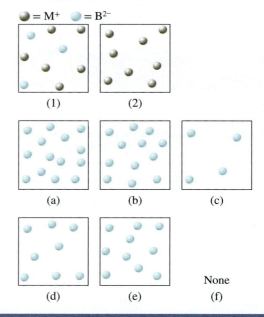

17.5 **Factors Affecting Solubility**

In this section, we examine the effect of several factors on solubility, including the common ion effect, pH, and the formation of complex ions.

The Common Ion Effect

The solubility product is an equilibrium constant, and precipitation of an ionic compound from solution occurs whenever the ion product exceeds the K_{sp} for that substance. In a saturated solution of AgCl, for example, the ion product $[Ag^+][Cl^-]$ is equal to K_{sp}. The solubility of AgCl in water can be calculated as follows, using the procedure introduced in Section 17.4:

$$1.6 \times 10^{-10} = [Ag^+][Cl^-]$$

In a solution in which AgCl is the only solute, $[Ag^+] = [Cl^-]$. Therefore:

$$1.6 \times 10^{-10} = s^2$$

and $s = 1.3 \times 10^{-5}$ M. Thus, the solubility of AgCl in water at 25°C is 1.3×10^{-5} M.

Now suppose we want to determine the solubility of AgCl in a solution already containing a solute that has an ion in common with AgCl. For example, consider dissolving AgCl in a 0.10-M solution of $AgNO_3$. In this case, the Ag^+ and Cl^- concentrations at equilibrium will not be equal. In fact, the Ag^+ ion concentration will be equal to 0.10 M *plus* the concentration contributed by AgCl. The equilibrium expression is:

$$1.6 \times 10^{-10} = [Ag^+][Cl^-] = (0.10 + s)(s)$$

Note that s still represents the concentration of Cl^- ion at equilibrium—and the *solubility* of AgCl. Because we expect s to be very small, we can simplify this calculation as follows:

$$(0.10 + s)\,M \approx 0.10\,M$$

Therefore:

$$1.6 \times 10^{-10} = 0.10s$$

and $s = 1.6 \times 10^{-9}$ M. Thus, AgCl is *significantly* less soluble in 0.10 M $AgNO_3$ than in pure water—due to the common ion effect. Figure 17.9 illustrates the common ion effect.

Sample Problem 17.10 shows how the common ion effect affects solubility.

Animation
Figure 17.9, Common Ion Effect.

> 🔊 **Student Hot Spot**
>
> Student data indicate you may struggle with the changes in solubility caused by the common ion effect. Access the eBook to view additional Learning Resources on this topic.

Student Note: The common ion effect is an example of Le Châtelier's principle [◄◄ Section 15.5].

SAMPLE PROBLEM (17.10)

Calculate the molar solubility of silver chloride in a solution that is 6.5×10^{-3} M in silver nitrate.

Strategy Silver nitrate is a strong electrolyte that dissociates completely in water. Therefore, the concentration of Ag^+ before any AgCl dissolves is 6.5×10^{-3} M. Use the equilibrium expression, the K_{sp} for AgCl, and an equilibrium table to determine how much AgCl will dissolve.

Setup The dissolution equilibrium and the equilibrium expression are:

$$AgCl(s) \rightleftharpoons Ag^+(aq) + Cl^-(aq) \qquad 1.6 \times 10^{-10} = [Ag^+][Cl^-]$$

Solution

$AgCl(s) \rightleftharpoons Ag^+(aq) \; + \; Cl^-(aq)$

		Ag^+	Cl^-
Initial concentration (M):	—	6.5×10^{-3}	0
Change in concentration (M):	—	$+s$	$+s$
Equilibrium concentration (M):	—	$6.5 \times 10^{-3} + s$	s

Substituting these concentrations into the equilibrium expression gives:

$$1.6 \times 10^{-10} = (6.5 \times 10^{-3} + s)(s)$$

We expect s to be very small, so:

$$6.5 \times 10^{-3} + s \approx 6.5 \times 10^{-3}$$

and:

$$1.6 \times 10^{-10} = (6.5 \times 10^{-3})(s)$$

Thus:

$$s = \frac{1.6 \times 10^{-10}}{6.5 \times 10^{-3}} = 2.5 \times 10^{-8}\ M$$

Therefore, the molar solubility of AgCl in $6.5 \times 10^{-3}\ M$ AgNO$_3$ is $2.5 \times 10^{-8}\ M$.

THINK ABOUT IT

The molar solubility of AgCl in water is $\sqrt{1.6 \times 10^{-10}} = 1.3 \times 10^{-5}\ M$. The presence of $6.5 \times 10^{-3}\ M$ AgNO$_3$ reduces the solubility of AgCl by a factor of ~500.

Practice Problem ATTEMPT Calculate the molar solubility of AgI in (a) pure water and (b) 0.0010 M NaI.

Practice Problem BUILD Arrange the following salts in order of increasing molar solubility in 0.0010 M AgNO$_3$: AgBr, Ag$_2$CO$_3$, AgCl, AgI, Ag$_2$S.

Practice Problem CONCEPTUALIZE The diagram on the left shows a saturated solution of a slightly soluble ionic compound. In the diagram on the right, enough of the nitrate salt of the cation has been added to increase the concentration of cations (yellow). How many anions (blue) must be included in the second diagram for it to correctly represent the solution after the addition?

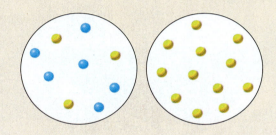

pH

The solubility of a substance can also depend on the pH of the solution. Consider the solubility equilibrium of magnesium hydroxide:

$$\mathrm{Mg(OH)_2}(s) \rightleftharpoons \mathrm{Mg^{2+}}(aq) + 2\mathrm{OH^-}(aq)$$

According to Le Châtelier's principle, adding OH$^-$ ions (increasing the pH) shifts the equilibrium from right to left, thereby decreasing the solubility of Mg(OH)$_2$. (This is actually another example of the common ion effect.) On the other hand, adding H$^+$ ions (decreasing the pH) shifts the equilibrium from left to right, and the solubility of Mg(OH)$_2$ *increases*. Thus, insoluble bases tend to dissolve in acidic solutions. Similarly, insoluble acids tend to dissolve in basic solutions.

To examine the effect of pH on the solubility of Mg(OH)$_2$, we first calculate the pH of a saturated Mg(OH)$_2$ solution:

$$K_{\mathrm{sp}} = (s)(2s)^2 = 4s^3$$

$$4s^3 = 1.2 \times 10^{-11}$$

$$s^3 = 3.0 \times 10^{-12}$$

$$s = 1.4 \times 10^{-4}\ M$$

At equilibrium, therefore:

$$[\mathrm{OH^-}] = 2(1.4 \times 10^{-4}\ M) = 2.8 \times 10^{-4}\ M$$

$$\mathrm{pOH} = -\log(2.8 \times 10^{-4}) = 3.55$$

$$\mathrm{pH} = 14.00 - 3.55 = 10.45$$

Figure 17.9

Common Ion Effect

We prepare a saturated solution by adding AgCl to water and stirring.

In the resulting saturated solution, the concentrations of Ag^+ and Cl^- are equal, and the product of their concentrations is equal to K_{sp}.

$$[Ag^+][Cl^-] = 1.6 \times 10^{-10}$$

Therefore, the concentrations are

$$[Ag^+] = 1.3 \times 10^{-5}\ M$$
$$\text{and } [Cl^-] = 1.3 \times 10^{-5}\ M$$

Neither AgCl nor a saturated solution of AgCl is purple. The color has been used in this illustration to clarify the process.

START

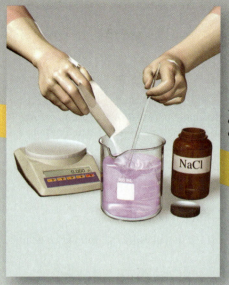

After filtering off the solid AgCl, we dissolve enough NaCl to make the concentration of Cl⁻ = 1.0 M.

Because the concentration of Cl⁻ is now larger, the product of Ag⁺ and Cl⁻ concentrations is no longer equal to K_{sp}.

$$[Ag^+][Cl^-] = (1.3 \times 10^{-5}\ M)(1.0\ M) > 1.6 \times 10^{-10}$$

In any solution saturated with AgCl at 25°C, the product of [Ag⁺] and [Cl⁻] must equal the K_{sp} of AgCl. Therefore, AgCl will precipitate until the product of ion concentrations is again 1.6×10^{-10}.

Note that this causes nearly all the dissolved AgCl to precipitate. With a Cl⁻ concentration of 1.0 M, the highest possible concentration of Ag⁺ is $1.6 \times 10^{-10}\ M$.

$$[Ag^+](1.0\ M) = 1.6 \times 10^{-10}$$
$$\text{therefore, } [Ag^+] = 1.6 \times 10^{-10}\ M$$

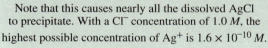

The amount of AgCl precipitated is exaggerated for emphasis. The actual amount of AgCl would be extremely small.

(See Visualizing Chemistry questions VC17.5–VC17.8 on page 903.)

What's the point?

When two salts contain the same ion, the ion they both contain is called the "common ion." The solubility of a slightly soluble salt such as AgCl can be decreased by the addition of a *soluble* salt with a common ion. In this example, AgCl is precipitated by adding NaCl. AgCl could also be precipitated by adding a soluble salt containing the Ag⁺ ion, such as $AgNO_3$.

In a solution with a pH of *less* than 10.45, the solubility of $Mg(OH)_2$ would increase. The dissolution process and the effect of additional H_3O^+ ions are summarized as follows:

$$Mg(OH)_2(s) \rightleftharpoons Mg^{2+}(aq) + 2\cancel{OH^-(aq)}$$

$$2H_3O^+(aq) + 2\cancel{OH^-(aq)} \longrightarrow 4H_2O(l)$$

$$Overall: \quad Mg(OH)_2(s) + 2H_3O^+(aq) \rightleftharpoons Mg^{2+}(aq) + 4H_2O(l)$$

If the pH of the medium were higher than 10.45, $[OH^-]$ would be higher and the solubility of $Mg(OH)_2$ would decrease because of the common ion (OH^-) effect.

The pH also influences the solubility of salts that contain a basic anion. For example, the solubility equilibrium for BaF_2 is:

$$BaF_2(s) \rightleftharpoons Ba^{2+}(aq) + 2F^-(aq)$$

and:

$$K_{sp} = [Ba^{2+}][F^-]^2$$

In an acidic medium, the high $[H_3O^+]$ will shift the following equilibrium to the left, consuming F^-:

$$HF(aq) + H_2O(l) \rightleftharpoons H_3O^+(aq) + F^-(aq)$$

As the concentration of F^- decreases, the concentration of Ba^{2+} must increase to satisfy the equality $K_{sp} = [Ba^{2+}][F^-]^2$ and maintain the state of equilibrium. Thus, more BaF_2 dissolves. The process and the effect of pH on the solubility of BaF_2 can be summarized as follows:

$$BaF_2(s) \rightleftharpoons Ba^{2+}(aq) + 2\cancel{F^-(aq)}$$

$$2H_3O^+(aq) + 2\cancel{F^-(aq)} \longrightarrow 2HF(aq) + 2H_2O(l)$$

$$Overall: \quad BaF_2(s) + 2H_3O^+(aq) \rightleftharpoons Ba^{2+}(aq) + 2HF(aq) + 2H_2O(l)$$

> **Student Hot Spot**
>
> Student data indicate you may struggle with the effect of pH on solubility. Access the eBook to view additional Learning Resources on this topic.

The solubilities of salts containing anions that do *not* hydrolyze, such as Cl^-, Br^-, and NO_3^-, are unaffected by pH.

Sample Problem 17.11 demonstrates the effect of pH on solubility.

SAMPLE PROBLEM (17.11)

Which of the following compounds will be more soluble in acidic solution than in water: (a) CuS, (b) AgCl, (c) PbSO$_4$?

Strategy For each salt, write the dissociation equilibrium equation and determine whether it produces an anion that will react with H_3O^+. Only an anion that is the conjugate base of a weak acid will react with H_3O^+.

Setup
(a) $CuS(s) \rightleftharpoons Cu^{2+}(aq) + S^{2-}(aq)$
S^{2-} is the conjugate base of the weak acid HS^-. S^{2-} reacts with H_3O^+ as follows:

$$S^{2-}(aq) + H_3O^+(aq) \longrightarrow HS^-(aq) + H_2O(l)$$

(b) $AgCl(s) \rightleftharpoons Ag^+(aq) + Cl^-(aq)$
Cl^- is the conjugate base of the strong acid HCl. Cl^- does not react with H_3O^+.

(c) $PbSO_4(s) \rightleftharpoons Pb^{2+}(aq) + SO_4^{2-}(aq)$
SO_4^{2-} is the conjugate base of the weak acid HSO_4^-. It reacts with H_3O^+ as follows:

$$SO_4^{2-}(aq) + H_3O^+(aq) \longrightarrow HSO_4^-(aq) + H_2O(l)$$

A salt that produces an anion that reacts with H_3O^+ will be more soluble in acid than in water.

Solution CuS and PbSO$_4$ are more soluble in acid than in water. (AgCl is no more or less soluble in acid than in water.)

THINK ABOUT IT

When a salt dissociates to give the conjugate base of a weak acid, H_3O^+ ions in an acidic solution *consume* a product (base) of the dissolution. This drives the equilibrium to the right (more solid dissolves) according to Le Châtelier's principle.

Practice Problem ATTEMPT Determine if the following compounds are more soluble in acidic solution than in pure water: (a) $Ca(OH)_2$, (b) $Mg_3(PO_4)_2$, (c) $PbBr_2$.

Practice Problem BUILD Other than those in Sample Problem 17.11 and those in Practice Problem A, list three salts that are more soluble in acidic solution than in pure water.

Practice Problem CONCEPTUALIZE If an ionic compound's solubility is affected by the presence of acid in solution, will its solubility necessarily also be affected by the presence of base? Explain.

Complex Ion Formation

A *complex ion* is an ion containing a central metal cation bonded to one or more molecules or ions. Complex ions are crucial to many chemical and biological processes. Here we consider the effect of complex ion formation on solubility. In Chapter 22 we discuss the chemistry of complex ions in more detail.

Student Note: Lewis acid-base reactions in which a metal cation combines with a Lewis base result in the formation of complex ions.

Transition metals have a particular tendency to form complex ions. For example, a solution of cobalt(II) chloride ($CoCl_2$) is pink because of the presence of the $Co(H_2O)_6^{2+}$ ions (Figure 17.10). When HCl is added, the solution turns blue because the complex ion $CoCl_4^{2-}$ forms:

$$Co^{2+}(aq) + 4Cl^-(aq) \rightleftharpoons CoCl_4^{2-}(aq)$$

Copper(II) sulfate ($CuSO_4$) dissolves in water to produce a blue solution. The hydrated copper(II) ions are responsible for this color; many other sulfates (e.g., Na_2SO_4) are colorless. Adding a few drops of concentrated ammonia solution to a $CuSO_4$ solution causes the formation of a light-blue precipitate, copper(II) hydroxide:

$$Cu^{2+}(aq) + 2OH^-(aq) \longrightarrow Cu(OH)_2(s)$$

Figure 17.10 (Left) An aqueous cobalt(II) chloride solution. The pink color is due to the presence of $Co(H_2O)_6^{2+}$ ions. (Right) After the addition of HCl solution, the solution turns blue because of the formation of the complex $CoCl_4^{2-}$ ions.
Ken Karp/McGraw Hill

Figure 17.11 (Left) An aqueous solution of copper(II) sulfate. (Center) After the addition of a few drops of concentrated aqueous ammonia solution, a light-blue precipitate of $Cu(OH)_2$ is formed. (Right) When more concentrated aqueous ammonia solution is added, the $Cu(OH)_2$ precipitate dissolves to form the dark-blue complex ion $Cu(NH_3)_4^{2+}$.

Ken Karp/McGraw Hill

The OH^- ions are supplied by the ammonia solution. If more NH_3 is added, the blue precipitate redissolves to produce a beautiful dark-blue solution, this time due to the formation of the complex ion $Cu(NH_3)_4^{2+}$ (Figure 17.11):

$$Cu(OH)_2(s) + 4NH_3(aq) \rightleftharpoons Cu(NH_3)_4^{2+}(aq) + 2OH^-(aq)$$

Thus, the formation of the complex ion $Cu(NH_3)_4^{2+}$ increases the solubility of $Cu(OH)_2$.

A measure of the tendency of a metal ion to form a particular complex ion is given by the **formation constant (K_f)** (also called the stability constant), which is the equilibrium constant for the complex ion formation. The larger K_f is, the more stable the complex ion is. Table 17.5 lists the formation constants of a number of complex ions.

The formation of the $Cu(NH_3)_4^{2+}$ ion can be expressed as:

$$Cu^{2+}(aq) + 4NH_3(aq) \rightleftharpoons Cu(NH_3)_4^{2+}(aq)$$

TABLE 17.5	Formation Constants of Selected Complex Ions in Water at 25°C	
Complex Ion	**Equilibrium Expression**	**Formation Constant (K_f)**
$Ag(NH_3)_2^+$	$Ag^+ + 2NH_3 \rightleftharpoons Ag(NH_3)_2^+$	1.5×10^7
$Ag(CN)_2^-$	$Ag^+ + 2CN^- \rightleftharpoons Ag(CN)_2^-$	1.0×10^{21}
$Cu(CN)_4^{2-}$	$Cu^{2+} + 4CN^- \rightleftharpoons Cu(CN)_4^{2-}$	1.0×10^{25}
$Cu(NH_3)_4^{2+}$	$Cu^{2+} + 4NH_3 \rightleftharpoons Cu(NH_3)_4^{2+}$	5.0×10^{13}
$Cd(CN)_4^{2-}$	$Cd^{2+} + 4CN^- \rightleftharpoons Cd(CN)_4^{2-}$	7.1×10^{16}
CdI_4^{2-}	$Cd^{2+} + 4I^- \rightleftharpoons CdI_4^{2-}$	2.0×10^6
$HgCl_4^{2-}$	$Hg^{2+} + 4Cl^- \rightleftharpoons HgCl_4^{2-}$	1.7×10^{16}
HgI_4^{2-}	$Hg^{2+} + 4I^- \rightleftharpoons HgI_4^{2-}$	2.0×10^{30}
$Hg(CN)_4^{2-}$	$Hg^{2+} + 4CN^- \rightleftharpoons Hg(CN)_4^{2-}$	2.5×10^{41}
$Co(NH_3)_6^{3+}$	$Co^{3+} + 6NH_3 \rightleftharpoons Co(NH_3)_6^{3+}$	5.0×10^{31}
$Zn(NH_3)_4^{2+}$	$Zn^{2+} + 4NH_3 \rightleftharpoons Zn(NH_3)_4^{2+}$	2.9×10^9
$Cr(OH)_4^-$	$Cr^{3+} + 4OH^- \rightleftharpoons Cr(OH)_4^-$	8×10^{29}

The corresponding formation constant is:

$$K_f = \frac{[Cu(NH_3)_4^{2+}]}{[Cu^{2+}][NH_3]^4} = 5.0 \times 10^{13}$$

The large value of K_f in this case indicates that the complex ion is very stable in solution and accounts for the very low concentration of copper(II) ions at equilibrium.

Recall that K for the sum of two reactions is the product of the individual K values [◀◀ Section 15.3]. The dissolution of silver chloride is represented by the equation:

$$AgCl(s) \rightleftharpoons Ag^+(aq) + Cl^-(aq)$$

The sum of this equation and the one representing the formation of $Ag(NH_3)_2^+$ is:

$AgCl(s) \rightleftharpoons Ag^+(aq) + Cl^-(aq)$	$K_{sp} = 1.6 \times 10^{-10}$
$\underline{Ag^+(aq) + 2NH_3(aq) \rightleftharpoons Ag(NH_3)_2^+(aq)}$	$K_f = 1.5 \times 10^7$
$AgCl(s) + 2NH_3(aq) \rightleftharpoons Ag(NH_3)_2^+(aq) + Cl^-(aq)$	

and the corresponding equilibrium constant is $(1.6 \times 10^{-21})(1.5 \times 10^7) = 2.4 \times 10^{-3}$. This is significantly larger than the K_{sp} value, indicating that much more AgCl will dissolve in the presence of aqueous ammonia than in pure water. In general, the effect of complex ion formation generally is to *increase* the solubility of a substance.

Finally, there is a class of hydroxides, called amphoteric hydroxides, which can react with both acids and bases. Examples are $Al(OH)_3$, $Pb(OH)_2$, $Cr(OH)_3$, $Zn(OH)_2$, and $Cd(OH)_2$. $Al(OH)_3$ reacts with acids and bases as follows:

$$Al(OH)_3(s) + 3H_3O^+(aq) \longrightarrow Al^{3+}(aq) + 6H_2O(l)$$

$$Al(OH)_3(s) + OH^-(aq) \rightleftharpoons Al(OH)_4^-(aq)$$

The increase in solubility of $Al(OH)_3$ in a basic medium is the result of the formation of the complex ion $Al(OH)_4^-$ in which $Al(OH)_3$ acts as the Lewis acid and OH^- acts as the Lewis base. Other amphoteric hydroxides react similarly with acids and bases.

Many of the equilibrium problems we encounter involve equilibrium constants, such as K_a, K_b, and K_{sp}, that are very small. Often, this enables us to neglect the unknown x in the denominator of the equilibrium expression, which simplifies the math necessary to solve the problem [◀◀ Section 16.5]. The solution of an equilibrium problem involving complex ion formation is complicated both by the magnitude of K_f and by the stoichiometry of the reaction. Consider the combination of aqueous copper(II) ions and ammonia to form the complex ion $Cu(NH_3)_4^{2+}$:

$$Cu^{2+}(aq) + 4NH_3(aq) \rightleftharpoons Cu(NH_3)_4^{2+}(aq) \qquad K_f = 5.0 \times 10^{14}$$

Let's say we wish to determine the molar concentration of free copper(II) ion in solution when 0.10 mole of $Cu(NO_3)_2$ is dissolved in a liter of 3.0 M NH_3. We cannot solve this with the same approach we used to determine the pH of a weak acid solution. Not only can we not neglect x, the amount of copper(II) ion consumed in the reaction, but having to raise the ammonia concentration to the fourth power in the equilibrium expression results in an equation that is not easily solved. Another approach is needed.

Because the magnitude of K_f is so large, we begin by assuming that *all* the copper(II) ion is consumed to form the complex ion. Then we consider the equilibrium in terms of the *reverse* reaction; that is, the dissociation of $Cu(NH_3)_4^{2+}$, for which the equilibrium constant is the reciprocal of K_f:

$$Cu(NH_3)_4^{2+}(aq) \rightleftharpoons Cu^{2+}(aq) + 4NH_3(aq) \qquad K = 2.0 \times 10^{-15}$$

Now we construct an equilibrium table and, because this K is so small, we can expect x (the amount of complex ion that dissociates) to be insignificant compared to the concentration of the complex ion and the concentration of ammonia [note that the

Student Note: Formation of a complex ion *consumes* the metal ion produced by the dissociation of a salt, increasing the salt's solubility simply due to Le Châtelier's principle [◀◀ Section 15.5].

Student Note: The term *free* is used to refer to a metal ion that is *not* part of a complex ion.

concentration of ammonia, which had been 3.0 M, has been diminished by $4 \times 0.10\ M$ due to the amount required to complex 0.10 mole of copper(II) ion]:

$$Cu(NH_3)_4^{2+}(aq) \rightleftharpoons Cu^{2+}(aq) + 4NH_3(aq)$$

Initial concentration (M):	0.10	0	2.6
Change in concentration (M):	$-x$	$+x$	$+4x$
Equilibrium concentration (M):	$0.10 - x$	x	$2.6 + 4x$

We can neglect x with respect to the concentrations of $Cu(NH_3)_4^{2+}$ and NH_3 ($0.10 - x \approx 0.10$ and $2.6 + 4x \approx 2.6$), and the solution becomes:

$$\frac{[Cu^{2+}][NH_3]^4}{[Cu(NH_3)_4^{2+}]} = \frac{x(2.6)^4}{0.10} = 2.0 \times 10^{-15}$$

and $x = 4.4 \times 10^{-18}\ M$. Note that because the formation constant is so large, the amount of copper that remains uncomplexed is extremely small. As always, it is a good idea to check the answer by plugging it into the equilibrium expression:

$$\frac{(4.4 \times 10^{-18})[2.6 + 4(4.4 \times 10^{-18})]^4}{0.10 - 4.4 \times 10^{-18}} = 2.0 \times 10^{-15}$$

Sample Problem 17.12 lets you practice applying this approach to a complex-ion formation equilibrium problem.

SAMPLE PROBLEM 17.12

In the presence of aqueous cyanide, cadmium(II) forms the complex ion $Cd(CN)_4^{2-}$. Determine the molar concentration of free (uncomplexed) cadmium(II) ion in solution when 0.20 mole of $Cd(NO_3)_2$ is dissolved in a liter of 2.0 M sodium cyanide (NaCN).

Strategy Because formation constants are typically very large, we begin by assuming that all the Cd^{2+} ion is consumed and converted to complex ion. We then determine how much Cd^{2+} is produced by the subsequent dissociation of the complex ion, a process for which the equilibrium constant is the reciprocal of K_f.

Setup From Table 17.5, the formation constant (K_f) for the complex ion $Cd(CN)_4^{2-}$ is 7.1×10^{16}. The reverse process:

$$Cd(CN)_4^{2-}(aq) \rightleftharpoons Cd^{2+}(aq) + 4CN^-(aq)$$

has an equilibrium constant of $1/K_f = 1.4 \times 10^{-17}$. The equilibrium expression for the dissociation is:

$$1.4 \times 10^{-17} = \frac{[Cd^{2+}][CN^-]^4}{[Cd(CN)_4^{2-}]}$$

The formation of complex ion will consume some of the cyanide originally present. Stoichiometry indicates that four CN^- ions are required to react with one Cd^{2+} ion. Therefore, the concentration of CN^- that we enter in the top row of the equilibrium table will be $[2.0\ M - 4(0.20\ M)] = 1.2\ M$.

Solution We construct an equilibrium table:

$$Cd(CN)_4^{2-}(aq) \rightleftharpoons Cd^{2+}(aq) + 4CN^-(aq)$$

Initial concentration (M):	0.20	0	1.2
Change in concentration (M):	$-x$	$+x$	$+4x$
Equilibrium concentration (M):	$0.20 - x$	x	$1.2 + 4x$

and, because the magnitude of K is so *small*, we can neglect x with respect to the initial concentrations of $Cd(CN)_4^{2-}$ and CN^- ($0.20 - x \approx 0.20$ and $1.2 + 4x \approx 1.2$), so the solution becomes:

$$\frac{[Cd^{2+}][CN^-]^4}{[Cd(CN)_4^{2-}]} = \frac{x(1.2)^4}{0.20} = 1.4 \times 10^{-17}$$

and $x = 1.4 \times 10^{-18}\ M$.

THINK ABOUT IT

When you assume that all the metal ion is consumed and converted to complex ion, it's important to remember that some of the complexing agent (in this case, CN^- ion) is consumed in the process. Don't forget to adjust its concentration accordingly before entering it in the top row of the equilibrium table.

Practice Problem A TTEMPT In the presence of aqueous ammonia, cobalt(III) forms the complex ion $Co(NH_3)_6^{3+}$. Determine the molar concentration of free cobalt(III) ion in solution when 0.15 mole of $Co(NO_3)_3$ is dissolved in a liter of 2.5 M aqueous ammonia.

Practice Problem B UILD Use information from Tables 17.4 and 17.5 to determine the molar solubility of chromium(III) hydroxide in a buffered solution with pH = 11.45.

Practice Problem C ONCEPTUALIZE Beginning with a saturated solution of AgCl, which of the following graphs best represents how the concentrations of free silver and chloride ions change as NH_3 is added to the solution?

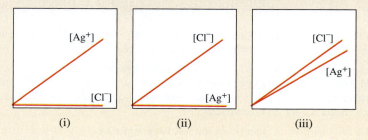

(i)   (ii)   (iii)   (iv)

CHECKPOINT – SECTION 17.5 Factors Affecting Solubility

17.5.1 Calculate the molar solubility of AgCl in 0.10 M $CaCl_2$.

a) $1.6 \times 10^{-10} M$ d) $1.3 \times 10^{-5} M$

b) $1.6 \times 10^{-9} M$ e) $2.6 \times 10^{-8} M$

c) $8.0 \times 10^{-10} M$

17.5.2 Which of the following substances will be more soluble in *acidic* solution than in pure water? (Select all that apply.)

a) $PbCO_3$ d) $Fe(OH)_3$

b) AgS   e) CaF_2

c) AgI

17.5.3 In which of the solutions would the slightly soluble salt MC_2 be *most* soluble, and in which would it be *least* soluble? (For clarity, water molecules and counter ions in soluble salts are not shown.)

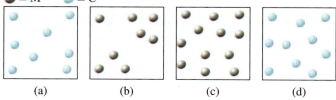

● = M^{2+} ○ = C^-

(a)   (b)   (c)   (d)

17.6 Separation of Ions Using Differences in Solubility

In chemical analysis, it sometimes is necessary to remove one type of ion from solution by precipitation while leaving other ions in solution. For instance, the addition of sulfate ions to a solution containing both potassium and barium ions causes $BaSO_4$ to precipitate out, thereby removing most of the Ba^{2+} ions from the solution. The other "product," K_2SO_4, is soluble and will remain in solution. The $BaSO_4$ precipitate can be separated from the solution by filtration.

Fractional Precipitation

Even when both products are insoluble, we can still achieve some degree of separation by choosing the proper reagent to bring about precipitation. Consider a solution that contains Cl^-, Br^-, and I^- ions. One way to separate these ions is to convert them to insoluble silver halides. According to their K_{sp} values, the solubility of the silver halides decreases from AgCl to AgI. Thus, when a soluble compound such as silver nitrate is slowly added to this solution, AgI begins to precipitate first, followed by AgBr, and then AgCl. This practice is known as *fractional precipitation.*

Sample Problem 17.13 describes the separation of only two ions (Cl^- and Br^-), but the procedure can be applied to a solution containing more than two different types of ions.

SAMPLE PROBLEM   17.13

Silver nitrate is added slowly to a solution that is 0.020 M in Cl^- ions and 0.020 M in Br^- ions. Calculate the concentration of Ag^+ ions (in mol/L) required to initiate the precipitation of AgBr without precipitating AgCl.

Strategy Silver nitrate dissociates in solution to give Ag^+ and NO_3^- ions. Adding Ag^+ ions in sufficient amount will cause the slightly soluble ionic compounds AgCl and AgBr to precipitate from solution. Knowing the K_{sp} values for AgCl and AgBr (and the concentrations of Cl^- and Br^- already in solution), we can use the equilibrium expressions to calculate the maximum concentration of Ag^+ that can exist in solution without exceeding K_{sp} for each compound.

Setup The solubility equilibria, K_{sp} values, and equilibrium expressions for AgCl and AgBr are:

$$AgCl(s) \rightleftharpoons Ag^+(aq) + Cl^-(aq) \qquad K_{sp} = 1.6 \times 10^{-10} = [Ag^+][Cl^-]$$

$$AgBr(s) \rightleftharpoons Ag^+(aq) + Br^-(aq) \qquad K_{sp} = 7.7 \times 10^{-13} = [Ag^+][Br^-]$$

Because the K_{sp} for AgBr is smaller (by a factor of more than 200), AgBr should precipitate first; that is, it will require a lower concentration of added Ag^+ to begin precipitation. Therefore, we first solve for $[Ag^+]$ using the equilibrium expression for AgBr to determine the minimum Ag^+ concentration necessary to initiate precipitation of AgBr. We then solve for $[Ag^+]$ again, using the equilibrium expression for AgCl to determine the *maximum* Ag^+ concentration that can exist in the solution without initiating the precipitation of AgCl.

Solution Solving the AgBr equilibrium expression for Ag^+ concentration, we have:

$$[Ag^+] = \frac{K_{sp}}{[Br^-]} \quad \text{and} \quad [Ag^+] = \frac{7.7 \times 10^{-13}}{0.020} = 3.9 \times 10^{-11} \, M$$

For AgBr to precipitate from solution, the silver ion concentration must exceed 3.9×10^{-11} M. Solving the AgCl equilibrium expression for the Ag^+ concentration, we have:

$$[Ag^+] = \frac{K_{sp}}{[Cl^-]}$$

and:

$$[Ag^+] = \frac{1.6 \times 10^{-10}}{0.020} = 8.0 \times 10^{-9} \, M$$

For AgCl *not* to precipitate from solution, the silver ion concentration must stay below 8.0×10^{-9} M. Therefore, to precipitate the Br^- ions without precipitating the Cl^- from this solution, the Ag^+ concentration must be greater than 3.9×10^{-11} M and less than 8.0×10^{-9} M.

THINK ABOUT IT

If we continue adding $AgNO_3$ until the Ag^+ concentration is high enough to begin the precipitation of AgCl, the concentration of Br^- remaining in solution can also be determined using the K_{sp} expression:

$$[Br^-] = \frac{K_{sp}}{[Ag^+]} = \frac{7.7 \times 10^{-13}}{8.0 \times 10^{-9}}$$

$$= 9.6 \times 10^{-5} \, M$$

Thus, by the time AgCl begins to precipitate, $(9.6 \times 10^{-5} \, M) \div (0.020 \, M) = 0.0048$, so less than 0.5 percent of the original bromide ion remains in the solution.

Practice Problem **ATTEMPT** Lead(II) nitrate is added slowly to a solution that is 0.020 M in Cl^- ions. Calculate the concentration of Pb^{2+} ions (in mol/L) required to initiate the precipitation of $PbCl_2$. (K_{sp} for $PbCl_2$ is 2.4×10^{-4}.)

Practice Problem **BUILD** Calculate the concentration of Ag^+ (in mol/L) necessary to initiate the precipitation of (a) AgCl and (b) Ag_3PO_4 from a solution in which $[Cl^-]$ and $[PO_4^{3-}]$ are each 0.10 M. (K_{sp} for Ag_3PO_4 is 1.8×10^{-18}.)

Practice Problem **CONCEPTUALIZE** The first two diagrams show saturated solutions of the sparingly soluble ionic compounds AX and BX_2. The third diagram shows a solution of soluble salts containing the cations A^+ and B^{2+}. (The anions of the soluble salts are not shown.) Which of the sparingly soluble compounds will precipitate first as NaX is added to the third solution?

  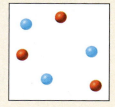

Qualitative Analysis of Metal Ions in Solution

The principle of selective precipitation can be used to identify the types of ions present in a solution. This practice is called *qualitative analysis.* There are about 20 common cations that can be analyzed readily in aqueous solution. These cations can be divided into five groups according to the solubility products of their insoluble salts (Table 17.4). Because an unknown solution may contain from 1 to all 20 ions, any analysis must be carried out systematically from group 1 through group 5. The general procedure for separating these 20 ions is as follows:

> **Student Note:** Note that these group numbers do *not* correspond to groups in the periodic table.

- *Group 1 cations.* When dilute HCl is added to the unknown solution, only the Ag^+, Hg_2^{2+}, and Pb^{2+} ions precipitate as insoluble chlorides. The other ions, whose chlorides are soluble, remain in solution.
- *Group 2 cations.* After the chloride precipitates have been removed by filtration, hydrogen sulfide is added to the unknown solution, which is acidic due to the addition of HCl. Metal ions from group 2 react to produce metal sulfides:

$$M^{2+}(aq) + H_2S(aq) + 2H_2O(l) \rightleftharpoons MS(s) + 2H_3O^+(aq)$$

  In the presence of H^+, this equilibrium shifts to the *left.* Therefore, only the metal sulfides with the *smallest* K_{sp} values precipitate under acidic conditions. These are Bi_2S_3, CdS, CuS, and SnS, (see Table 17.4). The solution is then filtered to remove the insoluble sulfides.
- *Group 3 cations.* At this stage, sodium hydroxide is added to the solution to make it basic. In a basic solution, the metal sulfide equilibrium shifts to the right and the more soluble sulfides (CoS, FeS, MnS, NiS, ZnS) now precipitate out of solution. The Al^{3+} and Cr^{3+} ions actually precipitate as the hydroxides $Al(OH)_3$ and $Cr(OH)_3$, rather than as the sulfides, because the hydroxides are less soluble. The solution is filtered again to remove the insoluble sulfides and hydroxides.
- *Group 4 cations.* After all the group 1, 2, and 3 cations have been removed from solution, sodium carbonate is added to the basic solution to precipitate Ba^{2+}, Ca^{2+}, and Sr^{2+} ions as $BaCO_3$, $CaCO_3$, and $SrCO_3$. These precipitates, too, are removed from solution by filtration.
- *Group 5 cations.* At this stage, the only cations possibly remaining in solution are Na^+, K^+, and NH_4^+. The presence of NH_4^+ ions can be determined by adding sodium hydroxide:

$$NaOH(aq) + NH_4^+(aq) \longrightarrow Na^+(aq) + H_2O(l) + NH_3(g)$$

The ammonia gas is detected either by its characteristic odor or by observing a wet piece of red litmus paper turning blue when placed above (not in contact with) the solution. To confirm the presence of Na^+ and K^+ ions, a flame test is usually used in which a piece of platinum wire (chosen because platinum is inert) is dipped into the original solution and then held over a Bunsen burner flame. Na^+ ions emit a yellow flame when heated in this manner, whereas K^+ ions emit a violet flame (Figure 17.12). Figure 17.13 summarizes this scheme for separating metal ions.

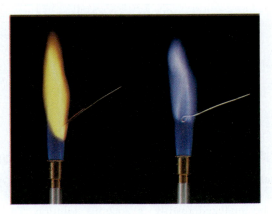

Figure 17.12 Flame tests for sodium (yellow flame) and potassium (violet flame).

Stephen Frisch/McGraw Hill

Animation

Chemical Analysis—flame tests of metals.

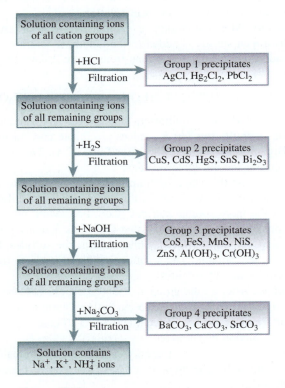

Figure 17.13 A flowchart for the separation of cations in qualitative analysis.

CHECKPOINT – SECTION 17.6   Separation of Ions Using Differences in Solubility

17.6.1 A solution is 0.10 M in Br^-, CO_3^{2-}, Cl^-, I^-, and SO_4^{2-} ions. Which compound will precipitate first as silver nitrate is added to the solution?

a) AgBr

b) Ag_2CO_3

c) AgCl

d) AgI

e) Ag_2SO_4

17.6.2 Barium nitrate is added slowly to a solution that is 0.10 M in SO_4^{2-} ions and 0.10 M in F^- ions. Calculate the concentration of Ba^{2+} ions (in mol/L) required to initiate the precipitation of $BaSO_4$ without precipitating BaF_2.

a) 1.7×10^{-6} M

b) 1.1×10^{-9} M

c) 1.7×10^{-4} M

d) 1.7×10^{-5} M

e) 1.1×10^{-8} M

Chapter Summary

Section 17.1

- The presence of a common ion suppresses the ionization of a weak acid or weak base. This is known as the ***common ion effect.***

- A common ion is added to a solution in the form of a *salt.*

Section 17.2

- A solution that contains significant concentrations of both members of a conjugate acid-base pair (weak acid–conjugate base or weak base–conjugate acid) is a *buffer solution* or simply a ***buffer.***

- Buffer solutions resist pH change upon addition of small amounts of strong acid or strong base. Buffers are important to biological systems.

- The pH of a buffer can be calculated using an equilibrium table or with the ***Henderson-Hasselbalch equation.***

- The pK_a of a weak acid is $-\log K_a$. When the weak acid and conjugate base concentrations in a buffer solution are equal, $pH = pK_a$.

- We can prepare a buffer with a specific pH by choosing a weak acid with a pK_a close to the desired pH.

Section 17.3

- The titration curve of a strong acid–strong base titration has a long, steep region near the equivalence point.

- Titration curves for weak acid–strong base or weak base–strong acid titrations have a significantly shorter steep region.

- The pH at the equivalence point of a strong acid–strong base titration is 7.00.

- The pH at the equivalence point of a weak acid–strong base titration is above 7.00.

- The pH at the equivalence point of a weak base–strong acid titration is below 7.00.

- Acid-base indicators are usually weak organic acids that exhibit two different colors depending on the pH of the solution. The ***endpoint*** of a titration is the point at which the color of the indicator changes. It is used to estimate the *equivalence point* of a titration.

- The indicator used for a particular titration should exhibit a color change in the pH range corresponding to the steep region of the titration curve.

Section 17.4

- The ***solubility product constant (K_{sp})*** is the equilibrium constant that indicates to what extent a slightly soluble ionic compound dissociates in water.

- K_{sp} can be used to determine ***molar solubility*** or ***solubility*** in g/L, and vice versa.

- K_{sp} can also be used to predict whether or not a precipitate will form when two solutions are mixed.

Section 17.5

- Solubility is affected by common ions, pH, and complex ion formation. The ***formation constant (K_f)*** indicates to what extent complex ions form.

- A salt that dissociates to give a strong conjugate base such as fluoride ion will be more soluble in acidic solution than in pure water.

- A salt that dissociates to give hydroxide ion will be more soluble at lower pH and less soluble at higher pH.

- The solubility of an ionic compound increases when the formation of a ***complex ion*** consumes one of the products of dissociation.

Section 17.6

- Ions can be separated using ***fractional precipitation.***

- Fractional precipitation schemes can be designed based on K_{sp} values.

- Groups of cations can be identified through the use of selective precipitation. This is the basis of ***qualitative analysis.***

Key Words

Buffer, 854	Endpoint, 872	Henderson-Hasselbalch equation, 856	Qualitative analysis, 893
Common ion effect, 854	Formation constant (K_f), 888		Solubility, 875
Complex ion, 887	Fractional precipitation, 892	Molar solubility, 875	Solubility product constant (K_{sp}), 875

Key Equations

17.1 $\text{pH} = \text{p}K_a + \log \dfrac{[\text{A}^-]}{[\text{HA}]}$	The pH of a buffer is calculated as the sum of pK_a of the weak acid component and base-10 log of the ratio of conjugate base to weak acid.
17.2 $\text{p}K_a = -\log K_a$	The pK_a of a weak acid is equal to minus the base-10 log of its K_a value.
17.3 $\text{pH} = \text{p}K_a + \log \dfrac{[\text{conjugate base}]}{[\text{weak acid}]}$	This is the more general form of the Henderson-Hasselbalch equation, which is used to calculate the pH of a buffer. Because the buffer components are dissolved in the same volume, the ratio of conjugate base to weak acid can be calculated using either molar concentrations or absolute molar amounts.

Buffers

The Henderson-Hasselbalch equation (Equation 17.3) enables us to calculate the pH of a buffer if we know the concentrations of the weak acid and conjugate base (or of the weak base and conjugate acid); and to determine the pH of a buffer after the addition of strong acid or base. It can also be used to determine the amounts or *relative* amounts of both members of a conjugate pair necessary to prepare a buffer of specified pH.

The most familiar form of the Henderson-Hasselbalch equation is:

$$pH = pK_a + \log\frac{[\text{conjugate base}]}{[\text{weak acid}]}$$

Although the equation contains the ratio of *molar concentrations* of conjugate base and weak acid, because both are contained within the same volume, volume cancels in the numerator and denominator of the log term. Therefore, we can also use *moles* or *millimoles* of conjugate base and weak acid, which is more convenient in many cases:

$$pH = pK_a + \log\frac{\text{mol conjugate base}}{\text{mol weak acid}} \quad \text{or} \quad pH = pK_a + \log\frac{\text{mmol conjugate base}}{\text{mmol weak acid}}$$

If we wish to prepare a buffer in a specific pH range, we must first select a suitable conjugate pair. Because the concentrations of weak acid and conjugate base cannot differ by more than a factor of 10, the log term in the Henderson-Hasselbalch equation can have values only from −1 through 1:

$$pH = pK_a \pm 1$$

Therefore, we must select a weak acid with a pK_a within one pH unit of the desired buffer pH. For example, to prepare a buffer with a pH between 4 and 5, we could use acetic acid ($HC_2H_3O_2$). [From Table 16.6, K_a for acetic acid is 1.8×10^{-5}, so its pK_a is $-\log(1.8 \times 10^{-5}) = 4.74$.]

To prepare a buffer with a specific pH, we solve the Henderson-Hasselbalch equation to determine the relative amounts of weak acid and conjugate base. The following flowchart illustrates this process for preparation of a buffer with pH = 4.15 using acetic acid ($pK_a = 4.74$) and sodium acetate ($NaC_2H_3O_2$):

$$4.15 \Rightarrow 4.74 + \log\left[\frac{[NaC_2H_3O_2]}{[HC_2H_3O_2]}\right]$$

$$-0.59 = \log\left[\frac{[NaC_2H_3O_2]}{[HC_2H_3O_2]}\right]$$

To eliminate the log term, we must take the antilog (10^x) of both sides of the equation:

$$10^{-0.59} = \frac{[NaC_2H_3O_2]}{[HC_2H_3O_2]}$$

This gives:

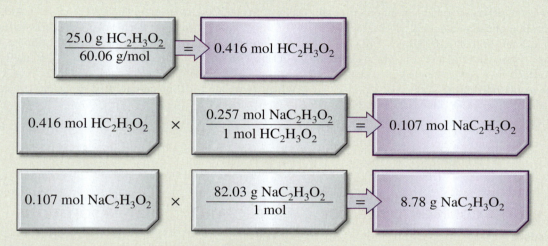

$$0.257 = \frac{[NaC_2H_3O_2]}{[HC_2H_3O_2]}$$

This means that we need 0.257 mole of sodium acetate for every mole of acetic acid. If the amount of one member of the conjugate pair is specified, this ratio enables us to calculate the amount of the other member. For instance, if we know that the buffer must contain 25.0 g acetic acid, we can determine the amount of sodium acetate necessary to achieve the desired pH. We will need the molar masses of both members of the conjugate pair [◀◀ Section 3.4]:

$$\frac{25.0 \text{ g } HC_2H_3O_2}{60.06 \text{ g/mol}} = 0.416 \text{ mol } HC_2H_3O_2$$

$$0.416 \text{ mol } HC_2H_3O_2 \times \frac{0.257 \text{ mol } NaC_2H_3O_2}{1 \text{ mol } HC_2H_3O_2} = 0.107 \text{ mol } NaC_2H_3O_2$$

$$0.107 \text{ mol } NaC_2H_3O_2 \times \frac{82.03 \text{ g } NaC_2H_3O_2}{1 \text{ mol}} = 8.78 \text{ g } NaC_2H_3O_2$$

Therefore, we would need 8.78 g sodium acetate. Because the pH of a buffer does not depend on volume, we could combine these amounts of acetic acid and sodium acetate in any volume that is convenient.

Key Skills Problems

17.1
Which of the acids in Table 16.6 can be used to prepare a buffer of pH 6.5? (Select all that apply.)

(a) hydrofluoric acid   (b) benzoic acid
(c) hydrocyanic acid   (d) phenol   (e) none

17.2
What molar ratio of sodium cyanide to hydrocyanic acid is necessary to prepare a buffer with pH = 9.72?

(a) 0.39:1   (b) 0.41:1   (c) 2.3:1
(d) 2.6:1   (e) 1:1

17.3
How many moles of sodium benzoate must be added to 175 mL of 0.955 M benzoic acid to prepare a buffer with pH = 5.05?

(a) 0.18   (b) 0.15   (c) 6.8   (d) 7.2   (e) 1.2

17.4
How much sodium fluoride must be dissolved in 250 mL of 0.98 M HF to prepare a buffer with pH 3.50?

(a) 23 g   (b) 2.2 g   (c) 15 g   (d) 92 g   (e) 0.98 g

Questions and Problems

Applying What You've Learned

One strategy used by oyster farmers to adjust ocean-water pH is to add sodium carbonate, Na_2CO_3, to the water. The added carbonate ion reacts with excess hydronium ion to produce hydrogen carbonate ion HCO_3^-.

$$CO_3^{2-}(aq) + H_3O^+(aq) \longrightarrow HCO_3^-(aq) + H_2O(l)$$

Problems:
(a) Determine the pH of an aqueous solution, originally 0.025 M in $NaHCO_3$, when enough Na_2CO_3 has been added to make the concentration of carbonate ion 0.018 M. [◄◄ Sample Problem 17.1] (b) Determine the pH of 1.0 L of this buffer after the addition of 125 mL of 0.080 M HCl. [◄◄ Sample Problem 17.2] (c) Calculate the solubility of calcium carbonte, $CaCO_3$, in water at 25°C, first assuming that none of the carbonate ion reacts with water; [◄◄ Sample Problem 17.7] and then taking into account the reaction of carbonate ion with water. [◄◄ Sample Problem 15.4] (d) Will the aqueous solubility of $CaCO_3$ increase or decrease as pH falls? [◄◄ Sample Problem 17.11] (e) Which compound will precipitate first as $Ca(NO_3)_2$ is added to a solution that is 2.0 M in CO_3^{2-} and 0.20 M in PO_4^{3-}? [◄◄ Sample Problem 17.13]

SECTION 17.1: THE COMMON ION EFFECT

Review Questions

17.1 Use Le Châtelier's principle to explain how the common ion effect affects the pH of a weak acid solution.

17.2 Describe the effect on pH (increase, decrease, or no change) that results from each of the following additions: (a) potassium acetate to an acetic acid solution, (b) ammonium nitrate to an ammonia solution, (c) sodium formate (HCOONa) to a formic acid (HCOOH) solution, (d) potassium chloride to a hydrochloric acid solution, (e) barium iodide to a hydroiodic acid solution.

17.3 Define pK_a for a weak acid. What is the relationship between the value of the pK_a and the strength of the acid?

17.4 The pK_a values of two monoprotic acids HA and HB are 5.9 and 8.1, respectively. Which of the two is the stronger acid?

Computational Problems

17.5 Determine the pH of (a) a 0.40 M CH_3COOH solution and (b) a solution that is 0.40 M CH_3COOH and 0.20 M CH_3COONa.

17.6 Determine the pH of (a) a 0.20 M NH_3 solution, and (b) a solution that is 0.20 M NH_3 and 0.30 M NH_4Cl.

SECTION 17.2: BUFFER SOLUTIONS

Visualizing Chemistry
Figure 17.1

VC 17.1 Which pair of substances can be dissolved together to prepare a buffer solution?

CH_3COOH	CH_3COONa	NaOH	HCl	NaCl
(i)	(ii)	(iii)	(iv)	(v)

a) i/ii
b) i/ii, i/iii, ii/iv
c) i/ii, i/iii, ii/iii, iv/v

VC 17.2 Consider the buffer shown in Figure 17.1, in which $[CH_3COOH] = 0.10\ M$, $[CH_3COO^-] = 0.10\ M$, and pH = 4.74. How would the pH of the buffer change if we doubled the volume by adding 100 mL of water?
a) The pH would increase.
b) The pH would decrease.
c) The pH would not change.

VC 17.3 Consider a buffer similar to the one shown in Figure 17.1, in which $[CH_3COOH] = 0.50\ M$ and $[CH_3COO^-] = 0.75\ M$. Which substance can we add more of without causing a drastic change in pH?
a) HCl   b) NaOH   c) Neither

VC 17.4 According to Figure 17.1, the reaction that occurs when strong acid is added to the buffer is:

$$H_3O^+(aq) + CH_3COO^-(aq) \longrightarrow CH_3COOH(aq) + H_2O(l)$$

Estimate the value of K for this reaction.
a) 1×10^{14} b) 6×10^{-10} c) 6×10^4

Review Questions

17.7 What is a buffer solution? What must a solution contain to be a buffer?

17.8 Using only a pH meter, water, and a graduated cylinder, how would you distinguish between an acid solution and a buffer solution at the same pH?

Computational Problems

17.9 Calculate the pH of a buffer system made up of 0.15 M NH_3/0.35 M NH_4Cl.

17.10 Calculate the pH of the following two buffer solutions: (a) 2.0 M CH_3COONa/2.0 M CH_3COOH, (b) 0.20 M CH_3COONa/0.20 M CH_3COOH. Which is the more effective buffer? Why?

17.11 The pH of a bicarbonate-carbonic acid buffer is 7.50. Calculate the ratio of the concentration of carbonic acid (H_2CO_3) to that of the bicarbonate ion (HCO_3^-).

17.12 What is the pH of a buffer system made up of 0.10 M Na_2HPO_4/0.15 M KH_2PO_4?

17.13 The pH of a sodium acetate–acetic acid buffer is 4.50. Calculate the ratio $[CH_3COO^-]/[CH_3COOH]$.

17.14 The pH of blood plasma is 7.40. Assuming the principal buffer system is HCO_3^-/H_2CO_3, calculate the ratio $[HCO_3^-]/[H_2CO_3]$. Is this buffer more effective against an added acid or an added base?

17.15 Calculate the pH of the 0.20 M NH_3/0.20 M NH_4Cl buffer. What is the pH of the buffer after the addition of 10.0 mL of 0.10 M HCl to 65.0 mL of the buffer?

17.16 Calculate the pH of 1.00 L of the buffer 1.00 M CH_3COONa/1.00 M CH_3COOH before and after the addition of (a) 0.080 mol NaOH and (b) 0.12 mol HCl. (Assume that there is no change in volume.)

Conceptual Problems

17.17 Which of the following solutions can act as a buffer: (a) KCl/HCl, (b) $KHSO_4$/H_2SO_4, (c) Na_2HPO_4/NaH_2PO_4, (d) KNO_2/HNO_2?

17.18 Which of the following solutions can act as a buffer: (a) KCN/HCN, (b) Na_2SO_4/$NaHSO_4$, (c) NH_3/NH_4NO_3, (d) NaI/HI?

17.19 A diprotic acid, H_2A, has the following ionization constants: $K_{a_1} = 1.1 \times 10^{-3}$ and $K_{a_2} = 2.5 \times 10^{-6}$. To make up a buffer solution of pH 5.80, which combination would you choose: $NaHA$/H_2A or Na_2A/$NaHA$?

17.20 A student is asked to prepare a buffer solution at pH 8.60, using one of the following weak acids: $HA(K_a = 2.7 \times 10^{-3})$, $HB(K_a = 4.4 \times 10^{-6})$, $HC(K_a = 2.6 \times 10^{-9})$. Which acid should the student choose? Why?

17.21 The following diagrams contain one or more of the compounds: H_2A, NaHA, and Na_2A, where H_2A is a weak diprotic acid. (1) Which of the solutions can act as buffer solutions? (2) Which solution is the most effective buffer solution? Water molecules and Na^+ ions have been omitted for clarity.

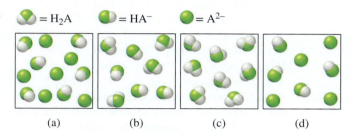

= H_2A = HA^- = A^{2-}

(a)　　(b)　　(c)　　(d)

17.22 The following diagrams represent solutions containing a weak acid HA($pK_a = 5.0$) and its sodium salt NaA. (1) Which solution has the lowest pH? Which has the highest pH? (2) How many different species are present after the addition of two H^+ ions to solution (a)? (3) How many different species are present after the addition of four OH^- ions to solution (b)?

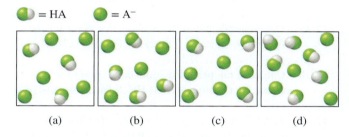

= HA   = A^-

(a)　　(b)　　(c)　　(d)

SECTION 17.3: ACID-BASE TITRATIONS

Review Questions

17.23 Briefly describe what happens in an acid-base titration.

17.24 Sketch titration curves for the following acid-base titrations: (a) HCl versus NaOH, (b) HCl versus CH_3NH_2, (c) CH_3COOH versus NaOH. In each case, the base is added to the acid in an Erlenmeyer flask. Your graphs should show the pH on the y axis and the volume of base added on the x axis.

17.25 Explain how an acid-base indicator works in a titration. What are the criteria for choosing an indicator for a particular acid-base titration?

17.26 The amount of indicator used in an acid-base titration must be small. Why?

Computational Problems

17.27 A 0.2688-g sample of a monoprotic acid neutralizes 16.4 mL of 0.08133 M KOH solution. Calculate the molar mass of the acid.

17.28   A 5.00-g quantity of a diprotic acid was dissolved in water and made up to exactly 250 mL. Calculate the molar mass of the acid if 25.0 mL of this solution required 11.1 mL of 1.00 M KOH for neutralization. Assume that both protons of the acid were titrated.

17.29   In a titration experiment, 12.5 mL of 0.500 M H_2SO_4 neutralizes 50.0 mL of NaOH. What is the concentration of the NaOH solution?

17.30   In a titration experiment, 20.4 mL of 0.883 M HCOOH neutralizes 19.3 mL of $Ba(OH)_2$. What is the concentration of the $Ba(OH)_2$ solution?

17.31   A 0.1276-g sample of an unknown monoprotic acid was dissolved in 25.0 mL of water and titrated with a 0.0633 M NaOH solution. The volume of base required to bring the solution to the equivalence point was 18.4 mL. (a) Calculate the molar mass of the acid. (b) After 10.0 mL of base had been added during the titration, the pH was determined to be 5.87. What is the K_a of the unknown acid?

17.32   A solution is made by mixing exactly 500 mL of 0.167 M NaOH with exactly 500 mL of 0.100 M CH_3COOH. Calculate the equilibrium concentrations of H^+, CH_3COOH, CH_3COO^-, OH^-, and Na^+.

17.33   Calculate the pH at the equivalence point for the following titration: 0.20 M HCl versus 0.20 M methylamine (CH_3NH_2).

17.34   Calculate the pH at the equivalence point for the following titration: 0.10 M HCOOH versus 0.10 M NaOH.

17.35   A 25.0-mL solution of 0.100 M CH_3COOH is titrated with a 0.200 M KOH solution. Calculate the pH after the following additions of the KOH solution: (a) 0.0 mL, (b) 5.0 mL, (c) 10.0 mL, (d) 12.5 mL, (e) 15.0 mL.

17.36   A 10.0-mL solution of 0.300 M NH_3 is titrated with a 0.100 M HCl solution. Calculate the pH after the following additions of the HCl solution: (a) 0.0 mL, (b) 10.0 mL, (c) 20.0 mL, (d) 30.0 mL, (e) 40.0 mL.

Conceptual Problems

17.37   Referring to Table 17.3, specify which indicator or indicators you would use for the following titrations: (a) HCOOH versus NaOH, (b) HCl versus KOH, (c) HNO_3 versus CH_3NH_2.

17.38   A student carried out an acid-base titration by adding NaOH solution from a burette to an Erlenmeyer flask containing an HCl solution and using phenolphthalein as the indicator. At the equivalence point, she observed a faint reddish-pink color. However, after a few minutes, the solution gradually turned colorless. What do you suppose happened?

17.39   The ionization constant K_a of an indicator HIn is 1.0×10^{-6}. The color of the nonionized form is red and that of the ionized form is yellow. What is the color of this indicator in a solution whose pH is 8.00?

17.40   The K_a of a certain indicator is 2.0×10^{-6}. The color of HIn is green and that of In^- is red. A few drops of the indicator are added to an HCl solution, which is then titrated against an NaOH solution. At what pH will the indicator change color?

17.41   The following diagrams represent solutions at various stages in the titration of a weak base B (such as NH_3) with HCl. Identify the solution that corresponds to (1) the initial stage before the addition of HCl, (2) halfway to the equivalence point, (3) the equivalence point, (4) beyond the equivalence point. Is the pH greater than, less than, or equal to 7 at the equivalence point? Water and Cl^- ions have been omitted for clarity.

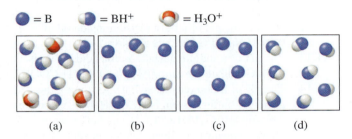

(a)   (b)   (c)   (d)

17.42   The following diagrams represent solutions at various stages in the titration of a weak acid HA with NaOH. Identify the solution that corresponds to (1) the initial stage before the addition of NaOH, (2) halfway to the equivalence point, (3) the equivalence point, (4) beyond the equivalence point. Is the pH greater than, less than, or equal to 7 at the equivalence point? Water and Na^+ ions have been omitted for clarity.

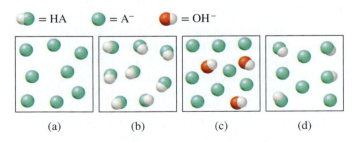

(a)   (b)   (c)   (d)

SECTION 17.4: SOLUBILITY EQUILIBRIA

Review Questions

17.43   Use $BaSO_4$ to distinguish between the terms *solubility* and *solubility product*.

17.44   Why do we usually not quote the K_{sp} values for soluble ionic compounds?

17.45 Write balanced equations and solubility product expressions for the solubility equilibria of the following compounds: (a) CuBr, (b) ZnC_2O_4, (c) Ag_2CrO_4, (d) Hg_2Cl_2, (e) $AuCl_3$, (f) $Mn_3(PO_4)_2$.

17.46 Write the solubility product expression for the ionic compound A_xB_y.

17.47 How can we predict whether a precipitate will form when two solutions are mixed?

17.48 Silver chloride has a larger K_{sp} than silver carbonate (see Table 17.4). Does this mean that AgCl also has a larger molar solubility than Ag_2CO_3?

Computational Problems

17.49 Calculate the concentration of the following ions: (a) $[I^-]$ in a saturated solution of AgI, (b) $[Al^{3+}]$ in $[Al(OH)_3]$ solution with $[OH^-] = 2.9 \times 10^{-9}\ M$.

17.50 From the solubility data given, calculate the solubility products for the following compounds: (a) SrF_2, 7.3×10^{-2} g/L, (b) Ag_3PO_4, 6.7×10^{-3} g/L.

17.51 The molar solubility of $MnCO_3$ is $4.2 \times 10^{-6}\ M$. What is K_{sp} for this compound?

17.52 The solubility of an ionic compound MX(molar mass = 346 g) is 4.63×10^{-3} g/L. What is K_{sp} for this compound?

17.53 The solubility of an ionic compound M_2X_3 (molar mass = 288 g) is 3.6×10^{-17} g/L. What is K_{sp} for this compound?

17.54 Using data from Table 17.4, calculate the molar solubility of BaF_2.

17.55 What is the pH of a saturated Iron(II) hydroxide solution?

17.56 The pH of a saturated solution of a metal hydroxide MOH is 9.68. Calculate the K_{sp} for this compound.

17.57 If 20.0 mL of 0.10 M $Ba(NO_3)_2$ is added to 50.0 mL of 0.10 M Na_2CO_3, will $BaCO_3$ precipitate?

17.58 A volume of 75 mL of 0.060 M NaF is mixed with 25 mL of 0.15 M $Sr(NO_3)_2$. Calculate the concentrations in the final solution of NO_3^-, Na^+, Sr^{2+}, and F^-. (K_{sp} for $SrF_2 = 2.0 \times 10^{-10}$.)

SECTION 17.5: FACTORS AFFECTING SOLUBILITY

Visualizing Chemistry
Figure 17.9

VC 17.5 Which of the following would cause precipitation of the largest amount of AgCl when 0.1 mole is dissolved in the saturated solution shown in Figure 17.9?
a) NaCl
b) CsCl
c) Both would cause precipitation of the same amount.

VC 17.6 Which of the following would cause precipitation of the largest amount of AgCl when 0.1 g is dissolved in the saturated solution shown in Figure 17.9?
a) NaCl
b) CsCl
c) Both would cause precipitation of the same amount.

VC 17.7 How would the concentration of silver ion in the last view of Figure 17.9 be different if we dissolved the sodium chloride *before* dissolving the silver chloride?
a) It would be higher.
b) It would be lower.
c) It would not change.

VC 17.8 How would the concentration of silver ion in the last view of Figure 17.9 be different if we added silver nitrate to the saturated solution of AgCl instead of sodium chloride?
a) It would be higher.
b) It would be lower.
c) It would not change.

Review Questions

17.59 How does the common ion effect influence solubility equilibria? Use Le Châtelier's principle to explain the decrease in solubility of $CaCO_3$ in an Na_2CO_3 solution.

17.60 The molar solubility of AgCl in $6.5 \times 10^{-3}\ M$ $AgNO_3$ is $2.5 \times 10^{-8}\ M$. In deriving K_{sp} from these data, which of the following assumptions are reasonable? (a) K_{sp} is the same as solubility. (b) K_{sp} of AgCl is the same in $6.5 \times 10^{-3}\ M$ $AgNO_3$ as in pure water. (c) Solubility of AgCl is independent of the concentration of $AgNO_3$. (d) $[Ag^+]$ in solution does not change significantly upon the addition of AgCl to $6.5 \times 10^{-3}\ M$ $AgNO_3$. (e) $[Ag^+]$ in solution after the addition of AgCl to $6.5 \times 10^{-3}\ M$ $AgNO_3$ is the same as it would be in pure water.

17.61 Give an example to illustrate the general effect of complex ion formation on solubility.

Computational Problems

17.62 How many grams of $CaCO_3$ will dissolve in 3.0×10^2 mL of 0.050 M $Ca(NO_3)_2$?

17.63 The solubility product of $PbBr_2$ is 8.9×10^{-6}. Determine the molar solubility in (a) pure water, (b) 0.20 M KBr solution, and (c) 0.20 M $Pb(NO_3)_2$ solution.

17.64   Calculate the molar solubility of AgCl in a 1.00-L solution containing 10.0 g of dissolved $CaCl_2$.

17.65   Calculate the molar solubility of $BaSO_4$ in (a) water and (b) a solution containing 1.0 M SO_4^{2-} ions.

17.66   Which of the following ionic compounds will be more soluble in acid solution than in water: (a) $BaSO_4$, (b) $PbCl_2$, (c) $Fe(OH)_3$, (d) $CaCO_3$?

17.67   Which of the following will be more soluble in acid solution than in pure water: (a) CuI, (b) Ag_2SO_4, (c) $Zn(OH)_2$, (d) BaC_2O_4, (e) $Ca_3(PO_4)_2$?

17.68   Compare the molar solubility of $Mg(OH)_2$ in water and in a solution buffered at a pH of 9.0.

17.69   Calculate the molar solubility of $Fe(OH)_2$ in a solution buffered at (a) a pH of 8.00 and (b) a pH of 10.00.

17.70   The solubility product of $Mg(OH)_2$ is 1.2×10^{-11}. What minimum OH^- concentration must be attained (e.g., by adding NaOH) to decrease the Mg concentration in a solution of $Mg(NO_3)_2$ to less than 1.0×10^{-10} M?

17.71   Calculate whether or not a precipitate will form if 2.00 mL of 0.60 M NH_3 is added to 1.0 L of 1.0×10^{-3} M $FeSO_4$.

17.72   If 2.50 g of $CuSO_4$ is dissolved in 9.0×10^2 mL of 0.30 M NH_3, what are the concentrations of Cu^{2+}, $Cu(NH_3)_4^{2+}$, and NH_3 at equilibrium?

17.73   Calculate the concentrations of Cd^{2+}, $Cd(CN)_4^{2-}$, and CN^- at equilibrium when 0.50 g of $Cd(NO_3)_2$ dissolves in 5.0×10^2 mL of 0.50 M NaCN.

17.74   If NaOH is added to 0.010 M Al^{3+}, which will be the predominant species at equilibrium: $Al(OH)_3$ or $Al(OH)_4^-$? The pH of the solution is 14.00. [K_f for $Al(OH)_4^- = 2.0 \times 10^{33}$.]

17.75   Calculate the molar solubility of AgI in a 1.0 M NH_3 solution.

Conceptual Problems

17.76   Both Ag^+ and Zn^{2+} form complex ions with NH_3. Write balanced equations for the reactions. However, $Zn(OH)_2$ is soluble in 6 M NaOH, and AgOH is not. Explain.

17.77   Explain, with balanced ionic equations, why (a) CuI_2 dissolves in ammonia solution, (b) AgBr dissolves in NaCN solution, and (c) $HgCl_2$ dissolves in KCl solution.

SECTION 17.6: SEPARATION OF IONS USING DIFFERENCES IN SOLUBILITY

Review Questions

17.78   Outline the general procedure of qualitative analysis.

17.79   Give two examples of metal ions in each group (1 through 5) in the qualitative analysis scheme.

Computational Problems

17.80   Solid NaI is slowly added to a solution that is 0.010 M in Cu^+ and 0.010 M in Ag^+. (a) Which compound will begin to precipitate first? (b) Calculate [Ag^+] when CuI just begins to precipitate. (c) What percent of Ag^+ remains in solution at this point?

17.81   Find the approximate pH range suitable for the separation of Fe^{3+} and Zn^{2+} ions by precipitation of $Fe(OH)_3$ from a solution that is initially 0.010 M in both Fe^{3+} and Zn^{2+}.

17.82   In a group 1 analysis, a student obtained a precipitate containing both AgCl and $PbCl_2$. Suggest one reagent that would enable the student to separate AgCl(s) from $PbCl_2$(s).

17.83   In a group 1 analysis, a student adds HCl acid to the unknown solution to make [Cl^-] = 0.15 M. Some $PbCl_2$ precipitates. Calculate the concentration of Pb^{2+} remaining in solution.

Conceptual Problems

17.84   Both KCl and NH_4Cl are white solids. Suggest one reagent that would enable you to distinguish between these two compounds.

17.85   Describe a simple test that would allow you to distinguish between $AgNO_3$(s) and $Cu(NO_3)_2$(s).

17.86   The first two diagrams show saturated solutions of the sparingly soluble ionic compounds AY and B_2Y, respectively. The third diagram shows a solution of soluble salts containing the cations A^{2+} and B^+. Determine which of the sparingly soluble ionic compounds will precipitate first as Na_2Y is added to the third solution.

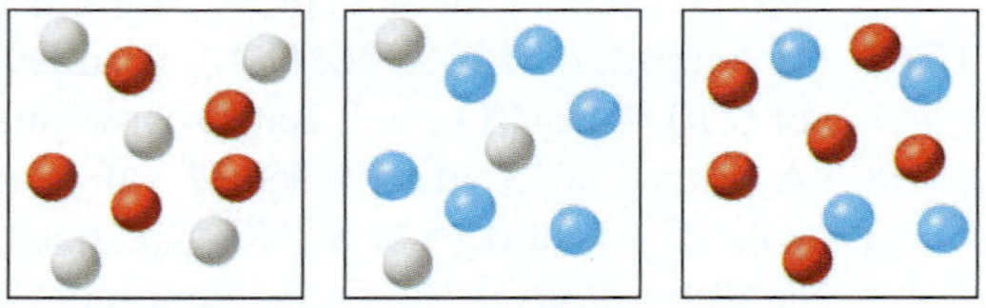

17.87   The first two diagrams show saturated solutions of the sparingly soluble ionic compounds AZ_2 and BZ_3, respectively. The third diagram shows a solution of soluble salts containing the cations A^{2+} and B^+. Determine which of the sparingly soluble ionic compounds will precipitate first as NaZ is added to the third solution.

ADDITIONAL PROBLEMS

17.88 Sketch the titration curve of a weak acid with a strong base like the one shown in Figure 17.4. On your graph, indicate the volume of base used at the equivalence point and also at the half-equivalence point, that is, the point at which half of the acid has been neutralized. Explain how the measured pH at the half-equivalence point can be used to determine K_a of the acid.

17.89 A 200-mL volume of NaOH solution was added to 400 mL of a 2.00 M HNO$_2$ solution. The pH of the mixed solution was 1.50 units greater than that of the original acid solution. Calculate the molarity of the NaOH solution.

17.90 The pK_a of butyric acid (HBut) is 4.7. Calculate K_b for the butyrate ion (But$^-$).

17.91 A solution is made by mixing exactly 500 mL of 0.167 M NaOH with exactly 500 mL 0.100 M HCOOH. Calculate the equilibrium concentrations of H$_3$O$^+$, HCOOH, HCOO$^-$, OH$^-$, and Na$^+$.

17.92 The titration curve shown here represents the titration of a weak diprotic acid (H$_2$A) versus NaOH. (a) Label the major species present at the marked points. (b) Estimate the pK_{a_1} and pK_{a_2} values of the acid. Assume that any salt hydrolysis is negligible.

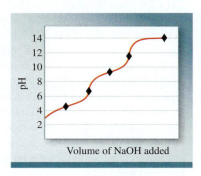

Volume of NaOH added

17.93 Cd(OH)$_2$ is an insoluble compound. It dissolves in excess NaOH in solution. Write a balanced ionic equation for this reaction. What type of reaction is this?

17.94 A student mixes 50.0 mL of 1.00 M Ba(OH)$_2$ with 86.4 mL of 0.494 M H$_2$SO$_4$. Calculate the mass of BaSO$_4$ formed and the pH of the mixed solution.

17.95 For which of the following reactions is the equilibrium constant called a solubility product?
(a) Zn(OH)$_2$(s) + 2OH$^-$(aq) $\rightleftharpoons$ Zn(OH)$_4^{2-}$(aq)
(b) 3Ca^{2+}(aq) + 2PO$_4^{3-}$(aq) $\rightleftharpoons$ Ca$_3$(PO$_4$)$_2$(s)
(c) CaCO$_3$(s) + 2H$^+$(aq) $\rightleftharpoons$
$\qquad\qquad$ Ca^{2+}(aq) + H$_2$O(l) + CO$_2$(g)
(d) PbI$_2$(s) $\rightleftharpoons$ Pb^{2+}(aq) + 2I$^-$(aq)

17.96 Water containing Ca^{2+} and Mg^{2+} ions is called *hard water* and is unsuitable for some household and industrial use because these ions react with soap to form insoluble salts, or curds. One way to remove the Ca^{2+} ions from hard water is by adding washing soda (Na$_2$CO$_3 \cdot$ 10H$_2$O). (a) The molar solubility of CaCO$_3$ is 9.3 × 10^{-5} M. What is its molar solubility in a 0.050 M Na$_2$CO$_3$ solution? (b) Why are Mg^{2+} ions not removed by this procedure? (c) The Mg^{2+} ions are removed as Mg(OH)$_2$ by adding slaked lime [Ca(OH)$_2$] to the water to produce a saturated solution. Calculate the pH of a saturated Ca(OH)$_2$ solution. (d) What is the concentration of Mg^{2+} ions at this pH? (e) In general, which ion (Ca^{2+} or Mg^{2+}) would you remove first? Why?

17.97 Equal volumes of 0.12 M AgNO$_3$ and 0.14 M ZnCl$_2$ solution are mixed. Calculate the equilibrium concentrations of Ag$^+$, Cl$^-$, Zn^{2+}, and NO$_3^-$.

17.98 Find the approximate pH range suitable for separating Mg^{2+} and Zn^{2+} by the precipitation of Zn(OH)$_2$ from a solution that is initially 0.010 M in Mg^{2+} and Zn^{2+}.

17.99 Calculate the solubility (in g/L) of Ag$_2$CO$_3$.

17.100 A volume of 25.0 mL of 0.100 M HCl is titrated against a 0.100 M CH$_3$NH$_2$ solution added to it from a burette. Calculate the pH values of the solution after (a) 10.0 mL of CH$_3$NH$_2$ solution has been added, (b) 25.0 mL of CH$_3$NH$_2$ solution has been added, (c) 35.0 mL of CH$_3$NH$_2$ solution has been added.

17.101 The molar solubility of Pb(IO$_3$)$_2$ in a 0.10 M NaIO$_3$ solution is 2.4 × 10^{-11} mol/L. What is K_{sp} for Pb(IO$_3$)$_2$?

17.102 When a KI solution was added to a solution of mercury(II) chloride, a precipitate [mercury(II) iodide] formed. A student plotted the mass of the precipitate versus the volume of the KI solution added and obtained the following graph. Explain the shape of the graph.

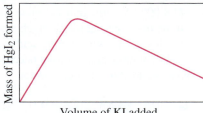

Volume of KI added

17.103 Which of the following compounds, when added to water, will increase the solubility of CdS: (a) LiNO$_3$, (b) Na$_2$SO$_4$, (c) KCN, (d) NaClO$_3$?

17.104 The pK_a of phenolphthalein is 9.10. Over what pH range does this indicator change from 95 percent HIn to 95 percent In$^-$?

17.105 Solid NaBr is slowly added to a solution that is 0.010 M in Cu$^+$ and 0.010 M in Ag$^+$. (a) Which compound will begin to precipitate first? (b) Calculate [Ag$^+$] when CuBr just begins to precipitate. (c) What percent of Ag$^+$ remains in solution at this point?

17.106 Cacodylic acid is $(CH_3)_2AsO_2H$. Its ionization constant is 6.4×10^{-7}. (a) Calculate the pH of 50.0 mL of a 0.10-M solution of the acid. (b) Calculate the pH of 25.0 mL of 0.15 M $(CH_3)_2AsO_2Na$. (c) Mix the solutions in parts (a) and (b). Calculate the pH of the resulting solution.

17.107 What reagents would you employ to separate the following pairs of ions in solution: (a) Na^+ and Ba^{2+}, (b) K^+ and Pb^{2+}, (c) Zn^{2+} and Hg^{2+}?

17.108 Look up the K_{sp} values for $BaSO_4$ and $SrSO_4$ in Table 17.4. Calculate the concentrations of Ba^{2+}, Sr^{2+}, and SO_4^{2-} in a solution that is saturated with both compounds.

17.109 In principle, amphoteric oxides, such as Al_2O_3 and BeO can be used to prepare buffer solutions because they possess both acidic and basic properties (see Section 16.11). Explain why these compounds are of little practical use as buffer components.

17.110 $CaSO_4$ ($K_{sp} = 2.4 \times 10^{-5}$) has a larger K_{sp} value than that of Ag_2SO_4 ($K_{sp} = 1.4 \times 10^{-5}$). Does it necessarily follow that $CaSO_4$ also has greater solubility (g/L)? Explain.

17.111 Describe how you would prepare a 1-L 0.20 M CH_3COONa/0.20 M CH_3COOH buffer system by (a) mixing a solution of CH_3COOH with a solution of CH_3COONa, (b) mixing a solution of CH_3COOH with a solution of NaOH, and (c) mixing a solution of CH_3COONa with a solution of HCl.

17.112 Phenolphthalein is the common indicator for the titration of a strong acid with a strong base. (a) If the pK_a of phenolphthalein is 9.10, what is the ratio of the nonionized form of the indicator (colorless) to the ionized form (reddish pink) at pH 8.00? (b) If 2 drops of 0.060 M phenolphthalein are used in a titration involving a 50.0-mL volume, what is the concentration of the ionized form at pH 8.00? (Assume that 1 drop = 0.050 mL.)

17.113 Of the reactions depicted, which best represents (a) what occurs when strong acid is added to a buffer solution, and (b) what occurs when strong base is added to a buffer solution?

⬤ = HA   ⬤ = A⁻   ◐ = H_3O^+ ⬤ = OH⁻   ◐ = H_2O

(a) ⬤ + ⬤ ⟶ ◐ + ⬤

(b) ⬤ + ◐ ⟶ ⬤ + ◐

(c) ⬤ + ◐ ⇌ ⬤ + ◐

(d) ⬤ + ◐ ⇌ ⬤ + ⬤

(e) ◐ + ⬤ ⟶ ◐ + ◐

17.114 The molar mass of a certain metal carbonate, MCO_3, can be determined by adding an excess of HCl acid to react with all the carbonate and then "back-titrating" the remaining acid with NaOH. (a) Write an equation for these reactions. (b) In a certain experiment, 20.00 mL of 0.0800 M HCl was added to a 0.1022-g sample of MCO_3. The excess HCl required 5.64 mL of 0.1000 M NaOH for neutralization. Calculate the molar mass of the carbonate and identify M.

17.115 Consider the ionization of the following acid-base indicator:

$$HIn(aq) + H_2O(l) \rightleftharpoons H_3O^+(aq) + In^-(aq)$$

The indicator changes color according to the ratios of the concentrations of the acid to its conjugate base. When $[HIn]/[In^-] \geq 10$, color of acid (HIn) predominates. When $[HIn]/[In^-] \leq 0.1$, color of conjugate base (In^-) predominates. Show that the pH range over which the indicator changes from the acid color to the base color is pH $= pK_a \pm 1$, where K_a is the ionization constant of the acid HIn.

17.116 One way to distinguish a buffer solution from an acid solution is to dilute both. (a) Consider a buffer solution made of 0.500 M CH_3COOH and 0.500 M CH_3COONa. Calculate its pH before and after it has been diluted 10-fold. (b) Compare the result in part (a) with the pH of a 0.500 M CH_3COOH solution before and after it has been diluted 10-fold.

17.117 (a) Referring to Figure 17.4, describe how you would determine the pK_b of the base. (b) Derive an analogous Henderson-Hasselbalch equation relating pOH to pK_b of a weak base B and its conjugate acid HB^+. Sketch a titration curve showing the variation of the pOH of the base solution versus the volume of a strong acid added from a burette. Describe how you would determine the pK_b from this curve.

17.118 $AgNO_3$ is added slowly to a solution that contains 0.1 M each of Br^-, CO_3^{2-}, and SO_4^{2-} ions. What compound will precipitate first and what compound will precipitate last?

17.119 The following diagrams represent solutions of MX, which may also contain one or both of the soluble salts, MNO_3 and NaX. (Na^+ and NO_3^- ions are not shown.) If (a) represents a saturated solution of MX, classify each of the other solutions as unsaturated, saturated, or supersaturated.

⬤ = M^+ ◯ = X^-

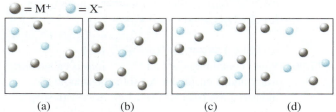

(a)   (b)   (c)   (d)

Engineering Problems

17.120 A 2.0-L kettle contains 116 g of boiler scale ($CaCO_3$). How many times would the kettle have to be completely filled with distilled water to remove all the deposit at 25°C?

17.121 Radiochemical techniques are useful in estimating the solubility product of many compounds. In one experiment, 50.0 mL of a 0.010 M $AgNO_3$ solution containing a silver isotope with a radioactivity of 74,025 counts per min per mL was mixed with 100 mL of a 0.030 M $NaIO_3$ solution. The mixed solution was diluted to 500 mL and filtered to remove all the $AgIO_3$ precipitate. The remaining solution was found to have a radioactivity of 44.4 counts per min per mL. What is the K_{sp} of $AgIO_3$?

Biological Problems

17.122 One of the most common antibiotics is penicillin G (benzylpenicillinic acid), which has the following structure:

It is a weak monoprotic acid:

$$HP \rightleftharpoons H^+ + P^- \quad K_a = 1.64 \times 10^{-3}$$

where HP denotes the parent acid and P^- the conjugate base. Penicillin G is produced by growing molds in fermentation tanks at 25°C and a pH range of 4.5 to 5.0. The crude form of this antibiotic is obtained by extracting the fermentation broth with an organic solvent in which the acid is soluble. (a) Identify the acidic hydrogen atom. (b) In one stage of purification, the organic extract of the crude penicillin G is treated with a buffer solution at pH = 6.50. What is the ratio of the conjugate base of penicillin G to the acid at this pH? Would you expect the conjugate base to be more soluble in water than the acid? (c) Penicillin G is not suitable for oral administration, but the sodium salt (NaP) is because it is soluble. Calculate the pH of a 0.12 M NaP solution formed when a tablet containing the salt is dissolved in a glass of water.

17.123 Barium is a toxic substance that can seriously impair heart function. For an X ray of the gastrointestinal tract, a patient drinks an aqueous suspension of 20 g $BaSO_4$. If this substance were to equilibrate with the 5.0 L of the blood in the patient's body, what would be $[Ba^{2+}]$? For a good estimate, we may assume that the K_{sp} of $BaSO_4$ at

body temperature is the same as at 25°C. Why is $Ba(NO_3)_2$ not chosen for this procedure?

17.124 Tris [tris(hydroxymethyl)aminomethane] is a common buffer for studying biological systems: (a) Calculate the pH of the tris buffer after mixing 15.0 mL of 0.10 M HCl solution with 25.0 mL of 0.10 M tris. (b) This buffer was used to study an enzyme-catalyzed reaction. As a result of the reaction, 0.00015 mole of H^+ was consumed. What is the pH of the buffer at the end of the reaction? (c) What would be the final pH if no buffer were present?

17.125 Calcium oxalate is a major component of kidney stones. Predict whether the formation of kidney stones can be minimized by increasing or decreasing the pH of the fluid present in the kidney. The pH of normal kidney fluid is about 8.2. [The first and second acid ionization constants of oxalic acid ($H_2C_2O_4$) are 6.5×10^{-2} and 6.1×10^{-5}, respectively. The solubility product of calcium oxalate is 3.0×10^{-9}.]

17.126 Histidine is one of the 20 amino acids found in proteins. Shown here is a fully protonated histidine molecule, where the numbers denote the pK_a values of the acidic groups:

(a) Show stepwise ionization of histidine in solution. (*Hint:* The H^+ ion will first come off from the strongest acid group followed by the next strongest acid group and so on.) (b) A dipolar ion is one in which the species has an equal number of positive and negative charges. Identify the dipolar ion in part (a). (c) The pH at which the dipolar ion predominates is called the isoelectric point, denoted by pI. The isoelectric point is the average of the pK_a values leading to and following the formation of the dipolar ion. Calculate the pI of histidine. (d) The histidine group plays an important role in buffering blood (the pH of blood is about 7.4). Which conjugate acid-base pair shown in part (a) is responsible for maintaining the pH of blood?

17.127 Amino acids are building blocks of proteins. These compounds contain at least one amino group ($-NH_2$) and one carboxyl group ($-COOH$). Consider glycine (NH_2CH_2COOH). Depending on the pH of the solution, glycine can exist in one of three possible forms:

Fully protonated: $^+NH_3-CH_2-COOH$
Dipolar ion: $^+NH_3-CH_2-COO^-$
Fully ionized: $NH_2-CH_2-COO^-$

Predict the predominant form of glycine at pH 1.0, 7.0, and 12.0. The pK_a of the carboxyl group is 2.3 and that of the ammonium group (NH_3^+) is 9.6.

Environmental Problems

17.128 For the past 300 million years, the pH of ocean water has been fairly steady at about 8.2. Today, it is 8.1. Determine what percentage increase in hydronium ion concentration this pH change represents.

17.129 The maximum allowable concentration of Pb^{2+} ions in drinking water is 0.05 ppm (i.e., 0.05 g of Pb^{2+} in 1 million grams of water). Is this guideline exceeded if an underground water supply is at equilibrium with the mineral anglesite ($PbSO_4$) ($K_{sp} = 1.6 \times 10^{-8}$)?

Multiconcept Problems

17.130 A sample of 0.96 L of HCl gas at 372 mmHg and 22°C is bubbled into 0.034 L of 0.57 M NH_3. What is the pH of the resulting solution? Assume the volume of solution remains constant and that the HCl is totally dissolved in the solution.

17.131 When lemon juice is added to tea, the color becomes lighter. In part, the color change is due to dilution, but the main reason for the change is an acid-base reaction. What is the reaction? (*Hint:* Tea contains "polyphenols," which are weak acids, and lemon juice contains citric acid.)

17.132 How many milliliters of 1.0 M NaOH must be added to 200 mL of 0.10 M NaH_2PO_4 to make a buffer solution with a pH of 7.50?

17.133 Which of the following solutions has the highest $[H_3O^+]$: (a) 0.10 M HF, (b) 0.10 M HF in 0.10 M NaF, (c) 0.10 M HF in 0.10 M SbF_5? (*Hint:* SbF_5 reacts with F^- to form the complex ion SbF_6^-.)

17.134 Distribution curves show how the fractions of a nonionized acid and its conjugate base vary as a function of the pH of the medium. Plot distribution curves for CH_3COOH and its conjugate base CH_3COO^- in solution. Your graph should show fraction as the y axis and pH as the x axis. What are the fractions and pH at the point where these two curves intersect?

17.135 A 1.0-L saturated silver carbonate solution at 5°C is filtered to remove undissolved solid and treated with enough hydrochloric acid to decompose the dissolved compound. The carbon dioxide generated is collected in a 19-mL vial and exerts a pressure of 114 mmHg at 25°C. What is the K_{sp} of silver carbonate at 5°C?

17.136 Draw distribution curves for an aqueous carbonic acid solution. Your graph should show fraction of species present as the y axis and pH as the x axis. Note that at any pH, only two of the three species (H_2CO_3, HCO_3^-, and CO_3^{2-}) are present in appreciable concentrations. Use the K_a values in Table 16.8.

17.137 Acid-base reactions usually go to completion. Confirm this statement by calculating the equilibrium constant for each of the following cases: (a) A strong acid reacting with a strong base. (b) A strong acid reacting with a weak base (NH_3). (c) A weak acid (CH_3COOH) reacting with a strong base. (d) A weak acid (CH_3COOH) reacting with a weak base (NH_3). (*Hint:* Strong acids exist as H_3O^+ ions and strong bases exist as OH^- ions in solution. You need to look up K_a, K_b, and K_w.)

17.138 Calculate x, the number of molecules of water in oxalic acid hydrate ($H_2C_2O_4 \cdot xH_2O$), from the following data: 5.00 g of the compound is made up to exactly 250 mL solution, and 25.0 mL of this solution requires 15.9 mL of 0.500 M NaOH solution for neutralization.

Standardized-Exam Practice Problems

Physical and Biological Sciences

Aqueous acid reacts with carbonate ions to produce carbonic acid, which produces carbon dioxide. A 1.0-L saturated silver carbonate solution at 5°C is treated with enough hydrochloric acid to consume all the carbonate in solution. The carbon dioxide generated is collected in a 19-mL vial and exerts a pressure of 114 mmHg at 25°C.

1. Which of the following reactions represents the overall process of acid and carbonate reacting to give carbon dioxide?

   a) $CO_3^{2-} + 2H^+ \longrightarrow H_2CO_3$
   b) $H_2CO_3 \rightleftharpoons H_2O + CO_2$
   c) $CO_3^{2-} + 2H^+ \longrightarrow H_2O + CO_2$
   d) $Ag_2CO_3 + 2H^+ \longrightarrow 2Ag^+ + CO_2 + H_2O$

2. What is the K_{sp} of silver carbonate at 5°C?

   a) 2.5×10^{-11}
   b) 5.4×10^{-8}
   c) 6.3×10^{-12}
   d) 2.7×10^{-8}

3. At 25°C, the K_{sp} of silver carbonate is 8.1×10^{-12}. Based on this and the answer to question 1, what can be said about the dissolution of silver carbonate?

   a) It is endothermic.
   b) It is exothermic.
   c) It is neither exothermic nor endothermic.
   d) It produces hydrogen gas.

4. Which of the following, if added to a saturated solution of Ag_2CO_3, would increase the solubility of Ag_2CO_3?

   a) Na_2CO_3
   b) $NaHCO_3$
   c) $AgNO_3$
   d) HNO_3

Answers to In-Chapter Materials

Answers to Practice Problems

17.1A 4.22. **17.1B** 9.11. **17.2A** 5.0. **17.2B** 0.33 mole.
17.3A Dissolve 0.6 mol CH_3COONa and 1 mol CH_3COOH in enough water to make 1 L of solution. **17.3B** 2.35–4.35.
17.4A (a) 5.04, (b) 8.76, (c) 12.2. **17.4B** (a) 8.3 mL, (b) 12.5 mL, (c) 27.0 mL. **17.5A** 5.91. **17.5B** Aniline. **17.6A** (a) Bromophenol blue, methyl orange, methyl red, or chlorophenol; (b) any but thymol blue, bromophenol blue, or methyl orange; (c) cresol red or phenolphthalein. **17.6B** Urea. **17.7A** (a) 1.3×10^{-5} M, 1.8×10^{-3} g/L; (b) 1.0×10^{-13} M, 1.5×10^{-11} g/L; (c) 4.0×10^{-5} M, 5.9×10^{-3} g/L. **17.7B** (a) 2.2×10^{-3} M, 0.53 g/L; (b) 1.3×10^{-4} M, 3.5×10^{-2} g/L; (c) 1.7×10^{-15} M, 8.8×10^{-13} g/L. **17.8A** (a) 1.5×10^{-14}, (b) 1.7×10^{-6}, (c) 1.3×10^{-15}. **17.8B** (a) 1.5×10^{-14}, (b) 4.8×10^{-29}, (c) 9.0×10^{-33}. **17.9A** (a) Yes, (b) yes, (c) no. **17.9B** 4.4×10^{-8} g, (b) 1.2 g, (c) 0.037 g. **17.10A** (a) 9.1×10^{-9} M, (b) 8.3×10^{-14} M.
17.10B $Ag_2S < AgI < AgBr < AgCl < Ag_2CO_3$. **17.11A** (a) Yes, (b) yes, (c) no. **17.11B** Any salts containing the CO_3^{2-} ion, the OH^- ion, the S^{2-} ion, or any anion of a weak acid such as the SO_3^{2-} ion or the F^- ion. **17.12A** 1.8×10^{-34} M. **17.12B** 6.8×10^{-2} M. **17.13A** 0.60 M. **17.13B** (a) 1.6×10^{-9} M, (b) 2.6×10^{-6} M.

Answers to Checkpoints

17.1.1 a, c, d. **17.1.2** b. **17.2.1** b, c, e. **17.2.2** a. **17.2.3** c. **17.2.4** b. **17.2.5** a, b. **17.2.6** d. **17.3.1** a, c, d, e. **17.3.2** c. **17.3.3** a. **17.3.4** a. **17.3.5** b. **17.3.6** e. **17.3.7** d. **17.3.8** a. **17.3.9** c. **17.4.1** a. **17.4.2** c. **17.4.3** e. **17.4.4** c. **17.4.5** b, c, d, e. **17.5.1** c. **17.5.2** a, b, d, e. **17.5.3** b, d. **17.6.1** d. **17.6.2** b.

CHAPTER 18

Entropy, Free Energy, and Equilibrium

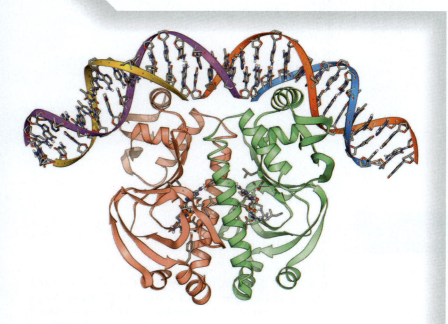

Ribbon diagram, also known as a Richardson diagram, of a gene activator protein.

LAGUNA DESIGN/Science Photo Library/Science Source

In This Chapter, You Will Learn

About the three laws of thermodynamics and how to use thermodynamic quantities to determine whether or not a process is expected to be spontaneous.

Before You Begin, Review These Skills

- System and surroundings [◄◄ Section 5.1]
- Hess's law [◄◄ Section 5.5]
- Equilibrium expressions [◄◄ Section 15.3]
- Chapter 5 Key Skills [◄◄ pages 238–239]

How Living Systems Must Obey the Laws of Thermodynamics

Under normal physiological conditions, polypeptide chains fold spontaneously into unique three-dimensional structures known as native proteins, which perform various biological functions. Because an unfolded polypeptide chain can assume many possible configurations while the corresponding native protein can have only one specific arrangement, the folding process is accompanied by a decrease in entropy of the system. (Note that solvent molecules, water in this case, can also play a role in affecting the entropy change.) In accord with the second law of thermodynamics, any spontaneous process must result in an increase in the entropy of the universe. It follows, therefore, that there must be an increase in the entropy of the surroundings that outweighs the decrease in the entropy of the system. The intramolecular attractions between amino acid residues cause the folding of the polypeptide chain to be exothermic. The energy produced by the process spreads out, increasing molecular motion in the surroundings—thereby increasing the entropy of the surroundings.

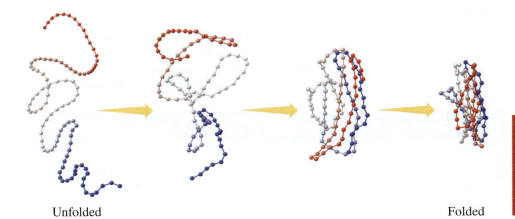

Unfolded   Folded

At the end of this chapter, you will be able to answer several questions regarding entropy changes [►► Applying What You've Learned, page 948].

18.1   Spontaneous Processes

Student Note: The conditions that most often are specified are *temperature, pressure,* and in the case of a solution, *concentration.*

An understanding of thermodynamics enables us to predict whether or not a reaction will occur when reactants are combined. This is important in the synthesis of new compounds in the laboratory, the manufacturing of chemicals on an industrial scale, and the understanding of natural processes such as cell function. A process that *does* occur under a specific set of conditions is called a ***spontaneous process.*** One that does *not* occur under a specific set of conditions is called ***nonspontaneous.*** Table 18.1 lists examples of familiar spontaneous processes and their nonspontaneous counterparts. These examples illustrate what we know intuitively: Under a given set of conditions, a process that occurs spontaneously in one direction does not also occur spontaneously in the opposite direction.

Processes that result in a decrease in the energy of a system often are spontaneous. For example, the combustion of methane is exothermic:

$$CH_4(g) + 2O_2(g) \longrightarrow CO_2(g) + 2H_2O(l) \qquad \Delta H° = -890.4 \text{ kJ/mol}$$

Thus, the energy of the system is lowered because heat is given off during the course of the reaction. Likewise, in the acid-base neutralization reaction:

$$H^+(aq) + OH^-(aq) \longrightarrow H_2O(l) \qquad \Delta H° = -56.2 \text{ kJ/mol}$$

heat is given off, lowering the energy of the system. Each of these processes is spontaneous, and each results in a lowering of the system's energy.

Now consider the melting of ice:

$$H_2O(s) \longrightarrow H_2O(l) \qquad \Delta H° = 6.01 \text{ kJ/mol}$$

In this case, the process is endothermic and yet it is also spontaneous at temperatures above 0°C. Conversely, the freezing of water is an *exothermic* process:

$$H_2O(l) \longrightarrow H_2O(s) \qquad \Delta H° = -6.01 \text{ kJ/mol}$$

Yet it is *not* spontaneous at temperatures above 0°C.

Based on the first two examples, and many others like them, we might conclude that exothermic processes tend to be spontaneous and, indeed, a negative ΔH does *favor* spontaneity. The last two examples, however, make it clear that the sign of ΔH *alone* is insufficient to predict spontaneity in every circumstance. For the remainder of this chapter, we examine the *two* factors that determine whether or not a process is spontaneous under a given set of conditions.

18.2   Entropy

To predict the spontaneity of a chemical or physical process, we need to know both the change in *enthalpy* [◄◄ Section 5.3] and the change in *entropy* associated with the process. We first encountered the concept of entropy in our discussion of solution formation [◄◄ Section 13.2]. We now look in more detail at what entropy is, and why it matters.

TABLE 18.1	Familiar Spontaneous and Nonspontaneous Processes
Spontaneous	**Nonspontaneous**
Ice melting at room temperature	Water freezing at room temperature
Sodium metal reacting violently with water to produce sodium hydroxide and hydrogen gas [◄◄ Section 7.7]	Sodium hydroxide reacting with hydrogen gas to produce sodium metal and water
A ball rolling downhill	A ball rolling uphill
The rusting of iron at room temperature	The conversion of rust back to iron metal at room temperature
Water freezing at −10°C	Ice melting at −10°C

A Qualitative Description of Entropy

Qualitatively, the **entropy (S)** of a system is a measure of how *spread out* or how *dispersed* the system's energy is. The simplest interpretation of this is how spread out a system's energy is in *space*. In other words, for a given system, the greater the volume it occupies, the greater its entropy. This interpretation explains how the process in Figure 18.1 occurs spontaneously despite there being no enthalpy change. Because they are moving, the gas molecules that were originally confined to one side of the container possess *motional energy*. Motional energy includes *translational* energy, in which the entire molecule moves through space [◄◄ Section 5.1]; *rotational energy,* in which the molecule spins about an axis running through its center of mass; and *vibrational energy,* in which atoms of a molecule move relative to one another. In the absence of a barrier preventing it, the motional energy of molecules will spread out to occupy a larger volume. The dispersal of a system's motional energy to occupy a larger volume when the barrier is removed constitutes an *increase* in the system's entropy. Just as spontaneity is favored by a process being exothermic, spontaneity is also favored by an increase in the system's entropy. Whether it is the enthalpy change, the entropy change, or both, for a process to be spontaneous, *something* must favor spontaneity.

A Quantitative Definition of Entropy

At this point, it is useful to introduce the mathematical definition of entropy proposed by Ludwig Boltzmann:

$$S = k \ln W \qquad \text{Equation 18.1}$$

where k is the Boltzmann constant (1.38×10^{-23} J/K) and W is the number of energetically equivalent different ways the molecules in a system can be arranged. To illustrate what this means, let's consider a simplified version of the process shown in Figure 18.1. Prior to the removal of the barrier between the left and right sides of the container, at any given instant, each molecule has a particular location, somewhere in the left side of the container. To narrow down the possible locations of the molecules, we imagine that each side of the container is divided into a number of equal smaller volumes called *cells*. In the simplest scenario, with just one molecule in the system, the number of possible locations of the molecule is equal to the number of cells. If the system contains *two* molecules, the number of possible arrangements is equal to the number of cells *squared*. (Note that a cell may contain more than one molecule.) Each time we increase the number of molecules by one, the number of possible arrangements increases by a factor equal to the number of cells. In general, for a volume consisting of X cells, and containing N molecules, the number of possible arrangements, W, is given by the equation:

$$W = X^N \qquad \text{Equation 18.2}$$

Figure 18.2 illustrates this for a simple case involving just two molecules. We imagine the container is divided into four cells each with volume v. Initially, both molecules are confined to the left side, which consists of two cells. With two molecules in two cells,

Student Note: The Boltzmann constant is equal to the gas constant, R (in J/K · mol), divided by Avogadro's constant, N_A.

Student Note: The number of possible arrangements is sometimes called the number of *microstates*.

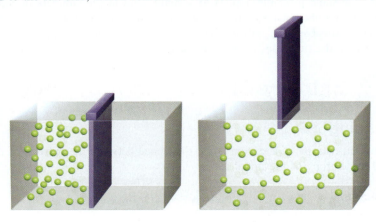

Figure 18.1 A spontaneous process. The rapidly moving gas molecules originally confined to one side of a container spread out to fill the whole container when the barrier is removed.

Figure 18.2 (a) Before the barrier is removed, the molecules are both in the left side of the container, which we imagine is divided into two cells of equal volume. There are four possible arrangements of two molecules in two cells. (b) When the barrier between the two sides of the container is removed, the volume (and the number of cells) available to the molecules doubles. The new number of possible arrangements is $4^2 = 16$, eight of which have the molecules in opposite sides of the container—the most probable outcome.

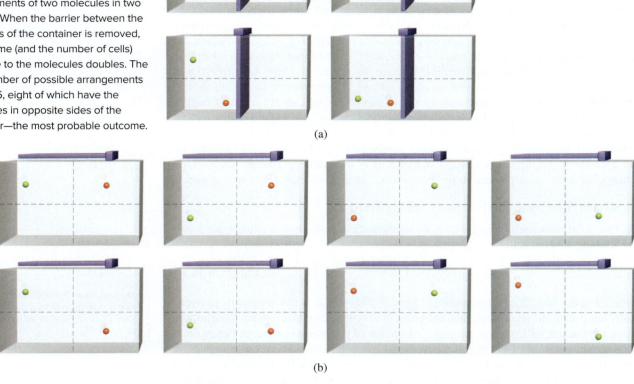

(a)

(b)

there are $2^2 = 4$ possible arrangements of the molecules [Figure 18.2(a)]. When the barrier is removed, doubling the volume available to the molecules, the number of cells also doubles. With four cells available, there are $4^2 = 16$ possible arrangements of the molecules. Eight of the sixteen arrangements have the molecules on opposite sides of the container [Figure 18.2(b)]. Of the other eight arrangements, four have both molecules on the left side [as shown in Figure 18.2(a)], and four have both molecules on the right side (not shown). There are three different states possible for this system.

1. One molecule on each side (eight possible arrangements)
2. Both molecules on the left (four possible arrangements)
3. Both molecules on the right (four possible arrangements)

Student Note: The state with the largest number of possible arrangements has the greatest entropy.

The most *probable* state is the one with the *largest number of possible arrangements.* In this case, the most probable state is the one with one molecule on each side of the container. The same principle applies to systems with larger numbers of molecules. Increasing the number of molecules increases the number of possible arrangements, but the most probable state will be the one in which the gas molecules are divided evenly between the two sides of the container.

18.3   Entropy Changes in a System

Calculating ΔS_{sys}

The change in entropy of a system is the difference between the entropy of the final state and the entropy of the initial state:

Equation 18.3 $\Delta S_{sys} = S_{final} - S_{initial}$

Using Equation 18.1, we can write an expression for the entropy of each state:

$$\Delta S_{\text{sys}} = k \ln W_{\text{final}} - k \ln W_{\text{initial}} = k \ln \frac{W_{\text{final}}}{W_{\text{initial}}}$$

Combining this result with Equation 18.2 gives:

$$\Delta S_{\text{sys}} = k \ln \frac{(X_{\text{final}})^N}{(X_{\text{initial}})^N} = k \ln \left(\frac{X_{\text{final}}}{X_{\text{initial}}} \right)^N = k N \ln \left(\frac{X_{\text{final}}}{X_{\text{initial}}} \right)$$

Because X is the number of cells, and the volume of each cell is v, the total volume is related to the number of cells by:

$$V = Xv \qquad \text{or} \qquad X = \frac{V}{v}$$

We substitute V_{final}/v for X_{final} and V_{initial}/v for X_{initial} to get:

$$\Delta S_{\text{sys}} = k N \ln \frac{V_{\text{final}}/v}{V_{\text{initial}}/v} = k N \ln \frac{V_{\text{final}}}{V_{\text{initial}}}$$

Finally, because the Boltzmann constant, k, is the gas constant, R, divided by Avogadro's constant:

$$k = \frac{R}{N_A}$$

and because the number of molecules, N, is the product of the number of moles, n, and Avogadro's constant, N_A:

$$N = n \times N_A$$

$$kN = \left(\frac{R}{N_A} \times n \times N_A \right) = nR$$

the equation becomes:

$$\Delta S_{\text{sys}} = nR \ln \frac{V_{\text{final}}}{V_{\text{initial}}} \qquad \textbf{Equation 18.4}$$

> **Student Note:** Remember that the difference between two logs is equal to the log of the corresponding quotient:
>
> $$\ln A - \ln B = \ln \frac{A}{B}$$
>
> and that $\ln A^x = x \ln A$ [▶▶ Appendix 1]

Sample Problem 18.1 shows how to use Equation 18.4 to calculate the entropy change for a process like the one shown in Figure 18.1, the expansion of an ideal gas at constant temperature.

SAMPLE PROBLEM 18.1

Determine the change in entropy for 1.0 mole of an ideal gas originally confined to one-half of a 5.0-L container when the gas is allowed to expand to fill the entire container at constant temperature.

Strategy This is the isothermal expansion of an ideal gas. Because the molecules spread out to occupy a greater volume, we expect there to be an increase in the entropy of the system. Use Equation 18.3 to solve for ΔS_{sys}.

Setup $R = 8.314$ J/K · mol, $n = 1.0$ mole, $V_{\text{final}} = 5.0$ L, and $V_{\text{initial}} = 2.5$ L.

Solution

$$\Delta S_{\text{sys}} = nR \ln \frac{V_{\text{final}}}{V_{\text{initial}}} = 1.0 \ \text{mol} \times \frac{8.314 \ \text{J}}{\text{K} \cdot \text{mol}} \times \ln \frac{5.0 \ \text{L}}{2.5 \ \text{L}} = 5.8 \ \text{J/K}$$

THINK ABOUT IT

Remember that for a process to be spontaneous, *something* must favor spontaneity. If the process is spontaneous but not exothermic (in this case, there is no enthalpy change), then we should expect ΔS_{sys} to be positive.

Practice Problem Ⓐ TTEMPT Determine the change in entropy (ΔS_{sys}), for the expansion of 0.10 mole of an ideal gas from 2.0 L to 3.0 L at constant temperature.

(Continued on next page)

Practice Problem BUILD To what fraction of its original volume must a 0.50-mol sample of ideal gas be compressed at constant temperature for ΔS_{sys} to be -6.7 J/K?

Practice Problem CONCEPTUALIZE Which equation is correct for calculating ΔS_{sys} for a gaseous reaction that occurs at constant volume?

$$\Delta S_{sys} = nR \qquad \Delta S_{sys} = nRT \ln \frac{P_{initial}}{P_{final}} \qquad \Delta S_{sys} = nR \ln \frac{P_{initial}}{P_{final}} \qquad \begin{array}{l}\text{There is not enough information} \\ \text{provided to determine this.}\end{array}$$

$$\qquad\quad \text{(i)} \qquad\qquad\qquad\quad \text{(ii)} \qquad\qquad\qquad\qquad \text{(iii)} \qquad\qquad\qquad\qquad\quad \text{(iv)}$$

Standard Entropy, $S°$

Student Note: For even the simplest of hypothetical systems, where there are only two possible positions for molecules ($X = 2$), most calculators cannot display a number as large as the result of Equation 18.2 for even as few as 500 molecules—much less for the *enormous* number of molecules present in any *real* sample. (If your calculator is like most, with $X = 2$, you can calculate the number of possible arrangements for $N \le 332$ molecules. Try it: $2 \wedge 332 = ?$ and $2 \wedge 333 = ?$)

Although Equation 18.1 provides a quantitative definition of entropy, we seldom use it or Equation 18.3 to calculate the entropy change for a real process because of the difficulty involved in determining W, the number of different possible arrangements (Equation 18.2) in a macroscopic system. Instead, for processes other than isothermal expansion or compression of an ideal gas (for which we can use Equation 18.4), we routinely determine entropy changes using tabulated values.

Using calorimetry [◄◄ Section 5.4], it is possible to determine the *absolute* value of the entropy of a substance, $S;$ something we cannot do with either energy or enthalpy. (Recall that while we can determine ΔU and ΔH for a process that a system undergoes, we cannot determine the absolute values of either U or H for a system [◄◄ Sections 5.2 and 5.3].) **Standard entropy ($S°$)** is the absolute entropy of a substance at 1 atm. (Tables of standard entropy values typically are the values at 25°C because so many processes are carried out at room temperature—although temperature is *not* part of the standard state definition and therefore must be specified.) Table 18.2 lists standard entropies of a few elements and compounds. Appendix 2 provides a more extensive listing. The units of entropy are J/K · mol. We use joules rather than kilojoules because entropy values typically are quite small. The entropies of substances (elements and compounds) are always positive (i.e., $S > 0$), even for elements in their standard states. (Recall that the standard *enthalpy* of formation, $\Delta H_f°$, for elements in their standard states is arbitrarily defined as zero, and for compounds it may be either positive or negative [◄◄ Section 5.6].) You will find that tables, including Appendix 2, contain *negative* absolute entropies for some aqueous ions. Unlike a substance, an individual ion cannot be studied experimentally. Therefore, standard entropies of ions are actually *relative* values, where a standard entropy of zero is arbitrarily assigned to the hydrated hydrogen (hydronium) ion. Depending on an ion's extent of hydration, its standard entropy may be positive or negative, relative to that of hydrogen (hydronium) ion.

Referring to Table 18.2, we can identify several important trends:

- For a given substance, the standard entropy is greater in the liquid phase than in the solid phase. [Compare the standard entropies of Na(s) and Na(l).] This

TABLE 18.2	Standard Entropy Values ($S°$) for Some Substances at 25°C		
Substance	$S°$ (J/K · mol)	Substance	$S°$ (J/K · mol)
$H_2O(l)$	69.9	C(diamond)	2.4
$H_2O(g)$	188.7	C(graphite)	5.69
Na(s)	51.05	$O_2(g)$	205.0
Na(l)	57.56	$O_3(g)$	237.6
Na(g)	153.7	$F_2(g)$	203.34
He(g)	126.1	Au(s)	47.7
Ne(g)	146.2	Hg(l)	77.4

results from there being greater molecular motion in a liquid, resulting in many possible arrangements of atoms in the liquid phase; whereas the positions of atoms in the solid are fixed.

- For a given substance, the standard entropy is greater in the gas phase than in the liquid phase. [Compare the standard entropies of Na(l) and Na(g) and those of $H_2O(l)$ and $H_2O(g)$.] This results from there being much greater molecular motion in a gas, resulting in many more possible arrangements of atoms in the gas phase than in the liquid phase—in part because the gas phase occupies a much greater volume than either of the condensed phases.

- For two monatomic species, the one with the larger molar mass has the greater standard entropy. [Compare the standard entropies of He(g) and Ne(g).]

- For two substances in the same phase, and with similar molar masses, the substance with the more complex molecular structure has the greater standard entropy. [Compare the standard entropies of $O_3(g)$ and $F_2(g)$.] The more complex a molecular structure, the more different types of motion the molecule can exhibit. A diatomic molecule such as F_2, for example, exhibits only one type of vibration, whereas a bent triatomic molecule such as O_3 exhibits three different types of vibrations. Each mode of motion contributes to the total number of available energy levels within which a system's energy can be dispersed. Figure 18.3 illustrates the ways in which the F_2 and O_3 molecules can rotate and vibrate.

- In cases where an element exists in two or more allotropic forms, the form in which the atoms are more mobile has the greater entropy. [Compare the

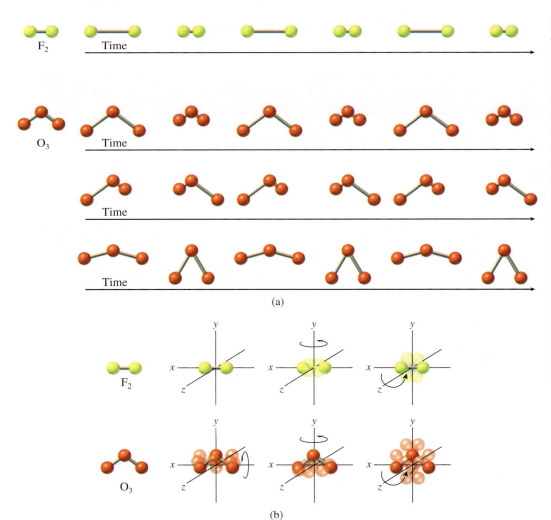

Figure 18.3 In addition to translational motion, molecules exhibit both *vibrations,* in which the atoms' positions relative to one another change, and *rotations,* in which the molecule rotates about its center of mass. (a) A diatomic molecule such as fluorine only exhibits one type of vibration. A bent, triatomic molecule such as ozone exhibits three types of vibration. (b) A diatomic molecule exhibits two different rotations, whereas a bent, triatomic molecule exhibits three different rotations. (Note that rotation of F_2 about the x axis would cause no change in the positions of either atom in the molecule.)

each of the possible molecular speeds within the range as a discrete energy level, we can see that at higher temperatures, there is a greater number of possible molecular speeds and, therefore, a greater number of energy levels available to the molecules in the system. With a greater number of available energy levels, there is a greater number of possible arrangements of molecules *within* those levels and, therefore, a greater entropy.

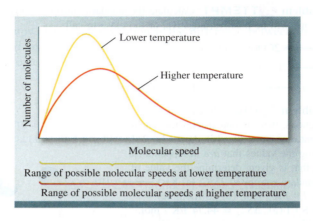

Animation
Figure 18.4, Factors That Influence the Entropy of a System.

Because the entropy of a substance in the gas phase is always significantly greater than its entropy in either the liquid or solid phase, a reaction that results in an increase in the number of gas molecules causes an increase in the system's entropy. For reactions that do not involve gases, an increase in the number of solid, liquid, or aqueous molecules also usually causes an entropy increase.

By considering these factors, we can usually make a reasonably good prediction of the sign of ΔS°_{rxn} for a physical or chemical process, without having to look up the absolute entropy values for the species involved. Figure 18.4 summarizes the factors that can be used to compare entropies and illustrates several comparisons.

In addition to melting, vaporization/sublimation, temperature increase, and reactions that increase the number of gas molecules, which can always be counted upon to result in an entropy increase, the process of *dissolving* a substance often leads to an increase in entropy. In the case of a molecular solute, such as sucrose (sugar), dissolving causes dispersal of the molecules (and, consequently, of the system's *energy*) into a larger volume—resulting in an increase in entropy. In the case of an ionic solute, the analysis is slightly more complicated. We saw in our discussion of solution formation [◄◄ Section 13.2] that the dissolution of ammonium nitrate (NH_4NO_3) is spontaneous, even though it is endothermic, because the system's entropy increases when the ionic solid dissociates and is dispersed throughout the solution. In general, this is the case for ionic solutes in which the charges on ions are small. (In the case of NH_4NO_3, they are +1 and −1.) However, when ions are dispersed in water, they become hydrated (surrounded by water molecules in a specific arrangement [◄◄ Figure 4.4]). This leads to a *decrease* in the entropy of the *water,* as hydration reduces the mobility of some of the water molecules by fixing them in positions around the dissolved ions. When the charges on ions are low, the increase in entropy of the solute typically outweighs the decrease in entropy of the water—resulting in an overall *increase* in the entropy of the system—as is the case with NH_4NO_3. By contrast, when highly charged ions such as Al^{3+} and Fe^{3+} are hydrated, the decrease in entropy of the water can actually outweigh the increase in entropy of the solute, leading to an overall *decrease* in entropy of the system. Table 18.3 lists the changes in entropy associated with the spontaneous dissolution of several ionic solids.

TABLE 18.3	Entropy Changes for the Dissolution (ΔS°_{soln})* of Some Ionic Solids at 25°C	
Dissolution Equation		**ΔS°_{soln} (J/K · mol)**
$NH_4NO_3(s) \longrightarrow NH_4^+(aq) + NO_3^-(aq)$		108.1
$AlCl_3(s) \longrightarrow Al^{3+}(aq) + 3Cl^-(aq)$		−253.2
$FeCl_3(s) \longrightarrow Fe^{3+}(aq) + 3Cl^-(aq)$		−266.1

*Note that the subscript "rxn" in ΔS°_{rxn} changes to "soln" to refer specifically to the *solution* process.

Sample Problem 18.3 lets you practice making qualitative predictions of the sign of ΔS°_{rxn}.

SAMPLE PROBLEM 18.3

For each process, determine the sign of ΔS for the system: (a) decomposition of $CaCO_3(s)$ to give $CaO(s)$ and $CO_2(g)$, (b) heating bromine vapor from 45°C to 80°C, (c) condensation of water vapor on a cold surface, (d) reaction of $NH_3(g)$ and $HCl(g)$ to give $NH_4Cl(s)$, and (e) dissolution of sugar in water.

Strategy Consider the change in energy/mobility of atoms and the resulting change in number of possible positions that each particle can occupy in each case. An increase in the number of arrangements corresponds to an increase in entropy and therefore a positive ΔS.

Setup Increases in entropy generally accompany solid-to-liquid, liquid-to-gas, and solid-to-gas transitions; the dissolving of one substance in another; a temperature increase; and reactions that increase the net number of moles of gas.

Solution ΔS is (a) positive, (b) positive, (c) negative, (d) negative, and (e) positive.

THINK ABOUT IT

For reactions involving only liquids and solids, predicting the sign of ΔS° can be more difficult, but in many such cases an increase in the total number of molecules and/or ions is accompanied by an increase in entropy.

Practice Problem ATTEMPT For each of the following processes, determine the sign of ΔS: (a) crystallization of sucrose from a supersaturated solution, (b) cooling water vapor from 150°C to 110°C, (c) sublimation of dry ice.

Practice Problem BUILD Make a qualitative prediction of the sign of ΔH°_{soln} for $AlCl_3(s)$ and the dissolution of $FeCl_3(s)$. See Table 18.3. Explain your reasoning.

Practice Problem CONCEPTUALIZE Consider the gas-phase reaction of A_2 (blue) and B_2 (orange) to form AB_3. What are the correct balanced equation and the sign of ΔS for the reaction?

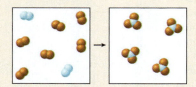

(a) $A_2 + B_2 \longrightarrow AB_3$, negative

(b) $2A_2 + 3B_2 \longrightarrow 4AB_3$, positive

(c) $2A_2 + 3B_2 \longrightarrow 4AB_3$, negative

(d) $A_2 + 3B_2 \longrightarrow 2AB_3$, negative

(e) $A_2 + 3B_2 \longrightarrow 2AB_3$, positive

Figure 18.4

Factors That Influence the Entropy of a System

Volume Change

Quantum mechanical analysis shows that the spacing between translational energy levels is inversely proportional to the volume of the container. Thus, when the volume is increased, more energy levels become available within which the system's energy can be dispersed.

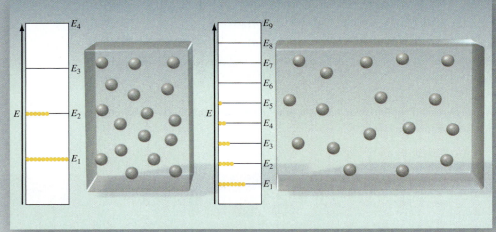

Temperature Change

At higher temperatures, molecules have greater kinetic energy—making more energy levels accessible. This increases the number of energy levels within which the system's energy can be dispersed, causing entropy to increase.

Molecular Complexity

Unlike atoms, which exhibit only translational motion, molecules can also exhibit rotational and vibrational motions. The greater a molecule's complexity, the greater the number of possible ways it can rotate and vibrate. The ozone molecule (O_3), for example, is more complex than the fluorine molecule (F_2) and exhibits more different kinds of vibrations and rotations. (See Figure 18.3.) This results in more energy levels within which the system's energy can be dispersed. The number and spacing of additional energy levels have been simplified to keep the illustration clear.

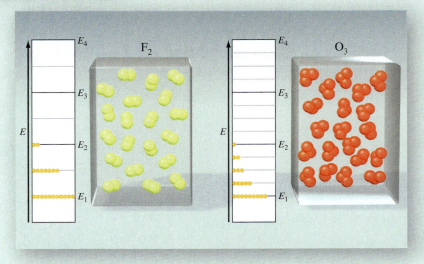

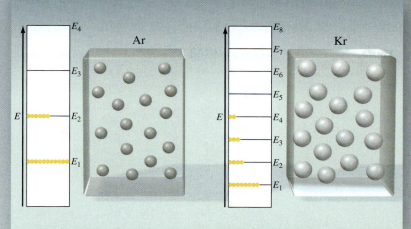

Molar Mass

The energy levels for a substance with a larger molar mass are more closely spaced. Kr, for example, has roughly twice the molar mass of Ar. Thus, Kr has roughly twice as many energy levels within which the system's energy can be dispersed.

Phase Change

Because of greater mobility, there are many more different possible arrangements (W) of molecules in the liquid phase than there are in the solid phase; and there are many, *many* more different possible arrangements of molecules in the gas phase than there are in the liquid phase. Entropy of a substance increases when it is melted ($s \rightarrow l$), vaporized ($l \rightarrow g$), or sublimed ($s \rightarrow g$).

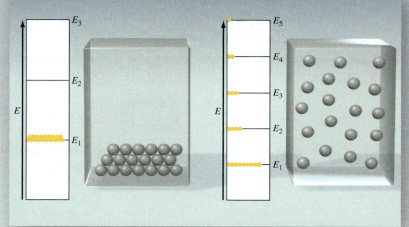

$$N_2O_4(g) \longrightarrow 2NO_2(g)$$

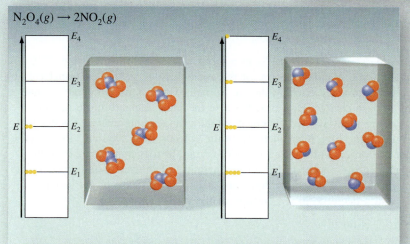

Chemical Reaction

When a chemical reaction produces more gas molecules than it consumes, the number of different possible arrangements of molecules (W) increases and entropy increases.

What's the point?

Although several factors can influence the entropy of a system or the entropy change associated with a process, often one factor dominates the outcome. Each of these comparisons shows a qualitative illustration of one of the important factors.

(See Visualizing Chemistry questions VC 18.1–VC 18.4 on pages 948–949.)

CHECKPOINT – SECTION 18.3   Entropy Changes in a System

18.3.1 For which of the following physical processes is ΔS negative? (Select all that apply.)

a) Freezing ethanol

b) Evaporating water

c) Mixing carbon tetrachloride and benzene

d) Heating water

e) Condensing bromine vapor

18.3.2 For which of the following chemical reactions is ΔS negative? (Select all that apply.)

a) $2O_3(g) \longrightarrow 3O_2(g)$

b) $4Fe(s) + 3O_2(g) \longrightarrow 2Fe_2O_3(s)$

c) $2H_2O_2(aq) \longrightarrow 2H_2O(l) + O_2(g)$

d) $2Li(s) + 2H_2O(l) \longrightarrow 2LiOH(aq) + H_2(g)$

e) $2NH_3(g) \longrightarrow N_2(g) + 3H_2(g)$

18.3.3 Identify the correct balanced equation and the sign of ΔS for the reaction shown here.

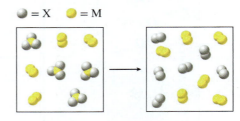

a) $4MX_3 + 4M_2 \longrightarrow 6X_2 + 6M_2$, ΔS positive

b) $4MX_3 \longrightarrow 6X_2 + 2M_2$, ΔS negative

c) $2MX_3 \longrightarrow 3X_2 + M_2$, ΔS positive

d) $2MX_3 + 2M_2 \longrightarrow 3X_2 + 3M_2$, ΔS positive

e) $2MX_3 + 2M_2 \longrightarrow 3X_2 + 3M_2$, ΔS negative

18.4   Entropy Changes in the Universe

Recall that the *system* typically is the part of the universe we are investigating (e.g., the reactants and products in a chemical reaction). The *surroundings* are everything else [◄◄ Section 5.1]. Together, the system and surroundings make up the *universe*. We have seen that the *dispersal* or *spreading out* of a system's energy corresponds to an increase in the system's entropy. Moreover, an increase in the system's entropy is one of the factors that determines whether or not a process is spontaneous. However, correctly predicting the spontaneity of a process requires us to consider entropy changes in both the system and the surroundings.

Consider the following processes:

- An ice cube spontaneously melts in a room where the temperature is 25°C. In this case, the motional energy of the air molecules at 25°C is transferred to the ice cube (at 0°C), causing the ice to melt. There is no temperature change during a phase change. However, because the molecules are more mobile and there are many more different possible arrangements in *liquid* water than there are in ice, there is an increase in the entropy of the system. In this case, because the process of melting is endothermic, heat is transferred *from* the surroundings *to* the system and the temperature of the surroundings decreases. The slight decrease in temperature causes a small decrease in molecular motion and a decrease in the entropy of the surroundings.

  ΔS_{sys} is positive.

  ΔS_{surr} is negative.

- A cup of hot water spontaneously cools to room temperature as the motional energy of the water molecules spreads out to the cooler surrounding air. Although the loss of energy from the system and corresponding temperature decrease cause a *decrease* in the entropy of the *system,* the increased temperature of the surrounding air causes an *increase* in the entropy of the *surroundings.*

  ΔS_{sys} is negative.

  ΔS_{surr} is positive.

Thus, it is not just the entropy of the *system* that determines if a process is spontaneous, the entropy of the *surroundings* is also important. There are also examples of spontaneous processes in which ΔS_{sys} and ΔS_{surr} are *both* positive. The decomposition of hydrogen peroxide produces water and oxygen gas, $2H_2O_2(l) \longrightarrow 2H_2O(l) + O_2(g)$. Because the reaction results in an increase in the number of gas molecules, we know that there is an increase in the entropy of the system. However, this is an exothermic reaction, meaning that it also gives off heat to the surroundings. An increase in temperature of the surroundings causes an increase in the entropy of the surroundings as well. (Note that there are no spontaneous processes in which ΔS_{sys} and ΔS_{surr} are both negative, which will become clear shortly.)

Calculating ΔS_{surr}

When an exothermic process takes place, the heat transferred from the system to the surroundings increases the temperature of the molecules in the surroundings. Consequently, there is an increase in the number of energy levels accessible to the molecules in the surroundings and the entropy of the surroundings increases. Conversely, in an endothermic process, heat is transferred from the surroundings to the system, decreasing the entropy of the surroundings. Remember that for constant-pressure processes, the heat released or absorbed, q, is equal to the enthalpy change of the system, ΔH_{sys} [◀◀ Section 5.3]. The change in entropy for the surroundings, ΔS_{surr}, is directly proportional to ΔH_{sys}:

$$\Delta S_{surr} \propto -\Delta H_{sys}$$

The minus sign indicates that a negative enthalpy change in the system (an *exothermic* process) corresponds to a positive entropy change in the surroundings. For an *endothermic* process, the enthalpy change in the system is a positive number and corresponds to a negative entropy change in the surroundings.

In addition to being directly proportional to ΔH_{sys}, ΔS_{surr} is inversely proportional to temperature:

$$\Delta S_{surr} \propto \frac{1}{T}$$

Combining the two expressions gives:

$$\Delta S_{surr} = \frac{-\Delta H_{sys}}{T} \qquad \textbf{Equation 18.7}$$

Student Hot Spot

Student data indicate you may struggle with entropy changes in system and surroundings. Access the eBook to view additional Learning Resources on this topic.

The Second Law of Thermodynamics

We have seen that both the system and surroundings can undergo changes in entropy during a process. The sum of the entropy changes for the system and the surroundings is the entropy change for the universe overall:

$$\Delta S_{univ} = \Delta S_{sys} + \Delta S_{surr} \qquad \textbf{Equation 18.8}$$

The *second law of thermodynamics* says that for a process to be spontaneous as written (in the forward direction), ΔS_{univ} must be positive. Therefore, the system may undergo a *decrease* in entropy, as long as the surroundings undergoes a larger *increase* in entropy, and vice versa. A process for which ΔS_{univ} is *negative* is not spontaneous as written.

In some cases, ΔS_{univ} is neither positive nor negative but is equal to zero. This happens when the entropy changes of the system and surroundings are equal in magnitude and opposite in sign and describes a specific type of process known as an *equilibrium* process. An *equilibrium process* is one that does not occur spontaneously in either the net forward or net reverse direction but can be made to occur by the addition or removal of energy from a system at equilibrium. An example of an equilibrium process is the melting of ice at 0°C. (Remember that at 0°C, ice and liquid water are in equilibrium with each other [◀◀ Section 11.6].)

With Equations 18.6 and 18.7, we can calculate the entropy changes for both the system and surroundings in a process. We can then use the second law of

Student Note: The equilibrium processes that we will encounter are phase changes.

thermodynamics (Equation 18.8) to determine if the process is spontaneous or non-spontaneous as written or if it is an equilibrium process.

Consider the synthesis of ammonia at 25°C:

$$N_2(g) + 3H_2(g) \longrightarrow 2NH_3(g) \qquad \Delta H^\circ_{rxn} = -92.6 \text{ kJ/mol}$$

From Sample Problem 18.2 (b), we have $\Delta S^\circ_{sys} = -199$ J/K · mol, and substituting ΔH°_{sys} (−92.6 kJ/mol) into Equation 18.7, we get:

$$\Delta S_{surr} = \frac{-(-92.6 \times 1000) \text{ J/mol}}{298 \text{ K}} = 311 \text{ J/K} \cdot \text{mol}$$

The entropy change for the universe is:

$$\Delta S^\circ_{univ} = \Delta S^\circ_{sys} + \Delta S^\circ_{surr}$$
$$= -199 \text{ J/K} \cdot \text{mol} + 311 \text{ J/K} \cdot \text{mol}$$
$$= 112 \text{ J/K} \cdot \text{mol}$$

Because ΔS°_{univ} is positive, the reaction will be spontaneous at 25°C. Keep in mind, though, that just because a reaction is spontaneous does not mean that it will occur at an observable rate. The synthesis of ammonia is, in fact, extremely slow at room temperature. Thermodynamics can tell us whether or not a reaction will occur spontaneously under specific conditions, but it does not tell us how fast it will occur.

The spontaneity that we have seen as favored by a process being exothermic is due to the spreading out of energy from the system to the surroundings; thus, the negative ΔH_{sys} corresponds to a positive ΔS_{surr}. It is this positive contribution to the overall ΔS_{univ} that actually favors spontaneity.

Sample Problem 18.4 lets you practice identifying spontaneous, nonspontaneous, and equilibrium processes.

SAMPLE PROBLEM 18.4

Determine if each of the following is a spontaneous process, a nonspontaneous process, or an equilibrium process at the specified temperature:
(a) $H_2(g) + I_2(g) \longrightarrow 2HI(g)$ at 0°C, (b) $CaCO_3(s) \longrightarrow CaO(s) + CO_2(g)$ at 200°C, (c) $CaCO_3(s) \longrightarrow CaO(s) + CO_2(g)$ at 1000°C, (d) $Na(s) \longrightarrow Na(l)$ at 98°C. (Assume that the thermodynamic data in Appendix 2 do not vary with temperature.)

Strategy For each process, use Equation 18.6 to determine ΔS°_{sys} and Equations 5.19 and 18.7 to determine ΔH°_{sys} and ΔS°_{surr}. At the specified temperature, the process is *spontaneous* if ΔS_{sys} and ΔS_{surr} sum to a positive number, *nonspontaneous* if they sum to a negative number, and an *equilibrium process* if they sum to zero. Note that because the *reaction* is the *system*, ΔS_{rxn} and ΔS_{sys} are used interchangeably.

Setup From Appendix 2:

(a) $S°[H_2(g)] = 131.0$ J/K · mol, $S°[I_2(g)] = 260.57$ J/K · mol, $S°[HI(g)] = 206.3$ J/K · mol; $[H_2(g)] = 0$ kJ/mol, $\Delta H^\circ_f[I_2(g)] = 62.25$ kJ/mol, $\Delta H^\circ_f[HI(g)] = 25.9$ kJ/mol.

(b), (c) In Sample Problem 18.2(a), we determined that for this reaction, $\Delta S^\circ_{rxn} = 160.5$ J/K · mol, $\Delta H^\circ_f[CaCO_3(s)] = -1206.9$ kJ/mol, $\Delta H^\circ_f[CaO(s)] = -635.6$ kJ/mol, $\Delta H^\circ_f[CO_2(g)] = -393.5$ kJ/mol.

(d) $S°[Na(s)] = 51.05$ J/K · mol, $S°[Na(l)] = 57.56$ J/K · mol; $\Delta H^\circ_f[Na(s)] = 0$ kJ/mol, $\Delta H^\circ_f[Na(l)] = 2.41$ kJ/mol.

Solution

(a) $\Delta S^\circ_{rxn} = [2S°(HI)] - [S°(H_2) + S°(I_2)]$
$= (2)(206.3 \text{ J/K} \cdot \text{mol}) - [131.0 \text{ J/K} \cdot \text{mol} + 260.57 \text{ J/K} \cdot \text{mol}] = 21.03 \text{ J/K} \cdot \text{mol}$

$\Delta H^\circ_{rxn} = [2\Delta H^\circ_f(HI)] - [\Delta H^\circ_f(H_2) + \Delta H^\circ_f(I_2)]$
$= (2)(25.9 \text{ kJ/mol}) - [0 \text{ kJ/mol} + 62.25 \text{ kJ/mol}] = -10.5 \text{ kJ/mol}$

$\Delta S_{surr} = \dfrac{-\Delta H_{rxn}}{T} = \dfrac{-(-10.5 \text{ kJ/mol})}{273 \text{ K}} = 0.0385 \text{ kJ/K} \cdot \text{mol} = 38.5 \text{ J/K} \cdot \text{mol}$

$\Delta S_{univ} = \Delta S_{sys} + \Delta S_{surr} = 21.03 \text{ J/K} \cdot \text{mol} + 38.5 \text{ J/K} \cdot \text{mol} = 59.5 \text{ J/K} \cdot \text{mol}$

ΔS_{univ} is positive; therefore, the reaction is spontaneous at 0°C.

(b), (c) $\Delta S^\circ_{rxn} = 160.5$ J/K · mol

$$\Delta H^\circ_{rxn} = [\Delta H^\circ_f(CaO) + \Delta H^\circ_f(CO_2)] - [\Delta H^\circ_f(CaCO_3)]$$

$$= [-635.6 \text{ kJ/mol} + (-393.5 \text{ kJ/mol})] - (-1206.9 \text{ kJ/mol}) = 177.8 \text{ kJ/mol}$$

(b) $T = 200°C$ and

$$\Delta S_{surr} = \frac{-\Delta H_{sys}}{T} = \frac{-(177.8 \text{ kJ/mol})}{473 \text{ K}} = -0.376 \text{ kJ/K} \cdot \text{mol} = -376 \text{ J/K} \cdot \text{mol}$$

$$\Delta S_{univ} = \Delta S_{sys} + \Delta S_{surr} = 160.5 \text{ J/K} \cdot \text{mol} + (-376 \text{ J/K} \cdot \text{mol}) = -216 \text{ J/K} \cdot \text{mol}$$

ΔS_{univ} is negative, so the reaction is nonspontaneous at 200°C.

(c) $T = 1000°C$ and

$$\Delta S_{surr} = \frac{-\Delta H_{sys}}{T} = \frac{-(177.8 \text{ kJ/mol})}{1273 \text{ K}} = -0.1397 \text{ kJ/K} \cdot \text{mol} = -139.7 \text{ J/K} \cdot \text{mol}$$

$$\Delta S_{univ} = \Delta S_{sys} + \Delta S_{surr} = 160.5 \text{ J/K} \cdot \text{mol} + (-139.7 \text{ J/K} \cdot \text{mol}) = 20.8 \text{ J/K} \cdot \text{mol}$$

In this case, ΔS_{univ} is positive; therefore, the reaction is spontaneous at 1000°C.

(d) $\Delta S^\circ_{rxn} = S°[Na(l)] - S°[Na(s)] = 57.56$ J/K · mol $- 51.05$ J/K · mol $= 6.51$ J/K · mol

$$\Delta H^\circ_{rxn} = \Delta H^\circ_f[Na(l)] - \Delta H^\circ_f[Na(s)] = 2.41 \text{ kJ/mol} - 0 \text{ kJ/mol} = 2.41 \text{ kJ/mol}$$

$$\Delta S_{surr} = \frac{-\Delta H_{rxn}}{T} = \frac{-(2.41 \text{ kJ/mol})}{371 \text{ K}} = -0.0650 \text{ kJ/K} \cdot \text{mol} = -6.50 \text{ J/K} \cdot \text{mol}$$

$$\Delta S_{univ} = \Delta S_{sys} + \Delta S_{surr} = 6.51 \text{ J/K} \cdot \text{mol} + (-6.50 \text{ J/K} \cdot \text{mol}) = 0.01 \text{ J/K} \cdot \text{mol} \approx 0$$

ΔS_{univ} is zero; therefore, the reaction is an equilibrium process at 98°C. In fact, this is the melting point of sodium.

THINK ABOUT IT

Remember that standard enthalpies of formation have units of kJ/mol, whereas standard absolute entropies have units of J/K · mol. Make sure that you convert kilojoules to joules, or vice versa, before combining the terms. The small difference between the magnitudes of ΔS_{sys} and ΔS_{surr} in part (d) results from thermodynamic values not being entirely independent of temperature. The tabulated values of $S°$ and ΔH°_f are for 25°C.

Practice Problem **A**TTEMPT   For each of the following, calculate ΔS_{univ} and identify the process as a spontaneous process, a nonspontaneous process, or an equilibrium process at the specified temperature: (a) $CO_2(g) \longrightarrow CO_2(aq)$ at 25°C, (b) $N_2O_4(g) \longrightarrow 2NO_2(g)$ at 10.4°C, (c) $PCl_3(l) \longrightarrow PCl_3(g)$ at 61.2°C. (Assume that the thermodynamic data in Appendix 2 do not vary with temperature.)

Practice Problem **B**UILD   (a) Calculate ΔS_{univ} and determine if the reaction $H_2O_2(l) \longrightarrow H_2O_2(g)$ is spontaneous, nonspontaneous, or an equilibrium process at 163°C. (b) The reaction $NH_3(g) + HCl(g) \longrightarrow NH_4Cl(s)$ is spontaneous in the forward direction at room temperature but, because it is exothermic, becomes less spontaneous with increasing temperature. Determine the temperature at which it is no longer spontaneous in the forward direction. (c) Determine the boiling point of Br_2. (Assume that the thermodynamic data in Appendix 2 do not vary with temperature.)

Practice Problem **C**ONCEPTUALIZE   The table at right shows the signs of ΔS_{sys}, ΔS_{surr}, and ΔS_{univ} for four processes. Where possible, fill in the missing table entries. Indicate where it is not possible to determine the missing sign and explain.

Process	ΔS_{sys}	ΔS_{surr}	ΔS_{univ}
1	−	−	
2	+		+
3	−	+	
4		−	+

The Third Law of Thermodynamics

Finally, we consider the third law of thermodynamics briefly in connection with the determination of standard entropy. We have related the entropy of a system to the number of possible arrangements of the system's molecules. The larger the number of possible arrangements, the larger the entropy. Imagine a pure, perfect crystalline substance at absolute zero (0 K). Under these conditions, there is essentially no molecular motion and, because the molecules occupy fixed positions in the solid, there is only one way to arrange the molecules. From Equation 18.1, we write:

$$S = k \ln W = k \ln 1 = 0$$

According to the ***third law of thermodynamics,*** the entropy of a perfect crystalline substance is *zero* at absolute zero. As temperature increases, molecular motion increases, causing an increase in the number of possible arrangements of the molecules and in the number of accessible energy levels, among which the system's energy can be dispersed. (See Figure 18.4.) This results in an increase in the system's entropy. Thus, the entropy of any substance at any temperature above 0 K is greater than zero. If the crystalline substance is impure or imperfect in any way, then its entropy is greater than zero even at 0 K because without perfect crystalline order, there is more than one possible arrangement of molecules.

The significance of the third law of thermodynamics is that it enables us to determine experimentally the *absolute* entropies of substances. Starting with the knowledge that the entropy of a pure crystalline substance is zero at 0 K, we can measure the increase in entropy of the substance as it is heated. The change in entropy of a substance, ΔS, is the difference between the final and initial entropy values:

$$\Delta S = S_{\text{final}} - S_{\text{initial}}$$

where S_{initial} is zero if the substance starts at 0 K. Therefore, the measured *change* in entropy is equal to the *absolute* entropy of the substance at the final temperature:

$$\Delta S = S_{\text{final}}$$

The entropy values arrived at in this way are called *absolute* entropies because they are *true* values—unlike standard enthalpies of formation, which are derived using an arbitrary reference. Because the tabulated values are determined at 1 atm, we usually refer to absolute entropies as *standard* entropies, $S°$. Figure 18.5 shows the increase in entropy of a substance as temperature increases from absolute zero. At 0 K, it has a zero entropy value (assuming that it is a perfect crystalline substance). As it is heated, its entropy increases gradually at first because of greater molecular motion within the crystal. At the melting point, there is a large increase in entropy as the

Figure 18.5 Entropy increases in a substance as temperature increases from absolute zero.

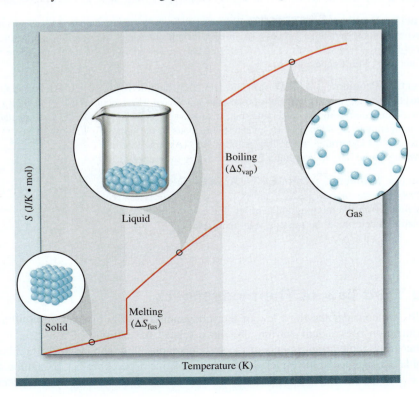

solid is transformed into the liquid. Further heating increases the entropy of the liquid again due to increased molecular motion. At the boiling point, there is a large increase in entropy as a result of the liquid-to-vapor transition. Beyond that temperature, the entropy of the gas continues to increase with increasing temperature.

CHECKPOINT – SECTION 18.4   Entropy Changes in the Universe

18.4.1 Using data from Appendix 2, calculate $\Delta S°$ (in J/K · mol) for the following reaction:

$$2NO(g) + O_2(g) \longrightarrow 2NO_2(g)$$

a) 145.3 J/K · mol

b) −145.3 J/K · mol

c) 59.7 J/K · mol

d) −59.7 J/K · mol

e) −421.2 J/K · mol

18.4.2 Using data from Appendix 2, calculate $\Delta S°$ (in J/K · mol) for the following reaction:

$$CH_4(g) + 2O_2(g) \longrightarrow CO_2(g) + 2H_2O(l)$$

a) 107.7 J/K · mol

b) −107.7 J/K · mol

c) 2.6 J/K · mol

d) 242.8 J/K · mol

e) −242.8 J/K · mol

18.4.3 The diagrams show a spontaneous chemical reaction. What can we deduce about ΔS_{surr} for this process?

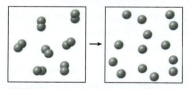

a) ΔS_{surr} is positive.

b) ΔS_{surr} is negative.

c) ΔS_{surr} is zero.

d) There is not enough information to deduce the sign of ΔS_{surr}.

18.4.4 The diagrams show a spontaneous chemical reaction. What can we deduce about ΔS_{surr} for this process?

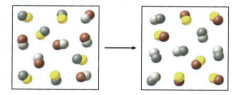

a) ΔS_{surr} is positive.

b) ΔS_{surr} is negative.

c) ΔS_{surr} is zero.

d) There is not enough information to deduce the sign of ΔS_{surr}.

18.5 Predicting Spontaneity

Gibbs Free-Energy Change, ΔG

According to the second law of thermodynamics, $\Delta S_{univ} > 0$ for a spontaneous process. What we are usually concerned with and usually *measure*, however, are the properties of the system rather than those of the surroundings or those of the universe overall. Therefore, it is convenient to have a thermodynamic function that enables us to determine whether or not a process is spontaneous by considering the system alone.

We begin with Equation 18.8. For a spontaneous process:

$$\Delta S_{univ} = \Delta S_{sys} + \Delta S_{surr} > 0$$

Substituting $-\Delta H_{sys}/T$ for ΔS_{surr}, we write:

$$\Delta S_{univ} = \Delta S_{sys} + \left(-\frac{\Delta H_{sys}}{T} \right) > 0$$

Multiplying both sides of the equation by T gives:

$$T\Delta S_{univ} = T\Delta S_{sys} - \Delta H_{sys} > 0$$

Now we have an equation that expresses the second law of thermodynamics (and predicts whether or not a process is spontaneous) in terms of only the *system*. We no longer need to consider the surroundings. For convenience, we can rearrange the preceding equation, multiply through by -1, and replace the $>$ sign with a $<$ sign:

$$-T\Delta S_{univ} = \Delta H_{sys} - T\Delta S_{sys} < 0$$

According to this equation, a process carried out at constant pressure and temperature is spontaneous if the changes in enthalpy and entropy of the system are such that $\Delta H_{sys} - T\Delta S_{sys}$ is less than zero.

To express the spontaneity of a process more directly, we introduce another thermodynamic function called the **Gibbs**[1] **free energy (G),** or simply **free energy:**

Equation 18.9	$G = H - TS$

Each of the terms in Equation 18.9 pertains to the system. G has units of energy just as H and TS do. Furthermore, like enthalpy and entropy, free energy is a state function. The change in free energy, ΔG, of a system for a process that occurs at constant temperature is:

Equation 18.10	$\Delta G = \Delta H - T\Delta S$

Equation 18.10 enables us to predict the spontaneity of a process using the change in enthalpy, the change in entropy, and the absolute temperature. At constant temperature and pressure, for processes that are spontaneous as written (in the forward direction), ΔG is negative. For processes that are not spontaneous as written but that are spontaneous in the reverse direction, ΔG is positive. For systems at equilibrium, ΔG is zero.

> **Student Note:** In this context, *free energy* is the energy available to do work. Thus, if a particular process is accompanied by a release of usable energy (i.e., if ΔG is negative), this fact alone guarantees that it is spontaneous, and there is no need to consider what happens to the rest of the universe.

- $\Delta G < 0$ The reaction is spontaneous in the forward direction (and nonspontaneous in the reverse direction).
- $\Delta G > 0$ The reaction is nonspontaneous in the forward direction (and spontaneous in the reverse direction).
- $\Delta G = 0$ The system is at equilibrium.

Often we can predict the sign of ΔG for a process if we know the signs of ΔH and ΔS. Table 18.4 shows how we can use Equation 18.10 to make such predictions.

Based on the information in Table 18.4, you may wonder what constitutes a "low" or a "high" temperature. For the example given in the table, 0°C is the temperature that divides high from low. Water freezes spontaneously at temperatures

TABLE 18.4	Predicting the Sign of ΔG Using Equation 18.10 and the Signs of ΔH and ΔS			
When ΔH Is	**And ΔS Is**	**ΔG Will Be**	**And the Process Is**	**Example**
Negative	Positive	Negative	Always spontaneous	$2H_2O_2(aq) \longrightarrow 2H_2O(l) + O_2(g)$
Positive	Negative	Positive	Always nonspontaneous	$3O_2(g) \longrightarrow 2O_3(g)$
Negative	Negative	Negative when $T\Delta S < \Delta H$ Positive when $T\Delta S > \Delta H$	Spontaneous at low T Nonspontaneous at high T	$H_2O(l) \longrightarrow H_2O(s)$ (freezing of water)
Positive	Positive	Negative when $T\Delta S > \Delta H$ Positive when $T\Delta S < \Delta H$	Spontaneous at high T Nonspontaneous at low T	$2HgO(s) \longrightarrow 2Hg(l) + O_2(g)$

---

1. Josiah Willard Gibbs (1839–1903). American physicist. One of the founders of thermodynamics. Gibbs was a modest and private individual who spent almost all his professional life at Yale University. Because he published most of his work in obscure journals, Gibbs never gained the eminence that his contemporary and admirer James Maxwell did. Even today, very few people outside of chemistry and physics have ever heard of Gibbs.

below 0°C, and ice melts spontaneously at temperatures above 0°C. At 0°C, a system of ice and water is at equilibrium. In general, though, the temperature that divides "high" from "low" depends on the individual reaction. To determine that temperature, we must set ΔG equal to 0 in Equation 18.10 (i.e., the equilibrium condition):

$$0 = \Delta H - T\Delta S$$

Rearranging to solve for T yields:

$$T = \frac{\Delta H}{\Delta S}$$

The temperature that divides high from low for a particular reaction can now be calculated if the values of ΔH and ΔS are known.

Sample Problem 18.5 demonstrates the use of this approach.

SAMPLE PROBLEM 18.5

According to Table 18.4, a reaction will be spontaneous only at high temperatures if both ΔH and ΔS are positive. For a reaction in which $\Delta H = 199.5$ kJ/mol and $\Delta S = 476$ J/K · mol, determine the temperature (in °C) above which the reaction is spontaneous.

Strategy The temperature that divides high from low is the temperature at which $\Delta H = T\Delta S$ ($\Delta G = 0$). Therefore, we use Equation 18.10, substituting 0 for ΔG and solving for T to determine temperature in kelvins; we then convert to degrees Celsius.

Setup

$$\Delta S = \left(\frac{476 \text{ J}}{\text{K} \cdot \text{mol}}\right)\left(\frac{1 \text{ kJ}}{1000 \text{ J}}\right) = 0.476 \text{ kJ/K} \cdot \text{mol}$$

Solution

$$T = \frac{\Delta H}{\Delta S} = \frac{199.5 \text{ kJ/mol}}{0.476 \text{ kJ/K} \cdot \text{mol}} = 419 \text{ K}$$

$$= (419 - 273) = 146°\text{C}$$

THINK ABOUT IT

Spontaneity is favored by a release of energy (ΔH being negative) and by an increase in entropy (ΔS being positive). When both quantities are positive, as in this case, only the entropy change favors spontaneity. For an endothermic process such as this, which requires the input of heat, it should make sense that adding more heat by increasing the temperature will shift the equilibrium to the right, thus making it "more spontaneous."

Practice Problem A TTEMPT A reaction will be spontaneous only at low temperatures if both ΔH and ΔS are negative. For a reaction in which $\Delta H = -380.1$ kJ/mol and $\Delta S = -95.00$ J/K · mol, determine the temperature (in °C) below which the reaction is spontaneous.

Practice Problem B UILD Given that the reaction $4Fe(s) + 3O_2(g) + 6H_2O(l) \longrightarrow 4Fe(OH)_3(s)$ is spontaneous at temperatures below 1950°C, estimate the standard entropy of $Fe(OH)_3(s)$.

Practice Problem C ONCEPTUALIZE Which of the following graphs best represents the relationship between ΔG and temperature for a process that is exothermic and for which ΔS is negative?

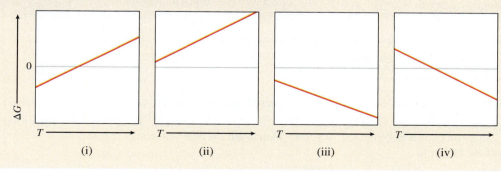

(i)   (ii)   (iii)   (iv)

Student Note: The introduction of the term $\Delta G°$ enables us to write Equation 18.10 as $\Delta G° = \Delta H° - T\Delta S°$

Standard Free-Energy Changes, $\Delta G°$

The *standard free energy of reaction* ($\Delta G°_{rxn}$) is the free-energy change for a reaction when it occurs under standard-state conditions—that is, when reactants in their standard states are converted to products in *their* standard states. The conventions used by chemists to define the standard states of pure substances and solutions are:

- Gases   1 atm pressure
- Liquids   Pure liquid
- Solids   Pure solid
- Elements   The most stable allotropic form at 1 atm and 25°C
- Solutions   1 molar concentration

To calculate $\Delta G°_{rxn}$, we start with the general equation:

$$a\text{A} + b\text{B} \longrightarrow c\text{C} + d\text{D}$$

The standard free-energy change for this reaction is given by:

Equation 18.11 $\Delta G°_{rxn} = [c\Delta G°_f(\text{C}) + d\Delta G°_f(\text{D})] - [a\Delta G°_f(\text{A}) + b\Delta G°_f(\text{B})]$

Equation 18.11 can be generalized as follows:

Equation 18.12 $\Delta G°_{rxn} = \Sigma n\Delta G°_f \text{ (products)} - \Sigma m\Delta G°_f \text{ (reactants)}$

where m and n are stoichiometric coefficients. The term $\Delta G°_f$ is the *standard free energy of formation* of a compound—that is, the free-energy change that occurs when 1 mole of the compound is synthesized from its constituent elements, each in its standard state. For the combustion of graphite:

$$\text{C(graphite)} + O_2(g) \longrightarrow CO_2(g)$$

the standard free-energy change (from Equation 18.12) is:

$$\Delta G°_{rxn} = [\Delta G°_f(CO_2, \text{gas})] - [\Delta G°_f(\text{C, graphite}) + \Delta G°_f(O_2, \text{gas})]$$

As with standard enthalpy of formation, the standard free energy of formation of any element (in its most stable allotropic form at 1 atm) is defined as zero. Thus:

$$\Delta G°_f(\text{C, graphite}) = 0 \quad \text{and} \quad \Delta G°_f(O_2, \text{gas}) = 0$$

Student Hot Spot

Student data indicate you may struggle with free-energy changes. Access the eBook to view additional Learning Resources on this topic.

Therefore, the standard free-energy change for the reaction in this case is equal to the standard free energy of formation of CO_2:

$$\Delta G°_{rxn} = \Delta G°_f(CO_2)$$

Appendix 2 lists the values of $\Delta G°_f$ at 25°C for a number of compounds.

Sample Problem 18.6 demonstrates the calculation of standard free-energy changes.

SAMPLE PROBLEM   18.6

Calculate the standard free-energy changes for the following reactions at 25°C:

(a) $CH_4(g) + 2O_2(g) \longrightarrow CO_2(g) + 2H_2O(l)$

(b) $2MgO(s) \longrightarrow 2Mg(s) + O_2(g)$

Strategy Look up the $\Delta G°_f$ values for the reactants and products in each equation, and use Equation 18.12 to solve for $\Delta G°_{rxn}$.

Setup From Appendix 2, we have the following values: $\Delta G°_f[CH_4(g)] = -50.8$ kJ/mol, $\Delta G°_f[CO_2(g)] = -394.4$ kJ/mol, $\Delta G°_f[H_2O(l)] = -237.2$ kJ/mol, and $\Delta G°_f[MgO(s)] = -569.6$ kJ/mol. All the other substances are elements in their standard states and have, by definition, $\Delta G°_f = 0$.

Solution

(a) $\Delta G_{rxn}^{\circ} = (\Delta G_f^{\circ}[CO_2(g)] + 2\Delta G_f^{\circ}[H_2O(l)]) - (\Delta G_f^{\circ}[CH_4(g)] + 2\Delta G_f^{\circ}[O_2(g)])$

$= [(-394.4 \text{ kJ/mol}) + (2)(-237.2 \text{ kJ/mol})] - [(-50.8 \text{ kJ/mol}) + (2)(0 \text{ kJ/mol})]$

$= -818.0 \text{ kJ/mol}$

(b) $\Delta G_{rxn}^{\circ} = (2\Delta G_f^{\circ}[Mg(s)] + \Delta G_f^{\circ}[O_2(g)]) - (2\Delta G_f^{\circ}[MgO(s)])$

$= [(2)(0 \text{ kJ/mol}) + (0 \text{ kJ/mol})] - [(2)(-569.6 \text{ kJ/mol})]$

$= 1139 \text{ kJ/mol}$

THINK ABOUT IT

Note that, like standard enthalpies of formation (ΔH_f°), standard free energies of formation (ΔG_f°) depend on the *state* of matter. Using water as an example, $\Delta G_f^{\circ}[H_2O(l)] = -237.2$ kJ/mol and $\Delta G_f^{\circ}[H_2O(g)] = -228.6$ kJ/mol. Always double-check to make sure you have selected the right value from the table.

Practice Problem **ATTEMPT** Calculate the standard free-energy changes for the following reactions at 25°C:

(a) $H_2(g) + Br_2(l) \longrightarrow 2HBr(g)$

(b) $2C_2H_6(g) + 7O_2(g) \longrightarrow 4CO_2(g) + 6H_2O(l)$

Practice Problem **BUILD** For each reaction, determine the value of ΔG_f° that is not listed in Appendix 2:

(a) $Li_2O(s) + 2HCl(g) \longrightarrow 2LiCl(s) + H_2O(g)$ $\Delta G_{rxn}^{\circ} = -244.88$ kJ/mol

(b) $Na_2O(s) + 2HI(g) \longrightarrow 2NaI(s) + H_2O(l)$ $\Delta G_{rxn}^{\circ} = -435.44$ kJ/mol

Practice Problem **CONCEPTUALIZE** For which of the following species is $\Delta G_f^{\circ} = 0$?

$Br_2(l)$	$I_2(g)$	$CO_2(g)$	$Xe(g)$
(i)	(ii)	(iii)	(iv)

Using ΔG and ΔG° to Solve Problems

It is the sign of ΔG, the free-energy change, *not* the sign of ΔG°, the standard free-energy change, that indicates whether or not a process will occur spontaneously under a given set of conditions. What the *sign* of ΔG° tells us is the same thing that the magnitude of the equilibrium constant (K) tells us [◄◄ Section 15.2]. A negative ΔG° value corresponds to a large K value (products favored at equilibrium), whereas a positive ΔG° value corresponds to a small K value (reactants favored at equilibrium).

Student Note: The sign of ΔG° does indicate whether or not a process is spontaneous when *all* reactants and products are in their standard states, but this is very seldom the case.

Like equilibrium constants, ΔG° values change with temperature. One of the uses of Equation 18.10 is to determine the temperature at which a particular equilibrium will begin to favor a desired product. For example, calcium oxide (CaO), also called quicklime, is an extremely valuable inorganic substance with a variety of industrial uses, including water treatment and pollution control. It is prepared by heating limestone (CaCO₃), which decomposes at a high temperature:

$$CaCO_3(s) \rightleftharpoons CaO(s) + CO_2(g)$$

The reaction is reversible, and under the right conditions, CaO and CO₂ readily recombine to form CaCO₃ again. To prevent this from happening in the industrial preparation, the system is never maintained at equilibrium; rather, CO₂ is constantly removed as it forms, shifting the equilibrium from left to right, thus promoting the formation of calcium oxide.

An important piece of information for the chemist responsible for maximizing CaO production is the temperature at which the decomposition equilibrium of CaCO₃ begins to favor products. We can make a reliable estimate of that temperature as follows. First we calculate $\Delta H°$ and $\Delta S°$ for the reaction at 25°C, using the data in Appendix 2. To determine $\Delta H°$, we apply Equation 5.19:

$$\Delta H° = [\Delta H°_f(CaO) + \Delta H°_f(CO_2)] - [\Delta H°_f(CaCO_3)]$$

$$= [(-635.6 \text{ kJ/mol}) + (-393.5 \text{ kJ/mol})] - (-1206.9 \text{ kJ/mol})$$

$$= 177.8 \text{ kJ/mol}$$

Next we apply Equation 18.6 to find $\Delta S°$:

$$\Delta S° = [S°(CaO) + S°(CO_2)] - S°(CaCO_3)$$

$$= [(39.8 \text{ J/K} \cdot \text{mol}) + (213.6 \text{ J/K} \cdot \text{mol})] - (92.9 \text{ J/K} \cdot \text{mol})$$

$$= 160.5 \text{ J/K} \cdot \text{mol}$$

From Equation 18.10, we can write:

$$\Delta G° = \Delta H° - T\Delta S°$$

and we obtain:

$$\Delta G° = (177.8 \text{ kJ/mol}) - (298 \text{ K})(0.1605 \text{ kJ/K} \cdot \text{mol})$$

$$= 130.0 \text{ kJ/mol}$$

Student Note: Be careful with units in problems of this type. $S°$ values are tabulated using joules, whereas $\Delta H°_f$ values are tabulated using kilojoules.

Because $\Delta G°$ is a large positive number, the reaction does *not* favor product formation at 25°C (298 K). And, because $\Delta H°$ and $\Delta S°$ are both positive, we know that $\Delta G°$ will be negative (product formation will be favored) at high temperatures. We can determine what constitutes a high temperature for this reaction by calculating the temperature at which $\Delta G°$ is zero:

$$0 = \Delta H° - T\Delta S°$$

or:

$$T = \frac{\Delta H°}{\Delta S°}$$

$$= \frac{(177.8 \text{ kJ/mol})(1000 \text{ J/kJ})}{0.1605 \text{ kJ/K} \cdot \text{mol}}$$

$$= 1108 \text{ K}(835°\text{C})$$

At temperatures higher than 835°C, $\Delta G°$ becomes negative, indicating that the reaction would then favor the formation of CaO and CO_2. At 840°C (1113 K), for example:

$$\Delta G° = \Delta H° - T\Delta S°$$

$$= 177.8 \text{ kJ/mol} - (1113 \text{ K})(0.1605 \text{ kJ/K} \cdot \text{mol})\left(\frac{1 \text{ kJ}}{1000 \text{ J}}\right)$$

$$= -0.8 \text{ kJ/mol}$$

Student Hot Spot

Student data indicate you may struggle with the effect of temperature on free-energy change. Acess the eBook to view additional Learning Resources on this topic.

At still higher temperatures, $\Delta G°$ becomes increasingly negative, thus favoring product formation even more. Note that in this example we used the $\Delta H°$ and $\Delta S°$ values at 25°C to calculate changes to $\Delta G°$ at much higher temperatures. Because both $\Delta H°$ and $\Delta S°$ actually change with temperature, this approach does not give us a truly accurate value for $\Delta G°$, but it does give us a reasonably good estimate.

Equation 18.10 can also be used to calculate the change in entropy that accompanies a phase change. At the temperature at which a phase change occurs (i.e., the melting point or boiling point of a substance), the system is at equilibrium ($\Delta G = 0$). Therefore, Equation 18.10 becomes:

$$0 = \Delta H - T\Delta S$$

or:

$$\Delta S = \frac{\Delta H}{T}$$

Consider the ice-water equilibrium. For the ice-to-water transition, ΔH is the molar heat of fusion (see Table 11.8) and T is the melting point. The entropy change is therefore

$$\Delta S_{\text{ice}\rightarrow\text{water}} = \frac{6010 \text{ J/mol}}{273 \text{ K}} = 22.0 \text{ J/K} \cdot \text{mol}$$

Thus, when 1 mole of ice melts at 0°C, there is an increase in entropy of 22.0 J/K · mol. The increase in entropy is consistent with the increase in possible arrangements from solid to liquid. Conversely, for the water-to-ice transition, the decrease in entropy is given by:

$$\Delta S_{\text{water}\rightarrow\text{ice}} = \frac{-6010 \text{ J/mol}}{273 \text{ K}} = -22.0 \text{ J/K} \cdot \text{mol}$$

The same approach can be applied to the water-to-steam transition. In this case, ΔH is the heat of vaporization and T is the boiling point of water.

Sample Problem 18.7 examines the phase transitions in benzene.

SAMPLE PROBLEM 18.7

The molar heats of fusion and vaporization of benzene are 10.9 and 31.0 kJ/mol, respectively. Calculate the entropy changes for the solid-to-liquid and liquid-to-vapor transitions for benzene. At 1 atm pressure, benzene melts at 5.5°C and boils at 80.1°C.

Strategy The solid-liquid transition at the melting point and the liquid-vapor transition at the boiling point are *equilibrium* processes. Therefore, because ΔG is zero at equilibrium, in each case we can use Equation 18.10, substituting 0 for ΔG and solving for ΔS, to determine the entropy change associated with the process.

Setup The melting point of benzene is 5.5 + 273.15 = 278.7 K and the boiling point is 80.1 + 273.15 = 353.3 K.

Solution

$$\Delta S_{\text{fus}} = \frac{\Delta H_{\text{fus}}}{T_{\text{melting}}} = \frac{10.9 \text{ kJ/mol}}{278.7 \text{ K}}$$

$$= 0.0391 \text{ kJ/K} \cdot \text{mol} \quad \text{or} \quad 39.1 \text{ J/K} \cdot \text{mol}$$

$$\Delta S_{\text{vap}} = \frac{\Delta H_{\text{vap}}}{T_{\text{boiling}}} = \frac{31.0 \text{ kJ/mol}}{353.3 \text{ K}}$$

$$= 0.0877 \text{ kJ/K} \cdot \text{mol} \quad \text{or} \quad 87.7 \text{ J/K} \cdot \text{mol}$$

THINK ABOUT IT

For the same substance, ΔS_{vap} is always significantly larger than ΔS_{fus}. The change in number of possible arrangements is always bigger in a liquid-to-gas transition than in a solid-to-liquid transition.

Practice Problem **A**TTEMPT The molar heats of fusion and vaporization of argon are 1.3 and 6.3 kJ/mol, respectively, and argon's melting point and boiling point are −190°C and −186°C, respectively. Calculate the entropy changes for the fusion and vaporization of argon.

Practice Problem **B**UILD Using data from Appendix 2 and assuming that the tabulated values do not change with temperature, (a) calculate $\Delta H^{\circ}_{\text{fus}}$ and $\Delta S^{\circ}_{\text{fus}}$ for sodium metal and determine the melting temperature of sodium, and (b) calculate $\Delta H^{\circ}_{\text{vap}}$ and $\Delta S^{\circ}_{\text{vap}}$ for sodium metal and determine the boiling temperature of sodium.

Practice Problem **C**ONCEPTUALIZE Explain why, in general, we can use the equation $\Delta S = \dfrac{\Delta H}{T}$ to calculate ΔS for a phase change but not for a chemical reaction.

CHECKPOINT – SECTION 18.5   Predicting Spontaneity

18.5.1 A reaction for which ΔH and ΔS are both negative is

  a) nonspontaneous at all temperatures.

  b) spontaneous at all temperatures.

  c) spontaneous at high temperatures.

  d) spontaneous at low temperatures.

  e) at equilibrium.

18.5.2 At what temperature (in °C) does a reaction go from being nonspontaneous to spontaneous if it has $\Delta H = 171$ kJ/mol and $\Delta S = 161$ J/K · mol?

  a) 270°C

  b) 670°C

  c) 1100°C

  d) 790°C

  e) 28°C

18.5.3 Using data from Appendix 2, calculate $\Delta G°$ (in kJ/mol) at 25°C for the reaction:

$$CH_4(g) + 2O_2(g) \longrightarrow CO_2(g) + 2H_2O(g)$$

  a) −580.8 kJ/mol

  b) 580.8 kJ/mol

  c) −572.0 kJ/mol

  d) −800.8 kJ/mol

  e) −818.0 kJ/mol

18.5.4 Calculate ΔS_{sub} (in J/K · mol) for the sublimation of iodine in a closed flask at 45°C:

$$I_2(s) \longrightarrow I_2(g)$$

$\Delta H_{sub} = 62.4$ kJ/mol.

  a) 1.4 J/K · mol

  b) 196 J/K · mol

  c) 1387 J/K · mol

  d) 0.196 J/K · mol

  e) 721 J/K · mol

18.5.5 What can be deduced about the spontaneity of the reaction represented in the diagrams?

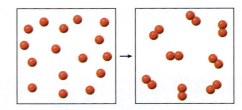

  a) It is spontaneous only at high temperatures.

  b) It is spontaneous only at low temperatures.

  c) It is spontaneous at all temperatures.

  d) It is nonspontaneous at all temperatures.

  e) There is not enough information to determine this.

18.6 Free Energy and Chemical Equilibrium

Student Note: Even for a reaction that *starts* with all reactants and products in their standard states, as soon as the reaction begins, the concentrations of all species change and standard-state conditions no longer exist.

Reactants and products in a chemical reaction are almost always in something *other* than their standard states—that is, solutions usually have concentrations other than 1 *M* and gases usually have pressures other than 1 atm. To determine whether or not a reaction is spontaneous, therefore, we must take into account the actual concentrations and/or pressures of the species involved. And although we can determine $\Delta G°$ from tabulated values, we need to know ΔG to determine spontaneity.

Relationship Between ΔG and $\Delta G°$

The relationship between ΔG and $\Delta G°$, which is derived from thermodynamics, is:

Equation 18.13	$\Delta G = \Delta G° + RT \ln Q$

Student Note: The Q used in Equation 18.13 is Q_c for reactions that take place in solution and Q_P for reactions that take place in the gas phase.

where R is the gas constant (8.314 J/K · mol or 8.314×10^{-3} kJ/K · mol), T is the absolute temperature at which the reaction takes place, and Q is the reaction quotient [◄◄ Section 15.2]. Thus, ΔG depends on two terms: $\Delta G°$ and $RT \ln Q$. For a given reaction at temperature T, the value of $\Delta G°$ is fixed but that of $RT \ln Q$ can vary because Q varies according to the composition of the reaction mixture.

Consider the following equilibrium:

$$H_2(g) + I_2(s) \rightleftharpoons 2HI(g)$$

Using Equation 18.12 and information from Appendix 2, we find that $\Delta G°$ for this reaction at 25°C is 2.60 kJ/mol. The value of ΔG, however, depends on the pressures of both gaseous species. If we start with a reaction mixture containing solid I_2, in which $P_{H_2} = 4.0$ atm and $P_{HI} = 3.0$ atm, the reaction quotient, Q_P, is:

$$Q_P = \frac{(P_{HI})^2}{(P_{H_2})} = \frac{(3.0)^2}{4.0} = \frac{9.0}{4.0}$$
$$= 2.25$$

Using this value in Equation 18.13 gives:

$$\Delta G = \frac{2.60 \text{ kJ}}{\text{mol}} + \left(\frac{8.314 \times 10^{-3} \text{ kJ}}{\text{K} \cdot \text{mol}} \right)(298 \text{ K})(\ln 2.25)$$
$$= 4.3 \text{ kJ/mol}$$

Because ΔG is positive, we conclude that, starting with these concentrations, the forward reaction will not occur spontaneously as written. Instead, the *reverse* reaction will occur spontaneously and the system will reach equilibrium by consuming part of the HI initially present and producing more H_2 and I_2.

If, on the other hand, we start with a mixture of gases in which $P_{H_2} = 4.0$ atm and $P_{HI} = 1.0$ atm, the reaction quotient, Q_P, is:

$$Q_P = \frac{(P_{HI})^2}{(P_{H_2})} = \frac{(1.0)^2}{(4.0)} = \frac{1}{4}$$
$$= 0.25$$

Using this value in Equation 18.13 gives:

$$\Delta G = \frac{2.60 \text{ kJ}}{\text{mol}} + \left(\frac{8.314 \times 10^{-3} \text{ kJ}}{\text{K} \cdot \text{mol}} \right)(298 \text{ K})(\ln 0.25)$$
$$= -0.8 \text{ kJ/mol}$$

With a negative value for ΔG, the reaction will be spontaneous as written—in the forward direction. In this case, the system will achieve equilibrium by consuming some of the H_2 and I_2 to produce more HI.

Sample Problem 18.8 uses $\Delta G°$ and the reaction quotient to determine in which direction a reaction is spontaneous.

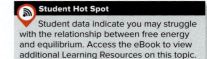

Student Hot Spot

Student data indicate you may struggle with the relationship between free energy and equilibrium. Access the eBook to view additional Learning Resources on this topic.

SAMPLE PROBLEM 18.8

The equilibrium constant, K_P, for the reaction:

$$N_2O_4(g) \rightleftharpoons 2NO_2(g)$$

is 0.113 at 298 K, which corresponds to a standard free-energy change of 5.4 kJ/mol. In a certain experiment, the initial pressures are $P_{N_2O_4} = 0.453$ atm and $P_{NO_2} = 0.122$ atm. Calculate ΔG for the reaction at these pressures, and predict the direction in which the reaction will proceed spontaneously to establish equilibrium.

(Continued on next page)

Strategy Use the partial pressures of N_2O_4 and NO_2 to calculate the reaction quotient Q_P, and then use Equation 18.13 to calculate ΔG.

Setup The reaction quotient expression is:

$$Q_P = \frac{(P_{NO_2})^2}{P_{N_2O_4}} = \frac{(0.122)^2}{0.453} = 0.0329$$

Solution

$$\Delta G = \Delta G^\circ + RT \ln Q_P$$

$$= \frac{5.4 \text{ kJ}}{\text{mol}} + \left(\frac{8.314 \times 10^{-3} \text{ kJ}}{\text{K} \cdot \text{mol}}\right)(298 \text{ K})(\ln 0.0329)$$

$$= 5.4 \text{ kJ/mol} - 8.46 \text{ kJ/mol}$$

$$= -3.1 \text{ kJ/mol}$$

Because ΔG is negative, the reaction proceeds spontaneously from left to right to reach equilibrium.

THINK ABOUT IT

Remember, a reaction with a positive ΔG° value can be spontaneous if the starting concentrations of reactants and products are such that $Q < K$.

Practice Problem **A** **TTEMPT** ΔG° for the reaction:

$$H_2(g) + I_2(s) \rightleftharpoons 2HI(g)$$

is 2.60 kJ/mol at 25°C. Calculate ΔG, and predict the direction in which the reaction is spontaneous if the starting concentrations are $P_{H_2} = 5.25$ atm and $P_{HI} = 1.75$ atm.

Practice Problem **B** **UILD** What is the minimum partial pressure of H_2 required for the preceding reaction to be spontaneous in the forward direction at 25°C if the partial pressure of HI is 0.94?

Practice Problem **C** **ONCEPTUALIZE** Consider the reaction in Sample Problem 18.8. Which of the following graphs best shows what happens to ΔG as the partial pressure of N_2O_4 is increased?

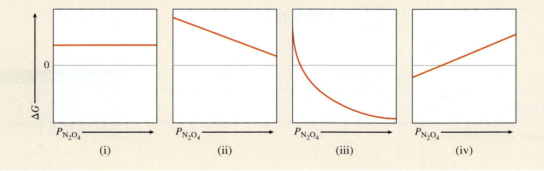

(i)   (ii)   (iii)   (iv)

Relationship Between ΔG° and K

By definition, $\Delta G = 0$ and $Q = K$ at equilibrium, where K is the equilibrium constant. Thus, $\Delta G = \Delta G^\circ + RT \ln Q$ (Equation 18.13) becomes:

$$0 = \Delta G^\circ + RT \ln K$$

Student Note: In this equation and in Equation 18.14, K is K_c for reactions that take place in solution and K_P for reactions that take place in the gas phase.

or:

Equation 18.14	$\Delta G^\circ = -RT \ln K$

According to Equation 18.14, then, the larger K is, the more negative ΔG° is. For chemists, Equation 18.14 is one of the most important equations in thermodynamics

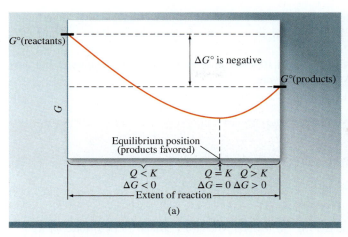

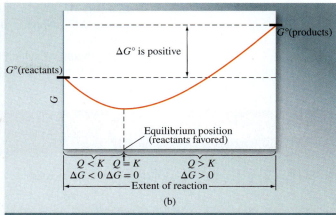

Figure 18.6 (a) $\Delta G° < 0$. At equilibrium, there is a significant conversion of reactants to products. (b) $\Delta G° > 0$. At equilibrium, reactants are favored over products. In both cases, the net reaction toward equilibrium is from left to right (reactants to products) if $Q < K$ and right to left (products to reactants) if $Q > K$. At equilibrium, $Q = K$.

because it enables us to find the equilibrium constant of a reaction if we know the change in standard free energy, and vice versa.

It is significant that Equation 18.14 relates the equilibrium constant to the standard free-energy change, $\Delta G°$, rather than to the actual free-energy change, ΔG. The actual free-energy change of the system varies as the reaction progresses and becomes zero at equilibrium. On the other hand, $\Delta G°$, like K, is a constant for a particular reaction at a given temperature. Figure 18.6 shows plots of the free energy of a reacting system versus the extent of the reaction for two reactions. Table 18.5 summarizes the relationship between the *magnitude* of an equilibrium constant and the *sign* of the corresponding $\Delta G°$. Remember this important distinction: It is the sign of ΔG and not that of $\Delta G°$ that determines the direction of reaction spontaneity. The sign of $\Delta G°$ only tells us the relative amounts of products and reactants when equilibrium is reached, *not* the direction the reaction must go to reach *equilibrium.*

For reactions with very large or very small equilibrium constants, it can be very difficult—sometimes impossible—to determine K values by measuring the concentrations of the reactants and products. Consider, for example, the formation of nitric oxide from molecular nitrogen and molecular oxygen:

$$N_2(g) + O_2(g) \rightleftharpoons 2NO(g)$$

At 25°C, the equilibrium constant, K_P, is:

$$K_P = \frac{(P_{NO})^2}{(P_{N_2})(P_{O_2})} = 4.0 \times 10^{-31}$$

The very small value of K_P means that the concentration of NO at equilibrium will be exceedingly low and, for all intents and purposes, impossible to measure

Animation
Chemical Equilibrium—equilibrium.

Student Note: The sign of $\Delta G°$ tells us the same thing that the magnitude of K tells us. The sign of ΔG tells us the same thing as the comparison of Q and K values [◄◄ Section 15.4].

TABLE 18.5	Relationship Between K and $\Delta G°$ as Predicted by Equation 18.14		
K	ln K	$\Delta G°$	**Result at Equilibrium**
> 1	Positive	Negative	Products are favored.
= 1	0	0	Neither products nor reactants are favored.
< 1	Negative	Positive	Reactants are favored.

directly. In such a case, the equilibrium constant is more conveniently determined using $\Delta G°$, which can be calculated either from tabulated $\Delta G_f°$ values or from $\Delta H°$ and $\Delta S°$.

Sample Problems 18.9 and 18.10 show how to use $\Delta G°$ to calculate K and how to use K to calculate $\Delta G°$, respectively.

SAMPLE PROBLEM   18.9

Using data from Appendix 2, calculate the equilibrium constant, K_P, for the following reaction at 25°C:

$$2H_2O(l) \rightleftharpoons 2H_2(g) + O_2(g)$$

Strategy Use data from Appendix 2 and Equation 18.12 to calculate $\Delta G°$ for the reaction. Then use Equation 18.14 to solve for K_P.

Setup

$$\Delta G° = (2\Delta G_f°[H_2(g)] + \Delta G_f°[O_2(g)]) - (2\Delta G_f°[H_2O(l)])$$
$$= [(2)(0 \text{ kJ/mol}) + (2)(0 \text{ kJ/mol})] - [(2)(-237.2 \text{ kJ/mol})]$$
$$= 474.4 \text{ kJ/mol}$$

Solution

$$\Delta G° = -RT \ln K_P$$
$$\frac{474.4 \text{ kJ}}{\text{mol}} = -\left(\frac{8.314 \times 10^{-3} \text{ kJ}}{\text{K} \cdot \text{mol}}\right)(298 \text{ K}) \ln K_P$$
$$-191.5 = \ln K_P$$
$$K_P = e^{-191.5}$$
$$= 7 \times 10^{-84}$$

THINK ABOUT IT

This is an extremely small equilibrium constant, which is consistent with the large, positive value of $\Delta G°$. We know from everyday experience that water does not decompose spontaneously into its constituent elements at 25°C.

Practice Problem Ⓐ**TTEMPT** Using data from Appendix 2, calculate the equilibrium constant, K_P, for the following reaction at 25°C:

$$2O_3(g) \rightleftharpoons 3O_2(g)$$

Practice Problem Ⓑ**UILD** K_f for the complex ion $Ag(NH_3)_2^+$ is 1.5×10^7 at 25°C. Using this and data from Appendix 2, calculate the value of $\Delta G_f°$ for $Ag(NH_3)_2^+(aq)$.

Practice Problem Ⓒ**ONCEPTUALIZE** Which of the following graphs best shows the relationship between $\Delta G°$ and equilibrium constant (K)?

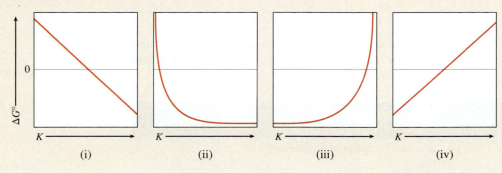

(i)   (ii)   (iii)   (iv)

SAMPLE PROBLEM 18.10

The equilibrium constant, K_{sp}, for the dissolution of silver chloride in water at 25°C:

$$AgCl(s) \rightleftharpoons Ag^+(aq) + Cl^-(aq)$$

is 1.6×10^{-10}. Calculate $\Delta G°$ for the process.

Strategy Use Equation 18.14 to calculate $\Delta G°$.

Setup $R = 8.314 \times 10^{-3}$ kJ/K · mol and $T = (25 + 273) = 298$ K.

Solution

$$\Delta G° = -RT \ln K_{sp}$$

$$= -\left(\frac{8.314 \times 10^{-3} \text{ kJ}}{\text{K} \cdot \text{mol}}\right)(298 \text{ K})\ln(1.6 \times 10^{-10})$$

$$= 55.9 \text{ kJ/mol}$$

THINK ABOUT IT

The relatively large, positive $\Delta G°$, like the very small K value, corresponds to a process that lies very far to the left. Note that the K in Equation 18.14 can be any type of K_c (K_a, K_b, K_{sp}, etc.) or K_P.

Practice Problem **A**TTEMPT Calculate $\Delta G°$ for the process:

$$BaF_2(s) \rightleftharpoons Ba^{2+}(aq) + 2F^-(aq)$$

The K_{sp} of BaF_2 at 25°C is 1.7×10^{-6}.

Practice Problem **B**UILD K_{sp} for $Co(OH)_2$ at 25°C is 3.3×10^{-16}. Using this and data from Appendix 2, calculate the value of $\Delta G_f°$ for $Co(OH)_2(s)$.

Practice Problem **C**ONCEPTUALIZE Fill in each blank with *positive, negative, zero,* or *impossible to determine:*

When $Q = K$, ΔG is _____. When $Q < K$, ΔG is _____. When $Q > K$, ΔG is _____.

CHECKPOINT – SECTION 18.6   Free Energy and Chemical Equilibrium

18.6.1 For the reaction:

$$A(aq) + B(aq) \rightleftharpoons C(aq)$$

$\Delta G° = -1.95$ kJ/mol at 25°C. What is ΔG (in kJ/mol) at 25°C when the concentrations are [A] = [B] = 0.315 M and [C] = 0.405 M?

a) −1.95 kJ/mol

b) 1.53 kJ/mol

c) −5.43 kJ/mol

d) 8.16 kJ/mol

e) −12.1 kJ/mol

18.6.2 Consider the reaction:

$$X(g) + Y(g) \rightleftharpoons Z(g)$$

for which $\Delta G° = -1.2$ kJ/mol at 25°C. Without knowing the pressures of the reactants and the product, we know that this reaction

a) is spontaneous as written.

b) is nonspontaneous as written.

c) favors reactants at equilibrium.

d) favors products at equilibrium.

e) is at equilibrium.

18.6.3 The $\Delta G°$ for the reaction:

$$N_2(g) + 3H_2(g) \rightleftharpoons 2NH_3(g)$$

is -33.3 kJ/mol at 25°C. What is the value of K_P?

a) 2×10^{-6}

b) 7×10^5

c) 3×10^{-70}

d) 1

e) 1×10^{-1}

18.6.4 The K_{sp} for iron(III) hydroxide [Fe(OH)$_3$] is 1.1×10^{-36} at 25°C. For the process:

$$Fe(OH)_3(s) \rightleftharpoons Fe^{3+}(aq) + 3OH^-(aq)$$

determine $\Delta G°$ (in kJ/mol) at 25°C.

a) -2.73×10^{-36} kJ/mol

b) -17.2 kJ/mol

c) 17.2 kJ/mol

d) -205 kJ/mol

e) 205 kJ/mol

18.7   Thermodynamics in Living Systems

Many biochemical reactions have a positive $\Delta G°$ value, yet they are essential to the maintenance of life. In living systems, these reactions are coupled to an energetically favorable process, one that has a negative $\Delta G°$ value. The principle of coupled reactions is based on a simple concept: we can use a thermodynamically favorable reaction to drive an unfavorable one. Suppose, for example, that we want to extract zinc from a zinc sulfide (ZnS). The following reaction will not work because it has a large positive $\Delta G°$ value:

$$ZnS(s) \longrightarrow Zn(s) + S(s) \qquad \Delta G° = 198.3 \text{ kJ/mol}$$

On the other hand, the combustion of sulfur to form sulfur dioxide is favored because of its large negative $\Delta G°$ value:

$$S(s) + O_2(g) \longrightarrow SO_2(g) \qquad \Delta G° = -300.1 \text{ kJ/mol}$$

By coupling the two processes, we can bring about the separation of zinc from zinc sulfide. In practice, this means heating ZnS in air so that the tendency of S to form SO$_2$ will promote the decomposition of ZnS:

$$
\begin{aligned}
ZnS(s) &\longrightarrow Zn(s) + \cancel{S(s)} & \Delta G° &= 198.3 \text{ kJ/mol} \\
\cancel{S(s)} + O_2(g) &\longrightarrow SO_2(g) & \Delta G° &= -300.1 \text{ kJ/mol} \\
\hline
ZnS(s) + O_2(g) &\longrightarrow Zn(s) + SO_2(g) & \Delta G° &= -101.8 \text{ kJ/mol}
\end{aligned}
$$

Coupled reactions play a crucial role in our survival. In biological systems, enzymes facilitate a wide variety of nonspontaneous reactions. In the human body, for example, food molecules, represented by glucose ($C_6H_{12}O_6$), are converted to carbon dioxide and water during metabolism, resulting in a substantial release of free energy:

$$C_6H_{12}O_6(s) + 6O_2(g) \longrightarrow 6CO_2(g) + 6H_2O(l) \qquad \Delta G° = -2880 \text{ kJ/mol}$$

In a living cell, this reaction does not take place in a single step; rather, the glucose molecule is broken down with the aid of enzymes in a series of steps. Much of the free energy released along the way is used to synthesize adenosine triphosphate (ATP) from adenosine diphosphate (ADP) and phosphoric acid (Figure 18.7):

$$ADP + H_3PO_4 \longrightarrow ATP + H_2O \qquad \Delta G° = 31 \text{ kJ/mol}$$

The function of ATP is to store free energy until it is needed by cells. Under appropriate conditions, ATP undergoes hydrolysis to give ADP and phosphoric acid, with a release of 31 kJ/mol of free energy, which can be used to drive energetically unfavorable reactions, such as protein synthesis.

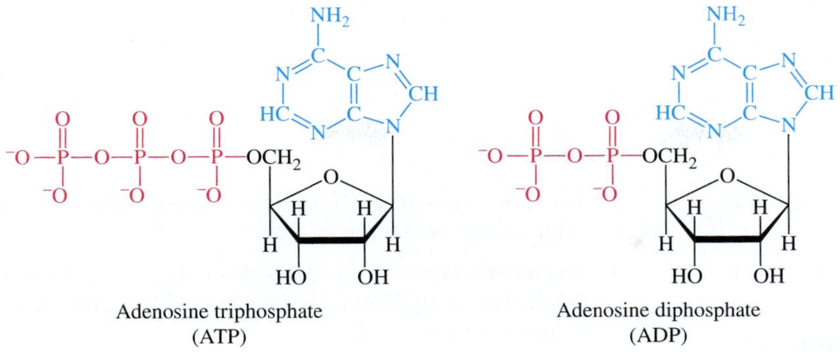

NH2
Adenosine triphosphate
(ATP)
Adenosine diphosphate
(ADP)

Figure 18.7 Structures of ATP and ADP.

Proteins are polymers made of amino acids. The stepwise synthesis of a protein molecule involves the joining of individual amino acids. Consider the formation of the dipeptide (a unit composed of two amino acids) alanylglycine from alanine and glycine. This reaction represents the first step in the synthesis of a protein molecule:

$$\text{alanine} + \text{glycine} \longrightarrow \text{alanylglycine} \qquad \Delta G° = 29\ \text{kJ/mol}$$

The positive $\Delta G°$ value means this reaction does not favor the formation of product, so only a little of the dipeptide would be formed at equilibrium. With the aid of an enzyme, however, the reaction is coupled to the hydrolysis of ATP as follows:

$$\text{ATP} + H_2O + \text{alanine} + \text{glycine} \longrightarrow \text{ADP} + H_3PO_4 + \text{alanylglycine}$$

The overall free-energy change is given by $\Delta G° = -31\ \text{kJ/mol} + 29\ \text{kJ/mol} = -2\ \text{kJ/mol}$, which means that the coupled reaction now favors the formation of product and an appreciable amount of alanylglycine will be formed under these conditions. Figure 18.8 shows the ATP-ADP interconversions that act as energy storage (from metabolism) and free-energy release (from ATP hydrolysis) to drive essential reactions.

Free energy
Glucose
ATP
Proteins
CO2 + H2O
ADP
Amino acids

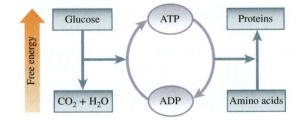

Figure 18.8 Schematic representation of ATP synthesis and coupled reactions in living systems. The conversion of glucose to carbon dioxide and water during metabolism releases free energy. The released free energy is used to convert ADP to ATP. The ATP molecules are then used as an energy source to drive unfavorable reactions such as protein synthesis from amino acids.

Chapter Summary

Section 18.1

- A *spontaneous process* is one that occurs under a specified set of conditions.

- A *nonspontaneous process* is one that does *not* occur under a specified set of conditions.

- Spontaneous processes do not necessarily happen quickly.

Section 18.2

- *Entropy (S)* is a thermodynamic state function that measures how dispersed or spread out a system's energy is.

Section 18.3

- Entropy change for a process can be calculated using standard entropy values or can be predicted qualitatively based on factors such as temperature, phase, and number of molecules.

- Whether or not a process is spontaneous depends on the change in *enthalpy* and the change in *entropy* of the system.

- Tabulated *standard entropy* values are absolute values.

Section 18.4

- According to the *second law of thermodynamics,* the entropy change for the universe is positive for a *spontaneous* process and zero for an *equilibrium process.*

- According to the *third law of thermodynamics,* the entropy of a perfectly crystalline substance at 0 K is zero.

Section 18.5

- The *Gibbs free energy (G)* or simply the *free energy* of a system is the energy available to do work.

- The *standard free energy of reaction* (ΔG°_{rxn}) for a reaction tells us whether the equilibrium lies to the right (negative ΔG°_{rxn}) or to the left (positive ΔG°_{rxn}).

- *Standard free energies of formation* (ΔG°_f) can be used to calculate standard free energies of reaction.

Section 18.6

- The free-energy change (ΔG) is determined using the standard free-energy change (ΔG°) and the reaction quotient (Q).

- The sign of ΔG tells us whether the reaction is spontaneous under the conditions described.

- ΔG° is related to the equilibrium constant, K. A negative ΔG° corresponds to a large K; a positive ΔG° corresponds to a small K.

Section 18.7

- In living systems, thermodynamically favorable reactions provide the free energy needed to drive necessary but thermodynamically unfavorable reactions.

Key Words

Key Equations

18.1 $S = k \ln W$	The entropy S of a system is equal to the product of the Boltzmann constant (k) and ln of W, the number of possible arrangements of molecules in the system.
18.2 $W = X^N$	The number of possible arrangements W is equal to the number of possible locations of molecules X raised to the number of molecules in the system N.
18.3 $\Delta S_{sys} = S_{final} - S_{initial}$	The entropy change in a system ΔS_{sys}, is equal to final entropy, ΔS_{final}, minus initial entropy, $\Delta S_{initial}$.
18.4 $\Delta S_{sys} = nR \ln \dfrac{V_{final}}{V_{initial}}$	For a gaseous process involving a volume change, entropy change is calculated as the product of the number of moles (n), the gas constant (R), and ln of the ratio of final volume to initial volume [$\ln(V_{final}/V_{initial})$].
18.5 $\Delta S^{\circ}_{rxn} = [cS^{\circ}(C) + dS^{\circ}(D)] - [aS^{\circ}(A) + bS^{\circ}(B)]$	Standard entropy change for a reaction (ΔS°_{rxn}) can be calculated using tabulated values of absolute entropies (S°) for products and reactants.
18.6 $\Delta S^{\circ}_{rxn} = \Sigma nS^{\circ}(\text{products}) - \Sigma mS^{\circ}(\text{reactants})$	ΔS°_{rxn} is calculated as the sum of absolute entropies for products minus the sum of absolute entropies for reactants. Each species in a chemical equation must be multiplied by its coefficient.
18.7 $\Delta S_{surr} = \dfrac{-\Delta H_{sys}}{T}$	Entropy change in the surroundings (ΔS_{surr}) is calculated as the ratio of minus the enthalpy change in the system ($-\Delta H_{sys}$) to absolute temperature (T).
18.8 $\Delta S_{univ} = \Delta S_{sys} + \Delta S_{surr}$	Entropy change in the universe (ΔS_{univ}) is equal to the sum of entropy change for the system (ΔS_{sys}) and entropy change for the surroundings (ΔS_{surr}).
18.9 $G = H - TS$	Gibbs free energy (G) is the difference between enthalpy (H) and the product of absolute temperature and entropy (TS).
18.10 $\Delta G = \Delta H - T\Delta S$	The change in free energy (ΔG) is calculated as the difference between change in enthalpy (ΔH) and the product of absolute temperature and change in entropy ($T\Delta S$).
18.11 $\Delta G^{\circ}_{rxn} = [c\Delta G^{\circ}_f(C) + d\Delta G^{\circ}_f(D)] - [a\Delta G^{\circ}_f(A) + b\Delta G^{\circ}_f(B)]$	Standard free-energy change for a reaction (ΔG°_{rxn}) can be calculated using tabulated values of free energies of formation (ΔG°_f) for products and reactants.
18.12 $\Delta G^{\circ}_{rxn} = \Sigma n\Delta G^{\circ}_f(\text{products}) - \Sigma m\Delta G^{\circ}_f(\text{reactants})$	ΔG°_{rxn} is calculated as the sum of free energies of formation for products minus the sum of free energies of formation for reactants. Each species in a chemical equation must be multiplied by its coefficient.
18.13 $\Delta G = \Delta G^{\circ} + RT \ln Q$	ΔG is calculated as the sum of ΔG° and the product of R, T, and ln of the reaction quotient, Q.
18.14 $\Delta G^{\circ} = -RT \ln K$	At equilibrium, Q is equal to K and ΔG is equal to zero. ΔG° is equal to minus the product of R, T, and ln of K.

Key Constant

Section 18.2 Boltzmann constant $k = 1.38 \times 10^{-23}$ J/K	The Boltzmann constant is equal to the gas constant, 8.314 J/K · mol, divided by Avogadro's constant, 6.022×10^{23}/mol. It is used to calculate the entropy of a system when the number of energetically equivalent different arrangements (W) is known.

In Chapter 15, you learned how to calculate reaction quotient (Q), and that by comparing its magnitude to that of the equilibrium constant (K), you can determine whether a reaction must proceed to the right or to the left in order to reach equilibrium. In Chapter 18, you learned how to calculate free-energy change (ΔG) and standard free-energy change ($\Delta G°$), and that the sign of ΔG indicates whether or not a process is spontaneous. It is useful to consider how these parameters are related, and that they give us essentially the same information. Equation 18.14 relates K and $\Delta G°$:

$$\Delta G° = -RT \ln K$$

According to this equation, if $\Delta G°$ is negative, the value of K is greater than 1. Recall that $K > 1$ tells us the equilibrium *lies to the right*. This is different from saying that it will *proceed* to the right. A reaction may have a large equilibrium constant and yet proceed to the *left* in order to reach equilibrium—if the starting conditions are such that Q is greater than K. For example, $K = 54.3$ for the following reaction at 430°C:

$$H_2(g) + I_2(g) \rightleftharpoons 2HI(g)$$

Because K is greater than 1, we would say that this equilibrium lies to the right. However, if we initially have only HI and no H_2 or I_2, clearly the reaction must *proceed* to the *left* in order for equilibrium to be established.

So what does it mean for an equilibrium to *lie* to the right? It means that if the starting conditions are *standard* conditions (the concentrations of all aqueous species are 1 M and the pressures of all gaseous species are 1 atm), products will predominate at equilibrium. This corresponds to the reaction proceeding to the right from *standard* initial conditions. Likewise, for an equilibrium that lies to the *left,* if the starting conditions are standard conditions, reactants will predominate at equilibrium—the reaction will proceed to the *left* from standard initial conditions.

We can also describe this in terms of spontaneity. The sign of $\Delta G°$ tells us whether or not a process or reaction is *spontaneous* under *standard* conditions. $\Delta G°$ being *negative* is simply another way to say that the equilibrium lies to the right. ΔG, on the other hand, tells us whether or not a process is spontaneous under *actual* conditions—which usually are not standard. ΔG and $\Delta G°$ are related by Equation 18.13:

$$\Delta G = \Delta G° + RT \ln Q$$

The sign of ΔG depends not only on the sign of $\Delta G°$, but also on the value of Q. The information provided by the sign of ΔG is the same as that provided by the *comparison* of Q and K.

By combining Equations 18.13 and 18.14, we can write an expression that indicates at a glance the relationship between the sign of ΔG and the relative magnitudes of Q and K.

Substituting the right side of Equation 18.14 for the $\Delta G°$ term in Equation 18.13 gives:

$$\Delta G = -RT \ln K + RT \ln Q$$

or:

$$\Delta G = -RT \ln Q - RT \ln K$$

And because $\ln A - \ln B = \ln \dfrac{A}{B}$ [▶◀ Appendix 1], we get:

$$\Delta G = RT \ln \dfrac{Q}{K}$$

This enables us to simply compare Q and K. When Q is greater than K, the log term in our final equation will be positive and ΔG will be positive—meaning that the reaction must proceed to the left to establish equilibrium. When Q is *less* than K, the log term and ΔG will both be *negative*—meaning that the reaction will proceed to the *right*.

Key Skills Problems

18.1
Which of the following must be negative for a process to be spontaneous as written?

(a) $\Delta G°$ (b) ΔG (c) K (d) Q (e) R

18.2
ΔG for a reaction is always negative when

(a) $\Delta G°$ is negative (b) $K < 1$ (c) $K > 1$
(d) $Q < K$ (e) $Q > K$

18.3
The diagram shown here depicts a system at equilibrium for the reaction $A_2 + B_2 \rightleftharpoons 2AB$.

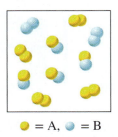

● = A, ● = B

$\Delta G°$ for the reaction is −4.8 kJ/mol. Which of the following diagrams depicts starting conditions for the reaction where ΔG is negative? Select all that apply.

(a)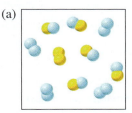

(b)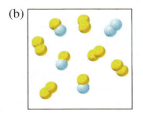

(c)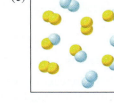

(d)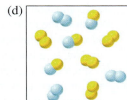

18.4
The reaction shown here has $\Delta G° = -1.83$ kJ/mol at 25°C. Under the starting conditions shown in the diagram, the reaction will proceed to the right. Which of the following could be the reaction depicted in the diagram? Select all that apply.

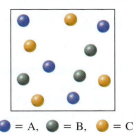

● = A, ● = B, ● = C

(a) $C \rightleftharpoons A + B$ (c) $2C \rightleftharpoons 2A + B$
(b) $2C \rightleftharpoons A + B$ (d) $2C \rightleftharpoons A + 2B$

Questions and Problems

Applying What You've Learned

When a folded protein in solution is heated to a high enough temperature, its polypeptide chain will unfold to become the denatured protein—a process known as "denaturation." The temperature at which most of the protein unfolds is called the "melting" temperature. The melting temperature of a certain protein is found to be 63°C, and the enthalpy of denaturation is 510 kJ/mol.

Problems:
(a) Estimate the entropy of denaturation, assuming that the denaturation is a single-step equilibrium process; that is, folded protein ⇌ denatured protein [◀◀ Sample Problem 18.7]. The single polypeptide protein chain has 98 amino acids. Calculate the entropy of denaturation per amino acid. (b) Assuming that ΔH and ΔS do not change with temperature, determine ΔG for the denaturation at 20°C [◀◀ Sample Problem 18.5]. (c) Assuming that the ΔG value from part (b) is $\Delta G°$ for the denaturation, determine the value of the equilibrium constant for the process at 20°C [◀◀ Sample Problem 18.9].

SECTION 18.1: SPONTANEOUS PROCESSES

Review Questions

18.1 Explain what is meant by a *spontaneous process*. Give two examples each of spontaneous and nonspontaneous processes.

18.2 Which of the following processes are spontaneous and which are nonspontaneous: (a) dissolving table salt (NaCl) in hot soup, (b) climbing Mt. Everest, (c) spreading fragrance in a room by removing the cap from a perfume bottle, (d) separating helium and neon from a mixture of the gases?

18.3 Which of the following processes are spontaneous and which are nonspontaneous at a given temperature?
(a) $NaNO_3(s) \xrightarrow{H_2O} NaNO_3(aq)$ saturated soln
(b) $NaNO_3(s) \xrightarrow{H_2O} NaNO_3(aq)$ unsaturated soln
(c) $NaNO_3(s) \xrightarrow{H_2O} NaNO_3(aq)$ supersaturated soln

SECTION 18.2: ENTROPY

Review Questions

18.4 Describe what is meant by the term *entropy*. What are the units of entropy?

18.5 What is the relationship between entropy and the number of possible arrangements of molecules in a system?

Conceptual Problems

18.6 Referring to the setup in Figure 18.2, determine the number of possible arrangements, W, and calculate the entropy before and after removal of the barrier if the number of molecules is (a) 10, (b) 50, (c) 100.

18.7 In the setup shown, a container is divided into eight cells and contains two molecules. Initially, both molecules are confined to the left side of the container. (a) Determine the number of possible arrangements before and after removal of the central barrier. (b) After the removal of the barrier, how many of the arrangements correspond to the state in which both molecules are in the left side of the container? How many correspond to the state in which both molecules are in the right side of the container? How many correspond to the state in which the molecules are in opposite sides of the container? Calculate the entropy for each state and comment on the most probable state of the system after removal of the barrier.

SECTION 18.3: ENTROPY CHANGES IN A SYSTEM

▶▶▶ **Visualizing Chemistry**
Figure 18.4

VC 18.1 Consider two gas samples at STP: one consisting of a mole of F_2 gas ($S° = 203.34$ J/K · mol) and one consisting of a mole of F gas ($S° = 158.7$ J/K · mol). What factors account for the difference in standard entropies of these two species?

Volume (i)	Molar mass (ii)	Temperature (iii)	Phase (iv)	Molecular complexity (v)

a) i, ii, iii, and iv
b) ii and v
c) ii, iv, and v

VC 18.2 Now consider the reaction $F_2(g) \longrightarrow 2F(g)$ at constant temperature and pressure. What factors contribute to the entropy increase associated with the reaction?

Volume increase (i)	Molar mass increase (ii)	Increased number of molecules (iii)	Phase change (iv)	Increase in molecular complexity (v)

  a) i and iii
  b) i, ii, and iii
  c) i, iv, and v

VC 18.3 Which of the following best describes why entropy always increases with temperature?
  a) As temperature increases, the number of molecules increases.
  b) As temperature increases, energy levels become more closely spaced.
  c) As temperature increases, the molecules become more energetic and can access more energy levels.

VC 18.4 Which of the following best explains why entropy typically increases with molar mass?
  a) As molar mass increases, the number of molecules increases.
  b) As molar mass increases, energy levels become more closely spaced.
  c) As molar mass increases, the molecules become more energetic and can access more energy levels.

Review Questions

18.8 How does the entropy of a system change for each of the following processes?
(a) A solid melts.
(b) A liquid freezes.
(c) A liquid boils.
(d) A vapor is converted to a solid.
(e) A vapor condenses to a liquid.
(f) A solid sublimes.
(g) A solid dissolves in water.

18.9 How does the entropy of a system change for each of the following processes?
(a) Bromine liquid vaporizes.
(b) Water freezes to form ice.
(c) Naphthalene, the key component of mothballs, sublimes.
(d) Sugar crystals form from a supersaturated solution.
(e) A block of lead melts.
(f) Iodine vapor condenses to form solid iodine.
(g) Carbon tetrachloride dissolves in liquid benzene.

18.10 Predict whether the entropy change is positive or negative for each of the following reactions. Give reasons for your predictions.
(a) $2KClO_4(s) \longrightarrow 2KClO_3(s) + O_2(g)$
(b) $H_2O(g) \longrightarrow H_2O(l)$
(c) $2Na(s) + 2H_2O(l) \longrightarrow 2NaOH(aq) + H_2(g)$
(d) $N_2(g) \longrightarrow 2N(g)$

18.11 State whether the sign of the entropy change expected for each of the following processes will be positive or negative, and explain your predictions.
(a) $PCl_3(l) + Cl_2(g) \longrightarrow PCl_5(s)$
(b) $2HgO(s) \longrightarrow 2Hg(l) + O_2(g)$
(c) $H_2(g) \longrightarrow 2H(g)$
(d) $U(s) + 3F_2(g) \longrightarrow UF_6(g)$

Computational Problems

18.12 Calculate ΔS_{sys} for (a) the isothermal expansion of 2.0 mol of an ideal gas from 10.0 L to 15.0 L, (b) the isothermal expansion of 1.5 mol of an ideal gas from 20.0 L to 22.5 L, and (c) the isothermal compression of 5.0 mol of an ideal gas from 100.0 L to 75.0 L.

18.13 Calculate ΔS_{sys} for (a) the isothermal compression of 0.0050 mol of an ideal gas from 112 mL to 52.5 mL, (b) the isothermal compression of 0.015 mol of an ideal gas from 225 mL to 22.5 mL, and (c) the isothermal expansion of 22.1 mol of an ideal gas from 122 L to 275 L.

18.14 Using the data in Appendix 2, calculate the standard entropy changes for the following reactions at 25°C:
(a) $S(rhombic) + O_2(g) \longrightarrow SO_2(g)$
(b) $MgCO_3(s) \longrightarrow MgO(s) + CO_2(g)$
(c) $2C_2H_6(g) + 7O_2(g) \longrightarrow 4CO_2(g) + 6H_2O(l)$

18.15 Using the data in Appendix 2, calculate the standard entropy changes for the following reactions at 25°C:
(a) $H_2(g) + CuO(s) \longrightarrow Cu(s) + H_2O(g)$
(b) $2Al(s) + 3ZnO(s) \longrightarrow Al_2O_3(s) + 3Zn(s)$
(c) $CH_4(g) + 2O_2(g) \longrightarrow CO_2(g) + 2H_2O(l)$

Conceptual Problems

18.16 For each pair of substances listed here, choose the one having the larger standard entropy value at 25°C. The same molar amount is used in the comparison. Explain the basis for your choice.
(a) $Li(s)$ or $Li(l)$, (b) $C_2H_5OH(l)$ or $CH_3OCH_3(l)$ (Hint: Which molecule can hydrogen-bond?), (c) $Ar(g)$ or $Xe(g)$, (d) $CO(g)$ or $CO_2(g)$, (e) $O_2(g)$ or $O_3(g)$, (f) $NO_2(g)$ or $N_2O_4(g)$.

18.17 Arrange the following substances (1 mole each) in order of increasing entropy at 25°C: (a) $Ne(g)$, (b) $SO_2(g)$, (c) $Na(s)$, (d) $NaCl(s)$, (e) $H_2(g)$. Give the reasons for your arrangement.

SECTION 18.4: ENTROPY CHANGES IN THE UNIVERSE

Review Questions

18.18 State the second law of thermodynamics in words, and express it mathematically.

18.19 State the third law of thermodynamics in words, and explain its usefulness in calculating entropy values.

Computational Problems

18.20   Calculate ΔS_{surr} for each of the reactions in Problem 18.14 and determine if each reaction is spontaneous at 25°C.

18.21   Calculate ΔS_{surr} for each of the reactions in Problem 18.15 and determine if each reaction is spontaneous at 25°C.

18.22   Using data from Appendix 2, calculate ΔS_{rxn}° and ΔS_{surr} for each of the reactions in Problem 18.10 and determine if each reaction is spontaneous at 25°C.

18.23   Using data from Appendix 2, calculate ΔS_{rxn}° and ΔS_{surr} for each of the reactions in Problem 18.11 and determine if each reaction is spontaneous at 25°C.

SECTION 18.5: PREDICTING SPONTANEITY

Review Questions

18.24   Define *free energy*. What are its units?

18.25   Why is it more convenient to predict the direction of a reaction in terms of ΔG_{sys} instead of ΔS_{univ}? Under what conditions can ΔG_{sys} be used to predict the spontaneity of a reaction?

18.26   What is the significance of the sign of ΔG_{sys}?

18.27   From the following combinations of ΔH and ΔS, predict if a process will be spontaneous at a high or low temperature: (a) both ΔH and ΔS are negative, (b) ΔH is negative and ΔS is positive, (c) both ΔH and ΔS are positive, (d) ΔH is positive and ΔS is negative.

Computational Problems

18.28   Calculate ΔG° for the following reactions at 25°C:
(a) $N_2(g) + O_2(g) \longrightarrow 2NO(g)$
(b) $H_2O(l) \longrightarrow H_2O(g)$
(c) $2C_2H_2(g) + 5O_2(g) \longrightarrow 4CO_2(g) + 2H_2O(l)$
(*Hint:* Look up the standard free energies of formation of the reactants and products in Appendix 2.)

18.29   Calculate ΔG° for the following reactions at 25°C:
(a) $2Mg(s) + O_2(g) \longrightarrow 2MgO(s)$
(b) $2SO_2(g) + O_2(g) \longrightarrow 2SO_3(g)$
(c) $2C_2H_6(g) + 7O_2(g) \longrightarrow 4CO_2(g) + 6H_2O(l)$
(See Appendix 2 for thermodynamic data.)

18.30   From the values of ΔH and ΔS, predict which of the following reactions would be spontaneous at 25°C: reaction A: $\Delta H = 10.5$ kJ/mol, $\Delta S = 30$ J/K · mol; reaction B: $\Delta H = 1.8$ kJ/mol, $\Delta S = -113$ J/K · mol. If either of the reactions is nonspontaneous at 25°C, at what temperature might it become spontaneous?

18.31   Find the temperatures at which reactions with the following ΔH and ΔS values would become spontaneous: (a) $\Delta H = -126$ kJ/mol, $\Delta S = 84$ J/K · mol; (b) $\Delta H = -11.7$ kJ/mol, $\Delta S = -105$ J/K · mol.

18.32   The molar heats of fusion and vaporization of ethanol are 7.61 and 26.0 kJ/mol, respectively. Calculate the molar entropy changes for the solid-liquid and liquid-vapor transitions for ethanol. At 1 atm pressure, ethanol melts at −117.3°C and boils at 78.3°C.

18.33   The molar heats of fusion and vaporization of mercury are 23.4 and 59.0 kJ/mol, respectively. Calculate the molar entropy changes for the solid-liquid and liquid-vapor transitions for mercury. At 1 atm pressure, mercury melts at −38.9°C and boils at 357°C.

18.34   Consider the formation of a dimeric protein:

$$2P \longrightarrow P_2$$

At 25°C, we have $\Delta H^{\circ} = 17$ kJ/mol and $\Delta S^{\circ} = 65$ J/K · mol. Is the dimerization favored at this temperature? Comment on the effect of lowering the temperature. Does your result explain why some enzymes lose their activities under cold conditions?

18.35   Use the values listed in Appendix 2 to calculate ΔG° for the following alcohol fermentation:

$$C_6H_{12}O_6(s) \longrightarrow 2C_2H_5OH(l) + 2CO_2(g)$$

Conceptual Questions

18.36   The reaction represented here is spontaneous under standard conditions. What can be deduced about the sign of ΔH° for the reaction?

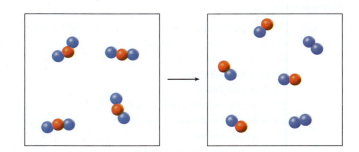

18.37   Which of the following graphs best represents the relationship between ΔG and temperature for a process that is (a) exothermic and for which ΔS is negative, (b) endothermic and for which ΔS is positive, and (c) exothermic and for which ΔS is positive?

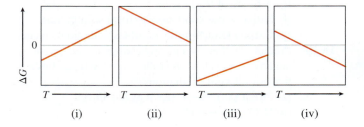

SECTION 18.6: FREE ENERGY AND CHEMICAL EQUILIBRIUM

Review Questions

18.38 Explain the difference between ΔG and $\Delta G°$.

18.39 Explain why Equation 18.14 is of great importance in chemistry.

18.40 Fill in the missing entries in the following table:

K	$\ln K$	$\Delta G°$	Result at equilibrium
< 1			
		0	
			Products are favored

Computational Problems

18.41 Calculate K_P for the following reaction at 25°C:

$$H_2(g) + I_2(g) \rightleftharpoons 2HI(g) \qquad \Delta G° = 2.60 \text{ kJ/mol}$$

18.42 For the autoionization of water at 25°C:

$$H_2O(l) \rightleftharpoons H^+(aq) + OH^-(aq)$$

K_w is 1.0×10^{-14}. What is $\Delta G°$ for the process?

18.43 Consider the following reaction at 25°C:

$$Fe(OH)_2(s) \rightleftharpoons Fe^{2+}(aq) + 2OH^-(aq)$$

Calculate $\Delta G°$ for the reaction. K_{sp} for $Fe(OH)_2$ is 1.6×10^{-14}.

18.44 Calculate $\Delta G°$ and K_P for the following equilibrium reaction at 25°C:

$$2H_2O(g) \rightleftharpoons 2H_2(g) + O_2(g)$$

18.45 (a) Calculate $\Delta G°$ and K_P for the following equilibrium reaction at 25°C:

$$PCl_5(g) \rightleftharpoons PCl_3(g) + Cl_2(g)$$

(b) Calculate ΔG for the reaction if the partial pressures of the initial mixture are $P_{PCl_5} = 0.0029$ atm, $P_{PCl_3} = 0.27$ atm, and $P_{Cl_2} = 0.40$ atm.

18.46 The equilibrium constant (K_P) for the reaction:

$$H_2(g) + CO_2(g) \rightleftharpoons H_2O(g) + CO(g)$$

is 4.40 at 2000 K. (a) Calculate $\Delta G°$ for the reaction. (b) Calculate ΔG for the reaction when the partial pressures are $P_{H_2} = 0.25$ atm, $P_{CO_2} = 0.78$ atm, $P_{H_2O} = 0.66$ atm, and $P_{CO} = 1.20$ atm.

18.47 Consider the decomposition of calcium carbonate:

$$CaCO_3(s) \rightleftharpoons CaO(s) + CO_2(g)$$

Calculate the pressure in atm of CO_2 in an equilibrium process (a) at 25°C and (b) at 800°C. Assume that $\Delta H° = 177.8$ kJ/mol and $\Delta S° = 160.5$ J/K · mol for the temperature range.

18.48 The equilibrium constant K_P for the reaction:

$$CO(g) + Cl_2(g) \rightleftharpoons COCl_2(g)$$

is 5.62×10^{35} at 25°C. Calculate $\Delta G_f°$ for $COCl_2$ at 25°C.

18.49 At 25°C, $\Delta G°$ for the process:

$$H_2O(l) \rightleftharpoons H_2O(g)$$

is 8.6 kJ/mol. Calculate the vapor pressure of water at this temperature.

18.50 Calculate $\Delta G°$ for the process

$$C(\text{diamond}) \rightleftharpoons C(\text{graphite})$$

Is the formation of graphite from diamond favored at 25°C? If so, why is it that diamonds do not become graphite on standing?

SECTION 18.7: THERMODYNAMICS IN LIVING SYSTEMS

Review Questions

18.51 What is a coupled reaction? What is its importance in biological reactions?

18.52 What is the role of ATP in biological reactions?

Computational Problems

18.53 Referring to the metabolic process involving glucose in Section 18.7, calculate the maximum number of moles of ATP that can be synthesized from ADP from the breakdown of 1 mole of glucose.

18.54 In the metabolism of glucose, the first step is the conversion of glucose to glucose 6-phosphate:

$$\text{glucose} + H_3PO_4 \longrightarrow \text{glucose 6-phosphate} + H_2O$$
$$\Delta G° = 13.4 \text{ kJ/mol}$$

Because $\Delta G°$ is positive, this reaction does not favor the formation of products. Show how this reaction can be made to proceed by coupling it with the hydrolysis of ATP. Write an equation for the coupled reaction, and estimate the equilibrium constant for the coupled process.

ADDITIONAL PROBLEMS

18.55 Predict the signs of ΔH, ΔS, and ΔG of the system for the following processes at 1 atm: (a) ammonia melts at −60°C, (b) ammonia melts at −77.7°C, (c) ammonia melts at −100°C. (The normal melting point of ammonia is −77.7°C.)

18.56 Calculate ΔG for the reaction:

$$H_2O(l) \rightleftharpoons H^+(aq) + OH^-(aq)$$

at 25°C for the following initial concentrations:
(a) $[H^+] = 1.0 \times 10^{-7}$ M, $[OH^-] = 1.0 \times 10^{-7}$ M
(b) $[H^+] = 1.0 \times 10^{-3}$ M, $[OH^-] = 1.0 \times 10^{-4}$ M
(c) $[H^+] = 1.0 \times 10^{-12}$ M, $[OH^-] = 2.0 \times 10^{-8}$ M
(d) $[H^+] = 3.5$ M, $[OH^-] = 4.8 \times 10^{-4}$ M

18.57 Which of the following thermodynamic functions are associated only with the first law of thermodynamics: S, U, G, and H?

18.58 A student placed 1 g of each of three compounds A, B, and C in a container and found that after 1 week no change had occurred. Offer some possible explanations for the fact that no reactions took place. Assume that A, B, and C are totally miscible liquids.

18.59 Consider the following reaction at 298 K:

$$2H_2(g) + O_2(g) \longrightarrow 2H_2O(l) \qquad \Delta H° = -571.6 \text{ kJ/mol}$$

Calculate ΔS_{sys}, ΔS_{surr}, and ΔS_{univ} for the reaction.

18.60 Consider the following facts: Water freezes spontaneously at $-5°C$ and 1 atm, and ice has a lower entropy than liquid water. Explain how a spontaneous process can lead to a decrease in entropy.

18.61 Ammonium nitrate (NH_4NO_3) dissolves spontaneously and endothermically in water. What can you deduce about the sign of ΔS for the solution process?

18.62 Calculate the equilibrium pressure of CO_2 due to the decomposition of barium carbonate ($BaCO_3$) at 25°C.

18.63 (a) Trouton's rule states that the ratio of the molar heat of vaporization of a liquid (ΔH_{vap}) to its boiling point in kelvins is approximately 90 J/K · mol. Use the following data to show that this is the case and explain why Trouton's rule holds true:

	T_{bp} (°C)	ΔH_{vap}(kJ/mol)
Benzene	80.1	31.0
Hexane	68.7	30.8
Mercury	357	59.0
Toluene	110.6	35.2

(b) Use the values in Table 11.6 to calculate the same ratio for ethanol and water. Explain why Trouton's rule does not apply to these two substances as well as it does to other liquids.

18.64 Referring to Problem 18.63, explain why the ratio is considerably smaller than 90 J/K · mol for liquid HF.

18.65 Which of the following are not state functions: S, H, q, w, T?

18.66 For reactions carried out under standard-state conditions, Equation 18.10 takes the form $\Delta G° = \Delta H° - T\Delta S°$. (a) Assuming $\Delta H°$ and $\Delta S°$ are independent of temperature, derive the equation:

$$\ln\frac{K_2}{K_1} = \frac{\Delta H°}{R}\left(\frac{T_2 - T_1}{T_1 T_2}\right)$$

where K_1 and K_2 are the equilibrium constants at T_1 and T_2, respectively. (b) Given that at 25°C K_c is 4.63×10^{-3} for the reaction:

$$N_2O_4(g) \rightleftharpoons 2NO_2(g) \qquad \Delta H° = 58.0 \text{ kJ/mol}$$

calculate the equilibrium constant at 65°C.

18.67 The sublimation of carbon dioxide at $-78°C$ is given by:

$$CO_2(s) \longrightarrow CO_2(g) \qquad \Delta H_{sub} = 25.2 \text{ kJ/mol}$$

Calculate ΔS_{sub} when 84.8 g of CO_2 sublimes at this temperature.

18.68 Under what conditions does a substance have a standard entropy of zero? Can an element or a compound ever have a negative standard entropy?

18.69 A student looked up the $\Delta G_f°$, $\Delta H_f°$, and $\Delta S°$ values for CO_2 in Appendix 2. Plugging these values into Equation 18.10, the student found that $\Delta G_f° \neq \Delta H_f° - T\Delta S°$ at 298 K. What is wrong with this approach?

18.70 Consider the following Brønsted acid-base reaction at 25°C:

$$HF(aq) + Cl^-(aq) \rightleftharpoons HCl(aq) + F^-(aq)$$

(a) Predict whether K will be greater or smaller than 1. (b) Does $\Delta S°$ or $\Delta H°$ make a greater contribution to $\Delta G°$? (c) Is $\Delta H°$ likely to be positive or negative?

18.71 At 0 K, the entropy of carbon monoxide crystal is not zero but has a value of 4.2 J/K · mol, called the residual entropy. According to the third law of thermodynamics, this means that the crystal does not have a perfect arrangement of the CO molecules. (a) What would be the residual entropy if the arrangement were totally random? (b) Comment on the difference between the result in part (a) and 4.2 J/K · mol. (*Hint:* Assume that each CO molecule has two choices for orientation, and use Equation 18.1 to calculate the residual entropy.)

18.72 Crystallization of sodium acetate from a supersaturated solution occurs spontaneously (see Figure 13.2). Based on this, what can you deduce about the signs of ΔS and ΔH?

18.73 Consider the thermal decomposition of $CaCO_3$:

$$CaCO_3(s) \rightleftharpoons CaO(s) + CO_2(g)$$

The equilibrium vapor pressures of CO_2 are 22.6 mmHg at 700°C and 1829 mmHg at 950°C. Calculate the standard enthalpy of the reaction. [*Hint:* See Problem 18.66(a).]

18.74 A certain reaction is spontaneous at 72°C. If the enthalpy change for the reaction is 19 kJ/mol, what is the *minimum* value of ΔS (in J/K · mol) for the reaction?

18.75 Predict whether the entropy change is positive or negative for each of these reactions:
(a) $Zn(s) + 2HCl(aq) \rightleftharpoons ZnCl_2(aq) + H_2(g)$
(b) $O(g) + O(g) \rightleftharpoons O_2(g)$
(c) $NH_4NO_3(s) \rightleftharpoons N_2O(g) + 2H_2O(g)$
(d) $2H_2O_2(l) \rightleftharpoons 2H_2O(l) + O_2(g)$

18.76 The reaction $NH_3(g) + HCl(g) \rightleftharpoons NH_4Cl(s)$ proceeds spontaneously at 25°C even though there is a decrease in entropy in the system (gases are converted to a solid). Explain.

18.77 Use the following data to determine the normal boiling point, in kelvins, of mercury. What assumptions must you make to do the calculation?

$$Hg(l): \Delta H_f^{\circ} = 0 \text{ (by definition)}$$
$$S^{\circ} = 77.4 \text{ J/K} \cdot \text{mol}$$
$$Hg(g): \Delta H_f^{\circ} = 60.78 \text{ kJ/mol}$$
$$S^{\circ} = 174.7 \text{ J/K} \cdot \text{mol}$$

18.78 The molar heat of vaporization of ethanol is 39.3 kJ/mol, and the boiling point of ethanol is 78.3°C. Calculate ΔS for the vaporization of 0.50 mole of ethanol.

18.79 A certain reaction is known to have a ΔG° value of −122 kJ/mol. Will the reaction necessarily occur if the reactants are mixed together?

18.80 Derive the equation:

$$\Delta G = RT \ln \frac{Q}{K}$$

where Q is the reaction quotient, and describe how you would use it to predict the spontaneity of a reaction.

18.81 Calculate ΔG° and K_P for the following processes at 25°C:
(a) $H_2(g) + Br_2(l) \rightleftharpoons 2HBr(g)$
(b) $\frac{1}{2}H_2(g) + \frac{1}{2}Br_2(l) \rightleftharpoons HBr(g)$
Account for the differences in ΔG° and K_P obtained for parts (a) and (b).

18.82 The reaction represented here is exothermic. What can be deduced about its spontaneity?

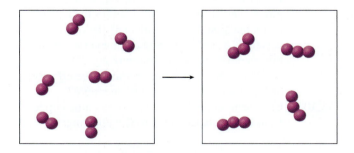

18.83 What can be deduced about the spontaneity of the reaction represented here?

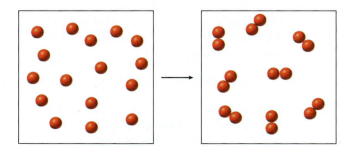

18.84 For a reaction with a negative ΔG° value, which of the following statements is false? (a) The equilibrium constant K is greater than one. (b) The reaction is spontaneous when all the reactants and products are in their standard states. (c) The reaction is always exothermic.

18.85 Consider the reaction:

$$N_2(g) + O_2(g) \rightleftharpoons 2NO(g)$$

Given that ΔG° for the reaction at 25°C is 173.4 kJ/mol, (a) calculate the standard free energy of formation of NO and (b) calculate K_P of the reaction. (c) One of the starting substances in smog formation is NO. Assuming that the temperature in a running automobile engine is 1100°C, estimate K_P for the given reaction. (d) As farmers know, lightning helps to produce a better crop. Why?

18.86 Heating copper(II) oxide at 400°C does not produce any appreciable amount of Cu:

$$CuO(s) \rightleftharpoons Cu(s) + \tfrac{1}{2}O_2(g) \qquad \Delta G^{\circ} = 127.2 \text{ kJ/mol}$$

However, if this reaction is coupled to the conversion of graphite to carbon monoxide, it becomes spontaneous. Write an equation for the coupled process, and calculate the equilibrium constant for the coupled reaction.

18.87 Consider the decomposition of magnesium carbonate:

$$MgCO_3(s) \rightleftharpoons MgO(s) + CO_2(g)$$

Calculate the temperature at which the decomposition begins to favor products. Assume that both ΔH° and ΔS° are independent of temperature.

18.88 Describe two ways that you could determine ΔG° of a reaction.

18.89 The activity series in Table 4.6 shows that reaction (a) is spontaneous whereas reaction (b) is nonspontaneous at 25°C:
(a) $Fe(s) + 2H^+(aq) \longrightarrow Fe^{2+}(aq) + H_2(g)$
(b) $Cu(s) + 2H^+(aq) \longrightarrow Cu^{2+}(aq) + H_2(g)$
Use the data in Appendix 2 to calculate the equilibrium constant for these reactions and hence confirm that the activity series is correct.

18.90 The rate constant for the elementary reaction:

$$2NO(g) + O_2(g) \longrightarrow 2NO_2(g)$$

is $7.1 \times 10^9/M^2 \cdot$ s at 25°C. What is the rate constant for the reverse reaction at the same temperature?

18.91 A 74.6-g ice cube floats in the Arctic Sea. The pressure and temperature of the system and surroundings are at 1 atm and 0°C, respectively. Calculate ΔS_{sys}, ΔS_{surr}, and ΔS_{univ} for the melting of the ice cube. What can you conclude about the nature of the process from the value of ΔS_{univ}? (The molar heat of fusion of water is 6.01 kJ/mol.)

18.92 Which of the following is not accompanied by an increase in the entropy of the system: (a) mixing of two gases at the same temperature and pressure, (b) mixing of ethanol and water, (c) discharging a battery, (d) expansion of a gas followed by compression to its original temperature, pressure, and volume?

18.93 Hydrogenation reactions (e.g., the process of converting C=C bonds to C—C bonds in the food industry) are facilitated by the use of a transition metal catalyst, such as Ni or Pt. The initial step is the adsorption, or binding, of hydrogen gas onto the metal surface. Predict the signs of ΔH, ΔS, and ΔG when hydrogen gas is adsorbed onto the surface of Ni metal.

18.94 Give a detailed example of each of the following, with an explanation: (a) a thermodynamically spontaneous process, (b) a process that would violate the first law of thermodynamics, (c) a process that would violate the second law of thermodynamics, (d) an irreversible process, (e) an equilibrium process.

18.95 The following diagram shows the variation of the equilibrium constant with temperature for the reaction:

$$I_2(g) \rightleftharpoons 2I(g)$$

Calculate $\Delta G°$, $\Delta H°$, and $\Delta S°$ for the reaction at 872 K. (*Hint:* See Problem 18.66.)

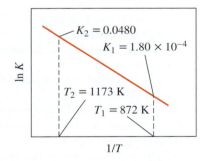

18.96 The standard enthalpy of formation and the standard entropy of gaseous benzene are 82.93 kJ/mol and 269.2 J/K · mol, respectively. Calculate $\Delta H°$, $\Delta S°$, and $\Delta G°$ for the given process at 25°C. Comment on your answers:

$$C_6H_6(l) \longrightarrow C_6H_6(g)$$

18.97 Consider the gas-phase reaction between A_2 (green) and B_2 (red) to form AB at 298 K:

$$A_2(g) + B_2(g) \rightleftharpoons 2AB(g) \qquad \Delta G° = -3.43 \text{ kJ/mol}$$

(a) Which of the following reaction mixtures is at equilibrium?
(b) Which of the following reaction mixtures has a negative ΔG value?
(c) Which of the following reaction mixtures has a positive ΔG value?

The partial pressures of the gases in each frame are equal to the number of A_2, B_2, and AB molecules times 0.10 atm. Round your results to two significant figures.

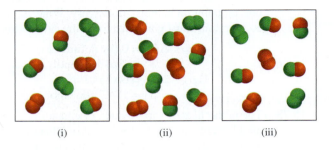

(i)   (ii)   (iii)

Engineering Problems

18.98 Large quantities of hydrogen are needed for the synthesis of ammonia. One preparation of hydrogen involves the reaction between carbon monoxide and steam at 300°C in the presence of a copper-zinc catalyst:

$$CO(g) + H_2O(g) \rightleftharpoons CO_2(g) + H_2(g)$$

Calculate the equilibrium constant (K_P) for the reaction and the temperature at which the reaction favors the formation of CO and H_2O. Will a larger K_P be attained at the same temperature if a more efficient catalyst is used?

18.99 The following reaction was described as the cause of sulfur deposits formed at volcanic sites:

$$2H_2S(g) + SO_2(g) \rightleftharpoons 3S(s) + 2H_2O(g)$$

It may also be used to remove SO_2 from powerplant stack gases. (a) Identify the type of redox reaction it is. (b) Calculate the equilibrium constant (K_P) at 25°C, and comment on whether this method is feasible for removing SO_2. (c) Would this procedure become more effective or less effective at a higher temperature?

18.100 Comment on the feasibility of extracting copper from its ore chalcocite (Cu_2S) by heating:

$$Cu_2S(s) \longrightarrow 2Cu(s) + S(s)$$

Calculate the $\Delta G°$ for the overall reaction if this process is coupled to the conversion of sulfur to sulfur dioxide and $\Delta G_f°(Cu_2S) = -86.1$ kJ/mol.

18.101 One of the steps in the extraction of iron from its ore (FeO) is the reduction of iron(II) oxide by carbon monoxide at 900°C:

$$FeO(s) + CO(g) \rightleftharpoons Fe(s) + CO_2(g)$$

If CO is allowed to react with an excess of FeO, calculate the mole fractions of CO and CO_2 at equilibrium. State any assumptions.

18.102 In the Mond process for the purification of nickel, carbon monoxide is combined with heated nickel to produce $Ni(CO)_4$, which is a gas and can therefore be separated from solid impurities:

$$Ni(s) + 4CO(g) \rightleftharpoons Ni(CO)_4(g)$$

Given that the standard free energies of formation of $CO(g)$ and $Ni(CO)_4(g)$ are -137.3 and -587.4 kJ/mol, respectively, calculate the equilibrium constant of the reaction at 80°C. Assume that ΔG_f° is temperature independent.

18.103 Water gas, a mixture of H_2 and CO, is a fuel made by combining steam with red-hot coke (a by-product of coal distillation):

$$H_2O(g) + C(s) \rightleftharpoons CO(g) + H_2(g)$$

From the data in Appendix 2, estimate the temperature at which the reaction begins to favor the formation of products.

18.104 (a) Over the years, there have been numerous claims about "perpetual motion machines," machines that will produce useful work with no input of energy. Explain why the first law of thermodynamics prohibits the possibility of such a machine existing. (b) Another kind of machine, sometimes called a "perpetual motion of the second kind," operates as follows. Suppose an ocean liner sails by scooping up water from the ocean and then extracting heat from the water, converting the heat to electric power to run the ship, and dumping the water back into the ocean. This process does not violate the first law of thermodynamics, for no energy is created—energy from the ocean is just converted to electric energy. Show that the second law of thermodynamics prohibits the existence of such a machine.

Biological Problems

18.105 The enthalpy change in the denaturation of a certain protein is 125 kJ/mol. If the entropy change is 397 J/K · mol, calculate the minimum temperature at which the protein would denature spontaneously.

18.106 The concentration of glucose inside a cell is 0.12 mM and that outside a cell is 12.3 mM. Calculate the Gibbs free-energy change for the transport of 3 moles of glucose into the cell at 37°C.

18.107 Active transport is the process in which a substance is transferred from a region of lower concentration to one of higher concentration. This is a nonspontaneous process and must be coupled to a spontaneous process, such as the hydrolysis of ATP. The concentrations of K^+ ions in the blood plasma and in nerve cells are 15 mM and 400 mM, respectively (1 m$M = 1 \times 10^{-3}$ M).

Use Equation 18.13 to calculate ΔG for the process at the physiological temperature of 37°C:

$$K^+(15 \text{ m}M) \longrightarrow K^+(400 \text{ m}M)$$

In this calculation, the ΔG° term can be set to zero. What is the justification for this step?

18.108 When a native protein in solution is heated to a high enough temperature, its polypeptide chain will unfold to become the denatured protein. The temperature at which a large portion of the protein unfolds is called the melting temperature. The melting temperature of a certain protein is found to be 46°C, and the enthalpy of denaturation is 382 kJ/mol. Estimate the entropy of denaturation, assuming that the denaturation is a two-state process; that is, native protein $\longrightarrow$ denatured protein. The single polypeptide protein chain has 122 amino acids. Calculate the entropy of denaturation per amino acid. Comment on your result.

18.109 The pH of gastric juice is about 1.00 and that of blood plasma is 7.40. Calculate the Gibbs free energy required to secrete a mole of H^+ ions from blood plasma to the stomach at 37°C.

18.110 The following reaction represents the removal of ozone in the stratosphere:

$$2O_3(g) \rightleftharpoons 3O_2(g)$$

Calculate the equilibrium constant (K_P) for this reaction. In view of the magnitude of the equilibrium constant, explain why this reaction is not considered a major cause of ozone depletion in the absence of human-made pollutants such as the nitrogen oxides and CFCs. Assume the temperature of the stratosphere is -30°C and ΔG_f° is temperature independent.

18.111 Carbon monoxide (CO) and nitric oxide (NO) are polluting gases contained in automobile exhaust. Under suitable conditions, these gases can be made to react to form nitrogen (N_2) and the less harmful carbon dioxide (CO_2). (a) Write an equation for this reaction. (b) Identify the oxidizing and reducing agents. (c) Calculate the K_P for the reaction at 25°C. (d) Under normal atmospheric conditions, the partial pressures are $P_{N_2} = 0.80$ atm, $P_{CO_2} = 3.0 \times 10^{-4}$ atm, $P_{CO} = 5.0 \times 10^{-5}$ atm, and $P_{NO} = 5.0 \times 10^{-7}$ atm. Calculate Q_P, and predict the direction toward which the reaction will proceed. (e) Will raising the temperature favor the formation of N_2 and CO_2?

Multiconcept Problems

18.112 The K_{sp} of AgCl is given in Table 17.4. What is its value at 60°C? [*Hint:* You need the result of Problem 18.66(a) and the data in Appendix 2 to calculate ΔH°.]

18.113 Consider two carboxylic acids (acids that contain the −COOH group): CH_3COOH (acetic acid, $K_a = 1.8 \times 10^{-5}$) and $CH_2ClCOOH$ (chloroacetic acid, $K_a = 1.4 \times 10^{-3}$). (a) Calculate $\Delta G°$ for the ionization of these acids at 25°C. (b) From the equation $\Delta G° = \Delta H° - T\Delta S°$, we see that the contributions to the $\Delta G°$ term are an enthalpy term ($\Delta H°$) and a temperature times entropy term ($T\Delta S°$). These contributions are listed here for the two acids:

	$\Delta H°$(kJ/mol)	$T\Delta S°$(kJ/mol)
CH_3COOH	−0.57	−27.6
$CH_2ClCOOH$	−4.7	−21.1

Which is the dominant term in determining the value of $\Delta G°$ (and hence K_a of the acid)? (c) What processes contribute to $\Delta H°$? (Consider the ionization of the acids as a Brønsted acid-base reaction.) (d) Explain why the $T\Delta S°$ term is more negative for CH_3COOH.

18.114 Many hydrocarbons exist as structural isomers, which are compounds that have the same molecular formula but different structures. For example, both butane and isobutane have the same molecular formula of C_4H_{10}. Calculate the mole percent of these molecules in an equilibrium mixture at 25°C, given that the standard free energy of formation of butane is −15.9 kJ/mol and that of isobutane is −18.0 kJ/mol. Does your result support the notion that straight-chain hydrocarbons (i.e., hydrocarbons in which the C atoms are joined along a line) are less stable than branch-chain hydrocarbons?

18.115 Use the data from Table 15.1 to calculate $\Delta G°$ for the decomposition of gaseous N_2O_4. (Note that for Equation 18.14 to give the correct value for $\Delta G°$, the K value for a gaseous reaction must be a K_P.) What is the significance of the sign of the calculated $\Delta G°$?

Standardized-Exam Practice Problems

Physical and Biological Sciences

In chemistry, the standard state for a solution is 1 M. This means that each solute concentration expressed in molarity is divided by 1 M. In biological systems, however, we define the standard state for the H^+ ions to be 1×10^{-7} M. Consequently, the change in the standard Gibbs free energy according to these two conventions will be different for reactions involving the uptake or release of H^+ ions, depending on which convention is used. We will therefore replace $\Delta G°$ with $\Delta G°'$, where the prime denotes that it is the standard Gibbs free-energy change for a biological process.

1. What is most likely the reason for the standard state for H^+ being defined as 1×10^{-7} M in biological systems?

   a) It is an arbitrary convention.
   b) Physiological conditions reduce all concentrations by a factor of 10^7.
   c) Physiological conditions increase all concentrations by a factor of 10^7.
   d) Biological systems have a pH of about 7.

2. Consider the reaction:

$$A + B \longrightarrow C + xH^+$$

   where x is a stoichiometric coefficient. Using the equation:

$$\Delta G = \Delta G° + RT \ln Q$$

   determine the relationship between $\Delta G°'$ and $\Delta G°$, keeping in mind that ΔG is the same for a process regardless of which convention is used.
   a) $\Delta G°' = \Delta G° + RT \ln (1 \times 10^7)$
   b) $\Delta G°' = \Delta G° + RT \ln (1 \times 10^{-7})^x$
   c) $\Delta G°' = \Delta G° + e^{-7/RT}$
   d) $\Delta G°' = \Delta G° + RT \times e^{-7}$

3. Determine the relationship between $\Delta G°'$ and $\Delta G°$ for the reverse process:

$$C + xH^+ \longrightarrow A + B$$

   a) $\Delta G°' = \Delta G° + RT \ln (1 \times 10^7)^x$
   b) $\Delta G°' = \Delta G° + RT \ln (1 \times 10^{-7})$
   c) $\Delta G°' = \Delta G° + e^{7/RT}$
   d) $\Delta G°' = \Delta G° + RT \times e^7$

4. NAD^+ and NADH are the oxidized and reduced forms of nicotinamide adenine dinucleotide, two key compounds in the metabolic pathways. For the oxidation of NADH:

$$NADH + H^+ \longrightarrow NAD^+ + H_2$$

   $\Delta G°$ is −21.8 kJ/mol at 298 K. Calculate $\Delta G°'$ for this process, and calculate ΔG using either the chemical or the biological convention when [NADH] = 1.5×10^{-2} M, $[H^+]$ = 3.0×10^{-5} M, [NAD] = 4.6×10^{-3} M, and P_{H_2} = 0.010 atm.
   a) 18.1 kJ/mol, 50.3 kJ/mol
   b) 21.8 kJ/mol, 29.6 kJ/mol
   c) 18.1 kJ/mol, 10.3 kJ/mol
   d) 21.8 kJ/mol, 10.3 kJ/mol

Answers to In-Chapter Materials

Answers to Practice Problems

18.1A 0.34 J/K. **18.1B** $\frac{1}{5}$. **18.2A** (a) 173.6 J/K · mol,
(b) −139.8 J/K · mol, (c) 215.3 J/K · mol. **18.2B** (a) $S°[K(l)] =$
71.5 J/K · mol, (b) $S°[S_2Cl_2(g)] = 331.5$ J/K · mol, (c) $S°[MgF_2(s)] =$
57.24 J/K · mol. **18.3A** (a) Negative, (b) negative, (c) positive.
18.3B The sign of $\Delta H°$ for both dissolution processes is negative.
Something must favor spontaneity; if not entropy change, then
enthalpy change. Because these processes both involve decreases
in the system's entropy, they must be exothermic, or they
could not be spontaneous. **18.4A** (a) $\Delta S_{univ} = -27.2$ J/K · mol,
nonspontaneous, (b) $\Delta S_{univ} = -28.1$ J/K · mol, nonspontaneous,
(c) $\Delta S_{univ} = 0$ J/K · mol, equilibrium. **18.4B** (a) $\Delta S_{univ} =$
5.2 J/K · mol, spontaneous, (b) 346°C, (c) 58°C. **18.5A** 3728°C.
18.5B 108 J/K · mol. **18.6A** (a) −106 kJ/mol, (b) −2935 kJ/mol.
18.6B (a) $\Delta G_f°[Li_2O(s)] = -561.2$ kJ/mol, (b) $\Delta G_f°[NaI(s)] =$
−286.1 kJ/mol. **18.7A** $\Delta S_{fus} = 16$ J/K · mol, $\Delta S_{vap} = 72$ J/K · mol.
18.7B (a) $\Delta H_{fus}° = 2.41$ kJ/mol, $\Delta S_{fus}° = 6.51$ J/K · mol, $T_{melting} =$
97°C. (b) $\Delta H_{vap}° = 105.3$ kJ/mol, $\Delta S_{vap}° = 96.1$ J/K · mol, $T_{boiling} =$
823°C. **18.8A** 1.3 kJ/mol, reverse reaction is spontaneous.
18.8B 2.5 atm. **18.9A** 2×10^{57}. **18.9B** −16.8 kJ/mol.
18.10A 32.9 kJ/mol. **18.10B** −454.4 kJ/mol.

Answers to Checkpoints

18.3.1 a, e. **18.3.2** b. **18.3.3** c. **18.4.1** b. **18.4.2** e. **18.4.3** b.
18.4.4 d. **18.5.1** d. **18.5.2** d. **18.5.3** d. **18.5.4** b. **18.5.5** b.
18.6.1 b. **18.6.2** d. **18.6.3** b. **18.6.4** e.

Design Icon Credits: Animation icon: ©McGraw Hill Education; Student Hot Spot icon: ©LovArt/Shutterstock.com; Satellite view of planet Earth: Stockbyte/Alamy

CHAPTER 19

Electrochemistry

Potato clocks are popular as science demonstrations and experiments.

Science Photo Library/Alamy Stock Photo

In This Chapter, You Will Learn

How chemical reactions can produce electric energy and how reaction conditions affect the energy production. You will also learn how electric energy can be used to drive chemical reactions that otherwise would not occur.

Before You Begin, Review These Skills

- Oxidation numbers [◂◂ Section 4.4]
- The reaction quotient, Q [◂◂ Section 15.2]
- Gibbs free energy [◂◂ Section 18.5]

Making Batteries out of Food

A very popular science demonstration and experiment is the powering of a small electronic device using potatoes. In fact, "potato clocks" have a sufficient following to have generated a significant market for potato-clock kits on Amazon.com; and might more properly be called potato *batteries*, as the the electricity they generate can be used to power any small electronic device.

A potato battery can be constructed by inserting two electrodes, typically one zinc and one copper, into each of two potatoes—a total of *four* electrodes. Electrical leads are used to connect the electrodes, as shown in the photo; and are connected to a small digital clock. The flow of electricity results from a difference in electrical potential between the electrodes. (A potato battery can be constructed using a single potato with just two electrodes, but the energy output is insufficient to power a clock.) Potato batteries have also been used to power LED lights; and both projects can also be done using other larder items, such as lemons and pickles. The critical component common to all of these is *acid*. In the case of the potato, the acid is phosphoric acid (H_3PO_4); in the case of the lemon, it is citric acid ($H_3C_6H_5O_7$); and in the case of the pickle, it is acetic acid ($HC_2H_3O_2$).

Constructing batteries using such common household items as potatoes or fruit requires an understanding of the principles of *electrochemistry*.

At the end of this chapter, you will be able to solve a series of problems related to the electrochemistry of potato batteries [▸▸ Applying What You've Learned, page 1000].

Student Note: Now is a good time to
review how oxidation numbers are
assigned [◄◄ Section 4.4].

19.1 Balancing Redox Reactions

In Chapter 4 we briefly discussed oxidation-reduction or "redox" reactions, those in
which electrons are transferred from one species to another. In this section we review
how to identify a reaction as a redox reaction and look more closely at how such
reactions are balanced.

A redox reaction is one in which there are *changes* in oxidation states, which
we identify using the rules introduced in Chapter 4. The following are examples of
redox reactions:

$$2KClO_3(s) \longrightarrow 2KCl(s) + 3O_2(g)$$
$$(+1)(+5)(-2) (+1)(-1) (0)$$

$$CH_4(g) + 2O_2(g) \longrightarrow CO_2(g) + 2H_2O(l)$$
$$(-4)(+1) (0) (+4)(-2) (+1)(-2)$$

$$Sn(s) + Cu^{2+}(aq) \longrightarrow Cu(s) + Sn^{2+}(aq)$$
$$(0) (+2) (0) (+2)$$

Equations for redox reactions, such as those shown here, can be balanced by inspection, the method of balancing introduced in Chapter 3 [◄◄ Section 3.3], but remember
that redox equations must be balanced for mass (number of atoms) *and* for charge
(number of electrons) [◄◄ Section 4.4]. In this section we introduce the *half-reaction
method* to balance equations that cannot be balanced simply by inspection.

Consider the aqueous reaction of the iron(II) ion (Fe^{2+}) with the dichromate
ion ($Cr_2O_7^{2-}$):

$$Fe^{2+} + Cr_2O_7^{2-} \longrightarrow Fe^{3+} + Cr^{3+}$$

Because there is no species containing oxygen on the product side of the equation,
it would not be possible to balance this equation simply by adjusting the coefficients of reactants and products. However, there are two things about the reaction
that make it possible to *add* species to the equation to balance it—without changing
the chemical reaction it represents:

- The reaction takes place in aqueous solution, so we can add H_2O as needed to
  balance the equation.
- This particular reaction takes place in acidic solution, so we can add H^+ as
  needed to balance the equation. (Some reactions take place in basic solution,
  enabling us to add OH^- as needed for balancing. We explain more about this
  shortly.)

After writing the unbalanced equation, we balance it stepwise as follows:

1. Separate the unbalanced reaction into **half-reactions.** A half-reaction is an oxidation or a reduction that occurs as part of the overall redox reaction:

Oxidation:	$Fe^{2+} \longrightarrow Fe^{3+}$
Reduction:	$Cr_2O_7^{2-} \longrightarrow Cr^{3+}$

2. Balance each of the half-reactions with regard to atoms other than O and H. In
   this case, no change is required for the oxidation half-reaction. We adjust the
   coefficient of the chromium(III) ion to balance the reduction half-reaction:

Oxidation:	$Fe^{2+} \longrightarrow Fe^{3+}$
Reduction:	$Cr_2O_7^{2-} \longrightarrow 2Cr^{3+}$

3. Balance both half-reactions for O by adding H_2O. Again, the oxidation in this case requires no change, but we must add seven water molecules to the product side of the reduction:

$$\text{Oxidation:} \qquad Fe^{2+} \longrightarrow Fe^{3+}$$
$$\text{Reduction:} \qquad Cr_2O_7^{2-} \longrightarrow 2Cr^{3+} + 7H_2O$$

4. Balance both half-reactions for H by adding H^+. Once again, the oxidation in this case requires no change, but we must add 14 hydrogen ions to the reactant side of the reduction:

$$\text{Oxidation:} \qquad Fe^{2+} \longrightarrow Fe^{3+}$$
$$\text{Reduction:} \quad 14H^+ + Cr_2O_7^{2-} \longrightarrow 2Cr^{3+} + 7H_2O$$

5. Balance both half-reactions for charge by adding electrons. To do this, we determine the total charge on each side and add electrons to make the total charges equal. In the case of the oxidation, there is a charge of +2 on the reactant side and a charge of +3 on the product side. Adding one electron to the product side makes the charges equal:

$$\text{Oxidation:} \qquad \underbrace{Fe^{2+}}_{} \longrightarrow \underbrace{Fe^{3+} + e^-}_{}$$
$$\text{Total charge:} \qquad +2 \qquad\qquad +2$$

In the case of the reduction, there is a total charge of $[(14)(+1) + (1)(-2)] = +12$ on the reactant side and a total charge of $[(2)(+3)] = +6$ on the product side. Adding six electrons to the reactant side makes the charges equal:

$$\text{Reduction:} \qquad \underbrace{6e^- + 14H^+ + Cr_2O_7^{2-}}_{} \longrightarrow \underbrace{2Cr^{3+} + 7H_2O}_{}$$
$$\text{Total charge:} \qquad\qquad +6 \qquad\qquad\qquad +6$$

6. If the number of electrons in the balanced oxidation half-reaction is not the same as the number of electrons in the balanced reduction half-reaction, multiply one or both of the half-reactions by the number(s) required to make the number of electrons the same in both. In this case, with one electron in the oxidation and six in the reduction, multiplying the oxidation by 6 accomplishes this:

$$\text{Oxidation:} \qquad\qquad 6(Fe^{2+} \longrightarrow Fe^{3+} + e^-)$$
$$6Fe^{2+} \longrightarrow 6Fe^{3+} + \boxed{6e^-}$$
$$\text{Reduction:} \qquad \boxed{6e^-} + 14H^+ + Cr_2O_7^{2-} \longrightarrow 2Cr^{3+} + 7H_2O$$

7. Finally, add the balanced half-reactions back together and cancel the electrons, in addition to any other identical terms that appear on both sides:

$$6Fe^{2+} \longrightarrow 6Fe^{3+} + \cancel{6e^-}$$
$$\cancel{6e^-} + 14H^+ + Cr_2O_7^{2-} \longrightarrow 2Cr^{3+} + 7H_2O$$
$$\overline{6Fe^{2+} + 14H^+ + Cr_2O_7^{2-} \longrightarrow 6Fe^{3+} + 2Cr^{3+} + 7H_2O}$$

A final check shows that the resulting equation is balanced both for mass and for charge.

Some redox reactions occur in basic solution. When this is the case, balancing by the half-reaction method is done exactly as described for reactions in acidic solution, but it requires two additional steps:

8. For each H^+ ion in the final equation, add one OH^- ion to each side of the equation, combining the H^+ and OH^- ions to produce H_2O.

9. Make any additional cancellations made necessary by the new H_2O molecules.

Sample Problem 19.1 shows how to use the half-reaction method to balance a reaction that takes place in basic solution.

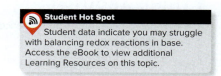

Student Hot Spot

Student data indicate you may struggle with balancing redox reactions in base. Access the eBook to view additional Learning Resources on this topic.

SAMPLE PROBLEM   19.1

Permanganate ion and iodide ion react in basic solution to produce manganese(IV) oxide and molecular iodine. Use the half-reaction method to balance the equation:

$$MnO_4^- + I^- \longrightarrow MnO_2 + I_2$$

Strategy The reaction takes place in basic solution, so apply steps 1 through 9 to balance for mass and for charge.

Setup Identify the oxidation and reduction half-reactions by assigning oxidation numbers:

$$MnO_4^- + I^- \longrightarrow MnO_2 + I_2$$
$$(+7)(-2)(-1)(+4)(-2)(0)$$

Solution

Step 1.   Separate the unbalanced reaction into half-reactions:

Oxidation: $I^- \longrightarrow I_2$

Reduction: $MnO_4^- \longrightarrow MnO_2$

Step 2.   Balance each half-reaction for mass, excluding O and H:

$$2I^- \longrightarrow I_2$$
$$MnO_4^- \longrightarrow MnO_2$$

Step 3.   Balance both half-reactions for O by adding H_2O:

$$2I^- \longrightarrow I_2$$
$$MnO_4^- \longrightarrow MnO_2 + 2H_2O$$

Step 4.   Balance both half-reactions for H by adding H^+:

$$2I^- \longrightarrow I_2$$
$$4H^+ + MnO_4^- \longrightarrow MnO_2 + 2H_2O$$

Step 5.   Balance the total charge of both half-reactions by adding electrons:

$$2I^- \longrightarrow I_2 + 2e^-$$
$$3e^- + 4H^+ + MnO_4^- \longrightarrow MnO_2 + 2H_2O$$

Step 6.   Multiply the half-reactions to make the numbers of electrons the same in both:

$$3(2I^- \longrightarrow I_2 + 2e^-)$$
$$2(3e^- + 4H^+ + MnO_4^- \longrightarrow MnO_2 + 2H_2O)$$

Step 7.   Add the half-reactions back together, cancelling electrons:

$$6I^- \longrightarrow 3I_2 + \cancel{6e^-}$$
$$\cancel{6e^-} + 8H^+ + 2MnO_4^- \longrightarrow 2MnO_2 + 4H_2O$$
$$\overline{8H^+ + 2MnO_4^- + 6I^- \longrightarrow 2MnO_2 + 3I_2 + 4H_2O}$$

Step 8.   For each H^+ ion in the final equation, add one OH^- ion to each side of the equation, combining the H^+ and OH^- ions to produce H_2O:

$$8H^+ + 2MnO_4^- + 6I^- \longrightarrow 2MnO_2 + 3I_2 + 4H_2O$$
$$+ 8OH^- + 8OH^-$$
$$\overline{(8H_2O) + 2MnO_4^- + 6I^- \longrightarrow 2MnO_2 + 3I_2 + (4H_2O) + 8OH^-}$$

Step 9.   Carry out any cancellations made necessary by the additional H_2O molecules:

$$4H_2O + 2MnO_4^- + 6I^- \longrightarrow 2MnO_2 + 3I_2 + 8OH^-$$

THINK ABOUT IT

Verify that the final equation is balanced for mass and for charge. Remember that electrons cannot appear in the overall balanced equation.

Practice Problem **A**TTEMPT Use the half-reaction method to balance the following equation in basic solution:

$$CN^- + MnO_4^- \longrightarrow CNO^- + MnO_2$$

Practice Problem **B**UILD Use the half-reaction method to balance the following equation in acidic solution:

$$Fe^{2+} + MnO_4^- \longrightarrow Fe^{3+} + Mn^{2+}$$

Practice Problem **C**ONCEPTUALIZE In Chapter 3, you learned to balance chemical equations by changing coefficients only—it was not permissible to add species to the equation. Explain why it is all right to add water and hydronium (or hydroxide) to the equations in this chapter as part of the balancing process.

CHECKPOINT – SECTION 19.1   Balancing Redox Reactions

19.1.1 Which of the following equations does *not* represent a redox reaction? (Select all that apply.)

a) $NH_3 + HCl \longrightarrow NH_4Cl$

b) $2H_2O_2 \longrightarrow 2H_2O + O_2$

c) $2O_3 \longrightarrow 3O_2$

d) $3NO_2 + H_2O \longrightarrow NO + 2HNO_3$

e) $LiCl \longrightarrow Li^+ + Cl^-$

19.1.2 MnO_4^- and $C_2O_4^{2-}$ react in basic solution to form MnO_2 and CO_3^{2-}. What are the coefficients of MnO_4^- and $C_2O_4^{2-}$ in the balanced equation?

a) 1 and 1

b) 2 and 1

c) 2 and 3

d) 2 and 6

e) 2 and 2

19.2   Galvanic Cells

When zinc metal is placed in a solution containing copper(II) ions, Zn is oxidized to Zn^{2+} ions whereas Cu^{2+} ions are reduced to Cu [◄◄ Section 4.4]:

$$Zn(s) + Cu^{2+}(aq) \longrightarrow Zn^{2+}(aq) + Cu(s)$$

The electrons are transferred directly from the reducing agent, Zn, to the oxidizing agent, Cu^{2+}, in solution. However, if we physically separate two half-reactions from each other, we can arrange it such that the electrons must travel through a wire to pass from the Zn atoms to the Cu^{2+} ions. As the reaction progresses, it generates a flow of electrons through the wire and thereby generates electricity.

The experimental apparatus for generating electricity through the use of a spontaneous reaction is called a ***galvanic cell.*** Figure 19.1 shows the essential components of a galvanic cell. A zinc bar is immersed in an aqueous $ZnSO_4$ solution in one container, and a copper bar is immersed in an aqueous $CuSO_4$ solution in another container. The cell operates on the principle that the oxidation of Zn to Zn^{2+} and the reduction of Cu^{2+} to Cu can be made to take place simultaneously in separate locations with the transfer of electrons between them occurring through an external wire. The zinc and copper bars are called ***electrodes.*** By definition, the ***anode*** in a galvanic cell is the electrode at which *oxidation* occurs and the ***cathode*** is the electrode at which *reduction* occurs. (Each combination of container, electrode, and solution is called a ***half-cell.***) This particular galvanic cell is known as a Daniell cell, named for its inventor.

The half-reactions for the galvanic cell shown in Figure 19.1 are:

Oxidation: $\qquad Zn(s) \longrightarrow Zn^{2+}(aq) + 2e^-$

Reduction: $Cu^{2+}(aq) + 2e^- \longrightarrow Cu(s)$

Animation
Electrochemistry—operation of a voltaic cell.

Student Note: A *galvanic* cell can also be called a *voltaic* cell. Both terms refer to a cell in which a spontaneous chemical reaction generates a flow of electrons.

Figure 19.1

Construction of a Galvanic Cell

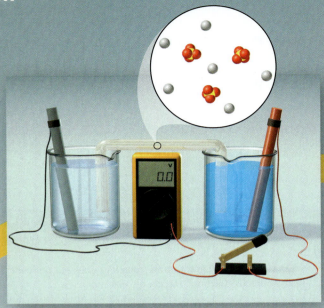

We add a salt bridge, a tube containing a solution of a strong electrolyte—in this case Na_2SO_4. Having this solution in electrical contact with the two solutions in the beakers allows ions to migrate toward the electrodes, ensuring that the two compartments remain electrically neutral.

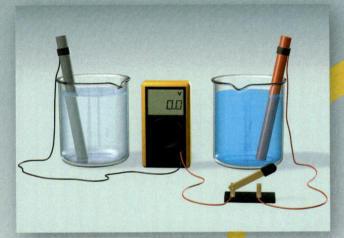

The two metal pieces are the electrodes in the galvanic cell. We connect the electrodes with a length of wire routed through a voltmeter and a switch so that we can complete the circuit when we have completed construction of the cell.

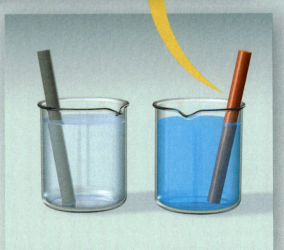

To make the reaction between zinc and copper more useful, we can construct a galvanic cell. In one beaker, we place a piece of zinc metal in a 1.00-*M* solution of Zn^{2+} ions. In the other, we place a piece of copper metal in a 1.00-*M* solution of Cu^{2+} ions.

As shown in Figure 4.6, when zinc metal (Zn) is immersed in a solution containing copper ions (Cu^{2+}), the zinc is oxidized to zinc ions (Zn^{2+}), and copper ions are reduced to copper metal (Cu).

$$Zn(s) + Cu^{2+}(aq) \longrightarrow Zn^{2+}(aq) + Cu(s)$$

This is an oxidation-reduction reaction in which electrons flow spontaneously from zinc metal to the copper ions in solution. The lightening of the blue color indicates that the concentration of Cu^{2+} has decreased. Copper metal is deposited on the solid zinc surface. Some of the zinc metal has gone into solution as Zn^{2+} ions, which do not impart any color in the solution.

before   after

When we close the switch, we complete the circuit; the voltmeter indicates the initial potential of the cell: 1.10 V.

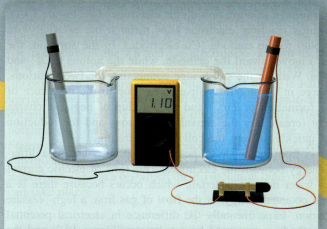

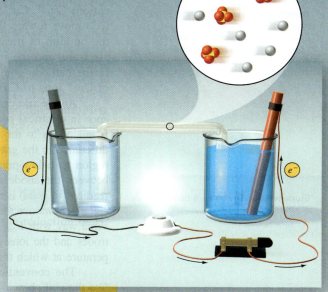

When we replace the voltmeter with a lightbulb, electrons flow from the zinc electrode (the anode) to the copper electrode (the cathode). The flow of electrons lights the bulb. Anions in the salt bridge migrate toward the anode; cations migrate toward the cathode.

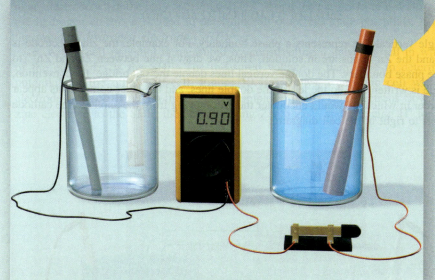

As the reaction proceeds, zinc metal from the anode is oxidized, increasing the Zn^{2+} concentration in the beaker on the left; and additional copper metal is deposited on the cathode, decreasing the Cu^{2+} concentration in the beaker on the right. As the concentrations of both ions change, the potential of the cell decreases. After allowing the reaction to proceed for a time, we can reinsert the voltmeter to measure the decreased voltage.

What's the point?

Zinc is a stronger reducing agent than copper, so there is a natural tendency for electrons to flow from zinc metal to copper ions. We can harness this flow of electrons by forcing the half-reactions to occur in separate compartments. Electrons still flow from Zn to Cu^{2+}, but they must flow through the wire connecting the two electrodes. The potential of the cell decreases as the reaction proceeds, as the reading on the voltmeter shows.

(See Visualizing Chemistry questions VC 19.1–VC 19.4 on page 1000.)

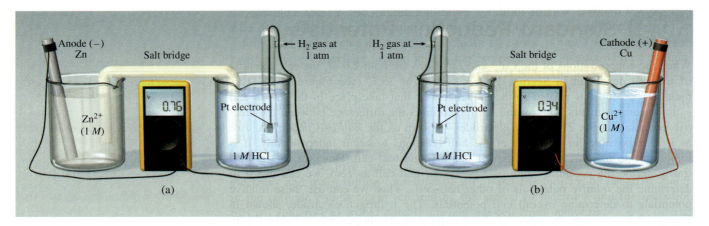

Figure 19.4 (a) A cell consisting of a zinc electrode and a hydrogen electrode. (b) A cell consisting of a copper electrode and a hydrogen electrode. Both cells are operating under standard-state conditions. Note that in (a) the SHE is the cathode, but in (b) it is the anode.

In this case, therefore, the zinc electrode is the anode (where oxidation takes place) and the SHE is the cathode (where reduction takes place). The cell diagram is:

$$Zn(s) \mid Zn^{2+}(1\ M) \parallel H^+(1\ M) \mid H_2(1\ atm) \mid Pt(s)$$

The measured potential for this cell is 0.76 V at 25°C. We can write the half-cell reactions as follows:

Anode (oxidation): $\quad\quad\quad\quad\quad\quad\quad Zn(s) \longrightarrow Zn^{2+}(1\ M) + 2e^-$

Cathode (reduction): $\quad 2H^+(1\ M) + 2e^- \longrightarrow H_2(1\ atm)$

Overall: $\quad\quad\quad\quad\quad\quad Zn(s) + 2H^+(1\ M) \longrightarrow Zn^{2+}(1\ M) + H_2(1\ atm)$

By convention, the standard cell potential, E°_{cell}, which is composed of a contribution from the anode and a contribution from the cathode, is given by:

Equation 19.1 $\quad\quad\quad\quad\quad E^\circ_{cell} = E^\circ_{cathode} - E^\circ_{anode}$

where $E^\circ_{cathode}$ and E°_{anode} are the standard reduction potentials of the cathode and anode, respectively. For the Zn-SHE cell, we write:

$$E^\circ_{cell} = E^\circ_{H^+/H_2} - E^\circ_{Zn^{2+}/Zn}$$

$$0.76\ V = 0 - E^\circ_{Zn^{2+}/Zn}$$

where the subscript "H^+/H_2" means "$2H^+ + 2e^- \longrightarrow H_2$" and the subscript "$Zn^{2+}/Zn$" means "$Zn^{2+} + 2e^- \longrightarrow Zn$." Thus, the standard reduction potential of zinc, $E_{Zn^{2+}/Zn}$ is −0.76 V.

The standard reduction potential of copper can be determined in a similar fashion, by using a cell with a copper electrode and a SHE [Figure 19.4(b)]. In this case, electrons flow from the SHE to the copper electrode when the circuit is completed; that is, the *copper* electrode is the cathode and the SHE is the *anode*. Reduction of Cu^{2+} ions causes the mass of the Cu electrode to increase.

The cell diagram is:

$$Pt(s) \mid H_2(1\ atm) \mid H^+(1\ M) \parallel Cu^{2+}(1\ M) \mid Cu(s)$$

and the half-cell reactions are:

Anode (oxidation): $\quad\quad\quad\quad\quad\quad H_2(1\ atm) \longrightarrow 2H^+(1\ M) + 2e^-$

Cathode (reduction): $\quad Cu^{2+}(1\ M) + 2e^- \longrightarrow Cu(s)$

Overall: $\quad\quad\quad\quad\quad H_2(1\ atm) + Cu^{2+}(1\ M) \longrightarrow 2H^+(1\ M) + Cu(s)$

Under standard-state conditions and at 25°C, the measured potential for the cell is 0.34 V, so we write:

$$E^\circ_{cell} = E^\circ_{cathode} - E^\circ_{anode}$$

$$= E^\circ_{Cu^{2+}/Cu} - E^\circ_{H^+/H_2}$$

$$0.34 \text{ V} = E^\circ_{Cu^{2+}/Cu} - 0$$

Thus, the standard reduction potential of copper, $E^\circ_{Cu^{2+}/Cu}$, is 0.34 V, where the subscript "Cu^{2+}/Cu" means "$Cu^{2+} + 2e^- \longrightarrow Cu$."

Having determined the standard reduction potentials of Zn and Cu, we can use Equation 19.1 to calculate the cell potential for the Daniell cell described in Section 19.2:

$$E^\circ_{cell} = E^\circ_{cathode} - E^\circ_{anode}$$

$$= E^\circ_{Cu^{2+}/Cu} - E^\circ_{Zn^{2+}/Zn}$$

$$= 0.34 \text{ V} - (-0.76 \text{ V})$$

$$= 1.10 \text{ V}$$

As in the case of ΔG° [◀◀ Section 18.5], we can use the sign of E° to predict whether a reaction lies to the right or to the left. A positive E° means that the redox reaction will favor the formation of products at equilibrium. Conversely, a negative E° indicates that reactants will be favored at equilibrium. We examine how E°_{cell}, ΔG°, and K are related in Section 19.4.

Table 19.1 lists standard reduction potentials, in order of decreasing reduction potential, for a number of half-cell reactions. To avoid ambiguity, all half-cell reactions are shown as reductions. A galvanic cell is composed of two half-cells, and therefore two half-cell reactions. Whether or not a particular half-cell reaction occurs as a reduction in a galvanic cell depends on how its reduction potential compares to that of the other half-cell reaction. If it has the greater (or more *positive*) reduction potential of the two, it will occur as a reduction. If it has the smaller (or more *negative*) reduction potential of the two, it will occur in the reverse direction, as an oxidation. (The half-reaction with the greater reduction potential has the greater potential to occur as a reduction.)

Consider the example of the cell described in Section 19.2. The two half-cell reactions and their standard reduction potentials are:

$$Cu^{2+} + 2e^- \longrightarrow Cu \qquad E^\circ = 0.34 \text{ V}$$

$$Zn^{2+} + 2e^- \longrightarrow Zn \qquad E^\circ = -0.76 \text{ V}$$

The Cu half-reaction, having a greater (more positive) reduction potential, is the half-reaction that will occur as a reduction:

$$Cu^{2+} + 2e^- \longrightarrow Cu$$

The Zn half-reaction has a smaller (less positive) reduction potential and will occur, instead, as an oxidation:

$$Zn \longrightarrow Zn^{2+} + 2e^-$$

Adding the two half-reactions gives the overall cell reaction:

$$Zn + Cu^{2+} \longrightarrow Zn^{2+} + Cu$$

Consider, however, what would happen if we were to construct a galvanic cell combining the Zn half-cell with an Mn half-cell. The reduction potential of Mn is −1.18 V:

$$Zn^{2+} + 2e^- \longrightarrow Zn \qquad E^\circ = -0.76 \text{ V}$$

$$Mn^{2+} + 2e^- \longrightarrow Mn \qquad E^\circ = -1.18 \text{ V}$$

Student Hot Spot

Student data indicate you may struggle with calculations involving cell potentials. Access the eBook to view additional Learning Resources on this topic.

Student Note: The electrode where reduction occurs is the cathode. Therefore, $E^\circ_{cathode} = 0.34$ V.

Student Note: The electrode where oxidation occurs is the anode. Therefore, $E^\circ_{anode} = -0.76$ V.

TABLE 19.1	Standard Reduction Potentials at 25°C*	
Half-Reaction		**$E°$(V)**
$F_2(g) + 2e^- \longrightarrow 2F^-(aq)$		+2.87
$O_3(g) + 2H^+(aq) + 2e^- \longrightarrow O_2(g) + H_2O(l)$		+2.07
$Co^{3+}(aq) + e^- \longrightarrow Co^{2+}(aq)$		+1.82
$H_2O_2(aq) + 2H^+(aq) + 2e^- \longrightarrow 2H_2O(l)$		+1.77
$PbO_2(s) + 4H^+(aq) + SO_4^{2-}(aq) + 2e^- \longrightarrow PbSO_4(s) + 2H_2O(l)$		+1.70
$Ce^{4+}(aq) + e^- \longrightarrow Ce^{3+}(aq)$		+1.61
$MnO_4^-(aq) + 8H^+(aq) + 5e^- \longrightarrow Mn^{2+}(aq) + 4H_2O(l)$		+1.51
$Au^{3+}(aq) + 3e^- \longrightarrow Au(s)$		+1.50
$Cl_2(g) + 2e^- \longrightarrow 2Cl^-(aq)$		+1.36
$Cr_2O_7^{2-}(aq) + 14H^+(aq) + 6e^- \longrightarrow 2Cr^{3+}(aq) + 7H_2O(l)$		+1.33
$MnO_2(s) + 4H^+(aq) + 2e^- \longrightarrow Mn^{2+}(aq) + 2H_2O(l)$		+1.23
$O_2(g) + 4H^+(aq) + 4e^- \longrightarrow 2H_2O(l)$		+1.23
$Br_2(l) + 2e^- \longrightarrow 2Br^-(aq)$		+1.07
$NO_3^-(aq) + 4H^+(aq) + 3e^- \longrightarrow NO(g) + 2H_2O(l)$		+0.96
$2Hg^{2+}(aq) + 2e^- \longrightarrow Hg_2^{2+}(aq)$		+0.92
$Hg_2^{2+}(aq) + 2e^- \longrightarrow 2Hg(l)$		+0.85
$Ag^+(aq) + e^- \longrightarrow Ag(s)$		+0.80
$Fe^{3+}(aq) + e^- \longrightarrow Fe^{2+}(aq)$		+0.77
$O_2(g) + 2H^+(aq) + 2e^- \longrightarrow H_2O_2(aq)$		+0.68
$MnO_4^-(aq) + 2H_2O(l) + 3e^- \longrightarrow MnO_2(s) + 4OH^-(aq)$		+0.59
$I_2(s) + 2e^- \longrightarrow 2I^-(aq)$		+0.53
$O_2(g) + 2H_2O(l) + 4e^- \longrightarrow 4OH^-(aq)$		+0.40
$Cu^{2+}(aq) + 2e^- \longrightarrow Cu(s)$		+0.34
$AgCl(s) + e^- \longrightarrow Ag(s) + Cl^-(aq)$		+0.22
$SO_4^{2-}(aq) + 4H^+(aq) + 2e^- \longrightarrow SO_2(g) + 2H_2O(l)$		+0.20
$Cu^{2+}(aq) + e^- \longrightarrow Cu^+(aq)$		+0.15
$Sn^{4+}(aq) + 2e^- \longrightarrow Sn^{2+}(aq)$		+0.13
$2H^+(aq) + 2e^- \longrightarrow H_2(g)$		0.00
$Pb^{2+}(aq) + 2e^- \longrightarrow Pb(s)$		−0.13
$Sn^{2+}(aq) + 2e^- \longrightarrow Sn(s)$		−0.14
$Ni^{2+}(aq) + 2e^- \longrightarrow Ni(s)$		−0.25
$Co^{2+}(aq) + 2e^- \longrightarrow Co(s)$		−0.28
$PbSO_4(s) + 2e^- \longrightarrow Pb(s) + SO_4^{2-}(aq)$		−0.31
$Cd^{2+}(aq) + 2e^- \longrightarrow Cd(s)$		−0.40
$Fe^{2+}(aq) + 2e^- \longrightarrow Fe(s)$		−0.44
$Cr^{3+}(aq) + 3e^- \longrightarrow Cr(s)$		−0.74
$Zn^{2+}(aq) + 2e^- \longrightarrow Zn(s)$		−0.76
$2H_2O(l) + 2e^- \longrightarrow H_2(g) + 2OH^-(aq)$		−0.83
$Mn^{2+}(aq) + 2e^- \longrightarrow Mn(s)$		−1.18
$Al^{3+}(aq) + 3e^- \longrightarrow Al(s)$		−1.66
$Be^{2+}(aq) + 2e^- \longrightarrow Be(s)$		−1.85
$Mg^{2+}(aq) + 2e^- \longrightarrow Mg(s)$		−2.37
$Na^+(aq) + e^- \longrightarrow Na(s)$		−2.71
$Ca^{2+}(aq) + 2e^- \longrightarrow Ca(s)$		−2.87
$Sr^{2+}(aq) + 2e^- \longrightarrow Sr(s)$		−2.89
$Ba^{2+}(aq) + 2e^- \longrightarrow Ba(s)$		−2.90
$K^+(aq) + e^- \longrightarrow K(s)$		−2.93
$Li^+(aq) + e^- \longrightarrow Li(s)$		−3.05

Increasing strength as oxidizing agent

Increasing strength as reducing agent

*For all half-reactions the concentration is 1 M for dissolved species and the pressure is 1 atm for gases. These are the standard-state values.

In this case, the greater (less negative) reduction potential is that of Zn, so the Zn half-reaction will occur as a reduction and the Zn electrode will be the cathode. The Mn electrode will be the anode. The cell potential is therefore:

$$E^\circ_{cell} = E^\circ_{cathode} - E^\circ_{anode}$$

$$= E^\circ_{Zn^{2+}/Zn} - E^\circ_{Mn^{2+}/Mn}$$

$$= (-0.76 \text{ V}) - (-1.18 \text{ V})$$

$$= 0.42 \text{ V}$$

The overall cell reaction is:

$$Mn + Zn^{2+} \longrightarrow Mn^{2+} + Zn$$

By using Equation 19.1, we can predict the direction of an overall cell reaction.

It is important to understand that the standard reduction potential is an *intensive* property (like temperature and density), not an *extensive* property (like mass and volume) [◄◄ Section 1.4]. This means that the value of the standard reduction potential does *not* depend on the amount of a substance involved. Therefore, when it is necessary to multiply one of the half-reactions by a coefficient to balance the overall equation, the value of E° for the half-reaction remains the same. Consider a galvanic cell made up of a Zn half-cell and an Ag half-cell:

$$Zn(s) \mid Zn^{2+}(1 \text{ } M) \parallel Ag^+(1 \text{ } M) \mid Ag(s)$$

The half-cell reactions are:

$$Ag^+ + e^- \longrightarrow Ag \qquad E^\circ = 0.80 \text{ V}$$

$$Zn^{2+} + 2e^- \longrightarrow Zn \qquad E^\circ = -0.76 \text{ V}$$

The Ag half-reaction, with the more positive standard reduction potential, will occur as a reduction, and the Zn half-reaction will occur as an oxidation. Balancing the equation for the overall cell reaction requires multiplying the reduction (the Ag half-reaction) by 2:

$$2(Ag^+ + e^- \longrightarrow Ag)$$

We can then add the two half-reactions and cancel the electrons to get the overall, balanced equation:

$$2Ag^+ + 2e^- \longrightarrow 2Ag$$

$$\underline{+Zn \longrightarrow Zn^{2+} + 2e^-}$$

$$2Ag^+ + Zn \longrightarrow 2Ag + Zn^{2+}$$

The standard cell potential can be calculated using Equation 19.1:

$$E^\circ_{cell} = E^\circ_{cathode} - E^\circ_{anode}$$

$$= E^\circ_{Ag^+/Ag} - E^\circ_{Zn^{2+}/Zn}$$

$$= 0.80 \text{ V} - (-0.76 \text{ V})$$

$$= 1.56 \text{ V}$$

Although we multiplied the Ag half-reaction by 2, we did *not* multiply its standard reduction potential by 2.

Table 19.1 is essentially an extended version of the activity series [◄◄ Section 4.4]. Sample Problem 19.2 illustrates how to use standard reduction potentials to predict the direction of the overall reaction in a galvanic cell.

Student Note: It is a common error to multiply E° values by the same number as the half-cell equation. Think of the reduction potential as the height of a waterfall. Just as water falls from a higher elevation to a lower elevation, electrons move from an electrode with a higher potential toward one with a lower potential. The amount of water falling does not affect the net change in its elevation. Likewise, the number of electrons moving from higher potential to lower potential does not affect the size of the change in potential.

SAMPLE PROBLEM (19.2)

A galvanic cell consists of an Mg electrode in a 1.0 M $Mg(NO_3)_2$ solution and a Cd electrode in a 1.0 M $Cd(NO_3)_2$ solution. Determine the overall cell reaction, and calculate the standard cell potential at 25°C.

Strategy Use the tabulated values of $E°$ to determine which electrode is the cathode and which is the anode, combine cathode and anode half-cell reactions to get the overall cell reaction, and use Equation 19.1 to calculate $E°_{cell}$.

Setup The half-cell reactions and their standard reduction potentials are:

$$Mg^{2+} + 2e^- \longrightarrow Mg \qquad E° = -2.37 \text{ V}$$
$$Cd^{2+} + 2e^- \longrightarrow Cd \qquad E° = -0.40 \text{ V}$$

Because the Cd half-cell reaction has the greater (less negative) standard reduction potential, it will occur as the reduction. The Mg half-cell reaction will occur as the oxidation. Therefore, $E°_{cathode} = -0.40$ V and $E°_{anode} = -2.37$ V.

Solution Adding the two half-cell reactions together gives the overall cell reaction:

$$Mg \longrightarrow Mg^{2+} + 2e^-$$
$$\underline{Cd^{2+} + 2e^- \longrightarrow Cd}$$
$$\textit{Overall:} \quad Mg + Cd^{2+} \longrightarrow Mg^{2+} + Cd$$

The standard cell potential is:

$$E°_{cell} = E°_{cathode} - E°_{anode}$$
$$= E°_{Cd^{2+}/Cd} - E°_{Mg^{2+}/Mg}$$
$$= (-0.40 \text{ V}) - (-2.37 \text{ V})$$
$$= 1.97 \text{ V}$$

THINK ABOUT IT

If you ever calculate a *negative* voltage for a galvanic cell potential, you have done something wrong—check your work. Under standard-state conditions, the overall cell reaction will proceed in the direction that gives a positive $E°_{cell}$.

Practice Problem **A**TTEMPT Determine the overall cell reaction and $E°_{cell}$ at (25°C) of a galvanic cell made of a Cd electrode in a 1.0 M $Cd(NO_3)_2$ solution and a Pb electrode in a 1.0 M $Pb(NO_3)_2$ solution.

Practice Problem **B**UILD A galvanic cell with $E°_{cell} = 0.30$ V can be constructed using an iron electrode in a 1.0 M $Fe(NO_3)_2$ solution, and either a tin electrode in a 1.0 M $Sn(NO_3)_2$ solution, or a chromium electrode in a 1.0 M $Cr(NO_3)_2$ solution—even though Sn^{2+}/Sn and Cr^{3+}/Cr have different standard reduction potentials. Explain and give the overall balanced reaction for each cell.

Practice Problem **C**ONCEPTUALIZE For a galvanic cell consisting of a Zn electrode immersed in a solution that is 0.10 M in $Zn^{2+}(aq)$ and a Cu electrode immersed in a solution that is 0.10 M in $Cu^{2+}(aq)$, which of the following graphs best represents the concentrations of metal ions as a function of time?

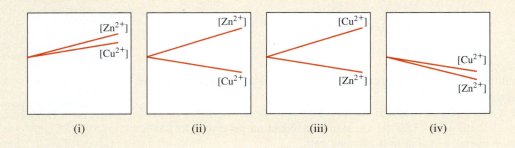

(i)   (ii)   (iii)   (iv)

Standard reduction potentials can also be used to determine what, if any, redox reaction will take place when reactants are combined in the same beaker—rather than divided into half-cells. Unlike a galvanic cell, in which we use reduction potentials to determine if the reaction proceeds in the *forward* or *reverse* direction, when the reaction is not divided into half-cells, we use reduction potentials to determine *if* the reaction will occur as written—or *not*.

Sample Problem 19.3 illustrates this technique.

SAMPLE PROBLEM 19.3

Predict what redox reaction will take place, if any, when molecular bromine (Br_2) is added to (a) a 1-*M* solution of NaI and (b) a 1-*M* solution of NaCl. (Assume a temperature of 25°C.)

Strategy In each case, write the equation for the redox reaction that *might* take place and use $E°$ values to determine whether or not the proposed reaction will actually occur.

Setup From Table 19.1:

$$Br_2(l) + 2e^- \longrightarrow 2Br^-(aq) \qquad E° = 1.07 \text{ V}$$
$$I_2(s) + 2e^- \longrightarrow 2I^-(aq) \qquad E° = 0.53 \text{ V}$$
$$Cl_2(g) + 2e^- \longrightarrow 2Cl^-(aq) \qquad E° = 1.36 \text{ V}$$

Solution

(a) If a redox reaction is to occur, it will be the oxidation of I^- ions by Br_2:

$$Br_2(l) + 2I^-(aq) \longrightarrow 2Br^-(aq) + I_2(s)$$

Because the reduction potential of Br_2 is greater than that of I_2, Br_2 will be reduced to Br^- and I^- will be oxidized to I_2. Thus, the preceding reaction *will* occur.

(b) In this case, the proposed reaction is the reduction of Br_2 by Cl^- ions:

$$Br_2(l) + 2Cl^-(aq) \longrightarrow 2Br^-(aq) + Cl_2(g)$$

However, because the reduction potential of Br_2 is smaller than that of Cl_2, this reaction will *not* occur. Cl_2 is more readily reduced than Br_2, so Br_2 is not reduced by Cl^-.

THINK ABOUT IT

We can use Equation 19.1 and treat problems of this type like galvanic cell problems. Write the proposed redox reaction, and identify the "cathode" and the "anode." If the calculated $E°_{cell}$ is positive, the reaction will occur. If the calculated $E°_{cell}$ is negative, the reaction will not occur.

Practice Problem ATTEMPT Determine what redox reaction, if any, occurs (at 25°C) when lead metal (Pb) is added to (a) a 1.0-*M* solution of $NiCl_2$ and (b) a 1.0-*M* solution of HCl.

Practice Problem BUILD Would it be safer to store a cobalt(II) chloride solution in a tin container or an iron container? Explain.

Practice Problem CONCEPTUALIZE A piece of nickel metal is added to a solution that is 1.0 *M* in three different chloride salts: $CoCl_2$, $NiCl_2$, and $SnCl_2$. Which of the following graphs best represents the concentrations of metal ions in solution as a function of time?

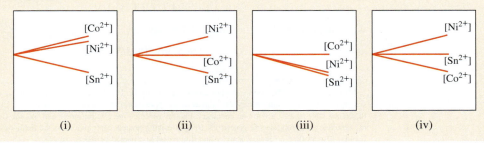

(i)   (ii)   (iii)   (iv)

CHECKPOINT – SECTION 19.3   Standard Reduction Potentials

19.3.1 Calculate E°_{cell} at 25°C for a galvanic cell made of a Cr electrode in a solution that is 1.0 M in Cr^{3+} and an Au electrode in a solution that is 1.0 M in Au^{3+}.

a) −2.03 V

b) 1.11 V

c) −1.11 V

d) 2.24 V

e) 0.76 V

19.3.2 Calculate E°_{cell} at 25°C for a galvanic cell made of a Cr electrode in a solution that is 1.0 M in Cr^{3+} and an Ag electrode in a solution that is 1.0 M in Ag^+.

a) 0.06 V

b) −0.06 V

c) 1.54 V

d) 3.14 V

e) 3.02 V

19.3.3 What redox reaction, if any, will occur at 25°C when Al metal is placed in a solution that is 1.0 M in Cr^{3+}?

a) $Al^{3+} + Cr \longrightarrow Al + Cr^{3+}$

b) $Cr^{3+} + Al \longrightarrow Cr + Al^{3+}$

c) $Al^{3+} + Cr^{3+} \longrightarrow Al + Cr$

d) $Al + Cr \longrightarrow Al^{3+} + Cr^{3+}$

e) No redox reaction will occur.

19.3.4 What redox reaction, if any, will occur at 25°C when Zn metal is placed in a solution that is 1.0 M in Sr^{2+}?

a) $Zn^{2+} + Sr \longrightarrow Zn + Sr^{2+}$

b) $Sr^{2+} + Zn \longrightarrow Sr + Zn^{2+}$

c) $Zn^{2+} + Sr^{2+} \longrightarrow Zn + Sr$

d) $Zn + Sr \longrightarrow Zn^{2+} + Sr^{2+}$

e) No redox reaction will occur.

19.4 Spontaneity of Redox Reactions Under Standard-State Conditions

We now look at how E°_{cell} is related to thermodynamic quantities such as ΔG° and K. In a galvanic cell, chemical energy is converted to electric energy. The electric energy produced in a galvanic cell is the product of the cell potential and the total electric charge (in coulombs) that passes through the cell:

$$\text{electric energy} = \text{volts} \times \text{coulombs}$$
$$= \text{joules}$$

The total charge is determined by the number of moles of electrons (n) that pass through the circuit. By definition:

$$\text{total charge} = nF$$

where F, the Faraday[1] constant, is the electric charge contained in 1 mole of electrons. One faraday is equivalent to 96,485.3 C, although we usually round the number to three significant figures. Thus:

$$1\,F = 96,500 \text{ C/mol } e^-$$

Because:

$$1\,J = 1\,C \times 1\,V$$

we can also express the units of faraday as:

$$1\,F = 96,500 \text{ J/V} \cdot \text{mol } e^-$$

---

1. Michael Faraday (1791–1867). English chemist and physicist. Faraday is regarded by many as the greatest experimental scientist of the nineteenth century. He started as an apprentice to a bookbinder at the age of 13, but became interested in science after reading a book on chemistry. Faraday invented the electric motor and was the first person to demonstrate the principle governing electric generators. Besides making notable contributions to the fields of electricity and magnetism, Faraday also worked on optical activity and discovered and named benzene.

The measured cell potential is the maximum voltage that the cell can produce. This value is used to calculate the maximum amount of electric energy that can be obtained from the chemical reaction. This energy is used to do electrical work ($w_{electrical}$), so:

$$w_{max} = w_{electrical}$$
$$= -nFE_{cell}$$

where w_{max} is the maximum amount of work that can be done. The negative sign on the right-hand side indicates that the electrical work is done *by* the system *on* the surroundings. In Chapter 18 we defined *free energy* as the energy available to do *work*. Specifically, the change in free energy, ΔG, represents the maximum amount of useful work that can be obtained from a reaction:

$$\Delta G = w_{max}$$

Therefore, we can write:

Equation 19.2 $\Delta G = -nFE_{cell}$

Both n and F are positive quantities and ΔG is negative for a spontaneous process, so E_{cell} must be positive for a spontaneous process. For reactions in which reactants and products are in their standard states, Equation 19.2 becomes:

Equation 19.3 $\Delta G° = -nFE°_{cell}$

Equation 19.3 makes it possible to relate $E°_{cell}$ to the equilibrium constant, K, of a redox reaction. In Section 18.6 we saw that the standard free-energy change, $\Delta G°$, for a reaction is related to its equilibrium constant as follows [◄◄ Section 18.6, Equation 18.14]:

$$\Delta G° = -RT \ln K$$

Therefore, if we combine Equations 18.14 and 19.3, we get:

$$-nFE°_{cell} = -RT \ln K$$

Solving for $E°_{cell}$ gives:

Equation 19.4 $E°_{cell} = \dfrac{RT}{nF} \ln K$

When $T = 298$ K and n moles of electrons are transferred per mole of reaction, Equation 19.4 can be simplified by inserting the values for R and F:

$$E°_{cell} = \frac{\left(8.314 \dfrac{J}{K \cdot mol}\right)(298 \text{ K})}{(n)\left(96{,}500 \dfrac{J}{V \cdot mol}\right)}$$

$$= \frac{0.0257 \text{ V}}{n} \ln K$$

And, by converting to the base-10 logarithm of K, we get:

Equation 19.5 $E°_{cell} = \dfrac{0.0592 \text{ V}}{n} \log K$ (at 25°C)

Thus, if we know any one of the three quantities $\Delta G°$, K, or $E°_{cell}$, we can convert to the others by using Equations 18.14, 19.3, and 19.5.

Sample Problems 19.4 and 19.5 demonstrate the interconversions among $\Delta G°$, K, and $E°_{cell}$. For simplicity, the subscript "cell" is not shown.

SAMPLE PROBLEM 19.4

Calculate the standard free-energy change for the following reaction at 25°C:

$$2Au(s) + 3Ca^{2+}(1.0\ M) \rightleftharpoons 2Au^{3+}(1.0\ M) + 3Ca(s)$$

Strategy Use $E°$ values from Table 19.1 to calculate $E°$ for the reaction, and then use Equation 19.3 to calculate the standard free-energy change.

Setup The half-cell reactions are:

$$\text{Cathode (reduction):} \quad 3Ca^{2+}(aq) + 6e^- \longrightarrow 3Ca(s)$$
$$\text{Anode (oxidation):} \quad 2Au(s) \longrightarrow 2Au^{3+}(aq) + 6e^-$$

From Table 19.1, $E°_{Ca^{2+}/Ca} = -2.87$ V and $E°_{Au^{3+}/Au} = 1.50$ V.

Solution

$$E°_{cell} = E°_{cathode} - E°_{anode}$$
$$= E°_{Ca^{2+}/Ca} - E°_{Au^{3+}/Au}$$
$$= -2.87\ \text{V} - 1.50\ \text{V}$$
$$= -4.37\ \text{V}$$

Next, substitute this value of $E°$ into Equation 19.3 to obtain $\Delta G°$:

$$\Delta G° = -nFE°$$

The overall reaction shows that $n = 6$, so

$$\Delta G° = -(6e^-)(96{,}500\ \text{J/V} \cdot \text{mol}\ e^-)(-4.37\ \text{V})$$
$$= 2.53 \times 10^6\ \text{J/mol}$$
$$= 2.53 \times 10^3\ \text{kJ/mol}$$

THINK ABOUT IT

The large positive value of $\Delta G°$ indicates that reactants are favored at equilibrium, which is consistent with the fact that $E°$ for the reaction is negative.

Practice Problem **A**TTEMPT Calculate $\Delta G°$ for the following reaction at 25°C:

$$3Mg(s) + 2Al^{3+}(aq) \rightleftharpoons 3Mg^{2+}(aq) + 2Al(s)$$

Practice Problem **B**UILD The hydrazinium ion, $N_2H_5^+$, reacts in acidic solution with molecular bromine to form nitrogen gas and aqueous bromide. $\Delta G°$ for this reaction is -5.02×10^5 J/mol. Calculate $E°_{red}$ for $N_2H_5^+$ and write the corresponding half-reaction.

Practice Problem **C**ONCEPTUALIZE Which of the following graphs best represents the relationship between $\Delta G°$ and $E°$ for a chemical reaction?

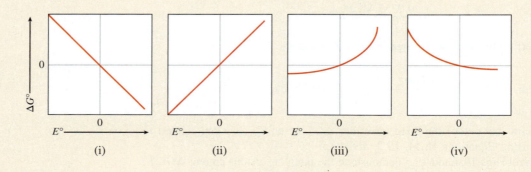

(i) (ii) (iii) (iv)

SAMPLE PROBLEM 19.5

Calculate the equilibrium constant for the following reaction at 25°C:

$$Sn(s) + 2Cu^{2+}(aq) \rightleftharpoons Sn^{2+}(aq) + 2Cu^+(aq)$$

Strategy Use $E°$ values from Table 19.1 to calculate $E°$ for the reaction, and then calculate the equilibrium constant using Equation 19.5 (rearranged to solve for K).

Setup The half-cell reactions are:

$$\text{Cathode (reduction):} \quad 2Cu^{2+}(aq) + 2e^- \longrightarrow 2Cu^+(aq)$$

$$\text{Anode (oxidation):} \quad Sn(s) \longrightarrow Sn^{2+}(aq) + 2e^-$$

From Table 19.1, $E°_{Cu^{2+}/Cu^+} = 0.15$ V and $E°_{Sn^{2+}/Sn} = -0.14$ V.

Solution

$$E°_{cell} = E°_{cathode} - E°_{anode}$$
$$= E°_{Cu^{2+}/Cu^+} - E°_{Sn^{2+}/Sn}$$
$$= 0.15 \text{ V} - (-0.14 \text{ V})$$
$$= 0.29 \text{ V}$$

Solving Equation 19.5 for K gives:

$$K = 10^{nE°/0.0592 \text{ V}}$$
$$= 10^{(2)(0.29 \text{ V})/0.0592 \text{ V}}$$
$$= 6 \times 10^9$$

THINK ABOUT IT

A positive standard cell potential corresponds to a large equilibrium constant.

Practice Problem A TTEMPT Calculate the equilibrium constant for the following reaction at 25°C:

$$2Ag(s) + Fe^{2+}(aq) \rightleftharpoons 2Ag^+(aq) + Fe(s)$$

Practice Problem B UILD Like equilibrium constants, $E°_{cell}$ values are temperature dependent. At 80°C, $E°_{cell}$ for the cell diagram shown is 0.18 V.

$$Pt \mid H_2(g) \mid HCl(aq) \parallel AgCl(s) \mid Ag(s)$$

The corresponding cell reaction is:

$$H_2(g) + 2AgCl(s) \rightleftharpoons 2Ag(s) + 2H^+(aq) + 2Cl^-(aq)$$

Calculate the equilibrium constant for this reaction at 80°C.

Practice Problem C ONCEPTUALIZE Which of the following graphs best represents the relationship between K and $E°$ for a chemical reaction?

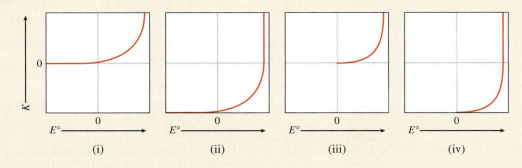

(i)   (ii)   (iii)   (iv)

CHECKPOINT – SECTION 19.4 Spontaneity of Redox Reactions Under Standard-State Conditions

19.4.1 Calculate K at 25°C for the following reaction:

$$Fe^{2+}(aq) + Ni(s) \rightleftharpoons Fe(s) + Ni^{2+}(aq)$$

a) 6×10^{-4}

b) 4×10^{-7}

c) 3×10^{6}

d) 1×10^{-13}

e) 2×10^{3}

19.4.2 Calculate $\Delta G°$ at 25°C for the following reaction:

$$3Cu^{2+}(aq) + Cr(s) \rightleftharpoons 3Cu^{+}(aq) + Cr^{3+}(aq)$$

a) 2.6×10^{2} kJ/mol

b) -2.6×10^{2} kJ/mol

c) 1×10^{45} kJ/mol

d) -86 kJ/mol

e) 86 kJ/mol

19.5 Spontaneity of Redox Reactions Under Conditions Other than Standard State

So far we have focused on redox reactions in which the reactants and products are in their standard states. Standard-state conditions, though, are difficult to come by and usually impossible to maintain. Just as there is an equation that relates ΔG to $\Delta G°$ [◄ Section 18.6, Equation 18.13], there is an equation that relates E to $E°$. We now derive this equation.

The Nernst Equation

Consider a redox reaction of the type:

$$aA + bB \longrightarrow cC + dD$$

From Equation 18.13:

$$\Delta G = \Delta G° + RT \ln Q$$

Because $\Delta G = -nFE$ and $\Delta G° = -nFE°$, the equation can be expressed as:

$$-nFE = -nFE° + RT \ln Q$$

Dividing the equation through by $-nF$, we get:

$$E = E° - \frac{RT}{nF} \ln Q \qquad\qquad \textbf{Equation 19.6}$$

where Q is the reaction quotient [◄ Section 15.2]. Equation 19.6 is known as the *Nernst[2] equation.* At 298 K, Equation 19.6 can be rewritten as:

$$E = E° - \frac{0.0257 \text{ V}}{n} \ln Q$$

---

2. Walther Hermann Nernst (1864–1941). German chemist and physicist. Nernst's work was mainly on electrolyte solutions and thermodynamics. He also invented an electric piano. Nernst was awarded the Nobel Prize in Chemistry in 1920 for his contribution to thermodynamics.

or using the base-10 logarithm of Q as:

$$E = E° - \frac{0.0592 \text{ V}}{n} \log Q \qquad \textbf{Equation 19.7}$$

During the operation of a galvanic cell, electrons flow from the anode to the cathode, resulting in product formation and a decrease in reactant concentration. Thus, Q increases, which means that E decreases. Eventually, the cell reaches equilibrium. At equilibrium, there is no net transfer of electrons, so $E = 0$ and $Q = K$, where K is the equilibrium constant.

The Nernst equation enables us to calculate E as a function of reactant and product concentrations in a redox reaction. For example, for the cell pictured in Figure 19.1:

$$Zn(s) + Cu^{2+}(aq) \longrightarrow Zn^{2+}(aq) + Cu(s)$$

The Nernst equation for this cell at 25°C can be written as:

$$E = 1.10 \text{ V} - \frac{0.0592 \text{ V}}{2} \log \frac{[Zn^{2+}]}{[Cu^{2+}]}$$

If the ratio $[Zn^{2+}]/[Cu^{2+}]$ is less than 1, log $([Zn^{2+}]/[Cu^{2+}])$ is a negative number, making the second term on the right-hand side of the preceding equation a positive quantity. Under this condition, E is *greater* than the standard potential $E°$. If the ratio is greater than 1, E is *smaller* than $E°$.

Sample Problem 19.6 shows how to use the Nernst equation.

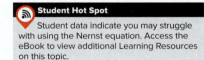

Student Hot Spot

Student data indicate you may struggle with using the Nernst equation. Access the eBook to view additional Learning Resources on this topic.

SAMPLE PROBLEM 19.6

Predict whether the following reaction will occur spontaneously as written at 298 K:

$$Co(s) + Fe^{2+}(aq) \longrightarrow Co^{2+}(aq) + Fe(s)$$

assuming $[Co^{2+}] = 0.15 \ M$ and $[Fe^{2+}] = 0.68 \ M$.

Strategy Use $E°$ values from Table 19.1 to determine $E°$ for the reaction, and use Equation 19.7 to calculate E. If E is positive, the reaction will occur spontaneously.

Setup From Table 19.1:

$$Cathode \ (reduction): \quad Fe^{2+}(aq) + 2e^- \longrightarrow Fe(s)$$
$$Anode \ (oxidation): \quad Co(s) \longrightarrow Co^{2+}(aq) + 2e^-$$

$$E°_{cell} = E°_{cathode} - E°_{anode}$$
$$= E°_{Fe^{2+}/Fe} - E°_{Co^{2+}/Co}$$
$$= -0.44 \text{ V} - (-0.28 \text{ V})$$
$$= -0.16 \text{ V}$$

The reaction quotient, Q, for the reaction is $[Co^{2+}]/[Fe^{2+}]$. Therefore, $Q = (0.15/0.68) = 0.22$.

Solution From Equation 19.7:

$$E = E° - \frac{0.0592 \text{ V}}{n} \log Q$$
$$= -0.16 \text{ V} - \frac{0.0592 \text{ V}}{2} \log 0.22$$
$$= -0.14 \text{ V}$$

The negative E value indicates that the reaction is *not* spontaneous as written under the conditions described.

(Continued on next page)

THINK ABOUT IT

For this reaction to be spontaneous as written, the ratio of $[Fe^{2+}]$ to $[Co^{2+}]$ would have to be enormous. We can determine the required ratio by first setting E equal to zero:

$$0 \text{ V} = -0.16 \text{ V} - \frac{0.0592 \text{ V}}{2} \log Q$$

$$-\frac{(0.16 \text{ V})(2)}{0.0592 \text{ V}} = \log Q$$

$$\log Q = -5.4$$

$$Q = 10^{-5.4} = \frac{[Co^{2+}]}{[Fe^{2+}]} = 4 \times 10^{-6}$$

For E to be positive, therefore, the ratio of $[Fe^{2+}]$ to $[Co^{2+}]$, the reciprocal of Q, would have to be greater than 3×10^5 to 1.

Practice Problem ATTEMPT   Will the following reaction occur spontaneously at 298 K if $[Fe^{2+}] = 0.60$ M and $[Cd^{2+}] = 0.010$ M?

$$Cd(s) + Fe^{2+}(aq) \longrightarrow Cd^{2+}(aq) + Fe(s)$$

Practice Problem BUILD   Consider the electrochemical cell in Practice Problem 19.5B, for which $E°_{cell} = 0.18$ V at 80°C:

$$H_2(g) + 2AgCl(s) \rightleftharpoons 2Ag(s) + 2H^+(aq) + 2Cl^-(aq)$$

If pH = 1.05 in the anode compartment, and $[Cl^-] = 2.5$ M in the cathode compartment, determine the partial pressure of H_2 necessary in the anode compartment for the cell potential to be 0.27 V at 80°C.

Practice Problem CONCEPTUALIZE   Consider a galvanic cell based on the reaction in Sample Problem 19.5. Which of the following graphs best represents what happens to the value of E as the Cu^{2+} ion concentration is increased?

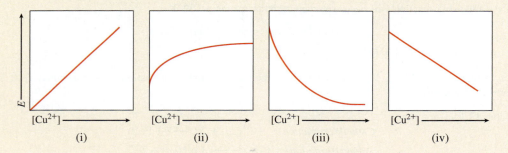

(i)   (ii)   (iii)   (iv)

Concentration Cells

Because electrode potential depends on ion concentrations, it is possible to construct a galvanic cell from two half-cells composed of the same material but differing in ion concentrations. Such a cell is called a ***concentration cell.***

Consider a galvanic cell consisting of a zinc electrode in 0.10 M zinc sulfate in one compartment and a zinc electrode in 1.0 M zinc sulfate in the other compartment (Figure 19.5). According to Le Châtelier's principle, the tendency for the reduction:

$$Zn^{2+}(aq) + 2e^- \longrightarrow Zn(s)$$

to occur increases with increasing concentration of Zn^{2+} ions. Therefore, reduction should occur in the more concentrated compartment and oxidation should take place on the more dilute side. The cell diagram is:

$$Zn(s) \mid Zn^{2+}(0.10 \text{ } M) \parallel Zn^{2+}(1.0 \text{ } M) \mid Zn(s)$$

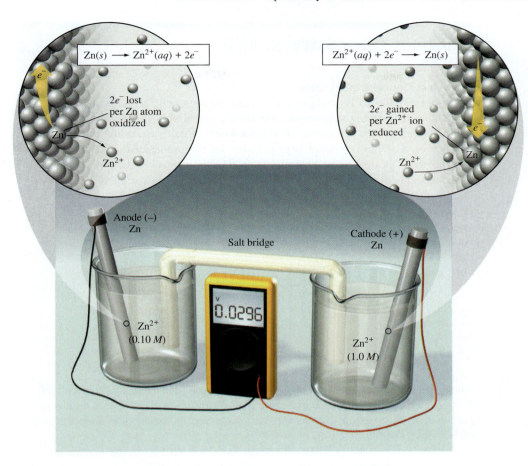

Figure 19.5 A concentration cell. Oxidation will occur in the container with the lower Zn^{2+} concentration. Reduction will occur in the container with the higher Zn^{2+} concentration.

and the half-cell reactions are:

Oxidation: $Zn(s) \longrightarrow Zn^{2+}(0.10\ M) + 2e^-$

Reduction: $Zn^{2+}(1.0\ M) + 2e^- \longrightarrow Zn(s)$

Overall: $Zn^{2+}(1.0\ M) \longrightarrow Zn^{2+}(0.10\ M)$

The cell potential is:

$$E = E° - \frac{0.0592\ \text{V}}{2}\log\frac{[Zn^{2+}]_{\text{dilute}}}{[Zn^{2+}]_{\text{concentrated}}}$$

where the subscripts "dilute" and "concentrated" refer to the 0.10 M and 1.0 M concentrations, respectively. The $E°$ for this cell is zero (because the same electrode-ion combination is used in both half-cells; i.e., $E°_{\text{cathode}} = E°_{\text{anode}}$), so:

$$E = 0 - \frac{0.0592\ \text{V}}{2}\log\frac{0.10}{1.0}$$

$$= 0.030\ \text{V}$$

The cell potential for a concentration cell is typically *small* and decreases continually during the operation of the cell as the concentrations in the two compartments approach each other. When the concentrations of the ions in the two compartments are equal, E becomes zero and no further change occurs.

Bringing Chemistry to Life

Biological Concentration Cells

A biological cell can be compared to a concentration cell for the purpose of calculating its membrane potential. Membrane potential is the electric potential that exists across the membrane of various kinds of cells, including muscle cells and nerve cells. This potential is responsible for the propagation of nerve impulses and heartbeat. A membrane potential is established whenever there are unequal concentrations of the same type of ion in the interior and exterior of a cell—and the membrane is permeable to the ion. For example, the concentrations of Na^+ ions in the interior and exterior of a nerve cell are 1.5×10^{-2} M and 1.5×10^{-1} M, respectively. Treating the situation as a concentration cell and applying the Nernst equation for a single ion, we can write:

$$E_{Na^+} = E_{Na^+}^{\circ} - \frac{0.0592 \text{ V}}{1} \log \frac{[Na^+]_{ex}}{[Na^+]_{in}}$$

$$= -(0.0592 \text{ V}) \log \frac{1.5 \times 10^{-1}}{1.5 \times 10^{-2}}$$

$$= 0.059 \text{ V} \quad \text{or} \quad 59 \text{ mV}$$

ECG strip

Charles D. Winters/Timeframe Photography/ McGraw Hill

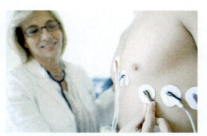

Patient with ECG electrodes

Science Photo Library/Alamy Stock Photo

where the subscripts "ex" and "in" denote "exterior" and "interior," respectively. We have set $E_{Na}^{\circ} = 0$ because the same type of ion is involved. Thus, an electric potential of 59 mV exists across the membrane due to the unequal concentrations of Na^+ ions.

When a nerve cell is stimulated, there is a large change in the membrane permeability, decreasing the membrane potential temporarily to about 34 mV. This sudden change in the potential is called the *action potential*. Once created, the action potential propagates along the nerve fiber until it reaches either a synaptic junction (the connection between nerve cells) or a neuromuscular junction (the connection between a nerve cell and a muscle cell). In the muscle cell of the heart, an action potential is generated during the heartbeat. This potential produces enough current to be detected by electrodes placed on the chest. The amplified signals can be recorded: either on a moving chart or displayed on an oscilloscope. The record, called an *electrocardiogram* (ECG, also known as an EKG, where K is from the German word *kardio* for heart), is a valuable tool in the diagnosis of heart disease.

In Chapter 17, we studied solubility equilibria and learned how to use tabulated K_{sp} values to determine solubilities of ionic compounds that are only very slightly soluble. The tabulated K_{sp} values are determined by measuring the concentration of one of a compound's constituent ions in a saturated solution of the compound. For example, to determine the K_{sp} of silver bromide (AgBr), we must measure the concentration of Ag^+, or the concentration of Br^-. Because the K_{sp} of silver bromide is 7.7×10^{-13}, the concentration of either ion in a saturated solution is 8.8×10^{-7} M. You may have wondered how such a small concentration is measured. It cannot be done using visible spectrophotometry, the method described in Chapter 4; not only because the concentration is so low, but also because a solution AgBr is colorless—meaning that it does not absorb visible light. In fact, concentrations of ions in these cases are measured using concentration cells.

Sample Problem 19.7 illustrates this process.

An electrochemical cell is constructed for the purpose of determining the K_{sp} of silver cyanide (AgCN) at 25°C. One half-cell consists of a silver electrode in a 1.00-M solution of silver nitrate. The other half-cell consists of a silver electrode in a saturated solution of silver cyanide. The cell potential is measured and found to be 0.470 V. Determine the concentration of silver ion in the saturated silver cyanide solution and the value of K_{sp} for AgCN.

Strategy Use Equation 19.7 to solve for the unknown concentration of silver ion. The half-cell with the higher Ag^+(1.0 M $AgNO_3$) concentration will be the cathode; the half-cell with the lower, unknown Ag^+ concentration (saturated AgCN solution) will be the anode. The overall reaction is Ag^+(1.0 M) $\longrightarrow$ Ag^+(x M).

Setup Because this is a concentration cell, $E°_{cell} = 0$ V. The reaction quotient, Q, is (x M)/(1.00 M); and the value of n is 1.

Solution

$$E_{cell} = E°_{cell} - \frac{0.0592\ \text{V}}{n} \log Q$$

$$0.470\ \text{V} = 0 - \frac{0.0592\ \text{V}}{1} \log \frac{x}{1.00}$$

$$-7.939 = \log \frac{x}{1.00}$$

$$10^{-7.939} = 1.15 \times 10^{-8} = \log \frac{x}{1.00}$$

$$x = 1.15 \times 10^{-8}$$

Therefore, $[Ag^+] = 1.15 \times 10^{-8}$ M and K_{sp} for AgCN $= x^2 = 1.3 \times 10^{-16}$.

THINK ABOUT IT

Remember that in a saturated solution of a salt that dissociates into two ions, the ion concentrations are equal to each other and each ion concentration is equal to the square root of K_{sp} [◀◀ Section 17.4].

Practice Problem Ⓐ**TTEMPT** An electrochemical cell is constructed for the purpose of determining the K_{sp} of copper(I) chloride, CuCl, at 25°C. One half-cell consists of a copper electrode in a 1.00-M solution of copper(I) nitrate. The other half-cell consists of a copper electrode in a saturated solution of copper(I) chloride. The measured cell potential is 0.175 V. Determine the concentration of copper(I) ion in the saturated copper(I) chloride solution and the value of K_{sp} for CuCl.

Practice Problem Ⓑ**UILD** The K_{sp} of copper(II) ferrocyanide ($Cu_2[Fe(CN)_6]$) is 1.3×10^{-16} at 25°C. Determine the potential of a concentration cell in which one half-cell consists of a copper electrode in 1.00 M copper(II) nitrate, and the other consists of a copper electrode in a saturated solution of $Cu_2[Fe(CN)_6]$. Ferrocyanide, ($[Fe(CN)_6]^{4-}$), is a *complex ion* [◀◀ Section 17.5].

Practice Problem Ⓒ**ONCEPTUALIZE** When the circuit in a silver chloride concentration cell is first completed, the concentration of silver ion in the cathode compartment is significantly higher than it is in the anode compartment. During operation of the cell, the concentrations become closer as the concentration decreases in the cathode compartment and increases in the anode compartment. If the cell continues to operate for long enough, will the concentration of silver ion in the anode compartment ever be greater than that in the cathode compartment? Explain.

CHECKPOINT − SECTION 19.5   Spontaneity of Redox Reactions Under Conditions Other than Standard State

19.5.1 Calculate E at 25°C for a galvanic cell based on the reaction:

$$Zn(s) + Cu^{2+}(aq) \rightleftharpoons Zn^{2+}(aq) + Cu(s)$$

in which $[Zn^{2+}] = 0.55\ M$ and $[Cu^{2+}] = 1.02\ M$.

a) 1.10 V

b) 1.11 V

c) 1.09 V

d) 0.0118 V

e) 1.03 V

19.5.2 Calculate the cell potential at 25°C of a galvanic cell consisting of an Ag electrode in 0.15 M AgNO$_3$ and an Ag electrode in 1.0 M AgNO$_3$.

a) 0.0 V

b) 0.049 V

c) −0.049 V

d) 0.024 V

e) −0.024 V

19.5.3 Calculate E°_{cell} at 80°C for a galvanic cell based on the reaction:

$$H_2(g) + 2AgCl(s) \rightleftharpoons 2Ag(s) + 2H^+(aq) + 2Cl^-(aq)$$

in which $[H^+] = 0.10\ M$, $[Cl^-] = 1.5\ M$, and $P = 1.25$ atm. ($E^{\circ}_{cell} = 0.18$ V at 80°C).

a) 0.12 V

b) 0.16 V

c) 0.18 V

d) 0.22 V

e) 0.24 V

19.5.4 Which of these would cause an increase in the cell potential of the electrochemical cell in question 19.5.2?

(i) Adding AgNO$_3$ to the cathode compartment

(ii) Adding saturated AgCl solution to the anode compartment

(iii) Adding NaCl to the anode compartment

a) i and iii

b) ii and iii

c) iii only

d) i, ii, and iii

e) None of these

To answer questions 19.5.5 and 19.5.6, consider the following cell diagram:

$$Pb(s)\ |\ Pb(NO_3)_2(0.60\ M)\ ||\ AgNO_3(0.40\ M)\ |\ Ag(s)$$

19.5.5 Determine the initial value of E_{cell} under the conditions shown in the cell diagram; and determine the initial value of E_{cell} if $[Ag^+]$ were increased by a factor of 4 at 25°C.

a) 0.91 V, 0.88 V

b) 0.91 V, 0.95 V

c) 0.93 V, 0.95 V

d) 0.93 V, 0.96 V

e) 0.92 V, 0.88 V

19.5.6 Which of the following would cause a decrease in the cell potential?

(i) Adding NH$_3$ to the cathode compartment

(ii) Adding NH$_3$ to the anode compartment

(iii) Adding NaNO$_3$ to the anode compartment

a) i only

b) i and iii

c) iii only

d) i, ii, and iii

e) None of these

19.6 Batteries

A *battery* is a galvanic cell, or a series of connected galvanic cells, that can be used as a portable, self-contained source of direct electric current. In this section we examine several types of batteries.

Dry Cells and Alkaline Batteries

The most common batteries, *dry cells* and *alkaline batteries,* are those used in flashlights, toys, and certain portable electronics such as CD players. The two are similar in appearance, but differ in the spontaneous chemical reaction responsible for producing a voltage. Although the reactions that take place in these batteries are somewhat complex, the reactions shown here approximate the overall processes.

A dry cell, so named because it has no fluid component, consists of a zinc container (the anode) in contact with manganese dioxide and an electrolyte (Figure 19.6). The electrolyte consists of ammonium chloride and zinc chloride in water, to which starch is added to thicken the solution to a paste so that it is less likely to leak. A carbon rod, immersed in the electrolyte in the center of the cell, serves as the cathode. The cell reactions are:

Anode: $\qquad\qquad\qquad Zn(s) \longrightarrow Zn^{2+}(aq) + 2e^-$

Cathode: $\ 2NH_4^+(aq) + 2MnO_2(s) + 2e^- \longrightarrow Mn_2O_3(s) + 2NH_3(aq) + H_2O(l)$

Overall: $\ \ Zn(s) + 2NH_4^+(aq) + 2MnO_2(s) \longrightarrow Zn^{2+}(aq) + Mn_2O_3(s) + 2NH_3(aq) + H_2O(l)$

The voltage produced by a dry cell is about 1.5 V.

An alkaline battery is also based on the reduction of manganese dioxide and the oxidation of zinc. However, the reactions take place in a *basic* medium, hence the name *alkaline* battery. The anode consists of powdered zinc suspended in a gel, which is in contact with a concentrated solution of KOH. The cathode is a mixture of manganese dioxide and graphite. The anode and cathode are separated by a porous barrier (Figure 19.7):

Anode: $\qquad\qquad Zn(s) + 2OH^-(aq) \longrightarrow Zn(OH)_2(s) + 2e^-$

Cathode: $\quad 2MnO_2(s) + 2H_2O(l) + 2e^- \longrightarrow 2MnO(OH)(s) + 2OH^-(aq)$

Overall: $\quad Zn(s) + 2MnO_2(s) + 2H_2O(l) \longrightarrow Zn(OH)_2(s) + 2MnO(OH)(s)$

Alkaline batteries are more expensive than dry cells and offer superior performance and shelf life.

Lead Storage Batteries

The lead storage battery commonly used in automobiles consists of six identical cells joined together in series. Each cell has a lead anode and a cathode made of lead dioxide (PbO_2) packed on a metal plate (Figure 19.8). Both the cathode and the anode

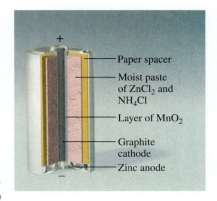

Figure 19.6 Interior view of the type of dry cell used in flashlights and other small devices.

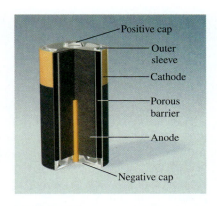

Figure 19.7 Interior view of an alkaline battery.

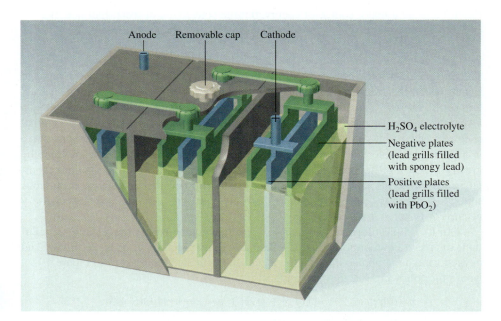

Figure 19.8 Interior view of a lead storage battery. Under normal operating conditions, the concentration of the sulfuric acid solution is about 38 percent by mass.

are immersed in an aqueous solution of sulfuric acid, which acts as the electrolyte. The cell reactions are:

Anode: $\qquad\qquad\qquad\qquad\qquad\qquad$ $Pb(s) + SO_4^{2-}(aq) \longrightarrow PbSO_4(s) + 2e^-$

Cathode: $\qquad$ $PbO_2(s) + 4H^+(aq) + SO_4^{2-}(aq) + 2e^- \longrightarrow PbSO_4(s) + 2H_2O(l)$

Overall: $\quad$ $Pb(s) + PbO_2(s) + 4H^+(aq) + 2SO_4^{2-}(aq) \longrightarrow 2PbSO_4(s) + 2H_2O(l)$

Under normal operating conditions, each cell produces 2 V. A total of 12 V from the six cells is used to power the ignition circuit of the automobile and its other electric systems. The lead storage battery can deliver large amounts of current for a short time, such as the time it takes to start the engine.

Unlike dry cells and alkaline batteries, the lead storage battery is rechargeable. Recharging the battery means reversing the normal electrochemical reaction by applying an external voltage at the cathode and the anode. (This kind of process is called *electrolysis,* which we discuss in Section 19.7.)

Lithium-Ion Batteries

Sometimes called "the battery of the future," lithium-ion batteries have several advantages over other battery types. The overall reaction that takes place in the lithium-ion battery is:

Anode: $\qquad\qquad\qquad$ $Li(s) \longrightarrow Li^+ + e^-$

Cathode: $\quad$ $Li^+ + CoO_2 + e^- \longrightarrow LiCoO_2(s)$

Overall: $\qquad$ $Li(s) + CoO_2 \longrightarrow LiCoO_2(s)$

Student Note: Lithium has the largest negative reduction potential of all the metals, making it a powerful reducing agent.

The overall cell potential is 3.4 V, which is relatively large. Lithium is also the lightest metal—only 6.941 g of Li (its molar mass) are needed to produce 1 mole of electrons. Furthermore, a lithium-ion battery can be recharged hundreds of times. These qualities make lithium batteries suitable for use in portable devices such as cell phones, digital cameras, and laptop computers.

Fuel Cells

Fossil fuels are a major source of energy, but the conversion of fossil fuel into electric energy is a highly inefficient process. Consider the combustion of methane:

$$CH_4(g) + 2O_2(g) \longrightarrow CO_2(g) + 2H_2O(l) + energy$$

To generate electricity, heat produced by the reaction is first used to convert water to steam, which then drives a turbine, which then drives a generator. A significant fraction of the energy released in the form of heat is lost to the surroundings at each step (even the most efficient power plant converts only about 40 percent of the original chemical energy into electricity). Because combustion reactions are redox reactions, it is more desirable to carry them out directly by electrochemical means, thereby greatly increasing the efficiency of power production. This objective can be accomplished by a device known as a ***fuel cell,*** a galvanic cell that requires a continuous supply of reactants to keep functioning. Strictly speaking, a fuel cell is not a battery because it is not self-contained.

In its simplest form, a hydrogen-oxygen fuel cell consists of an electrolyte solution, such as a potassium hydroxide solution, and two inert electrodes. Hydrogen and oxygen gases are bubbled through the anode and cathode compartments (Figure 19.9), where the following reactions take place:

Anode: $\qquad$ $2H_2(g) + 4OH^-(aq) \longrightarrow 4H_2O(l) + 4e^-$

Cathode: $\quad$ $O_2(g) + 2H_2O(l) + 4e^- \longrightarrow 4OH^-(aq)$

Overall: $\qquad\qquad$ $2H_2(g) + O_2(g) \longrightarrow 2H_2O(l)$

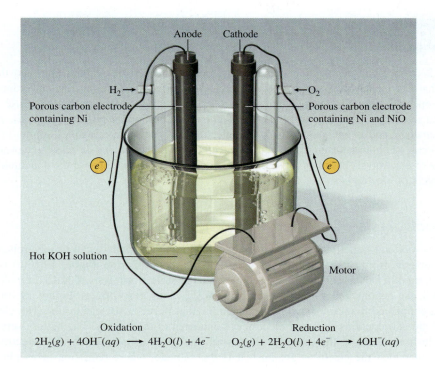

Anode   Cathode

$H_2 \rightarrow$

Porous carbon electrode
containing Ni

$\leftarrow O_2$

Porous carbon electrode
containing Ni and NiO

e^- e^-

Hot KOH solution

Motor

Oxidation
$2H_2(g) + 4OH^-(aq) \longrightarrow 4H_2O(l) + 4e^-$

Reduction
$O_2(g) + 2H_2O(l) + 4e^- \longrightarrow 4OH^-(aq)$

Figure 19.9 A hydrogen-oxygen fuel cell. The Ni and NiO embedded in the porous carbon electrodes are catalysts.

Using $E°$ values from Table 19.1, the standard cell potential is calculated as follows:

$$E°_{cell} = E°_{cathode} - E°_{anode}$$

$$= 0.40\ \text{V} - (-0.83\ \text{V})$$

$$= 1.23\ \text{V}$$

Thus, the cell reaction is spontaneous under standard-state conditions. Note that the reaction is the same as the hydrogen combustion reaction, but the oxidation and reduction are carried out separately at the anode and the cathode. Like platinum in the standard hydrogen electrode, the electrodes serve two purposes. They serve as electrical conductors, and they provide the necessary surfaces for the initial decomposition of the molecules into atomic species, which must take place before electrons can be transferred. Electrodes that serve this particular purpose are called "electrocatalysts." Metals such as platinum, nickel, and rhodium also make good electrocatalysts.

In addition to the H_2-O_2 system, a number of other fuel cells have been developed. Among these is the propane-oxygen fuel cell. The corresponding half-cell reactions are:

Anode: $\qquad\qquad C_3H_8(g) + 6H_2O(l) \longrightarrow 3CO_2(g) + 20H^+(aq) + 20e^-$

Cathode: $\quad 5O_2(g) + 20H^+(aq) + 20e^- \longrightarrow 10H_2O(l)$

Overall: $\qquad\qquad C_3H_8(g) + 5O_2(g) \longrightarrow 3CO_2(g) + 4H_2O(l)$

The overall reaction is identical to the burning of propane in oxygen.

Unlike batteries, fuel cells do not store chemical energy. Reactants must be constantly resupplied as they are consumed, and products must be removed as they form. However, properly designed fuel cells may be as much as 70 percent efficient, which is about twice as efficient as an internal combustion engine. In addition, fuel-cell generators are free of the noise, vibration, heat transfer, thermal pollution, and other problems normally associated with conventional power plants. Nevertheless, fuel cells are not yet in widespread use. One major problem is the expense of electrocatalysts that can function efficiently for long periods of time without contamination. One notable application of fuel cells is their use in space vehicles. Hydrogen-oxygen fuel cells provide electric power (and drinking water) for space flight.

19.7 Electrolysis

In Section 19.6, we mentioned that lead storage batteries are rechargeable and that recharging means reversing the electrochemical processes by which the battery ordinarily operates through the application of an external voltage. This process, the use of electric energy to drive a nonspontaneous chemical reaction, is called **electrolysis.** An **electrolytic cell** is one used to carry out electrolysis. The same principles apply to the processes in both galvanic and electrolytic cells. In this section, we discuss three examples of electrolysis based on those principles. We then examine some of the quantitative aspects of electrolysis.

Electrolysis of Molten Sodium Chloride

In its molten (melted) state, sodium chloride, an ionic compound, can be electrolyzed to separate it into its constituent elements, sodium and chlorine. Figure 19.10(a) is a diagram of a Downs cell, which is used for the large-scale electrolysis of NaCl. In molten NaCl, the cations and anions are the Na^+ and Cl^- ions, respectively. Figure 19.10(b) is a simplified diagram showing the reactions that occur at the electrodes. The electrolytic cell contains a pair of electrodes connected to the battery. The battery serves to push electrons in the direction they would not flow spontaneously. The electrode toward which the electrons are pushed is the cathode, where reduction takes place. The electrode away from which electrons are drawn is the anode, where oxidation takes place. The reactions at the electrodes are:

$$
\begin{aligned}
\textit{Anode (oxidation):} && 2Cl^-(l) &\longrightarrow Cl_2(g) + 2e^- \\
\textit{Cathode (reduction):} && 2Na^+(l) + 2e^- &\longrightarrow 2Na(l) \\
\hline
\textit{Overall:} && 2Na^+(l) + 2Cl^-(l) &\longrightarrow 2Na(l) + Cl_2(g)
\end{aligned}
$$

This process is a major industrial source of pure sodium metal and chlorine gas.

Using data from Table 19.1, we estimate $E°_{cell}$ to be -4 V for this process. The negative standard reduction potential indicates that for the process to occur as written, a minimum of approximately 4 V must be supplied by the battery to drive the reaction in the desired direction. In practice, an even higher voltage is required because of inefficiencies in the electrolytic process and because of overvoltage, a phenomenon we discuss later in this section.

Student Note:

Cell Type	Chemical Reaction	Electric Energy
Galvanic	Spontaneous	Produced
Electrolytic	Nonspontaneous	Consumed

Student Note: We can only estimate the value of $E°_{cell}$ because the values in Table 19.1 refer to the species in aqueous solution.

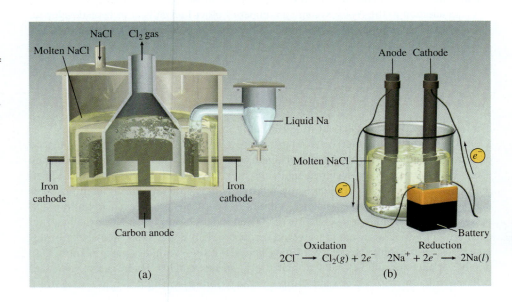

Figure 19.10 (a) A practical arrangement called a Downs cell for the electrolysis of molten NaCl (m.p. = 801°C). The sodium metal formed at the cathodes is in the liquid state. Because liquid sodium metal is lighter than molten NaCl, the sodium floats to the surface, as shown, and is collected. Chlorine gas forms at the anode and is collected at the top. (b) A simplified diagram showing the electrode reactions during the electrolysis of molten NaCl. The battery is needed to drive the nonspontaneous reaction.

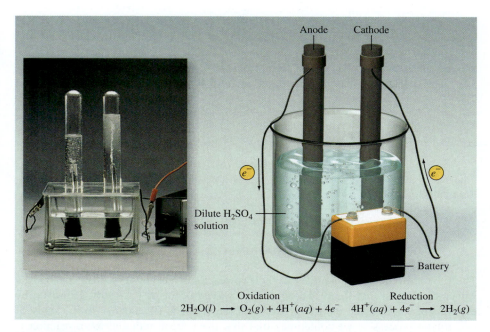

Oxidation
$$2H_2O(l) \longrightarrow O_2(g) + 4H^+(aq) + 4e^-$$

Reduction
$$4H^+(aq) + 4e^- \longrightarrow 2H_2(g)$$

Figure 19.11 Apparatus for small-scale electrolysis of water. The volume of hydrogen gas generated at the cathode is twice that of oxygen gas generated at the anode.

Stephen Frisch/McGraw Hill

Electrolysis of Water

Under ordinary atmospheric conditions (1 atm and 25°C), water will not spontaneously decompose to form hydrogen and oxygen gas because the standard free-energy change for the reaction is a large positive quantity:

$$2H_2O(l) \longrightarrow 2H_2(g) + O_2(g) \qquad \Delta G° = 474.4 \text{ kJ/mol}$$

However, this reaction can be made to occur in an electrolytic cell like the one shown in Figure 19.11. This cell consists of a pair of electrodes made of a nonreactive metal, such as platinum, immersed in water. When the electrodes are connected to the battery, nothing happens because there are not enough ions in pure water to carry much of an electric current. The reaction occurs readily in a 0.1 M H$_2$SO$_4$ solution, however, because there is a sufficient ion concentration to conduct electricity. Immediately, gas bubbles begin to appear at both electrodes.

The processes at the electrodes are:

Anode:	$2H_2O(l) \longrightarrow O_2(g) + 4H^+(aq) + 4e^-$
Cathode:	$4H^+(aq) + 4e^- \longrightarrow 2H_2(g)$
Overall:	$2H_2O(l) \longrightarrow O_2(g) + 2H_2(g)$

Note that there is no net consumption of the acid.

> **Student Note:** Remember that at 25°C pure water has only a very low concentration of ions [◄ Section 16.2]:
> $$[H^+] = [OH^-] = 1 \times 10^{-7} \, M$$

Electrolysis of an Aqueous Sodium Chloride Solution

An aqueous sodium chloride solution is the most complicated of the three examples of electrolysis considered here because NaCl(*aq*) contains several species that could be oxidized and reduced. The reductions that might occur at the cathode are:

$$2H^+(aq) + 2e^- \longrightarrow H_2(g) \qquad\qquad E° = 0.00 \text{ V}$$

$$2H_2O(l) + 2e^- \longrightarrow H_2(g) + 2OH^-(aq) \qquad E° = -0.83 \text{ V}$$

or:

$$Na^+(aq) + e^- \longrightarrow Na(s) \qquad\qquad E° = -2.71 \text{ V}$$

We can rule out the reduction of Na^+ ion because of the large negative $E°$ value. Under standard-state conditions, the reduction of H^+ is more apt to occur than the reduction of H_2O. However, in a solution of NaCl, the H^+ concentration is very low, making the reduction of H_2O the more probable reaction at the cathode.

The oxidation reactions that might occur at the anode are:

$$2Cl^-(aq) \longrightarrow Cl_2(g) + 2e^-$$

or

$$2H_2O(l) \longrightarrow O_2(g) + 4H^+(aq) + 4e^-$$

Referring to Table 19.1, we find:

$$Cl_2(g) + 2e^- \longrightarrow 2Cl^-(aq) \qquad E° = 1.36 \text{ V}$$

$$O_2(g) + 4H^+(aq) + 4e^- \longrightarrow 2H_2O(l) \qquad E° = 1.23 \text{ V}$$

The standard reduction potentials of the two reactions are not very different, but the values do suggest that the oxidation of H_2O *should* occur more readily. However, by experiment we find that the gas produced at the anode is Cl_2, not O_2. In the study of electrolytic processes, we sometimes find that the voltage required for a reaction is considerably higher than the electrode potentials would indicate. The ***overvoltage*** is the difference between the calculated voltage and the actual voltage required to cause electrolysis. The overvoltage for O_2 formation is quite high. Under normal operating conditions, therefore, Cl_2 gas forms at the anode instead of O_2.

Thus, the half-cell reactions in the electrolysis of aqueous sodium chloride are:

Anode (oxidation): $\qquad 2Cl^-(aq) \longrightarrow Cl_2(g) + 2e^-$

Cathode (reduction): $\qquad 2H_2O(l) + 2e^- \longrightarrow H_2(g) + 2OH^-(aq)$

Overall: $\qquad 2H_2O(l) + 2Cl^-(aq) \longrightarrow H_2(g) + Cl_2(g) + 2OH^-(aq)$

Student Note: In the electrolysis of aqueous solutions, the water itself may be oxidized and/or reduced.

As the overall reaction shows, the concentration of the Cl^- ions decreases during electrolysis and that of the OH^- ions increases. Therefore, in addition to H_2 and Cl_2, the useful by-product NaOH can be obtained by evaporating the aqueous solution at the end of the electrolysis.

Electrolysis has many important applications in industry, mainly in the extraction and purification of metals. We discuss some of these applications in Chapter 23.

Quantitative Applications of Electrolysis

The quantitative treatment of electrolysis was developed primarily by Faraday. He observed that the mass of product formed (or reactant consumed) at an electrode was proportional to both the amount of electricity transferred at the electrode and the molar mass of the substance being produced (or consumed). In the electrolysis of molten NaCl, for example, the cathode reaction tells us that one Na atom is produced when one Na^+ ion accepts an electron from the electrode. To reduce 1 mole of Na^+ ions, we must supply an Avogadro's number (6.02×10^{23}) of electrons to the cathode. On the other hand, stoichiometry tells us that it takes 2 moles of electrons to reduce 1 mole of Mg^{2+} ions and 3 moles of electrons to reduce 1 mole of Al^{3+} ions:

$$Na^+ + e^- \longrightarrow Na$$

$$Mg^{2+} + 2e^- \longrightarrow Mg$$

$$Al^{3+} + 3e^- \longrightarrow Al$$

In an electrolysis experiment, we generally measure the current (in amperes) that passes through an electrolytic cell in a given period of time. The relationship between charge (in coulombs) and the current is:

$$1 \, C = 1 \, A \times 1 \, s$$

That is, a coulomb is the quantity of electric charge passing any point in the circuit in 1 s when the current is 1 A. Therefore, if we know the current (in amperes) and how long it is applied (in seconds), we can calculate the charge (in coulombs). Knowing the charge enables us to determine the number of moles of electrons. And knowing the number of moles of electrons allows us to use stoichiometry to determine the number of moles of product. Figure 19.12 shows the steps involved in calculating the quantities of substances produced in electrolysis.

To illustrate this approach, consider an electrolytic cell in which molten $CaCl_2$ is separated into its constituent elements, Ca and Cl_2. Suppose a current of 0.452 A is passed through the cell for 1.50 h. How much product will be formed at each electrode? The first step is to determine which species will be oxidized at the anode and which species will be reduced at the cathode. Here the choice is straightforward because we have only Ca^{2+} and Cl^- ions. The cell reactions are:

Anode (oxidation):	$2Cl^-(aq) \longrightarrow Cl_2(g) + 2e^-$
Cathode (reduction):	$Ca^{2+}(l) + 2e^- \longrightarrow Ca(l)$
Overall:	$Ca^{2+}(l) + 2Cl^-(l) \longrightarrow Ca(l) + Cl_2(g)$

The quantities of calcium metal and chlorine gas formed depend on the number of electrons that pass through the electrolytic cell, which in turn depends on *charge*, or current × time:

$$\text{coulombs} = 0.452 \, A \times 1.50 \, h \times \frac{3600 \, s}{1 \, h} \times \frac{1 \, C}{1 \, A \cdot s} = 2.441 \times 10^3 \, C$$

Because 1 mol e^- = 96,500 C and 2 mol e^- are required to reduce 1 mole of Ca^{2+} ions, the mass of Ca metal formed at the cathode is calculated as follows:

$$\text{grams Ca} = (2.441 \times 10^3 \, C)\left(\frac{1 \, \text{mol} \, e^-}{96,500 \, C}\right)\left(\frac{1 \, \text{mol Ca}}{2 \, \text{mol} \, e^-}\right)\left(\frac{40.08 \, \text{g Ca}}{1 \, \text{mol Ca}}\right) = 0.507 \, \text{g Ca}$$

The anode reaction indicates that 1 mol of chlorine is produced per 2 mol e^- of electricity. Hence, the mass of chlorine gas formed is:

$$\text{grams Cl}_2 = (2.441 \times 10^3 \, C)\left(\frac{1 \, \text{mol} \, e^-}{96,500 \, C}\right)\left(\frac{1 \, \text{mol Cl}_2}{2 \, \text{mol} \, e^-}\right)\left(\frac{70.90 \, \text{g Cl}_2}{1 \, \text{mol Cl}_2}\right) = 0.897 \, \text{g Cl}_2$$

Sample Problem 19.8 applies this approach to electrolysis in an aqueous solution.

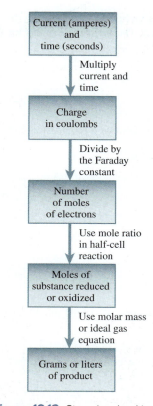

Figure 19.12 Steps involved in calculating amounts of substances reduced or oxidized in electrolysis.

SAMPLE PROBLEM (19.8)

A current of 1.26 A is passed through an electrolytic cell containing a dilute sulfuric acid solution for 7.44 h. Write the half-cell reactions, and calculate the volume of gases generated at STP.

Strategy As shown in Figure 19.12, we can use current and time to determine charge. We can then convert charge to moles of electrons, and use the balanced half-reactions to determine how many moles of product form at each electrode. Finally, we can convert moles to volume.

(Continued on next page)

Setup The half-cell reactions for the electrolysis of water are:

$$\text{Anode:} \qquad 2H_2O(l) \longrightarrow O_2(g) + 4H^+(aq) + 4e^-$$

$$\text{Cathode:} \quad 4H^+(aq) + 4e^- \longrightarrow 2H_2(g)$$

$$\text{Overall:} \qquad \overline{2H_2O(l) \longrightarrow O_2(g) + 2H_2(g)}$$

Remember that STP for gases means 273 K and 1 atm.

Solution

$$\text{coulombs} = (1.26 \text{ A})(7.44 \text{ h})\left(\frac{3600 \text{ s}}{1 \text{ h}}\right)\left(\frac{1 \text{ C}}{1 \text{ A} \cdot \text{s}}\right) = 3.375 \times 10^4 \text{ C}$$

At the anode:

$$\text{moles O}_2 = (3.375 \times 10^4 \text{ C})\left(\frac{1 \text{ mol } e^-}{96,500 \text{ C}}\right)\left(\frac{1 \text{ mol O}_2}{4 \text{ mol } e^-}\right) = 0.0874 \text{ mol O}_2$$

The volume of 0.0874 mol O_2 at STP is given by:

$$V = \frac{nRT}{P}$$

$$= \frac{(0.0874 \text{ mol})(0.08206 \text{ L} \cdot \text{atm/K} \cdot \text{mol})(273.15)}{1 \text{ atm}} = 1.96 \text{ L O}_2$$

Similarly, for hydrogen we write:

$$\text{moles H}_2 \, (3.375 \times 10^4 \text{ C})\left(\frac{1 \text{ mol } e^-}{96,500 \text{ C}}\right)\left(\frac{1 \text{ mol H}_2}{2 \text{ mol } e^-}\right) = 0.175 \text{ mol H}_2$$

The volume of 0.175 mol H_2 at STP is given by:

$$V = \frac{nRT}{P}$$

$$= \frac{(0.175 \text{ mol})(0.08206 \text{ L} \cdot \text{atm/K} \cdot \text{mol})(273.15)}{1 \text{ atm}} = 3.92 \text{ L H}_2$$

THINK ABOUT IT

The volume of H_2 is twice that of O_2 (see Figure 19.11), which is what we would expect based on Avogadro's law (at the same temperature and pressure, volume is directly proportional to the number of moles of gas: $V \propto n$) [◄◄ Section 10.2].

Practice Problem A TTEMPT A constant current of 0.912 A is passed through an electrolytic cell containing molten $MgCl_2$ for 18 h. What mass of Mg is produced?

Practice Problem B UILD A constant current is passed through an electrolytic cell containing molten $MgCl_2$ for 12 h. If 4.83 L of Cl_2 (at STP) is produced at the anode, what is the current in amperes?

Practice Problem **C**ONCEPTUALIZE Diagram (i) shows the ions in an aqueous solution of sodium chloride. Which of the other diagrams [(ii)–(iv)] could represent the system after electrolysis? (Water molecules are not shown.)

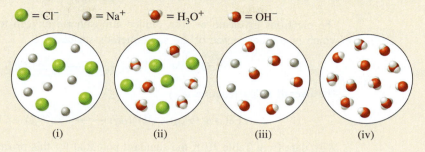

= Cl^- = Na^+ = H_3O^+ = OH^-

(i)   (ii)   (iii)   (iv)

CHECKPOINT – SECTION 19.7   Electrolysis

19.7.1 In the electrolysis of molten $CaCl_2$, a current of 1.12 A is passed through the cell for 3.0 h. What is the mass of Ca produced at the cathode?

a) 2.51 g

b) 1.26 g

c) 5.02 g

d) 10.0 g

e) 2.42×10^5 g

19.7.2 How long will a current of 0.995 A need to be passed through water (containing H_2SO_4) for 5.00 L of O_2 to be produced at STP?

a) 6.0 h

b) 8.2 h

c) 1.5 h

d) 3.0 h

e) 24.0 h

19.7.3 The diagram shows an electrolytic cell being powered by a galvanic cell. Identify each of the electrodes from left to right as an anode or a cathode.

a) Cathode, anode, anode, cathode

b) Cathode, anode, cathode, anode

c) Anode, cathode, anode, cathode

d) Anode, anode, cathode, cathode

e) Anode, cathode, cathode, anode

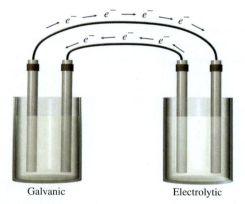

Galvanic   Electrolytic

19.8 Corrosion

The term *corrosion* generally refers to the deterioration of a metal by an electrochemical process. There are many examples of corrosion, including rust on iron, tarnish on silver, and the green layer that forms on copper and brass. In this section we discuss the processes involved in corrosion and some of the measures taken to prevent it.

The formation of rust on iron requires oxygen and water. Although the reactions involved are quite complex and not completely understood, the main steps are believed to be as follows. A region of the metal's surface serves as the anode, where the following oxidation occurs:

$$Fe(s) \longrightarrow Fe^{2+}(aq) + 2e^-$$

The electrons given up by iron reduce atmospheric oxygen to water at the cathode, which is another region of the same metal's surface:

$$O_2(g) + 4H^+(aq) + 4e^- \longrightarrow 2H_2O(l)$$

The overall redox reaction is:

$$2Fe(s) + O_2(g) + 4H^+(aq) \longrightarrow 2Fe^{2+}(g) + 2H_2O(l)$$

With data from Table 19.1, the standard potential for this process can be calculated as follows:

$$E°_{cell} = E°_{cathode} - E°_{anode}$$

$$= 1.23 \text{ V} - (-0.44 \text{ V})$$

$$= 1.67 \text{ V}$$

Note that this reaction occurs in an *acidic* medium; the H^+ ions are supplied in part by the reaction of atmospheric carbon dioxide with water to form the weak acid, carbonic acid (H_2CO_3).

The Fe^{2+} ions formed at the anode are further oxidized by oxygen as follows:

$$4Fe^{2+}(aq) + O_2(g) + (4 + 2x)H_2O(l) \longrightarrow 2Fe_2O_3 \cdot xH_2O(s) + 8H^+(aq)$$

This hydrated form of iron(III) oxide is known as rust. The amount of water associated with the iron(III) oxide varies, so we represent the formula as $Fe_2O_3 \cdot xH_2O$.

Figure 19.13 shows the mechanism of rust formation. The electric circuit is completed by the migration of electrons and ions; this is why rusting occurs so rapidly in saltwater. In cold climates, salts (NaCl or $CaCl_2$) spread on roadways to melt ice and snow are a major cause of rust formation on automobiles.

Other metals also undergo oxidation. Aluminum, for example, which is used to make airplanes, beverage cans, and aluminum foil, has a much greater tendency to oxidize than does iron. Unlike the corrosion of iron, though, corrosion of aluminum produces an insoluble layer of protective coating (Al_2O_3) that prevents the underlying metal from additional corrosion.

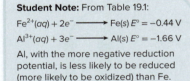

Student Note: From Table 19.1:

$Fe^{2+}(aq) + 2e^- \longrightarrow Fe(s)$ $E° = -0.44$ V

$Al^{3+}(aq) + 3e^- \longrightarrow Al(s)$ $E° = -1.66$ V

Al, with the more negative reduction potential, is less likely to be reduced (more likely to be oxidized) than Fe.

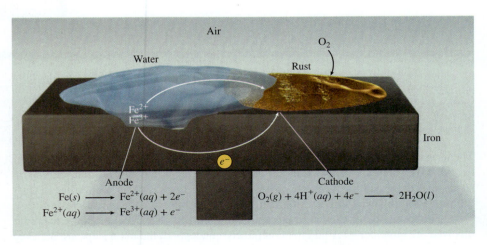

Figure 19.13 The electrochemical process involved in rust formation. The H^+ ions are supplied by H_2CO_3, which forms when CO_2 from air dissolves in water.

Coinage metals such as copper and silver also corrode, but much more slowly than either iron or aluminum:

$$Cu(s) \longrightarrow Cu^{2+}(aq) + 2e^-$$

$$Ag(s) \longrightarrow Ag^+(aq) + e^-$$

In ordinary atmospheric exposure, copper forms a layer of copper carbonate ($CuCO_3$), a green substance referred to as *patina,* that protects the metal underneath from further corrosion. Likewise, silverware that comes into contact with foodstuffs develops a layer of silver sulfide (Ag_2S).

A number of methods have been devised to protect metals from corrosion. Most of these methods are aimed at preventing rust formation. The most obvious approach is to coat the metal surface with paint to prevent exposure to the substances necessary for corrosion. If the paint is scratched or otherwise damaged, however, thus exposing even the smallest area of bare metal, rust will form under the paint layer. The surface of iron metal can be made inactive by a process called *passivation.* A thin oxide layer is formed when the metal is treated with a strong oxidizing agent such as concentrated nitric acid. A solution of sodium chromate is often added to cooling systems and radiators to prevent rust formation.

The tendency for iron to oxidize is greatly reduced when it is alloyed with certain other metals. For example, in stainless steel, an alloy of iron and chromium, a layer of chromium oxide forms that protects the iron from corrosion.

An iron container can be covered with a layer of another metal such as tin or zinc. A "tin" can is made by applying a thin layer of tin over iron. Rust formation is prevented as long as the tin layer remains intact. However, once the surface has been breached by a scratch or a dent, rusting occurs rapidly. If we look up the standard reduction potentials in Table 19.1, we find that when tin and iron are in contact with each other, tin, with its greater reduction potential, acts as a cathode. Iron acts as an anode and is therefore oxidized:

$$Sn^{2+}(aq) + 2e^- \longrightarrow Sn(s) \qquad E° = -0.14 \text{ V}$$

$$Fe^{2+}(aq) + 2e^- \longrightarrow Fe(s) \qquad E° = -0.44 \text{ V}$$

Zinc-plating, or **galvanization,** protects iron from corrosion by a different mechanism. According to Table 19.1:

$$Zn^{2+}(aq) + 2e^- \longrightarrow Zn(s) \qquad E° = -0.76 \text{ V}$$

so zinc is *more* easily oxidized than iron. Like aluminum, zinc oxidizes to form a protective coating. Even when the zinc layer is compromised, though, and the underlying iron is exposed, zinc is still the more easily oxidized of the two metals and will act as the anode. Iron will be the cathode, thereby remaining reduced.

Galvanization is one example of *cathodic protection,* the process by which a metal is protected by being made the cathode in what amounts to a galvanic cell. Another example is the use of zinc or magnesium bars to protect underground storage tanks and ships. When a steel tank or hull is connected to a more easily oxidized metal, corrosion of the steel is prevented.

Animation
Electrochemistry—chrome plating.

Chapter Summary

Section 19.1

- Redox reactions are those in which oxidation numbers change. *Half-reactions* are the separated oxidation and reduction reactions that make up the overall redox reaction.

- Redox equations can be balanced via the half-reaction method, which allows for the addition of H_2O to balance O, H^+ to balance H, and OH^- for reactions taking place in basic solution.

Section 19.2

- An electrochemical cell in which a spontaneous chemical reaction generates a flow of electrons through a wire is called a *galvanic cell.*

- Half-reactions in a galvanic cell take place in separate compartments called *half-cells.* Half-cells contain *electrodes* in solutions and are connected via an external wire and by a *salt bridge.*

- The electrode at which reduction occurs is called the *cathode;* the electrode at which oxidation occurs is called the *anode.*

- The difference in electric potential between the cathode and the anode is the *cell potential* (E_{cell}).

Section 19.3

- We use *standard reduction potentials* ($E°$) to calculate the standard cell voltage or *standard cell potential* ($E°_{cell}$).

- Half-cell potentials are measured relative to the *standard hydrogen electrode (SHE),* the half-reaction for which has an arbitrarily defined standard reduction potential of zero.

Section 19.4

- $E°_{cell}$ is related to the standard free-energy change ($\Delta G°$) and to the equilibrium constant, K. A positive $E°_{cell}$ corresponds to a negative ($\Delta G°$) value and a large K value.

Section 19.5

- E_{cell} under other than standard-state conditions is determined from $E°_{cell}$ and the reaction quotient, Q, using the *Nernst equation.*

- A *concentration cell* has the same type of electrode and the same ion in solution (at different concentrations) in the anode and cathode compartments.

Section 19.6

- *Batteries* are portable, self-contained sources of electric energy consisting of galvanic cells—or a series of galvanic cells.

- *Fuel cells* are not really batteries but also supply electric energy via a spontaneous redox reaction. Reactants must be supplied constantly for a fuel cell to operate.

Section 19.7

- *Electrolysis* is the use of electric energy to drive a nonspontaneous redox reaction. An electrochemical cell used for this purpose is called an *electrolytic cell.*

- The voltage that must actually be supplied to drive a nonspontaneous redox reaction is greater than the calculated amount because of *overvoltage.*

- Electrolysis is used to recharge lead storage batteries, separate compounds into their constituent elements, and separate and purify metals.

- We can calculate the amount of a substance produced in electrolysis if we know the current applied to the cell and the length of time for which it is applied.

Section 19.8

- *Corrosion* is the undesirable oxidation of metals.

- Corrosion can be prevented by coating the metal surface with paint, a less easily oxidized metal, or a more easily oxidized metal such as zinc.

- The use of a more easily oxidized metal is known as *cathodic protection,* wherein the metal being protected is made the cathode in a galvanic cell. *Galvanization* is the cathodic protection of iron or steel using zinc.

Key Words

Anode, 963	Corrosion, 993	Galvanic cell, 963	Overvoltage, 990
Battery, 984	Electrode, 963	Galvanization, 995	Salt bridge, 966
Cathode, 963	Electrolysis, 988	Half-cell, 963	Standard hydrogen electrode
Cell potential (E_{cell}), 966	Electrolytic cell, 988	Half-reaction, 960	(SHE), 967
Concentration cell, 980	Fuel cell, 986	Nernst equation, 978	Standard reduction potential ($E°$), 967

Key Equations

19.1 $E°_{cell} = E°_{cathode} - E°_{anode}$	Standard cell potential, $E°_{cell}$, is calculated by subtracting $E°_{red}$ of the anode from $E°_{red}$ of the cathode.
19.2 $\Delta G = -nFE_{cell}$	Free-energy change, ΔG, is calculated as the product of the number of electrons transferred in a redox reaction, n, the Faraday constant, F, and the cell potential, E_{cell}.
19.3 $\Delta G° = -nFE°_{cell}$	Standard free-energy change, $\Delta G°$, is calculated as the product of the number of electrons transferred in a redox reaction, n, the Faraday constant, F, and the standard cell potential, $E°_{cell}$.
19.4 $E°_{cell} = \dfrac{RT}{nF} \ln K$	Standard cell potential, $E°_{cell}$, is proportional to the natural log (ln) of the equilibrium constant, K.
19.5 $E°_{cell} = \dfrac{0.0592 \text{ V}}{n} \log K$ (at 25°C)	At 25°C, the relationship between $E°_{cell}$ and K is simplified.
19.6 $E = E° - \dfrac{RT}{nF} \ln Q$	At conditions other than standard, cell potential, E_{cell}, is proportional to the natural log (ln) of the reaction quotient, Q.
19.7 $E = E° - \dfrac{0.0592 \text{ V}}{n} \log Q$	At 25°C, the relationship between E_{cell} and Q is simplified.

Key Constant

Section 19.4 Faraday constant

$F = 96{,}485.3$ C/mol e^-	The Faraday constant is the charge (in Coulombs) on a mole of electrons. It is usually rounded to three significant figures, 96,500 C/mol e^-, and is used in the conversions among ΔG, E_{cell}, and Q; and among $\Delta G°$, $E°_{cell}$, and K.

Electrolysis of Metals

Electrolysis is used extensively in the processing and refining of metals. Converting between the current (and the time over which it is applied), and the amount of metal produced requires application of dimensional analysis [◄◄ Key Skills Chapter 1]. When we are given the current in amperes (A) and the time in seconds (s), we first convert to charge in coulombs (C):

| Current (A) | × | Time (s) | = | Charge (C) |

We then divide by Faraday's constant to get the number of moles of electrons:

$$\frac{\text{Charge (C)}}{96{,}500 \text{ C/mol } e^-} = \text{mol } e^-$$

Using the reduction half-reaction, we can determine the number of moles of electrons needed to reduce a mole of metal:

$$M^{n+} + ne^- \longrightarrow M \qquad \frac{1 \text{ mol metal}}{n \text{ mol } e^-}$$

Finally, we convert moles electrons to moles metal:

$$\text{mol } e^- \times \frac{1 \text{ mol metal}}{n \text{ mol } e^-} = \text{mol metal}$$

Consider the following example. A current of 2.09 A is applied to a solution containing chromium(III) nitrate for 2.10 h. We determine the number of moles of Cr deposited as follows:

$$2.09 \text{ A} \times 2.10 \text{ h} \times \frac{60 \text{ min}}{1 \text{ h}} \times \frac{60 \text{ s}}{1 \text{ min}} = 1.58 \times 10^4 \text{ C}$$

$$\frac{1.58 \times 10^4 \text{ C}}{96{,}500 \text{ C/mol } e^-} = 0.1637 \text{ mol } e^-$$

$$Cr^{3+} + 3e^- \longrightarrow Cr \qquad \frac{1 \text{ mol Cr}}{3 \text{ mol } e^-}$$

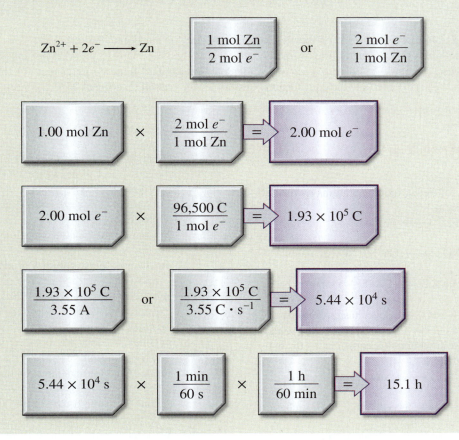

$$0.1637 \text{ mol } e^- \quad \times \quad \frac{1 \text{ mol Cr}}{3 \text{ mol } e^-} \quad = \quad 0.0546 \text{ mol Cr}$$

Conversely, we may know both the amount of metal to be produced and the current to be applied, and have to determine the amount of time required. If the goal is to produce 1.00 mol of zinc by electrolysis of a solution containing Zn^{2+} ions, and we apply a current of 3.55 A, we determine the time over which the current must be applied as follows.

The reduction half-reaction indicates that 2 moles of electrons are needed to reduce a mole of zinc:

$$Zn^{2+} + 2e^- \longrightarrow Zn \qquad \frac{1 \text{ mol Zn}}{2 \text{ mol } e^-} \quad \text{or} \quad \frac{2 \text{ mol } e^-}{1 \text{ mol Zn}}$$

$$1.00 \text{ mol Zn} \quad \times \quad \frac{2 \text{ mol } e^-}{1 \text{ mol Zn}} \quad = \quad 2.00 \text{ mol } e^-$$

$$2.00 \text{ mol } e^- \quad \times \quad \frac{96{,}500 \text{ C}}{1 \text{ mol } e^-} \quad = \quad 1.93 \times 10^5 \text{ C}$$

$$\frac{1.93 \times 10^5 \text{ C}}{3.55 \text{ A}} \quad \text{or} \quad \frac{1.93 \times 10^5 \text{ C}}{3.55 \text{ C} \cdot \text{s}^{-1}} \quad = \quad 5.44 \times 10^4 \text{ s}$$

$$5.44 \times 10^4 \text{ s} \quad \times \quad \frac{1 \text{ min}}{60 \text{ s}} \quad \times \quad \frac{1 \text{ h}}{60 \text{ min}} \quad = \quad 15.1 \text{ h}$$

Key Skills Problems

19.1

How much copper metal can be produced by electrolysis of a solution containing Cu^{2+} ions by a current of 1.85 A applied for exactly four hours?

(a) 0.276 mol   (b) 0.138 mol   (c) 0.552 mol
(d) 5.14×10^9 mol   (e) 2.00 mol

19.2

What mass of cadmium will be produced by electrolysis of a solution of $Cd(NO_3)_2$ when a current of 4.83 A is applied for six hours and 15 minutes?

(a) 63.3 g   (b) 127 g   (c) 253 g   (d) 31.6 g   (e) 57.2 g

19.3

Of the following aqueous solutions, identify the one that would yield the *smallest* mass of metal and the one that would yield the *largest* mass of metal when it is electrolyzed with a current of 2.00 A for 25 minutes.

$\qquad CuSO_4 \qquad AgNO_3 \qquad AuCl_3 \qquad ZnSO_4 \qquad Cr(NO_3)_3$

(a) $ZnSO_4$, $CuSO_4$ (b) $CuSO_4$, $ZnSO_4$ (c) $AgNO_3$, $Cr(NO_3)_3$
(d) $Cr(NO_3)_3$, $AuCl_3$ (e) $AuCl_3$, $AgNO_3$

19.4

When a current of 5.22 A is applied over 3.50 hours to a solution containing metal ions, 20.00 grams of metal are produced. Which of the following could be the metal ion in the solution?

(a) Zn^{2+} (b) Au^{3+} (c) Ag^+ (d) Ni^{2+} (e) Cu^{2+}

Questions and Problems

Applying What You've Learned

The half-reactions involved in the generation of electricity in a potato battery are:

$$\text{Cathode: } 2H^+(aq) + 2e^- \longrightarrow H_2(g)$$
$$\text{Anode: } Zn(s) \longrightarrow Zn^{2+}(aq) + 2e^-$$

Problems:
(a) Write the overall balanced equation for the reaction that occurs in the potato battery [◄◄ Sample Problem 19.1].
(b) Calculate the standard cell potential for the overall equation [◄◄ Sample Problem 19.2]. (c) What species is oxidized and what species is reduced [◄◄ Sample Problem 19.3]? (d) Calculate the standard free-energy change for the overall reaction [◄◄ Sample Problem 19.4]. (e) Calculate the equilibrium constant for the overall reaction at 25°C [◄◄ Sample Problem 19.5].

SECTION 19.1: BALANCING REDOX EQUATIONS

Problems

19.1 Balance the following redox equations by the half-reaction method:
(a) $H_2O_2 + Fe^{2+} \longrightarrow Fe^{3+} + H_2O$ (in acidic solution)
(b) $Cu + HNO_3 \longrightarrow Cu^{2+} + NO$ (in acidic solution)
(c) $CN^- + MnO_4^- \longrightarrow CNO^- + MnO_2$ (in basic solution)
(d) $Br_2 \longrightarrow BrO_3^- + Br^-$ (in basic solution)
(e) $S_2O_3^{2-} + I_2 \longrightarrow I^- + S_4O_6^{2-}$ (in acidic solution)

19.2 Balance the following redox equations by the half-reaction method:
(a) $Mn^{2+} + H_2O_2 \longrightarrow MnO_2 + H_2O$ (in basic solution)
(b) $Bi(OH)_3 + SnO_2^{2-} \longrightarrow SnO_3^{2-} + Bi$ (in basic solution)
(c) $Cr_2O_7^{2-} + C_2O_4^{2-} \longrightarrow Cr^{3+} + CO_2$ (in acidic solution)
(d) $ClO_3^- + Cl^- \longrightarrow Cl_2 + ClO_2$ (in acidic solution)
(e) $Mn^{2+} + BiO_3^- \longrightarrow Bi^{3+} + MnO_4^-$ (in acidic solution)

SECTION 19.2: GALVANIC CELLS

 Visualizing Chemistry
Figure 19.1

VC 19.1 In the first scene of the animation, when a zinc bar is immersed in an aqueous copper(II) sulfate solution, solid copper deposits on the bar. What reaction would take place if a *copper* bar were immersed in an aqueous *zinc* sulfate solution?
a) No reaction would take place.
b) Solid copper would still deposit on the bar.
c) Solid zinc would deposit on the bar.

VC 19.2 What causes the change in the potential of the galvanic cell in Figure 19.1 as the cell operates?
a) Changes in the sizes of the zinc and copper electrodes.
b) Changes in the concentrations of zinc and copper ions.
c) Changes in the volumes of solutions in the half-cells.

VC 19.3 Why does the color of the blue solution in the galvanic cell (Figure 19.1) fade as the cell operates?
a) Blue Cu^{2+} ions are replaced by colorless Zn^{2+} ions solution.
b) Blue Cu^{2+} ions are removed from solution by reduction.
c) Blue Cu^{2+} ions are removed from solution by oxidation.

VC 19.4 What happens to the mass of the copper electrode in the galvanic cell in Figure 19.1 as the cell operates?
a) It increases.
b) It decreases.
c) It does not change.

Review Questions

19.3 Define the following terms: *anode, cathode, cell voltage, electromotive force, standard reduction potential.*

19.4 Describe the basic features of a galvanic cell. Why are the two components of the cell separated from each other?

19.5 What is the function of a salt bridge? What kind of electrolyte should be used in a salt bridge?

19.6 What is a cell diagram? Write the cell diagram for a galvanic cell consisting of an Al electrode placed in a 1 M Al(NO$_3$)$_3$ solution and an Ag electrode placed in a 1 M AgNO$_3$ solution.

19.7   What is the difference between the half-reactions discussed in redox processes in Chapter 4 and the half-cell reactions discussed in Section 19.2?

SECTION 19.3: STANDARD REDUCTION POTENTIALS

Review Questions

19.8   Discuss the spontaneity of an electrochemical reaction in terms of its standard emf (E_{cell}°).

19.9   After operating the galvanic cell in Figure 19.1 for a few minutes, the cell potential decreases. Explain.

Computational Problems

19.10   Calculate the standard emf of a cell that uses the Mg/Mg^{2+} and Cu/Cu^{2+} half-cell reactions at 25°C. Write the equation for the cell reaction that occurs under standard-state conditions.

19.11   Calculate the standard emf of a cell that uses Ag/Ag$^+$ and Al/Al^{3+} half-cell reactions. Write the cell reaction that occurs under standard-state conditions.

Conceptual Problems

19.12   Predict whether Fe^{3+} can oxidize I$^-$ to I$_2$ under standard-state conditions.

19.13   Which of the following reagents can oxidize H$_2$O to O$_2$(g) under standard-state conditions: H$^+$(aq), Cl$^-$(aq), Cl$_2$(g), Cu^{2+}(aq), Pb^{2+}(aq), MnO$_4^-$(aq) (in acid)?

19.14   Consider the following half-reactions:

$$\text{MnO}_4^-(aq) + 8\text{H}^+(aq) + 5e^- \longrightarrow \text{Mn}^{2+}(aq) + 4\text{H}_2\text{O}(l)$$

$$\text{NO}_3^-(aq) + 4\text{H}^+(aq) + 3e^- \longrightarrow \text{NO}(g) + 2\text{H}_2\text{O}(l)$$

Predict whether NO$_3^-$ ions will oxidize Mn^{2+} to MnO$_4^-$ under standard-state conditions.

19.15   Predict whether the following reactions would occur spontaneously in aqueous solution at 25°C. Assume that the initial concentrations of dissolved species are all 1.0 M.
(a) Ca(s) + Cd^{2+}(aq) $\longrightarrow$ Ca^{2+}(aq) + Cd(s)
(b) 2Br$^-$(aq) + Sn^{2+}(aq) $\longrightarrow$ Br$_2$(l) + Sn(s)
(c) 2Ag(s) + Ni^{2+}(aq) $\longrightarrow$ 2Ag$^+$(aq) + Ni(s)
(d) Cu$^+$(aq) + Fe^{3+}(aq) $\longrightarrow$ Cu^{2+}(aq) + Fe^{2+}(aq)

19.16   Which species in each pair is a better oxidizing agent under standard-state conditions: (a) Br$_2$ or Au^{3+}, (b) H$_2$ or Ag$^+$, (c) Cd^{2+} or Cr^{3+}, (d) O$_2$ in acidic media or O$_2$ in basic media?

19.17   Which species in each pair is a better reducing agent under standard-state conditions: (a) Na or Li, (b) H$_2$ or I$_2$, (c) Fe^{2+} or Ag, (d) Br$^-$ or Co^{2+}?

SECTION 19.4: SPONTANEITY OF REDOX REACTIONS UNDER STANDARD-STATE CONDITIONS

Review Questions

19.18   Use the information in Table 2.1, and calculate the Faraday constant.

19.19   Write the equations relating ΔG° and K to the standard emf of a cell. Define all the terms.

19.20   Compare the ease of measuring the equilibrium constant electrochemically with that by chemical means.

Computational Problems

19.21   What is the equilibrium constant for the following reaction at 25°C?

$$\text{Mg}(s) + \text{Zn}^{2+}(aq) \rightleftharpoons \text{Mg}^{2+}(aq) + \text{Zn}(s)$$

19.22   The equilibrium constant for the reaction:

$$\text{Sr}(s) + \text{Mg}^{2+}(aq) \rightleftharpoons \text{Sr}^{2+}(aq) + \text{Mg}(s)$$

is 2.69×10^{12} at 25°C. Calculate E° for a cell made up of Sr/Sr^{2+} and Mg/Mg^{2+} half-cells.

19.23   Use the standard reduction potentials to find the equilibrium constant for each of the following reactions at 25°C:
(a) Br$_2$(l) + 2I$^-$(aq) $\rightleftharpoons$ 2Br$^-$(aq) + I$_2$(s)
(b) 2Ce^{4+}(aq) + 2Cl$^-$(aq) $\rightleftharpoons$ Cl$_2$(g) + 2Ce^{3+}(aq)
(c) 5Fe^{2+}(aq) + MnO$_4^-$(aq) + 8H$^+$(aq) $\rightleftharpoons$ Mn^{2+}(aq) + 4H$_2$O(l) + 5Fe^{3+}(aq)

19.24   Calculate ΔG° and K_c for the following reactions at 25°C:
(a) Mg(s) + Pb^{2+}(aq) $\rightleftharpoons$ Mg^{2+}(aq) + Pb(s)
(b) O$_2$(g) + 4H$^+$(aq) + 4Fe^{2+}(aq) $\rightleftharpoons$ 2H$_2$O(l) + 4Fe^{3+}(aq)
(c) 2Al(s) + 3I$_2$(s) $\rightleftharpoons$ 2Al^{3+}(aq) + 6I$^-$(aq)

19.25   Under standard-state conditions, what spontaneous reaction will occur in aqueous solution among the ions Ce^{4+}, Ce^{3+}, Co^{3+}, and Co^{2+}? Calculate ΔG° and K_c for the reaction.

19.26   Given that $E^\circ = 0.52$ V for the reduction Cu$^+$(aq) + e^- $\longrightarrow$ Cu(s), calculate E°, ΔG°, and K for the following reaction at 25°C:

$$2\text{Cu}^+(aq) \rightleftharpoons \text{Cu}^{2+}(aq) + \text{Cu}(s)$$

SECTION 19.5: SPONTANEITY OF REDOX REACTIONS UNDER CONDITIONS OTHER THAN STANDARD STATE

Review Questions

19.27   Write the Nernst equation, and explain all the terms.

19.28   Write the Nernst equation for the following processes at some temperature T:
(a) Mg(s) + Sn^{2+}(aq) $\rightleftharpoons$ Mg^{2+}(aq) + Sn(s)
(b) 2Cr(s) + 3Pb^{2+}(aq) $\rightleftharpoons$ 2Cr^{3+}(aq) + 3Pb(s)

Computational Problems

19.29   What is the potential of a cell made up of Zn/Zn^{2+} and Cu/Cu^{2+} half-cells at 25°C if $[Zn^{2+}] = 0.25\ M$ and $[Cu^{2+}] = 0.15\ M$?

19.30   Calculate $E°$, E, and ΔG for the following cell reactions.
(a) $Mg(s) + Sn^{2+}(aq) \rightleftharpoons Mg^{2+}(aq) + Sn(s)$
    $[Mg^{2+}] = 0.045\ M$, $[Sn^{2+}] = 0.035\ M$
(b) $3Zn(s) + 2Cr^{3+}(aq) \rightleftharpoons 3Zn^{2+}(aq) + 2Cr(s)$
    $[Cr^{3+}] = 0.010\ M$, $[Zn^{2+}] = 0.0085\ M$

19.31   Calculate the standard potential of the cell consisting of a Cd/Cd^{2+} half-cell and the SHE. What will the emf of the cell be if $[Cd^{2+}] = 0.96\ M$, $P_{H_2} = 1.50$ atm, and $[H^+] = 0.080\ M$?

19.32   What is the emf of a cell consisting of a Pb^{2+}/Pb half-cell and a $Pt/H^+/H_2$ half-cell if $[Pb^{2+}] = 0.10\ M$, $[H^+] = 0.050\ M$, and $P_{H_2} = 1.0$ atm?

19.33   Referring to the arrangement in Figure 19.1, calculate the $[Cu^{2+}]/[Zn^{2+}]$ ratio at which the following reaction is spontaneous at 25°C:

$$Cu(s) + Zn^{2+}(aq) \longrightarrow Cu^{2+}(aq) + Zn(s)$$

19.34   Calculate the emf of the following concentration cell:

$$Mg(s) \mid Mg^{2+}(0.24\ M) \parallel Mg^{2+}(0.53\ M) \mid Mg(s)$$

SECTION 19.6: BATTERIES

Review Questions

19.35   What is a battery? Describe several types of batteries.

19.36   Explain the differences between a primary galvanic cell—one that is not rechargeable—and a storage cell (e.g., the lead storage battery), which is rechargeable.

19.37   Discuss the advantages and disadvantages of fuel cells over conventional power plants in producing electricity.

Problems

19.38   The hydrogen-oxygen fuel cell is described in Section 19.6. (a) What volume of $H_2(g)$, stored at 25°C at a pressure of 155 atm, would be needed to run an electric motor drawing a current of 8.5 A for 3.0 h? (b) What volume (in liters) of air at 25°C and 1.00 atm will have to pass into the cell per minute to run the motor? Assume that air is 20 percent O_2 by volume and that all the O_2 is consumed in the cell. The other components of air do not affect the fuel-cell reactions. Assume ideal gas behavior.

19.39   Calculate the standard emf of the propane fuel cell (discussed at the end of Section 19.6) at 25°C, given that $\Delta G_f°$ for propane is -23.5 kJ/mol.

SECTION 19.7: ELECTROLYSIS

Review Questions

19.40   What is the difference between a *galvanic* cell and an *electrolytic* cell?

19.41   What is Faraday's contribution to quantitative electrolysis?

19.42   Define the term *overvoltage*. How does overvoltage affect electrolytic processes?

Computational Problems

19.43   The half-reaction at an electrode is:

$$Mg^{2+}(\text{molten}) + 2e^- \longrightarrow Mg(s)$$

Calculate the number of grams of magnesium that can be produced by supplying 1.00 F to the electrode.

19.44   Consider the electrolysis of molten barium chloride ($BaCl_2$). (a) Write the half-reactions. (b) How many grams of barium metal can be produced by supplying 0.50 A for 30 min?

19.45   Considering only the cost of electricity, would it be cheaper to produce a ton of sodium or a ton of zinc by electrolysis?

19.46   If the cost of electricity to produce magnesium by the electrolysis of molten magnesium chloride is $155 per ton of metal, what is the cost (in dollars) of the electricity necessary to produce (a) 10.0 tons of aluminum, (b) 30.0 tons of sodium, and (c) 50.0 tons of calcium?

19.47   One of the half-reactions for the electrolysis of water is:

$$2H_2O(l) \longrightarrow O_2(g) + 4H^+(aq) + 4e^-$$

If 0.076 L of O_2 is collected at 25°C and 755 mmHg, how many faradays of electricity had to pass through the solution?

19.48   How many faradays of electricity are required to produce (a) 0.84 L of O_2 at exactly 1 atm and 25°C from aqueous H_2SO_4 solution, (b) 1.50 L of Cl_2 at 750 mmHg and 20°C from molten NaCl, and (c) 6.0 g of Sn from molten $SnCl_2$?

19.49   Calculate the amounts of Cu and Br_2 produced in 1.25 h at inert electrodes in a solution of $CuBr_2$ by a current of 3.75 A.

19.50   In the electrolysis of an aqueous $AgNO_3$ solution, 0.67 g of Ag is deposited after a certain period of time. (a) Write the half-reaction for the reduction of Ag^+. (b) What is the probable oxidation half-reaction? (c) Calculate the quantity of electricity used (in coulombs).

19.51   A steady current was passed through molten $CoSO_4$ until 10.2 g of metallic cobalt was produced. Calculate the number of coulombs of electricity used.

19.52   A constant electric current flows for 3.75 h through two electrolytic cells connected in series. One contains a solution of $AgNO_3$ and the second a solution of $CuCl_2$. During this time, 2.00 g of silver is deposited in the first cell. (a) How many grams of copper are deposited in the second cell? (b) What is the current flowing (in amperes)?

19.53 What is the hourly production rate of chlorine gas (in kg) from an electrolytic cell using aqueous NaCl electrolyte and carrying a current of 1.500×10^3 A? The anode efficiency for the oxidation of Cl^- is 93.0 percent.

19.54 Chromium plating is applied by electrolysis to objects suspended in a dichromate solution, according to the following (unbalanced) half-reaction:

$$Cr_2O_7^{2-}(aq) + e^- + H^+(aq) \longrightarrow Cr(s) + H_2O(l)$$

How long (in hours) would it take to apply a chromium plating 1.0×10^{-2} mm thick to a car bumper with a surface area of 0.25 m^2 in an electrolytic cell carrying a current of 25.0 A? (The density of chromium is 7.19 g/cm^3.)

19.55 The passage of a current of 0.750 A for 25.0 min deposited 0.369 g of copper from a $CuSO_4$ solution. From this information, calculate the molar mass of copper.

19.56 A quantity of 0.300 g of copper was deposited from a $CuSO_4$ solution by passing a current of 3.00 A through the solution for 304 s. Calculate the value of the Faraday constant.

19.57 In a certain electrolysis experiment, 1.44 g of Ag were deposited in one cell (containing an aqueous $AgNO_3$ solution), while 0.120 g of an unknown metal X was deposited in another cell (containing an aqueous XCl_3 solution) in series with the $AgNO_3$ cell. Calculate the molar mass of X.

19.58 One of the half-reactions for the electrolysis of water is:

$$2H^+(aq) + 2e^- \longrightarrow H_2(g)$$

If 0.845 L of H_2 is collected at 25°C and 782 mmHg, how many faradays of electricity had to pass through the solution?

SECTION 19.8: CORROSION

Review Questions

19.59 Steel hardware, including nuts and bolts, is often coated with a thin plating of cadmium. Explain the function of the cadmium layer.

19.60 "Galvanized iron" is steel sheet that has been coated with zinc; "tin" cans are made of steel sheet coated with tin. Discuss the functions of these coatings and the electrochemistry of the corrosion reactions that occur if an electrolyte contacts the scratched surface of a galvanized iron sheet or a tin can.

19.61 Tarnished silver contains Ag_2S. The tarnish can be removed by placing silverware in an aluminum pan containing an inert electrolyte solution, such as NaCl. Explain the electrochemical principle for

this procedure. [The standard reduction potential for the half-cell reaction $Ag_2S(s) + 2e^- \longrightarrow 2Ag(s) + S^{2-}(aq)$ is -0.71 V.]

19.62 How does the tendency of iron to rust depend on the pH of the solution?

ADDITIONAL PROBLEMS

19.63 For each of the following redox reactions, (i) write the half-reactions, (ii) write a balanced equation for the whole reaction, (iii) determine in which direction the reaction will proceed spontaneously under standard-state conditions:
(a) $H_2(g) + Ni^{2+}(aq) \longrightarrow H^+(aq) + Ni(s)$
(b) $MnO_4^-(aq) + Cl^-(aq) \longrightarrow Mn^{2+}(aq) + Cl_2(g)$
    (in acid solution)
(c) $Cr(s) + Zn^{2+}(aq) \longrightarrow Cr^{3+}(aq) + Zn(s)$

19.64 The oxidation of 25.0 mL of a solution containing Fe^{2+} requires 26.0 mL of 0.0250 M $K_2Cr_2O_7$ in acidic solution. Balance the following equation, and calculate the molar concentration of Fe^{2+}:

$$Cr_2O_7^{2-} + Fe^{2+} + H^+ \longrightarrow Cr^{3+} + Fe^{3+}$$

19.65 As discussed in Section 19.5, the potential of a concentration cell diminishes as the cell operates and the concentrations in the two compartments approach each other. When the concentrations in both compartments are the same, the cell ceases to operate. At this stage, is it possible to generate a cell potential by adjusting a parameter other than concentration? Explain.

19.66 A sample of iron ore weighing 0.2792 g was dissolved in an excess of a dilute acid solution. All the iron was first converted to Fe(II) ions. The solution then required 23.30 mL of 0.0194 M $KMnO_4$ for oxidation to Fe(III) ions. Calculate the percent by mass of iron in the ore.

19.67 The composition of the metal used to mint pennies has varied over the years. Originally 100 percent copper, pennies have been minted from a copper-nickel alloy, a copper-zinc alloy (bronze), zinc-plated steel, and copper-plated zinc. Modern pennies (post 1982) are copper-plated zinc. A popular chemistry demonstration involves dissolving the zinc core of a penny in a strongly acidic aqueous solution. For the demonstration to work, the copper coating on the zinc must be breached before the penny is submerged in the solution. Explain why this is necessary.

19.68 The first diagram on page 1004 represents a galvanic cell constructed using an SHE and an M/M^{3+} half-cell, where M is a metal. Identify which of the other diagrams [(i)–(iv)] best represents a close-up view of the two electrodes as the cell operates (a) if M is Al and (b) if M is Au. (Tiny black spheres represent reduced metal.)

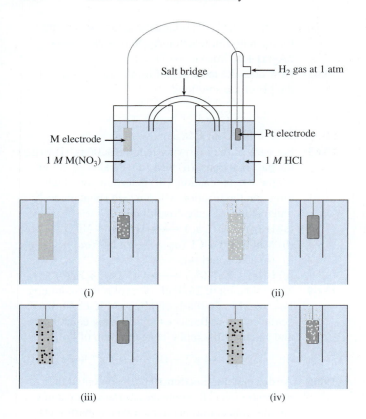

Salt bridge

H$_2$ gas at 1 atm

M electrode

Pt electrode

1 *M* M(NO$_3$)

1 *M* HCl

(i)   (ii)

(iii)   (iv)

19.69 Based on the following standard reduction potentials:

$$Fe^{2+}(aq) + 2e^- \longrightarrow Fe(s) \qquad E_1^\circ = -0.44 \text{ V}$$

$$Fe^{3+}(aq) + e^- \longrightarrow Fe^{2+}(aq) \qquad E_2^\circ = 0.77 \text{ V}$$

calculate the standard reduction potential for the half-reaction:

$$Fe^{3+}(aq) + 3e^- \longrightarrow Fe(s) \qquad E_3^\circ = ?$$

19.70 Complete the following table. State whether the cell reaction is spontaneous, nonspontaneous, or at equilibrium.

E	Δ*G*	**Cell Reaction**
> 0		
	> 0	
= 0		

19.71 From the following information, calculate the solubility product of AgBr:

$$Ag^+(aq) + e^- \longrightarrow Ag(s) \qquad E^\circ = 0.80 \text{ V}$$

$$AgCl(s) + e^- \longrightarrow Ag(s) + Cl^-(aq) \qquad E^\circ = 0.22 \text{ V}$$

19.72 Consider a galvanic cell composed of the SHE and a half-cell using the reaction $Ag^+(aq) + e^- \longrightarrow$ $Ag(s)$. (a) Calculate the standard cell potential. (b) What is the spontaneous cell reaction under standard-state conditions? (c) Calculate the cell potential when [H$^+$] in the hydrogen electrode is

changed to (i) 1.0×10^{-2} *M* and (ii) 1.0×10^{-5} *M* all other reagents being held at standard-state conditions. (d) Based on this cell arrangement, suggest a design for a pH meter.

19.73 A galvanic cell consists of a silver electrode in contact with 346 mL of 0.100 *M* AgNO$_3$ solution and a magnesium electrode in contact with 288 mL of 0.100 *M* Mg(NO$_3$)$_2$ solution. (a) Calculate *E* for the cell at 25°C. (b) A current is drawn from the cell until 1.20 g of silver has been deposited at the silver electrode. Calculate *E* for the cell at this stage of operation.

19.74 Calculate the equilibrium constant for the following reaction at 298 K:

$$Zn(s) + Cu^{2+}(aq) \rightleftharpoons Zn^{2+}(aq) + Cu(s)$$

19.75 Calculate the emf of the following concentration cell at 25°C:

$$Cu(s) \mid Cu^{2+}(0.080 \text{ M}) \parallel Cu^{2+}(1.2 \text{ M}) \mid Cu(s)$$

19.76 The cathode reaction in the Leclanché cell is given by:

$$2MnO_2(s) + Zn^{2+}(aq) + 2e^- \longrightarrow ZnMn_2O_4(s)$$

If a Leclanché cell produces a current of 0.0050 A, calculate how many hours this current supply will last if there is initially 4.0 g of MnO$_2$ present in the cell. Assume that there is an excess of Zn^{2+} ions.

19.77 For a number of years, it was not clear whether mercury(I) ions existed in solution as Hg$^+$ or as Hg$_2^{2+}$. To distinguish between these two possibilities, we could set up the following system:

$$Hg(l) \mid \text{soln A} \parallel \text{soln B} \mid Hg(l)$$

where soln A contained 0.263 g mercury(I) nitrate per liter and soln B contained 2.63 g mercury(I) nitrate per liter. If the measured emf of such a cell is 0.0289 V at 18°C, what can you deduce about the nature of the mercury(I) ions?

19.78 An aqueous KI solution to which a few drops of phenolphthalein have been added is electrolyzed using an apparatus like the one shown here:

David A. Tietz/Editorial Image, LLC

Describe what you would observe at the anode and the cathode. (*Hint:* Molecular iodine is only slightly soluble in water, but in the presence of I$^-$ ions, it forms the brownish-yellow color of I$_3^-$ ions.)

19.79 A piece of magnesium metal weighing 1.56 g is placed in 100.0 mL of 0.100 M AgNO$_3$ at 25°C. Calculate [Mg^{2+}] and [Ag$^+$] in solution at equilibrium. What is the mass of the magnesium left? The volume remains constant.

19.80 Describe an experiment that would enable you to determine which is the cathode and which is the anode in a galvanic cell using copper and zinc electrodes.

19.81 An acidified solution was electrolyzed using copper electrodes. A constant current of 1.18 A caused the anode to lose 0.584 g after 1.52×10^3 s. (a) What is the gas produced at the cathode, and what is its volume at STP? (b) Given that the charge of an electron is 1.6022×10^{-19} C, calculate Avogadro's number. Assume that copper is oxidized to Cu^{2+} ions.

19.82 In a certain electrolysis experiment involving Al^{3+} ions, 60.2 g of Al is recovered when a current of 0.352 A is used. How many minutes did the electrolysis last?

19.83 Consider the oxidation of ammonia:

$$4NH_3(g) + 3O_2(g) \longrightarrow 2N_2(g) + 6H_2O(l)$$

(a) Calculate the $\Delta G°$ for the reaction. (b) If this reaction were used in a fuel cell, what would the standard cell potential be?

19.84 When an aqueous solution containing gold(III) salt is electrolyzed, metallic gold is deposited at the cathode and oxygen gas is generated at the anode. (a) If 9.26 g of Au is deposited at the cathode, calculate the volume (in liters) of O$_2$ generated at 23°C and 747 mmHg. (b) What is the current used if the electrolytic process took 2.00 h?

19.85 In an electrolysis experiment, a student passes the same quantity of electricity through two electrolytic cells, one containing a silver salt and the other a gold salt. Over a certain period of time, the student finds that 2.64 g of Ag and 1.61 g of Au are deposited at the cathodes. What is the oxidation state of gold in the gold salt?

19.86 People living in cold-climate countries where there is plenty of snow are advised not to heat their garages in the winter. What is the electrochemical basis for this recommendation?

19.87 Given that:

$$2Hg^{2+}(aq) + 2e^- \longrightarrow Hg_2^{2+}(aq) \qquad E° = 0.92 \text{ V}$$

$$Hg_2^{2+}(aq) + 2e^- \longrightarrow 2Hg(l) \qquad E° = 0.85 \text{ V}$$

calculate $\Delta G°$ and K for the following process at 25°C:

$$Hg_2^{2+}(aq) \longrightarrow Hg^{2+}(aq) + Hg(l)$$

(The preceding reaction is an example of a *disproportionation reaction* in which an element in one oxidation state is both oxidized and reduced.)

19.88 Fluorine (F$_2$) is obtained by the electrolysis of liquid hydrogen fluoride (HF) containing potassium fluoride (KF). (a) Write the half-cell reactions and the overall reaction for the process. (b) What is the purpose of KF? (c) Calculate the volume of F$_2$ (in liters) collected at 24.0°C and 1.2 atm after electrolyzing the solution for 15 h at a current of 502 A.

19.89 A 300-mL solution of NaCl was electrolyzed for 6.00 min. If the pH of the final solution was 12.24, calculate the average current used.

19.90 A piece of magnesium ribbon and a copper wire are partially immersed in a 0.1 M HCl solution in a beaker. The metals are joined externally by another piece of metal wire. Bubbles are seen to evolve at both the Mg and Cu surfaces. (a) Write equations representing the reactions occurring at the metals. (b) What visual evidence would you seek to show that Cu is not oxidized to Cu^{2+}? (c) At some stage, NaOH solution is added to the beaker to neutralize the HCl acid. Upon further addition of NaOH, a white precipitate forms. What is it?

19.91 An aqueous solution of a platinum salt is electrolyzed at a current of 2.50 A for 2.00 h. As a result, 9.09 g of metallic Pt is formed at the cathode. Calculate the charge on the Pt ions in this solution.

19.92 Consider a galvanic cell consisting of a magnesium electrode in contact with 1.0 M Mg(NO$_3$)$_2$ and a cadmium electrode in contact with 1.0 M Cd(NO$_3$)$_2$. Calculate $E°$ for the cell, and draw a diagram showing the cathode, anode, and direction of electron flow.

19.93 Use the data in Table 19.1 to show that the decomposition of H$_2$O$_2$ (a disproportionation reaction) is spontaneous at 25°C:

$$2H_2O_2(aq) \longrightarrow 2H_2O(l) + O_2(g)$$

19.94 Consider the galvanic cell in Figure 19.1. When viewed externally, the anode appears negative and the cathode positive (electrons are flowing from the anode to the cathode). Yet in solution anions are moving toward the anode, which means that it must appear positive to the anions. Because the anode cannot simultaneously be negative and positive, give an explanation for this apparently contradictory situation.

19.95 Explain why most useful galvanic cells give voltages of no more than 1.5 to 2.5 V. What are the prospects for developing practical galvanic cells with voltages of 5 V or more?

19.96 A silver rod and a SHE are dipped into a saturated aqueous solution of silver oxalate (Ag$_2$C$_2$O$_4$), at 25°C. The measured potential difference between the rod and the SHE is 0.589 V, the rod being positive. Calculate the solubility product constant for silver oxalate.

19.97 Zinc is an amphoteric metal; that is, it reacts with both acids and bases. The standard reduction potential is −1.36 V for the reaction:

$$Zn(OH)_4^{2-}(aq) + 2e^- \longrightarrow Zn(s) + 4OH^-(aq)$$

Calculate the formation constant (K_f) for the reaction:

$$Zn^{2+}(aq) + 4OH^-(aq) \rightleftharpoons Zn(OH)_4^{2-}(aq)$$

19.98 Use the data in Table 19.1 to determine whether or not hydrogen peroxide will undergo disproportion in an acid medium:
$2H_2O_2 \longrightarrow 2H_2O + O_2$.

19.99 The magnitudes (but *not* the signs) of the standard reduction potentials of two metals X and Y are:

$$Y^{2+} + 2e^- \longrightarrow Y \quad |E°| = 0.34 \text{ V}$$
$$X^{2+} + 2e^- \longrightarrow X \quad |E°| = 0.25 \text{ V}$$

where the ‖ notation denotes that only the magnitude (but not the sign) of the $E°$ value is shown. When the half-cells of X and Y are connected, electrons flow from X to Y. When X is connected to a SHE, electrons flow from X to SHE. (a) Are the $E°$ values of the half-reactions positive or negative? (b) What is the standard emf of a cell made up of X and Y?

19.100 A galvanic cell is constructed as follows. One half-cell consists of a platinum wire immersed in a solution containing 1.0 M Sn^{2+} and 1.0 M Sn^{4+}; the other half-cell has a thallium rod immersed in a solution of 1.0 M Tl$^+$. (a) Write the half-cell reactions and the overall reaction. (b) What is the equilibrium constant at 25°C? (c) What is the cell voltage if the Tl$^+$ concentration is increased 10-fold? ($E°_{Tl^+/Tl} = −0.34$ V.)

19.101 Given the standard reduction potential for Au^{3+} in Table 19.1 and:

$$Au^+(aq) + e^- \longrightarrow Au(s) \quad E° = 1.69 \text{ V}$$

answer the following questions. (a) Why does gold not tarnish in air? (b) Will the following disproportionation occur spontaneously?

$$3Au^+(aq) \longrightarrow Au^{3+}(aq) + 2Au(s)$$

(c) Predict the reaction between gold and fluorine gas.

19.102 Calculate $E°$ for the reactions of mercury with (a) 1 M HCl and (b) 1 M HNO$_3$. Which acid will oxidize Hg to Hg$_2^{2+}$ under standard-state conditions? Can you identify which pictured test tube contains HNO$_3$ and Hg and which contains HCl and Hg?

Ken Karp/McGraw Hill

19.103 Because all alkali metals react with water, it is not possible to measure the standard reduction potentials of these metals directly as in the case of, say, zinc. An indirect method is to consider the following hypothetical reaction:

$$Li^+(aq) + \tfrac{1}{2}H_2(g) \longrightarrow Li(s) + H^+(aq)$$

Using the appropriate equation presented in this chapter and the thermodynamic data in Appendix 2, calculate $E°$ for Li$^+(aq) + e^- \longrightarrow$ Li(s) at 298 K. Use 96,485.338 C/mol e^- for the Faraday constant. Compare your result with that listed in Table 19.1.

19.104 A galvanic cell using Mg/Mg^{2+} and Cu/Cu^{2+} half-cells operates under standard-state conditions at 25°C, and each compartment has a volume of 218 mL. The cell delivers 0.22 A for 31.6 h. (a) How many grams of Cu are deposited? (b) What is the [Cu^{2+}] remaining?

19.105 Given the following standard reduction potentials, calculate the ion-product, K_w, for water at 25°C:

$$2H^+(aq) + 2e^- \longrightarrow H_2(g) \qquad E° = 0.00 \text{ V}$$
$$2H_2O(l) + 2e^- \longrightarrow H_2(g) + 2OH^-(aq) \quad E° = −0.83 \text{ V}$$

19.106 Fluorine is a highly reactive gas that attacks water to form HF and other products. Follow the procedure in Problem 19.103 to show how you can determine indirectly the standard reduction potential for fluorine. Compare your result with the value in Table 19.1.

19.107 Consider a Daniell cell like the one shown in Figure 19.1 operating under non-standard-state conditions. Suppose that the cell's reaction is multiplied by 2. What effect does this have on each of the following quantities in the Nernst equation: (a) E, (b) $E°$, (c) Q, (d) ln Q, (e) n?

19.108 A spoon was silver-plated electrolytically in an AgNO$_3$ solution. (a) Sketch a diagram for the process. (b) If 0.884 g of Ag was deposited on the spoon at a constant current of 18.5 mA, how long (in min) did the electrolysis take?

19.109 Comment on whether F$_2$ will become a stronger oxidizing agent with increasing H$^+$ concentration.

19.110 Explain why chlorine gas can be prepared by electrolyzing an aqueous solution of NaCl but fluorine gas cannot be prepared by electrolyzing an aqueous solution of NaF.

19.111 Calculate the pressure of H$_2$ (in atm) required to maintain equilibrium with respect to the following reaction at 25°C:

$$Pb(s) + 2H^+(aq) \rightleftharpoons Pb^{2+}(aq) + H_2(g)$$

given that [Pb^{2+}] = 0.035 M and the solution is buffered at pH 1.60.

Industrial Problems

19.112 Industrially, copper is purified by electrolysis. The impure copper acts as the anode, and the cathode is made of pure copper. The electrodes are immersed in a $CuSO_4$ solution. During electrolysis, copper at the anode enters the solution as Cu^{2+} while Cu^{2+} ions are reduced at the cathode. (a) Write half-cell reactions and the overall reaction for the electrolytic process. (b) Suppose the anode was contaminated with Zn and Ag. Explain what happens to these impurities during electrolysis. (c) How many hours will it take to obtain 1.00 kg of Cu at a current of 18.9 A?

19.113 Gold will not dissolve in either concentrated nitric acid or concentrated hydrochloric acid. However, the metal does dissolve in a mixture of the acids (one part HNO_3 and three parts HCl by volume), called *aqua regia*. (a) Write a balanced equation for this reaction. (*Hint:* Among the products are $HAuCl_4$ and NO_2.) (b) What is the function of HCl?

Engineering Problems

19.114 To remove the tarnish (Ag_2S) on a silver spoon, a student carried out the following steps. First, she placed the spoon in a large pan filled with water so the spoon was totally immersed. Next, she added a few tablespoonfuls of baking soda (sodium bicarbonate), which readily dissolved. Finally, she placed some aluminum foil at the bottom of the pan in contact with the spoon and then heated the solution to about 80°C. After a few minutes, the spoon was removed and rinsed with cold water. The tarnish was gone, and the spoon regained its original shiny appearance. (a) Describe with equations the electrochemical basis for the procedure. (b) Adding NaCl instead of $NaHCO_3$ would also work because both compounds are strong electrolytes. What is the added advantage of using $NaHCO_3$? (*Hint:* Consider the pH of the solution.) (c) What is the purpose of heating the solution? (d) Some commercial tarnish removers contain a fluid (or paste) that is a dilute HCl solution. Rubbing the spoon with the fluid will also remove the tarnish. Name two disadvantages of using this procedure compared to the one described previously.

19.115 A construction company is installing an iron culvert (a long cylindrical tube) that is 40.0 m long with a radius of 0.900 m. To prevent corrosion, the culvert must be galvanized. This process is carried out by first passing an iron sheet of appropriate dimensions through an electrolytic cell containing Zn^{2+} ions, using graphite as the anode and the iron sheet as the cathode. If the voltage is 3.26 V, what is the cost of electricity for depositing a layer 0.200 mm thick if the efficiency of the process is 95 percent? The electricity rate is $0.12 per kilowatt hour (kWh), where 1 W = 1 J/s and the density of Zn is 7.14 g/cm³.

19.116 The concentration of sulfuric acid in the lead-storage battery of an automobile over a period of time has decreased from 38.0 percent by mass (density = 1.29 g/mL) to 26.0 percent by mass (1.19 g/mL). Assume the volume of the acid remains constant at 724 mL. (a) Calculate the total charge in coulombs supplied by the battery. (b) How long (in hours) will it take to recharge the battery back to the original sulfuric acid concentration using a current of 22.4 A?

19.117 Lead storage batteries are rated by ampere-hours, that is, the number of amperes they can deliver in an hour. (a) Show that 1 Ah = 3600 C. (b) The lead anodes of a certain lead storage battery have a total mass of 406 g. Calculate the maximum theoretical capacity of the battery in ampere-hours. Explain why in practice we can never extract this much energy from the battery. (*Hint:* Assume all the lead will be used up in the electrochemical reaction, and refer to the lead storage-battery electrode reactions in Section 19.6.) (c) Calculate $E°_{cell}$ and $\Delta G°$ for the battery.

19.118 Compare the pros and cons of a fuel cell, such as the hydrogen-oxygen fuel cell, and a coal-fired power station for generating electricity.

Biological Problems

19.119 Oxalic acid ($H_2C_2O_4$) is present in many plants and vegetables. (a) Balance the following equation in acid solution:

$$MnO_4^- + C_2O_4^{2-} \longrightarrow Mn^{2+} + CO_2$$

(b) If a 1.00-g sample of plant matter requires 24.0 mL of 0.0100 M $KMnO_4$ solution to reach the equivalence point, what is the percent by mass of $H_2C_2O_4$ in the sample?

19.120 The ingestion of a very small quantity of mercury is not considered too harmful. Would this statement still hold if the gastric juice in your stomach were mostly nitric acid instead of hydrochloric acid? Explain.

19.121 Calcium oxalate (CaC_2O_4) is insoluble in water. This property has been used to determine the amount of Ca^{2+} ions in blood. The calcium oxalate isolated from blood is dissolved in acid and titrated against a standardized $KMnO_4$ solution as described in Problem 19.119. In one test it is found that the calcium oxalate isolated from a 10.0-mL sample of blood requires 24.2 mL of 9.56×10^{-4} M $KMnO_4$ for titration. Calculate the number of milligrams of calcium per milliliter of blood.

19.122 Cytochrome-c is a protein involved in biological electron transfer processes. The redox half-reaction is shown by the reduction of the Fe^{3+} ion to the Fe^{2+} ion:

$$\text{cyt } c(Fe^{3+}) + e^- \longrightarrow \text{cyt } c(Fe^{2+}) \qquad E° = 0.254 \text{ V}$$

Calculate the number of moles of cyt $c(Fe^{3+})$ formed from cyt $c(Fe^{2+})$ with the Gibbs free energy derived from the oxidation of 1 mole of glucose.

19.123   The nitrite ion (NO_2^-) in soil is oxidized to the nitrate ion (NO_3^-) by the bacterium *Nitrobacter agilis* in the presence of oxygen. The half-reactions are:

$$NO_3^- + 2H^+ + 2e^- \longrightarrow NO_2^- + H_2O \qquad E° = 0.42 \text{ V}$$
$$O_2 + 4H^+ + 4e^- \longrightarrow 2H_2O \qquad E° = 1.23 \text{ V}$$

Calculate the yield of ATP synthesis per mole of nitrite oxidized. (*Hint:* Refer to Section 18.7.)

Environmental Problems

19.124   In recent years, there has been much interest in electric cars. List some advantages and disadvantages of electric cars compared to automobiles with internal combustion engines.

19.125   The SO_2 present in air is mainly responsible for the phenomenon of acid rain. The concentration of SO_2 can be determined by titrating against a standard permanganate solution as follows:

$$5SO_2 + 2MnO_4^- + 2H_2O \longrightarrow 5SO_4^{2-} + 2Mn^{2+} + 4H^+$$

Calculate the number of grams of SO_2 in a sample of air if 7.37 mL of 0.00800 M $KMnO_4$ solution is required for the titration.

19.126   The zinc-air battery shows much promise for electric cars because it is lightweight and rechargeable:

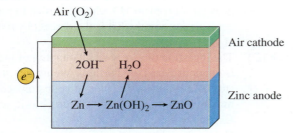

Air (O_2)

Air cathode

$2OH^-$ H_2O

$Zn \longrightarrow Zn(OH)_2 \longrightarrow ZnO$

Zinc anode

The net transformation is $Zn(s) + \frac{1}{2}O_2(g) \longrightarrow ZnO(s)$. (a) Write the half-reactions at the zinc-air electrodes, and calculate the standard emf of the battery at 25°C. (b) Calculate the emf under actual operating conditions when the partial pressure of oxygen is 0.21 atm. (c) What is the energy density (measured as the energy in kilojoules that can be obtained from 1 kg of the metal) of the zinc electrode? (d) If a current of 2.1×10^5 A is to be drawn from a zinc-air battery system, what volume of air (in liters) would need to be supplied to the battery every second? Assume that the temperature is 25°C and the partial pressure of oxygen is 0.21 atm.

19.127   A current of 6.00 A passes through an electrolytic cell containing dilute sulfuric acid for 3.40 h. If the volume of O_2 gas generated at the anode is 4.26 L (at STP), calculate the charge (in coulombs) on an electron.

19.128   A 9.00×10^2 mL amount of 0.200 M MgI_2 solution was electrolyzed. As a result, hydrogen gas was generated at the cathode and iodine was formed at the anode. The volume of hydrogen collected at 26°C and 779 mmHg was 1.22×10^3 mL. (a) Calculate the charge in coulombs consumed in the process. (b) How long (in min) did the electrolysis last if a current of 7.55 A was used? (c) A white precipitate was formed in the process. What was it, and what was its mass in grams? Assume the volume of the solution was constant.

19.129   When 25.0 mL of a solution containing both Fe^{2+} and Fe^{3+} ions is titrated with 23.0 mL of 0.0200 M $KMnO_4$ (in dilute sulfuric acid), all the Fe^{2+} ions are oxidized to Fe^{3+} ions. Next, the solution is treated with Zn metal to convert all the Fe^{3+} ions to Fe^{2+} ions. Finally, 40.0 mL of the same $KMnO_4$ solution is added to the solution to oxidize the Fe^{2+} ions to Fe^{3+}. Calculate the molar concentrations of Fe^{2+} and Fe^{3+} in the original solution.

Standardized-Exam Practice Problems

Physical and Biological Sciences

A galvanic cell is constructed by immersing a piece of copper wire in 25.0 mL of a 0.20 *M* $CuSO_4$ solution and a zinc strip in 25.0 mL of a 0.20 *M* $ZnSO_4$ solution. Cu^{2+} ions react with aqueous NH_3 to form the complex ion $Cu(NH_3)_4^{2+}$:

$$Cu^{2+}(aq) + 4NH_3(aq) \longrightarrow Cu(NH_3)_4^{2+}(aq)$$

1. Using the equation:

$$E = E° - \frac{0.0592\ V}{n}\log Q$$

   calculate the emf of the cell at 25°C.

   a) 0.0 V
   b) 1.10 V
   c) 0.90 V
   d) 1.30 V

2. What would happen if a small amount of concentrated NH_3 solution were added to the $CuSO_4$ solution?

   a) Nothing.
   b) Emf would increase.
   c) Emf would decrease.
   d) Not enough information to determine.

3. What would happen if a small amount of concentrated NH_3 solution were added to the $ZnSO_4$ solution?

   a) Nothing.
   b) Emf would increase.
   c) Emf would decrease.
   d) Not enough information to determine.

4. In a separate experiment, 25.0 mL of 3.00 *M* NH_3 is added to the $CuSO_4$ solution. If the emf of the cell is 0.68 V at equilibrium, calculate the formation constant (K_f) of $Cu(NH_3)_4^{2+}$.

   a) 9.4×10^{22}
   b) 1.1×10^{-23}
   c) 1.5×10^{-14}
   d) 1.5×10^{14}

Answers to In-Chapter Materials

Answers to Practice Problems

19.1A $2MnO_4^- + H_2O + 3CN^- \longrightarrow 2MnO_2 + 2OH^- + 3CNO^-$.
19.1B $MnO_4^- + 5Fe^{2+} + 8H^+ \longrightarrow Mn^{2+} + 5Fe^{3+} + 4H_2O$.
19.2A $Cd + Pb^{2+} \longrightarrow Cd^{2+} + Pb$, $E°_{cell} = 0.27$ V. **19.2B** In one cell, the iron electrode is the anode and the overall reaction is $Fe(s) + Sn^{2+}(aq) \longrightarrow Fe^{2+}(aq) + Sn(s)$. $E°_{cell} = E°_{Sn^{2+}/Sn} - E°_{Fe^{2+}/Fe} = (-0.14\ V) - (-0.44\ V) = 0.30$ V. In the other cell, the iron electrode is the cathode and the overall reaction is $3Fe^{2+}(aq) + 2Cr(s) \longrightarrow 3Fe(s) + 2Cr^{3+}(aq)$. $E°_{cell} = E°_{Fe^{2+}/Fe} - E°_{Cr^{3+}/Cr} = (-0.44\ V) - (-0.74\ V) = 0.30$ V. **19.3A** (a) No reaction, (b) $2H^+ + Pb \longrightarrow H_2 + Pb^{2+}$.
19.3B Cobalt has a more positive reduction potential than iron. Cobalt ion would, therefore, be *reduced* in the presence of iron metal; and the iron in an iron container would oxidize to Fe^{2+}. Metal that is oxidized goes from the solid phase to the aqueous phase, meaning that the container would effectively *dissolve*. Cobalt has a less positive reduction potential than tin. Tin metal would not be oxidized by Co^{2+} ion, and the container would remain intact. A cobalt(II) chloride solution would be more safely stored in a tin container than in an iron container. **19.4A** −411 kJ/mol. **19.4B** −0.23 V. **19.5A** 1×10^{-42}. **19.5B** 1.4×10^5. **19.6A** Yes, the reaction is spontaneous. **19.6B** 18 atm. **19.7A** $[Cu^+] = 1.11 \times 10^{-3}$ *M*, $K_{sp} = 1.2 \times 10^{-6}$. **19.7B** 0.16 V. **19.8A** 7.44 g Mg. **19.8B** 0.96 A.

Answers to Checkpoints

19.1.1 a, c, e. **19.1.2** c. **19.3.1** d. **19.3.2** c. **19.3.3** b. **19.3.4** e. **19.4.1** b. **19.4.2** b. **19.5.1** b. **19.5.2** b. **19.5.3** e. **19.5.4** d. **19.5.5** b. **19.5.6** a. **19.7.1** a. **19.7.2** e. **19.7.3** c.

CHAPTER 20

Nuclear Chemistry

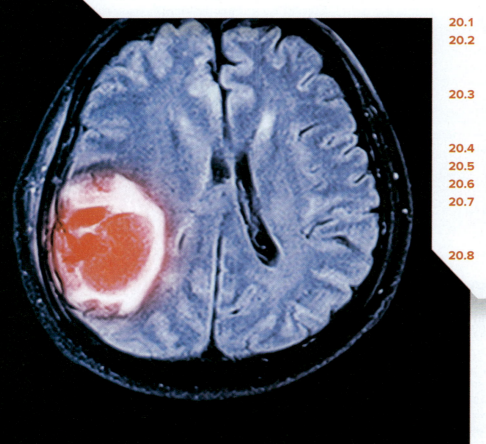

The CT scan shows a brain tumor that might be difficult or impossible to treat by conventional surgical methods. Nuclear medicine, including boron neutron capture therapy (BNCT), enables doctors to treat cancers of this type.

Puwadol Jaturawutthichai/Alamy Stock Photo

In This Chapter, You Will Learn

Some of the fundamentals of nuclear chemistry and how nuclear reactions are important to living systems and to society.

Before You Begin, Review These Skills

- Radioactivity [⏮ Section 2.2]
- Atomic number, mass number, and isotopes [⏮ Section 2.3]
- First-order kinetics [⏮ Section 14.3]
- Chapter 14 Key Skills [⏮ pages 712–713]

How Nuclear Chemistry Is Used to Treat Cancer

Brain tumors are some of the most difficult cancers to treat because the site of the malignant growth makes surgical excision difficult or impossible. Likewise, conventional radiation therapy using X rays or γ rays from outside the skull is usually not effective. An ingenious approach to this problem is *boron neutron capture therapy* (BNCT). This technique involves first administering a boron-10 compound that is selectively taken up by tumor cells and then applying a beam of low-energy neutrons to the tumor site. ^{10}B captures a neutron to produce ^{11}B, which disintegrates via the following nuclear reaction:

$$^{10}_{5}\text{B} + ^{1}_{0}\text{n} \longrightarrow ^{7}_{3}\text{Li} + ^{4}_{2}\alpha$$

The highly energetic particles produced by this reaction destroy the tumor cells in which the ^{10}B is concentrated. Because the particles are confined to just a few micrometers, they preferentially destroy tumor cells without damaging neighboring normal cells.

BNCT is a highly promising treatment and is an active area of research. One of the major goals of the research is to develop suitable compounds to deliver ^{10}B to the desired site. For such a compound to be effective, it must meet several criteria. It must have a high affinity for tumor cells, be able to pass through membrane barriers to reach the tumor site, and have minimal toxic effects on the human body.

This is one example of how *nuclear chemistry* is important in the treatment of cancer.

At the end of this chapter, you will be able to answer questions about isotopes used in nuclear medicine [⏭ Applying What You've Learned, page 1040].

20.1 Nuclei and Nuclear Reactions

With the exception of hydrogen ($_1^1H$), all nuclei contain protons and neutrons. Some nuclei are unstable and undergo *radioactive decay,* emitting particles and/or electromagnetic radiation [◄◄ Section 2.2]. Spontaneous emission of particles or electromagnetic radiation is known as **radioactivity.** All elements having an atomic number greater than 83 are unstable and are therefore radioactive. Polonium-210 ($_{84}^{210}Po$), for example, decays spontaneously to Pb by emitting an α particle.

Another type of nuclear process, known as **nuclear transmutation,** results from the bombardment of nuclei by neutrons, protons, or other nuclei. An example of a nuclear transmutation is the conversion of atmospheric $_7^{14}N$ to $_6^{14}C$ and $_1^1H$, which results when the nitrogen isotope is bombarded by neutrons (from the sun). In some cases, heavier elements are synthesized from lighter elements. This type of transmutation occurs naturally in outer space, but it can also be achieved artificially, as we explain in Section 20.4.

Radioactive decay and nuclear transmutation are *nuclear reactions,* which differ significantly from ordinary chemical reactions. Table 20.1 summarizes the differences.

To discuss nuclear reactions in any depth, we must understand how to write and balance nuclear equations. Writing a nuclear equation differs somewhat from writing equations for chemical reactions. In addition to writing the symbols for the various chemical elements, we must explicitly indicate the number of subatomic particles in *every* species involved in the reaction.

The symbols for subatomic particles are as follows:

$$_1^1H \text{ or } _1^1p \qquad _0^1n \qquad _{-1}^0e \text{ or } _{-1}^0\beta \qquad _{+1}^0e \text{ or } _{+1}^0\beta \qquad _2^4\alpha \text{ or } _2^4He$$

$$\text{proton} \qquad\qquad \text{neutron} \qquad\qquad \text{electron} \qquad\qquad\qquad \text{positron} \qquad\qquad\qquad \alpha \text{ particle}$$

In accordance with the notation introduced in Section 2.3, the superscript in each case denotes the mass number (the total number of neutrons and protons present) and the subscript is the atomic number (the number of protons). Thus, the "atomic number" of a proton is 1, because there is one proton present, and the "mass number" is also 1, because there is one proton but no neutrons present. On the other hand, the mass number of a neutron is 1, but its atomic number is zero, because there are no protons present. For the electron, the mass number is zero (there are neither protons nor neutrons present), but the atomic number is −1, because the electron possesses a unit negative charge.

The symbol $_{-1}^0e$ represents an electron in or from an atomic orbital. The symbol $_{-1}^0\beta$, on the other hand, represents an electron that, although physically identical to any other electron, comes from a nucleus (in a decay process in which a neutron is

TABLE 20.1 Comparison of Chemical Reactions and Nuclear Reactions

Chemical Reactions	Nuclear Reactions
1. Atoms are rearranged by the breaking and forming of chemical bonds.	1. Elements are converted to other elements (or isotopes).
2. Only electrons in atomic or molecular orbitals are involved in the reaction.	2. Protons, neutrons, electrons, and other subatomic particles such as α particles may be involved.
3. Reactions are accompanied by the absorption or release of relatively small amounts of energy.	3. Reactions are accompanied by the absorption or release of tremendous amounts of energy.
4. Rates of reaction are influenced by temperature, pressure, concentration, and catalysts.	4. Rates of reaction normally are not affected by temperature, pressure, or catalysts.

converted to a proton and an electron) and not from an atomic orbital. The *positron* has the same mass as the electron, but bears a charge of +1. The α particle has two protons and two neutrons, so its atomic number is 2 and its mass number is 4.

Student Note: An α particle is identical to a helium-4 nucleus and can be represented as either $^4_2\alpha$ or $^4_2$He.

In balancing any nuclear equation, we must balance the total of all atomic numbers and the total of all mass numbers for the products and reactants. If we know the atomic numbers and mass numbers of all but one of the species in a nuclear equation, we can identify the *unknown* species by applying these rules, as shown in Sample Problem 20.1.

SAMPLE PROBLEM 20.1

Identify the missing species X in each of the following nuclear equations:

(a) $^{212}_{84}$Po $\longrightarrow$ $^{208}_{82}$Pb + X

(b) $^{90}_{38}$Sr $\longrightarrow$ X + $^{0}_{-1}\beta$

(c) X $\longrightarrow$ $^{18}_{8}$O + $^{0}_{+1}\beta$

Strategy Determine the mass number for the unknown species, X, by summing the mass numbers on both sides of the equation:

$$\Sigma \text{ reactant mass numbers} = \Sigma \text{ product mass numbers}$$

Similarly, determine the atomic number for the unknown species:

$$\Sigma \text{ reactant atomic numbers} = \Sigma \text{ product atomic numbers}$$

Use the mass number and atomic number to determine the identity of the unknown species.

Setup (a) 212 = (208 + mass number of X); mass number of X = 4. 84 = (82 + atomic number of X); atomic number of X = 2.

(b) 90 = (mass number of X + 0); mass number of X = 90. 38 = [atomic number of X + (−1)]; atomic number of X = 39.

(c) Mass number of X = (18 + 0); mass number of X = 18. Atomic number of X = (8 + 1); atomic number of X = 9.

Solution (a) X = $^4_2\alpha$: $^{212}_{84}$Po $\longrightarrow$ $^{208}_{82}$Pb + $^4_2\alpha$

(b) X = $^{90}_{39}$Y: $^{90}_{38}$Sr $\longrightarrow$ $^{90}_{39}$Y + $^{0}_{-1}\beta$

(c) X = $^{18}_{9}$F: $^{18}_{9}$F $\longrightarrow$ $^{18}_{8}$O + $^{0}_{+1}\beta$

THINK ABOUT IT

The rules of summation that we apply to balance nuclear equations can be thought of as the *conservation of mass number* and the *conservation of atomic number*.

Practice Problem ATTEMPT Identify X in each of the following nuclear equations:

(a) $^{78}_{33}$As $\longrightarrow$ X + $^{0}_{-1}\beta$

(b) $^1_1$H + $^4_2$He $\longrightarrow$ X

(c) $^{258}_{100}$Fm $\longrightarrow$ $^{257}_{100}$Fm + X

Practice Problem BUILD Identify X in each of the following nuclear equations:

(a) X + $^{0}_{-1}\beta$ $\longrightarrow$ $^{244}_{94}$Pu

(b) $^{238}_{92}$U $\longrightarrow$ X + $^4_2$He

(c) X $\longrightarrow$ $^{14}_{7}$N + $^{0}_{-1}e$

Practice Problem CONCEPTUALIZE For each process, specify the identity of the product.

$\xleftarrow{\text{electron capture}}$ ^{222}Rn $\xrightarrow{\text{alpha emission}}$ $\xleftarrow{\text{electron capture}}$ ^{132}Cs $\xrightarrow{\text{beta emission}}$

20.1.1 Identify the species X in the following nuclear equation:

$$^{222}_{86}\text{Rn} \longrightarrow \text{X} + {}^4_2\alpha$$

a) $^{226}_{84}\text{Po}$

b) $^{226}_{88}\text{Ra}$

c) $^{212}_{84}\text{Po}$

d) $^{218}_{84}\text{Po}$

e) $^{218}_{88}\text{Ra}$

20.1.2 Identify the species X in the following nuclear equation:

$$^{15}_{8}\text{O} \longrightarrow \text{X} + {}^0_{-1}\beta$$

a) $^{15}_{9}\text{F}$

b) $^{14}_{8}\text{O}$

c) $^{16}_{9}\text{F}$

d) $^{15}_{7}\text{N}$

e) $^{14}_{7}\text{N}$

20.2 Nuclear Stability

The nucleus occupies a very small portion of the total volume of an atom, but it contains most of the atom's mass because both the protons and the neutrons reside there. In studying the stability of the atomic nucleus, it is helpful to know something about its density, because it tells us how tightly the particles are packed together. As a sample calculation, let us assume that a nucleus has a radius of 5×10^{-3} pm and a mass of 1×10^{-22} g. These figures correspond roughly to a nucleus containing 30 protons and 30 neutrons. Density is mass/volume, and we can calculate the volume from the known radius (the volume of a sphere is $\frac{4}{3}\pi r^3$, where r is the radius of the sphere). First we convert the picometer units to centimeters. Then, we calculate the density in g/cm³:

$$r = (5 \times 10^{-3} \text{ pm})\left(\frac{1 \times 10^{-12} \text{ m}}{1 \text{ pm}}\right)\left(\frac{100 \text{ cm}}{1 \text{ m}}\right) = 5 \times 10^{-13} \text{ cm}$$

$$\text{density} = \frac{\text{mass}}{\text{volume}} = \frac{1 \times 10^{-22} \text{ g}}{\frac{4}{3}\pi r^3} = \frac{1 \times 10^{-22} \text{ g}}{\frac{4}{3}\pi(5 \times 10^{-13} \text{ cm})^3}$$

$$= 2 \times 10^{14} \text{ g/cm}^3$$

This is an exceedingly high density. The highest density known for an element is 22.65 g/cm³, for iridium (Ir). Thus, the average atomic nucleus is roughly 9×10^{12} (or 9 *trillion*) times as dense as the densest element known!

The enormously high density of the nucleus means that some very strong force is needed to hold the particles together so tightly. From *Coulomb's law* we know that like charges repel and unlike charges attract one another. We would thus expect the protons to repel one another strongly, particularly when we consider how close they must be to each other. However, in addition to the repulsion, there are also short-range attractions between proton and proton, proton and neutron, and neutron and neutron. The stability of any nucleus is determined by the difference between coulombic repulsion and the short-range attraction. If repulsion outweighs attraction, the nucleus disintegrates, emitting particles and/or radiation. If attractive forces prevail, the nucleus is stable.

Patterns of Nuclear Stability

The principal factor that determines whether a nucleus is stable is the *neutron-to-proton ratio (n/p)*. For stable atoms of elements having low atomic number (≤ 20), the n/p value is close to 1. As the atomic number increases, the neutron-to-proton ratios of the stable nuclei also increase. This deviation at higher atomic numbers arises because

TABLE 20.2	Number of Stable Isotopes with Even and Odd Numbers of Protons and Neutrons		
Protons	Neutrons	Number of Stable Isotopes	
Odd	Odd	4	
Odd	Even	50	
Even	Odd	53	
Even	Even	164	

more neutrons are needed to counteract the strong repulsion among the protons and stabilize the nucleus. The following rules are useful in gauging whether or not a particular nucleus is expected to be stable:

1. There are more stable nuclei containing 2, 8, 20, 50, 82, or 126 protons or neutrons than there are containing other numbers of protons or neutrons. For example, there are 10 stable isotopes of tin (Sn) with the atomic number 50 and only 2 stable isotopes of antimony (Sb) with the atomic number 51. The numbers 2, 8, 20, 50, 82, and 126 are called magic numbers.
2. There are many more stable nuclei with even numbers of both protons and neutrons than with odd numbers of these particles (Table 20.2).
3. All isotopes of the elements with atomic numbers higher than 83 are radioactive.
4. All isotopes of technetium (Tc, $Z = 43$) and promethium (Pm, $Z = 61$) are radioactive.

Figure 20.1 shows a plot of the number of neutrons versus the number of protons in various isotopes. The stable nuclei are located in an area of the graph known as the

Student Note: The significance of these numbers for nuclear stability is similar to the numbers of electrons associated with the very stable noble gases (i.e., 2, 10, 18, 36, 54, and 86 electrons).

Student Note: Of the two stable isotopes of antimony mentioned in rule 1, both have even numbers of neutrons: $^{121}_{51}Sb$ and $^{123}_{51}Sb$.

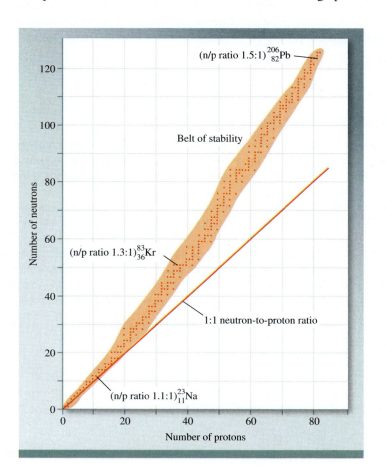

Figure 20.1 Plot of neutrons versus protons for various stable isotopes, represented by dots. The straight line represents the points at which the neutron-to-proton ratio is 1. The shaded area represents the belt of stability.

belt of stability. Most radioactive nuclei lie outside this belt. Above the belt of stability, the nuclei have higher neutron-to-proton ratios than those within the belt (for the same number of protons). To lower this ratio (and hence move down toward the belt of stability), these nuclei undergo the following process, called *β-particle emission:*

$$\frac{1}{0}n \longrightarrow \frac{1}{1}p + \frac{0}{-1}\beta$$

Beta-particle emission leads to an increase in the number of protons in the nucleus and a simultaneous decrease in the number of neutrons. Some examples are:

$$^{14}_{6}C \longrightarrow ^{14}_{7}N + ^{0}_{-1}\beta$$

$$^{40}_{19}K \longrightarrow ^{40}_{20}Ca + ^{0}_{-1}\beta$$

$$^{97}_{40}Zr \longrightarrow ^{97}_{41}Nb + ^{0}_{-1}\beta$$

Below the belt of stability, the nuclei have lower neutron-to-proton ratios than those in the belt (for the same number of protons). To increase this ratio (and hence move up toward the belt of stability), these nuclei may emit a positron:

$$^{1}_{1}p \longrightarrow ^{1}_{0}n + ^{0}_{+1}\beta$$

An example of positron emission is:

$$^{38}_{19}K \longrightarrow ^{38}_{18}Ar + ^{0}_{+1}\beta$$

Alternatively, a nucleus may undergo electron capture:

$$^{1}_{1}p + ^{0}_{-1}e \longrightarrow ^{1}_{0}n$$

Electron capture is the capture of an electron—usually a $1s$ electron—by the nucleus. The captured electron combines with a proton in the nucleus to form a *neutron* so that the atomic number decreases by 1 while the mass number remains the same. Electron capture has the same net effect on the nucleus as positron emission. Examples of electron capture are:

$$^{37}_{18}Ar + ^{0}_{-1}e \longrightarrow ^{37}_{17}Cl$$

$$^{55}_{26}Fe + ^{0}_{-1}e \longrightarrow ^{55}_{25}Mn$$

Nuclear Binding Energy

A quantitative measure of nuclear stability is the ***nuclear binding energy,*** which is the energy required to break up a nucleus into its component protons and neutrons. This quantity represents the conversion of mass to energy that occurs during an exothermic nuclear reaction.

The concept of nuclear binding energy evolved from studies of nuclear properties showing that the masses of nuclei are always less than the sum of the masses of the *nucleons,* which is a general term for the protons and neutrons in a nucleus. For example, the $^{19}_{9}F$ isotope has an atomic mass of 18.99840 amu. The nucleus has 9 protons and 10 neutrons and therefore a total of 19 nucleons. Using the known masses of the $^{1}_{1}H$ atom (1.007825 amu) and the neutron (1.008665 amu), we can carry out the following analysis. The mass of 9 $^{1}_{1}H$ atoms (i.e., the mass of 9 protons and 9 electrons) is:

$$9 \times 1.007825 \text{ amu} = 9.070425 \text{ amu}$$

and the mass of 10 neutrons is:

$$10 \times 1.008665 \text{ amu} = 10.08665 \text{ amu}$$

Therefore, the atomic mass of an $^{19}_{9}F$ atom calculated from the known numbers of electrons, protons, and neutrons is:

$$9.070425 \text{ amu} + 10.08665 \text{ amu} = 19.15708 \text{ amu}$$

This value is larger than 18.99840 amu (the measured mass of $^{19}_{9}F$) by 0.15868 amu.

The difference between the mass of an atom and the sum of the masses of its protons, neutrons, and electrons is called the ***mass defect.*** According to relativity theory, the loss in mass shows up as energy (heat) given off to the surroundings. Thus, the formation of $^{19}_{9}F$ is *exothermic*. According to *Einstein's mass-energy equivalence relationship* ($E = mc^2$, where E is energy, m is mass, and c is the velocity of light), we can calculate the amount of energy released. We start by writing:

Equation 20.1 $$\Delta E = (\Delta m)c^2$$

where ΔE and Δm are defined as follows:

$$\Delta E = \text{energy of product} - \text{energy of reactants}$$

$$\Delta m = \text{mass of product} - \text{mass of reactants}$$

Thus, the change in mass is:

$$\Delta m = 18.99840 \text{ amu} - 19.15708 \text{ amu}$$

$$= -0.15868 \text{ amu}$$

or:

$$\Delta m = (-0.15868 \text{ amu})\left(\frac{1 \text{ kg}}{6.0221418 \times 10^{26} \text{ amu}}\right)$$

$$= -2.6349 \times 10^{-28} \text{ kg}$$

> **Student Note:** When you apply Einstein's equation, $E = mc^2$, it is important to remember that mass defect must be expressed in kilograms for the units to cancel properly.
>
> $1 \text{ kg} = 6.0221418 \times 10^{26}$ amu

Because $^{19}_{9}F$ has a mass that is less than the mass calculated from the number of electrons and nucleons present, Δm is a negative quantity. Consequently, ΔE is also a negative quantity; that is, energy is released to the surroundings as a result of the formation of the fluorine-19 nucleus. We calculate ΔE as follows:

$$\Delta E = (-2.6349 \times 10^{-28} \text{ kg})(2.99792458 \times 10^8 \text{ m/s})^2$$

$$= -2.3681 \times 10^{-11} \text{ kg} \cdot \text{m}^2/\text{s}^2$$

$$= -2.3681 \times 10^{-11} \text{ J}$$

> **Student Note:** Remember that joule is a *derived* unit:
>
> $1 \text{ J} = 1 \text{ kg} \cdot \text{m}^2/\text{s}^2$
>
> [◀ Section 5.1].

This is the amount of energy released when one fluorine-19 nucleus is formed from 9 protons and 10 neutrons. The nuclear binding energy of the nucleus is 2.3681×10^{-11} J, which is the amount of energy needed to decompose the nucleus into separate protons and neutrons. In the formation of 1 mole of fluorine nuclei, for instance, the energy released is:

$$\Delta E = (-2.3681 \times 10^{-11} \text{ J})(6.0221418 \times 10^{23}/\text{mol})$$

$$= -1.4261 \times 10^{-13} \text{ J/mol}$$

$$= -1.4261 \times 10^{-10} \text{ kJ/mol}$$

> **Student Note:** Note that when we report a nuclear binding energy per mole, we give just the magnitude without the negative sign.

The nuclear binding energy, therefore, is 1.4261×10^{10} kJ for 1 mole of fluorine-19 nuclei, which is a tremendously large quantity when we consider that the enthalpies of ordinary chemical reactions are on the order of only 200 kJ. The procedure we have followed can be used to calculate the nuclear binding energy of any nucleus.

As we have noted, nuclear binding energy is an indication of the stability of a nucleus. When comparing the stability of any two nuclei, however, we must account for the fact that they have different numbers of nucleons. It makes more sense, therefore, to compare nuclei using the *nuclear binding energy per nucleon:*

$$\text{nuclear binding energy per nucleon} = \frac{\text{nuclear binding energy}}{\text{number of nucleons}}$$

Figure 20.2 Plot of nuclear binding energy per nucleon versus mass number.

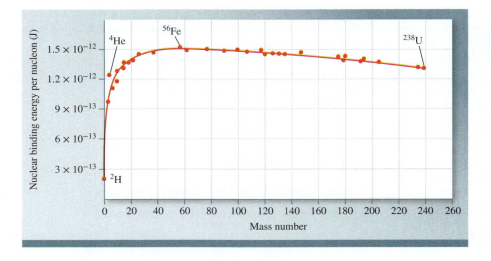

For the fluorine-19 nucleus:

$$\text{nuclear binding energy per nucleon} = \frac{2.3681 \times 10^{-11} \text{ J}}{19 \text{ nucleons}}$$

$$= 1.2464 \times 10^{-12} \text{ J/nucleon}$$

The nuclear binding energy per nucleon makes it possible to compare the stability of all nuclei on a common basis. In general, the greater the nuclear binding energy per nucleon, the more stable the nucleus. Figure 20.2 shows the variation of nuclear binding energy per nucleon plotted against mass number. As you can see, the curve rises rather steeply. The highest binding energies per nucleon belong to elements with intermediate mass numbers—between 40 and 100—and are greatest for elements in the iron, cobalt, and nickel region (elements in Groups 8, 9, and 10) of the periodic table. This means that the net attractive forces among the particles (protons and neutrons) are greatest for the nuclei of these elements.

Nuclear binding energy and nuclear binding energy per nucleon are calculated for an iodine nucleus in Sample Problem 20.2.

SAMPLE PROBLEM 20.2

The atomic mass of $^{127}_{53}\text{I}$ is 126.904473 amu. Calculate the nuclear binding energy of this nucleus and the corresponding nuclear binding energy per nucleon.

Strategy To calculate the nuclear binding energy, we first determine the difference between the mass of the nucleus and the mass of all the protons and neutrons, which yields the mass defect. Next, we must apply Einstein's mass-energy relationship $[\Delta E = (\Delta m)c^2]$.

Solution There are 53 protons and 74 neutrons in the iodine nucleus. The mass of 53 ^1_1H atoms is:

$$53 \times 1.007825 \text{ amu} = 53.41473 \text{ amu}$$

and the mass of 74 neutrons is:

$$74 \times 1.008665 \text{ amu} = 74.64121 \text{ amu}$$

Therefore, the predicted mass for $^{127}_{53}\text{I}$ is 53.41473 + 74.64121 = 128.05594 amu, and the mass defect is:

$$\Delta m = 126.904473 \text{ amu} - 128.05594 \text{ amu}$$

$$= -1.15147 \text{ amu}$$

$$= (-1.15147 \text{ amu})\left(\frac{1 \text{ kg}}{6.0221418 \times 10^{26} \text{ amu}}\right)$$

$$= -1.91206 \times 10^{-27} \text{ kg}$$

The energy released is:

$$\Delta E = (\Delta m)c^2$$
$$= (-1.91206 \times 10^{-27} \text{ kg})(2.99792458 \times 10^8 \text{ m/s})^2$$
$$= 1.71847 \times 10^{-10} \text{ kg} \cdot \text{m}^2/\text{s}^2$$
$$= 1.71847 \times 10^{-10} \text{ J}$$

Thus the nuclear binding energy is 1.71847×10^{-10} J. The nuclear binding energy per nucleon is obtained as follows:

$$\frac{1.71847 \times 10^{-10} \text{ J}}{127 \text{ nucleons}} = 1.35313 \times 10^{-12} \text{ J/nucleon}$$

Practice Problem **A**TTEMPT Calculate the nuclear binding energy (in joules) and the nuclear binding energy per nucleon of $^{209}_{83}\text{Bi}$ (208.980374 amu).

Practice Problem **B**UILD The nuclear binding energy for $^{197}_{79}\text{Au}$ is 1.2683×10^{-12} J/nucleon. Determine the mass of a $^{197}_{79}\text{Au}$ atom.

Practice Problem **C**ONCEPTUALIZE Which of the following graphs best represents the relationship between energy released (ΔE) and mass defect (Δm)?

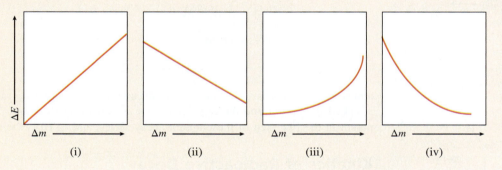

(i)   (ii)   (iii)   (iv)

CHECKPOINT – SECTION 20.2   Nuclear Stability

20.2.1 Determine the binding energy per nucleon in a ^{238}U nucleus (238.0507847 amu).

a) 2.891×10^{-10} J/nucleon

b) 1.215×10^{-12} J/nucleon

c) 1.186×10^{-12} J/nucleon

d) 2.823×10^{-10} J/nucleon

e) 3.212×10^{-24} J/nucleon

20.2.2 What is the energy associated with a mass defect of 2.000 amu?

a) 3.321×10^{-27} J

b) 2.989×10^{-7} J

c) 1.800×10^{17} J

d) 2.989×10^{-10} J

e) 3.309×10^{-45} J

20.2.3 What is the change in mass (in kg) for the following reaction? $\Delta H°$ for the reaction is -890.4 kJ/mol.

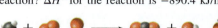

a) 1.479×10^{-21} kg

b) 1.479×10^{-24} kg

c) 1.645×10^{-35} kg

d) 4.932×10^{-33} kg

e) 9.907×10^{-15} kg

20.2.4 What type of radioactive decay will the isotopes ^{13}B and ^{188}Au most likely undergo?

a) Beta emission, positron emission

b) Beta emission, beta emission

c) Positron emission, beta emission

d) Positron emission, positron emission

Figure 20.3
Decay series for uranium-238. (Times shown are half-lives.)

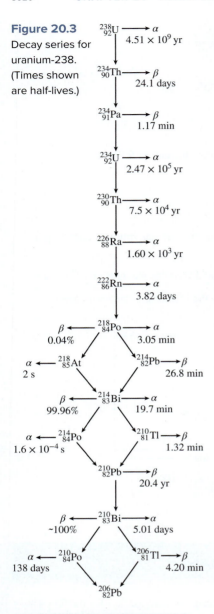

Figure 20.3
Decay series for uranium-238. (Times shown are half-lives.)

20.3 Natural Radioactivity

Nuclei that do not lie within the belt of stability, as well as nuclei with more than 83 protons, tend to be unstable. The spontaneous emission by unstable nuclei of particles or electromagnetic radiation, or both, is known as radioactivity. The main types of radioactivity are the emission of α particles (doubly charged helium nuclei, He^{2+}); the emission of β particles (electrons of nuclear origin); the emission of γ rays, which are very-short-wavelength (0.1 nm to 104 nm) electromagnetic waves; the emission of positrons; and electron capture.

The disintegration of a radioactive nucleus often is the beginning of a ***radioactive decay series,*** which is a *sequence* of nuclear reactions that ultimately result in the formation of a stable isotope. Figure 20.3 shows the decay series of naturally occurring uranium-238, which involves 14 steps. This decay scheme, known as the *uranium decay series,* also shows the half-lives of all the nuclei involved.

It is important to be able to balance the nuclear reaction for each of the steps in a radioactive decay series. For example, the first step in the uranium decay series is the decay of uranium-234 to thorium-234, with the emission of an α particle. Hence, the reaction is represented by:

$$^{238}_{92}U \longrightarrow {}^{234}_{90}Th + {}^{4}_{2}\alpha$$

The next step is represented by:

$$^{234}_{90}Th \longrightarrow {}^{234}_{91}Pa + {}^{0}_{-1}\beta$$

and so on. In a discussion of radioactive decay steps, the beginning radioactive isotope is called the *parent* and the product isotope is called the *daughter.* Thus, $^{238}_{92}U$ is the parent in the first step of the uranium decay series, and $^{234}_{90}Th$ is the daughter.

Kinetics of Radioactive Decay

All radioactive decays obey first-order kinetics. Therefore, the rate of radioactive decay at any time t is given by:

$$\text{rate of decay at time } t = kN$$

where k is the first-order rate constant and N is the number of radioactive nuclei present at time t. According to Equation 14.3, the number of radioactive nuclei at time zero (N_0) and time t (N_t) is:

$$\ln \frac{N_t}{N_0} = -kt$$

and the corresponding half-life of the reaction is given by Equation 14.5:

$$t_{1/2} = \frac{0.693}{k}$$

The half-lives and, therefore, the rate constants of radioactive isotopes vary greatly from nucleus to nucleus. Looking at Figure 20.3, for example, we find that $^{238}_{92}U$ and $^{214}_{84}Po$ are two extreme cases:

$$^{238}_{92}U \longrightarrow {}^{234}_{90}Th + {}^{4}_{2}\alpha \qquad t_{1/2} = 4.51 \times 10^9 \text{ yr}$$

$$^{214}_{84}Po \longrightarrow {}^{210}_{82}Pb + {}^{4}_{2}\alpha \qquad t_{1/2} = 1.6 \times 10^{-4} \text{ s}$$

These two rate constants, after conversion to the same time unit, differ by many orders of magnitude. Furthermore, the rate constants are unaffected by changes in environmental conditions such as temperature and pressure. These highly unusual features are not seen in ordinary chemical reactions (see Table 20.1).

Student Note: Most nuclear scientists (and some general chemistry books) use the symbol λ instead of k for the rate constant of nuclear reactions.

Student Note: Although Equation 14.3 uses concentrations of a reactant at times t and 0, it is the *ratio* of the two that is important, so we can also use the *number* of radioactive nuclei in this equation.

Animation
Nuclear Chemistry—radioactive half-life.

Dating Based on Radioactive Decay

The half-lives of radioactive isotopes have been used as "atomic clocks" to determine the ages of certain objects. Some examples of dating by radioactive decay measurements are described here.

The carbon-14 isotope is produced when atmospheric nitrogen is bombarded by cosmic rays:

$$^{14}_{7}N + ^{1}_{0}n \longrightarrow ^{14}_{6}C + ^{1}_{1}H$$

The radioactive carbon-14 isotope decays according to the equation:

$$^{14}_{6}C \longrightarrow ^{14}_{7}N + ^{0}_{-1}\beta$$

This reaction is the basis of radiocarbon or "carbon-14" dating. To determine the age of an object, we measure the *activity* (disintegrations per second) of ^{14}C and compare it to the activity of ^{14}C in living matter.

Sample Problem 20.3 shows how to use radiocarbon dating to determine the age of an artifact.

SAMPLE PROBLEM 20.3

A wooden artifact is found to have a ^{14}C activity of 9.1 disintegrations per second. Given that the ^{14}C activity of an equal mass of fresh-cut wood has a constant value of 15.2 disintegrations per second, determine the age of the artifact. The half-life of carbon-14 is 5715 years.

Strategy The activity of a radioactive sample is proportional to the number of radioactive nuclei. Thus, we can use Equation 14.3 with activity in place of concentration:

$$\ln \frac{^{14}C \text{ activity in artifact}}{^{14}C \text{ activity in fresh-cut wood}} = -kt$$

To determine k, though, we must solve Equation 14.5, using the value of $t_{1/2}$ for carbon-14 (5715 years) given in the problem statement.

Setup Solving Equation 14.5 for k gives:

$$k = \frac{0.693}{5715 \text{ yr}} = 1.21 \times 10^{-4} \text{ yr}^{-1}$$

Solution

$$\ln \frac{9.1 \text{ disintegrations per second}}{15.2 \text{ disintegrations per second}} = -1.21 \times 10^{-4} \text{ yr}^{-1}(t)$$

$$t = \frac{-0.513}{-1.21 \times 10^{-4} \text{ yr}^{-1}} = 4240 \text{ yr}$$

Therefore, the age of the artifact is 4.2×10^{3} years.

THINK ABOUT IT

Carbon dating cannot be used for objects older than about 60,000 years (about 10 half-lives). After that much time has passed, the activity of carbon-14 has fallen to a level too low to be measured reliably.

Practice Problem ATTEMPT A piece of linen cloth found at an ancient burial site is found to have a ^{14}C activity of 4.8 disintegrations per minute. Determine the age of the cloth. Assume that the carbon-14 activity of an equal mass of living flax (the plant from which linen is made) is 14.8 disintegrations per minute.

Practice Problem BUILD What would be the ^{14}C activity in a 2500-year-old wooden object? Assume that the ^{14}C activity of an equal mass of fresh-cut wood is 13.9 disintegrations per second.

Practice Problem CONCEPTUALIZE The Think About It box in Sample Problem 20.3 explains why carbon dating cannot be used to date objects that are older than 60,000 years. Explain why it also cannot be used to date objects that are only a few years old.

(Continued on next page)

Because some of the intermediate products in the uranium decay series have very long half-lives (see Figure 20.3), this series is particularly suitable for estimating the age of rocks found on Earth and of extraterrestrial objects. The half-life for the first step ($^{238}_{92}$U to $^{234}_{90}$Th) is 4.51×10^9 years. This is about 20,000 times the second largest value (i.e., 2.47×10^5 years), which is the half-life for $^{234}_{92}$U to $^{230}_{90}$Th. As a good approximation, therefore, we can assume that the half-life for the *overall* process (i.e., from $^{238}_{92}$U to $^{206}_{82}$Pb) is equal to the half-life of the first step:

$$^{238}_{92}U \longrightarrow \; ^{206}_{82}Pb + 8\,^4_2\alpha + 6\,^0_{-1}\beta \qquad t_{1/2} = 4.51 \times 10^9 \text{ yr}$$

In naturally occurring uranium minerals, we should and do find some lead-206 formed by radioactive decay. Assuming that no lead was present when the mineral was formed and that the mineral has not undergone chemical changes that would allow the lead-206 isotope to be separated from the parent uranium-238, it is possible to estimate the age of the rocks from the mass ratio of $^{206}_{82}$Pb to $^{238}_{92}$U. According to the preceding nuclear equation, 1 mol (206 g) of lead is formed for every 1 mol (238 g) of uranium that undergoes complete decay. If only half a mole of uranium-238 has undergone decay, the mass ratio ^{206}Pb/^{238}U becomes:

$$\frac{206 \text{ g/2}}{238 \text{ g/2}} = 0.866$$

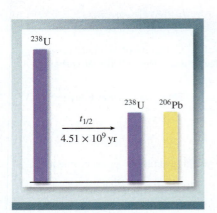

^{238}U

$t_{1/2}$

4.51×10^9 yr

^{238}U ^{206}Pb

Figure 20.4 After one half-life, half of the original uranium-238 has been converted to lead-206.

and the process would have taken a half-life of 4.51×10^9 years to complete (Figure 20.4). Ratios lower than 0.866 mean that the rocks are less than 4.51×10^9 years old, and higher ratios suggest a greater age. Interestingly, studies based on the uranium series, as well as other decay series, put the age of the oldest rocks and, therefore, probably the age of Earth itself, at 4.5×10^9, or 4.5 billion, years.

One of the most important dating techniques in geochemistry is based on the radioactive decay of potassium-40. Radioactive potassium-40 decays by several different modes, but the one relevant for dating is that of electron capture:

$$^{40}_{19}K + \,^0_{-1}e \longrightarrow \; ^{40}_{18}Ar \qquad t_{1/2} = 1.2 \times 10^9 \text{ yr}$$

The accumulation of gaseous argon-40 is used to gauge the age of a specimen. When a potassium-40 atom in a mineral decays, argon-40 is trapped in the lattice of the mineral and can escape only if the material is melted. Melting, therefore, is the procedure for analyzing a mineral sample in the laboratory. The amount of argon-40 present can be conveniently measured with a mass spectrometer. Knowing the ratio of argon-40 to potassium-40 in the mineral and the half-life of decay makes it possible to establish the ages of rocks ranging from millions to billions of years old.

Sample Problem 20.4 shows how to use radioisotopes to determine the age of a specimen.

SAMPLE PROBLEM 20.4

A rock is found to contain 5.51 mg of ^{238}U and 1.63 mg of ^{206}Pb. Determine the age of the rock ($t_{1/2}$ of ^{238}U = 4.51×10^9 yr).

Strategy We must first determine what mass of ^{238}U decayed to produce the measured amount of ^{206}Pb and then use it to determine the original mass of ^{238}U. Knowing the initial and final masses of ^{238}U, we can use Equation 14.3 to solve for t.

Setup

$$1.63 \text{ mg } ^{206}\text{Pb} \times \frac{238 \text{ mg } ^{238}\text{U}}{206 \text{ mg } ^{206}\text{Pb}} = 1.88 \text{ mg } ^{238}\text{U}$$

Thus, the original mass of ^{238}U was 5.51 mg + 1.88 mg = 7.39 mg. The rate constant, k, is determined using Equation 14.5 and $t_{1/2}$ for ^{238}U:

$$k = \frac{0.693}{4.51 \times 10^9} = 1.54 \times 10^{-10} \text{ yr}^{-1}$$

Solution

$$\ln \frac{5.51 \text{ mg}}{7.39 \text{ mg}} = -1.54 \times 10^{-10} \text{ yr}^{-1}(t)$$

$$t = \frac{-0.294}{-1.54 \times 10^{-10} \text{ yr}^{-1}} = 1.91 \times 10^9 \text{ yr}$$

The rock is 1.9 billion years old.

THINK ABOUT IT

This is slightly more complicated than the radiocarbon problem. We cannot use the measured masses of the two isotopes in Equation 14.3 because they are masses of different elements.

Practice Problem A TTEMPT Determine the age of a rock that contains 12.75 mg of ^{238}U and 1.19 mg of ^{206}Pb.

Practice Problem B UILD How much ^{206}Pb will be in a rock sample that is 1.3×10^8 years old and that contains 3.25 mg of ^{238}U?

Practice Problem C ONCEPTUALIZE Isotope X decays to isotope Y with a half-life of 45 days. Which diagram most closely represents the sample of X after 105 days?

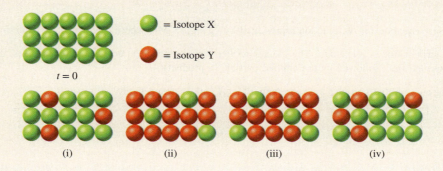

CHECKPOINT – SECTION 20.3   Natural Radioactivity

20.3.1 Determine the age of a rock found to contain 4.31 mg of ^{238}U and 2.47 mg of ^{206}Pb.

a) 7.19×10^9 years

b) 3.62×10^9 years

c) 1.18×10^8 years

d) 3.30×10^9 years

e) 1.78×10^8 years

20.3.2 Determine the ^{14}C activity in disintegrations per second (dps) of a wooden artifact that is 23,000 years old. (Assume that fresh-cut wood of equal mass has an activity of 15.2 dps.)

a) 16.3 dps

b) 15.2 dps

c) 0.935 dps

d) 11.5 dps

e) 4.31 dps

20.3.3 Iron-59 decays to cobalt via beta emission with a half-life of 45.1 days. From the diagram, determine how many half-lives have elapsed since the sample was pure iron-59.

a) 2

b) 3

c) 0.25

d) 8

e) 14

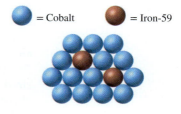

The scope of nuclear chemistry would be rather narrow if study were limited to natural radioactive elements. An experiment performed by Rutherford in 1919, however, suggested the possibility of producing radioactivity artificially. When he bombarded a sample of nitrogen with α particles, the following reaction took place:

$$^{14}_{7}\text{N} + ^{4}_{2}\alpha \longrightarrow ^{17}_{8}\text{O} + ^{1}_{1}\text{p}$$

An oxygen-17 isotope was produced with the emission of a proton. This reaction demonstrated for the first time the feasibility of converting one element into another, by the process of nuclear transmutation. Nuclear transmutation differs from radioactive decay in that transmutation is brought about by the *collision* of two particles.

The preceding reaction can be abbreviated as $^{14}_{7}\text{N}(\alpha,\text{p})^{17}_{8}\text{O}$. In the parentheses the bombarding particle is written first, followed by the emitted particle.

Sample Problem 20.5 shows how to use this notation to represent nuclear transmutations.

SAMPLE PROBLEM **20.5**

Write the balanced nuclear equation for the reaction represented by $^{56}_{26}\text{Fe}(\text{d},\alpha)^{54}_{25}\text{Mn}$, where d represents a deuterium nucleus.

Strategy The species written first is a reactant. The species written last is a product. Within the parentheses, the bombarding particle (a reactant) is written first, followed by the emitted particle (a product).

Setup The bombarding and emitted particles are represented by $^{2}_{1}\text{H}$ and $^{4}_{2}\alpha$, respectively.

Solution

$$^{56}_{26}\text{Fe} + ^{2}_{1}\text{H} \longrightarrow ^{54}_{25}\text{Mn} + ^{4}_{2}\alpha$$

THINK ABOUT IT

Check your work by summing the mass numbers and the atomic numbers on both sides of the equation.

Practice Problem **A**TTEMPT Write an equation for the process represented by $^{106}_{46}\text{Pd}(\alpha,\text{p})^{109}_{47}\text{Ag}$.

Practice Problem **B**UILD Write the abbreviated form of the following process:

$$^{33}_{17}\text{Cl} + ^{1}_{0}\text{n} \longrightarrow ^{31}_{15}\text{P} + ^{3}_{2}\text{He}$$

Practice Problem **C**ONCEPTUALIZE One of the major aspirations of alchemy, the historical precursor to chemistry, was the transmutation of common metals into gold—although transmutation as we know it was not a process that was known to alchemists. Explain why we still mine gold today, despite the fact that it can be produced via transmutation of mercury.

Particle accelerators made it possible to synthesize the so-called *transuranium elements,* elements with atomic numbers greater than 92. Neptunium ($Z = 93$) was first prepared in 1940. Since then, 24 other transuranium elements have been synthesized. All isotopes of these elements are radioactive. Table 20.3 lists the transuranium elements that have been reported and some of the reactions by which they have been produced.

Although light elements are generally not radioactive, they can be made so by bombarding their nuclei with appropriate particles. As we saw in the previous section,

TABLE 20.3	Preparation of the Transuranium Elements		
Atomic Number	**Name**	**Symbol**	**Preparation***
93	Neptunium	Np	$^{238}_{92}U + ^{1}_{0}n \longrightarrow ^{239}_{93}Np + ^{0}_{-1}\beta$
94	Plutonium	Pu	$^{239}_{93}Np \longrightarrow ^{239}_{94}Pu + ^{0}_{-1}\beta$
95	Americium	Am	$^{239}_{94}Pu + ^{1}_{0}n \longrightarrow ^{240}_{95}Am + ^{0}_{-1}\beta$
96	Curium	Cm	$^{239}_{94}Pu + ^{4}_{2}\alpha \longrightarrow ^{242}_{96}Cm + ^{1}_{0}n$
97	Berkelium	Bk	$^{241}_{95}Am + ^{4}_{2}\alpha \longrightarrow ^{243}_{97}Bk + 2^{1}_{0}n$
98	Californium	Cf	$^{242}_{96}Cm + ^{4}_{2}\alpha \longrightarrow ^{245}_{98}Cf + ^{1}_{0}n$
99	Einsteinium	Es	$^{238}_{92}U + 15^{1}_{0}n \longrightarrow ^{253}_{99}Es + 7^{0}_{-1}\beta$
100	Fermium	Fm	$^{238}_{92}U + 17^{1}_{0}n \longrightarrow ^{255}_{100}Fm + 8^{0}_{-1}\beta$
101	Mendelevium	Md	$^{253}_{99}Es + ^{4}_{2}\alpha \longrightarrow ^{256}_{101}Md + ^{1}_{0}n$
102	Nobelium	No	$^{246}_{96}Cm + ^{12}_{6}C \longrightarrow ^{254}_{102}No + 4^{1}_{0}n$
103	Lawrencium	Lr	$^{252}_{98}Cf + ^{10}_{5}B \longrightarrow ^{257}_{103}Lr + 5^{1}_{0}n$
104	Rutherfordium	Rf	$^{249}_{98}Cf + ^{12}_{6}C \longrightarrow ^{257}_{104}Rf + 4^{1}_{0}n$
105	Dubnium	Db	$^{249}_{98}Cf + ^{15}_{7}N \longrightarrow ^{260}_{105}Db + 4^{1}_{0}n$
106	Seaborgium	Sg	$^{249}_{98}Cf + ^{18}_{8}O \longrightarrow ^{263}_{106}Sg + 4^{1}_{0}n$
107	Bohrium	Bh	$^{209}_{83}Bi + ^{54}_{24}Cr \longrightarrow ^{262}_{107}Bh + ^{1}_{0}n$
108	Hassium	Hs	$^{208}_{82}Pb + ^{58}_{26}Fe \longrightarrow ^{265}_{108}Hs + ^{1}_{0}n$
109	Meitnerium	Mt	$^{209}_{83}Bi + ^{58}_{26}Fe \longrightarrow ^{266}_{109}Mt + ^{1}_{0}n$
110	Darmstadtium	Ds	$^{208}_{82}Pb + ^{62}_{28}Ni \longrightarrow ^{269}_{110}Ds + ^{1}_{0}n$
111	Roentgenium	Rg	$^{209}_{83}Bi + ^{64}_{28}Ni \longrightarrow ^{272}_{111}Rg + ^{1}_{0}n$
112	Copernicium	Cn	$^{208}_{82}Pb + ^{70}_{30}Zn \longrightarrow ^{277}_{112}Cn + ^{1}_{0}n$
113	Nihonium	Nh	$^{288}_{115}Mc \longrightarrow ^{284}_{113}Nh + ^{4}_{2}\alpha$
114	Flerovium	Fl	$^{244}_{94}Pu + ^{48}_{20}Ca \longrightarrow ^{289}_{114}Fl + 3^{1}_{0}n$
115	Moscovium	Mc	$^{243}_{95}Am + ^{48}_{20}Ca \longrightarrow ^{288}_{115}Mc + 3^{1}_{0}n$
116	Livermorium	Lv	$^{248}_{96}Cm + ^{48}_{20}Ca \longrightarrow ^{292}_{116}Lv + 4^{1}_{0}n$
117	Tennessine	Ts	$^{249}_{97}Bk + ^{48}_{20}Ca \longrightarrow ^{293}_{117}Ts + 4^{1}_{0}n$
118	Oganesson	Og	$^{249}_{98}Cf + ^{48}_{20}Ca \longrightarrow ^{294}_{118}Og + 3^{1}_{0}n$

*Some of the transuranium elements have been prepared by more than one method.

the radioactive carbon-14 isotope can be prepared by bombarding nitrogen-14 with neutrons. Tritium ($^{3}_{1}H$) is prepared according to the following bombardment:

$$^{6}_{3}Li + ^{1}_{0}n \longrightarrow ^{3}_{1}H + ^{4}_{2}\alpha$$

Tritium decays with the emission of β particles:

$$^{3}_{1}H \longrightarrow ^{3}_{2}He + ^{0}_{-1}\beta \qquad t_{1/2} = 12.5 \text{ yr}$$

Many synthetic isotopes are prepared by using neutrons as projectiles. This approach is particularly convenient because neutrons carry no charges and therefore are not repelled by the targets—the nuclei. In contrast, when the projectiles are positively charged particles (e.g., protons or α particles), they must have considerable kinetic energy to overcome the electrostatic repulsion between themselves and the target nuclei. The synthesis of phosphorus from aluminum is one example:

$$^{27}_{13}Al + ^{4}_{2}\alpha \longrightarrow ^{30}_{15}P + ^{1}_{0}n$$

A *particle accelerator* uses electric and magnetic fields to increase the kinetic energy of charged species so that a reaction will occur (Figure 20.5). Alternating

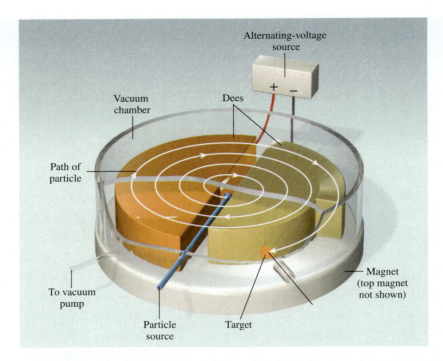

Figure 20.5 Schematic diagram of a cyclotron particle accelerator. The particle (an ion) to be accelerated starts at the center and is forced to move in a spiral path through the influence of electric and magnetic fields until it emerges at a high velocity. The magnetic fields are perpendicular to the plane of the dees (so-called because of their shape), which are hollow and serve as electrodes.

Figure 20.6 Section of a particle accelerator.

Monty Rakusen/Image Source

the polarity (i.e., + and −) on specially constructed plates causes the particles to accelerate along a spiral path. When they have the energy necessary to initiate the desired nuclear reaction, they are guided out of the accelerator into a collision with a target substance.

Various designs have been developed for particle accelerators, one of which accelerates particles along a linear path of about 3 km (Figure 20.6). It is now possible to accelerate particles to a speed well above 90 percent of the speed of light. (According to Einstein's theory of relativity, it is impossible for a particle to move *at* the speed of light. The only exception is the photon, which has a zero rest mass.) The extremely energetic particles produced in accelerators are employed by physicists to smash atomic nuclei to fragments. Studying the debris from such disintegrations provides valuable information about nuclear structure and binding forces.

CHECKPOINT – SECTION 20.4   Nuclear Transmutation

20.4.1 Identify the balanced nuclear equation for the reaction represented by $^{98}_{42}\text{Mo}(\text{d},\text{n})^{99}_{43}\text{Tc}$.

a) $^{98}_{42}\text{Mo} + ^{1}_{1}\text{H} \longrightarrow ^{99}_{43}\text{Tc}$

b) $^{98}_{42}\text{Mo} \longrightarrow ^{99}_{43}\text{Tc} + ^{2}_{1}\text{H} + ^{1}_{0}\text{n}$

c) $^{98}_{42}\text{Mo} + ^{1}_{0}\text{n} \longrightarrow ^{99}_{43}\text{Tc} + ^{2}_{1}\text{H}$

d) $^{98}_{42}\text{Mo} + ^{2}_{1}\text{H} + ^{1}_{0}\text{n} \longrightarrow ^{99}_{43}\text{Tc}$

e) $^{98}_{42}\text{Mo} + ^{2}_{1}\text{H} \longrightarrow ^{99}_{43}\text{Tc} + ^{1}_{0}\text{n}$

20.4.2 Identify the correct abbreviated form of the equation

$$^{125}_{53}\text{I} + ^{0}_{-1}\beta \longrightarrow ^{125}_{52}\text{Te} + \gamma$$

a) $^{125}_{53}\text{I}(\gamma,\beta)^{125}_{52}\text{Te}$

b) $^{125}_{53}\text{I}(\beta,\gamma)^{125}_{52}\text{Te}$

c) $^{125}_{52}\text{Te}(\beta,\gamma)^{125}_{53}\text{I}$

d) $^{125}_{52}\text{I}(\beta,\gamma)^{125}_{53}\text{Te}$

e) $^{125}_{52}\text{I}(\gamma,\beta)^{125}_{53}\text{Te}$

20.5 Nuclear Fission

Nuclear fission is the process in which a heavy nucleus (mass number > 200) divides to form smaller nuclei of intermediate mass and one or more neutrons. Because the heavy nucleus is less stable than its products (see Figure 20.2), this process releases a large amount of energy.

The first nuclear fission reaction to be studied was that of uranium-235 bombarded with slow neutrons, whose speed is comparable to that of air molecules at room temperature. Under these conditions, uranium-235 undergoes fission, as shown in Figure 20.7. Actually, this reaction is very complex: more than 30 different elements have been found among the fission products (Figure 20.8). A representative reaction is:

$$^{235}_{92}\text{U} + ^{1}_{0}\text{n} \longrightarrow ^{90}_{38}\text{Sr} + ^{143}_{54}\text{Xe} + 3^{1}_{0}\text{n}$$

Although many heavy nuclei can be made to undergo fission, only the fission of naturally occurring uranium-235 and of the artificial isotope plutonium-239 has any practical importance. Table 20.4 lists the nuclear binding energies of uranium-235 and its fission products. As the table shows, the binding energy per nucleon for uranium-235 is less than the sum of the binding energies for strontium-90 and xenon-143. Therefore, when a uranium-235 nucleus is split into two smaller nuclei, a certain amount of energy is released.

Let's estimate the magnitude of this energy. The difference between the binding energies of the reactants and products is $(1.23 \times 10^{-10} + 1.92 \times 10^{-10})$ J $- (2.82 \times 10^{-10})$ J, or 3.3×10^{-11} J per uranium-235 nucleus. For 1 mole of uranium-235, the energy released would be $(3.3 \times 10^{-11})(6.02 \times 10^{23})$, or 2.0×10^{13} J. This is an *extremely* exothermic reaction, considering that the heat of combustion of 1 ton of coal is only about 5×10^{7} J.

The significant feature of uranium-235 fission is not just the enormous amount of energy released, but the fact that more neutrons are produced than are originally captured in the process. This property makes possible a **nuclear chain reaction,** which is a self-sustaining sequence of nuclear fission reactions. The neutrons generated during the initial stages of fission can induce fission in other uranium-235 nuclei, which in turn produce more neutrons, and so on. In less than a second, the reaction can become uncontrollable, liberating a tremendous amount of heat to the surroundings. Figure 20.9 shows two types of fission reactions. For a chain reaction to occur, enough uranium-235 must be present in the sample to capture the neutrons. Otherwise, many of the neutrons will escape from the sample and the chain reaction will not occur. In this situation the mass of the sample is said to be *subcritical*. Figure 20.9 shows what happens when the amount of the fissionable material is equal to or greater than the **critical mass,** the minimum mass of fissionable material required to generate a self-sustaining nuclear chain reaction. In this case, most of the neutrons will be captured by uranium-235 nuclei, and a chain reaction will occur.

The first application of nuclear fission was in the development of the atomic bomb. How is such a bomb made and detonated? The crucial factor in the bomb's design is the determination of the critical mass for the bomb. A small atomic bomb is equivalent to 20,000 tons of TNT (trinitrotoluene). Because 1 ton of TNT releases about 4×10^{9} J of energy, 20,000 tons would produce 8×10^{13} J. Recall that 1 mole, or 235 g, of uranium-235 liberates 2.0×10^{13} J of energy when it undergoes fission. Thus, the mass of the isotope present in a small bomb must be at least:

$$(235 \text{ g})\left(\frac{8 \times 10^{13} \text{ J}}{2.0 \times 10^{13} \text{ J}}\right) \approx 1 \text{ kg}$$

An atomic bomb is never assembled with the critical mass already present. Instead, the critical mass is formed by using a conventional explosive, such as TNT, to force the

TABLE 20.4	Nuclear Binding Energies of ^{235}U and Its Fission Products
	Nuclear Binding Energy
^{235}U	2.82×10^{-10} J
^{90}Sr	1.23×10^{-10} J
^{143}Xe	1.92×10^{-10} J

Animation
Nuclear chemistry—Nuclear chain reaction.

Animation
Nuclear chemistry—Nuclear fission.

Figure 20.7

Nuclear Fission and Fusion

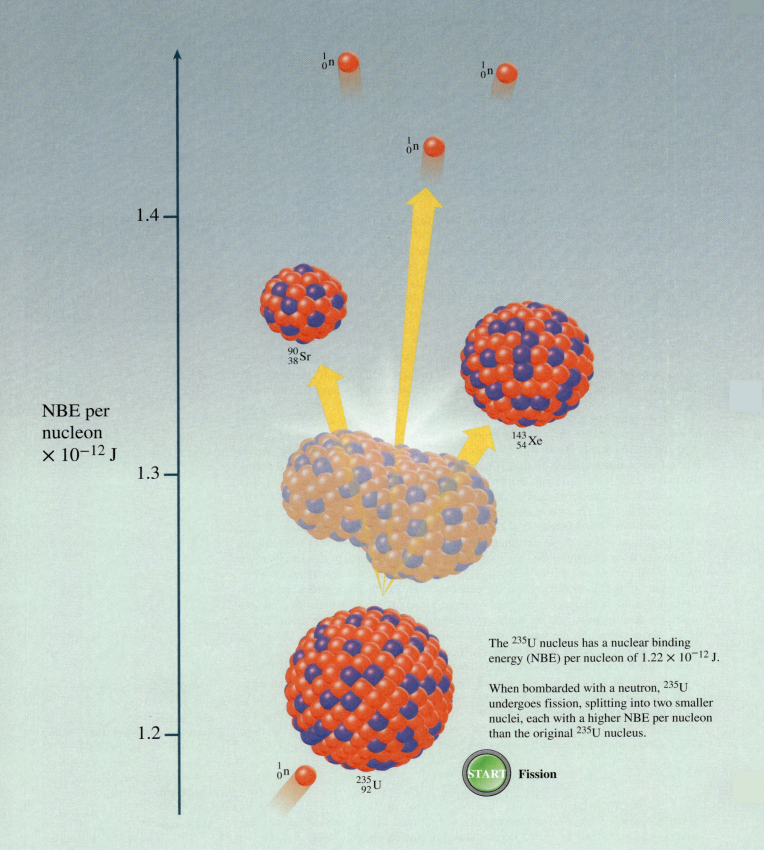

NBE per nucleon $\times 10^{-12}$ J

$^{90}_{38}\text{Sr}$

$^{143}_{54}\text{Xe}$

$^{1}_{0}\text{n}$

$^{235}_{92}\text{U}$

The ^{235}U nucleus has a nuclear binding energy (NBE) per nucleon of 1.22×10^{-12} J.

When bombarded with a neutron, ^{235}U undergoes fission, splitting into two smaller nuclei, each with a higher NBE per nucleon than the original ^{235}U nucleus.

START **Fission**

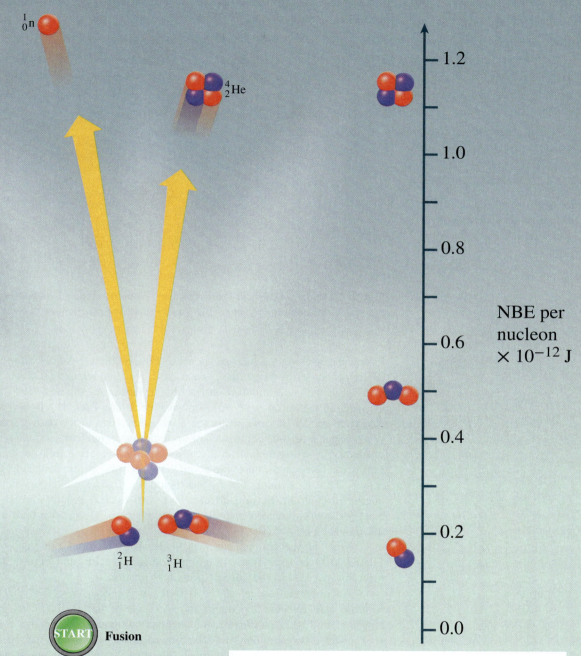

$^1_0$n

$^4_2$He

1.2

1.0

0.8

NBE per
nucleon
$\times 10^{-12}$ J

0.6

0.4

0.2

$^2_1$H $^3_1$H

0.0

START **Fusion**

The ^{2}H and ^{3}H nuclei have the following
nuclear binding energies per nucleon:

^{2}H: 0.185×10^{-12} J
^{3}H: 0.451×10^{-12} J

At very high temperatures, the ^{2}H and ^{3}H nuclei
undergo fusion to produce a ^{4}He nucleus and a
neutron. The ^{4}He nucleus has a significantly
higher NBE per nucleon: 1.13×10^{-12} J

What's the point?

Large nuclei, such as ^{235}U, can achieve greater nuclear stability
by splitting into smaller nuclei with greater NBE per nucleon.
Small nuclei achieve stability by undergoing fusion to produce
a larger nucleus with a greater NBE per nucleon. Note that
different scales are used to show the change in NBE per nucleon
for the two processes. There is a much greater change in NBE
per nucleon in the fusion process than in the fission process. As
with chemical reactions, nuclear reactions are favored when the
products are more stable than the reactants.

(See Visualizing Chemistry questions VC 20.1–VC 20.4 on page 1042.)

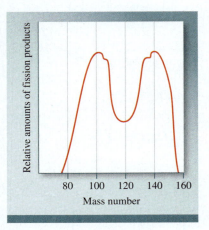

Figure 20.8 Relative yields of the products resulting from the fission of ^{235}U as a function of mass number.

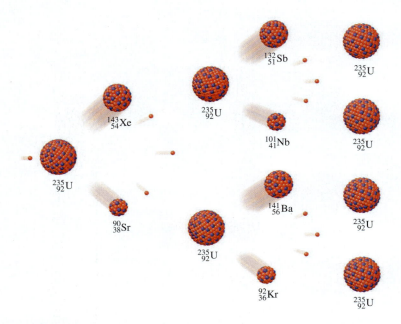

Figure 20.9 If a critical mass is present, many of the neutrons emitted during the fission process will be captured by other ^{235}U nuclei and a chain reaction will occur.

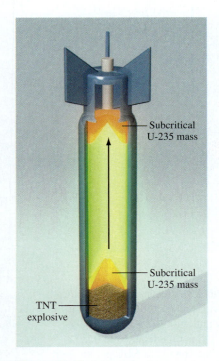

Figure 20.10 Schematic diagram of an atomic bomb. The TNT explosives are set off first. The explosion forces the sections of fissionable material together to form an amount considerably larger than the critical mass.

fissionable sections together, as shown in Figure 20.10. Neutrons from a source at the center of the device trigger the nuclear chain reaction. Uranium-235 was the fissionable material in the bomb dropped on Hiroshima, Japan, on August 6, 1945. Plutonium-239 was used in the bomb exploded over Nagasaki 3 days later. The fission reactions generated were similar in these two cases, as was the extent of the destruction.

A peaceful but controversial application of nuclear fission is the generation of electricity using heat from a controlled chain reaction in a nuclear reactor. Currently, nuclear reactors provide about 20 percent of the electric energy in the United States. This is a small but by no means negligible contribution to the nation's energy production. Several different types of nuclear reactors are in operation; we briefly discuss the main features of three of them, along with their advantages and disadvantages.

Most of the nuclear reactors in the United States are *light water reactors.* Figure 20.11 is a schematic diagram of such a reactor, and Figure 20.12 shows the refueling process in the core of a nuclear reactor.

An important aspect of the fission process is the speed of the neutrons. Slow neutrons split uranium-235 nuclei more efficiently than do fast ones. Because fission reactions are highly exothermic, the neutrons produced usually move at high velocities. For greater efficiency, they must be slowed down before they can be used to induce nuclear disintegration. To accomplish this goal, scientists use **moderators,** which are substances that can reduce the kinetic energy of neutrons. A good moderator must satisfy several requirements: It should be nontoxic and inexpensive (as very large quantities of it are necessary), and it should resist conversion into a radioactive substance by neutron bombardment. Furthermore, it is advantageous for the moderator to be a fluid so that it can also be used as a coolant. No substance fulfills all these requirements, although water comes closer than many others that have been considered. Nuclear reactors that use light water (H_2O) as a moderator are called *light* water reactors because $^{1}_{1}H$ is the lightest isotope of the element hydrogen.

The nuclear fuel consists of uranium, usually in the form of its oxide, U_3O_8. Naturally occurring uranium contains about 0.7 percent of the uranium-235 isotope, which is too low a concentration to sustain a small-scale chain reaction. For effective operation of a light water reactor, uranium-235 must be enriched to a concentration of 3 or 4 percent.

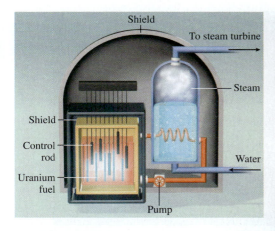

Figure 20.11 Schematic diagram of a nuclear fission reactor. The fission process is controlled by cadmium or boron rods. The heat generated by the process is used to produce steam for the generation of electricity via a heat exchange system.

Figure 20.12 Refueling the core of a nuclear reactor.
Steve Allen/Brand X Pictures

In principle, the main difference between an atomic bomb and a nuclear reactor is that the chain reaction that takes place in a nuclear reactor is kept under control at all times. The factor limiting the rate of the reaction is the number of neutrons present. This can be controlled by lowering cadmium or boron *control rods* between the fuel elements. These rods capture neutrons according to the equations:

$$^{113}_{48}\text{Cd} + ^{1}_{0}\text{n} \longrightarrow ^{114}_{48}\text{Cd} + \gamma$$

$$^{10}_{5}\text{B} + ^{1}_{0}\text{n} \longrightarrow ^{7}_{3}\text{Li} + ^{4}_{2}\alpha$$

where γ denotes gamma rays. Without the control rods, the reactor core would melt from the heat generated and release radioactive materials into the environment. Nuclear reactors have rather elaborate cooling systems that absorb the heat given off by the nuclear reaction and transfer it outside the reactor core, where it is used to produce enough steam to drive an electric generator. In this respect, a nuclear power plant is similar to a conventional power plant that burns fossil fuel. In both cases, large quantities of cooling water are needed to condense steam for reuse. Thus, most nuclear power plants are built near a river or a lake. Unfortunately, this method of cooling causes thermal pollution.

Another type of nuclear reactor uses D_2O, or *heavy* water, as the moderator, rather than H_2O. Deuterium absorbs neutrons much less efficiently than does ordinary hydrogen. Because fewer neutrons are absorbed, the reactor is more efficient and does not require enriched uranium. More neutrons leak out of the reactor, too, though this is not a serious disadvantage.

The main advantage of a heavy water reactor is that it eliminates the need for building expensive uranium enrichment facilities. However, D_2O must be prepared by either fractional distillation or electrolysis of ordinary water, which can be very expensive considering the amount of water used in a nuclear reactor. In countries where hydroelectric power is abundant, the cost of producing D_2O by electrolysis can be reasonably low. At present, Canada is the only nation successfully using heavy water nuclear reactors. The fact that no enriched uranium is required in a heavy water reactor allows a country to enjoy the benefits of nuclear power without undertaking work that is closely associated with weapons technology.

A ***breeder reactor*** uses uranium fuel, but unlike a conventional nuclear reactor, it produces more fissionable materials than it uses.

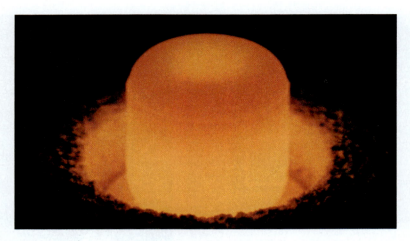

Figure 20.13 Radioactive plutonium oxide (PuO_2) has a red glow.

H.S. Photos/Alamy Stock Photo

When uranium-238 is bombarded with fast neutrons, the following reactions take place:

$$^{238}_{92}\text{U} + ^{1}_{0}\text{n} \longrightarrow ^{239}_{92}\text{U}$$

$$^{239}_{92}\text{U} \longrightarrow ^{239}_{93}\text{Np} + ^{0}_{-1}\beta \qquad t_{1/2} = 23.4 \text{ min}$$

$$^{239}_{93}\text{Np} \longrightarrow ^{239}_{94}\text{Pu} + ^{0}_{-1}\beta \qquad t_{1/2} = 2.35 \text{ days}$$

In this manner, the nonfissionable uranium-238 is transmuted into the fissionable isotope plutonium-239 (Figure 20.13).

In a typical breeder reactor, nuclear fuel containing uranium-235 or plutonium-239 is mixed with uranium-238 so that breeding takes place within the core. For every uranium-235 (or plutonium-239) nucleus undergoing fission, more than one neutron is captured by uranium-238 to generate plutonium-239. Thus, the stockpile of fissionable material can be steadily increased as the starting nuclear fuels are consumed. It takes about 7 to 10 years to regenerate the sizable amount of material needed to refuel the original reactor and to fuel another reactor of comparable size. This interval is called the *doubling time*.

Another *fertile* isotope is $^{232}_{90}\text{Th}$. Upon capturing slow neutrons, thorium is transmuted to uranium-233, which, like uranium-235, is a fissionable isotope:

$$^{232}_{90}\text{Th} + ^{1}_{0}\text{n} \longrightarrow ^{233}_{90}\text{Th}$$

$$^{233}_{90}\text{Th} \longrightarrow ^{233}_{91}\text{Pa} + ^{0}_{-1}\beta \qquad t_{1/2} = 22 \text{ min}$$

$$^{233}_{91}\text{Pa} \longrightarrow ^{233}_{92}\text{U} + ^{0}_{-1}\beta \qquad t_{1/2} = 27.4 \text{ days}$$

Uranium-233 ($t_{1/2} = 1.6 \times 10^5$ years) is stable enough for long-term storage.

Although the amounts of uranium-238 and thorium-232 in Earth's crust are relatively plentiful (4 ppm and 12 ppm by mass, respectively), the development of breeder reactors has been very slow. To date, the United States does not have a single operating breeder reactor, and only a few have been built in other countries, such as France and Russia. One problem is economics; breeder reactors are more expensive to build than conventional reactors. There are also more technical difficulties associated with the construction of such reactors. As a result, the future of breeder reactors, in the United States at least, is rather uncertain.

Many people, including environmentalists, regard nuclear fission as a highly undesirable method of energy production. Many fission products such as strontium-90 are dangerous radioactive isotopes with long half-lives. Plutonium-239, used as a nuclear fuel and produced in breeder reactors, is one of the most toxic substances known. It is an α-emitter with a half-life of 24,400 years.

Accidents, too, present many dangers. An accident at the Three Mile Island reactor in Pennsylvania in 1979 first brought the potential hazards of nuclear plants to public attention. In this instance, very little radiation escaped the reactor, but the plant remained closed for more than a decade while repairs were made and safety issues addressed. Then, on April 26, 1986, a reactor at the Chernobyl nuclear plant in Ukraine surged out of control. The fire and explosion that followed released much radioactive material into the environment. People working near the plant died within weeks as a result of the exposure to the intense radiation. The long-term effect of the radioactive fallout from this incident has not yet been clearly assessed, although agriculture and dairy farming were affected by the fallout. The number of potential cancer deaths attributable to the radiation contamination is estimated to be between a few thousand and more than 100,000.

In addition to the risk of accidents, the problem of radioactive waste disposal has not been satisfactorily resolved even for safely operated nuclear plants. Many suggestions have been made as to where to store or dispose of nuclear waste, including burial underground, burial beneath the ocean floor, and storage in deep geologic formations. But none of these sites has proved absolutely safe in the long run. Leakage of radioactive wastes into underground water, for example, can endanger nearby communities. The ideal disposal site would seem to be the sun, where a bit more radiation would make little difference, but this kind of operation requires space technology that is 100 percent reliable.

Because of the hazards, the future of nuclear reactors is clouded. What was once hailed as the ultimate solution to our energy needs in the twenty-first century is now being debated and questioned by both the scientific community and the general public. It seems likely that the controversy will continue for some time.

20.6 Nuclear Fusion

In contrast to the nuclear fission process, ***nuclear fusion,*** the combining of small nuclei into larger ones, is largely exempt from the waste disposal problem.

Figure 20.2 showed that for the lightest elements, nuclear stability increases with increasing mass number. This behavior suggests that if two light nuclei combine or fuse together to form a larger, more stable nucleus, an appreciable amount of energy will be released in the process. This is the basis for ongoing research into the harnessing of nuclear fusion for the production of energy.

Nuclear fusion occurs constantly in the sun. The sun is made up mostly of hydrogen and helium. In its interior, where temperatures reach about 15 million degrees Celsius, the following fusion reactions are believed to take place:

$$\ _{1}^{1}\text{H} + \ _{1}^{2}\text{H} \longrightarrow \ _{2}^{3}\text{He}$$

$$\ _{2}^{3}\text{He} + \ _{2}^{3}\text{He} \longrightarrow \ _{2}^{4}\text{He} + 2\ _{1}^{1}\text{H}$$

$$\ _{1}^{1}\text{H} + \ _{1}^{1}\text{H} \longrightarrow \ _{1}^{2}\text{H} + \ _{-1}^{0}\beta$$

Because fusion reactions take place only at very high temperatures, they are often called ***thermonuclear reactions.***

A major concern in choosing the proper nuclear fusion process for energy production is the temperature necessary to carry out the process. Some promising reactions are listed here:

Reaction	Energy Released
$\ _{1}^{2}\text{H} + \ _{1}^{2}\text{H} \longrightarrow \ _{1}^{3}\text{H} + \ _{1}^{1}\text{H}$	6.3×10^{-13} J
$\ _{1}^{2}\text{H} + \ _{1}^{3}\text{H} \longrightarrow \ _{2}^{3}\text{He} + 2\ _{0}^{1}\text{n}$	2.8×10^{-12} J
$\ _{3}^{6}\text{Li} + \ _{1}^{2}\text{H} \longrightarrow 2\ _{2}^{4}\text{He}$	3.6×10^{-12} J

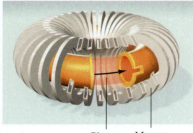

Figure 20.14 A magnetic plasma confinement design called a tokamak.

Figure 20.15 This small-scale fusion reaction was carried out at the Lawrence Livermore National Laboratory using one of the world's most powerful lasers, Nova.

Alliance Images/Alamy Stock Photo

Figure 20.16 Explosion of a thermonuclear bomb.

World History Archive/Alamy Stock Photo

These reactions must take place at extremely high temperatures, on the order of 100 million degrees Celsius, to overcome the repulsive forces between the nuclei. The first reaction is particularly attractive because the world's supply of deuterium is virtually inexhaustible. The total volume of water on Earth is about 1.5×10^{21} L. Because the natural abundance of deuterium is 0.015 percent, the total amount of deuterium present is roughly 4.5×10^{21} g, or 5.0×10^{15} tons. Although it is expensive to prepare deuterium, the cost is minimal compared to the value of the energy released by the reaction.

In contrast to the fission process, nuclear fusion looks like a very promising energy source, at least on paper. Although thermal pollution would be a problem, fusion has the following advantages: (1) the fuels are cheap and almost inexhaustible and (2) the process produces little radioactive waste. If a fusion machine were turned off, it would shut down completely and instantly, without any danger of a meltdown.

If nuclear fusion is so great, why isn't there even one fusion reactor producing energy? Although we possess the scientific knowledge to design such a reactor, the technical difficulties have not yet been solved. The basic problem is finding a way to hold the nuclei together long enough, and at the appropriate temperature, for fusion to occur. At temperatures of about 100 million degrees Celsius, molecules cannot exist, and most or all of the atoms are stripped of their electrons. This state of matter, a gaseous mixture of positive ions and electrons, is called *plasma*. The problem of containing this plasma is a formidable one. No solid container can exist at such temperatures, unless the amount of plasma is small, but then the solid surface would immediately cool the sample and quench the fusion reaction. One approach to solving this problem is to use *magnetic confinement*. Because plasma consists of charged particles moving at high speeds, a magnetic field will exert a force on it. As Figure 20.14 shows, the plasma moves through a doughnut-shaped tunnel, confined by a complex magnetic field. Thus, the plasma never comes in contact with the walls of the container.

Another promising design employs high-power lasers to initiate the fusion reaction. In test runs, a number of laser beams transfer energy to a small fuel pellet, heating it and causing it to *implode*—that is, to collapse inward from all sides and compress into a small volume (Figure 20.15). Consequently, fusion occurs. Like the magnetic confinement approach, laser fusion presents a number of technical difficulties that still need to be overcome before it can be put to practical use on a large scale.

The technical problems inherent in the design of a nuclear fusion reactor do not affect the production of a *hydrogen bomb,* also called a *thermonuclear* bomb. In this case, the objective is all power and no control. Hydrogen bombs do not contain gaseous hydrogen or gaseous deuterium; they contain solid lithium deuteride (LiD), which can be packed very tightly. The detonation of a hydrogen bomb occurs in two stages—first a fission reaction and then a fusion reaction. The required temperature for fusion is achieved with an atomic bomb. Immediately after the atomic bomb explodes, the following fusion reactions occur, releasing vast amounts of energy (Figure 20.16):

$$^{6}_{3}\text{Li} + ^{2}_{1}\text{H} \longrightarrow 2\,^{4}_{2}\alpha$$

$$^{2}_{1}\text{H} + ^{2}_{1}\text{H} \longrightarrow ^{3}_{1}\text{H} + ^{1}_{1}\text{H}$$

There is no critical mass in a fusion bomb, and the force of the explosion is limited only by the quantity of reactants present. Thermonuclear bombs are described as being "cleaner" than atomic bombs because the only radioactive isotopes they produce are tritium, which is a weak β-particle emitter ($t_{1/2} = 12.5$ years), and the products of the fission starter. Their damaging effects on the environment can be aggravated, however, by incorporating in the construction some nonfissionable material such as cobalt. Upon bombardment by neutrons, cobalt-59 is converted to cobalt-60, which is a very strong γ-ray emitter with a half-life of 5.2 years. The presence of radioactive cobalt isotopes in the debris or fallout from a thermonuclear explosion would be fatal to those who survived the initial blast.

20.7 Uses of Isotopes

Radioactive and stable isotopes alike have many applications in science and medicine. We have previously described the use of isotopes in the study of reaction mechanisms [◄◄ Section 14.5] and in dating artifacts (Section 20.3). In this section we discuss a few more examples.

Chemical Analysis

The formula of the thiosulfate ion is $S_2O_3^{2-}$. For some years, chemists were uncertain as to whether the two sulfur atoms occupied equivalent positions in the ion. The thiosulfate ion is prepared by treating the sulfite ion with elemental sulfur:

$$SO_3^{2-}(aq) + S(s) \longrightarrow S_2O_3^{2-}(aq)$$

When thiosulfate is treated with dilute acid, the reaction is reversed. The sulfite ion is re-formed, and elemental sulfur precipitates:

$$S_2O_3^{2-}(aq) \xrightarrow{H^+} SO_3^{2-}(aq) + S(s)$$

If this sequence is started with elemental sulfur enriched with the radioactive sulfur-35 isotope, the isotope acts as a "label" for S atoms. All the labels are found in the sulfur precipitate; none of them appears in the final sulfite ions. As a result, the two atoms of sulfur in $S_2O_3^{2-}$ are not structurally equivalent, as would be the case if the structure were:

$$\left[\ddot{\text{O}} - \ddot{\text{S}} - \ddot{\text{O}} - \ddot{\text{S}} - \ddot{\text{O}} \right]^{2-}$$

If the sulfur atoms were equivalent, the radioactive isotope would be present in both the elemental sulfur precipitate and the sulfite ion. Based on spectroscopic studies, we now know that the structure of the thiosulfate ion is:

$$\left[\begin{array}{c} \ddot{\text{S}} \\ \| \\ \ddot{\text{O}} - \text{S} - \ddot{\text{O}} \\ \| \\ \ddot{\text{O}} \end{array} \right]^{2-}$$

The study of photosynthesis is also rich with isotope applications. The overall photosynthesis reaction can be represented as:

$$6CO_2 + 6H_2O \longrightarrow C_6H_{12}O_6 + 6O_2$$

In Section 14.5, we learned that the ^{18}O isotope was used to determine the source of O_2. The radioactive ^{14}C isotope helped to determine the path of carbon in photosynthesis. Starting with $^{14}CO_2$, it was possible to isolate the intermediate products during photosynthesis and measure the amount of radioactivity of each carbon-containing compound. In this manner, the path from CO_2 through various intermediate compounds to carbohydrate could be clearly charted. Isotopes, especially radioactive isotopes that are used to trace the path of the atoms of an element in a chemical or biological process, are called *tracers.*

Isotopes in Medicine

Tracers are also used for diagnosis in medicine. Sodium-24 (a β-emitter with a half-life of 14.8 h) injected into the bloodstream as a salt solution can be monitored to trace the flow of blood and detect possible constrictions or obstructions in the circulatory system. Iodine-131 (a β-emitter with a half-life of 8 days) has been used to test the activity of the thyroid gland. A malfunctioning thyroid can be detected by giving the patient a drink of a solution containing a known amount of $Na^{131}I$ and measuring the radioactivity just above the thyroid to see if the iodine is absorbed at the normal rate. Another radioactive isotope of iodine, iodine-123 (a γ-ray emitter), is used to image the brain (Figure 20.17). In each of these cases, though, the amount of

Animation
Nuclear chemistry—Nuclear medical techniques.

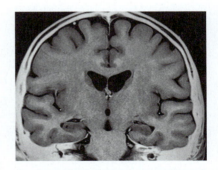

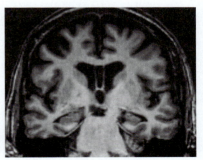

Figure 20.17 ^{123}I image of a normal brain (top) and the brain of an Alzheimer's victim (bottom).

(top): *Flowersandtraveling/Shutterstock*
(bottom): *Atthapon Raksthaput/Shutterstock*

Figure 20.18 Schematic diagram of a Geiger counter. Radiation (α, β, or γ rays) entering through the window ionizes the argon gas to generate a small current flow between the electrodes. This current is amplified and is used to flash a light or operate a counter with a clicking sound.

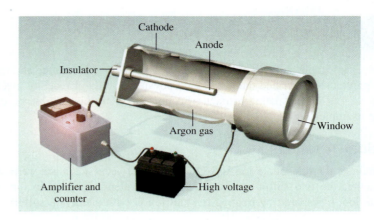

radioisotope used must be kept small to prevent the patient from suffering permanent damage from the high-energy radiation.

Technetium, the first artificially prepared element, is one of the most useful elements in nuclear medicine. Although technetium is a transition metal, all its isotopes are radioactive. In the laboratory it is prepared by the nuclear reactions:

$$\begin{align} ^{98}_{42}\text{Mo} + ^{1}_{0}\text{n} &\longrightarrow ^{99}_{42}\text{Mo} \\ ^{99}_{42}\text{Mo} &\longrightarrow ^{99m}_{43}\text{Tc} + ^{0}_{-1}\beta \end{align}$$

where the superscript "m" denotes that the technetium-99 isotope is produced in its *excited* nuclear state. This isotope has a half-life of about 6 hours, decaying by γ radiation to technetium-99 in its nuclear *ground* state. Thus, it is a valuable diagnostic tool. The patient either drinks or is injected with a solution containing ^{99m}Tc. By detecting the γ rays emitted by ^{99m}Tc, doctors can obtain images of organs such as the heart, liver, and lungs.

A major advantage of using radioactive isotopes as tracers is that they are easy to detect. Their presence even in very small amounts can be detected by photographic techniques or by devices known as counters. Figure 20.18 is a diagram of a Geiger counter, an instrument widely used in scientific work and medical laboratories to detect radiation.

20.8 Biological Effects of Radiation

In this section, we examine briefly the effects of radiation on biological systems. But first we must define the quantitative measures of radiation. The fundamental unit of radioactivity is the *curie* (Ci); 1 Ci corresponds to exactly 3.70×10^{10} nuclear disintegrations per second. This decay rate is equivalent to that of 1 g of radium. A *millicurie* (mCi) is one-thousandth of a curie. Thus, 10 mCi of a carbon-14 sample is the quantity that undergoes $(10 \times 10^{-3})(3.70 \times 10^{10}) = 3.70 \times 10^{8}$ disintegrations per second.

The intensity of radiation depends on the number of disintegrations as well as on the energy and type of radiation emitted. One common unit for the absorbed dose of radiation is the *rad* (radiation *a*bsorbed *d*ose), which is the amount of radiation that results in the absorption of 1×10^{-5} J per gram of irradiated material. The biological effect of radiation depends on the part of the body irradiated and the type of radiation. For this reason, the rad is often multiplied by a factor called the *RBE* (relative *b*iological *e*ffectiveness). The product is called a *rem* (*r*oentgen *e*quivalent for *m*an):

$$\text{number of rems} = \text{number of rads} \times 1\ \text{RBE}$$

Of the three types of nuclear radiation, α particles usually have the least-penetrating power. Beta particles are more penetrating than α particles, but less so than γ rays. Gamma rays have very short wavelengths and high energies. Furthermore, because they carry no charge, they cannot be stopped by shielding materials as easily

Animation
Nuclear Chemistry—alpha, beta, and gamma emission.

TABLE 20.5	Average Yearly Radiation Doses for Americans
Source	**Dose (mrem/yr)***
Cosmic rays	20–50
Ground and surroundings	25
Human body[†]	26
Medical and dental X rays	50–75
Air travel	5
Fallout from weapons tests	5
Nuclear waste	2
Total	**133–188**

*1 mrem = millirem = 1×10^{-3} rem.
[†]The radioactivity in the body comes from food and air.

as α and β particles. If α- or β-emitters are ingested or inhaled, however, their damaging effects are greatly aggravated because the organs will be constantly subject to damaging radiation at close range. For example, strontium-90, a β-emitter, can replace calcium in bones, where it does the greatest damage.

Table 20.5 lists the average amounts of radiation an American receives every year. For short-term exposures to radiation, a dosage of 50 to 200 rems will cause a decrease in white blood cell counts and other complications, while a dosage of 500 rems or greater may result in death within weeks. Current safety standards permit nuclear workers to be exposed to no more than 5 rems per year and specify a maximum of 0.5 rem of human-made radiation per year for the general public.

The chemical basis of radiation damage is that of ionizing radiation. Radiation (of either particles or γ rays) can remove electrons from atoms and molecules in its path, leading to the formation of ions and radicals. **Radicals** (also called *free radicals*) are molecular fragments having one or more unpaired electrons; they are usually short lived and highly reactive. When water is irradiated with γ rays, for example, the following reactions take place:

$$H_2O \xrightarrow{\text{radiation}} H_2O^+ + e^-$$

$$H_2O^+ + H_2O \longrightarrow H_3O^+ + \cdot OH$$
$$\text{hydroxyl radical}$$

The electron (in the hydrated form) can subsequently react with water or with a hydrogen ion to form atomic hydrogen, and with oxygen to produce the superoxide ion (O_2^-) (a radical):

$$e^- + O_2 \longrightarrow \cdot O_2^-$$

In the tissues, the superoxide ions and other free radicals attack cell membranes and a host of organic compounds, such as enzymes and DNA molecules. Organic compounds can themselves be directly ionized and destroyed by high-energy radiation.

It has long been known that exposure to high-energy radiation can induce cancer in humans and other animals. Cancer is characterized by uncontrolled cellular growth. On the other hand, it is also well established that cancer cells can be destroyed by proper radiation treatment. In radiation therapy, a compromise is sought. The radiation to which the patient is exposed must be sufficient to destroy cancer cells without killing too many normal cells and, it is hoped, without inducing another form of cancer.

Radiation damage to living systems is generally classified as *somatic* or *genetic*. Somatic injuries are those that affect the organism during its own lifetime. Sunburn, skin rash, cancer, and cataracts are examples of somatic damage. Genetic damage means inheritable changes or gene mutations. For example, a person whose chromosomes have been damaged or altered by radiation may have deformed offspring.

Bringing Chemistry to Life

Radioactivity in Tobacco

"SURGEON GENERAL'S WARNING: Smoking Is Hazardous to Your Health." Warning labels such as this appear on every package of cigarettes sold in the United States. The link between cigarette smoke and cancer has long been established. There is, however, another cancer-causing mechanism in smokers. The culprit in this case is a radioactive environmental pollutant present in the tobacco leaves from which cigarettes are made.

The soil in which tobacco is grown is heavily treated with phosphate fertilizers, which are rich in uranium and its decay products. Consider a particularly important step in the uranium-238 decay series:

$$\ce{^{226}_{88}Ra} \longrightarrow \ce{^{222}_{86}Rn} + \ce{^{4}_{2}\alpha}$$

The product formed, radon-222, is an unreactive gas. (Radon is the only gaseous species in the uranium-238 decay series.) Radon-222 emanates from radium-226 and is present at high concentrations in soil gas and in the surface air layer under the vegetation canopy provided by the field of growing tobacco. In this layer some of the daughters of radon-222, such as polonium-218 and lead-214, become firmly attached to the surface and interior of tobacco leaves. As Figure 20.3 shows, the next few decay reactions leading to the formation of lead-210 proceed rapidly. Gradually, the concentration of radioactive lead-210 can build to quite a high level.

During the combustion of a cigarette, tiny insoluble smoke particles are inhaled and deposited in the respiratory tract of the smoker and are eventually transported and stored at sites in the liver, spleen, and bone marrow. Measurements indicate a high lead-210 content in these particles. The lead-210 content is not high enough to be hazardous *chemically* (it is insufficient to cause *lead poisoning*), but it is hazardous because it is radioactive. Because of its long half-life (20.4 years), lead-210 and its radioactive daughters—bismuth-210 and polonium-210—continue to build up in the body of a smoker over the years. Constant exposure of the organs and bone marrow to α- and β-particle radiation increases the probability the smoker will develop cancer. The overall impact on health is similar to that caused by indoor radon gas.

Chapter Summary

Section 20.1

- Spontaneous emission of particles or radiation from unstable nuclei is known as **radioactivity.** Unstable nuclei emit α particles, β particles, **positrons,** or γ rays.

- **Nuclear transmutation** is the conversion of one nucleus to another. Nuclear reactions are balanced by summing the mass numbers and the atomic numbers.

Section 20.2

- Stable nuclei with low atomic numbers have neutron-to-proton ratios close to 1. Heavier stable nuclei have higher ratios. Nuclear stability is favored by certain numbers of nucleons including even numbers and "magic" numbers.

- The difference between the actual mass of a nucleus and the mass calculated by summing the masses of the individual nucleons is the **mass defect.**

- **Nuclear binding energy,** determined by using Einstein's equation $E = mc^2$, is a measure of nuclear stability.

Section 20.3

- Uranium-238 is the parent of a natural **radioactive decay series** that can be used to determine the ages of rocks. Radiocarbon dating is done using carbon-14.

Section 20.4

- **Transuranium elements** are created by bombarding other elements with accelerated neutrons, protons, α particles, or other nuclei.

Section 20.5

- **Nuclear fission** is the splitting of a large nucleus into two smaller nuclei and one or more neutrons. When the free neutrons are captured efficiently by other nuclei, a **nuclear chain reaction** can occur in which the fission process is sustained. The minimum amount of fissionable material required to sustain the reaction is known as the **critical mass.**

- Nuclear reactors use the heat from a controlled nuclear fission reaction to produce power. Fission is controlled, in part, by **moderators**—materials that limit the speed of liberated neutrons but that do not themselves undergo fission when bombarded with neutrons. The three important types of reactors are light water reactors, heavy water reactors, and **breeder reactors.** Breeder reactors produce more fissionable material than they consume.

Section 20.6

- **Nuclear fusion,** the type of reaction that occurs in the sun, is the combination of two light nuclei to form one heavier nucleus. Fusion reactions are sometimes referred to as **thermonuclear reactions** because they take place only at very high temperatures.

Section 20.7

- Radioactive isotopes are easy to detect and thus make excellent **tracers** in chemical reactions and in medical procedures.

Section 20.8

- High-energy radiation damages living systems by causing ionization and the formation of **radicals,** or *free radicals,* which are chemical species with unpaired electrons.

Key Words

Breeder reactor, 1031	Nuclear binding energy, 1016	Nuclear transmutation, 1012	Radioactivity, 1012
Critical mass, 1027	Nuclear chain reaction, 1027	Positron, 1013	Thermonuclear reaction, 1033
Mass defect, 1017	Nuclear fission, 1027	Radical, 1037	Tracers, 1035
Moderator, 1030	Nuclear fusion, 1033	Radioactive decay series, 1020	Transuranium elements, 1024

Key Equation

20.1 $\Delta E = (\Delta m)c^2$ Einstein's equation allows us to calculate the energy change associated with the loss of a given mass (Δm) in a nuclear process. Mass must be expressed in kg for units to cancel properly.

Questions and Problems

Applying What You've Learned

In addition to BNCT, another promising treatment for brain tumors is brachytherapy using iodine-125. In brachytherapy, "seeds" containing ^{125}I are implanted directly into the tumor. As the radioisotope decays, γ rays destroy the tumor cells. Careful implantation prevents the radiation from harming nearby healthy cells.

Problems:

(a) ^{125}I is produced by a two-step process in which ^{124}Xe nuclei are bombarded with neutrons to produce ^{125}Xe—a process called *neutron activation*. ^{125}Xe then decays by electron capture to produce ^{125}I, which also decays by electron capture. Write nuclear equations for the two steps that produce ^{125}I from ^{124}Xe, and identify the product of the electron-capture decay of ^{125}I [◄◄ Sample Problem 20.1]. (b) The mass of an ^{125}I nucleus is 124.904624 amu. Calculate the nuclear binding energy and the nuclear binding energy per nucleon [◄◄ Sample Problem 20.2]. (c) The half-life of ^{125}I is 59.4 days. How long will it take for the activity of implanted ^{125}I seeds to fall to 5.00 percent of its original value [◄◄ Sample Problem 20.3]? (d) Iridium-192 is another isotope used in brachytherapy. It is produced by a nuclear transmutation. Identify the target nucleus X, and write the balanced nuclear equation for the reaction represented by $^{191}X(n,\gamma)^{192}Ir$ [◄◄ Sample Problem 20.5].

Brachytherapy seeds (shown with a penny to illustrate their size).
David A. Tietz/Editorial Image, LLC

SECTION 20.1: NUCLEI AND NUCLEAR REACTIONS

Review Questions

20.1   How do nuclear reactions differ from ordinary chemical reactions?

20.2   What are the steps in balancing nuclear equations?

20.3   What is the difference between $_{-1}^{0}e$ and $_{-1}^{0}\beta$?

20.4   What is the difference between an electron and a positron?

Problems

20.5   Complete the following nuclear equations, and identify X in each case:

(a) $_{12}^{26}Mg + _{1}^{1}p \longrightarrow \alpha + X$

(b) $_{27}^{59}Co + _{1}^{2}H \longrightarrow _{27}^{60}Co + X$

(c) $_{92}^{235}U + _{0}^{1}n \longrightarrow _{36}^{94}Kr + _{56}^{139}Ba + 3X$

(d) $_{24}^{53}Cr + _{2}^{4}\alpha \longrightarrow _{0}^{1}n + X$

(e) $_{8}^{20}O \longrightarrow _{9}^{20}F + X$

20.6   Complete the following nuclear equations, and identify X in each case:

(a) $_{53}^{135}I \longrightarrow _{54}^{135}Xe + X$

(b) $_{19}^{40}K \longrightarrow _{-1}^{0}\beta + X$

(c) $_{27}^{59}Co + _{0}^{1}n \longrightarrow _{25}^{56}Mn + X$

(d) $_{92}^{235}U + _{0}^{1}n \longrightarrow _{40}^{99}Zr + _{52}^{135}Te + 2X$

SECTION 20.2: NUCLEAR STABILITY

Review Questions

20.7   State the general rules for predicting nuclear stability.

20.8   What is the belt of stability?

20.9   Why is it impossible for the isotope $_{2}^{2}He$ to exist?

20.10   Define *nuclear binding energy, mass defect,* and *nucleon.*

20.11   How does Einstein's equation, $E = mc^2$, enable us to calculate nuclear binding energy?

20.12   Why is it preferable to use nuclear binding energy per nucleon for a comparison of the stabilities of different nuclei?

Computational Problems

20.13   The radius of a uranium-235 nucleus is about 7.0×10^{-3} pm. Calculate the density of the nucleus in g/cm^3. (Assume the atomic mass is 235 amu.)

20.14   For each pair of isotopes listed, predict which one is less stable: (a) $_{3}^{6}Li$ or $_{3}^{9}Li$, (b) $_{11}^{23}Na$ or $_{11}^{25}Na$, (c) $_{20}^{48}Ca$ or $_{21}^{48}Sc$.

20.15   For each pair of elements listed, predict which one has more stable isotopes: (a) Co or Ni, (b) F or Se, (c) Ag or Cd.

20.16   In each pair of isotopes shown, indicate which one you would expect to be radioactive: (a) $_{10}^{20}Ne$ or $_{10}^{17}Ne$, (b) $_{20}^{40}Ca$ or $_{20}^{45}Ca$, (c) $_{42}^{95}Mo$ or $_{43}^{92}Tc$, (d) $_{80}^{195}Hg$ or $_{80}^{196}Hg$, (e) $_{83}^{209}Bi$ or $_{96}^{242}Cm$.

20.17 Given that:

$$H(g) + H(g) \longrightarrow H_2(g) \qquad \Delta H^\circ = -436.4 \text{ kJ/mol}$$

calculate the change in mass (in kg) per mole of H_2 formed.

20.18 Estimates show that the total energy output of the sun is 5×10^{26} J/s. What is the corresponding mass loss in kg/s of the sun?

20.19 Calculate the nuclear binding energy (in joules) and the binding energy per nucleon of the following isotopes: (a) $_3^7\text{Li}$ (7.01600 amu) and (b) $_{17}^{35}\text{Cl}$ (34.96885 amu).

20.20 Calculate the nuclear binding energy (in joules) and the binding energy per nucleon of the following isotopes: (a) $_2^4\text{He}$ (4.002603 amu) and (b) $_{74}^{184}\text{W}$ (183.950928 amu).

20.21 Given that the nuclear binding energy of ^{48}Cr is 1.37340×10^{-12} J/nucleon, calculate the mass of a single ^{48}Cr atom.

20.22 Given that the nuclear binding energy of ^{192}Ir is 1.27198×10^{-12} J/nucleon, calculate the mass of a single ^{192}Ir atom.

SECTION 20.3: NATURAL RADIOACTIVITY

Review Questions

20.23 Discuss factors that lead to nuclear decay.

20.24 Outline the principle for dating materials using radioactive isotopes.

Computational Problems

20.25 Fill in the blanks in the following radioactive decay series:

(a) $^{232}\text{Th} \xrightarrow{\alpha} \underline{\hspace{1cm}} \xrightarrow{\beta} \underline{\hspace{1cm}} \xrightarrow{\beta} {}^{228}\text{Th}$

(b) $^{235}\text{U} \xrightarrow{\alpha} \underline{\hspace{1cm}} \xrightarrow{\beta} \underline{\hspace{1cm}} \xrightarrow{\alpha} {}^{227}\text{Ac}$

(c) $\underline{\hspace{1cm}} \xrightarrow{\alpha} {}^{233}\text{Pa} \xrightarrow{\beta} \underline{\hspace{1cm}} \xrightarrow{\alpha} \underline{\hspace{1cm}}$

20.26 A radioactive substance undergoes decay as follows:

Time (days)	Mass (g)
0	500
1	389
2	303
3	236
4	184
5	143
6	112

Calculate the first-order decay constant and the half-life of the reaction.

20.27 The radioactive decay of Tl-206 to Pb-206 has a half-life of 4.20 min. Starting with 5.00×10^{22} atoms of Tl-206, calculate the number of such atoms left after 42.0 min.

20.28 A freshly isolated sample of ^{90}Y was found to have an activity of 9.8×10^5 disintegrations per minute at 1:00 P.M. on December 3, 2006. At 2:15 P.M. on December 17, 2006, its activity was measured again and found to be 2.6×10^4 disintegrations per minute. Calculate the half-life of ^{90}Y.

20.29 A wooden artifact has a ^{14}C activity of 18.9 disintegrations per minute, compared to 27.5 disintegrations per minute for live wood. Given that the half-life of ^{14}C is 5715 years, determine the age of the artifact.

20.30 In the thorium decay series, thorium-232 loses a total of six α particles and four β particles in a 10-stage process. What is the final isotope produced?

20.31 Consider the decay series $A \longrightarrow B \longrightarrow C \longrightarrow D$ where A, B, and C are radioactive isotopes with half-lives of 4.50 s, 15.0 days, and 1.00 s, respectively, and D is nonradioactive. Starting with 1.00 mole of A, and none of B, C, or D, calculate the number of moles of A, B, C, and D left after 30 days.

20.32 The activity of radioactive carbon-14 decay of a piece of charcoal found at a volcanic site is 11.2 disintegrations per second. If the activity of carbon-14 decay in an equal mass of living matter is 18.3 disintegrations per second, what is the age of the charcoal? (See Problem 20.29 for the half-life of carbon-14.)

20.33 The age of some animal bones was determined by carbon-14 dating to be 8.4×10^3 years old. Calculate the activity of the carbon-14 in the bones in disintegrations per minute per gram, given that the original activity was 15.3 disintegrations per minute per gram. (See Problem 20.29 for the half-life of carbon-14.)

20.34 Given that the half-life of ^{238}U is 4.51×10^9 years, determine the age of a rock found to contain 1.09 mg ^{238}U and 0.08 mg ^{206}Pb.

20.35 Determine the ratio of ^{238}U to ^{206}Pb in a rock that is 1.7×10^8 years old. (See Problem 20.34 for the half-life of ^{238}U.)

SECTION 20.4: NUCLEAR TRANSMUTATION

Review Questions

20.36 What is the difference between radioactive decay and nuclear transmutation?

20.37 How is nuclear transmutation achieved in practice?

Conceptual Problems

20.38 Write balanced nuclear equations for the following reactions, and identify X: (a) $\text{X}(p,\alpha)_6^{12}\text{C}$, (b) $_{13}^{27}\text{Al}(d,\alpha)\text{X}$, (c) $_{25}^{55}\text{Mn}(n,\gamma)\text{X}$.

20.39 Write the abbreviated forms for the following reactions:
(a) $^{14}_{7}N + ^{4}_{2}\alpha \longrightarrow ^{17}_{8}O + ^{1}_{1}p$
(b) $^{9}_{4}Be + ^{4}_{2}\alpha \longrightarrow ^{12}_{6}C + ^{1}_{0}n$
(c) $^{238}_{92}U + ^{2}_{1}H \longrightarrow ^{238}_{93}Np + 2^{1}_{0}n$

20.40 Write balanced nuclear equations for the following reactions, and identify X: (a) $^{80}_{34}Se(d,p)X$, (b) $X(d,2p)^{9}_{3}Li$, (c) $^{10}_{5}B(n,\alpha)X$.

20.41 Write the abbreviated forms for the following reactions:
(a) $^{40}_{20}Ca + ^{2}_{1}H \longrightarrow ^{41}_{20}Ca + ^{1}_{1}p$
(b) $^{32}_{16}S + ^{1}_{0}n \longrightarrow ^{32}_{15}P + ^{1}_{1}p$
(c) $^{239}_{94}Pu + ^{4}_{2}\alpha \longrightarrow ^{242}_{96}Cm + ^{1}_{0}n$

20.42 Describe how you would prepare astatine-211, starting with bismuth-209.

20.43 A long-cherished dream of alchemists was to produce gold from cheaper and more abundant elements. This dream was finally realized when $^{198}_{80}Hg$ was converted into gold by neutron bombardment. Write a balanced equation for this reaction.

SECTION 20.5: NUCLEAR FISSION

▶▶▶ Visualizing Chemistry
Figure 20.7

VC 20.1 The fission of ^{235}U can result in a variety of products, including those shown in Figure 20.7. Which of the following equations does *not* represent another possible fission process?
a) $^{1}_{0}n + ^{235}_{92}U \longrightarrow ^{137}_{52}Te + ^{97}_{40}Zr + 2^{1}_{0}n$
b) $^{1}_{0}n + ^{235}_{92}U \longrightarrow ^{142}_{56}Ba + ^{91}_{36}Kr + 3^{1}_{0}n$
c) $^{1}_{0}n + ^{235}_{92}U \longrightarrow ^{137}_{55}Cs + ^{90}_{37}Rb + 3^{1}_{0}n$

VC 20.2 How many neutrons are produced in the fission reaction shown?
$$^{1}_{0}n + ^{239}_{94}Pu \longrightarrow ^{109}_{44}Ru + ^{129}_{50}Sn + \_\_\_^{1}_{0}n$$
a) 1
b) 2
c) 3

VC 20.3 The fission of ^{235}U shown in Figure 20.7 is represented by the equation:
$$^{1}_{0}n + ^{235}_{92}U \longrightarrow ^{143}_{54}Xe + ^{90}_{38}Sr + 3^{1}_{0}n$$
How does the combined mass of products compare to the combined mass of reactants for this process?
a) The combined mass of products is smaller than the combined mass of reactants.
b) The combined mass of products is larger than the combined mass of reactants.
c) The combined mass of products is equal to the combined mass of reactants.

VC 20.4 The fusion of $^{2}_{1}H$ and $^{3}_{1}H$ shown in Figure 20.7 is represented by the equation:
$$^{2}_{1}H + ^{3}_{1}H \longrightarrow ^{4}_{2}He + ^{1}_{0}n$$

How does the combined mass of products compare to the combined mass of reactants for this process?
a) The combined mass of products is smaller than the combined mass of reactants.
b) The combined mass of products is larger than the combined mass of reactants.
c) The combined mass of products is equal to the combined mass of reactants.

Review Questions

20.44 Define *nuclear fission, nuclear chain reaction,* and *critical mass.*

20.45 Which isotopes can undergo nuclear fission?

20.46 Explain how an atomic bomb works.

20.47 Explain the functions of a moderator and a control rod in a nuclear reactor.

20.48 Discuss the differences between a light water and a heavy water nuclear fission reactor. What are the advantages of a breeder reactor over a conventional nuclear fission reactor?

20.49 No form of energy production is without risk. Make a list of the risks to society involved in fueling and operating a conventional coal-fired electric power plant, and compare them with the risks of fueling and operating a nuclear fission-powered electric plant.

SECTION 20.6: NUCLEAR FUSION

Review Questions

20.50 Define *nuclear fusion, thermonuclear reaction,* and *plasma.*

20.51 Why do heavy elements such as uranium undergo fission whereas light elements such as hydrogen and lithium undergo fusion?

20.52 How does a hydrogen bomb work?

20.53 What are the advantages of a fusion reactor over a fission reactor? What are the practical difficulties in operating a large-scale fusion reactor?

SECTION 20.7: USES OF ISOTOPES

Conceptual Problems

20.54 Describe how you would use a radioactive iodine isotope to demonstrate that the following process is in dynamic equilibrium:
$$PbI_2(s) \rightleftharpoons Pb^{2+}(aq) + 2I^-(aq)$$

20.55 Consider the following redox reaction:
$$IO_4^-(aq) + 2I^-(aq) + H_2O(l) \longrightarrow I_2(s) + IO_3^-(aq) + 2OH^-(aq)$$
When KIO_4 is added to a solution containing iodide ions labeled with radioactive iodine-128, all the radioactivity appears in I_2 and none in the IO_3^- ion. What can you deduce about the mechanism for the redox process?

20.56   Explain how you might use a radioactive tracer to show that ions are not completely motionless in crystals.

20.57   Each molecule of hemoglobin, the oxygen carrier in blood, contains four Fe atoms. Explain how you would use the radioactive $^{59}_{26}Fe$ ($t_{1/2}$ = 46 days) to show that the iron in a certain food is converted into hemoglobin.

SECTION 20.8: BIOLOGICAL EFFECTS OF RADIATION

Review Questions

20.58   List the factors that affect the intensity of radiation from a radioactive element.

20.59   What are *rad* and *rem,* and how are they related?

20.60   Explain, with examples, the difference between somatic and genetic radiation damage.

20.61   Compare the extent of radiation damage done by α, β, and γ sources.

ADDITIONAL PROBLEMS

20.62   How does a Geiger counter work?

20.63   Strontium-90 is one of the products of the fission of uranium-235. This strontium isotope is radioactive, with a half-life of 28.1 years. Calculate how long (in years) it will take for 1.00 g of the isotope to be reduced to 0.200 g by decay.

20.64   Nuclei with an even number of protons and an even number of neutrons are more stable than those with an odd number of protons and/or an odd number of neutrons. What is the significance of the even numbers of protons and neutrons in this case?

20.65   Tritium (^3H) is radioactive and decays by electron emission. Its half-life is 12.5 years. In ordinary water the ratio of ^1H to ^3H atoms is 1.0×10^{17} to 1. (a) Write a balanced nuclear equation for tritium decay. (b) How many disintegrations will be observed per minute in a 1.00-kg sample of water?

20.66   (a) What is the activity, in millicuries, of a 0.500-g sample of $^{237}_{93}Np$? (This isotope decays by α-particle emission and has a half-life of 2.20×10^6 years.) (b) Write a balanced nuclear equation for the decay of $^{237}_{93}Np$.

20.67   The following equations are for nuclear reactions that are known to occur in the explosion of an atomic bomb. Identify X.
(a) $^{235}_{92}U + ^1_0n \longrightarrow ^{140}_{56}Ba + 3^1_0n + X$
(b) $^{235}_{92}U + ^1_0n \longrightarrow ^{144}_{55}Cs + ^{90}_{37}Rb + 2X$
(c) $^{235}_{92}U + ^1_0n \longrightarrow ^{87}_{35}Br + 3^1_0n + X$
(d) $^{235}_{92}U + ^1_0n \longrightarrow ^{160}_{62}Sm + ^{72}_{30}Zn + 4X$

20.68   Calculate the nuclear binding energies (in J/nucleon) for the following species: (a) ^{10}B (10.0129 amu), (b) ^{11}B (11.009305 amu), (c) ^{14}N (14.003074 amu), (d) ^{56}Fe (55.93494 amu).

20.69   Write complete nuclear equations for the following processes: (a) tritium (^3H) undergoes β decay, (b) ^{242}Pu undergoes α-particle emission, (c) ^{131}I undergoes β decay, (d) ^{251}Cf emits an α particle.

20.70   The nucleus of nitrogen-18 lies above the stability belt. Write the equation for a nuclear reaction by which nitrogen-18 can achieve stability.

20.71   Astatine, the last member of Group 17, can be prepared by bombarding bismuth-209 with α particles. (a) Write an equation for the reaction. (b) Represent the equation in the abbreviated form as discussed in Section 20.4.

20.72   How are scientists able to tell the age of a fossil?

20.73   (a) Assuming nuclei are spherical in shape, show that the radius (r) of a nucleus is proportional to the cube root of mass number (A). (b) In general, the radius of a nucleus is given by $r = r_0 A^{1/3}$, where r_0, the proportionality constant, is given by 1.2×10^{-15} m. Calculate the volume of the ^{238}U nucleus.

20.74   Modern designs of atomic bombs contain, in addition to uranium or plutonium, small amounts of tritium and deuterium to boost the power of explosion. What is the role of tritium and deuterium in these bombs?

20.75   (a) Calculate the energy released when a U-238 isotope decays to Th-234. The atomic masses are as follows: U-238: 238.05078 amu; Th-234: 234.03596 amu; and He-4: 4.002603 amu. (b) The energy released in part (a) is transformed into the kinetic energy of the recoiling Th-234 nucleus and the α particle. Which of the two will move away faster? Explain.

20.76   Sources of energy on Earth include fossil fuels, geothermal power, gravity, hydroelectric power, nuclear fission, nuclear fusion, the sun, and wind. Which of these have a "nuclear origin," either directly or indirectly?

20.77   From the definition of curie, calculate Avogadro's number, given that the molar mass of ^{226}Ra is 226.03 g/mol and that it decays with a half-life of 1.6×10^3 years.

20.78   Tritium contains one proton and two neutrons. There is no significant proton-proton repulsion present in the nucleus. Why, then, is tritium radioactive?

20.79   The usefulness of radiocarbon dating is limited to objects no older than 60,000 years. What percent of the carbon-14, originally present in the sample, remains after this period of time?

20.80   The radioactive potassium-40 isotope decays to argon-40 with a half-life of 1.2×10^9 years. (a) Write a balanced equation for the reaction. (b) A sample of moon rock is found to contain 18 percent potassium-40 and 82 percent argon by mass. Calculate the age of the rock in years. (Assume that all the argon in the sample is the result of potassium decay.)

20.81   Name two advantages of a nuclear-powered submarine over a conventional submarine.

20.82   Both barium (Ba) and radium (Ra) are members of Group 2 and are expected to exhibit similar chemical properties. However, Ra is not found in barium ores. Instead, it is found in uranium ores. Explain.

20.83   As a result of being exposed to the radiation released during the Chernobyl nuclear accident, the dose of iodine-131 in a person's body is 7.4 mC $(1 \text{ mC} = 1 \times 10^{-3} \text{ Ci})$. Use the relationship rate = λN to calculate the number of atoms of iodine-131 to which this radioactivity corresponds. (The half-life of ^{131}I is 8.1 days.)

20.84   Which of the following poses a greater health hazard: a radioactive isotope with a short half-life or a radioactive isotope with a long half-life? Explain. [Assume the same type of radiation (α or β) and comparable energetics per particle emitted.]

20.85   In 1997, a scientist at a nuclear research center in Russia placed a thin shell of copper on a sphere of highly enriched uranium-235. Suddenly, there was a huge burst of radiation, which turned the air blue. Three days later, the scientist died of radiation exposure. Explain what caused the accident. (*Hint:* Copper is an effective metal for reflecting neutrons.)

20.86   A radioactive isotope of copper decays as follows:

$$^{64}\text{Cu} \longrightarrow {}^{64}\text{Zn} + {}_{-1}^{0}\beta \qquad t_{1/2} = 12.8 \text{ h}$$

Starting with 84.0 g of ^{64}Cu, calculate the quantity of ^{64}Zn produced after 18.4 h.

20.87   A 0.0100-g sample of a radioactive isotope with a half-life of 1.3×10^9 years decays at the rate of 2.9×10^4 disintegrations per minute. Calculate the molar mass of the isotope.

20.88   The half-life of ^{27}Mg is 9.50 min. (a) Initially there were 4.20×10^{12} ^{27}Mg nuclei present. How many ^{27}Mg nuclei are left 30.0 min later? (b) Calculate the ^{27}Mg activities (in Ci) at $t = 0$ and $t = 30.0$ min. (c) What is the probability that any one ^{27}Mg nucleus decays during a 1-s interval? What assumption is made in this calculation?

20.89   During the past two decades, syntheses of elements 110 through 118 have been reported. Element 110 was created by bombarding ^{208}Pb with ^{62}Ni, element 111 was created by bombarding ^{209}Bi with ^{64}Ni, element 112 was created by bombarding ^{208}Pb with ^{66}Zn, element 114 was created by bombarding ^{244}Pu with ^{48}Ca, element 115 was created by bombarding ^{243}Am with ^{48}Ca, element 116 was created by bombarding ^{248}Cm with ^{48}Ca, element 117 was created by bombarding ^{249}Bk with ^{48}Ca, and element 118 was created by bombarding ^{249}Cf with ^{48}Ca. Write an equation for each synthesis and predict the chemical properties of these elements.

Engineering Problems

20.90   Explain why achievement of nuclear fusion in the laboratory requires a temperature of about 100 million degrees Celsius, which is much higher than that in the interior of the sun (15 million degrees Celsius).

20.91   An electron and a positron are accelerated to nearly the speed of light before colliding in a particle accelerator. The resulting collision produces an exotic particle having a mass many times that of a proton. Does this result violate the law of conservation of mass? Explain.

20.92   Americium-241 is used in smoke detectors because it has a long half-life (458 years) and its emitted α particles are energetic enough to ionize air molecules. Using the given schematic diagram of a smoke detector, explain how it works.

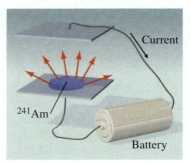

20.93   The constituents of wine contain, among others, carbon, hydrogen, and oxygen atoms. A bottle of wine was sealed about 6 years ago. To confirm its age, which of the isotopes would you choose in a radioactive dating study? The half-lives of the isotopes are: ^{14}C: 5715 years; ^{15}O: 124 s; ^{3}H: 12.5 years. Assume that the activities of the isotopes were known at the time the bottle was sealed.

20.94   To detect bombs that may be smuggled onto airplanes, the Federal Aviation Administration (FAA) will soon require all major airports in the United States to install thermal neutron analyzers. The thermal neutron analyzer will bombard baggage with low-energy neutrons, converting some of the nitrogen-14 nuclei to nitrogen-15, with simultaneous emission of γ rays. Because nitrogen content is usually high in explosives, detection of a high dosage of γ rays will suggest that a bomb may be present. (a) Write an equation for the nuclear process. (b) Compare this technique with the conventional X-ray detection method.

Biological Problems

20.95   The carbon-14 decay rate of a sample obtained from a young tree is 0.260 disintegration per second per gram of the sample. Another wood sample prepared from an object recovered at an archaeological excavation gives a decay rate of 0.186 disintegration per second per gram of the sample. What is the age of the object?

20.96 Why is strontium-90 a particularly dangerous isotope for humans? The half-life of strontium-90 is 29.1 years. Calculate the radioactivity in millicuries of 15.6 mg of ^{90}Sr.

20.97 After the Chernobyl accident, people living close to the nuclear reactor site were urged to take large amounts of potassium iodide as a safety precaution. What is the chemical basis for this action?

20.98 Cobalt-60 is an isotope used in diagnostic medicine and cancer treatment. It decays with γ-ray emission. Calculate the wavelength of the radiation in nanometers if the energy of the γ ray is 2.4×10^{-13} J/photon.

Environmental Problems

20.99 Identify two of the most abundant radioactive elements that exist on Earth. Explain why they are still present. (You may wish to consult a website such as that of the University of Sheffield and WebElements Ltd, UK, webelements.com.)

20.100 Nuclear waste disposal is one of the major concerns of the nuclear industry. In choosing a safe and stable environment to store nuclear wastes, consideration must be given to the heat released during nuclear decay. As an example, consider the β decay of ^{90}Sr (89.907738 amu):

$$^{90}_{38}\text{Sr} \longrightarrow \,^{90}_{39}\text{Y} + \,^{0}_{-1}\beta \qquad t_{1/2} = 28.1 \text{ yr}$$

The ^{90}Y (89.907152 amu) further decays as follows:

$$^{90}_{39}\text{Y} \longrightarrow \,^{90}_{40}\text{Zr} + \,^{0}_{-1}\beta \qquad t_{1/2} = 64 \text{ h}$$

Zirconium-90 (89.904703 amu) is a stable isotope. (a) Use the mass defect to calculate the energy released (in joules) in each of the preceding two decays. (The mass of the electron is 5.4857×10^{-4} amu.) (b) Starting with 1 mole of ^{90}Sr, calculate the number of moles of ^{90}Sr that will decay in a year. (c) Calculate the amount of heat released (in kJ) corresponding to the number of moles of ^{90}Sr decayed to ^{90}Zr in part (b).

Multiconcept Problems

20.101 The quantity of a radioactive material is often measured by its activity (measured in curies or millicuries) rather than by its mass. In a brain scan procedure, a 70-kg patient is injected with 20.0 mCi of ^{99m}Tc, which decays by emitting γ-ray photons with a half-life of 6.0 h. Given that the RBE of these photons is 0.98 and only two-thirds of the photons are absorbed by the body, calculate the rem dose received by the patient. Assume all the ^{99m}Tc nuclei decay while in the body. The energy of a γ-ray photon is 2.29×10^{-14} J.

20.102 In 2006, an ex-KGB agent was murdered in London. The investigation following the agent's death revealed that he was poisoned with the radioactive isotope ^{210}Po, which had apparently been added to his food. (a) ^{210}Po is prepared by bombarding ^{209}Bi with neutrons. Write an equation for the reaction. (b) The half-life of ^{210}Po is 138 days. It decays by α particle emission. Write the equation for the decay process. (c) Calculate the energy of an emitted α particle. Assume both the parent and daughter nuclei have zero kinetic energy. The atomic masses of ^{210}Po, ^{206}Pb, and $^{4}_{2}\alpha$ are 209.98286, 205.97444, and 4.00150 amu, respectively. (d) Ingestion of 1 μg of ^{210}Po could prove fatal. What is the total energy released by this quantity of ^{210}Po over the course of 138 days?

20.103 Alpha particles produced by radioactive decay eventually pick up electrons from their surroundings to form helium atoms. Calculate the volume (in mL) of He collected at STP when 1.00 g of pure ^{226}Ra is stored in a closed container for 125 years. (Assume that there are five α particles generated per ^{226}Ra as it decays to ^{206}Pb.)

Standardized-Exam Practice Problems

Physical and Biological Sciences

The radioactive isotope ^{238}Pu, used in pacemakers, decays by emitting an α particle with a half-life of 86 years. The energy of the emitted α particle is 9.0×10^{-13} J, which is the energy per decay. After 10 years, the activity of the isotope decreases by 8.0 percent. (Power is measured in watts or J/s.)

1. What is the rate constant for the decay?

   a) 2.6×10^{-10} s^{-1}
   b) 86 year^{-1}
   c) 2.7×10^{-9} s^{-1}
   d) 0.012 year^{-1}

2. Identify the equation for the decay process.

   a) $^{238}Pu \longrightarrow {}^{236}Pu + \alpha$
   b) $^{238}Pu + \alpha \longrightarrow {}^{234}U$
   c) $^{238}Pu \longrightarrow {}^{234}U + \alpha$
   d) $^{238}Pu \longrightarrow {}^{234}Pu + \alpha$

3. Assuming that all the α-particle energy is used to run the pacemaker, calculate the power output at $t = 0$ and $t = 10$ years. Initially, 1.0 mg of ^{238}Pu was present in the pacemaker.

   a) 0.081 mW, 0.075 mW
   b) 2.6×10^{-10} mW, 2.4×10^{-10} mW
   c) 1.0 mW, 0.92 mW
   d) 0.58 mW, 0.53 mW

4. How long will it take for the power output to fall to 80 percent of its original value?

   a) 28 years
   b) 69 years
   c) 86 years
   d) 43 years

Answers to In-Chapter Materials

Answers to Practice Problems

20.1A (a) $^{78}_{34}$Se, (b) $^{5}_{3}$Li, (c) $^{1}_{0}$n. **20.1B** (a) $^{244}_{95}$Am, (b) $^{234}_{90}$Th, (c) $^{14}_{6}$C. **20.2A** 2.6280×10^{-10} J; 1.2574×10^{-12} J/nucleon. **20.2B** 196.9665 amu. **20.3A** 9.3×10^{3} yr. **20.3B** 1.0×10^{1} dps. **20.4A** 6.6×10^{8} yr. **20.4B** 5.7×10^{-2} mg. **20.5A** $^{106}_{46}$Pd + $^{4}_{2}$He $\longrightarrow$ $^{1}_{1}$H + $^{109}_{47}$Ag. **20.5B** $^{33}_{17}$Cl(n,$^{3}_{2}$He)$^{31}_{15}$P.

Answers to Checkpoints

20.1.1 d. **20.1.2** a. **20.2.1** b. **20.2.2** d. **20.2.3** c. **20.2.4** a. **20.3.1** d. **20.3.2** c. **20.3.3** b. **20.4.1** e. **20.4.2** b.

Environmental Chemistry

Earth's atmosphere contains a region of increased ozone concentration, commonly known as the "ozone layer." In recent decades, pollutants have contributed to the destruction of ozone, potentially reducing the ozone layer's ability to protect life on Earth from harmful radiation.

Digital Vision/Getty Images

In This Chapter, You Will Learn

About some of the facets of environmental chemistry.

Before You Begin, Review These Skills

- Bond enthalpy [⏮ Section 8.9]
- Catalysis [⏮ Section 14.6]
- Molecular polarity [⏮ Section 9.2]
- Chapter 9 Key Skills [⏮ pages 454–455]

How Human-made Molecules Have Impacted the Environment

The presence of ozone (O_3) in the atmosphere is responsible for the absorption of light that otherwise would significantly alter life on Earth. The wavelengths of light that are absorbed by stratospheric ozone are known to cause cancer, genetic mutations, and the destruction of plant life. Ozone is destroyed naturally by the absorption of this short-wavelength light—and is regenerated by another solar process. The balance of ozone destruction and regeneration can be disrupted, however, by the presence of substances not found naturally in the atmosphere.

In 1973, F. Sherwood "Sherry" Rowland and Mario Molina, chemistry professors at the University of California–Irvine, began studying the chemical behavior of chlorofluorocarbons (CFCs) in the atmosphere. They discovered that although CFC molecules were extraordinarily stable in the troposphere, the very stability that made them attractive as coolants and propellants also allowed them to survive the gradual diffusion into the stratosphere where they would ultimately be broken down by high-energy ultraviolet radiation. Rowland and Molina proposed that chlorine atoms liberated in the breakdown of CFCs could potentially catalyze the destruction of large amounts of ozone in the stratosphere. The work of Rowland and Molina, along with that of other atmospheric scientists, provoked a debate among the scientific and international communities regarding the fate of the ozone layer—and the planet. The Montreal Protocol, once referred to by United Nations Secretary-General Kofi Annan as "perhaps the single most successful international agreement to date," is an international treaty designed to protect the ozone layer. Originally signed in 1987, it calls for a halt to the production of substances believed to contribute to the destruction of stratospheric ozone.

In 1995, Rowland and Molina, along with Dutch atmospheric chemist Paul Crutzen, were awarded the Nobel Prize in Chemistry for their elucidation of the role of human-made chemicals in the catalytic destruction of stratospheric ozone.

At the end of this chapter, you will be able to answer questions about an environmentally important molecule [⏭ Applying What You've Learned, page 1072].

21.1 Earth's Atmosphere

Earth is unique among the planets of our solar system in having an atmosphere that is chemically active and rich in oxygen. Mars, for example, has a much thinner atmosphere that is about 90 percent carbon dioxide. Jupiter has no solid surface; it is made up, instead, of 90 percent hydrogen, 9 percent helium, and 1 percent other substances.

It is generally believed that three billion or four billion years ago, Earth's atmosphere consisted mainly of ammonia, methane, and water. There was little, if any, free oxygen present. Ultraviolet (UV) radiation from the sun probably penetrated the atmosphere, rendering the surface of Earth sterile. However, the same UV radiation may have triggered the chemical reactions (perhaps beneath the surface) that eventually led to life on Earth. Primitive organisms used energy from the sun to break down carbon dioxide (produced by volcanic activity) to obtain carbon, which they incorporated in their own cells. The major by-product of this process, called *photosynthesis,* is oxygen. Another important source of oxygen is the *photodecomposition* of water vapor by UV light. Over time, the more reactive gases such as ammonia and methane have largely disappeared, and today our atmosphere consists mainly of oxygen and nitrogen gases. Biological processes determine to a great extent the atmospheric concentrations of these gases, one of which is reactive (oxygen) and the other unreactive (nitrogen).

Table 21.1 shows the composition of dry air at sea level. The total mass of the atmosphere is about 5.3×10^{18} kg. Water is excluded from this table because its concentration in air can vary drastically from location to location.

Figure 21.1 shows the major processes involved in the cycle of nitrogen in nature. Molecular nitrogen, with its triple bond, is a very stable molecule. However, through

TABLE 21.1	Composition of Dry Air at Sea Level
Gas	**Composition (% by Volume)**
N_2	78.09
O_2	20.95
Ar	0.93
CO_2	0.041
Ne	0.0015
He	0.000524
Kr	0.00014
Xe	0.000006

Figure 21.1 The nitrogen cycle. Although the supply of nitrogen in the atmosphere is virtually inexhaustible, it must be combined with hydrogen or oxygen before it can be assimilated by higher plants, which in turn are consumed by animals. Juvenile nitrogen is nitrogen that has not previously participated in the nitrogen cycle.

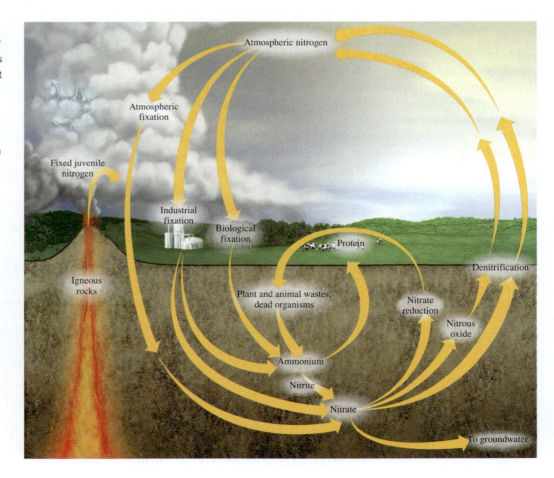

biological and industrial ***nitrogen fixation*** (the conversion of molecular nitrogen into nitrogen compounds), atmospheric nitrogen gas is converted into nitrates and other compounds suitable for assimilation by algae and plants. Another important mechanism for producing nitrates from nitrogen gas is lightning. The steps are:

$$N_2(g) + O_2(g) \xrightarrow{\text{electric energy}} 2NO(g)$$
$$2NO(g) + O_2(g) \longrightarrow 2NO_2(g)$$
$$2NO_2(g) + H_2O(l) \longrightarrow HNO_2(aq) + HNO_3(aq)$$

About 30 million tons of HNO_3 are produced this way annually. Nitric acid is converted to nitrate salts in the soil. These nutrients are taken up by plants, which in turn are ingested by animals. Animals use the nutrients from plants to make proteins and other essential biomolecules. Denitrification reverses nitrogen fixation to complete the cycle. For example, certain anaerobic organisms decompose animal wastes as well as dead plants and animals to produce free molecular nitrogen from nitrates.

The main processes of the global oxygen cycle are shown in Figure 21.2. This cycle is complicated by the fact that oxygen takes so many different chemical forms. Atmospheric oxygen is removed through respiration and various industrial processes (mostly combustion), which produce carbon dioxide. Photosynthesis is the major mechanism by which molecular oxygen is regenerated from carbon dioxide and water.

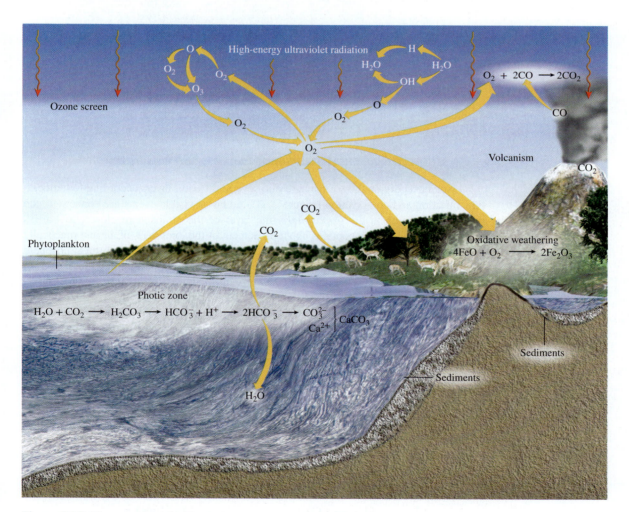

Figure 21.2 The oxygen cycle. The cycle is complicated because oxygen appears in so many chemical forms and combinations, primarily as molecular oxygen, in water, and in organic and inorganic compounds.

Figure 21.3 Regions of Earth's atmosphere. Notice the variation of temperature with altitude. Most of the phenomena shown here are discussed in the chapter.

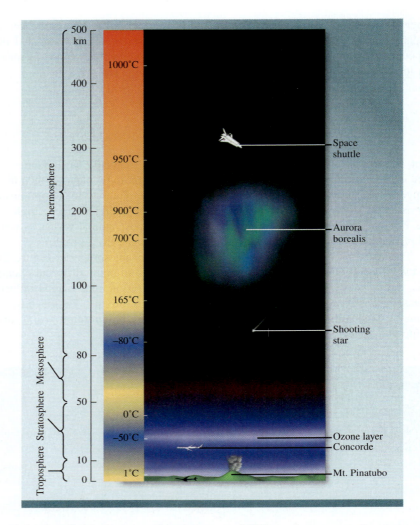

Scientists divide the atmosphere into several different layers according to temperature variation and composition (Figure 21.3). As far as visible events are concerned, the most active region is the ***troposphere,*** the layer of the atmosphere closest to Earth's surface. The troposphere contains about 80 percent of the total mass of air and practically all the atmosphere's water vapor. The troposphere is the thinnest layer of the atmosphere (10 km), but it is where all the dramatic events of weather—rain, lightning, hurricanes—occur. Temperature decreases almost linearly with increasing altitude in this region.

Above the troposphere is the ***stratosphere,*** which consists of nitrogen, oxygen, and ozone. In the stratosphere, the air temperature *increases* with altitude. This warming effect is the result of exothermic reactions triggered by UV radiation from the sun (see Section 21.3). One of the products of this reaction sequence is ozone (O_3), which serves to prevent harmful UV rays from reaching Earth's surface.

In the ***mesosphere,*** which is above the stratosphere, the concentration of ozone and other gases is low, and the temperature decreases again with increasing altitude. The ***thermosphere,*** or ***ionosphere,*** is the uppermost layer of the atmosphere. The increase in temperature in this region is the result of the bombardment of molecular oxygen and nitrogen and atomic species by energetic particles, such as electrons and protons, from the sun. Typical reactions are:

$$N_2 \longrightarrow 2N \qquad\qquad \Delta H° = 941.4 \text{ kJ/mol}$$
$$N \longrightarrow N^+ + e^- \qquad\qquad \Delta H° = 1400 \text{ kJ/mol}$$
$$O_2 \longrightarrow O_2^+ + e^- \qquad\qquad \Delta H° = 1176 \text{ kJ/mol}$$

In reverse, these processes liberate the equivalent amount of energy, mostly as heat. Ionized particles are responsible for the reflection of radio waves back toward Earth.

21.2 Phenomena in the Outer Layers of the Atmosphere

In this section we discuss two dazzling phenomena that occur in the outer regions of the atmosphere. One is a natural event. The other is a curious by-product of human space travel.

Aurora Borealis and Aurora Australis

Violent eruptions on the surface of the sun, called *solar flares,* result in the ejection of myriad electrons and protons into space, where they disrupt radio transmission and provide us with spectacular celestial light shows known as *auroras* (Figure 21.4). These electrons and protons collide with the molecules and atoms in Earth's upper atmosphere, causing them to become ionized and electronically excited. Eventually, the excited molecules and ions return to the ground state with the emission of light. For example, an excited oxygen atom emits photons at wavelengths of 558 nm (green) and between 630 and 636 nm (red):

$$O^* \longrightarrow O + h\nu$$

where the asterisk denotes an electronically excited species and $h\nu$ denotes the emitted photon [◀◀ Section 6.2]. Similarly, the blue and violet colors often observed in auroras result from the transition in the ionized nitrogen molecule:

$$N_2^{+*} \longrightarrow N_2^+ + h\nu$$

The wavelengths for this transition fall between 391 and 470 nm.

The incoming streams of solar protons and electrons are oriented by Earth's magnetic field so that most auroral displays occur in doughnut-shaped zones about 2000 km in diameter centered on the North and South Poles. *Aurora borealis* is the name given to this phenomenon in the Northern Hemisphere. In the Southern Hemisphere, it is called *aurora australis.* Sometimes, the number of solar particles is so immense that auroras are also visible from locations as far south as Olympia, Washington.

Sample Problem 21.1 shows how to determine the maximum wavelength capable of breaking a chemical bond.

Figure 21.4 Aurora borealis, commonly referred to as the northern lights.

Theo Allofs/Digital Vision/Getty Images

SAMPLE PROBLEM 21.1

The bond enthalpy of O_2 is 498.7 kJ/mol. Calculate the maximum wavelength (in nm) of a photon that can cause the dissociation of an O_2 molecule.

Strategy We want to calculate the wavelength of a photon that will break an O=O bond. Therefore, we need the amount of energy in one bond. The bond energy of O_2 is given in units of kJ/mol. The units needed for the energy of one bond are J/molecule. Once we know the energy in one bond, we can calculate the minimum frequency and maximum wavelength needed to dissociate one O_2 molecule.

Setup The conversion steps are

$$\text{kJ/mol} \longrightarrow \text{J/molecule} \longrightarrow \text{frequency of photon} \longrightarrow \text{wavelength of photon}$$

Solution First we calculate the energy required to dissociate one O_2 molecule:

$$\text{energy per molecule} = \frac{498.7 \times 10^3 \, \text{J}}{1 \, \text{mol}} \times \frac{1 \, \text{mol}}{6.022 \times 10^{23} \, \text{molecules}} = 8.281 \times 10^{-19} \frac{\text{J}}{\text{molecule}}$$

(Continued on next page)

The energy of the photon is given by $E = h\nu$ (Equation 6.2). Therefore:

$$\nu = \frac{E}{h} = \frac{8.281 \times 10^{-19}\ \text{J}}{6.63 \times 10^{-34}\ \text{J} \cdot \text{s}} = 1.25 \times 10^{15}\ \text{s}^{-1}$$

Finally, we calculate the wavelength of the photon, given by $\lambda = c/\nu$ (Equation 6.1), as follows:

$$\lambda = \frac{3.00 \times 10^8\ \text{m/s}}{1.25 \times 10^{15}\ \text{s}^{-1}} = 2.40 \times 10^{-7}\ \text{m} = 240\ \text{nm}$$

THINK ABOUT IT

In principle, any photon with a wavelength of 240 nm or *shorter* can dissociate an O_2 molecule.

Practice Problem **A**TTEMPT Calculate the wavelength (in nm) of a photon needed to dissociate an O_3 molecule:

$$O_3 \longrightarrow O + O_2 \qquad \Delta H° = 107.2\ \text{kJ/mol}$$

Practice Problem **B**UILD Which of the following gaseous species is dissociated by visible light: CS_2, F_2, HI, ClF, HCN? (See Table 8.6.)

Practice Problem **C**ONCEPTUALIZE Which of the following graphs best represents the relationship between the wavelength of light and the maximum bond enthalpy it can dissociate?

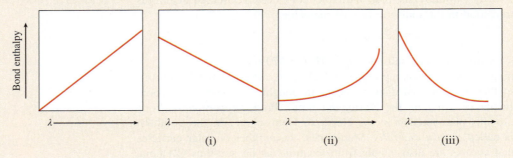

(i)   (ii)   (iii)

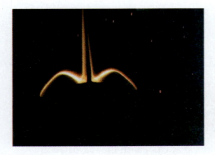

Figure 21.5 The glowing tail section of the space shuttle viewed from inside the vehicle.

Source: *NASA-JSC*

The Mystery Glow of Space Shuttles

A human-made light show that baffled scientists for several years was produced by space shuttles orbiting Earth. In 1983, astronauts first noticed an eerie orange glow on the outside surface of their spacecraft at an altitude about 300 km above Earth (Figure 21.5).

The light, which usually extended about 10 cm away from the protective silica heat tiles and other surface materials, was most pronounced on the parts of the shuttle facing its direction of travel. This fact led scientists to postulate that collision between oxygen atoms in the atmosphere and the fast-moving shuttle somehow produced the orange light. Spectroscopic measurements of the glow, as well as laboratory tests, strongly suggested that nitric oxide (NO) and nitrogen dioxide (NO_2) also played a part. It was believed that oxygen atoms interacted with nitric oxide adsorbed on (i.e., bound to) the shuttle's surface to form electronically excited nitrogen dioxide:

$$O + NO \longrightarrow NO_2^*$$

As the NO_2^* left the shell of the spacecraft, it emitted photons at a wavelength of 680 nm (orange):

$$NO_2^* \longrightarrow NO_2 + h\nu$$

Support for this explanation came inadvertently in 1991, when astronauts aboard *Discovery* released various gases, including carbon dioxide, neon, xenon, and nitric oxide, from the cargo bay in the course of an unrelated experiment. Expelled one at a time, these gases scattered onto the surface of the shuttle's tail. The nitric oxide caused the normal shuttle glow to intensify markedly, but the other gases had no effect on it.

What was the source of the nitric oxide on the outside of the spacecraft? Scientists believed that some of it may have come from the exhaust gases emitted by the shuttle's rockets and that some of it was present in the surrounding atmosphere. The shuttle glow did not harm the vehicle, but it did interfere with spectroscopic measurements on distant objects made from the spacecraft.

CHECKPOINT – SECTION 21.2   Phenomena in the Outer Layers of the Atmosphere

21.2.1 What maximum wavelength (in nm) of light is necessary to break the bond in a nitrogen molecule (N_2)? (The bond energy of N_2 is 941.4 kJ/mol.)

a) 127 nm

b) 211 nm

c) 236 nm

d) 942 nm

e) 941 nm

21.2.2 What process gives rise to the aurora borealis and the aurora australis?

a) Formation of oxygen and nitrogen molecules

b) Collisions between nitrogen atoms

c) Collisions between oxygen atoms

d) Emission of photons of visible light

e) Absorption of photons of visible light

21.3 Depletion of Ozone in the Stratosphere

Ozone in the stratosphere prevents UV radiation emitted by the sun from reaching Earth's surface. The formation of ozone in this region begins with the *photodissociation* of oxygen molecules by solar radiation at wavelengths below 240 nm:

$$O_2 \xrightarrow{\text{UV} < 240 \text{ nm}} O + O \qquad \textbf{Equation 21.1}$$

The highly reactive O atoms combine with oxygen molecules to form ozone as follows:

$$O + O_2 + M \longrightarrow O_3 + M \qquad \textbf{Equation 21.2}$$

where M is some inert substance such as N_2. The role of M in this exothermic reaction is to absorb some of the excess energy released and prevent the spontaneous decomposition of the O_3 molecule. The energy that is not absorbed by M is given off as heat. (As the M molecules themselves become de-excited, they release more heat to the surroundings.) In addition, ozone itself absorbs UV light between 200 and 300 nm:

$$O_3 \xrightarrow{\text{UV}} O + O_2 \qquad \textbf{Equation 21.3}$$

The process continues when O and O_2 recombine to form O_3 as shown in Equation 21.2, further warming the stratosphere.

If all the stratospheric ozone were compressed into a single layer at STP on Earth, that layer would be only about 3 mm thick! Although the concentration of ozone in the stratosphere is very low, it is sufficient to filter out (i.e., absorb) solar radiation in the 200- to 300-nm range (see Equation 21.3). In the stratosphere, it acts as our protective shield against UV radiation, which can induce skin cancer, cause genetic mutations, and destroy crops and other forms of vegetation.

The formation and destruction of ozone by natural processes is a dynamic equilibrium that maintains a constant concentration of ozone in the stratosphere. Since the mid-1970s scientists have been concerned about the harmful effects of CFCs on the ozone layer. Generally known by the trade name Freons, CFCs were first synthesized

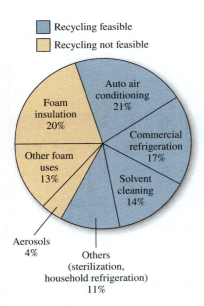

Figure 21.6 Uses of CFCs. Since 1978, the use of aerosol propellants has been banned in the United States.

in the 1930s. Some of the common ones are $CFCl_3$ (Freon 11), CF_2Cl_2 (Freon 12), $C_2F_3Cl_3$ (Freon 113), and $C_2F_4Cl_2$ (Freon 114). Because these compounds are readily liquefied, relatively inert, nontoxic, noncombustible, and volatile, they have been used as coolants in refrigerators and air conditioners, in place of highly toxic liquid sulfur dioxide (SO_2) and ammonia (NH_3). Large quantities of CFCs are also used in the manufacture of disposable foam products such as cups and plates, as aerosol propellants in spray cans, and as solvents to clean newly soldered electronic circuit boards (Figure 21.6). In 1977, the peak year of production, nearly 1.5×10^6 tons of CFCs were produced in the United States. Most of the CFCs produced for commercial and industrial use are eventually discharged into the atmosphere.

Because of their relative inertness, the CFCs slowly diffuse unchanged up to the stratosphere, where UV radiation of wavelengths between 175 and 220 nm causes them to decompose:

$$CFCl_3 \longrightarrow CFCl_2 + Cl$$
$$CF_2Cl_2 \longrightarrow CF_2Cl + Cl$$

The reactive chlorine atoms then undergo the following reactions:

Equation 21.4 $Cl + O_3 \longrightarrow ClO + O_2$

Equation 21.5 $ClO + O \longrightarrow Cl + O_2$

The overall result (the sum of Equations 21.4 and 21.5) is the net removal of an O_3 molecule from the stratosphere:

Equation 21.6 $O_3 + O \longrightarrow 2O_2$

The oxygen atoms in Equation 21.5 are supplied by the photochemical decomposition of molecular oxygen and ozone described earlier. The Cl atom plays the role of a catalyst in the reaction mechanism scheme represented by Equations 21.4 and 21.5 because it is not used up and therefore can take part in many such reactions. In fact, one Cl atom can destroy up to 100,000 O_3 molecules before it is removed by some other reaction. The ClO (chlorine monoxide) species is an intermediate because it is produced in the first elementary step (Equation 21.4) and consumed in the second step (Equation 21.5). The preceding mechanism for the destruction of ozone has been supported by the detection of ClO in the stratosphere in recent years (Figure 21.7). As can be seen, the concentration of O_3 decreases in regions that have high amounts of ClO.

Figure 21.7 The variations in concentrations of ClO and O_3 with latitude.

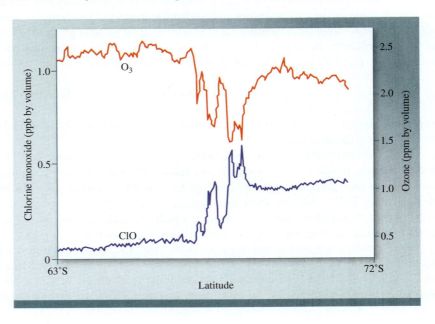

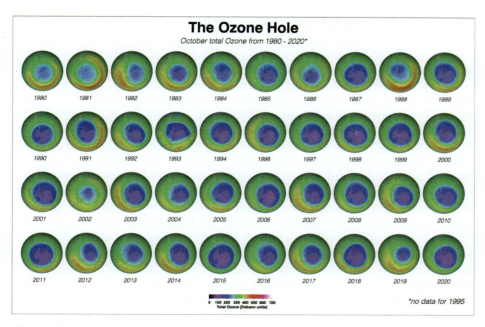

The Ozone Hole
October total Ozone from 1980 - 2020*

1980 1981 1982 1983 1984 1985 1986 1987 1988 1989

1990 1991 1992 1993 1994 1996 1997 1998 1999 2000

2001 2002 2003 2004 2005 2006 2007 2008 2009 2010

2011 2012 2013 2014 2015 2016 2017 2018 2019 2020

0 100 200 300 400 500 600 700
Total Ozone (Dobson units)

*no data for 1995

Figure 21.8 In recent decades, scientists have found that the ozone layer in the stratosphere over the South Pole has become thinner. This map, based on data collected over a number of years, shows the depletion of ozone in blue.

Source: *H.S. Photos/Alamy Stock Photo*

Another group of compounds that can destroy stratospheric ozone are the nitrogen oxides, generally denoted as NO_x. (Examples of NO_x are NO and NO_2.) These compounds come from the exhausts of high-altitude supersonic aircraft and from human and natural activities on Earth. Solar radiation decomposes a substantial amount of the other nitrogen oxides to nitric oxide (NO), which participates in the destruction of ozone as follows:

$$O_3 \longrightarrow O_2 + O$$
$$NO + O_3 \longrightarrow NO_2 + O_2$$
$$\underline{NO_2 + O \longrightarrow NO + O_2}$$
$$\textit{Overall:} \qquad 2O_3 \longrightarrow 3O_2$$

In this case, NO is the catalyst and NO_2 is the intermediate. Nitrogen dioxide also reacts with chlorine monoxide to form chlorine nitrate:

$$ClO + NO_2 \longrightarrow ClONO_2$$

Chlorine nitrate is relatively stable and serves as a "chlorine reservoir," which plays a role in the depletion of the stratospheric ozone over the North and South Poles.

Polar Ozone Holes

In the mid-1980s, evidence began to accumulate that an "Antarctic ozone hole" developed in late winter, depleting the stratospheric ozone over Antarctica by as much as 50 percent (Figure 21.8). In the stratosphere, a stream of air known as the "polar vortex" circles Antarctica in winter. Air trapped within this vortex becomes extremely cold during the polar night. This condition leads to the formation of ice particles known as polar stratospheric clouds (PSCs) (Figure 21.9). Acting as a heterogeneous catalyst, these PSCs provide a surface for reactions converting HCl (emitted from Earth) and chlorine nitrate to more reactive chlorine molecules:

$$HCl + ClONO_2 \longrightarrow Cl_2 + HNO_3$$

Figure 21.9 Polar stratospheric clouds containing ice particles can catalyze the formation of Cl atoms and lead to the destruction of ozone.

Lamont Poole/NASA

By early spring, the sunlight splits molecular chlorine into chlorine atoms:

$$Cl_2 + h\nu \longrightarrow 2Cl$$

which then attack ozone as shown earlier.

The situation is not as severe in the warmer Arctic region, where the vortex does not persist quite as long. Studies have shown that ozone levels in this region have declined between 4 and 8 percent in the past decade. Volcanic eruptions, such as that of Mount Pinatubo in the Philippines in 1991, inject large quantities of dust-sized particles and sulfuric acid aerosols into the atmosphere. These particles can perform the same catalytic function as the ice crystals at the South Pole. As a result, the Arctic hole is expected to grow larger for several years following an eruption.

Recognizing the serious implications of the loss of ozone in the stratosphere, nations throughout the world have acknowledged the need to drastically curtail or totally stop the production of CFCs. In 1978, the United States was one of the few countries to ban the use of CFCs in hair sprays and other aerosols. The Montreal protocol was signed by most industrialized nations in 1987, setting targets for cutbacks in CFC production and the complete elimination of these substances by the year 2000. Although progress has been made in this respect, many nations have not been able to abide by the treaty because of the importance of CFCs to their economies. Recycling could play a significant supplementary role in preventing CFCs already in appliances from escaping into the atmosphere. As Figure 21.6 shows, more than half of the CFCs in use are recoverable.

An intense effort is under way to find CFC substitutes that are not harmful to the ozone layer. One of the promising candidates is called hydrochlorofluorocarbon-123, or HCFC-123 (CF_3CHCl_2). The presence of the hydrogen atom makes the compound more susceptible to oxidation in the lower atmosphere, so it never reaches the stratosphere. Specifically, it is attacked by the hydroxyl radical in the troposphere:

$$CF_3CHCl_2 + OH \longrightarrow CF_3CCl_2 + H_2O$$

The CF_3CCl_2 fragment reacts with oxygen, eventually decomposing to CO_2, water, and hydrogen halides that are removed by rainwater. Unfortunately, the same hydrogen atom also makes the compound more active biologically than the CFCs. Laboratory tests have shown that HCFC-123 can cause tumors in rats, although its toxic effect on humans is not known. Another promising group of compounds that can substitute for CFCs are the hydrofluorocarbons (HFCs). Because they do not contain chlorine, HFCs will not promote the destruction of ozone even if they diffuse to the stratosphere. Examples of these compounds are CF_3CFH_2, CF_3CF_2H, CF_3CH_3, and CF_2HCH_3. In particular, CF_3CFH_2 is already widely used in place of CFCs in air conditioning and refrigeration applications.

Although it is unclear whether the CFCs already released to the atmosphere will eventually result in catastrophic damage to life on Earth, it is conceivable that the depletion of ozone can be slowed by reducing the availability of Cl atoms. Indeed, some chemists have suggested sending a fleet of planes to spray 50,000 tons of ethane (C_2H_6) or propane (C_3H_8) high over the South Pole in an attempt to heal the hole in the ozone layer. Being a reactive species, the chlorine atom would react with the hydrocarbons as follows:

$$Cl + C_2H_6 \longrightarrow HCl + C_2H_5$$

$$Cl + C_3H_8 \longrightarrow HCl + C_3H_7$$

The products of these reactions would not affect the ozone concentration. A less realistic plan is to rejuvenate the ozone layer by producing large quantities of ozone and releasing it into the stratosphere from airplanes. Technically this solution is feasible, but it would be enormously costly and it would require the collaboration of many nations.

Having discussed the chemistry in the outer regions of Earth's atmosphere, we focus in Sections 21.4 through 21.8 on events closer to us—that is, in the troposphere.

21.4 Volcanoes

Volcanic eruptions, Earth's most spectacular natural displays of energy, are instrumental in forming large parts of Earth's crust. The upper mantle, immediately under the crust, is nearly molten. A slight increase in heat, such as that generated by the movement of one crustal plate under another, melts the rock. The molten rock, called *magma,* rises to the surface and generates some types of volcanic eruptions (Figure 21.10).

An active volcano emits gases, liquids, and solids. The gases spewed into the atmosphere include primarily N_2, CO_2, HCl, HF, H_2S, and water vapor. It is estimated that volcanoes are the source of about two-thirds of the sulfur in the air. On the slopes of Mount St. Helens, which last erupted in 1980, deposits of elemental sulfur are visible near the eruption site. At high temperatures, the hydrogen sulfide gas given off by a volcano is oxidized by air:

Figure 21.10 2018 eruption of Hawaii's Kilauea volcano.

Arctic Images/Alamy Stock Photo

$$2H_2S(g) + 3O_2(g) \longrightarrow 2SO_2(g) + 2H_2O(g)$$

Some of the SO_2 is reduced by more H_2S from the volcano to elemental sulfur and water:

$$2H_2S(g) + SO_2(g) \longrightarrow 3S(s) + 2H_2O(g)$$

The rest of the SO_2 is released into the atmosphere, where it reacts with water to form acid rain (see Section 21.6).

The tremendous force of a volcanic eruption carries a sizable amount of gas into the stratosphere. There SO_2 is oxidized to SO_3, which is eventually converted to sulfuric acid aerosols in a series of complex reactions. In addition to destroying ozone in the stratosphere (see Section 21.3), these aerosols can affect climate. Because the stratosphere is above the atmospheric weather patterns, the aerosol clouds often persist for more than a year. They absorb solar radiation and thereby cause a drop in temperature at Earth's surface. However, this cooling effect is local rather than global, because it depends on the site and frequency of volcanic eruptions.

21.5 The Greenhouse Effect

Although carbon dioxide is only a trace gas in Earth's atmosphere, with a concentration of about 0.04 percent by volume (see Table 21.1), it plays a critical role in controlling our climate. The so-called **greenhouse effect** describes the trapping of heat near Earth's surface by gases in the atmosphere, particularly carbon dioxide. The glass roof of a greenhouse transmits visible sunlight and absorbs some of the outgoing infrared (IR) radiation, thereby trapping the heat. Carbon dioxide acts somewhat like a glass roof, except that the temperature rise in the greenhouse is due mainly to the restricted air circulation inside. Calculations show that if the atmosphere did not contain carbon dioxide, Earth would be 30°C cooler!

Figure 21.11 shows the carbon cycle in our global ecosystem. The transfer of carbon dioxide to and from the atmosphere is an essential part of the carbon cycle. Carbon dioxide is produced when any form of carbon or a carbon-containing compound is burned in an excess of oxygen. Many carbonates give off CO_2 when heated, and all give off CO_2 when treated with acid:

$$CaCO_3(s) \longrightarrow CaO(s) + CO_2(g)$$
$$CaCO_3(s) + 2HCl(aq) \longrightarrow CaCl_2(aq) + H_2O(l) + CO_2(g)$$

Carbon dioxide is also a by-product of the fermentation of sugar:

$$\underset{\text{glucose}}{C_6H_{12}O_6(aq)} \xrightarrow{\text{yeast}} \underset{\text{ethanol}}{2C_2H_5OH(aq)} + 2CO_2(g)$$

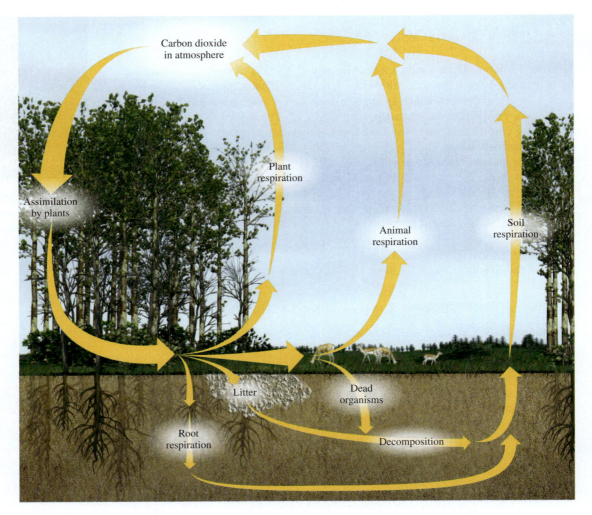

Figure 21.11 The carbon cycle.

Carbohydrates and other complex carbon-containing molecules are consumed by animals, which respire and release CO_2 as an end product of metabolism:

$$C_6H_{12}O_6(aq) + 6O_2(g) \longrightarrow 6CO_2(g) + 6H_2O(l)$$

Another major source of CO_2 is volcanic activity. Carbon dioxide is removed from the atmosphere by photosynthetic plants and certain microorganisms:

$$6CO_2(g) + 6H_2O(l) \longrightarrow C_6H_{12}O_6(aq) + 6O_2(g)$$

After plants and animals die, the carbon in their tissues is oxidized to CO_2 and returns to the atmosphere. In addition, there is a dynamic equilibrium between atmospheric CO_2 and carbonates in the oceans and lakes.

The solar radiant energy received by Earth is distributed over a band of wavelengths between 100 and 5000 nm, but much of it is concentrated in the 400- to 700-nm range, which is the visible region of the spectrum (Figure 21.12). By contrast, the thermal radiation emitted by Earth's surface is characterized by wavelengths longer than 4000 nm (the IR region) because of the much lower average surface temperature compared to that of the sun. The outgoing IR radiation can be absorbed by water and carbon dioxide, but not by nitrogen and oxygen.

All molecules vibrate, even at the lowest temperatures. The energy associated with molecular vibration is quantized, much like the electronic energies of atoms and molecules. To vibrate more energetically, a molecule must absorb a photon of a specific wavelength in the IR region. First, however, its dipole moment *must* change

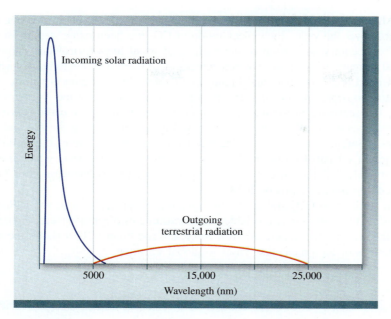

Figure 21.12 The incoming radiation from the sun and the outgoing radiation from Earth's surface.

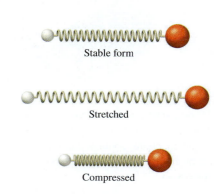

Figure 21.13 Vibration motion of a diatomic molecule. Chemical bonds can be stretched and compressed like a spring.

during the course of a vibration. [Recall that the dipole moment of a molecule is the product of the charge and the distance between charges (Equation 8.1)]. Figure 21.13 shows how a diatomic molecule can vibrate. If the molecule is homonuclear like N_2 and O_2, there can be no change in the dipole moment; the molecule has a zero dipole moment no matter how far apart or close together the two atoms are. We call such molecules IR-inactive because they *cannot* absorb IR radiation. On the other hand, all heteronuclear diatomic molecules are IR-active; that is, they all can absorb IR radiation because their dipole moments constantly change as the bond lengths change.

A *polyatomic* molecule can vibrate in more than one way. Water, for example, can vibrate in three different ways as shown in Figure 21.14. Because water is a polar molecule, any of these vibrations results in a change in dipole moment because there is a change in bond length. Therefore, an H_2O molecule is IR-active. Carbon dioxide has a linear geometry and is nonpolar. Figure 21.15 shows two of the four ways a CO_2 molecule can vibrate. One of them [Figure 21.15(a)] symmetrically displaces atoms from the center of gravity and will not create a dipole moment, but the other vibration [Figure 21.15(b)] is IR-active because the dipole moment changes from zero to a maximum value in one direction and then reaches the same maximum value when it changes to the other extreme position.

Upon receiving a photon in the IR region, a molecule of H_2O or CO_2 is promoted to a higher vibrational energy level:

$$H_2O + h\nu \longrightarrow H_2O^*$$
$$CO_2 + h\nu \longrightarrow CO_2^*$$

(the asterisk denotes a vibrationally excited molecule). These energetically excited molecules soon lose their excess energy either by collision with other molecules or by spontaneous emission of radiation. Part of this radiation is emitted to outer space and part returns to Earth's surface.

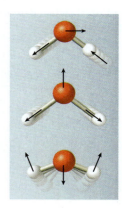

Figure 21.14 The three different modes of vibration of a water molecule. Each mode of vibration can be imagined by moving the atoms along the arrows and then reversing the direction of motion.

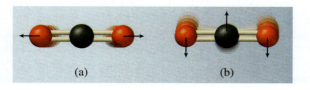

(a)   (b)

Figure 21.15 Two of the four ways a carbon dioxide molecule can vibrate. The vibration in (a) does not result in a change in dipole moment, but the vibration in (b) renders the molecule IR-active.

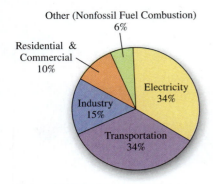

Figure 21.16 Sources of carbon dioxide emission in the United States. Note that not all the emitted CO_2 enters the atmosphere. Some of it is taken up by carbon dioxide "sinks," such as the ocean.

Figure 21.17 Yearly variation of carbon dioxide concentration at Mauna Loa, Hawaii, the source of the longest running record of atmospheric CO_2. The general trend clearly points to an increase of carbon dioxide in the atmosphere.

Although the total amount of water vapor in our atmosphere has not altered noticeably over the years, the concentration of CO_2 has been rising steadily since the turn of the century as a result of the burning of fossil fuels (petroleum, natural gas, and coal). Figure 21.16 shows the percentages of CO_2 emitted due to human activities in the United States in 2016, and Figure 21.17 shows the variation of carbon dioxide concentration over a period of years, as measured in Hawaii. In the Northern Hemisphere, the seasonal oscillations are caused by the removal of carbon dioxide by photosynthesis during the growing season and its buildup during the fall and winter months. The trend is toward an increase in CO_2. The current rate of increase is more than 1 ppm (> 1 part CO_2 per million parts air) by volume per year, which is equivalent to roughly 10^{10} tons of CO_2! According to the website co2now.org, the CO_2 level exceeded 410 ppm in April of 2018.

In addition to CO_2 and H_2O, other greenhouse gases, such as the CFCs, CH_4, NO_x, and N_2O, contribute appreciably to the warming of the atmosphere. Figure 21.18 shows the increase in temperature over the years, and Figure 21.19 shows the relative contributions of the greenhouse gases to global warming.

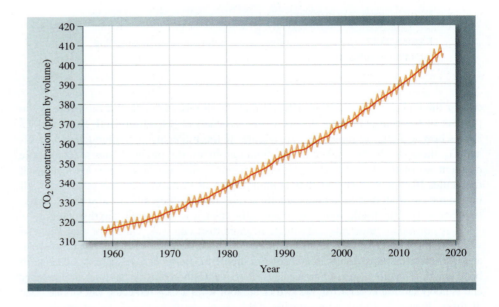

Figure 21.18 Temperature rise on Earth's surface from 1880 to 2017.

It is predicted by some climate models that should the buildup of greenhouse gases continue at its current rate, Earth's average temperature will increase by as much as 6°C in the twenty-first century. Although a temperature increase of a few degrees may seem insignificant, it is actually large enough to disrupt the delicate thermal balance on Earth and could cause glaciers and ice caps to melt. Consequently, the sea level would rise and coastal areas would be flooded. Predicting weather trends is extremely difficult, though, and there are other potentially moderating factors to take into account before concluding that global warming is inevitable and irreversible. For example, the ash from volcanic eruptions diffuses upward and can stay in the atmosphere for years. By reflecting incoming sunlight, volcanic ash can cause a cooling effect. Furthermore, the warming effect of CFCs in the troposphere is offset by its action in the stratosphere. Because ozone is a polar polyatomic molecule, it is also an effective greenhouse gas. A decrease in ozone brought about by CFCs actually produces a noticeable drop in temperature.

To combat the greenhouse effect, we must lower carbon dioxide emissions. This can be done by improving energy efficiency in automobiles and in household heating and lighting, and by developing nonfossil fuel energy sources, such as photovoltaic cells. Nuclear energy is a viable alternative, but its use is highly controversial due to the difficulty of disposing of radioactive waste and the fact that nuclear power stations are more prone to accidents than conventional power stations (see Chapter 20). The phasing out of CFCs, the most potent greenhouse gas, will help to slow down the warming trend. The recovery of methane gas generated at landfills and the reduction of natural gas leakages are other steps we could take to control CO_2 emissions. Finally, the preservation of the Amazon jungle, tropical forests in Southeast Asia, and other large forests is vital to maintaining the steady-state concentration of CO_2 in the atmosphere. Converting forests to farmland for crops and grassland for cattle may do irreparable damage to the delicate ecosystem and permanently alter the climate pattern on Earth.

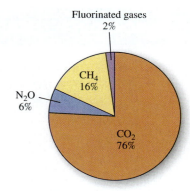

Figure 21.19 Contribution to global warming by various greenhouse gases. The concentrations of fluorinated gases and methane are much lower than that of carbon dioxide. However, because they can absorb IR radiation much more effectively than CO_2, they make significant contributions to the overall warming effect.

Student Note: Global climate change is the subject of the Academy Award–winning 2006 documentary, *An Inconvenient Truth,* presented by former vice president Al Gore.

SAMPLE PROBLEM 21.2

Which of the following qualify as greenhouse gases: CO, NO, NO_2, Cl_2, H_2, Ne?

Strategy To behave as a greenhouse gas, either the molecule must possess a dipole moment or some of its vibrational motions must generate a temporary dipole moment.

Setup The necessary conditions immediately rule out homonuclear diatomic molecules and atomic species.

Solution Only CO, NO, and NO_2, which are all polar molecules, qualify as greenhouse gases. Both Cl_2 and H_2 are homonuclear diatomic molecules, and Ne is atomic. These three species are all IR-inactive.

THINK ABOUT IT

CO_2, the best-known greenhouse gas, is *nonpolar.* It is only necessary for at least one of a molecule's vibrational modes to induce a *temporary* dipole for it to act as a greenhouse gas.

Practice Problem **A**TTEMPT Which of the following is a more effective greenhouse gas: CO or H_2O? Explain.

Practice Problem **B**UILD Both O_2 and O_3 exhibit molecular vibration. Despite this, only O_3 acts as a greenhouse gas. Explain.

Practice Problem **C**ONCEPTUALIZE Which of these molecules qualifies as a greenhouse gas?

Cl_2 SO_2 SO_3 CH_4 N_2 CS_2

CHECKPOINT – SECTION 21.5   The Greenhouse Effect

21.5.1 Which of the following can act as a greenhouse gas? (Select all that apply.)

a) CH_4

b) N_2

c) Rn

d) O_3

e) Xe

21.5.2 The greenhouse effect is

a) caused by depletion of stratospheric ozone.

b) entirely the result of human activity.

c) a natural phenomenon that has been enhanced by human activity.

d) the absorption of the sun's energy by molecules and atoms in the upper atmosphere.

e) responsible for the aurora borealis and the aurora australis.

21.6 Acid Rain

Every year acid rain causes hundreds of millions of dollars' worth of damage to stone buildings and statues throughout the world. The term *stone leprosy* is used by some environmental chemists to describe the corrosion of stone by acid rain (Figure 21.20). Acid rain is also toxic to vegetation and aquatic life. Many well-documented cases show dramatically how acid rain has destroyed agricultural and forest lands and killed aquatic organisms.

Precipitation in the northeastern United States has an average pH of about 4.3 (Figure 21.21). Because atmospheric CO_2 in equilibrium with rainwater would not be expected to result in a pH less than 5.5, sulfur dioxide (SO_2) and, to a lesser extent, nitrogen oxides from auto emissions are believed to be responsible for the high acidity of rainwater. Acidic oxides, such as SO_2, react with water to give the corresponding acids. There are several sources of atmospheric SO_2. Nature itself contributes much SO_2 in the form of volcanic eruptions. Also, many metals exist combined with sulfur in nature. Extracting the metals often entails *smelting*, or *roasting*, the ores—that is, heating the metal sulfide in air to form the metal oxide and SO_2. For example:

$$2ZnS(s) + 3O_2(g) \longrightarrow 2ZnO(s) + 2SO_2(g)$$

The metal oxide can be reduced more easily than the sulfide (by a more reactive metal or in some cases by carbon) to the free metal.

Figure 21.20 This photo of a badly eroded statue illustrates the damage done by acid rain.

Ryan McGinnis/Alamy Stock Photo

Figure 21.21 Mean precipitation pH in the United States in 1994. Most SO_2 comes from the midwestern states. Prevailing winds carry the acid droplets formed over the Northeast. Nitrogen oxides also contribute to acid rain formation.

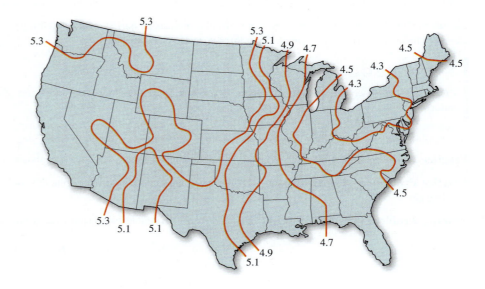

Although smelting is a major source of SO_2, the burning of fossil fuels in industry, in power plants, and in homes accounts for most of the SO_2 emitted to the atmosphere (Figure 21.22). The sulfur content of coal ranges from 0.5 to 5 percent by mass, depending on the source of the coal. The sulfur content of other fossil fuels is similarly variable. Oil from the Middle East, for instance, is low in sulfur, whereas that from Venezuela has a high sulfur content. To a lesser extent, the nitrogen-containing compounds in oil and coal are converted to nitrogen oxides, which can also acidify rainwater.

All in all, some 50 million to 60 million tons of SO_2 are released into the atmosphere each year! In the troposphere, SO_2 is almost all oxidized to H_2SO_4 in the form of aerosol, which ends up in wet precipitation or acid rain. The mechanism for the conversion of SO_2 to H_2SO_4 is quite complex and not fully understood. The reaction is believed to be initiated by the hydroxyl radical (OH):

$$OH + SO_2 \longrightarrow HOSO_2$$

The $HOSO_2$ radical is further oxidized to SO_3:

$$HOSO_2 + O_2 \longrightarrow HO_2 + SO_3$$

The sulfur trioxide formed would then rapidly react with water to form sulfuric acid:

$$SO_3 + H_2O \longrightarrow H_2SO_4$$

SO_2 can also be oxidized to SO_3 and then converted to H_2SO_4 on particles by heterogeneous catalysis. Eventually, the acid rain can corrode limestone and marble ($CaCO_3$). A typical reaction is:

$$CaCO_3(s) + H_2SO_4(aq) \longrightarrow CaSO_4(s) + H_2O(l) + CO_2(g)$$

Sulfur dioxide can also attack calcium carbonate directly:

$$2CaCO_3(s) + 2SO_2(g) + O_2(g) \longrightarrow 2CaSO_4(s) + 2CO_2(g)$$

There are two ways to minimize the effects of SO_2 pollution. The most direct approach is to remove sulfur from fossil fuels before combustion, but this is technologically difficult to accomplish. A cheaper but less efficient way is to remove SO_2 as it is formed. For example, in one process powdered limestone is injected into the power plant boiler or furnace along with the coal (Figure 21.23). At high temperatures, the following decomposition occurs:

$$\underset{\text{limestone}}{CaCO_3(s)} \longrightarrow \underset{\text{quicklime}}{CaO(s)} + CO_2(g)$$

Figure 21.22 Sulfur dioxide and other air pollutants being released into the atmosphere from a coal-burning power plant.
Larry Lee Photography/Corbis/Getty Images

Animation
Oil refining process.

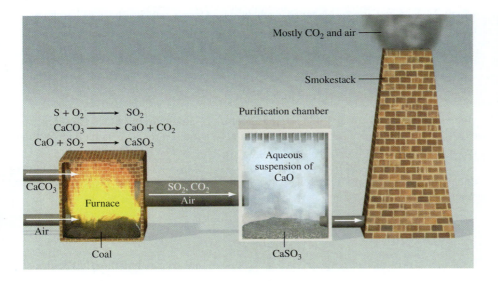

Figure 21.23 Common procedure for removing SO_2 from burning fossil fuel. Powdered limestone decomposes into CaO, which reacts with SO_2 to form $CaSO_3$. The remaining SO_2 is combined with an aqueous suspension of CaO to form $CaSO_3$.

Figure 21.24 Spreading calcium oxide (CaO) over acidified soil. This process is called liming.

NaMaKuKi/Shutterstock

The quicklime reacts with SO_2 to form calcium sulfite and some calcium sulfate:

$$CaO(s) + SO_2(g) \longrightarrow CaSO_3(s)$$

$$2CaO(s) + 2SO_2(g) + O_2(g) \longrightarrow 2CaSO_4(s)$$

To remove any remaining SO_2, an aqueous suspension of quicklime is injected into a purification chamber prior to the gases' escape through the smokestack. Quicklime is also added to lakes and soils in a process called *liming* to reduce their acidity (Figure 21.24). Installing a sulfuric acid plant near a metal ore refining site is also an effective way to cut SO_2 emission because the SO_2 produced by roasting metal sulfides can be captured for use in the synthesis of sulfuric acid. This is a very sensible way to turn what is a pollutant in one process into a starting material for another process!

21.7 Photochemical Smog

The word *smog* was coined to describe the combination of smoke and fog that shrouded London during the 1950s. The primary cause of this noxious cloud was sulfur dioxide. Today, however, ***photochemical smog,*** which is formed by the reactions of automobile exhaust in the presence of sunlight, is much more common.

Automobile exhaust consists mainly of NO, CO, and various unburned hydrocarbons. These gases are called *primary pollutants* because they set in motion a series of photochemical reactions that produce *secondary pollutants.* It is the secondary pollutants—chiefly NO_2 and O_3—that are responsible for the buildup of smog.

Nitric oxide is the product of the reaction between atmospheric nitrogen and oxygen at high temperatures inside an automobile engine:

$$N_2(g) + O_2(g) \longrightarrow 2NO(g)$$

Once released into the atmosphere, nitric oxide is oxidized to nitrogen dioxide:

$$2NO(g) + O_2(g) \longrightarrow 2NO_2(g)$$

Sunlight causes the photochemical decomposition of NO_2 (at a wavelength shorter than 400 nm) into NO and O:

$$NO_2(g) + h\nu \longrightarrow NO(g) + O(g)$$

Atomic oxygen is a highly reactive species that can initiate a number of important reactions, one of which is the formation of ozone:

$$O(g) + O_2(g) + M \longrightarrow O_3(g) + M$$

where M is some inert substance such as N_2. Ozone attacks the C=C linkage in rubber:

$$\begin{array}{c} R \\ \diagdown \\ C=C \\ \diagup \quad \diagdown \\ R \qquad R \end{array} + O_3 \longrightarrow \begin{array}{c} R \quad O \quad R \\ \diagdown \diagup \diagdown \diagup \\ C \qquad C \\ \diagup \quad | \quad | \quad \diagdown \\ R \quad O-O \quad R \end{array} \xrightarrow{H_2O} \begin{array}{c} R \\ \diagdown \\ C=O \\ \diagup \\ R \end{array} + \begin{array}{c} R \\ \diagdown \\ O=C \\ \diagdown \\ R \end{array} + H_2O_2$$

where R represents groups of C and H atoms. In smog-ridden areas, this reaction can cause automobile tires to crack. Similar reactions are also damaging to lung tissues and other biological substances.

Ozone can be formed also by a series of very complex reactions involving unburned hydrocarbons, nitrogen oxides, and oxygen. One of the products of these reactions is peroxyacetyl nitrate (PAN):

$$CH_3-\underset{\underset{O}{\|}}{C}-O-O-NO_2$$

PAN is a powerful lachrymator, or tear producer, and causes breathing difficulties.

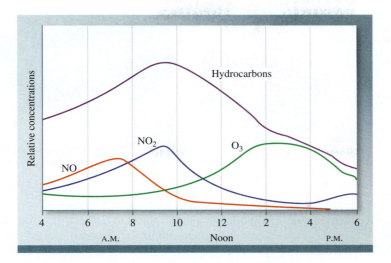

Figure 21.25 Typical variations with time in concentration of air pollutants on a smoggy day.

Figure 21.25 shows typical variations with time of primary and secondary pollutants. Initially, the concentration of NO_2 is quite low. As soon as solar radiation penetrates the atmosphere, though, more NO_2 is formed from NO and O_2. The concentration of ozone remains fairly constant at a low level in the early morning hours. As the concentration of unburned hydrocarbons and aldehydes increases in the air, the concentrations of NO_2 and O_3 also rise rapidly. The actual amounts depend on the location, traffic, and weather conditions, but their presence is always accompanied by haze (Figure 21.26). The oxidation of hydrocarbons produces various organic intermediates, such as alcohols and carboxylic acids, which are all less volatile than the hydrocarbons themselves. These substances eventually condense into small droplets of liquid. The dispersion of these droplets in air, called an *aerosol*, scatters sunlight and reduces visibility. This interaction also makes the air look hazy.

As the mechanism of photochemical smog formation has become better understood, major efforts have been made to reduce the buildup of primary pollutants. Most automobiles now are equipped with catalytic converters designed to oxidize CO and unburned hydrocarbons to CO_2 and H_2O and to reduce NO and NO_2 to N_2 and O_2 [◄◄ Section 14.6]. More efficient automobile engines and better public transportation systems would also help to decrease air pollution in urban areas. A recent technological innovation to combat photochemical smog is to coat automobile radiators and air conditioner compressors with a platinum catalyst. So equipped, a running car can purify the air that flows under the hood by converting ozone and carbon monoxide to oxygen and carbon dioxide:

$$O_3(g) + CO(g) \xrightarrow{\text{Pt}} O_2(g) + CO_2(g)$$

In a city like Los Angeles, where the number of miles driven in one day equals nearly 300 million, this approach would significantly improve the air quality and reduce the "high-ozone level" warnings frequently issued to its residents.

Figure 21.26 A smoggy day in a big city.

Hung Chung Chih/Getty Images

21.8 Indoor Pollution

Difficult as it is to avoid air pollution outdoors, it is no easier to avoid pollution indoors. The air quality in homes and in the workplace is affected by human activities, by construction materials, and by other factors in our immediate environment. The common indoor pollutants are radon, carbon monoxide, carbon dioxide, and formaldehyde.

The Risk from Radon

In a highly publicized case in 1984, an employee reporting for work at a nuclear power plant in Pennsylvania set off the plant's radiation monitor. Astonishingly, the source of his contamination turned out not to be the plant, but radon in his home!

Figure 21.27 (a) Sources of background radiation. (b) Radon occurrence in the United States.

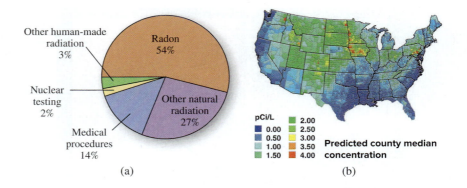

Other human-made radiation
3%

Radon
54%

Nuclear testing
2%

Other natural radiation
27%

Medical procedures
14%

pCi/L

0.00	2.00
0.50	2.50
1.00	3.00
1.50	3.50
	4.00

Predicted county median concentration

(a)   (b)

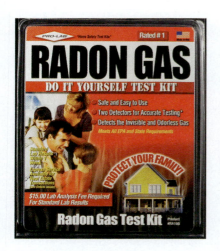

Figure 21.28 Home radon detector.

David A. Tietz/Editorial Image, LLC

A lot has been said and written about the potential dangers of radon as an air pollutant. Just what is radon? Where does it come from? And how does it affect our health?

Radon is a member of Group 18 (the noble gases). It is an intermediate product of the radioactive decay of uranium-238 (see Figure 20.3). All isotopes of radon are radioactive, but radon-222 is the most hazardous because it has the longest half-life—3.8 days. Radon, which accounts for slightly over half the background radioactivity on Earth, is generated mostly from the phosphate minerals of uranium (Figure 21.27).

Since the 1970s, high levels of radon have been detected in homes built on reclaimed land above uranium mill tailing deposits. The colorless, odorless, and tasteless radon gas enters a building through tiny cracks in the basement floor. It is slightly soluble in water, so it can be spread in different media. Radon-222 is an α-emitter. When it decays, it produces radioactive polonium-214 and polonium-218, which can build up to high levels in an enclosed space. These solid radioactive particles can adhere to airborne dust and smoke, which are inhaled into the lungs and deposited in the respiratory tract. Over a long period of time, the α particles emitted by polonium and its decay products, which are also radioactive, can cause lung cancer.

What can be done to combat radon pollution indoors? The first step is to measure the radon level in the basement with a reliable test kit. Short-term and long-term kits are available (Figure 21.28). The short-term tests use activated charcoal to collect the decay products of radon over a period of several days. The container is sent to a laboratory where a technician measures the radioactivity (γ rays) from radon-decay products lead-214 and bismuth-214. Knowing the length of exposure, the lab technician back-calculates to determine radon concentration. The long-term test kits use a piece of special polymer film on which an α particle will leave a "track." After several months' exposure, the film is etched with a sodium hydroxide solution and the number of tracks counted. Knowing the length of exposure enables the technician to calculate the radon concentration. If the radon level is unacceptably high, then the house must be regularly ventilated. This precaution is particularly important in recently built houses, which are well insulated. A more effective way to prevent radon pollution is to reroute the gas before it gets into the house (e.g., by installing a ventilation duct to draw air from beneath the basement floor to the outside).

Currently there is considerable controversy regarding the health effects of radon. The first detailed studies of the effects of radon on human health were carried out in the 1950s when it was recognized that uranium miners suffered from an abnormally high incidence of lung cancer. Some scientists have challenged the validity of these studies because the miners were also smokers. It seems quite likely that there is a synergistic effect between radon and smoking on the development of lung cancer. Radon decay products will adhere not only to tobacco tar deposits in the lungs, but also to the solid particles in cigarette smoke, which can be inhaled by smokers and nonsmokers. More systematic studies are needed to evaluate the environmental impact of radon. In the meantime, the Environmental Protection Agency (EPA) has recommended remedial action where the radioactivity level due to radon exceeds 4 picocuries (pCi) per liter of air. [A curie corresponds to 3.70×10^{10} disintegrations of radioactive nuclei per second; a picocurie is a trillionth of a curie, or 3.70×10^{-2} disintegrations per second (dps).]

SAMPLE PROBLEM 21.3

The half-life of Rn-222 is 3.8 days. Starting with 1.0 g of Rn-222, how much will be left after 10 half-lives?

Strategy All radioactive decays obey first-order kinetics, making the half-life independent of the initial concentration.

Setup Because the question involves an integral number of half-lives, we can deduce the amount of Rn-222 remaining without using Equation 14.3.

Solution After one half-life, the amount of Rn left is 0.5×1.0 g, or 0.5 g. After two half-lives, only 0.25 g of Rn remains. Generalizing the fraction of the isotope left after n half-lives as $(1/2)^n$, where $n = 10$, we write

$$\text{quantity of Rn-222 left} = 1.0 \text{ g} \times (1/2)^{10}$$

$$= 9.8 \times 10^{-4} \text{ g}$$

> **THINK ABOUT IT**
>
> An alternative solution is to calculate the first-order rate constant from the half-life and use Equation 14.3:
>
> $$\ln\frac{N_t}{N_0} = -kt$$
>
> where N is the mass of Rn-222. Try this and verify that your answers are the same. (Since most of the kinetics problems we encounter do not involve an integral number of half-lives, we generally use Equation 14.3 as the *first* approach to solving them.)

Practice Problem **A**TTEMPT The concentration of Rn-222 in the basement of a house is 1.8×10^{-6} mol/L. Assume the air remains static, and calculate the concentration of the radon after 2.4 days.

Practice Problem **B**UILD How long will it take for the radioactivity due to radon to fall to a level considered acceptable by the EPA if the starting activity is 2.25×10^3 dps/L?

Practice Problem **C**ONCEPTUALIZE The first diagram represents a sample of radon gas. Which of diagrams (i)–(iii) best represents the system after about 10 days?

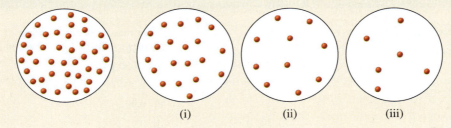

| (i) | (ii) | (iii) |

Carbon Dioxide and Carbon Monoxide

Both carbon dioxide (CO_2) and carbon monoxide (CO) are products of combustion.

In the presence of an abundant supply of oxygen, CO_2 is formed; in a limited supply of oxygen, both CO and CO_2 are formed. The indoor sources of these gases are gas cooking ranges, woodstoves, space heaters, tobacco smoke, human respiration, and exhaust fumes from cars (in garages). Carbon dioxide is not a toxic gas, but it does have an asphyxiating effect. In airtight buildings, the concentration of CO_2 can reach as high as 2000 ppm by volume (compared with 3 ppm outdoors). Workers exposed to high concentrations of CO_2 in skyscrapers and other sealed environments become fatigued more easily and have difficulty concentrating. Adequate ventilation is the solution to CO_2 pollution.

Like CO_2, CO is a colorless and odorless gas, but it differs from CO_2 in that it is highly poisonous. The toxicity of CO lies in its unusual ability to bind very strongly to hemoglobin, the oxygen carrier in blood. Both O_2 and CO bind to the Fe(II) ion in hemoglobin, but the affinity of hemoglobin for CO is about 200 times greater than it is for O_2. Hemoglobin molecules with tightly bound CO (called carboxyhemoglobin) cannot carry the oxygen needed for metabolic processes. At a concentration of 70 ppm, CO can cause drowsiness and headache; at higher concentrations, death may result when about half the hemoglobin molecules become complexed with CO. The best first-aid response to CO poisoning is to remove the victim immediately to an area with a plentiful oxygen supply or to give mouth-to-mouth resuscitation.

Formaldehyde

Formaldehyde (CH_2O) is a rather disagreeable-smelling liquid used as a preservative for laboratory specimens. Industrially, formaldehyde resins are used as bonding agents in building and furniture construction materials such as plywood and particle board. In addition, urea-formaldehyde insulation foams are used to fill wall cavities. The resins and foams slowly break down to release free formaldehyde, especially under acid and humid conditions. Low concentrations of formaldehyde in the air can cause drowsiness, nausea, headaches, and other respiratory ailments. Laboratory tests show that breathing high concentrations of formaldehyde can induce cancers in animals, but whether it has a similar effect in humans is unclear. The safe standard of formaldehyde in indoor air has been set at 0.1 ppm by volume.

Because formaldehyde is a reducing agent, devices have been constructed to remove it by means of a redox reaction. Indoor air is circulated through an air purifier containing an oxidant such as $Al_2O_3/KMnO_4$, which converts formaldehyde to the less harmful and less volatile formic acid (HCOOH). Proper ventilation is the best way to remove formaldehyde. However, care should be taken not to remove the air from a room too quickly without replenishment, because a reduced pressure would cause the formaldehyde resins to decompose faster, resulting in the release of *more* formaldehyde.

CHECKPOINT – SECTION 21.8   Indoor Pollution

21.8.1 What is the risk posed by radon gas in homes?

a) respiratory distress

b) nausea

c) dizziness

d) coma

e) lung cancer

21.8.2 What mass of a 1.0-g sample of radon remains after 30 days? (The half-life of radon is 3.8 days.)

a) 0.030 g

b) 0.0040 g

c) 0.99 g

d) 0.97 g

e) 0.0010 g

Chapter Summary

Section 21.1

- Earth's atmosphere is made up mainly of nitrogen and oxygen, plus a number of other trace gases. Molecular nitrogen in the atmosphere is incorporated into other compounds via **nitrogen fixation.**

- The regions of the atmosphere, from Earth's surface outward, are the **troposphere,** the **stratosphere,** the **mesosphere,** the **thermosphere,** and the **ionosphere.**

- The chemical processes that go on in the atmosphere are influenced by solar radiation, volcanic eruption, and human activities.

Section 21.2

- In the outer regions of the atmosphere, the bombardment of molecules and atoms by solar particles gives rise to the *aurora borealis* in the Northern Hemisphere and the *aurora australis* in the Southern Hemisphere. The glow on a space shuttle was caused by excitation of molecules adsorbed on the shuttle's surface.

Section 21.3

- Ozone in the stratosphere absorbs harmful UV radiation in the 200- to 300-nm range and protects life underneath. For many years, chlorofluorocarbons have been destroying the ozone layer.

Section 21.4

- Volcanic eruptions can lead to air pollution, deplete ozone in the stratosphere, and affect climate.

Section 21.5

- Carbon dioxide's ability to absorb infrared radiation enables it to trap some of the outgoing heat from Earth, warming its surface—a phenomenon known as the **greenhouse effect.** Other gases such as the CFCs and methane also contribute to the greenhouse effect. Global warming refers to the result of the *enhanced* greenhouse effect caused by human activities.

Section 21.6

- Sulfur dioxide, and to a lesser extent nitrogen oxides, generated mainly from the burning of fossil fuels and from the roasting of metal sulfides, causes acid rain.

Section 21.7

- *Photochemical smog* is formed by the photochemical reaction of automobile exhaust in the presence of sunlight. It is a complex reaction involving nitrogen oxides, ozone, and hydrocarbons.

Section 21.8

- Indoor air pollution is caused by radon, a radioactive gas formed during uranium decay; carbon monoxide and carbon dioxide, products of combustion; and formaldehyde, a volatile organic substance released from resins used in construction materials.

Key Words

Greenhouse effect, 1059

Ionosphere, 1052

Mesosphere, 1052

Nitrogen fixation, 1051

Photochemical smog, 1066

Stratosphere, 1052

Thermosphere, 1052

Troposphere, 1052

Questions and Problems

Applying What You've Learned

The research of Paul Crutzen, the third recipient of the Nobel Prize for Chemistry in 1995, involved the effect of nitric oxide (NO) on the destruction of stratospheric ozone. Unlike CFCs, which may take 50 to 100 years to diffuse into the upper atmosphere, nitric oxide is introduced directly to the stratosphere in the exhaust of high-altitude aircraft. Early in the 1970s, the United States considered construction of a large fleet of supersonic transport airplanes (SSTs), similar to the Concorde. Environmentalists argued, based in part on the work of Paul Crutzen, that to do so would significantly endanger the ozone layer.

Problems:

(a) The bond enthalpy of NO is 630.6 kJ/mol. Determine the maximum wavelength of light required to break the bond in an NO molecule [◄◄ Sample Problem 21.1]. (b) Can nitric oxide act as a greenhouse gas? Explain [◄◄ Sample Problem 21.2].

SECTION 21.1: EARTH'S ATMOSPHERE

Review Questions

21.1   Describe the regions of Earth's atmosphere.
21.2   Briefly outline the main processes of the nitrogen and oxygen cycles.
21.3   Explain why, for maximum performance, supersonic airplanes need to fly at a high altitude (in the stratosphere).
21.4   Jupiter's atmosphere consists mainly of hydrogen (90 percent) and helium (9 percent). How does this mixture of gases contrast with the composition of Earth's atmosphere? Why does the composition differ?

Problems

21.5   Referring to Table 21.1, calculate the mole fraction of CO_2 and its concentration in parts per million by volume.
21.6   Calculate the partial pressure of CO_2 (in atm) in dry air when the atmospheric pressure is 754 mmHg.
21.7   Describe the processes that result in the warming of the stratosphere.
21.8   Calculate the mass (in kg) of nitrogen, oxygen, and carbon dioxide gases in the atmosphere. Assume that the total mass of air in the atmosphere is 5.25×10^{21} g.

SECTION 21.2: PHENOMENA IN THE OUTER LAYERS OF THE ATMOSPHERE

Review Questions

21.9   What process gives rise to the aurora borealis and aurora australis?
21.10   Why can astronauts not release oxygen atoms to test the mechanism of shuttle glow?

Computational Problems

21.11   The highly reactive OH radical (a species with an unpaired electron) is believed to be involved in some atmospheric processes. Table 8.6 lists the bond enthalpy for the oxygen-to-hydrogen bond in OH as 460 kJ/mol. What is the longest wavelength (in nm) of radiation that can bring about the following reaction?

$$OH(g) \longrightarrow O(g) + H(g)$$

21.12   The green color observed in the aurora borealis is produced by the emission of a photon by an electronically excited oxygen atom at 558 nm. Calculate the energy difference between the two levels involved in the emission process.

SECTION 21.3: DEPLETION OF OZONE IN THE STRATOSPHERE

Review Questions

21.13   Briefly describe the absorption of solar radiation in the stratosphere by O_2 and O_3 molecules.
21.14   Explain the processes that have a warming effect on the stratosphere.
21.15   List the properties of CFCs, and name four major uses of these compounds.
21.16   How do CFCs and nitrogen oxides destroy ozone in the stratosphere?
21.17   What causes the polar ozone holes?
21.18   How do volcanic eruptions contribute to ozone destruction?
21.19   Describe ways to curb the destruction of ozone in the stratosphere.
21.20   Discuss the effectiveness of some of the CFC substitutes.

Computational Problems

21.21 Given that the quantity of ozone in the stratosphere is equivalent to a 3.0-mm-thick layer of ozone on Earth at STP, calculate the number of ozone molecules in the stratosphere and their mass in kilograms. (*Hint:* The radius of Earth is 6371 km and the surface area of a sphere is $4\pi r^2$, where r is the radius.)

21.22 Referring to the answer in Problem 21.21, and assuming that the level of ozone in the stratosphere has already fallen 6.0 percent, calculate the number of kilograms of ozone that would have to be manufactured on a daily basis so that we could restore the ozone to the original level in 100 years. If ozone is made according to the process $3O_2(g) \longrightarrow 2O_3(g)$, how many kilojoules of energy would be required?

Conceptual Problems

21.23 Both Freon-11 and Freon-12 are made by the reaction of carbon tetrachloride (CCl_4) with hydrogen fluoride. Write equations for these reactions.

21.24 Why are CFCs not decomposed by UV radiation in the troposphere?

21.25 The average bond enthalpies of the C—Cl and C—F bonds are 340 and 485 kJ/mol, respectively. Based on this information, explain why the C—Cl bond in a CFC molecule is preferentially broken by solar radiation at 250 nm.

21.26 Like CFCs, certain bromine-containing compounds such as CF_3Br can also participate in the destruction of ozone by a similar mechanism starting with the Br atom:

$$CF_3Br \longrightarrow CF_3 + Br$$

Given that the average C—Br bond energy is 276 kJ/mol, estimate the longest wavelength required to break this bond. Will this compound be decomposed in the troposphere only or in both the troposphere and stratosphere?

21.27 Draw Lewis structures for chlorine nitrate ($ClONO_2$) and chlorine monoxide (ClO).

21.28 Draw Lewis structures for HCFC-123 (CF_3CHCl_2) and CF_3CFH_2.

SECTION 21.4: VOLCANOES

Review Questions

21.29 What are the effects of volcanic eruptions on climate?

21.30 Classify the reaction between H_2S and SO_2 that leads to the formation of sulfur at the site of a volcanic eruption.

SECTION 21.5: THE GREENHOUSE EFFECT

Review Questions

21.31 What is the greenhouse effect? What is the criterion for classifying a gas as a greenhouse gas?

21.32 Why is more emphasis placed on the role of carbon dioxide in the greenhouse effect than on that of water?

21.33 Describe three human activities that generate carbon dioxide. List two major mechanisms for the uptake of carbon dioxide.

21.34 Deforestation contributes to the greenhouse effect in two ways. What are they?

21.35 How does an increase in world population enhance the greenhouse effect?

21.36 Is ozone a greenhouse gas? If so, sketch three ways an ozone molecule can vibrate.

21.37 What effects do CFCs and their substitutes have on Earth's temperature?

21.38 Why are CFCs more effective greenhouse gases than methane and carbon dioxide?

Computational Problems

21.39 The annual production of zinc sulfide (ZnS) is 4.0×10^4 tons. Estimate the number of tons of SO_2 produced by roasting it to extract zinc metal.

21.40 Calcium oxide or quicklime (CaO) is used in steelmaking, cement manufacture, and pollution control. It is prepared by the thermal decomposition of calcium carbonate:

$$CaCO_3(s) \longrightarrow CaO(s) + CO_2(g)$$

Calculate the yearly release of CO_2 (in kg) to the atmosphere if the annual production of CaO in the United States is 1.7×10^{10} kg.

SECTION 21.6: ACID RAIN

Review Questions

21.41 Name the gas that is largely responsible for the acid rain phenomenon.

21.42 List three detrimental effects of acid rain.

21.43 Briefly discuss two industrial processes that lead to acid rain.

21.44 Discuss ways to curb acid rain.

21.45 Water and sulfur dioxide are both polar molecules, and their geometry is similar. Why is SO_2 not considered a major greenhouse gas?

SECTION 21.7: PHOTOCHEMICAL SMOG

Review Questions

21.46 What is photochemical smog? List the factors that favor the formation of photochemical smog.

21.47 What are primary and secondary pollutants?

21.48 Identify the gas that is responsible for the brown color of photochemical smog.

21.49 The safety limits of ozone and carbon monoxide are 120 ppb by volume and 9 ppm by volume, respectively. Why does ozone have a lower limit?

21.50 Suggest ways to minimize the formation of photochemical smog.

21.51 In which region of the atmosphere is ozone beneficial? In which region is it detrimental?

Computational Problems

21.52 The gas-phase decomposition of peroxyacetyl nitrate (PAN) obeys first-order kinetics:

$$CH_3COOONO_2 \longrightarrow CH_3COO + NO_2$$

with a rate constant of $4.9 \times 10^{-4} \, s^{-1}$. Calculate the rate of decomposition (in M/s) if the concentration of PAN is 0.55 ppm by volume. Assume STP conditions.

21.53 On a smoggy day in a certain city, the ozone concentration was 0.42 ppm by volume. Calculate the partial pressure of ozone (in atm) and the number of ozone molecules per liter of air if the temperature and pressure were 20.0°C and 748 mmHg, respectively.

SECTION 21.8: INDOOR POLLUTION

Review Questions

21.54 List the major indoor pollutants and their sources.

21.55 What is the best way to deal with indoor pollution?

21.56 Why is it dangerous to idle a car's engine in a poorly ventilated place, such as the garage?

21.57 Describe the properties that make radon an indoor pollutant. Would radon be more hazardous if ^{222}Rn had a longer half-life?

Computational Problems

21.58 A volume of 5.0 L of polluted air at 18.0°C and 747 mmHg is passed through lime water [an aqueous suspension of $Ca(OH)_2$], so that all the carbon dioxide present is precipitated as $CaCO_3$. If the mass of the $CaCO_3$ precipitate is 0.026 g, calculate the percentage by volume of CO_2 in the air sample.

21.59 A concentration of 8.00×10^2 ppm by volume of CO is considered lethal to humans. Calculate the minimum mass of CO (in grams) that would become a lethal concentration in a closed room 17.6 m long, 8.80 m wide, and 2.64 m high. The temperature and pressure are 20.0°C and 756 mmHg, respectively.

ADDITIONAL PROBLEMS

21.60 As mentioned in the chapter, spraying the stratosphere with hydrocarbons such as ethane and propane should eliminate Cl atoms. What is the drawback of this procedure if used on a large scale for an extended period of time?

21.61 Briefly describe the harmful effects of the following substances: O_3, SO_2, NO_2, CO, $CH_3COOONO_2$ (PAN), Rn.

21.62 The equilibrium constant (K_P) for the reaction

$$N_2(g) + O_2(g) \rightleftharpoons 2NO(g)$$

is 4.0×10^{-31} at 25°C and 2.6×10^{-6} at 1100°C, the temperature of a running car's engine. Is this an endothermic or exothermic reaction?

21.63 Although the hydroxyl radical (OH) is present only in a trace amount in the troposphere, it plays a central role in its chemistry because it is a strong oxidizing agent and can react with many pollutants as well as some CFC substitutes. The hydroxyl radical is formed by the following reactions:

$$O_3 \xrightarrow{\lambda \, = \, 320 \, nm} O^* + O_2$$
$$O + H_2O \longrightarrow 2OH$$

where O* denotes an electronically excited atom. (a) Explain why the concentration of OH is so small even though the concentrations of O_3 and H_2O are quite large in the troposphere. (b) What property makes OH a strong oxidizing agent? (c) The reaction between OH and NO_2 contributes to acid rain. Write an equation for this process. (d) The hydroxyl radical can oxidize SO_2 to H_2SO_4. The first step is the formation of a neutral HSO_3 species, followed by its reaction with O_2 and H_2O to form H_2SO_4 and the hydroperoxyl radical (HO_2). Write equations for these processes.

21.64 The equilibrium constant (K_P) for the reaction $2CO(g) + O_2(g) \rightleftharpoons 2CO_2(g)$ is 1.4×10^{90} at 25°C. Given this enormous value, why doesn't CO convert totally to CO_2 in the troposphere?

21.65 How are past temperatures determined from ice cores obtained from the Arctic or Antarctica? (*Hint:* Look up the stable isotopes of hydrogen and oxygen. How does energy required for vaporization depend on the masses of H_2O molecules containing different isotopes? How would you determine the age of an ice core?)

21.66 The balance between SO_2 and SO_3 is important in understanding acid rain formation in the troposphere. From the following information at 25°C:

$$S(s) + O_2(g) \rightleftharpoons SO_2(g) \qquad K_1 = 4.2 \times 10^{52}$$
$$2S(s) + 3O_2(g) \rightleftharpoons 2SO_3(g) \qquad K_2 = 9.8 \times 10^{128}$$

calculate the equilibrium constant for the reaction:

$$2SO_2(g) + O_2(g) \rightleftharpoons 2SO_3(g)$$

21.67 The effective incoming solar radiation per unit area on Earth is 342 W/m². Of this radiation, 6.7 W/m² is absorbed by CO_2 at 14,993 nm in the atmosphere. How many photons at this wavelength are absorbed per second in 1 m² by CO_2? (1 W = 1 J/s.)

21.68 A glass of water initially at pH 7.0 is exposed to dry air at sea level at 20°C. Calculate the pH of the water when equilibrium is reached between atmospheric CO_2 and CO_2 dissolved in the water, given that Henry's law constant for CO_2 at 20°C is 0.032 mol/L · atm. (*Hint:* Assume no loss of water due to evaporation, and use Table 21.1 to calculate the partial pressure of CO_2. Your answer should correspond roughly to the pH of rainwater.)

21.69 Ozone in the troposphere is formed by the following steps:

$$NO_2 \longrightarrow NO + O \quad (1)$$
$$O + O_2 \longrightarrow O_3 \quad \quad (2)$$

The first step is initiated by the absorption of visible light (NO_2 is a brown gas). Calculate the longest wavelength required for step 1 at 25°C. (*Hint:* You need to first calculate ΔH and hence ΔE for step 1. Next, determine the wavelength for decomposing NO_2 from ΔE.)

21.70 Instead of monitoring carbon dioxide, suggest another gas that scientists could study to substantiate the fact that CO_2 concentration is steadily increasing in the atmosphere.

21.71 Describe the removal of SO_2 by CaO (to form $CaSO_3$) in terms of a Lewis acid-base reaction.

21.72 Which of the following settings is the most suitable for photochemical smog formation: (a) Gobi desert at noon in June, (b) New York City at 1 P.M. in July, (c) Boston at noon in January? Explain your choice.

21.73 As stated in the chapter, about 50 million tons of sulfur dioxide is released into the atmosphere every year. (a) If 20 percent of the SO_2 is eventually converted to H_2SO_4, calculate the number of 1000-lb marble statues the resulting acid rain can damage. As an estimate, assume that the acid rain only destroys the surface layer of each statue, which is made up of 5 percent of its total mass. (b) What is the other undesirable result of the acid rain damage?

21.74 Peroxyacetyl nitrate (PAN) undergoes thermal decomposition as follows:

$$CH_3(CO)OONO_2 \longrightarrow CH_3(CO)OO + NO_2$$

The rate constant is 3.0×10^{-4} s^{-1} at 25°C. At the boundary between the troposphere and stratosphere, where the temperature is about −40°C, the rate constant is reduced to 2.6×10^{-7} s^{-1}. (a) Calculate the activation energy for the decomposition of PAN. (b) What is the half-life of the reaction (in min) at 25°C?

21.75 What is ironic about the following cartoon?

21.76 Calculate the standard enthalpy of formation (ΔH_f°) of ClO from the following bond energies: Cl_2: 242.7 kJ/mol; O_2: 498.7 kJ/mol; ClO: 206 kJ/mol.

21.77 The carbon dioxide level in the atmosphere today is often compared with that in preindustrial days. Explain how scientists use tree rings and air trapped in polar ice to arrive at the comparison.

Engineering Problems

21.78 A 14-m by 10-m by 3.0-m basement had a high radon content. On the day the basement was sealed off from its surroundings so that no exchange of air could take place, the partial pressure of ^{222}Rn was 1.2×10^{-6} mmHg. Calculate the number of ^{222}Rn isotopes ($t_{1/2} = 3.8$ days) at the beginning and end of 31 days. Assume STP conditions.

21.79 In 1991, it was discovered that nitrous oxide (N_2O) is produced in the synthesis of nylon. This compound, which is released into the atmosphere, contributes *both* to the depletion of ozone in the stratosphere and to the greenhouse effect. (a) Write equations representing the reactions between N_2O and oxygen atoms in the stratosphere to produce nitric oxide (NO), which is then oxidized by ozone to form nitrogen dioxide. (b) Is N_2O a more effective greenhouse gas than carbon dioxide? Explain. (c) One of the intermediates in nylon manufacture is adipic acid [$HOOC(CH_2)_4COOH$]. About 2.2×10^9 kg of adipic acid is consumed every year. It is estimated that for every mole of adipic acid produced, 1 mole of N_2O is generated. What is the maximum number of moles of O_3 that can be destroyed as a result of this process per year?

Biological Problems

21.80 A person was found dead of carbon monoxide poisoning in a well-insulated cabin. Investigation showed that he had used a blackened bucket to heat water on a butane burner. The burner was found to function properly with no leakage. Explain, with an appropriate equation, the cause of his death.

21.81 Methyl bromide (CH_3Br, b.p. = 3.6°C) is used as a soil fumigant to control insects and weeds. It is also a marine by-product. Photodissociation of the C−Br bond produces Br atoms that can react with ozone similar to Cl, except more effectively.

Do you expect CH_3Br to be photolyzed in the troposphere? The bond enthalpy of the C$-$Br bond is about 293 kJ/mol.

21.82 As stated in the chapter, carbon monoxide has a much higher affinity for hemoglobin than oxygen does. (a) Write the equilibrium constant expression (K_c) for the following process:

$$CO(g) + HbO_2(aq) \rightleftharpoons O_2(g) + HbCO(aq)$$

where HbO_2 and HbCO are oxygenated hemoglobin and carboxyhemoglobin, respectively. (b) The composition of a breath of air inhaled by a person smoking a cigarette is 1.9×10^{-6} mol/L CO and 8.6×10^{-3} mol/L O_2. Calculate the ratio of [HbCO] to [HbO_2], given that K_c is 212 at 37°C.

Multiconcept Problems

21.83 The molar heat capacity of a diatomic molecule is 29.1 J/K · mol. Assuming the atmosphere contains only nitrogen gas and there is no heat loss, calculate the total heat intake (in kJ) if the atmosphere warms up by 3°C during the next 50 years. Given that there are 1.8×10^{20} moles of diatomic molecules present, how many kilograms of ice (at the North and South Poles) will this quantity of heat melt at 0°C? (The molar heat of fusion of ice is 6.01 kJ/mol.)

21.84 Assume that the formation of nitrogen dioxide:

$$2NO(g) + O_2(g) \longrightarrow 2NO_2(g)$$

is an elementary reaction. (a) Write the rate law for this reaction. (b) A sample of air at a certain temperature is contaminated with 2.0 ppm of NO by volume. Under these conditions, can the rate law be simplified? If so, write the simplified rate law. (c) Under the conditions described in part (b), the half-life of the reaction has been estimated to be 6.4×10^3 min. What would the half-life be if the initial concentration of NO were 10 ppm?

21.85 An electric power station annually burns 3.1×10^7 kg of coal containing 2.4 percent sulfur by mass. Calculate the volume of SO_2 emitted at STP.

21.86 The concentration of SO_2 in the troposphere over a certain region is 0.16 ppm by volume. The gas dissolves in rainwater as follows:

$$SO_2(g) + H_2O(l) \rightleftharpoons H^+(aq) + HSO_3^-(aq)$$

Given that the equilibrium constant for the preceding reaction is 1.3×10^{-2}, calculate the pH of the rainwater. Assume that the reaction does not affect the partial pressure of SO_2.

Standardized-Exam Practice Problems

Physical and Biological Sciences

Indoor air pollutants, including radon gas, can pose significant health risks. Radon is a radioactive gas found in the soils and rocks of Earth's crust. It finds its way into homes through cracks in the foundation or basement, and sometimes through the water supply. Radon is invisible and odorless, and becomes a health hazard when it is allowed to build up inside the home. The isotope of radon commonly found in homes is radon-222, a product of the uranium-238 decay chain, formed by the α decay of radium-226. Radon-222, in turn, decays by α emission.

1. Select the equation that correctly represents the α decay of radium-226.

 a) $^{226}Ra \longrightarrow {}^{226}Rn + \alpha$ c) $^{226}Ra + \alpha \longrightarrow {}^{226}Rn$
 b) $^{226}Ra \longrightarrow {}^{222}Rn + \alpha$ d) $^{226}Ra + \alpha \longrightarrow {}^{222}Rn$

2. How many neutrons does a radium-226 nucleus contain?

 a) 226   b) 88   c) 113   d) 138

3. What is the product when radon-222 decays by α emission?

 a) Radon-218   c) Polonium-218
 b) Radium-218   d) Polonium-226

4. What would the product be if radon-222 decayed by β emission?

 a) Francium-222   c) Radon-221
 b) Radon-223   d) Astatine-222

Answers to In-Chapter Materials

Answers to Practice Problems

21.1A 1120 nm. **21.1B** HI, F_2, ClF. **21.2A** H_2O. **21.2B** To act as a greenhouse gas, a molecule must be IR-active. To be IR-active, a molecule must undergo a change in dipole moment as the result of one or more of its vibrations. **21.3A** 1.2×10^{-6} mol/L. **21.3B** 52.8 days.

Answers to Checkpoints

21.2.1 a. **21.2.2** d. **21.5.1** a, d. **21.5.2** c. **21.8.1** e. **21.8.2** b.

Coordination Chemistry

Lead paint, known for its brightness and durability, was commonly used to paint homes, fences, and interior walls. Its use in homes was banned in 1978 because of the health risks associated with exposure to lead.

David Kay/Shutterstock

How Coordination Chemistry Is Used to Treat Lead Poisoning

Although the Consumer Product Safety Commission (CPSC) banned the residential use of lead-based paint in 1978, millions of children remain at risk for exposure to lead from deteriorating paint in older homes. Lead poisoning is especially harmful to children under the age of 5 years because it interferes with growth and development and it has been shown to lower IQ. Symptoms of chronic exposure to lead include diminished appetite, nausea, malaise, and convulsions. *Blood lead level* (BLL), expressed as micrograms per deciliter (μg/dL), is used to monitor the effect of chronic exposure. A BLL $\leq$ 10 μg/dL is considered normal; a BLL > 45 μg/dL requires medical and environmental intervention. At high levels (>70 μg/dL), lead can cause seizures, coma, and death.

Treatment for lead poisoning involves *chelation therapy,* in which a chelating agent is administered orally, intravenously, or intramuscularly. Chelating agents form strong coordinate-covalent bonds to metal ions, forming stable, water-soluble complex ions that are easily removed from the body via the urine. One of the drugs commonly used for this purpose is dimercaptosuccinic acid (DMSA), marketed under the name Chemet. Chelation therapy relies on *coordination chemistry.*

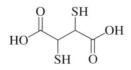

Dimercaptosuccinic acid (DMSA)

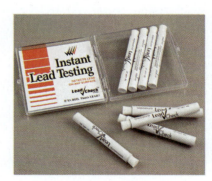

H.S. Photos/Alamy Stock Photo

David A. Tietz/Editorial Image, LLC

At the end of this chapter, you will be able to answer questions about some important coordination compounds [▶▶ **Applying What You've Learned, page 1101**].

22.1 Coordination Compounds

Coordination compounds contain *coordinate covalent* bonds [⏮ Section 8.8] formed by the reactions of metal ions with groups of *anions* or *polar molecules.* The metal ion in these kinds of reactions acts as a Lewis acid, accepting electrons, whereas the anions or polar molecules act as Lewis bases, donating pairs of electrons to form bonds to the metal ion. Thus, a coordinate covalent bond is a covalent bond in which one of the atoms donates *both* of the electrons that constitute the bond. Often a coordination compound consists of a *complex ion* and one or more *counter* ions. In writing formulas for such coordination compounds, we use square brackets to separate the complex ion from the counter ion.

Student Note: A complex ion is one in which a metal cation is covalently bound to one or more molecules or ions [⏮ Section 17.5].

$$\underbrace{K_2[PtCl_6]}$$

This compound consists of the complex ion $PtCl_6^{2-}$ and two K^+ counter ions.

Student Note: We can use the term *coordination complex* to refer to a compound, such as $Fe(CO)_5$, or to a complex ion.

Some coordination compounds, such as $Fe(CO)_5$, do not contain complex ions. Most but not all of the metals in coordination compounds are transition metals. Our understanding of the nature of coordination compounds stems from the classic work of Alfred Werner,[1] who prepared and characterized many coordination compounds. In 1893, at the age of 26, Werner proposed what is now commonly referred to as Werner's coordination theory.

Nineteenth-century chemists were puzzled by a certain class of reactions that seemed to violate valence theory. For example, the valences of the elements in cobalt(III) chloride and those in ammonia seem to be completely satisfied, and yet these two substances react to form a stable compound having the formula $CoCl_3 \cdot 6NH_3$. To explain this behavior, Werner postulated that most elements exhibit two types of valence: *primary valence* and *secondary valence.* In modern terminology, primary valence corresponds to the oxidation number and secondary valence to the coordination number of the element. In $CoCl_3 \cdot 6NH_3$, according to Werner, cobalt has a primary valence of 3 and a secondary valence of 6.

Today we use the formula $[Co(NH_3)_6]Cl_3$ to indicate that the ammonia molecules and the cobalt atom form a complex ion; the chloride ions are not part of the complex but are counter ions, held to the complex ion by Coulombic attraction.

Properties of Transition Metals

Transition metals are those that either have incompletely filled *d* subshells or form ions with incompletely filled *d* subshells (Figure 22.1). The Group 12 metals—Zn, Cd, and Hg—do not fit either of these criteria. As a result, they are *d*-block metals, but they are not actually transition metals. Incompletely filled *d* subshells give rise to

1	2	3	4	5	6	7	8	9	10	11	12	13	14	15	16	17	18
1 H																	2 He
3 Li	4 Be											5 B	6 C	7 N	8 O	9 F	10 Ne
11 Na	12 Mg											13 Al	14 Si	15 P	16 S	17 Cl	18 Ar
19 K	20 Ca	21 Sc	22 Ti	23 V	24 Cr	25 Mn	26 Fe	27 Co	28 Ni	29 Cu	30 Zn	31 Ga	32 Ge	33 As	34 Se	35 Br	36 Kr
37 Rb	38 Sr	39 Y	40 Zr	41 Nb	42 Mo	43 Tc	44 Ru	45 Rh	46 Pd	47 Ag	48 Cd	49 In	50 Sn	51 Sb	52 Te	53 I	54 Xe
55 Cs	56 Ba	57 La	72 Hf	73 Ta	74 W	75 Re	76 Os	77 Ir	78 Pt	79 Au	80 Hg	81 Tl	82 Pb	83 Bi	84 Po	85 At	86 Rn
87 Fr	88 Ra	89 Ac	104 Rf	105 Db	106 Sg	107 Bh	108 Hs	109 Mt	110 Ds	111 Rg	112 Cn	113 Nh	114 Fl	115 Mc	116 Lv	117 Ts	118 Og

Figure 22.1 The transition metals (shown in green). Note that although the Group 12 elements (Zn, Cd, Hg) are described as transition metals by some chemists, neither the metals nor their ions possess incompletely filled *d* subshells.

1. Alfred Werner (1866–1919). Swiss chemist. Werner started as an organic chemist but did his most notable work in coordination chemistry. For his theory of coordination compounds, Werner was awarded the Nobel Prize in Chemistry in 1913.

TABLE 22.1	**Electron Configurations and Other Properties of the Fourth-Period Transition Metals**								
	Sc	Ti	V	Cr	Mn	Fe	Co	Ni	Cu
Electron configuration									
M	$4s^2 3d^1$	$4s^2 3d^2$	$4s^2 3d^3$	$4s^1 3d^5$	$4s^2 3d^5$	$4s^2 3d^6$	$4s^2 3d^7$	$4s^2 3d^8$	$4s^1 3d^{10}$
M^{2+}	—	$3d^2$	$3d^3$	$3d^4$	$3d^5$	$3d^6$	$3d^7$	$3d^8$	$3d^9$
M^{3+}	[Ar]	$3d^1$	$3d^2$	$3d^3$	$3d^4$	$3d^5$	$3d^6$	$3d^7$	$3d^8$
Electronegativity									
	1.3	1.5	1.6	1.6	1.5	1.8	1.9	1.9	1.9
Ionization energy (kJ/mol)									
First	631	658	650	652	717	759	760	736	745
Second	1235	1309	1413	1591	1509	1561	1645	1751	1958
Third	2389	2650	2828	2986	3250	2956	3231	3393	3578
Radius (pm)									
M	162	147	134	130	135	126	125	124	128
M^{2+}	—	90	88	85	80	77	75	69	72
M^{3+}	81	77	74	64	66	60	64		

TABLE 22.2	**Physical Properties of Elements K to Zn**											
	1	2				Transition Metals						12
	K	Ca	Sc	Ti	V	Cr	Mn	Fe	Co	Ni	Cu	Zn
Atomic radius (pm)	235	197	162	147	134	130	135	126	125	124	128	138
Melting point (°C)	63.7	838	1539	1668	1900	1875	1245	1536	1495	1453	1083	419.5
Boiling point (°C)	760	1440	2730	3260	3450	2665	2150	3000	2900	2730	2595	906
Density (g/cm³)	0.86	4.51	3.0	4.51	6.1	7.19	7.43	7.86	8.9	8.9	8.96	7.14

several notable properties, including distinctive colors, the formation of paramagnetic compounds, catalytic activity, and the tendency to form complex ions. The most common transition metals are scandium through copper, which occupy the fourth row of the periodic table. Table 22.1 lists the electron configurations and some of the properties of these metals.

Most of the transition metals exhibit a close-packed structure in which each atom has a coordination number of 12. Furthermore, these elements have relatively small atomic radii. The combined effect of closest packing and small atomic size results in strong metallic bonds. Therefore, transition metals have higher densities, higher melting points and boiling points, and higher heats of fusion and vaporization than the main group and Group 12 metals (Table 22.2).

Transition metals exhibit variable oxidation states in their compounds. Figure 22.2 shows the oxidation states of the first row of transition metals. Note that all these metals can exhibit the oxidation state +3 and *nearly* all can exhibit the oxidation state +2. Of these two, the +2 oxidation state is somewhat more common for the heavier elements. The highest oxidation state for a transition metal is +7, exhibited by manganese ($4s^2 3d^5$). Transition metals exhibit their highest oxidation states in compounds that contain highly electronegative elements such as oxygen and fluorine—for example, V_2O_5, CrO_3, and Mn_2O_7.

Student Note: The oxidation state of O in each of these compounds is −2, making those of V, Cr, and Mn +5, +6, and +7, respectively [◀◀ Section 4.4].

Figure 22.2 Oxidation states of the first-row transition metals. The most stable oxidation numbers are shown in red. The zero oxidation state is encountered in some compounds, such as $Ni(CO)_4$ and $Fe(CO)_5$.

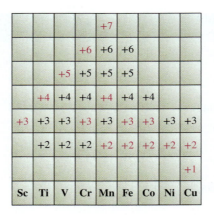

				+7				
			+6	+6	+6			
		+5	+5	+5	+5			
	+4	+4	+4	+4	+4	+4		
+3	+3	+3	+3	+3	+3	+3	+3	+3
	+2	+2	+2	+2	+2	+2	+2	+2
								+1
Sc	**Ti**	**V**	**Cr**	**Mn**	**Fe**	**Co**	**Ni**	**Cu**

Ligands

The molecules or ions that surround the metal in a complex ion are called **ligands** (Table 22.3). The formation of covalent bonds between ligands and a metal can be thought of as a Lewis acid-base reaction. (Recall that a Lewis base is a species that donates a pair of electrons [◄◄ Section 16.12].) To be a ligand, a molecule or ion must have at least one unshared pair of valence electrons, as these examples illustrate:

TABLE 22.3	Common Ligands
Name	**Structure**
Monodentate	
Ammonia	H—N̈—H with H below
Carbon monoxide	:C≡O:
Chloride ion	:C̈l:⁻
Cyanide ion	[:C≡N:]⁻
Thiocyanate ion	[:S̈—C≡N:]⁻
Water	H—Ö—H
Bidentate	
Ethylenediamine	$H_2N̈—CH_2—CH_2—N̈H_2$
Oxalate ion	oxalate structure
Polydentate	
Ethylenediaminetetraacetate ion (EDTA)	EDTA structure

Therefore, ligands play the role of Lewis bases. The transition metal, on the other hand, acts as a Lewis acid, accepting (and sharing) pairs of electrons from the Lewis bases.

The atom in a ligand that is bound directly to the metal atom is known as the **donor atom.** For example, *nitrogen* is the donor atom in the $[Cu(NH_3)_4]^{2+}$ complex ion:

$$
\begin{bmatrix}
& & H & & \\
& & | & & \\
& H\!-\!N\!-\!H & & \\
& H & H & & \\
& | & | & & \\
H\!-\!N\!-\!Cu\!-\!N\!-\!H & & \\
& H & H & & \\
& H\!-\!N\!-\!H & & \\
& & H & &
\end{bmatrix}^{2+}
$$

The **coordination number** in a coordination compound refers to the number of donor atoms surrounding the central metal atom in a complex ion. The coordination number of Cu^{2+} in $[Cu(NH_3)_4]^{2+}$ is 4. The most common coordination numbers are 4 and 6, although coordination numbers of 2 and 5 are also known.

Depending on the number of donor atoms a ligand possesses, it is classified as monodentate (1 donor atom), bidentate (2 donor atoms), or polydentate (> 2 donor atoms). Table 22.3 lists some common ligands. Figure 22.3 shows how ethylenediamine, sometimes abbreviated "en," forms two bonds to a metal atom.

Bidentate and polydentate ligands are also called **chelating agents** because of their ability to hold the metal atom like a claw (from the Greek *chele*, meaning "claw"). One example is EDTA (Figure 22.4), a polydentate ligand used to treat metal poisoning. Six donor atoms enable EDTA to form a very stable complex ion with lead. This stable complex enables the body to remove lead from the blood.

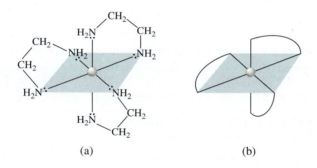

(a)   (b)

Figure 22.3 (a) Structure of a metal-ethylenediamine complex cation, such as $[Co(en)_3]^{2+}$. Each ethylenediamine molecule provides two N donor atoms and is therefore a bidentate ligand. (b) Simplified structure of the same complex cation.

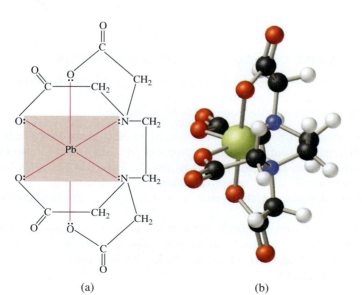

(a)   (b)

Figure 22.4 (a) EDTA complex of lead. The complex bears a net charge of 2− because each O donor atom has one negative charge and the lead ion carries two positive charges. Only the lone pairs that participate in bonding are shown. Note the octahedral geometry about the Pb^{2+} ion. (b) Molecular model of the Pb^{2+}-EDTA complex. The light green sphere is the Pb^{2+} ion.

Student Note: Remember that the oxidation number of a monoatomic ion is equal to the charge [◄◄ Section 4.4].

The oxidation state of a transition metal in a complex ion is determined using the known charges of the *ligands* and the known *overall* charge of the complex ion. In the complex ion $[PtCl_6]^{2-}$, for example, each chloride ion ligand has an oxidation number of -1. For the overall charge of the ion to be -2, the Pt must have an oxidation number of $+4$.

Sample Problem 22.1 shows how to determine transition metal oxidation states in coordination compounds.

SAMPLE PROBLEM (22.1)

Determine the oxidation state of the central metal atom in each of the following compounds: (a) $[Ru(NH_3)_5(H_2O)]Cl_2$, (b) $[Cr(NH_3)_6](NO_3)_3$, and (c) $Fe(CO)_5$.

Strategy Identify the components of each compound, and use known oxidation states and charges to determine the oxidation state of the metal.

Setup (a) $[Ru(NH_3)_5(H_2O)]Cl_2$ consists of a complex ion (the part of the formula enclosed in square brackets) and two Cl^- counter ions. Because the overall charge on the compound is zero, the complex ion is $[Ru(NH_3)_5(H_2O)]^{2+}$. There are six ligands: five ammonia molecules and one water molecule. Each molecule has a zero charge (i.e., each ligand is neutral), so the charge on the metal is equal to the overall charge on the complex ion.

(b) $[Cr(NH_3)_6](NO_3)_3$ consists of a complex ion and three NO_3^- ions, making the complex ion $[Cr(NH_3)_6]^{3+}$. Each of the six ammonia molecule ligands is neutral (i.e., each has a zero charge), making the charge on the metal equal to the overall charge on the complex ion.

(c) $Fe(CO)_5$ does not contain a complex ion. The ligands are CO molecules, which have a zero charge, so the central metal also has a zero charge.

Solution (a) $+2$ (b) $+3$ (c) 0

THINK ABOUT IT

To solve a problem like this, you must be able to recognize the common polyatomic ions and you must know their charges.

Practice Problem **A**TTEMPT Give oxidation numbers for the metals in (a) $K[Au(OH)_4]$ and (b) $K_4[Fe(CN)_6]$.

Practice Problem **B**UILD The oxidation state of cobalt in each of the following species is $+3$. Determine the overall charge on each complex: $[Co(NH_3)_4Cl_2]^?$, $[Co(CN)_6]^?$, $[Co(NH_3)_3Br_3]^?$, $[Co(en)_2Cl_2]^?$, $[Co(en)_3]^?$, $[Co(NH_3)_6]^?$, $[Co(NH_3)_5Cl]^?$.

Practice Problem **C**ONCEPTUALIZE Explain why the bonds between transition metals in very high oxidation states (>4) and nonmetals are covalent rather than ionic.

Nomenclature of Coordination Compounds

Now that we have discussed the various types of ligands and the oxidation numbers of metals, our next step is to learn how to name coordination compounds. The rules for naming ionic coordination compounds are as follows:

1. The cation is named before the anion, as in other ionic compounds. The rule holds regardless of whether the complex ion bears a net positive or a net negative charge. In the compounds $K_2[Fe(CN)_6]$ and $[Co(NH_3)_4]Cl$, for example, we name the K^+ and $[Co(NH_3)_4]^+$ cations first, respectively.

2. Within a complex ion, the ligands are named first, in alphabetical order, and the metal ion is named last.

3. The names of anionic ligands end with the letter *o*, whereas neutral ligands are usually called by the names of the molecules. The exceptions are H_2O (aqua), CO (carbonyl), and NH_3 (ammine). Table 22.4 lists some common ligands and their nomenclature.

4. When two or more of the same ligand are present, use Greek prefixes *di, tri, tetra, penta,* and *hexa,* to specify their number. Thus, the ligands in the cation $[Co(NH_3)_4Cl_2]^+$ are "tetraamminedichloro." (Note that prefixes are *not* used for the purpose of alphabetizing the ligands.)

5. The oxidation number of the metal is indicated in Roman numerals immediately following the name of the metal. For example, the Roman numeral III is used to indicate the +3 oxidation state of chromium in $[Cr(NH_3)_4Cl_2]^+$, which is called tetraamminedichlorochromium(III) ion.

6. If the complex is an anion, its name ends in *–ate.* In $K_4[Fe(CN)_6]$, for example, the anion $[Fe(CN)_6]^{4-}$ is called hexacyanoferrate(II) ion. Note that the Roman numeral indicating the oxidation state of the metal *follows* the suffix *–ate.* Table 22.5 lists the names of anions containing metal atoms.

> **Student Note:** When the name of a ligand already contains a Greek prefix, a different set of prefixes is used to denote the number of the ligand:
>
> | | |
> |---|---|
> | 2 | bis |
> | 3 | tris |
> | 4 | tetrakis |
>
> Two ethylenediamine ligands, for example, would be specified by bis(ethylenediamine).

> **Student Note:** Note that there is no space between the compound name and the parenthetical Roman numeral.

TABLE 22.4	Names of Common Ligands in Coordination Compounds
Ligand	**Name of Ligand in Coordination Compound**
Bromide (Br^-)	Bromo
Chloride (Cl^-)	Chloro
Cyanide (CN^-)	Cyano
Hydroxide (OH^-)	Hydroxo
Oxide (O^{2-})	Oxo
Carbonate (CO_3^{2-})	Carbonato
Nitrite (NO_2^-)	Nitro
Oxalate ($C_2O_4^{2-}$)	Oxalato
Ammonia (NH_3)	Ammine
Carbon monoxide (CO)	Carbonyl
Water (H_2O)	Aqua
Ethylenediamine	Ethylenediamine
Ethylenediaminetetraacetate	Ethylenediaminetetraacetate

TABLE 22.5	Names of Anions Containing Metal Atoms
Metal	**Name of Metal in Anionic Complex**
Aluminum	Aluminate
Chromium	Chromate
Cobalt	Cobaltate
Copper	Cuprate
Gold	Aurate
Iron	Ferrate
Lead	Plumbate
Manganese	Manganate
Molybdenum	Molybdate
Nickel	Nickelate
Silver	Argentate
Tin	Stannate
Tungsten	Tungstate
Zinc	Zincate

Sample Problems 22.2 and 22.3 apply these rules to the nomenclature of coordination compounds.

SAMPLE PROBLEM (22.2)

Write the names of the following coordination compounds: (a) $[Co(NH_3)_4Cl_2]Cl$ and (b) $K_3[Fe(CN)_6]$.

Strategy For each compound, name the cation first and the anion second. Refer to Tables 22.4 and 22.5 for the names of ligands and anions containing metal atoms.

Setup (a) The cation is a complex ion containing four ammonia molecules and two chloride ions. The counter ion is chloride (Cl^-), so the charge on the complex cation is +1, making the oxidation state of cobalt +3.

(b) The cation is K^+, and the anion is a complex ion containing six cyanide ions. The charge on the complex ion is −3, making the oxidation state of iron +3.

Solution (a) Tetraamminedichlorocobalt(III) chloride   (b) Potassium hexacyanoferrate(III)

THINK ABOUT IT

When the anion is a *complex* ion, its name must end in *–ate*, followed by the metal's oxidation state in Roman numerals. Also, do not use prefixes to denote numbers of counter ions.

Practice Problem **A**TTEMPT Give the correct name for (a) $[Co(NH_3)_4Br_2]Cl$, (b) $[Cr(H_2O)_4Cl_2]Cl$, and (c) $K_2[CuCl_4]$.

Practice Problem **B**UILD Give the correct name for (a) $Na_3[Fe(CN)_6]$, (b) $[Cr(en)_2Cl_2]Br$, and (c) $[Co(en)_3]Cl_3$. [*Hint:* For parts (b) and (c), read the Student Note (shown earlier) regarding the use of prefixes.]

Practice Problem **C**ONCEPTUALIZE Draw the structure of the en ligand.

SAMPLE PROBLEM (22.3)

Write formulas for the following compounds: (a) pentaamminechlorocobalt(III) chloride and (b) dichlorobis(ethylenediamine)platinum(IV) nitrate.

Strategy If you can't remember them yet, refer to Tables 22.4 and 22.5 for the names of ligands and anions containing metal atoms.

Setup (a) There are six ligands: five NH_3 molecules and one Cl^- ion. The oxidation state of cobalt is +3, making the overall charge on the complex ion +2. Therefore, there are two chloride ions as counter ions.

(b) There are four ligands: two bidentate ethylenediamines and two Cl^- ions. The oxidation state of platinum is +4, making the overall charge on the complex ion +2. Therefore, there are two nitrate ions as counter ions.

Solution (a) $[Co(NH_3)_5Cl]Cl_2$ (b) $[Pt(en)_2Cl_2](NO_3)_2$

THINK ABOUT IT

Although ligands are alphabetized in a compound's name, they do not necessarily appear in alphabetical order in the compound's formula.

Practice Problem **A**TTEMPT Write the formulas for (a) pentaaquabromoruthenium(II) nitrate, (b) potassium tetrabromodichloroplatinate(IV), and (c) sodium hexanitrocobaltate(III).

Practice Problem **B**UILD Write the formulas for (a) bis(ethylenediamine)oxalatovanadium(IV) chloride, (b) dibromobis(ethylenediamine)chromium(III) nitrate, and (c) tris(ethylenediamine)platinum(IV) sulfate.

Practice Problem **C**ONCEPTUALIZE Explain why coordination-compound nomenclature uses the prefixes bis, tris, and tetrakis, instead of simply using di, tri, and tetra.

22.1.1 Select the correct name for the compound $[Cu(NH_3)_4]Cl_2$.

a) Coppertetraammine dichloride

b) Tetraamminecopper(II) chloride

c) Tetraaminedichlorocuprate(II)

d) Dichlorotetraaminecopper(II)

e) Tetraaminedichlorocopper(II)

22.1.2 Select the correct name for the compound $K_3[FeF_6]$.

a) Tripotassiumironhexafluoride

b) Hexafluorotripotassiumferrate(III)

c) Hexafluoroiron(III) potassium

d) Potassium hexafluoroferrate(III)

e) Potassium ironhexafluorate

22.1.3 Select the correct formula for pentaaminenitrocobalt(III).

a) $[Co(NH_3)_5NO_2]^{3+}$

b) $[Co(NH_3)_5NO_2]^{2+}$

c) $Co(NH_3)_5NO_2$

d) $[Co(NH_3)_5](NO_2)$

e) $[Co(NH_3)_5](NO_2)_2$

22.1.4 Select the correct formula for tetraaquadichlorochromium(III) chloride.

a) $[Cr(H_2O)_4Cl_2]Cl_3$

b) $[Cr(H_2O)_4Cl_2]Cl_2$

c) $[Cr(H_2O)_4Cl_2]Cl$

d) $[Cr(H_2O)_4]Cl_3$

e) $[Cr(H_2O)_4]Cl_2$

22.2 Structure of Coordination Compounds

The geometry of a coordination compound often plays a significant role in determining its properties. Figure 22.5 shows four different geometric arrangements for metal atoms with monodentate ligands. In these diagrams we see that structure and the coordination number of the metal relate to each other as follows:

Coordination Number	Structure
2	Linear
4	Tetrahedral or square planar
6	Octahedral

In studying the geometry of coordination compounds, we sometimes find that there is more than one way to arrange the ligands around the central atom. Such compounds in which ligands are arranged differently, known as *stereoisomers,* have distinctly different physical and chemical properties. Coordination compounds may exhibit two types of stereoisomerism: *geometric* and *optical.*

Geometrical isomers are stereoisomers that cannot be interconverted without breaking chemical bonds. Geometric isomers come in pairs. We use the terms *cis* and *trans* to distinguish one geometric isomer of a compound from the other. *Cis* means that two particular atoms (or groups of atoms) are adjacent to each other, whereas *trans* means that the atoms (or groups of atoms) are on opposite sides in the structural

Student Note: In general, stereoisomers are compounds that are made up of the same types and numbers of atoms, bonded together in the same sequence, but with different spatial arrangements.

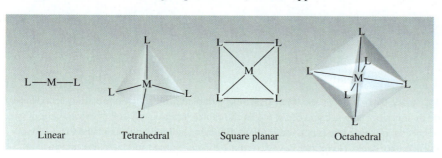

Figure 22.5 Common geometries of complex ions. In each case M is a metal and L is a monodentate ligand.

Linear   Tetrahedral   Square planar   Octahedral

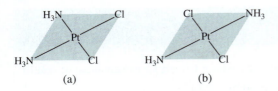

(a)   (b)

Figure 22.6 The (a) *cis* and (b) *trans* isomers of diamminedichloroplatinum(II). Note that the two Cl atoms are adjacent to each other in the *cis* isomer and diagonally across from each other in the *trans* isomer.

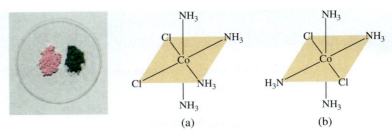

(a)   (b)

Figure 22.7 The (a) *cis* and (b) *trans* isomers of tetraamminedichlorocobalt(III) ion, $[Co(NH_3)_4Cl_2]^+$. The ion has only two geometric isomers.

(photo): *Ken Karp/McGraw Hill*

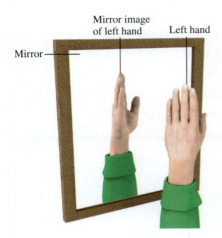

Figure 22.8 A left hand and its mirror image.

formula. The *cis* and *trans* isomers of coordination compounds generally have quite different colors, melting points, dipole moments, and chemical reactivities. Figure 22.6 shows the *cis* and *trans* isomers of diamminedichloroplatinum(II). Note that although the types of bonds are the same in both isomers (two Pt—N and two Pt—Cl bonds), the spatial arrangements are different. Another example is the tetraamminedichlorocobalt(III) ion, shown in Figure 22.7.

Optical isomers are nonsuperimposable mirror images. (*Superimposable* means that if one structure is laid over the other, the positions of all the atoms will match.) Like geometric isomers, optical isomers come in pairs. However, the optical isomers of a compound have *identical* physical and chemical properties, such as melting point, boiling point, dipole moment, and chemical reactivity toward molecules that are not *themselves* optical isomers. Optical isomers differ from each other, though, in their interactions with plane-polarized light, as we explain shortly.

The structural relationship between two optical isomers is analogous to the relationship between your left and right hands. If you place your left hand in front of a mirror, the image you see will look like your right hand (Figure 22.8). Your left hand and right hand are mirror images of each other. They are nonsuperimposable, however, because when you place your left hand over your right hand (with both palms facing down), they do not match. This is why a right-handed glove will not fit comfortably on your left hand.

Figure 22.9 shows the *cis* and *trans* isomers of dichlorobis(ethylenediamine) cobalt(III) ion and the mirror image of each. Careful examination reveals that the *trans* isomer and its mirror image are superimposable, but the *cis* isomer and its mirror image are not. Thus, the *cis* isomer and its mirror image are *optical isomers*.

Optical isomers are described as *chiral* (from the Greek word for "hand") because, like your left and right hands, chiral molecules are nonsuperimposable. Isomers that are superimposable with their mirror images are said to be *achiral*. Chiral molecules play a vital role in enzyme reactions in biological systems. Many drug molecules are chiral, although only one of a pair of chiral isomers is biologically effective [▶▶ Section 25.4].

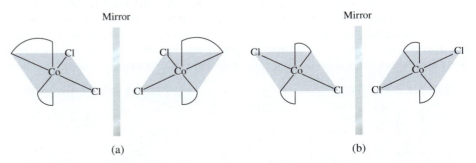

(a)   (b)

Figure 22.9 The (a) *cis* and (b) *trans* isomers of dichlorobis(ethylenediamine)cobalt(III) ion and their mirror images. If you could rotate the mirror image in (b) 90° clockwise about the vertical position and place the ion over the *trans* isomer, you would find that the two are superimposable. No matter how you rotate the *cis* isomer and its mirror image in (a), however, you cannot superimpose one on the other.

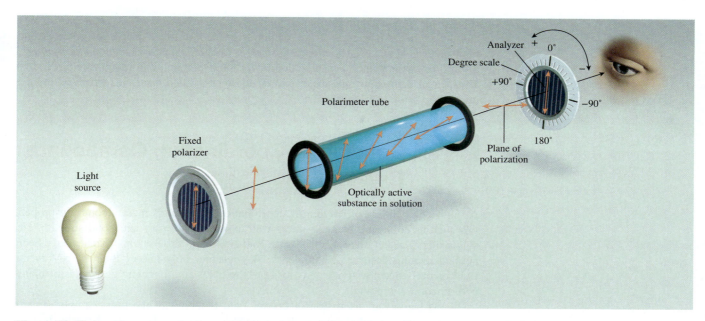

Figure 22.10 Operation of a polarimeter. Initially, the tube is filled with an achiral compound. The analyzer is rotated so that its plane of polarization is perpendicular to that of the polarizer. Under this condition, no light reaches the observer. Next, a chiral compound is placed in the tube as shown. The plane of polarization of the polarized light is rotated as it travels through the tube so that some light reaches the observer. Rotating the analyzer (either to the left or to the right) until no light reaches the observer again allows the angle of optical rotation to be measured.

Chiral molecules are said to be optically active because of their ability to rotate the plane of polarization of polarized light as it passes through them. Unlike ordinary light, which vibrates in all directions, plane-polarized light vibrates only in a single plane. We use a ***polarimeter*** to measure the rotation of polarized light by optical isomers (Figure 22.10). A beam of unpolarized light first passes through a Polaroid sheet, called the polarizer, and then through a sample tube containing a solution of an optically active, chiral compound. As the polarized light passes through the sample tube, its plane of polarization is rotated either to the right (clockwise) or to the left (counterclockwise). This rotation can be measured directly by turning the analyzer in the appropriate direction until minimal light transmission is achieved (Figure 22.11).

Figure 22.11 Polarized lenses. No light passes through the lenses when they are rotated so that their planes of polarization are perpendicular.

David A. Tietz/Editorial Image, LLC

If the plane of polarization is rotated to the right, the isomer is said to be ***dextrorotatory*** and the isomer is labeled *d;* if the rotation is to the left, the isomer is ***levorotatory*** and the isomer is labeled *l.* The *d* and *l* isomers of a chiral substance, called ***enantiomers,*** always rotate the plane of polarization by the same amount, but in opposite directions. Thus, in an equimolar mixture of two enantiomers, called a ***racemic mixture,*** the net rotation is zero.

22.3 Bonding in Coordination Compounds: Crystal Field Theory

A satisfactory theory of bonding in coordination compounds must account for properties such as color and magnetism, as well as stereochemistry and bond strength. No single theory as yet does all this for us. Rather, several different approaches have been applied to transition metal complexes. We consider only one of them here—crystal field theory—because it accounts for both the color and magnetic properties of many coordination compounds.

We begin our discussion of crystal field theory with the most straightforward case—namely, complex ions with octahedral geometry. Then we describe how it applies to tetrahedral and square-planar complexes.

Crystal Field Splitting in Octahedral Complexes

Crystal field theory explains the bonding in complex ions purely in terms of electrostatic forces. In a complex ion, two types of electrostatic interaction come into play. One is the attraction between the positive metal ion and the negatively charged ligand or the negatively charged end of a polar ligand. This is the force that binds the ligands to the metal. The second type of interaction is the electrostatic repulsion between the lone pairs on the ligands and the electrons in the *d* orbitals of the metals.

The *d* orbitals have different orientations [◀◀ Section 6.7], but in the absence of an external disturbance, they all have the same energy. In an octahedral complex, a central metal atom is surrounded by six lone pairs of electrons (on the six ligands), so all five *d* orbitals experience electrostatic repulsion. The magnitude of this repulsion depends on the orientation of the *d* orbital that is involved. Take the $d_{x^2-y^2}$ orbital as an example. In Figure 22.12, we see that the lobes of this orbital point toward the

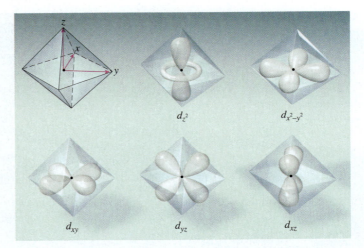

Figure 22.12 The five *d* orbitals in an octahedral environment. The metal atom (or ion) is at the center of the octahedron, and the six lone pairs on the donor atoms of the ligands are at the corners.

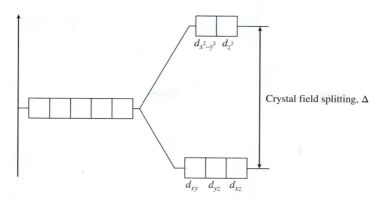

Figure 22.13 Crystal field splitting between *d* orbitals in an octahedral complex.

corners of the octahedron along the *x* and *y* axes, where the lone-pair electrons are positioned. Thus, an electron residing in this orbital would experience a greater repulsion from the ligands than an electron would in the d_{xy}, d_{yz}, or d_{xz} orbitals. For this reason, the energy of the $d_{x^2-y^2}$ orbital is increased relative to the d_{xy}, d_{yz}, and d_{xz} orbitals. The d_{z^2} orbital's energy is also greater, because its lobes are pointed at the ligands along the *z* axis. As a result of these metal-ligand interactions, the five *d* orbitals in an octahedral complex are split between two sets of energy levels: a higher level with two orbitals ($d_{x^2-y^2}$ and d_{z^2}) having the same energy, and a lower level with three equal-energy orbitals (d_{xy}, d_{yz}, and d_{xz}), as shown in Figure 22.13. The ***crystal field splitting (Δ)*** is the energy difference between two sets of *d* orbitals in a metal atom when ligands are present. The magnitude of Δ depends on the metal and the nature of the ligands; it has a direct effect on the color and magnetic properties of complex ions.

Color

In Chapter 6 we learned that white light, such as sunlight, is a combination of all colors. A substance appears black if it absorbs all the visible light that strikes it. If it absorbs no visible light, it is white or colorless. An object appears green if it absorbs all light but reflects the green component. An object also looks green if it reflects all colors except red, the *complementary* color of green (Figure 22.14).

What has been said of reflected light also applies to *transmitted* light (i.e., the light that passes *through* the medium, such as a solution). Consider the hydrated cupric ion ($[Cu(H_2O)_6]^{2+}$); it absorbs light in the orange region of the spectrum, so a solution of $CuSO_4$ appears blue to us. Recall from Chapter 6 that when the energy of a photon is equal to the difference between the ground state and an excited state, absorption occurs as the photon strikes the atom (or ion or compound), and an electron is promoted to a higher level. Using these concepts, we can calculate the energy change involved in the electron transition. The energy of a photon is given by:

$$E = h\nu$$

where *h* represents Planck's constant (6.63×10^{-34} J · s) and ν is the frequency of the radiation, which is (5.00×10^{14} s^{-1}) for a wavelength of 600 nm. Here $E = \Delta$, so we have:

$$\Delta = h\nu$$
$$= (6.63 \times 10^{-34} \text{ J} \cdot \text{s})(5.00 \times 10^{14} \text{ s})$$
$$= 3.32 \times 10^{-19} \text{ J}$$

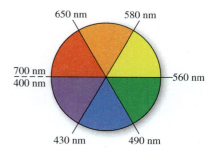

Figure 22.14 A color wheel with appropriate wavelengths. Complementary colors, such as red and green, are on opposite sides of the wheel.

Figure 22.15 (a) The process of photon absorption, and (b) a graph of the absorption spectrum of $[Ti(H_2O)_6]^{3+}$. The energy of the incoming photon is equal to the crystal field splitting. The maximum absorption peak in the visible region occurs at 498 nm.

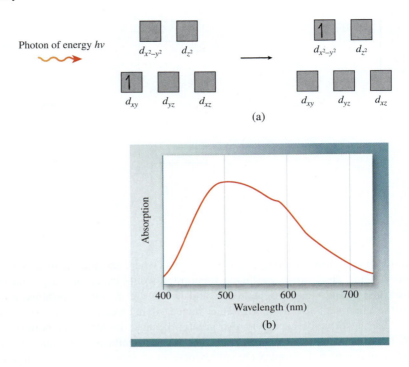

(a)

(b)

This value is very small, but it is the energy absorbed by only *one* ion. If the wavelength of the photon absorbed by an ion lies outside the visible region, then the transmitted light looks the same (to us) as the incident light—white—and the ion appears colorless.

The best way to measure crystal field splitting is to use spectroscopy to determine the wavelength at which light is absorbed. The $[Ti(H_2O)_6]^{3+}$ ion provides a straightforward example, because Ti^{3+} has only one $3d$ electron (Figure 22.15). The $[Ti(H_2O)_6]^{3+}$ ion absorbs light in the visible region of the spectrum (Figure 22.16). The wavelength corresponding to maximum absorption is 498 nm [Figure 22.15(b)]. To calculate the crystal field splitting energy, we start by writing:

$$\Delta = h\nu$$

Next, recall that:

$$\nu = \frac{c}{\lambda}$$

Figure 22.16 Colors of some of the first-row transition metal ions in solution. From left to right: Ti^{3+}, Cr^{3+}, Mn^{2+}, Fe^{3+}, Co^{2+}, Ni^{2+}, Cu^{2+}. The Sc^{3+} and V^{5+} ions are colorless.

Charles D. Winters/Timeframe Photography/McGraw Hill

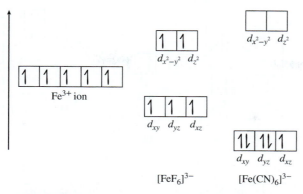

Figure 22.17 Energy-level diagrams for the Fe^{3+} ion and for the $[FeF_6]^{3-}$ and $[Fe(CN)_6]^{3-}$ complex ions.

where c is the speed of light and λ is the wavelength. Therefore:

$$\Delta = \frac{hc}{\lambda} = \frac{(6.63 \times 10^{-34} \text{ J} \cdot \text{s})(3.00 \times 10^8 \text{ m/s})}{(498 \text{ nm})(1 \times 10^{-9} \text{ m/nm})} = 3.99 \times 10^{-19} \text{ J}$$

This is the energy required to excite *one* $[Ti(H_2O)_6]^{3+}$ ion. To express this energy difference in the more convenient units of kJ/mol, we write:

$$\Delta = (3.99 \times 10^{-19} \text{ J/ion})(6.02 \times 10^{23} \text{ ions/mol})$$

$$= 240,000 \text{ J/mol}$$

$$= 240 \text{ kJ/mol}$$

Aided by spectroscopic data for a number of complexes, all having the same metal ion but different ligands, chemists calculated the crystal field splitting for each ligand and established the following **spectrochemical series,** which is a list of ligands arranged in increasing order of their abilities to split the d orbital energy levels:

$$I^- < Br^- < Cl^- < OH^- < F^- < H_2O < NH_3 < \text{en} < CN^- < CO$$

These ligands are arranged in the order of increasing value of Δ. CO and CN^- are called *strong-field ligands,* because they cause a large splitting of the d orbital energy levels. The halide ions and hydroxide ion are *weak-field ligands,* because they split the d orbitals to a lesser extent.

Magnetic Properties

The magnitude of the crystal field splitting also determines the magnetic properties of a complex ion. The $[Ti(H_2O)_6]^{3+}$ ion, having only one d electron, is always paramagnetic. However, for an ion with several d electrons, the situation is less immediately clear. Consider, for example, the octahedral complexes $[FeF_6]^{3-}$ and $[Fe(CN)_6]^{3-}$ (Figure 22.17). The electron configuration of Fe^{3+} is $[Ar]3d^5$, and there are two possible ways to distribute the five d electrons among the d orbitals. According to Hund's rule [◄◄ Section 6.8], maximum stability is reached when the electrons are placed in five separate orbitals with parallel spins. This arrangement can be achieved only at a cost, however, because two of the five electrons must be promoted to the higher-energy $d_{x^2-y^2}$ and d_{z^2} orbitals. No such energy investment is needed if all five electrons enter the d_{xy}, d_{yz}, and d_{xz} orbitals. According to Pauli's exclusion principle [◄◄ Section 6.8], there will be only one unpaired electron present in this case.

Figure 22.18 shows the distribution of electrons among d orbitals that results in low- and high-spin complexes. The actual arrangement of the electrons is determined

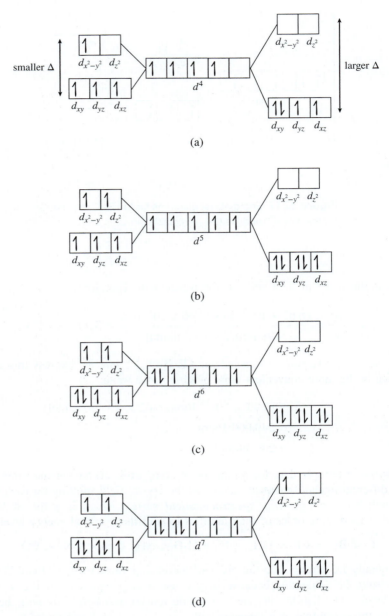

Figure 22.18 Orbital diagrams for the high-spin and low-spin octahedral complexes corresponding to the electron configurations of (a) d^4, (b) d^5, (c) d^6, and (d) d^7.

by the amount of stability gained by having maximum parallel spins versus the investment in energy required to promote electrons to higher d orbitals. Because F^- is a weak-field ligand, the five d electrons enter five separate d orbitals with parallel spins to create a high-spin complex. The cyanide ion is a strong-field ligand, though, so it is energetically preferable for all five electrons to be in the lower orbitals, thus forming a low-spin complex. High-spin complexes are more paramagnetic than low-spin complexes.

The actual number of unpaired electrons (or spins) in a complex ion can be found by magnetic measurements, and in general, experimental findings support predictions based on crystal field splitting. However, a distinction between low- and high-spin complexes can be made only if the metal ion contains more than three and fewer than eight d electrons, as shown in Figure 22.18.

Sample Problem 22.4 shows how to determine the number of spins in an octahedral complex.

SAMPLE PROBLEM 22.4

Predict the number of unpaired spins in the $[Cr(en)_3]^{2+}$ ion.

Strategy The magnetic properties of a complex ion depend on the strength of the ligands. Strong-field ligands, which cause a high degree of splitting among the d orbital energy levels, result in low-spin complexes. Weak-field ligands, which cause only a small degree of splitting among the d orbital energy levels, result in high-spin complexes.

Setup The electron configuration of Cr^{2+} is $[Ar]3d^4$; and en is a strong-field ligand.

Solution Because en is a *strong*-field ligand, we expect $[Cr(en)_3]^{2+}$ to be a low-spin complex. According to Figure 22.18, all four electrons will be placed in the lower-energy d orbitals (d_{xy}, d_{yz}, and d_{xz}) and there will be a total of two unpaired spins.

THINK ABOUT IT

It is easy to draw the wrong conclusion regarding high- and low-spin complexes. Remember that the term *high spin* refers to the number of spins (*unpaired electrons*), not to the energy levels of the d orbitals. The greater the energy gap between the lower-energy and higher-energy d orbitals, the greater the chance that the complex will be *low* spin.

Practice Problem ATTEMPT How many unpaired spins are in $[Mn(H_2O)_6]^{2+}$? (*Hint:* H_2O is a weak-field ligand.)

Practice Problem BUILD Transition metal ions can have as few as one d electron or as many as nine d electrons. Which numbers of d electrons (1–9) result in the same number of unpaired spins in both the high-spin and low-spin states?

Practice Problem CONCEPTUALIZE Transition metal complexes containing CN^- ligands are often yellow in color, whereas those containing H_2O ligands are often green or blue. Explain.

Tetrahedral and Square-Planar Complexes

So far we have concentrated on octahedral complexes. The splitting of the d orbital energy levels in tetrahedral and square-planar complexes, though, can also be accounted for satisfactorily by the crystal field theory. In fact, the splitting pattern for a tetrahedral ion is just the reverse of that for octahedral complexes. In this case, the d_{xy}, d_{yz}, and d_{xz} orbitals are more closely directed at the ligands and therefore have more energy than the $d_{x^2-y^2}$ and d_{z^2} orbitals (Figure 22.19). Most tetrahedral complexes are high-spin complexes. Presumably, the tetrahedral arrangement reduces the magnitude of the metal-ligand interactions, resulting in a smaller Δ value. This is a reasonable assumption because the number of ligands is smaller in a tetrahedral complex.

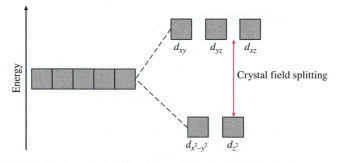

Figure 22.19 Crystal field splitting between d orbitals in a tetrahedral complex.

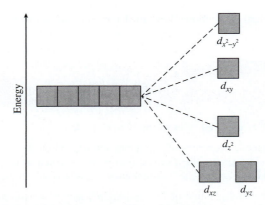

Figure 22.20 Energy-level diagram for a square-planar complex. Because there are more than two energy levels, we cannot define crystal field splitting as we can for octahedral and tetrahedral complexes.

As Figure 22.20 shows, the splitting pattern for square-planar complexes is the most complicated. The $d_{x^2-y^2}$ orbital possesses the highest energy (as in the octahedral case), and the d_{xy} orbital is the next highest. However, the relative placement of the d_{z^2} and the d_{xz} and d_{yz} orbitals cannot be determined simply by inspection and must be calculated.

CHECKPOINT – SECTION 22.3   **Bonding in Coordination Compounds: Crystal Field Theory**

22.3.1 How many unpaired spins would you expect the $[Mn(CO)_6]^{2+}$ ion to have?

a) 0

b) 1

c) 2

d) 3

e) 5

22.3.2 Which of the following metal ions can potentially form both low-spin and high-spin complexes? (Select all that apply.)

a) Ti^{2+}

b) Cu^+

c) Fe^{2+}

d) Ni^{2+}

e) Cr^{3+}

22.4   Reactions of Coordination Compounds

Complex ions undergo *ligand exchange* (or *substitution*) reactions in solution. The rates of these reactions vary widely, depending on the nature of the metal ion and the ligands.

In studying ligand exchange reactions, it is often useful to distinguish between the stability of a complex ion and its tendency to react, which we call *kinetic lability*. Stability in this context is a thermodynamic property, which is measured in terms of the species' formation constant K_f [◄◄ Section 17.5]. For example, we say that the complex ion tetracyanonickelate(II) is *stable* because it has a large formation constant ($K_f = 1 \times 10^{30}$):

$$Ni^{2+} + 4CN^- \rightleftharpoons [Ni(CN)_4]^{2-}$$

By using cyanide ions labeled with the radioactive isotope carbon-14, chemists have shown that $[Ni(CN)_4]^{2-}$ undergoes ligand exchange very rapidly in solution. The following equilibrium is established almost as soon as the species are mixed:

$$[Ni(CN)_4]^{2-} + 4{*}CN^- \rightleftharpoons [Ni({*}CN)_4]^{2-} + 4CN^-$$

where the asterisk denotes a ^{14}C atom. Complexes like the tetracyanonickelate(II) ion are termed *labile complexes* because they undergo *rapid* ligand exchange reactions. Thus, a thermodynamically *stable* species (i.e., one that has a *large* formation constant) is not necessarily unreactive.

A complex that is thermodynamically *unstable* in acidic solution is $[Co(NH_3)_6]^{3+}$. The equilibrium constant for the following reaction is about 1×10^{20}:

$$[Co(NH_3)_6]^{3+} + 6H^+ + 6H_2O \rightleftharpoons [Co(H_2O)_6]^{3+} + 6NH_4^+$$

When equilibrium is reached, the concentration of the $[Co(NH_3)_6]^{3+}$ ion is very low. This reaction requires several days to complete, however, because the $[Co(NH_3)_6]^{3+}$ ion is so inert. This is an example of an *inert complex*—a complex ion that undergoes very slow exchange reactions (on the order of hours or even days). It shows that a thermodynamically unstable species is not necessarily chemically reactive. The rate of reaction is determined by the energy of activation, which is high in this case.

Most complex ions containing Co^{3+}, Cr^{3+}, and Pt^{2+} are kinetically inert. Because they exchange ligands very slowly, they are easy to study in solution. As a result, our knowledge of the bonding, structure, and isomerism of coordination compounds has come largely from studies of these compounds.

22.5 Applications of Coordination Compounds

Coordination compounds are found in living systems and have many uses in the home, in industry, and in medicine. We briefly describe a few examples in this section.

Metallurgy

The extraction of silver and gold by the formation of cyanide complexes and the purification of nickel by converting the metal to the gaseous compound $Ni(CO)_4$ are typical examples of the use of coordination compounds in metallurgical processes.

Chelation Therapy

Previously we mentioned that chelation therapy is used in the treatment of lead poisoning. Other metals, such as arsenic and mercury, can also be removed using chelating agents.

Chemotherapy

Several platinum-containing coordination compounds, including cisplatin $[Pt(NH_3)_2Cl_2]$ and carboplatin $[Pt(NH_3)_2(OCO)_2C_4H_6]$, can effectively inhibit the growth of cancerous cells. The mechanism for the action of cisplatin is the *chelation* of DNA, the molecule that contains the genetic code. During cell division, the double-stranded DNA unwinds into two single strands, which must be accurately copied for the new cells to be identical to their parent cell. X-ray studies show that cisplatin binds to DNA by forming cross-links in which the two chlorides on cisplatin are replaced by nitrogen atoms in the adjacent guanine bases on the same strand of the DNA. (Guanine is one of the four bases in DNA [▶▶ Section 25.6, Figure 25.15].) This causes a bend in the double-stranded structure at the binding site. It is believed that this structural distortion is a key factor in inhibiting replication. The damaged cell is then destroyed by the body's immune system. Because the binding of cisplatin to DNA requires both Cl atoms to be on the same side of the complex, the *trans* isomer of the compound is totally ineffective as an anticancer drug.

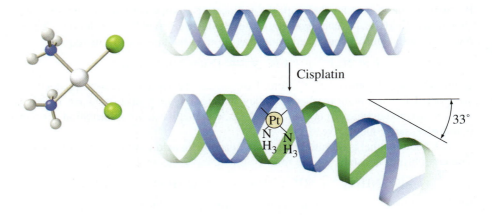

Cisplatin

33°

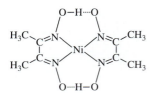

Figure 22.21 Structure of nickel dimethylglyoxime. The overall structure is stabilized by hydrogen bonds.

INGREDIENTS: Water, Butylene Glycol, Mineral Oil Alcohol, Propylene Glycol Dicaprylate/Dicaprate, Isoparaffin, Glyceryl Stearate, Tri-PPG-3 Myristyl Ether Citrate, Cetearyl Alcohol, Dimethicone, Methylparaben, Polysorbate 60, DMDM Hydantoin, Carbomer, Ethylparaben, Propylparaben, Disodium EDTA, Sodium Hydroxide, Xanthan Gum, Butylparaben, Titanium Dioxide

QUESTIONS? Call 1-877-543-8477,

David A. Tietz/Editorial Image, LLC

Chemical Analysis

Although EDTA has a great affinity for a large number of metal ions (especially 2+ and 3+ ions), other chelates are more selective in binding. Dimethylglyoxime, for example, forms an insoluble brick-red solid with Ni^{2+} and an insoluble bright-yellow solid with Pd^{2+}. These characteristic colors are used in qualitative analysis to identify nickel and palladium. Furthermore, the quantities of ions present can be determined by gravimetric analysis [◀◀ Section 4.6] as follows: To a solution containing Ni^{2+} ions, say, we add an excess of dimethylglyoxime reagent, and a brick-red precipitate forms. The precipitate is then filtered, dried, and weighed. Knowing the formula of the complex (Figure 22.21), we can readily calculate the amount of nickel in the original solution.

Detergents

The cleansing action of soap in hard water is hampered by the reaction of the Ca^{2+} ions in the water with the soap molecules to form insoluble salts or curds. In the late 1940s the detergent industry introduced a "builder," usually sodium tripolyphosphate, to circumvent this problem. The tripolyphosphate ion is an effective chelating agent that forms stable, soluble complexes with Ca^{2+} ions. Sodium tripolyphosphate revolutionized the detergent industry. Because phosphates are plant nutrients, however, wastewater containing phosphates discharged into rivers and lakes causes algae to grow, resulting in oxygen depletion. Under these conditions, most or all aquatic life eventually succumbs. This process is called *eutrophication*. Consequently, many states have banned phosphate detergents since the 1970s, and manufacturers have reformulated their products to eliminate phosphates.

Sequestrants

In addition to its use in medicine and chemical analysis, EDTA is used as a food additive to sequester metal ions. EDTA sequesters copper, iron, and nickel ions that would otherwise catalyze the oxidation reactions that cause food to spoil. EDTA is a common preservative in a wide variety of consumer products.

Bringing Chemistry to Life

The Coordination Chemistry of Oxygen Transport

Because of its central function as an oxygen carrier for metabolic processes, hemoglobin is probably the most studied of all the proteins. The molecule contains four folded long chains called subunits. Hemoglobin carries oxygen in the blood from the lungs to the tissues, where it delivers the oxygen molecules to myoglobin. Myoglobin, which is made up of only one subunit, stores oxygen for metabolic processes in the muscle.

The porphine molecule forms an important part of the hemoglobin structure. Upon coordination to a metal, the H^+ ions that are bonded to two of the four nitrogen atoms in porphine are displaced. Complexes derived from porphine are called porphyrins, and the iron-porphyrin combination is called the heme group. The iron in the heme group has an oxidation state of $+2$; it is coordinated to the four nitrogen atoms in the porphine group and also to a nitrogen donor atom in a ligand that is attached to the protein. The sixth ligand is a water molecule, which binds to the Fe^{2+} ion on the other side of the ring to complete the octahedral complex. This hemoglobin molecule is called deoxyhemoglobin and imparts a bluish tinge to venous blood. The water ligand can be replaced readily by molecular oxygen (in the lungs) to form red *oxy*hemoglobin, which is found in arterial blood. Each subunit contains a heme group, so each hemoglobin molecule can bind up to four O_2 molecules.

Animation
Organic and Biochemistry—Oxygen binding in hemoglobin.

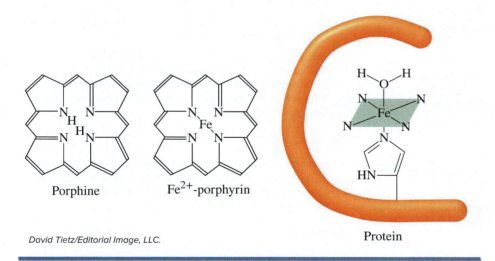

Porphine   Fe^{2+}-porphyrin

Protein

David Tietz/Editorial Image, LLC.

Chapter Summary

Section 22.1

- *Coordination compounds* contain coordinate covalent bonds between a metal ion (often a transition metal ion) and two or more polar molecules or ions.

- The molecules or anions that surround a metal in a coordination complex are called *ligands.*

- Many coordination compounds consist of a *complex ion* and a *counter* ion.

- Transition metals are those that have incompletely filled *d* subsells—or that *give rise* to ions with incompletely filled *d* subsells.

- Transition metals exhibit variable oxidation states ranging from +1 to +7.

- To act as a ligand, a molecule or ion must have at least one unshared pair of electrons. The atom that bears the unshared pair of electrons is the *donor atom.*

- Ligands are classified as *monodentate, bidentate,* or *polydentate,* based on the number of donor atoms they contain. Bidentate and polydentate ions are also known as *chelating agents.*

- The *coordination number* is the number of donor atoms surrounding a metal in a complex.

- Ionic coordination compounds are named by first naming the cation and then the anion. Complex ions are named by listing the ligands in alphabetical order, followed by the metal and its oxidation state (as a Roman numeral). When the complex ion is the anion, the anion's name ends in *–ate.*

Section 22.2

- The coordination number largely determines the geometry of a coordination complex.

- Coordination compounds containing different arrangements of the same ligands are *stereoisomers.* The two types of stereoisomerism are geometric and optical.

- *Geometrical isomers* contain the same atoms and bonds arranged differently in space.

- *Optical isomers* are nonsuperimposable mirror images. We call a *pair* of optical isomers *enantiomers.* The rotation of polarized light is measured with a *polarimeter.*

- Enantiomers rotate the plane of plane-polarized light in opposite directions. The enantiomer that rotates it to the right is called *dextrorotatory* and is labeled *d.* The enantiomer that rotates it to the left is called *levorotatory* and is labeled *l.* An equal mixture of a pair of enantiomers, called a *racemic mixture,* does not cause any net rotation of plane-polarized light.

Section 22.3

- Ligands in a coordination complex cause the energy levels of the *d* orbitals on a metal to split. The difference in energy between the lower and higher *d* orbital energy levels is called the *crystal field splitting (Δ).*

- The magnitude of Δ depends on the nature of the ligands in the complex. The *spectrochemical series* orders some common ligands in order of increasing *field strength.*

- *Strong*-field ligands give rise to a larger Δ value; *weak*-field ligands yield a smaller Δ value.

- Crystal field splitting sometimes changes the number of unpaired electrons, and therefore the magnetic properties, of a metal.

- Complexes containing transition metals with d^4, d^5, d^6, or d^7 configurations may be *high spin* or *low spin.* In high-spin complexes, the number of unpaired electrons is maximized because Δ is small; in low-spin complexes, the number of unpaired electrons is minimized because Δ is large.

Section 22.4

- Complex ions undergo ligand exchange in solution. The rate at which ligand exchange occurs is a measure of a complex's *kinetic lability* and does not necessarily correspond directly to the complex's *thermodynamic stability.*

Section 22.5

- Coordination chemistry is important in many biological, medical, and industrial processes.

Key Words

Chelating agent, 1083	Dextrorotatory, 1090	Levorotatory, 1090	Racemic mixture, 1090
Coordination compound, 1080	Donor atom, 1083	Ligand, 1082	Spectrochemical series, 1093
Coordination number, 1083	Enantiomers, 1090	Optical isomers, 1088	Stereoisomers, 1087
Crystal field splitting (Δ), 1091	Geometrical isomers, 1087	Polarimeter, 1089	

Questions and Problems

Applying What You've Learned

Elevated BLL and other heavy metal poisoning can be treated with one of several chelating agents, including DMSA and EDTA. EDTA is administered intravenously either as the sodium salt (Endrate) or as the calcium disodium salt (Versenate). Endrate is not approved for the treatment of lead poisoning because of its high affinity for calcium. It is approved, however, for treating *hypercalcemia,* a condition in which there is excess calcium in the blood—usually as a result of bone cancer. The accidental use of Endrate during treatment for lead poisoning resulted in the death of a 2-year-old girl in February of 2005. The girl's death was attributed to sudden cardiac arrest caused by the removal of too much calcium from her blood.

Chelation therapy works by the administration of a ligand, which binds to metal ions already in the body. Many drugs, including cisplatin [◄◄ Chapter 3, Applying What You've Learned], are themselves coordination compounds in which the central metal ion binds to electron-rich sites (such as oxygen or sulfur atoms) in biological molecules.

Problems:
(a) Determine the oxidation state of platinum in cisplatin, $[Pt(NH_3)_2Cl_2]$ [◄◄ Sample Problem 22.1]. (b) Give the systematic name for cisplatin [◄◄ Sample Problem 22.2]. (c) Write the formula for the compound potassium hexachloroplatinate(IV) [◄◄ Sample Problem 22.3].

SECTION 22.1: COORDINATION COMPOUNDS

Review Questions

22.1 What distinguishes a transition metal from a main group metal?

22.2 Why is zinc not considered a transition metal?

22.3 Explain why atomic radii decrease very gradually from scandium to copper.

22.4 Without referring to the text, write the ground-state electron configurations of the first-row transition metals. Explain any irregularities.

22.5 Write the electron configurations of the following ions: V^{5+}, Cr^{3+}, Mn^{2+}, Fe^{3+}, Cu^{2+}, Sc^{3+}, Ti^{4+}.

22.6 Why do transition metals have more oxidation states than other elements?

22.7 Give the highest oxidation states for scandium to copper.

22.8 Define the following terms: *coordination compound, ligand, donor atom, coordination number, chelating agent.*

22.9 Describe the interaction between a donor atom and a metal atom in terms of a Lewis acid-base reaction.

Conceptual Problems

22.10 Complete the following statements for the complex ion $[Co(en)_2(H_2O)CN]^{2+}$. (a) En is the abbreviation for _____. (b) The oxidation number of Co is _____. (c) The coordination number of Co is _____. (d) _____ is a bidentate ligand.

22.11 Complete the following statements for the complex ion $[Cr(C_2O_4)_2(H_2O)_2]^-$. (a) The oxidation number of Cr is _____. (b) The coordination number of Cr is _____. (c) _____ is a bidentate ligand.

22.12 Give the oxidation numbers of the metals in the following species: (a) $K_3[Fe(CN)_6]$, (b) $K_3[Cr(C_2O_4)_3]$, (c) $[Ni(CN)_4]^{2-}$.

22.13 Give the oxidation numbers of the metals in the following species: (a) Na_2MoO_4, (b) $MgWO_4$, (c) $Fe(CO)_5$.

22.14 What are the systematic names for the following ions and compounds?
(a) $[Co(NH_3)_4Cl_2]^+$ (c) $[Co(en)_2Br_2]^+$
(b) $[Cr(NH_3)_3Cl_3]$ (d) $[Co(NH_3)_6]Cl_3$

22.15 What are the systematic names for the following ions and compounds?
(a) $[cis\text{-}Co(en)_2Cl_2]^+$ (c) $[Co(NH_3)_5Cl]Cl_2$
(b) $[Pt(NH_3)_5Cl]Cl_3$

22.16 Write the formula for each of the following ions and compounds: (a) tetrahydroxozincate(II), (b) pentaaquachlorochromium(III) chloride, (c) tetrabromocuprate(II), (d) ethylenediaminetetraacetatoferrate(II).

22.17 Write the formula for each of the following ions and compounds: (a) dichlorobis(ethylenediamine)chromium(III), (b) pentacarbonyliron(0), (c) potassium tetracyanocuprate(II), (d) tetraammineaquachlorocobalt(III) chloride.

SECTION 22.2: STRUCTURE OF COORDINATION COMPOUNDS

Review Questions

22.18 Define the following terms: *stereoisomers, geometric isomers, optical isomers, plane-polarized light.*

22.19 Specify which of the following structures can exhibit geometric isomerism: (a) linear, (b) square planar, (c) tetrahedral, (d) octahedral.

22.20 What determines whether a molecule is chiral? How does a polarimeter measure the chirality of a molecule?

22.21 Explain the following terms: *enantiomers, racemic mixtures.*

Conceptual Problems

22.22 The complex ion $[Ni(CN)_2Br_2]^{2-}$ has a square-planar geometry. Draw the structures of the geometric isomers of this complex.

22.23 How many geometric isomers are in the following species: (a) $[Co(NH_3)_2Cl_4]^-$, (b) $[Co(NH_3)_3Cl_3]$?

22.24 Draw structures of all the geometric and optical isomers of each of the following cobalt complexes:
(a) $[Co(NH_3)_6]^{3+}$
(b) $[Co(NH_3)_5Cl]^{2+}$
(c) $[Co(C_2O_4)_3]^{3-}$

22.25 Draw structures of all the geometric and optical isomers of each of the following cobalt complexes:
(a) $[Co(NH_3)_4Cl_2]^+$, (b) $[Co(en)_3]^{3+}$.

SECTION 22.3: BONDING IN COORDINATION COMPOUNDS: CRYSTAL FIELD THEORY

Review Questions

22.26 Briefly describe crystal field theory.

22.27 Define the following terms: *crystal field splitting, high-spin complex, low-spin complex, spectrochemical series.*

22.28 What is the origin of color in a coordination compound?

22.29 Compounds containing the Sc^{3+} ion are colorless, whereas those containing the Ti^{3+} ion are colored. Explain.

22.30 For the same type of ligands, explain why the crystal field splitting for an octahedral complex is always greater than that for a tetrahedral complex.

Conceptual Problems

22.31 The absorption maximum for the complex ion $[Co(NH_3)_6]^{3+}$ occurs at 470 nm. (a) Predict the color of the complex, and (b) calculate the crystal field splitting in kJ/mol.

22.32 From each of the following pairs, choose the complex that absorbs light at a longer wavelength: (a) $[Co(NH_3)_6]^{2+}$, $[Co(H_2O)_6]^{2+}$; (b) $[FeF_6]^{3-}$, $[Fe(CN)_6]^{3-}$; (c) $[Cu(NH_3)_4]^{2+}$, $[CuCl_4]^{2-}$.

22.33 A solution made by dissolving 0.875 g of $Co(NH_3)_4Cl_3$ in 25.0 g of water freezes at $-0.56°C$. Calculate the number of moles of ions produced when 1 mole of $Co(NH_3)_4Cl_3$ is dissolved in water, and suggest a structure for the complex ion present in this compound.

22.34 Predict the number of unpaired electrons in the following complex ions: (a) $[Cr(CN)_6]^{4-}$, (b) $[Cr(H_2O)_6]^{2+}$.

22.35 Plastocyanin, a copper-containing protein found in photosynthetic systems, is involved in electron transport, with the copper ion switching between the +1 and +2 oxidation states. The copper ion is coordinated with two histidine residues, a cysteine residue, and a methionine residue in a tetrahedral configuration. How does the crystal field splitting (Δ) change between these two oxidation states?

SECTION 22.4: REACTIONS OF COORDINATION COMPOUNDS

Review Questions

22.36 Define the terms *labile complex* and *inert complex.*

22.37 Explain why a thermodynamically stable species may be chemically reactive and a thermodynamically unstable species may be unreactive.

Conceptual Problems

22.38 Oxalic acid ($H_2C_2O_4$) is sometimes used to clean rust stains from sinks and bathtubs. Explain the chemistry underlying this cleaning action.

22.39 The $[Fe(CN)_6]^{3-}$ complex is more labile than the $[Fe(CN)_6]^{4-}$ complex. Suggest an experiment that would prove that $[Fe(CN)_6]^{3-}$ is a labile complex.

22.40 Aqueous copper(II) sulfate solution is blue in color. When aqueous potassium fluoride is added, a green precipitate is formed. When aqueous potassium chloride is added instead, a bright-green solution is formed. Explain what is happening in these two cases.

22.41 When aqueous potassium cyanide is added to a solution of copper(II) sulfate, a white precipitate, soluble in an excess of potassium cyanide, is formed. No precipitate is formed when hydrogen sulfide is bubbled through the solution at this point. Explain.

22.42 A concentrated aqueous copper(II) chloride solution is bright green in color. When diluted with water, the solution becomes light blue. Explain.

22.43 In a dilute nitric acid solution, Fe^{3+} reacts with thiocyanate ion (SCN^-) to form a dark-red complex:

$$[Fe(H_2O)_6]^{3+} + SCN^- \rightleftharpoons H_2O + [Fe(H_2O)_5NCS]^{2+}$$

The equilibrium concentration of $[Fe(H_2O)_5NCS]^{2+}$ may be determined by how darkly colored the solution is (measured by a spectrometer). In one such experiment, 1.0 mL of 0.20 M $Fe(NO_3)_3$ was mixed with 1.0 mL of 1.0×10^{-3} M KSCN and 8.0 mL of dilute HNO_3. The color of the solution quantitatively indicated that the $[Fe(H_2O)_5NCS]^{2+}$ concentration was 7.3×10^{-5} M. Calculate the formation constant for $[Fe(H_2O)_5NCS]^{2+}$.

SECTION 22.5: APPLICATIONS OF COORDINATION COMPOUNDS

Review Question

22.44 Describe and give examples of the applications of coordination compounds.

ADDITIONAL PROBLEMS

22.45 How many geometric isomers can the following square-planar complex have?

22.46 $[Pt(NH_3)_2Cl_2]$ is found to exist in two geometric isomers designated I and II, which react with oxalic acid as follows:

$$I + H_2C_2O_4 \longrightarrow [Pt(NH_3)_2C_2O_4]$$
$$II + H_2C_2O_4 \longrightarrow [Pt(NH_3)_2(HC_2O_4)_2]$$

Comment on the structures of I and II.

22.47 What are the oxidation states of Fe and Ti in the ore ilmenite ($FeTiO_3$)? (*Hint:* Look up the ionization energies of Fe and Ti in Table 22.1; the fourth ionization energy of Ti is 4180 kJ/mol.)

22.48 As we read across the first-row transition metals from left to right, the +2 oxidation state becomes more stable in comparison with the +3 state. Why is this so?

22.49 Which is a stronger oxidizing agent in aqueous solution, Mn^{3+} or Cr^{3+}? Explain your choice.

22.50 Draw qualitative diagrams for the crystal field splittings in (a) a linear complex ion ML_2, (b) a trigonal-planar complex ion ML_3, and (c) a trigonal-bipyramidal complex ion ML_5.

22.51 The Cr^{3+} ion forms octahedral complexes with two neutral ligands X and Y. The color of CrX_6^{3+} is blue while that of CrY_6^{3+} is yellow. Which is a stronger field ligand, X or Y?

22.52 A student has prepared a cobalt complex that has one of the following three structures: $[Co(NH_3)_6]Cl_3$, $[Co(NH_3)_5Cl]Cl_2$, or $[Co(NH_3)_4Cl_2]Cl$. Explain how the student would distinguish between these possibilities by an electrical conductance experiment. At the student's disposal are three strong electrolytes—NaCl, $MgCl_2$, and $FeCl_3$—which may be used for comparison purposes.

22.53 The K_f for the formation of the complex ion between Pb^{2+} and $EDTA^{4-}$:

$$Pb^{2+} + EDTA^{4-} \rightleftharpoons [Pb(EDTA)]^{2-}$$

is 1.0×10^{18} at 25°C. Calculate $[Pb^{2+}]$ at equilibrium in a solution containing 1.0×10^{-3} M Pb^{2+} and 2.0×10^{-3} M $EDTA^{4-}$.

22.54 Explain the following facts: (a) Copper and iron have several oxidation states, whereas zinc has only one. (b) Copper and iron form colored ions, whereas zinc does not.

22.55 A student in 1895 prepared three coordination compounds containing chromium, with the following properties:

Formula	Color	Cl$^-$ Ions in Solution per Formula Unit
(a) $CrCl_3 \cdot 6H_2O$	Violet	3
(b) $CrCl_3 \cdot 6H_2O$	Light green	2
(c) $CrCl_3 \cdot 6H_2O$	Dark green	1

Write modern formulas for these compounds, and suggest a method for confirming the number of Cl$^-$ ions present in solution in each case. (*Hint:* Some of the compounds may exist as hydrates.)

22.56 Explain the difference between these two compounds: $CrCl_3 \cdot 6H_2O$ and $[Cr(H_2O)_6]Cl_3$.

22.57 From the standard reduction potentials listed in Table 19.1 for Zn/Zn^{2+} and Cu/Cu^{2+}, calculate $\Delta G°$ and the equilibrium constant for the reaction:

$$Zn(s) + 2Cu^{2+}(aq) \longrightarrow Zn^{2+}(aq) + 2Cu^+(aq)$$

22.58 Using the standard reduction potentials listed in Table 19.1 and the *Handbook of Chemistry and Physics,* show that the following reaction is favorable under standard-state conditions:

$$2Ag(s) + Pt^{2+}(aq) \longrightarrow 2Ag^+(aq) + Pt(s)$$

What is the equilibrium constant of this reaction at 25°C?

22.59 The Co^{2+}-porphyrin complex is more stable than the Fe^{2+}-porphyrin complex. Why, then, is iron the metal ion in hemoglobin (and other heme-containing proteins)?

22.60 What are the differences between geometric isomers and optical isomers?

22.61 Manganese forms three low-spin complex ions with the cyanide ion with the formulas $[Mn(CN)_6]^{5-}$, $[Mn(CN)_6]^{4-}$, and $[Mn(CN)_6]^{3-}$. For each complex ion, determine the oxidation number of Mn and the number of unpaired d electrons.

22.62 Hydrated Mn^{2+} ions are practically colorless (see Figure 22.16) even though they possess five $3d$ electrons. Explain. (*Hint:* Electronic transitions in which there is a change in the number of unpaired electrons do not occur readily.)

22.63 Which of the following hydrated cations are colorless: $Fe^{2+}(aq)$, $Zn^{2+}(aq)$, $Cu^+(aq)$, $Cu^{2+}(aq)$, $V^{5+}(aq)$, $Ca^{2+}(aq)$, $Co^{2+}(aq)$, $Sc^{3+}(aq)$, $Pb^{2+}(aq)$? Explain your choice.

22.64 Aqueous solutions of $CoCl_2$ are generally either light pink or blue. Low concentrations and low temperatures favor the pink form, whereas high concentrations and high temperatures favor the blue form. Adding hydrochloric acid to a pink solution of $CoCl_2$ causes the solution to turn blue; the pink color is restored by the addition of $HgCl_2$. Account for these observations.

22.65 Suggest a method that would allow you to distinguish between *cis*-$Pt(NH_3)_2Cl_2$ and *trans*-$Pt(NH_3)_2Cl_2$.

22.66 You are given two solutions containing $FeCl_2$ and $FeCl_3$ at the same concentration. One solution is light yellow, and the other one is brown. Identify these solutions based only on color.

22.67 The label of a certain brand of mayonnaise lists EDTA as a food preservative. How does EDTA prevent the spoilage of mayonnaise?

22.68 The compound 1,1,1-trifluoroacetylacetone (tfa) is a bidentate ligand:

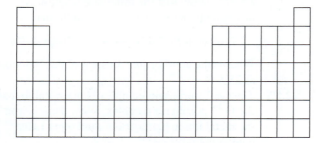

It forms a tetrahedral complex with Be^{2+} and a square-planar complex with Cu^{2+}. Draw structures of these complex ions, and identify the type of isomerism exhibited by these ions.

22.69 Locate the transition metal atoms and ions in the periodic table shown here. Atoms: (a) $[Kr]5s^24d^5$. (b) $[Xe]6s^24f^{14}5d^4$. Ions: (c) $[Ar]3d^3$ (ion with charge 4+). (d) $[Xe]4f^{14}5d^8$ (ion with charge 3+).

Biological Problems

22.70 Carbon monoxide binds to Fe in hemoglobin some 200 times more strongly than oxygen. This is the reason why CO is a toxic substance. The metal-to-ligand sigma bond is formed by donating a lone pair from the donor atom to an empty sp^3d^2 orbital on Fe. (a) On the basis of electronegativities, would you expect the C or O atom to form the bond to Fe? (b) Draw a diagram illustrating the overlap of the orbitals involved in the bonding.

22.71 Suffocation victims usually look purple, but a person poisoned by carbon monoxide often has rosy cheeks. Explain.

22.72 Copper is known to exist in the +3 oxidation state, which is believed to be involved in some biological electron-transfer reactions. (a) Would you expect this oxidation state of copper to be stable? Explain. (b) Name the compound K_3CuF_6 and predict the geometry and magnetic properties of the complex

ion. (c) Most of the known Cu(III) compounds have square-planar geometry. Are these compounds diamagnetic or paramagnetic?

22.73 Chemical analysis shows that hemoglobin contains 0.34 percent of Fe by mass. What is the minimum possible molar mass of hemoglobin? The actual molar mass of hemoglobin is about 65,000 g. How do you account for the discrepancy between your minimum value and the actual value?

22.74 In biological systems, the Cu^{2+} ions coordinated with S atoms tend to form tetrahedral complexes, whereas those coordinated with N atoms tend to form octahedral complexes. Explain.

22.75 Oxyhemoglobin is bright red, whereas deoxyhemoglobin is purple. Show that the difference in color can be accounted for qualitatively on the basis of high-spin and low-spin complexes. (*Hint:* O_2 is a strong-field ligand. See the Bringing Chemistry to Life box in Section 22.5.)

Multiconcept Problems

22.76 The formation constant for the reaction $Ag^+ + 2NH_3 \rightleftharpoons [Ag(NH_3)_2]^+$ is 1.5×10^7, and that for the reaction $Ag^+ + 2CN^- \rightleftharpoons [Ag(CN)_2]^-$ is 1.0×10^{21} at 25°C (see Table 17.5). Calculate the equilibrium constant and $\Delta G°$ at 25°C for the reaction:

$$[Ag(NH_3)_2]^+ + 2CN^- \rightleftharpoons [Ag(CN)_2]^- + 2NH_3$$

22.77 Commercial silver-plating operations frequently use a solution containing the complex $Ag(CN)_2^-$ ion. Because the formation constant (K_f) is quite large, this procedure ensures that the free Ag^+ concentration in solution is low for uniform electrodeposition. In one process, a chemist added 9.0 L of 5.0 M NaCN to 90.0 L of 0.20 M $AgNO_3$. Calculate the concentration of free Ag^+ ions at equilibrium. See Table 17.5 for K_f value.

22.78 Consider the following two ligand exchange reactions:

$$[Co(H_2O)_6]^{3+} + 6NH_3 \rightleftharpoons [Co(NH_3)_6]^{3+} + 6H_2O$$
$$[Co(H_2O)_6]^{3+} + 3en \rightleftharpoons [Co(en)_3]^{3+} + 6H_2O$$

(a) Which of the reactions should have a larger $\Delta S°$? (b) Given that the Co—N bond strength is approximately the same in both complexes, which reaction will have a larger equilibrium constant? Explain your choices.

22.79 (a) The free Cu(I) ion is unstable in solution and has a tendency to disproportionate:

$$2Cu^+(aq) \rightleftharpoons Cu^{2+}(aq) + Cu(s)$$

Use the information in Table 19.1 to calculate the equilibrium constant for the reaction. (b) Based on your results in part (a), explain why most Cu(I) compounds are insoluble.

Standardized-Exam Practice Problems

Physical and Biological Sciences

The following questions are not based on a passage.

1. What are the correct names for the complex ions $[Ni(H_2O)_6]^{2+}$ and $[Ni(NH_3)_6]^{2+}$?

   a) Hexaaquanickel(VI) ion and hexaamminenickel(VI) ion
   b) Nickel(II) hexaaqua ion and nickel(II) hexaammine ion
   c) Hexaaquanickel(II) ion and hexaamminenickel(II) ion
   d) Hexaaquanickelate and hexaamminenickelate

2. One of the complex ions in question 1 is green, and the other is purple. Based on the spectrochemical series, which is which?

   $$I^- < Br^- < Cl^- < OH^- < F^- < H_2O < NH_3 < en < CN^- < CO$$

   a) $[Ni(H_2O)_6]^{2+}$ is green and $[Ni(NH_3)_6]^{2+}$ is purple.
   b) $[Ni(H_2O)_6]^{2+}$ is purple and $[Ni(NH_3)_6]^{2+}$ is green.
   c) Both ions should be the same color because they contain the same metal.
   d) There is not enough information to determine which is which.

3. How many geometric isomers are possible for the complex ion $[Ni(H_2O)_4(NH_3)_2]^{2+}$?

   a) 1   b) 2   c) 4   d) 6

4. How many ions in solution would result from dissolving $[Ni(H_2O)_4(NH_3)_2]Cl_2$?

   a) 2   b) 3   c) 8   d) 11

Answers to In-Chapter Materials

Answers to Practice Problems

22.1A (a) K: +1, Au: +3; (b) K: +1, Fe: +2. **22.1B** (a) +1, −3, 0, +1, +3, +3, +2. **22.2A** (a) Tetraamminedibromocobalt(III) chloride, (b) tetraaquadichlorochromium(III) chloride, (c) potassium tetrachlorocuprate(II). **22.2B** (a) Sodium hexacyanoferrate(III), (b) dichlorobis(ethylenediamine)chromium(III) chloride, (c) tris(ethylenediamine)cobalt(III) chloride.
22.3A (a) $[Ru(H_2O)_5Br]NO_3$, (b) $K_2[PtBr_4Cl_2]$, (c) $Na_3[Co(NO_2)_6]$.
22.3B (a) $[V(en)_2C_2O_4]Cl_2$, (b) $[Cr(en)_2Br_2]NO_3$, (c) $[Pt(en)_3](SO_4)_2$.
22.4A 5. **22.4B** 1, 2, 3, 8, and 9.

Answers to Checkpoints

22.1.1 b. **22.1.2** d. **22.1.3** b. **22.1.4** c. **22.3.1** b. **22.3.2** c.

Organic Chemistry

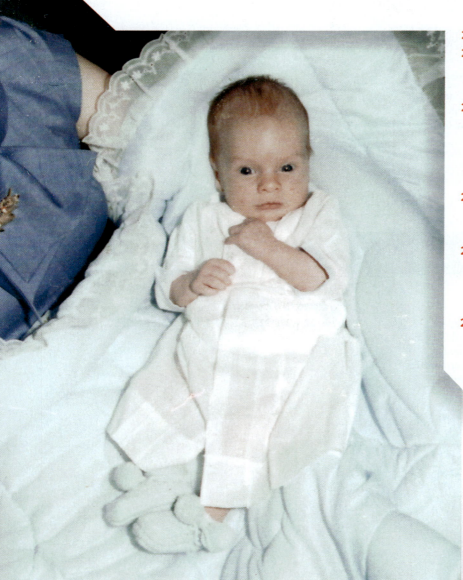

This baby, born in 1958, is one of the millions of healthy babies born in the United States during the late 1950s and early 1960s. During this period, thousands of babies in other countries suffered terrible birth defects as the result of the drug thalidomide.

Courtesy of Julia Burdge

In This Chapter, You Will Learn

Some of the basic concepts of organic chemistry and how the principles of chemical bonding contribute to the understanding of organic compounds and reactions.

Before You Begin, Review These Skills

- Lewis structures and formal charge [⏮ Sections 8.5 and 8.6]
- Resonance [⏮ Section 8.7]
- Molecular geometry and polarity [⏮ Sections 9.1 and 9.2]

The Importance of Organic Chemistry to the Development of New Drugs

Beginning in 1957, the drug thalidomide was marketed in 48 countries around the world as a sleeping pill and as an antinausea medicine for pregnant women suffering from morning sickness. By 1962, the drug was shown to have caused horrific birth defects and an untold number of fetal deaths. Thalidomide interferes with spinal cord and limb development, and more than 10,000 babies had been born with severe spinal cord abnormalities and malformed or absent limbs. Many of the victims were born to mothers who reportedly had taken just *one* thalidomide pill early in their pregnancies. At the time, thalidomide was not approved for use in the United States, but the otherwise worldwide tragedy did prompt the U.S. Congress to enact a new law to give the FDA more control over the testing and approval of new drugs.

Student Note: Thalidomide was not approved for use in the United States thanks in large part to the vigilance of one doctor at the FDA. She was troubled by inadequate research into the safety of the drug and steadfastly refused to approve the drugmaker's application.

In August of 1998, the FDA approved thalidomide for the treatment of *erythema nodosum leprosum* (ENL), a painful inflammatory skin condition associated with leprosy. This approval is controversial because of the drug's infamous history, but thalidomide has shown tremendous promise in the treatment of a wide variety of painful and debilitating conditions, including complications from certain cancers, AIDS, and some autoimmune disorders such as lupus and rheumatoid arthritis. Because of the dangers known to be associated with thalidomide, researchers are working on developing *analogues*—drugs that are chemically similar enough to have the same therapeutic benefits but chemically *different* enough *not* to have the undesirable and/or dangerous properties. Two such analogues that have been approved by the FDA are lenalidomide and pomalidomide, shown here:

At the end of this chapter, you will be able to answer several questions about the drugs thalidomide, lenalidomide, and pomalidomide [⏭ Applying What You've Learned, page 1151].

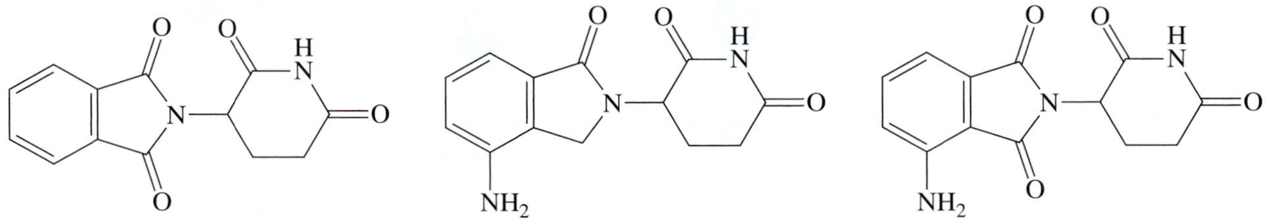

Thalidomide Lenalidomide Pomalidomide

Scientists who develop new drugs such as lenalidomide and pomalidomide must understand the principles and concepts of *organic chemistry*.

23.1   Why Carbon Is Different

Organic chemistry is usually defined as the study of compounds that contain carbon. This definition is not entirely satisfactory, though, because it would include such things as cyanide and cyanate complexes and metal carbonates, which are considered to be *inorganic.* A somewhat more useful definition of organic chemistry is the study of compounds that contain carbon and hydrogen, although many organic compounds also contain other elements, such as oxygen, sulfur, nitrogen, phosphorus, or the halogens, and many do not contain hydrogen. Examples of organic compounds include the following:

CH_4	C_2H_5OH	$H_2C_6H_6O_6$	CH_3NH_2	CCl_4
Methane	Ethanol	Ascorbic acid	Methylamine	Carbon tetrachloride

Early in the study of organic chemistry there was thought to be some fundamental difference between compounds that came from living things, such as plants and animals, and those that came from nonliving things, such as rocks. Compounds obtained from plants or animals were called *organic,* whereas compounds obtained from nonliving sources were called *inorganic.* In fact, until early in the nineteenth century, scientists believed that only nature could produce organic compounds. In 1829, however, Friedrich Wöhler prepared urea, a well-known organic compound, by combining the inorganic substances lead cyanate and aqueous ammonia:

$$Pb(OCN)_2 + 2NH_3 + 2H_2O \xrightarrow{\Delta} \underset{\text{Urea}}{2NH_2CONH_2} + Pb(OH)_2$$

Wöhler's synthesis of urea dispelled the notion that organic compounds were fundamentally different from inorganic compounds—and that they could only be produced by nature. We now know that it is possible to synthesize a wide variety of organic compounds in the laboratory; in fact, many thousands of new organic compounds are produced in research laboratories each year. In this chapter, we consider several types of organic compounds that are important biologically.

Because of its unique nature, carbon is capable of forming millions of different compounds. Carbon's position in the periodic table (Group 14, Period 2) gives it the following set of unique characteristics:

Student Note: To form an ion that is isoelectronic with a noble gas, a C atom would have to either gain or lose *four* electrons [◄◄ Section 7.5]—something that is energetically impossible under ordinary conditions. This is not to say that carbon *cannot* form ions. But in the vast majority of its compounds, carbon acquires a complete octet by *sharing* electrons—rather than by gaining or losing them.

- The electron configuration of carbon ($[He]2s^2 2p^2$) effectively prohibits *ion formation.* This and carbon's electronegativity, which is intermediate between those of metals and nonmetals, cause carbon to complete its octet by sharing electrons. In nearly all its compounds, carbon forms four covalent bonds, which can be oriented in as many as four different directions:

Methane	Formaldehyde	Carbon dioxide
(4 σ bonds)	(3 σ bonds & 1 π bond)	(2 σ bonds & 2 π bonds)

Boron and nitrogen, carbon's neighbors in Groups 13 and 15, respectively, usually form covalent compounds, too, but B and N form ions more readily than C.

- Carbon's small atomic radius allows the atoms to approach one another closely, giving rise to short, *strong,* carbon-carbon bonds and *stable* carbon compounds. In addition, carbon atoms that are *sp-* or *sp²*-hybridized approach one another closely enough for their singly occupied, unhybridized *p* orbitals to overlap effectively—giving rise to relatively strong π bonds [◄◄ Section 9.5]. Recall that elements in the same group generally exhibit similar chemical behavior [◄◄ Section 2.4]. Silicon atoms, however, are bigger than carbon atoms, so silicon atoms generally cannot approach one another closely enough for their unhybridized *p* orbitals to overlap significantly. As a result, very few compounds exhibit significant π bonding between Si atoms:

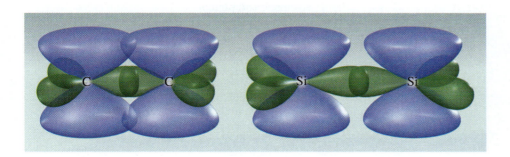

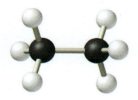

Ethane

- Carbon's valence electrons are in the second shell (*n* = 2), where there are no *d* orbitals. The valence electrons of silicon, on the other hand, are in the third shell (*n* = 3), where there are *d* orbitals, which can be occupied or attacked by lone pairs on another substance—resulting in a reaction. This reactivity makes silicon compounds far less stable than the analogous carbon compounds. Ethane (CH₃−CH₃), for example, is stable in both water and air, whereas disilane (SiH₃−SiH₃) is unstable—breaking down in water and combusting spontaneously in air.

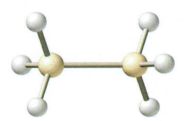

Disilane

These attributes enable carbon to form chains (straight, branched, and cyclic) containing single, double, and triple carbon-carbon bonds. Carbon's formation of chains is called **catenation.** This, in turn, results in an endless array of organic compounds containing any number and arrangement of carbon atoms. Each carbon atom in a compound can be classified by the number of other carbon atoms to which it is bonded. A carbon atom that is bonded to just one other carbon atom is called a *primary* carbon; one that is bonded to two other carbon atoms is called a *secondary* carbon; one that is bonded to three other carbon atoms is called a *tertiary* carbon; and one that is bonded to four other carbon atoms is called a *quaternary* carbon. These four types of carbon atoms are identified with the labels 1°, 2°, 3°, and 4°, respectively. Each of the four carbon types is labeled in the following structure:

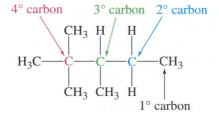

One of the important organic molecules that we encountered in Chapter 9 is benzene (C₆H₆). Organic compounds that are related to benzene, or that contain one or more benzene rings, are called ***aromatic*** compounds. Organic molecules that are *not* aromatic are called ***aliphatic*** compounds:

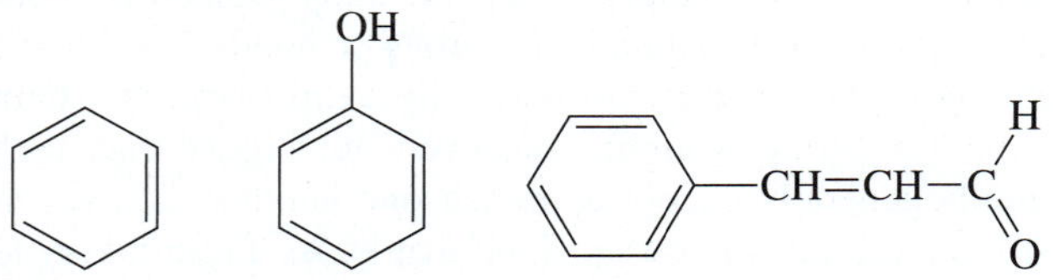

Benzene   Phenol   Cinnamaldehyde

Aromatic compounds

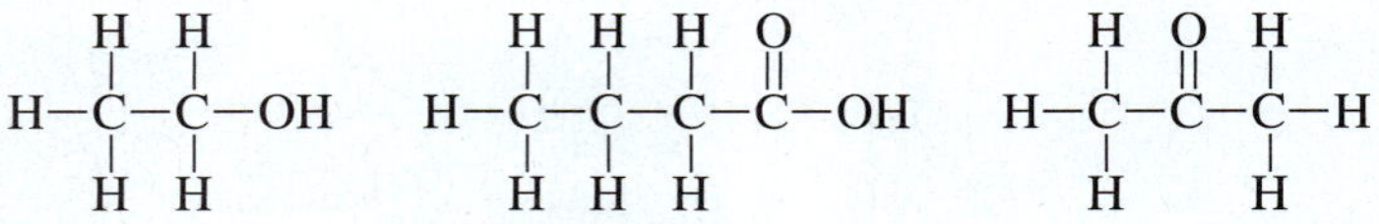

Ethanol   Butyric acid   Acetone

Aliphatic compounds

23.2 Organic Compounds

Organic molecules occur in seemingly limitless variety. In Chapter 2 we encountered *alkanes,* organic compounds consisting of only carbon and hydrogen and containing only single bonds. A variety of different types of organic compounds, each with their own characteristic properties, result from the following:

Student Note: A functional group is a group of atoms that determines many of a molecule's properties [◄◄ Section 2.7].

1. Carbon's ability to form chains by bonding with itself
2. The presence of elements other than carbon and hydrogen
3. Functional groups
4. Multiple bonds

Classes of Organic Compounds

Student Note: Recall that isomers are different compounds with the same chemical formula [◄◄ Section 9.2].

In this section, we discuss several types of organic compounds and how we represent them. Consider two isomers of C_3H_6O:

$$CH_3CH_2CH \quad\quad CH_3CCH_3$$

Propanaldehyde   Acetone

Although the two isomers contain exactly the same atoms, their different arrangements of atoms result in two very different compounds. The first is an *aldehyde* called propanal. The second is a *ketone* called acetone. Aldehydes and ketones are two classes of organic compounds. The classes of organic compounds that we discuss in this chapter are alcohols, carboxylic acids, aldehydes, ketones, esters, amines, and amides.

A class of organic compounds often is represented with a general formula that shows the atoms of the functional group(s) explicitly, and the remainder of the molecule using one or more R's, where R represents an alkyl group. An ***alkyl group*** is a portion of a molecule that resembles an alkane. In fact, an alkyl group is formed by removing one hydrogen atom from the corresponding alkane. The *methyl group* ($-CH_3$), for example, is formed by removing a hydrogen atom from methane (CH_4),

TABLE 23.1	Alkyl Groups	
Name	**Formula**	**Model**
Methyl	$-CH_3$	
Ethyl	$-CH_2CH_3$	
Propyl	$-CH_2CH_2CH_3$	
Isopropyl	$-CH(CH_3)_2$	
Butyl	$-CH_2CH_2CH_2CH_3$	
tert-butyl	$-C(CH_3)_3$	
Pentyl	$-CH_2CH_2CH_2CH_2CH_3$	
Isopentyl	$-CH_2CH_2CH(CH_3)_2$	
Hexyl	$-CH_2CH_2CH_2CH_2CH_2CH_3$	
Heptyl	$-CH_2CH_2CH_2CH_2CH_2CH_2CH_3$	
Octyl	$-CH_2CH_2CH_2CH_2CH_2CH_2CH_2CH_3$	

the simplest alkane. Methyl groups are found in many organic molecules. Table 23.1 lists some of the simplest alkyl groups. Table 23.2 gives the general formula for each of the classes of organic compounds discussed in this chapter and the Lewis structure of each functional group:

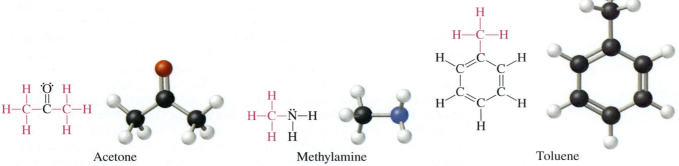

Acetone   Methylamine   Toluene

Student Note: Some functional groups have special names and some do not.

TABLE 23.2 General Formulas for Select Classes of Organic Compounds

Class	General Formula	Lewis Structure	Functional Group
Alcohol	ROH	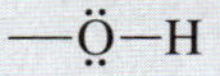 $-\ddot{\text{O}}-\text{H}$	Hydroxy group
Carboxylic acid	RCOOH	$-\overset{\overset{\ddot{\text{O}}}{\|}}{\text{C}}-\ddot{\text{O}}-\text{H}$	Carboxy group
Ester	RCOOR′	$-\overset{\overset{\ddot{\text{O}}}{\|}}{\text{C}}-\ddot{\text{O}}-\text{R}'$	Ester group
Aldehyde	RCHO	$-\overset{\overset{\ddot{\text{O}}}{\|}}{\text{C}}-\text{H}$	Carbonyl group
Ketone*	RCOR′	$-\overset{\overset{\ddot{\text{O}}}{\|}}{\text{C}}-\text{R}'$	Carbonyl group
Amine†	RNH₂	$-\overset{\underset{\text{H}}{\|}}{\ddot{\text{N}}}-\text{H}$	Amino group (primary, 1°)
Amine	RNR′H	$-\overset{\underset{\text{H}}{\|}}{\ddot{\text{N}}}-\text{R}'$	Amino group (secondary, 2°)
Amine	RNR′R″	$-\overset{\underset{\text{R}''}{\|}}{\ddot{\text{N}}}-\text{R}'$	Amino group (tertiary, 3°)
Amide	RCONH₂	$-\overset{\overset{\ddot{\text{O}}}{\|}}{\text{C}}-\overset{\underset{\text{H}}{\|}}{\ddot{\text{N}}}-\text{H}$	Amide group (primary, 1°)
Amide	RCONR′H	$-\overset{\overset{\ddot{\text{O}}}{\|}}{\text{C}}-\overset{\underset{\text{H}}{\|}}{\ddot{\text{N}}}-\text{R}'$	Amide group (secondary, 2°)
Amide	RCONR′R″	$-\overset{\overset{\ddot{\text{O}}}{\|}}{\text{C}}-\overset{\underset{\text{R}''}{\|}}{\ddot{\text{N}}}-\text{R}'$	Amide group (tertiary, 3°)

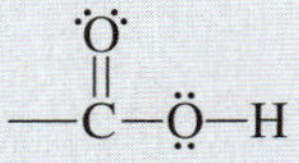

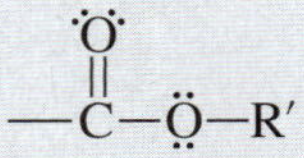

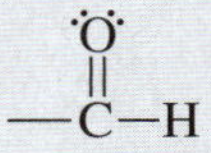

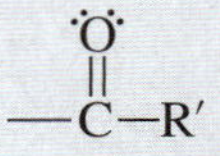

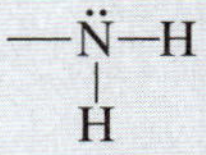

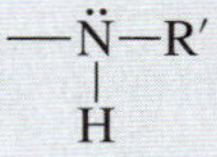

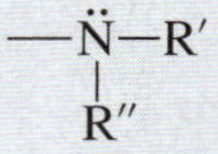

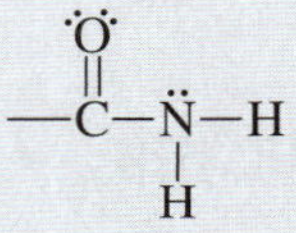

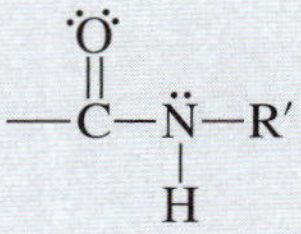

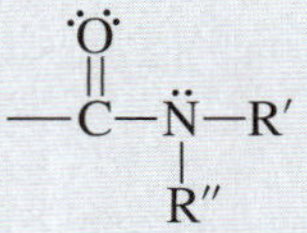

*R′ represents a second alkyl group that may or may not be identical to the first alkyl group R. Likewise, R″ represents a third alkyl group that may or may not be identical to R or R′.

†The designations 1°, 2°, and 3° refer to how many R groups are bonded to the N atom.

The functional groups in the types of compounds shown in Table 23.2 are the *hydroxy* group (in *alcohols*), the *carboxy* group (in *carboxylic acids*), the −COOR group (in *esters*), the *carbonyl* group (in *aldehydes* and *ketones*), the *amino* group (in *amines*), and the *amide* group (in *amides*). Functional groups determine many of the properties of a compound, including what types of reactions it is likely to undergo. Figure 23.1 shows ball-and-stick models and electrostatic potential maps of the hydroxy, carboxy, carbonyl, amino, and amide functional groups.

A compound consisting of an alkyl group and the functional group −OH is an *alcohol*. The identity of an individual alcohol depends on the identity of R, the alkyl group. For example, when R is the *methyl* group, we have CH_3OH. This is methyl alcohol or methanol, also known as wood alcohol. It is highly toxic and can cause blindness or even death in relatively small doses. When R is the *ethyl* group, we have

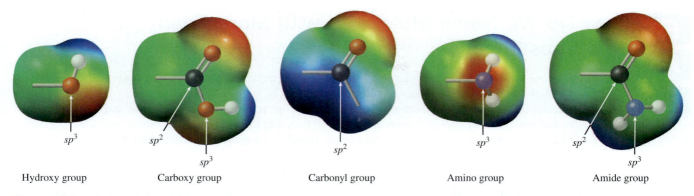

sp³ — Hydroxy group   sp² — Carboxy group   sp² — Carbonyl group   sp³ — Amino group   sp² — Amide group

sp³ (Carboxy group)   sp³ (Carbonyl group)   sp³ (Amide group)

Figure 23.1 Models and electrostatic potential maps of the hydroxy, carboxy, carbonyl, amino, and amide functional groups.

CH_3CH_2OH. This is ethyl alcohol or ethanol. Ethanol is the alcohol in alcoholic beverages. When R is the *isopropyl* group, we have $(CH_3)_2CHOH$. This is isopropyl alcohol. Isopropyl alcohol, what we commonly call "rubbing alcohol," is widely used as a disinfectant.

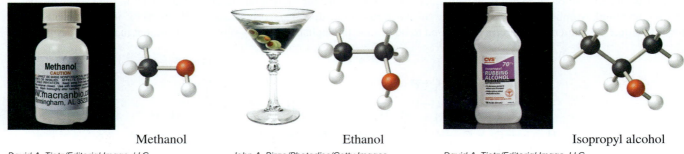

Methanol

Ethanol

Isopropyl alcohol

Naming Organic Compounds

Organic compounds are named systematically using International Union of Pure and Applied Chemistry (IUPAC) rules.

Root Names for Alkanes and Alkyl Groups

Number of Carbons	Name	Number of Carbons	Name	Number of Carbons	Name
1	Meth-	5	Pent-	8	Oct-
2	Eth-	6	Hex-	9	Non-
3	Prop-	7	Hept-	10	Dec-
4	But-				

Prefixes for Halogen Substituents

—F Fluoro   —Cl Chloro   —Br Bromo   —I Iodo

Alkanes

In Chapter 2, we encountered the names of some simple, straight-chain alkanes such as pentane:

$$\begin{array}{ccccccccc} & H & & H & & H & & H & & H \\ & | & & | & & | & & | & & | \\ H- & C & - & C & - & C & - & C & - & C & -H \\ & | & & | & & | & & | & & | \\ & H & & H & & H & & H & & H \end{array}$$

How Do We Name Molecules with More Than One Substituent?

A systematic name must identify a compound unambiguously. Therefore, a special system of rules must be followed to name molecules that contain more than one substituent.

In molecules that contain two or more identical substituents, the prefixes *di, tri, tetra, penta,* and so forth, are used to denote the number of substituents. Numbers are then used to denote their positions. (Note that two substituents may be bonded to the same carbon atom. In this case, the carbon's number is repeated with a comma between the numbers.)

2,3-Dimethylpentane   2,2-Dichlorohexanoic acid

In the case where two or more different substituents are present, the substituent names are alphabetized in the systematic name of the compound. Numbers are used to indicate the positions of the alphabetized substituents. If a prefix is used to denote two or more identical substituents, the prefix is *not* used to determine the alphabetization—only the substituent name is used:

4-Ethyl-2-methylhexane   4-Ethyl-2,2-dimethylhexane

To name substituted alkanes (i.e., those that have ***substituents,*** which are groups other than −H bonded to the carbons of the chain), we follow a series of steps:

1. Identify the longest continuous carbon chain to get the *parent name.*
2. Number the carbons in the continuous chain, beginning at the end closest to the substituent. Commonly encountered substituents include alkyl groups and halogens.
3. Identify the substituent and use a *number* followed by a dash and a *prefix* to specify its *location* and *identity,* respectively.

Step 1:   The longest continuous carbon chain contains five C atoms:

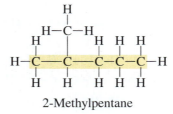

2-Methylpentane

(We could also identify the carbon chain as:

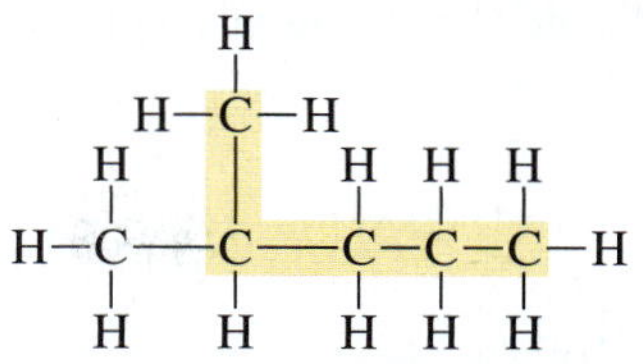

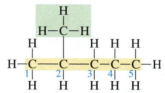

but the number of C atoms in the chain is the same either way.) The parent name of a five-carbon chain is *pentane*. (See root names.)

Step 2:   We number the carbon atoms beginning at the end nearest the substituent (shaded in green):

Step 3:   The substituent is a methyl group, $-CH_3$. It is attached to carbon 2. Therefore, the name is 2-methylpentane.

Sample Problem 23.1 lets you practice naming some simple organic compounds.

SAMPLE PROBLEM 23.1

Give names for the following compounds:

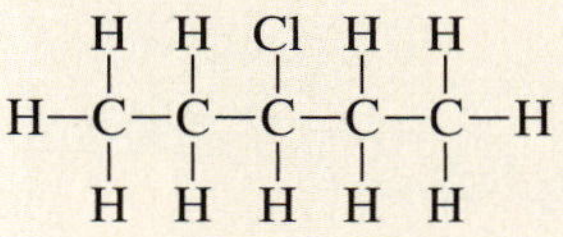

(a) 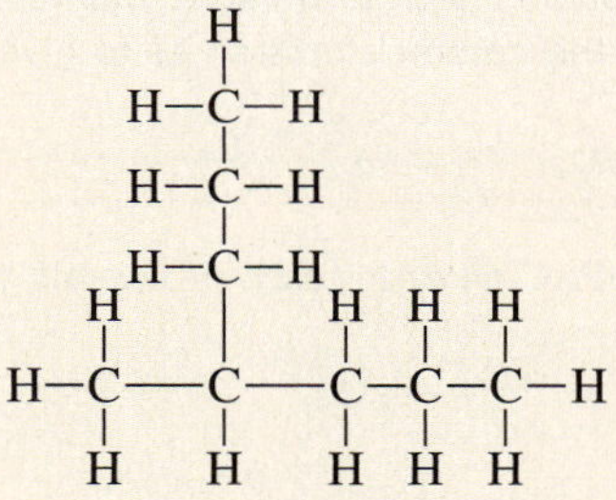   (b) 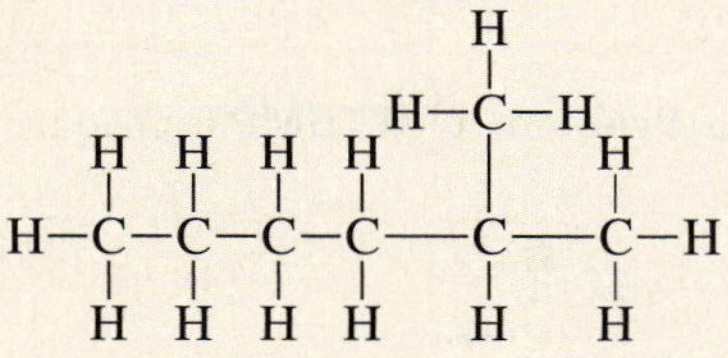   (c)

Strategy Use the three-step procedure for naming substituted alkanes: (1) name the parent alkane, (2) number the carbons, and (3) name and number the substituent. (Consult Table 2.7 for parent alkane names.)

Setup (a) This is a five-carbon chain. We can number the carbons starting at either end because the Cl substituent will be located on carbon 3 either way:

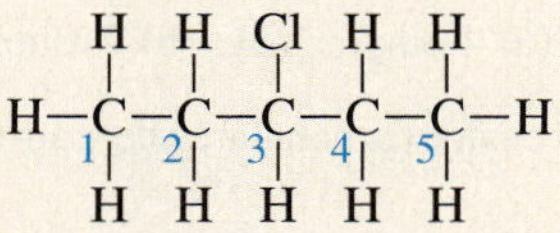

(b) This may look like a substituted pentane, too, but the longest carbon chain in this molecule is seven carbons long:

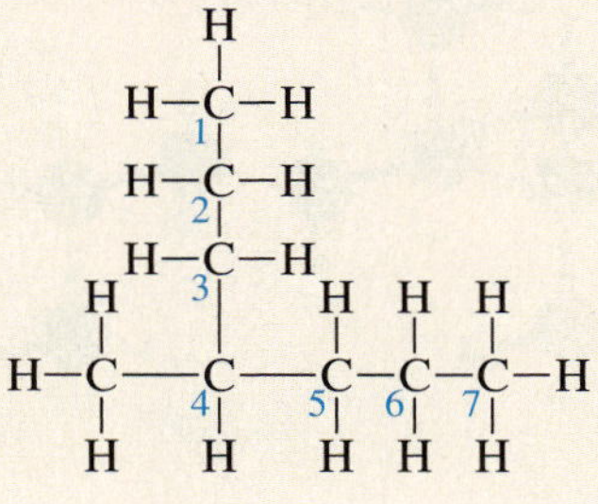

(Continued on next page)

Although Lewis structures appear to be flat and to contain 90° angles, the C atoms in this molecule are all sp^3-hybridized (four electron domains around each) and there is free rotation about the C—C bonds [◀◀ Section 9.5]. Thus, the molecule can also be drawn as:

$$
\begin{array}{c}
\text{H} \\
| \\
\text{H—C—H} \\
\text{H H H} \quad | \quad \text{H H H} \\
| \ \ | \ \ | \quad\quad | \ \ | \ \ | \\
\text{H—C—C—C——C——C—C—C—H} \\
1 \ \ \ 2 \ \ \ 3 \quad\ \ 4 \quad\ \ 5 \ \ \ 6 \ \ \ 7 \\
| \ \ | \ \ | \quad\quad | \ \ | \ \ | \\
\text{H H H} \quad\ \ \text{H} \quad\ \ \text{H H H}
\end{array}
$$

The substituent is a methyl group on carbon 4.

(c) This is a substituted *hexane:*

$$
\begin{array}{c}
\text{H} \\
| \\
\text{H—C—H} \\
\text{H H H H} \quad | \quad \text{H} \\
| \ \ | \ \ | \ \ | \quad\quad | \\
\text{H—C—C—C—C——C——C—H} \\
6 \ \ \ 5 \ \ \ 4 \ \ \ 3 \quad\ 2 \quad\ 1 \\
| \ \ | \ \ | \ \ | \quad\quad | \\
\text{H H H H} \quad\ \text{H} \quad\ \text{H}
\end{array}
$$

Solution (a) 3-chloropentane, (b) 4-methylheptane, (c) 2-methylhexane

THINK ABOUT IT

A common error is to misidentify the parent alkane. Double-check to be sure you have identified the *longest* continuous carbon chain in the molecule. Also be sure to number the carbon atoms so as to give the substituent the *lowest* possible number.

Practice Problem **A**TTEMPT Give the systematic IUPAC name for each of the following:

(a)
$$
\begin{array}{c}
\text{H} \\
| \\
\text{H—C—H} \\
| \\
\text{H—C—H} \\
\text{H H} \quad | \quad \text{H H} \\
| \ \ | \quad\quad | \ \ | \\
\text{H—C—C——C——C—C—H} \\
| \ \ | \quad\quad | \ \ | \\
\text{H H} \quad\ \text{H} \quad\ \text{H H}
\end{array}
$$

(b)
$$
\begin{array}{c}
\text{H H H H Br H} \\
| \ \ | \ \ | \ \ | \ \ \ | \ \ \ | \\
\text{H—C—C—C—C—C—C—H} \\
| \ \ | \ \ | \ \ | \ \ \ | \ \ \ | \\
\text{H H H H H H}
\end{array}
$$

(c)
$$
\begin{array}{c}
\text{H H H H} \\
| \ \ | \ \ | \ \ | \\
\text{H—C—C—C—C—H} \\
| \ \ | \ \ | \ \ | \\
\text{H Cl H H}
\end{array}
$$

Practice Problem **B**UILD Draw structures for (a) 4-ethyloctane, (b) 2-fluoropentane, and (c) 3-methyldecane.

Practice Problem **C**ONCEPTUALIZE How many carbons are there in the longest carbon chain in this molecule?

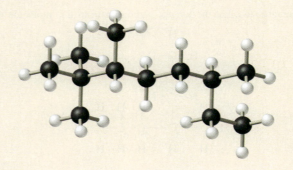

How Do We Name Compounds with Specific Functional Groups?

Alcohols

Identify the longest continuous carbon chain that includes the carbon to which the −OH group is attached. This is the parent alkyl group. Name it according to the number of carbons it contains, and change the –e ending to –ol. Number the C atoms such that the −OH group has the lowest possible number, and, when necessary, use a number to indicate the position of the −OH group:

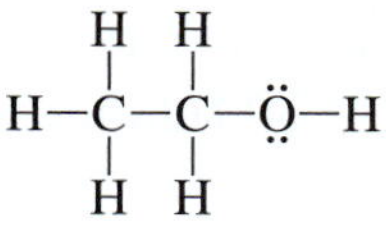

Ethanol 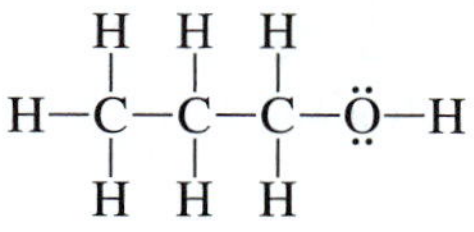   1-Propanol 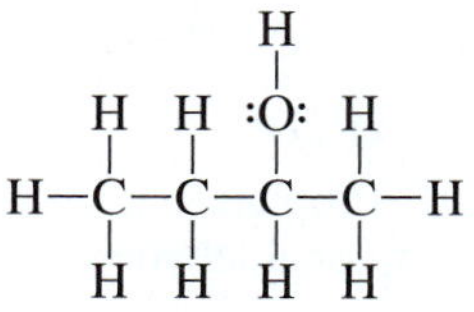   2-Butanol

When the chain that bears the −OH group also bears an alkyl substituent, the chain is numbered in the direction that gives the lowest possible number to the carbon attached to −OH:

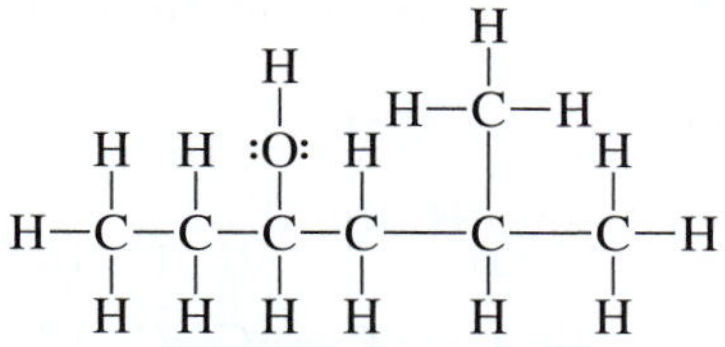

5-Methyl-3-hexanol

Carboxylic Acids

Identify the longest continuous carbon chain that includes the carboxy group. Name it according to the number of carbons it contains, and change the –e ending to –oic acid. Number the C atoms starting with the carbonyl carbon. Use numbers and prefixes to indicate the position and the identity of any substituents.

> **Student Note:** The carbonyl carbon is the one that is doubly bonded to oxygen.

   Many organic compounds have common names in addition to their systematic names. Common names for some of the carboxylic acids shown here are given in parentheses:

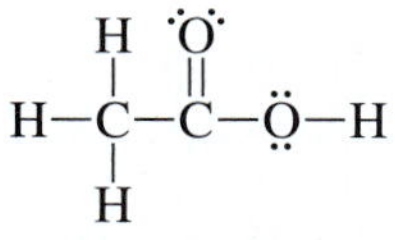

Ethanoic acid   Propanoic acid   Butanoic acid   5-Methylhexanoic acid
(acetic acid)   (propionic acid)   (butyric acid)

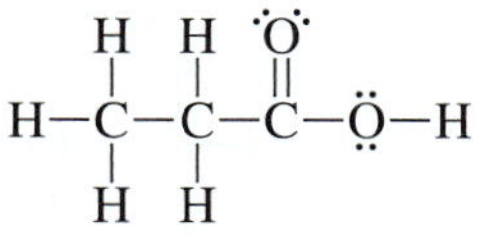

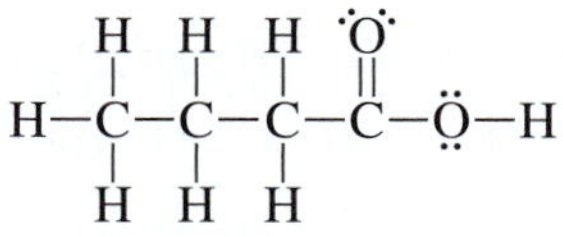

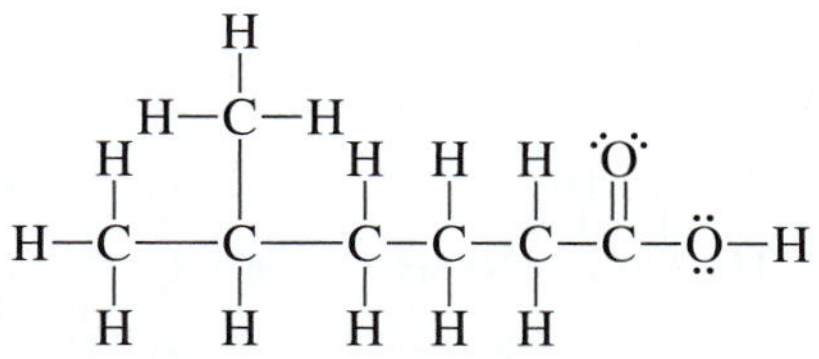

Esters

Name esters as derivatives of carboxylic acids by replacing the –ic acid ending with –ate:

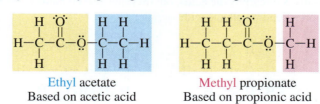

Ethyl acetate   Methyl propionate
Based on acetic acid   Based on propionic acid

(The first part of an ester's name specifies the substituent that replaces the ionizable hydrogen of the corresponding carboxylic acid.)

(Continued on next page)

Aldehydes

Identify the longest continuous carbon chain that includes the carbonyl group. Name it according to the number of carbons it contains, and change the –e ending to –al. Number the C atoms starting with the carbonyl carbon. Use numbers and prefixes to indicate the position and the identity of any substituents:

> **Student Note:** With carboxylic acids and aldehydes, because we always begin numbering the carbons at the carbonyl, it is not necessary to include the number of the carbonyl carbon (1) in the name.

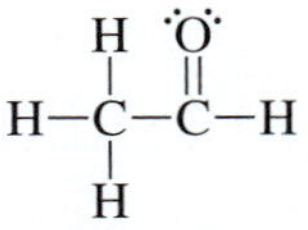

Ethanal (acetaldehyde)   Propanal (propionaldehyde)   Butanal 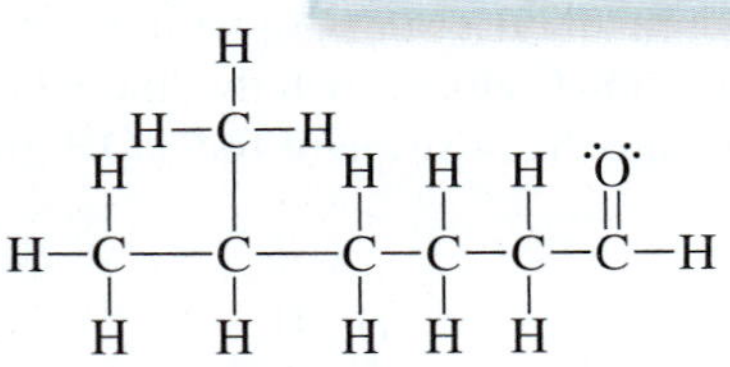   5-Methylhexanal

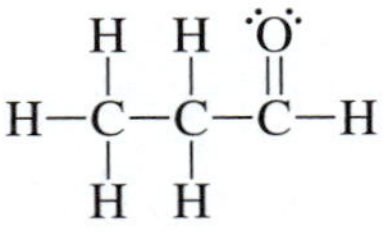

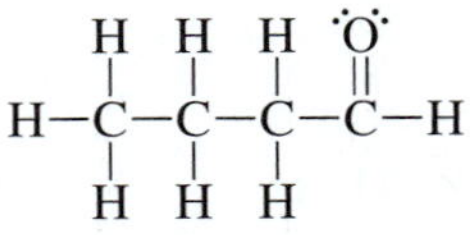

Ketones

Identify the longest continuous carbon chain that includes the carbonyl group. Name it according to the number of carbons it contains, and change the –e ending to –one. If necessary, number the C atoms to give the carbonyl carbon the lowest possible number. Use numbers and prefixes to indicate the position and the identity of any substituents:

> **Student Note:** Note that there is only one possible location for the carbonyl carbon in propanone (the middle carbon), making a number in the name unnecessary. If the carbonyl occurred on either of the other carbons, the compound would be an *aldehyde*, not a ketone.

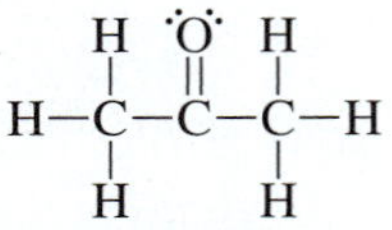

Propanone (acetone)   2-Butanone (ethyl methyl ketone)   5-Methyl-3-hexanone

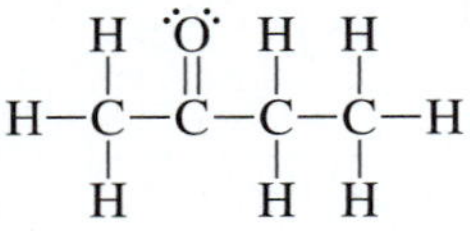

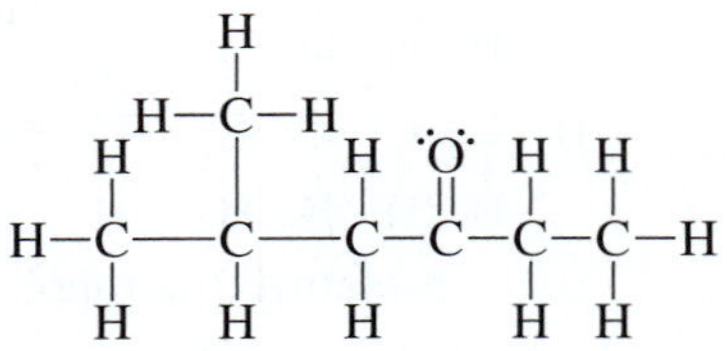

Primary Amines

Identify the longest continuous carbon chain that includes the carbon to which the $-NH_2$ group is bonded. Name it according to the number of carbons it contains, and change the –e ending to –amine. Number the C atoms starting with the carbon to which the $-NH_2$ group is bonded. Use numbers and prefixes to indicate the position and the identity of any substituents:

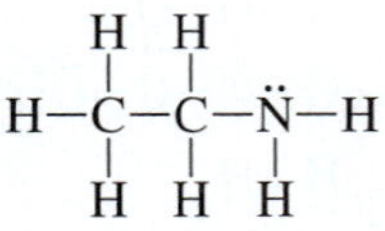

Ethanamine (ethylamine)   1-Propanamine (propylamine)   1-Butanamine (butylamine)   5-Methyl-1-hexanamine (5-methylhexylamine)

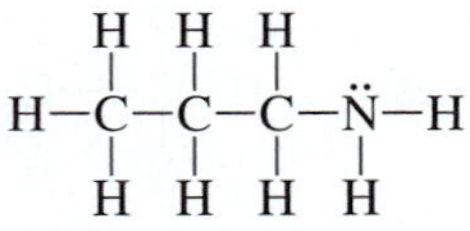

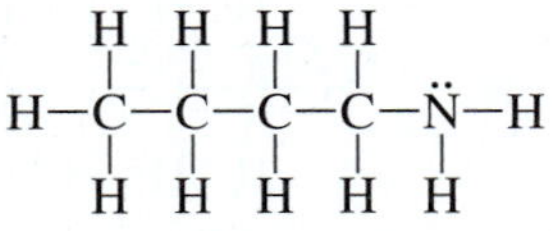

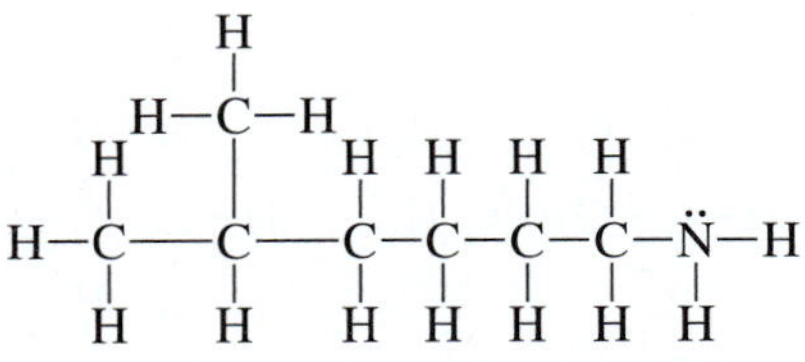

Primary Amides

Primary amides are named as derivatives of carboxylic acids, but they can also be named by replacing the –e ending of the corresponding alkane with –amide:

> **Student Note:** We do not introduce the nomenclature for secondary and tertiary amines and amides in this text.

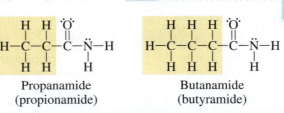

Propanamide (propionamide)   Butanamide (butyramide)

Many compounds contain more than one functional group. An *amino acid,* for example, contains both the amine group and the carboxy group:

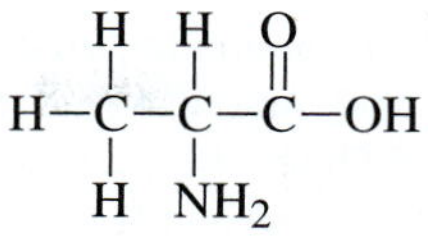

Alanine

Sample Problem 23.2 lets you practice identifying functional groups in molecules.

SAMPLE PROBLEM (23.2)

Many familiar substances are organic compounds. Some examples include aspartame, the artificial sweetener in the sugar substitute Equal and in many diet sodas; salicylic acid, found in some acne medicines and wart-removal treatments; and amphetamine, a stimulant used to treat narcolepsy, attention-deficit hyperactivity disorder (ADHD), and obesity. Identify the functional group(s) in each molecule:

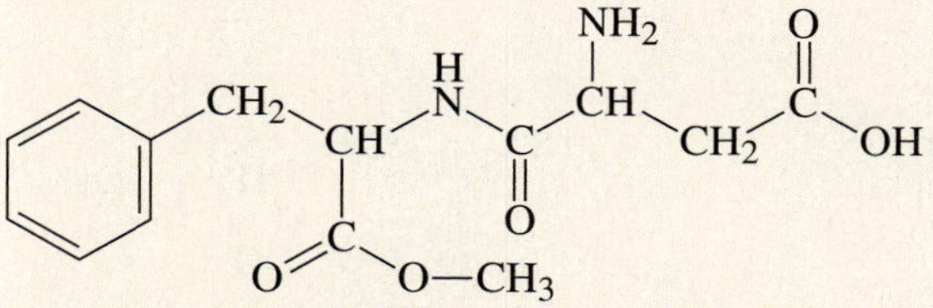

Aspartame
(a)

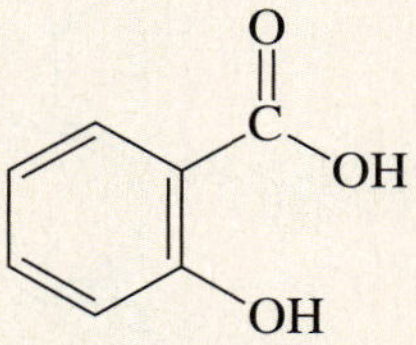

Salicylic acid
(b)

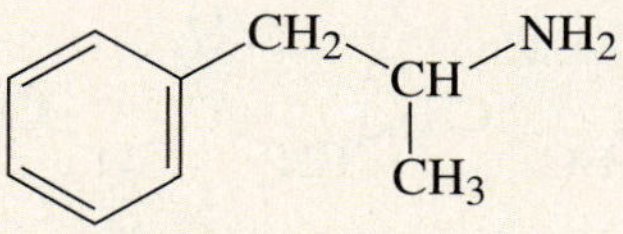

Amphetamine
(c)

Strategy Look for and identify the combinations of atoms shown in Table 23.2.

Setup (a) From right to left, aspartame contains a —COOH group, an —NH$_2$ group, a —CONHR group, and a —COOR group:

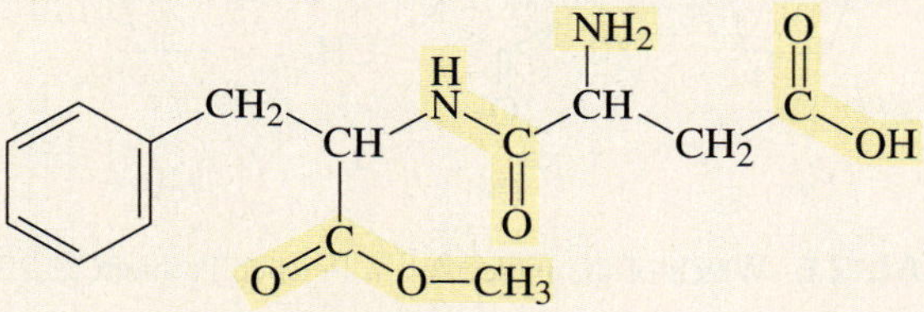

(b) Salicylic acid contains an —OH group and a —COOH group:

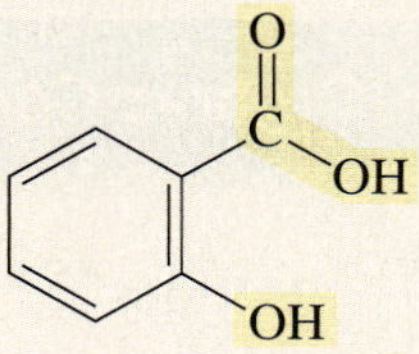

(c) Amphetamine contains an —NH$_2$ group:

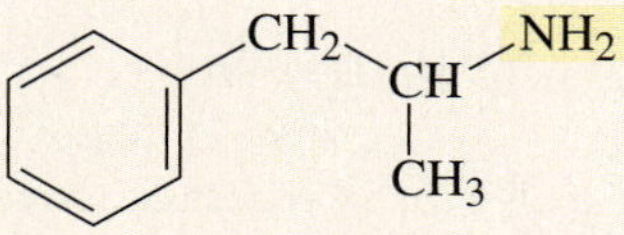

Solution (a) From right to left, aspartame contains a *carboxy* group, an *amino* group, an *amide* group, and a —COOR (ester) group.

(b) Salicylic acid contains a *hydroxy* group and a *carboxy* group.

(c) Amphetamine contains an *amino* group.

(Continued on next page)

THINK ABOUT IT

In part (b), the salicylic acid molecule contains a benzene ring and is therefore *aromatic*. When the hydroxy group is attached to a benzene ring, the resulting aromatic compound is a *phenol*, not an alcohol. Amphetamine has several legitimate medicinal uses, but it is also one of the most commonly misused drugs in the United States. Because it frequently is prescribed to adolescents to treat ADHD, much of it finds its way into high schools, where its misuse is a serious problem. A closely related compound that frequently makes headlines is *methamphetamine:*

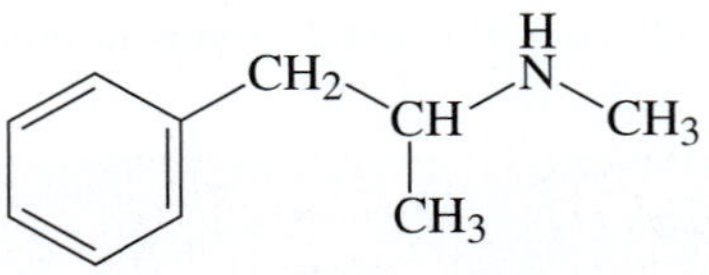

In August of 2005, an issue of *Newsweek* magazine devoted a cover story to methamphetamine and its abuse.

Practice Problem Ⓐ**TTEMPT** Identify the functional groups in each of the following molecules:

(a) Ethyl butyrate   (b) Aspirin   (c) Tartaric acid

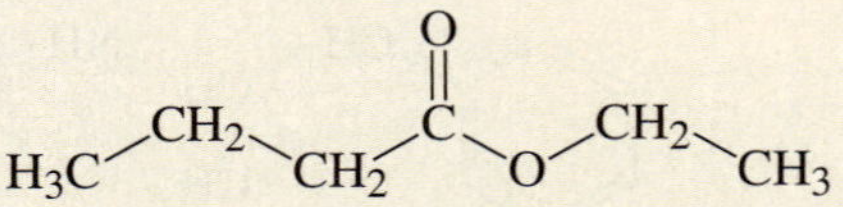

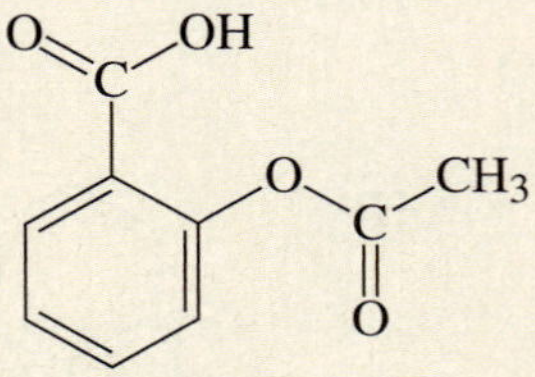

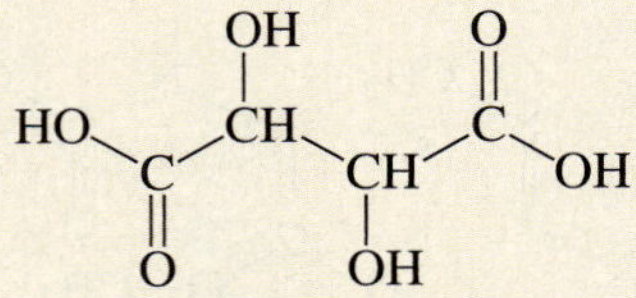

Practice Problem Ⓑ**UILD** Identify the functional groups in each of the following cyclic compounds:

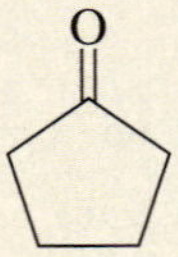

(a) 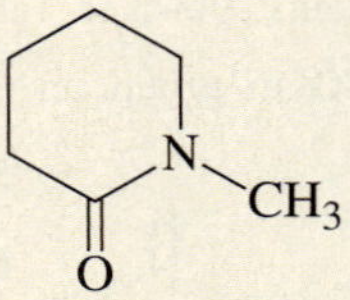   (b) 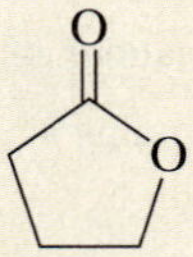   (c)

Practice Problem Ⓒ**ONCEPTUALIZE** Which of the molecules in Practice Problem 23.2B can form hydrogen bonds with itself?

CHECKPOINT – SECTION 23.2   Organic Compounds

23.2.1 Identify the name of the following compound:

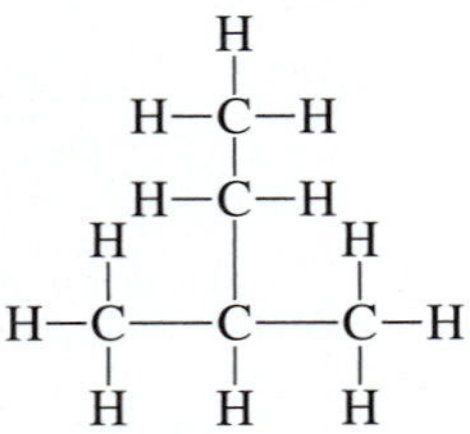

a) 2-Ethylpropane

b) 2-Ethylbutane

c) 2-Methylbutane

d) 2-Methylpentane

e) 3-Methylbutane

23.2.2 Identify the name of the following compound:

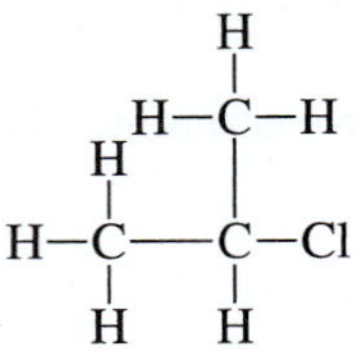

a) 1-Chloroethane

b) 2-Chloropropane

c) 2-Chloromethane

d) 1-Chloropropane

e) 3-Chloroethane

23.2.3 Identify the name of the following compound:

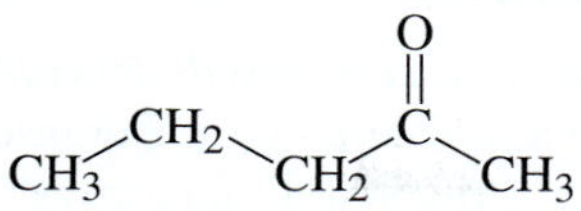

a) 4-Pentanone

b) 2-Pentanone

c) Ethylpentanal

d) Propylpentanal

e) Propylethanal

23.2.4 Identify the name of the following compound:

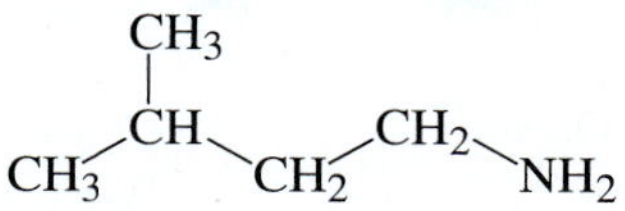

a) 3-Methyl-1-ethanamine

b) 2-Methyl-1-ethanamine

c) 1-Pentanamine

d) 2-Methyl-1-butanamine

e) 3-Methyl-1-butanamine

23.2.5 Identify the functional group(s) in the following molecule:

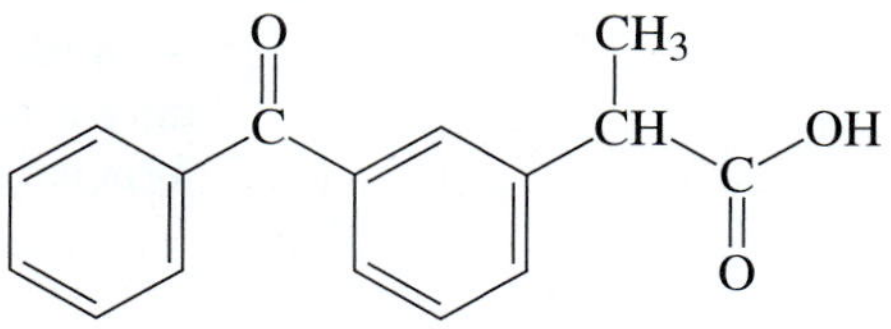

(Select all that apply.)

a) Hydroxy   c) Carbonyl   e) Amide

b) Carboxy   d) Amino

23.2.6 Identify the functional group(s) in the following molecule:

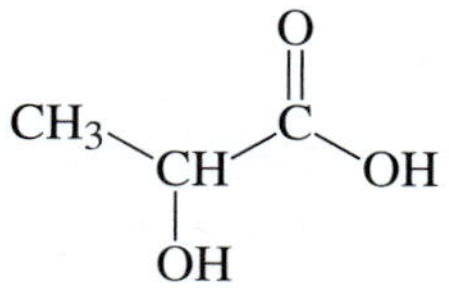

(Select all that apply.)

a) Hydroxy   c) Carbonyl   e) Amide

b) Carboxy   d) Amino

23.3 Representing Organic Molecules

You've learned previously how to represent molecules using *molecular* and *structural* formulas [◄◄ Section 2.7], as well as using Lewis structures [◄◄ Section 8.3]. In this section, we describe several additional ways to represent molecules—ways that are particularly useful in the study of organic chemistry.

The representation of organic molecules is especially important because the atoms in an organic molecule, unlike those in inorganic compounds, may be arranged in an enormous variety of different ways. For example, there are literally dozens of different ways that a compound containing five carbon atoms, one oxygen atom, and the necessary number of hydrogen atoms can be arranged, with each arrangement representing a unique organic compound. Here are 10 possibilities:

Student Note: The "necessary" number of H atoms is the number necessary to *complete the octet* of each C and O atom [◄◄ Section 8.3].

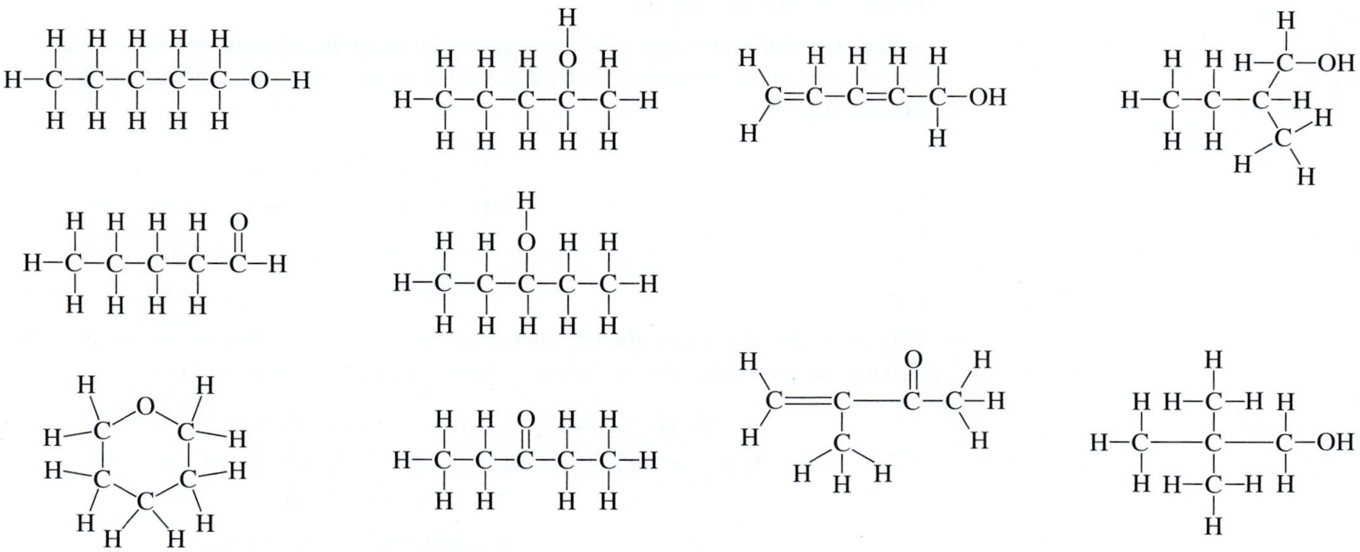

Condensed Structural Formulas

A **condensed structural formula,** or simply a **condensed structure,** shows the same information as a *structural formula,* but in a *condensed* form. For instance, the molecular formula, structural formula, and condensed structural formula of octane are as follows:

C_8H_{18} $CH_3CH_2CH_2CH_2CH_2CH_2CH_2CH_3$ $CH_3(CH_2)_6CH_3$

Molecular formula   Structural formula   Condensed structural formula

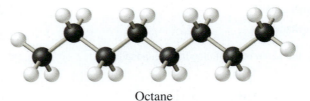

Octane

In the condensed structural formula, the identical adjacent groups of atoms (in this case $-CH_2-$ groups) are enclosed in parentheses and subscripted to denote their number.

In molecules where the carbon atoms do not form a single, unbranched chain, branches are indicated using additional parentheses in the condensed structural formula. For example, 2-methylheptane has the same molecular formula (C_8H_{18}) as octane. In effect, the chain is one C atom *shorter* than in octane and one of the hydrogen atoms on the second C atom has been replaced by a methyl group. The molecular formula, structural formula, and condensed structural formula of 2-methylheptane are as follows:

$\overset{\displaystyle CH_3}{\underset{\displaystyle |}{}}$

C_8H_{18} $CH_3CHCH_2CH_2CH_2CH_2CH_3$ $CH_3CH(CH_3)(CH_2)_4CH_3$

Molecular formula   Structural formula   Condensed structural formula

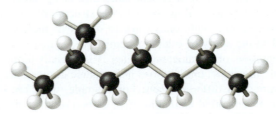

2-Methylheptane

Kekulé Structures

Kekulé structures are similar to Lewis structures except that they do not show lone pairs. For molecules that contain no lone pairs, such as octane, the Lewis and Kekulé structures are identical:

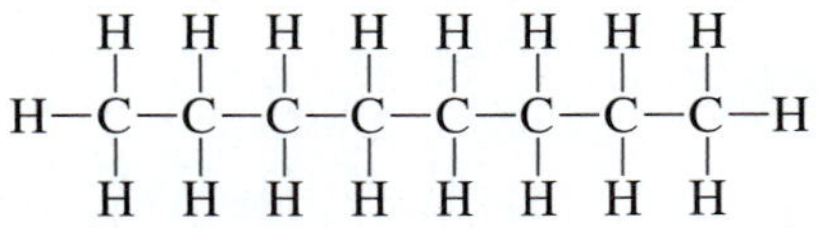

```
    H  H  H  H  H  H  H  H
    |  |  |  |  |  |  |  |
H―C―C―C―C―C―C―C―C―H
    |  |  |  |  |  |  |  |
    H  H  H  H  H  H  H  H
```
Octane

(Many of the structures already shown in this chapter are Kekulé structures.) The Kekulé structures for several familiar organic molecules are as follows:

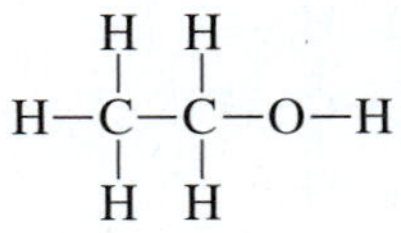

```
    H  H                 O              H  O
    |  |                 ||             |  ||
H―C―C―O―H        H―C―H        H―C―C―O―H
    |  |                                |
    H  H                                H
```
Ethanol 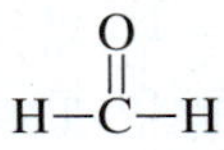   Formaldehyde 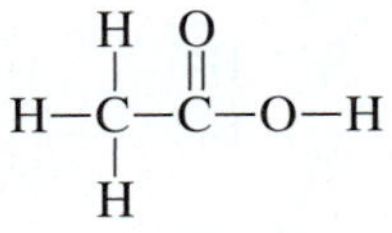   Acetic acid

Bond-Line Structures

Bond-line structures are especially useful for representing complex organic molecules. A **bond-line structure** consists of straight lines that represent carbon-carbon bonds. The carbon atoms themselves (and the attached hydrogen atoms) are not shown, but you need to know that they are there. The structural formulas and bond-line structures for several hydrocarbons [◄◄ Section 2.7] are as follows:

Name	Structural Formula	Bond-Line Structure
Pentane	$CH_3CH_2CH_2CH_2CH_3$	
Isopentane	$\overset{\displaystyle CH_3}{\underset{\displaystyle \vert}{CH_3CHCH_2CH_3}}$	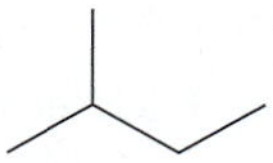
Neopentane	$CH_3\overset{\displaystyle CH_3}{\underset{\displaystyle \underset{\textstyle CH_3}{\vert}}{\overset{\vert}{C}}}CH_3$	

The end of each straight line in a bond-line structure corresponds to a carbon atom (unless another atom of a different type is explicitly shown at the end of the line). Additionally, there are as many hydrogen atoms attached to each carbon atom as are necessary to give each carbon atom a total of four bonds.

When a molecule contains an element *other* than carbon or hydrogen, those atoms, called **heteroatoms,** are shown explicitly in the bond-line structure. Furthermore, while the hydrogens attached to carbon atoms typically are not shown, hydrogens attached to heteroatoms *are* shown, as illustrated by the molecule ethylamine in the following:

Name	Structural Formula	Bond-Line Structure
Propanone (acetone)	$\overset{\displaystyle O}{\overset{\displaystyle \vert\vert}{CH_3CCH_3}}$	
Ethylamine	$CH_3CH_2NH_2$	
Tetrahydrofuran	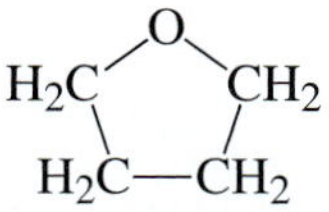 $\overset{\displaystyle H_2C\overset{O}{}CH_2}{H_2C-CH_2}$	

Although carbon and hydrogen atoms need not be shown in a bond-line structure, some of the C and H atoms *can* be shown for the purpose of emphasizing a particular part of a molecule. Often when we show a molecule, we choose to emphasize the *functional group(s),* which are largely responsible for the properties and reactivity of the compound [◄◄ Section 2.7].

Student Note: Study these structural formulas and bond-line structures, and those shown earlier, and make sure you understand how to interpret the bond-line structures.

Sample Problem 23.3 shows how to interpret bond-line structures.

SAMPLE PROBLEM (23.3)

Write a molecular formula and a structural formula (or *condensed* structural formula) for the following:

(a)    (b) 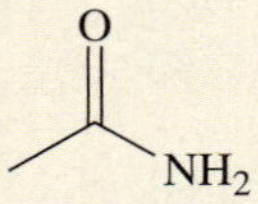   O ∥ NH₂

Strategy Count the C atoms represented and the heteroatoms shown. Determine how many H atoms are present using the octet rule.

Setup Each line represents a bond. (Double lines represent double bonds.) Count one C atom at the end of each line unless another atom is shown there. Count the number of H atoms necessary to complete the octet of each C atom:

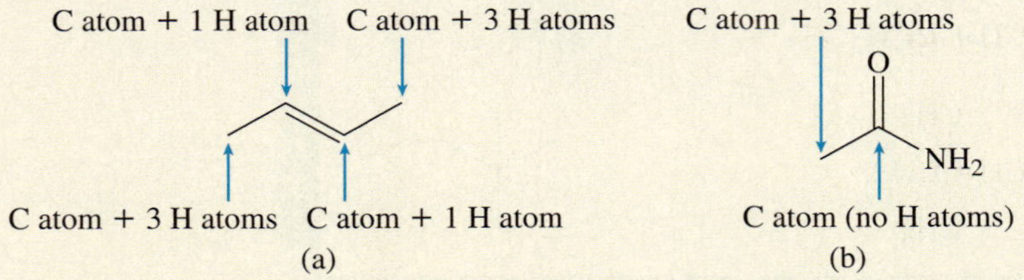

C atom + 1 H atom   C atom + 3 H atoms   C atom + 3 H atoms

O

NH₂

C atom + 3 H atoms   C atom + 1 H atom   C atom (no H atoms)
              (a)   (b)

Solution (a) Molecular formula: C_4H_8; structural formula: $CH_3(CH)_2CH_3$.

(b) Molecular formula: C_2H_5NO; structural formula: CH_3CONH_2.

THINK ABOUT IT

Make sure that each C atom is surrounded by four electron pairs: four single bonds, two single bonds and a double bond, or two double bonds. Remember that the single bonds to H typically are not shown in a bond-line structure—you have to remember that they are there and account for the H atoms when you deduce the formula.

Practice Problem **A**TTEMPT Write a structural formula for the compound represented by the following bond-line structure:

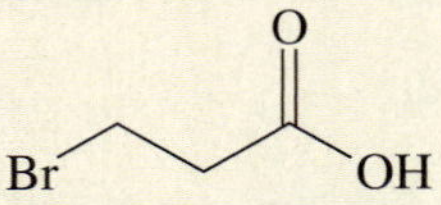

O ∥
Br   OH

Practice Problem **B**UILD Draw the bond-line structure for $(CH_3)_2C=CHNH_2$.

Practice Problem **C**ONCEPTUALIZE How many hydrogen atoms are there in the molecule represented here?

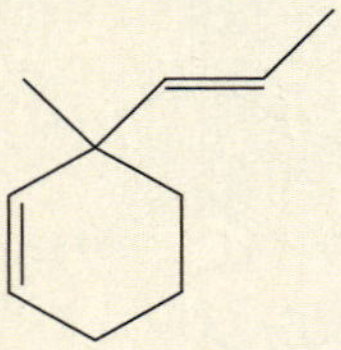

Resonance

Recall from Chapter 8 that many molecules and ions can be represented by more than one Lewis structure [◄◄ Section 8.7]. Furthermore, two or more equally valid Lewis structures that differ only in the position of their electrons are called *resonance structures*. For example, SO_3 can be represented by three different Lewis structures [◄◄ Sample Problem 8.10].

None of these Lewis structures represents the SO_3 molecule accurately. The bonds in SO_3 are actually *equivalent*—equal in length and strength, which we would not expect if two were single bonds and one were a double bond. Each individual resonance structure implies that electron pairs are *localized* in specific bonds or on specific atoms. The concept of resonance allows us to envision certain electron pairs as *delocalized* over several atoms [◄◄ Section 9.7]. Delocalization of electron pairs imparts additional stability to a molecule (or polyatomic ion), and a species that can be represented by two or more resonance structures is said to be *resonance stabilized.*

Chemists sometimes use curved arrows to specify the differences in positions of electrons in resonance structures. In SO_3, for example, the *"repositioning" of electrons* in the resonance structures can be indicated as follows:

Student Note: Remember that the real structure of the molecule is neither the first structure nor the second, but rather something in between that cannot be represented by a single structure [◄◄ Section 8.7].

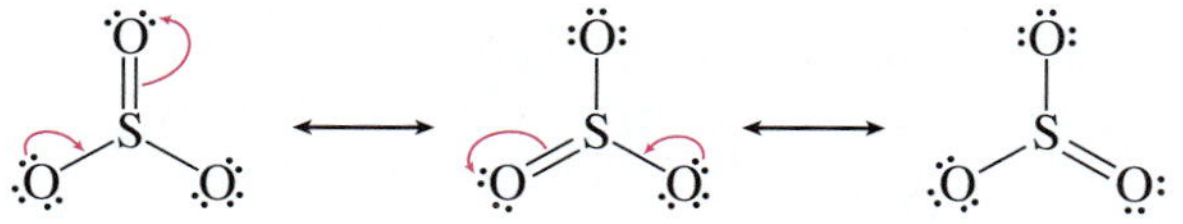

Resonance stabilization is observed in many organic species as well and affects chemical properties such as in the acidic behavior of ethanol (CH_3CH_2OH) and ethanoic acid (CH_3COOH), also commonly known as *acetic* acid. Each of these molecules has one *ionizable hydrogen atom* [◄◄ Section 2.7], enabling it to behave as a Brønsted acid [◄◄ Section 4.3]. However, the concentration of hydronium ions in a solution of ethanoic acid is hundreds of thousands of times higher than that in a comparable solution of ethanol. The reason for this large discrepancy is that the hydrogen atom on ethanoic acid is far more *easily* ionized than the one on ethanol. *Resonance* helps us explain why.

When a species loses its ionizable hydrogen atom, what remains is an anion. In the case of ethanoic acid and ethanol, the anions are:

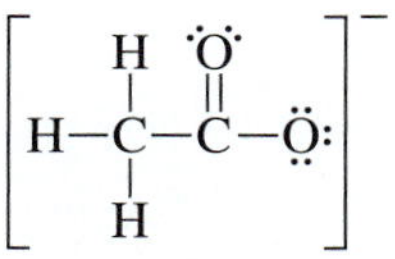

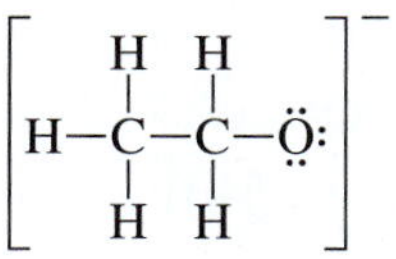

We can draw a second resonance structure for the anion produced by the ionization of ethanoic acid by repositioning the electron pairs as follows:

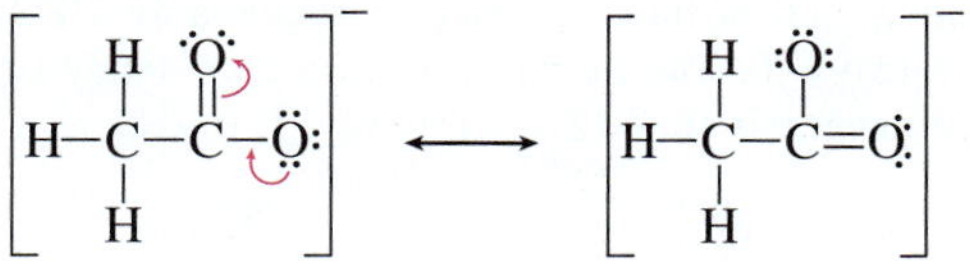

The negative charge on the anion resides on an oxygen atom. Which oxygen atom bears the charge depends on which resonance structure we look at. In essence, the greater the number of possible locations for the negative charge, the more stable the *anion*. And the more stable the anion, the more easily the ionizable hydrogen atom is lost, resulting in the production of more hydronium ions in solution.

Student Note: We determine where a charge resides in a polyatomic anion by calculating the *formal charge* on each atom [◄◄ Section 8.6].

It is not possible to draw additional resonance structures for the anion produced by the ionization of ethanol because there is nowhere else to put the lone pairs that reside on the O atom. They cannot be moved in between the O and C atoms because C can only have four electron pairs around it. (We were able to do this with CH_3COO^- because we could also move one of the pairs already around C to the other O atom. See the curved arrows in the preceding resonance structure.)

Sample Problem 23.4 illustrates the use of curved arrow notation and determination of resonance structures.

SAMPLE PROBLEM 23.4

Adenosine triphosphate (ATP) is sometimes called the "universal energy carrier" or "molecular energy currency." It contains two high-energy bonds (shown in red) that, when *hydrolyzed* (broken by the addition of water), release the energy necessary for cell function. Resonance stabilization of the hydrogen phosphate ion is one of the reasons the breakdown of ATP releases energy:

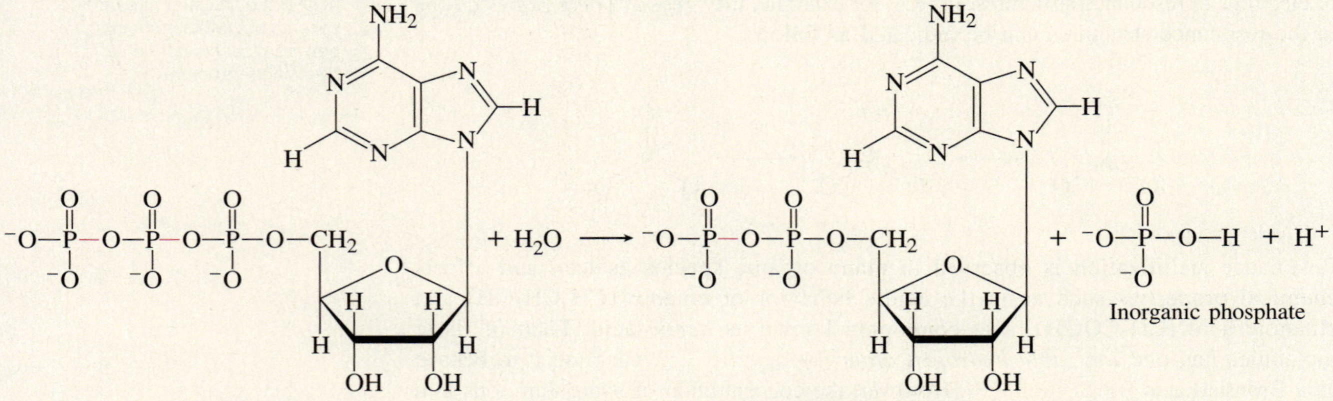

Adenosine triphosphate   Adenosine diphosphate

Draw all the possible resonance structures for the hydrogen phosphate ion (HPO_4^{2-}). Use curved arrows to indicate how electrons are repositioned, and determine the position(s) of the negative charges.

Strategy Draw a valid Lewis structure for HPO_4^{2-}, and determine whether and where electrons can be repositioned to produce one or more additional structures. Indicate the movement of electrons with curved arrows, and draw all possible resonance structures. Calculate the formal charge on each atom to determine the placement of charges.

Setup A valid Lewis structure for the hydrogen phosphate ion is:

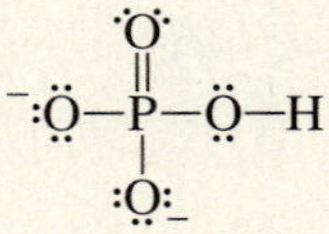

For the purpose of determining formal charges, P and O have five and six valence electrons, respectively.

Solution A lone pair can be moved from one of the oxygen atoms to create a double bond to the phosphorus, and a pair of electrons from the *original* double bond can be moved onto *that* oxygen atom. The net result is simply a repositioning of the double bond by moving two electron pairs. This can be done once more, giving a total of three resonance structures for HPO_4^{2-}. In each of the resonance structures, the formal charge on phosphorus is $[5 - (5)] = 0$. The formal charge on each singly bonded oxygen is $[6 - (1 + 6)] = -1$, and the formal charge on the doubly bonded oxygen is $[6 - (2 + 4)] = 0$:

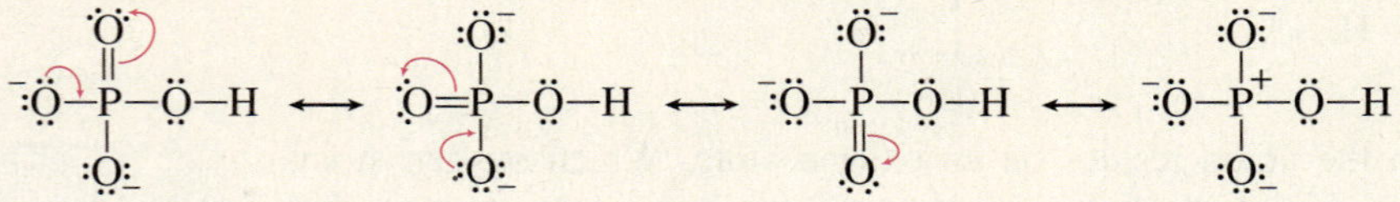

THINK ABOUT IT

ATP can also be hydrolyzed to give AMP (adenosine *mono*phosphate) and *pyrophosphate* ($P_2O_7^{4-}$). Pyrophosphate hydrolyzes, in turn, to give two hydrogen phosphate ions. The oxygen atoms that can help delocalize the negative charges are highlighted:

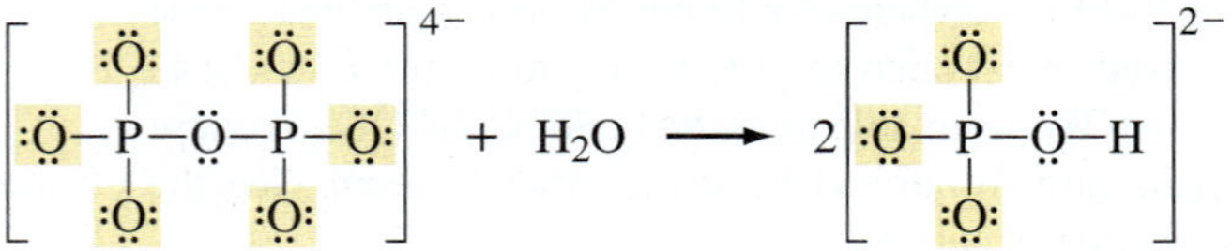

These structures can also be drawn with one double bond to each phosphorus atom, to minimize formal charges [◄◄ Section 8.8].

Practice Problem **A**TTEMPT Follow the curved arrows to draw a second resonance structure for the HCOO⁻ ion.

$$H-C-\overset{\displaystyle ::O::}{\underset{}{}}\overset{}{\ddot{O}}:^-$$

Practice Problem **B**UILD Given the following two resonance structures, draw the curved arrows on the first structure that will give rise to the second structure.

$$\left[:\ddot{O}-\overset{\displaystyle :\ddot{O}:}{\underset{:\ddot{O}:}{S}}-\ddot{O}: \right]^{2-} \longleftrightarrow \left[:\ddot{O}-\overset{\displaystyle :O:}{\underset{.\ddot{O}.}{S}}-\ddot{O}: \right]^{2-}$$

Practice Problem **C**ONCEPTUALIZE In which of the following examples do the curved arrows *not* correspond correctly to the repositioning of electrons needed to arrive at the resonance structure shown?

$$\left[:\ddot{O}-\ddot{N}=\ddot{O}: \right]^- \longleftrightarrow \left[:O=\ddot{N}-\ddot{O}: \right]^-$$
(i)

$$\left[:\ddot{O}-\ddot{N}=\ddot{O}: \right]^- \longleftrightarrow \left[:O=\ddot{N}-\ddot{O}: \right]^-$$
(ii)

$$\left[:\ddot{O}-\ddot{N}=\ddot{O}: \right]^- \longleftrightarrow \left[:O=\ddot{N}-\ddot{O}: \right]^-$$
(iii)

CHECKPOINT – SECTION 23.3   Representing Organic Molecules

23.3.1 Give the molecular formula of the compound represented by

a) $C_5H_{12}O$ c) $C_6H_{12}O$ e) $C_6H_{13}O$

b) C_6H_8O d) C_5H_6O

23.3.2 Give the molecular formula of the compound represented by

a) $C_5H_{12}O$ c) $C_6H_{12}O$ e) $C_4H_{13}O$

b) $C_6H_{13}O$ d) $C_6H_{14}O$

23.3.3 Which of the following pairs of species are resonance structures? (Select all that apply.)

a) $H-C=C-\overset{H}{\underset{H}{C}}-H \longleftrightarrow H-\overset{H}{\underset{H}{C}}-\overset{\ddot{O}}{C}-\overset{H}{\underset{H}{C}}-H$

b) $H-\overset{H}{\underset{H}{C}}-\overset{\ddot{O}}{C}-\ddot{O}:^- \longleftrightarrow H-\overset{H}{C}=\overset{:O:}{C}-\ddot{O}:^-$

c) $H-\overset{H}{\underset{H}{C}}-\overset{\ddot{O}}{C}-\ddot{O}:^- \longleftrightarrow H-\overset{H}{\underset{H}{C}}-\overset{\ddot{O}}{C}=\ddot{O}:^-$

d) $H-\overset{H}{\underset{H}{C}}-\overset{H}{\underset{:\ddot{O}:}{C}}-\overset{H}{\underset{H}{C}}-H \longleftrightarrow H-\overset{H}{\underset{H}{C}}-\overset{H}{\underset{H}{C}}-\overset{H}{\underset{H}{C}}-\ddot{O}-H$

e) $:\ddot{O}-\overset{\ddot{O}}{C}-\overset{\ddot{O}}{C}-\ddot{O}:^- \longleftrightarrow :O=\overset{:\ddot{O}:^-}{C}-\overset{:\ddot{O}:^-}{C}=O:$

23.3.4 Which of the following structural formulas represents a species that has two or more resonance structures? (Select all that apply.)

a) HCOOH   c) CH_2CH_2 e) CO_2

b) HCOO⁻   d) O_3

23.4 Isomerism

We first encountered the term *isomer* in the context of molecular geometry and molecular polarity [◄◄ Section 9.2]. Isomers are different compounds that have the same chemical formula. In this section, we will discuss the different types of isomerism (namely, constitutional isomerism and stereoisomerism) and their importance to organic chemistry.

Constitutional Isomerism

Also known as *structural* isomerism, **constitutional isomerism** occurs when the same atoms can be connected in two or more different ways. For example, there are three different ways to arrange the atoms in a compound with the chemical formula C_5H_{12}. Constitutional isomers have distinct names and generally have different physical and chemical properties. Table 23.3 lists the three constitutional isomers of C_5H_{12} along with their boiling points for comparison.

Animation
Organic and Biochemistry—Structural
isomers of hexane.

Stereoisomerism

Stereoisomers are molecules that contain identical bonds but differ in the orientation of those bonds in space. Two types of stereoisomers exist: geometrical isomers and optical isomers. *Geometrical isomers* occur in compounds that have restricted rotation about a bond. For instance, compounds that contain carbon-carbon double bonds can form geometrical isomers. Individual geometrical isomers have the same names but are distinguished by a prefix such as *cis* or *trans*. Dichloroethylene, ethylene in which two of the H atoms (one on each C atom) have been replaced by Cl atoms, exists as two geometrical isomers. The isomer in which the Cl atoms both lie on the same side (above or below, in this example) of the double bond is called the *cis* isomer.

TABLE 23.3	Constitutional Isomers of C_5H_{12}			
Name	Structural Formula	Bond-Line	Ball-and-Stick Model	BP (°C)
Pentane (*n*-pentane)				36.1
Methylbutane (isopentane)				27.8
2,2-Dimethylpropane (neopentane)				9.5

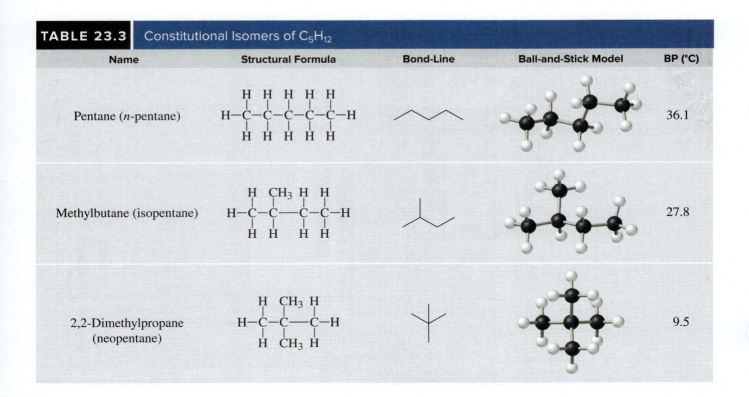

Cl   Cl   H   Cl   H   Cl

C=C   C=C   C=C

H   H   Cl   H   H   Cl

cis-Dichloroethylene   *trans*-Dichloroethylene   1,1-Dichloroethylene

Figure 23.2 Three isomers of dichloroethylene.

The isomer in which the Cl atoms lie on opposite sides of the double bond is called the **trans** isomer. (The compound in which both Cl atoms are attached to the same C atom is a *constitutional* isomer rather than a stereoisomer.) Figure 23.2 depicts the isomers of $C_2H_2Cl_2$.

Geometrical isomers usually have different physical and chemical properties. *Trans* isomers tend to be more stable and are generally easier to synthesize than their *cis* counterparts. The existence of *cis* isomers in living systems is a testament to how much better nature is at chemical synthesis than we are. As the information in the Bringing Chemistry to Life box at the end of this section demonstrates, geometrical isomerism is sometimes of tremendous biological significance.

Stereoisomers that are mirror images of each other, but are not superimposable, are called **optical isomers.** Consider the hypothetical organic molecule shown in Figure 23.3. It consists of an sp^3-hybridized carbon atom that is bonded to four different groups. Its mirror image appears identical to it, just as your right and left hands appear identical to each other. But, if you have ever tried to put a right-handed glove on your left hand, or vice versa, you know that your hands are not identical. Imagine rotating the molecule on the right so that its green and red spheres coincide with those of the molecule on the left. Doing so results in the yellow sphere of one molecule coinciding with the blue sphere on the other. These two molecules are mirror images of each other, but they are not identical.

Molecules with nonsuperimposable mirror images are called **chiral;** and a pair of such mirror-image molecules are called **enantiomers.** Most of the chemical properties of enantiomers and nearly all their physical properties are identical. Their chemical properties differ only in reactions that involve another chiral species, such as a chiral molecule or a receptor site that is shaped to fit only one enantiomer. Most biochemical processes consist of a series of chemically specific reactions that use chiral receptor sites to facilitate reaction by allowing only the specific reactants to fit (and thus react).

Animation
Organic and Biochemistry—Chiral molecules.

Student Note: The word *chiral* comes from the Greek word *cheir* for "hand." Chiral molecules may be right-handed or left-handed.

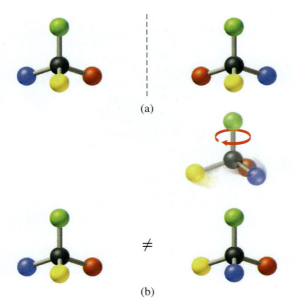

Figure 23.3 (a) Nonsuperimposable mirror images. (b) Despite being mirror images of each other, enantiomers are different compounds.

(a)

≠

(b)

In organic chemistry, it often is necessary to represent tetrahedral molecules (three-dimensional objects) on paper (a two-dimensional surface). By convention, this is done using solid lines to represent bonds that lie in the plane of the page, dashes to represent bonds that point behind the page, and wedges to represent bonds that point in front of the page:

Methane   1, 2-Dichloroethane

One property of chiral molecules is that the two enantiomers rotate the plane of plane-polarized light in opposite directions; that is, they are *optically active.* Unlike ordinary light, which oscillates in all directions, plane-polarized light oscillates only in a single plane. We use a polarimeter, shown schematically in Figure 23.4, to measure the rotation of polarized light by optical isomers. A beam of unpolarized light first passes through a Polaroid sheet, called the *polarizer,* and then through a sample tube containing a solution of an optically active, chiral species. As the polarized light passes through the sample tube, its plane of polarization is rotated either to the right or to the left. The amount of rotation can be measured by turning the analyzer in the appropriate direction until minimal light transmission is achieved.

Minimal transmission occurs when the plane of polarization of the light is perpendicular to that of the analyzer through which it is viewed. This effect can be demonstrated using two pairs of polarized sunglasses as shown in Figure 23.5. If the plane of polarization is rotated to the right, the isomer is said to be *dextrorotatory* and is labeled *d;* if it is rotated to the left, the isomer is called *levorotatory* and labeled *l.* *Enantiomers* always rotate the light by the same amount, but in opposite directions. Thus, in an equimolar mixture of both enantiomers, called a ***racemic mixture,*** the net rotation is zero.

Student Note: There are several conventions used to designate specific enantiomers. *Dextro-* and *levo-* prefixes refer to the direction of rotation of polarized light. *R* and *S,* the most commonly used designations, are assigned based on the "priority" assigned to each of the four groups attached to the chiral carbon—something that is beyond the scope of this text.

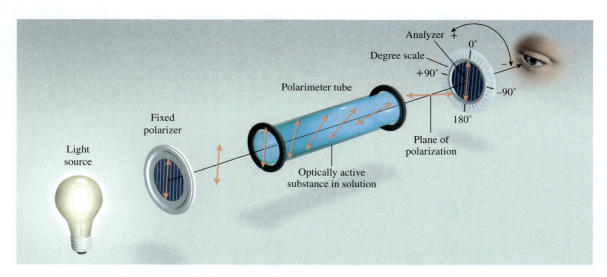

Figure 23.4 Schematic of a polarimeter.

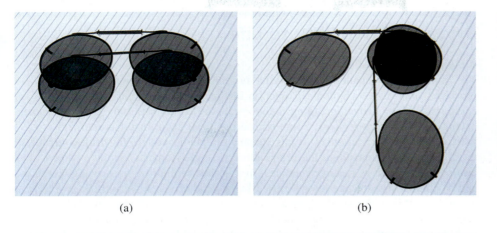

         (a)   (b)

Figure 23.5 Two pairs of polarized sunglasses. (a) When two polarized lenses overlap with their planes of polarization parallel to each other, light is transmitted. (b) When one pair is rotated so that its plane of polarization is perpendicular to the other, no light is transmitted through the overlapped lenses.

(both): *Charles D. Winters/Timeframe Photography/McGraw Hill*

Bringing Chemistry to Life

Plane-Polarized Light and 3-D Movies

We see in three dimensions because our eyes view the world from slightly different positions. Our brains synthesize a three-dimensional (3-D) picture based on the two different pictures sent to it by our eyes. Modern 3-D movies make use of this phenomenon to make it seem as though objects on the screen are actually moving toward the viewer.

Three-dimensional movies are filmed using two different cameras at slightly different angles to the action. Thus, there are actually *two* movies that must be shown to us simultaneously. To make sure that our two eyes receive two different perspectives, each movie is projected through a polarizer, which polarizes the two projections in directions perpendicular to each other (Figure 23.6).

If we were to watch the movie without the special glasses provided by the movie house, we would see the blurry combination of the two movies. However, the 3-D glasses consist of polarized lenses, with planes of polarization that are mutually perpendicular. The left lens, polarized in one direction, blocks the image that is polarized perpendicular to it. The right lens, polarized in the other direction, blocks the other image. Our eyes are "tricked" into seeing two different movies, which our brain combines to form one 3-D image. The results of this process can be quite impressive!

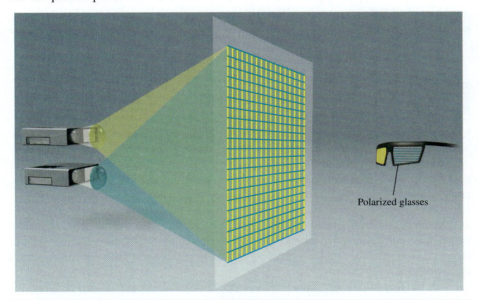

Polarized glasses

Figure 23.6 Two different versions of the same movie are projected onto the screen. Polarized lenses make it so that each eye sees only one version. The result is the perception of three-dimensional action.

Bringing Chemistry to Life

Biological Activity of Enantiomers

The importance of enantiomers in living systems cannot be overstated. Numerous processes important to biological function involve one enantiomer of a chiral compound. Many drugs, including thalidomide, are chiral with only one enantiomer having the desired *properties*. Some such drugs have been manufactured and marketed as racemic mixtures. It has become common, though, for drug companies to invest in *chiral switching,* the preparation of single-isomer versions of drugs originally marketed as racemic mixtures, in an effort to improve on existing therapies and to combat the revenue losses caused by generic drugs. A fairly high-profile example of this is the single-isomer drug Nexium, the so-called purple pill. AstraZeneca, makers of Nexium, held a patent for the drug Prilosec, originally a prescription heartburn medication that is now available over the counter. Prilosec is a racemic mixture of the chiral compound omeprazole. Prior to the 2002 expiration of its patent on Prilosec, AstraZeneca began producing and marketing Nexium, which contains only the therapeutically effective enantiomer (*S*)-omeprazole or *esomeprazole.*

Another example of chiral switching is that of the selective serotonin reuptake inhibitor (SSRI) antidepressant Celexa, which was introduced to the market in 1998 by Forest Laboratories. Celexa is a racemic mixture of (*R*)-citalopram oxalate and (*S*)-citalopram oxalate. While only the (*S*) enantiomer has therapeutic antidepressant properties, both enantiomers contribute to the side effects of the drug and therefore limit effectiveness and patient tolerance. In 2002, the FDA approved Lexapro, a new antidepressant derived from Celexa but from which the therapeutically ineffective (*R*) enantiomer has been removed. The benefits of isolating the active isomer include smaller required dosages, reduced side effects, and a faster and better patient response to the drug.

Although the intellectual property laws regarding single-isomer drug patents are somewhat ambiguous, chiral switching has enabled some pharmaceutical companies to extend the time that they are able to market their popular prescriptions exclusively. Strictly speaking, the FDA does not consider a single enantiomer of an already approved chiral drug to be a "new chemical entity," which is a requirement for obtaining a patent on a compound. Early in the history of chiral switching, however, there was some disagreement among patent examiners regarding what constituted a new chemical entity, and patents were granted on single-isomer drugs that might not be granted today.

23.5   Organic Reactions

Coulomb's law, which we first encountered in Section 7.4, measures the force of the attraction between opposite charges. The attraction between regions of opposite charge on neighboring species and the resulting movement of electrons are the basis for our understanding of many organic reactions. We begin by defining what *electrophiles* and *nucleophiles* are, two terms used frequently in organic chemistry.

An **electrophile** is a species with a positive or partial positive charge. Literally, an electrophile "loves electrons." Thus, an electrophile is attracted to a region of negative or partial negative charge. An electrophile may be a cation, such as H^+, or the positive portion of a polar molecule, such as the H atom in HCl. Electrophiles are *electron poor.*

A **nucleophile** is a species with a negative or partial negative charge. Literally, a nucleophile "loves a nucleus." A nucleophile is attracted to a region of positive or partial positive charge (i.e., an electrophile). A nucleophile may be an anion, such as Cl⁻, or the negative portion of a polar molecule, such as the Cl atom in HCl. Nucleophiles are *electron rich*. Electron-rich sites and electron-poor sites are attracted to one another.

Addition Reactions

The electrostatic potential maps of HCl and C_2H_4 shown in Figure 23.7 demonstrate that both molecules have regions of partial positive charge and partial negative charge. For example, HCl is a polar molecule, with H bearing a partial positive *charge,* due to the large difference in electronegativity between hydrogen and chlorine. Moreover, the carbon-carbon double bond in C_2H_4 consists of two pairs of shared electrons, one pair in a sigma bond and one pair in a pi bond [I◄◄ Section 9.5], making the double bond a region of partial negative charge. The partial positive charge on the H in the HCl molecule is an electrophile. The double bond in ethylene, a region of relatively high electron density, is a nucleophile.

A reaction takes place when the positive end of the HCl molecule approaches the double bond in ethylene. The pi bond *breaks,* and the electrons it contained move as indicated by the curved *arrows* shown in the following equation, forming a sigma bond between the H atom of the HCl molecule and one of the C atoms. As this new bond forms, two things happen:

1. Because there cannot be more than one bond to the H atom, the original bond between H and Cl breaks. Both of the electrons originally shared by H and Cl go with the Cl atom. The resulting intermediate species are shown in square brackets in the following equation (dashed lines represent bonds that are being formed):

The C atom on the right bears a positive charge after the valence electron it originally shared (in the pi bond with the other C atom) is removed from it completely. A species such as this, in which one of the carbons is surrounded by only six electrons, is called a **carbocation.**

Although carbon must obey the octet in any stable compound, some reactions involve transient, *intermediate* species in which a carbon atom may be electron deficient—having only three electron pairs around it.

Student Note: Although it ionizes completely in aqueous solution [I◄◄ Section 4.1], HCl exists as *molecules* in the gas phase.

Student Note: When an electrophile approaches another species and accepts electrons from it to form a bond, this is called *electrophilic attack.*

Student Note: Curved arrows are used to illustrate the mechanism by which an organic reaction occurs. Unlike their use in resonance structures, curved arrows in a reaction mechanism correspond to the actual *movement* of electrons.

Figure 23.7 The partial positive charge on H in HCl is attracted to the region of electron density in ethylene's double bond.

2. The C atom forming the new sigma bond to the H atom changes from sp^2-hybridized to sp^3-hybridized:

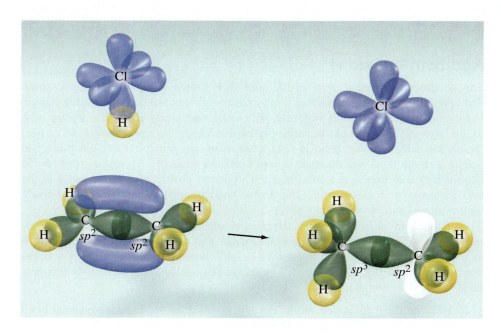

The carbon atom bearing the positive charge is still sp^2-hybridized at this point.

The chloride ion produced when both of the electrons originally shared by H and Cl go with Cl is a *nucleophile*. It is attracted to the newly formed positive charge and two of its electrons form a bond to the positively charged C atom as shown by the curved arrows.

The formation of a new sigma bond between the Cl and C atoms causes the hybridization of the second C to change from sp^2 to sp^3:

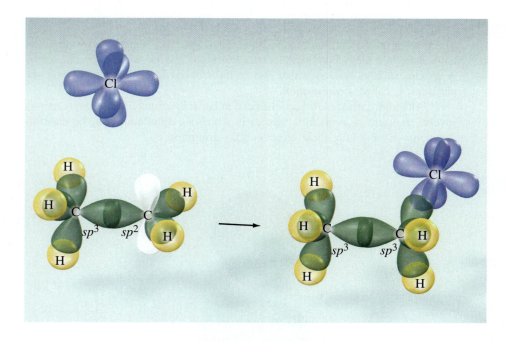

The overall reaction is called an *addition reaction*. Specifically, this is an **electrophilic addition** reaction because it begins with the electrophilic attack on HCl by the region of electron density in the double bond.

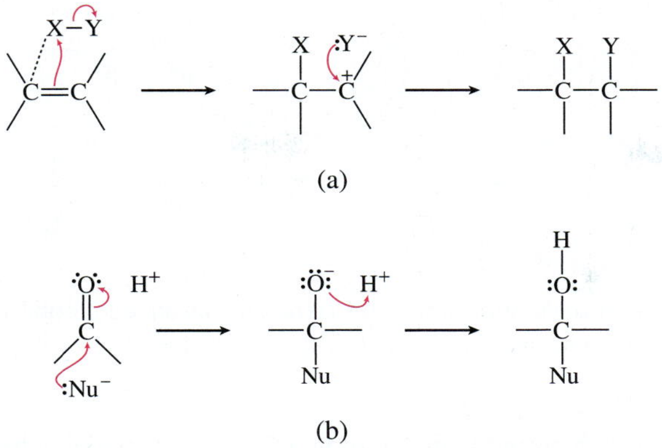

(a)

(b)

Figure 23.8 (a) Electrophilic addition reaction.
(b) Nucleophilic addition reaction. Curved arrows indicate
movement of electrons. (Nu represents a nucleophile.)

Addition reactions can also begin with nucleophilic attack, in which case the reaction is called a ***nucleophilic addition.*** In a nucleophilic addition reaction, a bond forms when a nucleophile donates a pair of electrons to an electron-deficient atom. Water, for example, reacts with carbon dioxide to produce carbonic acid. Although the addition reaction happens essentially all at once, it is helpful to think of the movement of electrons as the following stepwise process:

1.

One of the lone pairs on the O atom in water attacks the C atom, which is electron deficient because of the highly electronegative O atoms bonded to it. As the new bond forms, one of the original C—O pi bonds breaks. The electron pair from the broken pi bond is repositioned on the corresponding O atom, leaving the atom with a negative *charge*.

> **Student Note:** Nucleophilic attack in which only one electron is donated by the attacking species can also occur. These are called *radical* reactions.

2.

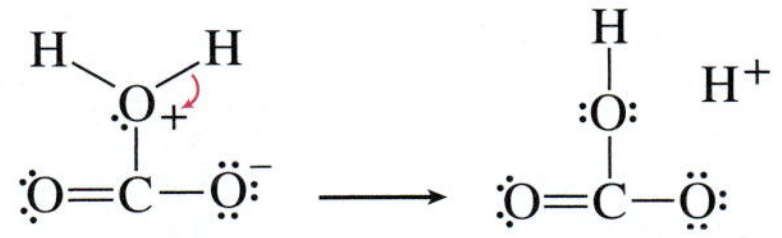

In essence, one of the O—H bonds in water breaks, with both electrons remaining with the O atom. The H atom separates from the water molecule as a proton.

3.

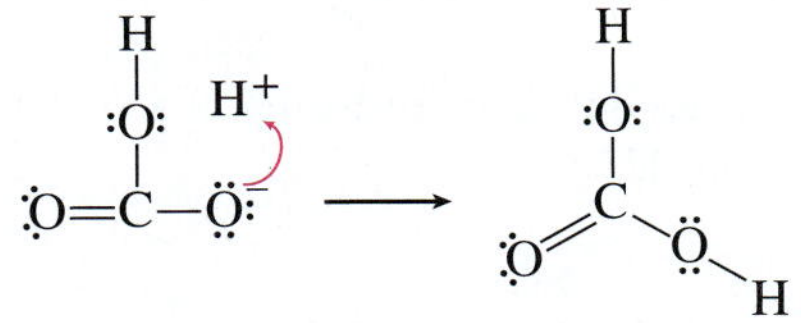

Finally, the negatively charged O atom acquires the proton lost by the water molecule that served as the nucleophile.

A summary of the mechanisms of electrophilic and nucleophilic addition reactions is given in Figure 23.8.

Substitution Reactions

Electrophilic and nucleophilic attack can also lead to ***substitution reactions*** in which one group is replaced by another group. *Electrophilic substitution* occurs when an electrophile attacks an aromatic molecule and replaces a hydrogen atom. Nucleophilic substitution occurs when a nucleophile replaces another group on a carbon atom. Figure 23.9 shows the general mechanisms for substitution reactions:

> **Student Note:** Specifically, this type of reaction is called an *electrophilic aromatic substitution* reaction.

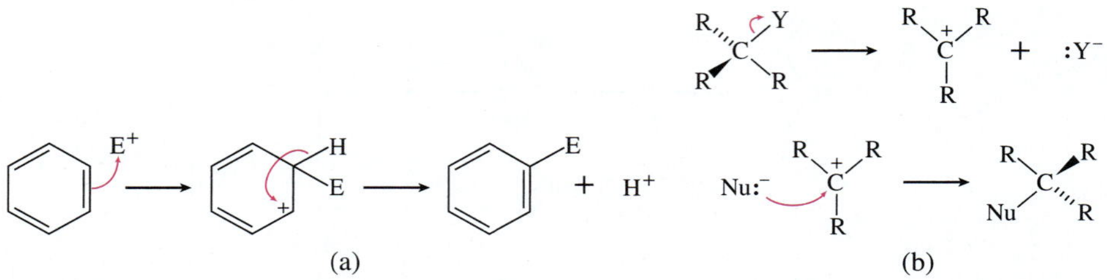

(a)   (b)

Figure 23.9 (a) Electrophilic substitution. Benzene attacks an electrophile (E^+). (b) Nucleophilic substitution reaction.

The nitration of benzene is an example of an electrophilic substitution reaction:

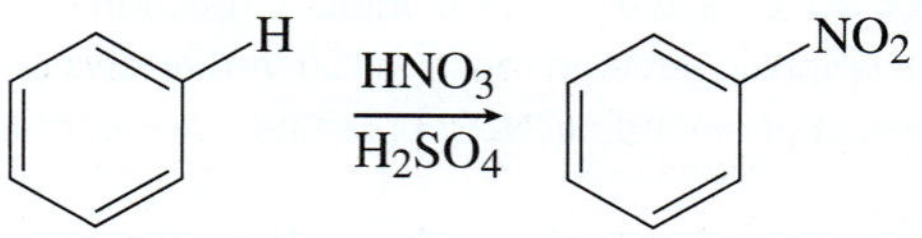

Nitric acid and sulfuric acid react to produce the nitronium ion ($^+NO_2$), which acts as the electrophile:

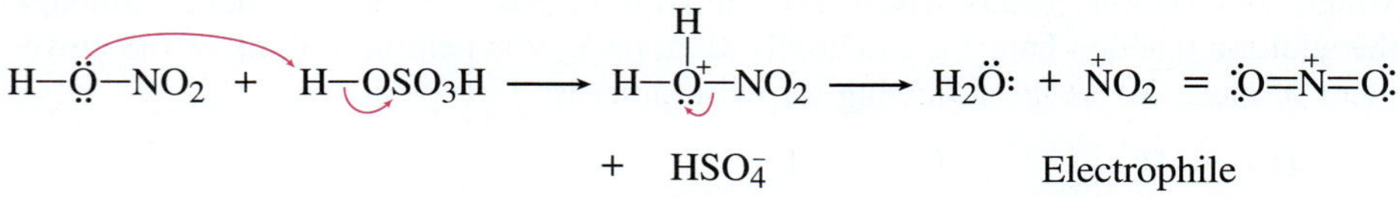

The positively charged nitronium ion is attracted to the electron-rich pi bonds of the benzene ring. A bond forms between one of the carbon atoms and the nitronium ion, breaking one of benzene's pi bonds:

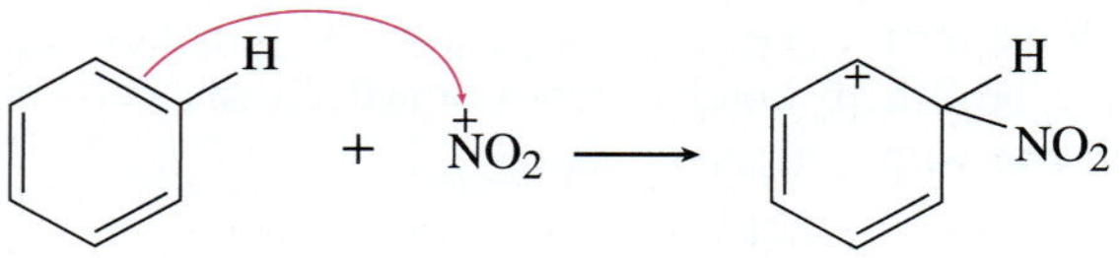

The resulting carbocation is stabilized by resonance:

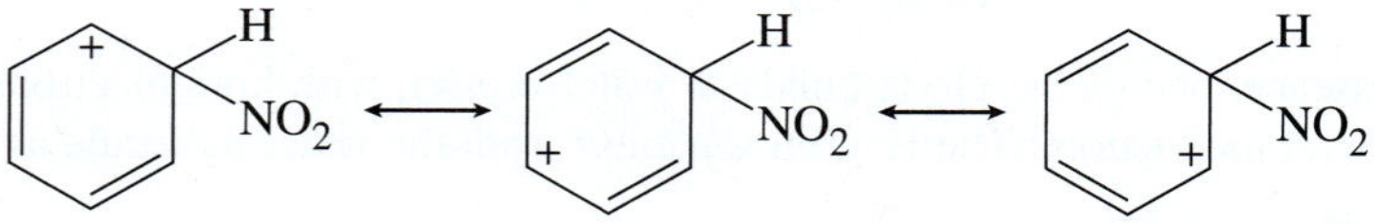

Finally, the electrons in the C—H bond move to the ring, restoring the original pi bond, and the hydrogen atom leaves as a proton, H^+:

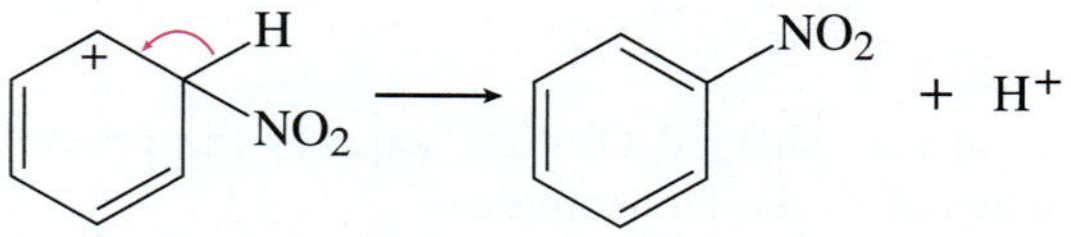

When necessary, one additional step can convert the $-NO_2$ group into the $-NH_2$ group:

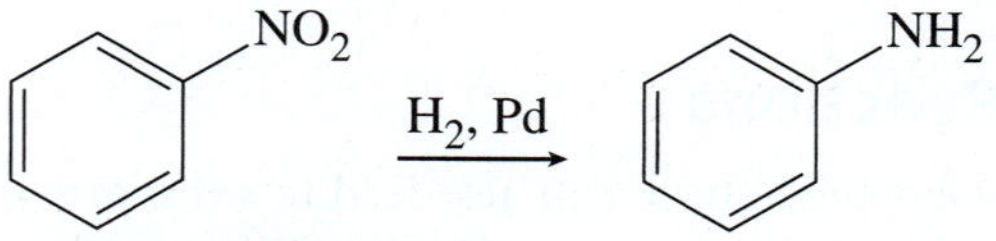

Student Note: The C atom in methyl bromide bears a partial positive charge because it is bonded to the somewhat more electronegative bromine atom [◀◀ Section 8.4].

A simple example of nucleophilic substitution is the reaction of an alkyl halide such as methyl bromide with a nucleophile such as the chloride ion. The chloride ion is attracted to the partial positive charge on the C atom in *methyl bromide*. A lone

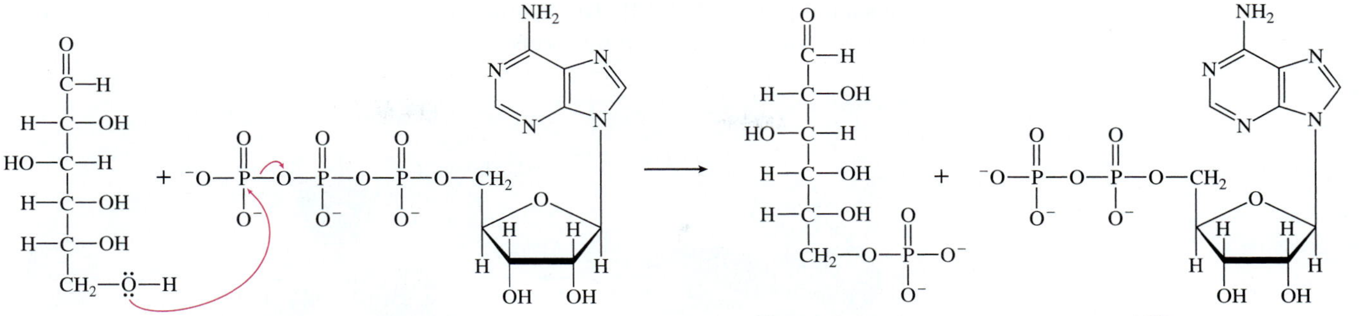

Figure 23.10 Reaction of glucose and ATP to produce glucose-6-phosphate and ADP.

pair on the chloride ion moves to form a sigma bond between the Cl atom and the C atom. And, because there can be no more than four electron pairs around the C atom, the original C—Br bond breaks, with both of the electrons originally shared by C and Br going to the Br atom. The result is a methyl chloride molecule and a bromide ion:

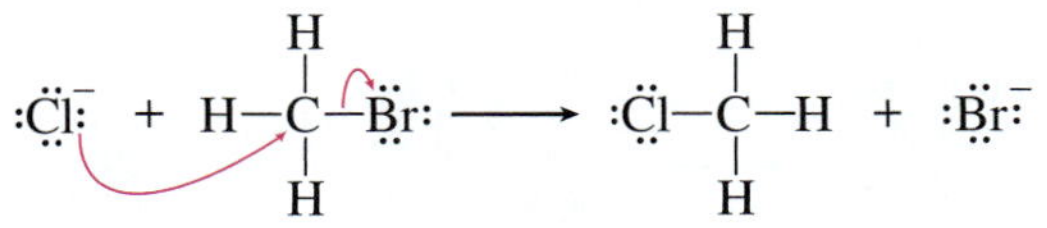

Nucleophilic substitutions are particularly important in living systems. The hydrolysis of ATP (adenosine triphosphate), which was described in Section 23.3, is an example of nucleophilic substitution. In hydrolysis, the oxygen atom in water acts as the nucleophile, attacking the electron-deficient phosphorus atom. A similar reaction happens between glucose and ATP, as shown in Figure 23.10. The P atom that is attacked is electron deficient because of the four highly electronegative O atoms bonded to it. As a bond forms between the attacking O and the P, one of the original P—O bonds breaks. The net result is the replacement of the original —H group on glucose with a —PO_4^{3-} group to give glucose-6-phosphate and ADP (adenosine *diphosphate*).

Student Note: Digestion of proteins also begins with a nucleophilic substitution reaction.

Bringing Chemistry to Life

S$_N$1 Reactions

Thalidomide is a chiral drug, but only one of its enantiomers has the desired therapeutic properties. The other enantiomer causes severe birth defects. Thalidomide was originally dispensed as a racemic mixture, giving patients equal amounts of both enantiomers. Unlike some chiral drugs, thalidomide cannot be administered as a single isomer to avoid the undesirable enantiomer. Within hours of administering one enantiomer of thalidomide, both enantiomers are found in roughly equal amounts in the blood. Although the mechanism by which one enantiomer of thalidomide is converted to the other is the subject of some debate, one way that enantiomers can be interconverted is via a nucleophilic substitution reaction.

Nucleophilic substitution reactions fall into two categories, called S$_N$1 reactions and S$_N$2 reactions. (The numbers 1 and 2 refer to a specific aspect of the *kinetics* of the reactions [◄◄ Chapter 14].) The nucleophilic substitution that converts one enantiomer to a mixture of both is an S$_N$1 reaction. An S$_N$1 reaction begins when one of the groups bonded to a carbon "leaves," leaving behind a carbocation [Figure 23.9(b)]. The hybridization of the carbon atom changes from sp^3 to sp^2 when the carbocation forms.

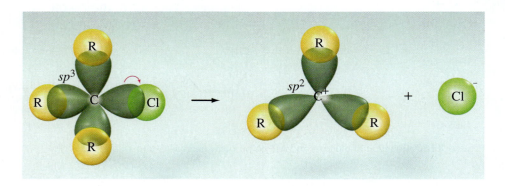

Student Note: sp^2-Hybridized orbitals form a trigonal *plane* [◄◄ Table 9.2].

Commonly encountered "leaving groups" include Cl^-, Br^-, and I^-. A carbocation is an unstable species and, being positively charged, is prone to nucleophilic attack. Because the carbocation is *planar* about the carbon that bears the positive charge, it is equally likely that a nucleophile will attack from either side:

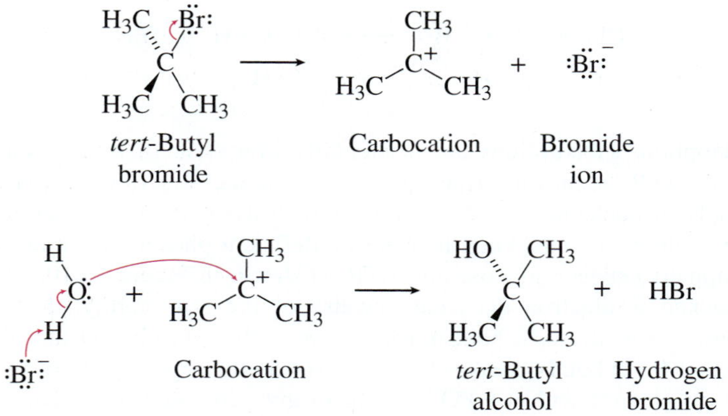

In the reaction of $(CH_3)_3CBr$ with H_2O, the product is the same regardless of whether the nucleophile (water) attacks from the front or from the back. When a molecule loses a leaving group from a chiral carbon, however, as would be the case with the thalidomide molecule, attack from one side of the carbocation will yield one enantiomer whereas attack from the other side will yield the other enantiomer. The result is a racemic mixture. The conversion of a single enantiomer to a racemic mixture of both enantiomers is called **racemization:**

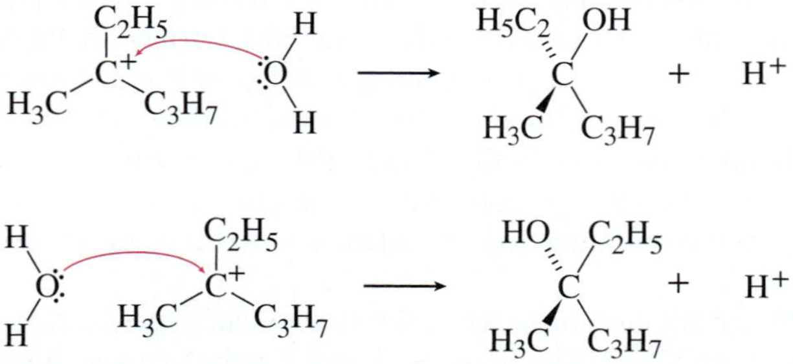

Nucleophilic attack on opposite sides of the carbocation leads
to two different products, which are mirror images of each other.

Sample Problem 23.5 shows how to draw mechanisms for addition and substitution reactions.

SAMPLE PROBLEM 23.5

Using curved arrows to indicate the movement of electrons, draw the mechanism for each of the following reactions: (a) nucleophilic addition of CN^- to CH_3CHO and (b) electrophilic substitution of benzene with $^+SO_3H$. (Draw all resonance structures for the carbocation intermediate.)

Strategy For nucleophilic addition, draw Lewis structures with the nucleophile close to the electron-poor atom where attack will occur (in this case, the carbonyl carbon). For electrophilic substitution, draw Lewis structures with the electrophile close to the site of attack on the benzene ring. Remember that nucleophiles are attracted to and react with electron-poor atoms whereas electrophiles are attracted to and react with electron-rich areas of the molecule. Using this information, and the octet rule, determine which electrons are likely to be involved in the reaction and indicate their repositioning with curved arrows.

Setup The nucleophile is CN^-. The site of attack is the carbonyl C in CH_3CHO, which is electron-poor because it is bonded to the more electronegative O atom.

The electrophile is $^+SO_3H$. The site of attack is the electron-rich, delocalized pi bonds of the benzene ring.

Solution

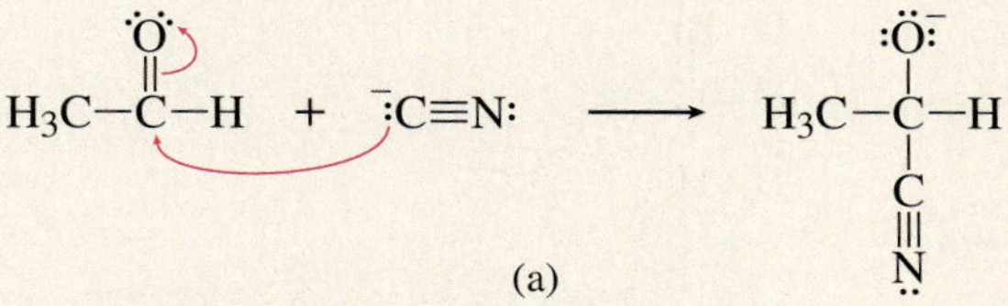

(a)

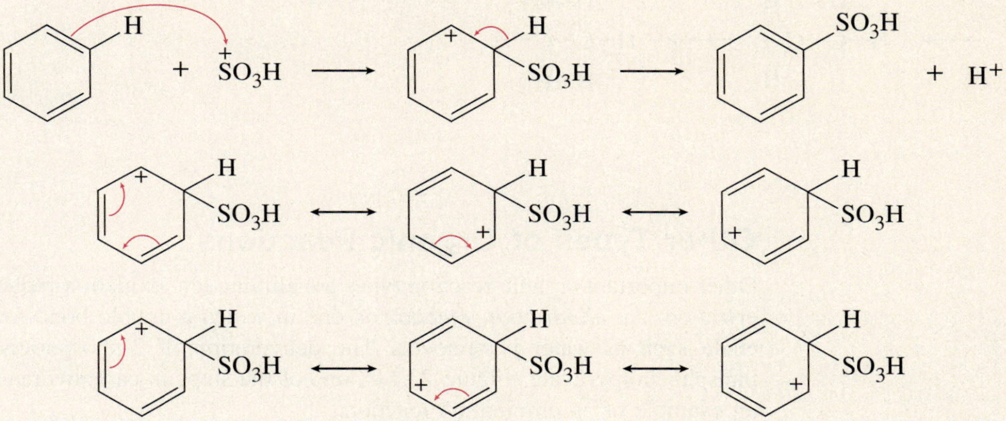

(b)

THINK ABOUT IT

The C atoms in benzene are all equivalent, so the choice of which C will bear the substituent in part (b) is arbitrary. All the following represent the same product:

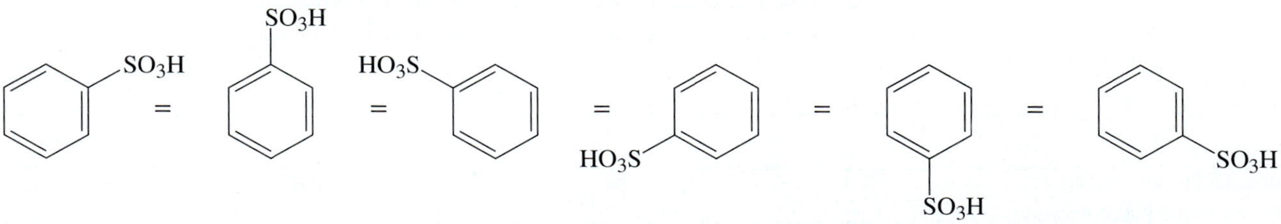

(Continued on next page)

Practice Problem **A**TTEMPT   Draw mechanisms for (a) nucleophilic addition of H^- to CH_3COCH_3 and (b) electrophilic substitution of benzene with Cl^+.

Practice Problem **B**UILD   Draw all the possible products that could result from the following electrophilic substitution reaction:

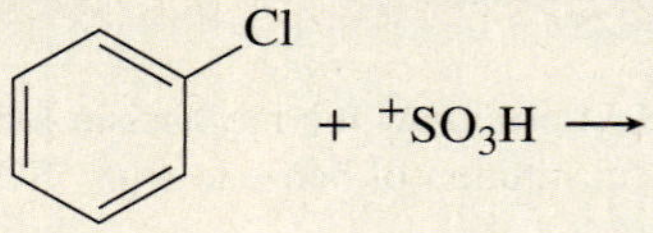

$$+ \ ^+SO_3H \longrightarrow$$

Practice Problem **C**ONCEPTUALIZE   In which of the following examples do the curved arrows correspond correctly to the proper movement of electrons in the mechanism of electrophilic addition?

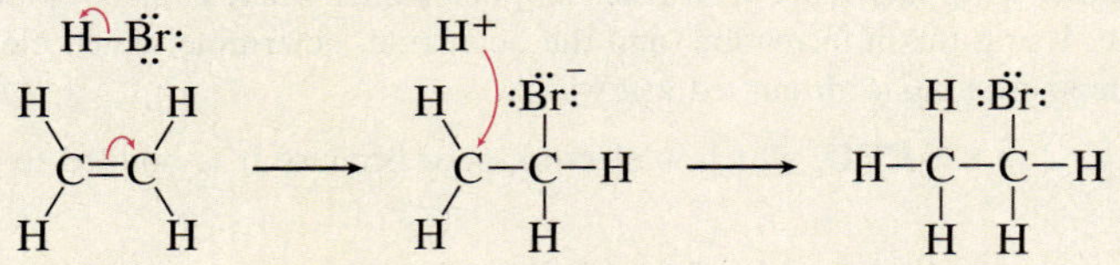

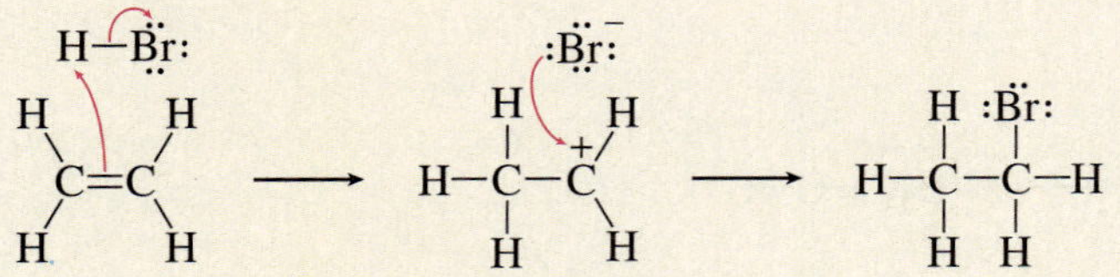

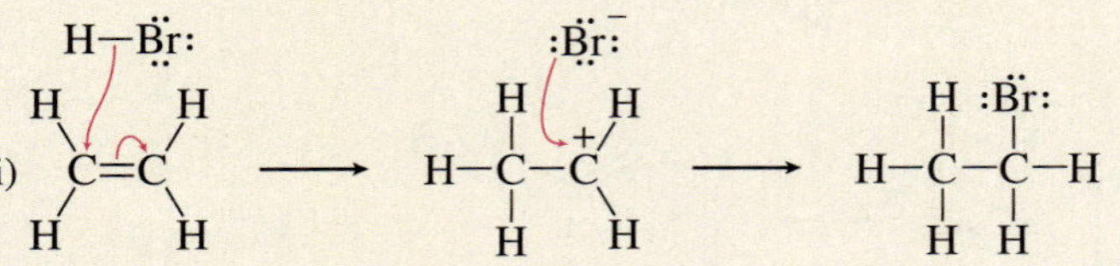

Other Types of Organic Reactions

Other important organic reaction types are elimination, oxidation-reduction, and isomerization. An **elimination reaction** is one in which a double bond forms and a molecule such as water is removed. The dehydration of 2-phosphoglycerate to form phosphoenolpyruvate (Figure 23.11), one of the steps in carbohydrate metabolism, is an example of an elimination reaction.

Oxidation-reduction reactions, as we learned previously [◄◄ Section 4.4], involve the *loss* and *gain* of electrons, respectively. Determining which species have lost or gained electrons may seem less straightforward with organic reactions than with the inorganic reactions we have encountered, but the following guidelines can help you decide when an organic molecule has been oxidized or reduced:

1. When a molecule gains O or loses H, it is *oxidized.*
2. When a molecule loses O or gains H, it is *reduced.*

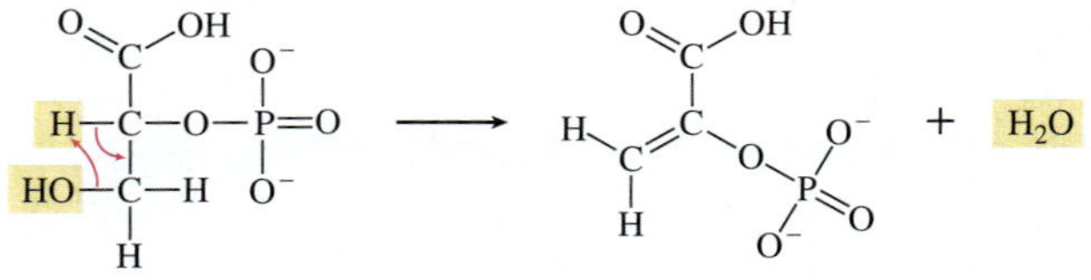

Figure 23.11 The highlighted atoms are those that constitute the eliminated water molecule.

An important biological example of oxidation is the enzyme-catalyzed reaction by which ethanol is converted to acetaldehyde in the liver:

$$CH_3CH_2OH \longrightarrow CH_3CHO$$

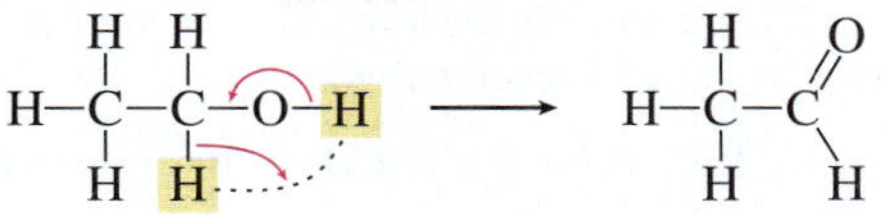

The ethanol molecule loses two H atoms in this reaction, so it is *oxidized*.

  Isomerization reactions are those in which one isomer is converted to another. The interconversion between the sugars aldose and ketose is an *example:*

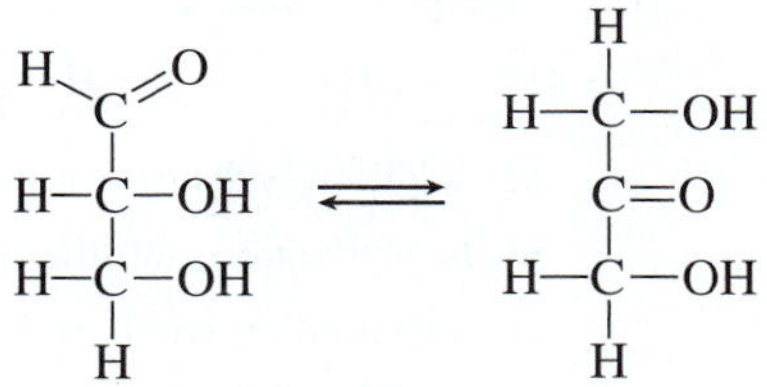

Aldose   Ketose

Student Note: Many of the organic reactions in living systems require *enzymes*, natural *catalysts*, in order to occur rapidly enough to be useful [◄◄ Chapter 14].

Bringing Chemistry to Life

The Chemistry of Vision
Vision, our ability to perceive light, is the result of an isomerization reaction. Our eyes contain millions of cells called rods that are packed with *rhodopsin,* an 11-*cis*-retinal molecule

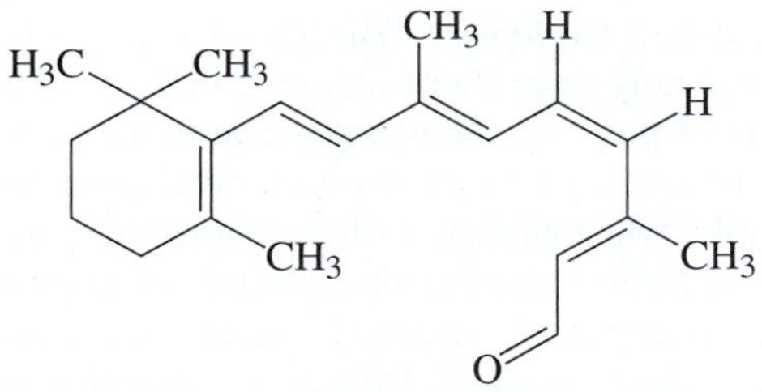

11-*cis*-retinal

bonded to a large protein. When visible light strikes rhodopsin, the retinal molecule isomerizes to the all-*trans* isomer:

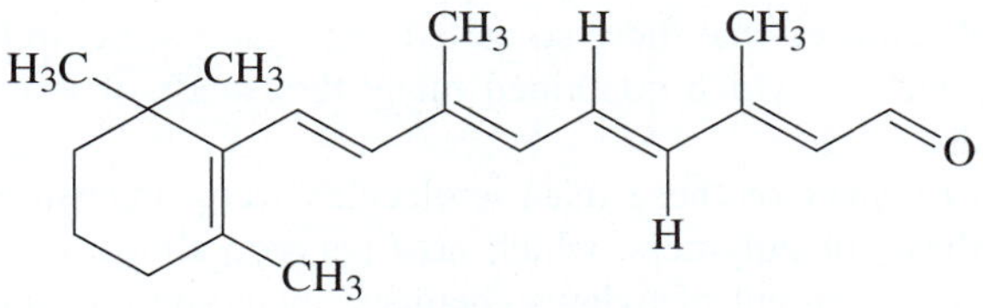

All-*trans*-retinal

  This isomerization causes such a significant change in the structure of retinal that it separates from the protein. These events trigger the electrical impulses that stimulate the optic nerve and result in the brain receiving the signals that we know as vision. The all-*trans*-retinal diffuses away from the protein and is converted back to 11-*cis*-retinal. The regenerated 11-*cis*-retinal can then rebind with the protein. Interestingly, the conversion of the all-*trans* isomer back to the 11-*cis* isomer, which is necessary for us to perceive light, is much slower than the light-induced conversion from *cis* to *trans*. This is why after looking at a bright light, you have a "blind spot" for a period of time.

CHECKPOINT – SECTION 23.5   Organic Reactions

23.5.1 Identify each species as a nucleophile or an electrophile.

$CH_3-\ddot{\underset{..}{O}}{:}^{-}$ $H-\ddot{\underset{..}{S}}{:}^{-}$ $\underset{\underset{CH_3}{|}}{\overset{\overset{CH_3}{|}}{H_3C-\overset{+}{C}}}$

(i)   (ii)   (iii)

a) Nucleophile, electrophile, electrophile

b) Nucleophile, nucleophile, nucleophile

c) Electrophile, nucleophile, electrophile

d) Nucleophile, nucleophile, electrophile

e) Electrophile, electrophile, electrophile

23.5.2 Identify each reaction as addition, substitution, elimination, or isomerization.

(i) $H_2C{=}CHCH_2CH_2CH_3 + Br_2 \longrightarrow BrCH_2-\underset{\underset{Br}{|}}{CHCH_2CH_2CH_3}$

(ii) $CH_3Br + OH^- \longrightarrow CH_3OH + Br^-$

(iii) $\underset{\overset{|}{H}}{\overset{\overset{|}{H_3C}}{}}C{=}C\underset{\overset{|}{CH_3}}{\overset{\overset{|}{H}}{}} \longrightarrow \underset{\overset{|}{H}}{\overset{\overset{|}{H_3C}}{}}C{=}C\underset{\overset{|}{H}}{\overset{\overset{|}{CH_3}}{}}$

a) Addition, substitution, isomerization

b) Isomerization, substitution, addition

c) Substitution, addition, isomerization

d) Addition, addition, isomerization

e) Substitution, substitution, addition

Student Note: Polymers typically have very high molar masses—sometimes thousands or even millions of grams.

Student Note: Some of the other properties that suggested a high-molar-mass solute were low osmotic pressure and negligible freezing-point depression. These are known as *colligative* properties [◄◄ Chapter 13].

Animation
Organic and Biochemistry—Natural and synthetic polymers.

23.6   Organic Polymers

Polymers are molecular compounds, either natural or synthetic, that are made up of many *repeating units* called *monomers*. The physical properties of these so-called macromolecules differ greatly from those of small, ordinary molecules.

The development of polymer chemistry began in the 1920s with the investigation of the puzzling behavior of some materials including wood, gelatin, cotton, and rubber. For example, when rubber, with the known empirical formula of C_5H_8, was dissolved in an organic solvent, the solution displayed several properties, including a higher than expected viscosity, which suggested that the dissolved compound had a very high molar *mass*. Despite the experimental evidence, though, scientists at the time were not ready to accept the idea that such giant molecules could exist. Instead, they postulated that materials such as rubber consisted of aggregates of small molecular units, like C_5H_8 or $C_{10}H_{16}$, held together by intermolecular forces. This misconception persisted for a number of years, until Hermann Staudinger[1] clearly showed that these so-called aggregates were, in fact, enormously large molecules, each of which contained many thousands of atoms held together by covalent bonds.

Once the *structures* of these macromolecules were understood, the way was open for the synthesis of polymers, which now pervade almost every aspect of our daily lives. About 90 percent of today's chemists, including biochemists, work with polymers. In this section, we discuss the reactions that result in polymer formation and some of the natural polymers that are important to biology.

Addition Polymers

Addition polymers form when monomers such as ethylene join end to end to make polyethylene. Reactions of this type can be initiated by a *radical*—a species that

---

1. Hermann Staudinger (1881–1963). German chemist. One of the pioneers in polymer chemistry. Staudinger was awarded the Nobel Prize in Chemistry in 1953.

contains an unpaired electron [◄◄ Section 8.8]. The mechanism of addition polymerization is as follows:

1. The radical, which is unstable because of its unpaired electron, attacks a carbon atom on an ethylene molecule. This is the *initiation* of the reaction.
2. This attack would result in the carbon atom in question having more than eight electrons around it. To keep the carbon atom from having too many electrons around it, the double bond breaks.
3. One of the electrons in the double bond, together with the electron from the radical, becomes a new bond between the ethylene molecule and the radical.
4. The other electron remains with the other carbon atom, generating a new radical species.
5. The new radical species, also unstable, attacks another ethylene molecule, causing the same sequence of events to happen again. The generation of a new, highly reactive radical species at each step is known as *propagation* of the reaction.
6. Each step lengthens the chain of carbon atoms and results in the formation of a new radical, continuing the propagation of the reaction. Reactions such as this are known as *chain* reactions. They continue until the system runs out of ethylene molecules or until the radical species encounters another radical species—resulting in *termination* of the reaction:

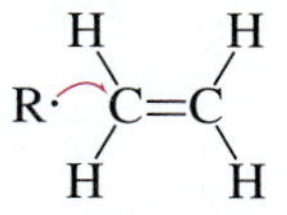

Initiation step

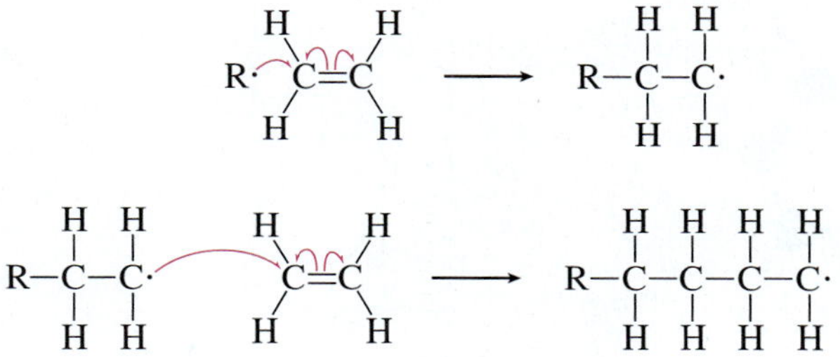

Propagation steps

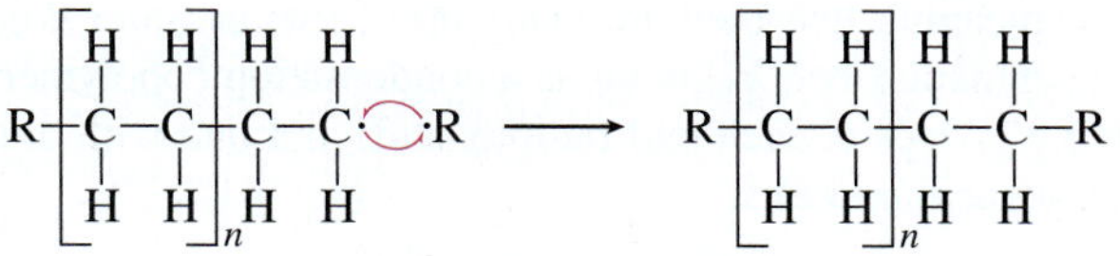

Termination step

Some familiar and important addition polymers are listed in Table 23.4.

Condensation Polymers

Reactions in which two or more molecules become connected with the elimination of a small molecule, often water, are called ***condensation reactions*** (Figure 23.12).

(a)

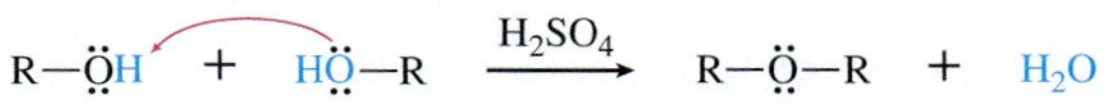

(b)

Figure 23.12 (a) Condensation reaction between an alcohol and a carboxylic acid to form an ester. (b) Condensation of two alcohol molecules in the presence of sulfuric acid to form an ether.

TABLE 23.4 Addition Polymers

Name	Monomer Unit	Structure	Uses
Polytetrafluoroethylene (Teflon)	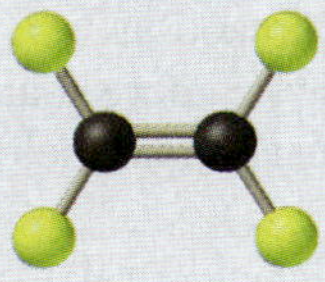	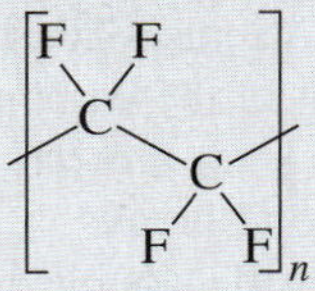 $\begin{bmatrix} F & F \\ C & \\ & C \\ F & F \end{bmatrix}_n$	Nonstick coatings
Polyethylene	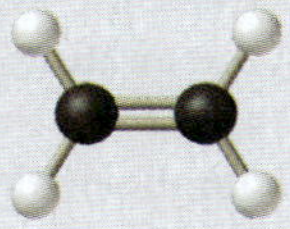	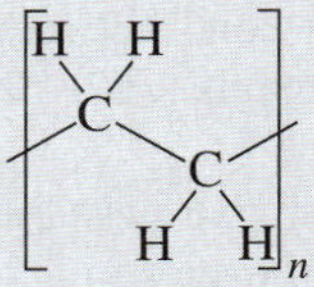 $\begin{bmatrix} H & H \\ C & \\ & C \\ H & H \end{bmatrix}_n$	Plastic bags, bottles, toys
Polypropylene	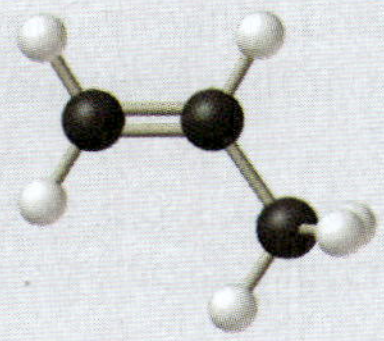	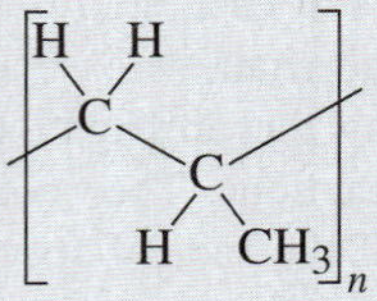 $\begin{bmatrix} H & H \\ C & \\ & C \\ H & CH_3 \end{bmatrix}_n$	Carpeting, bottles
Polyvinylchloride (PVC)	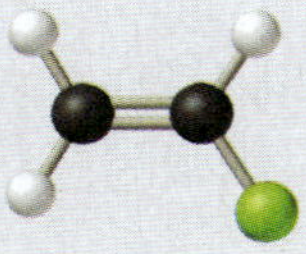	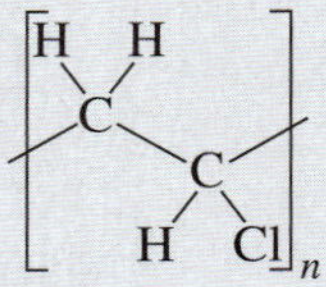 $\begin{bmatrix} H & H \\ C & \\ & C \\ H & Cl \end{bmatrix}_n$	Water pipes, garden hoses, plastic wrap
Polystyrene	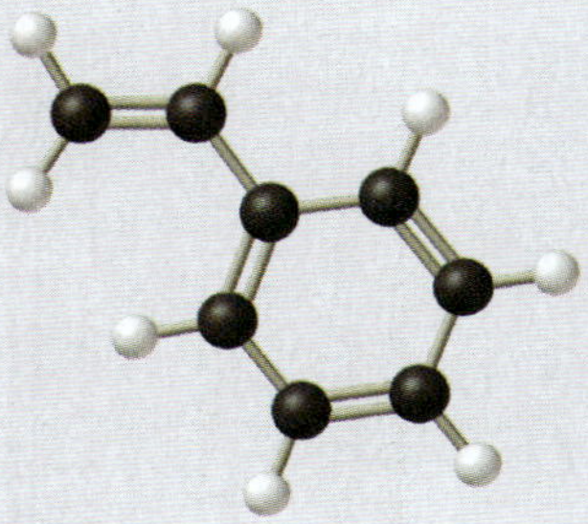	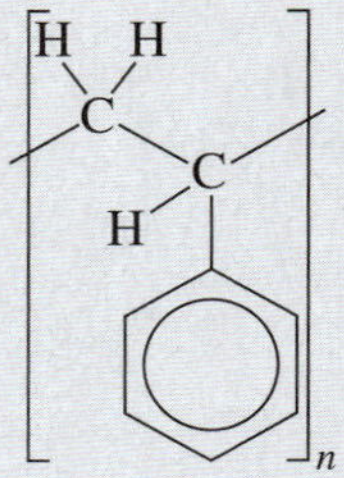	Packing material, insulation, furniture

Condensation polymers form when molecules with two different functional groups combine, with the elimination of a small molecule, often water. Many condensation polymers are **copolymers,** meaning that they are made up of two or more *different* monomers.

The first synthetic fiber, *nylon 66,* is a condensation copolymer of two molecules: one with carboxy groups at each end (adipic acid) and one with amine groups at both ends (hexamethylenediamine):

$$H_2N-(CH_2)_6-NH_2 \quad + \quad HOOC-(CH_2)_4-COOH$$

Hexamethylenediamine   Adipic acid

Condensation

$$H_2N-(CH_2)_6-N-\overset{O}{\overset{\|}{C}}-(CH_2)_4-COOH \quad + \quad H_2O$$
H

Further condensation reactions

$$-(CH_2)_4-\overset{O}{\overset{\|}{C}}-N-(CH_2)_6-N-\overset{O}{\overset{\|}{C}}-(CH_2)_4-\overset{O}{\overset{\|}{C}}-N-(CH_2)_6-$$
H   H   H

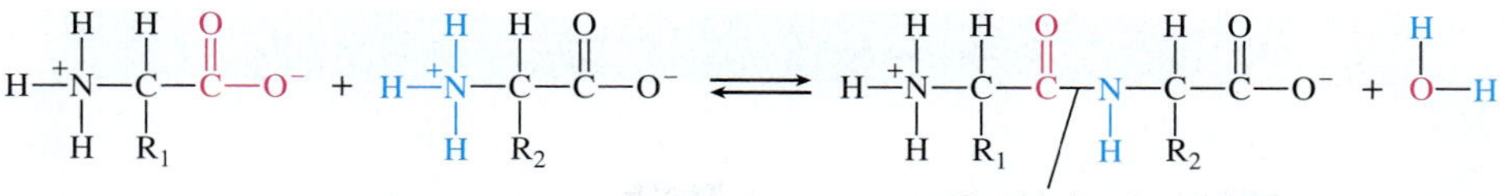

Figure 23.13 Formation of a peptide bond with elimination of water.

Nylon was first made by Wallace Carothers[2] at DuPont in 1931. The versatility of nylons is so great that the annual production of nylons and related substances now amounts to several billion pounds.

Biological Polymers

Naturally occurring polymers include *proteins, polysaccharides,* and *nucleic acids.*

   Proteins, polymers of amino acids, play an important role in nearly all biological processes. The human body contains an estimated 100,000 different kinds of proteins, each of which has a specific physiological function. An amino acid has both the carboxylic acid functional group and the amino functional group. Amino acids are joined together into chains when a condensation reaction occurs between a carboxy group on one molecule and an amino group on another molecule (Figure 23.13).

   The bonds that form between amino acids are called **peptide bonds.** Very long chains of amino acids assembled in this way are called *proteins,* while shorter chains are called *polypeptides.*

   Amino acids consist of a central carbon atom bonded to four different groups: an amino group, a carboxy group, a hydrogen atom, and an additional group (highlighted in Figure 23.14) consisting of carbon, hydrogen, and sometimes other elements such as nitrogen or sulfur. Proteins are made essentially from the 20 different amino acids shown in Figure 23.14. The identity of a protein depends on which of the 20 amino acids it contains and on the order in which the amino acids are assembled.

   Polysaccharides are polymers of sugars such as glucose and fructose. Starch and cellulose are two polymers of glucose with slightly different linkages—and very different properties. In starch, glucose molecules are connected by what biochemists call α linkages. This enables animals, including humans, to digest the starch in such foods as corn, wheat, potatoes, and rice:

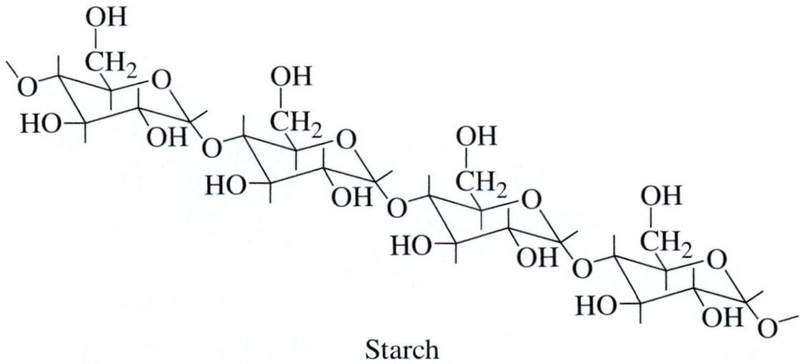

> **Student Note:** Peptide bonds are also called *amide bonds* or *amide linkages* because they contain the amide functional group.

Starch

---

2. Wallace H. Carothers (1896–1937). American chemist. Besides its enormous commercial success, Carothers's work on nylon is ranked with that of Staudinger in clearly elucidating macromolecular structure and properties. Depressed by the death of his sister and wrongly believing that his life's work had been a failure, Carothers committed suicide at the age of 41.

Figure 23.14 The 20 amino acids essential to living organisms. The shaded area represents the **R** group.

Name	Abbreviation	Structure
Alanine	Ala	$H_3C-\overset{\overset{\displaystyle H}{|}}{\underset{\underset{\displaystyle NH_3^+}{|}}{C}}-COO^-$
Arginine	Arg	$H_2N-\overset{\overset{\displaystyle H}{|}}{\underset{\underset{\displaystyle NH}{\|}}{C}}-N-CH_2-CH_2-CH_2-\overset{\overset{\displaystyle H}{|}}{\underset{\underset{\displaystyle NH_3^+}{|}}{C}}-COO^-$
Asparagine	Asn	$H_2N-\overset{\overset{\displaystyle O}{\|}}{C}-CH_2-\overset{\overset{\displaystyle H}{|}}{\underset{\underset{\displaystyle NH_3^+}{|}}{C}}-COO^-$
Aspartic acid	Asp	$HOOC-CH_2-\overset{\overset{\displaystyle H}{|}}{\underset{\underset{\displaystyle NH_3^+}{|}}{C}}-COO^-$
Cysteine	Cys	$HS-CH_2-\overset{\overset{\displaystyle H}{|}}{\underset{\underset{\displaystyle NH_3^+}{|}}{C}}-COO^-$
Glutamic acid	Glu	$HOOC-CH_2-CH_2-\overset{\overset{\displaystyle H}{|}}{\underset{\underset{\displaystyle NH_3^+}{|}}{C}}-COO^-$
Glutamine	Gln	$H_2N-\overset{\overset{\displaystyle O}{\|}}{C}-CH_2-CH_2-\overset{\overset{\displaystyle H}{|}}{\underset{\underset{\displaystyle NH_3^+}{|}}{C}}-COO^-$
Glycine	Gly	$H-\overset{\overset{\displaystyle H}{|}}{\underset{\underset{\displaystyle NH_3^+}{|}}{C}}-COO^-$
Histidine	His	$HC=C-CH_2-\overset{\overset{\displaystyle H}{|}}{\underset{\underset{\displaystyle NH_3^+}{|}}{C}}-COO^-$ (imidazole ring)
Isoleucine	Ile	$H_3C-CH_2-\overset{\overset{\displaystyle CH_3}{|}}{\underset{\underset{\displaystyle H}{|}}{C}}-\overset{\overset{\displaystyle H}{|}}{\underset{\underset{\displaystyle NH_3^+}{|}}{C}}-COO^-$

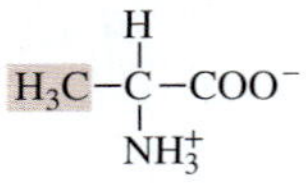

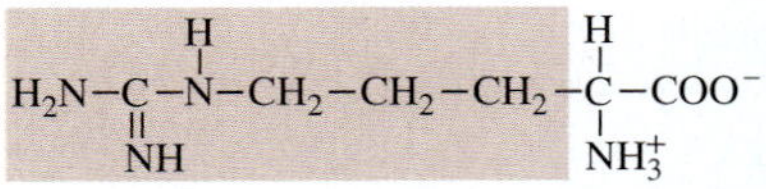

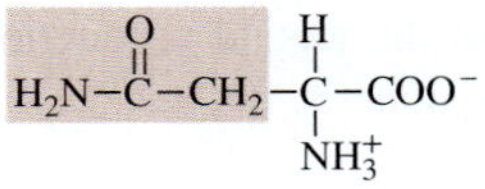

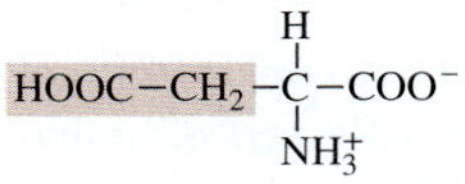

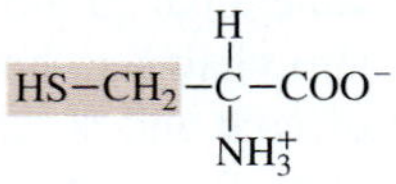

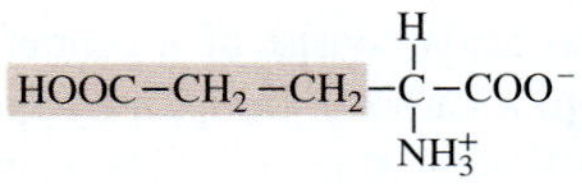

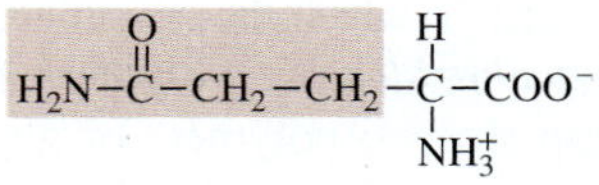

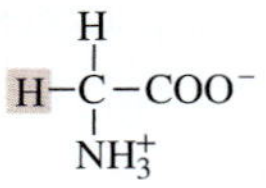

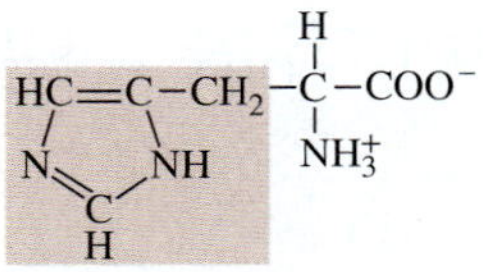

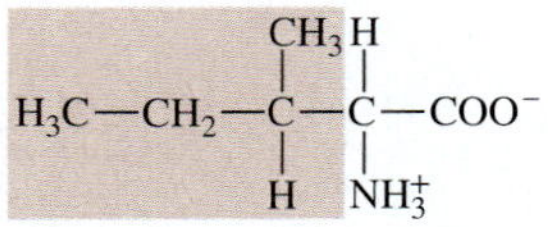

Figure 23.14 (Continued).

Name	Abbreviation	Structure
Leucine	Leu	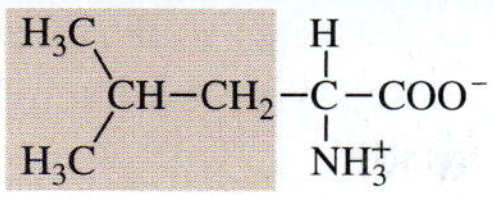
Lysine	Lys	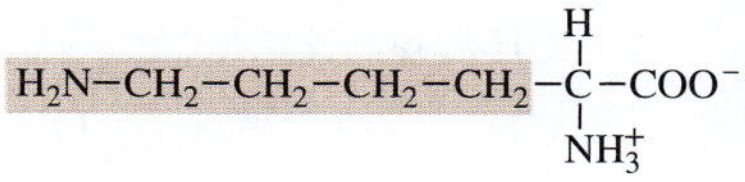
Methionine	Met	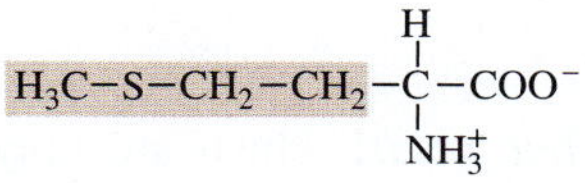
Phenylalanine	Phe	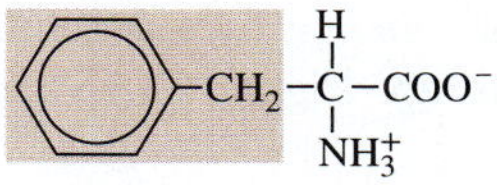
Proline	Pro	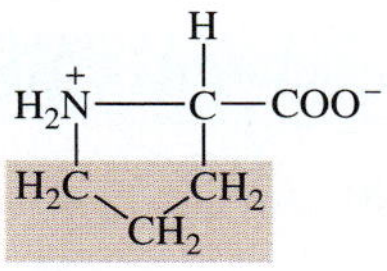
Serine	Ser	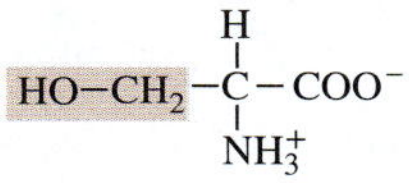
Threonine	Thr	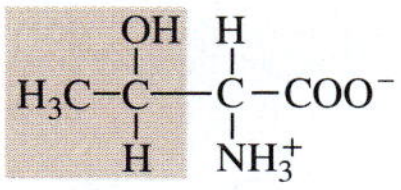
Tryptophan	Trp	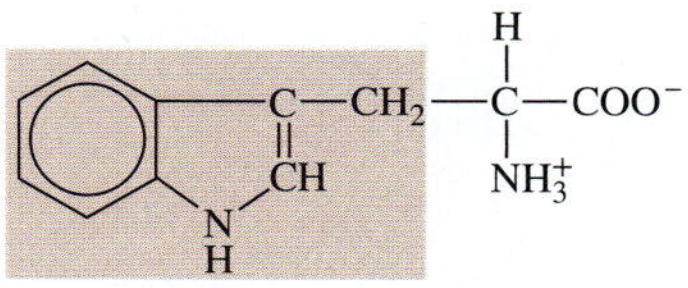
Tyrosine	Tyr	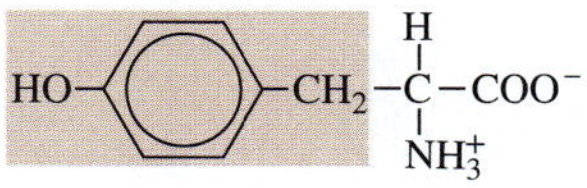
Valine	Val	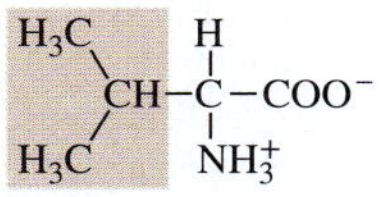

In cellulose, the glucose molecules are connected by β linkages. Digestion of cellulose requires enzymes that most animals do not have. Species that *do* digest cellulose, such as termites and ruminants (including cattle, sheep, and llamas), do so with the help of enzyme-producing symbiotic bacteria in the gut:

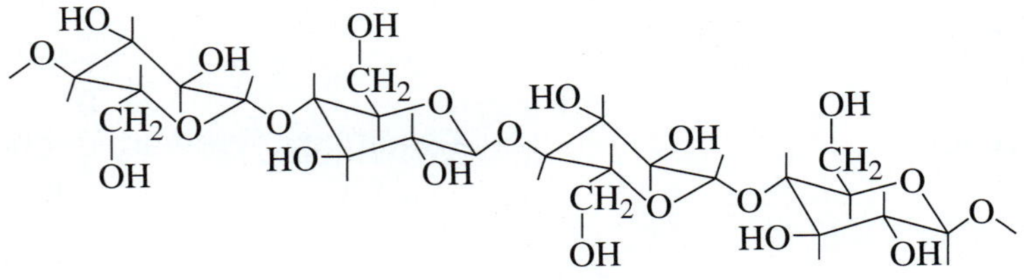

Cellulose

Animation
Organic and Biochemistry—Molecular structure in DNA.

Nucleic acids, which are polymers of *nucleotides,* play an important role in protein synthesis. There are two types of nucleic acids: *deoxyribonucleic acid (DNA)* and *ribonucleic acid (RNA).* Each *nucleotide* in a nucleic acid consists of a purine or pyrimidine *base,* a furanose sugar (*deoxyribose* for DNA; *ribose* for RNA), and a phosphate group. Figure 23.15 shows the building blocks of DNA and RNA. The

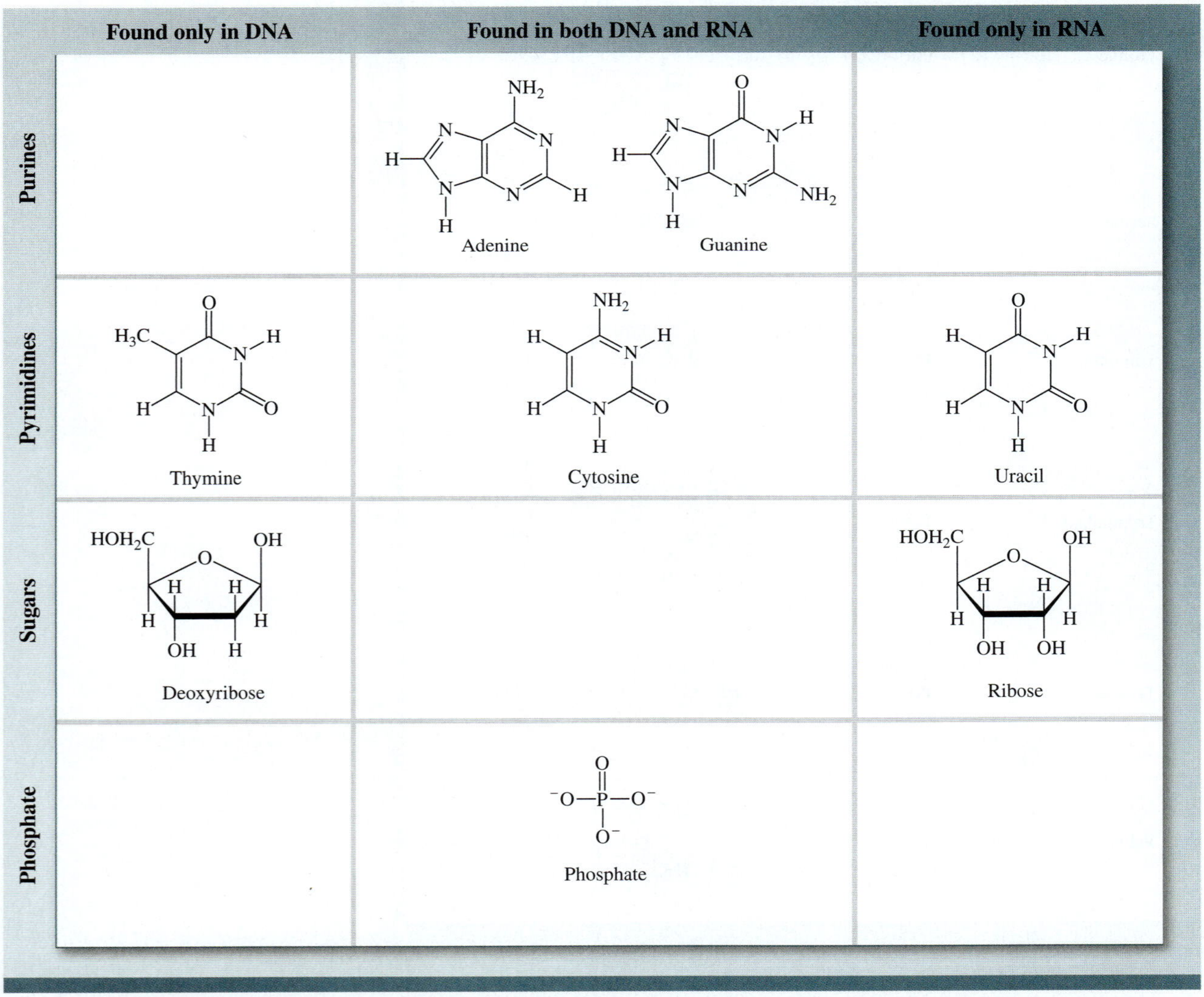

	Found only in DNA	Found in both DNA and RNA	Found only in RNA
Purines		Adenine Guanine	
Pyrimidines	Thymine	Cytosine	Uracil
Sugars	Deoxyribose		Ribose
Phosphate		Phosphate	

Figure 23.15 The components of the nucleic acids DNA and RNA.

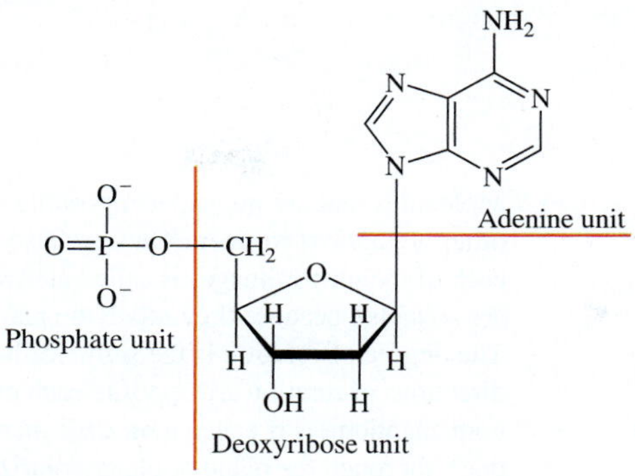

Figure 23.16 Structure of a nucleotide, one of the repeating units in DNA.

components of a nucleotide are linked together as shown in Figure 23.16. These molecules are among the largest known—they can have molar masses of up to tens of billions of grams. RNA molecules, on the other hand, typically have molar masses on the order of tens of thousands of grams. Despite their sizes, the composition of nucleic acids is relatively simple compared with that of proteins. Proteins consist of up to 20 different amino acids, whereas DNA and RNA consist of only four different nucleotides each.

Chapter Summary

Section 23.1

- Organic chemistry is the study of carbon-based substances. Although it was once thought that organic compounds could only be produced by nature, thousands of new organic compounds are now synthesized each year by scientists.

- Carbon's position in the periodic table makes it uniquely able to form long, stable chains—a process called *catenation.*

- *Aromatic* compounds contain one or more benzene rings. *Aliphatic* compounds are organic compounds that do not contain benzene rings.

Section 23.2

- *Alkyl groups* consist of just carbon and hydrogen. They are derived from the corresponding alkane by removing one hydrogen atom and are represented generically in the formulas of organic compounds with the letter R.

- Functional groups are specific arrangements of atoms that are responsible for the properties and reactivity of organic compounds. Common functional groups and their formulas include

Alcohol	ROH
Carboxylic acid	RCOOH
Aldehyde	RCHO
Ketone	RCOR′
Ester	RCOOR′
Amine	RNH_2, RNHR′, or RNR′R″
Amide	$RCONH_2$, RCONHR′, or RCONR′R″

- Many organic compounds contain more than one functional group. A *substituent* in an alkane is a group other than hydrogen that is bonded to the carbon chain.

- An *amino acid* contains both the carboxy group and the amino group.

Section 23.3

- *Condensed structural formulas* or *condensed structures* abbreviate a series of repeating units, such as $-CH_2CH_2CH_2CH_2-$, into a more compact form, such as $-(CH_2)_4-$. *Kekulé structures* are similar to Lewis structures but do not show the lone pairs on a molecule. *Bond-line structures* use straight lines to represent C—C bonds. They typically do not show the C atoms explicitly except for the purpose of emphasizing a particular functional group. H atoms are not shown in bond-line structures—again, except to emphasize a functional group. The number of H atoms bonded to a C atom must be inferred from the number of C—C bonds in the structure. *Heteroatoms,* atoms other than C and H, are always shown explicitly in a bond-line structure.

Section 23.4

- *Constitutional isomers* are molecules in which the same atoms are connected differently. *Stereoisomers* are molecules in which the same atoms are connected by the same bonds but the bonds are oriented differently. *Geometrical isomers* arise due to restricted rotation about a carbon-carbon double bond. *Cis* and *trans* isomers are geometrical isomers.

- Molecules that are nonsuperimposable mirror images of each other are *optical isomers.* They are also referred to as *chiral,* and each of the mirror images is called an *enantiomer.* Optical isomers are so called because they rotate the plane of plane-polarized light. The degree of rotation is the same for both enantiomers, but the directions of rotation are opposite each other. An equal mixture of both enantiomers is called a *racemic mixture.* A racemic mixture does not rotate the plane of plane-polarized light.

Section 23.5

- An *electrophile* generally is a positively charged ion that is attracted to electrons. A *nucleophile* is a negatively charged ion or a partially negatively charged atom in a polar molecule. Nucleophiles and electrophiles are attracted to each other.

- A *carbocation* is an intermediate species in which one of the carbon atoms is surrounded by only six electrons and bears a positive charge.

- *Electrophilic addition* reactions and *nucleophilic addition* reactions involve the addition of a molecule or an ion to another molecule.

- *Substitution reactions* occur when an electrophile replaces a hydrogen on an aromatic ring or when a nucleophile replaces a leaving group on a carbon atom.

- Carbocations are the intermediate species in *racemization,* a nucleophilic substitution reaction in which a single enantiomer is converted into a racemic mixture.

- An *elimination reaction* is one in which a double bond forms and a small molecule, such as water, is eliminated.

- *Isomerization reactions* convert one isomer into another.

Section 23.6

- *Polymers* are long chains of repeating molecular units called *monomers. Polysaccharides* are polymers of sugars. *Addition polymers* form when a radical species attacks a double bond, forming a new, longer radical species that attacks another double bond, and so on.

- An elimination reaction that joins two molecules is a *condensation reaction. Condensation polymers* form when molecules with two different functional groups undergo a condensation reaction.

- *Copolymers* are polymers that contain more than one type of monomer.

- *Proteins* and *polypeptides* are biological polymers in which the monomers are amino acids. Amino acids are joined by *peptide bonds,* which result from the condensation reaction between the amino group of one amino acid and the carboxyl group of another amino acid.

- *Nucleic acids* are polymers of *nucleotides.* The two types of nucleic acid are *deoxyribonucleic acid (DNA)* and *ribonucleic acid (RNA).* Each nucleotide in a nucleic acid consists of a purine or pyrimidine base, a furanose sugar (deoxyribose for DNA; ribose for RNA), and a phosphate group linked together.

Key Words

Addition polymer, 1142
Alcohol, 1112
Aldehyde, 1112
Aliphatic, 1110
Alkyl group, 1110
Amide, 1112
Amine, 1112
Amino acid, 1119
Aromatic, 1110
Bond-line structure, 1123
Carbocation, 1133
Carboxylic acid, 1112
Catenation, 1109
Chiral, 1129
Cis, 1128
Condensation polymer, 1144
Condensation reaction, 1143
Condensed structural formula, 1122
Condensed structure, 1122
Constitutional isomerism, 1128
Copolymer, 1144
Deoxyribonucleic acid (DNA), 1148
Electrophile, 1132
Electrophilic addition, 1134
Elimination reaction, 1140
Enantiomer, 1129
Ester, 1112
Geometrical isomer, 1128
Heteroatom, 1123
Isomerization reaction, 1141
Kekulé structure, 1122
Ketone, 1112
Monomer, 1142
Nucleic acid, 1148
Nucleophile, 1132
Nucleophilic addition, 1135
Nucleotide, 1148
Optical isomers, 1129
Peptide bond, 1145
Polymer, 1142
Polypeptide, 1145
Polysaccharides, 1145
Protein, 1145
Racemic mixture, 1130
Racemization, 1138
Ribonucleic acid (RNA), 1148
Stereoisomers, 1128
Substituent, 1114
Substitution reaction, 1135
Trans, 1129

Questions and Problems

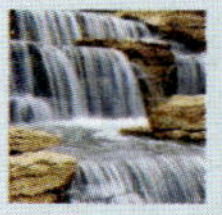

Applying What You've Learned

Although it was approved by the FDA in 1998, thalidomide is the most regulated prescription drug in history, because it is known to harm developing fetuses. The drug's manufacturer, Celgene Corporation, has developed the *System for Thalidomide Education and Prescribing Safety* (STEPS) program. In order for physicians to prescribe thalidomide to their patients, they must be registered in the STEPS program. Female patients must have a negative pregnancy test within 24 hours of beginning treatment and must undergo periodic pregnancy testing throughout treatment. Both female and male patients must comply with mandatory contraceptive measures, patient registration, and patient surveys. Moreover, new patients must view an informational video in which a thalidomide victim explains the potential dangers of the drug.

Problems:
(a) From the structures given at the beginning of the chapter, identify the functional groups in thalidomide [⏮ Sample Problem 23.2]. (b) Write molecular formulas for thalidomide, lenalidomide, and pomalidomide [⏮ Sample Problem 23.3]. (c) Thalidomide is converted to the drug pomalidomide by substitution of an amino group for one of the H atoms on the aromatic portion of the molecule. Using curved arrows, draw the mechanism for this reaction and all the resonance structures for the carbocation intermediate [⏮ Sample Problem 23.5].

SECTION 23.1: WHY CARBON IS DIFFERENT

Review Questions

23.1 Explain why carbon is able to form so many more compounds than any other element.

23.2 Why was Wöhler's synthesis of urea so important for the development of organic chemistry?

23.3 What are aromatic organic compounds? What are aliphatic organic compounds?

SECTION 23.2: ORGANIC COMPOUNDS

Review Questions

23.4 What are functional groups? Why is it logical and useful to classify organic compounds according to their functional groups?

23.5 Draw the Lewis structure for each of the following functional groups: alcohol, aldehyde, ketone, carboxylic acid, amine.

23.6   Name the classes to which the following compounds belong:
(a) C_4H_9OH
(b) C_2H_5CHO
(c) C_6H_5COOH
(d) CH_3NH_2

Conceptual Problems

23.7   Classify each of the following molecules as alcohol, aldehyde, ketone, carboxylic acid, or amine:
(a) $CH_3-CH_2-NH_2$

(b) $CH_3-CH_2-C\overset{\displaystyle O}{\underset{\displaystyle H}{\big|}}$

(c) $CH_3-\overset{\displaystyle }{\underset{\displaystyle O}{C}}-CH_2-CH_3$

(d) $H-\overset{\displaystyle O}{C}-OH$

(e) $CH_3-CH_2CH_2-OH$

23.8   Draw structures for molecules with the following formulas:
(a) CH_4O
(b) C_2H_6O
(c) $C_3H_6O_2$
(d) C_3H_8O

23.9   Name each of the following compounds:

(a)

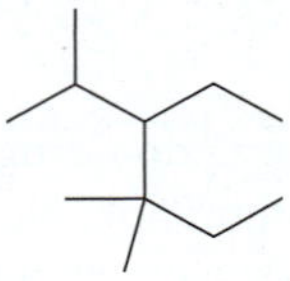

(b) $CH_3\overset{\displaystyle CH_3}{\underset{\displaystyle CH_3}{C}}CH_2CH_2CH_2\overset{\displaystyle OH}{CH}CH_3$

(c) $ClCHCH_2CH_2\overset{\displaystyle O}{\overset{\displaystyle \|}{C}}H$
      $\underset{\displaystyle CH_2CH_3}{|}$

23.10   Name each of the following compounds:
(a)

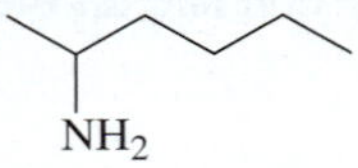

(b)

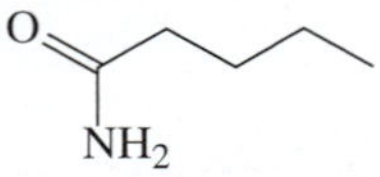

(c)

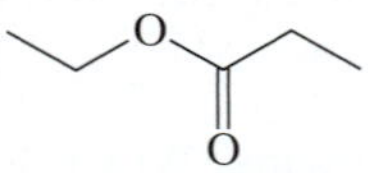

23.11   Give the name of the alkane represented by the model shown:

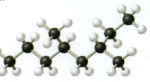

23.12   The molecular formula corresponding to the model is $C_5H_9ClO_2$. What is the name of the compound?

23.13   Write the structural formula for each of the following based on its systematic name. The names quoted in parentheses are so-called common names. Common names are not systematic and are more difficult, sometimes impossible, to connect with a unique structure.
(a) 2,2,4-Trimethylpentane ("isooctane")
(b) 3-Methyl-1-butanol ("isoamyl alcohol")
(c) Hexanamide ("caproamide")
(d) 2,2,2-Trichloroethanal ("chloral")

23.14   Write the structural formula for each of the following:
(a) 3,3-Dimethyl-2-butanone ("pinacolone")
(b) 3-Hydroxybutanal ("acetaldol")
(c) Ethyl pentanoate ("ethyl valerate")
(d) 6-Methyl-2-heptanamine ("isooctylamine")

23.15   Classify the oxygen-containing groups in the plant hormone abscisic acid:

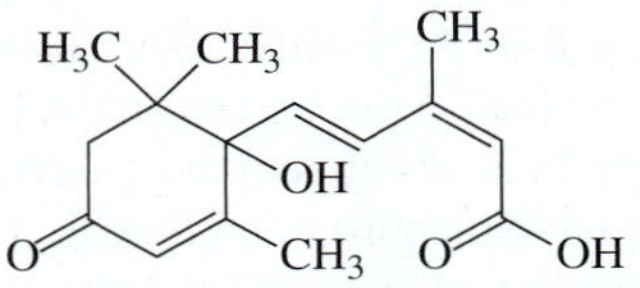

23.16   Identify the functional groups in the antipsychotic drug haloperidol:

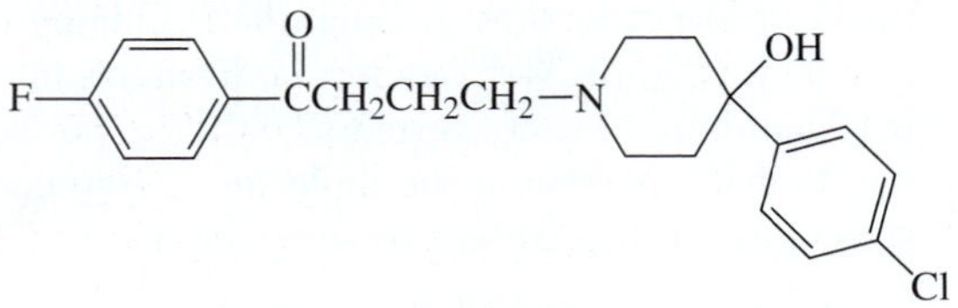

23.17   PABA was the active UV-absorbing compound in earlier versions of sunblock creams. What functional groups are present in PABA?

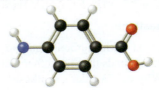

23.18 Lidocaine ($C_{14}H_{22}N_2O$) is a widely used local anesthetic. Classify its nitrogen-containing functional groups:

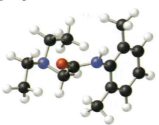

SECTION 23.3: REPRESENTING ORGANIC MOLECULES

Conceptual Problems

23.19 Write structural formulas for the following organic compounds: (a) 3-methylhexane, (b) 2,3-dimethylpentane, (c) 2-bromo-4-phenylpentane, (d) 3,4,5-trimethyloctane.

23.20 Write structural formulas for the following compounds: (a) 1,1,3-trichloro-2-propanol, (b) 3-methyl-3-pentanamine, (c) 3-bromo-1-chloro-2-butanone, (d) propyl-4-bromobutanoate.

23.21 (a) Convert $CH_3(CH_2)_4C(O)CH_2CO_2H$ to a Kekulé structure and to a bond-line structure.
(b) Convert the following to a condensed structure and to a bond-line structure:

$$CH_3CH_2CHCH_2COCCH_3$$

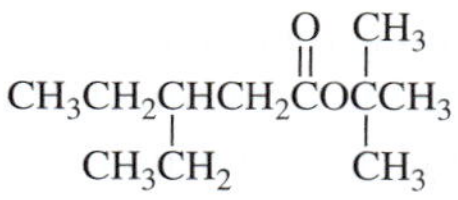

(c) Convert the following to a condensed structure and to a Kekulé structure:

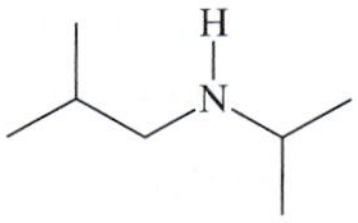

23.22 (a) Convert $(CH_3)_2C{=}CHCO_2H$ to a Kekulé structure and to a bond-line structure.
(b) Convert the following to a condensed structure and to a bond-line structure:

$$CH_3CH_2NCH_2CH_3$$
$$CH_3$$

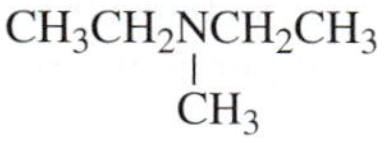

(c) Convert the bond-line structure of the general anesthetic isoflurane to a condensed structure and to a Kekulé structure:

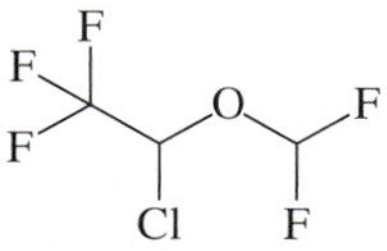

23.23 Convert each of the molecular models to a condensed structural formula, a Kekulé structure, and a bond-line structure.

(a) 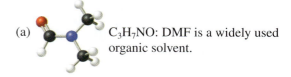 C_3H_7NO: DMF is a widely used organic solvent.

(b)  $C_6H_8O_7$: Citric acid contributes to the tart taste of citrus fruits.

(c) 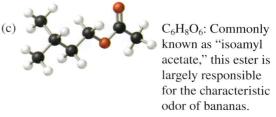 $C_6H_8O_6$: Commonly known as "isoamyl acetate," this ester is largely responsible for the characteristic odor of bananas.

23.24 Convert each of the molecular models to a condensed structural formula, a Kekulé structure, and a bond-line structure.

(a) 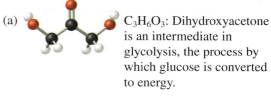 $C_3H_6O_3$: Dihydroxyacetone is an intermediate in glycolysis, the process by which glucose is converted to energy.

(b) C_4H_5Cl: Chloroprene gives neoprene on polymerization.

(c) 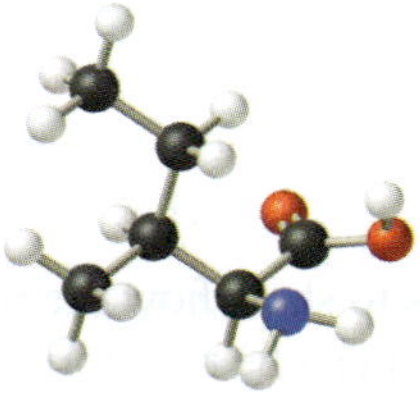 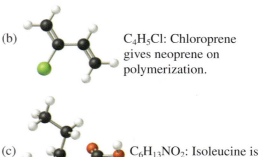 $C_6H_{13}NO_2$: Isoleucine is one of the 20 amino acids that occur in proteins.

23.25 Given a structural formula or bond-line structure, rewrite it in the other style.
(a) 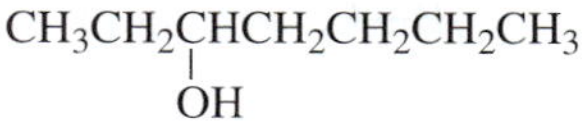 $CH_3CH_2CHCH_2CH_2CH_2CH_3$
OH

(b)

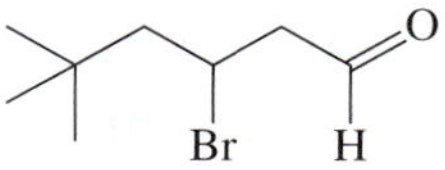

(c)

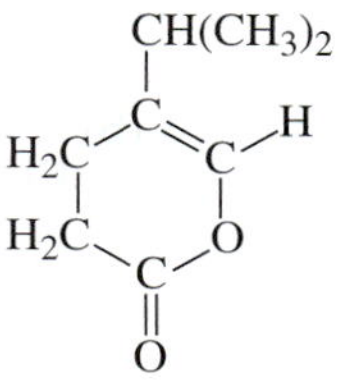

(c) Certain reactions of aldehydes and ketones begin with isomerization of the aldehyde or ketone to an *enol* isomer. Enols contain an —OH group attached to a carbon-carbon double bond. Write a chemical equation for the isomerization of acetone $[(CH_3)_2C{=}O]$ to its enol isomer.

23.52   Classify the following reactions according to whether they are addition, substitution, elimination, or isomerization.

(a)

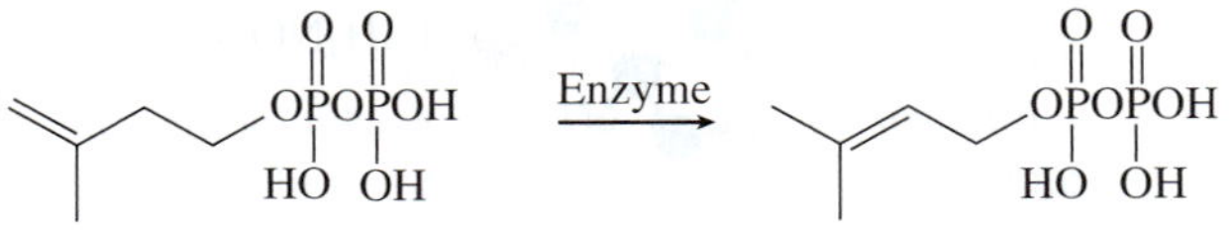

(b) $CH_3CH{=}CH_2 + Cl_2 \xrightarrow{\text{heat}} ClCH_2CH{=}CH_2 + HCl$

(c) $CH_3CH{=}CH_2 + Cl_2 \longrightarrow CH_3CHCH_2Cl$
$\qquad\qquad\qquad\qquad\qquad\quad |$
$\qquad\qquad\qquad\qquad\qquad\ Cl$

(d) $CH_3CHCH_2Cl + NaSH \longrightarrow CH_3CHCH_2SH + NaCl$
$\quad\ \ |$
$\quad\ Cl$
$\qquad\qquad\qquad\qquad\qquad\qquad\qquad |$
$\qquad\qquad\qquad\qquad\qquad\qquad\quad Cl$

(e)

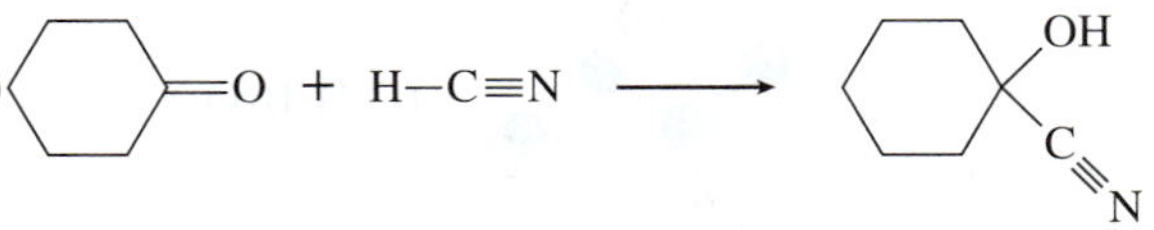

(f)

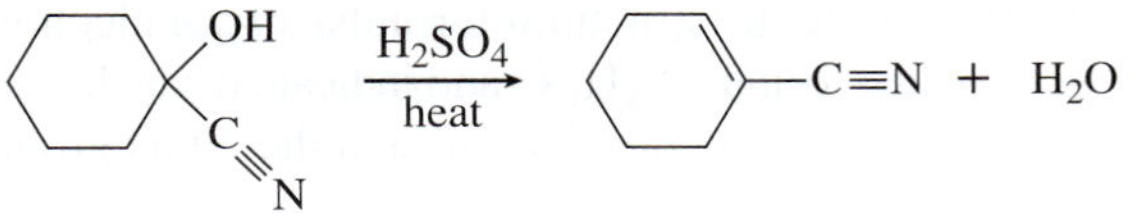

23.53   Which of the following are carbocations?

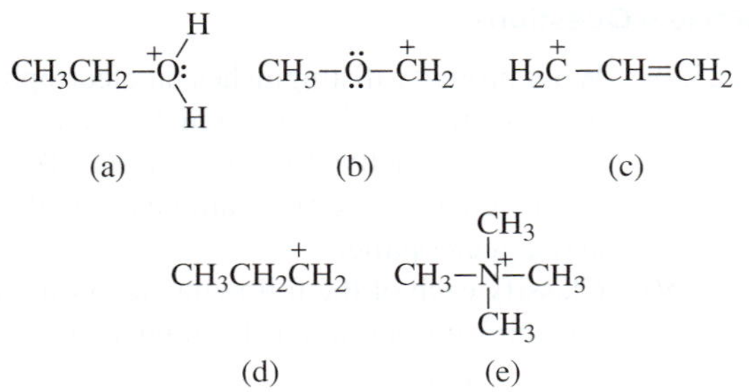

Conceptual Problems

23.54   Halogenated hydrocarbons are biodegraded in the natural environment by reactions catalyzed by dehalogenase enzymes. The reaction that takes place with 1,2-dichloroethane begins with nucleophilic substitution involving a carboxylate site of the enzyme.

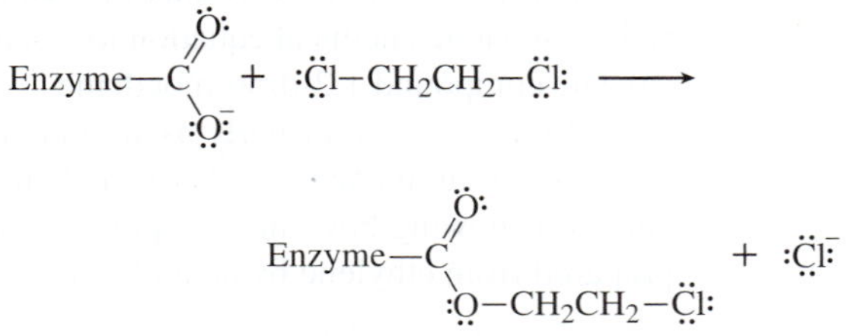

Expand this equation by adding curved arrows to show the movement of electrons.

23.55   A common reaction in carbohydrate biochemistry is the conversion of an aldose to a ketose. The glucose to fructose isomerization is a specific example; the equation illustrates the general case:

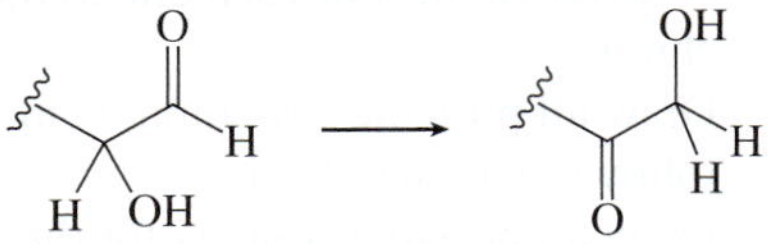

The first stage in the reaction is shown in the following equation. Enzymes facilitate the reaction, but for simplicity the overall change can be approximated with water molecules. Use curved arrows to show the flow of electrons in the equation.

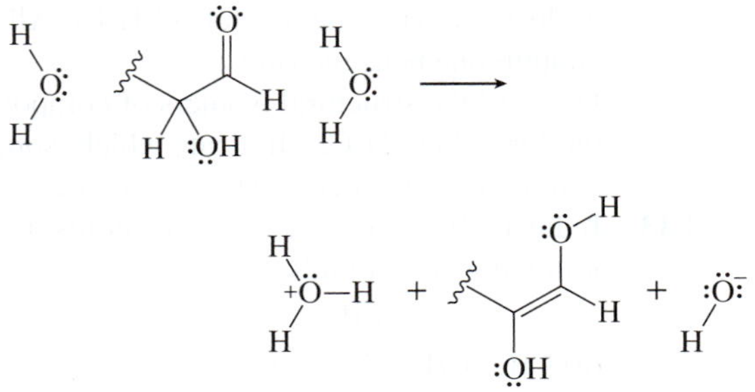

23.56   (a) Benzene reacts with *tert*-butyl cation $[(CH_3)_3C^+]$ by a two-step electrophilic aromatic substitution mechanism to yield *tert*-butylbenzene $[C_6H_5C(CH_3)_3]$. Write a chemical equation for each step in the mechanism and use curved arrows to track electron flow.
(b) Electrophilic addition of hydrogen chloride to styrene gives the product shown. Write the mechanism for this reaction including curved arrows.

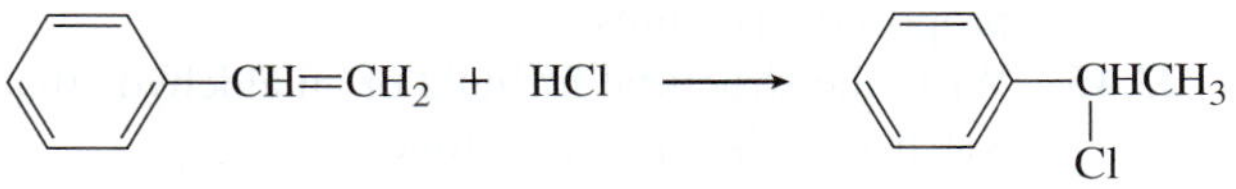

23.57   (a) Acetylide ion undergoes nucleophilic addition to aldehydes and ketones to give the species shown. Subsequent addition of water yields an acetylenic alcohol. Add curved arrows to the equations to show how the reaction occurs.

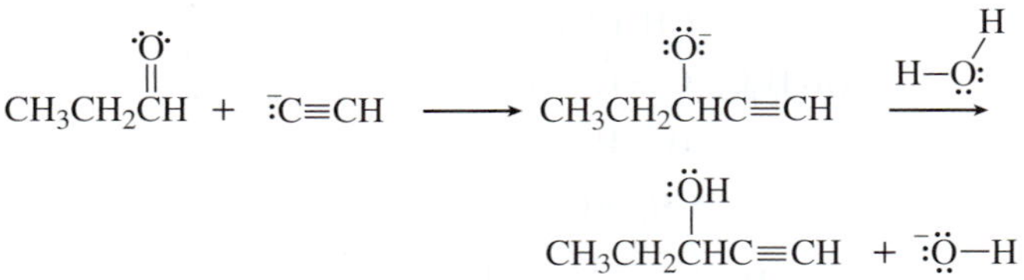

(b) A nucleophilic site on a molecule can substitute for a halogen elsewhere in the same molecule to form a ring. Use curved arrows to show how the following *cyclization* is related to a conventional nucleophilic substitution.

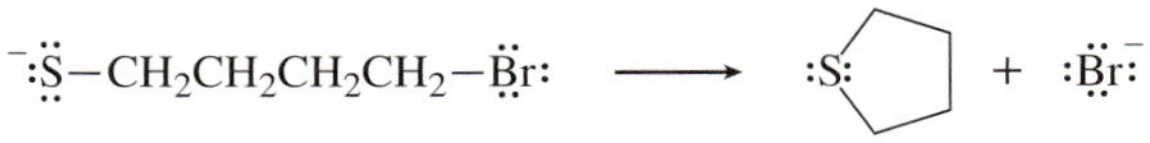

23.58 Complete the following reaction by including essential unshared electron pairs and formal charges. Use curved arrows to show electron flow.

$$(CH_3)_3B + \overset{\bigcirc}{O} \longrightarrow (CH_3)_3B-\overset{\bigcirc}{O}$$

23.59 The product of the reaction of chlorobenzene with $NaNH_2$ is a very reactive species called *benzyne*. Add necessary electron pairs and formal charges to the net ionic equation, and use curved arrows to show how benzyne is formed. To what general type of reaction does this belong?

$$H_2N^- + \text{[chlorobenzene ring structure]} \longrightarrow$$

$$H_2N-H + \text{[benzyne ring structure]} + Cl^-$$

23.60 Consider the following reactions of butanal:

$$\text{(i)} \quad CH_3CH_2CH_2\overset{O}{\overset{\|}{C}}H \xrightarrow[H_2O]{NaBH_4} CH_3CH_2CH_2CH_2OH$$

$$\text{(ii)} \quad CH_3CH_2CH_2\overset{O}{\overset{\|}{C}}H \xrightarrow[H_2O]{H_2CrO_4} CH_3CH_2CH_2\overset{O}{\overset{\|}{C}}OH$$

In which reaction is butanal oxidized? In which reaction is it reduced?

23.61 Esters can be prepared by the acid-catalyzed condensation of a carboxylic acid and an alcohol:

$$CH_3CH_2\overset{O}{\overset{\|}{C}}OH + CH_3OH \longrightarrow CH_3CH_2\overset{O}{\overset{\|}{C}}OCH_3 + H_2O$$

Is this an oxidation-reduction reaction? If so, identify the species being oxidized and the species being reduced.

23.62 A compound has the empirical formula $C_5H_{12}O$. Upon controlled oxidation, it is converted into a compound of empirical formula $C_5H_{10}O$, which behaves as a ketone. Draw possible structures for the original compound and the final compound.

23.63 Isopropanol is prepared by reacting propylene (CH_3CHCH_2) with sulfuric acid, followed by treatment with water. (a) Show the sequence of steps leading to the product. What is the role of sulfuric acid? (b) Draw the structure of an alcohol that is an isomer of isopropanol. (c) Is isopropanol a chiral molecule?

SECTION 23.6: ORGANIC POLYMERS

Review Questions

23.64 Define the following terms: *monomer, polymer, copolymer.*

23.65 Name 10 objects that contain synthetic organic polymers.

23.66 Calculate the molar mass of a particular polyethylene sample, $\text{+}CH_2-CH_2\text{+}_n$, where $n = 4600$.

23.67 Describe the two major mechanisms of organic polymer synthesis.

23.68 What are the steps involved in polymer formation by chain reaction?

23.69 Polysaccharides, proteins, and nucleic acids comprise the three main classes of biopolymers. Compare and contrast them with respect to structure and function. What are the building-block units for each? What are the key functional groups involved in linking the units together? In which biopolymer is there the greatest variety of building-block structure? In which is there the least?

Conceptual Problems

23.70 Teflon is formed by a radical addition reaction involving the monomer tetrafluoroethylene. Show the mechanism for this reaction.

23.71 Vinyl chloride ($H_2C=CHCl$), undergoes copolymerization with 1,1-dichloroethylene, ($H_2C=CCl_2$), to form a polymer commercially known as Saran. Draw the structure of the polymer, showing the repeating monomer units.

23.72 Deduce plausible monomers for polymers with the following repeating units:
(a) $\text{+}CH_2-CH_2\text{+}_n$,

(b) $\text{[}-CO-\bigcirc-CONH-\bigcirc-NH-\text{]}_n$

23.73 Deduce plausible monomers for polymers with the following repeating units:
(a) $\text{+}CH_2-CH=CH-CH_2\text{+}_n$
(b) $\text{+}CO\text{+}CH_2\text{+}_6NH\text{+}_n$

23.74 Draw the structures of the dipeptides that can be formed from the reaction between the amino acids glycine and alanine.

23.75 Draw the structures of the dipeptides that can be formed from the reaction between the amino acids glycine and lysine.

23.76   From among the given nucleotides, identify those that occur naturally in RNA and those that occur in DNA. Do any *not* occur in either RNA or DNA?

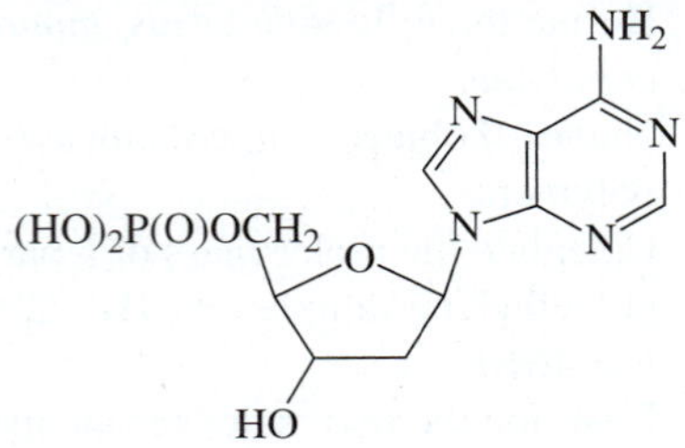

(a)

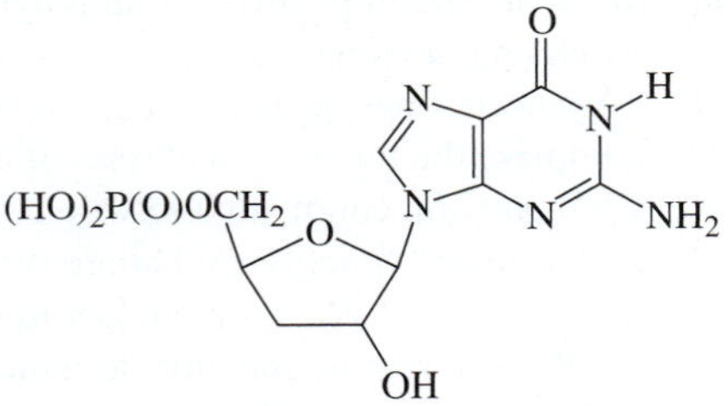

(b)

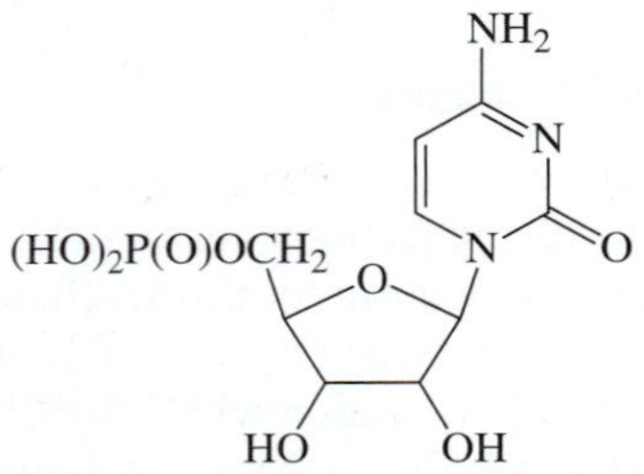

(c)

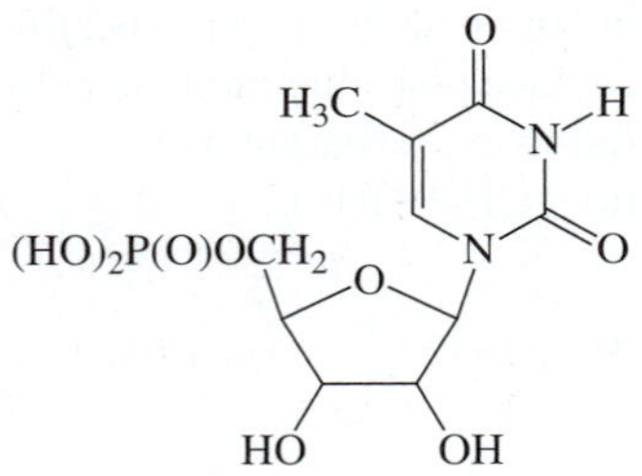

(d)

ADDITIONAL PROBLEMS

23.77   Write structural formulas for all the constitutionally isomeric C_4H_9 alkyl groups. Check your answers with Table 23.1, and note the names of these groups.

23.78   *Ethers* are compounds (excluding esters) that contain the C—O—C functional group. An acceptable way to name them is to list the two groups attached to oxygen in alphabetical order as separate words, followed by the word *ether*. If the two groups are the same, add the prefix *di* to the name of the alkyl group. Thus, $CH_3OCH_2CH_3$ is "ethyl methyl ether" and $CH_3CH_2OCH_2CH_3$ is "diethyl ether."

Write structural formulas and provide names for all the constitutionally isomeric ethers in which only the C_4H_9 alkyl groups from Problem 23.77 are attached to oxygen. Which of these ethers is potentially chiral?

23.79   Carbon dioxide reacts with sodium hydroxide according to the following equation:

$$CO_2 + 2NaOH \longrightarrow Na_2CO_3 + H_2O$$

The overall reaction is the result of two separate reactions.

Reaction I:

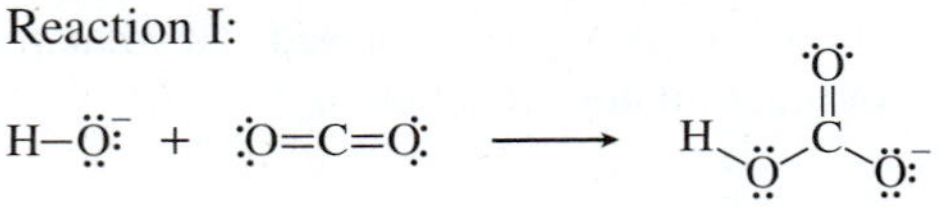

Reaction II:

(a) Use curved arrows to track the flow of electrons in reaction I.
(b) Use curved arrows to track the flow of electrons in reaction II.
(c) Classify reaction I as electrophilic addition, nucleophilic addition, electrophilic substitution, or acid-base.
(d) Classify reaction II according to the choices in part (c).

23.80   *Alkynes* are hydrocarbons that contain a carbon-carbon triple bond.
(a) Write structural formulas for all the isomeric alkynes of molecular formula C_5H_8.
(b) Are any of the alkynes chiral?
(c) Are any of the alkynes stereoisomeric?

23.81   Among the many alkenes of molecular formula C_6H_{12}, only one is chiral.
(a) Write a structural formula for this alkene.
(b) Place substituents on the tetrahedral carbons so as to represent the two enantiomers of this alkene.

23.82   Match each molecular model with the correct line-wedge-dash structure.

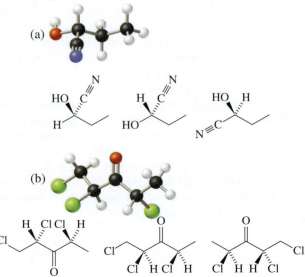

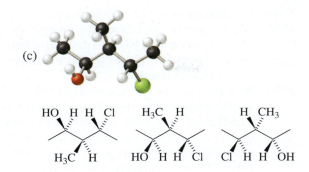

(c)

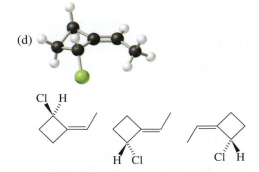

(d)

23.83 In each of the following pairs, specify whether the two structural formulas represent constitutional isomers, *cis,trans*-stereoisomers, enantiomers, resonance structures, or are simply a different representation of the same structure.

(a) [structure] and [structure]

(b) [structure] and [structure]

(c) [structure] and [structure]

(d) [structure] and [structure]

23.84 In each of the following pairs, specify whether the two structural formulas represent constitutional isomers, *cis,trans*-stereoisomers, enantiomers, resonance structures, or are simply a different representation of the same structure.

(a) HÖ ... ÖH and HÖ ... ÖH
H_2N: H   H :NH_2

(b) HÖ ... ÖH and HÖ ... ÖH
H_2N: H H_2N: H

(c) HÖ ... ÖH and HÖ ... ÖH
H_2N: H H_2N: H

(d) HÖ ... ÖH and HÖ ... Ö⁻
H_2N: H H_3N H

23.85 The electrophile in the following aromatic substitution is $(CH_3)_2CH^+$, and the mechanism involves the intermediate shown. Write two other resonance structures for this intermediate.

[reaction scheme]

$$H{-}\text{(ring)}{-}CH_3 \;+\; (CH_3)_2CHCl \xrightarrow{\text{catalyst}}$$

$$(CH_3)_2CH{-}\text{(ring)}{-}CH_3 \;+\; HCl$$

via: $(CH_3)_2CH{-}\text{(ring)}^+{-}CH_3$

23.86 Two isomeric alkenes are formed in the dehydration of 2-methyl-2-butanol. What are these two alkenes? Are they constitutional isomers or stereoisomers?

23.87 Electrophilic addition of HCl to *cis*-2-butene gave 2-chlorobutane, which was determined *not* to be optically active when examined with a polarimeter.

$$\underset{H}{\overset{H_3C}{>}}C{=}C\underset{H}{\overset{CH_3}{<}} \;+\; HCl \longrightarrow CH_3\underset{Cl}{CH}CH_2CH_3$$

Which of the following is the better explanation for the lack of optical activity in the 2-chlorobutane formed in this reaction?
(a) 2-Chlorobutane is not chiral.
(b) Two enantiomers of 2-chlorobutane were formed in equal amounts.

23.88 Reactions such as the following have been used in carbohydrate synthesis since the nineteenth century.

[structure with OH, OH, O groups]
$$\xrightarrow[\text{NaCN}]{\text{HCN}}$$
[product structure with OH groups and CN]

Classify this procedure according to reaction type.
(a) Electrophilic addition
(b) Electrophilic substitution
(c) Nucleophilic addition
(d) Nucleophilic substitution

23.89   Excluding compounds that have rings, there are three hydrocarbons that have the molecular formula C_4H_6. Write their structural formulas, and specify the hybridization of each carbon in these isomers.

23.90   Give the structures of the two tertiary amines that are isomers of $CH_3CH_2CH_2CH_2CH_2NH_2$.

23.91   In which class of compounds, esters or amides, do you think electron donation into the carbonyl group is more pronounced? Explain.

Ester   Amide

23.92   (a) A plane of symmetry in a molecule is a plane passing through the middle of the molecule that bisects the molecule into two mirror-image halves. If a molecule has a plane of symmetry, it cannot be chiral and cannot be optically active. Which of the following have at least one plane of symmetry? Do any have more than one?

1,1-Dichlorocyclopropane

cis-1,2-Dichlorocyclopropane

trans-1,2-Dichlorocyclopropane

(b) Specify the relationships (constitutionally isomeric or stereoisomeric) among these compounds.

ENGINEERING PROBLEMS

23.93   Kevlar is a copolymer used in bulletproof vests. It is formed in a condensation reaction between the following two monomers:

Sketch a portion of the polymer chain showing several monomer units. Write the overall equation for the condensation reaction.

23.94   Describe the formation of polystyrene.

23.95   Nylon can be destroyed easily by strong acids. Explain the chemical basis for the destruction. (*Hint:* The products are the starting materials of the polymerization reaction.)

23.96   Nylon was designed to be a synthetic silk.
(a) The average molar mass of a batch of nylon 66 is 12,000 g/mol. How many monomer units are there in this sample? (b) Which part of nylon's structure is similar to a polypeptide's structure? (c) How many different tripeptides (made up of three amino acids) can be formed from the amino acids alanine (Ala), glycine (Gly), and serine (Ser), which account for most of the amino acids in silk?

Biological Problems

23.97   The α-amino acids found in proteins are based on the structural formula:

$$\underset{\underset{NH_2}{|}}{RCHCO_2H}$$

From the C_4H_9 alkyl groups in Problem 23.77, find the ones that correspond to "R" among the amino acids listed in Figure 23.14. Match the name of the amino acid with the name of the group.

23.98   How many different tripeptides can be formed by lysine and alanine?

23.99   The amino acid glycine can be condensed to form a polymer called polyglycine. Draw the repeating monomer unit.

Multiconcept Problems

23.100   The combustion of 20.63 mg of compound Y, which contains only C, H, and O, with excess oxygen gave 57.94 mg of CO_2 and 11.85 mg of H_2O. (a) Calculate how many milligrams of C, H, and O were present in the original sample of Y. (b) Derive the empirical formula of Y. (c) Suggest a plausible structure for Y if the empirical formula is the same as the molecular formula.

23.101   All alkanes give off heat when burned in air. Such *combustion* of alkanes is exothermic, and the sign of $\Delta H°$ is negative. The general equation for the combustion of the alkanes of molecular formula C_5H_{12} is

$$C_5H_{12} + 8O_2 \longrightarrow 5CO_2 + 6H_2O$$

The values of $\Delta H°$ for the combustion of the three pentane isomers are

$$CH_3CH_2CH_2CH_2CH_3 = -3536 \text{ kJ/mol}$$
$$(CH_3)_2CHCH_2CH_3 = -3529 \text{ kJ/mol}$$
$$(CH_3)_4C = -3515 \text{ kJ/mol}$$

(a) What do these data tell you about the effect of chain branching on the relative potential energies and stabilities of these isomers?

(b) Assume this effect is general to predict which one of the 18 C_8H_{18} isomers should be the most stable.

23.102 (a) Use VSEPR to predict the geometry at the carbon shown in bold in the carbocation, radical, and anion derived from the *tert*-butyl group. Is the arrangement of bonds to this carbon linear, tetrahedral, trigonal planar, or trigonal pyramidal?

$$^+C(CH_3)_3 \qquad \cdot C(CH_3)_3 \qquad :\overline{C}(CH_3)_3$$

| *tert*-Butyl | Cation | Radical | Anion |

(b) Which of these species is the most nucleophilic? Which is the most electrophilic?
(c) Predict which species reacts with water to give $HC(CH_3)_3$.
(d) Write a chemical equation for the reaction in part (c), and use curved arrows to show the flow of electrons.

Standardized-Exam Practice Problems

Verbal Reasoning

In 1960, Canadian-born doctor and pharmacologist Frances Kelsey was hired by the FDA to review applications for the approval of new drugs. Her first assignment was an application by the William S. Merrell Company of Cincinnati, Ohio, requesting approval of thalidomide as a sedative and anti-emetic for pregnant women suffering from morning sickness. Kelsey rejected the application, citing the manufacturer's failure to prove the drug's safety, and requested that further studies be done. Although the company initially complied with her request, Kelsey, still unsatisfied with the data, refused the application a second time. Eventually, Merrell appealed to Kelsey's superiors to pressure her to approve the drug. By early 1961, though, a study in England reported that repeated use of the drug might have serious nervous-system side effects. In addition, Kelsey's research in pharmacology early in her career made her question the effects of the drug on a developing fetus. She continued to resist the pressure to approve the drug and was vindicated late in 1961 when a German scientist, Dr. Widukind Lenz, reported that use of thalidomide by pregnant women might be the cause of the epidemic of *phocomelia*, a malformation of limbs in newborn babies, in Germany.

U.S. law at the time allowed drugs that had not yet been approved by the FDA to be distributed to physicians for "experimental use." Nevertheless, only 17 thalidomide babies are known to have been born in the United States. If not for the diligence and dedication of Dr. Frances Kelsey, there would almost certainly have been thousands more. In recognition of her incalculable service to the people of the United States, Dr. Kelsey was given the President's Award for Distinguished Federal Civilian Service, the highest award ever given to a civilian. President John F. Kennedy presented the medal to Kelsey at a ceremony at the White House in August 1962. As a result of the thalidomide tragedy, new laws were enacted placing a greater burden on drug manufacturers to prove the safety and efficacy of their drugs. In addition, the concept of "informed consent" was introduced, preventing the distribution of unapproved drugs to unsuspecting patients.

1. The main point of the passage is that Frances Kelsey

   a) was born in Canada.
   b) worked for the FDA.
   c) prevented the thalidomide tragedy in the United States.
   d) was awarded the Distinguished Federal Civilian Service medal.

2. According to the passage, Dr. Kelsey refused to approve thalidomide for distribution in the United States because

   a) her superiors pressured her not to approve it.
   b) she knew that it had caused thousands of birth defects in Europe.
   c) 17 babies had been born with birth defects as a result of the drug.
   d) she was not satisfied with the manufacturer's studies on the safety of the drug.

3. Based on the passage, what was likely the reason that 17 thalidomide babies were born in the United States?

   a) Physicians in the United States gave the drug to their patients without FDA approval.
   b) Pregnant women arrived from Europe and gave birth in the United States.
   c) U.S. citizens obtained the drug from overseas sources.
   d) Some women in the United States failed to heed the manufacturer's warning about use during pregnancy.

4. Based on the passage, the author most likely considers Kelsey to be

   a) a celebrity.
   b) an activist.
   c) a hero.
   d) a role model.

Answers to In-Chapter Materials

Answers to Practice Problems

23.1A (a) 3-Ethylpentane, (b) 2-bromohexane, (c) 2-chlorobutane.

23.1B (a) 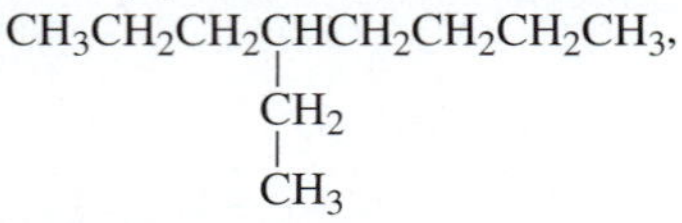 $CH_3CH_2CH_2CHCH_2CH_2CH_2CH_3$,

$$CH_2$$
$$CH_3$$

(b) 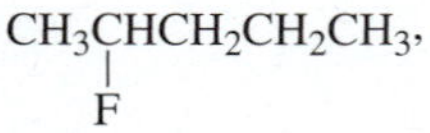 $CH_3CHCH_2CH_2CH_3$,

$$F$$

(c) 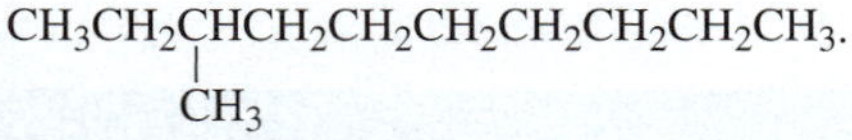 $CH_3CH_2CHCH_2CH_2CH_2CH_2CH_2CH_3$.

$$CH_3$$

23.2A (a) Ester, (b) carboxy group; ester, (c) 2 hydroxy groups, 2 carboxy groups.

23.2B (a) Ketone, (b) 3° amide, (c) ester.

23.3A CH_2BrCH_2COOH. **23.3B**

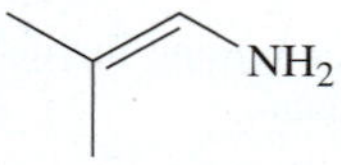

23.4A

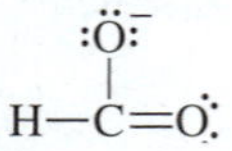

23.4B

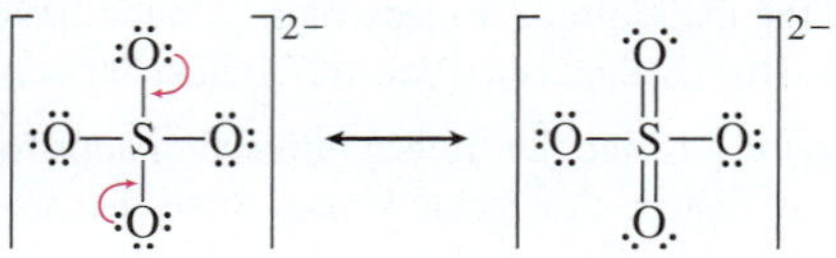

23.5A (a)

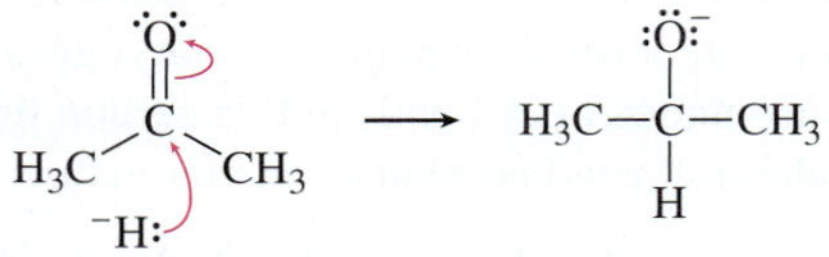

(b)

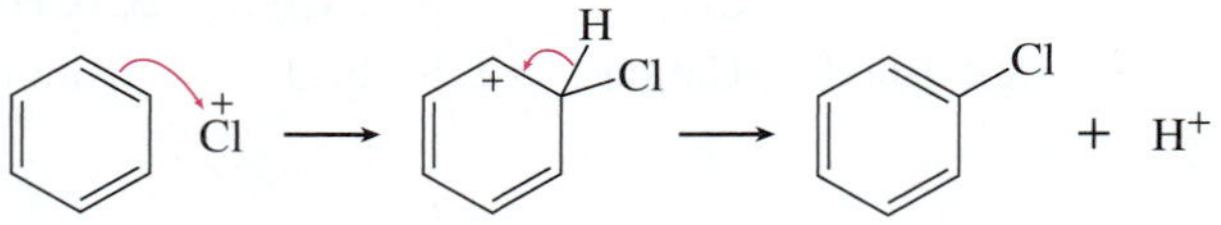

23.5B

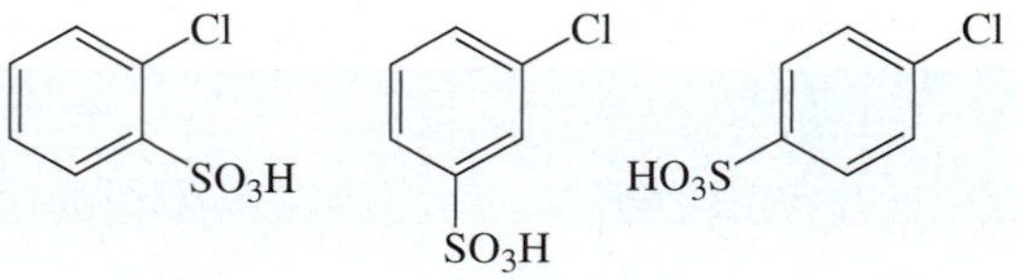

Answers to Checkpoints

23.2.1 c. **23.2.2** b. **23.2.3** b. **23.2.4** e. **23.2.5** b, c. **23.2.6** a, b.
23.3.1 b. **23.3.2** d. **23.3.3** e. **23.3.4** b, d. **23.5.1** d. **23.5.2** a.

Appendix 1

Mathematical Operations

Scientific Notation

Chemists often deal with numbers that are either extremely large or extremely small. For example, in 1 g of the element hydrogen there are roughly

$$602{,}200{,}000{,}000{,}000{,}000{,}000{,}000$$

hydrogen atoms. Each hydrogen atom has a mass of only

$$0.00000000000000000000000166 \text{ g}$$

These numbers are cumbersome to handle, and it is easy to make mistakes when using them in arithmetic computations. Consider the following multiplication:

$$0.0000000056 \times 0.00000000048 = 0.0000000000000000002688$$

It would be easy for us to miss one zero or add one more zero after the decimal point. Consequently, when working with very large and very small numbers, we use a system called *scientific notation*. Regardless of their magnitude, all numbers can be expressed in the form

$$N \times 10^n$$

where N is a number between 1 and 10 and n, the exponent, is a positive or negative integer (whole number). Any number expressed in this way is said to be written in scientific notation.

Suppose that we are given a certain number and asked to express it in scientific notation. Basically, this assignment calls for us to find n. We count the number of places that the decimal point must be moved to give the number N (which is between 1 and 10). If the decimal point has to be moved to the left, then n is a positive integer; if it has to be moved to the right, n is a negative integer. The following examples illustrate the use of scientific notation:

1. Express 568.762 in scientific notation:

$$568.762 = 5.68762 \times 10^2$$

Note that the decimal point is moved to the left by two places and $n = 2$.

2. Express 0.00000772 in scientific notation:

$$0.00000772 = 7.72 \times 10^{-6}$$

Here the decimal point is moved to the right by six places and $n = -6$.

Keep in mind the following two points. First, $n = 0$ is used for numbers that are not expressed in scientific notation. For example, 74.6×10^0 ($n = 0$) is equivalent to 74.6. Second, the usual practice is to omit the superscript when $n = 1$. Thus, the scientific notation for 74.6 is 7.46×10 and not 7.46×10^1.

Next, we consider how scientific notation is handled in arithmetic operations.

Addition and Subtraction

To add or subtract using scientific notation, we first write each quantity—say N_1 and N_2—with the same exponent, n. Then we combine N_1 and N_2; the exponents remain the same. Consider the following examples:

$$(7.4 \times 10^3) + (2.1 \times 10^3) = 9.5 \times 10^3$$
$$(4.31 \times 10^4) + (3.9 \times 10^3) = (4.31 \times 10^4) + (0.39 \times 10^4)$$
$$= 4.70 \times 10^4$$
$$(2.22 \times 10^{-2}) - (4.10 \times 10^{-3}) = (2.22 \times 10^{-2}) - (0.41 \times 10^{-2})$$
$$= 1.81 \times 10^{-2}$$

Multiplication and Division

To multiply numbers expressed in scientific notation, we multiply N_1 and N_2 in the usual way, but *add* the exponents together. To divide using scientific notation, we divide N_1 and N_2 as usual and subtract the exponents. The following examples show how these operations are performed:

$$(8.0 \times 10^4) \times (5.0 \times 10^2) = (8.0 \times 5.0)(10^{4+2})$$
$$= 40 \times 10^6$$
$$= 4.0 \times 10^7$$
$$(4.0 \times 10^{-5}) \times (7.0 \times 10^3) = (4.0 \times 7.0)(10^{-5+3})$$
$$= 28 \times 10^{-2}$$
$$= 2.8 \times 10^{-1}$$
$$\frac{6.9 \times 10^7}{3.0 \times 10^{-5}} = \frac{6.9}{3.0} \times 10^{7-(-5)}$$
$$= 2.3 \times 10^{12}$$
$$\frac{8.5 \times 10^4}{5.0 \times 10^9} = \frac{8.5}{5.0} \times 10^{4-9}$$
$$= 1.7 \times 10^{-5}$$

Basic Trigonometry

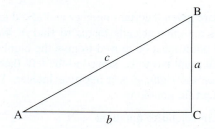

In the triangle shown, A, B, and C are angles (C = 90°) and a, b, and c are side lengths. To calculate unknown angles or sides, we use the following relationships:

$$a^2 + b^2 = c^2$$

$$\sin A = \frac{a}{c}$$

$$\cos A = \frac{b}{c}$$

$$\tan A = \frac{a}{b}$$

Logarithms

Common Logarithms

The concept of logarithms is an extension of the concept of exponents, which is discussed on page A-1. The common, or base-10, logarithm of any number is the power to which 10 must be raised to equal the number. The following examples illustrate this relationship:

Logarithm	Exponent
$\log 1 = 0$	$10^0 = 1$
$\log 10 = 1$	$10^1 = 10$
$\log 100 = 2$	$10^2 = 100$
$\log 10^{-1} = -1$	$10^{-1} = 0.1$
$\log 10^{-2} = -2$	$10^{-2} = 0.01$

In each case, the logarithm of the number can be obtained by inspection.

Because the logarithms of numbers are exponents, they have the same properties as exponents. Thus, we have

Logarithm	Exponent
$\log AB = \log A + \log B$	$10^A \times 10^B = 10^{A+B}$
$\log \dfrac{A}{B} = \log A - \log B$	$\dfrac{10^A}{10^B} = 10^{A-B}$

Furthermore, $\log A^n = n \log A$.

Now suppose we want to find the common logarithm of 6.7×10^{-4}. On most electronic calculators, the number is entered first and then the log key is pressed. This operation gives us

$$\log 6.7 \times 10^{-4} = -3.17$$

Note that there are as many digits *after* the decimal point as there are significant figures in the original number. The original number has two significant figures, and the "17" in -3.17 tells us that the log has two significant figures. The "3" in -3.17 serves only to locate the decimal point in the number 6.7×10^{-4}. Other examples are

Number	Common Logarithm
62	1.79
0.872	-0.0595
1.0×10^{-7}	-7.00

Sometimes (as in the case of pH calculations) it is necessary to obtain the number whose logarithm is known. This procedure is known as taking the antilogarithm; it is simply the reverse of taking the logarithm of a number. Suppose in a certain calculation we have pH = 1.46 and are asked to calculate $[H^+]$. From the definition of pH (pH = $-\log [H^+]$), we can write

$$[H^+] = 10^{-1.46}$$

Many calculators have a key labeled $\log^{-1}$ or INV log to obtain antilogs. Other calculators have a 10^x or y^x key (where x corresponds to -1.46 in our example and y is 10 for base-10 logarithm). Therefore, we find that $[H^+] = 0.035\ M$.

Natural Logarithms

Logarithms taken to the base e instead of 10 are known as natural logarithms (denoted by ln or $\log_e$); e is equal to 2.7183. The relationship between common logarithms and natural logarithms is as follows:

$$\log 10 = 1 \qquad\qquad 10^1 = 10$$

$$\ln 10 = 2.303 \qquad e^{2.303} = 10$$

Thus,

$$\ln x = 2.303 \log x$$

To find the natural logarithm of 2.27, say, we use ln key to get

$$\ln 2.27 = 0.820$$

For a calculator that does not have an ln key, we can proceed as follows:

$$2.303 \log 2.27 = 2.303 \times 0.356$$
$$= 0.820$$

Sometimes we may be given the natural logarithm and asked to find the number it represents. For example,

$$\ln x = 59.7$$

On many calculators, we simply use the e key:

$$e^{59.7} = 8 \times 10^{25}$$

The Quadratic Equation

A quadratic equation takes the form

$$ax^2 + bx + c = 0$$

If coefficients a, b, and c are known, then x is given by

$$x = \frac{-b \pm \sqrt{b^2 - 4ac}}{2a}$$

Suppose we have the following quadratic equation:

$$2x^2 + 5x - 12 = 0$$

Solving for x, we write

$$x = \frac{-5 \pm \sqrt{(5)^2 - 4(2)(-12)}}{2(2)}$$

$$= \frac{-5 \pm \sqrt{25 + 96}}{4}$$

Therefore,

$$x = \frac{-5 + 11}{4} = \frac{3}{2}$$

and

$$x = \frac{-5 - 11}{4} = -4$$

Successive Approximation

In the determination of hydrogen ion concentration in a weak acid solution, use of the quadratic equation can sometimes be avoided using a method known as *successive approximation*. Consider the example of a 0.0150 M solution of hydrofluoric acid (HF). The K_a for HF is 7.10×10^{-4}. To determine the hydrogen ion concentration in this solution, we construct an equilibrium table and enter the initial concentrations, the expected change in concentrations, and the equilibrium concentrations of all species:

$$HF(aq) \rightleftharpoons H^+(aq) + F^-(aq)$$

Initial concentration (M):	0.0150	0	0
Change in concentration (M):	$-x$	$+x$	$+x$
Equilibrium concentration (M):	$0.0150 - x$	x	x

Using the rule that x can be neglected if the initial acid concentration divided by the K_a is greater than 100, we find that in this case x cannot be neglected ($0.0150/7.1 \times 10^{-4} \approx 21$). Successive approximation involves first neglecting x with respect to initial acid concentration

$$\frac{x^2}{0.0150 - x} \approx \frac{x^2}{0.0150} = 7.10 \times 10^{-4}$$

and solving for x:

$$x^2 = (0.0150)(7.10 \times 10^{-4}) = 1.07 \times 10^{-5}$$

$$x = \sqrt{1.07 \times 10^{-5}} = 0.00326 \ M$$

We then solve for x again, this time using the calculated value of x on the bottom of the fraction:

$$\frac{x^2}{0.0150 - x} = \frac{x^2}{0.0150 - 0.00326} = 7.10 \times 10^{-4}$$

$$x^2 = (0.0150 - 0.00326)(7.10 \times 10^{-4}) = 8.33 \times 10^{-6}$$

$$x = \sqrt{8.33 \times 10^{-6}} = 0.00289 \ M$$

Note that the calculated value of x decreased from 0.00326 to 0.00289. We now use the new calculated value of x on the bottom of the fraction and solve for x again:

$$\frac{x^2}{0.0150 - x} = \frac{x^2}{0.0150 - 0.00289} = 7.10 \times 10^{-4}$$

$$x^2 = (0.0150 - 0.00289)(7.10 \times 10^{-4}) = 8.60 \times 10^{-6}$$

$$x = \sqrt{8.60 \times 10^{-6}} = 0.00293 \ M$$

This time the value of x increased slightly. We use the new calculated value and solve for x again.

$$\frac{x^2}{0.0150 - x} = \frac{x^2}{0.0150 - 0.00293} = 7.10 \times 10^{-4}$$

$$x^2 = (0.0150 - 0.00293)(7.10 \times 10^{-4}) = 8.57 \times 10^{-6}$$

$$x = \sqrt{8.57 \times 10^{-6}} = 0.00293 \ M$$

This time we find that the answer is still 0.00293 M, so there is no need to repeat the process. In general, we apply the method of successive approximation until the value of x obtained does not differ from the value obtained in the previous step. The value of x determined using successive approximation is the same value we would get if we were to use the quadratic equation.

Appendix 2

Thermodynamic Data at 1 atm and 25°C*

Inorganic Substances			
Substance	ΔH_f° (kJ/mol)	ΔG_f° (kJ/mol)	S° (J/K · mol)
Ag(s)	0	0	42.7
Ag^+(aq)	105.9	77.1	73.9
AgCl(s)	−127.0	−109.7	96.1
AgBr(s)	−99.5	−95.9	107.1
AgI(s)	−62.4	−66.3	114.2
$AgNO_3$(s)	−123.1	−32.2	140.9
Al(s)	0	0	28.3
Al^{3+}(aq)	−524.7	−481.2	−313.38
Al_2O_3(s)	−1669.8	−1576.4	50.99
As(s)	0	0	35.15
AsO_4^{3-}(aq)	−870.3	−635.97	−144.77
AsH_3(g)	171.5		
H_3AsO_4(s)	−900.4		
Au(s)	0	0	47.7
Au_2O_3(s)	80.8	163.2	125.5
AuCl(s)	−35.2		
$AuCl_3$(s)	−118.4		
B(s)	0	0	6.5
B_2O_3(s)	−1263.6	−1184.1	54.0
H_3BO_3(s)	−1087.9	−963.16	89.58
H_3BO_3(aq)	−1067.8	−963.3	159.8
Ba(s)	0	0	66.9
Ba^{2+}(aq)	−538.4	−560.66	12.55
BaO(s)	−558.2	−528.4	70.3
$BaCl_2$(s)	−860.1	−810.66	125.5
$BaSO_4$(s)	−1464.4	−1353.1	132.2
$BaCO_3$(s)	−1218.8	−1138.9	112.1
Be(s)	0	0	9.5
BeO(s)	−610.9	−581.58	14.1

* The thermodynamic quantities of ions are based on the reference states that $\Delta H_f^\circ[H^+(aq)] = 0$, $\Delta G_f^\circ[H^+(aq)] = 0$, and $S^\circ[H^+(aq)] = 0$.

Substance	ΔH_f° (kJ/mol)	ΔG_f° (kJ/mol)	S° (J/K · mol)
$Br_2(l)$	0	0	152.3
$Br_2(g)$	30.7	3.14	245.13
$Br^-(aq)$	−120.9	−102.8	80.7
$HBr(g)$	−36.2	−53.2	198.48
$C(graphite)$	0	0	5.69
$C(diamond)$	1.90	2.87	2.4
$CCl_4(g)$	−95.7	−62.3	309.7
$CCl_4(l)$	−128.2	−66.4	216.4
$CO(g)$	−110.5	−137.3	197.9
$CO_2(g)$	−393.5	−394.4	213.6
$CO_2(aq)$	−412.9	−386.2	121.3
$CO_3^{2-}(aq)$	−676.3	−528.1	−53.1
$HCO_3^-(aq)$	−691.1	−587.1	94.98
$H_2CO_3(aq)$	−699.7	−623.2	187.4
$CS_2(g)$	115.3	65.1	237.8
$CS_2(l)$	87.3	63.6	151.0
$HCN(aq)$	105.4	112.1	128.9
$CN^-(aq)$	151.0	165.69	117.99
$(NH_2)_2CO(s)$	−333.19	−197.15	104.6
$(NH_2)_2CO(aq)$	−319.2	−203.84	173.85
$Ca(s)$	0	0	41.6
$Ca(g)$	179.3	145.5	154.8
$Ca^{2+}(aq)$	−542.96	−553.0	−55.2
$CaO(s)$	−635.6	−604.2	39.8
$Ca(OH)_2(s)$	−986.6	−896.8	83.4
$CaF_2(s)$	−1214.6	−1161.9	68.87
$CaCl_2(s)$	−794.96	−750.19	113.8
$CaSO_4(s)$	−1432.69	−1320.3	106.69
$CaCO_3(s)$	−1206.9	−1128.8	92.9
$Cd(s)$	0	0	51.46
$Cd^{2+}(aq)$	−72.38	−77.7	−61.09
$CdO(s)$	−254.6	−225.06	54.8
$CdCl_2(s)$	−389.1	−342.59	118.4
$CdSO_4(s)$	−926.17	−820.2	137.2
$Cl_2(g)$	0	0	223.0
$Cl(g)$	121.7	105.7	165.2
$Cl^-(aq)$	−167.2	−131.2	56.5
$HCl(g)$	−92.3	−95.27	187.0
$Co(s)$	0	0	28.45
$Co^{2+}(aq)$	−67.36	−51.46	155.2

(Continued)

Substance	ΔH_f° (kJ/mol)	ΔG_f° (kJ/mol)	S° (J/K · mol)
$CoO(s)$	−239.3	−213.38	43.9
$Cr(s)$	0	0	23.77
$Cr^{2+}(aq)$	−138.9		
$Cr_2O_3(s)$	−1128.4	−1046.8	81.17
$CrO_4^{2-}(aq)$	−863.16	−706.26	38.49
$Cr_2O_7^{2-}(aq)$	−1460.6	−1257.29	213.8
$Cs(s)$	0	0	82.8
$Cs(g)$	76.50	49.53	175.6
$Cs^+(aq)$	−247.69	−282.0	133.05
$CsCl(s)$	−442.8	−414.4	101.2
$Cu(s)$	0	0	33.3
$Cu^+(aq)$	51.88	50.2	−26.4
$Cu^{2+}(aq)$	64.39	64.98	−99.6
$CuO(s)$	−155.2	−127.2	43.5
$Cu_2O(s)$	−166.69	−146.36	100.8
$CuCl(s)$	−134.7	−118.8	91.6
$CuCl_2(s)$	−205.85		
$CuS(s)$	−48.5	−49.0	66.5
$CuSO_4(s)$	−769.86	−661.9	113.39
$F_2(g)$	0	0	203.34
$F(g)$	80.0	61.9	158.7
$F^-(aq)$	−329.1	−276.48	−9.6
$HF(g)$	−271.6	−270.7	173.5
$Fe(s)$	0	0	27.2
$Fe^{2+}(aq)$	−87.86	−84.9	−113.39
$Fe^{3+}(aq)$	−47.7	−10.5	−293.3
$FeO(s)$	−272.0	−255.2	60.8
$Fe_2O_3(s)$	−822.2	−741.0	90.0
$Fe(OH)_2(s)$	−568.19	−483.55	79.5
$Fe(OH)_3(s)$	−824.25		
$H(g)$	218.2	203.2	114.6
$H_2(g)$	0	0	131.0
$H^+(aq)$	0	0	0
$OH^-(aq)$	−229.94	−157.30	−10.5
$H_2O(g)$	−241.8	−228.6	188.7
$H_2O(l)$	−285.8	−237.2	69.9
$H_2O_2(g)$	−136.1	−105.5	232.9
$H_2O_2(l)$	−187.6	−118.1	109.6
$Hg(l)$	0	0	77.4
$Hg^{2+}(aq)$		−164.38	
$HgO(s)$	−90.7	−58.5	72.0

Substance	ΔH_f° (kJ/mol)	ΔG_f° (kJ/mol)	S° (J/K · mol)
$HgCl_2(s)$	−230.1		
$Hg_2Cl_2(s)$	−264.9	−210.66	196.2
$HgS(s)$	−58.16	−48.8	77.8
$HgSO_4(s)$	−704.17		
$Hg_2SO_4(s)$	−741.99	−623.92	200.75
$I_2(g)$	62.25	19.37	260.57
$I_2(s)$	0	0	116.7
$I(g)$	106.8	70.21	180.67
$I^-(aq)$	−55.9	−51.67	109.37
$HI(g)$	25.9	1.30	206.3
$K(s)$	0	0	63.6
$K^+(aq)$	−251.2	−282.28	102.5
$KOH(s)$	−425.85		
$KCl(s)$	−435.87	−408.3	82.68
$KClO_3(s)$	−391.20	−289.9	142.97
$KClO_4(s)$	−433.46	−304.18	151.0
$KBr(s)$	−392.17	−379.2	96.4
$KI(s)$	−327.65	−322.29	104.35
$KNO_3(s)$	−492.7	−393.1	132.9
$Li(s)$	0	0	28.0
$Li(g)$	159.3	126.6	138.8
$Li^+(aq)$	−278.46	−293.8	14.2
$LiCl(s)$	−408.3	−384.0	59.30
$Li_2O(s)$	−595.8		
$LiOH(s)$	−487.2	−443.9	50.2
$Mg(s)$	0	0	32.5
$Mg(g)$	150	115	148.55
$Mg^{2+}(aq)$	−461.96	−456.0	−117.99
$MgO(s)$	−601.8	−569.6	26.78
$Mg(OH)_2(s)$	−924.66	−833.75	63.1
$MgCl_2(s)$	−641.8	−592.3	89.5
$MgSO_4(s)$	−1278.2	−1173.6	91.6
$MgCO_3(s)$	−1112.9	−1029.3	65.69
$Mn(s)$	0	0	31.76
$Mn^{2+}(aq)$	−218.8	−223.4	−83.68
$MnO_2(s)$	−520.9	−466.1	53.1
$N(g)$	470.4	455.5	153.3
$N_2(g)$	0	0	191.5
$N_3^-(aq)$	245.18		
$NH_3(g)$	−46.3	−16.6	193.0

(Continued)

Substance	ΔH_f° (kJ/mol)	ΔG_f° (kJ/mol)	S° (J/K · mol)
$NH_4^+(aq)$	−132.80	−79.5	112.8
$NH_4Cl(s)$	−315.39	−203.89	94.56
$NH_3(aq)$	−80.3	−26.5	111.3
$N_2H_4(l)$	50.4		
$NO(g)$	90.4	86.7	210.6
$NO_2(g)$	33.85	51.8	240.46
$N_2O_4(g)$	9.66	98.29	304.3
$N_2O(g)$	81.56	103.6	219.99
$HNO_2(aq)$	−118.8	−53.6	
$HNO_3(l)$	−173.2	−79.9	155.6
$NO_3^-(aq)$	−206.57	−110.5	146.4
$Na(s)$	0	0	51.05
$Na(l)$	2.41	0.50	57.56
$Na(g)$	107.7	77.3	153.7
$Na^+(aq)$	−239.66	−261.87	60.25
$NaOH(aq)$	−469.6	−419.2	49.8
$Na_2O(s)$	−415.9	−376.56	72.8
$NaCl(s)$	−410.9	−384.0	72.38
$NaI(s)$	−288.0		
$Na_2SO_4(s)$	−1384.49	−1266.8	149.49
$NaNO_3(s)$	−466.68	−365.89	116.3
$Na_2CO_3(s)$	−1130.9	−1047.67	135.98
$NaHCO_3(s)$	−947.68	−851.86	102.09
$Ni(s)$	0	0	30.1
$Ni^{2+}(aq)$	−64.0	−46.4	−159.4
$NiO(s)$	−244.35	−216.3	38.58
$Ni(OH)_2(s)$	−538.06	−453.1	79.5
$O(g)$	249.4	230.1	160.95
$O_2(g)$	0	0	205.0
$O_3(aq)$	−12.09	16.3	110.88
$O_3(g)$	142.2	163.4	237.6
$P(white)$	0	0	44.0
$P(red)$	−18.4	13.8	29.3
$PCl_3(l)$	−319.7	−272.3	217.1
$PCl_3(g)$	−288.07	−269.6	311.7
$PCl_5(g)$	−374.9	−305.0	364.5
$PO_4^{3-}(aq)$	−1284.07	−1025.59	−217.57
$P_4O_{10}(s)$	−3012.48		
$PH_3(g)$	9.25	18.2	210.0
$HPO_4^{2-}(aq)$	−1298.7	−1094.1	−35.98
$H_2PO_4^-(aq)$	−1302.48	−1135.1	89.1
$Pb(s)$	0	0	64.89

Substance	ΔH_f° (kJ/mol)	ΔG_f° (kJ/mol)	S° (J/K · mol)
$Pb^{2+}(aq)$	1.6	−24.3	21.3
$PbO(s)$	−217.86	−188.49	69.45
$PbO_2(s)$	−276.65	−218.99	76.57
$PbCl_2(s)$	−359.2	−313.97	136.4
$PbS(s)$	−94.3	−92.68	91.2
$PbSO_4(s)$	−918.4	−811.2	147.28
$Pt(s)$	0	0	41.84
$PtCl_4^{2-}(aq)$	−516.3	−384.5	175.7
$Rb(s)$	0	0	69.45
$Rb(g)$	85.8	55.8	170.0
$Rb^+(aq)$	−246.4	−282.2	124.27
$RbBr(s)$	−389.2	−378.1	108.3
$RbCl(s)$	−435.35	−407.8	95.90
$RbI(s)$	−328	−326	118.0
$S(rhombic)$	0	0	31.88
$S(monoclinic)$	0.30	0.10	32.55
$SO(g)$	5.01	−19.9	221.8
$SO_2(g)$	−296.4	−300.4	248.5
$SO_3(g)$	−395.2	−370.4	256.2
$SO_3^{2-}(aq)$	−624.25	−497.06	43.5
$SO_4^{2-}(aq)$	−907.5	−741.99	17.15
$H_2S(g)$	−20.15	−33.0	205.64
$HSO_3^-(aq)$	−627.98	−527.3	132.38
$HSO_4^-(aq)$	−885.75	−752.87	126.86
$H_2SO_4(l)$	−811.3		
$SF_6(g)$	−1096.2		
$Si(s)$	0	0	18.70
$SiO_2(s)$	−859.3	−805.0	41.84
$Sr(s)$	0	0	54.39
$Sr^{2+}(aq)$	−545.5	−557.3	−39.33
$SrCl_2(s)$	−828.4	−781.15	117.15
$SrSO_4(s)$	−1444.74	−1334.28	121.75
$SrCO_3(s)$	−1218.38	−1137.6	97.07
$U(s)$	0	0	50.21
$UF_6(g)$	−2147	−2064	378
$Zn(s)$	0	0	41.6
$Zn^{2+}(aq)$	−152.4	−147.2	−106.48
$ZnO(s)$	−348.0	−318.2	43.9
$ZnCl_2(s)$	−415.89	−369.26	108.37
$ZnS(s)$	−202.9	−198.3	57.7
$ZnSO_4(s)$	−978.6	−871.6	124.7

Organic Substances

Substance	Formula	ΔH_f° (kJ/mol)	ΔG_f° (kJ/mol)	S° (J/K · mol)
Acetic acid(*l*)	CH_3COOH	−484.2	−389.45	159.8
Acetaldehyde(*g*)	CH_3CHO	−166.35	−139.08	264.2
Acetone(*l*)	CH_3COCH_3	−246.8	−153.55	198.7
Acetylene(*g*)	C_2H_2	226.6	209.2	200.8
Benzene(*l*)	C_6H_6	49.04	124.5	172.8
Butane(*g*)	C_4H_{10}	−124.7	−15.7	310.0
Ethanol(*l*)	C_2H_5OH	−276.98	−174.18	161.0
Ethane(*g*)	C_2H_6	−84.7	−32.89	229.5
Ethylene(*g*)	C_2H_4	52.3	68.1	219.5
Formic acid(*l*)	$HCOOH$	−409.2	−346.0	129.0
Glucose(*s*)	$C_6H_{12}O_6$	−1274.5	−910.56	212.1
Methane(*g*)	CH_4	−74.85	−50.8	186.2
Methanol(*l*)	CH_3OH	−238.7	−166.3	126.8
Propane(*g*)	C_3H_8	−103.9	−23.5	269.9
Sucrose(*s*)	$C_{12}H_{22}O_{11}$	−2221.7	−1544.3	360.2

Appendix 3

Solubility Product Constants at 25°C

Compound	Dissolution Equilibrium	K_{sp}
Bromides		
Copper(I) bromide	$CuBr(s) \rightleftharpoons Cu^+(aq) + Br^-(aq)$	4.2×10^{-8}
Lead(II) bromide	$PbBr_2(s) \rightleftharpoons Pb^{2+}(aq) + 2Br^-(aq)$	6.6×10^{-6}
Mercury(I) bromide	$Hg_2Br_2(s) \rightleftharpoons Hg_2^{2+}(aq) + 2Br^-(aq)$	6.4×10^{-23}
Silver bromide	$AgBr(s) \rightleftharpoons Ag^+(aq) + Br^-(aq)$	7.7×10^{-13}
Carbonates		
Barium carbonate	$BaCO_3(s) \rightleftharpoons Ba^{2+}(aq) + CO_3^{2-}(aq)$	8.1×10^{-9}
Calcium carbonate	$CaCO_3(s) \rightleftharpoons Ca^{2+}(aq) + CO_3^{2-}(aq)$	8.7×10^{-9}
Lead(II) carbonate	$PbCO_3(s) \rightleftharpoons Pb^{2+}(aq) + CO_3^{2-}(aq)$	3.3×10^{-14}
Magnesium carbonate	$MgCO_3(s) \rightleftharpoons Mg^{2+}(aq) + CO_3^{2-}(aq)$	4.0×10^{-5}
Silver carbonate	$Ag_2CO_3(s) \rightleftharpoons 2Ag^+(aq) + CO_3^{2-}(aq)$	8.1×10^{-12}
Strontium carbonate	$SrCO_3(s) \rightleftharpoons Sr^{2+}(aq) + CO_3^{2-}(aq)$	1.6×10^{-9}
Chlorides		
Lead(II) chloride	$PbCl_2(s) \rightleftharpoons Pb^{2+}(aq) + 2Cl^-(aq)$	2.4×10^{-4}
Mercury(I) chloride	$Hg_2Cl_2(s) \rightleftharpoons Hg_2^{2+}(aq) + 2Cl^-(aq)$	3.5×10^{-18}
Silver chloride	$AgCl(s) \rightleftharpoons Ag^+(aq) + Cl^-(aq)$	1.6×10^{-10}
Chromates		
Lead(II) chromate	$PbCrO_4(s) \rightleftharpoons Pb^{2+}(aq) + CrO_4^{2-}(aq)$	2.0×10^{-14}
Silver(I) chromate	$Ag_2CrO_4(s) \rightleftharpoons 2Ag^+(aq) + CrO_4^{2-}(aq)$	1.2×10^{-12}
Fluorides		
Barium fluoride	$BaF_2(s) \rightleftharpoons Ba^{2+}(aq) + 2F^-(aq)$	1.7×10^{-6}
Calcium fluoride	$CaF_2(s) \rightleftharpoons Ca^{2+}(aq) + 2F^-(aq)$	4.0×10^{-11}
Lead(II) fluoride	$PbF_2(s) \rightleftharpoons Pb^{2+}(aq) + 2F^-(aq)$	4.0×10^{-8}

(Continued)

Compound	Dissolution Equilibrium	K_{sp}
Hydroxides		
Aluminum hydroxide	$Al(OH)_3(s) \rightleftharpoons Al^{3+}(aq) + 3OH^-(aq)$	1.8×10^{-33}
Calcium hydroxide	$Ca(OH)_2(s) \rightleftharpoons Ca^{2+}(aq) + 2OH^-(aq)$	8.0×10^{-6}
Chromium(III) hydroxide	$Cr(OH)_3(s) \rightleftharpoons Cr^{3+}(aq) + 3OH^-(aq)$	3.0×10^{-29}
Copper(II) hydroxide	$Cu(OH)_2(s) \rightleftharpoons Cu^{2+}(aq) + 2OH^-(aq)$	2.2×10^{-20}
Iron(II) hydroxide	$Fe(OH)_2(s) \rightleftharpoons Fe^{2+}(aq) + 2OH^-(aq)$	1.6×10^{-14}
Iron(III) hydroxide	$Fe(OH)_3(s) \rightleftharpoons Fe^{3+}(aq) + 3OH^-(aq)$	1.1×10^{-36}
Magnesium hydroxide	$Mg(OH)_2(s) \rightleftharpoons Mg^{2+}(aq) + 2OH^-(aq)$	1.2×10^{-11}
Strontium hydroxide	$Sr(OH)_2(s) \rightleftharpoons Sr^{2+}(aq) + 2OH^-(aq)$	3.2×10^{-4}
Zinc hydroxide	$Zn(OH)_2(s) \rightleftharpoons Zn^{2+}(aq) + 2OH^-(aq)$	1.8×10^{-14}
Iodides		
Copper(I) iodide	$CuI(s) \rightleftharpoons Cu^+(aq) + I^-(aq)$	5.1×10^{-12}
Lead(II) iodide	$PbI_2(s) \rightleftharpoons Pb^{2+}(aq) + 2I^-(aq)$	1.4×10^{-8}
Silver iodide	$AgI(s) \rightleftharpoons Ag^+(aq) + I^-(aq)$	8.3×10^{-17}
Phosphates		
Calcium phosphate	$Ca_3(PO_4)_2(s) \rightleftharpoons 3Ca^{2+}(aq) + 2PO_4^{3-}(aq)$	1.2×10^{-26}
Iron(III) phosphate	$FePO_4(s) \rightleftharpoons Fe^{3+}(aq) + PO_4^{3-}(aq)$	1.3×10^{-22}
Sulfates		
Barium sulfate	$BaSO_4(s) \rightleftharpoons Ba^{2+}(aq) + SO_4^{2-}(aq)$	1.1×10^{-10}
Calcium sulfate	$CaSO_4(s) \rightleftharpoons Ca^{2+}(aq) + SO_4^{2-}(aq)$	2.4×10^{-5}
Lead(II) sulfate	$PbSO_4(s) \rightleftharpoons Pb^{2+}(aq) + SO_4^{2-}(aq)$	1.8×10^{-8}
Mercury(I) sulfate	$Hg_2SO_4(s) \rightleftharpoons Hg_2^{2+}(aq) + SO_4^{2-}(aq)$	6.5×10^{-7}
Silver sulfate	$Ag_2SO_4(s) \rightleftharpoons 2Ag^+(aq) + SO_4^{2-}(aq)$	1.5×10^{-5}
Strontium sulfate	$SrSO_4(s) \rightleftharpoons Sr^{2+}(aq) + SO_4^{2-}(aq)$	3.8×10^{-7}
Sulfides		
Bismuth sulfide	$Bi_2S_3(s) \rightleftharpoons 2Bi^{3+}(aq) + 3S^{2-}(aq)$	1.6×10^{-72}
Cadmium sulfide	$CdS(s) \rightleftharpoons Cd^{2+}(aq) + S^{2-}(aq)$	8.0×10^{-28}
Cobalt(II) sulfide	$CoS(s) \rightleftharpoons Co^{2+}(aq) + S^{2-}(aq)$	4.0×10^{-21}
Copper(II) sulfide	$CuS(s) \rightleftharpoons Cu^{2+}(aq) + S^{2-}(aq)$	6.0×10^{-37}
Iron(II) sulfide	$FeS(s) \rightleftharpoons Fe^{2+}(aq) + S^{2-}(aq)$	6.0×10^{-19}
Lead(II) sulfide	$PbS(s) \rightleftharpoons Pb^{2+}(aq) + S^{2-}(aq)$	3.4×10^{-28}
Manganese(II) sulfide	$MnS(s) \rightleftharpoons Mn^{2+}(aq) + S^{2-}(aq)$	3.0×10^{-14}
Mercury(II) sulfide	$HgS(s) \rightleftharpoons Hg^{2+}(aq) + S^{2-}(aq)$	4.0×10^{-54}
Nickel(II) sulfide	$NiS(s) \rightleftharpoons Ni^{2+}(aq) + S^{2-}(aq)$	1.4×10^{-24}
Silver sulfide	$Ag_2S(s) \rightleftharpoons 2Ag^+(aq) + S^{2-}(aq)$	6.0×10^{-51}
Tin(II) sulfide	$SnS(s) \rightleftharpoons Sn^{2+}(aq) + S^{2-}(aq)$	1.0×10^{-26}
Zinc sulfide	$ZnS(s) \rightleftharpoons Zn^{2+}(aq) + S^{2-}(aq)$	3.0×10^{-23}

Dissociation Constants for Weak Acids and Bases at 25°C

Weak Acids				
Name	**Formula**	K_{a_1}	K_{a_2}	K_{a_3}
Acetic	$HC_2H_3O_2$ (CH_3COOH)	1.8×10^{-5}		
Acetylsalicylic	$HC_9H_7O_4$	3.0×10^{-4}		
Ascorbic	$H_2C_6H_6O_6$	8.0×10^{-5}	1.6×10^{-12}	
Benzoic	$HC_7H_5O_2$ (C_6H_5COOH)	6.5×10^{-5}		
Carbonic	H_2CO_3	4.2×10^{-7}	4.8×10^{-11}	
Chloroacetic	$HC_2H_2O_2Cl$ $(CH_2ClCOOH)$	1.4×10^{-3}		
Chlorous	$HClO_2$	1.1×10^{-2}		
Citric	$H_3C_6H_5O_7$	7.4×10^{-4}	1.7×10^{-5}	4.0×10^{-7}
Dichloroacetic	$HC_2HO_2Cl_2$ $(CHCl_2COOH)$	5.5×10^{-2}		
Formic	$HCHO_2$ $(HCOOH)$	1.7×10^{-4}		
Hydrocyanic	HCN	4.9×10^{-10}		
Hydrofluoric	HF	7.1×10^{-4}		
Hydrosulfuric	H_2S	9.5×10^{-8}	$\sim 1 \times 10^{-19}$	
Nitrous	HNO_2	4.5×10^{-4}		
Oxalic	$H_2C_2O_4$	6.5×10^{-2}	6.1×10^{-5}	
Phenol	C_6H_5OH	1.3×10^{-10}		
Phosphoric	H_3PO_4	7.5×10^{-3}	6.2×10^{-8}	4.8×10^{-13}
Phosphorous	H_3PO_3	5×10^{-2}	2×10^{-7}	
Sulfuric	H_2SO_4	very large	1.3×10^{-2}	
Sulfurous	H_2SO_3	1.3×10^{-2}	6.3×10^{-8}	
Trichloroacetic	$HC_2O_2Cl_3$ (CCl_3COOH)	2.2×10^{-1}		
Trifluoroacetic	$HC_2O_2F_3$ (CF_3COOH)	3.0×10^{-1}		

Weak Bases		
Name	**Formula**	**K_b**
Ammonia	NH_3	1.8×10^{-5}
Aniline	$C_6H_5NH_2$	3.8×10^{-10}
Ethylamine	$C_2H_5NH_2$	5.6×10^{-4}
Methylamine	CH_3NH_2	4.4×10^{-4}
Pyridine	C_5H_5N	1.7×10^{-9}
Urea	H_2NCONH_2	1.5×10^{-14}

Glossary

A

absolute temperature scale. A scale based on −273.15°C (absolute zero) being the lowest point. (10.2)

absolute zero. Theoretically the lowest obtainable temperature: −273.15°C or 0 K. (10.2)

absorbance. Negative base ten logarithm of transmittance. (4.5)

absorption spectrum. Plot of absorbance as a function of wavelength of incident light. (4.5)

accuracy. The closeness of a measurement to the true or accepted value. (1.5)

acid. See *Arrhenius acid*, *Brønsted acid*, and *Lewis acid*.

acid ionization constant (K_a). The equilibrium constant that indicates to what extent a weak acid ionizes. (16.5)

actinide series. Series of elements that has partially filled $5f$ and/or $6d$ subshells. (6.9)

activated complex. A transient species that forms when molecules collide in an effective collision. Also known as the *transition state*. (14.4)

activation energy (E_a). The minimum amount of energy to begin a chemical reaction. (14.4)

activity series. A list of metals arranged from top to bottom in order of decreasing ease of oxidation. (4.4)

actual yield. The amount of product actually obtained from a reaction. (3.7)

addition polymer. A large molecule that forms when small molecules known as monomers join together. (23.6)

addition polymerization. Process by which monomers combine to form polymers without the elimination of small molecules, such as water. (12.1)

adhesion. The attractions between unlike molecules. (11.2)

alcohol. A compound consisting of an alkyl group and the functional group OH. (23.2)

aldehyde. A compound containing a hydrogen atom bonded to a carbonyl group. (23.2)

aliphatic. Describes organic molecules that do not contain the benzene ring. (23.1)

alkali metal. An element from Group 1, with the exception of H (i.e., Li, Na, K, Rb, Cs, and Fr). (2.4)

alkaline earth metal. An element from Group 2 (Be, Mg, Ca, Sr, Ba, and Ra). (2.4)

alkane. A hydrocarbon having the general formula C_nH_{2n+2}, where $n = 1, 2, \ldots$ (2.7)

alkyl group. A portion of a molecule that resembles an alkane. (23.2)

allotrope. One of two or more distinct forms of an element. (2.7)

alloy. Homogeneous mixture of two or more metals. (24.2)

alpha particle. A helium ion with a positive charge of +2. (2.2)

alpha ray. See *alpha particle*.

amalgam. A substance made by combining mercury with one or more other metals. (24.2)

amide. An organic molecule that contains an amide group. (23.2)

amine. An organic molecule that contains an amino group. (23.2)

amino acid. A compound that contains both an amino group and a carboxy group. (23.2)

amorphous solid. A solid that lacks a regular three-dimensional arrangement of atoms. (11.5)

amphoteric. Describes an oxide that displays both acidic and basic properties. (7.7, 16.2)

amplitude. The vertical distance from the midline of a wave to the top of the peak or the bottom of the trough. (6.1)

angular momentum quantum number (ℓ). Describes the shape of the atomic orbital. (6.6)

anion. An ion with a negative charge. (2.6)

anisotropic. Dependent upon the axis of measurement. (12.3)

anode. The electrode at which oxidation occurs. (19.2)

antibonding molecular orbital. A molecular orbital that is higher in energy than the atomic orbitals that combined to produce it. (9.6)

aqueous. Dissolved in water. (3.3)

aromatic. Describes organic compounds that are related to benzene or that contain one or more benzene rings. (23.1)

Arrhenius acid. Substance that increases H^+ concentration when added to water. (2.7, 4.3)

Arrhenius base. Substance that increases OH^- concentration when added to water. (4.1, 4.3)

Arrhenius equation. An equation that gives the dependence of the rate constant of a reaction on temperature: $k = Ae^{-E_a/RT}$. (14.4)

atactic. Describes polymers in which the substituents are oriented randomly along the polymer chain. (12.1)

atom. The basic unit of an element that can enter into chemical combination. (2.2)

atomic ion. Atom that has lost or gained one or more electrons, giving it a positive or negative charge. (2.6)

atomic mass. The mass of the atom given in atomic mass units (amu). (2.5)

atomic mass unit (amu). A mass exactly equal to one-twelfth the mass of one carbon-12 atom. (2.5)

atomic number (Z). The number of protons in the nucleus of each atom of an element. (2.3)

atomic orbital. The wave function of an electron in an atom. (6.5)

atomic radius. Metallic: One-half the distance between the nuclei in the two adjacent atoms of the same element in a metal. Covalent: One-half the distance between the nuclei of the two identical atoms in a diatomic molecule. (7.4)

atomic weight. The average atomic mass. (2.5)

Aufbau principle. The process by which the periodic table can be built up by successively adding one proton to the nucleus and one electron to the appropriate atomic orbital. (6.8)

autoionization of water. Ionization of water molecules to give H^+ and OH^- ions. (16.2)

Avogadro's law. The volume of a sample of gas (V) is directly proportional to the number of moles (n) in the sample at constant temperature and pressure: $V \propto n$. (10.2)

Avogadro's number (N_A). The number of atoms in exactly 12 g of carbon-12: 6.022×10^{23}. (3.4)

axial. Describes the two bonds that form an axis perpendicular to the trigonal plane. (9.1)

B

band theory. A theory wherein atomic orbitals merge to form energy bands. (24.3)

barometer. An instrument used to measure atmospheric pressure. (10.1)

base. A compound that dissolves in water to produce hydroxide ions. See also *Arrhenius base*, *Brønsted base*, and *Lewis base*.

base ionization constant (K_b). The equilibrium constant that indicates to what extent a weak base ionizes. (16.6)

battery. A portable, self-contained source of electric energy consisting of galvanic cells or a series of galvanic cells. (19.6)

Beer-Lambert law. Equation relating absorbance to molar absorptivity, solution concentration, and distance that light travels through a solution. (4.5)

beta particle. An electron. (2.2)

beta ray. See *beta particle*.

bimolecular. Describes a reaction in which two reactant molecules collide. (14.5)

binary compound. A substance that consists of just two different elements. (2.7)

blackbody radiation. The electromagnetic radiation emitted from a heated solid. (6.2)

body-centered cubic cell. A unit cell with one atom at the center of the cube and one atom at each of the eight corners. (11.3)

boiling point. The temperature at which vapor pressure equals atmospheric pressure. (11.6)

bond angle. The angle between two adjacent A—B bonds. (9.1)

bond enthalpy. The enthalpy change associated with breaking a particular bond in 1 mole of gaseous molecules. (8.9)

bond order. A number based on the number of electrons in bonding and antibonding molecular orbitals that indicates, qualitatively, how stable a bond is. (9.6)

bond-line structure. A structure in which straight lines represent carbon-carbon bonds. (23.3)

bonding molecular orbital. A molecular orbital that is lower in energy than the atomic orbitals that combined to produce it. (9.6)

Born-Haber cycle. The cycle that relates the lattice energy of an ionic compound to quantities that can be measured. (8.2)

Boyle's law. The pressure of a fixed amount of gas at a constant temperature is inversely proportional to the volume of the gas: $V \propto 1/P$. (10.2)

Bragg equation. An equation relating the wavelength of X rays, the angle of diffraction, and the spacing between atoms in a lattice. (11.4)

breeder reactor. A nuclear reactor that produces more fissionable material than it consumes. (20.5)

Brønsted acid. A substance that donates a proton (H^+). (4.3, 16.1)

Brønsted base. A substance that accepts a proton (H^+). (4.3, 16.1)

buffer. A solution that contains significant concentrations of both members of a conjugate pair (weak acid/conjugate base or weak base/conjugate acid). (17.2)

C

calorimetry. The measurement of heat changes. (5.4)

capillary action. The movement of liquid up a narrow tube as the result of adhesive forces. (11.2)

carbocation. A species in which one of the carbons is surrounded by only six electrons. (23.5)

carbon nanotube. A tube made of carbon atoms with dimensions on the order of nanometers. (12.5)

carboxylic acid. An organic acid that contains a carboxy group. (23.2)

catalyst. A substance that increases the rate of a chemical reaction without itself being consumed. (14.6)

catenation. The formation of long carbon chains. (23.1)

cathode. The electrode at which reduction occurs. (19.2)

cation. An ion with a positive charge. (2.6)

cell potential (E_{cell}). The difference in electric potential between the cathode and the anode. (19.2)

ceramics. Polymeric inorganic compounds that share the properties of hardness, strength, and high melting points. (12.2)

chalcogens. Elements in Group 16 (O, S, Se, Te, and Po). (2.4)

Charles's law. The volume of a fixed amount of gas (V) maintained at constant pressure is directly proportional to absolute temperature (T): $V \propto T$. Also known as Charles's and Gay-Lussac's law. (10.2)

chelating agent. A polydentate ligand that forms complex ions with metal ions in solution. (22.1)

chemical change. A process in which one or more substances are changed into one or more new substances. (1.4)

chemical energy. Energy stored within the structural units (molecules or polyatomic ions) of chemical substances. (5.1)

chemical equation. Chemical symbols used to represent a chemical reaction. (3.3)

chemical formula. Chemical symbols and numerical subscripts used to denote the composition of the substance. (2.7)

chemical property. Any property of a substance that cannot be studied without converting the substance into some other substance. (1.4)

chemistry. The study of matter and the changes it undergoes. (1.1)

chiral. Describes molecules with nonsuperimposable mirror images. (23.4)

cholesteric. Describes molecules that are parallel to each other within each layer, but where each layer is rotated with respect to the layers above and below it. (12.3)

cis. Describes the isomer in which two substituents both lie on the same side of a double bond. (23.4)

Clausius-Clapeyron equation. A linear relationship that exists between the natural log of vapor pressure and the reciprocal of absolute temperature. (11.2)

closed system. A system that can exchange energy (but not mass) with the surroundings. (5.2)

cohesion. The attraction between like molecules. (11.2)

colligative properties. Properties that depend on the number of solute particles in solution but do not depend on the nature of the solute particles. (13.5)

collision theory. The reaction rate is directly proportional to the number of molecular collisions per second. (14.4)

colloid. A dispersion of particles of one substance throughout another substance. (13.7)

combination reaction. A reaction in which two or more reactants combine to form a single product. (3.7)

combined gas law. The equation that relates the parameters (pressure, volume, and absolute temperature) of an ideal gas in one state to the parameters of the sample in another state. (10.2)

combustion. Burning in air. (3.3, 4.4)

combustion analysis. An experimental determination of an empirical formula by a reaction with oxygen to produce carbon dioxide and water. (3.5)

common ion effect. The presence of a common ion suppresses the ionization of a weak acid or weak base. (17.1)

complex ion. Charged species consisting of a central metal cation bonded to two or more anions or polar molecules. (17.5)

composite material. A material made from two or more substances with different properties that remain separate in the bulk material. (12.2)

compound. A substance composed of atoms of two or more elements chemically united in fixed proportions. (1.2)

concentration. Amount of solute relative to the volume of a solution or to the amount of solvent in a solution. (4.5)

concentration cell. A cell that has the same type of electrode and the same ion in solution (at different concentrations) in the anode and cathode compartments. (19.5)

condensation. The phase transition from gas to liquid. (11.2)

condensation polymer. A large molecule that forms when small molecules undergo condensation reactions. (12.1, 23.6)

condensation reaction. An elimination reaction in which two or more molecules become connected with the elimination of a small molecule, often water. (23.6)

condensed structural formula. Shows the same information as a structural formula, but in a condensed form. (23.3)

condensed structure. Chemical structure simplified by using abbreviations for repeating structural units. (23.3)

conduction band. The antibonding band. (12.6)

conductor. A substance through which electrons move freely. (24.3)

conjugate acid. The cation that remains when a Brønsted base accepts a proton. (16.1)

conjugate base. The anion that remains when a Brønsted acid donates a proton. (16.1)

conjugate pair. The combination of a Brønsted acid and its conjugate base (or the combination of a Brønsted base and its conjugate acid). (16.1)

constitutional isomers. Compounds with the same chemical formula but different structures. (23.4)

conversion factor. A fraction in which the same quantity is expressed one way in the numerator and another way in the denominator. (1.6)

coordinate covalent bond. A covalent bond in which one of the atoms donates both electrons. (8.8)

coordination compound. A compound that contains coordinate covalent bonds between a metal ion (often a transition metal ion) and two or more polar molecules or ions. (22.1)

coordination number. The number of donor atoms surrounding the metal atom in a complex ion. (11.3, 21.1)

copolymer. A polymer made of two or more different monomers. (12.1, 23.6)

corrosion. The undesirable oxidation of metals. (19.8)

Coulomb's law. The force (F) between two charged objects (Q_1 and Q_2) is directly proportional to the product of the two charges and inversely proportional to the distance (d) between the objects squared. (7.4)

covalent bond. A shared pair of electrons. (8.3)

covalent bonding. Two atoms sharing a pair of electrons. (8.3)

covalent radius. Half the distance between adjacent, identical nuclei in a molecule. (7.4)

critical mass. The minimum amount of fissionable material required to sustain a reaction. (20.5)

critical pressure (P_c). The minimum pressure that must be applied to liquefy a substance at its critical temperature. (11.6)

critical temperature (T_c). The temperature above which the gas phase cannot be liquefied, no matter how great the applied pressure. (11.6)

cross-link. A bond that forms between a functional group off the backbone of the polymer chain that interacts with another functional group of a second polymer strand creating a new covalent bond and causing the polymer to be stronger and more rigid. (12.1)

crystal field splitting (Δ). The difference in energy between the lower and higher d-orbital energy levels. (22.3)

crystalline solid. A solid that possesses rigid and long-range order; its atoms, molecules, or ions occupy specific positions. (11.3)

D

***d* orbital.** Atomic orbital in which the angular momentum quantum number (ℓ) is 2. (6.7)

Dalton's law of partial pressures. The total pressure exerted by a gas mixture is the sum of the partial pressures exerted by each component of the mixture. (10.5)

dative bond. A covalent bond in which one of the atoms donates both electrons. (8.8)

de Broglie wavelength. A wavelength calculated using the following equation: $\lambda = h/mu$. (6.4)

decomposition reaction. A reaction in which one reactant forms two or more products. (3.7)

degenerate. Having an equal energy. (6.8)

delocalized. Spread out over the molecule or part of the molecule, rather than confined between two specific atoms. (9.7)

density. The ratio of mass to volume. (1.3)

deoxyribonucleic acid (DNA). Biological polymer arranged in a double-helix shape, consisting of two long chains of nucleotides in which the sugar is deoxyribose. (23.6)

deposition. The phase change from gas to solid. (11.6)

dextrorotatory. The term used to describe the enantiomer that rotates the plane-polarized light to the right. (22.2)

diagonal relationships. Similarities in chemical properties of elements that are in different groups but that are positioned diagonally to one another in the periodic table. (7.7)

diamagnetic. A species without unpaired electrons that is weakly repelled by magnetic fields. (9.6)

diatomic molecule. A molecule that contains two atoms. (2.6)

diffusion. The mixing of gases. (10.6)

dilution. The process of preparing a less concentrated solution from a more concentrated one. (4.5)

dimensional analysis. The use of conversion factors in problem solving. (1.6)

diode. An electronic device that restricts the flow of electrons in a circuit to one direction. (12.6)

dipole moment (μ). A quantitative measure of the polarity of a bond. (8.4)

dipole-dipole interactions. Attractive forces that act between polar molecules. (11.1)

diprotic acid. An acid with two ionizable protons. (4.3)

dispersion forces. Attractive forces that act between all molecules, including nonpolar molecules, resulting from the formation of instantaneous dipoles and induced dipoles. Also called *London dispersion forces*. (11.1)

displacement reaction. Reaction in which two reactants trade components (double displacement) or where a component of a reactant is removed (single displacement). (4.4)

disproportionation reaction. Occurs when an element undergoes both oxidation and reduction in the same reaction. (4.4)

dissociation. The process by which an ionic compound, upon dissolution, breaks apart into its constituent ions. (4.1)

donor atom. The atom that bears the unshared pair of electrons. (22.1)

doping. The addition of very small quantities of an element with one more or one fewer valence electron than the natural semiconductor. (12.6)

double bond. A multiple bond in which the atoms share two pairs of electrons. (8.3)

dynamic equilibrium. Occurs when a forward process and reverse process are occurring at the same rate. (11.2)

E

effective collision. A collision that results in a reaction. (14.4)

effective nuclear charge (Z_{eff}). The actual magnitude of positive charge that is "experienced" by an electron in the atom. (7.3)

effusion. The escape of a gas from a container into a vacuum. (10.6)

elastomer. A material that can stretch or bend and then return to its original shape as long as the limits of its elasticity are not exceeded. (12.1)

electrode. A piece of conducting metal in an electrochemical cell at which either oxidation or reduction takes place. (19.2)

electrolysis. The use of electric energy to drive a nonspontaneous redox reaction. (19.7)

electrolyte. A substance that dissolves in water to yield a solution that conducts electricity. (4.1)

electrolytic cell. An electrochemical cell used for electrolysis. (19.7)

electromagnetic spectrum. Consists of radio waves, microwave radiation, infrared radiation, visible light, ultraviolet radiation, X rays, and gamma rays. (6.1)

electromagnetic wave. A wave that has an electric field component and a magnetic field component. (6.1)

electron. A negatively charged subatomic particle found outside the nucleus of all atoms. (2.2)

electron affinity (*EA*). The energy released (the negative of the enthalpy change, ΔH) when an atom in the gas phase accepts an electron. (7.4)

electron configuration. The distribution of electrons in the atomic orbitals of an atom. (6.8)

electron density. The probability that an electron will be found in a particular region of an atom. (6.5)

electron domain. A lone pair or a bond, regardless of whether the bond is single, double, or triple. (9.1)

electron spin quantum number (m_s). The fourth quantum number that differentiates two electrons in the same orbital. (6.6)

electron-domain geometry. The arrangement of electron domains (bonds and lone pairs) around a central atom. (9.1)

electronegativity. The ability of an atom in a compound to draw electrons to itself. (8.4)

electrophile. A region of positive or partial positive charge. (23.5)

electrophilic addition. An addition reaction that begins when an electrophile approaches a region of electron density. (23.5)

electrostatic energy. The potential energy that results from the interaction of charged particles. (5.1)

element. A substance that cannot be separated into simpler substances by chemical means. (1.2)

elementary reaction. A reaction that occurs in a single collision of the reactant molecules. (14.5)

elimination reaction. A reaction in which a double bond forms and a molecule such as water is removed. (23.5)

emission spectrum. The light emitted, either as a continuum or in discrete lines, by a substance in an excited electronic state. (6.3)

empirical formula. The chemical formula that conveys with the smallest possible whole numbers the ratio of combination of elements in a compound. (2.6)

enantiomers. Molecules that are mirror images of each other but cannot be superimposed. (22.2, 23.4)

endothermic process. A process that absorbs heat. (5.1)

endpoint. The point at which the color of the indicator changes. (4.6, 17.3)

energy. The capacity to do work or transfer heat. (5.1)

enthalpy (*H*). A thermodynamic quantity defined by the equation $H = U + PV$. The change in enthalpy, ΔH, is equal to the heat exchanged between the system and surroundings at constant pressure q_P. (5.3)

enthalpy of reaction (ΔH_{rxn}). The difference between the enthalpy of the products and the enthalpy of the reactants. (5.3)

entropy (*S*). A thermodynamic state function that describes how dispersed a system's energy is. (13.2, 18.2)

enzymes. Biological catalysts. (14.6)

equatorial. The three bonds that are arranged in a trigonal plane in a trigonal bipyramidal geometry. (9.1)

equilibrium. A state in which forward and reverse processes are occurring at the same rate. (15.1)

equilibrium constant (K_c). A number equal to the ratio of the equilibrium concentrations of products to the equilibrium concentrations of reactants, with each concentration raised to the power of its stoichiometric coefficient. (15.2)

equilibrium expression. The quotient of product concentrations and reactant concentrations, each raised to the power of its stoichiometric coefficient. (15.2)

equilibrium process. A process that can be made to occur by the addition or removal of energy but does not happen on its own. (18.4)

equilibrium vapor pressure. The pressure exerted by the molecules that have escaped to the gas phase, once the pressure has stopped increasing. (11.2)

equivalence point. The point in a titration where the reaction (e.g., neutralization) is complete. (4.6)

ester. An organic molecule containing a —COOR group. (23.2)

evaporation. The phase change from liquid to gas at a temperature below the boiling point. (11.2)

excess reactant. The reactant present in a greater amount than necessary to react with all the limiting reactant. (3.7)

excited state. A state that is higher in energy than the ground state. (6.3)

exothermic process. A process that gives off heat. (5.1)

extensive property. A property that depends on the amount of matter involved. (1.4)

F

***f* orbital.** Atomic orbital in which the angular momentum quantum number (ℓ) is 3. (6.7)

face-centered cubic cell. A cubic unit cell with one atom on each of the six faces and one atom at each of the eight corners. (11.3)

family. The elements in a vertical column of the periodic table. (2.4)

first law of thermodynamics. Energy can be converted from one form to another, but cannot be created or destroyed. (5.2)

first-order reaction. A reaction whose rate depends on the reactant concentration raised to the first power. (14.3)

formal charge. Method of electron "bookkeeping" in which shared electrons are divided equally between the atoms that share them. (8.6)

formation constant (K_f). The equilibrium constant that indicates to what extent complex-ion formation reactions occur. (17.5)

formula mass. The mass of a formula unit. (3.1)

formula weight. See *formula mass*.

fractional precipitation. The separation of a mixture based upon the components' solubilities. (17.6)

free energy. The energy available to do work. (18.5)

free radical. A molecule with an odd number of electrons. (8.8)

freezing point. The temperature liquid and solid phases exist in equilibrium. Also known as *melting point*. (11.6)

frequency (ν). The number of waves that pass through a particular point in 1 s. (6.1)

fuel cell. A voltaic cell in which reactants must be continually supplied. (19.6)

fullerenes. Molecules with elongated and elliptical cages of 70 and 80 carbon atoms. (12.5)

functional group. The part of a molecule characterized by a special arrangement of atoms that is largely responsible for the chemical behavior of the parent molecule. (2.7)

fusion. The phase transition from solid to liquid (melting). (11.6)

G

galvanic cell. An electrochemical cell in which a spontaneous chemical reaction generates a flow of electrons. (19.2)

galvanization. The cathodic protection of iron or steel using zinc. (19.8)

gamma rays. High-energy radiation. (2.2)

gas constant (R). The proportionality constant that appears in the ideal gas equation. (10.3)

gas laws. Equations that relate the volume of a gas sample to its other parameters: temperature (T), pressure (P), and number of moles (n). (10.2)

geometrical isomers. Molecules that contain the same atoms and bonds arranged differently in space. (22.2, 23.4)

Gibbs free energy (G). The energy available to do work. (18.5)

glass. Commonly refers to an optically transparent fusion product of inorganic materials that has cooled to a rigid state without crystallizing. (11.5)

Graham's law. The rates of diffusion and effusion are inversely proportional to the square root of the molar mass of the gas. (10.6)

graphene. Two-dimensional sheet of sp^2-hybridized carbon atoms. (12.5)

gravimetric analysis. An analytical technique based on the measurement of mass. (4.6)

greenhouse effect. Describes the trapping of heat near Earth's surface by gases in the atmosphere, particularly carbon dioxide. (21.5)

ground state. The lowest energy state of an atom. (6.3)

group. The elements in a vertical column of the periodic table. (2.4)

H

half-cell. One compartment of an electrochemical cell containing an electrode immersed in a solution. (19.2)

half-life ($t_{1/2}$). The time required for the reactant concentration to drop to half its original value. (14.3)

half-reaction. The separated oxidation and reduction reactions that make up the overall redox reaction. (4.4, 19.1)

half-reaction method. Balancing an oxidation-reduction equation by separating the oxidation and the reduction, balancing them separately, and adding them back together. (4.4)

Hall process. Electrolytic reduction of aluminum from anhydrous aluminum oxide (corundum). (24.7)

halogens. The elements in Group 17 (F, Cl, Br, I, and At). (2.4)

heat. The transfer of thermal energy between two bodies that are at different temperatures. (5.1)

heat capacity (C). The amount of heat required to raise the temperature of an object by 1°C. (5.4)

Heisenberg uncertainty principle. It is impossible to know simultaneously both the momentum (p) (defined as mass times velocity, $m \times u$) and the position (x) of a particle with certainty. (6.5)

Henderson-Hasselbalch equation. $pH = pK_a + \log ([\text{conjugate base}]/[\text{weak acid}])$. (17.2)

Henry's law. The solubility of a gas in a liquid is proportional to the pressure of the gas over the solution. (13.4)

Henry's law constant (k). The proportionality constant that is specific to the gas-solvent combination and varies with temperature. (13.4)

Hess's law. The change in enthalpy that occurs when reactants are converted to products in a reaction is the same whether the reaction takes place in one step or in a series of steps. (5.5)

heteroatom. Any atom in an organic molecule other than carbon or hydrogen. (23.3)

heterogeneous catalysis. A catalysis in which the reactants and the catalyst are in different phases. (14.6)

heterogeneous mixture. A mixture in which the composition varies. (1.2)

heteronuclear. Containing two or more different elements. (2.7)

high-temperature superconductor. A material that shows no resistance to the flow of electrons at an unusually high temperature. (12.7)

homogeneous catalysis. A catalysis in which the reactants and the catalyst are in the same phase. (14.6)

homogeneous mixture. A mixture in which the composition is uniform. Also called a *solution*. (1.2)

homonuclear. Containing atoms of only one element. (2.7)

Hund's rule. The most stable arrangement of electrons in orbitals of equal energy is the one in which the number of electrons with the same spin is maximized. (6.8)

hybridization. The mixing of atomic orbitals. (9.4)

hydrate. A compound with a specific number of water molecules within its solid structure. (2.6)

hydration. The process by which water molecules surround solute particles in an aqueous solution. (4.2)

hydrocarbon. A compound containing only carbon and hydrogen. (2.7)

hydrogen bonding. A special type of dipole-dipole interaction that occurs only in molecules that contain H bonded to a small, highly electronegative atom, such as N, O, or F. (11.1)

hydrogen displacement. A redox reaction in which the hydrogen in a compound is replaced by a metal cation and reduced to hydrogen gas. (4.4)

hydronium ion. A hydrated proton (H_3O^+). (4.3)

hydrophilic. Water-loving. (13.7)

hydrophobic. Water-fearing. (13.7)

hypertonic. Describes a solution with a higher concentration of dissolved substances than plasma. (13.5)

hypothesis. A tentative explanation for a set of observations. (1.1)

hypotonic. Describes a solution that has a lower concentration of dissolved substances than plasma. (13.5)

I

ideal gas. A hypothetical sample of gas whose pressure-volume-temperature behavior is predicted accurately by the ideal gas equation. (10.3)

ideal gas equation. An equation that describes the relationship among the four variables P, V, n, and T. (10.3)

ideal solution. A solution that obeys Raoult's law. (13.5)

indicator. A substance that has a distinctly different color in acidic and basic media. (4.6)

initial rate. The instantaneous rate at the beginning of a reaction. (14.2)

inorganic compounds. Compounds that do not contain carbon or that are derived from nonliving sources. (2.7)

instantaneous dipole. A fleeting nonuniform distribution of electron density in a molecule without a permanent dipole. (11.1)

instantaneous rate. The reaction rate at a specific time. (14.1)

insulator. A substance that does not conduct electricity. (24.3)

integrated rate law. ln ([A]$_t$/[A]$_0$) = $-kt$. (14.3)

intensive property. A property that does not depend on the amount of matter involved. (1.4)

intermediate. A chemical species that is produced in one step of a reaction mechanism and consumed in a subsequent step. (14.5)

intermolecular forces. The attractive forces that hold particles together in the condensed phases. (11.1)

International System of Units. See *SI units*. (1.3)

ion. Atom or molecule that has lost or gained one or more electrons giving it a positive or negative charge. (2.6)

ion pair. Ions in solution that are held together by electrostatic forces. (13.5)

ion-dipole interactions. Coulombic attractions between ions and polar molecules. (11.1)

ion-product constant. The product of hydronium ion and hydroxide ion concentrations in an aqueous solution. (16.2)

ionic bonding. An electrostatic attraction that holds oppositely charged ions together in an ionic compound. (8.2)

ionic compound. Substance consisting of ions held together by electrostatic attraction. (2.6)

ionic equation. Chemical equation in which all strong electrolytes are shown as ions. (4.2)

ionic radius. The radius of a cation or an anion. (7.6)

ionizable hydrogen atom. A hydrogen atom that can be lost as a hydrogen ion, (H^+). (2.7)

ionization. The process by which a molecular compound forms ions when it dissolves. (4.1)

ionization energy (*IE*). The minimum energy required to remove an electron from an atom in the gas phase. (7.4)

ionosphere. The outermost layer of the atmosphere—also known as the thermosphere. (21.1)

isoelectronic. Describes two or more species with identical electron configurations. (7.5)

isoelectronic series. A series of two or more species that have identical electron configurations but different nuclear charges. (7.6)

isolated system. A system that can exchange neither energy nor mass with the surroundings. (5.2)

isomerization reaction. A reaction in which one isomer is converted to another. (23.5)

isotactic. Polymers in which all the substituents (i.e., the R groups) are in the same relative orientation (i.e., on the same side of the polymer chain). (12.1)

isotonic. Equal in concentration and osmotic pressure. (13.5)

isotopes. Atoms that have the same atomic number (*Z*) but different mass numbers (*A*). (2.3)

isotropic. Independent of the axis of measurement. (12.3)

J

joule. SI unit of energy, 1 kg · m^2/s^2. (5.1)

K

Kekulé structure. A structure similar to a Lewis structure in which the lone pairs may not be shown. (23.3)

Kelvin temperature scale. A temperature scale offset from the Celsius scale by 273.15. One kelvin (1 K) is equal in magnitude to one degree Celsius (1°C). (10.2)

kelvin. The SI base unit of temperature. (1.3)

ketone. An organic compound consisting of two R groups bonded to a carbonyl. (23.2)

kinetic energy. The energy that results from motion. (5.1)

kinetic molecular theory. A theory that explains how the molecular nature of gases gives rise to their macroscopic properties. (10.6)

L

lanthanide (rare earth) series. A series of 14 elements that have incompletely filled $4f$ subshells or that readily give rise to cations that have incompletely filled $4f$ subshells. (6.9)

lattice. A three-dimensional array of cations and anions. (2.6)

lattice energy. The amount of energy required to convert a mole of ionic solid to its constituent ions in the gas phase. (8.2)

lattice points. Positions occupied by atoms, ions, or molecules in a unit cell. (11.3)

lattice structure. The arrangement of the particles in a crystalline solid. (11.3)

law. A concise verbal or mathematical statement of a reliable relationship between phenomena. (1.1)

law of conservation of energy. First law of thermodynamics stating that energy can be neither created nor destroyed. (5.1)

law of conservation of mass. An alternate statement of the first law of thermo-dynamics stating that matter can be neither created nor destroyed. (2.1)

law of definite proportions. Different samples of a given compound always contain the same elements in the same mass ratio. (2.1)

law of mass action. For a reversible reaction at equilibrium and a constant temperature, the reaction quotient, Q, has a constant value, K (the equilibrium constant). (15.2)

law of multiple proportions. Different compounds made up of the same elements differ in the number of atoms of each kind that combine. (2.1)

Le Châtelier's principle. When a stress is applied to a system at equilibrium, the system will respond by shifting in the direction that minimizes the effect of the stress. (15.5)

levorotatory. The term used to describe the enantiomer that rotates the plane-polarized light to the left. (22.2)

Lewis acid. A species that can accept a pair of electrons. (16.12)

Lewis base. A species that can donate a pair of electrons. (16.12)

Lewis dot symbol. An elemental symbol surrounded by dots, where each dot represents a valence electron. (8.1)

Lewis structure. A representation of covalent bonding in which shared electron pairs are shown either as dashes or as pairs of dots between two atoms, and lone pairs are shown as pairs of dots on individual atoms. (8.3)

Lewis theory of bonding. A chemical bond involves atoms sharing electrons. (8.3)

ligand. A molecule or anion that can form coordinate bonds to a metal to form a coordination complex. (22.1)

limiting reactant. The reactant that is completely consumed and determines the amount of product formed. (3.7)

line spectra. The emission or absorption of light only at discrete wavelengths. (6.3)

liquid crystal. A substance that exhibits properties of both a liquid, such as the ability to flow and to take on the shape of a container, and those of a crystal, such as a regular arrangement of particles in a lattice. (12.3)

localized. Describes electrons that are shared between two specific atoms and that cannot be repositioned to generate additional resonance structures. (9.7)

London dispersion forces. Attractive forces that act between all molecules, including nonpolar molecules, resulting from the formation of instantaneous dipoles and induced dipoles. Also called simply *dispersion forces*. (11.1)

lone pair. A pair of valence electrons that are not involved in covalent bond formation. (8.3)

M

magnetic quantum number (m_ℓ). Describes the orientation of an orbital in space. (6.6)

main group elements. Elements in the s– and p– blocks of the periodic table. (7.2)

manometer. A device used to measure the pressure of gases relative to atmospheric pressure. (10.1)

mass. A measure of the amount of matter in an object or sample. (1.3)

mass defect. The difference between the actual mass of a nucleus and the mass calculated by summing the masses of the individual nucleons. (20.2)

mass number (A). The number of neutrons and protons present in the nucleus of an atom of an element. (2.3)

matter. Anything that occupies space and has mass. (1.1)

Meissner effect. The exclusion of magnetic fields. (12.7)

melting point. The temperature liquid and solid phases exist in equilibrium. Also known as *freezing point*. (11.6)

mesosphere. The region located above the stratosphere in which the concentration of ozone and other gases is low and temperature decreases again with increasing altitude. (21.1)

metal. Element with a tendency to lose electrons, located left of the zigzag line on the periodic table. (2.4)

metallic radius. Half the distance between the nuclei of two adjacent, identical metal atoms. (7.4)

metalloid. An element with properties intermediate between those of metals and nonmetals. (2.4, 7.4)

metallurgy. The preparation, separation, and purification of metals. (24.2)

mineral. A naturally occurring substance with a characteristic chemical composition and specific physical properties. (24.1)

miscible. Mutually soluble in any proportions. (13.2)

mixture. A combination of two or more substances in which the substances retain their distinct identities. (1.2)

moderator. A material that limits the speed of liberated neutrons but does not itself undergo fission when bombarded with neutrons. (20.5)

molality (m). The number of moles of solute dissolved in 1 kg (1000 g) of solvent. (13.3)

molar absorptivity. Proportionality constant relating absorbance to solution concentration and distance that light travels through a solution. (4.5)

molar concentration. Molarity. The number of moles of solute per liter of solution. (4.5)

molar enthalpy of sublimation (ΔH_{sub}). The energy, usually expressed in kilojoules, required to sublime 1 mole of a solid. (11.6)

molar heat of fusion (ΔH_{fus}). The energy, usually expressed in kJ/mol, required to melt 1 mole of a solid. (11.6)

molar heat of vaporization (ΔH_{vap}). The amount of heat required to vaporize a mole of substance at its boiling point. (11.6)

molar mass ($\mathscr{M}$). The mass in grams of 1 mole of the substance. (3.4)

molar solubility. The number of moles of solute in one liter of saturated solution (mol/L). (17.4)

molarity (M). Molar concentration. The number of moles of solute per liter of solution. (4.5)

mole (mol). The amount of a substance that contains as many elementary entities (atoms, molecules, formula units, etc.) as there are atoms in exactly 0.012 kg (12 g) of carbon-12. (3.4)

mole fraction (χ_i). The number of moles of a component divided by the total number of moles in a mixture. (10.5)

molecular equation. A chemical equation written with all compounds represented by their chemical formulas. (4.2)

molecular formula. A chemical formula that gives the number of atoms of each element in a molecule. (2.7)

molecular geometry. The arrangement of bonded atoms. (9.1)

molecular mass. The sum of the atomic masses (in amu) of the atoms that make up a molecule. (3.1)

molecular orbital. An orbital that results from the interaction of the atomic orbitals of the bonding atoms. (9.6)

molecular orbital theory. A theory that describes the orbitals in a molecule as

bonding and antibonding combinations of atomic orbitals. (9.6)

molecular weight. Average molecular mass. (3.1)

molecularity. The number of molecules involved in a specific step in a reaction mechanism. (14.5)

molecule. A combination of two or more atoms in a specific arrangement held together by chemical bonds. (2.7)

monatomic ion. An ion that contains only one atom. (2.6)

monomer. A small molecule that can be linked in large numbers to form a large molecule (polymer). (12.1, 23.6)

monoprotic acid. An acid with one ionizable proton. (4.3)

multiple bond. A chemical bond in which two atoms share two or more pairs of electrons. (8.3)

N

n-type semiconductor. Semiconductors in which an electron-rich impurity is added to enhance conduction. (12.6, 24.3)

nanotechnology. The development and study of extremely small-scale materials and objects. (12.5)

nematic. Describes an arrangement in which molecules are all aligned parallel to one another but with no organization into layers or rows. (12.3)

Nernst equation. An equation relating the emf of a galvanic cell with the standard emf and the concentrations of reactants and products. (19.5)

net ionic equation. Chemical equation from which spectator ions have been removed. (4.2)

neutralization reaction. A reaction between an acid and a base. (4.3)

neutron. An electrically neutral subatomic particle with a mass slightly greater than that of a proton. (2.2)

newton (N). The SI unit of force. (10.1)

nitrogen fixation. The conversion of molecular nitrogen into nitrogen compounds. (21.1)

noble gas core. A representation in an electron configuration that shows in brackets the most recently completed noble gas. (6.9)

noble gases. Elements in Group 18 (He, Ne, Ar, Kr, Xe, and Rn). (2.4)

node. A collection of points at which electron density in an atom is zero. (6.4)

nonconductor. A substance that does not conduct electricity. (12.6)

nonelectrolyte. A substance that dissolves in water to yield a solution that does not conduct electricity. (4.1)

nonmetal. Element with a tendency to gain electrons, located in the upper right portion of the periodic table. (2.4)

nonpolar. Having a uniform distribution of electron density. (8.4)

nonspontaneous process. A process that does not occur under a specified set of conditions. (18.1)

nonvolatile. Having no measurable vapor pressure. (13.5)

normal boiling point. Temperature at which a substance boils at 1 atm pressure. (11.6)

nuclear binding energy. The energy required to separate the nucleons in a nucleus. (20.2)

nuclear chain reaction. A self-sustaining reaction sequence of fission reactions. (20.5)

nuclear fission. The splitting of a large nucleus into smaller nuclei and one or more neutrons. (20.5)

nuclear fusion. The combination of two light nuclei to form one heavier nucleus. (20.6)

nuclear transmutation. The conversion of one nucleus to another. (20.1)

nucleic acid. Macromolecule formed by polymerization of nucleotides. (23.6)

nucleons. Protons and neutrons. (2.3)

nucleophile. A region of negative or partial negative charge. (23.5)

nucleophilic addition. An addition reaction that begins when a nucleophile donates a pair of electrons to an electron-deficient atom. (23.5)

nucleotide. A structural unit consisting of a sugar (ribose or deoxyribose) bonded to both a cyclic-amine base and a phosphate group. (23.6)

nucleus. The central core of the atom that contains the protons and neutrons. (2.2)

O

octet rule. Atoms will lose, gain, or share electrons to achieve a noble gas electron configuration. (8.3)

open system. A system that can exchange mass and energy with its surroundings. (5.2)

optical isomers. Nonsuperimposable mirror images. (22.2, 23.4)

ore. A mineral deposit concentrated enough to allow economical recovery of a desired metal. (24.1)

organic compounds. Compounds containing carbon and hydrogen, sometimes in combination with other elements such as oxygen, nitrogen, sulfur, and the halogens. (2.7)

osmosis. The selective passage of solvent molecules through a porous membrane from a more dilute solution to a more concentrated one. (13.5)

osmotic pressure (π). The pressure required to stop osmosis. (13.5)

overvoltage. The difference between the electrode potential and the actual voltage required to cause electrolysis. (19.7)

oxidation. Loss of electrons. (4.4)

oxidation number. The charge an atom would have if electrons were transferred completely. (4.4)

oxidation state. The charge an atom would have if electrons were transferred completely. (4.4)

oxidation-reduction reaction. A chemical reaction in which electrons are transferred from one reactant to another. (4.4)

oxidizing agent. A species that accepts electrons. (4.4)

oxoacid. An acid consisting of one or more ionizable protons and an oxoanion. (2.7)

oxoanion. A polyatomic anion that contains one or more oxygen atoms bonded to a central atom. (2.6)

P

p orbital. Atomic orbital in which the angular momentum quantum number (ℓ) is 1. (6.7)

p-type semiconductor. A semiconductor in which an electron-poor impurity is added to enhance conduction. (12.6, 24.3)

paramagnetic. A species with unpaired electrons that are attracted by magnetic fields. (9.6)

partial pressure (P_i). The pressure exerted by a component in a gas mixture. (10.5)

pascal (Pa). The SI unit of pressure. (10.1)

Pauli exclusion principle. No two electrons in an atom can have the same four quantum numbers. (6.8)

peptide bond. The bond that forms between amino acids. (23.6)

percent by mass. The ratio of the mass of an individual component to the total mass, multiplied by 100 percent. (13.3)

percent composition by mass. The percent of the total mass contributed by each element in a compound. (3.2)

percent dissociation. The percentage of dissolved molecules (or formula units, in the case of an ionic compound) that separate into ions in solution. (13.6)

percent ionic character. The ratio of experimentally-measured dipole moment to calculated dipole moment—multiplied by 100%. (8.4)

percent ionization. Quantitative description of the degree to which an electrolyte exists as ions in solution. (13.6)

percent yield. The ratio of actual yield to theoretical yield, multiplied by 100 percent. (3.7)

period. A horizontal row of the periodic table. (2.4)

periodic table. A chart in which elements having similar chemical and physical properties are grouped together. (2.4)

pH. A scale used to measure acidity. pH = $-\log$ [H$^+$]. (16.3)

phase change. Occurs when a substance goes from one phase to another phase. (11.6)

phase diagram. Summarizes the conditions (temperature and pressure) at which a substance exists as a solid, a liquid, or a gas. (11.7)

photochemical smog. Air pollution resulting from the interaction of sunlight with nitrogen monoxide and carbon monoxide from automobile exhaust. (21.7)

photoelectric effect. A phenomenon in which electrons are ejected from the surface of a metal exposed to light of at least a certain minimum frequency. (6.2)

photon. A quantum of light. (6.2)

physical change. A process in which the state of matter changes but the identity of the matter does not change. (1.4)

physical property. A property that can be observed and measured without changing the identity of a substance. (1.4)

pi (π) bond. A bond that forms from the interaction of parallel p orbitals. (9.5)

pOH. A scale used to measure basicity. pOH = $-\log$ [OH$^-$]. (16.3)

polar. Having a nonuniform electron density. (8.4)

polar covalent bond. Bonds in which electrons are unequally shared. (8.4)

polarimeter. Device used to measure the angle of rotation of plane-polarized light caused by an optically active compound. (22.2)

polarized. A molecule in which a dipole moment has been induced. (11.1)

polyamide. A polymer in which the monomers are connected by amide linkages. (12.1)

polyatomic. Molecules containing more than two atoms. (2.7)

polyatomic ion. Molecule that has lost or gained one or more electrons giving it a positive or negative charge. (2.6)

polyester. A polymer in which the monomers are connected by ester linkages. (12.1)

polymer. Molecular compounds, either natural or synthetic, that are made up of many repeating units called monomers. (12.1, 23.6)

polypeptide. Short chains of amino acids. (23.6)

polyprotic acid. An acid with more than two ionizable protons. (2.7, 4.3)

polysaccharides. Biological polymer consisting of sugars. (23.6)

positron. A subatomic particle with the same mass as an electron, but with a positive charge. (20.1)

potential energy. The energy possessed by an object by virtue of its position. (5.1)

precipitate. An insoluble solid product that separates from a solution. (4.2)

precipitation reaction. A chemical reaction in which a precipitate forms. (4.2)

precision. The closeness of agreement of two or more measurements of the same quantity. (1.5)

pressure. The force applied per unit area. (10.1)

principal quantum number (n). Designates the size of the orbital. (6.6)

product. A substance that forms in a chemical reaction. (3.3)

protein. A polymer of amino acids. (23.6)

proton. A positively charged particle in the nucleus of an atom. (2.2)

pyrometallurgy. Metallurgic processes carried out at high temperatures. (24.2)

Q

qualitative analysis. The determination of the types of ions present in a solution. (17.6)

qualitative property. A property of a system that can be determined by general observation. (1.4)

quantitative property. A property of a system that can be measured and expressed with a number. (1.4)

quantum. The smallest quantity of energy that can be emitted (or absorbed) in the form of electromagnetic radiation. (6.2)

quantum numbers. Numbers required to describe the arrangement of electrons in an atom. (6.6)

R

racemic mixture. An equimolar mixture of enantiomers that does not rotate the plane of plane-polarized light. (22.2, 23.4)

racemization. The conversion of a single enantiomer to a racemic mixture of both enantiomers. (23.5)

radiation. The emission and transmission of energy through space in the form of waves. (2.2)

radical. A chemical species with an odd number of electrons. (20.8)

radioactive decay series. A sequence of nuclear reactions that ultimately result in the formation of a stable isotope. (20.3)

radioactivity. The spontaneous emission of particles or radiation from unstable nuclei. (2.2, 20.1)

Raoult's law. The partial pressure of a solvent over a solution, P_1, is given by the vapor pressure of the pure solvent, P_1°, times the mole fraction of the solvent in the solution, χ_1. (13.5)

rate constant (k). The proportionality constant in a rate law. (14.1)

rate law. An equation relating the rate of reaction to the concentrations of reactants. (14.2)

rate of reaction. The change in concentration of reactants or products per unit time. (14.1)

rate-determining step. The slowest step in a reaction mechanism. (14.5)

reactant. A substance that is consumed in a chemical reaction. (3.3)

reaction mechanism. Series of steps by which a chemical reaction occurs. (14.5)

reaction order. The sum of the powers to which all reactant concentrations appearing in the rate law are raised. (14.2)

reaction quotient (Q_c). A fraction with product concentrations in the numerator and reactant concentrations in the denominator, each raised to its stoichiometric coefficient. (15.2)

redox reaction. A chemical reaction in which electrons are transferred from one reactant to another. (4.4)

redox titration. Titration in which reactants undergo a redox reaction. (4.6)

reducing agent. A species that can donate electrons. (4.4)

reduction. A gain of electrons. (4.4)

resonance structures. Two or more equally valid Lewis structures for a single molecule that differ only in the positions of electrons. (8.7)

reversible process. A process in which the products can react to form reactants. (15.1)

ribonucleic acid (RNA). Biological polymer consisting of nucleotides in which the sugar is ribose. (23.6)

root-mean-square speed (u_{rms}). The molecular speed that is inversely proportional to the molecular mass. (10.6)

S

s orbital. Atomic orbital in which the angular momentum quantum number (ℓ) is 0. (6.7)

salt bridge. An inverted U tube containing an inert electrolyte solution, such as KCl or NH_4NO_3, that maintains electrical neutrality in an electrochemical cell. (19.2)

salt hydrolysis. The reaction of a salt's constituent ions with water to produce either hydroxide ions or hydronium ions. (16.10)

salt. An ionic compound made up of the cation from a base and the anion from an acid. (4.3)

saturated solution. A solution that contains the maximum amount of a solute that will dissolve in a solvent at a specific temperature. (13.1)

scientific method. A systematic approach to experimentation. (1.1)

second law of thermodynamics. The entropy of the universe increases in a spontaneous process and remains unchanged in an equilibrium process. (18.4)

second-order reaction. A reaction whose rate depends on the concentration of one reactant raised to the second power or on the product of the concentrations of two different reactants each raised to the first power. (14.3)

semiconductor. A substance that normally does not conduct electricity but that will conduct at elevated temperatures or when combined with a small amount of certain other elements. (12.6, 24.3)

semipermeable membrane. A membrane that allows the passage of solvent molecules but blocks the passage of solute molecules. (13.5)

shielding. The partial obstruction of nuclear charge by core electrons. (7.3)

SI units. International System of Units. A system of units based on metric units. (1.3)

sigma (σ) bond. A bond in which the shared electron density is concentrated directly along the internuclear axis. (9.5)

significant figures. Meaningful digits in a measured or calculated value. (1.5)

simple cubic cell. The basic repeating unit in which there is one atom at each of the eight corners in a cube. (11.3)

single bond. A pair of electrons shared by two atoms. (8.3)

sintering. A method used to form objects by heating a finely divided substance. (12.2)

smectic. Containing molecules ordered in two dimensions. The molecules are aligned parallel to one another and are further arranged in layers that are parallel to one another. (12.3)

solubility. The maximum amount of solute that will dissolve in a given quantity of solvent at a specific temperature. (4.2, 13.1, 17.4)

solubility product constant (K_{sp}). The equilibrium constant that indicates to what extent a slightly soluble ionic compound dissolves in water. (17.4)

solute. The dissolved substance in a solution. (4.1)

solution. A homogeneous mixture consisting of a solvent and one or more solutes. (4.1)

solvation. The process by which solute molecules are surrounded by solvent molecules. (13.2)

solvent. A substance in a solution that is present in the largest amount. (4.1)

specific heat (s). The amount of heat required to raise the temperature of 1 g of a substance by 1°C. (5.4)

spectator ion. An ion that does not participate in the reaction and appears on both the reactant and product side in the ionic equation. (4.2)

spectrochemical series. A list of ligands arranged in increasing order of their abilities to split the d orbital energy levels. (22.3)

spontaneous process. A process that occurs under a specified set of conditions. (18.1)

standard atmospheric pressure. The pressure that would support a column of mercury exactly 760 mm high at 0°C. (10.1)

standard enthalpy of formation (ΔH_f°). The heat change that results when 1 mole of a compound is formed from its constituent elements in their standard states. (5.6)

standard enthalpy of reaction (ΔH_{rxn}°). The enthalpy of a reaction carried out under standard conditions. (5.6)

standard entropy (S°). The absolute entropy of a substance at 1 atm. (18.3)

standard free energy of formation (ΔG_f°). The free-energy change that occurs when 1 mole of the compound forms from its constituent elements in their standard states. (18.5)

standard free energy of reaction (ΔG_{rxn}°). The free-energy change for a reaction when it occurs under standard-state conditions. (18.5)

standard hydrogen electrode (SHE). A half-cell based on the half-reaction $2H^+(1\ M) + 2e^- \longrightarrow H_2(1\ atm)$, which has an arbitrarily defined standard reduction potential of zero. (19.3)

standard reduction potential (E°). The potential associated with a reduction half-reaction at an electrode when the ion concentration is 1 M and the gas pressure is 1 atm. (19.3)

standard solution. A solution of precisely known concentration. (4.6)

standard temperature and pressure (STP). 0°C and 1 atm. (10.3)

state functions. Properties that are determined by the state of the system, independent of how the state was achieved. (5.2)

state of a system. The values of all relevant macroscopic properties, such as composition, energy, temperature, pressure, and volume. (5.2)

stereoisomers. Molecules that contain identical bonds but differ in the orientation of those bonds in space. (22.2, 23.4)

stoichiometric amounts. Quantities of reactants in the same relative amounts as those represented in the balanced chemical equation. (3.6)

stoichiometric coefficients. The numeric values written to the left of each species in a chemical equation to balance the equation. (3.3)

stratosphere. The region of the atmosphere located above the troposphere and consisting of nitrogen, oxygen, and ozone. (21.1)

strong conjugate acid. The conjugate acid of a weak base. Acts as a weak Brønsted acid in water. (16.7)

strong conjugate base. The conjugate base of a weak acid. Acts as a weak Brønsted base in water. (16.7)

strong electrolyte. An electrolyte that ionizes or dissociates completely. (4.1)

structural formula. A chemical formula that shows the general arrangement of atoms within the molecule. (2.7)

structural isomer. Molecules that have the same chemical formula but different arrangements of atoms. (9.2)

sublimation. The phase change from solid to gas. (11.6)

substance. Matter with a definite (constant) composition and distinct properties. (1.2)

substituent. A group other than $^-$H bonded to the carbons of an organic molecule. (23.2)

substitution reaction. One group is replaced by another group by electrophilic or nucleophilic attack. (23.5)

superconducting transition temperature (T_c). The temperature below which an element, compound, or material becomes superconducting. (12.7)

superconductor. A substance with no resistance to the flow of electrons. (12.7)

supercooling. A phenomenon in which a liquid can be temporarily cooled to below its freezing point. (11.6)

supercritical fluid. A fluid at a temperature and pressure that exceed T_c and P_c. (11.6)

supersaturated solution. A solution that contains more dissolved solute than is present in a saturated solution. (13.1)

surface tension. The amount of energy required to stretch or increase the surface of a liquid by a unit area. (11.2)

surroundings. The part of the universe not included in the system. (5.1)

suture. Thread used by physicians to close wounds and/or incisions. (12.4)

syndiotactic. Polymers in which the substituents alternate positions along the polymer chain. (12.1)

system. The specific part of the universe that is of interest to us. (5.1)

T

tacticity. Describes the relative arrangements of chiral carbon atoms within a polymer. (12.1)

termolecular. Involving three reactant molecules. (14.5)

theoretical yield. The maximum amount of product that can be obtained from a reaction. (3.7)

theory. A unifying principle that explains a body of experimental observations and the laws that are based on them. (1.1)

thermal energy. The energy associated with the random motion of atoms and molecules. (5.1)

thermochemical equation. A chemical equation that includes the enthalpy change. (5.3)

thermochemistry. The study of the heat associated with chemical reactions and physical processes. (5.1)

thermodynamics. The scientific study of the interconversion of heat and other kinds of energy. (5.2)

thermonuclear reaction. Generally refers to a fusion reaction. (20.6)

thermoplastic. Polymers that can be melted and reshaped. (12.1)

thermosetting. Polymers that assume their final shape as part of the chemical reaction that forms them. (12.1)

thermosphere. The outermost layer of the atmosphere—also known as the ionosphere. (21.1)

third law of thermodynamics. The entropy of a pure crystalline solid is zero at absolute zero (0 K). (18.4)

titration. The gradual addition of a solution of known concentration to another solution of unknown concentration until the chemical reaction between the two solutions is complete. (4.6)

tracers. Radioactive isotopes that are used to trace the path of the atoms of an element in a chemical or biological process. (20.7)

trans. The isomer in which two substituents lie on opposite sides of a double bond. (23.4)

transition element. An element that has—or readily forms one or more ions that have—an incompletely filled *d* subshell (Groups 3 to 11). (2.4)

transition metals. The elements in Groups 3 to 11. (2.4)

transition state. A transient species that forms when molecules collide in an effective collision. Also known as an *activated complex*. (14.4)

transmittance. Ratio of the intensity of light transmitted through a sample (I) to the intensity of the incident light (I_0). (4.5)

transuranium elements. Elements with atomic numbers greater than 92, created by bombarding other elements with accelerated neutrons, protons, alpha particles, or other nuclei. (20.4)

triple bond. A multiple bond in which the atoms share three pairs of electrons. (8.3)

triple point. The point at which all three phase boundary lines meet. (11.7)

triprotic acid. An acid molecule with three ionizable protons. (4.3)

troposphere. The layer of the atmosphere closest to Earth's surface. (21.1)

Tyndall effect. The scattering of visible light by colloidal particles. (13.7)

U

unimolecular. Describes a reaction involving one reactant molecule. (14.5)

unit cell. The basic repeating structural unit of a crystalline solid. (11.3)

unsaturated solution. A solution that contains less solute than it has the capacity to dissolve. (13.1)

V

valence band. The bonding band. (12.6)

valence bond theory. Atoms share electrons when an atomic orbital on one atom overlaps with an atomic orbital on the other. (9.3)

valence electrons. The outermost electrons of an atom. (7.2)

valence-shell electron-pair repulsion (VSEPR). A model that accounts for electron pairs in the valence shell of an atom repelling one another. (9.1)

van der Waals equation. An equation relating the volume of a real gas to the other parameters, P, T, and n. (10.7)

van der Waals forces. The attractive forces that hold particles together in the condensed phases that include dipole-dipole interactions (including hydrogen bonding) and dispersion forces. (11.1)

van't Hoff factor (i). The ratio of the actual number of particles in solution after dissociation to the number of formula units initially dissolved in solution. (13.5)

vaporization. The phase change from liquid to gas at the boiling point. (11.2)

viscosity. A measure of a fluid's resistance to flow. (11.2)

visible spectrophotometry. Measurement of absorption of visible light for qualitative and quantitative analysis. (4.5)

volatile. Describes a substance that has a high vapor pressure. (11.2, 13.5)

W

wavelength (λ). The distance between identical points on successive waves. (6.1)

weak acid. An acid that ionizes only partially. (16.5)

weak base. A base that ionizes only partially. (16.6)

weak conjugate acid. A conjugate acid of a strong base. Does not react with water. (16.7)

weak conjugate base. A conjugate base of a strong acid. Does not react with water. (16.7)

weak electrolyte. A compound that produces ions upon dissolving but exists in solution predominantly as molecules that are not ionized. (4.1)

X

X-ray diffraction. A method of using X rays to bombard a crystalline sample to determine the structure of the crystal. (11.4)

Z

zeroth-order reaction. A constant rate, independent of reactant concentration. (14.3)

Answers

To Odd-Numbered Problems

Chapter 1

1.5 (a) Hypothesis. (b) Law. (c) Theory. **1.7** (a) C and O. (b) F and H. (c) N and H. (d) O. **1.13** (a) K. (b) Sn. (c) Cr. (d) B. (e) Ba. (f) Pu. (g) S. (h) Ar. (i) Hg. **1.15** (a) Homogeneous mixture. (b) Element. (c) Compound. (d) Homogeneous mixture. (e) Heterogeneous mixture. (f) Homogeneous mixture. (g) Heterogeneous mixture. **1.17** (a) Element. (b) Compound. (c) Compound. (d) Element. **1.23** 13.56 g/mL. **1.25** (a) 35°C. (b) −11°C. (c) 39°C. (d) 1011°C. (e) −459.67°F. **1.27** 69.1 mL. **1.29** (a) 388.36 K. (b) 3.10×10^2 K. (c) 6.30×10^2 K. **1.31** The picture on the right best illustrates the measurement of the boiling point of water using the Celsius and Kelvin scales. A temperature on the Kelvin scale is numerically equal to the temperature in Celsius plus 273.15. **1.37** (a) Qualitative. (b) Quantitative. (c) Quantitative. (d) Qualitative. (e) Quantitative. **1.39** (a) Physical change. (b) Chemical change. (c) Physical change. (d) Chemical change. (e) Physical change. **1.41** 99.9 g; 20°C; 11.35 g/cm³. **1.47** (a) 0.0152. (b) 0.0000000778. (c) 0.000001. (d) 1600.1. **1.49** (a) 1.8×10^{-2}. (b) 1.14×10^{10}. (c) -5×10^4. (d) 1.3×10^3. **1.51** (a) One. (b) Three. (c) Three. (d) Four. (e) Three. (f) One. (g) One or two. **1.53** (a) 1.28. (b) 3.18×10^{-3} mg. (c) 8.14×10^7 dm. **1.55** Tailor Z's measurements are the most accurate. Tailor Y's measurements are the least accurate. Tailor X's measurements are the most precise. Tailor Y's measurements are the least precise. **1.57** (a) 1.10×10^8 mg. (b) 6.83×10^{-5} m³. (c) 7.2×10^3 L. (d) 6.24×10^{-8} lb. **1.59** 5.2595×10^5 min. **1.61** (a) 81 in/s. (b) 1.2×10^2 m/min. (c) 7.4 km/h. **1.63** 602 km/h. **1.65** 3.7×10^{-3} g Pb. **1.67** (a) 1.85×10^{-7} m. (b) 1.4×10^{17} s. (c) 7.12×10^{-5} m³. (d) 8.86×10^4 L. **1.69** 6.25×10^{-4} g/cm³. **1.71** 0.88 s. **1.73** (a) 2.5 cm. (b) 2.55 cm. **1.75** (a) Chemical. (b) Chemical. (c) Physical. (d) Physical. (e) Chemical. **1.77** (a) 8.08×10^4 g. (b) 1.4×10^{-6} g. (c) 39.9 g. **1.79** 53.91 cm³. **1.81** 10.50 g/cm³. **1.83** 11.4 g/cm³. **1.85** −40°F = −40°C. **1.87** 4.8×10^{19} kg NaCl = 5.3×10^{16} tons NaCl. **1.89** The density of the crucible is equal to the density of pure platinum. **1.91** (a) 75.0 g Au. (b) A troy ounce is heavier than an ounce. **1.93** (a) 0.5%. (b) 3.1%. **1.95** Gently heat the liquid to see if any solid remains after the liquid evaporates. Also, collect the vapor and then compare the densities of the condensed liquid with the original liquid. **1.97** The volume occupied by the ice is larger than the volume of the glass bottle. The glass bottle would break. **1.99** 277.4 s = 4 min 37.35 s. **1.101** (a) Homogeneous. (b) Heterogeneous. **1.103** 6.0×10^{12} g Au; 2.6×10^{14}. **1.105** 7.3×10^{21} kg Si. **1.107** Density = 7.20 g/cm³; $r = 0.853$ cm. **1.109** It would be more difficult to prove that the unknown substance is an element. Most compounds would decompose on heating, making them easy to identify. **1.111** 1.1×10^2 yr. **1.113** 2.54×10^6 g Cu. **1.115** 9.5×10^{10} kg CO₂. **1.117** 2.3×10^4 kg NaF/yr; 99% NaF wasted. **1.119** 5×10^2 mL/breath. **1.121** %Error (°F) = 0.1%; %Error (°C) = 0.3%. **1.123** 4.0×10^{-19} g/L.

Key Skills
1.1 d **1.2** e **1.3** a **1.4** b

Chapter 2

2.3 $\dfrac{\text{ratio of S to O in compound 1}}{\text{ratio of S to O in compound 2}} = \dfrac{1.002}{0.668} \approx 1.5$

2.5 $\dfrac{\text{ratio of F to S in } S_2F_{10}}{\text{ratio of F to S in } SF_4} = \dfrac{2.962}{2.370} = 1.25{:}1 = 6{:}5{:}4;$

$\dfrac{\text{ratio of F to S in } SF_6}{\text{ratio of F to S in } SF_4} = \dfrac{3.555}{2.370} = 1.5{:}1 = 3{:}2.$ **2.7** $0.667{:}1 = 2{:}3.$

2.15 0.12 mi. **2.21** 145. **2.23** $^{15}_{7}$N: protons = 7, electrons = 7, neutrons = 8; $^{33}_{16}$S: protons = 16, electrons = 16, neutrons = 17; $^{63}_{29}$Cu: protons = 29, electrons = 29, neutrons = 34; $^{84}_{38}$Sr: protons = 38, electrons = 38, neutrons = 46; $^{130}_{56}$Ba: protons = 56, electrons = 56, neutrons = 74; $^{186}_{74}$W: protons = 74, electrons = 74, neutrons = 112; $^{202}_{80}$Hg: protons = 80, electrons = 80, neutrons = 122. **2.25** (a) $^{186}_{74}$W. (b) $^{201}_{80}$Hg. (c) $^{76}_{34}$Se. (d) $^{239}_{94}$Pu. **2.27** (a) 20. (b) 32. (c) 78. (d) 198. **2.35** Metallic character (a) increases as you progress down a group of the periodic table and (b) decreases from the left to right across the periodic table. **2.37** Na and K; N and P; F and Cl. **2.39** Iron: Fe, period 4, upper-left square of Group 18; Iodine: I, period 5, Group 17; Sodium: Na, period 3, Group 1; Phosphorus: P, period 3, Group 15; Sulfur: S, period 3, Group 16; Magnesium: Mg, period 3, Group 2. **2.45** 207.2 amu. **2.47** 192.2 amu. **2.49** ^{6}Li = 7.5%, ^{7}Li = 92.5%. **2.55** Na⁺: 11 protons, 10 electrons; Ca²⁺: 20 protons, 18 electrons; Al³⁺: 13 protons, 10 electrons; Fe²⁺: 26 protons, 24 electrons; I⁻: 53 protons, 54 electrons; F⁻: 9 protons, 10 electrons; S²⁻: 16 protons, 18 electrons; O²⁻: 8 protons, 10 electrons; N³⁻: 7 protons, 10 electrons. **2.57** (a) Na₂O. (b) FeS. (c) Co₂(SO₄)₃. (d) BaF₂. **2.59** Ionic: LiF, BaCl₂, KCl; Molecular: SiCl₄, B₂H₆, C₂H₄. **2.61** (a) Potassium dihydrogen phosphate. (b) Potassium hydrogen phosphate. (c) Hydrogen bromide. (d) Hydrobromic acid. (e) Lithium carbonate. (f) Potassium dichromate. (g) Ammonium nitrite. (h) Hydrogen iodate (in water, iodic acid). (i) Phosphorus pentafluoride. (j) Tetraphosphorus hexoxide. (k) Cadmium iodide. (l) Strontium sulfate. (m) Aluminum hydroxide. **2.63** (a) RbNO₂. (b) K₂S. (c) NaHS. (d) Mg₃(PO₄)₂. (e) CaHPO₄. (f) PbCO₃. (g) SnF₂. (h) (NH₄)₂SO₄. (i) AgClO₄. (j) BCl₃. **2.65** (a) Mg(NO₃)₂. (b) Al₂O₃. (c) LiH. (d) Na₂S. **2.75** (a) Polyatomic, elemental form, not a compound. (b) Polyatomic, compound. (c) Diatomic, compound. **2.77** Elements: N₂, S₈, H₂; Compounds: NH₃, NO, CO, CO₂, SO₂. **2.79** (a) CN. (b) CH. (c) C₉H₂₀. (d) P₂O₅. (e) BH₃. **2.81** C₃H₇NO₂. **2.83** (a) Nitrogen trichloride. (b) Iodine heptafluoride. (c) Tetraphosphorus hexoxide. (d) Disulfur dichloride. **2.85** (a) NF₃: nitrogen trifluoride. (b) PBr₅: phosphorus pentabromide. (c) SCl₂: sulfur dichloride. **2.87** Acid: compound that produces H⁺; Base: compound that produces OH⁻; Oxoacids: acids that contain oxygen; Oxoanions: the anions that remain when oxoacids lose H⁺ ions; Hydrates: ionic solids that have water molecules in their formulas. **2.89** (c) Changing the electrical charge of an atom usually has a major effect on its chemical properties. The two electrically neutral carbon isotopes should have nearly identical chemical properties. **2.91** P^{3-}. **2.93** NaCl is an ionic compound; it doesn't consist of molecules. **2.95** (a) Molecule and compound. (b) Element and molecule. (c) Element. (d) Molecule and compound. (e) Element. (f) Element and molecule. (g) Element and molecule. (h) Molecule and compound. (i) Compound, not molecule. (j) Element. (k) Element and molecule. (l) Compound, not molecule. **2.97** It establishes a standard mass unit that permits the measurement of masses of all other isotopes relative to carbon-12. **2.99** $^{11}_{5}B$, protons = 5, neutrons = 6, electrons = 5, net charge = 0; $^{54}_{26}Fe^{2+}$, protons = 26, neutrons = 28, electrons = 24, net charge = +2; $^{31}_{15}P^{3-}$, protons = 15, neutrons = 16, electrons = 18, net charge = −3; $^{196}_{79}Au$, protons = 79, neutrons = 117, electrons = 79, net charge = 0; $^{222}_{86}Rn$, protons = 86, neutrons = 136, electrons = 86, net charge = 0. **2.101** (a) Li^+. (b) S^{2-}. (c) I^-. (d) N^{3-}. (e) Al^{3+}. (f) Cs^+. (g) Mg^{2+}. **2.103** Group 17, binary: HF, hydrofluoric acid; HCl, hydrochloric acid; HBr, hydrobromic acid; HI, hydroiodic acid. Group 17, oxoacids: $HClO_4$, perchloric acid; $HClO_3$, chloric acid; $HClO_2$, chlorous acid; HClO, hypochlorous acid; $HBrO_3$, bromic acid; $HBrO_2$, bromous acid; HBrO, hypobromous acid; HIO_4, periodic acid; HIO_3, iodic acid; HIO, hypoiodous acid. Examples of oxoacids containing other Group 13–16 elements are: H_3BO_3, boric acid; H_2CO_3, carbonic acid; HNO_3, nitric acid; HNO_2, nitrous acid; H_3PO_4, phosphoric acid; H_3PO_3, phosphorous acid; H_3PO_2, hypophosphorous acid; H_2SO_4, sulfuric acid; H_2SO_3, sulfurous acid. Binary acids formed from other Group 13–16 elements: H_2S, hydrosulfuric acid. **2.105** ^4_2He: protons = 2, neutrons = 2, neutrons/protons = 1.00; $^{20}_{10}Ne$: protons = 10, neutrons = 10, neutrons/protons = 1.00; $^{40}_{18}Ar$: protons = 18, neutrons = 22, neutrons/protons = 1.22; $^{84}_{36}Kr$: protons = 36, neutrons = 48, neutrons/protons = 1.33; $^{132}_{54}Xe$: protons = 54, neutrons = 78, neutrons/protons = 1.44. **2.107** Cu, Ag, and Au are fairly chemically unreactive. This makes them especially suitable for making coins and jewelry that you want to last a very long time. **2.109** MgO and SrO. **2.111** (a) 2:1. (b) 1:2. (c) 2:1. (d) 5:2. **2.113** The mass of fluorine reacting with hydrogen and deuterium would be the same. The ratio of F atoms to hydrogen (or deuterium) atoms is 1:1 in both compounds. This does not violate the law of definite proportions. When the law of definite proportions was formulated, scientists did not know of the existence of isotopes. **2.115** (a) Br. (b) Rn. (c) Se. (d) Rb. (e) Pb. **2.117** Mg^{2+}, HCO_3^-, $Mg(HCO_3)_2$, Magnesium bicarbonate; Fe^{3+}, NO_2^-, $Fe(NO_2)_3$, Iron(III) nitrite; Mn^{2+}, ClO_3^-, $Mn(ClO_3)_2$, Manganese(II) chlorate; Co^{2+}, PO_4^{3-}, $Co_3(PO_4)_2$, Cobalt(II) phosphate; Hg_2^{2+}, I^-, Hg_2I_2, Mercury(I) iodide; Cu^+, CO_3^{2-}, Cu_2CO_3, Copper(I) carbonate. **2.119** 1.908×10^{-8} g. The predicted change (loss) in mass is too small a quantity to measure. Therefore, for all practical purposes, the law of conservation of mass is assumed to hold for ordinary chemical processes. **2.121** Chloric acid, nitrous acid, hydrocyanic acid, and sulfuric acid. **2.123** (a) Yes. (b) Acetylene: any formula with C:H = 1:1 (CH, C_2H_2, etc.); Ethane: any formula with C:H = 1:3 (CH_3, C_2H_6, etc.). **2.125** (a) $cA^{1/3} = r$ (c is a constant). (b) 5.1×10^{-44} m³. (c) 3.4×10^{-15}; yes.

Key Skills
2.1 d **2.2** c **2.3** c **2.4** a

Chapter 3

3.3 (a) 50.48 amu. (b) 92.02 amu. (c) 64.07 amu. (d) 84.16 amu. (e) 34.02 amu. (f) 342.3 amu. (g) 17.03 amu. **3.5** (a) 16.04 amu. (b) 46.01 amu. (c) 80.07 amu. (d) 78.11 amu. (e) 149.9 amu. (f) 174.27 amu. (g) 310.2 amu. **3.9** 78.77% Sn; 21.23% O. **3.11** (d) Ammonia, NH_3. **3.13** 39.89% Ca, 18.50% P, 41.41% O, 0.20% H. **3.15** (a) 60; 14; 0.75.

(b) 24; 1; 3.33. (c) 15; 1; 25. **3.21** (a) $KOH + H_3PO_4 \longrightarrow K_3PO_4 + H_2O$. (b) $Zn + AgCl \longrightarrow ZnCl_2 + Ag$. (c) $NaHCO_3 \longrightarrow Na_2CO_3 + H_2O + CO_2$. (d) $NH_4NO_2 \longrightarrow N_2 + H_2O$. (e) $CO_2 + KOH \longrightarrow K_2CO_3 + H_2O$. (f) $3KOH + H_3PO_4 \longrightarrow K_3PO_4 + 3H_2O$; $Zn + 2AgCl \longrightarrow ZnCl_2 + 2Ag$; $2NaHCO_3 \longrightarrow Na_2CO_3 + H_2O + CO_2$; $NH_4NO_2 \longrightarrow N_2 + 2H_2O$; $CO_2 + 2KOH \longrightarrow K_2CO_3 + H_2O$ **3.23** (a) Potassium and water react to form potassium hydroxide and hydrogen. (b) Barium hydroxide and hydrochloric acid react to form barium chloride and water. (c) Copper and nitric acid react to form copper nitrate, nitrogen monoxide, and water. (d) Aluminum and sulfuric acid react to form aluminum sulfate and hydrogen. (e) Hydrogen iodide reacts to form hydrogen and iodine. **3.25** (a) $2N_2O_5 \longrightarrow 2N_2O_4 + O_2$. (b) $2KNO_3 \longrightarrow 2KNO_2 + O_2$. (c) $NH_4NO_3 \longrightarrow N_2O + 2H_2O$. (d) $NH_4NO_2 \longrightarrow N_2 + 2H_2O$. (e) $2NaHCO_3 \longrightarrow Na_2CO_3 + H_2O + CO_2$. (f) $P_4O_{10} + 6H_2O \longrightarrow 4H_3PO_4$. (g) $2HCl + CaCO_3 \longrightarrow CaCl_2 + H_2O + CO_2$. (h) $2Al + 3H_2SO_4 \longrightarrow Al_2(SO_4)_3 + 3H_2$. (i) $CO_2 + 2KOH \longrightarrow K_2CO_3 + H_2O$. (j) $CH_4 + 2O_2 \longrightarrow CO_2 + 2H_2O$. (k) $Be_2C + 4H_2O \longrightarrow 2Be(OH)_2 + CH_4$. (l) $3Cu + 8HNO_3 \longrightarrow 3Cu(NO_3)_2 + 2NO + 4H_2O$. (m) $S + 6HNO_3 \longrightarrow H_2SO_4 + 6NO_2 + 2H_2O$. (n) $2NH_3 + 3CuO \longrightarrow 3Cu + N_2 + 3H_2O$. **3.27** (d) $3A + 2B \longrightarrow 2C + D$. **3.33** 5.8×10^3 light-yr. **3.35** 9.96×10^{-15} mol Co. **3.37** 3.01×10^3 g Au. **3.39** (a) 4.664×10^{-23} g/Si atom. (b) 9.273×10^{-23} g/Fe atom. **3.41** 2.450×10^{23} atoms Cu. **3.43** 2 atoms of lead. **3.45** 409 g/mol. **3.47** 3.01×10^{22} C atoms, 6.02×10^{22} H atoms, 3.01×10^{22} O atoms. **3.49** 39.3 g S. **3.51** 5.97 g F. **3.53** (a) CH_2O. (b) KCN. **3.55** $C_8H_{10}N_4O_2$. **3.57** 6.12×10^{21} molecules. **3.59** $C_9H_{16}O_4$; $C_9H_{16}O_4$; 57.43% C, 8.57% H, 34.00% O. **3.63** $C_{10}H_{20}O$. **3.65** $C_3H_7O_2NS$. **3.67** (a) Diagram (b). (b) Diagram (a). **3.71** 1.01 mol Cl_2. **3.73** 2.0×10^1 mol CO_2. **3.75** (a) $2NaHCO_3 \longrightarrow Na_2CO_3 + CO_2 + H_2O$. (b) 78.3 g $NaHCO_3$. **3.77** 255.9 g C_2H_5OH; 0.324 L. **3.79** 0.294 mol KCN. **3.81** $NH_4NO_3(s) \longrightarrow N_2O(g) + 2H_2O(g)$. (b) 2.0×10^1 g N_2O. **3.83** 71.48 g O_2. **3.89** MnO_2 is the limiting reactant; 81.5 g Cl_2 are produced. **3.91** 31.31 g $CO(NH_2)_2$ and 55.76 g NH_4Cl are produced; NH_3 is consumed completely; and 1.134 g $COCl_2$ remain. **3.93** The reaction produces 32.12 g H_2O. The resulting solution contains 155.3 g K_2SO_4 and 12.59 g unreacted H_2SO_4. **3.95** (a) 7.05 g O_2. (b) 92.9%. **3.97** 3.85×10^3 g C_6H_{14}. **3.99** 8.55 g S_2Cl_2; 76.6%. **3.101** $O_2 + 4NO_2 \longrightarrow 2N_2O_5$. The limiting reactant is NO_2. **3.103** 6 mol NH_3 produced; 1 mol H_2 left. **3.105** (a) Combustion. (b) Combination. (c) Decomposition. **3.107** Diagram (b). **3.109** (a) 0.212 mol O. (b) 0.424 mol O. **3.111** Cl_2O_7. **3.113** 700 g. **3.115** (a) 4.3×10^{22} Mg atoms. (b) 1.6×10^2 pm. **3.117** 0.0011 mol chlorophyll. **3.119** (a) 4.24×10^{22} K^+ ions, 4.24×10^{22} Br^- ions. (b) 4.58×10^{22} Na^+ ions, 2.29×10^{22} SO_4^{2-} ions. (c) 4.34×10^{22} Ca^{2+} ions, 2.89×10^{22} PO_4^{3-} ions. **3.121** 6.022×10^{23} amu = 1 g. **3.123** 16.00 amu. **3.125** (e) 0.50 mol Cl_2. **3.127** $PtCl_2$ and $PtCl_4$. **3.129** (a) Compound X: MnO_2; Compound Y: Mn_3O_4. (b) $3MnO_2 \longrightarrow Mn_3O_4 + O_2$. **3.131** Mg_3N_2, magnesium nitride. **3.133** 28.97 g/mol. **3.135** $BaBr_2$. **3.137** 32.17% NaCl, 20.09% Na_2SO_4, 47.75% $NaNO_3$. **3.139** (a) $C_3H_8(g) + 5O_2(g) \longrightarrow 3CO_2(g) + 4H_2O(l)$. (b) 482 g CO_2. **3.141** (a) $Zn(s) + H_2SO_4(aq) \longrightarrow ZnSO_4(aq) + H_2(g)$. (b) 64.2%. (c) We assume that the impurities are inert and do not react with the sulfuric acid to produce hydrogen. **3.143** (a) $C_3H_8(g) + 3H_2O(g) \longrightarrow 3CO(g) + 7H_2(g)$. (b) 909 kg H_2. **3.145** 1.85×10^5 kg CaO. **3.147** CH_2O. **3.149** (a) C_3H_7NO. (b) $C_6H_{14}N_2O_2$. **3.151** 30.20% C, 5.069% H, 44.57% Cl, 20.16% S. **3.153** (a) 6.532×10^4 g. (b) 7.6×10^2 g HG. **3.155** $C_3H_2ClF_5O$; 184.50 g/mol. **3.157** 6.1×10^5 tons H_2SO_4. **3.159** $C_2H_3NO_5$. **3.161** (a) \$0.47/kg. (b) 0.631 kg K_2O. **3.163** 3.1×10^{23} molecules/mol.

Key Skills
3.1 b **3.2** c **3.3** e **3.4** a

Chapter 4

4.7 Diagram (c). **4.9** (a) Strong electrolyte. (b) Nonelectrolyte. (c) Weak electrolyte. (d) Strong electrolyte. **4.11** (a) Nonconducting. (b) Conducting. (c) Conducting. **4.13** Since HCl dissolved in water conducts electricity, $HCl(aq)$ must actually exist as $H^+(aq)$ cations and $Cl^-(aq)$ anions. Since HCl dissolved in benzene solvent does not conduct electricity, we must assume that the HCl molecules in benzene solvent do not ionize, but rather exist as un-ionized molecules. **4.17** Diagram (d). **4.19** (a) Insoluble. (b) Insoluble. (c) Soluble. (d) Soluble. **4.21** (a) $2Ag^+(aq) + 2NO_3^-(aq) + 2Na^+(aq) + SO_4^{2-}(aq) \longrightarrow Ag_2SO_4(s) + 2Na^+(aq) + 2NO_3^-(aq)$; $2Ag^+(aq) + SO_4^{2-}(aq) \longrightarrow Ag_2SO_4(s)$. (b) $Ba^{2+}(aq) + 2Cl^-(aq) + Zn^{2+}(aq) + SO_4^{2-}(aq) \longrightarrow BaSO_4(s) + Zn^{2+}(aq) + 2Cl^-(aq)$; $Ba^{2+}(aq) + SO_4^{2-}(aq) \longrightarrow BaSO_4(s)$. (c) $2NH_4^+(aq) + CO_3^{2-}(aq) + Ca^{2+}(aq) + 2Cl^-(aq) \longrightarrow CaCO_3(s) + 2NH_4^+(aq) + 2Cl^-(aq)$; $Ca^{2+}(aq) + CO_3^{2-}(aq) \longrightarrow CaCO_3(s)$. **4.23** (a) No precipitate forms. (b) $Ba^{2+}(aq) + SO_4^{2-}(aq) \longrightarrow BaSO_4(s)$. **4.31** (a) Brønsted base. (b) Brønsted base. (c) Brønsted acid. (d) Brønsted acid and Brønsted base. **4.33** (a) $HC_2H_3O_2(aq) + KOH(aq) \longrightarrow KC_2H_3O_2(aq) + H_2O(l)$; *Ionic*: $HC_2H_3O_2(aq) + K^+(aq) + OH^-(aq) \longrightarrow C_2H_3O_2^-(aq) + K^+(aq) + H_2O(l)$; *Net ionic*: $HC_2H_3O_2(aq) + OH^-(aq) \longrightarrow C_2H_3O_2^-(aq) + H_2O(l)$. (b) $H_2CO_3(aq) + 2NaOH(aq) \longrightarrow Na_2CO_3(aq) + 2H_2O(l)$; *Ionic*: $H_2CO_3(aq) + 2Na^+(aq) + 2OH^-(aq) \longrightarrow 2Na^+(aq) + CO_3^{2-}(aq) + 2H_2O(l)$; *Net ionic*: $H_2CO_3(aq) + 2OH^-(aq) \longrightarrow CO_3^{2-}(aq) + 2H_2O(l)$. (c) $2HNO_3(aq) + Ba(OH)_2(aq) \longrightarrow Ba(NO_3)_2(aq) + 2H_2O(l)$; *Ionic*: $2H^+(aq) + 2NO_3^-(aq) + Ba^{2+}(aq) + 2OH^-(aq) \longrightarrow Ba^{2+}(aq) + 2NO_3^-(aq) + 2H_2O(l)$; $2H^+(aq) + 2OH^-(aq) \longrightarrow 2H_2O(l)$ or $H^+(aq) + OH^-(aq) \longrightarrow H_2O(l)$. **4.41** (a) $2Sr \longrightarrow 2Sr^{2+} + 4e^-$, Sr is the reducing agent; $O_2 + 4e^- \longrightarrow 2O^{2-}$, O_2 is the oxidizing agent. (b) $2Li \longrightarrow 2Li^+ + 2e^-$, Li is the reducing agent; $H_2 + 2e^- \longrightarrow 2H^-$, H_2 is the oxidizing agent. (c) $2Cs \longrightarrow 2Cs^+ + 2e^-$, Cs is the reducing agent; $Br_2 + 2e^- \longrightarrow 2Br^-$, Br_2 is the oxidizing agent. (d) $3Mg \longrightarrow 3Mg^{2+} + 6e^-$, Mg is the reducing agent; $N_2 + 6e^- \longrightarrow 2N^{3-}$, N_2 is the oxidizing agent. **4.43** H_2S (−2), S^{2-} (−2), HS^- (−2) < S_8 (0) < SO_2 (+4) < SO_3 (+6), H_2SO_4 (+6). **4.45** (a) +1. (b) +7. (c) −4. (d) −1. (e) −2. (f) +6. (g) +6. (h) +7. (i) +4. (j) 0. (k) +5. (l) −1/2. (m) +5. (n) +3. **4.47** (a) +1. (b) −1. (c) +3. (d) +3. (e) +4. (f) +6. (g) +2. (h) +4. (i) +2. (j) +3. (k) +5. **4.49** If nitric acid is a strong oxidizing agent and zinc is a strong reducing agent, then zinc metal will probably reduce nitric acid when the two react; that is, N will gain electrons and the oxidation number of N must decrease. Since the oxidation number of nitrogen in nitric acid is +5, the nitrogen-containing product must have a smaller oxidation number for nitrogen. The only compound in the list that doesn't have a nitrogen oxidation number less than +5 is N_2O_5. This is never a product of the reduction of nitric acid. **4.51** Molecular oxygen is a powerful oxidizing agent. In SO_3, the oxidation number of the element bound to oxygen (S) is at its maximum value (+6); the sulfur cannot be oxidized further. The other elements bound to oxygen in this problem have less than their maximum oxidation number and can undergo further oxidation. Only SO_3 does not react with molecular oxygen. **4.53** (a) Decomposition. (b) Displacement. (c) Decomposition. (d) Combination. **4.59** 232 g KI. **4.61** 6.00×10^{-3} mol $MgCl_2$. **4.63** (a) 1.16 M. (b) 0.608 M. (c) 1.78 M. **4.65** (a) 136 mL. (b) 62.2 mL. (c) 47 mL. **4.67** Dilute 323 mL of the 2.00 M HCl solution to a final volume of 1.00 L. **4.69** Dilute 3.00 mL of the 4.00 M HNO_3 solution to a final volume of 60.0 mL. **4.71** (a) $BaCl_2$: 0.300 M Cl^-; NaCl: 0.566 M Cl^-; $AlCl_3$: 3.606 M Cl^-. (b) 1.28 M $Sr(NO_3)_2$. **4.73** 2.325 M. **4.81** 0.215 g AgCl. **4.83** 0.165 g NaCl; $Ag^+(aq) + Cl^-(aq) \longrightarrow AgCl(s)$. **4.85** (a) 42.78 mL. (b) 158.5 mL. (c) 79.23 mL.

4.87 1.74 g. **4.89** 0 g (no insoluble product). **4.91** (a) First combination; (b) second combination; (c) second combination. **4.93** Diagram (b) = H_3PO_4; Diagram (c) = HCl; Diagram (d) = H_2SO_4. **4.95** (a) Redox. (b) Precipitation. (c) Acid-base. (d) Combination. (e) Redox. (f) Redox. (g) Precipitation. (h) Redox. (i) Redox. (j) Redox. **4.97** (d) 0.20 M $Mg(NO_3)_2$ (highest concentration of ions). **4.99** 773 mL. **4.101** (a) Weak electrolyte. (b) Strong electrolyte. (c) Strong electrolyte. (d) Nonelectrolyte. **4.103** (a) $C_2H_5ONH_2$ molecules. (b) K^+ and F^- ions. (c) NH_4^+ and NO_3^- ions. (d) C_3H_7OH molecules. **4.105** 1146 g/mol. **4.107** 1.28 M. **4.109** 43.4 g $BaSO_4$. **4.111** 1.72 M. **4.113** (1) Electrolysis to ascertain if hydrogen and oxygen were produced, (2) the reaction with an alkali metal to see if a base and hydrogen gas were produced, and (3) the dissolution of a metal oxide to see if a base was produced (or a nonmetal oxide to see if an acid was produced). **4.115** 1.09 M $Ca(NO_3)_2$. **4.117** Diagram (a) showing Ag^+ and NO_3^- ions. The reaction is $AgOH(aq) + HNO_3(aq) \longrightarrow H_2O(l) + AgNO_3(aq)$. **4.119** (a) Check with litmus paper, combine with carbonate or bicarbonate to see if CO_2 gas is produced and combine with a base and check for neutralization with an indicator. (b) Titrate a known quantity of acid with a standard NaOH solution. (c) Visually compare the conductivity of the acid with a standard NaCl solution of the same molar concentration. **4.121** No. The oxidation number of all oxygen atoms is zero. **4.123** (a) $HI(aq) + KOH(aq) \longrightarrow KI(aq) + H_2O(l)$, evaporate to dryness. (b) $2HI(aq) + K_2CO_3(aq) \longrightarrow 2KI(aq) + CO_2(g) + H_2O(l)$, evaporate to dryness. **4.125** (a) Combine any soluble magnesium salt with a soluble hydroxide, and filter the precipitate. (b) Combine any soluble silver salt with any soluble iodide salt, and filter the precipitate. (c) Combine any soluble barium salt with any soluble phosphate salt, and filter the precipitate. **4.127** (a) Add Na_2SO_4. (b) Add KOH. (c) Add $AgNO_3$. (d) Add $Ca(NO_3)_2$. (e) Add $Mg(NO_3)_2$. **4.129** Reaction 1: $SO_3^{2-}(aq) + H_2O_2(aq) \longrightarrow SO_4^{2-}(aq) + H_2O(l)$; Reaction 2: $SO_4^{2-}(aq) + Ba^{2+}(aq) \longrightarrow BaSO_4(s)$. **4.131** Cl_2O (+1), Cl_2O_3 (+3), ClO_2 (+4), Cl_2O_6 (+6), Cl_2O_7 (+7). **4.133** $[Na^+] = 0.5295$ M, $[NO_3^-] = 0.4298$ M, $[OH^-] = 0.09968$ M, $[Mg^{2+}] \approx 0$ M. **4.135** 1.41 M $KMnO_4$. **4.137** (a) The precipitate $CaSO_4$ formed over Ca preventing the Ca from reacting with the sulfuric acid. (b) Aluminum is protected by a tenacious oxide layer with the composition Al_2O_3. (c) These metals react more readily with water: $2Na(s) + 2H_2O(l) \longrightarrow 2NaOH(aq) + H_2(g)$. (d) The metal should be placed below Fe and above H. (e) Any metal above Al in the activity series will react with Al^{3+}. Metals from Mg to Li will work. **4.139** 56.2% NaBr. **4.141** (a) 1.40 M Cl^-. (b) 4.96 g Cl^-. **4.143** (a) Acid: H_3O^+; base: OH^-.

(b) Acid: NH_4^+; base: NH_2^-.

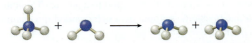

4.145 When a solid dissolves in solution, the volume of the solution usually changes. **4.147** Electric furnace method: $P_4(s) + 5O_2(g) \longrightarrow P_4O_{10}(s)$ (redox), $P_4O_{10}(s) + 6H_2O(l) \longrightarrow 4H_3PO_4(aq)$ (acid-base); wet process: $Ca_5(PO_4)_3F(s) + 5H_2SO_4(aq) \longrightarrow HF(aq) + 3H_3PO_4(aq) + 5CaSO_4(s)$ (acid-base and precipitation). **4.149** (a) $CaF_2(s) + H_2SO_4(aq) \longrightarrow CaSO_4(s) + 2HF(g)$; $2NaCl(s) + H_2SO_4(aq) \longrightarrow Na_2SO_4(aq) + 2HCl(g)$. (b) The sulfuric acid would oxidize the Br^- and I^- ions to Br_2 and I_2. (c) $PBr_3(l) + 3H_2O(l) \longrightarrow 3HBr(g) + H_3PO_3(aq)$. **4.151** (a) $4KO_2(s) + 2CO_2(g) \longrightarrow 2K_2CO_3(s) + 3O_2(g)$. (b) −1/2.

(c) 34.4 L air. **4.153** 4.99 grains. **4.155** (a) $Pb(NO_3)_2(aq)$ + $Na_2SO_4(aq) \longrightarrow PbSO_4(s) + 2NaNO_3(aq)$; *net ionic:* $Pb^{2+}(aq)$ + $SO_4^{2-}(aq) \longrightarrow PbSO_4(s)$. (b) 6.34×10^{-5} M. **4.157** $Cu^{2+}(aq)$ + $S^{2-}(aq) \longrightarrow CuS(s)$, 2.31×10^{-4} M Cu^{2+}. **4.159** (a) Nonelectrolytes: CH_3CH_2OH, H_2O; weak electrolyte: $HC_2H_3O_2$; strong electrolytes: $K_2Cr_2O_7$, H_2SO_4, $Cr_2(SO_4)_3$, K_2SO_4.

(b) *Ionic equation:* $3CH_3CH_2OH(g) + 4K^+(aq) + 2Cr_2O_7^{2-}(aq)$ + $8H^+(aq) + 2HSO_4^-(aq) \longrightarrow 3HC_2H_3O_2(aq) + 4Cr^{3+}(aq) + 8SO_4^{2-}$ $(aq) + 4K^+(aq) + 11H_2O(l)$; *net ionic equation:* $3CH_3CH_2OH(g)$ + $2Cr_2O_7^{2-}(aq) + 8H^+(aq) + 2HSO_4^-(aq) \longrightarrow 3HC_2H_3O_2(aq)$ + $4Cr^{3+}(aq) + 8SO_4^{2-}(aq) + 11H_2O(l)$.

(c) $3CH_3CH_2OH(g)$ + $2K_2Cr_2O_7(aq)$ + $8H_2SO_4(aq) \longrightarrow 3HC_2H_3O_2(aq)$ + $2Cr_2(SO_4)_3(aq)$ + $2K_2SO_4(aq)$ + $11H_2O(l)$.

| −2 +1 −2 +1 −2 +1 | +1 +6 −2 | +1 +6 −2 | +1 0 +1 −2 | +3 +6 −2 | +1 +6 −2 | +1 −2 |
| −2 +3 −2 +2 −2 +1 | +2 +12 −14 | +2 +6 −8 | +1 0 +3 −4 | +6 +18 −24 | +2 +6 −8 | +2 −2 |

(d) 8.5×10^{-4} M. (e) 15 mL. (f) $[K^+] = 1.7 \times 10^{-3}$ M; $[Cr_2O_7^{2-}] = 8.5 \times 10^{-4}$ M.

Key Skills
4.1 b **4.2** c **4.3** c **4.4** d

Chapter 5

5.7 Law of conservation of energy. **5.9** Energy is needed to break chemical bonds, while energy is released when bonds are formed. **5.13** −46 J. **5.15** 925 J (work done *on* the system). **5.17** (a) Diagram (ii). (b) Diagram (ii). (c) Diagram (ii). **5.25** (a) 0 J. (b) −9.5 J. (c) −18 J. **5.27** 4.51 kJ/g. **5.29** 4.80×10^2 kJ. **5.31** 595.8 kJ/mol. **5.35** 728 kJ. **5.37** 50.7°C. **5.39** 26.3°C. **5.41** 2.36 J/g · °C. **5.43** Metal A. **5.47** (a) 150 kJ/mol, (b) 600 kJ/mol, (c) −150 kJ/mol, (d) 300 kJ/mol, (e) −300 kJ/mol. **5.49** 0.30 kJ/mol. **5.51** −238.7 kJ/mol. **5.53** −438.6 kJ/mol. **5.55** −300 kJ/mol. **5.57** 180 kJ/mol. **5.59** 105 kJ/mol. **5.63** $CH_4(g)$ and $H(g)$. **5.65** $\Delta H_f^\circ[H_2O(l)]$. **5.67** (a) −41.2 kJ/mol. (b) −1046.4 kJ/mol. **5.69** (a) −724 kJ/mol. (b) -1.37×10^3 kJ/mol. (c) -2.01×10^3 kJ/mol. **5.71** −3924 kJ/mol. **5.73** −175.3 kJ. **5.75** −71.58 kJ/g B_5H_9. **5.77** (a) $\Delta H_f^\circ[Br_2(l)] = 0$. $\Delta H_f^\circ[Br_2(g)] > 0$. (b) $\Delta H_f^\circ[I_2(s)] = 0$. $\Delta H_f^\circ[I_2(g)] > 0$. **5.79** $2Ag(s) + \frac{1}{2}O_2(g) \longrightarrow$ $Ag_2O(s)$, $\Delta H_f^\circ[Ag_2O] = \Delta H_{rxn}^\circ$; $Ca(s) + Cl_2(g) \longrightarrow CaCl_2(s)$, $\Delta H_f^\circ(CaCl_2) = \Delta H_{rxn}^\circ$; calorimetry can be used to measure the enthalpy changes. **5.81** In a chemical reaction, the same elements and the same numbers of atoms are always on both sides of the equation. This provides a consistent reference which allows the energy change in the reaction to be interpreted in terms of the chemical or physical changes that have occurred. In a nuclear reaction, the same elements are not always on both sides of the equation and no common reference point exists. **5.83** −44.35 kJ/mol. **5.85** 0.492 J/g·°C. **5.87** −350.7 kJ/mol. **5.89** $1.92/gal ethanol. **5.91** 5.60 kJ/mol. **5.93** Reaction (a). **5.95** (a) 0 J. (b) −9.1 J. **5.97** 5.35 kJ/°C. **5.99** Burning graphite in oxygen will form both CO and CO_2. **5.101** −277.0 kJ/mol. **5.103** 104 g. **5.105** $w = 0$, $\Delta U = -5153$ kJ/mol. **5.107** 96.21%. **5.109** 58.1°C. **5.111** (a) As heat is added to water at 25°C, the temperature increases until the boiling point is reached. (b) At 1 atm of pressure, water at 100°C will remain at that temperature until the added heat has converted all the liquid to a gas. (c) The temperature of a system can change without heat being added if a chemical reaction occurs.

	q	w	ΔU	ΔH
(a)	−	0	−	−
(b)	−	−	−	−
(c)	+	−	?	+
(d)	+	0	+	+
(e)	+	−	0	0

5.113 **5.115** 23.6°C. **5.117** The first reaction, which is exothermic, can be used to promote the second reaction, which is endothermic. Thus, the two gases are produced

alternately. **5.119** -3.60×10^2 kJ/mol Zn or -3.60×10^2 kJ/2 mol Ag^+. **5.121** 4.1 cents. **5.123** −9.78 kJ/mol. **5.125** 3.0×10^9 atomic bombs. **5.127** (a) Although we cannot measure ΔH_{rxn}° for this reaction, the reverse process is the combustion of glucose. We could easily measure ΔH_{rxn}° for this combustion by burning a mole of glucose in a bomb calorimeter. (b) 1.1×10^{19} kJ. **5.129** 5.8×10^2 m. **5.131** Water has a larger specific heat than air. Thus, cold, damp air can extract more heat from the body than cold, dry air. By the same token, hot, humid air can deliver more heat to the body. **5.133** (a) $2LiOH(aq)$ + $CO_2(g) \longrightarrow Li_2CO_3(aq) + H_2O(l)$. (b) 1.1 kg CO_2, 1.2 kg LiOH. **5.135** (a) $CaC_2(s) + 2H_2O(l) \longrightarrow Ca(OH)_2(s) + C_2H_2(g)$. (b) 1.51×10^6 J. **5.137** (a) Glucose: 31 kJ, sucrose: 33 kJ. (b) Glucose: 15 m, sucrose: 16 m. **5.139** -5.2×10^6 kJ. **5.141** Since the humidity is very low in deserts, there is little water vapor in the air to trap and hold the heat radiated back from the ground during the day. Once the sun goes down, the temperature drops dramatically. 40°F temperature drops between day and night are common in desert climates. Coastal regions have much higher humidity levels compared to deserts. The water vapor in the air retains heat, which keeps the temperature at a more constant level during the night. In addition, sand and rocks in the desert have small specific heats compared with water in the ocean. The water absorbs much more heat during the day compared to sand and rocks, which keeps the temperature warmer at night. **5.143** (a) $3N_2H_4(l) \longrightarrow 4NH_3(g) + N_2(g)$. (b) −336.5 kJ/mol. (c) $N_2H_4(l) + O_2(g) \longrightarrow N_2(g) + 2H_2O(l)$, $\Delta H_{rxn}^\circ = -622.0$ kJ/mol; $4NH_3(g) + 3O_2(g) \longrightarrow 2N_2(g) + 6H_2O(l)$, $\Delta H_{rxn}^\circ = -1529.6$ kJ/mol. (d) Ammonia. **5.145** 0.44 J/g · °C. The metal could be either Fe or Ni.

Key Skills
5.1 e **5.2** a **5.3** c **5.4** b

Chapter 6

6.5 (a) 3.5×10^3 nm. (b) 5.30×10^{14} Hz. **6.7** 3.26×10^7 nm, microwave. **6.9** 7.0×10^2 s. **6.15** 2.82×10^{-19} J. **6.17** (a) 4.6×10^7 nm, not in the visible region. (b) 4.3×10^{-24} J/photon. (c) 2.6 J/mol. **6.19** 1.29×10^{-15} J. **6.21** (a) 1.2×10^2 photons. **6.23** Infrared photons have insufficient energy to cause the chemical changes. **6.25** A "blue" photon (shorter wavelength) is higher energy than a "yellow" photon. For the same amount of energy delivered to the metal surface, there must be fewer "blue" photons than "yellow" photons. Thus, the yellow light would eject more electrons since there are more "yellow" photons. Since the "blue" photons are of higher energy, blue light will eject electrons with greater kinetic energy. **6.29** 3.027×10^{-19} J. **6.31** $\nu = 1.60 \times 10^{14}$ Hz, $\lambda = 1.88 \times 10^3$ nm. **6.33** 5. **6.35** Analyze the emitted light by passing it through a prism. **6.37** Excited atoms of the chemical elements emit the same characteristic frequencies or lines in a terrestrial laboratory, in the Sun, or in a star many light-years distant from Earth. **6.41** 0.565 nm. **6.43** 9.96×10^{-32} cm. **6.51** 1.6×10^{-11} m. **6.53** $\Delta u \geq 4.38 \times 10^{-26}$ m/s. This uncertainty is far smaller than can be measured. **6.57** $\ell = 1$, $m_\ell = -1$, 0, and 1; $\ell = 0$, $m_\ell = 0$. **6.59** $4s$, $4p$, $4d$, and $4f$ subshells; 1, 3, 5, and 7 orbitals, respectively. **6.65** (a) $n = 2$, $\ell = 1$, $m_\ell = 1$, 0, or -1. (b) $n = 3$, $\ell = 0$, $m_\ell = 0$. (c) $n = 5$, $\ell = 2$, $m_\ell = 2$, 1, 0, -1, or -2. **6.67** A $2s$ orbital is larger than a $1s$ orbital and exhibits a node. Both have the same spherical shape. The $1s$ orbital is lower in energy than the $2s$. **6.69** In H, energy depends only on n, but for all other atoms, energy depends on n and ℓ. **6.71** (a) $2s$. (b) 3. (c) Equal. (d) Equal. (e) $5s$. **6.73** (a) Orbital b. (b) Orbitals a and d. (c) None. **6.77** (a) Two. (b) Six. (c) Ten. (d) Fourteen. **6.79** $3s$: two; $3d$: ten; $4p$: six; $4f$: fourteen; $5f$: fourteen. **6.81** (a) is wrong because the magnetic quantum number m_ℓ can have only whole-number values. (b) is wrong because the magnetic quantum number m_ℓ can only have the value 0 when the angular momentum quantum number ℓ is 0. (c) is wrong because the magnetic quantum number m_ℓ can only have the value 0 when the angular momentum quantum number ℓ is 0. (e) is wrong because the electron spin quantum number m_s can have only half-integral values. **6.83** B: 1; Ne: 0; P: 3; Sc: 1; Mn: 5; Se: 2; Kr: 0; Fe: 4; Cd: 0; I: 1; Pb: 2. **6.85** S^+. **6.95** $[Kr]5s^2 4d^5$. **6.97** Ge: $[Ar]4s^2 3d^{10} 4p^2$; Fe: $[Ar]4s^2 3d^6$; Zn: $[Ar]4s^2 3d^{10}$; Ni: $[Ar]4s^2 3d^8$; W: $[Xe]6s^2 4f^{14} 5d^4$; Tl: $[Xe]6s^2 4f^{14} 5d^{10} 6p^1$. **6.99** Part (b) is correct in the view of contemporary quantum theory. Bohr's explanation of emission and absorption line spectra appears to have universal validity. Parts (a) and (c) are artifacts of Bohr's early planetary model of the hydrogen atom and are not considered to be valid today. **6.101** (a) 4. (b) 6. (c) 10. (d) 1. (e) 2. **6.103** (a) Metal A: 3.4×10^{-19} J; metal B: 5.6×10^{-19} J; metal C: 6.6×10^{-19} J; metal C has the highest binding energy. (b) Electrons will be ejected from metals A and B. **6.105** He: $n = 3 \longrightarrow 2$: $\lambda = 164$ nm; $n = 4 \longrightarrow 2$: $\lambda = 121$ nm; $n = 5 \longrightarrow 2$: $\lambda = 108$ nm; $n = 6 \longrightarrow 2$: $\lambda = 103$ nm. H: $n = 3 \longrightarrow 2$: $\lambda = 656$ nm; $n = 4 \longrightarrow 2$: $\lambda = 486$ nm; $n = 5 \longrightarrow 2$: $\lambda = 434$ nm; $n = 6 \longrightarrow 2$: $\lambda = 410$ nm. All the Balmer transitions for He^+ are in the ultraviolet region, whereas the transitions for H are all in the visible region.

6.107 (a) $\underset{1s^2}{\uparrow\downarrow}$ $\underset{2s^2}{\uparrow\downarrow}$ $\underset{2p^5}{\uparrow\downarrow \;\; \uparrow\downarrow \;\; \uparrow}$. (b) [Ne] $\underset{3s^2}{\uparrow\downarrow}$ $\underset{3p^3}{\uparrow \;\; \uparrow \;\; \uparrow}$.

(c) [Ar] $\underset{4s^2}{\uparrow\downarrow}$ $\underset{3d^7}{\uparrow\downarrow \;\; \uparrow\downarrow \;\; \uparrow \;\; \uparrow \;\; \uparrow}$.

6.109 (a) False. (b) False. (c) True. (d) False. (e) True. **6.111** (a) He, $1s^2$. (b) N, $1s^2 2s^2 2p^3$. (c) Na, $1s^2 2s^2 2p^6 3s^1$. (d) As, $[Ar]4s^2 3d^{10} 4p^3$. (e) Cl, $[Ne]3s^2 3p^5$. **6.113** (b) and (d) are allowed transitions. All of the transitions in Figure 6.11 are possible as long as ℓ for the final state differs from ℓ of the initial state by 1.

6.115 $\dfrac{1}{\lambda_1} = \dfrac{1}{\lambda_2} + \dfrac{1}{\lambda_3}$. **6.117** 1.06 nm. **6.119** (a) 2.29×10^{-6} nm. (b) 6.0×10^{-2} kg. **6.121** $\lambda = 0.382$ pm, $\nu = 7.86 \times 10^{20}$ s^{-1}. **6.123** In the photoelectric effect, light of sufficient energy shining on a metal surface causes electrons to be ejected (photoelectrons). Since the electrons are charged particles, the metal surface becomes positively charged as more electrons are lost. After a long enough period of time, the positive surface charge becomes large enough to start attracting the ejected electrons back toward the metal with the result that the kinetic energy of the departing electrons becomes smaller. **6.125** 17.4 pm. **6.127** $\lambda = 0.596$ m; microwave/radio region. **6.129** 483 nm. **6.131** 2.2×10^5 J. **6.133** (a) We note that the maximum solar radiation centers around 500 nm. Thus, over billions of years, organisms have adjusted their development to capture energy at or near this wavelength. The two most notable cases are photosynthesis and vision. (b) Astronomers record black-body radiation curves from stars and compare them with those obtained from objects at different temperatures in the laboratory. Because the shape of the curve and the wavelength corresponding to the maximum depend on the temperature of an object, astronomers can reliably determine the temperature at the surface of a star from the closest matching curve and wavelength. **6.135** 3.3×10^{28} photons. **6.137** 4.10×10^{23} photons.

Key Skills
6.1 b **6.2** d **6.3** c **6.4** d

Chapter 7

7.17 Selenium, $1s^2 2s^2 2p^6 3s^2 3p^6 4s^2 3d^{10} 4p^4$. **7.19** (a) and (d); (b) and (e); (c) and (f). **7.21** (a) Group 1. (b) Group 15. (c) Group 18. (d) Group 10. **7.25** (a) $\sigma = 2$ and $Z_{eff} = +4$. (b) $2s$, $Z_{eff} = +3.22$; $2p$, $Z_{eff} = +3.14$. The values are lower than those in part (a) because the $2s$ and $2p$ electrons actually do shield each other somewhat. **7.33** 8.40×10^6 kJ/mol. **7.35** Na > Mg > Al > P > Cl. **7.37** Fluorine. **7.39** Left to right: S, Se, Ca, K. **7.41** The atomic radius is largely determined by how strongly the outer-shell electrons are held by the nucleus. The larger the effective nuclear charge, the more strongly the electrons are held and the smaller the atomic radius. For the second period, the atomic radius of Li is largest because the $2s$ electron is well shielded by the filled $1s$ shell. The effective nuclear charge that the outermost electrons feel increases across the period as a result of incomplete shielding by electrons in the same shell. Consequently, the orbital containing the electrons is compressed and the atomic radius decreases. **7.43** K < Ca < P < F < Ne. **7.45** The Group 13 elements (such as Al) all have a single electron in the outermost p subshell, which is well shielded from the nuclear charge by the inner electrons and the ns^2 electrons. Therefore, less energy is needed to remove a single p electron than to remove a paired s electron from the same principal energy level (such as for Mg). **7.47** 496 kJ/mol is paired with $1s^2 2s^2 2p^6 3s^1$. 2080 kJ/mol is paired with $1s^2 2s^2 2p^6$, a very stable noble gas configuration. **7.49** Cl. **7.51** Alkali metals have a valence electron configuration of ns^1 so they can accept another electron in the ns orbital. On the other hand, alkaline earth metals have a valence electron configuration of ns^2. Alkaline earth metals have little tendency to accept another electron, because it would have to go into a higher energy p orbital. **7.57** Fe. **7.59** Be^{2+} and He; N^{3-} and F^-; Fe^{2+} and Co^{3+}; S^{2-} and Ar. **7.61** (a) Cr^{3+}. (b) Sc^{3+}. (c) Rh^{3+}. (d) Ir^{3+}. **7.65** (a) Cl. (b) Na^+. (c) O^{2-}. (d) Al^{3+}. (e) Au^{3+}. **7.67** The Cu^+ ion is larger than Cu^{2+} because it has one more electron. **7.73** $-199.7°C$. **7.75** Since ionization energies decrease going down a column in the periodic table, francium should have the lowest first ionization energy of all the alkali metals. As a result, Fr should be the most reactive of all the Group 1 elements toward water and oxygen. The reaction with oxygen would probably be similar to that of K, Rb, or Cs. **7.77** The Group 11 elements are much less reactive than the Group 1 elements. The Group 11 elements are more stable because they have much higher ionization energies resulting from incomplete shielding of the nuclear charge by the inner d electrons. The ns^1 electron of a Group 1 element is shielded from the nucleus more effectively by the completely filled noble gas core. Consequently, the outer s electrons of Group 11 elements are more strongly attracted by the nucleus. **7.79** (a) $Li_2O(s) + H_2O(l) \longrightarrow 2LiOH(aq)$. (b) $CaO(s) + H_2O(l) \longrightarrow Ca(OH)_2(aq)$. (c) $SO_3(g) + H_2O(l) \longrightarrow H_2SO_4(aq)$.

7.81 BaO. As we move down a column, the metallic character of the elements increases. **7.83** (a) Br. (b) N. (c) Rb. (d) Mg. **7.85** $O^{2-} < F^- < Na^+ < Mg^{2+}$. **7.87** O^+ and N; S^{2-} and Ar; N^{3-} and Ne; As^{3+} and Zn; Cs^+ and Xe. **7.89** (d). **7.91** Fluorine is a yellow-green gas that attacks glass; chlorine is a pale yellow gas; bromine is a fuming red liquid; iodine is a dark, metallic-looking solid. **7.93** F. **7.95** H^-. Since H^- has only one proton compared to two protons for He, the nucleus of H^- will attract the two electrons less strongly compared to He. **7.97** Li_2O, lithium oxide, basic; BeO, beryllium oxide, amphoteric; B_2O_3, diboron trioxide, acidic; CO_2, carbon dioxide, acidic; N_2O_5, dinitrogen pentoxide, acidic. **7.99** 0.66. **7.101** 77.5%. **7.103** (a) Matches bromine (Br_2). (b) Matches hydrogen (H_2). (c) Matches calcium (Ca). (d) Matches gold (Au). (e) Matches argon (Ar). **7.105** X must belong to Group 14; it is probably Sn or Pb because it is not a very reactive metal (it is certainly not reactive like an alkali metal). Y is a nonmetal since it does not conduct electricity. Since it is a light yellow solid, it is probably phosphorus (Group 15). Z is an alkali metal since it reacts with air to form a basic oxide or peroxide. **7.107** (a) $IE_1 = 3s^1$ electron, $IE_2 = 2p^6$ electron, $IE_3 = 2p^5$ electron, $IE_4 = 2p^4$ electron, $IE_5 = 2p^3$ electron, $IE_6 = 2p^2$ electron, $IE_7 = 2p^1$ electron, $IE_8 = 2s^2$ electron, $IE_9 = 2s^1$ electron, $IE_{10} = 1s^2$ electron, $IE_{11} = 1s^1$ electron. (b) Each break ($IE_1 \longrightarrow IE_2$ and $IE_9 \longrightarrow IE_{10}$) represents the transition to another shell ($n = 3 \longrightarrow 2$ and $n = 2 \longrightarrow 1$).

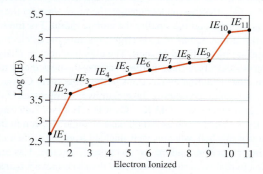

7.109 LiH (lithium hydride), CH_4 (methane), NH_3 (ammonia), H_2O (water), and HF (hydrogen fluoride); $LiH + H_2O \longrightarrow LiOH + H_2$; $CH_4 + H_2O \longrightarrow$ no reaction at room temperature; $NH_3 + H_2O \longrightarrow NH_4^+ + OH^-$; $H_2O + H_2O \longrightarrow H_3O^+ + OH^-$; $HF + H_2O \longrightarrow H_3O^+ + F^-$. **7.111** (a) $2KClO_3(s) \longrightarrow 2KCl(s) + 3O_2(g)$. (b) $N_2(g) + 3H_2(g) \longrightarrow 2NH_3(g)$ (industrial); $NH_4Cl(s) + NaOH(aq) \longrightarrow NH_3(g) + NaCl(aq) + H_2O(l)$. (c) $CaCO_3(s) \longrightarrow CaO(s) + CO_2(g)$ (industrial); $CaCO_3(s) + 2HCl(aq) \longrightarrow CaCl_2(aq) + H_2O(l) + CO_2(g)$. (d) $Zn(s) + H_2SO_4(aq) \longrightarrow ZnSO_4(aq) + H_2(g)$. (e) Same as (c), (first equation). **7.113** Examine a solution of Na_2SO_4, which is colorless. This shows that the SO_4^{2-} ion is colorless. Thus, the blue color is due to $Cu^{2+}(aq)$. **7.115** Z_{eff} increases from left to right across the table, so electrons are held more tightly. (This explains the electron affinity values of C and O.) Nitrogen has a zero value of electron affinity because of the stability of the half-filled $2p$ subshell (that is, N has little tendency to accept another electron). **7.117** Once an atom gains an electron forming a negative ion, adding additional electrons is typically an unfavorable process due to electron-electron repulsions. 2nd and 3rd electron affinities do not occur spontaneously and are therefore difficult to measure. **7.119** 2A. There is a large jump from the second to the third ionization energy, indicating a change in the principal quantum number n. **7.121** (a) SiH_4, GeH_4, SnH_4, PbH_4. (b) RbH should be more ionic than NaH. (c) $Ra(s) + 2H_2O(l) \longrightarrow Ra(OH)_2(aq) + H_2(g)$. (d) Be (diagonal relationship).

7.123 Li: $Z_{eff} = 1.26$, $Z_{eff}/n = 0.630$; Na: $Z_{eff} = 1.84$, $Z_{eff}/n = 0.613$; K: $Z_{eff} = 2.26$, $Z_{eff}/n = 0.565$. As we move down a group, Z_{eff} increases. This is what we would expect because shells with larger n values are less effective at shielding the outer electrons from the nuclear charge. The Z_{eff}/n values are fairly constant, meaning that the screening per shell is about the same. **7.125** Nitrogen. Lithium forms a stable nitride (Li_3N). **7.127** (c) Carbon. **7.129** 6.94×10^{-19} J/electron. If there are no other electrons with lower kinetic energy, then this is the electron from the valence shell. UV light of the *longest* wavelength (lowest energy) that can still eject electrons should be used. **7.131** (a) [Ne]. (b) [Ne]. (c) [Ar]. (d) [Ar]. (e) [Ar]. (f) $[Ar]3d^6$. (g) $[Ar]3d^9$. (h) $[Ar]3d^{10}$. **7.133** The binding of a cation to an anion results from electrostatic attraction. As the +2 cation gets smaller (from Ba^{2+} to Mg^{2+}), the distance between the opposite charges decreases and the electrostatic attraction increases. **7.135** (a) It was determined that the periodic table was based on atomic number, not atomic mass. (b) Argon: 39.95 amu; potassium: 39.10 amu. **7.137** The electron configuration of titanium is: $[Ar]4s^23d^2$. K_2TiO_4 is unlikely to exist because of the oxidation state of Ti (+6). Ti in an oxidation state greater than +4 is unlikely because of the very high ionization energies needed to remove the fifth and sixth electrons. **7.139** 343 nm; ultraviolet.

Key Skills
7.1 e **7.2** d **7.3** e **7.4** d

Chapter 8

8.3 (a) $\cdot$Be$\cdot$ (b) $\cdot$K (c) $\cdot$Ca$\cdot$ (d) $\cdot\ddot{G}a\cdot$ (e) $\cdot\ddot{O}\cdot$ (f) $:\ddot{B}r\cdot$ (g) $\cdot\ddot{N}\cdot$ (h) $:\ddot{I}\cdot$ (i) $\cdot\ddot{A}s\cdot$ (j) $\cdot\ddot{F}\cdot$

8.5 (a) $:\ddot{I}\cdot$ (b) $\left[:\ddot{I}:\right]^-$ (c) $\cdot\ddot{S}\cdot$ (d) $\left[:\ddot{S}:\right]^{2-}$ (e) $\cdot\ddot{P}\cdot$ (f) $\left[:\ddot{P}:\right]^{3-}$ (g) $\cdot$Na (h) Na^+ (i) $\cdot Mg\cdot$ (j) Mg^{2+} (k) $\cdot\ddot{A}l\cdot$ (l) $\left[:\ddot{A}s:\right]^{3+}$ (m) $\cdot\ddot{P}b\cdot$ (n) $\left[:Pb\right]^{2+}$

8.19 860 kJ/mol. **8.21** (a) Decreases the ionic bond energy. (b) Triples the ionic bond energy. (c) Increases the bond energy by a factor of 4. (d) Increases the bond energy by a factor of 2.

8.23

(a) $Na\cdot + :\ddot{F}\cdot \longrightarrow Na^+:\ddot{F}:^-$

(b) $2K\cdot + \cdot\ddot{S}\cdot \longrightarrow 2K^+:\ddot{S}:^{2-}$

(c) $\dot{B}a + \cdot\ddot{O}\cdot \longrightarrow Ba^{2+}:\ddot{O}:^{2-}$

(d) $\dot{A}l\cdot + \cdot\ddot{N}\cdot \longrightarrow Al^{3+}:\ddot{N}:^{3-}$

8.33 (a) BF_3, boron trifluoride, covalent. (b) KBr, potassium bromide, ionic. **8.37** 0.057 **8.39** Cl–Cl < Br–Cl < Si–C < Cs–F. **8.41** (a) Covalent. (b) Polar covalent. (c) Ionic. (d) Polar covalent. **8.43** C–H ($\Delta EN = 0.4$) < Br–H ($\Delta EN = 0.7$) < F–H ($\Delta EN = 1.9$) < Li–Cl ($\Delta EN = 2.0$) < Na–Cl ($\Delta EN = 2.1$) < K–F ($\Delta EN = 3.2$). **8.45** Greatest percent ionic character will be in the compound consisting of the yellow element in period 5 and the blue element in period 2 because they have the greatest difference in electronegativities.

8.47

(a) $:\ddot{F}-\ddot{O}-\ddot{F}:$

(b) $:\ddot{F}-\ddot{N}=\ddot{N}-\ddot{F}:$

(c) H—Si—Si—H with H atoms on each Si

(d) $^-\ddot{O}-H$

(e) H—C—C—Ö:$^-$ with :O: double bonded and :Cl: below

(f) H—C—N$^+$—H with H atoms

8.49
(a) $\overset{..}{O}=C$ with two $\overset{..}{Br}:$ (b) H—$\overset{..}{Se}$—H (c) H—$\overset{..}{N}$—H with $\overset{|}{\overset{..}{O}}$—H

(d) $:\overset{..}{Cl}$—$\overset{\overset{:O:}{|}}{P}$—$\overset{..}{Cl}:$ with $\overset{|}{\underset{..}{Cl}}:$ (e) H—C—C—$\overset{..}{Br}:$ with H,H,H,H (f) $:\overset{..}{Cl}$—$\overset{..}{N}$—$\overset{..}{Cl}:$ with $\overset{|}{\underset{..}{Cl}}:$

8.51 (a) $\overset{..}{O}=\overset{+}{N}=\overset{..}{O}$ (b) $\overset{..}{S}=C=\overset{..}{N}:^-$ or $^-:\overset{..}{S}$—C≡N:
(c) $^-:\overset{..}{S}$—$\overset{..}{S}:^-$ (d) $:\overset{..}{F}$—$\overset{..}{Cl}\overset{+}{:}$—$\overset{..}{F}:$

8.53 (a) Neither oxygen atom has a complete octet, and the left-most hydrogen atom shows two bonds (4 electrons). Hydrogen can hold only two electrons in its valence shell.

(b) H—C—C—$\overset{..}{O}$—H (with H, H, :O: arrangement)

8.57 (a) H—C with $:\overset{..}{O}:^-$ and $:\overset{..}{O}:$ ⟷ H—C with $:O:$ and $:\overset{..}{O}:^-$

(b) $\overset{H}{\overset{}{}}C=\overset{+}{N}$ with $:\overset{..}{O}:^-$ and $:\overset{..}{O}:$ ⟷ $\overset{H}{}C$—$\overset{+}{N}$ with $:\overset{..}{O}:^-$ and $:\overset{..}{O}:$ ⟷ $\overset{H}{}C$—$\overset{+}{N}$ with $:O:$ and $:\overset{..}{O}:^-$

8.59 H—$\overset{..}{N}$=$\overset{+}{N}$=$\overset{..}{N}:^-$ ⟷ H—$\overset{..}{N}$—$\overset{+}{N}$≡N: ⟷ H—$\overset{+}{N}$≡$\overset{+}{N}$—$\overset{..}{N}:^{2-}$

8.61 $\overset{..}{O}$=C=$\overset{..}{N}:^-$ ⟷ $^-:\overset{..}{O}$—C≡N: ⟷ $:\overset{+}{O}$≡C—$\overset{..}{N}:^{2-}$

8.63 (structure left) H—C ... ⟷ (structure right) H—C ... (adenine resonance structures)

8.71 No. $:\overset{..}{Cl}$—Be—$\overset{..}{Cl}:$ To make an octet on Be is not plausible,

$^+:\overset{..}{Cl}$=$\overset{2-}{Be}$=$\overset{+}{\overset{..}{Cl}}:$ **8.73** No. Cl—Sb—Cl structure (with Cl, Cl, Cl, Cl) **8.75** Coordinate covalent bond,

$:\overset{..}{Cl}$—Al—$\overset{..}{Cl}:$ + $:\overset{..}{Cl}:^-$ ⟶ $:\overset{..}{Cl}$—$\overset{..}{Al}^-$—$\overset{..}{Cl}:$ **8.77** Completed octet

on S: $\left[:\overset{..}{O}$—$\overset{\overset{:O:}{||}}{S}$—$\overset{..}{O}:\right]^{2-}$; zero formal charge on S: $\left[:\overset{..}{O}$—$\overset{\overset{:O:}{|}}{\underset{:O:}{S}}$—$\overset{..}{O}:\right]^{2-}$

8.81 303.0 kJ/mol. **8.83** (a) −2759 kJ/mol. (b) −2855.4 kJ/mol.
8.85 −651 kJ/mol. **8.87** (a) Exothermic, (b) endothermic,
(c) not enough information to determine.
8.89 Ionic: NaF, MgF_2, AlF_3. Covalent: SiF_4, PF_5, SF_6, ClF_3.
8.91 KF is an ionic compound. It is a solid at room temperature made up of K^+ and F^- ions. It has a high melting point, and it is a strong electrolyte. Benzene, C_6H_6, is a covalent compound that exists as discrete molecules. It is a liquid at room temperature. It has a low melting point, is insoluble in water, and is a nonelectrolyte.
8.93 $^-\overset{..}{N}$=$\overset{+}{N}$=$\overset{..}{N}:^-$ ⟷ $:N$≡$\overset{+}{N}$—$\overset{..}{N}:^{2-}$ ⟷ $^{2-}:\overset{..}{N}$—$\overset{+}{N}$≡N:
8.95 (a) $AlCl_4^-$, (b) AlF_6^{3-}, (c) $AlCl_3$. **8.97** CF_2 would be very unstable because carbon does not have an octet. LiO_2 would not be stable because the lattice energy between Li^+ and superoxide O_2^-

would be too low to stabilize the solid. $CsCl_2$ requires a Cs^{2+} cation. The second ionization energy is too large to be compensated by the increase in lattice energy. PI_5 appears to be a reasonable species. However, the iodine atoms are too large to have five of them "fit" around a single P atom. **8.99** (a) False. (b) True. (c) False. (d) False.
8.101 −67 kJ/mol. **8.103** N_2, since it has a triple bond.
8.105 CH_4 and NH_4^+; N_2 and CO; C_6H_6 and $B_3N_3H_6$.

8.107 $\left[:\overset{..}{O}$—$\overset{\overset{:\overset{..}{O}:}{|}}{P}$—$\overset{..}{O}:\right]^{3-}$ $\left[:\overset{..}{O}$—$\overset{\overset{:O:}{||}}{P}$—$\overset{..}{O}:\right]^{3-}$

H—$\overset{..}{O}$—$\overset{\overset{:\overset{..}{O}:}{|}}{Cl}$—$\overset{..}{O}:$ H—$\overset{..}{O}$—$\overset{\overset{\overset{.}{O}}{|}}{Cl}$=$\overset{..}{O}:$

$\overset{..}{O}$=$\overset{\overset{\overset{.}{O}}{||}}{S}$—$\overset{..}{O}:$ $:O$=$\overset{\overset{\overset{.}{O}}{||}}{S}$=O:
$:\overset{..}{O}$=$\overset{\overset{\overset{.}{O}}{||}}{S}$—$\overset{..}{O}:$ $:O$=$\overset{\overset{\overset{.}{O}}{||}}{S}$=O:

8.109 H—$\overset{..}{N}:^-$ + H—$\overset{..}{O}:$ ⟶ H—$\overset{..}{N}$—H + $^-\overset{..}{O}$—H (with H below each)

8.111 The central iodine atom in I_3^- has *ten* electrons surrounding it: two bonding pairs and three lone pairs. The central iodine has an expanded octet. Elements in the second period such as fluorine cannot have an expanded octet as would be required for F_3^-.

8.113 $:N$≡$\overset{+}{N}$—$\overset{..}{N}$=$\overset{+}{N}$=$\overset{..}{N}:^-$ ⟷ $^-\overset{..}{N}$=$\overset{+}{N}$=$\overset{..}{N}$—$\overset{+}{N}$≡N: ⟷
$:N$≡$\overset{+}{N}$—$\overset{..}{N}^-$—$\overset{+}{N}$≡N:

8.115 Form (a) is the most important structure with no formal charges and all satisfied octets. Form (b) is likely not as important as (a) because of the positive formal charge on O. Forms (c) and (d) do not satisfy the octet rule for all atoms and are likely not important. **8.117** The arrows indicate coordinate covalent bonds.

(Al dimer structure with Cl atoms)

This dimer does not possess a dipole moment.
8.119 (a) −9.2 kJ/mol. (b) −9.2 kJ/mol.
8.121 (a) $^-:C$≡$O:^+$ (b) $:N$≡$O:^+$ (c) $^-:C$≡N: (d) $:N$≡N:
8.123 True. Each noble gas atom already has completely filled ns and np subshells. **8.125** (a) 114 kJ/mol. (b) The bond in F_2^- is weaker.
8.127 (a) $:\overset{..}{N}$=$\overset{..}{O}$ ⟷ $^-:N$=$\overset{..}{O}:^+$ The first structure is the most important. (b) No. **8.129** 347 kJ/mol. **8.131** EN(O) = 3.2 (Pauling 3.5); EN(F) = 4.4 (Pauling 4.0); EN(Cl) = 3.5 (Pauling 3.0). **8.133** C—C: 347 kJ/mol; N—N: 193 kJ/mol; O—O: 142 kJ/mol. Lone pairs appear to weaken the bond. **8.135** 2×10^2 kJ/mol. **8.137** (1) You could estimate the lattice energy of the solid by trying to measure its melting point. Mg^+O^- would have a lattice energy (and, therefore, a melting point) similar to that of Na^+Cl^-. This lattice energy and melting point are much lower than those of $Mg^{2+}O^{2-}$. (2) You could determine the magnetic properties of the solid. An Mg^+O^- solid would be paramagnetic while $Mg^{2+}O^{2-}$ solid is diamagnetic. See Chapter 9 of the text.
8.139

H—$\overset{..}{N}$—C—C—$\overset{..}{O}$—H (with :O: above C, H—C—H below, aromatic ring structure below)

8.141

$$H-\overset{\overset{\displaystyle H}{|}}{\underset{\underset{\displaystyle H}{|}}{C}}-\ddot{N}=C=\ddot{O} \longleftrightarrow H-\overset{\overset{\displaystyle H}{|}}{\underset{\underset{\displaystyle H}{|}}{C}}-\overset{+}{N}\equiv C-\ddot{O}\mathord{:}^-$$

8.143 (a) $:\!\ddot{F}-\overset{\overset{\displaystyle :\ddot{C}l:}{|}}{\underset{\underset{\displaystyle :\ddot{C}l:}{|}}{C}}-\ddot{C}l\mathord{:}$ (b) $:\!\ddot{F}-\overset{\overset{\displaystyle :\ddot{F}:}{|}}{\underset{\underset{\displaystyle :\ddot{C}l:}{|}}{C}}-\ddot{C}l\mathord{:}$

(c) $H-\overset{\overset{\displaystyle :\ddot{F}:}{|}}{\underset{\underset{\displaystyle :\ddot{F}:}{|}}{C}}-\ddot{C}l\mathord{:}$ (d) $:\!\ddot{F}-\overset{\overset{\displaystyle :\ddot{F}:}{|}}{\underset{\underset{\displaystyle :\ddot{F}:}{|}}{C}}-\overset{\overset{\displaystyle :\ddot{F}:}{|}}{\underset{\underset{\displaystyle :\ddot{F}:}{|}}{C}}-H$

8.145 $:\!\ddot{C}l-\ddot{O}-\overset{\overset{\displaystyle \cdot\cdot}{\overset{\displaystyle \ddot{O}}{\|}}}{\underset{\underset{\displaystyle +}{}}{N}}-\ddot{O}\mathord{:}^-$

8.147

(a) $H\overset{\diagdown H}{} \quad H^+ \longleftrightarrow H^+ \quad \overset{H\diagdown}{}H \longleftrightarrow \overset{H^+}{H-\!\!\!-\!\!\!-H}$ (b) -413 kJ/mol.

Key Skills

8.1 a **8.2** b, e **8.3** c **8.4** a

Chapter 9

9.7 (a) Trigonal pyramidal. (b) Tetrahedral. (c) Tetrahedral.
(d) Seesaw. **9.9** (a) Tetrahedral. (b) Trigonal planar. (c) Trigonal
pyramidal. (d) Bent. (e) Bent. **9.11** (a) Linear. (b) Tetrahedral.
(c) Trigonal bipyramidal. (d) Trigonal pyramidal. (e) Tetrahedral.
9.13 Carbon at the center of H_3C-: electron domain geometry =
tetrahedral, molecular geometry = tetrahedral; carbon at center
of $-CO-OH$: electron-domain geometry = trigonal planar,
molecular geometry = trigonal planar; oxygen in $-O-H$: electron
domain geometry = tetrahedral, molecular geometry = bent.
9.17 (a) Polar. (b) Nonpolar. **9.19** Only (c) is polar. **9.29** (a) sp^3.
(b) sp^3. **9.31** In BF_3, B is sp^2-hybridized. In NH_3, N is sp^3-hybridized.
In F_3B-NH_3, B and N are both sp^3-hybridized. **9.33** sp^3d.
9.37 (a) sp. (b) sp. (c) sp. **9.39** sp. **9.41** Nine σ bonds and nine π
bonds. **9.43** 36 σ bonds and 10 π bonds. **9.49** In order for the two
hydrogen atoms to combine to form a H_2 molecule, the electrons
must have opposite spins. If two H atoms collide and their electron
spins are parallel, no bond will form. **9.51** $Li_2^+ = Li_2^- < Li_2$.
(Both Li_2^+ and Li_2^- have bond order of ½. Li_2 has a bond order of 1.)
9.53 B_2^+, with a bond order of ½. (B_2 has a bond order of 1.)
9.55 The Lewis diagram has all electrons paired (incorrect) and a
double bond (correct). The MO diagram has two unpaired electrons
(correct), and a bond order of 2 (correct). **9.57** O_2: bond order = 2,
paramagnetic; O_2^+: bond order = 2.5, paramagnetic; O_2^-: bond order =
1.5, paramagnetic; O_2^{2-}: bond order = 1, diamagnetic. **9.59** The two
shared electrons that make up the single bond in B_2 both lie in pi
molecular orbitals and constitute a pi bond. The four shared electrons
that make up the double bond in C_2 all lie in pi molecular orbitals and
constitute two pi bonds. **9.63** The left symbol shows three delocalized double bonds (correct). The right symbol shows three localized
double bonds and three single bonds (incorrect).

9.65 (a) $\ddot{O}=\overset{+}{N}-\ddot{O}\mathord{:}^- \longleftrightarrow :\!\ddot{O}-\overset{+}{N}=\ddot{O}$

(b) sp^2. (c) Sigma bonds join the nitrogen atom to the fluorine and
oxygen atoms. There is a pi molecular orbital delocalized over the
N and O atoms. **9.67** The central oxygen atom is sp^2-hybridized. The
unhybridized $2p_z$ orbital on the central oxygen overlaps with the
$2p_z$ orbitals on the two terminal atoms. **9.69** $:\!\ddot{B}r-Hg-\ddot{B}r\mathord{:}$ Linear.
You could establish the geometry of $HgBr_2$ by measuring its dipole

moment. **9.71** Bent; sp^3. **9.73** (a) $[Ne_2](\sigma_{3s})^2(\sigma_{3s}^*)^2(\pi_{3p_y})^2(\pi_{3p_z})^2(\sigma_{3p_x})^2$.
(b) 3. (c) Diamagnetic. **9.75** (a) 180°. (b) 120°. (c) 109.5°. (d) 109.5°.
(e) 180°. (f) 120°. (g) 109.5°. (h) 109.5°. **9.77**

(a) $\overset{\overset{\displaystyle :\ddot{F} \quad \ddot{F}:}{\diagdown \,B\, \diagup}}{\underset{\underset{\displaystyle :\ddot{F}:}{|}}{}}$, planar. (b) $\left[:\!\ddot{O}-\overset{\overset{\displaystyle \cdot\cdot}{\overset{\displaystyle \ddot{O}}{\|}}}{Cl}-\ddot{O}\mathord{:}\right]^-$ $\begin{matrix} \\ :\ddot{O}\mathord{:}^- \end{matrix}$, nonplanar.

(c) $H-C\equiv N\mathord{:}$; polar. (d) $:\!\ddot{F}\overset{\overset{\displaystyle :\ddot{O}}{}}{\diagdown\,\diagup}\ddot{F}\mathord{:}$, polar. (e) $\overset{\overset{\displaystyle :\ddot{N}}{}}{\overset{\cdot}{O}\diagdown\,\diagup\ddot{O}\mathord{:}}$,
greater than 120°. Experimental value is around 135°.
9.79 (a) Nonpolar. (b) Polar. **9.81** Only ICl_2^- and $CdBr_2$ are linear.
9.83 (a) sp^2. (b) The molecule on the right is polar. **9.85** (a) Polar.
(b) Nonpolar. **9.87** (a) Trigonal bipyramidal, square planar,
octahedral. (b) Octahedral. **9.89** The molecule is linear and
symmetric about the molecular axis, so we do not expect the
molecule to possess a dipole moment. **9.91** C has no d orbitals
but Si does ($3d$). Thus, H_2O molecules can add to Si in hydrolysis
(valence-shell expansion). **9.93** The carbons are all sp^2-hybridized.
The nitrogen double bonded to carbon in the ring is sp^2-hybridized.
The other nitrogens are sp^3-hybridized. **9.95** F_2 has 8 electrons in
bonding orbitals and 6 electrons in antibonding orbitals, giving
it a bond order of 1. F_2^- has 8 electrons in bonding orbitals and
7 electrons in antibonding orbitals, giving it a bond order of 1/2.
9.97 As the molecule vibrates with one of its bending modes, it
deviates from its equilibrium linear geometry, producing a transient
dipole moment. The CO_2 molecule can also vibrate by an asymmetric
shift in the positions of the atoms along the molecular axis, causing an
asymmetric distribution of charge that also creates a transient
dipole moment. In both cases, the transient dipole moments
disappear as the molecule relaxes to its equilibrium geometry.
9.99 $[He_2](\sigma_{2s})^2(\sigma_{2s}^*)^2(\pi_{2p_y})^2(\pi_{2p_z})^2(\sigma_{2p_x})^2$; CO is isoelectronic with CN^-.
9.101 In the Lewis structure, all the electrons are paired. From
molecular orbital theory, the electrons would be arranged as follows:
$[He_2](\sigma_{2s})^2(\sigma_{2s}^*)^2(\sigma_{2p_x})^2(\pi_{2p_y})^2(\pi_{2p_z})^2(\pi_{2p_y}^*)^1(\pi_{2p_z}^*)^1$. This shows two
unpaired electrons. To pair these electrons, energy is required to flip
the spin of one, thus making the Lewis structure an excited state.
9.103 Tetrahedral. **9.105** (a) $C_6H_8O_6$. (b) Hybridization of each of the
five central O atoms is sp^3; hybridization of the sixth (double bonded)
peripheral O atom is sp^2. Three C atoms (those with double bonds)
are sp^2-hybridized; the other three C atoms are sp^3-hybridized.
(c) The geometry about each of the five central O atoms is bent.
The geometry about each of the three sp^2 C atoms is trigonal planar.
The geometry about each one of the three sp^3 C atoms is tetrahedral.
9.107 (a) Although the O atoms are sp^3-hybridized, they are locked
in a planar structure by the benzene rings. The molecule is
symmetrical and therefore is not polar. (b) 20 σ bonds and 6 π bonds.
9.109 (a) $:\!\overset{-}{C}\equiv\overset{+}{O}\mathord{:}$ The electronegativity difference between O and C
suggests that electron density should concentrate on the O atom, but
assigning formal charges places a negative charge on the C atom.
Therefore, we expect CO to have a small dipole moment. (b) CO is
isoelectronic with N_2, bond order 3. This agrees with the triple bond
in the Lewis structure. (c) Since C has a negative formal charge,
it is more likely to form bonds with Fe^{2+}. ($OC-Fe^{2+}$ rather than
$CO-Fe^{2+}$). **9.111** The $S-S$ bond is a normal 2-electron shared
pair covalent bond. Each S is sp^3-hybridized, so the $X-S-S$
angle is about 109°. **9.113** Rotation about the sigma bond in
1,2-dichloroethane does not destroy the bond, so the bond is free
to rotate. Thus, the molecule is nonpolar because the $C-Cl$ bond
moments cancel each other because of the averaging effect
brought about by rotation. The π bond between the C atoms in
cis-dichloroethylene prevents rotation (in order to rotate, the
π bond must be broken). Therefore, the molecule is polar.

9.115 $S_8(s) + 16SO_3(g) \longrightarrow 24SO_2(g)$. S_8: 0, sp^3; SO_3: +6, sp^2; SO_2: +4, sp^2. 4.99 kg SO_3. 5.99 kg SO_2.

Key Skills
9.1 c **9.2** d **9.3** b **9.4** a, e

Chapter 10

10.13 0.493 atm, 0.500 bar, 375 torr, 5.00×10^4 Pa. **10.15** 13.1 m.
10.17 7.3 atm. **10.21** 45.9 mL. **10.23** 587 mmHg. **10.25** 31.8 L.
10.27 b. **10.29** (a) Diagram (d). (b) Diagram (b). **10.35** 6.6 atm.
10.37 2.15 atm. **10.39** 0.70 L. **10.41** 38.5 L. **10.43** 2.3×10^{-2} atm.
10.45 92.45 g/mol. **10.47** 2.1×10^{22} N_2 molecules, 5.6×10^{22} O_2
molecules, 2.7×10^{20} O_2 molecules. **10.49** 2.98 g/L. **10.51** SF_4.
10.53 1590°C. **10.55** 3.70×10^2 L. **10.57** 88.9%. **10.59** $2Y_3 +$
$3XY \longrightarrow 3XY_3$, 50.0 mL. **10.61** 94.7%. Assuming the impurity
(or impurities) do not produce CO_2. **10.63** $C_2H_5OH(l) + 3O_2(g) \longrightarrow$
$2CO_2(g) + 3H_2O(l)$; 1.79×10^3 L air. **10.67** $\chi_{purple} = 0.222$,
$\chi_{red} = 0.278$, $\chi_{yellow} = 0.500$; $P_{red} = 0.275$ atm, $P_{yellow} = 0.496$ atm;
$P_{total} = 0.991$ atm. **10.69** 349 mmHg. **10.71** 19.8 g Zn. **10.73** $P_{N_2} =$
217 mmHg, $P_{H_2} = 650$ mmHg. **10.75** (a) Box 2. (b) Box 2.
10.83 $u_{rms}(N_2) = 472$ m/s, $u_{rms}(O_2) = 441$ m/s, $u_{rms}(O_3) = 360$ m/s.
10.85 RMS = 2.8 m/s, average speed = 2.7 m/s. The root-mean-
square value is always greater than the average value, because
squaring favors the larger values compared to just taking the
average value. **10.87** 43.8 g/mol, CO_2. **10.89** (a) The molar mass
of the yellow gas is less than the molar mass of the black gas.
(b) The molar mass of the red gas is less than the molar mass of the
blue gas. **10.95** No. **10.97** Ne. **10.99** C_6H_6. **10.101** (a) $P_{ii} = 4.0$ atm,
$P_{iii} = 2.67$ atm. (b) 2.67 atm, $P_A = 1.33$ atm, $P_B = 1.33$ atm.
10.103 (a) $2KClO_3(s) \longrightarrow 2KCl(s) + 3O_2(g)$. (b) 6.21 L.
10.105 (a) $C_3H_8(g) + 5O_2(g) \longrightarrow 3CO_2(g) + 4H_2O(g)$. (b) 11.4 L
CO_2. **10.107** 0.166 atm O_2, 0.333 atm NO_2. **10.109** (a) First, the
total pressure (P_{Total}) of the mixture of carbon dioxide and hydrogen
must be determined at a given temperature in a container of known
volume. Next, the carbon dioxide can be removed by reaction with
sodium hydroxide. The pressure of the hydrogen gas that remains
can now be measured under the same conditions of temperature
and volume. Finally, the partial pressure of CO_2 can be calculated.
(b) The most direct way to measure the partial pressures would be
to use a mass spectrometer to measure the mole fractions of the
gases. The partial pressures could then be calculated from the mole
fractions and the total pressure. Another way to measure the partial
pressures would be to realize that helium has a much lower boiling
point than nitrogen. Therefore, nitrogen gas can be removed by
lowering the temperature until nitrogen liquefies. Helium will
remain as a gas. As in part (a), the total pressure is measured first.
Then, the pressure of helium can be measured after the nitrogen is
removed. Finally, the pressure of nitrogen is simply the difference
between the total pressure and the pressure of helium.
10.111 33.1% Na_2CO_3. **10.113** 0.400 g H_2O. **10.115** C_6H_6.

10.117 (a) $8\left(\frac{4}{3}\pi r^3\right)$. (b) $\dfrac{4N_A\left(\frac{4}{3}\pi r^3\right)}{1 \text{ mole}}$. The volume actually
occupied by a mole of molecules with radius r is $N_A\left(\frac{4}{3}\pi r^3\right)$.

10.119 146 pm. **10.121** The partial pressure of carbon dioxide is
higher in the winter because carbon dioxide is utilized less by
photosynthesis in plants. **10.123** 42.6 K. **10.125** Radon, because
it is radioactive so that its mass is constantly changing (decreasing).
The number of radon atoms is not constant. **10.127** 53.4%.
10.129 $P_{NO_2} = 0.53$ atm, and $P_{N_2O_4} = 0.45$ atm. **10.131** The nitrogen

sample will have the greatest volume. **10.133** The law of conservation
of energy (or the first law of thermodynamics) states that energy
cannot be created or destroyed. While individual gas particles
constantly exchange energy with other particles and with the
container walls, the overall energy remains constant. This does not
violate energy conservation. **10.135** (a) Of the substances listed in
Figure 10.23, Cl_2 and NH_3 have normal boiling points that are significantly higher than the others, so we may conclude that the intermolecular attractions in these two substances are relatively large. These
strong attractions lead to a measured molar volume that is less than
the molar volume of an ideal gas. (b) For Ar, H_2, He, N_2, Ne, and O_2,
the normal boiling points are relatively low, indicating relatively weak
intermolecular attractions. Since the attractions are weak, it is the
excluded volume that dominates the deviation from nonideal behavior,
and excluded volume causes the actual volume to be higher than
expected ideally. **10.137** $\chi_{CO} = 0.544$. **10.139** Warm air rises because
of its buoyancy. This buoyancy is a direct result of the decreased
density of the warm air relative to the surrounding air. On a molecular
level, the decreased density of the warm air can be accounted for by
considering that the molecules in the warm air move with more speed
and thus more kinetic energy at higher temperatures. These more-
energetic molecules are able to open up a larger "bubble" of volume
within the surrounding air than they would otherwise, thus making
the density less than the surrounding air. **10.141** (a) 0.112 mol CO/min.
(b) 2.0×10^1 min. **10.143** 7.8×10^3 L NH_3. **10.145** (a) 3.4 g Hg.
(b) Yes. (c) *Physical:* The sulfur powder covers the Hg surface,
thus retarding the rate of evaporation. *Chemical:* Sulfur reacts
slowly with Hg to form HgS. HgS has no measurable vapor
pressure. **10.147** 1.8×10^2 mL. **10.149** 50.1%. **10.151** 4.20 L.
10.153 20.9 kg O_2; 1.58×10^4 L O_2. **10.155** (a) $u_{mp} = 421$ m/s,
$u_{rms} = 515$ m/s. The most probable speed (u_{mp}) will be 81.6% of the
root-mean-square speed (u_{rms}) at a given temperature. (b) 1200 K.
10.157 The air inside the egg expands with increasing temperature.
The increased pressure can cause the egg to crack. **10.159** (a) 2.1.
(b) 5% by volume. **10.161** 445 mL. **10.163** (a) 0.86 L. (b) The
advantage in using the ammonium salt is that more gas is produced
per gram of reactant. The disadvantage is that one of the gases
is ammonia. The strong odor of ammonia would *not* make the
ammonium salt a good choice for baking. **10.165** (a) 229 K = −44°C.
(b) 72.4%. **10.167** 2.3×10^3 L. **10.169** (a) $CaO(s) + CO_2(g) \longrightarrow$
$CaCO_3(s)$, $BaO(s) + CO_2(g) \longrightarrow BaCO_3(s)$. (b) 10.5% CaO,
89.5% BaO. **10.171** 101.0 J.

Key Skills
10.1 d **10.2** c **10.3** b **10.4** e

Chapter 11

11.9 Butane would be a liquid in winter (boiling point −44.5°C), and
on the coldest days even propane would become a liquid (boiling
point −0.5°C). Only methane would remain gaseous (boiling point
−161.6°C). **11.11** (a) Dispersion. (b) Dispersion and dipole-dipole.
(c) Dispersion and dipole-dipole. (d) Dispersion and ionic.
(e) Dispersion. **11.13** (e) CH_3COOH. **11.15** 1-Butanol has greater
intermolecular forces because it can form hydrogen bonds.
11.17 (a) Xe, it is larger and therefore has stronger dispersion forces.
(b) CS_2, it is larger and therefore has stronger dispersion forces.
(c) Cl_2, it is larger and therefore has stronger dispersion forces.
(d) LiF, it is an ionic compound, and the ion-ion attractions are much
stronger than the dispersion forces between F_2 molecules. (e) NH_3, it
can form hydrogen bonds and PH_3 cannot. **11.19** (a) Dispersion and
dipole-dipole, including hydrogen bonding. (b) Dispersion only.
(c) Dispersion only. (d) Covalent bonds. **11.21** The compound

with $-NO_2$ and $-OH$ groups on adjacent carbons can form hydrogen bonds with itself (intramolecular hydrogen bonds). Such bonds do not contribute to intermolecular attraction and do not help raise the melting point of the compound. The other compound, with the $-NO_2$ and $-OH$ groups on opposite sides of the ring, can form only intermolecular hydrogen bonds; therefore, it will take a higher temperature to escape into the gas phase. **11.33** 321.2 mmHg. **11.35** 84.2°C. **11.37** (a) None; (b) iii. **11.39** Ethylene glycol has two $-OH$ groups, allowing it to exert strong intermolecular forces through hydrogen bonding. Its viscosity should fall between ethanol (1 OH group) and glycerol (3 OH groups). **11.41** Liquid X has a larger ΔH_{vap} than does liquid Y. **11.49** Simple cubic: one sphere; body-centered cubic: two spheres; face-centered cubic: four spheres. **11.51** 6.20×10^{23} atoms/mol. **11.53** 458 pm. **11.55** XY_3. **11.57** 0.220 nm. **11.59** ZnO. **11.63** Molecular solid. **11.65** Molecular: Se_8, HBr, CO_2, P_4O_6, and SiH_4; covalent: Si and C. **11.67** Diamond: each carbon atom is covalently bonded to four other carbon atoms. Because these bonds are strong and uniform, diamond is a very hard substance. Graphite: the carbon atoms in each layer are linked by strong bonds, but the layers are bound by weak dispersion forces. As a result, graphite may be cleaved easily between layers and is not hard. In graphite, all atoms are sp^2 hybridized; each atom is covalently bonded to three other atoms. The remaining unhybridized $2p$ orbital is used in pi bonding, forming a delocalized molecular orbital. The electrons are free to move around in this extensively delocalized molecular orbital, making graphite a good conductor of electricity in directions along the planes of carbon atoms. **11.89** 2.72×10^3 kJ. **11.91** (a) Other factors being equal, liquids evaporate faster at higher temperatures. (b) The greater the surface area, the greater the rate of evaporation. (c) Weak intermolecular forces imply a high vapor pressure and rapid evaporation. **11.93** Two phase changes occur in this process. First, the liquid is turned to solid (freezing), then the solid ice is turned to gas (sublimation). **11.95** When steam condenses to liquid water at 100°C, it releases a large amount of heat equal to the enthalpy of vaporization. Thus, steam at 100°C exposes one to more heat than an equal amount of water at 100°C. **11.99** Initially, the ice melts because of the increase in pressure. As the wire sinks into the ice, the water above the wire refreezes. Eventually the wire actually moves completely through the ice block without cutting it in half. **11.101** (a) Ice would melt. (If heating continues, the liquid water would eventually boil and become a vapor.) (b) Liquid water would vaporize. (c) Water vapor would solidify without becoming a liquid. **11.103** (d). **11.105** Covalent. **11.107** CCl_4. **11.109** 24.2°. **11.111** 760 mmHg. **11.113** It has reached the critical point; the point of critical temperature (T_c) and critical pressure (P_c). **11.115** Crystalline SiO_2. Its regular structure results in a more efficient packing. **11.117** (a) and (b). **11.119** 233 pm. **11.121** (a) K_2S. Ionic forces are much stronger than the dipole-dipole forces in $(CH_3)_3N$. (b) Br_2. Both molecules are nonpolar, but Br_2 has a larger mass. **11.123** SO_2 will behave less ideally because it is polar and has greater intermolecular forces. **11.125** 62.4 kJ/mol. **11.127** Smaller ions can approach polar water molecules more closely, resulting in larger ion-dipole interactions. The greater the ion-dipole interaction, the larger is the heat of hydration. **11.129** (a) 30.7 kJ/mol. (b) 192.5 kJ/mol. It requires more energy to break the bond than to vaporize the molecule. **11.131** (a) Decreases. (b) No change. (c) No change. **11.133** $CaCO_3(s) \longrightarrow CaO(s) + CO_2(g)$. Initial state: one solid phase; final state: two solid phase components and one gas phase component. **11.135** (a) Pumping allows Ar atoms to escape, thus removing heat from the liquid phase. Eventually the liquid freezes. (b) The slope of the solid-liquid line of cyclohexane is positive. Therefore, its melting point increases with

pressure. (c) These droplets are super-cooled liquids. (d) When the dry ice is added to water, it sublimes. The cold CO_2 gas generated causes nearby water vapor to condense, hence the appearance of fog. **11.137** The time required to cook food depends on the boiling point of the water in which it is cooked. The boiling point of water increases when the pressure inside the cooker increases. **11.139** (a) Extra heat produced when steam condenses at 100°C. (b) Avoids extraction of ingredients by boiling in water. **11.141** The fuel source for the Bunsen burner is most likely methane gas. When methane burns in air, carbon dioxide and water are produced. The water vapor produced during the combustion condenses to liquid water when it comes in contact with the outside of the cold beaker. **11.143** 6.019×10^{23}. **11.145** 127 mmHg. **11.147** 55°C. **11.149** 0.833 g/L. The hydrogen-bonding interactions in HF are relatively strong, and since the ideal gas equation ignores intermolecular forces, it underestimates significantly the density of HF gas near its boiling point. **11.151** Fluoromethane. Of the three compounds, only fluoromethane has a permanent dipole moment. **11.153** (a) Two triple points: diamond/graphite/liquid and graphite/liquid/vapor. (b) Diamond. (c) Apply high pressure at high temperature. **11.155** (a) ~2.3 K. (b) ~10 atm. (c) ~5 K. (d) No. **11.157** Ethanol mixes well with water. The mixture has a lower surface tension and readily flows out of the ear channel. **11.159** Ratio = $e^{-401} \approx 0$. **11.161** The molecules are all polar. The F atoms can form H-bonds with water and other $-OH$ and $-NH$ groups in the membrane, so water solubility plus easy attachment to the membrane would allow these molecules to pass the blood-brain barrier. **11.163** When water freezes it releases heat, helping keep the fruit warm enough not to freeze. Also, a layer of ice is a thermal insulator.

Key Skills
11.1 a, d, e **11.2** a, d **11.3** b **11.4** d

Chapter 12

12.3 The monomer must have a triple bond.

12.5
$$\left[\begin{array}{c} H \\ | \\ -C- \\ | \\ CH_3 \end{array} \begin{array}{c} C_2H_5 \\ | \\ C \\ | \\ CH_3 \end{array} \begin{array}{c} H \\ | \\ -C \\ \end{array}\begin{array}{c} \\ = \\ \end{array} \begin{array}{c} \\ C- \\ | \\ H \end{array}\right]_n \quad \text{or} \quad \left[\begin{array}{c} H \\ | \\ -C- \\ | \\ CH_3 \end{array} \begin{array}{c} C_2H_5 \\ | \\ C \\ | \\ CH_3 \end{array} \begin{array}{c} H \\ | \\ -C \\ | \\ H \end{array}\begin{array}{c} \\ = \\ \end{array}\begin{array}{c} \\ C- \\ | \\ H \end{array}\right]_n$$

12.9 (1) Produce the alkoxide: $Sc(s) + 2C_2H_5OH(l) \longrightarrow Sc(OC_2H_5)(alc) + 2H^+(alc)$ ("alc" indicates a solution in alcohol); (2) Hydrolyze to produce hydroxide pellets: $Sc(OC_2H_5)(alc) + 2H_2O(l) \longrightarrow Sc(OH)_2(s) + 2C_2H_5OH(alc)$; (3) Sinter pellets to produce ceramic: $Sc(OH)_2(s) \longrightarrow ScO(s) + 2H_2O(g)$. **12.11** Bakelite is best described as a thermosetting composite polymer. **12.15** No. These polymers are too flexible, and liquid crystals require long, relatively rigid molecules. **12.19** As shown, it is an alternating condensation copolymer of the polyester class. **12.21** Metal amalgams expand with age; composite fillings tend to shrink. **12.25** sp^2. **12.27** Dispersion forces; dispersion forces. **12.31** (a) n-type. (b) p-type. **12.35** $Bi_2Sr_2CuO_6$. **12.37** Plastic polymer: covalent bonds, disulfide (covalent) bonds, H-bonds and dispersion forces. Ceramics: ionic and network covalent bonds. **12.39** Two are $+2$ ([Ar]$3d^9$), one is $+3$ ([Ar]$3d^8$). The $+3$ oxidation state is unusual for copper. **12.41** It is amphoteric, since it reacts with both acid and base. **12.43** The green light has a shorter wavelength (higher energy) than red, so the LED in the exit sign has the greater band gap. **12.45** Fluoroapatite is less soluble than hydroxyapatite, particularly in acidic solutions. Dental fillings must also be insoluble.

Chapter 13

13.9 "Like dissolves like." Naphthalene and benzene are nonpolar, whereas CsF is ionic. **13.11** $O_2 < Br_2 < LiCl < CH_3OH$. **13.15** (a) 8.47%. (b) 17.7%. (c) 11%. **13.17** (a) 0.0610 m. (b) 2.04 m. **13.19** (a) 1.7 m. (b) 0.87 m. (c) 7.0 m. **13.21** 3.2×10^2 m. **13.23** 18.3 M; 27.4 m. **13.25** $\chi(N_2) = 0.677$, $\chi(O_2) = 0.323$. Due to the greater solubility of oxygen, it has a larger mole fraction in solution than it does in the air. **13.33** 45.9 g. **13.35** 1.0×10^{-5} mol/L. **13.37** According to Henry's law, the solubility of a gas in a liquid increases as the pressure increases ($c = kP$). The soft drink tastes flat at the bottom of the mine because the carbon dioxide pressure is greater and the dissolved gas is not released from the solution. As the miner goes up in the elevator, the atmospheric carbon dioxide pressure decreases and dissolved gas is released from his stomach. **13.39** 3.3 atm. This pressure is only an estimate since we ignored the amount of CO_2 that was present in the unopened container in the gas phase. **13.41** The dissolution of the red solute is exothermic. The dissolution of the green solute is endothermic. The numerical value of ΔH_{soln} is greater for the red solute, since changing the temperature produces a greater difference in solubility. **13.57** 30.8 mmHg. **13.59** 88.6 mmHg. **13.61** 187 g. **13.63** 0.59 m. **13.65** $-5.4°C$. **13.67** Boiling point: 102.8°C, freezing point: $-10.0°C$. (b) Boiling point: 102.0°C, freezing point: $-7.14°C$. **13.69** Both NaCl and $CaCl_2$ are strong electrolytes. Urea and sucrose are nonelectrolytes. The NaCl or $CaCl_2$ will yield more particles per mole of the solid dissolved, resulting in greater freezing-point depression. Also, sucrose and urea would make a mess when the ice melts. **13.71** 2.47. **13.73** 9.16 atm. **13.75** (a) $CaCl_2$. (b) Urea. (c) $CaCl_2$. $CaCl_2$ is an ionic compound and is therefore an electrolyte in water. Assuming that $CaCl_2$ completely dissociates, the total ion concentration will be $3 \times 0.35 = 1.05$ m, which is larger than the urea (nonelectrolyte) concentration of 0.90 m. **13.77** 0.15 m $C_6H_{12}O_6 > 0.15$ m $CH_3COOH > 0.10$ m $Na_3PO_4 > 0.20$ m $MgCl_2 > 0.35$ m NaCl. **13.79** (a) Na_2SO_4. (b) $MgSO_4$. (c) KBr. **13.81** a. **13.85** 4.3×10^2 g/mol; $C_{24}H_{20}P_4$. **13.87** 1.75×10^4 g/mol. **13.89** 342 g/mol. **13.91** 15.7%. **13.95** (a) Fat soluble. (b) Water soluble.

13.97 1.2×10^2 g/mol;

$$H_3C-\overset{O-H\cdots O}{\underset{O\cdots H-O}{C}}C-CH_3$$

13.99 7.81×10^3 g/mol **13.101** 3.5. **13.103** Water soluble. **13.105** Fat soluble. **13.107** Reverse osmosis involves no phase changes and is usually cheaper than distillation or freezing. 34 atm. **13.109** (a) Solubility decreases with increasing lattice energy. (b) Ionic compounds are more soluble in a polar solvent. (c) Solubility increases with enthalpy of hydration of the cation and anion. **13.111** 1.43 g/mL; 37.0 m. **13.113** NH_3 can form hydrogen bonds with water; NCl_3 cannot. **13.115** 3%. **13.117** 12.3 M. **13.119** 14.2%. **13.121** 1.9 m. **13.123** Boiling point: c < a = d < b; freezing point: c < a = d < b; van't Hoff factor: d < a = c < b **13.125** (a) 0.099 L. (b) 9.9. **13.127** About 0.4 molal. **13.129** Assume a solution volume of 1 L and write Henry's law as $n = kP$. Substituting kP for n in the ideal gas equation gives $V = kRT$. This equation shows that the volume of a gas that dissolves in a given amount of solvent depends on the temperature, not the pressure of the gas. **13.131** 1.8×10^2 g/mol. **13.133** (a) At reduced pressure, the solution is supersaturated with CO_2. (b) As the escaping CO_2 expands it cools, condensing water vapor in the air to form fog. **13.135** 33 mL, 67 mL. **13.137** Egg yolk contains lecithins that solubilize oil in water (see Figure 13.18). The nonpolar oil becomes soluble in water because the nonpolar tails of lecithin dissolve in the oil, and the polar heads of the lecithin molecules dissolve in polar water (Like dissolves like). **13.139** $\Delta P = 2.05 \times 10^{-5}$ mmHg; $\Delta T_f = 8.9 \times 10^{-5}°C$; $\Delta T_b = 2.5 \times 10^{-5}°C$; $\pi = 0.889$ mmHg. **13.141** 32 m. This is an extremely high concentration. **13.143** The saturated NaCl solution is hypotonic, relative to physiological conditions. Consequently, by osmosis, water gradually moves across the semipermeable membrane into the NaCl compartment in the pill. The increase in volume of the NaCl compartment pushes the elastic membrane to the right, causing the drug to exit through the small holes in the rigid wall at a constant rate. **13.145** (a) Runoff of the salt solution into the soil increases the salinity of the soil. If the soil becomes hypertonic relative to the tree cells, osmosis would reverse, and the tree would lose water to the soil and eventually die of dehydration. (b) Assuming the collecting duct acts as a semipermeable membrane, water would flow from the urine into the hypertonic fluid, thus returning water to the body. **13.147** 0.295 M; $-0.55°C$. **13.149** (a) 2.14×10^3 g/mol. (b) 4.50×10^4 g/mol. **13.151** 282.5 g/mol; $C_{19}H_{38}O$. **13.153** 168 m.

Key Skills

13.1 a, c **13.2** d **13.3** e **13.4** a

Chapter 14

14.5 (a) Rate $= \dfrac{\Delta[H_2]}{\Delta t} = -\dfrac{\Delta[I_2]}{\Delta t} = \dfrac{1}{2}\dfrac{\Delta[HI]}{\Delta t}$.

(b) Rate $= -\dfrac{1}{5}\dfrac{\Delta[Br^-]}{\Delta t} = -\dfrac{\Delta[BrO_3^-]}{\Delta t} = -\dfrac{1}{6}\dfrac{\Delta[H^+]}{\Delta t} = \dfrac{1}{3}\dfrac{\Delta[Br_2]}{\Delta t}$.

14.7 (a) 0.066 M/s. (b) 0.033 M/s. **14.15** 8.1×10^{-6} M/s. **14.17** The reaction is first order in A and first order overall; $k = 0.213$ s^{-1}. **14.19** (a) 2. (b) 0. (c) 2.5. (d) 3. **14.21** First order; $k = 1.19 \times 10^{-4}$ s^{-1}. **14.27** 30 min. **14.29** (a) 0.034 M. (b) 17 s; 23 s. **14.31** 4.5×10^{-10} M/s; 5.4×10^6 s. **14.33** (a) 4:3:6. (b) The relative rates would be unaffected; each absolute rate would decrease by 50%. (c) 1:1:1. **14.41** 3.0×10^3 s^{-1}. **14.43** 51.8 kJ/mol. **14.45** 1.3×10^2 kJ/mol. For maximum freshness, fish should be frozen immediately after capture and kept frozen until cooked. **14.47** 1.3×10^2 kJ/mol. **14.49** Diagram (a). **14.59** (a) Second order. (b) The first step is the slower (rate-determining) step. **14.61** Mechanism I can be discarded. Mechanisms II and III are possible. **14.71** (i) and (iv). **14.73** Temperature, energy of activation, concentration of reactants, and a catalyst. **14.75** Temperature must be specified. **14.77** 0.035 s^{-1}. **14.79** 272 s. **14.81** Since the methanol contains no oxygen-18, the oxygen atom must come from the phosphate group and not the water. The mechanism must involve a bond-breaking process like:

$$CH_3-O\overset{O}{\underset{O-H}{\overset{\|}{P}}}-O-H$$

14.83 Most transition metals have several stable oxidation states. This allows the metal atoms to act as either a source or a receptor of electrons in a broad range of reactions. **14.85** (a) Rate $= k[CH_3COCH_3][H^+]$. (b) 3.8×10^{-3} $M^{-1} \cdot s^{-1}$. (c) $k = k_1k_2/k_{-1}$. **14.87** (I) Fe^{3+} oxidizes I^-: $2Fe^{3+} + 2I^- \longrightarrow 2Fe^{2+} + I_2$; (II) Fe^{2+} reduces $S_2O_8^{2-}$: $2Fe^{2+} + S_2O_8^{2-} \longrightarrow 2Fe^{3+} + 2SO_4^{2-}$; overall reaction: $2I^- + S_2O_8^{2-} \longrightarrow I_2 + 2SO_4^{2-}$. ($Fe^{3+}$ undergoes a redox cycle: $Fe^{3+} \longrightarrow Fe^{2+} \longrightarrow Fe^{3+}$.) The uncatalyzed reaction is slow because both I^- and $S_2O_8^{2-}$ are negatively charged, which makes their mutual approach unfavorable.

14.89 (a) (i) Rate $= k[A]^0 = k$,

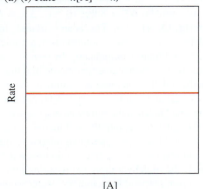

(ii) The integrated rate law is: $[A] = -kt + [A]_0$,

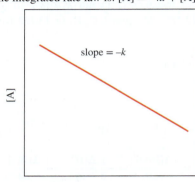

(b) $t_{1/2} = \dfrac{[A]_0}{2k}$. (c) $t = 2t_{1/2}$. **14.91** There are three gases present and we can measure only the total pressure of the gases. To measure the partial pressure of azomethane at a particular time, we must withdraw a sample of the mixture, analyze, and determine the mole fractions. Then, $P_{azomethane} = P_T\chi_{azomethane}$.

14.93

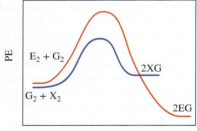

Reaction progress

14.95 (a) A catalyst works by changing the reaction mechanism, thus lowering the activation energy. (b) A catalyst changes the reaction mechanism. (c) A catalyst does not change the enthalpy of reaction. (d) A catalyst increases the forward rate of reaction. (e) A catalyst increases the reverse rate of reaction.

14.97 At very high $[H_2]$: $k_2[H_2] >> 1$, rate $= \dfrac{k_1[NO]^2[H_2]}{k_2[H_2]} = \dfrac{k_1}{k_2}[NO]^2$.

At very low $[H_2]$: $k_2[H_2] << 1$, rate $= \dfrac{k_1[NO]^2[H_2]}{1} = k_1[NO]^2[H_2]$.

The result from Problem 14.80 agrees with the rate law determined for low $[H_2]$. **14.99** Rate $= k[N_2O_5]$; $k = 1.0 \times 10^{-5}$ s^{-1}. **14.101** The red bromine vapor absorbs photons of blue light and dissociates to form bromine atoms: $Br_2 \longrightarrow 2Br\cdot$. The bromine atoms collide with methane molecules and abstract hydrogen atoms: $Br\cdot + CH_4 \longrightarrow$ $HBr + \cdot CH_3$. The methyl radical then reacts with Br_2, giving the

observed product and regenerating a bromine atom to start the process over again: $\cdot CH_3 + Br_2 \longrightarrow CH_3Br + Br\cdot$, $Br\cdot + CH_4 \longrightarrow$ $HBr + \cdot CH_3$, and so on. **14.103** (a) 1.13×10^{-3} M/min. (b) 6.83×10^{-4} M/min; 8.8×10^{-3} M. **14.105** (a) Rate $= k[X][Y]^2$. (b) 0.019 M^{-2}s^{-1}. **14.107** (a) The activation energy of reaction B is larger than that of reaction A. (b) $E_a \approx 0$. Orientation factor is not important.

14.109 (a) $\dfrac{\Delta[B]}{\Delta t} = k_1[A] - k_2[B]$. (b) $[B] = \dfrac{k_1}{k_2}[A]$.

14.111 (a) Three. (b) Two. (c) The third step. (d) Exothermic. **14.113** 0.40 atm. **14.115** (a) Catalyst: Mn^{2+}; intermediates: Mn^{3+}, Mn^{4+}; first step is rate determining. (b) Without the catalyst, the reaction would be a termolecular one involving 3 cations: Tl^+ and two Ce^{4+}. The reaction would be slow. (c) The catalyst is a homogeneous catalyst because it has the same phase (aqueous) as the reactants. **14.117** 4.1×10^2 kJ/mol.

14.119 $n = 0$, $t_{1/2} = C\dfrac{1}{[A]_0^{-1}} = C[A]_0$;

$n = 1$, $t_{1/2} = C\dfrac{1}{[A]_0^0} = C$;

$n = 2$, $t_{1/2} = C\dfrac{1}{[A]_0}$.

14.121 (a) $k = 0.0247$ yr^{-1}. (b) 9.8×10^{-4}. (c) 187 yr. **14.123** 5.7×10^5 yr. **14.125** Second order, $k = 0.42/M \cdot$ min. **14.127** (a) 2.5×10^{-5} M/s. (b) 2.5×10^{-5} M/s. (c) 8.3×10^{-6} M. **14.129** Lowering the temperature would slow all chemical reactions, which would be especially important for those that might damage the brain.

14.131 $\overline{M} = \dfrac{2M[P]_0}{[P]_0 + [P]_0e^{-kt}} = \dfrac{2M}{1 + e^{-kt}}$. The rate constant, k, can be determined by plotting $\left(\dfrac{2M - \overline{M}}{\overline{M}}\right)$ versus t. The plot will give a straight line with a slope of $-k$. **14.133** Second order; $k = 2.4 \times 10^7$ M^{-1}s^{-1}. **14.135** (a) 0.0350 min^{-1}. (b) 110 kJ/mol. (c) Since all the steps are elementary steps, we can deduce the rate law simply from the equations representing the steps. The rate laws are: initiation: rate $= k_i[R_2]$; propagation: rate $= k_p[M][M_1]$; termination: rate $= k_t[M'][M'']$. The reactant molecules are the ethylene monomers, and the product is polyethylene. Recalling that intermediates are species that are formed in an early elementary step and consumed in a later step, we see that they are the radicals $M'\cdot$, $M''\cdot$, and so on. (The R$\cdot$ species also qualifies as an intermediate.)

Key Skills
14.1 d **14.2** e **14.3** b **14.4** a

Chapter 15

15.9 (a) $\dfrac{[N_2][H_2O]^2}{[NO]^2[H_2]^2}$. (b) 1.4×10^2. **15.11** 1.08×10^7. **15.13** (a) 6, (b) 6, and (c) 9. **15.21** 2.40×10^{33}. **15.23** 3.5×10^{-7}. **15.25** (a) 8.2×10^{-2}. (b) 0.29. **15.27** $K_P = 9.52$, $K_c = 4.87 \times 10^2$. **15.29** 7.09×10^{-3}. **15.31** 5.6×10^6. **15.33** $K_P = 9.6 \times 10^{-3}$, $K_c = 3.9 \times 10^{-4}$. **15.35** 4.0×10^{-6}. **15.37** (a) $A + C \rightleftharpoons AC$. (b) $A + D \rightleftharpoons AD$. **15.39** The equilibrium pressure is less than the original pressure. **15.41** 0.173 mol H_2. **15.43** $[H_2] = [Br_2] = 1.53 \times 10^{-4}$ M; $[HBr] = 0.227$ M. **15.45** $P_{COCl_2} = 0.408$ atm; $P_{CO} = P_{Cl_2} = 0.352$ atm. **15.47** $P_{CO} = 1.96$ atm; $P_{CO_2} = 2.54$ atm. **15.49** The forward reaction will not occur. **15.55** (a) The equilibrium would shift to the right. (b) The equilibrium would be unaffected. (c) The equilibrium would be unaffected. **15.57** (a) No effect. (b) No effect. (c) Shift to the left. (d) No effect. (e) Shift to the left. **15.59** (a) Shift to the right. (b) Shift to the left. (c) Shift to the right. (d) Shift to the left. (e) A catalyst has no effect on equilibrium position. **15.61** No change. **15.63** (a) Shift to the right. (b) No effect. (c) No effect. (d) Shift to the left. (e) Shift to the right. (f) Shift to the left. (g) Shift to the right. **15.65** (a) $2O_3(g) \rightleftharpoons 3O_2(g)$, $\Delta H_{rxn}^{\circ} = -284.4$ kJ/mol. (b) Equilibrium would shift to the left. The number of O_3 molecules would increase and the number of O_2 molecules would decrease. **15.67** (a) $P_{NO} = 0.24$ atm; $P_{Cl_2} = 0.12$ atm. (b) $K_P = 0.017$. **15.69** (b). **15.71** (a) 8×10^{-44}. (b) A mixture of H_2 and O_2 can be kept at room temperature because of a very large activation energy. **15.73** (a) 1.7. (b) $P_A = 0.69$ atm, $P_B = 0.81$ atm. **15.75** 4.0. **15.77** $P_{H_2} = 0.28$ atm; $P_{Cl_2} = 0.051$ atm; $P_{HCl} = 1.67$ atm. **15.79** 5.0×10^1 atm. **15.81** -3. **15.83** 6.28×10^{-4}. **15.85** (a) 1.16. (b) 53.7%. **15.87** There is a temporary dynamic equilibrium between the melting of ice cubes and the freezing of water between the ice cubes. **15.89** [H_2]: 0.07 M; [I_2]: 0.18 M; [HI]: 0.83 M. **15.91** (c); N_2O_4(colorless) $\longrightarrow$ $2NO_2$(brown) is consistent with the observations. The reaction is endothermic, so heating darkens the color. Above 150°C, the NO_2 breaks up into colorless NO and O_2: $2NO_2(g) \longrightarrow 2NO(g) + O_2(g)$. An increase in pressure shifts the equilibrium back to the left, restoring the color by producing NO_2. **15.93** (a) Color deepens. (b) Increases. (c) Decreases. (d) Increases. (e) Unchanged. **15.95** 3.5×10^{-2}. **15.97** (a) 1.8×10^{-16}. (b) [H^+][OH^-]: 1.0×10^{-14}; [H^+] = [OH^-]: 1.0×10^{-7} M. **15.99** [NH_3]: 0.042 M; [N_2]: 0.086 M; [H_2]: 0.26 M. **15.101** (a) $K_P = \dfrac{\left(\dfrac{4x^2}{1+x}\right)P}{1-x} = \dfrac{4x^2}{1-x^2}P$.

(b) If P increases, the fraction $\dfrac{4x^2}{1-x^2}$ (and therefore x) must decrease. Equilibrium shifts to the left to produce less NO_2 and more N_2O_4 as predicted. **15.103** $K_P = K_c$: (d), (g); cannot write a K_P: (c), (f). **15.105** (a) 0.49 atm. (b) 23%. (c) 3.7%. (d) Greater than 0.037 mol. **15.107** Potassium is more volatile than sodium. Therefore, its removal shifts the equilibrium from left to right. **15.109** $P_{SO_2Cl_2} = 3.58$ atm, $P_{SO_2} = P_{Cl_2} = 2.71$ atm. **15.111** 0.038. **15.113** (a) 1.0×10^{-6} atm. (b) 2.6×10^{-16} atm. (c) Endothermic. (d) Lightning; the electrical energy promotes the endothermic reaction. **15.115** 3.3×10^2 atm. **15.117** (a) $K_P = 2.6 \times 10^{-6}$; $K_c = 1.1 \times 10^{-7}$. (b) 2.2 g; 22 mg/m³; yes. **15.119** (a) Shifts to the right. (b) Shifts to the right. (c) No change. (d) No change. (e) No change. (f) Shifts to the left. **15.121** Panting decreases the concentration of CO_2 because CO_2 is exhaled during respiration. This decreases the concentration of carbonate ions, shifting the equilibrium to the left. Less $CaCO_3$ is produced. Two possible solutions would be either to cool the chickens' environment or to feed them carbonated water. **15.123** (a) A catalyst speeds up the rates of the forward and reverse reactions to the same extent. (b) A catalyst would not change the energies of the reactant and product. (c) The first reaction is exothermic. Raising the temperature would favor the reverse reaction, increasing the amount of reactant and decreasing the amount of product at equilibrium. The equilibrium constant, K, would decrease. The second reaction is endothermic. Raising the temperature would favor the forward reaction, increasing the amount of product and decreasing the amount of reactant at equilibrium. The equilibrium constant, K, would increase. (d) A catalyst lowers the activation energy for the forward and reverse reactions to the same extent. Adding a catalyst to a reaction mixture will simply cause the mixture to reach equilibrium sooner. The same equilibrium mixture could be obtained without the catalyst, but we might have to wait longer for equilibrium to be reached. If the same equilibrium position is reached, with or without a catalyst, then the equilibrium constant is the same. **15.125** (a) -115 kJ/mol. (b) We start by writing the van't Hoff equation at two different temperatures.

$$\ln K_1 = \frac{\Delta H^{\circ}}{RT_1} + C, \ln K_2 = \frac{\Delta H^{\circ}}{RT_2} + C, \ln K_1 - \ln K_2 = \frac{-\Delta H^{\circ}}{RT_1} - \frac{-\Delta H^{\circ}}{RT_2},$$

$\ln \dfrac{K_1}{K_2} = \dfrac{\Delta H^{\circ}}{R}\left(\dfrac{1}{T_2} - \dfrac{1}{T_1}\right)$. Assuming an endothermic reaction, $\Delta H^{\circ} > 0$ and $T_2 > T_1$. Then, $\dfrac{\Delta H^{\circ}}{R}\left(\dfrac{1}{T_2} - \dfrac{1}{T_1}\right) < 0$, meaning that $\ln \dfrac{K_1}{K_2} < 0$ or $K_1 < K_2$. A larger K_2 indicates that there are more products at equilibrium as the temperature is raised. This agrees with Le Châtelier's principle that an increase in temperature favors the forward endothermic reaction. The opposite of the above discussion holds for an exothermic reaction. (c) 434 kJ/mol.

Key Skills
15.1 e **15.2** a **15.3** b **15.4** d

Chapter 16

16.3 (a) Both. (b) Both. (c) Acid. (d) Both. (e) Acid. (f) Both. (g) Base. (h) Base. (i) Acid. (j) Acid. **16.5** (a) NO_2^-. (b) HSO_4^-. (c) HS^-. (d) CN^-. (e) $HCOO^-$. **16.7** a, b. **16.15** (a) 4.0×10^{-13} M. (b) 6.0×10^{-10} M. (c) 1.2×10^{-12} M. (d) 5.7×10^{-3} M. **16.17** (a) 2.05×10^{-11} M. (b) 3.07×10^{-8} M. (c) 5.95×10^{-11} M. (d) 2.93×10^{-1} M. **16.19** (a), (c), and (e). **16.23** 7.1×10^{-12} M. **16.25** (a) 3.00. (b) 13.89. **16.27** (a) 3.8×10^{-3} M. (b) 6.2×10^{-12} M. (c) 1.1×10^{-7} M. (d) 1.0×10^{-15} M. **16.29** 2.5×10^{-5} M. **16.31** 0.40 g. **16.33** pH < 7, [H^+] > 1.0×10^{-7} M, solution is acid; pH > 7, [H^+] < 1.0×10^{-7} M, solution is basic; pH = 7, [H^+] = 1.0×10^{-7} M, solution is neutral. **16.35** Strong acids—any four of: HCl, HBr, HI, HNO_3, H_2SO_4, $HClO_3$, and $HClO_4$; Strong bases—any four of: LiOH, NaOH, $Ca(OH)_2$, $Sr(OH)_2$, KOH, RbOH, CsOH, and $Ba(OH)_2$. **16.37** Since the ionization of strong acids and bases is complete, these reactions are not treated as equilibria but rather as processes that go to completion. **16.39** (a) -0.009. (b) 1.46. (c) 5.82. **16.41** (a) 7.4×10^{-7} M. (b) 1.8×10^{-5} M. (c) 0.056 M. **16.43** (a) pOH = -0.093; pH = 14.09. (b) pOH = 0.36; pH = 13.64. (c) pOH = 1.07; pH = 12.93. **16.45** (a) 1.1×10^{-3} M. (b) 5.5×10^{-4} M. **16.53** 2.59. **16.55** 5.31. **16.57** (a) 9.8%. (b) 84%. (c) 0.97%. **16.59** 1.3×10^{-6}. **16.61** 4.8×10^{-9}. **16.63** 8.4×10^{-5} M. **16.65** c. **16.67** (a) Strong base. (b) Weak base. (c) Weak base. (d) Weak base. (e) Strong base. **16.71** 6.97×10^{-7}. **16.73** 11.98. **16.75** (a) has the highest K_b value, (b) has the lowest K_b value. **16.79** $K_b(CN^-) = 2.0 \times 10^{-5}$; $K_b(F^-) = 1.4 \times 10^{-11}$; $K_b(CH_3COO^-) = 5.6 \times 10^{-10}$; $K_b(HCO_3^-) = 2.4 \times 10^{-8}$. **16.81** (a) A^-. (b) B^-. **16.85** pH (0.040 M HCl) = 1.40; pH (0.040 M H_2SO_4) = 1.31. **16.87** [H_3O^+] = [HCO_3^-] = 1.4×10^{-4} M; [CO_3^{2-}] = 4.8×10^{-11} M. **16.89** 1.00. **16.91** (a) Diagram c. (b) Diagrams b and d. **16.95** (a) $H_2SO_4 > H_2SeO_4$. (b) $H_3PO_4 > H_3AsO_4$. **16.97** The conjugate bases are $C_6H_5O^-$ from phenol and CH_3O^- from methanol. The $C_6H_5O^-$ is stabilized by resonance:

The CH_3O^- ion has no such resonance stabilization. A more stable conjugate base means an increase in the strength of the acid. **16.103** 4.82. **16.105** 5.39. **16.107** (a) Neutral. (b) Basic. (c) Acidic. (d) Acidic.

16.109 HZ < HY < HX. **16.111** pH > 7. **16.115** (a) Al_2O_3 < BaO < K_2O. (b) CrO_3 < Cr_2O_3 < CrO. **16.117** $Al(OH)_3(s) + OH^-(aq) \longrightarrow$ $Al(OH)_4^-(aq)$. This is a Lewis acid-base reaction. **16.121** $AlCl_3$ is a Lewis acid with an incomplete octet of electrons and Cl^- is the Lewis base donating a pair of electrons. **16.123** CO_2, SO_2, and BCl_3. **16.125** (a) $AlBr_3$ is the Lewis acid; Br^- is the Lewis base. (b) Cr is the Lewis acid; CO is the Lewis base. (c) Cu^{2+} is the Lewis acid; CN^- is the Lewis base. **16.127** $CH_3COO^-(aq)$ and $HCl(aq)$; this reaction will *not* occur to any measurable extent. **16.129** pH = 1.70; percent ionizations = 2.3%. **16.131** c. **16.133** (a) For the forward reaction, NH_3 is both the acid and the base. For the reverse reaction, NH_4^+ and NH_2^- are the acid and base, respectively. (b) NH_4^+ corresponds to H_3O^+; NH_2^- corresponds to OH^-. For the neutral solution, $[NH_4^+] = [NH_2^-]$.

16.135 $K_a = \dfrac{[H^+][A^-]}{[HA]}$; $[HA] \approx 0.1\ M$, $[A^-] \approx 0.1\ M$. Therefore,

$K_a = [H^+] = \dfrac{K_w}{[OH^-]}$ and $[OH^-] = \dfrac{K_w}{K_a}$. **16.137** 1.7×10^{10}. **16.139** (a) H^- (base$_1$) + H_2O (acid$_2$) $\longrightarrow$ OH^- (base$_2$) + H_2 (acid$_1$). (b) H^- is the reducing agent and H_2O is the oxidizing agent. **16.141** 6.02. **16.143** PH_3 is a weaker base than NH_3. **16.145** (a) HNO_2. (b) HF. (c) BF_3. (d) NH_3. (e) H_2SO_3. (f) HCO_3^- and CO_3^{2-}. **16.147** A strong acid: diagram (b); a weak acid: diagram (c); a very weak acid: diagram (d). **16.149** $Cl_2(g) + H_2O(l) \rightleftharpoons HCl(aq) + HClO(aq)$; $HCl(aq) + AgNO_3(aq) \rightleftharpoons AgCl(s) + HNO_3(aq)$. In the presence of OH^- ions, the first equation is shifted to the right: H^+ (from HCl) + $OH^- \longrightarrow H_2O$. Therefore, the concentration of HClO increases. (The "bleaching action" is due to ClO^- ions.) **16.151** 11.80. **16.153** Loss of the first proton from a polyprotic acid is always easier than the subsequent removal of additional protons. The ease with which a proton is lost (i.e., the strength of the acid) depends on the stability of the anion that remains. An anion with a single negative charge is more easily stabilized by resonance than one with two negative charges. **16.155** Magnesium. **16.157** 7.2×10^{-3} g. **16.159** 1.000. **16.161** (a) The pH of the solution of HA would be lower. (b) The electrical conductance of the HA solution would be greater. (c) The rate of hydrogen evolution from the HA solution would be greater. **16.163** 1.4×10^{-4}. **16.165** 2.7×10^{-3} g. **16.167** (a) NH_2^- (base) + H_2O (acid) $\longrightarrow$ $NH_3 + OH^-$; N^{3-} (base) + $3H_2O$ (acid) $\longrightarrow NH_3 + 3OH^-$. (b) N^{3-}. **16.169** 21 mL. **16.171** When the smelling salt is inhaled, some of the powder dissolves in the basic solution. The ammonium ions react with the base as follows: $NH_4^+(aq) + OH^-(aq) \longrightarrow NH_3(aq) + H_2O$. It is the pungent odor of ammonia that prevents a person from fainting. **16.173** 2.8×10^{-2}. **16.175** The F^- ions replace OH^- ions during the remineralization process: $5Ca^{2+} + 3PO_4^{3-} + F^- \longrightarrow Ca_5(PO_4)_3F$ (fluorapatite). Because F^- is a weaker base than OH^-, fluorapatite is more resistant to attacks by acids compared to hydroxyapatite. **16.177** 4.41. **16.179** 5.2×10^{-10}. **16.181** 4.26.

Key Skills

16.1 b **16.2** d **16.3** d **16.4** b

Chapter 17

17.5 (a) 2.57. (b) 4.44. **17.9** 8.89. **17.11** 0.024. **17.13** 0.58. **17.15** 9.25; 9.18. **17.17** (c) and (d). **17.19** Na_2A/NaHA. **17.21** (a) Solutions (a), (b), and (c). (b) Solution (a). **17.27** 202 g/mol. **17.29** 0.25 M. **17.31** (a) 1.10×10^2 g/mol. (b) 1.6×10^{-6}. **17.33** 5.82. **17.35** (a) 2.87. (b) 4.56. (c) 5.34. (d) 8.78. (e) 12.10. **17.37** (a) Cresol red or phenolphthalein. (b) Most of the indicators in Table 17.3, except thymol blue and, to a lesser extent, bromophenol blue and methyl orange.

(c) Bromophenol blue, methyl orange, methyl red, or chlorophenol blue. **17.39** Yellow. **17.41** (a) Diagram (c). (b) Diagram (b). (c) Diagram (d). (d) Diagram (a). The pH at the equivalence point is below 7 (acidic). **17.49** (a) $[I^-] = 9.1 \times 10^{-9}\ M$. (b) $[Al^{3+}] = 7.4 \times 10^{-8}\ M$. **17.51** 1.8×10^{-11}. **17.53** 3.3×10^{-93}. **17.55** 9.50. **17.57** Yes. **17.63** (a) 0.013 M or $1.3 \times 10^{-2}\ M$. (b) $2.2 \times 10^{-4}\ M$. (c) 3.3×10^{-3} M. **17.65** (a) $1.0 \times 10^{-5}\ M$. (b) $1.1 \times 10^{-10}\ M$. **17.67** (b), (c), (d), and (e). **17.69** (a) 0.016 M or $1.6 \times 10^{-2}\ M$. (b) $1.6 \times 10^{-6}\ M$. **17.71** A precipitate of $Fe(OH)_2$ will form. **17.73** $[Cd^{2+}] = 1.1 \times 10^{-18}\ M$, $[Cd(CN)_4^{2-}] = 4.2 \times 10^{-3}$ $[CN^-] = 0.48\ M$. **17.75** $3.5 \times 10^{-5}\ M$. **17.77** (a) $Cu^{2+}(aq) + 4NH_3(aq) \rightleftharpoons [Cu(NH_3)_4]^{2+}(aq)$. (b) $Ag^+(aq) + 2CN^-(aq) \rightleftharpoons [Ag(CN)_2]^-(aq)$. (c) $Hg^{2+}(aq) + 4Cl^-(aq) \rightleftharpoons [HgCl_4]^{2-}(aq)$. **17.81** Greater than 2.68 but less than 8.11. **17.83** 0.011 M. **17.85** Chloride ion will precipitate Ag^+ but not Cu^{2+}. So, dissolve some solid in H_2O and add HCl. If a precipitate forms, the salt was $AgNO_3$. **17.87** BZ_3. **17.89** 1.3 M. **17.91** $[H^+] = 3.0 \times 10^{-13}\ M$; $[HCOOH] = 8.8 \times 10^{-11}\ M$; $[HCOO^-] =$ 0.0500 M; $[OH^-] = 0.0335\ M$; $[Na^+] = 0.0835\ M$. **17.93** $Cd(OH)_2(s) +$ $2OH^-(aq) \rightleftharpoons Cd(OH)_4^{2-}(aq)$; this is a Lewis acid-base reaction. **17.95** (d). **17.97** $[Ag^+] = 2.0 \times 10^{-9}\ M$; $[Cl^-] = 0.080\ M$; $[Zn^{2+}] =$ 0.070 M; $[NO_3^-] = 0.060\ M$. **17.99** 0.035 g/L. **17.101** 2.4×10^{-13}. **17.103** (c). **17.105** (a) AgBr. (b) $1.8 \times 10^{-7}\ M$. (c) 0.0018%. **17.107** (a) Add sulfate. (b) Add sulfate. (c) Add iodide. **17.109** They are insoluble in water. **17.111** (a) Mix 500 mL of 0.40 M CH_3COOH with 500 mL of 0.40 M CH_3COONa. (b) Mix 500 mL of 0.80 M CH_3COOH with 500 mL of 0.40 M NaOH. (c) Mix 500 mL of 0.80 M CH_3COONa with 500 mL of 0.40 M HCl. **17.113** (a) Figure (b). (b) Figure (a). **17.115** pH = $pK_a \pm 1$. **17.117** (a) The pK_b value can be determined at the half-equivalence point of the titration (half the volume of added acid needed to reach the equivalence point). At this point in the titration pH = pK_a, where K_a refers to the acid ionization constant of the conjugate acid of the weak base. The Henderson-Hasselbalch equation reduces to pH = pK_a when [acid] = [conjugate base]. Once the pK_a value is determined, the pK_b value can be calculated as follows: $pK_a + pK_b = 14.00$.

(b) $pOH = pK_b + \log \dfrac{[BH^+]}{[B]}$. The titration curve would look very much like Figure 17.4 of the text, except the y-axis would be pOH and the x-axis would be volume of strong acid added. The pK_b value can be determined at the half-equivalence point of the titration (half the volume of added acid needed to reach the equivalence point). At this point in the titration, the concentrations of the buffer components, [B] and [BH$^+$], are equal, and hence pOH = pK_b. **17.119** (a) Saturated. (b) Unsaturated. (c) Supersaturated. (d) Unsaturated. **17.121** 3.0×10^{-8}. **17.123** $[Ba^{2+}] = 1.0 \times 10^{-5}\ M$. $Ba(NO_3)_2$ is too soluble to be used for this purpose. **17.125** Decreasing the pH would increase the solubility of calcium oxalate and should help minimize the formation of calcium oxalate kidney stones. **17.127** At pH = 1.0: $^+NH_3-CH_2-COOH$; at pH = 7.0: $^+NH_3-CH_2-COO^-$; at pH = 12.0: $NH_2-CH_2-COO^-$. **17.129** Yes. **17.131** The ionized polyphenols have a dark color. In the presence of citric acid from lemon juice, the anions are converted to the lighter-colored acids. **17.133** (c). **17.135** 8.8×10^{-12}. **17.137** (a) 1.0×10^{14}. (b) 1.8×10^9. (c) 1.8×10^9. (d) 3.2×10^4.

Key Skills

17.1 e **17.2** d **17.3** e **17.4** a

Chapter 18

18.7 (a) With barrier: 16; without barrier: 64. (b) 16; 16; 32; both particles on one side: $S = 3.83 \times 10^{-23}$ J/K; particles on opposite sides: $S = 4.78 \times 10^{-23}$. The most probable state is the one with the larger entropy; that is, the state in which the particles are on opposite

sides. **18.13** (a) -0.031 J/K. (b) -0.29 J/K. (c) 1.5×10^2 J/K.
18.15 (a) 47.5 J/K · mol. (b) -12.5 J/K · mol. (c) -242.8 J/K · mol.
18.17 (c) $<$ (d) $<$ (e) $<$ (a) $<$ (b). **18.21** (a) 291 J/K · mol; spontaneous.
(b) 2.10×10^3 J/K · mol; spontaneous. (c) 2.99×10^3 J/K · mol;
spontaneous. **18.23** (a) $\Delta S_{sys} = -75.6$ J/K · mol; $\Delta S_{surr} = 185$ J/K · mol;
spontaneous. (b) $\Delta S_{sys} = 215.8$ J/K · mol; $\Delta S_{surr} = -609$ J/K · mol;
not spontaneous. (c) $\Delta S_{sys} = 98.2$ J/K · mol; $\Delta S_{surr} = -1.46 \times$
10^3 J/K · mol; not spontaneous. (d) $\Delta S_{sys} = -282$ J/K · mol;
$\Delta S_{surr} = 7.20 \times 10^3$ J/K · mol; spontaneous. **18.29** (a) -1139 kJ/mol.
(b) -140.0 kJ/mol. (c) -2935 kJ/mol. **18.31** (a) All temperatures.
(b) Below 111 K. **18.33** $\Delta S_{fus} = 99.9$ J/K · mol; $\Delta S_{vap} = 93.6$ J/K · mol.
18.35 -226.6 kJ/mol. **18.37** (a) (i), (b) (iv), (c) (iii). **18.41** 0.35.
18.43 79 kJ/mol. **18.45** (a) $\Delta G_{rxn}^{\circ} = 35.4$ kJ/mol; $K_P = 6.2 \times 10^{-7}$.
(b) 44.6 kJ/mol. **18.47** (a) 1.6×10^{-23} atm. (b) 0.535 atm.
18.49 3.1×10^{-2} atm or 23.6 mmHg. **18.53** 93 ATP molecules.
18.55 $\Delta H_{fus} > 0$, $\Delta S_{fus} > 0$. (a) $\Delta G_{fus} < 0$. (b) $\Delta G_{fus} = 0$. (c) $\Delta G_{fus} > 0$.
18.57 U and H. **18.59** $\Delta S_{sys} = -327$ J/K · mol; $\Delta S_{surr} = 1918$ J/K · mol;
$\Delta S_{univ} = 1591$ J/K · mol. **18.61** ΔS must be positive ($\Delta S > 0$).
18.63 (a) Benzene: $\Delta S_{vap} = 87.8$ J/K · mol; hexane: $\Delta S_{vap} =$
90.1 J/K · mol; mercury: $\Delta S_{vap} = 93.7$ J/K · mol; toluene: $\Delta S_{vap} =$
91.8 J/K · mol; Trouton's rule is a statement about ΔS_{vap}°. In most
substances, the molecules are in constant and random motion in
both the liquid and gas phases, so $\Delta S_{vap}^{\circ} \approx 90$ J/K · mol. (b) Ethanol:
$\Delta S_{vap} = 111.9$ J/K · mol; water: $\Delta S_{vap} = 109.4$ J/K · mol. In ethanol
and water, there are fewer possible arrangements of the molecules
due to the network of H-bonds, so ΔS_{vap}° is greater. **18.65** q and w
are *not* state functions. **18.67** 249 J/K. **18.69** Equation 18.10
represents the standard free-energy change for a reaction, and not
for a particular compound like CO_2. The correct form is: $\Delta G^{\circ} =$
$\Delta H^{\circ} - T\Delta S^{\circ}$. For a given reaction, ΔG° and ΔH° would need to be
calculated from standard formation values (graphite, oxygen, and
carbon dioxide) first, before plugging into the equation. Also,
ΔS° would need to be calculated from standard entropy values.
C(graphite) $+ O_2(g) \longrightarrow CO_2(g)$. **18.71** (a) 5.76 J/K · mol.
(b) The fact that the actual residual entropy is 4.2 J/K · mol
means that the orientation is not totally random. **18.73** 174 kJ/mol.
18.75 (a) Positive. (b) Negative. (c) Positive. (d) Positive. **18.77** 625 K.
We assume that ΔH° and ΔS° do not depend on temperature.
18.79 No; a negative ΔG° tells us that a reaction has the potential
to happen, but gives no indication of the rate. **18.81** (a) $\Delta G^{\circ} =$
-106.4 kJ/mol; $K_P = 4 \times 10^{18}$. (b) $\Delta G^{\circ} = -53.2$ kJ/mol; $K_P =$
2×10^9. The K_P in (a) is the square of the K_P in (b). Both ΔG° and
K_P depend on the number of moles of reactants and products specified
in the balanced equation. **18.83** Because the reaction results in a
smaller number of molecules, the sign of ΔS is negative. Because
the reaction involves the forming of bonds, the sign of ΔH is also
negative. When both ΔS and ΔH are negative, the reaction is
spontaneous only at low temperatures. **18.85** (a) 86.7 kJ/mol.
(b) 4×10^{-31}. (c) 3×10^{-6}. (d) Lightning supplies the energy necessary
to drive this reaction, converting the two most abundant gases in the
atmosphere into NO(g). The NO gas dissolves in the rain, which
carries it into the soil where it is converted into nitrate and nitrite by
bacterial action. This "fixed" nitrogen is a necessary nutrient for
plants. **18.87** $T > 673.2$ K. **18.89** (a) 7.6×10^{14}. (b) 4.1×10^{-12}.
The activity series is correct. The very large value of K for reaction
(a) indicates that *products* are highly favored, whereas the very small
value of K for reaction (b) indicates that *reactants* are highly
favored. **18.91** $\Delta S_{sys} = 91.1$ J/K; $\Delta S_{surr} = -91.1$ J/K; $\Delta S_{univ} = 0$;
the system is at equilibrium. **18.93** ΔG must be negative; ΔS must
be negative; ΔH must be negative. **18.97** (a) (iii). (b) (i). (c) (ii).
18.99 (a) Disproportionation redox reaction. (b) 8.2×10^{15};
this method is feasible for removing SO_2. (c) Less effective.

18.101 $\chi_{CO} = 0.45$; $\chi_{CO_2} = 0.55$; We assumed that ΔG° calculated
from ΔG_f° values was temperature independent. **18.103** 976 K =
703°C. **18.105** 42°C. **18.107** 8.5 kJ/mol, since we are dealing with
the same ion (K^+). **18.109** 38 kJ. **18.111** (a) 2CO $+$ 2NO $\longrightarrow$
$2CO_2 + N_2$. (b) The oxidizing agent is NO; the reducing agent is
CO. (c) $K_P = 3 \times 10^{120}$. (d) $Q_P = 1.2 \times 10^{14}$; to the right. (e) No.
18.113 (a) CH_3COOH: 27 kJ/mol; $CH_2ClCOOH$: 16 kJ/mol.
(b) The *system's* entropy change dominates. (c) The breaking and
making of specific O—H bonds. Other contributions include solvent
separation and ion solvation. (d) The CH_3COO^- ion, which is
smaller than CH_2ClCOO^-, can participate in hydration to a greater
extent, leading to solutions with fewer possible arrangements.
18.115 $\Delta G^{\circ} = 5.4$ kJ/mol. The sign of ΔG° is positive, indicating
that this reaction, under standard conditions, is not spontaneous.
This agrees with the value of K_P (< 1), which also indicates that the
reaction lies to the *left*.

Key Skills
18.1 b **18.2** d **18.3** b, c, d **18.4** b

Chapter 19

19.1 (a) $2H^+ + H_2O_2 + 2Fe^{2+} \longrightarrow 2Fe^{3+} + 2H_2O$. (b) $6H^+ +$
$2HNO_3 + 3Cu \longrightarrow 3Cu^{2+} + 2NO + 4H_2O$. (c) $3CN^- + 2MnO_4^- +$
$H_2O \longrightarrow 3CNO^- + 2MnO_2 + 2OH^-$. (d) $6OH^- + 3Br_2 \longrightarrow$
$BrO_3^- + 3H_2O + 5Br^-$. (e) $2S_2O_3^{2-} + I_2 \longrightarrow S_4O_6^{2-} + 2I^-$.
19.11 Al(s) $+ 3Ag^+(1.0\ M) \longrightarrow Al^{3+}(1.0\ M) + 3Ag(s)$; $E_{cell}^{\circ} =$
2.46 V. **19.13** $Cl_2(g)$ and $MnO_4^-(aq)$. **19.15** (a) Spontaneous. (b) Not
spontaneous. (c) Not spontaneous. (d) Spontaneous. **19.17** (a) Li.
(b) H_2. (c) Fe^{2+}. (d) Br^-. **19.21** 3×10^{54}. **19.23** (a) 2×10^{18}.
(b) 3×10^8. (c) 3×10^{62}. **19.25** $Co^{3+}(aq) + Ce^{3+}(aq) \longrightarrow$
$Co^{2+}(aq) + Ce^{4+}(aq)$; $\Delta G^{\circ} = -20$ kJ/mol; $K = 4 \times 10^3$.
19.29 1.09 V. **19.31** $E_{cell}^{\circ} = 0.40$ V; $E_{cell} = 0.33$ V. **19.33** 6.0×10^{-38}.
19.39 1.09 V. **19.43** 12.2 g. **19.45** Zinc. **19.47** 0.012 F. **19.49** 5.56 g
Cu; 14.0 g Br_2. **19.51** 3.34×10^4 C. **19.53** 1.84 kg Cl_2/h.
19.55 63.3 g/mol. **19.57** 27.0 g/mol. **19.63** (a) Half-reactions:
$H_2(g) \longrightarrow 2H^+(aq) + 2e^-$, $Ni^{2+}(aq) + 2e^- \longrightarrow Ni(s)$.
Balanced equation: $H_2(g) + Ni^{2+}(aq) \longrightarrow 2H^+(aq) + Ni(s)$. The
reaction will proceed to the left. (b) Half-reactions: $5e^- + 8H^+(aq)$
$+ MnO_4^-(aq) \longrightarrow Mn^{2+}(aq) + 4H_2O$; $2Cl^-(aq) \longrightarrow Cl_2(g) +$
$2e^-$. Balanced equation: $16H^+(aq) + 2MnO_4^-(aq) + 10Cl^-(aq)$
$\longrightarrow 2Mn^{2+}(aq) + 8H_2O(l) + 5Cl_2(g)$. The reaction will proceed
to the right. (c) Half-reactions: $Cr(s) \longrightarrow Cr^{3+}(aq) + 3e^-$,
$Zn^{2+}(aq) + 2e^- \longrightarrow Zn(s)$. Balanced equation: $2Cr(s) +$
$3Zn^{2+}(aq) \longrightarrow 2Cr^{3+}(aq) + 3Zn(s)$. The reaction will proceed to
the left. **19.65** A small nonzero emf will appear if the temperatures
of the two half-cells are different. **19.67** (a) Acid does not oxidize
copper. The copper coating on a penny must be breached for the
acid to reach the zinc, which dissolves as the acid oxidizes Zn(s)
to $Zn^{2+}(aq)$. **19.69** -0.037 V. **19.71** 1.6×10^{-10}. **19.73** (a) 3.14 V.
(b) 3.13 V. **19.75** 0.035 V. **19.77** Mercury(I) is Hg_2^{2+}. **19.79** 1.44 g Mg;
$[Ag^+] = 7 \times 10^{-55}\ M$; $[Mg^{2+}] = 0.0500\ M$. **19.81** (a) H_2, 0.206 L.
(b) 6.09×10^{23} e^-/mol e^-. **19.83** (a) -1356.8 kJ/mol. (b) 1.17 V.
19.85 $+3$. **19.87** $\Delta G^{\circ} = 6.8$ kJ/mol; $K = 0.064$. **19.89** 1.4 A.
19.91 $+4$. **19.93** $H_2O_2(aq) + 2H^+(aq) + 2e^- \longrightarrow 2H_2O(l)$, $E_{cathode}^{\circ} =$
1.77 V. $H_2O_2(aq) \longrightarrow O_2(g) + 2H^+(aq) + 2e^-$, $E_{anode}^{\circ} = 0.68$ V.
$E_{cell}^{\circ} = E_{cathode}^{\circ} - E_{anode}^{\circ} = 1.09$ V. The decomposition is spontaneous.
19.95 Cells of higher voltage require very reactive oxidizing and
reducing agents, which are difficult to handle. (From Table 19.1
of the text, we see that 5.92 V is the theoretical limit of a cell made
up of Li^+/Li and F_2/F^- electrodes under standard-state conditions.)
Batteries made up of several cells in series are easier to use.
19.97 $K_f = 2 \times 10^{20}$. **19.99** (a) E_{red}° for X is negative. E_{red}° for Y

is positive. (b) 0.59 V. **19.101** (a) Gold does not tarnish in air because the reduction potential for oxygen is not sufficiently positive to result in the oxidation of gold. (b) Yes. (c) $2Au + 3F_2 \longrightarrow$ $2AuF_3$. **19.103** -3.05 V. **19.105** 1×10^{-14}. **19.107** (a) Unchanged. (b) Unchanged. (c) Squared. (d) Doubled. (e) Doubled. **19.109** As $[H^+]$ increases, F_2 does become a stronger oxidizing agent. **19.111** 4.4×10^2 atm. **19.113** (a) $Au(s) + 3HNO_3(aq) + 4HCl(aq)$ $\longrightarrow HAuCl_4(aq) + 3H_2O(l) + 3NO_2(g)$. (b) The function of HCl is to increase the acidity and to form the stable complex ion, $AuCl_4^-$. **19.115** $217. **19.117** (a) $1A \cdot h = 1A \times 3600s = 3600$ C. (b) 105 A·h. This ampere hour cannot be fully realized because the concentration of H_2SO_4 keeps decreasing. (c) $E_{cell}^\circ = 2.01$ V; $\Delta G^\circ = -388$ kJ/mol. **19.119** (a) $2MnO_4^-(aq) + 16H^+(aq) + 5C_2O_4^{2-}(aq) \longrightarrow$ $2Mn^{2+}(aq) + 10CO_2(g) + 8H_2O(l)$. (b) 5.40%. **19.121** 0.232 mg Ca^{2+}/mL blood. **19.123** 5 mol ATP/mol NO_2^-. **19.125** 0.00944 g SO_2. **19.127** 1.60×10^{-19} C/e^-. **19.129** $[Fe^{2+}] = 0.0920$ M. $[Fe^{3+}] = 0.0680$ M.

Key Skills
19.1 b **19.2** a **19.3** d **19.4** d

Chapter 20

20.5 (a) $^{23}_{11}Na$. (b) ^1_1p or ^1_1H. (c) ^1_0n. (d) $^{56}_{26}Fe$. (e) $^0_{-1}\beta$. **20.13** 2.72×10^{14} g/cm^3. **20.15** (a) Ni. (b) Se. (c) Cd. **20.17** -4.85×10^{-12} kg/mol H_2. **20.19** (a) 6.30×10^{-12} J; 9.00×10^{-13} J/nucleon. (b) 4.78×10^{-11} J; 1.37×10^{-12} J/nucleon. **20.21** 7.963×10^{-26} kg.

20.25 (a) $^{232}_{90}Th \xrightarrow{\alpha} ^{228}_{88}Ra \xrightarrow{\beta} ^{228}_{89}Ac \xrightarrow{\beta} ^{228}_{90}Th$.

(b) $^{235}_{92}U \xrightarrow{\alpha} ^{231}_{90}Th \xrightarrow{\beta} ^{231}_{91}Pa \xrightarrow{\alpha} ^{227}_{89}Ac$.

(c) $^{237}_{93}Np \xrightarrow{\alpha} ^{233}_{91}Pa \xrightarrow{\beta} ^{233}_{92}U \xrightarrow{\alpha} ^{229}_{90}Th$.

20.27 4.88×10^{19} atoms. **20.29** 3.09×10^3 yr. **20.31** No A remains, 0.25 mole of B, no C is left, 0.75 mole of D. **20.33** 5.5 dpm. **20.35** 43:1. **20.39** (a) $^{14}N(\alpha,p)^{17}O$. (b) $^9Be(\alpha,n)^{12}C$. (c) $^{238}U(d,2n)^{238}Np$. **20.41** (a) $^{40}Ca(d,p)^{41}Ca$. (b) $^{32}S(n,p)^{32}P$. (c) $^{239}Pu(\alpha,n)^{242}Cm$. **20.43** $^{198}_{80}Hg + ^1_0n \longrightarrow ^{199}_{80}Hg \longrightarrow ^{198}_{79}Au + ^1_1p$. **20.55** The fact that the radioisotope appears only in the I_2 shows that the IO_3^- is formed only from the IO_4^-. **20.57** Add iron-59 to the person's diet, and allow a few days for the iron-59 isotope to be incorporated into the person's body. Isolate red blood cells from a blood sample and monitor radioactivity from the hemoglobin molecules present in the red blood cells. **20.63** 65.3 yr. **20.65** (a) $^3_1H \longrightarrow ^3_2He + ^0_{-1}\beta$. (b) 70.5 dpm. **20.67** (a) $^{93}_{36}Kr$. (b) ^1_0n. (c) $^{146}_{57}La$. (d) ^1_0n. **20.69** (a) $^3_1H \longrightarrow ^3_2He + ^0_{-1}\beta$. (b) $^{242}_{94}Pu \longrightarrow ^4_2\alpha + ^{238}_{92}U$. (c) $^{131}_{53}I \longrightarrow ^{131}_{54}Xe + ^0_{-1}\beta$. (d) $^{251}_{98}Cf \longrightarrow ^{247}_{96}Cm + ^4_2\alpha$. **20.71** (a) $^{209}_{83}Bi + ^4_2\alpha \longrightarrow ^{211}_{85}At + 2^1_0n$. (b) $^{209}_{83}Bi(\alpha, 2n)^{211}_{85}At$. **20.73** (a) $r = r_0A^{1/3}$, where r_0 is a proportionality constant. (b) 1.7×10^{-42} m^3. **20.75** (a) 1.83×10^{-12} J. (b) The α particle will move away faster because it is smaller. **20.77** 6.1×10^{23} atoms/mol. **20.79** 0.070%. **20.81** The nuclear submarine can be submerged for a long period without refueling; Conventional diesel engines receive an input of oxygen. A nuclear reactor does not. **20.83** 2.8×10^{14} iodine-131 atoms. **20.85** A small-scale chain reaction (fission of ^{235}U) took place. Copper played the crucial role of reflecting neutrons from the splitting uranium-235 atoms back into the uranium sphere to trigger the chain reaction. Note that a sphere has the most appropriate geometry for such a chain reaction. In fact, during the implosion process prior to an atomic explosion, fragments of uranium-235 are pressed roughly into a sphere for the chain reaction to occur (see Section 20.5 of the text). **20.87** 2.1×10^2 g/mol. **20.89** Using A for element 110, D for element 111, E for element 112, G for element 114, J for element 115, L for element 116, M for element 117, and Q for element 118: $^{208}_{82}Pb + ^{62}_{28}Ni \longrightarrow ^{270}_{110}A$;

$^{209}_{83}Bi + ^{64}_{28}Ni \longrightarrow ^{273}_{111}D$; $^{208}_{82}Pb + ^{66}_{30}Zn \longrightarrow ^{274}_{112}E$; $^{244}_{94}Pu + ^{48}_{20}Ca \longrightarrow ^{289}_{114}G + 3^1_0n$; $^{243}_{95}Am + ^{48}_{20}Ca \longrightarrow ^{291}_{115}J$; $^{248}_{96}Cm + ^{48}_{20}Ca \longrightarrow ^{296}_{116}L$; $^{249}_{97}Bk + ^{48}_{20}Ca \longrightarrow ^{297}_{117}M$; $^{249}_{98}Cf + ^{48}_{20}Ca \longrightarrow ^{297}_{118}Q$; A and D are transition metals. E resembles Zn, Cd, and Hg. G is in the carbon family, J is in the nitrogen family, and L is in the oxygen family. M is a halide and Q is a noble gas and likely a metalloid. **20.91** Since the new particle's mass exceeds the sum of the masses of the electron and positron, the process violates the law of conservation of mass. But, it does not violate Einstein's more general law of mass-energy conservation, $\Delta E = \Delta mc^2$. The large mass of the new particle reflects the fact that the process is extremely endothermic. **20.93** Only ^3H has a suitable half-life. The other half-lives are either too long or too short to determine the time span of 6 years accurately. **20.95** 2.77×10^3 yr. **20.97** Normally the human body concentrates iodine in the thyroid gland. The purpose of the large doses of KI is to displace radioactive iodine from the thyroid and allow its excretion from the body. **20.99** U-238, $t_{1/2} = 4.5 \times 10^9$ yr and Th-232, $t_{1/2} = 1.4 \times 10^{10}$ yr. They are still present because of their long half-lives. **20.101** 0.49 rem. **20.103** 3.4 mL.

Chapter 21

21.5 3.3×10^{-4}; 330 ppm. **21.7** In the stratosphere, the air temperature rises with altitude. This warming effect is the result of exothermic reactions triggered by UV radiation from the sun. **21.11** 260 nm. **21.21** 4.0×10^{37} molecules; 3.2×10^{12} kg O_3. **21.23** $CCl_4 + HF \longrightarrow HCl + CFCl_3$ (Freon-11); $CFCl_3 + HF \longrightarrow HCl + CF_2Cl_2$ (Freon-12). **21.25** $E = 479$ kJ/mol. Solar radiation preferentially breaks the $C-Cl$ bond. There is not enough energy to break the $C-F$ bond.

21.27 $:\ddot{C}l-\ddot{O}-\overset{+}{N}-\ddot{O}^-$ with $:\underset{||}{O}:$; $:\ddot{C}l-\ddot{O}\cdot$. **21.39** 2.6×10^4 tons. **21.47** Primary pollutants, such as automobile exhaust consisting mainly of NO, CO, and various unburned hydrocarbons, set in motion a series of photochemical reactions that produce secondary pollutants. It is the secondary pollutants, chiefly NO_2 and O_3, that are responsible for the buildup of smog. **21.49** While carbon monoxide is a primary pollutant, it is the secondary pollutants, such as ozone, that are responsible for the buildup of smog. **21.51** Most automobiles now are equipped with catalytic converters designed to oxidize CO and unburned hydrocarbons to CO_2 and H_2O and to reduce NO and NO_2 to N_2 and O_2. More efficient automobile engines and better public transportation systems would help to decrease air pollution in urban areas. A recent technological innovation to combat photochemical smog is to coat automobile radiators and air conditioner compressors with a platinum catalyst. So equipped, a running car can purify the air that flows under the hood by converting ozone and carbon monoxide to oxygen and carbon dioxide. **21.53** 4.1×10^{-7} atm; 1×10^{16} molecules/L. **21.59** 378 g. **21.61** O_3: greenhouse gas, toxic to humans, attacks rubber; SO_2: toxic to humans, forms acid rain; NO_2: forms acid rain, destroys ozone; CO: toxic to humans; PAN: a powerful lachrymator, causes breathing difficulties; Rn: causes lung cancer. **21.63** (a) Its small concentration is the result of the high reactivity of the OH radical. (b) OH has an unpaired electron; free radicals are always good oxidizing agents. (c) $OH + NO_2 \longrightarrow HNO_3$. (d) $OH + SO_2 \longrightarrow HSO_3$; $HSO_3 + O_2 + H_2O \longrightarrow H_2SO_4 + HO_2$. **21.65** Most water molecules contain oxygen-16, but a small percentage of water molecules contain oxygen-18. The ratio of the two isotopes in the ocean is essentially constant, but the ratio in the water vapor evaporated from the oceans is temperature-dependent, with the vapor becoming slightly enriched with oxygen-18 as temperature increases. The water locked up in ice cores provides a historical record of this oxygen-18 enrichment, and thus ice cores contain information about past global temperatures. **21.67** 5.1×10^{20} photons. **21.69** 394 nm. **21.71** The lone pair on the S in SO_2 functions as the Lewis base and the Ca in CaO functions as a Lewis acid. **21.73** (a) 6.2×10^8. (b) The CO_2 liberated from limestone contributes to global warming. **21.75** The use of the aerosol liberates CFCs that destroy the ozone layer. **21.77** The size of tree rings can be related to CO_2 content, where the number of rings indicates the age of the tree. The amount of CO_2 in ice can be directly measured from portions of polar ice in different layers obtained by drilling. The "age" of CO_2 can be determined by radio-carbon dating and other methods. **21.79** (a) $N_2O + O \longrightarrow 2NO$; $2NO + 2O_3 \longrightarrow 2NO_2 + 2O_2$; overall: $N_2O + O + 2O_3 \longrightarrow 2NO_2 + 2O_2$. (b) N_2O is a more effective greenhouse gas than CO_2 because it has a permanent dipole. (c) 3.0×10^{10} mol. **21.81** Yes. Light of wavelength 409 nm (visible) or shorter will break the C−Br bond. **21.83** 1.6×10^{19} kJ; 4.8×10^{16} kg. **21.85** 5.2×10^8 L.

Chapter 22

22.11 (a) +3. (b) 6. (c) Oxalate ion ($C_2O_4^{2-}$). **22.13** (a) Na: +1; Mo: +6. (b) Mg: +2; W: +6. (c) Fe: 0. **22.15** (a) *cis*-Dichlorobis(ethylenediamine) cobalt(III). (b) Pentamminechloroplatinum(IV) chloride. (c) Pentamminechlorocobalt(III) chloride. **22.17** (a) $[Cr(en)_2Cl_2]^+$. (b) $Fe(CO)_5$. (c) $K_2[Cu(CN)_4]$. (d) $[Co(NH_3)_4(H_2O)Cl]Cl_2$. **22.23** (a) Two. (b) Two.

22.25 (a) 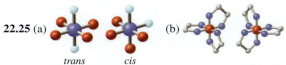 (b)

trans   *cis*

22.31 (a) Orange. (b) 255 kJ/mol. **22.33** Two moles. $[Co(NH_3)_4Cl_2]Cl$. Refer to Problem 22.25 (a) for a diagram of the structure of the complex ion. **22.35** Δ would be greater for the higher oxidation state. **22.39** Use a radioactive label such as $^{14}CN^-$ (in NaCN). Add NaCN to a solution of $K_3Fe(CN)_6$. Isolate some of the $K_3Fe(CN)_6$ and check its radioactivity. If the complex shows radioactivity, then it must mean that the CN^- ion has participated in the exchange reaction. **22.41** $Cu(CN)_2$ is the white precipitate. It is soluble in KCN(aq), due to formation of $[Cu(CN)_4]^{2-}$, so the concentration of Cu^{2+} is too small for Cu^{2+} ions to precipitate with sulfide. **22.43** 1.4×10^2. **22.45** 3. **22.47** Ti^{3+}; Fe^{3+}. **22.49** Mn^{3+} is $3d^4$ and Cr^{3+} is $3d^5$. Therefore, Mn^{3+} has a greater tendency to accept an electron and is a stronger oxidizing agent. The $3d^5$ electron configuration of Cr^{3+} is a stable configuration. **22.51** Y. **22.53** 0.0 *M*. **22.55** (a) $[Cr(H_2O)_6]Cl_3$, number of ions: 4. (b) $[Cr(H_2O)_5Cl]Cl_2 \cdot H_2O$, number of ions: 3. (c) $[Cr(H_2O)_4Cl_2]Cl \cdot 2H_2O$, number of ions: 2. Compare the compounds with equal molar amounts of NaCl, $MgCl_2$, and $FeCl_3$ in an electrical conductance experiment. The solution that has similar conductance to the NaCl solution contains (c); the solution with the conductance similar to $MgCl_2$ contains (b); and the solution with conductance similar to $FeCl_3$ contains (a). **22.57** $\Delta G° = -1.8 \times 10^2$ kJ/mol; $K = 6 \times 10^{30}$. **22.59** Iron is much more abundant than cobalt. **22.61** $[Mn(CN)_6]^{5-}$: Mn is +1, one unpaired *d* electron. $[Mn(CN)_6]^{4-}$: Mn is +2, one unpaired *d* electron. $[Mn(CN)_6]^{3-}$: Mn is +3, two unpaired *d* electrons. **22.63** Complexes are expected to be colored when the highest occupied orbitals have between one and nine *d* electrons. Zn^{2+}, Cu^+, and Pb^{2+} are d^{10} ions. V^{5+}, Ca^{2+}, and Sc^{3+} are d^0 ions. **22.65** Dipole moment measurement. Only the *cis* isomer has a dipole moment. **22.67** EDTA sequesters metal ions (like Ca^{2+} and Mg^{2+}), which are essential for the growth and function of bacteria. **22.69** (a) Tc. (b) W. (c) Mn^{4+}. (d) Au^{3+}. **22.71** The purple color is caused by the build-up of deoxyhemoglobin. When either oxyhemoglobin or

deoxyhemoglobin takes up CO, the carbonylhemoglobin takes on a red color, the same as oxyhemoglobin. **22.73** 1.6×10^4 g hemoglobin/ mol Fe. The discrepancy between our minimum value and the actual value can be explained by realizing that there are four iron atoms per mole of hemoglobin. **22.75** Oxyhemoglobin absorbs higher energy light than deoxyhemoglobin. Oxyhemoglobin is diamagnetic (low spin), while deoxyhemoglobin is paramagnetic (high spin). These differences occur because oxygen (O_2) is a strong-field ligand. **22.77** 2.2×10^{-20} *M*. **22.79** (a) 2.7×10^6. (b) Free Cu^+ ions are unstable in solution. Therefore, the only stable compounds containing Cu^+ ions are insoluble.

Chapter 23

23.7 (a) Amine. (b) Aldehyde. (c) Ketone. (d) Carboxylic acid. (e) Alcohol. **23.9** (a) 3-Ethyl-2,4,4-trimethylhexane. (b) 6,6-Dimethyl-2-heptanol. (c) 4-Chlorohexanal. **23.11** 3,5-Dimethyloctane. **23.13** (a) $(CH_3)_3CCH_2CH(CH_3)_2$. (b) $HO(CH_2)_2CH(CH_3)_2$. (c) $CH_3(CH_2)_4C(O)NH_2$. (d) Cl_3CCHO.

23.15

A = Carbonyl (Ketone)
B = Carboxy (Carboxylic acid)
C = Hydroxy (Alcohol)

23.17 Primary amino / Carboxy

23.19 (a) $CH_3CH_2CHCH_2CH_2CH_3$ with CH_3 branch (b) $CH_3CHCHCH_2CH_3$ with CH_3 branch

(c) $CH_3CHCH_2CHCH_3$ with Br and C_6H_5 branches (d) $CH_3CH_2CHCHCH_2CH_2CH_3$ with CH_3, CH_3 branches

23.21 (a) Kekulé: H−C−C−C−C−C−C−C−C−O−H (with H substituents and O carbonyls)

Bond-line: (structure with two C=O groups and OH)

(b) Condensed: $(C_2H_5)_2CHCH_2CO_2C(CH_3)_3$,

Bond-line: (structure)

(c) Condensed: $(CH_3)_2CHCH_2NHCH(CH_3)_2$,

Kekulé: H−C−C−C−N−C−C−H (with H and CH substituents)

23.23 (a) Condensed structural: $(CH_3)_2NCHO$,

Kekulé:

$$H-\overset{\overset{\displaystyle H}{|}}{\underset{\underset{\displaystyle H-\overset{\overset{\displaystyle H}{|}}{\underset{\underset{\displaystyle H}{|}}{C}}-H}{|}}{C}}-\overset{\overset{\displaystyle O}{\|}}{N}-\overset{}{C}-H$$

Bond-line:

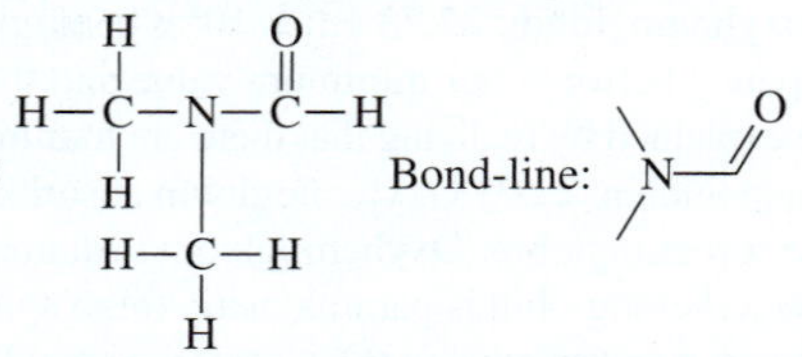

(b) Condensed structural: $(CH_2COOH)_2C(OH)COOH$,

Kekulé:

$$H-\overset{\overset{\displaystyle H}{|}}{\underset{\underset{\displaystyle OH}{|}}{C}}-\overset{\overset{\displaystyle OH}{|}}{\underset{\underset{\displaystyle OH}{|}}{C}}-\overset{\overset{\displaystyle H}{|}}{\underset{\underset{\displaystyle OH}{|}}{C}}-H$$

with $O=C$, $C=O$, $C=O$

Bond-line:

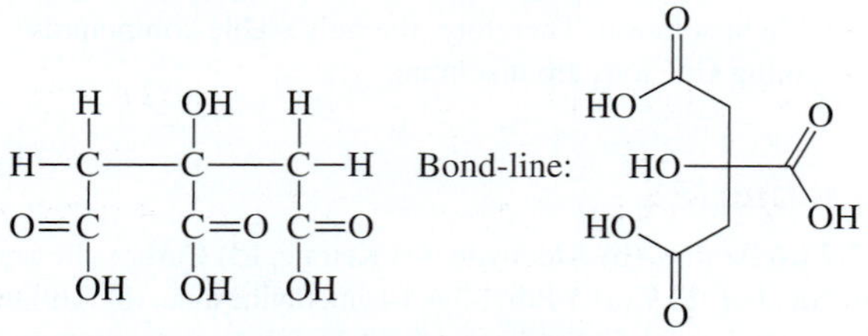

(c) Condensed structural: $(CH_3)_2CH(CH_2)_2OC(O)CH_3$,

Kekulé:

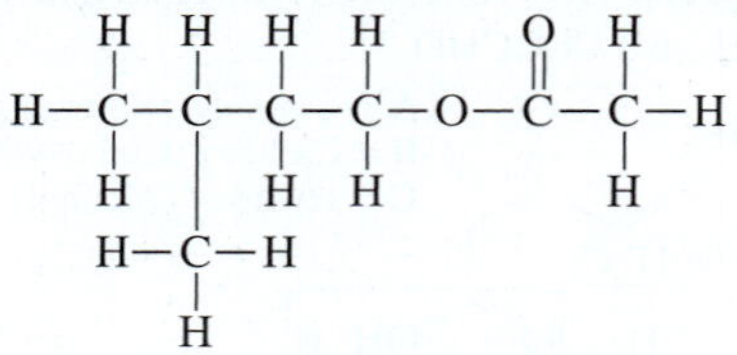

Bond-line:

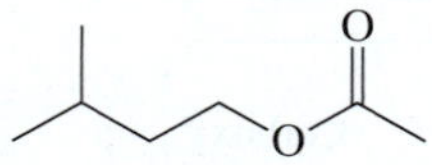

23.25 (a) OH

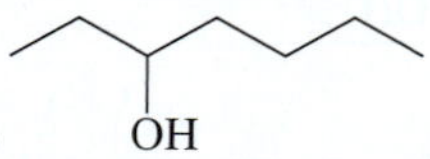

(b) $(CH_3)_3CCH_2CHCH_2\overset{\overset{\displaystyle O}{\|}}{C}H$ with Br

(c) O

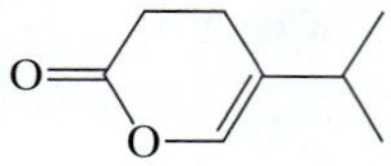

23.27 (a) $CH_3-C\equiv N: \longleftrightarrow CH_3-C\overset{+}{=}\ddot{N}:^-$

(b)

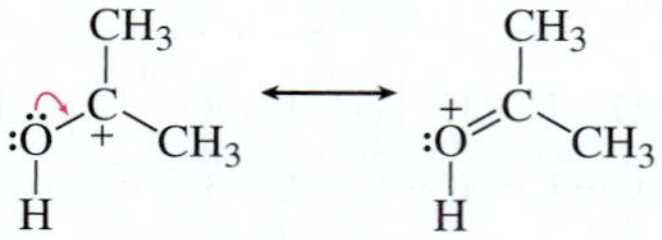

(c)

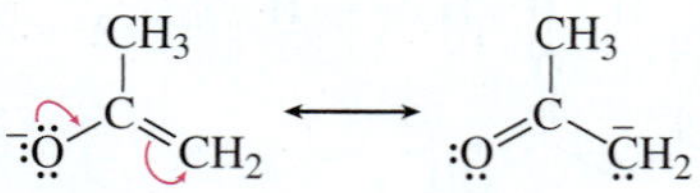

23.29 (a)

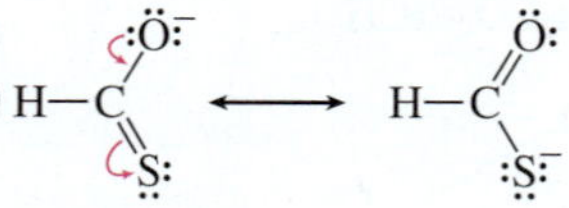

(b) $CH_3-\ddot{O}-CH_2 \longleftrightarrow CH_3-\overset{+}{\ddot{O}}=CH_2$

(c)

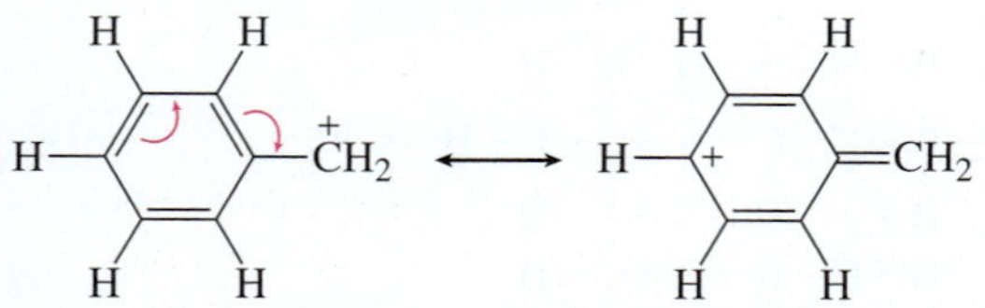

23.37

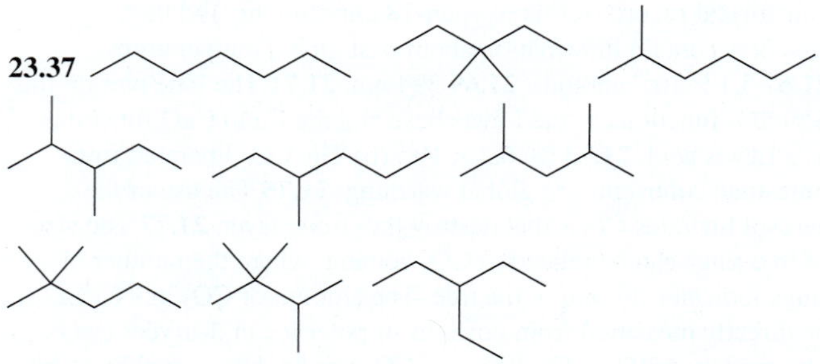

23.39 Br... Br... Br... Br... (with Br)

23.41 Cl

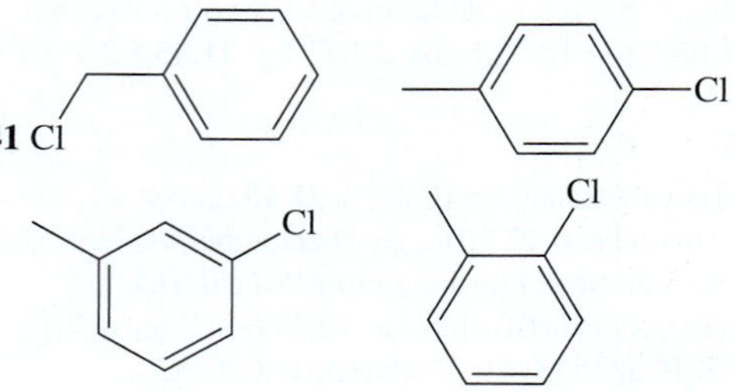

23.43 (a) $CH_3-CH_2-\overset{\overset{\displaystyle CH_3}{|}}{\underset{\underset{\displaystyle *}{}}{C}}H-\overset{*}{C}H-\overset{\overset{\displaystyle O}{\|}}{C}-NH_2$ with NH_2 (b)

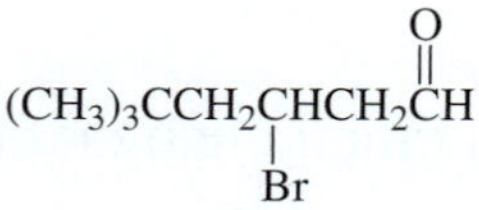

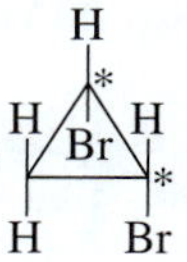

23.45 $CH_3CH_2\overset{\overset{\displaystyle O}{\|}}{C}-H$

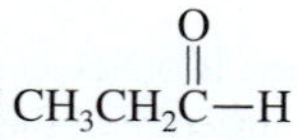

23.47 (a) 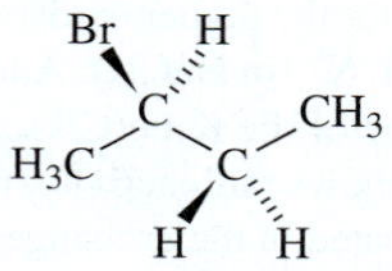 (b) 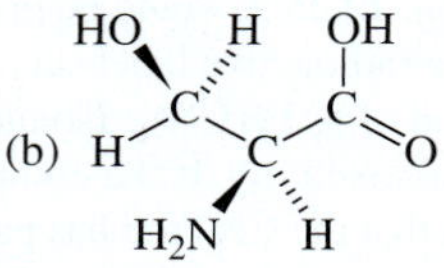 (c) 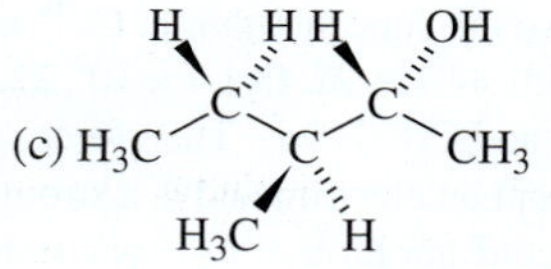 (d)

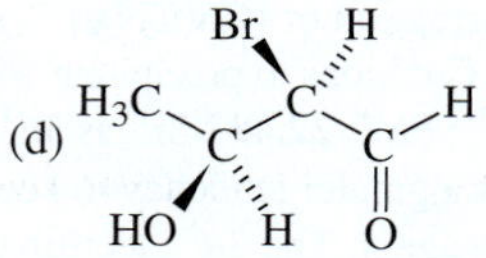

23.55

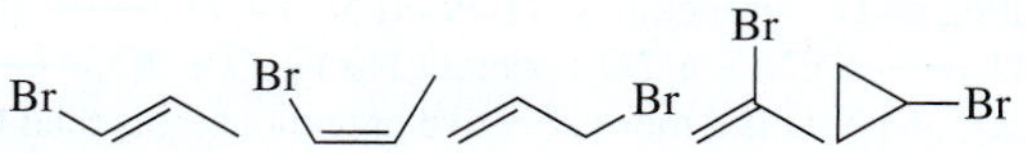

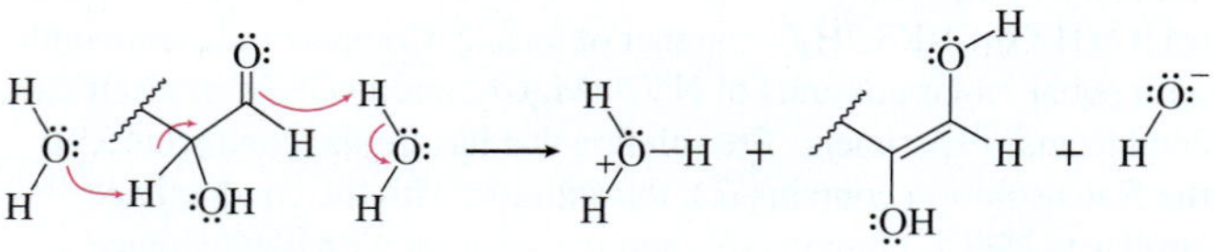

23.57 (a)

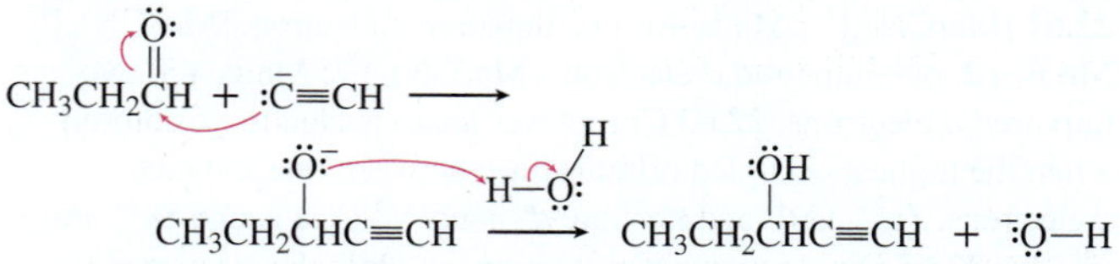

$CH_3CH_2\overset{\overset{\displaystyle O}{\|}}{C}H + :\bar{C}\equiv CH \longrightarrow$

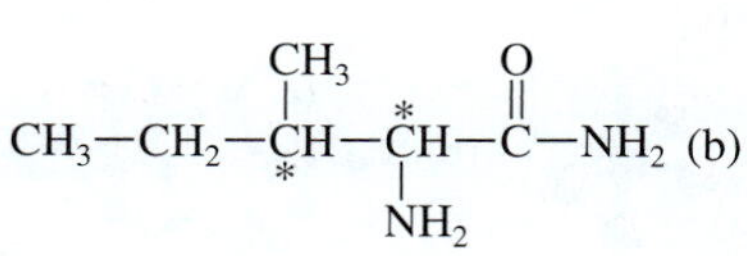

$CH_3CH_2CHC\equiv CH \longrightarrow CH_3CH_2CHC\equiv CH + {}^-\ddot{O}-H$

(b) $^-\ddot{S}-CH_2CH_2CH_2CH_2-\ddot{B}r: \longrightarrow :S: + :\ddot{B}r:^-$

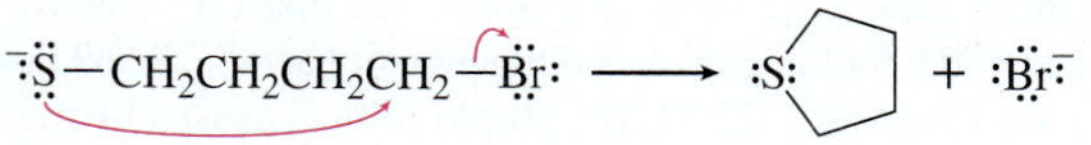

23.59

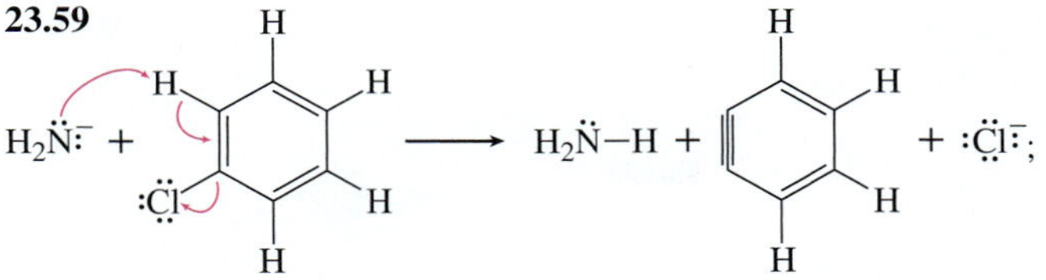

elimination reaction. **23.61** No.

23.63 (a) Sulfuric acid is a catalyst.

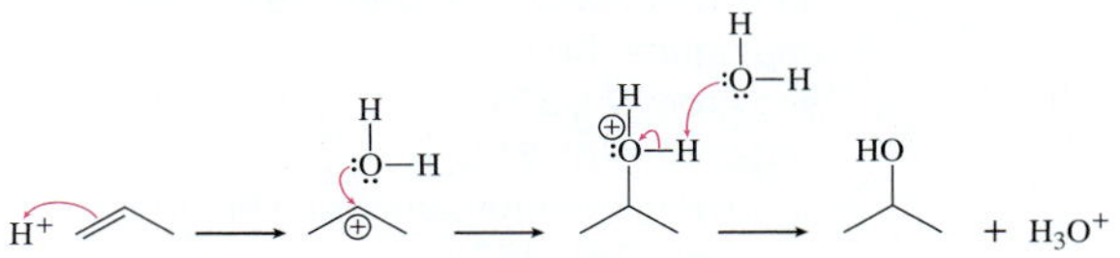

(b) *n*-Propanol: OH   (c) No.

23.71

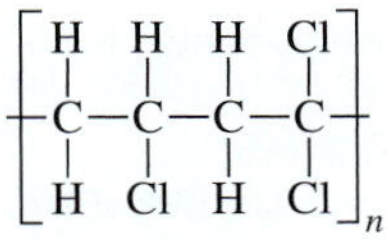

23.73 (a) $CH_2=CHCH=CH_2$

(b)

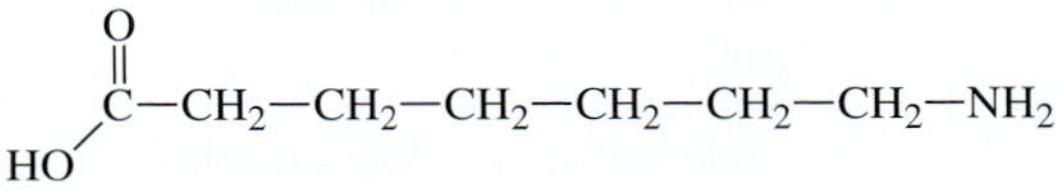

23.75

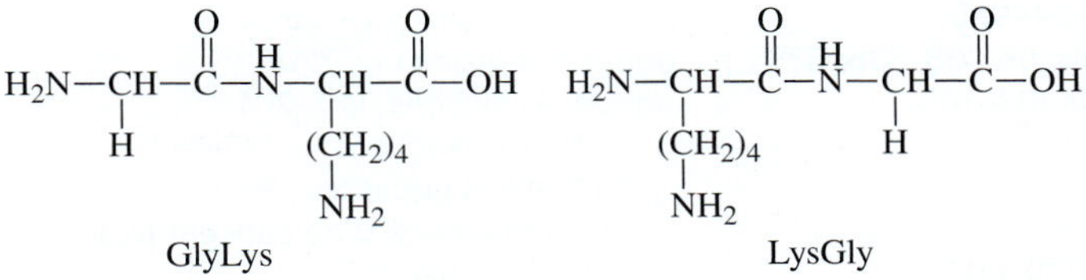

GlyLys   LysGly

23.77

$CH_3CH_2CH_2CH_2-$ $CH_3CHCH_2CH_3$ CH_3CHCH_2- CH_3C-

Butyl   *sec*-Butyl   Isobutyl   *tert*-Butyl

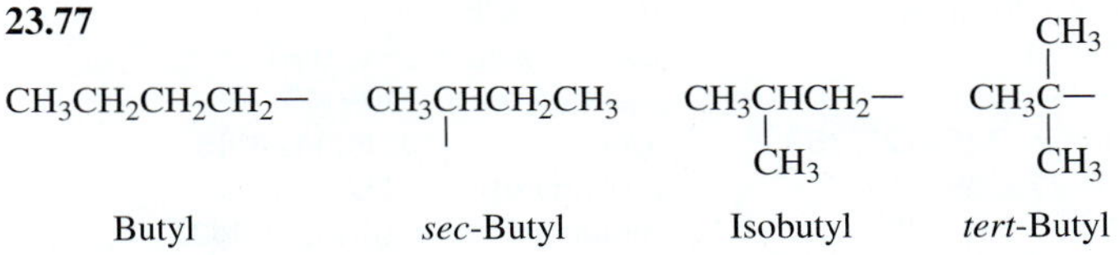

23.79 (a)

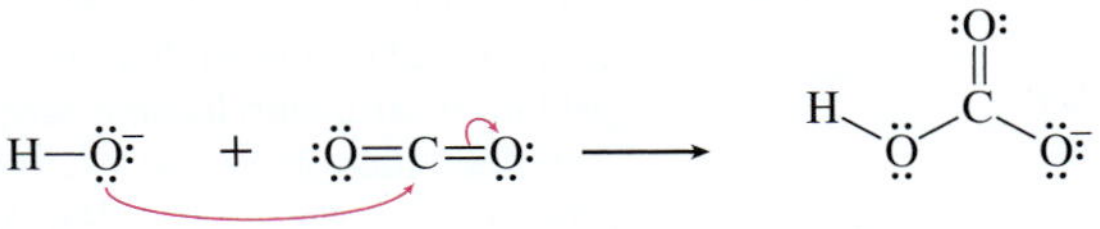

(b)

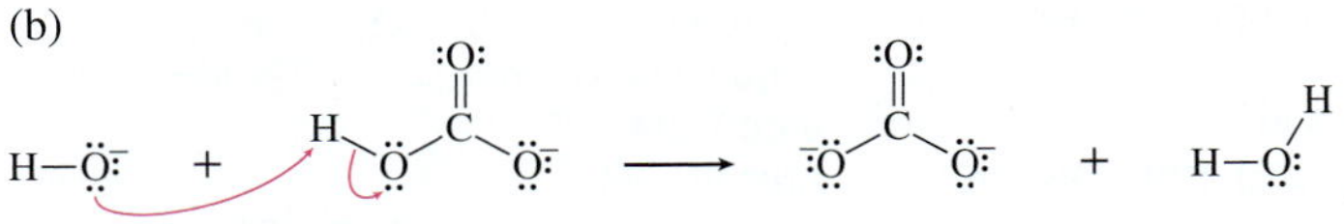

(c) Nucleophilic addition. (d) Acid-base.

23.81 (a) 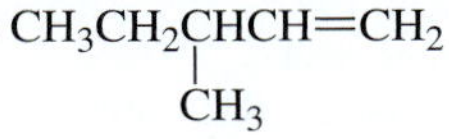 $CH_3CH_2CHCH=CH_2$ with CH_3

(b)

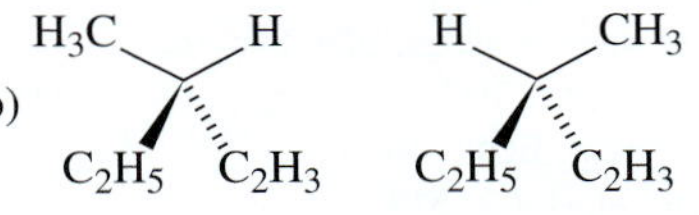

23.83 (a) *Cis/trans* stereoisomers. (b) Constitutional isomers. (c) Resonance structures. (d) Different representations of the same structure.

23.85

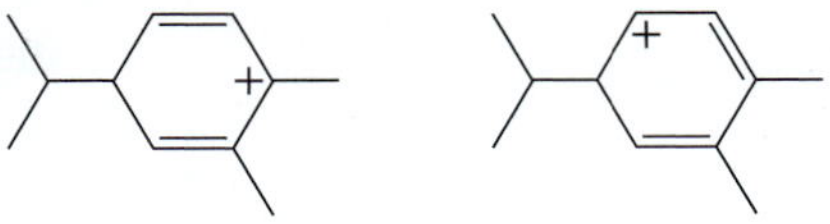

23.87 (b).

23.89 $\overset{sp^3}{CH_3}\overset{sp^3}{CH_2}\underset{sp}{C}\equiv\underset{sp}{CH}$ $\overset{sp^3}{CH_3}\underset{sp}{C}\equiv\underset{sp}{C}\overset{sp^3}{CH_3}$ $H_2C=CHCH=CH_2$ all sp^2

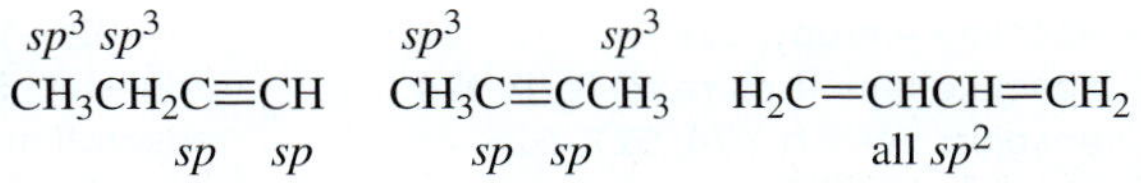

23.91 Since N is less electronegative than O, electron donation in the amide would be more pronounced.

23.93

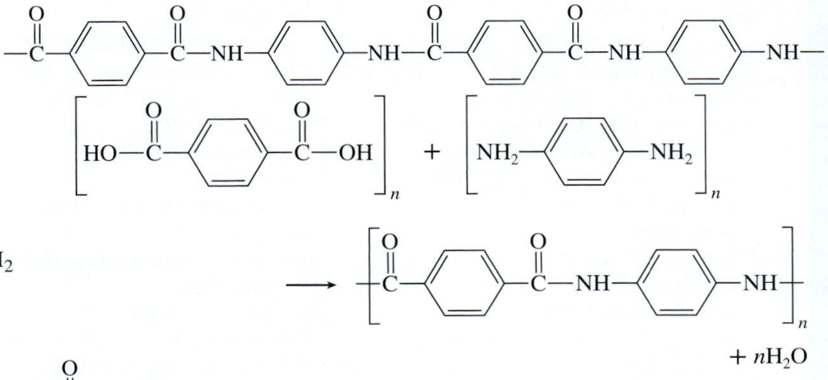

23.95 The N atom in the amide bond is protonated, then nucleophilic addition of water to the carbonyl cleaves the amide bond to produce the original acid and amine;

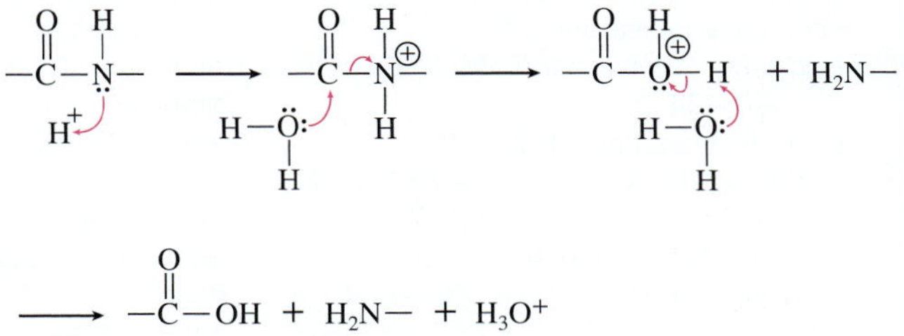

23.97 $CH_3CHCH_2CHCO_2H$ with CH_3 and NH_2 $CH_3CH_2CHCHCO_2H$ with CH_3 and NH_2

Leucine   Isoleucine

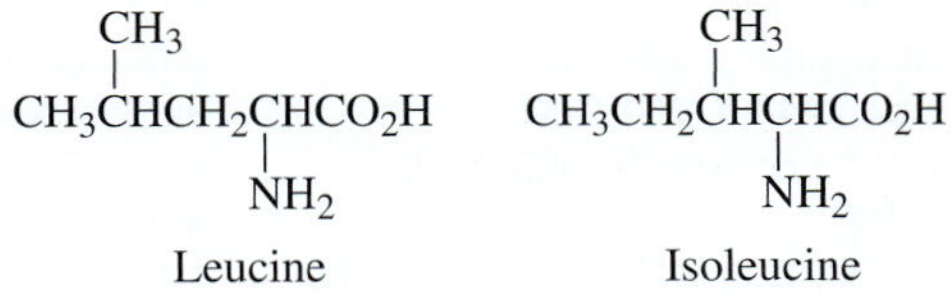

23.99

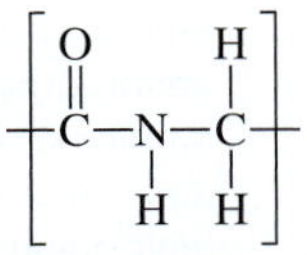

23.101 (a) The more negative ΔH implies stronger alkane bonds; branching decreases the total bond enthalpy (and overall stability) of the alkane. (b) The least highly branched isomer (*n*-octane).

Index

A

absolute entropy, 928
absolute temperature scale, 11, 474
absolute zero, 11, 474, 927–928
absorbance, 178
absorption, 647
absorption spectrum, 178
accuracy, 24–25
acetic acid, 76, 143, 417, 803, 959, 1125
acetone, 68, 1110
acetylene, 434–440, 592
acetylsalicylic acid, 120–121
achiral isomers, 1088, 1089
acid(s), 142–145, 786–849. *See also*
    specific processes and types
  in acid-base reactions, 152–157
  Brønsted, 153–155
  as catalysts, 706
  defined, 71, 142, 153
  formulas for, 145
  identification of, 145
  molecular structure and strength of,
    822–825
  naming of, 71
  percent ionization of, 808–810
  pH of, 791–797
acid-base indicators, 871–874
acid–base indicators, 181
acid-base neutralization, 155–156,
  180–184
acid-base reactions, 152–157
acid-base titrations, 180–184, 862–874
  acid-base indicators in, 181, 871–874
  salt hydrolysis and, 838–839
  strong acid-strong base, 862–864
  strong acid-weak base, 862, 869–871
  weak acid-strong base, 862, 865–869
acidic oxides, 345
acidic salt solutions, 827–829
acidification, ocean, 851, 900
acid ionization constants, 803–812,
  815–817
acid rain, 200, 786, 787, 840, 1064–1066
actinides, 294–295, 317–319
action potential, 982
activated complex, 689
activation energy, 688–689, 704
active metals, 163, 338–339
active transport, 796
activity series, 162–163
actual yield, 119
addition polymerization, 584
addition polymers, 584–590, 1143
addition reactions, 1133–1135, 1139–1140

adenine, 75
adenosine triphosphate (ATP), 942, 943,
  1126–1127, 1137
adhesion, 538
adsorption, 647, 704–705
aerobic capacity, 727
aerosols, 646, 1067
air bags, 487–488
air pollution
  in acid rain, 200, 786, 787, 840,
    1064–1066
  indoor, 1038, 1067–1070, 1076
  in photochemical smog, 1066–1067
alcohol(s), 72, 1110, 1112–1113
  naming of, 1117
  rubbing, 626, 1113
  wood, 1112
alcohol dehydrogenase (ADH), 665, 708
aldehyde(s), 1110, 1112, 1118
aldehyde dehydrogenase (ALDH),
  665, 708
aliphatic compounds, 1110
alkali metals, 56, 163, 338–339
alkaline batteries, 984–985
alkaline earth metals, 56, 163, 338–339
alkanes, 72, 73, 1110, 1113–1115
Alka-Seltzer, 118–119
alkenes, 583, 592
alkyl group, 1110–1114
allotropes, 67–68, 320, 603
alpha (α) particles, 1020, 1036–1037
alpha (α) particles, 49–51
alpha (α) rays, 49, 50
altitude, and hemoglobin, 726–727, 767
altitude sickness, 514, 726, 767
aluminum, 994–995
aluminum oxide, 63
Alzheimer's disease, 1035
amalgam, dental, 600–601
amide(s), 1112, 1118
amide group, 1112–1113
amide linkages, 532, 590–591, 1145
amine(s), 1112, 1118
amine group, 1112, 1119
amino acid(s), 532–533, 1119, 1145–1146
amino acid residue, 532
amino group, 1112–1113
ammonia, 143
ammonium chloride, 63, 827–828
ammonium nitrate, 622–623
amorphous solids, 543, 556–557
ampere (A), 10, 48
amperes (A), 991
amphetamine, 1119–1120
amphoteric hydroxides, 832–833

amphoteric oxides, 345
amphoteric species, 790
amplitude, of waves, 256–257
analogues, 1107
angstrom (Å), 374
angstroms (Å), 51
angular momentum quantum number, 279,
  280, 281–282
anhydrous compounds, 66
aniline, 813
anions, 60–64
  in coordination compounds, 1080,
    1084–1086
  polyatomic, 61–62, 145
anisotropic liquid crystals, 596–597
Annan, Kofi, 1049
anodes, 47, 963–973
Antabuse (disulfiram), 708
antacids, 796–797
anthrax, 387
antibonding molecular orbitals,
  441–442, 605
antifreeze, 664
Antizol (fomepizole), 665
aqueous equilibrium, 770
aqueous solutions, 140–201
  acid-base reactions in, 152–157
  in chemical equations, 96
  concentration of (*See* concentration
    of solutions)
  defined, 618
  electrolysis of, 989–990
  electrolytes in, 140–146
  ionic compounds in, 142–146
  molarity of, 169–172
  molecular compounds in, 142–143,
    145–146
  oxidation of metals in, 162–163
  oxidation-reduction (redox) reactions
    in, 158–168
  precipitation reactions in, 146–152
  properties of, 142–146
  quantitative analysis of, 179–186
argon, 314–315
Aristotle, 44
aromatic compounds, 1110, 1120
Arrhenius, Svante, 3, 154
Arrhenius acid, 154
Arrhenius base, 154
Arrhenius equation, 690–695
artificial hearts, 601
artificial heart valves, 601
artificial joints, 602
artificial skin, 601
ascorbic acid, 43, 819